CW00968226

Power Plant Instrumentation and Control Handbook

Power Plant Instrumentation and Control Handbook

A Guide to Thermal Power Plants

Second Edition

Swapan Basu
Systems & Controls Kolkata, India

Ajay Kumar Debnath
Systems & Controls Kolkata, India

Academic Press is an imprint of Elsevier
125 London Wall, London EC2Y 5AS, United Kingdom
525 B Street, Suite 1650, San Diego, CA 92101, United States
50 Hampshire Street, 5th Floor, Cambridge, MA 02139, United States
The Boulevard, Langford Lane, Kidlington, Oxford OX5 1GB, United Kingdom

Library of Congress Cataloging-in-Publication Data
A catalog record for this book is available from the Library of Congress

British Library Cataloguing-in-Publication Data
A catalogue record for this book is available from the British Library

ISBN 978-0-12-819504-8

For information on all Academic Press publications
visit our website at https://www.elsevier.com/books-and-journals

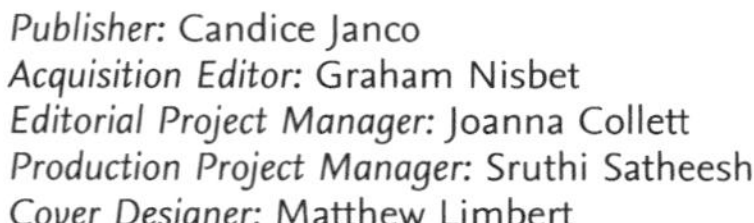

Publisher: Candice Janco
Acquisition Editor: Graham Nisbet
Editorial Project Manager: Joanna Collett
Production Project Manager: Sruthi Satheesh
Cover Designer: Matthew Limbert

Typeset by SPi Global, India

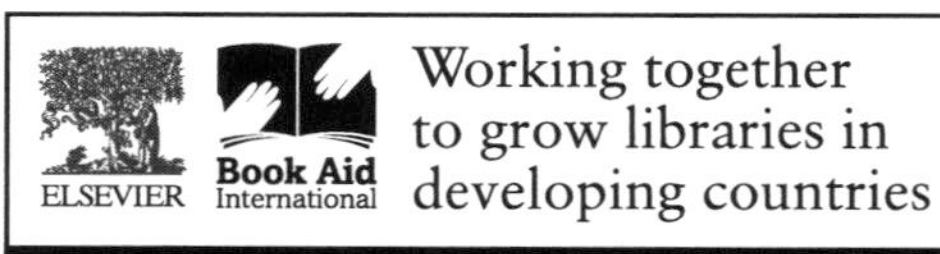

Contents

3. Plant P&ID (Process) Discussions

4. General Instruments

5. Special Instrument

6. Final Control Element

9. Turbo Generator Control System

10. Coordinated Control System

14. Plant Safety Lifecycle and Safety Integrated Level

Preface

Technical books with theoretical and practical approaches on several subsystems of thermal power plant instrumentation and control are available plentiful. However, this book stands out unique as it provides a balance between two extreme lines of thinking, by giving a comprehensive approach toward measurements and controls of thermal power plants.

The book serves as a handbook primarily for professionals who work in thermal power plant instrumentation and control systems. It's an invaluable resource for budding engineers who are beginning their careers in the field. Professionals in other disciplines will also benefit greatly from this book's scope and practical approach. Lastly, it serves as an essential reference tool for engineering students pursuing a career in thermal power plant instrumentation and control systems.

Instead of complicated, advanced mathematical deductions and far-reaching topics, the authors provide physical explanations in a narrow focus that create a solid understanding of these systems The text incorporates abridged descriptions of each subject, relevant figures, and tables that create a simple, clear understanding of these complex topics. In all subjects, detailed specification of the instruments, subsystems, and systems, are included so the book will continue to be a reference tool that will be relevant for years to come.

The text explores future trends in subcritical, supercritical and ultra-supercritical power plants. In this revision additional chapters are included to provide essential information on processes and control systems of these types of thermal power plants. With the introduction of standards like ISA 84.00.00 and IEC 61508/61511, it is important to implement a safety lifecycle for all plants. This revised edition includes a new chapter about safety lifecycles and SIL details that are pertinent to fossil fuel power plants. To keep pace with trends in developmental work in the field of electronics, control, and communication engineering, the authors include information on the means and methods of system integrations, such as with HART, wireless HART Fieldbus systems, OPC servers, etc. This revision also includes details about the latest developments in double and triple offset valves. Applications of artificial intelligence and fuzzy logic in power plant instrumentation are explored in this edition, as well as, hot topics such as IoT and IIoT.

This invaluable information comes from years of study and research by the authors and decades of global experience in thermal power plant instrumentation engineering. The authors wish to thank the companies that entrusted them to work in this specialized area of engineering. The authors hope their lifework developed through research and experience will benefit future engineers who can be of service to mankind by providing scarce pollution-free energy for future generations.

Swapan Basu
Ajay Kumar Debnath

Acknowledgments

First, the authors express their gratitude to the professors of their engineering institution, erstwhile Bengal Engineering and Science University, Shibpur (now IIEST), and their power plant and instrumentation gurus: The late Samir Kumar Shome (former DCL) and the late Makhan Lal Chakraborty (former DCL) for their brilliant instruction in this area. Authors are extremely indebted to Dr. Shankar Sen (former professor of BESU) for his unwavering encouragement to develop the book. The authors are thankful to the International Electrotechnical Commission ***for granting permission*** to use its figures from IEC 61508 and 61511 in the book. and would like to acknowledge as follows:

While developing the book, the authors were inspired by information and suggestions from their former colleagues M/S D.K. Sarkar, J.K. Sarkar, D.J. Gupta, S. Chakraborty, A. Thakur, and Arijit Ghosh. The authors also express sincere gratitude to M/S A. Bhattachariya (Kolkata), A. Sarkar (Norway), A. Tendulkar (Mumbai), N. Kirloskar (Pune), and S. Mohanty (Gurgaon) for their support and sharing of technical information. The authors would like to thank the writers of books and internet resources that stimulated and considerably helped us. Finally, we thank the entire team of Elsevier, the publisher that nurtured this book to its fruition.

Last but not the least, authors would like to thank their children Idai, Piku, Arijita and Arijit for their continuous inspiration and support. Authors would like to convey special thanks to their wives Bani Basu and Syama Debnath for managing the family show with care and encouragement to the authors who had to refuse all other project work to dedicate their time for to this book. The authors sincerely acknowledge that without all of this support, it would have been impossible to publish the book.

Chapter 1

Introduction

Chapter Outline

1 INTRODUCTION

The authors of this book have been associated with the instrumentation and control systems of modern power plants for more than two decades while working with a leading consulting firm. They are still in touch with modern technology by associating with the engineering and consultancy activities of ongoing projects. We wanted to document their extended experience in the form of a reference book so that professional engineers, working engineers in power plants, and students could benefit from the knowledge gathered during their tenure.

There are so many valuable and good books available on a variety of subjects related to power plants regarding boilers, turbines, generators, and their subsystems, but it is very difficult to get a single book or single volume of a book to cater to the equipment, accessories, or items along with the instrumentation and control systems associated with them. In this book, there is a very brief description of the system and equipment along with diagrams for a cursory idea about the entire plant. Up-to-date piping and instrumentation diagrams (P&IDs) are included to better understand the tapping locations of the measuring and control parameters of the plant.

Various types of instruments, along with sensors, transmitters, gauges, switches, signal conditioners/converters, etc., have been discussed in depth in dedicated chapters, whereas special types of instruments are covered in separate chapters. Instrument data sheets or specification sheets are included so that beginners may receive adequate support for preparing the documents required for their daily work.

The control system in Chapters 8–10 incorporates the latest control philosophy that has been adopted in several power stations.

This book mainly emphasizes subcritical boilers, but a separate appendix is provided on supercritical boilers because of their economic and low-pollution aspects, which create a bigger demand and need than conventional subcritical boilers.

It is hoped that this book may help students and/or those who perform power plant-oriented jobs.

2 FUNDAMENTAL KNOWLEDGE ABOUT BASIC PROCESS

Power plant concepts are based on the laws of thermodynamics, which depict the relationship among heat, work, and various properties of the systems. All types of energy transformations related to various systems (e.g., mechanical, electrical, chemical, etc.) may fall under the study of thermodynamics and are basically founded on empirical formulae and system and/or process behavior. A thermodynamic system is a region in space on a control volume or a mass under study toward energy transformation within a system and transfer of energy across the boundaries of the system.

2.1 Ideas Within and Outside the System

1. *Surrounding*: Space and matter outside the thermodynamic system.
2. *Universe*: Thermodynamic system and *surroundings* put together.

Power Plant Instrumentation and Control Handbook. https://doi.org/10.1016/B978-0-12-819504-8.00001-9

3. *Thermodynamic systems*:
   a. *Closed*: Only energy may cross the boundaries with the mass remaining within the boundary.
   b. *Open*: Transfer of mass takes place across the boundary.
   c. *Isolated*: The system is isolated from its surroundings and no transfer of mass or energy takes place across the boundary.
4. *State*: It is the condition detailed in such a way that one state may be differentiated from all other states.
5. *Property*: Any observable characteristics measurable in terms of numbers and units of measurement, including physical qualities such as pressure, temperature, flow, level, location, speed, etc. The property of any system depends only on the state of the system and not on the process by which the state has been achieved.
   a. *Intensive*: Does not depend on the mass of the system (e.g., pressure, temperature, specific volume, and density).
   b. *Extensive*: Depends on the mass of the system (i.e., volume).
6. *Specific weight*: The weight density (i.e., weight per unit volume).
7. *Specific volume*: Volume per unit mass.
8. *Pressure*: Force exerted by a system per unit area of the system.
9. *Path*: The thermodynamic system passes through a series of states.
10. *Process*: Where various changes of state take place.
11. *Cyclic process*: The process after various changes of state complete their journey at the same initial point of the state.

2.1.1 Zeroeth Law of Thermodynamics

"If two systems are both in thermal equilibrium with a third system, they are in thermal equilibrium with each other." Thermal equilibrium displays no change in the thermodynamic coordinates of two isolated systems brought into contact; thus, they have a common and equal thermodynamic property called *temperature*. With the help of this law, the measurement of temperature was conceived. A thermometer uses a material's basic property, which changes with temperature.

2.1.1.1 Energy

"The definition in its simplest form is capacity for producing an effect." There are a variety of classifications for energy.

1. Stored energy may be described as the energy contained within the system's boundaries. There are various forms, such as:
   a. Potential
   b. Kinetic
   c. Internal
2. Energy in transition may be described as energy that crosses the system's boundaries. There are various types, such as:
   a. Heat energy (thermal energy)
   b. Electrical energy
   c. Work

2.1.1.2 Work

"Work is transferred from the system during a given operation if the sole effect external to the system can be reduced to the rise of a weight." This form of energy is transferred from one system to another system originally at different temperatures. It may take place by contact and without mass flow across the boundaries of the two systems. This energy flows from a higher temperature to a lower temperature; it is energy in transition only and not the property. The unit in the metric system is kcal and is denoted by Q.

2.1.1.3 Specific Heat

Specific heat is defined as the amount of heat required to raise the temperature of a substance of unit mass by one degree. There are two types of specific heat:

1. At constant pressure and denoted as C_p
2. At constant volume and denoted as C_v

Heat energy is a path function and the amount of heat transfer can be given by the following:

$$ {}_1Q_2 = \text{Integration from } T_1 \text{ to } T_2 \text{ of } mC_n dT, $$
$$ \text{i.e.,} \int_{T_1}^{T_2} (mC_n dT), $$

where 1 and 2 are two points in the path through which change takes place in the system, m is the mass, C_n is the specific heat and maybe C_p, dT is the differential temperature, and T_1 and T_2 are the two temperatures at point 1 and 2 of the path.

2.1.1.4 Perfect Gas

A particular gas that obeys all laws strictly under all conditions is called a *perfect gas*. In reality, no such gas exists; however, by applying a fair approximation, some gases are considered as perfect (air and nitrogen) and obey the gas laws within the range of pressure and temperature of a normal thermodynamic application.

2.1.2 Boyle's Law and the Charles Law

2.1.2.1 Boyle's Law—Law I

The volume of a given mass of a perfect gas varies inversely as the absolute pressure when temperature is constant.

2.1.2.2 Charles Law—Law II

The volume of a given mass of a perfect gas varies directly as the absolute temperature, if the pressure is constant.

2.1.3 General and Combined Equation

From a practical point of view, neither Boyle's Law nor Charles Law is applicable to any thermodynamic system because volume, pressure, and temperature, etc., all vary simultaneously as an effect of others. Therefore, it is necessary to obtain a general and combined equation for a given mass undergoing interacting changes in volume, pressure, and temperature:

$$v \propto T/p, \text{ when } T \text{ is constant (Boyle's Law)}$$

$$v \propto T, \text{ when } p \text{ is constant (Charles Law).}$$

Therefore, $v \propto T/p$ when both pressure and temperature vary.

or

$$v = k \cdot T/p,$$

where k is a constant that depends on the temperature scale and properties of gas, or

$$pv = mRT,$$

where m is the mass of gas and R is a constant. This depends on the temperature scale and properties of gas: p = absolute pressure of gas in kgf/m^2, v = volume of gas in m^3, m = mass of gas in kg, and T = absolute temperature of gas in K. Therefore $R = pV/mT$ = kgf/m^2 × m^3/kg × K = kgf m/kg/K.

R = 30.26 kgf m/kg/K for nitrogen
R = 29.27 kgf m/kg/K for air
R = 26.50 kgf m/kg/K for oxygen
R = 420.6 kgf m/kg/K for hydrogen

2.1.3.1 Universal Gas Constant

After experiments were performed, it was revealed that for any ideal gas, the product of its characteristic gas constant and molecular weight is a constant number equal to 848. Therefore, by virtue of this revelation, 848 kgf m/kg/K is called the universal gas constant.

For example: MR = molecular weight in kg × R

MR = 29.00 × 29.27 ≈ 848 for air
MR = 2.016 × 420.6 ≈ 848.5 for hydrogen
MR = 28.016 × 30.26 ≈ 847.6 for nitrogen
MR = 32 × 26.5 ≈ 848 for oxygen

2.1.4 Avogadro's Law/Hypothesis—Law III

This states that the molecular weights of all the perfect gases occupy the same volume under the same conditions of pressure and temperature.

2.1.5 First Law of Thermodynamics

When a system undergoes a cyclic change, the algebraic sum of work transfers is proportional to the algebraic sum of heat transfers, or work or heat is mutually convertible one into the other.

Joules' experiments on this subject led to an interesting and important observation showing that the net amount of heat in kcal to be removed from the system was directly proportional to the net amount of work done in kcal on the system.

It is the convention that whenever work is done by the system, the amount of work transfer is considered as +ve, and when work is done on the system, the amount of work transfer is considered as −ve.

2.1.5.1 Internal Energy

There exists a property of a system called energy E, such that a change in its value is the algebraic sum of the heat supplied and the work done during any change in state.

$$dE = \partial Q - \partial W$$

This is also described as corollary 1 of the First Law of Thermodynamics.

Energy E may include many types of energies, such as kinetic, potential, electric, magnetic, surface tension, etc., but these values, negligible considering the thermodynamic system, are ignored and only the energy due to change in temperature is considered. This type of energy is called internal energy and is denoted by U.

2.1.5.2 Adiabatic Work

Whenever the change of state takes place without any heat transfer, it is called an *adiabatic process*. The equation can be written as follows:

$$\Delta U = W_{ad}; \ W_{ad} \text{ is the adiabatic work done}$$

It can be established that change in internal energy ΔU is independent of process path. Thus, it is evident that adiabatic work W_{ad} would remain the same for all adiabatic paths between the same pair of end states.

2.1.6 Law of the Conservation of Energy

"In an isolated system, the energy of the system remains constant." This is known as the second corollary of the First Law of Thermodynamics.

2.1.6.1 Constant Volume Process

The volume of the system is constant. Work done being zero, due to heat addition to the system, there would be an increase in internal energy or vice versa.

2.1.6.2 Constant Pressure or Isobaric Process

In this process, the system is maintained at constant pressure and any transfer of heat would result in work done by the system or on the system.

2.1.6.3 Enthalpy

The sum of internal energy and pressure volume product (i.e., $U+pV$) is known as *enthalpy* and is denoted by H. As both U, p, and V are known as system properties, enthalpy is also a system property.

2.1.6.4 Constant Temperature of the Isothermal Process

The system is maintained at a constant temperature by any means and an increase in volume would result in a decrease in pressure and vice versa.

2.1.7 Second Law of Thermodynamics

There is a limitation of the First Law of Thermodynamics, as it assumes a reversible process. In nature, there is actually a directional law, which implies a limitation on the energy transformation other than that imposed by First Law of Thermodynamics.

Whenever energy transfers or changes from one system to another are equal, there is no violation of the First Law of Thermodynamics; however, that does not happen in practice. Thus, there must exist some directional law governing the transfer of energy.

2.1.8 Heat Engine

A heat engine is a cyclically operating system across whose boundary is a cyclically operating system across which only heat and work flow. This definition incorporates any device operating cyclically and its primary purpose is transformation of heat into work.

Therefore if a boiler, turbine, condenser, and pump are separately considered in a power plant, they do not stand included in the definition of heat engines because each individual device in the system does not complete a cycle (Fig. 1.1).

When put together, however, the combined system satisfies the definition of a heat engine. Referring to Fig. 1.2, the heat enters the boiler and leaves at the condenser. The difference between these equals work at the turbine and pump. The working medium is water and it undergoes a cycle of processes. Passing through the boiler and transforming to steam, it goes to the turbine and then to the condenser where it changes back into water and goes to the feed pump, and finally to the boiler again to its initial state.

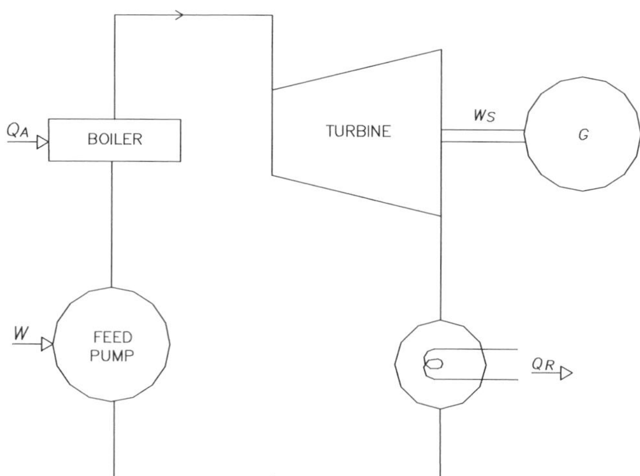

FIG. 1.1 Power plant as a basic heat engine.

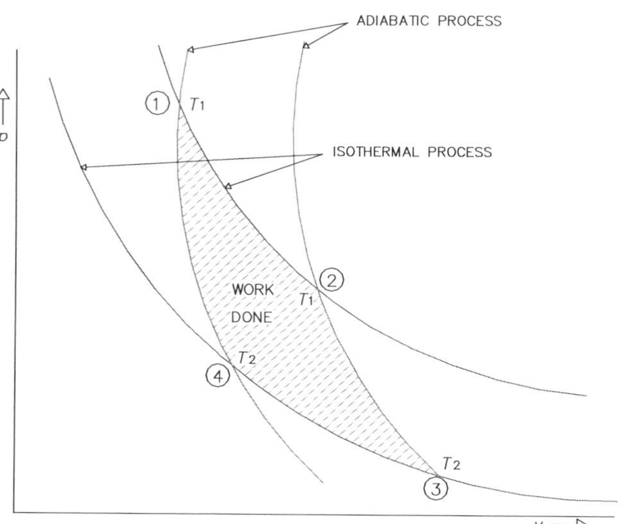

FIG. 1.2 *p-v* diagram of a Carnot (reversible) cycle.

2.1.8.1 Kelvin Planck Statement of the Second Law of Thermodynamics

It is impossible to construct an engine that, while operating in a cycle, produces no other effect except to extract heat from a single reservoir and do the equivalent amount of work. Thus, it is imperative that some heat be transferred from the working substance to another reservoir, or cyclic work is possible only with two temperature levels involved and the heat is transferred from a high temperature to a heat engine and from a heat engine to a low temperature.

2.1.8.2 Clausius Statement of the Second Law of Thermodynamics

"It is impossible for heat energy to flow spontaneously from a body at lower temperature to a body at higher temperature."

2.2 Recapitulation: Various Cycles: Carnot, Rankine, Regenerative, and Reheat

2.2.1 Reversible Cycle: Carnot

Here, a reversible cycle was proposed by Sadi Carnot, the inventor of this, in which the working medium receives heat at one temperature and rejects heat at another temperature. This is achieved by two isothermal processes and two reversible adiabatic processes, shown in the simplified schematic in Fig. 1.2.

A given mass of gas (system) is expanded isothermally from point 1 at temperature T_1 to point 2 (after receiving heat q_1 from an external source). So, work is done by the system. The system is now allowed to expand further to point 3 at temperature T_2 through a reversible adiabatic process, meaning no exchange of heat or transfer except work is done due to expansion.

Now the system at point 3 is allowed to reject heat q_2 to a sink at temperature T_2 isothermally up to point 4 by compressing (i.e., doing work on the system). At point 4, the system is again compressed up to point 1, the starting point, through a reversible adiabatic process (i.e., without any heat transfer). Now because the system has completed a cycle and returned to its initial state, its internal energy remained the same, as per the First Law of Thermodynamics. Now, $q_1 - q_2 = W =$ work done.

2.2.2 Application of Carnot Cycle in a Power Plant

The previous schematic in Fig. 1.2 is a classical demonstration of the Carnot cycle. The water-steam flow cycle of a steam power plant is shown in Fig. 1.3.

Here, the isothermal process or heat transfers take place in the boiler at temperature T_1 and in the condenser at temperature T_2. In these two operations, the fluid is undergoing a change in phase; in other words, in the boiler, water is transformed to steam at temperature T_1 and in the condenser, steam is transformed into water at temperature T_2.

The reversible adiabatic expansion is performed at the turbine and reversible adiabatic compression takes place in the (boiler) feed pump.

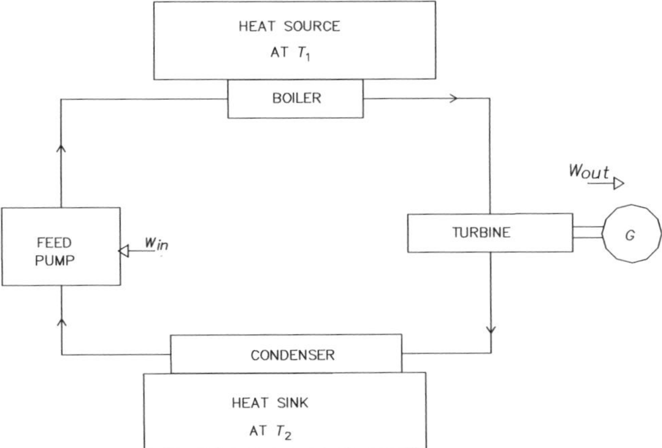

FIG. 1.3 Water-steam simplified flow cycle of a power plant.

2.2.3 Carnot Theorem or Corollary 2

No engine working between two temperatures can be more efficient than the reversible engine working between the same two temperatures or the Carnot engine (hypothetical). Among all engines operating between fixed temperatures, it is the most efficient.

2.2.4 Properties of Steam

Water is introduced into the boiler by a feed pump at a certain pressure and temperature, adding some energy to the system. At the boiler, heat is added to raise the temperature at a saturation temperature corresponding to that initial pressure. This is called "sensible heat," as the rise in temperature is evident. When the saturation stage is attained, further addition of heat would change the phase of water to steam without a temperature rise but a sensible change in volume. This stage would continue until dry saturation steam is available. As there is no change in temperature, the heat added is called "latent heat" and is denoted by L.

2.2.4.1 Steam Table

Normally, the properties of steam include different parameters such as pressure, temperature, volume, enthalpy, entropy, etc., and their interrelations are experimentally determined and presented in a tabular form. These values are referred to and required values are obtained from reference tables instead of calculating from the equations, which are very complex.

2.2.4.2 Wet Steam

Wet steam may be described as steam with a mixture of liquid water and water vapor suspended in it. The fraction of steam present in the mixture by weight is called the *dryness* fraction of steam.

2.2.4.3 Superheated Steam

Superheated steam behavior is like a perfect gas; the volume of a given mass can be determined by the Charles Law (i.e., p is constant). All the properties of superheated steam are normally found in reference steam tables, the figures of which were found by performing experiments to explain variations in specific heat and other influencing factors.

2.2.4.4 Entropy

It can be proved that the integral value of change in heat transfers divided by temperature in a cyclic path is equal to zero.

$$\text{Cyclic}\int (\partial q/T)_{\text{rev}} = 0$$

or

$$(\partial Q/T) = dS,$$

where S is called entropy, or a change in entropy during a reversible process can be written as follows:

$$\int_1^2 (\partial Q/T)_{\text{rev}} = \int_1^2 dS = (s_2 - s_1) = \Delta S$$

For unit mass, $\int_1^2 (\partial Q/T)_{\text{rev}} = \int_1^2 dS = \Delta S.$

2.2.4.4.1 Corollary 5 Corollary 5 of the Second Law of Thermodynamics indicates that there exists a property called entropy of a system such that for a reversible process from point 1 to point 2 in a process path, its change is given as

$$\int_1^2 (\partial Q/T)_{\text{rev}} \text{ for a unit mass}$$

Therefore, it is evident that entropy is not a path function but a point function and a change of entropy can be shown as:

$$ds = (dU + pdV)/T$$

or, in another way,

$$T\,ds = dU + pdV$$

This equation is very important as it is evident that the relationships among all parameters are thermodynamic properties and not path functions such as heat or work. It is interesting that the equation

$$T\,ds = dU + pdV$$

is applicable to both reversible and irreversible processes, but

$$\partial Q = T\,ds \text{ and } \partial Q = dU + pdV$$

are only applicable to reversible processes.

2.2.5 Temperature-Entropy Diagram

As it is known that ${}_1Q = \int_{s_1}^{s_2} T\,ds$, it can be graphically realized as the area under the curve with temperature and entropy as the coordinates as seen in Fig. 1.4. Fig. 1.5 also graphically

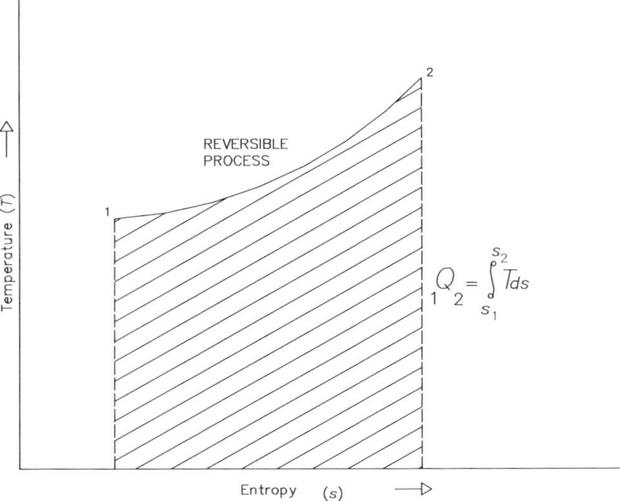

FIG. 1.4 Temperature-entropy diagram of reversible process.

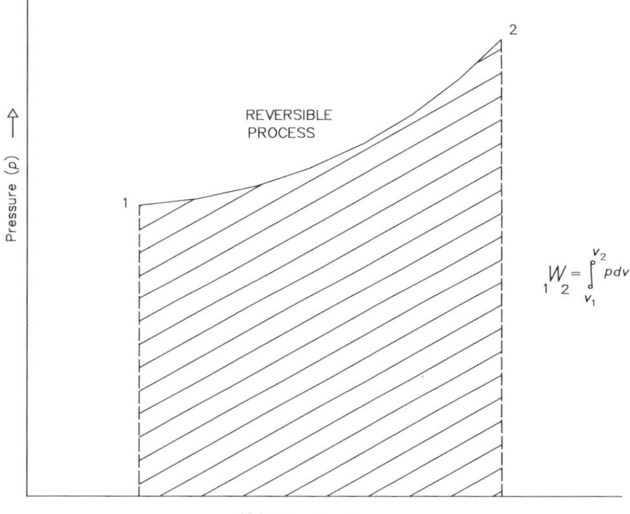

FIG. 1.5 Pressure volume diagram of reversible process.

represents the work done in a separate set of pressure and volume coordinates; for example, work done in these coordinates is

$${}_1W_2 = \int_{v_1}^{v_2} p\,dv$$

By the First Law of Thermodynamics:

$$\text{Cyclic}\int \partial Q = \int dW$$

(i.e., heat transferred to the system is equal to the work done by the system). From the previous equation, a very important conclusion can be drawn: the "enclosed area for a reversible cyclic process represents work done by heat transfers on both p-V as well as T-s coordinates." Thus, in the Carnot cycle represented on the *p-V* or *T-s* coordinates, the enclosed area denotes work done or heat transfers. From various logical derivations and approximations, it can be said that for an irreversible process, the entropy change is not equal to $(\partial Q/T)$, but more than $(\partial Q/T)$; in other words, the (ds) isolated system is ≥ 0, which is known as Corollary 6 of the Second Law of Thermodynamics.

2.2.6 Entropy of Different Phases of Water and Steam

2.2.6.1 Entropy of Water

By definition, $ds = dq/T = C_p \cdot dT/T$; therefore, $(s_2 - s_1) = \int_{T_1}^{T_2} C_p dT/T = C_p \log_e T_2/T_1$ If 0°C or 273 K is chosen as the datum for entropy, then the entropy of water at any temperature T would be $s = C_p \log_e T/273$ and the entropy of water at saturation temperature T_s is $s_w = C_{pw} \log_e T_s/273$.

2.2.6.2 Entropy of Steam

Heat required to convert a unit mass of water to a unit mass of dry saturated steam is the latent heat of vaporization and is denoted by L. Therefore, $s_L = L/T_s$, or, the entropy of vaporization of wet steam is $xS_L = xL/T_s$, where x = the dryness fraction of steam; in other words, it is the fraction of dry saturation steam to the total mass of the steam. Entropy of dry saturated steam is given by the following:

$$s = s_w + s_L = C_{pw} \log_e T_s/273 + xL/T_s.$$

2.2.6.3 Entropy of Superheated Steam

For a unit mass of dry saturated steam to get superheated to temperature T_{sup} at constant pressure, the entropy excursion may be given as follows:

$$s_{sup} - s_s = \int_{T_s}^{T_{sup}} C_p \cdot dT_{sup}/T_s = C_p \log_e T_{sup}/_s.$$

Therefore, the entropy of superheated steam may be expressed as follows:

$$s_{sup} = C_{pw} \log_e T_s/273 + L/T_s + C_p \log_e T_{sup}/T_s.$$

These equations are very cumbersome and are not used much because these entropy values can be found in reference steam tables.

2.2.7 Temperature-Entropy Diagram of Steam

From the equation $s_w = C_{pw} \log_e T_s/273$, different values of saturation temperature are plotted against values of entropy at different pressures (see Fig. 1.6).

In this figure, the portion of the graph from point 1 to 2 is considered the water or liquid line. From point 2 to point 3, the path is a straight horizontal line at constant saturation temperature T_s denoting the water and vapor mixture phase. At point 3, the dry saturation stage is achieved. From point 3, if the process follows path 3-4, then different values of dry saturated temperatures are available at lower saturation pressure up to point 4. These two lines or paths when plotted for higher pressure corresponding to a higher saturation temperature would finally merge at point C, which is called the critical point. Here, the saturation temperature is 374.065°C and the pressure is 225.415 kgf/cm^2. At this point, water transforms into the gaseous phase (i.e., dry saturation steam) directly without passing through the two-phase system, and the latent heat of vaporization is zero.

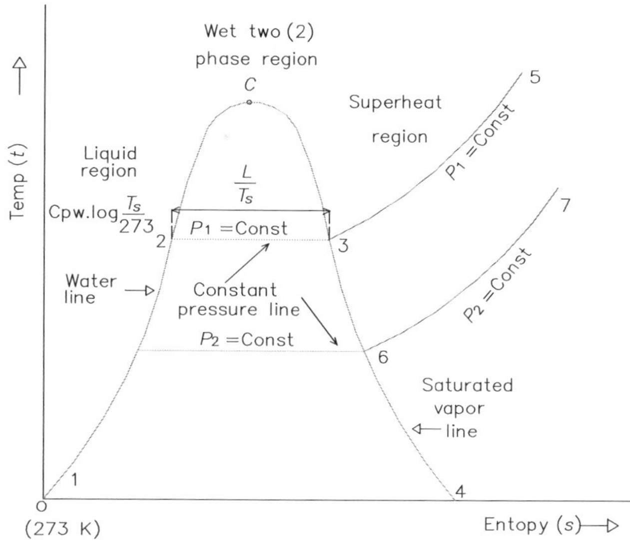

FIG. 1.6 Temperature-entropy diagram of steam.

In path 3-4, at any point, if the steam is further heated at constant pressure, the process will follow path 3-5 or 6-7 up to the temperatures of superheated steam corresponding to the heat added. After this, the region is denoted as a superheat region.

2.2.7.1 Pressure-Volume Diagram

The pressure-volume diagram corresponding to the temperature-entropy diagram is illustrated in Fig. 1.7.

The critical point C is at 225.415 kgf/cm^2. Liquid, wet, and superheat regions are depicted; 1-2 and extension up to point C is the water line. Line 3-4 and extension up to point C is the dry saturation line. Constant pressure heating is represented by 1-2-3-5.

2.2.7.2 Steam Generators/Boilers

Steam generators or boilers represent devices for generating steam for various applications:

1. A power-generation plant with the help of steam turbines.

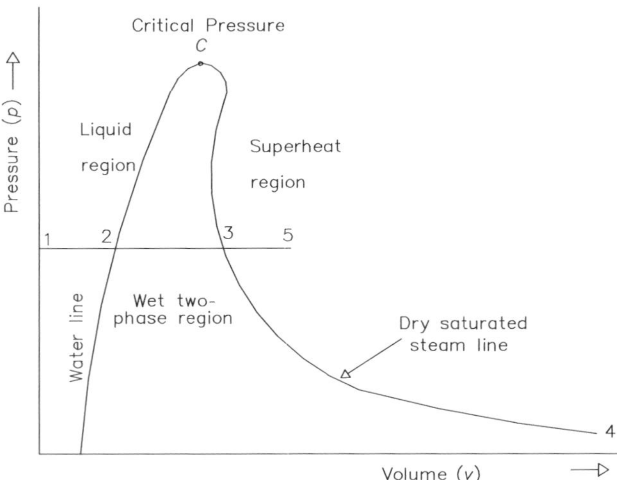

FIG. 1.7 Pressure-volume diagram of steam.

2. An industrial or process plant, for example, textile, bleaching, steel, etc.
3. Heating steam as in an HVAC system.

Boilers are designed to transmit heat through the burning of fuel (e.g., coal, oil, natural gas, etc.). The basic requirements to be satisfied are:

1. Safe handling of water.
2. Safe handling and delivery of steam at the desired quality and quantity.
3. Efficient heat transfer from an external heat source.
4. Ability to cater to large and rapid load changes.
5. Minimum leakage.
6. Minimum refractory material use.

2.2.7.3 Boiler Classifications

Boilers are classified mainly by:

1. Utilization.
2. Tube content, shape, and position.
3. Furnace position and firing.
4. Heat source/fuel type.
5. Circulation of water.

2.2.7.3.1 Use Boilers are primarily stationary and mobile. Stationary boilers are used for:

1. Power plants.
2. Utility or process plants.
3. HVAC plants.

Mobile boilers are used for:

1. Marine vessels.
2. Locomotive engines.

2.2.7.3.2 Tube Contents There are two types of tubes: fire and water. Fire tubes contain hot gases inside tubes surrounded by water. These types are of limited use. Water tubes contain water and steam inside the tube with surrounding hot gases. All large plants have this type of boiler. Tubes may be bent, straight, or sinuous and can be positioned in a horizontal, vertical, or inclined way.

2.2.7.3.3 Furnace Position and Firing A furnace can be externally or internally fired. For an internally fired system, the furnace region is completely surrounded by water tubes (also called *water walls*). The firing system may be front-fired, opposed-fired, downshot, corner-fired, etc.

2.2.7.3.4 Heat Source A heat source may be the combustion of:

1. Solid fuels such as coal, ignite, bagasse, etc.
2. Liquid fuels such as high-speed diesel oil, fuel oil, coal tar, etc.
3. Gaseous fuels such as natural gas, hot waste gas as a byproduct of some other plant, etc.
4. Electrical energy.
5. Nuclear energy.

2.2.7.3.5 Forced or Natural Circulation Circulation of water in a majority of applications is done naturally where a natural convection current is produced by applying heat. In forced circulation systems, separate pumps are provided for complete or partial circulation. The Rankine cycle (complete expansion cycle) is considered the standard cycle for comparing steam power plants comprised of boilers, turbines, condensers, etc. (Fig. 1.8). Fig. 1.9 illustrates the

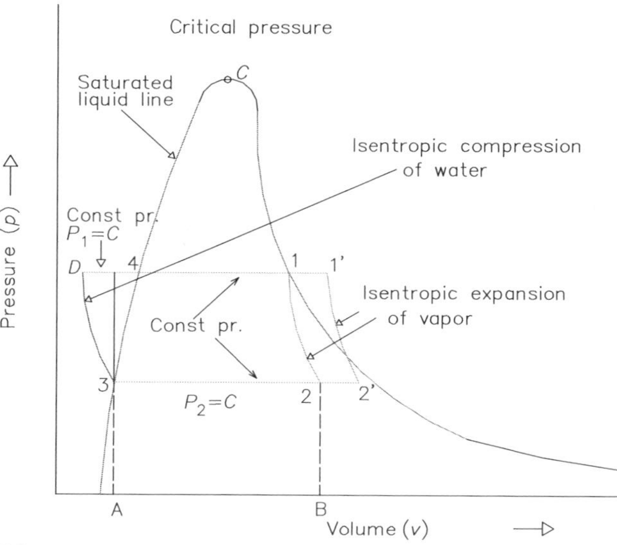

FIG. 1.8 Pressure-volume diagram of steam in a Rankine cycle.

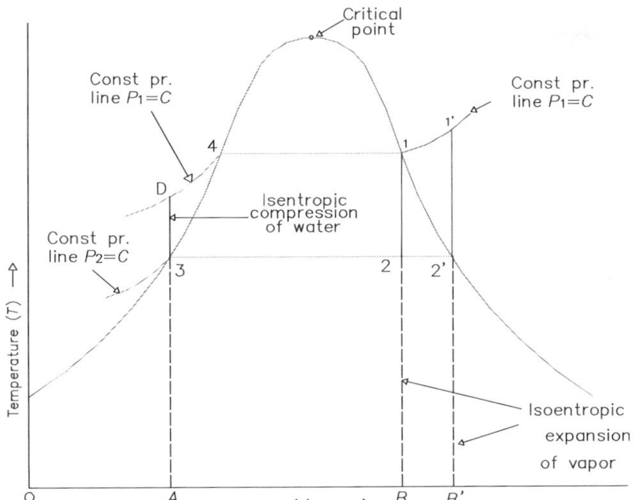

FIG. 1.9 Temperature-entropy diagram of steam in a Rankine cycle.

process with the various components of steam power plants on both *p-v* and *T-s* plots for unit mass.

The boiler delivers steam at point 1 as dry saturated steam or at point 1′ as superheated steam and then to the turbine with the assumption of no heat loss due to transportation through pipelines. The steam expands isentropically in the ideal engine (turbine) to point 2 or 2′. After that, the steam passes to the condenser without any heat loss between the turbine and condenser. Steam at point 2 or 2′ is condensed to completely saturated water at point 3 at pressure p_2. This saturated water is compressed isentropically to pressure p_1 represented by the process path 3D by different pumps. From this, the boiler receives water at pressure p_1 but at a lower temperature, and then heat is added to raise the temperature at T_4 and further transforms it to steam at constant pressure (and temperature). It is clear now that D-4-1 (or 1′ for superheated steam) is the process carried out in the boiler.

When an ideal engine receives steam at higher pressure and rejects it at lower temperature after isentropic expansion, the efficiency would refer to the engine alone; this efficiency is called the Rankine engine efficiency.

2.3 Regenerative Cycle/Heater/Extraction System

2.3.1 Regenerative Cycle

The regenerative cycle is illustrated in Fig. 1.10. Before going to the boiler, the condensate, also known as feed water (FW), after the boiler feed pump (BFP) discharge is heated at various points to avoid irreversible mixing of cold condensate with hot boiler water, which causes loss of cycle efficiency. Various methods are adopted to do this reversibly by interchange of heat within the system, thereby improving cycle thermal efficiency. This method is called *regenerative feed heating* and the cycle is called the *regenerative cycle*. This is implemented by extracting or bleeding small quantities of steam from suitable points throughout the turbine stages utilizing the heat contents of an extracted or bled steam. The vessels where the exchange of heat takes place are called *heaters*. Here, the steam totally condenses in the heater shell and is allowed to pass to the next lower pressure heater shell to maintain its own level and to prevent ingress of water into the turbine from the high level in the heater (TWDPS). The outlet water leaves the heater with a higher temperature than the inlet water.

In different cylinders or turbine stages, a numbers of extraction outlets are used for regeneration or heating FW through a number of heaters with a suitable temperature and pressure—gland steam coolers (GSC), low-pressure heaters (LPH), and high-pressure heaters (HPH)—to ultimately match the boiler FW inlet temperature. Extraction steam is also provided from the turbine for deaeration of FW and in many plants for a separate BFP driven by a steam turbine in addition to a motor-driven feed pump.

The condensate from the condenser hot well first passes through the GSC to gain heat or temperature and then proceeds to the steam ejector (or a vacuum pump) to gain further heat/temperature (not shown in Fig. 1.10).

In GSCs, all the gland steams are collected from glands provided at different casings of the turbine to prevent leakage of pressurized steam to the atmosphere in high-pressure stages and air into the turbine in subatmospheric pressure stages. The heat contained is utilized for condensate heating. An ejector is provided to such air ingress in the condenser to help maintain the vacuum therein by ejecting steam at a very high velocity. Both these vessels get the initial steam from the auxiliary steam (AS) header at no load or a low-load condition of the turbine and switch over to cold reheat (CRH) steam or extraction steam as necessary.

LPH 1 is normally installed in the steam chest between the low-pressure turbine (LPT) exhaust and the condenser to reduce the load on the condenser and heat gained by the condensate after leaving the ejector.

LPH 2 gets condensate from the LPH 1 outlet and extraction steam from LPT at a slightly higher pressure called Ex 2. Similarly, LPH 3 receives condensate from the LPH 2 outlet and extraction steam from the LPT at a pressure higher than Ex 2 (called Ex 3). Next comes the extraction steam for the deaerator from the intermediate-pressure turbine (IPT) exhaust, which is called Ex 4 or the fourth extraction. It serves two purposes: heating of condensate from the LPH 3 outlet and a very important service called deaeration of condensate. In some power plants, after LPH 3, there is another LP heater (LPH 4), which then receives steam from Ex 4. The deaerator then receives steam from the IPT exhaust, which is called Ex 5 or the fifth

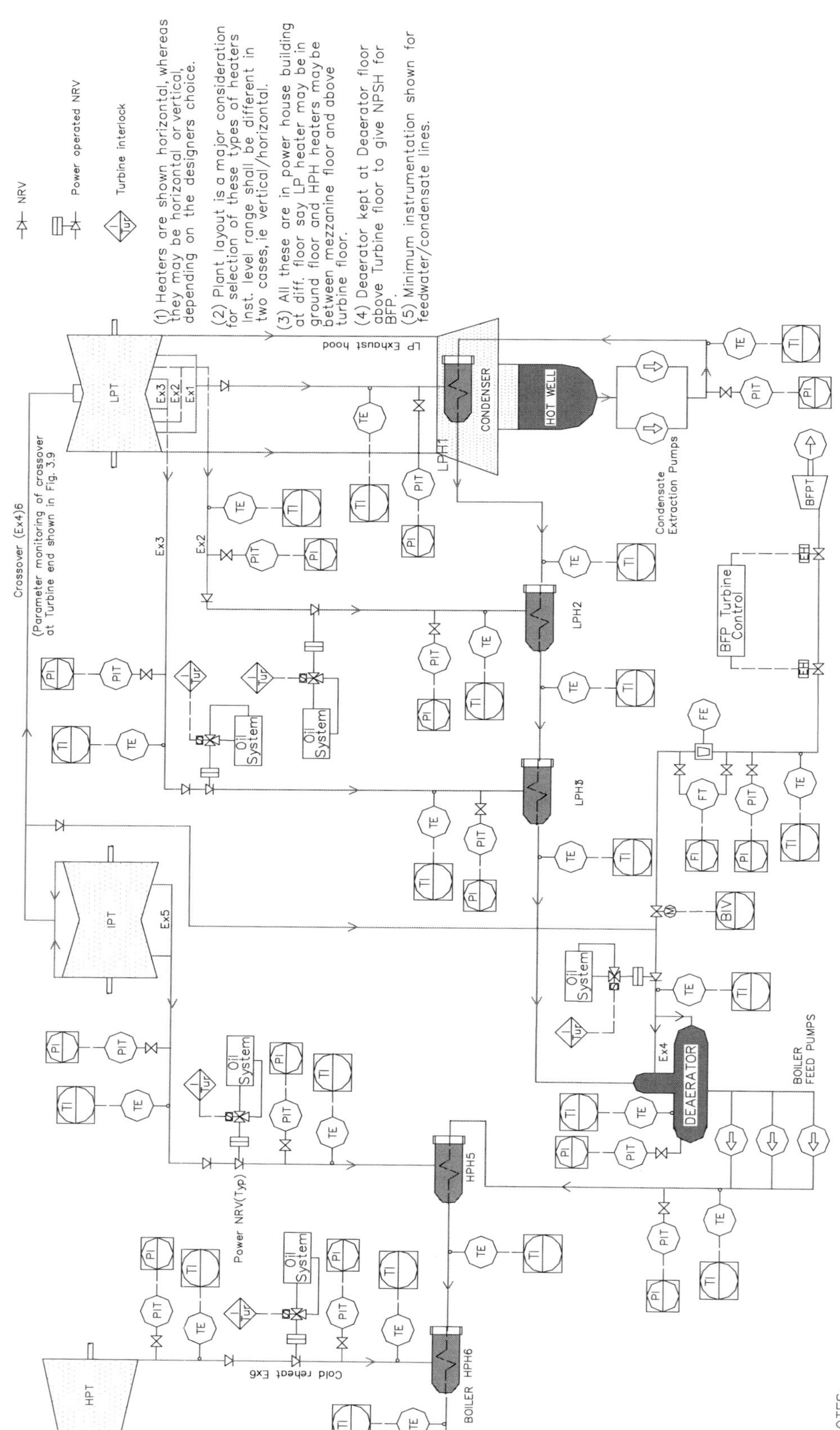

NRV
Power operated NRV
Turbine interlock
(1) Heaters are shown horizontal, whereas they may be horizontal or vertical, depending on the designers choice.
(2) Plant layout is a major consideration for selection of these types of heaters Inst. level range shall be different in two cases, ie vertical/horizontal.
(3) All these are in power house building at diff. floor say LP heater may be in ground floor and HPH heaters may be between mezzanine floor and above turbine floor.
(4) Deaerator kept at Deaerator floor above Turbine floor to give NPSH for BFP.
(5) Minimum instrumentation shown for feedwater/condensate lines.
Crossover (Ex4)6
(Parameter monitoring of crossover at Turbine end shown in Fig. 3.9
LPT
Ex3
Ex2
Ex1
LP Exhaust hood
CONDENSER
HOT WELL
LPH1
Condensate Extraction Pumps
BFPT
BFP Turbine Control
LPH2
LPH3
Oil System
FE
FT
FI
PIT
PI
TE
TI
IPT
Ex5
BIV
Ex4
DEAERATOR
BOILER FEED PUMPS
HPH5
HPH6
HPT
Cold reheat Ex6
Power NRV(Typ)
TO BOILER
NOTES:
(a) In LP heaters dotted coils shown for condensate in HP heaters dotted coil shown for feedwater.
(b) In addition, heating aerator has another function to deaerate.

FIG. 1.10 Extraction steam/regenerative cycle/flow/schematic diagram.

extraction. After the deaerator, the condensate goes to the BFP or boiler feed.

Booster pump suction may depend on the size of the plant, and has been renamed *boiler feed water*. The BFP discharge FW then goes to HPH 5 (or HPH 6), then to HPH 6 (or HPH 7), and then HPH 7 or 8 (if any) before finally proceeding to the boiler through an economizer. HPH 5 is provided with the heating steam from the intermediate extraction of IPT called Ex 5. HPH 6 is provided with the heating steam from the HPT exhaust or the CRH line called Ex 6, or the sixth extraction.

2.3.2 Various Valves and Their Operations

2.3.2.1 Main Steam Stop Valve

The boiler outlet steam passes through the stop valve before going to the consumer/end user; this is called the *main steam stop valve* (MSSV or MSV). The primary purpose of this vital accessory is to isolate the boiler by interrupting the steam circuit during start up, shut down, or in case of an emergency. Normally, this valve is motor operated. For a bigger power plant, a small bypass valve is provided to facilitate easy opening of the MSV. During start up, the pressure upstream of the MSV increases while the pressure downstream is almost zero. The differential pressure across the valve and the valve size is very high for high-capacity plants, and the operating thrust/torque required by the actuator is also very high while the valve opens from a fully closed position. To circumvent the situation, a small bypass valve, which opens first with less thrust/torque (line size is small), is provided. As the pressure downstream builds, pressure equalization takes place between the upstream and downstream side and the MSV can then open, requiring less thrust/torque. During normal plant operation, the valve remains in full open condition.

2.3.2.2 Nonreturn (Check) Valve

This valve allows the fluid to flow forward under pressure, but checks the fluid flow in the reverse direction. The valve plug moves up from the seat when pressure applied from the bottom of the plug is higher than that at the top of the plug. It will remain in this position as long as the differential pressure multiplied by the plug area is higher than the spring force applied to the plug to keep it in the shut-off position. In the reverse condition, when pressure downstream (top of plug) is higher than upstream (bottom of plug), the plug moves down by the force of the differential pressure aided by the spring force and sits tightly on the seat to arrest any flow. Nonreturn or check valves are provided in every flow path, irrespective of steam or water service, wherever there is a chance of return flow under any operating condition. The valve is normally self-actuated—that is, no external power is required. In the extraction line, one check valve is a TWDPS requirement. In some instances, these may be power-assisted.

2.3.2.3 Start-Up Vent Valve

This type of valve is in the main steam header and, as the name implies, is required for the start-up period only. The valve regulates service, and through it steam is allowed to vent out of the system to the atmosphere, as required automatically or manually for purging and/or heating of the pipelines. During the initial start-up stage, the line is drained with the help of the startup drain valve, similar to what was discussed previously. These valves are generally motor-operated.

2.3.2.4 Safety (Pop-Up) Valve

These valves are of immense importance to the safety of the plant and its personnel. Whenever there is a pressure build-up in the pipeline beyond the limit, the valve should operate or pop up to release steam to the atmosphere until the pressure comes down to a safe value. Although a loss of energy and mass of working fluid occurs, it is inevitable during any untoward situation rendered uncontrollable by a normal control system. There are various types of safety valves: electromatic (or relief), spring-loaded, dead weight, fusible plug, etc.

2.3.2.5 Electromatic Safety (or Relief) Valve

This is a pilot solenoid-operated valve that is energized automatically from the pressure switch at a very high set point, which allows working fluid to operate the actuator of the safety valve. It can be operated through remote manual command as well from the control room/tower.

2.3.2.5.1 Spring-Loaded Safety Valve Normally, this valve operates as a last resort to the safety system against high pressure. Under normal plant operation, the spring tension is high enough to hold the valve plug on its seat to ensure a closed position until a very high pressure set point is reached. At this point and above, the force against the spring lifts the plug over its seat to allow extra steam to escape, unless the steam pressure comes down to a normal value. The discharge capacity should be selected so that it is equal to the evaporative capacity to avoid frequent build-up of pressure (actuation of this valve). Other types of safety valves are no longer in use, hence, they are not discussed.

2.3.2.6 Blowdown Valve

This type of valve removes sludge, sediment, and other impurities collected at the bottom-most location in the water flow path. It also helps to drain the system completely. There are two types of blowdown valves: continuous and intermittent.

2.3.2.6.1 Continuous Blowdown Valve This valve opens continuously to maintain the dirty material level at a minimum value. The opening of the valve is varied as per requirements with a predefined control signal. The motorized actuator can also receive manual commands; the method of control is the operator's prerogative.

2.3.2.6.2 Intermittent Blowdown Valve This type of valve blows down dirty water as necessary. Its operation may be predefined, based on a cyclic or time-framed full open/close signal or a manual command from the operator with a motorized actuator. Boiler drum conductivity may be one of the parameters to operate this valve in automatic control—not uncommon in medium- to large-size boilers in utility stations.

2.3.2.7 Drain Valve

During the start up of the plant after a prolonged shutdown or a cold start up, the pipelines and various equipment need to be warmed before loading the boiler. To achieve this, the heating steam is admitted phase by phase in a very slow manner to avoid dissimilar heat causing an expansion of various casings and pipes. While heating metal works, the steam gets condensed and collected at the bottom of the pipeline with a siphon-type design at various strategic locations. At the bottom, condensed water is drained through this valve with a motorized actuator when the level in the drain pipe reaches a predefined value to avoid frequent operation. Level switches are provided for automatic operation. The operator is provided a manual command.

2.3.3 Steam Trap

This type of element drains out the condensed water from steam pipes and jackets used for heating, thus resulting in partial condensation, and simultaneously arresting the steam inside from escaping (hence the name). Generally, there are two types of steam traps available: float or bucket and thermal expansion. The operation is self-contained and mechanical; it does not require any external power, therefore, it is not discussed further.

2.3.4 Steam Separator

As the name implies, it separates water particles suspended in generated steam from the boiler and carries the flow of steam to the turbine or engine. To work properly, it should be located far enough away from the steam generator to separate water particles from the steam for most of the transportation line. A drum-type boiler is in the drum where the water particles drop into the water section.

The steam path of the steam separator is guided by a series of baffles. The water particles are heavier and have greater inertia. Because of this, after striking the baffles they fall by gravitational force to the bottom of the vessel. The dry steam is practically unaffected and gets transported out. The collected water is then drained through the drain line.

2.4 Reheat Cycles in Utility Boiler—Hot and CRH Lines

2.4.1 Reheat Cycle in Utility Boiler

The steam from the high-pressure turbine (HPT) outlet or exhaust is returned to the boiler to reheat the steam at a temperature (generally) equal to the original main steam temperature (Fig. 1.11). Reheating is done to avoid wet steam in the turbine blade, which causes erosion because of water particles in the wet steam. The international standard moisture limit for steam in a turbine is ~10%. Water particles in the turbine are against TWDPS for the preceding reason.

The modern power plant concept is based on multiple turbine cylinders with the rotors coupled to a single shaft. Figs. 1.11 and 1.12 show high-pressure steam, popularly called *main steam*, from the boiler that enters the HPT where it is isentropically expanded from pressure p_1 to p_2 (path 1-2 in the T-s diagram) and removed as high-pressure exhaust. This is normally called *cold reheat*.

This high-pressure exhaust steam is then readmitted to the reheater part of the boiler for reheating, after which the changed steam is at a temperature equal to the main steam and a pressure equal to p_3.The reheated steam from the boiler is known as hot reheat (HRH). It then reenters the turbine intermediate-pressure (IP) cylinder and expands isentropically up to pressure p_4 (point 4 in the T-s diagram). The temperature is maintained at the outlet of the reheater header by providing heaters at various stages, but before entering each heater there are desuperheaters (spray type). These help to avoid overheating and a rise in temperature of the reheated steam to achieve precise control of it by spraying adequate water through control valves. However, in practice reheater temperature is controlled primarily by the burner tilt for tangential firing and by operation of a bypass damper or readmission of cold flue gas near the furnace hopper in front/opposed/downshot fired boilers. Spray control is used mostly in emergencies by setting its controller set point slightly higher than the normal reheat temperature controller set point.

By reheating, more work is done, that is, more output from the turbo generator as shown in the T-s diagram (Fig. 1.12). Without reheat, the steam cycle would follow the path 1-2-5_{Sup}; however, with reheating, the path followed is 1-2-3-4-5_{Rh}. The extra work done is in the area vertically under line 2-3, that is, the area enclosed by the path 2-3-5_{Rh}-5_{Sup}-2. In modern and/or high-capacity power plants, twin cylinder IP modules (as shown in Fig. 1.11) are used for various reasons, although there was only a single cylinder in an earlier design.

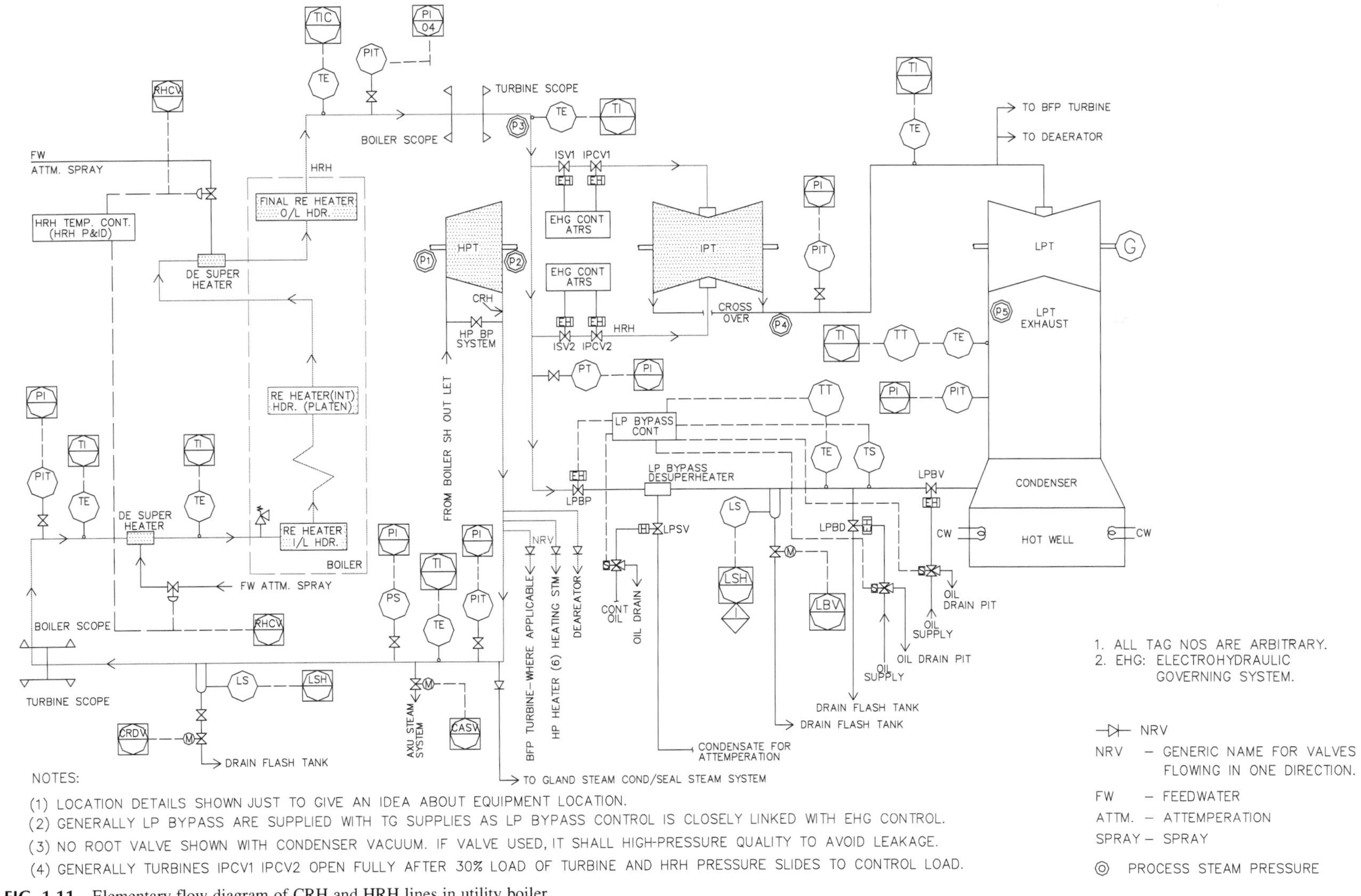

FIG. 1.11 Elementary flow diagram of CRH and HRH lines in utility boiler.

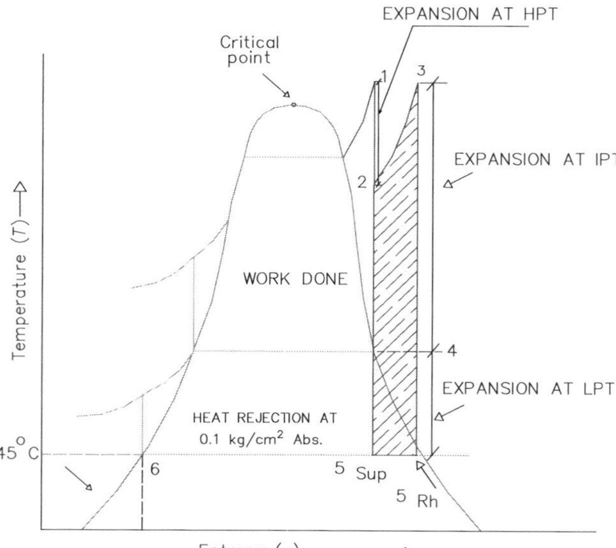

FIG. 1.12 Reheat steam cycle in modern power plant in *T-s* axis.

The CRH header at the HP turbine outlet supplies steam to the following plant components:

1. Heating steam for HPH 6.
2. Turbine-type drive of BFP (if any) during start up.
3. Gland seal system.
4. Deaerator during start up.

There is another important system, the HP bypass system, which enables the main steam to bypass the turbine to meet the CRH line after suitable temperature and pressure conditioning. If there is turbine tripping (or outage), the steam generation of the boiler cannot stop immediately. As boiler start up is a very time-consuming process, steam generation is kept uninterrupted and diverted to a bypass line by closing the main steam line isolating/regulating valves. This also happens during turbine start up. There is a pressure-reducing valve (PRV) that reduces the steam pressure equal to a simulated pressure set point generated from the control system. After pressure reduction, the steam is cooled by spray water from the FW line through a control valve; thus, a simulated CRH steam condition is generated. The spray water pressure is also regulated so as to prevent excess water and over pressure in bypass line (Fig. 3.4). This is necessary because the boiler is running at the existing load and the reheater must get steam at the CRH condition to keep from overheating.

2.5 Gas Turbine Types (Frames)/Black Startup

A gas turbine (GT) is a type of internal combustion engine, also called a *combustion turbine*. Theoretically, basic GTs can be categorized into two classes: open and closed cycle. An open-cycle turbine has its main operating fluid and gas/air taken from the atmosphere and returned to the atmosphere after residual heat rejection. In this turbine, fuel is burned with the help of air within the system and combustion products along with the rest of the air form the working fluid. Some part of the exhaust gas may be retained to preheat the inlet air and balance the gas allowed to return to the atmosphere. This system is called a *semiclosed cycle*.

In a closed-cycle system, the main operating fluid (e.g., steam in a steam power plant) is not allowed to leave the system and the transfer of heat (work between the system and surroundings) takes place. In other words, it may be categorized as a "hot air engine." Fig. 1.13 illustrates this type of mechanical system.

The GT's operating principle, for either an open- or closed-cycle type, is based on the thermodynamic cycle known as the Brayton cycle. In this type of system, the atmospheric air is the operating fluid, which is compressed to accommodate a sufficient amount of air in a given limited-volume combustion chamber to assist in the burning of fuel. When combustion takes place—that is, energy is added to the gas stream in the high-pressure environment of the combustor—the air quickly heats to a high temperature and tends to expand abruptly. The products of combustion are then forced into the turbine section, guided through a set of nozzles mounted throughout the periphery of the rotor, and passed over the adjacent turbine blades. The expansion takes place within the turbine at various stages. The consequent high velocity of the gas flow causes the turbine to spin, which powers the compressor and shaft with mechanical energy. The thermal energy transferred to the turbine comes from a reduction in temperature and pressure and comes out as exhaust gas. The total work developed at the turbine by expansion after subtraction of the amount consumed by the compressor (which normally ranges from 50% to 60%) is then available for power generation. In a Brayton cycle, the total work developed is proportional to the absolute temperature of the operating fluid passing through a device such as the turbine. It is therefore a natural choice to operate the turbine at the highest operable

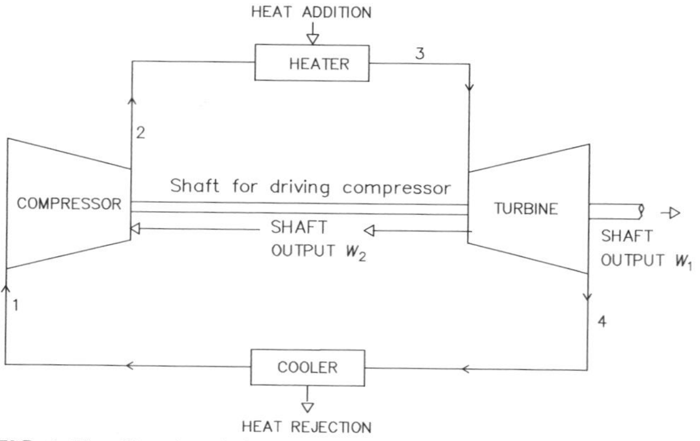

FIG. 1.13 Closed cycle in GT schematic diagram.

temperature within the limits posed by the metals subjected to these temperatures. For this purpose, the inlet blades are arranged to be cooled by air or steam (if working in a combined cycle with a heat-recovery steam generator, HRSG) and will be discussed briefly in the latter part of this chapter.

With the tremendous development of GT technology and other factors, the closed-loop cycle has become obsolete and a subject of theoretical interests only. However, the heating of clean air, and it acting like working fluid instead of hot products of combustion, is still in use, but in an open-loop configuration only. The system is briefly discussed in Section 2.5.9. There is also research that is developing a closed-cycle GT based on helium or supercritical carbon dioxide as the working fluid and utilizing nuclear/solar energy as the heat source.

For a smaller engine, the general GT design criterion is that the rotational speed of the shaft must be high enough to maintain blade-tip velocity because the maximum pressure ratio achieved by the GT compressor depends on the blade-tip velocity. The maximum power and efficiency is in turn proportional to the maximum pressure ratio of the engine. To summarize, the maximum powers (close to its own rating) and efficiencies of various machines can be attainable if the blade-tip velocity is constant, which means if the diameter of a rotor is doubled, the rotational speed must be half its previous value. For example, if a very large jet engine operates at around 10,000 rpm, a comparatively small GT has to run at a much higher speed, such as 100,000 rpm, to attain its maximum (near rated) output and efficiency.

GTs are used in many fields: aviation, power generation, marine vessels, and even in road transport systems. With the advent of aircraft, the GT is broadly classified and developed as a jet engine. When used as an engine for aircraft, GTs are generally called jet engines, not GTs. In this type of GT, the available energy left after driving the compressor and associated components, in the form of high-pressure gas and a huge volume of atmospheric air, is allowed to accelerate, which provides the formation of jet and, consequently, the thrust necessary for desired aircraft operation. Two basic types of jet engines are presently available in aviation technology.

Jet engines optimized to produce thrust from direct impulse of the exhaust gases are called "turbojets." The other type of jet engine has a large fan (driven by the GT shaft) at the air intake of the engine, which supplies a huge amount of air to produce extra thrust in addition to thrust produced by the exhaust gases. This is called a "turbofan" or "fan jet." However, further discussion about aviation engines is beyond the scope of this book.

GT thermal efficiency is lower than the thermal efficiency of comparable diesel/reciprocating engines. Thermal efficiency of GTs within a 30 MW rating varies from 35% to 40%. This fact results in ~20% higher fuel consumption in a GT working in single-cycle (without heat recovery) mode than that of a comparable reciprocating engine. GT thermal efficiency is proportional to GT output; hence, the thermal efficiency of small GTs, for example, within a 5 MW capacity, normally is not >30%. As far as the capital cost is concerned, the initial investment for a GT engine within a 30 MW range is ~20% higher than in reciprocating engines of similar rating.

In the electric power generation field, two modern basic categories have emerged: heavy duty industrial GTs, which are specifically designed for stationary duty, and aeroderivative GTs, which are derived from jet engines. Industrial GTs are different from aeroderivatives both in structure and in service, but both are used in electric power generation. For the industrial GTs, the frames, bearings, and different stage blades are heavier compared with aeroderivative GT blades. The size of industrial GTs varies widely, ranging from small mobile plants to large power-generating plants.

Heavy-duty industrial GTs are also known as frame GTs. These are meant entirely for the station-mounted electrical power-generating units with typical average efficiencies of 40%, if installed as a standalone unit.

The efficiency may increase to the typical figure of 60% when waste heat with a very high temperature from the GT exhaust is utilized by an HRSG to power a conventional steam turbine. This is widely known as a combined cycle plant. The waste heat can also be recovered in other ways, such as by heating, cooling, or refrigeration through suitable equipment in a cogeneration configuration.

The industrial GT has been designed to extract power from the shaft to drive an alternator or AC generator. Normally, in this system, the GT exit pressure is kept very close to the atmospheric pressure with some margin to enable the hot exhaust to reach the desired destination. The typical compression ratio in this type of GT is 16:1. The electrical power output capacity from the GTs typically ranges from 40 to 350 MW. It is best used as a base load power plant for continuous running.

Lower capacity GTs, usually up to 40 MW or below, may be used for power generation as well as for a mechanical drive, for example, compressors for long-distance gas pipelines, air compressors for blast furnaces, for maintaining well pressure in the petroleum industry, or to enable different process plants to work at an elevated pressure environment.

The output power can be extracted from the GT in many forms, such as shaft power to drive trains, ships, etc. The exhaust gas pressure is similar to the atmospheric pressure discussed previously.

The advantages and disadvantages of industrial GTs include:

1. Rugged, less expensive, less maintenance time, more availability, fewer intervals between overhauls.

2. Less efficient, heavy structure.

The aeroderivative GTs are naturally lightweight and thermally efficient with a decreased start-up time. The compression ratio may be raised to 30:1, compared with 16:1 for industrial GTs. The capacity rating of aeroderivative GTs is available up to 50 MW. This type of machine is slightly more efficient and more costly than the standard industrial GTs. Aeroderivative GTs, although expensive, are used in electrical power generation to take up variable load because of their quick start/shut down; they also handle load changes more smoothly than industrial machines. Because they are lightweight, they are used in marine vessels.

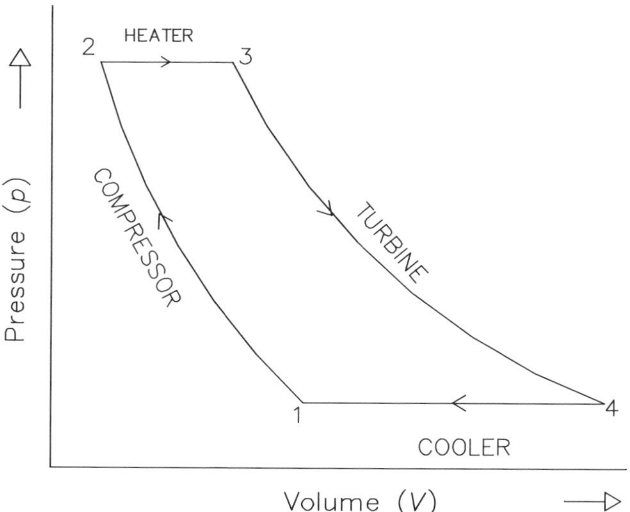

FIG. 1.15 *p-v* diagram of basic closed cycle in GT.

2.5.1 GT Basic Closed Loop Cycle

Figs. 1.14 and 1.15 depict the working principle of this type of GT. Gases passing through an ideal GT undergo three thermodynamic processes. Point 1 denotes the entrance of cold gas/air to the compressor and work is done on the system to raise the pressure of the system, which may be considered isentropic compression. Point 2 denotes the starting of heat added to the system while passing through the heater, thus raising the temperature. This is considered isobaric (constant pressure) combustion. Point 3 denotes entry to the turbine where expansion of gas/air takes place ultimately to a lower exhaust pressure as isentropic expansion and shaft power develop. Part of it is used to run the compressor and the rest is used to run the generator. Point 4 denotes the entry of hot exhaust to the cooler where heat is rejected to reach the initial condition (point 1).With the above assumptions and design parameters, the process may be conceived as an ideal and reversible cycle.

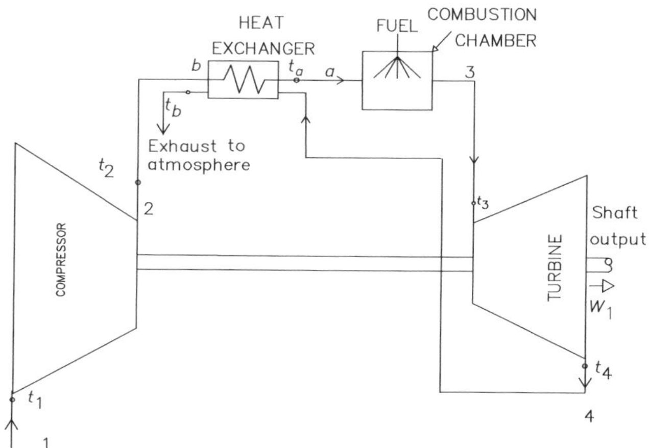

FIG. 1.16 Schematic diagram of GT cycles with heat exchangers.

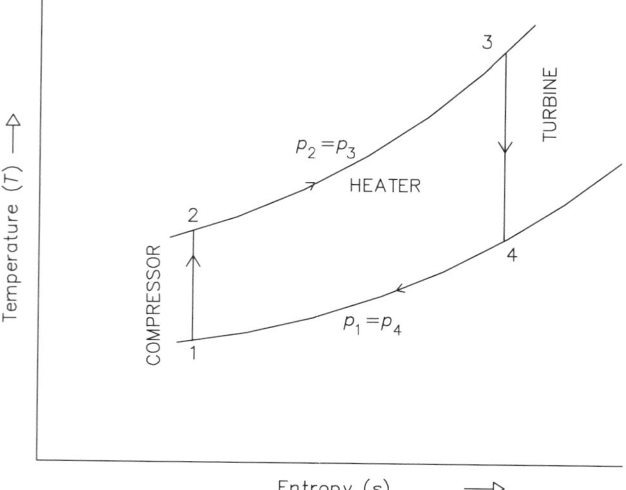

FIG. 1.14 *T-s* diagram of basic closed cycle in GT.

2.5.2 GT Basic Open Loop Cycles

2.5.2.1 GT Cycles With Heat Exchangers/ Regenerator

In this type of GT, the fuel-saving system is envisioned as seen in Fig. 1.16. The hot exhaust gas is utilized to preheat the cold compressed air by passing through a heat exchanger before going into the atmosphere. With a suitable design, it is possible to raise the temperature of the cold compressed air from t_2 to $t_a=t_4$ and lower the temperature of gas leaving the turbine t_4 to $t_b=t_2$, as shown in Figs. 1.16 and 1.17.

Therefore, it is apparent that the heat transfer has been taking place at each interval of the heat exchanger with a very low, practically negligible temperature difference. With the above assumptions and design parameters, the process may be conceived as an ideal and reversible cycle. The external heat would be less than the amount rejected by the exhaust gas at the heat exchanger.

2.5.2.2 GT Cycles With Intercooling and Reheating

A system with a regenerator improves the cycle's thermal efficiency, but not the work ratio. The work ratio would only be improved by decreasing the compressor work, increasing the turbine shaft work, or both at the same time.

From Figs. 1.18 and 1.19, suppose the process starts at the compressor inlet with atmospheric pressure and temperature at point 1 to point 5 if compression work takes place at a single stage. If the compression work is done at two stages, then the path will be 1-2 and 3-4 instead of 1-5 with cooling taking place at an intermediate constant pressure, p_x. From the series of constant pressure lines, it is evident that the vertical distance between any two such lines on the left side is less than the right side distance. So, the vertical distance 3-4 is less than that of 2-5; the work done on the two-stage compressors is less than that of a single-stage compressor.

By suitable design, it may be possible to cool the second-stage cooler inlet temperature at atmospheric condition, that is, $T_3 = T_1$.This is called perfect intercooling. If the reheat part is analyzed, the vertical distance to the right is more than those with less reheating, which means work done by the turbine with reheating is more than that without reheating. From the diagram it is also clear that the total area (work done) with intercooling and reheating is more than that without them. The shaded areas in Fig. 1.19 are the additional work available by using the above-mentioned plan.

2.5.3 GT With Single and Double Shaft (Turboshaft)

Generally, GTs can be classified as a single-shaft or double-shaft configuration. A brief description of these two types is incorporated in the following sections.

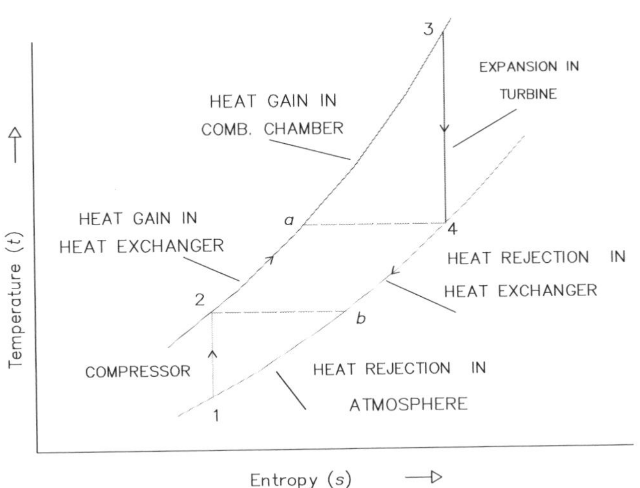

FIG. 1.17 GT cycle in *T-s* axis with heat exchanger.

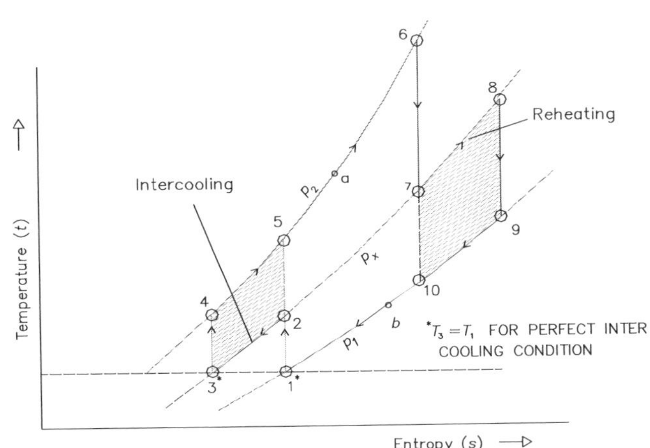

FIG. 1.19 GT cycles with intercooling and reheating in *T-s* axis.

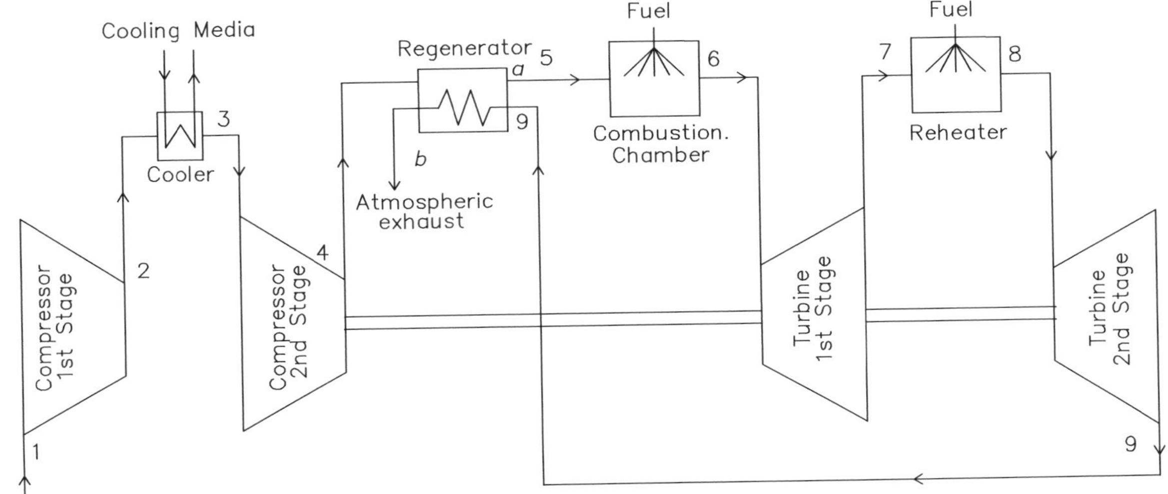

FIG. 1.18 Schematic diagram of GT cycles with intercooling and reheating.

2.5.3.1 GT With Single Shaft

A single-shaft GT is normally used when the connected load does not need significant speed variation during the operation range, for example, the alternator or generator. In this configuration, the compressors, along with the turbine and generator, are connected as a single continuous shaft, which means all would be running at the same speed.

2.5.3.2 GT With Double Shaft (Turboshaft)

In this configuration, the turbine part is mechanically separated into two parts: an HPT and an LPT. Here, the compressor rotor and the HPT form a common shaft. The work developed at HPT (also a high speed turbine; often referred to as a "gas generator" or "compressor turbine") by converting thermal energy through kinetic energy is utilized solely to drive the compressor rotor. On the other hand, the LPT (also a low speed turbine; often called the "power turbine" or "free-wheeling turbine") is coupled with the output shaft, which may be a generator or any other load, such as an aerodynamic drive, etc.

The two turbine shafts are separated mechanically but are "aerodynamically" coupled by the hot gases exiting from the compressor turbine/gas generator and entering the power turbine. This type of GT is often called a "turboshaft engine," which is used to drive compression trains such as gas-pumping stations, natural gas liquefaction plants, marine vessels, etc. This configuration is used to increase speed and power output flexibility. The design of modern helicopters often utilizes the application of this type of GT, and it is preferred because the compressor turbine/gas generator turbine spins separately from the power turbine.

The advantages of a GT with double shaft include:

1. The speed of both turbines can be varied to meet the prevailing demands independent of each other.
2. The starting torque for the load requirement is less than the single shaft as the power turbine is mechanically decoupled from the compressor turbine.

2.5.4 GT Firing Temperature and Pressure Ratio

The two most important parameters for determining the characteristics of a system in GTs based on the Brayton Cycle are the turbine firing temperature and the pressure ratio. With the advancement of technology, the modern trend is to operate the turbine at a higher firing temperature and pressure ratio. The combined effect of these two factors is responsible for higher efficiency and specific power, turbine exhaust temperature of the GT, etc. The higher exhaust temperature may range from 425°C to 600°C for small to larger GTs, respectively. Such high exhaust temperature allows further use of this heat energy through the HRSG in a combined cycle plant.

2.5.4.1 Turbine Firing Temperature

The firing temperature is the highest temperature attained in the system; it is called the turbine inlet temperature or first-stage nozzle outlet temperature. As previously stated, this temperature is proportional to the total power developed by the GT; hence, the higher temperature is a natural trend for a GT.

Restrictions of higher firing temperature and solutions are as follows:

1. Oxides of nitrogen (NO_X) are part of the pollutant gases emitted by the GT. They depend on the combustion temperature (similar to the turbine inlet or firing temperature), so care is taken to restrict them. There are certain methods of firing control by regulating air flow (primary and secondary air, SA) at different locations in the combustion chamber, which assist in maintaining lower temperature.
2. The other restriction for achieving higher temperature is the operating limits of the wetted parts of the turbine materials used. Technocommercial consideration of the metal selection dictates the temperature.

This problem is also solved by cooling the first-stage nozzles so that the products of combustion entering the turbine blades cool after leaving the nozzle trails. For GTs without HRSG or meant for aerodynamic service, this cooling system employs air injection through the ports of the hollow nozzle walls. The pressurized cooling air enters the hot gas stream and instantly mixes to reduce the entrant gas temperature.

For GTs with HRSG (as in a combined cycle plant), the nozzle cooling utilizes the steam as a cooling medium. In this method, dilution of hot gas with air mixing does not take place. Instead, a fraction of comparatively low-temperature CRH steam from the steam turbine is diverted to travel inside the hollow portion of the adjacent nozzle walls in a closed loop. It exits out back to join the HRH steam line after gaining heat in the IP section of the steam turbine for further cyclic requirement. This type of scheme is illustrated in Fig. 1.20.

2.5.4.2 Turbine Pressure Ratio

This parameter is the ratio of turbine inlet and outlet pressure, which should be the same as the compressor outlet and inlet pressure in an ideal cycle, although the actual pressure ratio is less than what it should be because of loss in the combustion system. The minimum GT pressure ratio is $7\,kg/cm^2$ for even the smallest available set. For larger machines, the ratio is higher.

2.5.5 Various Sections of GT

Like reciprocating/piston engines, GT engines take the same four steps to accomplish their task, but in four distinctly different sections as stated below (see Fig. 1.21):

1. Inlet (air) section.
2. Compressor section with diffuser.
3. Combustion section (combustor).
4. Turbine (and exhaust) section.

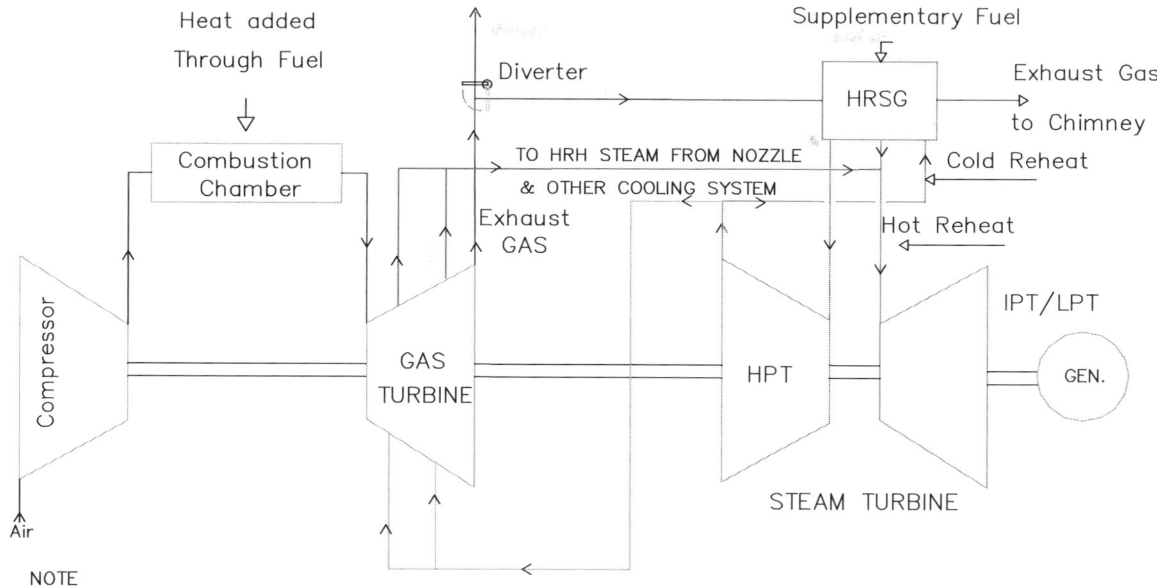

FIG. 1.20 Schematic diagram of combined cycle plant with GT nozzle cooling.

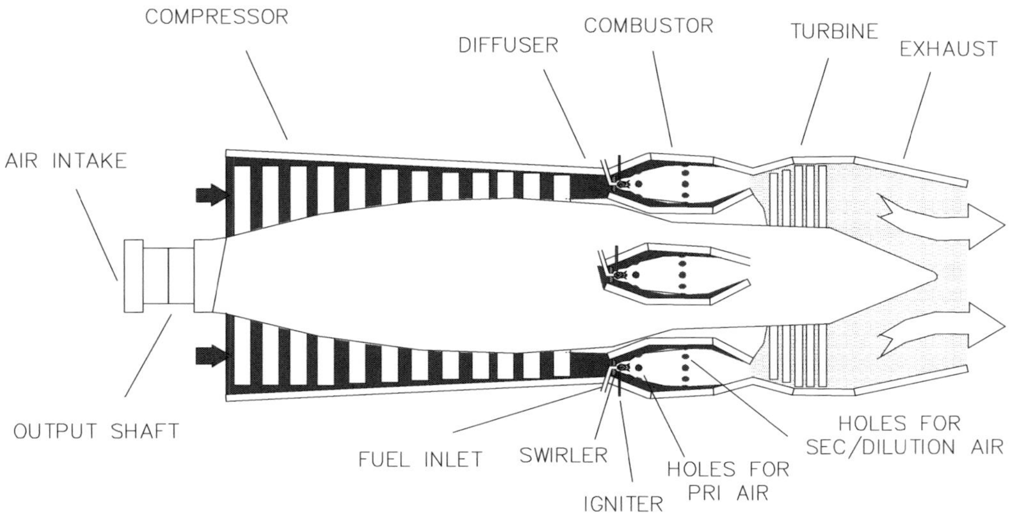

FIG. 1.21 Typical location of different sections of a GT.

2.5.5.1 GT Inlet Sections

A considerable mass of air must be supplied to the turbine for the complete system to properly function. This mass of air is supplied by the compressor through the inlet section. It is essential that the air inlet section (ducting) must provide clean, smooth, and continuous air flow to the heat engine to ensure long engine life by preventing erosion, corrosion, and mechanical damage to different GT parts. Mechanical damage may occur through inadvertent suction of small engine parts (nuts and bolts and washers, etc.) and even flying creatures like bats, birds, etc.

2.5.5.2 GT Compressor Section With Diffuser

In a GT (which is a rotary machine), compression is accomplished through aerodynamic activity as the air passes through the different stages of the compressor, contrary to the piston engine, which uses confinement. The compressor has many stages of blades and vanes to suit the outlet pressure by using the pressure ratio. It also has inlet and outlet guide vanes for different purposes. The inlet guide vanes (IGVs) allow the required air flow to the first-stage compressor blades at the "best" angle. The purpose of the outlet guide vanes is to straighten the air so the combustor is provided air flow in the proper direction.

It is very important that the compressor is fed with smooth air flow to minimize losses due to friction, air leakage, and turbulence to achieve higher efficiency. Pressure builds up in each stage of the moving blades and stator vanes, and is added up in the last stage according to the required pressure ratio.

A diffuser is provided at the end of the compressor where pressurized air enters through the compressor outlet guide vanes. The diffuser, as a divergent duct, converts the maximum part of the velocity component of the flowing air into static pressure. This effect results in a point of highest static pressure and lowest velocity in the entire engine where the outlet of the diffuser and the inlet of the combustor meet. Apart from rendering this aerodynamic service, the diffuser section is also used for the following minimum facilities:

1. Engine structural support.
2. Mounting of fuel nozzles for combustors.
3. Support for the rear compressor bearings and seals.
4. Oil passages for the rear compressor and front turbine bearings.
5. Place for bleed air take-off (if any).

2.5.5.3 GT Combustor With Transition Section

The pressurized air from the diffuser then enters the combustor section (the combustion chamber where combustion of fuel with the presence of air takes place). There are a number of burners called liners/cans located within the annular space between the inner and outer combustion cases. They are uniformly distributed throughout the periphery of the round (sectional) combustion chamber and positioned in such a way that the flame produced must avoid direct contact with combustor metal parts. Flame formation takes place at the front end or primary zone of the combustor section with the support of primary air (PA; ~25%–30% of total air flow). The SA, the balance of total air (often referred to as dilution air), is injected after the flame-forming primary zone for various purposes:

1. To control the flame pattern and stoichiometric ratio.
2. To cool the combustor metals.
3. To dilute the temperature of the products of combustion gases at the turbine inlet.
4. To increase mass flow according to GT capacity by power requirement.

The latter part or the rear end of the combustor is called the transition section, which is a very convergent-shaped duct. It assists in accelerating the hot gas stream against the reduction of static pressure before entering the turbine section.

2.5.5.4 Turbine Section of GT

The turbine converts thermal energy into kinetic/mechanical energy by expansion of the hot and high-pressure gases to a lower temperature and pressure through many stages, each of which consists of a circumferential row of stationary vanes or nozzles followed by a row of rotating blades. Turbine stage stator vanes and moving blades are placed just in opposite order to that of the compressor. In the turbine, the stator vanes increase hot gas velocity; then part of the energy is imparted to the rotor blades.

The efficiency of the turbine depends mostly on the energy conversion capabilities of the vanes/blades and having the least amount of cross air flow around them instead of a guided and desired gas path. The turbine section of the GT engine has to drive the compressor and all engine accessories, meet the mechanical and electrical losses, and produce usable shaft power output to drive the generator or other load.

The next and last section is the exhaust through which the turbine discharges hot gas. Some amount of energy remains in the exhaust gas and is allowed to discharge in the atmosphere or to the HRSG plant or to produce jet thrust (for aviation services), depending on the design concept.

2.5.6 *Black Starting of GT*

GT plant black start-up occurs when the plant becomes isolated from the surrounding electric power system and thus must bootstrap itself into an operating state. The degree of difficulty in making a black start depends on the size of the plant and the power needed to drive all the auxiliary equipment, which must be operating before the plant can start generating electric power.

Fuel supply for the black start comes from the main station's existing storage tank, which must have sufficient stock to supply the main plant and its facility during an emergency. However, it is important that the engine of the black start facility has a separate fuel storage tank big enough for 8 h of continuous running at full load. It provides operational flexibility in the event of main supply failure and should add little to the overall plant cost.

Black starting a large GT generator with a fuel system is done using a fuel flow control valve to control the GT speed during and after start up to a predetermined limit at a speed higher than rated and higher than the normal limit speed. In a black start mode, a load rate limit is applied to this fuel flow control valve operation when the actual load change step exceeds a predetermined limit amount. It also applies the auxiliary plant loads sequentially to the GT set to a predetermined sequence after successful start up of the same.

The predetermined load limit is adjustable to the allowed load pickup limits of the GT. The predetermined load step limit differs for successive load steps in accordance with the expected size of the actual load steps. A control signal output limit is normally applied to the fuel valve control system and the control signal output limit is raised in this mode setting to permit step load increases to the turbine. Provisions are made for tracking the load step to allow reset of the fuel limit for each step and track ahead for the subsequent step. A temperature limit control is also incorporated into the fuel flow valve control.

There is a full remote control and monitoring system for the automatic alarm annunciation system, remote control, and synchronizing facility with data logging of the plant from the station as well as the local service depot. The priority would be restoration of power supplies to essential infrastructure (e.g., lighting loads, sewage works, pumping stations). In the extremely rare occurrence of a total grid failure, it is expected that the transmission system would be restored within approximately 8 h. The black start facility would provide full load within 60 s of start up. In addition, under extreme circumstances, the facility can operate for up to 72 h continuously. A suitable control system should provide an immediate and quick start, synchronization, and transient start performance of the black start plant. The connected grid system requirement should test the black start plant on a regular basis, ensuring reliable operation and testing the switching systems.

The electrical power, when finally restored to the external grid, is usually restored gradually to help secure the stability of the system. For example, disconnection of electric power for a long period during summer or winter may demand enormous amounts of power at the time of resumption for air conditioning units and heating devices.

2.5.7 Different Steps to Implement Black Start

As previously discussed, black start is the process of restoring power by bringing all the connected power plants back into operation in the absence/failure of the external electric power supply system network. To provide a black start, it is necessary that the power stations have a small diesel generating (DG) set that can be used to start larger GTs, which in turn can then be used to start the main thermal power station generators. The hydroelectric or hydel generating set can also be used to source a black start, as it requires less electric power to start up by itself. A hydel power station requires a very small amount of initial power to start (just sufficient to excite the generator field coils and open the power-operated fluid intake gates) and can put a large block of power online very quickly to allow start up. Providing such a large standby capacity at each station to facilitate a black start is extremely uneconomical, and power thus generated must be provided to other stations over the designated tie lines. The main power station with steam turbines and fossil-fueled power stations may then use the GT output power, which can supply $\sim 10\%$ of unit capacity for the BFP and forced draft (FD), induced draft (ID), and fuel-related equipment.

A typical black start sequence might be like this:

1. A small battery set starts a small DG set (or hydel generating station if available). Multiple DG sets can be used in parallel to power the emergency auxiliaries and the starting up mechanism and may be used as standby/emergency generators. They were specially designed to handle the high harmonic load. The electrical power from the DG set is used to bring the GT station into operation.
2. Essential electrical connections from the GT to other areas are resumed.
3. Power from the GT is used to start any main unit providing base load.
4. The power from the base load unit is used to restart all the other power plants in the entire system.
5. For a comparatively bigger electric grid, black start in a particular station may not be useful as it would unnecessarily take much longer to complete the process. In this situation, it would be wise to start different power stations (or "islands," each supplying local load areas) with individual black start facility and then synchronize and reconnect these stations to form the complete grid.

2.5.8 Different Systems to Implement Black Start

For simplicity, the DG set is the stored energy source for all types of black starts.

2.5.8.1 Self-Contained Black Start

A self-contained unit consists of an AC generator with GT as the prime mover, and is installed with all the auxiliaries required to operate independently to provide a black start. The system has the following auxiliaries/subsystems as well as both AC and DC drives as backup (as applicable):

1. Lube oil pump.
2. Compressor.
3. Fuel oil pump.
4. DG set with self-starter motor (battery power requirement).
5. Hydraulic turning gear mechanism for the GT rotor.

This system is able to provide a black start even when only battery power and liquid fuel are available. Fig. 1.22 is an example of this system.

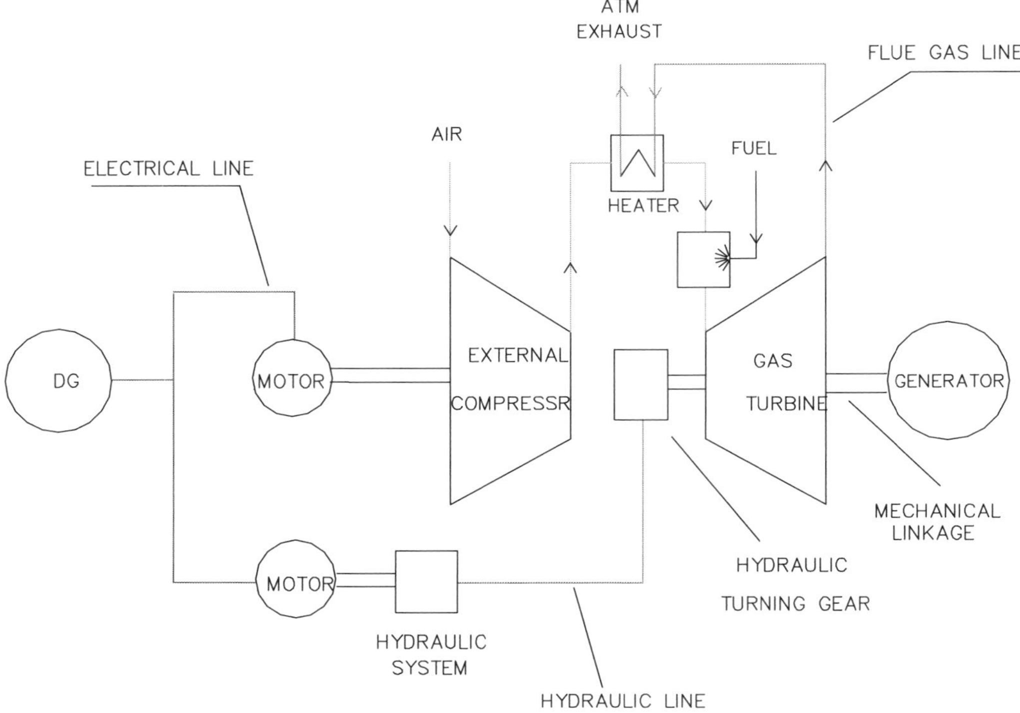

FIG. 1.22 Schematic diagram of GT black start (self-contained).

2.5.8.2 Black Start Through Variable Frequency Drive or Load Commutated Inverter

This system has the following auxiliaries/subsystems:

1. DG set to provide an auxiliary bus, which would supply emergency power, variable frequency drive (VFD), and the GT generator rotor's exciter coil.
2. The VFD/load commutated inverter (LCI) output to drive the GT rotor.
3. Other auxiliary equipment as stated in Section 2.5.8.1 to run DG sets.

The DG set must be suitable to handle a high harmonic load from the VFD if provided. Fig. 1.23 is an example of this system. There are two ways the VFD/LCI output can spin the GT rotor, and they are described below.

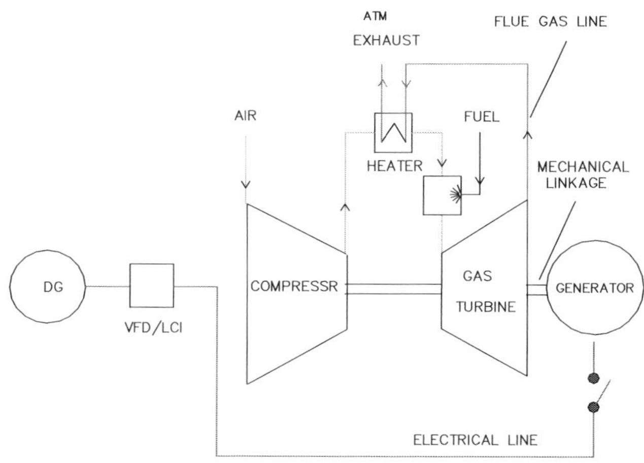

FIG. 1.23 Schematic diagram of GT black start with VFD/LCI for stator winding.

2.5.8.2.1 VFD/LCI to Drive the GT Directly In this system, the VFD/LCI receives its power supply from the DG set(s) and its output drives the rotor of the GT generator as a synchronous motor to rotate the GT shaft from the initial static position up to running speed. Considerations for this system include:

1. Amortisseur or starting windings to support the initial starting torque when the GT acts as a motor. The amortisseur windings are in the synchronous motors, but while the generator acts as a motor, they must be properly sized for high combined inertia loads such as in the gear box, compressor, and turbines.
2. A proper timing sequence must be designed for injecting the rotor field current so that synchronous motor action starts after starting the torque.
3. A proper timing sequence to withdraw the VFD/LCI output near the synchronous speed when used as a synchronous generator.

Amortisseur windings are additional windings in the rotating member in the form of bars; they occur in the rotor of synchronous motors. These windings are made of copper bars short circuited at both ends, embedded in the head of the pole, close to the face of the pole. They are similar to the skewed, short-circuited bars seen in the rotor of the squirrel cage induction motors. During the static start-up condition, the amortisseur windings are used to start the machine under its own power as an induction motor. The synchronous motor is not self-starting; hence, the motor starts initially as an induction motor through the action of these amortisseur windings and takes it unloaded to almost synchronous speed. When sufficient speed has been attained, the excitation to the rotor main winding of the synchronous machine is switched on when the rotor is "pulled in" by the synchronous torque. The motor then runs at the synchronous speed as a synchronous machine. During loaded condition, the function of these windings is to minimize the effect of load fluctuations by dampening the torsional oscillations in the rotor, and for that they are also known as damper windings.

The LCI uses technology involving load-commutated and phase-controlled power thyristors. The AC power supply from the DG set is applied to a converter, which acts as a rectifier and connects the output to an inverter through a DC link inductor (known as load the LCI). This is normally used for controlling the speed of the synchronous motor and incorporates thyristor full-wave bridges; it does not need forced commutation at the inverter stage. The automatic thyristor turn-off is achieved with a synchronous motor as the load, if it has a leading phase angle with respect to the load voltage. For a given load, sufficiently increasing the field will produce a leading power factor. These can be referenced in the appropriate textbook.

With the LCI drive, there is natural commutation of the thyristors of the inverter bridge(s) by the load, and this is available over a significant range of motor speed. The overall LCI operation is simple and reliable because of the phase-controlled thyristors and elimination of the forced commutation, which involves a capacitor discharge method to provide reverse biasing of the conducting thyristors. The output power directly connects the stator winding of the GT generator to act as a synchronous motor. The torque of the machine is controlled by the current through a DC link inductor, which controls the machine speed as required.

The inverter output frequency must be high enough to obtain a sinusoidal wave form; otherwise, there is a risk of having a square waveform at lower frequencies. The higher frequency, on the other hand, would necessitate a lower turn-off time of the thyristors, which must be lower than the commutation time of the thyristors to obtain reliable functioning of the system. There is an upper limit of the frequency beyond which the commutation would not take place with the undesirable triggering of the thyristors.

The minimum combination of thyristors has a 6/6 pulse at the converter and inverter assembly, respectively, with other variations up to a 24/12 pulse system. The synchronous machine is excited by a field winding supplied with current from an exciter (may be of several types), as discussed in detail in Chapter 10, Section 3.

2.5.8.3 Black Start Through DG With Automatic Voltage Regulator Directly Connected to the GT Generator (Without VFD/LCI)

Fig. 1.24 is an illustration of this system. Here, the DG output is directly connected to the turbine generator before the DG is started. There is no need for VFD/LCI as the DG would act as a frequency convertor. The automatic voltage regulator (AVR) supplied with the DG set generates an output voltage approximately proportional to the frequency/speed as the DG accelerates from a standstill condition. The DG set must be properly selected to support the start-up and acceleration of the GT generator with amortisseur windings. For such a scheme, two separate DG sets help with the lighting/emergency load and the black start of the GT. This system may be cheaper as the cost of two generators without the static convertor (VFD/LCI) should be less than the cost of one large generator plus a static convertor.

2.5.8.4 Black Start Through Hydraulic Drive

This system is similar to what is described in Section 2.5.8.1. In this system, the GT uses an electric motor (induction) to spin the turbine through a hydraulic coupling (for achieving variable speed) up to the rated speed, thus eliminating the use of an external compressor. A soft starter to start the motor may be required for a higher-capacity plant. A GT set handles the starting load and the emergency auxiliary powers. Fig. 1.25 illustrates this system.

2.5.8.5 Black Start Through Electric Drive (Induction Motor)

This method is similar to a black start with a hydraulic drive, but without the hydraulic coupling. Here, the GT uses a suitably sized induction starter motor to spin the turbine through a VFD/LCI (for achieving variable speed) up to the rated speed without requiring an external compressor. Additionally, with this method there is a limiting inrush current at the instance of switching on. For example, a 15 MW GT set would require a 200 kW induction motor (typical value) and the inrush current may be quite high if started directly online. Modern VFD/LCI usually acts as a load from the DG set output at a power factor close to unity and with minimum harmonics. This would control the voltage amplitude and frequency to produce the slow initial turning and then the acceleration required by the connected turbine through mechanical coupling (Fig. 1.26).

2.5.9 GT With Compressed Air Energy Storage Facility

Previously in this chapter, GTs using a compressor or compressed air supply differently were discussed. For example, in Section 2.5.3.2, the two-shaft configuration has a power turbine without the compressor, and there is a separate compressor turbine that drives the compressor only and its hot gas exhaust is further expanded in the power turbine. In

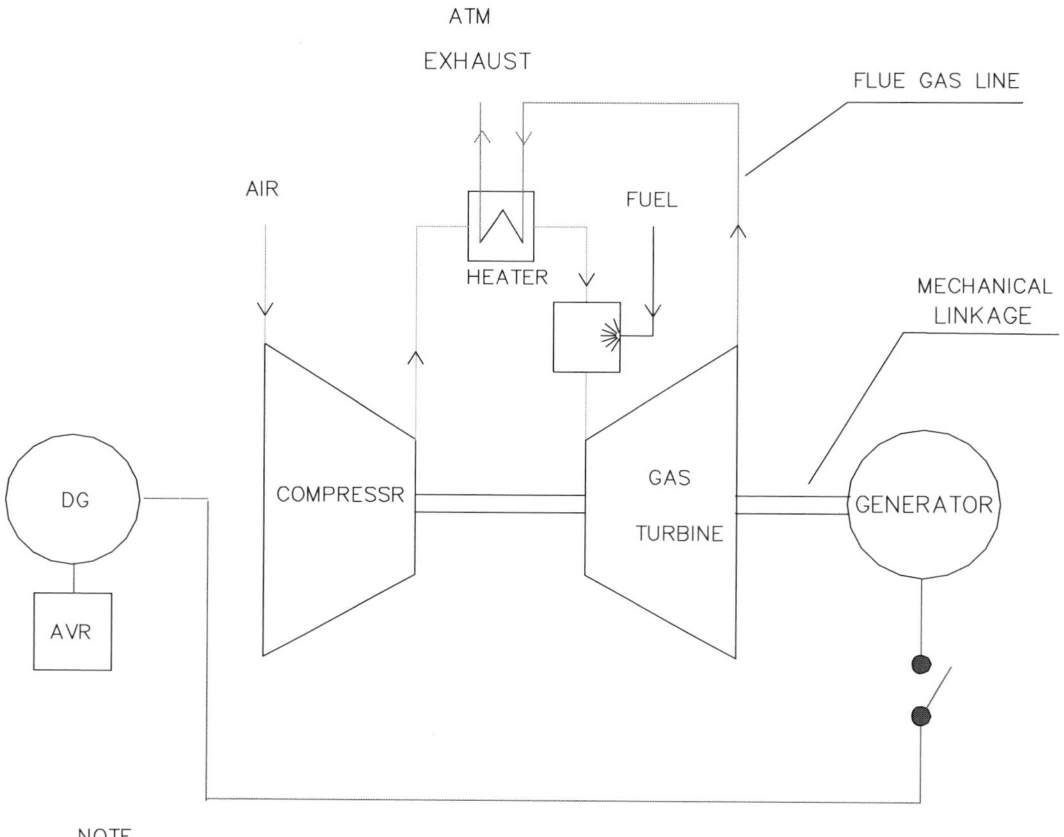

FIG. 1.24 Schematic diagram of GT black start through DG with AVR.

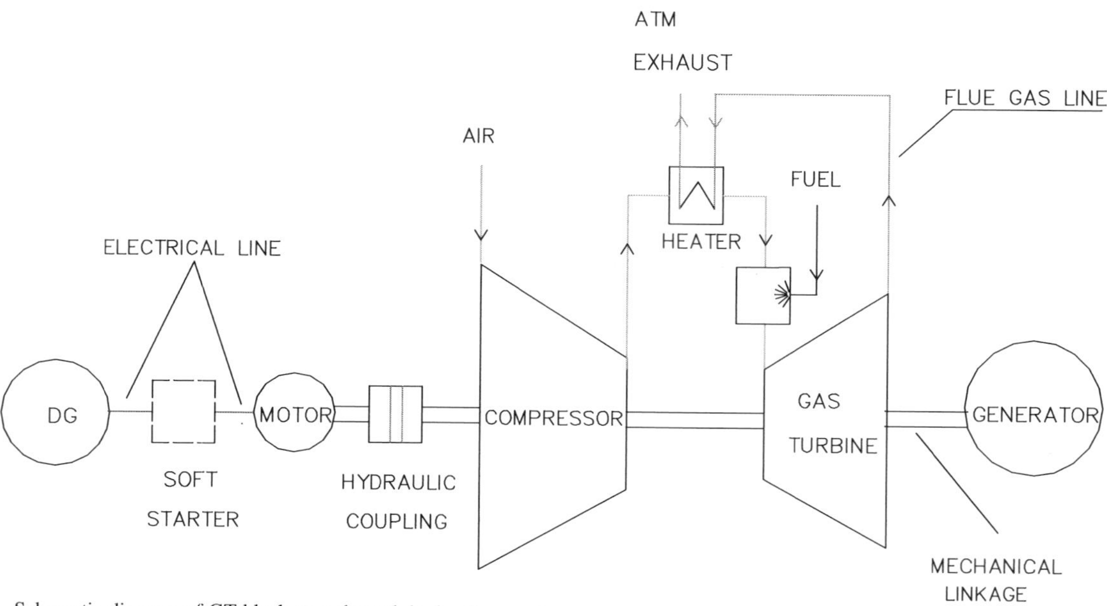

FIG. 1.25 Schematic diagram of GT black start through hydraulic coupling.

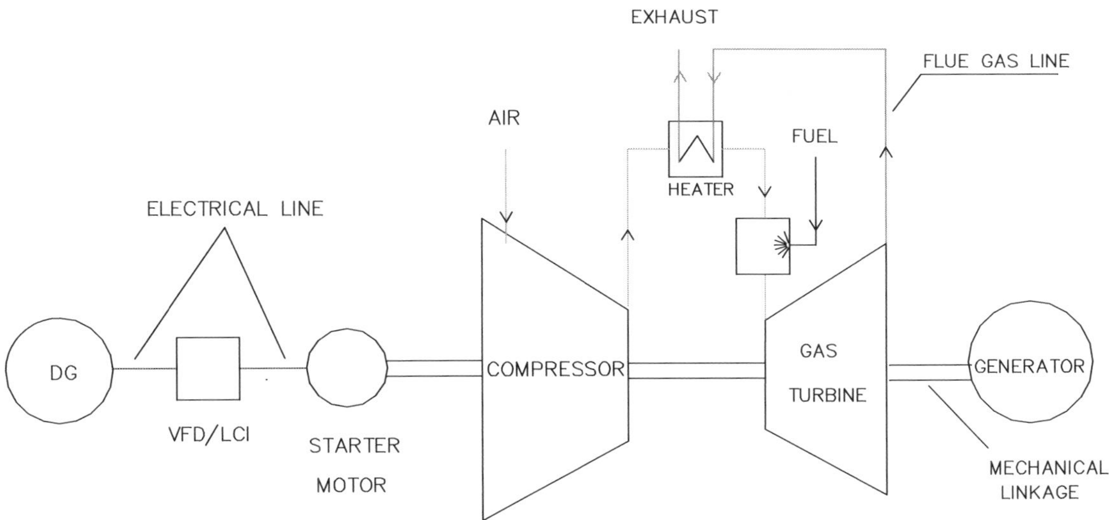

FIG. 1.26 Schematic diagram of GT black start through VFD/LCI and starter motor.

Section 2.5.8.1, a self-contained GT black start also had a separate starting compressor. More developments have been made toward improving the efficiency of the GT. This is done by separating the compressor and the turbine with a compressed air source and not directly from the compressor. The compressor operates independently with the power supply whenever available, and the compressed air is stored in a separate reservoir. This new design operates more cheaply than earlier configurations.

The compressed air operates the turbine when required. As previously mentioned, the compressor takes about 50%–60% of the total developed power from the GT, with an overall improved efficiency. Another advantage is that the energy can be stored when electric power is available during low demand at a low cost and used during high demand. This can be done because the GT has a separate compressed air source.

2.5.10 GT Emissions

As a product of combustion, the oxides are the main polluting agents, along with volatile organic compounds (VOC) and particulate matter (PM). The main components of oxides are NO_X, CO, and SO_x, out of which the SO_x percentage is comparatively low as GTs operate on desulfized fuel. Using liquid fuel contributed through ash and metallic additives produces PM. Formation of thermal NO_X is dependent on the operating flame temperature, which is increased at higher loads. Formation of CO and VOC is the result of incomplete combustion, which takes place during low load operation in the GT. The desired condition of pollutant gas emission limits is shown in Fig. 1.28.

2.5.10.1 NO_X Control in GT

The control measure of NO_X in a GT is similar to that of a steam generator. As previously discussed, the power developed and the efficiency from the Brayton Cycle depend on the firing temperature from the turbine inlet temperature. The GT is designed for a high temperature at the operating load; thus NO_X (thermal) control reduces the flame temperature only and lets the temperature rise further before the hot gas reaches the turbine inlet. The other source of NO_X generation is called "fuel NO_X" because the source is from fuel only and is not dependent on the temperature but is related to the availability of oxygen (O_2). The O_2 reacts with the gaseous state of the nitrogen compounds (NCH and NH_3) to generate NO in the air-rich condition. Under fuel-rich conditions, these nitrogen compounds, because they are unstable, produce N_2 gas only.

2.5.10.1.1 NO_X Control of GT Through Lean Air/Fuel Ratio Control A portion of the total air is mixed with the fuel before the flame develops; this is called PA and would be present at ~25%–30%. This is called the lean fuel/air ratio, which is much less than the stoichiometric air requirement. The flame produced by this incomplete combustion results in a lower flame temperature, thereby suppressing thermal NO_X generation. The balance portion of the air flow or SA is added after to achieve complete combustion.

2.5.10.1.2 NO_X Control of GT Through Lean Premixed Air/Fuel Combustion Another way to control NO_X is premixed combustion. Here, the major part of the fuel is in a gaseous state and part is total compressed air (typically 50%–60%, depending on the combustor design). These are mixed together and injected around the surface of the combustor so that local high-temperature zones are avoided. A small fraction of fuel is injected through the central part of the combustor where the igniter is located. In the absence of the stoichiometric air requirement, the fuel produces a flame but with a lower temperature, and secondary combustion takes place where the premixed air/fuel is injected and later when balanced air flow is injected near the rear part of the combustor. This process creates the flame hot spot temperature and less generation of NO_X. It demands a specially designed mixing chamber and a combustor and turbine as well. The NO_X level can be substantially reduced to ~9 ppm using this method, as claimed by a number of reputed manufacturers.

2.5.10.1.3 NO_X Control of GT Through Selective Catalytic Reduction Selective catalytic reduction (SCR) is another method of controlling the NO_X percentage in the exhaust hot gas as a part of postcombustion emission control. In this process, ammonia, with a suitable catalytic agent, is sprayed over the exhaust gas to react with NO_X, resulting in N_2 and H_2O. Depending on the proper selection and exhaust gas condition, SCR can increase NO_X elimination by ~80%–90%. Generally, there are three types of SCR systems, named for the temperature ranges in which they operate:

Low temperature: Works between 150°C and 200°C and is located downstream of the HRSG exhaust duct. This is not suitable for GT installation without HRSG.

Moderate temperature: Works in the range between 200°C and 425°C and is located between the GT and HRSG or may be within the HRSG where the two temperatures (of the hot exhaust gas and the catalyst operating range) match each other. This is also not suitable for GT installation without HRSG.

High temperature: Works between 425°C and 600°C and is located just at the GT exhaust, irrespective of whether the installation includes HRSG or not.

While SCR application has a number of advantages, there are also a few drawbacks. First, the SCR is considerably more expensive. Second, there can be health hazards caused by the presence of ammonia, which demands in situ availability and may leak after prolonged operation.

2.5.11 GT as External Combustion Engine

As previously discussed, most GTs operate as internal combustion engines, but they can also work as external combustion engines. This system may be described as a turbine version of a hot air engine. Another way to describe the system is either as an externally fired GT or as an indirectly fired GT. The process is somewhat similar to the closed-loop cycle discussed earlier, but uses fresh air at the compressor inlet instead of recycling the turbine exhaust (Fig. 1.27).

The advantage of external combustion is twofold. Here, the heat is added to the system through the heat exchanger; hence, only clean hot air with no combustion products travels through the power turbine. This means the turbine blades are not subjected to combustion products, so they can use cheaper quality fuels of low caloric value. Nonconventional types of fuels such as powered biomass (sawdust, for example) or conventional pulverized fossil fuels may also be used. Due to indirect heat transfer, the thermal

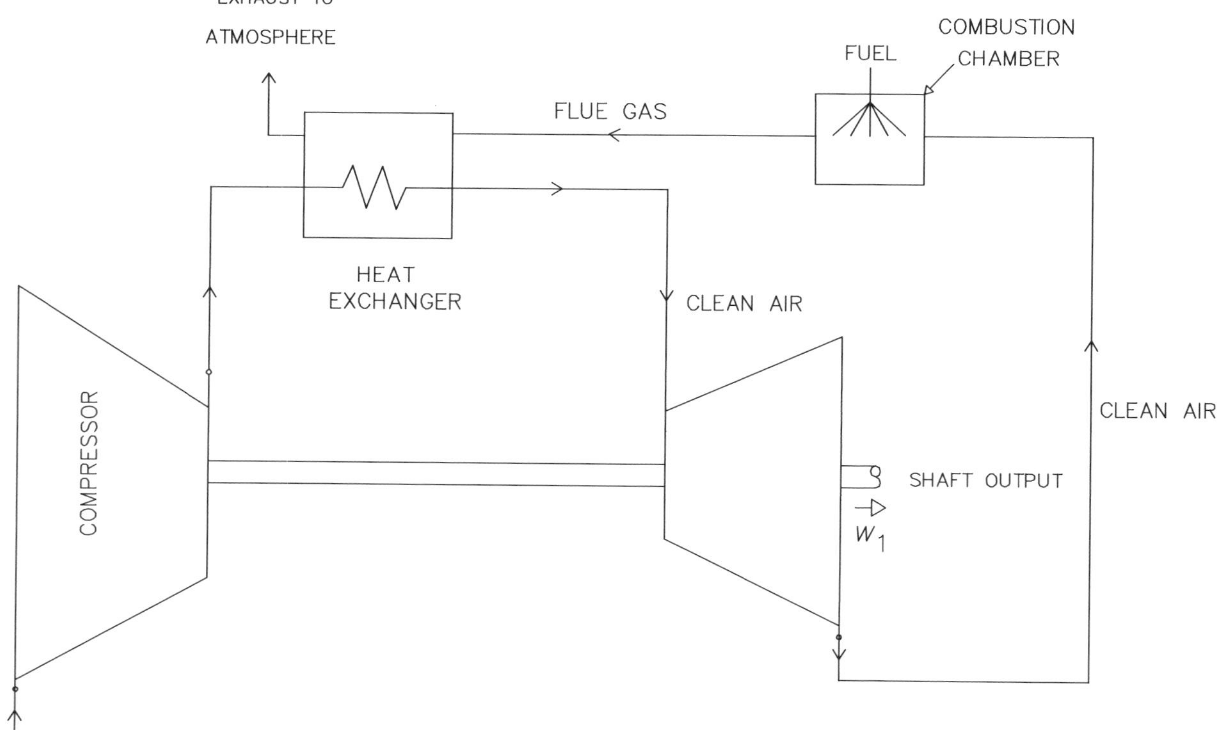

FIG. 1.27 Schematic diagram of GT with external combustor and heat exchanger.

efficiency of the external combustion engine is lower than that of the direct type of internal combustion engine.

2.5.12 GT Fuels

This is one of the most advantageous aspects of a GT. The GT is considered a multifuel engine and can operate on almost all the commercially available fuels such as gas, diesel, biodiesel, kerosene, natural gas, biogas, propane, and even powdered solid fuel such as coal and biomass, with the external combustor. Some of these fuels, such as solid fuels, diesel, or kerosene, also require something easily ignitable to start.

GTs operate on gaseous fuel or liquid fuel, and they also can be run on both fuels simultaneously because they are not restricted by their mass flow ratio. However, there are restrictions regarding the quality of the liquid fuel used. The presence of vanadium and sulfur cause high-temperature corrosion of the turbine blades, which ultimately results in loss of engine performance. There are specified limits and different standards for harmful ingredient content (typically 1% for sulfur and 0.5% for vanadium), which avoids deterioration of blade metals. Beyond these restrictions, it is common industrial practice not to use any residual fuel or any kind of cheaper distillates.

2.5.13 GT Control Systems

Other than control loops for auxiliaries, there are four main control loops associated with the GT control systems:

1. Speed/load.
2. Temperature.
3. Fuel flow.
4. Air flow.

2.5.13.1 Speed/Load Control Systems

This control system has the load reference as its set point duly corrected by rotor speed deviation. Actual shaft output is the measured variable; it generates the requirement or fuel demand as the controller output.

2.5.13.2 Temperature Control System

The purpose of this control system is to control the combustor exit or turbine inlet temperature, as discussed earlier, to save turbine metals as well as to limit the emission of CO and NO_X. The relationship between temperature and emission of these gases is depicted in Fig. 1.28. The temperature acts as a measured variable and is compared with the fixed set value of temperature. The output forms the

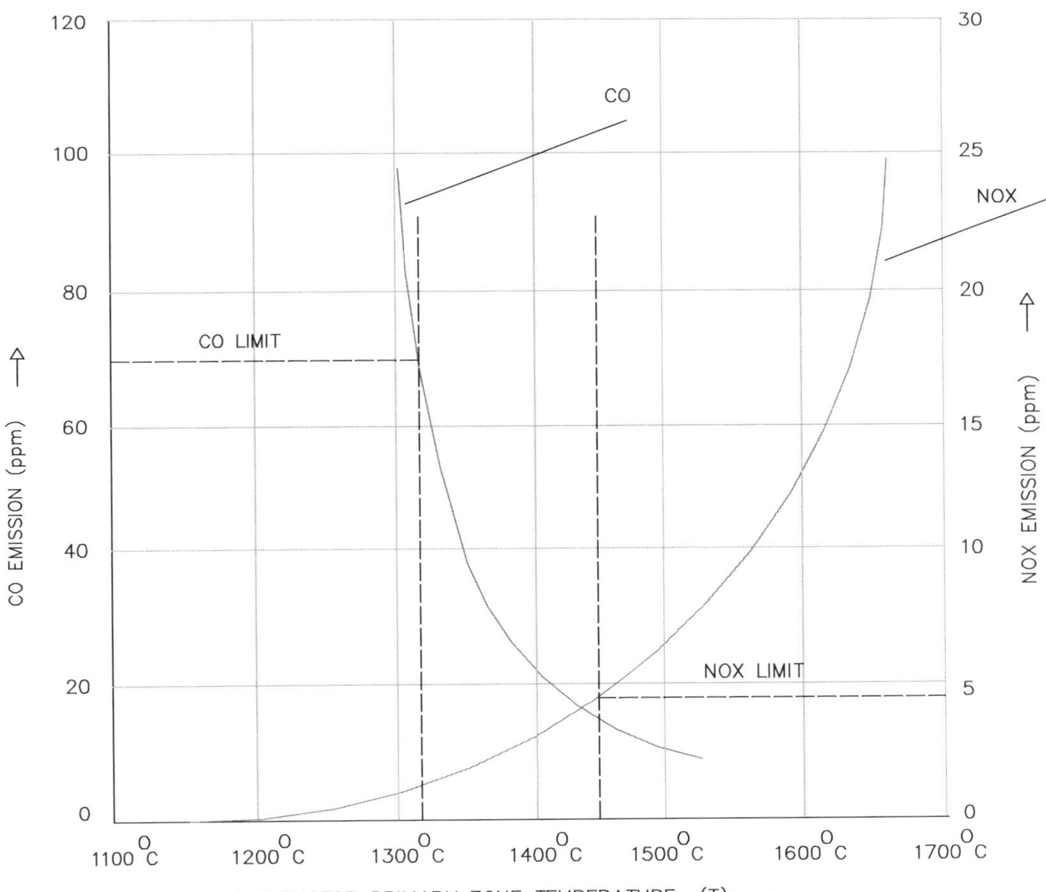

FIG. 1.28 Emission of pollutant gases presence with respect to temperature.

temperature control signal, which influences the "fuel flow control system" without any direct final control element. The controller output also determines the air flow requirement in that loop.

2.5.13.3 Fuel Flow Control System

The fuel flow demand from the speed/load control system is compared with the temperature control system output signal for a low value selection, and the selected lower signal determines the ultimate fuel flow demand. The final control element is the flow control valve, which regulates actual fuel flow to the GT.

2.5.13.4 Air Flow Control System

The air flow demand comes from the temperature control system output signal and acts as a set point against the actual air flow as a measured variable. The controller output adjusts the air flow through a final control element, for example, the IGVs of the compressor. This air flow controls the desired gas temperature along with combustion control, as a major part of the inlet air flow is injected at the rear portion of the combustor, which acts as a cooling agent. Normally, this temperature set point is kept lower than the GT rated value by ~1% (typically). For a particular GT, the air flow is proportional to the rotor speed. Variation from the maximum rated speed would change the characteristics of the air flow within the compressor.

2.5.14 Effects of Atmospheric Condition on GT Operation

Atmospheric condition plays a vital role in the performance of the GT. As stated earlier, the air flow is proportional to the rotor speed where flow signifies volume flow and not the mass flow. This means the compressor supplies a speed-dependent volume of air to the turbine without any relation to air mass or density. The turbine output depends on the mass of air passing through it; hence, at higher air density the air mass flow would be higher for the same volume of air and, consequently, the power output would be more when air is available at a lower density from the atmosphere.

2.5.15 Influencing Factors of GT Efficiency and Performance

The efficiency of a GT is primarily defined by the specific fuel consumption of the engine at a given set of conditions. The performance requirement, on the other hand, is mainly determined by the amount of power developed at the output shaft of the GT for a given set of conditions, which includes the standard day conditions with the temperature and pressure duly specified. In general, these data are a datum line with which a variety of GTs can be compared. The majority of GTs are rated at 15°C and 1.033 kg/cm^2. The type of operation for which the engine is designed also dictates the performance requirement of a particular GT. There are a number of influencing factors that affect both the efficiency and performance of a GT, and they are summarized in the following list:

1. *Air mass flow rate* through the turbine determines the machine performance. Any constraint hampering the desired flow condition would jeopardize the overall performance of the machine.
2. *Pressure ratio of the compressor* affects both the performance and the efficiency of the overall GT.
3. *Hot gas temperatures at the turbine inlet* affect both the performance and the efficiency of the overall GT.
4. *Individual accessory/component efficiencies* influence both the performance and the efficiency of the overall GT.
5. *Air/gas leakage* also influences performance and efficiency.

2.6 Recovery Boilers: Introduction

The HRSG is the most popular type of recovery boiler that uses supplementary fuels (Fig. 1.29). Some utility/process plants, especially in industrial applications, produce large amounts of excess heat with exhaust (e.g., flue gas) beyond what can be efficiently used in the process. HRSGs are boilers that reuse or recover this extra energy from hot exhaust gases of combustion chambers from, for example, generation plants, GTs, DGs/engine, etc., for generating steam. HRSGs are found in many combined cycle power plants. A very general schematic diagram is shown in Fig. 1.30 for an HRSG used in a combined cycle power plant. Exhaust gases in the

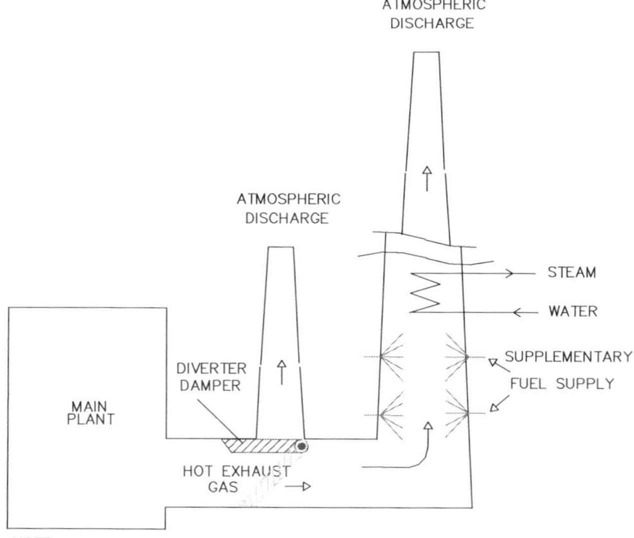

FIG. 1.29 Schematic diagram of heat recovery steam generators.

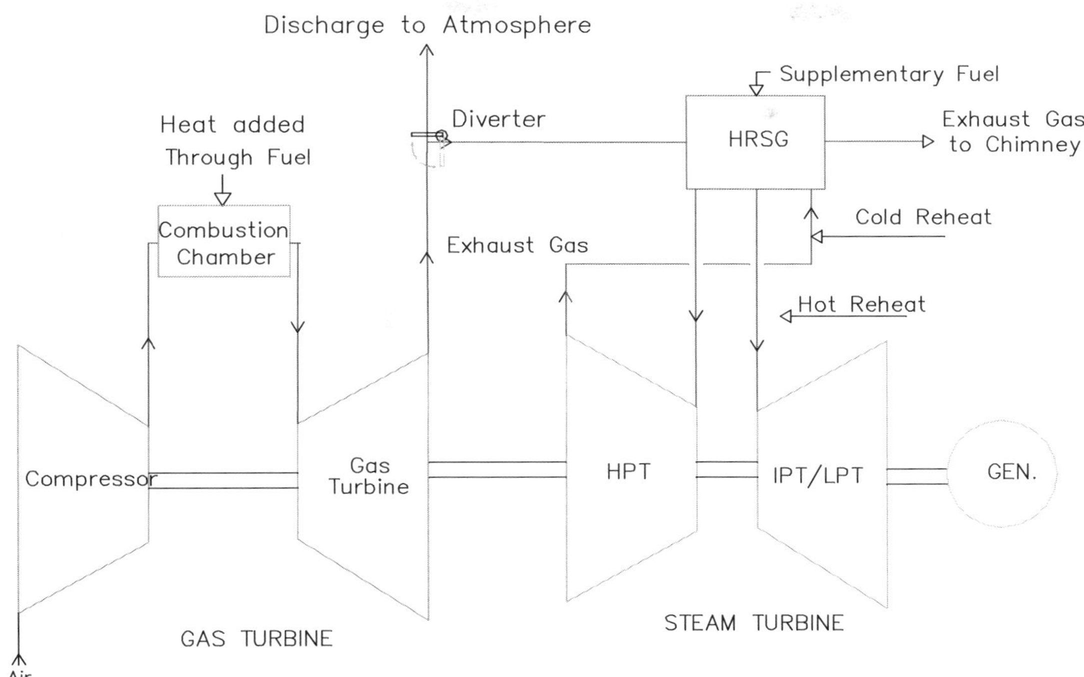

FIG. 1.30 Schematic diagram of HRSG utilized in combined cycle.

main plant (e.g., GT) of ~400–650°C are again allowed to pass through and heat another bank of tubes mounted in the exhaust.

High-pressure and high-temperature water is circulated through the tubes, which produces steam through heat transfer from the hot exhaust gases. One advantage to an HRSG is that it separates the caustic compounds in the flue gases from the occupants and equipment that use the waste heat. In most plants, a flapper damper (or "diverter") is employed to vary flow across the heat transfer surfaces of the heat exchanger to maintain a specific design temperature for the hot water or steam generation rate.

The foremost requirement for an HRSG is that the hot exhaust gases must have sufficient reusable heat to produce steam at the required condition. HRSGs may be designed for either convective or radiant heat sources with horizontal or vertical shell boilers or water tube boilers and be suitable for individual applications ranging from gases from power plant furnaces, incinerators, GTs, to DG/engine exhausts. Where additional steam or pressurized hot water is needed, it may be necessary to provide supplementary heat to the exhaust gas with a duct burner.

While designing the system, it is important that gas-exit temperatures are maintained at a predetermined level to prevent reaching the dewpoint and properly introducing soot blowing to achieve acceptable thermal efficiency. Today, even the small CHP stations would normally incorporate an HRSG/waste-heat boiler. Problems may arise from the source of waste heat (exhaust gas). Carryover from some types of furnaces can cause strongly bonded deposits and carbon from heavy oil-fired engines.

2.7 Process Boiler: Steam Supply at Different Pressures Compared With Steam Turbine Operation for Utility Purposes

Different industries use steam at various pressures and temperatures to make steel, textiles, chemicals, dairy, paper, fertilizer, etc. At the same time, captive power plants have been built to meet industry standards, that is, a dual requirement of power and heating and/or process work steam. As the normal turbine exhaust steam pressure is too low to be used for further heating, a back-pressure turbine was introduced to make both ends meet. With suitably designed initial and exhaust pressure of the turbine, it is possible to generate the required power and heating/process steam for a particular industry.

2.7.1 Back-Pressure Turbine

As shown in Fig. 1.31, steam is generated at a moderately high temperature and pressure to suit the turbine power

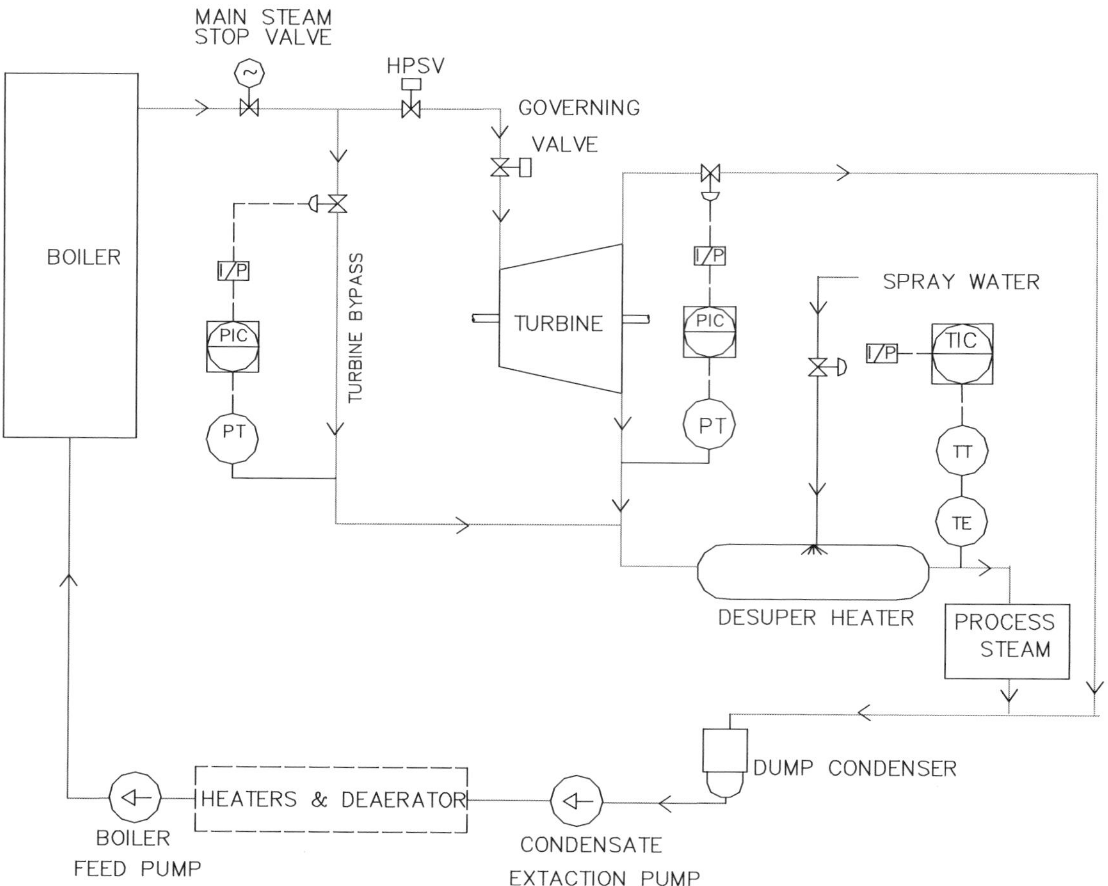

FIG. 1.31 Schematic diagram of back-pressure turbine.

requirement. The exhaust steam is normally superheated and not usable in most of the processes for various reasons. These reasons include: (1) a problem in controlling its temperature because it varies with the initial superheat, and (2) the rate of heat transfer from superheated steam to the heating surface is lower than that of saturated steam. To circumvent this situation, the exhaust steam temperature is lowered by using a desuperheater. By spraying water over the superheated steam, the water vaporizes and the steam cools to make saturated steam. This steam is then used by the process heater and comes out totally condensed as water. The exhaust pressure is controlled to avoid variation in saturated steam temperature.

If the back-pressure turbine is the only power unit, then the quantity of exhaust steam is controlled solely by the load on the turbine. For a low quantity, the main steam is allowed to pass through a pressure-reducing valve to the desuperheater to fulfill the process heating requirement. On the other hand, for a larger quantity, an extra portion of exhaust steam is bypassed to another location.

If the back-pressure turbine is running in parallel with other machines, then the output solely depends on the heat load and control of the supply steam pressure is necessary so that exhaust steam pressure is maintained at a fairly constant value. A power station with this type of configuration requires steam conditioning with the help of a steam turbine bypass system to accomplish a fast and smooth start and stop as well as additional protection for the equipment if a turbine trips or there is any other emergency situation. One source of protection is the turbine bypass valve (BPV), which opens very quickly to prevent safety valves from popping up and prevents wastage of steam and heat. Downstream temperature control is also important; it has to match the process application with parameters close to the set point.

The additional but normal tasks done by the bypass system are listed below:

1. The steam to process requirement is supplied through the bypass system when the turbine is out of operation.
2. The turbine operates with insufficient load to fulfill the process steam requirement; that is, it makes up the shortfall of steam supply from the steam turbine compared to system demand.

3. For the back-pressure turbine connected to a generator, process steam is a priority requirement that may demand continuous operation of the bypass system for an allocated period of time.

2.7.2 Pass Out Turbine

There are some cases where the power available from a back-pressure turbine is less than required by the plant (Fig. 1.32). This may be due to a low heating requirement, a high back pressure, or a combination of both. The problem may be overcome in a two-stage turbine, where the main incoming steam expands through the high-pressure stage and supplies the heating steam from the exhaust. The balance steam passes through the low-pressure stage of the turbine to satisfy the power requirement. For any further heating steam requirement at any other pressure and temperature, a number of stages may be incorporated in the turbine for optimum power and heating output.

2.8 Pressure-Reducing and Desuperheating Station: Purpose and Importance for Process Boilers/Initial Heating Up and Other Purposes

2.8.1 Pressure-Reducing and Desuperheating Station

As the name implies, the pressure-reducing and desuperheating station (PRDS) is a unit that conditions steam from a main system to an AS source of supply by reducing steam pressure and temperature using a pressure-regulating valve and a water spray valve. In a process plant, steam is required at different pressure and temperature conditions demanded by different equipment, and the main purpose of this PRDS is to provide a steam supply to satisfy these needs.

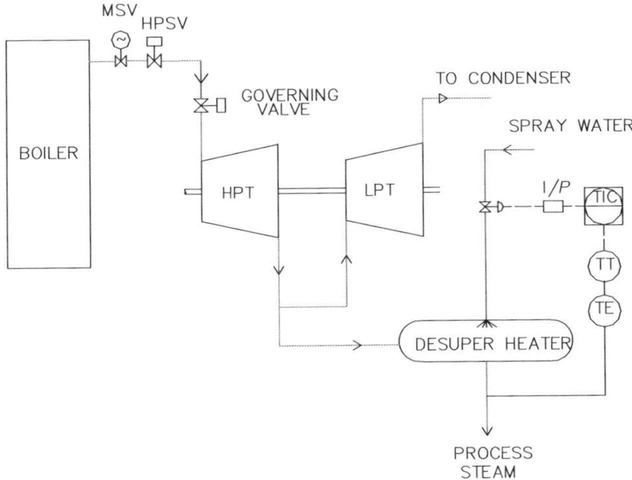

FIG. 1.32 Schematic diagram of pass-out turbine.

2.8.2 Objective of the System

In the PRDS, there is a steam header with fixed pressure and temperature (the parameter depends on the plant requirement) for a particular area where steam is supplied to various subsystems within the plant. This process-related steam is called AS. The basic purpose of the PRDS includes:

- Maintaining constant pressure and temperature at the steam header, irrespective of the source of supply for the AS.
- Supplying AS to plant auxiliaries at a constant pressure and temperature.
- Having more than one PRDS in a plant with a different pressure and temperature, depending on the area and process parameters of plant auxiliaries.

2.8.3 System Description

For the process boiler, there are several auxiliary plants and auxiliaries that require steam at different conditions. In addition, there is normally a turbo generator to provide electric power as a captive power plant. High-pressure steam required for the steam turbine therefore must be reduced to the desired pressure set point. This is achieved by a PRV that opens by a controller [may be a single-loop controller or a part of the main distributed control system (DCS) depending on the size and complexity of the plant] to a desired pressure set point. The inlet and outlet of the PRV (Fig. 1.33) are controlled with two remotely operated motorized isolation valves (post indicating valves) for isolating the control valve during maintenance.

During this time, the system can be made operational by remotely throttling (inching) the BPV, which heats up the line before starting the PRV operation by crack opening it. After pressure reduction is accomplished at the PRV, the steam is then allowed to pass through the desuperheating station to lower the temperature of the steam to the required value. The attemperation water is sprayed at high velocity through nozzles and at adequate quantity on the desuperheater. The desired value or set point is compared with a PID controller to generate the output signal, which determines the position of the desuperheating control valve (DSCV) by attemperation water flow to the desuperheater. At the inlet and outlet of the control valve (DSCV), there are two isolating remotely operated motorized isolation valves (TIVs) used to isolate the control valve during maintenance. During this time the system can be made operational by remotely throttling the BPV. In Fig. 1.33, two PRDSs are illustrated. The number of such stations depends on the requirement of the plant's auxiliaries.

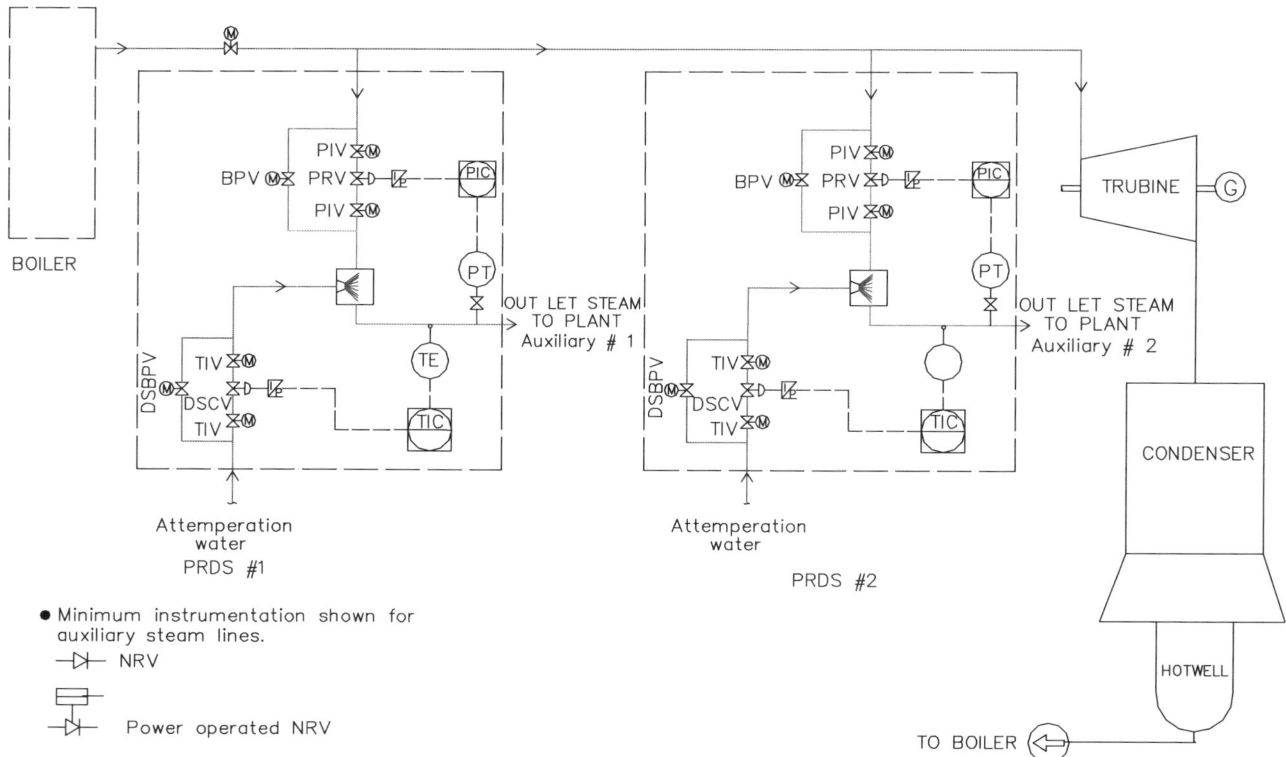

FIG. 1.33 Pressure-reducing and desuperheating station.

2.9 Vacuum and Dump Condenser

Whenever steam is involved in a plant, there should be a condenser to prevent venting and reuse of the process medium. There are two types of condensers: vacuum and dump. Modern thermal power plants use the former while the process plants and CHP plants use the latter. As discussed in Chapter 2, Section 3, the main function of the condenser is to condense the exhaust steam coming out of the turbine or process plant into water and recycle, as depicted in the Rankine cycle. Steam is condensed in a vessel by extracting heat through cooling and converting to water.

2.9.1 Vacuum Condenser

The very name implies that this type of condenser works under vacuum so that maximum boiler heat output is converted into work by the turbine. The typical operating parameters are 0.1 kg/cm^2 absolute, that is, ~0.9 kg/cm^2 vacuum at 45.4°C. The volume shrinkage is so huge that vacuum is created almost instantly. For instance, 1 kg of steam occupying 15,000 L at 0.1 kg/cm^2 pressure becomes only 1.0 L of water. For more details, see Chapter 2, Section 3.

2.9.2 Dump Condenser

Dump condensers are mainly installed to reuse working fluid, that is, the bulk quantity of steam that might have been wasted by venting. In many power plants, dump condensers are provided, in addition to vacuum condensers, to divert the excess steam during turbine start up, huge load throw off, turbine tripping, or bypass.

In certain plant applications such as combined cycle plants, trash to steam plants, etc., the steam surface (dump) condenser is required to condense the steam that has bypassed the steam turbine. In the bypass scenario, the steam turbine is usually not functioning. The steam from the steam-generating devices bypasses the steam turbine and is admitted to the condenser at a suitable pressure and temperature. Dump condensers can be furnished as a system that may consist of a steam pressure/temperature reducing station and a condensate recovery system, which includes level controls, a condensate pump, and an electrical control panel for automatic operation.

Dump condensers are normally used for process plants and CHP plants when demand requirements, be it steam or turbo generator oriented, may vary and alter the steam pressure condition, which may cause safety valves to pop up. In general, plants such as combined cycle, cogeneration, and refuse-recovery plants where the steam generation is required to be continued while steam turbine maintenance is performed simultaneously, use dump condensers. However, in power plants where process steam is not required, the dump condenser is avoided due to space and extra cost constraints. The operating parameters of a dump condenser depend on the type of plant, which may not be under vacuum condition.

2.9.2.1 Air-Cooled Dump Condenser

Some air-cooled dump condensers are not used for condensing steam; instead they are used to cool hot gases before releasing to the atmosphere during an emergency. For example, plants utilizing residual heat through an HRSG boiler may need emergency cooling if the HRSG is taken out of service for maintenance or it breaks down. There must be some way to absorb the input heat after diverting the heating medium before passing through the chimney. In that situation, a fast air-cooled dump condenser that can immediately be put into service from a standstill condition to operate with full capacity may be used.

2.9.2.2 Water-Cooled Dump Condenser

The water-cooled dump condensers are used when there is a continuous requirement of steam condensation, for example, back-pressure turbine exhaust, power plants with high-pressure and low-pressure flash tanks, or process plants utilizing steam for heating purposes (Figs. 1.31 and 1.34).

2.9.2.3 System Description of Different Types of Dump Condensers

The basic functions of all water-cooled dump condensers are the same: condense low-pressure steam and convert it to water by taking away the latent heat and a small amount of heat energy with a slight degree of superheat. The pressure of the condensing vessel, which is at saturated condition, varies from plant to plant usage and design criteria.

2.9.2.3.1 System Description of Dump Condensers in CHP Plants In a CHP plant, or cogeneration (cogen) plant, the main purpose is to provide steam to the process plant. The generation of electric power is the secondary purpose and the steam turbine (back pressure) passes all the steam flow as required for the process at an exhaust pressure designed to suit the process demands. Ideally, there should be no flow through the steam condensers when everything is running as scheduled and steam is not recovered from the process outlet. Normally, the boilers generate sufficient steam in the base load operation with the help of supplementary firing, which may be reduced or removed in case of lower demand to avoid dumping of steam.

The problem arises when equipment limitations or outages alter system demand. This may occur any time with no prior warning and cause abrupt and large changes in steam demand, depending on the type of disturbance. When, in case of emergency, the base load steam supply increases above the requirements even after total withdrawal of supplementary firing, the excess steam has to be diverted to the dump condensers for immediate removal from the system. Dump condenser steam temperature control is very important to ensure that vapor locking does not occur during the dumping operation.

2.9.2.3.2 System Description of Dump Condensers in Thermal Power Plants Small-capacity thermal power plants normally do not require a dump condenser, as the main condenser handles the steam condensing at very high vacuum. The 500 MW TPS or larger capacity plants with high- and low-pressure flash tanks use a dump condenser (Fig. 1.34) High-pressure flash tanks are connected to the drain lines of HPHs for emergency dumping of steam in case of increased levels; LPHs and flash tanks are connected in the same way. Whenever the level of high-pressure flash tanks increases, it drains to the low-pressure flash tanks, which in turn drain to a condenser when levels increase. The high-pressure flash tank's steam vent line is connected to move fluid to the dump condenser's steam chest area through a suitable nozzle connection to create necessary design pressure.

3 PROCESS PARAMETERS AND RANGES

Parameters are the basic physical and chemical characteristics or properties by which the state/condition of a matter or mass can be described. Some parameters are essential for various control systems, calculations, etc., such as flow or level. There are various process parameters, for example, pressure, temperature, flow, level, etc., that are required for measurement, control, data acquisition for analysis, storage and historical archives, and alarm annunciation systems as well as many other subsystems. All these systems are provided to ensure safe and uninterrupted operation, maximum efficiency with minimum cost, inventory control, equipment life expectancy, etc. Some parameters are necessary for plant guaranty and acceptance tests with special types of instruments at various strategic measuring points. It is not always possible to directly measure or control properties, so it becomes necessary to deal with variables such as pressure, temperature, flow, level, humidity, viscosity, density, etc.

Power plant operation is mainly based on the laws of thermodynamics, which demand instruments for parameters such as pressure, temperature, fluid flow for steam and water, and level of and pressure drops across miscellaneous tanks and vessels. Because a turbine is a rotating machine, it requires a different set of parameters such as speed, vibration, eccentricity, expansion, valve position, etc., as well as conventional parameters. For a generator, parameters are electrical and include voltage, current, power (MW/M VAR), frequency, etc.

Some analytical parameters are also important for checking the condition of the working fluid, that is, dissolved oxygen, pH, conductivity, hydrazine, etc., for water and dissolved silica, dissolved hydrogen, conductivity etc., for steam. Another set of special analytical parameters is for environmental pollution control. These include smoke/particulate emissions, SO_X, NO_X, carbon dioxide, carbon monoxide, oxygen percentage, etc., in flue gas. Typical values of a 210 MW thermal power plant parameters are included in Tables 1.1–1.3.

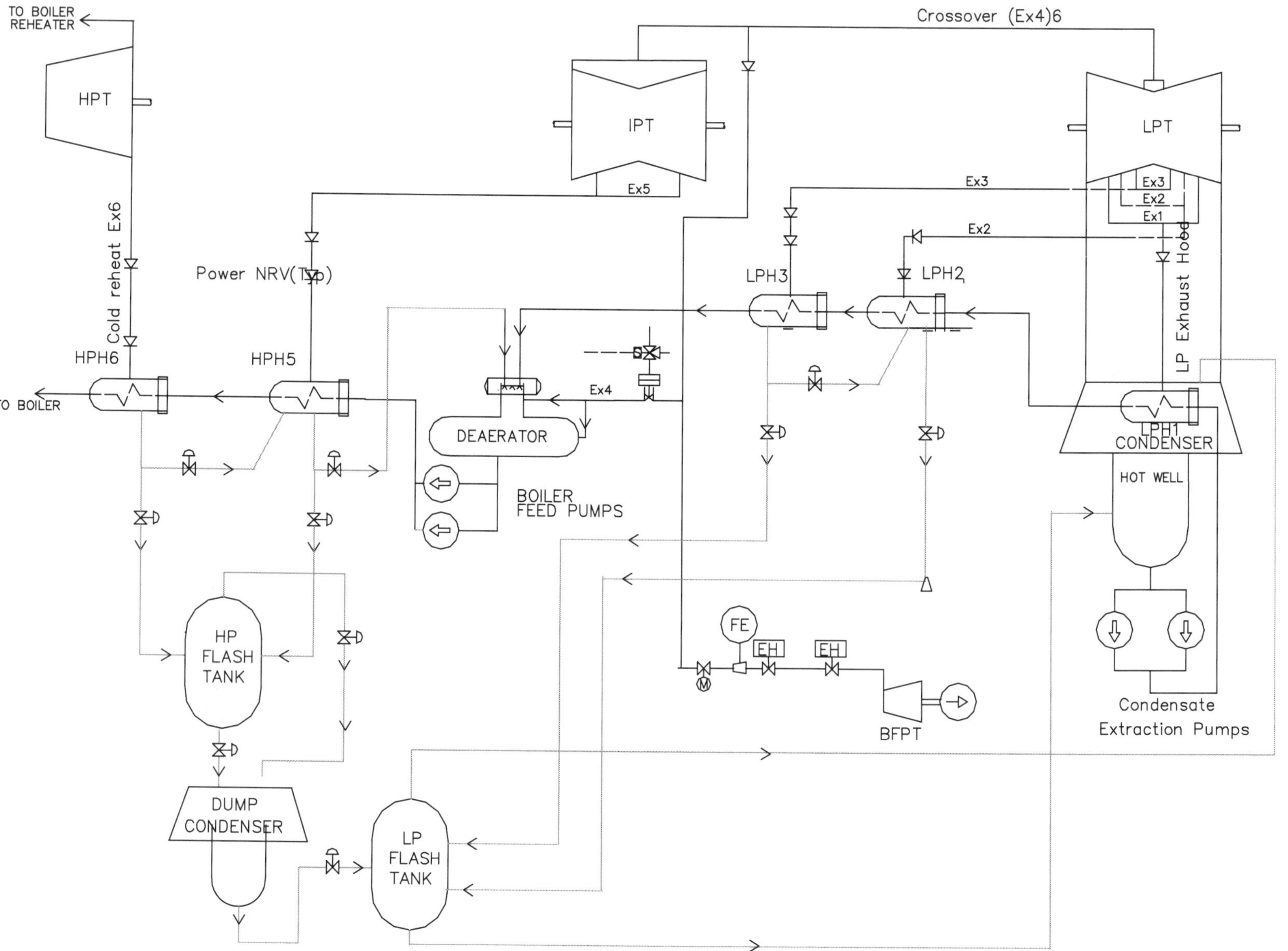

NOTE:
(a) In LP heaters dotted coils shown for condensate & in HP heaters dotted coil shown for Feed water.
(b) Instrumentation are not shown

FIG. 1.34 Dump condenser/HP and LP flash tank flow/schematic diagram.

TABLE 1.1 Typical Pressure Values

Parameters	100% BMCR Turbine VWO	94% BMCR 210 MW	80% BMCR 168 MW	60% BMCR 126 MW	Range	Unit
Steam at boiler outlet	158	–	–	–	200	kg/cm^2
Steam at turbine inlet	150	–	–	–	200	kg/cm^2
Steam at condenser	0.1	0.1	0.1	0.1	0–1	kg/cm^2 a
Steam at HPH 6	37	–	–	–	60	kg/cm^2
Steam at HPH 5	–	–	–	–	–	–
Steam at deaerator	10				15	kg/cm^2
Steam at reheater inlet (CRH)	38	37	29	22	60	kg/cm^2
Steam at reheater outlet (HRH)	35	34	26	19	60	kg/cm^2
Furnace	−3	−3	−3	−3	±25	mm of wcl

BMCR, boiler maximum continuous rating.

TABLE 1.2 Temperature Values

Parameters	100% BMCR Turbine VWO	94% BMCR 210 MW	80% BMCR 168 MW	60% BMCR 126 MW	Range	Unit
Steam at boiler outlet	540	540	540	540	600	°C
Steam at HP turbine inlet	535	535	535	535	600	°C
Steam at reheater inlet (CRH)	335	330	323	315	450	°C
Steam at reheater outlet (HRH)	540	540	540	540	600	°C
Economizer inlet FW	244	241	231	219	300	°C

BMCR, boiler maximum continuous rating.

TABLE 1.3 Typical Flow Values

Parameters	100% BMCR Turbine VWO	94% BMCR 210 MW	80% BMCR 168 MW	60% BMCR 126 MW	Range	Unit
Main steam	690	650	552	410	800	Te/Hr
HRH steam	600	571	458	351	800	Te/Hr

BMCR, boiler maximum continuous rating.

3.1 Purpose of Parameter Measurements

Measurements of various parameters are made for different purposes. With the advent of modern and state-of-the art technology, the measurement system is keeping pace with requirements for running plants in a safe and economic way. First and foremost is the plant and operator's safety. Next is the efficient running of the plant so that technocommercial viability is established. The parameters required for different calculations, for example, plant efficiency, metal stress evaluation, etc., are measured and transmitted to the appropriate software packages. Some measurements are made for diagnostic or postmortem analysis. To circumvent single transmitter/sensor/switch failure, a previously decided plan of action should be enacted (one out of two or two out of three redundancy) depending on the criticality or necessity of the process parameter.

3.1.1 Measurements for Plant Safety

Measurements of plant safety are associated with the parameters whose values must be kept at a particular level; if they go beyond the maximum or minimum limit, there may be a catastrophic result. For example, the normal operating parameter for a furnace pressure is (−)3 mm (g) of water column (wcl). If the pressure goes beyond 200 mm toward the positive side, there is the possibility of an explosion; if it goes toward the negative side, there is the possibility of an implosion. It is too expensive to manufacture a furnace to withstand this higher capacity.

Another example is the boiler drum level. Normally, the drum level is maintained at a level below the center line of the drum by ~100 mm. The design parameter does not permit the level to go beyond 150 mm toward the positive or negative side. If there is a positive side increase, water could enter the particles inside the turbine. On the negative side, a limit is imposed to avert drum burnout from loss of adequate water to absorb the heat or uneven heat distribution, which causes stress in the drum metal.

The measurements are normally on/off switches that give binary signals to form start/stop/trip logic (to be implemented in dual or triple redundant hardware) for almost all the equipment/drive or a central system, which decides the sequential or manually intervened start/stop/trip signals for the equipment/drive as per the present condition of the plant. The other important signals related to this parameter are

Loss of both/all FD or ID fans.
Loss of all flame-sensing (ultraviolet or flickering) scanners.
Loss of all fuel (no oil and coal flow).

3.1.2 Measurement for Efficient Running/ Control of the Plant

These measurements are made to control the important parameters that enable the plant to have a continuous and trouble-free operation; this is done with an automatic control system called the DCS. The signals are continuous and generally analog (4–20 mA DC) or digital (through a fieldbus or remote terminal unit).These are called control parameters.

The plant is a thermal power station, so the measurement of steam temperatures is very important as the work done by the turbine depends on the difference between the turbine inlet and exhaust temperature of each stage. The inlet temperature of the HPT is controlled by spraying an appropriate quantity of hot water at the boiler end, and that of the IP turbine is controlled by tilting the burner assembly or SH/RH bypass damper (as per manufacturer) beyond its limit by spraying hot water as a last resort. The turbine low-pressure exhaust temperature, although not controlled through DCS, is controlled by the condenser cooling water from natural resources or by using a suitably designed cooling tower system.

The boiler load is maintained by controlling the main steam pressure at the turbine inlet. This control subsystem is called the master pressure control and is used to adjust the fuel input to the furnace. The set point of the controlled main steam pressure may be a fixed value or a variable (sliding pressure), depending on the configuration of the boiler, turbine, and other associated equipment. The final control elements (the actuator/drive, through which actual final control action is executed) for fuel are the flow control valve for oil and/or the mill/pulverizer (for coal) feeder speed.

The fuel, whether oil, natural gas, or coal, is heated before going to the furnace for efficient burning by a suitable temperature controller. There are separate controls for oil, which is heated by the steam coil and/or mat heaters in the oil day tank and storage tank. After that, the oil is heat traced up to the burner with a provision for short and long recirculation. For coal, the mill outlet temperature is the measured variable with a fixed set value of the controller, and the outputs are sent to the hot air and cold air dampers to maintain the mill temperature. The mill air flow to carry the coal depends on the boiler load and is done by either the above-mentioned hot air damper or the PA fan vane controls, depending on the manufacturer.

Air flow control plays a very important role in the efficient burning of fuels. The exact air flow depends on the quantity and quality of fuel input, which is guided by the stoichiometric ratio. More air flow (for example, 20%) is necessary to ensure complete combustion. This is achieved by measuring the residual oxygen and carbon monoxide percentage contained in the outgoing flue gas; the control system will take corrective action for arranging extra air to the furnace. FD fans provide the combustion air with the help of a variable blade pitch or inlet vane, depending on the manufacturer.

Another very important control is the furnace pressure (draft) control, which is accomplished by ID fans. The pressure transmitters for control purposes are located in strategic positions for representative measurement. Furnace pressure is maintained slightly lower (approximately −3 mm of wcl) than atmospheric pressure to arrest the flame and flue gas within the chamber, thus avoiding operational and environmental hazards. For the boiler drum, level control is very important for the reasons stated earlier. Drum levels, as measured by redundant level transmitters, are fed to the controller along with main steam flow and FW flow signals to receive the feedback of flow imbalance. There is a fixed set value against which the controller output drives the final control element such as a control valve in the main feed line or the scoop tube of hydraulic coupling to change the speed of the BFP. When the feed control valve controls the drum level, the scoop tube controls the differential pressure across the valve for smooth control. The problem with the latter control strategy is that there is a continuous energy loss across the valve, which can be eliminated if the valve is kept wide open at higher load with the pump changing speed as per requirement.

Level transmitters are provided for each HPH and LHP for its own shell level control. The flow control valves are in the drain line of each heater to the next lower pressure heater or vessel, for example, HPH 6 to HPH 5, HPH 5 to the deaerator, or LPH 3 to LPH 2, etc. Deaerator level control is achieved in two ways, and which way is chosen depends on the manufacturer's recommendation. The measuring level transmitter outputs are sent to the controller with a fixed set point. According to one philosophy, the main condensate inlet valve after the condensate extraction pump's discharge opens or closes as per the controller output, that is, if the deaerator level decreases, the inlet valve will open and vice versa. Here, condenser hot well level control is accomplished by dumping deaerator water into a storage tank called the *condensate surge tank* through a separate line in case of a high level. In turn, the deaerator level decreases and then the condensate extraction pump discharge valve opens more so that the hot well level becomes normal. For a hot well low level, condensate from the storage tank is transferred to the condenser hot well whose level increases to normal.

Another philosophy for deaerator level control is that whenever the level rises, the condensate is dumped into the condensate surge tank. When the level decreases, condensate from the storage tank is transferred to the condenser hot well, whose level increases. Its controller output then directs the extra condensate to a deaerator through the main condensate inlet valve to replenish its level. In both philosophies, the two valves operate in such a way that one valve will operate when the other valve has reached a full closed condition in either the auto or manual position of the controller.

Measuring transmitters for condenser hot well level control are selected based on the suitability in the vacuum service condition where the operating pressure is 0.1 kg/cm^2 absolute. The output of transmitters after selection/averaging is sent to the controller against a fixed set point. The controller output may be directed to the main condensate inlet valve after condensate extraction pumps discharge the line, as per the latter deaerator level control philosophy. If the deaerator level control assumes the latter control philosophy, then the level controller output is sent to the dump control valve in the surge tank if there is a high level; if the level decreases, then the condensate make-up valve to the condenser opens. The two valves operate in such a way that one valve will operate when the other valve has reached a full closed condition in either auto or manual position of the controller. Here, there is another valve in the recirculation line that is fully open at start up and goes on closing as the load increases, and at ~40% load it is fully closed.

Although the preceding control and plant safety parameters are required for this category of measurement, there are other parameters measured for indication, recording, or data storage purposes. These parameters may be needed for various calculations, posttrip analysis, formation of process graphics, management data preparation, noncritical alarm annunciation signal generation, storing with critical alarms, etc.

3.2 Type of Instruments and Their Selection: Discussion

In a modern power plant, various types of instruments are required that need careful study and ultimately an informed decision is necessary before selecting them. These instruments may include a pressure transmitter or a temperature element/transmitter, etc. A brief discussion about these instruments may help beginners with power plant requirements.

As far as range is concerned, the general notion is that the normal operating value will be approximately two-thirds of the value of the selected range. For example, for a parameter with an operating pressure of 67 kg/cm^2, the range will be 100 kg/cm^2. The trim materials or the wetted parts of an instrument also guide the selection as per the physical or chemical nature of the operating fluid, such as temperature, pH, slurry service or not, whether of an abrasive nature, etc. Selection of an instrument is also dependent on the characteristics of the instrument, such as static and dynamic characteristics. Static characteristics of process variables do not change, such as accuracy, reproducibility, sensitivity, etc., whereas dynamic characteristics of process variables do change, such as responsiveness, fidelity, etc.

3.2.1 Pressure Elements/Gauges/Switches/Transmitters

In a large-capacity modern power plant, the operating pressure varies from 0.1 kg/cm^2 absolute to ~300 kg/cm^2 g. The simplest device used to measure pressure is the manometer, which is used for measuring very low pressure. Gauges are used for higher pressure application. Switches are used for alarm generation and binary control purposes, for example, starting, tripping, and interlocked operation of different drive motors and safety relays or equipment. Some sensing elements use the amount of expansion or displacement characteristic of the material when subjected to pressure; these are called elastic deformation pressure elements, and they change shape under pressure. For gauges, switches, and even for a pneumatic pressure transmitter, the expansion or displacement characteristic is utilized by different elements: bellows, diaphragms, bourdons, helices, capsules, and spirals, depending on the pressure maximum and minimum ranges. Some typical values are indicated in Table 1.4.

Elements such as the ring balance type and wound resistance in varying shapes are now obsolete. For electronic transmitters, different types of elements are used: capacitance, dual inductance, strain gauges, piezoresistive, piezoelectric, silicon resonance, twin resonance, nitinol wires, etc. In capacitance-type sensor elements, a diaphragm is used as a primary element with measuring pressure on one side and atmospheric pressure on the other. The diaphragm moves toward the lower pressure side, and this deflection is sensed as change in capacitance and ultimately into a two-wire 4–20 mA DC signal. For a strain gauge, when pressure is applied, the deformation due to stress is converted into an electrical signal. For other operating principles, the change in electrical characteristics due to physical change after applying pressure is sensed and converted into a 4–20 mA DC signal. For smart transmitters, superimposed digital signals proportional to input pressure are also available in the same output terminals. For differential pressure transmitters, the same philosophy applies: two process pressures are connected to the transmitter with higher pressure tapping the impulse line on the high side and lower pressure tapping the impulse line on the low side.

TABLE 1.4 Pressure-Sensing Elements and Corresponding Ranges

Element Type	Minimum Range	Maximum Ranges (kg/cm^2)
Diaphragm	50 mm of wcl	28
Bellows	125 mm of wcl	56
Capsule	25 mm of wcl	3.5
Bourdon tubes	0.85 kg/cm^2	7000
Spiral	1.05 kg/cm^2	280
Helix	3.5 kg/cm^2	700

3.2.2 Flow Measurement

There are various was to measure flow: flow switches, flow transmitters, and flow gauges. With flow gauges, sight flow glass is used to see that fluid is passing without any calibration. Flow gauges are also called flow meters or rotameters, and there are positive displacement types such as ovalgear meters, nutating disc meters, etc. These meters indicate flow rate as well as cumulative flow value and even have a compatible electrical signal of 4–20 mA.

Flow switches and flow transmitters are also used for flow measurement. There are various types such as flapper, target, and diaphragm. Switch contacts are available in flow meters, but normally are not used for control purposes. Flow transmitters may be a differential pressure transmitter (DPT) type with two pressure impulse lines connected across flow elements or a differential producer. There are many types of differential producers, such as flow nozzles, orifice plates, and Venturi tubes, and their use depends on the service condition (discussed in Chapter 4, Section 4). The differential pressure produced is proportional to the square of flow as deducted from Bernoulli's theorem and the transmitter output is calibrated accordingly.

Other types of flow elements are also available that create restriction in a particular section of a specially designed flow path for open-channel flow measurement. The flow elements are weir notches (V-shaped or rectangular notches) or a Parshall flume. When passing through this type of restriction or the flow path (Parshall flume), the liquid level increases near the path inlet and this increased level is proportional to the flow. Normally, this type of level is measured by an ultrasonic level sensor and transmitter with the built-in software providing flow output.

Magnetic flow meters are based on the electromagnetic property of a conductor. Here, the electromotive force (EMF) generated is directly proportional to flow velocity. The only criterion for the flowing media is that the conductivity must be >0.5 μS/cm. There are other types of flow meters such as vortex, coriolis, wedge, thermal mass, etc., but they are not normally used in power plant flow measurement.

3.2.3 Level Gauges/Switches/Transmitters

Level gauges are normally called gauge glass, which is located near the vessel. A calibrated glass tube of suitable thickness and material (borosilicate or toughened) is placed vertically with the help of upper and lower limbs (pipes/tubes) covering the level to be observed. Almost all heaters, tanks, and boiler drums are provided with gauge glass for a direct reading. Other types of gauge glass are available for boiler drums with bicolor indicators showing steam as red and water as green.

TABLE 1.5 Temperature Ranges of Different Thermocouples in Use

Types of Thermo Couples	Iron-Constantan (Type J),	Copper-Constantan (Type T)	Chromel-Alumel (Type K)	Platinum-Platino Rhodium (13%) (Type R)	Platinum-Platino Rhodium (10%) (Type S)
Temperature range in °C	−210 to 1200	−200 to 400	250–1372	−50 to 1768	−50 to 1768
Used for measurement parameters			Steam temp. above 300°C	Furnace temperature ~1000°C	Furnace temperature ~1000°C

Level switches of various types are available including conductivity, capacitance, float, magnetic float, displacer, ultrasonic, paddle, and radio frequency. For an open tank, conductivity switches may be used when the fluid is highly or moderately conductive, and capacitance switches may be used for low conductive fluid. Level ranges can go up to 5–10m. Displacer and float types are also used, but the range is limited to 0–2m for top-mounted switches. Side-mounted float types may be used at any level. Ultrasonic or radio frequency types may be used for a high range up to 30m. Paddle-type switches are used only for solid level service, that is, coal or ash in the bunker.

For a pressurized tank, side-mounted magnetic float switches may be used at any level, but they are limited to a pressure rating of ~25kg/cm^2. For very high pressure, a special type of conductivity switch is available called a Hydrastep, which is used for boiler tripping from very high and low drum levels.

Level transmitters, when operated on the differential pressure principle at the same pressure or DPT, are used for open tanks or pressurized tanks, respectively. These types are used for level measurements for drums, all heaters, deaerators, condenser hot wells, and condenser surge tanks. Displacer transmitters are also used for pressurized tanks, whereas conductivity, capacitance, ultrasonic, and radio frequency level transmitters are used for open tank measurement.

3.2.4 Temperature Measurement

Temperature gauges and switches use the volume expansion characteristics of fluid after heat or temperature is applied. There is a bulb containing transmission fluid, which is placed at the sensing point. The transmission fluid may be an inert gas such as nitrogen or liquid such as mercury, alcohol, etc. The receiving side may be bellows, a diaphragm, a bourdon, a helix, a capsule, or a spiral-like, pressure-sensing element. A bimetallic element is also used that will bend in one direction when heat is applied to two metals of dissimilar coefficient of expansion that are joined together.

Unlike the pressure transmitter where the sensing element and processing circuitry are located in the same enclosure, temperature sensors/elements are located separately at the sensing point inside the process. Normally, the sensors/elements are isolated from the process fluid, which may have a high velocity of an abrasive and/or hazardous nature. This isolation is done using a thermowell, which takes care of process-side problems and is normally threaded and/or welded to the process pipe or vessel. Inside the thermowell, there is a female thread to accommodate the sensors/elements.

Sensors/elements consist of a resistance temperature detector (RTD), a thermocouple (THC), a thermistor, etc. An RTD uses the property of changing resistance of a wire with respect to changes in temperature. The resistance in a medium increases when the temperature rises. It is used in the temperature range of (−)250°C to 850°C. There are three types of RTD: copper (53 Ω at 0°C); platinum (100 or 1000 Ω at 0°C), also a platinum resistance thermometer (PRT); and nickel (100 or 500 Ω at 0°C). Normally, PRTs are widely used in power plants to measure water and air temperature < 300°C.

Thermistors also show a similar quality, but in the reverse direction. The resistance of the element decreases when the temperature rises. Thermocouples use the thermoelectric property of EMF in millivolts between two open ends when two dissimilar metal wires are solidly connected at the end and subjected to a higher temperature than the other open ends. Various types of thermocouples are used: iron-constantan (type J), copper-constantan (type T), chromel-alumel (type K), platinum-platino rhodium [two types with 10% (S type) and 13% (R) type) of platinum in the alloy] (Table 1.5).

BIBLIOGRAPHY

[1] P.L. Ballaney, Heat Engines, Vol. I (Mechanical Engineering), sixth ed., Khanna Publishers, India, 1973.

[2] F.J. Brooks, GE Gas Turbine Performance Characteristics [GE Power System *GER-3567H*(10/00)], GE Power Systems, Schenectady, NY, 2000. 16 pages.

[3] Product Bulletin CEPSI 2008/GTj of Siemens Gas Turbine (SGT-800).
[4] Design Features/Concepts-Transcanada.
[5] Update: Advanced Gas Turbine Technology (February 2004 Modern Power Systems)—GE
[6] DG Set System, Bureau of Energy Efficiency, Ministry of Power, Government of India (Chapter 9).
[7] Excerpts From Wired For Success by Randy Rundle, Krause Publications, November 1, 1995
[8] Island operation tests in a SCC-800 CHP plant—an example from an operating plant in Sweden, in: Paper Presented by Göran Tjellander, Siemens, Sweden.
[9] Lesson 40: Load-Commutated Current Source Inverter (CSI)—Version 2 EE IIT, Kharagpur.
[10] CHP—guide to steam conditioning, in: Paper Presented by M/s CCI.
[11] Excerpts From Energy Solutions Center, DG Consortium 2004 (Chapter 4/4-3)

Chapter 2

Main Equipment

Chapter Outline

Power Plant Instrumentation and Control Handbook. https://doi.org/10.1016/B978-0-12-819504-8.00002-0

1 OVERVIEW OF MAIN EQUIPMENT TYPES, FUNCTION, AND DESCRIPTION

The modern large thermal power plants have evolved through years of research work, experience, and state-of-the-art technologies but the basic equipment/subsystems are more or less the same, although with much sophistication and technological development to increase efficiencies.

Five fundamental elements, as shown in Fig. 2.1, represent a power plant as a basic heat engine:

(i) Boiler
(ii) Turbine
(iii) Condenser
(iv) Boiler feed pump (BFP)
(v) Generator

There are some other subsystems such as regenerative heating cycle or offsite packages such as water treatment plant (DM plant and pretreatment plant), coal handling plant, ash handling plant, oil handling plant, CW and auxiliary cooling water (ACW) system, condensate polishing unit (CPU), etc.

This section briefly describes the functions of the above elements with important auxiliaries.

1.1 Boiler

A boiler is a combination of several items that presents a means for combustion to provide heat energy to be transferred to water until it becomes heated steam; it is then used for transferring the heat to the turbine. Water is used as the working fluid for transferring heat to a process, as it is not expensive. However, tremendous expansion after water changes to steam means it needs careful handling. The main subsystems or the system auxiliaries are divided into three principal categories: the fuel and air/draft system, the feed water system, and the steam system. Each has various supplementary equipment to make it suitable for use in large and modern power plants.

1.1.1 Fuel and Air/Draft System

The *fuel and air/draft system* includes all necessary auxiliary equipment required for fuel to provide the necessary heat. The amount of fuel and air needed for steam generation is automatically controlled as per steam demand. Mills/pulverizers are required for solid fuels, pumping and heating units are required for liquid fuels, and compressors/boosters are required for gaseous fuels with a suitable type of burners and air supply arrangements. Supplementary equipment/items include:

(i) *Mills and pulverizers*

For solid fuels, the type of mills/pulverizers and associated feeders depends on the fuels used. They are supplied with the main equipment to achieve higher thermal efficiencies and for spontaneous ignition through burning of finely powdered fuel.

Pumps and Heating Units

For liquid fuels, supply pumps are needed and heating units are additionally required for a viscous type of fuel such as HFO/LSHS, etc.

(ii) *Fuel burners*

This category covers both oil and solid fuel with various designs, for example, low NO_x burners. The burner type also varies with the type of atomizer, etc.

(iii) *Air flow control elements*, for example, forced draft (FD) air fans, primary air (PA) fans, air dampers or registers for fuel air, auxiliary/secondary air, overfire air, etc.
(iv) *Draft system or flue gas system* includes induced draft (ID) air fans and gas dampers for evacuating combustion products from the furnace.
(v) *Soot Blowing System* blows hot steam on a regular basis to effect a periodic cleaning of soot buildup, which damages boiler metals and causes inefficient operation, on the radiant furnace surfaces, boiler tube banks, economizers, and air heaters.

1.1.2 Feed Water System

The feed water system provides necessary piping, heat exchangers to preheat a water recirculation system (if so designed), and feed flow control to match steam demand. Pressurized and preheated water supplied to the boiler as a working fluid to be converted into steam is known as *feed water*.

1.1.3 Steam System

The steam system include evaporation of water up to superheating, reheating of cold steam (for a hot reheat system), and collection and transfer of the steam through a piping system to the turbine with excess steam pressure being regulated/vented through suitable valves automatically.

1.2 Turbine

The turbine accepts steam from the boiler with high temperature and pressure, then converts the heat energy into mechanical energy for driving the generator with some vital subsystems indicated below.

1.2.1 Turbine Oil System

This very important subsystem supplies the lube oil to all the bearings, hydraulic control oil, jacking oil, and actuator power oil through individual oil coolers, strainers, redundant pumps, etc.

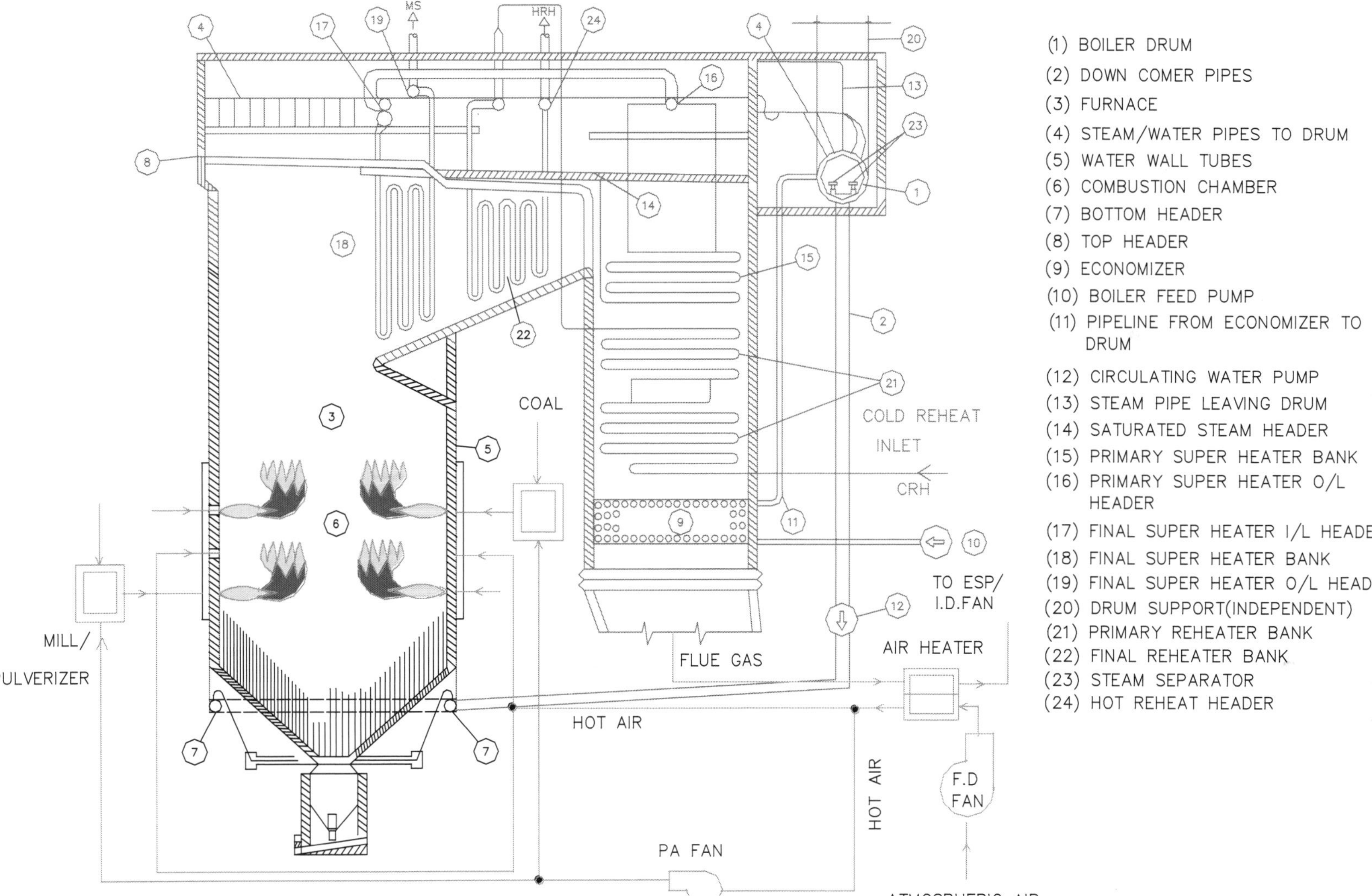
MS
HRH
COAL
COLD REHEAT
INLET
CRH
TO ESP/
I.D.FAN
MILL/
PULVERIZER
FLUE GAS
AIR HEATER
HOT AIR
HOT AIR
F.D
FAN
PA FAN
ATMOSPHERIC AIR
(1) BOILER DRUM
(2) DOWN COMER PIPES
(3) FURNACE
(4) STEAM/WATER PIPES TO DRUM
(5) WATER WALL TUBES
(6) COMBUSTION CHAMBER
(7) BOTTOM HEADER
(8) TOP HEADER
(9) ECONOMIZER
(10) BOILER FEED PUMP
(11) PIPELINE FROM ECONOMIZER TO DRUM
(12) CIRCULATING WATER PUMP
(13) STEAM PIPE LEAVING DRUM
(14) SATURATED STEAM HEADER
(15) PRIMARY SUPER HEATER BANK
(16) PRIMARY SUPER HEATER O/L HEADER
(17) FINAL SUPER HEATER I/L HEADER
(18) FINAL SUPER HEATER BANK
(19) FINAL SUPER HEATER O/L HEADER
(20) DRUM SUPPORT(INDEPENDENT)
(21) PRIMARY REHEATER BANK
(22) FINAL REHEATER BANK
(23) STEAM SEPARATOR
(24) HOT REHEAT HEADER

FIG. 2.1 Schematic diagram: solid fuel firing system.

1.2.2 Turning Gear/Barring Gear

Turning/barring gear is basically a gear arrangement provided to rotate the TG rotor shaft at a very low speed before the turbine starts up and after unit shutdown at low speed (about 1% rated speed) until the temperature comes down sufficiently.

1.2.3 HP/LP Bypass System

These subsystems are supplied to protect the boiler reheater from a burnout when no steam is passing through the turbine with the boiler generating steam during start-up, running, or tripped condition.

1.2.4 Governor and Isolation Valves

These valves are provided for both high and intermediate pressure turbines. Both are hydraulically operated, and are special valves with high controllability and isolating capability, respectively.

1.2.5 Gland Sealing Systems

Many parts of the turbine operate at such a high pressure that it cannot prevent leakage to the full extent. On the other hand, the lower-pressure parts operate at subatmospheric regions and so air ingress cannot be totally eliminated. Turbine glands are provided with this subsystem to seal the glands in such a way that leakage steam from the high side is controlled at a certain pressure by a control valve, which allows extra steam from its outlet to the low-pressure side glands to arrest air ingress by maintaining certain pressure.

1.3 Condenser

The function of the condenser is to condense the steam coming from the turbine exhaust. A high vacuum is maintained to extract steam and then cooled vis-a-vis condensed through an exchange of heat with the cooling water from a separate source. The subsystems comprised are indicated below.

1.3.1 Steam Ejectors or External Vacuum Pumps, Vacuum Breaker

Steam ejectors or vacuum pumps are provided for maintaining the high vacuum at the condenser. Starting ejectors are meant for the initial operation of the condenser to achieve vacuum quickly whereas running ejectors are for regular operation. A vacuum breaker is provided for speedy breaking of the condenser vacuum in an emergency to arrest the turbine from overspeeding in case of problems on the generator side.

1.3.2 Circulating Water Pump System or Cooling Towers

The circulating water (CW) is supplied to the condenser by external means. The open-loop configuration envisages transferring of sea or river water through the circulating water pump (CWP) and back with a higher temperature after the heat exchange. The close loop configuration envisages cooling towers where the hot return cooling water is sprayed from a height to exchange heat with the ambient air.

1.3.3 Condensate Extraction Pumps

These pumps are connected to the condenser hotwell to take out the condensate and deliver it to the rest of the water circuit of the whole cycle.

1.3.4 Heat Exchangers (Drain/Gland Steam Cooler, LP Heaters)

The CEP discharge line is connected to several heat exchangers, as stated above, in series up to the deaerator and is called "condensate."

1.4 Boiler Feed Pumps

The outlet of the deaerator is connected to the BFP suction and from there the working fluid is called "feed water (FW)" and reaches the boiler proper with the help of boiler feed pumps. It impels the FW with requisite pressure dictating the operating point of steam generation. BFPs are always supplied in a redundant configuration to avert interruption of operation.

Normally one motor-driven BFP is kept for start-up and emergency situations and one or two turbine-driven BFPs (may also be motor) are kept for operating up to a full load.

1.5 Generator and Exciter

Having received the mechanical energy from the turbine, the generator set is the ultimate equipment to supply electric power with important auxiliaries, as briefly indicated below.

1.5.1 Hydrogen Cooling System

Generators of modern thermal power plants get very efficient cooling by circulating pure hydrogen, whose thermal conductivity is almost seven times that of air. A safe level of hydrogen purity is essential to prevent a potentially explosive mixture of hydrogen and air. The other is the revenue loss as a drop in hydrogen purity would cause additional windage losses and the reduction in generator efficiency is the ultimate consequence.

1.5.2 Hydrogen and Seal Oil System

Hydrogen and air are a very explosive mixture and hence they need to be sealed to arrest hydrogen leakage from the generator itself. The same is done through oil sealing applied near the bearings at a higher pressure than that of hydrogen. It is a full-fledged subsystem with redundant pumps, coolers, and strainers.

1.5.3 Excitation System

Being a synchronous machine, the generator needs separate magnetic field excitation. The increase or decease in excitation current dictates the generator output voltage. Different types of excitation systems are discussed in the relevant Section 4.2 of this chapter.

1.5.4 Stator Cooling Water System

The generator stator windings are supposed to carry huge current, which means high copper loss and the resultant heat is dissipated by flowing of ultrapure water through the specially designed hollow stator conductor.

1.5.5 Generator Inerting System

It is extremely important that the introduction and removal of hydrogen are done carefully to and from the generator casing while commissioning and decommissioning of the system. To avoid the explosive mixture of air and hydrogen during commissioning, air is at first purged from the generator casing by an inert gas (carbon dioxide has been the dominant choice for years). However, argon or nitrogen are also gaining in popularity. The hydrogen is then introduced slowly and going on replacing the purge gas. While decommissioning the turbogenerator, the purge sequence is reversed.

2 STEAM GENERATOR: BOILER

The steam generation plant or boiler forms one of the fundamental units of thermal power plants, which is the practical and sophisticated representation of the basic heat engine following the Rankin cycle. The term "boiler" can be used for subcritical steam generators. However, at pressures above that such as super- or ultracritical application, the term "boiler" is not applicable. This is because there is no boiling at all and the term "steam generators" is the most appropriate one.

A boiler is a combination of several items that presents a means for combustion providing heat energy to be transferred to water until it becomes heated steam. The furnace contains accessories for producing heat by burning fuel. The heat as an intermediate step transferred to water to make saturated steam at a temperature depends on the pressure above the boiling water. The rate of steam generation depends on the furnace temperature in relation to the firing rate. The saturated steam is then further heated to a higher temperature to reduce suspended water particles, making a given volume of steam to carry more heat in relation to producing more work; this is popularly called superheated steam. The hot steam under very high pressure is then used for transferring the heat to the turbine. By this, the higher temperature difference does not allow the tendency of the steam to condense as an effect of pressure and temperature decrease during work done plus friction and conduction loss with the comparatively colder walls of the steam path plus loss across the control valves (HPCV and IPCV). Water is used as the working fluid for transferring heat as it is not expensive. The huge volume change after water is converted to steam means it needs careful handling.

The product of combustion with residual heat then passes through the reheater and economizer to preheat the feed water before it reaches the boiler and then through the air heater to preheat the air, making it suitable for combustion.

The very name "super/ultracritical steam generators" implies that they operate at super/ultra critical steam pressure ranging around 230–254 kg/cm^2 (g) where actual boiling has already ceased to occur just over the critical pressure of 22.64 MPa or 222.36 kg/cm^2 (g) (374°C), which means the steam generator has no water and steam interface. The steam with very high pressure and temperature passes below the critical point while passing through the high-pressure turbine doing work. Therefore the term "boiler" may not be used for a super- or ultracritical pressure steam generator, as no boiling takes place in reality. This process results in less fuel consumption and therefore generates fewer greenhouse effects. Increases in pressure and temperature in SC and USC systems increase the efficiency of the Rankin cycle. It has now become possible to increase cycle efficiency from 30% up to 50% in USC with a future target of about 52%. The corresponding decrease in CO_2 is expected to the extent of 30%. The details are included in the Appendix E of this book and in the new Chapter 13 (2nd Edition).

2.1 Boiler/Steam Generator Subsystems

The boiler system comprises three main subsystems: the fuel/air/draft system, the feed water system, and the steam system. A brief discussion on each system follows.

2.1.1 Fuel System

The *fuel system* includes all necessary auxiliary systems/equipment required to make fuel available for providing the necessary heat. The equipment required in the fuel system depends on the type of fuel used in the system.

The amount of fuel and air needed for steam generation is automatically controlled as per steam demand.

Mills/pulverizers with primary air fans are required for solid fuels so as to convert them into fine powder form and make them transportable pneumatically. Feeders with feed control systems are required to convey the solid fuels from the bunker to the mills/pulverizers.

Fig. 2.1 shows the process.

The drawing shows only a typical view with the interconnection among the furnace, the mill/pulverizer, the PA fan, the FD fan, etc. There are many options, for example, the PA fan may be mill-wise or it may have suction from the atmosphere, etc. The draft system is a very important subsystem for maintaining the furnace at a subatmospheric pressure (−3 to −5 mm of wcl) and for other important purposes, as will be discussed later.

Pumping and heating units and suitable atomizing media are required for liquid fuels while compressor/boosters are required for gaseous fuels with a suitable type of burners and air supply arrangements. Fig. 2.2 is presented for reference.

Fuel burners suitable for each fuel system with an air compartment and atomizing arrangement are also provided for liquid fuel.

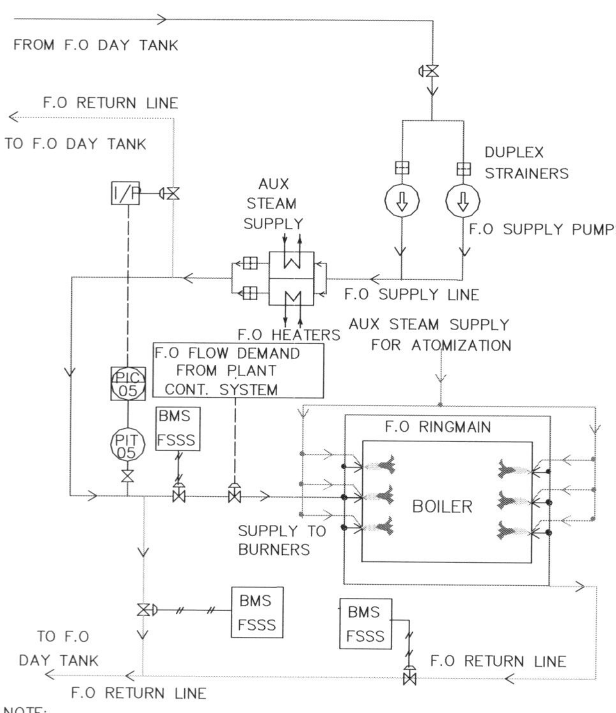

FIG. 2.2 Schematic diagram: fuel oil firing system.

2.1.2 Feed Water System (Ref. Fig. 2.3*)*

In normal running condition, feed water is supplied through ~BFP from two sources with a major part as the *condensate*, which is nothing but condensed steam from the condenser and many other heat exchangers used in the processes. The minor part (3%–5%) is the continuous and/or intermittent *makeup water* (DM water), which comes from outside the boiler proper.

It also includes *a blow down and chemical dosing system*.

2.1.2.1 Blow Down and Chemical Dosing System

This system enables the blow down of feed water to drain if needed when feed water is determined to contain dissolved solids that may corrode the boiler metal, hinder proper heat exchanging by scale formation (discussed in Chapter 5 in detail) also resulting in localized overheating and finally causing boiler tube failure, encourage foaming causing carryover of water to turbine, etc. Blow down is done in a drum-type boiler only as the same is done from bottom part of the drum or a separate blow down tank which do not exist in other kind of boiler.

Chemical dosing is done as a continuous process to retain the harmful content of the feed water under control to avoid/minimize the blow down requirement causing both heat and DM water loss. In doing that, the water-steam circuit is maintained as slightly alkaline (around a typical pH value between 8.5 and 9.4) in order to minimize the iron transport and to reduce corrosion. Chemicals dosed are phosphate at a high-pressure zone that is the boiler drum/blow down tank for pH control; sodium sulfite or hydrazine for oxygen content control; ammonia at LP heater O/L, BFP suction for pH control by measuring and controlling conductivity, etc. Pumps with automatic stroke control are provided.

2.1.3 Steam System (Ref. Fig. 2.4*)*

The steam produced in the boiler is collected and transferred to the turbine by appropriate piping. Throughout the system, excess steam pressure is regulated/vented automatically through suitable valves. Steam is superheated to get more work and to avoid condensation after work done at high-pressure turbine parts; a reheater is provided to superheat the HPT exhaust for the same reason.

To avoid congestion, desuperheaters are not shown and only the main superheaters and reheaters are shown with their tentative locations. The drum would be replaced with a separator and a storage vessel for a super/ultracritical SG plant.

2.2 Deaerator

Though not a direct part of the boiler, this provides mechanical-type deaeration for the removal of the dissolved gases (CO_2 and O_2) that are highly desirable for a boiler's

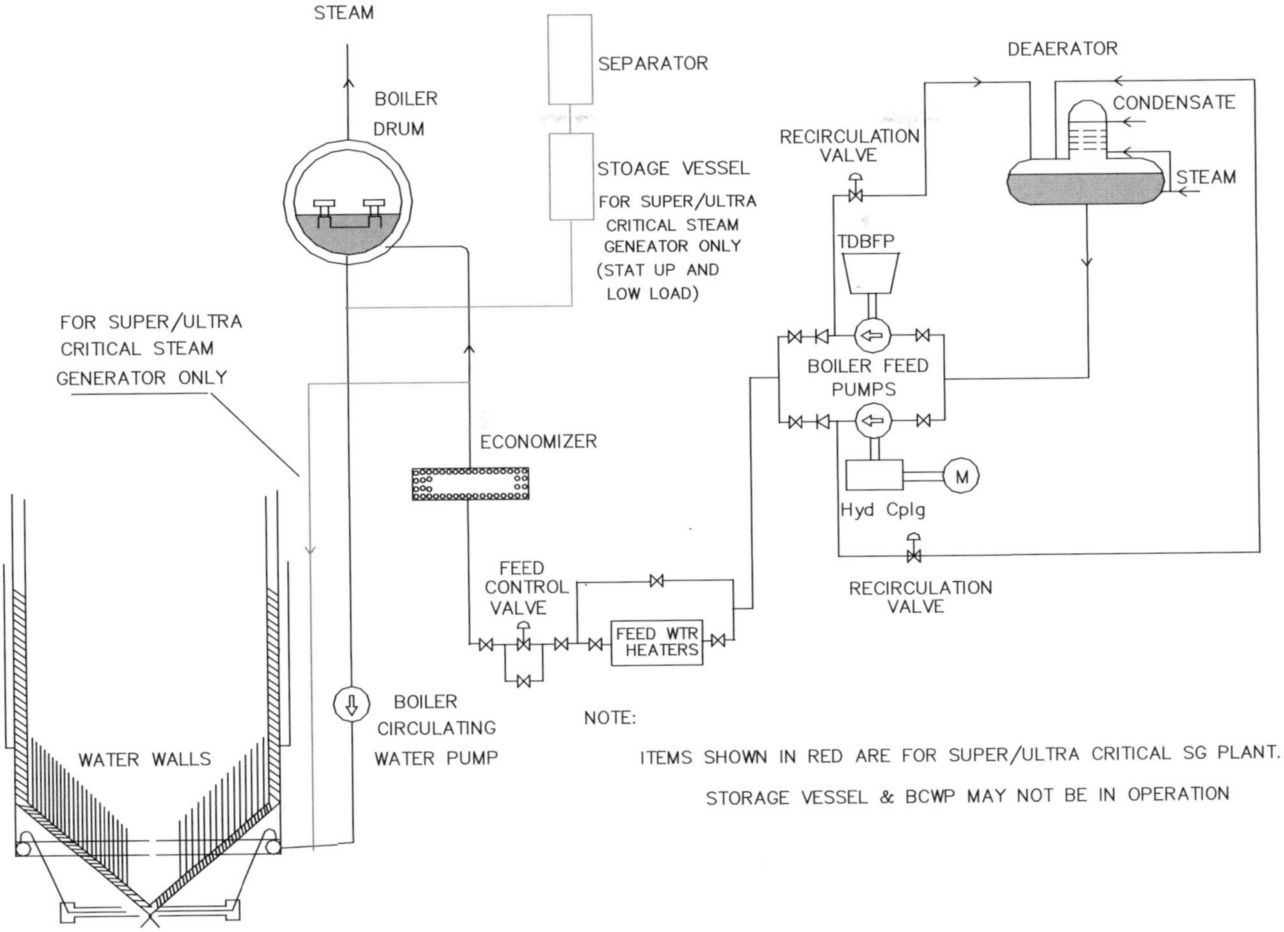

STEAM
BOILER
DRUM
SEPARATOR
STOAGE VESSEL
FOR SUPER/ULTRA
CRITICAL STEAM
GENEATOR ONLY
(STAT UP AND
LOW LOAD)
DEAERATOR
CONDENSATE
STEAM
RECIRCULATION
VALVE
TDBFP
BOILER FEED
PUMPS
M
Hyd Cplg
RECIRCULATION
VALVE
FOR SUPER/ULTRA
CRITICAL STEAM
GENERATOR ONLY
ECONOMIZER
FEED
CONTROL
VALVE
FEED WTR
HEATERS
WATER WALLS
BOILER
CIRCULATING
WATER PUMP
NOTE:
ITEMS SHOWN IN RED ARE FOR SUPER/ULTRA CRITICAL SG PLANT.
STORAGE VESSEL & BCWP MAY NOT BE IN OPERATION

FIG. 2.3 Schematic diagram: typical feed water system.

MS
HRH
CRH
FLUE GAS
LEGEND: –
(1) BOILER DRUM
(2) DOWN COMER PIPES
(3) FURNACE
(4) STEAM/WATER PIPES TO DRUM
(5) WATER WALL TUBES
(6) COMBUSTION CHAMBER
(7) BOTTOM HEADER
(8) TOP HEADER
(9) ECONOMIZER
(10) BOILER FEED PUMP
(11) PIPELINE FROM ECONOMIZER TO DRUM
(12) CIRCULATING WATER PUMP
(13) STEAM PIPE LEAVING DRUM
(14) SATURATED STEAM HEADER
(15) PRIMARY SUPER HEATER BANK
(16) PRIMARY SUPER HEATER O/L HEADER
(17) FINAL SUPER HEATER I/L HEADER
(18) FINAL SUPER HEATER BANK
(19) FINAL SUPER HEATER O/L HEADER
(20) DRUM SUPPORT(INDEPENDENT)
(21) PRIMARY REHEATER BANK
(22) FINAL REHEATER BANK
(23) STEAM SEPARATOR
(24) HOT REHEAT HEADER

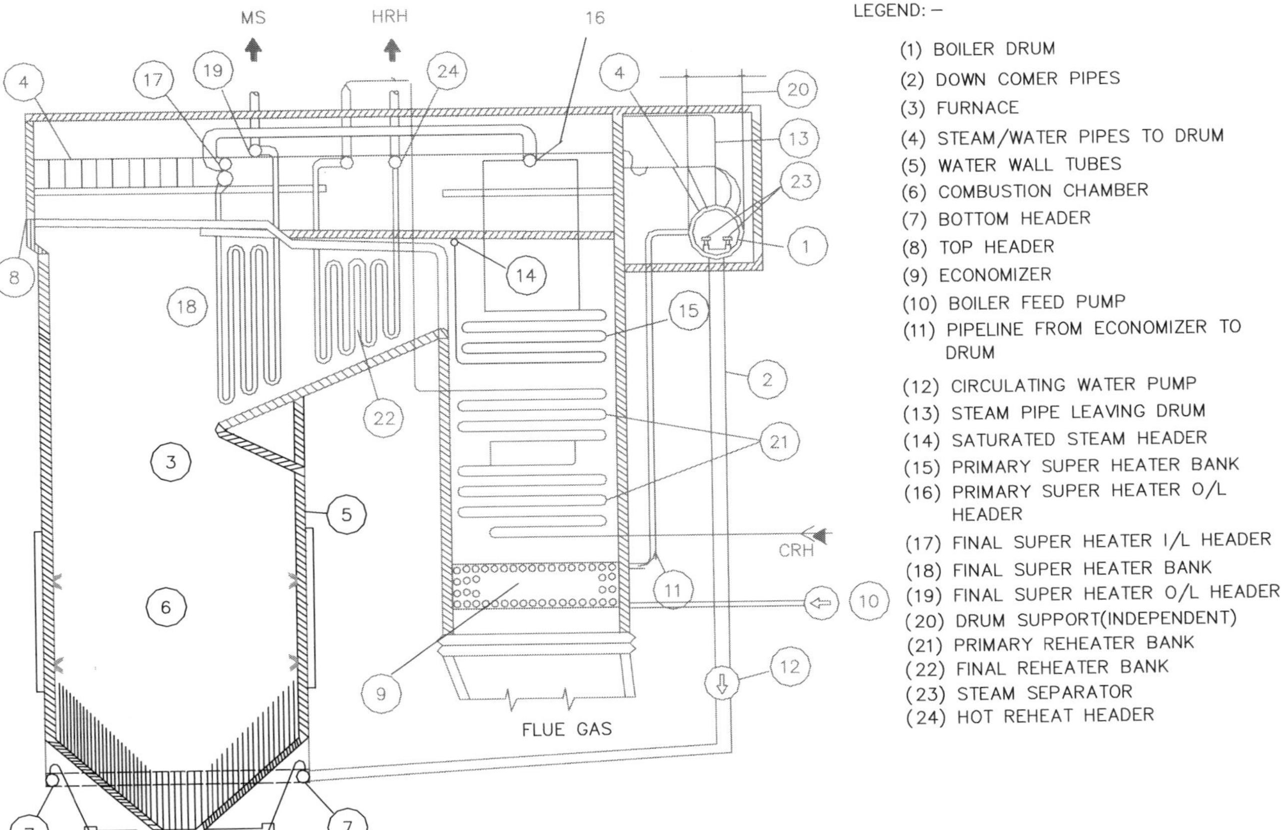

FIG. 2.4 Schematic diagram: steam system.

safe operation. The same is accomplished prior to dose chemical oxygen scavengers (hydrazine) by heating feed water by steam to reduce the concentration of oxygen and carbon dioxide from the feed water. Being most economical, it can reduce the oxygen content up to 0.005 mg/L. Steam is preferred for deaeration because steam is essentially free of O_2 and CO_2. Readily available steam adds the heat required to complete the process.

2.3 Soot Blowing System

It was already discussed in brief in Section 1 of this chapter.

2.4 Boiler Circulating Water System

In some plants, the BCW pumps are deployed for better circulation of feed water in the boiler with its suction connected to the steam drum and the discharge connected to the boiler waterwall header. Furnace height may get reduced for having BCWP. For a super/ultracritical SG plant, this pump is necessary for start-up and low load operation. In normal load operation, it is kept off through interlocked operation.

2.5 Mills and Pulverizers

For solid fuels, mills and pulverizers are supplied with the main equipment so as to achieve higher thermal efficiencies through burning of finely powdered fuel. The type of mills and pulverizers and associated feeders depends on the type of fuels.

2.6 Pumps and Heating Units

For liquid fuels, supply pumps are needed; heating units are additionally required for viscous fuel, namely for heavy fuel oil (HFO), low sulfur heavy stock (LSHS), etc.

2.7 Fans

Air is the most desired constituent after the fuel for the combustion process; the types of fans used in the modern thermal power plants depend on the type of fuels.

2.7.1 Forced Draught Fans

The air is supplied through the forced draught (FD) fans and air ducts. Before reaching the combustion chamber, the air is preheated through the air heater to increase the overall boiler efficiency. Furnaces usually operate on negative pressure for various reasons. Automatically controlled dampers are part of this system to control the desired quantity of air maintaining the differential pressure of supply air and furnace.

2.7.2 Induced Draught Fans

Induced draught (ID) fans are provided to maintain a negative pressure in the furnace by sucking the products of combustion from the furnace with a slight positive pressure at the discharge end near the bottom of the chimney/stack. This positive pressure assisted by the "stack effect" of the heated chimney (120°C approximately) causes the flue gas to leave the chimney. Automatically controlled dampers are also a part of the SG plant that maintains the approximate furnace draught within the safe margin.

2.7.3 Primary Air Fans

The function of the primary air (PA) fans is to transport the fine grains of solid fuels with the required quantity and temperature with automatically controlled dampers through proportioning of cold PA with hot PA obtained by heating a portion of PA with the help of air heaters.

2.7.4 Seal Air Fans

Seal air (SA) fans provide air at a slightly higher than PA pressure to seal all the dust-laden equipment, namely the mills/pulverizers, associated dampers, relevant parts of the feeders, etc.

2.7.5 Scanner Air Fans

The function of scanner air (ScA) fans is to cool the "flame scanner" heads meant for detecting the presence and quality of fuel flame.

2.8 Fans: Functions of ID, FD FANS, Effect on Control System, Control Devices

Sections 1 and 2 of this chapter briefly discuss the vital roles played by different fans in the SG plant; additional information is incorporated below.

2.8.1 Fans

The coal (solid fuel)-fired thermal power plant normally consumes approximately 5%–8% of the electricity it produces as the auxiliary load. This auxiliary power requirement broadly includes mills/pulverizers; ID, FD, PA, and other fans; motorized BF pumps, CW pumps, BCW pumps, etc.

Auxiliary drives are designed to operate reliably under all operating conditions. By applying various methods, this allows continuously variable load operation matching with demand.

Air is the most desired constituent after fuel for the combustion process. Many types of fans are used in modern thermal power plants, depending on the type of fuels. The supply of air and the withdrawal of the combustion product are handled by the fans.

Speed control technology is an alternative way to mechanical control system for reducing the auxiliary power requirement. Of the above drives, fans are devices used to produce a gas (including air) flow and are mainly associated with the SG package. Different types of fans available in industries such as axial (-flow) fans and radial fans are mostly used in a thermal power plant while cross-flow fans and mixed-flow fans are used elsewhere.

2.8.1.1 Axial Fans (Ref. Fig. 2.5A)

Axial fans have blades where outlet air flows in the direction along the shaft or axis around which the blades rotate; hence, they are so named.

This type of fan is used in a wide variety of applications ranging from small cooling fans for computers/electronics to the giant fans used in very large power plants such as ID fans. Axial fans are characterized by low pressure output with application from a low to high flow range.

Generally, axial fans are classified according to their type of wheel blade, as indicated below:

(i) K-wheel type: fixed blades, simple construction.
(ii) A-wheel type: adjustable blades (in standstill condition), highly efficient.
(iii) C-wheel type: adjustable blades (even in running condition), highly efficient, and air volume is highly variable.

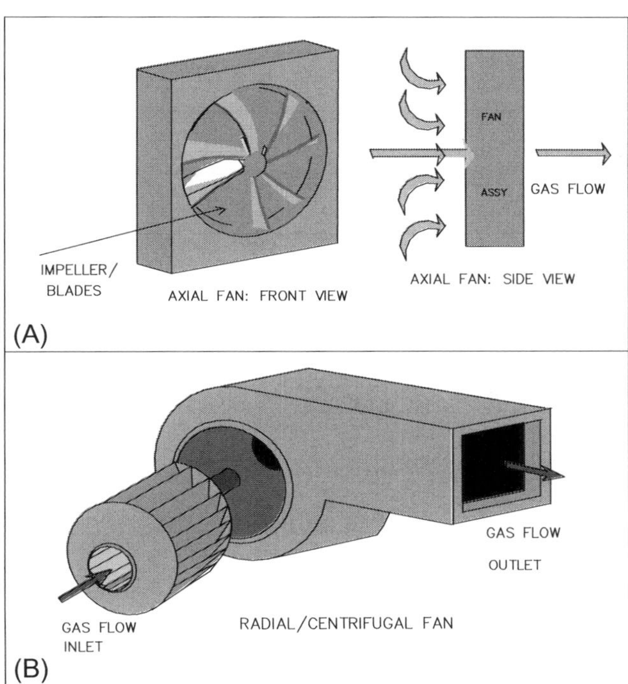

FIG. 2.5 Schematic diagram: axial and radial fans.

2.8.1.1.1 Control Device of Axial Fans Axial fans are provided with two basic type of drives, that is, fixed and variable speed, with control devices as indicated below:

(a) Fixed-speed drives

(i) Variable-pitch control

A variable-pitch controlled fan is used for precise control of static pressure inside the duct or large chamber. The fan wheel rotates at constant speed with blades arranged to revolve around a control-pitch hub to change its angle of attack, thus enabling change of the axial flow.

Outlet Damper Control

The flow/pressure control is achieved by continuous positioning of the fan outlet damper; here, system resistance is the ultimate controlling agency that, when increased, lowers the efficiency.

Inlet Damper Control

Here, the flow/pressure control is achieved by continuous positioning of the parallel inlet damper of the fan. If properly installed, it provides an inlet swirl in the direction of the fan rotation, which would cause a reduction of fan power consumption due to prerotation of the air's incoming course.

(ii) Inlet vane control

The flow/pressure control is achieved by adjusting the pitch of the vanes located at the fan inlet. This is another method that offers improved efficiency compared to the outlet damper control.

(b) Variable-speed drives with hydraulic coupling

A variable speed fan has many number of discharge head versus flow characteristic curves at each speed similar to a fixed speed fan having only one speed. The curve moves up or down with the same shape when the speed is increased or decreased, respectively.

Similarly, the efficiency also attains its optimum value at a certain point in the discharge head-flow curve for each speed, whereas only one point is available in the case of a fixed-speed fan. Because the speed can be varied, the best efficiency point can also be moved to suit the requirement.

The speed variation can be achieved though hydraulic coupling and variable frequency drive (VFD), though its application is not suitable for very high power requirements.

Hydraulic coupling envisages a hydraulic connection between the electric motor and the fan (or pump) in an oil chamber. The oil level in that chamber determines the speed of the driven equipment. The servodrive deployed for the oil level control is called the "Scoop Tube" and is widely used for ID fans and BFP speed control vis a vis suction/discharge pressure control; Chapter 6 may be referred to for detailed discussion.

2.8.1.2 Radial/Centrifugal Fans (Ref. Fig. 2.5B)

The radial/centrifugal fans have an impeller as the moving component, which is nothing but a set of blades or ribs suitably placed around a central shaft; they are noisier. They are sometimes called "Squirrel cages" or "Scroll fans."

In centrifugal fans, the air flows through the action of deflection and centrifugal force. When the impeller rotates, air is allowed to enter the fan around the open space between the impeller and the shaft and forced to proceed at a right angle to the shaft at the outlet. With respect to axial fans, gas comes out here at higher pressure with less flow.

Radial/centrifugal fans also are classified according to their type of wheel blade:

- **(i)** F-wheel type: Forward curved blades. Highly efficient.
- **(ii)** B-wheel type: Backward curved blades. Highly efficient. Air volume change is negligible with changing pressure. Low noise and low energy consumption.
- **(iii)** P-wheel type: Backward and straight blades. Highly efficient and air volume change is negligible with changing pressure. Self-cleaning type.
- **(iv)** T-wheel type: Radial and straight blades. Self-cleaning type.

2.8.1.2.1 Control Device of Radial Centrifugal Fans For radial/centrifugal fans, in general, two basic type of drives—fixed speed and variable speed—are applicable, for which control devices are provided with radial fans.

In fact, all control devices for axial fans are applicable for radial fans also, except the fan blade pitch control (meant for axial fans only).

2.8.2 Functions of ID and FD Fans

The word "draught" means the force needed to draw something gaseous, which may be the outcome of the differential pressure (DP) created or may be the current of gas causing the flow. In any furnace, the first thing to support combustion is to supply adequate air along with fuel by forced draught fans and to extract all the products of combustion with the help of induced draught fans. Induced draught fans together with natural draught fans contribute to proper and efficient combustion and easy supply as well as evacuation of gases.

2.8.2.1 Functions of FD Fans

2.8.2.1.1 Forced Draught Fans (FD Fan) The combustion air is supplied through the FD fan and air ducts. Before reaching the furnace chamber, the air is preheated through the air heater in order to increase the overall efficiency of the boiler. Normally, radial/centrifugal fans are deployed for FD fan service, which would offer more discharge pressure and flow, as is expected. The FD fans supply the total air or most of the combustion air required, which is divided into many categories such as fuel air, auxiliary air, overfire air, PA (if suction taken from FD), etc. They are controlled to meet different types of process variables, but the total air flow as a whole is maintained so as to keep the oxygen percentage level in the flue gas outlet at the preset value.

2.8.2.1.2 Induced Draught Fans (ID) After burning the fuel in the furnace, the product of combustion with the residual heat is guided to pass through the economizer to preheat the feed water and then through the air heater to preheat the air to make it suitable for combustion. Furnaces usually have a negative pressure for various reasons and automatic control devices are part of this system.

2.9 Air Flue Gas Path Equipment

The air and flue gas path covers numerous pieces of equipment such as both the primary and final superheaters (SH), the reheaters (RH), the air heaters (AH), the economizers, the selective catalytic reduction (SCR) plant, the electrostatic precipitator (ESP) as a part of the flue gas dust collector system, the flue gas desulfurization (FGD) plant, etc. The pressure parts are taken care of in clause no. 2.4 of this chapter.

2.9.1 Air Heaters

Economizers and air heaters in combination are often referred to as a heat trap. Details on economizers are discussed in clause no. 2.4 of this chapter. The air heater is the last element in the flue gas path where the residual heat contained in the flue gas is utilized in air drying for better, more spontaneous, and more efficient combustion of fuel as well as lower stack outlet temperature as per local pollution control board stipulation. The effect on fuel savings is in the range of 5%–10% by installing air heaters. The preheated air for the combustion process increases boiler efficiency practically at all loads.

For bigger capacity modern power plants, tubular or regenerative air heaters are best suited and are used for higher gas and air pressure as per the recent trend.

2.9.1.1 Classification of Air Heaters

The principle of operation dictates the classification of air heaters, which are generally of two kinds: recuperative or regenerative air heaters.

2.9.1.1.1 Recuperative Air Heaters They are mostly of a *tubular type* (may be *plate type* as well to some extent) where the heat input may be from the boiler flue gas or a separate source, namely other furnaces or steam coil, etc. Here, the heat exchange takes place directly from the source on one side of the heater surface to the air on the other side of the surface.

Tubular-type air heaters consist of bundles of vertical tubes through which the flue gas flows upward or downward. Cold air flows horizontally across the external surface of the tubes. No of traverse is multiplied by putting baffles in the air path for proper heat exchanges. In another design, the air flows through horizontal tubes and exchanges of heat take place with the hot flue gas flowing upward or downward across the tube bundle. For efficient heat exchanging, there may be a number of separate horizontal tube bundles, one above the other.

2.9.1.1.2 Regenerative Air Heaters These are normally two types—*rotary and stationary*—with the heat source being the boiler flue gas or the separate other furnace, etc.

The *rotary type* is depicted in Fig. 2.6A.

Here, the heat exchange takes place indirectly from the hot flue gases to the air through some intermediate medium acting as heat storage. The main advantages of rotary regenerators are economy of space, greater thermal efficiency, and lower capital costs. One major disadvantage of this type is leakage, which can be as high as 10%.

Stationary plate-type regenerative air heaters deploy the same basic heat transfer principles for the rotating-plate regenerative preheater. Fig. 2.6B incorporates a sketchy idea for the same. The heat-absorbing element or plates in this type of regenerative air preheater are stationary against the rotating air ducts so that the heat-absorbing plates are alternately exposed and the heat exchange is accomplished.

Hot flue gas enters from the top of the preheater and heat is absorbed by the unexposed portion of the static plates (not blocked by the rotating air outlet ducts passing above the plates); it flows downward toward the exit. Due to the slow rotational movement of the air ducts, the air inside of it absorbs heat from the static plate they pass over the hot sections previously exposed. The incoming air is thus heated and flows upward toward the exit.

2.9.1.1.3 Steam Coil Air Preheater and Corrosion Control The steam coil air preheater (SCAPH) is located just before the air heater in the air path so that the air temperature increases while entering the air heater. Here, steam is passed through the coil whereas air passes through the shell to exchange heat from the auxiliary steam to heat the air. During the cold rainy season, or in some cases of start-up, this is used to heat air before admitting to the air heater. This system is useful to avoid acid corrosion in the air heater. Here, the average of gas outlet and air inlet temperatures are taken as the measured variables and controlled against a calculated preset temperature value, which will not allow the acid condensation as discussed above.

Corrosion is a common phenomenon that occasionally takes place when the air heater metal temperature comes down below a certain level (the acid dew point), particularly at the flue gas outlet/air inlet end. The flue gas containing *sulfur* from different types of fuels, moisture, etc., is responsible; very low concentrations of even a few ppm of sulfur combined with moisture in flue gas. This can damage metal components severely when the temperature of the surface exposed to the flue gas falls below the H_2SO_4 dewpoint temperature. Under those conditions, corrosion occurs because of the presence of a thin film of acidic electrolytes over the surface, giving rise to localized and uniform corrosion. To assess the corrosion aggressiveness of the flue gas, several techniques have been introduced:

(i) By *chemical analysis*, the SO_3 content in the flue gas can easily be determined.
(ii) By using *electrical conductivity* readings, evaluation of the acid dewpoint of the gas and the rate of acid buildup below the dewpoint can be made.
(iii) By *weight-loss measurements* taken through inserting probes in the gas stream or by determining the iron concentration in the corrosion products.

These experimental approaches are limited by the short-term characteristics of the readings and by the indirect nature of the variables obtained in the first two techniques as well as by the average corrosion values obtained during long-term specimen exposure in the third technique.

Two or three common methods are applied for the above:

(i) A portion of total air is bypassed so that heat exchange cannot take place to farther lower the metal temperature.
(ii) Arranging the recirculation of hot air from the air heater outlet to the force draught fan inlet so that the air temperature increases while entering the air heater.
(iii) Using a *steam coil air preheater*, which will not allow the acid condensation as discussed above.

2.9.2 Dust Collector Units

This major part of the air and flue gas path does not take part in the thermal cycle but plays a vital role in pollution control. It separates the fly ash from the flue gas before going to the chimney.

2.9.2.1 Mechanical Dust Collector Units

Mechanical dust collectors used in some smaller plants and cement plants utilize three basic forces—centrifugal, inertial, and gravitational—to separate clean gas through admitting dust-laden gas tangentially through the whirl vanes. Particles are subject to centrifugal force and the dragging force. Clean gas is dragged to the vortex sink and goes out as clean gas. For the dust, the inertial, gravity, and centrifugal forces compel the dust to be injected to the dust hopper below, as shown in Fig. 2.7A, the mechanical dust collector portion of the main drawings titled, "Fly ash separators: mechanical type and ESP."

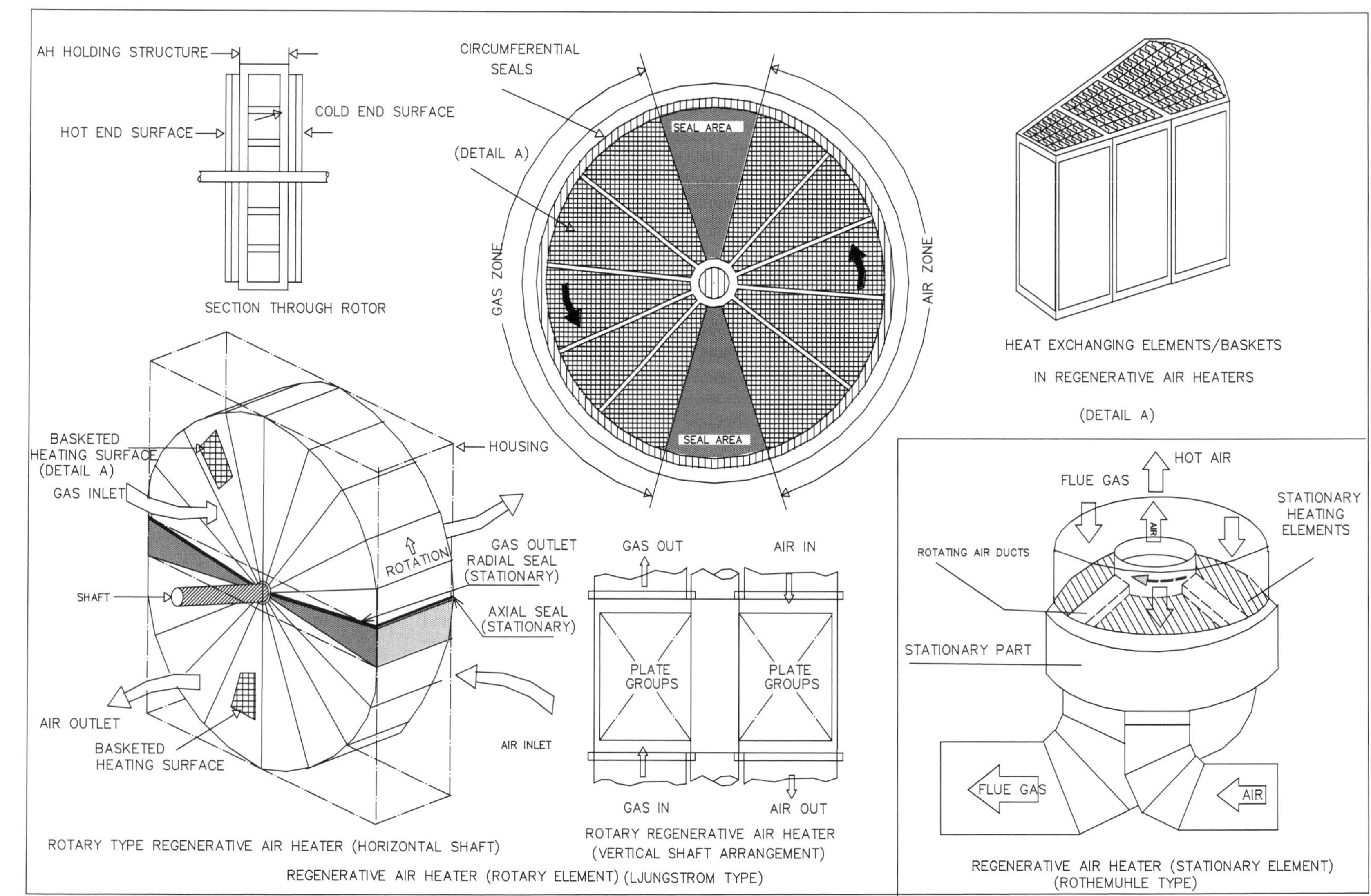

FIG. 2.6 Different types of regenerative air heaters.

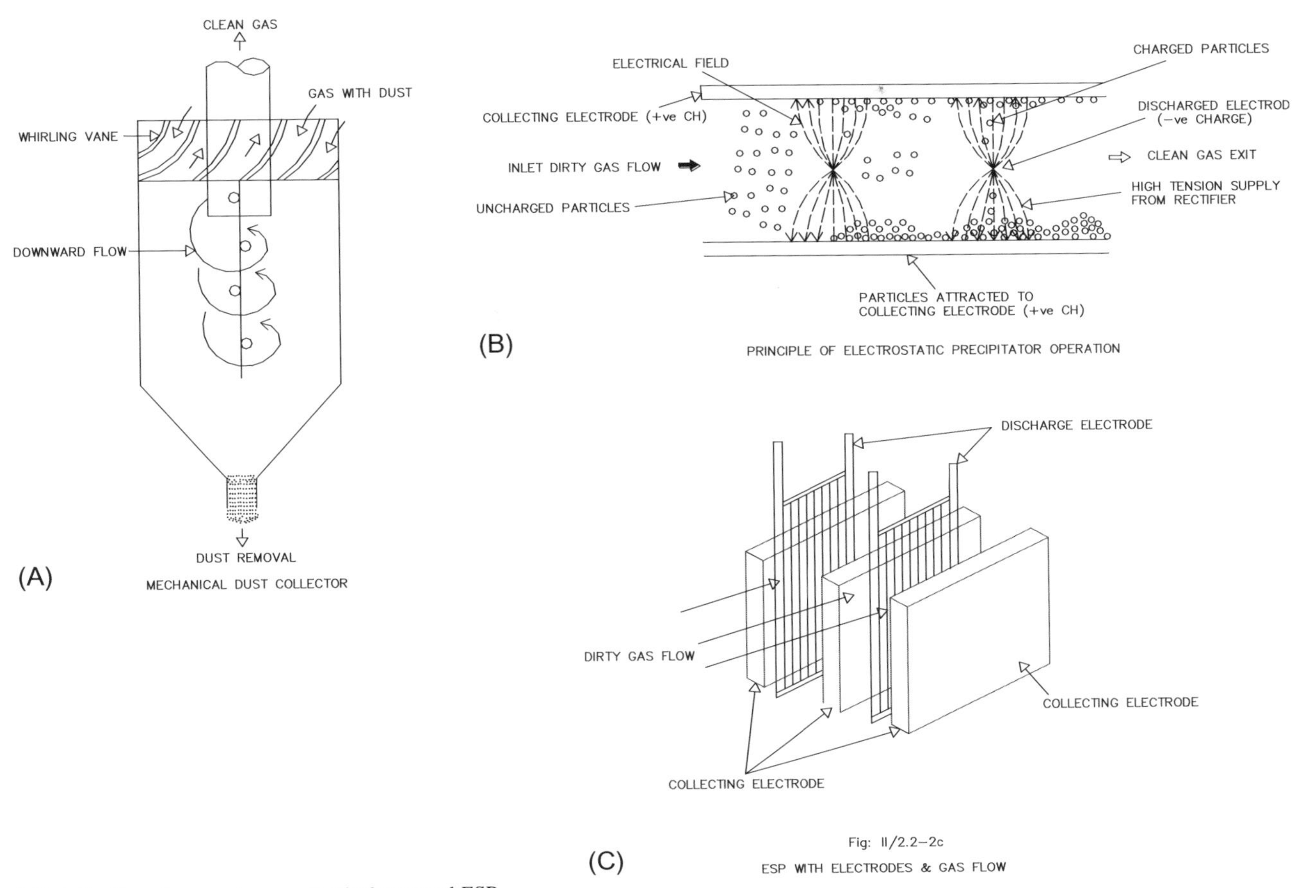

FIG. 2.7 Fly ash separators: mechanical type and ESP.

Among various other methods of dust removal such as fabric filter/bag dust filter, wet scrubber, Venturi scrubber, etc., ESPs in modern power plants are mostly used.

2.9.2.2 Electrostatic Precipitator

In order to collect the dust particles, the carrier gas is ionized by a high electric field. As shown in Fig. 2.7B and C of the ESP portion of the main drawings titled, "Fly ash separators: mechanical and ESP," flue gas is passed through the discharge electrode (basically a row of thin vertical wires) and the collector electrode (a number of large flat metal plates placed vertically) with a small distance between them but is subjected to a highly regulated electric field created by several thousand volts. The gas stream flows through the row of wires connected to the negative terminal of the DC supply and then passes through the collecting plates connected to the positive terminal of the DC supply.

The applied DC voltage is high enough to create electric discharge (corona) and ionize the gas around the electrodes. Negative ions thus emanated flow toward the positive plates and the gasborne particles are charged coming in contact while passing through. These negatively ionized particles are then attracted to positively charged grounded collector plates. The layer does not collapse due to high electrostatic pressure with the current, though small, flowing into the collected layer.

Dust is collected at the hopper below by rapping the electrodes. This is popular because it consumes much less power with a lower current while offering highly effective dust separation coupled with low draft loss combined with no effect on the gas composition. The collection efficiency depends on the collection area or sizing, the hopper design, the gas velocity (for high sulfur content gases use higher velocity, for example, 1.5 m/s (some manufacturers use the term gas volume)), the retention time (the time particles spend in the presence of the collecting electrode at design velocity, the number of fields (for reliability, ESP is divided into independent fields (3, 4, 5, 6)), field strength, power input, automatic voltage control (to enable ESP operating at peak voltage), etc.

2.9.2.3 Flue Gas Desulfurization Plant

Any type of fuel contains more or less sulfur in both organic and inorganic forms. In solid fossil fuels, such as all types of coals, for example, brown coals/lignites, bituminous or anthracite coals, etc., and in many unrefined oils and HFOs, organic sulfur is present in the form of various types of sulfides, bisulfides, etc. These organic compounds are also found in natural gases and raw fuel gases. Also, sulfur compounds are generally found as hydrogen sulfide (H_2S) and carbonyl sulfide (COS). Inorganic sulfur is available as sulfates (Na_2SO_4, $CaSO_4$, $FeSO_4$) and pyritic sulfur (FeS_2). For more details, see Section 10.3 of Chapter 8.

2.9.2.4 Flue Gas Denitrification Plant

There are several methods of flue gas denitrification that have been adopted to eliminate the nitrogen oxides (NOx) from the hot flue gas generated as a byproduct of combustion. A SCR plant is one of several denitrification methods and is the most widely used. This plant also constitutes a significant part of the pollution control measure. SCR is generally located in between the economizer and the AH where ammonia is injected into a catalytic chamber. For more details, see Fig. 2.2-4 and Section 10.3 of Chapter 8. The other methods are also briefly discussed in the same chapter.

Different subplants in a flue gas path with various combinations are shown in Fig. 2.8.

2.9.3 Stoichiometric Air-Fuel Ratio and Excess Air

The judicious amounts of fuel and oxygen are the main constituents of a stable combustion process. Theoretically, for complete combustion, the requirement of oxygen has a specific value for burning a given amount of fuel. The perfect or ideal fuel ratio is called the *Stoichiometric* air-fuel ratio. During the stoichiometric combustion process, no fuel or air is left over. But unfortunately in practice, burning processes are never ideal. Therefore, the additional oxygen, which comes from atmospheric air, than that required for ideal combustion has to be supplied for complete combustion that is to burn the fuel completely to ensure safety, better efficiency and pollution control. However, the excess air also accounts for some heat loss and fanning loss. For incomplete combustion, which means an inadequate supply of air to the burner, there are many undesirable results that take place, such as unburnt fuel with carbon monoxide, soot, smoke, etc., from the boiler. The outcome would be reduced boiler efficiency, reduced heat transfer (for deposition of soot in the pipe surface), pollution (for belching smoke), flame instability, and ultimately a potential condition for explosion due to the sporadic deposit of unburnt fuel. The term *excess air* refers to this additional amount of air requirement more than that of the theoretical value. An exact mixing proportion for air and fuel needs to be established for a stoichiometric or ideal combustion. To determine the excess air at which the combustion system operates, the value of the *Stoichiometric* air-fuel ratio for that particular fuel is required. Generally, 20% excess air flow is maintained for pulverized coal-fired boilers whereas 5% excess air flow is quite common for natural gas-fired boilers. However, in the case of modern control systems with fast response and technical advancement, strong endeavors are ongoing to bring down this "excess air" by operating the combustion near the Stoichiometric point using O_2 trim control in conjunction with CO analyzer signal injection. Gas turbine operation calls for a very high value, up to 300% excess air because excess air is utilized for cooling purposes as well.

Boilers normally operate at an excess air level not only to ensure efficient running and safe conditions but also to provide protection from the nonavailability of sufficient oxygen with the variations in fuel composition and operation snag in the fuel-air control system.

It should be kept in mind that excess air would not take part in the reaction but would take away a portion of the heat generated by the combustion of fuel after it leaves the air heater, thereby reducing boiler efficiency. Too much excess air may also disturb the flame stability. All these aspects may be taken into consideration while deciding on the percentage of excess air to be provided.

2.10 Firing System

There are different types of firing systems, depending on the manufacturer's design. There are many manufacturers with two major types of firing systems: fixed burner fired or a tilting type with corner fired burners. The detailed control philosophy is included in Sections 4 and 5.2 for mill to be put suitably of Chapter 8.

2.10.1 Type of Firing Systems

2.10.1.1 Fixed-Type Firing

This type is generally used in the Babcock design and also has different options. The following are those types (Ref. Fig. 2.8):

- **(i)** Front or wall firing. In this type, the burners are placed on the front wall or the side walls of the furnace. Each set of burners connects to one pulverizer to maintain balance. The air required for each burner enters the furnace adjacent to the burners.
- **(ii)** Opposed firing. Here, the burners are placed in opposite walls in equal numbers and in symmetrical disposition.

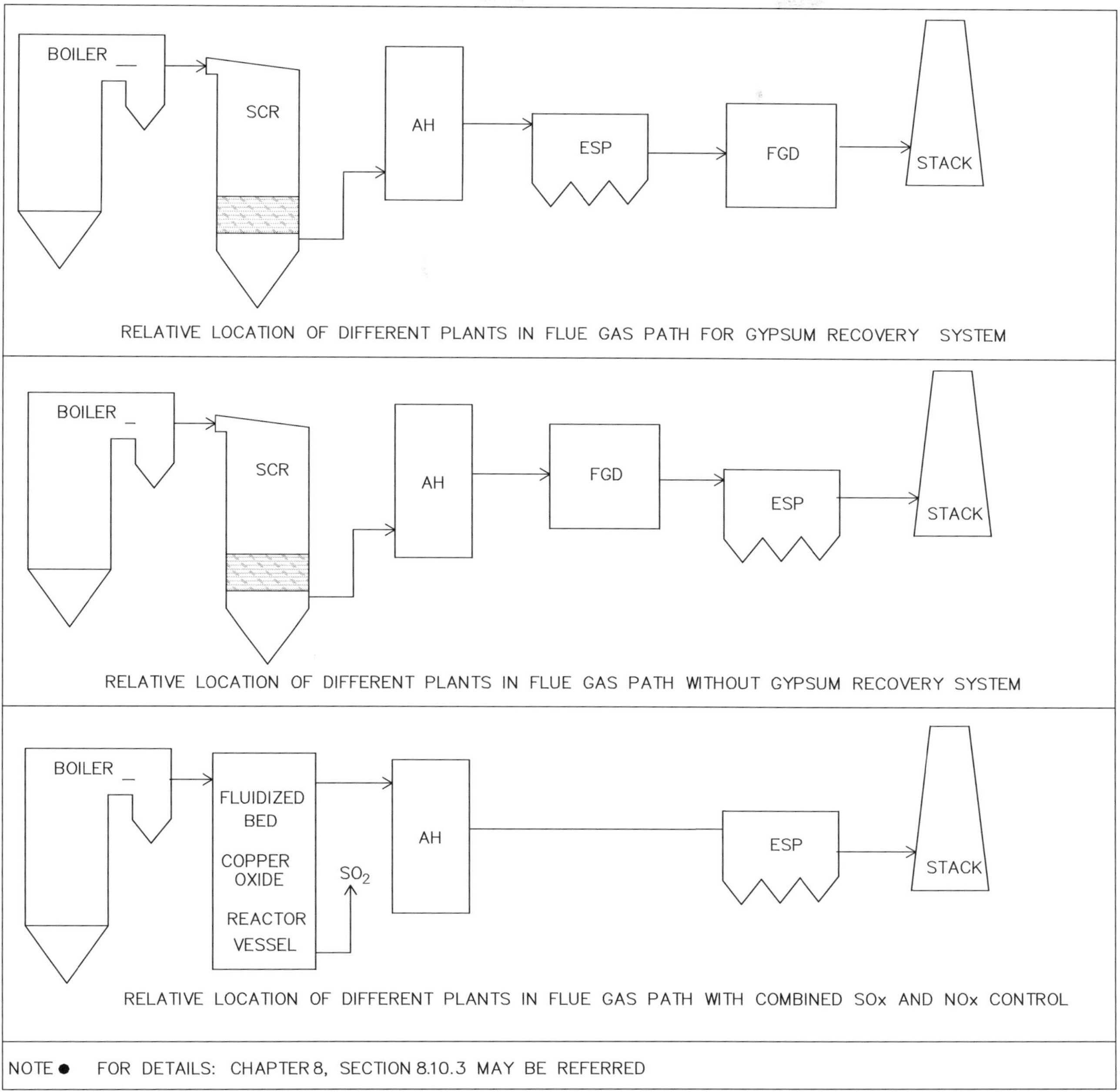

FIG. 2.8 Relative location of different plants in flue gas system.

(iii) Downshot firing. In this type of furnace, the burners are placed in such a way that the flames are pointed downward. This type is normally used with anthracitic coals or other type of fuels that take more time to burn.

2.10.1.2 Corner or Tangential Firing

Here, the coal feeds into the furnace at the four corners of the furnace. This produces a cyclonic effect in the furnace, ensuring a proper mixing of air and fuel that is essential for proper combustion. Due to this effect, a fireball is created with the burners firing tangential to it. The combustion air also enters into the furnace tangentially to the fireball above and below these burners. Each pulverizer supplies coal to one level so that the firing is always balanced. In most of the tangential firing systems, the burners can tilt 30 degrees up or down, which causes the fire ball to move up and down, thus aiding in controlling steam temperature and combustion conditions. These are called tangential tilting burners.

2.10.2 Combustion Air

The air supply for the combustion of fuels is provided in two parts.

- Around 25%–30% of the air, called primary air, is for drying the coal and transporting the coal. This enters the furnace along with the coal through the burners. PA header air pressure is controlled by the PA fan and the flow is controlled by mill-wise dampers.
- The remaining air enters the furnace adjacent to the burners from air headers or wind boxes, which are air chambers on the furnace walls; this is called the secondary air. An elaborate dedicated set of control loops (part of the total system) is provided for the control of total air. For Babcock boilers, FD header air pressure is controlled by FD fans and the flow of air through the air registers/secondary air dampers to the burners. For CE boilers, total air is controlled by FD fans and the distribution of this air to the respective burners is governed by secondary air damper control.
- The burner tips and the air entry points have splitters, swirlers, and vanes to create turbulence at the burner tip. These arrangements are required with an eye to target the most popular term in firing fuels, that is, the three Ts:
    - o. *Turbulence*: Ensure proper mixing of the air and coal particles.
    - o. *Time*: Provides sufficient time for the particle to burn.
    - o. *Temperature*: Make the ignition temperature available to entering coal particles.

The total effect is to prevent the flame from being blown away from the burner while at the same time containing the flame inside the furnace.

2.10.3 Start-Up Procedure

Because the ignition point of coal is around 300°C, there should be another ignition source to start the firing. This takes place in two stages.

- The first stage is to use an igniter that provides the electric spark for supplying the initial ignition energy to ignite small gas or light diesel oil (LDO) burners. This is like starting the firing in domestic gas burners.
- This in turns lights up other higher capacity oil burners, usually firing HFO. Once these oil burners are burning and the furnace is hot enough, only then is permission for coal firing given. Because there is sufficient ignition energy available, the coal particles easily catch fire. Once the coal flame is sufficiently strong and stabilized, the oil burners are cut off.

However, an alternative system, the direct firing of HFO by a high energy arc (HEA) igniter capable of supplying the required energy for HFO ignition, is also available.

The type of burner system that is to be adopted depends on the type of coal that is to be fired. Manufacturer's preference/license agreements are also responsible in deciding the type of burners.

2.10.4 Mills/Pulverizers/Feeders/Primary Air/Burners

In a solid fossil-fired thermal power plant, the mills/pulverizers are one of the most important pieces of equipment to make solid fuels (coal or lignite/brown coal, for example) in the correct form or shape for catching the flame appropriately. Details about the mills/pulverizers are discussed below and their control systems are discussed in Sections 4 and 5 of Chapter 8.

Coal mills, also known as coal pulverizers, are basically coal crushers. Inside this equipment, there are different grinding arrangements used to reduce the higher coal size into a smaller size for further processing or transportation by air for burning. Coal is pulverized/crushed/ground into a fine powder and is capable of being burnt in the same way as a gaseous fuel burns. This method enables the solid fuel to take part in combustion in a very efficient way. Although there is a very fine demarcation between pulverizers and mills, they can be taken as the same so far as getting the powdered fuel is concerned. To be more precise, pulverizers often mean using pressure or collision-oriented force to crush materials. On the other hand, mills are known to apply grinding force to achieve the same end product.

The advantage of efficient combustion has automatically prompted users to go in for coal mills, but there are some disadvantages as well that could not be eliminated but instead reduced to a lower acceptable level. The main and significant disadvantage of considerable vibrations needs to be taken care of through proper design improvement and using an appropriate bearing arrangement with a lubrication system. The pulverizing or milling agent may be any of the following four types.

A vertical roller mill or bowl-type mill for coal utilizes heavy rollers that have also been hydraulically loaded in order to crush the coal through application of pressure by means of a continuously rotating table passing underneath the rollers. In this type of coal mill, hot air is introduced through an opening at the bottom of the mill and is used to transport the pulverized fuel from the top, through a cyclone separator, to the corresponding set of burners.

A *ring-roll and a ball-race mill* or a *ball-type mill* for coal are basically the same type of coal mill, except that a ring-roll mill uses rollers and rings and a ball-race mill uses balls and races or raceways. In this type of coal mill, the ring or race is rotating while the different sizes of heavy metal balls or rollers are stationary; springs force the two together to crush the coal.

An *impact mill* or a *hammer type* is a high-speed type of coal mill. In impact mills, the coal is fed by means of an inlet box at the top of the mill. Crushing occurs in two steps: first, the coal is crushed using primary beater tools at the top of the rotor, then grinding tools are used to crush the coal again at the side of the rotor.

Ball-tube mills are constructed mainly of a rotating tube that contains cast alloy balls. The raw coal is inserted into either side of the mill through two hollow trunnions. In order to pulverize the coal, the tube is rotated and the balls are used as grinding media in order to deform and/or break apart the coal through the use of friction.

2.10.4.1 Selection of Equipment

Selection of the mill/pulverizer plays an important role for providing efficient fuel flow vis-à-vis combustion control. The selection considerations of equipment include the cost, the type of fuel, efficiency, capacity and space requirements, power consumption, noises and vibrations, sealing arrangements, capabilities toward separating and taking out harmful materials such as magnetic components, etc.

2.10.4.2 Advantages of Using a Mill/Pulverizer

There are manifold functions that are performed by the mill/pulverizer. The performance of a power plant depends greatly on that of the mill/pulverizer. The following items are some of those features:

(a) Faster response toward load changing as pulverized fuel (PF) burns easily and quickly.
(b) Transportation of PF through primary air is easier.
(c) Increased thermal efficiency due to less carbon loss. There is a provision of low excess air requirement but that also depends on the optimization through proper application of the control system.
(d) Different sizes and varieties of solid fuels are acceptable when this plant is used.
(e) It is permissible to burn PF in combination with other gases and oils.

2.10.4.3 Functions and Controls

The milling/pulverizing plant with associated auxiliaries is installed to perform multifarious activities. The auxiliaries are the mill/pulverizer, feeders, primary air (PA) fans, seal air (SA) fans, scanner air (ScA) fans, fuel burners, igniters, etc.

2.10.4.3.1 The Mill/Pulverizers Hot primary air is supplied to dry and convey pulverized feed from mill/pulverizers to the burners. This is done by proper mixing of hot and cold primary air with the help of an automatic temperature control loop discussed in Sections 5.1 and 5.2 (Mill outlet temperature control for ball tube mill) of Chapter 8. The PA fan suction may be taken from the atmosphere and heated through a PA heater or from the hot FD fan after the air heaters.

Another automatic control loop unique to ball-tube mills is differential pressure across the mill to be maintained at a constant value, assuring raw feed storage capacity inside the mill. It is discussed in Section 5.3.2 (Mill drum level control loop) of Chapter 8.

2.10.4.3.2 The Feeders The feeders with variable speeds are provided to feed the raw fuel.

For mills/pulverizers other than ball-tube mills, the speed varies in proportion with the boiler load demand. The raw feed already available inside the mill/pulverizer acts as the storage for meeting the immediate requirement and then the proper feeder speed takes care of the situation by bringing back storage status within acceptable limits. The automatic feeder speed control loop is discussed in Section 4 (Fuel flow control loop) of Chapter 8.

For ball-tube mills, the speed does not vary with the boiler load directly. Here, the primary air flow varies with the boiler load and the feeder speed varies to maintain the mill DP/level (feed plus ball) as a constant storage of raw feed. It is discussed in detail in Section 5.3.2 (Mill drum level control loop) of Chapter 8.

2.10.4.3.3 The Primary Air *The primary air* does various jobs for the milling/pulverizing plant for making the concept of a direct firing system possible. Those functions are: (i) transportation of PF to the burners. The quantity varies as the boiler load varies; (ii) drying of raw feed by hot PA for effective conversion into fine particles; and (iii) PA in PF/PA mixture produced at the classifier outlet provides the initial O_2 for combustion.

For all types of mills/pulverizers, the PA flow varies in proportion with the boiler load demand. PA flow control is taken care of in the fuel flow control loop, discussed in Sections 4 and 5 (for ball-tube mills) of Chapter 8 for mills/pulverizers. It is to be noted that a minimum line velocity through the coal conduits has to be maintained to keep the pulverized but solid fuel in suspension. Due to the lack of proper fuel-air line velocity, there is a possibility of the occurrence of a dangerous situation, such as solid fuel particles may fall out of suspension and deposit in the pipe lines into separate fuel slugs. The untoward deposition may turn into a fire hazard or even an explosion. With mill outlet temperature being an important factor along with other factors such as pipe size, length of pipes, altitude, etc., they influence the line velocity and thus are required to be maintained. PA inlet pressure control for ball-tube mills is discussed in Section 5.3.3.2 of Chapter 8.

2.10.4.3.4 The Burners The burners provides the arrangement for further mixing of the PF and primary air mixture with the secondary air having the desired quantity as needed for a perfect or a near-perfect condition of combustion and deliver the whole mixture to the furnace.

In some types of burners, a part of the secondary air (called fuel air) is mixed with the PF and primary air mixture and the flow rate varies proportionally with the boiler load. The other part of the secondary air (called auxiliary air) is utilized to control the wind box (WB) to furnace differential pressure. This arrangement of using only a part of the secondary air with the PF and primary air mixture is to ensure a fuel-rich initial combustion, a precondition for a low NO_x firing atmosphere. The further combustion takes place with the help of the other parts of the secondary air.

2.10.4.3.5 The Igniters Igniters are required to light the first flame in the furnace with the help of oil and gas, as available. Normally, they are spark devices for gas and light oil. HEA or equivalent igniters are used for lighting the HFO directly. When sufficient ignition energy is available by burning gas or oil for enough time and quantity to raise the temperature, the adjacent PF burner can be brought into service, considering that the flame now would be self-sustaining with the main fuel. When oil is considered as the light up fuel, it must be atomized to minute particles to create a large surface area to come into contact with the oxygen available with the air so as to assure complete and speedy burning of oils. The same philosophy is applicable for PF where the particle sizes must be small enough to create a larger surface area for achieving the ideal condition for a proper combustion. Igniters operate in open loop control logics, including checking of start permissive signals, starting sequences, and shutdown and tripping sequences in association with respective flame scanners.

2.10.4.3.6 Seal Air Fans Seal air (SA) fans are used for sealing the fuel dust and PA mixture from coming out of the milling/pulverizing plant. The seal air pressure is kept slightly higher than the PA pressure and supplied to all probable joints/flanges of this plant. Redundant fans ensure an uninterrupted supply of pressurized seal air. Open loop safety logics are provided for these fans.

2.10.4.3.7 Scanner Air Fans Scanner air (ScA) fans are not actually part of the process for this plant. These fans supply the cooling air to the flame scanners (important for furnace safety) provided to check the health of the flames of the individual burners or of two adjacent burners. Scanner lenses need cooling media against the high temperature of the furnace. Open loop safety logics are provided for these fans.

2.10.4.4 Ignition Support

Ignition supports are not only required to light the initial flame in the furnace but also to provide the oil support with the help of oil or gas as and when necessary. Ignition support is not necessary for dry and good quality main fuel as the flame is self-sustaining. The problem arises when the fuel quality is poor or the number of mills/boiler loads is not sufficient to hold the flame, or in the rainy season when a higher temperature may be necessary for sustained furnace flame. For these eventualities, a control system must be ready to support supplementary firing, if needed.

2.10.5 F.O. Systems

Fig. 2.9 in its simplest form may be referred to get a brief idea about the fuel oil circuit without going into the actual instrumentation system, which would be available in the corresponding P and I Diagram.

For an HFO firing system, proper care is taken to maintain the temperature in relation to viscosity throughout the pipeline with steam tracing or an electrical heating arrangement, if necessary. For LDO, this problem does not arise for its low viscous characteristics. The FO system also demands a selection of suitable valves and strainers for smooth and clean handling of the fluid.

Basically, there are three steps that are provided as a complete system in large thermal power plants. These are briefly discussed below.

2.10.5.1 Oil Unloading System

The fuel oil unloading system includes the unloading of oil from the oil tankers (may be roadway truck or railway wagon) through flexible hoses to the adjacent redundant storage tanks to cater to the requirement of both start-up and emergency support. With HFO being a viscous liquid, its temperature has to be maintained properly so as to attain the required viscosity/fluidity.

2.10.5.2 Storage and Transfer System

The storage and transfer system for HFO incorporates the storage tanks with suitable heaters and temperature sensors for the control system to maintain the viscosity to enable the storage pumps to transfer the oil to the day tank to cater to the day-to-day needs. Normally, two types of heaters are provided. A "floor mat" type of heater is located near the lower portion of the inside tank to regulate the temperature and the other one is located near the outlet, which is called a suction heater to ensure the flowability of HFO connected to the suction of the transfer pump. The pipe lines are also electrically heated or steam traced through the jacket, along with the conveying pipe length. The storage pumps are accompanied by heaters and strainers so as to keep system fluidity as desired.

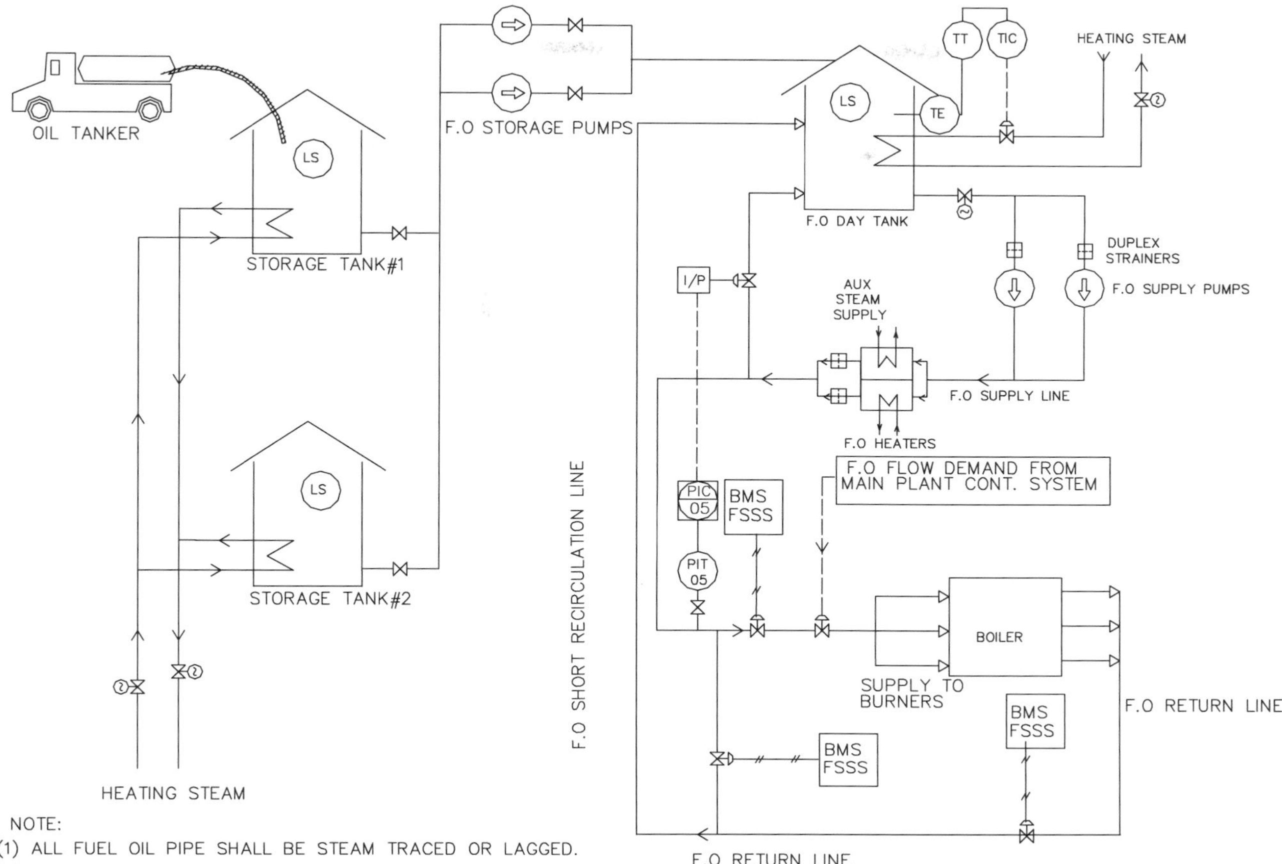

NOTE:
(1) ALL FUEL OIL PIPE SHALL BE STEAM TRACED OR LAGGED.
(2) FOR LDO LINES STEAM TRACING OR HEATING IS NOT NECESSARY.
(3) MINIMUM INSTRUMENTS ARE SHOWN.

FIG. 2.9 Fuel oil flow diagram.

2.10.5.3 Pumping and Heating Unit

This unit is a subsection that starts with the day tank outlet, supply oil to the oil burners and the return oil line up to the suction line of the day tank. This part incorporates redundant heaters (for HFO) and pumps to improve fluidity and then pressurize the oil sufficiently so as to reach the far end of the burner elevations and back to the tank (part is not required for the combustion).

Whenever the power plant is in operation, the oil flows in the above route, even if there is no oil requirement. The idea is to keep the oil system in readiness for any eventuality such as a sudden load shedding or a "run back" situation when a few mills/pulverizers are to be cut off within a very short time and the resultant flame might become unstable, badly needing the instant oil support. Oil pressure is maintained through a pressure control valve in the bypass line diverting more or less oil to the day tank. For details, see Section 10.4 of Chapter 8.

The FO trip valve (FOTV) is provided at the common inlet of the burners so as to initiate a shutdown of the oil supply instantly at an emergency situation, such as tripping of the boiler. The downstream line at the common outlet of the burners is provided with another quick closing - ON-OFF valve to facilitate oil recirculation in the plant's running condition. In certain thermal power plants, an oil leak test valve (OLTV) is provided to check whether the HOTV or the oil burners and downstream pipelines are leaking. The details about all these ON-OFF valves are discussed in Section 12.1 of Chapter 8 on BMS/FSSS.

The oil route is somewhat different for a unit under a long shutdown condition. There is a small ON-OFF valve or a regulating valve called a short recirculation valve (SRV) at the HOTV inlet and connected to the day tank. In some plants, it opens only when the unit goes on a prolonged shutdown. In some other units, it is always open to about 10% in the normal operation of the unit and opens fully in case of a long shutdown.

2.10.6 *Tangential Tilt Burners (Ref* Fig. 2.10*)*

A furnace with this type of coal burner has the advantage of raising or lowering the burning zone by the burner-tilting mechanism, as depicted in Fig. 2.10. The burners are corner fired, located in four corners, and focused in such a way that all are tangent to a common imaginary circle.

When firing starts, a huge fireball with swirling action forms. As soon as the burner is made to move vertically upward or downward, the fire ball vis-a vis the flame envelope would also move in that direction. It changes the pattern of heat transfer to the superheater and reheater banks, causing a good variation in the amount of radiation heat the reheater receives and providing an efficient method of controlling the steam temperature. This is the premier means to regulate reheater temperature control in combustion engineering (CE) manufactured boilers (Fig. 2.11).

The system so far erases the requirement for spray water and confines the use for fine tuning and emergency purposes only. In some control schemes, the system is used for controlling the furnace outlet temperature as well.

The steam temperature control system thus becomes genetically different in these two types of furnaces, that is, one with a tangentially fired tilting burner and the other having a fixed burner.

2.10.7 *Reheat Temperature Control by Tangential Tilting Burner Assembly*

The firing system in any furnace of the power plant boiler having reheater banks affects the steam temperature of both the superheater and the reheater. It is, therefore, very difficult to control the steam temperature by a single control system, as a corrective action for one steam heater bank may have an adverse effect on the other bank. To circumvent the situation, separate control strategies have been considered by the control engineers in association with the equipment designer.

Reheat steam, having a low operating pressure, is subjected to different thermodynamic conditions than that of superheater steam. There may be problems arising from injecting colder spray water and an unwarranted effect on the overall efficiency target of the power plant. For example, carryover of water particles may conflict with TWDPS stipulation. In fact, for that reason, the spray arrangement is made to inject at cold reheat and never in HRH.

The modern control system for reheater steam temperature thus normally avoids water spraying and uses a primary control system that would take care of the control of reheat steam, by either adjusting through *tilting of burners* or by apportioning of hot *flue gas* flow through the super heater and reheater banks. The provision of a spray water injection system is also kept as a last resort when the above concerned final control element has reached the extreme position and further influence is not achievable or the temperature has shot up to a particular predefined value when emergency spray is necessary. This type of control arrangement is normally provided for the boiler where the final SH bank is located in the radiation zone.

2.10.8 *Reheat Temperature Control by Gas dampers*

For boilers having *gas dampers* as final control elements for reheater steam temperature, take the task of apportioning the flue gas flow through both the superheater and reheater banks by adjusting the opening of these gas dampers.

2.10.8.1 Gas Dampers in Main Flue Gas Path

In some boilers, two separate and dedicated sets of dampers are provided in the flue gas path of both the superheater and the reheater banks controlling individual flue gas flow. As these dampers are located across the main path, they control the total flue gas volume, therefore proper care has to be taken so that both the dampers do not get a closed signal at the same time. In that situation, the flue gas path will be constricted so badly that overpressurization may damage the structure itself.

To obviate the untoward situation, the control philosophy would be such that when one particular set of dampers is fully opened, only then is the other set allowed to throttle.

Emergency spray may be necessary here as the control through gas dampers is a slow response system for large units and a sudden rise in temperature following a rise in furnace heat input to cope up with the load demand.

2.10.8.2 Gas Dampers in Bypass Flue Gas Path

Some boilers are provided with SH and RH bypass gas dampers for controlling reheat steam temperature. For any corrective action issued by the controller, the dampers operate in the opposite direction; for example, at low RH steam temperature, the RH gas bypass damper would close, whereas the SH gas bypass damper would open.

As the water spray is an emergency or secondary control system in case of reheat temperature control, the spray valves are normally shut unless the temperature at the reheat outlet reaches a predetermined higher set value. The spray valves are provided at the inlet of the reheater so that after spraying a proper quantity, the temperature is lower than the previous value. After gaining heat input from the flue gas, the outlet temperature is brought back to the normal control region vis-a-vis a set value where the primary control system and final control elements may be back in action.

In case of turbine tripping, the reheater flow would be collapsed. In that case, the spray valves must immediately be closed so that cold spray water cannot enter the turbine in any way.

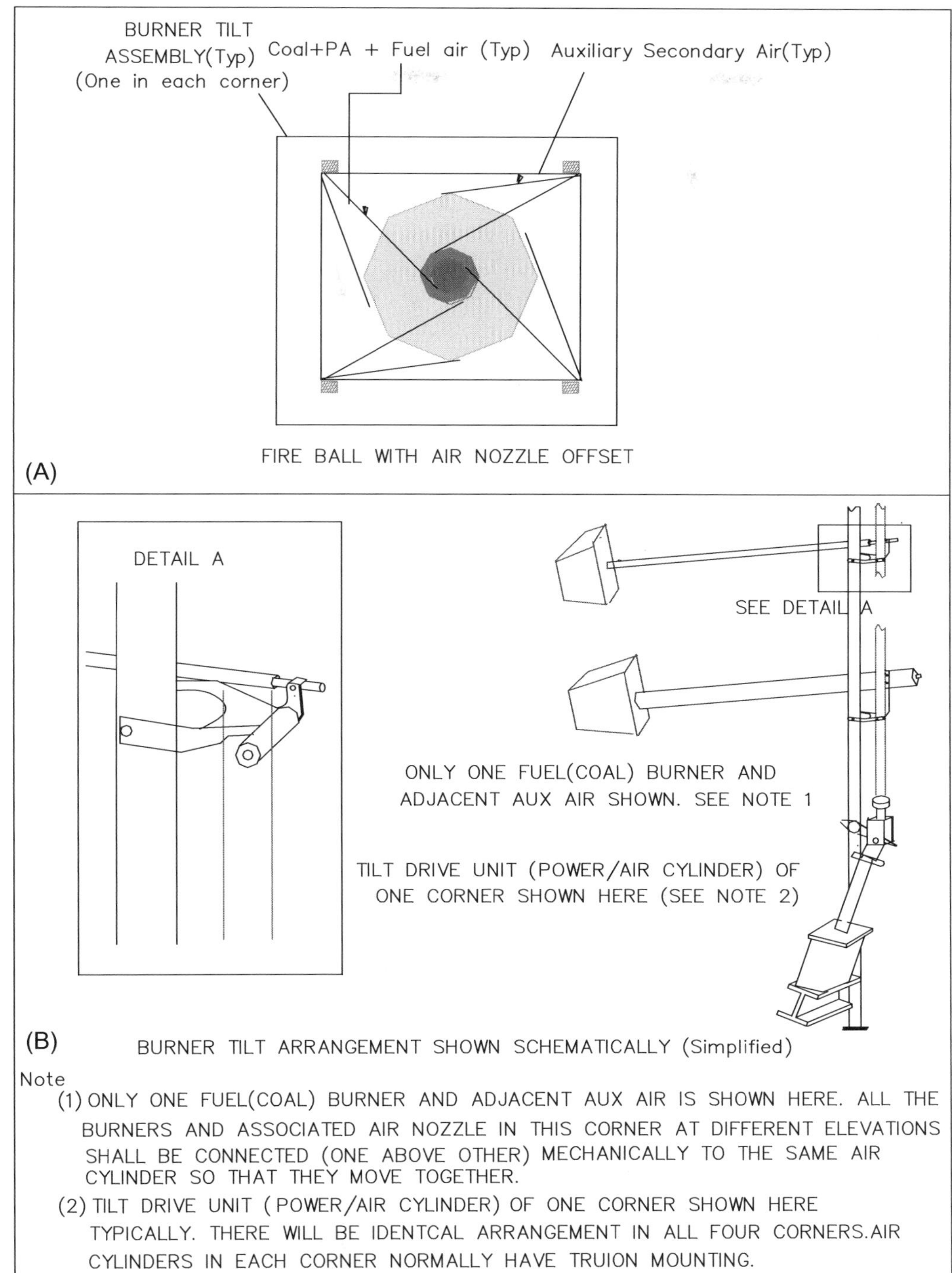

FIG. 2.10 Tangential fired coal/oil/gas tilting burners.

2.10.9 Reheat Temperature Control by Gas Recirculation Damper

With boilers with a furnace having an oil-firing system or combined oil and coal firing systems, some manufacturers provide a gas recirculation damper at a suitable location of the flue gas path through which a part of the flue gas is recirculated back into the furnace.

In general, when the gas is reintroduced near the vicinity of the primary burning zone and utilized for superheated steam of the reheater outlet temperature, it is designated

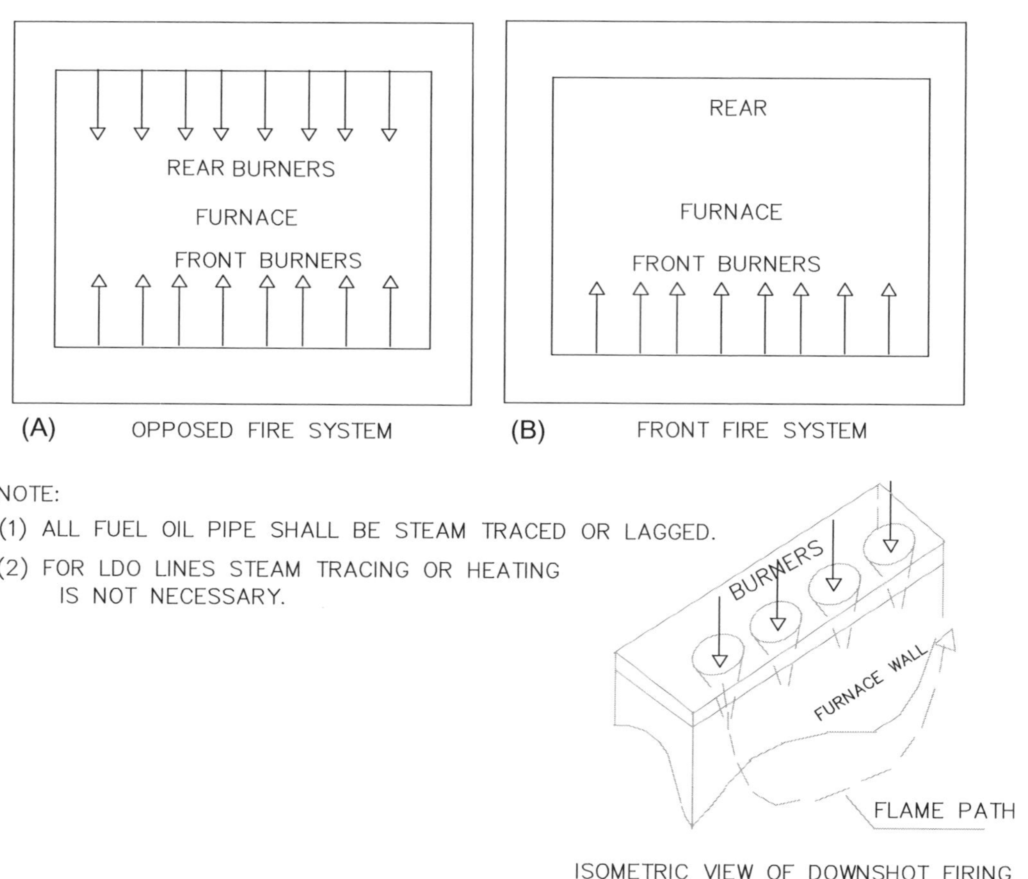

FIG. 2.11 Opposed fire/front fire/downshot firing system.

as gas recirculation; the other choice is to reintroduce the gas at the furnace outlet and the system is often referred to as gas tempering. The selections of both take off and injection points of flue gas recirculation may vary according to the design of the manufacturer, depending on the type of fuel, the size of the plant, etc.

Take off points may be from the economizer outlet, the air heater outlet or the ID fan discharge.

Reintroduction points may also be to any of the following locations: (i) Bottom part (through the bottom ash hopper) of the furnace, which is closer to the burner or in the near vicinity of the initial burning zone; (ii) the furnace outlet (first pass) or the final SH inlet; and (iii) in the combustion air supply line and sometimes a combination of the air line and bottom hoppers for the control of temperature and NOx. Flue gas recirculation is one of the most efficient methods of altering the pattern of heat absorption/distribution inside the furnace to effect the controlling means of reheater outlet steam temperature and the reduction of NOx generation as an effect of lower furnace temperature; these details are discussed in Section 10.3 of Chapter 8. By controlling the gas recirculation flow, the steam temperature of both the superheater and the reheater may also be controlled. Normally, this effect is applied to the reheat temperature control and the aftereffects are taken care of by the SH temperature control through spray water. The temperature excursions are totally dependent on the recirculation gas flow when the point of take off and reintroduction are fixed for a particular furnace. For instance, the flow of gas recirculation requirement increases at low load in achieving a reheat temperature of the desired level. On the contrary, the flow of gas recirculation is brought to the lower values when the load increases. Flue gas recirculation affects little of the total heat absorption or the total flue gas mass, which could otherwise reduce the boiler efficiency.

2.10.9.1 Flue Gas Recirculation Introduced in the Furnace

When introduction of flue gas is made in the flue gas side, this system experiences:

(i) An increase in total flue gas mass flow inside the furnace and boiler part up to the take off point, which assists with more heat transfer to the superheater, the reheater, and the economizer.

(ii) A reduction of overall flue gas temperatures due to readmission of colder flue gas. Enhanced heat transfer would enable flue gas temperatures after different types of heater stages to decrease more.

(iii) Less generation of SO_x and NO_x for colder flue gas injection near the burner, resulting in a low temperature around the burners.

When the gas recirculation take off point pressure is less than the introduction point, gas recirculation fans are essential to force the part of the return flue gas passage back to the comparatively higher pressure furnace through a long way. The gas recirculation damper is also provided and the reverse flow of very hot flue gas must be arrested by it to protect the fans in case of failure (stopping/tripping) of the fan itself.

If the take off point is made from the ID fan discharge, no additional fans are required.

2.10.9.2 Flue Gas Recirculation Introduced in the Combustion Air

Another typical flue gas recirculation take off point is from the boiler economizer outlet zone or from the outlet duct upstream of the air heater. There are two methods of this type of flue gas recirculation system:

(i) The flue gas is reintroduced through a separate line and a hot gas fan to the combustion air duct at a suitable location where the recirculated flue gas and the combustion air are thoroughly mixed by some means in the duct.

(ii) This system eliminates the need for a separate and extra hot gas fan and mixing device. This method, known as induced flue gas recirculation, utilizes the FD fan to pull (induce) flue gas from the boiler exhaust duct into the combustion air at the fan inlet. The fan also serves as the mixing device and hence this is considered a very cost-effective method. The control dampers in the flue gas recirculation path are provided to achieve the desired degree of control over the operating boiler load range. However, this system may have some relation with the FD fan design/sizing and hence may be suitably taken care of.

2.10.10 Boiler Tube Failure and Metal Temperature

The pressure parts of a boiler consist of a bunch of tubes and connecting headers containing both water and steam. These tubes are often found to be affected with various kinds of failures having its pressure boundary is damaged by a leak or rupture. They experience a loss of metals, which may make them prone to be broken due to this untoward thinning of the tube metal wall if a boiler inspection is made at regular intervals.

If the boiler tube failure is not properly addressed with the right instrumentation, inspection, and analysis, repeat or multiple failures would occur in a boiler having the same failure mechanism and root cause. There are some important factors responsible for boiler tube failures that can be categorized as the root cause related to the following items: (i) working fluid chemistry; (ii) superheater (SH) and reheater (RH) temperature control; (iii) combustion control; and (iv) problems in the firing area.

Some measuring procedures or methods have been established with the help of the right instruments at the right locations and times using radiography, hydrostatic tests, and temperature measurements, which help to prevent or reduce this type of failure. A proper life expectancy calculation is also necessary to enhance the residual life of the boiler tube metals and structures.

2.10.11 Major Reasons for Boiler Tube Failure

The following are the major categories that are responsible for tube failure: (i) tube inside or water-side corrosion; (ii) tube outside or fireside corrosion; (iii) tube outside erosion; (iv) stress rupture; (v) high-temperature graphitization; and (vi) fatigue.

2.10.11.1 Tube Inside or Water-side Corrosion

The problem is most likely to develop during boiler start-up cycles. This type of failure is observed at the inner side of the pipe with wide transgranular cracks, which typically occur adjacent to external attachments and weldments such as buck stays, seal plates, scallop bars, etc. A combination of some damaging factors leads to the breakdown of the protective magnetite of the inside surface of the boiler tube. Corrosion of the tube surface then starts in the absence of this protective layer.

This type of corrosion may typically be caused by:

(a) *Caustic corrosion*, which may be caused by:
- Random build up by FW-borne deposits or preboiler corrosion products at locations of high heat flux.
- Concentration of sodium hydroxide due to chemicals carried over by the feed water.

(b) *Pitting or localized corrosion*, which is normally affected by:
- FW with high acidic or oxygen concentrations.
- Existence of close-fitting surfaces and deposits where differences in oxygen concentration can be produced.

(c) *Stress corrosion cracking* (SCC)

The combined effect of high-tensile stresses and the presence of corrosive fluids is responsible for this type of failure. Stress corrosion cracking (SCC) failures are distinguished by the brittle type of crack with a noticeable thickness. These failure locations are also observed near the source of higher external stresses such as attachments. The corrosive fluid may come as carryover from the steam drum into the superheater or may get contaminated during boiler acid cleaning without the faulty superheater protection. Normally, austenitic (stainless steel) superheater tubes are subjected to SCC and can propagate in the tube wall from the inner wall surface with transgranular or intergranular cracks.

(d) *Hydrogen embrittlement*

Hydrogen embrittlement is a most common type of failure effected by the combined outcome of excessive deposition on inside tube surfaces and the boiler FW having a low pH value. The digression of water chemistry or acidic contaminants, that is, low pH, resulting through, for example, condenser leakage particularly with a saltwater cooling medium, may cause further deposition and concentrate on the build up already taken place. The resultant deposit initiates corrosion, which releases atomic hydrogen that propagates into the tube metal wall and reacts with the carbon present in the steel. This phenomenon is called decarburization and causes intergranular separation or intergranular microcracking. The final damage done is the loss of ductility or embrittlement of the tube material, leading to tube failure with severe rupture.

2.10.11.2 Tube Outside or Fireside Corrosion

There are quite a few types of tube outside or fireside corrosions, namely, (i) superheater fireside ash corrosion, (ii) high-temperature oxidation, (iii) waterwall fireside corrosion, and (iv) fireside corrosion fatigue.

2.10.11.2.1 Superheater Fireside Ash Corrosion Superheater fireside ash corrosion is associated with the ash characteristics of the fuel. The boiler design also constitutes an important factor to this type of tube failure. It is normally the outcome of coal firing, although certain types of oil firing also can contribute. The *very high temperatures* of combustion gases and metals passing through different types of steam heaters in the convection passes are of immense importance. The molten ash is considered highly corrosive and damage takes place when certain portions of coal/oil ash remain in a molten state on the steam heater tube surfaces. Ash characteristics are considered in the boiler design when establishing the size, geometry, and materials used.

The ultimate damage is outer side tube metal loss and consequently an increase in tube stress. Affected tubes normally are covered with spots or have a pock-marked appearance when scale and corrosion products are removed.

2.10.11.2.2 High-Temperature Oxidation High-temperature oxidation takes place at any point when the *metal outside surface temperature exceeds* the oxidation limit of the tube material. The appearance of this type of tube failure is fairly similar to that of fireside ash corrosion and they are often confused. Experts can distinguish the actual root cause between the two types with the help of tube analysis and evaluation of scales and deposits of both metal inside and outside.

2.10.11.2.3 Waterwall Fireside Corrosion Waterwall fireside corrosion is associated with the reducing or substoichiometric atmospheres taking place during the process of burning and the products of combustion that pass through the external surfaces of waterwall tubes. The damage causes external tube metal loss with a consequent thinning of the tube metal wall as well as increased tube strain.

2.10.11.2.4 Fireside Corrosion Fatigue Fireside corrosion fatigue is the combined effect of thermal fatigue and corrosion. The outside surfaces of the tube often experience thermal fatigue stress cycles, which may take place from different causes such as soot blowing, normal shedding of slag, of normal cyclic operation of the boiler. The *thermal effect of load variation or cycling* in combination with the cyclic stress of material over a long period of time may become the source of a crack on the less-elastic external tube scales, exposing the actual tube metal wall to repeated corrosion.

Because this type of tube failure is developed over a considerably longer period of time, the tube surfaces tend to appear to resemble what is described as "elephant hide," "alligator hide," or craze cracking. A series of cracks on the tube surfaces thus developed propagate into the tube wall. The long-term appearance is like a series of circumferential cracks over the tube length. This type of tube failure is normally observed on furnace wall tubes of coal-fired once-through boiler designs, but are also occasionally found on tubes in drum-type boilers.

2.10.11.3 Tube Outside or Fireside Erosion

The erosion of tube external surfaces occurs as a result of abrasion from the impingement of various gases and solids on the tube and causes loss of materials of the outside diameter. The erosion medium can be any of the following abrasive materials in the combustion chamber: (i) fly ash, (ii) coal particles, (iii) falling slag, and (iv) soot blowing steam.

It has been observed that when the soot blower steam is the primary cause, the erosion may be accompanied by thermal fatigue erosion. As the phenomenon results in metal loss of the tube's outside diameter, the damage would be visible on the impact side of the tube. Ultimate failure results from rupture due to increasing strain as the tube material erodes.

2.10.11.4 Stress Rupture

Stress ruptures short of tube failures may be categorized into three types, as indicated and discussed below.

2.10.11.4.1 Short-Term Overheating This type of tube failure often happens during the boiler start-up for the lack of following proper start-up procedure. Short-term overheat failures are the result of an extremely *high tube metal temperature* due to low or inadequate flow of cooling media through the water and/or steam tubes. Tube metal temperatures may even reach as high as that of combustion gas temperatures of 1000°C or greater, paving the way for the untoward situation. The cause may be a very simple thing such as accumulation of condensation products in the superheater tubes during boiler start up, which restricts the steam to flow to the desired minimum value and results in a ductile rupture of the tube metal where the fracture surface becomes a thin edge.

The typical physical appearance of the failure may be recognized by what is popularly termed the "fish mouth" opening in the tube.

2.10.11.4.2 High Temperature Creep or Long-Term Overheat Creep is the process where metals exposed to high temperatures and sustained stress over long periods of time will gradually deform and eventually fail. Long-term, overheat-related failures are applicable when tubes are subjected to high temperatures over a long period, say months or years. Steam superheater tubes are subject to operation under high temperature and may fail due to the "creep effect" after prolonged operation at the end of their creep life. The waterwall tubes may also fail due to prolonged overheating. Problems such as deposits, scale, etc., causing restrictions in the waterwall tubes may result in very high tube temperatures. Contrasting the physical appearance of the failed tube to that of short-term overheat, here the longitudinal split is narrow and the swelling is minimum. The build up or the external scale deposited on the tube surface after prolonged operation also develops secondary cracking.

Superheater and reheater tubes containing steam normally operate at temperatures of 550°C or above. Research reveals that working temperatures at more than 482°C cause metal tube failure due to the phenomenon called creep rupture. Oxide layers are formed on the internal surface of these tubes, which restrains heat transfer through the wall and causes the tube metal temperature to increase over the operating time.

Heater banks for superheaters, reheaters, and economizers are made of various materials such as carbon steels (CS) in the cooler sections or stainless steel grades in the hottest outlet sections. The allowable stresses in superheater designs using alloy materials for the hottest sections are based on a finite creep-rupture life set by the ASME code. It is imperative that quite a few numbers of these tubes would be sure to rupture due to the creep effect, even if some other factor does not cause the premature failure of these superheater tubes.

For alloy superheater tubes, stress and temperature are not usually constant during a tube's service life. When a tube enters service, the metal in contact with the internal steam begins to form a layer of magnetite (Fe_3O_4) in the form of scales. This oxide grows thicker as the service life increases and its growth over time is dependent *on metal temperature*. This oxide layer acts as a baffle plate to heat transfer and as its thickness increases, the metal temperatures must also increase to maintain a constant outlet steam temperature. Typically, tube metal temperatures increase more than 1°C per scale thickness of every one-thousandth of an inch, or 0.03 mm of ID oxide formation. In addition to the metal temperature increase, tube wall thinning due to OD metal loss for problems such as erosion, corrosion, or other environmental damage may take place over time. The resulting effect—the loss or thinning of the tube wall—would obviously turn into increased stress in the affected tubes operating at a fairly constant operating fluid pressure.

There are some methods that calculate the *creep life prediction* of SH metal tubes through measuring the real time tube metal temperature and scale thickness vis a vis tube stress.

2.10.11.4.3 Dissimilar Metal Welds Dissimilar metal welds (DMWs) describe the butt weld where two dissimilar metals are welded. For example, in a power plant, an austenitic material like stainless steel is welded to a metal-like ferrite alloy. The peculiarity of this type of failure is that the damage due to failure at DMW locations takes place only on the ferrite side of the butt weld. Many factors are responsible for this type of failure: (i) high stresses at the austenitic-to-ferrite interface due to dissimilar expansion of two materials, (ii) excessive external loading stresses, (iii) excessive thermal cycling, (iv) creep characteristics of the ferrite material, and (v) variation or continuous high operating temperatures.

The symptoms of failure are characterized normally by no or least warning of tube damage condition. As discussed earlier, the damage occurs at the ferrite side of the weld throughout the circumference of the weld fusion line. This type of failure may lead to catastrophic damage in nature as the entire tube will fail across the cross-section of the tube.

2.10.11.5 High-Temperature Graphitization

This type of failure is caused mainly due to the *high temperature of the metal tubes* of the plant while operating for a long time. Boiler tubes made of carbon steel or a similar type of steel with a higher carbon content may experience a unique type of damage causing degradation of the steel metals, which is popularly called graphitization. The term degradation is used as these materials, if exposed to excessive temperatures, would cause disbanding of the iron carbide molecules in the steel and graphite nodules are formed as a consequence. The ultimate effect of this phenomenon is lower strength of materials and tube failure with a thick-edge, brittle-type fracture.

2.10.11.6 Fatigue

Fatigue as is well known for its extensive damaging effect and caused by the cyclical stresses in the component attributable to various factors such as mechanical vibration, thermal cycle, and corrosion fatigue.

2.10.11.6.1 Mechanical Vibration Fatigue The tendency of this type of fatigue is to affect the area of the materials near the vicinity of high stress or constraint. Damage may be noticeable as cracks around the outer side of the tube diameter. Mechanical vibration fatigue damage is initiated from the external source causing stresses. The stresses due to externally sourced vibration may be transmitted from different origins, such as:

(i) High-frequency, low-amplitude stresses associated with flue gas flow or soot blowers.
(ii) Low-frequency, high-amplitude stresses associated with boiler load cycling.

As stated above, the areas prone to fatigue failure are near the location of constraint, for example, tube attachments, welds, penetrations, supports, or similar tube assemblies/locations.

2.10.11.6.2 Thermal Fatigue Thermal stresses are responsible for thermal fatigue and have been defined as "stresses in metal resulting from nonuniform temperature distribution."

Thermal fatigue is a continuous process of developing a form of failure that builds up within the metal components where the process is subjected to alternate heating and cooling, resulting in expansion and contraction of the metals. Whenever there is a difference of temperature between the metal surface and the inner metal, better to say mid-metal, the thermal stress develops. It will be quite significant when the subject metal thickness is appreciable. Another reason for developing thermal stresses when the surrounding material/constraints hinder the metal expansion/contraction.

Expansion of two pieces of metal at different rates (due to different coefficients of expansion) may also contribute to stress development. For example, weld attachments where thermal expansion or contraction of one or both pieces of metal result in tensile stresses on the tube interior/exterior.

These cyclic thermal stresses out of alternate temperature excursion cause fatigue similar to that of mechanical stresses and cracks appear. The reason for cracking is because of the rapid change of temperature in the material that causes thermal expansion or contraction. Under thermal fatigue, cracks can initiate and propagate, and eventually failure occurs. This type of failure is quite common in thermal power plants where turbulent mixing and flow of fluids cause fast and unwanted thermal transients in boiler tubes. To minimize the thermal stress and at the same time to achieve the desired load condition at the earliest possible time, the boiler manufacturers must provide suggestive guidelines on the optimum time frame of heating and cooling for proper start-up and shut down of boilers.

As the thermally induced stresses are due to the cyclic operation of the boiler fuel firing start and shutdown, it is to be noted that the stresses occur in every firing cycle with varying magnitudes, depending on the procedure of firing start and shutdown.

The causes are due to the resistance offered by the different boiler pressure part metals within the boiler frame to the movement experienced by the thermal expansions and contractions, popularly termed thermal shock. These metals such as SH/RH/boiler drum/water tube are operated at high temperatures and time-to-time load variations, or in other words, thermally induced stress cycling creates metal fatigue. Though the phenomenon is also called "shock," which means a sudden impact, the actual effect does not take place suddenly but over a period of time. Each start-up and shutdown also causes metal fatigue and reduces metal life expectancy to some extent, depending on the time and type of the same. The resulting effect is ultimately tube failure. Most of the failures of this type occur after operating the boiler and quite a few starts/stops over a certain period of time, or sometimes after a short interval depending on the severity of operation glitches and the malfunctions of control systems. Usually, it is observed that a considerably longer time occurs before the actual damage detection becomes evident.

As it appears in reality, these types of failures are not likely to be very dangerous in nature, so far as damages to the equipment are concerned but the same may be a very costly and serious affair toward the repair charge and the revenue loss against idle downtime. Failures of this type, known as "tube failure," may be described as the failures of the tube metal to prevent leaking through sectional cracking. They may also be of different types, for example, leaks between the tube-to-tube sheet joints, cracked tube sheet ligaments, or broken stays.

2.10.11.6.3 Corrosion Fatigue Corrosion Fatigue is applicable to fire side corrosion fatigues and already discussed in Section 2.10.11.2.4 (Table 2.1).

2.10.12 Measurement of Boiler Tube Metal Temperatures

Boiler tube metal temperatures are very important parameters and hence their measurements have become a necessity as far as boiler tube failure is concerned. During analysis of tube failure, it must be noticed that many of the failure factors are related to and/or caused by high temperature or temperature fluctuation.

Several methods are adopted to measure the metal temperature of tubes in the gas path. There are even methods of calculations to arrive at the metal temperature from measuring other related parameters. These metal temperature values, whether measured or calculated, are normally taken for initial understanding of the boiler behavior under running conditions. This initial observation period may extend up to the first 1–2 years.

In the method where temperature sensors (normally thermocouples) are used, there are strategic locations for installing start-up thermocouples and Permanent thermocouples. Start-up thermocouples are meant for the hottest locations, where it is assumed or expected that the thermocouples would be burnt out after some period of time. The other locations are so chosen such that the flue gas temperatures at that area are not so high as to cause burning out of the thermocouples. During this initial period, the interrelation among the measured values of the start-up and permanent thermocouples is drawn so that the inferential values of the start-up thermocouples can be made use of for further desirable purposes, even after the same metal temperature measuring devices are burnt out. In the long run, the numbers of thermocouples available for this purposes become very limited and hence the inferential values of burnt out thermocouple measuring points play a vital role in assessing tube failure.

Another method of calculating the waterside tube metal temperature is from the corresponding water temperature of a particular tube. It is a well-known fact that the temperature rise of a metal tube will be much quicker than that of the water flowing through it. As the temperature measurements are nothing but an index of internal kinetic energy which commensurate with the quicker vibration of free electrons associated with the heat imparted to the tube metal. On the contrary, water does not possess such free electrons, and held within by the molecular bonds and thus vibrating movements are restricted. Considering the heat input to the boiler tube is equal to the heat transmitted to the water inside the tube and causing the temperature rise in both metal tube and water for a particular time period. If the temperature rise of the water is measured, the tube's temperature rise vis-a-vis the final temperature can be calculated, assuming a suitable mass of water and metal tube.

2.10.12.1 Measuring Location of Boiler Tube Metal Temperatures

Quite a few measuring locations are selected for both permanent and start-up purposes. The locations are at different elevations for mainly water tube walls (normally for the 500 MW and above TPS), superheaters, and reheaters. The drum mid-metal temperatures around the periphery are also measured to avert catastrophic disaster in the case of low drum level. Data have been acquired for both laboratory stress analysis and field test programs, from which the necessary action is to be explored to take care of the power output requirement along with the requirements of the boiler manufacturer's recommendation to protect the structural integrity of the boiler.

Following are the approximate or typical numbers of metal temperatures measured at different strategic locations of the heat transfer elements of a 500 MW TPS:

Location	Numbers	Alarm Value
Platen SH outlet	45 nos.	575°C
Final SH outlet	40 nos.	590°C
Final RH outlet	25 nos.	612°C
Primary SH outlet	10 nos.	460°C
Drum	10 nos.	

In some power plants, economizer metal temperatures are also measured. During the start-up and shutdown, the trend of these measurements is very helpful for operators, especially during start-up, which is normally monitored in DCS.

2.10.12.2 Problems and Solutions of Measuring Boiler Tube Metal Temperatures

As environmental or ambient temperatures are very high in these areas, the tube metal temperatures are extremely difficult to measure because the object temperature is totally different from that of the environment where the sensor is located. The sensor or probe is supposed to respond to the tip temperature, and hence it is extremely important that proper contact, to eliminate or reduce thermal resistance, must be made with the object where the measurement can be conveniently accomplished, utilizing only a small area. It is always better if an appropriate (and externally insulated

TABLE 2.1 Summary Table for Tube Leakage

Leakage Type	Symptoms	Reasons	Remarks
acid attack	Corrosive attack on internal tube. Irregular pitting	Poor boiler chemical cleaning and inadequate postcleaning. Passivation	Like Swiss cheese
Caustic attack	Localized loss of ID resulting in stress and strain in the tube wall	Excess deposition on ID tube surface—loss of cooling media flow. Local deposit boiling and concentration of boiler water chemicals	Finally resulting caustic condition
Dissimilar metal weld (DMW)	Material failure at ferrite side of the weld along weld fusion line without warning	When austenitic (SS) joins with ferrites due to high stress at austenitic interface, difference in expansion properties, creep of ferrite due to loading stress, etc.	
Erosion	OD metal loss on impact side of the tube	Impingement of external surface may be due to fly ash or soot blowing steam which may cause thermal fatigue	Rupture due to increased strain
Fireside corrosion fatigue	Series of cracks on OD propagates in to wall	Combination of corrosion and thermal fatigue. OD experience thermal stress cycle in addition to subjecting material to cyclic stress	"Elephant hide," "Alligator hide," "Craze cracking"
Graphitization	Brittle with thick edge fracture	Long-term operation at relatively high temperature to damage CS with more carbon. Formation of graphite nodules	Dissociation of iron carbide in steel
High temperature oxidation	External tube metal loss, excess strain in tube	Local high temperature causes high temperature oxidation. Similar and confusing with fireside ash corrosion. Best differentiated by tube analysis	
Hydrogen embrittlement	Loss of ductility and embrittlement finally brittle, catastrophic rupture	Deposition on Tube ID and Low pH. Under deposit corrosion atomic hydrogen is released which reacting with carbon in the steel causes inter granular separation	Water chemistry upset may be due to condenser leakage
Long term overheat	Small swelling and longitudinal split is narrow. Tube often has scale build-ups	Long-term[a] overheating at SH, RH, and stress due to high temperature. Also, water walls fail due to this	
Mechanical failure	OD-initiated localized drack due to high stress and constraints	Cyclic stress in component. Thermal fatigue affects mechanical damage, with external stress	
Oxygen pitting	Aggressive localized corrosion and loss of tube wall	Presence of excess air may be due to pump air leakage or failure of operation of preboiler water treatment equipment. Mostly in economizer FW inlet and nondrainable surfaces	Mainly occurs during long idle period
Short-term overheat	Ductile rupture of tube metal characterized by "fish mouth"	Occurs during start-up, when metal tube temperature may be extremely elevated due to lack of steam/water, or SH tubes not cleared of condensation at start-up	
Stress corrosion cracking (SCC)	Thick-wall, brittle-type crack, in location with external stress	Mainly in austenitic SH material, due to high tensile stress coupled with corrosive fluid presence	
SH fireside ash corrosion	External tube wall loss and increased strain in tube	Function of ash characteristic and boiler design. Mainly in coal (may be for oil). Certain constituent of ash in molten state causes high corrosion	
Waterside corrosion fatigue	ID-initiated wide transgranular cracks to external attachment	Combination of thermal fatigue and corrosion. Breakdown of protective magnetite on ID of the tube	Found in seal plat, scallop bars
Waterwall fireside corrosion	External tube metal loss, thinning and increase strain	When reducing atmosphere is created due to combustion., mainly in recovery boiler	Also in burning zone of fossil fuel boiler

[a] *Overheat spread over months and year.*

Courtesy Babcock Library Document.

from the environment) heat-sinking compound be used that would ensure maximum contact vis-a-vis minimum interference. There are methods that follow the above guidelines to avoid such harsh environmental problems, as discussed below:

(i) Using a thermopad, which is nothing but a block of metal suitably shaped and curved to match the tube curvature. The thermopad is meant for pressing the sensor with the tube metal surface with a proper fixing arrangement through nuts and bolts or welding.

(ii) A skewer-shaped, pointed-tip sensor is a good practical choice, but the measurement accuracy is always under question. A spring-loaded arrangement would facilitate good contact. A pocket sort of arrangement by drilling or scraping the surface material would ensure better accuracy.

(iii) A thermocouple made of fine wire diameter may be used and laying should be such that the extension wires are drawn along up to a certain length of the tube surface, which ensures good contact with the material body.

(iv) Using a noncontact type measurement, that is, radiation thermometry, would also enable avoiding the cumbersome contact type. The instrument used here is popularly called a radiation pyrometer, and is widely used for industrial measurement purposes. This is a noninvasive type of measurement and so avoids the problem of interference.

There are certain disadvantages of radiation thermometry that must be considered. These may be the background radiation reflected from the surface or a radioactive property, that is, emissivity of the surface, etc.; they will need suitable correction. Emissivity may change drastically with surface roughness, surface composition, object body temperature, etc.

2.11 Boiler Drum, Pressure Parts With Locations

2.11.1 Boiler Accessories

Fig. 2.12 illustrates the very important accessories and equipment of a typical boiler plant of a modern fossil-fired thermal power station of considerably high capacity. The functions of these items are described in a very brief way to offer an overall idea about the plant.

Item 1 is the boiler *drum* where *boiler feed water* is fed from the BFP (item 10) via the *economizer* (item 9) for onward distribution by the *down comer pipes* (item 2) throughout the *waterwall* (item 4) surrounding the *furnace* (item 3) vis-a-vis the boiler. The economizer is a bunch of water tubes through which cold water from the BFP is passed for gaining some heat from the flue gas before going out of the main boiler structure. The waterwall is the series of tubes connected together like a wall along the periphery of the boiler/furnace to ultimately form a leak-proof enclosure where the exchange of heat takes place. However, the whole surface area of the waterwall is externally insulated by a suitable material to arrest heat loss to the surroundings and then clad for personnel convenience. Water getting heat input from the flue gas (item 6) goes upward as hot water of lower density and ultimately starts vaporizing near the uppermost part of the tubes. This mixture of saturated steam and water is then brought back to the upper part of the drum for separation of steam from water through a steam separator (item 23).

The formation of steam in the once-through boiler is a little different and is discussed in Appendix E. Saturated steam from the drum is then passed through the *primary (/platen/) superheater* (PSH) (item 15) bank and the *final superheater* (FSH) (item 18) bank for getting dry superheated steam ready from the *final superheater outlet header* (item 19).

In some plants, the feed water from the boiler drum is distributed to the waterwall by the help of a separate set of BCW *pumps* (item 12) known as a forced circulation system, whereas the natural circulation system does not need those pumps.

Larger plants have steam reheaters that are meant for heating the steam once again from the high-pressure turbine exhaust called cold reheat (CRH) steam for reuse in the intermediate pressure turbine for higher output. Here, as shown, CRH steam from the turbine enters the *primary reheater* (item 21) bank and then to the *final reheater* bank (item 22) for getting dry superheated steam ready from the *final reheater outlet header* (item 24); it is then called hot reheat (HRH) steam. In an ultrasupercritical steam generating plant, more than one reheating system is provided, operating at a different pressure for achieving a more efficient plant.

There may be additional intermediate superheaters and reheaters to match the steam parameters for external usage. The incoming boiler feed water may not be coming directly from the BFP but via the high-pressure water heaters, the feed control (valve) station, etc.

There are plants where the boiler drum is eliminated with appropriate design and the saturation point takes place at critical pressure (225.415 kgf/cm^2) and temperature 374.065°C. The latent heat of vaporization is zero at this point, or, in other words, water transforms from the liquid phase to the gaseous phase without going through the two-phase condition. These units are called *supercritical steam generators*, as the term "boiler" is a misnomer in this case with no boiling taking place at all. Brief details are included separately elsewhere in this book.

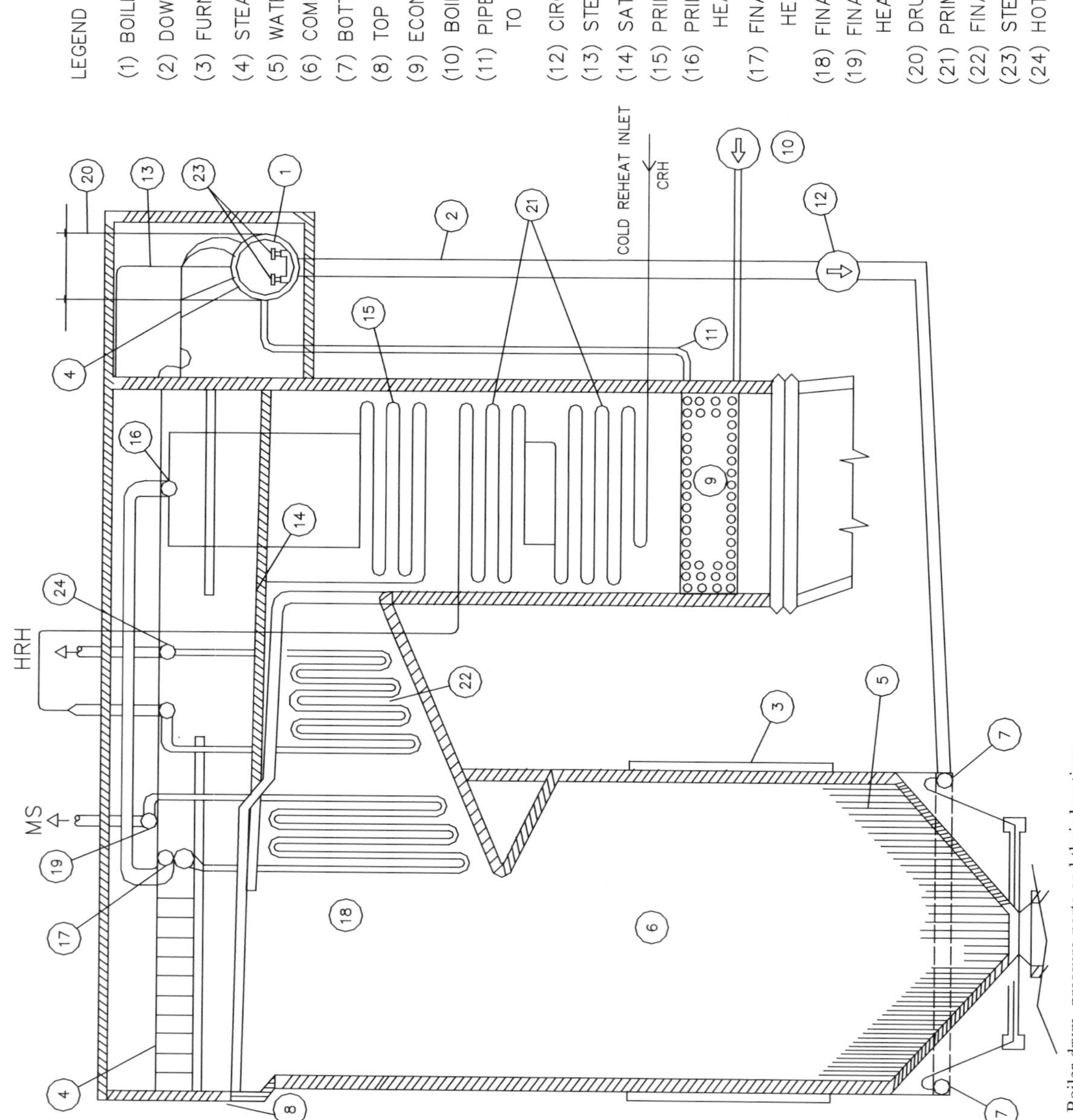

FIG. 2.12 Boiler drum, pressure parts and their locations.

2.11.2 Drum Internals (*Ref.* Fig. 2.13)

The boiler drum serves many purposes, but the two main functions are to separate steam and water coming from the waterwalls and include an arrangement to provide better quality of outgoing steam and water inside the drum ready for circulation. Apart from that, the boiler drum is also considered a high-pressure intermediate steam storage unit, though the water level vis-a-vis the quantity of drum water is not comparable with respect to the total steam requirement of the plant, capable of repeatedly and frequently receiving waterwall tube fluid, basically a mixture of saturated steam and water, flowing at high temperatures and pressure different than those of the drum wall. The priority is to design the drum interior with a view to minimizing thermal stressing of the drum wall of thickness in the approximate and typical range of 100–150 mm, depending on the size of the boiler and other parameters.

The baffles are provided so as to ensure a change of direction of the incoming saturated steam-water mixture and facilitate gravity separation in the inner/open drum space as well as assist the uniform distribution of the drum metal temperature to avoid thermal stress during the start-up/on-load condition. The separators, utilizing the whirling action of the fluid path, assist in the elimination of water particles from the steam. The steam quality is aided by the screen driers provided at the top of the drum's internal structure. They are used to heat the wet steam from the separator and remove the remaining moisture to achieve a dry quality steam at the drum outlet.

Feed water lines are connected to the boiler drum at the bottom portion submerged in water, an arrangement that enables the comparatively cold fluid to avoid direct contact with the drum metal during admission, thus assuring minimum thermal stress.

Provisions are also made to connect the high-pressure chemical dosing system in order to minimize the scales, sludge, and corrosive reactions of feed water-borne objectionable compounds.

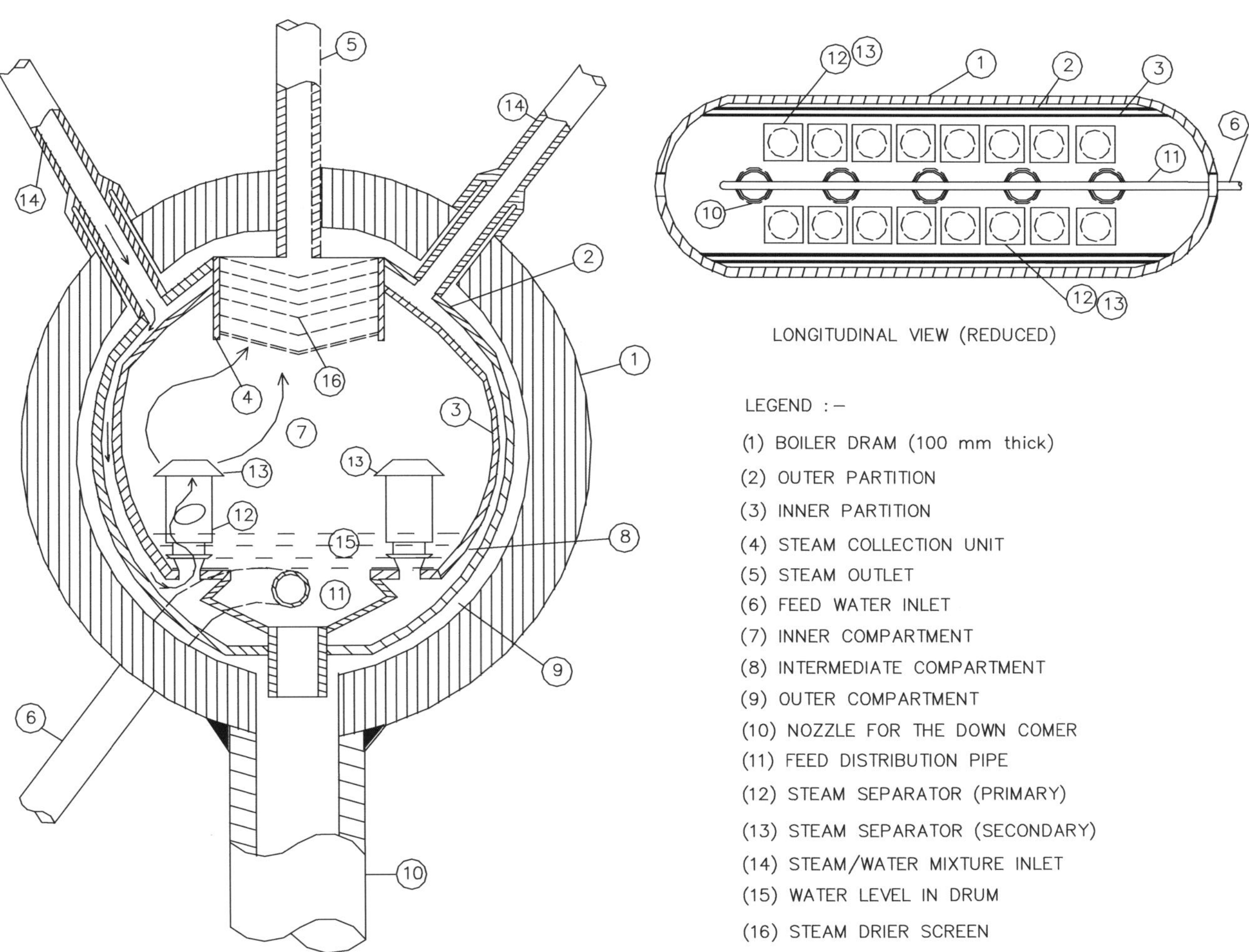

FIG. 2.13 Internals of the boiler drum.

The blowdown connections are provided to facilitate both continuous and intermediate removal of a portion of the drum water in order to maintain the water quality vis-a-vis the concentration of feed water-borne solids.

The length of the drum normally depends on the furnace width, but for high capacity units, the steam-water separators and purifying equipment dictate the overall design of the drum. Fig. 2.13 illustrates a sketchy view of the major part of the drum internal arrangements. Baffles or partitions are spaced throughout the entire length so that the interior of the drum is divided into three distinctly separate compartments: the outer (item 9), the intermediate (item 8), and the inner (item 7). Partitions are so made as to form a sealed fluid tight installation isolating the intermediate compartment. The fluid mixture coming from the furnace waterwall tubes enters the drum directly into the isolated intermediate compartment through the inlet nozzles (item 13). After entering the intermediate compartment, most of the steam-water mixture flows through steam separators (item 12) and into the inner compartment. The purpose of the outer compartment is to provide thermal isolation between the drum wall and the water-steam mixture passage in the intermediate compartment and steam in the inner compartment. A portion of the steam-water mixture from the intermediate compartment is allowed to percolate to the outer compartment through holes in the partition between the intermediate and outer compartments. It is done in order to minimize the pressure difference across the outer partition and thereby permit only a limited amount of fluid flow to the outer compartment. The inner and outer compartments are connected to each other near the bottom of the drum, with the inner compartment filled with a controlled level of water described below. There is created a suction head near the multiple downcomer nozzles (item 10) at the bottom of the drum due to the high velocity of water flowing out, thereby ensuring continuous removal of water from the intermediate compartment.

The FW is introduced to the drum by means of the feed distribution pipe (item 11), which is located within the inner compartment of the drum and extends longitudinally, as depicted in the transverse sectional drawing of Fig. 2.13. The level of FW is established initially and thereafter is maintained continuously at a preset value (item 11) by means of external feed water pumps at a pressure matching the drum pressure, taking care of the transmission and other losses. The temperature of the inlet feed water is also important and so is raised by heat exchange through regenerative feed heaters before introduction to the drum through the feed inlet pipe (item 6).

As the modern steam generating units operate at very high fluid pressures in the range of 150–200 kg/cm^2, it is imperative that their pressure components must have adequate wall thickness to sustain such pressures over the estimated life. These drums normally have a length in the range of 20–30 m with a diameter in the range of 1–2.5 m. Under such conditions, the wall thickness of the drum must be more than sufficient in the range of 100–250 mm.

It is obvious that the higher wall thickness is a necessity so far as pressure criteria is concerned, but then the problem comes is the thermal stress of the drum at a saturation temperature corresponding to the drum pressure. While starting up a high-pressure boiler, the foremost consideration demands that a uniform temperature must be maintained throughout the length of the drum so that excessive thermal stresses are avoided concerning the drum wall. It can be accomplished by increasing the pressure of the plant slowly and thereby the temperature of the drum uniformly. This procedure requires considerable time for starting up a large boiler plant, which may not be acceptable to the user from a revenue point of view. Nowadays, the boiler drum designs are made toward the increasing rate of temperature change so as to achieve a reduced time required for starting up and shutting down a boiler, up to say 150°C/h from the initial moderate temperature value of 40°C/h. However, it is quite evident that operating at the quick start/stop mode, the boiler drum wall experiences severe thermal stresses against economic advantage and the frequent occurrence of such start-ups/shutdowns has, in some instances, resulted ultimately in the failure of the drum.

2.11.3 Type of Superheaters and Reheaters (Ref. Fig. 2.14*)*

It has been observed that saturated steam, if introduced in the turbine, the work done by the turbine results in a loss of energy by the steam associated with partial condensation of steam. The amount of work done by the turbine could have been more had it not been limited by the amount of steam converted to moisture in spite of a pressure drop across the turbine stage. It is also true that a turbine generally converts the heat of superheat to work done without forming moisture; the heat contained therein is recoverable in totality. It is therefore economic that the admission of steam must be with such a degree of superheat that there must not be moisture formation in any stage of the turbine so as to ensure the maximum amount of work done in the turbine.

SHs and RHs are the heat exchanger banks that are used to elevate the steam temperature above saturated temperature at that pressure, which is effectively the degree of superheat. Before entering SHs/RHs, it is a precondition that the steam quality must be dry saturated. Any addition of heat would obviously increase the steam temperature, depending upon the heat input, the mass of steam, the temperature difference between the steam inlet and the heating

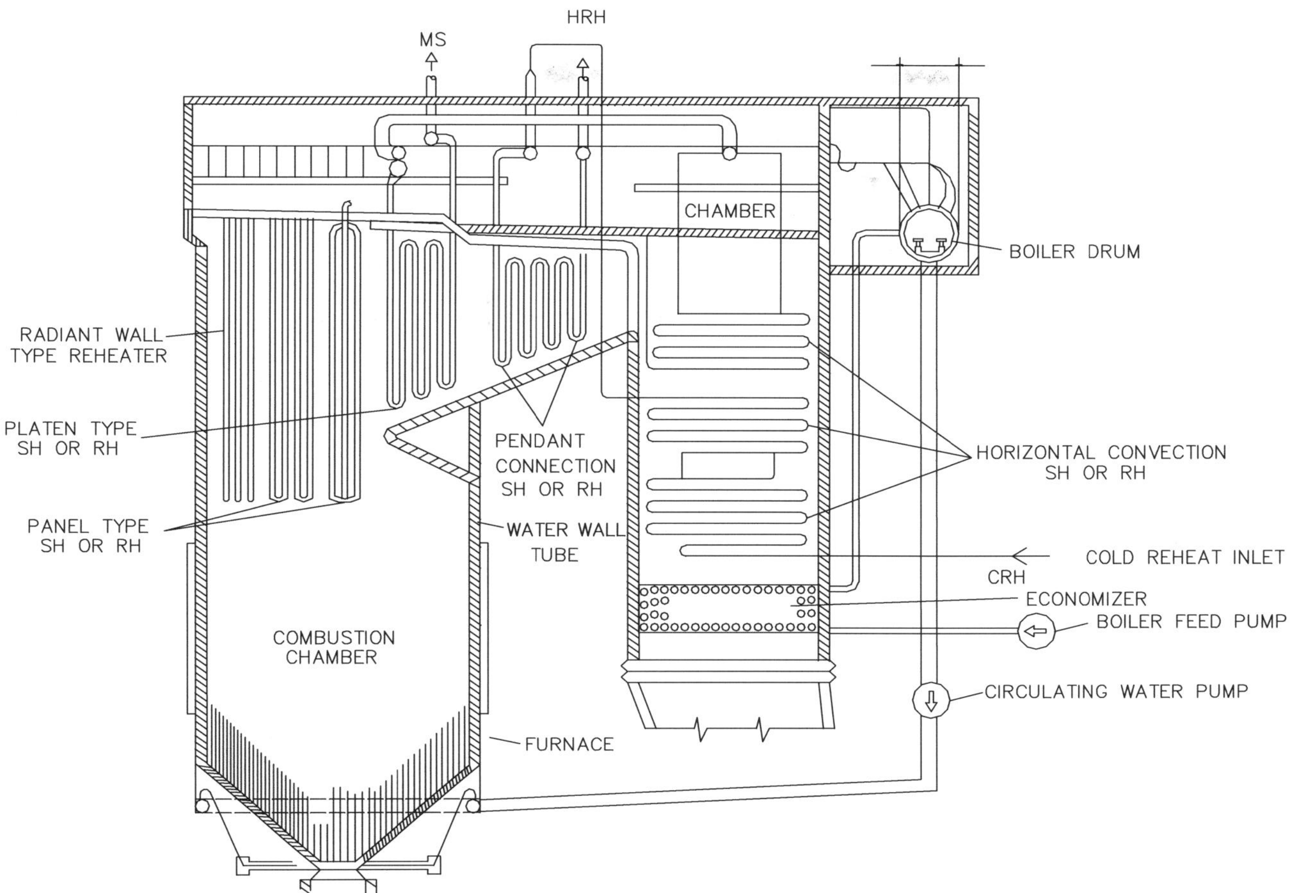

FIG. 2.14 Type and locations of superheater and reheaters.

media, the specific heat and duration of heat exchanging, etc. The type of SHs and RHs depends on the particular requirement of the plant where they would be installed. Reheaters are also superheaters but the name is so because it accepts the steam from the HP turbine exhaust or cold reheat (CRH) steam, which is again heated and sent to the IP turbine to avoid condensation therein. There are two basic types of SH/RH: *convection and radiant superheaters*.

The *convection heat exchanger* is that where the heat transfer is taken place by convection or movement/mass transfer of fluids. The convection design is the original form of SHs and was initially adopted for relatively low temperature applications. They are placed in the flue gas path so that direct contact is made between the flue gas and the SH tube metal surfaces through which the steam is forced to flow. This type of SH/RHs is placed in the location where heat transfer by radiation is negligible with respect to the other type. Hence, the percentage of heat absorbed in the furnace decreases and more heat is conveyed through the flue gas for absorption in the superheater region in case of increasing boiler load. This means that when the load increases, the temperature at the convection SH/RH outlets is more than that of the decreasing loads.

In the convection zone, the SH/RH is installed in a manner such that counterblow of steam and gas (heating media) is established, having the steam inlet at the bottom whereas the gas inlet at the top of the boiler that is from the opposite side ensuring a maximum mean temperature difference between the two opposite flowing fluid and minimum heating surface in the primary zone.

The *radiation heat exchanger* is where the heat is transferred through radiation or electromagnetic waves transmitted by the source. They are located in such a place where the flue gas when travelling need not be in direct contact with them, which means the they receive a negligible amount of heat transfer through convection. The fact regarding radiation energy is that it depends on the heat absorbed in the furnace surfaces, which does not bear the direct proportional relation with the boiler load rather than at a lesser rate when load increases. The steam outlet temperature of radiant SH/RHs, contrary to the convection type, shows decreasing characteristics as the steam flow at higher load also increases but receives less percentage of heat by radiation. Higher outlet temperature demands more surface area of the heaters, namely wall, pendant, or platen-type superheaters and reheaters. Those are basically radiant heat exchangers and necessarily must be mounted near the combustion zone where the exchange of heat shall be by radiation.

For very high outlet temperatures, both locations as well as types of superheaters and reheaters are considered. In general, all high-capacity thermal power stations have both types of SH/RHs. The combination of these two opposite characteristic curves and proper coordination of them provide the considerably flat superheat outlet temperature over a wide range of the load variation, especially in the higher percentage of loads.

The final SH/RH outlet temperatures are maintained within a very close limit of set values so as to avoid overheating of superheater metals and turbine expansion, causing reduced clearance due to high steam temperature. Low steam temperature causes moisture carryover and damages the turbine blades. The outlet temperatures mainly vary due to load changes but also due to the formation of insulating media such as deposition of ash and slag over the SH/RH tubes.

2.11.4 Economizers

The economizer is a bunch of horizontal water tubes usually located in the back pass or second pass (after the primary superheaters and reheaters) of the boiler through which the feed water enters the boiler to absorb the residual heat from the exhaust flue gas and transfer it to the steam production cycle. This process helps in improving overall thermal efficiency (and thus justifying the name attributed), but on the other hand, it reduces the stack outlet temperature for better environmental conditions, as per different government stipulations. It has been observed that loss (through cooling) of flue gas temperature by every 22.2°C increases the boiler efficiency by approximately 1%. This also means comparatively less fuel would be required to produce the same amount of steam, resulting in fewer carbon emissions. According to the location, it is necessarily a convection type of heat exchanger providing a forced and once-through flow path of FW, effectively making it an FW heater inside the boiler or steam generator.

The economizer tubes may be of different types such as bare tubes, finned tubes (spiral or long finned), etc., but bare tubes are preferred, keeping in mind the problem associated with ash fouling. However, the design of the economizer has always been guided by economic considerations, the stack outlet temperature, the provision of air preheaters, etc. For a given temperature of the stack outlet, the higher gas temperature at the economizer outlet calls for a lower size of economizer and a bigger size of air preheaters. In many cases, the size of the economizer is restricted so as to make way for higher air temperature vis a vis the air preheater size for better fuel conditioning and for efficient combustion in turn.

It is preferable that the flue gas from the boiler flows down across the economizer tubes and for the water to go up after entering the boiler at the bottom. This counterblow design minimizes the requirement of surface area as well as the draft loss experienced by the flue gas. The concept of upward direction FW flow eliminates the possibility of unstable water flow, assists uniform flue gas distribution, and realizes the idea of steaming economizers, which makes the provision of generating some steam in the outlet section even.

Steam or vapor formation in the economizer is a common phenomenon (practically limited to about 20% of full-load FW flow) and also advantageous to some extent rather than in the boiler surface. In addition to availing of waste heat from the flue gas, it also assists the increased life expectancy of the boiler by preventing the admission of cold feed water. For preheated FW through the economizer, the temperature differential is much less and consequently would cause less thermal stress on the boiler metals, and thus a longer operating life.

2.11.5 Boiler Water Circulation

It is needless to mention that water-tube boilers are only used in the current large thermal power plants, the type being a shell-type boiler where water is circulated inside the tubes. The flue gas or the heat source is passing through externally to raise the water temperature to the desired level. The water-tube boilers are much more advantageous for very high pressure usage as the design employs multiple small diameter tubes to minimize the hoop stress. There are two types of methods utilized for circulating the water inside the water-tube boiler: natural circulation or forced circulation.

2.11.5.1 Natural Water Circulation

A large share of water-tube boilers operate on the principle of natural water circulation, which is also described as thermosiphoning as the water is circulated by the thermal effect. The reason behind the natural circulation can be explained in two ways: (i) In natural circulation boilers, the circulation of water depends on the difference between the low density of an ascending concoction of hot water and steam and a descending mass of relatively colder and steam-free water with higher density. (ii) Natural circulation is also caused by convection currents due to the application of heat to water through the tube metal wall, which results in uneven heating of the total water course contained in the boiler.

The boiler drum, waterwalls, and tubes constitute the multipassage section of the water circuit. Some large-diameter insulated water pipes are connected to the drum, taken out of the boiler, and reconnected to the waterwall tubes through water drums or headers located at the bottom part of the boiler, depending on the arrangement of the steam and water spaces. These pipes from the steam drum to the water drum or header are popularly called downcomers and form the colder leg of the circulation path, as heat is not added to the same.

On the other hand, the water in waterwall tubes gets heated from combustion products from the hotter leg. As the water tubes approaches the boiler drum, the water temperature rises and steam bubbles are formed. Comparatively cold FW is introduced into the boiler drum and because the density of the cold water is greater, it comes down through the downcomer toward the lower part, pushing up or displacing the low density fluid, that is, the combination of hot water and saturated steam. This sets up a continuous circulation and thus steam is being collected and naturally separated from the hot water in the steam drum, and are taken off. The size and number of downcomers varies with the plant capacity and type of boiler.

The natural circulation mostly is dependent on the height of the water tube, also called a riser, and the downcomer and the difference of the mean fluid density in them.

To maintain the same level of steam output at higher design pressures, the distance between the lower part and the steam drum—the height of the downcomer—must be increased, or some means of enhanced circulation must be introduced.

The natural circulation is normally effective up to 150 kg/cm^2 pressure, which assists sufficient fluid velocity vis a vis heat transfer.

The methods to achieve natural circulation may also be of two types: the free or the accelerated type. For a free natural circulation boiler, the water tubes are installed almost in a horizontal direction, with only a slight vertical inclination. When the water tubes are installed at a much greater angle of inclination, the rate of water circulation would increase proportionately. Therefore, boilers in which the tubes follow a steep inclination from the steam drum to the bottom water headers may be called the accelerated type.

2.11.5.2 Forced Water Circulation

For very high-pressure drum plants, this problem could be eliminated by forced or controlled water circulation. This system introduces the BCW pump in between the steam drum and the waterwall, which enables the cold feed water from the drum to get extra pressure head to overcome the limitations experienced through natural circulation. The pump ensures sufficient velocity for the desired heat transfer, making the system independent of the density difference of water/steam and the frictional loss due to mainly water flow through the tubes. Water circulation is assured even before the introduction of burners due to the additional pump head.

The advantage of the recirculation pump is that the frictional loss through the tubes may be considered more by using small-diameter tubes to minimize hoop stress at very high pressures. More small-diameter tubes offer many plus points such as less tube weight, reduced tube thickness, increased heat transfer area, lower thermal stress, etc.

2.11.5.3 Water Circulation in Once-through/ Supercritical or Ultracritical Steam Generators

In a *once-through* boiler, there is no steam drum and hence the service of BCW pump does not arise (except for start-up purposes). Theoretically, the boiler here should be merely a piece of tubing through which water enters as input and exits as steam as output.

2.12 Miscellaneous Boiler Equipment

2.12.1 Main Steam Stop Valve

The boiler outlet steam passes through the stop valve before going to the consumer/user end, called the *main steam stop valve (MSSV or MSV)*. The main purpose of this vital accessory is to isolate the boiler by interrupting the steam circuit during start-up or shutdown, or in case of emergency. Normally this valve is motor-operated. For a bigger plant, a small bypass valve is provided to facilitate easy opening of the main steam stop valve. During start-up, the pressure upstream of the MSV becomes greater and greater while the pressure downstream is almost zero. The differential pressure across the valve and the valve size itself being very high for high-capacity plants, the operating thrust/torque required is also very high while opening the valve from fully closed position. To circumvent the situation, the small bypass valve opens first with less thrust/torque (the line size being small) and upon pressure building up at the downstream, the MSV can then open, requiring less thrust/torque. During normal operation of the plant, the valve remains in full open condition.

2.12.2 Nonreturn (Check) Valve

The function of this valve is to allow the fluid flow to the forward direction under pressure but check the fluid flow in the reverse direction. Here, the construction of the valve plug is such that it moves up from the seat when pressure applied from the bottom of the plug is higher than that of the top of the plug and remains in that position as long as the differential pressure multiplied by the plug area is higher than the spring force applied to the plug for keeping the valve in the shut-off position. In the reverse condition when pressure at the downstream (top of plug) is higher than upstream (bottom of plug), the plug moves down by force due to the nature of the differential pressure aided by the spring force and sits tightly on the seat to arrest any flow.

Nonreturn or check valves are provided in every flow path, irrespective of steam or water service, wherever there is a chance of return flow under any operating condition. The valve is normally self-actuated so that no external power is required.

2.12.3 Start-Up Vent Valve

These type of valves are provided in the main steam header and, as the name implies, are required for the start-up period only. The valves are of regulating service, that is, steam is allowed to vent out of the system to the atmosphere as required; this is determined automatically or manually for purging and/or heating of the pipe lines. These valves are generality motor-operated.

2.12.4 Safety (Pop Up)Valve

These valves are of immense importance as far as plant and personnel safety are concerned. Whenever there is a pressure build up in the pipe line beyond the limit, the valve should operate or pop up to release the steam to the atmosphere until the pressure comes down to a safe value. Though it is a loss of energy and mass of working fluid, it is inevitable during any untoward situation.

There are various types of safety valves: electromatic safety (or relief) valve, spring-loaded safety valve, dead weight safety valve, fusible plug, etc.

2.12.5 Electromagnetic Safety (or Relief) Valve

This is a pilot solenoid operated valve that gets energized from the pressure switch at a very high set point and allows the working fluid to operate the actuator of the safety valve. It operates through remote manual command as well as from the control room/tower.

2.12.6 Spring-Loaded Safety Valve

Normally, this valve operates as a last resort to the safety system against high pressure. Under normal operation of the plant, the spring tension is high enough to hold the valve plug on its seat to ensure a closed position until the very high pressure set point is reached. At this point and above, the force against the spring is more to lift the plug over its seat and allow extra steam to escape, unless the steam pressure comes down to the normal value. The discharge capacity should be selected such that the same is equal to the evaporative capacity to avoid frequent build up of pressure vis-a-vis actuation of this valve. Other types of safety valves are not in use nowadays and hence are not discussed.

2.12.7 Blow Down Valve

This type of valve is provided to remove sludge, sediments, and other impurities collected at the bottom-most location in the water flow path. It also provides the means of complete draining of the system. They are of two types: continuous and intermittent blow down valve.

2.12.8 Continuous Blow Down Valve

The function of this valve is to open continuously so as to maintain the dirty material level to a minimum value. The opening of the valve is varied as per requirements with a predefined control signal. The motorized actuator gets manual commands as the operator's prerogative in addition to automatic controls.

2.12.9 Intermittent Blow Down Valve

It is a valve for blowing down the dirty water as per the need. Its operation may be predefined based on a cyclic or time-framed full open/close signal or a manual command from the operator with the motorized actuator.

2.12.10 Drain Valve

During the start-up of the plant after a prolonged shut down or a cold start-up, the pipelines and various equipment need to be warmed up before loading the boiler. To achieve this, heating steam is admitted phase by phase in a very slow manner to avoid dissimilar heating vis-a-vis expansion of various casings and pipes. While heating metal works, the steam gets condensed and collected at the bottom of the pipeline (with a siphon sort of design at different strategic locations) before being drained out through these valves with the motorized actuator. The valve opens when the level in the drain pipe reaches a predefined value through level switches for automatic operation along with the operator's manual action.

2.12.11 Steam Trap

The function of this type of element is provided to drain the condensed water from the steam pipes and jackets used for heating. This results in partial condensation and at the same time arrests the steam inside from escaping in a self-contained operation, being a mechanical type without any external power. Generally, float/bucket/thermal expansion traps are available.

2.12.12 Steam Separator

As the name implies, its function is to separate water particles suspended in generated steam from the boiler and carried by the flow of the steam to the turbine. To serve the purpose properly, its location calls for a minimum distance from the steam generator so as not to allow water for the rest of the transportation line. In the *drum-type boiler*, it is provided in the drum itself where the water particles drop in the water section.

Here, the steam path is guided to impinge on a series of baffles where water particles, being a heavier material, have a higher inertia and fall below after it strikes the baffles by gravitational force at the bottom of the vessel for collection and draining out. The dry steam, practically unaffected by this arrangement, gets out for transportation.

2.12.13 Swing Check Valve

This type of valve is a variation of the nonreturn or check valve and is normally provided at the turbine end. The specialty of this valve compared to the normal nonreturn or check valve is that it requires some external power while opening and after that the forward flow thrust keeps it in full open position without any external power until the flow is more than some predetermined value. At very low/reverse flow, the valve gets closed due to spring and other action automatically.

The valve plug is a circular lid connected to the valve body at a particular point around which it can swing almost 90 degrees. It sits tightly on a ring seat, matching the main pipe diameter, due to the spring action and gravitational force and at low flow and reverse flow thrust due to the untoward situation of the sudden loss of forward flow. It is normally installed in the horizontal pipeline and due to the swinging action, it remains in the horizontal position at open condition, allowing f the forward flow to pass through and take a vertical position at the closed condition, inhibiting any flow in the reverse direction.

3 TURBINE TYPE

Basically, three types turbines are classified by the types of fluid they handle: steam turbine (steam), gas turbine (gas/hot air), and the Hydel turbine (water). Gas turbines are briefly discussed in Section 2.4 of Chapter 1. The Hydel turbine operates on the principle of high velocity of water coming from a higher elevation impinging on the turbine impeller. The steam turbine is the one that utilizes high pressure and temperature steam as the prime moving force.

3.1 Steam Turbines: Turbine Types and Classification

The steam turbine operates wholly on the dynamic nature of steam. Here, steam pressure is allowed to fall in a narrow passage (namely the nozzle) and results in increased velocity due to the conversion of some heat energy. The high velocity steam impinges on the moving part (called the blade) of the turbine with a change of direction of motion, therefore causing a change in momentum and ultimately in force. The blades, being the moving type, experience a motion due to this force and mechanical work is done. It is known that the maximum work done is achievable when the outlet velocity is half the inlet velocity. This process of expansion and direction changing may take place once or a number of times in succession.

3.1.1 Impulse Turbine

A simple impulse turbine deploys only one set of nozzles where complete expansion from the steam chest pressure to the exhaust (condenser) pressure is performed along with the highest velocity (near 1000 m/s) at the outlet of nozzles vis-a-vis the inlet of the moving blade. The result is a very high rotational speed of the turbine requiring a high reduction gear. The outlet velocity is also very high, causing energy loss of 10%–12% of the initial kinetic energy entering the turbine. This type of turbine is also called the *De Laval turbine*.

In this type of turbine, the blade inlet and outlet pressures remain the same and the work done is solely due to a change in velocity. Fig. 2.15 depicts the basic functioning of the same.

3.1.2 Pressure Compounded Impulse Turbine (*Ref.* Fig. 2.16)

The above type of problem of the turbine high speed (rotational) is eliminated by contemplating the single impulse stage in several stages in a series connected to the same output rotor.

Here, the exhaust steam from the first stage moving blades (rotor) enters the nozzles of the second stage and so on. The nozzles in each stage are fixed to the diaphragm of the stationary part of the turbine. Each stage can be conceived as a set of nozzles and moving blades and undertakes the splitting up action of the total pressure drop by them. The expansion of steam takes place only in nozzles of each stage. Because the pressure drop is divided into several stages, the steam velocity at the nozzle outlet vis-a-vis the rotor blade inlet is reduced and thereby the rotor speed is reduced. In fact, the rotor speed can be reduced as per requirement by introducing a number of stages to suit and thereby also reducing the leaving loss to 1%–2%. This type of turbine is also called the *Rateau turbine*.

3.1.3 Velocity Compounded Impulse Turbine (*Ref.* Fig. 2.17)

This type of turbine consists of only one set of nozzles at the inlet and a number of moving blades (rotor) and fixed blades (casing) in each stage. The entire expansion takes place in

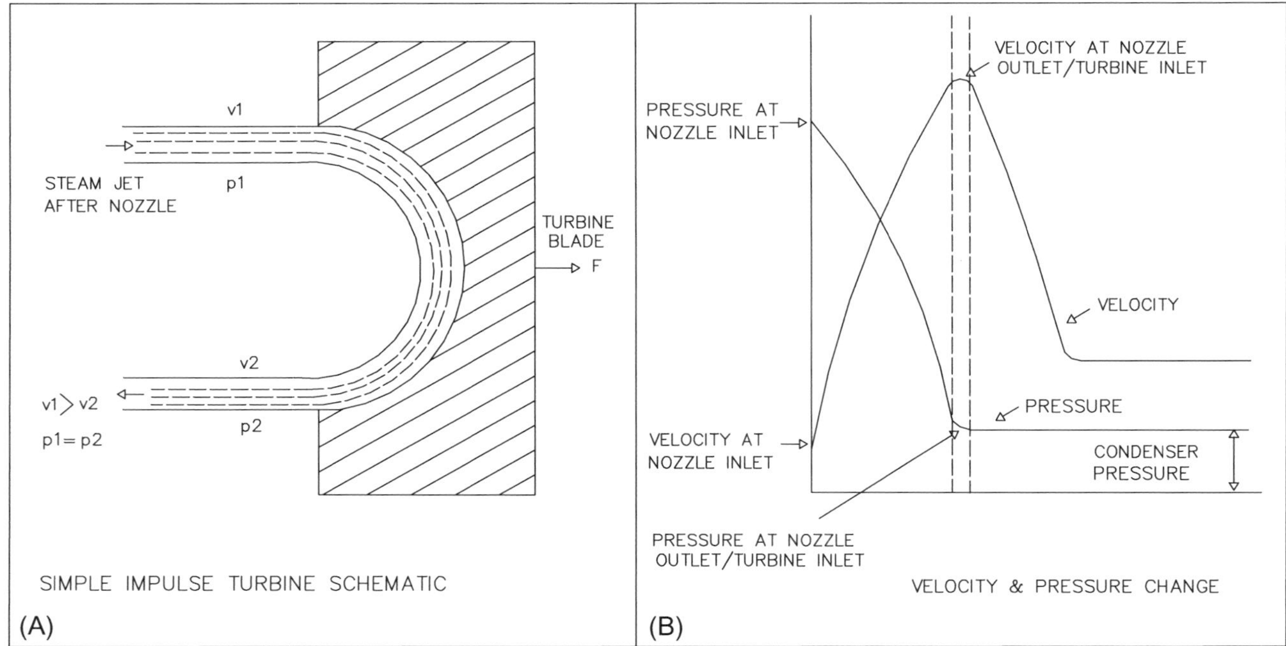

FIG. 2.15 Simple impulse turbine schematic and pressure/velocity changes.

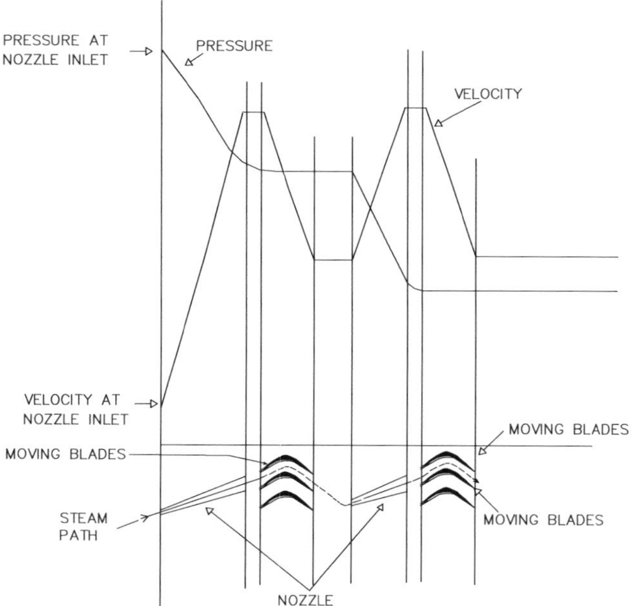

FIG. 2.16 Pressure compounded impulse turbine schematic and pressure/velocity changes.

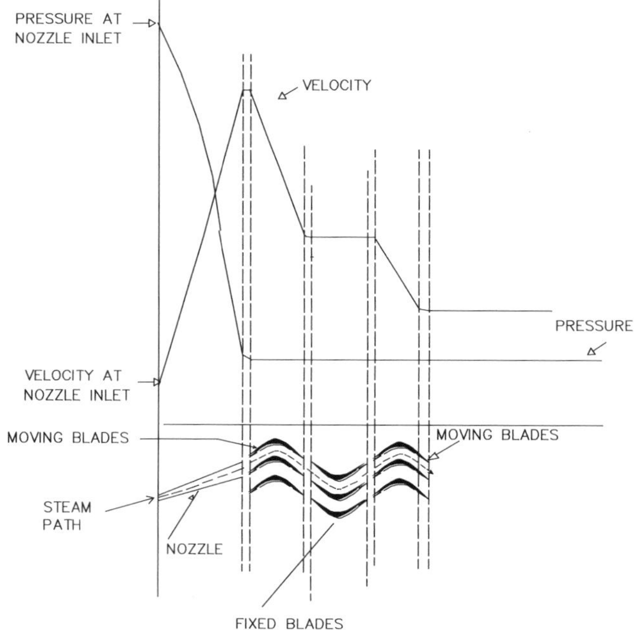

FIG. 2.17 Velocity compounded impulse turbine schematic and pressure/velocity changes.

the nozzles and the high velocity (of steam) breaks into several parts in each stage. The fixed blades are provided only to cause change of direction of the steam jet for the next stage; this is also called the *Curtis turbine*.

Velocity compounding may be achieved with only one rotor using a reentry turbine. The total enthalpy transfer performed in the nozzle itself and the resulting high-velocity steam pass through the moving blades and enter a reversing chamber, which guides the steam to change its direction to pass through another set of moving blades located at a different angular position from the first set of moving blades at different locations; the velocity of the steam decreases gradually, giving low speed and high efficiency.

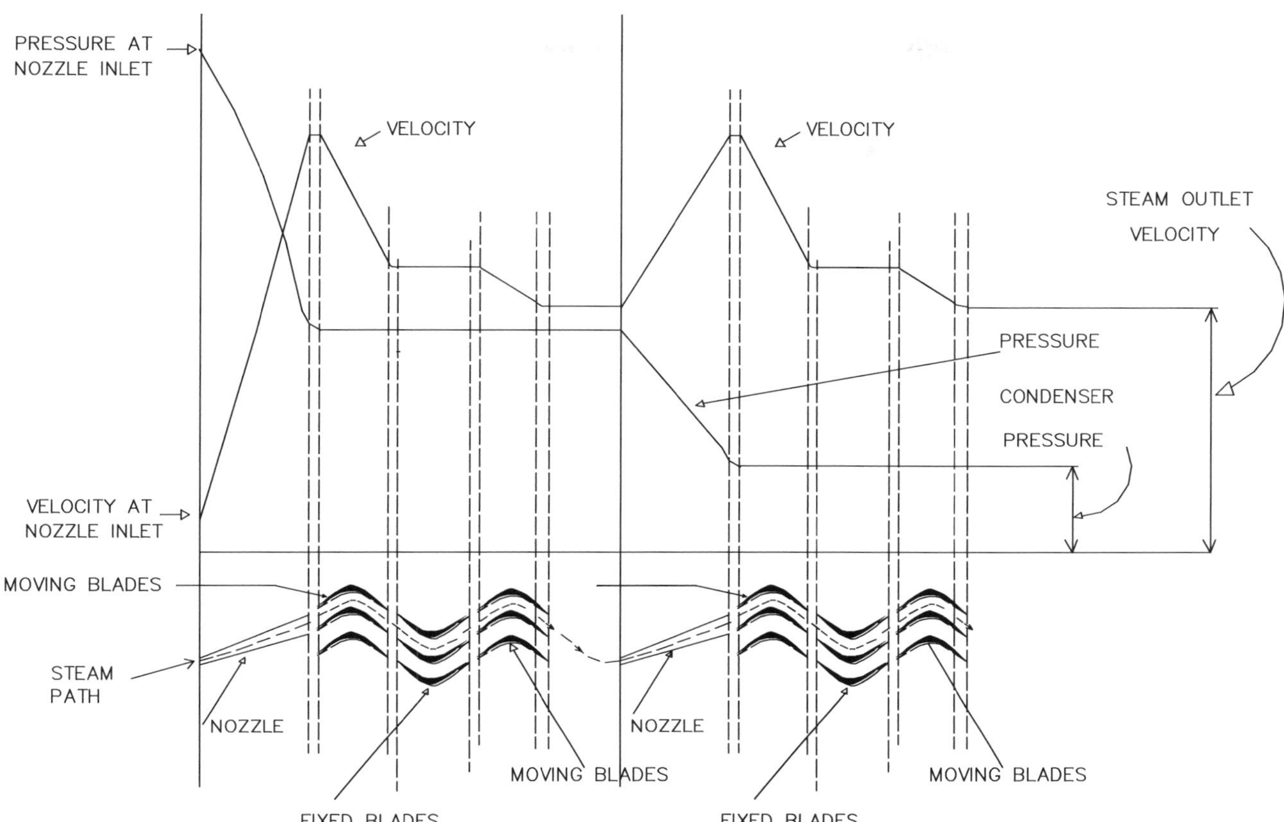

FIG. 2.18 Pressure velocity compounded turbine schematic and pressure/velocity changes.

3.1.4 Pressure Velocity Compounded Turbine (*Ref.* Fig. 2.18)

In this type of turbine, both pressure and velocity compounding are combined in different stages. Each stage is identical to a velocity compounded turbine. The number of stages determines the total drop of pressure to take place in those stages. This type of turbine was very popular for simple construction compared to other types, but as the efficiency is low, it is rarely used nowadays.

3.1.5 Impulse-Reaction Turbine (*Ref.* Fig. 2.19)

In the reaction turbine, as the name implies, the rotor is rotated mainly by the reactive force in place through the direct impulse or thrust alone. Here, both types of applications are done by providing many moving blades in the rotor and the same number of fixed blades in the stator.

The *fixed blades* in this type of turbine act as nozzles as there are no typical nozzles provided as such. The blades are projected radially from the stator inner periphery; they are shaped and mounted in such a manner that the spaces between the blades are formed just like the shape of nozzles, as may be seen in the cross-sectional view. Steam is directed onto the rotor by the fixed vanes or nozzles of the stator. In contrast with only a few nozzles, steam enters into the whole circumference of the inlet chamber and leaves the stator nozzles and fills the rotor chamber all along. Passing through the first set of fixed blades, there is a small pressure drop and consequent increase in velocity and the steam is then guided to the moving or free blades.

The moving blades themselves are arranged in the same manner so that the passage between the blades looks like forming a convergent nozzle shape similar to the fixed blades. This particular design allows a small pressure drop and hence an increase in kinetic energy. The steam entering into the first set of moving blades imparts torque to the shaft due to a change in direction at the outlet and hence a change in momentum thereby impulse to the moving blades. This pressure drop gives rise to a reaction in the opposite direction to that of added velocity.

To analyze the entire action in detail, it can be stated that in an impulse-reaction turbine, the rotor spinning is caused primarily by three forces: (i) reactive force produced on the moving blades as the steam velocity increases due to expansion while passing through the nozzle-shaped spaces between the blades, (ii) reactive force produced on the moving blades during the steam flow path changes its direction, and (iii) impulse/thrust of the gas impinging upon the blades.

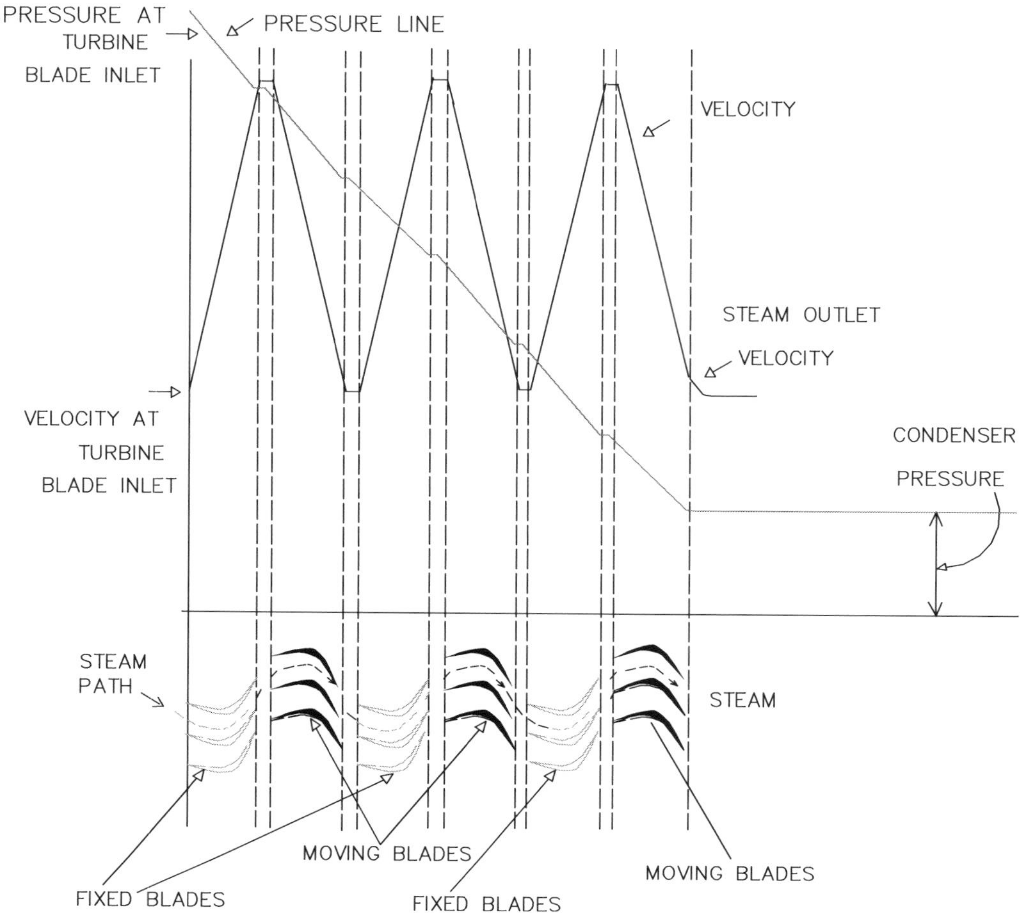

FIG. 2.19 Impulse-reaction turbine schematic and pressure/velocity changes.

Thus, as previously noted, a reaction turbine is moved primarily by reactive force but also to some extent by direct impulse/thrust. Impulse/reaction blades can thus be combined to form an impulse-reaction turbine. To summarize, this turbine combines the rotational forces of the previously described turbines; that is, it derives its rotational force or torque from both the impulse of the steam striking the turbine blades and the reactive force of the steam with increased velocity and changing direction. The resultant motive force is the vector sum of these impulse and reaction forces. As a small pressure drop occurs across both the stator and the rotor, there is a series of continuous and smooth pressure drops throughout the turbine length. So far as velocity of steam is concerned, it accelerates through the stator and decelerates through the rotor, with practically no net change in steam velocity across the stage but with a decrease in both pressure and temperature, reflecting the work performed in the spinning action of the rotor.

Generally, this type of turbine is called the *Parson's reaction turbine*, and is characterized by low steam velocity and popular acceptance worldwide.

3.2 Basic Turbine Type: HP, IP, LP Cylinders

The modern power plant concept is based on multiple turbine cylinders/rotors that are coupled to a single shaft (Figs. 2.20 and 2.21).

As the name implies, the high-pressure steam from the boiler enters the HP (high-pressure) cylinder, where it is isentropically expanded from pressure p_1 to p_2 (point 1 and 2 in the T-s diagram) and taken out as HP exhaust and normally termed *cold reheat*. This HP exhaust steam is then readmitted to the *reheater* part of the boiler for reheating and the steam condition is changed from point 2 and 3 in the T-s diagram. Reheating is done to avoid wet steam in the turbine blade, which otherwise would cause erosion as the standard moisture limit followed internationally for the steam in the turbine is around 10%. The reheated steam from the boiler, popularly known as *hot reheat*, then enters the turbine IP (intermediate pressure) cylinder and expands isentropically up to pressure p_4 (point 4). In a modern power plant and/or high capacity power plants, a twin cylinder IP module is also provided for various reasons such as better balancing of heavy mass,

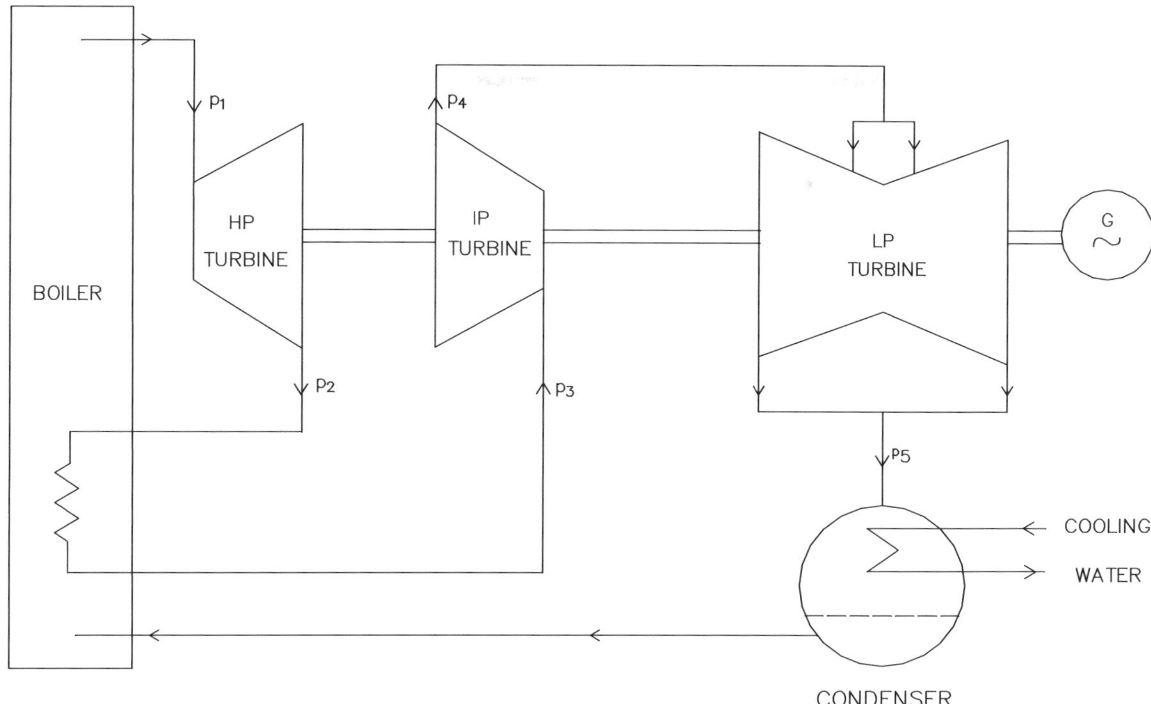

FIG. 2.20 Schematic diagram of turbine stages.

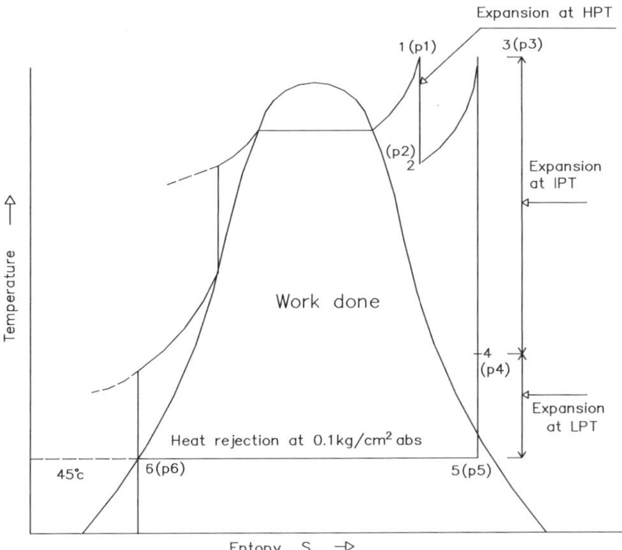

FIG. 2.21 Reheat steam cycle in modern power plant with reheat steam in T-s axis.

etc. After doing work, steam is then taken out from the IP cylinder as IP exhaust and readmitted to the LP (low pressure) cylinder, which is normally a twin cylinder configuration with steam admission near the middle portion of the cylinder with a symmetrical distribution arrangement. Here, inlet steam enters at pressure p_4 and further expands isentropically at a point 5 of pressure nearly about 0.1 kg/cm^2 absolute. Exhaust from the LP cylinder is then directed to the condenser, which is actually responsible for maintaining this very low pressure. Here condensation vis-a-vis conversion of a huge volume of low-pressure steam into water takes place through heat rejection to an external cooling agency. In the T-s diagram, the path is shown from point 5 to point 6, which is actually the saturated water line, and finally taken out of the turbine scope with the help of the condensate extraction pump (CEP).

The design of the turbine part so far as number of stages is concerned takes care of many aspects, namely, plant capacity/efficiency, thrust balancing of the turbine, etc. In different cylinders, the number of extraction outlets is used for the regeneration cycle. To avoid irreversible mixing of cold condensate with hot boiler water, which causes a loss of cycle efficiency, extraction steams are used to heat this cold condensate from the hot well. Before going to the boiler, this condensate, also known as feed water, is heated at various points by extraction steam with a suitable temperature and pressure through heat exchangers called gland steam heaters (GSH), low-pressure heaters (LPH), and high-pressure heaters (HPH) to ultimately match the boiler feed water inlet temperature. Extraction steam is also provided from the turbine for deaeration of feed water and in many plants for a separate steam turbine-driven BFP in addition to a motor-driven feed pump.

3.3 Turbine Oil Systems

There are three main oil systems so far as turbine subsystems are concerned: lubricating oil, control oil, and jacking oil. The type of oil used nowadays for all these purposes is mostly fire-resistant synthetic oil.

3.3.1 *Lubricating Oil (Ref.* Fig. 2.22*)*

Fig. 2.22 depicts the basic idea and purpose of the lubricating oil meant for all the turbine bearings with dirt-free lubricating oil (lube oil or LO) at the proper temperature, pressure, and quantity under all operating conditions. Even during the posttrip coast-down period, the supply of the right quantity of lube oil is extremely essential. The turbine rotor shaft, having high speed rotation and heavy mass, is supported on different types of journal and thrust bearings at different places of each cylinder. These bearings do require lubrication and cooling media to minimize the frictional effect and dissipation of heat generated due to friction. LO, serving both purposes, is necessarily a closed-loop system consisting of the oil tank, pump, oil cooler, etc. As the duty of this oil system is of utmost importance, the turbine shaft-driven oil pump, known as the main oil pump (MOP), provides the necessary supply pressure during normal operating conditions. There are separate AC and DC motor-driven auxiliary oil pumps (AOP) to provide LO during the start-up and shut-down periods. The temperature of all the bearing's outlet oil is measured in addition to the bearing metal at some places. The oil temperature is controlled automatically by water flow variation to the oil cooler. The supply pressure is maintained at a constant value by providing a self pressure regulating valve (PRV). However, oil from the emergency DC oil pump discharge is connected to the system, bypassing the pressure control part, and in some plants, even the oil cooler, considering its emergency service requirement. A dual filter is provided for removal of dirt in oil.

The AC motor-driven auxiliary oil pumps (AOP 1 and 2, for example) start automatically if the pressure falls below certain predetermined values due to any problem in the oil circuit under running conditions or during start-up/shut down for an AOP failure/problem.

Under normal operating condition, if the LO pressure falls further below some predetermined value, the turbine trip circuits are energized after a short delay and the TG trips. The DC motor-driven pump known as the emergency DC oil pump, rated at about 65% duty, is also made ready as an automatic standby to start after a further fall of oil pressure mainly due to the failure of the AC motor-driven pumps.

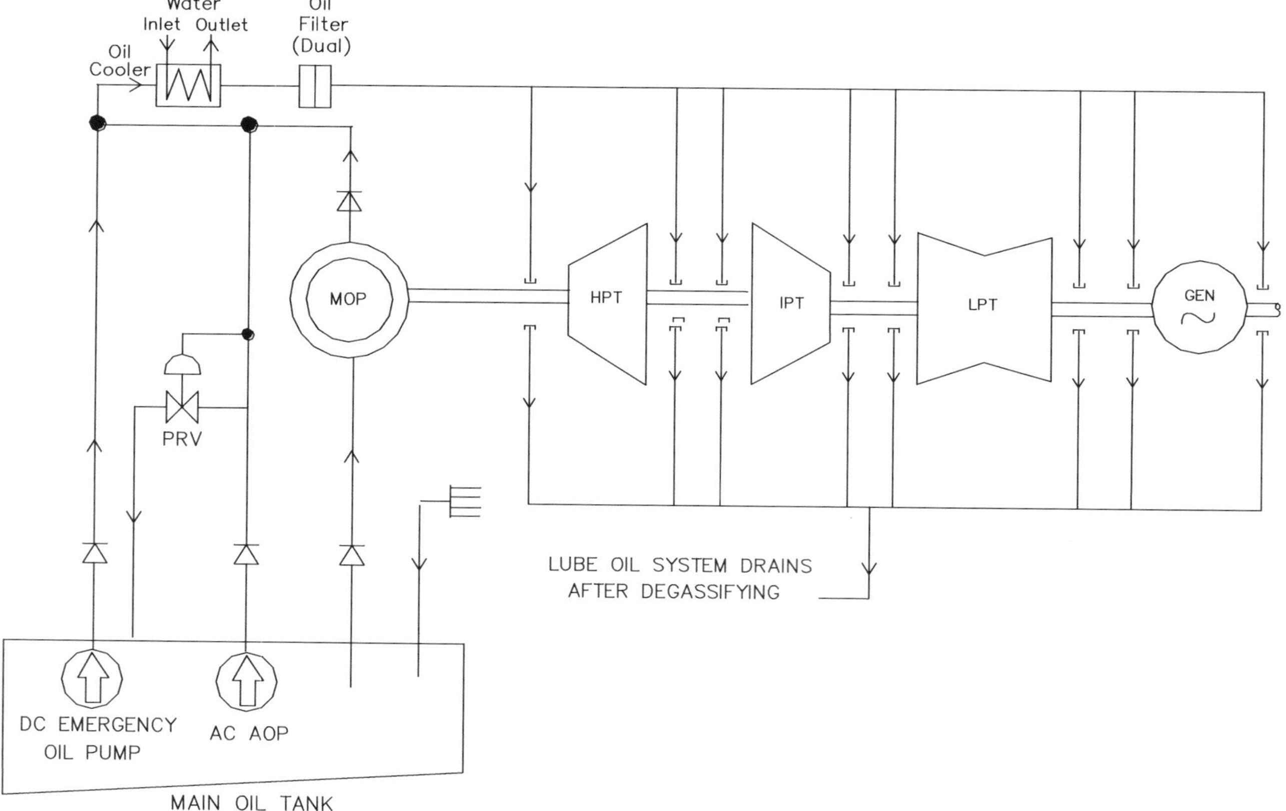

FIG. 2.22 Schematic diagram of turbine lube oil system.

There is a very important mechanism called the *barring gear*, also known as the turning gear, in the turbine subsystem. This mechanism is meant for providing rotation of the turbo generator (TG) rotor shaft at a very low speed as a part of the start-up procedure before actual running of the TG set and also after stoppages of the unit during the shut-down procedure.

When the unit is going on shutdown or is tripped, the steam inlet valve is closed and the turbine coasts down to a dead stop or standstill. In this condition, when the turbine stops completely. If it remains in one position for a long time, the heat inside the turbine casing will have the tendency to concentrate in the casing's top half portion and will obviously make the rotor shaft's top half portion hotter than the bottom half. The rotor shaft therefore could warp or bend even to the range of a few microns and would result in an eccentric shaft; it can only be detectable by high-precision eccentricity meters. It is obvious that however small the magnitude of the eccentricity thus formed, it would be enough to cause harmful vibrations to the entire turbo generator unit at the time it restarts. To prevent this, an automatic interlocking system makes the rotor shaft continue turning at low speed (around 1% of the rated speed) by the motor-driven barring (turning) gear unless and until the temperature drops down sufficiently to deserve a complete stoppage. The gearing system being used to achieve a very high reduction ratio and to turn a heavy mass requires a continuous supply of lube oil in the gearbox to assure reduced frictional loss, low noise, and cooling.

3.3.2 *Control Oil (Ref.* Fig. 2.23*)*

This system is meant for supplying oil to the control circuit of the turbine load/speed controller and hydraulic power oil for operating the on/off interceptor and modulating the duty hydraulic governor valves known as the *electro-hydraulic governing (EHG) system.* This part of the control and hydraulic power circuit can be described as the blood supply to the EHG system, the brain of the turbine subsystem.

Oil coming from the MOP and AC motor-driven auxiliary oil pump discharge lines is connected together with a common self PRV to maintain the control oil supply pressure at a fairly constant value.

Controlled oil pressure is supplied to all the electro hydraulic converters that receive electrical control signals from the *electro-hydraulic control (EHC) system.* The output is then transmitted to the respective servomotors of the various governing valves of the HP and IP turbines as the input signal with the power oil from the control oil delivery header.

The *trip circuit* is also fed from this header to accomplish automatic tripping of the turbine under extreme conditions to avoid unsafe and damaging operation of the TG set.

There is another circuit supplied from the above system called the *automatic turbine tester (ATT).* The operation of different valves is simulated through this system but without actually operating the valves or operating for a very small time so that the normal operation of the main power plant so far as generation is concerned is not jeopardized.

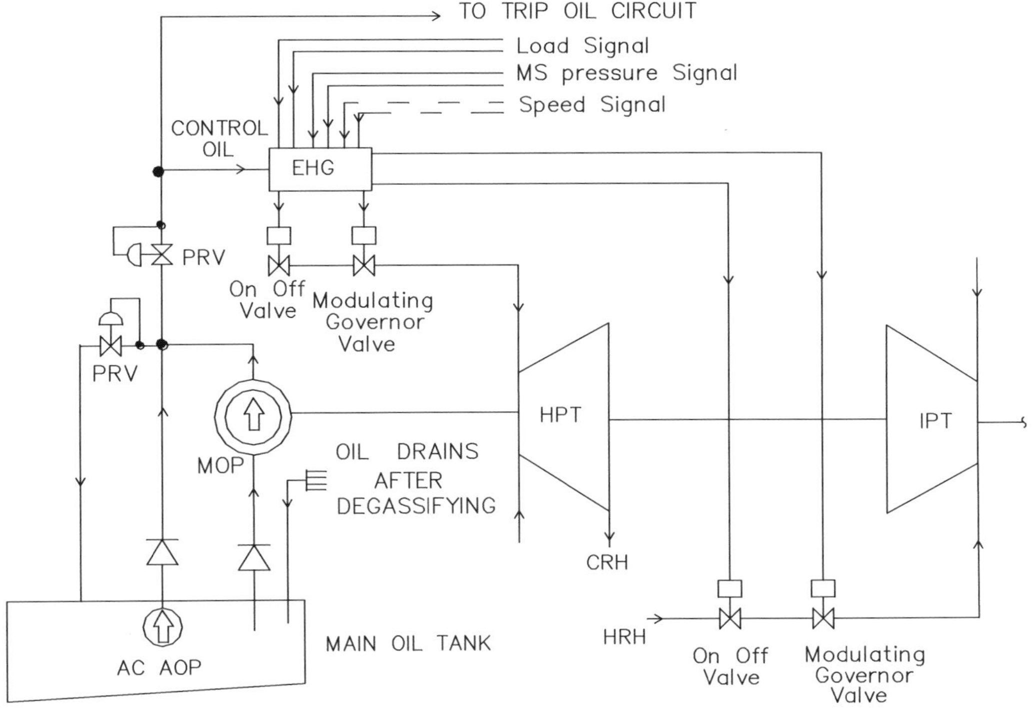

FIG. 2.23 Schematic diagram of turbine control oil system.

The automatic interlocked start and stop operation of the AOPs is already taken care of in the part of the lube oil system. The DC-operated emergency LOY pump has got nothing to do with this circuit as the control system is inactive as the turbine has already tripped. Then, this pump is to start if the AOPs fail to deliver the adequate oil pressure required for emergency services.

3.3.3 Jacking Oil (Ref. Fig. 2.24)

While starting the TG set after long stoppages, there is practically no space between the lower part of the rotor shaft and the bearing part on which the rotor rests. The heavy mass of the rotor may damage itself and the bearing part due to high static friction just before starting even the lube oil is supplied as it cannot enter the space in between. To prevent this untoward situation, another oil circuit has been envisaged with very high pressure at about 160–200 kg/cm^2 by a separate pump called the *jacking oil pump (JOP)*. It takes its suction from the lube oil header or main oil tank. This high-pressure oil is injected from the bottom of each journal bearing through a special arrangement to virtually float the rotor shaft before turning from the standstill condition. The thin oil film thus created due to a very small elevated / floated rotor condition helps to reduce the damaging frictional effect and then after, the turbine starts turning at a low speed by the turning/barring gear. Running under this condition for a preset time and with other prerequisites being satisfied, for example, the turbine gets steam admitted for generating power, etc., then it stops and the normal lubricating system resumes.

3.4 Extraction and Gland Sealing System

Before going to the boiler, the condensate/feed water is heated at various points to avoid an irreversible mixing of cold condensate with hot boiler water, which causes a loss of cycle efficiency. Different methods are adopted to do so reversibly *by an interchange of heat within the system* and thereby improving the cycle efficiency. The method is called *regenerative feed heating* and the cycle is called

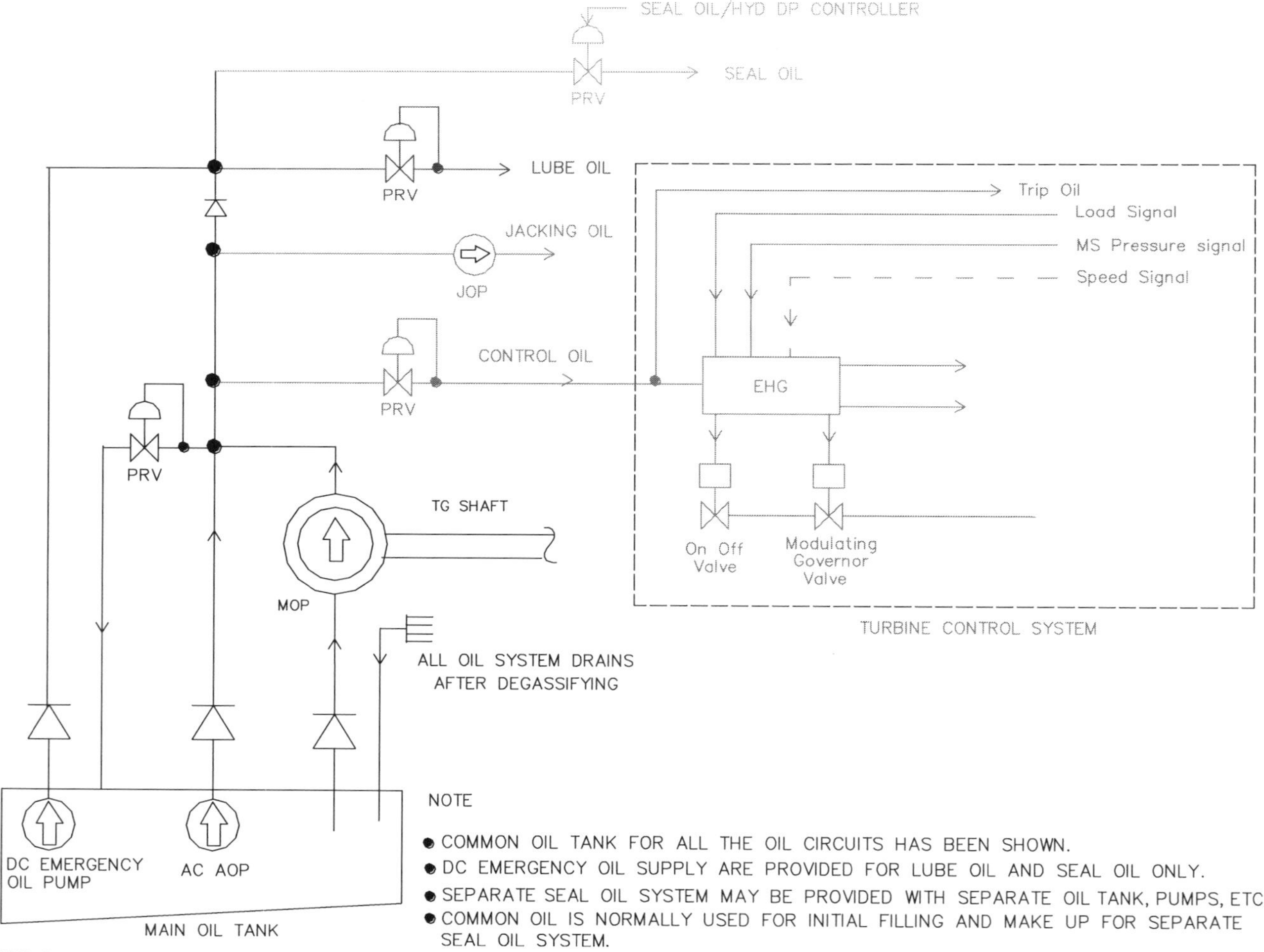

FIG. 2.24 Simplified schematic diagram of combined turbine control/lube and seal oil system.

the *regenerative cycle*. This is implemented by extracting or bleeding small quantities of steam from suitable points throughout the turbine stages, utilizing the heat content of the extracted or bled steam. The vessels where the exchange of heat takes place are called regenerative heaters. Here, the steam totally condenses and the outlet water leaves the heater with a higher temperature.

In different turbine stages, there are a number of extraction outlets with suitable temperature and pressure used for regenerating feed heating through a number of heaters called the GSH, LPH, and HPH to ultimately match the boiler feed water inlet temperature. Extraction steam is also provided from the turbine for deaeration of feed water at the deaerator and separate turbine-driven BFPs (if any).

The GSH gets heating steam comes out from the gland seals provided to prevent leakage of steam from the high/intermediate pressure turbines (HPT, IPT) casings. A special type of glands called *labyrinth* glands is the common choice to minimize leakage, but as there must be some amount of steam leakage, those are collected through the header and passed on to GSH where they are allowed to mix with the main condensate to get a common condensate temperature at the outlet higher than the inlet.

3.5 Condenser and Evacuation System

3.5.1 Function

The modern power plant using the *Rankin cycle* has many components, but the main and common component is the *condenser*. The main function of the condenser is to condense the exhaust steam coming out of the turbine into the water and reuse as depicted in the Rankin cycle. The usage of the condenser has many more advantages, such as:

- **(i)** Maximization of turbine efficiency by maintaining proper vacuum.
- **(ii)** Getting through an expansion of steam by lowering the steam exhaust temperature (saturation) much below the atmospheric saturation temperature through lowering the exhaust pressure, that is, increased vacuum.
- **(iii)** Reduction of steam flow with the given plant output by increasing the work area in the T-S diagram, and getting more work done.

The complete function is achieved in a separate vessel called the condenser, which is directly connected to the turbine exhaust and causes exhaust steam with the help of exchanging heat with the external cooling water to condense; hence, it is called the condenser.

3.5.2 What is Vacuum

A vacuum is a reading of pressure below atmospheric pressure. In an enclosed vessel, the vacuum can be created by partial withdrawal of the gaseous fluid occupying the given volume of the vessel. Generally, it involves lot of compression work on the gas but in the case of water vapor, it is achieved easily and with much less work if it is condensed by cooling. The volume shrinkage is so huge that a vacuum is created almost instantly. For example, at 0.1 kgf/cm^2 pressure (normal condenser pressure), 1 kg of steam occupies 15,000 L against 1.0 L of water only. Much less work is needed to transfer the condensate from the vacuum to positive pressure.

3.5.3 Type of Condenser

In general, condensers may be categorized into two types; direct (jet/contact) and indirect (surface) cooling. In *the jet/contact type*, cooling water and steam mix directly with each other whereas in the *surface condenser*, there is a barrier between the steam and cooling water and both have a separate circuit without mixing, and are most commonly used in modern power plants. The exhaust steam after becoming condensate at the outlet is driven through the appropriate pumping to the boiler as feed water for the next cycle.

The cooling water known as *CW* may be sourced from a once-through (from the ocean, river, or lake) or a closed-loop (using a cooling tower, spray ponds, etc.) system.

3.5.4 Operation of Condenser

The surface condenser is the type most commonly used in modern power plants and hence it will be the only one taken into consideration. Here, the main heat transfer takes place at the outside of the tube, thereby cooling vis-a-vis condensing of the saturated steam, and hence heating of the CW inside the tube. The water inlet temperature to the condenser determines the operating temperature/pressure of the condenser. As the temperature decreases, the condenser pressure will also decrease and thus a decrease in pressure will increase the plant output and efficiency.

The surface condenser operates under high vacuum and therefore is liable to suck noncondensable gases within itself. These types of gases may comprise mostly air due to leakage into the cycle from the components operating at that high vacuum system condition. The gases may result from decomposition of water into hydrogen and oxygen by thermal and chemical reactions and must be driven out of the system.

3.5.5 Evacuation of Condenser (*Ref.* Fig. 2.25)

Evacuation of the condenser is a must condition for the following reasons:

- **(i)** If gases leak into the system from outside or generate within the system, the total pressure in the condenser will increase and thus decrease turbine efficiency and output.

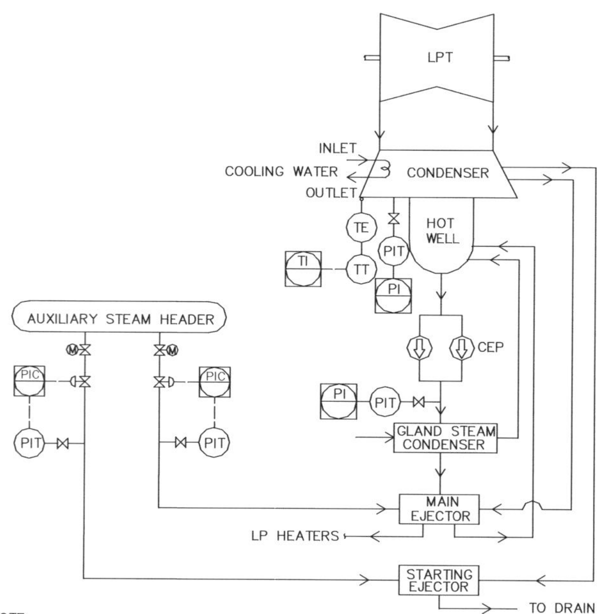

NOTE:
(1) LOCATION DETAILS SHOWN IS JUST TO GIVE AN IDEA ABOUT EQUIPMENT LOCATION.
(2) NOTE NO ROOT VALVE SHOWN WITH CONDENSER VACUUM. IF VALVE USED, IT SHALL HIGH PRESSURE QUALITY TO AVOID LEAKAGE.

FIG. 2.25 Condenser and evacuation system flow diagram.

(ii) Undesirable gases will cloud the outer surfaces of the condenser tubes and will seriously hamper the heat transfer of the exhaust steam to the CW, resulting in high temperature vis-a-vis pressure lowering turbine efficiency and output.

(iii) Free oxygen is corrosive in nature and always dangerous to the metal piping and components. It must be removed from the condenser to improve the life expectancy of the cycle components, including steam generators.

For the above purposes, *air ejectors or vacuum pumps* are used, which are the devices instrumental for discharging any gas or air into the atmosphere.

3.5.6 Evacuation Methods

Two methods are adopted for the evacuation process of the condenser, thus venting out noncondensable gases:

3.5.6.1 Steam Jet Air Ejectors (Ref. Fig. 2.26)

Steam jet ejectors have a low initial cost and are reliable pieces of equipment for producing a vacuum. The ejector design has basic advantages such as absence of moving parts and simplicity of operation.

The conventional steam jet ejector has four basic parts: the steam chest, the nozzle(s), the mixing chamber, and the diffuser. Fig. 2.26 illustrates basic ejector operation. High pressure motivating fluid (here, steam from auxiliary PRDS) enters at the inlet of a converging-diverging nozzle and comes out with a very high velocity. This causes rapid expansion in the ejector steam chest, thus resulting in high vacuum. The suction line being connected to the condenser through the pipeline sucks noncondensable gases or air at the steam chest, thus causing removal of those elements from the condenser. The suction fluid enters and mixes with the motivating fluid in the mixing chamber and both are then recompressed through the diffuser.

The nozzle described above may be of different designs, and some of them are:

3.5.6.1.1 Single-Nozzle Ejector Single-nozzle ejectors are used for either critical or noncritical flow and are suitable for use only with one set of design conditions.

3.5.6.1.2 Multiple Nozzle Ejector Multiple-nozzle ejectors are unique in both design and performance and are said to be effective in steam savings of 10%–20% when compared with single-nozzle units designed for the same conditions.

3.5.6.1.3 Types The standard steam-jet air ejector available is a two-stage, twin-element type; however, a one-element type is also available.

The standard type two-stage ejector is arranged with an intercooler and an aftercooler, which are common to the two elements. The air ejector type is selected according to the amount of condensed water produced by condensers.

3.5.6.2 Spindle-Operated Ejector

Spindle-operated ejectors are suggested for services whenever suction or discharge pressures vary. During operation, a pneumatically driven tapered spindle moves in and out of the nozzle orifice to control motive fluid flow. The flow can otherwise be controlled by a separate valve as well.

3.5.6.2.1 Liquid Ring Vacuum Pump (Ref. Fig. 2.27*)* The operating principle of this type uses a liquid compressant to compress the evacuated noncondensable gases or air and discharge them to the atmosphere.

These type of pumps are becoming more and more important in modern power plants. The simplified design and operating principle provides many advantages over other types of rotary gas pumps. Liquid ring vacuum pumps can be used for wide applications on a very large scale.

Operating Principle of Liquid Ring Vacuum Pumps: Fig. 2.27 shows a cross-section across the length of the liquid ring vacuum pump. It operates on the principle of a rotary liquid piston. The shaft and impeller assembly are mounted eccentrically relative to the pump casing, which means the centers of the shaft/impeller assembly and pump casing are different, as shown in the figure. As the impeller rotates, the service liquid (water in this case) continuously

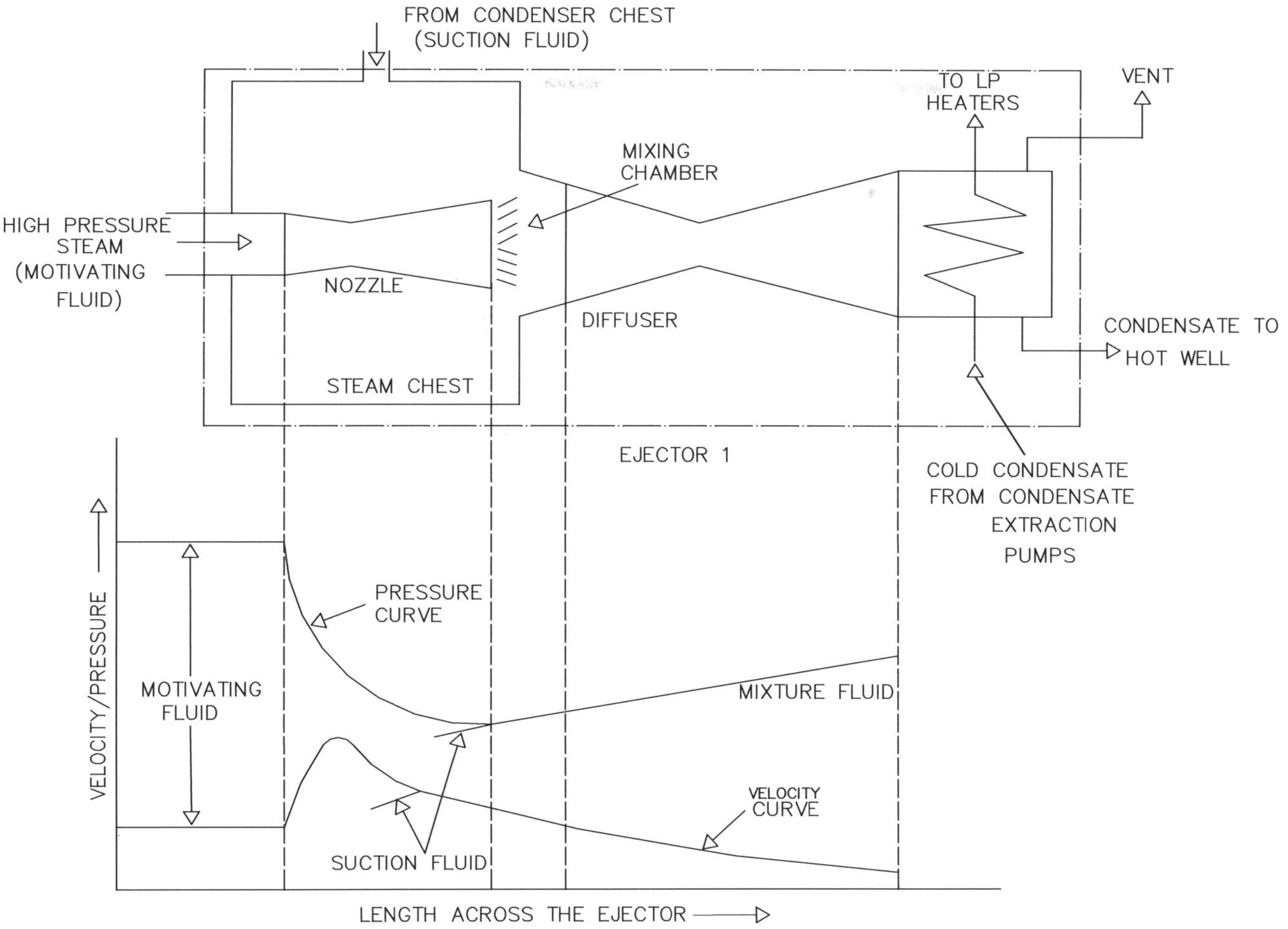

NOTE:
(1) LOCATION DETAILS SHOWN IS JUST TO GIVE AN IDEA ABOUT EQUIPMENT LOCATION.

FIG. 2.26 Ejector operation and pressure/velocity excursion curve.

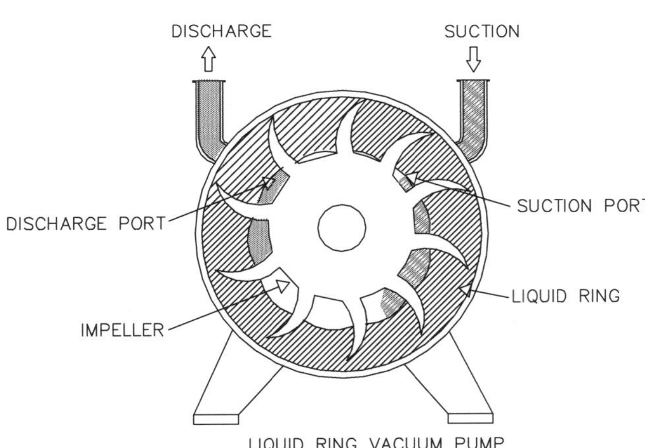

FIG. 2.27 Cross-sectional view of liquid ring vacuum pump.

supplied to the pump is forced outward by centrifugal force to form a liquid ring revolving concentric to the pump casing. By virtue of the eccentric position of the impeller, the liquid ring will move toward the periphery or away from the shaft, resulting in a liquid piston action displacing the air or gases between the spaces of the impeller blades.

Air/gas is drawn into the open space through the suction port. As soon as the suction port is crossed, the service liquid is forced back into the spaces between the impeller blades, as depicted in the diagram, starts compressing the air or gases. When the spaces between the impeller blades reach the discharge point, the liquid ring will force the air so compressed between the blades toward the discharge port.

3.5.6.2.2 Dry Vacuum Pump *Dry vacuum pumps* are devices that need no working fluid. There are two types of pumps namely, using the mechanism of volume reduction or the mixing of lower-pressure gas with higher-pressure discharge gas (as in a root blower).

Unlike a liquid ring vacuum pump, dry vacuum pumps normally run hot because there is no liquid to absorb the heat of compression. Consequently, this increases the temperature of the process gas. To counteract this, some designs provide for precompression by recycling discharge gas that has been cooled.

3.5.6.2.3 *Once-Through Oil Vacuum Pump*

These pumps are a sliding-vane type that use once-through oil (OTO) to seal clearances and lubricate moving parts. The vanes are in slots of the rotor and are mounted eccentrically to the pump chamber.

During rotation, centrifugal forces push the vanes out of the slot and against the walls of the chamber, creating pockets whose size varies similarly to those of the liquid ring vacuum pump. Because of this variation, suction draws process gas into the pump from the vessel being evacuated and compression takes place as the vane rotates toward the discharge side of the device, decreasing the area and forcing the gas and lube oil against the discharge valve opens slightly above atmosphere.

In the condensers, there is an air-cooler section provided to support the removal of noncondensable gases. It consists of a bundle of tubes, and they are baffled to collect noncondensables. The volume and size of the air removal equipment is reduced for less volume of noncondensables due to the cooling effect.

3.5.6.3 Mode of Evacuation Methods: Hogging and Holding

Before exhaust steam is admitted, all the noncondensable gases/air must be removed from the condenser. There are two modes of operating the noncondensable/air removal equipment: hogging and holding.

In *hogging mode*, large volumes of noncondensable gases/air are removed very quickly, thereby reducing the condenser pressure below atmospheric to a certain predetermined level.

Holding mode begins to operate when the desired pressure is achieved after successful operation of the hogging mode to remove all the remaining noncondensables/air.

3.5.6.4 Performance

The two-stage, twin-element air ejector can extract a specified amount of dry air with its single element and maintain a very high vacuum on the order of 735 mmHg. This air ejector is designed to extract saturated steam and air from the condenser about 2.3 times the volume of air.

The cooler is divided into the intercooler and the aftercooler, which are mounted on the outlets of two stages of the ejector, respectively. It is designed to condense steam. The second-stage ejector extracts air with a small amount of saturated steam and discharges it into the atmosphere. High-velocity steam jets from the nozzle entrain the air. The diffuser raises the entrained air pressure. The nozzle is usually made of stainless steel while the extraction chamber and diffuser are of steel castings or steel plates; however, the material may differ according to the condition of the steam handled.

3.5.6.5 Starting Air Ejector and Priming Air Ejector

When starting the turbine, the special type of air ejector employed to rapidly discharge a large amount of air from the condenser is called the *starting air ejector*.

Another type of air ejector is used to prime the cooling water system of the condenser; it is called the *priming air ejector* and is discussed in Section 3.7 of this chapter.

3.6 Start-Up and Thermal Stress

Steam with very high pressure and temperature is to pass through the turbine for getting power output and that is why it is very important to take care of the thermal stresses and must be kept under normal value during *start-ups* and normal operation. The start-up procedure is more important as hot steam would cause the cold turbine metals stressed during initial entry of steam especially for large thermal power plants (around 210 MW and above) where the temperature rising slopes during start-up and large load excursions must be made limited to restrain the unsteady thermal stresses in the various parts of HP and IP. It has long been observed that cyclic thermal stresses or low-cycle fatigue more than the design limits along with other sources can cause rotors to develop a dangerous condition called cracking.

It is basically the temperature difference between the hot steam (outer surfaces) and the cold middle metal body that causes the thermal stress. With a set of particular initial steam temperatures, the flow rate is regulated at the optimum values for such time that enables start-up of the turbine in the shortest period for maintaining the thermal stress within the permissive limit. This procedure is extremely important, as the thermal stresses beyond limits act upon the life expectancy of the steam turbine while the start-up time of the plant is also jeopardized.

3.6.1 Types of Start-Up Procedures

There are three types of start-up procedure: the *cold, warm, and hot start-ups*. A cold start-up is the most important as the temperature difference is the maximum in this case and needs careful operation.

A cold start-up of a steam turbine is applicable after a very long period of shutdown when the turbine case chest, valves, and rotor temperature are around the ambient temperature.

Technically, cold starts are considered when the metal temperature is less than 200°C. Warm start-ups are applicable when the plant goes for a two-shift operation or resumes operation after the weekend shutdown when the temperatures of different metals are not so less by properly entrapping the steam within the turbine. The hot start-ups are naturally meant for when there is a need for restarting the plant immediately after an emergency trip out.

3.6.2 *Damage Mechanisms*

It has been observed that the damage caused by cold starts is about three times that caused by warm or hot start-ups. It has also been found that the start-up stress becomes less toward the depth of the metal with a proportionate build up of cracking growth.

Two major damage mechanisms are responsible for most of the turbine component failures, causing lapses in safety measures and costly maintenance arrangements. These two damage mechanisms are *creep* and *low cycle fatigue*, which are mainly associated with high process temperature and cyclic operation of the plants.

3.6.2.1 Creep

This type of failure is a mechanical phenomenon observed on metal bodies caused by very high temperature process fluid. The damage is more aggravated as rotor metals work under the action of very high centrifugal forces. Turbine metals are mainly a special type of steel that undergoes dislocations on grain boundaries and results in permanent deformation of the affected area.

3.6.2.2 Low Cycle Fatigue

Low cycle fatigue is the thermal effect on thick metals when the process temperature varies and accordingly, stresses develop. The operation of the thermal power plants always demands changes in load at any moment, resulting in stress development. As the start and stop operations are also associated with changes in temperature, the turbine metals are sure to experience stress induction.

3.6.2.3 Combined Effect of Creep and Low Cycle Fatigue

The thick turbine metal component parts such as HP/IP cylinders, governor control and isolation valve bodies, high pressure castings, etc., are affected by the combined effect of creep and low cycle fatigue.

These two major damage mechanisms resulting from operation through high temperature and a longer run time engender the cumulative latent destruction of material components with a reduction in physical properties and distortion/deformation of internal properties. The ultimate outcome is obviously the less efficient running of the turbine until it is noticed.

3.6.3 *Relation of Stress, Temperature, and Start-Ups*

The temperature rising slopes (*heating ramps*) must be controlled by controlling the steam flow rate in such a way as to experience minimum thermal stresses on the turbine. It is also important to avoid high differential expansion between the rotor and stator case that could tend to make contact between them because of their different thermal inertia. The most critical part of the plant so far as the thermal stresses are concerned is the turbine shaft in contact with the highest temperature steam, that is, downstream of the first stage of the turbine. The start-up control system must therefore be capable of keeping the controlled variables within the safe allowable limits.

3.6.3.1 Stress and Temperature Curves With Relation to Time

Fig. 2.28 shows the different temperatures for the HP turbine, including the three lower curves representing the internal, mean, and external rotor temperatures, while the two upper curves represent the temperature of the superheated steam at the turbine inlet and the temperature of the first (impulse) stage outlet steam, that is, downstream of HPT nozzles where the steam first comes into contact with the rotor. During the turbine start up, the steam flow and the heat transfer coefficient are obviously very low; therefore, the rotor temperatures increase slowly. After a certain period of time, the HP valves are opened more rapidly and the rotor surface temperature travels nearer to both the HPT inlet and the first-stage outlet steam temperature. The temperature and time values as shown in the graphical representation are indicative only and vary according to the capacity and design of the plant. At the end of the transient period, all temperatures are more or less equal and the TG set is all set to take up the load.

The possibilities of a considerable reduction in start-up time, however, depend on state-of-the-art knowledge and improvements of the steam turbine operation. With the warm-up procedure being the first step of the start-up procedure, it needs much more attention. For warm up to take place, the turbine needs the admission of steam having a temperature above the corresponding metal temperature. The differential temperature (dT) determines the time taken by the warm-up process while the approximate or nearest value of the same is absolutely dependent on the time by which the machine was out of operation. For example, a longer outage would obviously demand a lower dT.

The plant start-up process includes different steps such as waiting for the proper admission steam quality, warm up of the turbine metals such as valves and stator/rotor blades, turbine step-by-step acceleration to synchronizing speed and then switching over to power generation mode, etc. All these steps are time consuming and the total time

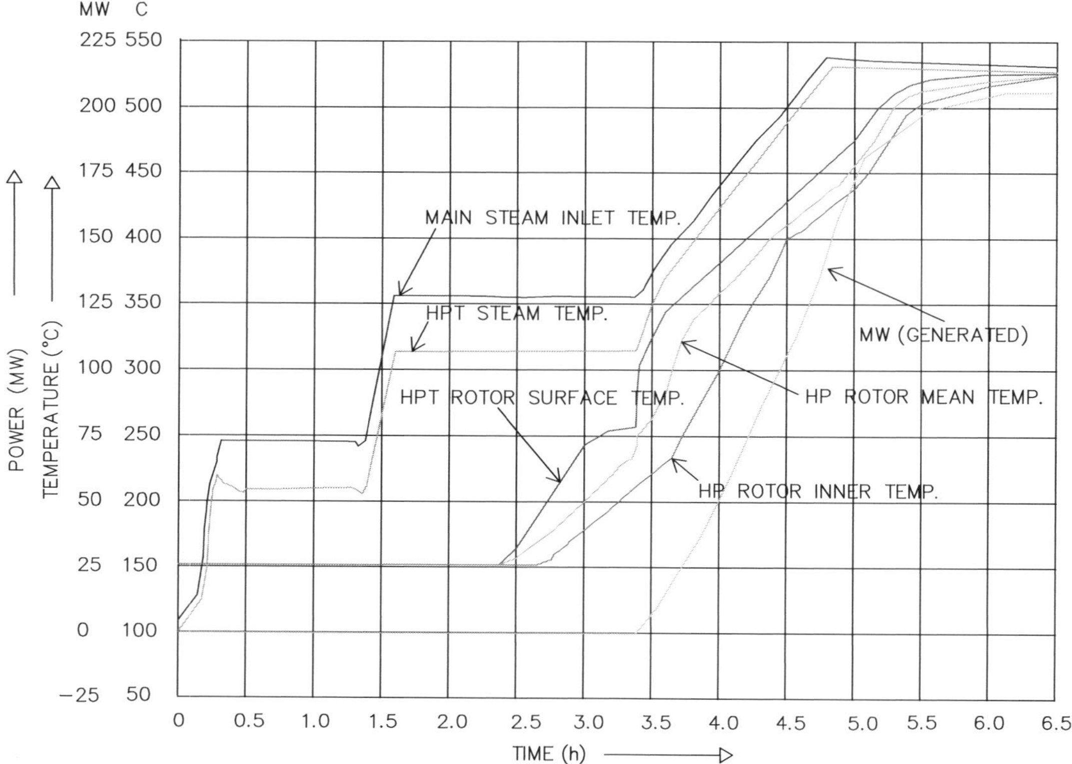

FIG. 2.28 Change in main steam inlet, HPT steam, rotor temperatures and MW (Gen) with time for normal start-up.

can be reduced if the possibility of doing some steps in parallel could be explored, which may provide operational flexibility and improvement without sacrificing the turbine health and life expectancy.

Fig. 2.29 shows the corresponding thermal stresses at the rotor surfaces of HPT and IPT. The maximum values, which actually determine the lifetime consumption over the start-up cycle, would be around a few thousand kilograms per square centimeter (kg/cm^2), and from this point, the turbo generator starts to take up the load.

The above approximate excursion figures, however, indicate the start-up procedure in a moderately unadventurous or conventional way to accommodate a sufficient safety factor toward unpredictable situations that could crop up during start-up of the plant. In other words, more importance is put on plant safety and not on efficient running of the plant.

Improvement of the start-up transient behavior can be made with a view to reducing the start-up time by either keeping the stress the same as before but taking the generator load earlier or minimizing the stress values. A brief discussion is incorporated below.

3.6.3.2 Stress and Temperatures Curves With Relation to Longer Soaking Time

After the initial heating ramp, the period called *heat soaking* is coming into the picture, which means the heating is continued for that period of time unless the desired temperature is achieved.

Fig. 2.30 shows the tentative excursions/trends of the parameters during start-up when an arrangement is provided to give the turbine more time for soaking.

Here, the main objective is to reduce the maximum stress value, meaning less lifetime consumption, without increasing the start-up time with respect to normal start-up transient values.

In order to achieve the same objective, it is necessary to reduce the thermal shock at the very commencement of the steam turbine picking up the generator load. This can be obtained by allowing for a longer soak time for the turbine, that is, once the turbine has reached full speed, it is kept running without taking any load so that the steam flow can heat the rotor for an extended period greater than that of the time period shown earlier. The turbine start up is therefore initiated earlier than the normal start-up transient, and the turbine is kept idling for say 1 h.

The load pick-up phase is then started, and the rate of load change is to be steered in order to get a flat stress curve. After a desired load level is achieved, the rate of change is increased to the maximum allowable level. The resulting temperature and stress plots are shown in Figs. 2.30 and 2.31.

Here, two thermal stress peaks are observed, with each of them contributing fatigue and thereby lifetime consumption. If the peak values are analyzed, it can be seen that the first

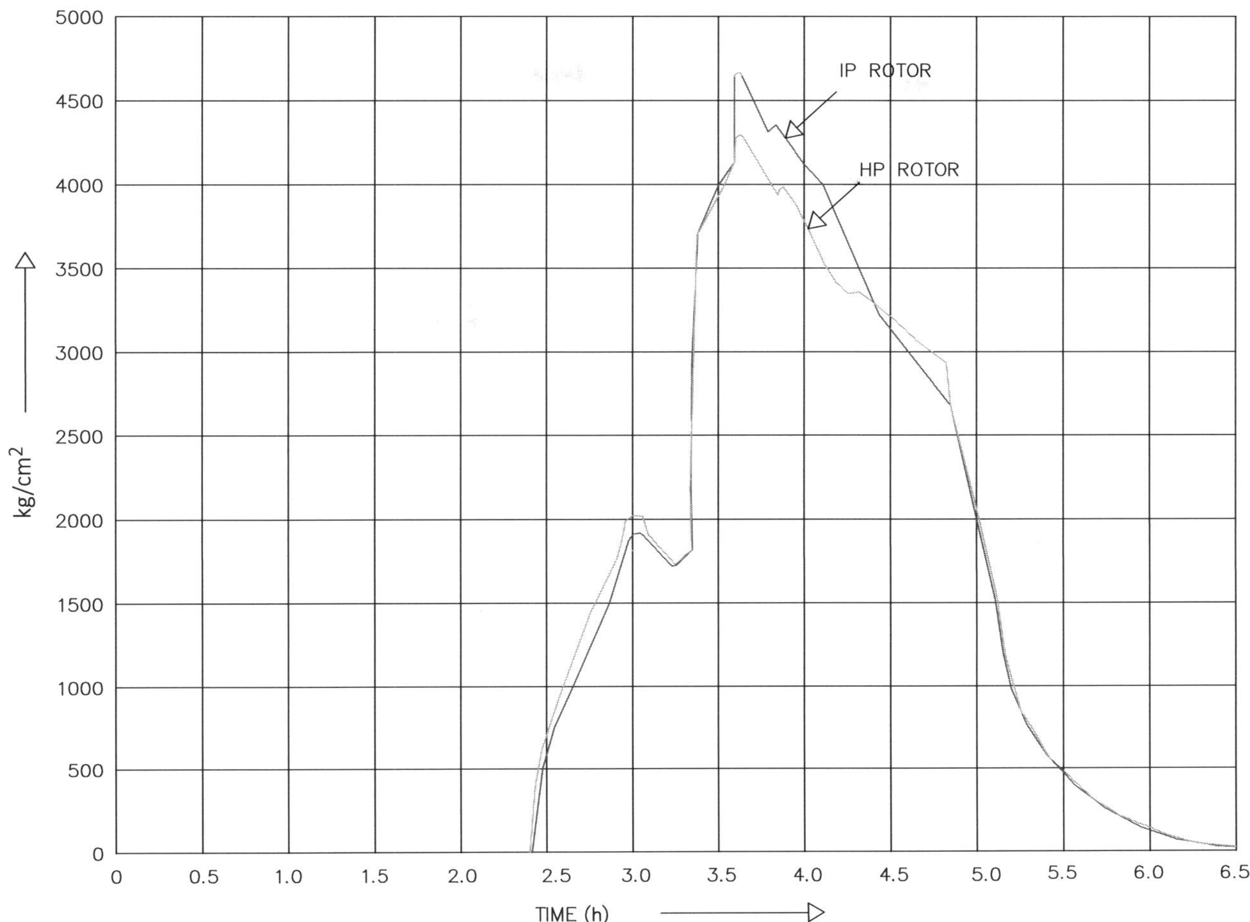

FIG. 2.29 HP and IP turbine rotor thermal stress development for normal start-up.

peak value may be slightly higher than the *elastic limit*, and thus correspond to a very low additional lifetime consumption. While the second peak is observed to be higher than the first peak, it is, however, well below the corresponding stress value shown earlier for normal start-up. The start-up time is also reduced by a considerable value vis-a-vis fuel savings.

3.6.3.3 Stress and Temperatures Curves With Relation to Fast Start-Up

Fig. 2.32 shows the tentative excursions/trends of the parameters during start-up when an arrangement is provided to give the turbine less time for soaking. Fig. 2.33 shows the tentative excursions/trends of the parameters during that type of start-up.

Here, the start-up time is minimal and the soaking period is much less as the generator load is connected earlier but with an upper limit of the stress peak the same as that of the value experienced with a normal start-up time. The benefit is thus the same lifetime consumption with early load delivery as compared to the normal start-up time. The net advantage is much less start-up time vis-a-vis maximum fuel savings in comparison with the normal start-up transient values without compromising the stress factor, that is, lifetime consumption.

It is to be noted that the thermal stress peaks are basically due to the thermal shock at the beginning of the start-up phase. Once this peak is attained, the lifetime consumption is the same regardless of how fast the stress goes back to zero.

3.7 Miscellaneous Turbine Auxiliaries

Some very important components will only be discussed in this section due to space limitations.

3.7.1 *Turbine Bearings*

Basically, journal-type bearings are used to take care of the weight of the TG rotor shafts and allow them to rotate freely while there is a thrust-type bearing provided to absorb the impact of the axial thrust produced during the rotation of the huge mass of the shaft.

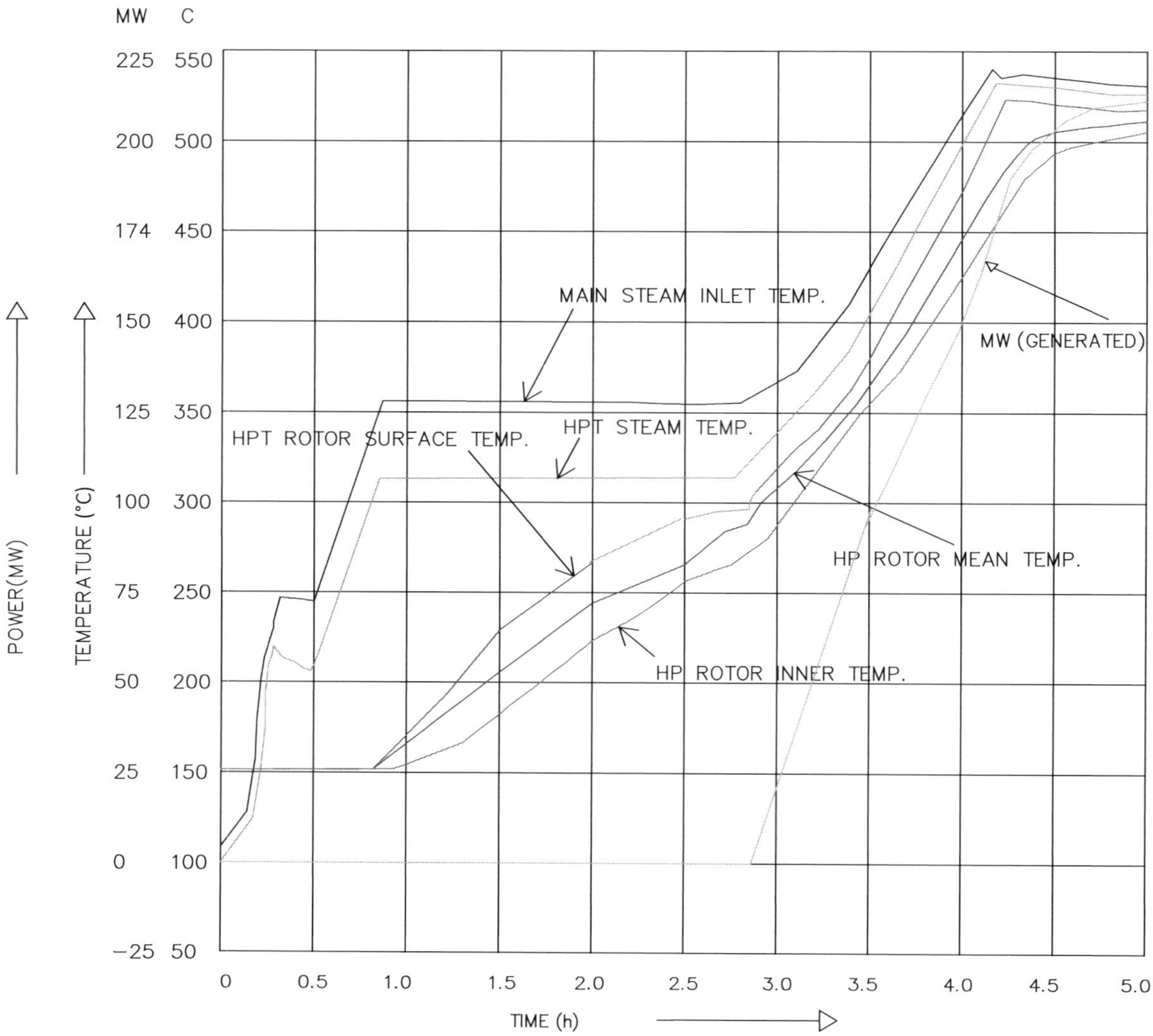

MW C
225 550
200 500
174 450
150 400
125 350
100 300
75 250
50 200
25 150
0 100
−25 50
POWER(MW)
TEMPERATURE (°C)
MAIN STEAM INLET TEMP.
MW (GENERATED)
HPT ROTOR SURFACE TEMP.
HPT STEAM TEMP.
HP ROTOR MEAN TEMP.
HP ROTOR INNER TEMP.
0 0.5 1.0 1.5 2.0 2.5 3.0 3.5 4.0 4.5 5.0
TIME (h)

FIG. 2.30 Change in main steam, HP turbine steam, rotor temperatures, and MW with longer soaking period.

3500
3000
2500
2000
1500
1000
500
0
kg/cm2
IP ROTOR
HP ROTOR
0 0.5 1.0 1.5 2.0 2.5 3.0 3.5 4.0 4.5 5.0
TIME (h)

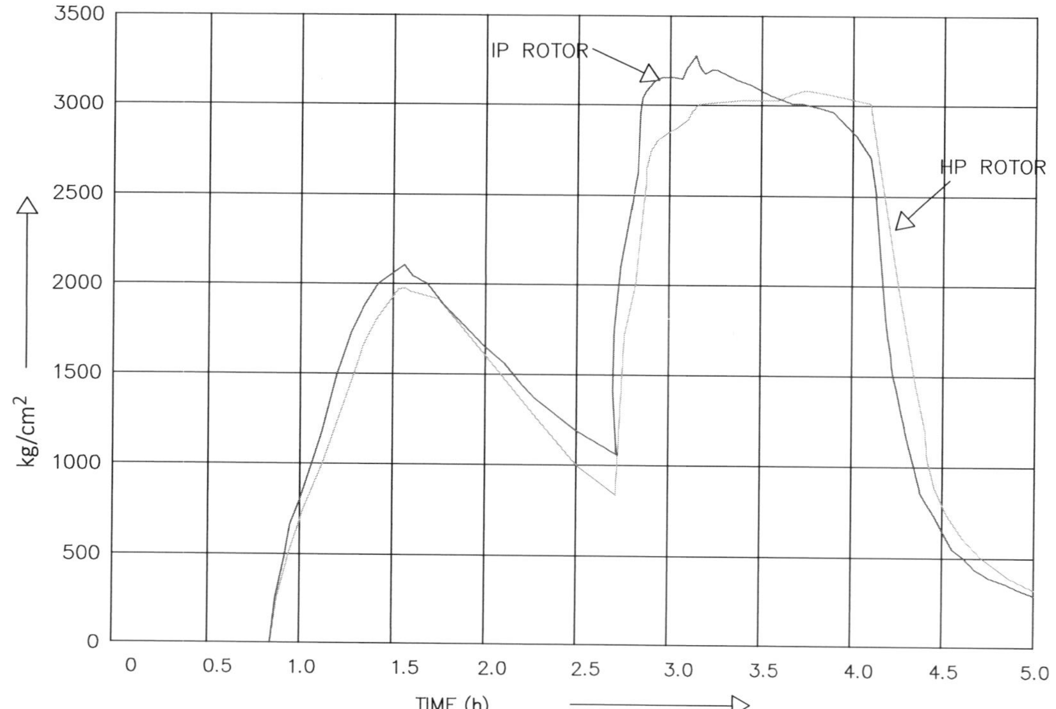

FIG. 2.31 HP and IP turbine rotor stress development during start-up with longer soaking period.

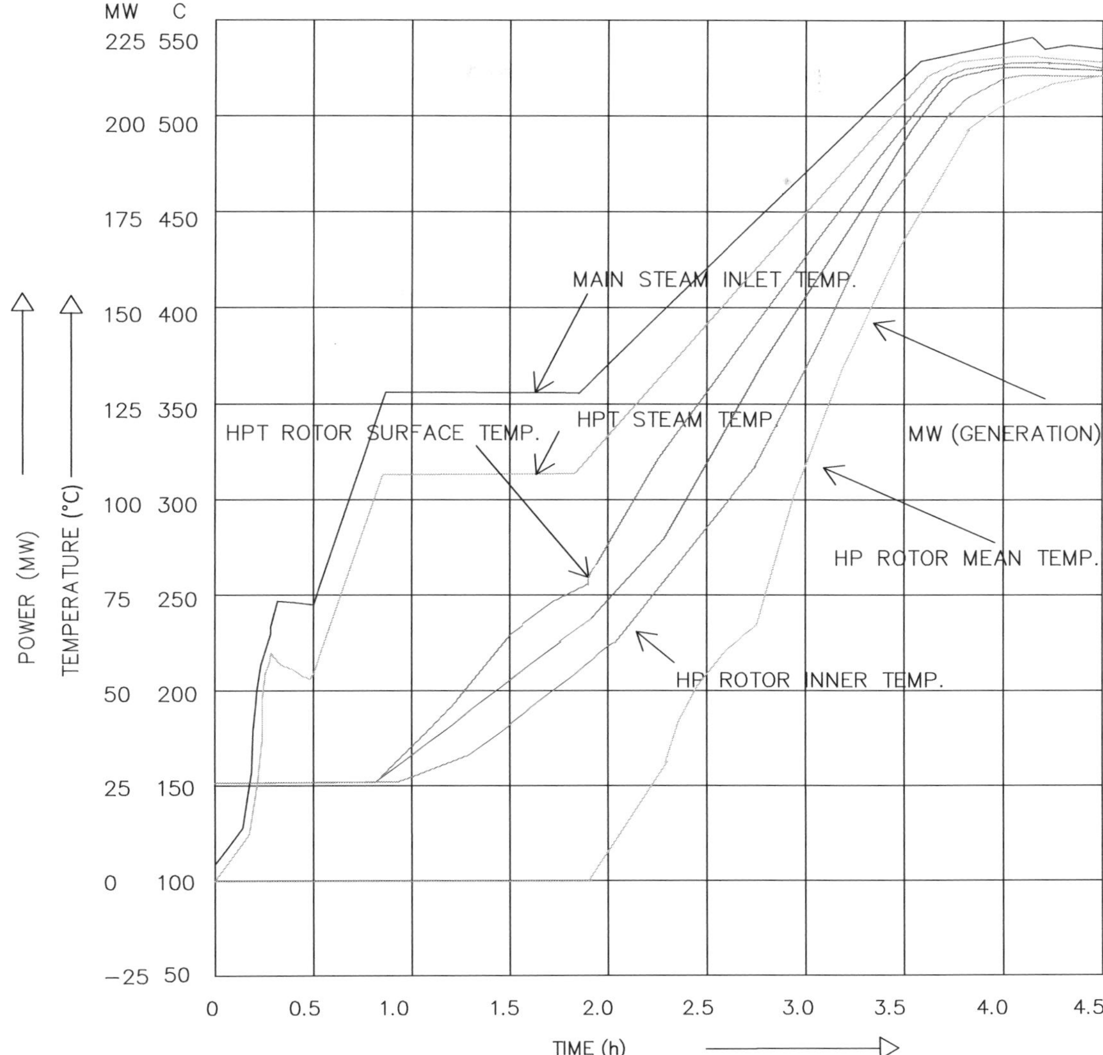

FIG. 2.32 Change in main steam, HP turbine steam, rotor temperatures, and MW with fast start-up.

3.7.1.1 Main Bearing

The main bearing consisting of a bearing proper and a shell is a spherical surface, seat-supporting type. This is an elliptical shape that, compared to conventional bearings (reference Fig. 2.34A and B), offers a much greater effect in stabilizing the shaft during turbine operation. Large-capacity rotors use tilting pad-type bearings that provide a greater shaft-stabilizing function. Lubricating oil with the proper volume is fed to each bearing by an orifice installed in the feed-oil tube.

The lower half of the shell that fits into the groove of the bearing stand is constructed so that its position can be adjusted by an adjusting plate. The upper and lower halves are bolted together to complete the shell, which is directly bolted to the horizontal coupling surface of the bearing stand to securely retain the bearing in position.

3.7.1.2 Thrust Bearing

The thrust bearing consists of the bearing proper and a thrust-receiving surface. The thrust-receiving surface is a taper-land type, which possesses great resistance against pressure. Several radial oil grooves divide the surface into segments, the surfaces of which are radially and circumferentially precision-machined and properly inclined. Various types of thrust bearings are shown in Fig. 2.34A–F.

Lubricating oil is supplied from the inner thrust-receiving surface to the oil grooves, and forced into the inclined surface by a rotating collar bearing to form a powerful pressure-resisting surface.

Nowadays, old-style elliptical or four-lobed bearings are replaced (and hence the description is not included) by a tilting pad bearing for adjustment facilities of shaft positions. The advantages of tilting pad bearings are, for example,

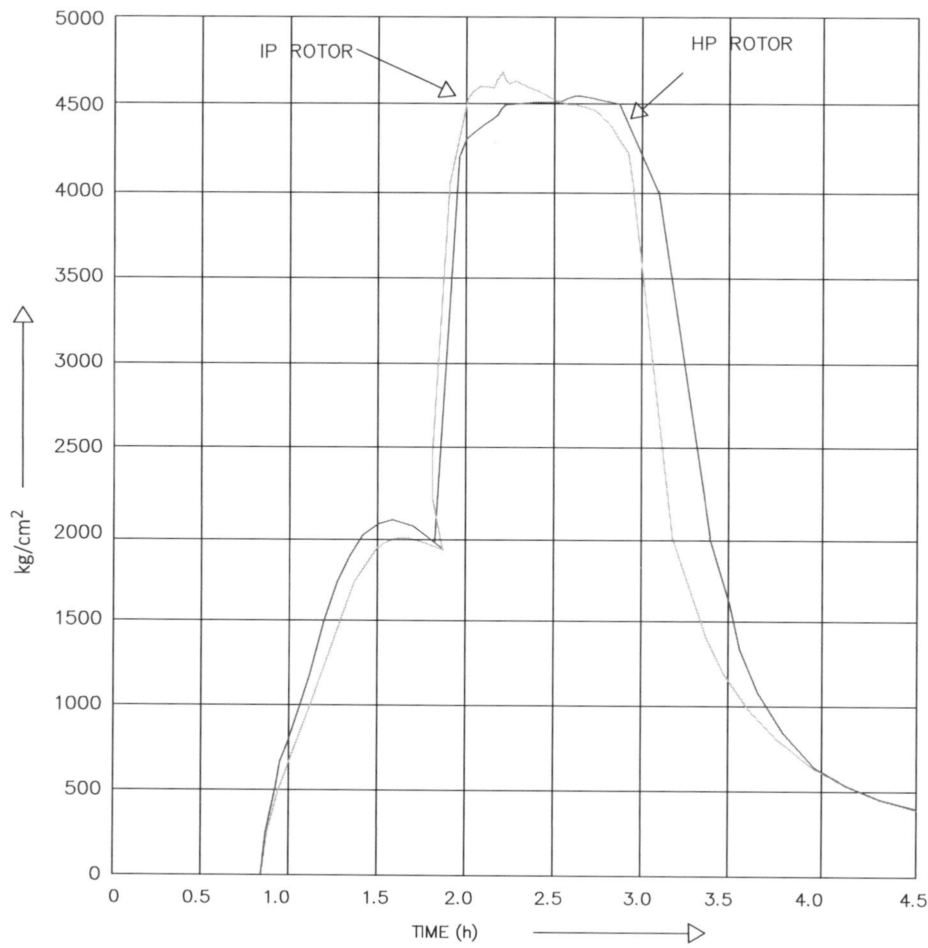

FIG. 2.33 HP and IP turbine rotor thermal stress developments during start-up with fast start-up.

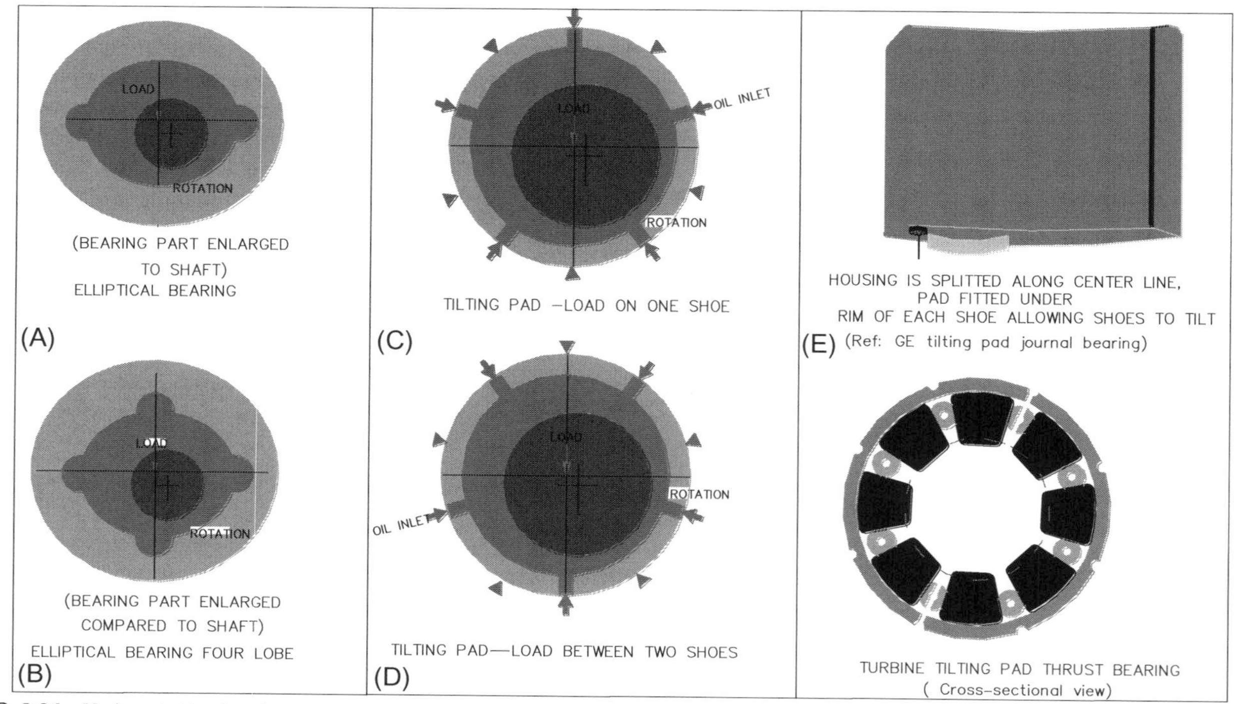

FIG. 2.34 Various turbine bearing types.

(i) total avoidance of oil whirl and dynamic instability, (ii) lesser amplitude of shaft vibration, (iii) increased safety and operability, and (iv) easier thermo element placement. The thrust bearing (reference Fig. 2.35F) is a split type having adjustable sectional shoes or pads inserted in its back to facilitate adjusting the thrust-bearing gap. During operation, the rotating part of the bearing, through the action of viscous drag, draws fresh oil into the pad area.

Due to fluid pressure, the pad tilts (thus the name) slightly and narrow constriction is formed between the pad and the rest of the bearing surface. Behind this constriction, pressurized fluid builds up in the shape of a converging wedge and separates the moving parts. This self-renewing wedge or film of oil, as incompressible, takes the load, provides shock absorption, helps control vibration, reduces power losses, and extends bearing life over a wide range of speeds, loads, and temperatures. The tilt of the pad adjusts itself with bearing load and speed. Various design details ensure continued replenishment of the oil to avoid overheating and pad damage. Normally, babbit metals and polymer-coated (self-lubricated) tilting pad bearings are used.

The thrust bearing especially equipped with a thermostat for measuring the temperature of the front and rear thrust bearing metal as a means of sensitivity monitoring any abnormal condition of the thrust-bearing surface. Also, it is equipped with a protective device for automatically stopping the turbine in case of an unforeseen accident.

3.7.2 Gland Packing, Gland Steam Condenser (GSC), and Gland Exhaust Fan

The gland steam system incorporates the gland sealing system, basically a subsystem of the turbine shaft packing,

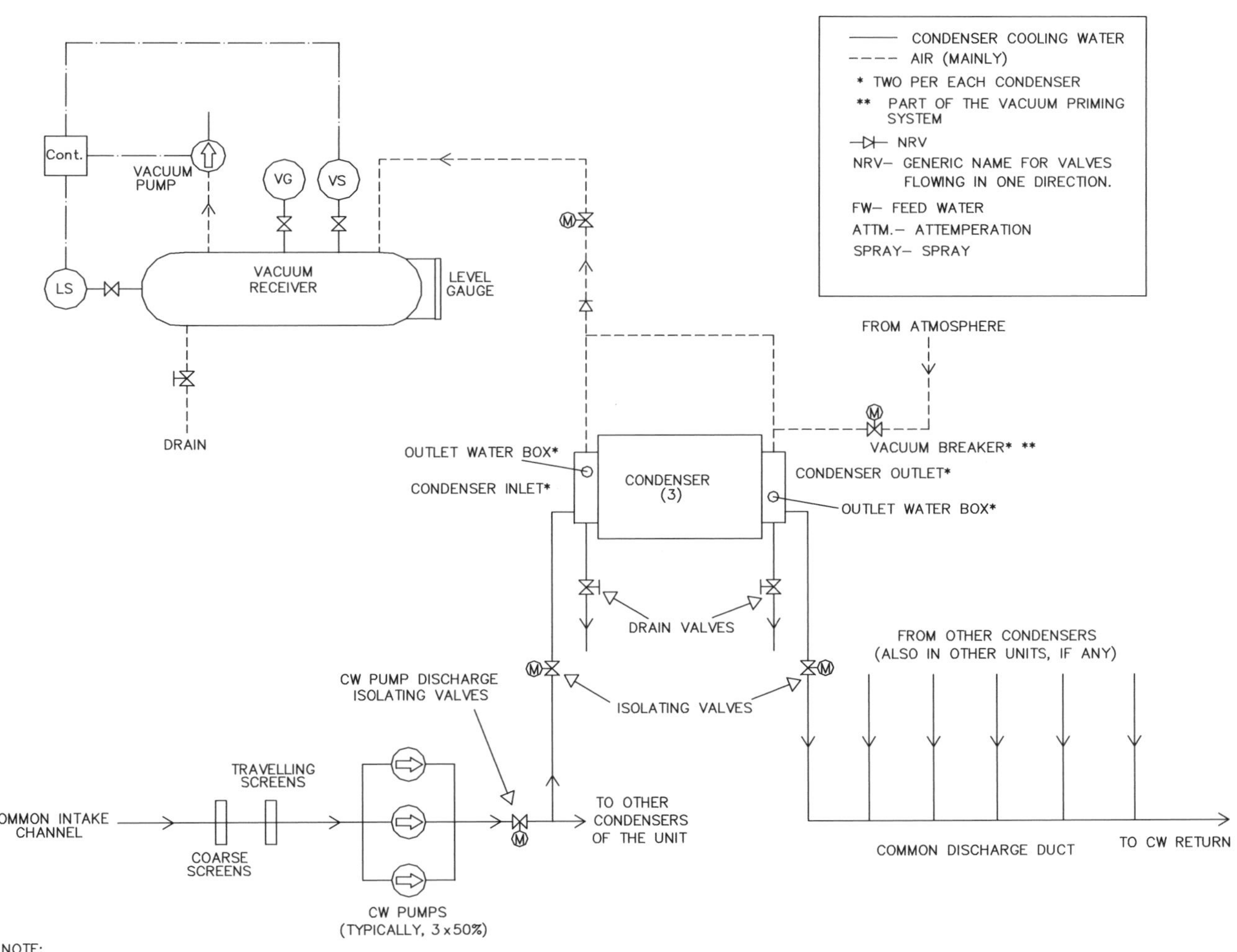

FIG. 2.35 Vacuum breaker and priming system.

to prevent the leakage of HP stage steam from or air into the LP stage of the turbine casing along the rotor shaft. The system not only prevents leakage but also provides arrangements for the removal of gland steam and air, which cannot be totally leakproof, from the turbine glands and sends it the GSC. The steam is condensed as water drain and sent to the condenser hot well. Leakage air is removed from the GSC to the atmosphere. The GSC maintains a partial vacuum in the glands to prevent steam leakage.

3.7.2.1 Gland Packing

Labyrinth gland packing: Labyrinth packing is used for the gland part of the turbine and for the inner surface of the nozzle diaphragm. It consists of several circular segments fitted into the grooves of their mating parts and equipped labyrinth packing should contact each other, thus reducing the local heating of the rotor.

The labyrinth packing for high-temperature parts consists of an alloy steel labyrinth body inserted with Cr-Mo steel strips; for low-temperature parts, that consists of a labyrinth body and strips integrally formed on nickel silver or phosphor bronze stock.

3.7.2.2 Gland Steam Condenser and Gland Exhaust Fan

When the gland seal is a steam-seal type, a gland condenser and a gland exhaust fan are installed. Steam leakage and air from the glands are led to the gland condenser, where the steam is condensed and the air and gases are discharged into the atmosphere by the gland exhaust fan. Turbine condensation is used as cooling water for the condensate to reclaim heat.

3.7.3 Priming System for Water Box and CW Pumps

Fig. 2.35 Vacuum Breaker and Priming System illustrates the basic functioning of the condenser vacuum system.

It is the system responsible for drawing the CW into the condenser along with CW pumps. As the CW intake and discharge points are normally situated at a distant location, it is necessary to suck the air in the long pipe line and create a vacuum that helps draw the CW through the pipeline. Removing the air from the system to lift water into the condenser before the circulating pumps are started is called *initial priming*.

Once the flow is established, it continues to run as a siphon effect with the aid of the CW pump discharge pressure to cater to the pipeline frictional pressure drop.

The *priming system* also incorporates *continuous priming* necessary to remove the air liberated from the water during normal operation of the condenser vis-a-vis CW. The vacuum equipment is often also used to remove air pockets from water boxes. Priming valves are often used between water boxes and the vacuum system to protect the vacuum system from being flooded with cooling water from the main condenser. It is also common to include a vacuum "reservoir" tank to dampen the effects of air surges.

It is obvious that cooling water temperature will have less impact on priming equipment than the main condenser vacuum equipment. A closed-cycle cooling water loop is usually considered as it will not force the use of exotic materials for the various cooler/heat exchangers. The system may also be considered for a once-through service water arrangement in case of smaller vacuum pumps and a clean water source is available. This will eliminate the need for a recirculated service water system and the associated maintenance.

A priming valve or stack must be installed at high points in the primed system to allow the air or noncondensable gases to be driven out of the system without excessive water being carried over into the vacuum system. A valve for the priming service should also be there to isolate the primed system from the vacuum system at a high water level in the primed system.

Priming valve wetted parts or trim materials, comes in contact with any carryover, should be designed to be compatible with the CW primed. The use of a drop-out receiver tank ahead of the vacuum pumps can help to protect the vacuum pumps from corrosive water spray.

For a continuously primed system, the flow of noncondensable gases is relatively small and hence it may be prudent to run the vacuum pumps intermittently to maintain the evacuation of a vacuum control tank. The added volume of a control tank prevents frequent cycling of the pump.

3.7.3.1 Priming Devices

3.7.3.1.1 Priming Air Ejector Such an ejector is normally a single-stage, steam-jet air ejector , which was already described in Section 3.7 of this chapter.

3.7.3.1.2 Liquid Ring Vacuum Pump These are the same type as described already in Section 3.7 of this chapter.

3.7.3.1.3 Dry Vacuum Pump The dry vacuum pumps are also widely used other than steam air ejectors and liquid ring vacuum pumps. The operating principle of dry vacuum pumps does not utilize any liquid such as oil or water as a sealing fluid. The steam is sucked, compressed, and forced to leave the condenser chamber. SCREW/LOBE/CLAW-type pumps are deployed for this application. Dry vacuum pumps are capable of creating a vacuum range that varies from atmosphere to 2.0 mm of (wcl).

3.7.3.1.4 Once Through Oil Vacuum Pump A rotary vane-type once-through (OT) oil vacuum pump utilizes oil as sealant/lubricant fluid for extracting clean, dry, nonreactive gases. These types of vacuum pumps are usually air-cooled, direct-drive 1750 rpm pumps and provide a vacuum range from atmosphere to 70.0 mm of wcl. Regular maintenance such as oil changes and periodic vane/filter changes are required to keep maximum serviceability.

3.7.3.2 Selection of the Priming Devices

The following parameters are important while selecting the priming devices.

(i) *Suction Pressure*—This is determined by the amount of head required to prime the system and the pressure drops between the vacuum pump and the system.
(ii) *Capacity*—This is determined by volume, the lowest system operating pressure, the temperature of water being handled in the condenser, and the time allowed to reduce the system pressure.

3.7.4 Vacuum Breaker

This is a special type of valve (zero leakage type) in the condenser priming circuit. The purpose of a vacuum breaker valve is to quickly allow air into the vacuum space of the condenser as well as to lower the pressure turbine exhaust hood to reduce the roll-down time of the TG set.

As a usual practice, the vacuum breaker valve is located on the steam turbine (LPT) or on the top of the condenser shell itself.

3.7.4.1 Main Objectives of Providing Vacuum Breaker

Following are the main as well as associated objectives in providing a vacuum breaker: (i) reduce or break the vacuum at the earliest, (ii) turbine speed to decrease as a fallout of the above action; otherwise, there could have been a huge acceleration of the steam turbine upon loss of load by the generator, (iii) avert the possibility of turbine rotor vibration going up, and (iv) decrease the loss of turbine lubricating oil and hydrogen seal oil pressure.

3.7.4.2 Capacity/Sizing of Vacuum Breaking Valves

There are many considerations for determining the capacity or sizing of a vacuum breaking valve. Among those are important criteria such as total volume of steam space within the LP turbine casing and the condenser, and the time required to reduce the vacuum.

The opening time must always be taken into consideration because in the event of a turbine over peed running condition, a reduction in turbine speed is directly influenced by the vacuum breaking valve size and opening time.

3.7.4.3 Types of Vacuum Breaking Valves

Vacuum breaking valves should follow the specifications laid down by the original equipment manufacturer unless and otherwise driven by some special consideration.

Globe and butterfly valves are the natural choice.

The sealing of the vacuum breaking valve is very important in reducing any potential threat of air leakage into the condenser; for example, resilient metal seals provide two basic characteristics such as elasticity and plasticity to ensure drop tight sealing. The valves should strictly be rated for vacuum service with a tight shut-off leakage class and preferably with an additional water seal system to prevent air in-leakage.

3.7.4.4 Location of Vacuum Breaking Valves

Vacuum breaking valves are located on the condenser itself due to demand from the design criteria to get close proximity to the condenser shell, considering other safety parameters. For example, personnel protection and anticipated/calculated noise levels must also be taken into account during the design stage of the intake piping of the vacuum breaking valve.

3.7.5 Online Condenser Tube Cleaning System

Online cleaning of condenser tubes is necessary for large modern power plants because they handle natural water, especially for the OT CW system; they are also subject to mineral fouling. At ambient temperature, biological growth inside the tubes is inevitable. All these phenomena will hamper the CW flow and lower the condenser/plant efficiency.

3.7.5.1 Principle of Operation (Ref. Fig. 2.36)

Suitable sizes of rubber sponge balls are injected at relatively high velocity into the CW flow circuit at the condenser inlet. The sponge balls are circulated through the condenser tubes, where they mechanically remove fouling biological growth, debris, and scales that build up on the tube surfaces. At the upstream of the inject point, there shall be a strainer/filter so as to prevent balls from escaping the desired route and forced to pass through the condenser itself. At the condenser outlet, a ball strainer is provided for the collection of sponge balls and reinjected at the inlet to continuously maintain the cleaning process. This continuous automatic operation eliminates the need for costly plant shutdowns to conduct manual condenser cleaning.

3.7.5.2 Basic Objectives

It is provided to ensure optimal heat transfer to maintain design plant parameters and thereby raising plant output to original design specifications. The cleaning system shall also lower fuel consumption/cost and the condenser life

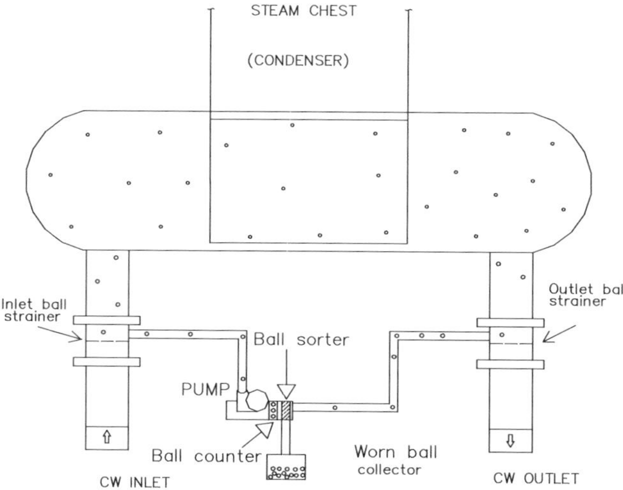
STEAM CHEST
(CONDENSER)
Inlet ball strainer
Outlet ball strainer
Ball sorter
PUMP
Ball counter
Worn ball collector
CW INLET
CW OUTLET
NOTE:
(1) LOCATION DETAILS SHOWN IS JUST TO GIVE AN IDEA ABOUT EQUIPMENT LOCATION.
(2) NOTE NO ROOT VALVE SHOWN WITH CONDENSER VACUUM. IF VALVE USED, IT SHALL HIGH PRESSURE QUALITY TO AVOID LEAKAGE.

FIG. 2.36 Schematic diagram of on line condenser tube cleaning system.

expectancy will be better if equipped with this system. Online cleaning of condenser tubes eliminates regular shutdowns toward manual tube cleaning, reduction in expense relating to use of chlorination.

3.7.5.3 Materials of Construction

Normally, these materials are stainless steel or rubber-lined carbon steel, depending on service requirements.

3.7.6 Condensate Polishing Unit

Functional details on the CPU are discussed in the following subsection: Section 4.0.4, Chapter 5; details given in the box at the end of this section may also be referenced.

3.7.6.1 Requirement of Condensate Polishing Unit (Ref. Fig. 2.37)

Boiler pressure is the key factor for determining the need for a condensate polishing system. With a steam generator plant with low steam pressures (below 40 kg/cm^2), the requirement of the CPU is usually avoided. Hard scale formation and boiler metal corrosion can easily be prevented through FW treatment before sending to the boiler proper. Adding suitable chemicals such as phosphate, etc., is more or less sufficient to remove contaminants from the condensate stream. The requirement of a CPU becomes a necessity for medium-pressure (40–170 kg/cm^2) and high-pressure (above 170 kg/cm^2) boilers. In once-through boilers, the CPU is absolutely essential.

It has also been observed that medium-pressure plants can manage to avoid a CPU by controlling FW contamination through intermittent and continuous blowdown of the boiler drum water. Condensate "scavenging," or cation exchange, is also often used to make corrosion-free condensate from the turbine. Where the contamination levels are not that high, partial CPU is also observed to avoid system pressure loss across the CPU. Very high-pressure systems are often "zero-solids" systems requiring condensate polishing to satisfy the water quality requirements of the major contaminant ions.

The condensate gets contaminated with different types of impurities while passing through pipelines, heat exchangers, and other equipment in the steam cycle. A serious type of contamination comes from seawater (used as condenser cooling water) leakage (having a high level of total dissolved solids (TDS) and chlorides). A CPU uses ion-exchange resins for eliminating impurities from the condensate. The high-pressure thermal cycle requirement demands a very high quality water/steam as the fluid medium. CPUs are essential because the condensate constitutes the major portion of the feed water. The risk of persistent contamination of the return condensate also calls for its application, thus minimizing the make-up water quantity. The CPU is a stand-alone system comprising hardware, drives, and instrumentation items required for its operation.

3.7.6.2 Presence of Corrosive Materials

Oxides of metals such as copper, nickel, and iron play major roles in corroding the pipe works throughout the cycle and need special attention. Their presence is felt in the condensate generally in two forms: particulate/suspended matter or dissolved salts. By rigorous and extensive *filtration*, the effect of these contaminating agents can be arrested.

3.7.6.2.1 Particulate or Suspended Matter Reusing contaminated condensate (with organic carbon components, dissolved solids, and materials inflicting turbidity) in feed water requires a thorough purging to match the specified quality. The condensate first passes through the ACF, which removes crude and suspended solids. There may be a block with the arrangement for providing a reverse osmosis process to further clean the condensate. Loaded with hot water membranes for high-temperature operation, the appropriate subsystem reduces incoming total organic compound (TOC) levels and eliminates a proportion of the dissolved contaminants in the water.

3.7.6.2.2 Dissolved Salts These are caused basically due to leakage from the CW of natural resources such as riverwater or seawater. *Ion exchanging* agents demineralize or neutralize dissolved salts, eliminating impurities that badly affect the condensate quality.

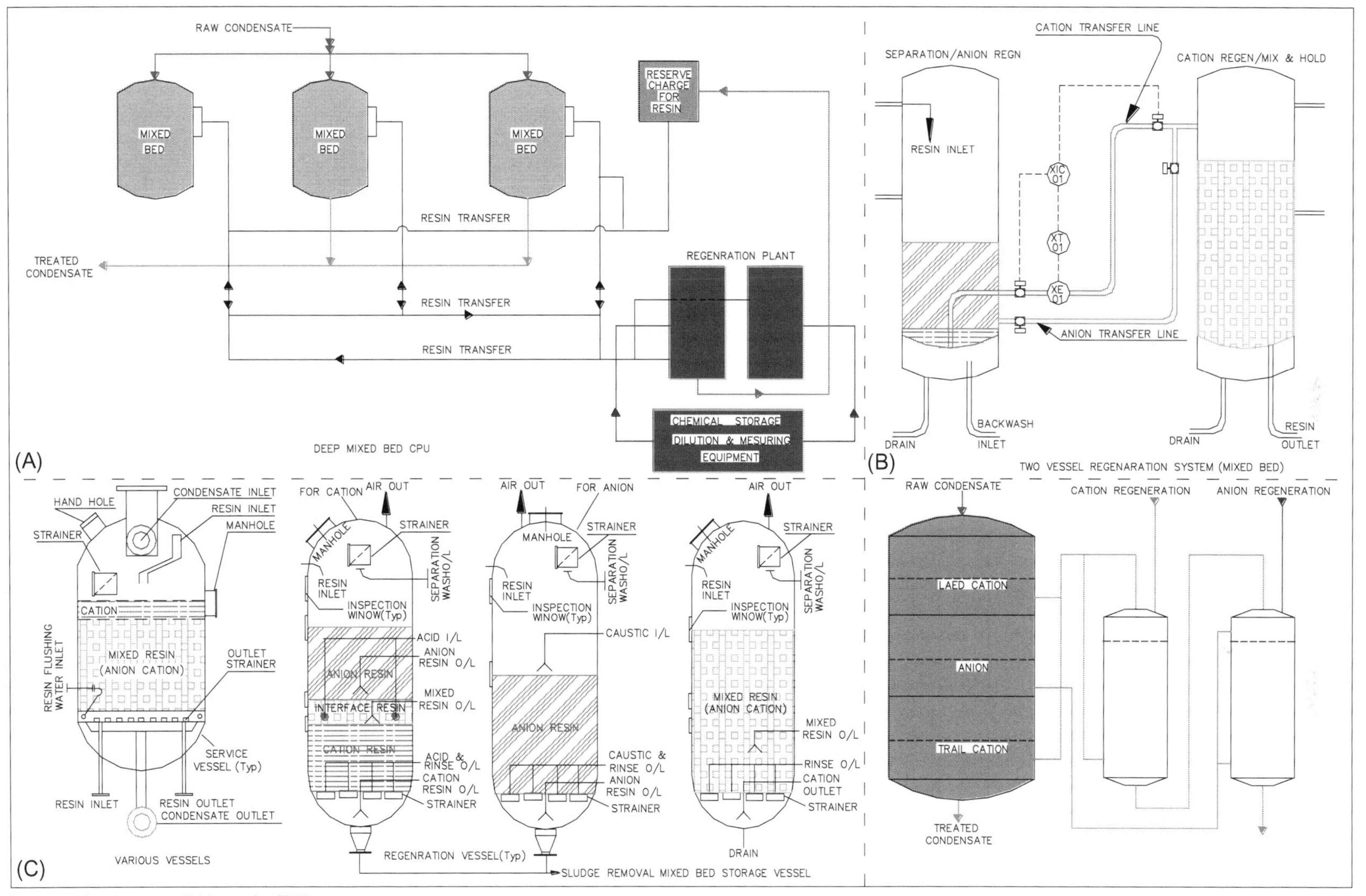

FIG. 2.37 Condensate polishing unit (CPU).

3.7.6.3 Types of Ion Exchangers

There are several types of ion exchangers provided as required by the SG plant/type of condenser cooling circuit.

3.7.6.3.1 Mixed Bed Exchanger The most common ion-exchanging system observed in CPUs is the application of an mixed bed (MB) exchanger, mainly for plants with low contamination levels. MB exchangers consist of a strong acid cation (SAC) exchange resin and a strong base anion exchange resin. Very high quality demineralized water is expected from the mixed bed exchangers because ion leakage from either the cation or the anion resin is readily cancelled by the other resin. Arrangements are made for maintaining a high flow rate (much higher compared to general DM plants). A resin bed depth of approximately 1 m is provided to keep the DP across the bed at techno-commercially acceptable levels. Generally, the mixed-bed CPU consists of several vessels, for example, three numbers operating in parallel. This is done to facilitate high flow and maintenance. Used resins are transferred to a different mode of operation for clean up and regeneration. Disposable mixed-bed resins are also used to avoid the regeneration process. A higher level of dissolved salts/mineral contamination suggests separate cation/anion exchangers and mixed-bed ion exchangers as well.

3.7.6.3.2 Lead Cation Resin followed by Mixed Bed of Strong Cation/Anion Resins When the main feed water undergoes all volatile treatment (AVT) by using ammonia, hydrazine, or other volatile amines to provide a reducing environment to control pH and corrosion to save the metals in the SG plant, the resins in the condensate polisher exhaust at a faster pace. This is because when AVT is used in a steam-condensate cycle, the amine is carried over and transported along with steam, building up amine levels ranging from 0.2 to 1.0 ppm in the cycle when steam condenses. This amine is readily exchanged onto the cation resin during condensate polishing. Eventually, the cation resin becomes exhausted to the amine form and obviously the run time would be shorter if the pH value is higher.

The improved technique for preventing the early exhaust time length of the mixed-bed type polishing unit is to allow the condensate to mix with a hydrogen form cation resin (lead cation resin) to remove the amine before it come into contact with the MB resins. This type of CPU is in general a variation of the MB system and may be termed a lead cation resin followed by MB of strong cation/anion resins. Corrosion products are also removed by this lead cation bed, eliminating the contamination of solids in the MB. The added advantage here is that lead cation regeneration can be done more frequently on an "as and when required" basis. The exhaustion time length being longer, the frequency of MB regeneration reduces as a consequence.

A process having two ion-exchanging beds, that is, cation and mixed bed in series, would obviously increase the differential pressure across the polishing unit and in turn increase the permanent loss/pumping costs. The continuous improvement on the bed size and bed design of different resins endeavors the optimization of the pressure loss features.

3.7.6.3.3 Cation-Anion-Cation Stacked Bed Exchanger This type of ion exchangers is the product of a relatively recent development with a single tank having individual compartments containing separate layers of cations, anions, and cation resins. The resins, particularly of the second and third layers, are never mixed as used in the MB type; each resin goes to its own external regeneration vessel. The advantage here is that the lead cation resin can be kept isolated until an ammonia break in the AVT systems is detected by the sensor at a suitable location. Leakage from the lead cation is taken care of in the final/third cation resin.

3.7.6.3.4 Simple Cation Bed Exchanger If the process permits, one may select a cost-effective alternative such as simple sodium or a hydrogen cycle CPU. This type is techno-commercially viable, taking into consideration the operating cost savings compared to the cost of chemicals and energy requirements.

3.7.6.3.5 Regeneration of Resins It has been observed that the external regeneration of the resins outside the CPU is comparatively better toward maintenance. Condensate contamination by the regenerating agents itself can be noticeably avoided through proper isolation of the regenerating media from the recirculating water circuit, with the added advantage of less time requirement by which a CPU is forced to remain offline. The only interruption is to take out the used resin and put to the external system for regeneration and then introduce regenerated resin to the condensate polishing circuit, enabling one regeneration system for multiple CPUs.

3.7.6.3.6 Capacity of the Polishing Unit The capacity of the polishing unit may vary depending on many factors. Plants with partial condensate polishing capacity are often found in large conventional thermal power plants with coal/solid fossil firing systems. However, in that case, the sizing of the make-up water treatment plant necessitates proper attention to feed the main plant during start-up periods apprehending most likeliness of the presence of contaminants at that time. For OT boilers, 100% CPUs are a must where a drum is not available for chemical injection.

Some other parameters are also spelled out to facilitate working out the capacity of the plant, for example (typically), the condensate flow corresponding to the maximum

TG output at 3%–4% make up, 0.1 kg/cm^2 (abs) back pressure, and considering all HP heaters out of service.

3.7.6.3.7 Quality of the Polishing Unit The quality of condensate at the CPU outlet with reference to its inlet determines the quality of the CPU. The running period in service is also important and to be considered for designing the polishing unit.

Some typical parameters are indicated below to get some idea about the water quality expected from the polishing unit (Table 2.2):

During unit start-up conditions, the quality of the CPU inlet would obviously be different; may be considered, for example, TDS 2000 ppb, Silica 150 ppb, and Crude 1000 ppb.

During a condenser tube-leakage condition, the quality of the CPU inlet would also vary depending on the type of condenser cooling system (Fig. 2.38).

3.8 Turbine Supervisory Instrumentation

For more details, Fig. 5.2 may be referenced.

All these very important measurements are used for getting valuable data for safe running, posttrip analysis, archive, etc., through sophisticated instruments and state-of-the-art technology. They are mainly: (i) eccentricity of the TG shaft near the HP turbine nondriving end, (ii) thrust position near the thrust bearing, (iii) key phasor position near the HP turbine shaft nondriving end, (iv) differential expansion of the TG shaft with respect to the corresponding turbine casing, (v) speed and acceleration, (vi) shell or casing expansion, (vii) vibration (absolute and relative) at different locations, (viii) turbine governor valve position, and (ix) all the bearing temperatures (Fig. 2.39).

The above lists mainly the type of instruments. There may be a number of similar instruments in different locations (depending on several factors) for which the detailed instruments list in Section 2.0 of Chapter 5 and Fig. 5.2 may be referenced.

3.8.1 Eccentricity

Eccentricity is the measurement of the rotor bow at slow speed through a suitable sensor. The type of sensor deployed to measure the eccentricity of the shaft is essentially the noncontact pickup, sometimes called a displacement or eddy current or even a proximity pickup. It may be caused by any or a combination of a misalignment, a fixed mechanical bow (bend)/temporary thermal bow, or a bow due to gravitational force. It also may be used to calculate the shaft attitude angle, which is an indicator of rotor stability.

3.8.2 Thrust Position

Thrust position is the measurement of axial movement of the rotor with respect to a fixed position when the machine is running under heavy load. Excessive horizontal movement of the rotor can result in catastrophic damage in mere seconds. It is common to use two or more noncontact sensors for emergency shutdown or tripping purposes prior to reaching a dangerous level.

3.8.3 Phase or Key Phasor or Phase Angle

Phase or phase angle is defined as the angle between the reference point in the shaft and the imbalance, called the

TABLE 2.2 Expected Water Quality at CPU Outlet

Measuring Parameters	Measuring Unit	Inlet of (CPU)	Outlet of (CPU)
Sodium, ppb	ppb	10	2
Iron	ppb	50	5
Silica	ppb	30	5
Chloride	ppb	10	2
Ammonia	ppb	500	[a]
pH	pH	9.2	[a]
Conductivity	μ mhos/cm	0.3	≤0.1
TDS (total dissolved solids) (excluding ammonia)	ppb	100	25
Suspended solids (crude)[b]	ppb	50	5

[a]*These values would vary depending on the boiler metallurgy.*
[b]*A sticky coating or an incrustation of filth or refuse.*

CPU Technology

1. Factor influencing performance of CPU: Discussed in clause nos 4.0.4 of Chapter 5 .
2. Major components associated with CPU are
   a. *Operating Vessel/ Service Vessel*
   b. *Separator /Cat Ion regeneration Vessel*
   c. *Anion Regeneration Vessel*
   d. *Mixed bed Resin /Storage tank*
   e. *Transfer pumps and recovery Vessel*
   f. *Acid & Caustic handling Tanks/ Measuring tanks, Vacuum pumps.*
   g. *Effluent Pumps*
   h. *Instrumentation & controls (discussed in Clause 4.0.4 of Chapter 5)*
3. Basic Principle of Regeneration : Commonly used Mixed bed systems, normally there are three service vessels, two working and other is stand by. Following are the major steps for Regenerations.
   a. Mixed resins are transported from *Service vessel* to *Separator /Cat Ion regeneration Vessel* in the form slurry with water supplied by *Transfer pumps*.
   b. Empty *Service vessel* is resin from *Mixed bed Resin/storage tank*.
   c. Mixed bad resin in *Separator/Cat Ion regeneration Vessel* is scrubbed with air & back wash with water to clean and separate resins—light Anion Resin at the top and heavier Cat ion resin at the bottom.
   d. With the help of water from the *Transfer pump* Anions are transported hydraulically to *Anion Regeneration Vessel* Before regeneration a portion from the *Separator /Cat Ion regeneration Vessel* is kept for interface tank to ensure that there is no Anion present to improve better performance.
   e. Cat ion resins are regenerated with dilute acid solution (measured quantity of acid with the help of *Measuring tank* and *Vacuum pump*) piped to *Separator /Cat Ion regeneration Vessel* at the top. Provisions are kept for Sulphation also. *For plants with NH3 Cycle or the condenser tube leakage exists Sulphation shall be done at regular interval.*
   f. Once regeneration is done acid solution is drained and resin is rinsed with DM water
   g. Anion is very similarly regenerated with the help of Caustic solution.
   h. After regeneration both Cat ion & Anion resins are transported hydraulically to *Mixed bed Resin /Storage tank* as slurry. The resins are mixed by air and rinsed with DM water. Then they are backwashed to separate the ions again to transport to next service vessel. All of Anion and half of Cat ion are transferred and mixed in the service vessel.
   i. Conductivities at stages are checked to ensure proper regeneration.
   j. Vessel details and various inlet /outlet lines have been elaborated in Fig. 2.37C
4. TRIPOL Process (A registered Trade mark): In this method instead of Mixed bed three layers as shown in Fig. 2.37D Lead Cat Ion_ Anion, Trail Cat ion. There are separate resin movement lines to avoid cross contaminations. The resin volume can be altered to suit the operating condition.

FIG. 2.38 CPU Technology details.

heavy spot on the rotor. This measurement is a secondary parameter, mainly utilized for diagnostic and analysis purpose in combination with the vibration data so as to work out the approximate size and location of the heavy spot as well as the critical speed of the machine. The reference point or mark is usually a keyway/notch on the shaft. The above-mentioned measurement is used for establishing the relationship of one vibration signal to another vibration signal, and is expressed in degrees. The once per revolution phase reference relates the static coordinate system of the vibration sensor to the rotating coordinate system of the machine. The output signal is generally the measure of shaft speed and is utilized as a reference for measuring the vibration phase lag angle. This measurement is a very important tool for diagnosing the machinery problem. It is not usually continuously monitored, but is available as a diagnostic aid in vibration analysis, mainly for the balancing. In general, it is an indication or warning suggesting any problem related to shaft cracks, shaft misalignment, or rotor damage/mass loss. The noncontact sensor probe is typically used to observe a once-per-turn event such as a key or a notch. Phase, phase angle, or key phasor is, as already mentioned, a tool to measure the relationship of how one vibration signal relates to another vibration signal and the resultant signal after the analysis is generally used to calculate the location

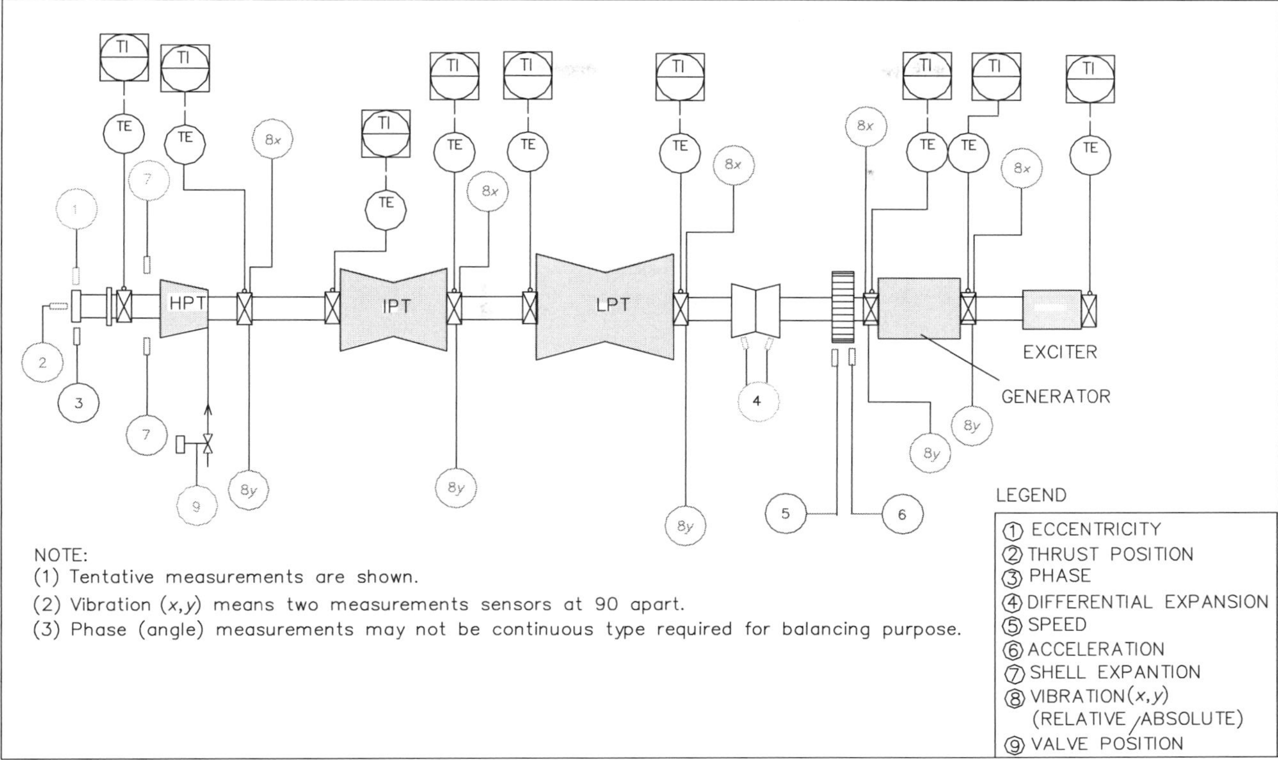

FIG. 2.39 Turbine supervisory instrumentation.

for placing the balance or counterweight. This parameter is not required to be displayed continuously but is monitored periodically to determine changes in the rotor balance condition, deviations in system stiffness such as a cracked shaft, etc. The installation requires locating or installing a once-per-turn event such as a key/notch that the sensor would detect only one time per revolution of the shaft. The installation of the notch demands special arrangements to cut the notch, but it is convenient for the instrument to view a notch for the installation/adjustment purpose. On the contrary, keys are easier to apply using glues/epoxies but are subject to coming off due to high centrifugal forces.

3.8.4 Measurement of Expansion

As the steam temperatures of the thermal power plant vary a lot at different operating points of the turbine, namely start-up and running/tripping/shutdown conditions, the turbine metals of different areas expand in dissimilar ways according to the corresponding steam temperatures. This measurement helps the operator to watch and ensure that the measured values are within limits. The unwanted expansion of the turbine casing or rotor shaft would lead the plant to experience disastrous consequences and should trip the machine prior to that limiting condition.

3.8.4.1 Shell or Case Expansion

Case expansion is the relation of the shell or case in relation to its foundation. Turbine generator sets are tied down at one end and allowed to thermally expand at the other end. The LVDT type of sensors or any other type of position transmitters is used to measure this thermal growth, which would be of considerable length on the order of a few centimeters for large machines. In some plants, there are provisions for monitoring the thermal expansion of the end position of both sides so as to ensure that the case expansions take place evenly throughout the length of the machine.

3.8.4.2 Differential Expansion

Differential expansion on a turbine is the relative expansion measurement of the rotor's axial thermal growth with respect to that of the case. Differences in thermal growth of the rotor to case growth if becomes less so as to touch each other, it would lead to rubbing and catastrophic failure of the machine and personnel injury also. LVDT sensors are used to measure this parameter.

3.8.5 Speed

The above type of probe or a magnetic pickup or Hall probe is the sensor for measuring speed. For this measurement, some provisions are made in the turbine shaft, for example, a key, notch, or toothed gear so that the sensor can detect the revolution of the shaft. Generally, three sensors are used for a two out of three voting and used for control system, measurement, plant tripping purpose. Zero speed detection for turning gear engagement could be taken from this arrangement also.

3.8.6 Measurement of Vibration

Vibration measurement is basically the radial motion of the rotor shaft. The output signal gives all information about the problem of the rotating machine such as misalignment, cracked shaft, oil whirl, unbalance, or any other dynamic snags. Vibration measurement in a rotating machine provides the most important data for diagnostic purposes. With the limited number of measuring locations due to high cost and minimum availability of space, modern technology has evolved with the proper diagnostic system by several reputed manufacturers to guide plant management speedily and accurately about the assessment of a machine's condition.

Vibration measurement may be provided in a single plane or a two-plane configuration. The latter arrangement envisages the sensors at 90 degrees apart and perpendicular to the shaft cross-section.

3.8.6.1 Shaft Relative Vibration

All the bearings provided with the TG unit are basically the journal type, except the thrust type. Due to the interaction of various forces acting onto the shaft journal, it would move with respect to the bearing housing. It has been observed that the vibration components of higher frequencies are absorbed/damped through the oil film considerably and may be ignored. The vibration sensors are provided to monitor the relative motion between the shaft journal and the bearing housing. The type of sensor deployed to measure the relative vibration of the shaft is essentially the noncontact pickup. These are sometimes called a displacement or eddy current or even a proximity pickup. This type of sensor measures not only the relative vibration but also the relative position of the shaft with respect to the bearing housing to ascertain the clearances between them. These sensors measure the dynamic motion (radial vibration) of the rotor shaft relative to the bearing housing. This parameter will detect common machinery problems such as rubs, imbalances, bearing stability, etc. Eddy probes also provide a measurement of the shaft position within the bearing clearance, giving warnings of harmful preloads and misalignment.

3.8.6.2 Absolute Vibration

For machines with a light casing-to-rotor weight ratio, vibration is readily transmitted to the bearing housing. This measurement is also utilized to calculate the absolute vibration of the shaft with the help of the relative vibration of the shaft with respect to bearing housing. An accelerometer or velocity transducer is a cost-effective vibration sensor for these machines.

3.8.7 Turbine Governing Valve Position

Valve position is measured with an LVDT or any other type of position transmitter observing the lift of the governing valve position. It is an indicator of approximate turbine load when the TG set is not running under sliding pressure control mode.

3.8.8 Temperature

The temperature of bearings is a measure of the health of the bearings in running condition by sensing how hot a bearing is while operating. Several reasons can contribute to overheating of a particular bearing such as partial failure of the lubricating system such as inadequate pressure and/or flow, overloading on the machine, misalignment of the TG shaft, etc. These sensors may be T/Cs or RTDs, but normally RTDs are preferred for better performance in this temperature region.

4 GENERATOR

The generator in a thermal power plant is a very significant subsystem driven by the steam turbine and is necessarily a synchronous generator or an alternator. It is very important electrical equipment and is categorized as a rotating machine converting mechanical energy into electrical energy when operated as a generator and vice versa when operated as a motor. In general, they are called synchronous machines because these machines operate with a constant speed very much synchronized with the frequency of the alternating voltage appearing at the terminals of the same. Synchronous generators are also used in hydroelectric, Hydel plants, and gas turbines.

The basic generator is based on two major parts; one being the *rotor*, that is, the rotating part and the other is the static part, or the *stator*. The rotor is so designed and wire wound as to form a DC magnetic field and made to rotate along the inner periphery of the hollow stator; this is often called a rotor magnetic field, a magnetic field, or even only the field.

When the field rotates by external agency (a steam turbine for instance), it produces a change of the magnetic field with respect to the static stator winding and thus excites them so as to induce voltages in the stator windings vis-a-vis the generator output terminals as well. Direct current is required to flow through the field winding to create constant magnetic field and is much smaller compared to the stator winding.

Modern generators are normally self-excited or with a static excitation system. Fig. 9.16 shows, in a simplified way, how a self-excited system works, with some of the power output from the rotor used to power the field coils. The static excitation system, Fig. 2.40C, on the other hand envisages generator-produced electric power to excite the field residual magnetism (depending on the magnetic material) when the generator is turned off. The same can be achieved with a permanent magnet located on the rotor. When the rotor starts to rotate with no load connected to the generator terminal, the initial weak magnetic field generates a weak voltage in the stator windings. This weak voltage by some means increases the field excitation current and thereby further increases the voltage in the stator windings until the perpetual action enables the generator to develop the full terminal voltage. For self-excited generators, it is important to note that the external load is not connected to it during start-up as the effect of the load current would prevent development of the proper terminal voltage. The hollow part of the stator is mounted around the rotor with a very small air gap between them. Both the rotor and stator are made of a ferromagnetic material wrapped in a set of coils within the slots distributed throughout the circumference. The stator windings are connected as a three-phase AC voltage system. More details are available in subsequent sections of this chapter and Section 10 of Chapter 9.

4.1 Basic Generator Details

Generators comprise two main subsystems, the *rotor* or *field* (rotating part) and the *stator* (the static part), which was previously called armature.

4.1.1 Rotor (Field)

This part of the machine is to rotate freely (driven by a prime mover) and to produce the magnetic field either by a permanent magnet or by direct current flowing through a coil fixed onto it called the field winding. The coil is wrapped over a pair or multiple pair of magnetic cores made of ferromagnetic material that acts as a magnetic pair of poles producing an electromagnetic field around them when unidirectional current flows. They are designed so as to create a constant magnetic field and also to interact with the magnetic field produced by stator winding itself when current flows through it. The field winding direct current is much smaller compared to the stator winding. The number of pole pairs may be one or more (up to three pair) but all the coils are connected in series so that each electromagnetic pole strength is necessarily the same. There are two kinds of rotors, for example, (i) salient pole rotors, which are mainly used for slow-speed machines such as hydraulic turbo generators, etc. and (ii) smooth cylindrical rotors used for GTs and SGTs.

The main function of the rotor is to produce the magnetic field to excite the machine so as to induce voltages in the stator windings and ultimately at stator output terminals.

Modern generators are normally self-excited (Fig. 9.16) where some of the power from the rotor is used to power the field coils. The rotor iron retains some amount of residual magnetism (depending on the magnetic material) when the generator is turned off. The same can be achieved with a permanent magnet located on the rotor. When the rotor starts to rotate with no load connected to the generator terminal, the initial weak magnetic field generates a weak voltage in the stator windings. This weak voltage by some means increases the field excitation current and thereby further increases the voltage in the stator windings until the perpetual action enables the generator to develop the full terminal voltage. For self-excited generators, it is important to note that the external load is not connected to it during start-up as the effect of the load current would prevent developing the proper terminal voltage.

4.1.1.1 Field Flashing

Field flashing of the rotor is often performed when there is a lack of adequate residual magnetism to enable the machine to develop full voltage. This system envisages injection of direct current (DC) into the rotor from a separate source for a very short period of time, which is why the name "field flashing" is popularly used. The DC source may be fed directly from the station battery or from the suitable AC/DC converter, taking connection from an AC source. On many occasions, small alternators also may need field flashing for initial starting.

4.1.2 Stator (Armature)

The stator is the static part of the machine, mounted around the rotor. The stator is also made of a ferromagnetic material wrapped in a set of coils within the slots distributed throughout the circumference. The stator windings are connected to a three-phase AC voltage system. The winding of the stator is star (or wye/Y) connected because of the following reasons: (i) Star phase voltage $= 58\%$ $(1/\sqrt{3})$ of line voltage. Naturally, the insulation thickness would be less than that required for line voltage, hence comparatively more space is available for a higher conductor size to carry out more current than what is required [line current required is $\sqrt{3}(1.732)$ times more compared to delta connected stator], hence higher power. (ii) In star connection distorting line to neutral harmonics do not appear as they cancel each other. The neutral point is always connected to ground with the lowest possible neutral to ground line resistance.

It is worth mentioning that the efficiency increases with the power rating of the generator, for example, the efficiency for a 10 MW may be $\sim 90\%$ whereas for a 100 MW it must be more.

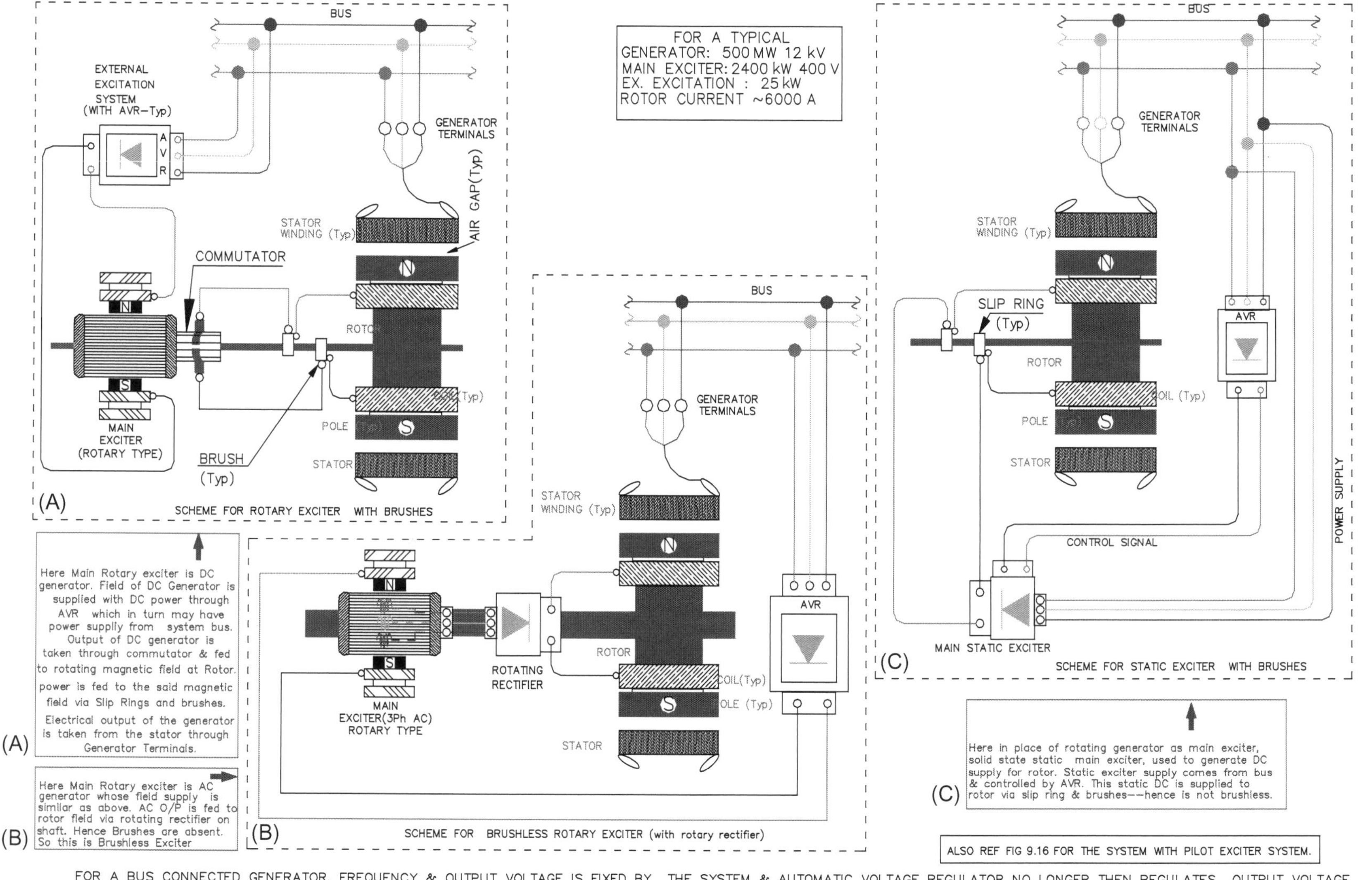

FIG. 2.40 Schematic diagram of different excitation systems.

4.1.3 Working Principle

For the generator, mechanical energy is converted into electrical energy. The turbine, may it be steam, gas, or Hydel, supplies the mechanical energy by way of applying torque and hence the subsequent rotation of the shaft. Due to the relative motion between the magnetic field produced by the rotor-mounted magnetized poles, the intensity of the magnetic field passing through the stator windings would change as a function of time. When the rotor field is energized, electromotive force (EMF) is induced in the generator stator, which is proportional to the rotor field current and speed and capable of supplying the electrical output to the external load. The frequency of this voltage would also be proportional to the rotor speed. The moment the generator output terminal is connected to the supply bus (or the grid), the rotor would be locked to the synchronous torque of the network system and the rotor speed would solely be governed by the frequency of the grid. The speed would now be the synchronous speed and as the frequency of the generator voltage depends directly on the speed of the machine, both the grid and generator voltage frequencies are exactly the same. The stator coils are distributed throughout the circumference in such a manner that three separate bunches of coils are formed in various combinations (series, parallel, or both) and called three phases. The arrangement is made in the way that the three phases of the stator windings are 120 degrees electrically apart and thus the EMF induced there would obviously be alternating three-phase voltages with sine wave characteristics.

The field on the rotor is often called the exciter and the DC current through the same is thus called the excitation current. The actual power handled by the stator is huge compared to that of the rotor. It is also worthwhile to note that in the case of the generator operating in an island concept, that is, in isolation from the grid or the electrical network system, the field excitation current would be responsible for the generated voltage. However, when the generator is connected to the grid with a number of other generators connected forming an infinite bus, the excitement of the field excitation current would control the reactive power generated due to the action of increased armature reactions, etc.

Generators are designed to produce moderately high voltage, which is a compromise between the high insulation provision and handling of high electric current. With generators up to 50 MVA, the voltage is 6.6 kV; higher-sized generators above 1000 MW may have voltages up to 32 kV.

4.1.4 Cooling of Stator and Rotor

Generators/alternators are supposed to deliver electric power at a certain output voltage and current. For large thermal power stations, those values may be very high, especially the current, which is reduced afterward by the use of the generator transformer by raising the output voltage, thereby reducing the current flowing through the grid. Therefore, the generator windings should be capable of sourcing the high-density currents that in turn produce enormous heating effects. To overcome/reduce this undesirable heating, the *stator windings* are made of hollow copper conductors through which the pure quality cooling water circulates. With this cooling water, the problem often experienced is the formation of scale depositions on the internal walls of the hollow conductors. These deposits lessen water flow as well as heat transfer; both are detrimental to generator stator cooling while also reducing the load delivering capacity and forcing frequent downtime. To get minimum scale deposition, maximum heat transfer, and the lowest possible electrical conductivity of the cooling water, a small portion of the normal stator cooling water flow (approximately 1%–10%) is bypassed for rejuvenation through an MB demineralization (DM) plant and returned to the original stream. Another way is to take advantage of the station shutdown and go for chemical cleaning of the hollow conductors to remove the scale. However, cleaning operations are a long-standing process and invite corrosion of metals of hollow conductors.

The rotor is also subjected to carrying high field excitation current and the heat generated due to a high centrifugal force. For that reason, pressurized hydrogen is introduced inside the generator housing, which enables cooling of both the stator core and rotor elements and taken out from a suitable point for reentry after heat release through an external cooling arrangement.

4.1.5 Hydrogen Cooling Including Seal Oil System

In the early days, electric generators were air-cooled. To cope with the huge power demand, the size of the generators/alternators became increasingly large. As the larger machines generate more heat from various sources such as stator windings, rotor windings, windage losses, and the stator core iron that require more cooling, the use of air became obsolete due to limitations in dissipating the huge wasted heat. For larger machines, windage losses and increased noise generation also demand removal of the air-cooling system.

Helium was considered earlier as a substitute for air but it could improve the cooling only five times better with around five times higher cost compared to air. Hydrogen (H_2) was selected in spite of its disadvantages. Nowadays, hydrogen-cooled generators are used for almost all the large power plants. H_2 as a coolant has the following qualities that put it ahead of other coolants:

- **(i)** High relative specific heat, which is approximately 14–16 times better than air so far as (H_2)-cooling effect is concerned. This property enables it to absorb a large amount of heat compared to other fluids.
- **(ii)** Hydrogen has a density of 1/14 that of air at the same temperature and pressure, thereby reducing windage losses sand noise. This characteristic offers minimum windage loss as hydrogen has the lowest molecular weight. Due to the very high rotational speed of the rotor, the friction (also known as drag) loss, an example of aerodynamic resistance, becomes predominant; this is the energy loss specifically termed windage loss. This loss is directly proportional to the velocity square and density of the atmosphere of the rotating mass.
- **(iii)** The highest thermal conductivity of all gases, which means it conducts the absorbed heat efficiently and transfers more quickly. This means that it behaves as an ideal coolant, extracting the heat generated around 7–10 times better than air.
- **(iv)** Higher electrical insulation characteristic than air when pressurized.
- **(v)** Negative oxidizing agent.
- **(vi)** vii) Offers better measurement/detection facility for the hydrogen sensors.

Considering the above points, it is obvious that, in a generator with a hydrogen cooling system, the loss would be overwhelmingly reduced and it would be much more cost-effective compared to an air-cooled machine.

Against all these plus points, hydrogen has the biggest disadvantage in that it, unlike air, poses a fire hazards. Hydrogen has a wide flammability range; however, experiments assert that it does not support combustion when its purity in air is more than 85% or less than 3%. Proper handling of hydrogen with purity at a very high level (as the low range is not applicable) can almost eliminate fire hazards, which means there is very little if any air in the generator casing to mix with the hydrogen. In that situation, the possibility of an explosion is very remote, even if there was a spark. Fig. 9.15 in Chapter 9 shows the H_2 flammability range and justifies the need for maintaining the purity of hydrogen. The system of maintaining hydrogen purity ensures no leaking of either air or hydrogen from mixing with each other. Care is also necessary to prevent air from contaminating the high-purity hydrogen while filling the generator casing after displacing the air. To facilitate that, carbon dioxide (CO_2) is used as an intermediate media for filling/displacing as much of the air inside the generator casing before and after running the turbine.

4.1.5.1 The Hydrogen System

Almost pure hydrogen gas in the range of 99% is supplied to the generator with a pressure around 2–3 kg/cm^2 g (may be up to 4 kg/cm^2 g) to prevent air leakage into the generator casing.

The hydrogen is circulated within the generator by fans mounted on each end of the rotor. While circulating around the generator, the hydrogen gets heated. For cooling, it is forced to pass through redundant water coolers. The cold hydrogen at the outlet of the coolers is returned back to the generator casing, forming a closed cycle.

Automatic control is provided for maintaining the hydrogen temperature at the generator inlet. The coolers or heat exchangers for hydrogen cooling are mounted normally on the side/top of the generator frame. Many power stations have captive hydrogen generation plants along with provisions for using gas cylinders in case of emergency or shutdown of the H_2 generation plant.

The H_2, being at a higher pressure, still tends to leak in the H_2 side of the seal (discussed later) where from return seal oil is collected and sent back to seal oil reservoir. A small amount of H_2 gas (and air also coming from the seal oil) is continuously vented to the atmosphere for purging/ scavenging and hence results in low pressure inside the generator if not replenished. But the fact is that most of the H_2 loss results from generator casing leaks, causing the pressure to decrease. To circumvent that situation, a pressure controller may be provided to maintain the pressure by allowing a small amount of high purity hydrogen gas into the generator casing through a control valve as a continuous feed system. In some plants, manual batch feeding is also done.

A purity monitor is provided to detect hydrogen purity during operation. This is more required in view of the seal oil borne air if liberated (detailed later) within the generator casing contaminates hydrogen purity. Monitoring the purity is preferred at multiple strategic locations in the generator to ensure that the purity is maintained to prevent a possible explosion or fire.

In case the hydrogen purity drops below a certain level (typically 85%–90%) as recommended by the prevailing standard or manufacturers, then the TG set may be tripped.

For filling H_2 in the generator for the first time or after a long shutdown, the air is first completely replaced with CO_2 and then charged with hydrogen to replace CO_2 to the desired purity value. When a long shut down is envisaged, vice versa. With CO_2 being a heavy gas, its entry and exit are made at the bottom of the generator while that of H_2 is made at the top. The H_2 purity monitors usually monitor the purity of the gas at the top of the enlargement tanks, where the air/H_2 mixture is vented through the scavenging system and inside the generator casing.

4.1.5.2 Moisture in Hydrogen Cooling System

In its delivered form, whether from an external or internal source, hydrogen gas has a dewpoint of around −50°C. The hydrogen dewpoint is normally maintained at less than 0°C for this type of requirement. Problems may crop up when the generator is in shutdown condition and moisture levels increase beyond this point. The retaining rings crack if moisture (water particles) is deposited on them. There are generators that are cooled by hydrogen with dew points of −30°C or lower.

4.1.6 The Seal Oil System

With hydrogen being hazardous, its leakage along the rotor shaft is prevented by a sealing system with the help of oil with pressure slightly more than hydrogen.

This differential pressure is maintained by a dedicated control loop, which is discussed in detail in Section 9 of Chapter 9. It receives makeup oil normally from the bearing lubricating oil system. The seal oil system defines the arrangement provided to arrest not only the escape of hydrogen from the generator casing but also the ingress of air into the said casing near the rotor shaft projected outside the casing at both ends.

The generator-rotor ends where the air is supposed to make ingress is sealed to keep the hydrogen inside the generator. Special seals on both the shaft ends ensure adequate sealing against the axial flow of gas toward the air side. Oil is used as the sealing medium to ensure tight sealing, and is injected on the shaft around the entire circumference of the shaft through the channels and orifices inside the seals. As the seal oil is at a higher pressure than the hydrogen inside the generator casing, a portion of the oil flows along the shaft to the air side of the generator and some of the oil flows along the shaft into the H_2 side of the generator. The oil used as the seal oil is generally the same lubricating oil as that used for the bearings, although in some plants, the seal oil system is separate from the lube oil system.

The oil is supposed to be in contact with air while transferred from the lube oil tank, and also to pass through the bearing drains. This air makes froths or small bubbles and somehow can be entrained in the oil. When oil is injected on the generator shaft, that air could be liberated from the oil that flows into the hydrogen side of the seal area. That unwanted air must be removed; otherwise, there is every possibility that it somehow continues to collect inside the generator casing, thus contaminating the hydrogen purity.

The system provided to remove air from the seal oil into the H_2 side of the generator seal is called a *scavenging/* purging system. The seal oil that flows into the hydrogen side of the seal area is generally passed to a vessel known as an enlargement tank.

It is expected that the entrained air liberated from the seal oil and along with a small amount of hydrogen leaked through the labyrinth seal, and accumulate there. This small amount of gas (hydrogen and air) is permitted to escape from the enlargement tank through a vent. The vent line must be long enough so that the mixture is discharged to a safe area away from any ignition source.

If the seal oil flow rates increase, then the entrained air liberated from the seal oil will also increase which means more contamination and decreased purity. So, it is very important to keep a provision for monitoring seal oil flow rates to detect an increase for alarms and trip signals. With only one common seal oil flow meter for both generator seals, it may not be possible to detect the defective seal, but the H_2 purity meter in each enlargement tank would indicate any problem with the decreased value of the purity percentage. More details in Chapter 9, Section 8.

4.2 Generator Excitation: Types With Advantages and Disadvantages

As discussed earlier, the generator requires direct current to have its magnetic field energized. The DC field current is arranged from a separate source called the field exciter. Fig. 2.40 and Fig. 9.16 may be referenced.

Generally, three types of excitation systems are used: (i) direct current excitation, (ii) rotating-type excitation (Fig. 2.40A and B), and (iii) static-type excitation (Fig. 2.40C).

4.2.1 Direct Current Excitation

In this type of DC excitation system, mainly used in the earlier stages, the DC generator armature is mounted around the same shaft of the main generator whereas the field is formed by a permanent magnet mounted on the rotor shaft. The DC voltage as generated is fed to the field excitation terminals with proper control arrangement.

4.2.2 Rotating-Type Excitation System

This type is categorized into two types: a brush system and a brushless system. The most important difference between them is the technique used in transmitting the DC field excitation current to generate the magnetic fields.

The brush excitation system can be mounted on the same shaft as the AC generator armature or can be housed separate from, but adjacent to, the generator. When housed separately, the excitation system is rotated by the generator rotor through a drive belt.

The distinguishing feature of brush excitation is that static brushes are used to transmit the DC exciting current to the rotating generator field. The current transmission is accomplished via rotating slip rings that are in contact with the spring-loaded brush for pressing against the slip ring for assured contact. Two slip rings are solidly mounted on the generator shaft and both of them are fully insulated from the shaft and each other. Out of the two rings, one is meant for the positive terminal of the field winding and the other is for the negative terminal.

For the brushless excitation system, Section 10 of Chapter 9 and Fig. 9.16 and Fig. 2.40B may be referred to for details.

4.2.3 Static Type Excitation System

A static excitation system (Fig. 2.40B) for generating the magnetic fields involves various options. These types may be classified as per the source of the excitation power taken from the AC voltage, for example, (i) Main generator winding, which is also called self-excitation. Auxiliary winding provided with the main generator winding is also another type of self-excitation. (ii) Station auxiliary system, which is also called separate excitation. A static excitation system contains no moving parts. Here the generator output is taken through a suitable transformer, controlled power converters including thyristors rectifiers, reactors, etc., for AC to DC voltage conversion and then supplied to the field windings through the brush/slip ring assembly.

Here, at the start, the residual magnetic field of the rotor may be used for start-up, but due to the strong magnetic field around the high voltage stator, this natural property may disappear or may be weak. Then, an external source of DC supply is necessary for the initial excitation of the field windings. The initial excitation may be obtained from the station or any other storage batteries. This action is achieved by a system called a field flushing system where the DC voltage is taken from either station (or separate) batteries, a system comprising solid-state components, or a DC generator with a separate excitation/self-excitation system.

4.2.4 Generator Field Excitation: Advantages and Disadvantages

4.2.4.1 Advantages and Disadvantages: Brush and Brushless Excitation

Synchronous generators (or motors) of small capacity still use the brush/slip ring electrical sliding contacts for rotor-mounted field excitation connections, as was used generally in the earlier phase of the invention of electrical machines.

The brush/slip ring electrical connections are nowadays not used for large and medium machines for various reasons. The same can be easily understood from the following: (i) unsuitability of use under an explosive or hazardous environment due to formation of electrical sparks at the making and breaking of contact, (ii) failure of electrical contact may occur due to contamination caused by dust mixed with oil, grease, or any other sticky material, and (iii) maintenance is very often necessary due to continuous wear and tear, which causes increased downtime and costs for undertaking replacement.

4.2.4.2 Advantages and Disadvantages: Permanent Magnets

Normally, high-performance permanent magnets are used for synchronous generators (or motors). They are made of magnetic materials such as the derivatives of neodymium and samarium-cobalt, etc., which offer characteristics such as high coercivity.

4.2.4.2.1 Advantages (i) Does not require an additional DC supply for the field excitation circuit, (ii) stable and secure during normal operation, (iii) higher power efficiency is available, (iv) generally, the brushless field excitation system (Fig. 9.16) utilizes the service o f permanent magnet, which makes it simpler and maintenance-free, and (v) with a suitable choice of high-performance permanent magnet materials, the air gap may be designed to a comparatively higher value. Air-gap depth tolerance may improve around 20% over other types, and then the magnetic leakage becomes predominant. This facility offers fewer constraints associated with the expansion of generator casings/structures.

4.2.4.2.2 Disadvantages (i) High-performance permanent magnet materials are expensive and the source is also limited. (ii) Flux density of high-performance permanent magnets is limited in spite of qualities such as high coercivity. They are also subject to the influence of the higher flux density of what is produced by the torque current MMF and combines as vector quantity. This resultant flux density undesirably magnetizes the permanent magnet, which may lead to higher air-gap flux density and result in magnetic core saturation constraints. (iii) Air-gap flux density is uncontrolled and may cause overvoltage and poor control reliability. (iv) Permanent magnetic fields always exist and may pose safety issues such as electrocution during erection, assembly, or maintenance. The use of permanent magnets is said to not be environmentally friendly.

5 BOILER FEED PUMP AND CONDENSATE EXTRACTION PUMP: ASSOCIATED MEASUREMENTS

A BFP may be considered as the heart in the human body and one of the five fundamental elements (as shown in Fig. 1.1) in a modern power plant. CEPs draw the condensate under vacuum from the condenser hotwell with some positive pressure at the discharge. This water, after passing through several heaters and a deaerator, reaches the suction of the BFP and is then pumped at a very high pressure so as to reach the boiler after passing through a number of HPH. Fig. 2.41 may be referred to for easy understanding of the system.

Centrifugal pumps are normally employed for both services. A centrifugal pump is a rotary machine with a set of

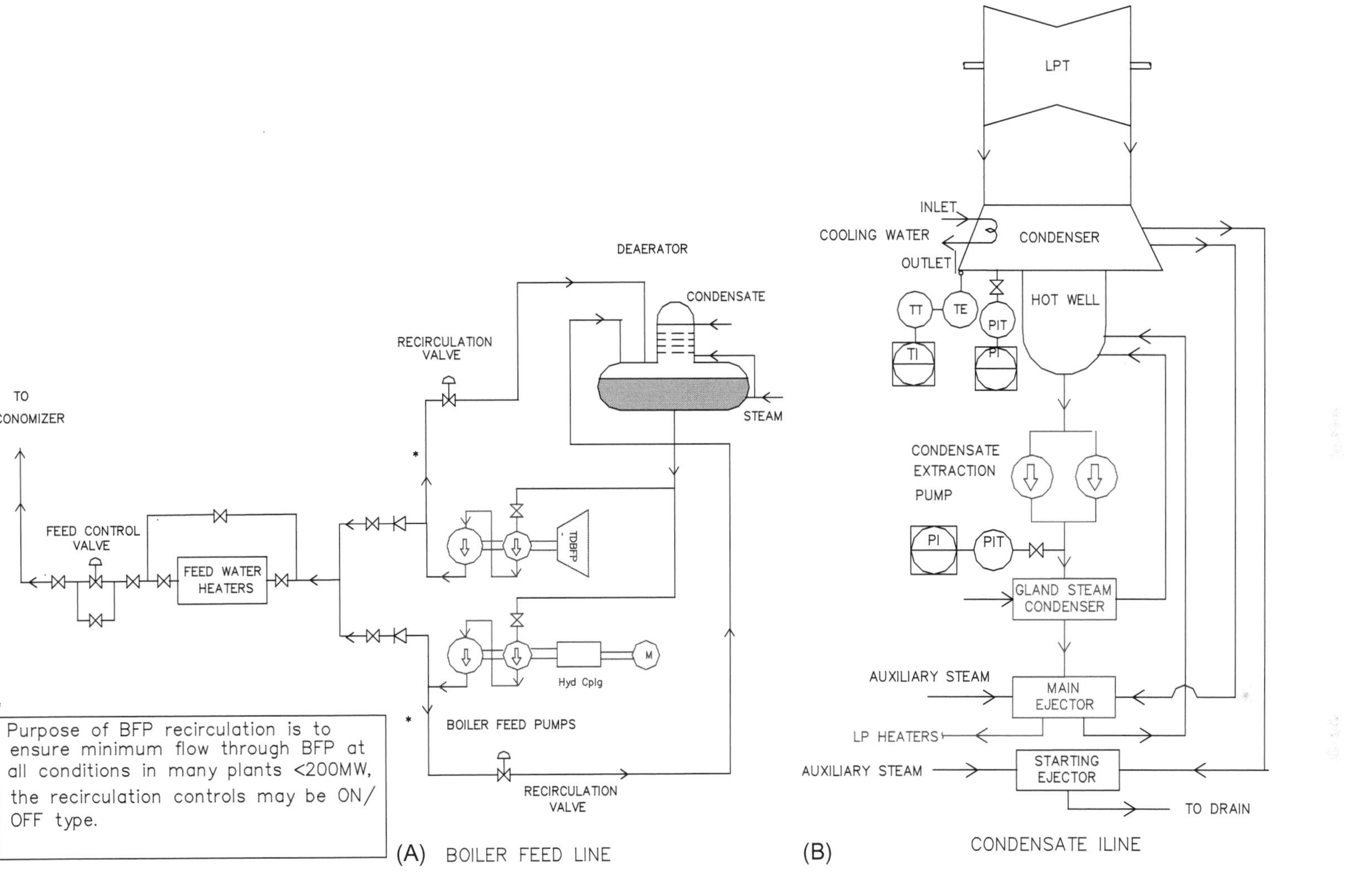

FIG. 2.41 Boiler feed pump and condensate extraction pump with feed water and condensate line.

impellers fixed on the rotating shaft to propel the fluid flow by the addition of mechanical energy to a fluid. The fluid enters the pump impeller along or near the rotating axis and is pushed by the impeller. At the outlet, the fluid comes out flowing radially or at a right angle to the axis of inlet flow, from where it exits into the downstream piping network.

Construction-wise, a centrifugal pump has three basic components: the impeller, diffuser, and a volute chamber or casing.

A multistage centrifugal pump consists of more than one set of impellers. During centrifugal operation of the pump, the suction zone of the impeller experiences a negative pressure that enables the fluid to enter the pump at the inlet flange. The fluid thus rushes to enter the first or a low-pressure stage impeller from the suction line. The water gains some positive pressure after coming out from the first stage impeller and is then guided to pass through the diffuser. Here, vanes are strategically arranged to force the water toward the second-stage impeller. The passage of this stage determines the capacity of the pump. The diameter of the stage is related to the centrifugal force and its relevant stage pressure.

Similarly, in the second-stage impeller, water gets more pressurized and enters the next stage. After the final stage, the water enters the volute chamber/casing, where the design is arranged to remove the water through the water outlet at the appropriate system pressure. The impellers may be mounted on the same shaft or on different shafts. If higher pressure is needed at the outlet, the impellers are connected in series, whereas for higher flow, the impellers are connected in parallel.

5.1 Boiler Feed Pumps

The function of BFPs is to provide water to the boiler. BFPs are very vital constituents of the power plant, as their failures cause the unavailability of the same in many a cases.

BFPs are subject to various types of operation such as running at low load during start-up, normal coast-down transient, immediate shutdown during plant trip condition, frequent load transients during normal load operation, etc. It is a prerogative that the designers would take care of the BFP safe passage while passing through their critical speeds during every occasion as may arise. BFPs not only supply the feed water to the boiler but also control the same if provided with a variable speed setting arrangement. In fact, BFPs can be used to deliver both the desired flow at the desired pressure by adjusting the speed through various methods.

5.1.1 Selection Criteria of BFP

(i) Flow of the feed water

Feed water flow is one of the most important parameters along with discharge pressure, which depends on the rate of the boiler outlet steam flow with some capacity margin as per standard. Normally, the margin is about 30% over the maximum flow requirement of the steam generator.

(ii) Discharge pressure of the BFP

Feed water discharge pressure is another important parameter. The BFP discharge pressure is obviously dictated by the boiler outlet steam pressure and friction losses in the pipelines. If the flow control is accomplished by a set of control valves in the discharge line, the pressure would have to be increased by another 5–10 kg/cm^2 toward the drop across said valve. There are a set of discharge pressure versus flow characteristic curves for every speed move upward and downward when the speed rises and falls respectively. The desired operating point can be selected by the right selection of speed. There is another characteristic curve provided by the manufacturer that indicates the minimum discharge pressure in every above curve (that is for a particular speed) beyond which the flow should not be increased for pump safety (Fig. 2.42).

(iii) Temperature of the feed water

Feed water temperature is an important parameter for the selection of BFP. Pumps are usually available with a temperature range of around 110–130°C. Am external water-cooling facility is provided for pumps to handle higher temperatures.

(iv) Net positive suction head (NPSH) required for BFP

This parameter is required for pump protection. This is the minimum absolute pressure needed at the suction or inlet of the pump above which it can operate safely. It must be ensured that the available suction head must be greater than the NPSH required to avoid cavitation inside the pump. The deaerator in one way assists the suction head to rise as its feed storage tank is normally located at an elevation of 20–30 ms. It is connected directly to the BFP suction and the water level in the feed storage tank always adds to the pump safety margin.

5.1.2 Types of Boiler Feed Pumps

A multistage centrifugal boiler feed water pump is mainly used in steam power plants. This type of pump is available in two broad classifications: equidirectional impellers and opposite directional impellers. There are several advantages and disadvantages associated with them, of which the most important points are efficiency and thrust balancing.

5.1.2.1 Boiler Feed Pumps With Equidirectional Impellers

Impellers in this design are set in the same direction thoughout the barrel of the pump. These pumps are compact and simple with more efficiency as compared to the other type of pump. Here, the fluid coming out of each impeller

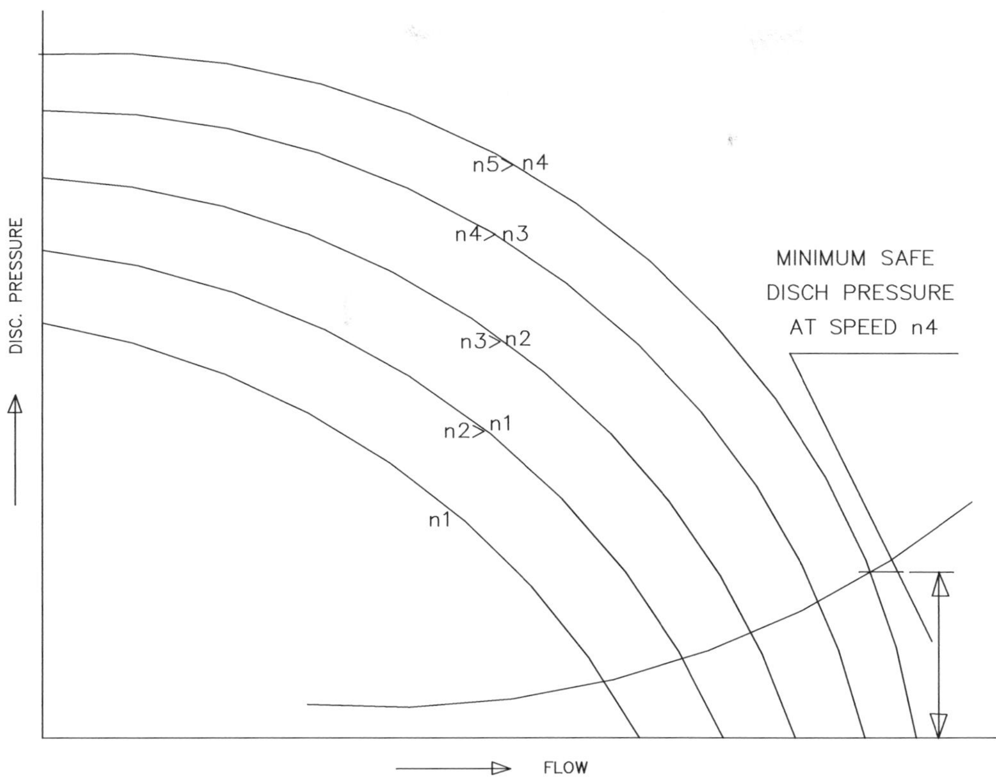
n5>n4
n4>n3
n3>n2
n2>n1
n1
MINIMUM SAFE
DISCH PRESSURE
AT SPEED n4
DISC. PRESSURE
FLOW
NOTE:
** IN CASE OF SUPERCRITICAL BOILERS & LARGE MW SUBCRITICAL BOILERS (WITH DIRECT
CONTROL FROM DRUM LEVEL CONTROL) FEED CONTROL STATION WILL CONSIST OF
LO LOW FEED CONTROL VALVE (30%) & bYPASS VALVE FOR FULL OPEN CONDITION.

FIG. 2.42 Boiler feed pump: discharge pressure versus flow characteristic curves.

at the outlet is passed to the inlet of the subsequent impeller. BFPs with this arrangement have a very high axial thrust, as the individual axial thrust of every impeller is added as they are unidirectional.

To overcome this problem, a *balancing drum*, also called a *balance chamber*, is provided to balance the axial thrust so as to reduce the stress exerted on the thrust bearing. It is evident that the balancing drum in this case is subjected to a total differential pressure due to all stages. The drum diameter and clearance must be greater because of the higher axial thrust that needs to be balanced. All these factors are responsible for balance chamber (balancing drum) leak-off flow to be greater than the other type. This leak-off flow is certainly to be increased in the course of time as all the clearances begin to increase due to wear and tear of all seals and rings. This leak-off flow is not contributing any energy conversion and hence efficiency would be reduced.

5.1.2.2 Boiler Feed Pumps With Opposite Directional Impellers

This design envisages impellers to be divided into two halves and set in the opposite direction. The outlet of the first group of stages is connected to the inlet of the next group of stages that lies at the opposite location of the pump through two crossover waterways. The number of impellers in each group of stage is just half the total number of impellers, if even. For odd numbers of impellers, the first group would consist of half the total number of impellers less one. Whatever the case, there would be some pressure drop due to this flow crossover arrangement and the system comes down to reduced efficiency. In contrast with the other type of pump design, here the axial thrust on the pumps is always well balanced. For obvious reasons, the sizing of the balance chamber (balancing drum) is less critical.

5.1.3 Boiler Feed Pumps With Associated Measurements

The following typical and minimum measurements are very important those are associated with (i) suction pressure, (ii) DP across BFP suction strainer, (iii) feed water flow, (iv) discharge pressure, (v) discharge temperature, (vi) feed water flow for all the desuperheaters, (vii) HP heaters outlet feed water temperature, (viii) HP heaters outlet feed water pressure, and (ix) feed water temperature before economizer inlet.

For large BFPs, some of the recommended integral instrumentation includes the following as a minimum, for example, (i) shaft vibration (both ends), (ii) bearing housing vibration (both ends), (iii) bearing temperature (at all bearings), (iv) axial rotor position, and (v) balance water flow. (See Fig. 3.18 for easy representation of the system and the purpose of measurements.)

5.2 Condensate Extraction Pumps

The CEP is a very important constituent of the total feed cycle, as shown in Fig. 2.41. Drawing condensate water from the hotwell of the condenser under vacuum, it delivers condensate water to the next phase of the feed cycle. It acts as a booster pump of the BFP by providing adequate pressure, which is also utilized to overcome the pressure drops across various LP heaters and pipelines up to the BFP suction.

As the condenser hotwell is directly connected to the suction flange of the CEP, it is customary that a portion of the pump delivery is recirculated to assure the minimum suction head by way of controlling the condenser hotwell level, especially during the low load operation. This is done in such a way that minimum flow through GSC is also ensured. There are also variations in system design, especially in larger plants where the CEP and GSC recirculation loop is independent of the hotwell level. The level in the hotwell would therefore be regulated independently by make up/dump valves.

A special design is adopted for the CEP shaft and other mechanical parts, enabling the volute chamber to hang in the sump and the bearings outside the sump.

The condensate available from the CEP discharge header is utilized to supply the following major services by tapping off from the main line before connected to the GSC/air ejector, which are (i) turbine exhaust hood spray and (ii) gland sealing system desuperheating.

5.2.1 Selection Criteria of Condensate Extraction Pumps

The selection criteria are the same as that of the BFPs, as indicated in Section 5.1.1 of this chapter.

5.2.2 Type of Condensate Extraction Pumps

CEP, as per the general layout of the power plant condenser hotwell, is usually a multistage centrifugal pump having a vertical shaft or, in other words, a vertical centrifugal pump. They are also referred to as cantilever pumps. The number of stages is dependent on the pump discharge pressure vis-a-vis the process requirement. These types of pumps are used for the process requiring a high flow capacity and pressure at the pump delivery. The advantage of handling high flow capacity is that they are able to operate with a specific minimum suction head.

CEP types vary in the shape of the housing of the impellers/diffusers, in general. The options are given below, although function-wise, they are basically the same: (i) bowl type and (ii) canister type.

Each bowl or canister contains a set of horizontal impellers around its inside periphery and diffuser with integral guide vanes, which direct the fluid to reach the inlet of the next stage bowl or canister located just at the top of the previous one. These vertical arrays of bowls or canisters determine the number of stages and are arranged in the pump barrel.

The CEPs may be categorized by the number of suction facilities, such as a single/double suction impeller. For a double suction impeller, the suction head requirement/safety factor improves.

5.2.3 Condensate Extraction Pumps With Associated Measurements

The following typical and minimum measurements associated with the CEP are very important: (i) condenser hotwell level, (ii) DP across CEP suction strainer, (iii) CEP discharge pressure, (iv) CEP discharge temperature, (v) CEP discharge flow, and (vi) GSC outlet temperature (Fig. 3.22).

6 DEAERATORS AND HEATERS

A deaerator is a very vital piece of equipment in thermal power plants that justifies its very name, that is, its main function is to remove air and other objectionable gas ingress from the main working fluid. The feed cycle heaters are also called regenerative feed heaters. The conception of providing feed heaters comes from increasing the thermal efficiency of the Rankin cycle on which steam power plants operate, thereby saving fuel and lowering the air pollution through flue gas emissions. The plain and simple thought behind putting feed heaters in the cycle is to increase the average temperature of the working media during the process of heat addition and bringing down the fluid temperature while the process is on heat rejection.

6.1 Deaerator

The deaerator, though, is primarily utilized for deaeration, that is, removal of gases from the condensate water. It also functions as a heater located in between the LP and HP heaters in the feed water cycle, as a storage unit of feed water, and also ensures the required NPSH for BFPs.

6.1.1 Reason for Deaeration

It has been experienced that the presence of contaminants such as dissolved oxygen, carbon dioxide, and other non-condensable gases in FW supplied to the boiler is the main wrongdoer affecting the equipment and pipe work metals directly through corrosion. For example, oxygen is the prime accused for corrosion attacks in hotwell tanks, FW pipelines, BFPs, and, of course, in boiler metals. Moreover, the presence of dissolved oxygen (DO) is the root cause for rapid localized corrosion in boiler tubes. It is aggravated by the low pH in the working fluid, which means the acidic behavior of water resulting from the presence of contaminants such as CO_2. CO_2 dissolves in water and produces corrosive carbonic acid, which is detrimental to the boiler metals for causing acid attacks. Measurement of feed water pH levels would be indicative for taking appropriate action toward the same. The bad effect of corrosion is quite visible by the pitting type damage, where deep penetration and perforation can often occur expeditiously without notice, even if the metal loss may be nominal. Thus it is extremely necessary to remove all dissolved gases from the boiler-bound FW to keep the boiler in a healthy condition with an increased life span.

Out of two major ways, one method is to apply chemical, but this is costly and may harm the metal/process if overdosed. The mechanical method, that is, deaeration, has proved more economical and thermally efficient compared to the former. Although elimination of DO is achieved by the mechanical method, in practice, it is usually a combination of both systems.

As stated earlier, the pH level in the condensate/feed water is indicative regarding the corrosion possibilities, which, if maintained above 8.5–9, the presence of both the above-mentioned damaging elements can be avoided. Above that pH level, carbon dioxide and oxygen are both not traceable. That is the reason for using the chemical method also by applying sodium sulfite, hydrazine or tannin as oxygen scavenger after the deaeration.

Normally, the FW temperature is raised to near but less than the saturation temperature by supplying the heat energy through direct mixing with the steam taken from the fourth extraction line or the crossover line from IPT to LPT. Operation at higher temperatures than this saturation pressure is avoided due to the close proximity of the saturation temperature, which may increase the probability of cavitation inside the BFP. However, this situation can be tackled by installing the feed tank at a very high level to add some pressure above the boiler saturation pressure when the water reaches the BFP suction. By increasing the temperature, more oxygen scavenging is attainable; it is to be kept in mind that this temperature must be less than the saturation temperature. Hence, the only way to get more operating temperature is to increase the saturation temperature of the fluid, which demands the increased deaerator shell pressure. Therefore, the units intending to less usage of chemicals for feed water dissolved oxygen content, pressurized deaerator is the only alternative solution.

6.1.2 Working Principles of Deaerator

Two different scientific theories paved the way for the working principles of the deaerator. The first theory is Henry's Law, which states that, "Gas solubility in a solution decreases as the gas partial pressure above the solution decreases." The second theory states that "Gas solubility in a solution decreases as the temperature of the solution rises and approaches saturation temperature." In fact, when a liquid attains its saturation temperature, the solubility of a gas in it is zero, but this can be achieved only when the liquid is subjected to turbulence by properly stirring the fluid to ensure complete deaeration. Both these theories are utilized in the deaerator operation for removal of dissolved gases from the boiler FW, which at the same time gains some temperature from the heating media as well.

Fig. 2.43 and Fig. 3.24 may be referred to for easy understanding of the system.

Inside the deaerator, an atmosphere of steam is maintained by injecting the same in the upper part, called the deaeration area, and the condensate is arranged to pass through a special type of spraying system such that it comes out in the shape of thin films or otherwise breaks the water into mist form/minute droplets. This system provides a high surface area-to-mass ratio and mixed/heated rapidly in touch with steam through faster heat transfer from the steam to the water and cause the whole of fluid inside deaerator be under saturated condition. The important point to note is that raising the water temperature and releasing the gases must take place within a very short residence period in the deaerator. This releases the dissolved gases, which are then carried with the excess steam to be vented to the atmosphere. Steam is also injected in the lower part of the deaerator, called the feed tank, for raising the saturated temperature of the system. The gases are thus released from the condensate and in turn leave the upper chamber through the vent from the deaerator. The combined effect of heating and removal of released oxygen and other gases enhances the gas liberation and lowers the gas concentration. Water after deaeration is then available in the feed tank. A thick layer of steam is maintained above the feed tank storage water to ensure that gases are not reabsorbed. A very efficient type of deaerator can eliminate dissolved oxygen concentration equal to or even less than 0.005 cc/L (7 ppb). So far as the concentration of carbon dioxide is concerned, some manufacturers claim complete elimination of the same.

6.1.3 Types of Deaerator (*Ref.* Fig. 2.44)

Deaerator water handling systems are of two types for providing greater surface area with their individual techno-commercial advantages and disadvantages. Those are: (i) tray

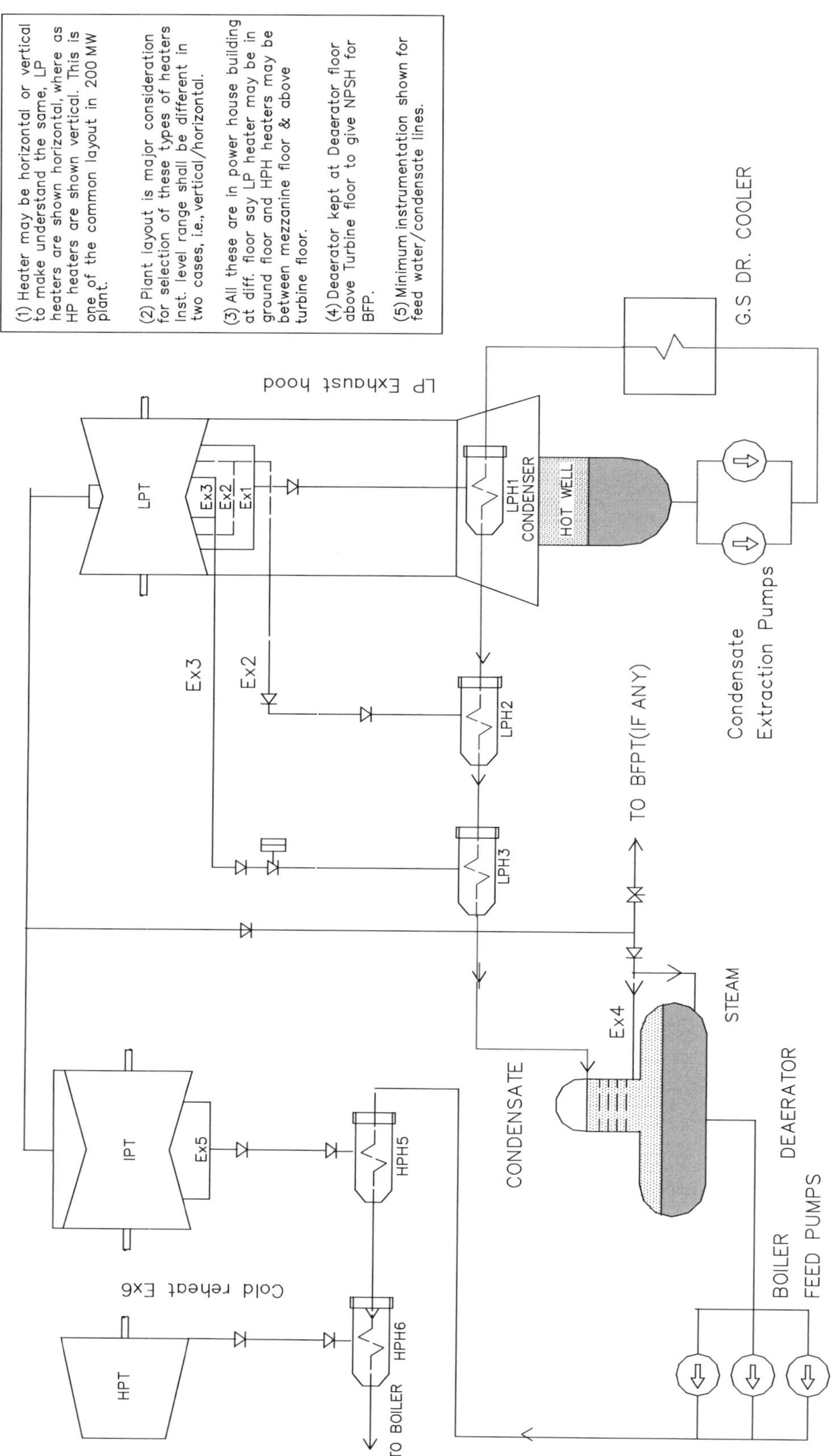

(1) Heater may be horizontal or vertical to make understand the same, LP heaters are shown horizontal, where as HP heaters are shown vertical. This is one of the common layout in 200 MW plant.
(2) Plant layout is major consideration for selection of these types of heaters Inst. level range shall be different in two cases, i.e., vertical/horizontal.
(3) All these are in power house building at diff. floor say LP heater may be in ground floor and HPH heaters may be between mezzanine floor & above turbine floor.
(4) Deaerator kept at Deaerator floor above Turbine floor to give NPSH for BFP.
(5) Minimum instrumentation shown for feed water/condensate lines.
LP Exhaust hood
LPT
Ex3
Ex2
Ex1
LPH1
CONDENSER
HOT WELL
G.S DR. COOLER
Condensate
Extraction Pumps
Ex3
Ex2
LPH2
TO BFPT(IF ANY)
LPH3
Ex4
STEAM
CONDENSATE
DEAERATOR
IPT
Ex5
HPH5
Cold reheat Ex6
HPT
HPH6
TO BOILER
BOILER
FEED PUMPS
NOTE: In LP heaters dotted coils shown for condensate & in HP heaters dotted coil shown for Feed water.

FIG. 2.43 Basic schematic diagram of deaerator operation.

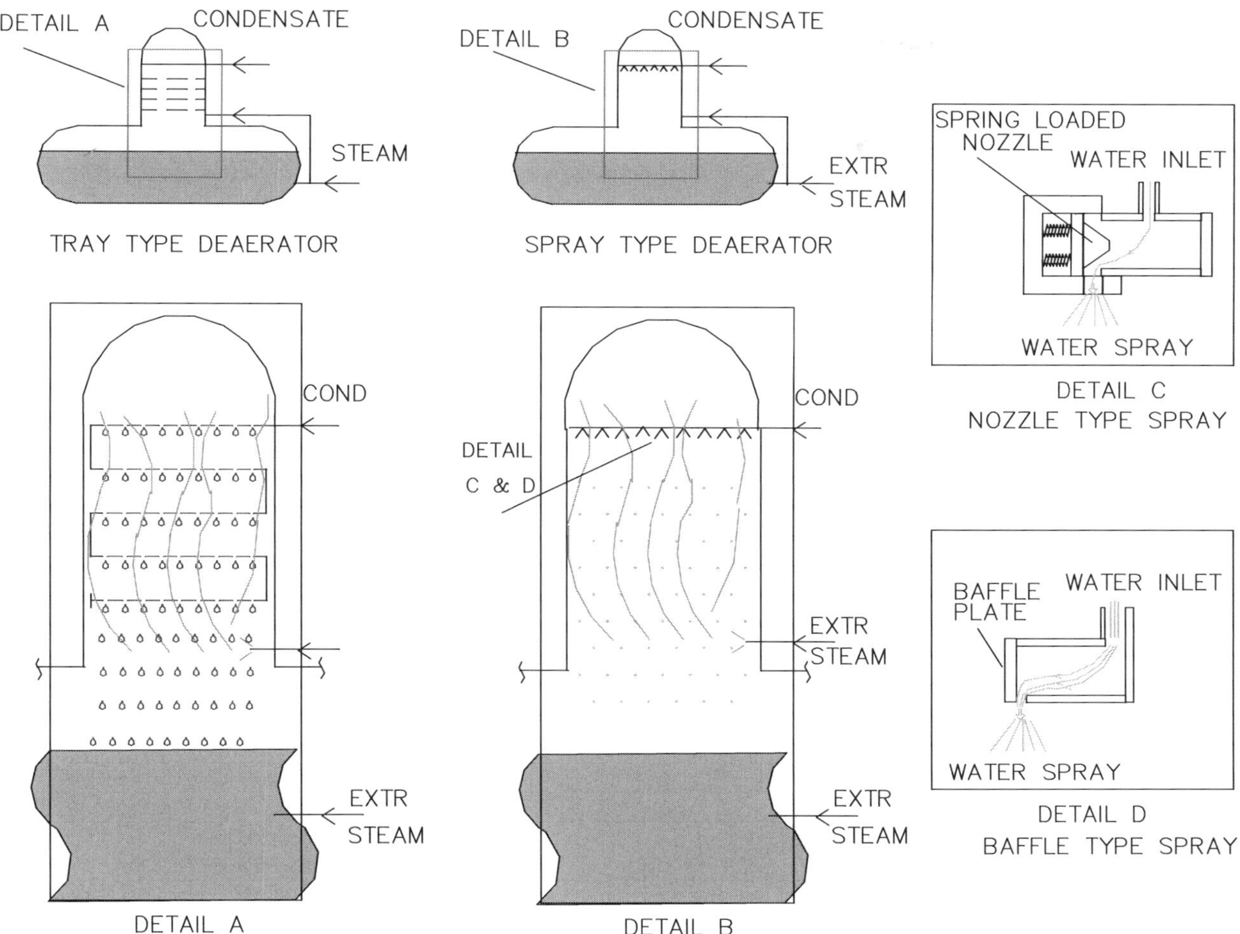

FIG. 2.44 Basic schematic diagram of different deaerator type.

type used for power plants with very high turndown ratio and (ii) spray type normally used for process plants.

6.1.3.1 Tray Type of Deaerator

This type of deaerator deploys a series of trays with small holes at the bottom. When pressurized water is passed along these trays, part of it comes out through the holes and is converted to small droplets and/or mist. They provide more surface area to come in contact with comparatively high-temperature steam to get heated to a higher temperature and release noncondensable gases, which vent to the atmosphere. The condensate and the major part of the steam that gets condensed fall in the feed tank, where the level is maintained in coordination with the rest of the condensate cycle.

Extraction steam may also be injected into the feed tank water side, depending on the system design. This is done so as to raise the condensate temperature in addition to the above arrangement. Pressure fluctuations in this type of deaerator may damage the tray assembly, so the deaerator pressure control needs special attention to keep it within limit.

6.1.3.2 Spray Type of Deaerator

This type of deaerator deploys a number of spray nozzles provided with the manifold where the condensate water enters. Another type may be arranged in such a way where the condensate water is guided to impinge like a jet against a baffle plate so as to come out as if water spray is achieved. With a number of such units, the water is converted into fine water particles, providing more surface to come in contact to get heated with steam quickly and release noncondensable gases. A heating arrangement through extraction steam may also be provided, depending upon system design.

6.2 Regenerative Feed Water Heaters

The temperature of the condensate from the CEP outlet is low, hence it is routed through several LPH and HPH to increase the water (both condensate and feed water) temperature with the idea to improve efficiency. This is accomplished by taking comparatively high-temperature extraction stream from different stages of the turbine at multiple points (of the Rankin cycle) through an exchange of heat. This very idea is termed *regeneration* and the vessel used as the heat exchanger for the transfer of heat from steam to water is called an *FW heater*. Theoretically, there are two main types of FW heaters: (i) *Open* FW heater where the steam directly mixes with the water and (ii) *Closed* FW heater where the steam does not mix with the water but an exchange of heat takes place through the metallic barrier between them.

Theoretically, both types are available but in practice, modern thermal power plants are equipped with closed FW heaters, except the deaerator.

6.2.1 Open Type Regenerative Feed Heaters

The open FW heater is fundamentally a mixing vessel where the extraction steam from the turbine combines with the water directly. In an ideal heat exchanging condition, the water leaves the heater as a saturated liquid at the vessel's inside pressure. It has been observed that the effect of adding one regenerative heat exchanger in the form of an open FW heater system improves the thermal efficiency simply by preheating/raising the temperature of the water before it enters the boiler. It has, however, also been observed that the power output is reduced.

6.2.2 Closed Type Regenerative Feed Heaters

Closed FW heaters, on the contrary, do not allow direct mixing. In a power plant, they are basically "shell and tube" heat exchangers where the FW temperature increases by gaining the heat from the extracted steam via the tube metal. The extraction steam enters the shell, which contains a bundle of metal tubes through which water passes through. The extraction steam then condenses at the outlet of the shell after losing temperature to the FW. Here, the two heat exchanging media are at different pressures as they do not mix with each other. The P and I diagrams relating to the feed water and condensate lines of a steam power plant with closed FW heaters may be referenced. The P and I Diagrams Fig. 3.19, 3.22, 3.24, 3.28, 3.29, and 3.31 depict typical instrumentation systems and flow paths.

In an ideal regenerative Rankin cycle with all closed heat exchangers (including drain cooler, gland steam cooler, etc.) and FW heaters, the condensate, steam drains from the heaters are cascaded to lower pressure heater/vessel and ultimately to the condenser. The boiler main steam expands in the turbine to an intermediate pressure to get work done. While the main part of the steam in the turbine continues to expand to the next lower stage, a small part of the steam is extracted from a suitable point of the turbine so as to make a thermal balance. The extraction steam is then directed to the FW heaters. A similar process goes on until the steam is expanded to the condenser pressure with several other extractions to heat various other closed FW heaters. The water (condensate and feed water), on the other hand, gets more and more heated through its journey from CEP discharge to the boiler inlet. The normal route of the steam condensate drain from any FW heater is toward the next lower pressure heater, but is diverted to the deaerator in the case of HP heaters and flash tanks/condensers, depending upon the situation. Some systems allow the condensate from the closed FW heater to a higher-pressure point in the Rankin cycle by suitably pumping the same.

The heater vent lines are also shown with three types of possibilities to vent the steam in heaters, if required. Those are during start-up, normal operation, and emergency. The HP heater vents are sent back to the system through the deaerator in case of the first two events, but in an emergency, the vents are diverted to the LPH or sump pit to suit the emergency condition.

The LP heater vents are sent back to the system through the LP flash tank in the case of the first two events; during an emergency, the vents are made open to the atmosphere through pressure relief valves.

6.2.3 Broad Comparison Between Open and Closed Regenerative Feed Water Heaters

In comparison with open FW heaters, it can be stated that the closed FW heaters are more expensive because of their complex design and internal arrangements. Because the two streams are not allowed to mix inside the heater, there is no need to provide a separate feed pump for each closed FW heater to raise the water pressure to make it suitable for the next stage.

Regarding plant efficiency, it has been observed that the thermal efficiency of the closed type is more or less equal to that of the open FW heater. As far as the power output is concerned, however, it is noticed that the same is less compared to that of the open FW heater. For closed heaters, certain eventualities such as tube rupture may take place. This calls for emergency drains, and may be taken care of suitably by diverting them to high-capacity process chambers such as hotwells and deaerators. The number of LPHs is in general the same for all high-capacity power plants, that is, LPH 1, 2, and 3. But the number of HPHs varies as per the capacity and even there may be a pair of identical HPHs in each stage and while any of the pair can be isolated in case of any mechanical or operational snag.

6.2.4 Operation Description of Closed Regenerative Feed Heaters

The bled steam is taken off from the suitable extraction point of the HP/LP turbine and then utilized to heat the comparatively high-pressure feed water or condensate. The steam is, however, ultimately condensed to a subcooled liquid called (HP/LP) heater condensate. As the two streams

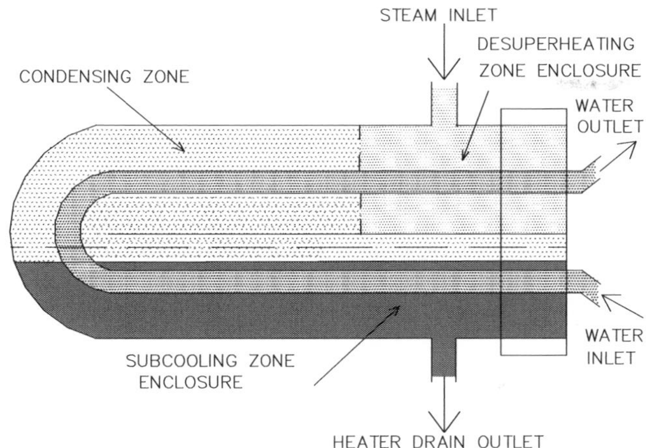

FIG. 2.45 Basic schematic diagram of HP/LP heater operation.

are separated mechanically, only a finite temperature difference is the necessary condition for heat transfer to take place. The closed FW heater is basically a counterflow heat exchanger and the pressurized fluid entering at a lower temperature is heated to the saturation temperature of the bled steam. There are three distinct heat transfer zones that can be seen in typical closed FW heaters, as shown in Fig. 2.45. Those heat zones are mentioned below:

(i) *Desuperheating zone enclosure*: Here, the extraction steam first enters the heater and is cooled (desuperheated) while raising the temperature of the FW. It leaves this heat transfer zone to a temperature approaching or almost equal to the saturation temperature reached by the conduction cooling by the water within the bundle of feed water pipes.

(ii) *The condensing zone*: It is the largest heat transfer zone within the heater shell, as the already made saturated steam condenses due to further cooling and gives up its latent heat to cooling water, thereby raising its temperature.

(iii) *The subcooling zone*: This is an enclosed zone that is a separate shrouded area within the heater shell, further cooling the condensed steam called heater drain while heating the incoming FW. The design of this heater sizing ensures that the subcooled condensate is reduced to within around 4–6°C (typically) above the incoming FW temperature.

7 CW AND ACW SYSTEM FUNCTION AND DESCRIPTION

The requirement of the condenser in a thermal power plant was discussed in Section 2.8 of Chapter 1 and Section 3.5 of this chapter. It is evident that when the condenser internal design is made to allow the condensate to be colder, the exhaust steam pressure is also reduced according to the saturation condition, and the efficiency of the Rankin cycle increases.

The prime object of a CW system is to cool the LP turbine exhaust steam to convert it to condensate. Typically, the cooling water causes the steam to condense at a temperature of about 45.4°C, and that creates an absolute pressure in the condenser of about 0.1 kg/cm^2. The great volume reduction that takes place when steam condenses to liquid creates the very low vacuum, which enhances the action of taking out the steam.

The limiting factor of the condenser functioning is the inlet CW temperature dictated by the prevailing climatic conditions at the power plant's location, which are lower during the winter season and cause more condensation in the turbine. Electrical output may decrease in plants operating in hot climates during the summer if their source of condenser CW becomes warmer.

The CW system provides for supplying water not only to the condenser for condensing the LP turbine exhaust steam, but also for secondary cooling of the ACW systems of different auxiliaries of the boiler/turbine. CW pumps are normally of the vertical wet pit type. A CW system utilizing seawater concrete volute type CW pumps is also used.

The ACW system is thus a subsystem of the CW system and its main function is to cool the closed/clarified cooling water (CCW) system catering to the cooling water requirements of all the auxiliary equipment such as the turbine lube oil coolers, the generator coolers of H_2 and the seal oil system, conductor cooling, etc.

7.1 Circulating Water System

There are two major classifications of the CW system adopted as per the location/design of the plant: (i) once-through/open and (ii) closed cycle or recirculating type using a cooling tower (CT).

7.1.1 Once-Through Circulating Water System

Here, cooling water is supplied directly to the condenser when the same is available in abundance near the plant, such as the river/seawater for coastal power stations.

The OT system is always accompanied by the desilting arrangement and also by traveling water screens of appropriate mesh size to arrest the entry of debris and biological species from the water source. It is installed at the intake section of the CW system. When sea/riverwater is utilized as cooling water, suitably sized debris filters have to be provided upstream of the condenser cooling circuit so that fouling with the condenser tubes is avoided. A time-based control system of traveling water screens with backwash facilities is provided with DP between upstream and downstream of traveling water screens as a controlling factor for its operation.

7.1.1.1 Once-Through Circulating Water System Components

A pump house consisting o f reservoir and pumps with sufficient redundancy draws raw water from the sea/river/lake and discharges the same as cooling media directly for the condenser, as depicted in Fig. 2.46. The CW pump house is normally located near the source to avoid additional earthwork toward making of canal. All statuses are indicated and alarmed (for abnormality) through level switches/transmitters, etc. The CWPs are provided with sufficient numbers to facilitate uninterrupted supply of cooling water to the main power plant(s). Individual pump discharges are connected to a separate header with interlocked operation through low discharge pressure and CWP tripping status. The discharge pressure is calculated on the basis of the frictional loss due to the long run of the pipelines and the cooling water inlet head near the condenser. Desilting and traveling water screens are provided to prevent unwanted materials entering the condenser cooling circuits.

7.1.2 Closed Cycle Type or Recirculating Type Circulating Water System

Previously, once-through or open systems were allowed with the upper limit of the maximum temperature increase of about 20°C compared to the intake temperature, but there are restrictions in many countries toward acceptance of this system. For example, nowadays, as per stipulations dated Feb. 1, 1999, by the Ministry of Environment and Forests (*MoEF*), Government of India on the power plants installed after Jan. 6, 1999, on the freshwater sources to meet their

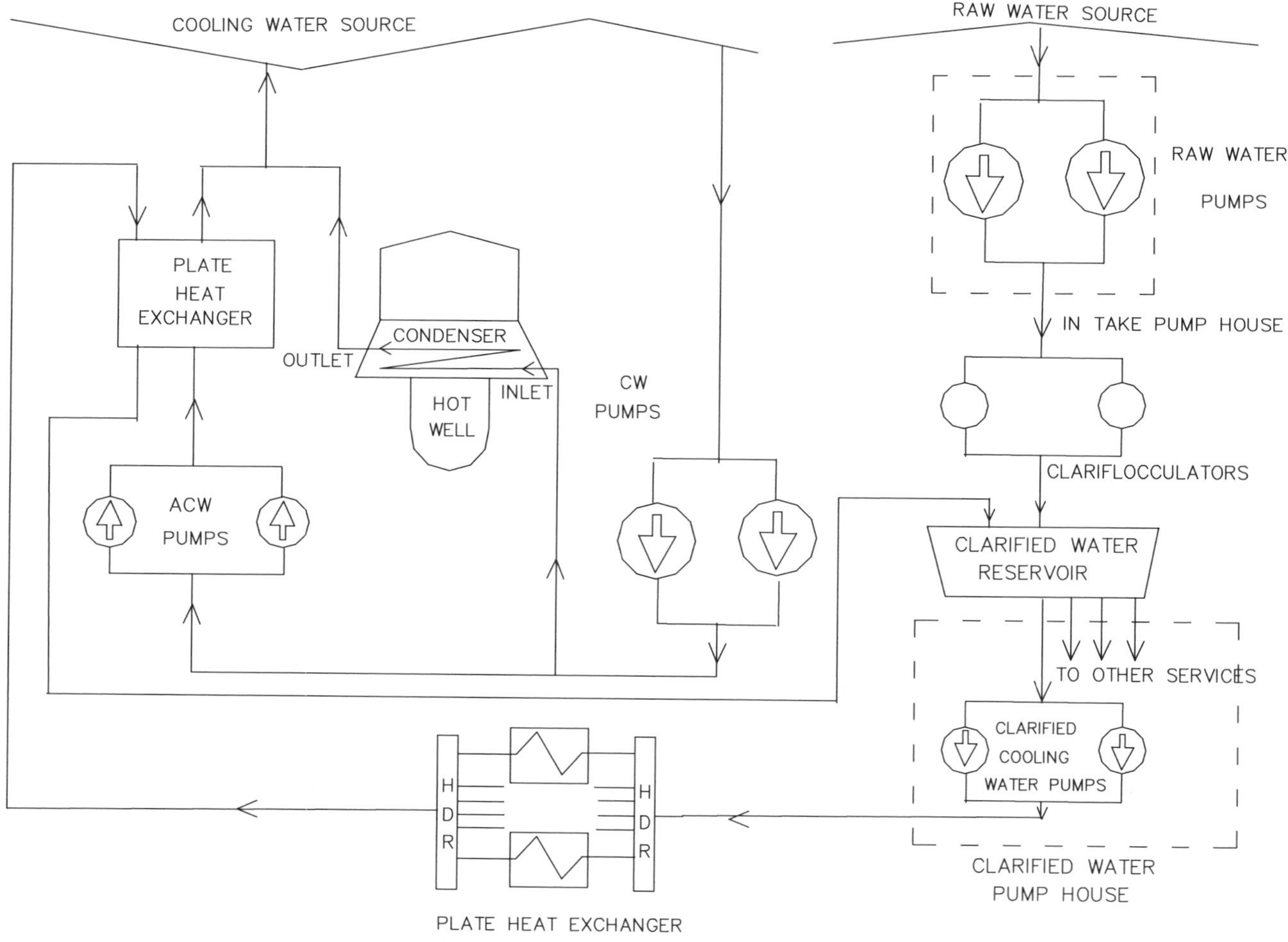

FIG. 2.46 Basic schematic/flow diagram of CW/ACW system without cooling tower.

water requirements are not permitted for installation of an OT condenser cooling system considering the thermal pollution aspects of the source waterbody. This means all inland power plants in countries under this restriction now have to provide cooling towers. Seawater-based cooling towers are also adopted at coastal sites, considering the techno-economic-environmental compulsions.

A closed cycle or recirculating CW system utilizes the services of clarified water with cooling towers when the service water is taken from freshwater sources such as lakes, reservoirs, and even rivers or canals (Fig. 2.47). The freshwater is then clarified through a clariflocculator and transferred to the CW reservoir with the help of clarified water pumps. Once the system starts, the transfer takes place for make-up purposes only. The closed CW circuit includes this reservoir receiving cold water from the CTs and CW pumps that supply cold water from the CT reservoir into the CW pipelines to pass it through the condensers and plate heat exchangers of the auxiliary cooing system. Hot water from the condensers and plate heat exchangers is returned back to the cooling towers through discharge ducts/pipe headers.

7.1.2.1 Types of Closed Cycle Type or Recirculating Type CW Systems

The two types of cooling towers normally available are the mechanical induced or natural draft cooling tower (NDCT). The selection is mainly dependent upon techno-economic considerations involving capital costs, operating expenses, and specific site-related aspects.

7.1.2.1.1 Mechanical Induced Draft Cooling Tower

As a general rule, mechanical induced draft cooling towers (IDCTs) are preferred for power plants located near the coal source point (pit head) as a result of low running costs due to the lower costs (low coal transportation charge) of power generation. The mechanical IDCT may have a single or a double inlet, and may be a crossflow or a counterflow type with the fans located on top of the tower.

7.1.2.1.2 Natural Draft Cooling Tower NDCTs, on the other hand, are preferred for power plants located near the load centers, which means far off from the coal source point as this process works without any rotating equipment,

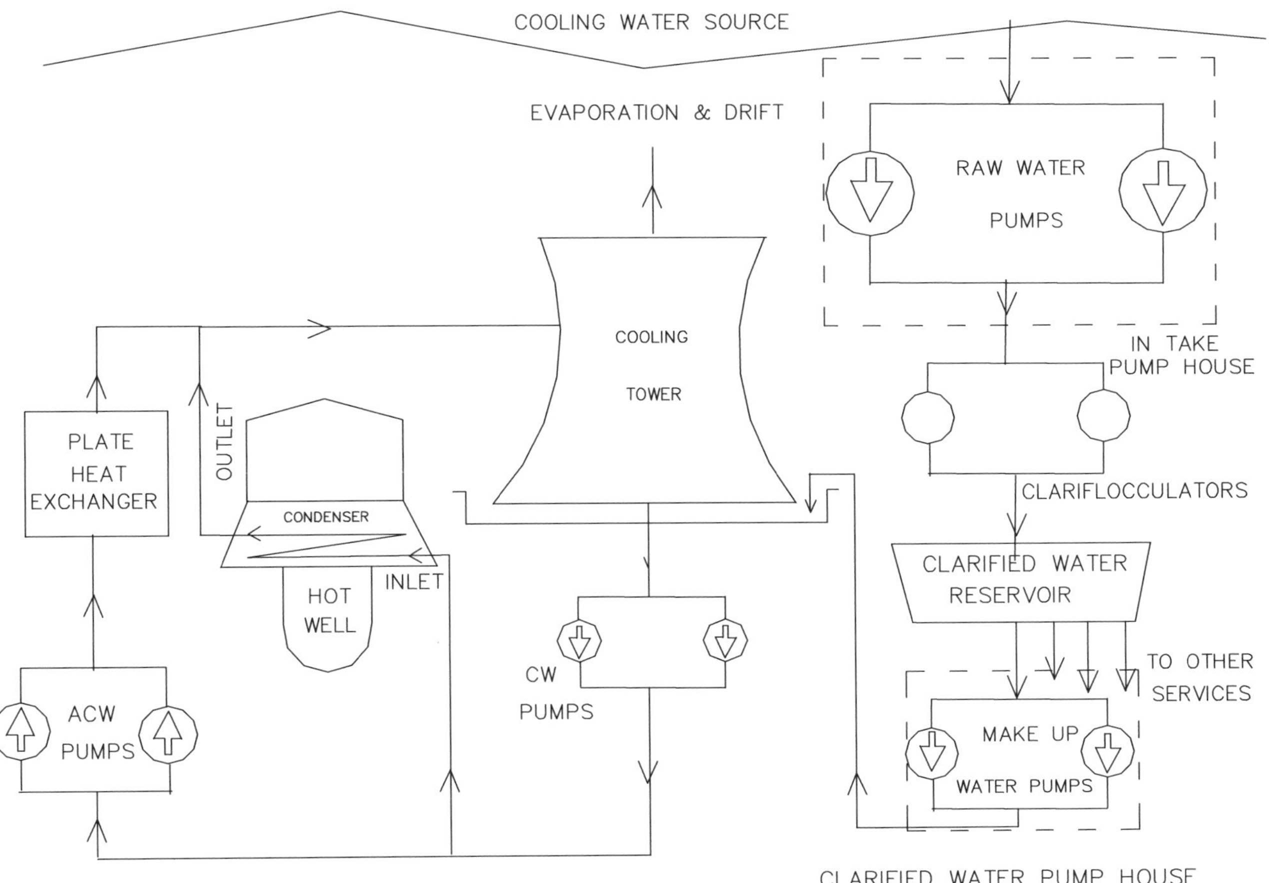

FIG. 2.47 Basic schematic/flow diagram of CW/ACW system with cooling tower.

namely cooling fans, thereby saving on costly power. The air flow rate through the NDCT depends upon the density difference between the ambient air and the relatively hot and humid light air inside the CT. For a plant site where the duration of summer ambient temperatures is considerably greater along with low relative humidity, an adequate density difference would not be available for suitable performance of an NDCT, and an IDCT is a much better choice.

7.2 Auxiliary Cooling Water and Closed Cooling Water Systems

The *ACW* system is to facilitate an uninterrupted supply of ACW to the coolers for cooling the hot return lines of system auxiliary coolers (*CCW system*) such as H_2 and seal oil coolers, etc. These two systems cater to all the cooling requirements other than the condenser cooling circuit.

7.2.1 Auxiliary Cooling Water System and Components

The ACW system is connected to the CW delivery line near the condenser with a common discharge header to supply the secondary sides of the CCW system coolers.

This system consists of redundant ACW pumps and plate-type heat exchangers with necessary valves, temperature/pressure measurements, and interlocked operation.

7.2.2 Closed/Clarified Cooling Water System

The CCW system takes care of the cooling water requirements of all the auxiliary equipment

such as turbine lube oil/control oil coolers, H_2 and seal oil coolers, stator winding water coolers, BFP auxiliaries, condensate pump bearings, all sample coolers for both continuous measurement and grab sampling, air compressor auxiliaries, etc. For the gas turbine (GT) auxiliaries, the CCW is provided with an individual ACW system because the pressure requirements of a GT cooling water system are generally high when compared to that of a steam turbine.

The primary side of this CCW system is actually the fluid utilizing passivated demineralized (DM) water called CCW as the cooling medium, which will be circulated in a closed circuit. The CCW basic schemes are included in Figs. 2.46 and 2.47 referenced above for both systems, that is, with or without CT.

7.2.2.1 Closed/Clarified Cooling Water System Components

The main components of this system are redundant CCW pumps, plate heat exchangers, and overhead expansion tanks. The CCW takes away the rejected heat from all auxiliary coolers and circulates in a closed circuit through the plate heat exchanger by redundant CCW pumps. An overhead expansion tank of adequate capacity is provided to ensure positive suction to the CCW pumps as well as to allow the expansion of water in the closed circuit. There is every possibility of leakage losses in the CCW system in gland packing at pumps, in flanged connections, at plate heat exchanger seals, etc. Make-up (MU) water is normally provided from the CEP discharge through interlocked operation. A solenoid-operated level control valve is provided to ensure the expansion tank level. During the initial fill for the system, water is supplied through MU water pumps with a chemical feed system to add chemicals to ensure adequate pH value, which is called the passivation of DM water.

7.3 CW Make-Up and Treatment System

The CW quality in terms of desired cycles of concentration (*COC*) is maintained by a blowdown operation from the cold water side of the CW system (CWP discharge line). It is used normally for low-grade applications such as ash handling (wet) systems, coal dust suppression, etc.

It is quite natural to lose water due to blowdown, evaporation, drift, etc. needs MU water replenishment supplied by CW make-up water pumps installed in the clarified water pump house.

Keeping in mind the CW discharge water temperature limitations for coastal power plants, the take of point of cooling tower blowdown is in general located near the colder side of the CW system or, in other words, from the discharge header of the CW pumps and the remaining blowdown is discharged back to sea with a lower temperature.

Scale and corrosion can result despite the application of different chemicals at the desired treatment levels, as a result of improper blowdown. For example, excessive blowdown means that suspended solids, scale-forming minerals, etc., are removed, making the CW more corrosive.

Excessive blowdown may also cause removal of the chemicals already applied for treatment, which may either result in less corrosion protection or increased chemical costs. On the other hand, if the blowdown is less than the desired value, scale-forming minerals may build up in the system to the level that could make it difficult for the treatment chemicals to overcome the tendency to form scale deposits.

In a number of plants, blowdown is not taken into consideration as a part of the chemical treatment system, but is included as a part of the overall treatment of the water-cleaning arrangement.

The COC, as stated above, determines the presence of unwanted chemicals in water without going in for chemical treatment. The meaning of COC is actually the number of times the constituents in the raw MU water are concentrated in a CT as water is evaporated. The COC is controlled solely by blowdown and is evaluated by the TDS present in the water or by measuring the conductivity of water. For any

CT system, insufficient blowdown as stated earlier would cause the COC level to go higher than the recommended value. This can lead to the formation of scale in spite of maintaining the proper chemical treatment levels.

7.3.1 Prevention of Microbiological Growth and Treatment System

For preventing microbiological growth in the CW system, a chlorine dosing is provided. A shock dosing of chlorine is carried out in the fore bay of the CW pump house. Apart from chlorine dosing in the CW, the dosing of sulfuric acid, an inhibitor of scale/corrosion/biocides, as required is done for their control and organic fouling in the CW and ACW systems. Side stream (SS) filters are provided in a clarified water-based, closed-cycle cooling water system to reduce the turbidity in the CW on account of suspended solids in the make-up water and atmospheric dust ingress through the cooling tower.

7.4 CW Make-Up and Treatment System Using Seawater

Seawater is used as the cooling media of the CW system, that is, for condenser cooling purposes near coastal areas. Both options—OT or closed-cycle cooling using CT—are applicable here. For this type of CW system, the thermal pollution effect on seawater dictates the selection of the system type, along with techno-commercial analysis based on the distance between the power station and the sea coast vis-a-vis the cost of pumping plus pipelines to transmit seawater. Prime considerations for CTs with seawater makeup are the CT drift and salt contamination in the environment.

7.4.1 Drift Problem With Cooling Tower (CT) Using Seawater

Elaborate study on the problem associated with CT drift it has been experienced that the sea water may contain very high concentration of TDS up to 50,000 ppm which is apprehended to be followed by corrosion of connected equipment.

Proper action needs to be taken to avoid the long-distance drift of high concentration saltwater while designing the CW system, such as deploying low-pressure spray nozzles used in the CT, etc.

The manufacturing cost of CT for a typical seawater application, operating at a COC ranging around 1.5, would be higher to accommodate the increase in CT size, which may go up to 10 percent bigger than a similar capacity of a CW system utilizing freshwater. The reason behind the same is the high salt concentration borne by seawater lowers the vapor pressure of same water vis-a-vis reducing the evaporative cooling rate depending on the degree of salt concentration. Approach temperatures (the difference between a wet bulb reading of the environment and the CT outlet cold water temperature) used for tower design must consider the effect on tower performance by using saltwater. For practical purposes, a 20% increase in approach temperature may be considered for cooling tower design for a seawater-based system compared to a freshwater-based one.

Keeping in mind the discharge water temperature limitations for coastal power plants, the cooling tower blowdown is affected from the colder side of the CW circuit or, in other words, from the discharge header of the CW pumps.

7.5 Instrumentation Requirement for a CW System

Other than the interlock and protection system, the necessary instrumentation for a CW system should include, (i) a vibration monitoring system for CW pumps and motors, (ii) bearing temperature (for a larger motor), (iii) chlorine leakage detection with a chlorine absorption and neutralizing system (to come in service automatically on the detection of chlorine leakage more than the alarm level, (iv) CW flow, (v) CW hot inlet temperature, and (vi) CW cold outlet temperature, wet bulb temperature

7.6 Auxiliary and Associated Subsystems of a CW/ACW/CCW System

The following important auxiliary and associated subsystems are provided to take care of various hazards related to CW/ACW /CCW systems. These are indicated to have a very precise but broad idea, for example, (i) a sludge transfer system from the stilling chamber near the source point, (ii) different dosing systems such as alum, lime, polyelectolyte, etc., (iii) a chlorination plant, (iv) a service/potable water system, (v) a firefighting system, and (vi) an effluent treatment plant.

8 DEMINERALIZING (DM) PLANT FUNCTION AND DESCRIPTION (FIG. 2.48)

Water is the system working fluid in any steam TG set be it coal fired thermal or nuclear power plant or combined cycle plant. Besides that, water is indispensible for various applications such as condenser/equipment cooling, chemical dosing, ash disposal, service/potable applications, etc., with different types of demands made available by the proper type of treatment accomplished in a particular subsection or plant. The main source of water known as raw water is drawn from rivers, lakes, or canals to suit the availability and/or geographical location of the plant.

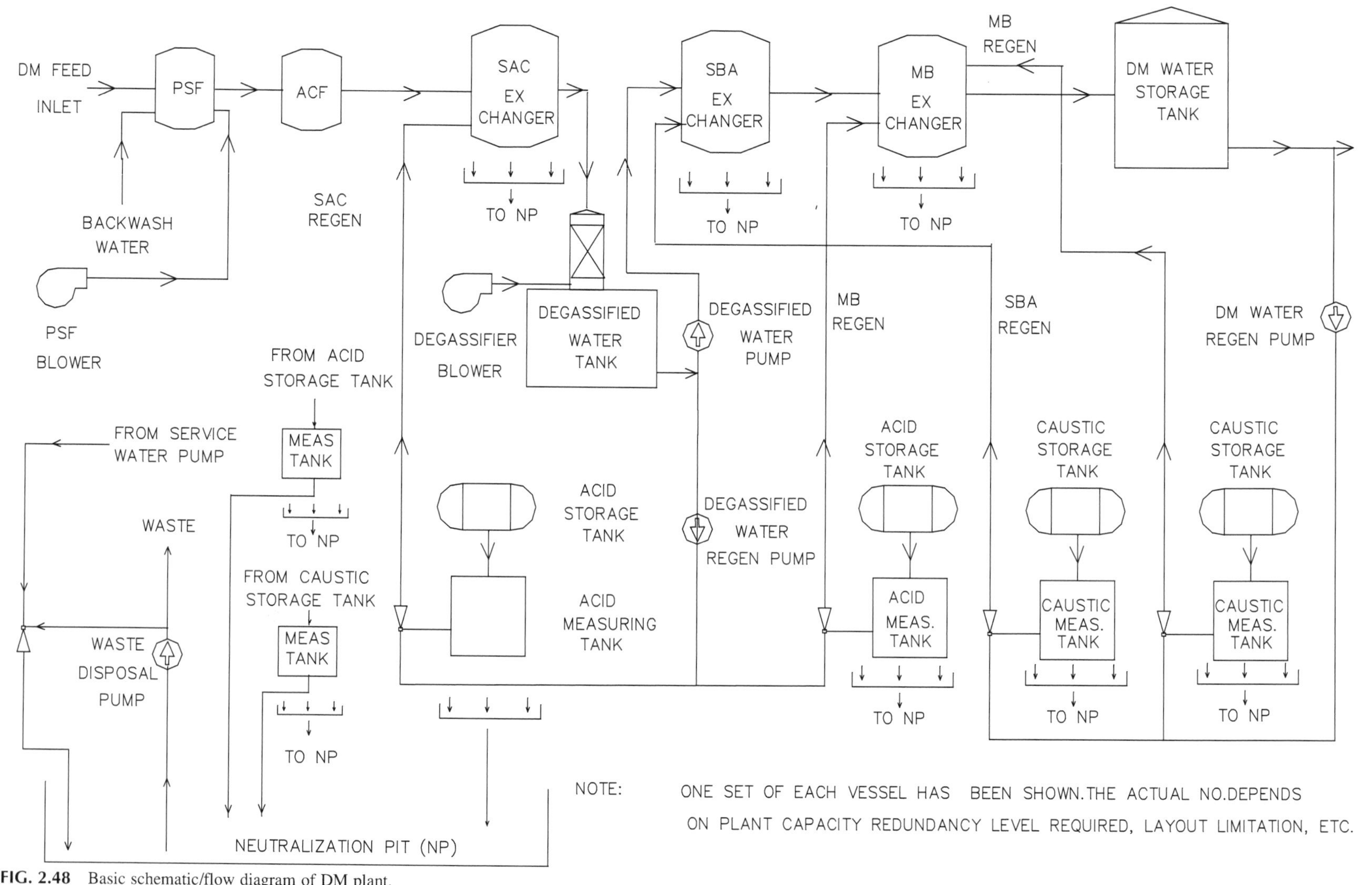

FIG. 2.48 Basic schematic/flow diagram of DM plant.

The applications or requirements of particular types of treated water used in a thermal power plant are:

(i) The main system working fluid requirement is DM water from the DM plant during initial filling and for regular make-up for the Rankin heat cycle. DM water is necessary for the CPU and the regeneration of ion exchanger units of the DM plant, as shown in Fig. 2.49.

(ii) Service water, cooling tower make-up water, potable water after further treatment (filtration and disinfection), equipment cooling, etc., require clarified water.

(iii) Ultrapure water for generator stator conductor cooling.

8.1 Necessity of a DM Plant and System Requirements

Waters available from any source may be mixed with various types of dissolved salts and dissociate in water, forming free positive (cations) and negative ions (anions). They are considered impurities for the boiler working fluid as they influence the operating efficiency and cause loss/failure of metal after prolonged use.

As discussed elsewhere, overheating is caused by the buildup of deposits mainly formed by the presence of calcium and magnesium ions (in the form of carbonates and bicarbonates). This may lead to catastrophic tube failures and decreased efficiency. These ions are known as hardness ions and must be removed from the water before its usage as boiler FW. For that reason, DM water is required to be FW and the quality is maintained by introducing the CPU after the condenser to combat the acquired contamination during cyclic travel.

The DM plant receives water input from the pretreatment plant, which also supplies the plant's clarified/filtered water. Drinking water is normally generated from filtered water after chlorination for disinfection. The overall water treatment plant, including pretreatment and post-treatment (DM plant), is shown in Fig. 2.49.

The DM plant consisting of an activated carbon filter (ACF) and ion-exchanging systems accomplishes the task of efficient removal of suspended solids, micro/macrobiotic matter, dissolved ions, carbon dioxide, silica, hardness, etc., from the water source.

DM plant types depend on the quality of raw water to provide an appropriate treatment system so that the final product is acceptable for plant requirements and environmental restrictions, as stipulated by the local pollution control board/regulation.

Seawater is converted by the reverse osmosis (RO) process (one-stage or two-stage RO as per application requirements) and then further treatment of the same to DM quality.

8.2 Inlet Water Quality Requirement for DM Plant Operation

In Section 6 of this chapter, the route of raw water was shown (Figs. 2.46 and 2.47), starting from the source up to the DM plant inlet. The route includes the raw water reservoir, raw water pumps, and the pretreatment (PT) plant (chlorination, aeration, and clariflocculation through conventional gravity clarifiers/reactor clarifiers) as a standard. The pretreatment plant may also provide an ultrafiltration facility, which separates particles on the basis of their molecular sizes for minimizing the effect of raw water containing high levels of colloidal silica as well as high hardness levels.

The quality of the filtered water to be taken for the DM plant for further processing dictates the chemical dosing requirements of the plant. The plant is normally designed for continuous and simultaneous operation of all the streams if more than one stream running is envisaged. Pretreatment plant output is termed clarified or filtered water and supplied to the DM plant, including other important services such as the CCW system, CT make up, potable water plant, service water system, firefighting system, etc.

8.3 Requirement of DM Water for Other Systems

DM water in a thermal power plant is required for a few purposes such as FW, DM plant regeneration, CPU regeneration, and the HP/LP dosing system.

8.4 System Operation of DM Plant

DM plant operation is nothing but elimination of the unwanted mineral salts from water by exchanging of both strong and aggressive cations and anions with weak ones that are less harmful or more acceptable ions to the process metals, as depicted in Fig. 2.48. The inlet water to the plant is guided to pass through suitable filters and ion exchangers, also called polishers, that may be of several types, such as SAC, weak acid cation (WAC), strong base anion (SBA), weak base anion (SBA), and MB.

The basic functions of different ion exchangers may be described very briefly as:

- SAC resins neutralize the cations of strong bases and converts neutral salts into their corresponding acids.
- SBA resins can neutralize the anions of strong acids and convert neutral salts into their corresponding bases.
- WAC resins are able to neutralize strong bases.
- WBA resins are able to neutralize strong acids.

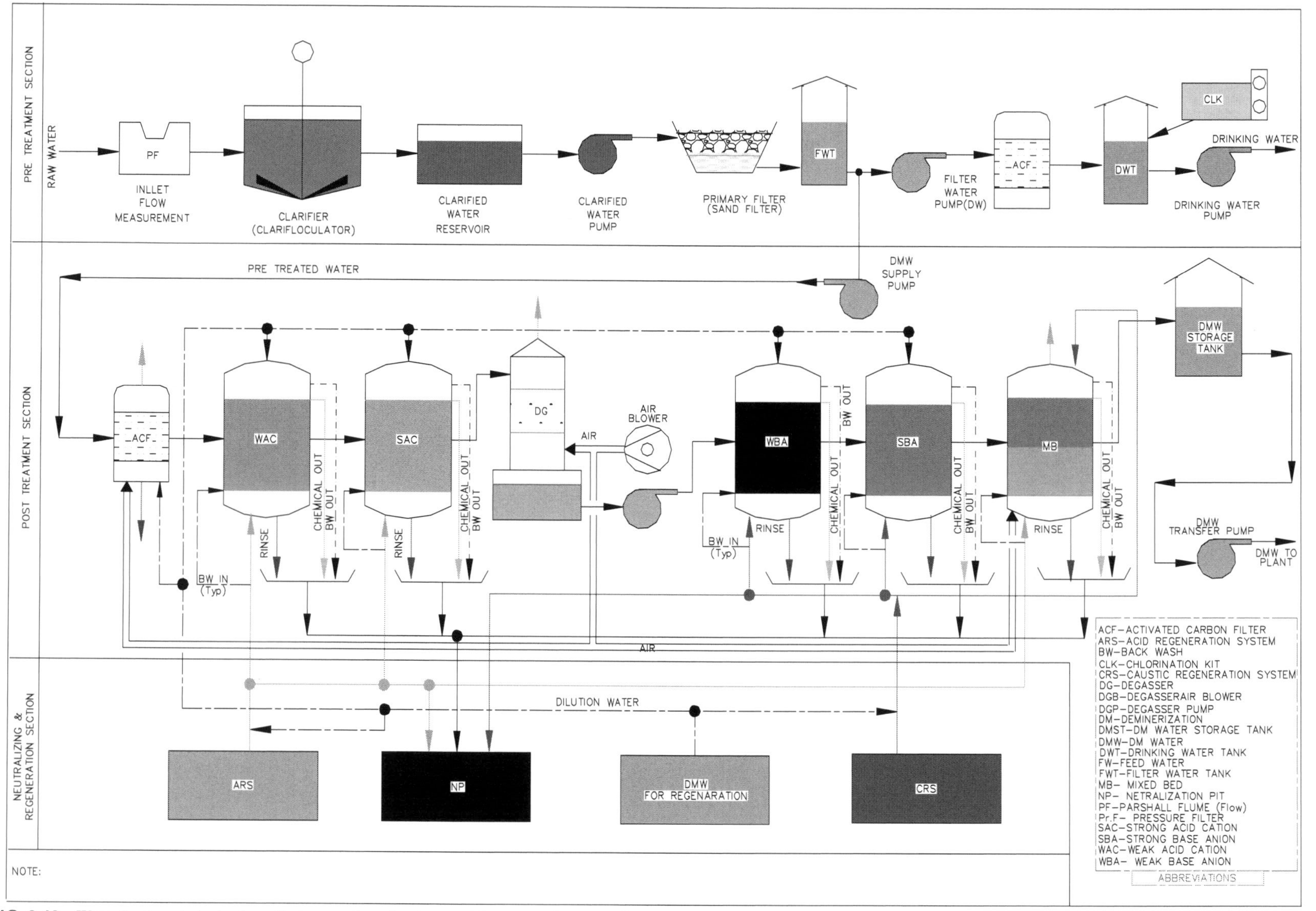

FIG. 2.49 Water treatment plant schematic (pre- and posttreatment with regeneration system).

These two resins (WAC and WBA) are used for dealkalization, partial demineralization, or full demineralization (in combination with strong resins).

When water passes first through a bed of SAC resin replacement of the metal cations, namely magnesium (Mg) and calcium (Ca), and releasing of the hydrogen ions would lower the solution pH making the outlet water acidic. The anions, namely sulfate (SO_4) and chloride (Cl), can then be removed with an SBA resin because the entering acidic water (H^+ ions) would react with the hydroxide ions (OH^-) of the SBA resins. This would cause the anions in the incoming water to be absorbed in the resin and (H^+ ions) with (OH^-) would react to produce water (H_2O), and the water becomes purer. In some plants, WBA resins are preferred over SBA resins because they require less regenerant chemical. The DM plant can take any type of inlet water with a TDS level of up to 500 ppm maximum.

8.4.1 Components and Their Function in DM Plant Operation

Normally, 2 × 100% or 3 × 50% DM streams are provided for the DM plant. The final product is transferred to DM storage tanks to cater to the intermittent requirements of various system MUs. Each stream consists of the following components, along with the common facility for the regeneration process: (i) pressure sand filter (PSF), (ii) ACF, (iii) ion exchange units (with regeneration arrangement), SAC and/or WAC, (iv) SBA and/or WBA, (v) MB, (vi) decarbonators and degassifiers, and (vii) drain neutralization system.

The ion exchange units containing the resin beds are normally vertical, rubber-lined, welded-plate steel construction and fabricated as per IS: 2825 or the equivalent. The vessels are rubber lined with a suitable material of appropriate thickness, for example, 4.5 mm to protect the metal of the vessels.

Brief discussions are incorporated below for each component with their dedicated functions.

8.4.1.1 Pressure Sand Filter

Pressure sand bed filters (PSF), often called rapid sand bed filters, consist of a pressure vessel that may normally be vertical or horizontal (in rare occasions) depending on the layout of the plant. The filter vessels are generally of welded mild steel construction lined with rubber/epoxy from inside. A minimum 50% free board is provided over the filtering bed depth to enable efficient backwash. Graded silica quartz sand and anthracite supported by layers of graded underbed consisting of pebbles and gravels are provided with the water inlet at the top. Incoming water is distributed uniformly throughout the cross-section of the filter to ensure that there are no preferred fluid paths where the sand may be washed away and the filter action is jeopardized. The bottom drainage system is kept to collect filtered water.

The sand grain size selection is important as smaller sand grains affect the increased surface area and consequently a higher decontamination at the water outlet which, on the other hand, demands extra pumping energy to drive the fluid through the bed. The optimum grain sizes are generally selected in the range 0.5–1.50 mm. The recommended bed depth is around 0.5–2.0 m, regardless of the application, of which the ratio of quartz sand and anthracite is around 7:50.

Raw water flows downward through the filter bed and the suspended matter is retained on the sand surface and between the sand grains immediately below the surface.

Rapid pressure sand bed filters are typically operated with a feed pressure of 1–4 kg/cm^2. The differential pressure (DP) across a clean sand bed is usually insignificantly low. The DP gradually builds up for a given flow rate as particulate solids are captured on the bed, which may not be uniform with depth. For obvious reasons, build ups would be more at a higher level with the concentration gradient decaying rapidly.

This type of filter captures particle sizes down to very small ones. In fact, there is no true cut-off size below which particles would not be arrested. Interestingly, the shape of the characteristic curve of efficiency versus filter particle size is a U-shape with high rates of particle capture for the smallest and largest particles, with a plunge in between for mid-sized particles.

When the pressure loss or flow is unacceptable, sensed by a pressure drop across the PSF of around 0.5 kg/cm^2, the filter is taken out of service and cleaning of the filter is effected by flow reversal or the bed is backwashed or pressure-washed to remove the accumulated particles. Backwashing of pressure filters is normally done once in 24 h online.

During backwash, the sand becomes fluidized with a volume expansion of about 30%, mixing and rubbing together with particulate solids to drive them off with the backwash fluid. The fluidizing flow requirement is typically 5–30 m^3/h/m^2 of filter bed area, depending on the depth of the bed, for a short period, that is, for a few minutes only. Filter backwash is taken to a common inlet chamber of raw water pumps. A periodic top-up of sand in the bed is required as the backwashing process causes sand loss.

To assist in cleaning the bed, the backwash operation is often preceded by air scouring through the underdrain system agitating the sand with a scrubbing action, which loosens the intercepted particles. The filter is now ready to be put back into service. For a 500 MW TPS, the typical backwashing flow rate would be between 25 and 30 m^3/h/m^2 and the air flow rate would be 50 m^3/h/m^2 of filter bed area.

8.4.1.2 Activated Carbon Filter

Carbon is used to absorb impurities and thousands of different chemicals and is considered the most powerful absorbent. Activated carbon has a slight electropositive charge deliberately added to it, which makes it attractive to different chemicals/impurities. When water passes through the positively charged carbon surface, the negative ions of the pollutants are drawn to the surface of the carbon granules.

8.4.1.2.1 Principle of Operation of ACF The two physical mechanisms through which ACF filtration works are: (i) adsorption for removal of organic compounds in water and (ii) catalytic reductions, which remove the residual disinfectants intentionally applied before such as chlorine and chloramines. Catalytic reduction takes advantage of the attraction between pollutants' negatively charged ions and positively charged activated carbon.

8.4.1.2.2 Factors Influencing Efficient Filtration The efficiency of ACF depends on three major factors: (i) more carbon would provide better results, (ii) more time of the water in contact with the filter by lowering the flow rate of water would lead to more absorption, and (iii) smaller size means greater surface area for a particular mass of carbon and the removal rate improves.

8.4.1.2.3 Types of Activated Carbon Filter There are two types of ACFs: granular activated carbon or powdered block carbon. The power plant application prefers the latter for its higher contaminant removal ratio, although both are effective.

8.4.1.2.4 Materials Used for Activated Carbon Filter Normally, carbon is used for this purpose and prepared from the raw materials after burning, and the char is utilized as carbon. Those are: (i) wood, (ii) coconut shells, and (iii) bituminous coal.

Of the above materials, carbon from coconut shells as used in ACF yields the best results against the higher cost.

8.4.1.2.5 Specification of Activated Carbon Filter The ACF is usually specified by these parameters: (i) upper ceiling of the flow velocity range with a typical value of 15 m/h, (ii) bed depth (typically1500 mm) and the minimum free board (around 75%) over the bed, (iii) the type of anticorrosive lining or coating applied to the vessel to prevent corrosion, (iv) cycle of backwash requirement using filtered water with a typical value of once every 24 h, and (v) minimum size of the particles up to which value the ACF could remove. The typical value ranges from 50 μm (least effective) down to 0.5 μm (most effective).

8.4.1.2.6 Advantages/Disadvantages of Using ACF
ACF normally eliminates/reduces volatile organic chemicals, including pesticides and herbicides. It also works well against chlorine, benzene radon, solvents, and trihalomethane compounds including turbidity, particulates, parasites (Giardia and Cryptosporidium, etc.). It is also very cost effective.

ACF requires very little maintenance; however, it is very important to ensure filter replacement on a regular basis to ensure proper filtration at all times. It is important to note that bad tastes and odors could return to the water, indicating the need for filter replacement.

Regarding the disadvantages of ACF, it in general is not very useful at removing dissolved inorganic contaminants or hardness/scale-causing metal compounds/salts, that is, heavy metals. However, high-quality ACF can be deployed for the removal of some iron, manganese, and hydrogen sulfide with moderate effectiveness.

ACF usually is not very effective regarding the removal of sediment/particulate material. It is always preferred that a sediment filter be installed before ACF so as to prolong the life by eliminating gross contaminants that could have clogged the activated carbon surface, thereby reducing the surface area vis-a-vis the efficiency of the system. The carbon block filters are generally better then granulated ACF at removing sediment.

8.4.1.2.7 Filter for Seawater Application Seawaters or certain other types of water can contain high pollutants (TDS more than 500 ppm) and/or carry antimony, arsenic, asbestos, barium, beryllium, cadmium, chromium, copper, fluoride, mercury, nickel, nitrates/nitrites, selenium, sulfate, thallium, and certain radio nuclides. This requires their removal through a reverse-osmosis water filter system or a distiller.

For production of demineralized water from seawater having a very high TDS in and around 40,000 ppm, a two-stage RO plant followed by a mixed-bed exchanger is employed. For some applications, a manganese greensand filter may also be useful.

8.4.1.3 Ion Exchange Units

The ion-exchange units constitute a very important part of the DM plant. The water from the ACF is forced to pass through ion-exchange units for exchanging detrimental cations and anions with weak/harmless ones. In most cases, the term "ion exchange" is used to denote the processes of decontamination and purification of aqueous and other ion-containing solutions with some special type of ion-exchanging materials. Ion exchange can, in general, be described as the exchange of ions between electrolytes and a complex material and a reversible process. An ion exchanger can be regenerated with desirable ions by washing with a solution having an excess presence of these ions.

The basic ion-exchange column consists of a resin bed of solid mass that is retained in the column with inlet and outlet screens, service and regeneration flow distributors, etc. Piping and valves are required to direct flow, and instrumentation is required to control all the sequence timing and other controls. The complete system is typically operated in a time cycle consisting of the following steps:

(i) Service (exhaustion)—A water solution containing ions is passed through the ion-exchange column or bed until the resins are exhausted.
(ii) Backwash—The bed is washed (generally with DM/filtered water) in the reverse direction of the service cycle in order to expand and resettle the resin bed.
(iii) Regeneration—Done by passing a concentrated (strong mineral acid/base) solution of the ion originally associated with it through the resin bed.
(iv) Rinse—Extra regenerant is removed from the exchanger by passing water through it.

Several types of ion exchangers are available such as functionalized porous or gel polymer resins, zeolites, clay, and soil humus. Two major categories of ion exchangers are available: cation exchangers (exchange cation/positive ions) or anion exchangers (exchange anion/negative ions).

There is also a special type of exchanger that has the unique characteristic of being able to exchange both cations and anions simultaneously; it is known as an amphoteric resin. It exploits the possibility of achieving useful radiochemical separations of both cation and anion exchange functions of the resin. As there are some limitations in this method, mixed-bed ion exchangers containing a mixture of anion and cation exchange resins are used with more efficient action.

Different Types of Ion Exchangers

The mass of solid ion exchange particles is either naturally available inorganic zeolites or synthetic organic resins, which are presently used widely because of the manufacturing advantage to make tailor-made characteristics to suit a specific application. Polyelectrolytes of high molecular weight are the best choice for an organic ion exchange resin. They exchange their mobile ions for ions of a similar charge from the medium in contact with them. Each resin has a distinct characteristic that roughly indicates the maximum quantity of exchanges that may take place per unit of the resin.

Both anion/cation resins are produced from the same basic organic polymers. They differ in the ionizable group attached to the hydrocarbon network. A particular functional group dictates resin's chemical characteristics and can be broadly classified as strong/weak cation/anion exchangers.

8.4.1.3.1 Cation Exchanger Resins For the water purification process in a DM plant, cation exchange resins exchange positively charged hydrogen mobile ions (H^+) for the positively charged ions (magnesium/calcium) from polluted water. For example, a resin with hydrogen ions would exchange itself for magnesium ions from solution with the following reaction:

$$2(R-SO_3)H + MgSO_4 \Leftrightarrow (R-SO_3)_2Mg + H_2SO_4$$

where (R-SO_3) is the resin; R indicates the organic portion of the resin (organic compound—any compound of carbon and another element or a radical); and SO_3 is the immobile portion of the ion active group (sulfonic acid group).

Two resin sites are needed for magnesium ions with a plus 2 valence (Mg^{2+}). Similarly, trivalent ferric ions would require three resin sites. The reaction as indicated above means the ion exchange process is reversible. The degree of the reaction to proceed to the right would depend on the resin's preference or selectivity. For example, the reaction shows that the resin group has more preference for magnesium ions compared to its preference for hydrogen ions, which is measured by the selectivity coefficient. The maximum service flow rate for a demineralizer range generally falls in the range from 15 to 25 m of resin. Flow rates of more than 25 $m^3/h/m^2$ of resin may cause increased sodium and silica leakage with waters.

Strong Acid Cation Exchanger Resins: The very name "strong acid resins" indicates that their chemical behavior is similar to that of a strong acid. The resins are highly ionized in both the acid (R-SO3H) and salt (R-SO3Na) forms. They can convert a metal salt to its corresponding acid, as per the following reaction:

$$2(R-SO_3)H + CaCl_2 \Leftrightarrow 2(R-SO_3)\,Ca + 2HCl$$

SAC resins derive their functionality from sulfonic acid groups (HSO_3^-). When used in demineralization, SAC resins remove nearly all raw water cations, replacing them with hydrogen ions by a few reactions as shown below:

$$\begin{array}{llcll} 2Na & 2HCO_3 & & 2Na & 2H_2CO_3 \\ Mg & SO_4 & +2(R-SO_3)H \Leftrightarrow 2(R-SO_3) & Mg & +\ H_2SO_4 \\ Ca & 2Cl & & Ca & 2HCl \end{array}$$

The hydrogen and sodium forms of strong acid resins are highly dissociated and the exchangeable Na^+ and H^+ are readily available for exchange over the entire pH range. Consequently, the exchange capacity of strong acid resins is independent of solution pH. These resins are used in the hydrogen form for complete deionization and used in the sodium form for water softening (calcium and magnesium removal).

The specification normally indicates as: High capacity, strongly acidic, sulfonated polystyrene, macroporous cation exchange resins in bed to be supplied for a strong cation-exchange unit.

Weak Acid Cation Exchanger Resins: WAC resin consists of the ionizable group is a carboxylic acid (COOH) as opposed to the sulfonic acid group (SO_3H) used in strong acid resins. These resins behave similarly to weak organic acids that are weakly dissociated. When operated in the hydrogen form, WAC resins remove cations that are associated with alkalinity, producing carbonic acid as shown:

$$\begin{matrix} 2Na \\ Mg \\ Ca \end{matrix}\; 2HCO_3 + 2(R-COO)H \Leftrightarrow 2(R-COO)\begin{matrix} 2Na \\ Mg \\ Ca \end{matrix} + 2H_2CO_3$$

These reactions are also reversible and permit the return of the exhausted WAC resin to the regenerated form. WAC resins are not able to remove all the cations in most water supplies. In some plants, to achieve full demineralization systems, SAC and WAC resins are utilized in a series combination (as shown in Fig. 2.49) for an economic solution by using more efficient WAC resin along with the full exchange capabilities of the SAC resin. They alone are used primarily for softening and dealkalization of high-hardness, high-alkalinity waters. The degree of dissociation of a weak acid resin is strongly influenced by the solution pH and therefore the resin capacity depends in part on solution pH.

Weak acid resins exhibit a much higher affinity for hydrogen ions than that of strong acid resins. This characteristic allows for regeneration to the hydrogen form with significantly less acid than is required for strong acid resins. Almost complete regeneration can be accomplished with stoichiometric amounts of acid with high regeneration efficiency compared to SAC resins, thereby reducing the waste acid and minimizing disposal problems.

8.4.1.3.2 Anion Exchanger Resins In the DM plant application, the anion exchange resins exchange negatively charged hydroxide mobile ions (OH^-) for the negatively charged ions (such as chloride, sulfate, nitrate, etc.) from the polluted water. For example, a resin with hydroxide ions available for exchange would exchange for chloride ions from solution. The reaction can be put like this:

$$(R-Q)(OH) + HCl \rightarrow (R-Q)Cl + H_2O$$

where (R-Q) is the anion resin. R indicates the organic portion of the resin. Q is the immobile portion of the ion active from the *quaternary ammonium functional group* (discussed below).

The reaction as indicated above means the ion exchange process is reversible. The degree of the reaction to proceed to the right would depend on the resin's preference or selectivity. This is measured by the selectivity coefficient.

Strong Base Anion Exchanger Resins: Like strong acid resins, strong base resins are highly ionized and can be used over the entire pH range. These resins are used in the hydroxide (OH) form for water deionization. They will react with anions in solution and can convert an acid solution to pure water.

SBA resins derive their functionality from quaternary ammonium functional groups.

Quaternary Ammonium Compound: This may be described generally as any of a group of compounds in which a central nitrogen atom is joined to four organic radicals and one acid radical. More specifically, it may be a type of ionic compound that can be regarded as derived from ammonium compounds (NH_4) by replacing the hydrogen atoms with organic groups.

$$R2 \quad \begin{matrix} R1 \\ N \\ R4 \end{matrix} \quad R3 \; +$$

(Quaternary ammonium cation)

The R (organic) groups may be the same or different alkyl or aryl groups. The R groups may be connected also.

Ammonium cations, also known as quats, are positively charged polyatomic ions of the basic structure NR_4^+. The quaternary ammonium cations are permanently charged, independent of the pH of the solution to which they belong. Quaternary ammonium salts/compounds are salts of quaternary ammonium cations with an anion.

Two types of quaternary ammonium groups, referred to as Type I and Type II, are used. Type I sites have three methyl groups:

$$R \;\begin{matrix} CH_3 \\ N \\ CH_3 \end{matrix}\; CH_3 \; +$$

In a Type II resin, one of the methyl groups is replaced with an ethanol group. The Type I resin is more stable and effective regarding removal of the weakly ionized acids compared to the Type II resin. Type II resins exhibit a more efficient resin as far as regeneration is concerned.

When in the hydroxide form, SBA resins remove all commonly encountered anions, as shown below:

$$\begin{matrix} 2H_2CO_3 \\ H_2SO_4 \\ 2HCl \end{matrix} + 2(R-Q)OH \Leftrightarrow 2(R-Q)\begin{matrix} 2HCO_3 \\ SO_4 \\ 2Cl \end{matrix} + 2H_2O$$

As these reactions are reversible, the regeneration of resin is done with a strong alkali, for example, caustic soda to get in the hydroxide form.

A typical specification for SBA is generally stated like this: Isoporous/macroporous, Type I in (OH) form, to be regenerated with caustic soda. The attrition loss of the anion resins will be limited to around 3%–4% per year for the first three years of operation.

Weak Base Anion Exchanger Resins: A weak base resin or a WBA resin derive its functionality from primary ($R\text{-}NH_2$), secondary (R-NHR′), or tertiary ($R\text{-}NR_2'$) amine groups. WBA resins readily remove sulfuric, nitric, and hydrochloric acids, as represented by the following reaction:

$$\begin{matrix} 2HNO_3 \\ H_2SO_4 \\ 2HCl \end{matrix} + 2(R-Q)OH \Leftrightarrow 2(R-Q)\begin{matrix} NO_3 \\ SO_4 \\ 2Cl \end{matrix} + 2H_2O$$

Regarding the degree of ionization, like WAC, WBA is also strongly influenced by pH. Consequently, weak base resins exhibit minimum exchange capacity above a pH of 7.0. These resins just absorb strong acids but are almost inactive toward the splitting purpose of salts.

Anion resin is much lighter than cation resin. Therefore, the backwash flow rates for anion exchange resins are much lower than those for cation resins.

8.4.1.3.3 Mixed Bed Exchangers The MB exchanger is a very important item used in both DM plants and. The difference is that in the CPU, the MB exchanger is normally used as the only exchanger whereas in the DM plant, it is used after cation and anion exchangers in series. The regeneration process is also different in those two systems. Here, both cation and anion resins are mixed and used together in a single vessel, but the ion-exchange process is repeated many times when water is flowing through the MB resin. This repeated ion-exchange process causes the water quality to develop a very high purity; hence, the process is termed "polishing."

More sophisticated equipment with mixed-bed systems with a counterflow arrangement is employed to yield purer water than conventional cation-anion demineralizers; obviously, this is a costlier system. It also calls for more intense operator attention than standard systems when a complicated regeneration sequence is adopted.

8.4.1.3.4 Regeneration of Resins Ion-exchange reactions are stoichiometric, which means they are predictable based on chemical relationships as well as *reversible*. Here, the ion exchangers exchange one ion for another, hold it temporarily, and then release it to a regenerant solution (Figs. 2.48 and 2.49).

After a prolonged service operation, the resins become exhausted after any of the three methods adopted: time lapsed (timer used), DP across exchanger more than set value and concentration of chemical at the bed less than set value.

This step is followed by backwash to expand and resettle the resin bed. The backwash step enables the removal of suspended solids accumulated on the bed during the service step and the elimination of chemicals that may have formed during this period. The backwash flow fluidizes the bed, releases the trapped particles, and reorients the resin particles according to size.

These ions are finally removed with the effluent during the regeneration process step, where a strong solution of regenerant containing the ions originally attached (H^+ or OH^- ions) to the resin is passed over the bed. After this step, the rinse step follows by passing water through it to remove excess regenerant from the exchanger.

Cation Regeneration: Strong acid resins may have some preference for Mg or Ca over hydrogen. Despite this preference, the resin can be converted back to the hydrogen form by contact with a concentrated solution of sulfuric acid (H_2SO_4):

$$(R\text{-}SO_3)_2Ca + H_2SO_4 \Leftrightarrow 2(R\text{-}SO_3H) + CaSO_4$$

This step is known as regeneration. In general terms, the higher the preference a resin exhibits for a particular ion, the greater the exchange efficiency in terms of resin capacity for removal of that ion from solution. However, that will cause increased consumption of chemicals for regeneration.

The regeneration ensures the resin is converted back to the hydrogen form (regenerated) by contact with a strong acid solution, or the resin can be rearranged to the sodium form with a sodium chloride solution. Hydrochloric acid (HCl) may also be used for regeneration. Cation regeneration pumps normally draw suction from degassed water storage tanks.

The total concentration of the strong acids in the cation effluent is a measure of the effectiveness of the resin in use. It is expressed in terms of free mineral acidity (FMA). During the normal service period, the FMA content is stable for most of the time. The FMA is usually slightly lower than the theoretical value because a small amount of sodium leakage takes place through the cation exchanger. The amount of sodium leakage depends on the regenerant level, the flow rate, and the presence of sodium compared to other cations in the inlet water. Sodium leakage increases as the ratio of sodium to total cations increases, as regenerant level decreases, and as flow rate increases.

The FMA in the effluent drops sharply when a cation exchange unit nears exhaustion, meaning immediate withdrawal of the service run of the exchanger and regeneration of the resin with an acid solution should begin. Sulfuric acid is normally used due to its availability/cost advantage, though improper use of sulfuric acid can cause irreversible fouling of the resin with calcium sulfate.

Sulfuric acid is usually applied at a high flow rate and an initial concentration of 2% or less. The acid concentration is gradually increased to 6%–8% to complete regeneration.

Some installations also use hydrochloric acid for regeneration, necessitating the use of special construction materials in the regenerant system.

Counterflow Cation Exchanger Regeneration: After regeneration, some ions, mostly sodium ions, still remain unaffected in the bottom of the resin bed, though the upper portion of the bed has been exposed to fresh regenerant and hence is highly regenerated. During service operation, cations are readily exchanged in the upper portion of the bed for the first few periods and then move down through the resin as the bed starts becoming exhausted. Sodium ions that remain in the bed after regeneration diffuse into the service water before leaving the exchanger. This sodium leakage entering the anion exchange unit produces caustic, raising the pH and conductivity of the effluent.

In a counterflow method, the regenerant flows in the opposite direction of the service flow. For this, the most highly regenerated resin is located where the service water leaves the exchanger. The highly regenerated resin removes the low level of contaminants that has escaped removal in the top of the bed. This results in higher water purity than cocurrent designs can produce.

The same philosophy can be adopted in the case of the MB exchanger.

Anion Regeneration: The anion exchange resins are generally regenerated with sodium hydroxide (48% w/v rayon grade in lye form as per the (IS: 252) solution of suitable strength.

The need for regeneration is sensed by the output of continuous measurement of conductivity and monitoring the silica analyzers' output reading of anion effluent water quality. In some instances, conductivity probes are placed in the resin bed above the underdrain collectors to detect resin exhaustion before silica breakthrough into the treated water occurs.

The equipment configuration and sequence of operation is just similar to the regeneration of cation power water pumps to be provided for anion units shall draw water from DM water storage tanks.

The water used for each step of anion resin regeneration should be free from hardness to prevent precipitation of hardness salts in the alkaline anion resin bed.

Continuous conductivity and silica analyzers are mostly used to monitor anion effluent or outlet water quality and detect the need for regeneration. In some instances, conductivity probes are placed in the resin bed above the underdrain collectors to detect resin exhaustion before silica breakthrough into the treated water occurs.

It may appear that complete removal of the cations and anions is possible, but in reality some leakage takes place. That is mostly sodium leakage from the cation exchanger, which is converted to sodium hydroxide in the anion exchanger; this makes the effluent pH of the anion outlet slightly alkaline. The caustic produced in the anions causes a small amount of silica leakage. The extent of silica leakage from the anions depends on the quality of the inlet water and regenerant dosage being used.

Demineralization using strong anion resins removes silica and other dissolved solids. Effluent silica and conductivity are important parameters to dictate the service run period of the system.

When silica breakthrough occurs near the end of the service run, the effluent silica level increases sharply with the momentary decrease of the conductivity level. This is because of exhaustion of hydroxide ions and conversion of sodium into sodium silicate (low conductivity) rather than sodium hydroxide. As anion resin nears complete exhaustion, the more conductive mineral ions break through, causing a subsequent increase in conductivity.

On detection of the end of a service run, the unit must immediately change over to the next step of operation. If the plant is allowed to remain in service mode after the breakpoint, the silica level in the effluent can increase beyond that of the influent water, due to the concentrating of silica that already occurred inside the anion exchanger during the service run.

Sodium hydroxide (4%) or caustic solution is used for regeneration of SBA exchangers. It is usually heated to 50°C or to the temperature specified by the resin manufacturer to improve upon the better removal of silica from the resin bed. Silica removal is also enhanced by a resin bed preheat step before the introduction of warm caustic.

Mixed Bed Regeneration: During MB regeneration, the resin is separated into distinct cation and anion fractions. The procedure is somewhat different from that of individual cation and anion regeneration. Here, the two resins are first separated by backwashing, which causes the lighter anion resins to settle on top of the cation resin. Regenerant acid is introduced through the bottom part of the vessel and caustic is introduced through the top part of the vessel. Two different regenerant streams coming from opposite sides meet at the interface between the cation and anion resin and are made to discharge through a collector near the resin meeting plane. After regeneration and the rinse step, air/water is blown for intermixing both resins, followed by rinsing again. The unit is then ready for service.

8.4.1.4 Decarbonators and Degassers

This part of the DM plant is included mainly as an economic measure. Decarbonators and degassers are beneficial to the next part of the system where they are installed. The effluent coming from the cation exchanger contains carbonic acid as a product of reaction, which again needs to be neutralized by the anion exchanger, meaning an excess amount of caustic for regeneration.

Outlet water from a cation exchanger is split into small droplets by sprays and trays in a decarbonator. The water is then allowed to flow through a stream of oppositely flowing air. By this, carbonic acid in the cation effluent dissociates into water and CO_2, which is separated from water by the forced draft or blower type air stream. This reduces the burden of the anion exchanger. CO_2 is removed to acceptable limits such as 10–20 ppm. The minor problem in this method is that the water effluent from a decarbonator is saturated with oxygen.

In a vacuum degasser, which is more effective in removing CO_2, the process is slightly different. The water droplets are allowed to pass through a packed column operated under vacuum. Carbon dioxide is removed from the water due to its decreased partial pressure in a vacuum. The removal of CO_2 can be attained up to the level of less than 2 ppm. The additional advantage here is that most of the oxygen is also removed from the water. The capital and running costs toward a vacuum degasser are greater than those of other decarbonators.

8.4.1.5 Drain Neutralization System

The drain neutralization system is provided to neutralize the waste drains from all the above-mentioned vessels handling chemicals, including the acid/caustic measuring tanks. The neutralizing pit accommodates neutralization of all waste by dosing a suitable additive from dedicated acid/caustic measuring tanks. The capacity is normally estimated for storing all waste water from one cation, anion, and mixed-bed unit with around 50% margin with RCC construction in a twin compartment. The mixing is done by recirculation of wastes and dosing materials.

The neutralized waste is transferred by separate pumps to the effluent treatment plant.

8.4.2 Advantages and Limitations

High-quality water can be produced by several other methods, including distillation. The cost of a DM plant is comparatively much less than any other system.

As the system utilizes ion exchanging, the inlet water quality requirements are to be maintained for efficient operation of the plant. Those include filtered water containing minimum iron and chlorine (considered as resin foulants and degrading agents) and fewer organic materials (considered as a threat as they attack and foul anion resins). Colloidal or nonreactive forms of silica are not removed using a DM plant. Even if the hot alkaline boiler water dissolves the colloidal material, it forms simple silicates that may enter the boiler in a soluble form. As such, they can form deposits on tube surfaces and volatilize into the steam.

8.5 Expected Quality of Various Components

All the components are expected to provide specific contributions with the following guidelines so that the performance of the components can be compared for necessary action. Following are the expected values (typical) of the measuring analyzers provided at the outlet of each vessel (Table 2.3).

8.6 Controls and Instrumentation

8.6.1 Controls

A DM plant demands a totally sequential operation covering total interlock, protection, monitoring, alarm, data logging, fault analysis, etc., to ensure operability, maintainability, and reliability. The hardware may be a programmable logic controller (PLC)-based or a

microprocessor-based distributed control system (DCS) with remote I/O cabinets wherever required, depending upon distance/location and the number of streams/inputs.

8.6.2 Instrumentation

The necessary instrumentation, namely flow measurements, analyzers such as pH, conductivity, silica, traces of CO_2, level/limit switches, gauges, etc., is included in the complete package of the DM plant.

TABLE 2.3 Expected Parameter Values at Vessel Outlet

Vessel Outlet	PSF	ACF	SAC	Degassifier	SBA	MB
Reactive silica					<0.1 ppm as SiO_2	<0.01 ppm as SiO_2
Conductivity					<5 micromho/cm at 25°C	<0.1 micromho/cm at 25°C
pH					>7.5	6.8–7.2
Sodium			<1 ppm as $CaCO_3$			
CO_2				<5 ppm		
Free chlorine		Nil				

9 COAL HANDLING: BASIC SYSTEM FUNCTION AND DESCRIPTION

A coal-handling plant (CHP) in a thermal power plant is a front-end facility whose main function is to receive coal and transfer it to the coal bunker with a proper coal size acceptable to mills.

Coal source, quality/size, transportation mode, main plant capacity, topography, etc., are a few factors that dictate the CHP design concept.

The coal supplied directly from the mine to the different destinations such as coal preparation plant/coal washeries, SG plants, etc., is called *run-of-mine* (ROM) coal and contains rocks, contaminant minerals contributed during the mining process, for example, machine parts, used consumables, and parts of ground excavation tools (shovels, etc.). ROM coal also includes a large variability of moisture and particle sizes depending on the types and location of the mines.

9.1 Influencing Factors of CHP Concept Design (Fig. 2.50)

9.1.1 Coal Source, Quality, and Size

The source and quality of coal are almost synonymous, as the quality of coal primarily depends on the source. The source may vary in different ways, such as mined coal from the adjacent area or an indigenous source. The second one is imported coal, which, of course, is costlier but expected to be of better quality.

The coal quality is for assessing different hardware and the elements of the CHP mainly depend on gross calorific value (GCV), the Hardgrove Grindability Index (HGI), and the moisture content.

Coal quality can be assessed by certain other parameters such as the amount of impurities that come with the coal and determine the coal GRADE. Those things are complex materials, namely ash and elements such as sulfur, sodium, and phosphorus, products that are either not good for better heat transfer (and hence low efficiency) or are detrimental to the public health when they emerge from the chimney. Chlorides, nitrates, sulfates, etc., cause corrosion in the boilers.

The general accepted quality/grade may be grossly expressed by heat available from the coal, for example:

Grade	UHV[a] Range (kcal/kg)
A	>6200
B	>5600 to <6200
C	>4940 to <5600
D	>4200 to <4940
E	>3360 to <4200
F	>2400 to <3360

[a]*Useful heat value.*

According to the major coal grades, as far as the type of availability of coal is concerned, lignite (brown coal) represents the bottom-most position against anthracite at the top

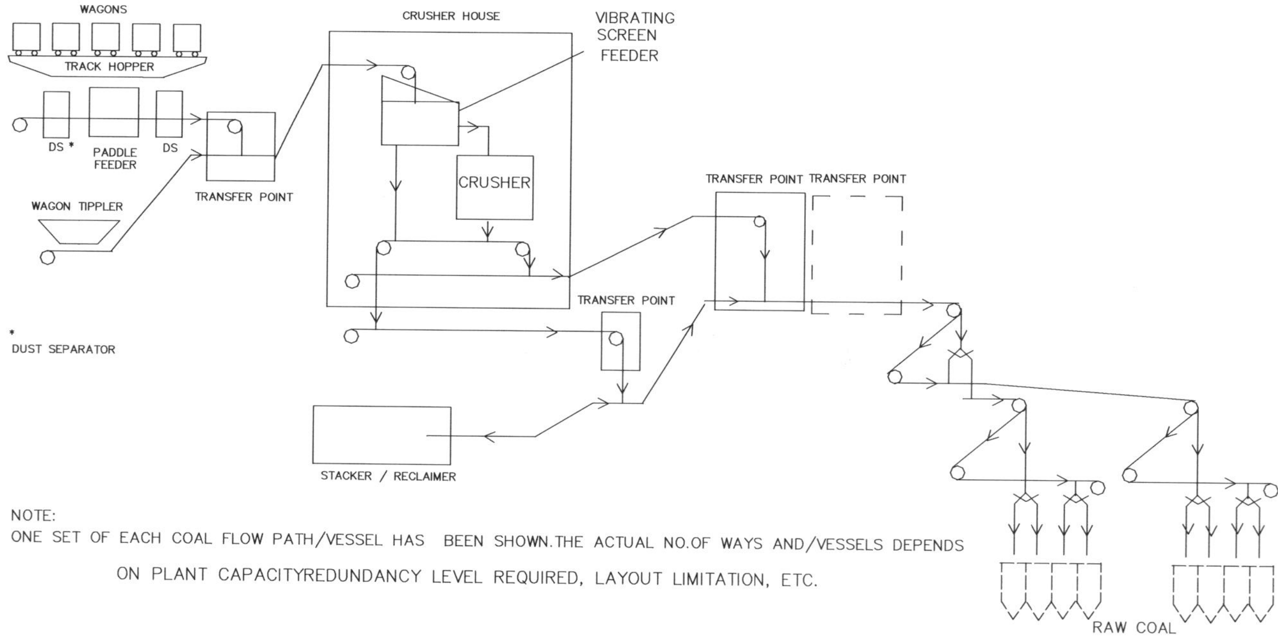

FIG. 2.50 Basic schematic/flow diagram of coal handling plant.

position. The subbituminous and bituminous coal are ranked in between. Each rank may be further subdivided according to the closed range UHV.

Most of the minerals and traces of different elements converted into ash after burning coal are nothing but oxides of different materials, for example, oxides of iron, magnesium, etc. Those minerals that do not constitute ashes break down into gaseous compounds and go out with the flue gas, including, for example, sulfur oxides commonly referred to as SOx.

The mineral content present in coal dictates the type of ash produced after burning. The melting point or the fusion temperature of ash influences the design of the furnace and other parts of the boilers as well.

The following types of test analyses are done to assess the type of coal available in the thermal plant for power generation (Table 2.4):

Blending of coals: Nowadays, many operation people in power plants in India accept the concept of utilizing the facilities for blending of indigenous and imported coal in view of the shortage of Indian coal. Sometimes, environmental reasons may also necessitate coal blending. Blending can be done in many ways. One method is to lay indigenous and imported coal in alternate layers on the belts while conveying the coal to bunkers; they then get mixed while falling into the bunkers. The other method is to stock indigenous and imported coal in layers in the stockyard.

The CHP broadly covers the subsystems such as unloading, screening, stacking, crushing, storing, reclaiming, and transfer to bunkers.

9.2 Unloading

Normally, the coal received at the thermal power station is unloaded by means of a wagon tippler or track hopper or by a combination of both, depending on the type of coal rakes used for transportation of coal to a particular station. There are many variations of transporting wagons and the track hopper at the receiving end has to be suitable with the particular wagon type, as per prior discussion with the railway authorities while designing the unloading system. Either a wagon tippler or a track hopper is well performing for a properly designed unloading system, as both the systems are proven and time-tested. Some factors influence the selection, including past experience, preference, groundwater level, cost of underground installation, economic considerations, etc. Technically, both systems are workable. The above-mentioned unloading systems are briefly stated in the following part.

TABLE 2.4 Analysis and Assessment of Coal Quality

1	Proximate Analysis (in %)	Ultimate Analysis (in %)
2	Moisture 2.6	Nitrogen: 1.5
3	Hydrogen 5.0.	Ash: 15.11
4	Volatile matter 32.2	Carbon: 67.6
5	Fixed carbon 50	Oxygen: 9.6
6		Sulfur: 1.2
7		Calorific value: 6700 kcal/kg
8		Air-dried loss: 0.3(%)
9		Forms of sulfur (%): Sulfate 0.01 Pyritic 0.66 Organic 0.55

9.2.1 Track Hopper Unloading System

Here, the coal is received through wagon rakes with a side or bottom opening arrangement. Previously, wagons with manually operated side opening panels were normally provided. Now, improved wagons with pneumatically or hydraulically operated discharge arrangements from the bottom are possible and are becoming more popular. If a power station is situated near the mine, then some clients prefer the bottom-discharge wagon with a merry-go-round system and for a track unloading hopper. The received coal is then unloaded in an underground RCC track hopper. Paddle feeders are provided under the track hopper to scoop the coal and feed it onto underground relevant conveyors. A belt-weighing arrangement is provided on these conveyors for measurement of the coal flow rate.

The track unloading hopper system by merry-go-round railways was introduced at a later date compared to its counterpart, but still found many customers, to its credit. Many of the existing plants still have track hopper unloading systems.

9.2.2 Wagon Tippler Unloading System

Railways in many countries do not favor supplying coal by bottom discharge wagons, but instead choose wagon tipplers by default. Coal received from wagons is unloaded in underground RCC hoppers by means of rotary wagon tipplers. Side arm chargers are employed for the placement of wagons on the tippler table and removal of the empty wagon from the tippler table after tippling.

Apron feeders are employed under each wagon tippler for extracting coal from the wagon tippler hopper and feeding it onto underground reclamation conveyors. Belt weigh scales are provided on these conveyors for measurement of the coal flow rate. Provisions are kept for shunting locomotives for placing the rakes in position for the side arm charger to handle and begin unloading operation.

9.3 Coal Crushing

It has been stressed that the coal supplier should initially crush the materials to a maximum size such as 300 mm, but they may be something else depending on the agreement or coal tie up. To circumvent the situation, the CHP keeps a crushing provision so that coal bunkers receive the materials at a maximum size of about 20–25 mm.

The unloaded coal in the hoppers is transferred to the crusher house through belt conveyors with different stop-overs in between such as the penthouse, transfer points, etc., depending on the CHP layout.

Suspended magnets for the removal of tramp iron pieces and metal detectors for identifying nonferrous materials are provided at strategic points to intercept unacceptable materials before they reach the crushers. There may be arrangements for manual stone picking from the conveyors, as suitable. Crushed coal is then sent directly to the stockyard.

A coal-sampling unit is provided for uncrushed coal. Online coal analyzers are also available, but they are a costly item. Screens (vibrating grizzly or rollers) are provided at the upstream of the crushers to sort out the smaller sizes as stipulated, and larger pieces are guided to the crushers.

Appropriate types of isolation gates, for example, rod or rack and pinion gates, are provided before screens to isolate one set of crushers/screens to carry on maintenance work without affecting the operation of other streams.

Vibrating grizzly or roller screens are provided upstream of the crushers for less than 25 (typical) mm coal particles bypass the crusher and coal size more than 25 mm then fed to the crushers. The crushed coal is either fed to the coal bunkers of the boilers or discharged to the coal stockyard through conveyors and transfer points, if any.

Coal crusher classification: In general, coal crushers are categorized into three types, as stated below:

(a) *Primary coal crusher*: These crushers are used for bigger coal sizes and have different types such as the coal jaw crusher, the coal hammer crusher, and the ring granulator.

(b) *Secondary coal crusher*: These are used when the supplied coal is big enough to be handled by a single crusher. The primary crusher converts the feed size to be acceptable to the secondary crusher.

9.3.1 Coal Jaw Crusher

This is used for crushing and breaking large coal in the first step of coal crushing plant applied most widely in coal crushing industry. Jaw crushers are designed for primary crushing of hard rocks without rubbing and with minimum dust. Jaw crushers may be utilized for materials such as coal, granite, basalt, river gravel, bauxite, marble, slag, hard rock, limestone, iron ore, magazine ore, etc., within a pressure resistance strength of 200 MPa. Jaw crushers are characterized for different features such as a simple structure, easy maintenance, low cost, high crushing ratio, and high resistance to friction/abrasion/compression with a longer operating lifespan.

9.3.1.1 Jaw Crusher Operating Principles

Fixed and movable jaw plates are the two main components. A motor-driven eccentric shaft through suitable hardware makes the movable jaw plate travel in a regulated track and hit the materials in the crushing chamber comprising a fixed-jaw plate to assert compression force for crushing.

9.3.2 Coal Hammer Crusher

A coal hammer crusher is developed for materials having pressure-resistance strength over 100 Mpa and humidity not more than 15%. A hammer crusher is suitable for mid-hard and light erosive materials such as coal, salt, chalk, gypsum, limestone, etc.

9.3.2.1 Coal Hammer Crusher Operating Principles

Hammer mills are primarily steel drums that contain a vertical or horizontal cross-shaped rotor mounted with pivoting hammers that can freely swing on either end of the cross. While the material is fed into the feed hopper, the rotor placed inside the drum is spun at a high speed. Thereafter, the hammers on the ends of the rotating cross thrust the material, thereby shredding and expelling it through the screens fitted in the drum.

9.3.3 Ring Granulator

Ring granulators are used for crushing coal to a size acceptable to the mills for conversion to powdered coal. A ring granulator prevents both the oversizing and undersizing of coal, helping the quality of the finished product and improving the workability. Due to its strong construction, a ring granulator is capable of crushing coal, limestone, lignite, or gypsum as well as other medium-to-hard friable items. Ring granulators are rugged, dependable, and specially designed for continuous high capacity crushing of materials. Ring granulators are available with operating capacities from 40 to 1800 tons/h or even more with a feed size up to 500 mm. Adjustment of clearance between the cage and the path of the rings takes care of the product gradation as well as compensates for wear and tear of the machine parts for maintaining product size. The unique combination of impact and rolling compression makes the crushing action yield a higher output with a lower noise level and power consumption. Here, the product is almost of uniform granular size with n adjustable range of less than 20–25 mm. As the crushing action involves minimum attrition, thereby minimum fines are produced with improving efficiency.

9.3.3.1 Ring Granulator Operating Principle

A ring granulator works on n operating principle similar to a hammer mill, but the hammers are replaced with rolling rings. The ring granulator compresses material by impact in association with shear and compression force. It comprises a screen plate/cage bar steel box with an opening in the top cover for feeding. The power-driven horizontal main shaft passes from frame side to frame side, supporting a number of circular discs fixed at regular intervals across its length within the frame. There are quite a few bars running parallel to the main shaft and around the periphery that pass through these discs near their outer edges. The bars are uniformly located about the center of the main rotating shaft. There are a series of rings in between the two consecutive disc spaces, mounted on each bar. They are free to rotate on the bars irrespective of the main shaft rotation. The entire cage assembly, located below the rotor assembly, can be set at a desired close proximity to the rings by screw jack mechanism adjustable from outside the crusher frame. The rotor assembly consisting of the shaft, discs, rings, etc., is fixed as far as the main shaft center line is concerned. This main shaft carries in roller bearings from the box sides. The movable cage frame arrangement is provided so as to set its inner radius marginally larger than that of the ring running periphery. When coal is fed from the top, the rings also rotate along with the shaft and around their own center line along the bars, which drags coal lumps and crushes them to the desired size. After the coal has been crushed by the coal crusher, a vibrating screen grades the coal by size and the coal is then transported via belt conveyor. In this process, a dewatering screen is optional to remove water from the product.

9.4 Coal Stacker and Reclaimer at Stockyard

Crushed coal is sent to the stockyard when coal bunkers are full. Stacking/reclaiming of coal is done by a bucket wheel stacker-cum-reclaimer moving on rails. The stacker-reclaimer stacks coal on either side of the yard conveyor. During stacking operations, the coal is fed from yard conveyors on a boom conveyor.

Coal is reclaimed by the stacker-reclaimer and fed to the coal bunkers when direct unloading from rakes is not done. During the reclaim operation, a boom conveyor discharges coal on the reversible yard conveyor for feeding coal to bunkers through a combination of conveyors and transfer points. There may be an emergency reclaim hopper (ERH) to reclaim coal by dozers when a stacker reclaimer is not in operation. The ERH may also be used for coal blending if provision permits. The coal stockpile storage capacity depends on the location of the plant and the coal source.

Magnetic separators/detectors at appropriate locations are provided again for removal of metals (ferrous/nonferrous) and then coal sampling unit for checking for required size of crushed coal before transferring to bunkers being now reclaimed.

9.5 Dust Control System and Ventilation system

Dust control systems have now become essential requirements of CHP for suppressing the dust emissions at different locations, that is, transfer points, feeders, crushers, etc. Dust control is achieved by a dedicated dust suppression and extraction system. There are two widely accepted methods: dry fog or plain water dust suppression systems.

(i) *Dry Fog Dust Suppression System* is normally provided for the wagon tippler/hopper complex, crusher receipt and discharge points, all transfer points, and the boom belt discharge of stacker-reclaimer.
(ii) *Plain Water Dust Suppression System* normally caters to the track hopper and wagon tippler area, as well as the stock pile area (with a swiveling nozzle system; this is also very much required from the fire point of view).

Dust suppression for all the working areas/locations of the CHP may be available from the mechanical ventilation system/pressurized ventilation system that suits the plant requirements.

The *pressurized ventilation* system is capable of pressurizing the inside of some closed areas slightly above atmospheric pressure to prevent dust ingress from outside.

The control rooms, office rooms and remote terminal unit (RTU) room, etc., are to be provided with an *air-conditioning* system while the MCC/switchgear rooms of the coal handling plant may be provided with a pressurized ventilation system. The rest of the areas may be provided with mechanical ventilation.

A dust extraction system is generally provided for locations, for example, the bunker floor, the crusher house, etc. The dust extraction system may be the Venturi scrubber system. One independent dust extraction system for each stream shall be provided.

9.6 Other Important Accessories

A few other important accessories apart from those discussed above do take part in the CHP process and some of them are briefly described below.

(i) *Apron Feeders*: Apron feeders are of robust construction and specially designed for handling ROM coal as normally supplied. An additional feature is to have the capability to ward off the choking problem of wet coal during the rainy season. A dribble conveyor is generally provided below the apron feeder for subsequent clean-up activities.

(ii) *Vibrating Grizzly Screens*: This type of screen is capable of segregating the required size of coal along with coal dust and any other contaminations such as muck and mud carried through the coal during the rainy season. Below each screen, separate hoppers/chutes are provided through which the segregated materials are directly fed to the belt conveyors/feeders.

(i) *Vibrating feeders*: Vibrating feeders are generally provided to feed materials into the primary crusher homogeneously and continuously. At the same time, it acts as a screen, thus reducing the crusher load. A vibrating feeder can continuously and evenly transfer materials to the crusher.

(ii) *Conveyors* (with motors, gear boxes, couplings, and pulleys) are like the arteries in a living body and transfer the coal throughout the wide expanse of the power plant.

Yard and boom conveyors are special types of conveyors used for the stacker-reclaimer subsystem.

(iii) *Side arm charger*: Side arm chargers are used for indexing forward the rake of loaded wagons, placing decoupled wagons on the tippler table, and withdrawing from the empty wagons.

(iv) *Paddle feeder*: Paddle feeders are provided to scoop out coal at the guaranteed capacity in both forward and reverse motions.

(v) *Electric hoists/manual hoists* are provided for handling the equipment during maintenance.

i. *Coal sampling unit*: The coal sampling unit is included to provide suitable "samples" conforming to ASTM-D-2234 or any other applicable standards for laboratory coal analysis.

ii. *Magnetic separator and suspended magnet*: The magnets are installed normally before and after the crusher. The strength of the magnet at the specified mounting height is typically 1000 gauss at the center point of the belt. Mounting heights may vary depending on the type of location, with a typical example at 450 mm before the crusher and 400 mm after the crusher. This is measured between the top of the conveyor belt or the bottom of the falling material trajectory (already discharged from the head pulley) and the surface of the magnetic separator belt.

iii. *Metal detectors* are installed to detect different materials such as aluminum, brass, copper, stainless steel, manganese steel, bars, scraps, etc.

Chutes and Hoppers: Chutes are in general used to direct the flow of bulk solids, for example, from one conveyor belt to another in the same or other direction. The minimum valley angle of chutes shall be 60 degrees from horizontal. In case of a vertical chute, the valley angle would be more than 80 degrees. The hoppers are, on the other hand, used for intermediate storage of bulk solids and the flow of material is controlled by some type of feeder. The materials flow from the hopper to the feeder at a continuous and regulated rate. Hoppers and chutes are made of good-quality abrasive-resistant steel such as LSLAS07 or an equivalent material with more than minimum thickness, as specified.

9.7 Instrumentation and Control

The sequential logic control system for the CHP is normally accomplished by a PLC or a microprocessor-based DCS to take care of the huge numbers of inputs/outputs (I/Os) and associated tasks such as interlock and protection, monitoring, alarm and data logging, etc.

The control system enables the selection of any coal flow path beginning from the receiving section to the final destination, along with associated equipment such as wagon tipplers/track hoppers, crushers, screens, feeders, etc., from the CHP control room.

A separate PLC-based system may be provided for the stacker-reclaimer, the wagon tipplers, and the dust extraction/suppression subsystems. It can then be hooked up to the main CHP control system via a suitable link. Local start/stop push button stations and deinterlock switches are mounted near each piece of equipment for start/stop during test/maintenance of the system.

A communication facility between the CHP control room and all the important strategic points such as the wagon tippler/track hopper control room, the stacker-reclaimer control cabin, the bunker floor, the main unit control room (UCR), etc.

The following minimum number of field/process switches along with measuring instruments are provided:

(i) Belt sway, pull chord, chute blockage switches.

(ii) Zero speed switch at the tail end for each conveyor.

(iii) Level switches in dust suppression system water tanks.

(iv) RTDs for conveyor drive/crusher motors for HT motors.

(v) Electronic static weigh measurement for the quantity of coal and the wagon wise on the wagon tippler table before and after tippling.

(vi) Vibration monitoring system for crushers/drives.

(vii) The electronic belt weighing system for the following locations:

- **(a)** Unloading conveyors to determine coal receipt rates.
- **(b)** Boom conveyor of the stacker-reclaimer to know the coal reclaim rates.
- **(c)** Belt conveyors to know the quantity of coal reclaimed/blended.
- **(d)** Conveyors feeding bunkers to know the fuel feed rates to coal bunkers.

9.8 Brief Details of Conveyor Safety Switches

9.8.1 Pull Chord Switches

These are rope-operated emergency switches mounted at equidistant (normally 10 m) places on either side along the conveyor that are used to stop the conveyor in case of emergency. When pulled, the switch remains in the latched position unless it is manually reset to avoid accidental restart. The rope is tied to the end rings of the lever and is carried opposite to each other. When the rope is pulled from any side, the switch is operated. Normally, the lever is at 45 degrees on either side, so when pulled snap acting stay put switch operates with moderate torque. Normally, these are available in an IP 55/65 enclosure (Flameproof/ Class II enclosures). A contact rating of 10/15 A @500 V AC in change over contact.

9.8.2 Belt Sway Switch

To protect the conveyor from damage due to misalignment, these switches are used. Belt sway switches can be mounted at equidistant (normally 25 m) places on either side along the conveyor. In case of excess sway, the edges pushes the operating lever (45 degrees on either side when angle changes to, say, 30 degrees) to actuate the switch. These may be the self-resettable type. Normally, these are available in an IP 55/65 enclosure (Flameproof /Class II enclosures). Contact rating of 10/15 A.

9.8.3 Zero Speed Switch

Monitoring of the speed is very essential in automation systems. In speed-control modulating loops, continuous speed monitoring is required, whereas in the CHP, a zero speed switch is required for conveyor operation. Also, under/over speed switches can be utilized. Noncontact = sensors are used for speed monitoring. These are mounted in the vicinity of the rotating device with the metallic flag. Thus, pulses are generated as the flag passes, and the same pulse is sent to the conditioning device. When the pulse count goes beyond the set point, the output relay operates to give contact. These speed settings are generally adjustable from 5 to 5000 rpm. These may be the self-resettable type. Normally these are available in an IP 55/ 65 enclosure (flameproof or class II enclosures). Contact rating of 10/15 A @500 V AC in change over contact or 1 NO/1 NC.

10 ASH HANDLING: BASIC SYSTEM FUNCTION AND DESCRIPTION

In coal-based thermal power plants, huge amounts of coal burn with a considerable amount of ash (roughly 200 tons/h of ash production is very likely in a 2 × 250 MW plant) as a part of the continuous process. This requires immediate ash removal to ward off associated hazards. The actual quantity, however, depends on several factors such as GCV, ash content, and other analysis values of the coal received. Fig. 2.51 depicts the overall idea of an ash handling system.

Ash is basically the unburnable part of coal and different metallic oxides such as SiO_2, Al_2O_3, Fe_2O_3, CaO, and MgO. It is most often sand and clay somehow mixed with coal. Commercially available, good-quality coal normally ranges from 10% to 15% ash against a 35%–45% range from bad-quality coal. After burning, the ash is removed from the combustion chamber. An ESP in the flue gas path is provided for arresting and collecting the flying ash, which is produced in the form of very fine grains and flies away with flue gas, hence the name. The other part, which is comparatively coarser, cannot fly and gets collected in the hopper provided at bottom of the furnace; it is therefore called "bottom ash."

Small quantities of ash are also collected in the air preheater, economizer, and stack. The total system starting from collection to disposal of ash is taken care of in a separate plant subsystem called the ash-handling plant (AHP). Size, percentage contribution, and location of various kinds of ashes in thermal power plants are shown in Fig. 2.55B. Out of the total ash in the boiler, more than 80% is fly ash. Ash from the APH and economizer contributes around 1% each of the total ash. Around 20% of total ash is bottom ash.

Previously most of the coal-fired power plants used a wet slurry system to convey all types of ash and accumulated in ash ponds. Starting in the 1970s, almost all coal-fired power plants began phasing out pond-based systems, adopting dry ash disposal that included a bottom-ash disposal system (with an option for semidry ash disposal).

The main reason for switching over to a dry fly-ash system is its demand as a raw material of cement plants, asphalt, etc., and the capacity constraints of ash ponds.

In some countries, there is an unambiguous stipulation regarding bottom ash disposal, for example in India, as per the Ministry of Environment and Forests (MoEF). The bottom ash is normally disposed of in slurry form in almost all the power stations. In some of the power stations, the semiwet type of bottom ash disposal is installed by use of hydrobins; the ash collected is then transferred directly for end use. Dry bottom-ash collection and disposal can also be adopted if an end user is available.

For a dry fly-ash extraction system, a full capacity system is to be provided to facilitate the utilization of fly ash in the dry form. In addition, a wet slurry system or high concentration slurry disposal system (HCSD) is also to be provided to cater to 100% ash. It is used to dispose of the balance of the unutilized fly ash until its full utilization is achieved.

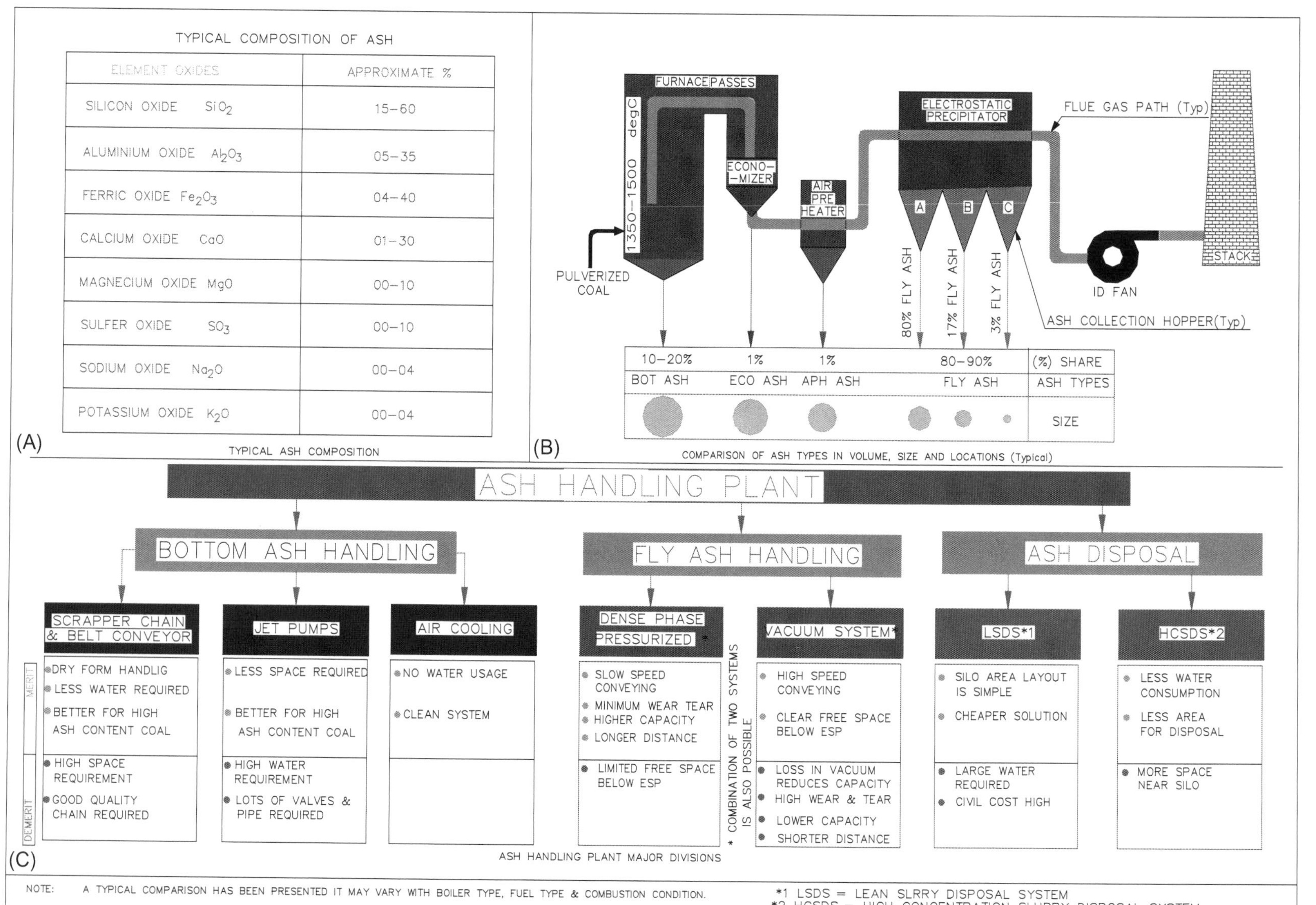

FIG. 2.51 Ash handling system overview (typical).

However, if full and continuous utilization of fly ash is foreseen from the inception of the

power plant, the wet slurry disposal of fly ash may not be installed.

10.1 Properties of Ash

10.1.1 Physical Properties

The color of ash varies from tan to different shades of gray (mainly depending on the presence of remaining carbon particles as a result of incomplete combustion. The lighter shades of gray generally indicate a higher quality of ash and a lower carbon content). Tan is normally contributed by lignite or subbituminous coal. Ash in its powdery form (fly ash or crushed/ground bottom ash) is normally found in a spherical shape with a glassy appearance. However, the particle sizes vary in accordance with the type/grade/rank of coal used. The specific gravity of ash may vary in the range of 2.0–3.0.

10.1.2 Chemical Properties

The chemical properties of ash depend on the type of coal burned as well as the handling and storage of the same. Other than metallic oxides, other very significant chemical properties of ash is percentage of unburnt carbon (measured by the loss on ignition method) and taken as a guideline to the suitability of its use as a substitute of cement in concrete (Table 2.5).

The percentage depends on the coal type, for example, bituminous, subbituminous, lignite, the geographical source, etc. Anthracite coal is rarely used in thermal power plants and hence is not considered.

10.2 Influencing Factors of AHP Concept Design

Some major factors such as station capacity/total coal requirement, GCV of design coal Ash content of worst coal, etc. have combined effect in determining the total ash quantum to be handled, size of AHP equipment and concept design of AHP in a power plant.

10.3 Mode of Ash Disposal

There exist two elemental processes called a dry ash system and a wet slurry system; another system called a semiwet system also exists that combines the other two. The subsequent section incorporates a brief discussion on them.

The AHP of a coal-fired thermal power station normally consists of the following methods adopted for different categories of ash.

10.3.1 Bottom Ash System

Bottom ash, which is the coarser/heavier part (about 10%–20%) of the total ash produced, falls at the bottom part of the furnace/combustion chamber due to gravity; it is collected from underneath. Hot ash gathered at the bottom of the hoppers forms a clinker as a hard deposit when it is cooled. BA collection and removal systems may be any of the processes—wet, semiwet, dry—as per their suitability considering the type of ash/coal and the techno-economic aspect.

10.3.1.1 Wet Bottom Ash System (Fig. 2.52)

Here, bottom ash is collected and extracted from a water impounded specially shaped storage refractory lined hopper facilitating intermittent removal. The hoppers are provided with a water seal trough and dip plates to make them suitable for arresting any air ingress during expansion of the furnace at higher temperatures. The shape of the hopper depends on many factors, such as the type of furnace, amount of bottom ash produced, etc. Provisions are there for overflow and make up of water on a continuous basis so that the bottom ash hopper temperature is maintained at 60°C.

Ash extraction is done through operation of the inclined feed gate at the bottom of the hopper with an appropriate hydraulic/pneumatic actuator. High-pressure water supplied to the jet pump acts as the motive water to pump the bottom ash slurry to the dewatering bins/slurry sump. The clinker grinder at each outlet converts the clinker to smaller particles for making slurry for further transportation. When the feed gate opens and the jet pump operates at the hopper outlet, the bottom ash slurry is made to arrive at the common ash (fly ash and bottom ash) slurry sump or dewatering bins, as per the design concept, to the final transport destination/ash pond.

10.3.1.2 Semiwet Bottom Ash System With Submerged Flight Conveyor

Semiwet bottom ash is removed with the help of hydrobins or dewatering bins, Then, the collected ash is disposed of directly from the hydrobins in the semidry form for end use.

A hydrobin is essentially a form of dewatering bin system. This system extracts bottom ash from the ash

TABLE 2.5 Oxide Forms of Various Metallic Elements Present in the Ash

Metal	Silica	Aluminum	Iron	Calcium	Magnesium	Sulfur	Sodium	Potassium
Oxides	SiO_2	Al_2O_3	Fe_2O_3	CaO	MgO	SO_3	Na_2O	K_2O
Ash %	15–60	5–35	4–40	1–30	0–10	0–10	0–4	0–4

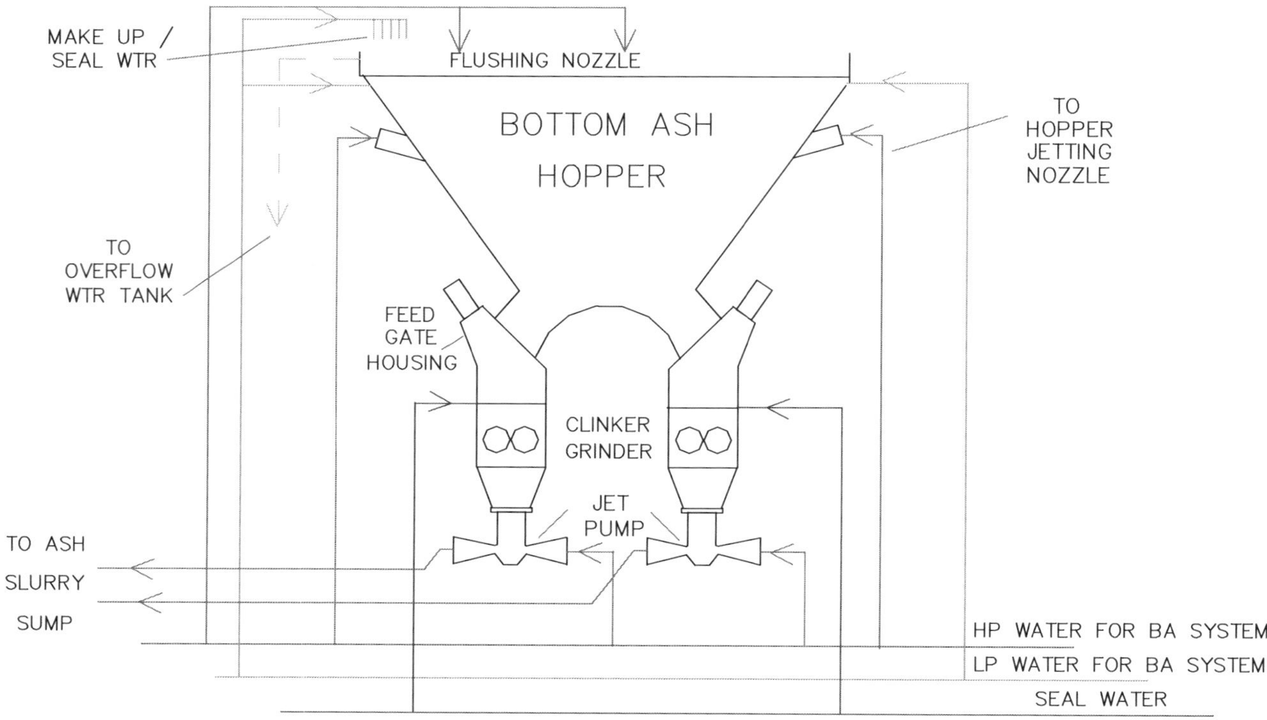

FIG. 2.52 Schematic/flow diagram of ash handling plant-wet bottom ash disposal system.

hopper as a slurry through a slurry pump to the dewatering bin, located outside, for ash water separation by dewatering the conveying supply water from the incoming slurry. Submerged flight conveyor (SFC) systems are also used to replace existing bottom ash equipment with a submerged mechanical chain and flight conveyor. Ash accumulates in the upper trough, which is filled with water to quench and cool the ash. Horizontal flights move the accumulated ash along the trough and up a dewatering ramp. At the top of the ramp, the ash falls through a discharge chute to a truck or bunker. The bottom ash in the bunker is picked up once or twice a day with a front-end loader and put into trucks; it may also be conveyed to hydrobins for water recovery.

The hydro/dewatering bin collects and dewaters the bottom-ash solids to approximately 15%–20% moisture. The clarified water thus available is stored in a surge tank, recirculated, and reused in the conveying cycle. The dewatered and semiwet ash is then loaded into trucks for delivery to the place of proper utilization. The dewatering hydrobin system renders advantages such as elimination of the ash storage ponds, low initial cost, and avoid shutdowns for conversion. The disadvantages are the running cost and usage of huge quantities of water.

The coarse ash collected from the economizer hoppers is connected to the bottom ash hopper top (above the maintained water level) by means of an adequately sized sloping pipe (for transporting slurry by gravity) in case of the jet pump system. However, if the calcium content is high in economizer ash or the BA hopper storage capacity is not enough, then economizer ash is disposed of separately.

10.3.1.3 Semiwet Bottom Ash System With Submerged Scrapper Conveyor

For plants where intermittent extraction is not acceptable in the plant design, continuous removal of bottom ash is stipulated. The system here comprises a dry, refractory-lined, bottom-ash hopper (BAH) of adequate capacity and proper shape. The hoppers are provided with a water seal trough and dip plates to make them suitable for arresting any air ingress during expansion of the furnace at higher temperatures. BAH protects the underneath ash extractor conveyor from the direct impact of hot ash. Hot ash is smoothly discharged to the water-filled slag bath of the special type ash conveyer called the submerged scrapper conveyor (SCC) below the BAH through horizontal hydropneumatically operated discharge gates.

The discharge chute at the hopper outlet remains dipped in the water of the slag bath of the SSC, ensuring sealing of the furnace bottom. The design of the BAH storage capacity would permit sufficient maintenance time for the SSC, depending on several factors such as the type of furnace, the type and amount of bottom ash produced, etc. (Fig. 2.53). Here, a temperature of 60°C within the SSC is always maintained by supplying high-pressure water to the slag bath.

The SCC collects continuously falling ash from the furnace into a water-filled trough to the hopper and the ash in turn is conveyed along the trough by submerged flight bars made of proper quality steel. It is then elevated up an inclined section to allow the water to drain back into the trough. After dewatering at the sloping portion, the moist ash is discharged to a clinker crusher for onward disposal through hydraulic sluice ways or belt conveyors.

The bottom ash is then converted into a slurry form with the help of the clinker grinder. Using the advantage of gravity flow if possible, the ash slurry is sent to a bottom-ash slurry sump through the trench. The bottom-ash slurry is then required to be suitably pumped, if gravity flow is not possible, from the bottom-ash slurry sump to the main ash slurry sump. This system comprises dry BAHs, horizontal hopper outlet gates, SCCs, clinker grinders, bottom-ash slurry sumps, etc. With a SCC system, the ash from the economizer is mainly evacuated and conveyed continuously in a wet form, and the ash slurry is led to the ash slurry pump house through trenches.

10.3.1.4 Dry Bottom Ash System

Bottom ash is collected in a refractory-lined hopper to withstand high temperatures; this is treated as a transition chute acting as a temporary volume chamber. The ash accumulated at the bottom is evacuated by a special extractor connected through an isolating gate/door for maintenance/other purposes. Arrangements are made so that air can percolate through the extractor with a steel plate conveyor and enter the combustion chamber due to the negative pressure that exists therein. Air travels in a counterflow direction along the surface of the ash, which rests on the steel plates, and encourages a reburning process of the still-glowing ash. This reduces the unburnt carbon level (almost complete combustion of coal), releasing additional thermal energy to the air before entering the combustion chamber. More fly ash rather than bottom ash is produced for better usage of the total ash production. The system does not conflict much with the combustion process and the character of the exhaust gas as the controlled amount of added combustion air would not be more than a tiny percentage of the total air flow. This process cools the ash, which is better for handling and protection of associated equipment without

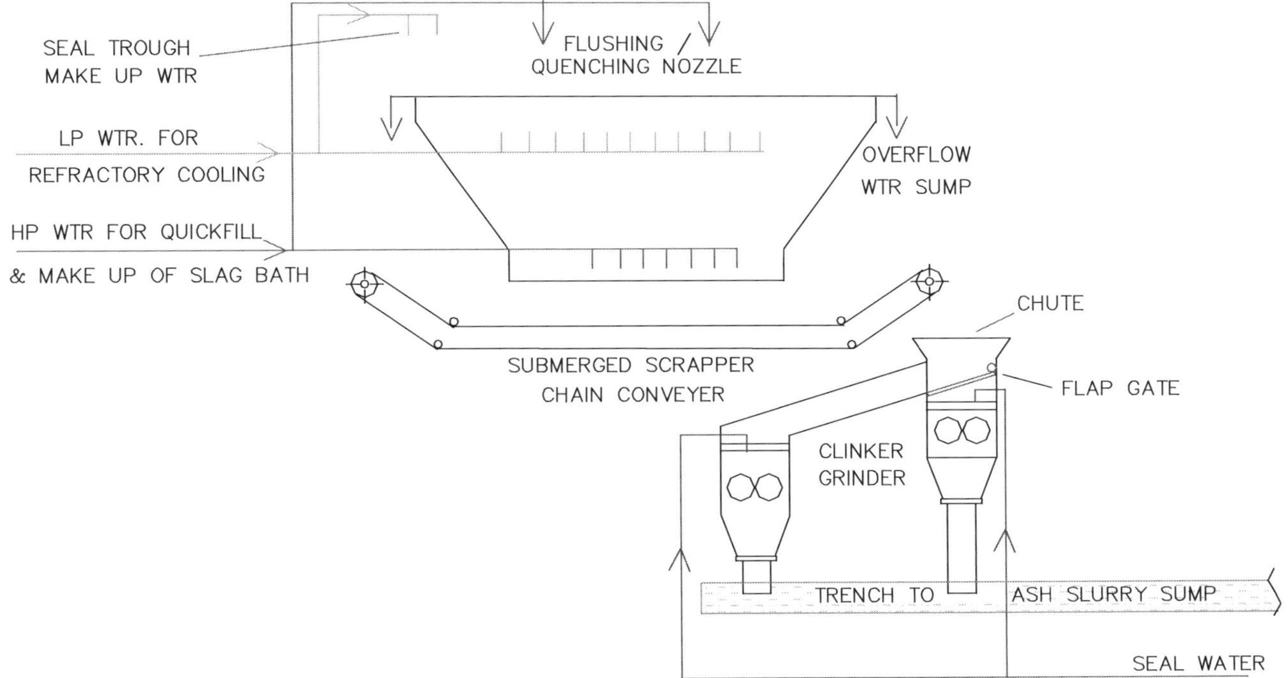

NOTE: ONE SET OF EACH ASH FLOW PATH/VESSEL HAS BEEN SHOWN.THE ACTUAL NO.OF ACCESSORIES AND VESSELS WOULD DEPEND ON PLANT CAPACITY, REDUNDANCY LEVEL REQUIRED, LAYOUT LIMITATION, ETC.

FIG. 2.53 Schematic/flow diagram of ash handling plant-bottom ash hopper disposal system.

requiring any additional fans. Large pieces are crushed small enough for pneumatic conveying.

The design criteria of the extractor takes care of the harsh conditions, for example, exposure to high temperatures, impact loads caused due to the fall of large and heavy clinkers, etc.

The cooled ash at the outlet of the extractor is then discharged into a crusher for reducing the clinker size, making it convenient for transport to a silo.

As the above system renders the complete elimination of all water in the bottom-ash process, the corrosion problems are also eliminated. The boiler becomes more efficient for recovery of most of the heat energy from the bottom ash.

The vacuum pneumatic conveying of dry bottom ash is the most popular system. The process exploits a vacuum design to convey ash to the point of shipment for external use. Because of the complete elimination of water, this system provides an increased lifetime as well as reduced cost and time.

The extractor may be a SFC (moving belt system) or a vibrating deck that moves ash from under the boiler to a crusher, where it is fed to a secondary conveyor.

Dry bottom ash handling system is probably more Environment friendly & economic (from capital cost as well as maintenance cost point of view). As short comparison of dry system with wet system has been presented in Table 2.6:

10.4 Fly Ash Handling System

Out of the three general types of furnaces provided in a coal-fired boiler, the dry-bottom furnace is the most common type, where about 80 percent of all the ash is fly ash. The other two types, the wet bottom and cyclone furnace, yield about 50% and 25% fly ash, respectively.

TABLE 2.6 Comparison of Dry and Wet System (AHP)

Dry System	Wet Systems	Remarks
No cooling water consumption	High cooling water required	
No additional water treatment is necessary	Additional water treatment for water recovery/ disposal is necessary	Cost saving in dry system
Thermal energy recovered	Significant energy loss	Energy gain in dry system
Reduction of Unburnt Carbon	Corrosion damage	
Improve boiler efficiency[a]	Steam may be produced and escape in to furnace	

[a]*Nearly 0.2% efficiency.*

Fly ash is useful in many applications because it is a pozzolan, which means it is a kind of siliceous or aluminosiliceous material. When in a finely powered form, it would combine with calcium hydroxide (from lime, Portland cement, or kiln dust) in the presence of water to form cementitious compounds. Fly ash can be used for landfilling and brickmaking also.

Two distinctly separate methods are available for a fly ash handling system, that is, a dry ash and wet ash type. Those are discussed briefly in the following subsections.

10.4.1 Dry Fly Ash Handling System

This type of fly ash handling system is applicable for ESP and air preheaters (APH), where dry fly ash is extracted from the relevant hoppers through a vacuum or pneumatic pressure conveying system as felt suitable. In some cases, a combination of both vacuum and pressurized systems is used.

The extraction and transportation of dry fly ash from the ESP is accomplished in two phases:

(i) ESP collection hoppers to the intermediate surge hoppers or collector tanks by a vacuum/pneumatic conveying system.
(ii) Intermediate surge hoppers to the storage silo, normally located at a distance from the main plant area, by conveying pneumatically.

Extraction and transportation of dry fly ash from the APH and duct hopper is accomplished pneumatically and connected to intermediate surge hoppers or collector tanks of ESPs. Alternatively, ash from APHs and duct hoppers can be handled in the wet form, and the ash slurry can be conveyed to the ash slurry pump house through trenches.

The advantages and disadvantages of both systems are shown in Fig. 2.55C, and are discussed below.

1. Pressurized system (Fig. 2.54)

Normally, for long-distance and high-capacity conveying, a system is used where the wear and tear of the accessories are less compared to the vacuum system. This system normally leaves very little space below the ESP hopper, which could be a disadvantage. There are normally two types of pressurized systems:

a. Dense phase: It can be used for conveying large-capacity (even up to 200 T/H) material over a long distance (~2 km).
b. Lean phase: In this method, both high-capacity (180 T/H) as well as long-distance (as high as 2.5 km) conveying is achievable.

2. Vacuum system (Fig. 2.54): the transportation is fast but the conveying capacity is less (~80 T/H), and it can be used for conveying materials over a moderate distance of ~1.5 km. The layout is simpler, but the capacity will be reduced if the vacuum fails.

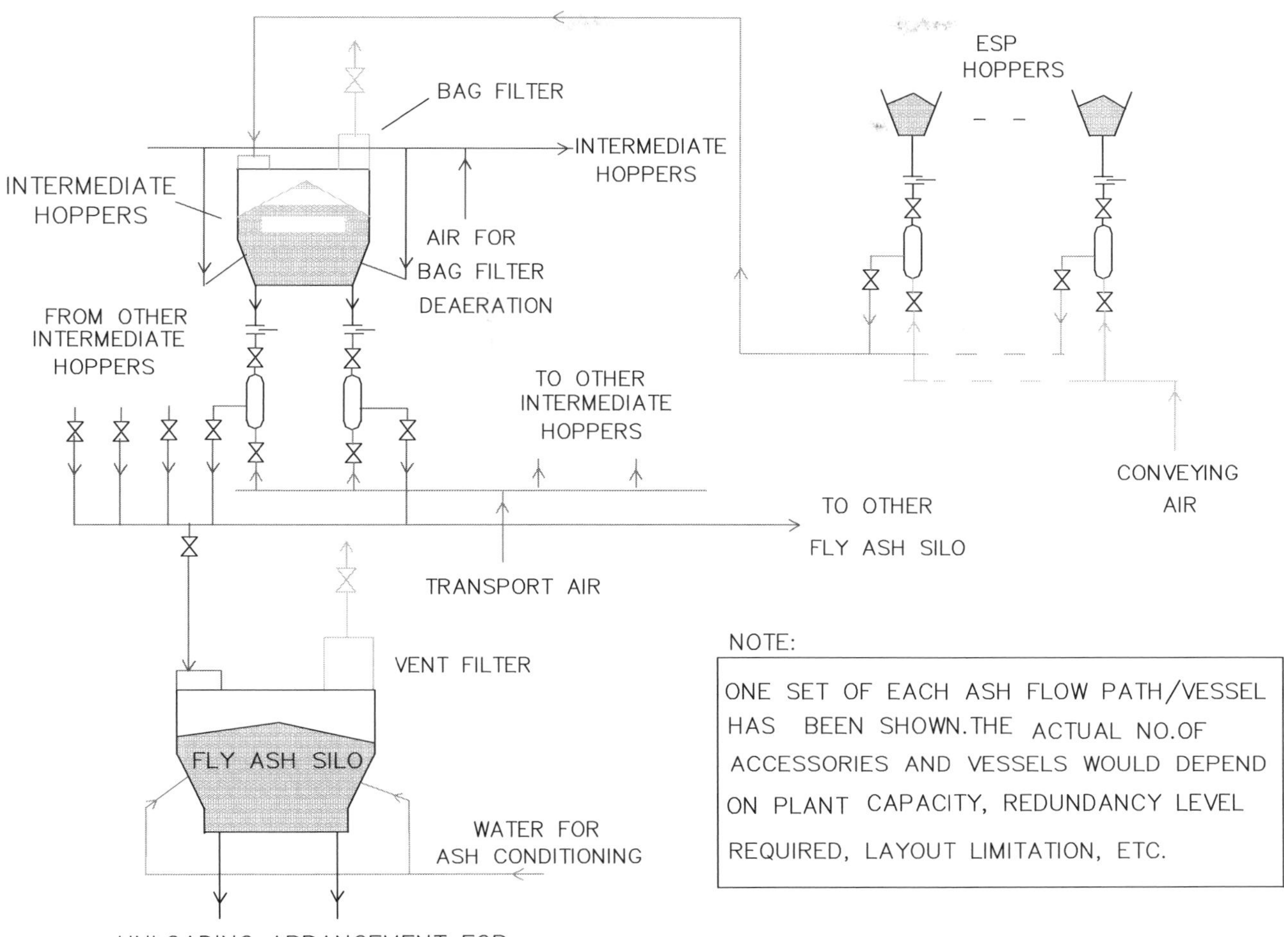

FIG. 2.54 Basic schematic/flow diagram of ash handling plant-pressurized dry fly ash hopper disposal system.

10.4.2 Wet Fly Ash Handling System

Here, fly ash mixed with water inside the wetting units turns into slurry before being pumped out with the help of jets or suitable pumps for transportation to a common slurry pit.

The wet disposal system can be a medium slurry disposal type or a HCSD type. The ash for the HCSD type can be taken from a silo or intermediate surge hopper, depending on the site layout.

10.5 Ash Water System

The entire water requirement of the ash-handling system may be met from the return header of a once-through CW system for a plant without a cooling tower (CT) facility.

For plants with CTs, AHP water is taken from the CT blowdown water of the station and the recovery water from the ash pond through decanting. For the latter case, a connection from the raw water is also provided for initial, fast fill and emergency makeup purposes. For equipment sealing/cooling, clarified water is provided.

An AHP water system consists of HP and LP water pumps, economizer ash water pumps, ash water sumps, etc.

10.5.1 Bottom-Ash Water System

Both HP and LP water pumps are utilized in a bottom-ash handling system.

Bottom-ash high-pressure water supply pumps are used to extract bottom ash from the furnace/economizer. When the bottom-ash handling system deploys jet pumps, HP pumps operate sequentially and in on-off mode, supplying to the jet pumps, hopper flushing, seal trough and gate housing flushing, etc.

For AHP with SSC systems, continuous operation of HP pumps is envisioned. In an SSC system, HP pumps supply water for quenching, trench jetting, seal trough flushing, gate cooling, sump stirring, etc.

Bottom-ash low-pressure water supply pumps are used for systems with jet pumps for refractory cooling, hopper cooling water to maintain hopper water temperature at 60°C, make up/fill of hopper water, seal trough, slurry sump, etc.

LP pumps, on the other hand, for systems with SSC, supply water for refractory cooling, cooling water for the upper trough of the SSC maintaining water temperature at 60°C, make up/fill of seal trough, ash slurry sump, cooling water to inspection windows, wash water to grinder, etc.

10.5.2 Fly-Ash Water System

Fly-ash high-pressure water supply pumps are required for wetting heads, air washers, fly ash slurry/trench jetting, combined ash slurry sump make up/stirring, etc.

Separate seal/cooling water pumps are provided for gland sealing of slurry pumps, vacuum pumps, and the sealing water requirement of clinker grinders and cooling of compressors.

With water being a very important medium for a power or process plant, every effort is taken to preserve it. For example, a BAH cooling water overflow may be recirculated after treating in a settling/surge tank and water recovery from the ash pond through decanting for wet ash disposal.

10.6 Ash Disposal System

According to the dry and wet systems, the disposal of ash in a plant is also tuned to match with the same. Those are briefly discussed below.

10.6.1 Wet-Ash Disposal System

Fly ash (of a wet system) and bottom ash are made to form slurry and channelized to accumulate into a common ash slurry sump. The same slurry sump is also utilized to dump the sludge from the pretreatment plant clarifier. The coalesced slurry is then propelled to the final disposal area or ash pond by low-speed centrifugal pumps suitable for ash slurry service.

The requirement of ash disposal at a long distance or higher elevation demands a number of ash slurry pumps in series and a booster station in case of an excessively high head requirement.

Alternatively, a HCSD system is employed, which uses the slurry concentration between 55% and 70%, depending on specific slurry rheology (viscoelastic properties of mud, polymers, and plastics, etc).

The wet disposal of fly ash is required during any emergent condition when dry disposal is out of service or during the commissioning or initial period of a running plant until full utilization of fly ash is accomplished.

10.6.2 Dry Ash Disposal System (Fig. 2.55)

The dry fly ash from the ESP and APH hoppers is transported to intermediate surge hoppers. Fly ash collected in intermediate surge hoppers is pneumatically conveyed

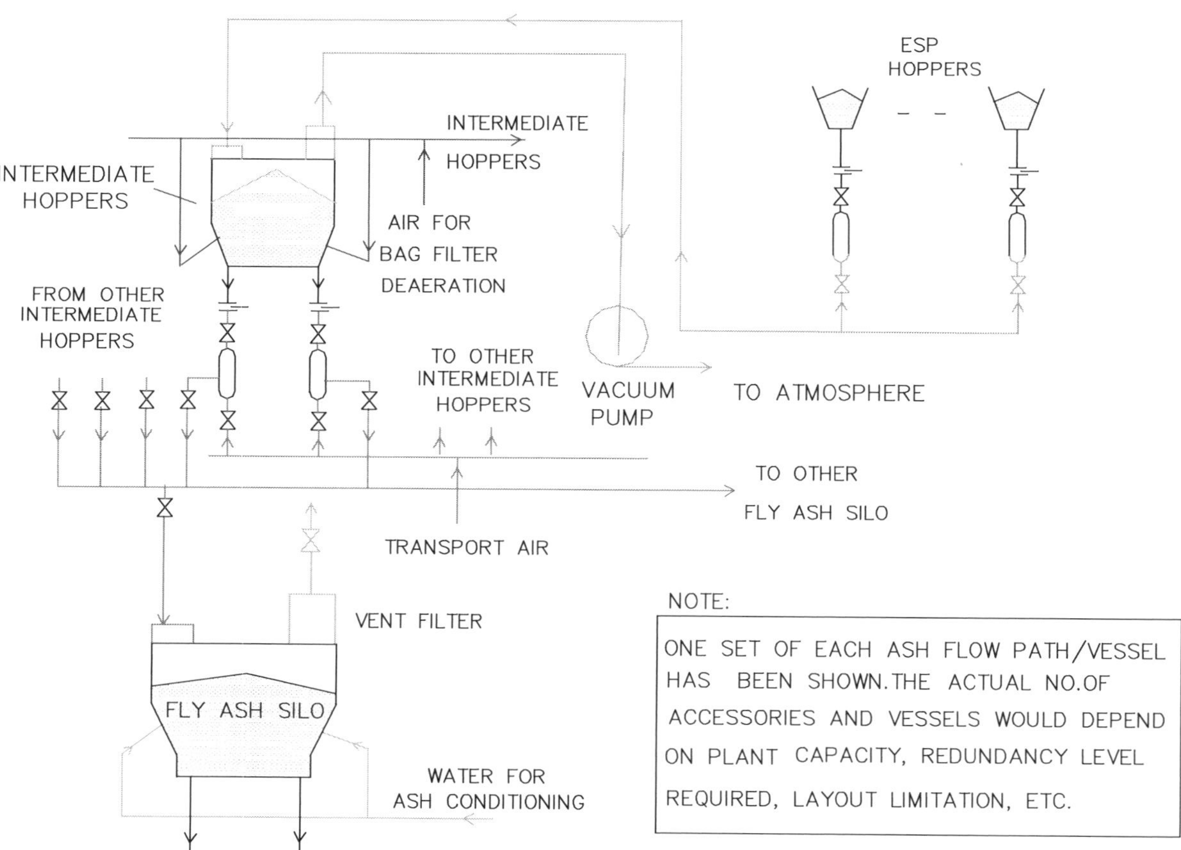

FIG. 2.55 Basic schematic/flow diagram of ash handling plant-dry fly ash hopper disposal system (vacuum).

to storage silos in a separate area where the same can be transported to elsewhere outside the plant for further usage in road construction, cement plants (as raw materials), plantations, etc.

BIBLIOGRAPHY

[1] Combustion-Fossil Power Systems: By Combustion Engineering Inc.

[2] Steam/Its Generation and Use: By M/s Babcock and Wilcox Company (New York, NY, 1978).

[3] Tech Sheet #113: By Heat Exchange Institute (HEI), Cleveland, Ohio.

[4] Product Bulletin of DEKKER Vacuum Technologies, Inc.

[5] Cooling Tower Details: By Bureau of Energy Efficiency, India.

[6] 0637567/4.2 SCA/APP.10.1.5/PSD Report, Dec. 16, 2006, by M/s Golder Associates.

[7] Sulfur: By M/s Zevenhoven and Kilpinen: June 1, 2006.

[8] Fast Start-up of a Combined-Cycle Power Plant: A Simulation Study with Modelica: By M/s Francesco Casella1, Francesco Pretolani 2.

[9] Boiler Tube Analysis: By M/s Babcock Wilcox (Power Generation Group).

[10] CIBO Energy Efficiency Handbook.

[11] POWER-GEN Europe June 7–9, 2011, Fiera, Milano-Milan-Italy.

[12] Tilting Pad Thrust Bearings, Factors Affecting Performance and Improvements With Directed Lubrication—Paper 13: Presented: By M/s M.K. Bielec and A.J. Leopardj.

[13] M.G. Say, Performance and Design of AC Machines, Pitman.

[14] Dr. S.K. Sen, Rotating Electrical Machines, Khanna Publishers.

[15] HPT Multistage Barrel Casing Boiler Feed Pump: Product Bulletin of M/s Sulzer, United States.

[16] J. Szargut, Influence of regenerative feed water heaters on the operational costs of steam power plants and HP plants, Int. J. Thermodyn. 8 (3) (2005) 137–141. ISSN: 1301-9724.

[17] Product Bulletin of M/s Roth Pump Company: Rock Island, United States.

[18] The Deaerator Principle: By M/s Spirax Sarco: Product Introduction.

[19] Engineering Thermodynamics—A Graphical Approach: By Israel Urieli (latest update: Nov. 18, 2013).

[20] Product Bulletin of M/s Clyde Bergemann Power Group/Delta/Ducon Conveying Technology Co. Inc., on Ashcon /Drycon System.

[21] Product Bulletin of M/s United Conveyor Corporation (UCC), Waukegan, Illinois, United States on Submerged Flight Conveyor.

[22] Product Bulletin of M/s Allen-Sherman-Hoff, a Division of Diamond Power International Inc.

[23] Report on Cooling Water Options for the New Generation of Nuclear Power Stations in the UK: Published by Environment Agency, Bristol.

[24] Standard Design Criteria/Guidelines for Balance of Plant of 2 × (500 MW or Above) Thermal Power Project: By M/s Central Electricity Authority, India.

[25] Chapter C of An Introduction to Coal Quality By Stanley P. Schweinfurth. The National Coal Resource Assessment Overview: Edited by Brenda S. Pierce and Kristin O. Dennen US Geological Survey Professional Paper 1625-F , US Geological Survey, Reston, Virginia, United States.

[26] Fly Ash handling System: F.L. Smidth: Catalog.

[27] Ash Handling System: NTPC Specification and Write up: Internet Document.

[28] Purolite Technical Bulletin: CPU: Internet Document.

Chapter 3

Plant P&ID (Process) Discussions

Chapter Outline

Power Plant Instrumentation and Control Handbook. https://doi.org/10.1016/B978-0-12-819504-8.00003-2

1 INTRODUCTION (P&ID PROCESS)

1.1 P&ID Basics

Probably, P&ID is the single type of drawing in a power and/or process plant that is of immense importance to all the disciplines involved, for example, mechanical (for equipment type and details), piping (piping detail), electrical (for a list of motors and their rating), civil and structural (for structural design, number of such equipment needing a special foundation), and instrumentation (starting point of all subsequent design). The main aim of P&ID is to describe the process involved as well as its associated instrumentation and control. Therefore, in this chapter, discussions will be on the processes of various systems in power generation and associated instrumentation, including developments in SC, USC, A-USC, CFBC, CCS, and oxyfuel combustion.

1.2 Instrumentation Symbols in P&ID

The symbols and legends for P&ID are elaborated in Fig. 3.1. Mostly, ISA S 5.1 (ISA norms) is used, but KKS representation is also used.

1.3 Piping Representation in P&ID

Normally, a pipe numbering system includes not only the pipe tag but also the *size* (important in line instruments) and *material* of the pipe. The pipe material code and schedule will help the reader to decide on the pressure rating and material of the instrument.

1.4 Process Parameter in P&ID

For designing a P&ID or while marking instrumentation on the process flow diagram (PFD), it is necessary to furnish the details discussed above. In the absence of process details in P&ID, the same is to be read in conjunction with the heat balance diagram. In the heat balance diagram, generally *four* parameters are indicated in a box with four cells detailed later. Normally, SI units should be used but in power station practice, certain units are obvious so those units are used, for example, FD discharge pressure shall be represented by xxx mm wc in place of kg/m^2 (SI unit for pressure). There are a few other process parameters such as normal *level*, high-low level, oxygen in the flue gas or conductivity in the steam, etc. It is necessary to mark these in P&ID wherever required.

1.5 Equipment in P&ID

As with instrumentation, it is conventional to represent all major equipment and valves with suitable symbols and associated tag numbers. In many P&IDs, for example, 2 × 3.5 MW MOTOR DRIVEN BFP marked below BFP. It is always better to put short specifications for major equipment at least in P&IDs.

1.6 Discussion on P&ID

Discussions on P&ID shall cover mainly the objectives and functions of the system, the basic system description, the major system equipment, major parameter monitoring in the system, the major control in the system, redundancy in measurement, and miscellaneous other points and relations with other systems.

The discussions are based on generalized terms, so various parameter values and the major equipment quantity and size will be generalized in nature and may not match any specific power station.

1.7 Redundancies for Transmitters (Sensor)

There are a few parameters in power plants where some redundancies are considered at the sensor level. Also, in certain cases, redundancies are considered at the controller as well as at the output display stages (Chapter 7). Some sensor redundancy schemes are high select, low select, one of two selections, and two of three covered here.

1.7.1 Transmitter (Sensor) Redundancy Considerations

In a sensor/transmitter redundancy, more than one sensor/transmitter is used for measurement of the same parameters to take care of random failure. For systematic failure, one needs to use sensors/transmitters based on different technologies, for example, for measurement of the drum level, one set of measurement with a DP transmitter and another based on Hydrate. However, on account of configurations, it is advisable *not* to consider the left drum level transmitter as redundant to that on the right. This is because there may be some level difference between the LHS and RHS of

LEGEND	DESCRIPTION	LEGEND	DESCRIPTION	LEGEND	DESCRIPTION	LEGEND	DESCRIPTION
	FIELD INSTRUMENT		NOZZLE		ELECTRIC SIGNAL		BALL VALVE
	PANEL INSTRUMENT		VERABAR FLOW ELEMENT		HYDRAULIC SIGNAL SOFT SIGNAL	N	BUTTERFLY VALVE
	DCS SIGNAL/TAG		SIMPLEX/DUPLEX THERMOCOUPLE		DCS OR DEDICATED CONTROL SYSTEM		CONTROL VALVE (GLOBE TYPE Typ.)
	BACK OF PANEL INST.	EP	ELECTRO PNEUMATIC ACTUATOR		CAPILLARY	f(x)	FINAL CONTROL ELEMENT
	INTERLOCK SYSTEM	EH	ELECTRO HYDRAULIC ACTUATOR		SOLENOID VALVE	REF.	PREFERENCE TO DWG (Drawing No. ref.)
A	ANNUNCIATION	M	ELECTRIC ACTUATOR		3 WAY SOLENOID VALVE (SINGLE COIL)	REF.	PREFERENCE FROM DWG (Drawing No. ref.)
XXXX	DEDICATED CONT.SYSTEM XXXX (SEE LEGEND)		PROCESS PIPE		GATE VALVE /ROOT VALVE		
	ORIFICE PLATE		PNEUMATIC SIGNAL		GLOBE VALVE		

COMMON ABBREVIATIONS USED IN POWER PLANT CONTROL SYSTEMS

LEGEND	CONTROL SYSTEM
ATRS	AUTOMATIC TURBINE RUN UP SYSTEM
ATT	AUTOMATIC TURBINE TESTING SYSTEM
BMS	BURNER MANGEMENT SYSTEM
CEM	CONTINUOUS EMISSION MONITORING SYSTEM
DAS	DATA ACQUISITION SYSTEM
DCS	DIGITAL CONTROL SYSTEM
EHG	ELECTRO HYDRALIC GOVERNOR CONTROL
FSSS	FURNCAE SAFE GUARD AND SUPERVISORY SYSTEM
GCS	GENERATOR CONTROL SYSTEM
HPBP	HIGH PRESSURE BY PASS SYSTEM
HSC	HYDROGEN SEAL OIL DP CONTROL
LPBP	LOW PRESSURE BY PASS SYSTEM
MIS	MANAGEMENT INFORMATION SYSTEM
OLCS	OPEN LOOP CONTROL SYSTEM
PLC	PROGRAMMABLE LOGIC CONTROL SYSTEM
SADC	SECONDARY AIR DAMPER CONTROL.
SER	SEQUENCE OF EVENT RECORDER
SOE	SEQUENCE OF EVENT
SPM	STACK PARTICULATE MONITORING
SSC	SEAL STEAM PRESSURE CONTROL
TPS	TURBINE PROTECTION SYSTEM
TSE	TURBINE STRESS EVALUATOR
TSI	TURBINE SUPERVISORY INSTRUMENTS

COMMONLY USED INSTRUMENTS & VALVE TAG NUMBERING SCHEME IN P&ID

LETTER	Ist PLACE FOR	2nd PLACE FOR	3rd PLACE FOR	4th PLACE FOR
A	ANALYSIS	ALARM	ALARM	ALARM
B	BURNER			
C	CONDUCTIVITY/CONTROL	CONTROL		
D	DENSITY/DIFFERENTIAL	DIFFERENTIAL/DISOLVE		
E	ELECTRICAL	ELEMENT	ELEMENT	ELEMENT
F	FLOW	FLAME		
G	USER CHOICE			HI
H	HAND /HYDROGEN	HYDRAZINE	HI	
I	CURRENT	INDICATING	INDICATING	
J	POWER			
K	TIME			
L	LEVEL		LO	LO
M	MOISTURE		MONITOR	
N	NITROGEN			
O	OXYGEN		OXYGEN	
P	PRESSURE	PRESSURE/p H		
Q	QUANTITY			
R	RAD ACTION/ RESIDUAL	RECORDING		
S	SPEED	SWITCH	SWITCH	TRANSMITTER
T	TEMPERATURE	TRANSMITTER	TRANSMITTER	
U	MULTI VARIABLE	UNIT		
V	VIBRATION	VOLT/VALVE		
W	WEIGHT			
X	UNCLASSIFIED			
Y	USER CHOICE			
Z	POSITION			

FIG. 3.1 Symbol and legend for P&ID.

the drum: (i) due to drum positioning (static difference) and (ii) there may be a difference in the level between the LHS and RHS of the drum levels due to nonuniform heating and steaming at two sides. Similarly, for boilers having steam outlets at two ends, there may be differences in temperature reading between the LHS and RHS. So, outlet temperatures of the SH (superheater) or RH (reheater) at the LHS and RHS should *not* be considered redundant.

In those cases, at best the average of LHS and RHS signals can be used to control the parameter but a close watch shall be maintained so that there is no spurious tripping due to deviation between the two sides.

1.7.2 Transmitter Monitoring and Inhibiting Selection

In intelligent control systems, out of limit for transmitter (<4 OR >20 mA) and open circuit and short circuit for thermocouples are easy and common. Also, smart transmitters have a diagnostic system that can detect faults and isolate itself. When a transmitter fault is detected, (then by selection) only good transmitters will be selected with alarms for faulty transmitters. In case of a transmitter connected via a fieldbus, such detections are more explicit and well reported in the system.

1.7.3 Redunancy for One of Two Selections

There can be three kinds of selections in one of two, discussed below.

1.7.3.1 Two Transmitters (Sensor) in High (or Low Selection)

These modes of selections are shown in Fig. 3.2A and B. In auto mode: normally, high or low select output shall be selected, but if out of two transmitters (sensor), one is detected as faulty (by the transmitter diagnostics or by an out-of-range detector), then it will be inhibited, so naturally the other will get selected. However, manually, any one of the two transmitters or high or low select output can be selected. Faults in sensors/transmitters will be alarmed.

1.7.3.2 One of Two Transmitters (Sensors) Selection With Average

The selection method is shown in Fig. 3.2C, (with soft average selection in DCS). In auto mode, normally the average output shall be selected, but when any one is detected as faulty, it will be inhibited so naturally the other will get selected. Manually, any one of the two transmitters or the average output can be selected. Faults in sensors/transmitters will be alarmed.

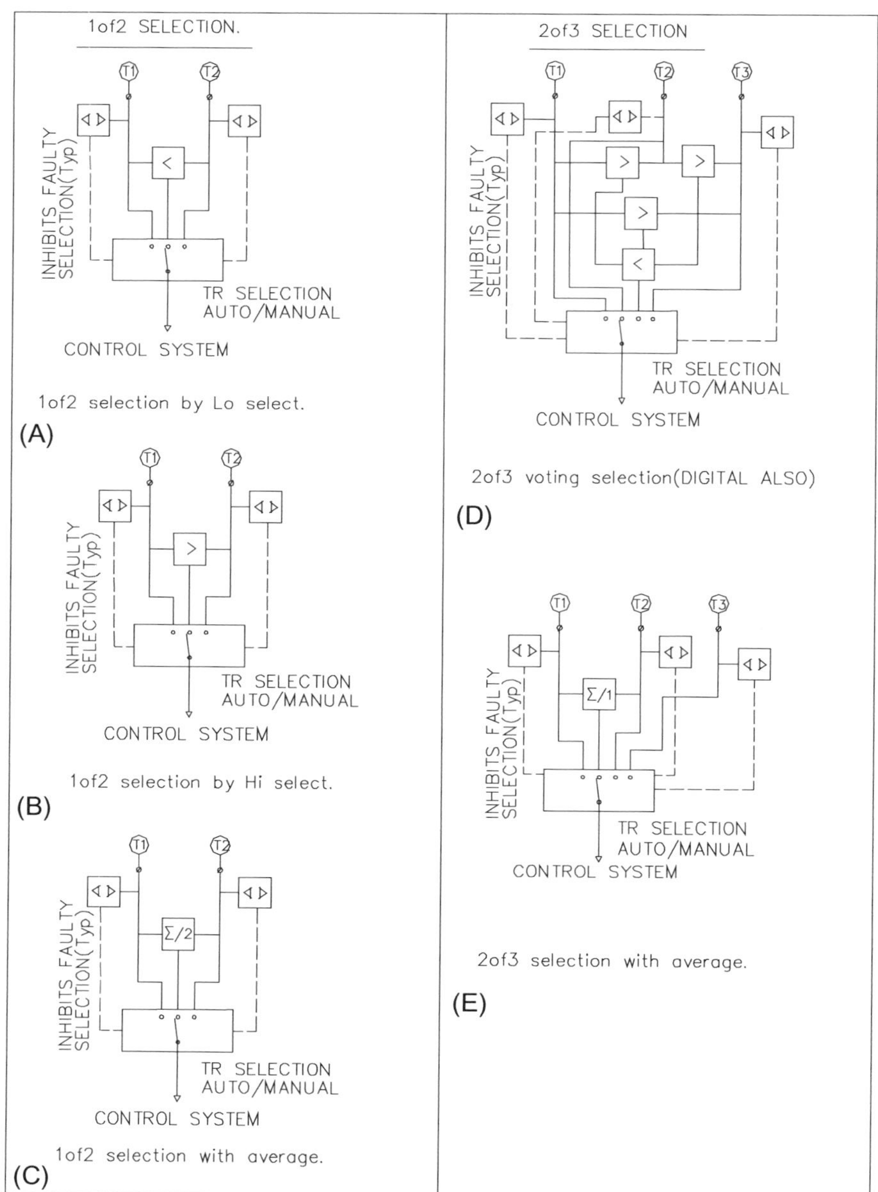

FIG. 3.2 Redundant instrument selection method.

1.7.4 Redundancy—Three Transmitters (Sensors) for Selection

For a few very important measurements, as listed below, three transmitters (sensors) are deployed in redundant mode: steam pressure at turbine inlet (steam header pressure—in process cum power plant), steam pressure at turbine first stage, furnace pressure, drum Level (in each LHS and RHS—so, six transmitters will be deployed-otherwise one of two that is, four transmitters), feed water flow (uncompensated), main steam flow (uncompensated), condenser vacuum, deaerator level (large plants) and secondary air flow.

1.7.4.1 Three Transmitters (Sensors) in 2 of 3 Selection

For connections, refer to Fig. 3.2E. In auto mode, normally the average output shall be selected, but if one transmitter (sensor) is detected as faulty, then it will be inhibited, so naturally the average will be on the other two transmitters.

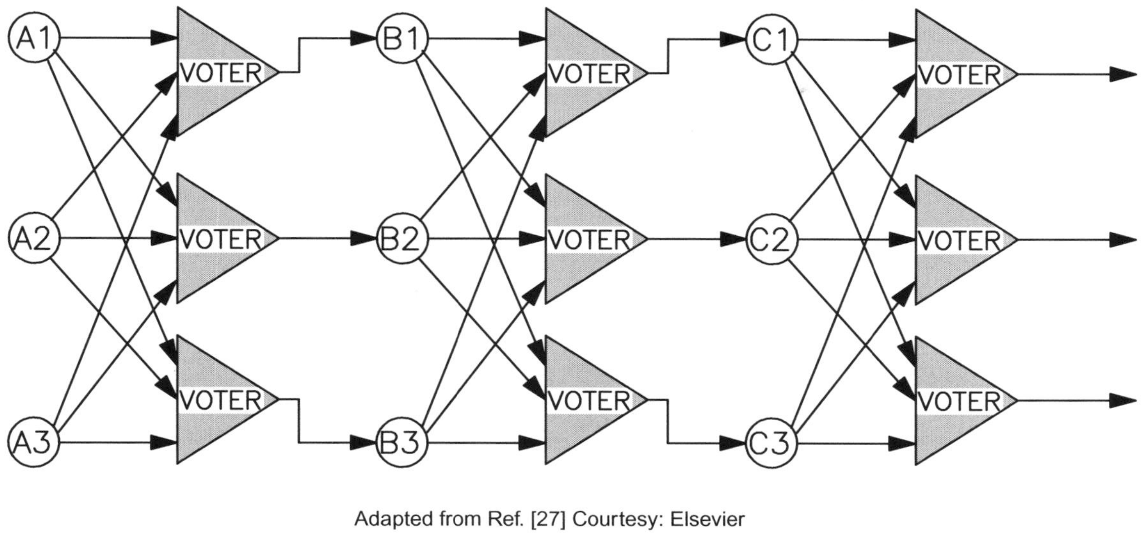

FIG. 3.3 Triple modular redundancy.

In manual mode, any one of the three transmitters or average outputs can be selected. Faults in sensors/transmitters will be alarmed.

1.7.4.2 Three Transmitters (Sensors) in 2 of 3 Voting Logic

As shown in Fig. 3.2D, three transmitter signals are initially voted through high selection between two transmitters, that is, high selection between one and two, one and three, and two and three. The output of these three high selections is fed to the low selection. Each of the transmitters, like other systems, is checked for health. The faulty transmitter is automatically voted out. In auto mode, the finally voted transmitter shall be selected. In manual mode, it is possible to select any of the three transmitters (if not faulty) or the voted transmitter.

1.7.5 Triple Modular Redundancy

Refer to Fig. 3.3 for connection details. Many times, in controllers in burner management systems (BMS) or GT controls, this type of redundancy is adapted. For details on redundancy and fault tolerance, Chapters 1 and 11 of the author's book [27] may be referenced.

1.8 Analytical Instruments and Control

In the P&IDs discussed below, there shall be a number of steam and water analysis (SWA) systems as well as dosing controls. Mainly, the sampling points are marked and control points are indicated in P&IDs.

2 MAIN STEAM (P&ID)

2.1 Objective and Function of the System

The basic purpose shall include the following major functions:

- A source for thermal energy supply to the turbine for mechanical work at the turbine.
- It is the source for many other steam supply steams such as auxiliary steam-boiler auxiliary steam (BAS) and turbine auxiliary steam (TAS) (CRH can be used).
- It has the ability to prevent overpressurization of the steam source by popping up of the safety valve to release excess steam.
- During start-up, shutdown, and huge power shedding by the turbine, the main steam bypasses the HP turbine and goes to the cold reheat line after pressure and temperature reduction.

2.2 System Description—Main Steam (Fig. 3.4)

2.2.1 Process and Piping—Main Steam

The main steam starts its journey from the outlet of the final superheater. In most of the boilers, the superheater outlet comes from the left and right side of the boilers. In those cases, after running for some distance, they may join at a point outside the boiler structure before entering the turbine building. Here, a few points as listed below need due consideration:

- Generally, piping up to the boiler outlet flange(s) is a part of the boiler itself and in this part all safety valves, boiler start-up vents, and drain valves are installed.
- From the boiler outlet up to the turbine inlet flange(s) is in the scope of piping. This scope part is discussed because the design engineer may need to coordinate among various agencies. When the main steam flow is measured, normally the flow element is placed in the common part of the piping.
- Pipe numbering: A *typical* depiction about pipe tag numbers is shown in Fig. 3.4, and is elaborated below.
- Process parameter in the P&ID diagram: It is customary to indicate the major parameters of the process. When parameters are not mentioned, a heat balance diagram may be used. So, in order for readers to understand, the same process parameter is shown here.
- P: pressure (in suitable units), E: enthalpy (KJ—or suitable unit—not common in P&ID), T: temperature (in degrees Celsius), and F: flow (In suitable units). There are a few other parameters such as level (normal level) that are marked at the place of the storage unit in P&ID.
- After the main steam line leaves the boiler house, a few outlets may be taken out, such as tapping for auxiliary steam, etc.
- Generally, after entering the turbine building, the main purpose of the MS line is to go to the high pressure turbine (HPT) to do some work at the HPT. However, based on plant configuration, there may be some bypass routes (HP bypass). Again, depending on the turbine (depending on the manufacturer), the type of turbine inlets may be one, two, four, etc. Generally, for HPTs, in most cases it is a single flow turbine with a single or double entry, as shown in Fig. 3.4 (double balanced entry). For example, a Siemens-KWU turbine has a single flow HPT. However, there are cases where the turbine is double flow, having four inlets. Small turbines may have a single entry.
- Here, a box has been shown as the boiler SH outlet, but it is quite a big one, as the boiler has a few meters length and breadth. Therefore, there may be some temperature difference between the two ends (Figs. 3.5 and 3.6).

2.2.2 *Bypass Path—Main Steam*

In many plants, there are some bypass route of this MS line to the turbine so that the boiler may run in the absence of a turbine. There are a number of such cases.

2.2.2.1 Process Plant Bypass Path—Main Steam (Fig. 3.7)

In case of process cum power plants, normally there is a bypass route for the MS to bypass the turbine and connect the same to another steam header of lower pressure. Also, part may be sent to the turbine and part to the bypass line. This is done to meet the following criteria

1. In process plants, there are requirements of steam at various pressures, such as HP pressure, medium pressure, and/or low pressure. The system is designed in such a way that these MP and LP steam pressures are the same as the extraction/exhaust of the turbine, for some economic benefits can be achieved.
2. In a case when the turbine is not available and still there is the requirement of medium/low pressure steam in the plant, then through the bypass lines with pressure reducing and desuperheating stations (PRDS), such requirements can be met.
3. As shown in a typical schematic in Fig. 3.7, the HPMP PRDS station is kept to get MP steam when the turbine is not in operation. Similarly, by MPLP, the PRDS ensures the LP steam supply in case of nonavailability of the turbine.

2.2.2.2 HP Bypass Main Steam

In medium to large plants (say, power plants >120 MW), there is one bypass line to bypass the HPT, so this is called the HP bypass system. This is used during turbine start-up, turbine shutdown, and for large power throw off by the turbine. Because the boiler has higher inertia, for certain cases of turbine trip and/or large power load throw off by the turbine, automatically the HP bypass comes into operation sensing pressure at the HP steam line. The set point of the HP bypass is set at a point higher than the set point of the inlet pressure. As mentioned in Chapter 8, the turbine trip automatically starts the HP bypass line through the protection input/quick opening criteria.

2.2.3 *Sampling Lines*

There will be sampling lines from the MS line at the SH outlet as well as the MS line at turbine inlet for online analytical instruments as well as for laboratory analysis. These are detailed in Chapter 5 SWA.

2.3 Major System Equipment—Main Steam

Major system equipment shall include the following:

1. Superheater outlet header(s): It is the header formed at the boiler where various subheaders from the final superheater join; it normally comes out from two sides of the boiler to form the MS header.
2. Start-up vent and drain valve (BSVV, BSDV 1 and 2 in the P&ID): Normally, there are two sets of remote-operated, semiautomatic, motorized valves used during the start-up of the boiler to vent and drain at the initial stage.
3. Boiler safety valve: In the superheater header, there is at least one mechanically operated safety valve to exhaust the excess steam in case of overpressurization. When the

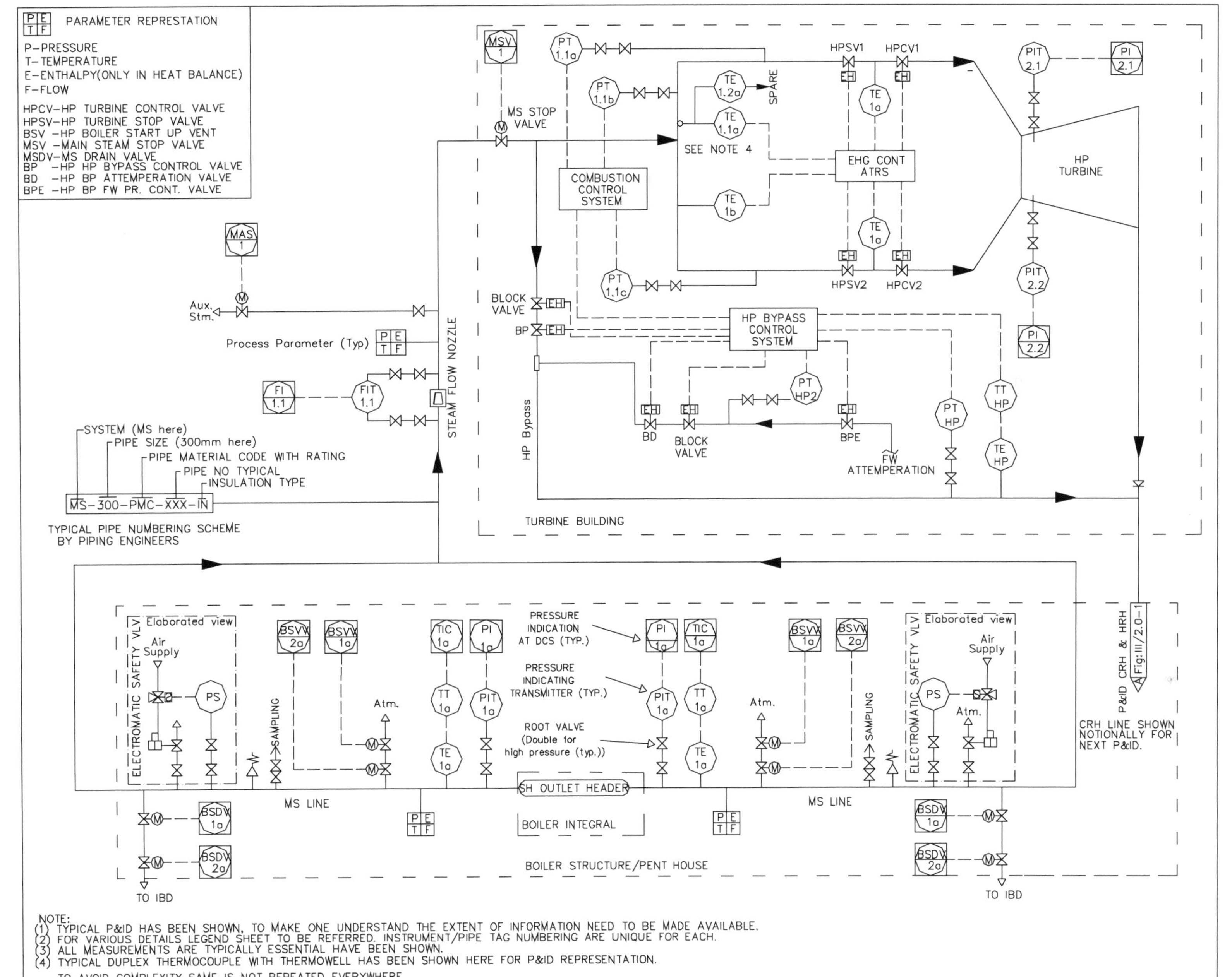

FIG. 3.4 Main steam P&I diagram.

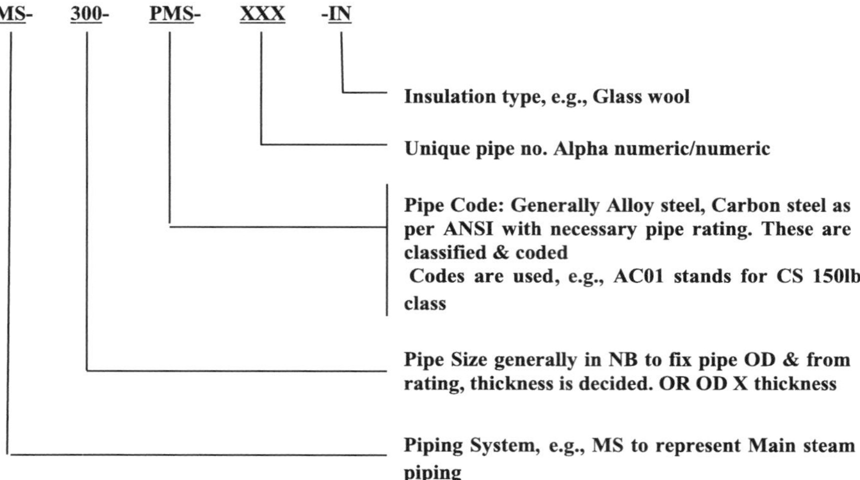

FIG. 3.5 Typical pipe numbering system.

P	E
T	F

FIG. 3.6 P&ID process parameter representation.

plants were commissioned, the safety pop-up function was carried out to set and check the pressure set point through spring.

4. Boiler electromatic safety valve: Apart from the mechanical safety valve, there is an electromatic safety valve operated with the help of a pressure switch, as in Fig. 3.4.
5. Main steam stop valve: Normally, this is a motor-operated remote semiautomatic valve to isolate the boiler steam supply to the turbine. Depending on the line size, this valve may have integral bypass (Chapter 6).
6. HP bypass system: This is meant to bypass the (excess) steam at the HPT inlet to the cold reheat line (CRH) formed from the HPT exhaust. There is a dedicated control system (Section 11 of Chapter 8) to regulate the pressure at a point slightly higher than that of the turbine inlet pressure. Another valve BD is meant to regulate the supply of feed water (FW) to reduce the temperature to match the CRH temperature. A BPE valve regulates the pressure of the FW header used to attemperate HP steam.
7. HPSV(s): These are basically on-off type isolation and stop valves at the entry to the HPT. These are controlled through an electrohydraulic governing control system. In some cases, it can be cracked open for heat soaking during start-up in some turbines to meet the starting criteria. Normally, these are combined with HPCV in a single body.
8. HPCV(s): A regulating HPCV is controlled by an electrohydraulic control (and/or hydraulic control) system of the turbine to regulate the quantity of steam entry to the HPT for varying the load in the turbine. The operation of these valves varies with constant pressure and sliding pressure modes of operations, as detailed in Chapter 9.

2.4 Major Parameters in Main Steam (MS)

2.4.1 Process Parameters

A few parameters that are *typical* for 200 MW and 500 MW have been depicted below. Normally, these are fixed by the thermal engineer while designing the system based on data from two major manufacturers: steam generator (SG) and turbine generator (TG) manufacturers. Normally, 60%–100% HP bypass is considered. On account of—large distance between—boiler and turbine, there will be pressure loss as big as 4–6 kg/cm^2 for larger plants. On account of higher permanent pressure loss, in many power plants of 500 MW and above, the steam flow is not measured directly. Instead, the turbine first-stage pressure is considered and/or the turbine first-stage pressure together with the HP bypass steam flow are used to compute the boiler load index (Table 3.1).

2.4.2 Monitoring of Process Parameter—Main Steam

The following main steam parameters (with typical range) are important to be monitored at the control room in almost all power plants.

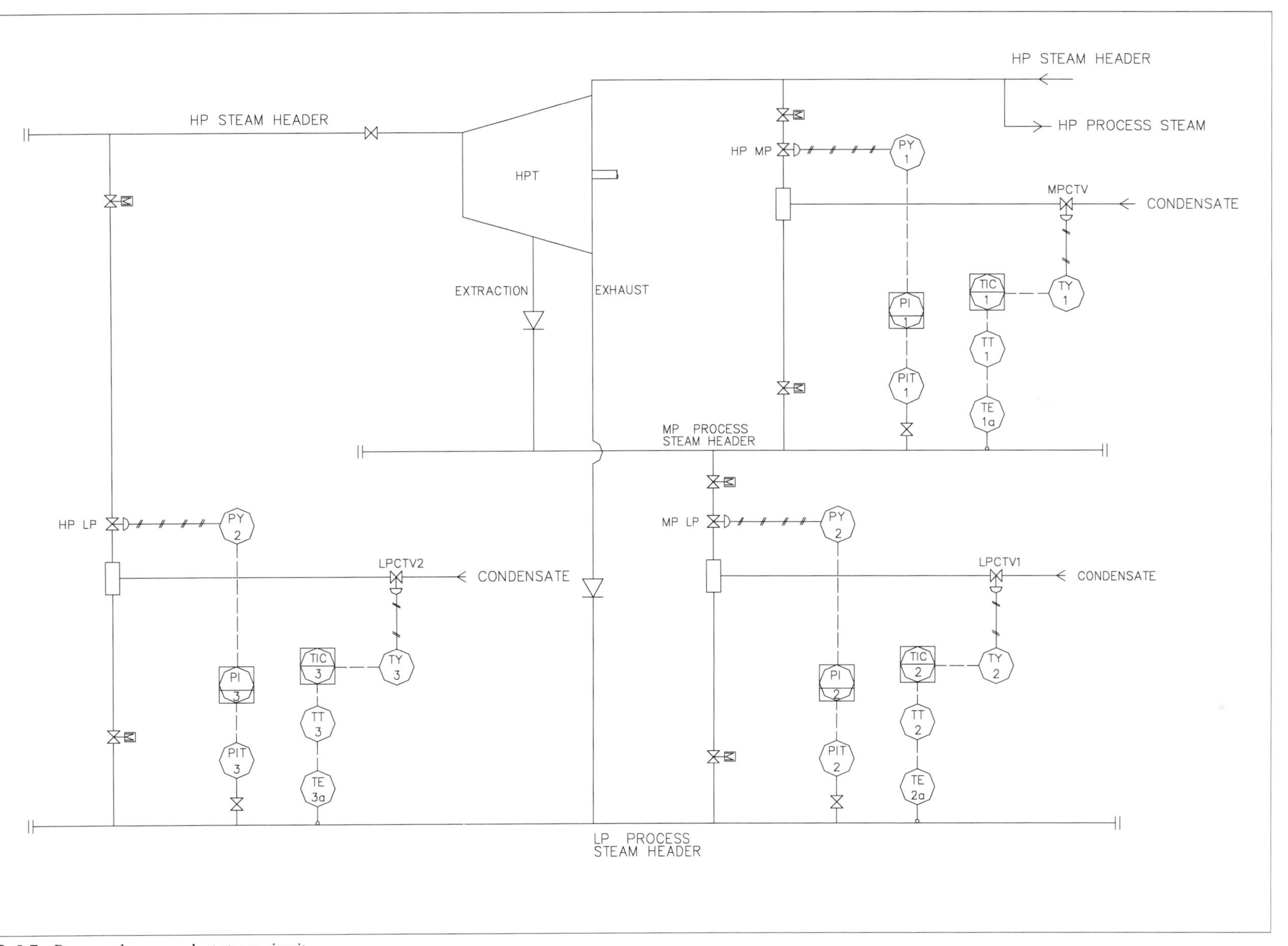

FIG. 3.7 Power and process plant steam circuit.

TABLE 3.1 MS Parameters

Steam Parameter	Boiler Outlet		Turbine Inlet		Remarks
Size	200 MW	500 MW	200 MW	500 MW	
Pressure	135–140	165–170	130–135	160–165	For smaller plants: as low as 30–40 kg/cm^2
Temp.	540	540	<540~536	<540~536	For smaller plants: as low as 450C but superheated steam at that pressure
Flow	~700–800	1600	~700–800	1600	Depends on MW and pressure design

Units: *P* in kg/cm^2, *T* in °C, *F* in t/h.

1. *Main steam pressure and temperature* at the boiler outlet. (MS temperature—control parameter). In case of a double outlet, these measurements need to be carried out at each end by smart indicating pressure transmitters. Typical range for 200 MW is 0–150 (160) kg/cm^2 whereas the same for 500 MW will be 0–200 kg/cm^2. However, in supercritical (SC)/ultrasupercritical (USC) plants, the MS pressure will be >200 bar, hence the range will be selected accordingly. Generally, for temperature measurements, ISA type K thermocouples are used for measuring. Depending on the actual parameter of the boiler, the range for the instrument (temperature transmitter) is selected. A typical range is 0%–600°C.
2. *Main steam flow*—Generally, flow elements (if used) are placed on the main steam line away from both boilers as well as the turbine end. This is done so as to ensure the straight length requirement. As already discussed, for larger plants, in order to avoid large permanent pressure loss, measuring the steam flow it is computed. Because the pressure is high, a flow nozzle along with smart differential pressure transmitters (DPTs) are used for the head-type measurement. Depending on the flow metering standard, the range shall be zero to several thousands of millimeters of the water column (typical safe value to choose the DP range as 0–10,000 mm wcl). However, the unit is expressed in terms of kg/h or t/h. Typical flow ranges are 0–900 t/h for 200MW and 0–1800 t/h for 500 MW plant.
3. *Turbine inlet steam pressure*: (steam header pressure in case of process plants, or where there are many boiler supplying to a header, and there are many consumers drawing steam from the same header). Generally, a fixed set points is used to control this parameter. In larger plants, especially in cases of once-through (OT) boilers used mainly in SC/USC plants, a sliding pressure operation within a load band of operation of say a 70%–90% load and fixed set points are used beyond the range of 70%–90% load. The HPBP set point in those cases also follows the main steam pressure set point *plus by a fixed value*.
4. *Temperature measurement at the inlet and outlet of HPSV and HPCV* are generally recommended by the TG manufacturer in order to compute thermal soaking by turbine metals, required for automatic starting and run up of the machine as well as for the thermal stress evaluator, etc. Because the final temperature reaches ~540°C, the ISA type K thermocouples are best suited for this application.
5. *Turbine first-stage steam pressure*: After the steam enters the turbine, the pressure at the first-stage pressure is measured.
6. *A few other parameters*: There are a few parameters that are not part of MS (but of CRH—hence, discussed there) that are measured at the CRH line, required for HP bypass control.

2.5 Controls in—Main Steam

There are some important controls associated with the MS system:

1. *Main steam temperature control*: The main steam temperature at the superheater outlet header(s) of the boiler is an important control parameter. After start-up, once the temperature of the MS reaches the desired value (say 540°C), it needs to be kept constant throughout all operating points of the plant by using the FW to attemperate. Initially, the fixed set point (adjustable) is kept much lower as there is insufficient steam flow. In cases of SC/USC-OT boilers, this is controlled by the fuel/feed water flow ratio.
2. *Turbine inlet (steam header pressure) pressure*: This parameter is the key parameter that needs to be kept. Based on this, the combustion in the boiler is regulated. Also, this parameter would be used to operate

the HP bypass valve (of course with a higher set point). Steam flow or load index for the boiler is also used as the feed forward signal for combustion control.

3. *EHG control of turbine*: This is the most important control system for turbine. It is very much interactive control with so many other parameters.
4. HP bypass control: During start-up/shutdown (and sudden load throw off of the turbine), this control plays the most important role in bypassing the requisite quantity of steam.
5. *Start-up vent and drain valve*: These are controlled (interlocked) for start-up of the boiler. These have remote-operated semiautomatic operation.
6. *Electromatic safety valve*: This is one interlock operated valve to pop up in case of higher pressure than the safety limit in the MS header near the boiler.

2.6 Redundancy in Measurement—Main Steam

- For critical parameters such as turbine inlet pressure "two of three redundancy," is quite common in larger power plants. In several power plants, dedicated one of two redundant pressure transmitters are used for HP bypass control, but in many places the turbine inlet pressure measurement is done with more redundancy and the same is used for HP bypass control also.
- It is recommended to utilize at least one of two redundancies for MS temperature at the SH outlet on each side (as applicable). For smaller power plants, one in each side is also deployed for measurement of the MS temperature at the SH outlet.
- In larger power plants such as those in the 200 MW/500 MW, deployment of two of three redundancies in DPTs for steam flow measurements is not uncommon. However, one of two DPTs is recommended because this is an important parameter.
- Turbine first stage MS pressure is also an important measurement, so this may call for at least one of two redundancies.

2.7 Miscellenous Points—Main Steam

The main steam system is very much related to many other systems and controls, thus these parameters need to be available in both Monitors dedicated for SG and TG. In a conventional layout, this is available at the central desk.

3 REHEAT STEAM (P&IDS): COLD AND HOT REHEAT

3.1 Reheat Steam System

The reheat steam has two parts: cold reheat and hot reheat.

3.2 Cold Reheat Steam System

3.2.1 Objectives and Functions of the System

The exhaust from the HP turbine is the starting point of the cold reheat (CRH) steam. The HPT exhaust stands for that piping section, where, the steam outlet from both the HPT and the HP bypass system comes and joins. Therefore, it is obvious that CRH steam may be generated even when the HPT may not be in operation and, as a combination of both and depending on the load, the CRH steam pressure at the HPT outlet may vary. Therefore, the HP bypass system needs this demanding need of variable pressure. Basic purposes shall include the following major functions:

- To return the HPT exhaust steam to the boiler for reheating to do further work at the turbine as hot reheat.
- It is the source for many other steam supply steams such as auxiliary steam—BAS and TAS.
- CRH is considered as an extraction from the HP turbine, (say as sixth) as an extraction steam for the regenerative heater.
- In many plants, CRH steam is used to heat the deaerator (through PRDS) at a lower load when extraction (cross-over) steam is unavailable.
- Sometimes (depending on the heat balance diagram), it goes to the gland steam condenser (GSC)/sealing steam system.
- It is provided with a facility to prevent overpressurization of steam by popping up of the safety valve to release excess steam.

3.2.2 System Description—Cold Reheat

The description given below shall be read in conjunction with Fig. 3.8.

CRH steam starts its journey from the outlet HPT and/or HP bypass (HPBP) station outlet with the aim to reach the reheater inlet at the boiler. Depending on the boiler turbine configuration, there may be a single or double CRH line. Naturally, in the case of two separate lines, measurement duplication for pressure and temperature would be necessary. However, separate monitoring of CRH pressure and temperature at the HP bypass outlet and at the HPT outlet is common. In many cases, the CRH steam line after running for some distance in a single pipe with bifurcation near the entry to the reheater inlet header from left and right side of the boilers. CRH lines have desuperheaters in CRH lines just at or near the entry point of the CRH at each reheat inlet header. In the CRH line, there are a few low elevation points from where it is drained to drain the flash tank to prevent water from entering the boiler. To prevent overpressurization, the necessary safety valves are provided in the line.

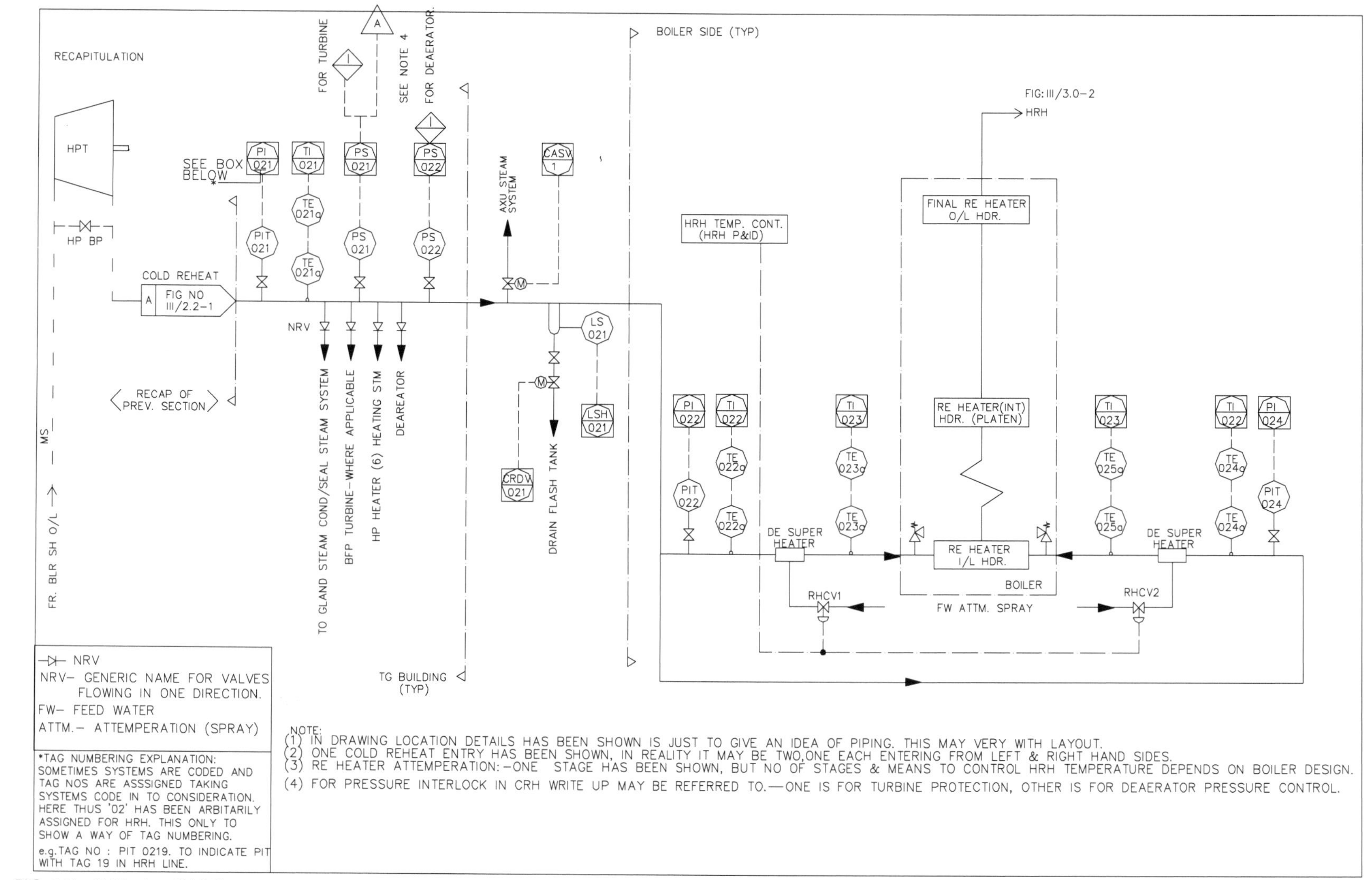

FIG. 3.8 Cold reheat P&I diagram.

3.2.3 Major System Equipment—Cold Reheat Steam

The major system equipment shall include the following:

- Reheat inlet header: It is the header where the cold reheat line(s) from the turbine sides enters the boiler. From here, there are several subheaders that go to form the reheater, which gets heated from the flue gas.
- Inlet safety valve: In order to prevent overpressurization, there is one safety valve at the reheater inlet of the boiler.
- Drain lines: There are a number of drain points during the journey of the CRH line. At each of the lowest points, there are U traps for draining (with level monitoring) to drain the flash tank.
- Desuperheater: For several boilers, there are stages of attemperations for controlling the temperature of the HRH at the RH outlet. So in those cases, the first-stage attemperation is kept at the reheater inlet header. There shall be another in the intermediate stage of the reheater.
- Miscellaneous supply lines: There are a number of lines that come out from the CRH line (with a remote-operated motorized valve): Auxiliary steam, HPH6 heating steam supply, gland steam, etc.

3.2.4 Major Parameter Measuring Monitoring in—CRH Steam

3.2.4.1 Process Parameter—CRH (Typical Values Only)

In the following table, we try to put forward a few process parameters at the turbine and boiler end (Table 3.2).

Depending on the actual piping route and sizing, a typical pressure drop between the boiler to the turbine and vice versa (each side) is considered to be 1–1.5 kg/cm^2. For larger plants, empirically a ~4 kg/cm^2 pressure drop can be considered from the CRH at the HPT to the HRH. It is to be noted that the quantity of CRH steam going to the boiler will be less than the MS flow because as there will be some consumption points for example, regenerative heater and/or for Auxiliary steam depending on mass and heat balance diagram for the plant.

3.2.4.2 Monitoring of Process Parameter—CRH Steam

The following CRH steam parameters (with typical ranges) are important to be monitored at the control room.

1. *CRH steam pressure and temperature at* the *HPT/HP bypass system outlet*. For these parameter measurement points, Fig. 3.8 may be referred to. The typical calibrated range for larger power plants is 0–50 kg/cm^2 (easy to adjust in the smart indicating pressure transmitters deployed). In case of a double outlet, the measurement needs to be carried out at each end. Generally, ISA type K thermocouples (with a temperature transmitter with typical range 0–400/450°C) are used to measure the temperature.
2. HP *bypass steam flow*—In many power plants, this parameter is measured to compute the Boiler Load Index. A flow nozzle with DPT (the range may be zero to several thousands of millimeters of water column) is used for flow measurement.
3. *Pressure monitoring outlet of HPT*: It is important to monitor the CRH pressure at the turbine end so that the pressure does not go too high. If it goes too high, then it will affect the pressure at the IPT as well as the LPT exhaust, that is, the condenser vacuum. For this reason, a block valve in the HP bypass line is common practice so that in case of high pressure, the system flow may be blocked (Fig. 3.4). Also, in an attemperation line, a separate pressure-reducing valve is introduced so that higher pressure at the FW cannot cause a pressure rise at the HP bypass line. Also, in an attemperation line, there is a block valve (Fig. 3.4). A pressure switch is placed to monitor pressure against the high set point. This will generate an alarm, and in some cases this pressure switch will cause a tripping of the turbine. There may be another pressure switch by which the CRH steam can be brought into operation for deaerator heating.
4. *CRH level monitoring*: At all major low points, there will be the monitoring of the level by float/displacement level switches in line with the recommendations of the turbine water damage prevention system (TWDPS) to prevent the possibility of water entry into the turbine.

TABLE 3.2 CRH Parameters

Parameter	Reheater Inlet		HPT/HPBP Outlet		Remarks
Size	200 MW	500 MW	200 MW	500 MW	
Pressure*	30–40	47–49.5	31–41	45–47	*Typical value, and may vary
Temp.*	316–340	355–360	316–340	355–360	
Flow	635–720	~1450	635–720	~1450	

Units: *P* in kg/cm^2, *T* in °C, *F* in t/h.

5. *Reheater inlet pressure and temperature measurement*: The measurement of the CRH steam pressure and temperature is done at the reheater inlet (for boiler performance). When there is a desuperheater at the reheater inlet as shown in Fig. 3.8, the temperature is monitored at both the inlet and outlet of the desuperheater or attemperator. Instrument types shall be the same as discussed in clause no. 3.2.4.2.1.

3.2.5 Controls in—CRH Steam

There are a few controls associated with the CRH system.

1. *HP bypass temperature control*: (Fig. 3.4 and Section 11 of Chapter 8): In the case of more than one HPBP control, for each set of HP bypass controls, there will be one spray control valve and perhaps one common pressure-regulating valve to reduce the FW pressure before the FW spray line.
2. *Reheater temperature control first-stage attemperation*: Generally, for reheater control, the spray water control at the CRH line inlet to the reheater is used during emergencies only. HRH temperature is regulated by the burner tilt control or gas bypass damper and/or the gas recirculation damper.
3. *CRH line level control*: At the low points in the CRH line, there will be the drain pots whose levels are monitored and interlocked to drain to the flash tank with alarm.
4. *CRH supply line valve—remote operated*: There are a few motorized valves—remote operation (semiautomatic) in the CRH supply lines to auxiliary steam, HPH VI, etc., as shown in Fig. 3.8.

3.2.6 Redundancy in Measurement—CRH Steam

Redundancy for the critical parameters such as HP bypass temperature control one of two (Fig. 3.8 redundancy not shown), turbine tripping from high pressure in the CRH line (two of three) is common.

3.2.7 Micellenous Points—CRH Steam

Like the main steam, CRH is very much related with many other systems and controls; hence, in a conventional layout, this is available at the central desk.

3.3 Hot Reheat Steam System

3.3.1 Objectives and Functions of the System

Basic purposes shall include the following major functions:

- A source for thermal energy supply to the intermediate-pressure and low-pressure turbines (IPT and LPT) to carry out the mechanical work at the turbine. There may be more than one reheater—hence, more than one HRH system in the thermal cycle (for example, USC plant design). It is worth noting that the maximum amount of work is done at the IPT and LPT, so the condenser is maintained at vacuum in utility power plants (with the exception of back-pressure condenser turbine systems).
- It has the ability to prevent overpressurization of the steam source by popping up a safety valve to release excess steam.
- During start-up, shutdown, and large load throw off, the HRH steam bypasses the IPT and LPT and goes to the condenser after pressure and temperature reduction with the help of the LP bypass (LPBP) system, as applicable.
- After passing through IPT, HRH goes to the LPT and is known as cross-over steam (also referred to as Extraction IV). A part of this cross-over steam is used to heat the deaerator during normal operation. For the plants having a steam turbine-driven boiler feed pump (BFPT), this cross-over steam is the source of energy for BFPT. Cross-over steam is also used for turbine sealing.

3.3.2 System Description—Hot Reheat (Fig. 3.9)

HRH steam starts its journey from the outlet of the final reheater. In many boilers, the reheater outlet headers come from the left and right side of the boilers. In those cases, after running for some distance, they join at a point outside the boiler structure before entering the turbine building. After running in a single header, they may be split into two or four headers to enter the IPT stage of the turbine. HRH also has a bypass path known as the LP bypass (LPBP) to bypass the IPT and LPT, as already mentioned in Section 3.3.1. Because in most cases, the electrohydraulic governing system (EHG) and LPBP go hand in hand, the LPBP is normally in the scope of the turbine supplier. There are a number of temperature protections associated with the LPBP to prevent the condenser from high pressure.

3.3.3 Major System Equipment—Hot Reheat Steam

Major system equipment shall include the following:

1. Reheat outlet header: It is the header from where the hot reheat steam comes out to enter the IPT. From the inlet header, a number of tube banks/panels form the reheater, which discharges at this header. It is worth noting that between the inlet header and the outlet header of the reheat, there may be a desuperheater where water is sprayed to control the reheater temperature as well as an outlet safety valve to prevent overpressurization.
2. The IPT stop valve (ISV 1, 2): At the turbine inlet, there is a stop valve in each entry line to the IPT to isolate the

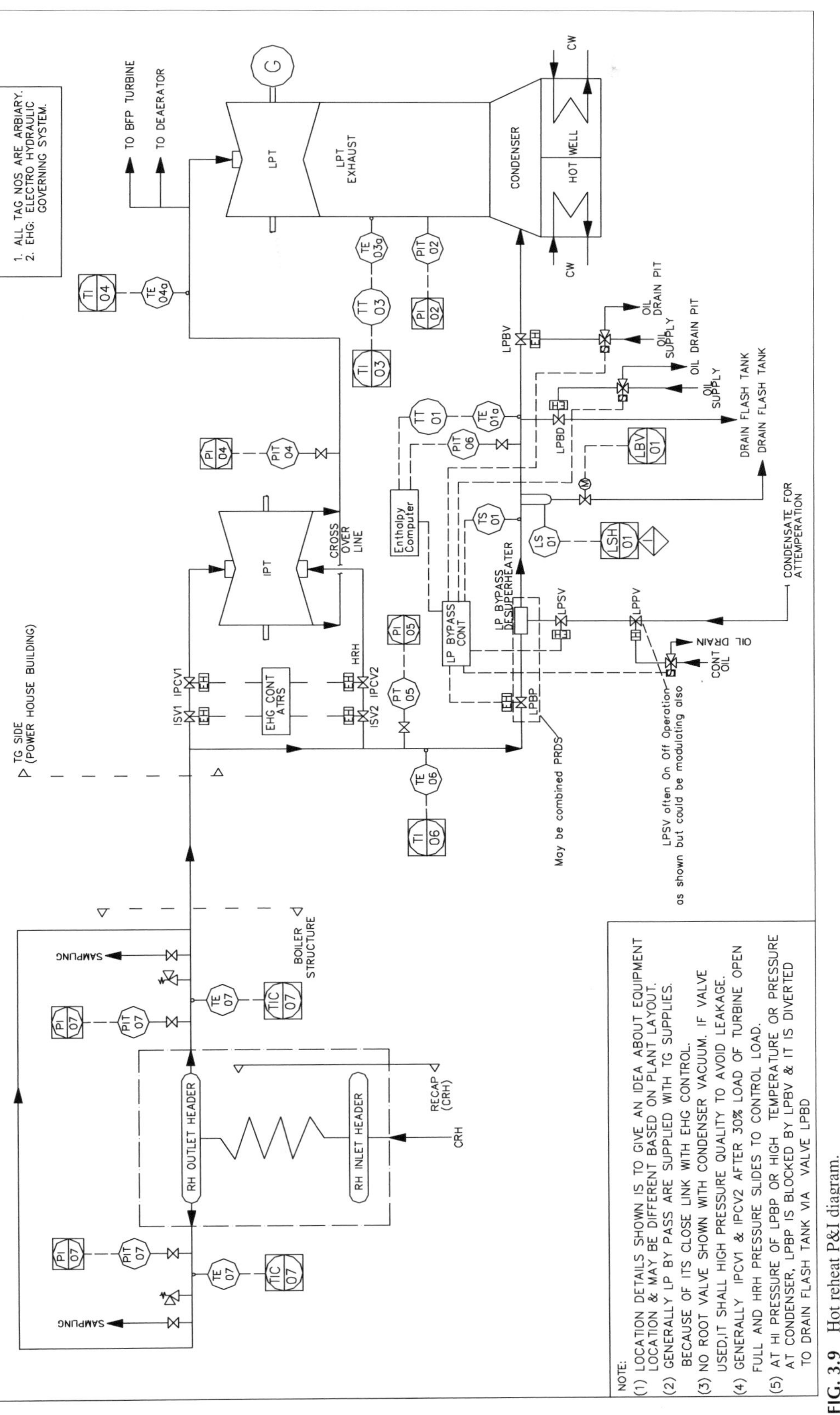

FIG. 3.9 Hot reheat P&I diagram.

steam entry to the IPT (hence, LPT). This is one hydraulic valve under the control of the EHG system of the turbine. Depending on the turbine start-up system, these valves can also be opened partially through the EHG control.

3. IPT control valve IPCV(1, 2): These are regulating valves that, when throttled, will regulate the steam admission to the IPT (as well as the LPT). Up to a 30% load of the turbine these valve throttle to control the load to the turbine sections. After that, these are kept wide open and the steam admission is to be guided by the HRH flow from the boiler. There could be some other variations also.
4. Cross-over line: It refers to the steam exhaust from the IPT for direct admission to the LPT. In certain cases, some extractions are taken from this line to supply steam to the deaerator and the BFP driving turbine (BFPT) during steady-state operation.
5. LP bypass system LPBP and LPSV: This is meant to bypass excess steam to the condenser by bypassing the IPT and LPT stages of the turbine. The pressure reduction in this line is done by the LPBP, the LP bypass valve. Now, there are two distinct situations here. When the LPBP is running in tandem with the turbine, then the LPBP pressure set point is to be derived from the HPT first-stage pressure (because during this time, the CRH pressure is variable, hence HRH pressure). In case the turbine is not running but the LPBP is in operation, during this time the LPBP needs to follow the HPBP; hence its set point may be derived from the HPBP (pressure-reducing valve position). Based on the above set point(s), the HRH pressure is reduced. The steam at reduced pressure is further cooled in a desuperheater, looking at the enthalpy *not* the temperature because the temperature at the desuperheater outlet after the LPBP is very near (or at) the saturation temperature. So, fixed enthalpy is to be maintained. In some of the designs, the control valve and desuperheater are built in one assembly. The spray control in this valve (LPSV) may not exactly behave like a standard control valve at times. It behaves as an on-off valve so as to ensure that the steam admitted to the condenser is not exceeding a preset value. In case the temperature exceeds that value for some reason, the same is not admitted to the condenser. It is instead sent to the drain flash tank, and if necessary the LPBP may be closed in extreme situations. A temperature element and a pressure instrument with suitable redundancy may be deployed to compute the enthalpy. Normally, fixed enthalpy is maintained (Fig. 3.9).
6. LPBV and LPBD: The purpose of the LPBV is to block the LP bypass steam flow to the condenser in case of high pressure and/or temperature. This is an interlock-operated valve generally connected with the LP bypass control. The LPBD is also an interlock-operated valve to allow passing the LP bypass steam to the drain flash tank in case the LPBV has blocked the flow. This is a typical scheme followed by some manufacturers.
7. LPPV: This valve is kept in the line to regulate the pressure of the condensate to the LPSV. Often, it has an on-off control but regulating/modulating controls are also seen.
8. Drain line: The LP bypass line going to the condenser has a U drain point to drain to the flash tank.

3.3.4 Major Parameter Measuring Monitoring in—HRH Steam

3.3.4.1 Process Parameter—HRH (Typical Values Only)

In the following table, a few process parameters at the turbine and boiler end have been listed to get a feel of the system (Table 3.3).

Cross-over steam (also referred to as EX4) has load-dependent pressure and temperature. Typically, the parameters for cross-over and LP exhaust, respectively, are 7 kg/cm^2 with temperature >340°C. LPT exhaust steam is ~0.081 kg/cm^2 with temperature about 40–45°C.

3.3.4.2 Monitoring of Process Parameter—HRH Steam

The following HRH steam parameters (with typical ranges) are important to be monitored at the control room.

TABLE 3.3 HRH Parameters

Parameter	HRH Outlet		IPT/LPBP Inlet		Remarks
Size	200 MW	500 MW	200 MW	500 MW	
Pressure*	29–39	46–47.5	28–38	45–46	*Typical value, may vary in actual case
Temp.*	540	540	540	540	
Flow	635-720	~1450	635-720	~1450	

Units: *P* in kg/cm^2, *T* in °C, *F* in t/h.

1. *HRH steam pressure and temperature at* the *reheater outlet*. The HRH at b the oiler outlet temperature is an important control parameter for the boiler. For larger power plants, a typical calibrated range for HRH pressure and temperature may be 0–50 kg/cm^2 and T 400–450°C, respectively.
2. *HRH steam pressure and temperature at* the *IPT/LP bypass inlet*. Similar to what is stated above. The HRH pressure at the IPT/LP bypass inlet is a control parameter for LP bypass control.
3. *Pressure temperature monitoring at cross-over steam*. Pressure and temperature measurement on this line at the turbine end is important. The typical range for a smart pressure transmitter is 0–10 kg/cm^2 while the Pt100RTD can be deployed for measurement of temperatures having a range of 0–450°C.
4. *LPT exhaust pressure and temperature*. Measurement of both these parameters is of great importance because this is basically a measurement of the condenser vacuum and condenser temperature, which are directly related to the plant's heat cycle performance. Referring to Fig. 3.9, it is to be noted that no root valve has been shown for pressure measurement, as this is in a vacuum. If any valve is to be provided, then it shall be of high-pressure class so that there is no leakage. A smart vacuum/absolute pressure transmitter in the short span of 200 mm wcl (0–0.2 kg/cm^2 A) shall be used. Whereas for temperature, the Pt100 RTD shall be deployed to measure a temperature in the range of 0–80°C.
5. *Temperature and pressure outlet of* the *LP bypass*. With reference to clause no. 3.3.3.6, the temperature alone at the outlet of the LP bypass (desuperheater outlet) is measured, but is not taken as a control parameter. Both temperature and pressure are measured to compute the enthalpy. For the temperature measurement, a Pt100RTDs/copper constantan (ISA type T) is used. For pressure measurement, a smart transmitter may be used. For protection in addition to temperature element, temperature switches are used to trip the LP bypass block valve and divert the steam to the drain flash tank. These are additional protections seen in many designs.

3.3.5 Controls in—HRH Steam

There are a few controls associated with constantan HRH system:

1. *HRH steam temperature control*. HRH steam temperature at the reheater outlet header(s) of the boiler is an important control parameter. After start-up, once the temperature of the HRH reaches the desired value (say 540°C in case of a large power plant), it is kept constant throughout. Initially, the fixed set point (manually adjustable) is kept much lower as there is insufficient steam flow. Later, when the steam flow is established, the set point is raised or may be set based on load (say, air flow). The temperature control means for HRH vary from boiler to boiler. In some designs, (corner fire) burner tilt is the primary control element whereas in wall-fire boilers, the gas bypass damper or gas recycling damper may be used as the main control element. Spray water is used during emergencies only. In some designs, spray water is used as the main control. However, in the case of SC boilers, the situations are a little different. Because here the FW:fuel ratio is most important, so the RH spray water is not really used for emergency purposes, but may be put into operation from the beginning.
2. *EHG control of turbine*: This is the most important control system for the turbine. In the case of IPT, the control valve is controlled up to 30% of the turbine load. After that, it is made wide open from the EHG, and during that time the steam pressure slides.
3. *LP bypass control*: During start-up/shutdown (and sudden load throw off of the turbine), this play the most important role in bypassing the requisite quantity of steam. Here, because the steam goes to the condenser, there is a block valve to protect the condenser during high temperature.

3.3.6 Redundancy in Measurement—HRH Steam

- The HRH boiler outlet temperature measurement is an important parameter, so the one of two redundancy is in general utilized in utility plants (for each side).
- LP bypass: The LPBP control system is normally supplied by turbine manufacturers with a sensor redundancy of one of two sensors for pressure as well as for temperature measurements. In many plants, there are two sets of such controls, and naturally the same is to be repeated for each control. In larger plants, redundancy in the temperature switch for turbine protection is also seen.
- The LPT exhaust temperature and pressure parameters are important and in larger utility stations, one of two redundancy is utilized.

3.3.7 Micellenous Points—HRH Steam

Like the main steam, the HRH parameters are at the central desk.

4 EXTRACTION STEAM (P&IDS): BLEED STEAM

4.1 Extraction Steam System

Extraction steam, also known as bleed steam, is the system for which the regeneration cycle has been made possible. In order to avoid thermal shock, the condensate and feed water are preheated by extraction steam before admission to the boiler.

4.2 Objectives and Functions of the System

Out of the total steam admitted to the turbine, the major part is utilized for doing the mechanical work in the turbine, whereas another part, instead of doing mechanical work, is utilized for preheating the condensate and feed water. Generally, the extraction lines are numbered from low pressure to high pressure (for example, the first three extraction lines from the LPT are numbered I, II, and III to heat the LP heaters 1, 2, and 3, respectively). However, reverse numbering systems are also seen. The number, location, and design parameters of each of these extraction lines purely depend on the heat balance diagram developed for the plant. The following are the basic purposes of extraction steam:

- In a typical subcritical turbine plant, say 200/500 MW, there are about three extraction lines from the LPT to the supply heating steam to three LP heater (LPH) shells to heat the condensate passing through the coil.
- Generally, from the cross-over pipe (exhaust line IPT—going to LPT), extraction steam (numbered Extraction IV) is supplied for deaerator heating as well as for deaeration function. The deaerator is therefore sometimes referred to as Heater 4. In addition to this, wherever there are turbine-driven BFPs, the steam supply for the same is taken from the cross-over steam.
- Normally, one extraction is taken from IPT for the steam supply to HP Heater (HPH) 5 feed water heating. However, there are designs where more than one extraction is taken from the IPT turbine. Based on design considerations, there may also be a single or double HPH.
- Feed water Heater 6 is generally heated by cold reheat (HPT exhaust). In certain designs, it has been found and is common that there is also extraction steam from the HPT. In some designs, it has been found that some extraction steam is also taken from the HPT for heating HP Heater 7.

4.3 System Description—Extraction Steam

The description given below for extraction stream needs to be studied in conjunction with Fig. 3.10 (with six extractions numbered from the LPT side).

The main function of each extraction line is preheating the condensate and feed water in the regenerative heaters. Each line is provided with at least one nonreturn valve so that under any circumstances of high level of the heater, water cannot ingress to the turbine. In extraction lines, there are power-operated NRVs that can be closed during certain operating conditions of the turbine. Also, by making heaters out, the turbine storage hence load shoot up can be made—as is done in super critical plants. It has been noted in some plants that there are also motorized valves in these lines for isolation purposes. The pressure in extraction lines is uncontrolled (meaning it depends on the turbine load). The various extraction lines are:

- *Extraction 1*: It starts from the bleeding line at a point in the LPT where the pressure is slightly higher than the LPT exhaust line and is used for heating the condensate at the coil of LPH 1 (may be located at the condenser neck).
- *Extraction 2*: This also is from the LPT with a pressure temperature higher than that of Ex. 1 for heating the condensate at the coil of LPH 2 located in the ground floor of the powerhouse building. However, in certain designs where there are seven heaters, this may be at the condenser also.
- *Extraction 3*: This also is from the LPT but has a pressure temperature higher than that of Ex. 2. It is used for heating the condensate at the coil of LPH 3 located in the ground floor of the powerhouse building.
- *Extraction 4*: This is a cross-over line coming out as exhaust from the IPT going to the LPT as steam input.
  - It goes for heating and deaeration or removal of gases (mainly dissolved oxygen) from the condensate at the deaerator, which is located at the deaerator floor in the powerhouse building.
  - Plants with BFPTs get the steam supply from this extraction line. This steam is admitted to the BFPT via the steam control valve to control the speed of the BFP (However there shall be at least one motor-operated BFP for start-up purposes.)
- *Extraction 5*: This line comes from the IPT before the cross-over line, having a higher pressure and temperature; it is used to heat the feed water in the heating coil of HPH 5.
- *Extraction 6*: This cold reheat steam line is used to heat the FW in the coil of HPH 6. When pressure in CRH line is > set point, this steam is utilized for BFPT and for deaerator until the Ex. 4 line is established (TG load of 55%).
- There are options for additional extraction from HPT also, for example, Ex.7.

4.4 Major System Equipment—Extraction Steam

The major system equipment in each of the extraction lines shall include the following:

- Nonreturn Valve: In each extraction line, there will be one nonreturn valve that allows extraction steam to flow into the heater and to prevent reverse flow.

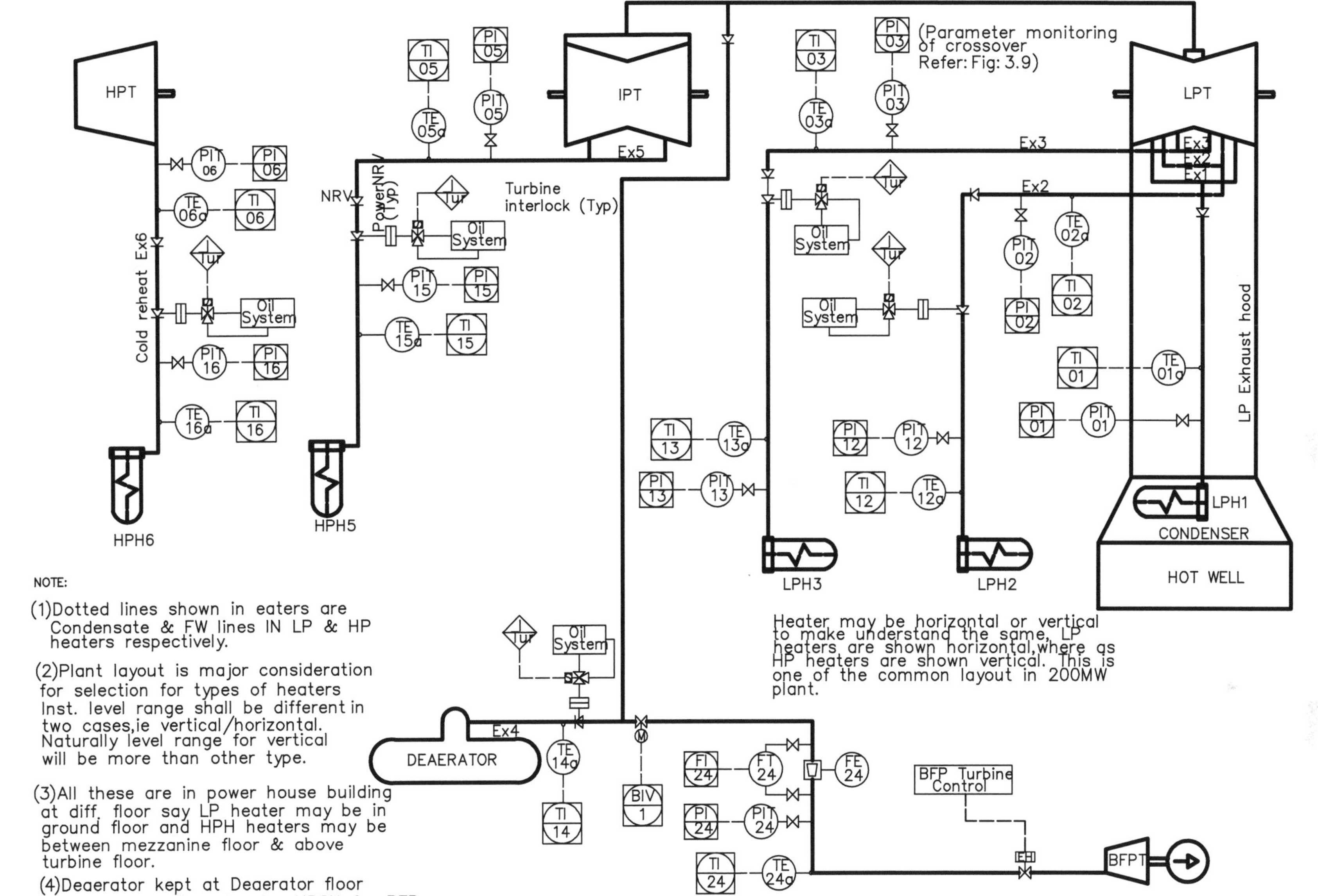
HPT
IPT
LPT
Ex5
Ex3
Ex2
Ex1
Ex4
Cold reheat Ex6
LP Exhaust hood
(Parameter monitoring of crossover Refer: Fig: 3.9)
Turbine interlock (Typ)
NRV
PowerNRV (Typ)
Oil System
Tur
PI 05
TI 05
PIT 05
TE 05a
PI 06
PIT 06
TE 06a
TI 06
PIT 16
PI 16
TE 16a
TI 16
PIT 15
PI 15
TE 15a
TI 15
TI 03
TE 03a
PI 03
PIT 03
TE 02a
TI 02
PIT 02
PI 02
TI 01
TE 01a
PI 01
PIT 01
TI 13
TE 13a
PI 13
PIT 13
PI 12
PIT 12
TI 12
TE 12a
HPH6
HPH5
LPH3
LPH2
LPH1
CONDENSER
HOT WELL
Heater may be horizontal or vertical to make understand the same, LP heaters are shown horizontal,where as HP heaters are shown vertical. This is one of the common layout in 200MW plant.
DEAERATOR
TE 14a
TI 14
BIV 1
FI 24
FT 24
FE 24
PI 24
PIT 24
TI 24
TE 24a
BFP Turbine Control
BFPT
NOTE:
(1)Dotted lines shown in eaters are Condensate & FW lines IN LP & HP heaters respectively.
(2)Plant layout is major consideration for selection for types of heaters Inst. level range shall be different in two cases,ie vertical/horizontal. Naturally level range for vertical will be more than other type.
(3)All these are in power house building at diff. floor say LP heater may be in ground floor and HPH heaters may be between mezzanine floor & above turbine floor.
(4)Deaerator kept at Deaerator floor above Turbine floor to give NPSH for BFP
(5)Deaerator also deaerates gases in addition to its heating function.

FIG. 3.10 Extraction steam P&ID.

- Power-operated nonreturn valve: Except in the extraction 1 line, all other extraction lines will have one additional power-operated nonreturn valve for easing out the turbine operation from the remote.

4.5 Major Parameter Monitoring in—Extraction Steam

4.5.1 Process Parameter—Extraction Steam (Typical Turbine Continuous Rating—TMCR Values Indicated)

See Table 3.4.

4.5.2 Monitoring of Process Parameter—Ex. Steam (Fig. 3.10)

Because the point of the consumption point (except for the heater located at the condenser) is away from the source at the turbine, normally the steam parameters are monitored at both ends. Also, these are required to monitor the performance of the turbine as well as the consuming equipment (heaters and BFPTs).

1. *EX1 steam pressure and temperature (from LPT to LPH1)*: A smart absolute pressure transmitter needs to be used for this purpose. The range shall be typically 0–1.5 kg/cm². For temperature measurement, an RTD (with a smart transmitter if used) can be deployed. The range of measurement shall be ~0–100°C.
2. *EX2 steam pressure and temperature (from LPT to LPH2)*: Measurements are the same as above, with typical pressure temperature ranges are of 0–2.5 kg/cm² and ~0–250°C.
3. *EX3 steam pressure and temperature (from LPT to LPH3)*: Measurements are similar to that in Ex2. Typical pressure range could be 5 kg/cm². Typical temperature range of measurement shall be ~0–350°C.
4. *EX4 (steam to deaerator and BFPT) steam pressure and temperature*: Similar to what has been discussed above. At the turbine end, smart transmitters in the range of 0–10 kg/cm² would suffice. However, at the consumer end, the same range would suffice for extraction steam, but because the deaerator/BFPT has a steam supply from other sources such as auxiliary steam, CRH, hence may have to select another transmitter to cater to those ranges. The instrument for temperature is the same as discussed above and a typical temperature range is ~0–350°C to 0–400°C. In the case of a BFPT, the steam flow is measured. Now, in many cases it has been found in the specifications that there will be separate steam lines (with separate governing systems) from CRH (and auxiliary steam until the time CRH not made available) and also steam from EX4. In that case, separate steam flow measurements are deployed. The steam flow may be in the range of 50–60 t/h naturally. A flow nozzle with smart DP transmitters is to be deployed.
5. *EX5 steam pressure and temperature (from IPT to HPH5)*: Similar to the discussions in Clause no.

TABLE 3.4 Extraction Steam Parameters

Ex No.	User	Press. Bar (A)	Temp. (°C)	Flow (t/h)	Remarks
For (TMCR) 200MW (typical)					
1	LPH 1	0.5–0.6	82	55	
2	LPH 2	1.2	<150	25–30	
3	LPH 3	2.5–3.0	200–220	25	
4	Deaerator + BFPT	7.0	300–320	85–90	Cross-over
5	HPH 5	15–16	425–450	70	
6	HPH 6	30–35	300–320	80	CRH
For (TMCR) 500MW (typical)					
1	LPH 1	0.64	87	79.6	
2	LPH 2	1.6	160	46	
3	LPH 3	3.0–4.0	220–230	47	
4	Deaerator + BFPT	10.0	350	160	Cross-over
5	HPH 5	22	470	95	
6	HPH 6	38–42	340–50	115	CRH

4.5.2.3. Typical pressure range could be 0–20 kg/cm² and typical temperature range of ~0–500°C.

6. *EX6 (CRH) steam pressure and temperature (from HPT exhaust to HPH6)*: Similar to the discussions in the above section. Typical pressure and temperature range would be 0–40 (50) kg/cm² and ~0–450°C, respectively.

4.6 Controls in—Extraction Steam

There is no direct modulating control associated with this system except the control for power-operated NRVs. Steam admission to BFPT has control (as part of the BFPT).

4.7 Redundancy in Measurement—Extraction Steam

Deaerator pressure measurement is critical as it is used to control the auxiliary steam and the CRH steam (Ex. steam is uncontrolled, varies with load). For that purpose, redundancy in the one of two mode is quite common.

4.8 Miscellaneous Points—Extraction Steam

Extraction steam measurements are more related to the turbine system, so these parameters are monitored at turbine end monitors. However, some of these measurements are repeated at the monitors pertinent to the control and monitor regenerative cycle, say the central monitor in the control room layout.

5 AUXILIARY STEAM (P&IDS)

5.1 Auxiliary Steam System

As the name suggests, auxiliary steam (AS) is the "steam supply" to help run various auxiliaries. In a power plant, the demand for auxiliary steam is there from initial start-up until the end when the unit is running at its maximum value (rated capacity). It is worth noting that the requirement is higher during start-up of the boiler and turbine and gradually, consumption decreases as the unit is loaded to its full capacity. When a unit is started, it starts with heavy fuel oil (HFO with a high energy arc igniter), so for heating of the same as well as for atomizing steam, auxiliary steam is necessary. Even if the unit is initiated with light diesel oil (LDO), then team is also necessary to heat the HFO so that it can be taken up at any time. Now, when the unit starts loading, the HFO is cut off slowly so the atomizing steam is not necessary. However, the heating steam is necessary, but the requirement is less only to keep the HFO in the liquid state through circulation. Similarly, during unit/turbine start-up, the steam requirement at the starting ejector (wherever applicable) is much more than the normal ejector. Also, for deaerator and gland steam, there is the requirement of auxiliary steam during start-up, but the requirement is reduced as the unit/turbine is loaded (normal ejectors in operation and deaerator is heated by extraction steam).

5.2 Objectives and Functions of the System

There will be a common steam header having a fixed pressure and temperature, (the parameter depends on the design of the power plant where steam is supplied to various consumers within the plant. These major functions of auxiliary steam (AS) shall include the following:

- To maintain constant pressure and temperature at the steam header, irrespective of the source of supply for the auxiliary steam.
- To supply auxiliary steam to the boiler and turbine auxiliaries at a constant pressure and temperature.
- In the boiler side, the auxiliary steam is supplied mainly to:
  - Fuel oil heating—storage/day tank heating, Line heating so that highly viscous HFO is in a flowable condition.
  - While HFO is fired, it is atomized with the help of auxiliary steam.
  - Steam supply to the steam coil air preheater (SCAPH) when in use.
  - Supply of steam for soot blowing (SB steam).
- On the turbine side, auxiliary steam is supplied mainly to:
  - Gland steam supply—during the start-up stage, the auxiliary steam is used to seal gland leakage.
  - Deaerator steam pegging with auxiliary steam.
  - Steam supply to the air ejector.
  - Supply of steam for BFP turbine (as applicable).

5.3 System Description—Auxiliary Steam

It is important to note that there exist various types and designs of the auxiliary steam system for various power plants. There are many power stations with lower MW size, or in process cum power plants, where two separate headers are developed for auxiliary steam. One supplies AS to the turbine, and is hence known as TAS while the other supplies AS to the boiler, and is hence known as BAS. Nowadays, in major utility stations with higher MWs, a common auxiliary steam header is formed. Depending on the source, there are two types in some plants: Main steam (MS) is taken as the sole source to form auxiliary steam header. In some designs, MS is used as one of the sources during start-up and CRH is used as the main source for the formation of auxiliary steam to avoid an unnecessarily high pressure drop, hence a loss of energy. All these are shown in detail in Figs. 3.11–3.14. It is worth noting that in some of the Chinese designs, it has been

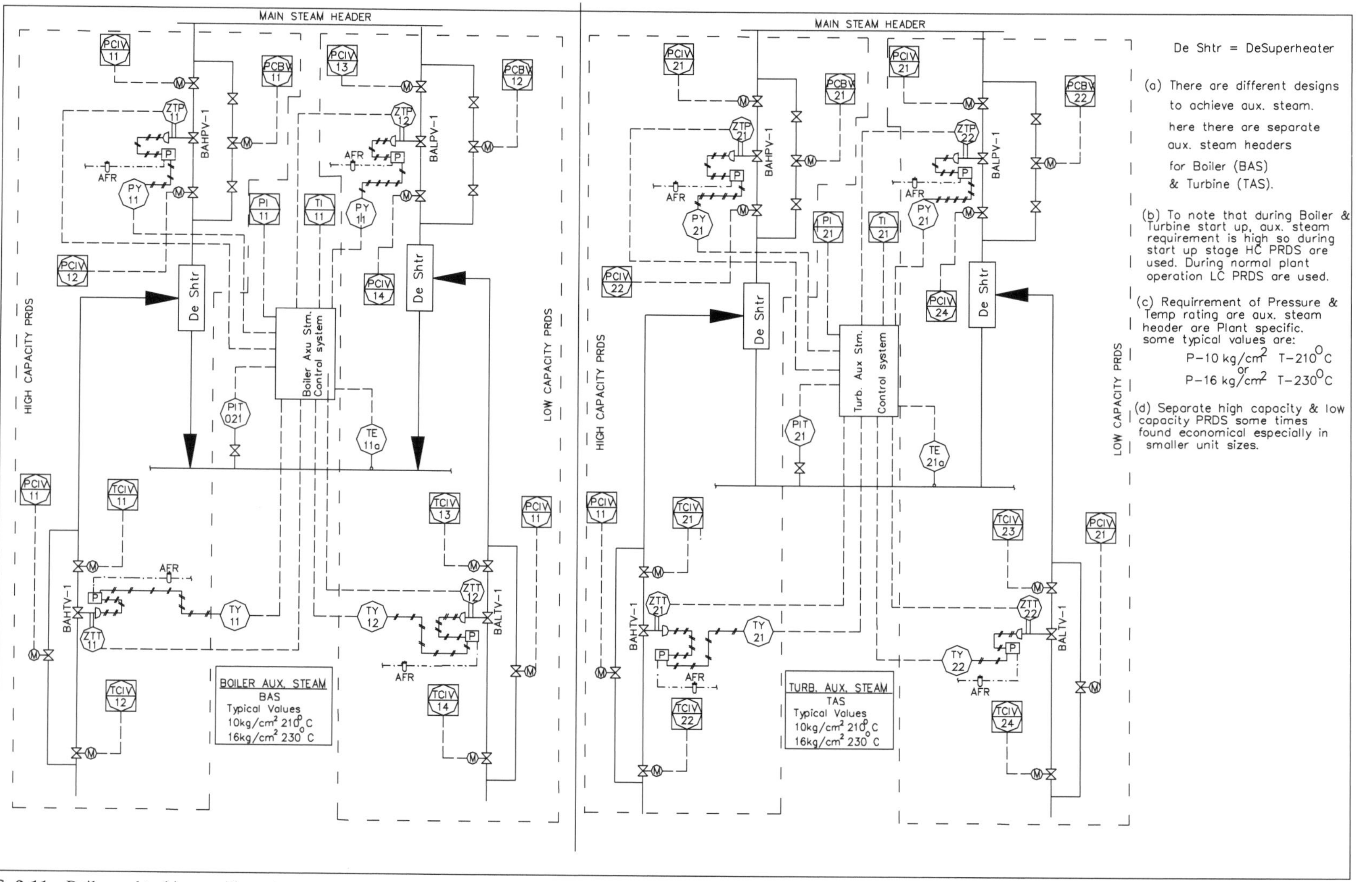

FIG. 3.11 Boiler and turbine auxiliary steam.

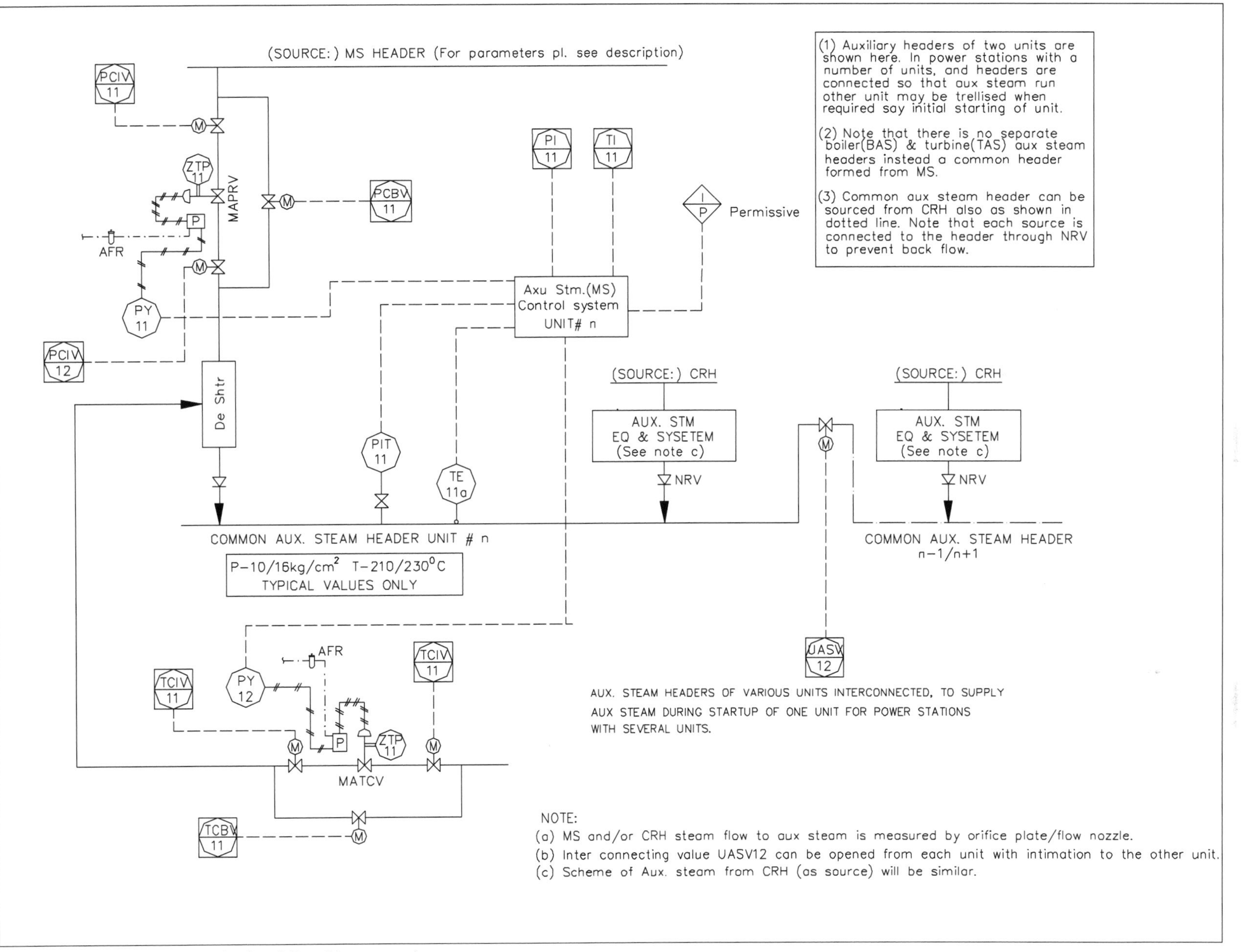

FIG. 3.12 Common aux, steam header from MS and interconnections.

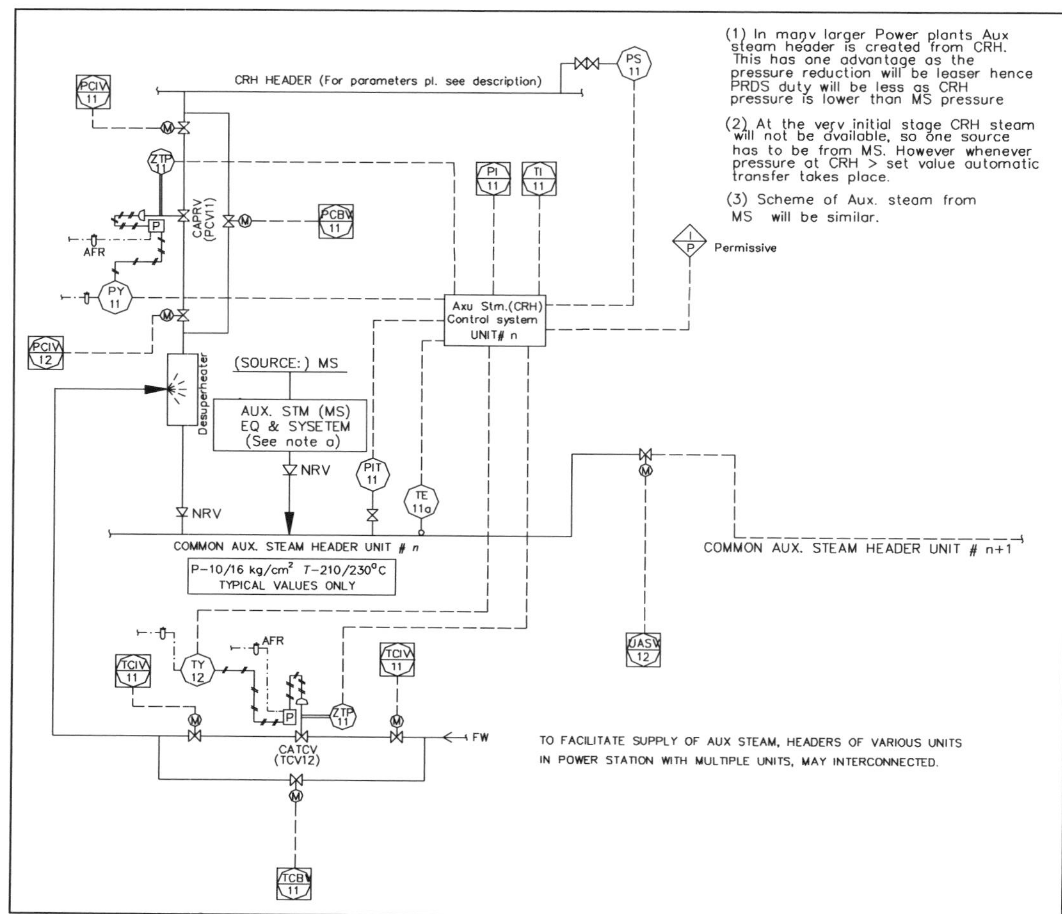

FIG. 3.13 Common aux, steam header from CRH and interconnections.

found that they use only pressure reduction to form an auxiliary steam header and temperature reduction is done according to the requirement of the consumers.

5.3.1 System Description—General (Figs. 3.11–3.14)

For discussions, we generally refer to instruments with tag numbers, as shown in Fig. 3.11; however these general discussions are equally applicable to other figures, except tag numbers may be different). High-pressure steam (MS or CRH) is reduced to the desired pressure set point with the help of a pressure-reducing control valve (*PRV*) whose opening is controlled by a controller (DCS) having the desired pressure set point. At the inlet (PCIV-11) and outlet (PCIV 12) of the control valve (for example, BAHPV-1 in Fig. 3.11), there are two remote-operated motorized isolation valves to isolate the control valve (for example, during maintenance). During this maintenance, the system can be made operational by remote manual throttling (inching) the bypass valve (PCBV-11). The purpose of the bypass valve is to heat the line before starting the PRV operation by cracking open the bypass valve.

After pressure reduction, the steam is passed through one desuperheating station to bring down the temperature of the steam to the desired point. The water supply to the desuperheater is controlled by controlling the feed water flow to the desuperheater. The desired temperature set point is compared in a PID controller (DCS) to generate control of the signal that regulates the FW flow to the desuperheater. The associated isolation and bypass operations are the same as explained above.

5.3.2 System Description—Separate BAS/TAS (Fig. 3.11)

In this type of design, as the figure suggests, there will be two separate auxiliary steam headers: the BAS and the TAS. In these types of systems, generally the MS is passed through the PRDS to generate BAS and TAS headers separately. As stated earlier, on account of different flow requirements of AS during start-up and at higher load, it

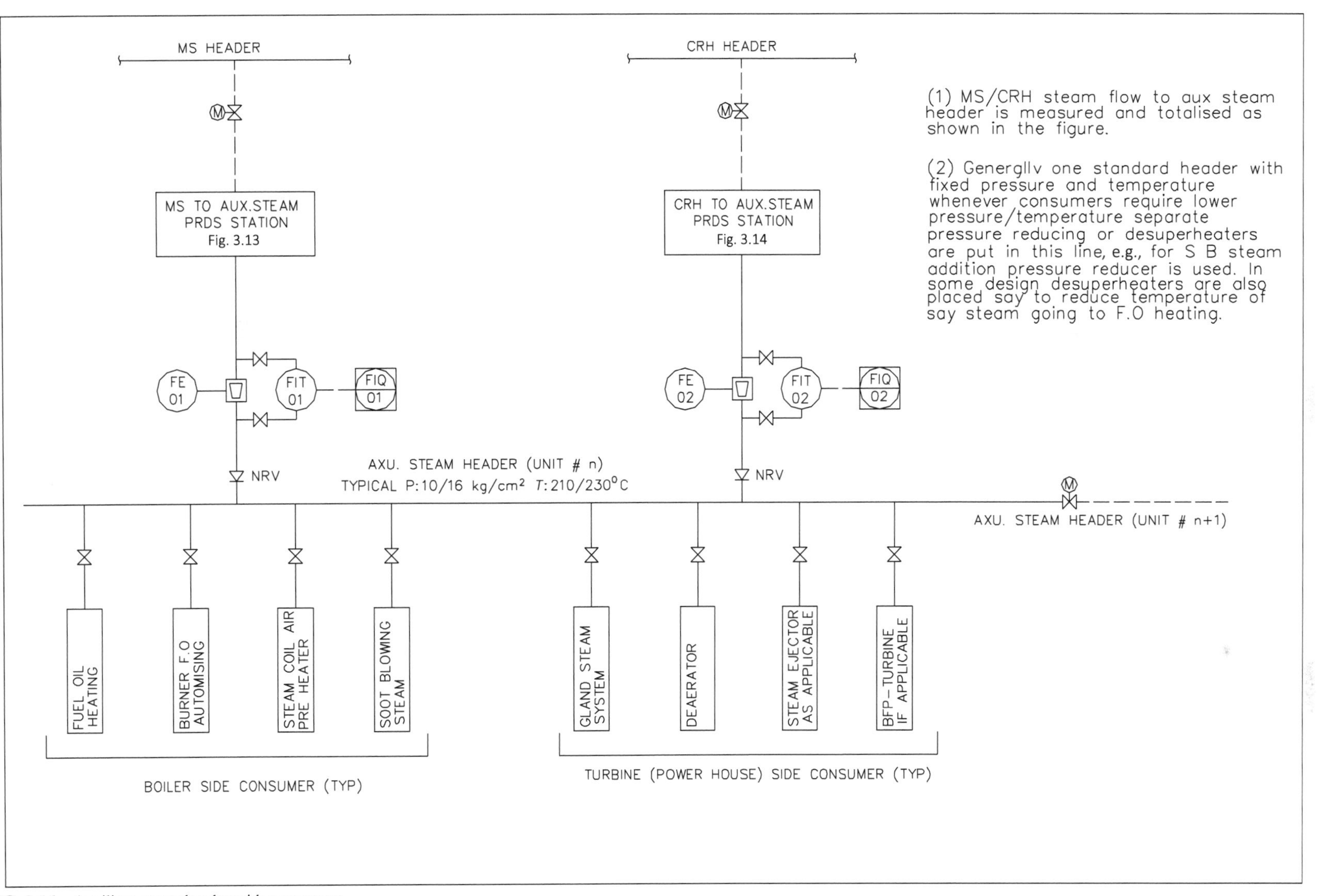

FIG. 3.14 Auxiliary steam header with consumers.

may not be possible for a single set of control valves to meet both requirements. Hence, there may be one high-capacity (HC) PRDS and one low-capacity (LC) PRDS, as indicated in the figure under reference.

Obviously, the question may arise as to why two separate headers.

For stations only meant for power generation, this configuration may not be desirable because there will be two sets of PRDS stations and each station with an HC and LC PRDS. Had there been one common header, then there may not be the requirement for separate sets of HC and LC PRDS stations. Because when the boiler is starting up (say, the pressurizing stage) it has a higher AS requirement, the turbine may not need steam at all, especially where there is a bypass in the system. So, combining two headers to form a common header could be a better option. However, this configuration may be beneficial in process cum power plants where the steam header consumers may be near the boiler (may be near the turbine side also) and the turbine is far off. Therefore, it is seen that from the layout point of view, this could be a better option to minimize energy loss (pressure drop) due to distance.

5.3.3 System Description—Common as Header From Main Steam (Fig. 3.12)

In this system, the common auxiliary steam header is created from the main steam with the help of PRDS stations, whose operations have been elaborated in clause no. 5.3.1. It is worth noting that in this type of PRDS station, there may not be any requirement to have separate HC and LC PRDS. From the main steam, auxiliary steam can be derived only after the boiler main steam pressure reaches a particular set point, say 14 kg/cm². Until that value is reached, auxiliary steam cannot be developed from the unit, and in those cases either the AS is imported from another unit as shown in the drawing for a station having multiple units or an auxiliary boiler needs to be arranged.

Because the MS pressure and temperatures are much higher (especially >200 MW units) than that for AS, it is economical to use CRH steam as the source for AS, especially when the power cycle has an HP bypass.

5.3.4 System Description—Common as Header From CRH Steam (Fig. 3.13)

This design is exactly the same as discussed in Section 5.3.3; only here, though, the AS is derived from CRH steam. Auxiliary steam can be derived only after the CRH steam pressure reaches a particular set point; otherwise, auxiliary steam cannot be developed from CRH. However, at the initial stage, the AS is derived from MS. It is therefore essential that there shall be some interlock operation for such a changeover (ref cl. No. 5.2.1).

5.3.5 System Description—Common Header With Auto switchover (Figs. 3.12 and 3.13)

During unit start-up, as stated earlier, the auxiliary steam needs to be imported from another unit and/or from an auxiliary boiler. When the unit is lit up and the main steam pressure is around a set value, say 17–18 kg/cm², the main steam pressure is put into the service. Initially, isolating valves are opened and the bypass valve is cracked open to warm the line. Then, the bypass valve is closed, the required set value is set in the controller for the pressure-reducing valve, and the valve starts opening when in auto mode, now depending on AS flow requirement imported supply may be cut off. After the unit is stabilized and the cold reheat pressure is > say 15 kg/cm², the CRH is put into the service. Here again in the CRH line to the AS, isolated valves are opened and the bypass valve is used for warming up purposes. The CRH Line PRV to AS is set to the desired value, say 16 kg/cm² (a *little higher* than the same in the MS line to the AS line PRV), and put in auto mode. Naturally, the CRH line PRV to AS will start opening to the full open condition and the AS header pressure > the set point for the MS line PRV set value. This will cause the MS line PRV to AS close and the CRH line PRV will be in operation automatically. Two separate temperature control valves are used to keep the AS header temperature at the desired value.

5.3.6 System Description as Header Consumers (Fig. 3.14)

In this diagram, all the probable consumers of AS for both the boiler and turbine sides have been shown. Also, a connection with another unit (in case of a multiunit power station) has been shown with the help of one motorized valve. This valve can be operated from either of the units. Whenever this valve is opened, generally the other unit is informed through the main.

5.4 Major System Equipment—Auxiliary Steam

The major system equipment auxiliary steam line shall include the following:

1. Pressure-reducing control valve: In each supply line to the AS header, there will be one pressure-reducing control valve to reduce the supply pressure to the desired value of the pressure at the AS header. Depending on the type of valve and duty, there may be separate HC and LC valves also, or a common one.
2. Temperature control valve: For cooling, there will be one desuperheater to which feed water will be supplied through one temperature control valve. The opening of the valve shall regulate the temperature by regulating the FW flow as a set point from the control system.

Depending on the type of valve and duty, there may be separate HC and LC valves also, or a common one.

3. Desuperheater: After one pressure-reducing control valve, there will be one desuperheater. Here, steam comes across the cooling FW in the form of the atomizing stage. It is also possible to use combined PRDS in place of a separate desuperheater.

5.5 Major Process Parameters and Measuring Monitoring in—AS

5.5.1 Process Parameter

The parameters in the auxiliary lines are purely functions of the design of the plant configuration and the requirements of various auxiliaries of the boiler and turbine (Table 3.5).

5.5.2 Monitoring of Process Parameter—Auxiliary Steam

Auxiliary steam parameter monitoring and control are done from the control. In the case of power station units having connected AS headers, the AS parameters are monitored in all concerned units. For details, Figs. 3.11–3.14 may be referred to.

1. *Auxiliary steam header pressure measurement*: This is one of the main parameters, and it is measured with the help of a smart pressure transmitter with a range of 0–20 kg/cm^2.
2. *Auxiliary steam header temperature measurement*: This is one of the main parameters, and it is measured with the help of RTD (Pt 100 at 0°C—typical). Along with RTD, a smart temperature transmitter with the range of 0–250/300°C may be used.
3. *Control valve position signal*: The position feedback signal of each of the control valves is measured. The valve position transmitters may be a discrete electrical transmitter with a 0%–100% scale with a typical output of 4–20 mA. Nowadays, smart actuators have built-in smart position feedback.
4. *MS/CRH header pressure measurement*: It is necessary to monitor the MS and CRH pressure before choosing one of them for the auxiliary steam header (Refer to discussion in clause no. 5.3.5). This is done by a smart transmitter with an interlock and or by a pressure switch having the set as discussed in clause no. 5.3.5.

TABLE 3.5 Typical Auxiliary Steam Parameters

MW	Pressure in kg/cm^2	Temp. in °C	Remarks
200	13	210	
500	16	230	

5.6 Controls in—Auxiliary Steam (Figs. 3.11–3.14)

1. *Auxiliary steam pressure control*: It is the most important control of the auxiliary steam. With the help of this control, the header pressure is kept constant. Depending on the AS flow, there may be a separate HC and/or LC control valve. In the case of a common header from the CRH, this control has an automatic transfer of source from MS to CRH.
2. *Auxiliary steam temperature control*: This control is used to keep the AS header at a constant temperature.
3. *Miscellaneous control*: There are several consumers who require auxiliary steam at a pressure and temperature lower than that available at the AS header. So, there will be a number of pressure and temperature controls for these. For example, the soot-blowing steam requirement is at a pressure lower than, say, the AS header pressure, so the necessary PRV is put to reduce the pressure.

5.7 Redundancy in Measurement—Auxiliary Steam

- In many higher MW utility stations, units have redundancy in the measurement of the auxiliary header pressure measurement. In temperature measurements, there is also redundancy in measurement.

5.8 Micellenous Points—Auxiliary Steam

Auxiliary steam measurements are more related to both the boiler as well as the turbine system, so these parameters are monitored at the central/common system monitor. Also, these measurements are repeated at the station system monitoring points.

6 FEED WATER STEAM (P&IDS)

6.1 Feed Water (FW) System

Feed water can be defined as return condensate + make up water.

Feed water is the high-pressure deaerated (removal of gases such as oxygen and carbon dioxide) water supplied to the boiler from which steam is generated. In the case of SC or USC boilers, the feed water has additional significance. In SC/USC boilers, there is no drum (storage vessel), so feed water flow has direct impact on fuel flow, that is, both feed water flow as well as fuel flow have to be precisely regulated to maintain the feed water-to-fuel flow ratio

FIG. 3.15 Benson point (definition).

Point/ load at which boiler control is switched from drum type to Once through is referred to/ considered as BENSON POINT/LOAD. At this point Separator-storage vessel levels almost vanishes). Typically this load is 25% BMCR. This information will be referred to later in this section.

necessary for proper operation of the plant. It is also important because the majority of the attemperation water is supplied from the feed water. There are several design alternatives available.

6.2 Objectives and Functions of the System

The feed water system provides treated high-pressure water to the boiler. Basic purposes shall include the following major functions (Fig. 3.15):

- To supply high-pressure water (free from dissolved gases such as O_2, CO_2, etc.) to the boiler during start-up, normal, and emergency operations. To supply attemperation water to the desuperheater sprays to control the superheater and reheater (emergency) temperature control systems and the desuperheater spray for HP bypass control.
- Ensure minimum flow of water through recirculation control of the BFP.
- Regulate the flow of feed to all types of boilers. In the case of a subcritical boiler, this is regulated to maintain the boiler drum level. In the case of a once-through (OT) drum level loop is missing instead there will be Separator/Storage tank level control, which is quite different in control philosophy from Drum level control.
- It also accepts chemical dosing to scavenge the dissolved gas further as well as to increase the pH value of the feed water.
- In the case of SC boilers, a precise ratio must be maintained with the fuel flow so as to avoid overheating, etc.

6.3 System Description—Feed Water (FW) System Variations

On account of variations in the BFP and its drives and feed control stations, there are some variations in feed water systems from plant to plant. The characteristic of a pump (Chapter 2) is nonlinear. Depending on the system resistance, an operating point is chosen. As is seen in the curve, when the flow is increased, the pressure falls. In smaller plants, this may be acceptable but in the case of a medium-sized plant, this is not acceptable because it will affect the net flow as well as the performance of the system. So the pump speed is chosen as a variable to change the flow without affecting the discharge pressure. The speed of the BFP can be varied in different ways, for example, by a hydraulic coupling scoop tube or by a variable frequency drive (VFD) for motor-driven BFPs. Also, many plants use a turbine-driven BFP, where the speed is varied with the opening of the BFP turbine steam control valve. It is well known that a drop across any valve at high flow is a loss of energy, so in higher MW plants, feed control stations at higher loads are removed and instead the feed flow is controlled by the speed control of the BFP. All these have been tried to cover with the help of various Figs. 3.16–3.20. In the case of an SC boiler through a low feed flow control valve, normally a minimum 30% flow is ensured during start-up, and until the *Benson point* is reached, the separator (somewhat like drum in subcritical boiler) level is controlled by the boiler water circulation pump. After a certain load, the feed water valve is wide open and the feed flow is precisely controlled by BFD speed control, as per the set point from boiler master demand with due consideration to maintain feed water-to-fuel ratio so as to maintain superheater temperature.

6.3.1 System Description—General (Fig. 3.16A and B)

Feed water is supplied from the deaerator, and is located at higher elevations (deaerator floor) to provide the necessary net positive suction head (NPSH) required for BFP at the ground floor level. After the feed pump, the FW is passed through a series of shell- and tube-type regenerative feed water heaters. Feed water is passed through the tubes whereas the steam is passed through the shell. After the heater, the main FW is passed through a feed control station that comprises a set of control valves. Feed control valves are meant to regulate and maintain the required feed water flow to the boiler. In the case of a drum boiler, these valves help to maintain the drum level at the desired (set) point. After the feed control station, the FW reaches the economizer of the boiler. A part of the FW is used for the purpose of attemperations, for example, at desuperheaters, HP bypass control. etc. Sometimes, the emergency reheat attemperation is taken from the booster stage output also, as shown in Fig. 3.17. Normally, for superheater temperature control, the spray flow is taken from the HP heater outlet to avoid thermal shock (because this attemperation water will meet the superheated steam at the desuperheater). However, for HP bypass, the attemperation flow is taken from the BFP discharge, but why? This is done probably to have less attemperation water for HP bypass (energy loss) and this also helps to ultimately avoid condenser loading. Also, the HP bypass steam does not go to the turbine

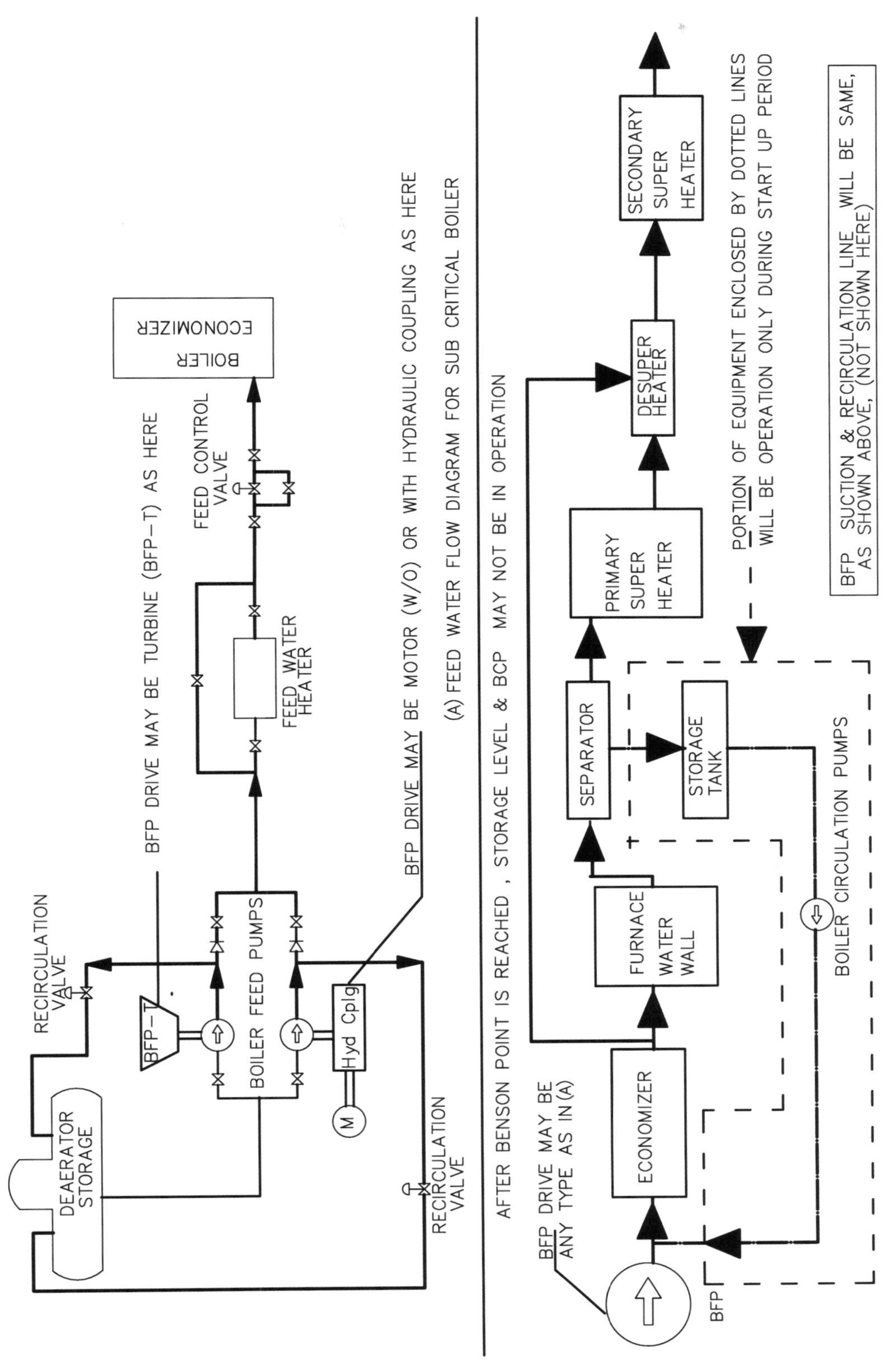

FIG. 3.16 Typical feed water flow diagram.

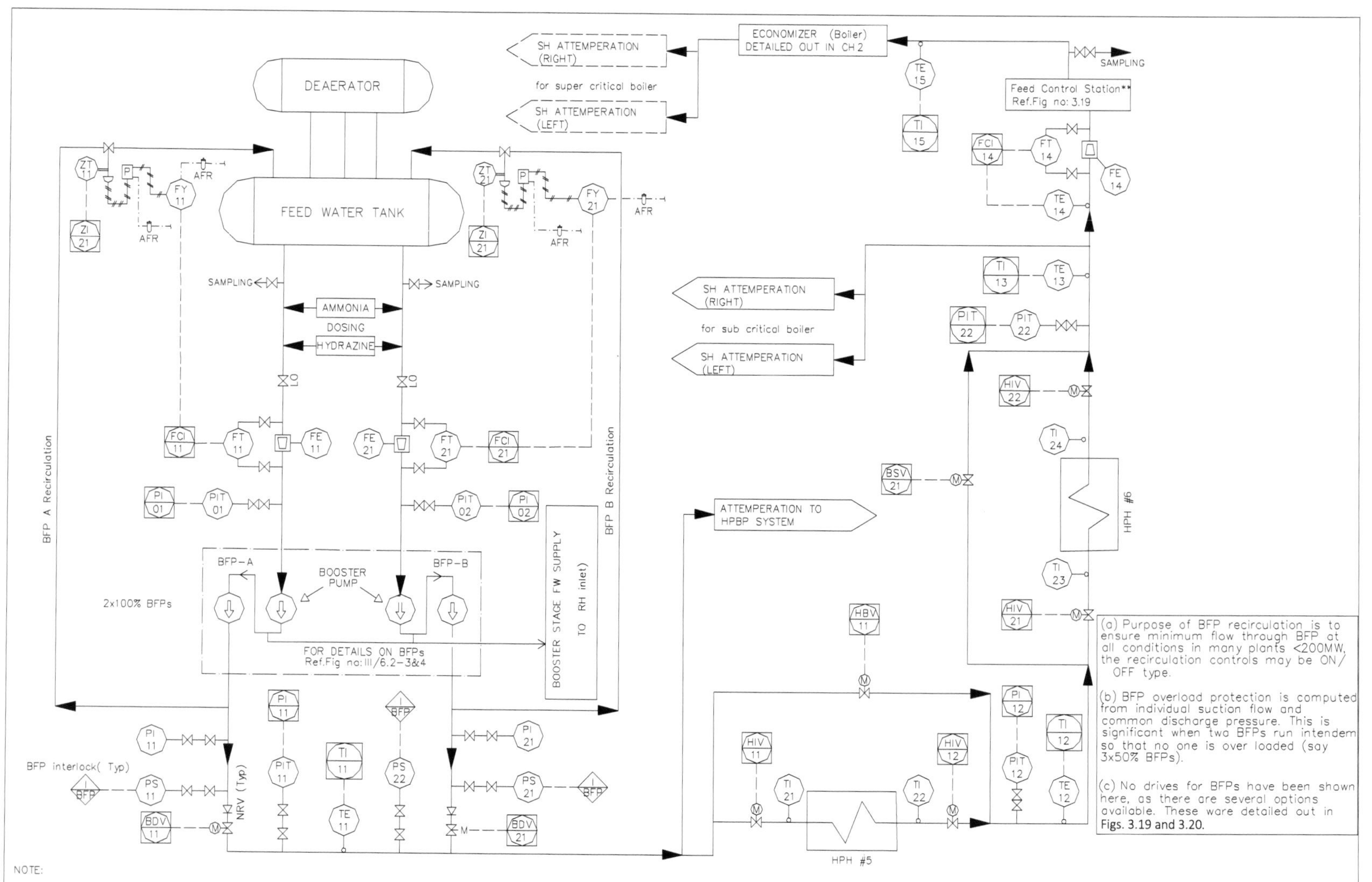

FIG. 3.17 Feed water system (overall P&ID).

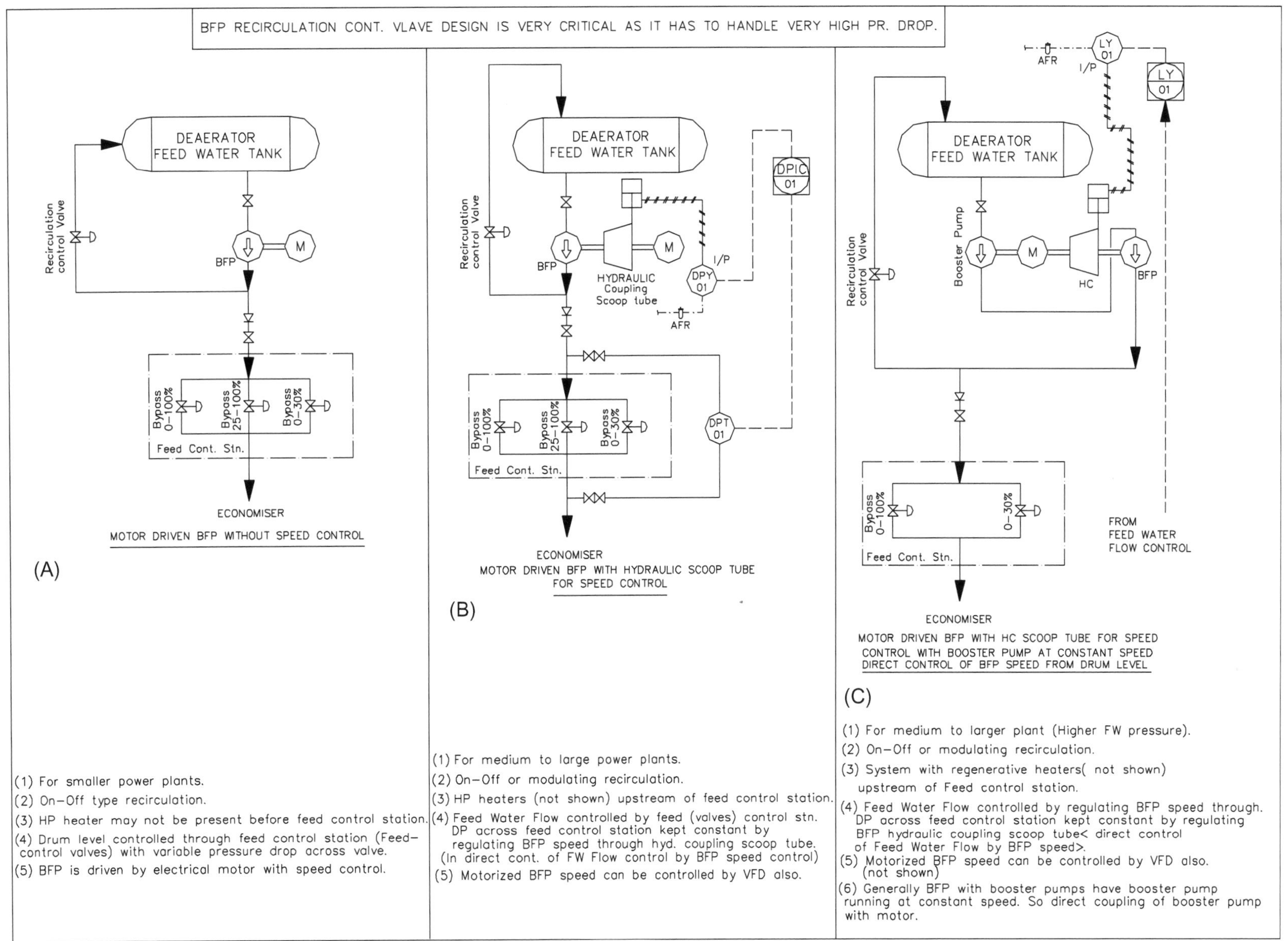

FIG. 3.18 FW system motor operated BFP design variations.

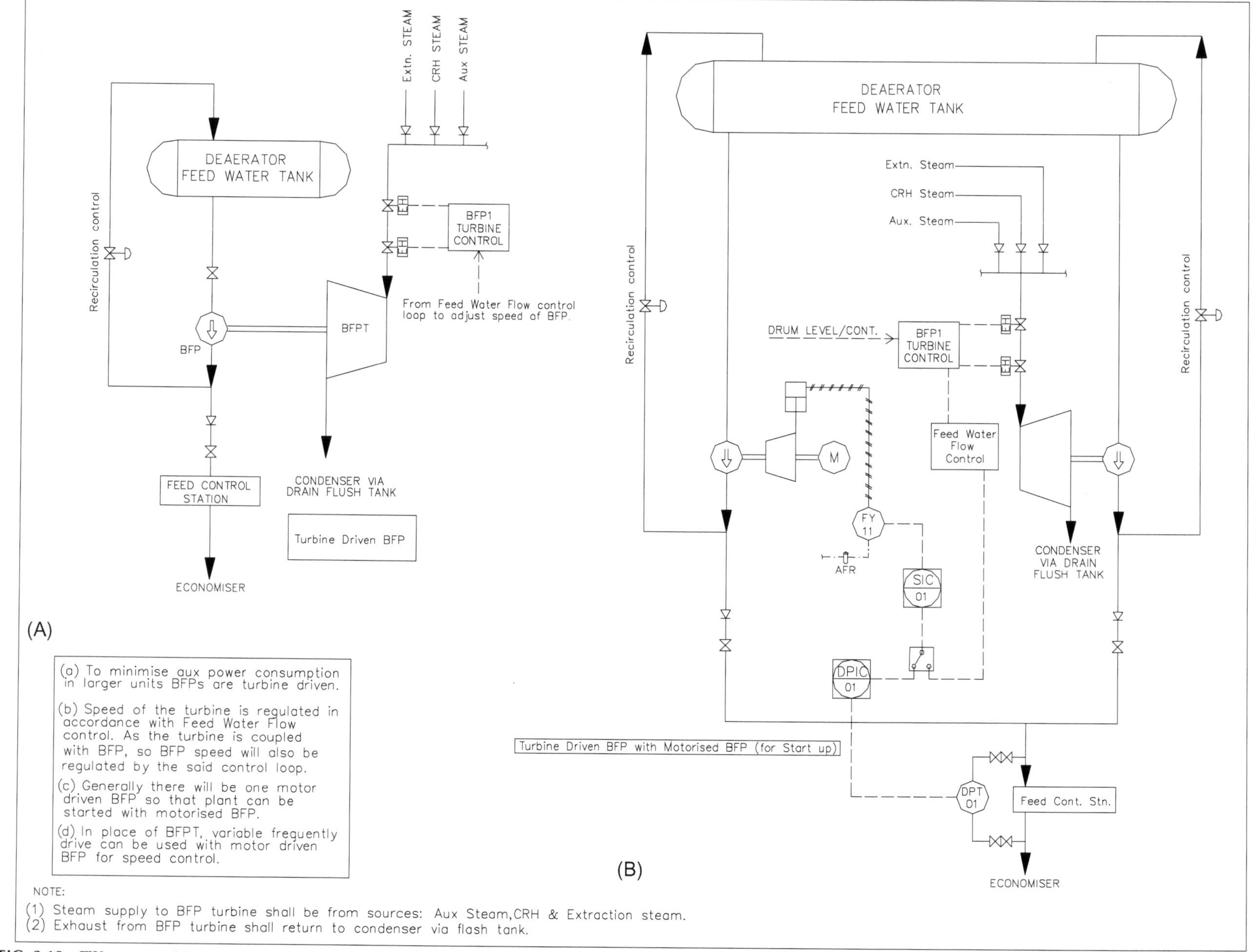
Extn. STEAM
CRH STEAM
Aux STEAM
DEAERATOR
FEED WATER TANK
Recirculation control
BFP1
TURBINE
CONTROL
BFPT
BFP
From Feed Water Flow control loop to adjust speed of BFP.
FEED CONTROL STATION
CONDENSER VIA DRAIN FLUSH TANK
Turbine Driven BFP
ECONOMISER
(A)
(a) To minimise aux power consumption in larger units BFPs are turbine driven.
(b) Speed of the turbine is regulated in accordance with Feed Water Flow control. As the turbine is coupled with BFP, so BFP speed will also be regulated by the said control loop.
(c) Generally there will be one motor driven BFP so that plant can be started with motorised BFP.
(d) In place of BFPT, variable frequently drive can be used with motor driven BFP for speed control.
NOTE:
(1) Steam supply to BFP turbine shall be from sources: Aux Steam,CRH & Extraction steam.
(2) Exhaust from BFP turbine shall return to condenser via flash tank.
DEAERATOR
FEED WATER TANK
Extn. Steam
CRH Steam
Aux. Steam
Recirculation control
Recirculation control
DRUM LEVEL/CONT.
BFP1
TURBINE
CONTROL
Feed Water
Flow
Control
M
FY
11
AFR
SIC
01
DPIC
01
CONDENSER
VIA DRAIN
FLUSH TANK
Turbine Driven BFP with Motorised BFP (for Start up)
DPT
01
Feed Cont. Stn.
ECONOMISER
(B)

FIG. 3.19 FW system turbine operated BFPs.

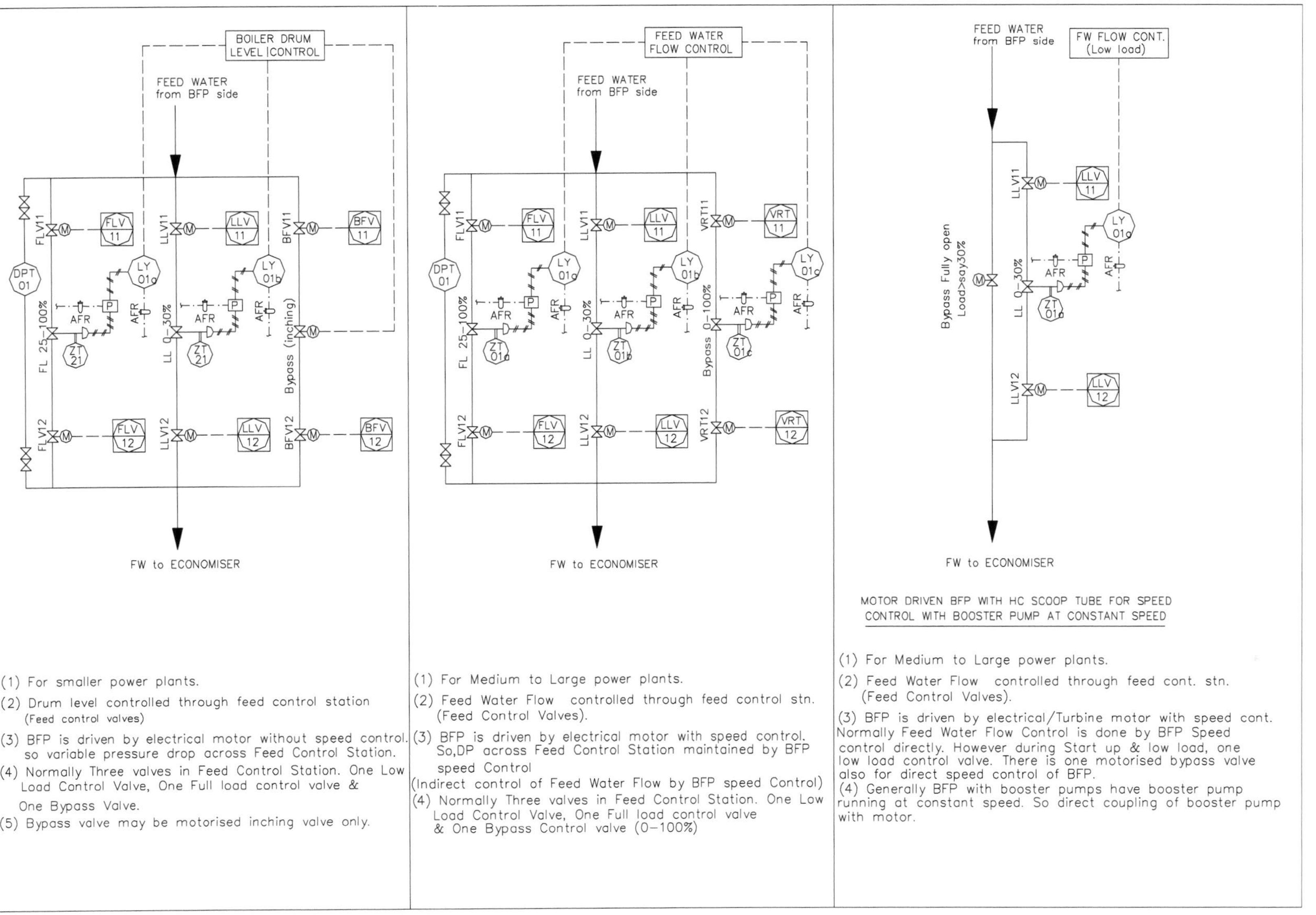

FIG. 3.20 FW system feed control station design.

(it is instead returned to the boiler for reheating, hence the fear from water carryover is not there). However, in the case of large *SC boilers*, it is customary to take the SH temperature control spray flow from the economizer outlet because in those case, the MS temperature is higher (to avoid thermal shock). However, in certain CFBC SC boilers, superheater spray flow at various stages (of multistage SH temperature control) of feed water flows is also taken from the HP heater outlet. Here one should note that these spray flow at "secondary stages" actually act as *trim* control, whereas the spray flow from the economizer outlet at "primary stage" is the main control where the maximum spray flow takes place. Also, in the case of SC boilers, steaming takes place in water walls/separators. During the *initial start-up* period, the water is separated from the steam at the separator and the water is collected at the storage tank whose level is maintained by regulating the boiler circulation pump (BCP) flow, with the help of a control valve (at the downstream of the BCP). During the process of water to steam at the separator inlet, the feed control valve is throttled to maintain the feed water flow rate of 25%–30% boiler maximum continuous rating (BMCR) through the economizer. Then, as the firing increases with the load, the steaming at the separator decreases (as it takes place in water walls), and at a certain value of steam flow, the BCP is stopped and there will be hardly any level in the storage tank. As shown in Fig. 3.16A, the equipment enclosed by the dotted lines becomes nonfunctional at higher loads, but the separator will be in service at all times. Therefore, unlike the drum boiler, there is no storage device in the once-through (OT) boiler design. Therefore, it is needless to say how the feed water-to-fuel ratio is important for proper operation of the boiler.

It is worth noting that a part of each BFP discharge (before the discharge valve) is recirculated back to the deaerator. This is BFP recirculation. *Recirculation control is necessary to ensure the minimum FW flow through the BFP under all conditions.* Initially, when there is no requirement of main FW, that time this recirculation ensures that total flow (main FW + recirculation) is > approximately 30% of the full flow of BFP. On account of the churning effect inside the heavy duty pump, a lot of heat will be generated, so some minimum water flow is necessary to take away this heat. The flow for the individual BFP suction is measured, and a set point of ~30% of flow is set to operate each BFP recirculation valve. When the FW flow through the BFP is more than the set point, the recirculation valve closes. In smaller and medium-sized power plants, these controls are the on-off type whereas in large plants >200 MW, these controls are the modulating type. However, there is no sacrosanct rule for this as a modulating control can also be used in lower plant sizes as well. From the discussion above on the SC boiler, it is seen that during the initial start-up, 30 BMCR feed water flow is ensured through the economizer. Therefore in the case of the SC and USC boilers, the requirement of BFP recirculation is less important, except for own starting/testing of the pump.

6.3.2 System Description—FW System Overall (Fig. 3.17)

In this figure, two stages of the BFP are shown: the booster stage and the main BFP. The booster pump is generally located at the ground floor, having its suction from the deaerator located at a higher elevation to ensure the required NPSH. The discharge of the booster pump acts as the suction of the BFP. The BFP may be located at the operating floor, having some piping between two pumps where even the flow elements are placed. However, it is better to place the flow element in the suction line from the deaerator, as shown in the drawing under reference from the straight length availability point of view. In the drawing under reference, 2 × 100% BFPs have been shown—this is common for utility stations of higher sizes. There could be some other options, for example, 3 × 50% also. This means that there will be two 50% pumps running and a third one on standby. The third one will cut in automatically in case of tripping on any pump. Based on reliability studies and pump performance data, a decision regarding the number of pump options may be made. For the option where two BFPs are running in tandem, there is the possibility that one may get overloaded. To prevent that, a BFP protection circuit is developed by measuring the FW flow to the individual BFP and the associated discharge pressure. This is not really applicable for one pump running at a time. Generally, in RH, attemperation spray is tapped from the booster stage output and HP bypass water is taken from BFP discharge. In many cases, the RH attemperation spray flow is insignificant and may be used during an emergency, yet in the case of SC boilers, that may not be so.

6.3.3 System Description—Feed Water Sampling System

At the suction line to each BFP chemical such as *ammonia* and *hydrazine* are dosed to increase pH value of FW and to scavenge dissolved gas respectively. There will be sampling lines from BFP suction lines and at the FW at the economizer outlet for analysis (Chapter 5).

6.3.4 System Description—Feed Water System Motor-Operated BFP (Fig. 3.18)

In this part, various aspects of BFPs and types of speed controls shall be discussed.

1. Fig. 3.18A: In smaller plants (sizes say <60 MW), BFPs are driven by a motor at constant speed. To ensure the minimum flow through the BFP, however, there will be on-off BFP recirculation valve to return the FW to

the deaerator. Now, the drum level is controlled by modulating the opening of the feed control valves. So, at different FW flows, there will be a different drop across the feed control valves, so these valves shall be selected giving due consideration to this aspect. In the feed control station, there are three control valves: one low load (say 0%–30%), another full load (25%–100%), and a bypass valve (0%–100%; may be motorized for smaller plants).

2. Fig. 3.18B: A hydraulic coupling scoop tube is a very popular means to regulate the speed of pumps/fans (for example, ID fans). A constant speed motor is coupled to the pump/fan by means of a hydraulic coupling (*turbo coupling* is represented in the drawing with the symbol of the turbine). Depending on the degree of coupling (filling), the speed of the driven item (BFP in this case) is varied. This coupling filling (degree) is varied according to the scoop tube position [Section 3.1.4.1 of Chapter 6]. A *VFD* is another way of regulating the speed of the motor-driven BFP [Section 3.1.4.2 of Chapter 6]. The speed of the turbine drive BFP (TDBFP) is regulated by regulating the steam flow to the turbine. There are different modes of controls for BFP speed:
   - **(a)** In one type, the differential pressure (DP) across the feed control valve station is kept constant. In this method, the feed flow is controlled by regulating the opening of the feed control valve (primary control). Naturally, there will be a variation in DP across the feed control valves. With the help of another controller, the DP across the feed control valve is kept constant by regulating the speed of the BFP.
   - **(b)** In this method, the feed flow is directly controlled by regulating the speed of the BFP; hence, the BFP speed is the primary control. It is worth noting that the direct variation speed of the BFP is done at BMCR >25%. At a low load, this method is not effective because for a low load, there exists one low-load feed control valve to regulate the FW flow. After the boiler load >25%–30%, the bypass valve to the feed control valve opens fully and the FW flow is controlled directly by the BFP speed regulation. It is recommended that the transfer of control from the feed control station to the BFP speed control be done manually from a plant safety point of view (Fig. 3.18C).
   - **(c)** In the case of OT boilers, the feed control valve comes into operation during start-up to make sure that the minimum (~30% BMCR) feed water is sent to the boiler. After a 30% load, the bypass valve is fully open, the feed control valve closes, and the BFP speed is controlled to regulate the flow to the boiler as well as to maintain the feed water-to-fuel ratio at the desired point.

3. Fig. 3.18C: This is the same as that discussed in connection with Fig. 3.18B:—with the difference here that there is a booster pump in addition to the BFP. Here, only to note that the speed of the booster pump is constant so, motor is directly coupled. Whereas the BFP is coupled to the motor via hydraulic coupling, as detailed above.

6.3.5 System Description—FW System TD BFP (Fig. 3.19)

In many units, especially in larger units to reduce auxiliary power consumption and for better steam utilization, a steam turbine is used to drive the BFP in the place of a motor. Extraction steam is the main source of steam supply to this boiler feed pump turbine (BFPT). In addition to the extraction steam, auxiliary steam and CRH steam supply lines are also taken so that it can run when the extraction steam supply is not available. Like a normal steam turbine, the governor control valve is regulated to control the speed of the turbine; hence, the BFP speed is in accordance with the control demand (ref. clause no. 6.3.4.2a and b and Fig. 3.19A). The exhaust steam from the BFPT is returned to the condenser via the drain flush tank. In power plants with BFPTs, there shall be at least one motor-driven BFP to cope with the very initial start-up program (Fig 3.19B) and/or for a quick auto cut in of the BFP in case of another BFP trip.

6.3.6 System Description—FW Feed Control Station (Fig. 3.20)

The feed control station regulates the feed water flow to the boiler as per demand. During start-up and low load (flow), the downstream pressure of the feed control station is very negligible (for a drum boiler, until drum pressurized. It is only the water head due to boiler drum height), and because the BFP flow is low, naturally the head will be high. As a consequence, at low flow there is a very high DP across the feed control station. Now at a higher load, the drum is pressurized, also as the FW demand is comparatively more and BFP discharge pressure is comparatively lower (that is, at higher load lower pressure drop across the control valve). Thus, the process conditions at the feed control station under two different conditions are completely divergent. Any single valve handling such a situation has to have a very special design. For this reason, in feed control stations there are separate control valves for low load and high load. There is also one bypass valve to cater to an emergency if one valve is in trouble and/or has been taken for maintenance. So the bypass control valve design is also very stringent. A variable resistance control valve could be a better option.

1. Fig. 3.20A: Here, there are three valves. The first is the low load valve to cater to the flow of say 0%–30%. Overlapping with the same, there is another full load control valve catering to 25%–100% full flow.

In smaller units, to economize the cost the bypass valves are sometimes chosen as motorized valves for inching (throttling) operations during an emergency.

2. Fig. 3.20B: This is the same as the system discussed in clause no. 6.3.5. The only difference is that unlike a bypass valve being a motorized inching valve, it is a special valve called a variable resistance trim (VRT) valve or a drag valve. Some special designs cater to the requirement of 0%–100% load condition of the FW.

   Here are a few points worth noting regarding the changeover of control valves:

   a. Changeover from a low feed control valve to a high load feed control valve could be done manually or automatically. Both systems exist and there are two schools of thought on the issue because many are of the opinion that if such changeover is done automatically, there may be some bump in the system. However, now with advancements in control systems, valves, and actuator technology, auto changeovers are not uncommon.
   b. Even under a higher load, there may be fluctuations in the DP across the feed control valves. In order avoid this, the DP is regulated by the speed control of the BFP, as discussed above. However, for smaller power plants with constant-speed BFP, control valves need to face fluctuating DP across the valve.
3. Fig. 3.20C: As discussed earlier, in order to prevent energy loss across the feed control valve, in current power plant designs the low load control valve (0%–30%) is kept to cater to the start-up requirements for both the OT boiler as well as for the drum boilers. To control the feed flow, a low load feed control valve is used (because at these conditions, the speed control of the BFP may not be a good proposition). Whenever the FW flow (load) >25%–30% (that is, the control valve is almost fully open), then the motorized bypass valve gets the signal to open the valve slowly. As the big bypass valve starts opening, the flow increases and the low feed control valve closes. Also, the feed flow control is transferred from the feed control station to the speed control of the BFP. At low load, the DP control to regulate the BFP speed may be there also, but when the control is directly transferred to the speed control of the BFP, the DP control (if it is at all there) will be nonfunctional. Here, it is to be noted that in most of the present-day designs, the bypass valve is considered a motorized valve that also has an inching operation facility.

6.4 Major System Equipment—Feed Water System

Major system equipment in the feed water line shall include the following:

1. *Deaerator*: A deaerator has two basic functions: get rid of dissolved gas and as a heater. Mechanical deaeration is the preliminary step to get rid of O_2 and other dissolved gases such as CO_2. Depending on the deaeration method, the deaerator can be divided into the tray type or the spray type. While deareating feed water with the help of steam, it also acts as a heater.
2. *Boiler feed pump*: The BFP is really the heart of not only this system but of the entire cycle. BFPs are in general a centrifugal pump with low tolerance. They have multistage impellers and call for a minimum 30% flow of hot water through it. For C&I engineers, the recirculation system and recirculation valve are very important. In some designs, mainly for larger units, booster pumps are used at the upstream of the BFP, that is, two-stage pressure raising. As discussed earlier, there could be variations in speed controls and drives for BFPs.
3. *HP heaters*: In the feed water system, there may be two to three shell and tube FW heaters (HPH 5 and 6) and these are heated with the help of extraction steam in the shell. After heating the feed water in the tube, the extraction steam is condensed in the heater, where it is finally again taken to the system.
4. *Feed control station*: The feed control station basically consists of a number of feed control valves along with associated isolation valve, as discussed in clause no. 6.3.5.
5. *Economizer*: It is a kind of heat exchanger where the heat of the flue gas is utilized to heat the feed water. Refer to clause nos. 9.4.6 in this chapter and Chapter 2.

6.5 Major Parameter Measuring Monitoring in—Feed Water

6.5.1 Process Parameter—Feed Water (Typical Values Only)

Some typical values have been put forward (Table 3.6).

6.5.2 Monitoring of Process Parameter—Feed Water System

For these measurements, Figs. 3.17–3.20 may be referred to.

1. *Individual BFP suction flow measurement*: At the suction of each of the BFPs, the flow elements (for example, flow nozzle) are placed to measure (by smart DPT with a typical range of 0–10000 mm wcl) the suction flow (typical range in t/h: 200 MW: 0–800; 500 MW: 0–1800) to each BFP (in designs, with the booster pump measurement point may be at the booster pump discharge). Please note that the flow range given is for 2 × 100% BFP configuration; naturally, the same will be halved for 3 × 50% BFP configuration. This BFP suction flow is not the main feed water flow

TABLE 3.6 Feed Water Parameters (Typ.)

Location	Pr. (kg/cm²)	Temp. (°C)	Flow (t/h)[a]	Remarks
200 MW				
BFP SUCTION	7.0+Dea*	180	800	*Water head
BFPDISCH.	180–190	180	800	
HPH 5 O/L	175–185*	215	800	*~5 kg/cm² drop
HPH 6 O/L	170–180*	248	800	*~5 kg/cm² drop
500 MW (drop across heater may be more)				
BFP SUCTION	9.0+Dea*	190	1700	*Water head
BFPDISCH	200	190	1700	
HPH 5 O/L	190*	230	1700	*~5 kg/cm² drop
HPH 6 O/L	~195*	270	1700	*~5 kg/cm² drop

[a]*Spray considered after HP heater (Reheater emergency spray flow).*

measurement (discussed in clause no. 6.5.2.8), but is used for BFP performance, Generating signal for BFP recirculation control and for the cases more than one BFP running (say in case of 3 × 50%), it helps to compute overload protection signal for BFP.

2. *Individual BFP suction pressure measurement*: At the suction of each of the BFPs, the pressure is monitored with a smart transmitter with a typical range of 0–15 kg/cm² to ensure that the required NPSH is achieved.
3. *BFP recirculation*: It is necessary to monitor the minimum FW flow through the BFP for controlling BFP recirculation. In many smaller plants, a separate flow monitor or a flow switch with an adjustable set point is deployed.
4. *BFP discharge pressure measurement*: The discharge pressure measurement at BFPs is for various purposes as discussed below.
   a. *Each pump discharges before the discharge valve*: Pressure gauge to see the pressure locally.
   b. *Each pump discharges before the discharge valve*: A pressure switch to stop the pump if the discharge pressure shoots up near the shut-off pressure for any discharge valve that fails to open.
   c. *BFP discharge header pressure transmitter*: Discharge pressure is measured by smart transmitters in the range of 0–200 (/250) kg/cm². This signal in conjunction with individual suction flow will be used to compute the protection signal for the BFP.
   d. *BFP discharge header pressure switch*: One pressure switch is kept to see if the discharge pressure drops below the set point. In that case, the standby pump may be automatically started.
5. *BFP discharge temperature measurement*: Temperature at the discharge is measured with the help of RTD (Pt 100) to monitor the water temperature. Typical range for the measurement will be 0–300°C.
6. *Balance leak-off flow measurement*: In some designs, the BFP balance leak-off Flow monitored BFP protections.
7. *Local temperature measurement across the FW heater*: Input and output of each of the FW heater are measured locally with the help of local temperature gauges to see the temperature gain across each heater. Typical range for the measurement will be 0–300°C.
8. *HPH 5 and 6 outlet temperature and pressure measurement*: Pressure and temperature at the outlet of each HPH is measured to check the pressure drop across the heater and the temperature gain. As the approximate pressure drop across the heater is 5 kg/cm² and temperature gain is ~30°C, so, the smart pressure transmitter will have a range of 0–200(/250) kg/cm² and a temperature range of 0–300°C for measurement with RTD (Pt 100).
9. *Differential pressure measurement across the feed control station*: The significance of the measurement was discussed in clause no. 6.3.5. The DP across the valve is measured with the help of a smart DP transmitter used in the range of 15 kg/cm².
10. *Feed water flow measurement*: The feed water flow (range discussed above) going to the boiler is measured with the help of a flow nozzle and a smart DPT (typical range of 0–10,000 mm wcl). This measurement is very important the for boiler drum level control (drum type). In the case of a SC once-through (OT) boiler also is extremely important and major controlling parameter. It is worth noting here that the location of this measurement is purely a function of the plant layout. Putting this flow element ahead of the feed water

control station is better in one sense in that it may theoretically require less straight length. However, sometimes the piping layout from the HPH to the feed control station may be a constraint. Again, when it is placed after the feed control station, naturally the straight length requirement will be more (because of control valves) and it may be very difficult without flow conditioners [26]. Also, the heater type (horizontal/vertical) has an effect on the selection of location. As discussed above, in the case of SC boilers, generally the spray flow is taken after the economizer, and the flow measurement is also done at that place.

11. *Temperature measurement at the economizer inlet*: The temperature at the economizer inlet is important as this is the terminal point for the boiler and is necessary not only for monitoring but also for SG performance calculations. RTD (pt 100) may be deployed for this purpose. Typical range will be 0–300°C. In some designs, this parameter is used in combustion control also.
12. *Analytical instruments and control*: The feed water system is also an extremely important place to ensure the purity of water going to the boiler. For scavenging dissolved oxygen and to maintain water pH, ammonia and hydrazine dosing take place in the FW system. This is done at the BFP suction. From an analysis point of view, the sampling points at I/L and O/L of the deaerator and economizer inlet are very important (Chapter V).

6.6 Controls in—Feed Water System (Fig. 3.20)

This system is also very important from a control point of view, as it is connected to many of the boiler control systems directly and/or indirectly.

1. Chemical dosing control: There is automatic (medium/large utility stations) control of chemicals at the suction of BFPs, as detailed in clause no. 6.1 of Chapter 5. There will be two chemicals for dosing:
   - e. *Ammonia*: Ammonia is dosed to increase the pH value of the feed water. This is done by measuring the pH in the feed water at the economizer inlet of the BFP discharge.
   - f. *Hydrazine*: Hydrazine is dosed to scavenge dissolved gas (O_2). This is done by measuring the residual hydrazine at the economizer inlet (to allow some retention path) or by measuring the dissolved oxygen.
2. *BFP recirculation control*: This control is to ensure a minimum FW flow through the BFP. In smaller/medium plants, the same may be done by a flow switch of the on-off type for control of the recirculation valve, whereas for larger utility stations, modulating control is deployed utilizing a BFP recirculation control valve.

   (In one design it has been found that the BFP recirculation line is connected to the deaerator without any valve, meaning constant recirculation. This may call for a loss in the system.)
3. *Attemperation control*: As seen in the P&ID, there are a number of attemperation lines (superheater, HP bypass, reheater, etc.) that have come out from the FW. So, a number of temperature controls by control valves in the water line will be there, as detailed in the respective system not repeated here.
4. *Discharge pressure control*: There are a number of controls by PS and PIT located at the BFP discharge. These are:
   - a. Individual discharge line pressure switch to prevent the pump getting closer to the shut-off pressure for some reason.
   - b. A pressure switch at the common header is used to cut in the idle in the pump in case of low header pressure (say due to tripping of a BFP).
   - c. In cases where more than one pump is running in tandem, then to prevent overload, protection signals being generated/computed from individual BFP flow and discharge pressure as discussed earlier.
5. *Feed control station*: This is the heart of the feed water system as it is responsible for the feed flow/drum level control. These valves are modulated in accordance with the demand signal from the feed flow controller. In case of an OT boiler, the same valve ensures minimum feed water flow through the economizer, as discussed earlier in clause no. 6.3.6.
6. *Drum level control*: It is part of the feed flow control.
7. *Separator storage tank level control*: Part of the start-up control for the OT boiler and is mainly concerned with the BCP flow, not the feed flow directly.
8. *DP control across feed control station*: The purpose of this control is to limit the differential pressure across the feed control station—discussed in detail in clause no. 6.3.6.

6.7 Redundancy in Measurement—Feed Water System

- The feed flow at the HP heater outlet: generally in utility stations, two of three (or one of two) redundancies are deployed.
- DP control across feed water control station: Because this is used to control the speed of the BFP, wherever applicable, and is directly related to feed water flow control, so generally in utility stations two of three (or one of two) redundancies are deployed.
- BFP common discharge pressure: For this measurement, also one of two redundancies is common in utility stations.

- Economizer inlet temperature: For this measurement, also one of two redundancies is common in utility stations.

6.8 Miscellaneous Points—Feed Water System

Feed water measurements are more related to a boiler system, so these parameters are monitored at the central/common system toward the boiler side monitor.

7 CONDENSATE SYSTEM (P&IDS)

7.1 Condensate System

After work is done at the turbine, the exhausted steam is cooled in the condenser to form a condensate system. In addition to that, various leakages, drains, and make up finally coming to the condenser system constitute the total condensate system. Exhausted steam is cooled at the condenser, then put into the cycle by the condensate extraction pump (CEP). Because it also receives water from various other sources, plus possible leakage in the circulating water (CW) system of the condenser, hence there is every possibility of contamination of condensate. Therefore, the sampling and analysis of the water quality at the condensate is very important. The condensate also contains higher dissolved oxygen, which is mechanically removed at the deaerator where the condensate completes its journey starting from the condenser, passing into the cycle with a new name-feed water.

7.2 Objectives and Functions of the System

Basic purposes shall include the following major functions:

- To supply low-pressure water to the deaerator at all points of operation.
- To supply attemperation water to the desuperheater sprays to the LP bypass system (as applicable) and the desuperheater sprays to the auxiliary steam system.
- It is the source of supply to various systems such as the chemical feed system, pump sealing, gland steam flow seal, flash tank, and turbine exhaust hood cooling.
- Ensure a minimum flow of water through the CEP and the GSC.
- Regulate the flow of condensate to the cycle vis-a-vis the deaerator.
- Thermal cycle make up and dump take place in this system.
- It also constitutes the source of potential energy through water storage at the hot well as well as at the deaerator.

7.3 System Description—Condensate System

Major variations in this system come from the number of heaters in the regenerative cycle. For example, some designs call for more than three LP heaters. Depending on the layout, these heaters may be horizontal or vertical. In the modern days for large power stations, the vacuum in the condenser is created and maintained by vacuum pumps; however, steam ejectors (with AS supply) are also not uncommon. Basically the function and purpose of this vacuum system is to take out air/gases from the condenser and bring the same to the vacuum.

7.3.1 System Description—General (Fig. 3.21)

From the condenser (at vacuum), the condensate is pumped by the CEPs. At the suction of each of the CEPs, there will be a removable filter (with a DP switch to monitor the filter) to remove dirt from the condensate before it enters the CEP. Because CEPs have suctions at the vacuum, the necessary sealing arrangement at CEPs is important. There could be redundancy in the CEPs with an automatic start-up facility in case of failure or malfunctioning of any CEP. In drawing $2 \times 100\%$ CEP have been considered. In the case of a trip of one pump and/or a defect in one pump, the other can start automatically. Each of the two pumps has one NRV and discharge valve, and they are connected to a common header. The common header goes to the GSC to heat the condensate utilizing the heat in the GSC (when ejectors are used, the condensate also passes through the main ejector(s) to gain heat, similar to that in the GSC). One flow element (FE 02) upstream of the GSC is utilized for recirculation control of the CEP and GSC (Fig. 3.23). In certain cases, the condenser may be partitioned, so sample lines need to be taken from both sides before connecting to the sample pump through suitable valves. Normally, the valves are open so that representative samples are available; in case of portioned operation, it can be collected from the operating side. There is a make-up line with a control valve to take regular controlled make up (Fig. 3.22).

7.3.2 System Description—Cond. System (LP Heaters Fig. 3.23)

Immediately after the GSC, there are a number of condensate outlets for consumers such as LP bypass spray, auxiliary steam spray, chemical feed system, vacuum breaker spray, pump sealing water system, etc. Also, there is the CEP and GSC recirculation line with a control valve RCV01 to ensure the minimum ($\sim>30\%$) condensate recirculation flow through the CEP and the GSC. There may be several other schemes for these recirculation controls (Section 7.3.4). A sample line is taken after CEP discharge for measurement of the conductivity and pH to assess the

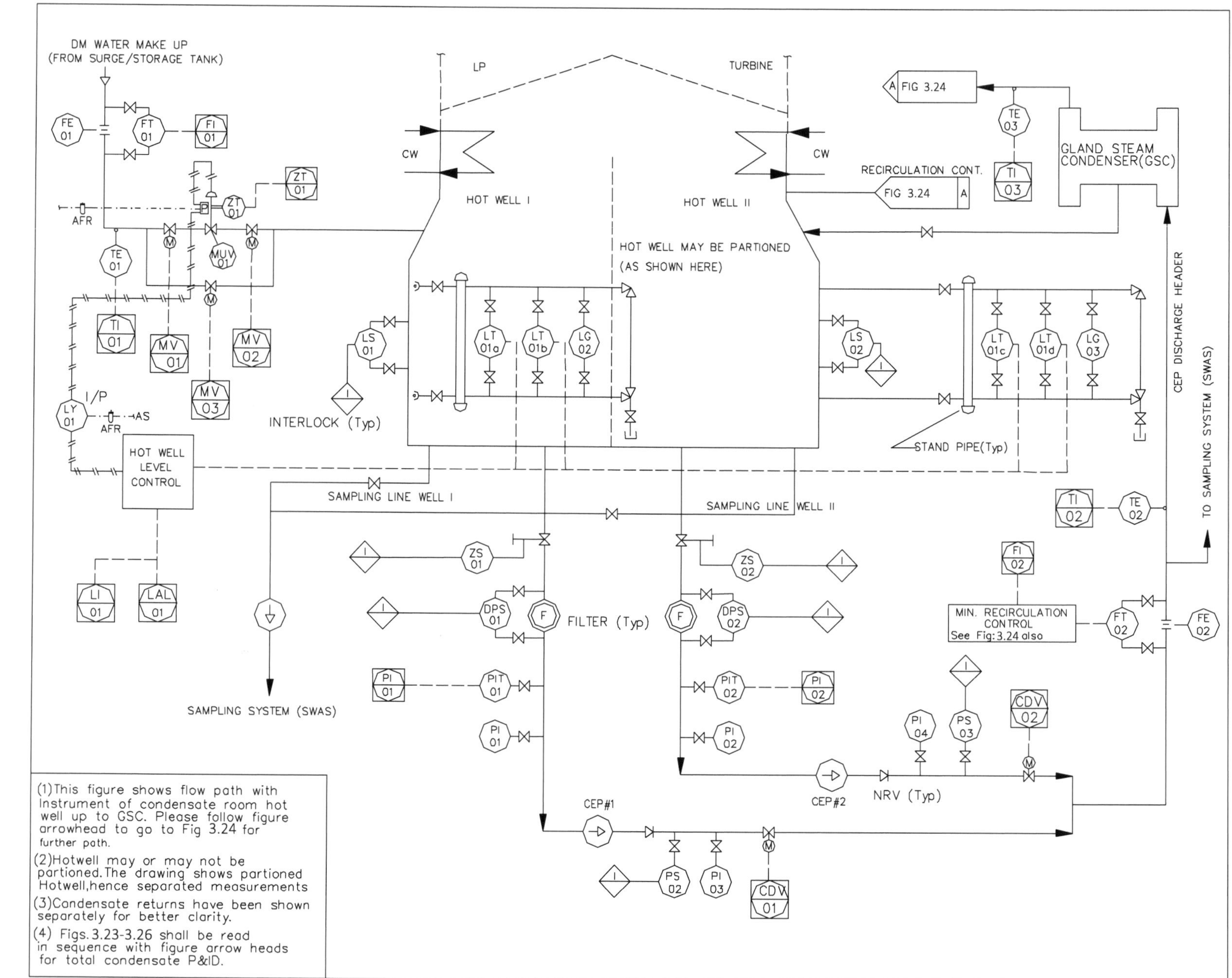

FIG. 3.21 Condensate P&ID hot well to GSC.

MAKE UP & DUMP: It is worth noting that, any thermal power plant, always need a make up to meet the requirement of leakage loss, evaporation loss etc. Apart from that sudden load throw off may require dumping condensate from the system.

FIG. 3.22 Condensate make up and dump systems.

condensate quality. As discussed earlier, there will be a dump line to dump the excess condensate. This dump line with the control valve DPV01 has been shown before the main condensate control valve MCV 01. After the main condensate valve, the condensate passes through the drain cooler and the LP (three) heaters before reaching its final destination, the deaerator. The LP Heater 1 is normally is located at the neck of the condenser. Each of the drain coolers and LP heaters have inlet/outlet isolation valves and a bypass valve so that at any time, each can be isolated and/or bypassed. The main condensate flow and quality are measured in the condensate line going to the deaerator. In the direct mode of control, the hot well level is maintained by the make up and dump valve whereas the main condensate control valve regulates the level in the deaerator (Section 7.3.4).

7.3.3 System Description—Condensate Return (Fig. 3.24)

The temperature of various condensate returns, drains for example, heater drain etc. are important factor to be considered, so that these can be directly be admitted to condenser. If the temperature is high, then it may cause an increase in temperature inside the condenser; hence, the condenser pressure will increase, meaning a lowering of cycle efficiency. It is therefore general practice to return all these drains to the condenser via the flash tank (HP for HP drains, LP for LP drains). Whenever the level in the flash tank increases, it opens the drain control valve through the level interlock to drain the same to the condenser. There is pressure interlock associated with the flash tank so that whenever the pressure inside the flash tank increases through the pressure interlock, the spray control valve to the flash tank will open to spray water to the flash tank to lower the flash tank temperature/ pressure.

7.3.4 System Description—Cond. Level Controls (Fig. 3.25)

In the P and ID (Figs. 3.21 and 3.23), the following loops have been considered, as shown in Fig. 3.25C.

- *The minimum recirculation control of* the *CEP and GSC* will be done independently by measuring the condensate flow through the CEP and GSC to ensure minimum flow. If the main condensate flow increases then from the set point, the loop will close the valve automatically.
- *The hot well level control* loop regulates the DM make up and dump control. This is also straightforward—when there is excess level in the hot well (say, load throw off), it dumps the excess condensate to, say, the surge tank. Similarly, whenever there is a dearth of condensate in the hot well, the control system will open the make-up line control valve for condensate make up.
- *The deaerator level control* regulates the main condensate flow to maintain the deaerator level.

7.3.4.1 In Fig. 3.25A and B, a Few Other Alternatives Are Shown

(A) In Fig. 3.25A, the hot well level control loop regulates both the minimum recirculation control valve as well as the main condensate control valve in opposite mode. Whenever the hot well level goes high, it will start opening the MCV to regulate the hot well level and begin sending condensate to the deaerator. Whenever the main control valve MCV opening $<$ certain value, meaning (FW TANK) level starts going down hence it will cause, Make up at Hot well causing Howell level to rise, and as a consequence, MCV starts opening and (because RCV and MCV operate in opposite direction), RCV will start closing. As a result of all these, the deaerator (FW TANK) level starts rising. If it raised beyond the desired point, then the excess will be dumped to the surge tank via DPV. This mode is a rather indirect way to maintain system balance and levels in the hot well and deaerator. On account of the capacity difference between the hot well and the deaerator (feed water tank), this method is sometimes effective.

(B) In Fig. 3.25B, these control loops are also alternatives to what has been shown in the main P&ID. This is also an indirect type of control and is quite similar to what has been discussed in (A) above. The main difference is here the CEP and GSC recirculation is independent, based on actual flow measurement (similar to that in P&ID). Here, the condenser hot well level is maintained by regulation of the main condensate flow, MCV. On the other hand, the deaerator level regulates the make up to the hot well or dump to the condensate storage (DM) tank.

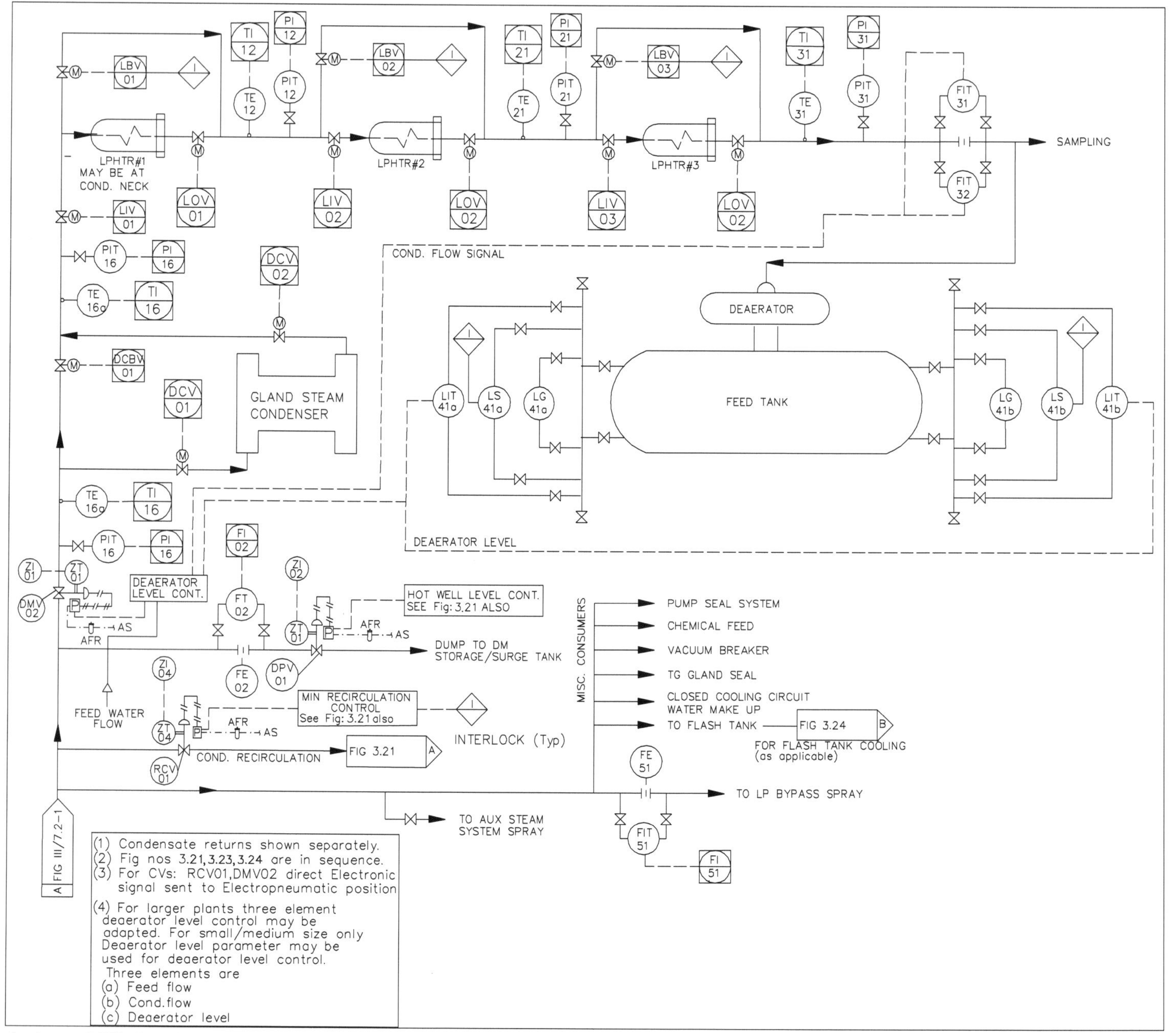

FIG. 3.23 Condensate P&ID through heaters.

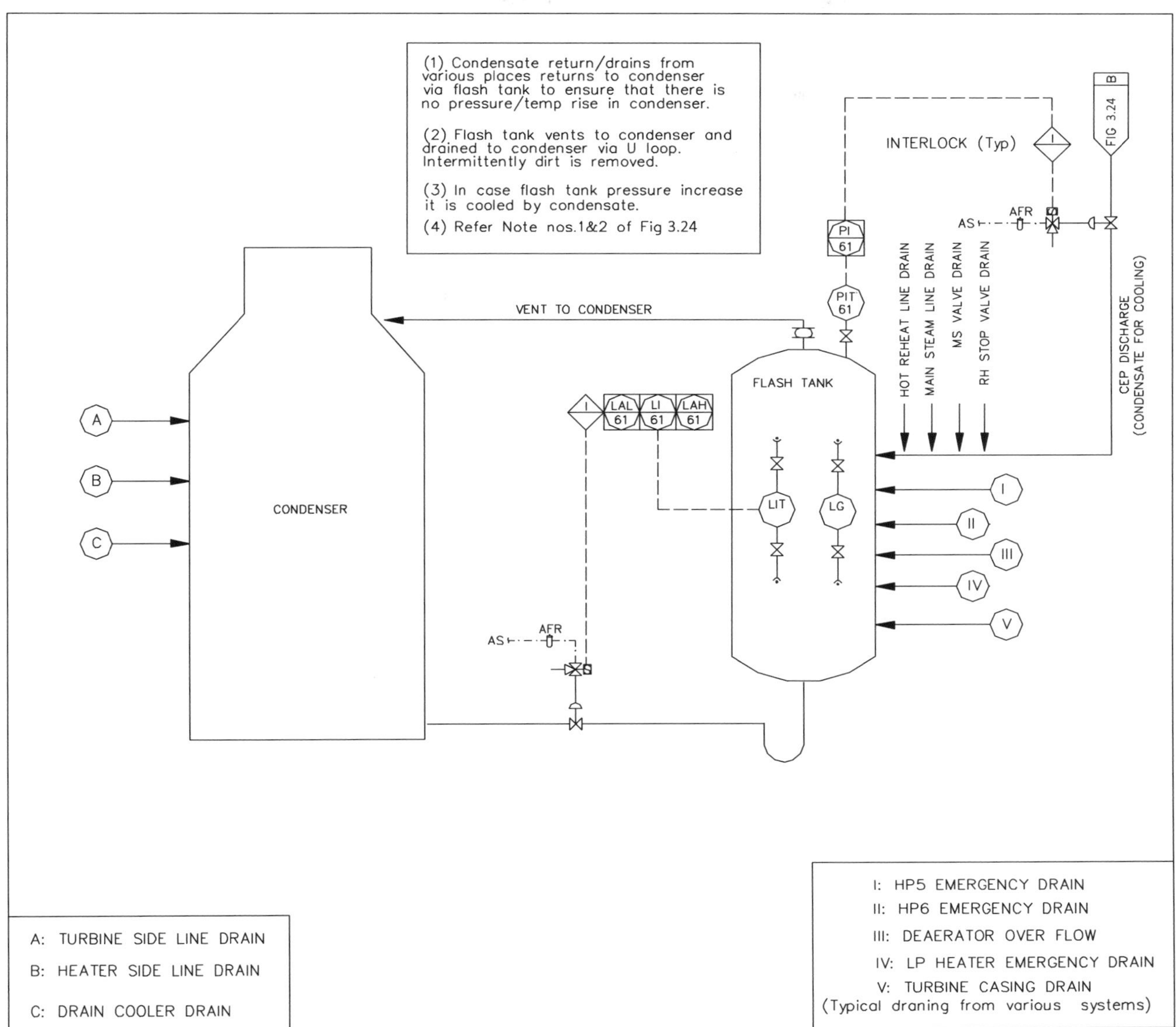

FIG. 3.24 Condensate P&ID condensate return.

7.3.5 System Description—Condensate Polishing Unit (Fig. 3.26)

A: CONDENSATE POLISHING why? And how much?

In larger units >200 MW of superthermal power stations plants, it is customary to deploy a condensate polishing unit. In heat exchangers where steam is condensed, it picks up contamination from the environment. Leakage at the heat exchangers and condenser are other sources of contamination. Air leakage and CO2 leakage cause the formation of oxides of Fe, Cu, Ni, and Si. *For larger power stations*, it is *essential* (especially for OT boilers) and *economical* to go for a CPU for better-quality water for the boiler as well as a longer turbine life. This condensate polishing may be partial as shown in the above picture, where part of the condensate directly goes to the GSC (main cycle) part goes to the main cycle only after passing through CPU. What percentage of the condensate shall pass through the CPU is part of process design and the water quality at the plant site; hence it is not dealt with here. Section 3.7.6 of Chapter 2 may be referred to.

B: CONDENSATE POLISHING Objective:

The major objective for using the CPU shall include but not be limited to the following:

- Better water quality and cycle clean up.
- Less blow down, hence less make up.
- Better heat transfer at the boiler.

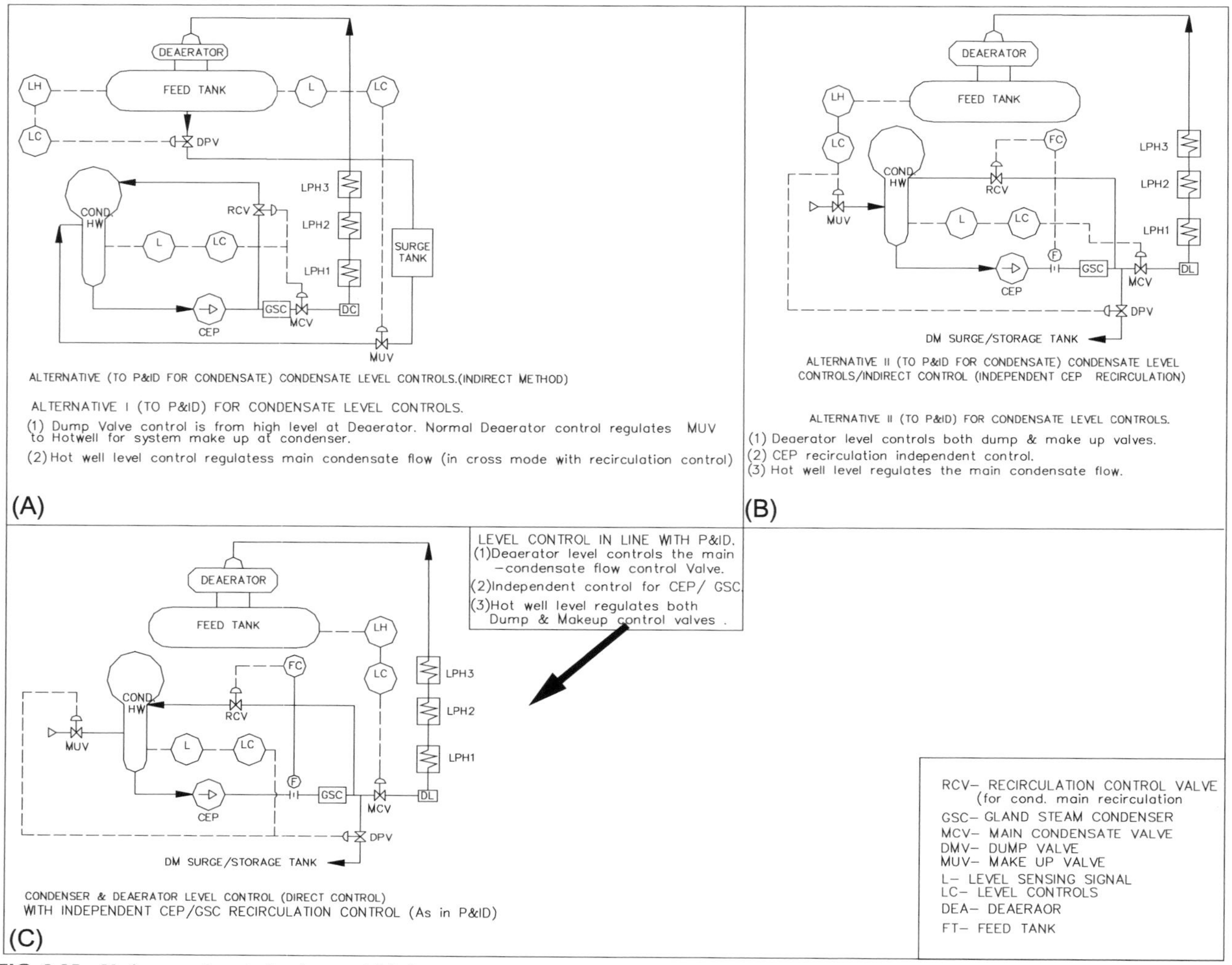

FIG. 3.25 Various condensate level control Modes.

- Quick start-up and reaching full load.
- Longer turbine life.
- Less pressure on the DM plant.

C: What is CONDENSATE POLISHING UNIT and types (Chapter 2, Section 3.7.6)

The CPU is basically a series of ion exchange units, and it may comprise:

- Strong acid cation (SAC).
- Weak acid cation (WAC).
- Mixed bed (MB).
- Strong base anion (SBA).
- Weak base anion (WBA).

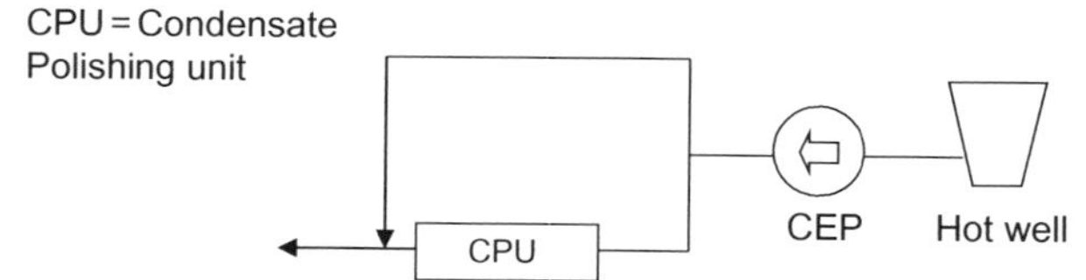

FIG. 3.26 Condensate line with CPU in service (Also refer to clause no. 3.7.6 and Fig. 2.37 of Chapter 2 for details).

The difference with the normal DM plant is the flow rate. In the case of the CPU, it has to have a *high flow rate*. Depending on the type chosen, regeneration may be in situ or external.

D: The instrumentation involved in this system is similar to that in the DM plant.

7.4 Major System Equipment—Condensate System

The major system equipment in the condensate line shall include the following:

1. *Condenser*: The purpose of the condenser is to condense the turbine exhaust hood steam by taking away the heat of vaporization with the help of the cooling water circuit in the condenser. This condensate collected is stored in the hot well of the condenser. In conjunction with the flash tank, it provides collection points condensate drain from various other points of the system also. The condenser

may be the plate type, the direct contact type, or the shell and tube type. Similarly, the cooling system may be the open type (once-through) where water is taken from water bodies, for example, a river, and returned to the source at a distance. In some cases, these may be closed types where the cooling water is cooled with the help of a cooling tower.

2. *Flash tank*: It is basically a storage device kept between the condenser and the condensate drains to ensure that under no circumstances does the high temperature go in to the condenser directly to raise the temperature of the latter. As discussed in Section 7.3.3, it ensures that low temperature condensate drains.
3. *LP heaters*: In a condensate system, there may be three (in P&ID LPH 1, 2, 3) to four shell and tube heaters. The condensate in the tube of the heaters is heated by the LP stage turbine extraction steam in the shell. Here, heaters are numbered 1, 2, and 3, taking the lowest heater as the one that receives the steam from the lowest extraction pressure, and so on. However, this numbering system may be the other way round.
4. *Drain cooler*: It extracts heat from the various drains (from the turbine), it receives to heat the condensate, and it is cooled down before sending the same to the condenser.
5. *Gland steam condenser*: As the name suggests, it condenses various leakage steams from the turbine glands. During start-up, there will be the requirement of auxiliary steam to seal the turbine glands, but as the load increases, there will be steam leaking from the glands of the higher pressure stages of the turbine. These are collected at this place, and thereby adding heat to the condensate as it passes through it in the tube.
6. *Condensate extraction pump*: The function of the CEP is to pump the condensate from the hot well to the various heaters other consumers and deaerator. It is a centrifugal pump with a constant-speed motor. Because many outlets are taken before the condensate line goes to the deaerator, the quantity of water handled in the CEP is more than that of the BFP, but with less header pressure (than BFP).

7.5 Major Parameter Measuring Monitoring in—Condensate

7.5.1 Process Parameter—Condensate (Typical Values Only)

See Table 3.7.

7.5.2 Monitoring of Process Parameter—Condensate System

Monitoring of parameters such as the condenser vacuum, level at the hot well, deaerator, etc., at the control room is extremely important (Figs. 3.21, 3.23, and 3.24).

1. *Make-up water flow measurement*: In almost all power stations, the flow of make-up water at the condenser is measured to monitor the actual make up to the system for checking of system performance and leakage in the system. For this one orifice plate, in conjunction with

TABLE 3.7 Condensate Parameters

Location	Pr.	Temp.	Flow	Remarks
200 MW				
CEP SUCTION (COND VAC)	0.37[a] (0.1)	38	800	[a]Water head
CEP DISCH.	22	40	800	
LPH 2 INLET	21	62	800[a]	Less for various desuperheating waters
LPH3 INLET	19	87	800[a]	
LPH3 OUTLET	17	125		
500 MW (drop across heater may be more)				
CEP SUCTION (COND VAC)	0.45[a] (0.1)	40	1700	[a]Water head
CEP DISCH.	32	40	1700	
LPH 2 INLET	30	85	1700	Less for various desuperheating waters
LPH3 INLET	27	115	1700	
LPH3 OUTLET	24	135		

Values given above are to get an idea, and may vary with actual. Units: *P* in kg/cm^2 A, *T* in °C, *F* in t/h.
[a]*Spray for AS, LP bypass considered before LP heaters.*

DP transmitter is used. Under normal conditions, the make shall be *3%–4% of normal flow* for the plant. However, in order to take care of sudden changes in load, the range of the flow measurement is kept wider, depending on the mass balance and heat balance for the plant. Typical (actual may vary) flow range shall be 250 t/h in 200 MW and 450 t/h for a 500 MW plant.

2. *Condensate level at hot well*: The condensate level at the hot well is a very important parameter. In order to avoid too many holes at the condenser body, normally a stand pipe on each side of the condenser, as shown in the P&ID, is used. Various tapping points for level measurements are taken from the stand pipe. LSs are shown intentionally direct tapping from condenser, to give an idea to the reader that direct tapping is also possible. LGs and LTs shall have ranges covering the allowed very high and very low level of the condenser. In many cases for control purposes, dedicated level transmitters with a narrow range are used. As the condenser may be partitioned, it is recommended that the number of transmitters in each side of the condenser will be the same. Apart from the closed-loop control system (CLCS), there are a number of interlock open-loop control systems (OLCS) also. Hot well level > min set value is taken as start permissive for initial starting of CEP to ensure required NPSH (Nominal Pump Suction Head). Similarly, in the event of the hot well level < very low set point, the running pump shall trip.
3. *Differential pressure across suction filter*: Differential pressure across the suction filter of the CEP is monitored and in the case of a high DP, an alarm is generated to take action for cleaning the filter.
4. *CEP suction pressure and valve open*: In order to ensure that the CEP has flooded suction pressure at the suction is measured with the help of a pressure transmitter with a typical range of 0–1 kg/cm^2. Additionally, the suction valve open signal is also taken as an interlock/start permissive for the pump.
5. *CEP/GSC minimum flow recirculation*: An orifice plate with a DPT is used to measure the flow. The flow range shall be >40% of the BMCR. The control was already described above.
6. *CEP discharge pressure measurement*: The discharge pressure measurement at the CEP is for various purposes, as discussed below.
   a. *Each pump discharge before discharge valve*: Pressure gauge to see pressure locally.
   b. *Each pump discharge before discharge valve*: Pressure switch to stop the pump if discharge pressure shoots up near the shut-off pressure for any reason (say) discharge valve failed to open.
   c. *CEP discharge header pressure transmitter*: Discharge pressure is monitored by smart transmitters in the range 0–30(/50) kg/cm^2.
   d. *CEP discharge header pressure switch*: One pressure switch at a common discharge header is used for automatic cutting in of the standby pump.
7. *Temperature measurement at CEP discharge and GSC O/L*: Temperature at the CEP discharge and GSC O/L are measured with the help of an RTD (Pt 100) to monitor water temperature and for calculation of the condensate temperature (heat) gain in each of the heaters. Typical range for measurement shall be 0–150°C.
8. *LP bypass condensate flow*: Depending on the system design and requirements of the turbine supplier, the spray flow going to the LP bypass system is measured in some of the systems, but it is not a mandatory measurement.
9. *Condensate dump flow measurement*: On account of system disturbances, (say) sudden load throw off, it may be necessary to dump a part of the system water to the DM storage/surge tank. Like the make-up system, here also the orifice plate in conjunction with the DP transmitter is deployed to check the overall performance of the system following a load throw off. Typical range for the measurement could be 250 t/h in 200 MW and 450 t/h for a 500 MW plant.
10. *Local temperature measurement across LP heater*: The input and output of each of the LP heaters is measured locally by local temperature gauges to see the temperature gain across each heater. Typical range for measurement shall be 0–200°C. These may be an integral part of the heater, and these are not shown in the drawing. Also, a terminal temperature difference measurement is necessary for calculation of heater performance, for which RTDs may be deployed.
11. *Drain cooler, LPHs outlet temperature, and pressure measurement*: Pressure and temperature at the outlet of each LPH and drain cooler is measured to check the pressure drop across the heater and the temperature gain. The temperature and pressure ranges shall be almost the same (ΔP and ΔT across the heater will be ~2–5 kg/cm^2 and ~25°C, respectively). The smart pressure transmitter in the range of 0–25(/35) kg/cm^2 and RTDs in the range of 0–150°C may be deployed for measurements.
12. *Deaerator (feed tank) level*: As discussed earlier for the deaerator (FW tank) level measurement, a stand pipe shall be used on either side of the deaerator. Various tapping points for LGs, LSs, and LTs (with ranges covering allowable very high and very low level of deaerator) are taken from stand pipes on either side. In many cases for control purposes, dedicated level transmitters with a narrow range are used. It is recommended that the number of transmitters in each side of

the deaerator shall be the same. Apart from the CLCS, there are a number of OLCS controls also. The FW tank level > minimum set value is taken as start permission for initial starting of the pump (BFP). Similarly, in the event of FW tank level < very low set point, the running BFP(s) shall trip. In larger plants for deaerator level control, a three-element control is used. In three-element control, the three parameters are the deaerator level, the condensate flow, and the FW flow.

13. *Condensate flow measurement*: The main condensate flow going to the deaerator is measured with the help of an orifice plate (or flow nozzle) and a smart DPT (typical range 0–8000 mm wcl). The approximate flow measuring range shall be 0–800 t/h (200 MW) and 0–1800 t/h (for 500 MW). This measurement is very important for deaerator (FW tank) level control as well as mass/heat balance calculations. It is recommended to have one temperature compensation for the same from a nearby temperature measurement system.
    a. *Sampling points*: Monitoring of water quality at the condensate is very important. Major sampling points are at each side of the condenser, CEP discharge, and deaerator inlet and outlet.

From these sampling points, the sample may be taken for manual analysis and/or be sent for SWA (Section 4 of Chapter 5).

7.6 Controls in—Condensate System (Figs. 3.21 and 3.23–3.25)

1. *CEP and GSC recirculation control*: This has been discussed in Section 7.4.3 (Fig. 3.25A).
2. *Condenser hot well level control*: As is seen in the P&ID, condensate hot well level is maintained by regulating two sets of valves. This loop with two alternatives has already been discussed and is not repeated.
3. *Deaerator (FW tank) level control*: In the P&ID, the deaerator (FW tank) level is controlled by regulating the main condensate flow control valve to the deaerator. In larger power plants, in order to achieve better response, three elements may be used. In alternative and/or indirect systems discussed in Section 7.3.4.1 (Fig. 3.25), the deaerator level signal is used to regulate the make-up and dump valves as well as the deaerator level.
4. *Flash tank control*: As discussed above, there two controls for this.
    a. *Level control*: The level in the flash tank is maintained, and in the case of a higher level, it is allowed to drain to the condenser.
    b. *Pressure control*: There is one line from the CEP discharge taken to spray cold condensate to the flash tank so that its temperature and pressure do not rise. Whenever the pressure inside the flash tank crosses the set point, water is sprayed so that at all times pressure and temperature are regulated within limits.

7.7 Redundancy in Measurement—Condensate System

- Hot well level: generally, one of two redundancy is deployed for measuring the level on each side of the hot well. However, in the case of a partitioned hot well, it is necessary to have *at least* one transmitter on each side of the hot well.
- Deaerator (FW tank) level measurement: Also, one of two redundancy is deployed for measurement of the deaerator (FW tank) level in each side.

7.8 Miscellaneous Points—Condensate System

Condensate measurements are more related to the turbine system, so these parameters are monitored at the central/common system monitor toward the turbine monitor.

8 HEATER DRAIN AND VENT (P&IDS)

8.1 Heater Drain and Vent System

In shell and tube heaters, after transferring the heat to the condensate (LPHs) or feed water (HPHs), the extraction steam at the shell condenses to the water, which therefore needs to be returned to the system at the condensate/deaerator the through heater drain and vent system.

8.2 Objectives and Functions of the System

The basic purpose of the system shall include the following major functions:

- To condense the extraction steam and to return to the system.
- Heater draining is necessary to prevent induction of water to the turbo generator by draining the water accumulated to the preceding heater, deaerator, and/or condenser.
- By proper selection of drain control valves, possible water hammering between the FW heater and condenser is prevented.
- Maintenance of proper level in the heaters so that the heat transfer to the condensate (LPH)/FW (HPH) from the extraction steam is ensured.
- To return the noncondensing gases to the condenser.

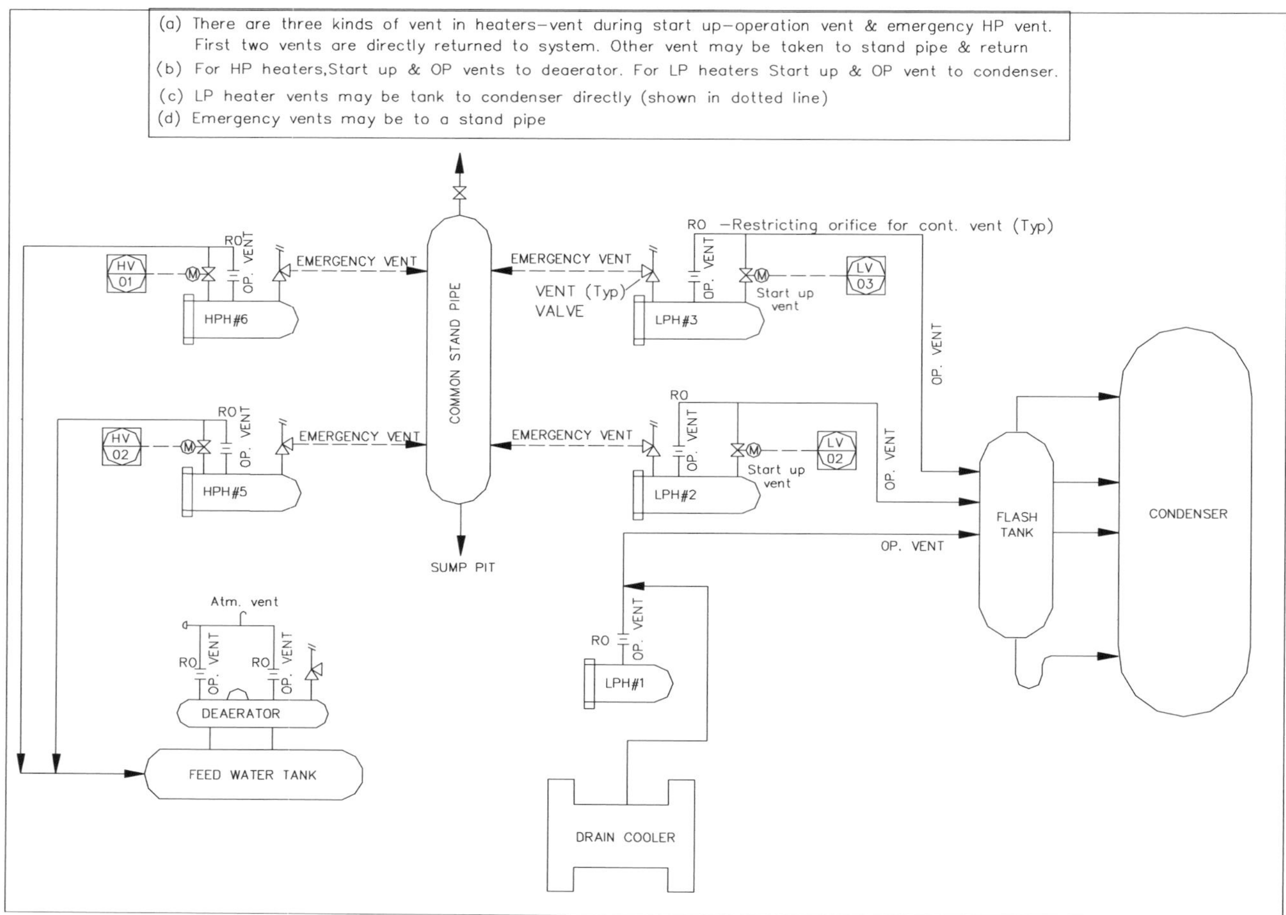

FIG. 3.27 Heater vents (HPH and LPH).

8.3 System Description—Heater Drain and Vent System

In each of the heaters, there are a number of drain and vent paths, depending on the operating points. Discussions are presented with the vent system first.

8.3.1 System Description—Heater Vent (Fig. 3.27)

Excepting LPH 1, all heaters have three vent lines: Start-up vent, operating vent, and pressure vent/safety valve vent (pressure safety valve).

1. Start-up vent: During start up, the heaters are vented through the motorized valve operated by the plant DCS in automatic/semi automatic mode. In cases of HP heaters, these lines return to the deaerator for possible high temperature and pressure. On the other hand, the start-up vent from LP heater 2 and 3 the start-up vent are controlled in the same manner, connected to the condenser.
2. Operating vent: Whenever heaters are in operation, there will be some continuous vent from the heaters. Generally, these constant vents are done through a restriction orifice (RO) that is sized with a lower beta ratio (flow coefficient) so that there is good amount of pressure drop across it. HP and LP heaters have operating vents connected to the deaerator and the condenser, respectively. The drain cooler vents to the condenser through the operating vent line of LP heater 1, as shown in the drawing. Depending on the system design, the vent lines may be connected to the condenser directly or else may be connected via flash tank, as shown, to avoid raising the temperature at the condenser.
3. Safety valve vent (pressure): Each of the heaters has one safety valve vent or pressure vent. Each heater (except LPH 1) has one spring-operated pressure safety valve. When the pressure in the heater is greater than the set point set by the set spring, it will pop up to the vent. These pressure vents may be connected to a common header with vent, and the drain line connected to the common drain/sump pit line.

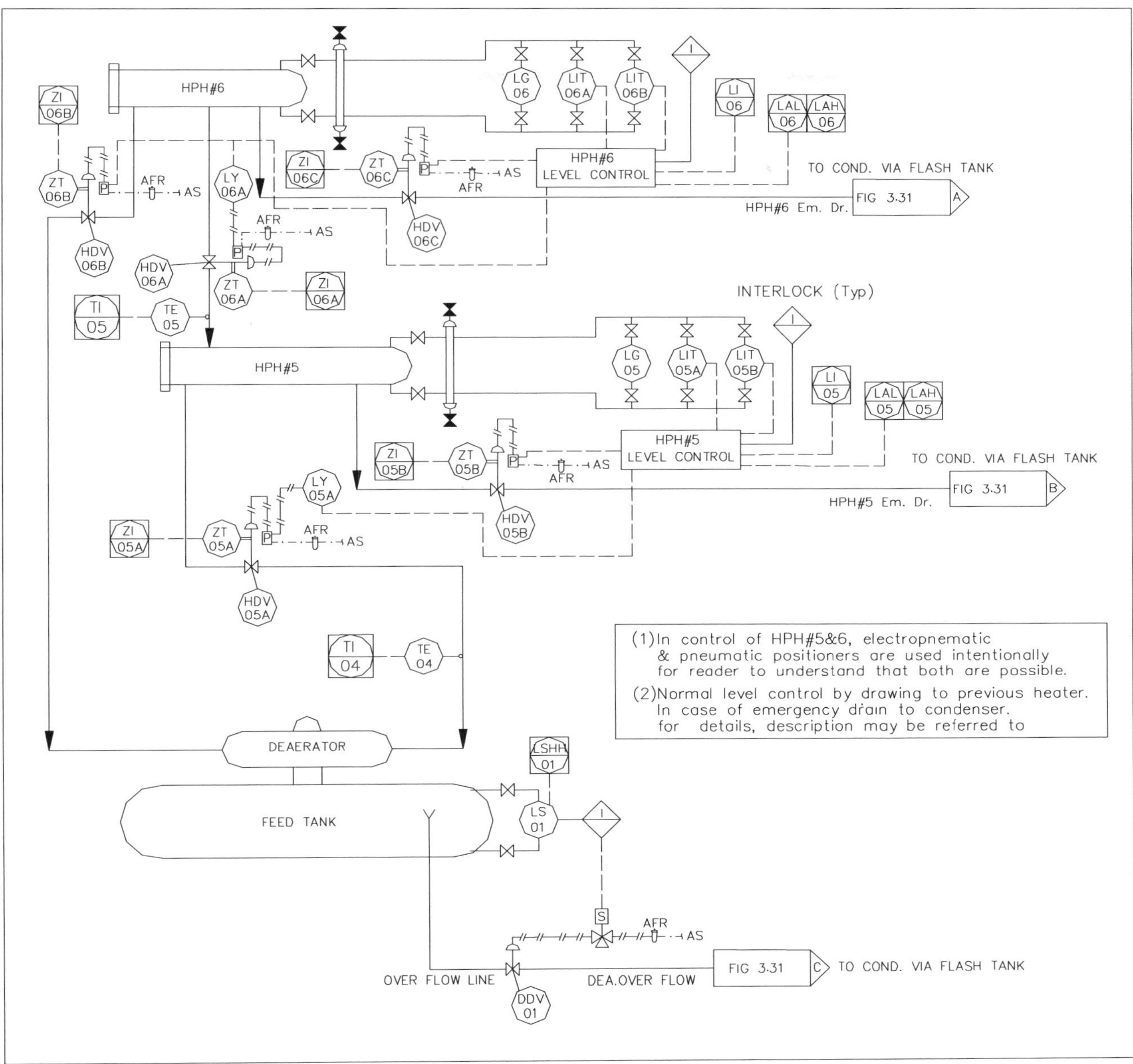

FIG. 3.28 HP heater drain system.

8.3.2 System Description—Heater Drain (Figs. 3.27 and 3.28)

As we see in the regenerative cycle, the major storage spaces are in the condenser and FW tank (deaerator). The basic philosophy behind the heater drains is that under normal conditions, the heater will drain to the preceding heater, so that there is not much change in operating condition between the draining fluid and where it is drained. However, there are situations when it is not possible for any heater to accept drain from its successor (say, heater 5 to accept the same from heater 6), or maybe the heater is bypassed. In those cases, the LPH heaters may be drained to the condenser or HPH may be drained to the deaerator in case the concerned heater is situated after the deaerator (for example, HPH 6 to deaerator). So, there will be at least two paths for each heater (Exception: LPH1, which always drains through the drain cooler and HPH 6 may have three draining paths instead of the two discussed, for example, normal drain to HPH 5, the deaerator, or in extreme cases, to the condenser via the flash tank). Here, one must remember that whenever *drain to condenser is referred, it is normally through the LP flash tank and may not be directly to the condenser*. Although not shown in the P&ID under reference, in some designs there are HP drain flash tanks (HPDFT) to accept drain from HPHs to avoid problems due to temperature. These HPDFTs on high level drains to LPDFT. All such flash tanks have provisions for cooling (Fig. 3.24) the inside temperature through the pressure interlock.

1. HP heater drain (Fig. 3.28):
   a. *HPH 6 normal level control*: In HP heater 6, there are three draining paths. Under normal situations, it shall drain to HPH 5 through HDV06A to HPH 5 to maintain a normal level at the desired set point. Now, in case there is some problem in HPH 5 (or it is bypassed) say, the level is already high in HPH 5 (level interlock in HPH 5), then through DCS HDV 06A shall be closed and HDV06B drains to the Deaerator, shall open to maintain the level in HPH 6. The same will also happen when the extraction pressure to HPH 5 is low (to prevent water ingress to the turbine extraction line through a probable high level in HPH 5). In case of a very high level in the deaerator, HDV06B may be closed and the drain may be sent to the condenser (through LPDFT).
   b. *HPH 6 high level control*: For HPH 6, there will be a high set point also. When the level in HPH 6 crosses that high set point value, then HDV06C will open to drain the water to the LPDFT and to the condenser. One of the probable causes of such a happening may be rupture of the heater tube (applicable for all heaters) of HPH 6 to cause a sharp rise in the heater level.
   c. *HPH 5 normal level control*: In HP heater 5, there are two draining paths. Under normal situations, it shall drain to the deaerator through HDV05A to maintain a normal level at the desired set point. In case of a very high level in the deaerator, HDV05A may be closed to drain to the condenser via LPDFT.
   d. *HPH 5 high level control*: For HPH 5, there will be a high set point also. Similar to HPH 6, when the level in HPH 5 crosses that high set point value, then HDV05B will open to drain the water to the condenser via LPDFT.
   e. *Deaerator overflow*: In case of deaerator overflow interlocked with a very high deaerator level (level switch), it will be drained to the condenser by DDV 01 (Fig. 3.29).
2. LP heaters drain (Fig. 3.30):
   a. *LPH 3 normal level control*: Under normal situations, it shall drain LPH 3 through LDV03A to LPH 2 to maintain a normal level. In case of a problem in LPH 2 (or it is bypassed) or the level is already high in LPH 2 (through level interlock in LPH2), LPH 3 would be should not be drained to the condenser via control valve LDV03B. This is the same thing as discussed above when the extraction pressure to LPH 2 is low—to prevent water ingress to the turbine extraction line through a high level in LPH 2.
   b. *LPH 3 high level control*: When the level in LPH 3 crosses that high set point value (say due to tube rupture), then LDV03B will open to drain the water to the condenser.
   c. *LPH 2 normal level control*: Under normal situations it shall drain to LPH 1 through LDV01A to maintain a normal level. In case of a high level in LPH 1, LDV02A may be closed and it will drain to the condenser.
   d. *LPH2 high level control*: Like LPH 3, when the level in LPH 2 crosses that high set point value, then LDV02B will open to drain the water to the condenser.
   e. *LPH 1*: The level in LPH is monitored but it is directly connected to the drain cooler, which in turn drains continuously to the condenser.

8.4 Major System Equipment—Heater Drain and Vent System:

Major system equipment (Chapter 2) shall include the following:

1. *Condenser*: The purpose of the condenser is to provide collection points for the condensate drain from the HP and LP heaters, so far as the heater drains and vents are concerned. HP heater drains are normally connected via the flash tank.
2. *Flash tank*: It is basically a storage device kept between the condenser and various drain points to ensure that under no circumstances does a high-temperature fluid go into the condenser to raise the temperature of the latter.
3. *HP and LP heaters*: There may be three or four LP heaters meant to heat the condensate and there are three to four HP heaters meant to heat the feed water, by extraction steam lines from the turbine as already discussed earlier. Depending on the manufacturer and plant layout, these heaters may be horizontal or vertical. In order to prevent water ingression into the turbine through the extraction lines, it is necessary to ensure that the level in these heaters never crosses the high-level set points (as per TWDP). Also, to have proper heat transfer to the condensate (LPH) or feed water (HPH), it is necessary to maintain the level at the normal set point.

8.5 Major Parameter Measuring Monitoring in—Heater Drain and Vent System

8.5.1 Process Parameter—Heater Drain and Vent System (Typical Values Only)

The parameters in the drain lines are purely a function of the heat and mass balance diagram for the plant (Table 3.8).

The situation is not always very congenial! This is because of the fact that HPH 5 or 6 Pressure & Temperature is quite high when compared the same in condenser. For these reasons in many designs HPDFT is introduced & HPH emergency drains may be connected to this. In larger size power stations there may be two nos. HPHs (viz. HPH 5A&B or HPH6A&B)

FIG. 3.29 Note on HP heater drains.

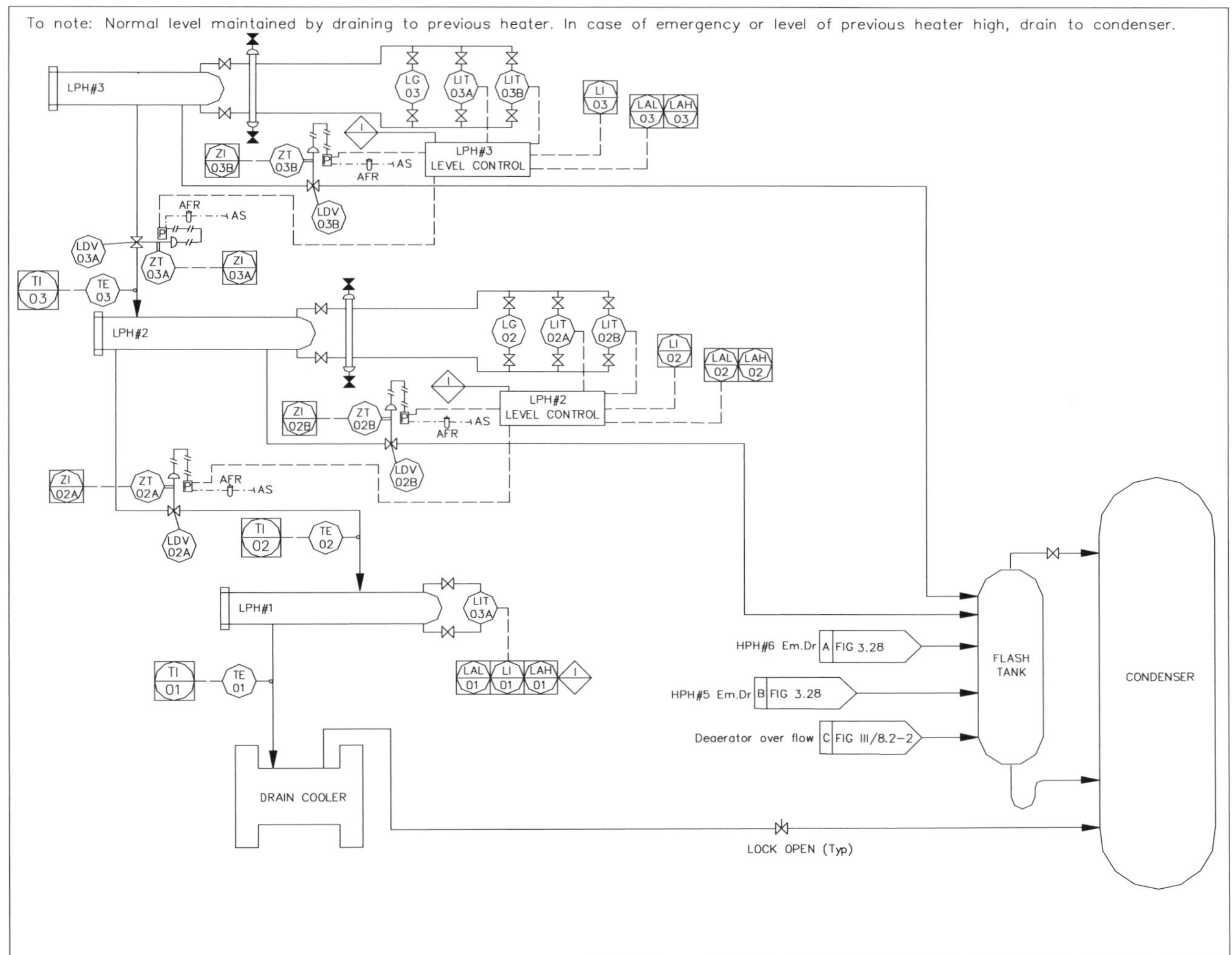

FIG. 3.30 LP heater drain system.

8.5.2 Monitoring of Process Parameter—Condensate System

From the TWDP point of view, the monitoring of levels at the heaters are quite important. Also, to have an idea about the heater performance, the drain outlet temperature for each of the heaters is equally important (Figs. 3.28 and 3.30).

1. *HP HEATER 5 and 6 or LP HEATER 2 and 3 level*: Because the heaters have two set points for control, in larger plants for better resolution, some choose two different ranges for level transmitters meant for normal and high-level control. In some design, the same set of redundant instruments is utilized. Generally, one of two redundancy is chosen in large plants for level instruments.
2. *LP HEATER 1 level*: The level at LPH 1 is monitored without any control.
3. *Drain line temperature*: Temperature at each drain line is monitored to check the transfer against each of the heaters.

TABLE 3.8 Heater Drain Parameters (Typical)

Ex No.	Heater No.	Pressure	Temperature.	Flow	Remarks
Typical value TMCR for 200 MW					
1	LPH 1	0.5–0.6	50	115	
2	LPH 2	1.2	82	50–55	
3	LPH 3	2.5–3.0	140–160	25	
5	HPH 5	15–16	~170	150	
6*	HPH 6	30–35	200	80	*(CRH)
Typical value TMCR for 500 MW					
1	LPH 1	0.64	52	175	
2	LPH 2	1.6	90	95	
3	LPH 3	3.0–4.0	140–160	47	
5	HPH 5	22	200	210	
6*	HPH 6	38–42	220–230	115	*(CRH)

Unit: Pressure in Bar (A), Temp in °C, Flow in t/h)

9 AIR AND FLUE GAS SYSTEM (P&IDS) (OUTLINE OF PULVERIZER INSTRUMENTATION)

9.1 Air and Flue Gas System

Air and flue gas, the draft Section of the boiler, play the most vital role in the performance and firing of the boiler. It encompasses not only both the air and fuel (mill) supply, but also covers the burning of fuel in the furnace and the flue gas produced thereby. There are classifications of various airs in the boiler: primary air, secondary or combustion air, scanner air, and seal air. Each of these types of air has a specific function to perform, and their main purposes are listed below:

Primary air (PA): It is the air that comes to the furnace along with the fuel such as pulverized coal. Because it is to carry pulverized coal, it has pressure greater than that of secondary air. Primary air is supplied by PA fan(s). A PA system may be two types. The first is a hot PA system, where the major part of the air at the suction of the PA fan is preheated through a air heater (AH). In a hot PA system, the fan capacity (mass) is less than that of cold system (of same size). The second is a cold PA system, where the PA fan handles the cold primary air, that is, not heated by AH (before the PA fan); naturally, for the same size, it has a higher capacity. Primary air is responsible for:

- Carrying pulverized coal to the furnace.
- Removal of moisture from coal so that it can burn spontaneously in the furnace.
- Keeping the temperature of the coal pulverizer within the specified limit so that there is no chance of fire in the mill (pulverizer).

Secondary air/combustion air (SA/CA): As the name suggests, combustion air is the main air (more than that for PA) supplied to the furnace required for combustion. This air is supplied to the furnace through the wind box/air register. The distribution of this secondary air is done as per the location of the fuel burner firing (say, elevation-wise in case of tangential tilt, or TT, burners). This distribution is controlled through the secondary air damper control or the air register control system.

- *Seal air system*: There are separate fans for this purpose to supply seal air to the coal mills and the coal firing system for sealing the coal from coming out. This is done not only to avoid dust pollution, but also as a safeguard against accumulation of fuel. Naturally, the seal air pressure will be higher than the PA pressure so that it can seal.
- *Scanner air system*: This air is responsible for cooling the various flame scanners and burners in some cases.

9.2 Objectives and Functions of the System

The air and flue gas system distinctly consists of three different sections: the flue gas path, the secondary (combustion) air path, and the primary air path. The basic purposes of these paths are listed below:

1. Flue gas:
   a. On burning fuel in the presence of air, flue gas is produced. It is very hot, and the basic purpose of this gas is to transfer heat to the various pressure parts in the boiler such as the primary and secondary

superheaters, the reheaters, E the economizer, etc., to heat the water and steam.

b. The transfer of heat to the various parts of the boiler may take place through radiation (Platen superheater) or partly by convection. Naturally, wherever the heat transfer is through convection, the flue gas flow plays a vital role in heat transfer.

c. Another function of flue gas is to preheat the air reaching the furnace (secondary air) and/or mill (primary air) so that spontaneous ignition of fuel can take place.

d. After all the heat transfer, the flue gas finally goes out through the chimney. It is very important to see that the flue gas at the chimney outlet does not cause pollution with gases, including NO_x, SO_x, CO_2, etc., that are outside the prescribed limit set by law.

e. Apart from pollution control, it is also important to see that there is not much unburnt carbon, CO, etc., to avoid a fire hazard. Also, excess air in the flue gas or chimney outlet means more energy loss in the form of heat. It is therefore the objective of the flue gas system to ensure that optimum burning is taking place.

f. From the point of view of plant operation and pollution control, it is important to ensure that the flue gas leaving the chimney is not at a high temperature ($<$120°C); otherwise, it could be used for steam temperature control by flue gas recirculation (FGR). Because in a rotary air heater heat transfer is more than the recuperating type, hence the flue gas temperature at the chimney outlet in the former case will be much less. Thus, gas recirculation for temperature control and for recuperating AH is common in plants with rotary air heaters (except for NO_x control).

2. Secondary air (combustion air):

a. To supply adequate oxygen for burning the fuel in the furnace.

b. In an air heater, the SA absorbs the heat from the flue gas to heat itself to assist in the instantaneous combustion of fuel in the furnace.

c. Always a little excess air is maintained to ensure complete combustion, so that the furnace atmosphere is safe for operation to circumvent the chances of accumulation of CO to cause an explosion. Again, the excess air shall be maintained at an optimum level so that unnecessary heat/energy is not lost through excess air.

d. Depending on the design of the boiler, the burners may be at four corners and at different elevations (for example, a TT burner boiler), they may be arranged in the wall at the front, or they may be on oppose fire mode (wall fire boiler). The main question here is whether the total quantity of SA must be distributed in such a way that all the burners get the optimum quantity of SA. While the FD fan inlet impeller/damper (of the fan speed) regulates the *total* combustion air flow, the *distribution* of air for each burner/over fire air etc. is regulated by the secondary air damper control (SADC)/air register.

3. Primary air:

a. This is sent to the mills with the help of the primary air fans. Depending on the boiler design, these PA fans may be common to all the mills, say 2 × 100% PA fans servicing all (say, six) mills. In these cases, the header pressure is maintained at a fixed value by regulating the flow through the fan (for example, the inlet vane control), and the individual mill gets air supply through the respective hot air (and cold air—for temperature control) damper. Whereas in other designs, there are as many PA fans as mills. So individual mill air flow is controlled by regulating the individual PA fan control.

b. As discussed earlier, the PA may be a hot or *cold* system. In a cold PA system, the PA fans take the suction from the atmosphere or cold discharge header of the FD fans, after the PA fans, a part of the air from PA fan, is heated in AH, sent to the mill after mixing with the other part of the PA flow which is not heated. In a hot PA system, the PA fans normally have their suction from secondary air, after mixing the hot SA (through AH) with cold SA.

c. Normally, in a cold PA system, the hot air damper (HAD) is regulated to meet the total air demand for the mill, whereas the cold air damper (CAD) is modulated mainly for mill temperature control. On the contrary, in a hot PA system, with several PA fans, each mill air flow is regulated by an individual fan flow control (for example, an inlet vane) and the HAD and CAD can be utilized for mill temperature control. In the case of a tube mill with a hot PA system, instead of controlling the individual side outlet temperature, the lower temperature of both sides is taken as the control parameter to regulate the HAD and CAD. However, an interlock from a very high temperature of an individual side may be provided to avoid a fire hazard in the mill.

d. To carry the pulverized coal to the furnace for burning.

9.3 System Description—Air and Flue Gas

9.3.1 System Description—Flue Gas (Figs. 3.31 and 3.32)

1. Flue gas is generated at the furnace chamber where the pressure is kept a little less than the atmospheric pressure (for example, [−] 5 mm wcl). In the case of

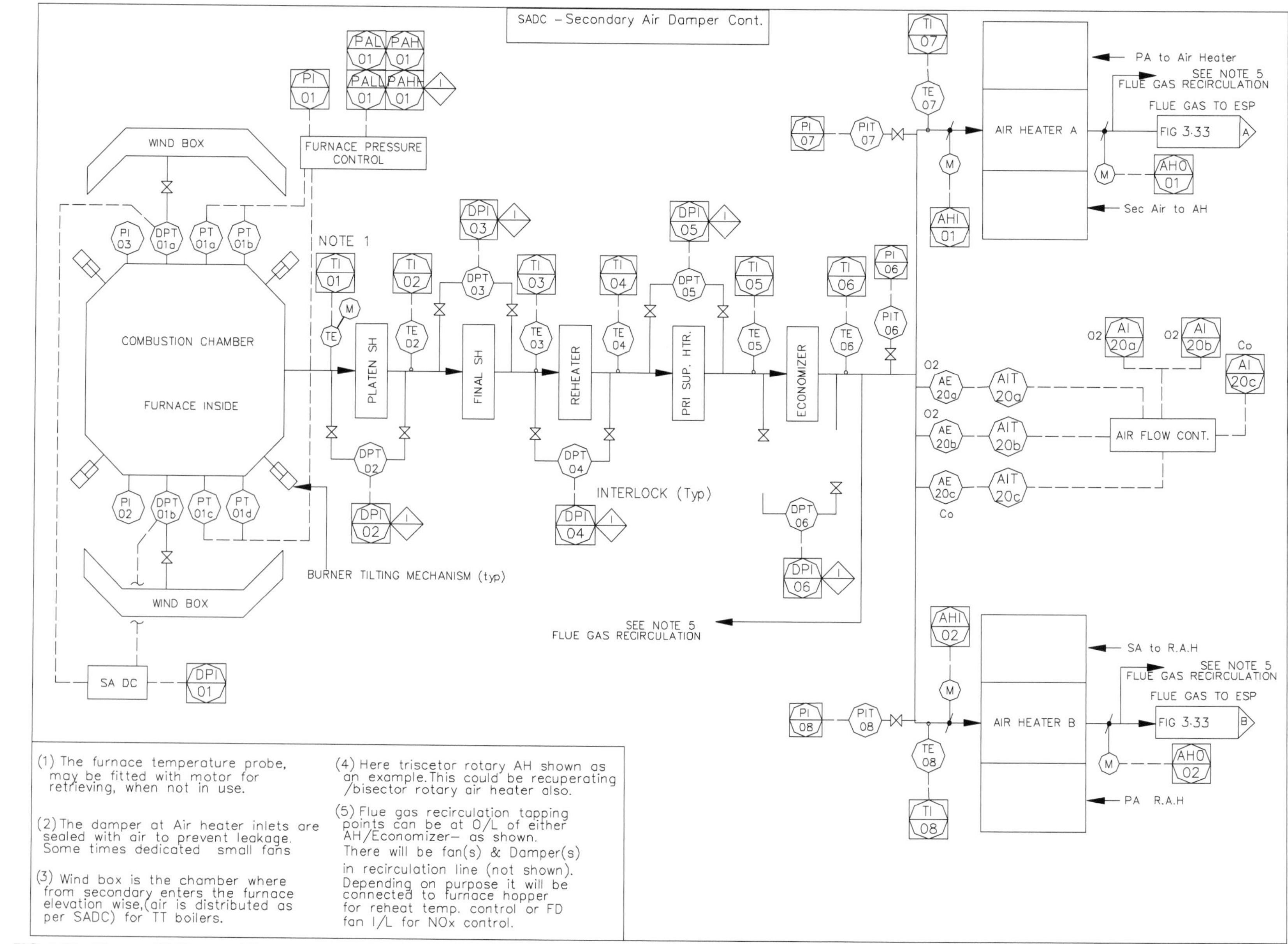

FIG. 3.31 Flue gas P&ID up to AH.

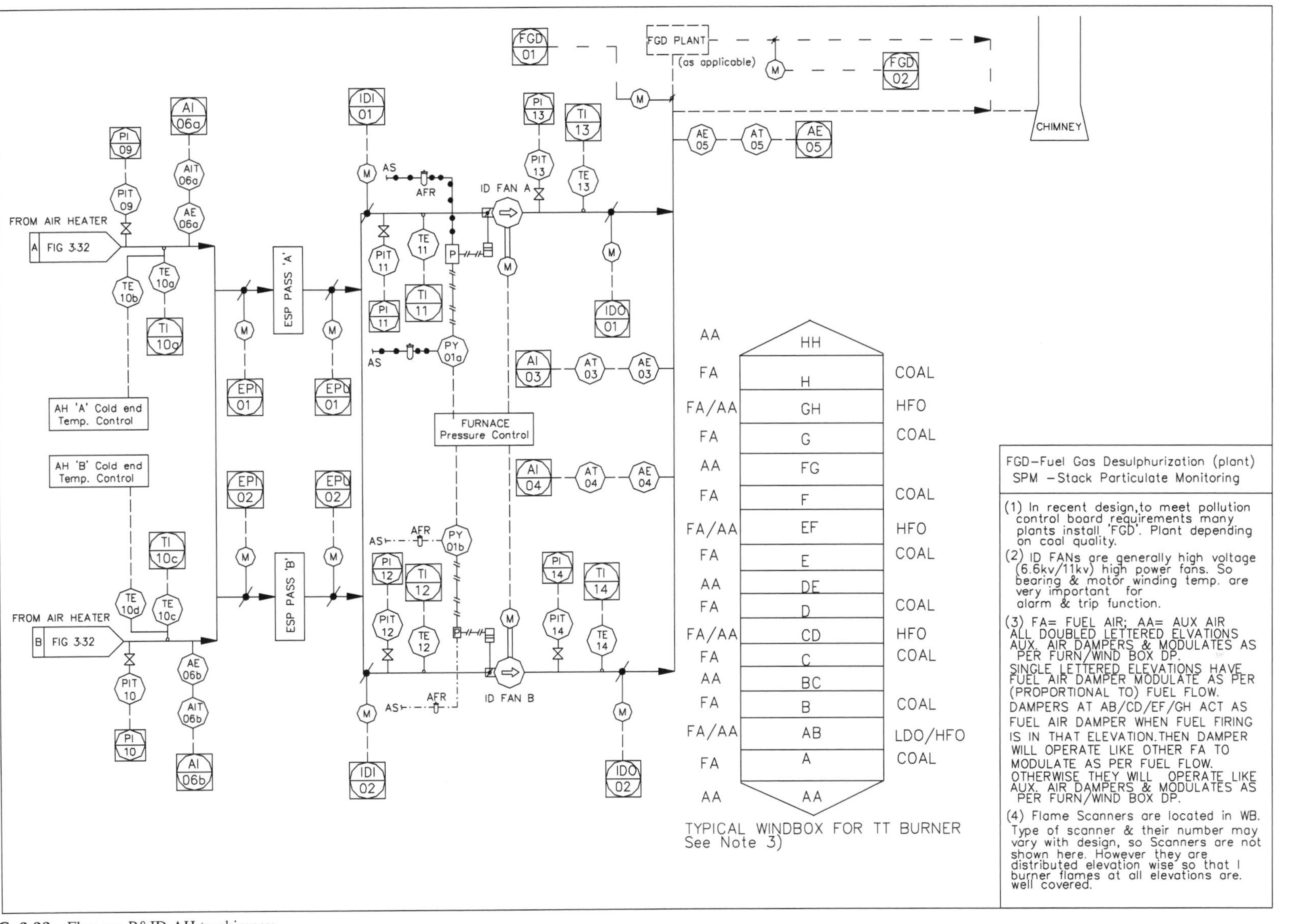

FIG. 3.32 Flue gas P&ID AH to chimney.

TT burners, the fireballs are generated here. Depending on the burner tilt position, the fireball may go up or down. The furnace wall is basically formed by water tubes, better referred to as water walls. Some amount of heat absorption takes place. The exact heat absorption in the furnace and various parts of the boiler is design-specific.

2. Because firing takes place at the furnace, from there a part of the heat radiates to the reheater (RH) or superheater (SH), which hangs like a pendant (also called a pendant superheater). Naturally, the heat transfer to steam from the flue gas is because of radiation and does not depend on flue gas quantity. Here, it is interesting to note that various boiler designs have different configurations of the superheater and reheater in the radiation and convection zones. In the P&ID, one typical design of the corner fire boiler has been considered. Thus, depending on the boiler configuration and design, the means for steam temperature control varies. In the case of a tangential burner, the tilt mechanism of the burner is utilized to regulate the RH steam temperature, which has an effect on SH temperature control. Again, in the case of a wall-fired boiler, and/or circular corner firing (CCF) with twin fire vortex (Mitsubishi design), gas dampers (may be with gas recirculation dampers) are used to change flue gas distribution between RH and SH to regulate reheat temperature control (influencing on SH temperature control). In the case of a CCF, the tilt can be used as a secondary means of RH temperature control.
3. As the flue gas rises along the height of the furnace, it passes over the SH and RH. Near the end of the first pass or in the second pass (based on boiler design), it passes over the primary SH (horizontal). In these SHs and RHs, heat transfer from the flue gas to the steam is by radiation and convection. Major heat transfer to SH/RH is through convection. The convection heat transfer rate is a direct function of the flue gas quantity, so the total heat absorption increases with boiler output. Therefore, the quantity of flue gas and the quantity of steam flowing through the coils of these SHs and RHs are dependent variables, that is, related to each other. For this reason, the set point of steam temperature control at the low load is not set at a fixed point (say 540°C). As during low load, the steam flow through the coils of the super heater will be low. Naturally it will take away less heat transferred by flue gas, so it needs to be cooled down by water. Firing shall also be regulated; otherwise, the superheater coil will be damaged. Therefore, a set point in the steam temperature control loop may be derived from air flow, which is an indication of load. For OT boilers, feed flow is also a function for temperature control.
4. In the second pass, the flue gas comes across the economizer, where it heats the feed water passing through the coil of the economizer. Meanwhile, by transferring heat, the temperature of the flue gas has come down (for example, ~350°C, 200 MW).
5. In an air heater (AH), the flue gas exchanges heat with air. Here again there are variations in designs of air heaters, as discussed in Section 9.4. In the drawing, a trisector Ljungstrom air heater has been considered. In the same AH, the flue gas heats both the PA as well as the SA. Instead of the trisector, the Ljungstrom AH could be a bisector (flue gas and either PA *or* SA). So in that case, separate AHs will be required for the PA and SA. In the case of a hot PA system, (because PA takes its suction from hot SA air), obviously, the air heater used will be a bisector air heater. Ljungstrom air heaters are rotary AH whereas other designs are the recuperating or stationary type. Air leakages as well as heat transferring efficiency both are high in the case of Ljungstrom/rotary air heaters when compared with other versions. For isolating the AH for maintenance, isolating dampers have been provided at the inlet as well as outlets of the air and flue gas paths.
6. Normally, an air heater is the last stage of heat exchange for flue gas and finally goes out through the chimney with the help of an induced draft fan (ID fan). However, in certain boiler designs where further heat recovery is done, by recycling a part of the flue gas to the furnace. This method was found to be quite popular to raise the temperature of the reheat steam temperature in wall-fired boilers. Gas recirculation may be used for NO_x control also.
7. *Flue gas recirculation*: A part of the flue gas at the outlet of the air heater or economizer (in some designs, it has been found that flue gas has been recirculated taking tapping at ID fan outlet to eliminate the need for separate fan) is recirculated back to the furnace with the help of fans, dampers, and ducts. It is introduced near the vicinity of the burning zone, that is, near the furnace hopper. With this, the total heat absorption may not be affected much but the pattern of heat absorption in the boiler is changed. In this method, with an increase in the percentage of FGR, heat absorption at the furnace is reduced whereas heat absorption in the reheater, primary superheater, economizer, etc., increases significantly. There is another point where the flue gas could be introduced for recirculation. It is the point near the furnace (first pass) exit. On account of the introduction of recirculated flue gas at this point, with an increase in the percentage of FGR, the flue gas temperature at the furnace exit as well as heat absorption at the secondary superheater will decrease. However, heat absorption at the economizer and the primary superheater would increase, with an increase in the percentage of FGR. (This may signify that less attemperation may be required in second-stage

attemperation, in case of two-stage superheater temperature control.) There could be another advantage of FGR. Because part of the heat is recovered from the flue gas, the temperature of the flue gas going to the ESP would get reduced in the course of time, leading to a better performance of the ESP. This method is also known as gas tempering.

8. *NO_x formation and elimination*: FGR plays some role in reducing thermal NO_x formation. For effective NO_x control, normally, this is used in conjunction with low NO_x burner (LBN), and/or over fire air dampers or additional air dampers in stages for some boilers (for example, Mitsubishi design). There are mainly two major sources of nitrogen for the formation of NO_x. One is molecular N_2 in air used for combustion and the other is bonded nitrogen in fuel. At high temperature, the N_2 present in air forms NO_x, more popularly known as thermal NO_x to pollute the atmosphere. NO_x formation depends mainly on temperature, turbulence (stoichiometric conditions), and time (resident combustion time)—the three Ts. To minimize this, a decrease in O_2 at the combustion zone and reducing the combustion gas temperature are necessary. Low NO_x burners (LNB) are designed in such a way as to mix fuel and air so that there will be a longer and more branched flame and the peak flame temperature is reduced so that there will be less NO_x formation. In the A-PM burner of Mitsubishi design, there will be a central weak flame and surrounding that there will be concentrated flame. According to the manufacturer, this would give good low NO_x performance with stable burner performance. It also helps to have less O_2 at the hottest combustion part—delayed mixing. In FGR technology, 20%–30% flue gas at ~350–400°C (may be from the air heater outlet) is recirculated and mixed with combustion air. This dilution decreases the flame temperature to result in less thermal NO_x formation. However, this method is more effective in oil- and natural gas-fired boilers. For coal-fired utility plants, gas recirculation is effective for temperature controls, but may not be *that* effective for NO_x control (Courtesy: EPA on utility boiler) if used in isolation. For this FGR, the increased rating for the ID fan and/or the damper and separate fan may be necessary. However, cost-effective designs are also available where separate fans and wind boxes are not necessary; instead, the flue gas is induced at the FD fan inlet. However, with LBN and FGR, NO_x reduction >70% can be achieved, provided the boiler design uses the "burner out of service (BOOS)," "over fire air (OFA)," or "additional air (AA)." In the OFA/AA port, staged combustion air downstream of the combustion zone is introduced. In case of tangential tilt burner designed boilers as such has inherently less NO_x formation. Separate over fire air (SOFA) and closely coupled over fire air (CCOFA) dampers are very common in reducing the NO_x level without the use of any chemicals. In some designs (for example, Mitsubishi design) AA is used in stages above the burning zone. In a nutshell, the fuel-enriched condition near the flame zone, with a comparatively lower temperature, is followed by the air-enriched zone (for complete combustion) by OFA/AA in a staged manner. In TT burners with SOFA, the CCOFA reduction level can even reach nearly 90%. Chemically, NOx is reduced by the selective catalytic reduction (SCR) method.
9. After the air heater, the flue gas passes through the electrostatic precipitator (EP), whose main function is to eliminate dust from the flue gas before it is discharged to the atmosphere through the chimney by the ID fan. In the initial days, mechanical dust collectors were in use, but now EPs are used for better efficiency. Because these are placed after air heaters, all these EPs are cold EPs.
10. Finally, the ID fan takes out the flue gas through the chimney. NO_x reduction was already discussed above, and hence is not repeated again. In the places where coal has a high sulfur content, there flue gas desulfurization (FGD) plants are installed. These are detailed in subsequent sections.
11. There will be inlet and outlet isolation dampers for ID fans, which are also used for no load starting of the fan, similar to that discussed in clause no. 9.3.2.2 (Figs. 3.33–3.37).

9.3.2 System Description—Secondary Air System (Figs. 3.34, 3.36, and 3.37)

1. Secondary air (combustion air) to the boiler is supplied by the FD fan. There may be inlet cones at the suction of these fans, which are nothing but restrictions in the system and could be used as a flow element to measure the total secondary air through each fan. The flow through each of the FD fans can be regulated or modulated by an (aero-foiled) impeller or inlet damper (Fig. 3.37B). These regulations are related to the total air control of the boiler. As discussed in clause no. 9.3.1.8 for NO_x control, the FGR line may be connected here.
2. There is also one discharge damper for each FD fan for isolation purposes. For "*no load*" starting of large high-voltage motors for fans (for example, FD/ID/PA fans), the isolating damper is kept closed.
3. Whenever the atmospheric temperature is low and/or contains more moisture, then it is customary to heat the air prior to its entry to the air heater. This is mainly applicable for the Ljungstrom rotary air heater. This preheating of the air is done at the steam-coiled air

FIG. 3.33 SCAPH cold end temperature control.

Why SCAPH? WHAT IS COLD END TEMPERATURE CONTROL?

Out of two types of air heater i) Tubular ii) Rotary type, rotary type has better heat transfer (but also higher leakage~ 10%), therefore the flue gas after transferring heat to SA or PA, gets more cooled. Now if fuel has sulphur and there will be SO_2 in flue gas which on account of leakage air gets converted to SO_3. In presence of moisture at cold end (where cold air is going in and cold flue gas is getting out of air heater) it will form H_2SO_4 which will cause corrosion. In order to prevent this SCAPH is used to heat up cold moisture laden air. This temperature is controlled with the help of regulating steam flow to the coil. This is cold end temperature control—detailed out with control loop under boiler control system. GENERALLY NOT APPLICABLE FOR TUBULAR AIR HEATER

$SO_2 + O_2 = SO_3 \quad SO_3 + H_2O = H_2SO_4$

preheater (SCAPH, Fig. 3.33). This is a coil heater where through the coil, auxiliary steam is passed to heat the air outside. The SCAPH has an inlet, outlet, and bypass damper to isolate the system as well as to bypass the system. The same type of SCAPH is utilized in both the SA and PA (cold PA system) systems.

4. After the FD fan, the SA goes to the AH either through SCAPH or directly. The AH is of two types: tubular air heater and rotary air heater, as discussed earlier. In the air heater, the air is heated with the help of flue gas coming from the economizer. In the secondary air side, also the air heater has inlet/outlet isolation dampers.
5. After the air heater, the secondary air heater goes straight to the wind box/air register, where there are a number of dampers meant to distribute air to various burners according to their operational conditions and/or to the over fire air port. In a hot PA system, a part of the secondary air (both hot and cold) goes to supply air for the primary air system, as shown in Fig. 3.36.

9.3.3 System Description—Primary Air System (Figs. 3.34, 3.36, and 3.37)

1. Primary air to the boiler is supplied by PA fan(s) to carry the pulverized coal to the boiler and perform other functions, as detailed in clause no. 9.2.3. Like FD fans, the PA flow can be measured by an inlet cone or Pitot tube. The flow control of the PA fan can be done by modulating the inlet damper (Figs. 3.35 and 3.37B) or the fan speed. For common PA fans to all mills, the PA header pressure will be chosen as the control parameter to regulate the flow through each fan. In the case of common PA fans, discharge from both PA fans forms a common header, as shown in Figs. 3.35 and 3.36A. From this common cold PA header, one line of cold air goes to the individual mill as cold air through CADs and other lines (to AH directly or to AH via SCAPH) supply hot air through HADs. As hot air is mainly responsible for carrying coal to the boiler, there will be an influence from the fuel flow control loop for the mill air flow loop.
2. Similar discussions as detailed in clause nos. 9.3.2.2, 9.3.2.3, and 9.3.2.4 are also applicable here for PA fans.
3. The configuration shown in Fig. 3.36A is for supplying to individual mills. This is normally adopted for TT boilers with bowl mills.
4. In the case of a hot PA system, a cold header and a hot air header are formed after the FD fan discharge header; after mixing both airs, the same is directed to individual PA fans, as shown in Figs. 3.36B and 3.38.

9.3.4 System Description—Scanner Air and Seal Air

1. Scanner air system (Refer: Fig. 3.34): A scanner air fan, with its suction from the discharge headers of FD fans, supplies scanner air to cool the flame scanners. Normally, a scanner air fan has a duplex suction filter (with DP monitoring for timely changeover of the filter) to ensure a pure air supply. In modern designs, the same air is used to cool the hot secondary air dampers also. Each scanner air fan has a motor-operated outlet damper with a BMS interlock. In the absence of scanner air, the boiler will be tripped and to ensure continuous scanner air supply, the differential pressure between the scanner air and furnace is monitored for interlock.
2. Seal air system (Fig. 3.35): The basic function of seal air is to seal the dust from the mills and associated devices at various positions. The seal air fan normally takes its suction from the cold PA header in the case of a cold PA system. Whereas in the case of a hot PA system, it takes its suction from the discharge headers of FD fans. Normally, it has a duplex suction filter (with DP monitoring for timely changeover of the filter) to ensure a pure air supply. Each seal air fan has a motor-operated outlet damper with interlock to trip the mill in the absence of seal air. In order to ensure continuous air supply,

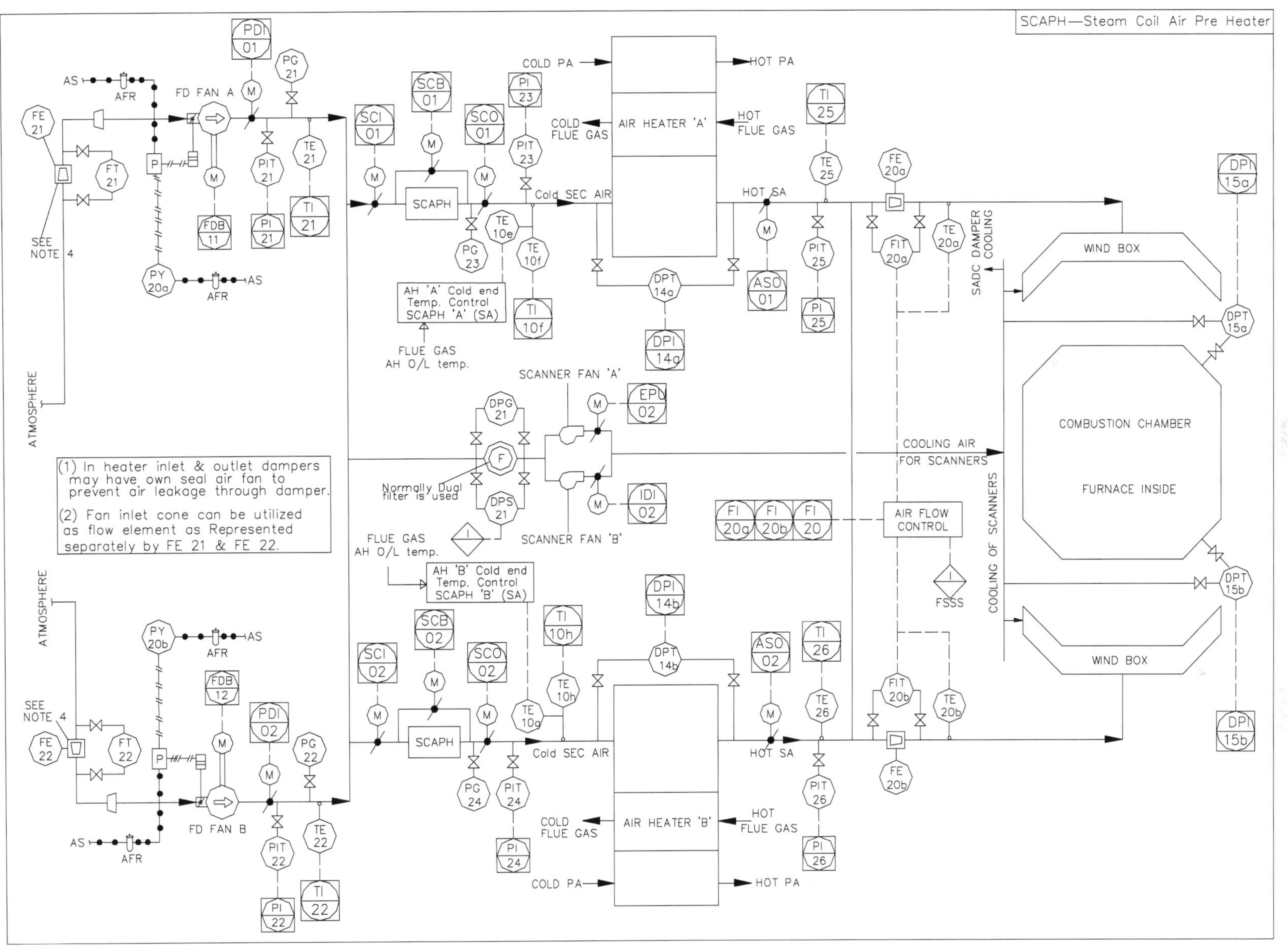

FIG. 3.34 Secondary air P&ID.

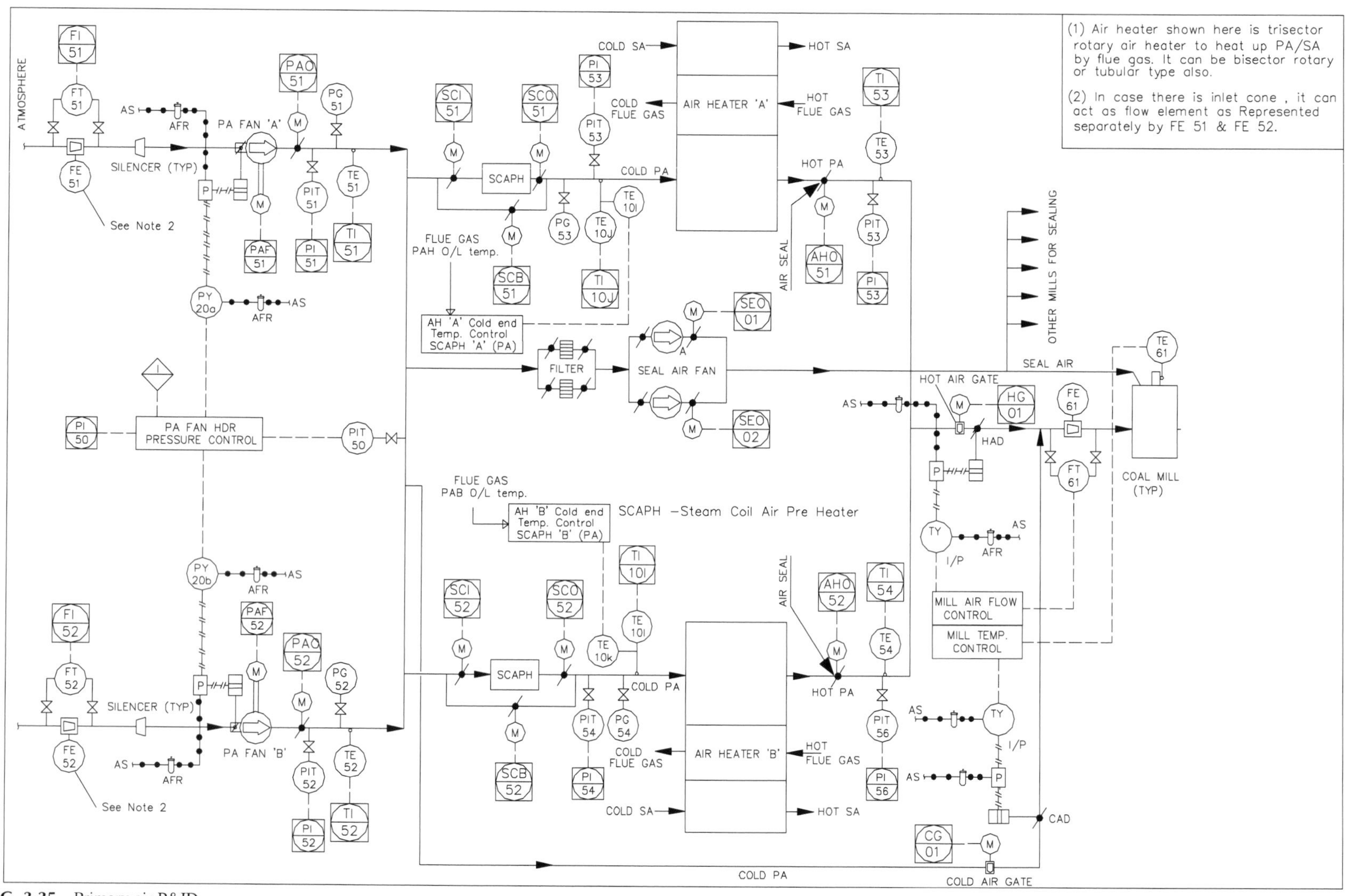

FIG. 3.35 Primary air P&ID.

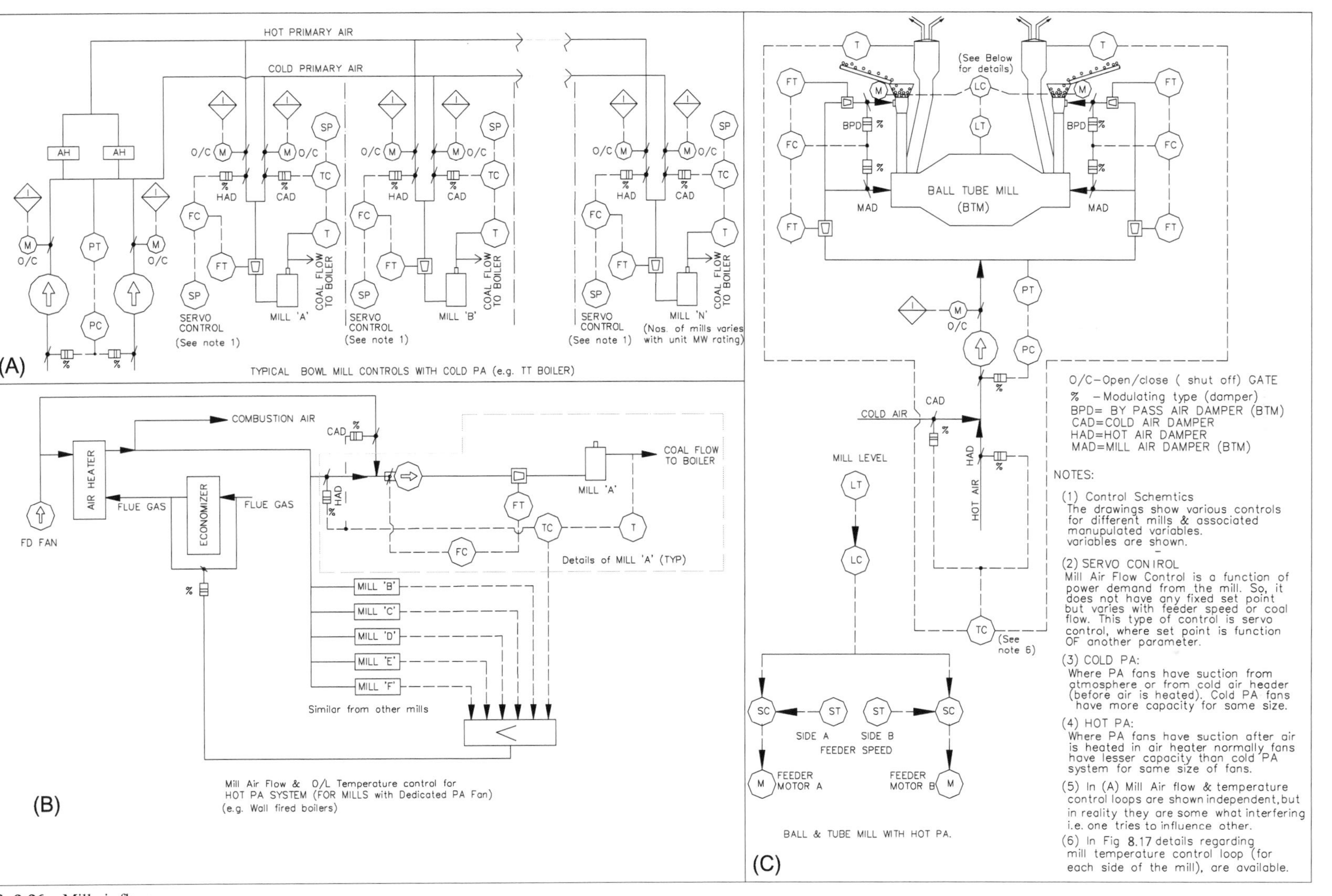

FIG. 3.36 Mill air flow.

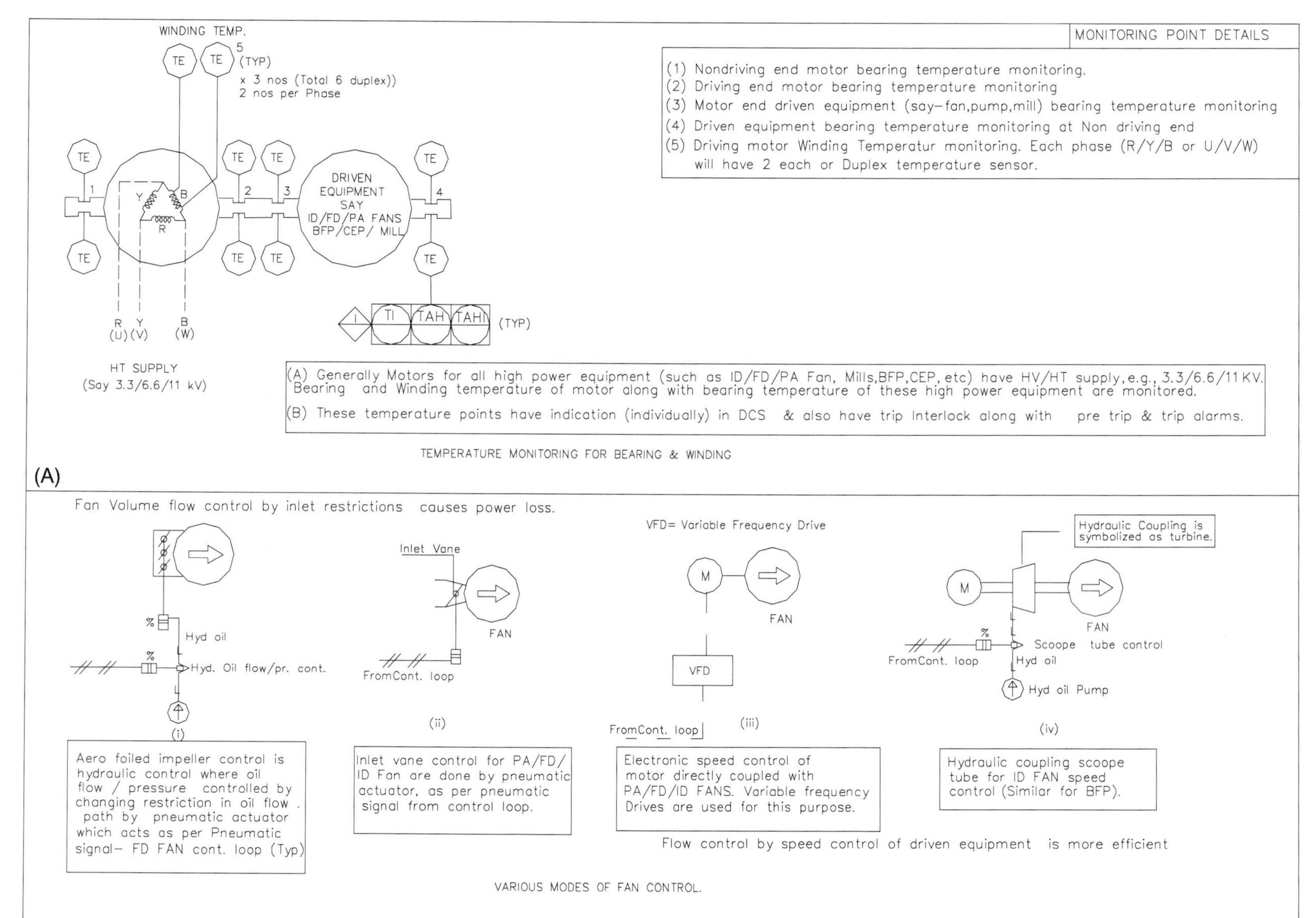

FIG. 3.37 Various temp. monitoring and type of controls for fans.

No Load starting of High Tension (HT) motor!
Big fans, pumps, mills in general are driven by HT motors (e.g., 3.3 /6.6/11 kV). Now if these motors start, with in full load starting, inrush current is very high say ~5-6 times Full load current, so to avoid such high inrush current, if* permitted, *it is customary to start these drives under no load condition , by closing the inlet/outlet damper /valves as applicable. This kind of starting is called No load starting. However in certain cases say for 2X100 %BFP, HT drive may have to start in full load. In case of tripping of one running BFP, other BFP needs to start in full load to avoid starvation of boiler, in case of failure of running BFP.

FIG. 3.38 HT motor starting—no load.

the differential pressure between across the seal air fan is monitored (not shown).

9.3.5 System Description—Monitoring and Controls (Fig. 3.37A and B)

The discussions in this section in connection with large fans and associated drives (motors) are equally *applicable* for all large equipment (though not discussed separately in respective sections) such as mills, BFPs, CW pumps, etc.

1. In most of the cases of medium to large power plants, the power consumed by various equipment such as ID, FD, PA, and mills (also BFP) is high, so motors to drive this equipment must have high tension (HT)/voltage (HV) supplies to limit the current. Naturally, this heavy equipment needs to be monitored for bearing and winding temperature (of the motors) by RTD (rarely by T/C), as shown in Fig. 3.37A, to ensure the protection and smooth operation of the heavy equipment. Similarly, the temperature at each of the three phases of the motor windings is monitored by the duplex (or two numbers) RTDs. Generally, 0–150°C is the standard range. All these RTDs are connected to the DCS for generation of a pretrip alarm as well as a trip whenever the temperature exceeds limits to avoid equipment/motor damage. Also, it is common practice to monitor the bearing vibration of this heavy equipment as well as the speed of big fans.
2. In Fig. 3.37B, various modes of control of fans have been depicted.
   a. In some fans, say in PA/ID fans, the flow control is regulated by adjustment of the *inlet vane* in the same manner as discussed below (see Chapter 6, Sections 3.1–3.3 for detailed discussion).
   b. In some of the fans, say an FD fan, there are air-foiled impellers that are regulated to regulate the air flow to the fan. Normally, on account of heavy design, these *impellers* are regulated by hydraulic actuators. The pressure flows to the hydraulic actuators are controlled by controlling the opening/restriction in the hydraulic line. Such opening control is done by a pneumatic actuator connected to the main control loop (Chapter 6, Section 3.4).
   c. *Speed control* of the fans (Pumps—say BFPs) is another method of regulating the flow through each of these fans. These speed controls can be either be by direct speed control (VFD) of the fan motor electronically (Section 3.1.4.2 of Chapter 6) or by an indirect but reliable method by a hydraulic coupling scoop tube (Section 3.1.4.1 of Chapter 6).

9.4 Major System Equipment—Air and Flue Gas Path

The major system equipment in the air and flue gas path is discussed below:

1. *Combustion chamber*: It is the chamber where the actual combination of O_2 with combustibles takes place. In fuel H_2, C and S are the major combustibles, out of which S is negligible compared to the other two. Hence, at this chamber the following reaction in the simple form takes place with the generation of huge heat: $C + O_2 = CO_2 + \text{Heat}$; $2H_2 + O_2 = 2H_2O + \text{Heat}$. The furnace chamber size depends on the three Ts—temperature, turbulence, and time, hence they vary with system design. The furnace chamber walls are water cooled for reduced maintenance as well as to reduce the gas temperature entering the convection zone. So, all walls are generally designed with drainable and weldable panels. The two major firing systems are horizontal firing and tangential firing.
   a. For horizontal/wall firing: Here, coal and PA are introduced tangentially into the coal nozzle, and the fuel is mixed with combustion air in the air register individually. In front horizontal firing, the burners are located in the row horizontally on the front wall. Whereas in opposed horizontal firing, the burners are located in the row horizontally both in the front and rear wall of the furnace. In circulating corner firing, there will be two sets of fireballs as there will be firing in a number of elevations. Like

tangential tilt firing discussed below, here it is also possible to move the fireball up and down.

b. Tangential tilt firing: When pulverized fuel and PA are sent to the furnace from the four corners (tangential to an imaginary circle) in a horizontal plane (at the elevation of burning), huge turbulence takes place and intensive mixing occurs. So, on account of firing at different elevations, there will be a fireball. In this mode, the burners can be tilted with respect to the horizontal plane. Here, secondary air dampers are away from the burners, and a mix up with combustion air takes place in the chamber. Thus tangential burners form a fireball and when these burners tilt, the fireball moves up and down (causing changes in heating effect)

2. *Superheater and reheater*:
   a. Superheater: In the superheater, the saturated steam gains heat to become superheated steam. In the case of a TT boiler, a large part of the heat transfer in superheaters is through radiation whereas for horizontal wall firing, the heat transfer is mainly by convection. Again in this type of design, it has been found that radiant superheat as a function of load as dropping in nature. As a result of these two, the overall change in the SH outlet temperature is a bit steady but, with load, increasing in nature. For this reason, attemperation flow in this type of boiler increases little with load. However, in a TT boiler, the attemperation water flow at higher load is not much at all. After leaving the furnace chamber, the flue gas first comes across the radiant superheater, where the major heat is transferred in radiation mode. In the case of a TT boiler, the final superheater is a vertical pendant Platen type above the furnace. Whereas in the case of wall-firing boilers, the final superheater is convection. The primary super eater in both cases is convection, and it may be located at the vertical pass in the rear side. In the second pass, there is a horizontal primary superheater. The superheater overall absorbs about 25% heat to raise the temperature from as low as 350°C to 540°C in medium to large boilers. Interstage attemperations are common.
   b. Reheater: In a TT boiler, the reheater is in the radiant zone. Reheat steam enters the reheater radiant wall inlet header and finally into the pendant reheater. In the TT boiler, reheaters are mostly vertical but horizontal reheaters are also found. In a wall horizontal firing (HF) boiler, the low temperature reheater may be located in the second pass in horizontal mode. Heat absorption is around 20% to raise the temperature from as low as 270° C to 540°C. The reheat temperature in HF firing is controlled by gas damper regulation (gas sharing between the superheater and reheater). In TT boilers, the reheat temperature control is affected by the burner tilt controls. However, in both systems there is a provision for emergency spray at the reheater inlet (Section 9.3.1).
3. *Economizer*: In order to improve boiler efficiency by extracting heat from the flue gas after the superheater, an economizer is deployed. The exact location of the economizer depends on the overall boiler design, but generally it is located ahead of the air heater in the second pass. Designing water and flue gas in counter flow—gas flowing downward across the tube, whereas water (at temperature appreciably lower than saturation temperature to prevent steaming in economizer) flow from bottom to top.
4. *Air heater*: The air heater has two functions, cooling the flue gas temperature before discharging to the atmosphere and raising the air temperature of the incoming combustion air and primary air (so that coal is dried and transportation of pulverized fuel is possible—in case of lignite this becomes the primary function). Two principal types of air heaters are in use—one tubular recuperative type and the other a regenerative rotary Ljungstrom type. In the former case, air flows horizontally across the tube and gas flows vertically. Whereas in the Ljungstrom air heater, sensible heat in the flue gas transferred to air through a regenerative heat transfer surface (which turns continuously through gas and air) of the rotor, which is turned by a motor at a speed of 1–3 rpm. Here, the gas flows upward and the air flows downward (Chapter 2). The efficiency and leakage between the two types of AH were already discussed earlier and are not repeated. The air heater cold end temperature control in the Ljungstrom air heater is very important.
5. Gas recirculation fan: A gas recirculation fan draws flue gas from the economizer/air heater outlet to recirculate the same to the furnace near the vicinity of the burning zone/furnace hopper (when considered for temperature control) or to the air circuit (for NOx control). The duty of the fan is severe on account of the high dust quantity as well as the temperature fluctuations it has to face.
6. *Selective catalytic reduction*: SCR is a means to convert NO_x with the help of a catalyst into N_2 and water. H_2O, aqueous ammonia, urea, etc., are added to the flue gas stream to remove NOx. It is located after the economizer.
7. *Dust removal*: Mechanical Dust collector: In this method, dust particulate from the flue gas path is

removed by centrifugal, inertial, and gravity forces (Section 2.2 of Chapter 2).

8. *Electrostatic precipitator*: (Section 2.2 of Chapter 2) In order to collect the dust particles, the carrier gas is ionized by high electric fields discharge electrode and collector electrode. The gas is passed between the electrodes. The gas ions in turn charge the dust particles to be collected at the collecting electrodes. Collected dust is collected at the hopper below by rapping the electrodes.
9. *Induced draft fan (ID FAN)*: The ID fan exhausts the combustion product from the boiler. In doing so, it creates a slight negative pressure in the furnace of say 5–12 mm wcl. This suction pressure gives the name "balanced draft." In conjunction with FD fan supply. As ID fans are located at the downstream of the particulate removal system, they are meant for clean service. For ID fan selection, due consideration is given to taking care of higher wear resistance due to the dust burden, corrosive ash particles, and high temperature it has to handle. In cases of FGD plants associated with the flue gas system, it acts as a booster fan. The capacity control of ID fans is done generally by inlet vane control (sometimes as coarse control—say in larger capacity fans), or speed control (sometime as fine control—say in larger capacity fans where both the controls are deployed). Speed control in turn is done by hydraulic coupling/VFD—electronic control. Speed controls are always smoother and give better performance.
10. *Flue gas desulfurization plant*: FGDs are installed to reduce the level of SO_2 in the flue gas discharged to the atmosphere. Limestone scrubbing (Chapter 2) is the most common method followed in FGD plants. It is a catalytic chemical process for the removal of SO_2. In this system, sulfur compounds combine with lime (CaO) or limestone ($CaCO_3$) to form slurry.
11. *Forced draft fan (FD fan)*: Total/combustion air necessary for the boiler is supplied with the help of FD fans. Normally these shall be sized, the stoichiometric air plus excess air. In the hot PA system, the fan size shall be more than in the cold PA system. A radial aerofoil (centrifugal) fan with impeller control or axial fans with pitch controls are used in general. Inlet silencers and screens are used for noise and protection of the fan.
12. Steam coil air preheater: In order to raise the cold end temperature to prevent acid formation, SCAPH is used to heat the air (both SA and PA) before it enters the regenerative air heater. This is used during start-up, low load operation, and with low (or highly humid) ambient temperature. This is located between the FD/PA fan and the air heater, and is supplied with steam from the auxiliary steam. The pressure loss is nominal: ~25 mm wcl.
13. *Wind box*: In corner-fired boilers, pulverized coals are delivered to a row of fuel nozzles in four corners arranged at different levels (one each corresponding to each pulverizer or section of pulverizer, for a tube mill). There are a number of secondary air dampers in each corner in each elevation. These are the auxiliary air damper and the fuel air damper, and these are regulated according to SADC for distribution of SA at various elevations. In wall-fired boilers, an air register serves the purpose of a wind box in distributing SA.
14. *Scanner air fan*: This is a low-voltage fan meant to cool flame scanners and burners. It takes the suction from the cold secondary air duct.
15. *Primary air fan (PA fan)*: These are large high-pressure fans meant to dry and transport coal from the pulverizers to the furnace. For controlling the volume through the PA fan, an inlet vane control is quite common. Inlet silencers and screens are used for noise and protection of the fan.
16. *Seal air fan*: This is a low-voltage fan meant to seal pulverizers and associated equipment. It takes its suction from the cold PA header.

9.5 Major Parameter Measuring Monitoring in—Air and Flue Gas

9.5.1 Process Parameter—Air and Flue Gas (Typical Value)

See Tables 3.9 and 3.10.

TABLE 3.9 Flue Gas Parameter: Pr. in mm wcl, Temp.: °C

Location	Pr.	Temp.	Remarks
Furnace	(−)5	1200	
Platen SH	(−)6	1150	
Reheater	(−)13	800	
Primary SH	(−)33	558	
Economizer	(−)55	338	
AH	(−)188	125	O_2 may be ~5% and CO may 100 ppm
ESP	(−)215	125	
ID fan	150*	120	*At full load, condition pressure may drop to ~40–50 mm wcl

TABLE 3.10 Air Parameter

Location	Pr.	Temp.	Remarks
FD fan disch.	174	Ambient	
SA at AH O/L	100	320	
SA AT wind box	75	318	
PA fan disch.	850	Ambient	
PA at AH O/L	798	319	
Mixed PA at mill	675	270	~90 t/h flow approx

9.5.2 *Monitoring of Process Parameter—Air and Flue Gas System*

Air and flue gas parameters are of immense importance, as they are directly related with the firing system vis-a-vis with the boiler performance. The major details are:

1. *Furnace/combustion chamber*: (a) Combustion at the furnace best occurs when it is maintained at a particular point that is slightly negative (should never be positive because then hot gas would come out of the inspection window). It is customary to have one in situ pressure gauge near the operating floor. The pressure transmitters deployed are the draft transmitter type (with a narrow range for control, typically ± 25 mm wcl). Another set of draft transmitters with a wide range (covering very high and very low trip set points) is used for monitoring purposes. The typical range may be wide range ± 250 mm wcl (tolerance ± 10–15 mm wcl). At times, dedicated pressure switches are used for alarm and trip but from a safety point of view, redundant transmitters are preferred as per IEC recommendations. In many places, it has been seen that wide and narrow band dedicated recorders (may be video graphic recorder) are installed. (b) A few differential pressures such as furnace to wind box DP, furnace scanner air DP, etc., are monitored by the DP transmitter as well as by the DP switch (especially for medium/large utility stations). Typical ranges for these are 0–100–120 mm wcl and 0–450 mm wcl, respectively. The furnace to wind box DP measurement is also very important, especially for corner-fired boilers (Ref. clause no. 9.6.2.3). (c) Temperature in the combustion chamber in some boilers is measured with the help of a retractable (motor-operated) temperature probe (ISA type S). The temperature span may be 0–1500°C for medium/large power plants.
2. *Superheater, reheater, and economizer*: Inlet and outlet temperature for each of Platen, final and primary super heater, reheater(s) and Economizer is common. For these ISA Type K T/C, with or without transmitters, are utilized for this purpose. ISA type R or S type thermocouples are preferred for combustion chamber temperature measurements in mid to large plants, as the temperature range may be 0–200°C in the Platen and final superheater zone. A temperature range of 0–700/500°C would suffice for temperature measurement in a low-temperature superheater, economizer, etc. It is worth noting that the differential pressure across each of the heaters may be monitored to alert the operator to high DP for soot blowing. In some designs, I/L and O/L pressures are measured in place of DP.
3. Air heater (flue gas): In order to check how much heat is transferred, the inlet and outlet flue gas temperatures are measured by ISA type K. Temperatures of 0–400°C and 0–200°C would do for I/L and O/L, respectively. On account of good linearity and less influenced by vibration K type T/C (also averaging is easier than RTD) is preferred. Air side temperatures are also monitored. Air heater outlet pressure is monitored by a smart transmitter with a selected span that may be 0–(−)200/250 mm wcl. Measurement of O_2 is common at the AH inlet and outlet to compare and monitor air leakage. This is especially so for rotary air heater higher leakage. AH inlet O_2 analyzers are generally a zirconia probe type for a better time response (as this signal may be taken to oxygen trim control). The standard range that could be selected is 0%–10% by vol. CO is also measured at the AH inlet for finer control of the air flow loop. The selected range may be 0–200 ppm.
4. ID fan: Pressure and temperature measurements at the inlet and outlet of the ID fan are common. Depending on the ID fan sizing, the pressure ranges at the inlet and outlet pressure transmitters are chosen, whereas a temperature in the range of 0–150 will do and RTD (Pt 100 at 0°C) will do. At the outlet of the ID fan, NO_x and SO_x are monitored in many plants. These along with particulate monitoring are required to monitor, as per pollution control board requirements for medium and large power plants. The speed control of the ID fan by the HC scoop tube or by VFD is a general means of control. Sometimes inlet vanes are also regulated as course control. Vane control means power loss.
5. ID/FD/PA fan monitoring and control: Refer to clause no. 9.3.5.1.
6. FD fan/PA fan: At the inlet of the FD/PA fans, in some cases there are inlet cones to create negative suction. With the help of the same FD/PA fan, the flow can be measured. DPT and compensating temperature elements such as RTD are used. At the discharge header, the pressure is measured by a local pressure gauge as well as by a pressure transmitter. The range is difficult to give as it depends on fan capacity, head, etc. At the discharge, the temperature is also measured by RTD. FD fan throughputs are regulated by impeller/damper

pitch control, whereas the inlet vane may be regulated to control the air flow through the PA fan.

7. SCAPH and AH: At the SCAPH outlet or air heater (AH), inlet temperature measurements in SA/PA lines are important as these are used for low temperature control of AH. For this, the ISA TYPE K T/C may be used. The pressure loss through SCPH is low, and is monitored by a smart transmitter (a span of say 0–200 mm wcl). At the AH outlet, pressure and temperature are also monitored with the help of ISA K T/C or RTD and smart transmitter (span of say 0–200 mm wcl), respectively.
8. Wind box (secondary air flow): Air flow at this point is done to compute the total SA flow to the boiler. An aerofoil (a special restriction) or a Pitot (/piccolo) tube as the flow element in conjunction with a mart DPT is normally deployed. For corner-fired boilers, a furnace to wind box DP differential pressure is used to regulate the dampers (that is, auxiliary air flow at various elevations (doubled lettered elevations). A typical TT boiler and burner arrangement is shown in Fig. 3.32. Normally, all doubled lettered elevations will have an auxiliary air damper that shall be modulated as per the furnace wind box DP. A fuel (coal) air damper at single lettered elevations will be modulated as per fuel flow (for example, feeder speed). At double lettered elevation where there are HFO and/or LDO (lower level) burners, the control is a little different. Normally, these elevations' air dampers will be operated as per the wind box to furnace DP when there is no oil firing. However, *whenever* burners in these elevations are firing, then only these dampers will be regulated *proportionally* as per fuel firing in the elevation and not as per furnace wind box DP to cater to firing requirements.
9. PA before mills: PA fan discharge header pressure is monitored by smart transmitters, and this parameter may be a control parameter in many cases for PA fan flow control. The total PA flow to each mill (hot and cold) is measured with the help of flow element and smart DPT (or by PA fan DP and then characterization). The flow is duly compensated for temperature. In each of the cold and hot PA lines, pressure and temperature are measured respectively by smart transmitters and RTDs.
10. Scanner air and seal air fan: Measurement of the discharge pressure of these fans is done. Also, scanner air to wind box and seal air to mill DP are monitored by smart DPTs to ensure proper flow.

9.6 Controls in—Air and Flue Gas System (Figs. 3.31, 3.32, and 3.34–3.36)

There are a number of vital controls in the air and flue gas system.

9.6.1 Controls in Flue gas systems

1. *Furnace pressure control*: Furnace pressure is a very vital control for proper firing and plant operation. Pressure inside the furnace is regulated by controlling the gas flow through the ID fan by either regulating the inlet vane and/or the fan speed. Additionally, a furnace pressure signal is used for alarm and trip.
2. *Flue gas O_2 and CO*: The flue gases O_2 and CO are involved in total air flow control to the boiler for optimizing the air flow control (oxygen trimming) to the boiler.
3. *Air heater outlet temperature*: The flue gas temperature at the air heater outlet is measured for low temperature control of the air, heater that is, the SCAPH control.
4. *Soot blowing*: Apart from these controls, the DP across each of the heaters may be considered as the initiating signal for SB control.

9.6.2 Controls in Air Systems

1. Air flow control: The duly temperature compensated, total secondary air flow measured at the wind box and/or at the inlet cone is used to regulate the total air flow for combustion by regulating impeller control/blade pitch of FD fans. Refer to clause no. 9.6.1.2 also.
2. SCAPH steam control: In order to maintain the low temperature end of the AH at a specified value, the steam supply to the SCAPH is regulated as per the actual temperature of both the air and flue gas at the AH cold end. For better measurements, the average of many sensors placed in large ducts is used for controlling.
3. SADC/Air register control: Fig. 3.32 and the discussions in clause no. 9.5.2.8 above may be referred to for details. For wall-fired boilers, the air register is responsible for air distribution.
4. PA header control: The discharge header of PA fans is maintained at a fixed value, and this involves regulation of the PA fan inlet vane.
5. Mill outlet temperature control: The mill outlet temperature control is done by regulating the mainly (except where there is a dedicated PA fan for each mill where both HAD and CAD are used) by CAD control. In some wall-fired boilers, an economizer bypass damper control is deployed for this purpose. For tube mills with a hot PA system, the individual side of the mill outlet temperature is not controlled; instead both sides are used to control the O/L temperature (Fig. 8.17B).
6. Mill air flow: In accordance with the fuel demand, the air flow to the mill is regulated by regulating the HAD (In the case of a dedicated PA fan for each mill, the vane control of the fan does the same function). In SC/USC boilers, this control is very much connected with the feed flow, as discussed in connection with feed

flow control. In a tube mill, the situation is a little different; Fig. 8.14 may be referenced.

9.7 Redundancy in Measurement

- *Two of three* redundancy is common in *furnace pressure* (sometimes redundancy on each side), *air flow, and oxygen (trim control) in flue gas.*
- *One of two* redundancy is common in control loops such as *PA header pressure, PA flow to the mill, mill outlet temperature, and SADC-WIND BOX-furnace DP.*

Now these redundancies are a function of investment also, so all such redundancies may not be applicable for smaller plants.

9.8 Miscellaneous Points—Air and Flue Gas System

1. The air and flue gas system measurements and control are related to the boiler system; hence, they are monitored at the boiler monitor normally kept on the left side of the operator facing the panel/monitor. In a conventional/backup panel, these instruments form the left most panel in general.
2. *Pulverizer instrumentation*: A few points listed below may be noted.
   a. As detailed in clause no. 9.3.5.1, for pulverizers, the bearing (for example, upper and lower radial bearing, radial and thrust worm bearing, lower thrust bearing, etc.) temperature, and winding (of motor) temperatures are also monitored by RTDs.
   b. *Measurement of mill DP* is also very important to monitor the mill jamming and flowthrough of the mill. In fact, for a tube mill, the DP across the mill is measured for mill level (also, the mill level is measured by a sound system). This mill level in the tube mill is a control parameter to regulate the coal feeder speed. In a tube mill, the coal feeder is not regulated directly as per fuel demand (as is done in Raymond and ball mills); instead, the air flow through the mill is regulated as per fuel demand.
   c. *Mill interlock and control*: The mill is normally interlocked with the boiler operation through the BMS/FSSS. In a tube mill, for each side of the mill, two types of air go into the mill. One is air flow through the mill and another is bypass air flow. Bypass air flow has two distinct functions: to increase velocity at lower load and preheat the coal from the conveyor. For this reason, bypass air flow going to the coal feeding pipe has some higher value initially, but with an increase in load as the air flow through the mill increases, the bypass air flow decreases. Thus, in a tube mill from the *damper position of* the *air flow through the mill signal*, the coal flow to the boiler can be computed. In a corner-fired boiler, normally each side of the tube mill corresponds to one elevation catering to four burners located at the corner of that elevation. In a horizontal-fired boiler, depending on the design, each mill caters to a few burners in the same elevation. As already discussed, the coal flow is a function of PA flow through the mill. In SC/USC boilers, the coal feed to the boiler is also influenced by feed flow. Therefore, these two loops, that is, fuel flow and feed flow, cannot be seen in isolation.

10 COGENERATION P&ID (GAS TURBINE HRSG)

10.1 Cogeneration/Combined Cycle Plants

In a *cogeneration (CoGen)* plant, in addition to the electric power generated by the gas turbine, the exhaust flue gas (from the GT) heat may be utilized to produce steam for other industrial purposes. In *a combined cycle* (CC) plant, in addition to electricity generation by GT, the exhaust gas from the GT, having sufficient heat (and/or with supplementary firing), is used as a source of heat in a heat recovery steam generator (HRSG) to generate steam to be used further in a steam turbine for Electricity Generation.

A few underlining advantages of the combined cycle follow:

A. With integration of an environmentally clean gasification plant (discussed in the appendix), it is a low-cost installation that takes less time to install.
B. Greater fuel flexibility coupled with higher thermal efficiency.
C. A high degree of operation flexibility such as base load, daily start/stop, and a mid-sized plant.
D. Highly reliable, available, and easy to start.
E. Comparatively more environmentally friendly.

For combined cycle plants, two basic configurations are possible.

I. A single-shaft combined cycle plant consists of one gas turbine, one steam turbine, one generator, and one HRSG. Here, the GT, steam turbine, and generators are coupled to one single shaft. This is simple in control and useful for base load operation. This configuration is discussed in clause no. 10.7.
II. In a *multiple* shaft configuration, a number of GTs and HRSGs may supply steam to the steam turbine via a common header, and a steam turbine generator (STG) can function independently of a particular GT function.

The control and operation of the gas turbine part is most important and demands special attention for equipment-specific control. For this reason, in most of these plants, the GT controls are included as a part of the GT/CoGen package (for example, MARK V control in a GE cogeneration plant) with the necessary interface for balance of plant (BOP) control, which may be supplied by an external agency (for example, an instrument vendor). To enhance the reliability and availability, these control systems use fault-tolerant redundancies, elaborated in Fig. 3.3. Before going into any further detail, it is necessary to discuss briefly the Brayton cycle based on which gas turbine operation. Explanations are mainly based on the GE gas turbine.

1. Thermodynamic principles for gas turbine operation—Brayton cycle: A short schematic drawing for a simple cycle gas turbine is shown in Fig. 3.39A. At ambient conditions, air enters the compressor at "A," where the same is compressed to a higher pressure without heating. Now because of the compression, there will be an increase in temperature, so at high pressure and temperature, air is discharged from the compressor to the combustion chamber at "B." In the combustion chamber, the fuel is also injected at a constant pressure for combustion. In the combustion chamber at the primary combustion zone, high temperatures are reached. The combustion chamber provides a means for mixing, burning, dilution, and cooling (so a volume change). It leaves the combustion chamber at a mixed average temperature to reach the turbine at C. In a gas turbine, the energy of the hot gas is converted into mechanical work. In the nozzle section, the hot gas expands, hence the heat energy is converted to kinetic energy in subsequent sections; energy is transformed to rotate the buckets and converted to work. Out of the total mechanical output, a part (nearly 50%) is utilized to drive the compressor, and balance is available to meet the mechanical load to drive the generator to produce electricity. For initial starting, some starting devices are deployed.
2. The associated P-V and T-S diagrams are shown in Fig. 3.39B and C, respectively.

Path A to B: This is the pressurization of air in the compressor, hence the associated rise in temperature, as shown in the diagrams referred to above.

Path B to C: This is the combustion part, meaning the addition of heat in the system at constant pressure. Naturally, in the P-V diagram, it is a horizontal line (no change in pressure but change in volume due to combustion) while in the T-S diagram, there is a gain in S with an associated rise in temperature also.

Path C to D: This part represents the expansion occurring in the turbine. Therefore, the volume increases and the pressure decreases in the P-V diagram with an associated fall in temperature in the T-S diagram.

Path D to A on the Brayton cycle is the constant pressure cooling, the gas exhausted to atmosphere in case of simple cycle.

3. As stated earlier, the air enters the compressor at ambient conditions, which may vary with time as well as with the place. Therefore, it is necessary to have a standard condition, which is normally 1.013 bar at 15° C with 60% RH. All calculations are done on this basis.
4. Two parameters that characterize the Brayton cycle are the firing temperature and the pressure ratio (pressure at B and A), which should be the same between C and D but in reality it is not due to loss. The firing temperature is generally as the reference temperature at the turbine inlet, as defined by ISO (document 2314). An increase in firing temperature increases the thermal efficiency. As the pressure ratio increases, the specific power output falls. The effect of these two parameters on MW output per kg per second is shown in Fig. 3.39E. These also affect the efficiencies. A combined cycle (Fig. 3.39D) has a less pronounced effect on efficiency, compared with that in the simple cycle.
5. The inlet guide vane in GT controls the air to the gas turbine to control temperature.
6. All first-stage nozzles are cooled either by air or steam so that the operating limit of the temperature for the materials used is not exceeded.
7. The following factors affect gas turbine performance.
   - Air temperature and site elevation.
   - Air humidity.
   - Inlet and exhaust losses.
   - Fuel type.
   - Fuel heating.
   - Diluents injection (NO_x Cont).
   - Inlet cooling.
   - Air extraction (IGCC).

10.2 Objectives and Functions of the System

Cogeneration/combined cycle plants have distinct functions, so these are discussed separately.

1. *Gas turbine*:
   a. Compressor: To supply air to the combustion chamber with higher pressure and temperature. This is connected to the GT shaft.
   b. Combustion chamber: Fuel is injected here for combustion with compressed air, and supply flue gas to GT increased volume.
   c. GT: The flue gas generated at the combustion chamber undergoes expansion at the gas turbine, and also the temperature is decreased for development of mechanical power to drive the same shaft-connected generator.

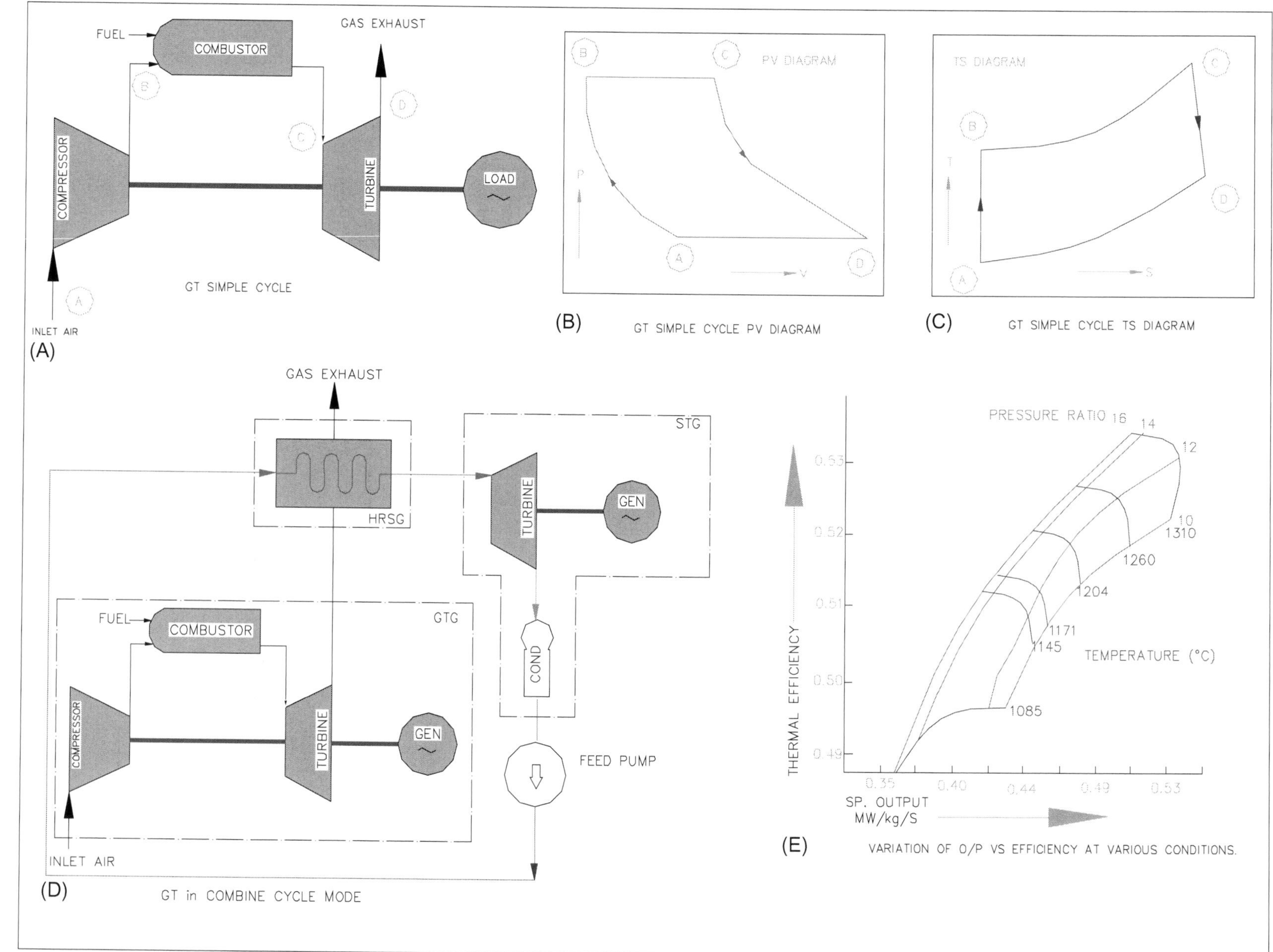

FIG. 3.39 Thermodynamics of GT in various modes.

d. The exhaust gas from GT in a simple cycle, having less heat, is allowed to escape to the atmosphere through the stack. In a CoGen/CC plant, the exhaust gas is partly/or fully sent to the HRSG via a diverting damper (discussed below).

2. Heat recovery steam generator
   a. To extract heat from hot flue gas from the GT to heat water to steam and exhaust the colder flue gas to the atmosphere via the chimney.
   b. Depending on the HRSG configuration, there may be supplementary firing to increase the throughput of the HRSG.
3. Diverting damper:
   a. The basic function of this is to allow a certain quantity of hot flue gas from the GT (as required) to the HRSG for utilization of heat in the exhaust gas. It also allows balance gas to be exhausted to the atmosphere.

10.3 System Description

10.3.1 System Description—Simple Cycle (Fig. 3.40A)

1. Out of many applications where the GT can be used, the discussion here is restricted to GT for power generation.
2. Through the inlet guide vane, the gas turbine brings the air into the compressor, which happens to be connected to the gas turbine proper in the same shaft. The compressor and combustion chambers discussed above are responsible for supplying hot flue gas to the GT.
3. On account of this gas expansion, there will be active mechanical power available to drive the shaft-connected generator to produce electricity. The GT requires cranking power from outside during start-up through a starting device, which could be a diesel engine, an electric motor with a torque converter, or even a steam turbine. For larger plants, a generator with variable power can be used. Gas turbines that are equipped with diesel engine starting devices are capable of starting at blackout conditions (black start). Naturally, in these cases the lube oil supply is established by pump(s) operated by DC (battery) power supplies, which would also power an oil-forwarding pump during black start-up.
4. The inlet guide vane, water, or steam injection is regulated to control emission and operating conditions.
5. In a simple cycle with an exhauster upward, may not require purging, and safe lighting can be done. Whereas for CoGen/CC plants, as a safety measure, purging will be necessary. At the purging speed, the GT is kept for a while so that there are 5–6 times air change is done through the combustion chamber and gas path; the typical purging time could be 1–10 min. When purging is complete, the turbine is decelerated for ignition to start at a specified ignition speed when ignition is reliable, with less system fatigue.
6. It is worth noting that there are sequences to be followed for ignition and firing as per the integral control system offered by the manufacturer.
7. The starting and ignition sequence shall mainly include (for example, GE Mark V):
   a. Ignition power supply made available.
   b. Detection of flame by flame scanner opposite the igniter. To establish completion of firing.
   c. Fuel regulated to warming up.
   d. In case of unsuccessful firing within a specified time, the purge sequence is repeated and attempts will be made to start a second time.
   e. After a warm-up period, the fuel flow is increased, allowing the GT to accelerate. At about 40%–50% speed, the gas turbine will be subject to a programmed rate of acceleration and speed up accordingly.
   f. From a speed of 40% to 80%–85%, the GT efficiency will be high enough to become self-sustaining. At 80% speed, the inlet vane of the compressor may be opened to the full speed. Near running speed synchronization may be initiated.
8. The fuel flow is controlled in accordance with the speed, load, and exhaust gas temperature (capability limit). Also, the inlet vane and injection control of water (or steam) is done to regulate emissions. It is worth noting that the ambient air temperature has a direct influence on fuel flow as well as start-up operation.
9. Generally, the GT is controlled by an automatic start/stop sequence and control and protection are integrated in one package (for example, GE Mark V).
10. NO_x emission control may be accomplished by injection of steam and water into the combustor. This amount of water/steam flow is a function of ambient humidity, fuel flow, and type. The steam flow injection quantity is more than water. Another method of NO_x control is dry low NO_x technology, where premixing of the air fuel is done and combustion takes place at the desired temperature at the desired air-fuel ratio.
11. The cooling down sequence is another important aspect of the gas turbine. A normal cooling down sequence may last from 5 h to 45–48 h. However, at any time, the cooling down sequence may be interrupted for restart (Fig. 3.41).

10.3.2 System Description—Cogeneration/Combined Cycle Plant (Figs. 3.40B and 3.42)

In Fig. 3.40B, an HRSG and a steam turbine have been added schematically. In this case, the exhaust flue gas from

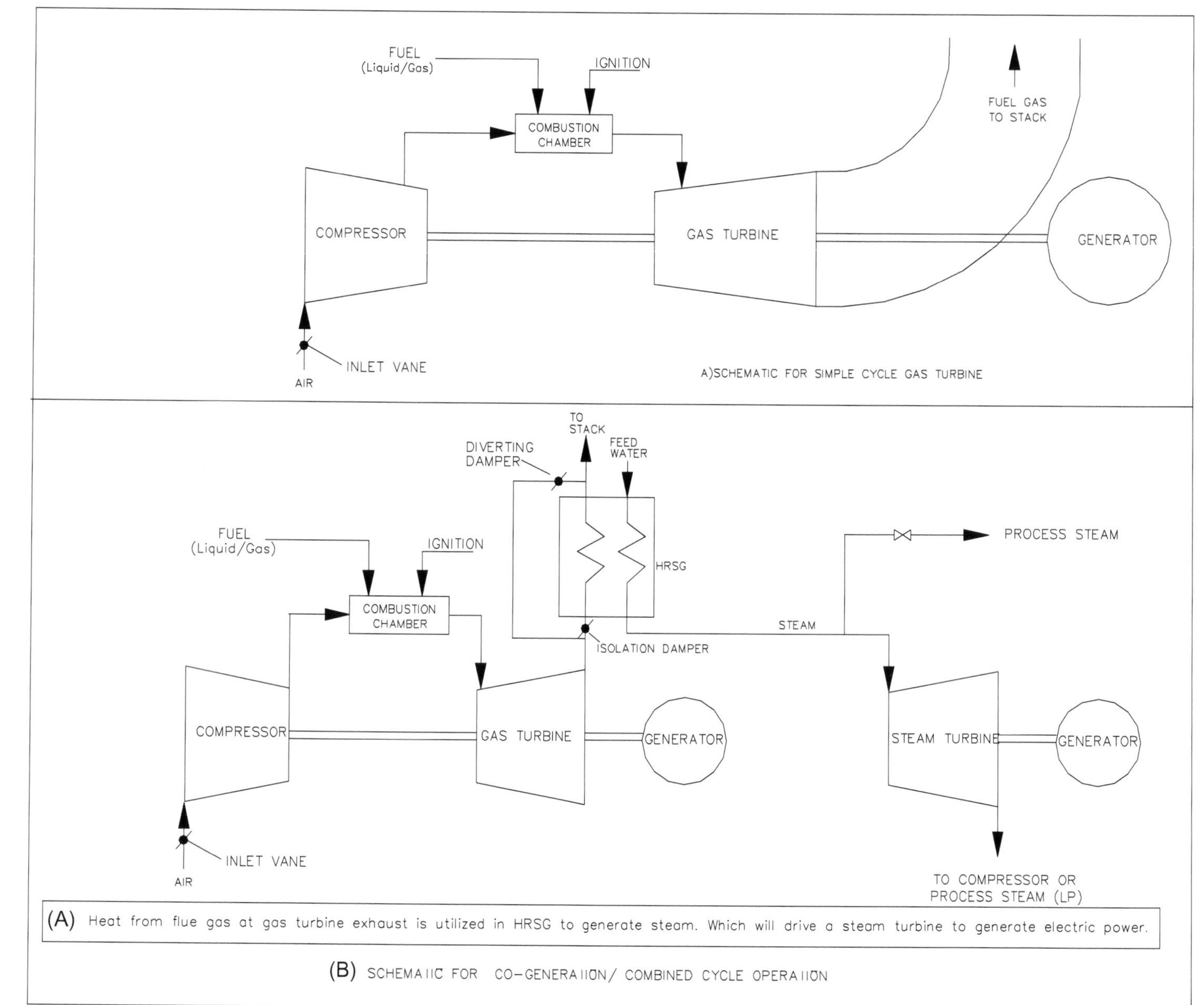

FIG. 3.40 GT in simple cycle and cogeneration (/combined cycle).

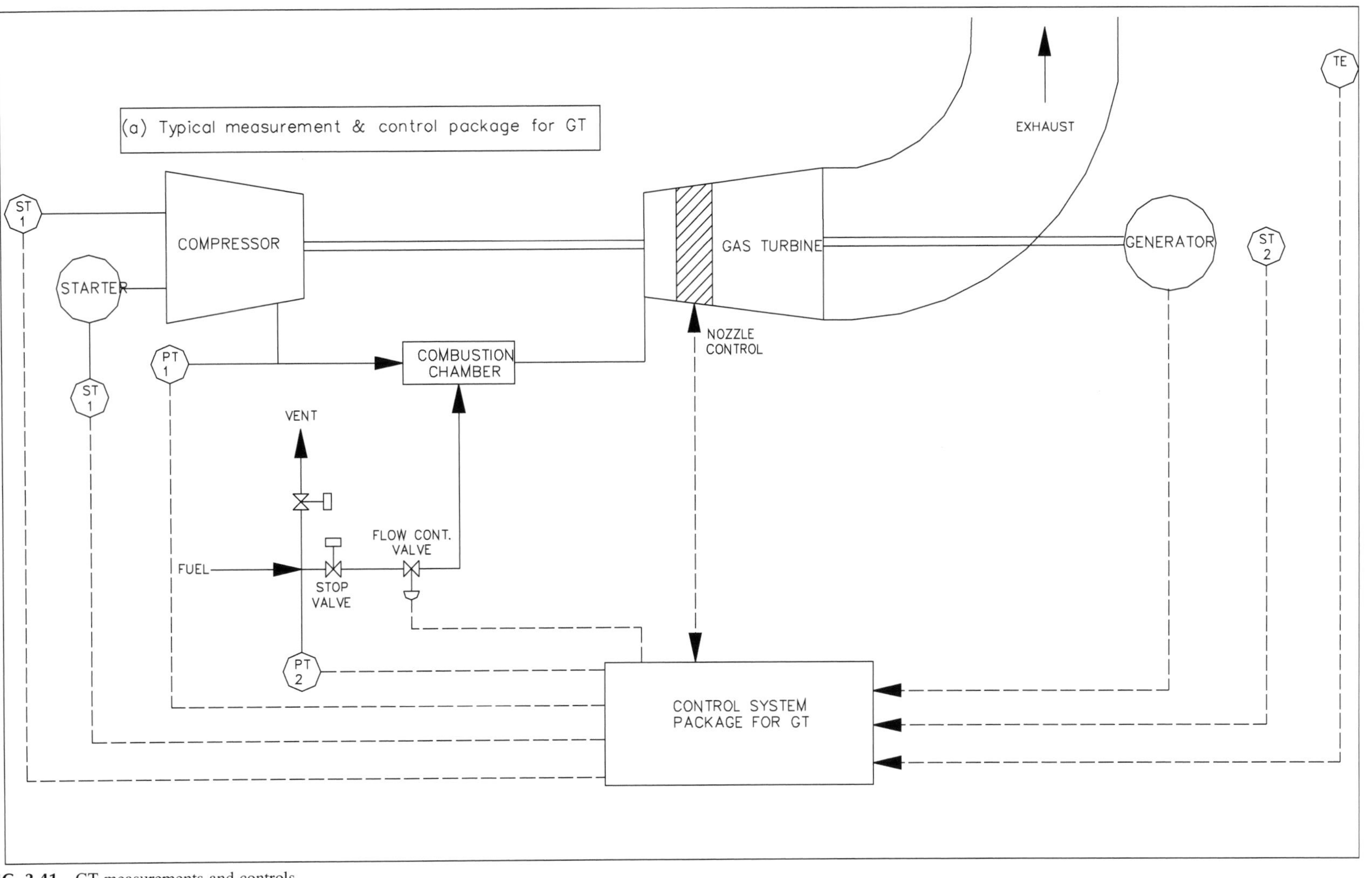

FIG. 3.41 GT measurements and controls.

the GT is sent to the HRSG for heat recovery and generation of steam from the feed water sent to the HRSG.

1. After releasing heat at the HRSG, the exhaust gas now comes out of the HRSG through the stack.
2. At the HRSG, steam is generated, which is utilized as process steam and/or sent to the STG. In a combined cycle plant, this steam expands and produces mechanical power to drive the connected generator to produce electricity. In a cogeneration plant, the steam generated at the HRSG is utilized for various processes in industrial plant (for example, an alumina/steel or paper plant). It is possible to have both a CC and a CoGen plant together from a set of GT(s), as detailed in Fig. 3.42.
3. The steam generation by HRSG can be controlled by a diverter damper, as shown in Figs. 3.40 and 3.42. On account of some limitations in the HRSG, a part of the exhaust gas only may be allowed to enter to HRSG balance may be exhausted through chimney. So there may be two stacks, as shown in Fig. 3.42.
4. Now the exhaust gas leaving the GT is in the range of ~500–550°C, which can be raised to 700–800°C by supplementary firing so that high-pressure (HP) superheated steam is available. The HP steam temperature can be regulated by attemperation control, as shown in Fig. 3.42.
5. The schematic shown in Fig. 3.42 is basically a CC plant, but a part of the steam is exported for process use, as seen in a CoGen plant. Medium-pressure steam is sent to the LP turbine (LPT). This steam comes from the HPT exhaust. Now, if there is an MP/LP steam header for (say) industrial use, then a part of the HP steam after PRDS could be sent to the MP/LP header so that even if the steam turbine does not run, the MP/LP steam header will not starve. Naturally, from the LP header, the steam supply may be sent to the LPT.
6. After expanding in the LPT, the steam is sent to the condenser. Then alike normal condensate is sent to Deaerator. In order to have better performance, the condensate may be heated by a GSC and heater. At the deaerator, dissolved gases are eliminated. Then the same is sent to the HP drum by the BFP as shown.
7. The hot flue gas is cooled in the HRSG nearly to 100–120°C, and sent to the atmosphere through the chimney.
8. BLACK START: A black start is the procedure for recovery of power after a total/partial shutdown of the electric supply. This is generally provided by a smaller peripheral black-start generating plant. Out of the various choices available, open cycle gas turbines, diesel engines, or ignition engines, are generally used. For gas turbines cranking power in the tune of 25% to 30% speed of unfired GT is given from here. The additional power may be used for gas path purging prior to the start-up/ignition necessary for CoGen/CC plants.

10.3.3 system Description—Heat Recovery Steam Generator

1. An HRSG essentially recovers the heat from the exhaust of the prime mover such as a gas turbine in a CoGen/CC plant.
2. Waste heat boilers may be horizontal/vertical shells or water tube boilers to produce steam (and hot water) by radiant or convection heat transfer.
3. As the name signifies, the prime function in this type of boiler is to utilize waste gas to the best but many of these HRSGs have supplementary firing also to increasing the heating capacity as well as for better efficiency.
4. Greater heat recovery is the primary object of the HRSG, keeping in mind that the final exhaust gas temperature is kept and maintained above the dewpoint.

10.4 Gas Turbine Control System

The control system for gas turbines is designed to cater to all the requirements of the system, a self-sufficient system. Both gas as well as liquid fuel are used in GTs, and the fuel flow is regulated in accordance with the speed and load of the GT. The temperature control (to fuel flow) limit and the inlet air flow control by the inlet vane are also important controls. In multistage combustion, the air-fuel ratio and fuel discharge pressure controls are also important. The control system also encompasses the NO_x control system, the interlock and protection system, and the start-up, shutdown, and cooling sequences. For generator synchronization, turbine speed hence frequency, Voltage control are necessary like any other generator. The control system also includes the operator interface through a color graphic monitor with a keyboard and printer. Some of the features found in a typical GT control system (for example, the GE Mark V) include the single failure alarm, protection backup, multiple failures to cause safe shutdown, fault-tolerant sensors and processors, fault-tolerant hardware and software, system diagnostics, alarms, and separation between the control processor and the communication/data processor.

10.5 Major Parameter Measuring Monitoring in—Cogeneration Plant—Gas Turbine

The parameters in the GT area are most important. A few such parameters are as follows:

- SPEED and FREQUENCY
- EXHAUST TEMPERATURE
- INLET VANE
- FUEL FLOW (GAS and LIQUID) and PRESSURE
- WATER and STEAM FLOW
- ACTUATOR STROKE

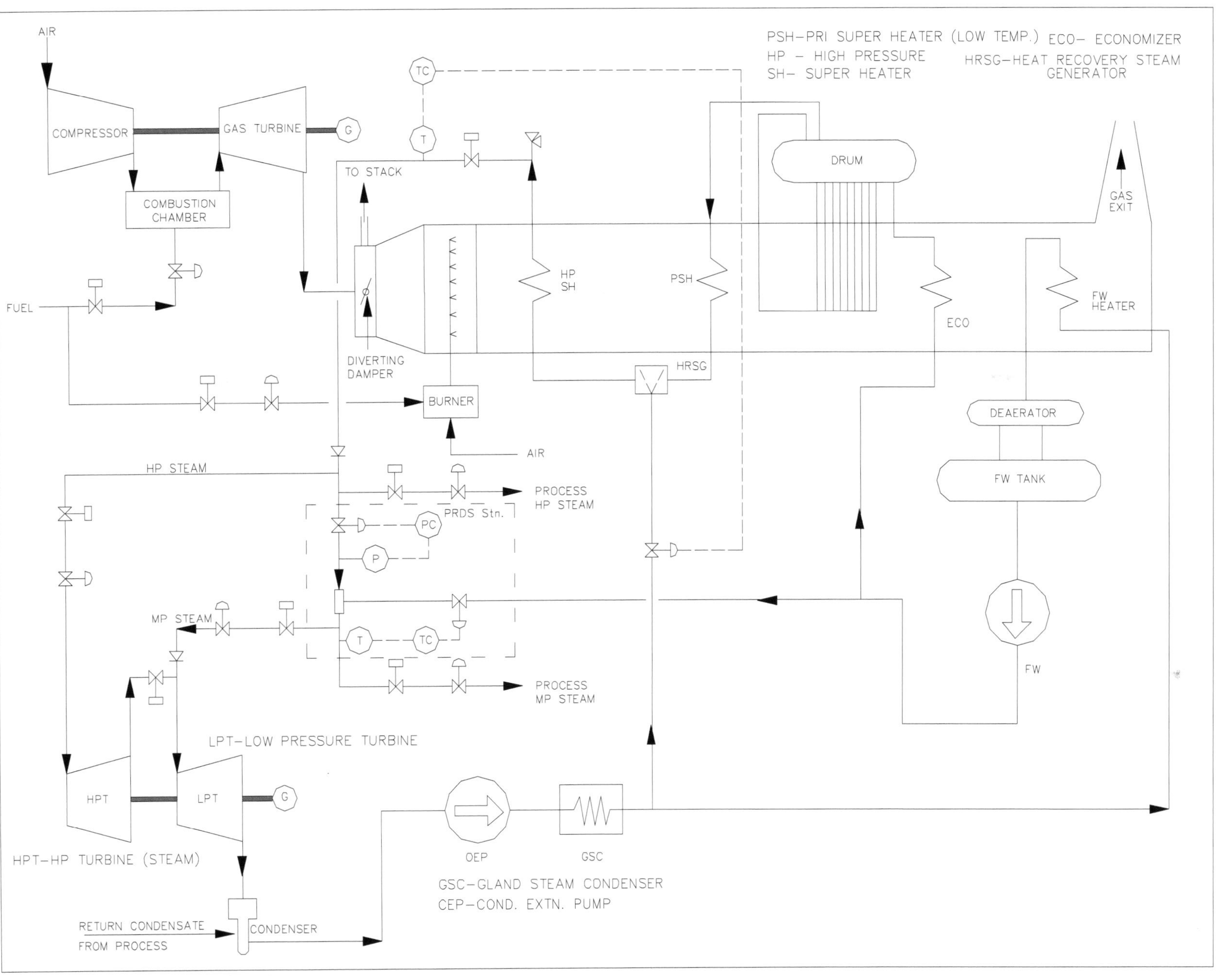

FIG. 3.42 CoGen (/comb. cycle) typical scheme.

- VIBRATION
- FLAME (SCANNER)
- FIRE (DETECTOR)
- LUBE and CONTROL OIL PRESSURE
- LUBE OIL TEMPERATURE
- DIFFERENT PR ACROSS FILTER
- NOZZLE OUTPUT
- WATER FLOW
- LOW PRESSURE AND TEMPERATURE
- CONTROL PRESSURE
- SERVO SOLENOIDS
- CONT. PROCESSOR
- PROT. PROCESSOR
- SOFTWARE

The HRSG and steam turbo generator (STG) also have sets of parameters for monitoring and control; these are similar to what has been discussed in various sections of this chapter, and are not repeated. However, it shall be kept in mind that these parameters are also equally important for overall performance of the plant.

10.6 Redundancy in—Gas Turbines

Because the controls and protections of the GT are extremely important, triple modular redundancies (TMR), as indicated in Fig. 3.3, are normally adopted for PLCs. Two of three redundancies in measurement of the following parameters is quite common.

- SPEED
- GENERATOR OUTPUT
- FUEL FLOW
- ACTUATOR STROKE

10.7 Single-Shaft Combined-Cycle Configuration

GE S107H and S109H are the typical examples of the above configuration, based on which the following short discussions have been presented to get an idea about the single-shaft configuration. A typical arrangement is shown in Fig. 3.43.

1. This configuration is excellent where the installation and operation of both the GT and ST are concurrent. The configuration is simple but efficient.
2. A steam turbine is installed between the GT and the generator with a solid coupling between the GT and ST with a direct-coupled generator with the steam turbine.
3. Single thrust bearing in the gas turbine compressor, and thrust bearing mounted between the steam turbine and generator.

...

DEVELOPED BASED ON SINGLE SHAFT STAG EQUIPMENT OF GE; COURTESY: GE

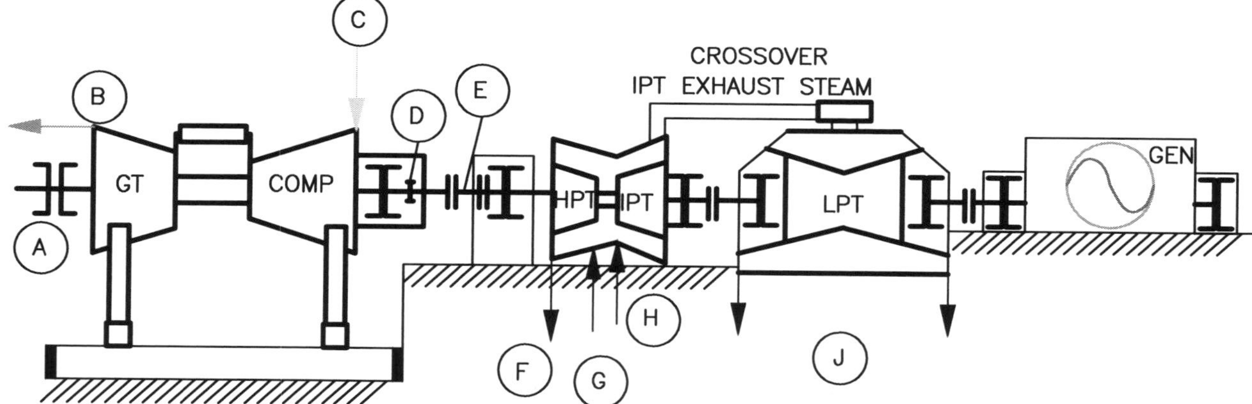

A: BEARING (Typ); B: GT EXHAUST GAS; C: AIR INLET; D: THRUST BEARING; E: SOLID COUPLING (Typ);

F: HPT EXHASUT STEAM; G: HPT INLET STEAM; H: IPT INLET STEAM; J: LPT DUAL EXHAUST

COMP: COMPRESSOR; GEN: GENERATOR; GT: GAS TURBINE; HPT: HIGH PRESSURE STEAM TURBINE;

IPT: INTERMEDIATE PRESSURE STEAM TURBINE; LPT: LOW PRESSURE STEAM TURBINE;

STG: STEAM TURBO GENERATOR

FIG. 3.43 Single shaft GT and STG.

4. Common lube oil system.
5. Control and operational aspect: The speed and load control are done through control of the gas turbine fuel flow valve. Steam turbine control valves are closely coordinated. These valves open after sufficient pressures and temperatures are developed. Then they are ramped open. At this stage, the valve open the pressure slides with the changes in load.
6. Frequency control is done with the help of the GT control. During overspeed, the HP and LP control valves start closing at a speed nearing 103% of the rated speed. They fully close at 105% of rated speed. IP control valves lag 2% behind.
7. For further details, see "Advanced Technology Combined Cycles" by M/S R.W. Smith, P. Polukort, C.E. Maslak, C.M. Jones, and B.D. Gardiner of GE Power Systems.

11 MISC. OTHER SYSTEMS

In this section, a few P&IDs that form the integral part of the turbine, generator, and boiler systems, as listed in Table 3.11, are taken up.

In each case, the objectives and functions of the system shall be discussed.

11.1 Turbogenerator-Related P&IDs.

11.1.1 System Description and Functions

11.1.1.1 Turbine Lube Oil (LO) System (Fig. 3.44)

1. With the turbo generator being a rotating system, there will be some friction at the bearings of the turbine. As a result, there will be generation of heat at the bearing. Lubrication oil (LO) is used to take away the heat from the bearing. The major functions are:
    a. To keep the bearing temperature <100–110°C.
    b. To take away heat by conduction—typically the oil leaving the bearing will be around 70°C.
    c. Less friction between the oil film and white metal.

TABLE 3.11 Various Auxiliary Systems

Turbine	Generator	Boiler
Lube oil	Seal oil	Heavy FO
Jacking oil	H_2 cooling	Light FO and atomizing air
Control oil	Stator water	Atomizing steam
Seal steam		Burner cooling and ignitor
Evacuation		Boiler circulation

2. Hot LO is cooled in an oil cooler outside the bearing housings.
3. While circulating in the bearing, the LO carries lot of dirt also; these are eliminated with the help of a filter.
4. There are a number of lubrication oil pumps. These are auxiliary oil pumps (*AOP*) that are driven by an AC motor; emergency oil pumps (*EOP*) driven by a DC motor; and a shaft -riven main oil pump (*MOP*). Generally, these pumps are centrifugal pumps with sufficient capacity. During initial start-up and shutdown, when a shaft-driven oil pump is not available, these AOPs are used. Depending on the unit size, the number and power of the AOPs is decided.
5. As the name suggests, an EOP is run during shutdown if there is system AC failure so as to save the turbine by circulating oil under this emergency situation.
6. During normal running, the oil supply to the various bearings is catered to by the main oil pump, which is coupled to the main shaft of the turbine and runs with the running of the turbine to supply LO. Pressure- and capacity-wise, the MOP is higher than AOPs, so when the MOP runs, the AOPs are cut out with an increase in LO pressure above a desired value. It is worth noting that in many places, there are oil-driven booster pumps (for example, the GE turbine lube oil system) with a discharge pressure around 1–2 kg/cm^2 that acts as suction to the main oil pump, which may deliver oil around 14–15 kg/cm^2.
7. All oil pumps such as *AOP*, the *main oil pump* (with/without a booster pump) and the EOP have suction from the main oil tank where the level is monitored, as discussed. Before supplying to the bearing, LO passes through a LO cooler and filter to ensure that dirt-free oil reaches the bearing at a temperature less than the desired value. This oil forms a header, and from there, the oil is supplied to each of the bearings individually, as shown in Fig. 3.44.
8. There may be another set of pumps called *a turning gear oil pump*, whose basic purpose is to supply oil to the bearings during turning gear operation of the turbine. It may share the same suction as used by the EOP. When the turbine runs at higher speed, this pump is put out of service automatically.
9. There exists autostarting of AOPs and EOP. These can be tested on a periodic basis, even when the system is running, from the local lube oil pump panel.
10. The pressure and temperature of LO entering the bearing is monitored by a pressure transmitter and temperature element, respectively. There are a number of pressure switches that are effective for changeover of AOP to EOP as required. Similarly cut off AOP in the event starting of MOP.

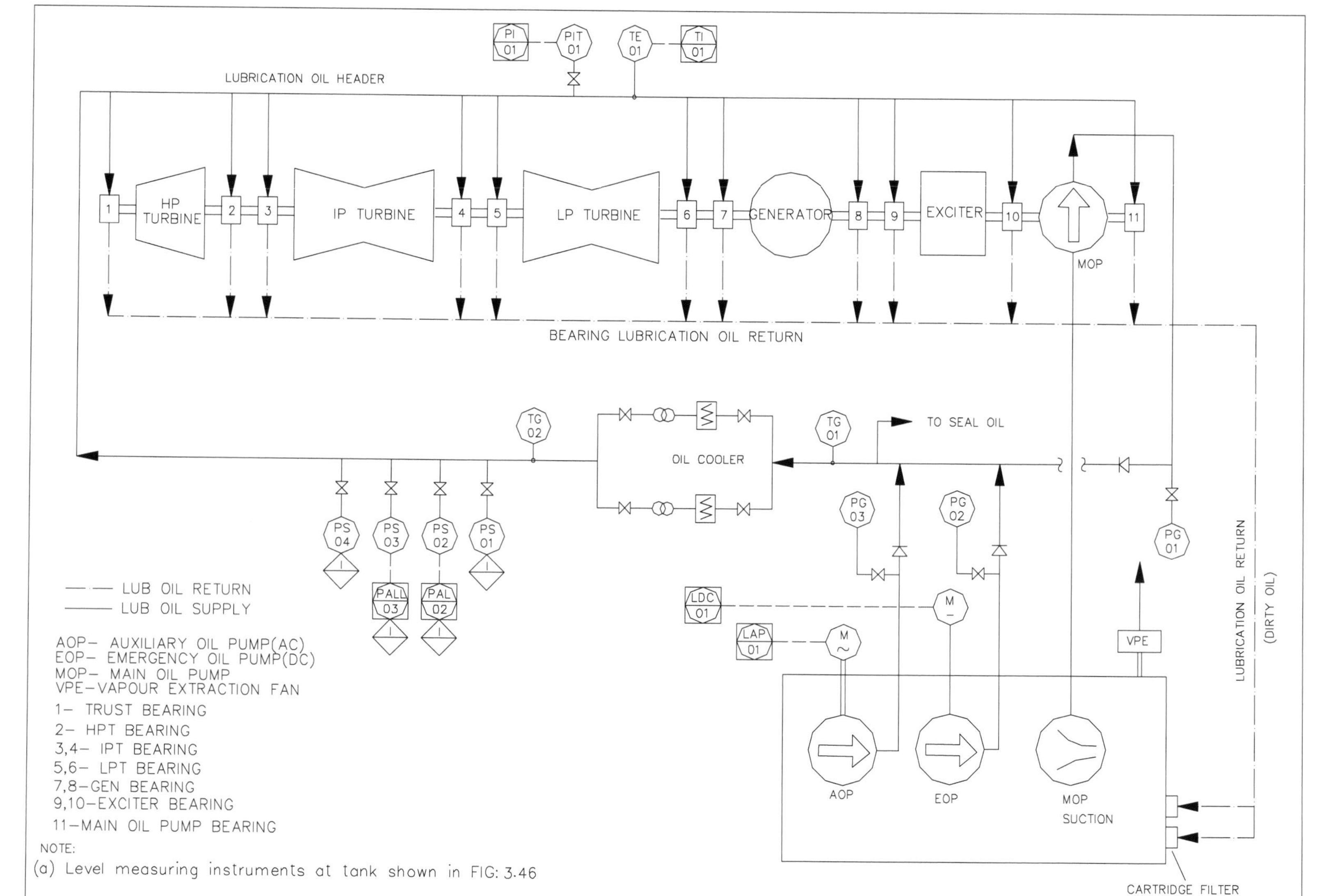

FIG. 3.44 Turbine lubrication oil system.

a. Initially, when AOP is the started and if the pressure falls below a set point, then another AOP (if any) will be started; when both AC pumps fail, then only will the EOP be started. When the turbine reaches the desired speed and the MOP can deliver oil, then the pressure will be greater than a set point when AOP will be stopped automatically. AOP delivers oil at around 2–3 kg/cm^2, so the pressure range could be 0–5 kg/cm^2.
b. If the LO pressure is less than a certain value, the turbine shall be tripped. Apart from these, there will be a low LO pressure alarm.
c. The oil tank level is monitored by level switches and level transmitters.
d. Performance of the oil cooler is monitored by at DCS by RTD (local temperature gauges).

11. After bearing cooling, the dirty oil goes to the return line and the oil tank.
12. Oil pressure is a criterion in automatic turbine run-up system (ATRS) functioning.
13. A part of the LO system may be taken as seal oil.
14. Normally, there is a local Panel and a gauge board pertinent to a turbine lube oil system. The local panel may be located near the turbine in the turbine operating floor, whereas the gauge board may be placed at the front pedestal.

11.1.1.2 Turbine Jacking Oil System (Fig. 3.45)

1. During start-up and shutdown of the turbine, the jacking oil system comes into operation. Before admitting steam, the turbine is turned at a slow speed by turning gear. During the start up of the turbine, jacking oil is necessary. Also, when the turbine is shutting down, then the jacking oil system comes into operation so that the hot shaft does not rest on the white metal of the bearing. As discussed earlier, there is a need to have a separation between the white metal in the bearing and shaft with the help of oil film. At low speed, say <75 rpm or so, this separation is achieved by jacking oil. During this time, oil at a very high pressure, say >200 kg/cm^2 (> 350 for higher MW units), is injected from the bottom to lift the shaft. The oil is kept injected until an oil film is developed.
2. Separate jacking oil pumps (JOPs) taking the suction from the main oil reservoir are deployed. In many places, these pumps are kept at the bearing pedestal. On account of the very high pressure (sometimes >350 kg/cm^2), there is a high importance of safety relief valves for these pumps.
3. Pressure and temperature at the jacking oil header are monitored. The normal range may be 0–350 kg/cm^2 for a medium to high MW power plant. At the main reservoir the level is monitored at the DCS with the help of

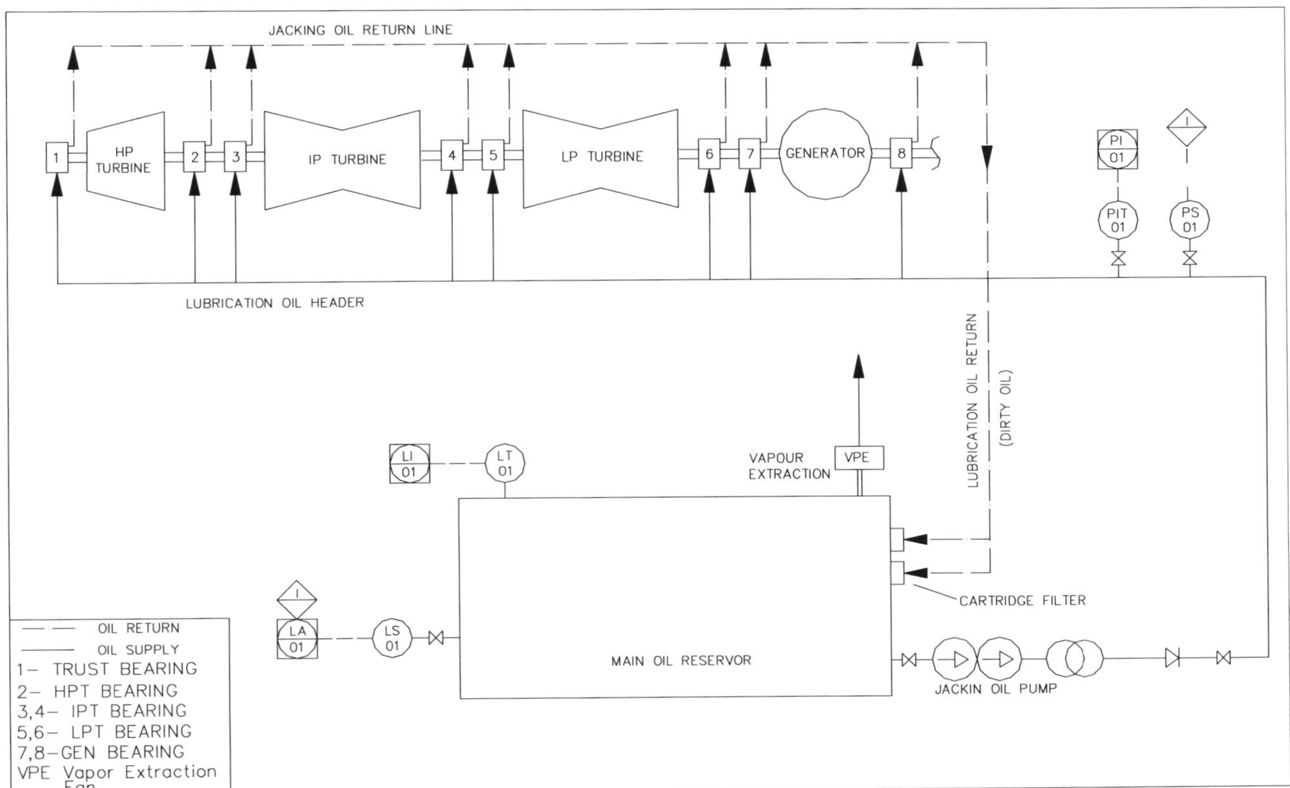

FIG. 3.45 Jacking oil scheme.

a level transmitter. A level switch is used for alarm and interlock.

4. As discussed earlier, jacking oil is used during low running of the turbine. During start-up and shutdown, the turbine rotor is rotated at a slow speed, ~100 rpm, by the turning gear. Basically, the turning gear is driven by a motor, and is coupled to the main turbine through a series of reduction gear. Turning gear is used to rotate the turbine rotor slowly during start-up and shutdown of the turbine to avoid uneven heating/cooling (to avoid shaft deformation). This is turning gear operation. During this time, the turning gear oil pump may be used to supply oil to the bearings.
5. When the turbine speed crosses, say 200 rpm, the turning gear stops operating. Jacking oil only operates during the time the turning gear is in operation.

11.1.1.3 Turbine Control Oil System (Fig. 3.46)

1. The control oil system is the same oil system used to operate the HP IP stop and control valves (the IP control valve may need to open for a load >30% MCR) along with many other control valves, such as sets of valves in the bypass system (say, an LP bypass system). So, the operation and performance of any control actuator pertinent to the turbine system depend mainly on the control oil system.
2. Basically, the control oil may have two components. The first is the high-pressure power oil needed to operate the large valves at high speed (mainly acting as controlled power supply) while the other is relatively lower pressure acting as the signal/control oil to operate pilot valves, solenoids, etc.
3. Because these oil pipe lines are exposed to or near the vicinity of high-pressure steam pipes, there will be

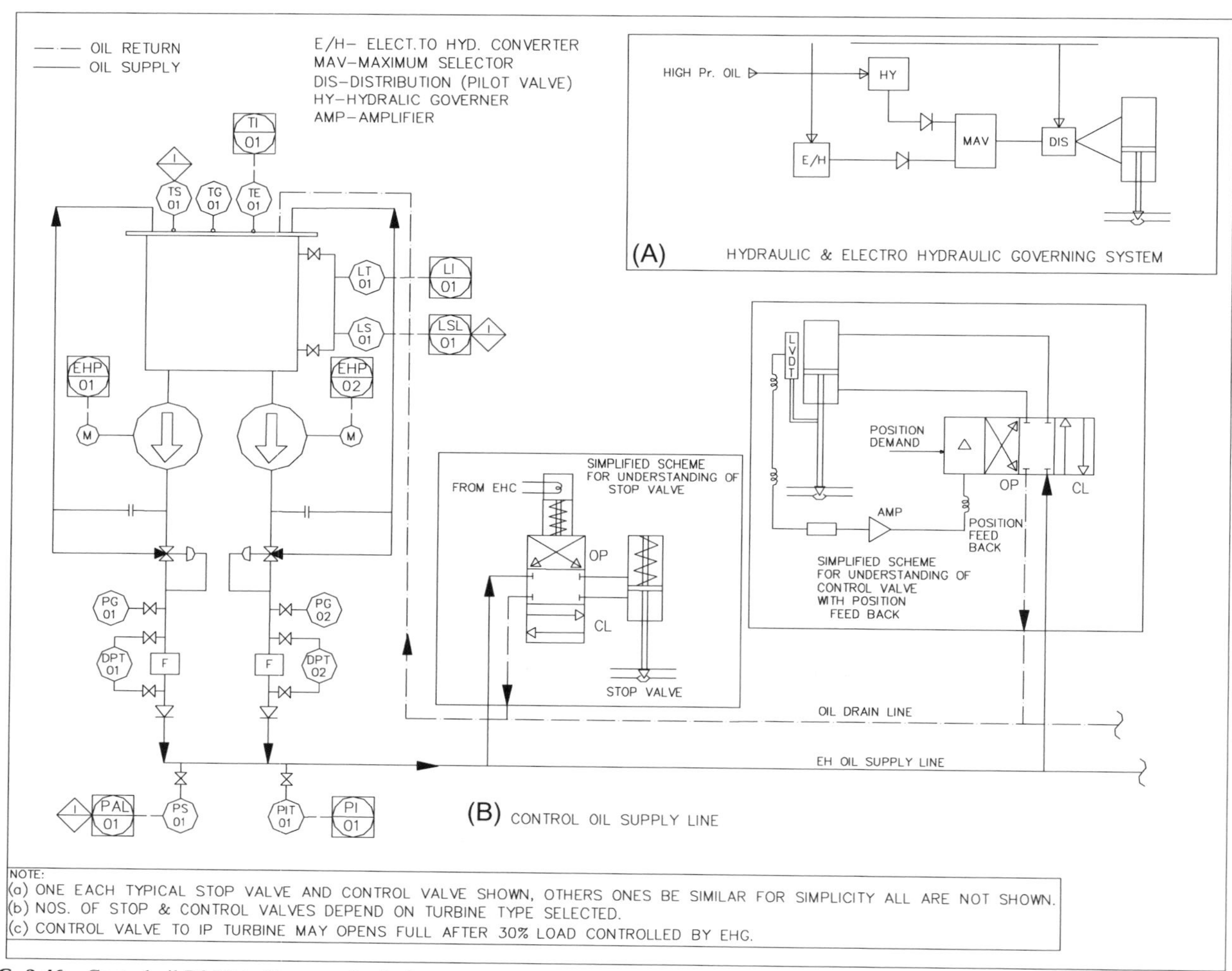

FIG. 3.46 Control oil P&ID (with gov. selection).

many fire hazards. Therefore, fire resistance fluid (FRF) is used for this purpose. Long service life as well as good fluidity is the characteristic of this fluid.

4. Depending on the power plant size, the pressure range of the control oil pumps may be as high as >160 and 30 kg/cm^2 in the lower side. There are several control oil lines formed from this FRF line. These are termed primary oil, secondary oil, trip oil, etc. Each of them has a separate function. Based on these oil pressures, pilot actuators drain/pressurize the main actuator with the main power fluid. The valves normally open against a spring. So whenever the pressurizing fluid drains, these valves close very fast (especially the stop valve) because of the spring force. While in cases of control valves there will be 3 position X 4 way solenoids to take care of the proportional valve movement in both directions.
5. The hydraulic circuit for the governing system is a little complex, but not very difficult to understand as long as the flow path is followed properly. However, in its *simplest* form, the same is depicted in Fig. 3.46. Control oil from the EH tank is pumped and passed through a filter before being sent to the actuators. The line pressure is maintained by a self-regulating pressure regulator. In some cases, a pressure control loop may be deployed for this.
6. Here, one three-position X four-way solenoid has been shown. With no power drain and supply line is blocked. The two output ports of the solenoid are blocked. When the coil is energized, the solenoid assumes an "OP" position so that the supply line is connected to the bottom half of the double acting cylinder, which pushes the piston against the spring force and the valve opens (Although not shown in the drawing, there will be feedback from the position signal so that during start-up, even the stop valve can be opened at a desired position). When the coil is deenergized (or the close coil is energized), the solenoid valve assumes the "CL" position. As a result, the supply line is connected to the top part of the cylinder, and the bottom part is drained. Naturally, the actuator closes the valve very quickly by the action of oil pressure as well as by the spring force, so that closing time is minimal. When any of the two positions is assumed by the actuator, power to the coil of the solenoid may be cut and the solenoid assumes the initial position where the actuator is locked.
7. The action of the control valve actuator valve is also similar. As long as there will be errors between position demand and position feedback, the solenoid will be departed from the initial position of rest (where the two ports connected with the actuator are blocked). When the opening demand is more than the actual open position, it will cause the solenoid to move to that position so that the top side is connected to the drain and the bottom side is connected to the supply (as shown in the left position in Fig. 3.46), and thus the control valve opens. Now, if the opening crosses the desired demand position, the reverse action will take place, so as to drain a part of the oil in the bottom part of the actuator, (and supply oil introduced in the top part) valve will start closing. Thus, we see that this opening and closing of the valve takes place proportionally to position the demand from the governing control system loop.
8. Each turbine has its own governor system comprising a set of valves operating in tandem to give the system unique characteristics specific for the machine assembly. Operational relation and operating points for these valves make the EHG system uniquely suitable for the particular turbine. One should bear in mind that this uniqueness is more related to the operating parameter and valve actuator characteristics. However, there are several basic features of the governing systems for all turbines, for example, speed control/load control, etc. On account of the implementation part of these basic features in hydraulic systems, the performance of the governing systems changes.
9. Most of the modern control systems have both hydraulic as well as electrohydraulic governing control systems. As shown in Fig. 3.46B, either system may selected by Max selection.
10. A few major characteristics of this governing system include:
    a. Capability to handle safely full-load rejection without overspeed, and to maintain system (grid) frequency.
    b. Speed droop characteristic to share the load with other machines.
    c. Up to low load, generally the IPCV is throttled whereas after, say a 30% load, the IPCV opens full, rendering the governor in HP control mode for better thermal efficiencies. In cases of sliding pressure operation of the turbine, the HPCV up to a certain load throttles, then opens almost full and the pressure at the turbine inlet is not fixed but regulated in coordination with the boiler up to an upper limit of the load. Beyond this upper limit of load, again the HPCV throttles (to have cushion in control) (Fig. 3.47).

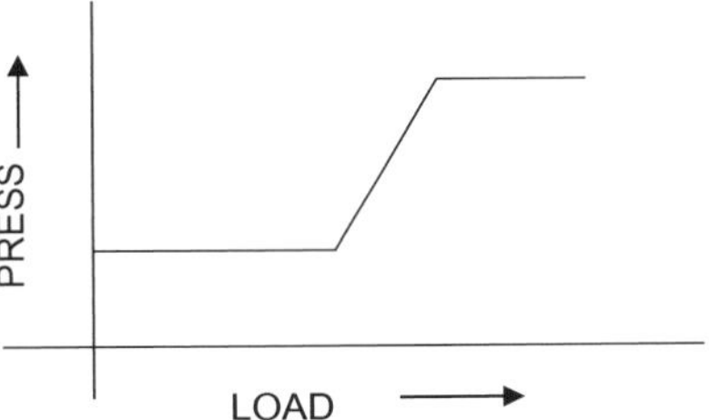

FIG. 3.47 PR versus LOAD.

d. In many large machines (say, KWU-Siemens), the LP bypass is directly related to the EHG system and the control oil circuit of both systems may be common. Naturally, in these cases, the LP bypass control system forms a part of the EHG package. Also, there are systems where a separate control system (as part of the DCS) has been provided for LP bypass control.

e. Chapter 9 discusses the issue in detail.

11.1.1.4 Turbine Seal Steam System (Figs. 3.48 and 3.49)

1. The majority of steam turbines use at both ends of the shaft with rings, to prevent any leakage causing reduction of turbine efficiency. As per API, gland sealing is required for all new turbines.
2. This system enables the turbine to be sealed where the shaft comes out of the casing. Sets of labyrinth packing are employed along the rotor where the shaft exits the casing; details are shown in Fig. 3.49B and D. These labyrinths create many chambers, causing a pressure drop. The differential pressure between the atmosphere and the inside turbine causes air leakage into the turbine on the LP side, whereas on the HP side, the steam escapes.
3. During start-up, shutdown and low-load operation, the TAS assists labyrinth packing in sealing the turbine shaft and preventing air leakage. During normal running conditions, the steam leakage from the HPT and IPT side would be utilized to seal the LP side leakage and any excess leakage steam would be dumped. In some designs, the seal steam line for the LPT sealing is passed through the attemperator to match the process parameter (Fig. 3.48B). In some other designs, steam from the HPT is drained to the condenser and the LPT is sealed from the leakage steam from the IPT to avoid the need for an attemperator (for example, Siemens KWU design).
4. As shown in Figs. 3.48 and 3.49C and D, at the gland steam at the farthest end of each gland, there will be a mixture of steam and air. This gland steam is taken to the GSC where air is removed by the gland steam extractor fan. The steam chamber and reheater stop valve leakages are also taken to the GSC, whose pressure is monitored by a pressure switch.
5. Auxiliary steam (AS) is supplied to the seal steam header, after reducing the pressure and temperature, as shown in Fig. 3.49A.
6. Normally, the seal steam header pressure is controlled by a dedicated control system, which may be a part of the turbine integral control. There will be redundant pressure transmitters (typical range −100 to +900 mm wcl) and two sets of valves, the supply steam control valve for the seal steam header and the leak-off steam control valve from the seal steam header," as detailed in Fig. 9.12.

11.1.1.5 Turbine Evacuation System (Fig. 3.50)

1. Thermal efficiency increases when the condenser is kept at vacuum, so the importance of this evacuation system cannot be overestimated. In many systems, evacuation is one of the major subgroups in a turbine control system.
2. Mainly, there are two pieces of equipment that are responsible for maintenance of the condenser vacuum: a vacuum pump and a steam ejector.
3. In a vacuum pump driven by electric motor draws the air steam mixture from the top of the condenser creating vacuum in the condenser. The air-steam mixture is cooled and separated. The air is released to the atmosphere and the condensed part is drained.
4. The other option is a steam ejector; in a steam ejector, the steam (auxiliary steam) is passed through a nozzle and allowed to expand in one chamber. Now, when the HP steam passing through a nozzle gets a chance to expand in the chamber, it creates a vacuum in the chamber; this chamber is connected to the condenser so that a condenser vacuum is created. Out of the mixed air and steam, air escapes to the atmosphere whereas the steam is cooled by the passing condensate, which absorbs some heat from the steam, allowing it to condense.

11.1.1.6 Generator Seal Oil System (Fig. 3.51)

1. Many of the larger generators are hydrogen-cooled. When hot hydrogen comes in contact with oxygen/air, it is vulnerable to explosion.
2. Seal oil (may be taken from LO oil) is used to seal an H_2 leakage. Seal oil is used at both ends, that is, the air side as well as the H_2 side.
3. There are two sets of seal oil systems. Each system comprises sets of seal oil pumps, heaters/coolers, filters, one header, and a return oil tank.
4. Because seal oil is necessary at all operating conditions, there are seal oil pumps driven by both AC and DC motors.
5. Seal oil pumps take their suction from the return oil tank. Discharge of the pump is passed through a heating/cooling circuit to keep the temperature at the desired point. This is done by measuring the temperature with a temperature gauge. Seal oil at the desired temperature is then passed through a set of filters. The DP across the filter is monitored and changeover takes place in case of high DP. There will be one recirculation line to keep the oil running at all times. The recirculation line may be returned to the return oil tank.

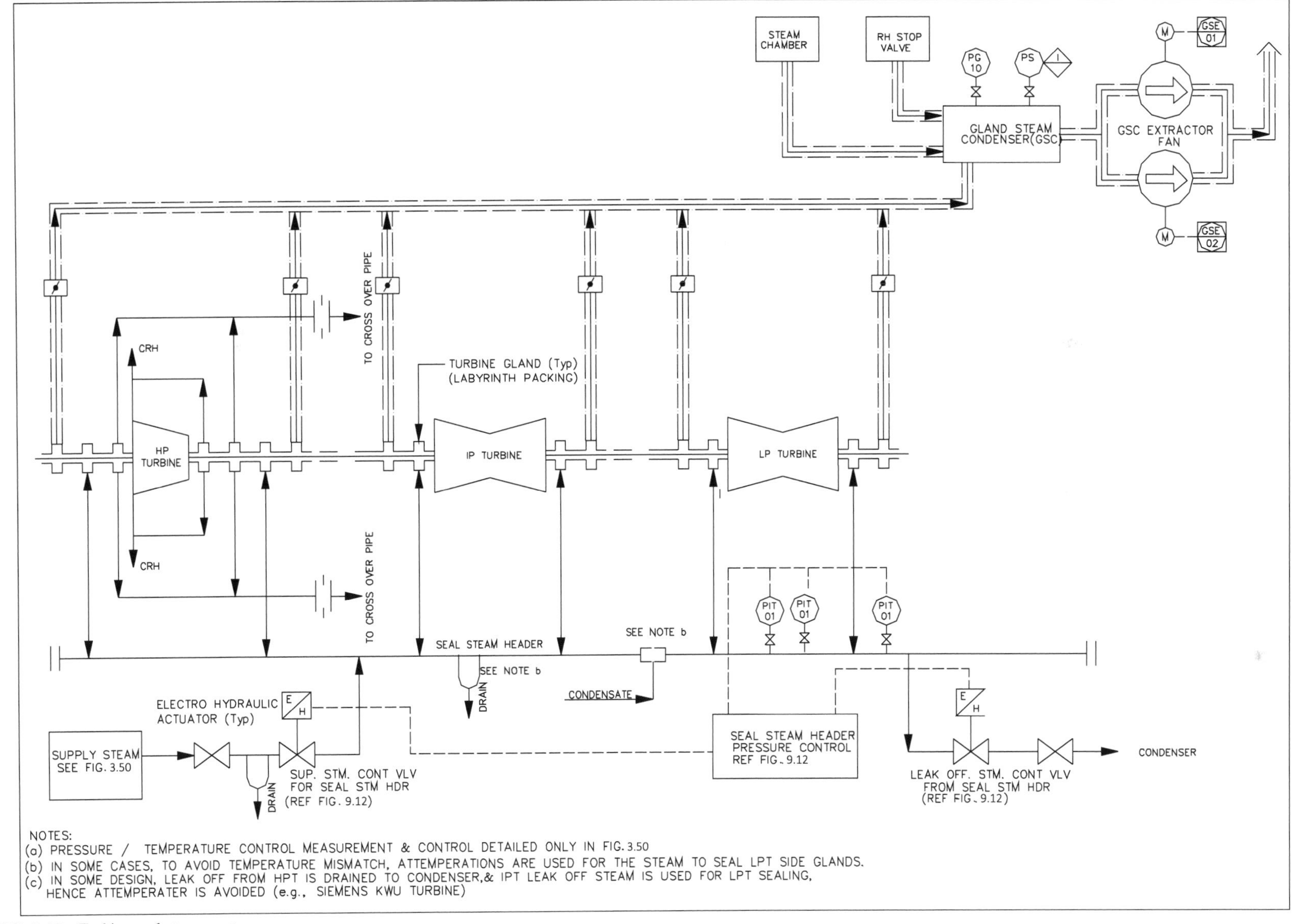

FIG. 3.48 Turbine seal steam system.

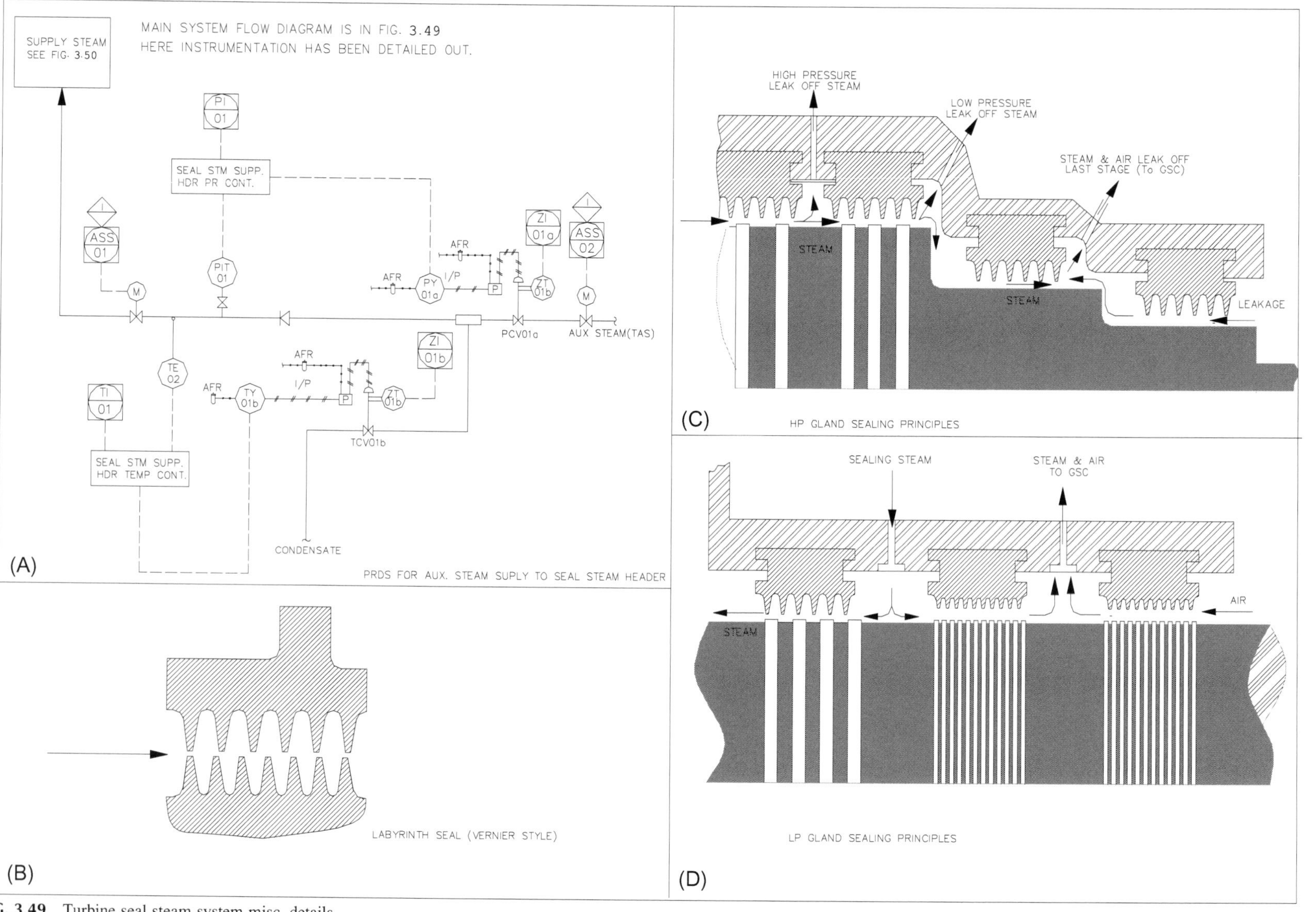

FIG. 3.49 Turbine seal steam system misc. details.

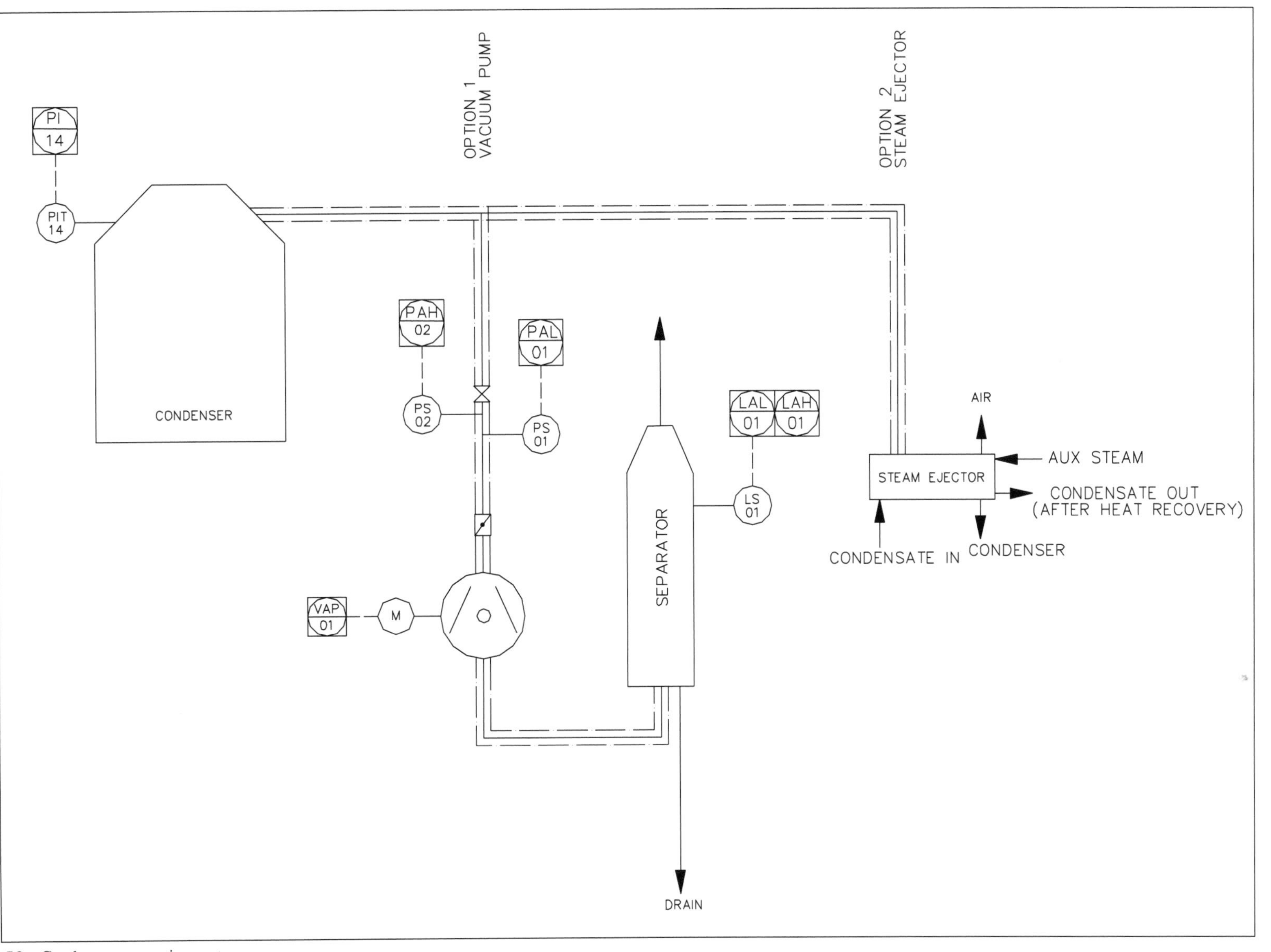

FIG. 3.50 Condenser evacuation system.

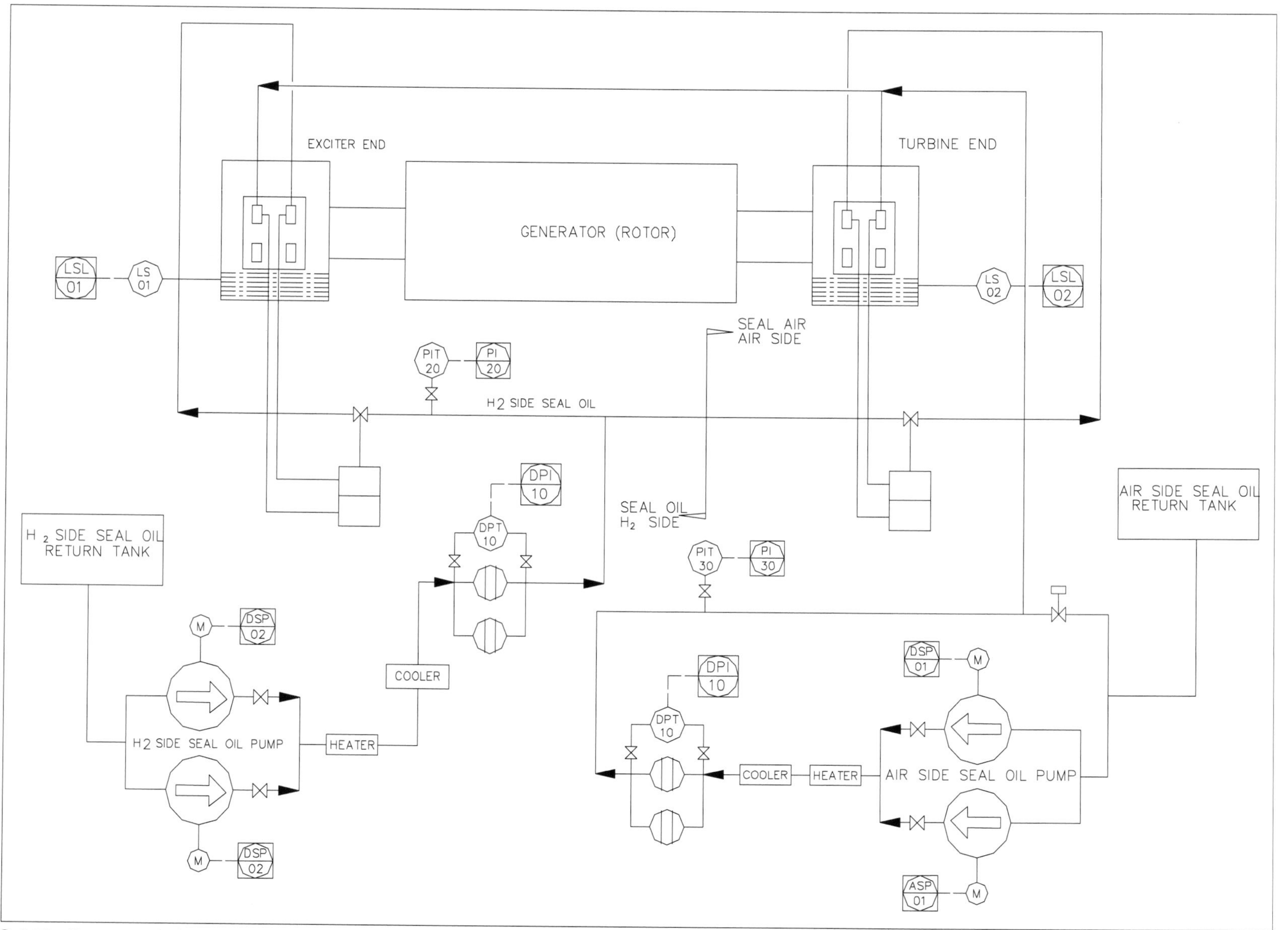

FIG. 3.51 Generator seal oil system.

6. The seal oil line after the filter forms the seal oil header. There will be two such headers, one for the air-side seal oil and the other for the H_2-side seal oil header.
7. Each of these seal oil headers supplies seal oil to the generator at both the turbine end as well as the exciter end.
8. One thing worth noting is that the H_2-side seal oil pressure always maintains a differential pressure with respect to air-side seal oil. This is done by measuring the seal oil pressure at both ends and connecting these pressures at two sides to two sides of a double-acting cylinder so that the H_2 side seal oil valve opens more or less to maintain the required differential pressure with respect to that at the air side, as shown in Fig. 3.51.
9. Seal oil—H_2 differential pressure is one of the most important parameters for generator operation. It is measured with redundant sensors, and interlocked with generator operation. Generally, seal oil is kept around 1.2 kg/cm^2 above H_2 pressure.
10. One of the critical things of a seal oil system is to remove the entrained H_2 and other gases from the seal oil before the seal oil is again sent to seal the shaft. This is done at the main tank through vacuum operation in a spray nozzle to break the seal oil into fine spray to remove gases. Also, after the shaft, sealing oil goes through a detraining section. As it passes through the detraining section, there will be pressure loss, so to take care of this, a DC pump may have to be run.

11.1.1.7 Generator H_2 System (Fig. 3.52)

1. On account of better thermal conductivity, low density, and higher specific heat, H_2 has been chosen for rotor cooling of many larger generators in place of air cooling. Air-cooled large generators are also not uncommon. For low-rating generators, H_2 cooling may not be applicable.
2. In many power stations, a separate H_2 generation plant is kept for H_2 supply. Also, there are provisions for H_2 supply from cylinders for emergencies.
3. There shall be a number of H_2 driers from where H_2 fills the space in the rotor. An H_2 drying operation is very important, and leak detectors are deployed to monitor the dryness of H_2. Moist H_2, being at a lower pressure, is sent to the dryer, whereas dry H_2 is sent with higher pressure.
4. H_2 quality and H_2 pressure are very important parameters and are always monitored in the main DCS.
5. H_2 is not filled directly into the generator, as there will be a chance of H_2 coming in contact with air. Normally, before filling H_2 in o the rotor which is filled with air. At first, the air is replaced with CO_2 by upward displacement of the air. So, the CO_2 cylinders shown in the P&ID. Fig. 3.52.When the rotor is completely filled with CO_2, it is ensured by measuring the CO_2 in the oxygen analyzer. After ensuring the purity of the CO_2 in the rotor, H_2 filling is started. Being heavier than H_2, CO_2 is filled by downward displacement of CO_2. Proper filling of H_2 is ensured by measuring the H_2 in CO_2
6. One of the reasons that favors H_2 as a cooling medium could be that the generator is completely sealed from a safety point of view, meaning sealed from dust, humidity, salt, etc. Also, no oxygen atmosphere is inside, so high voltage insulation is safe from stator activity (corona).

11.1.1.8 Stator-Cooling Water System (Fig. 3.53)

An effective means for heat dissipation of stator winding is very important in improving the performance of the generator. Larger generators have small diameter copper tubes carrying water around the stator winding. These are cooled by sending DM quality process water through them.

1. This process water is stored in a tank with a suitable heat exchange circuit so that a constant temperature is maintained. Stator-cooling water pumps take their suction from the tank. The discharge of the pumps is passed through a water cooler to maintain the desired temperature. After passing through the filter, stator-cooling water is passed through the hollow copper tubes around the stator winding. Water leaving the generator returns to the tank.
2. The level in the tank is monitored (with high and low alarms) at DCS by a set of level instruments. Also pressure, temperature, and flow of the stator-cooling water line is monitored. The inlet/outlet temperature of the stator-cooling water is a very important parameter to be monitored.
3. Corrosion is one of the major problems, as copper oxide flakes are formed inside the tube to choke the flow of water. For this, either aerated water >2 ppm DO is used to form a CuO layer or deaerated water with DO < 50 ppb is used to form a Cu_2O layer to prevent corrosion. As any corrosion is an electrochemical process, electrochemical corrosion potential may be used as a means to detect corrosion due to changes in water chemistry.
4. Water leakage is a great threat for generator operation, as it can damage insulation. There are suitable leak detectors (for example, one from GE).

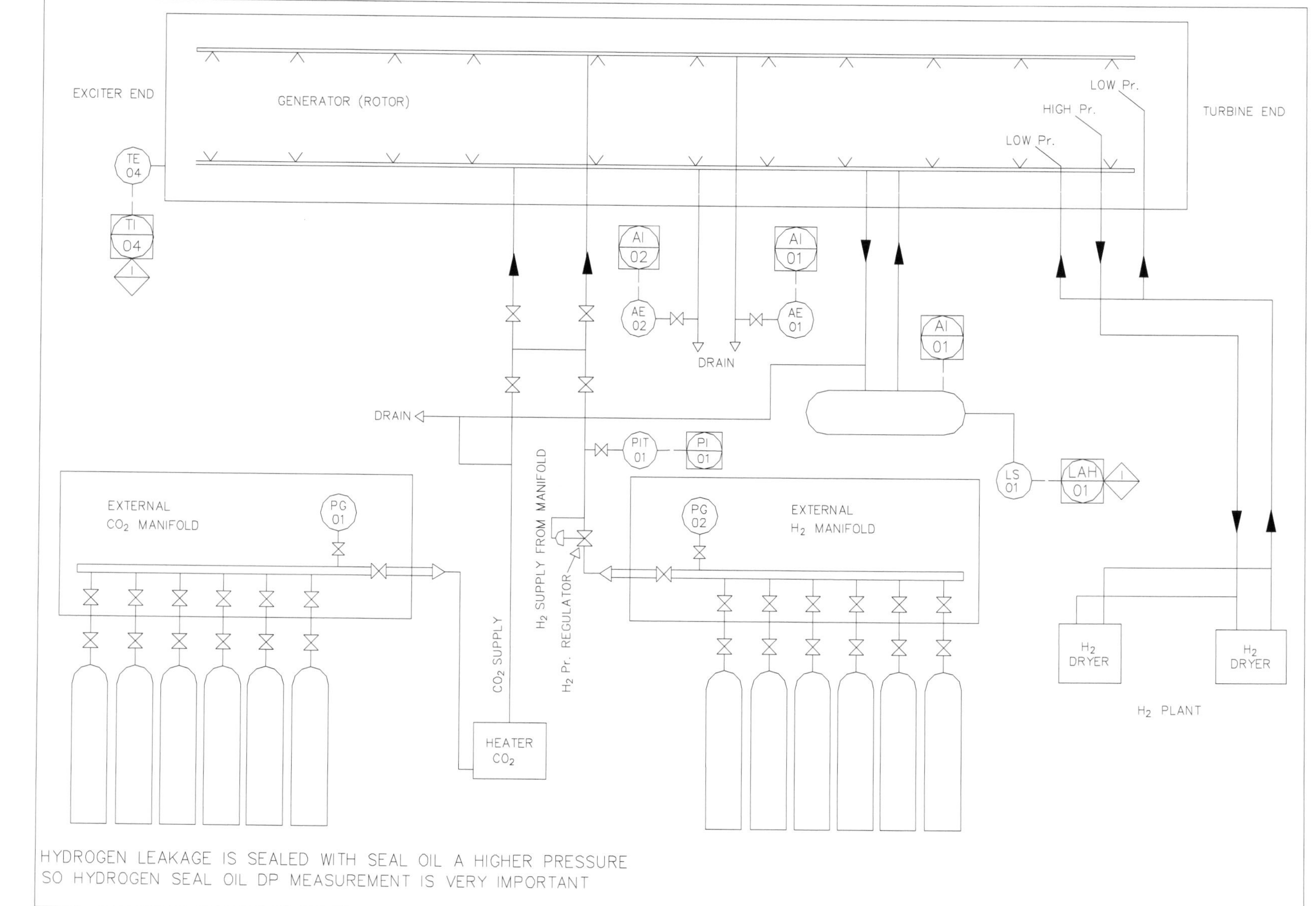

FIG. 3.52 Generator H_2 cooling system.

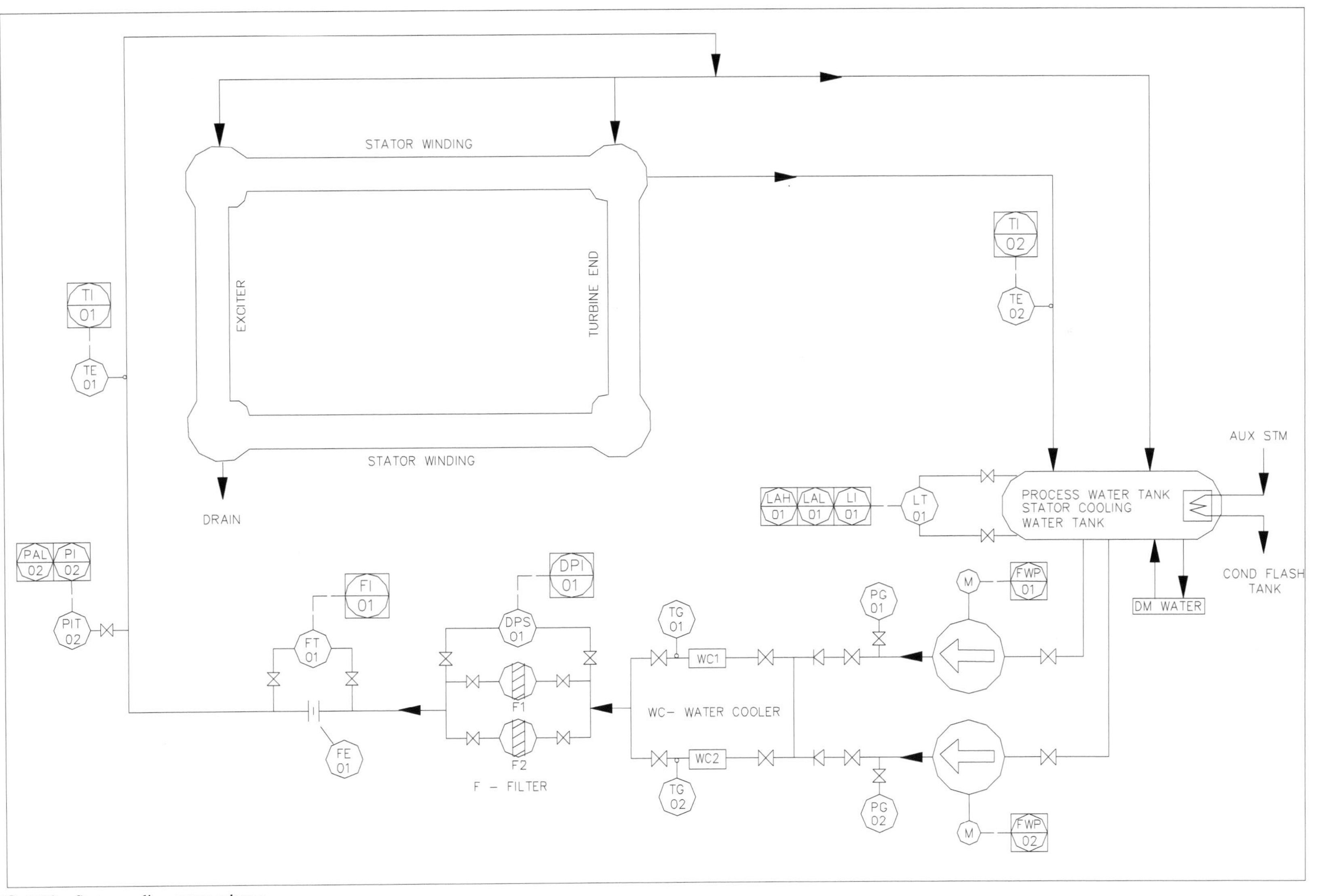

FIG. 3.53 Stator cooling water scheme.

11.2 Steam Generator Related P&IDs.

11.2.1 System Description and Functions

11.2.1.1 SG Heavy (Fuel) Oil (HFO) System

See Fig. 3.54 and Table 3.12.

1. In an oil-fired boiler, these can be used as the main fuel. For a PF boiler, HFO is used for supporting fuel, low load operation, and secondary fuel for coal flame stabilization. In some boilers, this may be directly used as the initial starting fuel (with a high energy arc igniter) whereas in an oil-fired boiler, this will be the primary fuel.
2. In a thermal power plant proper, there re two sections for HFO:
   a. The HFO heating and pumping section may be common for various units in a power station. Normally, this is located outside the unit powerhouse, with a separate control system and panel with a software link to the main DCS.
   b. FO preparation and firing section—this is mainly connected with oil equipment in the operating floor of the boiler and controlled from the DCS.
3. In power HFO, storage is normally done at two places: the HFO storage tank (main storage facility) way outside the powerhouse building, and the HFO day tank (a day's consumption for the unit) very near the powerhouse.
4. HFO is quite viscous and needs heating in achieving flowability. It is therefore essential that all the storage tanks have floor coil heaters (steam heating) and heating at the outlet nozzle. This is necessary so that it could be pumped and be made to flow. In addition to these, the entire flow path of the HFO is heat traced (either by electric or steam tracing). In modern plants, there is a continuous recirculation line so as to keep the oil always in flowable condition to provide quick fuel support.
5. The HFO shall be drawn from the HFO tank by a motor-operated, pressurizing pump, which normally is a rotary positive displacement horizontal pump with a relief valve. The pumps through discharge valves supply the heating unit, which is heated utilizing auxiliary steam. Discharge header pressure is measured and controlled by the recirculation valve to recirculate excess HFO back to the tank. In order to eliminate dirt, the pumps shall have a suction filter, with DP monitoring across it for quick changeover of the filter.
6. Temperature at the outlet of the heater is measured and controlled by regulating the supply of auxiliary steam to the heater to see that the temperature never crosses The flashpoint (actually some degree below the flashpoint) at the same time is above a set value to ensure better flowability.
7. The main oil path, after the strainer (filter) flows through the flow meter to measure the amount sent to the boiler (the difference between supply flow and return flow computes the oil consumption). A bypass to the flow meter is kept for flow meter maintenance purposes.
8. Immediately after the flow meter is the trip valve, required to cut off the fuel supply to the boiler in the event of a boiler trip.
9. After this, the oil follows the path to the oil guns. For TT boilers, there are two paths, as shown in Fig. 3.54. In one path, it goes to the boiler operating floor; in fact, up to the burner. The other is a constant quantity recirculation back to the tank to ensure that the FO is in running condition at all times. Normally, 10%–5% recirculation is kept. This is a long recirculation line to trace almost the entire oil path. This is a scheme followed by a number of manufacturers, if not by all. There is a short recirculation path also where the HFO is recirculated from the pump outlet back to the tank, as shown in Fig. 3.54.
10. For a TT boiler, the FO is divided into four parts to supply the four corners (for a corner-fired boiler). At each corner, the oil line goes up to the highest oil elevation. Also from each of these corners, there is a return line, as shown in Fig. 3.54. These return lines meet to form the main return line, which flows back to the tank via one flow meter and HFORTV.
11. The pressure control valve regulates the pressure of the oil at the boiler front. Depending on the boiler demand, the flow to the boiler will change, and accordingly, the opening of the pressure-regulating valve is necessary so that the supply flow changes so as to take care of changed flow conditions. This is done by measuring the pressure in the header at the boiler front (PIC 10). The boiler front oil header pressure and temperature are very important for boiler operation. Some of the important parameters are: oil pressure okay to start for oil firing, oil Pressure low trip and alarm, and FO temp > set value. On account of the high viscosity selection of valves, the control valves are critical. The ball valve is a good choice, as is shown in the P&ID.
12. Leak test: In certain boilers (seen in Chinese design), there is a valve parallel to the trip valve known as the leak test valve. This, in conjunction with the trip valve, the return valve, the control valve, and associated interlocks, checks the leakage in the entire line as well as leaks for the trip valve. Details are available in Section 12.1 of Chapter 8.

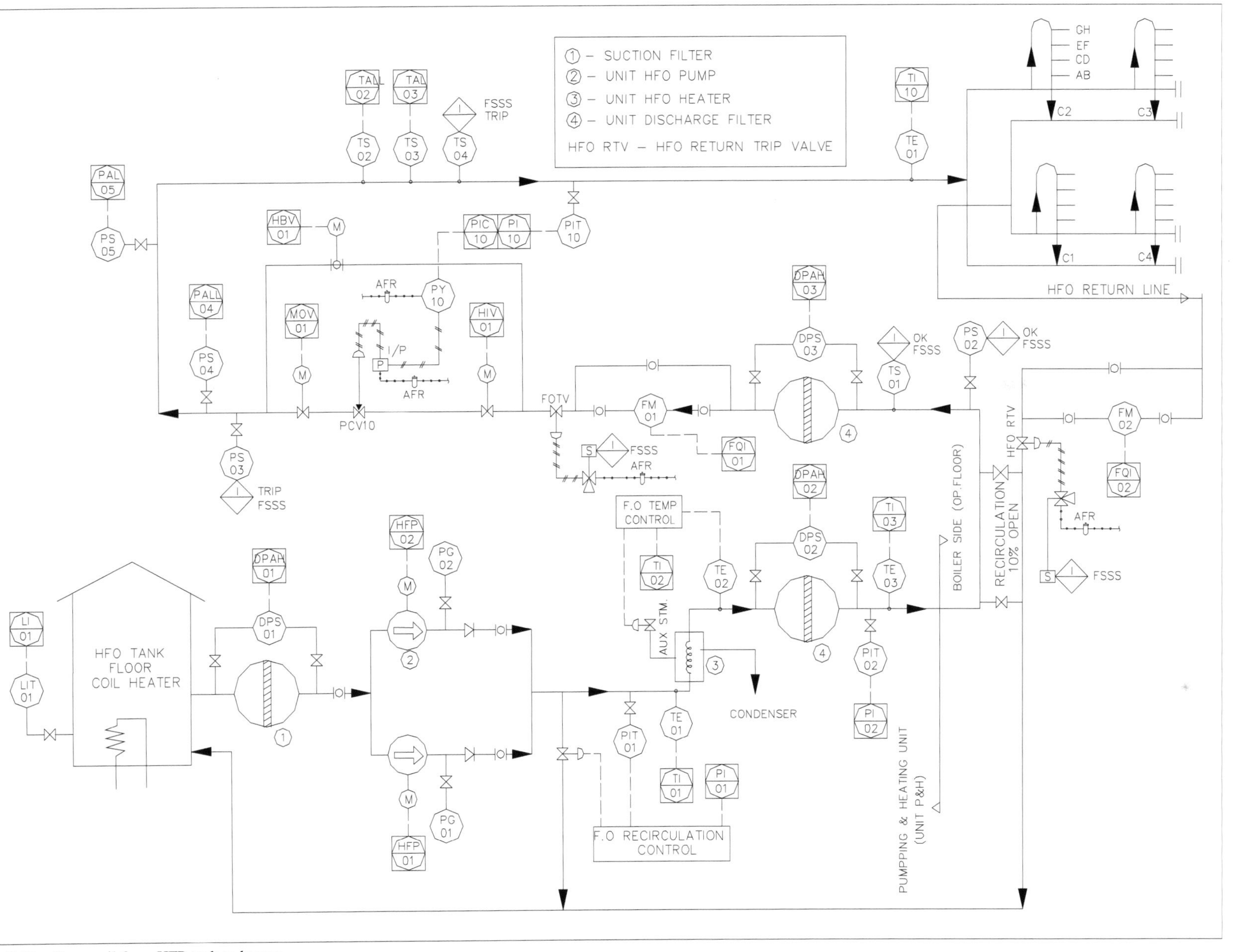

FIG. 3.54 Heavy oil from HFP tank to burner.

TABLE 3.12 HFO Parameters

SL	Characteristic	HFO	HPS	LSHS
1	Calorific value (kcal/Kg)	10,000	10,000	10,700
2	Flash point °C	66	66	75
3	Viscosity (CST) at 50°C	370	500	50
4	Specific gravity	0.95	0.9	0.86
5	Sulfur content (%wt)	4.5	4.5	1

11.2.1.2 Light Oil and Atomizing Air System (Fig. 3.55)

1. Light diesel oil
   - **A.** In many PF boilers, LDO is used for firing the boiler when no auxiliary steam is available or as the initial start-up fuel.
   - **B.** The LDO shall be drawn from the LDO tank by a motor-operated pressurizing pump, which normally is a rotary positive displacement horizontal pump with a relief valve. The discharge header pressure is measured and controlled by the recirculation valve (it could be a self-regulating upstream control valve) to recirculate the excess LDO back to the tank. In order to eliminate dirt, the pumps shall have a suction filter with DP monitoring across it for quick changeover of the filter.
   - **C.** In case of LDO, long recirculation may not be applicable as it does not have a flowability problem. However, in many designs, especially in Chinese designs, long recirculation lines similar to the HFO line have been noted.
   - **D.** Oil normally flows through the flow meter. A bypass to the flow meter is kept in the event that the flow meter is taken out for maintenance.
   - **E.** Immediately after the flow meter is the trip valve required to cut off the fuel supply to the boiler in the event of a boiler trip.
   - **F.** Leak test: As discussed in clause no. 11.2.1.5.14, a leak test is also applicable for LDO and similar leak test valves are used to carry out the leak test, as discussed in connection with the HFO.
   - **G.** The boiler front oil header pressure is very important for boiler operation. Some of the important parameters are oil pressure, okay to start oil firing, and oil pressure low trip and alarm.
2. Atomizing air
   1. In order to burn the LDO/HFO at the desired high rate, it is necessary to atomize the oil, that is, disperse the oil in a boiler as oil mist. In an LDO, air is used as the atomizing medium.
   2. Service air from the air header may be utilized for this purpose. The pressure shall be more than the LDO. This is monitored by the pressure switch in the air line.
   3. The atomizing air line is taken to all corners (or all burners) and elevations wherever the LDO line goes.

11.2.1.3 Atomization: Steam (Fig. 3.56)

1. HFO is also atomized at the boiler before firing for spontaneous firing as well as for achieving a high rate of firing.
2. Auxiliary steam is used for atomization. There is one control valve to regulate the steam header pressure so that the atomizing steam pressure is kept a little higher than the HFO. The pressure at the steam atomization line is monitored for low alarm and trip.
3. Atomization Types: The basis of atomization is the mechanical type to get a small droplet so that it vaporizes quickly. There are basically two types of atomization—internal mix and external mix. In the internal mix, the lower oil pressure is augmented by the high pressure of the atomizing medium. In the case of an external mix, the atomizing medium pressure need not be always higher than oil, as the interaction takes place just before the emission from the nozzle. There is another means for emulsification by water not normally seen in any power plant.

11.2.1.4 SG Burner Cooling, Ignitor, and Flame Detection System

1. Oil burners are cooled by either steam or air, depending on the design.
2. The ignition system mainly consists of the following:
   - **a.** Igniter, HEA igniter.
   - **b.** Warm-up oil.
   - **c.** Flame-detection system.
   - **d.** Control system.

11.2.1.5 Boiler Circulation System (Fig. 3.57)

1. Boiler circulation is necessary to cool the water walls, the economizer, etc. The circulation systems of the OT and drum boilers are different.
2. For a drum boiler, there are two types of circulation: natural and forced. In natural circulation, the circulation is thermally induced, without the use of any pump. In forced circulation, in addition to thermally induced circulation, it is also assisted by a pump called the BCP.
3. From the economizer, water reaches the drum, where through the downcomer it reaches the furnace bottom and passes through the water tubes in the furnace wall.

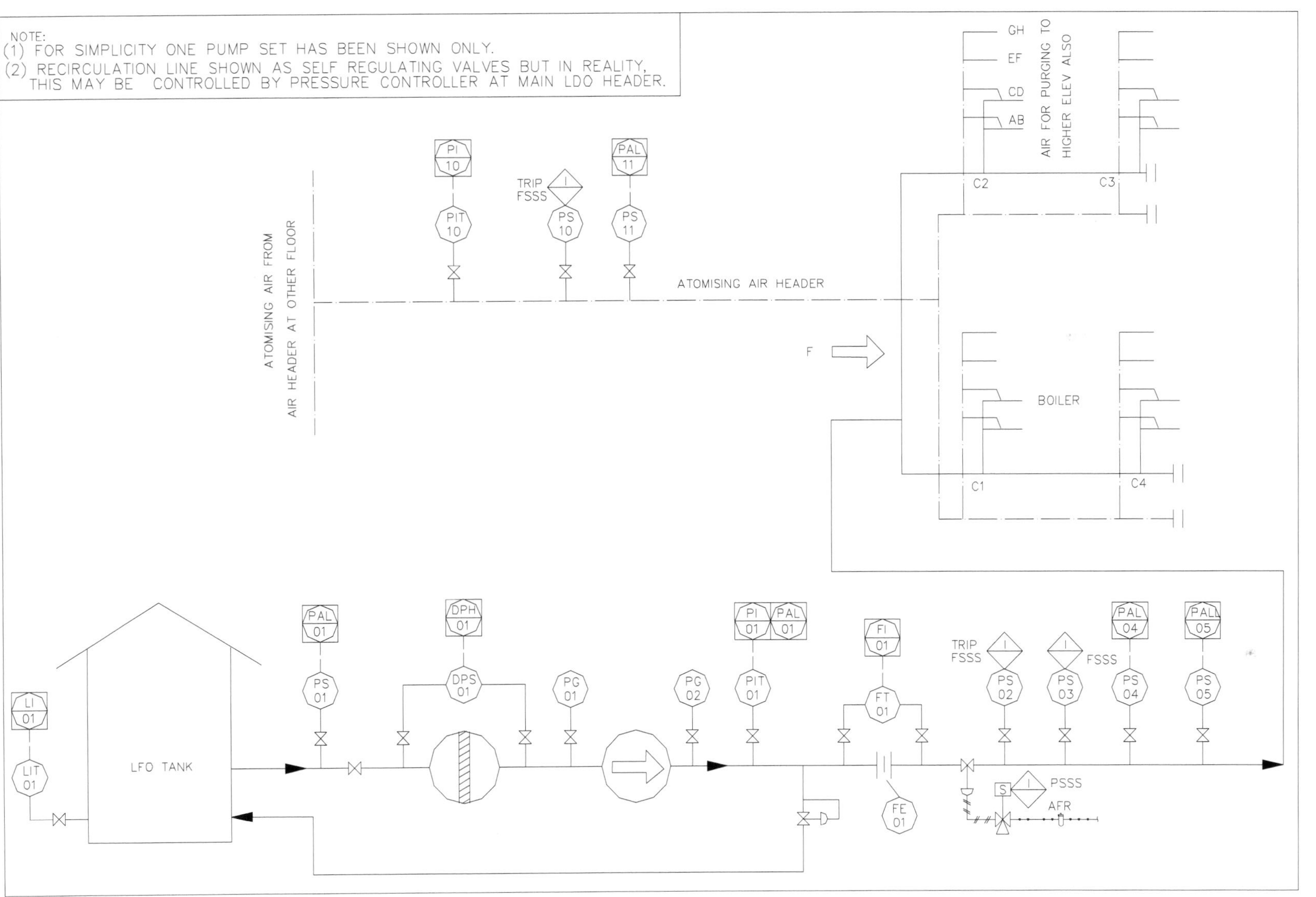

FIG. 3.55 Light oil from LDO tank to burner and atomizing air.

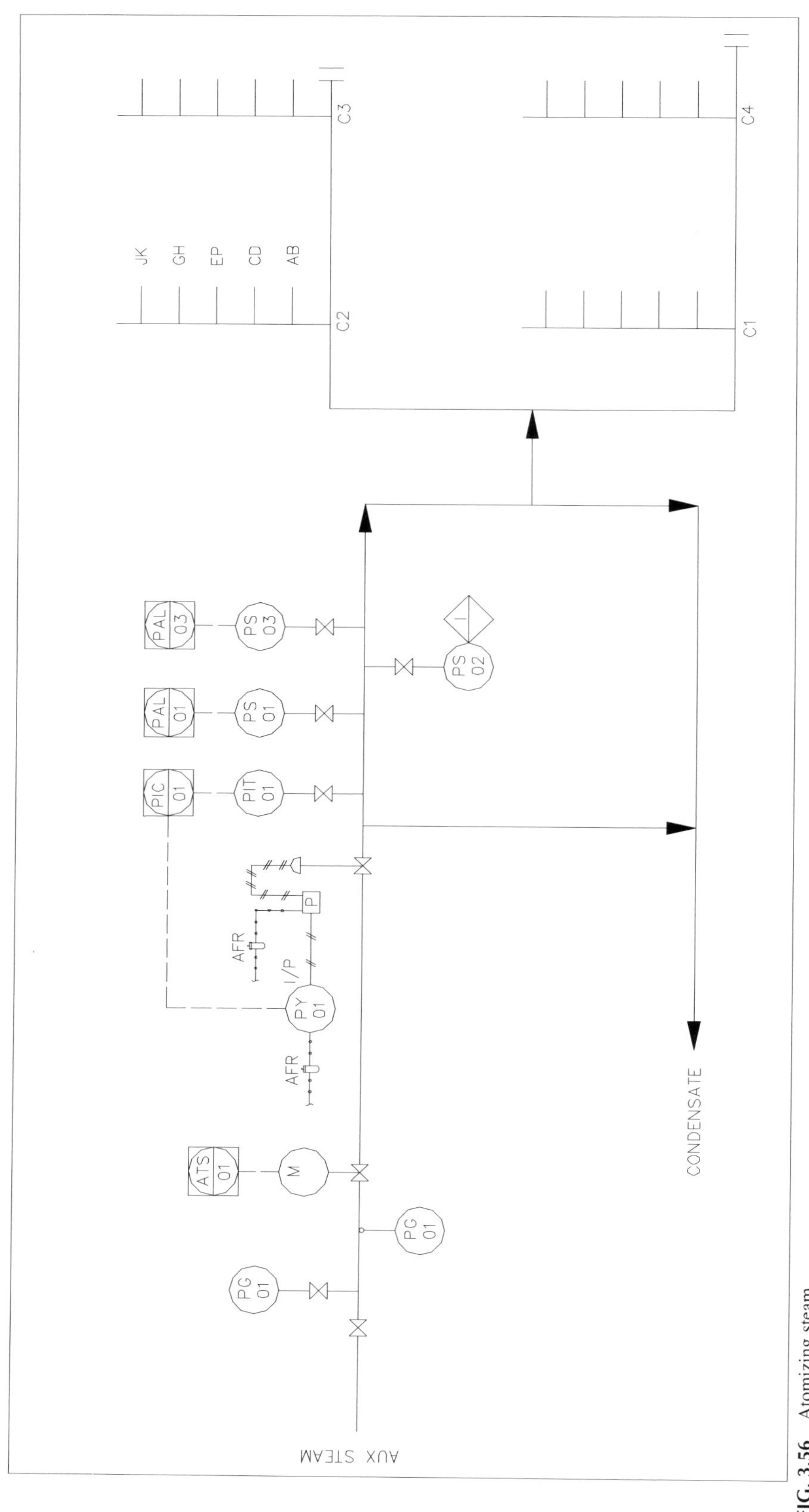

FIG. 3.56 Atomizing steam.

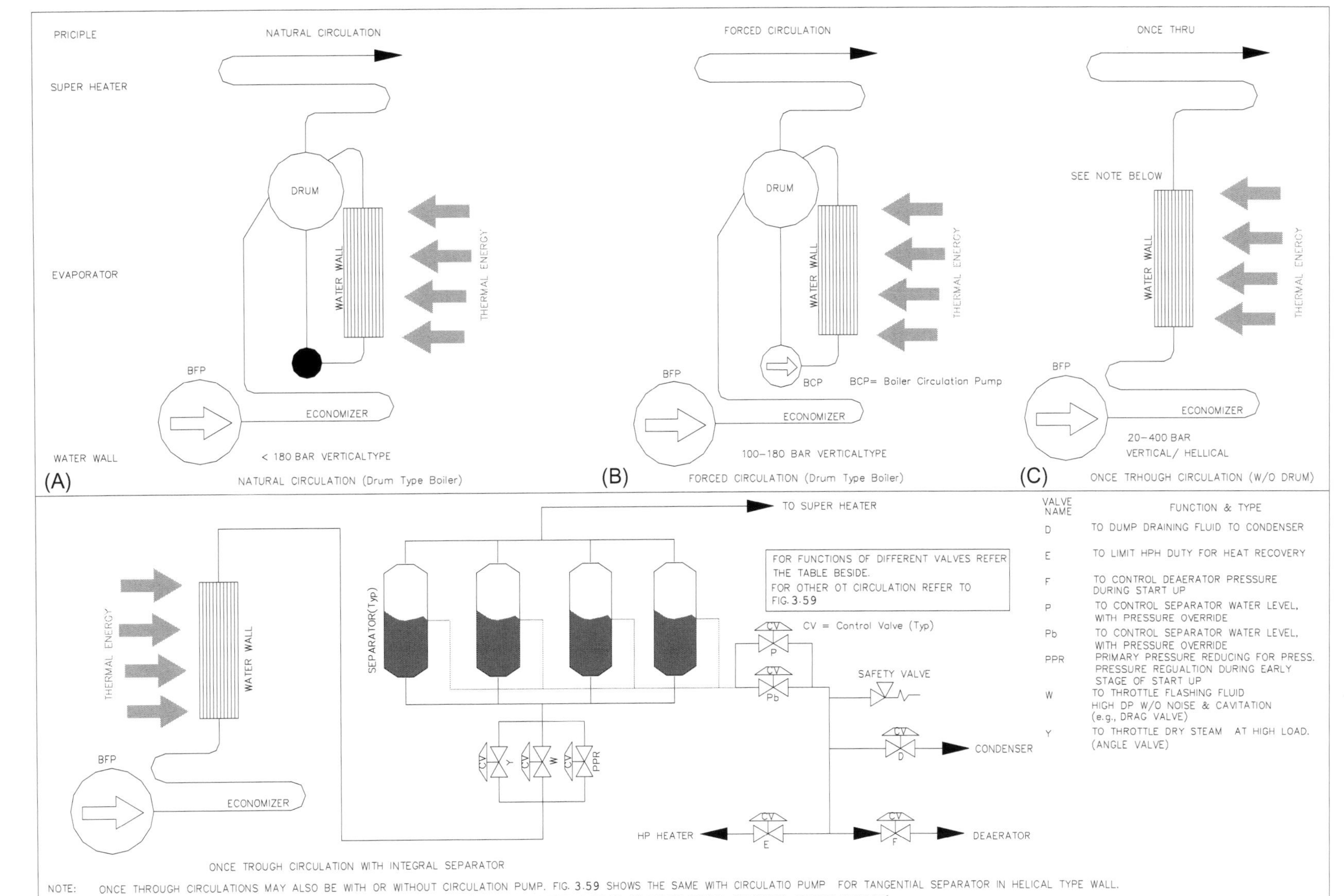

FIG. 3.57 Boiler circulation types.

When flue gas passes across the furnace water wall tubes, it heats the water inside the tube; naturally, its density decreases and would try to go up. These water tubes/overtubes carry the steam/water mixer to the top, that is, to the drum through the riser (because while carrying the steam/water mixture through the tubes, it gets heated by the flue gas). As a consequence of the same, water at the lower temperature (hence a higher density) will try to fill the gap. So because of the difference in density between the cooler water in the downcomer and the steam/water mixture in the riser tube, a natural circulation in the system is established due to the convection current. Here, it is important that the flow through the tube must be sufficient to cool the tube and prevent the tubes from overheating by the flue gas; the greater the heat absorption, the higher the circulation.

4. In forced circulation, as the name suggests, an external mechanical force is produced by pumps located at the bottom, as shown in Fig. 3.57B. In subcritical boilers, this type of circulation is found in cases where the boiler pressure is high with a lower static head energy (mainly for higher-rated boilers).
5. In a subcritical boiler, the water absorbs the latent heat to convert itself to steam. As the pressure is increased, the amount of latent heat required to convert water to steam is reduced. When the pressure is further increased, the latent heat of vaporization is zero. The pressure at which such a phenomenon occurs is called *critical pressure*.
6. At pressure > critical pressure ~225 kg/cm^2 (temperature ~374°C), there is no boiling, and water becomes steam directly. This kind of boiler is the SC boiler. In case the pressure and temperature parameters are in the range of 300 bar and 620°C, respectively, then the type of boiler may be referred to as an USC boiler.
7. Supercritical boilers are once-through (OT) in design, meaning that there will not be any stored vessel or drum associated with the boiler. The criticality associated with OT boilers is a design of the furnace steam/water evaporator circuitry, the start-up system, and the firing and heat recovery area (HRA). OT boilers are of two types: one with recirculation or a BCP (as in Fig. 3.58) and the other is without the pump.
8. In OT boilers, before the fuel is fired, a minimum fluid mass flow rate is established in the evaporator tube to protect from overheating. This is ensured normally by the BCP and by keeping the feed water control valve at a minimum (~25%–30% flow rate) opening. This is a basic requirement. In this region of start-up of OT boiler is similar to drum type boiler control with its storage tank. This is the Benson load.
9. Before the flow rate reaches the Benson point (25%–30%), the recirculation flow guarantees the safety. After the Benson point, the water-cooled wall can be cooled by enough mass flow rate.
10. Steam and water are separated at the separator, where the steam is put into the superheater and from the separator storage tank, the BCP sends the water back to the economizer. During start-up, if the circulation is improper, then there may be a serious temperature difference and damage to the boiler.
11. Safe operation of water circulation is highly dependent on BCP flow control as well as separator storage tank level control. Normally, the storage level is maintained by BCP flow control; when the storage level is getting high, then the excess flow is controlled by draining to the flash tank.
12. When the unit load is near and above the Benson point, the water level in the separator mostly disappears. Normally, the Benson point is passed in a rising speed. At a certain point of steam flow, the BCP is stopped, and automatic feed water flow control takes over. It is worth noting that the feed water-to-fuel flow ratio is very important for OT boilers. Steam temperature is maintained mainly by firing rate with auxiliary control from a spray water flow.
13. Temperature is a function of load. Normally, a little degree of superheat is kept at the separator. At the low load combustion rate and as the feed rate decreases, there may be saturated steam in the separator.
14. The OT design so far is mainly based on the Benson spiral design with a circulation pump. However, there are other designs also, for example, the Benson vertical design, where circulation without a circulation pump is possible. Also, the Foster Wheeler multipass design does not call for a separate circulation pump, but in this case the separator design is different (source: Foster Wheeler once-through utility boiler technology).

12 TRENDS IN POWER GENERATION PROCESS

12.1 Increased Pressure, Temperature Operation

Despite efforts to increase the use of renewable energy in the world, especially in Europe and many Third-World countries, fossil fuels account for the largest share of total electricity generation capacity. As a consequence, there will be a large contribution of CO_2 emissions. In order to cut CO_2 emissions, the main aim of plant owners is to increase generation efficiency to cut CO_2 emissions. The average efficiency of coal-fired generation worldwide is expected

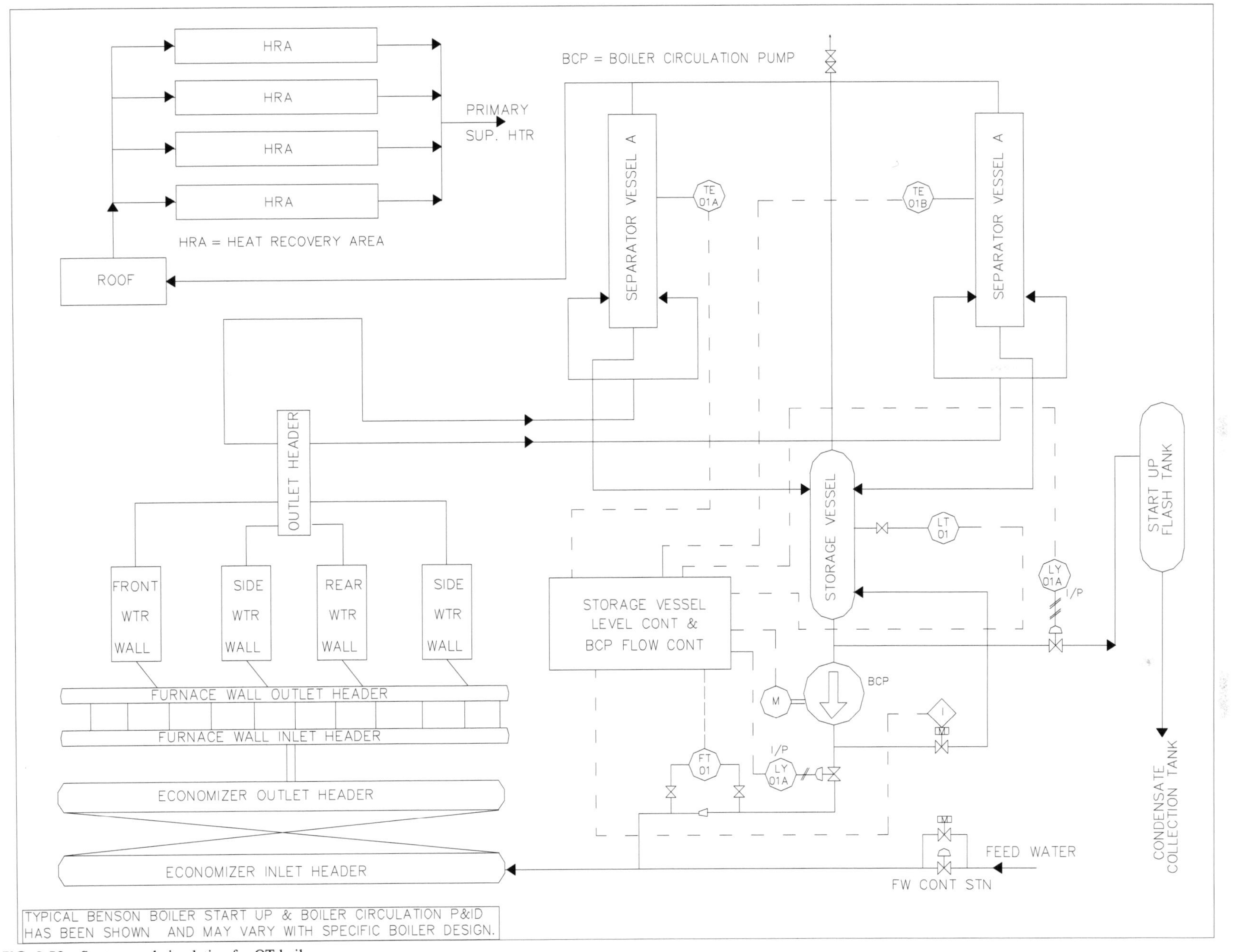

FIG. 3.58 Start-up and circulation for OT boiler.

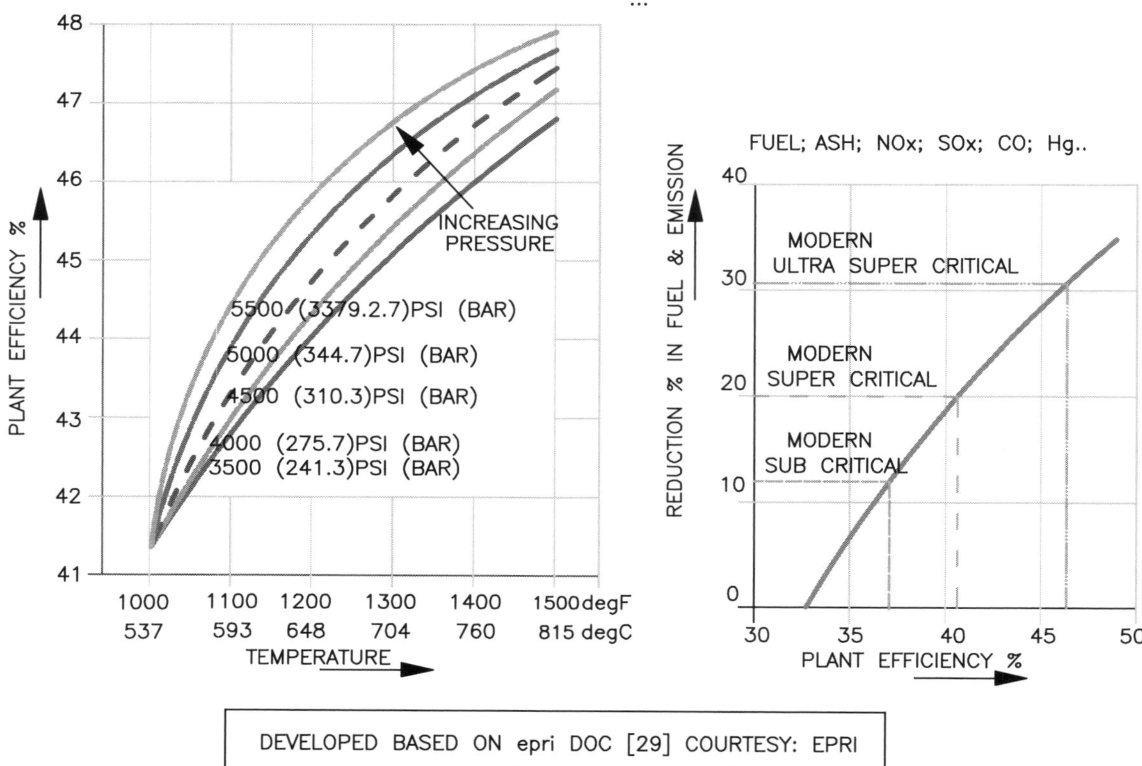

FIG. 3.59 Pressure temperature efficiency and emission.

to reach more than 40% by 2035 when subcritical plants may be phased out. The main EU policy documents relevant to advanced fossil-fuel power generation are [28]: Large Combustion Plant Directive (LCPD) (Directive 2001/80/EC); Industrial Emissions Directive (IED) (Directive 2010/75/EU); CHP Directive (Directive 2004/8/EC); Renewable Energy Sources Directive (RED) (Directive 2009/28/EC); and Emissions Trading Directive (2009/29/EC).

12.1.1 Subcritical, Supercritical, Ultrasupercritical, and Approach to AUSC

As seen in clause no. 11.2.1.5.6, in each case, the pressure and temperature of steam to drive the turbine increases. This means a smaller fuel requirement, hence fewer CO_2 emissions. According to the European commission, the new target is advanced ultrasupercritical (AUSC) operation with superheater temperatures nearing 700°C so as to attain an efficiency up to 50%. As shown in Fig. 3.59—developed based on an EPRI document [29]—there will be an increase in both the pressure and temperature of steam for the approach to AUSC.

Even with IGCC (process explained in appendix), with hybrid *fuel cells* (*a fuel cell* is an electrochemical energy conversion device that produces electricity, water, and heat) could possibly reach 60% efficiency, with zero emissions. The gasification process produces syngas containing mostly hydrogen, carbon monoxide. etc. These can be further processed by directly feeding to the fuel cell. Because heat is generated in the fuel cell, it can be recovered in the power cycle. The major constraints come here from materials in use. People are now focusing on nickel-based alloys (263/613/740H and others, for example, HR6W, HR35) and some additional alloys (Chapter 13). Also, people are considering using Oxyfuel in place of air and gas in place of coal (United States) (Table 3.13).

TABLE 3.13 CO_2 **Reduction (1000 MW—70% Annual Operation) [30]**

Plant Type	Rate of CO_2 Generation	Annual CO_2 Release (approx)
Subcritical	900 g/kWh	5.5 million tons
A-USC	800 g/kWh	5.0 million tons
A-USC with IGCC	~700 g/kWh	4.4 million tons
IGFC	~ 590 g/kWh	3.6 million tons

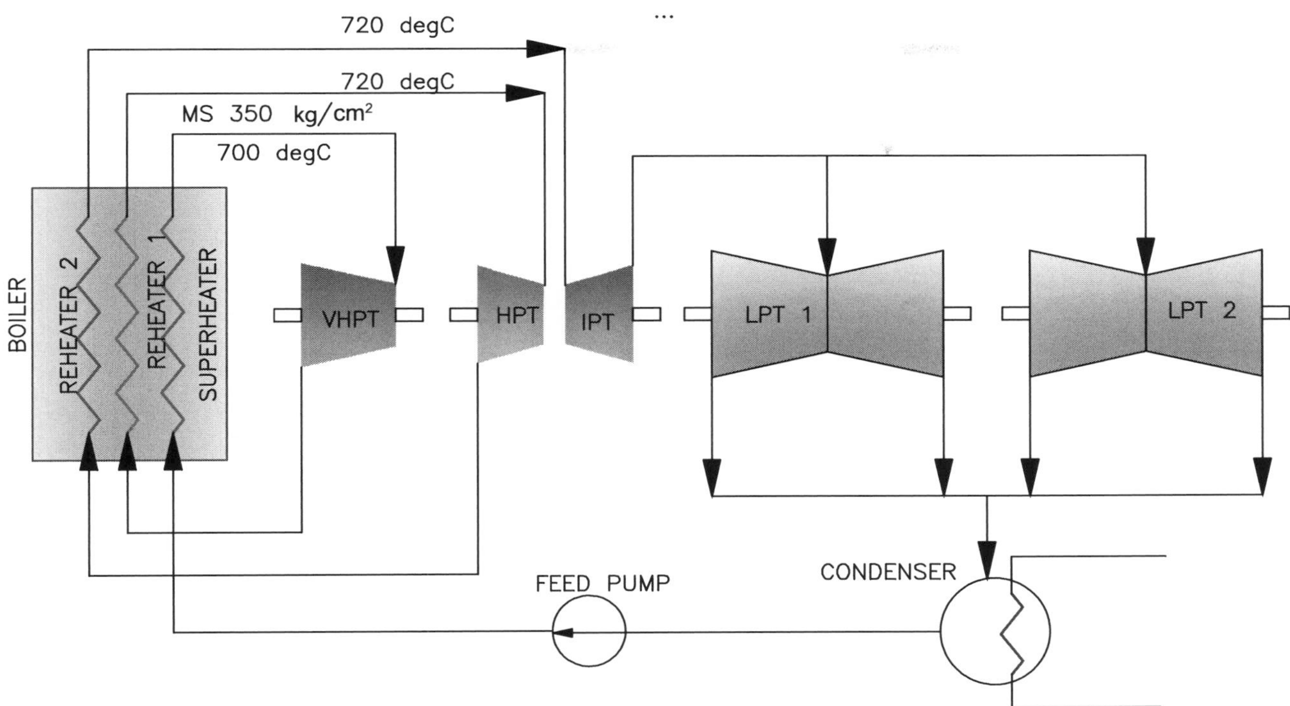

FIG. 3.60 A-USC plant-power cycle.

12.1.2 AUSC Power cycle

A typical cycle for A-USC is depicted in Fig. 3.60.

A-USC boilers have an increased steam temperature near 700°C with pressure as high as 300–350 bar or more for the realization of better efficiency. Although not shown in Fig. 3.60, normally there would be HP (4 × 25% for example, Siemens/Babcock) and LP (2 × 50%) bypass systems. As there are no drastic changes in the peripheral equipment, the A-USC boiler can be retrofitted to an existing power-generating facility [30]. A-USC power generation and CO_2 emissions can be reduced by approximately 12% from existing USC power generation while the efficiency can be improved by nearly 10% with respect to USC.

12.2 Various Ways for Reducing Emissions and Increasing Efficiency

Circulating fluidized bed (CFB) combustion (Section 12.4) is also an important system toward increasing plant efficiency with the additional advantage of fuel flexibility, heat integration, and low emissions. It is also well suited for Oxyfuel combustion. It allows the use of low-quality coal and biomass with carbon capture and storage (CCS).

Dry ash Handling and air recirculation not only provide ways to reduce water consumption but also heat utilization. Also, the plant can reuse 100% fly and bottom ash for road making, land filling, making fly ash bricks, and in cement industries. Normally, bottom ash is available at 900°C, but in some plants, bottom ash is cooled by air and such hot air is channeled to the boiler so that the bottom ash can be discharged at ambient temperatures. On account of this, the efficiency of the boiler also improves vig. Aditya Birla group captive plant in India where dry bottom ash system was supplied by Magaldi Italy [courtesy: Power Magazine Aug'2018—www.powermag.com]. In addition to the A-USC, the IGCC fuel cell is another area to increase the coal-to-electricity conversion efficiency. The most noticeable advantage is the power generation in ultraclean power without interruption.

12.3 Carbon Capture and Storage and Oxyfuel Combustion

CCS refers to a technology through which up to 90% of the carbon dioxide (CO_2) emissions produced from the use of fossil fuels in electricity generation and industrial processes can be captured to prevent CO_2 from entering our atmosphere. After capture, CO_2 is dehydrated and compressed. Liquid CO_2 is transported through a pipeline or by ship for safe storage at a carefully selected geological rock formation that is typically located several kilometers below the surface of the Earth.

Basically, there are three methods to accomplish CO_2 capture:

1. *Precombustion capture*: O_2 is extracted from the atmosphere in an air-separation unit (ASU). Oxygen is fed to

the gasifier where coal is baked at around 700°C to produce what is known as syngas (containing CO + H_2 along with CO_2 and steam). In a reactor, water is added to syngas to produce H_2 and CO_2. While the former gas is used for electricity generation, CO_2 is captured and transported. This is not feasible for retrofitting projects.

2. *Postcombustion capture*: This can be used for green field projects as well as for retrofitting projects. In this method, flue gas is passed through a series of filters to strip out CO2. Separated CO_2 is transported.
3. *Oxyfuel combustion*: Oxyfuel (Fig. 5.25) firing in a boiler is a low-risk technology coupled with a high-efficiency technique used for CO_2 capture in a power plant. It is can be used for a new power plant or to retrofit applications. The technology is adaptable for both pulverized coal (PC) and circulating fluidized bed (CFB) boilers. The FGR is used to regulate the temperature. In this system, the ASU is utilized to get pure oxygen. Pure oxygen aims to produce a flue gas stream with nearly 80% CO_2 concentration, depending on the carbon content of the fuel. The gas is then passed through a CO_2 purification unit where pollutants such as oxides of sulfur and nitrogen are removed.
4. *Storage*: As a part of storage, the transported liquid CO_2 is injected into the porous rock formations a few kilometers below the surface. Where suitable geological data are available, a depleted oil field may be used for this storage purpose.

12.4 CFBC Technology

Bubbling and circulating fluidized bed combustion (CFBC) are two commonly used firing technologies used in power generation. A typical CFBC is indicated in Fig. 3.61. In circulating bed solid circulated through cyclone for heat

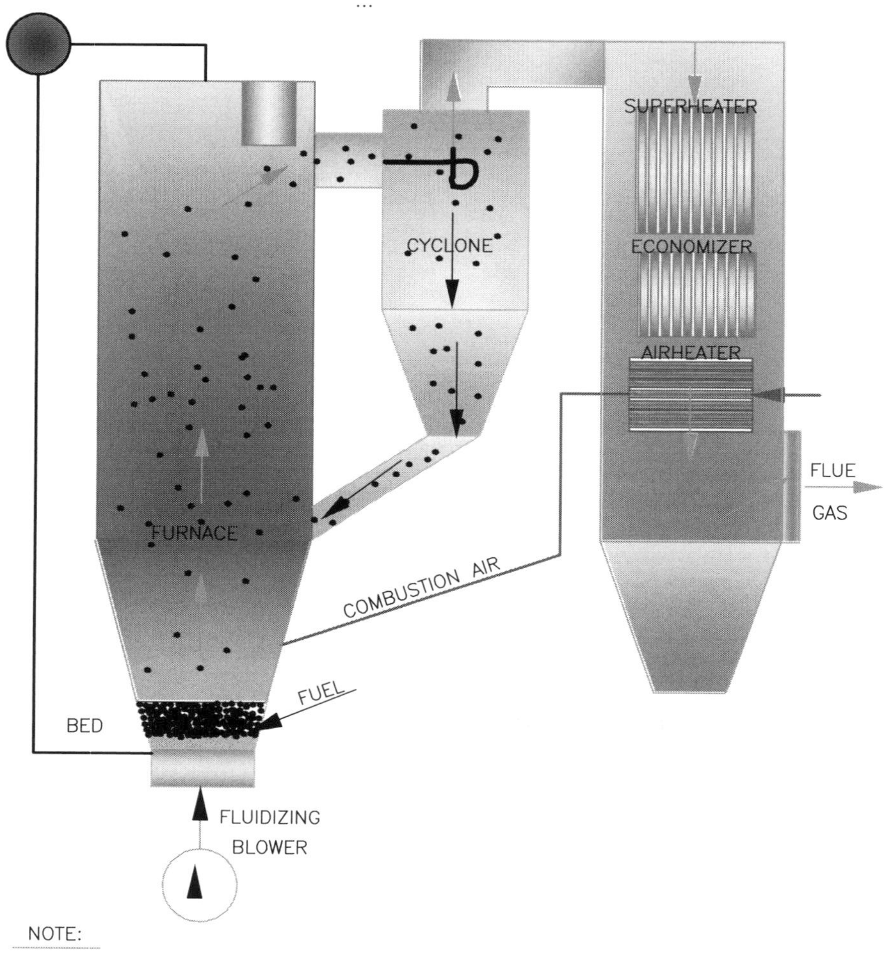
SUPERHEATER
CYCLONE
ECONOMIZER
AIRHEATER
FLUE
GAS
FURNACE
COMBUSTION AIR
FUEL
BED
FLUIDIZING
BLOWER
NOTE:
IN SOME DESIGN (COURTESY: FOSTER WHEELER) CFBC USC BOILER, SUPERHEATING
TAKES PLACE INTREX LOCATED AT SOLID RETURN PATH. NOT ON THE TOP AS SHOWN.

FIG. 3.61 CFBC boiler.

transfer. There is a bed of inert material at the bottom of the furnace. Fuel is spread over the bed, and air is supplied at high pressure from the bottom by a blower to lift the bed material and the fuel in suspension. Combustion takes place in the suspended condition. Secondary air fans provide preheated combustion air. Fine particles of partly burned coal, ash, and bed material are carried along with the flue gases to the upper areas of the furnace and then into a cyclone. On account of the cyclone, the heavier particles separate from the gas, fall to the hopper of the cyclone, and are returned to the furnace. On account of this solid recirculation, this type of fluidized boiler is referred to as a circulating fluidized bed combustion boiler.

One of the major benefits from a CFBC is that in this technology, a wide range of fuel can be used with lower air pollution. Good quality coal can be used in PC boilers and rejected quality coal as well as many other kinds of fuel such as biomass can be utilized in a CFBC. Also, CFBC technology offers almost pollution-free combustion without the need for back end FGD. Limestone can be added to control SO_2 and low NO_x combustion or selective noncatalytic reduction (SNCR) can be utilized for low NO_x. Added to this, there will be a reduction of CO_2 emissions. Heat utilization is excellent and that made it possible for USC steam generation in the CFBC. Some of the major benefits of the CFBC are:

- Wide range of solid and liquid fuels and high heat utilization through the blower.
- Carbon burnout of 98%–99% depending on fuel type.
- Part load operation as low as 40%, with a rate of load change up to 3%–7% per minute.
- Optimized for subcritical steam, SC, USC, and combined heat and power (CHP).
- Ensures safe temperature control for efficient combustion.
- Ensures in situ desulfurization control (limestone injection).
- Efficiently controlled combustion process suppressing the formation of thermal NO_x.
- In most cases, external SO_x/NO_x emission control is not required.
- Avoids blockage associated with low melting ashes.
- Use of standardized materials leads to cost-competitive solutions.
- A cleaner carbon footprint through biomass cocombustion or main combustion capabilities.
- Low-temperature CFB combustion coupled with USC steam technology.

In some design, for example, the Foster Wheeler design, an integrated recycle heat exchanger (INTREX) provides additional advantage for oxyfuel combustion. Unlike PC boilers, superheating is not done at the top but at the high-efficiency heat exchanger called the INTREX and protects superheater coil from corrosive flue gas and lower grade coil materials can be used for example, the KOSPO plant in South Korea [Courtesy: Power Magazine August 2018—www.powermag.com].

BIBLIOGRAPHY

[1] R.K. Kaporia, S. Kumar, K.S. Kasana, An analysis of a thermal power plant working on a Rankine cycle: a theoretical investigation, J. Energy S. Afr. 19 (1) 2008.

[2] Boiler Feed Pump: Sulzer: Pump Catalog.

[3] Boiler Superheater Reheaters: Book: Steam/Its Generation and Use: Bcock and Wilcox.

[4] R.E. Athey, E. Spencer, Deaerating Condenser Boosts Combined-Cycle Plant Efficiency, Graham Manufacturing Co.

[5] Economizers and Air Heater: Book: Steam/Its Generation and Use: Book: Babcock and Wilcox.

[6] Feed Water System: IRI Power Plant: Industrial Resources Inc.

[7] Fuel-Burning System: Book: Combustion Fossil Power Systems: Combustion Engineering Inc.

[8] D.L. Chase, P.T. Kehoe, GE Combined Cycle Product Line and Performance: G.E, New York.

[9] T. Kamei, T. Tomura, Y. Kato, Latest Power Plant Control System: Hitachi Documents.

[10] J. Yan, J. Wanggui, H. Ping, S. Varanasi, Main Electric Power and Balance of Plant Systems and Equipment of Qinshan Phas III CANDU Nuclear Power Plant: East China Electric Power Design Institute, Atomic Energy of Canada Limited.

[11] Mark VIe Control System Product Description: GEI-100600A: G.E. Document.

[12] MS Supply and Feed Water System: Canteach.candu.org (Internet): Training docu., 1996.

[13] PIP PIC001: Process Industry Practices.

[14] Pipe Symbols: web.cecs.pdx.edu/~graig/eas199B/PipeSymbols1: Internet ref.

[15] Power Generation Condensate Polishing Unit: PALL Power generation USA: Catalog.

[16] Preliminary Safety Analysis Report: Lungmen Unit 1 and 2.

[17] Process Diagram master2: 30678_12_ch12_p251-270. qxd: Internet ref.

[18] Pulverized Fuel System Fires: Babcock and Wilcox: System Write Up.

[19] Pulverizer and Pulverized Coal System: Combustion Fossil Power Systems: Combustion Engineering Inc.

[20] Simulation of Combined Cycle Cogeneration Power Plant: W. Pongprasert, S. Kerdsuwan: WIRC King Mongkut Institute of Technology Thailand.

[21] Standard Tech. Specification for Main plant Package of Subcritical Thermal Power Plant (500 MW and Above): Central Electricity Authority New Delhi: New Delhi, India, September 2008.

[22] Steam and power Conversion System: AP1000 Design Control Document.

[23] Steam Generation: Book: Combustion Fossil Power Systems: Combustion Engineering Inc.

[24] Steam Generator Auxiliaries: Book: Combustion Fossil Power Systems: Combustion Engineering Inc.

[25] Thermodynamic Analysis of Rankine-Kalina Combined Cycle: R.S. Murugan, P.M.V. Subbarao: IIT New Delhi, India.

[26] Plant Flow Measurement and Control Hand Book: Swapan Basu, Elsevier BV.

[27] Plant Hazard Analysis and Safety Instrumentation Systems: Swapan Basu, Elsevier BV.

[28] Advanced Fossil Fuel Power Generation; THEMATIC RESEARCH SUMMARY; European Commission; https://setis.ec.europa.eu/energy-research/sites/default/files/library/ERKC_%20TRS_Advanced_fossil_fuel_power_generation.pdf.

[29] J. Shingledecker, R. Purgert, Steam Turbine Materials for Advanced Ultrasupercritical (AUSC) Coal Power Plants, Electric Power Research Institute, https://netl.doe.gov/File%20Library/Events/2014/crosscutting/Crosscutting_20140522_1030B_EPRI.pdf.

[30] Development of Materials for Use in A-USC Boilers; Mitsubishi Heavy Industries, Mitsubishi Heavy Industries Technical Review, 52(4), 2015.

[31] A Steam Generator for 700C TO 760C Advanced Ultrasupercritical Design and Plant Arrangement: What Stays the Same and What Needs to Change; P.S. Weitzel; Babcock and Wilcox Barberton, Ohio; The Seventh International Conference on Advances in Materials Technology for Fossil Power Plants, 2013, http://www.babcock.com/products/-/media/3d7d9be9f78e4b6189355a80b7ed499f.ashx.

Chapter 4

General Instruments

Chapter Outline

1 INTRODUCTION

Modern thermal power plants are equipped with state-of-the-art instrumentation systems. Advancements in science and technology have transformed instrumentation into a continuously developing system, including both measurement and control systems. With the help of measuring instruments, the human civilization learned about the environment, the world, and beyond. By controlling instruments, the quality of necessities and enjoyment of life are not only enhanced, but improved. Modernization and progress in the industrial sector depends on precise, modern cost-effective instruments. The power plant instrumentation system comprises various types of measurements. In this chapter, discussions are restricted to general types of instruments for pressure, temperature, flow, and level.

1.1 Pressure Measurements

The simplest device for measuring pressure and differential pressure (DP) is the manometer or differential manometer. This is used to locally view low and extremely low pressure. Gauges are also used for local viewing purposes as well as for remote viewing through an extended impulse line for higher pressure applications. In some cases, low-pressure measurements were also taken remotely. Switches are used for alarm generation and binary control purposes, for example, starting, tripping, and interlocked operation of different drive motors, and safety relays or equipment. Some sensing elements use the amount of expansion or displacement characteristic of the material when subjected to pressure. This is called the elastic deformation pressure element, which practically changes its shape under pressure.

For gauges, switches, and even for pneumatic pressure transmitters, this characteristic is utilized by different elements, such as bellows, diaphragm, Bourdon, helix, capsule, and spiral, depending on the pressure maximum and minimum ranges. Different types of instruments deployed for each kind of measurement are listed in Table 4.1. There were many other element types available, for example, ring balance type, wound resistance in varying shapes, etc.

Power Plant Instrumentation and Control Handbook. https://doi.org/10.1016/B978-0-12-819504-8.00004-4

TABLE 4.1 Type of Pressure Gauge Switch Elements

Type of Measurement	Type of Instruments					
Pressure Gauges/Switches	**Bellow**	**Bourdon Tube**	**Capsule**	**Diaphragm**	**Helix**	**Spiral**
Maximum range kg/cm^2	56	7000	3.5	28	700	280
Minimum range mm of wcl or kg/cm^2	125 mm of wcl	0.85 kg/cm^2	25 mm of wcl	50mm of wcl	3.5 kg/cm^2	1.05 kg/cm^2

TABLE 4.2 Type of Temperature Gauge/Switch Elements

Temperature Gauges/Switches	Mercury in Glass	Alcohol, Toluene in Glass	Bimetallic	Gas/Liquid Filled, Pressure Spring Type
Range °C	−30 to 510	−180 to 650	−180 to 425	−250 to 550

1.2 Temperature Gauges and Switches

Temperature gauges and switches use the volume-expansion characteristics of fluid after an application of heat or temperature. A bulb containing transmission fluid is placed at the sensing point. The transmission fluid may be inert gas like nitrogen, liquid or mercury. The receiving side may be bellows, diaphragm, Bourdon tube gauge, helix, capsule, or a spiral-type pressure sensing element. A bimetallic element is also used, which bends in one direction when heat is applied to an assembly of two metals of a dissimilar coefficient of expansion that are joined together. Types of temperature gauge and switch elements are listed in Table 4.2.

Pressure spring-type temperature gauges are mainly used for a remote indication/switch where the indicator/dial case is brought at a distance from the sensing point through a capillary tube. Here, the fluid-filled bulb is inserted in the process, and a definite amount of volume expansion at a particular temperature pressurizes the internal system fluid, which is sensed by the pressure spring located inside the indicator/dial case. The pressure spring is a hollow spring of a simple C-shaped Bourdon tube or spiral, although a helix is more often used. A combination of liquid and vapor is also used as a filling liquid, but because vapor does not expand uniformly like liquid and gas, the scale becomes nonlinear at higher temperatures. The range is thereby limited to −180°C to 300°C.

1.3 Elements for Remote Pressure Transmitters

The operating pressure in modern power plants varies widely from 0.1 kg/cm^2 abs to approximately 350 kg/cm^2 g. To take care of this variation, different types of instruments are used with various types of sensing elements for electronic transmitters such as capacitance, dual inductance, strain gauges, piezoresistive, silicon resonance, twin resonance (nitinol) wires, etc. In capacitance elements, a diaphragm is used as the primary element with a higher measuring pressure on one side and atmospheric pressure (for gauge pressure measurement) or vacuum (for vacuum/absolute pressure measurement) on the other side. The diaphragm moves toward the lower pressure side and this displacement is sensed as a change in ratio of two capacitances (placed in such a way that if one increases the other would decrease) and ultimately converted into a two-wire 4–20 mADC signal. For strain gauge, when pressure is applied, the deformation due to stress is converted into an electrical signal. For other operating principles, the change in electrical characteristics due to physical change after applying pressure is sensed and converted into a 4–20 mADC signal. For smart transmitters (highway-addressable remote transmission, HART), superimposed digital signals proportional to input pressure are also available in the same output terminals. Other smart transmitters utilize a full digital signal in Fieldbus systems such as Profibus and Foundation Fieldbus.

The same philosophy applies with a differential pressure transmitter (DPT): two process pressure sides are connected to the transmitter with a higher pressure-tapping impulse line to the high side and lower pressure-tapping impulse line to the low side.

1.4 Elements for Remote Temperature Transmitters

With pressure transmitters, the sensing elements and processing circuitry are located in the same enclosure, whereas the temperature sensors/elements are located

TABLE 4.3 Type of THC and Temperature Ranges

Types of THCs	Iron-Constantan (Type J),	Copper-Constantan (Type T)	Chromel-Alumel (Type K)	Platinum-Platino Rhodium (13%) (Type R)	Platinum-Platino Rhodium (10%) (Type S)
Temperature Range in °C	−210 to 1200	−200 to 400	250–372	−50 to 1768	−50 to 1768
Used for measurement of parameters	Mill outlet/ bearing temperature		Steam temperature above 300°C	Furnace temperature ~1000°C	Furnace temperature ~1000°C

In all cases types are "ISA type."

separately at the sensing point inside the process. The sensors/elements are isolated from the process fluid, which may have a high velocity of abrasive and/or hazardous nature, with the help of a thermowell. This takes care of process-side complications and is normally threaded and/or welded to the process pipe or vessel. Inside the thermowell there is a female thread to accommodate the sensors/elements with a male thread.

Sensors/elements are of various types, such as a resistance temperature detector (RTD), thermocouple (THC) and thermistor. An RTD utilizes the property of changing "resistance" of an electrically conducting wire with respect to change in temperature: the resistance of a conducting wire increases when the temperature rises. The operation of the resistance thermometer depends on the electrical/ electronic circuitry. The temperature range covered by different types of RTDs is (−)250 to 850°C. There are three types of RTDs normally in use: copper (53 Ω at 0°C), platinum (100 Ω or 1000 Ω at 0°C), also called the platinum resistance thermometer (PRT), and nickel (100 Ω or 500 Ω at 0°C). PRTs are widely used in power plants to measure water and air temperature <400°C.

Thermistors also show the similar type of quality but in the reverse direction, that is, the resistance of the element decreases when the temperature rises. THCs utilize the thermoelectric property of generating electromotive force (emf) in millivolts (mV) between two open ends when two dissimilar metal wires are solidly connected at the end and subjected to higher temperatures than the other open ends. When the joined ends are heated or subjected to/inserted into a higher temperature, a small but measurable voltage (mV) is generated between the two open and cold ends, depending on the type of wires selected. This varies with hot ends and cold ends. Various types of THCs are in regular use: iron-constantan (type J), copper-constantan (type T), chromel-alumel (type K), platinum-platino rhodium [two types with 10% (S type), and 13% (R type) platinum in the alloy]. Table 4.3 shows examples of THC temperature ranges.

1.5 Temperature Transmitters

Temperature transmitters are used to convert the mV output from the temperature elements to 4–20 mADC with a superimposed digital signal for the smart version. The exact type/model of temperature transmitter is selected as per the type of element. The transmitters may be element head mounted, local/field mounted, or element head or cabinet/panel back mounted. The mV output of temperature elements (for THCs) depends on the temperature difference between the hot and cold junctions. Therefore, there must be a cold junction compensation facility at the transmitter end to take care of the ambient temperature variation at the location of the transmitter. The cable from element to transmitter needs to be the same specification as the THC element. This is called an extension cable; one with similar characteristics (as a cost-saving measure) is called a compensating cable.

As for RTDs, the output is resistance in ohms, and the variable signal cable resistance due to variation in ambient temperature would make the measurement unpredictable. Hence, various methods have been applied to obviate this problem by using a three-wire or four-wire compensation system. Detailed discussions are in Clause 3.2. Temperature ranges of THCs are in Table 4.4.

1.6 Flow Measurement

1.6.1 Sight Flow Glass Indicators

Sight flow glass indicators are used to see that fluid is flowing through the pipeline without any calibration. Flow gauges are provided with flow calibrations. Flow meters and rotameters are types of flow gauges. Oval gear meters and nutating disk meters are flow meters that use positive displacement. These meters are available with indication of flowrate as well as cumulative flow value, and have a compatible electrical signal of 4–20 mA.

TABLE 4.4 Type of Uncommon THCs and Temperature Ranges

Types of THCs	Platinum-Rhodium (30 and 6%) Alloy (Type B)	Rhenium-Tungsten 5 and 26% (Type C)	Nickel-Alloy, Molybdenum (18%), and Cobalt (0.8%) (Type M)	Platinum-Platino Rhodium (Type R)	Platinum-Platino Rhodium (Type S)
Range in °C	50–1800	0–320	0–1400	−50 to 768	−50 to 1768

1.6.2 Flow Switches

These are also available in various types, such as flapper, target, and diaphragm. Flow meters with switch contacts are also available. They are typically used for alarms, data acquisition systems, or open-loop control purposes.

1.6.3 Flow Elements or Differential Producers

These are special types of instruments that are installed for restriction inside a flowing channel like a pipe or duct, and, as a result, a pressure drop proportional to the square of flow is experienced across the element. This pressure drop is taken to the DPT for further use. Many types of flow elements or differential producers are available: flow nozzles, orifice plates, Venturi tubes, aerofoils, and piccolo tubes. Usage depends on the service condition. They are discussed in Clause 4.1.

Other types of flow elements that create restriction in a particular section of a specially designed flow path for open-channel flow measurement are also available. The flow elements are weir/notches (V or rectangular notches), or a Parshall flume. When passing through this type of restriction or the flow path itself, the liquid level increases near its inlet. This increased level is proportional to the flow. Normally, this type of level is measured by an ultrasonic level sensor and transmitter with built-in software providing output in terms of flow.

1.6.4 Flow Transmitters

Flow transmitters can have DPTs with two pressure impulse lines connected across the flow element or differential producer. The flow is calibrated so that the output is proportional to the square root of DP, thus produced across the restriction as deducted from Bernoulli's theorem.

Magnetic flow meters are based on the electromagnetic property of an electrical conductor. Here, the emf generated is directly proportional to velocity from the flow as the area of cross section is constant near the flow meter. The most important criterion for the flowing media is that the conductivity must be more than 0.5 μSiemens/cm. The velocity range is 0.5–15 m/s, but the practical higher limit is ~5 m/s.

There are other special types of flow meters such as the vortex, Coriolis mass flow meter, swirl, thermal mass, etc., but normally they are not used in power plant flow measurement. The Coriolis mass flow meter has limited use in oil flow services.

1.7 Level Measurement

1.7.1 Level Gauge

The level gauge, known as "gauge glass," is located near the vessel. This is a vital instrument for boiler drum-level indication without which, the Boiler Inspectorate will not allow the boiler to run. A calibrated glass tube of suitable thickness and material, generally borosilicate or toughened glass tubes, are placed vertically with the help of upper and lower limbs (pipes/tubes) connected to the process vessel that cover the level to be observed. Almost all of the heaters, tanks, and boiler drums are provided with gauge glass for a direct reading. Other types of special gauge glass are available for boiler drums. Those types of gauge glasses have bicolor indication with steam colored red, and water as green.

1.7.2 Level Switch

Level switches of various types include; Conductivity, capacitance, float, magnetic float, displacer, ultrasonic, paddle, radio frequency, etc. For an open tank, a conductivity type (for fluid that is highly or moderately conductive); capacitance type (for low-conductive fluid); and displacer and float type for top-mounted switches. Side-mounted float types are used at any level. Ultrasonic or radio frequency types can be used for a high range up to 30 m. Paddle type switches are used only for solid level service, for example, coal or ash in the bunker.

For a pressurized tank, side-mounted, magnetic float type switches may be used at any level up to a process pressure of 25 kg/cm^2. For particularly high pressure, Hydrastep or other similar switch types are used for boiler tripping for espeially high and low drum levels.

1.7.3 Level Transmitter

Level transmitters operate on DP principles similar to pressure or DPT, which are used for open tanks or pressurized tanks, respectively. These types are used for drums, all heaters, deaerators, condenser hotwells, condenser surge tanks, etc., for level measurements. Displacer type transmitters are also used for pressurized tanks, whereas conductivity, capacitance, ultrasonic, and radio frequency type level transmitters are used for open tank measurements.

2 PRESSURE MEASUREMENT: VARIOUS MEASURING POINTS AND RANGE SELECTION

Initial introduction of pressure measurements are already incorporated in Clause 1, Chapter 1, Clauses 3.1 and 3.2. Various measuring points in the power plant (typical value of a 250 MW plant) are given in Table 4.5 along with the operating/maximum continuous rating (MCR) value and instrument ranges. Typically, the calibrated ranges are selected so that the operating or MCR value is ~67% of the full range. But there may be some cases where the operating or MCR value is closer to the calibrated range, for example, the control range of an instrument.

2.1 Pressure Transmitter: Working Principle, Specification, Supplier, and Special Features

Pressure transmitters are devices that convert low-level electrical outputs from pressure-sensing elements to higher level signals that can be transmitted over a long distance for further processing and use in various systems. Various sensing technologies have been utilized to measure the pressure of liquids and gases. Previous chapters (Clause 1 and Chapter 1, Clause 3.2) discussed the types of transmitters that deployed the sensors and the types with working principles, namely capacitance, dual inductance, strain gauges (resistive), silicon resonance, piezoelectric, twin resonance wire, or nitinol wires.

The purpose of deploying pressure transmitters is to measure the pressure of various points of interest and to express the values in different units. Gauge pressure is the most common type of pressure measurement relative to the local atmospheric pressure. Absolute pressure is a measurement relative to a perfect vacuum that is −1.0 atm of pressure and whose physical values in different units are 1.033 kg/cm^2, 14.7 lb/sq. in., 1.013 bar, or 760 mmHg. In practice, these values are taken as 1.0 kg/cm^2 (or 10 m of water column head, wcl), 14.22 lb/sq. in., 1.0 bar, or 736 mmHg. Vacuum pressures are lower than atmospheric pressure and can be expressed either in an absolute pressure unit or in a negative gauge pressure. However, sealed gauge pressure is relative to 1 atm of pressure at sea level. DP reflects the difference between two input pressures. Compound pressure instruments can display both positive and negative pressures. Pressure transmitters are calibrated in different engineering units as per the prevalent standards of the user countries such as in kilograms per square centimeter, pounds per square inch, pascals or kilo pascals, bars or millibars, inches or centimeters of mercury, or inches or feet or meter or millimeter of water.

Pressure transmitters can be classified according to display types and other features made available by the manufacturer. Only a simple visual indication by a needle is provided in the analog meters. Digital display units present numeric or application-specific values in engineering units. These displays may be of different types such as liquid plasma, light emitting diode (LED), liquid crystal displays (LCD), and other types of multiline displays. Temperature compensation is provided in smart pressure transmitters as a built-in feature. In some pressure transmitters integral audible or visual alarms and/or output switches with relay contacts or soft logic forms are also available. Explosion-proof or intrinsically safe pressure transmitters are suitable for installation in hazardous areas with limited application in power plants.

Various types of electrical signals are available in pressure transmitters as output, such as analog voltage, and analog current, but the current output (4–20 mA) is always preferred because it is suitable for long transmission coverage and low-attenuation characteristics. Superimposed digitized output signals on pressure transmitters (smart-transmitter type) are normally encoded with frequency modulation (FM) or some other modulation scheme, such as sine wave or pulse train. The universally and widely accepted communication protocols that handle the pressure transmitter outputs include HART, the Ethernet, Profibus, and Foundation Fieldbus.

There are many criteria for performance specifications of a pressure transmitter, that include: working pressure, vacuum range, accuracy, response time, deadband, and operating temperature. Working pressure is the maximum, allowable pressure at which pressure transmitters are designed to operate. Devices should not exceed 75% of their maximum rated range. Vacuum range, another important measurement, covers the lowest vacuum pressure and the highest vacuum pressure. Accuracy is described as the degree of closeness of the measured value to the true value with respect to the span and expressed in percentage; in other words, it is the difference between the true value and the indication expressed as a percentage of span. In cases where the accuracy differs between the middle span and the first and last quarters of the scale, the largest percentage error is reported. Here accuracy is the overall accuracy including repeatability. Basic pressure transmitters can be force collector types or others.

TABLE 4.5 Typical Measuring Points for Pressure Transmitters

SR. No.	Service Description	Operating/MCR Value	Calibrated Range
1	Forced draft fan discharge pressure	250 mm wcl	0–400 mm wcl
2	Secondary air heater outlet pressure	180 mm wcl	0–300 mm wcl
3	Primary air heater inlet header pressure	1000 mm wcl	0–1500 mm wcl
4	Primary air heater outlet pressure	925 mm wcl	0–1400 mm wcl
5	Scanner air fan discharge pressure	400 mm wcl	0–600 mm wcl
6	Flue gas pressure at secondary air heater inlet	−55 mm wcl	−75 to 0 mm wcl
7	Flue gas pressure at secondary air heater outlet	−145 mm wcl	−150 to 0 mm wcl
8	Flue gas pressure at primary air heater inlet	−60 mm wcl	−140 to 0 mm wcl
9	Flue gas pressure at primary air heater outlet	−140 mm wcl	−200 to 0 mm wcl
10	Flue gas pressure at ESP inlet	−175 mm wcl	−250 to 0 mm wcl
11	Flue gas pressure at ESP outlet	−200 mm wcl	−300 to 0 mm wcl
12	Indirect fan discharge pressure	50 mm wcl	0–75 mm wcl
13	Primary air pressure to mill	900 mm wcl	0–1500 mm wcl
14	Furnace pressure	3 mm	−250 to +250 mm
15	Hot air duct pressure	270 mm	0–400 mm wcl
16	Seal air fan discharge pressure	1200 mm wcl	0–2000 mm wcl
17	Mill DP	550 mm	0–800 mm wcl
18	Feed control station inlet feed water pressure	185 kg/cm^2	0–250 kg/cm^2
19	Economizer inlet feed water pressure	175 kg/cm^2	0–250 kg/cm^2
20	Main steam outlet pressure	155 kg/cm^2	0–200 kg/cm^2
21	Reheater outlet pressure	38 kg/cm^2	0–60 kg/cm^2
22	Reheater inlet pressure	40 kg/cm^2	0–60 kg/cm^2
23	FO pressure at flow control valve outlet	10 kg/cm^2	0–15 kg/cm^2
24	FO pressure at FO pump discharge	15 kg/cm^2	0–20 kg/cm^2
25	LDO pressure at LDO pump discharge	10 kg/cm^2	0–15 kg/cm^2
26	Auxiliary steam pressure for SCAPH	15 kg/cm^2	0–20 kg/cm^2
27	Auxiliary steam pressure for atomization	10 kg/cm^2	0–15 kg/cm^2
28	Auxiliary steam pressure for soot blower	35 kg/cm^2	0–50 kg/cm^2
29	HPT steam chest pressure	150 kg/cm^2	0–200 kg/cm^2
30	HPT governor valve outlet pressure	150 kg/cm^2	0–200 kg/cm^2
31	CRH steam pressure at HPT exhaust	42 kg/cm^2	0–60 kg/cm^2
32	HP; first stage steam pressure	145 kg/cm^2	0–200 kg/cm^2
33	RH (IPT) steam chest pressure	35 kg/cm^2	0–50 kg/cm^2
34	IPT RH ESV outlet steam pressure	35 kg/cm^2	0–50 kg/cm^2
35	LPT inlet steam pressure	5.13 kg/cm^2	0–10 kg/cm^2
36	LPT exhaust steam pressure	1.08 kg/cm^2	0–1.50 kg/cm^2
37	Turbine gland steam header pressure	3 kg/cm^2	0–5 kg/cm^2
38	Gland steam supply pressure to LPT	0.5 kg/cm^2	0–1 kg/cm^2

TABLE 4.5 Typical Measuring Points for Pressure Transmitters—cont'd

SR. No.	Service Description	Operating/MCR Value	Calibrated Range
39	Condenser shell pressure	0.1 kg/cm^{2}abs	0–1 kg/cm^2 abs
40	Condenser air suction pressure	0.1 kg/cm^2 abs	0–1 kg/cm^2 abs
41	Condenser air extraction pump suction pressure	0.1 kg/cm^2 abs	0–1 kg/cm^2 abs
42	LPH 1 shell pressure	0.5 kg/cm^2 abs	0–1.5 kg/cm^2 abs
43	LPH 2 shell pressure	0.5 kg/cm^2 abs	0–1.5 kg/cm^2 abs
44	LPH 3 bled steam line pressure	2 kg/cm^2	0–3 kg/cm^2
45	L.PH 4 bled steam line pressure	5.7 kg/cm^2	0–10 kg/cm^2
46	Deaerator bled (IP extraction) steam pressure	10 kg/cm^2	0–15 kg/cm^2
47	Deaerator header steam pressure	10 kg/cm^2	0–15 kg/cm^2
48	Deaerator low-load steam supply pressure	3.5 kg/cm^2	0–5 kg/cm^2
49	Deaerator storage tank steam pressure	10 kg/cm^2	0–15 kg/cm^2
50	BFP suction pressure	13 kg/cm^2	20 kg/cm^2
51	Booster pump discharge feed water pressure	0–40 kg/cm^2	0–60 kg/cm^2
52	BFP pressure at discharge valve inlet	210 kg/cm^2	320 kg/cm^2
53	BFP discharge header pressure	205 kg/cm^2	320 kg/cm^2
54	HPH 7 outlet feed water pressure	195 kg/cm^2	0–250 kg/cm^2
55	Boiler drum steam pressure	170 kg/cm^2	0–250 kg/cm^2
56	Primary super heater inlet pressure	165 kg/cm^2	0–250 kg/cm^2
57	IPT bled steam line pressure	21 kg/cm^2	0–30 kg/cm^2
58	HPH 7 (from CRH) steam line pressure	42 kg/cm^2	0–60 kg/cm^2
59	DP across feed control station	7 kg/cm^2	0–10 kg/cm^2
60	Condensate pressure at condenser extraction pump discharge	27 kg/cm^2	0–40 kg/cm^2
61	Condensate pressure at drain cooler inlet	25 kg/cm^2	0–40 kg/cm^2
62	Condensate pressure at LPH 2 outlet	23 kg/cm^2	0–30 kg/cm^2
63	Condensate pressure at LPH 4 outlet	19 kg/cm^2	0–30 kg/cm^2
64	FRF (turbine oil) supply pump discharge pressure	68 kg/cm^2	0–100 kg/cm^2
65	FRF supply filter outlet pressure	66 kg/cm^2	0–100 kg/cm^2
66	CW pump discharge pressure	6 kg/cm^2	0–10 kg/cm^2
67	CW at condenser inlet pressure	5 kg/cm^2	0–10 kg/cm^2
68	CW at condenser outlet pressure	4.5 kg/cm^2	0–10 kg/cm^2
69	Generator stator coolant pump discharge pressure	7 kg/cm^2	0–10 kg/cm^2
70	Jacking oil pump discharge pressure	210 kg/cm^2	0–320 kg/cm^2
71	Lube oil supply pressure to TG bearings	2kg/cm^2	0–3 kg/cm^2
72	Condensate make-up bus pressure	7 kg/cm^2	0–10 kg/cm^2
73	Intake pump discharge pressure	10 kg/cm^2	0–16 kg/cm^2
74	ACW supply bus pressure	7 kg/cm^2	0–10 kg/cm^2
75	Service water bus pressure	7 kg/cm^2	0–10 kg/cm^2
76	DMCCW suction bus pressure	10 kg/cm^2	0–16 kg/cm^2

Continued

TABLE 4.5 Typical Measuring Points for Pressure Transmitters—cont'd

SR. No.	Service Description	Operating/MCR Value	Calibrated Range
77	DMCCW supply bus pressure at heat exchange inlet	7 kg/cm^2	0–10 kg/cm^2
78	DMCCW supply bus pressure at heat exchange outlet	7 kg/cm^2	0–10 kg/cm^2
79	DMCCW supply bus pressure at heat exchange outlet	7 kg/cm^2	0–10 kg/cm^2
80	Stator water cooling supply bus pressure	7 kg/cm^2	0–10 kg/cm^2
81	Auxiliary cooling water pump discharge pressure	6 kg/cm^2	0–10 kg/cm^2
82	Cooling water pressure to compressor	6 kg/cm^2	0–10 kg/cm^2
83	Instrument air compressor discharge pressure	8 kg/cm^2	0–12 kg/cm^2
84	Service air compressor discharge pressure	8 kg/cm^2	0–12kg/cm^2
85	Instrument air receiver pressure	8 kg/cm^2	0–12 kg/cm^2
86	Service air receiver pressure	8 kg/cm^2	0–12kg/cm^2
87	Instrument air pressure at burner area	7 kg/cm^2	0–10 kg/cm^2
88	Instrument air pressure at TG area	7 kg/cm^2	0–10 kg/cm^2
89	DP across gas recycling damper	15 mm wcl	0–20 mm wcl

BFP, boiler feed pump; CRH, cold reheat; CW, circulating water; ESP, exhaust steam pressure; ESV, emergency stop valve; FO, fuel oil; FRF, fire-resistant fluids; HP, high pressure; HPH, high-pressure heater; HPT, high-pressure, turbine; IPT, intermediate pressure turbine; LDO, light diesel oil; LPH; low-pressure heater; RH, reheat; TG, turbo generator; DMCCW, demineralized water clarified cooling water; ACW, auxiliary cooling water; SCAPH, steam coil air preheater.

2.1.1 *Force Collector Types*

Devices that respond to variations of applied pressure with mechanical deflection are Bourdon tubes, capsules, diaphragms, or bellows, which consist of a highly elastic or flexible element . Devices based on the movements of sealed pistons or cylinders are also available. All of these are categorized as force collector types. The electronic pressure sensors generally use a force collector device to measure strain (or deflection) as a result of applied force (pressure) over an area.

Piezoresistive devices sense shifts of electrical charges within a resistor. Piezoelectric pressure transmitters measure dynamic and quasi-static pressures. They operate in charge mode, which generates a high-impedance charge output, and voltage mode, which uses an amplifier to convert the high-impedance charge into a low-impedance output voltage. Thin-film devices consist of an extremely thin layer of material, usually titanium nitride or polysilicon, deposited on a substrate. Pressure transmitters that use microelectromechanical systems (i.e., variable capacitance) and vibrating elements are also available. More details about principles of operation of the force collector type sensors that use mechanical deflection or strain are discussed below.

2.1.1.1 Capacitive Type

Capacitive type sensors use a diaphragm and pressure cavity to create a variable capacitance to detect deflection due to applied pressure. A diaphragm is used as a primary element for measuring (high-process side) pressure on one side and atmospheric pressure (for pressure) on the other side (low-process side for DP) or vacuum (for vacuum/absolute pressure). The diaphragm moves toward the lower pressure side, and this deflection is sensed as change in capacitance. Ultimately, the transmitter part converts it into a compatible output signal. Common technologies use metal, ceramic, and silicon diaphragms. Generally, these technologies are applied to low and medium pressures (absolute, differential, and gauge).

2.1.1.2 Piezoresistive Strain Gauge

The piezoresistive effect of bonded or formed strain gauges detect strain/deformation due to applied pressure. The most common types are silicon (monocrystalline), polysilicon thin film, bonded metal foil, thick film, and sputtered thin film. This is one of the most commonly used sensing technologies for general purpose pressure measurement. Generally, these technologies are suited to measure absolute, gauge, vacuum, and DPs. The piezoresistive effect describes the variation of electrical resistance of a material due to applied mechanical stress. In contrast to the piezoelectric effect, the piezoresistive effect only causes a change in resistance; it does not produce an electric potential. The piezoresistive effect of metal sensors is only due to the change of the sensor geometry resulting from applied mechanical stress.

Metal piezoresistors, that is, strain gauges, are successfully used in a wide range of applications. The piezoresistive effect of semiconductor materials can be several orders of magnitudes larger than the geometrical piezoresistive effect in metals, and is present in materials like germanium, polycrystalline silicon, amorphous silicon, silicon carbide, and single crystal silicon. The piezoresistive effect of semiconductors is widely used for various sensor devices using many types of semiconductor materials.

The resistance of silicon changes due to applied stress in two ways: from the stress-dependent change of geometry and from the stress-dependent resistivity of the material. For integrated circuits used in analog and digital hardware, silicon is the choice of material and the use of piezoresistive silicon devices is of great interest. Easy integration of stress sensors with various electronic circuits makes it technically viable for vast applications, that is, silicon enabled a wide range of products using the piezoresistive effect. Many industrial instrumentation devices, such as pressure and acceleration sensors, utilize the piezoresistive effect in silicon. Piezoresistors are resistors made from a piezoresistive material and are usually used to measure mechanical stress. They are the simplest form of piezoresistive devices.

2.1.1.3 Inductance/Reluctance (Electromagnetic)

Transmitters using these types of sensors work like capacitive type sensors, which also use a diaphragm and pressure cavity, except capacitive types create a variable inductance/reluctance to detect deflection due to applied pressure. The diaphragm moves toward the lower pressure side and this deflection is sensed as a change in inductance/reluctance. Ultimately the transmitter part converts it into a compatible output signal. Other electromagnetic sensors measure the displacement of a diaphragm/magnetic object by means of changes in the linear variable differential transformer, the Hall effect, or by eddy current principle. The latter two are normally not used for pressure measurement.

2.1.1.4 Piezoelectric

Transmitters employing this type of sensor utilize certain materials, such as quartz, which helps the piezoelectric effect to measure the strain on the sensing mechanism due to pressure. Highly dynamic pressure measurement uses this technology.

Piezoelectricity is a property in crystals and certain ceramics that helps generate an electric field or electric potential proportional to perpendicularly applied mechanical stress. The applied stress induces voltage across the material if the two ends are not short circuited, or if they are measured by an instrument with high input impedance.

The generation of an electric potential when stress is applied exhibits the direct piezoelectric effect. Conversely, the generation of stress and/or strain when an electric field is applied exhibits the reverse piezoelectric effect. The piezoelectric effect is reversible; for example, lead zirconate titanate crystals exhibit a maximum shape change of about 0.1% of the original dimension.

2.1.2 Other Pressure Measurements (Normally Not Used in Power Plants)

2.1.2.1 Optical

The working principle utilizes the property of physical change of an optical fiber to detect strain due to applied pressure. This technology is employed for challenging service applications where the measuring point is located at a highly remote place, or the process temperature is high, or systems require high inherent electromagnetic immunity.

2.1.2.2 Potentiometric

Here the strain caused by applied pressure is detected by the motion of a wiper along a resistive mechanism.

The specifications and data sheet applicable to DPTs is seen in Table 4.6.

2.2 Pressure Switch: Working Principle, Specification, Supplier, and Special Features

A pressure switch is a type of switch that makes or breaks electrical contact when a certain set pressure is reached at input. The input is the process pressure connected through the impulse line. The process is air, gas, or liquid and would provide ON/OFF switching with adequate current-carrying capacity. The switch can be designed to make contact either on pressure rise or on pressure fall.

Pressure or DP switches are simple electromechanical devices operating on basic principles of deformation/deflection of sensing elements, movement transmitting/multiplying levers/gears, and springs to provide opposing forces.

Process pressure, when applied through the impulse line to the sensing element, creates a force that overcomes the pretensioned spring as per the setpoint, and then moves a balancing arm to effect a tiny movement required to actuate a micro switch (es).

There are three essential and basic elements, various combinations of which form hundreds of variants to suit a particular requirement of industrial applications. These include the following:

1. Sensing elements like Bourdon tubes, bellows, diaphragms (metallic or elastomeric), or a diaphragm sealed piston.
2. A spring to determine the range setpoint.
3. A variety of snap-acting micro switches.

TABLE 4.6 Data Sheet of Pressure Transmitters

SL. No.	Item	Specification	Special Feature
1	Make/model	ABB/Emersion/Honeywell/YEC	
2	Type/output	Smart two wire/4-20 mADC/HART/Profibus/FFbus	
3	Quantity		
4	Accessories	Mounting bracket, nuts bolts	
5	Body	Plated carbon steel/316 SS	
6	Enclosure	IP 66(NEMA4)/IP 67/FM/CSA	
7	Connection	1/2″ NPT for process	
8	Range		
9	Power supply	10.2–32/45 V DC	
10	Adjustment	Span and zero	
11	Sensor	Inductance/capacitance/piezoresistive	
12	Sensing material and medium	316 SS/Hastelloy C	
13	Terminals	1/2″ NPT	
14	Electrical Connection	1/2″ NPT	
15	Damping	Adjustable electrical damping	
16	Output indication	Local integral display	
17	Span adjustability	1:30	
18	Performance	Accuracy: ±0.075% Stability: ±10% of URL Vibration effect: ±10% of URL	

2.2.1 Bourdon Tube-Operated Sensor

This type of pressure switch operates on the primary principle of electromechanical devices employing Bourdon tubes. The essential elements include the Bourdon tube for the input section and the micro switch(es) for the output section. These are easily constructed utilizing time-proven German "C"/coiled Bourdon tubes of 316 SS/phosphor bronze/beryllium copper material.

When process pressure is applied to the input, that is, the sensing element, it creates a force inside the Bourdon tube and it bends outward proportional to the applied pressure similar to a pressure gauge. One or two micro switches are fixed (as per the pressure setting) in the instrument and are actuated by the amplified movement of the Bourdon end and remain actuated until the pressure decreases.

2.2.2 Bellows Sensor

This is the basic pressure switch actuated by a seamless bellows with an adjustable/fixed nonadjustable switching differential. A micro switch is actuated by the movement of the expansion of the bellows through a linkage, thereby, providing independent adjustable setpoints. Multiple setpoints are available, each with their own setting scale, spring, and switch. The auxiliary mechanism permits switching differential adjustment between 10% and 15% minimum to ~50% without disturbing the setpoint.

The materials are wetted parts (i.e., the bellows) such as phosphor bronze bellows with brass wetted parts, 316L SS bellows with 316 SS wetted parts, Monel bellows with Monel Wetted Parts, etc. The enclosures are normally die cast with aluminum pressure and weatherproof to IP 66 with nitrile gasket.

2.2.2.1 Diaphragm Sensor

A diaphragm sensor is typically used as a DP switch with pressures from two different sources (one may be atmospheric for the pressure switch) of a particular process that are connected to the two inputs. They are placed across a sensing diaphragm, which are metallic or elastomeric. The difference of pressure creates a force that overcomes the force of a pretensioned spring, and moves a balancing arm to effect the small required movement to actuate micro switch(es). High and low pressures are applied on either side of the specially contoured diaphragm and immediately they eliminate the errors due to the difference in area, which is a common problem encountered in two-element pressure differential switches.

Metallic diaphragms may be made of 316 SS, Monel, or Hastelloy C and elastomer diaphragms may be made of nitrile, EPDM, Viton, or silicone. This type of pressure switch comprises a thin diaphragm responsive to pressure change with the help of a rigid ring that secures it. The other parts are similar to bellows type sensors.

2.2.2.2 Diaphragm Sealed Piston

This type of switch is used for corrosive type process fluid. The sensing element essentially comprises a time-proven seal diaphragm with a piston for the actuation of the micro switch(es). It is versatile because it comes with various combinations of features. Diaphragm chemical seals can be either direct or remote mounted with a capillary and threaded/flanged connection, duly evacuated, and filled with suitable inert-type filling liquid. The wetted part material of the switch, including the diaphragm, can be 316 SS or Monel. The diaphragm can be flange connected for low process pressure or welded to the pressure housings to avoid a risk against leakage under extreme or specific conditions.

2.2.3 Accessories

The following accessories may be required as optional items:

1. Damping coil, snubber for eliminating effects of process pulsations for increased instrument life.
2. Breather/drain for flameproof enclosures.
3. Blow out disk for weatherproof enclosure.
4. Pigtail siphons for steam services.
5. Three-way cock/isolation valves/valve manifold for process isolation and zero checking.

2.2.4 ON/OFF Differential/Deadband of Switch Contact

This is the difference of process value (pressure in this case) between the points at which the switch actuates and then turns off. For example, for rising pressure a switch actuates at a high setpoint of 2.0 kg/cm^2 and remains actuated as long as the pressure is >2.0 kg/cm^2. If the pressure comes down, the switch does not turn off at exactly 2.0 kg/cm^2 but at 1.8 kg/cm^2. This difference of pressure (0.2 kg/cm^2) is called as ON/OFF differential. As another example, for falling pressure the switch actuates at a low setpoint of 1.0 kg/cm^2 and remains actuated as long as the pressure is <1.0 kg/cm^2. If the pressure increases, the switch does not turn off at exactly 1.0 kg/cm^2 but at 1.3 kg/cm^2. This difference of pressure (0.3 kg/cm^2) is the ON/OFF differential.

The switches are available with a fixed or an adjustable ON/OFF differential or deadband. The ON/OFF differential for each micro switch depends on the micro switch movement differential and the spring tension setting.

Pressure switches can be set as follows:

1. Normal or adequate for system/equipment/pump start permissive.
2. Low or High for alarm annunciation systems
3. Very Low or Very High for system/equipment/pump tripping.

A few parameters are indicated below with approximate operating/ set value for a typical 250 MW thermal power plant. Specifications for DP switches are shown in Table 4.7.

2.3 Pressure Gauge: Working Principle, Specification, Supplier, and Special Features

Gauges are used for local and remote viewing through extended impulse lines for higher pressure application. Gauge-type pressure measuring instruments may be categorized broadly by the magnitude of the pressure they measure. Examples include: pressure gauges, vacuum and compound gauges. Many types of gauges have been developed to measure pressure and the vacuum.

The manometer is another instrument used for local viewing. It is usually limited to measuring pressures close to atmospheric or of low magnitude. It is the simplest type of pressure gauge, requiring no hardware and using a graduated glass tube in a straight, vertical, or inclined configuration or in a U-shape for a hydrostatic liquid column.

A vacuum gauge is used to measure the pressure in a system below atmospheric pressure in a vacuum. The ranges can be divided into low vacuum, high vacuum, and ultrahigh vacuum.

As already discussed in Chapter 1, Clause 3.2 and Clauses 2.1 and 2.2, some sensing elements use the amount of expansion or displacement characteristics of the material when subjected to pressure. This is called an elastic deformation pressure element because it nearly changes its shape under pressure. For gauges, this characteristic is used by bellows, diaphragms, Bourdons, helixes, capsules, and spirals, depending on the pressure maximum and minimum ranges.

Although pressure is an absolute quantity, typical pressure measurements are usually made relative to atmospheric air pressure. In other cases, measurements are made relative to an absolute vacuum, to another reference like standard temperature and pressure (STP), or to normal temperature and pressure (NTP).

The principle of operation of the sensor element is the same as that of the pressure switch, as discussed in Clause

TABLE 4.7 Specification Sheet of Pressure Switches

SL. No.	Item	Specification	Special Feature
1	Make/model	Reputed manufacturer	
2	Range		
3	Sensor type	Bourdon/diaphragm/bellows	
4	Setpoint	1 or 2 setpoints	
5	Sensing material	316 SS/Monel/elastomer	
6	Repeatability	± 0.5% FSR	
7	Medium	Gas/air/water/steam/oil	
8	Switch unit	2 SPDT	
9	Rating	5 Amps, 230 V AC	
10	Connection size	1/2″ BSP. If necessary, suitable adapter will be used	
11	Casing/casing material	Weather/explosion proof, IP 65, die cast aluminum	
12	Electrical connection	1″ ET/3/4″ ET complete with cable gland	
13	Dead band	Adjustable with narrow/wide band	
14	Quantity	As per BOM	

2.2, in addition to that an indicator needle moving or rotating on a graduated scale supported by sector gear, sector gear axle pin, and indicator needle axle. This has a spur gear that engages the sector gear and extends through the face to drive the indicator needle. Due to the short distance between the lever arm link boss and the pivot pin, and the difference between the effective radius of the sector gear and that of the spur gear, any motion of the sensor element is greatly amplified. A small motion of the tube results in a large motion of the indicator needle. A hair spring is provided to preload the gear train to eliminate gear lash and hysteresis.

Along with conventional gauges, digital gauges are also available. They provide analog and/or digital output, data logging capability, high accuracy of about 0.08% (0.05% for TEST gauges), alarm contacts, NEMA-4x/IP 66 enclosure, back-lit display.

Specifications for pressure gauges are seen in Table 4.8.

3 TEMPERATURE MEASUREMENT: VARIOUS MEASURING POINTS AND RANGE SELECTION

Initial introduction of temperature measurements are already mentioned in Chapter 1, Clauses 3.1 and 3.2, and Clauses 1.2, 1.4, and 1.5 of this chapter. Various measuring points in the power plant (typical value of a 250 MW plant) are given in chart form (Table 4.9) along with operating/MCR values and instrument ranges. Normally, the calibrated ranges are selected so that the operating or MCR value is ~67% (but usually not beyond 75%) of the full range. There may be some cases where the operating or MCR value is closer to the calibrated range.

3.1 Temperature Element: Types and Classification, Immersion Length, and Connection Type

3.1.1 Temperature Elements

Temperature sensors/elements (only electrical sensing is discussed) are located separately at the sensing point inside or near the process, and outputs are used for further conditioning and converting to standard output, that is, 4–20 mADC, and/or superimposed digitized signal. Typically, the sensors/elements are isolated from the harsh system conditions, which may have a high velocity of corrosive/abrasives and/or a hazardous nature. This isolation is done by a thermowell of suitable material, which takes care of the process-side complications and is normally welded or flanged to the process through a stub. The element is then connected to the thermowell, which may be threaded for low pressure, welded for medium high pressure, or both threaded and welded for extremely high pressure services. Figs. 4.1A and B and 4.2A and B illustrate how different types of thermowells are mounted on the process and accommodate the elements.

TABLE 4.8 Data Sheet of Pressure Gauges

SL. No.	Item	Specification	Special Feature
1	Make/model		
2	Range		
3	Case	In die cast aluminum alloy moisture proof; case epoxy powder-coated black	
4	Bourdon material	Ph br, 17-4 SS/AISI 316 SS	
5	Movements	AISI 304 SS	
6	Process connection	1/2″ NPT(M) with bottom entry	
7	Mounting	Direct	
8	Dial size	150 mm	
9	Quantity	As per BOM	
10	Accessories	Snubber for pulsating services; siphon for steam service, isolating valves, 2VM	

There are various types of sensors/elements, namely, RTDs, THCs, and thermistors, the application of which depends mainly on the process temperature.

3.1.1.1 RTDs

The electrical resistance of a medium increases when the temperature rises. RTDs utilize this property of changing resistance of an electrically conducting wire with respect to change in temperature. RTDs are widely used in the temperature range of (−)250°C to 850°C. There are three types of RTDs normally in use: copper (53 Ω at 0°), platinum or PRT (100 Ω or 1000 Ω at 0°C) and nickel (100 Ω or 500 Ω platinum at 0°). Compared to copper and nickel RTDs, PRTs are in maximum use in power plants to measure water, oil, flue gas, and air temperatures less than ~400°C.

RTDs, as used for practical and industrial measurements, are available in three designs, one with a coil element, one that is wire-wound, and the other has a thin film element (Fig. 4.3A). They are able to measure temperature because of the physical principle of the positive temperature coefficient (PTC) of electrical resistance of metals. That means for higher temperatures, the electrical resistance between the end terminals of the element or device will increase.

- Coil element RTDs have a wire coil that expands freely as a "strain-free" design because it can withstand temperatures up to 850°C. The small platinum coil sensor looks like a filament of an incandescent lamp. The mandrel is a hard-fired ceramic oxide tube with equally spaced bores for inserting coil. These RTDs have been largely replaced by wire-wound types.
- Wire-wound RTD design is based on wire wrapping around an insulating mandrel with a round or flat shape. The coefficient of thermal expansion of both the mandrel and the sensor materials matches to avoid mechanical strain, which causes inaccuracy. The lead wire selection ensures that no emf is generated to cause measurement distortion. Temperature range of this type may go up to 660°C.
- Thin-film element RTDs are formed by depositing a thin layer (1–10 nm) of platinum (mostly) on a ceramic substrate and then coated with epoxy or glass for mechanical protection. This type of element is not as stable compared with the other types. The normal temperature range is 300°C, but may go up to 500°C by taking care of the "strain gauge" effect (mismatching of thermal coefficients).

The relationship between temperature and resistance is nearly linear, and over a wide range of temperatures RTDs have an infinitesimal response time of around a millisecond. They show extremely accurate behavior with resolution and measurement uncertainties (accuracy figured altogether) of ±0.1°C or better in special designs and are considered one of the most precise temperature sensors available. The other main advantage of using RTDs is stable output for a long period of time. They can be recalibrated easily and provide accurate readings even in relatively narrow temperature spans. Platinum is the preferred material for the most accurate measurement, because in its pure form, the temperature coefficient of resistance is almost linear with an accuracy of ±0.1°C, which can be readily achieved at moderate cost. Better accuracy is also possible, but instrument costs escalate to attain smaller error levels.

In a general industrial application, RTDs are encapsulated in a small tablet form within a metal sheath, which

TABLE 4.9 Typical Measuring Points for Temperatures (250 MW TPS)

SR. No.	Service Description	Operating/MCR Value (°C)	Calibrated Range (°C)
1	Mill outlet temperature	90	120
2	Mill inlet PA temperature	290	350
3	Mill and mill motor bearing temperature	85	150
4	ID fan lube oil temperature	35	50
5	ID fan and ID fan motor bearing temperature	85	150
6	I.D/FD/PA fan/mill motor winding temperature	110	150
7	ID fan hydraulic coupling working chamber oil temperature	35	50
8	FD fan lube oil temperature	35	50
9	FD fan and FD fan motor bearing temperature	85	150
10	PA fan and PA fan motor bearing temperature	85	150
11	FO temperature at FO pump discharge	100	200
12	FO temperature after pump and heating unit	120	200
13	FO temperature at burner front	120	200
14	FO temperature at return line	120	200
15	Auxiliary steam temperature for atomization	210	300
16	Flue gas temperature before primary superheater	800	1000
17	Flue gas temperature before final superheater	1150	1500
18	Flue gas temperature before final reheater	1050	1500
19	Flue gas temperature before primary reheater	800	1000
20	Flue gas at economizer inlet	550	1000
21	Flue gas temperature economizer outlet	345	500
22	Flue gas temperature at sec. air heater inlet	340	500
23	Flue gas temperature at secondary air heater outlet	130	200
24	Flue gas temperature at ID fan outlet	120	150
25	Flue gas temperature at primary air heater inlet	350	500
26	Flue gas temperature at primary air heater outlet	130	200
27	Secondary air heater inlet temperature		60
28	Secondary air heater outlet temperature	320	500
29	PA heater inlet header temperature		60
30	PA heater outlet temperature	320	450
31	Feed control station inlet feed water temperature	200	300
32	Economizer inlet feed water temperature	250	400
33	Main steam outlet temperature	540	600
34	HRH outlet temperature	540	600
35	CRH outlet temperature	330	600
36	Reheater inlet steam temperature after de-superheater	315	400

TABLE 4.9 Typical Measuring Points for Temperatures (250 MW TPS)—cont'd

SR. No.	Service Description	Operating/MCR Value (°C)	Calibrated Range (°C)
37	Drum outlet steam temperature	362	500
38	Steam at first stage supplemental heater inlet temperature	450	600
39	Drum outlet steam temperature	362	500
40	Steam at first stage supplemental heater inlet temperature	450	600
41	Steam at first stage supplemental heater outlet temperature	500	600
42	Steam at final supplemental heater outlet temperature	550	700
43	Superheater/reheater metal temperature		1200
44	Auxiliary steam temperature for SCAPH temperature control	200	300
45	Auxiliary steam temperature for soot blower	390	500
46	Main Steam at HPT turbine inlet temperature	545	1000
47	HRH Steam at IPT turbine inlet temperature	545	1000
48	Gland seal supply steam temperature (hot start)	450–530	1000
49	Gland seal supply steam temperature (cold start)	250	1000
50	FRF cooler outlet DMCW temperature	30	50
51	FRF cooler outlet FRF temperature	30	100
52	FRF cooler inlet DMCW temperature	50	
53	FRF temperature at FRF tank	40	100
54	BFP and BFP motor bearing temperature	85	150
55	BFP motor winding temperature	110	150
56	BFP fluid coup. working oil inlet temperature	35	60
57	BFP fluid coup. working oil outlet temperature	45	100
58	BFP fluid coup. lube oil inlet temperature	35	60
59	TG bearing drain oil temperature	55	100
60	TG lube oil cooler inlet temperature	50	60
61	TG lube oil cooler outlet temperature	35	60
62	Seal oil supply temperature	35	60
63	Generator casing cold hydrogen temperature		
64	Generator casing hot hydrogen temperature		
65	Stator coolant from generator temperature		
66	Stator coolant from generator temperature		
67	HP/LP housing temperature	60	100
68	CEP fan bearing temperature	50	75
69	Generator transformer oil temperature		
70	Generator transformer winding temperature		
71	Unit transformer oil temperature		
72	Unit transformer winding temperature		
73	Station transformer oil temperature		

Continued

TABLE 4.9 Typical Measuring Points for Temperatures (250 MW TPS)—cont'd

SR. No.	Service Description	Operating/MCR Value (°C)	Calibrated Range (°C)
74	Station transformer winding temperature		
75	Generator rotor temperature		
76	L.PH 1 outlet temperature	70	100
77	LPH 2 outlet temperature	84	120
78	LPH 3 outlet temperature	100	150
79	LPH 4 outlet temperature (if any)	120	150
80	Deaerator condensate outlet temperature	158	200
81	Deaerator steam temperature	365	500
82	HPH 6 outlet temperature	185	300
83	HPH 7 outlet temperature	210	300
84	HPH8 outlet temperature (if any)	235	300
85	Condenser shell inside temperature	40	100
86	Condenser extraction pump discharge temperature	47	100
87	LPH 1 bled steam temperature	82	150
88	LPH 2 bled steam temperature	135	200
89	LPH 3 bled steam temperature	190	300
90	LPH 4 bled steam temperature (if any)	250	400
91	Deaerator bled (IP extraction) steam temperature	365	500
92	HPH 6 bled steam temperature	475	600
93	HPH 7 bled steam temperature	330	500
94	HPH 8 bled steam temperature (if any)	390	600
95	BFP suction temperature	165	200

BFP, boiler feed pump; CEP, condensate extraction pump; CRH, cold reheat; FD, forced draft; FO, fuel oil; FRF, fire-resistant fluids; HP, high pressure; HPH, high-pressure heater; HPT, high-pressure, turbine;, ID, indirect draft;, IPT, intermediate pressure turbine; LP, low pressure; LPH; low-pressure heater; PA, primary air; RH, reheat; TG, turbo generator; DMCCW, demineralized water clarified cooling water; ACW, auxiliary cooling water; SCAPH, steam coil air preheater.

is further protected by a probe called a thermowell (discussed above). As per the theory of operation, they belong to the electrical component group and require electrical current to sense the change in resistance by a calibrated standard instrument.

The disadvantage of employing RTDs in comparison to THCs (described later) is mainly the smaller, overall temperature measuring range. Inherently, they are not made with rugged construction and are failure prone to higher vibration environments .The initial cost is also higher than other temperature-measuring sensors.

Because it is an electrical component, the RTD has lead wires that are required to be connected to the receiving device, such as a temperature transmitter. The lead wires have some inherent resistance that varies with the change in ambient temperature, which would introduce measurement error. The magnitude of error is greater with the long lead lengths required for remote temperature measurement in large-capacity power plants. There are many measuring systems employing three- and four-wire leads from the two end terminals of an RTD to minimize or limit such errors, or eliminate them altogether (Fig. 4.3B–D).

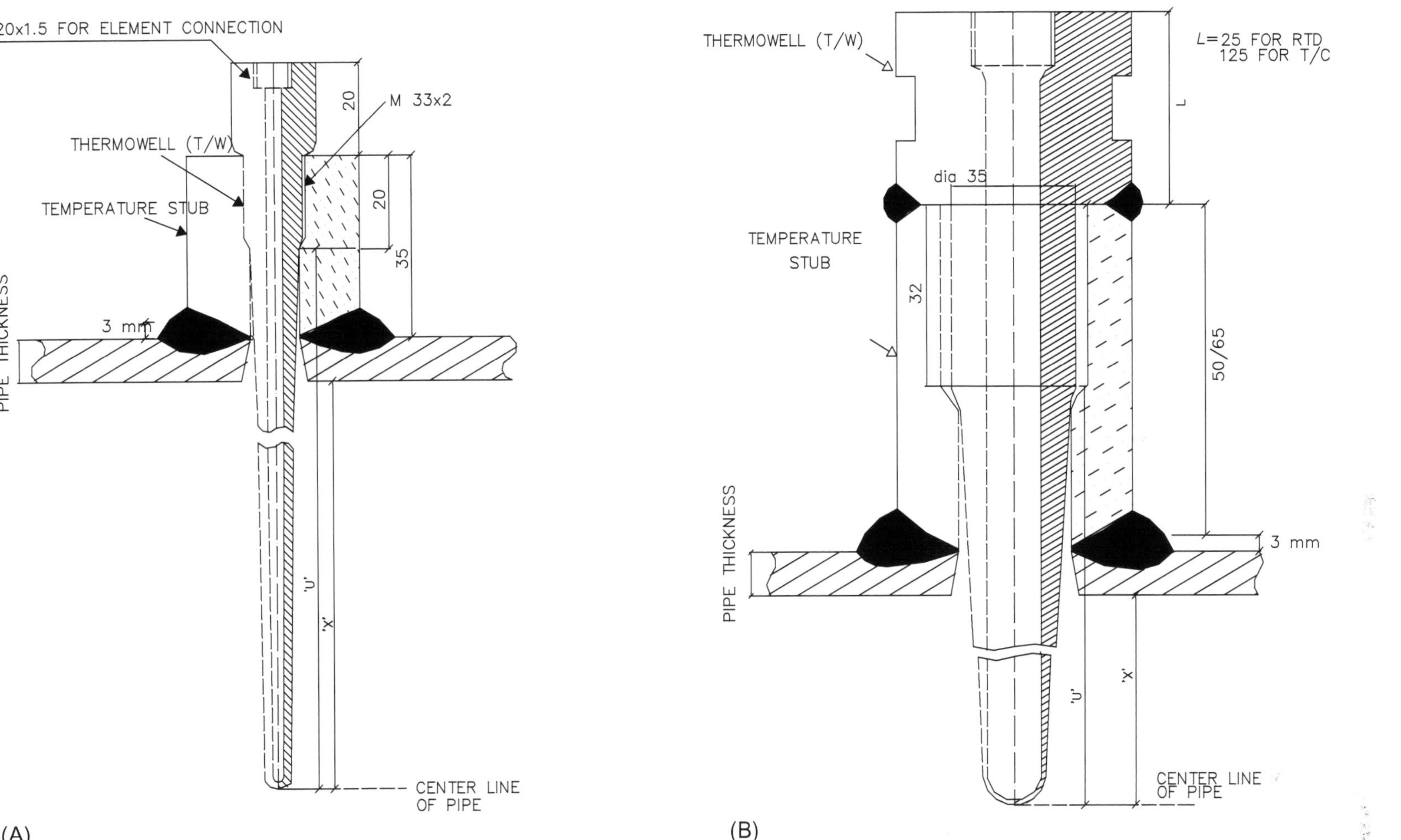

TAPPING THERMOWELL/ELEMENT AND T/C & RTD WITH SCREWED T/W DETAILS

TAPPING THERMOWELL/ELEMENT AND T/C & RTD WITH WELDABLE T/W DETAILS

NOTE:

(1) ALL DIMENSION ARE IN mm AND APPROXIMATE AND TENTATIVE.

(2) MATERIAL OF THERMO WELL CONSTRUCTION SHALL BE SS316 OR DEPEND ON PROCESS CONDITION.

(3) FOR PIPE OF 300 dia OR MORE ,INSERTION SHOULD NOT BE MORE THAN 150. FOR PIPE OF 300 dia OR LESS ,IMMERSION LENGTH IN INCREMENTS OF 25 TO BE SO SPECIFIED THAT TIP OF WELL REACH CENTER LINE OF PIPE +/−13. IMMERSION SHALL NOT BE LESS THAN 75 IN A PIPE OF 100dia OR SMALLER.

(4)THIS STANDARD IS NOT APPLICABLE FOR GAS AND AIR APPLICATION

(5) 'U' IS THE IMMERSION LENGTH, AND 'X' IS THE INSERTION LENGTH

FIG. 4.1 Thermowell/element details (steam, liquid).

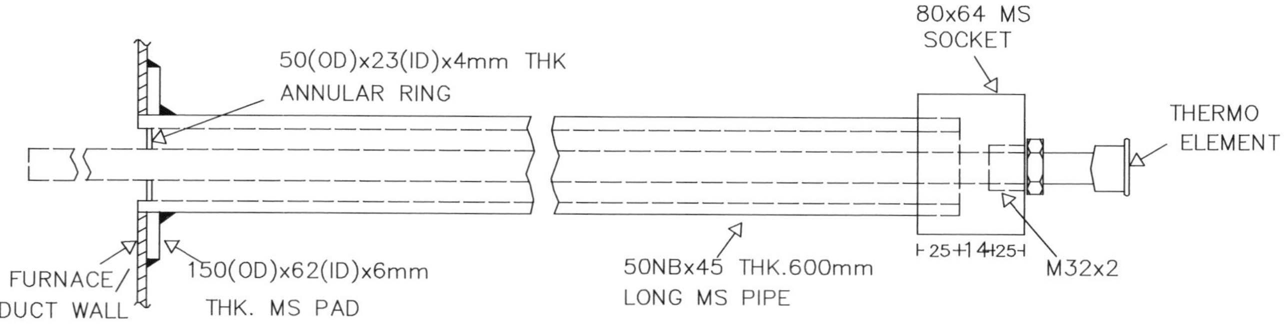

(A)

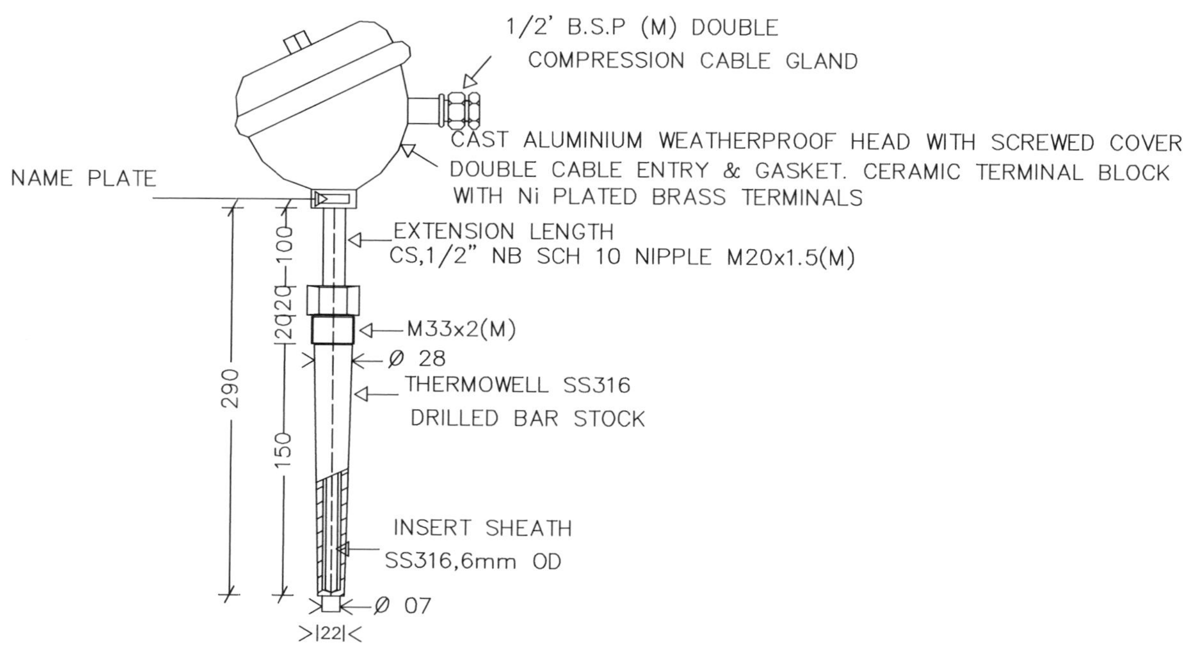

(B)

FIG. 4.2 Thermowell/element details (air, flue gas).

A common way to circumvent the long-lead resistance error problem is to place temperature transmitters near the measuring point. With this system, lead resistance would be minimized and their change in response to varying ambient temperatures is negligible. The temperature transmitters convert the sensor output to an analog current (4–20 mADC as a standard) and/or serial digital signal that can be transmitted over long distances by wire to the receiving device.

Another error of a comparatively minor nature, and often ignored, is the insulation resistance, which is also a function of temperature. At a reasonably higher temperature, the shunt resistance of the insulator may contribute an error into the measurement. This error is taken care of by calculating the effect of thermal properties of the insulator into the measuring circuit. There are different types of insulator material used, such as powdered magnesia (MgO), alumina (Al_2O_3), and similar compounds. These materials are carefully dried and sealed when encapsulated within the sheath along with the sensor element. ASTM E 1652-00 standard is related to insulation resistance testing to help determine the performance of these sealed elements.

PRTs are a costly element, but here are other provisions for RTDs that are less costly. They can be made cheaply in copper and nickel, although they have limitations in use due to practical problems. For example, the nickel RTDs show nonlinear characteristics, and therefore, can only be employed for restricted ranges. The problems with the copper RTDs are well known, such as oxidation of the wound sensor wire.

Platinum is the preferred material for precision measurement because in its pure form the temperature coefficient of resistance is nearly linear; enough so that temperature measurements within ±0.1°C can be readily achieved with moderately priced devices. Better resolution is possible, but equipment costs escalate rapidly at smaller error levels.

3.1.1.2 THCs

In 1821, renowned German scientist Thomas Johann Seebeck discovered that when any conductor is subjected to a thermal gradient, it will generate an emf. This is now known as the thermoelectric effect or the Seebeck effect. To measure this voltage, another conductor to the "hot" end has to be connected and the other terminal has to be brought back to the cold end. This additional conductor will also experience the temperature gradient and will develop an emf of its own in the same direction as that of the former one. As the two conductors of the same polarity are connected to form a hot junction, the emf generated in them will cause them to oppose each other. Fortunately, the magnitude of the generated emf also depends on the type of metal used.

Therefore, if a dissimilar metal is used to complete the circuit, the two conductors generate different emfs and a small difference in emf between the two cold junction terminals is measured. That difference increases with temperature, and is between 1 and 70 μV/°C) for standard metal combinations.

THCs utilize the thermoelectric property of generating emf in mVs between two open cold ends when two dissimilar metal wires are solidly connected at the end and subjected to higher temperatures than the other open ends. In other words, a THC is a junction between two different metals that produces an emf (mV) related to a temperature difference. THCs are a widely used type of temperature sensor for measurement of indication, recording, controls, etc.

THCs are of rugged construction, less costly, interchangeable, and can measure a wide range of temperatures. The major disadvantage of THCs is a limitation in measurement accuracy. Normally it is extremely difficult to achieve an accuracy of <1°C. THCs regularly used include: iron-constantan (type J), copper-constantan (type T), chromel-alumel (type K), and platinum-platino rhodium [two types with 10% (S type) and 13% (R type) of platinum in the alloy]. The thermal characteristic curve temperature is in mVs, and the accuracy and range of these THCs are depicted in Fig. 4.3E. From this figure it can be inferred that ISA type K shows a linear characteristic over a large temperature range and produces more mV/°C change in temperature. For this reason, ISA type K gets used more often in power plant applications.

Any junction of dissimilar metals will produce an electric potential (mV) related to temperature, but it may not be practical to use all of them as THCs, which must have a predictable and repeatable relationship between temperature and millivolt. For practical measurement of temperature, THCs are made of junctions of specific alloys with the previously mentioned qualities. Different alloys are used for different temperature ranges as seen in Table 4.10. Other important properties such as resistance to corrosion/oxidation may also be considered while choosing a type of THC; for example, iron-related alloys are prone to formation of oxides and are easily corroded.

3.1.1.2.1 Voltage-Temperature Relationship in THC

The nonlinear relationship between the temperature difference (ΔT) and the output voltage (mV) of a THC can be approximated by a polynomial:

$$\Delta T = \sum_{n=0}^{N} a_n v^n$$

The coefficients a_n are given for n from 0 to between 5 and 13 depending upon the metals. In some cases, better accuracy is obtained with additional nonpolynomial terms A database of voltage as a function of temperature and coefficients for computation of temperature from voltage and vice versa for many types of THCs is available online.

In modern equipment this equation is usually implemented in a digital controller or stored in a reference table; whereas, the older devices use analog circuits. In some electronic instruments linearization of nonlinear outputs of the THC is accomplished to improve the precision and accuracy of measurements.

3.1.1.2.2 THC Connection to Instruments If the measuring tapping point is away from the receiving instrument or device, an intermediate connection, as mentioned earlier, should be made by extension cable wires of the same materials as the main THC wire to avoid any intermediate junction, but this cable is costly. The alternative method is to use a less expensive compensating cable similar to the extension cable. Calibrations of THCs are usually standardized against a reference temperature of 0°C or 273 K. Receiving end instruments or devices integrally incorporate the temperature measurement of the environment and take care of any change in measuring temperature with respect to a reference temperature of

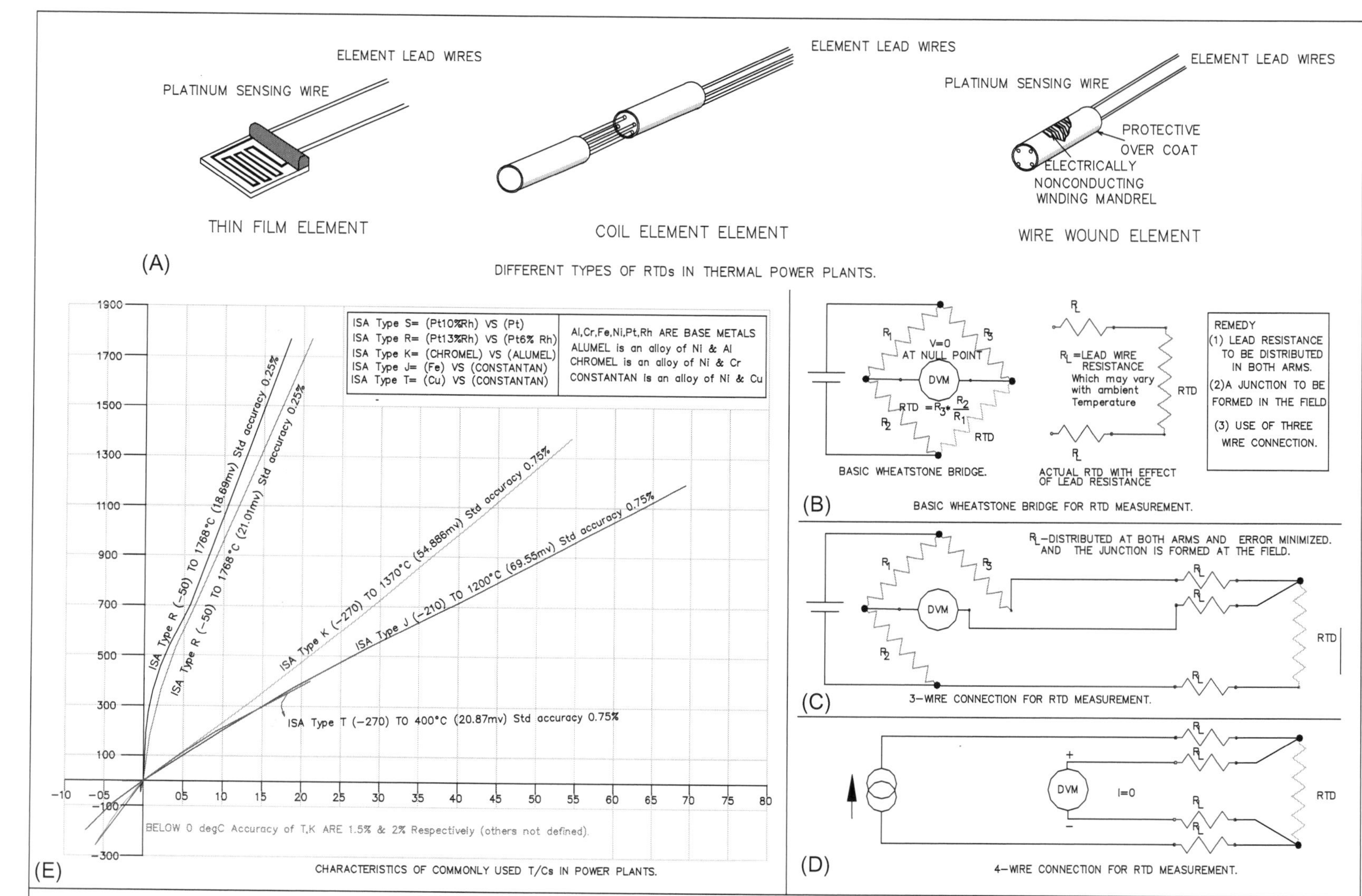

FIG. 4.3 Temperature element theory and practice.

TABLE 4.10 Polynomial Coefficients for K Type THC (0–500°C)

n	1	2	3	4	5	6	7	8	9
a^n	25.08355	7.860106 $\times 10^{-2}$	(−) 2.503131 $\times 10^{-1}$	8.315270 $\times 10^{-2}$	(−) 1.228034 $\times 10^{-2}$	9.804036 $\times 10^{-4}$	(−) 4.413030 $\times 10^{-5}$	1.057734 $\times 10^{-6}$	(−) 1.052755 $\times 10^{-8}$

0°C. Practical instruments use electronic methods to adjust the effect of varying temperatures at the instrument terminals. Therefore, cold junction compensations are necessary to eliminate the effect of variable ambient temperatures if the THC cold-end terminals are terminated in open ambient conditions for further connection to receiving instruments by cables other than extension/compensating cables.

THCs are widely used in thermal power plant applications for almost all high-temperature measurements as well as for kilns, gas turbine exhausts, diesel engines, and other industrial processes.

3.1.1.2.3 Cold Junction Compensation THCs measure the temperature difference between two points, not absolute temperature. As mentioned earlier, THCs are standardized against the reference temperature, which is 0°C. To measure a single temperature, one of the junctions (normally the cold junction) is maintained at a known reference temperature other than °C, for example, the control room of an air-conditioned environment. The other (hot) junction is at the temperature to be sensed. With a junction of known temperature, the mV temperature difference between the cold and reference junctions at °C can be obtained from the table and the appropriate correction applied. This is known as cold junction compensation. It is worth noting for understanding of cold junction compensation methods, that the emf (or the mV) is not generated at the junction of the two metals of the THC, but rather along that portion of the length of the two dissimilar metals that is subjected to a temperature gradient.

There are several methods of cold junction compensation:

- One method depicts termination of extension wires from various THCs in the vicinity (as a cost-saving measure) in a local junction box maintained at a constant temperature, thereby incorporating an artificial cold junction using a thermally sensitive device such as a thermistor or diode to measure the temperature of the cold junction box. From the junction box, less costly copper cable can be drawn up to the input terminals at the instrument with special care taken to minimize any temperature gradient between terminals. The voltage from a known cold junction can be incorporated in the measuring circuit. If temperature gradients are the same (which they should be), for these two copper wires from a THC, the millivolt generated by them due to difference in cold and reference junctions will cancel each other.
- Another method uses a THC at the cold junction box and an extension wire up to the measuring instrument, and adds this voltage to all other THC output of that cold junction box by suitable methods.

Temperature ranges of different THCs are given in Table 4.11 (but may be extended to upper and lower sides as depicted in Fig. 4.3E).

Table 4.12 describes properties of several different THC types. Within the tolerance columns, T represents the temperature of the hot junction, in °C. For example, a THC with a tolerance of $\pm 0.0025 \times T$ would have a tolerance of ±2.5°C at 1000°C.

THC junction types and response times are depicted in Fig. 4.4.

TABLE 4.11 Temperature Ranges of Different THCs

Types of THCs	Iron-Constantan (Type J)	Copper-Constantan (Type T)	Chromel-Alumel (Type K)	Platinum-Platino Rhodium (13%) (Type R)	Platinum-Platino Rhodium (10%) (Type S)
Temperature range in °C	−0 to 700	−185 to 300	0–1100	0–1600	0–1600
Used for measurement parameters	Flue gas	Air and oil	Steam, air/flue gas Temperature >300°C	Furnace Temperature ~1000°C	Furnace Temperature ~1000°C

TABLE 4.12 Properties and Color Codes of THCs

Type	Temp Range (Continuous) (°C)	Temp Range (Short Term) (°C)	Tolerance Class One (°C)	Tolerance Class Two (°C)	IEC Color Code	BS Color Code	ANSI Color Code
K	0 to +1100	−180 to +1300	*1	**1			
J	0 to +700	−180 to +800	*2	**2			
N	0 to +1100	−270 to +1300	*1	**1			
R	0 to +1600	−50 to +1700	*3	**3			Not defined
S	0 to 1600	−50 to +1750	*3	**3			Not defined
B	+200 to +1700	0 to +1820	Not available	**4	No standard use copper wire	No standard use copper wire	Not defined
T	−185 to +300	−250 to +400	*4	**5			
E	0 to 800	−40 to 900	*5	**6			

Tolerances class one (°C): 0 to +800 and −40 to +900: *1, ±1.5 between −40 and 375°C and ±0.004 × T between 375 and 1000°C; *2, ±1.5 between −40 and 375°C and ±0.004 × T between 375 and 750°C; *3, ±1.0 between 0 and 1100°C and ±[1 + 0.003 × (T − 1100)] between 1100 and 1600°C; *4, ±0.5 between −40 and 125°C and ±0.004 × T between 125 and 350°C; *5, ±1.5 between −40 and 375°C and ±0.004 × T between 375 and 800°C. Tolerances class two (°C): **1, ±2.5 between −40 and 333°C and ±0.0075 × T between 333 and 1200°C; **2, ±2.5 between −40 and 333°C and ±0.0075 × T between 333 and 750°C; **3, ±1.5 between 0 and 600°C and ±0.0025 × T between 600 and 1600°C; **4, ±0.0025 × T between 600 and 1700°C; **5, ±1.0 between −40 and 133°C and ±0.0075 × T between 133 and 350°C; **6, ±2.5 between −40 and 333°C and ±0.0075 × T between 333 and 900°C.

3.1.1.3 Thermistors

Thermistors were discovered first in 1833 by Michael Faraday while working on the semiconducting behavior of silver sulfide. Faraday noticed that the resistance of silver sulfide decreased dramatically as temperature increased. Today thermistors are used as temperature elements, but were difficult to produce earlier and applications were limited. Commercial production of thermistors did not begin until 1930.

A thermistor is a type of resistor whose resistance varies with temperature; that is, thermistors show qualities similar to RTDs. But when used as a temperature element, thermistor characteristics are in the reverse direction, that is, the resistance of the element decreases when the temperature rises. Thermistors also differ from RTDs with respect to the material used; a thermistor is generally made of ceramic or polymer, while RTDs are made from pure metals. Regarding temperature response, RTDs are better as they are useful over larger temperature ranges, whereas thermistors have a temperature in the range of −90 to 130°C.

Assuming a first-order approximation and that the relationship between resistance and temperature is linear, then:

$$\Delta R = k\Delta T$$

where ΔR is the change in resistance, ΔT is the change in temperature, and k is the first-order temperature coefficient.

Thermistors can be classified into two types, depending on the sign of k. If k is positive, the resistance increases with increasing temperature, and the device is called a PTC thermistor, or posistor. If k is negative, the resistance decreases with increasing temperature, and the device is called a negative temperature coefficient (NTC) thermistor. Resistors that are not thermistors are designed to have a k as close to zero as possible (smallest possible k), so that their resistance remains nearly constant over a wide temperature range.

- Thermistors are used as resistance thermometers in low-temperature measurements of the order of 10 K, that is, −263.2°C.
- Thermistors can be used as inrush current-limiting devices in power supply circuits. They present a higher resistance initially, which prevents large currents from flowing at turn-on, and then heat up and become much lower resistance to allow higher current flow during normal operation. These thermistors are usually much larger than measuring type thermistors, and are purposely designed for this application.
- Thermistors are regularly used in automotive applications. For example, they monitor things like coolant temperature and/or oil temperature inside the engine and provide data to the ECU and, indirectly
- Thermistors are also commonly used in modern digital thermostats and to monitor the temperature of battery packs while charging.

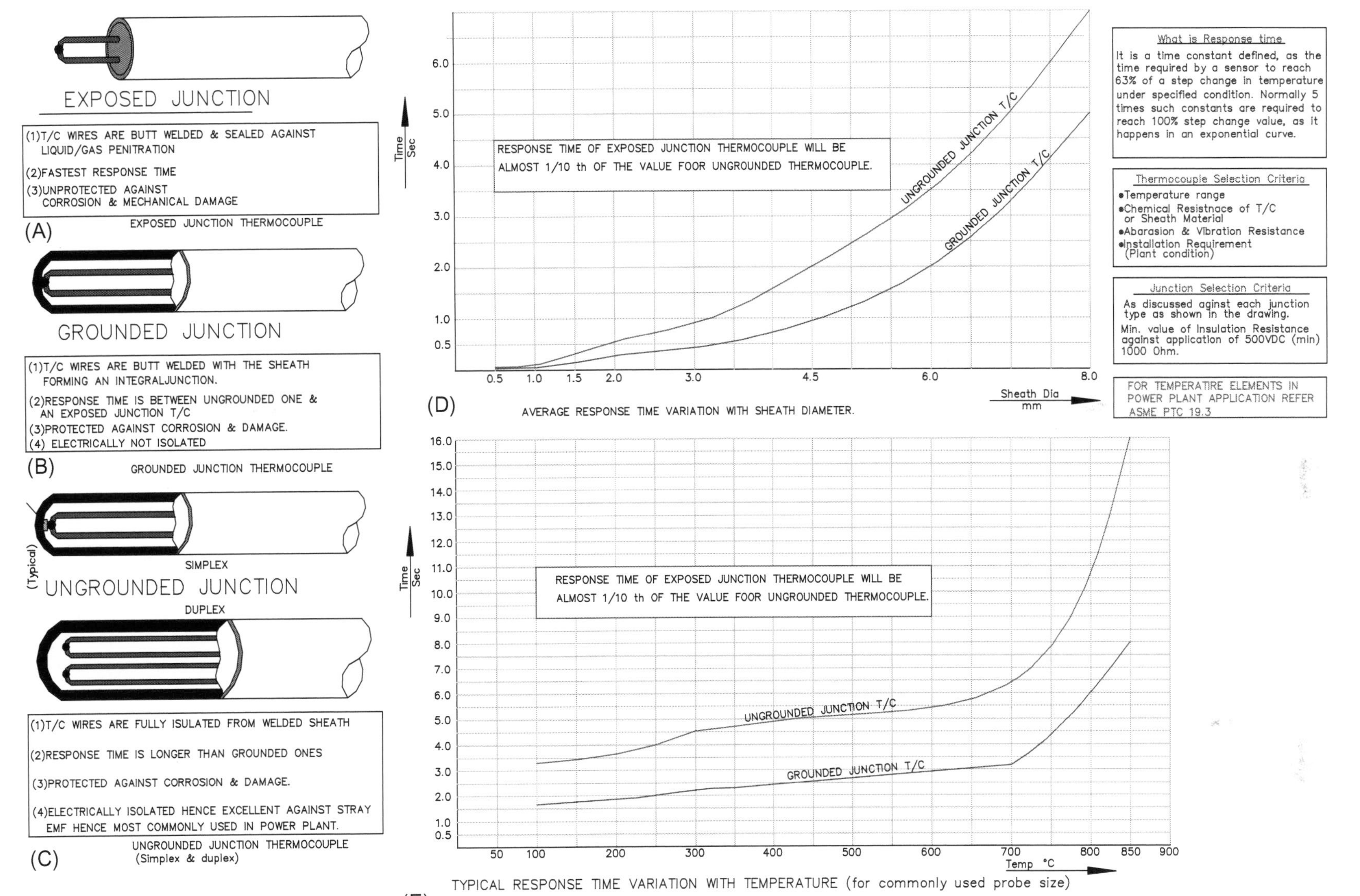

FIG. 4.4 Junction types and response time for THCs.

- PTC thermistors or posistors can be used as current-limiting devices for circuit protection as replacements for fuses. Current through the device causes a small amount of resistive heating. If the current is large enough to generate more heat than the device can lose to its surroundings, the device heats up, causing its resistance to increase and causing even more heating. This creates a self-reinforcing effect that drives the resistance upward, reducing the current and voltage available to the device.

3.1.2 Immersion Length

Immersion length is the length of the temperature element's portion that comes into contact with the measuring fluid. Insertion length is the length of the temperature element's portion that is inserted into the measuring fluid beyond the pipe or vessel wall. Thus the immersion length consists of the insertion length, plus pipe or vessel wall thickness, plus the length of temperature stub up to which the fluid can fill the gap between the stub and thermowell (Fig. 4.1A and B).

Normally the insertion length should be such that the tip of the element would be around the center (that is the radius of the pipe) when process fluid passes through a pipe. There is some flexibility in the industrial arena; the tip can be away by a maximum ± 13 mm from the center so that it can still measure the temperature of the fluid with the tip in the maximum velocity zone. However, for a bigger pipe, the maximum limit of insertion is 150 mm to avoid vibration and damage of the element due to abrasion and erosion. For ducts or vessels with low pressure, the insertion length may be more to suit the application. The minimum insertion length is normally 75 mm. For smaller pipes, the minimum length is maintained with the element mounted in the elbow or expander with suitable length or on a stub fixed at an angle of $\sim 45°$ with the pipe.

3.1.3 Connection Types

See Fig. 4.2B for a better understanding of connection details. The elements with a built-in sheath are provided for primary mechanical protection to the fragile elements. The sheath has a male thread for connection with the thermowell with rugged construction normally manufactured from solid bar stock with suitable material to suit the service condition. For low pressure, however, the thermowell may be fabricated. Some extension length to the element/sheath is provided after it comes out of the piping insulation to avoid high temperatures through conduction near the electrical connection of the element for remote instrumentation. There is a connection facility on the top of the arrangement called an element head. The head is weatherproof made of quality material with a screwed cover and suitable gasket. The cover is protected from free falling by a chain fixed to the body of the element. The output wires of the elements are connected firmly at the ceramic terminal block consisting of nickel- or cadmium-plated terminals clearly marking the terminal with universally accepted signs or symbols.

3.2 Temperature Transmitter: Working Principle, Specification, Supplier, and Special Features

The temperature transmitter is an instrument or device used to sense a weak, low-level temperature in the form of either resistance or millivolts, convert it to a universal output (i.e., 4–20 mA) and/or digital signals (namely HART/Fieldbus signal) representative of the sensed temperature, and transmit for remote instrumentation. Similar to the HART/Fieldbus signal, this universal output is a high-level analog signal in the form of a current signal, and much more suitable for long-distance transmission through a noisy environment. It is also ready for direct interface with many instrumentation items, such as an indicator, recorder, PLC, DCS, and SCADA system. Current and digital signals are immune to resistance in field-connecting cables or extension wires and stray electrical noise rather than resistance or voltage. This allows accurate signals to be received from a far-away sensor. The 4–20 mA output or digital signal and DC power share the same wire pair. It is not necessary to run power wires to every sensor. Temperature transmitters also linearize the temperature input signal, making them a good and acceptable low-cost signal conditioners.

In large power plants, temperature transmitters are used to control process parameters as well as for plant protection, real-time indication, hard or soft copy recording, analyzing, and historical archiving, by sensing a temperature of the process and transmitting the information to a remote location for different subsystems envisaged by the end user. The temperature transmitter output can be communicated over the loop to a control room, or the output can be communicated to other process subsystem devices so that the process can be monitored and controlled. The temperature transmitter may be located on the top of the sensor (head mounted) or in the field near the sensor, both of which provide minimum effect of variation in ambient temperature. The other location may be in the back of the control panel or rack mounted with due consideration of ambient temperature.

Temperature transmitter types are grouped in two broad categories for RTDs and THCs. There are several methods for making each of these devices.

3.2.1 Transmitter (RTD Input)

The RTD is basically a coil element/wire-wound/thin-film resistance with two end terminals. There are various types that use two-, three-, or four-wire measuring systems (Fig. 4.3B–D) depending upon the accuracy level desired and the expected cost. The two-wire system has one wire from each end and the three-wire system has one wire from each end and another wire from any end that goes to the measuring circuit. A two-wire system does not provide correct output due to variation in ambient temperature, as the resistance of the lead wires (both sides) changes unpredictably. The three-wire system is a better way to take care of this problem. The four-wire system has two wires from each end that go to the measuring circuit, which to a great extent eliminates the problem of variation in ambient temperature. In all of the wire measuring systems, the RTD is ultimately connected to a bridge circuit to generate a voltage signal.

There is another four-wire measuring system that is more accurate, because it almost completely eliminates the ambient temperature variations. The transmitter injects a precise and controlled current into the temperature sensor by two wires, and with the other two wires, the resultant voltage drop across the temperature sensor is used to measure resistance in a circuit, which provides high input impedance so that minute current flows. With almost zero level current, the lead resistance change will introduce minimum effect, and voltage drop across the RTD will be read as an open circuit voltage, even at the remote location. The voltage is then converted into a digital format using an analog-to-digital converter provided by a microprocessor. The microprocessor converts the measured voltage into a digital value representative of temperature. The temperature transmitter generally includes housing and a temperature probe, which attaches to the housing. To monitor a process temperature, the transmitter includes a sensor, such as an RTD or a THC. An RTD changes resistance in response to a change in temperature. By measuring the resistance of the RTD, the temperature can be calculated.

3.2.2 Transmitter (THC Input)

A thermocouple is placed in the process fluid and provides an output voltage (in mV) in response to a temperature change in the process fluid. Typically, the temperature transmitter is located in the field location and converts the millivolt input signal to 4–20 mADC transmitted to a control room. The connection from sensor to transmitter is made by either a THC extension cable or by a compensating cable for the reasons discussed in Clause 3.1 of this chapter.

Specifications for temperature transmitters are given in Tables 4.13 and 4.14.

3.3 Temperature Switches

As described in Chapter 1, Clause 3.2.4, most of the temperature switches utilize the volume expansion characteristics of fluid after application of heat or the temperature in turn. The bulb containing transmission fluid is placed at the sensing point .The transmission fluid may be an inert gas like nitrogen or a liquid like mercury or alcohol. The receiving side may be bellows, diaphragm, Bourdon, helix, capsule, or a spiral-type pressure sensing element.

Bimetallic elements are also used. They will bend in one direction when heat is applied to an assembly of two metals of dissimilar coefficients of expansion joined together. One of these is Invar, a nickel alloy that hardly expands, and another nickel alloy is used that expands considerably. They are welded together and rolled to the desired thickness to form a bimetallic element. The modern type uses a multiple helix or coil within a coil arrangement to accommodate space limitation. Temperature sensors/elements provided within the bulb, are located separately at the sensing point inside the process, and isolated from the process fluid to avoid contact from a high velocity of abrasive and/or hazardous nature. This isolation is achieved through a thermowell, which takes care of the process-side complications and is normally threaded and/or welded to the process pipe or vessel. Inside the thermowell, there is a female thread to accommodate the sensors/elements.

Because the temperature switch is a mechanism that is responsive to temperature changes, the movement from expansion of fluid or bending of dissimilar metal coupled together is amplified to open or close a switch by using electrical contacts when a predetermined minimum or maximum temperature is sensed by the responsive element. Temperature-sensitive switches are used for thermal protection of the process or to keep an electrical circuit from overheating . If a device gets too hot or cold (i.e., fuel oil required for burning), the temperature-sensitive switch can be used to generate an alarm or trip the system as per the requirement/severity of the process.

A closed system containing an equilibrium of liquid and vapor of a suitable, volatile chemical or gas is used to sense the process temperature. If the fluid is liquid or vapor, the expected volume expansion of fluid is converted into pressure as the volume of the system is fixed. This pressure within the system is proportional to the applied temperature and is applied to the bellows or diaphragm, Bourdon, helix, capsule, spiral, etc., which transmit a force, proportional to the temperature, to an operating bar or beam. The beam is restrained by an adjustable spring. When the force on the bar/beam overcomes the spring tension, it moves and operates a switch or switches. On reduction of the applied temperature, the force applied to the bar/beam also falls

TABLE 4.13 Data Sheet of Temperature Transmitters (RTD)

SL. No.	Item	Specification	Special Feature
1	Make/ model/type	ABB/YEC/Honeywell/Emersion	
2	Type of input	30WIRE, RTD (Pt.−100)	
3	Process connection	1/2″ Electrical connection	
4	Enclosure	Die cast aluminum, IP 65	
5	Local indicator	(Process/sensor or loop current value) LCD, five-digit, seven-segment sign and floating point Seven-digit alphanumeric characters 6 mm, 14 segments Response time 0.5 s, measuring error ±0.1%	
6	Output	4–20 mADC (maximum load resistance = 750 Ω) with superimposed digitized signal (HART protocol or Fieldbus signals such as Profibus and Foundation Fieldbus	
7	Power supply	8.5–30 V DC	
8	Monitoring	Sensor and self-diagnostic	
9	Signal refresh time	<0.4 s	
10	Long-term stability	<0.1% PA	
11	Test	Calibration test	
12	Electrical isolation	1.5 kV AC	
13.	Ambient temperature	−40 to 85°C	
14	Humidity	100 RH	
15	Range	As applicable	

and the beam is restored to its original position by the spring and the switch resets.

The switches are grouped into two categories, capillary and rigid stem. The rigid stem type is mounted directly to the process and the capillary type is mounted separately to a nearby wall or local panel.

It's important that the process temperature should not exceed the specified maximum working temperature. Normally accuracy is better than ±0.5%, which is the same as that of the master gauge used. As these switches are meant for outdoor installations, manufacturers make them available with sufficient protections against aggressive environments, for example, explosive gas-laden air, dust, temperature, and water. The instrument is mounted rigidly to avoid shock and vibration so that spurious actuation does not take place.

The case or enclosure must be weatherproof or explosion proof to suit the site condition. The following standards are normally followed in India:

1. Weatherproof to IP 66 IS 2147.
2. Flameproof cum weatherproof (die cast aluminum) CMRI approved to GR.IIA and IIB of IS 2148 1981 and weatherproof to IP 66. GR.IIC.
3. Hydrogen service may be required for a generator cooling circuit.

The corresponding international code, corresponding to CMRI certificates, is also applicable .

Regarding switch configuration and contact rating, the contents of Clause 2.2 are applicable; Table 4.15 is a data sheet of temperature switches.

TABLE 4.14 Data Sheet of Temperature Transmitters (THC)

SL. No.	Item	Specification	Special Feature
1	Make/model/type	ABB/YEC/Honeywell/Emersion	
2	Type of input	Type K/T/R/S/J	
3	Process connection	1/2″ Electrical connection	
4	Enclosure	Die cast aluminum, IP 65	
5	Local indicator	(Process/sensor or loop current value) LCD, five-digit, seven segment sign and floating point. Seven-digit alphanumeric characters 6 mm, 14 segments Response time 0.5 s, measuring error ±0.1 %	
6	Output (Cold junction compensated)	4–20 mADC (maximum load resistance = 750 Ω) with superimposed digitized signal (HART protocol or Fieldbus signals such as Profibus and Foundation Fieldbus	
7	Power supply	8.5–30 V DC	
8	Monitoring	Sensor and self-diagnostic	
9	Signal refresh time	<0.4 s	
10	Long-term stability	<0.1% PA	
11	Test	Calibration test	
12	Electrical isolation	1.5 kV AC	
13	Ambient temperature	−40 to 85°C	
14	Humidity	100 RH	
15	Range	As applicable	

3.4 Temperature Gauges

A temperature gauge, commonly called a thermometer, is a device that indicates the temperature of the process being monitored. The display can be a pointer on a graduated dial or a digital display. THCs, which generate millivolts depending on the temperature, are also used for digital thermometers. By measuring temperature, the thermal condition of a homogenous system is available. Measurement of temperature normally requires direct or indirect (through thermowell) contact between the process and the sensor in such a manner that the sensor responds quickly to the temperature of the process.

Temperature gauges like temperature switches utilize the volume expansion characteristics of fluid after application of heat or the temperature in turn. A bulb containing fluid (transmission fluid) is placed at the sensing point. The transmission fluid may be inert gas like nitrogen or liquid like mercury or alcohol. The receiving side may be bellows, a diaphragm, Bourdon, helix, capsule, or a spiral-like pressure-sensing element or simply a glass tube with a graduated scale. Bimetallic elements are also used that bend in one direction when heat is applied to an assembly of two metals of dissimilar coefficients of expansion joined together, as stated in the Clause 3.1. Suitable measures have been provided to correct errors from changes in casing and capillary expansion due to ambient temperature change. To minimize the capillary expansion effect, an Invar rod is inserted inside the capillary to reduce the fluid volume. The capillary is well protected against mechanical damage.

The operating principles are the same as those for temperature switches.

3.4.1 Mercury in Glass

It is the simplest type of temperature gauge there is a sensing glass bulb, basically a small volume reservoir chamber, filled with mercury and placed inside the process. After being heated, the mercury expands much more than the glass (about six times) and is forced to rise along the glass tube. The glass tube is a small tube with a fine bore joined at the opposite end of the bulb. For each particular temperature, the mercury rises to a certain point. The expansion path is then calibrated with a reference temperature bath.

TABLE 4.15 Data Sheet of Temperature Switches

SL. No.	Item	Specification	Special Feature
1	Make/model		
2	Type	Capillary/rigid stem	
3	Process/electrical connection	M33X2 for thermowell, M20X1.5 for element 1/2″ electrical connection	
4	Enclosure	Die cast aluminum, IP 65(as applicable)	
5	Sensing material	316 SS bulb for fluid-filled type Nickel alloys (Invar and other) for bimetallic type	
6	Setpoint	Low/high (adjustable)—1 or 2 setpoints	
7	Repeatability	±1% FSR	
8	Switch unit	2 SPDT	
9	Rating	5 Amps, 230 V AC	
10	Test	Calibration test	
11	Range	As applicable	
12	Medium	Air/water/steam/oil	
13	Ambient temperature	−40 to 85°C	
14	Humidity	100 RH	

The glass tube is not marked but the graduations are engraved on metal scales with both tube and scale enclosed in a single casing. Mercury in some glass thermometers is calibrated for complete immersion and some with partial immersion. The fluid can be alcohol instead of mercury, which has the coefficient of linear expansion even six times more than mercury.

3.4.2 *Mercury or Alcohol in Steel*

The principle of operation is exactly same as in Clause 3.1, but has a different final output. In the switch, the output movement of the element is to operate a switch assembly but in the gauge, the output movement is connected to a pointer which moves on a calibrated and graduated scale

3.4.3 *Bimetallic Gauges*

This is one of the most dependable temperature-measuring instruments. It employs a suitable metal or metal alloy with the required coefficient of linear expansion. One of them is Invar, which expands little upon application of heat. Welded together with another metal (alloy) that has a better coefficient of linear expansion and rolled to the desired thickness, the combination is used in bimetallic gauges. Alloys with widely differing coefficient of linear expansion values are used for short temperature ranges, whereas for longer temperature ranges, metals (alloys) of closer coefficient of linear expansions are used. The term used to describe the thermal activity of a bimetal is flexibility. The actual movement of a bimetallic strip is proportional to its flexibility. One end of a straight strip is fixed, and the other end of the strip deflects or bends, which varies proportionally with the temperature, to the square of its length and inversely to the thickness. It can be used in a helical configuration, and where the angular deflection is determined by the same factors. The pointer attached to the free end is then allowed to show its movement on a properly calibrated scale.

The modern bimetallic gauges/thermometers use multiple helical arrangements or coils within coils. This configuration allows the use of long bimetallic elements in a small space. The fixed end is normally located at the bottom, whereas the free end is located on a shaft to which the pointer is attached. The industrial gauges have a thicker stem. Alarm contacts with secondary pointers for indicating maximum and minimum temperatures settings are also available. A typical specification sheet of temperature gauges is indicated in Table 4.16.

4 FLOW MEASUREMENT, VARIOUS MEASURING POINTS, VARIOUS TYPES, AND RANGE SELECTION

Flow measurement in a power plant is a crucial parameter. Whenever the word "total" is used, the mass flow comes into

TABLE 4.16 Data Sheet of Temperature Gauges

SL. No.	Item	Specification	Special Features
1	Make/model		
2	Range	As applicable	
3	Type	Alcohol/mercury in glass/steel Bimetallic	
4	Process/electrical connection	M33X2 for thermowell,M20X1.5 for element 1/2″ electrical connection	
5	Style	Capillary/rigid stem	
3.	Medium	Gas/air/water/steam	
4	Accuracy	±0.5% FSD	
5	Ambient temperature	−40 to 80°C	
6	Humidity	100 RH	
7	Case	In die cast aluminum alloy moisture proof Case epoxy powder-coated black	
8	Dial size	150 mm	
9	Quantity	As applicable	

the picture. Everything like ultimate power output, efficiency of the power plant, total enthalpy, etc., depends on the mass flow of steam, coal, and oil. Various kinds of flow meters are shown in Fig. 4.5 with selection guidelines from permanent pressure loss, discharge coefficient variations, etc.

4.1 Units of Measurement

There are two basic types of flow measurement: volumetric and mass flowrates. Both gas and liquid flow can be measured in volumetric as well as in mass flowrates. It is well known that a mass of a particular homogeneous material can be calculated by multiplying density of the material with the volume of that material. That means these measurements can be converted between one another if the material's density is known. For liquid, the density is almost independent of the liquid conditions like pressure, although it varies slightly with the temperature. For gas services, the density is a dependent variable of both pressure and temperature and, to a lesser extent, the gas composition.

For liquids, various units are used depending on the application and industry, and also on the country of the users. In most countries the metric (MKS) system is used: tonne/hour (h) or /h (or second) for mass flow and liters/h or cubic meter/h for volume flow. In the United States or UK the flow unit is ton/h or gallons per minute (GPM).

The scenario is different for gases as they are compressible fluids and change volume when undergoing a change in pressure and/or temperature. For a definite amount of mass, a volume of gas under one set of pressure and temperature conditions is not equal to the same gas under different conditions. Therefore, whenever a gas flowrate is involved, references must be indicated to denote the pressure and temperature at which condition this particular volume is applicable. Generally three norms are followed: STP, NTP, ISA.

- *STP*: Defined by the International Union of Pure and Applied Chemistry (IUPAC) as air at 0°C (273.15 K, 32°F) and 10^5 Pa (1 bar). It is commonly used in the imperial and American systems of units with air at 60°F (520°R) and 14.696 psia (15.6°C, 1 atm).
- *NTP*: Defined as air at 20°C (293.15 K, 68°F) and 1 atm (1.0325 bar; 101.325 kN/m^2, 101.325 kPa, 14.7 psia, 0 psig, 29.92 in Hg, 760 torr). Density 1.204 kg/m^3 (0.075 lb/ft^3).
- ISA: Defined to 101.325 kPa (1 atm, 1.0325 bar), 15°C and 0% humidity.

$$[1\,\text{Pa} = 10^{-6}\,\text{N/mm}^2 \times 10^{-5}\text{bar} = 0.1020\,\text{kPa/m}^2 \\ = 1.02 \times 10^{-4}\,\text{m}\,\text{H}_2\text{O} = 9.869 \times 10^{-6}\,\text{atm} \\ = 1.45 \times 10^{-4}\,\text{psi}\left(1\,\text{bf/in}^2\right)]$$

The following important measuring points are normally provided for a comprehensive audit of the flow measurements in a typical 250 MW thermal power plant with two-stage superheater attemperation, flue gas recycling from a chimney inlet to furnace hopper, and emergency reheater attemperation (Table 4.17).

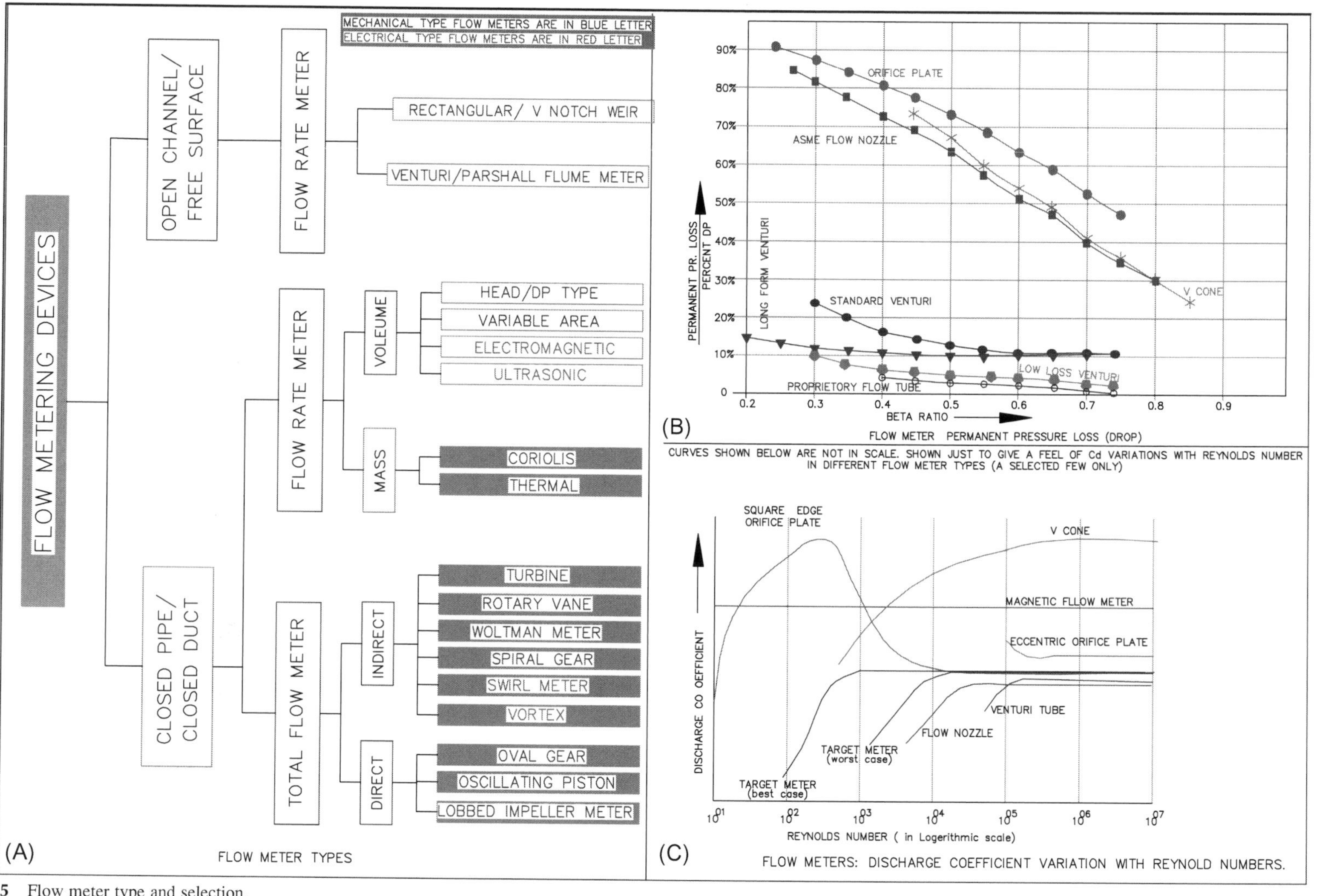

FIG. 4.5 Flow meter type and selection.

TABLE 4.17 Typical Flow Measurement Points (250 MW TPS)

SL. No.	Service Description	Operating/MCR Value	Calibrated Range
1	Total air flow	1800 T/h	2600 T/h
2	Secondary air flow to burner	320 T/h	0–450 T/h
3	Primary air flow to mill	320 T/h	0–450 T/h
4	Flue gas recycling flow	250 T/h	0–350 T/h
5	Main steam flow	750 T/h	0–1000 T/h
6	Feed water flow before feed control station	800 T/h	0–1200 T/h
7	RH steam flow	700 T/h	0–1000 T/h
8	RH attemperation flow	40 T/h	0–60 T/h
9	First stage attemperation flow	140 T/h	200 T/h
10	Second stage attemperation flow	100 T/h	0–150 kg/cm^2
11	Condensate flow after gland steam cooler outlet	800 T/h	0–1200 T/h
12	Gland steam supply header flow	2 T/h	0–3 T/h
13	RFW tank to condenser (in surge)	50 T/h	0–75 T/h
14	Condenser to RFW tank (out surge)	120 T/h	0–175T/h
15	LPH 3 drain pump discharge flow	35 T/h	0–49 T/h
16	LPH 4 condensate to deaerator flow	0–60 T/h	0–90 T/h
17	Raw water flow	3500 M^3/h	0–5000 M^3/h
18	CCW flow to the TG package	2250 M^3/h	0–3200 M^3/h
19	Cooling tower makeup flow	1400 M^3/h	0–2000 M^3/h
20	Stator coolant flow at generator inlet	0–55 T/h	0–80 T/h

CCW, component cooling water; LPH, low-pressure heater; RFW, reserve feed water; RH, reheater; TG, turbo generator.

4.1.1 Pressure-Based Measurements (Head Type Instruments)

In modern and large-capacity power plants, flow measurements are normally achieved through flow elements based on the Bernoulli's principle, thereby necessitating a differential producer as briefly indicated in Chapter 1. The flow transmitter then converts the DP to a 4–20 mADC signal corresponding to flow with a superimposed digitized signal (HART) or other signal through different types of fieldbuses.

Various types of flow meters have been developed, based on Bernoulli's principle, by measuring the DP across a well-designed constriction called a "differential producer." They are popularly called pressure-based instruments. The velocity from the volume flow is proportional to the square root of the DP thus produced. This DP is not permanent. At the down edge of the element, ~60%–50% of this DP is recovered while the rest of the 40%–50% is unavailable as permanent pressure loss. The following clauses are examples of these flow elements.

4.1.1.1 Venturi Meter

This constricts the flow in such a fashion that the cross section through the travel gradually decreases from the original one and then becomes steady for a certain length, and again increases downstream up to the original cross section. This method is widely used to measure flowrate in the transmission of gas through pipelines, and has been used since the Roman Empire. The coefficient of discharge of Venturi meters ranges from 0.93 to 0.97.The permanent loss is about 40% of the DP produced and eventually the second lowest among this category. This type of meter can be used for both circular and rectangular cross sections. Also there are two categories/versions of this device: classic long form and short form.

4.1.1.2 Flow Nozzle

This is a modified, shorter version of the Venturi meter that is suitable for circular cross sections with converging-diverging profiles without any straight portion in between. Details are discussed in Clause 4.1.

4.1.1.3 Orifice Plate

This is a plate with a hole through it, placed in the flow path, which allows only fluid to pass through it. The flow path has a smaller cross-sectional area than the original pipe/conduit area, and gets constricted. This helps to develop the DP across the plate and measurements of corresponding flowrate can be obtained. It is basically a crude form of Venturi meter, but with higher permanent pressure from energy losses. There are three types of orifices: concentric, eccentric, and segmental. Krell's orifice is also available with a circular target at the center and annular flow encircling it. There are various types of tapping styles available with all these elements.

4.1.1.4 Dall Tube

This is another version of a shortened Venturi meter. Dall tubes are widely used for measuring the flowrate of medium and large pipes when the pressure drop is lower compared with an orifice plate. It has the lowest permanent pressure loss compared with other differential producers.

4.1.1.5 Pitot Tube

This is a DP-measuring instrument used to measure additional stagnation fluid pressure equivalent to flow velocity with respect to the static pressure by impinging the flow straight to the other side or back of the tube wall after entering through the hole. Bernoulli's equation is used to calculate the dynamic pressure and hence, fluid velocity.

4.1.1.6 Piccolo Tube

This follows the same principle as the pitot tube. Here there are multiple holes in a vertical line facing the flow. Having multiple holes, there is less probability of being affected by choking in a dust-laden service.

4.1.1.7 Multihole Pressure Probes

Multihole pressure probes (also called impact probes) are based on the same theory as pitot tubes. Depending on the type of application, this probe consists of three or more holes on the measuring tip arranged in a specific pattern. More holes allow the instrument to measure the direction of the flow velocity in addition to its magnitude (after appropriate calibration). Three holes arranged in a line allow the pressure probes to measure the velocity vector in two dimensions. Introduction of more holes, e.g., five holes arranged in a "plus" formation, allow measurement of the 3D velocity vector.

4.1.1.8 Aerofoil

This follows the operating principle of the Venturi meter. The construction of the element depicts multiple Venturi meters covering the total cross section of the flow path and they are connected/welded to each other by sealing the path in between so the flow path is divided and passes through each individual Venturi-like passage.

4.1.1.9 V Cone Flow Device

This design incorporates a contour-shaped cone at the center of the pipe with annular passages for the flow passage. The V cone flow element is a particularly accurate (as high as 0.5%) flow device under a head-type flow measurement for both dirty as well as clean fluid measurements. It has better permanent pressure loss compared with orifice plates. Details are discussed in Clause 4.1.7.

4.1.2 Mechanical Flow Meter

A few examples of mechanical flow meters include: variable area meter, piston meter/rotary piston, oval gear meter, turbine flow meter, paddle wheel meter, nutating disk meter, Woltmann meter, single and multiple jet meter. More details are available in Clause 4.2.

4.2 Flow Elements: Selection and Sizing

Flow elements are basic type of instruments that provide an obstacle in the flow path. Some flow measurements, normally for the closed pipe/duct, work on the principle of DP produced across flow elements. Some flow elements work on the principle of level rise over the flow elements for the open channel flow- measuring system.

4.2.1 Flow Measurement by DP Method

The downstream pressure after an obstacle will always be lower than the upstream pressure, producing a pressure drop across the obstacle cum flow element. All these types of flow elements are differential producers and require a DPT to measure the pressure drop and convert it to a compatible linear signal. Flow elements like orifice, nozzle, and Venturi meters constitute a part of this category and ae based on the Bernoulli equation (Fig. 4.6). To have a better understanding it is necessary to look at the Bernoulli equation.

The Bernoulli equation can be expressed as follows:

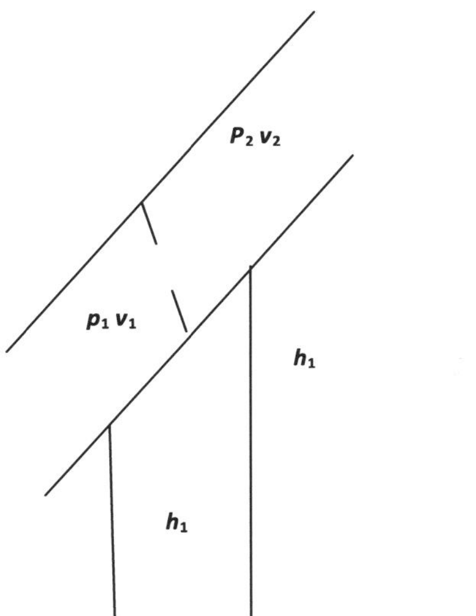

FIG. 4.6 Bernoulli's equation explanation.

$$p_1 + 1/2\rho v_1^2 + h_1 = p_2 + 1/2\rho v_2^2 + h_2 \quad (4.1)$$

where p is the pressure, ρ is the density, v is the flow velocity, and h is the elevation heights.

For simplicity's sake, a horizontal flow (neglecting minor elevation differences between measuring points) is assumed by eliminating the upstream and downstream elevation heights h_l and h_2.

The Bernoulli equation can then be modified to the following:

$$p_1 + 1/2\rho v_1^2 = p_2 + 1/2\rho v_2^2 \quad (4.2)$$

The continuity equation can be utilized in this connection by further assuming uniform velocity profiles in the upstream and downstream flow. The equation as expressed is as follows:

$$q = v_1 A_1 = v_2 A_2 \quad (4.3)$$

where q is the flowrate and A is the flow area.

Combining Eqs. (4.2), (4.3), assuming $A_2 < A_1$, gives the "ideal" E:

$$q = A_2\left[2(p_1 - p_2)/\rho\left\{1-(A_2/A_1)^2\right\}\right]^{1/2} \quad (4.4)$$

From the above equation it is evident that for a given area and area ratio, that is, geometry of the pipe/tube /duct, the flowrate can be determined by measuring the pressure difference $p_1 - p_2$.

The actual flowrate, however, is less than the theoretical flowrate q due to various geometrical conditions.

The ideal equation (4.4) thus can be modified with a discharge coefficient:

$$q_{act} = C_d X(q)$$

or

$$q_{act} = C_d A_2\left[2(p_1 - p_2)/\rho\left(1-(A_2/A_1)^2\right)\right]^{1/2} \quad (4.5)$$

where C_d is the discharge coefficient and q_{act} is the actual flowrate.

The discharge coefficient C_d is a function of the actual minimum jet size, flow element orifice opening, and the area ratio $= A_{vc}/A_2$ where A_{vc} = area in "vena contracta." Vena Contracta is the minimum jet area that appears physically just downstream the restriction. The viscous effect is usually expressed in terms of the nondimensional parameter Reynolds Number (*Re*). It is obvious that the velocity of the fluid will be maximum and the pressure will be minimum at vena contracta following the Bernoulli and continuity equations. At the end of this created flow pattern disturbance, the unrest settles down with a steady flow after a short distance from the vena contracta. From this point the velocity of fluid decreases to the same level as before the obstruction (considering the same cross-sectional area of the flow path before and after the flow element). The pressure, however, regains a value near but lower than the pressure before the obstacle and incurs a permanent pressure loss to the flow, which bears a relation in percentage to the measuring pressure drop $(p_1 - p_2)$ across the flow element.

Eq. (4.5) can be modified with diameters as follows:

$$q_{act} = C_d\pi/4d^2\left[2(p_1 - p_2)/\rho\left(1-\beta^4\right)\right]^{1/2} \quad (4.6)$$

where d is the orifice, Venturi, or nozzle bore/throat diameter, D is the upstream and downstream pipe diameter, $\beta = d/D$ diameter ratio or beta ratio, and $\pi = 3.14$.

Mass flow for fluids can be obtained by simply multiplying Eq. (4.6) with the density:

$$m_{act} = C_d\pi/4d^2\rho\left[2(p_1 - p_2)/\rho\left(1-\beta^4\right)\right]^{1/2} \quad (4.7)$$

Consideration must be given regarding pressure reduction and change in density of the fluid while measuring the mass flow in gases. The above formula is only applicable with the limitations of relatively small changes in pressure and density.

4.2.2 The Venturi Meter

As already discussed in Clause 4, the Venturi meter constricts the flow by gradual change in cross-section throughout the meter travel. The cross-section decreases from the original inlet and then becomes steady for a certain length and again increases downstream up to the original cross section (Fig. 4.7).

This long-used method is widely used to measure flowrate in the transmission of gas through pipelines. The discharge coefficient, C_d, of the Venturi meter is comparatively high to other similar types of flow elements and ranges from 0.93 to 0.975. The permanent pressure loss is ~40% of the

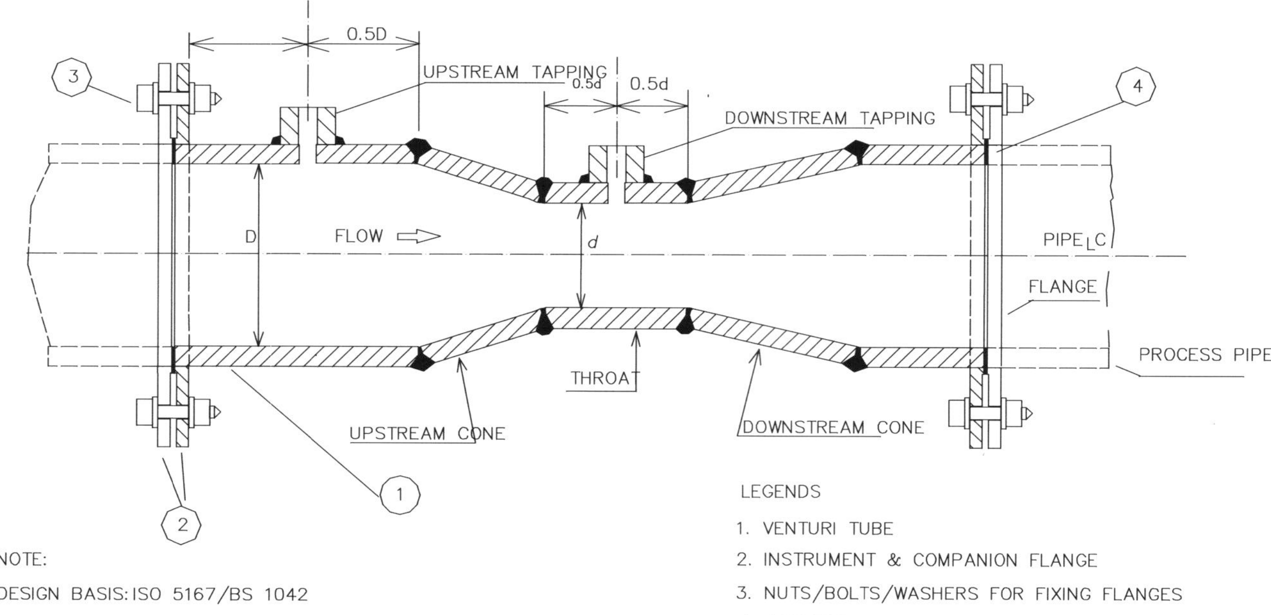

FIG. 4.7 Longitudinal cross-sectional view of Venturi tube.

DP produced and eventually is just higher than the lowest flow meter, that is, the Dall tube. These types can be used both for circular and rectangular cross sections. In the Venturi meter the fluid is accelerated throughout the inlet due to its converging nature as a cone with an angle of ~15°–20° and the pressure difference between the upstream side of the cone and the throat (minimum cross-sectional area) is measured and provides a signal for the rate of flow.

The fluid slows down in a cone with a smaller angle (5°–7°) where most of the kinetic energy is converted back to pressure energy. Because of the cone and the gradual reduction in the area there is no vena contracta. High-pressure recovery makes the Venturi meter suitable for the system or process where small pressure heads are available and/or only low permanent pressure loss is admissible .

A discharge coefficient of $C_d = 0.975$ can be indicated as standard, but the value varies noticeably at low values of the Reynolds number (Table 4.18).

Venturi tube meters are used in thermal power plants for medium temperature and low-pressure gas flow services, for example, flue gas, primary and secondary air flow, mill air flow, etc. Some manufacturers use Aerofoil, which is a modified version of the Venturi tube as described earlier.

Additional features of Venturi type flow elements include:

1. The Venturi tube is practically suitable for all the services like clean, dirty, and viscous liquid and some slurry services.
2. Straight length before and after the element required varies from 5 to 20 times diameter pipe length (upstream) and 5–10 times diameter pipe length (downstream) depending on the layout and installation of items (regarded as constriction) in the vicinity of the element's upstream and downstream pipe length.

TABLE 4.18 Different Values of Discharge Coefficient C_d (for Flow Nozzle) in Relation to Different Values of β Ratio and Reynolds Number (Re)

Beta Ratio $\beta = d/D$	Reynolds Number	Reynolds Number	Reynolds Number	Reynolds Number
0.2	0.968	0.988	0.994	0.995
0.4	0.957	0.984	0.993	0.995
0.6	0.95	0.981	0.992	0.995
0.8	0.94	0.978	0.991	0.995

3. Typical accuracy is 1% or even lower than full scale range.
4. Rangeability is as high as 4 is to 1.
5. Viscosity effect is high.
6. Relative cost is medium.

4.2.3 Flow Nozzle

As briefly discussed earlier, the flow nozzle is a modified version of the Venturi tube with a comparatively shorter length (long radius nozzle and Venturi nozzle with longer lengths are also used occasionally) flow element (Figs. 4.8 and 4.9). Flow nozzles are used in thermal power plants for high-temperature and high-pressure steam and water flow services, for example, main steam, hot and cold reheat steam, auxiliary steam, feed water and low pressure, temperature services like various condensate flow, etc.

The differences compared with the Venturi tubes include:

1. Flow nozzles are suitable for circular cross section only.
2. Flow nozzles are also designed with a converging-diverging profile but any straight portion in between is not required.

Flow nozzles are used to determine the flowrate of the fluid through pipes with these three different types:

1. ISA 1932 nozzle: This type of flow element was developed in 1932 by the International Organization for Standardization (ISO). The ISA 1932 nozzle is commonly used outside the United States.
2. Long radius nozzle: This type is a variation of the ISA 1932 nozzle.
3. Venturi nozzle: This is a hybrid type of flow element with an inlet convergent clause similar to the ISA 1932 nozzle and a divergent section similar to a Venturi tube flow meter.

Flow elements like the pitot tube (Fig. 4.10) and the Parshall flume (Fig. 4.11; both discussed later in this chapter) are also used for air and open channel water flow measurements, respectively, although they are not popular in power plant applications.

Listed below are additional features of Venturi type flow elements:

1. The flow nozzle is recommended for both clean and dirty liquids.
2. Straight length before and after the element required varies from 10 to 30 times diameter pipe length (upstream) and 5–10 times diameter pipe length (downstream) depending on the layout and installation of items (regarded as constriction) in the vicinity of the element's upstream and downstream pipe length.
3. Typical accuracy is 1%–2% of full-scale range
4. Rangeability is 4 to 1.
5. Viscosity effect is high.
6. Relative cost is medium.
7. Relative pressure loss is medium.

4.2.4 Orifice Plate

The orifice plate consists of a flat circular plate with an outer diameter greater than the inner diameter of the measuring fluid pipe and a thickness of $\geq$5 mm as per the line pressure and material used. A circular (or a circular segmental) hole is drilled in it which may not be centrally located, for example, an eccentric orifice plate. There is a pressure tapping point upstream from the orifice plate and another just downstream. There are generally four methods of placing the tapping points. The value of discharge coefficient C_d of the meter depends upon the position of tapping points.

1. *Flange tapping*: Tapping locations are on the orifice holding flanges, 1 in. upstream and 1 in. downstream the inlet face of the orifice.
2. *Vena contracta tapping*: Tapping location 1 pipe diameter (actual inside) upstream and 0.3–0.8 pipe diameter downstream the face of the orifice.
3. *D and D/2 tapping*: Tapping location 1 pipe diameter (actual inside) upstream and 0.5 pipe diameter downstream the face of the orifice.
4. *Pipe tapping*: Tap location 2.5 times nominal pipe diameter upstream and 8 times nominal pipe diameter downstream the face of the orifice.

Carrier ring type orifice plates have been contemplated for small diameter flow pipes where normal flange tapping (i.e., 1 in. upstream and 1 in. downstream the inlet face of the orifice) may be difficult to achieve. The carrier ring is a pair of thick plates/disks suitably grooved to hold the orifice plate in place in between them and make room for drilling flange tapping points. Fig. 4.12 indicates the longitudinal cross-sectional view of the carrier ring type orifice plate.

The discharge coefficient C_d varies considerably with changes in beta ratio (β) and the Reynolds number (Table 4.19). A discharge coefficient of $C_d = 0.60$ may be standard, but the value varies noticeably at low values of the Reynolds number (Fig. 4.5C).

The pressure recovery is limited for an orifice plate and the permanent pressure loss depends primarily on the area ratio. For an area ratio of 0.5, the head loss is about 70%–75% of the orifice DP. Orifice plate connections are suitable for flange connections other than carrier ring configurations. The flanges may be the weld type with a weldable neck (Fig. 4.13) or it may be a common type connected to the main process pipe through the counter flange.

There are classifications of orifice plate, such as square-edge orifice (most commonly used), segmental orifice, and eccentric orifice plate—as shown in Fig. 4.14.

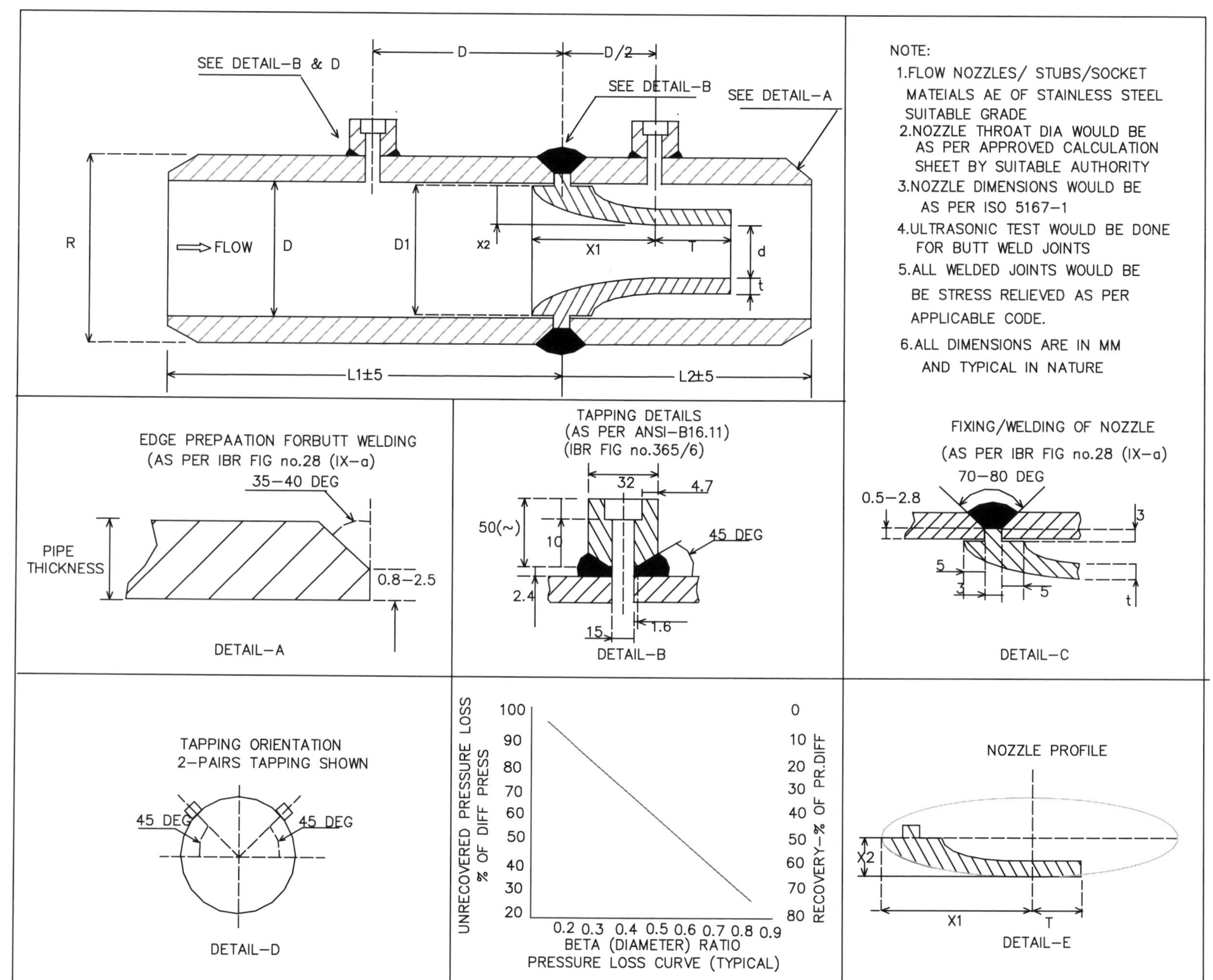

FIG. 4.8 Longitudinal cross-sectional view of flow nozzle assembly.

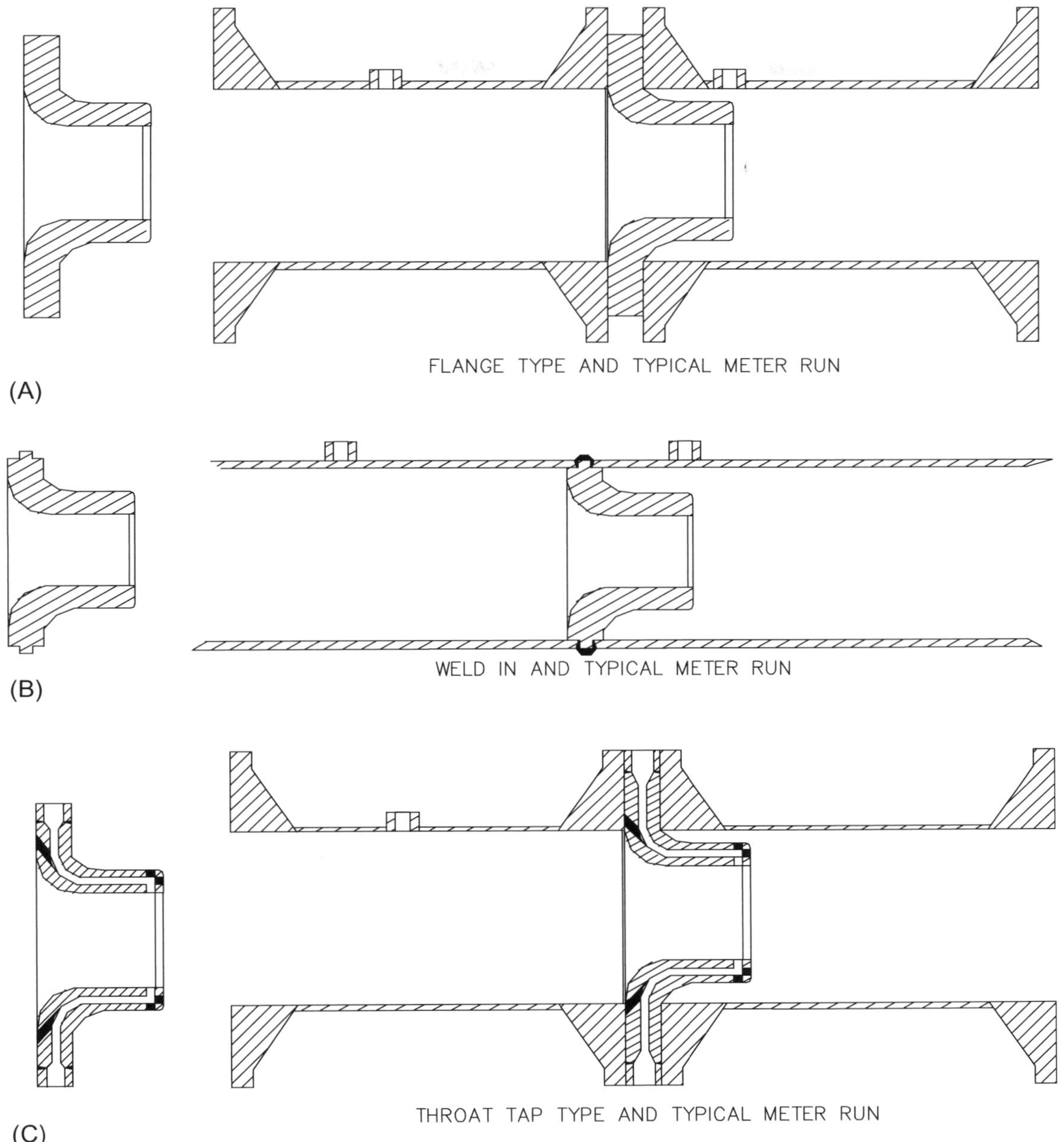

FIG. 4.9 Flange or weldable type with D, D/2, or throat tapping of flow nozzle.

Additional features of orifice plate type flow elements include:

1. The orifice meter is recommended for clean and dirty liquids and some slurry services.
2. Straight length before and after the element required varies from 10 to 35 times diameter pipe length (upstream) and 5–20 times diameter pipe length (downstream) depending on the layout and installation of items (regarded as constriction) in the vicinity of the element's upstream and downstream pipe length.
3. Typical accuracy is 2%–4% of full-scale range.
4. Rangeability is 4 to 1.
5. Viscosity effect is high.
6. Relative cost is medium.
7. Relative pressure loss is medium.

The following standards are followed:

1. American Society of Mechanical Engineers (ASME) 2001: Measurement of fluid flow using small bore precision orifice meters, ASME MFC-14M-2001.
2. International Organization of Standards (ISO 5167-1-2003): Measurement of fluid flow by means of pressure

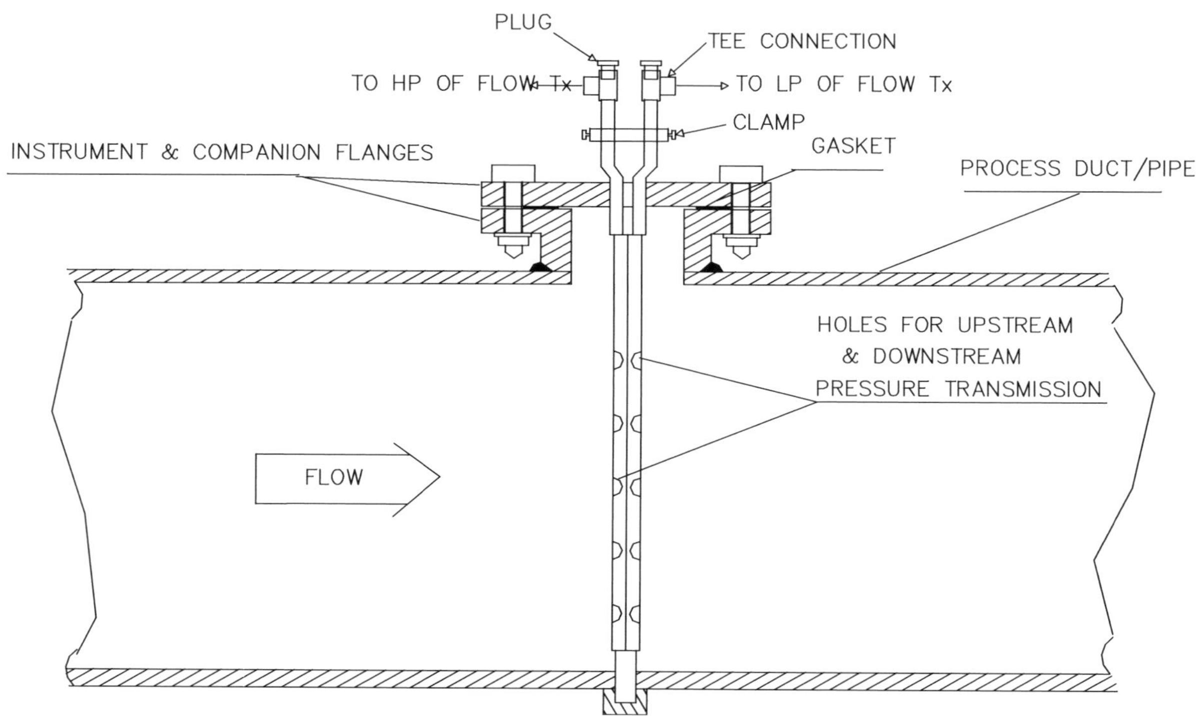

FIG. 4.10 Schematic diagram of the construction of a pitot tube.

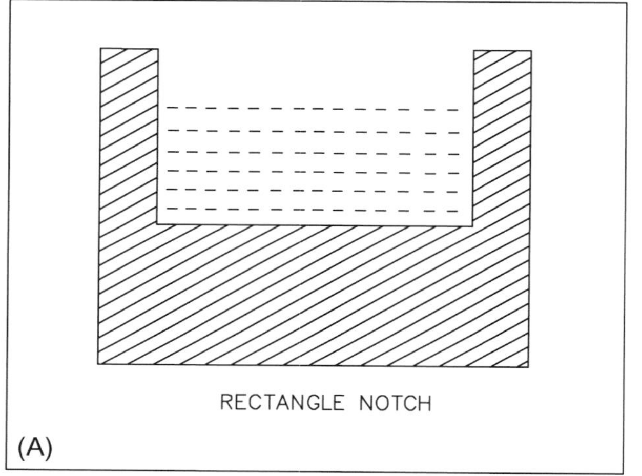

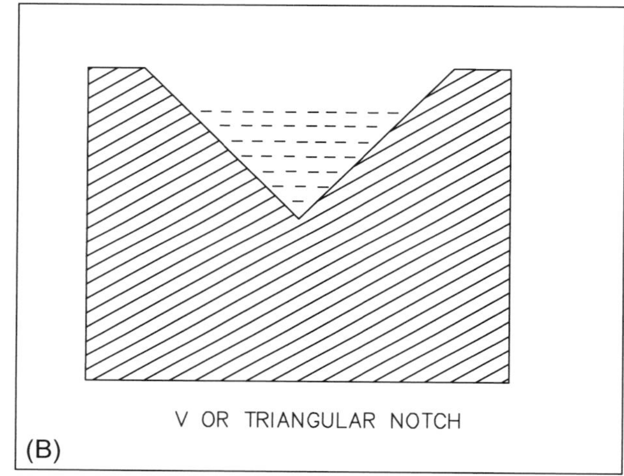

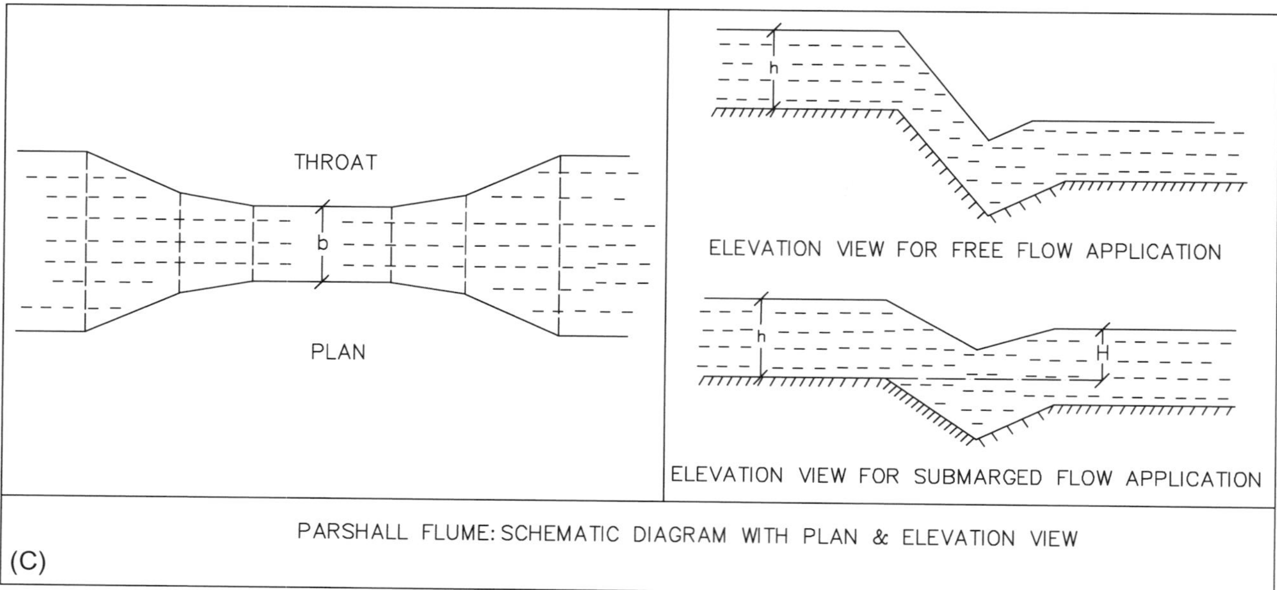

FIG. 4.11 Open channel flow-measuring elements.

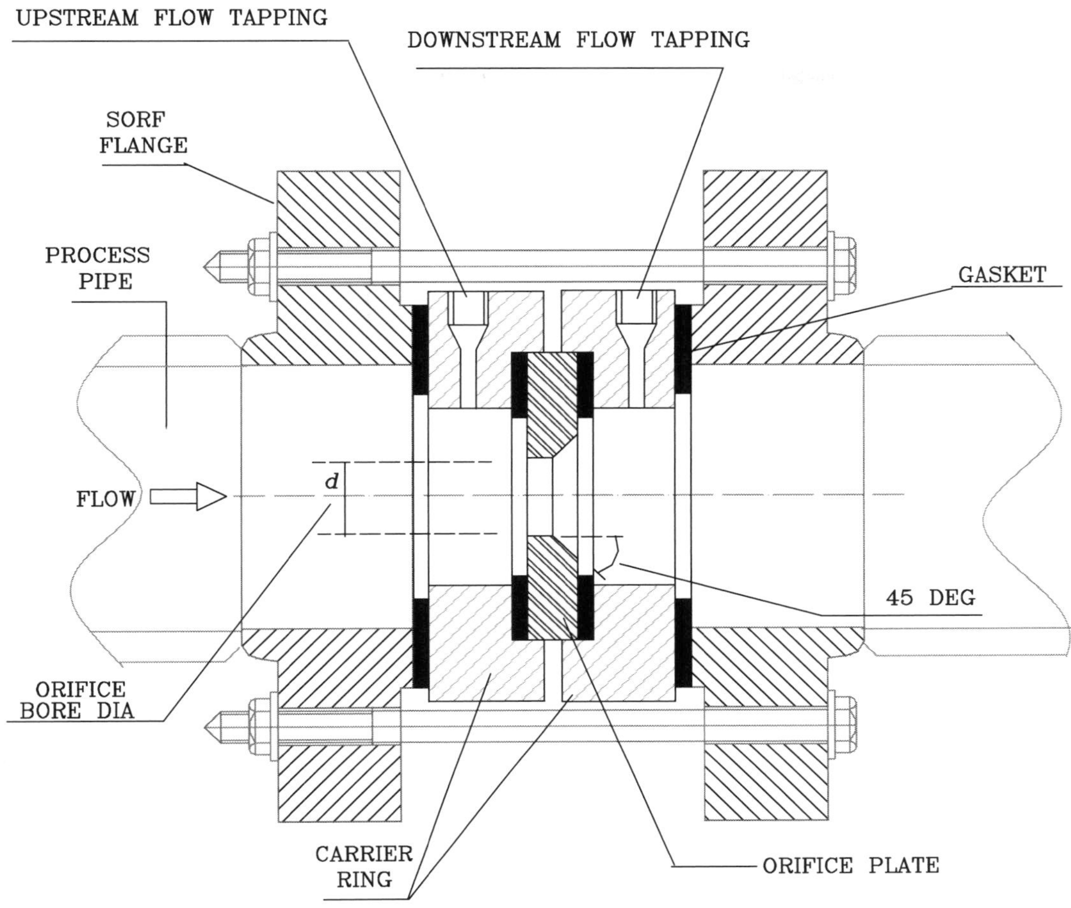

FIG. 4.12 Longitudinal cross-sectional view of carrier ring type orifice plate assembly.

differential devices. Part 1: Orifice plates, nozzles, and Venturi tubes inserted in circular cross-section conduits running full.

3. International Organization of Standards (ISO 5167-1) Amendment 1 1998: Measurement of fluid flow by means of pressure differential devices. Part 1: Orifice plates, nozzles, and Venturi tubes inserted in circular cross-section conduits running full. Reference number: ISO 5167-1:1991/Amd. 1:1998(E).
4. ASME B16.36–1996 Orifice Flanges, ASME FED 01 Jan 1971 Fluid Meters—Their Theory and Application, Sixth Edition.
5. BS 1042 and BS EN 5167-1.

4.2.5 Krell's Bar Orifice Plate

Krell's bar orifice plate was designed by an eminent manufacturer in West Germany (Hartmann and Braun) for the measurement of fluid flow in ducts or pipes in a large cross-sectional area. The principle of operation is similar to that of an averaging pitot tube (Annubar) or a pitot tube. The element design claimed to be more suitable for flow measurement of gaseous fluid containing dust.

Its construction (Fig. 4.15) is opposite of the conventional orifice plate but similar to Annubar with a much simpler design. There is a small diameter blind circular plate installed near the central axis of the flow path temporarily restricting the flow. The HP and LP impulse pipes/tubes are welded on two sides of this plate so that their open end is located at the center of the plate.

The fluid velocity becomes zero when it impinges on the plate and is converted into equivalent pressure. The upstream impulse pipe senses the static head plus the dynamic (or velocity) head, whereas the downstream impulse pipe senses the static head only. The DPT connected to this impulse lines generates the appropriate flow signal.

TABLE 4.19 Different Values of Discharge Coefficient C_d (for Orifice Plates) in Relation to Different Values of β Ratio and Reynolds Number (*Re*)

Beta Ratio $\beta = d/D$	Reynolds Number	Reynolds Number	Reynolds Number	Reynolds Number
0.2[a]	0.60	0.595	0.594	0.594
0.4	0.61	0.603	0.598	0.598
0.6	0.62	0.608	0.603	0.603
0.8	0.64	0.614	0.609	0.609

[a]*Normally not applicable for measuring purposes, but for the permanent pressure reduction break down or restricting or restriction) orifice plate.*

Here the bore of both impulse pipes is comparatively large and thereby problems with dust choking are minimal. The DP produced by this type of element is more than that of Annubar, resulting in high resolution and accuracy.

Additional features for Krell's bar orifice plate-type flow elements include the following:

1. The meter is recommended for clean and dirty liquids and some slurry services.
2. Straight length before and after the element required varies from 10 to 35 times diameter pipe length (upstream) and 5–20 times diameter pipe length (downstream) depending on the layout and installation of items (regarded as constriction) in the vicinity of the element's upstream and downstream pipe length.
3. Typical accuracy is ~2% of full-scale range.
4. Relative cost is low.
5. Relative pressure loss is medium.

4.2.6 Dall Tube

The Dall tube is a shortened version of a Venturi meter guided by Bernoulli's equation. The pressure drop in this type of instrument is lower than an orifice plate. The tube consists of a short inlet section followed by an abrupt reduction of inlet diameter called an "inlet shoulder." That's followed by the converging inlet cone and divergent outlet cone, which are separated from each other by a gap, or a slot, called a throat. With these flow meters the flowrate in a Dall tube is determined by measuring the pressure drop across the two tappings with one upstream the edge of the inlet shoulder and the other at the throat. The permanent pressure losses in these meters are lower than orifice meters or, as per some manufacturers, even lower than the Venturi tubes. For medium and large pipe sizes, Dall tubes are normally suitable for measuring the flowrate.

4.2.7 Pitot Tube

A pitot tube is a pressure-measuring instrument used to measure fluid flow velocity by determining the stagnation pressure. Bernoulli's equation is used to calculate the dynamic pressure and fluid velocity The operating principle is similar that for a Krell's bar orifice plate. See Fig. 4.10 for construction of the pitot tube.

4.2.8 V Cone Flow Device

The V cone flow element, as shown in Fig. 4.16A, is also an especially accurate (as high as 0.5%) flow device under the head type flow measurement. It can be used for both dirty and lean fluid measurements. In comparison to the orifice it has better permanent pressure loss. The design incorporates a contour-shaped cone at the center of the pipe with annular passages, which direct the flow without impacting it against an abrupt surface thus avoiding wear of edges of the cone by dirty fluids. Because of this feature, recalibration of V cones is rarely required.

Extremely short vortices are formed as the flow passes the cone and creates a low-amplitude, high-frequency signal proportional to flow. Pressure loss here is much lower than that of the orifice plate. These flow devices show high accuracy, repeatability, and rangeability with low permanent pressure loss and short straight-length requirement.

This is a DP type flow element, so one tapping (static pressure) is taken at slightly upstream the cone with the other tapping located downstream the face of the cone itself. The central hole of the cone is connected to a pipe for transmitting the downstream pressure to the flow transmitter.

This type of flow element offers excellent accuracy, repeatability, and rangeability compared with other flow elements and covers the pipe size from 15 to 3000 mm

4.2.9 Sizing of Flow Elements (Typical)

Various types of flow-element sizing calculation software are available following the standards as stipulated by different organizations as mentioned in the Bibliography. However, a sample calculation is given below to show how sizing is normally done (Table 4.20).

4.2.10 Flow Measurement by Level Excursion Method (Ref. Fig. 4.11)

Several types of hydraulic structures are used, for example, notches (rectangular and V or triangular notch), weirs, and

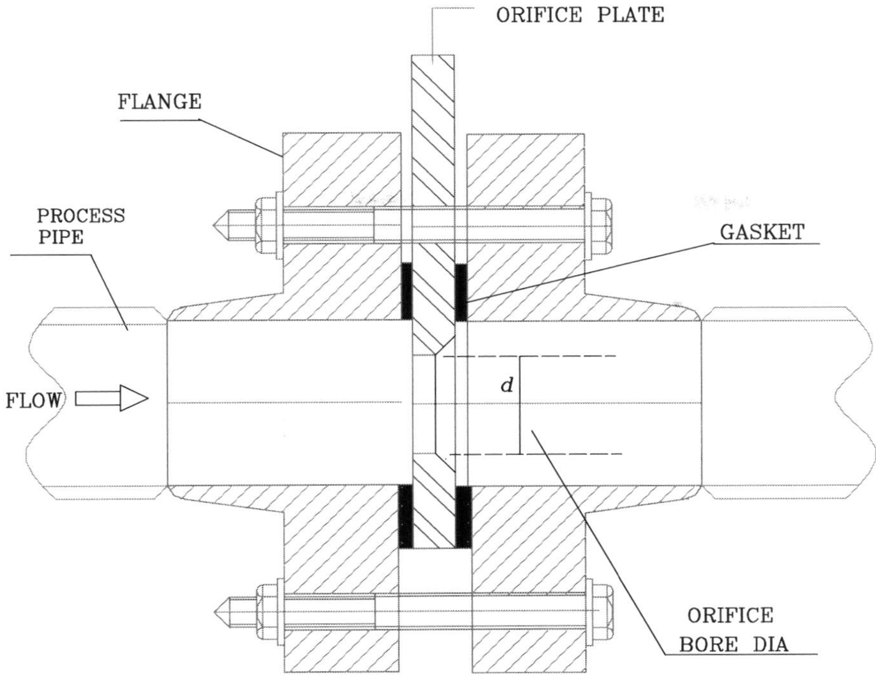

NOTE
DESIGN AS PER ISO5167–1/BS–1042
ORIFICE WOULD BE MADE OF STAINLESS STEEL OF SUITABLE GRADE
CARRIER RING WOULD BE MADE OF STEEL SUITABLE FOR THE POCESS
DIFFERENT TYPES OF TAPPING IS APPLICABLE VIZ D–D/2,CORNER OR FLANGE TAPPING

FIG. 4.13 Longitudinal cross-sectional view of orifice plate assembly with weldneck type flanges.

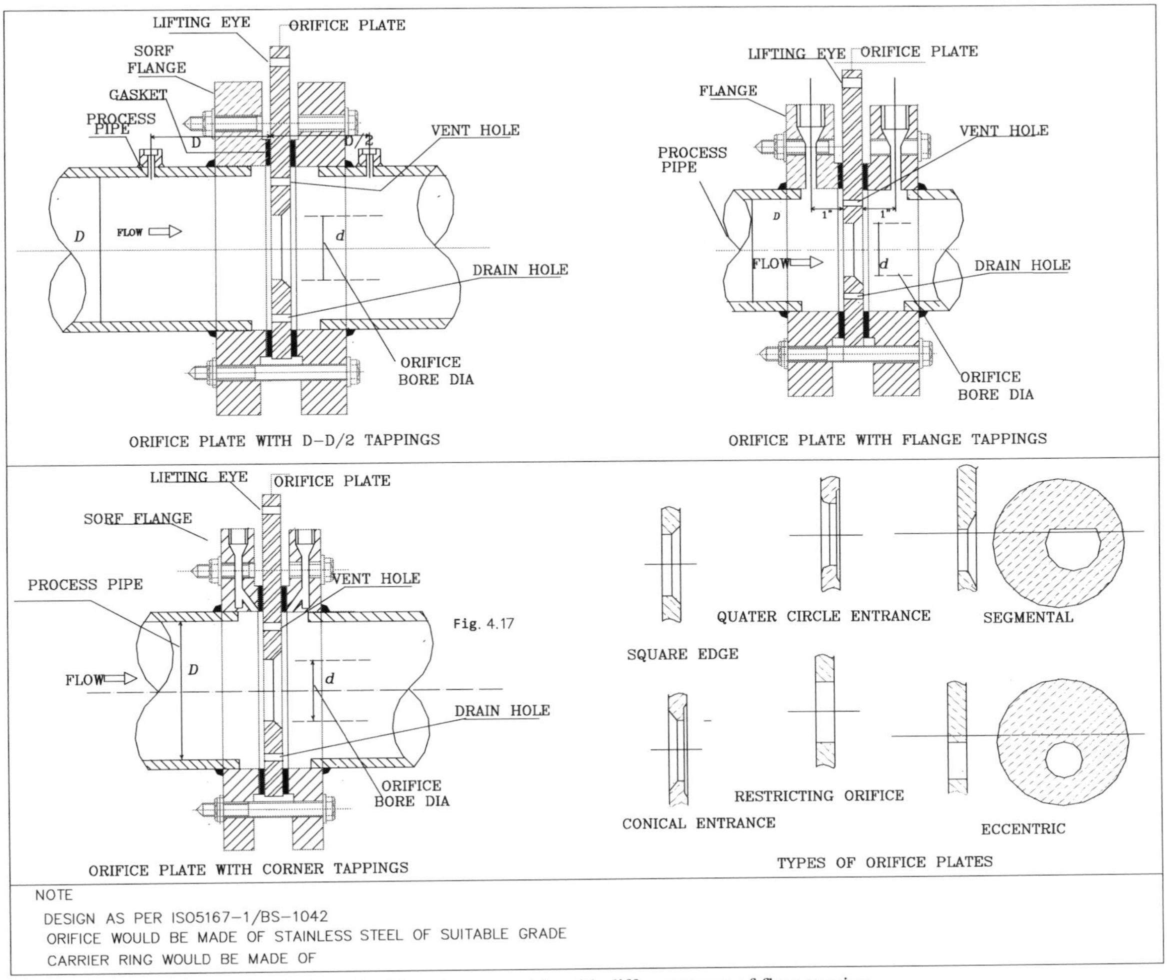

LIFTING EYE
ORIFICE PLATE
SORF FLANGE
GASKET
PROCESS PIPE
D
D/2
VENT HOLE
FLOW
d
DRAIN HOLE
ORIFICE BORE DIA
ORIFICE PLATE WITH D–D/2 TAPPINGS

LIFTING EYE
ORIFICE PLATE
FLANGE
PROCESS PIPE
VENT HOLE
D
FLOW
d
DRAIN HOLE
ORIFICE BORE DIA
ORIFICE PLATE WITH FLANGE TAPPINGS

LIFTING EYE
ORIFICE PLATE
SORF FLANGE
PROCESS PIPE
VENT HOLE
Fig. 4.17
D
FLOW
d
DRAIN HOLE
ORIFICE BORE DIA
ORIFICE PLATE WITH CORNER TAPPINGS

SQUARE EDGE
QUATER CIRCLE ENTRANCE
SEGMENTAL
CONICAL ENTRANCE
RESTRICTING ORIFICE
ECCENTRIC
TYPES OF ORIFICE PLATES

NOTE
DESIGN AS PER ISO5167–1/BS–1042
ORIFICE WOULD BE MADE OF STAINLESS STEEL OF SUITABLE GRADE
CARRIER RING WOULD BE MADE OF

FIG. 4.14 Longitudinal cross-sectional view of orifice plate assembly with different types of flow tapping.

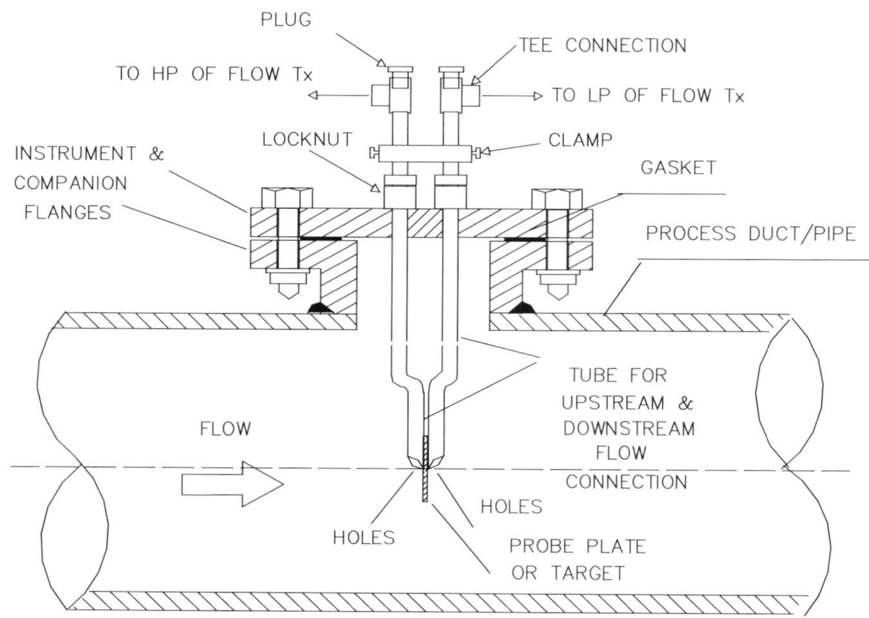

FIG. 4.15 Schematic diagram of the construction of Krell's bar.

flumes (such as rectangular, trapezoidal, or Parshall), for the measurement of flow in the open channel. In the existing channel, notches or weirs can be used because the flow measurement has to be done in the open area where there is a chance of siltation. These elements are removable and cleaning can be done for both the channel and element to restore measurement accuracy. Whenever there is a flow change, the level will change proportionally at the upper vicinity of the elements, which can be measured normally by an ultrasonic level transmitter with suitable software to convert the level signal into flow.

Notches or weirs are made of thin metal plates, and the general form is either triangular or rectangular in shape. The minimum requirements are

1. Width should be not less than 150 mm.
2. The contraction from either side should be greater than 100 mm.

Parshall flumes are a special shape and size and have to be erected during the construction of the main flow channel and cannot be put in the existing one. Level changes as flow changes and ultrasonic level transmitters with suitable software are utilized.

Parshall flume calculation is based on the ISO 9826 (1992) and ASTM D1941 (1991) standards or USBR (1997) standard. Parshall flumes must be constructed in strict compliance with specifications as per these standards as far as shape and their dimensions are concerned. The standard is valid for both free-flowing and submerged Parshall flumes. A free-flowing Parshall flume can be observed physically by the decrement in water head/level at the throat. For submerged flow, the downstream water flows back into the throat over the original head/level making it impossible to identify actual decrement had it been for the free-flow system. Therefore, after developing the empirical formula, the flow calculation of the submerged type of system requires two head measurements: one in the approach channel and one in the throat. On the other hand, only the upstream head/level measurement is required for the free-flow measuring system.

Flumes are also designed to force a transition from subcritical to super critical flow as is done with weirs. For Parshall flumes, the transition is caused by designing flumes to incorporate a constriction at the throat and a sharp drop in head/level in the channel bottom. This transition causes flow to pass through a critical depth in the flume throat, and there is a direct relationship between water depth and velocity or the flowrate. It is physically difficult to measure critical depth in a flume because its exact location is difficult to determine and may vary with flowrate. Through mass conservation, the upstream depth is related to the critical depth. Therefore, flowrate can be determined by measuring the upstream depth, which is a highly reliable measurement.

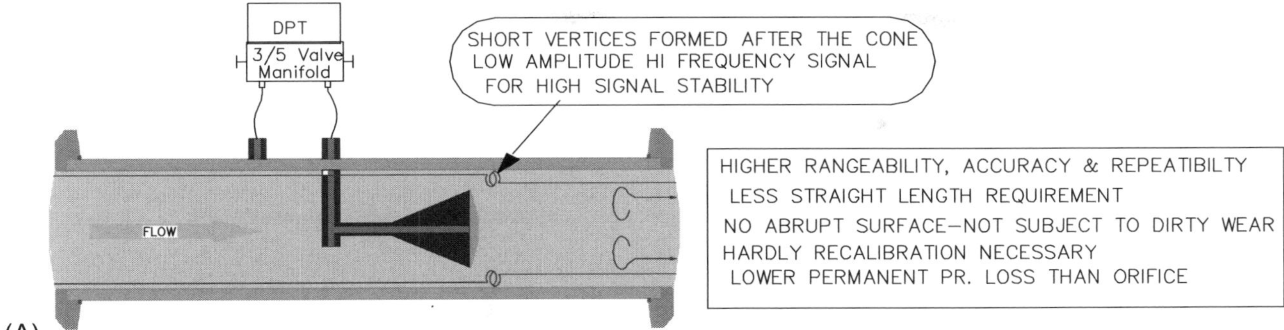

PERMANENT PRESSURE LOSS CURVE SHOWN IN Fig. 4.5B

DISCHARGE CO EFFICIENT VARIATION CURVE SHOWN IN Fig. 4.5C

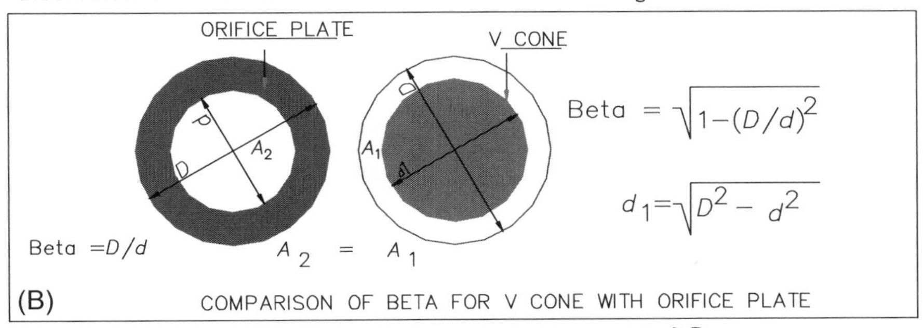

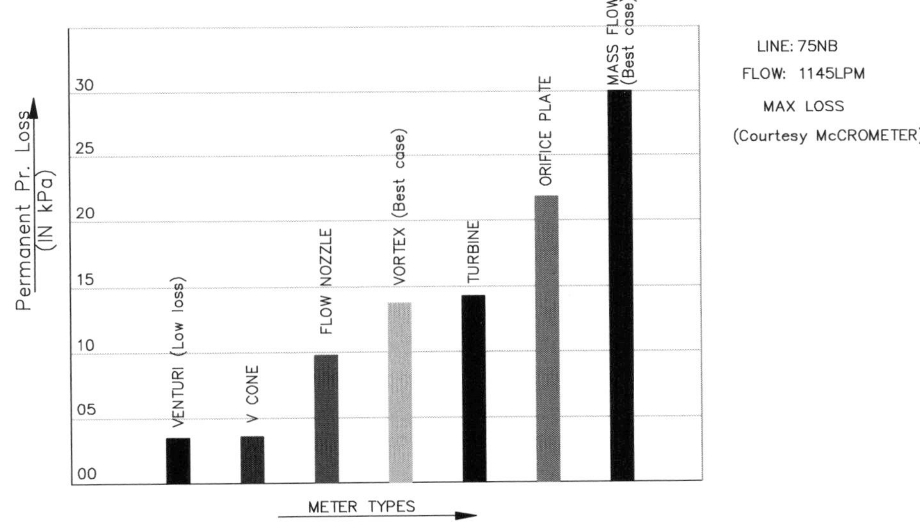

FIG. 4.16 V cone flow device.

For free flow $Q = Ch^n$, where Q is in m^3/s and h is in m. For submerged flow, ISO Eq. (12) is rewritten as $Q = C\,h^n - Q_e$, where C and n are found from the above equation based on width (b).

Q_e accounts for the effects of submergence, where m = meters, s = seconds, b = throat width [m], C = Parshall flume constant [empirical units], h = measured upstream head [m], and H = measured downstream head [m].

This is only needed if the flume is submerged or to determine mathematically if the flume is submerged. Usually, one can see if there is a hydraulic jump, but determining the ratio H/h is a quantitative method, where H/h = submergence ratio, n = Parshall flume power constant [unit less], Q = flowrate (discharge) through flume [m^3/s], and Q_e = reduction in flowrate due to submergence [m^3/s].

TABLE 4.20 Orifice Plate Bore Calculation as per BS 1042 and BS EN 5167-1

Date:	
Tag no. .... G.A. Drawing no....................................	
Service of fluid	DM
Pipe ID	188.40 mm = 0.18840 m (206 OD × 8.8 mm)
Pipe material	Copper
Material of orifice plate	316 SS
Operating temperature (t)	40°C
Upstream gauge pressure (Pg)	3.00 kg/cm^2 = 294198.3 Pa
Barometric pressure (Pb)	1.02 kg/cm^2 = 100027.4 Pa
Operating pressure (Pa)	394225.76 Pa
Specific gravity (σ)	0.992
Flow range (Q_{vr})	360 T/h (6000 LPM)
Dynamic viscosity at operating temperature (Mew)	0.00653170 poise = 0.00065317 Pa/s
Type of tapping	Flange
DP (h)	5000 mm wcl = 49033.068 Pa
Upstream pipe diameter at operating temperature: $D = 0.1884641$ m Approximate $CE\beta^2 = 4.Q_v/[\varepsilon \pi D^2 \sqrt{(2.h/\rho)}] = 0.36345487$ (assuming $\varepsilon = 1$)$Nu = \mu$ in Pa/s/ρ in kg/m^3 = 0.00000066 m^2/s^2 Reynolds number: Re at continuous flow, $Re\ D = 4\ Q_v \cdot 0.7/(\pi.D\ Nu) = 724046.08$ Using Table 4.2 of Clause 1.4, 1992, p. 7 and interpolating from ReD column, $\beta = 71909476$ Using the equation given in 8.3.2.1 of Clause 1.1, 1992: For $\beta = 71909476$, discharge coefficient $C_d = 60272277$ $E = 1.168324$ Calculating $CE\beta^2$ from C, E, and β just obtained: $CE\beta^2 = 0.36412735$ By repeating above steps, finally $\beta = 0.71859276$, $C = 0.60275702$, and $CE\beta^2 = 0.36345483$ Bore diameter at operating temperature, $d = 0.13542891$ m Bore diameter at 20°C, $d_{20} = 0.13538017$ m = 135.3802 mm Vent hole diameter = 5.0 mm Modified bore diameter = 135.2786 mm Pressure loss (Cl.8.4 of Clause.1.1: 1992) = 2334.31 mm wcl = 46.69% of DP	

4.3 Flow Transmitters and Meters

The DPT used as a flow transmitter has already been discussed in Chapter 1, Clauses 3.1 and 3.2. The nonlinear relationship has an impact on the DPT operating range and requires the electronic DPTs to linearize the signal before transmitting it to the control system. Whatever the operating and ambient conditions, state-of-the art technology in the measurement of flowrate incorporates electronic devices that can correct for varying pressure and temperature, that is, density conditions, nonlinearities, and for the characteristics of the flowing media such as viscosity.

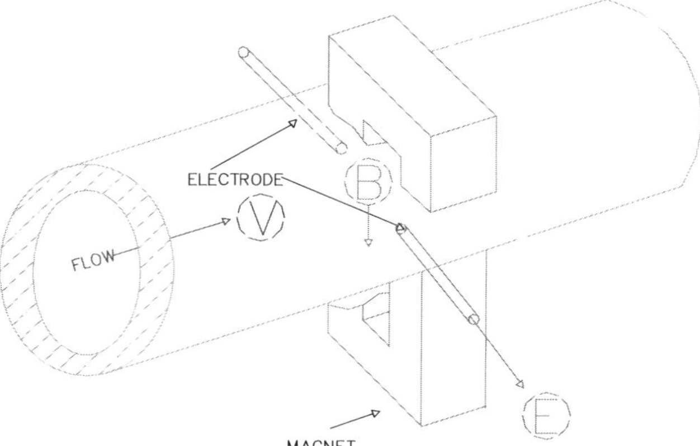

FIG. 4.17 Schematic diagram of magnetic flow meter.

4.3.1 Different Types of Flow Transmitters

4.3.1.1 Magnetic Flow Meters

One of the most common flow meters, apart from DPT and mechanical types, is the magnetic type. A magnetic flow meter is applied to the metering tube, which results in a potential difference proportional to the flow velocity perpendicular to the flux lines. The principle of operation is based on the Faraday's law of electromagnetic induction (Fig. 4.17).

The meter comprises a tube of nonconducting material, for example, a rubber-lined nonmagnetic steel tube. Mounted on the tube wall is a pair of electrodes opposite to each other that make contact or touch the flowing media within the tube. The arrangement is surrounded by a magnet with its magnetic field at right angles to the electrode and the flow direction. The flowing media must be an electrical conductor with minimum conductivity of 5 μSiemens/cm (Table 4.21).

As the moving fluid (regarded as electrical conductor) flows through the nonconducting tube and through the magnetic field, voltage is induced, which is sensed by the electrode. As per Faraday's law of electromagnetic induction, the magnitude of the induced voltage depends on the following:

1. Strength of the magnetic field.
2. Distance between the electrode.
3. Velocity of conducting fluid flowing through the tube.

If the first two conditions are kept constant, the voltage induced will be entirely proportional to the velocity of the flowing media. If the area of the tube is fixed and known, the volume flow can be calibrated with respect to the induced voltage. The electronic part of a smart flow transmitter will convert the voltage to a signal like 4–20 mADC and superimposed digital signal (HART) or digital signal through Profibus, Foundation Fieldbus, etc. The velocity range covered by these meters can be 1.0 m/s up to 10 m/s without hampering the accuracy.

TABLE 4.21 Data Sheet for Magnetic Flow Meters

SL. No.	Item	Specification
1	Make	ABB/Krone L/Foxboro
2	Model/type of transmitter	Remote/head mounted
3	Line size	As per client (15–2600 mm)
4	Display	3 line, engineering unit
5	Metering tube	Stainless tube
6	Earthing electrode	Fitted
7	Maximum detectable velocity	15M/SC
8	Maximum working pressure	6 bar
9	MOC	Electrode: 316 SS/Hastelloy C/titanium/tantalum/platinum iridium/nickel/tungsten carbide Lining: elastomer/Teflon (PTFE)/ PFA/neoprene/polyurethane
10	Test	3 point
11	Protection class	IP 68
12	Power supply	240V AC
13	Process connection	Flat face carbon steel, ANSI
14	Ambient temperature	−10 to 60°C
15	Totalizer	Integral with membrane keypad
16	Accuracy	1%
17	Range	As per client (0.005–190,000 M3/H)
18	Cable	5–100 m or more
19	Power supply	230 V AC/24 V DC
20	Output	4–20m A (single/dual)/HART/RS 422/423

4.3.1.2 Ultrasonic Type

4.3.1.2.1 Doppler, Transit Time, and Ultrasonic Type

This type of ultrasonic measurement employs the difference of the transit time of ultrasonic pulses propagating toward and against flow direction. The average velocity of the fluid and the speed of sound can be calculated along the path of the ultrasonic beam by the transit time difference obtained and their product. Using the two transit times t_{up} and $t_{down,}$ and the distance between receiving and transmitting transducers L, and the inclination angle α, the equations can be written as follows:

$$v = L/2\sin\alpha \times \left[\left(t_{up} - t_{down}\right)/t_{up} \times t_{down}\right]$$

$$c = L/2 \times \left[\left(t_{up} + t_{down}\right)/t_{up} \times t_{down}\right]$$

where U is the average velocity of the fluid along the sound path and c is the speed of sound.

Ultrasonic flow meters are used for the measurement of natural gas flow, water (any fluid) flow in a closed pipe, ducts, etc. The expected speed of sound for a given sample of fluid can also be calculated and can be compared to the speed of sound empirically measured by an ultrasonic flow meter (Fig. 4.18A).

4.3.1.2.2 Doppler Frequency Shift Method

Flow of fluid can also be measured by the help of the Doppler "frequency shift" method. This method uses the characteristics of an ultrasonic beam affected by the movement of a fluid when it is passing through the fluid. An arrangement is made so that while passing an ultrasonic beam through the flowing media it will bounce back off a reflective plate at the other end and then after the direction of the beam is reversed and received by the sensor. As the frequency of the transmitted beam is affected by the flowing fluid in the closed pipe, by comparing the frequency of the upstream beam with the downstream beam, the fluid flow through the closed pipe can be measured. The volume flow can be calculated by the mismatch between the upstream beam and the downstream beam frequencies. To make the flow independent of the cross-sectional area of the pipe, the wide-beam sensor is normally selected. It is mandatory that the flow stream contain sonically reflective materials to act as reflecting plate as described above. The flowing media containing solid particles or entrained air bubbles are suitable for the flow meter to work exploiting the Doppler principle of frequency shifting (Fig. 4.18B).

4.3.1.2.3 Level Excursion Type

The flow elements used for this type of measurement were discussed earlier. The level is measured by an ultrasonic level transmitter, which senses the actual level, and not the velocity of the fluid, from the bottom line of the fluid flowing through the channel upstream just near the flow element. An ultrasonic level transmitter is equipped with the appropriate software to convert the level signal into the flow signal according to the flow element design.

4.3.1.3 Coriolis Flow Meters

A Coriolis flow meter is basically a mass flow meter for measuring the amount or quantity or the mass of fluid, rather than volume, flowing through a closed pipe or tube. Using

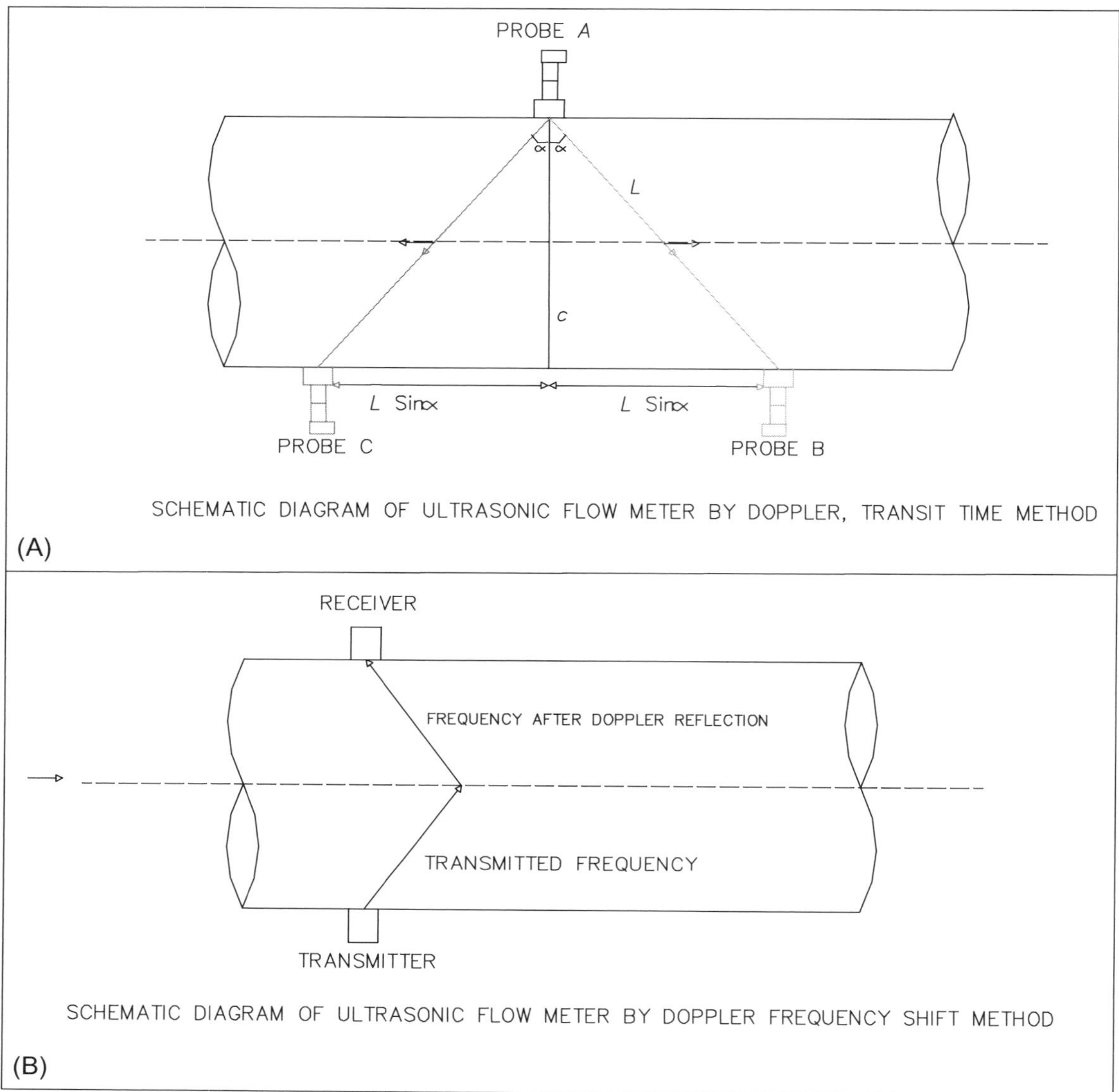

FIG. 4.18 Schematic view of Doppler and transit time/frequency shift type flow meters.

the Coriolis Effect, which causes a laterally vibrating tube to distort, a direct measurement of mass flow can be obtained in a Coriolis flow meter. The added advantage is that a direct measurement of the density of the fluid can also be obtained with this instrument. Coriolis measurements can be decidedly accurate irrespective of the type of gas or liquid that is measured; the same measurement tube can be used for any gas and liquid without individually calibrating each instrument. As in a conventional mass flow meter, the fluid passes through the Coriolis flow meter containing a pair of smooth pipes or tubes without any moving parts, avoiding regular cleaning and maintenance (Fig. 4.19).

4.3.1.3.1 Coriolis Principles If a moving mass is subjected to an oscillation, perpendicular to the motion, depending on the mass flow, a Coriolis force occurs. Because of this force, sensor(s) at the inlet and outlet are subjected to this force differently. At the inlet, because of mass addition, it tries to oppose the vibration, while at the outlet side, mass going out tries to add to the vibration. Naturally, there will be a twisting motion in the tubes. The resultant phase shift in tubes' oscillation is a measure of mass flow. Whenever mass of a vibrating system changes, there will be change in vibration frequency. When measuring density, keeping volume flow the same, the change in frequency of vibration will be a measure of density (volume flow is the same and meter mass does not change). To eliminate thermal influence on the tube, that is, volume change of the meter itself, an RTD is used to measure the temperature (Fig. 4.20).

There are two parallel tubes through which the fluid is passed. A mechanical vibrator is arranged to cause vibration

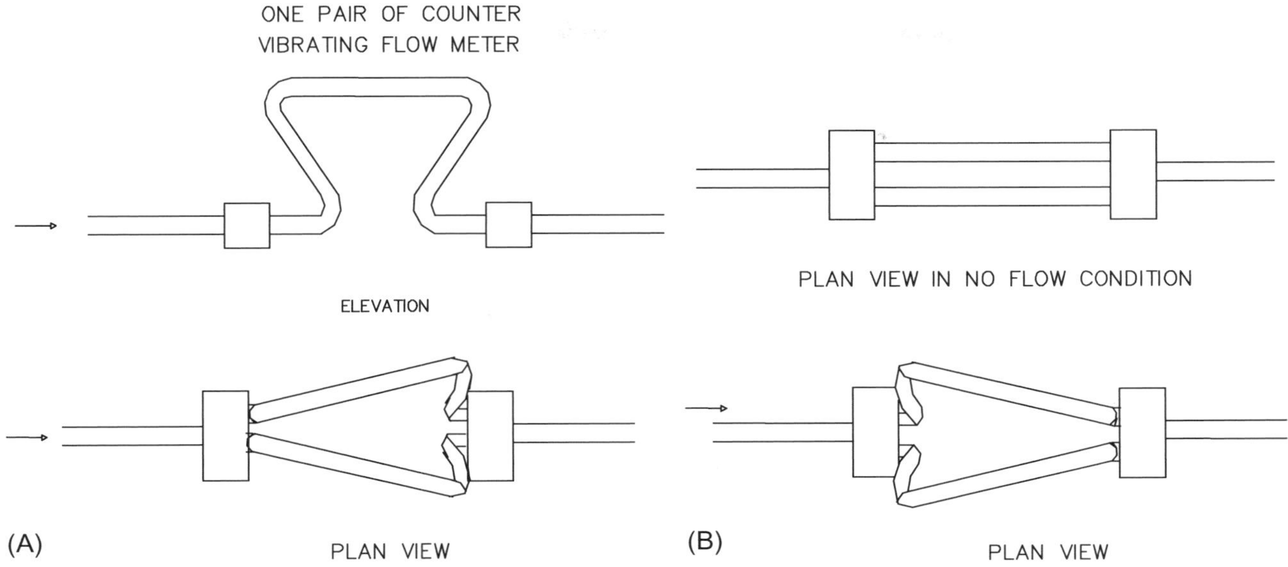
ONE PAIR OF COUNTER
VIBRATING FLOW METER
ELEVATION
PLAN VIEW IN NO FLOW CONDITION
(A)
PLAN VIEW
(B)
PLAN VIEW
ALTERNATE POSITION OF FLOW TUBES VIBRATING FLUID AT FLOWING CONDITION

FIG. 4.19 Schematic diagram of Coriolis flow meter.

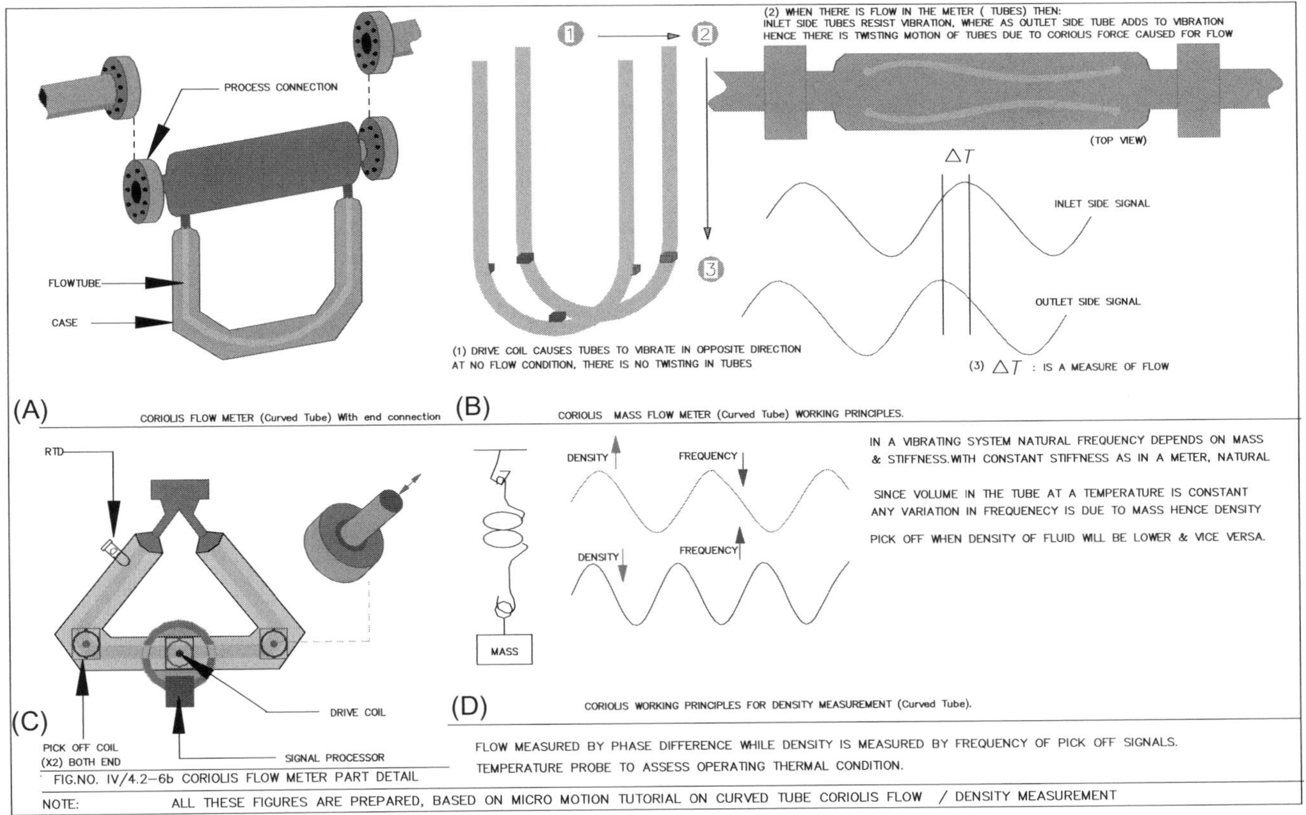
PROCESS CONNECTION
FLOWTUBE
CASE
(A)
CORIOLIS FLOW METER (Curved Tube) With end connection
(2) WHEN THERE IS FLOW IN THE METER (TUBES) THEN:
INLET SIDE TUBES RESIST VIBRATION, WHERE AS OUTLET SIDE TUBE ADDS TO VIBRATION
HENCE THERE IS TWISTING MOTION OF TUBES DUE TO CORIOLIS FORCE CAUSED FOR FLOW
1
2
3
(TOP VIEW)
ΔT
INLET SIDE SIGNAL
OUTLET SIDE SIGNAL
(1) DRIVE COIL CAUSES TUBES TO VIBRATE IN OPPOSITE DIRECTION
AT NO FLOW CONDITION, THERE IS NO TWISTING IN TUBES
(3) ΔT : IS A MEASURE OF FLOW
(B)
CORIOLIS MASS FLOW METER (Curved Tube) WORKING PRINCIPLES.
RTD
DRIVE COIL
(C)
PICK OFF COIL
(X2) BOTH END
SIGNAL PROCESSOR
FIG.NO. IV/4.2-6b CORIOLIS FLOW METER PART DETAIL
DENSITY
FREQUENCY
DENSITY
FREQUENCY
MASS
IN A VIBRATING SYSTEM NATURAL FREQUENCY DEPENDS ON MASS
& STIFFNESS.WITH CONSTANT STIFFNESS AS IN A METER, NATURAL
SINCE VOLUME IN THE TUBE AT A TEMPERATURE IS CONSTANT
ANY VARIATION IN FREQUENECY IS DUE TO MASS HENCE DENSITY
PICK OFF WHEN DENSITY OF FLUID WILL BE LOWER & VICE VERSA.
(D)
CORIOLIS WORKING PRINCIPLES FOR DENSITY MEASUREMENT (Curved Tube).
FLOW MEASURED BY PHASE DIFFERENCE WHILE DENSITY IS MEASURED BY FREQUENCY OF PICK OFF SIGNALS.
TEMPERATURE PROBE TO ASSESS OPERATING THERMAL CONDITION.
NOTE: ALL THESE FIGURES ARE PREPARED, BASED ON MICRO MOTION TUTORIAL ON CURVED TUBE CORIOLIS FLOW / DENSITY MEASUREMENT

FIG. 4.20 Coriolis principles.

in these two tubes. So the instrument is less sensitive to external vibrations, the parallel tubes vibrate to counter each other. The amplitude of the vibration is kept small, although it can be felt by touching the flow tubes. The vibrating frequency of the tubes solely depends on the size of the mass flow meter, although the ranges may vary from manufacturer to manufacturer in the domain of about 100–1000 times per second. When no fluid is flowing, the vibration of the two tubes is symmetrical. The two arms vibrate with the same frequency overall, but when the mass flows, the two vibrations shift to a sort of asynchronous situation. The two vibrations are shifted in phase with respect to each other, and the degree of phase shift is a measure for the amount of mass that is flowing through the tubes.

4.3.1.4 Vortex Flow Meter

Vortex flow meters are a direct type of flow-measuring device with operating fundamentals based on the von Karman Vortex Street principle. In this instrument, a bluff body or shedder bar is arranged at the center of flow path. When the fluid passes through it, disturbances in the flow called vortices are alternately formed on both sides of the shedder. These vortices are shed immediately after they are formed due to the flow stream forming a vortex street as per the Karman Vortex Street principle (Fig. 4.21).

The frequency of the vortex formation and shedding depends on several factors: velocity of the fluid (υ), width of the shedder (d), and Reynolds number (Re). The relationship of velocity with flow and frequency can be given as $f = S \times \upsilon/d$, where S is the Strouhal number.

The Strouhal number (S) is a dimensionless number that defines the quality of the vortex flowrate measurements and bears a relationship with Reynolds number as shown in graph of Fig. 4.21B, which is moderately constant in a long stretch of Reynolds number.

Reynolds number can be expressed as $Re = \upsilon \times D/\ni$, where D is the inner pipe diameter and $\ni$ is the kinematic viscosity. The frequency associated with vortex creation or shedding is therefore a dependent variable of fluid velocity only as L is constant and S is fairly constant for a particular wide range of Reynolds numbers and independent of the fluid density and viscosity.

The sensor type employed is often a piezoelectric crystal, the details of which were discussed earlier in the chapter. The local pressure near the sensor, located strategically, changes every time a vortex is created or in other sheds. The pressure changes are sensed by the piezoelectric crystal and converted into a small

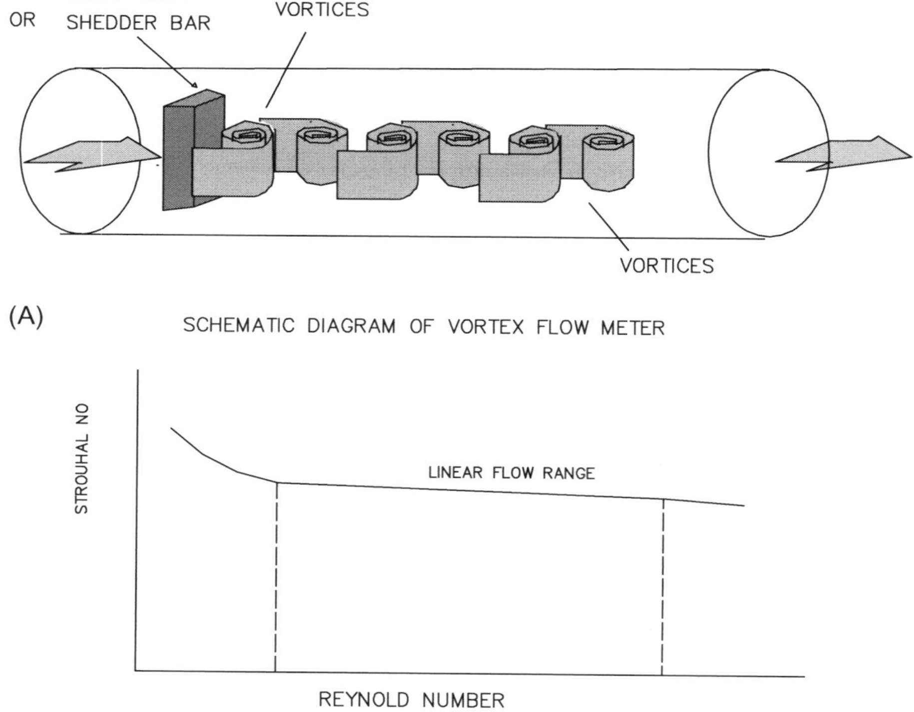

FIG. 4.21 Schematic diagram of vortex type flow meter.

electrical signal or voltage pulse representing the vortex shedding frequency. Because the frequency of the voltage pulse is also proportional to the fluid velocity, a volumetric flowrate is calculated using the cross-sectional area of the flow meter. The frequency is measured, and the flowrate is calculated by the flow meter electronics using the above equation.

4.3.1.5 Swirl Flow Meters

Another direct type of flow-measuring device is the swirl flow meter. When the flow enters the instrument through the inlet guide body, it is forced to swirl or rotate. A vortex core is created in the center of this rotation and is forced again into a secondary spiral-shaped rotation by the backflow. The frequency of this secondary rotation is fairly proportional to the flowrate. When the internal geometry design is optimal, the relationship is assumed to be linear over a wide flow range. The sensor type employed is a piezoelectric crystal and its principle of operation is similar to the vortex flow meters.

4.3.1.6 Mass Flow Meters

This device works according to the thermal measuring principle of a hot-film anemometer. This measuring method determines the gas mass flow directly with a result that it is totally independent of pressure and temperature influences. The mass flow meter is used in a process plant as well as in the food and beverage industries for flow measurement of gases and gas mixtures (Table 4.22).

TABLE 4.22 Data Sheet for Thermal Mass Flow Meters

SL. No.	Item	Specification
1	Make	ABB/Krone Marshal/ Foxboro
2	Model	–
3	Line size	As per client (25–200 mm)
4	Operating pressure	40 bar
5	Response time	<0.5 s
6	Process medium	As per customer
7	Type of transmitter	Head mounted
8	Maximum working temperature	−25 to 300°C
9	MOC of sensor	Ceramic or metal (316 SS)
10	Accuracy	<0.9% (for air/ nitrogen) <1.8% (for other gases)
11	Protection class	IP 65
12	Power supply	230V AC
13	Process connection	DIN/ANSI flanged
14	Ambient temperature	−25 to 70°C
15	Output	4–20 m A
16	Power supply	230 V AC /24 V DC
17	Range	As per client

Thermal mass flow instruments employs different ways to detect the mass flow dependent of cooling a heated resistor as a measuring signal. In a hot-film anemometer with differential temperature control, the heated platinum resistor is maintained at a constant over temperature, by controlling power input to the resistor, in relation to an unheated platinum sensor inside the gas flow. The power required for heating the platinum resistor to maintain the over temperature bears a direct relation to the flowrate and the material properties of the gas. With a known (and constant) gas composition, the mass flow can be determined electronically by evaluating the heater current versus the mass flow curve without additional pressure and temperature compensation. Another way to determine the mass flow is to use the constant power method. The temperature difference is measured, which results from a constant heating power and depends on the amount of heat dissipated by the gas mass flow. Together with the standard density of the gas, this directly results in the standard volume flow, and an accuracy smaller than 1% of the measuring value is achieved (Fig. 4.22).

The measuring systems are made up of a transducer and a pipe component. The transducer comprises the sensor unit and an electronic transmitter circuit, and it directly delivers an electrically isolated 0/4–20 mA output signal. It is designed as a flange-mounted insertion sensor and is installed in the pipe component in a defined way. The pipe component is available in nominal sizes ranging from 25 to 200 mm and in various designs. It is also possible to install the transducer in square ducts or pipes of any size by using a suitable adapter.

4.3.1.6.1 Typical Applications

- Gas flow measurement in thermal power plant
- Gas burner control
- Gas measurement for air separation systems
- Hydrogen measurement in processes

4.3.2 Mechanical Flow Meter

There are several types of mechanical flow meters, such as a variable area meter, piston meter/rotary piston, oval gear meter, turbine flow meter, paddle wheel meter, nutating disk

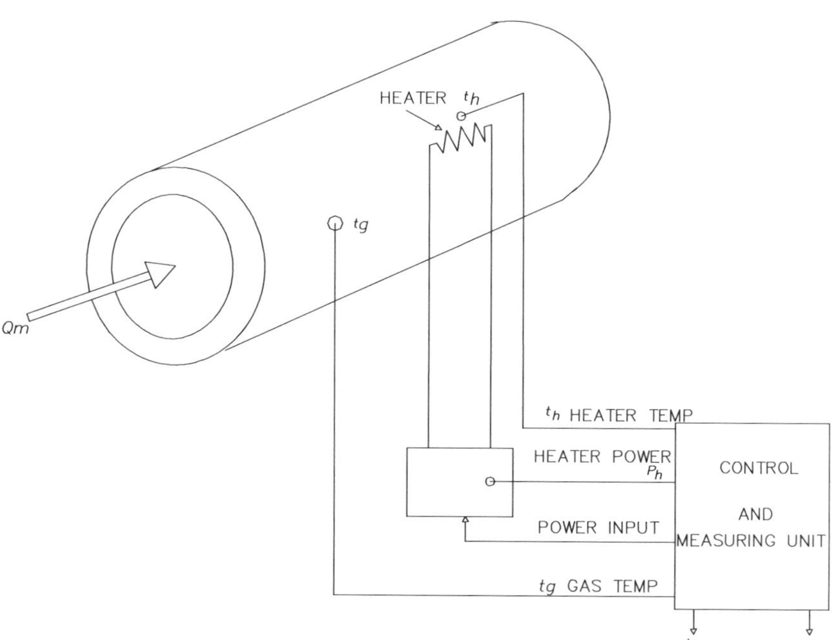

FIG. 4.22 Mass flow meter schematic diagram.

meter, Woltmann meter, and single- and multi-jet meters. They will be discussed briefly in this section, though all of them are not used in a thermal power plant.

4.3.2.1 Variable Area Meter

The variable area meter, also known as a rotameter, a vertically tapered tube, with the small diameter facing down. It's normally made of glass, with a float inside that is pushed up by the force from the upward fluid flow and pulled down by gravitational force. When the flowrate increases, greater viscous and pressure forces (due to increased DP acting on the float area) cause it to rise until it becomes stationary at a location in the tube that is wide enough to restore the DP at the value before rise in flow took place. The two forces balance each other to keep the float stationary at a new position. Floats are normally made of stainless steel in many different shapes; spheres and spherical ellipses are the most common. Some are designed to spin visibly in the fluid stream to aid the user in determining whether the float is stuck or not. Rotameters are available for a wide range of liquids but are most commonly used with water or air, for example, in demineralized (DM) water. They can be made to reliably measure flow down to 1% accuracy. Remote flow transmissions are also available with this type with the necessary transducer, which is magnetically coupled with the float.

Another type of variable area meter is called a valve-type area meter. This type of instrument applies the movement of the spindle containing a plug or a piston of a self-positioning valve. For each flow, there will be particular position of the valve plug or a piston to maintain the DP at a constant value. The valve position is then calibrated against the flow. Remote flow transmissions are also available with this type with the necessary transducer, which is magnetically coupled with the movement (Fig. 4.23).

4.3.2.2 Piston Meter/Rotary Piston

Piston meters are also known as rotary piston or semi-positive displacement meters. They are under the category of positive displacement type meter and most common water flow-measuring devices and are used for meter sizes up to 40 mm. The piston meter operates on a piston rotating within a chamber of known volume. Every rotation of the piston ensures that a fixed amount of water goes out after passing through the piston chamber. The number of rotations through a gear mechanism is counted, calibrated, and displayed.

4.3.2.3 Oval Gear Meter

The oval gear meter is a also a type of positive displacement meter. The operating principle of the oval gear meter is based on two or more oval-shaped gears configured to rotate always at right angles to one another. It has two sides connected to the inlet and outlet ports. The teeth of the two gears always mesh, so fluid is not allowed to pass through the

Almost all international standard transmitter suppliers have this type of transmitter to enhance the compensation approach for better accuracies. In conventional methods, three separate measuring transmitters/devices were deployed, e.g., DP transmitter, static/absolute transmitter, and temperature elements (with transmitters), that feed these signals to a flow computer in PLC/DCS to get compensated flow as per the following equations: $Q_m = \sqrt{(dp*P/T)}$ where dp = DPT output, P and T stand for pressure and temperature compensation signal. However, in this dynamic compensation of flow, which is influenced by factors like discharge coefficient (C), velocity approach factor (E_v). gas expansion factor (Y_1), and density factor (ρ) are taken in to consideration as per the following formula (EN ISO 5167/AG3). $\underline{Q_m = C*E_v*Y_1*d^2*\sqrt{(dp*\rho)}}$

C:Discharge coefficient is influenced by Reynolds number (velocity profile), which in turn depends on viscosity, density, and velocity of the medium. Re= $(v*D*\rho)/\upsilon$, where v = velocity, D = internal pipe diameter, ρ = density, and υ = viscosity.

Y_1:Gas expansion factor is applicable for compressible fluids and depends on Beta (β = diameter ratio), isentropic exponent, DP and static pressure of fluid. There are separate formulas for each type of flow elements (such as flow nozzle, orifice plate. etc. (any standard flow-metering book may be referred to for formulae).

E_v: Velocity approach factor depends on β as per the following formula $Ev= \{1/\sqrt{(1-\beta^4)}\}$. Naturally the material of the pipe and flow element and their thermal expansion has influence. Dynamic compensation would result in a better accuracy approach.

P: Medium density directly influences flow rate as stated in the beginning. It is a function of T and P so for gas AGA 8 standard or steam table can be used for superheated and saturated steam, and for liquid ρ as function T may be used.

Normally a PC-based software approach is used while configuring these transmitters. Apart from this Fieldbus or HART 5 or 7 configuration tools can be used.

Even though these multivariable transmitters/sensors individually can offer very high accuracy 0.075% of FSD, actual mass flow measurement accuracy will be little less. The sensor for pressure could be absolute, gage pressure. Normally standard Pt RTD, and/or standard THCs can be used from external sources. The response time could be in the tune of 200–300 mS. Almost all the transmitters have advanced diagnostic features to monitor impulse line block, heat trace, CPU failure, hardware failure, and wrong configuration.

FIG. 4.23 Multivariable transmitter for flow measurement.

center of the meter. At one position of the inlet side gear, the teeth of the gear shut the fluid flow inlet because the elongated gear is protruding into the measurement chamber. At that time, at the outlet side of the meter, a cavity is formed that holds a fixed amount of fluid in the measurement chamber.

As the inlet fluid continues to push the gear arrangement, both gears rotate. The fluid in the measurement chamber is then allowed to be released into the outlet port. At the same time the fluid entering the inlet port will be driven into the measurement chamber, which is now open. The teeth on the outlet side then close keeping the fluid from entering back into the outlet fluid. This cycle continues as the gears rotate and allow a fixed amount of fluid to pass through the measurement chambers, and the number of rotations is counted, suitably calibrated, and displayed. Remote transmission through magnetic coupling or other suitable devices is also available. This type of instrument is used in thermal power plants for fuel oil, light diesel oil/high speed diesel, or naphtha for problems encountered while using segmental orifice plates.

4.3.2.4 Turbine Flow Meter

The operating principle of the turbine flow meter is based on the rotation of the turbine wheel whose blades are set in the flow path of a fluid stream. When the incoming fluid through the inlet port impinges on the turbine blades, a force is developed that causes the blades to move. The blades are fixed to the rotor and create torque and cause rotational motion. The speed of the meter is proportional to fluid velocity after a steady rotational speed is attained. In this type of meter, remote transmission through magnetic coupling or other suitable devices are also available. Turbine flow meters are used for the measurement of natural gas, air, and liquid flow.

4.3.2.5 Nutating Disk Meter

This type of instrument is commonly used for measuring liquid and water flow. The liquid, enters at the inlet side of the meter and impinges on the nutating disk. The disk is eccentrically mounted within the measuring chamber. Because the bottom and the top of the disk remain in contact with the measuring chamber, the disk starts "wobbling" or nutating with respect to the vertical axis due to the impact of the fluid. The inlet and outlet chambers are separated by a partition. For every nutation, a fixed amount of liquid passes through the chamber, and the total number of nutations for a particular period directly indicates the volume of the liquid that has passed through the meter. This instrument provides reliable measurements within 1%.

4.4 Flow Switch: Working Principle

Flow switches are also available of various types, such as flapper or paddle, target, disk, vane, diaphragm, shuttle, piston, etc. Switch contacts are available in flow meters also and normally provided for alarm, data acquisition systems, or open loop control for interlock and protection purposes. Other types of switches are also available employing the principle of thermal mass flow or vortex type flow

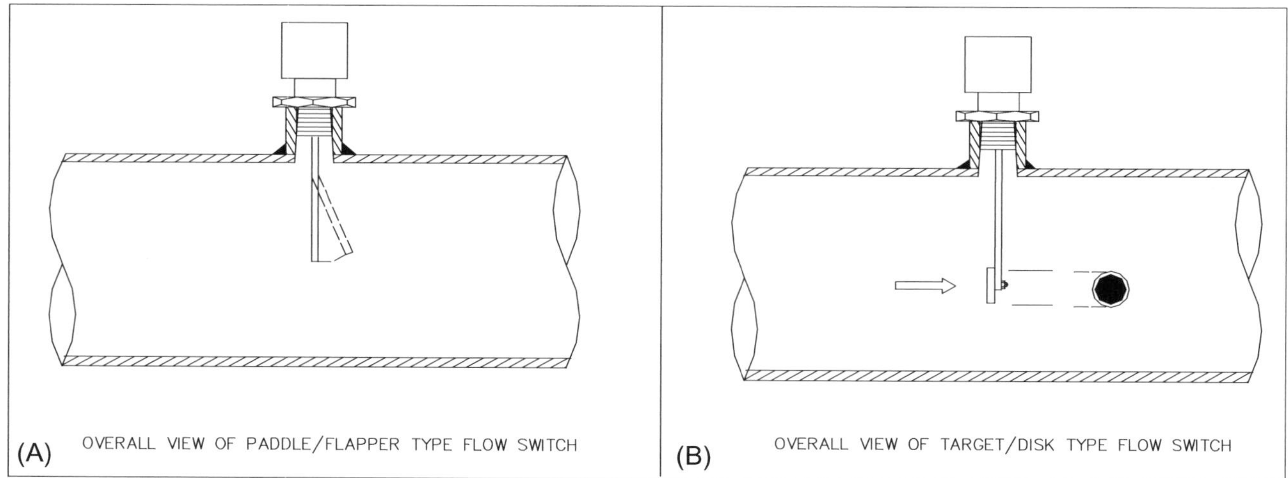

FIG. 4.24 Overall view of paddle/flapper/target/disk type flow switch.

measurement with the help of an electronic unit to calculate and provide alarm or tripping contacts.

4.4.1 Flapper or Paddle Type

The flow monitoring switch comprises a paddle/flapper system hanging through a hinge with the flat surface facing the flow direction. To this end a permanent magnet is attached; above this magnet is a reed contact, located outside the flow of fluid. A second magnet with opposing poles generates the force required to reset the switch so that it goes back to its original/normal starting point, that is, at the no-flow position. The paddle/flapper assembly is covered by a gaiter or bellows of suitable material to protect it from a process laden with hazardous components or particles (Fig. 4.24A).

When the flow increases, the force behind the flow compels the paddle/flapper assembly so that the paddle swings away. This causes the position of the magnet to shift with respect to its relation to the reed contact, activating the contact. If the flow value reduces, the flapper/paddle moves back to its starting position, activating the reed contact again. The force required to push the magnet back to the original position is provided by the two magnets repelling each other. Leaf springs are not usually employed for restoring original paddle/flapper position to get rid of its sensitiveness to pressure peaks. Magnetic forces are considerably more stable in the long term.

Flow switches should be mounted in a horizontal section of cooling water pipe when there is a straight horizontal of at least five pipe diameters on either side of the flow switch to avoid turbulent flow condition. The inlet line should allow free movement of the paddle/flapper in the pipe without rubbing the side. The paddle/flapper must be at the right angle to the flow and the arrow mark on the body must be in the same direction as the fluid flow.

4.4.2 Target or Disk or Vane Type Flow Switch

When there is a fluid flow through a pipe/tube with a velocity that exceeds a particular value, the fluid velocity becomes sufficient to push any object upward if suitably hinged at the top or cause rotation (in case of vane). The flow switch of this type consists of a valve body with a target or disk to sense the velocity of the flowing liquid. The force, thus developed against a target, disk, or vane causes it to lift out of the flow path. Disk, target, or, vane assemblies of different sizes and shapes for different flow requirements are utilized. They are arranged in an attraction sleeve just at the top of the target/disk/vane assembly. The sleeve then overcomes the pull of the spring and is arranged to move toward a magnet with the switch assembly located outside the sleeve pipe. The moment the sleeve swings into the field of the magnet, the magnet is attracted, actuating the switch. Setpoint is adjustable and is achieved by adjusting the spring at the top of sleeve pipe (Fig. 4.24B).

The switch resets when the disk assembly comes down due to fall of flow. Then the spring pulls back the magnet bringing the switch back in to a low- or no-flow condition. With a vane-operated flow switch, the operating principle of the switch is same except that a cam mechanism converts the angular movement of the vane to a suitable movement so the sleeve can operate the magnet by the switch.

4.4.3 Diaphragm Type Flow Switch

The diaphragm type flow switch works by an acting force due to DP on a set of diaphragms (see Fig. 4.25A).

Diaphragm type flow switches are similar in construction to DP switches. The only difference is that there is an opening between the inlet and outlet port letting fluid flow through the switch, whereas it is completely closed in DP switches. The opening acts as an orifice and a DP is developed, which is dependent on liquid flow. The

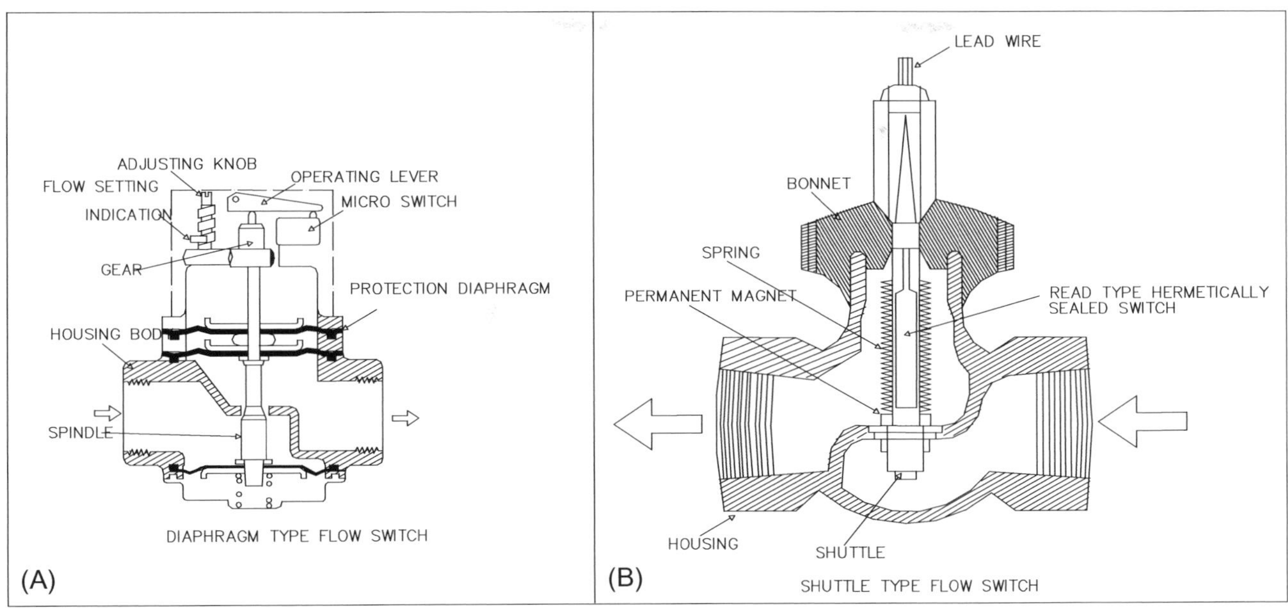

FIG. 4.25 Cutout view of diaphragm and shuttle type flow switch.

precondition of installing these switches is that the inlet port should be on the same line as the outlet port. When this is done, the liquid flow normally takes a zigzag route inside the switch body. This phenomenon leads to a large pressure drop, which is established when a liquid passes through the flow switch. Due to the action of DP across the self-created orifice of the switch housing, and the compression spring attached to the diaphragm assembly through a spindle, the spindle begins to rise or go down depending on the magnitude of flow. The higher value of flowrate would create a higher pressure drop and cause the spindle to move downward and upward if the flowrate goes down. The top attachment of the spindle with a projected lever is arranged to operate the micro switch. Setpoint adjustment facility at various flowrates is available. This can be achieved with the help of a knob provided at the gear assembly. This type of switch is sensitive to specifically low liquid flowrates with a minimum of acting components and is capable of withstanding high internal pressures.

4.4.4 Shuttle Type Flow Switch

Shuttle type flow switches work on the principle of an acting force due to DP or the velocity of the liquid acting on a disk. A shuttle is attached to the disk at the bottom and a magnet is just above the disk (Fig. 4.25B). The upper part of the shuttle body (spindle) moves up against a compression spring or gravitational force. The hollow spindle or stem incorporates a hermetically sealed reed switch within the shuttle.

As liquid flow increases over to the actuation setting, the shuttle is displaced. Upon displacement by fluid flow, this shuttle actuates the reed switch within the unit stem; the shuttle returns when flow decreases and sits on the port seat. The adjustable versions of this flow switch incorporate an internal, adjustable bypass vane by a setpoint adjuster, which is controlled externally as described in the following clause (Fig. 4.25B).

4.4.5 Piston Type Flow Switch

In this type of switch, a piston is positioned in the flow path. The piston incorporates a permanent magnet within it and moves along with the piston. When the DP is high enough to press the piston to the upside against the spring force, the magnet actuates the hermetically sealed reed switch placed outside the mechanical arrangement but coupled magnetically. The piston-metering land diameter dictates the bypass clearance and the actuation point is set (Fig. 4.26).

Fig. 4.26A shows the operation of the fixed setpoint type described above. Fig. 4.26B shows the operation of an adjustable setpoint type where an externally adjustable rotatable vane is provided in the main flow path to determine the bypass flow and the DP across the adjustable vane via the piston for actuation.The magnet-carrying piston is placed in the bypass line and it moves as the DP is adjusted.

4.4.6 Thermal Mass Flow Type

This type of flow switch features thermal dispersion technology. Here, the temperature difference is minimum for a full-flow condition and continues to increase as flow reduces for a cooling effect of the heated RTD. Changes in flow velocity directly affect the extent to which heat dissipates and, in turn, the magnitude of the temperature

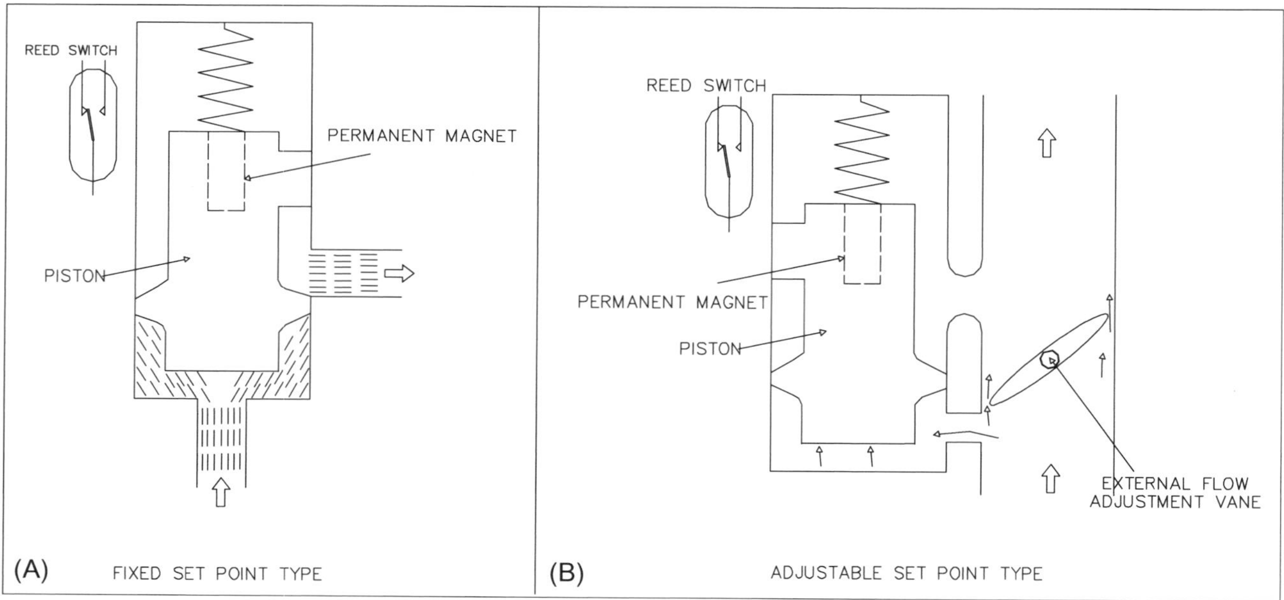

FIG. 4.26 Cutout view of piston type flow switch.

difference between the RTDs. An electronic control circuit converts the RTD temperature difference into a DC voltage signal. Both of these signals are provided at output terminals and are used to drive two adjustable setpoint alarm circuits. Both alarms are independently field configurable for flow, liquid level/interface, or temperature operations.

4.4.7 Flow Switch With Velocity-Based Actuation

The principle of operation is the same as the vortex type flow transmitter as they both operate on the vortex principle. The obstruction (shedder body) placed in the flow of the liquid sheds vortices downstream at a frequency proportional to the velocity of the liquid. This pattern of vortices is named after the von Karman Vortex Street. A piezoelectric sensor detects the vortices and generates voltage pulses, which are proportional to the liquid velocity by the flowrate. The instruments may be mounted in any position; the local output meter unit can be rotated. The instruments are supplied with two alarm outputs or with an analog output, thereby operating as a flow switch.

Typical specification sheets for different flow switches are indicated in Tables 4.23–4.27.

4.5 Flow Gauge

Flow meters are normally used to assess the performance of a system, and for calculations of the energy/power or heat rate/efficiency of a plant requiring continuous measurement of flow, and the use of electronic flow sensors rather than flow gauges without requiring any power. There are many different methods for measuring flow, and the direct form of those measuring instruments are flow gauges. Generally, flow gauges are called sight flow gauge glasses, which work without any external power supply. Some of the flow gauges can also provide electronic flow sensors for remote transmission (flow meters). Self-contained flow meters or sight flow gauge glasses, on the contrary, rely only on the dynamic characteristics of flow to provide a visual indication of it. Different manufacturers employ different design criteria, which change from one to another, but the basic concept of flow gauge operation depends solely on the principle of dynamic pressure.

Sight flow gauge glasses are the simplest type of flow gauges. They are is basically made of toughened glass or acrylic pipe or tubes, suitably protected from mechanical damage, through which the fluid passes and some sort of mechanism is provided to move a pointer on a graduated scale to indicate the flowrate. They provide a reliable, fast, and low-cost solution to measuring fluid flow. Flow gauges may be of the following type:

4.5.1 Sight Flow Glass

These types are only for viewing flow or no-flow conditions without any calibrated scale attached to it.

4.5.1.1 Full View Flow

This type is simply a piece of pipe or tube of transparent material through which the fluid passes and the flow is visible from outside just for a confirmatory flow or no-flow signal. The piece of pipe or tube diameter is normally more than the connecting process pipe or tube diameter to avoid any pressure loss due to the instrument.

TABLE 4.23 Data Sheet of Flapper/Paddle Type Flow Switches

SL. No.	Item	Specification	Special Feature
1	Make/model	Magnetrol	
2	Range	33–500 LPM (Eq. water)	
3	Setpoint	Low or high (adjustable)	
4	Sensing material	Brass/316 SS /PVC (for corrosive liquid)	
5	Repeatability	±1%–2% FSR	
6	Medium	Liquid/air/gas	
7	Switch unit	2 SPDT	
8	Rating	5 Amps, 230 V AC	
9	Line size	15–300 mm	
10	Maximum velocity	10 m/s	
11	Process connection	Threaded (25 mm)/Flange (50 mm)	
12	Casing/casing material	Weather proof, IP 65, die cast aluminum	
13	Electrical connection	1/2″ ET/3/4″ ET complete with cable gland	
14	Maximum line pressure	10 kg/cm^2(brass), 30 kg/cm^2 (SS)	
15	Maximum process temperature	100°C (brass), 170°C kg/cm^2 (SS)	
16	Mounting	Horizontal/vertical	

TABLE 4.24 Data Sheet of Target/Disk/Vane Type Flow Switches

SL. No.	Item	Specification	Special Feature
1	Make/model	Magnetrol	
2	Range	2–50 LPM (Eq. water)	
3	Setpoint	Low or high (adjustable)—one setpoint	
4	Sensing material	Brass/316 SS /PVC (for corrosive liquid)	
5	Repeatability	±1%–2% FSR	
6	Medium	Liquid/air/gas	
7	Switch unit	2 SPDT	
8	Rating	5 Amps, 230 V AC	
9	Line size	10–50 mm	
10	Maximum velocity	10 m/s	
11	Process connection	Threaded (25 mm)/Flange (50 mm)	
12	Casing/casing material	Weather proof, IP 65, die cast aluminum	
13	Electrical. connection	1/2″ ET/3/4″ ET complete with cable gland	
14	Maximum line pressure	16 kg/cm^2	
15	Maximum process temperature	110°C	
16	Mounting	Horizontal/vertical	

TABLE 4.25 Data Sheet of Shuttle Type Flow Switches

SL. No.	Item	Specification	Special Feature
1	Make/model	Magnetrol	
2	Range	2–50 LPM (Eq. water)	
3	Setpoint	Low or high (adjustable)—one setpoint	
4	Sensing material	Bronze/316 SS/Teflon (for corrosive liquid)	
5	Repeatability	±1% FSR	
6	Medium	Liquid/air/gas	
7	Switch unit	2 SPDT	
8	Rating	5 Amps, 230 V AC	
9	Line size	25–75 mm	
10	Maximum velocity	10 m/s	
11	Process connection	Threaded/flange	
12	Casing/casing material	Weather proof, IP 65, die cast aluminum	
13	Electrical connection	1/2″ ET/3/4″ ET complete with cable gland	
14	Maximum line pressure	30 kg/cm^2	
15	Maximum process temperature	150°C	
16	Mounting	Horizontal/vertical	

TABLE 4.26 Data Sheet of Piston Type Flow Switches

SL. No.	Item	Specification	Special Feature
1	Make/model	Magnetrol	
2	Range	0.5–500 LPM (Eq. water)	
3	Setpoint	Low or high (adjustable)—one setpoint	
4	Sensing material	Bronze/316 SS/Teflon (for corrosive liquid)	
5	Repeatability	±1% FSR	
6	Medium	Liquid/air/gas	
7	Switch unit	2 SPDT	
8	Rating	5 Amps, 230 V A.C	
9	Line size	8–80 mm	
10	Maximum velocity	2 m/s	
11	Process connection	Threaded/flange	
12	Casing/casing material	Weather proof, IP 65, die cast aluminum	
13	Electrical connection	1/2″ ET/3/4″ ET complete with cable gland	
14	Maximum line pressure	50 kg/cm^2	
15	Maximum process temperature	120°C	

TABLE 4.27 Data Sheet of Diaphragm Type Flow Switches

SL. No.	Item	Specification	Special Feature
1	Make/model	Magnetrol	
2	Range	10–50 LPM (Eq. water)	
3	Setpoint	Low or high (adjustable)—one setpoint	
4	Sensing material	Nitrile butadiene rubber (NBR)/Bronze/SS	
5	Repeatability	±1% FSR	
6	Medium	Water/liquid (noncorrosive)	
7	Switch unit	2 SPDT	
8	Rating	5 Amps, 230 V AC	
9	Port size	10, 15, and 20 mm	
10	Maximum velocity	2 m/s	
11	Process connection	Threaded/flange	
12	Casing/casing material	Weather proof, IP 65, die cast aluminum	
13	Electrical connection	1/2″ ET/complete with cable gland	
14	Maximum line pressure	12 kg/cm^2, 1–6 kg/cm^2 (op.)	
15	Maximum process temperature	5–60°C	

4.5.1.2 Rotary Type

Rotary type sight flow indicators are used in both transparent solutions and gases. These instruments are equipped with a rotating wheel and are suitable for installation in any position. However, they are also applicable for pipelines with dark opaque solutions where the motion of the wheel is easily detected. These indicators are most suitable for distant viewing purposes.

4.5.1.3 Drip, Ball, and Flapper Types

These are all direct type indicators without calibrated scales.

4.5.2 Variable Area Type

The principle of operation is already described in Clause 4.2.2.1. The rotameter or valve type area meters are generally used as an indicating instrument, but with help of attaching different types of followers with electronic transducer, the remote transmission is used.

4.5.3 Variable Orifice Type

A tapered shaft and a spring-loaded piston are the main components of this type of flow gauge. When there is no fluid flow, because of the action of the actuating spring the piston remains in its seat or initial position. As fluid enters from the inlet side, pressure acts on the piston against the spring and pushes upward to open the orifice formed between the inner diameter (ID) of the piston and outer diameter (OD) of the tapered shaft. For each flow, the piston assumes a particular position to allow the process flow through the corresponding orifice area. At that new position, the spring compression force is equal to the new DP produced by the new flow multiplied by the piston area. The position of the piston is now in equilibrium, when then provides an indication of flow. Flow can be measured directly by attaching a pointer to the piston position to a calibrated scale marked on the flow meter's transparent outer case. For remote transmission, the piston has an embedded magnet that moves a follower that has an electronic transducer.

5 LEVEL MEASUREMENT

Level measurements are of great importance for any industry or plant because water is required everywhere, but this type of measurement is specifically important for thermal power plants. All the regenerative heaters, gland coolers, ejectors, deaerators, and most importantly, boiler drums, require level measurements.

Some typical level measuring points for a 250 MW thermal power plant are given in Table 4.28.

5.1 Level Transmitters: Working Principle

The level measurement can be broadly categorized in two parts: continuous or point values. Continuous level

TABLE 4.28 Typical Level Measuring Points

SL. No.	Service Description	Operating/MCR Value	Calibrated Range
1	Boiler drum level (NWL-CL = −100 mm)	±5 mm wcl	−400 to +500 mm wcl
2	HP Heater 7 level (horizontal type) (NWL-CL = −500 mm)	NWL	500 mm wcl ±50 mm wcl (control)
3	HP Heater 6 level (horizontal type) (NWL-CL = −490 mm)	NWL	±500 mm wcl ±50 mm wcl (control)
4	Deaerator level (NWL-CL= +600 mm)	NWL	−750 mm wcl
5	LP Heater 4 level (horizontal type) (NWL-CL = −350 mm)	NWL	−400 mm wcl −100 mm wcl (control)
6	LP Heater 3 level (horizontal type) (NWL-CL = −250 mm)	NWL	−300 mm wcl
7	Gland steamer cooler level	1000 mm wcl	0–1500 mm wcl
8	Cooling water makeup head tank	1800 mm wcl	0–2500 mm wcl
9	DMCW, makeup tank level	1400 mm wcl	0–2000 mm wcl
10	Condensate storage tank level	1800 mm wcl	0–2500 mm wcl
11	Clarified water reservoir level	1400 mm wcl	0–2000 mm wcl
12	CW pit level	3000 mm wcl	0–4000 mm wcl
13	Condenser hotwell level	600mm	0–850 mm wcl
14	RFW tank level	4000 mm wcl	0–5000 mm wcl
15	Ammonia dosing tank level	1500 mm	0–2000 mm wcl
16	MU dosing tank level	2000 mm	0–3000 mm wcl
17	Acid dosing tank level	1500 mm	0–2000 mm wcl
18	Anti-scalant dosing tank level	1500 mm	0–2000 mm wcl
19	Phosphate solution tank level	700 mm	0–1000 mm wcl
20	Sodium solution tank level	700 mm	0–1000 mm wcl
21	Hydrazine dosing tank level	700 mm	0–1000 mm wcl
22	HFO storage tank level	1500 mm	0–2000 mm wcl
23	LFO storage tank level	1500 mm	0–2000 mm wcl

CW, circulating water; HFO, heavy fuel oil; LFO, light fuel oil; LP, low pressure; MU, morpholine; RFW, reserve feed water; DMCCW, demineralized water clarified cooling water; NWL, normal water level; HP, high pressure.

instruments measure levels within a specified range and determine the accurate amount (taking the accuracy figure into consideration) of the substance in a certain place. There are two major types of level measurement: direct and indirect. The first one employs a variation of levels directly to obtain a measurement. The other one measures levels indirectly by using a dependable variable that changes with the level. There are several level transmitters and they are grouped in the indirect category. Examples include pressure transmitters, DPTs, and transmitters employing ultrasonic sound waves, radio waves, radioactive signals, bubbler method, load cells, capacitance change, conductivity change, etc.

5.1.1 *Pressure or DP Type Level Transmitter*

Pressure or DPT transmitters have been used to measure liquid levels of open or pressurized tanks, respectively, with proper calibration taking care of the density value of the subject liquid. These types are used for level measurements of boiler drums, all heaters, deaerators, condenser hotwells, condenser surge tanks, and all dosing tanks.

For open tanks, a pressure tapping at a suitable location is connected through the impulse line, to the high-pressure port of the DPT with the low-pressure port open to atmosphere. The pressure (or DP) sensed by the DPT is proportional to the level with a known liquid density (ρ) and force due to (earth's) gravity (g). With proper calibration, the potential transform can now be used as a level transmitter.

Regarding the pressurized tank or vessel, single-pressure tapping does not serve this purpose as it senses the pressure of the vessel and the pressure of the liquid level. To balance out the vessel pressure, another pressure tapping for sensing it is connected, through the impulse line, to the low-pressure port of a DPT so that the output will only be proportional to the liquid level only with known ρ and g.

5.1.2 Displacer Type Level Transmitters

These types of transmitters utilize the buoyancy effect of a submersible matter in a liquid. A cylindrical solid metal body called a displacer is made to hang from a mechanical arrangement to sense the force exerted by the displacer submerged in the liquid of the vessel. The displacer may be located in the tank or in an extended part of the vessel according to the design to create the same level as in the vessel forming the sensing part of the transmitter. With the changes of the liquid level, the displacer is subjected to weight loss and in turn force exerted by the displacer will vary.

With the help of suitable linkage, the force exerted by the displacer is converted into torque and ultimately into 4–20 mADC and superimposed digitized signal.

5.1.3 Ultrasonic Type Level Transmitters

Ultrasonic type level measurements are used when making contact with the fluid or solid materials is not permitted and these are called noncontact type level measurements. Services, such as measuring the level of deep wells/hazardous areas/lack of tapping facility/highly viscous liquids, bunkers with bulk solids, water treatment applications/open channel measurement, etc., are accomplished through ultrasonic transmitters. The sensors are equipped with hardware capable of emitting high-frequency (20–200 kHz) sound waves that are reflected back to the sensor and detected by its receiving section.

As the velocity of sound is dependent on wave propagation, ultrasonic level sensor outputs are also liable to be contaminated by the changing velocity of sound due to environmental moisture, temperature, and pressures. Modern transmitters are capable of taking corrective actions to the same to improve measurement accuracy.

It is extremely important that sensors be mounted properly to get the best response from reflected sound waves. To avoid unfaithful return resulting in inaccurate output, the bunker, hopper, bin, tank, and vessel should be made relatively free of obstacles such as welding, brackets, or ladders around the measuring point. The modern measurement systems are equipped with a proper "intelligent" echo-processing arrangement to take care of this, but the "line of sight" of the transducer to the target is blocked.

There are several other conditions like turbulence, froth/foam/fumes, vapors of water (steam), or other chemicals, and even the changes in the concentration of the process material affect the output of ultrasonic sensors. Turbulence and froth/foam prevent the sound wave from being properly reflected to the sensor; fumes and vapors distort or absorb the sound wave. The amount of energy in the reflected sound waves depends on the concentration of the subject material and any variation therein will affect output response of the sensor. To circumvent those effects, deployment of stilling wells and wave guides are used as a common feature.

It is to be noted here that in this type of measurement, the ultrasonic sensor is designed to be used for both transmitting and receiving sound waves, and it is subject to a period of mechanical vibration popularly known as "ringing." This vibration should cease before the echoed signal can be processed. The net result is a distance from the face of the transducer within which the sensor cannot detect an object and is technically termed as blind or dead zone or blanking zone, typically 150 mm to 1 m, depending on the range of the transducer.

The low-level signal output from the sensor is then processed in the transmitter, which may be located a few meters away in an accessible place. The transmitter converts the input into 4–20 mADC and superimposed digitized signal in terms of level.

5.1.4 Conductivity Type Level Transmitters

Conductivity type level sensors, consisting of a probe with two electrodes, are used with power sources that have low-voltage and current-limiting capabilities. The voltage is applied across the two electrodes. The power supply is selected according to the conductivity of the liquid. The higher voltage versions will obviously be designed for the less conductive (higher resistance) mediums and vice versa. The electric circuit is completed via the conductive liquid contacting both the longest probe and a shorter probe. Conductivity type sensors are extremely safe because of low voltages and currents. They are intrinsically safe for use in hazardous areas. These sensors are simple and easy to install, but are affected by buildup. They will stop working properly if buildup insulates the electrodes from the conductive medium. Therefore, proper and regular maintenance is required to keep them running.

Conductivity type level sensors are suitable for the point level detection of a wide range of conductive liquids, such as water, and are well suited for highly corrosive liquids, such as acids like hydrochloric acid, nitric acid, sulfuric acid and caustic soda, ferric chloride, etc. For those corrosive conductive liquids, the sensor's electrodes should be constructed from 316 SS, Hastelloy B or C, or titanium as per the severity of the corrosiveness of the process fluid and insulated with spacers, separators, or holders of ceramic, polyethylene, and Teflon-based materials. Because corrosive liquids become more aggressive as temperature and pressure increase, suitable measures should be considered while specifying the instruments.

It is interesting to note that, in most of the services like water and wastewater wells, associated ladders, pumps, and other metal installations provide a ground return. However, in chemical tanks, insulated tanks, concrete tanks, and other nongrounded wells, the sensor may be supplied with an earth/ground rod for proper functioning of the instrument.

The low-level signal output from the sensor is then processed in the transmitter, which may be located a few meters away in an accessible place. The transmitter converts the input into 4–20 mADC and superimposed digitized signal in terms of level.

5.1.5 Capacitance or RF Type Level Transmitters

Capacitance type level transmitters are widely used in sensing the presence of various types of liquids, slurries, and solids. The sensors are equipped with hardware capable of emitting radio frequency signals acting in a capacitive circuit with the process material as the dielectric. Dual-probe capacitance type level sensors are also used to sense the interface between two immiscible liquids where dielectric constants are substantially different such as an "oil-water interface" application. This method of level measurement is frequently and popularly referred to as the RF type because of the application of radio frequency signals.

The advantage of this type of sensor is that it is rugged, simple, easily maintainable, and has no moving parts. Suitable designs for high temperature and pressure applications are also available. However, a problematic area does exist regarding the danger of high-static voltage discharge out of the rubbing and movement of low dielectric materials. This problem can be eliminated with a better design and proper grounding. An appropriate choice of materials as wetted parts in the probe may reduce, if not eliminate, the problems caused by abrasion and corrosion.

Another problem is that the output is also affected by buildup. Point level sensing of adhesives and high-viscosity materials such as oil and grease can result in the buildup of material on the probe; however, this can be minimized by using a self-tuning sensor. Therefore, proper and regular maintenance is required to keep this sensor running.

Capacitance type probes cannot be factory calibrated, instead these are calibrated in place so suitable care need to be taken. Splashguards or stilling wells and other appropriate devices may be necessary where process materials are liable to produce foam or froth and process applications are designed to control splashing or turbulence.

There use of capacitance probes is limited in tall bins storing bulk solids. Long probes and cables may be subjected to mechanical tension and friction due to the weight and abrasive nature of the bulk powder in the bins, which can frequently cause cable breakage.

Capacitance type level sensors can be designed to sense levels of process materials with a wide selection of dielectric constants ranging from 1.1 (coke and fly ash), 50 (sludge and slurries such as dehydrated cake), 88 (sewage slurry), 90 (liquid chemicals such as quicklime), or more. They are widely used in thermal power plants for level measurements of solids in ash hoppers and coal bunkers and liquids in cooling towers, DM plant tank levels, and CW storage wells/pond, etc.

5.1.6 Air Bubbler Type Level Transmitters

Air bubbler type level transmitters are used for open type vessels only. The pressure created by a liquid column is used in the bubbler method of level measurement. A pipe is installed vertically in the vessel with its open end downward covering the minimum level. The other end of the pipe/tube is connected to a regulated air supply and a pressure transmitter. A fixed flow of air is passed through the tube. The air supply pressure in the tube is adjusted in such a way so as to keep it slightly higher than the pressure exerted by the process liquid column. This can be accomplished by regulating the air pressure until the bubbles are seen slowly leaving the open end of the pipe or tube. Pressure in the tube is proportional to the depth (and of course the density) of the liquid over the outlet of the tube.

Air bubbler systems contain no moving parts, making them suitable for measuring the level of water with large quantities of suspended solids, sewage, drainage water, etc. The bubbler tube is the only part of the sensor that makes contact with the liquid, which must be compatible for handling with the material under measurement. This technique of level measurement may be a good and natural selection for classified "hazardous areas" if a pneumatic pressure transmitter is used. This type of measurement is not normally used in thermal power plant applications but discussed in view of its popular use in other industries.

In air bubbler systems provision should be made so that high-pressure air can be supplied through a bypass valve to clean the inside of the pipe/tube from solids that may clog the bubbler pipe/tube. This technique is inherently "self-cleaning." It is highly recommended for liquid level measurement applications where ultrasonic, float, or microwave techniques have proved undependable.

5.1.7 Magnetostriction and Guided Wave Radar

Both of these systems utilize a wave guide to measure the level in open as well as pressurized tanks. In power plant applications these types of instruments are quite popular for regenerative heater level measurements. Basic principles and data sheets of these instruments are reviewed in Clause 5.4, so they are not repeated here.

5.2 Level Switch: Working Principle

The point-level or discrete-level sensors only indicate whether the substance is above or below the sensing point. In general, those types are called level switches. Generally the discrete-level type detect levels that are excessively high or low. There are many types of level switches used in thermal power plants.

5.2.1 Magnetic and Mechanical Float Type Switch

The float type level switches can have different actuating principles of operation: magnetic, mechanical, cable, etc. They are involved in the opening or closing of a mechanical switch, either through direct contact with the switch (micro switch) or magnetic operation of a reed type switch.

In magnetically actuated float sensors, a permanent magnet sealed inside a float rises or falls to the actuation level, thereby the switching takes place through a magnetically coupled reed switch in the switch assembly. With a mechanically actuated float, switching occurs as a result of the direct movement of a float against a miniature (micro) switch. For both magnetic and mechanical float level switches, different factors like temperature, specific gravity/density (ρ), buoyancy, viscosity, etc., affect the selection of the stem and the float. For example, larger floats may be used with liquids with specific gravities as low as 0.5 while still maintaining buoyancy. The type of float used depends on the process fluid. The choice of float material is also influenced by temperature-induced changes in specific gravity and viscosity changes that directly affect buoyancy.

To protect the float-type sensors a shield can be provided so that the float can operate in a wide variety of liquids, including corrosives. Otherwise chemically compatible floats must be used for organic solvents. Float-style sensors should not be used with high-viscosity (thick) liquids, sludge, or liquids that adhere to the stem or floats or materials that contain contaminants such as metal chips. In these cases, a suitable protective below or gaiter must be used. A special application of float type sensors is the determination of interface level in oil-water separation systems. Two floats can be used with each float sized to match the specific gravity of the oil on one hand, and the water on the other. Another special application of a stem-type float switch is the installation of temperature or pressure sensors to create a multiparameter sensor. Magnetic float switches are popular for their simplicity, dependability, and low cost. Float switches can be side mounted and/or top mounted. Top-mounted switches may have multiple floats for multiple setpoints.

5.2.2 Gamma Ray Type Switch

A nucleonic level switch or gamma ray type switch measures the material level by the attenuation of gamma rays passing through a process vessel. The technique is used to sense the coal level in a bunker or level of molten steel in the process of a steel plant. The sensor case is arranged with a source of radiation, such as Cobalt 60 or Cesium 137 on one side and an appropriate detector on the other side. As the level of process material rises in the vessel, less of the gamma radiation is detected by the sensor. This technique falls under the noncontact type measurement category and can be both a continuous and point measurement.

5.2.3 Displacer Type Level Switch

This type of switch utilizes the buoyancy effect of submersible matter in a liquid. and the operating principle was discussed in Clause 5.1.2. A cylindrical solid metal body used as a displacer is provided for one setpoint. Multiple displacers are used in the same attachment for multiple setpoints and switches.

5.2.4 Vibration or Tuning Fork Type Level Switch

In this type of switch, a tuning fork is energized by a piezo crystal and vibrates at a resonant frequency of ~380 Hz. The receiver side incorporates another crystal to detect this frequency. When the tuning fork is covered by liquid, the resonant frequency is reduced by a dampening effect. This change is detected by the integral electronics and converted into a switch signal.

The piezo crystal driving the vibrating tuning fork is either screwed together or glued as per manufacturer's design. What makes the piezo crystal an unusual type of switch is that the output is unaffected by changing temperature, pressure, conductivity, density, or viscosity. Suitable design may take care of heavy buildup on forks. Typical applications are in storage and processing (coal level in the feeder bunker/hopper).

With proper selection of vibration frequency and suitable sensitivity adjustments, the following types of solids can be sensed and switching contacts will be available by this instrument:

1. Very fine powders (bulk density: 0.02–0.2 g/cm^3).
2. Fine powders (bulk density: 0.2–0.5 g/cm^3).
3. Granular solids (bulk density: 0.5 g/cm^3 or greater).
4. Highly fluidized powders.
5. Electrostatic materials.

Single-probe vibrating level sensors are ideal for bulk powder levels. Because only one sensing element contacts the powder, bridging between two probe elements is eliminated and media buildup is minimized. The vibration of the probe tends to eliminate buildup of material on the probe element. Vibrating level sensors are not affected by dust, static-charge buildup from dielectric powders, or changes in conductivity, temperature, pressure, humidity, or moisture content. They are less costlier than other similar types of instruments.

5.2.5 Rotating Paddle Type Level Switch

Rotating paddle level sensors are a popular selection in various type of bunkers and hoppers in thermal power plants, cement plants, and other material-handling plants. They have been used for many years with proven and established techniques for bulk solid-point level indication. The principle of operation uses a paddle wheel with a low speed that is rotated by a motor with a gear arrangement that turns the output shaft with the necessary speed. When the solid level reaches the paddle, it will stall because the solid materials will not allow the paddle to rotate anymore. The motor will still try to rotate with increased torque until a mechanical switch called a torque switch is operated. The buildups may pose a problem if the process material becomes sticky because of high moisture content in the atmosphere around the bunker or hopper.

Special paddle designs with low-torque motors are used for materials with low specific density, such as fly ash or bentonite. The paddle can be constructed from a variety of materials, but, to avoid buildup on the paddle, sticky material cannot be used.

5.2.6 RF Admittance Type Level Switch

An RF admittance level sensor employs a rod probe and RF source to detect the change in admittance. To eliminate the effects of the changing capacitance of the cable to ground, a shielded coaxial cable is used through which the power is supplied to the probe .The dielectric constant changes with the change in level around the probe. This changes the admittance due to corresponding, naturally formed capacitance, and this change is measured and converted into change in level.

Another type is available in which the probe electronic unit generates an RF signal and creates a field around the probe. The RF environment absorption changes around the electrode when the level changes and is reflected in the changes of generator supply current. When the level rises there is electrical field loss. This is detected by a change in generated current. This change is amplified by a coat guard amplifier (unaffected by external field and deposition over the probe). Typical specification sheets of different flow switches are indicated in Tables 4.29 and 4.30).

TABLE 4.29 Data Sheet of Float Type Level Switches

SL. No.	Item	Specification	Special Feature
1	Make/model	Magnetrol/Leveltrol	
2	Material	Body, external float chamber, and other wetted parts; ANSI 316 SS	
3	Spring	AISI 304 SS	
4	Flange	C.S. ANSI 150 RF	
5	Switch cover	Cast aluminum weather proof, as per IP 66; anti-corrosive painted	
6	Spring pipe	316 SS	
7	Signalization	Hermetically sealed glass encapsulated reed switch/snap or micro switch	
8	Contact form	2 NO + 2 NC	
9	Contact rating	5 A to 230 V, AC 50 Hz	
10	Repeatability	±2%	
11	ON/OFF differential	45 ± 5 mm (typical)	
12	Hydraulic test	Pressure: 10 kg/cm^2	
13	Maximum working temperature	100°C	

TABLE 4.30 Data Sheet of Displacer Type Level Switches

SL. No.	Item	Specification	Special Feature
1	Make/model	Magnetrol/Bestobell	
2	Material	Body, external float chamber, and other wetted parts; ANSI 316 SS	
3	Spring	AISI 304 SS	
4	Flange	C.S. ANSI 150 RF	
5	Switch cover	Cast aluminum weather proof, as per IP 66; anti-corrosive painted	
6	Number of setpoints	One/two/three	
7	Signalization	Hermetically sealed glass encapsulated reed switch/snap or micro switch	
8	Contact form	2 NO + 2 NC	
9	Contact rating	5 A to 230 V, AC 50 Hz	
10	Repeatability	±1%	
11	ON/OFF differential	45 ± 5 mm	
12	Hydraulic test	Pressure: 20 kg/cm^2	
13	Maximum working temperature	100°C	
14	Quantity	As per BOM	

5.3 Level Gauge: Working Principle

Level gauges form both direct and indirect types of level measurements. Direct type level gauges are do not involve any secondary device, such as the simplest and oldest type, such as bob and tape and other instruments such as sight level gauges, float and pulley types, displacer types, etc. The indirect type level gauges, on the contrary, require secondary device for processing input signals, which sense the effect of changing levels, for example, conductivity type level gauges and remote bicolor gauges.

5.3.1 Direct Level Gauges

There are several types of direct level measurements as indicated in the following clauses.

5.3.1.1 Sight Level Gauge Glass

As the name implies, the level can be seen in a graduated glass tube directly connected to the vessel. Normally the glass tube is located near the vessel intended for level measurement. The vessel is provided with two tapping points, covering the entire range of measurement, which are connected to the glass tube by the gauge glass assembly with an integral ball check valve for automatic isolation in case any breakage or damage takes place at the instrument end. As the bottom portion of both the vessel and measuring glass are connected, the liquid level in the measuring glass will be the same (except for very high pressure and temperature services where density plays a considerable role for a mismatch) as that of the vessel. Measurement is a simple matter of directly reading the position of the liquid level on the scale of the sight glass tube. See Fig. 4.27 for a general view of a sight level gauge glass and operation of a ball check valve shown for a boiler drum level service (this also applicable to other services).

In generally, there are two types of sight level gauge glasses: transparent and reflex. For colored liquid, the transparent type is sufficient to show the level, but for colorless liquid the transparent type may confuse the viewer, so the reflex type level gauges are used.

For reflex type gauges, the sight glass plate is grooved in a horizontal saw tooth fashion at an angle of 45° throughout its height. Fig. 4.28 illustrates the operating principle for reflex type gauges with a cross-sectional view of sight level gauge glass.

When the incident light ray reaches the reflex grooves of the glass plate, it is totally reflected if there is steam, vapor, or gas present in the region. As a result, the gaseous space looks like a bright silvery bar. When there is liquid, the incident ray is deviated and will not come back the same way. As a result, the liquid space looks like a black bar.

In some cases when the sight glass is required to be isolated from the corrosive liquid service, there are mica sheets in between the glass and liquid with the same effect as seen in general transparent type and reflex type gauges (Fig. 4.29).

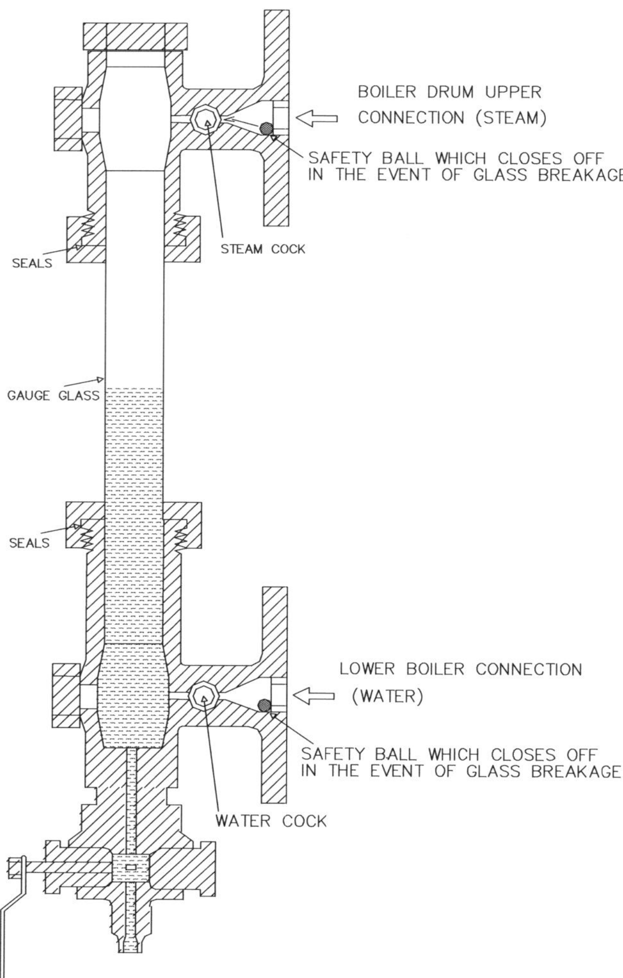

FIG. 4.27 Sight level gauge glass: mountings and accessories.

For transparent type gauges, the incident light ray from a rear-side lamp source reaches the sight glass at an angle of 90°, and there is no special effect of the viewing liquid and gaseous state; these gauges can easily be used for colored liquid. For colorless liquid, the axis of the rear-side lamp source and that of the viewer's eyes are different. The incident light ray from a rear-side lamp source is guided through the two mica sheets at an angle less than 90° and passes the medium between them. For the gaseous space, the light is guided straight forward and passes through the gas and both mica sheets without any deviation. In liquid space, the light ray passes the liquid medium and the mica sheets at an angle so they are refracted away due to a different index of refraction. The liquid level is like a black bar and the gas is like a bright, silvery bar.

5.3.1.2 Bicolor Level Gauges

Bicolor level gauges are used for boiler drum level measurement in thermal power plants where the steam and water interface is distinctively seen. There are a number of port holes throughout the range of measurement for the viewer to observe the approximate water level of the boiler drum. When put into operation, the viewer sees a green light through the port holes located in the water zone and a red light through port holes in the steam zone. With a long glass gauge, that is, without any discrete port holes, a light part green and part red appears on the bezel.

The principle of operation is based on the theory of an optical phenomenon. It is well known that different colors (in this case red and green) have a different index of refraction when passing obliquely through different media like glass, water, and steam, as in this case (Fig. 4.30).

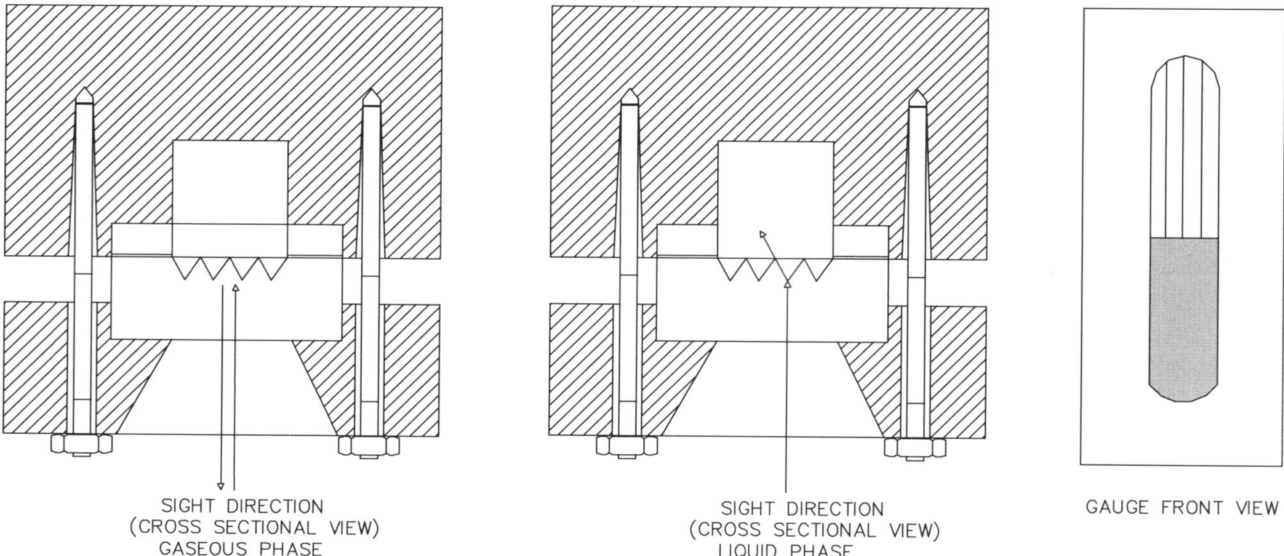

FIG. 4.28 Reflex type sight level gauge glass: schematic view.

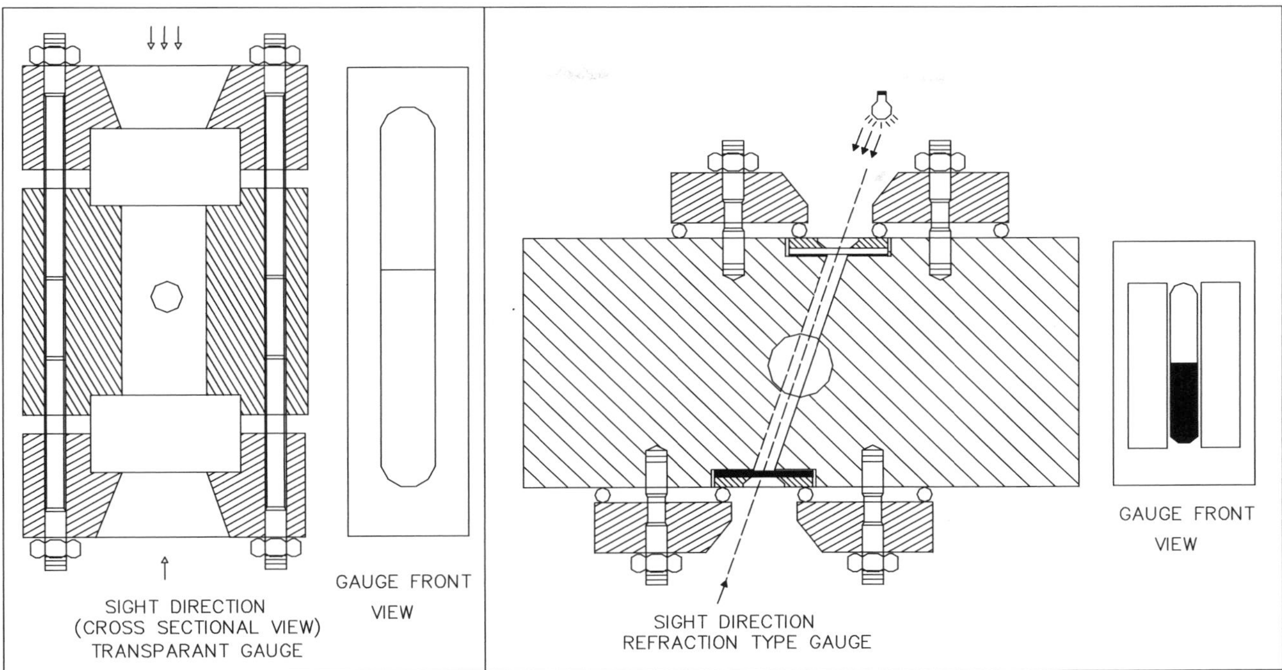

FIG. 4.29 Sight level gauge glass with mica sheets: schematic view.

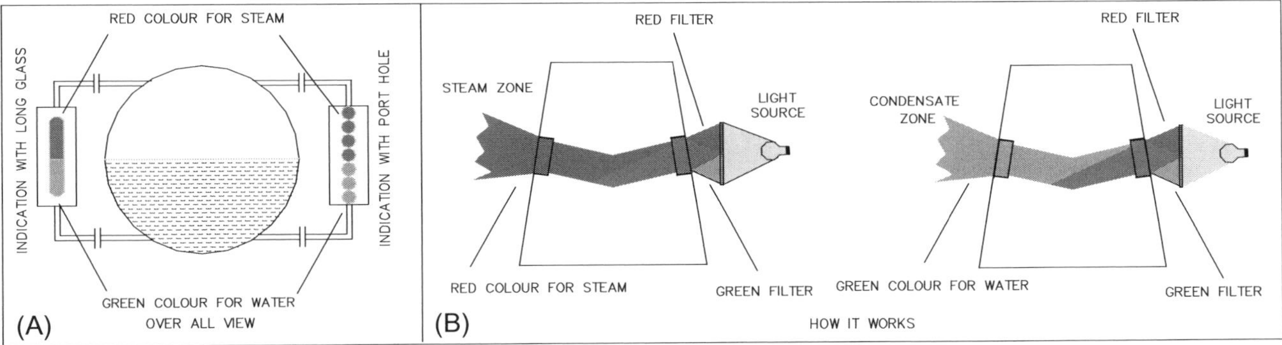

FIG. 4.30 Bicolor gauge glass: schematic view.

To achieve the desired result, the gauge body has a trapezoid section with glasses placed on its nonparallel faces. A lamp or an illuminator with special red and green filters is fitted on the gauge at the opposite side with respect to the observer. This special illuminator, as shown in Fig. 4.30, conveys light through the filters obliquely to the back glasses of the level gauge. These filters allow crossing only to red and green rays, which reach the media inside the level body through the back glass.

When the gauge port hole part contains steam, green rays are refracted to deviate considerably and are not allowed to reach the observer side; then only red light, whose rays are smoothly deviated by steam, passes through the entire internal hole and reaches the observer. When the gauge port hole part contains water, red rays are refracted to deviate considerably and are lost inside the internal part of the level gauge. Green rays are allowed can be seen in the observer's side glass.

5.3.1.3 Float Type Gauge

There are many types of float-operated level mechanisms for continuous direct liquid level gauges. The main and primary device is a float, which by reason of its buoyancy, follows the change in level of the liquid. The other mechanism transfers the float's movement to a pointer movement on a calibrated scale. The float can be shaped like a hollow metal sphere or disk, like synthetic material floats, or cylindrical-shaped like ceramic floats.

5.3.1.4 Float Pulley Type Gauge

Here the float movement is transferred to the pointer through a string or cable that is wound around a pulley or drum to which the indicating pointer is attached. At the other end of the cable, a firmly suspended weight keeps the string/cable under tension.

5.3.1.5 Float and Rotary Shaft Type Gauge

In this type of gauge, the float is attached to a shaft that transfers the motion of the float to an indicator. The main disadvantage is that a wide range of level measurement is not permitted. In spite of that, this gauge type has a mechanical advantage, which may be exploited for control and transmitter purposes.

5.3.1.6 Float and Magnet Type or Magnetic Level Gauge

A slight variation of the previous float and rotary shaft type gauge is the float and magnet type gauge. In a magnetic level gauge, there is a permanent magnet sealed inside a float, which rises or falls according to the level, thereby actuating a magnetically coupled pointer/indicator outside the vessel. The pointer/indicator freely moves on a calibrated scale showing the level of the fluid inside the vessel.

5.3.1.7 Displacer Type Gauge

It is similar to the buoyant float described earlier. With changes in the liquid level, the displacer is covered by the liquid. The more the displacer is submerged, the greater the force created by the displacer because of its buoyancy. This force is then transferred through a twisting or bending shaft. For every new liquid level, there is a new force on the shaft, causing it to assume a new position. The displacer float is more sensitive to small changes than the buoyant float and less subject to mechanical friction.

5.3.2 Indirect Type Level

Many indirect types of level gauges are available, but the conductivity type level gauge is the most popular.

5.3.2.1 Conductivity Type Level Gauge

These gauges are primarily meant for boiler drum level indication in thermal power plants. This is an indirect type of instrument that detects the electrical conductivity of steam and water by placing electrodes at different levels throughout the range of measurement. They are highly reliable and best suited for noticeably high temperature and pressure conditions (Fig. 4.31).

The electrodes are connected to the electronic section located in the gauge or indicator, which measures the conductivity between the electrodes and the metallic wall of stand pipe or the water column connected to the boiler drum through two pipes with double valves in each line and vent/drain valves. Each electrode is connected to a separate detection and indication channel. As the conductivity in the media like steam and water are widely varied, the electronic unit can easily distinguish the difference between the two phases.

The number of electrodes is solely dependent on the range or height of the level to be measured. The approximate distance between the electrodes is normally 50 mm. Although the instrument is a type of discrete or step measurement, the advantages are no moving parts, no calibration, easy installation, and no periodic maintenance. Alarm contacts are extremely reliable from this unit and usually connected to the master fuel tripping circuit and annunciation system.

5.3.2.2 Pressure Type Level Gauge

The simplest method is to use a pressure gauge with an impulse pipe that has a tapping point at zero level for corrosive material. A transmitting fluid, for example, air, can be used between liquid and the gauge. An air trap or diaphragm box provide the necessary isolation arrangement.

The bubbler methods, as discussed in Clause 5.1.6, is also a proven type of level measurement. For a pressurized vessel, a differential manometer may be used as discussed for DPTs in Clause 5.1.1.

5.4 Magnetostrictive and Guided Wave Radar Level Instruments

5.4.1 Introduction

One of the main ways to improve the efficiency of a plant is to reduce fuel costs. Optimum level control of feed water is one important way to increase the heat and the heat rate and therefore, the fuel flow. Regenerative heater-level controls in power plants are the major areas where magnetostrictive and guided wave radar (GWR) instruments are mostly applied. These instruments are discussed in this clause (Fig. 4.32).

5.4.2 Magnetostrictive Instruments

Prior to discussing the magnetostrictive instruments, it is necessary to have some knowledge about magnetostriction and magnetostrictive materials, because they work on the principle of magnetostriction. That is, with like heat it is possible to elongate and shorten a material with a magnetic/electric field. Ferromagnetic materials can be made longer and shorter by application and withdrawal of magnetic fields. Magnetostrictive materials can be magnetostrictive metals, alloys, or ferrite magnetostrictive materials.

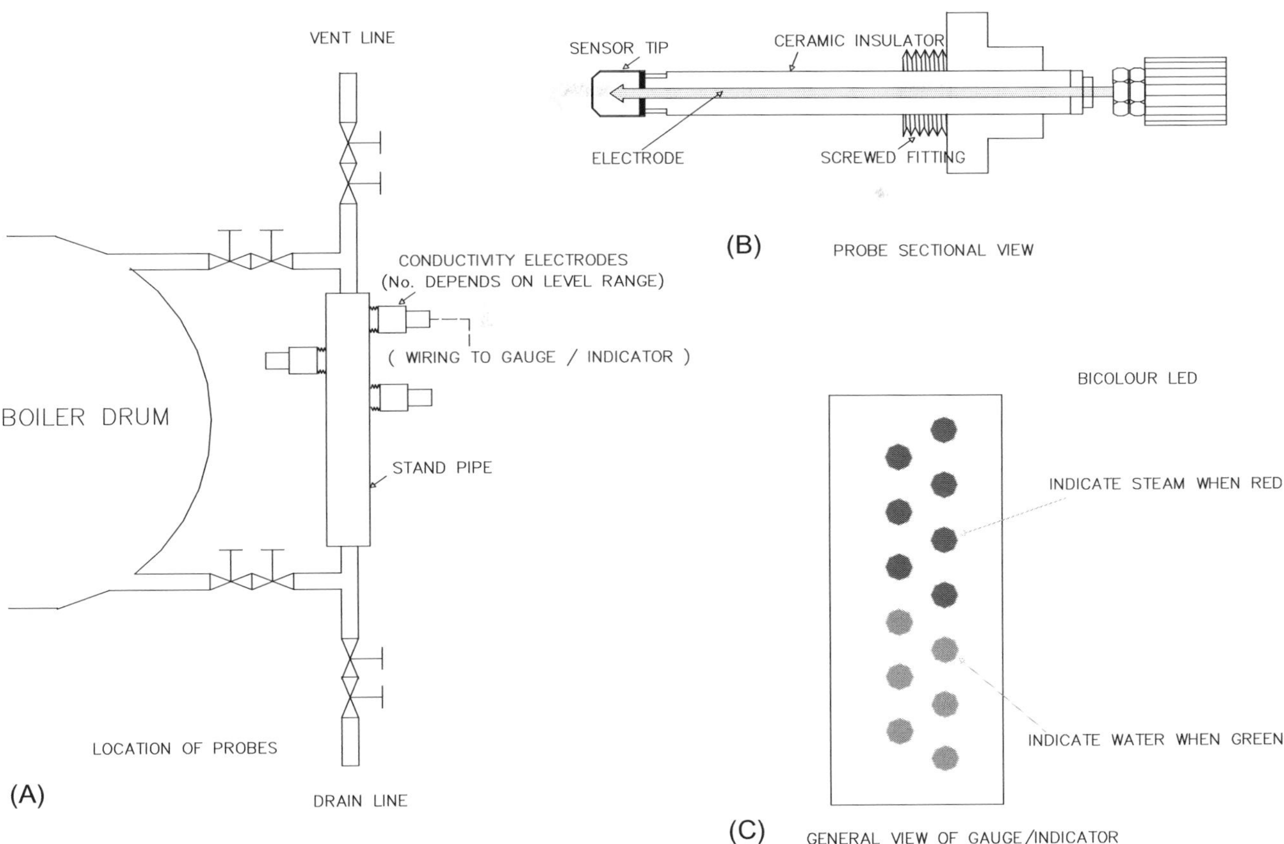

FIG. 4.31 Conductivity type level gauge: different views.

5.4.2.1 Working Principles

These instruments utilize magnetostrictive theory to measure liquid levels. Fig. 4.32A illustrates a wave guide made up of magnetostrictive material. The measurement is initiated by an interrogative current pulse through the wave guide/wire (magnetostrictive material). This generates an axial magnetic field (as shown in blue) along the length of the probe. There is a float with a magnet inside with its magnetic field at a right angle to the axial field as discussed earlier (due to current pulse). According to the level the float moves up and down and has a permanent inside (the float) magnet with its own magnetic field. The interaction of these two magnetic fields generates a mechanical torsion wave from the point of the float (or position transducer) in the wave guide or wire. This wave travels in both the directions. At the probe bottom end there is a damper that dampens the wave, whereas the other side (i.e., toward the probe head) reaches a piezoelectric sensor connected to the wire. Thus, at the head, the mechanical torsion wave is converted into an electrical pulse by a piezoelectric transducer located in the sensor housing. The time lapse between the starting pulse and receiving pulse is computed in the probe electronics to calculate the distance and the level. These kinds of probes can withstand harsh environmental conditions and they can be installed and commissioned easily (Table 4.31 for specification and standards for magnetostriction instruments).

5.4.3 *GWR*

There are two kinds of radar devices used in level measurements: free space (as used in tank gauging) and guided. The guided type is used in power plant applications. Normally microwaves in the range of 1.5–26 GHz are used for radar instruments. The lower end, that is, ~1.5 GHz, frequency range is used for GWR and the higher range is suitable for free space radar. This is based on the simple principle of reflected pulse. If T = total time to travel from device to liquid and back and electromagnetic waves travel at the speed of light C, then the distance (level) is $L = C \times T/2$.

The reflection property of the material depends on the dielectric number of the medium. In GWR transmitters direct energy pulses via wire/wave guide to the medium whose level needs to be measured. A major part of the energy is reflected back and measured by the electronics at the probe head; the time difference is a measure of the level. Normally process conditions such as vapor, temperature, and pressure do not affect radar instrument operation.

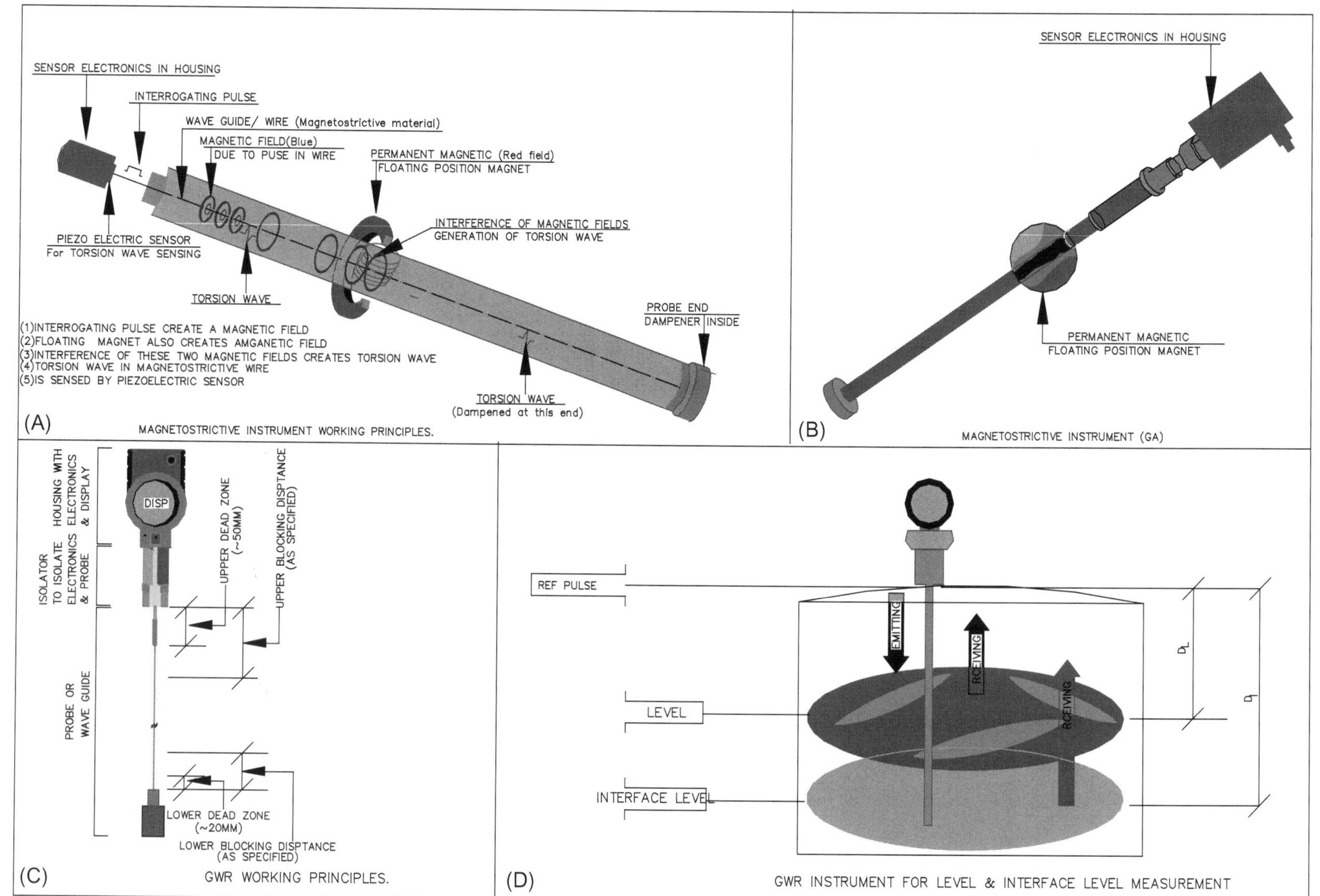

FIG. 4.32 Magnetostrictive and GWR instruments.

TABLE 4.31 Data Sheet for Magnetostriction Instruments

SL. No.	Specifying Point	Standard/Available Data	To Be Specified
1	Type	Magnetostrictive	
2	Measuring range	Up to 15 m is possible—consultation with details shown in manufacturer's catalog	To be specified
3	Tube material	316 SS	
4	Mounting	Threaded or flange	To be specified
5	Probe OD	Normally 16 mm (5/8″)	
6	Tr. Housing	Aluminum/SS	To be specified
7	Enclosure	IP 68 or better	
8	Out put	1 or 2 4–20 mA DC; also HART, Foundation Fieldbus	To be specified
9	Resolution	0.5% FSD	
10	Accuracy	0.01% FSD with repeatability ~0.005%	
11	Display	LED/LCD (backlit) and diagnostic display in alphanumeric characters	
12	Power supply	230 V AC/24 V DC as the case may be	
13	Ambient conditions	*T*: −40 to 70°C reheat (100%)	To be specified

TABLE 4.32 Data Sheet for GWR

SL. No.	Specifying Point	Standard/Available Data	To Be Specified
1	Type	GWR—rigid rod, flexible cable coaxial, etc.	To be specified
2	Measuring range	Up to 50 m possible—consultation with details shown in manufacturer's catalog	To be specified (as required)
3	Tube material	304/316L SS or Hastelloy C	
4	Isolation material	Teflon/ceramic	To be specified
5	Pressure temperature (°C) and pressure (bar)	As the case may be; also clarify if dynamic vapor compensation necessary	To be specified
6	Pressure connection/wave guide coupler	Threaded or flange	To be specified
7	Probe attachments	Centering disk, centering weight, etc.	
8	Transition housing	Aluminum/SS	To be specified
9	Enclosure	IP 68 or better	
10	Certification	International certification/ SIL requirement	To be specified if applicable
11	Output	1 or 2 4–20 mADC, also HART, and Foundation Fieldbus	To be specified
12	Resolution	1.6 mm available	To be specified
13	Accuracy	3/5 mm available	To be specified
14	Display	LED/LCD (backlit) and diagnostic display in alphanumeric characters	
15	Transition zone	Consult manufacturer's catalog	
16	Power supply	230 V AC/24 V DC, as the case may be	
17	Ambient conditions	*T*: −40 to 70°C reheat (100%)	To be specified

This is true as long as the vapor is not a polar gas whose dielectric number varies with process pressure and temperature, and the speed of microwave pulse may also vary.

It has been found that steam vapor at high pressure and temperature with a dielectric number >1.0 may affect the performance of the radar. However, there are probes with dynamic vapor compensation options used to compensate for changes in vapor space dielectric to get better accuracies. In power plants, GWR is applied for measurement of regenerative heater levels.

Fig. 4.32C depicts different parts of the probe. Electronics, displays, and adjustment devices are usually housed at the top of the probe. An isolation material, such as Teflon, is suitable for low temperature (i.e., ~150°C), which isolates the probe from its electronics to help eliminate high-temperature exposure to the electronics. In a high-temperature application, steam may damage the probe so ceramics may have to be used. It also houses the microwave frequency generators and receivers. The entire probe length cannot be effectively used for measurements because it has upper and lower blocking and a dead zone, as shown in Fig. 4.32C. This instruments also can be utilized for interface level measurements (Table 4.32 for specifications and standards of GWRs.)

BIBLIOGRAPHY

[1] J.P. Holman, Experimental Methods for Engineers (Pressure Measurement), sixth ed., 1994, (Chapter 6).

[2] Electronic Pressure Measurement Basics, applications and instrument selection: By M/s Eugen Gaßmann, Anna Gries (technical collaboration of M/s WIKA Alexander Wiegand SE & Co. KG.). First published in Germany in the series Die Bibliothek der Technik Original title: Elektronische Druckmesstechnik © 2009 by Süddeutscher Verlag onpact GmbH.

[3] Product Bulletin for Sight Glass Level Gauges of M/s PHONIX MESSTECHNIK GmbH.

[4] Pubblicazione B09-01-Rev.00. Eng: By M/s CESARE BONETTI S.p. A. Product Catalog.

[5] E. Oberg, F. Jones, H. Horton, H. Ryffel, C. McCauley (Eds.), Machinery's Handbook, 29th ed., 2012. Hardcover Published. ISBN 9780831129002, 2800 pp.

[6] Product Catalog of M/s Omega Engineering, Inc.

[7] Product Catalog-2600 Series Pressure Transmitters of M/s ABB Ltd.

[8] Product Catalog-600T Series Temperature Transmitters of M/s ABB Ltd.

[9] Product Catalog-for Multivariable Transmitters (Mass Flow) of M/s ABB Ltd.

[10] Product Catalog-for Thermal Mass Flowmeter of M/s ABB Ltd.

[11] Product Catalog-for Vortex Flowmeter of M/s ABB Ltd.

[12] Product Catalog-for Magnetic Flowmeter of M/s ABB Ltd.

[13] Product Catalog-for Ultrasonic Level & Flowmeter of M/s ABB Ltd.

[14] Product Catalog-for Coriolis Type Mass Flowmeter of M/s ABB Ltd.

[15] Product Catalog-for WEDGE Type Flowmeter of M/s ABB Ltd.

[16] Turbine Supervisory Instrumentation: By M/s SKF cmcp.

[17] Product Bulletin of M/s Burns Engineering, Inc., USA.

[18] Temperature Measurement: By Middle East Technical University Open Course Ware.

[19] Temperature Measurement: By Missouri University of Science and Technology, http://web.mst.edu/~cottrell/ME240/Resources/Temperature/Temperature.pdf

[20] Industrial Instrumentation (C-8\N-IND\BOOK1-6): By M/s IDC Technologies (India), http://www.idc-online.com/technical_references/pdfs/instrumentation/Industrial_Instrumentation%20-%20Flow.pdf.

[21] Product Catalog of M/s Imperil Flange and Fittings Company Inc (For Flow Elements).

[22] LMNO Engineering, Research, and Software, Ltd., USA (For Parshall Flume). Sharp Crested Weirs for Open Channel Flow Measurement, Course No: C02-022 Credit: 2 PDH Harlan H. Bengtson, PhD, P.E, http://www.cedengineering.com/upload/Sharp-Crested%20Weirs.pdf.

[23] Product Catalog for Ultrasonic Type Flow Measurement: By Fuji Electric Systems Company Ltd., Japan.

[24] Product Catalog for Diaphragm Style Flow Switch: By courtesy of M/s Steven Engineering, Inc.

[25] Product Catalog for Shuttle Type Flow Switch: By courtesy of M/s Clark Solutions, Inc.

[26] Product Information for Glass Level Gauges: By M/s KSR Kubler Niveau-Messtechnik AG.

[27] RF Level Measurement Handbook, 1999: By M/s Princo Instruments Inc. ELS 2005—Symposium on Unconventional Electrical Machines Suceava 22–23 September 2005. Method for Level Measurement in Hydroenergetic Accumulations Mariana MILICI, Leon MANDICI, Dan MILICI, http://www.els.usv.ro/pagini/past_editions/ELS%202005/B1-06.pdf.

[28] The Principles of Level Measurement October 1, 2000: By M/s Gabor Vass, Princo Instruments, Inc, http://www.sensorsmag.com/sensors/leak-level/the-principles-level-measurement-941.

[29] Product Catalog for RF Capacitance Level Controls (Form 1100): By M/s SOR Inc.

[30] MaxiFlo™ Product Catalog: By M/s Seil Enterprise Co: Seoul.

[31] Guided wave radar for level and interference: Vega America Inc.: Catalog.

[32] Guided wave radar Transmitter: www.smar.com: RD 400 catalog The Heat rate imperative: Special Application: Magnetrol Technical brocheure.

[33] Guided wave radar levelTransmitter: MT 5000 ABB Limited Catalog.

[34] Magnetostricktive Level Transmitter: AT 2000: ABB Limited Catalog.

[35] Advanced Differential Pressure Flow meter Technology: Macro-meter: ABLE Instruments & Controls Limited UK: Catalog.

[36] Coriolis Flow meter:Micromotion Tutorial: Micromotion- Emerson Process management.

[37] Industrial Flow measurement basic and Practice: ABB Limited: Technical book, ABB Automation Products GmbH, 2011.

[38] Coriolis Flow meter: Endress+Huaser: Catalog.

[39] A Flow Orientation: Flow & Level: Volume 4: Omega Instruments USA: Catalog.

Chapter 5

Special Instrument

Chapter Outline

1 SPECIAL INSTRUMENTS: INTRODUCTION

1.1 General

Apart from the conventional instruments in Chapter 4, there are a number of special instruments that have been included in the book, as follows:

- Vibration measurements in various fans, pumps, motors, mills, etc. (mechanical system).
- Turbovisory instruments (mechanical system, for turbine).
- Gas analysis (chemical system).
- Steam and water analysis (chemical system).
- Sample conditioning (for chemical analyzers).

Power Plant Instrumentation and Control Handbook. https://doi.org/10.1016/B978-0-12-819504-8.00005-6

- Blowdown and dosing control (chemical control).
- Pollution-related analyzer.

1.2 Vibration Measurement

- Vibration is the oscillatory motion of the object, and it can be linear or rotational.
- Relative position (proximity sensor), velocity (velocity sensor), and acceleration (accelerometer) are the sensors used to measure these parameters.

1.3 Turbovisory Instruments

Turbovisory instruments are provided to determine the deteriorating condition of the machine to ensure the safety of the machine as well as the operating personnel. Major turbovisory instruments shall include pedestal vibration, bearing vibration, shaft vibration, eccentricity, speed, valve position, metal temperature, axial shift, and casing expansion-differential expansion.

1.4 Gas Analysis

Gas analysis in the power plant is utilized for pollution control as well as the main equipment such as the SG and TG efficiency controls.

1.5 Steam and Water Analysis

The chemistry of the water and steam plays a great role in the performance and availability of the boiler and turbo generator. In case these chemical parameters are not properly monitored, this would cause corrosion and accelerate the deterioration of the plant performance due to depositions and corrosions. Conductivity, cat col. conductivity, pH, dissolved O_2, residual hydrazine, H_2 in steam, SiO_2 phosphate, and chloride analytical instruments are used.

1.6 Sample Conditioning

Prior to sending the sample to costly analytical instruments for analysis, it may be necessary to condition the sample. This is applicable for both gas analyzers (as applicable) and SWAS instruments.

1.7 Blowdown and Dosing Control System

In any power plant, there will be blowdown (continuous and intermittent) of the feed water (FW) to keep the quality of the FW.

To maintain the quality of the FW, it is necessary to dose some chemicals such as ammonia, hydrazine, phosphate, etc.

1.8 Pollution-Related Analysis

In order to meet the stringent pollution control requirements nowadays, it is necessary to install gas analyzers such as particulate monitoring, SO_X, NO_X, etc.

2 VIBRATION AND TURBOVISORY INSTRUMENTS

2.1 General IDEA of Vibration

- Vibration is an oscillatory motion of the object, set in a rotating machine on account of an imbalance of forces generated and/or acted upon the machine. Vibration frequencies are greatly related to the shaft speed and generally are an integral multiple of the shaft speed. The complexity in vibration measurement arises from the distribution of disturbances in a mechanical system.
- If an external (imbalance) force P_i is applied in Fig. 5.1A, then it will be set to vibrate, which will decay. The output will be given by:
  $P_i = md^2x/dt^2 + c\ dx/dt - k\ x$ where c is the damping factor (say friction) and k is the spring constant. Neglecting damping, the natural frequency will be given by $\omega = (k/m)^{1/2}$.
- An external imbalance force may be impressed to cause vibration. Oscillatory motion in its simplest form is

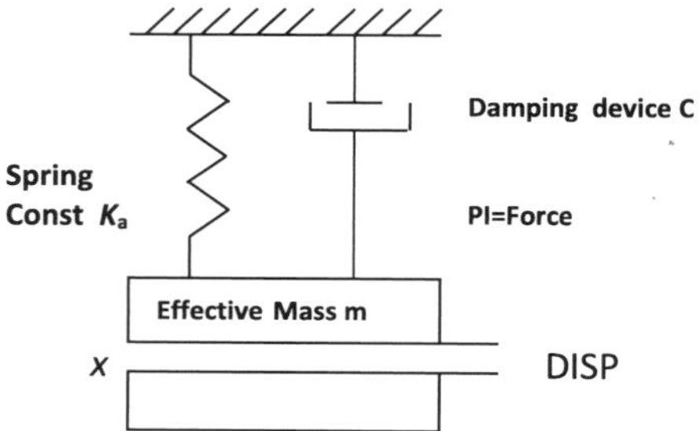

(A)

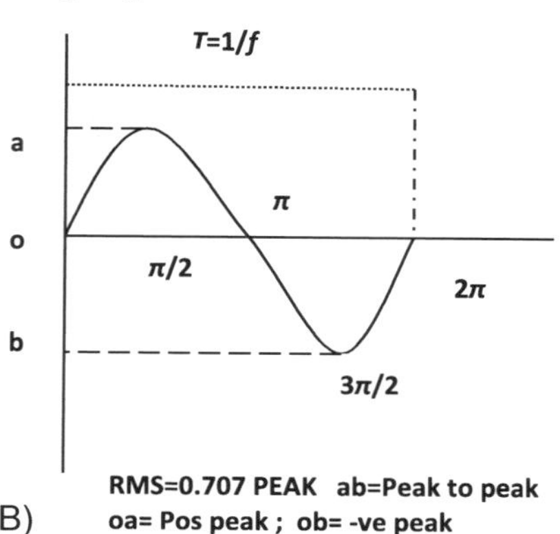

(B)

FIG. 5.1 (A) Simple vibration. (B) Oscillatory motion.

shown in Fig. 5.1B. It is seen that this is periodic in nature, having a time period of $T = 1/f$ where f is the frequency.

- The peak value is the extreme swing with respect to the datum in the positive and negative side, and peak = 1.414 RMS (root mean square value). It is worth noting that this RMS to peak conversion is true for the sinusoidal signal, but it will be wrong to multiply 1.414 to the RMS value of vibration some gets in his meter, to get peak value. So, it is necessary to ensure in peak or RMS—which form the meter output is given.
- The frequency of vibration is also very important. Generally, 10–1000 Hz is the range of frequencies encountered in a turbine system and drive vibration in power plant machines. In practice, it is quite common to measure the vibration in indirect form, for example, the transducer that operates on velocity can be used to get the displacement by simple integration.

2.2 Importance of Vibration Measurement

- To determine the natural frequency of vibration as well as to find out the damping ratio associated with it.
- The measurements of frequency and force of vibration in a rotating machine are necessary to develop a system to isolate it from vibration.
- At the time of manufacturing, all large rotating machines are properly balanced so as to minimize the chance of vibrations. The vibration frequency and force of vibrations are also predicted. However, in reality for various reasons (installation problems, linkage problems), the actual data may vary with the design data, so it is important to monitor vibrations in the machine to compare limiting data.
- Machine safety and prevention from failure/damage.

2.3 Basic Building Blocks for Vibration Measuring System

- Vibration transducer: Complete with a preamplifier, the transducer converts the motion (vibration) to an electrical signal. These pickups could be to measure displacement, velocity, or acceleration. The transducer pickups could be capacitance/eddy current/LVDT, or the piezoresistive type.
- Signal conditioning: The function of this is to convert the signals to a compatible electrical form so that they can be transmitted, displayed, or recorded. Nowadays, signal conditioning systems also perform the function of computation, communication, and many other software functions through software algorithms that are helpful for analysis, for example, FFT or software links for computer integration. Normally, they are complete with a power supply and monitoring system.
- Display/recording: There can be a simple monitor to display the vibration data. It can be a small subsystem connected to its own system by, say, a device net or local proprietary network for display and recording by the display monitor and printer. Normally, these are suitable for communication with the main DCS, by RS 232C/Ethernet, etc.
- The choice of pickup types can be categorized very roughly as follows:
  - Displacement measurements are done for very low frequency vibrations.
  - For moderately to moderately high frequency vibration, a velocity pickup is the choice.
  - When there is very high frequency vibration, and the vibrating force is measurable, an accelerometer can be used.

2.4 Vibration Measurement Points (Turbine Discussed Separately)—Condition Monitoring

In power plants, other than TG, there are a few other large drives where bearing vibrations are measured with suitable pickups (velocity type). A typical list with location is presented in Table 5.1.

In case the BFP is driven by a turbine, naturally the number of vibration pickups will be more and will be similar to what has been described for a small turbine.

2.5 Turbovisory Instruments: Measuring Points

2.5.1 *Steam Turbine (Fig. 5.2 and Table 5.2)*

Most of the possible measurements have been listed, and all may not be applicable for a single machine. Depending on

TABLE 5.1 Bearing Vibrations for Large Fans, Pumps and Motors

Equipment	No of Point	Location	Remarks
ID FAN	4–6	NDE and DE OF EACH OF FAN AND MOTOR	Depending on drive type
FD FAN	4		
PA FAN			
MILL	2	AT MOTORS	
BFP	4–6	NDE and DE OF EACH OF FAN AND MOTOR	Depending on configuration
CW PUMP	4		

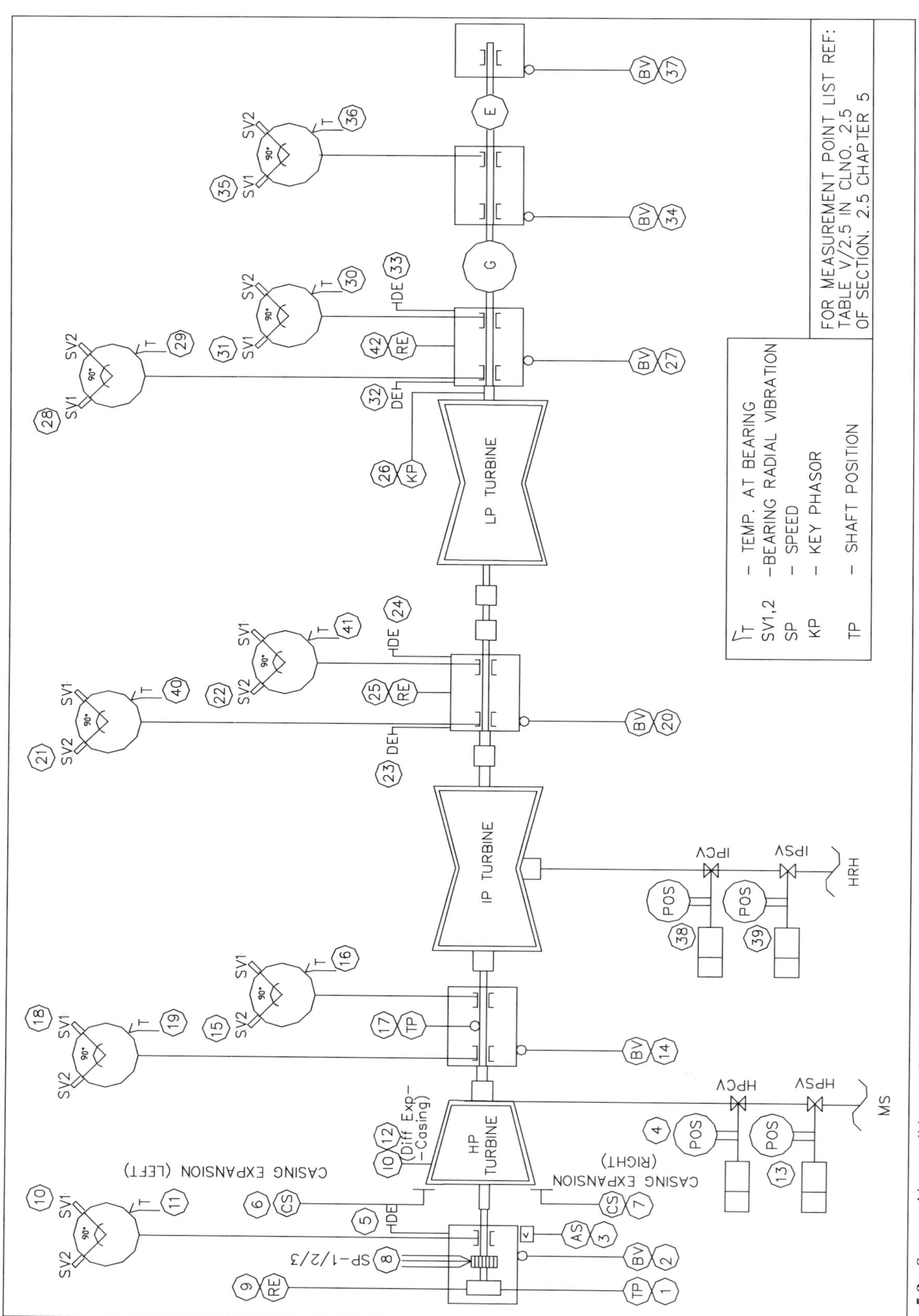

FIG. 5.2 Steam turbine condition monitoring.

TABLE 5.2 Steam Turbine Condition Monitoring

Sl	Meas. No.	Symbol	Measurement	Location	Purpose
1	3	AS	Absolute expansion	Front bearing pedestal	Absolute expansion (Location: fixed end shown one orientation only)
2	11	T	Bearing temp	Thrust bearing (HPT front)	White metal temperature of bearing
3	19	T	Bearing temp	HPT rear bearing	White metal temperature of bearing
4	16	T	Bearing temp	IPT front bearing	White metal temperature of bearing
5	40	T	Bearing temp	IPT rear bearing	White metal temperature of bearing
6	41	T	Bearing temp	LPT front bearing	White metal temperature of bearing
7	29	T	Bearing temp	LPT rear bearing	White metal temperature of bearing
8	30	T	Bearing temp	Gen front bearing	White metal temperature of bearing
9	36	T	Bearing temp	Gen rear bearing	White metal temperature of bearing
10	10	SV 1,2	Bearing vibration	Thrust bearing (HPT front)	Shaft vibration at bearing at 90°–*X*-*Y* axis
11	18	SV 1,2	Bearing vibration	HPT rear bearing	Shaft vibration at bearing at 90°–*X*-*Y* axis
12	15	SV 1,2	Bearing vibration	IPT front bearing	Shaft vibration at bearing at 90°–*X*-*Y* axis
13	21	SV 1,2	Bearing vibration	IPT rear bearing	Shaft vibration at bearing at 90°–*X*-*Y* axis
14	22	SV 1,2	Bearing vibration	LPT front bearing	Shaft vibration at bearing at 90°–*X*-*Y* axis
15	28	SV 1,2	Bearing vibration	LPT rear bearing	Shaft vibration at bearing at 90°–*X*-*Y* axis
16	31	SV 1,2	Bearing vibration	Gen front bearing	Shaft vibration at bearing at 90°–*X*-*Y* axis
17	35	SV 1,2	Bearing vibration	Gen rear bearing	Shaft vibration at bearing at 90°–*X*-*Y* axis
18	2	BV	Block vibration	Bearing pedestal HPT-Front	Bearing Bl. vibration (Abs)
19	14	BV	Block vibration	Bearing pedestal IPT-Front	Bearing Bl. vibration (Abs)
20	20	BV	Block vibration	Bearing pedestal LPT-Front	Bearing Bl. vibration (Abs)
21	27	BV	Block vibration	Bearing pedestal LPT-Rear	Bearing Bl. vibration (Abs)
22	34	BV	Block vibration	Bearing pedestal Gen-Rear	Bearing Bl. vibration (Abs)
23	37	BV	Block vibration	Bearing pedestal Exc'r-Rear	Bearing Bl. vibration (Abs)
24	6	CS	HP casing expansion left	Near HPT Fr. Pedestal	Expansion of HP casing (Left)
25	7	CS	HP casing expansion right	Near HPT Fr. Pedestal	Expansion of HP casing (Right)
26	5	DE	Diff. expansion	HPT Fr Brg Pedestal	Expansion of Rotor WITH RESPECT TO Near Bearing Pedestal
27	23	DE	Diff. expansion	IPT Rr Brg Pedestal	Expansion of Rotor WITH RESPECT TO Near Bearing Pedestal
28	24	DE	Diff. expansion	IPT Rr Brg Pedestal	Expansion of Rotor WITH RESPECT TO Near Bearing Pedestal
29	32	DE	Diff. expansion	LPT Rr Brg Pedestal	Expansion of Rotor WITH RESPECT TO Near Bearing Pedestal
30	33	DE	Diff. expansion	LPT Rr Brg Pedestal	Expansion of Rotor WITH RESPECT TO Near Bearing Pedestal
31	12	IO	Diff. expansion (casing)	HP casing	Diff. expansion inner and outer HP casing
32	26	KP	Key phasor	LPT Rr Brg Pedestal	Angle between heavy spot in rotor and reference mark

Continued

TABLE 5.2 Steam Turbine Condition Monitoringcont'd

Sl	Meas. No.	Symbol	Measurement	Location	Purpose
33	9	RE	Rotor Eccentricity	HPT Fr Brg Pedestal	Eccentricity—sagging of rotor due to uneven heating
34	25	RE	Rotor eccentricity	LPT Fr Brg Pedestal	Eccentricity—sagging of rotor due to uneven heating
35	42	RE	Rotor eccentricity	LPT Rr Brg Pedestal	Eccentricity—sagging of rotor due to uneven heating
36	1	TP	Shaft position	HPT Fr Brg Pedestal	Axial shift of the shaft (possibility—depends on turbine orientation. Redundancy asked for by manufacturer)
37	17	TP	Shaft position	IPT Fr Brg Pedestal	Axial shift of the shaft (possibility—depends on turbine orientation. Redundancy asked for by manufacturer)
38	8	SP	Speed	HPT Fr Brg Pedestal	Turbine speed most important parameter (so 2/3 redundancy common)
39	4	PoS	Valve position	HPCV	HP control valve position
40	13	PoS	Valve position	HPSV	HP stop valve position
41	38	PoS	Valve position	IPCV	IP control valve position
42	39	PoS	Valve position	IPSV	IP stop valve position

design and operational philosophy, the manufacturer recommends the most suitable measurements for the machine. However, measurements such as speed, bearing, temperature, and vibration measurements, axial shift, valve position, etc., are so important that these are seen in almost all machines.

2.5.1.1 Gas Turbine (GT) (Fig. 5.3 and Table 5.3)

However, the manufacturer sometimes also recommends a few supplementary measurements (e.g., GE's recommendation) for larger-frame machines.

- Casing Vibration: For machines with significant motion between the bearing housing and free space. This is a seismic vibration measurement, that is, an absolute vibration of the casing. The gas turbine which is not compliant on casing this is an optional measurement. Probes may be placed on the machine casing (radial bearing housing).
- Eccentricity: With a large machine with a heavy mass that can cause sagging, sensors shall be deployed to measure the residual bow during slow-speed operation.

2.6 Vibration and Turbovisory Measurement Issues

1. Speed Measurement SP (Fig. 5.4A): In fans, pump standard speed measuring instruments such as tachometers with transducers or photoelectric transducers are used. A toothed wheel may be used as a typical turbine speed measuring scheme, as shown in Fig. 5.4. Out of various types of sensors for turbine speed measurement, the Hall probe sensor is also popular (e.g., the Siemens turbine). Other types of transducers are proximity (namely, GE), etc. In general, especially for mid- to large-range turbines, redundant speed sensors are used. These speed sensors are directly connected to the governing control (EHG) and protection circuits, separately so that these are independent. The protection and measuring circuits may be in the triple modular redundancy (TMR, Chapter 3) configuration to ensure high reliability of protection. In some systems, separate electronic overspeed detection systems are available (e.g., GE-BN). Acceleration/deceleration measurements are also important for the operator during rolling of the turbine. The following points may be noted to understand the importance of this measurement:

a. It is worth noting that for a steam turbine, in its whole life, it is allowed to go into overspeed for only a *limited number* of times (as per the manufacturer's data), so the importance of speed measurements is extremely high.

b. In a turbine, the gap between the stator and rotor is very less (for larger turbine, gap is narrower), so, any overspeed can result in havoc to the turbine.

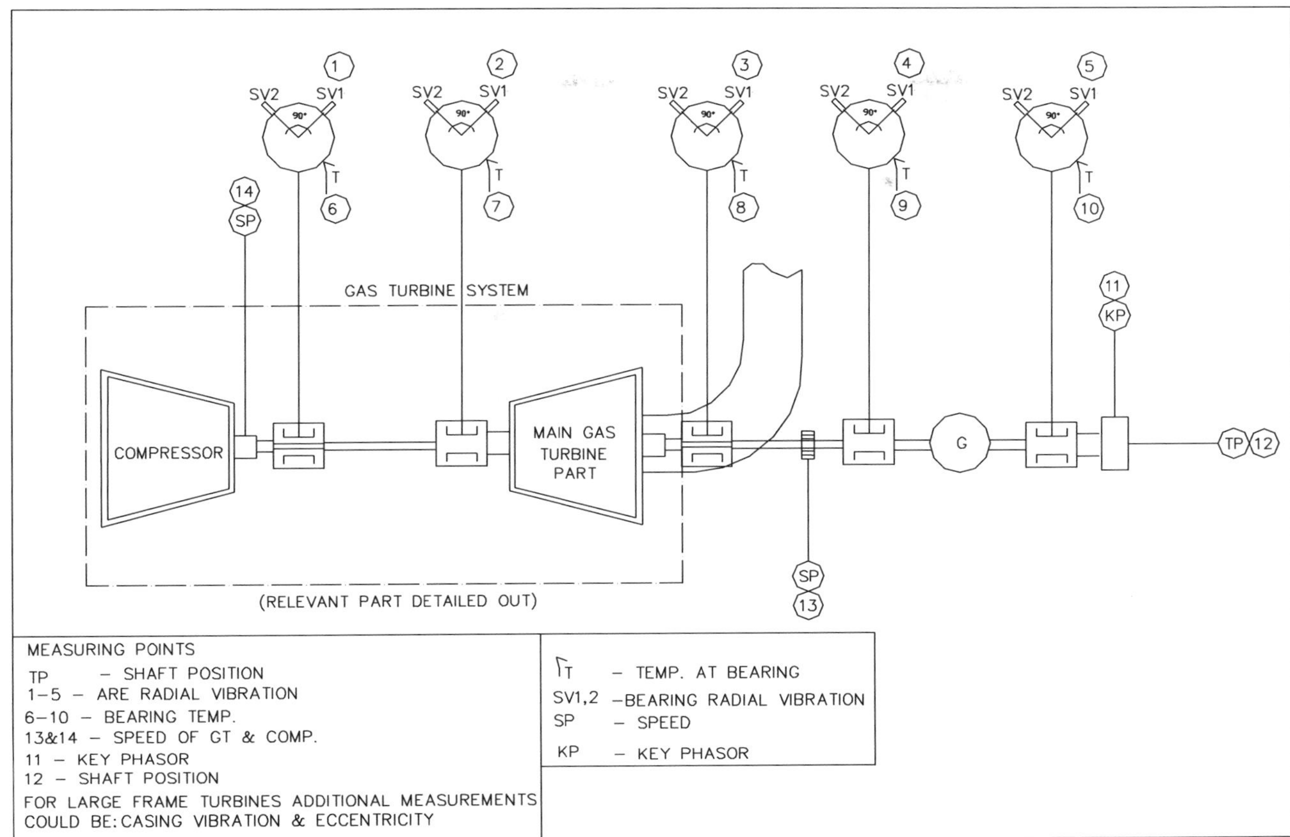

FIG. 5.3 Turbovisory instruments in gas turbine.

TABLE 5.3 Gas Turbine Condition Monitoring

Sl.	Meas. No.	Symbol	Measurement	Location	Purpose
1	1	SV 1,2	Bearing vibration	Comp end bearing	Shaft vibration at bearing at 90°–*X-Y* axis
2	2	SV 1,2	Bearing vibration	GT front. bearing	Shaft vibration at bearing at 90°–*X-Y* axis
3	3	SV 1,2	Bearing vibration	GT rear. bearing	Shaft vibration at bearing at 90°–*X-Y* axis
4	4	SV 1,2	Bearing vibration	Gen front bearing	Shaft vibration at bearing at 90°–*X-Y* axis
5	5	SV 1,2	Bearing vibration	IPT rear bearing	Shaft vibration at bearing at 90°–*X-Y* axis
6	6	T	Bearing temp	Comp end bearing	White metal temperature of bearing
7	7	T	Bearing temp	GT front. bearing	White metal temperature of bearing
8	8	T	Bearing temp	GT rear. bearing	White metal temperature of bearing
9	9	T	Bearing temp	Gen front bearing	White metal temperature of bearing
10	10	T	Bearing temp	IPT rear bearing	White metal temperature of bearing
11	13	SP	Speed	Compressor shaft	Compressor speed
12	14	SP	Speed	GT shaft	GT speed
13	11	KP	Key phasor	Thrust bearing pedestal	Angle between heavy spot in rotor and reference mark
14	12	TP	Shaft position	Thrust bearing pedestal	Thrust position

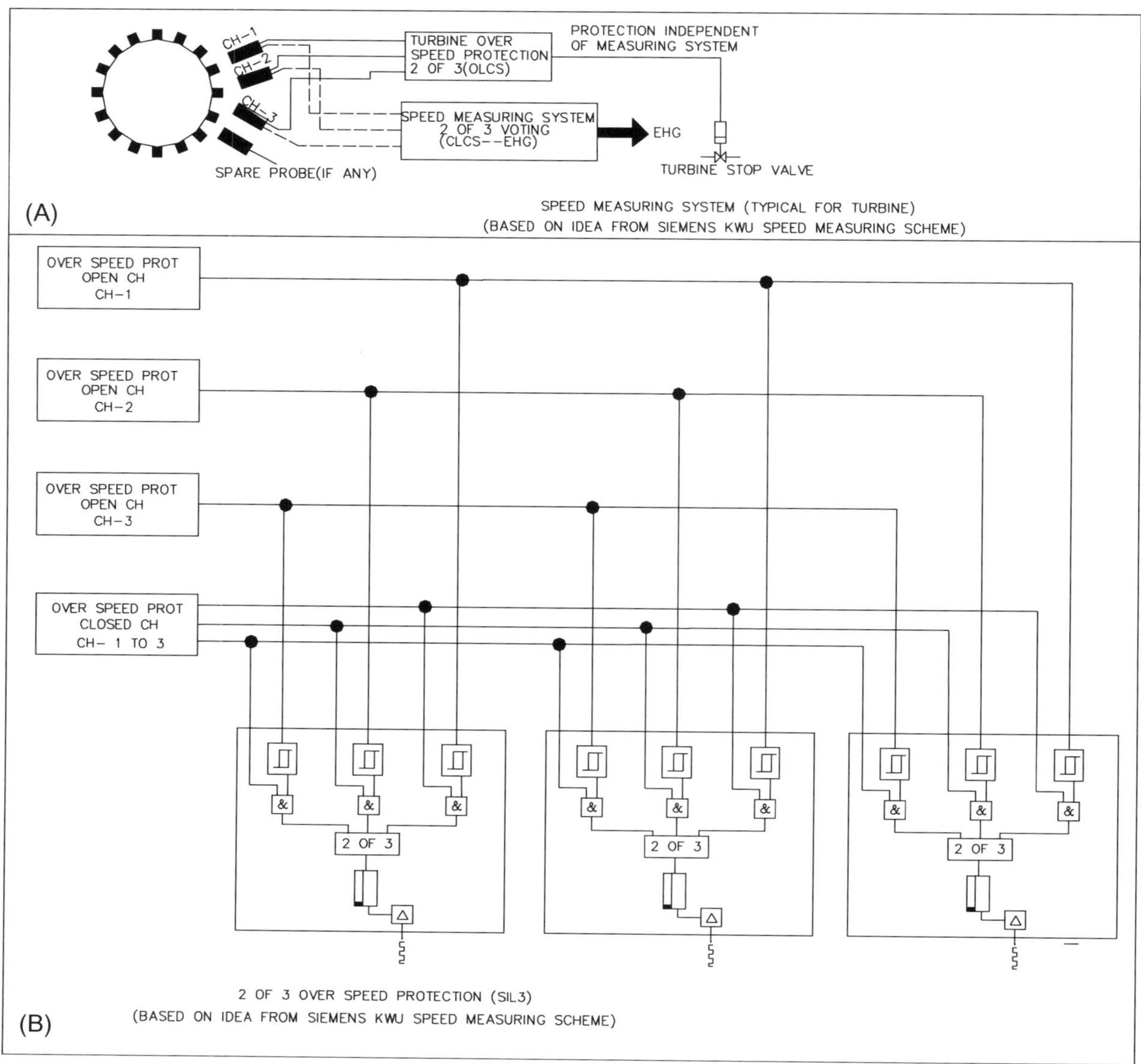

FIG. 5.4 Speed Measuring system.

c. During every start up, the operator needs to hold the speed during the *heat-soaking* period, whereas the operator has to cross quickly through critical speed (shaft critical frequency).

d. The turbine has to run at synchronized speed, as the generator (which is synchronized with the power system grid at a given frequency) is directly coupled with the turbine. However, the scenario is a little different for industrial turbines (e.g., BFPT), which are connected to the driving equipment through gear trains. However, in those cases, speed control is also important.

e. During an actual overspeed test, the governor accurately verifies the operation, so any overspeed will cause a protection trip of the turbine.

2. Each manufacturer has its own electronic overspeed detection system (*EODS*). In an EODS with *SIL3* approval (e.g., KWU), a signal from each channel is sent to the voting circuit as the open channel, whereas the same is sent through two of three voting circuits (as closed channel), giving three outputs to each of the protection channels—total of three protection channels (each two of three) for initiating a trip condition. Fig. 5.4B may be referenced. For TMR details, see Fig. 3.3.

3. Bearing vibration measurement (shaft vibration) SV1, 2 (Fig. 5.5A): As the vibration probes are placed radial to the vibration housing, as shown in Fig. 5.4A, they are also referred to as a radial vibration measurement. Here two probes are placed radially at 90°, as shown. Vibration probes may be in one plane or at two

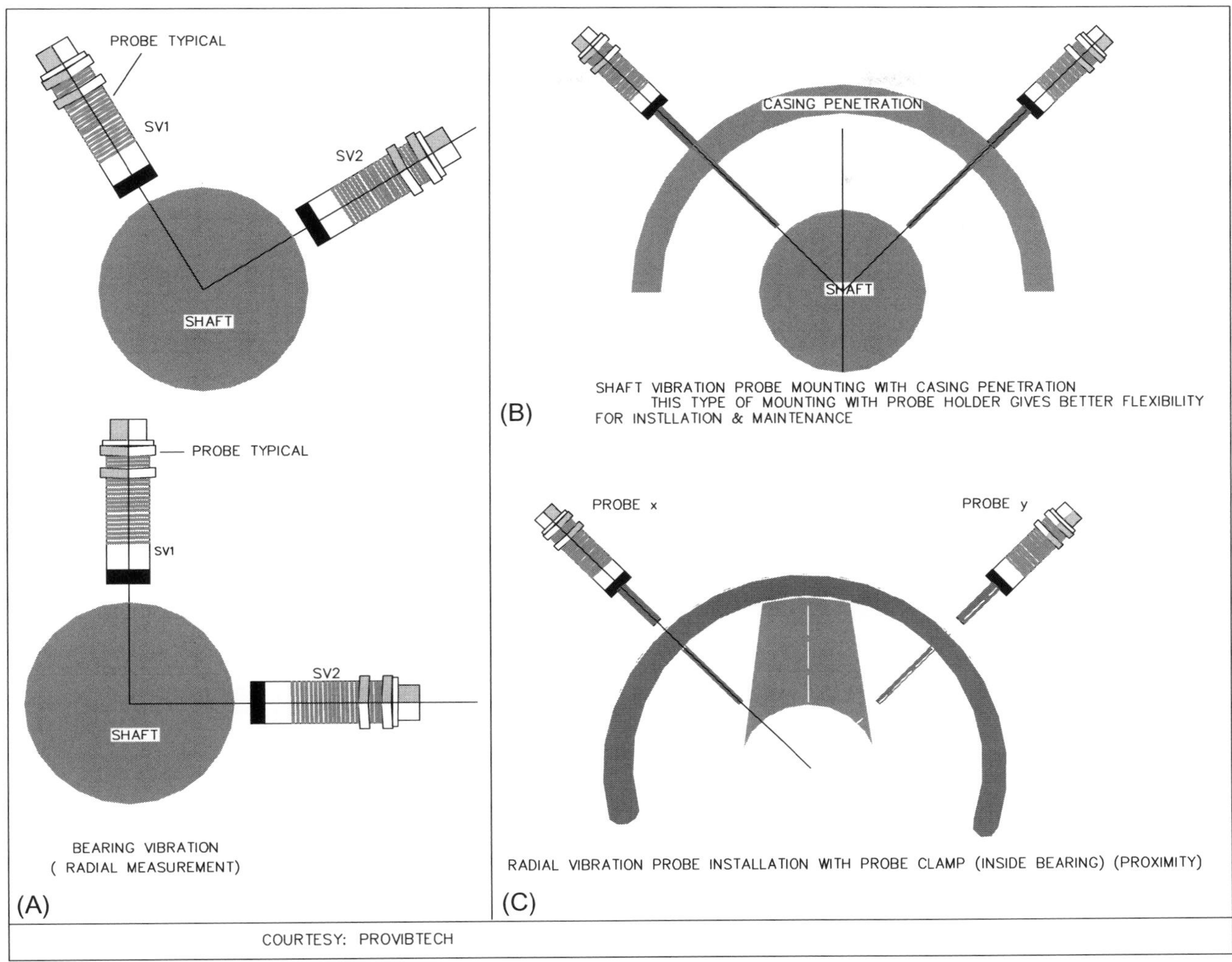

FIG. 5.5 Vibration measurement for bearing (shaft vibration).

different planes. Probes are put in place by a hole (and probe holder) in the bearing housing. Cap. Radial measurements give first-hand information. Regarding any unbalance, crack, misalignment, etc.

If the turbine or any other machine is allowed to run for long with excess vibration, then it may be damaged due to rubbing, fatigue, etc. In a turbine, it can be caused due to water induction. This monitoring helps in quick detection to save the machine.

4. Bearing block vibration (BV): Bearing block vibration, that is, absolute vibration, is measured to detect the absolute vibration of a machine casing with respect to free space; hence, seismic probes are used (Section 2.6.9). When there exists a light casing-to-rotor weight ratio, the vibration is transferred to the bearing housing. In order to analyze the signals at high frequencies, accelerometers are deployed. Otherwise, a velocity pick may be sufficient.
5. Phase measurement (KP) (Fig. 5.6): In the shaft, a key way is kept or a hole is drilled. A magnetic pickup, optical, or a similar type of sensor is put to measure the angle between the hole or key way and the heavy spot on the shaft. This is important for balancing the machine. Also, it can provide an early indication of misalignment and/or a crack in the shaft, change in blade disposition, loss of mass, etc.
6. Axial shift/thrust position (TP) (Fig. 5.7): This is one of the most important trips for the turbine. The turbine trip signal shall be generated from the probe through the necessary voting circuits, after each of them crosses the set point. This may be similar to shaft position measurement, with the difference that here the main aim is to see the wear of the thrust bearing by monitoring the movement of the thrust collar between the active and inactive thrust shoes, and their wear and tear. The measurement is carried out by one or two sets of probes on

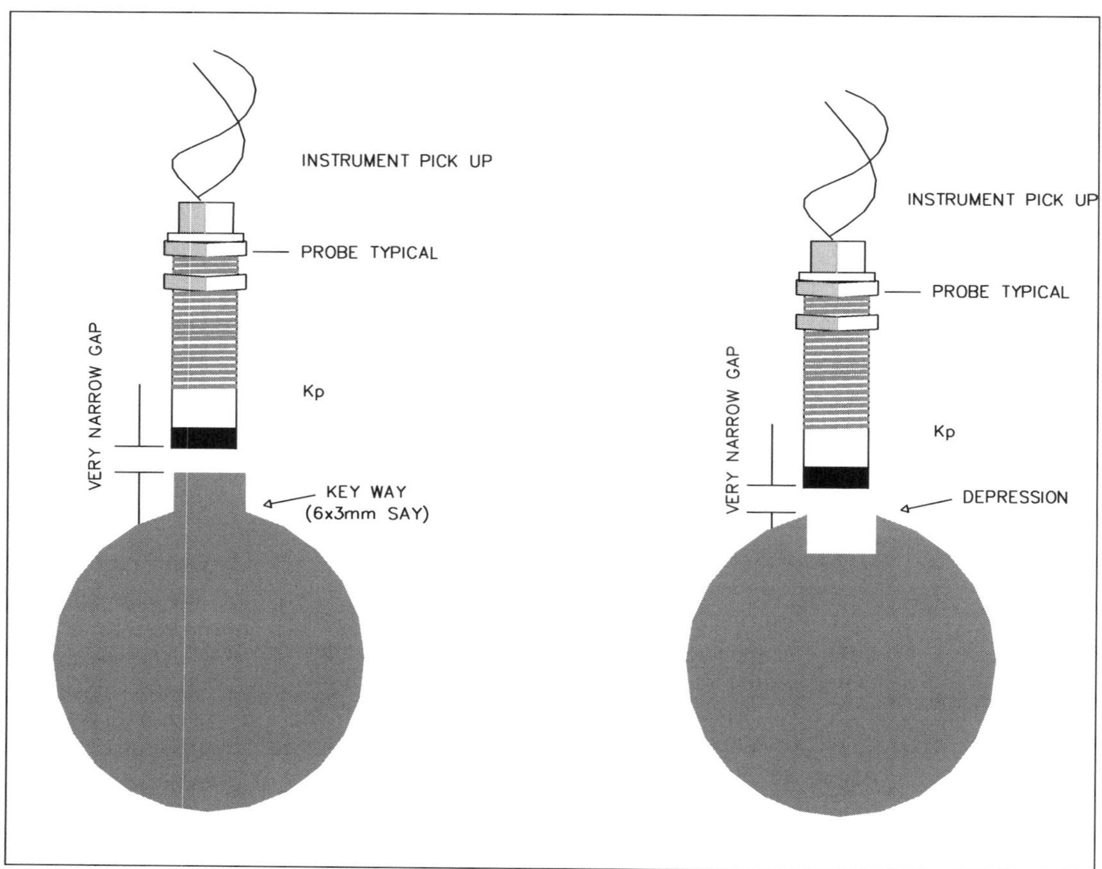

FIG. 5.6 Key phasor.

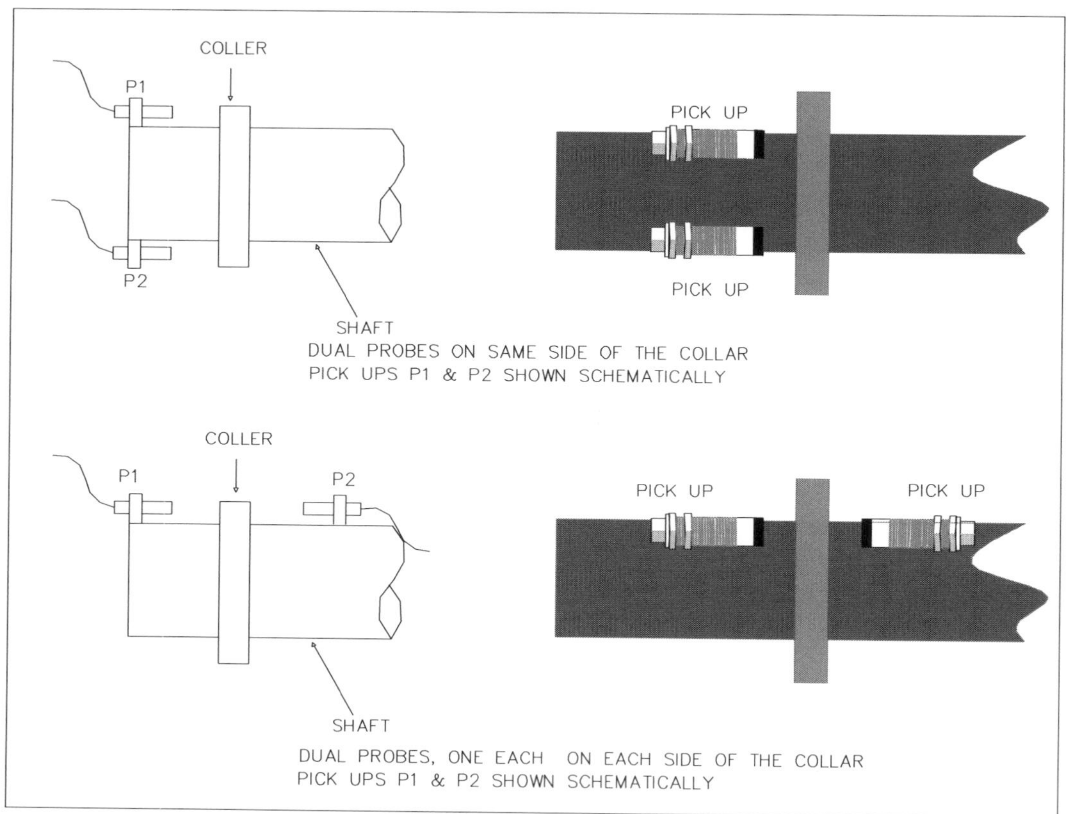

FIG. 5.7 Axial shift/thrust position.

the same or opposite sides of the collar (two of three is not uncommon in large plants). It measures the movement of the shaft in the axial direction with respect to some fixed reference, which could be the thrust bearing support structure or another nearby casing member. The thrust collar location with respect to the thrust bearing represents the shaft axial shift observed by the probe. The thrust position or the axial shift/position measured by the proximity probe is in the range of a few mm. Because thrust position measurements are taken within a short range of thrust bearing (normally within 300 mm or as recommended by the manufacturer), one needs to be careful about thermal expansion. Also, the instrument range shall be suitably chosen for the primary function to monitor the wear of the thrust bearing causing axial shift.

Axial shift indicates a number abnormality in the TG operation such as:

- Some improper operation in the turbine, for example, some problem in one of the piping such as extraction piping.
- Excess wear at the thrust bearing, which may cause severe damage to the turbine through rubbing in the seal.

7. HP turbine casing expansion (CS): The generator end of the HP casing is fixed in the axial direction. So, the HP casing due to thermal expansion/contraction has a chance only in the front end. It is measured at the supporting end in the front where little axial displacement is allowed without misalignment. From start-up to operation and shutdown, there will be a large variation in temperature to cause expansion. This parameter helps the operator know about the thermal growth on the HP casing. Installation of this is quite critical. The main body part is fixed with the foundation whereas the sensor probe plunger moves by a spring-loaded roller that is pressed against a support attached to the casing, as schematically shown in Fig. 5.8. One term called "cocked case" is popular with this measurement. This occurs when the turbine slider hangs up or sticks on one side of the foundation and continues to grow in the other direction. This has been elaborated in Fig. 5.8. In this connection, Section 2.12.3 shall be referred to also. To have a check on the same, it is necessary to use two LVDTs in place of one.

   This measurement is important as it helps to detect excessive resistance of the front casing support to casing thermal expansion, and resistance is marked by a jerky movement of the HP casing due to expansion.
8. Bearing temperature measurement (T) (Fig. 5.9): Misalignment, insufficient clearance in bearing, excess bearing loading, or improper lubrication at the bearing are the major reasons in any machines such as fans, pumps, HT motors, and turbines. Therefore, it is considered an important measurement not only in TG but also for other big drives. In many turbines, a thermocouple of a suitable ISA type is used while others generally use RTD for this purpose. These temperature elements may be brazed with the white metal of the bearing. In some places, a few number of points temperatures are monitored together to study the conditions of the turbine—such as metal temperature, lube oil temperature, exhaust hood temperature, etc. for these, dedicated temperature scanners with links to DCS may be deployed
9. Absolute shaft vibration measurement (Fig. 5.10): This measurement is used to find the motion of the shaft in

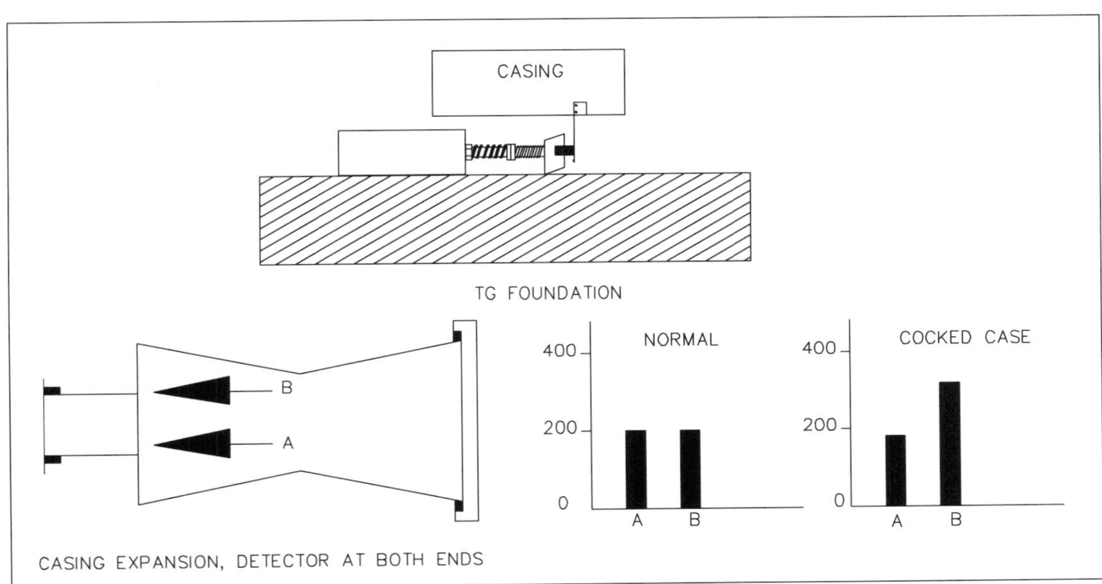

FIG. 5.8 HP casing expansion.

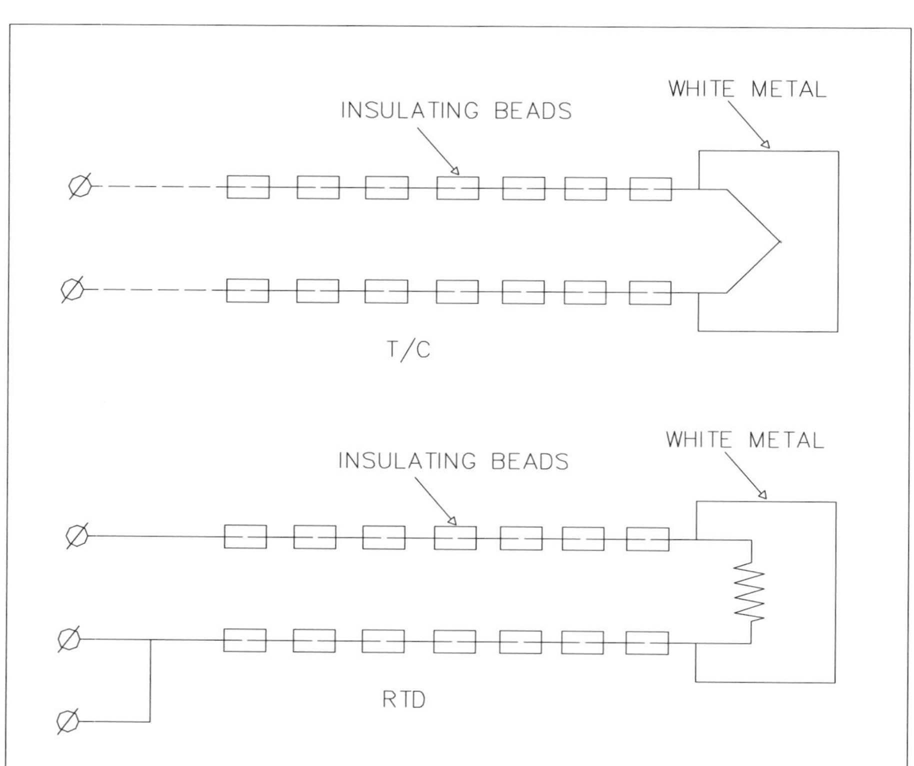

FIG. 5.9 Bearing temperature measurement.

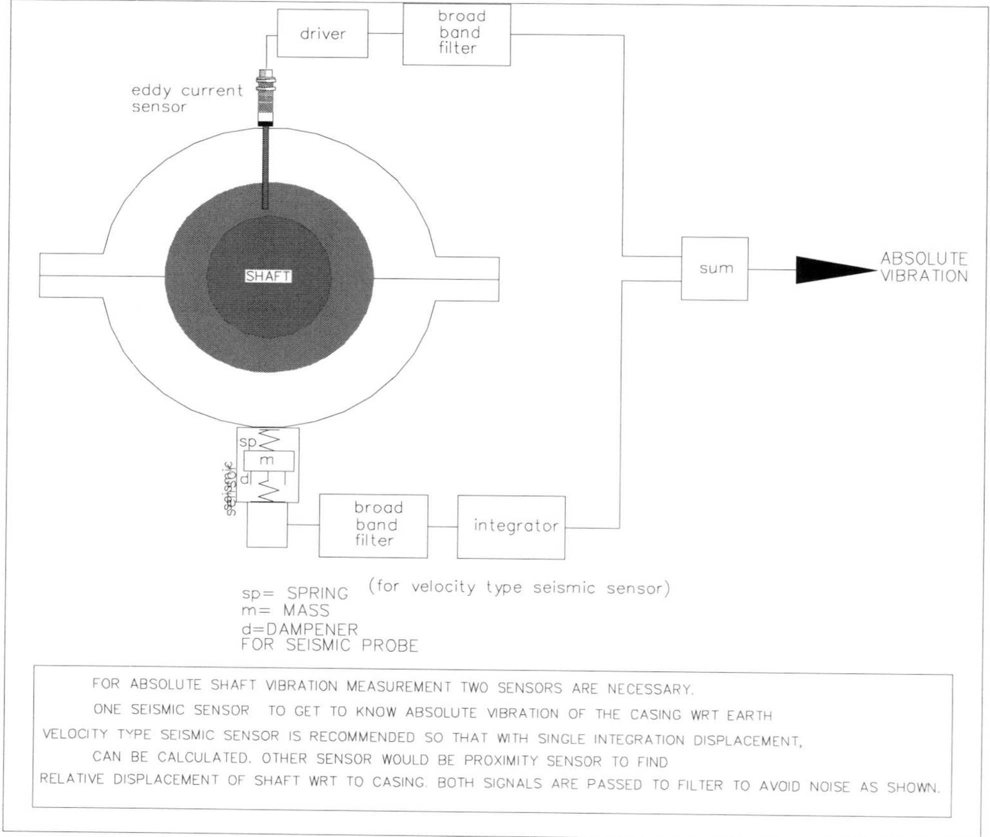

FIG. 5.10 Absolute shaft vib measurement.

free space. The rotor is much heavier (~5–6 times) than that of the casing. A seismic probe is used to measure the absolute vibration of the casing (with respect to free space). There will also be another probe to measure the relative motion of the shaft with respect to casing. In order to eliminate noise, suitable filters are used. Then these signals are summed to get the absolute shaft measurement.

10. Rotor eccentricity (RE) (Fig. 5.11): An eccentricity measurement indicates the deviation of the rotor's physical center and theoretical one. The rotor is a very heavy mass on the shaft, so naturally there will be the possibility of sagging or bowing of the rotor. This sag/bow of the rotor is measured by monitoring the rotor eccentricity. Because the hottest part is the HP turbine rotor, the highest possibility of a sag/bow would take place at this part, that is, the susceptibility to thermal deformation would be high. As is clear from Fig. 5.11, eccentricity is the maximum difference between the shaft deflections at the highest cases. The measurement of rotor eccentricity is possible only when the shaft is rotating. Therefore, this measurement

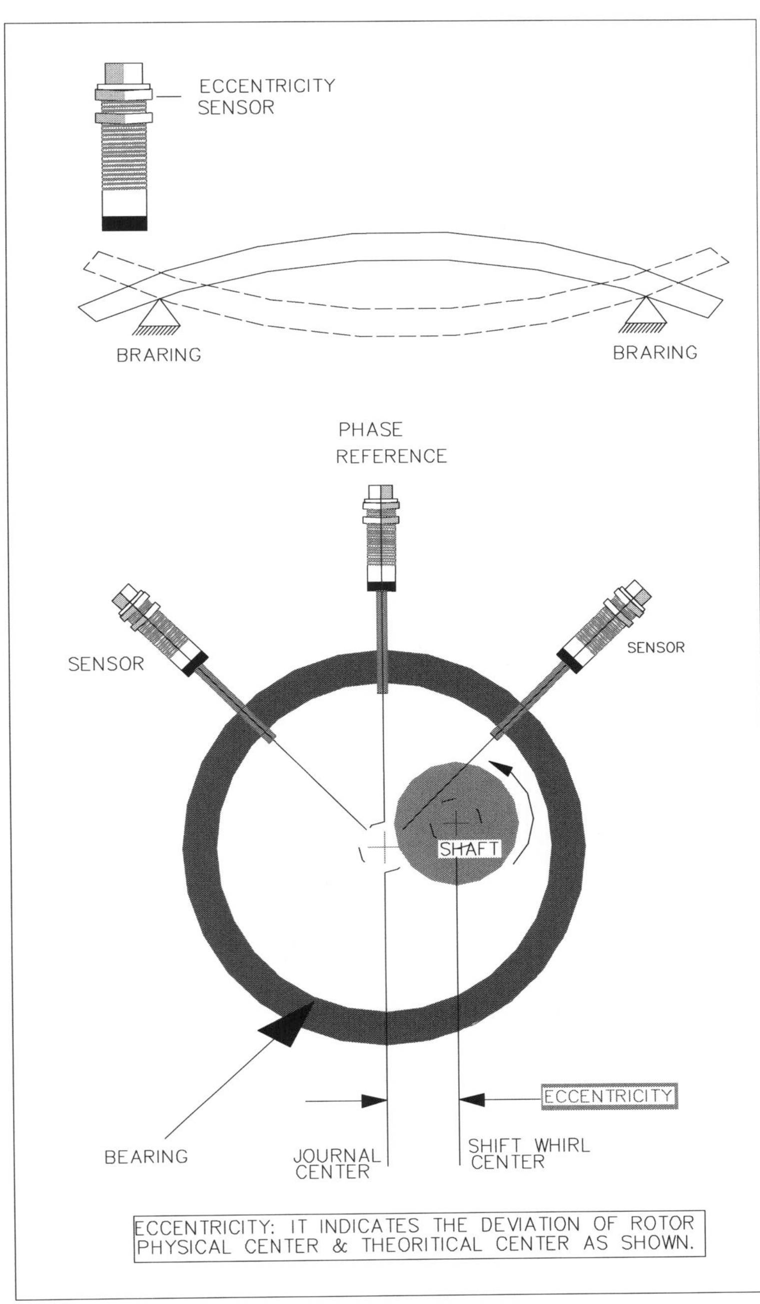

FIG. 5.11 Eccentricity measurement.

is carried out at the low speed of the turbine. This measurement helps the operator in bringing up the speed without rub or damage.

Detection of excess deformation at the time of low speed operation and ensuring that the turbine run up will take place after the sagging is straightened up are both important for rotor eccentricity.

11. Differential expansion (DE) (Fig. 5.12): The thermal expansion/contraction of the lighter hollow rotor is different from the thick casing, so the latter will change at a much slower rate than the former. This causes the need to measure the differential expansion between the rotor and the stator casing. This is more predominant during the cold start-up of large turbines. Therefore, the operator with the help of this measurement has to make sure that the casing has expended sufficiently so as to avoid rubbing by the rotor. Differential expansion measures the axial position of the rotor with respect to the machine casing at some distance from the thrust bearing where the axial shift of the rotor is least. The measurement is typically made with a proximity probe transducer mounted to the casing to measure the axial position of the rotor provided with a collar or special ramp shape given to the rotor, as is shown in the drawings
12. Valve position (PoS) (Fig. 5.13): Valve position measurement is done to know the opening/closing percentage of the valves, mainly the throttling/control valve. However, for fulfilling certain criterion during heat soaking, stop valves are also cracked open. An eddy current noncontact proximity rotary position transducer and/or any other type could be used for this purpose.

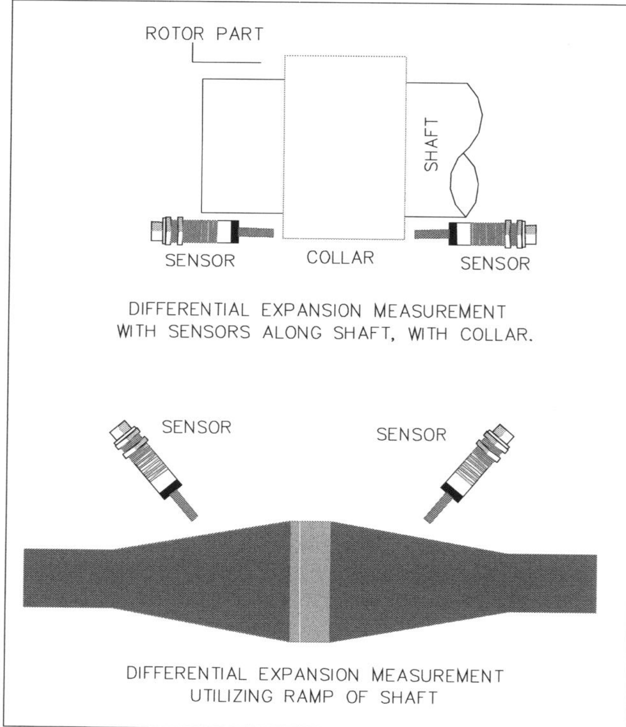

FIG. 5.12 Differential expansion measurement.

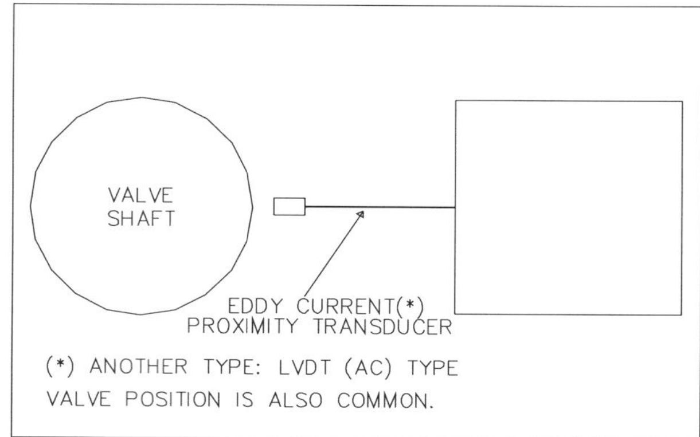

FIG. 5.13 Valve position measurement.

2.7 Frequency Range

As indicated earlier, the general frequency range encountered in a vibration is from 10 to 1000 Hz. However, in some large turbines, lower frequencies of vibrations are also common. Also, near critical speed, there is the possibility of low-frequency vibration, so for turbines, a low-frequency vibration measurement is very important. With a change of vibration frequency there will be changes in the type of sensor, for example, displacement to acceleration type. However, by mathematical computation, it is possible to get the desired vibration.

2.8 Measuring Sensor/Transducers

Generally, displacement, velocity, and acceleration are used to measure vibration and most of the other types of turbovisory instruments. However, for speed there are a few other types such as a tachogenerator, the photoelectric type, and Hall effect sensor types.

1. Displacement sensor/transducer: A displacement sensor can be used when there is a fixed reference, for example, displacement of the shaft with respect to the bearing housing. Sometimes, displacement is computed by integrating the output from a velocity sensor. When the displacement amplitude is measured, then the sensor is chosen so that its natural frequency is less than the lowest frequency of the vibrating medium in the question. Here, the total distance traveled by a vibrating body or another piece of equipment move (say shaft) is measured (normally in μm).
2. Velocity sensor/transducer: In this method, the velocity of vibration of a moving part (the speed at any instant of time) is utilized. Normally, RMS value of vibration in mm/s is chosen for the measurement. In this type of sensor/transducer, Faraday's law is used to generate

voltage (e) = $-N\ d\emptyset/dt$, where N represents the number of turns of the sensor coil and $d\emptyset/dt$ is proportional to the velocity of vibration. The transducers used is a moving coil, a moving magnet, or a change of reluctance, with relation to a fixed coil. Thus, the velocity of vibration is the speed at which the vibrating part is moving at any instant, and the output voltage is proportional to velocity vibration. Output is mV (mm/s). In order to measure a very low frequency (say 10 Hz) of vibration, the natural frequency of the transducer will also be lower $<$ say 8 Hz. At a vibration frequency above the natural frequency when the coil would remain stationary with respect to Earth's reference, the instrument is called a seismic transducer.

3. Accelerometer: In most of the acceleration measurements, Newton's second law $P = m^*f$ (where P = force, m = mass, and f = acceleration) is used. In this method, either spring compliance shows displacement of the mass relative to the casing is proportional to the acceleration is used or piezoelectric transducers are used (where charged developed is proportional to the force applied). In the modern turbines, piezoelectric transducers are used for measurement of pedestal vibration.

2.9 Factors Influencing Vibration Measurements

There are a few factors that influence measurements:

A. Cross-axis sensitivity. The sensitivity of an instrument is greatly influenced by the vibration occurring perpendicular to the sensitive axis of the instrument. It is recommended to keep within 10%.
B. Capacitance effect: In a piezoelectric sensor, the charge produced due to force is measured. It is therefore very sensitive to the stray capacitances coming from the cable and preamplifier, etc. Also, variations in relative humidity may have an effect on the same. For these reasons, the use of charge amplifiers and special cables may be recommended by the manufacturer.
C. Sensor loading: The sensitivity of a vibration sensor falls with less mass, whereas the addition of mass would create a mass loading of the vibration system and the problem of frequency response would be predominant. Therefore, it should be chosen carefully.
D. Coupling: The sensor and the surface at which the sensor is coupled shall be very stiff to avoid any spurious vibrations at low frequency.
E. External influence: Temperature, external magnetic field, etc., have effects on vibration reading. Also, damping has an effect on the measurement. A system with lower damping will take more time to come to the equilibrium after a disturbance.

2.10 Interpretation of Vibration Measurements

The vibration-measuring instruments are of a different category such as monitoring, machine diagnostic, and portable. Also, reading types are different as some are time peak overall, Average overall, FFT overall. Etc. The vibration signal may not be a simple sinusoidal wave, but may be like what is shown in Fig. 5.14B.

RMS = 0.707 peak is not valid for curve B. The main intention of the discussion is to conclude that there may be a mismatch between (say) the computed value of the RMS reading by an instrument (say) and a peak-detecting diagnostic instrument. It is always advisable to use an oscilloscope to compare where these can be checked.

2.11 Vibration Measurement

A few common sources of vibration are discussed below in Table 5.4.

- In this section, an effort will be made to discuss various kinds of vibration measuring systems. The discussions will be made on the basis of instrument types so as to cover other types of measurements also. For example,

FIG. 5.14 Vibration interpretation.

TABLE 5.4 A Few Common Sources of Vibration

Cause	Indication	Remarks
Bent shaft	High axial	Unsteady (In TG)
Misalignment	High axial	Axial amplitude high, higher of vertical or horizontal reading
Unbalance	Radial steady	Most common vibration cause
Gear Mesh	Radial high	Typical acceleration due to high frequency
Resonance	At machine critical speed amplitude high	OEM guide line for quickly passing the speed
Looseness	Proportional to looseness	Often coupled with misalignment

a proximity eddy current sensor can be used both for shaft vibration at bearing as well eccentricity measurements.

- Monitoring of the shaft vibration in the orthogonal axis *X–Y* axis (radial) is common. However, this alone is insufficient. So a casing vibration measurement is also necessary. Casing vibration is more useful when the rotor mass is sufficient to transfer vibration energy to the bearing. Phase measurement is helpful in analyzing the signal.
- For casing vibration, all three types can be used. A general guideline is to select the sensor type according to the machine RPM. So, for casing vibration, displacement is the better choice when M/C RPM < 600, whereas for 600 < M/CRPM < 60,000 a velocity sensor is the better choice. However acceleration with in integrator is often used.

2.11.1 Proximity Probes

An eddy current proximity sensor is most popular, but there is another type (mutual inductance used as a switching device).

- Eddy Current Sensing Principle: In this system, as shown in Fig. 5.15A and B, there is a transducer coil at the tip of the probe. Other than the coil, the sensor consists of an oscillator, power to the drive transducer oscillator, and a detector (demodulator). The oscillator generates the high frequency (1–3 MHz) signal through the transducer coil. This causes a changing magnetic field in the nearby vibrating target. As a consequence, a high-frequency magnetic field is generated on the target. Therefore, as per Faraday's law, an induced voltage is generated in the target body to cause the eddy current to flow. An eddy current in the conducting shaft would change the amplitude of the high-frequency signal from the probe. The energy loss due to the eddy current changes the oscillator operating point. This change is detected at the detector or pickup coil. When the target is at a fixed position, there will be a DC at the detector, meaning a new unchanging amplitude of the high-frequency signal due to the fixed target. Now, when the target changes position, the amplitude of the high-frequency signal changes in proportion to the air gap between the sensor and target. So, because of eccentricity or shaft vibration, when the gap changes and the amplitude changes in the carrier frequency current (original signal) caused by the eddy current, is detected by detector after being demodulated. A pickup coil signal is amplified and given as output. This type of sensor gives a signal for vibration amplitude or displacement.
- Change of reluctance sensing principle (mutual coupling): This type is another possibility, but it is not

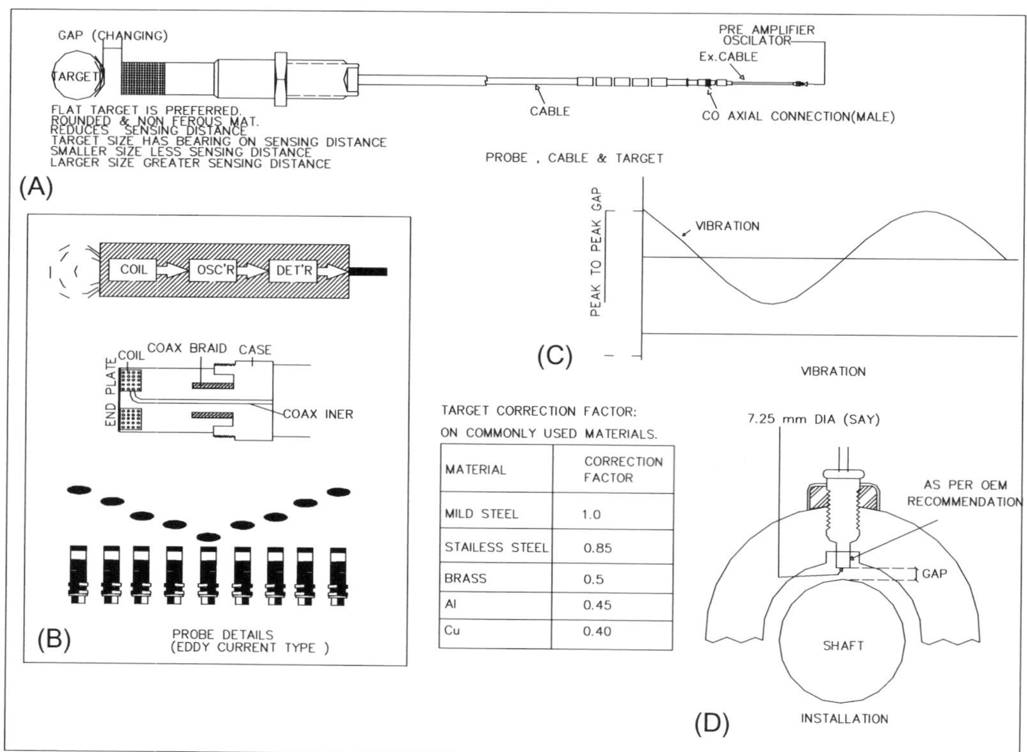

MATERIAL	CORRECTION FACTOR
MILD STEEL	1.0
STAILESS STEEL	0.85
BRASS	0.5
Al	0.45
Cu	0.40

FIG. 5.15 Proximity Probe.

popular. It consists of two coils: primary coils (being excited by, say, 10–20 KHz) and secondary coils. These two coils are placed across a metallic core. The second coil gets the voltage on account of the mutual coupling between them. Now, these arrangements are placed on one side of the diaphragm, forming a part of the path for magnetic flux. The whole assembly is placed near the target, which when vibrates changes the air gap between the target and the sensor, hence between the secondary coil. So the voltage generated in the second coil is a function of the air gap and is, in fact, proportional to the air gap.

2.11.1.1 Eddy Current Proximity Probes (Fig. 5.15)

1. On account of the eddy current, there will be a small voltage in the pickup coil. As long as the gap is not changing, the coil voltage will not change. Therefore, a DC voltage proportional (when the shaft rotates without vibration/eccentricity) to the average gap between the target and the sensor is available at the output due to the demodulation and signal-conditioning action of the sensor output circuit.
2. When the shaft vibrates (or there is eccentricity), then there will be an AC voltage (due to loss of the eddy or change in mutual inductance) on account of vibration (or eccentricity). This voltage (at the detector after demodulation) will be proportional to the amplitude vibration or eccentricity.
3. Only AC voltage is measured to note the vibration, eliminating the DC voltage after demodulation.
4. Depending on the coil size, the range could be 10 mils (0.25 mm) to 2.0 mm (80 mils).
5. Some of the special features preferred in selecting the systems shall include but are not limited to the following:
   - *Interchangeability* of probes, cables, and extension cables in the system.
   - *Compliance* with requirements of API670 Standard/ ISO 10816/ISO 7919/European CE marking/IEC, etc. Also For ATEX, CSA etc. For Hazardous application possible.
   - *Immunity* to radio frequency interference (RFI) and electromagnetic interference (EMI).
   - Some manufacturers offer a click lock (finger tight connector), for example, GE-B&N.
   - *High temperature* Withstand capability as high as 260°C.
   - *Sensor face* Can be plastic or stainless steel (having the capability to withstand more harsh environmental conditions). Sometimes full epoxy encapsulation to protect against shock, contamination.
   - Protection against: reverse polarity, short circuit.
   - In a vibration sensor, mild steel is considered the standard target; if other materials are used then a correction factor can be used to ascertain the actual range (nominal sensing range* correction factor for material = actual sensing range (Fig. 5.15—Courtesy Rockwell).
   - For ordering: Need to specify: length for overall probe, cable, extension cable, approving authority, accessories and contact type, and rating for switches only.
   - Application Note: Used for measurement of shaft/ bearing vibration radial vibration, differential expansion transducer (TG), eccentricity (TG), axial shift (TG)/thrust or axial position, and speed/zero speed/reverse rotation protection.

2.11.1.2 Eddy Current Proximity Probes Data Sheet

See Table 5.5.

2.11.2 Velocity-Seismic Probes

In a velocity probe, the velocity of vibration is measured so it involves the measurement of both the amplitude as well as the frequency.

2.11.2.1 Velocity Probe-Seismic Probe Working Principle

Seismic velocity-transducer working, on Faraday's induction principle, has a fixed conductor length, magnetic field strength so, the generated voltage is proportional to the velocity at which magnetic flux is cut in the coil. As shown in Fig. 5.16A, one permanent magnet is attached to the casing, whereas the coil assembly (coil with core considered as mass—natural frequency of the assembly is lower than the frequency of machine vibration—power station in the tune of 10 Hz) is suspended by a spring with dampening. At these frequencies, it will almost remain stationary in the space. It is called the seismic type. Now, when the instrument casing is fixed with the bearing housing (pedestal), and there is vibration in the pedestal, there will be relative motion of the magnet with respect to the fixed point, that is, the coil. Therefore there will be voltage induced in the coil, and this voltage is directly proportional to the velocity of relative motion, that is, the output is proportional to the velocity of the casing vibration. Displacement can be obtained by integrating the output. Sensitivity is expressed in terms of mV/mm/s.

2.11.2.2 Velocity-Seismic Probe Issues (Fig. 5.16B–D)

1. The instrument is a self-excited system.
2. After a particular frequency, the change in output drops because the transducer would operate near a damped natural frequency.

TABLE 5.5 Eddy Current Proximity Sensor (DS)

Sl	Specifying Point	Standard/Available Data	To Be Specified
1	No of input	1 per sensor standard	
2	Power supply	24 VDC standard	
3	Sensitivity	Varies with supplier 1–2 mV/V input	
4	Output resistance	<50 Ohm	
5	Probe length and total length	0.5–10 M (with extension wire)	To specify as per application
6	Extension cable	2.0–9 M (available)	
7	Gap/range	0.5–2.5 mm	
8	Extension cable capacitance	Supplier specific: ~70 pF/M (Typ)	
9	Extended temp range	Standard available: 120°C	
10	Min target size	15 mm flat or 50 mm diameter shaft for continuous, (or minimum one side face equal to the probe diameter for switches)	
11	Probe diameter	5–12 mm diameter, (may be > for switches)	
12	Probe TIP	Plastic (PPS) with casing SS, stainless steel, with epoxy painting	
13	Probe cable	Manufacturer standard	
14	Compliance and certification	Manufacturer's standard—some international authorities could be: ABS, EMC, IEC (ATEX, CSAETC for hazardous approval)	
15	Environmental condition	Available in the market: T=177°C OP range 100°C RH: 93% (IEC standard test)	
16	Field wiring	Standard triad cable	
17	Enclosure	IP 67/65, etc.	
18	Output	Voltage for continuous monitoring (linear)for switch: contact type, number, and rating	
19	Approval	As per project standard	To specify

3. It is recommended to use a magnetic screen to avoid influence due to the magnetic field of the main generator. To avoid erroneous results, physical installation of the transducer is very important.
4. A large portion of vibrations occurring at the rotor due to misalignment or imbalance is transmitted to the bearing housing or pedestal. This instrument the same can monitor and protect the machine. The orientation of the probe is also quite important.
5. The orientation of probe is very important for accurate results, and proper location consultation with the manufacturer/OEM is preferred.
6. With respect to the phase angle measurement, there will be a phase difference between the output and case velocity. This is a function of frequency, so for analysis, frequency should be taken into consideration.
7. Some of the special features preferred are as follows:
   a. Available two-wire configuration with linear output.
   b. These are less sensitive to impact and impulsive excitation, but can be affected by an external magnetic field, so better to use a magnetic screen.
   c. No external supply is necessary and finds applications in portable configuration.
   d. Up to a 10 HZ frequency range is available in the lower side.
   e. Compliance with requirements of API670 Standard/ISO 10816/ISO 7919/European CE marking/IEC, etc. Also For ATEX, CSA, etc. For Hazardous application possible.
   f. Orientation as shown in Fig. 5.16C is available.
   g. Transducers available to cater to temperatures up to 200°C.

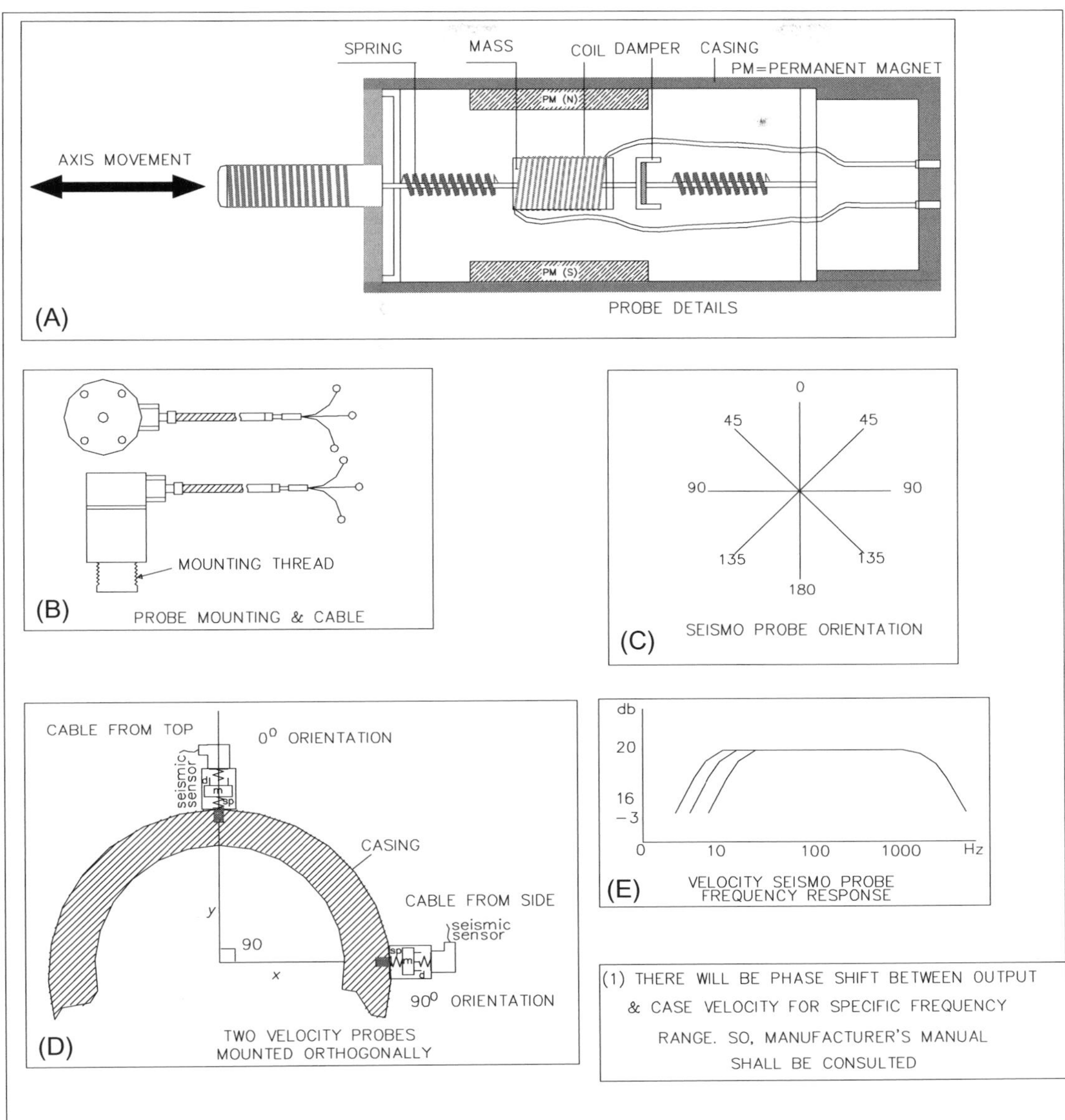

FIG. 5.16 Seismic velocity probe.

h. For ordering, the desired information is the transducer frequency range; orientation of the mounting with mounting drawings, mounting bases, and other accessories; electrical connection details; and approval authorities.

i. Application: Casing vibration (Table 5.6).

2.11.3 Acceleration-Seismic Probes

A seismic sensor acceleration probe is mainly used for high-frequency vibration measurement. Normally, compressive force is measured with the help of a known proof mass (seismic mass) to infer the acceleration. The output may be integrated to know velocity and displacement.

2.11.3.1 Acceleration Sensor—Working Principle

(a) When the body of the accelerometer is subjected to vibration, the mass mounted on the crystal wants to stay still in space due to inertia, so it compresses and stretches the piezoelectric crystal. According to Newton's law, $P = m^*f$ where P is the force for mass m having acceleration f. Again, spring compliance shows that relative displacement of mass with respect to case, hence is proportional to acceleration. Transducers based on this principle show that at the low-frequency end, the displacement produced is too low, hence, the output is very low, which may be suppressed by the background noise. So this type is not suitable for the measurement of vibrations with low frequencies.

TABLE 5.6 Velocity Sensor (Seismic) (DS)

Sl	Specifying Point	Standard/Available Data	To Be Specified
1	Frequency response	0-(−) 3 dB for minimum to 15 kHz	
2	Frequency range	±2.5–10 kHz	Choice of span
3	Electrical connection type and orientation	Terminal block/coaxial connector (copper/silver-plated brass) and top or side	
4	Mounting angle	0/(+/_) 45/135/180, etc.	Choice of angle
5	Mounting base	Circular/rectangular/isolated plate	
6	Agency approval	API670 Standard/ISO 10816/ISO 7919/European CE marking/IEC, etc.	To specify as per project standard
7	Mounting	Screwed/As per base selected	
8	Operating temperature	Up 200°C	
9	Humidity	95% Rh	
10	Casing/adapter	Anodized Al.	
11	Sensitivity	25 mv/mm/s to 4 mv/mm (freq > 5 Hz)	
12	Probe dimension	Manufacturer standard	
13	Output	Voltage for continuous monitoring (linear)/For switch: Contact type, number, and rating	
14	Lead wire	100/300 Meter available	
15	Transverse sensitivity	±10%	
16	Field wiring	Standard triad cable	
17	Enclosure	IP 67/65, etc.	
18	Hazardous class	To specify (Intrinsic safety ATEX, etc.)	As applicable

(b) *Piezoelectric sensor*: This the most popular type of accelerometer to measure the compressive force due to vibration. When force is applied (or applied force changes) to a certain kind of crystalline substance, for example, quartz, then between the two surfaces of the properly shaped crystal, charges are developed (or changed). These developed charges are proportional to the applied force, which in turn is proportional to the acceleration. The output through process of integration one can get velocity and displacement. Double integration poses a serious problem on the signal-to-noise (S/N) ratio vis-a-vis the frequency range, so the velocity is normally computed. As shown in Fig. 5.17B, there are several designs available such as the shear type, the bending type, and the compressive force type. Here, the force is applied to the tripod (for three directions) where the piezoelectric crystals are embedded. When the accelerometer is attached to the vibrating object, the piezoelectric elements experience vibrating force. A seismic body is placed against a spring so the seismic body will experience a changing compressive force. This produces a varying charge at the quartz. This signal is processed by a charge amplifier and other electronics to produce an output, as shown in Fig. 5.17D. The system is self-generating. In view of this, it can be considered as a variable capacitance with a variable voltage. Now, the charge $q = P^*a^*K$ where q is the charge developed due to force P, K is a constant for displacement (m/N), and a is a material constant, depends on the material, the geometry of the crystal, and the position of the electrodes. Now, force P in a spring is proportional to displacement X_i, so, $q = K_q{}^*X_i$, where K_q is the overall constant.

The commonly used quartz material is *lead zirconate titanate* ($Pb[Zr_xTi_{1-x}]O_3$ $0<x<1$), also called *PZT*. PZT is temperature-dependent for charge sensitivity and capaci tance value. It is a self-generating device with a high source impedance. Signal conditioning is done by a charge amplifier, as typically shown in Fig. 5.17C. The output voltage is proportional to the displacement, hence is also linear with force or acceleration (Fig. 5.18).

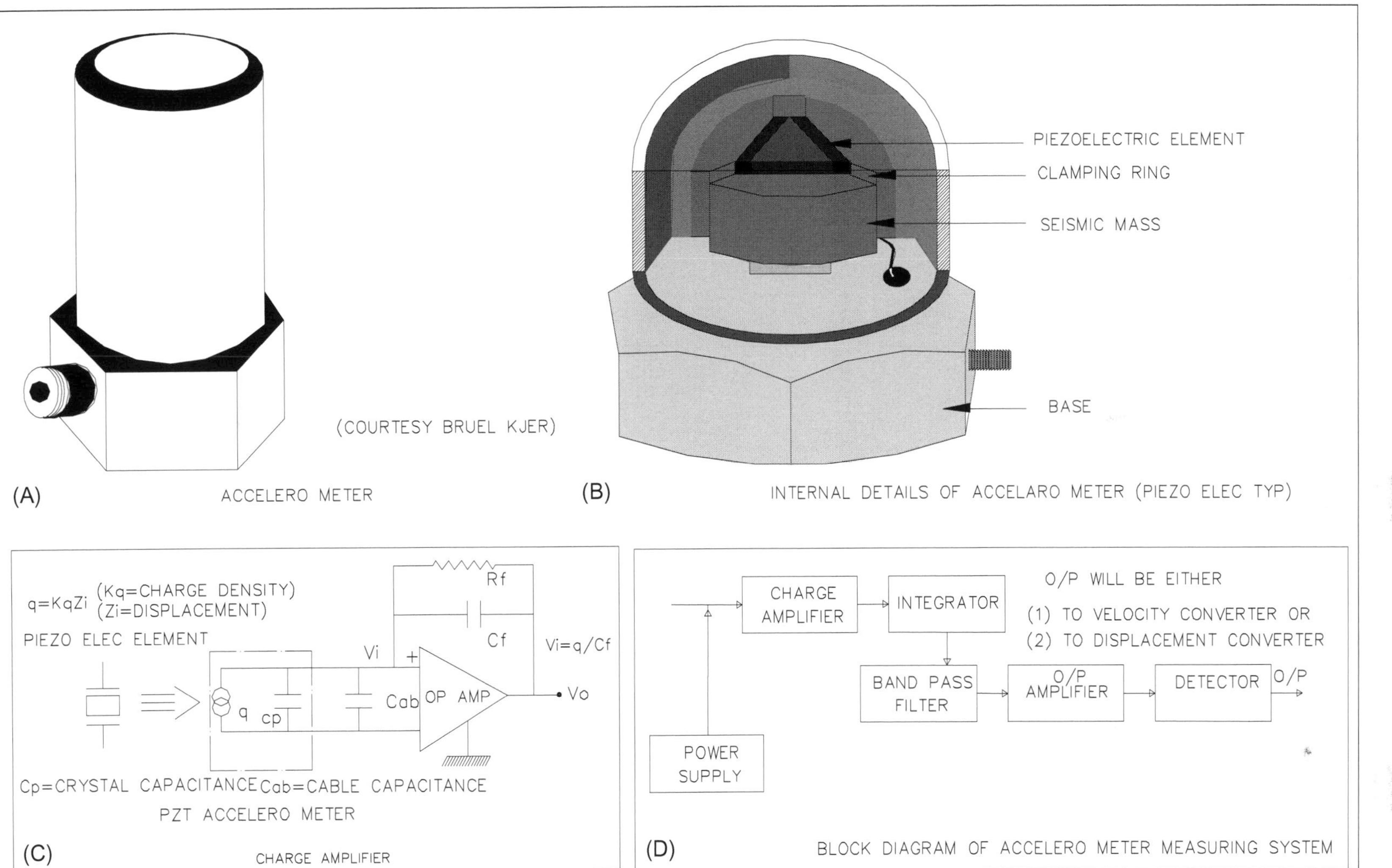

FIG. 5.17 Accelerometer measurement details.

Charge Amplifier: Charge amplifier (shown in fig 5.17) is used to minimize capacitance effect of cable. Since operational amplifier has very high input impedance, current through the feedback circuit is due to V_i to produce output V_o. With integrating operational feedback, output is integral of input current, which is the charge, so it is named as charge amplifier. In negative feedback OP Amp amplifier, $V_o/V_i = -Z_f/Z_i$ Where $Z_f=C_f$ and $Z_i=C_p$ parallel C_{cab} or $Z=C= C_p + C_{cab}$ so $V_o/V_i = -C/C_f$ ($z=1/\omega C$)

Therefore $V_o = -C/C_f * V_i$ or $V_o = -C/C_f * q/C$ or $V_o = -q/C_f$ or $V_o = K_q X_i/C_f$ (since $q=K_q * X_i$)

Thus it is seen that

- ***O/P voltage is linearly related to displacement caused due to acceleration.***
- ***O/P changes instantaneously with I/P without distortion***
- ***Sensitivity, K_p, & time constant are independent of Cable capacitance & C_p***
- ***Disadvantage : It has lower S/N ratio and natural frequency is lost due loss of stiffness. Also in practice the feedback Capacitance has some Resistance (not considered in the calculation.)*** **which would have some minor impact on the amplifier performance. Normal value of C_f 10 to 10^6 pF & Rf 10^{10},to 10^{14} Ω could be a better choice**

FIG. 5.18 Charge amplifier.

Sensitivity is expressed in terms of charge per unit acceleration, pC/g or mV/g.

2.11.3.2 Accelerometer-Seismic Probe (Fig. 5.17A–D)

1. Three nos. Zirconate Titanate elements as Piezoelectric material may be placed in Delta as shown in the drawing. Also, Sections 2.11.2.2.5 and 2.11.2.2.6 are applicable here.
2. Some of the special features preferred in selecting the systems shall include but are not limited to the following:
   a. Transducers in a two-wire configuration with linear output are available with a standard power supply 24 VDC, and a high-temperature version up to 200°C.
   b. Minimum frequency range: Up to 15 Hz frequency range in the lower side.
   c. Compliance with requirements of API670 Standard/ISO 10816/ISO 7919/European CE marking/IEC, etc. Also For ATEX, CSA, etc. For hazardous application possible.
   d. For ordering, the desired information is mounting threads and stud details, agency approval, cable length with details, accessories, and electrical connection details.
   e. Application: Casing vibration in acceleration and velocity (Table 5.7).

TABLE 5.7 Accelerometer (DS)

Sl	Specifying Point	Standard/Available Data	To Be Specified
1	Frequency response	±2 dB for minimum to 15 kHz	
2	Frequency range	±2.5 to 15 kHz	Choice of span
3	Acceleration range	490 M/s^2 for frequency range up to 15 kHz	
4	Linearity and noise	±1% within range, noise: 0.039 M/s^2	
5	Transverse sensitivity	<4%–5%	
6	Magnetic sensitivity:	0.4 g/T	
7	Temperature	−55 to 250°C (high temp)	
8	Humidity	>93%	
9	Shock peak to peak	5000–20,000 g	
10	Capacitance	1200 pF (a typical value)	
11	Cable length (M)	Up to 350	
12	Grounding	Isolated from case to case	
13	Mounting stud	Manufacturer standard	
14	Cable connection	Manufacturer standard	

2.11.4 *Application Note*

(a) Bearing vibration (radial vibration): Turbine, BFPT, ID/FD/PA; the proximity type may be a *choice* also.

TABLE 5.7 Accelerometer (DS)cont'd

Sl	Specifying Point	Standard/Available Data	To Be Specified
15	Case material	304 L SS/Manufacturer standard	
16	Accessories	Mounting base and others	To specify
17	Agency approval	API, CE marking, hazardous, etc.	As applicable

(b) Casing vibration: turbine, generator, BFPT, etc., seismic, velocity, or accelerometer for specific parameter to be measured.

(c) *A typical* guideline for the choice of sensor based on frequency is given in Table 5.8.

2.11.5 Machine Health/Condition Monitoring System

2.11.5.1 The Concept of Monitoring and Analysis

In power plants, any severe condition of the turbine can cause a complete trip of the plant. Also, failures of other major auxiliaries such as the FD/ID/PA fans or BFPs would result in a drastic reduction or complete stoppage of generation. Higher protection strategies such as SIL 3 or API 670 standards are prescribed to protect the machines. With a suitable machine conditioning and analysis system, it is possible to predict such failures and in most of the cases, preventive action can be undertaken to avoid such a catastrophic situation. Some of the features of condition monitoring are:

- Complete mechanical health monitoring and feedback to be integrated with the main control system with a suitable display. Apart from vibration, other parameters such as differential expansion, eccentricity, speed, key phasor, etc., can be accommodated.
- Generation of a predictive alarm based on analysis along with a normal alarm and trip based on protection standards (such as API 670/SIL 1 or SIL3).
- Depending on applicability, internal/external intrinsic safety can be provided.

TABLE 5.8 Sensor Choice Guidelines

Displacement	Velocity	Acceleration
0–200 Hz (Eddy current)	2–20 kHz (electromagnetic)	0 to >20 kHz (piezoelectric)

- Highly accurate and highly sensitive transducer (e.g., 0.1 mV/monitoring gives the ability to record and identify the vibration and allied parameter measuring unit or better).
- Communication gateway and software support (e.g., MODBUS, Ethernet).

2.11.5.2 Condition Monitoring System Components

Machine health/condition monitoring systems mainly consist of several functional subsystems that mainly includes an instrument rack, power supplies, a rack interface module, Communication module, monitoring modules, a relay module, a display module (PC compatibility), a safety barrier (if any), supporting software, and networking capability (if any) (Figs. 5.19 and 5.20).

1. *Instrument rack*: It is the rack to house various modules for connection:
   - **a.** Standard available: for example, 19″ EIA rail mounted; possible with enclosure protection NEMA 4× (IP66) for field applications and layout as per manufacturer standard.
   - **b.** I/O module mounting: one after another in racks with a back plane connection.
2. *Power supply*: Major features of power supply units are:
   - **a.** Standard AC (/DC) supplies, compatible with standard instrumentation power supplies of the country in voltage as well as frequency. Dual supply possible, with one as primary and the other the back up with changeover facility.
   - **b.** Line protection and noise filter are standard. Also, self-testing facilities of power supplies to ensure correct voltage.
3. *Rack interface function* (the first interface of the system):
   - **a.** Functionally: Interface to the configuration (with configuration software to configure the rack), display, and monitoring software (for data retrieval). Many racks provide LEDs for healthy condition (OK) and transmitting/receiving mode: T_x/R_x etc. (*courtesy: B&N module*).
   - **b.** With the help of a self-diagnostic feature, it is possible to carry out self-diagnosis as well as monitoring for the entire rack (may be in addition to the self-monitoring feature available in different modules).
   - **c.** The entire rack configuration can be done from the module with the help of any portable configuring unit with a security level (may be hardware as well as software) to access and configure.
   - **d.** It can provide permanent connectivity also. There may be a dedicated module for this function (*courtesy: B&N module*).

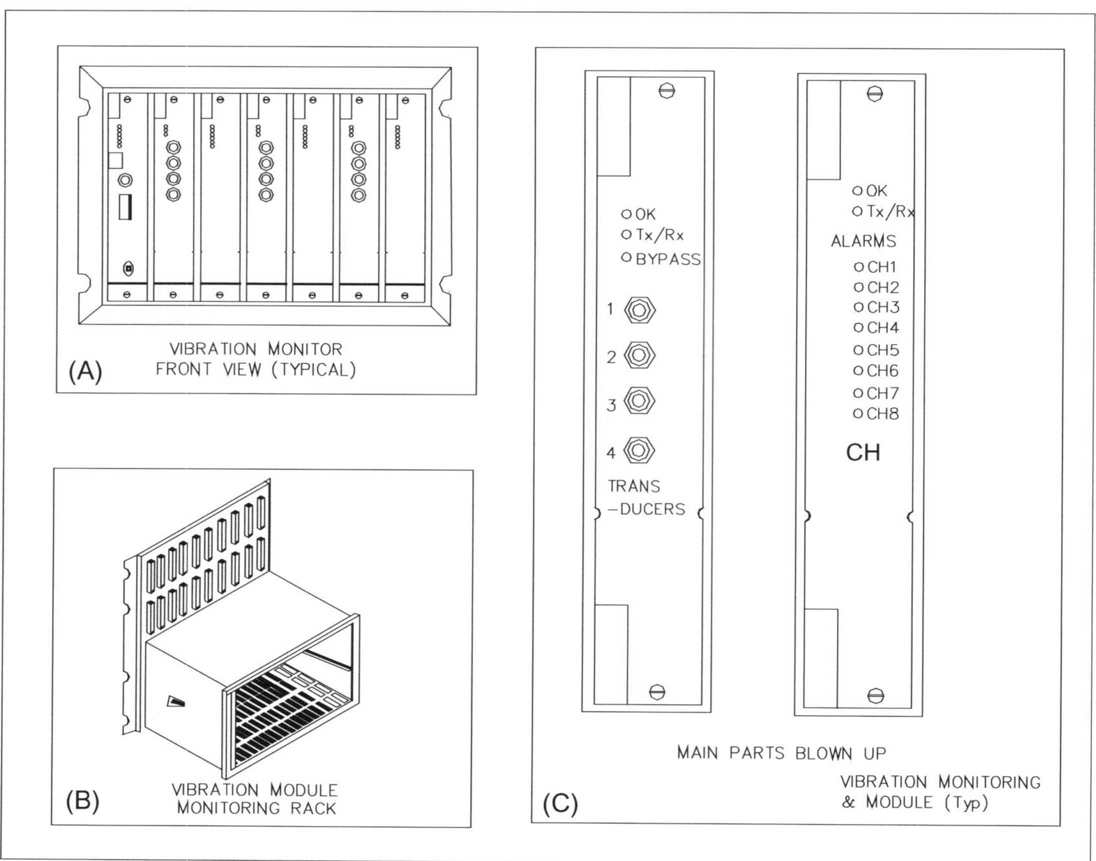

FIG. 5.19 Vibration monitoring system.

No of modules, I/O channel, Communication module and their relative disposition in the instrument rack are purely based on manufacturer's design, so in reality these may vary WRT to description here! This is general guideline only, based on data from a reputed manufacturer.

FIG. 5.20 Condition monitoring system guideline.

4. *Communication module*: Independent of the monitoring system and other functions, the gateway in the rack is designed to digitally transfer the vibration data and status to the outside world, for example, to DCS. The function of this gateway module is to support a number of industry standard links and protocols (for example, RS 485 and MODBUS protocol, Courtesy: 6600 Series of Entek-Rockwell). With issues related to communication capability, a few points to be considered include: *baud rate* and *distance*, communication channel *traffic mode*, and *rush of traffic*. Support for Ethernet, fieldbus, or serial links is important.
5. *Monitoring module*: Many channel types provide some advanced datasets, for example, gap voltage/filtered voltage (in addition to the amplitude of vibration) etc. Some of the modules may provide buffered output (say, 4–20 mA) at the module. Some of the modules are capable of providing data such as the frequency response of the channel. These are purely as per the manufacturer's standard (e.g., 3500/40 M of Bently Nevada has four channels to include different parameters or 3500/42 M provide buffered output—*Courtesy B&N*). It has been noticed in many of the standard power plant specification, which calls for measurement of both velocity as well as displacement. Naturally, one type of the probe Seismic or proximity type need to be provided, and other may be found out by signal conditioning!
6. *Relay module*: A relay module provides the necessary relay contacts that are necessary to generate alarms and a trip. These relay modules, in conjunction with supporting software, can provide simple relay output to a very complex logic sequence. The relay module

For example "FS" category of Bently Nevada 3500/53 module have SIL3 (Refer Chapter 14) protection. Similarly 3500/42, 40M, etc. have both SIL for case vibration & SIL3 certifications for over speed. – For these one needs To consult manufacturer's catalog.

FIG. 5.21 Note on SIL (refer Chapter 14) for condition monitoring.

may vary with the manufacturer. Now, because these relays may be used to trip equipment, there may be built-in redundancy. Generally, a TMR relay module is deployed for this service to get the SIL3 protection system (TUV certified or API 670). When called for, authentic *certification* from a concerned authority on the module is necessary (especially for trip function) (Fig. 5.21).

1. *Display module*: Currently, several display choices are available. Some of these displays are LED/backlit LCD/backlit high-resolution dot matrix VGA/laptop or computer displays. Except for laptop displays, the others can be mounted on the front of the monitor or put at a remote place. These displays can be: configuring details, status of channels (bar graph display with text). Based on the link used, there could be distance limitation, and/or there could be a separate work station or laptop/PC with supporting display software and suitable authority approval for hazardous applications.
2. *Safety barrier and isolation*: For the applications where transducers are located at hazardous locations, there is a necessity to have an intrinsic barrier located at the safe zone (e.g., GT and/or cogeneration or IGCC). Generally for such cases, external barriers are used. Even in a normal power plant, it is necessary to have some galvanic isolation.
3. *Supporting software system*: Almost all the monitoring systems available in modern days are all an intelligent type with the manufacturer's own system software for system support and communicating with the main DCS, for example, the Plant WEB of the Emerson 6600 series monitoring system of Entek (Rockwell). Major functions covered in software shall include module configuration, display management (including data capture), control configuration, alarm management, self-diagnostic functions, prediction control software, communication software, etc. A power predictive analysis tool, priority assignment, history data management, graphic support, etc., are advanced features of system software support.
4. *Networking capability*: Some of the monitoring systems available on the market can support networking. Predictive intelligence of the monitoring system in a networking environment lets the operator track the condition of the machine at all operating conditions. Bachmann Germany, VIBguard, B&N Trendmaster Pro online are examples of intelligent condition monitoring systems, and they can support fieldbus systems.

2.11.6 Machine Vibration Analysis System

Vibration analysis is a useful tool to perform a number of functions:

1. Usefulness: Evaluating machine conditions, machine fault diagnosis, monitoring and trending of machine conditions over time to predict faults.
2. Various Methods are: (1) Casing vibration with limits, so that by analysis, the condition can be monitored. (2) Radial vibration to detect faults in the rolling element of the bearing to detect fault condition by software for analysis. (3) FFT: A mathematical algorithm for transforming the vibration in the time domain to the frequency domain. Amplitude versus frequency is useful for fault recognition. In the time domain, unless the fault has gone to the advanced stage, changes may not be noticed properly. Whereas in the frequency domain, the signal not only registers the problem in its early growth, but also predicts the fault. It is used to detect faults such as misalignment, unbalance, or an eccentric component. For such an analytical approach, the FFT and bearing condition spectrum (BCS) need to be implemented on a periodic basis (e.g., monthly). Digital signal processors (DSP) are used to analyze the various vibration parameters that are collected periodically.
3. Standards: standards such as VDI 2056 are available to assist this analysis program.
4. Portable FFT analyzers are used in many plants where on a daily basis the data are collected and stored. Later, the same may be used to analyze and predict the fault.

Table 5.9 gives the data sheet (typical) for machine condition monitoring.

2.12 Other Turbovisory Instruments

The basic scheme for steam and gas turbine supervisory instruments (TSI) has been presented in Figs. 5.2 and 5.3, respectively.

2.12.1 STG TSI Specification

While specifying the TSI for STG, the following requirements may be looked into

1. TSI are a part of the steam turbine condition monitoring system. Depending on the application, it shall be one complete package to include sensors, transmitters, converters, a limit value monitoring (alarm/trip) amplifier

TABLE 5.9 Machine Condition Monitoring-Analysis System (DS-Typ.)

Sl	Specifying Point	Standard/Available Data	To Be Specified
1	Instrument rack size	Based on manufacturer's data, specify if no depth limitation	
2	Orientation of modules	Top/bottom, or in a wider rack based on layout space available	To specify
3	Termination	External marshaling/back of rack	
4	Power supply	Single/dual, AC, DC voltage, frequency to be specified. (To specify additional power)	
5	Configuration	(1) Method of configuration, rack interface module if any. (2) Configuration means, from front or any other place, any security protection (3) Security level	
6	Communication	Types of links, protocol, fieldbus support, Ethernet support both wired or wireless. If any other, network support to specify. Specify the redundancy in communication channel if any	
7	Module types	Whether single- or multimeasurement type. Number of channels allowed	
8	Output	Analog output if any, also from which channel and available. Also data output as mentioned above	
9	Status indication	Healthy status LED, T_x, R_x on during communication. Alarm LED, transducer LED	
10	Relay module	Number of channels, status indication of each	
11	Software	Configuration, display support, analysis software, FFT, and other mathematical support software	
12	Grounding	Isolated from case to case.	
13	Display options	Type and area of display, lines of display, information for display	
14	Standard (as applicable)	SIL 3/SIL1, API 670, etc., to be specified	
15	Authority certificate	TUV For SIL 3, and others for hazardous area if any	
16	Intrinsic Safety/Isolation	If required, specify and method. External, internal, etc.	
17	Environmental condition	Ambient temperature, humidity etc.	To Specify
18	Other special feature		To Specify

and another signal conditioning and analysis module, power (dual) supplies, and all special cables and junction boxes as the terminal point.

2. A list of TSI for medium to large sizes is presented in Table 5.2. The list is a typical one; normally, it (with a redundancy issue) is decided based on the manufacturer's recommendation in consultation with the owner.
3. TSI systems normally have facilities to communicate with the main DCS.
4. Also, a metal temperature measurement, a duplex (or two simplex) RTD 100Ω at 0°C, or an ISA Type K (e.g., in the KWU turbine, thermocouples are used mainly) is included in this package.
5. The commonly used sensor types are: proximity sensors (clause no. 2.11.1.2), seismic probes (clause nos. 2.11.2 and 2.11.3), a linear variable differential transformer (LVDT) for casing expansion and valve position monitoring (clause nos. 2.12.3 and 2.12.4), and a Hall probe for speed measurement.
6. Systems are intelligent enough to cater to the following requirements:
   - **a.** Data storage for analysis.
   - **b.** Online/spectral/harmonic analysis.
   - **c.** FFT analysis.
   - **d.** Onsite calibration and configuring facility.
   - **e.** Software support for configuration, display, and communication.
7. Sometimes, there are separate system monitoring cum analysis systems with a suitable diagnostic system (for preventive actions) specified for the utility station. These intelligent systems may operate independently but have the capability to communicate with the main

control system and TSI systems. In addition to the various criteria discussed in Section 2.12.1.6, the system normally also supports the following features:

a. Reliability: The offered system shall meet the requirements specified in API 670 or better and certification from a competent authority (e.g., TUV). Depending on applications, an SIL3 certification would be necessary.

b. Overall system accuracy: Typical values of 0.01 mv/measuring unit as per the project standard.

c. Some systems have an intelligent module to accommodate a number of channels with a mix of various kinds of measurement in one module (Provib tech PT2060/10). Major features of these modules shall include: DSP, programmability, and communication for seamless networking. Also, for the interlock and alarm, there will be relay modules with contact output of suitable ratings.

2.12.2 Smaller STG

Utility Turbines or BFP Turbine TSI SPECIFICATION: Number of measurements for smaller turbines and/or BFP-T could be:

1. Axial shift-thrust position (Minimum of two sensors).
2. Triple-redundant speed measurement.
3. Overspeed protection.
4. Differential expansion.
5. Casing overall expansion.
6. Control valve position.
7. Bearing radial vibration (orthogonal sensors).
8. Casing vibration (if recommended).
9. Bearing and casing metal temperature.
10. Possibility to integrate with DCS.

2.12.3 Casing Expansion

With the basics discussed in Section 2.6.7, detailed discussions on casing expansion, measurement principles, and installation are elaborated here:

1. The turbine shell is anchored at one end to the foundation, whereas the other end is allowed to expand due to thermal expansion. As shown in Fig. 5.22A and B, the LVDT is used to measure the casing thermal growth. The body of the LVDT is designed to be attached to the turbine foundation, whereas the spring-loaded roller is pressed against a bracket and attached to the turbine case. The roller tip of the plunger is allowed to ride freely without

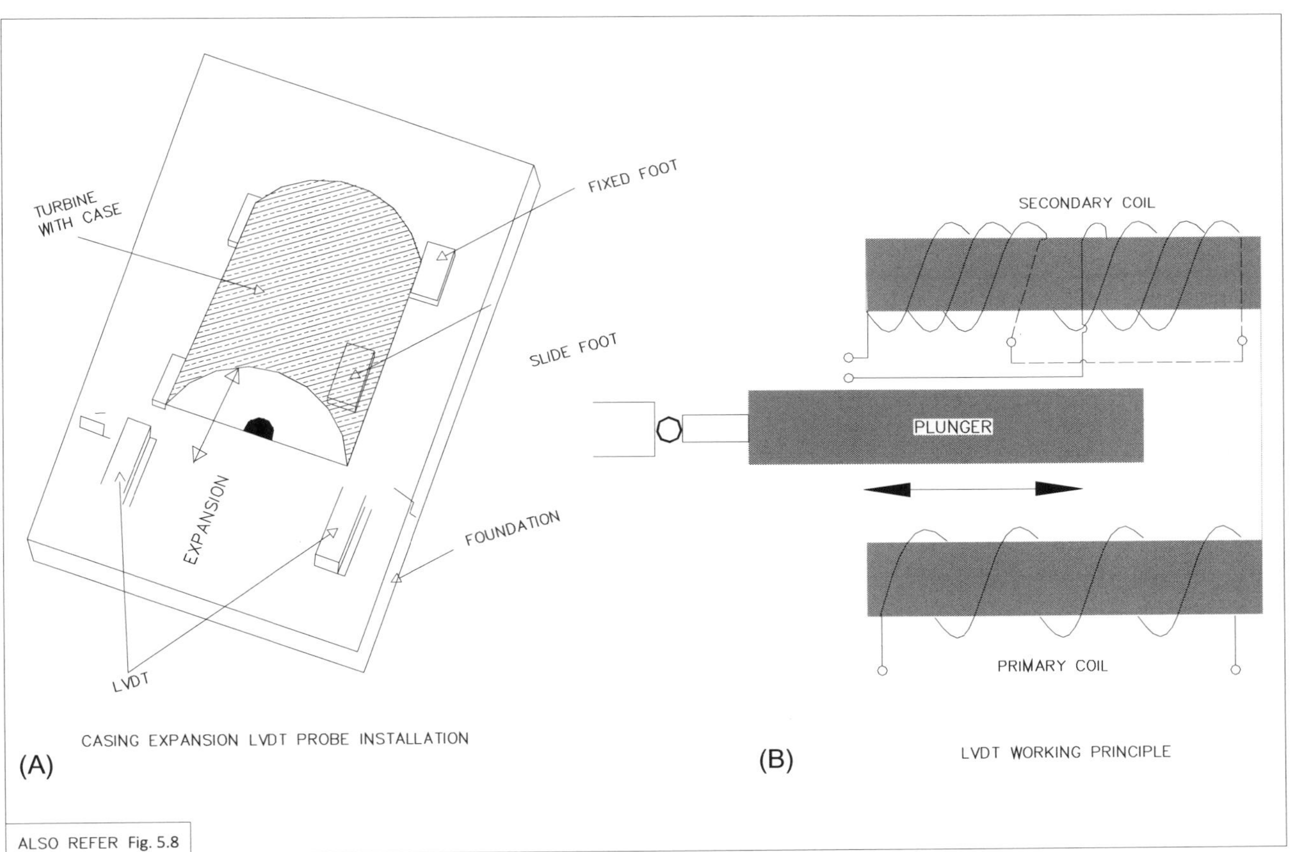

FIG. 5.22 Casing expansion LVDT probe installation.

creating any problems with the turbine operation. As the plunger goes into the LVDT body, the signal output will grow. Depending on the direction of installation of the LVDT, the thermal expansion of the casing signal (+ve or −ve) in the LVDT will increase (+ve signal will increase).

2. LVDTs are electromagnetic devices having three coils, as shown in Fig. 5.22B; would on a hollow tube. The primary coil is excited with supply voltage, which, on account of induction, will induce voltage on the secondary coil with movement of the plunger in the hollow tube. Secondary coils are so wound that as the plunger is in the center, the induced voltage in both the coils is the same in magnitude but opposite in sign so that the net voltage across the secondary coil (in differential mode) is zero. As the plunger goes in either direction, there will be an increase in one of the secondary coils while the same on the other will decrease and we get a voltage across the secondary coil. The LVDT can be the AC or DC type.
3. Casing expansion sensor specification (standard specification for guidance only)
   a. Operating range: 0–25/50/100 mm.
   b. Core magnetic material (Ni-Fe core common).
   c. Scale factor: 0.14/0.34/0.4 V/mm.
   d. Linearity: 0.5% FSD.
   e. Stability: 0.12% FSD.
   f. Environmental condition: (−)18 to 80°C 0–93 Rh.
   g. Ordering information: linear range/scale factor, spring action.

2.12.4 Valve Position

The correct valve position is necessary for efficient operation of the steam turbine. Potentiometric and LVDT are the common methods deployed to generate such signals. Potentiometric is the standard one where a standard resistor can be supplied with a constant voltage. And as the valve changes the position, the slider across the resistance will change. So there will be a change in output voltage. With the help of a V/I converter, the standard 4–20 mA DC can be achieved. Another type of valve position transducer used, is the AC LVDT type discussed above, where the rod inside the LVDT body moves as per the valve position.

The LVDT valve position sensor specification for TSI application (a standard specification for guidance only):

1. Type: single or dual ACLVDT.
2. LVDT internal core: magnetism material (Ni-Fe core common).
3. Core connecting rod material/length: to be specified (Al with length 6″ to 15″ standard available).
4. Linear range: to be specified (available 25 mm to 500 mm).
5. Linearity: 0.25% or better.
6. Environmental condition: $T = (-)50$ to 150°C 93% RH.
7. Ordering information: application, length of rod, and linear range.

2.12.5 Application Notes on Installation

There are a few issues that may be considered for installation. When vibration probes are placed with a clamp, it is possible to adjust the penetration. In case of thrust measurement, it is possible to place probes on the same and opposite sides of fixed structures for better flexibility. Differential expansion is most important during start up and shut down, so it is necessary to ensure the availability of these then. The bow due to eccentricity can be a fixed mechanical bow (within limits), temporary thermal bow, or a gravity bow. This deviation has to be very minimal for start up, so there will be two proximity pickups set orthogonally to measure the extremes of the bow. Based on eccentricity, the operator decides the timing of slow rolling and heating.

2.12.6 System Configuration (Fig. 5.23)

Normally, the turbine condition monitoring system forms a part of the common system network of the TG package. The monitoring and analysis systems are also intelligent and could reside in the same network as that of the turbine control system (TCS) or could be another standalone system with a link for monitoring and communication. This is similar to what has been detailed in clause no. 2.11.5.2. A typical TCS configuration and its integration with the DCS and condition monitoring system is shown in Fig. 5.23. TCS and main DCS may be integrated, and then these can be in the same network through the Ethernet or through a fieldbus, along with other machine conditioning systems as shown in Fig. 5.23.

3 GAS ANALYZERS

3.1 Gas Analysis Requirements and Types

1. *Basics*: Basically, in a thermal power plant, the chemical energy of the fuel is finally converted to electrical energy through the generator. Overall, the conversion efficiency in subcritical is 3035%. Now, by utilizing various techniques, for example, Oxyfuel (Fig. 5.24) technology, SC/USC/A-USC plants with IGCC or CFBC (Chapter 3) even up to ~55%–60% efficiency are achievable. All these developments in technologies also helped to bring down the pollution due to CO_2, NO_x, SO_x, etc., to a great extent (e.g., with Oxyfuel technology:, NO_x is almost zero) due to high conversion efficiency. Monitoring and analysis of flue gas plays an important role not only in optimizing combustion and increasing efficiency of the boiler, but also in maintaining the outside air quality within limits. In modern power plants, in addition to the dedusting plants (e.g., ESP), desulfurization and denitrification plants are also installed. These new plants in addition to controlling pollution, also offer commercial benefits, for

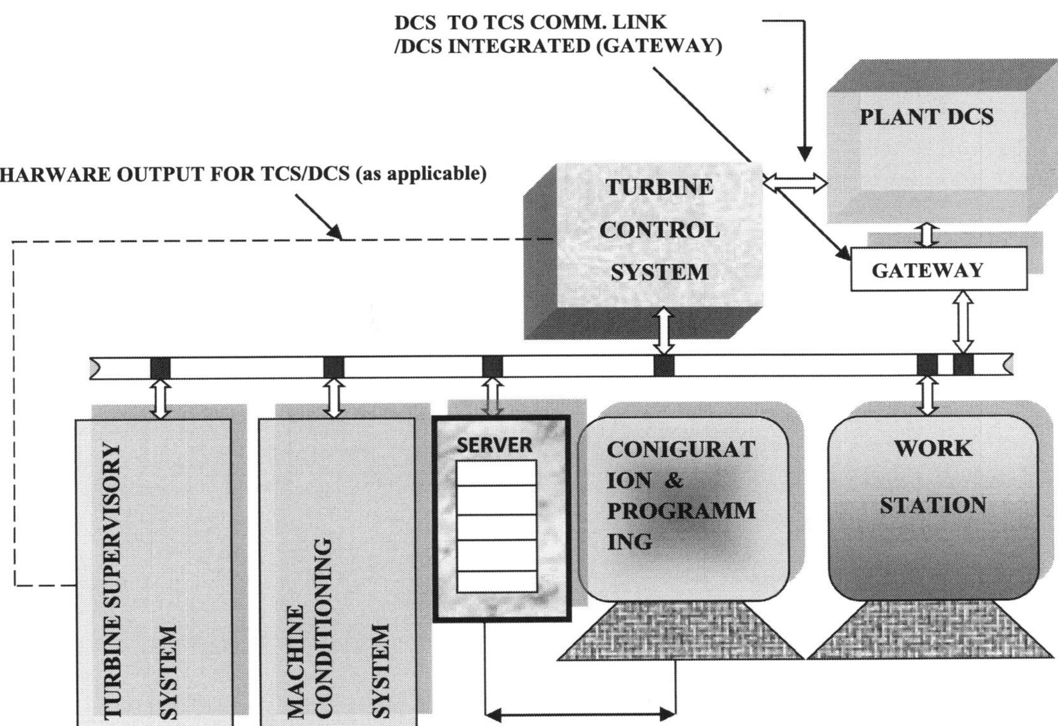

FIG. 5.23 Turbine supervisory system configuration (typical).

OXY FUEL In oxy fuel combustion fuels are burnt in an environment of Oxygen (in place of Air, using air separator unit to remove nitrogen—hence less NO_x) and recycled combustion gas. Exhaust flue gas is passed through an acid condenser and CO_2 purification unit to make capture ready CO_2 for sequestration of CO2. This type of design deploys a separate combustor unit where fuel is burnt with O_2 (after air separator) & recycled flue gas. (Also Refer Section 12.4.3 of Chapter 3)

FIG. 5.24 Oxy fuel.

example, production of gypsum from the desulfurization plant. Also, in order to get the maximum benefit in terms of efficiency, it is necessary to optimize the vol% of oxygen (excess air) in the flue gas. Therefore, it can be argued that gas analysis in power plants is necessary for combustion control, to prevent hazards (e.g., CO in ESP), and for maintenance of the atmospheric air quality. Apart from these types, there are a few other types of gas measurements that are involved in power plants, for example, H_2 purity measurement in the turbo generator is required to get better efficiency. Therefore the basic objectives of gas analyzers are improvement of efficiency, elimination of man-machine hazards (CO in ESP), and protection of the environment.

2. Flue gas purification: Optimization of oxygen, use of low NO_x burner, etc., are a few methods undertaken in modern power plants to improve efficiencies and reduce pollution. These are precombustion steps. Post-combustion steps are:

- Dedusting: Dust collectors, a bag dust filter, and electrostatic precipitator (ESP) are the normal way to dedust flue gas. Out of them, the ESP is the most popular. Here, the action of force on charged particles in an electric field is used to remove dust (Chapter 2).
- Desulfurization: Desulfurization plants to remove SO_x, based on liquid purification are mostly used. The core of such a plant is spraying of limestone water to form gypsum and to remove SO_2 (details available in Section 10.3, Chapter 2).
- Denitrification plant: Selective catalytic reduction (SCR) process is commonly used to remove NO_x. Here, it is worth noting that during the removal of NO_x, ammonia is formed and the exact metering of NH_3 is important. Details are discussed later in the book.
- International Standard: Coal-based power plants rank at the top in air pollution. An environmental impact assessment (EIA) showed that on account of pollution created in large combustion plants, there have been serious impacts of various constituent matters on the air quality. For thermal power stations, suspended particulates, oxides of sulfur (SO_x), oxides of nitrogen (NO_x), and carbon dioxide (CO_2) are the major

pollutants. The Clear act 1990 has raised concerns over trace concentration metals in flue gas from power plants and other places. Now, traces of heavy metals, particularly mercury, are also monitored. International standards specify the emission limits and the methodology for monitoring the pollution-constituting items. In Europe in 2001, Directive 2001/80/EC came into effect concerning large combustion plants. The European Committee for Standardization (CEN, Comité for Eurpéenne de Normalization) put CEN into use. Some of the features include stricter measures for particulates and limits for SO_2 and NO_x, monitoring of Hg (solid fuel), TOC have been introduced and Mandatory use of EN 14181 and EN ISO 14956.

- EN 14181: "Stationary source emission—Quality assurance of automated measuring systems." This gives guidelines concerning features of automated measuring instruments, their installation, and maintenance. The quality assurance levels for such instruments are QAL1: suitability of the measuring instruments for the task. As per ENISO 14956, QAL2: installation, calibration, uncertainty, and drift control.
- ENISO 14956: Deals with the definition of suitability of an automated measuring system and procedural details.
- Some of the other international standards/certifying followed are: UK MCERT, United States EPA requirements, and Japan JIS.

Limiting value and associated range of instruments, meant to monitor these air polluting gases in thermal power plants, vary with standards as well as with type of fuel (solid/liquid gas, etc.). To get an idea, the following table is presented (Table 5.10).

3. Various gas compounds: Major gaseous compounds generated during combustion of fuel (coal) are:
   - CO: CO is generated mainly due to incomplete combustion and can be minimized by regulating excess air, but this also accounts for system loss. So, it is always desirable to keep the stoichiometric ratio at a value where CO generation is minimum. Practical achievable minimum value could be around 50 mg/M^3.
   - CO_2: Formed on account of complete combustion of carbon in excess air. Apparently, this is not harmful, but has a greenhouse gas effect and hence need to be lowered by capturing and/or CO_2 sequestration (seizure). There have been several methods to reduce CO_2 emissions. CO_2 generation in SC and USC boilers is much lower than subcritical boilers. Oxyfuel combustion is also a very popular method. In Oxyfuel combustion, fuel is burnt with oxygen (with O_2 separator slightly auxiliary power sensitive) in the presence of recirculated flue gas. In this cycle, CO_2 is made ready for CO2 capture (Chapter 3).
   - NO_x: Oxides of nitrogen are formed during burning of fuel in air. Use of a low NO_x burner is an approach to reduce the formation of NO_x. The addition of separate over fire air (SOFA)*, or closely coupled over fire air (CCOFA) dampers injects air above the combustion zone for complete combustion while there will be low air in the burner zone (< stoichiometric ratio), so that the flame temperature is low hence less thermal NO_x as well as NO_x formation due to bonded nitrogen in fuel (Chapters 2 and 8). The arch-fired burner approach (low volatile coal) of FWC, and the ultralow NO_x approach of ALSTOM reduces drastically the NO_x formation. The post-combustion methodology to remove NO_x from flue gas is selective catalytic reduction(SCR), where NH_3 reacts with NO and NO_2, so NH_3 monitoring needs to be done in addition to monitoring of NO_x.
   - SO_x: The sulfur content of coal forms SO_x (mainly SO_2) when coal is burnt. However, a part of SO_2 is oxidized to form SO_3, which may combine with water vapor to form H_2SO_4—cold end corrosion at AH. Postcombustion FGD plants are deployed (Chapter 8) to get rid of SO_x.
   - Apart from the above currently after the Clean Air Act, trace elements such as mercury and other heavy metals in flue gas are also monitored. Also NH_3 as stated above is monitored (Fig. 5.25).

TABLE 5.10 Gas Components Limit and Range

Item	Limit Value*	Expected Range	Unit	Remarks
CO	150/100(GT)	0–125/625	.mg/M^3	*Lower for fuel: liquid and gas
SO_2	200	0–600	.mg/M^3	
NO_x	200	0–300/1500	.mg/M^3	*For GT 50 and gaseous fuel 100
Par#	97–100	0–200	.mg/NM^3	# Par = Particulate
Hg	0.1		μg/m^3	

Top three are in line with 13 FEPL Germany whereas the bottom two are as per EPA.

CO_2 capture & storage (CCS) means basically Striping of CO_2, Transport & Storage of CO_2 for future sequestration. Sequestration can be done by geo engineering, mixing with oil storage in the sea (as done in North Sea), or by chemical means of formation of carbonate of metals from oxides of metals. Refer Section 12.4 of Chapter 3 for details.

FIG. 5.25 CO_2 capture and storage (CCS).

Oxygen analysis: Now the air supply to the boiler ideally should be the amount that is required exactly for complete combustion, that is, where the stoichiometric ratio $\lambda = 1$. However, depending on the burner design, there may be some places where the fuel may escape the main firing zone and remain without getting completely burnt. This will give rise to the chances of CO formation, which is explosive especially when CO containing flue gas passes through the ESP, where there will be electric discharges. So, in order to assure safe conditions, excess air is maintained for assuring complete combustion. Now, there are two contradictory situations. If air is not in excess, there is a chance of incomplete combustion and associated hazards. Also, if excess air is greater, then there will be loss of heat due to excess air going out with the flue gas. Also, more excess air means more NO_x formation at high temperatures. It has been found that at the value of the stoichiometric ratio λ little (slightly) > 1 that is, where CO is minimum is the optimum point. Also, at this point the concentration change of CO is maximum (<100 ppm). For this reason, oxygen (analyzer) trim control and fine tuning through the CO analyzer are done.

Calculation for excess air in percentage is rather easy:

$$\text{Excess Air in\%} = \{(O_2\,\text{in\%})/(21 - O_2\,\text{in\%})\} \times 100$$

In this connection, Fig. 5.26B may be referred to. Here, one point to be noted is that *although the area of max efficiency has been shown around the vertical line ($\lambda = 1$), the operating point can never be on the left side of the vertical line as in that case there shall be a deficiency of O_2. This is unacceptable, so it has to be only on the right side of the vertical line*. This strategy is implemented in oxygen trim control, which is realized by a zirconia O_2 probe for its fast response time. On account of the following reasons, it is recommended that the zirconia probe may be located at the economizer outlet in the flue gas path.

- Air heater leakage, especially in a Ljungstrom (regenerative) air heater, is high, so this may not get a representative value, meaning the probe cannot be located.
- An oxygen probe can be placed near the combustion zone, but on account of high temperatures, the proposition is very costly and the instrument availability will be restricted. However, to find air leakage in n air heater (AH), O_2 is measured at AH O/L and I/L.

4. *Gas analyzing types*: Analyzers used for gas analysis could be
   - The extractive principle where a gas sample is taken from the process flow. Here, the analyzers are kept at a little far-off place. Samples are drawn from the duct by the steel (better to use stainless steel 316/320) sampling probe of nominal bore around 20 mm. The number of such probes in a duct depends on the duct size and gas stratification. In such cases of multiple probes, there shall be an automatic method to select the probes one after another. Prior to admitting the same to the analyzer, the sample is conditioned and prepared. The gas is drawn up to the conditioning system with the help of a stainless steel tube—a 10 mm diameter could be optimum. This preparation means drying the gas by cooling so that moisture could be removed. For dust removal, various kinds of filters are utilized. Ceramic filters are used to remove coarse dust while membrane filters are used for finer particles. Sometimes, water mists are used to remove dust. Standard cooler and/or special coolers such as an electric cooler to bring down the temperature to around 25°C are used.

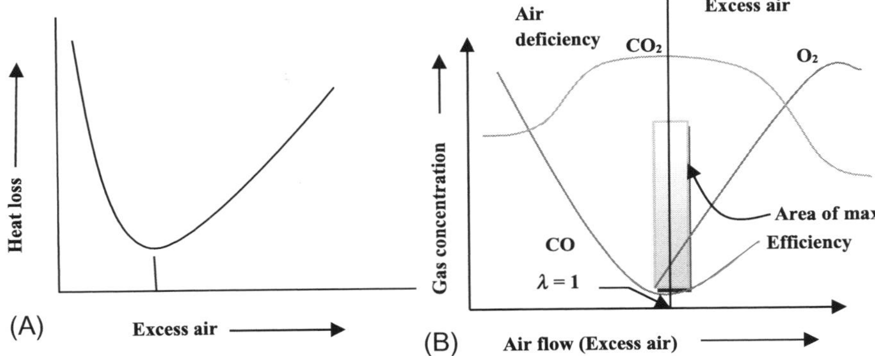

FIG. 5.26 (A) Heat loss versus excess air. (B) Optimal operating point.

Take-off points are also very important (e.g., take-off points should be pointing downward toward the duct so that the dust due to gravity can flow into the duct). The choice of sampling tube (after the probe) diameter is also crucial; when the diameter is too small, then the sample volume change will be less, and one can expect to get a better representative sample but there will be chances of frequent choking of the tube by dust-laden flue gas. On the contrary, if the diameter is bigger, then the chances of getting choked are less, but the time for volume change may be more–hence less representative samples will be available. In case of extraction instruments, a number of analyzers can be connected to one sampling line. Also, in most of the cases of an extractive system, there would be a provision for cleaning the probe in periodic intervals. Depending on the principle of sample extraction, a typical response time is 10–20 s.

- In Situ: In this, the principle analyzer is directly put at the point of measurement. This is done when a fast response (5–10 s) is desired, for example, a zirconia probe for O_2 online infrared analyzers and opacity monitors.

5. Measuring points of flue gas analyzers in a power plant (Table 5.11):

Gas analysis in gas turbine: CO, CO_2, NO_x, and SO_2 SPM are commonly measured in gas turbine plants. As per EPA, measurements are done at the GT exhaust/HRSG inlet as well as the HRSG outlet. In the case of natural gas, SO_2 and SPM will be minimal. In natural gas, hydrocarbons are also measured.

6. *Generator cooling gas analysis*: The efficiency of a generator depends highly on its cooling system. H_2 Has been chosen as the cooling medium for generator rotors on account of H_2 shows higher thermal conductivity, has a higher heat capacity, a low frictional loss for rotating parts, and a high dielectric breaking strength.

However, a fire hazard is associated with the use of H_2, as it forms an explosive mixture when it comes into contact with

TABLE 5.11 List of Gas Analyzers in Power Plant (Typ)

Location	Mea	Range (Typ)	Unit	Remarks/(Purpose)
Econ. O/L	O_2	0–10	vol%	Accuracy 0.5% FSD (E) 5% Reading (I)/(Zirconia type used for O_2 trim control)
AH O/L	O_2	0–10/0–5*	vol%	Accuracy 0.5% FSD (E) 5% Reading (I)/EX (e.g., paramagnetic for AH leakage)
	CO	0–300/500	mg/M^3	Accuracy: 2% reading/for fine tuning of O_2 trim control (I) usually in situ
	NO_x	0–1800	mg/M^3	1% FSD accuracy; denitrification control, as applicable
Stack	NO_x	0–1500	mg/M^3	Accuracy: >5% reading/(pollution control)/(I)
	SO_2	0–400	mg/M^3	Accuracy: >5% reading/(pollution control)/(I)
	SPM	0–20	mg/M^3	Accuracy: >5% reading/(pollution control)/ (I) (0%–100% for opacity)
Denitri'on plant*	NO_x	0–1500	mg/M^3	*Denitrification plant as applicable Accuracy: 1% FSD/pollution control
	NH_3	0–10/25	mg/M^3	*Denitrification plant as applicable Accuracy 2%/NH_3 slip monitoring from denitrification plant(I)
Desulfurization plant*	SO_2	0–4000	mg/M^3	.* Desulfurization plant as applicable Accuracy: 1% FSD/pollution control
ESP I/L	CO	0–300/500	mg/M^3	2% reading/ESP protection(I)
	NH_3	0–10	mg/M^3	Accuracy 2%/denitrification control wherever applicable and ESP protection(I)
Coal stack/silo	CO	0–5%	Vol	Accuracy: 1%–2% FSD; Type based on field condition; smoldering fire/explosion

Symbol: I = in situ type; EX = extractive type; FSD = full scale division.Wherever in situ type is possible, it is marked with (I).

air. The requirement of analyzers and purity value is guided by IEC 60842 and EN 60034. In power plants, the H_2 purity analyzers are used to measure the H_2 in air and in carbon dioxide. CO_2 purity in air is also measured.

3.2 Analyzer Selection Criteria

1. *Type selection*: whether extractive or in situ.
2. *Selection of analysis method*: Same parameters may be analyzed by a number of method; O_2 can be measured by a zirconia probe or by the paramagnetic/thermomagnetic method. Similarly, SO_2 may be analyzed by infrared as well as by ultraviolet. So for this, due consideration should be given to the range, installation facility, proneness of the instrument in the application, and international certification.
3. *Desired accuracy*.
4. *Calibration method and frequency* of calibration needed.
5. Analysis frequency.
6. Sampling method (as applicable).
7. Data handling.
8. Display and recording details.
9. Maintenance issue.
10. Cost.

3.3 Component and Analysis Types

Table 5.12 (analysis types) gives a list of various measuring methods a system designer needs to chose the correct one as per the above discussions. Gas analysis with the cheaper electrochemical cell method may not be very accurate and is less popular in utility stations

3.4 Absorption Principle for Analysis

1. *IR method of detection (typical wavelength: 2.5–10 μm)*: Working principles: Hetero atomic molecules such as CO, CO_2,SO_2, and NO_x show the property of absorption in some regions in the IR spectrum. These gas components show distinct behavior in their *absorption pattern to a particular frequency*, SO the selection of a proper optical filter based on the frequency of absorption is very important. Typical schemes are shown in Fig. 5.27. Infrared radiation from a common hot filament source is passed through the optical filter sensitive to the desired IR frequency. Now, this filtered radiation is passed through both reference sources as well as through the cell/chamber containing the sample gas. If the sample contains the desired gas, the energy absorption will be different than that in the reference cell, containing noninfrared absorbing gas (N_2). This causes half the detector cell to have a different pressure on the pressure-sensitive diaphragm to cause deflection of the diaphragm (the higher the gas concentration, the higher the deflection), hence the response of the condensing microphone pickup. In order to have AC pickup in the condenser microphone, the beam is chopped. This scheme is shown in Fig. 5.27A. There **is** a variation of this kind of pickup where the sample gas is passed through a fixed volume chamber at a preset time interval and the chamber is then sealed with the sample gas trapped inside. The infrared radiation is pulsed into the chamber via an infrared transparent window. The pulsating pressure change is measured by the microphone as a frequency change to produce the signal. For this, Fig. 5.27B may be referred to. In both cases, the pickup signal is amplified and rectified to detect

TABLE 5.12 Gas Analysis Types

Component	Location	Types of Measurement	Remarks
CO_2	Flue gas	Infrared (IR)/Electrochemical cell (EC) Thermal conductivity (TC)	IR preferred. TC are also in use
CO_2	Generator	IR/EC/TC	TC are used mostly
CO	Flue gas	IR/EC	IR preferred
H_2	Generator	TC	
H_2	Main steam	TC (Katharometer)	
SO_2/SO_x	Flue gas	UV/IR	
NO_x	Flue gas	Chemiluminescence/IR	
NO_2/NO	Flue gas	Chemiluminescence/IR/UV	
O_2	Flue gas	Para- or thermomagnetic/zirconia	Zirconia for control loop
NH_3	Flue gas	Absorption spectroscopy (near IR)	Also possible by chemiluminescence method.

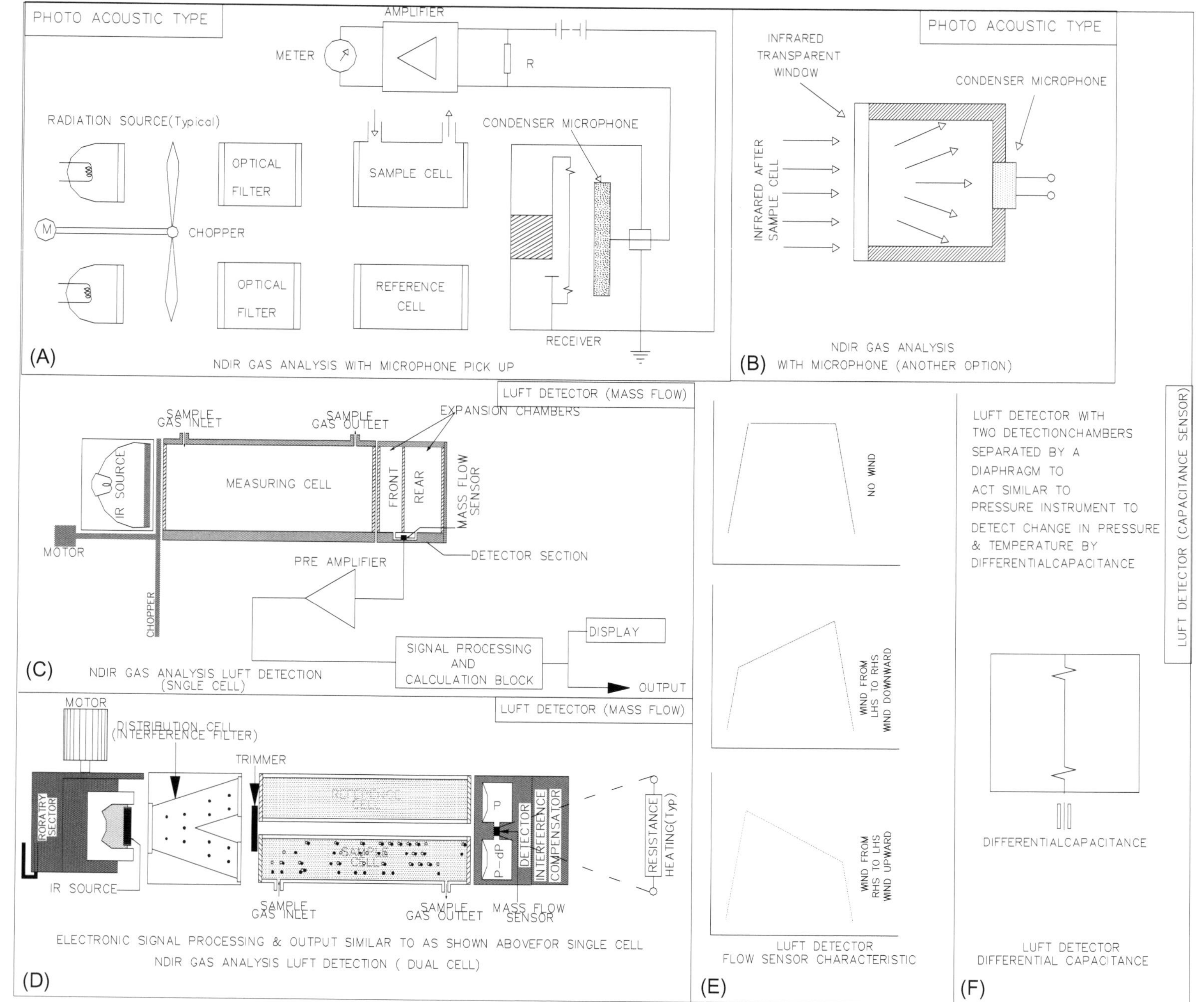

FIG. 5.27 NDIR principle for gas analysis.

the presence of the desired gas in the detector as a ratio of the two signals. The above method of detection is generally referred to the photoacoustic method. The other kind of detector, based on optopneumatic detection, is popularly known as the LUFT detector (named after inventor Kurl Luft). These are shown in Fig. 5.27C–F. In the detection part, there may be two chambers and if they are connected, then the thermal mass flow sensor (a resistor) may be placed between the two detection chambers. In case the two detection chambers are not connected, then on account of pressure variation, there will be deflection in the diaphragm, which will cause a change in differential capacitance like the normal capacitive pressure sensor/transmitter. Fig. 5.27C shows a single measuring chamber through which the sample gas is passed. An infrared source passes infrared radiation through the chamber. Based on gas concentration, the infrared radiation of the required frequency will be absorbed by the measuring chamber. In order to have AC sensing, the infrared radiation is chopped by the chopper. So, when the radiation is allowed to pass through, it will send energy (less absorbed by sample gas), which will increase the temperature and pressure in the front detection chamber. Naturally, to balance the pressure, it will push the pneumatic mass to the rear chamber through the opening containing the thermal mass flow detector (resistance) to detect the pneumatic mass flow. The higher the concentration of desired gas in the sample, the less will be the mass flow as more energy will be absorbed. In order to have an AC pulse, as stated earlier, radiation is interrupted when naturally, when there is no ration, there will be flow extra pneumatic mass flow from rear to front detection chamber to balance the pressure. In place of a single chamber, there could be two chambers, as shown in Fig. 5.27D. The reader's attention is drawn to Fig. 5.27E, Where it is seen that in the case of no wind, the characteristic curve is flat, whereas during one-way flow (say, left to right) the temperature variation is in one direction. Naturally, in a reverse flow, the variation will be in the other direction, meaning that on account of two-way flow (by chopping the infrared radiation) there will be a better response (change in temperature) of the mass flow detector. So, this variation in temperature is measured in the LUFT detector as gas concentration. In case the two chambers are not connected but sealed by a diaphragm, the pressure variation will cause a variation in differential capacitance, as shown in Fig. 5.27F

The aforesaid type of detection is possible in an extractive analyzer.

In in situ instruments, the transmitter and detector are placed across the duct. In an in situ analyzer system, the transmitter and receiver are placed opposite each other in the duct/stack, as shown in Fig. 5.28. There is one IR source at the transmitter that beams the IR through a chopper and lens to the flue gas in the duct to reach the receiver located at the opposite end of the duct/stack. IR after passing through the flue gas, reaches the receiver where two beams of exactly the same wavelength are created by a beam splitter (which receives the radiation through the lens in the receiver). The receiver employs a gas cell and optical filter to filter out interfering signals (placed ahead of the splitter). There are

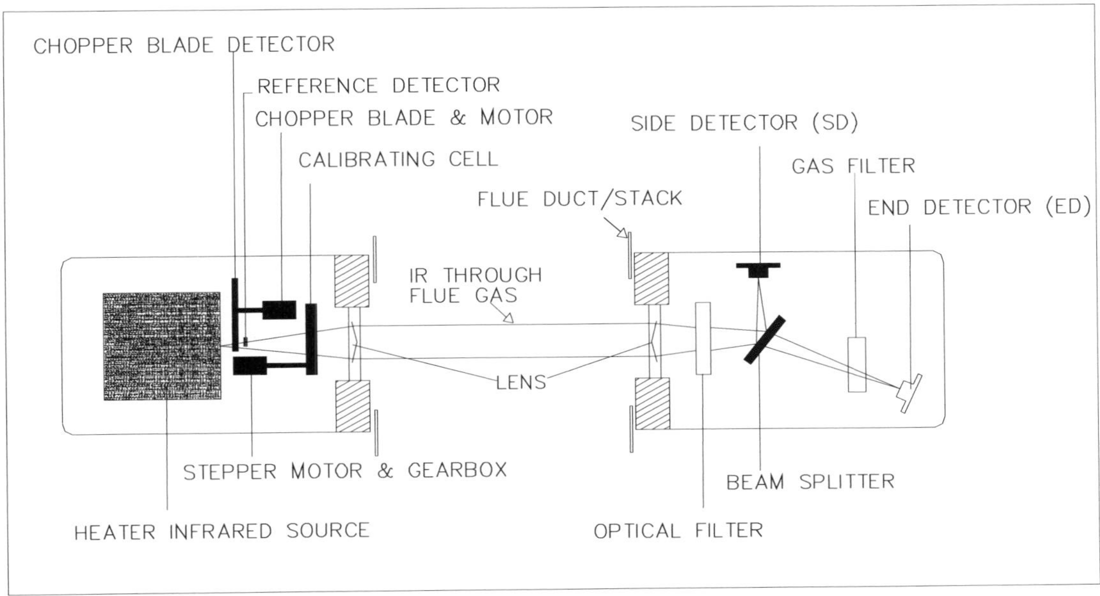

FIG. 5.28 IR gas analysis (in situ).

two solid-state detectors—a side detector (SD) and an end detector (ED)—to detect the radiation (after absorption by the desired gas in the flue gas). The SD detects the IR coming from the flue gas; the remaining part is passed through the gas cell containing the desired gas of known concentration. So, from the SD and ED, a ratio signal can be created that in electronics is compared with the calibrated ratio to detect the concentration of the gas component in the flue gas.

2. *UV method of detection (typical wavelength: 200-600 nm)*: SO_2, NO, NO_2 In flue gas can be measured in this principle. The radiation spectrum required for measuring gas as well as for reference is generated in filter wheel box, from the UV lamp shown in Fig. 5.29. There are interference filter in the filter wheel box. Two interfering filter having different characteristic are put in to the beam path for accurate measurement. The beam splitter directs filter radiation to both reference chamber as well as sampling chamber. UV Detector (diode array or spectrophotometric detector) behind each of these chambers detects the radiation to calculate the quotient received by detectors. In situ instruments with integral control and calibration unit are also available. In resonance absorption spectrometer principles, NO,NO_2, and NH_3 can be measured, without the need of NO converter. By this principle, NH_3 slip in SCR can be measured.

3. *Laser type absorption detection*: A few analyzers in the market use of single line molecular absorption spectroscopy utilizing a diode LASER emitting near infrared light. LDS 6 of Siemens is an example of the same. This has been shown in Fig. 5.30 this is in situ multichannel analyzer (courtesy Siemens) utilizing Laser beam.

3.5 Chemiluminescence Type Analysis System (Ref. Fig. 5.39)

This type of analysis system may be deployed to detect NO_x, NH_3, etc. When NO reacts with ozone (O_3) to produce NO_2, a part (10%) of electronically excited NO_2 reverts back to NO and gives away illumination. The light emission is directly proportional to the concentration of NO present in the gas, (because each NO molecule produces light). The volume of sample gas and excess ozone are carefully regulated. The emitted light is measured by photomultiplier and electronics. As is clear from the above discussion, it is basically for NO, so for measurement of NO2/NO_x suitable converter is necessary to convert NO_x to NO concentration.

This conversion is normally achieved by passing the sample over a heated catalyst to convert all oxides of nitrogen to NO, before the main reaction chamber. In some instruments, there is provision for performing the automatic switching of the catalyst in and out of the sample path so that the resulting signal may be used to measure NO only by indirectly measuring NO_2 produced by the reaction with ozone. Similar technique may be used to perform measurement of Ammonia (NH_3) but by oxidizing reaction (e.g., $NH_3 + O_2 \Longrightarrow NO + H_2O$) before the main reaction chamber.

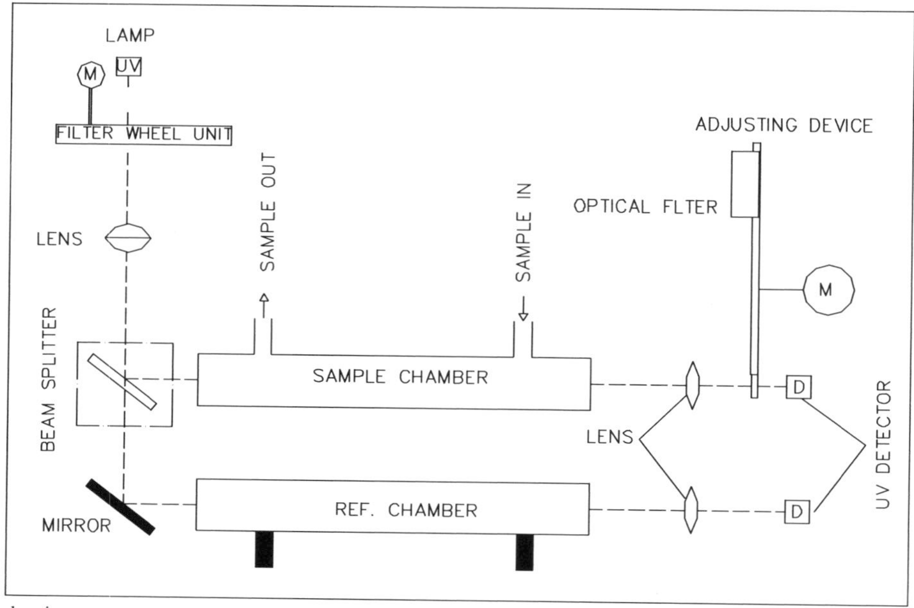

FIG. 5.29 UV gas anlaysis.

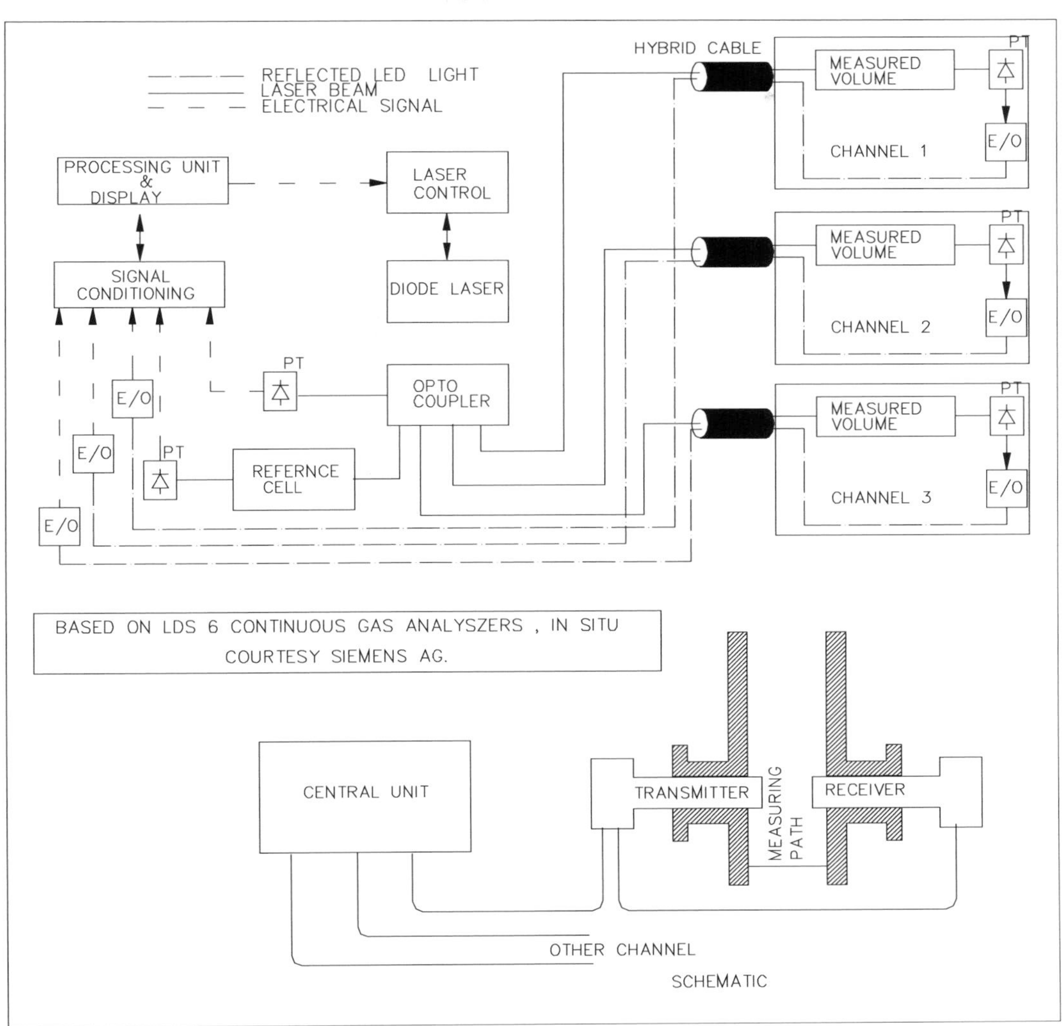

FIG. 5.30 LASER gas analysis.

3.6 Paramagnetic and Thermo Magnetic Oxygen Detection System

While most of the common gases are Diamagnetic (repelled out of magnetic field), Oxygen is strongly paramagnetic. Magnetic susceptibility of O_2 Principle is used to measure O_2 present in the sample, to give an indication of O_2 concentration by volume.

Paramagnetic: Normally a set of two glass sphere (filled with N_2 or other diamagnetic gas) is kept suspended by a thin wire with a metal piece acting as a mirror, and kept at balance in a magnetic field as shown in Fig. 5.31. A light source focuses on the metal piece which reflects the light to two nos. of photo diodes (connected in reverse mode) placed to receive the reflected light. At balanced condition both the photodiodes receive same amount of light. Thus any movement of the metal piece can be detected by the photo diodes due to variation in amount of light. When O_2 molecules with magnetic susceptibility, starts flowing, displaces the spheres, which are moved away from the magnetic field. So the metal piece/mirror rotates, as a result, the light detected by the two photo detectors are different. Photodiodes are connected to a differential amplifier which would cause unbalanced current flows through the feedback coil mounted on suspension, and on account of interaction of this magnetic field due to current in the coil, and permanent magnet the sphere would be brought back to the original balanced position. So the feedback circuit current is proportional to the concentration of O_2 in the sample. Manufactures like Fuji Japan, Siemens, ABB etc. have instrument of this type.

Thermo magnetic (refer Fig. 5.32): The magnetic susceptibility of O_2 also depends on temperature. In this analyzer there will be one (or two) platinum wire(s) which is heated by external source and is connected to a wheat stone bridge. A magnet is placed across *one* platinum wire. Now

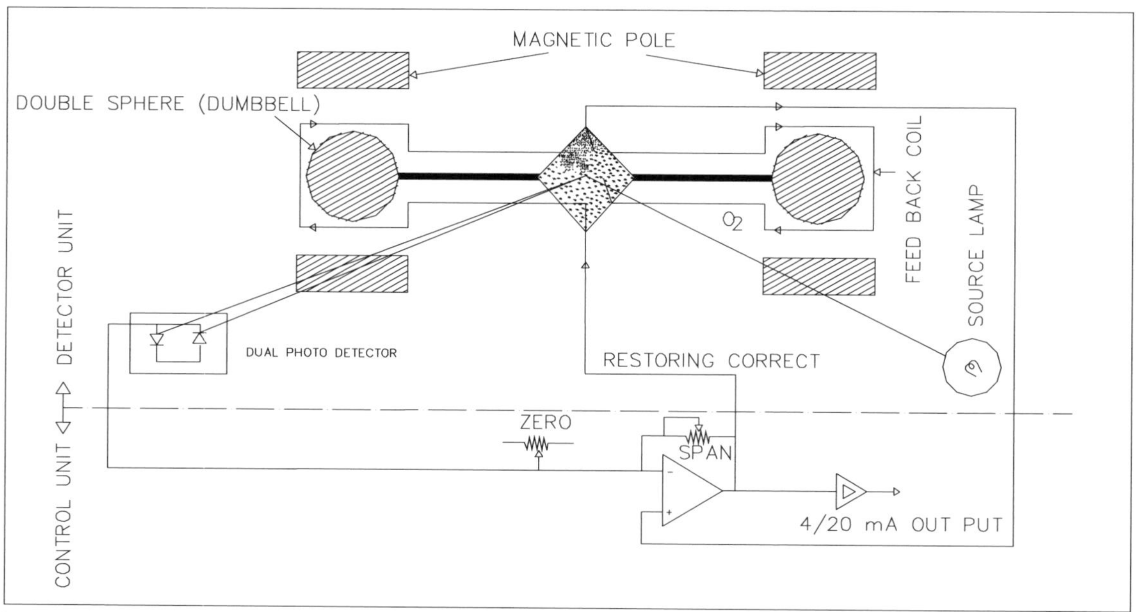

FIG. 5.31 Paramagnetic oxygen analyzer.

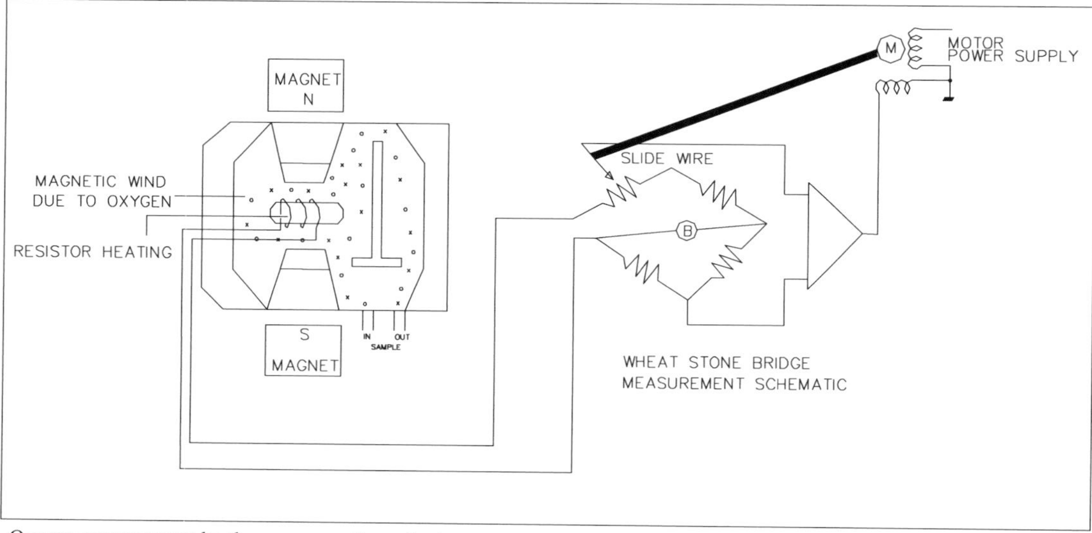

FIG. 5.32 Oxygen measurement by thermo magnetic method.

when sample gas containing O_2 is introduced, On account of Oxygen's affinity toward magnet, it will be attracted toward the magnet, hence will pass over the hot Platinum wire. As soon as it passes over the hot magnet it loses susceptibility, and moves away. This place will be filled in by the fresh sample. Thus a magnetic wind will be established. This magnetic wind takes away the heat from the Platinum wire. Naturally the amount of heat taken away is proportional to Oxygen volume which is measured by change of resistance in a Wheatstone bridge discussed above (which may compare this resistance change with respect to other Platinum wire which is not subject to magnetic field when two wire s are used). ABB has analysis system based on this principle. A similar system utilizes two thermistors and are located in such a way that the magnetic wind cools down one in the inlet side of the magnetic wind and the other is located at the outlet side of the magnetic wind and gets heated by the same. They are connected to a Wheatstone bridge. Additional sensor elements are used to measure gas heat capacity and viscosity to calculate the oxygen percentage accurately by the microprocessor-based transmitter to arrive at the resulting signal proportional to the oxygen concentration. This type is available with APX (Panametrics product) under GE Industrial and Sensing.

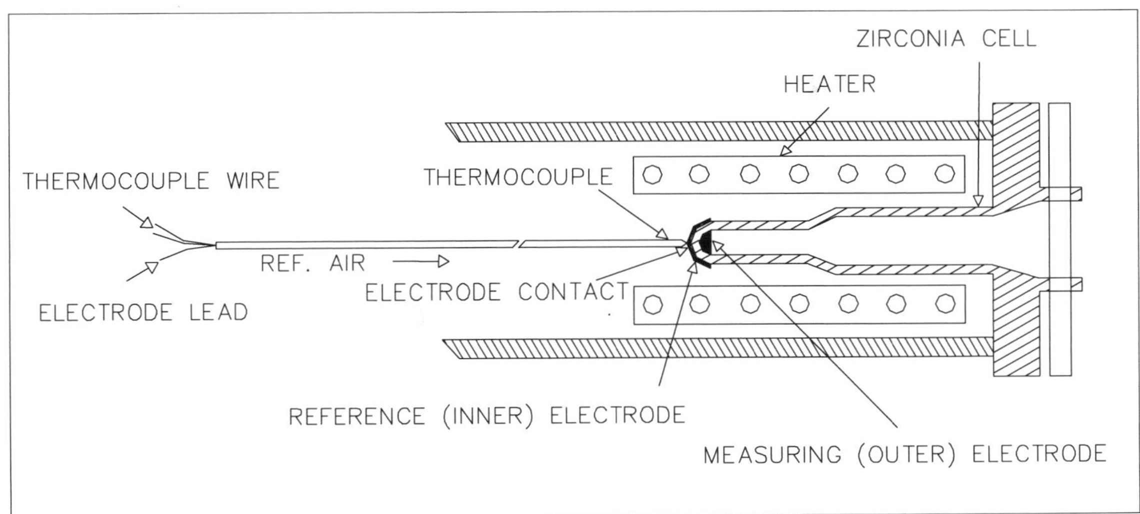

FIG. 5.33 Zirconia probe for oxygen measurement.

3.7 Zirconia Oxygen Detection System (Fig. 5.33)

An in situ oxygen detection method based on the "Nernst" relation for zirconium oxide is used in a zirconia probe oxygen analyzer (used for boiler air flow O_2 trim control, it's a better response time). When heated at around 700° C, zirconium oxide (on account of oxygen vacancy in the ceramic lattice) is selectively conductive to oxygen ions. When two gases differing in O_2 concentration are on opposite sides of the zirconium oxide membrane, a DC voltage is generated as per the established Nernst relation (Fig. 5.34).

As is seen from the above that the generated voltage is not linear with O_2 concentration, hence a linearization circuit is necessary. Also, as the log of zero is infinite, the sensor cannot measure zero concentration, but instead a very low concentration. Again, from the equation it is seen that it has dependence on temperature, which needs to be maintained precisely so as to get an accurate result. The reference air (around 500–1000 mL at 1 bar) can be supplied by a pump or regulated instrument air supply through an inlet port in the probe terminal head. Arrangements are also provided for suitable venting of reference air in the return path. In most of the cases, heaters in the probes are used to maintain the temperature. High-temperature versions of these probes do not require heaters, and they can be put close to the combustion point, which is desirable. A calibration gas inlet port is there to check the probe with test gas.

NERNST EQUATION: $E = (RT/nF)\ln(P_1/P_2)$
Where E= *measurable Voltage DC;* R = *universal Gas constant (8.314 Joules/mole K;* T=*Temperature in K;* n= *no of electrons transferred;* F= *Faraday constant;* P_1= *Partial pressure of ref gas(air~ 20.9%)* P_2=O_2 *partial Pressure in sample gas. Empirically the relationship has been simplified to*
$E(mV) = F\,(=0.0496) * T * (\log_{10} P1/P2) \pm C(mV)$ ***where C = cell constant.***

FIG. 5.34 Nernst equation—discussions.

3.8 Thermal Conductivity Gas Detection System—Katharometer (Fig. 5.35)

In a Katharometer, the concentration of a particular gas in the sample is measured, utilizing thermal conductivity of the particular gas. Generally, the thermal conductivity of the gas is inversely related with its molecular weight, as is seen from Table 5.13, Comparison of thermal conductivity of gases. H_2 has the highest conductivity with the least molecular weight, whereas SO_2 has high molecular weight but less thermal conductivity.

When current is passed through the electrical conductor surrounded by gas in a chamber, the temperature will rise until equilibrium is reached. At this point, the energy supplied to the conductor matches the thermal loss due to the surroundings, provided the loss due to radiation, end conduction, etc., is the minimum. Now the thermal conductivity difference between the measuring gas and reference gas would cause the difference in heat loss—hence the temperature difference between the two resistances. It is known that the electrical resistance is proportional to the temperature. Therefore, there will be an imbalance in the Wheatstone bridge circuit causing an imbalanced current, which is measured by the DC amplifier and another circuit. Thus, there will be 4–20 mA current proportional to the percentage concentration in a known volume of gas being measured. In a Katharometer, matched platinum wire filaments are used as resistance in the Wheatstone bridge. H_2 and CO_2 are the major gases measured in this principle.

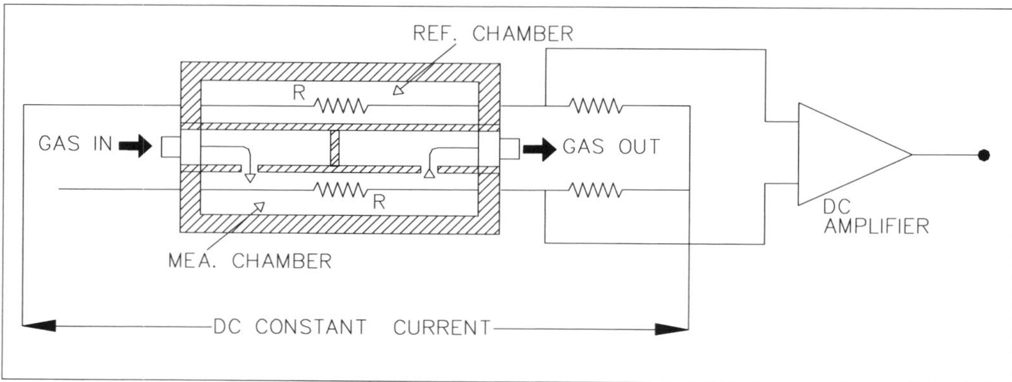

FIG. 5.35 Thermal conductivity analyzer.

TABLE 5.13 Comparison of Thermal Conductivity of Gases

Gas	Comparative Thermal Conductivity to Air at 0°C	Reference Gas
SO_2	——▶	
CO_2	———▶	N_2, AIR, Ar
CO	—————▶	
NO	—————▶	
N_2O	————▶	
H_2	——————————————————————▶	N_2 (CO_2)

3.9 Combustible Analyzer

Combustibles such as $CO+H_2$ are analyzed using the catalytic combustion method. In this method, Pt is used as a catalytic sensor for detecting $CO+H_2$. This is useful for detecting incomplete combustion. Combustibles are measured in terms of 0–2000 ppm level.

3.10 Gas Chromatograph

For gas turbines, special instrumentation such as a gas chromatograph (GC) provides a cost-effective analytical package. GC is an analytical instrument that measures the content of various components in a sample.

Principle: In order to achieve an optimal gas turbine and get an assured load rating, it is essential to know the composition details of the fuel gas (mainly natural gas is preferred). The quality of gas depends on the source as well as dilutions at various stages. Accurate determination of fuel gas composition allows for optimal adjustments of the air-to-fuel ratio, enabling the combustion turbine to operate at its most cost-effective, efficient point while reducing NO_x emissions and ensuring savings through avoiding fines. Process gas chromatographs are used to determine the gas composition to derive the calorific value, density, and Wobbe index mentioned above. Typical application and calculation details are depicted in Fig. 5.36.

The sample solution injected into the instrument enters a gas stream (helium or nitrogen), popularly called carrier gas. It transports the sample into a separation tube known as the column. The constituent components are separated inside the column. The detector measures the quantity of the components that exit the column. When a sample with an unknown concentration is measured, a standard sample with known concentration is first injected and then the standard sample peak retention time (or the appearance time) and area are compared to the test sample to calculate the concentration.

In the power industry, where emissions are closely regulated, operating cleaner and more efficiently also means a tremendous savings by avoiding potential costly fines. GCs provide the individual component concentrations, a process gas chromatograph also provides fuel gas heating values, allowing optimal adjustment of the air-to-fuel ratio. With a correct stoichiometric air-fuel ratio, total fuel and air will be consumed during the combustion process, burning cleanly and enabling the turbine to operate in its most efficient, cost-effective manner. ISO standard ISO 6974 is the prevailing standard. In the chromatograph, gases are separated into methane, ethane, and CO_2. Normally, two-minute and four-minute cycle times for C_1 through C_6^+

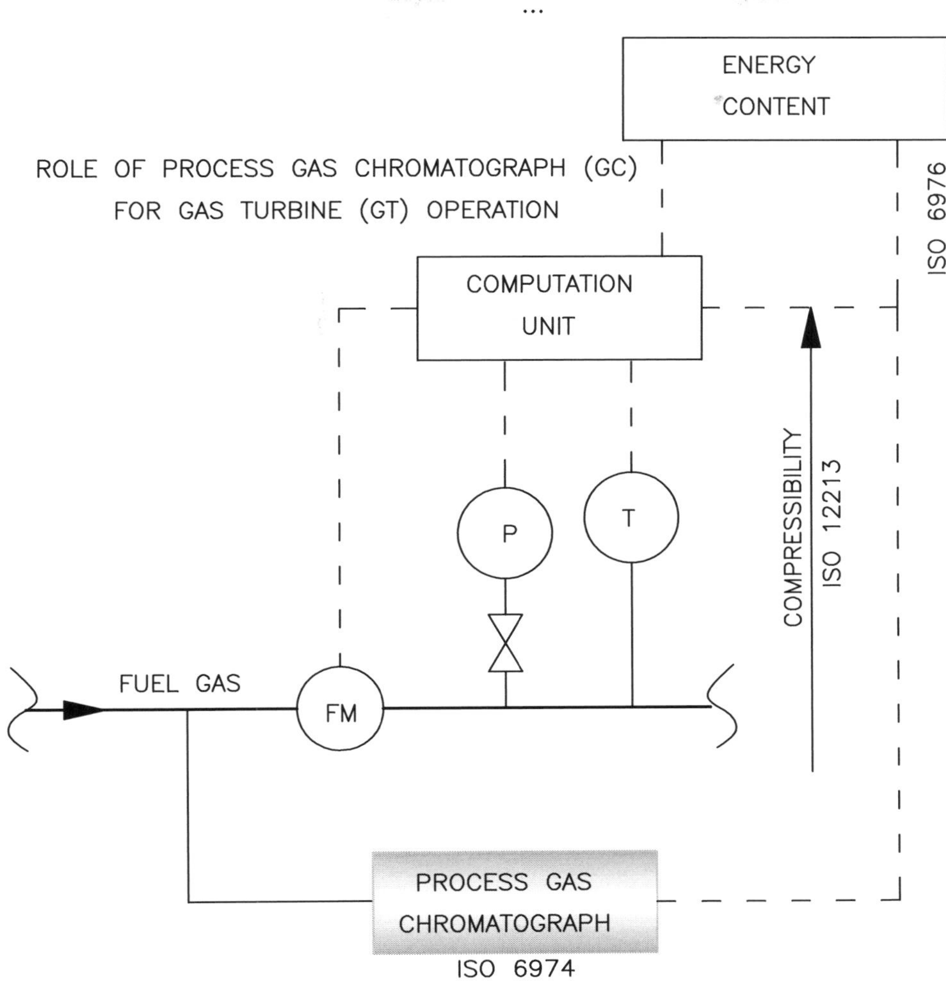

FIG. 5.36 Fuel gas energy calculation by GC.

composition and physical property calculations. Most process gas chromatographs are fully compatible with a modern Ethernet/fieldbus for communication to DCS. For turbine applications, dual-detector process gas chromatographs are used (e.g., the Emerson process gas chromatograph). Many process gas chromatographs also include monitoring of N_2 and H_2S gases.

- *Gas composition*: *The main aim is to operate* the plant at the stoichiometric point, so that all C is converted to CO_2, S to SO_2, and H_2 to H_2O so that there is no unburnt constituent in the exhaust gas. This also helps in reducing NO_x to the minimal value without the use of a high-cost low NO_x burner. A gas chromatograph helps in accurately determining the gas composition so that the burner can operate efficiently at the stoichiometric point.
- *Wobbe Index (WI)*: Gas burners get damaged from varying flame dynamics. This can be controlled by using WI, which relates the energy content of the gas per unit volume at NTP divided by the square root of specific gravity. Gas chromatographs can calculate WI to provide reliable data for GT operation. For a low heating value (LHV) in kJ/m^3 or BTU/scf and specific gravity S:

$$\mathrm{WI} = \frac{\mathrm{LHV}}{\sqrt{S}} \tag{5.1}$$

- *Hydrocarbon dewpoint*: When the natural gas temperature falls below the hydrocarbon dewpoint, then heavier hydrocarbon components start dropping, resulting in a catastrophic (flashback) situation for the GT. Gas chromatographs can calculate the hydrocarbon dewpoint of fuel gas at various pressures and help to mitigate damage due to this.

3.11 Extractive Multianalysis System

In this type of analysis system, a number of gas components are analyzed utilizing a single probe. The sampling system needs to face many challenging demands for the process conditions such as changing pressure, temperature, and dust loading conditions. In power stations, such instruments are used to monitor gas components such as CO, CO_2/NO, N_2O, NO_2, and NO_x/SO_2/O_2. A few typical features of these types of instruments have been presented below from some well-reputed manufacturers.

1. Measuring gas components with ranges: CO: 0–125/600 mg/M^3; CO_2: 0%–40% V; NO/NO_2/N_2O/NO_x: 0–100/1000 (1500) mg/M^3; SO_2: 0–75/1000 mg/M^3; O_2: 0%–10% V.
2. Number of gas components: 3–6 available.
3. International certification: As per the applicable standard (as discussed in clause no. 3.1), e.g., QAL 1/2/3 for European systems.
4. Analysis technology such as a photometer, UV type, NDIR, paramagnetic, flame ionization detection, etc., to measure the gas components.
5. Number of analyzers: Normally, with 3–4 analyzers in one single system.
6. Overall accuracy 1%–2% full-scale division (depends on individual instrument accuracy, as discussed later).
7. Analyzer performance specification: To be specified (some guidelines have been given in this book against individual analyzer type).
8. Sample probe and sampling system:
   a. Dust retention factor >99% with heating so as to avoid clogging.
   b. Probe material/length: high temperature stainless steel/length as required.
   c. Temperature withstand capability: say, up to 1000°C.
   d. Sampling: gas cooler, feed pump/extraction sample line, heating.
9. Power supply system: say, 230 VAC 1 Ph 50 Hz (110 VAC 1 Ph 60 Hz) or DC power supply as applicable/available.
10. Output: analog/digital, Alarm for each channel.
11. Contact rating: As applicable (e.g., 30/50 VA, etc.).
12. Calibration method: currently autocalibration methods are available for most cases, from the analyzer front panel.
13. Display type: Digital readout/graphical.
14. Built-in system software: To handle the sampling system, other system software for networking as applicable. In-house graphic generation, history logging, data storage, sampling of data, etc.
15. Interface: HART/Profibus/fieldbus/MODBUS, Ethernet, etc.
16. IT compatibility: Whether industrial computer compatible/web browser access, etc.
17. Examples; ABB ACX/AO 2000 Rosemount: X Stream, etc. (Fig. 5.37).

3.12 In Situ (Modified) Multianalysis System

Here, one probe to diffuse the flue gas into the sensor is put at the end of the probe. There are a number of stainless steel filters to eliminate the high dust of the flue gas. The sensor used is the NDIR type, for example, Codel G-CEM 4000.

A few typical features of the type of instrument are presented below:

1. Measuring gas components with ranges: CO/NO/NO_2/N_2O/NO_x/SO_2: 0–3000/6000 ppm; CO_2: 0%–25% V; Measuring unit options: % or mg/M^3.
2. Number of gas components: Up to 5.
3. Overall accuracy 1%–2% full-scale division.
4. International certification: As per the applicable standard.
5. Analyzer performance specification: To specify (data sheets).

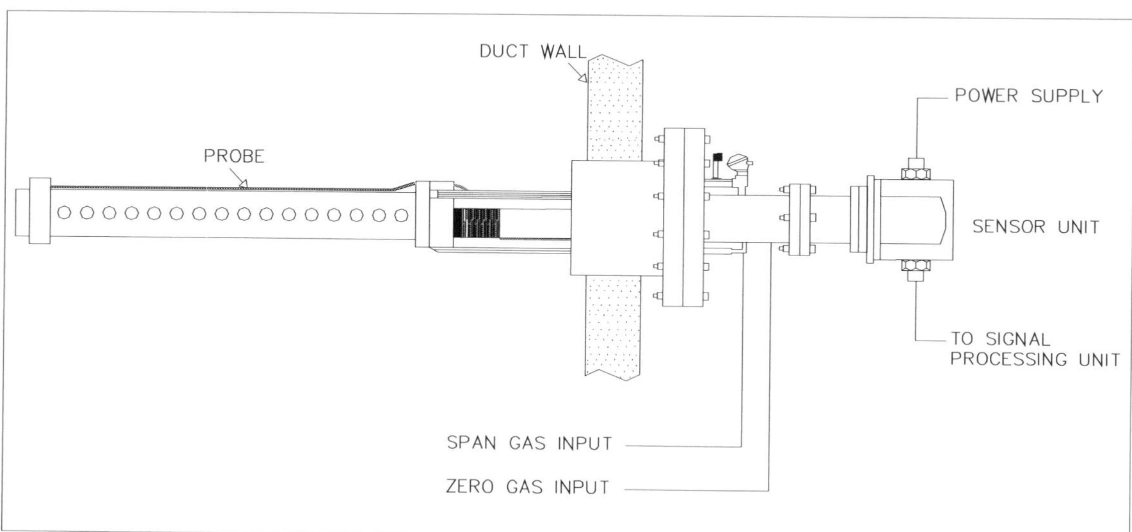

FIG. 5.37 Multigas analysis probe.

6. Sample probe and sampling system: Dust retention >99%, probe length 1–2 M with diffusion cell 1 M and temperature up to 1000°C.
7. Power supply system: say, 230 VAC 1 Ph 50 Hz (110 VAC 1 Ph 60 Hz) or DC supply as applicable/available.
8. Output: analog/digital, alarm 4–20 mA + SPST contact for alarm.
9. Contact rating: as applicable (e.g., 30/50 VA).
10. Calibration method: autocalibration with span/zero gas.
11. Display type: digital readout/graphical.
12. Built-in system software: For networking as applicable. In-house graphic generation, history logging, data storage, sampling of data, etc.
13. Interface: HART/Profibus/fieldbus/MODBUS, Ethernet etc.
14. Special feature: Pressure and temperature measurement for normalization.

3.13 In Situ Multichannel Analysis System

As shown in Fig. 5.38.

There are several in situ single-line analyzers connected to a central analysis system used to measure a number of gas components in the flue gas. Each set of transmitter/receiver units form a channel. A Siemens LDS 6 analyzer is an example of such a kind of analyzer. Power plant ammonia traces can be measured in a similar way.

Specification: Typical and generalized specification for guidance only (Fig. 5.39 and Tables 5.14 and 5.15).

3.14 Opacity/Particulate Monitoring (Fig. 5.40A and B)

In order to keep the environment clean, the particulates such as unburnt coal, fly ash, etc., coming out of the stack need to be monitored properly. Essentially, it is desirable to maintain 0.15 mg/M^3 dust in the exhaust gas. The particulates in flue gas are measured either in mg/M3 or as percent opacity. Some of the measuring principles of particulates are discussed below:

1. Working principles
   a. *Transmissiometer (light extinction)*: The instrument measures the loss of light intensity (visible range) across a particulate-laden flue gas, as a function of dust concentration (Fig. 5.41).

      In the instrument, a collimated beam of visual light is put across the gas path in the duct or stack, directed toward the receiver at the opposite end. The receiver measures the loss of intensity as discussed to convert it into an output signal. For better accuracy, a dual-pass transmissiometer is used with a mirror on the opposite side of the stack. For a particulate-monitoring instrument, RED/near IR light is used instead of visual light.
   b. *Optical scintillation*: In this method, the receiver measures the modulation of light due to the dust particles. The greater the dust concentration, the greater will be the momentary interruption of the light beam, hence the variation in amplitude of the light received. After manual gravimetric sampling, it can be calibrated as mg/M^3.

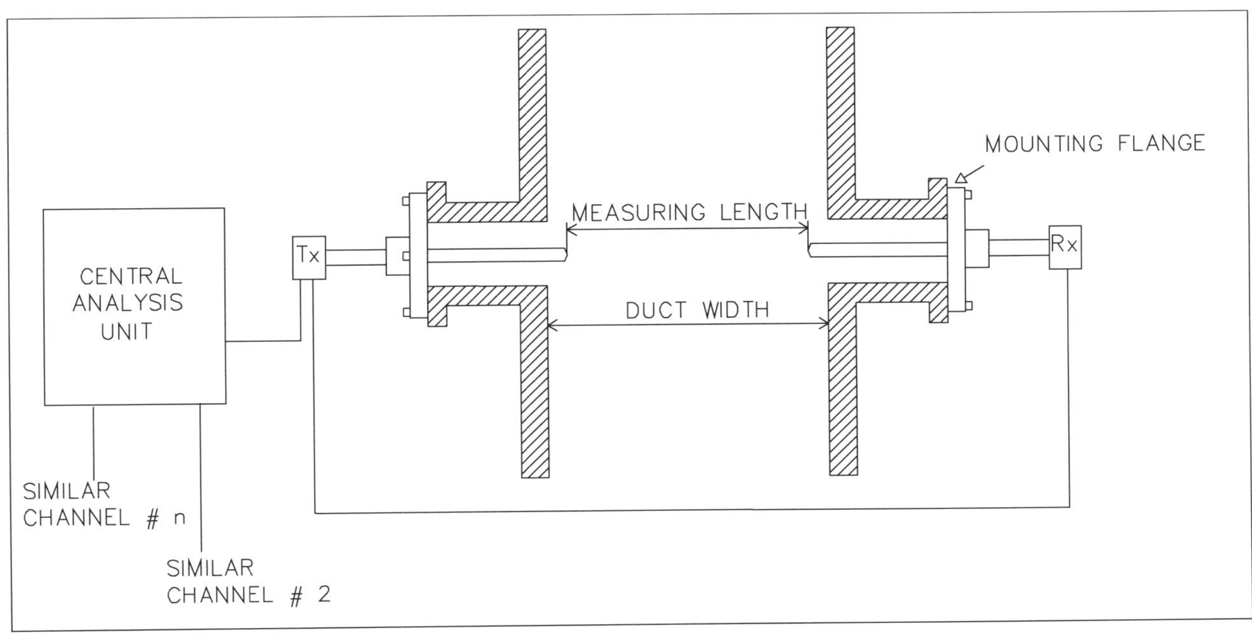

FIG. 5.38 Multichannel in situ analyzer.

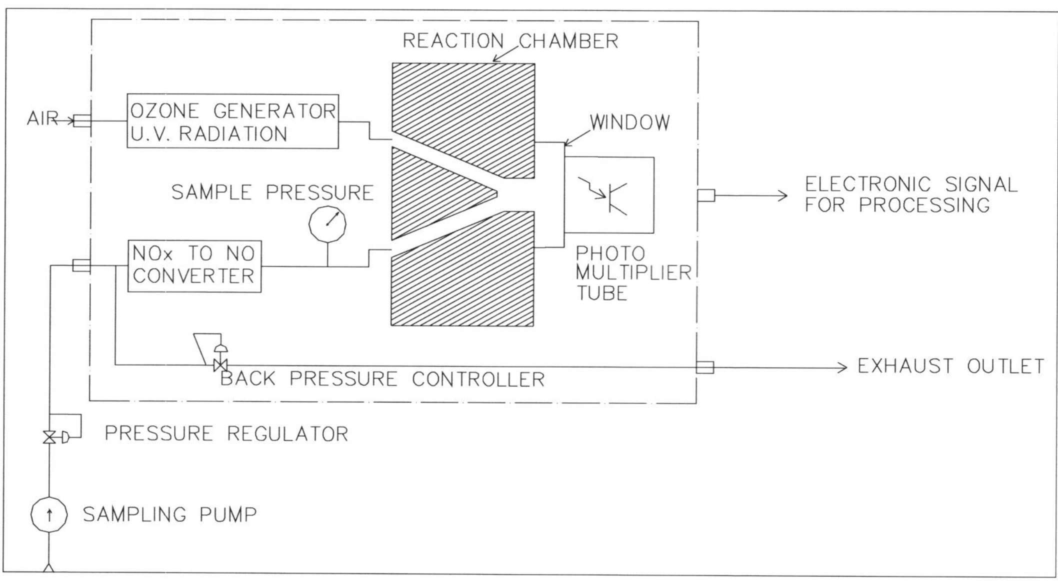

FIG. 5.39 NO_x analyzer (chemiluminescence).

TABLE 5.14 NO_x **Analyzer (Chemiluminescence) DS**

Sl	Specifying Point	Standard/Available Data	To Be Specified
1	Working principles	Chemiluminescence (Che)/IR/UV (NO, NO_2)	
2	Span	0–25/1500 mg/M3/(IR) 0–5000/6000 ppm (Chem)	Selectable range may be specified
3	Drift (zero/span)	Per week: Z: 2% FSD/S: <0.5%	
4	Accuracy	<0.1 ppm/1–2 (IR) %FSD	
5	Response time	Typically 60–200 s as per type	
6	Output	4–20 mA DC + Hi/Hi-Hi Al	
7	Contact details	SPST/SPDT 50 VA or 50 V 1A	
8	International certification	As per country requirement (clause no. 3.1.2)	To be specified
9	Probe details	Length (1–2 M) mat: SS (Norm)	To be specified
10	Sampling	As applicable to specify pump, flow rate, pressure temperature	To be specified
11	Display	Backlit LCD 32 channel alphanumeric/4.5 digit display	
12	Connectivity	RS 232/485/MODBUS/Profibus/HART/Ethernet, etc.	
13	Power supply	230 VAC/24 VDC as the case	
14	Housing Prot'n/material	IP rating (67)/Epoxy painted Al	To be specified
15	Ambient condition	Temperature/Humidity	To be specified
16	Accessories (Sampling if any)	Mounting flange/air and other requirement if any for cleaning, sampling system (as applicable)	To be specified
17	Software detail	For data storage, networking, sampling etc. as applicable	To be specified depending on need

TABLE 5.15 SO_2 **Analyzer**

Sl	Specifying Point	Standard/Available Data	To Be Specified
1	Working principles	IR/UV/Thermal Conductivity	
2	Span	0–25 ppm/0–75 mg/M^3	Selectable range may be specified
3	Drift (zero/span)	Per week: Z: 2% FSD/S: <0.5%	
4	Accuracy	1%–2% FSD IR (2%)	
5	Response time	Typically 200 s	
6	Output	4–20 mA DC + Hi/Hi-Hi Al contact	
7	Contact details	SPST/SPDT 50 VA or 50 V 1 A	
8	International certification	As per country requirement (clause no. 3.1)	To be specified
9	Probe details	Length (1–2 M) mat: SS (Norm)	To be specified
10	Sampling system	As applicable to specify pump, flow rate, pressure temperature condition	To be specified
11	Display	Backlit LCD 32 channel alphanumeric/4.5 Digit	
12	Connectivity	RS232/485; MODBUS/Profibus/HART/Ethernet, etc.	
13	Power supply	230 VAC/24 VDC	
14	Housing/Material	IP rating (67)/Epoxy painted Al.	To be specified
15	Ambient	Temperature/Humidity	To be specified
16	Accessories (sampling if any)	Mounting flange/air and other requirement if any for cleaning, sampling system (as applicable)	To be specified
17	Software detail	For data storage, networking, sampling, etc. as applicable	To be specified depending on need

N.B.: ABB/Rosemount/Honeywell/Siemens/YEL are some of the standard suppliers.

c. *Light scattering*: Light is both absorbed as well as scattered by the particles in the path of the light. The amount of such scattering is a function of dust concentration as well as the size, shape, and color of the particles. A light-scatter instrument measures the amount of light in a particular direction (forward/backward) and the output is proportional to the dust concentration, that is, the particulate matter in the flue gas. The light scattered by the particles is focused by the receiving optics. Receiving the light by the receiver/detector generates a current proportional to the dust concentration. The angle of source to the receiving optics and the characteristic of the optics determine the volume of gas to be covered.

d. *Beta attenuation technique*: In this method, beta rays are passed through the material. Part will be absorbed and reflected while another part will be passed. Attenuation of intensity is a measure of particles present in the path. Basically, the sampling instruments measure the energy absorbed from the beta particle as the beta particle passes through the particulate collected at the filter. The two major items are a beta source and detector. From a safety point of view, a good half-life C14 is chosen as the source. The detector does the beta count as the beta particle passes through the particulate in the filter. The output is mass concentration. These instruments also measure the volume of gas extracted so as to give a reading of mg/M^3.

2. *Some suppliers*: Rosemount/Lear Siegler/Durag
3. *Opacity instrument*: Most in situ opacity instruments work on a double-pass transmissiometer (Table 5.16).
4. Dust measuring instrument (Table 5.17)

3.15 Ammonia (in Flue Gas) Analyzer

When measured, generally the measurement of NH_3 in power plants is done as part of a multianalyzer system, either extractive (ABB Limas11HW) or in situ (e.-g., Siemens LDS6) (Table 5.18).

3.16 Mercury in Stack Gas Analyzer

As a part of continuous emission monitoring (CEM), there are two main popular types of mercury analyzers.

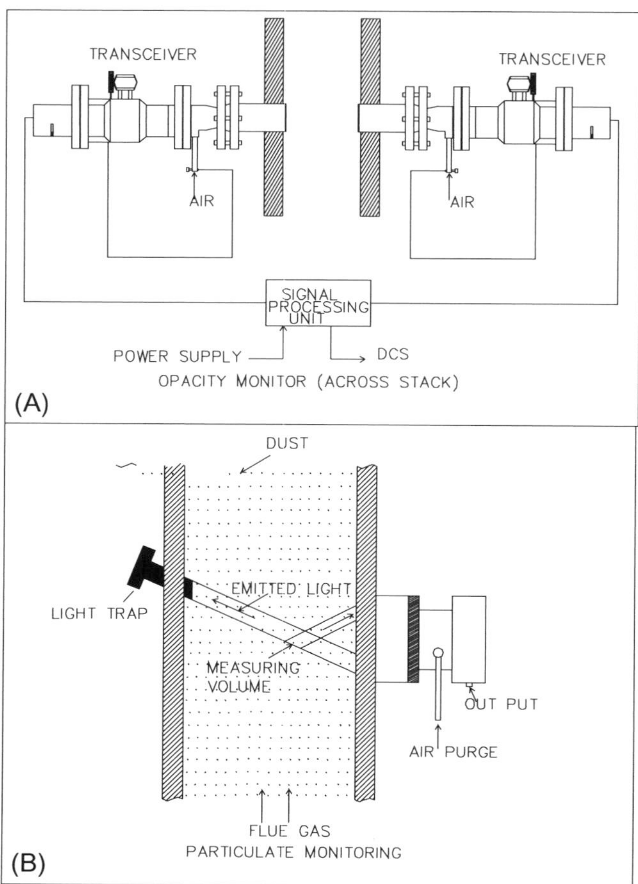

FIG. 5.40 Opacity monitor and particulate monitoring.

- *Atomic absorption spectroscopy (AAS)*: The system includes a thermocatalytic converter positioned inside the sampling probe. The purpose of the same is to reduce the ionic mercury compounds to elemental mercury vapor. The mercury is finally measured in a UV atomic absorption photometer.
- *Atomic fluorescence spectroscopy (AFS)*: This system utilizes the phenomenon of releasing stored energy in atoms in the form of light (there are ways), which is known as fluorescence. *AFS* measures this emitted light, the wavelength of which indicates the identity of the atoms. The intensity of the fluoresced light is directly proportional to the concentration of atoms and the intensity of the excitation source. Cold vapor atomic fluorescence (CVAF) spectroscopy is specific to mercury and is without interference from SO_2 or HCl (Table 5.19).

TABLE 5.16 Opacity Instrument

Sl	Specifying Point	Standard/Available Data	To Be Specified
1	Light source details	Solid-state light modulation, high intensity LED (~650 nM). Expected life (~5 years) with aging comp'n	To Specify selected feature.
2	Span	Full scale in % or in mg/NM^3	To be specified
3	Drift (zero/span)	0.1% opacity/month a typical value	
4	Accuracy	1% FSD or 0.2% opacity	
5	Response time	Typically 10–15 s	
6	Output	4-20 mA DC + Hi/HiHi Al contact	To be specified
7	Contact details	SPST/SPDT 50VA 50V 1A	
8	International certification	Ref. clause no. 3.1 for country requirement	To be specified
9	Path length/gas condition	Length: 0.5 to 12/15 M Flue gas temperature > 400°C	To be specified
10	Calibration	Automatic and manual zero/Span Cal. Also to comp build up	
11	Display	$4_{1/2}$ LED with diagnostic display	
12	Connectivity	RS 232/485/MODBUS/HART/Profibus/Ethernet, etc.	
13	Power supply	230 VAC/24 VDC	
14	Housing/Mat'l	IP rating (65)/Epoxy painted Al.	To be specified
15	Ambient	T: Up to 50°C RH (0.93%)	To be specified
16	Accessories sampling if any	Mounting flange/air purge system/service module	To be specified to
17	Software detail	For data storage, networking, remote control/cal, etc., as applicable	To be specified
18	Special feature	Service module for diagnostics	Optional

If the intensity received at the detector is I & the reference intensity is I_0 then,

Transmittance T= I/I_0, while opacity = 1-T, & optical density (D)= log(1/T)

<u>After manual gravimetric sampling, opacity can be related to dust concentration in mg/M^3.</u>

FIG. 5.41 Transmittance.

TABLE 5.17 Dust Measuring Instrument

Sl	Specifying Point	Standard/ Available Data	To Be Specified
1	Working principle	Forward scattering of light (or LASER)	
2	Span	0–10/200 mg/ M^3(or NM^3)	To be specified
3	Accuracy	2% FSD	
4	Response time	~10 s	
5	Output	4–20 mA DC + Hi/ HiHi Al contact	To be specified
6	Contact details	SPST/SPDT 50VA 50V 1A	
7	International certification	Clause no. 3.1 for country requirement	To be specified
8	Length/temp	0.5–5 M/Temp >200°C	To be specified
9	Function test	Automatic	
10	Air purge	Integral, to be provided	
11	Display	LED with diagnostic	
12	Connectivity	RS 232/485/ MODBUS/HART/ Profibus/Ethernet, etc.	
13	Power supply	230 VAC/24 VDC	
14	Housing material	IP rating (65)/Epoxy painted Al.	To be specified
15	Ambient	T: Up to 50°C RH (0.93%)	
16	Accessories	Mounting flange/air purge system/ service module (as applicable), shutter	To be specified
17	Software detail	For data storage, networking, remote control/cal, etc., as applicable	To be specified
18	Special feature	Diagnostic module/ LASER protection	Optional

TABLE 5.18 Ammonia Analyzer (DS)

Sl	Specifying Point	Standard/Available Data	To Be Specified
1	Working principle	Absorption spectroscopy (UV/LASER)/ Chemiluminescence	
2	Span	0–10/25 0–200 ppm	To be specified
3	Accuracy	2% measured value	
4	Output	4–20 mA DC + Hi/HiHi Al contact	To be specified
5	Contact details	SPST/SPDT 50VA 50V 1A	
6	International certification	Clause no. 3.1 for country requirement	To be specified
7	Sampling details	SS 2–4 M probe + heater/ cooler	If applicable
8	Display	5 digit LED + Diagnostic display	To be specified
9	Response time laser protection	<3 s/class I LASER protection	For LASER type
10	Connectivity	RS 232/485/MODBUS/ HART/Profibus/Ethernet, etc.	
11	Power supply	230 VAC 50 Hz/110 VAC 60 Hz	
12	Housing material	IP rating (65)/Epoxy painted Al.	To be specified
13	Ambient	*T*: Up to 45°C RH (0.93%)	To be specified
14	Accessories	Mounting flange/ sampling system (if applicable), shutter	To be specified
15	Software detail	For data storage, networking, remote control/cal, etc., as applicable	To be specified
16	Sample gas	In situ and extraction type	To be specified
17	Special feature	Diagnostic module/ LASER protection	Optional
18	Some suppliers	ABB/Maihak/Siemens	

3.17 Oxygen Analyzers

3.17.1 Oxygen Analyzer Types

The two mainly used types of oxygen in flue gas analyzers are *paramagnetic* and *zirconia*. Because oxygen in the flue gas is vital parameter, dedicated analyzers are used in power plants.

1. Zirconia probe: Some special features of the instrument are:

   There is another type as shown in Fig. 5.42, where a diffusion-based sampling system is used to draw the sample to the zirconia call located outside the probe

TABLE 5.19 Mercury in Stack Gas Analyzer (DS)

Sl	Specifying Point	Standard/Available Data	To Be Specified
1	Working principle	Absorption UV spectroscopy after converting Hg vapor or fluorescence	
2	Span	0–45/500 μg/NM^3	To be specified
3	Accuracy	1%–2% measured value	
4	Output	4-20 mA DC + Hi/HiHi Al contact	To be specified
5	Contact details	SPST/SPDT 50 VA 50 V 1A	
6	International certification	Clause no. 3.1 for country requirement	To be specified
7	Sampling details	SS 2–4 M probe + heater/cooler	As applicable
8	Display	Five-digit LED/LCD (backlit) + diagnostic display	To be specified
9	Connectivity	RS 232/485/MODBUS/HART/Profibus/Ethernet, etc.	
10	Power supply	230 VAC/24 VDC	
11	Housing material	IP rating (65)/Epoxy painted Al.	To be specified
12	Ambient	*T*: Up to 45°C RH (0.93%)	To be specified
13	Accessories	Mounting flange/sampling system (if applicable), shutter	To be specified
14	Software detail	For data storage, networking, remote control/cal, etc., as applicable	To be specified
15	Process temp	<250°C	To be specified
16	Some suppliers	Durag/Gasmet/Opsis/Applitek	

to reduce particles into the analyzer. A heater block at the inlet of the sampling line reduces the chances of corrosion and helps to draw the sample in as well as to maintain temperature (e.g., analyzers from GE, Ametek).

a. For a zirconia probe, there is a requirement of reference air and this can be supplied by a pump or from instrument quality air.

b. Output is temperature-dependent; hence, it is regulated by a heater. In the type shown in Fig. 5.33, the zirconia cell (available from ABB, Emerson, etc.) is a part of the probe inserted into the flue gas duct. In a zirconia probe, ceramic filters are built in to reduce the particle content. For the high-temperature version, ceramic aluminous porcelain, inconel (<1000°C), and an alumina (<1900°C) sheath are used. Flame arrester porous ceramic filters are common in this type of instrument. Some of the analyzers have inherent intrinsic safety features also.

c. Zirconia probe mounting is shown in Fig. 5.43.

d. Also, the analyzer output is logarithmic, so output linearization is necessary.

e. As per the flue gas duct size, the probe length needs to be specified to get a representative signal (especially for the type shown in Fig. 5.43).

f. On account of warm-up time, the probe has some start-up time. A few diagnostic features such as power failure, thermocouple open circuit, heater open circuit, etc., may be considered for specification of the instrument.

g. In some of the types, by using a catalyst, combustibles can be measured as an optional feature to monitor incomplete combustion.

2. Paramagnetic type

Some features of paramagnetic oxygen analyzers are no reference gas is necessary (less running cost), an elaborate sampling system (with routine maintenance), linear output, and better accuracy. Zero and span gas necessary for analyzer calibration and as the strong magnetic property of Oxygen is utilized hence interfering gas has very little effect.

3.17.2 Zirconia Oxygen Analyzer

See Table 5.20.

3.17.3 Paramagnetic Oxygen Analyzer

See Table 5.21.

3.18 CO Analyzer

3.18.1 CO Analyzer Feature

Most of the CO analyzers used in power plants are based on the nondestructive infrared (NDIR) type. CO analysis may be part of a composite CEM or by a dedicated instrument.

1. Multiple component gas analyzers features normally considered include: Sampling (Pr, temperature, probe material, heating and drying considerations); autoprobe cleaning; central processing and calculation; periodic auto zero span calibration facility; optional feature: combustible monitoring; display and operations console

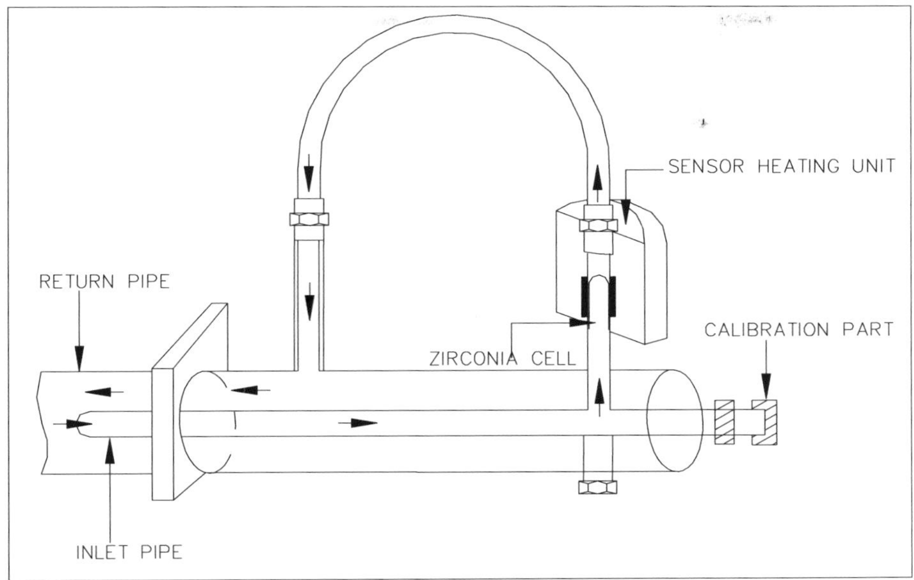

FIG. 5.42 Zirconia O_2 analyzer by diffusion sampling.

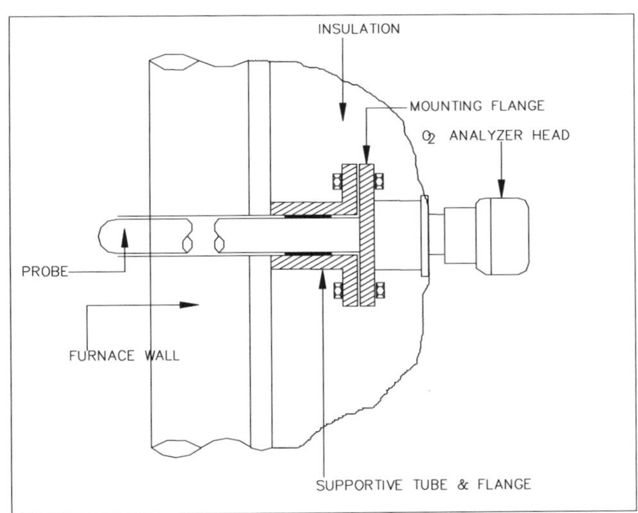

FIG. 5.43 Zirconia probe mounting.

with analysis facility; and dedicated software with a networking facility.

2. General points for single and/or multiple component gas analyzers shall mainly include: international certification; output types and alarm contacts details; diagnostic features; operator friendliness; installation and mounting facilities; connectivity (links, HART, Fieldbus), and networking facilities.

3.18.2 CO Analyzer Specification

See Table 5.22.

3.19 CO_2 Analyzer

Both Thermal conductivity as well as NDIR Types are used. (Features in clause no. 3.18.1 may be considered here also—as applicable.) (Table 5.23).

3.20 H_2 (Purity) Gas Analyzer in TG

3.20.1 H_2 (Purity) Analyzer in Turbo-Generator

Normally, this is done by a complete system in place of an individual instrument.

1. Turbogenerator efficiency is directly linked with the purity of the hydrogen gas used for cooling the generator. If there is a drop in H_2 purity, there will be a drop in generator efficiency.
2. Two sets of analyzers are required: H_2 in CO_2 and H_2 in air.
3. For safety, international standard IEC 842 and EN 60034-3 are followed. Intrinsic safety and two-stage alarms are not uncommon. When H_2 concentration is $<$ set point (e.g., 95%) an alarm is generated.

3.20.2 H_2 Analysis System (Complete Analysis System)

See Table 5.24.

3.21 Discussions on Flue Gas Measurements

As indicated earlier, the major concerns are to increase efficiencies and reduce emissions. In gas analysis, also there has been some scope of improvement. Some of these are enumerated below:

TABLE 5.20 Zirconia Oxygen Analyzer (DS)

Sl	Specifying Point	Standard/Available Data	To Be Specified
1	Measuring range	0–2.5 5 by volume available but for power plant 0%–10% by volume	
2	Probe length (M)	0.5–2.0	To specify
3	Process temperature limit	600–1900°C available	To specify
4	Accuracy	±1% or reading	
5	Response time	Typically 10–15 s	To specify
6	Output	One or two 4–20 mA DC +Hi/Hi Hii alarm contact, HART + diagnostic alarm	To be specified
7	Contact details	SPST/SPDT 50VA 50V 1A	
8	International certification	As per country requirement (clause no. 3.1 of this chapter)	To be specified
9	Probe materials	Depending on temperature (clause no. 3.17.1.1a in this chapter)	To specify
10	Calibration	Automatic and manual zero/span cal. Possible in some instrument	
11	Display possibility	LED/LCD display with diagnostic display in alphanumeric character microprocessor-based	
12	Connectivity	RS 232/485/MODBUS/ HART/Profibus/Ethernet, etc.	
13	Power supply	230 VAC/24 VDC as the case may be	
14	Housing Prot'n	IP rating (65)	To be specified
15	Ambient Cond.	*T*: Up to 50°C RH (0.93%)	To be specified
16	Accessories	Mounting flange/air source, flame arrestor, filter	To be specified
17	Software detail	For data storage, networking, remote control/cal, etc., as applicable	To be specified
18	Zirconia type	In situ/diffusion sampling/ fast response/high-temperature version, etc.	To specify
19	Special feature	Special cable and cabling details, etc.	Any other

TABLE 5.21 Paramagnetic Oxygen Analyzer (DS)

Sl	Specifying Point	Standard/Available Data	To Be Specified
1	Measuring range	0–25 5 by volume available but for power plant 0%–10% by volume	
2	Probe length (M)	0.5–2.0	To specify
3	Process temperature	600–1900°C available	To specify
4	Accuracy	±0.1% O_2/1% FSD	
5	Response time	~20 s	To specify
6	Output	One or two 4–20 mA DC +Hi/Hi Hii alarm, HART + diagnostic alarm	To be specified
7	Contact details	SPST/SPDT 50 VA 50 V 1A	
8	International certification	As per country requirement (clause no. 3.1)	To be specified
9	Sampling system	316 SS or better for probe material. SS tubing for sample line, sample pressure, temperature, and flow rate limit spec.	To specify
10	Calibration	Zero/span cal. with respective gas	
11	Display type	LED/LCD display with diagnostic display	
12	Connectivity	RS 232/485/MODBUS/ HART/Profi Bus/Ethernet, etc.	
13	Power supply	230 VAC/115 VAC as the case may be	
14	Housing Prot'n	IP rating (65)	To be specified
15	Ambient Cond.	*T*: Up to 50°C RH (0.93%)	To be specified
16	Accessories	Complete sampling system, zero span gas, mounting flange, etc.	To be specified
17	Software detail	For data storage, networking, remote control/cal, etc., if applicable	To be specified depending on need
18	Zero/Span drift	<2%	To specify
19	Special feature	Other special feature if any	Optional

TABLE 5.22 CO Analyzer (DS)

Sl	Specifying Point	Standard/Available Data	To Be Specified
1	Measuring range	0–50/6000 ppm/ 0–75 mg/M^3	
2	Probe/*Path* length (M)	0.5–2.0 M (for sampling type)/6–8 M path length (in situ)	To specify
3	Temperature	<1600°C (material as per temp)	To specify
4	Accuracy	±2% measurement	
5	Response time	<10 s	To specify
6	Output	One or two 4–20 mA DC + Hi/Hi-Hi alarm contact, HART + diagnostic alarm	To be specified
7	Contact details	SPST/SPDT 50VA 50 V 1A	
8	International certification	As per country requirement (clause no. 3.1)	To be specified
9	Zero/span drift	<2% Zero/0.5%Span/ week	
10	Calibration	Automatic and manual zero/span cal. Possible in some instrument	
11	Display possibility	LED/LCD display with diagnostic display in alphanumeric character microprocessor-based	
12	Connectivity	RS 232/485/MODBUS/ HART/F. Fieldbus/Profi Bus/Ethernet, etc.	
13	Power supply	230 VAC/24 VDC as the case may be	
14	Housing/ Material	IP rating (65)/Aluminum	To be specified
15	Ambient	*T*: Up to 50°C RH (0.93%)	To be specified
16	Accessories	Mounting flange, special cable if any	To be specified
17	Software detail	For data storage, networking, remote control/cal, etc., as applicable	To be specified
18	Other gases (as applicable)	To specify also number of such gases	To specify
19	Special feature	Integrated flow/ Temperature measurement	Optional

TABLE 5.23 CO_2 Analyzer (DS)

Sl	Specifying Point	Standard/Available Data	To Be Specified
1	Measuring range	0–5/25 ppm, 0–50 mg/ M^3, 0%–30% vol	
2	Probe/path length (M)	0.5–2.0 M	To specify
3	Temperature	<1600°C (material as per temp)	To specify
4	Accuracy	±2% measurement	
5	Response time	<10s (@1 lit/min flow)	To specify
6	Output	One or two 4–20 mA DC + Hi/Hi-Hi alarm contact, HART + diagnostic alarm	To be specified
7	Contact details	SPST/SPDT 50 VA 50 V 1A	
8	International certification	As per country requirement (clause no. 3.1)	To be specified
9	Zero/span drift	<2% Zero/1% Span/ week	
10	Calibration	Automatic and manual zero/span	
11	Display possibility	LED/LCD display with diagnostic display in alphanumeric character microprocessor-based	
12	Connectivity	RS 232/485/MODBUS/ HART/Profibus/Ethernet, etc.	
13	Power supply	230 VAC/24 VDC as the case may be	
14	Housing/ Material	IP rating (65)/Aluminum	To be specified
15	Ambient	*T*: Up to 50°C RH (0.93%)	To be specified
16	Accessories	Mounting flange, sampling system, special cable if any	To be specified
17	Software detail	For data storage, networking, remote control/cal, etc., as applicable	To be specified depending on need
18	Other gases (as applicable)	Gas mixture to be specified especially for TG application	To specify
19	Special feature	Diagnostic feature	Optional

TABLE 5.24 H_2 Analysis system (DS)

Sl	Specifying Point	Standard/Available Data	To Be Specified
1	Measuring range	80%–100% volume H_2 in air 0%–100% volume H_2 in purge gas 0%–100% volume air in purge gas	Range selection switch
2	Gas inlet condition	Pr in kPa, temp:°C and flow rate L/min	To specify
3	Cabinet	IP 54 cabinet for cont and operation	To specify
4	Accuracy	±0.25% FSD	To specify
5	Response time	<10 s	To specify
6	Output	One or two 4–20 mA DC + alarm contacts with adj set point + diagnostic alarm	To be specified
7	Contact details	SPST/SPDT 50 VA 50 V 1A	
8	International certification	As per country requirement (clause no. 3.1). also for intrinsic safety if any	To be specified
9	Zero/span drift	<1% Zero/0.5% Span/week	
10	Calibration	Automatic and manual zero/Span cal. Possible in some instruments.	
11	Display	LED/LCD display with diagnostic display	
12	Connectivity	RS 232/485/MODBUS/HART/Profibus/Ethernet, etc.	
13	Power supply	230 VAC/24 VDC	
14	Housing/Mat'l	IP rating (65)/Aluminum	To be specified
15	Ambient	*T*: Up to 50°C RH (0.93%)	To be specified
16	Accessories	Mounting flange, special cable if any	To be specified
17	Software detail	For data storage, networking, remote control/cal, etc., as applicable	Need based specification

3.21.1 *Oxyfuel*

The oxygen-fired pulverized coal power plant technology assists in removing completely (reported: 90% CO_2 removal efficiency) the CO_2 generated in the combustion process with very little air pollutant. Also, the oxygen-fired PC cycle is the simplest (with no special chemicals and/or process for CO_2 separation), and demands the least modification. Also, membrane separation for an O_2-fired PC (also CFBC with a wider range of fuels) power plant shows good advancement in this area. In short, oxygen-fired PC provides improved boiler efficiency (especially when gas recirculation is used for sensible heat recovery) and very low emissions. The A-USC of the Babcock boiler design has already implemented the same.

3.21.2 *Continuous Emission Monitoring (CEM)*

CEM in gas analysis is extremely helpful in monitoring flue gas for emissions. Normally, NO_x (NO, NO_2), SO_x (SO_2), NH_3, HgO, and CO_2 combustibles are covered under CEM. Major monitoring points are *boiler outlet*: NO_x, SO_2, CO, O_2; *denitrification plant* O/L: NO_x, SO_2 and NH_3 slip; flue gas *desulfurization outlet*: NO_x, SO_2, CO, CO_2, H_2O, Hg; and *GT and CC with HRSG*: CEM at GT outlet to stack, GT outlet to HRSG, and HRSG outlet.

3.21.3 *Flue Gas Flow*

For large stack monitoring applications, the EPA demands a flue gas flow measurement for continuous emission rate monitoring systems (CERM). Also, automatic measuring systems of EU flow measurement are necessary and should meet standard EN 15267-1-3 and EN 14181. Large stack diameters and lack of straight length pose serious challenges to this flow measurement. Added to these, there will be high temperature and high dust content in the gas. Out of various techniques of flow measurement such as coriolis, DP, electromagnetic, turbine, and thermal mass flow, each of them has some advantages and disadvantages. For further reading, the author's book [121] may be referenced.

4 STEAM AND WATER ANALYSIS SYSTEM

4.1 Basics of Steam and Water Chemistry With Measurements and Controls (Normal Utility Station/Boiler)

Modern power plants, especially steam turbines, have extremely low tolerance for any deposition and contamination. Water chemistry plays a vital role to achieve maximum efficiency with minimum down time. Flow accelerated corrosion (FAC), which is electrochemical corrosion and is distinct from erosion, is extremely important in large power plants, especially with respect to pressure drops in the system. Let us first look at steam and water chemistry before going into further details on steam and water analysis systems (SWAS). Contamination in feed water mainly comes from: (1) make-up water (DM plant malfunctioning), (2) CW from condenser leakage (severe for seawater cooling), (3) conditioning chemicals such as NH_3/N_2H_4 (residual to be removed) and/or injection of oxygen in an oxygenated system (discussed later), (4) corrosion products (Fe, Cu, etc.) formed in the FW system, and (5) recycling of boiler water (mainly SiO_2) (Fig. 5.44).

[*Note that oxygenated systems, for example, oxygen injection for SC, USC, and A-USC plants, are discussed separately later in this chapter.*]

1. *Corrosion and deposition*: Basically, these two phenomena are related to each other. Corrosion produces solid *metallic oxides and hydroxides*, which are carried away by the system flow to be deposited at another place. Also, high concentrations of deposits cause corrosion. The basic issues on these have been covered here so as to get a fair idea about the necessity for various SWAS measurements, dosing, and blowdowns in power plants. Most of the common metals used in power plants are susceptible to corrosion in a favorable environment to form metallic oxides by reacting with atmospheric oxygen and/or water. Carbon steel (CS) is the common material used in the boiler and the medium is water, so there is a possibility of a high degree of corrosion. At higher temperatures, in an alkaline medium, insoluble iron hydroxides are formed. Also mild steel readily corrodes in water and the rate of corrosion is being regulated by pH value. In a typical iron corrosion, *magnetite* will be formed: $3Fe + 4H_2O = Fe_2O_3 + 4H_2$. This black iron oxide, as a typical product of corrosion, inhibits further excess corrosion by forming a magnetite layer. Also, cu and cu alloys form a protective layer of *cuprous oxide*. Some metals such as Al, Cr, and Ti can form protective oxide layers when they come in contact with air and moisture. For this reason, cast Al is used as the casing of many instruments. The stability of these protective layers depends highly on the pH value. To maintain pH value, ammonia is added, but it corrodes copper alloys. So, the pH value shall be maintained at a moderate value. Dissolved oxygen (DO) may cause pitting. This is especially important at closed heaters and the economizer, where temperature rises rapidly. At higher temperatures, even a small concentration of DO could cause severe corrosion and economizer tube failure due to pitting. Variation of pH is another big cause for corrosion. Some other boiler corrosions are acidic attack, chlorine corrosion, stress corrosion, scale formation on the heating surface, and caustic corrosion.

Deposition could be caused by carryover from, say, the softener, ion exchange, condensate system leakage, etc. Boiler depositions take place (i) At drum (as applicable), the hottest area where steam is formed, that is, where nucleate boiling occurs. (ii) In the whole boiler system, there exists a thermal gradient so at the hottest point again the deposition takes place. (iii) The region of low flow. Deposition in the tube of the superheater can lead to tube failure or leakage due to overheating. Carryover can cause damage to the turbine blade, may cause sticking of the stop/governor valve, or may lead to catastrophic damage to the turbine due to overspeed. Silica is an important factor. CW is one of the sources of silica. It needs controlling at the DM plant also. Operating conditions may lead to chemical damage to the system such as foaming, synthetic contamination, etc.

2. *Sources of contamination*: The major source of water contamination in power plants is the leakage of the circulating water (CW) circuit in the condenser and other heat exchangers. The chemical quality of CW influences the scale formation and corrosion in the boiler. Seawater containing Ca/Mg ions reacts in the boiler to form mineral acids. Early indications of such failure are possible by cation conductivity at the CEP discharge. Make-up water is another source of contamination. One of the major equipment/systems installed in medium and large power plants (to improve the quality of water) is a full or partial condensate polishing unit (CPU). For large supercritical, ultrasupercritical, and advanced supercritical plants, a CPU is essential; they are also used in large subcritical plants. Maintenance of water and steam purity is extremely important to prevent:
   - **a.** Corrosion in water and steam systems in the boilers and turbines.
   - **b.** Formation of any deposit and/or scale in the tubes and pipes and turbine deposition.
3. *Hydrogen embrittlement*: Sometimes referred to as hydrogen-induced cracking, this is the process by which metals, mainly high strength steel, become brittle on exposure to hydrogen. At high temperatures, the solubility of hydrogen increases, allowing it to diffuse into the metal. These hydrogen atoms recombine in the void of the metal space to form a hydrogen molecule and create pressure from the inside. When this pressure is high enough to overcome the tensile strength of the metal, cracks start. These may come into the metal during the manufacturing process or any of the processes such as cathodic protection, phosphating, pickling, etc., or during operation, for example, decarburization (when carbon steel is exposed to high $T > 500°C$ and $P > 100\ kg/cm^2$). Decarburization at the steam side may be $3Fe + 4H_2O = Fe_3O_4 + 4H_2$, resulting in hydrogen damage of the steel. For this reason, H_2 in the main steam is monitored by a Katharometer.

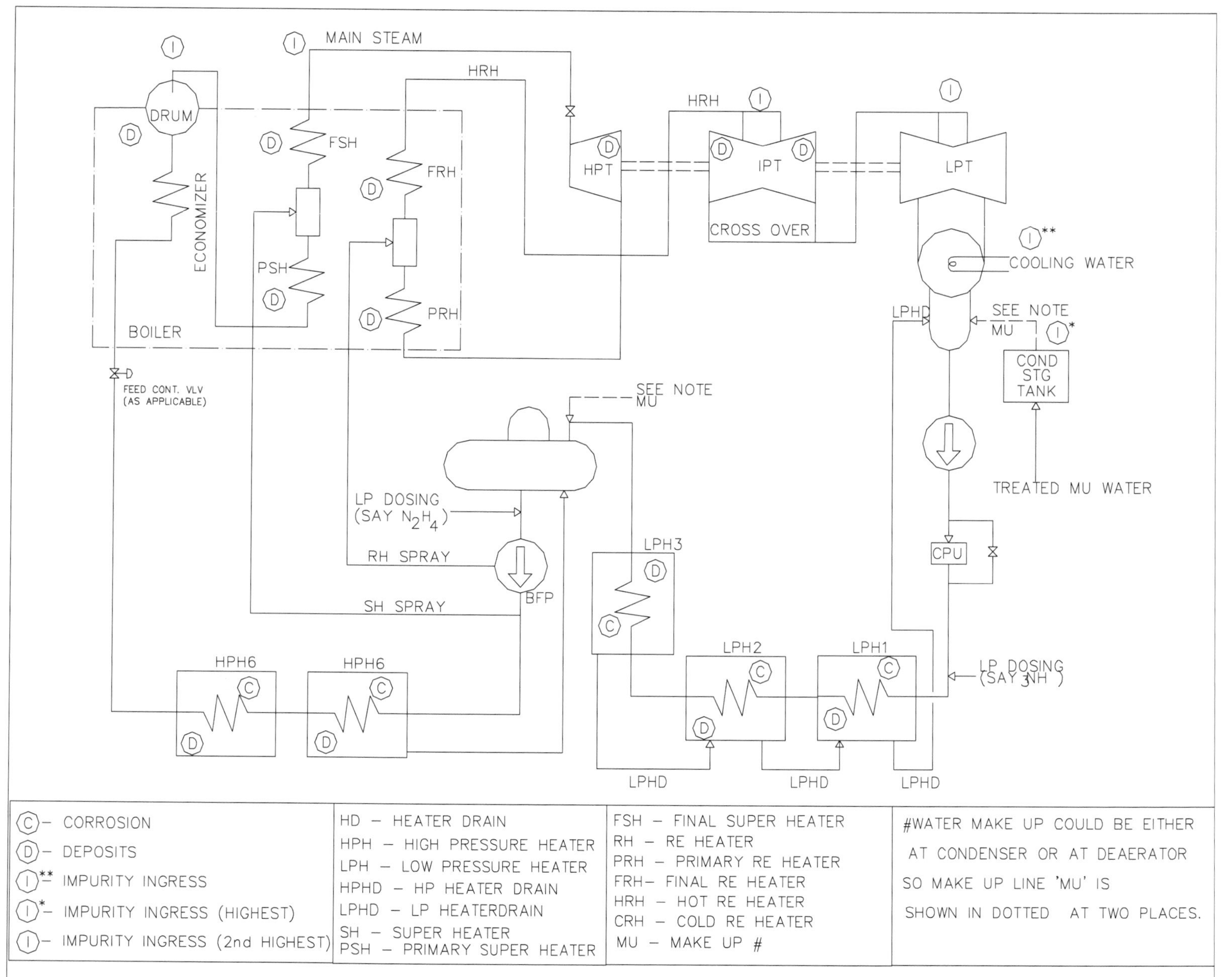

FIG. 5.44 Typical corrosion and deposits in power cycle.

4. *Treatment requirement*: It is necessary to treat the boiler water to produce a protective coat on the tubes required for good quality water and steam. For this, it is necessary that the water shall be *low controlled alkaline* with a *very low level of dissolved oxygen* and a *very low presence of ions such as* Cl^-. *Silica* from the CW is another damaging factor as it deposits on the turbine. Silica needs to be controlled at the DM plant also. The percentage of impurity allowed in the boiler water/steam is a function of the boiler operating pressure, so it is more *critical* for high-pressure boilers.

   Two basic ways to regulate corrosion in water are **a**ll **v**olatile **t**reatment (AVT) and oxygenated treatment (OT). Two types of AVT are: (i) AVT (R) using ammonia (NH_3) and as a reducing agent such as hydrazine (N_2H_4) and (ii) AVT (O) same as AVT (R) minus the reducing agent, that is, N_2H_4. Ammonia is a commonly used cheaper material to maintain the pH in the boiler water. N_2H_4 is used to scavenge dissolved oxygen and protect Cu alloys from ammonia. It may be good for the Cu alloy to form a thick harder magnetite layer. In OT, ammonia is used to control pH and remove a little oxygen, but a slightly oxidizing environment is maintained to promote formation of an oxidizing layer on the metal surface. In subcritical or supercritical plants (with CPU), oxygen treatment (OT) is done after start up. This treatment feeds oxygen to the condensate and FW to arrest corrosion as well as internal scale formation—hence, less chemical cleaning. Normally after start up during operation such OT is done to reduce internal scale formation at operating temperature (also ref. Appendix E). Oxygen may enter the system from air in leakage through the LP turbine and condenser. Also, oxygen may come from make-up water and/or from the return condensate. Oxygen removal is done by mechanical deaeration, chemical injection, and catalytic deaeration. Deaeration of water at the deaerator is done to remove dissolved oxygen, carbon dioxide, and other noncondensable gases. It is also done to heat the water to the temperature where the solubility of undesirable gases to a minimum. It is possible to bring DO to 3–40ppb and almost zero (free) carbon dioxide. The rest of DO is scavenged by chemical means such as hydrazine (or sodium sulfite/ ascorbic acid/diethyl hydroxylamine) for example, $N_2H_4+O_2=N_2+2H_2O$.

   On account of health hazards, many do not prefer hydrazine. Instead, many use DEHA, which has nearly 40 times less toxicity when tested in animals. Diethylhyrdoxylamine (DEHA) has unique properties such as volatility and passivating steel surface. Coupled with low toxicity, this makes DEHA a good oxygen scavenger in modern power plants. Some of the features of DHA shall include: Rapid scavenging of O_2 at FW temperature and pH control; passivate internal surfaces of the boiler; volatility with steam similar to neutralizing amine with the ability for distilled of boiler and protects steam and condensate equipment. Under heat it yields two neutralizing amines to raise the pH value Low toxicity hence less health hazard.

   Reaction with oxygen: $4(C_2H_5)_2NOH + 9O_2 = CH_3COOH + 2N_2 + 6H_2O$.

   In hydroxide, the alkalinity of boiler acetic acid does not pose any problems and acetates are formed. The only shortcoming is its sluggishness in the reaction kinetics of oxygen capture at lower temperature. This is circumvented by use of coscavenger hydroquinone (HQ) used with DEHA.

   A direct relation between pH and conductivity, as shown in Fig. 5.45, is important to note. At a high-pressure boiler, phosphate dosing is done. All these exercises are done to minimize depositions. Otherwise, silica, sodium salt, etc., are carried over with steam to the turbine where with a fall in pressure and temperature, the solubility falls and get deposited at the turbine.
5. *Treatment requirements for high-pressure, high-temperature plants*: For chemistry control in OT boilers, the following considerations are important:
   - No phase separating device such as a boiler drum.
   - FW, boiler water, and steam are a continuous fluid stream.
   - Sliding pressure operation (fixed pressure is less susceptible to corrosion).
   - FW treatments are AVT and OT in supercritical plants.

As discussed, AVT (R) forms a thick layer which may give rise to an increase in metal temperature, especially in supercritical boilers; hence, they require frequent cleaning. Also, hydrazine has an environmental effect. AVT (O) excludes the use of any oxidizing agent so the electrochemical potential (ECP) will be higher. It also favors formation of hematite (which is less soluble) on top of magnetite. This layer is thinner and minimizes orifice fouling. On the other hand, AVT (O) provides more ECP, which forms a hematite layer that is stable and there will be less iron oxide in the FW. Major systems use OT (e.g., Hitachi design for 1000MW boiler) but OT also has a little disadvantage as oxygen when coupled with anions will be highly corrosive.

6. *Sodium (Na) measurement*: Generally, sodium is present as its chloride and sulfate forms and can come to the system through: (a) condenser tube leakage; (b) Na^+ is the first cation break when a bed is exhausted (CPU/ DM.); and (c) sodium triphosphate is added at the drum to remove scaling effect could be a source also.

It is important to maintain the sodium balance between the condensate and the steam. Excess sodium would result in the problem of deposition in superheater tubes and turbine blades. In case of OT boilers, and in almost all modern large

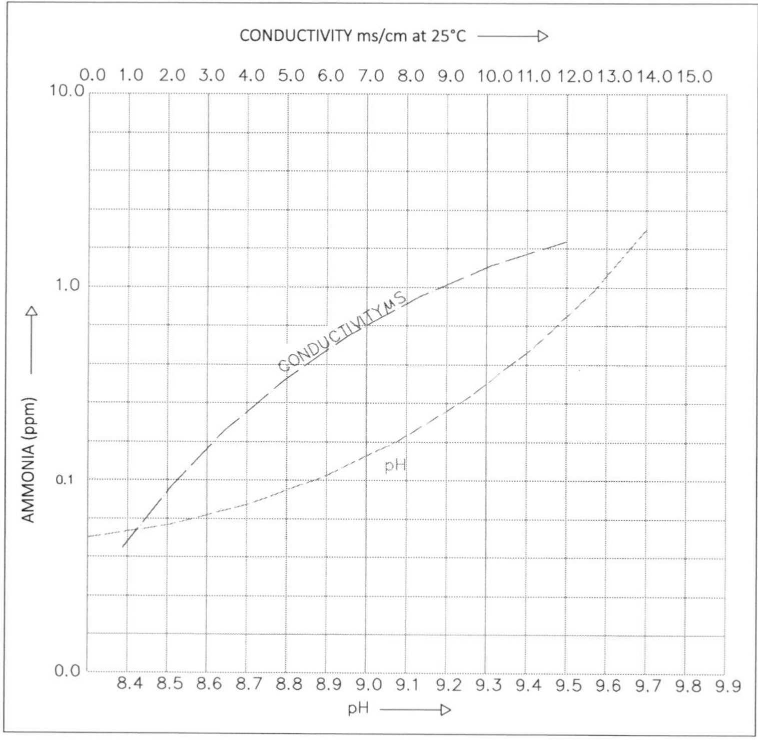

FIG. 5.45 Variation of pH and conductivity with respect to ammonia.

power plants, to have Na^+ monitoring which is sensitive to changing pH and temperature.

7. *SWAS measuring points dosing control and blowdown*: A detailed picture of the overall system is presented in Fig. 5.46—Power cycle Sampling points, Dosing points, and Blowdown points. In the tables, NA stands for "Not Applicable" (that is, parameters not measured/relevant).
   a. Notes on the list of measurements for SWAS:
      i. Parameters depend on operating condition of the boiler (on pressure), so these data are indicative of typical values only.
      ii. Sp. conductivity depends on dissolved solids and the presence of NH_3 and CO_2. It can change the value significantly. It also depends on pH.
      iii. Selectable ranges in instruments are indicated by "/" in the upper. Range.
      iv. Symbols used: $: depends on operating condition and local decision; EPT: equilibrium phosphate treatment. (Amongst the various phosphate treatment means EPT is very effective to maintain desired boiler pH.)
      v. pH value varies little with AVT/OT treatment of water.

Ammonia LP dosing systems are shown in two places in the drawing because: In plants without CPU, ammonia (LP) dosing has been shown at the *CEP discharge*. For plants with CPU, ammonia (LP) dosing has been done at the *CPU outlet* (Tables 5.25–5.31).

   b. Measuring point No. 8 not in USE (Tables 5.32–5.37)
   c. Typical parameters for supercritical boilers:
      DM WATER: Sp. conductivity: <0.1 μS (0.1–0.2 alarm), TOC < 300 ppb, SiO_2: <10 ppb (>10 alarm), Na < 5 ppb, Fe < 20 ppb.

HOT WELL OUTLET: Sp. cond'ty: 2.5–10 Cat. Cond'ty < 0.2 μS (0.2–0.5 alarm) pH 9.0–9.6, Fe < 20 ppb (Table 5.38).

MAIN STEAM/REHEAT STEAM

Sp. conductivity (μS) 2.5–7.0, Cat. conductivity (μS) < 0.2 pH 9.0–9.6 Na/K (ppb) < 5, SiO_2 (ppb) < 20, Fe (ppb) < 20.

8. Chemical dosing control and blowdown

Numbers indicated here are as per the number shown in Fig. 5.46.

a. *LP dosing control*: (1) Ammonia dosing: To control pH of the condensate. (2) Hydrazine dosing (AVT-R): At BFP suction to scavenge dissolved oxygen.
b. *HP dosing control*: 3 Phosphate dosing: Drum for controlling Sodium.
c. *Blowdown*: Point 12: From Boiler Drum.

4.2 SWAS—HRSG

Some measuring points at HRSGs for combined cycle/cogeneration plants (Tables 5.39 and 5.40).

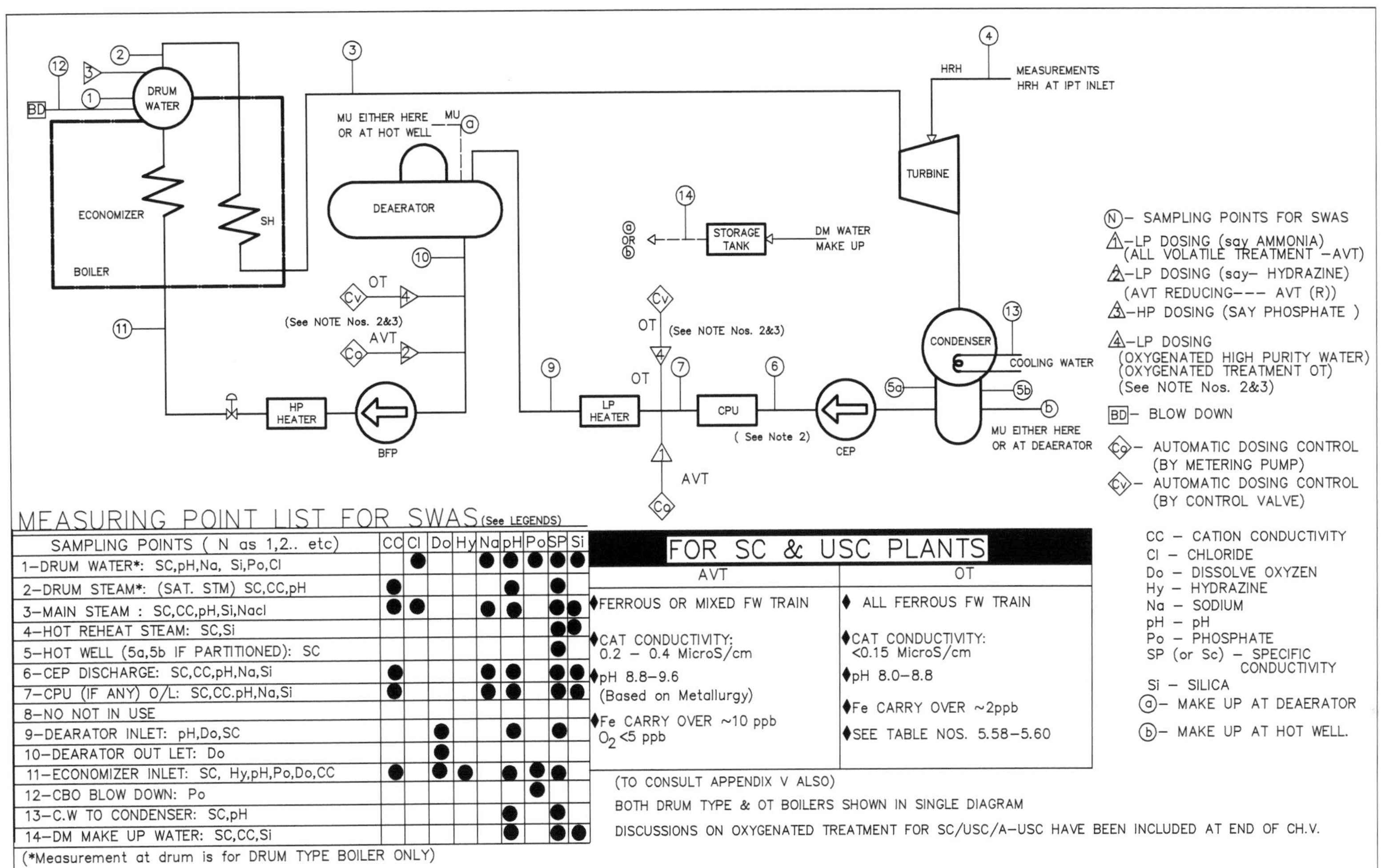

MEASURING POINT LIST FOR SWAS (See LEGENDS)

SAMPLING POINTS (N as 1,2.. etc)	CC	Cl	Do	Hy	Na	pH	Po	SP	Si
1–DRUM WATER*: SC,pH,Na, Si,Po,Cl		●			●	●	●	●	●
2–DRUM STEAM*: (SAT. STM) SC,CC,pH	●					●		●	
3–MAIN STEAM : SC,CC,pH,Si,Nacl	●	●			●	●		●	●
4–HOT REHEAT STEAM: SC,Si								●	●
5–HOT WELL (5a,5b IF PARTITIONED): SC								●	
6–CEP DISCHARGE: SC,CC,pH,Na,Si	●				●	●		●	●
7–CPU (IF ANY) O/L: SC,CC.pH,Na,Si	●				●	●		●	●
8–NO NOT IN USE									
9–DEARATOR INLET: pH,Do,SC			●			●		●	
10–DEARATOR OUT LET: Do			●						
11–ECONOMIZER INLET: SC, Hy,pH,Po,Do,CC	●		●	●		●	●	●	
12–CBO BLOW DOWN: Po							●		
13–C.W TO CONDENSER: SC,pH						●		●	
14–DM MAKE UP WATER: SC,CC,Si						●		●	●

(*Measurement at drum is for DRUM TYPE BOILER ONLY)

FOR SC & USC PLANTS	
AVT	OT
♦FERROUS OR MIXED FW TRAIN	♦ ALL FERROUS FW TRAIN
♦CAT CONDUCTIVITY: 0.2 – 0.4 MicroS/cm	♦CAT CONDUCTIVITY: <0.15 MicroS/cm
♦pH 8.8–9.6 (Based on Metallurgy)	♦pH 8.0–8.8
♦Fe CARRY OVER ~10 ppb O_2 <5 ppb	♦Fe CARRY OVER ~2ppb
	♦SEE TABLE NOS. 5.58–5.60

(TO CONSULT APPENDIX V ALSO)

BOTH DRUM TYPE & OT BOILERS SHOWN IN SINGLE DIAGRAM

DISCUSSIONS ON OXYGENATED TREATMENT FOR SC/USC/A-USC HAVE BEEN INCLUDED AT END OF CH.V.

(1) IN CASE THERE IS PARTITION HOT WELL, SEPARATE SAMPLING POINT TO BE CHOSEN FOR EACH SIDE OF HOT WELL AS SHOWN IN SAMPLING POINT NOS 5a & 5b

(2) ON ACCOUNT OF DEMAND FOR HIGH QUALITY FEED WATER, FOR SUPERCRITICAL, USC & A-USC, PLANTS, 100% CPU IS ESSENTIAL. SUB CRITICAL PLANTS MAY ALSO HAVE CPU. OXYGENATED TREATMENT WHICH REDUCES Fe CARRY OVER, CORROSION & SCALE FORMATION ETC. REQUIRES HIGH QUALITY WATER DEMANDING REQUIREMENT FOR CPU.

(3) FOR OT OXYGENATED HIGH PURITY WATER(O2>30ppb) IS USED. O2, H2O2 OR AIR CAN BE USED AS OXIDANTS. DOSING IS CONTROLLED BY CONTROL VALVE OPENING BASED ON FLOW & DO LEVEL.

FIG. 5.46 Power cycle sampling, dosing and blowdown points.

TABLE 5.25 Measuring point (MP): 1: Service: Drum Water

Parameter	Limit/Target/Range	Unit	Remarks
Chloride (Cl)	<200/<20/0–200	µg/kg or/L	$OP
pH	8.8–9.2		
Phosphate (PO)	<1*/Range:0–10	mg/L	*EPT
Silica (Si)	≪200/<20/0100	µg/kg or/L	
Sp. Cond'ty (SP)	/3/0–10	µS/cm	At 25°C
Sodium (Na) *	<100/5–10/0–10/100	µg/kg or/L	*$

TABLE 5.26 MP: 2: Service: Drum Steam

Parameter	Limit/Target/Range	Unit	Remarks
	Cat. Cond'ty (CC)		
	<0.8/<0.08/0–1/10	µg/kg or/L	Multirange for cont.
pH	8.8–9.2		
Sp. Cond'ty (SP)	/2–3*/0–10	µS/cm	At 25°C (*$)

TABLE 5.27 MP: 3: Service: Main Steam to HPT

Parameter	Limit/Target/Range	Unit	Remarks
Cat. Cond'ty (CC)	<3/<0.08/0–1*	µS/cm	
Chloride (Cl)	<100/<2/0–20/200	µg/kg or/L	*$
pH	>9*		*AVT treated
Silica (Si)	≪100/<5/0–50/100	µg/kg or/L	
Sp. Cond'ty (SP)	/2–3/0–10	µS/cm at	At 25°C
Sodium (Na)	<100/5–10/0–10/100	µg/kg or/L	*$

TABLE 5.28 MP: 4: Service: HRH Steam to IPT

Parameter	Limit/Target/Range	Unit	Remarks
Silica (Si)	≪100/<5/0–50/100	µg/kg or/L	
Sp. Cond'ty (SP)	–/2–3/0–10	µS/cm at	At 25°C

TABLE 5.29 MP 5: Service: Hot Well (Two Sets if Partitioned)

Parameter	Limit/Target/Range	Unit	Remarks
Sp. Cond'ty (SP)	11/3/0–20/50	µS/cm at	At 25°C

TABLE 5.30 MP: 6: Service: Condensate at CEP Discharge

Parameter	Limit/Target/Range	Unit	Remarks
Cat. Cond'ty (CC)	<3/<0.2/0–1	µS/cm	
pH*	>8.8–9.3		*AVT mainly
Silica (Si)	≪100/<20/0–50/100	µg/kg or/L	
Sp. Cond'ty (SP)	~15/3/0–10	µS/cm at	At 25°C
Sodium (Na)	<100/5–10/0–10/100	µg/kg or/L	*$

TABLE 5.31 MP: 7: Service: Condensate Polishing Unit (If Any) O/L**

Parameter	Limit/Target/Range	Unit	Remarks
Cat. Cond'ty (CC)	<3/<0.08/0–1/10	µS/cm	
pH	>8.8–9.3(9.2*)		*AVT mainly

TABLE 5.31 MP: 7: Service: Condensate Polishing Unit (If Any) O/L**cont'd

Parameter	Limit/Target/ Range	Unit	Remarks
Silica (Si)	≪100/<5/0–50/100	μg/kg or/L	
Sp. Cond'ty (SP)	~1/0.08/0–1	μS/cm at	At 25°C
Sodium (Na)	<200/<2/0–10/500	μg/kg or/L	*$

TABLE 5.32 MP: 9: Service: Deaerator Inlet

Parameter	Limit/Target/ Range	Unit	Remarks
DO	<200/<50/0–50/100	μg/kg or/L	
pH	8.6–9.0		
Sp. Cond'ty (SP)	15/2.5/0 –/550	μS/cm at	At 25°C

TABLE 5.33 MP: 10: Service: Deaerator Outlet

Parameter	Limit/Target/ Range	Unit	Remarks
DO	<100/<10/0–50/100	μg/ kg or/L	Value also depends on deaeration type

TABLE 5.34 MP: 11: Service: Feed Water at Economizer Inlet

Parameter	Limit/Target/Range	Unit	Remarks
Cat. Cond'ty (CC)	<3/<0.08/0–1*	μS/cm	
DO	<100*/<5–20/0–20/200	μg/kg or/L	* Time-dependent and possible for OT
Hydrazine (HY)	–/<10–20/0–50	ppb	
pH	>9*		*AVT treated
Phosphate (PO)	/<5–20/0–50/100	μg/kg or/L	
Sp. Cond'ty (SP)	<12/2–3/0–10	μS/cm at	At 25°C

Continued

TABLE 5.35 MP: 12: Service: CBD

Parameter	Limit/Target/ Range	Unit	Remarks
Phosphate (PO)	<20/0–50/100	μg/kg or/L	

TABLE 5.36 MP: 13: Service: CW to Condenser

Parameter	Limit/ Target/ Range	Unit	Remarks
pH	<8.3*		*Above this acid may be injected
Sp. Cond'ty (SP)	1.2	mS/cm at	At 25°C

TABLE 5.37 MP: 14: Service: DM Make Up

Parameter	Limit/Target/ Range	Unit	Remarks
Cat. Cond'ty (CC)	–/<0.1/0V1	μS/cm	
Silica (Si)	–/<20/0V50/100	μg/kg or/L	Silica (Si)
Sp.Cond'ty (SP)	–/<0.1/0V1	mS/cm at	At 25°C

TABLE 5.38 Typical Feed water Chemical Corrosion Parameter *(Alarm)*

Parameter	AVT	OT	Remarks
Sp. conductivity (μS)	4–10	2.0.5–7	
Cat. conductivity (μS)	0.2 (0.0.2–0.5)	<0.2 (>0.2–0.5)	
pH	9.2–9.6 (<9.2, >9.6)	8.5–9.0	
Na/K (ppb)	<5 (>5)	<5	
SiO_2 (ppb)	<20 (>20)		
Oxygen (ppb)	<10 (>50)	50–150	
Hydrazine (ppb)	Trace		
Fe (ppb)	<20		

TABLE 5.39 Chemical Measuring Points for HRSG (Typical)

Sampling Points	CC	DO	Na	pH	Si	SP	Remarks
Condenser hot well						X	For symbols see tables above
Condensate. pump discharge	X	X				X	
LP/IP evaporator water				X		X	
LP steam	X				X		
HP evaporator water				X	X	X	
HP/IP steam	X		X		X		
Other steam				X		X	

4.3 Condensate Polishing Unit

For larger plants and especially for supercritical plants, this is essential. During normal operating conditions with a steady flow rate, corrosion may be at a minimum due to chemical dosing. When the unit is in shutdown, the corrosion rate increases rapidly (especially due to oxygen

TABLE 5.40 Cause and Effect of Some Impurities in Water

Item	Effect	Remedy	Remarks
CO_2	Corrosive to form carbonic acid	Deaeration, and neutralized in alkaline medium	
Dissolve O_2	Corrosion, pitting, failure of boiler tubes, turbine blade	Deaeration-mechanical means, hydrazine dosing chemical means	
Sedimentation and turbidity	Sludge, scale carryover	Clarification, filtration at pretreatment plant. Also condensate blowdown	<5 ppm
Organic matter	Deposits, clogging, etc.	Chemical treatment at DM plant after pretreatment	
Hardness	Scale formation, thermal inefficiency, low heat transfer	Softening plant, internal treatment at boiler	
Na	Formation of carbonates, corrosion, even embrittlement	Deaeration of MU water, return condensate and ion exchange deionization and acid treatment of treatment of MU water	
SO_4	Scale formation	Deionization	Treatment through CPU as well
Cl_2	Priming, salt deposits at boiler tube and turbine	Deionization	Treatment through CPU as well
Si	Hard scale and deposit at turbine blade	Deionization	Treatment through CPU as well

entry). So, during start up, a lot of blowdown may be necessary, resulting in a loss of time and water. A CPU helps to rapidly start the system. Traces of soluble impurities entering the system get concentrated within the system in a high-pressure boiler (drum). In SC and USC, a single phase fluid exists, so all the boiler fluid enters the turbine. These may deposit on turbine blades, leading to inefficiency and mechanical damage. So, for once-through boilers in a supercritical plant, a 100% CPU is essential. Depending on operating conditions, the CPU may be partial (say 50%) or 100%. A CPU in the condensate line would cause a pressure drop at very high flow (~1550 m^3/h for a 660 MW plant is typical), hence system loss. The 3 × 50% vessel combination in CPU is the most popular. Specifications for water quality at the CPU outlet for fossil fuel SC units were conductivity <0.1 μS/cm at 25°C and Na < 5 μg/kg; even better water qualities are available now. Major factors influencing CPU selections are:

Flow rate, quality of the water to be treated, regeneration procedures and conditions, and CPU O/L water quality and quantity. Operationally, the CPU is sequential so for sequential operation, a dedicated PLC or open-loop control system is envisioned for CPU controls.Tthe modes of regeneration selection are: manually; time-lapsed (timer); flow quantity; and differential (pressure) instruments.

Similarly, the major controls are *sequence* selection (MAN/Semi AUTO/FullAUTO); local/remote *selection*; *a*uto/*m*anual regeneration mode selection; *transfer* initiate/interrupt; *and regeneration* initiate.

The plant operations could be from a manual—local control panel (LCP), remote in semiautomatic mode (e.g., regeneration selection manually then subsequent steps in auto), or a remote fully automatic including sequence selections. Control systems may be linked to DCS or direct integrated DCS controls may be used. Now, let us look out for various analytical instrument details.

4.4 SWAS Measurement Systems

4.4.1 Conductivity Measurement

Conductivity measurements are carried out in all power plants to get an idea about the soluble salts in the condensate and feed water. In a power station, two types of conductivity measurements are done: specific conductivity and cation conductivity.

- For measurement of conductivity by an analyzer, it is necessary to fix up the cell constant (Table 5.41).

TABLE 5.41 Conductivity Selection Guide Line

Conductivity Range (μS/cm)	0–20	1–100	10–200	10–1000	100–20,000
Cell constant	0.01	0.05	0.1	1	10
Remarks	Each sensor set has a selectable two-cell constant in the range 1–10, but this is thumb rule only, not fixed				

- Standard solution of 7.4 g KCl per liter at 20°C for cell constant ~1.0 is used whereas 0.74 g KCl per liter at 20°C for cell constant ~0.1 is used. These are approximate values.
- Size, shape, position, etc., of the measuring electrode in a conductivity cell determines the cell constant of the conductivity cell.
- Measurement of conductivity depends greatly on temperature. As conductivity in a solution increases with a rise in temperature, it is therefore necessary to put in temperature compensation.
- In a conductivity measurement with a direct electrical contact, the polarization effect also affects the reading. Polarization depends on the surface of the electrode and the frequency of AC excitation. Use of a carbon electrode is very helpful in eliminating the polarization effect.
- The conductivity measuring electrodes should be located at the places away from the high-temperature zone, vibration, and shocking zone.

4.4.1.1 Conductivity Analyzer Description

1. The arious kinds of conductivity cells are mainly *the insertion type, screw type, flow cell type, retractable type* with valve, etc. In the central SWAS panel, the flow cell type is the most popular. However, for a condenser hot well application, many use the insertion type to measure the conductivity online to avoid the use of a vacuum pump to bring the sample to the panel.
2. In case of insertion, dip type, etc., the length of insertion needs to be specified and in all types the connection size and style need to be specified.
3. The most-used materials from major manufacturers are:
   a. Electrode: carbon/graphite, stainless steel 316, titanium, and platinized (commonly used in power

plants but it may have problems associated with mechanical cleaning).

   b. Body: epoxy resin, stainless steel, or brass-epoxy resin flow cell is commonly used.
   c. Monitor material: ABS, epoxy-painted aluminum, carbon steel, etc., (manufacturer standard).

4. Mostly, the sensor is connected to the transmitter by a cable, which forms the part of the instrument to be supplied. Cable length desired is to be coordinated with the manufacturer and to be specified.
5. The maximum allowable operating pressure and temperature shall be specified or selected from the manufacturer's data sheet.
6. Temperature Senor: For temperature compensation, a Pt 100 RTD is generally used. Temperature compensations are automatic, and the compensation range is to be specified. In SWAS, the sample is generally conditioned to 25°C.
7. Apart from contact-type conductivity instruments, there are toroidal conductivity instruments also. In this type, there will be one driving coil and one pick-up coil. The driving coil is excited by AC voltage, and due to induction there will be current in the pick-up coil. The amount of pick up depends on the solution conductivity. This is shown in Fig. 5.47. Because in this method direct contact is not necessary, it is good to measure conductivity in oily and sticky materials. In SWAS, this may not find much application.
8. Cell constants 0.01, 0.1, and 1.0 μS/cm are generally used.
9. A self-diagnostic feature in the transmitter is common for all systems. Both remote and local alarms can be generated from the diagnostic program.
10. Most of the transmitters are the smart programmable type with an integral display. Most of them can perform on-off control functions and provide requisite alarm contacts.

4.4.1.2 Conductivity Analysis Working Principles

A conductivity cell consists of two electrodes made of carbon or metal. In contact with the electrolyte (sample), the two electrodes are excited by an AC source, as shown in Fig. 5.48. The electrolyte basically offers resistance to this current flow. Naturally, the amount of current flows through the sample are proportional to the conductivity (inverse of resistivity) of the sample. The measurement of conductivity depends on the cell constant, which is a function of the length, surface, and geometry of the electrode. The cell constant (c) is the current response of a sensor on account of sensor dimensions and geometry. Its unit is cm^{-1}. The relation between specific conductivity K and the cell constant is $K = c/R$, where $R =$ the resistance across electrodes. Because R is temperature-dependent, K is

What is Conductivity?

When current is passed through an electrolyte then it generally obeys Ohm's law, and the ability of the electrolyte to conduct electricity is the conductance of the electrolyte. Conductivity of any electrolyte is due to presence of ion in the electrolyte, so it cannot detect the material nor really the concentration. In certain cases, the concentration of an electrolyte in solution can be determined (compared) by conductivity, if the composition of the solution is known (as some times done in DM plants). Its unit is Siemens/cm (or mho/cm) For convenience expressed in terms of μS/cm

Why cat Ion Conductivity? What is it?

The disadvantage of using specific conductance is that some gases common to steam (such as carbon dioxide and ammonia) ionize in water solution. Even at extremely low concentrations, they interfere with measurement of dissolved solids by increasing conductivity. Low concentration of dissolve solid may give rise low conductivity, but this may be increased due to dissolve gases such as CO_2 or NH_3. Now when this is passed through Cat col. (acting in cat ion exchange principle) can help to detect dissolved solids. Now when the chemical reaction inside Cat col. will be when there is dissolved Ammonia:

R-H +NH_4OH R-NH_3 +H_2O -------- (1)

Whereas, the reaction when it contains salt it will be

R-H +NaCl R-Na + HCl-------------(2)

Cations replace NH_4 ions and outcome is water i.e. conductivity of NH_4 gets eliminated as it is pure water. On the contrary in (2) Outcome is HCl thus in this case conductivity almost three times than imparted by NaCl i.e. increased highly. So in normal operation if after cation conductivity reading is increased it suggests that there is contamination (due to say leakage).

FIG. 5.47 Concept of conductivity.

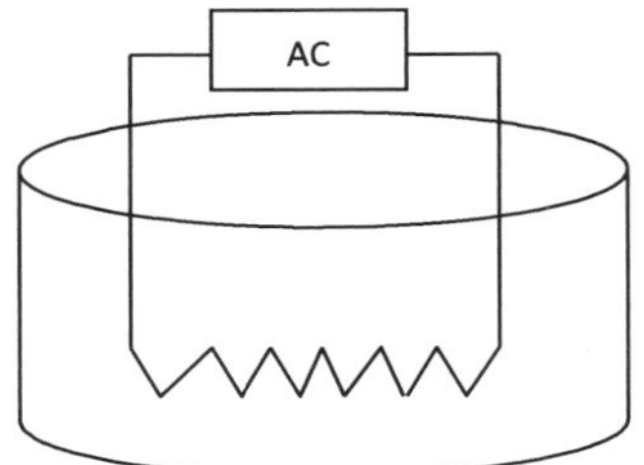

FIG. 5.48 Conductivity analysis working principles.

dependent on temperature. For this, conductivity is generally expressed at a specific temperature, normally 25°C. Also, Fig. 5.47 may be referred to for various types of conductivity instruments (Fig. 5.49).

Major accessories for the analysis system (Tables 5.42 and 5.43):

4.4.1.3 Conductivity—pH

As discussed in Section 4, there is a direct relationship between conductivity and pH. The same relation has been utilized in instruments like pH calculation by differential conductivity instruments. One such kind is Polymetron 9123. Here,conductivity before and after the cation is measured, and based on a built-in algorithm, the pH calculation is done

4.4.2 pH Measurement

See Fig. 5.50.

pH concept:

In 1909, Soresen coined the term. Here, **p** stands for the mathematical symbol of negative logarithm, H stands as the chemical symbol for hydrogen. The formal definition of $pH = -\log_{10} [H]^+$. From chemical dissociation principles, it is known that in any aqueous solution, the concentration (molar) of the hydrogen ion multiplied by the concentration of hydroxide ions is constant; $K_w = [H^+][OH^-]$ where K_w is the dissociation constant and dependent on temperature. $K_w = 1$ when $T = 25°C$. So, normally pH is defined at 25° C. Acid water concentration, with hydrogen ion concentration 1.0×10^{-4} has a pH value of 4 as per the definition of pH. The simplified Nernst relationship equation is:

$E_g = E^o{}_g + \{(2.303RT)/n \times F\} \times \log_{10} a$ where E_g = sum of reference potential and liquid junction; $E^o{}_g$ = potential when $a = 1$, where a = hydrogen ion activity, for hydrogen number of mole $n = 1$, T = Abs temp in K, $R = 1.986$ Cal/mol degree, F = faraday (col per mol). Finally $E = E^0 + 0.059$ V pH.

A few terms listed below and associated with pH measurements are important.

- *Reference electrode*: A reference electrode maintains a stable constant potential independent of the solution under test, but it depends on the concentration of the KCl solution. Mostly, Ag wire with an AgCl coating in close contact with the KCl solution is the reference electrode
- *Liquid junction* (LJ: Fig. 5.51B): The KCl electrolyte of the reference electrode has to have contact with the solution under test to complete the electrical path. This is done through a salt bridge or liquid junction, which is a porous plug that restricts the flow of the electrolyte from the reference electrode to maintain constant potential. The type of liquid junction has an influence on the response time of the electrode. Liquid junctions again come in different shapes such as flushed, notched, or recessed. It is recommended to consult the manufacturer for their use in the particular application. One of the major problems associated with pH is the plugging of the liquid junction (to cause the drift in reading), especially if there are suspended particles. If the pores are smaller, particles cannot accumulate in the pore (as they cannot get into the pore) and diffusion takes place. Double junctions help to avoid plugging from KCl, and also prevent poisoning.
- *Liquid junction potential*: Diffusion of KCl makes the electrical contact. A liquid junction potential exists when two dissimilar electrolyte solutions come in contact (through a salt bridge), but are not allowed to mix. On account of differences in concentration, there will be diffusion, which drives the ion from one side to the other. Because K+ and cl− have different mobility, during diffusion there will be potential because charge separation increases. This is the liquid junction potential. When dissimilar electrolytes come in contact, there will be a potential liquid junction, only the aim is to keep the same low and constant.
- *Glass membrane shape and types*: The glass membrane has various shapes to offer various properties. These are detailed in Fig. 5.51E. There are different types of membrane glass for accurate measurement of pH at different applications. The resistance of the measuring electrode can change in the electrode response. This pH electrode resistance depends on the composition of glass, its thickness, and shape.
- *Electrode selection*: The length of time necessary to get a stable reading from the instrument when a sensor is changed from the solution to another (or if temperature varies) is the response time, which depends highly on the electrode selection. Also, a liquid junction directly affects the flow rate of the electrolyte from the reference electrode to the sample, so the liquid junction potential varies. Therefore, the electrode selection is of immense importance.
- *Calibrating buffer*: A calibration buffer has to be of known pH at 25°C. While selecting the buffer solution, it is necessary to select the accurate/correct one. It is customary to use two known buffers for calibration, as there

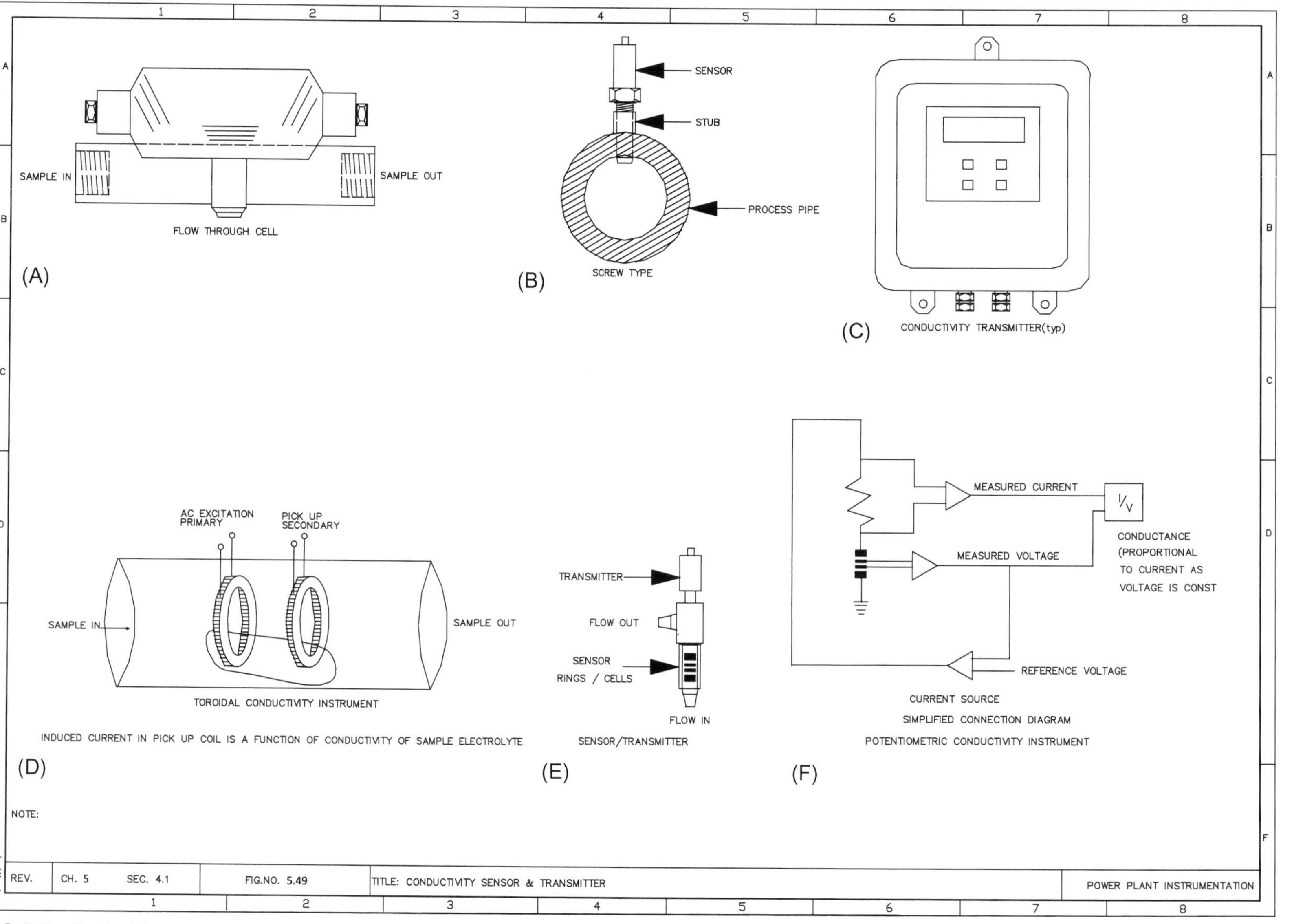

FIG. 5.49 Conductivity sensor and transmitter.

TABLE 5.42 Accessories for Conductivity Analyzer

• Temp. sensing probe	• Coaxial and other cable
• Electrode holder	• Junction box
• Flow cell	• Receptacle, cab gland, and other electrical. Accessories
• Cat column (if any)	• Any other as required

TABLE 5.43 Conductivity Analyzer (DS)

Sl	Specifying Point	Standard/ Available Data	To Be Specified
1	Type	Insertion/screw/ dip/flow cell/ retractable type	To specify
2	Output	1 or 2 4–20 mA Dc + alarm contacts	To be specified
3	Temperature compensation	Automatic temperature compensation	To specify
4	Accuracy/ linearity/ repeatability	±1% FSD/0.1 FSD/0.3FSD	
5	Process condition	To specify + Pr. and Temp Cond.	To be specified
6	Contact details	SPST/SPDT 50 VA 50 V 1A	
7	International certification	As per country requirement (clause no. 3.1)	To be specified
8	Materials	Manufacturer's standard	To specify
9	Calibration	Automatic and manual zero/span	
10	Display	LED/LCD (backlit) and diagnostic	
11	Connectivity	RS 232/485/ MODBUS/HART/ Profibus/Ethernet, etc.	
12	Power supply	230 VAC/24 VDC	
13	Housing	IP rating (65)	To be specified
14	Ambient	*T*: Up to 50°C RH (0.93%)	To be specified
15	Accessories	Discussed separately above	To be specified
16	Process connection	To specify based on project std.	1½″ or ¾″ NPT (typ)
17	Software detail	For data storage, networking, remote control (FTP)/cal, etc.	To be specified depending on need
18	Special feature	Programming feature, RFI effect standard	To be considered

Continued

are variations in commercial buffers. The difference between the two buffers should be minimal. Some pH meters have the ability to recognize the buffer, so it is better to consult the manufacturer for buffer pH values.

- *Electrode storage and use*: In case dry electrodes are supplied by the manufacturer, then the same need to be conditioned as per the manufacturer's recommendation or placed in an HCl acid solution for 12 h. pH electrodes should not be allowed to dry out, but instead may be kept in DM water at a temperature close to the temperature it would be subjected to.
- *Scale and slope control*: It is recommended that these shall be done in line with the manufacturer's recommendation. In modern smart instruments, this can be done at the meter also (slope is the mV response of the sensor per pH of the sample).
- *Na^+ error*: Ideally, the pH versus mV shall be a straight line, but at higher pH values, it is not. This is because ideally, the glass electrode needs to respond to H^+ but it is also affected by Na^+. As pH increases, the H^+ starts decreasing and Na^+ starts getting prominence. It is more so in boiler water, which contains Na^+ in some form. It is recommended to calibrate in an alkaline buffer.
- *Stirring*: For a well-buffered solution, stirring is not required but for poorly buffered solution stirring yields faster response.

4.4.2.1 pH Analyzer—Description

1. pH is measured by measuring the potential between the measuring electrode and the reference electrode. For industrial use, manufacturers offer a variety of pH sensors and transmitters. Table 5.44 gives a selection guide for material selection.

4.2. PH MEASUREMENT:

What is pH? What is pH scale?

Acidity and alkalinity of a solution is indicated by pH. It may be defined as the logarithm of the reciprocal of hydrogen-ion concentration in gram atoms per liter. pH= (-)$log_{10}[H^+]$ - $log_{10}(1/[H^+])$.

pH scale is defined between 0-14 division with pH 7 (neutral distilled water) pH values > 7 is basic and pH <7 is acidic in nature .HCl(1), H_2SO_4 (1.2) being in the extreme end in the acid side with NH_4OH(11)& NaOH(13) in the other (basic) end of pH table with distilled water at the middle(7)In power plant boiler feed water pH is kept little alkaline. To note that pH has direct influence on conductivity and measurement is temperature dependent..

FIG. 5.50 pH measurement.

2. There are pH electrodes with solution ground by a rod of different materials such as 316 SS, titanium, or hastelloy.
3. The liquid junctions could be of wood, Teflon, ceramic/ glass frit (ceramic), etc.
4. Sensors can be flow cell, inline, insertion, retractable or submersible. In the insertion type, the insertion length needs to be specified such as 1/2/3-in. insertion length. Flow cell materials need to be specified.
5. Specifying the operating temperature is very important, as the pH is temperature-dependent.
6. All transmitters are provided with temperature compensation. There may be variations in the mode of compensation. Temperature-compensating elements can be Pt 100 RTD, BalcoRTD (alloy of Ni and Fe at 7:3), identified temperature at 25°C, or another type.
7. Most of the sensors are provided with integral cables to connect the transmitter with the sensor. There are a few options available for types of connection and cable types as well as length to allow the designer to suitably locate the transmitter away from the sensor. However, in a central sample table, such options do not have much impact.
8. Smart transmitters with a self-diagnostic feature to generate remote and local alarms are common these days.
9. Generally, all these transmitters are programmable and capable of doing on-off control/alarms and have a local display.

4.4.2.2 pH Analyzer Working Principles (Fig. 5.51C)

pH Measurement is made with a glass electrode, and the EMF is measured with respect to the reference electrode. The H^+ sensitive electrode consists of a thin glass membrane onto which a glass tube is sealed. This glass electrode consists of an inert glass tube with a pH-sensitive glass tip, in various shapes, blown onto it by glass blowers. The thickness of the glass determines its resistance and affects its output. Sealed inside the tube is a solution of potassium chloride buffered at pH ~7. Silver wire coated with a silver chloride tip is inserted into the solution, acting as a measuring electrode. Through a complex mechanism, an electrical potential, directly proportional to the pH of the liquid, develops at each glass liquid surface. The overall potential of the measuring electrode is the sum of the internal potential at the glass surface plus the potential at the glass membrane surface outside (out of this, the inside potential is known as its known liquid). So, the measuring potential varies proportionately with the pH of the liquid. The reference electrode actually used to complete the circuit in measurement and potential considered as reference. The reference electrode is a silver wire coated with silver chloride in contact with the KCl solution (or may e gel) held in a glass or plastic tube. A porous liquid junction (defined earlier) is an integral part of the reference electrode, and it provides an electrical connection between the reference electrode and the liquid whose pH is to be measured.

The reference is required to develop a potential independent of the solution under a test, but dependent on the concentration of KCl.

1. For operation of the reference electrode without loss, it has to be hydrated. So, a salt bridge is necessary. This salt bridge is a porous plug so that restrictively, KCl can flow out of the reference electrode to establish electrical contact and maintain a constant potential for the reference electrode.
2. Now, when there is a difference in concentration between two electrolytes of different concentrations, there will be diffusion through the porous plug-liquid junction. As the mobility of Cl^- and K^+ are different, they will move through the liquid junction differently and there will be a charge difference, hence the liquid junction potential.
3. Whenever potential is talked about, it is with respect to a fixed point (say ground). Thus the net potential developed as per the Nernst equation (for measuring purposes) will be $E = E_{IR} + E_S + E_J + E_E - K^*T$ pH where all Es indicate potentials: E_{IR} = the internal reference electrode,. E_S = at the selective junction, E_J = at l the liquid junction, E_E = at the external electrode, and T is the temperature in Kelvin, K constant = 0.1984. Here, all the terms of Es are independent of the pH of the test

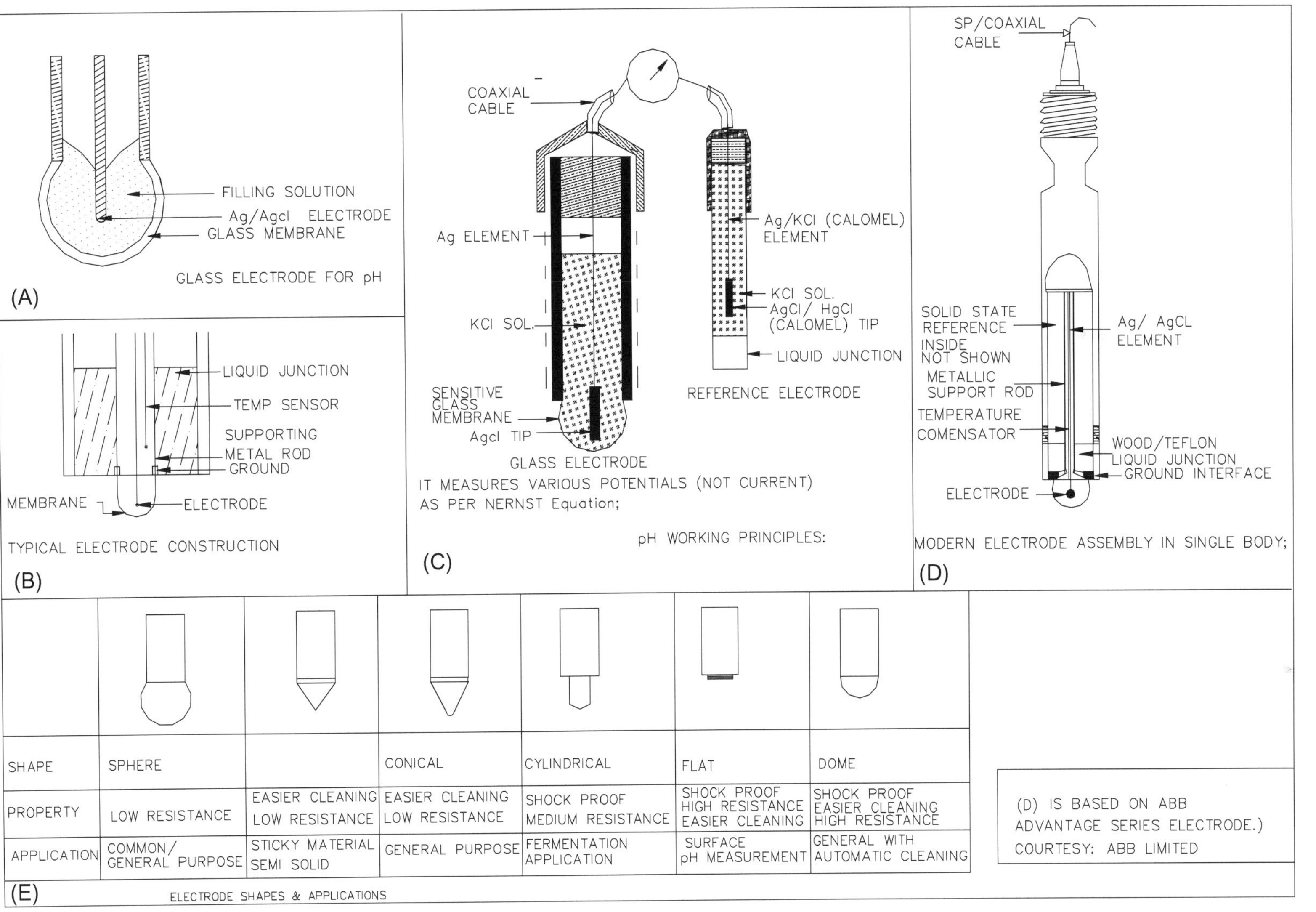

FIG. 5.51 pH sensor types.

TABLE 5.44 Sensor Material Selection

Flat glass	Coat resistance glass	Metallic, for example, platinum/antimony	Totally solid inner chamber charged with KCl, has solid-state ref. electrode

TABLE 5.45 Accessories for pH Analyzer

Temperature Sensing Probe	Flow Cell (as Applicable)	Electrode Holder
Coaxial and other Sp. cable	Junction box	Receptacle and other elec. accessories

solution. So, *E* is measured as a proportional signal to pH with dependence on temperature.

Major accessories for the analysis system (Table 5.45).

pH analyzer specification/data sheet (standard 0–14, special ranges, e.g., 9–12) (Table 5.46).

4.4.3 Dissolved Oxygen

"Dissolved oxygen (DO)" is responsible for oxidative corrosion, mainly in the form of pitting in the boiler. A deaerator is used to mechanically drive away gaseous substances from the condensate and the DO is monitored at the deaerator outlet.

4.4.3.1 Dissolved Oxygen Analyzer—Description

1. Dissolved oxygen measurements in power plants can be done by (i) *the* amperometric *principle* (some major suppliers: ABB/Rosemount/Polymetron) and (ii) *optical* (some major suppliers: Polymetron, YSI).
2. In an amperometric instrument, the diffusion of oxygen depends on temperature, so a temperature element is used with the system to compensate for temperature variations (Fig. 5.52). Some manufacturers use a means to minimize the distortion on the membrane to keep the pressure equal to the sample pressure to eliminate *the* little influence of pressure on the measurement. Many use a constant head unit before the flow cell.
3. A flowthrough cell is common for DO measurements. Some amperometric sensors are the disposable type, which also makes for easy maintenance. The majority of the DO analysis systems consist of sensors and smart transmitters with a self-diagnostic feature and support for various protocols/fieldbus systems. Normally, sensors and transmitters are connected by a special (or integral) cable forming a part of the instrument.

TABLE 5.46 pH Instrument (DS) (To Specify Range)

Sl	Specifying Point	Standard/ Available Data	To Be Specified
1	Type	Insertion/ submersible/ flow cell	To specify
2	Electrode cleaning	Autocleaning (as applicable)	To specify
3	Electrode type	From the above discussion	
4	Output	1 or 2 4–20mA DC +alarm Contacts	To be specified
5	Temp. comp.	Automatic	To specify
6	Accuracy	±0.02 pH/ 0.01pH*	*Repeatability
7	Stability	0.02 pH/week	
8	Resolution	0.01 pH	
9	Contact details	SPST/SPDT 50 VA 50V 1A	
10	International certification	As per country requirement (clause no. 3.1)	To be specified
11	Materials	Manufacturer's standard	To specify
12	Calibration	Automatic and manual zero/ span	
13	Display	LED/LCD (backlit) and diagnostic	Display
14	Connectivity**	RS 232/485/ MODBUS/ HART/**	
15	Power supply	230 VAC/24 VDC	
16	Housing	IP rating (IP65)	To be specified
17	Ambient	*T*: Up to 50°C RH (0.93%)	To be specified
18	Accessories	Discussed separately above	To be specified
19	Process connection	To specify based on project standard	1½″ or ¾″ NPT (typ)
20	Software detail	For data storage,	To be specified

TABLE 5.46 pH Instrument (DS) (To Specify Range)cont'd

Sl	Specifying Point	Standard/ Available Data	To Be Specified
		networking, remote control(FTP)/ cal, etc.	depending on need

** or Fieldbus such as PROFIBUS, FOUNDATION FIELDBUS, etc.

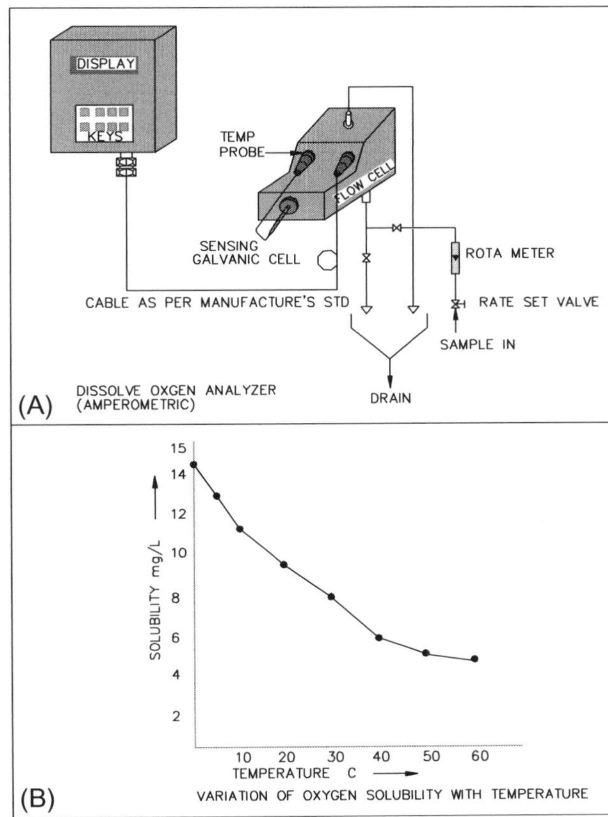

FIG. 5.52 Dissolved oxygen analyzer.

4. Typical sample parameter: Pr: ~0.3 kg/cm^2; *F*: 100–500mL, depends on the sample drawing system. *T*: 5–50°C. (Consult manufacturer's data sheet) Sample particulate: <8μm.
5. In an amperometric sensor (galvanic cell), the cathode is *silver* whereas the anode will be *gold* (or *lead* for disposable sensors, e.g., ABB).
6. An optical system utilizes luminescence lifetime detection. In some instruments, a single blue light LED is used in conjunction with a luminophore while some use both red and blue lights to measure the DO level, as in Fig. 5.54.
7. Description of the analyzers:
   a. Amperometric type: Sensors are mounted in the flow cell with a temperature element, as show, in Fig. 5.52. In a gold electrode, a Teflon membrane may be stretched over it (for example, Rosemount). A silver cathode is used.
   b. Optical type: In the optical type, there will be sources of light (e.g., LED). In some cases, only a blue light is used with the luminophore and color filter, whereas in other cases, two sources of light-red and blue—are used with the dye layer.
   c. Transmitter: A small signal from the sensors is amplified and conditioned at the smart transmitters with integral display and various connectivity discussed above.

4.4.3.2 Amperometric DO Analysis Working Principles (Fig. 5.53)

1. This galvanic cell consists of a gold (or lead) anode (covered with stretched Teflon) and a silver cathode. When oxygen diffuses, on account of the voltage applied, the reactions are:
   (a) Cathode: $O_2+2H_2O+4e^- \rightarrow 4OH^-$
   (b) Anode: $Au \rightarrow Au^+ + e^-$ or $Pb \rightarrow Pb^{2+} + 2e^-$
2. So, on account of the above reactions, there will be a current directly proportional to the concentration of DO in the sample.

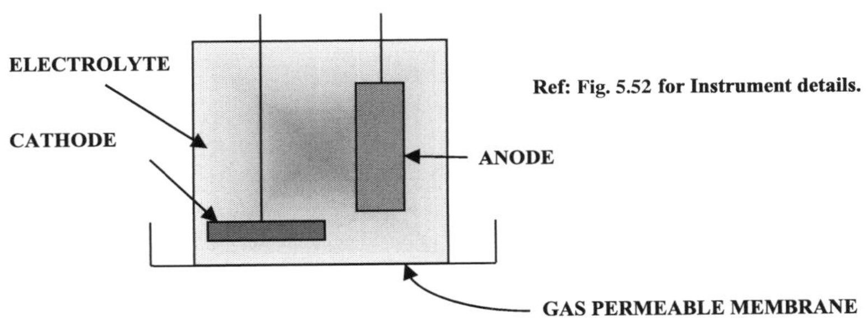

Ref: Fig. 5.52 for Instrument details.

FIG. 5.53 Polarographic do cell.

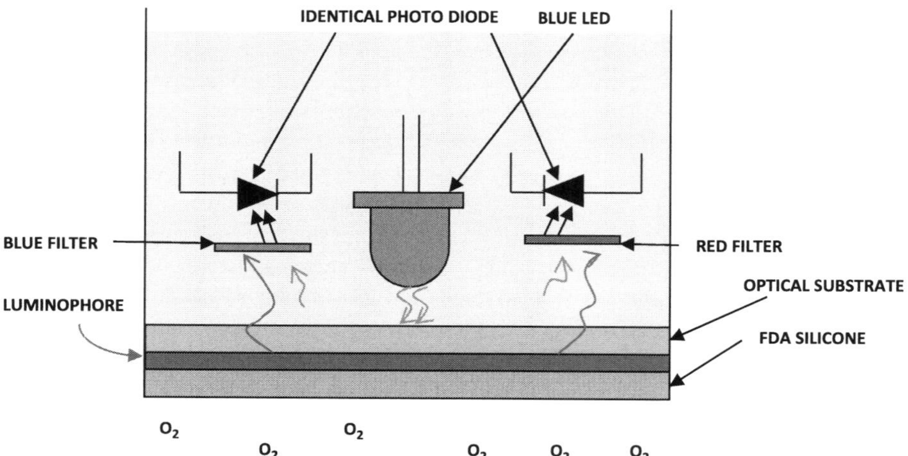

FIG. 5.54 Optical do sensor.

4.4.3.3 Dissolved Oxygen Analyzer Working Principle (Optical Type)

1. Oxygen as a triplet molecule (the triplet state is the most stable oxygen molecule) is able to efficiently quench the fluorescence and phosphorescence of certain luminophores. This effect (first described by Kautsky in 1939) is called "dynamic fluorescence quenching." the degree of fluorescence quenching is related to the frequency of collisions, which means the concentration, pressure, and temperature of the oxygen present in the sample.

   The optical DO sensor utilizes this property and provides the arrangement where a blue light is directed to an oxygen-active compound called **a** *luminophore* that has been stabilized in an oxygen-permeable polymer. The blue light causes electrons in the luminophore to be excited to a higher energy level, and to return to their original level after the emission of red light. When the luminophore comes in contact with elemental oxygen, the O2 molecules absorb the energy, resulting in reduced intensity of the red light emissions, that is, the fluorescence is quenched by oxygen. When the polymer-sensing surface is exposed to water, oxygen diffuses into the sensing surface according to the amount ("partial pressure") of oxygen in the water. In this method, one LED light source is used and there are two identical photo diodes, one with a red filter and another with a blue filter, to detect red and blue light, respectively (Fig. 5.54).
2. There will be some initial difference in reading of these two detectors when there is no DO. Let this difference be the base value. When oxygen diffuses into the system and absorbs red light, then the difference between the two detectors will be different from the base value, which is a measure of the concentration of dissolved oxygen. In reality, this is done with precision measurement of the phase difference of the two detectors and from there, the DO concentration is calculated.
3. Sometimes, a red dye layer is used. Once it is used with red light, when it does not cause any reaction on the dye and is taken as reference. The same dye is then flashed with blue light (Tables 5.47 and 5.48).

4.4.4 Residual Hydrazine

Hydrazine helps to maintain a protective magnetite layer on steel surfaces, so use of hydrazine as an oxygen scavenger in modern power plants is quite common for AVT (R) type treatment.

4.4.4.1 Hydrazine Analyzer—Description

1. Hydrazine measurements can be done on the basis of the *wet chemical colorimetric principle* (analyzers of this kind are available from suppliers like Rosemount) *as well as by measurement of potential in electrochemical cell-amperometric sensing* (analyzers of this kind are available from ABB, Emerson, Polymetron, etc.). Each of these analyses involves the addition of a chemical reagent to obtain the desired result. The system response

TABLE 5.47 Accessories for Dissolved Oxygen Analyzer

Temperature-sensing probe (Pt 100)	Autocalibration	Cable gland and other electrical accessories
Spare SENSOR for disposable type	Constant head unit	

TABLE 5.48 Dissolved Oxygen Analyzer (DS)

Sl	Specifying Point	Standard/ Available Data	To Be Specified
1	Type	Amperometric/ Optical	
2	Measuring range	0–20/200/ 20000 μg/kg	To be specified
3	Electrode	Ag cathode, Au/Pb anode	
4	Output	1 or 2 4–20 mA DC + alarm contact	To be specified
5	Response time	Typically <1 min	To specify
6	Accuracy	±5% reading	
7	Electrolyte life	Manufacturer's standard	Shelf life
8	Resolution	0.1 μg/kg	
9	Contact Details	SPST/SPDT 50VA 50V 1A	
10	International certification	As per country requirement (clause no. 3.1)	To be specified
11	Materials	Manufacturer's standard	To specify
12	Calibration	Automatic and manual zero/ span cal.	
13	Display	LED/LCD (backlit) and diagnostic*	*Alphanumeric
14	Connectivity	RS 232/485/ MODBUS/ HART/FB**	To specify
15	Power supply	230 VAC/24 VDC as the case may be	
16	Housing protection	IP rating (55)	To be specified
17	Ambient condition	*T*: Up to 50°C RH (0.93%)	To be specified
18	Accessories	Discussed separately above	To be specified
19	Software detail	For data storage, networking, remote control(FTP)/ cal, etc., as applicable	To be specified depending on need
20	Special feature	Salinity and pressure correction, cable length, etc.	To be considered

Continued

** FB = Fieldbus like PROFIBUS FOUNDATION FIELDBUS.

time for a hydrazine analyzer is on the order of 3–10 min, depending on the type selected. In both systems, there will be standard solutions as part of the instrument.

2. Colorimetric analyzers can be multistream, and they utilize a pump for reagent delivery. In the electrochemical cell analyzer, reagents such as caustic soda to adjust pH are mixed with the sample through a porous disc and a constant head unit to ensure the required pressure. Channel selection and calibration in most of the cases are automatic.
3. A self-diagnostic feature is common for all hydrazine analysis systems.
4. Sample parameter: Pr: ~0.3 kg/cm^2; *F*:100–500 mL, depends on sample drawing system. *T*: 5–50°C (consult manufacturer's data sheet). Sample particulate: <60 μm.
5. Analyzer sections: There are two sections: 1. *Analysis module*: (Colorimetric type): It is the section of the analyzer where the actual analysis takes place in a sample cell, present between the light source (LED) and the photo detector. 2. *Transmitter and control module*: A small signal from the sensors is amplified and conditioned at smart transmitters with integral display, multistream selection, and various connectivities.

4.4.4.2 Electrochemical Hydrazine Analyzer Working Principles (Fig. 5.55)

1. It consists of an outside jacket filled with silver oxide or gel, into which there is one porous ceramic cylinder. There will be a silver wire acting as a cathode wound round the outer surface of the ceramic cylinder. The sample is admitted from the bottom of this cylinder. Before sending the sample to the sensor, it is necessary to pass it through a constant head unit to ensure the desired pressure and flow. While the sample is admitted in the system, caustic soda, an alkali reagent, is mixed with the sample through a porous disc to raise the pH level for enhanced reaction in the alkaline environment (pH 10.2–10.5). From the bottom of the cylinder, the sample flows out from the top. Platinum wire acting as the anode is introduced in the cylinder at the center.

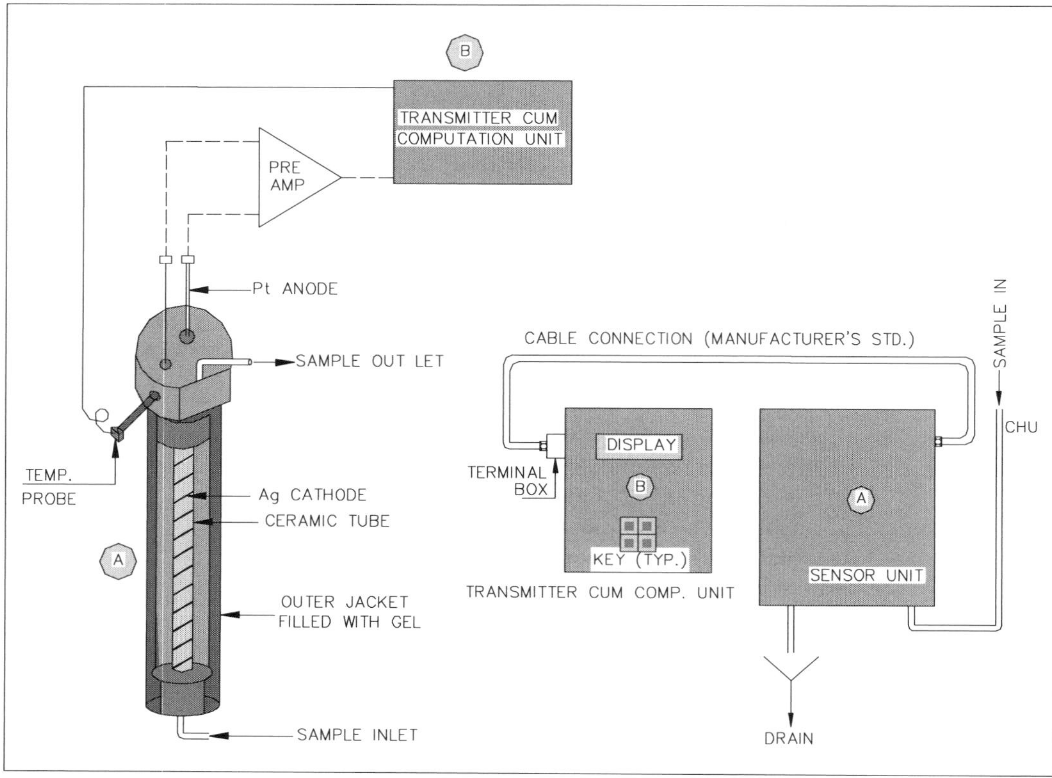

FIG. 5.55 Hydrazine analyzer system.

An electrical path between the two electrodes is complete via ionic transport through the porous ceramic cylinder. The resultant current is proportional to the hydrazine presence in the solution. The measurement is temperature sensitive, so a temperature element is inserted.

The reactions take place at the electrodes as follows:

$$\text{Pt anode}: N_2H_4 + 4OH^- \Leftrightarrow N_2 + 4H_2O + 4e^-$$

$$\text{Ag cathode}: 4H_2O + 4e^- \Leftrightarrow 2H_2 + 4OH^-$$

2. Some of the instruments sometimes use three electrodes (e.g., the Hach instrument). The third electrode is a reference electrode kept at a constant potential difference with the anode. The measuring principle based on the three electrode amperometric is also an electrochemical method. Here, the platinum anode used is called the working electrode while the stainless steel cathode is called the counter-electrode. A small polarization voltage is applied between the anode and cathode. The anode-cathode potential is kept constant with respect to a third Ag/AgCl electrode called the reference electrode. The system avoids interference effects, namely voltage drift, etc., resulting from changes in water composition that may appear in a two-electrode system.

Hydrazine is oxidized at the surface of the anode, that is, the platinum working electrode. The resulting current is directly proportional to the hydrazine concentration.

The sample is made alkaline with a pH of 10.2–10.5 by a dosing reagent such as the analyzer with two electrodes. The chemical reaction is the same as stated above. A temperature sensor for automatic compensation is provided. In some systems, Teflon bids driven by the sample are arranged to circulate on the surface of the anode for cleaning any deposits.

4.4.4.3 Colorimetric Hydrazine Analyzer Working Principle

This system is similar to the system discussed in connection with silica and phosphate analyzers in the next section. The sample along with reagents are mixed; after the reaction time is over, the same is put into the measuring cell. The measuring cell is placed between the LED light source (sending light with wavelength 450–480 nM) and the photo detector. The amount of light absorption is proportional to the amount of hydrazine present. At first, a reference measurement is done to take care of various interfering factors,

TABLE 5.49 Accessories for Hydrazine Analyzer

Standard Solution	Sample Tubes and Fittings	Cable Gland and Elec. Accessories
Reagent (built-in)	Temperature sensor (comp)	

namely the own color of the sample and different reagents, turbidity, refractive index variations, etc. Then, after reagent mixing and color development, another measurement with the sample is taken. The concentration is measured with the absorbance calculated based on the difference between the two measurements and the stored calibration parameters. For a high range measurement, a mixing pump is required to mix the sample with distilled water (dilution process). There shall be separate pump sets for reagents and samples.

4.4.4.4 Accessories of Analyzer

See Table 5.49.

4.4.4.5 Analyzer Data Sheet

See Table 5.50.

4.4.5 Miscellaneous Special Analysis System

4.4.5.1 Miscellaneous Special Analysis System

A few special analyzers with associated data sheets are discussed.

4.4.5.2 Silica and Phosphate Analyzers—General Description

1. Silica and phosphate measurements are done on the basis of the *wet chemical colorimetric principle (determining the intensity of a color and relating the same with the concentration of solution is the colorimetry principle)*. Each of these analyses involves the addition of a number of chemical reagents and a number of reactions at various stages. Therefore, it is a semicontinuous (rather than a batch-processing) system and each cycle may take around 10–15 min.
2. Normally, these analyzers (single/multichannels up to six channels) come in the form of a system having a panel (with space for reagents). Normally, the analysis system comprises three main sections: the basic analysis system (wet part), reagent supply lines, and a control and display unit (dry part), complete with a programmable microprocessor/embedded electronics for programming, processing, control, and diagnostics.
3. The reagent systems are complete with the necessary pumps, valves, and sample chamber. Samples are drawn

TABLE 5.50 Hydrazine Analyzer (DS)

Sl	Specifying Point	Standard/ Available Data	To Be Specified
1	Measuring principle	Colorimetric/ electrochemical cell	
2	Measuring range	0–100/200/ 1000 ppb with selection	To be specified
3	Number of streams	Single or Single/ 2/4/6 (colorimetric)	To specify
4	Reagent and Std sol	Typically one and Std sol.	To specify
5	Output	1 or 2 4–20 mA DC each stream (or time scale) output + alarm contact	To be specified
6	Accuracy/ Repeatability	±5% reading/± 01.5% reading	
7	Contact details	SPST/SPDT 50VA 50V 1A	
8	International certification	As per country requirement (clause no. 3.1)	To be specified
9	Materials	Manufacturer's standard	To specify
9	Calibration	Auto and man zero/ span cal. (frequency and timing as per manufacture)	
10	Display possibility	LED/LCD (backlit) with diagnostic display, microprocessor-based	
11	Connectivity	RS 232/485/ MODBUS/HART/ FB**	
12	Power supply	230 VAC/24 VDC	
13	Housing protection	IP rating (55)	To be specified
14	Ambient	*T*: Up to 50°C RH (0.93%)	To be specified
15	Software detail	For data storage, networking, remote control(FTP)/cal, etc.	To be specified
16	Accessories	Discussed separately above	To be specified

Continued

TABLE 5.50 Hydrazine Analyzer (DS)cont'd

Sl	Specifying Point	Standard/ Available Data	To Be Specified
17	Special feature	Reagent consumption, filling, drain etc. Type, material and number of electrodes. Temperature probe	To be considered

**FB = Fieldbus like Profibus Foundation Fieldbus.

by a peristaltic pump,/pressurizing system with a controlling solenoid valve.

4. Normally, channel selection and calibration are automatic. Each channel suitable continuous and alarm output are provided.
5. The sample parameter requirement varies depending on the sample lines.
6. Sections of the analyzer: In the analysis module, the actual analysis takes place in the sample cell, placed between the light source (LED) and photo detector. Reagents are supplied by peristaltic pumps described in clause no. 4.4.5.2.3 above. The display and control module described in clause no. 4.4.5.2.2 does display and control functions such as stream selection, mixing of reagents, cleaning, etc.

4.4.5.3 Silica Analyzer Working Principles

1. Heteropoly blue colorimetric analysis (molybdenum method) is used to detect the reactive silica trace. Sample pumps and solenoid valves are used to draw the sample and regulate the flow of the same. The sample cell is placed between a light source and photo detector to measure the amount of absorption of wavelength light in the range of 450–820 nM, depending on ranges.
2. Acidic molybdate solution is first mixed with the sample to form molybdosilicic/(molybdophosphoric) acid by reacting with traces of silica (phosphate).
3. Citric acid/surfactant reagent is added to mask the acid formed (as discussed above) and react with excess molybdate (to prevent molybdate from producing an interfering blue color). The surfactant, a wetting agent, minimizes air bubble formation on the sample cell walls. Light absorbance at this stage through this solution is measured, which gives the reference signal at zero silica condition. The color formed at this point is identical to the final color of a sample without any silica content and compensates for any background turbidity color of the sample, changes in light source, or contamination of the sample cell walls.
4. In the next stage, amino acid is added to reduce molybdosilicic to a blue-colored solution. Now, the blue color is the measure of the amount of Si in the sample. This is measured physically by the light (say, ~800nm) absorption method. This absorption is compared with reference absorption (clause no. 4.4.5.3.3).
5. Silica content in the sample is related to the ratio of absorption by reference and by the sample. This relation is not linear but is proportional to the log of the ratio of the reference sample. Standard solution is used to determine the constant of this relation.
6. Instrument calibrations could be programmable and repeated after a fixed cycle (e.g., twice a week to three weeks), which varies with the manufacturer.

4.4.5.4 Silica Analyzer Accessories

Silica analyzer accessories, in addition to cable glands and electrical accessories, are listed in Table 5.51.

4.4.5.5 Silica Analyzer Data Sheet

See Table 5.52.

4.4.5.6 Phosphate Analyzer Working Principles

In a power plant steam/water cycle, inorganic phosphorus is present in the form of phosphate, normally referred to as orthophosphate or soluble reactive phosphorus (SRP). It is available chiefly as ions of HPO_4^{2-} and with a small percentage present as PO_4^{3-}. These are called orthophosphate ions and result from the dissociation of orthophosphoric acid (H_3PO_4).

1. This analyzer is based on molybdophosphoric acid colorimetry (as suggested in ASTM D 515). Upon mixing the reagent with the sample, a color change indicates the presence of orthophosphate. Here, ascorbic acid is used to measure the orthophosphate. Surfactant is added to minimize bubble formation at the surface wall. Light absorption is measured through the solution in the sample, and is considered as a reference absorbance value similar to what was discussed above for silica. Molybdate reagent is then added to react with the orthophosphate to form heteropoly acid. For reducing the

TABLE 5.51 Silica Analyzer Accessories

All Reagents Required (For Desired Period)	Standard Solution	Tubes and Fittings
Cleaning solution and air source	Chilling unit, if any	Temperature probe for comp

TABLE 5.52 Silica Analyzer (DS)

Sl	Specifying Point	Standard/ Available Data	To Be Specified
1	Measuring range	0–20/500/5000 ppb multiple ranges available, upon selection by used	To be specified
2	Number of streams	Single/2/4/6	To specify
3	Reagent and standard solution	Typically four and standard solution at regular pressure	To specify
4	Output	4–20 mA DC + alarm contact/stream	To be specified
5	Accuracy	±0.5 μg/L/1% reading	
6	Repeatability	±01.5% reading	
7	Contact details	SPST/SPDT 50VA 50V 1A	
8	International certification	As per country requirement (clause no. 3.1)	To be specified
9	Materials	Manufacturer's standard	To specify
10	Calibration	Auto/man zero/span cal. Frequency and time period, manufacturer standard	
11	Display	LED/LCD (backlit) with diagnostic*	*Alpha numeric
12	Connectivity	RS 232/485/ MODBUS/HART/ FB**	
13	Power supply	230 VAC/24 VDC	
14	Housing Protection	IP rating (65)	To be specified
15	Ambient	*T*: Up to 50°C RH (0.93%)	To be specified
16	Software detail	For data storage, networking, remote control(FTP)/cal, etc., as applicable	To be specified
17	Accessories	Discussed separately above	
18	Special feature	Reagent consumption, filling, drain, etc.	To be specified

** FB = Fieldbus like Profibus/Foundation Fieldbus.

heteropoly acid, ascorbic acid is added; now, the molybdenum blue color is formed. After complete reaction, the light absorbance is measured. This increased absorption is due to the orthophosphate. So, the ratio of the final absorption and reference absorption is proportional to the orthophosphate present in the sample. Reactions:

$$\text{Phosphate} + \text{Molybdate} \Longrightarrow \text{Phosphomolybdic Acid}$$

$$\text{Phosphopmolybdic Acid} + \text{Ascorbic Acid}$$
$$\Longrightarrow \text{Reduced Phosphomolybdate Complex Blue}$$

2. If the vanadate molybdate reagent is used, it forms orthophosphate nanadomolybdophosphate to give a yellow color. The rest of the method is similar as discussed above.
3. Instrument calibrations could be programmable and repeated after a fixed cycle (e.g., twice a week to three weeks), and they vary with the manufacturer.
4. If, A_{SD} and A_Z denote the absorption by standard solution at full strength and zero corresponds to the standard PO_4 (μg/L), the slope will be

$$\text{Slope} = \text{standard } PO_4\ (\mu g/L)/(A_{SD} - A_Z)$$

so,

$$PO_4\ (\mu g/L) = A_{SA} \times \text{Slope} - \text{Instrument constant (reagent blank)}$$

where A_{SA} = absorption by sample solution.

5. *Phosphate analyzer accessories*: Similar as given in clause no. 4.4.5.4 (Table 5.53).

4.4.6 Sodium Analyzer

1. The measurement of Na is based on a direct potentiometric technique utilizing a highly sensitive ion specific electrode (ISE). The potential difference between the reference electrode and the ISE is a directly proportional logarithm of the Na concentration. This instrument is sensitive to changes in temperature as well as variations in pH. So, this ISE glass electrode composition contains sodium and is sensitive to H^+ but less sensitive to Na^+. The buffer solution contains a fixed level of sodium ions. Therefore, the difference in Na^+ in both sides of the glass electrode would result in potential.
2. On account of the high sensitivity to H^+ ions of the measuring electrode, it is necessary to maintain the sample with constant pH (preferably around pH 11) by adding a pH-adjusting reagent for measurement. Therefore, pH conditioning liquids are included in the measuring

TABLE 5.53 Phosphate Analyzer (DS)

Sl	Specifying Point	Standard/ Available Data	To Be Specified
1	Measuring range	0–1/5/15 ppm ranges with selection	To be specified
2	Number of streams	Single/2/4/6	To specify
3	Reagent and standard solution	Typically two and standard solution	To specify
4	Output	4–20 mA DC + alarm contact/stream	To be specified
5	Accuracy	±1% FSD 4% reading	
6	Repeatability	±01.5% Reading	
7	Contact details	SPST/SPDT 50VA 50V 1A	
8	International certification	As per country requirement (clause no. 3.1)	To be specified
9	Materials	Manufacturer's standard	To specify
10	Calibration	Auto/manual zero/ span dal. Frequency and time period, manufacturer's standard	
11	Display	LED/LCD (backlit) with diagnostic*	*Alpha numeric
12	Connectivity	RS 232/485/ MODBUS/HART/ FB**	
13	Power supply	230 VAC/24 VDC	
14	Housing protection	IP rating (65)	To be specified
15	Ambient	*T*: Up to 50°C RH (0.93%)	To be specified
16	Accessories	Discussed separately above	To be specified
17	Software detail	For data storage, networking, remote control(FTP)/cal, etc., as applicable	To be specified
18	Special feature	Reagent consumption, filling, drain, etc.	To be specified

**FB = Fieldbus like Profibus/Foundation Fieldbus.

system through a three-way valve. This is done before sending the sample to the measuring cell to ensure an accurate and repeatable Na^+ measurement (Fig. 5.56).

3. It is necessary to reactivate the analyzer from time to time so that it is not a sleeping element when measuring ultrapure water for a long time. For this, the instruments are programmed with automatic reactivation, just like the automatic calibration program. For this in the liquid handling section of these analyzers, standard solution and reactivate solution are included.
4. Two sections of the instrument are: (i) electronic control section with display and a transmission unit and (ii) a liquid handling section.
5. As the measurement is temperature-sensitive, normally the standard solution is kept in the range of the temperature of the sample line. This is done with the help of a heat exchanger in the standard solution, so that during calibration it is almost at the same temperature of the sample.

4.4.6.1 Sodium Analyzer Working Principle (Fig. 5.56):

1. On account of the ion sensitivity of the electrode, a potential measurement is done after pH conditioning. The sample is injected into an overflow tank to ensure a constant head. This is done to mix with the pH conditioning reagent as it flows down to the measuring cell. pH conditioning is done to ensure a constant pH value. There are three wells to house the thermistor (temperature compensation), ISE, and reference electrodes. Electrodes are connected to a preamplifier where the difference potential is measured and sent to the transmitter, which also receives a signal from the thermistor for temperature compensation.
2. Reactivation: After a fixed time, reactivating fluid is put into the well of the measuring cell to make it sensitive all times. After the reactive fluid is sent, it is rinsed with the sample so that it is sensitive (otherwise, the response time will at a later date be very sluggish).
3. For calibration, a standard solution can be injected into the system as shown in Fig. 5.56 (the standard solution may be heated).
4. Sample parameter requirement: Same as discussed for DO (clause no. 4.4.3.1.6) (Tables 5.54 and 5.55).

4.4.7 Chloride Analyzer

1. Chloride analyzers in power plant applications utilize ion selective (chloride) electrodes (ISE).
2. This analysis system also requires temperature compensation.
3. For calibration, the standard solutions are injected in the system, as shown in Fig. 5.57, through two solenoid valves.

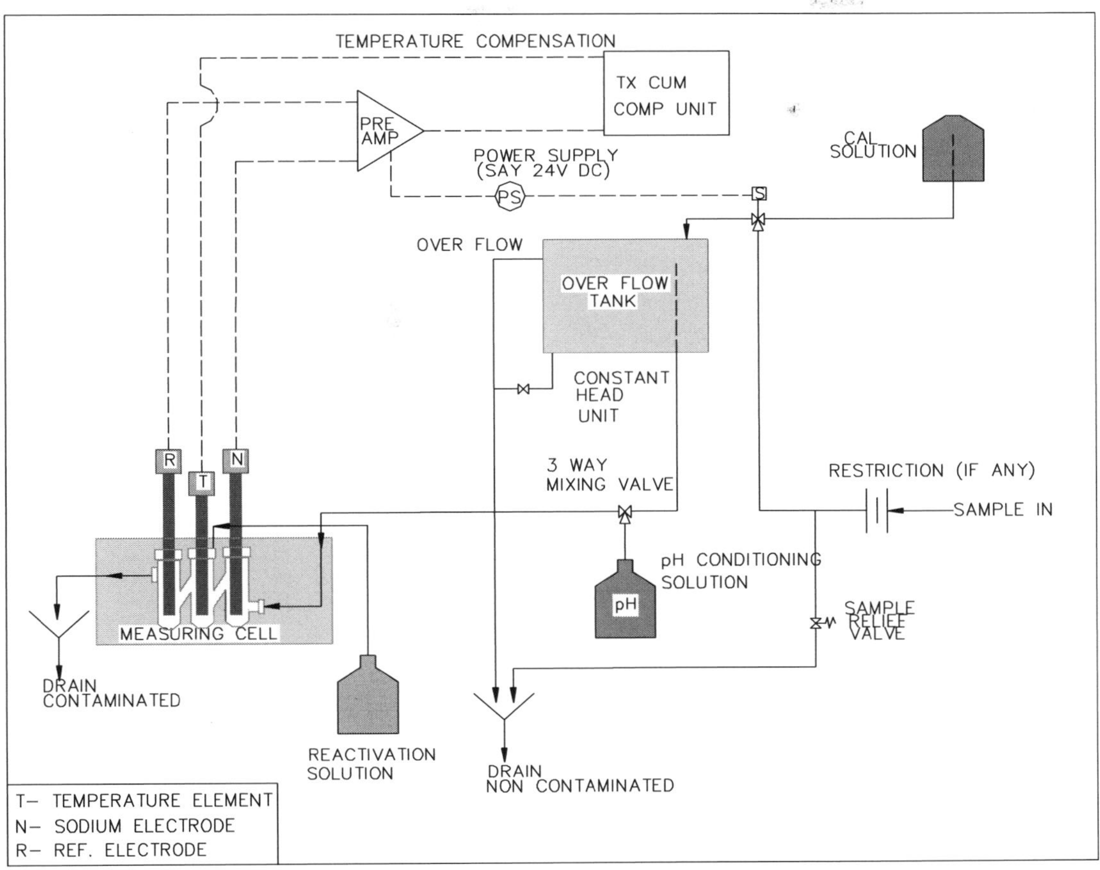

FIG. 5.56 Sodium analyzer working principles.

4. Calibration and measuring cycles are preprogrammed and are automatic.
5. Ammonium acetate, used as the reagent, may be drawn through a multichannel peristaltic pump and mixed. Reagent and standard solutions are available in a complete system.
6. Sample requirements: Same as given for a DO analyzer (clause no. 4.4.3.1.6).
7. There are two sections: a liquid handling section and an electronic section.
8. Electrodes are placed in a temperature-controlled flow cell where there is a heater to eliminate the effects of temperature variations in the sample and the ambient.
9. Electronics are placed on the back side of the instrument. There will be a front display and a key panel to operate the analyzer in manual as well as in automatic mode.

TABLE 5.54 Sodium Analyzer Accessories

All Reagent Solution in Same Enclosure	Standard Solution	Tube Fittings and Drain Lines
Reagent for 3 M trouble-free operation	Heat exchanger	Grab sampling system
Flow regulator restrictor	Cable gland and other electronic accessories	

4.4.7.1 Chloride Analyzer Working Principles

There are a number of methods to determine chlorine ions: ion sensitive electrode (ISE), culometric, and colorimetric.

Culometric method: In the culometric method, very low levels of silver ions are generated by passing a constant electrical current between the cathode (silver wire) and another electrode. Silver ions in turn react with the chloride ions of the sample solution to form silver chloride and change the sample conductivity measured by the voltage between the electrodes.

Colorimetric type: Similar to the silica/phosphate measurement, this type also uses the time-based mixing of reagents for the development of the color of the solution after reaction. At first, a measurement is taken as a reference to eliminate interfering factors such as the sample or reagent's own color, turbidity, refractive index variations,

TABLE 5.55 Sodium Analyzer (DS)

Sl	Specifying Point	Standard/Available Data	To Be Specified
1	Measuring range	0–10/100/10000 ppb selectable	To be specified
2	No of stream	Single/2/4/6	To specify
3	Reagent and standard solution	Typically one and standard solution at regular temp.	To specify
4	Output	4–20 mA DC + alarm contact/stream	To be specified
5	Response time	Typically 3–6 min	
6	Accuracy	±5% FSD	
7	Repeatability	±1.5% reading	
8	Contact details	SPST/SPDT 50VA 50V 1A	
9	International certification	As per country requirement (clause no. 3.1)	To be specified
10	Materials	Manufacturer's standard	To specify
11	Calibration	Automatic and manual zero/span cal. Frequency and time period as per manufacturer standard	
12	Display	LED/LCD (backlit) with diagnostic*	*Alphanumeric
13	Connectivity	RS 232/485/MODBUS/HART/FB**	
14	Power supply	230 VAC/24 VDC as the case may be	
15	Housing protection	IP rating (65)	To be specified
16	Ambient cond.	*T*: Up to 50°C RH (0.93%)	To be specified
17	Accessories	Discussed separately above	To be specified
18	Software detail	For data storage, networking, remote control(FTP)/cal, etc., as applicable	To be specified depending on need
19	Various keys	Menu key, up/down key, selection key, A/m	To specify
20	Special feature	Reagent consumption, filling, drain, Battery back up, if any, etc. Type and material of electrode, temp probe	To be considered

**FB = Fieldbus like Profibus, Foundation Fieldbus.

etc. The second measurement is taken after the reaction is over and the concentration is measured by the absorbance of light by the color sensed by the light detector and the sensors combination. The final result is computed on the basis of the difference between the two measurements and the stored calibration parameters. This semicontinuous method takes nearly 15 min to give a result.

Ion-selective electrode (Fig. 5.57): This system uses a *chloride* sensitive solid-state polycrystalline membrane at the electrode tip designed for the detection of chloride ions (Cl^-) in aqueous solution. The voltage that develops across the membrane is in relation to an Ag/AgCl double junction reference electrode (with a typical outer filling solution such as KNO_3 or CH_3COOLi). The two electrodes may be separate or may be combined in a common probe assembly. The ISE may be affected by numerous analytical interferences, which can be minimized or eliminated by adding the appropriate chemical reagents to the sample. For better results at low concentrations of chloride, a standard method for making measurements with ion-selective electrodes is to add ionic strength adjuster (ISA) solutions (e.g., $NaNO_3$) to each of the standard solutions and samples. Temperature variations affect electrode potentials and hence either maintained at a fixed value by heater or temperature compensation is applied. The pH range is wide (2–12 pH). The slope of measurement at 25°C is about 54 ± 5 mV/decade.

1. Here, the sample is fed through to the overflow tank (Fig. 5.57) so as to ensure constant pressure. On the way to the measuring cell, the sample passes through a set of solenoid valves, which will be used during

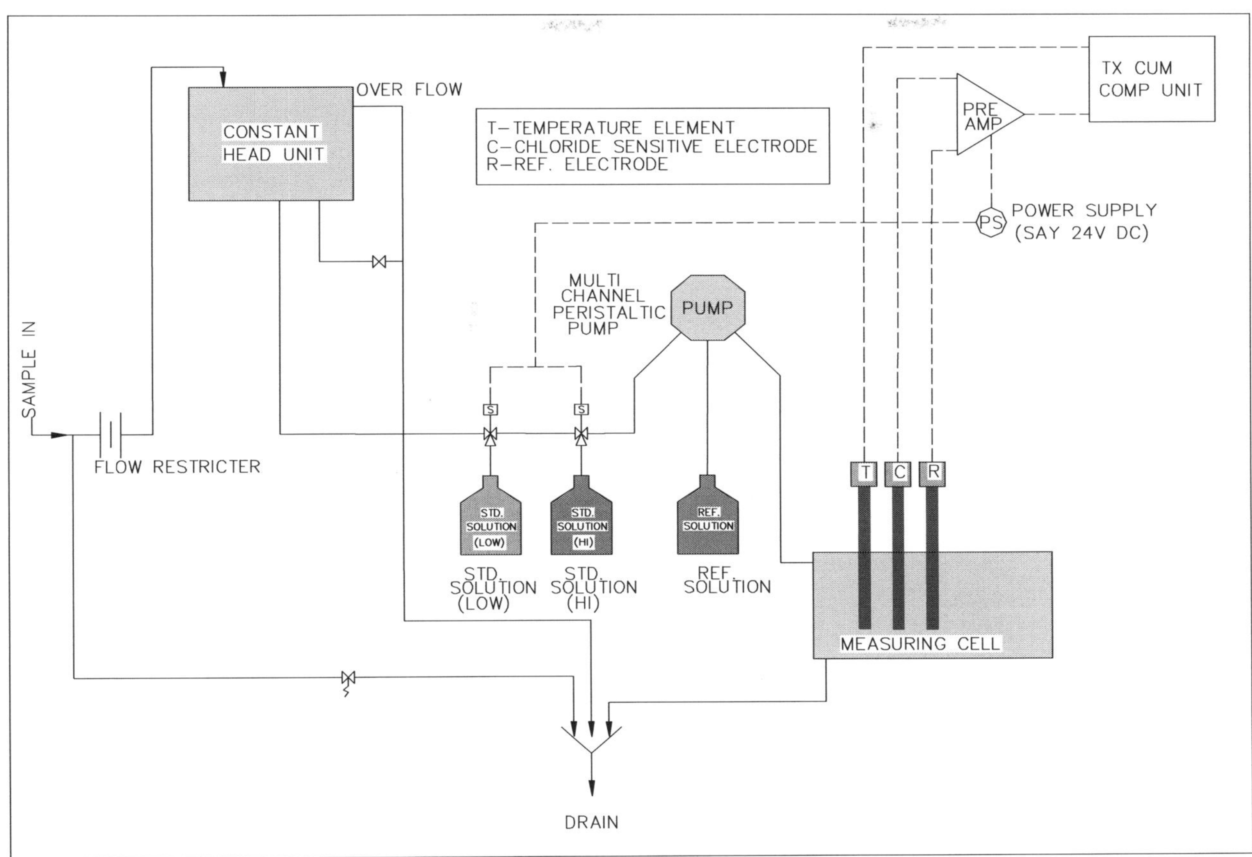

FIG. 5.57 Chloride analyzer system.

TABLE 5.56 Chloride Analyzer Accessories

Complete Reagent	Junction Box	Compensating Temperature Probe
Standard solution	Solenoid valves	Tube fittings, electronic accessories

calibration to admit standard solutions in low and high ranges. Reagent ammonium acetate is mixed through a multichannel peristaltic pump.

2. There are three wells to house the thermistor (temp. comp.), ISE, and reference electrodes.
3. The difference in potential between the two electrodes is a function of chloride ions.
4. Calibration is controlled by the built-in microprocessor (Tables 5.56 and 5.57).

5 SAMPLE CONDITIONING SYSTEM

For optimum operation of a power plant, monitoring of the steam and water chemistry is extremely important. For such

TABLE 5.57 Chloride Analyzer (DS)

Sl	Specifying Point	Standard/ Available Data	To Be Specified
1	Measuring range	0–10/100/ 1000 ppm selectable	To be specified
2	Number of streams	Single	
3	Reagent and standard solution	Typically one and two standard solutions	To specify
4	Output	4–20 mA DC +alarm contact/stream	To be specified
5	Response time	Typically 3–6 min	
6	Accuracy	±5% FSD/±1.5%–2% Reading*	*Repeatability

Continued

TABLE 5.57 Chloride Analyzer (DS)cont'd

Sl	Specifying Point	Standard/ Available Data	To Be Specified
8	Contact details	SPST/SPDT 50VA 50V 1A	
9	International certification	As per country requirement (clause no. 3.1)	To be specified
10	Materials	Manufacturer's standard	To Specify
11	Calibration	Automatic and manual zero/ span cal. frequency and time period	Manufacturer standard
12	Display	LED/LCD (backlit) with diagnostic*	*Alphanumeric
13	Connectivity	RS 232/485/ MODBUS/ HART/FB**	
14	Power supply	230 VAC/24 VDC as the case may be	
15	Housing	IP rating (65)	To be specified
16	Ambient cond.	*T*: Up to 50°C RH (0.93%)	To be specified
17	Accessories	Discussed separately above	To be specified
18	Software detail	Same as that for Na^+ (Table 5.55, clause no. 18)	
19	Special feature	Same as that for Na^+ (Table 5.55, clause no. 20)	

**FB = Fieldbus like Profibus/Foundation Fieldbus.

monitoring, it is essential that samples are taken from the strategic points in the plants, and to condition the samples before sending to the analyzers. Today, all sample points pertinent to SWAS throughout the main power plant and associated places are drawn to a central place, sometimes referred to as the "SWAS room," where the samples are conditioned and analyzed (exception: conditioned hot well conductivity, which *may be* monitored at the field to avoid complexity of sample drawings). Normally, all samples are brought at a common pressure and temperature (same platform): $P = \sim 2$ kg/cm^2 and $T = 25°C$.

1. Sample conditioning: Depending on the sample pressure and temperature condition, there may be two kinds of sample conditioning: Primary and secondary. Secondary sample conditioning is necessary for *all* the samples, even for those where primary sample conditioning has been done. In sample conditioning systems, pressure reduction is done after sample cooling, that is, coolers are placed in the upstream of the pressure regulator.
2. Sample line flow rate (min) is guided by a grab sample plus the number of analyzers connected with the sample line in question (with a minimum sample flow requirement of each analyzer). As a guideline for sample cooler selection as well as line size selection, the following conditions may be kept in mind. Approximate analyzer flow rate as per manufacturer recommendation only otherwise 50–70 CC/min (typical value); Grab sample max 500 CC/min (typical value); Sample line flow velocity ~2 m/s (typical value).

5.1 Primary and Secondary Sample Condition Devices

5.1.1 Primary Sample Conditioning

This is necessary for the samples that have either $P > 100$ kg/cm^2 *OR* $T > 100°C$. Primary sample conditioning's major components shall include mainly those items listed below. These may be placed in a rack at a suitable place in the field and may not be in the "SWAS Room" to avoid congestion.

1. Sample isolation globe valve.
2. Breakdown/restriction orifice.
3. Primary sample cooler.
4. Sample line relief valve.
5. Cooler shell relief valve.
6. Pressure reducing valve ($P > 100$ kg/cm^2).
7. Coolant sight flow indicator.
8. Tundishes/sink.

5.1.2 Secondary Sample Conditioning

All samples shall have secondary sample conditioning. Major components needed for each sample line for secondary sample conditioning shall include mainly those listed below.

1. Isolation globe/needle (low-temperature) valve.
2. Blowdown valve.
3. Secondary cooler
4. Shell relief valve for cooler.
5. Sample line relief valve.
6. Sample filter.
7. Pressure regulator.
8. Coolant water isolation valve.

9. Coolant flow sight flow indicator.
10. Solenoid block valve (with facility for manual operation).
11. Pressure gauge.
12. Pressure switch.
13. Temperature gauge.
14. Temperature switch.
15. Rate set valve (one for each analyzer).
16. Rota meter (one for each analyzer).
17. Chiller unit (+ temperature controller).
18. Three-way ball valve (for grab sampling and draining).
19. Cat column (for cat conductivity analyzer).

5.2 Primary Sampling P&ID

5.2.1 Primary Sample Conditioning P&ID

Three distinct cases have been considered here (Fig. 5.58).

5.2.2 When $P > 100\ kg/cm^2$ and $T > 100°C$

If the sampling point is located at an inaccessible place or far from the primary sample conditioning rack, it is recommended to use additional isolation valves. As a safety measure, relief valves have been considered, and a breakdown orifice is used to further reduce the pressure. The primary cooler takes the supply of the coolant from the equipment cooling water (DM water supply). Generally, the primary coolers are designed to bring the outlet temperature to ~45–50°C. In mid to large plants, the following sampling points would come under this: drum water/drum steam (as applicable), blowdown, FW at the economizer inlet, MS to HPT, and HRH to IPT.

5.2.3 When $P < 100\ kg/cm^2$ and $T > 100°C$

It is same as what was discussed above, only the PRV may not be necessary. The deaerator outlet is one of the sampling points falling under this category.

5.2.4 When $P < 100\ kg/cm^2$ and $T < 100°C$

Primary sample conditioning may not be applicable. CEP discharge, CPU outlet, DM MU water, CW to conditioner, etc. may fall under this category.

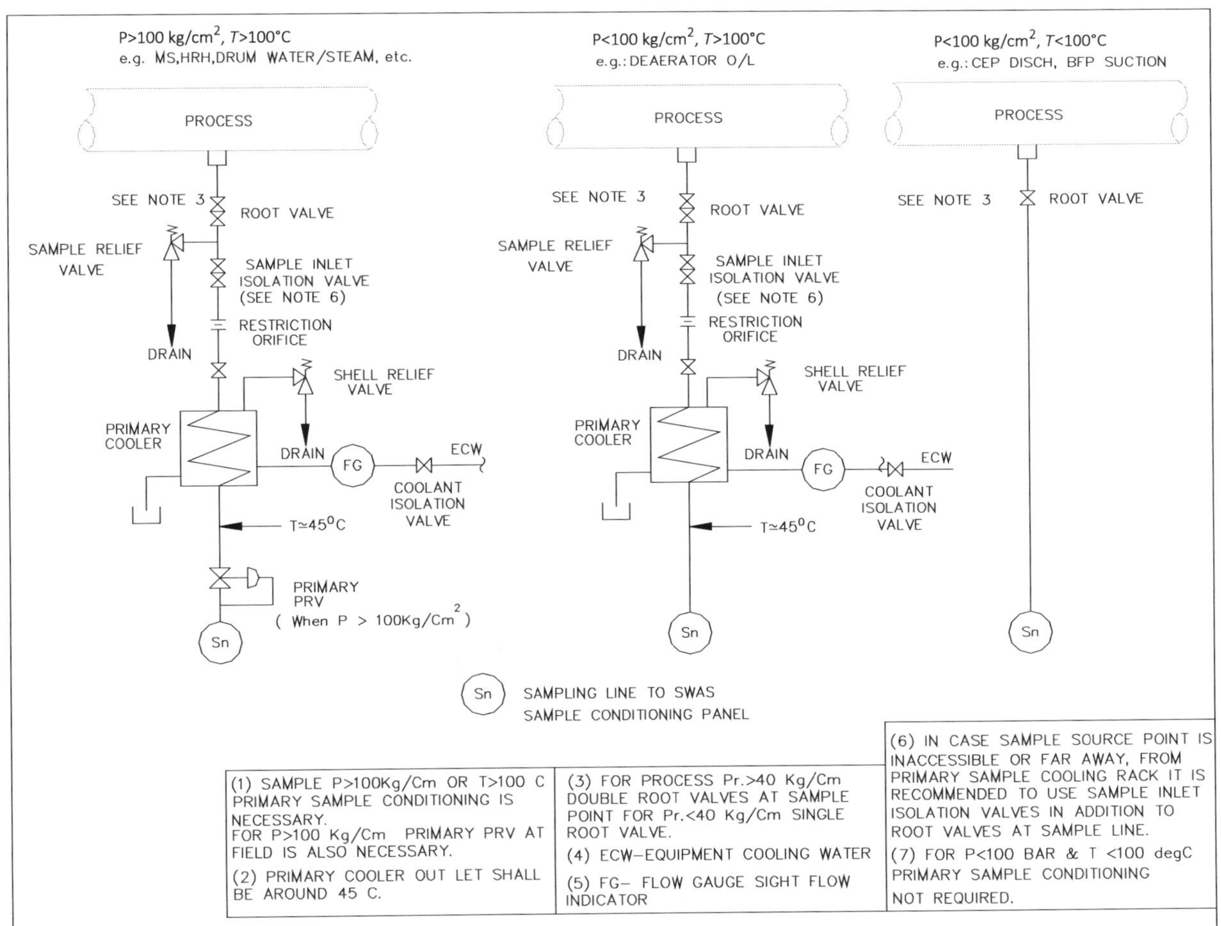

FIG. 5.58 Primary sample conditioning.

5.3 Secondary Sample Conditioning P&ID (Fig. 5.59)

Generally, secondary sample conditioning is done in the sample conditioning panel (SCP) of the SWAS. After primary sample conditioning, or directly (as applicable), the sample line (designated "Sn" for the nth sample line) in Fig. 5.58 is connected to the SCP via bulkhead fitting. For this reason, in Fig. 5.59, the primary conditioning circuit has been shown in dotted lines. Normally, the closed blowdown valve shown is used to drain/blowdown the line. The secondary cooler (or isothermal bath) is cooled by a chiller to ensure the sample O/L temperature is $<25\pm 1°$ C. A self-regulating, pressure-reducing valve (PRV) at the downstream of the filter ensures that the sample pressure at the outlet of the PRV is $2Kg/cm^2$. A solenoid-operated (with manual intervention) block valve (temperature shutoff valve) at the downstream of the PRV blocks the sampling flow, when the temperature or pressure exceeds the set limit, through the interlock. In the grab sample line, there is a back pressure valve to ensure suitable back pressure at all times in the sample line. The needle valves associated with rotameters adjust the sample flow to the analyzers, so these are also called rate set valves. Normally, analyzers are placed in a separate panel located beside the sample conditioning panel. In a cat conductivity analyzer, the cat column only is located in the sample conditioning panel.

5.4 Sample Conditioning Components

All pipe fitting and other wetted parts are generally chosen as 316 SS to avoid any contamination. Discussions put forward below are brief and generalized in nature; in case of any special requirement, the designer needs to take care of the same suitably.

5.4.1 Sample Isolation Valve (Both Primary and Secondary Sample Conditioning)

1. Type: Globe-type forged construction.
2. Pressure rating: As per the ANSI B16.34 standard to match upstream the sample line pressure at the maximum sample temperature.
3. Material: SS (ASTM A 182).
4. Flow: As per system design. Guide line: ref. clause no. 5.2.
5. Body size: As per sample line design.
6. Plug and spindle: Single piece. Guided plug.
7. Accessories: Suitable handle.

5.4.2 Blowdown Valve

Same as above with special features, such as being a slow-opening type and the throttle area shall be separate from the seating area.

5.4.3 Breakdown Orifice

1. Material: 316 SS.
2. Breakdown ratio: As per requirements.
3. Standard: BS/ISO.
4. Beta ratio: Project standard (<0.5).
5. Mounting: Welded/flanged of suitable rating and standards.

5.4.4 Primary Cooler

1. Type: Shell and submerged helical (coil counterflow design).
2. Coil material: 316 SS/Monel/Iconel.
3. Shell material: 304 SS.
4. Standard: ASTMD 1192 or equivalent.
5. Inlet temperature: As per sample, the temperature may be $>600°C$.
6. Special feature: Capable of withstanding simultaneously high temperature and pressure of the sample.
7. Preferred outlet temp: ~45–50°C.
8. Pressure rating: Suitable as the sample pressure rating as per ANSI.
9. Flow: Sample line flow (clause no. 5.2).
10. Capacity of cooling: Within 5°C of cooling water used.
11. Cooling water: Equipment cooling water of the plant (DM water).
12. Relief and check valve; Shell and cooling water header relief valve and cooling water outlet check valve.
13. Construction: Removable shell, and no welding in the sample coil.
14. Accessories: Shell relief valve of suitable rating.
15. Location: Sampling rack at a suitable location outside or field.

5.4.5 Sample Relief Valve (Both for Primary and Secondary Sample Processing)

1. Application: Primary sample conditioning: High pressure and temperature and secondary sample conditioning: moderate pressure.
2. Pressure setting: From outside.
3. Connection: Threaded/flange (ANSI/DIN, etc.).
4. Pressure/temperature rating: More than the corresponding sample cooler rating.
5. Set pressure: Sample pressure +20%.
6. Material: Body SS 316, Disc: SS 410/431.
7. Spring material: X12CrNi177 (Typical).

5.4.6 Primary Pressure Reducing Valve

1. Type: Piston type self-regulating for downstream pressure cont. PRV.
2. Pressure rating: As per ANSIB16.34.
3. Temperature rating: $>600°C$.
4. Material: 316 SS body and internal/wetted parts.

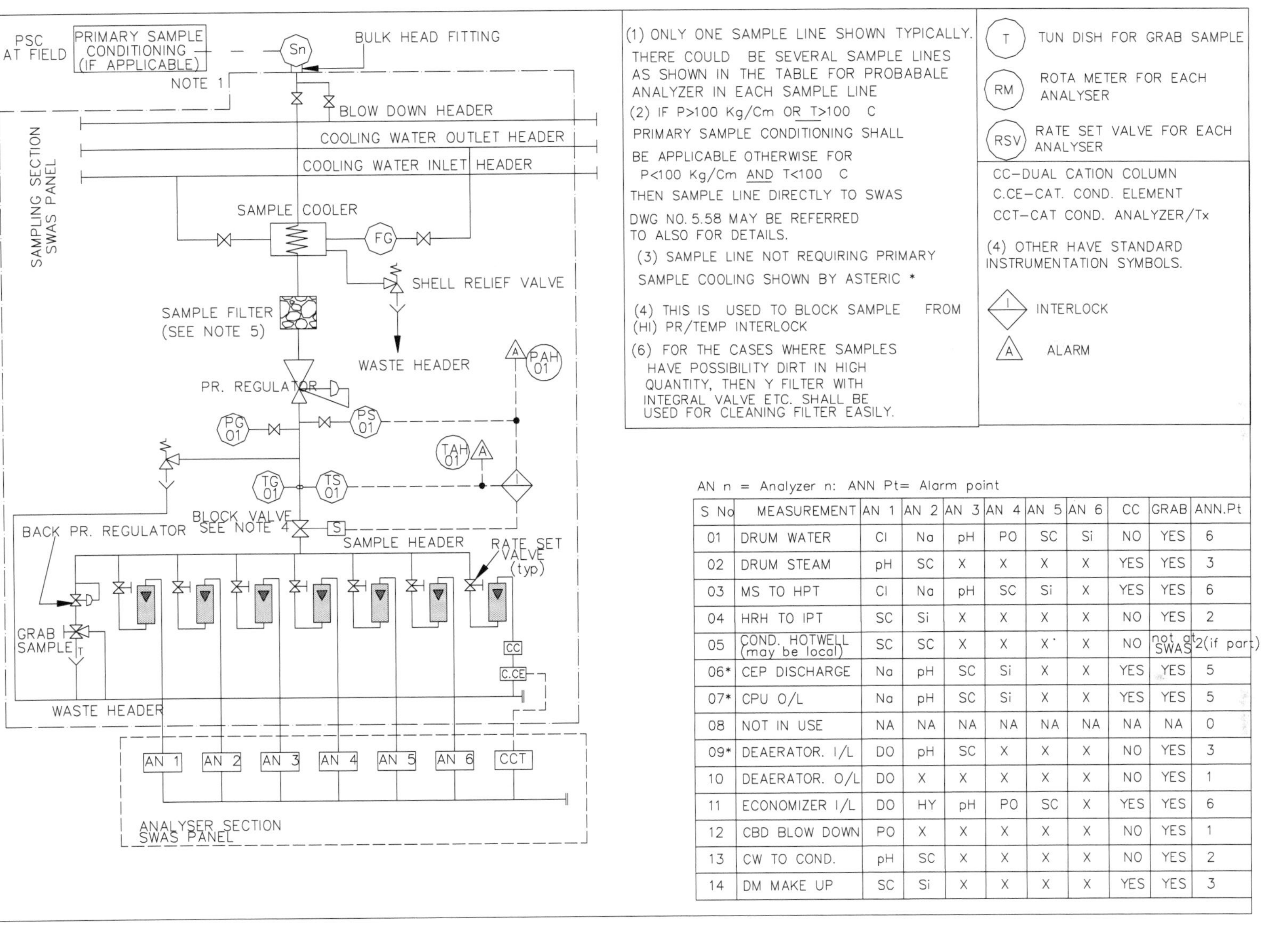

S No	MEASUREMENT	AN 1	AN 2	AN 3	AN 4	AN 5	AN 6	CC	GRAB	ANN.Pt
01	DRUM WATER	Cl	Na	pH	PO	SC	Si	NO	YES	6
02	DRUM STEAM	pH	SC	X	X	X	X	YES	YES	3
03	MS TO HPT	Cl	Na	pH	SC	Si	X	YES	YES	6
04	HRH TO IPT	SC	Si	X	X	X	X	NO	YES	2
05	COND. HOTWELL (may be local)	SC	SC	X	X	X	X	NO	not at SWAS	2(if part)
06*	CEP DISCHARGE	Na	pH	SC	Si	X	X	YES	YES	5
07*	CPU O/L	Na	pH	SC	Si	X	X	YES	YES	5
08	NOT IN USE	NA	NA	NA	NA	NA	NA	NA	NA	0
09*	DEAERATOR. I/L	DO	pH	SC	X	X	X	NO	YES	3
10	DEAERATOR. O/L	DO	X	X	X	X	X	NO	YES	1
11	ECONOMIZER I/L	DO	HY	pH	PO	SC	X	YES	YES	6
12	CBD BLOW DOWN	PO	X	X	X	X	X	NO	YES	1
13	CW TO COND.	pH	SC	X	X	X	X	NO	YES	2
14	DM MAKE UP	SC	Si	X	X	X	X	YES	YES	3

FIG. 5.59 SWAS sampling and analysis.

5. Output pressure setting: Wide range from outside.
6. Output pressure regulation: Desired output pressure setting is an aspect of system design but $\ll$100 kg/cm^2 (e.g., 35 kg/cm^2).
7. Flow: Sample line flow (clause no. 5.2).
8. Size: Design as per sample line size.

5.4.7 Coolant Flow Indicator

Sight flow as discussed in Section 4.5 of Chapter 4. For further details, refer to [123].

5.4.8 Secondary Cooler

1. Type: Shell and submerged helical (coil counterflow design).
2. Coil material: 316 SS/Monel/Iconel.
3. Shell material: 304 SS.
4. Standard: ASTMD 1192 or equivalent.
5. Inlet temperature: 45 ± 5°C.
6. Outlet temperature: $\sim 25 \pm 1$°C.
7. Pressure rating: As per ANSI, to suit pressure after primary conditioning.
8. Flow: Sample line flow (clause no. 5.2).
9. Capacity of cooling: Within 5°C of CW.
10. Cooling water: Chilled water, for example, 20°C.
11. Relief and check valve; shell and cooling water header relief valve and cooling water outlet check valve.
12. Construction: removable shell, and no welding in the sample coil.
13. Accessories: Shell relief valve of suitable rating.
14. Location: Sampling rack beside the SCP.

5.4.9 Sample Filter

1. Type: Cartridge.
2. Body: SS 316.
3. Element removal: Easy removal.
4. Particle size: Up to 40 μm or better.
5. Location: Secondary sample conditioning before PRV.

5.4.10 Secondary Pressure Reducing Valve

1. Type: Piston/spring-loaded diaphragm, self-regulating for downstream. Pressure cont. PRV.
2. Pressure rating: As per ANSIB16.34 for valves.
3. Material: 316 SS body and internal/wetted parts.
4. Output pressure setting: From outside.
5. Temperature rating: >Highest sample temp at the highest sample pressure.
6. Output pressure regulation: 2 kg/cm^2 sample output for analyzer.
7. Flow: Sample line flow (clause no. 5.1).
8. Size: Sample line size design.

5.4.11 Solenoid Valve

1. Type: Single coil normally open.
2. Number of ways: Two.
3. Coil voltage: 240/110 VAC or 24 VDC as per project requirement.
4. VA rating: To be specified (~10–15 W).
5. Insulation class: F.
6. Material: SS 316.
7. Size: Sample line size.
8. Pressure/temperature rating: As per system design.
9. Operation: Interlocked (to prevent high pressure/temperature to reach analyzers) with manual intervention to close manually from the SWAS panel.

5.4.12 Pressure Gauge

4″ dial pressure gauge (sample conditioning panel/rack front). Section 2.3 of Chapter 4.

5.4.13 Pressure Switch

Blind type with set point and dead band adjustment from front of switch (inside sample conditioning panel/rack). Section 2.2 of Chapter 4.

5.4.14 Temperature Gauge

4″ Dial bimetallic temperature gauge (sample conditioning panel/rack front) as discussed in Section 3.4 of Chapter 4.

5.4.15 Temperature Switch

Blind bimetallic temperature switch with set point and dead band adjustment from gauge front of switch (inside sample conditioning panel/rack). Section 3.3 of Chapter 4.

5.4.16 Rotameter

One for each analyzer in all the sample lines. Short borosilicate glass with 316 SS float type (normally located one after another in the front of the sample conditioning panel). Section 4.3 of Chapter 4. For further details, see [123].

5.4.17 Rate set Valve

This is a needle valve the comes as an accessory with the rotameters to set the desired flow to the individual analyzer. The valve shall be made of SS 316.

5.4.18 Chilling System and Isothermal Bath

1. Function: To cool each sample by chilled water at 20°C.
2. Type: Direct expansion with refrigerant tube in the shell.
3. Capacity: Redundant 100% capacity.
4. Performance: To maintain sample temperature at 25 ± 1°C at maximum flow samples, at their maximum temperature condition. Chilled water output 20°C of the quantity as demanded by the system concerned.

5. Condenser: Water cooling temperature at $40 \pm 2°C$.
6. Condenser design standard: ASME design code.
7. Compressor: Low noise, easily accessible, and complete with motor along with all electrical protection for the same. It shall also include auxiliaries such as a lubrication system, discharge valve, etc.
8. Chiller pump: Chiller water, pump motor, etc., in continuous operation.
9. Other auxiliaries: Electrical panel, storage tank, etc., complete with all necessary instruments such as a temperature gauge, level gauge, etc. These are available as part of the package.

5.4.19 Grab Sample and Three-Way Valve

The grab sample is collected via a three-way ball valve of SS 316 construction. A sufficient flexible tube shall be there to grab the samples. The grab sample flow shall be directed to the sampling trough, which drains to the waste header. The function of the three-way valve is to allow the sample to drain prior to collection. Also, through this, the entire line can be flushed.

5.4.20 Back Pressure Regulator

The back pressure regulator maintains the upstream pressure constant during fluctuation of the flow in the sample header at the time sample collection,. This in conjunction with the pressure regulator, provides a stable pressure and flow condition to the analyzers connected with the sample header.

5.4.21 Resin Column (For Cat Conductivity)

1. Function: Cation column.
2. Type: Duplex cat column.
3. Connection: Quick release connector (SS) for easy removal/change.
4. Accessories: Regeneration panel.

5.5 Sample Pipes, Tubes, and Fittings (Typical)

1. Standard: ISA SP 77.70 (latest).
2. Power piping: ANSI B 31.1 PTC 19.11.
3. High-pressure (HP) and high-temperature (HT) Tubing: SS $^3/_8$ in. (10mm) OD ASTM A213 Bulkhead (BH) fitting type 316H, 16 BWG tubing.
4. Secondary sampling: SS $^1/_4$ in. (6mm) OD ASTM A213 BH fitting type 316H, 16 BWG tubing.
5. Blowdown header: 80/160, As per Pr. Flange 2″ ANSI B16.5 Class 300/600lb. RF.
6. Closed cooling water pipe: 2in. (50mm) SS ASTM A312 Type 304 Scheme 40 Flange 2″ ANSI B16.5 Class 150lb. RF.
7. Drain header: 2in.(50mm) SS ASTM A312 Type 304 Scheme 40 Flange 2″ANSI B16.5 Class 150lb. RF.

5.6 Complete SWAS Rack/Panels

1. Sample conditioning rack: To house various components for sample conditioning. This may be an open rack and/or a closed enclosure with a corridor in between.
2. Analyzer panel/rack: The purpose of this is to house all analyzers.
3. Instrumentation cum electrical panel-A closed panel to house electrical components, recorder/controller power supply system, etc.

5.6.1 Sample Conditioning Panel/Rack

1. Type: Open rack/closed corridor, with easy access for adjustments.
2. Material: CRCA sheet ~2.5mm thick shall be deployed.
3. Mounting: Floor mounting with suitable foundation on base channel (e.g., 100mm). The mounting shall be without vibration, if necessary with a suitable antivibration pad.
4. Enclosure class—As applicable, depending on the type, shall be NEMA 12 minimum.
5. Surface preparation painting: All surfaces shall be cleaned, shop blasted, and painted with enamel painting. Also, other surface finishing as per manufacturer's standard. Exterior and interior paint color shall be as per project requirement (e.g., brilliant white paints for interior color). The paints shall be the anticorrosive type.
6. Finish: All sharp welds and corners shall be ground smooth. Both internal and exterior shall be free from scale, cutting, welding spatters, etc.
7. Accessibility and maintainability: Easy access without obstruction, and easy removal maintainability. There shall not be any crossing of tubes or obstruction from front or back of the analyzer and/or sampling component.
8. Completeness: Complete in all respects at the factory booth in tubing/piping as well as for electrical connections.
9. Location: SWAS Room (SCP and analyzer panel shall be in close proximity).
10. Headers: There shall be a number of headers such as cooling water headers and wastewater headers. All these shall be properly dressed and mounted in the sample conditioning panel. Suitable drains, sinks, etc., shall be provided.
11. Arrangement: All the monitoring instruments like the pressure gauge, temperature gauge, rotameter, etc., shall be mounted at the front of the panel. Also, rate set valves shall be in the front. Various headers, coolers, etc., shall be arranged suitably inside the panel.

5.6.2 Analyzer Panel

Various features discussed in clause nos. 5.6.1.2–5.6.1.9 above are applicable for this panel also.

1. Type: Freestanding enclosed panel.
2. Analyzer mounting: As far as possible, front surface mounting.
3. Arrangement: The electrical connections and electrical devices arrangement shall be as per logical functions. All drains, electrical, and other connections in the rear side of the panel. The sample line bulkhead is from the top rear side.
4. There will be a common drain line to collect the liquids leaving the analyzers.

5.6.3 Instrumentation Cum Electrical Panel

Various features discussed clause nos. 5.6.1.2–5.6.1.8 above are applicable for this panel also.

1. Type: Freestanding enclosed panel.
2. Location: SWAS room.
3. Major components: Power distribution breaker, recorders, controllers, analyzers, etc.
4. Devices: Apart from the electrical devices stated above, there may be a recorder and annunciator in this panel. Also, isolators, etc., exchanging signals with DCS.
5. Analyzer Output: Analyzers normally have 4–20m A DC, which may be connected to the recorder in this panel while the output may be sent to the DCS system of the main plant. Many analysis systems offer high/low alarms that can be annunciated in this panel and/or could be sent to DCS as a binary signal. In case the contact is insufficient, relays may be used, or contacts may not be sent to DCS as, from analog signal alarms can be generated in DCS.

Photograph of SWAS Panel (Courtesy: Forbes Marshall Pune India).

SWAS PANEL PHOTOGRAPH 1 Courtesy: Forbes Marshall Pune India.

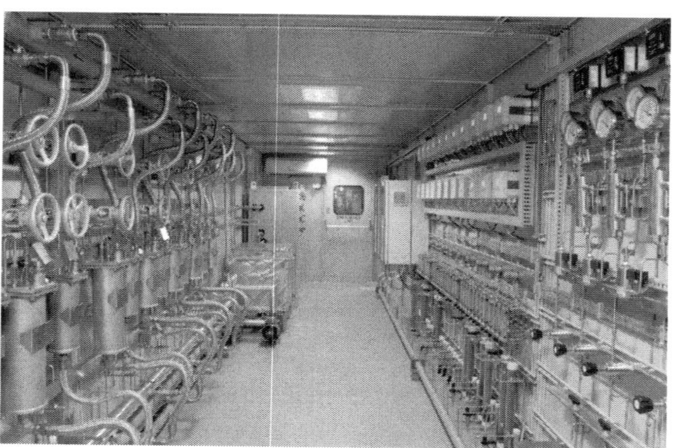

SWAS PANEL PHOTOGRAPH 2 Courtesy: Forbes Marshall Pune India.

Analyzer Panel for NTPC Sipat 3x660 MW Project

6 BLOW DOWN AND DOSING CONTROL

6.1 Blowdown

6.1.1 Functional Requirements

As the steam is generated from the boiler, the total dissolved solid (TDS) starts rising, and would form deposits and scales at boiler parts (e.g., at the tube). When these are carried by the steam, they will form deposits elsewhere, for example, the turbine. Therefore, whenever the TDS level crosses the limit set by the manufacturer, then blowdown is called for. Therefore, the basic purpose of blowdown is to regulate the boiler parameters within the prescribed limits to minimize carryovers, corrosion, and scale formation while driving out suspended particles. So, *TDS*, *alkalinity*, *silica*, and *suspended solids* are the major factors based on which blowdowns are carried out.

6.1.2 Range

The blowdown quantity depends directly on the maintenance of the water chemistry in the system concerned (also on the quality of water available). Apart from this, factors such as boiler design, turbine requirements and boiler operating conditions largely determine the blowdown quantity (as a percentage of feed water quantity). The range of blowdown lies somewhere between *2% and 20% of total feed water quantity*. ASME's "a consensus on operating practices for boiler blowdown" may also be referred to.

6.1.3 Blowdown Types

Blowdown may be manual and intermittent or continuous/automatic. Manual short time blowdowns are taken from the lowest point of the drum. In a continuous blowdown system, a part of the feed water is always discarded. It offers advantages for large heat recovery.

6.1.4 Blowdown Control

In an automatic blowdown system, the blowdown and its rate are regulated in relation to the concentration of dissolved solids present to maintain the proper water chemistry. Various means for automatic control of blowdown are:

1. Automatic blowdown control using Conductivity: (Fig. 5.60A)
    a. Objective: To blowdown drum water in relation to dissolved solids. Because TDS, alkalinity, silica, etc., directly affect the conductivity, the conductivity may be taken as a parameter to control the blowdown.
    b. Loop: In this method, a conductivity probe is used to measure the boiler conductivity and compare the same with the set point, which is set based on the allowed total contamination level. However, here one thing needs to be kept in mind: the rate of blowdown shall maintain a relation with the amount of feed water input to the boiler. So, the actual conductivity signal, for error generation, is duly multiplied by the feed water quantity, that is, by the boiler load index to take care of changes in the load. The error thus generated is fed to the blowdown PI controller, which generates a demand signal for the amount of blowdown. That is, the blowdown shall be varied with the load in the boiler.
2. As shown in Fig. 5.60B, there is another way that blowdown can be controlled. A TDS test is carried out before blowdown. Based on the test result, the blowdown valve opening is controlled by a timer to keep the valve open for a specified time. This is a very easy method but not accurate. For utility stations of moderate size, this may not be suitable. Also, it cannot take care of the load fluctuations.
3. A combination of timer, conductivity analyzer, and control is also possible. The "boiler boss" of the Rosemount is an example of the same.

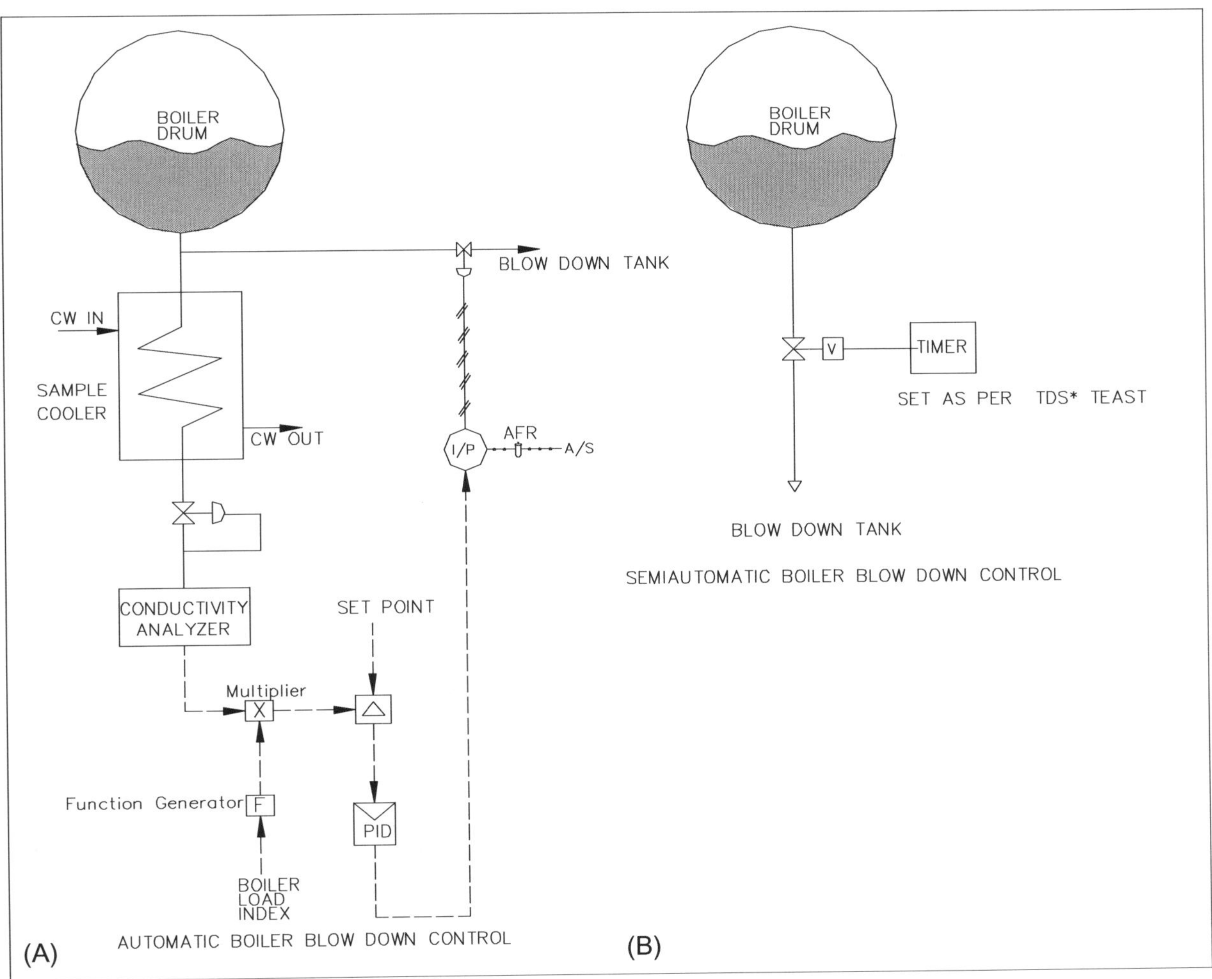

FIG. 5.60 Boiler blowdown control.

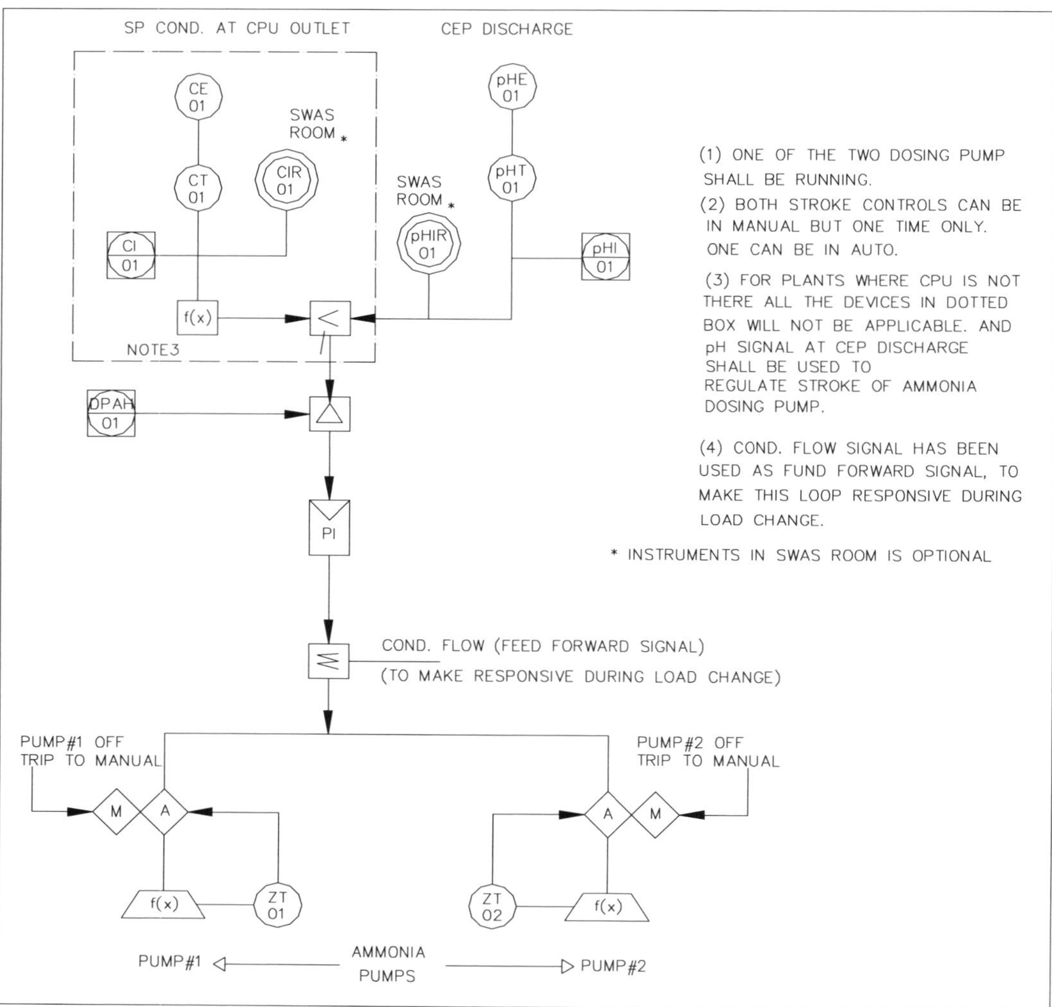

FIG. 5.61 Ammonia dosing control.

6.2 Dosing Control

In power stations, dosing is done at two places: the low-pressure condensate and the boiler drum, that is, in the high-pressure dosing control.

6.2.1 LP Dosing

In low-pressure condensate, ammonia and hydrazine are dosed at the CEP discharge and at the suction of the BFP, respectively. However, at plants with OT, N_2H_4 dosing is absent (clause no. 4.1.4 of this chapter). In plants where there will be CPU dosing, this may be done at the CPU outlet in place of the CEP discharge. The purpose of ammonia dosing is to control the pH of the water. The pH is kept slightly alkaline to avoid corrosion from chloride and other salts, whereas the dissolved oxygen is scavenged by dosing hydrazine. Very low ppb of dissolved oxygen can cause severe corrosion by pitting. Therefore, slight residual hydrazine is kept in the system.

1. Ammonia dosing control: (Fig. 5.61)
   - **a.** Objective: This loop is used to regulate the pH in the system by dosing ammonia. As shown in Fig. 5.45, there is a direct relation between the pH and sp. conductivity, so sp. conductivity has also been used through the function generator.
   - **b.** Loop: The pH set point is compared with the lower of the actual pH and derived pH from sp. conductivity or directly from the pH instrument, as detailed out in the loop (with notes). The error thus generated is fed to the PI controller to control the stroke of the running ammonia-dosing pump. In order to make the loop more responsive, the condensate flow has been taken as the feed forward signal.
   - **c.** Interlock: (I) Normally, only one pump will be in operation, so the A/M station pertaining to the operating pump shall be in auto. (II) When any pump is not in operation, the corresponding A/M station shall trip to manual. (III) In case of DCS, the A/M station shall be the soft A/M station in the operator's console.

2. Hydrazine control: (Fig. 5.62)
   a. Objective: This loop is used to scavenge dissolved oxygen to the extent possible. This loop may not be applicable for plants with OT (clause no. 4.1.4).
   b. Loop: DO here has been expressed as a function of residual hydrazine. The lower of this signal and the actual residual hydrazine is considered as the measuring signal, which is compared with the set point for residual hydrazine to get the PI output from the controller to regulate the stroke of the hydrazine-dosing pump. The output of the controller is summed with the feed water flow (may be BFP suction flow), which acts as a feed forward signal to make the loop responsive at the changing load.
   c. Interlock: Same as clause no. 6.2.1.1c above.

6.2.2 HP Dosing Control (Fig. 5.63)

The alkalinity in the system is maintained with phosphate dosing. Sodium compounds have influence on phosphate dosing. The alkalinity of the drum water is well maintained when it is done in conjunction with pH based on NaOH. The phosphate either itself or with NaOH could give an elevated pH.

1. Phosphate dosing control
   a. Objective: This loop is used to operate the boiler at the desired pH in presence.
   b. Loop: In the loop shown, the actual value of pH and PO is compared in separate controllers to generate demand for each parameter. PO demand in conjunction with the same from the drum water pH controller is used to regulate the stroke of the phosphate-dosing pump. The output of the controller is summed with feed water flow as the load of the system, which acts as a feed forward signal to make the loop responsive at the changing load. The pH and feed water signals are put into the loop through scaling and bias units to normalize the signal.
   c. Interlock: Same as clause no. 6.2.1.1c above.

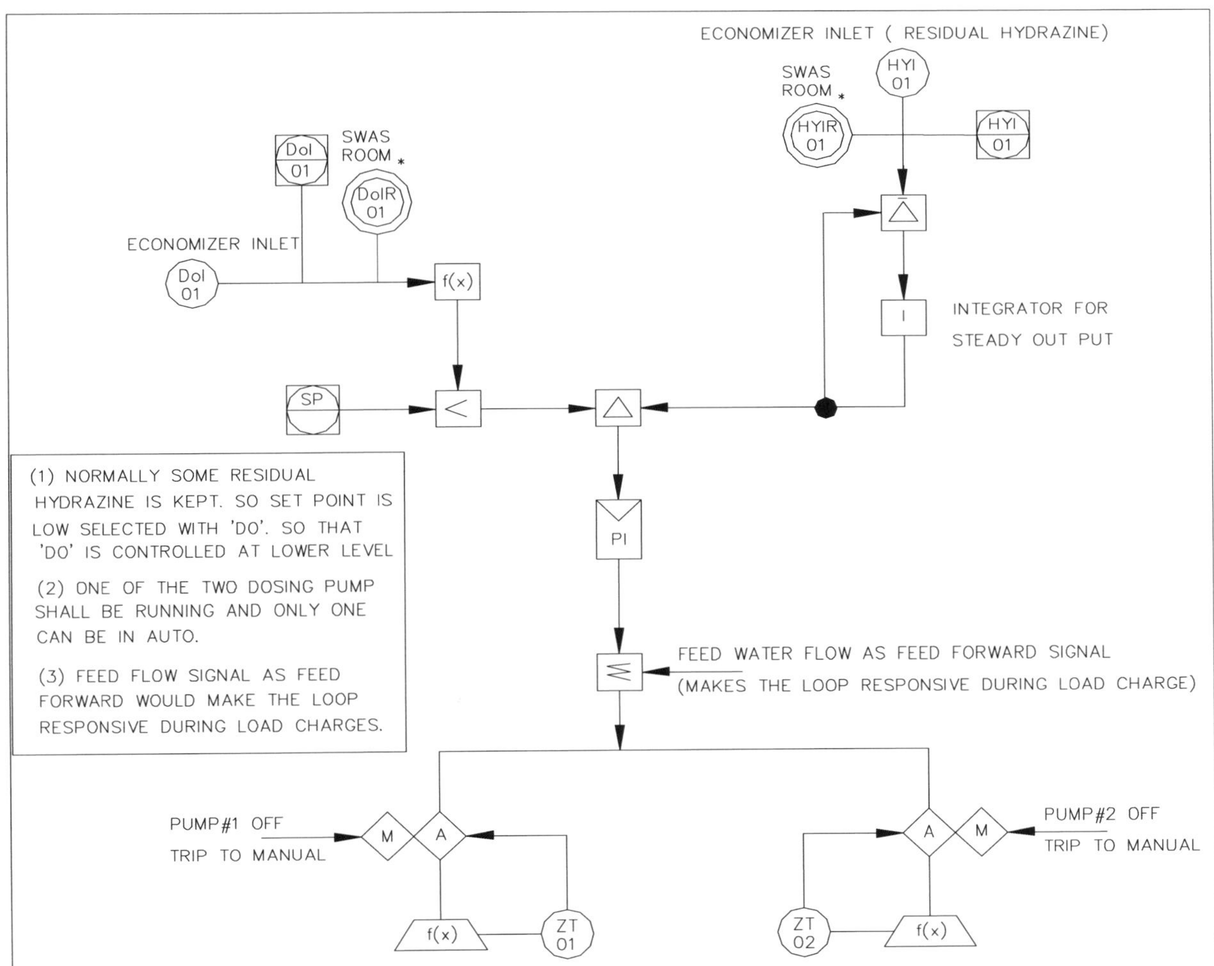

FIG. 5.62 Hydrazine dosing control.

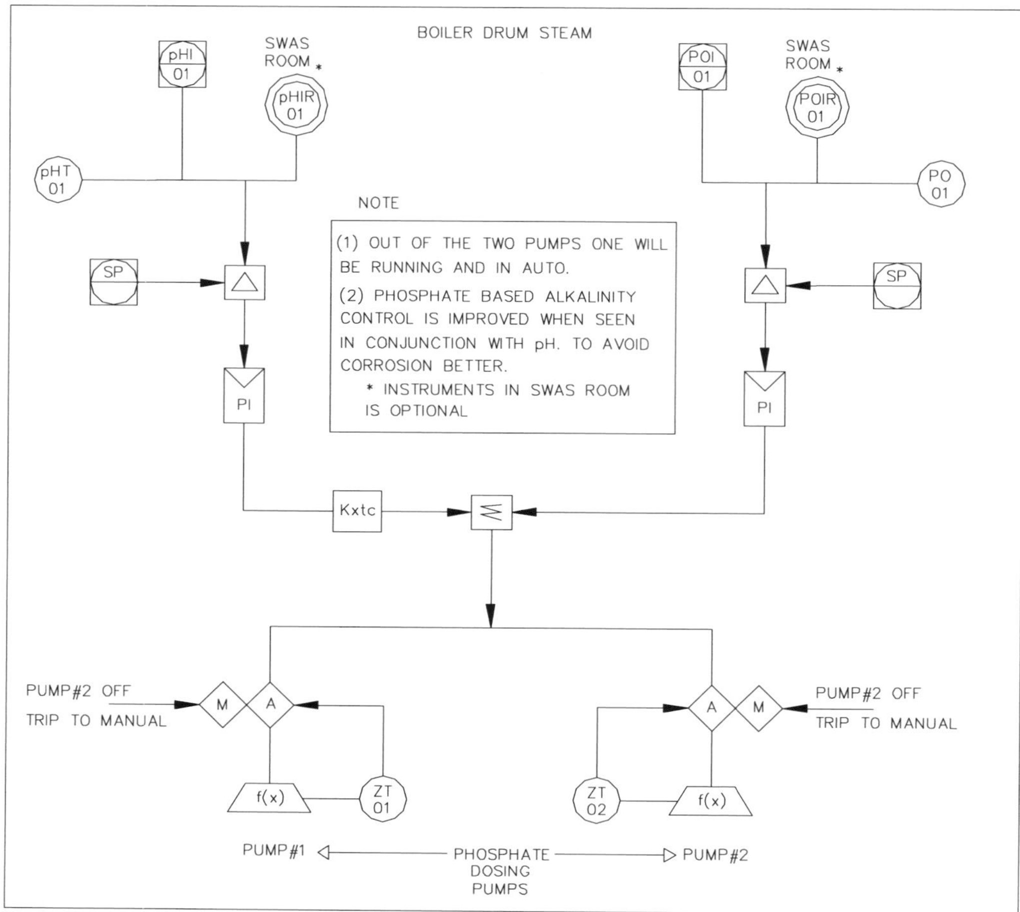

FIG. 5.63 Phosphate dosing control.

7 ANALYZERS FOR AIR POLLUTION MONITORING & CONTROL (NO_X CONTROL)

7.1 Pollution Control

Pollution on account of various gases along with dust emitted from the boiler and other from liquid pollution from various effluents.

7.1 Gas analyzer for pollution: This has been covered in Section 3.

7.2 Effluent and treatment: A power plant produces a lot of waste products from water treatment plants, regeneration and resin water, sludge, cooling tower blowdown, boiler blowdown, and various drains.

7.3 Based on local and national environmental regulations, there are certain guideline such as those from the EPA. The steam electric power generating effluent guidelines—postponement of certain compliance dates 2017 rule postponing certain compliance dates in steam electric power generating effluent guidelines (40 CFR Part 423). Related information is available at final rule 2015. The final rule sets new or additional requirements for wastewater streams from the various processes and byproducts, such as flue gas desulfurization, fly ash, bottom ash, flue gas mercury control, and gasification of fuels such as coal and petroleum coke. Effluent data and emission data for different plant types and fuel types have been elaborated in Tables 5 and 6 of document [124], for further reading.

It is important to know the typical sources of such contaminants.

7.3.1 Sources of liquid contaminants: Sources of contaminants in aqueous discharge from power plants in general shall include but are not limited to the following:

- Water treatment plant—regenerants, for example, acid and alkali, backwash, and others.
- Cooling water—chlorination and concentration factor.
- Boiler blowdown/drains and turbine drains—oil; metal oxides; dissolved additives.
- Ash lagoons—suspended solids; dissolved metals.

- Coal stocks drains—suspended solids; dissolved metals; acidity.
- Transformers and transmission—oil.
- Air heater washing—acidic waste; particulates; metals.

7.3.2 Apart from normal power plant discharge, there are auxiliary plants that are also associated with the main power plant, for example, FGD, SCR, and carbon capture and storage (CCS). These plants are necessary to limit gaseous emissions. These plants also contribute contaminants in aqueous discharges.

- FGD (Wet Limestone—Gypsum Process): High concentration of $CaCl_2$, traces of metals (Cd, Hg, Ni, Mo, etc.) from coal-ashes, particulates from gypsum, limestone, etc.
- SCR for NO_x removal: Ammonium salts, ammonia, and sulfates.
- CCS: Ammonium salts, amines, and associated products.

7.4 Various treatments needed shall include but are not limited to aeration, sludge treatment, chlorination, and neutralization of pH to meet the discharge requirements stipulated by various standards. There is a need for treatment of FGD water by raising the pH (around 9.5) as well as the addition of a coagulant, etc. in order to. Fluoride discharge depends on the $CaCl_2$ concentration and pH value. Similarly, there are limits for ammoniacal nitrogen in SCR discharge, for example, 10 mg/L ammonia. For this, technologies such as absorption/precipitation, microbiological digestion, and chemical oxidation are deployed. Postcombustion carbon capture is carried out by the absorber and stripper. Amine process and chilled ammonia are two major systems used for CCS. Thus, one can conclude that wastewater treatment technologies must be effective as well as reliable.

7.5 Acid and alkalis are added to the system for pH adjustments and monitoring of pH, conductivity, dissolved oxygen, ammonia, etc., is carried out.

7.6 Ammonia analyzer in effluent treatment plants: Ammonia is produced as the decay of organic nitrogen compounds and excess ammonia causes an imbalance in hydrological systems. Microprocessor-based ammonia ion-selective probe analyzers are used to monitor ammonia in effluent treatment plants, for example, Model 8232 of ABB. Prior to analysis, there is a need for sample conditioning such as regulation of flow, temperature, and adjustments of chemical composition. The instrument system responds to the partial pressure of ammonia in the sample solution and the potential developed is proportional to the logarithmic (base 10) relation with NH3 concentration. Ammonia obeys Henry's Law in dilute solutions, so the partial pressure of ammonia in the solutions is directly proportional to its concentration [125]. The measurement is directly influenced by temperature.

7.7 Total solids and monitoring of TDS/TSS: TDSs are inorganic salts and small amounts of organic matter present in solution in water. However, total suspended solids (TSS) are different from TDS, and these solids can be trapped by a no-ash glass fiber filter disc of approximately 0.45 mm pore size filter. The sum of TDS and TSS can be thought to be the total solids present in the water.

1. TDS measurement: TDS meters actually measure conductivity and calculate TDS with the help of a conversion factor (with a custom curve feature). There are two basic assumptions in measuring TDS from conductivity: All dissolved solids produce conductivity, and solutions having the same TDS have equal conductivity. However, these assumptions are not always true and for this reason it is necessary to measure conductivity and to calculate TDS. The instrument corrects the measured conductivity to 25°C and the conversion must be done through the custom curve feature, for example, the Rosemount 5081-C instrument.
2. Turbidity and TSS measurement: In contrast to the conductivity measurement for TDS, the turbidity and TSS monitor directs a beam of infrared light into the sample with the help of an LED. The light beam is scattered by particles in the sample, and the scattered light intensity is measured by a photo detector at 90° to the light beam. The scattered light detected is directly proportional to the turbidity of the sample. Analysis and signal conditioning are done by built-in electronics. As there is a known relationship between the amount of solids in suspension and the turbidity of a sample, the turbidity reading can be used to provide a real-time estimate of the level of suspended solids in the sample, for example, an ABB ATS430 turbidity meter.

8 OXYGENATED TREATMENT DISCUSSIONS

In addition to the discussions presented in clause no. 4.1.4, discussions on OT are presented here to have a closer look into this.

8.1 Basic Discussions on Treatment Issue

In OT, not only is a reducing agent eliminated, but oxygen is added along with NH_3. Naturally, there will be a slight rise in the DO level. As stated earlier, these are used mainly in plants without boiler drums, supercritical, USC and A-USC plants with CPU. One of the major reasons why OT is preferred is FAC. What is FAC? Flow-accelerated/flow-assisted corrosion (FAC) is metal loss that occurs in steel equipment due to the dissolution of the normal protective oxide layer into a the turbulent stream. Thus, on account of continuous loss of the oxide layer, there will loss of metal thickness. The loss of metal is dependent on the water chemistry, material composition, and hydrodynamics. CS materials are especially susceptible to FAC. Preventing the FAC of the water supply system with OT has an incomparable advantage. Some of the other preferable reasons behind the use of OT shall include but are not limited to the following:

- Long-term protection of the FW system due to a protective hematite layer.
- CPU operating period will be longer on account of low condensate NH_3.
- FAC will be minimum.
- Iron transportation will be highly reduced (reduction by 90%).
- Overall, the bio transfer factor (BTF) will be minimum [126].
- Crud will be generated less often [126].
- Less frequency for chemical cleaning.
- Outage time will be less and there will be faster start up.

However, on account of comparatively higher DO, researchers doubt that dissolved oxygen will have an effect on the growth and exfoliation of the oxide scale, and the oxygen oxidation mechanism is put forward, especially in SHs and RHs. The effect of OT on oxide growth and the exfoliation of superheater tubes is not entirely clear. So, the study of the affect of OT on the growth and exfoliation of the oxide scale (there has been significant controversy) will be the direction of scale research [127]. Now, let us take up the recommendations by the Electrical Power Research Institute (EPRI) for a once-through boiler. Some of the recommended figures from EPRI are elaborated upon in Tables 5.58 and 5.59.

Also the turbidity should be controlled within 2NTU. These are typical standard parameter specifications. As already stated in previous sections, it is very important to control the silica levels in the steam to avoid silicate scaling, which may contribute to turbine capacity and efficiency losses. It is quite challenging to maintain the feed water pH mentioned above, especially when the cat conductivity starts rising from 0.15 μS/cm to, say, 0.3 μS/cm. It is worth noting that during start up, it may not be possible to maintain the parameters listed in the tables above. Typical start-up parameters may be as shown in Table 5.60. Here, it is worth noting that the CPU has a great role to play during start up and commissioning.

According to the experts [128], the important issues that need to be addressed when implementing OT include:

- At what point during the start-up and commissioning process should the chemistry regime be switched from AVT to OT to prevent frequent switching back and forth between a reducing and an oxidizing environment?
- What would be the "detrimental effects" of going from an oxidizing atmosphere to a reducing (or close to reducing) atmosphere, for temporary periods?
- How can these "detrimental effects" be quantified and addressed during design and equipment procurement?

Thus, it is clear that there will be some pros and cons of OT. This is because of the fact that on account of a comparatively higher DO, some people doubt that this high DO may be responsible for the growth and exfoliation of the oxide scale, especially in SH and RHs. The effect of OT to oxide growth and exfoliation of superheater tubes is not entirely clear. So, the study of the affect of OT on the growth and exfoliation of the oxide scale (there has been significant controversy) will be the direction of scale research [127].

TABLE 5.58 EPRI: Recommended Parameter for Main Steam and Reheat Steam

Parameter	Target Value	Alarm Action Required
Cation conductivity (μS/cm)	≤0.15	0.3–0.6
Silica (ppb)	≤10	20–40
Sodium (ppb)	≤2	4–8
Chloride (ppb)	≤2	4–8
Sulfate	≤2	4–8
TOC (ppb)	≤100	>100

TABLE 5.59 Recommended Parameters for Feed Water Chemistry (EPRI)

Parameter	Recommended Values
Cation conductivity (μS/cm)	≤0.15
pH	8.08.5
Dissolved oxygen at economizer outlet (ppb)	30–150
Iron (ppb)	≤2*
Ammonia (ppb)	0.02–0.07

*The sodium and silica content should be ≤2 and ≤10 ppb, respectively.

TABLE 5.60 Typical Parameter Specification for Start Up [126, 128]

Parameter With Unit During Start Up	Typical Value
Cat conductivity (μS/cm)	<0.5
pH	≥9
Dissolved oxygen (ppb)	<100
Iron (ppb)	<20
Sodium (ppb)	<10
Silica (ppb)	<30
Turbidity NTU	<5

Having gained some knowledge about OT, it is better to explore how oxygen is injected into the system. In this connection, the notes in Fig. 5.46 may be referenced. In a typical OT system, NH_3 is added at the CPU outlet, as shown in Fig. 5.46. The dosing is done by adjusting the stroke of the dosing pump. The pH at the CPU outlet will be the controlling parameter. Also to take care about the load feed water flow should also be included in the dosing loop similar to that shown in Fig. 5.61. The cycle oxygen injection is regulated with the help of a set of flow-regulating valves whose opening is a regulated automatic controller (may be a part of the DCS). Dissolved oxygen at the outlet of the deaerator or the CEP is the control parameter. In order to take care of the cycle load, the feed water flow could be used as an anticipatory signal in the control loop, as discussed above.

BIBLIOGRAPHY

[1] 3300 Rotary Position Transducer (RPT): Bently Nevada.
[2] 3500 System Overview: Asset condition Monitoring: Bently Nevada.
[3] 6600 Series Machinery Protection Monitors: Entek-Rockwell Automation.
[4] Application Note—Power generation: Provitech (200 MW STG): www.provitech.com.
[5] CSI 6000 machinery Health Monitor: Emerson Process Management.
[6] Gas Turbine Vibration Monitoring—an overview: Mel Maalouf application Engineer Bently Nevada Products (GE energy).
[7] Inductive Proximity Sensors: Rockwell Automation.
[8] Machinery Protection System: GE Energy: 3500 series.
[9] Piezoelectric Accelerometer and vibration Preamplifier Handbook: Bruel and Kjaer: Denmark 1976.
[10] Piezoelectric Accelerometer Product data: Bruel and Kjaer.
[11] Power Generation Handbook: P. Kiameh.
[12] Principles of Vibration Measurement and Analysis: Bruel and Kjaer.
[13] Real-Time Condition Monitoring System Using Vibration Analysis for Turbine Bearing: A. Bandopahyay, S.K. Dasmondal, and B. Pal: ER and DCI Calcutta.
[14] Signal Conditioning: Turbine Speed: e.on Kraftwerke.
[15] STI Field Application Note—Turbine Supervisory Instrumentation: STI Vibration monitoring inc.
[16] Turbine Run Monitoring Signals for turbine protection: e.on Kraftwerke.
[17] Turbine Supervisory Instrumentation: SKF (CMCP): Field Application Note.
[18] Turbine Supervisory Instrumentation Selection Guide: Rockwell Automation: Technical literature (Internet document).
[19] Understanding Discrepancies n Vibration Amplitude reading between Different Instruments: S. Sabin, A. Hamad, R. Chitwood, and M. Qureshi: GE Energy.
[20] Understanding of Vibration Analysis: IN-PLANT SERVICES, INC: GREENVILLE, SC.
[21] Vibration (P.H. Sydenham): Instrumentation Reference Book: Edited by Walt Boyes.
[22] Vibration Application guide Metrix Vibration Condition: Monitoring and Protection.
[23] Vibration Measuring Instruments: Dr. B.K. Sridhara: Mysore, India.
[24] FCC Flue gas Emission Control Option: P.K. Niccum, E. Gbordzoe, S. Lang FCC: NPR A2002 Annual meeting March 2002.
[25] Advance Cemas NDIR (Multicomponent Measuring System): ABB limited: Catalog.
[26] Advance Optima High performance measuring technology Limus 11UV:ABB Analytical: ABB Catalog.
[27] Advance Optima the Integrated Analyzer System Solution: ABB Analytical: ABB Catalog.
[28] Advanced Pollution Instrumentation: Teledyne advanced Pollution Instrumentation: Product Data Sheet.
[29] B&W NO_x Reduction System and Equipment at Moss Landing Owner Station: B.Becker (B&W), D. Tonn (B&W), N. Stephenson (Cormetech United States), K Speer (Moss Landing United States).
[30] Continuous Emission Monitoring: Codel International Limited UK: Product Data Sheet.
[31] Continuous Gas Analyzers in Situ: Siemens: LDS 6 Catalog.
[32] Cross Duct Co Analyzer: Codel International Limited UK: Product Data Sheet.
[33] DEFOR Extractive UV Gas Analyzer: SICK Sensor Intelligence: Catalog.
[34] Environmental Assessment Report: Jhajjar Thermal Power Project: January 2009.
[35] EPA CICA Fact sheet: EPA United States: EPA-452/F-03-034.
[36] Flue gas Oxygen Analyzers: GE Sensing and Inspection Technologies.
[37] High-Temperature Zirconia Oxygen Probe: ABB limited: ZGP2.
[38] Improving Performance of N G Fired Assets in Power Plants With Process Chromatographs: Emerson Process Management: Application Note.
[39] Modular Continuous Gas Analyzers: ABB Limited: AO2000 Series.
[40] NDIR TYPE INFRARED GAS ANALYZER (FOUR-COMPONENT ANALYZER): Yoko gawa Electric Corporation Japan: Product Data Sheet.
[41] NO_x Analyzer: Rosemount Analytical: Catalog.
[42] Opacity Dust Density Monitor: Emerson Process Management: Catalog.
[43] Opacity Monitor: Lear Siegler Australia: Catalog.

[44] Paramagnetic Oxygen Analyzer: Fuji Electric Systems Co Ltd.: Catalog.
[45] Particulate Measurement System: PCME CO UK: QAL181.
[46] Principles of Operation: MEECO Inc.
[47] Putting Combustion Optimization to Work: N. Spring Sr. Editor:2009.
[48] Reduction of NO_x through Combustion Optimization: J. Cahill: Emerson Process Management.
[49] Specific Ion Analyzer: Honeywell: Specification.
[50] Standalone Type Infrared Gas Analyzer: Fuji Electric Co Ltd: Catalog.
[51] Superior Technology, Quality and Reliability From World Leader in Hydrogen Measurement: ABB Limited UK.
[52] Use of process analyzers in fossil fuel-fired power plants Solutions from Siemens: Power Industry—Siemens: Write up.
[53] Conductivity Sensor: Emerson Process Management: Product Data Sheet.
[54] A measure of success in the fight against corrosion: N. Kirloskar, Forbes Marshall: Power Engineering (PEI).
[55] Advantage series Two wire two electrode cond. T_x: ABB limited: Conductivity and pH.
[56] After Cation Conductivity (ACC) and Degassed Cation Conductivity (DCC) analyzer: Luchan Enterprises Co., Ltd.: Catalog.
[57] Alternatives to Hydrazine in Water Treatment at Thermal Power Plants: S. Tsubakizaki, M. Takada, H. Gotou, K. Mawatari, N. Ishihara, and R. Kai: Mitsubishi Heavy Industries Technical Review Vol. 46 2 (June 2009).
[58] An Introduction to steam boiler and steam raising: NEM Business Solutions (Internet document): cip.ukcenter.com.
[59] Analytical Solutions: Honeywell: System Writeup.
[60] Automatic Blowdown control Medium and high Pressure blowdown system: Forbes Marshall India: Application Note.
[61] Boiler Feedwater/Condensate Chemistry Control: KANUPP-IAEA Training.
[62] Boiler Steam and Water Sampling System: LECOL.
[63] Characteristics of Boiler Feed Water: Lenntech: Application Note.
[64] Chloride Analyzer: Sherwood UK:926.
[65] Clean Coal Technologies: K Ravikumar Director: Bharat Heavy Electricals Ltd.: 9th Assochem Energy Summit.
[66] Conductivity in Power Plants: Kemtron: Application Note.
[67] Conductivity Instrument: Polymetron: Product data sheet.
[68] Controlling Chemistry During Start Up and Commissioning of Once-Through Supercritical Boilers: K. Kirschenheiter, M. Chuk, C. Layman, and K. Sinha-Bechtel Corporation: December 2008.
[69] Dissolve Oxygen Analyzer: ABB Limited:4600.
[70] Dissolve Oxygen Analyzer: HACH Ultra: Hach 9182.
[71] DO Sensor: Rosemount Analytical: 499A.
[72] Dual Timer Automatic Boiler Blowdown Control System: Rosemount Analytical: Boiler Boss 1000.
[73] Electrochemistry Theory and Practice: Dr. A. Bier: Hach Ultra.
[74] Enhanced Phosphate Treatment for Drum Recirculating Boilers: J. Stodola (Ontario hydro), M.D. Silbert Marvin Silbert and Associates.
[75] First for Liquid Analysis:ABB Limited:Industrial Liquid Process Analyzer.
[76] Handbook of Industrial Water Treatment: GE Power and Water: Water and Process Technologies.
[77] Hydrazine Analyzer:Hach Ultra Switzerland: 9186.
[78] Handbook of Industrial Water Treatment:GE Power and Water: Water and Process Technologies.
[79] Hydrazine Analyzer: Hach Ultra Switzerland: 9186.
[80] Importance of Good Boiler Feedwater Treatment: Vogt Power International: Technical Bulletin (HRSG).
[81] Install an Automatic Blowdown Control System: Energy Efficiency and Renewable Energy: DOE United States.
[82] Instrumentation for Monitoring and Control of cycle chemistry for the steam-water circuits of fossil-fired and combined-cycle power plants: The International Association for the Properties of Water and Steam: Technical Guidance Document.
[83] Investigating a Proven Alternative to Phosphate Treatment for Low-Pressure HRSG Units Operating with Demineralized Water: G.J. McGiffney, BetzDearborn: GE Energy.
[84] Lecol Boiler SWAS: Luchan Enterprises Co., Ltd.: Catalog.
[85] Modern Power Plant Practices Vol F: Book—British Electricity International: Elsevier.
[86] Monitor for Fluoride, Ammonia, Nitrate and High Level Chloride: Honeywell Online Water Chemistry: 8230 Series.
[87] Online Hydrazine Analyzer: Emerson Process Management: CFA3018.
[88] Online Hydrazine Analyzer: Emerson Process Management: Product Data Sheet.
[89] Online Water Chemistry Measurement for Power Plants: Honeywell Online Water Chemistry.
[90] Online Water Quality Parameters as Indicators of Distribution System Contamination: J. Hall, A.D. Zaffiro, R.B. Marx, P.C Kefauver, E. Radhakrishnan, R.C. Haught, and J.G. Hermann.
[91] Optical DO Measurement Principles: Forbes Marshall India: VISIFERM Sensor.
[92] Optical DO Measurement Sensor: Polymetron 9123: Catalog.
[93] pH/ORP Sensors for Process Monitoring: ABB Limited: TB(X) 5 Series.
[94] pH Calculated by Differential Conductivity: Polymetron 9123: Catalog.
[95] Phosphate Analyzer: ABB Limited: Navigator 601.
[96] Phosphate Analyzer: ABB Limited: Navigator 601: Hach.
[97] Principles and Guidelines of pH Measurement: S. Rupert: Intech March 2005.
[98] Sample Sequencer:Sentry Equipment Corp. United States: Catalog.
[99] Silica Analyzer: ABB Limited: Navigator 600.
[100] Silica Analyzer: Hach: Catalog 60,000-18.
[101] Silica Analyzer: Polymetron: Product Data Sheet.
[102] Silica Monitor: ABB Limited: 8240.
[103] Single Channel Sodium Analyzer: Polymetron.
[104] Sodium Analyzer: Hach: 9245.
[105] Sodium Monitoring in the Water and Steam Cycle of Power Plants: E. Basset, S. Conan, P. Dudouit, P. Guillou, E.L. Hostis, R. Qarbi, G. Stehle: Hach Ultra Analytic SA Switzerland.
[106] Sodium Monitoring on Power Plant: ABB Limited: Tech. Note.
[107] Specification for SWAS: http://specswas.blogspot.in/: Nov. 30, 2009.
[108] Standard Design Criteria/Guideline for Balance of plant (2×500 MW TPP): Central Electricity Authority: New Delhi 2010.
[109] Steam and Water Analysis Brochure: Emerson Process Management: Writeup (Sentry United States).
[110] Steam and Water Analysis System Overview: Forbes Marshall India: System Write Up.
[111] Steam and Water Sampling: Sentry Equipment Corp. United States.
[112] Steam Generation in Power Plants: Hach Ultra: Application Note Power No. 12.

[113] Theory and Practices of pH Measurement: Emerson Process Management: December 2010.
[114] Total Chlorine Analysis System: Emerson Process Management: TCL Prod. Data Sheet.
[115] Two Wire Transmitters for pH, Conductivity, Oxygen and Chloride: Emerson Process Management: Product Data Sheet.
[116] Water Chemistry Aspects in Supercritical Unit: V Chandrashekharan: en. Sipat NTPC India.
[117] Water Quality—Cooling Tower Make-up Water Quality: tom baker: en.allexpert.com.
[118] Water Treatment: Honeywell Analytical Solutions.
[119] Water Treatment Hand Book by Degrimont.
[120] A Steam Generator for 700C TO 760C Advanced Ultrasupercritical Design and Plant Arrangement: What Stays the Same and What Needs to Change; P.S. Weitzel; Babcock and Wilcox Barberton, Ohio, United States; The Seventh International Conference on Advances in Materials Technology for Fossil Power Plants; October 2013; http://www.babcock.com/products/media/3d7d9be9f78e4b6189355a80b7ed499f.ashx.
[121] Plant Flow measurement and control handbook: Swapan Basu; Elsevier; July 2018; https://www.elsevier.com/books/plant-flow-measurement-and-control-handbook/basu/978-0-12-812437-6.
[122] Boiler Feed Water Oxygenated Treatment in Power Plants in China; Z. Li, W. Huang, S. Cao, and H. Zhang; PPChem; 2014; Waesseri GMBH; http://www.ppchem.net/errata/PPChem_2014_16_05_294-304_new.pdf.
[123] Plant Flow Measurement and Control Handbook; Swapan Basu; Elsevier; https://www.elsevier.com/books/plant-flow-measurement-and-control-handbook/basu/978-0-12-812437-6.
[124] Environmental, Health, and Safety Guidelines—Thermal Power Plant; World Bank Group; International Finance Corporation; December 2008 https://www.ifc.org/wps/wcm/connect/9a362534-bd1b-4f3a-9b42-a870e9b208a8/Thermal+Power+Guideline+2017+clean.pdf?MOD=AJPERES.
[125] Ammonia Probe 8002; Operating Instructions; ABB limited; IM/8002 Issue 11; https://library.e.abb.com/public/edc0e88a781be867c12574780028ed6f/IM_8002_11.pdf.
[126] WATER CHEMISTRY ASPECTS FOR SUPERCRITICAL UNIT; V Chandrasekharan; NTPC; Sipat India.
[127] Boiler Feed Water Oxygenated Treatment in Power Plants in China; Z. Li, W. huang, S. Cao, and H. Zhang; Power . PPChem 16(5); 2014; http://www.ppchem.net/errata/PPChem_2014_16_05_294-304_new.pdf.

Chapter 6

Final Control Element

Chapter Outline

1 VALVES AND ACTUATORS

1.1 Introduction

A control loop consists of process, sensor(s) for measurement, a controller subsystem and the final control element (FCE). Optimal command of process control depends on the performance of all control components including proper performance of the FCE. In almost all the cases the basic purpose of the FCE is to regulate flow (even in heating current flow). FCE are of various kinds of devices such as the damper, impeller, vane, blade pitch, VFD/coupling control (pump/blower/fan control), and control valve. Currently, there are a number of factors such as *high competition, low environmental impact, economic factors*, etc. that compel designers to develop control systems with a *high degree of performance, safety, and reliability*. By too often placing importance on the control loop, the FCE becomes a weak link. However, for overall better control performance, the FCE performance has to be good both in manual as well as in auto mode. There are several issues related to the control system, e.g., gain, deadband, system lag, etc. Out of these, gain is directly related with the FCE, viz. control valve. The FCE gain should be chosen critically and precisely (e.g., within a range of 2%–3%), otherwise, if it is higher then for a small change in the process, it will be multiplied by the gain to change the valve position. The performance is judged by the ability of the FCE to control process efficiently in response to the demand created by the controller, without any *backlash*, *deadband*, and *system lag*. In this definition, one important aspect is missed: *Gain*. The gain will be optimum so that the system can reach the stable position without much oscillation (due to overshoot/undershoot). Whenever the FCE is talked about, it basically consists of the actuator and the valve (or damper/impeller, etc.). While time based parameters like *backlash*, *deadband* and *system lag* are a contribution of the actuator, gain is the function of the control valve/damper, etc. From the discussion above, it is clear that control valves need to offer constant gain, which is a time independent proportional response to controller output, while the actuator is responsible for the time-based parameter. Ideally, the linear (first order) control system is considered, but in reality there will be deviation. Fig. 6.1 indicates various parts of the control valve.

1.2 Control Valve and Actuator: General

The most important issues related to control valves include

- control valve size
- trim characteristic
- MOC of various valve components
- selection of the correct actuator with or without a positioner
- proper accessories

Power Plant Instrumentation and Control Handbook. https://doi.org/10.1016/B978-0-12-819504-8.00006-8

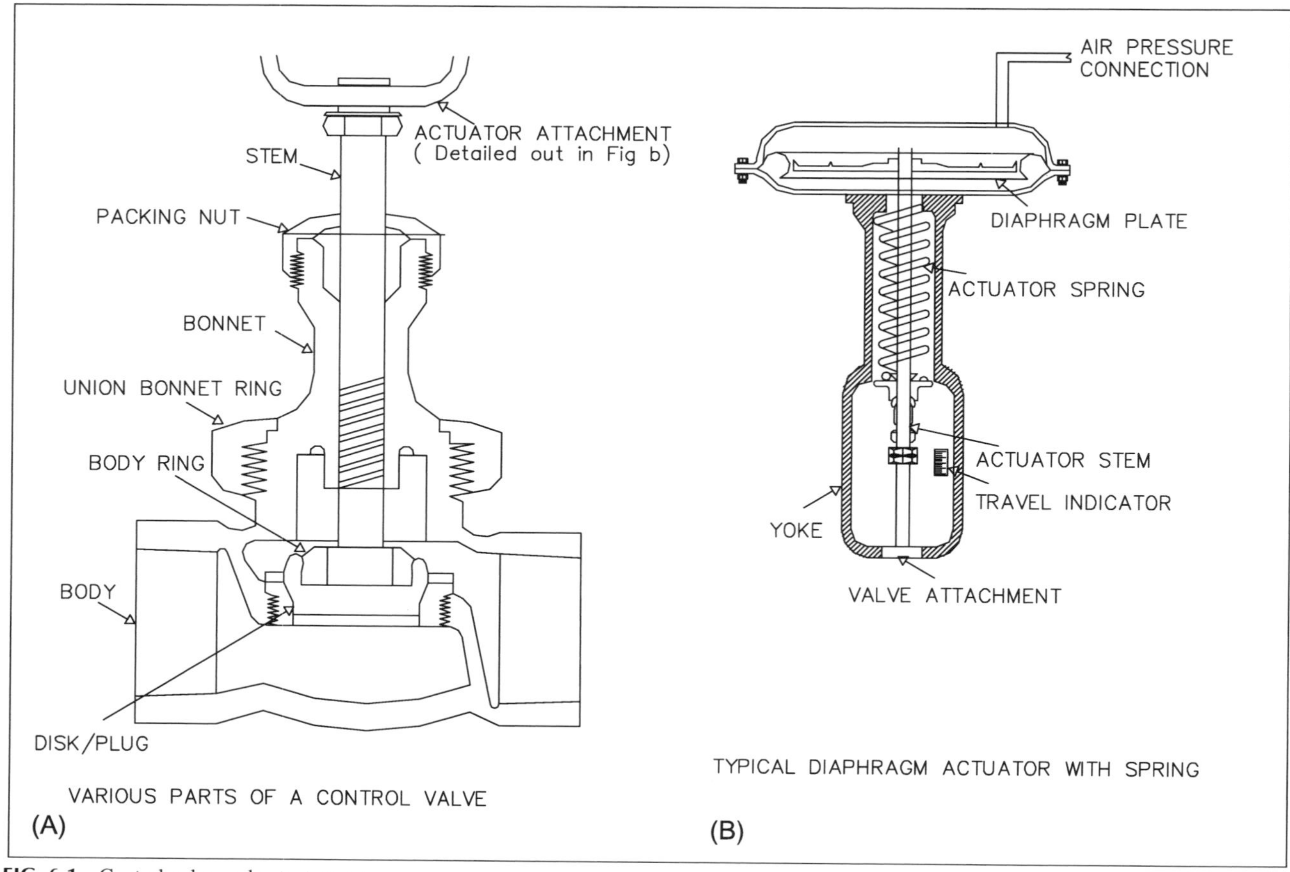

FIG. 6.1 Control valve and actuator parts.

Improper selection of the control valve *type* will result in the *wrong gain* for the system; improper *characteristics* will also result in *erroneous gain*, degrading the control loop performance. A properly selected actuator will result in very little backlash, deadband, etc. Often, it has been found that when the valve/FCE starts/stops, it does so with jerks. This is due to stiction, which is the resistance offered to start a motion. Because of stiction, the position of the air cylinder overshoots erratically and may move with jerks. This happens when the static friction *exceeds* dynamic friction. When there is stiction, there will be *overshoot and oscillation*. Valve fluctuation not only causes the poor performance, but may result in degradation of plant performance coupled with mechanical damage, wear and tear for the main equipment as well as for the valve itself. When a product is to be selected, then the total system should be taken into account so that it offers safe trouble-free performance. For further details, refer to "*Control Valve Technical Specification*" published by the ISA. Basically, all *control valves sizing* is based on *ISA S75.01 standard* (*IEC 534-2-1* and *543-2-2 standards* are also very close to ISA standard). Sizing catalogs published by manufacturers are also helpful. Here, a short discussion on the same has been presented. The terminology related to control valves and actuators is shown in Table 6.1.

1.3 Control Valve Sizing

ISA S20.50 provides the standard control valve datasheet, along with the necessary instructions to fill in the same properly.

The sizing and selection of the control valve is incomplete without any discussion on the control valve characteristics presented later in this chapter. Major issues related to control valve specifications mainly include: (1) Pr. class, trim characteristic, valve type, etc. for valve design; and (2) gas, oil, steam, and water along with service conditions for the process fluid condition. Symbols are preferred per ISA guidelines (Table 6.2).

In the table, various constants are defined with a pressure base of 101.3 kPa/1.013 bar and all pressures are in absolute

TABLE 6.1 Commonly Used Control Valve Terminology

Terminology	Definition	Remarks
Actuator	It is an externally powered device that supplies the motive force required for control valve movement. It can be electrical, hydraulic or pneumatic	
Bonnet	Part of the control valve that holds the packing box, and stem seal, and can guide the stem movement	
Bonnet extension	Bonnet with a greater dimension between the bonnet flange and packing box. Used in hot/cold service for better heat transfer	
Capacity	Rate of flow through a valve under stated conditions.	ISA
Control valve assembly	This includes all of the components mounted on the valve body that are required for its operation. Valve body assembly, actuator, positioner, air set, air lock relay, limit/torque switches, position transducer, etc. form a part of it	
Deadband	The range through which input can be varied upon reversal of direction without any observable change in the output signal, that is, the amount of controller output without observable change in the process variable	
Diaphragm pressure range	High and low pressure applied to the diaphragm to produce rated valve plug travel with atmospheric pressure in the valve body often referred to as *bench set.*	
Enthalpy	It is the sum of internal energy and the product of volume and associated pressure	
Entropy	In a thermodynamic process, it is the measure of energy that cannot be transformed into mechanical work	
Flow characteristic	The relation between the fluid flow through the valve and the percent of rate travel (0%–100%)	
Flow characteristic inherent	The flow characteristic when constant DP is maintained across the control valve	
Flow characteristics EQ percentage	Inherent flow characteristics of the valve that cause EQ percentage change in flow in the existing flow for equal change in travel	See Cl.1.3.1
Flow characteristics linear	An inherent flow characteristic that can be represented ideally by a straight line in a rectangular plot of flow vs. percent travel	See Cl.1.3.1
Flow characteristics quick opening	An inherent flow characteristic where there is maximum flow for minimum percent travel	See Cl.1.3.1
Hysteresis	The maximum difference in output for any single input, while input is increasing and decreasing in a calibration cycle	
Installed valve gain	The ratio of change flow through the valve to the change in percent travel of the valve under actual process conditions	
London pressure	It is the amount of pressure required to position a pneumatic actuator	
Process gain	The ratio of the change in the controlled process variable to the corresponding changes in the controller output	Cont. Valve
Maximum flow rate	This flow condition is consistent with plant or equipment operational maximum flow condition. The maximum flow condition is generally the governing case for the required maximum C_v	ISA Guide PP 486–487
Minimum flow rate	This flow rate is consistent with the plant turndown requirement or equipment turndown capability. The minimum flow condition generally subjects the control valve to the highest differential pressure condition. The minimum flow conditions are generally the governing case for the required trim performance	
Normal flow rate	This flow condition generally referred to as the design flow or material balance flow	
Rangeability	Ratio of maximum to minimum flow where inherent flow characteristics do not change the stated limit	

Continued

TABLE 6.1 Commonly Used Control Valve Terminology—cont'd

Terminology	Definition	Remarks
Rated C_v	Value of flow capacity at full open position	
Seat	The area of contact between the closure member and the associated mating surface that would cause complete shutoff of the valve. It is related with valve leakage. The net *contact force* determines the *seat load*	
Time constant	The time parameter is defined as the time interval between the first detectable change and the output reaches 63% of the final steady state value due to a step change in the input	
Valve cage	A part of a valve trim that surrounds the closure member	
Valve plug	The whole closure assembly often referred to as a valve plug	
Valve port	The orifice of the control valve that controls the flow through it. (There may be a difference in size between the body size and the port size)	
Valve trim	The internal component of a valve that modulates the flow of the controlled fluid. In a globe control valve, this should include, a seat ring (sitting surface for closure member), cage, stem along with the closure member	ISA

(Abs) pressure. A few factors such as F_L, x_T, C_d (defined in clause 1.2.1.4), etc. are available in manufacturer's table for representative sizing coefficient (guideline available in ISA standard). Valve sizing basis is mainly as per guidelines from ISA S75.01 and IEC534-2-1 and 2 are followed.

1.3.1. Sizing for Liquid Services (Table 6.2 May Be Referred to for Symbols)

1. Sizing formula for volume flow: $C_v = q/N\,F_p(\Delta P/G_f)^{1/2}$, where $N = 0.0865$ (when q in m^3/h, p in KPa); $N = 0.865$ when q in m^3/h, p in bar); and $N = 1$ if all in FPS.
2. Sizing formula mass flow $C_v = w/N\,F_p(\Delta P \times \gamma)^{1/2}$, where $N = 2.73$ (when w in kg/h, p in kPa and γ in kg/m^3); $N = 27.3$ (when w in kg/h, p in bar and γ in kg/m^3)and $N = 63.3$ if all in FPS.
3. Based on above C_v the valve size is selected from the manufacturer's table against a particular valve design selected for the service. Then, F_p is determined to finally get an accurate C_v value. Pipe geometry F_p can be calculated as per the formula below if it is not available from manufacturer's data:

$$F_p = \left[1 + \sum K/N\left(C_v/d^2\right)^2\right]^{-1/2}$$

where $N = 0.00214$ if d/D (mm); $N = 890$ if d or D (in.)

TABLE 6.2 Legends for Commonly Used Control Valves Symbols (Selected)

Sym	Meaning	Sym	Meaning	Sym	Meaning
C_v	Valve flow coefficient	G_g	Gas specific gravity	M	Molecular wt.
d	Nominal valve size	P_1	Up stream Abs static Pr.	N	Numerical const. for unit conversions
D	Pipe ID	P_2	Downstream Abs static Pr.	q	Vol. flow q_{max} choked flow at upstream Pr.
F_F	Liquid critical ratio factor	P_c	Abs critical pressure	T_1	Temperature in K
F_k	Ratio of Sp heat *factors*	P_v	Vap Pr.-Abs	w	Mass flow rate
F_L	Rated Pr. recovery factor	ΔP	Pressure drop ($P_1 - P_2$)	x	Ratio $\Delta P/P_1$
F_{Lp}	Combined Pr. recovery factor and geometry factor with fittings attached	ΔP_{max}	Max allowable Pr. drop (liquid) $\Delta P_{max\ LP}$ same but valve with fittings	x_T/ x_{TP}	Rated pressure drop factor without attachment. If there is attachment, then it is X_{TP}
F_p	Pipe geometry factor	γ_1	Sp. wt. at inlet condition	Y	Expansion factor
G_f	Liquid specific gravity	k	Ratio of Sp. heat	Z	Compressibility factor

$$\sum K = K_1 + K_2 + KB_1 - KB_2$$

where K_1 and K_2 are resistance coefficient of upstream and downstream fittings. KB_1 and KB_2 are Burnoullie's coefficient for the inlet and outlet. If the inlet and outlet pipe size of the control valve is the same, then KB_1/KB_2 [$KB = 1 - (d/D)^4$] will be equal, hence, may be omitted.

For inlet reducer $K_1 = 0.5[1 - (d^2/D^2)]^2$

For outlet reducer $K_2 = [1 - (d^2/D^2)]^2$

4. To check the choke flow condition: Max. flow with allowable drop is determined by $q_{max} = N \times F_L \times C_v \times [(P_1 - F_F P_v)/G_f]^{1/2}$ (where $N = 0.865$ when q in m^3/h, p in bar and $N = 1$ if all in FPS) and $F_F = 0.96\text{–}0.28(P_v/P_c)^{1/2}$. If there are attached fittings, then FL should be replaced with F_{LP}/F_P where $F_{LP} = [\{(K_1 + K_{B1})/N\}\{C_v^2/d^4\} + (1/F_L)^2]^{-1/2}$ when $N = 0.00214$ if d/D (mm) $N = 890$ if d/D in.

Note: The liquid pressure recovery factor F_L is a function of the ratio between the valve pressure drop and pressure differential between the inlet and pressure at vena contracta.

$F_L = [(\Delta P)/(P_1 - P_{vc})]^{1/2}$ values of different FL for various valve designs are available in the manufacturer's table for the particular design (ISA standard for valve sizing).

Allowable $\Delta P_{max} = F_L^2(P_1 - F_F \times P_v)$ $F_F = 0.96\text{–}0.28(P_v/P_c)^{1/2}$ if $\Delta P_{max} < \Delta P$, then there is a choke flow condition. So, C_v is to be recalculated by replacing ΔP with ΔP_{max} to avoid cavitation (Fig. 6.2).

1.3.2. Cavitation ($P_2 > P_v$): Related to the Pressure Recovery Factor for a Control Valve in Liquid Services

1. While controlling the flow, a control valve creates a pressure drop in the fluid. As the pressure drops, the velocity head increases due to restriction. In the flow path, at a point the fluid reaches maximum velocity, its pressure is minimum. This point is called vena contracta where pressure is minimum with the highest velocity. If this minimum static pressure goes below *liquid vapor pressure, then at the given temperature*, liquid bubbles will be formed. Past the vena contracta, the fluid decelerates and pressure recovery takes place as shown in Fig. 6.3. As the static pressure recovers and goes above the liquid vapor pressure, the bubbles will immediately collapse back into liquid. This phenomenon is known as *cavitation*. The collapse of bubbles produces a high-energy implosion that creates a high-impact force and metal fractures, which causes erosion.

In a globe valve, $P_1 - P_{vc}$ is about 125% of ΔP if P_{vc} (Fig. 6.3) $> P_v$ (vap pr.) this will not occur.

2. The major negative effects of cavitation include: Restriction of fluid flow, generation of high noise, erosion/pitting of metal surfaces, vibration/hammering, and damage to the plug/cage and guide.
3. The effect of cavitation is quite destructive, even on the appearance. The time required for such destruction

Relative Valve capacity : $C_d = C_v/d^2$ is useful factor in establishing the effect of geometry of different valve design which affects the pressure loss for a given flow. Thus it indicates pressure recovery in different style and design of valves. Larger the C_d value higher will be Pr. Recovery , naturally more possibility for choke flow.

FIG. 6.2 Relative valve capacity factor.

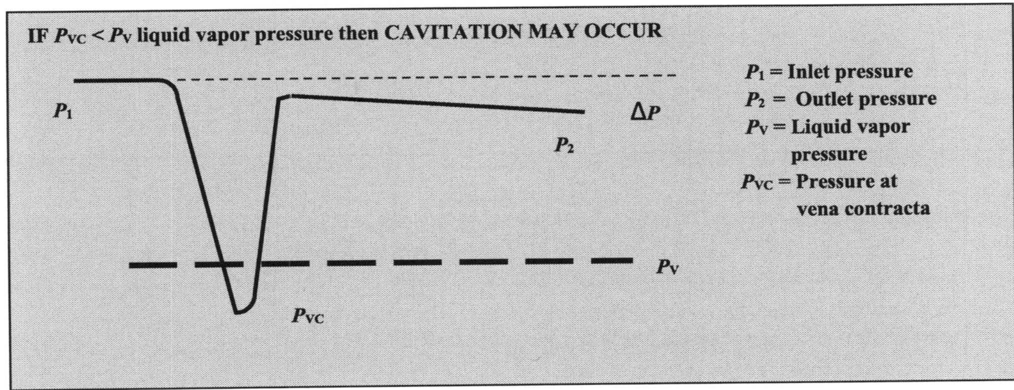

FIG. 6.3 Pressure profile in a control valve.

depends on the severity of cavitation as well as on the material used. Because inlet pressure is proportional to the energy available for cavitation, higher inlet pressure will cause more damage because of cavitation. When the calculated FL is more than the rated FL for the valve, the cavitation effect will be greater. Some of the special design trim such as Self-Drag®, Flash-Flo®, etc. that are available in the market are meant to fight cavitation.

4. All anticavitation trim design valves are designed and tested with water as the medium, but may not be the same for different liquids used in the industry. Because water has defined vapor pressure and collapses instantaneously, the effect of cavitation in water is more severe when compared with hydrocarbons. However, in the case of a power plant, such problems are mainly in the water application.
5. Choked flow: The cavitation restricts the flow. The vapor bubbles formed at vena contracta try to restrict the flow and increase until the flow is completely choked. Normally, when ΔP is increased, flow increases. Whereas, in choked flow condition, liquid flow will not increase even if ΔP is increased. Until the point that cavitation starts, $(\Delta P)^{1/2}$ is proportional to the flow. From the beginning of cavitation, with an increase of $(\Delta P)^{1/2}$, the flow curve starts bending, and at a choked flow condition, it is completely horizontal. Because it is almost impossible to define the point of damage, the critical pressure drop factor FL (FLP) is used to rate the trim type.
6. Cavitation avoidance: A few suggested methods to eliminate cavitation:
   - **a.** Materials: Use of *hardened* materials can avoid erosion due to cavitation. Some of the materials like *Stellite* 6B, *Stellite* 6 have high anticavitation resistance, and excellent corrosion resistance materials, but are rather costly. Whereas, *stainless steel 410 and stainless steel type 17-4 PH* have good anticavitation resistance with excellent corrosion resistance at a moderate cost.
   - **b.** Revised process conditions: (1) reduce operating temperature (avoid vapor formation) and (2) install the control valve with a chance of cavitation at the lowest point with low ΔP. The control valve is to be placed as close as possible to the pump.
   - **c.** Valve selection: In case of cavitation, valves may be selected that are designed to have *less* pressure recovery factor, that is, to have required F_L less than rated F_L by proper selection of valve. Also, trim can be designed to cause pressure drop in stages.
   - **d.** Installation: A number of control valves in a series.
   - **e.** Choked flow *reduced trim design* is recommended according to the ISA technical guide.

1.3.3. Sizing for Compressible Fluid Services (Table 6.2 May Be Referred to for Symbols)

1. Sizing formula for *volume flow with specific gravity of gas*:

$C_v = q/C_1 \times (x/G_g \times T_1 \times Z)^{1/2}$ where $C_1 = N \times F_p \times P_1 \times Y$.

N (for normal condition of $T_N = 0°C$) = 3.94 (with q in m^3/h P in kPa) OR = 394 (with q in m^3/h P in bar) when temperature in K.

N (for STD condition of $T_S = 15.5°C$) = 4.17 (with q in m^3/h P in kPa) OR = 417 (with q in m^3/h P in bar) when temperature in K.

N (for STD condition of $T_N = 60°F$) = 1360 (with q in scfh P in psia) temperature in R.

The expansion factor Y is important for compressible fluid valve sizing and the Y factor is affected by (1) pressure drop ratio x, (2) internal geometry of the valve, (3) Reynolds number, and (4) factor k.

The ratio of the port area to the body area, x_T, is responsible for this, valve geometry, and pressure drop ratio.

$Y = 1 - (x/3 \times F_k \times x_T)$ where $F_k = k/1.4$ here x_T is the pressure drop ratio to produce critical flow (with $F_k = 1$). Critical pressure drop occurs when $x > F_k \times x_T$ (or x_{TP} see symbol table). At that condition: $Y = 1 - 1/3 = 0.667$.

It signifies that at constant P_1, the flow will not increase even if ΔP is increased, due to critical pressure conditions. This means it should be such that $Y > 0.667$.

2. Sizing formula for volume flow with the molecular weight of gas (compressible fluid).

$C_v = q/C_2 \times (x/M \times T_1 \times Z)^{1/2}$ where $C_2 = N \times F_p \times P_1 \times Y$

N (for Normal condition of $T_N = 0°C$) = 21.2 (with q in m^3/h P in kPa) OR = 2120 (with q in m^3/h P in bar) when temperature in K.

N (for STD condition of $T_S = 15.5°C$) = 22.4 (with q in m^3/h P in kPa) OR = 2240 (with q in m^3/h P in bar) when temperature in K.

N (for STD condition of $T_N = 60°F = 7320$ (with q in scfh P in psia) temperature in R.

Y factor is the same as what was discussed above in 1.2.2.1.

3. Sizing formula for mass flow with molecular weight (compressible fluid)

$C_v = w/C_3 \times (x \times M/T_1 \times Z)^{1/2}$ where $C_3 = N \times F_p \times P_1 \times Y$

$N = 0.948$ (with w in kg/h P in kPa) OR = 94.8 (with w in kg/h P in bar) when temperature in K.

$N = 19.3$ (with q in lb/h P in psia) temperature in R.

Y factor should be the same as discussed above in 1.2.2.1.

4. Sizing formula for mass flow with specified weight γ_1 of gas:

$C_v = w/C_4 \times (x \times P_1 \times \gamma_1)^{1/2}$ where $C_4 = N \times F_p \times Y$

where $N = 2.73$ (when w in kg/h, p in kPa and γ in kg/m^3); $N = 27.3$ (when w in kg/h, p in bar and γ in kg/m^3) and $N = 63.3$ if all in FPS.

Y factor should be the same discussed above in 1.2.2.1.

5. Determination of X_{TP}: If the control valve is with the attached fittings, then this factor can be determined from $X_{TP} = X_T/FP_2[1 + (X_T \times K/N)(C_v^2/d^4)]^{-1}$ $K = K_1 + KB_1$ where $N = 0.00241$ when d or D in mm, or $N = 1000$ when d or D in in. This is determined to accurately calculate C_v.

1.3.4. Flashing liquid: ($P_2 < P_v$). Whenever the downstream pressure of the control valves becomes equal or below the vapor pressure of the liquid, then the liquid will flash, that is, dissipates into vapor. If $P_2 < P_v$. This is somewhat similar to cavitation but not exactly the same, because here, downstream pressure will not recover and the bubbles will not collapse as in cavitation. However, this phase change may result in high velocity, which causes erosion of the metal. While sizing the valve for flashing application, it is necessary to size the valve at choked flow condition. The downstream pipe should be large enough to accommodate additional volume. It is necessary to calculate vapor fraction; the velocity of this vaporized mixture (ISA control valve sizing standard). Due to erosion, the thickness of the body will be reduced (<acceptable code) so the body material will be suitably chosen. *Chrome molybdenum* (*WC6*) steel may be used; the body may be *angular*. *Cage control trim/multiple orifices* may be used. In flashing services with $\Delta P < 3.5$ kg/cm^2, *hardened trims* may suffice. In power plants, while sizing the steam drain valves and the heater drain valves (especially draining to condenser), due consideration is given to this approach. Reduced trim design could be a better option (Fig. 6.4).

1.3.5. After the C_v calculation, find the body size by matching the calculated C_v with the characteristic curve of the valve to see that the calculated C_v meets the criterion set by ISA or better. As per ISA technical specification, trim capacity will meet the following:

1. An equal percent trim will open the maximum 95% travel to meet the maximum flow.
2. A linear/quick opening should open 90% travel to meet the maximum flow.
3. In general, designs are done conservatively, so that later, a small capacity increase does not necessitate change of the control valve to accommodate the flow (e.g., max. flow at 85% opening).

1.3.6. Reduced trim applications: Per ISA guidelines, control valves with reduced trim are considered for the following cases:

- Pressure drop >5170 kPa
- Choked flow
- Flashing flow >5% wt. of liquid
- Future capacity increase
- Erosive fluid
- Gas flow with outlet velocity 0.3 mach (99 M/S)
- Noise >85 dBA (measured 1 m from source)

1.4 Control Valve Characteristics

1. The relationship between the control valve capacity and valve travel is known as flow characteristics of the control valve. Control valve characteristics, rangeability, gain, and the turndown ratio all are interrelated. All of these "describe the personality of the valve," according to author B.G. Liptak. These are important in that all these parameters, in conjunction with different elements in a control loop, decide the performance of the entire loop.
2. One control loop with various elements has been shown in Fig. 6.5. Where K_a = actuator gain, K_c = controller gain K_p = process gain, K_s = sensor gain K_v = valve gain, the overall gain of the loop is $K = K_a \times K_c \times K_p \times K_s \times K_v$. As long as the overall gain is constant,

***ISA recommendation for "Cavitation, Choked Flow and Flashing*: ISA S75.01 -ISA guide and Seller's valve Cavitation index data shall be used for determination of severity of Cavitation, choked flow or Flashing condition. Following techniques for Cavitation resistant valves may be followed (REF; p.191 ISA GUIDE)**

- **"Reduce the pressure in multiple stage"**
- **"Direct flow away from valve body, pipe wall"**
- **"Break the flow in many small streams"**
- **"Force the flow through multiple turns or tortuous path"**

FIG. 6.4 ISA recommendation for flashing.

it is stable. Out of all of these, it is clear that if the controller gain is independent of the load, then as long as $K_p \times K_v$ is constant, the system is stable because the overall gain is constant. This means that changes in the process have a direct impact on the control valve gain (assuming the controller is not changing with load change—nonadaptive). This introduction establishes that the control valve gain is related to the control valve characteristics. Before any discussion on control valve characteristics, it is necessary to examine control valve gain. When the control *valve gain* varies at a *constant rate with load* (flow), then the valve gain is called *Equal Percentage*. If the control valve gain *drops with load/* flow then, it is a *quick-opening type*. If there is *no variation* of valve gain with flow/load (i.e., the load and gain is related in *constant gradient* relationship), then it is *linear*. If the valve gain shows another nonlinear characteristic with the load, then it may be identified as a square root, parabolic, hyperbolic, etc. as shown in Fig. 6.6A.

The valve gain is the slope of the characteristics of a curve, i.e $\Delta Q/\Delta T_r$. The valve gain is influenced by the valve size, pressure drop, valve characteristics, and by flow. In Fig. 6.6B, the pump characteristics along with the system resistance illustrate how the control valve needs to change its opening to provide the necessary system resistance. The pump discharge pressure drops with the increase in the load or flow. As the ΔP of valve increases with the flow, the valve gain drops with the load. This suggests that after installation, the valve gain changes. This is the installed valve gain. When the actual flow is plotted against the corresponding valve opening, the installed valve characteristics are obtained, which is quite different from the theoretical flow where constant ΔP has been considered in testing.

3. At the time of tuning of the controller, it is done at a nominal load of the loop with the assumption that the loop gain (except adaptive gain) will not change with the load. In practice, the loop gain does not change with

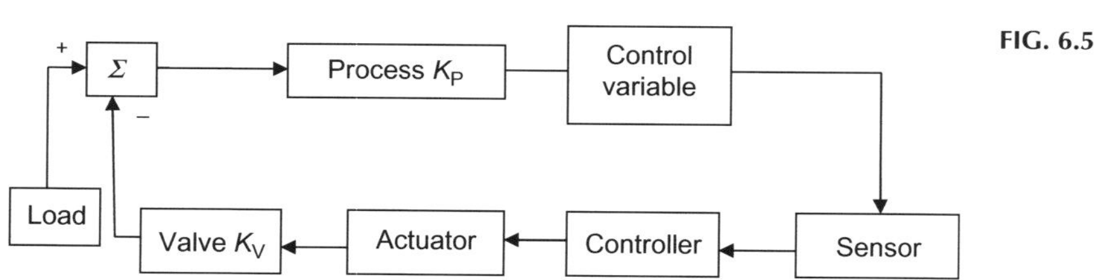

FIG. 6.5 Control loop.

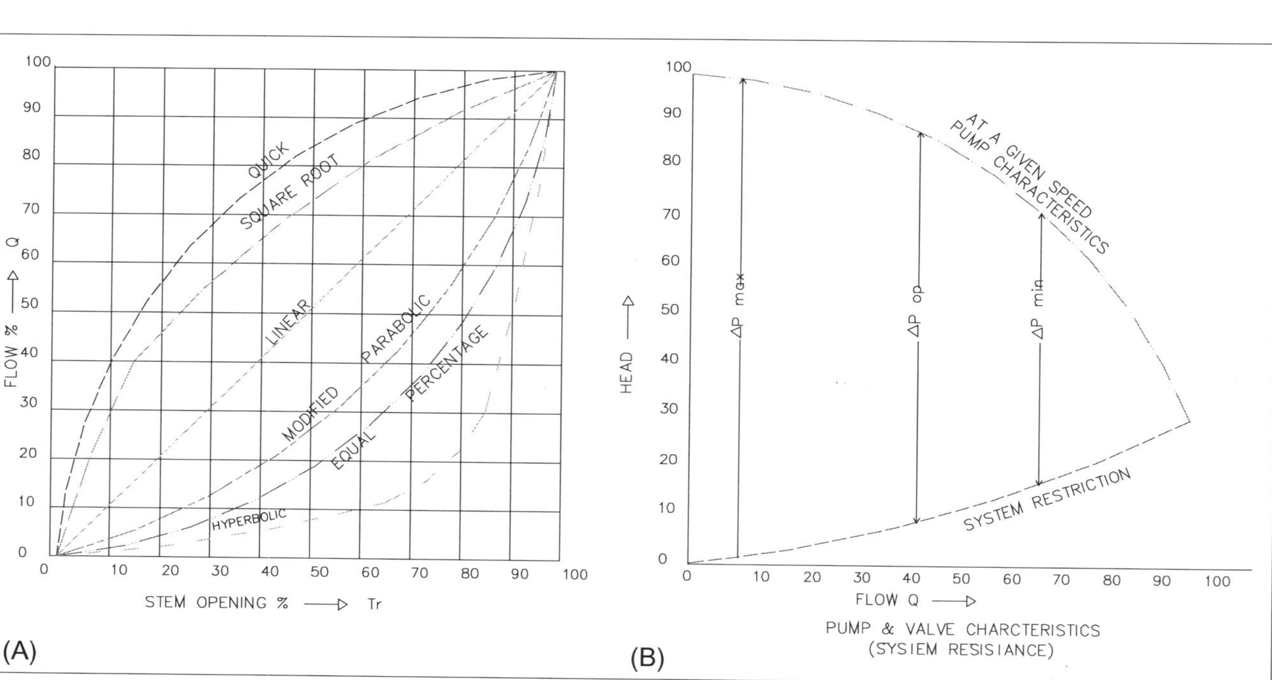

FIG. 6.6 Valve characteristics and system resistance.

the process load, so something must be done to compensate for the deviation from the theoretical, that is, the control valve characteristics are chosen to match the process type. Proper selection of the control valve for specific application is not easy and requires experience. There may be a number of schools of thought in selecting the valve characteristics for miscellaneous application, which may vary somewhat with respect to Tables 6.3–6.5. A rule of thumb for selection of the valve characteristic for various applications has been presented below in line with some standard suppliers Fig. 6.7.

TABLE 6.3 Control Valve For Level Application

ΔP	Load Condition	Characteristic
Constant	All Load	Linear
ΔP at Max load >20% of Min load ΔP	ΔP decreasing with load increasing	Linear
ΔP at Max load <20% of Min load ΔP	ΔP decreasing with load increasing	Equal percentage
ΔP at Max load >200% of Min load ΔP	ΔP Increasing with load increasing	Quick opening
ΔP at Max load <200% of Min load ΔP	ΔP Increasing with load increasing	Linear

TABLE 6.4 Control Valve in Flow Application (*Location of Control Valve and Flow Signal to the Controller*)

Flow Signal to the Controller	Control Valve Location	Recommended Characteristic	
		Wide Set Point Range	Small Flow Set Point Range[a]
Proportional to flow	Series or bypass	Linear	Equal percentage
Proportional to flow square	Series	Linear	Equal percentage
	Bypass	Equal percentage	Equal percentage

[a]*With large ΔP change with increasing load.*

TABLE 6.5 Control Valve For Pressure Application (*Process Fluid and Time Constant is Important*)

Fluid	Time Constant	Valve ΔP	Recommended Inherent Characteristic
LIQUID	Fast	Irrespective of ΔP	Equal percentage
LIQUID	Slow	$\Delta P_{Q\ max} <$ 0.2 $\Delta P_{Q\ min}$	Equal percentage
LIQUID	Slow	$\Delta P_{Q\ max} >$ 0.2 $\Delta P_{Q\ min}$	Linear
GAS	Fast	Irrespective of ΔP	Equal percentage
GAS	Slow	$\Delta P_{Q\ max} <$ 0.2 $\Delta P_{Q\ min}$	Equal percentage
GAS	Slow	$\Delta P_{Q\ max} >$ 0.2 $\Delta P_{Q\ min}$	Linear

Fast: *Short pipe length (liquid flow in pipe).* Slow: *Large volume (long distribution line, tank and receiver).*

a. *For Temperature Application*: Equal percentage characteristics, so far, is quite good for the temperature application having a higher time constant.

b. *Quick Opening Valves* are mainly used for ON/OFF service in a batch or semicontinuous process where there may be a requirement of large flow.

c. *Rangeability (R) and Turndown ratio (T)*: These two terms are similar.

Rangeabilty: = (max. controllable flow)/(min. controllable flow).

Turndown ratio = (normal max. flow)/(min. controllable flow). Generally, the normal max. flow is considered as 0.7 max. controllable flow (as in the flow element calculation, the average flow is considered 0.7 times max. flow), so $T = 0.7$ R.

4. *Small and large capacity valves–split range*: In a power plant, there are a few applications where it is necessary to use low capacity as well as high capacity (HC) control valves in tandem.

Auxiliary steam is another example. During the startup period, the flow demand of auxiliary steam is greater than during a normal run at a higher load. During the startup time, the valve has to face comparatively low DP with a higher demand of flow. On the other hand, when the unit is running normally, the demand for auxilary steam is low, but DP may be higher. For these reasons, low- and high-capacity valves

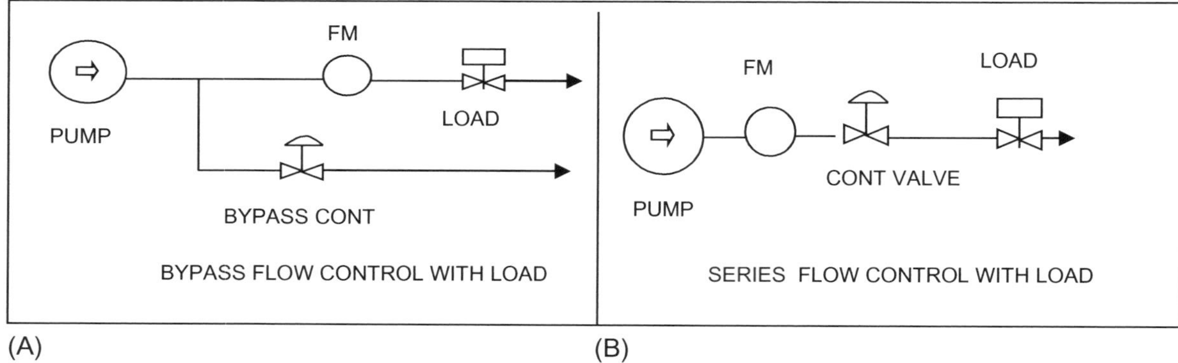

FIG. 6.7 Pump and valve configuration types.

are to be used. *How are the valves switched?* Fig. 6.8A explains that during the switch from large capacity to small capacity, there is a large fluctuation in case the switchover is abrupt. For this reason, sometimes the loop is shifted into manual, and the large capacity valve is slowly closed. The low capacity (LC) valve is opened. This can be done by using the split range arrangement as shown in Fig. 6.8B. When the LC valve is almost fully open, the HC valve is near the closing position. Suitable deadband between the two can also be done for better control. By using the floating control method, when the HC valve is slightly opened, the LC valve, which has a small capacity, closes to meet the demand. The LC valve may be closed by interlock after the HC valve achieves a certain position. The switching method could be done by electronic circuitry, which is easier to achieve, but it can also be achieved pneumatically. As shown in Fig. 6.8C, there is another kind of split range control that can be encountered in a power plant. In this system, one valve is a reverse-acting type, and the other is a direct-acting type to meet the same control strategy.

1.5 Seat Leakage Classification

1. Classifications are denoted by the quantity of fluid passing through a valve assembly when the valve is in the fully closed position with pressure differential, (static pressure) and temperature as specified. Control valves are classified according to ANSI B16.104. Control valves are designed for throttling services, yet they are expected to offer some type of shutoff.
2. The shutoff characteristic depends on many factors such as the valve type, the seat material, guiding, the actuator thrust, the type of fluid (the higher the flowability, the more chances for leakage), pressure drop across the valve, etc. A brand new control valve will conform to the leakage specification of the supplier, but after extended use the leakage value may change greatly. This could be due to the temperature gradient between the trim and body, or strain in pipe/valve etc. Also, when the valve operates at a temperature below the specified temperature, it would also cause leakage (Tables 6.6 and 6.7).
   a. Valve leakage classification: Class I–VI.
   b. Special name Class IV: Metal-to-metal kind of leakage that is expected from a metal plug and a metal seat.
   c. Special name Class VI: Soft seat where the plug or the seat, or both are made from soft composition material such as Teflon.

1.6 Actuator and its Sizing

Before going into much detail, a few fundamental things will be examined. How the valve orientation and the actuator orientation influence the actuator sizing is shown.

1. Basics of Actuator (Sizing): (Fig. 6.9) In a spring diaphragm actuator, any valve position is achieved by making the total force on the stem equal to zero. For the subsequent discussions, the following letters have been assigned to various variables such as: "A" = the area of the diaphragm, "P" = the air pressure "K" = the spring constant, "X" = the spring displacement, "ΔP" = the valve process pressure drop at the shutoff pressure. The general practice is to consider the "maximum upstream pressure" as ΔP unless it is ensured by a back pressure downstream, "A_v" = effective valve area (orifice). During the discussion, forces toward downward is arbitrarily chosen as positive.
   a. Fig. 6.9A explains that, in this case, the process flow comes from the *bottom* of the plug, and goes to the outlet from the *top*, that is, the process flow is *assisting* in the opening of the valve. Again, air supply is from the *top*, so air will *push* the plug to *close* the valve by compressing the spring in the actuator. It is clear that the force of air is acting in a positive direction, (direct acting) and it is *opposed* by the spring compression force, *and* force due to process flow:

$$P_a \times A - \Delta P \times A_v - K \times X = 0 \text{ or } P_a \times A = \Delta P \times A_v + K \times X \quad \text{(a)}$$

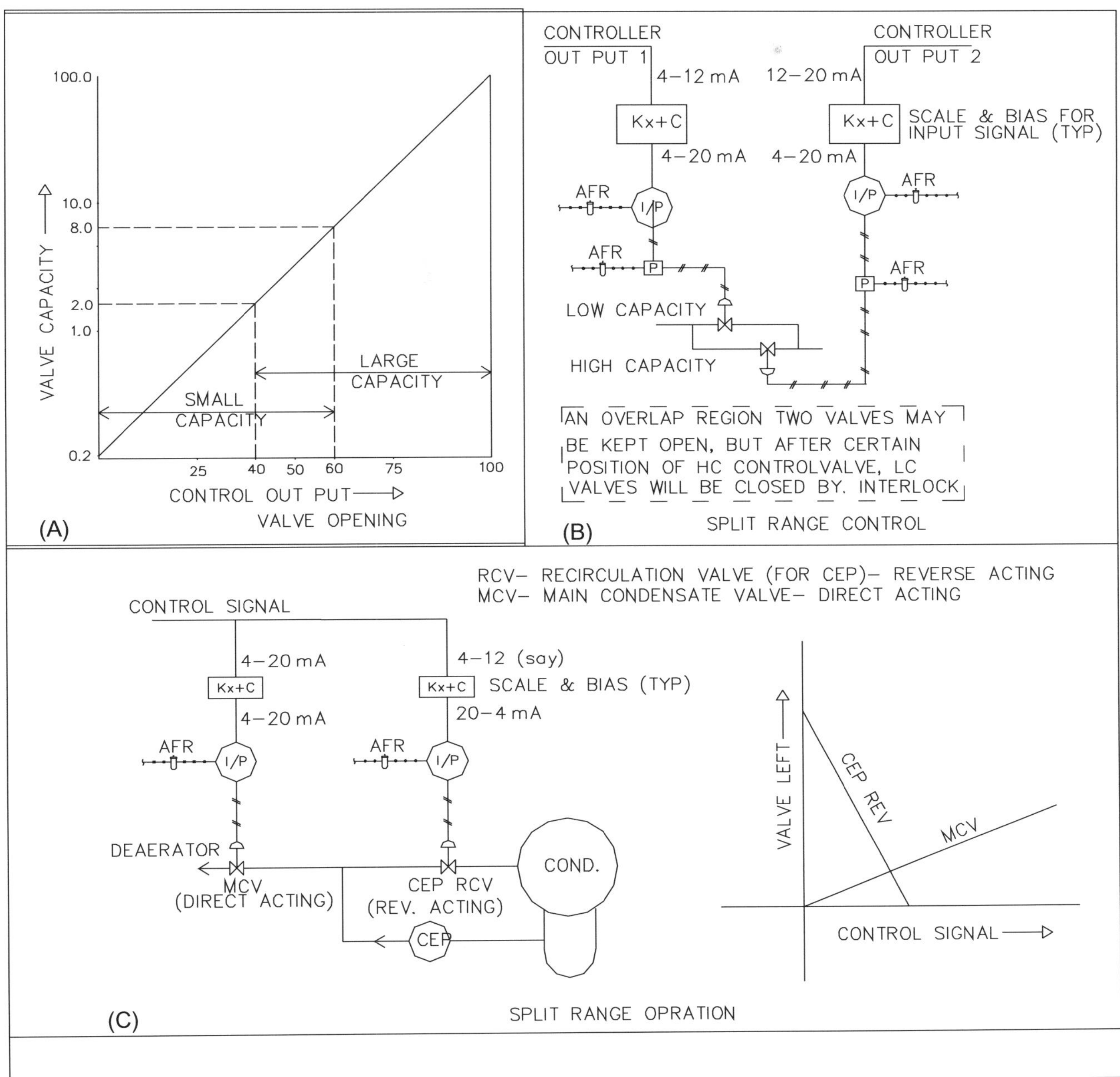

FIG. 6.8 Small and large capacity valves—split range.

So, the minimum force required to overcome is the *sum* of the two opposing forces. What happens if the orientation is changed?

b. Fig. 6.9B shows that, in this case, the process flow comes from the top of the plug, and goes to the outlet from the bottom, that is, the process flow is opposing the opening of valve. Again, air supply is from the bottom, so air will pull the plug to open the valve by compressing the spring in the actuator. The force of air is acting in a negative direction, (reverse acting), and it is opposed by a spring compression force, and a force due to process flow, so:

$$P_a \times A + \Delta P \times A_v + K \times X = 0 \text{ or } P_a \times A = \Delta P \times A_v + K \times X \quad \text{(b)}$$

Again, minimum force is required to overcome the *sum of the two* opposing forces. This is so, because, in this case, *both* the process flow and the actuator orientation has been changed.

TABLE 6.6 Control Valve Leakage Class (Leakage as a Percentage of Rated Full Capacity)

Class	Leakage%	Test Medium	Service ΔP/Temp.	Typical Construction
I		NO SHOP TEST		Same as Class II,III,IV
II	0.5	Air 45–60 psig	50 psid/3.4 bar in temp. range 50–125°F, whichever is lower	(1) Balanced single port, single graphite piston ring, metal seats, low seat load (2) Balanced double port, metal seats, high seat load
III	0.1	Same as above		(1) Balanced double port, soft seats, low seat load (2) Balanced single port, single graphite piston ring, lapped metal seat, medium seat load
IV	0.01	Same as above		(1) Balanced/unbalanced single seat, Teflon or multiple graphite piston ring, lapped metal seat, medium seat load.
V	$<5 \times 10^{-4}$ mL/min/in. orifice dia. per psi DP	Water at 100 psig or operating pressure	Service DP 50–125°F	(1) Unbalanced single port, soft/lapped metal seat, high seat load (2) Balanced single port, Teflon piston ring, soft seat, low seat load
VI	See below	Air/N_2 < 50 psig or operating pressure at 50 125°F		Unbalanced single port, low seat load

TABLE 6.7 Class VI Leakage Values

Port Diameter		Bubble Per Minute	mL Per Minute	Port Diameter		Bubble Per Minute	mL Per Minute
in.	mm			in.	mm		
1	25	1	0.15	4	102	11	1.7
1½	38	2	0.3	6	152	27	4
2	51	3	0.45	8	203	45	6.75
2½	64	4	0.6	10	254	63	9
3	76	6	0.9	12	305	81	11.5

c. Fig. 6.9C shows that, in this case, the process flow comes from the *bottom* of the plug, and goes to the outlet from the *top*, that is, the process flow is *assisting* in the opening of valve. Again, air supply is from the *bottom*, so the air will *pull* the plug to open the valve by compressing the spring in the actuator. Here, it is clear that the force of air acting is in a negative direction, (reverse acting) and it is *opposed* by the spring compression force, but *assisted* by the force due to the process flow, so:

$$P_a \times A - \Delta P \times A_v + K \times X = 0 \text{ or } P_a \times A = \Delta P \times A_v - K \times X \quad \text{(c)}$$

Here, the minimum force required is to overcome is the *difference* forces for process flow and spring compression hence will be less than that in cases in (a) and (b).

d. In Fig. 6.9D, it is found that, in this case, the process flow comes from the *top* of the plug, and goes to the outlet from the *bottom*, that is, the process flow is *opposing* the opening of the valve. Air supply is from top, so air will *push* the plug to *close* the valve by compressing the spring in the actuator. Here, it is clear that the force of air acting is in positive direction, (direct acting), and it is opposed by the spring compression force, and force due to the process flow, so:

$$P_a \times A + \Delta P \times A_v - K \times X = 0 \text{ or } P_a \times A = K \times X - \Delta P \times A_v \quad \text{(d)}$$

Here, the minimum force required is to overcome the *difference* of forces for spring compression and process flow,

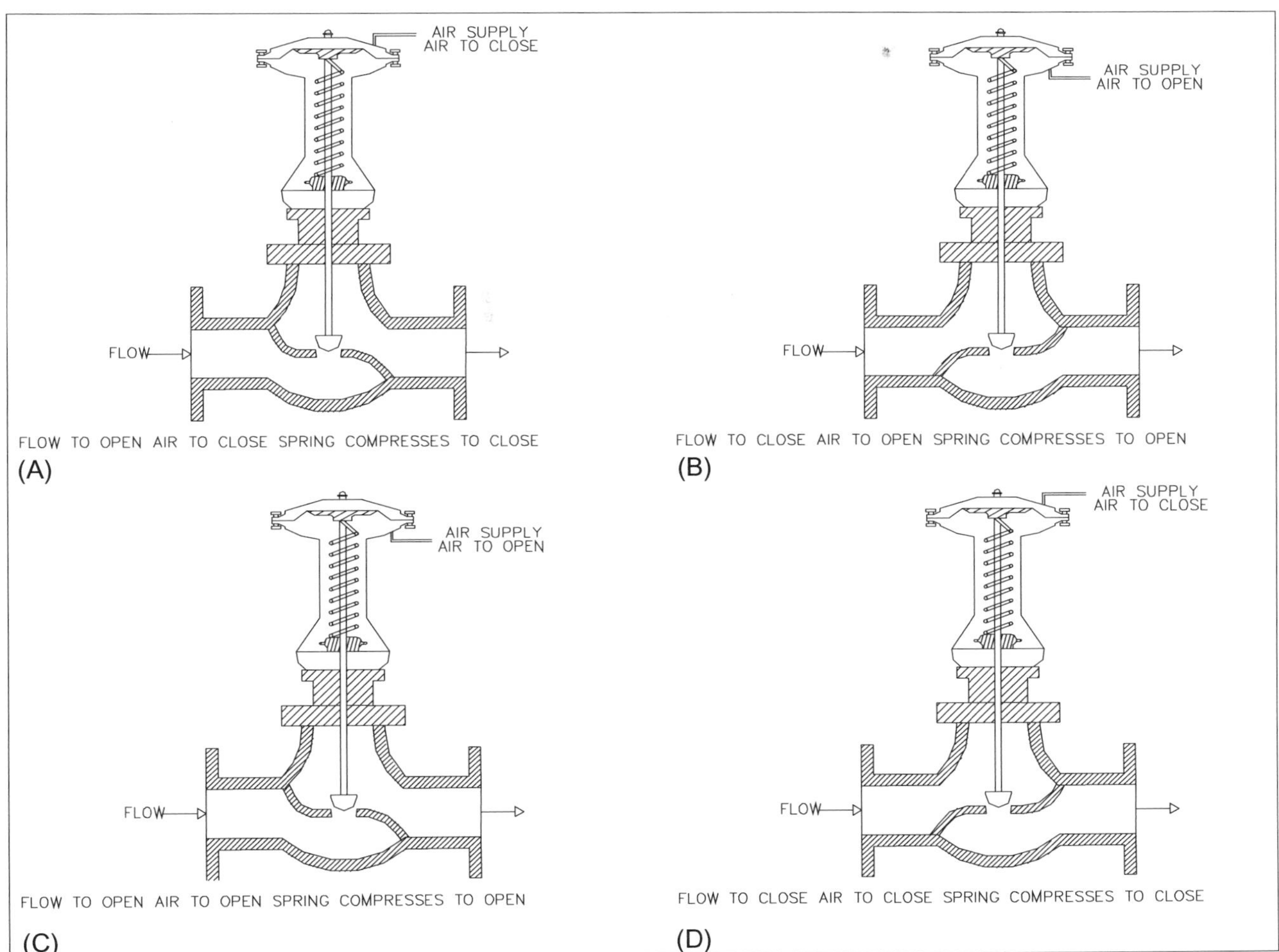

FIG. 6.9 Basics of actuators.

hence, it will be less than the minimum force required in equations (a) and (b). Based on these fundamentals, the actuator sizing has been discussed later.

1.7 Materials of Construction and Associated Tables for Pressure and Temperature Ratings

Data presented are mainly as per ASME B16.34.

1. In the following discussions, major control valves in the main power plants have been considered. Based on these guidelines, it will be possible to select the materials for control valves for other applications. A typical list of control valves in the main power plants is presented to enable the reader to have an idea on the subject. The materials shown are suggested as typical only for guidance.

The reader is advised to go through the necessary standard for pressure and temperature ratings and necessary derating at different conditions. See Table 6.8 for details of major control valves in power plant.

2. Valve body and trim materials composition with temperature limit: In Tables 6.9 and 6.10, a few materials that are common in power plant applications have been discussed. Consult a standard book on control valves for information about other materials that are available. Major compositions and temperature ratings that are discussed are per international standard.
3. Seat Materials: Some of the common seat materials are as follows with temperature limits (Table 6.11).
4. Valve/flange pressure temperature ratings: The pressure that each material (presented above) is capable of withstanding needs to be derated as per the operating temperature. This is done on the basis of pressure class selected for the valve as well as the operating temperature. For example, the working pressure for a valve (150 lb.) of ASTM A216-WCB at 200°F (93.33°C) is 260 psi (18.2798 kg/cm^2). Whereas with a valve of the same rating when used in a temperature of 600°F (315.5°C), the maximum working pressure would be 140 lb (9.8 kg/cm^2). This means that corresponding to each material

TABLE 6.8 List of Control Valves in a Power Plant With Suggested Materials (Quantity of Valves in Each Service Depends on the System Design)

Control Valve	Body Material	Trim Material (SS)
Lo/Hi range feed control valves (also bypass valve)	ASTM A 217GR C5	17-4 PH
SH or RH or other spray control valve	ASTM A 217GR C5	440C
HP or LP heater normal level control valve	ASTM A 217GR WC6	17-4 PH
HP or LP heater Hi/emergency level control valve	ASTM A 217GR C5	17-4 PH
BFP recirculation control valve	ASTM A 217GR C5	440C
CEP/GSC min flow control valve	ASTM A 217GR C5	17-4 PH
Aux steam or CRH to deaerator valve	ASTM A 216GR WCB	Type 316 Stellite
Normal or emergency makeup to condenser valve	SS Type 316	Type 316
Condensate flow to deaerator valve	ASTM A 216GR WCB	Type 316 Stellite
Deaerator drain to condenser valve	ASTM A 217GR C5	17-4 PH
Drain cooler level control valve	ASTM A 217GR WC6	17-4 PH
Steam Pr. valve for atomizing	ASTM A 216GR WCB	Type 316 Stellite
Control valves in PRDS for Aux. steam	ASTM A 216GR WCB	Type 316 Stellite
CBD tank level control valve	ASTM A 217GR C5	Type 316 Stellite
Condenser spill valve	ASTM A 217GR WC6	17-4 PH
SCAPH steam control valves	ASTM A 216GR WCB	17-4 PH
SCAPH drain valve	ASTM A 217GR C5	17-4 PH
Gland sealing supply valve	ASTM A 217GR WC6	17-4 PH
F.O heating and pressurizing valves	ASTM A 216GR WCB	Type 316

TABLE 6.9 Valve Body Material Composition (A Few Materials Commonly Used for Control Valves in a Power Plant) (in %) and Temperature Limit (in °C)

Casting	Forging	Other Major Composition (%)	Lo Temp Limit (°C)	Hi Temp Limit (°C)	Remarks
A216 Gr WCB	A105	C = 0.2–0.24; Cr = 0.2; Mn = 1.0–1.35; Mo = 0.06; Ni = 0.2; Si = 0.15–0.3	−29	535	Moderate temp and corrosion noncritical
A217 Gr C5	A182 F5	C = 0.15(M); Cr = 4.0–6.0; Mn = 0.3–0.6; Mo = 044–0.65; Ni = 0.5(M); Si = 0.5(M)	−29	590	Moderate corrosion
A217 Gr WC6	A182 F11-CL2	C = 0.1–0.15; Cr = 1.0–1.5; Mn = 0.3–0.8; Mo = 044–0.65	−29	535	Minimize graphitization
A217 Gr WC9	A182 F22-CL3	C = 0.15(M); Cr = 2.0–2.5; Mn = 0.3–0.6; Mo = 0.87–1.13; Si = 0.5	−29	565	Elevated temp and strength than F11

TABLE 6.10 Valve Trim Material Composition (in %) and Temperature Limit (in °C)

Type	Other Major Composition (%)	Lo Temp Limit (°C)	Hi Temp Limit (°C)	Remarks
316 SS—Casting: A351CF8M FORGING: A182 F316	C = 0.08; Cr = 18.0–21.0; Mn = 1.5 (M) Mo = 2.0–3.0 Ni = 9.0–12.0 Si = 1.5 (M)	−250	815 (316°C as trim)	Hi temp and corrosion resistance
316L SS—Casting: A351CF3M FORGING: A182 F316L	C = 0.035; Cr = 16.0–18.0; Mn = 2.0 (M) Mo = 2.0–3.0 Ni = 10.0–14.0 Si = 1.0 (M)	−250	454	
TYPE 17-4PH ASTM A564 Gr 630UNS S17400	C = 0.07(M); Cr = 15.0–17.5 Mn = 1.0 (M) Ni = 3.0–5.0 Si = 1.0 (M)	−29	343 (427°C as cage)	Less impact on loading
TYPE 440C	C = 0.95–1.2 Cr = 16.0–18.0 Mn = 1.0 Mo = 0.75 Si = 1.0(M)	−45	315 (427°C as cage)	

TABLE 6.11 Commonly Used Seat Materials with Temperature Limits

Materials	Lower Limit in °C	Upper Limit in °C
High density polyethylene (HDPE)	(−) 54	85
Neoprene	(−) 40	82
Nitrile rubber	(−) 29	93
Nylon	(−) 51	93
PTFE	(−) 268	230

specification and pressure rating (class of the valve), there would be a maximum pressure that the material can withstand at a particular temperature range. As stated above, there may be a necessity to derate to match the particular operating temperature. These can be found in ASME B16.34 Tables (in this case Table 2-1.1). It is therefore recommended to refer to ASME B16.34 (or relevant standard of the country for such ratings) or to contact flange/valve suppliers.

1.8 Special Considerations

There should be some special considerations for various applications:

1. High pressure service: For high pressure services, a forged body is used for greater strength. Erosion resistant materials should be selected for these services. A special seal and higher stem diameters are some of the other considerations.

2. High DP: Due to high DP, a high force may develop across the valve and internal parts to cause instability of the valve. Because of high DP, high velocity droplets may lead to erosion. For erosion, aberration, cavity "*SS 440C*" may be used. Also, in globe valves, there may be excessive stem thrust. For the "flow to close" type, high DP may damage the seat, whereas a "flow to open" type valve will open against the actuator force. For high DP, there may be sharp drop in outlet temperature, which may cause the material to become brittle, so due consideration should be given.

3. High temperature services: For these services where $T > 230°C$, the material temperature limitations discussed in 1.6 above must be taken into consideration. Also, temperature limitation due to packing needs should be considered. The packing material limitation: Teflon: 230°C, metallic packing: 480°C, graphite packing: 399°C.

1.9 Control Valve Noise

From a pollution standpoint, noise is the third greatest menace after air and water pollution. From human health condition standpoint, people are quite concerned about the same. Also noise is a kind of loss to the system. Valve noise is expressed in terms of dB. Valve noise, more precisely, aerodynamic noise, is defined in terms of sound pressure level (SPL). Sound pressure level in dB has been defined as follows:

$$\text{dB} = 20 \log_{10} (\text{measuring sound pressure level}/2 \times 10^{-4}\,\text{microbar}).$$

$= 20 \log_{10}$ (measuring sound pressure level$/2 \times 10^{-5}$ N/M^2).

1.9.1 Sources of Noise

There are three types of noise encountered in a control valve:

1. Mechanical vibration: Mechanical vibration is caused by the response of the valve's internal components to the turbulence through the valve. Turbulent strike and vortex shedding cause such vibrations. This could also blamed on improper design. This kind of reverberation indicates tonal frequency vibrations. Problems occur when these induced vibration frequencies match with the natural frequency of trim. There would be resonance. This can result in easy trim failure due to fatigue. Generally, guided valves have fewer chances to have such noise. Should it happen at all, they can be checked by changing the valve design or reducing the clearance between the guide and the plug, etc.
2. Aerodynamic noise: Aerodynamic noise arises as a result of the direct conversion of mechanical energy of flow into acoustic energy as the turbulent flow passes through the valve orifice. This noise is much more predominant when compared to the others by mechanical vibration. The point at which sonic speed is reached at vena contracta is a function of the valve design and F_L (rated pressure recovery factor). IEC-534-8-3 is the standard that is mostly followed to predict aerodynamic noise. This is typically shown in Fig. 6.10.

The basic process in such noise prediction consists of the following process:

a. As per process, condition to calculate the trim outlet velocity and find the noise source strength. The sound power generated in the fluid inside the valve and pipe is because of throttling pressure.

IEC Standard five step guidelines:

STEP 1: Determination of stream power at Vena contracta: If mass flow rate m is used in the equation in lace of actual mass it is given by $W_m = \frac{1}{2} * m * Uvc^2$ (velocity at vena contracta) — in watts.

STEP 2: Conversion of noise power at valve outlet: Major mechanical power at the vena contracta is converted to heat, but part is converted to noise. So, noise at vena contracta: $W_a = \eta * W_m$ where η empirically derived acoustic efficiency factors.

STEP 3: Sound pressure level at downstream: When the sound power is in the flowing medium is known, it will be converted to sound pressure level in the fluid downstream. From W_a actual pressure disturbance is determined , Once the actual pressure disturbance (pd) due to the downstream noise is known, It is compared to the standard pressure reference for the threshold of hearing ($p_o = 2 \times 10^{-5}$ Pa) to get Sound pressure level SPL.

STEP 4: Determination of a weighted sound pressure level outside pipe to determine how much of the sound pressure level gets transmitted through the pipe wall to the outside: Determining the sound transmission loss through the pipe wall is a very complicated. Statistical method is deployed (*discussed below*).

STEP 5: Translate Pressure level standard observer location: Standard acoustic theory can be used to determine how much of this sound pressure level gets transmitted to observer located at the standard location (1 m).

FIG. 6.10 IEC guidelines for noise determination.

b. To calculate the part of sound energy generated at the valve that is propagated to downstream.
c. To calculate how the pipe wall attenuates the noise as it passes from inside to outside, that is, transmission loss due to piping.
d. To estimate the weighted sound pressure level at a distance of 1m from the pipe.
3. Hydrodynamic noise: The major sources of hydrodynamic noise are: turbulent flow, cavitation, flashing, and mechanical vibration for previous issues. The problem associated with hydrodynamic noise comes from erosion as well as corrosion. IEC-534-8-4 standard may be used to measure the same.
4. Brief discussions on noise predictions (based on Masoneilan noise calculation manual):
   a. Aerodynamic noise:
      i. To calculate the total stream power at vena contracta *Wm*: Use kinetic energy equation stream power due to mass flow "*m*" is
         $Wm = ½ \times mU^2$ where U is the velocity at the vena contracta calculated using first law of thermodynamics.
      ii. The acoustic efficiency is calculated, for each flow for the regimes discussed below, $\eta = M_{vc}^{3.6}$ where M_{vc} = Mach number at vena contracta and peak frequency $f_p = 0.2U/d_{vc}$ where d_{vc} is the orifice diameter at vena contracta. This varies with flow regimes discussed below:
      iii. There are five defined flow regimes for a particular valve depending on the inlet pressure P_1, outlet pressure P_2, fluid physical parameter, and F_L for the valve. These flow regimes include: Subsonic, sonic with turbulent flow mixing, no compression but with flow shear, shock cell turbulent flow interaction, and constant acoustical efficiency.
      iv. Thus in each regime, efficiency is defined and calculated. Thus the sound power is calculated from the total power with the help of the efficiency in each regime.
      v. Only a portion of the sound power is propagated to the downstream, and this is designated by r_w, which is a function of the valve style. Typical values are as follows (Table 6.12):

TABLE 6.12 Factor for Noise Calculations

Valve Type	r_w
Globe/rotary globe	0.25
Butterfly/ball	0.5
Expander	1

      vi. The aim is to convert the sound power to SPL, which is done by
         Average SPL across the pipe cross section = $10\log_{10}(3.2 \times 10^9 W_a \rho_2 c_2)/(d_p^2)$ where Wa = sound power, ρ2 = downstream vena contracta gas density, c2 = downstream vena contracta sound speed, and dp = downstream pipe ID. Downstream noise propagation is given by
         $L_g = 16\ \log_{10}[1/(1 - M_2)]$ where M_2 is the Mach number at the outlet.
      vii. The net sound level at the pipe wall is given by $L_{pAe} = 5 + L_{pi} + TL + L_g$

where L_{pi} pipe internal sound pressure level: TL is transmission loss, which is dependent on flow peak frequency. These transmission loss regimes are determined from TL versus peak frequency relation.

      viii. Noise at 1 m from the pipe wall.

SPL = $L_{pAe} - 10\ \log_{10}[(dp + 2tp + 2)/(dp + 2tp)]$ where tp is pipe thickness.

In the above calculations, only basic steps have been discussed in line with the standard elaborated in Fig. 6.10. Standards and handbooks/standard curve are also available for detailed calculations.

   b. Hydrodynamic noise prediction. The main factor for hydrodynamic noise is associated with cavitation. There are two situations; Choked flow and nonchoked flow conditions. Again, for flashing fluid there will be a variety of situations. Noise prediction is rather complex. For details, refer to standard IEC-534-8-4.
5. Noise control means: The noise can be controlled by source treatment and path treatment.
   a. Source treatment for noise control
      i. Use of a small properly spaced jet: The size of the fluid jet affects the noise generation. (1) By reducing the size, the efficiency of the conversion of the mechanical energy to sound is minimized. Also it shifts the energy to high frequency and curtails the transmission through the pipeline. (2) A small size and proper spacing reduce the shock eddies.
      ii. Adiabatic flow with friction: The flow area of the valve trim is gradually increased in the downstream section to compensate for expansion and, more or less, constant velocity. Thus a sharp fall in enthalpy is arrested.
      iii. Multipath, multistage, and a combination of the two are the major source treatment. LO DB, cage style are the source treatment towards this.
      iv. For high DP $\Delta P/P_1 > 0.8$, a series restriction approach to split the pressure drop between the control valve and restrictor at downstream is effective.

   v. For hydrodynamic noise, which is mainly associated with cavitation, if the cavitation is predicted and eliminated then noise is drastically reduced.

b. Path treatment: A few methods used in this approach include:
   i. Silencer: This is an effective way, but there are a few problems associated with it such as low-flow velocity, which makes this ineffective in high capacity services. The operating condition may be too severe to use a silencer.
   ii. Pipe schedule increase; meaning higher wall thickness. It is a good treatment, but it needs to be maintained throughout the length of the downstream because noise, once generated, does not dissipate quickly (the system may be costly).
   iii. Pipe insulation: It is also effective, but has the same limitation as discussed above. Also, the insulation must be properly wrapped without any gap/void.

1.10 End Connection for Control Valve

In power plants, all three types of end connections of control valves are common.

1. Screw type: Screw types are common in small valves, and they are economical. The threads are female threads with different standards such NPT, BS, etc. Generally, these connections are restricted to low-pressure application and up to a size of 50 mm NB. Maintenance may not be easy.
2. Flange connection (Fig. 6.11): In a flange connection the valve is connected with pipe by mating the flange set with a gasket in between, and the two flanges are bolted. These can be used for a wide range of pressure and temperature (up to 800°C). From a maintenance point of view, this connection is a natural choice for easy removal of the valve. There is no difficulty in matching both end dimensions as long as it follows the international standards such as DIN, ANSI, BS, JIS, etc. In these standards, corresponding to the flange's nominal size and pressure class for a particular material, there

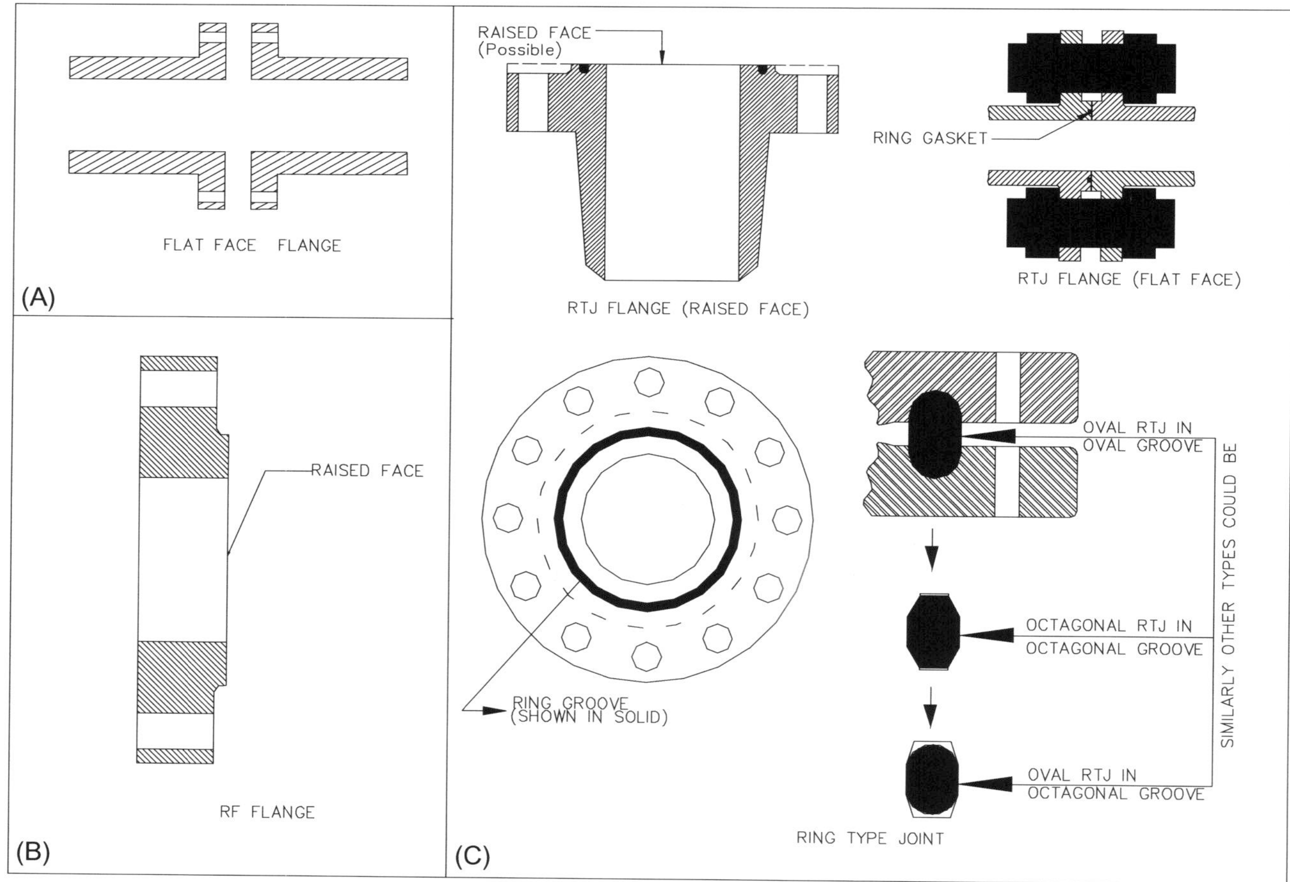

FIG. 6.11 Flange type valve end connection.

will be standard tables specifying all other required details such as bolt diameter, pitch, thickness, etc. Consult the standard to for details. There are three kinds of flanges for connections:

(a) Flat face (FF): Full face of contact of the two matching flanges with a gasket in between them. This is used for low pressure applications.

(b) Raised face (RF): A raised face that is in contact with the raised face of the matching flange with a gasket in between. The inner diameter of the raised face matches the valve opening with the outside diameter somewhere near, but less than a bolt diameter. The raised face has a concentric groove to accommodate the gasket and prevent a gasket blowout. A pressure range of up to 400 bar, and a temperature of up to 815°C can be withstood in this type of flange. However, for each material and pressure class the temperature pressure table should be consulted.

(c) Ring type joint (RTJ) flange is more or less the same as the RF type except there is one U-shaped groove and metallic gasket of elliptic/octagonal shape. It is used to put the ring type gasket in the groove when the bolts are tightened as shown in Fig. 6.11C. Pressure sealing of up to 1000 bar is possible but rarely used for a high temperature.

3. Weld type (Fig. 6.12): Weldable valves are leak tight, but difficult to remove. Also, care must be taken during welding so that the high temperature does not damage the trim, etc. There are two types of welding connections as shown in Fig. 6.12.

a. Socket weld ends are prepared on each side of the valve by boring a socket-type space where the pipe slips into the socket and rests there, and then it is welded as shown in Fig. 6.12A. A connection in a pipe of up to 50 mm NB utilizes this type.

b. Butt welding connections are used generally for line > 65 mm NB. For a butt weld, end preparation is done at each end of the valve, that is, beveled. Similarly, the pipe is also beveled, and welding is done as shown in Fig. 6.12B.

1.11 Control Valve Face-to-Face Dimension

ISA has developed the standard ISA S 75.*nn* (*nn* stands for various values, e.g., ISA S75.03) to specify the face-to-face dimensions for various kinds based on their pressure class, connection type, etc. Viz. face-to-face dimensions for flanged globe style control valve for classes 125,250, 300, and 600, etc. (ISA S75.03).

This means from the end-to-end dimension of the valve. The basic purpose of such a standard is to provide valve face-to-face dimensions for flanged and flangeless control valves regardless of the equipment manufacturer. This is helpful for the piping engineer to obtain the required data (see ISA standard and manufacturer's handbook for information). For the end connection of each valve type, size, and pressure class, there will be a unique face-to-face dimension viz. ISA S75.03 for *flanged globe style* valve of size 2" NPS. The *pressure class 300* with an *RTJ flange connection*, *"face-to-face dimension"* should be *282 mm (11.12 in.)*.

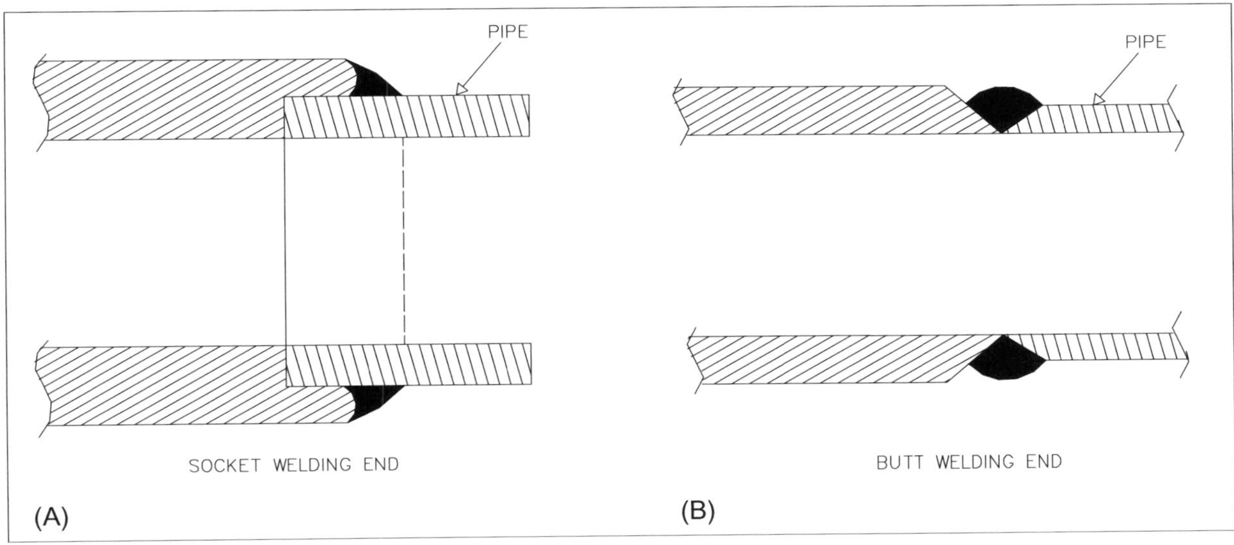

FIG. 6.12 Weld end connection.

1.12 Nominal Pipe Size and Pressure Rating

Since the control valve is an inline device, the nominal pipe size and associated pressure ratings are very important. Pressure ratings are normally specified either in terms of *ANSI (Class-in PSI)* or as per *ISO 7268* as "pressure nominal" *(PN-in Bar)*. Similarly, pipe sizes are defined either in terms of nominal pipe size (*in in.*) for the North American standard, or in ISO 6708 in diameter nominal (*in mm*). Common standards and their equivalence have been presented below (normally found in a power plant) (Tables 6.13 and 6.14).

2 CONTROL VALVE TYPES

2.1 Introduction

Types of valves and applications: It is a major task to select the proper control valve that is best suited for a particular application. In a power plant, the problem is slightly less than the same determination for a process plant. Depending on the application, some special types of valves, e.g., drag valve, triple offset valves are used. In power plants, there are many applications where there are chances of cavitation and/or flashing fluid applications. The term C_d ($= C_v/d^2$) as defined in clause 1.2.1.4 is very useful to see the pressure recovery factor. Generally, higher C_d means higher pressure recovery value. Hence one needs to be careful. Refer to Table 6.15 for a selection guide.

In this section, some details about the valves are discussed with special reference to globe valves, which are mainly used in power plants for control purposes. Prior to that, a few details about other important types of valves will be discussed. *Pressure ratings and sizes are explored in terms of NPS, ISO and ANSI. Equivalence between NPS/ ISO and ANSI/ISO is shown in Clause 1.11.*

2.1.1 Ball and Butterfly Valves

These valves belong to the *quarter turn* family, and are mainly used as ON/OFF valves, but have some modulating application. Also, both show good, tight shutoff characteristics. *Triple and double offset butterfly valves* are costly and are used in special applications like turbine drain/steam isolation service in power plants. These valves have been discussed at the end of this Section 2.4.

2.1.1.1. Ball Valves (Fig. 6.13A). As the name suggests, a ball valve has a sphere with a hole in the middle. When the hole is in line with the pipe axis, there is full flow, but when it is at 90° with the axis of the pipe, the flow is zero. Ball valve action has been detailed out schematically in Fig. 6.13B. So within 90° it has to complete full flow. Due to high pressure recovery factor it has chances for cavitation and noise. Also, the opening vs. actuator rotation is nonlinear. For all these reasons, it is not popular as control valve in power plant application (except for some special use). The ball valve can offer tight shutoff characteristic ANSI CL IV (metal seat) and VI (soft seat).

1. Ball valve types: There are mainly three kinds of ball valves. these are
   - **a.** *Full bore ball* valve: In this, the ball is oversized so that the hole in the ball matches with the pipe ID, so that full flow can pass.
   - **b.** *Reduced bore ball* valve: In this design, the flow through the valve is one pipe size smaller than valve's pipe size.
   - **c.** *Characterized type*: This type of ball valve should be the V or U notch ball valve. It is used in paper industries.
   - **d.** *Cage type*: Here the ball is positioned by a cage.
2. Conventional ball valves may be two-way and three-way with TEE port and/or L port, and are to be used as diverting, mixing valves as well as selective use of path (standard book).

TABLE 6.13 ANSI Class Rating and Pressure Nominal (at Standard Temperature)

ANSI Class Rating	Pressure in lb	Pressure Nominal (PN)	Remarks
150	290	20	Pressure nominal is the rating designator followed by a designation number approx. pressure rating in Bar 1 bar = 1 × 1.105 pa = 14.5038 psi So PN 20 stands 20 bar = 14.5 × 20 = 290 lb
300	750	50	
400	1000	64	
600	1500	100	
900	2250	150	
1500	3750	250	
2500	6250	420	

TABLE 6.14 Nominal Pipe Size (NPS) and Diameter Nominal (DN) Equivalence (in Power Plants)

NPS (in.)	DN (mm)	NPS (in.)	DN (mm)	NPS (in.)	DN (mm)
1/4	8	2½	65	20	500
3/8	10	3	80	24	600
1/2	15	4	100	28	700
3/4	20	6	150	32	800
1	25	8	200	48	1200
1¼	32	10	250	56	1400
1½	40	12	300	60	1500
2	50	16	400	72	1800

3. Ball valves are available from 12 mm to 1 m with pressure rating from 150 to 2500 lb class. With temperature 260°C.
4. Ball valves data sheet in power plant application. (Typical one may vary with application)
   a. Type: Conventional full bore.
   b. Size: 15–50 mm (Typical).
   c. No. of ways: 2/3 (in HVAC).
   d. O ring: Viton (Hi Temp.—Kalrez etc.).
   e. Material: Body: SS 316/CI/Bar stock brass. Ball: SS 316 or better Seat: Buna N, natural rubber.
   f. Max design Pr./Temp.: As per application.

2.1.1.2. Butterfly valve: As stated earlier, there is similarity in operation of a butterfly valve with a ball valve. Instead of a ball, there is a metal disk mounted on a rod. When the disk is parallel with the pipeline, there will be full flow, whereas,

TABLE 6.15 Valve Selection Guide (Triple and Double Offset Valves are not Included Here, Because These Special Types are Dealt With Separately at the End of This Section)

Feature	Ball	Butterfly	Digital	Globe (1)	Globe (2)	Saunders
Max Pressure class common (lb)	2500	600	2500	2500	2500	150
Service	Mainly on/off	Throttling	Throttling	Throttling	Throttling	Throttling
Min/max size (mm)	6/1200	50/1800	6/250	6/750	6/750	6/1800
Min/Max Temp °C	(−)55/350	(−)30/550	(−)30/650	VL/675	VL/675	(−)50/250
Max cap C_d $C_d = C_v/d^2$	25–40	40	13	12	15	20
Press. recovery F_L	0.6	0.65		0.85–0.9	0.85–0.9	
Characteristic	Good	Fair	V. Good	V. Good	V. Good	Fair
Hi DP (>15 bar)	Low (4)	Low (4)	Hi PD possible	Hi PD possible	Hi PD possible	NA
ANSI leakage CL	IV	VI	V	IV	IV	V
Abrasive service	Ceramics	NA	Bad	Good	Good	Good
Anticavitation	Fair	Fair	Fair	Good	Go od	NA
Flashing service	Bad	Bad	Fair	Good	Good	NA
Viscous service	Good	Good	Fair	Good	Good	Good
Comparative cost	0.8	0.7	2.4	***1(Ref)***	1.25	0.65
Special feature	(a) Hi capacity (b) Good rangeability (c) Tight shutoff (d) Light weight	(a) Hi capacity (b) Light weight	(a) Precise/Hi speed control	(a) Good sealing (b) Easy trim change (c) Hi Capacity (d) Good rangeability (e) Low noise		(a) Self cleaning (b) Good for water treatment service

(1) Globe valve single port.
(2) Globe valve double port.
(3) Data used in the table were taken from reputed valve manufacturers.
(4) Generally low DP, but high DP is also possible.

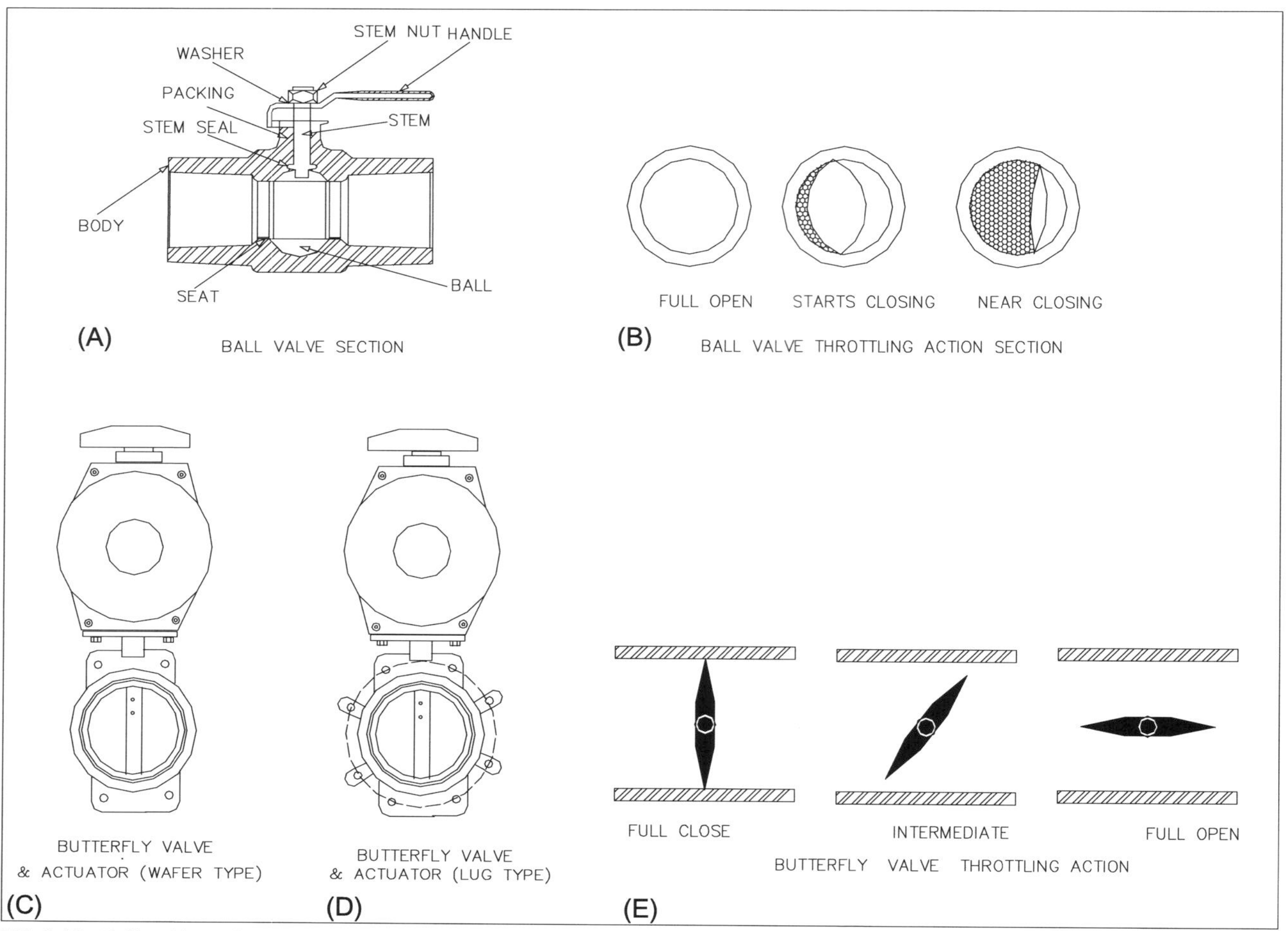

FIG. 6.13 Ball and butterfly valves.

when it is at 90°, the flow stops as shown in Fig. 6.13E. Generally, it also turns a quarter to give the full flow.

1. In the disc hard facing materials are applied. Deepening on design and seat configuration it can be *resilient BV* having a rubber seat with pressure withstand the capability of ~25 kg/cm^2 (e.g., Keystone Dubex RMI) and size up to 1200 mm. The other type is a *high performance butterfly valve*. These are high shutoff, lined valves with excellent throttling capabilities, at low torque with pressure capability as high as 50 kg/cm^2.
2. Depending on the type of mounting/installation, it can be in two categories. (1) *Wafer design* (Fig. 6.13C) provides unidirectional flow by using suitable seals and a flat valve face on both sides. It is put between two sets of flanges. (2) *Lug type design* (Fig. 6.13D) has a different mounting when compared with the wafer type. This design has threaded inserts on both sides of the body. This design is also placed between two flanges. Butterfly valves have a wide range of temperature designs up to 530°C.
3. In the CW system of a power plant, the butterfly valve is applied.
4. Butterfly valves datasheet in a power plant application. (Typically varies with application.)
   a. Type: Double flanged, double/triple eccentric resilient seat.
   b. Nominal size in mm: 300–200 (Typical).
   c. Material: Body: Ductile iron (epoxy lining possible) Seat: PTFE , natural rubber.
   d. Max design pressure/temperature: Depends on the application; Pr. $<$ PN 25 temp $<$ 100°C.
   e. Leakage: ANSI CL VI (Cl IV if the metal seat is selected).
   f. Connection: ISO/ANSI/DIN flange.

2.1.2 Other Miscellaneous Valves

In this part, a brief discussions shall be presented on a few other types of valves used in power plants. Some of these are used with actuators (e.g., diaphragm) and some are without actuators (e.g., rate set valves in SWAS, or simple isolation gate valve, etc. without any actuator). However, linear actuators will be used for the diaphragm and gate valves are discussed below.

2.1.2.1. Diaphragm valves (Fig. 6.14A–C): These valves are also known as "Saunders Valves." Corrosive fluid and slurries are the medium, where diaphragm valves find their applications.

1. The valve consists of a moving or elastic diaphragm that is moved up or down to regulate fluid flow. The valve stem moves the diaphragm to restrict the flow as shown in Fig. 6.14A–C.
2. In context of a power plant, these valves are extensively used in water treatment plants; 25–100 mm valve sizes are mostly used.
3. Body materials may be ductile iron, SS 316, etc. Whereas. the diaphragm materials are Teflon, Buna N or Neoprene.
4. Normally, these valves have quick opening characteristics.
5. Depending on the construction, there can be some variations such as:
   a. Straight through.
   b. Weir design where there is wear towards which the diaphragm is compressed to create the restriction.
   c. Dual range, which has two compressing units for better flow characteristics.
6. Diaphragm valves datasheet in power plant application. (A typical one may vary with Application.)
   a. Type: Weir/straight through.
   b. Size: 15–100 mm (Typical).
   c. Material: Body: SS 316(CF8)/DI CI/bar stock brass diaphragm: Buna N, natural rubber, EPDM.
   d. Max. design pressure/temperature: Depends on the application; PN 10/16 temp: 160°C.
   e. Connection: screwed, flanged.
   f. Body lining: natural rubber.

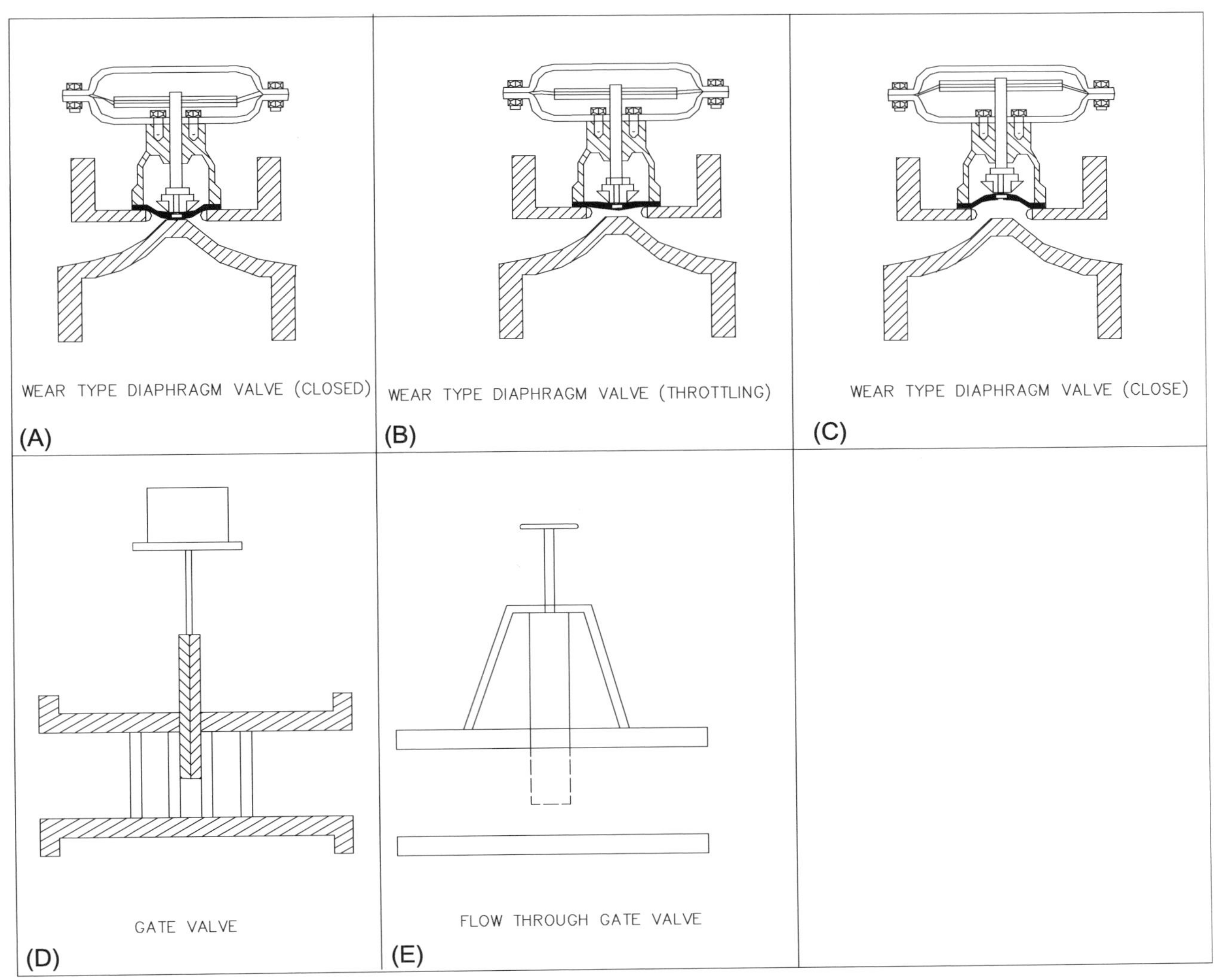

FIG. 6.14 Diaphragm and gate valve.

2.1.2.2. Gate valve: It is the most commonly used valve in industry. Generally, it is not used as a control valve.

1. Gate valves find most of their application in stop/block valves. Because in a partial opening position it may give rise to vibration. Because it offers practically no resistance to flow during open condition, hence, it offers the lowest pressure drop.
2. There are two basic designs of disc/gate: *parallel and tapered or wedge gate*. Wedge-type gates find their application as a gate damper at a coal chute, outlet damper, etc. Wedges can be: (a) solid wedge with very little flexibility; it cannot cope if there is any distortion. (b) Flexible wedge; there will be a slot around it so that it can withstand changes in temperature. (c) A split wedge is when two pieces meet at a close position.
3. According to the bonnet design, gate valves can be: (a) Screw design: inexpensive bronze valves. (b) Union joint: Same as (a), with flexibility to disassemble. (c) Bolted joint (most popular one). (d) Pressure-seal type for high pressure service. Extensively used for high pressure, high temperature application. (e) Welded joints: High pressure services.
4. Alike V-notch ball valve; gate valves that can also be with a V insert.
5. There is another kind "position disc" type as shown in Fig. 6.15. In an open position, the rotating disc's holes match with holes in a stationary disc. It has a hole to allow full flow, whereas in a closed position, it is at 90° to stop the flow; it can also be used to control the flow.
6. Gate valves datasheet in a power plant application:
   a. Type: Parallel slide knife edge.
   b. Nominal size in mm: 15–500 (mechanical steam service application).
   c. Material: Body: Ductile iron/SS 316/CS in some application with hard-facing (epoxy lining possible) Seat: metal-to-metal or PTFE.
   d. Max. design pressure/temperature: Depends on application; Up to 4500 lb class available (e.g., L and T).
   e. Leakage: ANSI CL II or better with soft seat possible.
   f. Connection: Screw/flange/weldable.

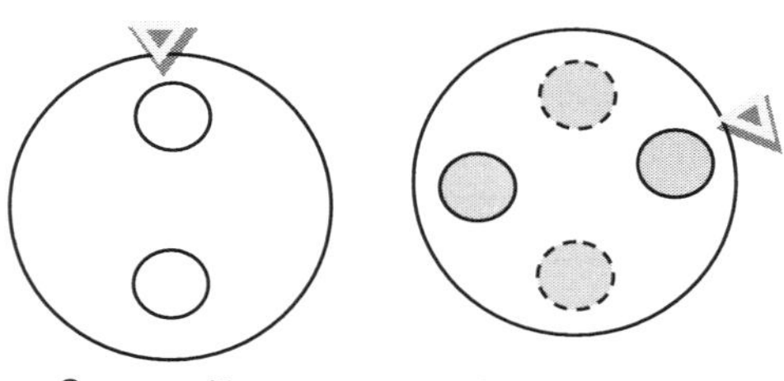

FIG. 6.15 Disc positioning gate valve in two positions.

2.1.3 Globe Control Valves

In most of the process industry, globe control valves enjoy the most popularity. However, globe design valves do not offer good tight shutoff characteristics. Issues like: simplicity *design and sizing*, relatively lower *dead time*, low *noise*, cavitation affect simple *linear* actuators (e.g., spring diaphragm/piston cylinder/electrical actuator). Special *design* variations for use in abrasive, corrosive, high pressure/high temperature applications (for almost all fluid types) and supports flow *characteristics* (viz. linear/EQ%/QC/Parabolic, etc.) make this valve type very popular. However, higher cost, low speed and low stiffness are major limitations for this type.

2.1.3.1. Parts of the control valve: Various parts of the control valves and actuators have been depicted in Fig. 6.1A and B. The valve stem is coupled with the stem of the actuator. Various parts shown are typical ones it varies with the types of valves.

2.1.3.2. Various kinds of globe valve design: The major pressure drop in the control valve takes place in the trim part of the control valve. Depending on the flow characteristics discussed, the shape of the plug varies.

These are shown in Fig. 6.16A–C for EQ%, linear, and quick opening. Most of the modern valves have contoured plug, ported plug, etc. to achieve the desired flow characteristics. A valve mainly comprises a body assembly, which includes bonnet, top and bottom closure; and trim, which includes plug stem guide bushing, etc. Body configuration determines the trim style to meet the process requirement.

1. Single seated valves with discussion on cage type trim: Available in numerous forms, this type of globe valves is the most popular because of (1) simple construction possible to manufacture from bar stock, forged, split body, etc., (2) lesser susceptibility to vibration, and (3) good tight shutoff characteristics with soft seating/metal-to-metal seating. However, actuator sizing is critical because of an unbalanced force over the entire port at high pressure. For proper performance and alignment with the seat, it is necessary to guide the plug. The various kinds of guide for the valve plugs: Stem guided (*when the guide bushing in the bonnet act on valve stem*), top guided (*plug is aligned by a single guide bushing in the bonnet* as shown in Fig. 6.17A, or top and bottom guided (*plug is aligned by guide bushing in the bonnet as well as at the bottom flange* as shown in Fig. 6.17B for a double-seat valve. However it is possible to have a single seated valve with top and bottom guide. Currently, the cage guide (*The outside diameter is close to the inside wall of the cylinder cage with opening.*) is finding popularity because it provides better controllability in complex applications. Various kinds of cage guides have been depicted in Fig. 6.16H and I. Balanced and unbalanced plugs are shown in Fig. 6.16F and G. As shown, windows of the wall of the cylinder determine flow

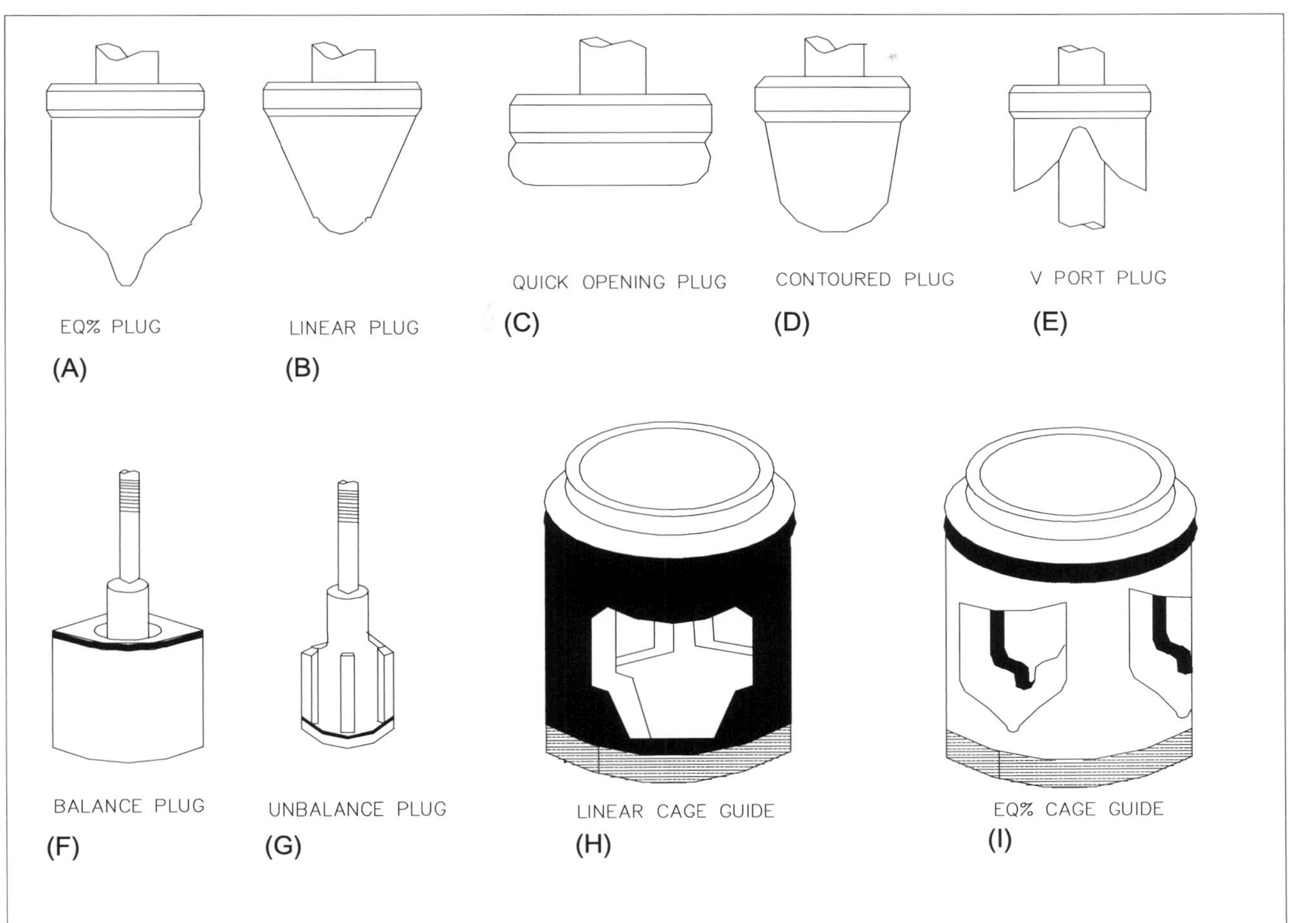

FIG. 6.16 Types of plugs and cages (globe valve).

characteristics. In this type, the interchange of the cage will change the flow characteristic without changing the plug of the seat. By interchange of the cage, special trim designs such as low noise/low cavity trims are possible. The other design has the plug characterized as per required flow types and the cage has a large opening to allow the flow. The plug surface next to the ring is responsible to achieve the flow characteristic. Unobstructed flow area changes in size and shape as per plug design. In cage guides, it is possible to modify various trim parts easily for achieving various requirements, such as low noise, reduction in cavitation, etc. In the case of balanced plugs, requirements of force for the actuator will be less. Whereas, for unbalanced plugs, shutoff will be better because of the unbalanced force. In a cage type valve design, both are possible. A balanced plug cage guide is popular as it gives an advantage of a lower actuator force. In this, the sliding piston plug seals the upper portion of the plug and the cage wall, and seals the high pressure end leakage. Downstream pressure acts on both sides at any position. As a result, the actuator has to overcome reduced unbalanced force; sizing will be lower. *A single seated valve, with the cage-style trim design in angle valves could be used in feedwater and in heater drain valves.*

2. *Double ported/seated valves*: These valves are heavier and larger and have lower shutoff characteristics. The leakage is Class II or Class III. As shown in Fig. 6.17B a double ported valve has one advantage, which is it can be converted from push-to-close to push-to-open type by opening the bottom portion of the bonnet and inverting the assembly. As seen from Fig. 6.17B, the force on the upper plug always tends to cancel the same on the bottom plug. So, this is, in that sense the balanced one, but because of dynamics of hydrostatic force, it is not totally balanced, hence, considered as a semi-balanced valve. Thus lower size actuators can be used. Normally, these are used for >100 mm valve sizes.

3. Special design valves: There are a few special valve designs available in the market for power plant application. These include:

a. High temperature (courtesy: Fisher *Control Valve Handbook*): Because of high temperature applications, the

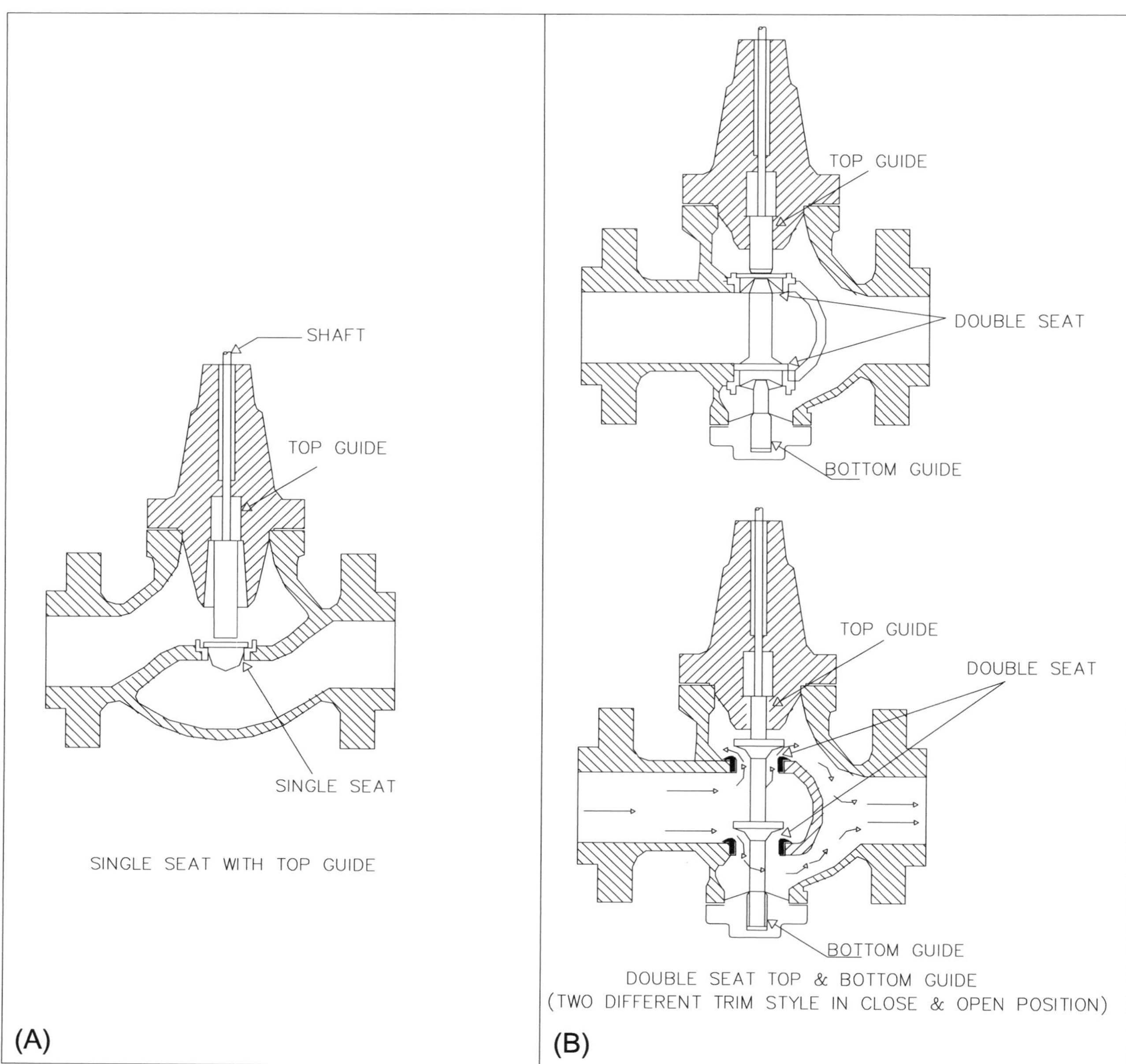

FIG. 6.17 Globe valve single and double seat.

boiler feedwater application, superheater control, and bypass control, applications in power plants require special consideration. These valves may have a special *bonnet* design discussed later and special materials of construction (detailed in Table 6.16). Also, standard gasket, elastomer, etc. are not suitable. These valves use metal-to-metal seat. Graphite packing material may be used (Fig. 6.18).

b. Lo dB trim valve: Lo dB trim valves have been designed for severe service application in compressible fluids. The design reduces the unbalance force to minimum requiring low actuator torque. There are various types of

TABLE 6.16 Material Guidance

Temperature T (°C)	Material Used	Source
$T > 538$	Cr-Mo steel	Courtesy: Fisher Control Valve
$T = 593$	ASTM A217 Gr WC9	
$T < 816$	ASTM CF8M	

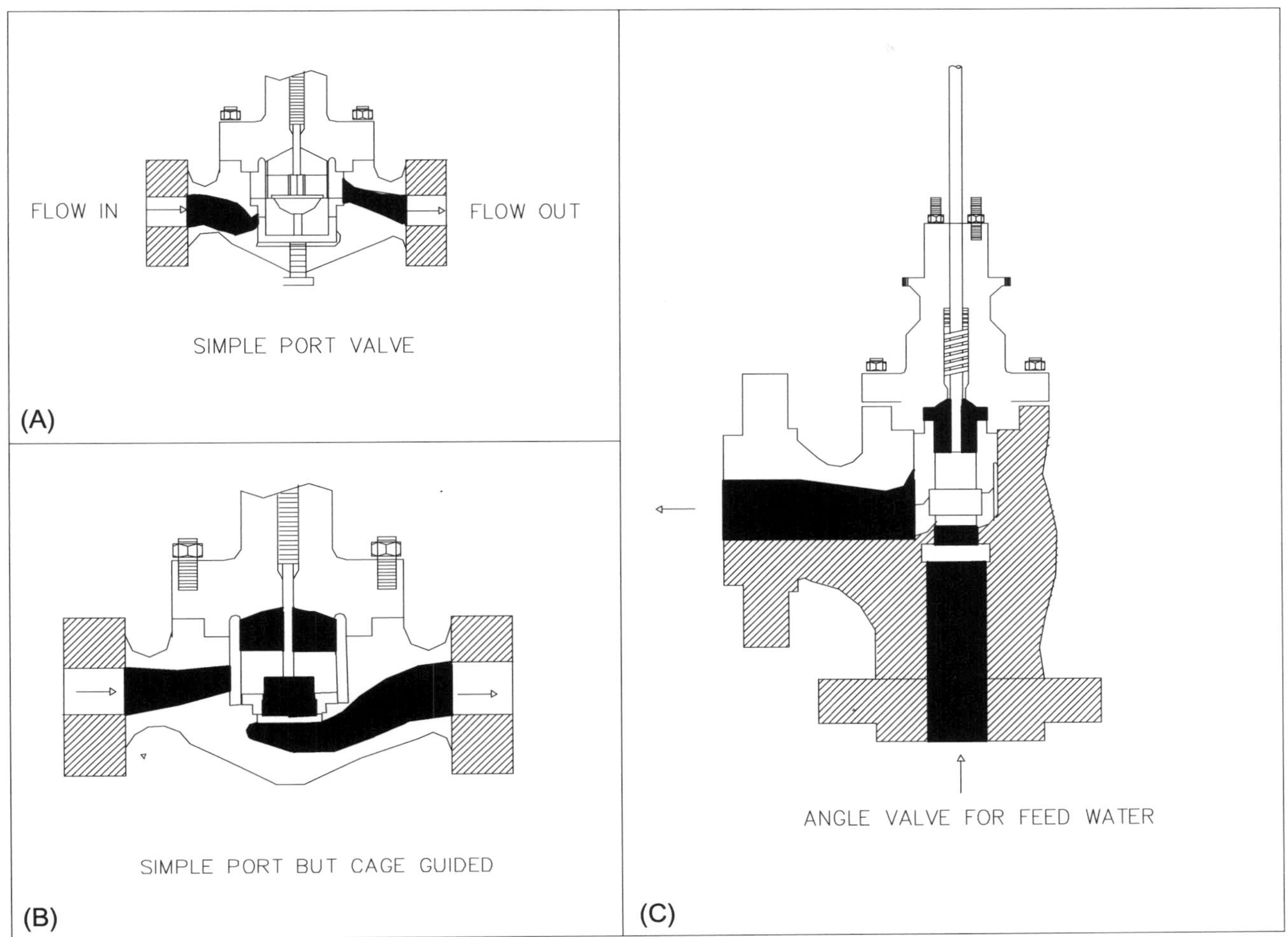

FIG. 6.18 Single Port Globe Control Valve.

designs such as single Lo dB trim with or without a diffuser, and a multistage Lo dB design that are meant for anticavitation. There is a rotary globe valve with Lo dB trim. Dresser Masoneilan have this type of design (Fig. 6.19A).

c. Variable resistance trim VRT (courtesy: Dresser Masoneilan) provides a good solution for cavitation protection in a high-pressure liquid application like boiler feedwater control valves. Here, process fluid is passed through a series of axially stacked plates with drilled holes, and positioned to provide the required level of cavitation protection with required flow rate.

d. Drag valve (courtesy: CCI): A drag-type trim is ideally suited for minimizing noise, at the required flow capacity. The valve design is a multistage design. Based on leakage requirements, both hard and soft seat can be used. This valve is suitable for boiler feed pump recirculation control where there is a requirement for high pressure breakdown without erosion/cavitation or vibration; leak tight at required flow rate (Fig. 6.19B).

e. Reduced trim: There are applications that require a larger valve body, but with reduced trim. As per ISA Guide (PP112-161), a reduced trim design may be considered for applications like: *Pressure drop* in excess of 5170 kPa, *Gas/Vapor outlet velocity* >0.3 Mach, high *noise* >85 dBA, *choked* flow, *flashing* exceeding 5%Wt. of liquid being vaporized, *erosive* fluid and future *capacity increase* anticipated.

2.1.3.3. Control valve bonnet: The bonnet mainly serves the following functions:

- It is the pressure retaining to the closure assembly; the plug stem/rotary shaft moves through it.
- At the bonnet, the actuators are mounted.
- Provide seal valve trim against process fluid.

This pressure retaining part and its design calculations shall be based on international standard e.g., ASME. In most of the cases, the bonnet is a bolted flange type as shown in Fig. 6.18A and B. In these designs with the cage trim, the bonnet applies force to prevent a leak from the valve body through the bonnet flange, as well as the seat ring. There are mainly three kinds of bonnets: Standard bonnet, extended bonnet and bellow seal bonnet.

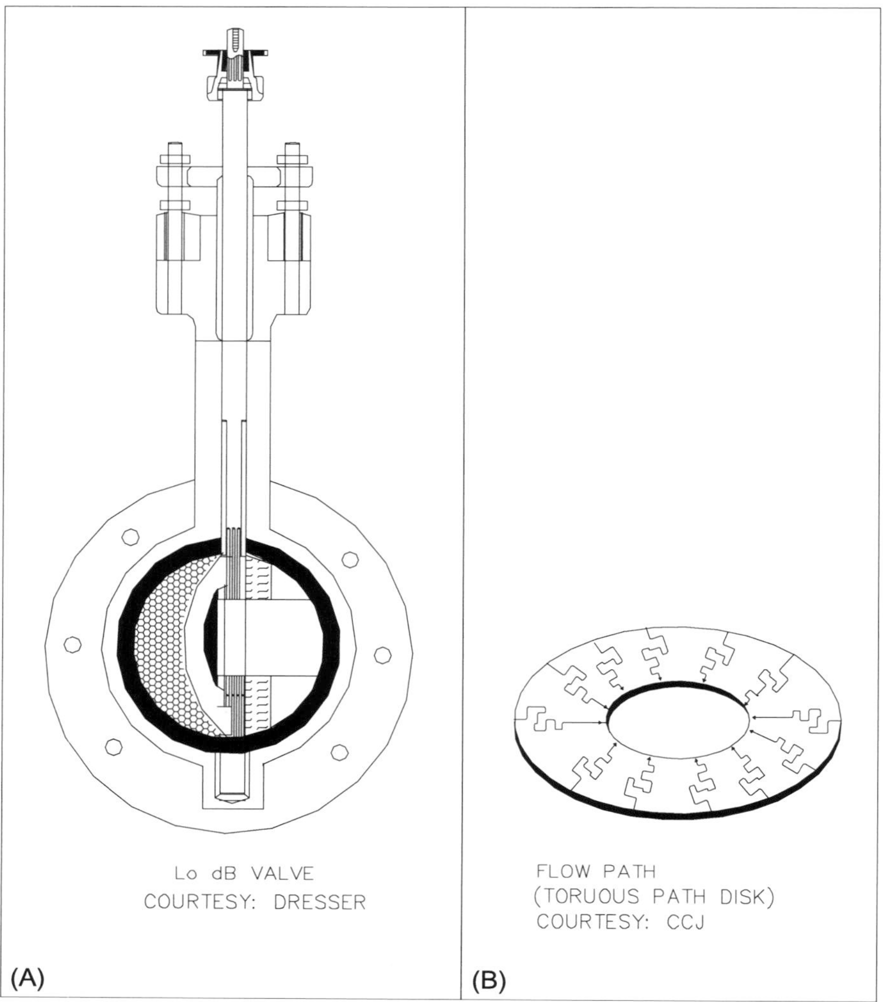

FIG. 6.19 Special globe control valve design.

1. *Standard bonnet*: This type, with standard seal packing and gasket materials can withstand pressure class ANSI 150-2500 and temperature −30 to 230°C (Even up to 300°C with special packing materials). However, as per ISA recommendations beyond 0–230°C, an extended bonnet should be used.
2. *Extended bonnet*: This type of bonnet could be used for extreme temperature conditions as packing far away from the extreme temperature application point (so that the temperature is within the range that is acceptable for the packing materials). At times, fans are used on the bonnet to dissipate heat. Stainless steel (<800°C) is preferred to CS (<400°C) due to lower thermal conductivity for heat dissipation.
3. *Bellow seal bonnet*: These are used for toxic fluid, volatile/radioactive fluid, etc.

2.1.3.4. Control valve packing: In order to prevent any leakage through the stem, there is a stuffing box at the upper part of the bonnet. A stuffing box is fixed with the valve by a flange connection in most of the cases. A stuffing box consists of a packing follower, a packing retainer, etc. The characteristics required for packing materials include: Nonreactivity with process fluid, the ability to seal stem, low-friction, the capability to withstand process temperature, and good service life. Teflon, laminated/filament graphite and Asbestos are common packing materials.

1. Teflon: This is plastic material with low friction. It is available in a V ring, a Chevron ring or cup, and a cone ring. It is resistant to almost all process fluids. It is suitable for a temperature range of (−40 to 230°C). In an extended bonnet design, the temperature can be raised up to 400°C. (Frosting in stem is not good as it may cause cracking in the packing). It is better suited with the smooth finish of the stem. PTFE V ring: A coil-type system usually used for $P < 20$ bar $T < 93$° C. Performance and service life, is good when coupled

with low friction. Seal PTFE packing with a live spring load is suitable for environmental application $P < 50$ bar $T < 230°C$. In nonenvironmental cases, it is used for high pressure temperature application. This type gives good performance, long life and friction is also low. Kalrez (with PTFE) is suitable for environmental and high temperature application.

2. Graphite: Suitable for a high temperature application up to 650°C, however an extension bonnet is better for temperature >400°C. It has long service life and thermal conductivity, but has medium to high friction. The valve manufacturers use laminated or filament graphite rings. It is necessary to ensure that no air is trapped. There are also environmental seals in graphite. This type gives good performance and long life.
3. Asbestos: Asbestos rings are used as packing materials. For high temperature application, reinforcement wires are used. Asbestos packing is used with lubrication. Currently, it is largely replaced by Teflon packing because prolonged inhalation of asbestos fibers can cause serious illnesses including malignant lung cancer. The European Union has banned all use of asbestos.

2.1.3.5. Steam conditioning control valves (PRDS) in power plant applications: In power plants and steam generation plants there are a number of pressure reduction and desuperheating stations (PRDS). What is a steam conditioning application? This system deals with the simultaneous reduction of pressure and temperature of the inlet steam to the required level. For example, main steam at 150 bar 540°C may have to be reduced to steam at 16 bar at 210° C. Because there is simultaneous reduction of both of the vital parameters forged type valves are best suited for the application. In these applications, valves are available up to 4500 lb class. Some of the advantages of the combined PRDS valves are: Higher rangeability, less spray water noise, less space for installation, and because of a turbulent expansion zone, better mixing. However, the characteristics of these valves vary greatly with the applications. There are some special trim designs that can accommodate rapid change in temperatures following a turbine trip. Precise control of temperature is important to improve the heating efficiency. It is necessary to inject a controlled quantity of spray water with the proper selection of desuperheater. Due to complex thermal and flow dynamic variables involved in the process of desuperheating, there are a number of physical, thermal, and geometrical factors on which the selection/success of the desuperheater depends. Factors affecting the performance may include:

- *Orientation of spray*: Proper placement of the desuperheater has great impact on the performance. Vertical orientation with the flow going up is a good selection for better mixing. Downstream fittings also play a role in the performance.
- *Spray water temperature*: The higher the temperature of the spray water the better the heat transfer. This is because with an increased temperature there will be an improvement of the *drop size distribution, latent heat of vaporization, the rate of vaporization,* etc.
- *Spray water quantity*: This directly affects the time for vaporization, and because heat transfer is time dependent, it would depend on the quantity of spray water.
- *Steam velocity*: Velocity directly affects the residence time, hence the mixing. If the velocity is too high, then there is not enough time to mix. On the other hand, if the velocity is too low, water droplets will try to drop out in want of turbulence. The recommended maximum/minimum velocity ranges could be 46-76m/sec and 9m/sec, respectively. This is in line with recommendations from renowned manufacturers.
- *Pipe size*: The pipe size affects the velocity of penetration and the coverage area for the flow.

The quantity of spray water can be calculated from the quantity of inlet/outlet steam flow, and enthalpies of inlet and outlet steam as well as from the enthalpy of water. Based on the quantity of spray water needed, Cv for the d-superheater is calculated. There are a number desuperheater designs available. These are:

- Fixed geometry nozzle: Simple mechanically atomized desuperheater with single/multiple spray nozzles. Generally used for constant load operation rangeability 5:1. These are fixed by a suitable flange on the branch pipe.
- Variable geometry nozzle: This is also mechanically atomized desuperheating system, only it has variable geometry activated by back pressure. Due to this configuration, it has rangeability of 20:1.
- Self-contained design: This is another kind of mechanically atomized design. It has variable geometry and is activated by back pressure.
- Steam atomized design used mainly for the systems with low velocity steam. The atomizing steam pressure is nearly twice that of inlet steam.

2.1.4 Triple and Double Offset Butterfly Valve

In a normal butterfly valve, the disc is at the center of the body. The major advantage of a butterfly valve is that it can be used for large line sizes using much less material. Normal butterfly valves have a disc diameter that is slightly higher than that of a seat diameter, so that it can give tight shutoff by squeezing the seat. Obviously, this is possible when the seat material is made of softer material for squeezing. Therefore the valves cannot be used in high pressure and extreme temperature (cryogenic/high temperature) conditions. On the other hand, globe valves with a metal plug and seat can also be utilized for these applications. In order to eliminate these application limitations, triple offset (TOV)/double offset butterfly valves have been developed so that all advantages of butterfly valves (e.g., high flow, high shutoff) can also be utilized for high pressure and extreme temperature conditions. Before going into detail, it is better to explore working principles of TOVs and double offset valves. However, TOVs are not only complex in design manufacture, they are expensive. So, to make a balance between cost and technical requirements double offset valves have been developed.

2.1.4.1. Triple offset and double offset butter fly valve working principles.

1. *Triple offset*

From the initial discussions above, it has been seen that in case of a globe valve with a metal seat and plug, it is possible to have good shutoff at high pressure and extreme temperature conditions. Referring to SK1, it shows that in a globe valve, cone-type plug seats are projected on conical seats with the help of linear motion. Unless the valve is completely closed, there is no surface-to-surface contact. At the closed position, additional force is applied to ensure a tight shutoff. In the case of a butterfly valve (BF), the same principle can be adapted. The difference is that in the case of a BF valve in place of a plug, there will be a disc, and instead of linear motion, there will be a quarter-turn rotational motion. In TOV, a similar principle can be applied. Let us consider a vertical disc as shown in SK2 of Fig. 6.20. Conceptually, the same has been detailed in SK2 of Fig. 6.21, where it has been shown how the disc is rotated at the center SK2 of Fig. 6.20, so that it fits into the seat when it is fully closed. Referring to SK3 of Fig. 6.20, it shows that when the disc is rotated, the upper portion gives way, but the bottom part of the disc rubs against the seat. In fact, it goes out of seat. This is possible only for a soft seat, but impossible for a metal seat (to support high pressure and extreme temperature). For this, a concept of offsets was created. The first offset is that bottom part of the is perpendicular to the bottom seat surface, so that distance of left and right side of bottom surfaces are same. In this method, when the disc is rotated, the rubbing at the bottom surface would be minimized but not eliminated as shown in SK5 of Fig. 6.20. This is possible only when the point of rotation is offset from the disc central axis, i.e., the shaft (point of rotation) is also offset with respect to the center line of the disc.

All three offsets have been marked by dotted lines in SK6 of Fig. 6.20 as well as SK5 of Fig. 6.21. From these, it is clear that the conic shape is at an angle with the pipe center line, and the conic shape is not integrated. Due to the nonintegrity-type design, it is very difficult to design and manufacture. So, TOVs are costly, too.

2. *Double offset valve (DOV):* In order to circumvent the problematic issues of TOVs, double offset valves were created. They can be easier to design and manufacture. and are comparatively less costly. In a double offset, the third offset is dispensed with, that is, in place of an oblique, cone shape, an integrated cone shape is used. In double offset the two offsets are:

- The shaft center axis is offset with respect to the centerline of the disc.
- The shaft center axis is offset with respect to the center axis of the valve.

In the discussion above Sk5 of Fig. 6.20, there will be a small rubbing at the seat. So for the metallic seat to withstand high pressure/extreme temperature it has to accept some rubbing. This means it has to be metal, and at the same time, elastic. For that issue a seamless tube plate in the form of a ring in the valve body is used as a seat, so that it is elastic in nature to take care of small rubbing [based on details from Vanessa data]. Other materials such as PTFE can also be used. From the discussion, it appears that basically, a double offset valve is a modified version of a triple offset valve. The double offset value is easier to design and manufacture and it is less costly, and at the same time provides a tight shutoff even with high pressure and extreme temperature.

Both the types are depicted in Sk7 of Fig. 6.20.

Now let us look into the details of triple/double offset valves.

2.1.4.2. Technical features of TOVs and DOVs:

A TOV provides a good solution when a tight seal is required, with a wide range of flow. Because of the compact, quarter-turn design, and lightweight structure, TOVs can be installed and operated easily, and they require less pipe bracing. Also, the replaceable seal ring allows quick and easy repair. So, TOVs are selected for specific applications because of their sealing features as well as cost savings over other metal-seated valve types (with large flow). In this connection, double offset valves offer more cost savings. Major features of TOVs are as follows:

1. No rubbing 90° rotation with metal-to-metal torque seating to ensure zero leakage.
2. It can offer bidirectional flow in line with international standards e.g., ISO 5208, EN 12266-1 and API 598. Sizes available: DN 80 to DN 1800.
3. For extreme temperature application, a bolted extension bonnet is possible.
4. A heavy-duty bearing to withstand high pressure and wear.
5. Standards: Normally TOVs, DOVs follow international standards listed below:
   - Design: API 609, ASME B16.34, EN 593; EN 12516.
   - Testing: API 598.
   - Pressure/temperature rating (with materials): ASME B16.34.
   - Face-to-face dimension: API 609; ASME B16.10; EN 558; ISO 5752.
   - End connection types and standards: ASME B16.5 B16.25/B16.47 EN12627.

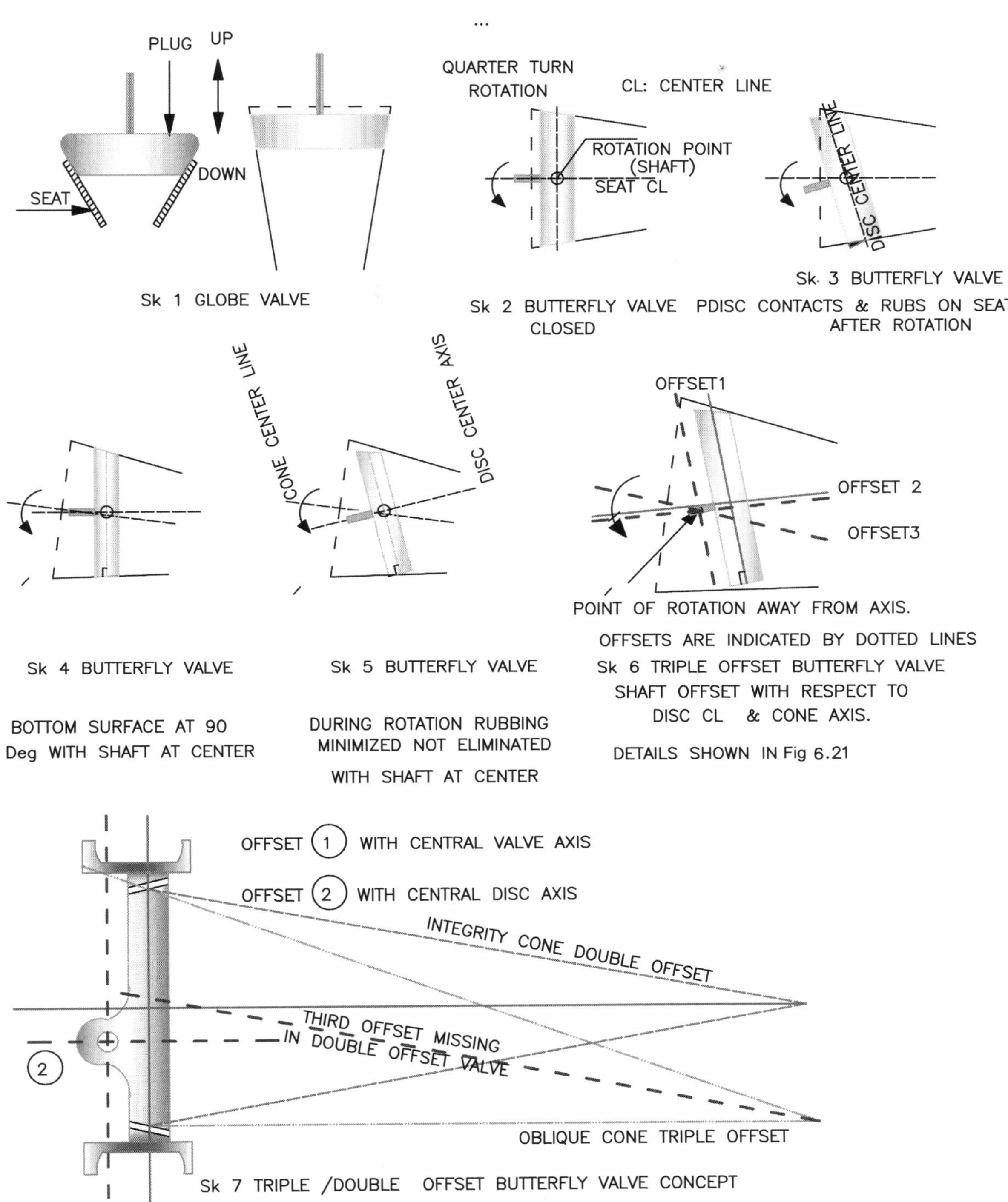

FIG. 6.20 Triple/double offset butterfly valve working principle.

2.1.4.3. Specification of TOVs/DOVs

Basically, there is not much difference in specification of triple/double offset valves except the differences discussed above. A short specification of triple/double offset valves have been presented in Table 6.17. Like other butterfly valves, these are also available both in wafer type and lug type.

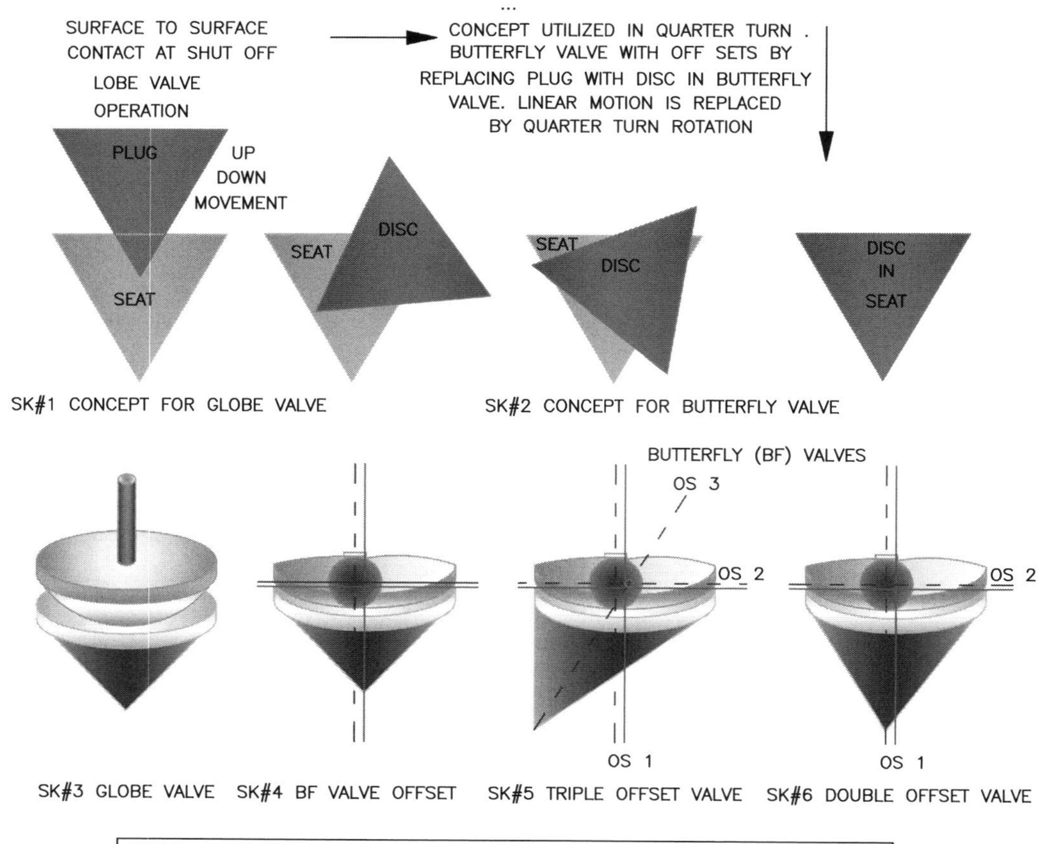

NOTES:
OS 1: CONIC AXIS IS OFFSET WITH RESPECT TO CENTER LINE OF PIPE
OS 2: DRIVE SHAFT IS OFFSET WITH RESPECT TO AXIS OF THE VALVE
OS 3: ROTATIONAL POINT IS SHIFTED FROM CENTER TO SELECT OPTIMIZED ANGLE SO THAT THERE IS NO RUBBING AT SEAT.
IN TRIPLE OFFSET VALVE ALL THREE OFFSETS ARE PRESENT. HOWEVER, MANUFACTURING IS DIFFICULT DUE TO THIRD OFFSET. HENCE COST IS HIGH.
IN DOUBLE OFFSET 3rd OFFSET IS MISSING AND SPECIAL SEAT TYPES ARE USED.

FIG. 6.21 Concept of triple and double offset butterfly valve.

TABLE 6.17 Specification for TOVs/DOVs

Sl.	Specifying Point	Standard/Available Data	To Specify
1	Body material	ASTM A 216/217/351/352 Gr WCB/WCC/LCB CF3 CF8M/EN 10213 1.4408	
2	Disc material	ASTM A216/217/351/352 Gr.WCB/WCC/LCB CF8M/EN 10213 1.4408 or ASTM A182 Type F316/EN 10222-5 1.4401	
3	Laminar seal	SS410+ Graphite	
4	Solid seal ring	UNS S20910 (Nitronic 50®) hard faced (courtesy: Vanessa)	
5	Bonnet	ASTM A351 CF8M/EN 10213 1.4408	
6	Shaft	ASTM A479 Type SS 410/Type XM19 (Nitronic 50®)	
7	Gasket	SS 316+ graphite	
8	Bearing	316 SS	
9	Packing	Graphite	
10	Application	TOVs/DOVs find their applications in process plants. In power plants, stream drains are major applications	

3 DAMPERS AND MISCELLANEOUS OTHER FINAL CONTROL ELEMENTS

3.1 Introduction

Dampers/louvers, vane, impeller, etc. in power plants are used to control the flow of gas through fans and/or ducts. There are a few other kinds of dampers like a butterfly damper and a guillotine damper. In this section, dampers and other control means such as a blade pitch as in axial fans, and other controls such as a hydraulic coupling scoop tube and variable frequency drive (VFD) to cover miscellaneous FCEs. Various dampers used are multiple blades/louvers and radial vanes.

3.1.1 Control Dampers

A control damper basically consists of a frame, blade, axles, linkage seal normal parallel dampers (Fig. 6.22B) are similar to butterfly valves. and it gives the flow directed toward a parallel opening. In an opposed blade configuration, flow direction is straight, similar to what has been shown in Fig. 6.23A. Opposed blade dampers give better control over the entire range of control, whereas, a parallel blade is better suited for control, e.g., 70%–100% flow. These control dampers are generally installed in the duct with a flange as shown in Fig. 6.22C). There is a special kind of control damper known as an Airfoil control damper that is used when extreme low leakage is desirable for a system with medium- to high-pressure and velocity. In power plant applications, some consideration should be given while designing the control damper and damper actuator. For example, a multiple burner tilt actuator requires extremely high torque, whereas long stroke length is necessary for a damper's actuator in air application.

1. Dampers/vanes connected with various types of fans need special attention. In power plants, the large fans like FD/ID/PA fans are generally centrifugal fans (outlet flow at 90° with inlet flow) or axial fan, which have a flow that is parallel to the shaft that rotates the blades. Air flow through these fans can be controlled by an inlet vane, fan pitch control, or fan speed (discussed later).

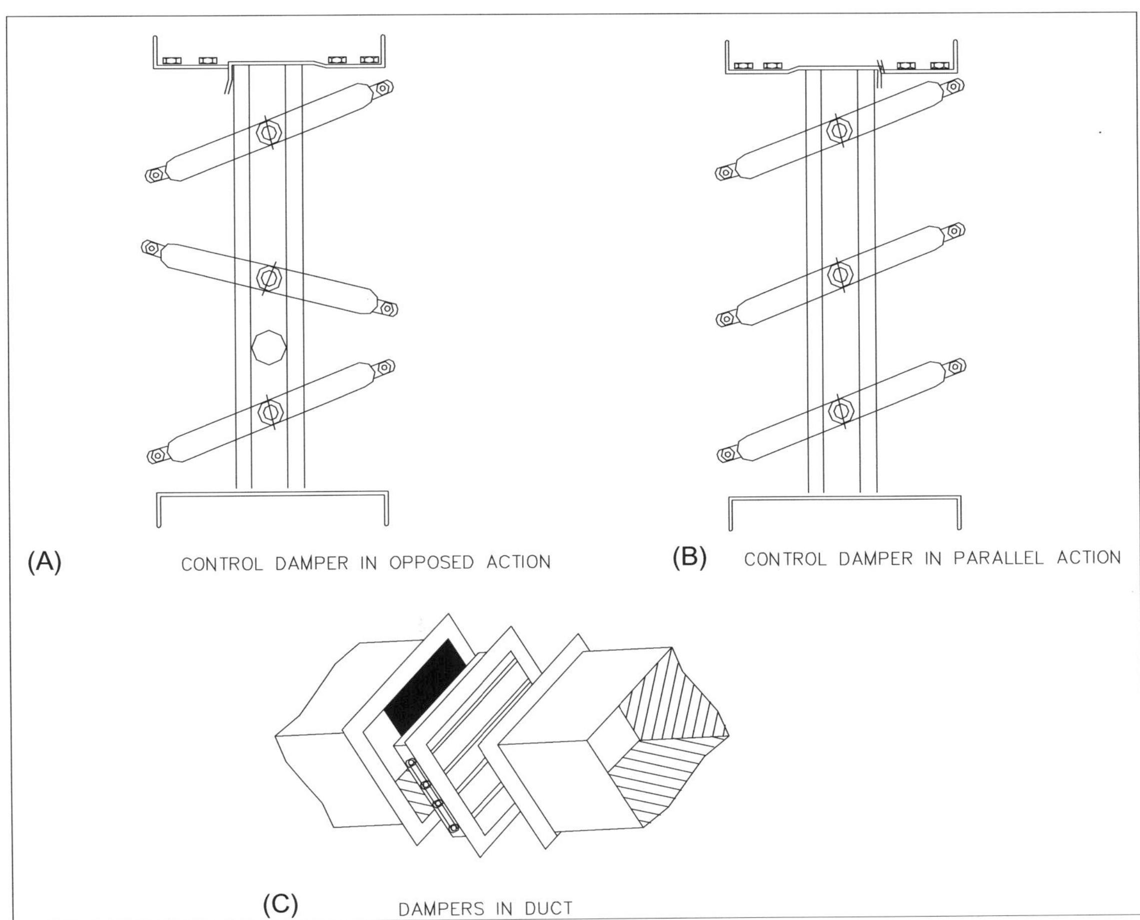

FIG. 6.22 Control damper.

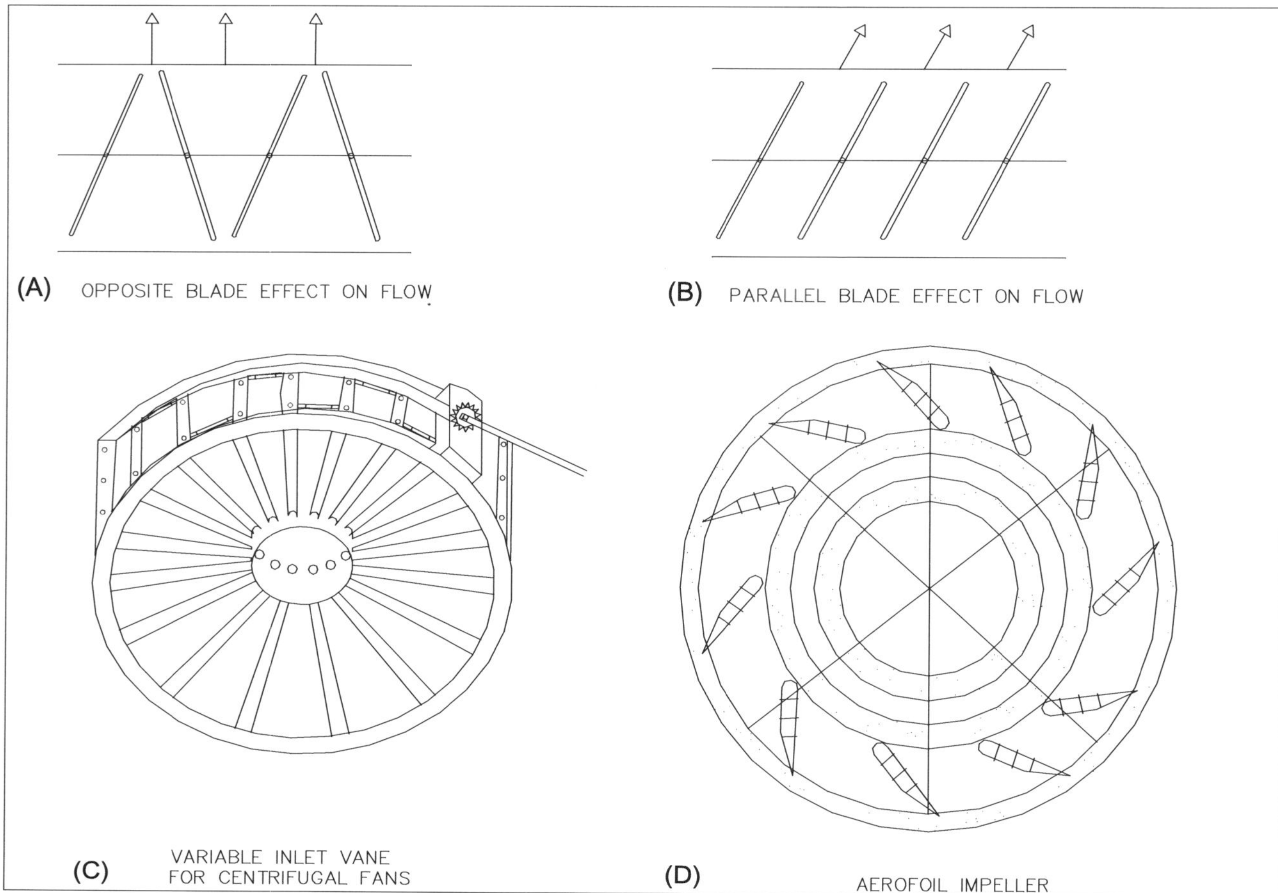

FIG. 6.23 Dampers, variable vanes, and impellers.

3.1.2 Vane/Damper Control

There will be a damper that is sized to handle the maximum air velocity. Outlet dampers used to throttle air flow can be a parallel type as well as an opposed type as shown in Fig. 6.23A and B. When the outlet dampers are throttled, resistance will cause a drop in fan pressure (i.e., energy loss), so this is not a good choice for utility boilers. Generally, outlet dampers in utility boilers are used as ON/OFF isolating dampers. There can be an inlet box damper for the fans in all types of boilers, but mainly used in clean air application. Parallel blade inlet box dampers have less power efficiency than variable angle inlet vanes. So, in a utility boiler, variable inlet vanes are popular. These inlet vanes could be the nested type (used when there is no duct and space constraints). Another type is an external variable inlet vane.

3.1.2.1. Variable inlet vane (Fig. 6.23C): This control is used for *centrifugal fans with a fixed speed* to control air flow through the fan. This can also be applied to *fixed speed, fixed pitch axial fans, (fans with* higher pressure and velocity). In most of the cases, these are external to the fan so they are accessible to linkage, for maintenance and flexibility in the mounting location. Variable inlet vanes can alter the fan performance by preswirling the gas before it enters the impeller (in contrast to the loss of power associated with conventional dampers). When the vanes are fully open, resistance is extremely low. As the vanes close (change in angle), the air enters the system in the direction of the rotation as the impeller. This causes a reduction in the total pressure generated across the impeller and a reduction in power. So, the variable inlet vane is an energy saving scheme when compared with conventional dampers. Depending on the operating condition and damper size, there are two options available:

- Cantilever blade: Used in high temperature flue gas containing corrosive and abrasive materials. The vortex vanes are designed to offer the least resistance. In corrosive application, these are used with an inlet cone, and the vanes send the air to the eye of the impeller for accurate and better control. Single and double vane designs are available.
- Center hub supported blade (Fig. 6.23C): This option is used for comparatively clean air having higher velocity

and pressure. In a variable vane controlled system, the fan selection is generally done with full pressure and flow with no vane in place. So, when a vane is in place, the maximum performance is brought down. It is worth noting that the variable inlet vane control is extremely sensitive at a *lower load*, meaning that with a slight change in the angle there will be a larger flow change, whereas, the same characteristic is reversed at a higher load (nonlinear).

3.1.3 Blade Pitch Control (Fig. 6.23D)

The blade pitch control is applied to the axial fan with a fixed speed to control the air flow. The blade angle is adjusted, with the help of a hydraulic actuator (fixed with the shaft), from minimum to maximum position, to ensure maximum possible efficiency. It is worth noting that in these cases, some important and critical items include: LO system, sliding bearing, and suitable oil control sealing (especially a stationary hydraulic control valve and rotating hydraulic cylinder). The flow change with a change in the blade angle is proportional, so linear characteristics are possible. The response time of this type of control is quite good, in the tune of <30 s (better is possible). This design is in use for FD (may be also for PA/ID/FGD) fans. Aerofoil-shaped blades are fixed to a hub mounted for axial sliding motion on a drive shaft. The other end of the link is connected to the driving shaft of the hydraulic actuator. As the hub moves axially, a rotational movement is transmitted through the pin and link arrangements to the fan blades to vary their pitch. This force is balanced at the actuator by a spring force. This way of controlling the gas flow is linear with the angle of the impeller.

3.1.4 Speed Control (Applicable for Both Fans and Pumps)

The most energy efficient way to control flow through a fan is speed control. Because the fan speed control is only possible with the help of an external system, any inefficiency of the system should be introduced into the overall efficiency of the system. Also, cost is another factor, which increases as the unit size increases. However, the electrical method may not be that expensive. There are various methods for control of fan speed viz. two-speed motor fluid, drive speed, control, and variable speed drive. Out of the three methods, the last two are used extensively.

3.1.4.1. Hydraulic coupling for speed control: Hydraulic coupling working on a hydrodynamic principle consists of one pump more termed as *impeller* and a reaction element; a *turbine* generally known as a *rotor* enclosed in suitable casings, as shown in Fig. 6.24B. For understanding the scheme, a schematic has been presented in Fig. 6.24A). Both the impeller and rotor face each other with an air gap. An impeller with a shaft is connected to the prime mover motor of constant speed. A rotor shaft is connected to the driven equipment (e.g., ID fan or BFP) through suitable arrangement. Oil is filled in the coupling. There is a fusible plug that blows off and drains in case of sustained overloading. There is no mechanical coupling between these shafts. The power is transmitted from the drive to the driven equipment by virtue of a fluid filled coupling. The rotation of the impeller imparts kinetic energy to the fluid in a closed circuit, and this energy is converted in the rotor into mechanical energy to rotate it as shown in Fig. 6.24B and C. Fluid flow in a closed circuit from the impeller to the rotor is through the air gap at the outer periphery, whereas, the same from the rotor to impeller through the inner periphery. In order to establish a flow of fluid from impeller to rotor, there should be a *head difference*, hence, there will be a *difference in the speed due to the slip developed.* Essentially this difference in speed is known as a slip; an important factor and inherent characteristic of the coupling. As the slip *increases* more and more fluid, hence, *more torque* is transmitted. The system is not linear in the power transmission, but in proportion to the *cube of the speed of the driving machine. Both* sides develop torque in proportion to the *square of the speed of the respective driven member.* One advantage of the system is that it can develop high output torque at low output shaft speed, *even* when the speed of the drive motor is high. Efficiency is *inversely* proportional to slip. In the speed control of the fan (ID fan) or pump (BFP), one sliding scoop tube enters through a clearance in the coupling rotating casing to vary the oil quantity. According to the position of the scoop tube, the fluid above it is drained out and as a result the amount of coupling changes. The speed of the driven machine depends on this and is controlled accurately. This system offers smooth control of the speed. From the control loop electronic signal after being converted to a pneumatic signal regulates the stroke of the power cylinder used to position the scoop tube (Fig. 6.24A–C).

3.1.4.2. VFD for speed control: In Fig. 6.26, the basic building block for a VFD has been shown. AC motor speed(S) is given by: $S = (120 \times f)/(P)$ where f is the frequency of the supply and P is the number of poles. Out of the two parameters responsible for motor speed, P is not changed generally, as it calls for a physical change in the motor and rewinding, but it is easier to change the frequency of supply.

As long as the f/P ratio is maintained, the rated torque can be developed. Fig. 6.25 shows the voltage frequency characteristics. This indicates that whenever speed of an induction motor is controlled both frequencies as well as voltage are controlled in order to get a different torque. This ratio is varied in the input section, which consists of an isolation transformer, a rectifier circuit, and a DC bus section comprising suitable filter, di/dt arrestor, etc. The isolation transformer separates the system from the input supply,

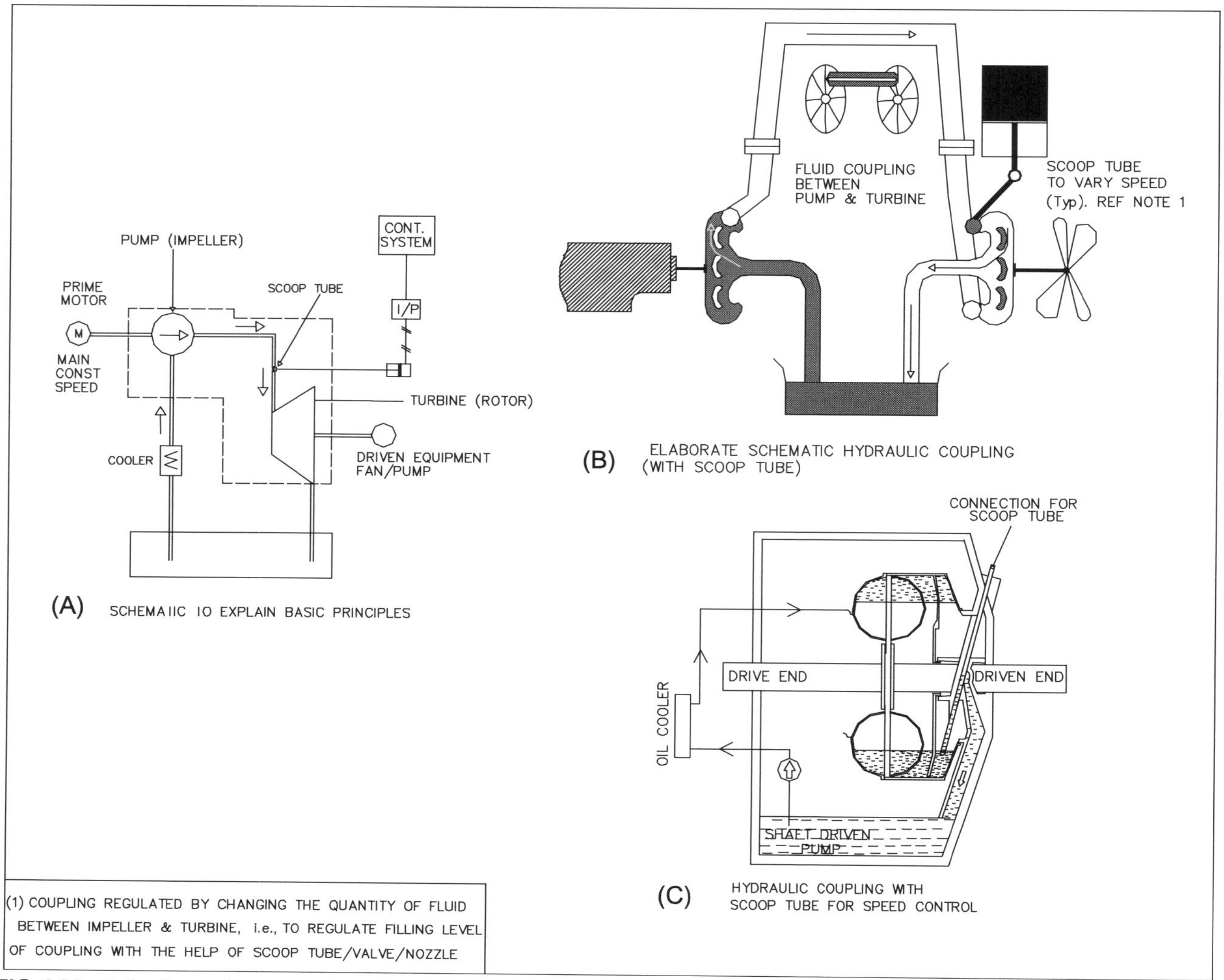

FIG. 6.24 Hydraulic coupling scoop tube.

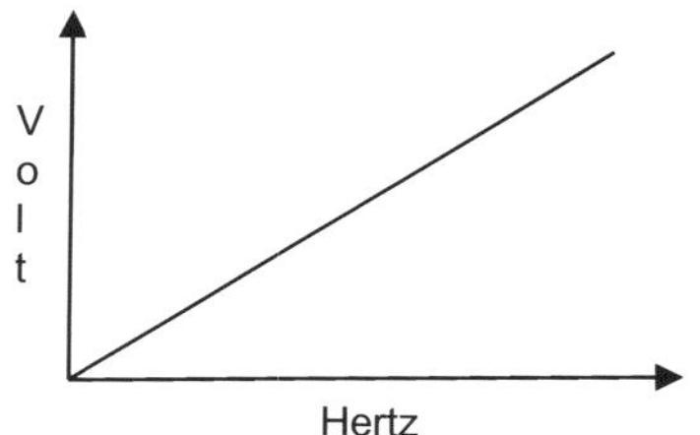

FIG. 6.25 Torque development.

and also helps to develop a multiphase rectifier circuit as shown in Fig. 6.26. This shows 12 diodes in a multiphase rectifier circuit to convert the supply AC (50/60 Hz) to a DC circuit. There is a filter to smoothen out the DC voltage. In fact, the multiphase rectifier circuit is used to get better DC voltage. The smoothener DC is fed to the inverter section. The inverter DC voltage is transformed into AC voltage (refer to a standard book on Power Electronics for inverter action), with the help of Silicon Controlled Rectifier (SCR) or currently, with the help of an IGBT. (Insulated Gate Bipolar Transistor is an advanced three-terminal power semiconductor switching device working on minority carrier. It has high input impedance combined with the capability to handle high bipolar current. It combines the advantages of MOSFET and BJT; it is a voltage-controlled device.) IGBT is used to turn ON/OFF DC voltage to provide bipolar pulses of equal magnitude. In the control board, the set point is compared, and it regulates to turn on the waveform positive half or negative half of the power device. The longer the device is on, the higher the output voltage, hence, the frequency is higher. The shorter the power device is on, the lower the output voltage, hence, the frequency is lower. The power device is made ON/OFF by a carrier frequency also known as a switching frequency. The higher the switching frequency, the more the resolution of each of PWM, as is the smoother waveform of the AC signal. The typical frequency is 3–4 kHz. Fig. 6.27A

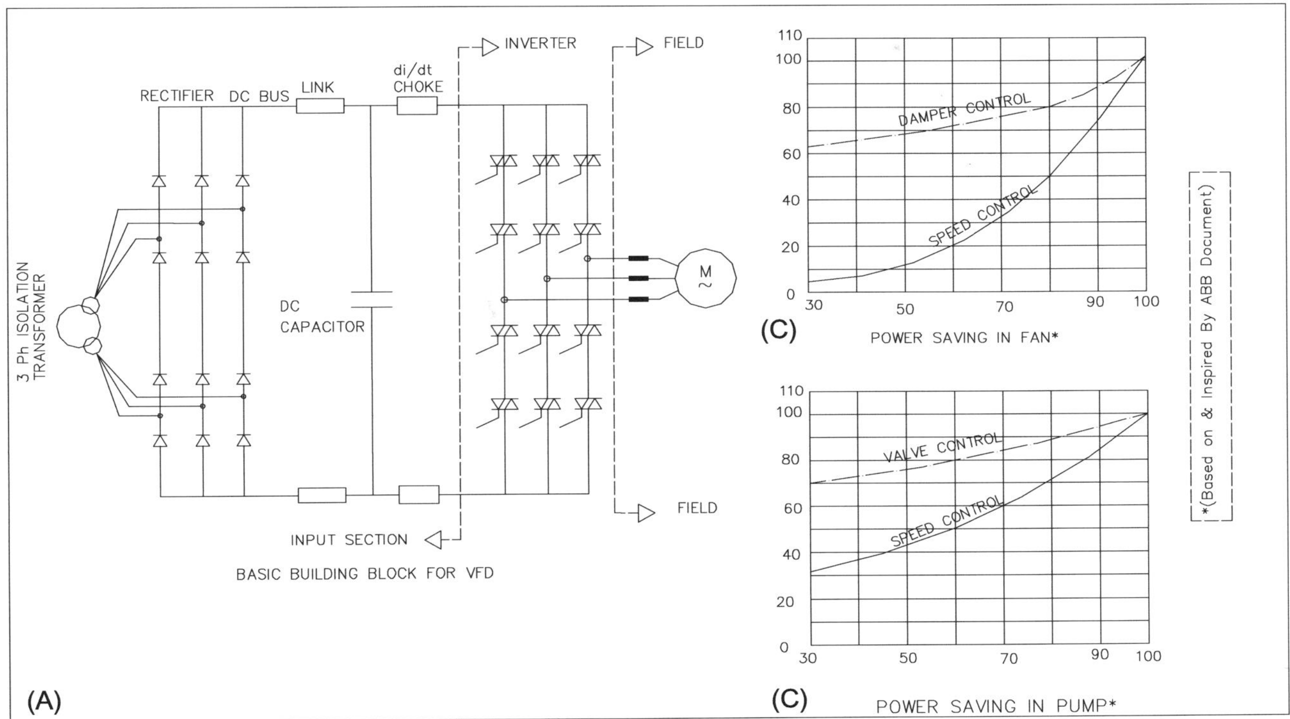

FIG. 6.26 Basic block diagram for VFD (3ph) and power saving diagram.

and B shows the inverter cells and the voltage levels. The typical waveforms have been depicted in Fig. 6.27C and D. Here, each inverter cell is to generate a voltage level. The typical energy savings in fan and pumps are shown as a percentage in Fig. 6.26B and C. It shows that the power saving is more prominent in case of a fan at lower loads.

As the discussion on speed controls ends, it is time to look into the details on actuators for FCEs.

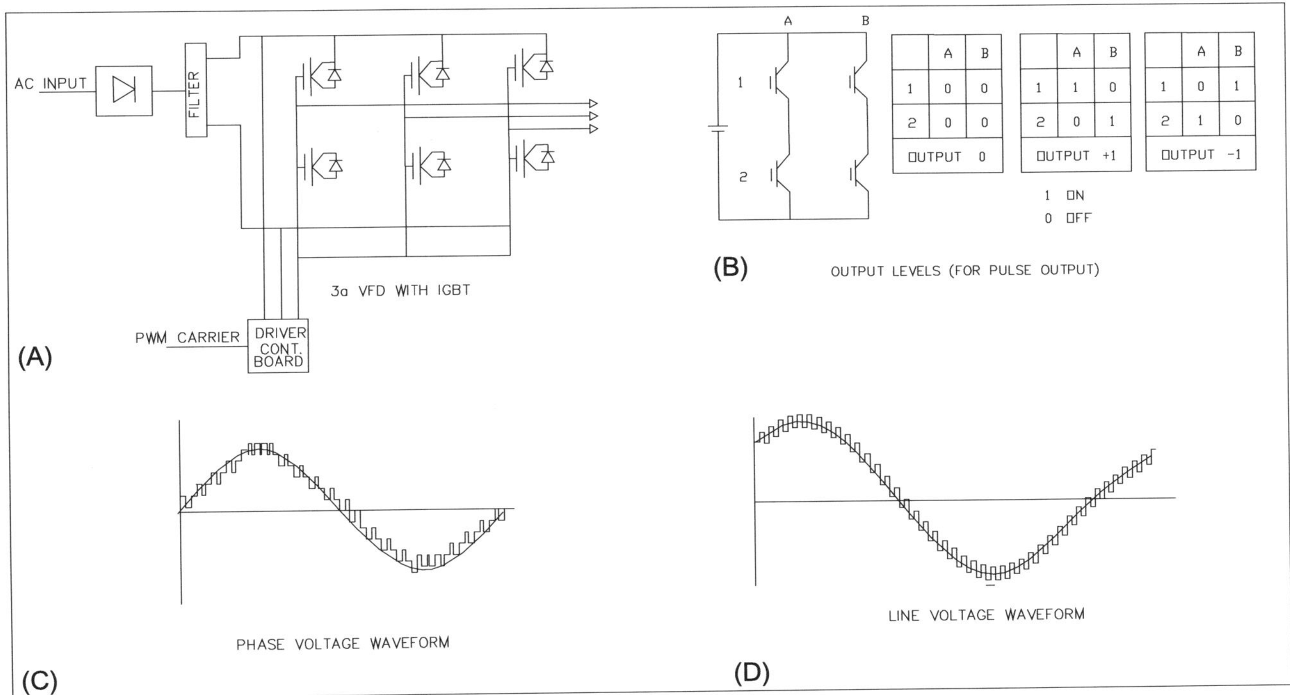

FIG. 6.27 Inverter circuit and waveform.

4 ACTUATORS

4.1 Introduction

The actuator's part in the FCE is to accept an external signal from the control system, and in response to the same, regulate the motion of FCE to modify its position to achieve the desired control. So, it has a twofold function; respond to the external signal and accommodate various accessories to achieve the desired movement.

In a power plant, all three types of actuators, viz. electrical, hydraulic, and pneumatic are in use. In many plants, the electrical actuator with power amplifiers is used in regulating functions to overcome excess inertia problems associated with the electrical actuators. However, electrical actuators are more often used in ON/OFF functions in power plant applications. In many control applications such as turbine (HP/IP) stem control valves, HP/LP bypass valves, inlet vane control or aerofoil, and hydraulic coupling scoop tube controls of fans, etc., hydraulic actuators are more popular. However, these actuators normally come as a part of the main equipment supply. A pneumatic actuator, which is also the recommendation of ISA, is more common in power plant applications. Actuators can be classified as a linear actuator and a rotary actuator. An electrical actuator such as servomotors are rotary actuators, whereas pneumatic power cylinders are a linear type. In power plants, the majority of hydraulic and pneumatic actuators are linear (rotary actuator for butterfly valve/damper).

4.1.1 Actuator Sizing

The force required for an actuator varies with the valve and actuator orientation. Discussion focuses on various other forces and then the sizing of the actuator.

Forces for actuator sizing:

1. Unbalance force at the plug: This results from the fluid pressure (ΔP) at shutoff. Generally, maximum upstream pressure is considered here. The unbalance area is chosen carefully, based on the configuration. Even for a balanced trim, there will be some unbalance force. This data are obtainable from manufacturer: As a rule of thumb, the following may be considered:

Unbalance area (UA) for balance trim may be: $(\pi/4)$ [(Cage ID)^2 – (Seat ID)^2].

UA for unbalanced trim could be $(\pi/4)$ (Seat ID)^2.

2. Seat load (SL): It is calculated from the shutoff requirement of the valve. A list of the recommended seat loads needs to meet the requirement for ANSI/FCI 70.2 or /IEC534-4 leakage tests. The data given below is based on standard manufacturer data. These data are available from all manufacturers and the same need to be consulted. Based on standard DP, the seat load calculation basis has been presented below.
   Leakage Class I: As required because no test is recommended.
   Leakage Class II: 20 lb per lineal in. of port circumference.
   Leakage Class III: 40 lb per lineal in. of port circumference.
   Leakage Class IV: 40 lb per lineal in. of port circumference Size $<$4-3/8 in. port diameter.
   Leakage Class IV :80 lb per lineal in. of port circumference Size $>$4-3/8 in. port diameter.
   Leakage Class VI: 300 lb per lineal in. of port circumference.
   For optimum performance of the boiler feedwater service seat load to be considered, 1000 lb per lineal in. for any DP.
3. Packing friction (PF): A table presented below for finding the packing frictional force for commonly used packing materials such PTFE graphite, etc. (Table 6.18).
4. Miscellaneous other forces: There may be several other additional forces like seal friction, etc. The manufacturer needs to share this information with the system designer.

Plug seal friction force can be calculated:

$$\text{Plug seal force (SF)} = 2\pi(\text{Cage ID}) + (0.02355)\left[(\text{cage ID})^2 - (\text{Seal Groove})^2\right] \times \Delta P$$

TABLE 6.18 Packing Material Selection

Packing Material	Temp Range (°C)	Sealing	Life	Packing Friction (kg)[a]	Remarks
Single PTFE	(–)46–230	Good	Long	9–45	Depending on pressure class and stem length
Double PTFE	(–)46–230	Good	Long	13–70	-DO-
Graphite	(–)198–538	Good	Long	45–450	(1) -DO- (2) High pressure withstand capability

[a] *This is to be chosen in consultation with the manufacturer.*

5. Actuator Output
   a. Actuator output (flow to close valve) = allowable ΔP $\times$ UA + SF + SL + PF
   b. Actuator output (flow to open valve) = allowable ΔP $\times$ UA + SL + PF (unbalance trim)

Here, remember to ensure that the pressure class of the valve is higher than the allowed pressure drop.
   c. As discussed in Clause 1.5 of Chapter VI, that

Direct actuator output = A (effective diaphragm area) $\times$ (Air pressure − Final spring pressure, i.e., spring pressure when travel reaches full travel valve).

Reverse Actuator output = A (effective diaphragm area) $\times$ (Air pressure − Initial spring pressure, i.e., spring pressure when valve starts moving).

In case of a rotary-type actuator torque, the sum of all torque is calculated. Power cylinders for the dynamic torques are calculated based on the mounting type and style. Then from the linkage arrangement, the force is calculated, and finally, from the above calculation, the power cylinder area is calculated.

4.1.2 Some ISA Guidelines on Actuators

Control valve technical specification: Courtesy of Control Valves: Practical Guides for Measurement and Control by G. Borden and P.G. Friedmann, published by ISA 1998:

After an examination of actuator sizing, the focus turns to some of the guidelines set forth by the ISA (At the end of each guideline, the reference is shown in brackets.) to check data from the supplier for the complete selection of the actuator.

4.1.2.1. Actuator systems should be a pneumatic type, any other than pneumatic is applied only as an exception (ISA guide p. 101). Here, one point to be noted is that in power stations, there are a few control valves where it is convention to use hydraulic actuators, viz. turbine (HP/IP) control valves, all control valves in HP/LP bypass systems, some actuators for fan controls boiler system, etc. In general, they form a part of main equipment supply.

4.1.2.2. A valve body size greater than 2″ and a pressure drop in excess of 4000 kPa diaphragm actuator shall not be used. (IAS guide p. 101).

4.1.2.3. Supply air pressure shall be <930 kPa g (IAS guide p. 101). (It may be better to specify a little lower value while specifying air supply pressure and to include an air filter regulator (AFR).)

4.1.2.4. Material for pneumatic tubing fitting, valve shall be at minimum 316 SS (carbon steel, copper, bronze, and SS 304 shall not be used) (IAS guide p. 101).

4.1.2.5. Valve travel position total accuracy shall be <2% and overshoot during testing (shop or field) shall not be greater than 2% (IAS guide p. 101).

4.1.2.6. Air failure position shall be accomplished without the aid of process pressure (IAS guide p. 101).

4.1.2.7. All applications need to be verified for the actual stroke speed requirement. Boosters may be used to stroke time requirements, but stroke movement shall be stable at 20%, 50%, 80% control signal steps (IAS guide p. 470).

4.1.2.8. Bolted yoke assembly shall not be used (IAS guide p. 470).

4.1.3 Pneumatic Actuator Selection

Refer to Table 6.19 for the selection of pneumatic actuators:

4.1.4 Piston Cylinders

As shown in Fig. 6.28, in this arrangement the piston actuator can be used in linear as well as in a rotary application with suitable linkage, feedback spring and cam, etc. In a control application, piston cylinders are used with a positioner. At a steady state condition of the pilot valve, the air supply ports to both sides of the main piston are closed, and exhaust ports are also not connected. Based on the control signal, the diaphragm will push or pull the arm, and as a result, the pilot valve will be moved up or down. When the diaphragm is pushed down due to a higher control air pressure signal, the pilot valve piston is also pushed down to the extent that is proportional to the additional control air pressure. As a result of this, the exhaust port of the lower side of the pilot valve will remain closed, and the air supply port will open, while the exhaust port of the upper side will open proportionately, and the air supply port will remain closed, as seen in Fig. 6.28B. During this period, the air supply will be connected to the lower part of the piston. This will cause the upper part of the piston cylinder to become pressurized, and the lower part drains. As a result, the lever will move down to change the position of FCE. When there is displacement of the lever arm, there will be feedback to the main beam via linkage, lever, feedback spring (getting compressed), etc. as shown in Fig. 6.28A and B. This displacement of FCE will stop when the force due to air pressure on the beam will be balanced by the feedback spring. The reverse will happen when there will be lower pressure in control air pressure, as shown in Fig. 6.28C. Through a suitable linkage arrangement, the same actuator can be used for a rotary application. For zero and range adjustment there are separate screws for calibration. There various mounting arrangements of the power cylinders. One type is straight mounting/fixed mounting, another type is trunnion mounting. Trunnion mounting is generally used when the load moves in an arc as in various damper controls, to transfer the complete thrust. The cylinder sizing is also influenced by the type of mounting possible at the site.

4.1.5 Safe Failure

There are a number of options available before the designer to choose fail safe mode.

Generally three types are chosen. These are fail open (FO), fail close (FC), and stay put (SP) (which means to

TABLE 6.19 Salient Points for Pneumatic Actuator Selection

Feature	Diaphragm (Linear)	Piston (Linear)	Rotary Diaphragm	Rotary Piston
Thrust	Moderate	High	Higher torque possible	Higher torque possible
Speed	Slow	Faster		Faster
Stroke	Short	Longer	Short	Longer
Fail safe	Possible without additional accessories	No fail safe without accessories	Possible without accessories	No fail safe without accessories
Design	Simple compact	Simple, mounting may be special	Complex with mounting sometimes special	
Air pressure	Low to moderate	Higher	Moderate	Higher
Control	Excellent with/without control device	Excellent with control device	Good control with suitable device	Good control with suitable device
Other features	Lacks stiffness	Higher stiffness	Easily reversible	
Application	Control linear valves	Linear valves and damper	Rotary valves, e.g., ball, butterfly	Rotary valves of higher sizes

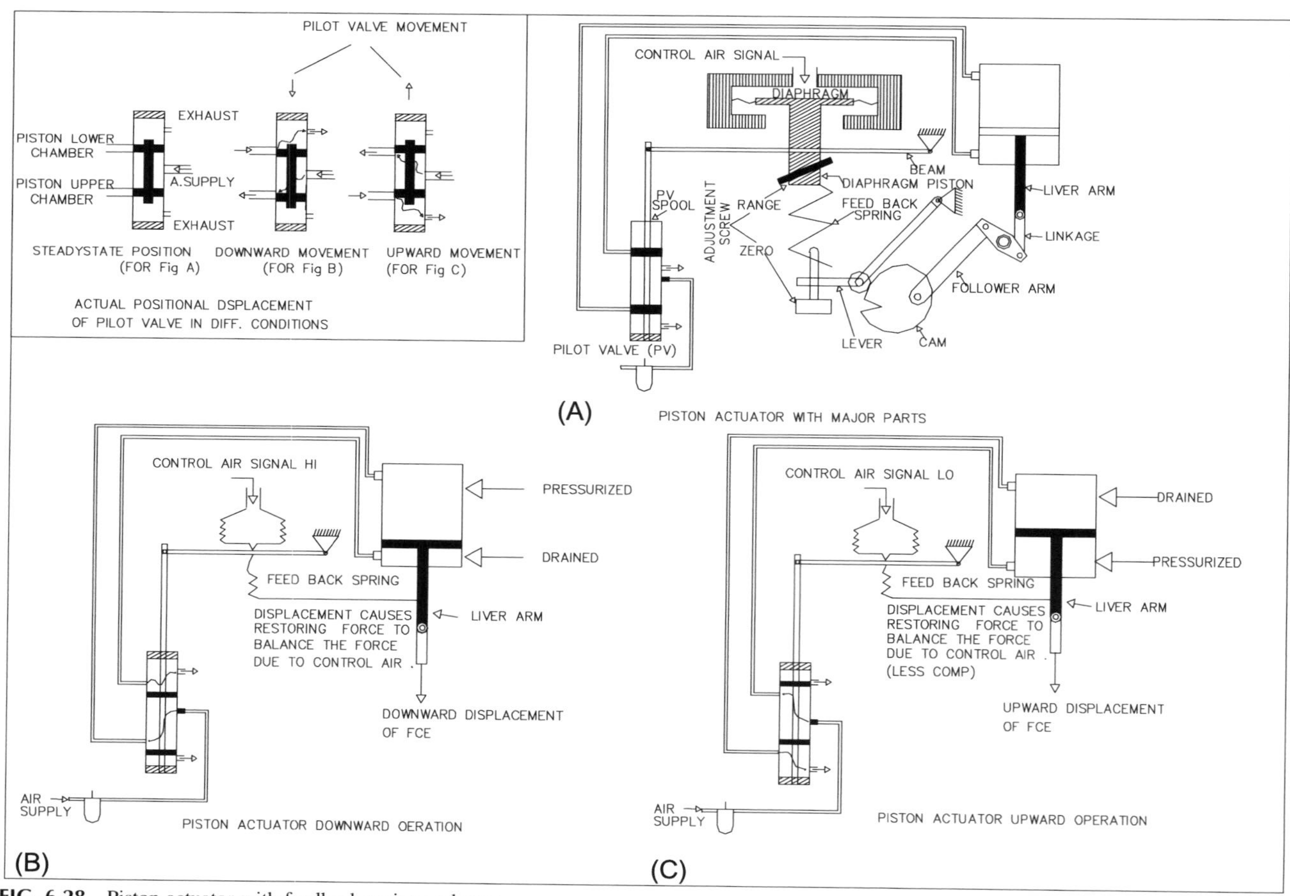

FIG. 6.28 Piston actuator with feedback spring and cam.

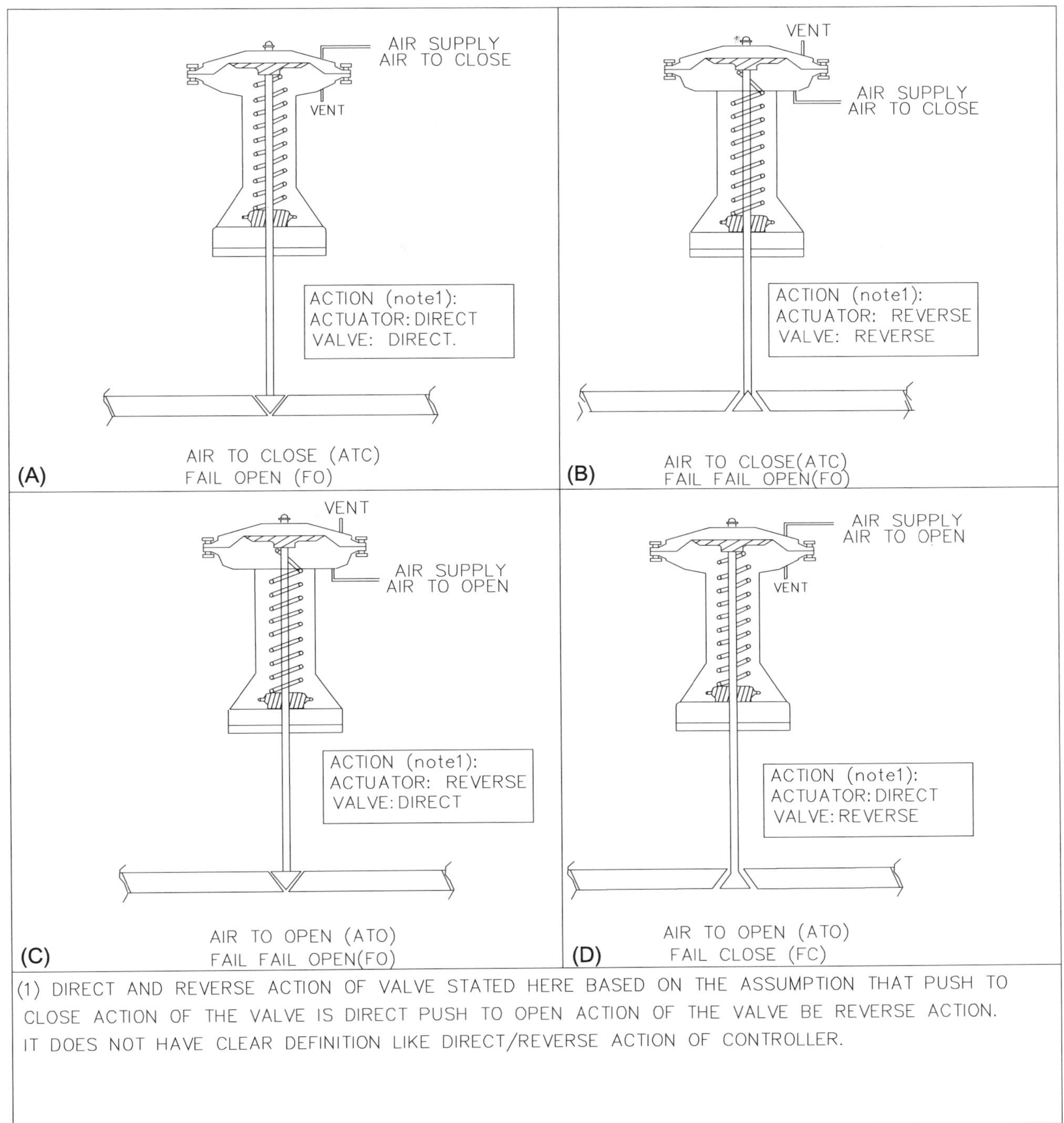

FIG. 6.29 Fail safe condition for linear actuators.

maintain the old position prior to the failure) or fail lock (FL). The fail lock condition can be attended to by use of some accessories. As shown in Fig. 6.29A–D, four combinations are possible.

4.1.5.1. Direct action: air to close ATC: FO In Fig. 6.29A, air pressure is applied from the top of the diaphragm, so when increasing the pressure, the stem lowers. It is referred to as direct acting and it will compress the spring to close the valve (direct acting valve), that is, the air is required to close (ATC). This is FO, because in case of air supply failure, the valve gets naturally open due to a spring action.

4.1.5.2. Reverse action: ATC: FO In Fig. 6.29B, air pressure is applied from the bottom of the diaphragm, so by increasing the pressure, the stem rises. It is referred to as reverse acting, and will compress the spring to FO because during air supply failure, the valve opens due to a spring action, and returns to its original position.

4.1.5.3. Reverse action: Air to open (ATO): FC In Fig. 6.29C, air pressure is applied from the bottom of the diaphragm, so when increasing the pressure, the stem rises. It is referred to as reverse acting, and will compress the spring to open the valve (direct acting valve), that is, air is required to open (ATO). This is FC because during air supply failure, the valve closes due to spring action, and returns to its original position.

4.1.5.4. Direct action: ATO: FC In Fig. 6.29D, air pressure is applied from the top of the diaphragm, so when pressure increases, the stem lowers. It is referred to as direct acting, and it will compress the spring to open the valve (reverse acting valve), that is, air is required to open (ATO). This is FC because during air supply failure, the valve closes due to spring action, and returns to its original position.

4.1.5.5. Rotary actuators: These actuators work similarly to the FC/FO. In power plants, many dampers are required to be done like the same; however in most cases, piston actuators and/or electrical actuators are used.

4.1.5.6. Fail Lock: In power plants, many of the valves and dampers are required to attain FL condition (especially in a modulating application), that is, stay put condition. In case of air failure, to achieve a FL condition, it is necessary that air pressure at the diaphragm (or air pressure on each side of the double acting piston cylinder) be locked at the value where it was prior to air failure. This is done with the help of an air lock relay.

1. Air lock relay: It is a pressure sensing valve, as shown in Fig. 6.30A. With the help of a pressure adjusting screw, the spring pressure can be set. As long as the pressure on the diaphragm is above the desired pressure value. it can compress the spring to keep the valve open. As soon as

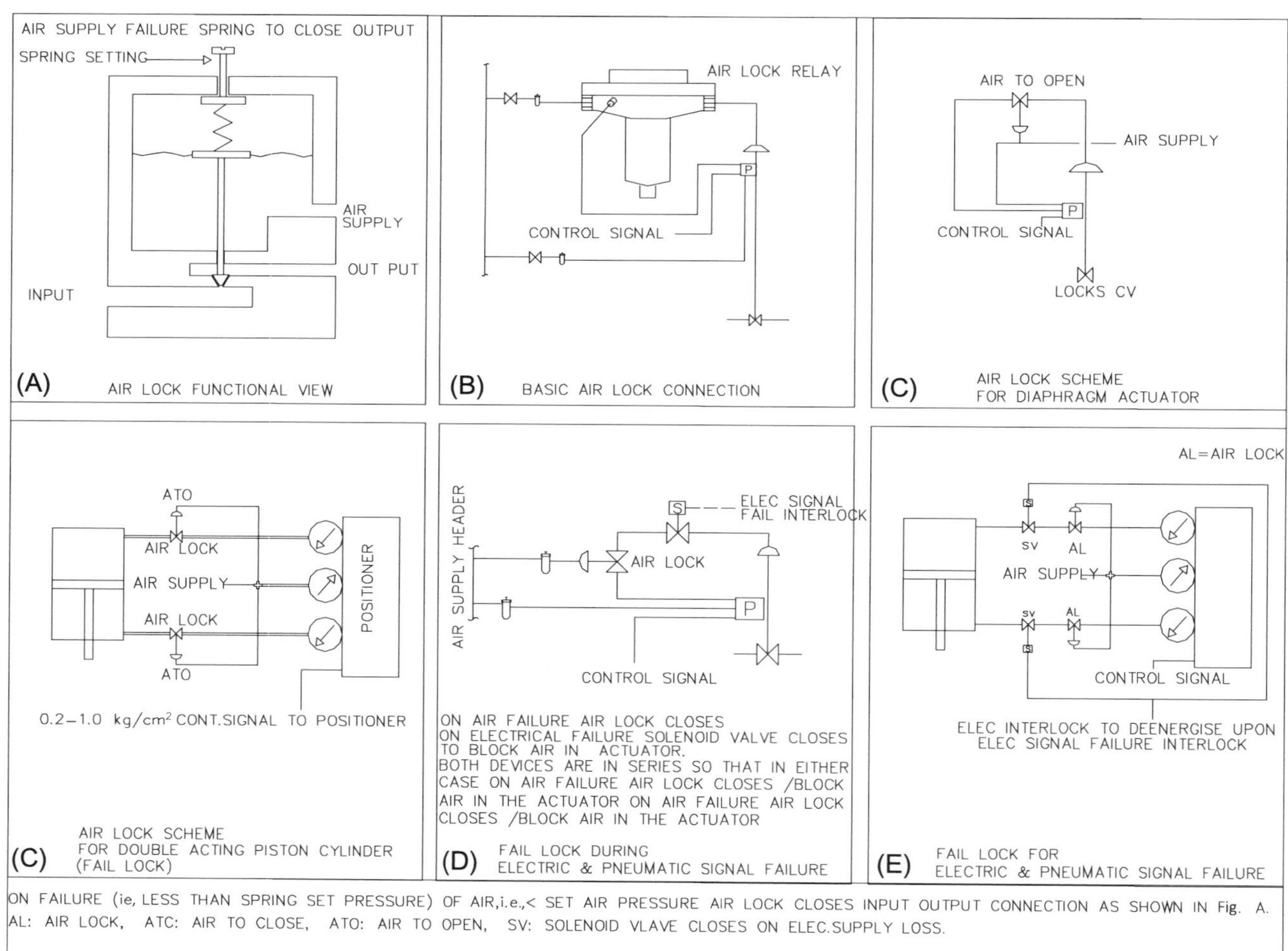

FIG. 6.30 Air lock: misc. fail lock condition.

the air pressure falls below the spring compression force, the spring will try to go back to its original position and the valve will close, isolating the input and output ports, that is, isolate the valve, or close the valve port so that it maintains its position prior to air failure. The air lock relay has a minimum of three ports such as the input/output ports, and supply port. There are a few other versions available to be used with solenoid valves viz. 3/5 port versions from Rotex.

2. Air lock (AL) connection and functions (Fig. 6.30B–D): The air lock relay is placed between the positioner (where in use) output and FCE input. The air supply to the positioner and the air lock is supplied through a separate AFR. However, the supply to these AFRs is from the same nearest air header as shown. Upon air failure, AL closes the path to the FCE so that it maintains the position to achieve FL. In case of a double acting piston cylinder, two such ALs are used.
3. FL during an electric signal failure is achieved by using a separate solenoid valve in a series with AL as shown in Fig. 6.30E and F. The electrical signal failure may be generated from the DCS; may be electric power failure or output port failure. Upon generation, the electrical signal failure interlock solenoid valve is closed, so that air pressure in the actuators remain unchanged to ensure FL.
4. It is worth noting that it is not possible to keep this locked position for long unless some accumulator, etc. is used. However, measures discussed above could be an immediate solution so that there is no bump in the system. Also, if the valve is at a partial open position, the exact position retention is not predictable for certain, so there may be some deviation, especially with time.
5. Electro-hydraulic actuator: Because of a fast response, high thrust, excellent throttling ability, with better stiffness, electro-hydraulic actuators are used extensively in power plants. In spite of having a few major advantages discussed above, there are a few limitations such as high cost, a heavier actuator, its performance depends on the temperature of the operating oil, the system is complex, and tuning is not easy. These actuators are used in turbine generation (TG) system,. HP bypass system, and in some cases, for impeller control of fans.
6. Electric actuators: Electric actuators are a combination of servomotor with worm gear to prevent reversal of drive direction by an unbalanced load. Generally, three-phase AC motors with a specialty to deliver higher torque (ranging up >80,000 NM) are deployed. However, single phase AC, or DC motors have also been used infrequently, These can be used for both linear as well as rotary drives. Electric actuators are compact in design and can deliver high torque. A slower speed (because of inertia), and the limit in duty cycle make it less popular in modulating applications, but with the latest technology, the majority of these problems have been overcome. Based on functions and movement mechanisms, actuators can be classified as follows:

- Multiturn actuator: Used for gate valves, damper applications.
- Part-turn actuators: Used for butterfly and ball valves, and damper application. The torque delivery may be for part revolution, e.g., 90°.
- Linear actuators: These are mainly used for globe valve, etc. However because of the low cost and the simplicity of a pneumatic actuator, it has an edge over its electrical counterpart.

Some essential parts of electrical actuators:

1. Worm gear: Worm gear is a gear in the form of a screw to prevent damage due to unbalanced forces. Worm gear and worm wheels have axis at 90° to each other, so power transmission is at 90°, which makes them different from spur gear. In a single start worm, for a 360° turn of a worm gear, the wheel moves one turn. Worm gears are mainly used for large gear/speed reductions in the range of 20:1, and even up to 300:1. Worm gears possess a unique feature such that the worm, when connected to the drive motor, can easily turn the gear, but the gear finds it difficult to turn the worm. The reason is that the angle on the worm is very shallow, and when the gear connected to the FCE tries to rotate it, the frictional force between the gear and the worm does not allow any worm movement. It is a self-locking system, and reverse movement is prevented.
2. Limit switches: When the dampers/valves reach the end position, the actuator needs to be stopped. For this purpose, limit switches are used. This is limit seating and actuator has to move with definite torque to ensure positive seating. Limit switches can be used for intermediate position, also.
3. Torque switch: This prevents the actuator overload due to over torque, as a consequence it may stop at any intermediate position whenever torque exceeds the set torque value. This also helps positive seating of valve/damper, and at the same time helps the actuator from being overloaded, that is, same for the actuator during jamming.
4. Motor control center (MCC), motor starter and integral control: The control action of the actuator needs to be controlled with the help of contactors/thyristors. There are a number of options available, and with the advent of fieldbus technology the choices are enhanced. A few of the control strategies are discussed below:
   a. MCC control in combination with PLC/DCS: The power circuit (with provision for MCC operation)

is a part of MCC, normally the interlock and remote operation manual command and/or auto interlock being cooked at PLC/DCS. In one MCC, there are draw out/nondraw out type modules to accommodate the electrical control for a number of motors (LT motor). Normally. there will be bus bars R,Y,B, phase, and neutral. In Fig. 6.31, a short specification for the bus bars is given. It should be noted that because this is a typical diagram, the data mentioned, e.g., 415 V 50 Hz, etc. are examples. They are only to indicate that for specification, all details like voltage, frequency, number of phase, fault level, as well as bus bar material (e.g., Cu or Al) need to be mentioned to enable the manufacturer to design the system.

b. Intelligent MCC: Currently, there are MCCs that are intelligent (e.g., in sum of ABB), so that these can be directly hooked up with PLCs/DCSs for exchanging data, that is, supervision control. These provide cost-effective protection and control for LT motors. It also helps to save the cost of cable and discrete elements in MCC. RS interface, status LEDs availability of local interface make them quite popular.

c. Built-in starter with the actuator: There are cases where there is a built-in starter module on the head of the actuator to accommodate contactors and external connections to PLC/DCS for manual/auto command as discussed above.

d. Smart actuators: Smart actuators could be pneumatic or electric. In modern days, the electrical actuators are available not only with a starter, but with the necessary control circuit that communicates with DCS or the rest of the system through the fieldbus system. Refer to Clause 4.8 for details.

e. Referring to Fig. 6.31, a short description of the system has been presented below. In the circuit, a typical three-phase AC servo motor has been shown that will reverse its direction of motion whenever any two phases are altered. In this case, *R* and *Y* phase has been reversed with the help of X_o (*F*) and X_c (*R*). The assignment of forward (*F*) and reverse (*R*) is purely arbitrary, only intended name two relays. Here, there are two selections, local and remote, which has shown that the actuator can be operated from two places. In reality, there may be many other selections, such as local, remote, MCC, unit "*n*" (viz. a valve in common systems like a CW system that may be operated from unit 1 or 2), etc. There may be a selection for test (to test the contactor operation) and normal operation, etc. In cases of series interlock (one assigned as safety interlock, which if any cannot be bypassed even during local operation (e.g., fuel shutoff valve cannot be opened before some interlocks are satisfied say PR. OK). Whereas, in a remote circuit this may be any safety interlock and/or sequential interlock in a series. Similarly, the "REM Close" or "REM Open" could be a combination of manual command and/or auto interlock (viz. whenever *one* FD stops its damper shall close). As discussed above, the FCE will stop operating as soon as the end position is reached or torque is > the set value. In this circuit X_o (*F*), and X_c (*R*) have been locked with self contact so it is in ON/OFF service, as FCE cannot be cutoff at an intermediate point unless STOP PB has been operated. If such self locking contact is removed, then the circuit could be used in modulating (*inching*) service and in that case STOP PB may be eliminated, because in the case of inching operation (without self-locking contact), the actuator will operate as long as opening/closing command is available (by the push button/or from the auto circuit). As a standard practice, a thermal switch has been put in a series in a power circuit.

5. Special considerations for electrical actuators in modulating services: In power plants, electrical actuators are quite common, in fact it is preferred for isolation services, e.g., isolating dampers for fans or main steam stop valves, etc. In modulating services, FCEs need to be operated almost all the time depending on the output of the controllers, and as a result of this there is the possibility of *overheating* of the gear and the motor. Another problem associated with an electrical actuator is its inability to go to the *desired position (full open/close)* upon failure of the motive power. This is because the spring size needed would be much higher considering the power delivered by the electric motor. However, it is inherently stay put, which is an advantage in many services. With the advancement of technology, some of these problems have been overcome. These include:

- Failure to go to the desired positions have been achieved by the super capacitor* that can not only go to the end position, but any desired position can be achieved.
- With the use of accurate sensors, and advanced technology it is possible to circumvent the problems discussed above and overall resolution, coupled with high speed of response is possible. For example, with power amplifiers, it is possible to have a much better response time by circumventing an inertia problem. Avoidance of dead time (in a pneumatic actuator, it is sometimes necessary to overcome the static friction, and in case of small changes in a modulating loop this is significant) and a small amount of overshoot is a great advantage of an electric actuator over its pneumatic counterpart.
- Enclosures available in IP 67/68.
- Intelligent actuators are possibilities.

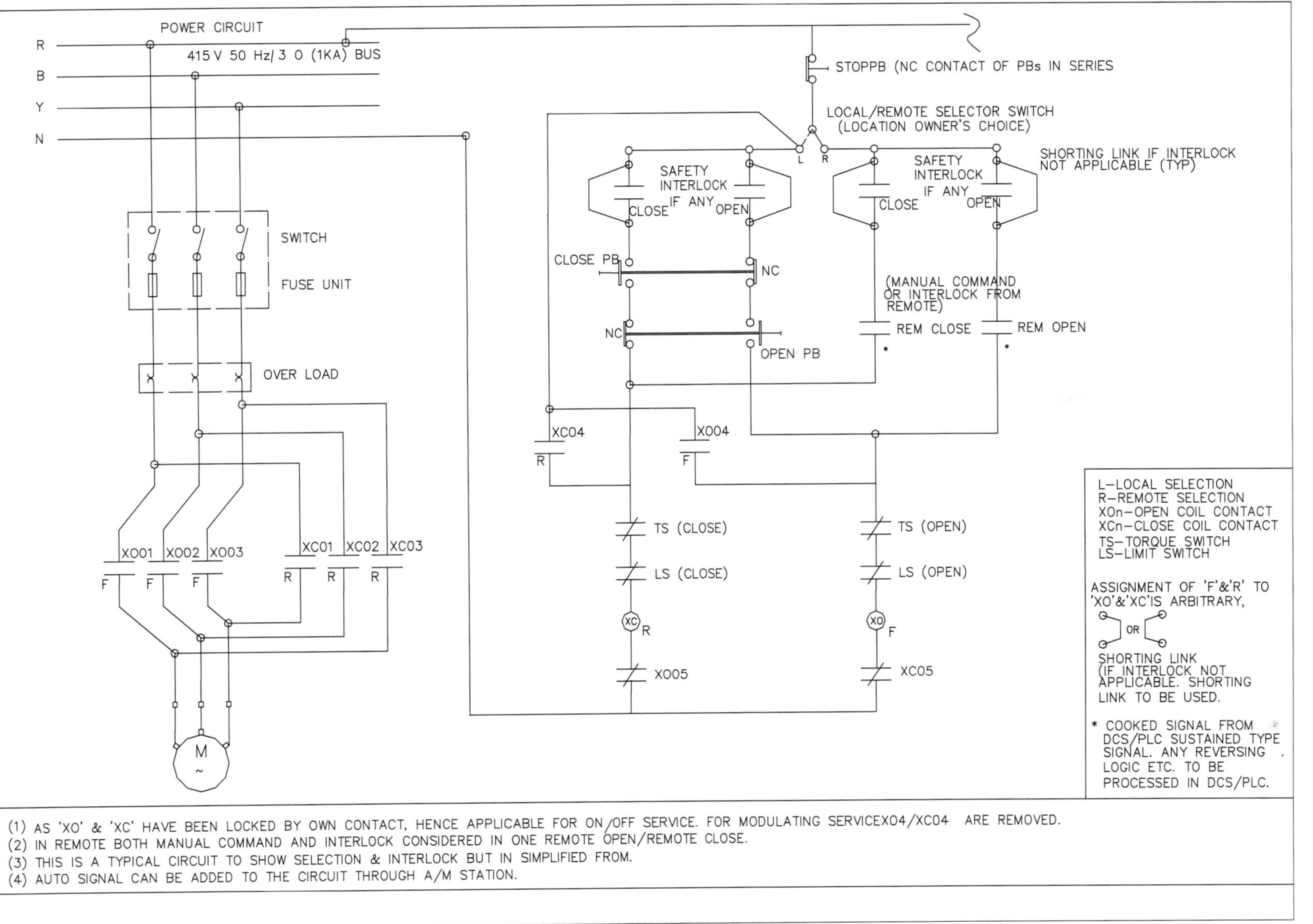

FIG. 6.31 Electrical schematic for electrical actuator.

**** Super Capacitor : This is one of the latest device (in this millennium) often referred to as electrical Double Layer Capacitor. Unlike Normal capacitor it has electrolyte and electrodes like battery. Ordinary capacitors are specified in Micro, Nano, Pico Farads. Super capacitors are specified in Farad. It has energy density in the range of 1–10 Wh/kg. For high voltage connections these capacitors are used in series. It could to used in Actuator control to hold energy in case of main power failure. These can be recharged and discharged many times. Unlimited cycle time, Rapid charging and low impedance are the major advantages of the device.***

FIG. 6.32 Super capacitor.

- Self-diagnostic features.
- Low-power device that could be used even in conjunction with solar power (Fig. 6.32).

4.1.6 Intelligent Actuator

Intelligent actuators have built-in intelligence to work in tandem with other devices in a third-party network based on open systems like fieldbus, Profibus, etc. The heart of such actuators is the microprocessor, which allows improved positioning and communication with the external world.

Due to the intelligence, it can provide electronic positioning with the help of the PID controller, and it can operate on external analog/digital setpoint. Salient features:

1. System can provide field supervisory system, and can provide additional data for system optimization.
2. Self-diagnostic feature for critical applications.
3. Built-in PID control for external set point tracking and critical control.
4. Easily adaptable of control loop configuration.
5. Retrofitting where additional hardware would otherwise take more space.
6. Huge saving in the cost of cable when connected in two-wire system of connections.
7. In some configurations, a number of such actuators are connected to a valve control master station, which actually communicates with an external DCS deployed to control a number of valves, dampers, etc. This is shown in Fig. 6.33A. In this configuration, complex control algorithms of the field units reside in the master station which exchange data and command with DCS. Cable cost is significantly reduced, but major control is in the field.
8. Intelligent actuators are also available to support open connect systems such as foundation fieldbus as shown in Fig. 6.33B. As per foundation fieldbus connection

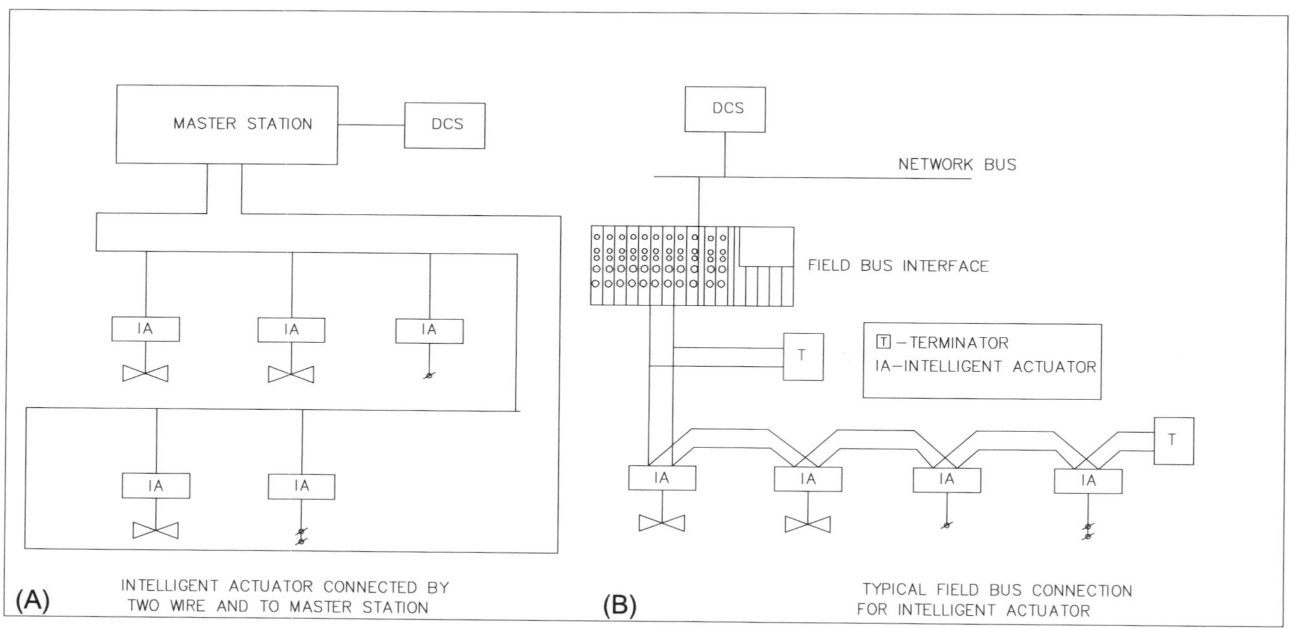

FIG. 6.33 Intelligent actuator connections.

methodology, there will be a terminator at the two ends. Also, systems can support PROFIbus, Devicenet, MODBUS, etc. to connect the intelligent actuators to open systems.

5 ACCESSORIES

5.1 Introduction

Major accessories required for satisfactory operation of actuators have been discussed here. Table 6.20 elaborates on the functionalities of these items.

Based on the function desired, the designer chooses the required ones for the application.

5.1.1 Air Filter Regulator

In a plant, there will be one common instrument air line with pressure maintained nearly at 7 kg/cm^{2}g (Table 6.21). Whereas, the air supply requirements for control valves, power cylinders, I/P converters vary. AFR is used to supply air to the pneumatic device at the required pressure. The other main function of AFR is to remove oil, moisture, and other contaminants from the air supply to save the costly pneumatic instruments/actuators. It also helps to remove pressure surges as well as protects the devices from low temperature or high dew point. Various parts of the AFR are detailed in Fig. 6.34A.

5.1.2 Air Lock Relay (Fig. 6.30A)

In many applications, to lock the actuator of the FCE at the last position where it was when air supply failure occurs (e.g., rupture in the line) (Table 6.22). In case of a double acting cylinder, there will be two numbers, such air lock relays where in other applications, e.g., diaphragm actuator, etc. One such will suffice. Schematic of air lock has been depicted in Fig. 6.30A with application details in Fig. 6.30B–F.

5.1.3 E(I)/P Converter (Fig. 6.34A and B)

This converts the electrical control signal in to a pneumatic one. In the case of an intelligent positioner, this forms part of the positioner. Out of various types of E(I)/P converters available in the market, one that works on a force balance principle is the simplest and most popular. Mounting details and working principles have been shown in Fig. 6.34A and B. There is a plunger coil near a permanent magnet, with two poles on the opposite side of the fulcrum. When the

TABLE 6.20 Functions of Various Major Accessories for FCE

Accessories	Function	Remarks
Air filter regulator (AIR SET)	To supply air to the instrument, the final control element, and its accessories at the desired pressure. It filters out the air, and help to remove dust, and moisture, if any	Often called air set: Selection of the appropriate range for service is essential
Air lock relay	On air failure, it holds the last position of FCE	Fail lock operation
E(I)/H or E(I)/P converter	Converter used with FCE to convert a standard electrical control signal to hydraulic/pneumatic standard control signal	E/H converter discussed in connection with turbine (Chapter 9)
Hand wheel	To provide local manual control, override control.	*Side*/top mounted
Limit switches	To indicate the end and intermediate position. Also used to stop FCE at desired position	Essential for ON/OFF services
Positioner	Primary function is to ensure that FCE accurately reaches the desired position, which is demanded by the controller/control system	It has a few other functions explained while discussing positioner
Position indicator	Local indication of FCE, generally mechanical (scale-pointer type)	For local indication
Position transmitter	Device used to transmit FCE position at remote place (e.g., voltage, 4–20 m ADC)	(*) 0.2–1.0 kg/cm^2 also used
Solenoid valve	Solenoid valve is used as an accessory, generally for ON/OFF operation. For interlock and/or to attend the fail lock position in case of electrical signal failure or failure of modulating services, also	
Torque switch	To stop electrical actuators of FCE when allowable torque limit is slightly exceeded	Safety for electrical actuators
Volume booster	To amplify the pneumatic signal in volume, and/or pressure. Mainly used when a number of pneumatic actuators are to be operated through one pneumatic signal	

TABLE 6.21 Air Set (DS)

Sl	Specifying Point	Standard/Available Data	To Be Specified
1	Fluid	Compressed instrument quality air	
2	Inlet pressure (bar)	15–20	
3	Outlet pressure (bar)	0.4–2.1 or 0.7–4.2 or 2.8–8.8 adjustable	Sq. head screw/hand wheel
4	Temperature max °C	(−)40–100°C	
5	Filter size	5 μm (in general)	
6	Filter element	Sintered bronze/polypropylene	
7	Body	Die cast aluminum	
8	Seals	Neoprene	
9	Mounting	Pipeline, panel on actuator assembly with bracket	
10	Accessories	Over pressure vent/drain port	
11	Connection size	6/15 mm	To be specified
12	Pressure gauge	Yes (in dual scale) as accessories	To be specified

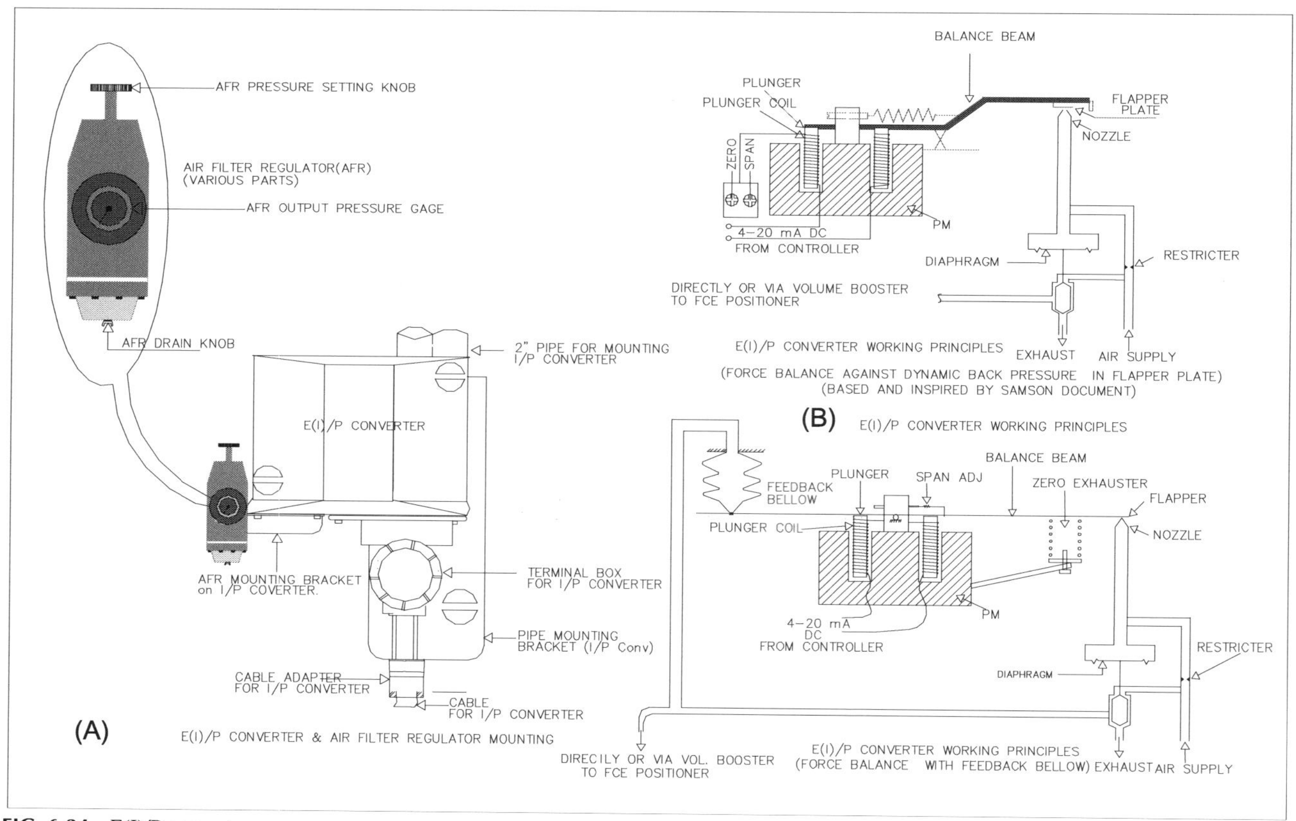

FIG. 6.34 E(I)/P converter.

TABLE 6.22 Air Lock Relay (DS)

Sl	Specifying Point	Standard/Available Data	To Be Specified
1	Fluid	Compressed instrument quality air	
2	Pressure (bar)	Supply pressure: 7 bar (e.g.)	
3	Temperature	(–)30–90°C	
4	Diaphragm	Reinforced rubber	
5	Trim	Brass with rubber sheet	
6	Body/ connection	Brass, aluminum/¼″ or 6 mm	
7	Switching point	Adjustable in 1.2–3.0 kg/cm^2	

system is in a balanced condition, the balance beam is in equilibrium to give a pneumatic output proportional to current into the device. When the current input to the plunger coil increases, the coil on the right side of the plunger is pulled towards the right side, so the flapper is pushed towards the nozzle. As a result, there will be an increase in back pressure and pressure in the diaphragm, which will try to close the exhaust, and the output air pressure of the E/P converter increases. This also causes an increase in feedback below to bring the balance beam to its equilibrium position once the forces are balanced. There will be zero adjustment to the bellow or the span adjusting spring. There could be other possibilities where the plunger coil force is balanced by the dynamic back pressure as shown in Fig. 6.34B (in the upper part of the drawing, *based on SAMSON document*) (Table 6.23).

5.1.4 Hand Wheel (Fig. 6.35A and B)

This FCE accessory is used to operate the valve manually at site (field). The mounting of the hand wheel may be at top and/or side. When mounted sideways, a suitable link and a clutch arrangement is necessary for it to operate. Normally, hand wheels have a neutral position, which allows the actuator to be operated remotely. Generally, a clockwise

TABLE 6.23 E(I)/P Converter (DS)

Sl	Specifying Point	Standard/Available Data	To Be Specified
1	Input	0/4–20 mA DC (for split operation 4–12 or 12–20 mA DC may be available (e.g., Model no. TEIP of ABB)	To specify
2	Output	0.2–1.0 kg/cm^2 or 3–15 psi, or 0.2–1.0 bar intermediate/higher ranges like 6–30 psi, etc. are also possible (e.g., Model no. TEIP of ABB)	To specify
3	Ex proof approval	Possible if applicable to indicate the desired certification standard	To specify
4	Characteristics	Linear—direct/reverse	
5	Air supply/ capacity/ consumption	1.4–5.4 bar (*Supply pressure*)/~2.0–8 Nm^3/h (*Capacity*)/0.16 Nm^3/h (*Consumption*)	All figures mentioned here are typical only
6	Encl. class	IP 65/NEMA 4	To specify
7	Hysteresis/ sensitivity	0.3%/0.1% of FSD	
8	Accuracy	<0.5% FSD	
9	Influence variables	Air supply/vibration/mounting position	Manufacturer std.
10	Connection	6 mm/¼″ NPT, etc.	
11	Materials of construction	Die cast aluminum/plastic manufacturer Std.	
12	Accessories	Mounting bracket for 2″ pipe mounting/wall mounting	
13	Environmental	(–)20–80°C 97% humidity	
14	Special feature	Construction ruggedness, vibration effect, lightweight, and dynamic response	Split range if desired (to specify)

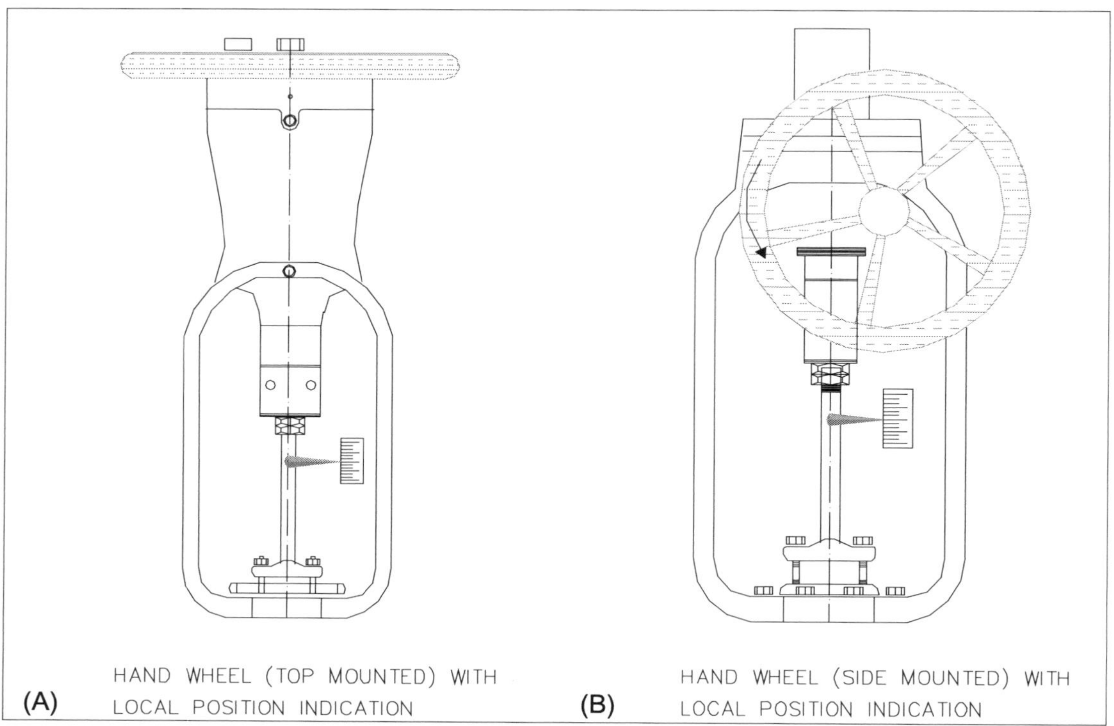

FIG. 6.35 Hand wheel and local position indicator.

TABLE 6.24 Hand Wheel (DS)

Sl	Specifying Point	Standard/Available Data	To Be Specified
1	Material	Ductile/cast iron	
2	Mounting	Yolk attached to the valve body	
3	Maximum travel	Valve specific	
4	Max thrust	Valve specific	

rotation of the hand wheel closes the valve and a counterclockwise rotation opens the valve (Table 6.24).

5.1.5 Limit Switches

As stated earlier, these FCE accessories are essential for electrical actuators to stop the motor at the end point. Also, they are necessary for the FCE for ON/OFF service to know the FCE position for all types of actuators. It is necessary to specify a few data for the particular application. In case of intermediate position, such position in percentage will be specified (Table 6.25).

5.1.6 Positioner

In its crude form, the positioner is basically a proportional controller with integration action to nullify offset. Its job is to ensure that the FCE takes the desired position as per the position demand from the control system. The positioner helps to avoid errors caused by packing friction, or any other jamming. Split-range operation is possible with the help of the positioner. A properly designed position will have optimum gain to achieve the desired position without overshoot and improve control loop performance. There are two types of positioners: (1) pneumatic positioner and (2) smart pneumatic positioner.

5.1.6.1. Pneumatic positioner: As shown in Fig. 6.36, when the input signal changes, for instance, it increases, the position of the diaphragm will then be pushed down. As a result, through the pilot valve, the supply line to the valve will get connected to the air supply line; the exhaust line will remain isolated. This will cause more air pressure on the diaphragm to push the valve to close with the additional force. As the valve closes, the positioner feedback cam will be pushed up to compress the spring to balance the force due to the air pressure. This will continue until the position spring force balances the input signal pressure multiplied by the sensing diaphragm area, so that both forces balance at the new assured position of the valve. So during this time, the supply line of the valve is connected to the air supply line pushing more and more air in the system to increase the pressure on the diaphragm actuator

TABLE 6.25 Limit Switch (DS)

Sl	Specifying Point	Standard/Available Data	To Be Specified
1	Function	Accurate poisoning of actuator	
2	Encl. class	IP 67/NEMA 4X, etc.	To specify
3	Ex proof approval	Classification as per NFPA/ATEX	If applicable
4	Contact type	Single pole double throw/COC, etc.	
5	Contact Mat. and type	Silver plated Cu/self-cleaning type	Snap acting
6	Contact rating	AC 3 A @240 V 1 Ph 50 Hz 3 A DC 0.3 A@220 V, etc.	
7	Insulation resistance	>100 MΩ	
8	Actuator strength	4-5 times operating force <1 min	In a direction
9	Mechanical resistance	High shock and vibration resistance	
10	Life	>1 million operation	
11	Environmental	(−)20–50°C 97% humidity	
12	Termination	Suitable for 2.5 mm^2 cable	
13	Speed of operation	Fast, but not damaging the actuator	
14	Manufacturer	The safety of the actuator depends on the reputed manufacturer	

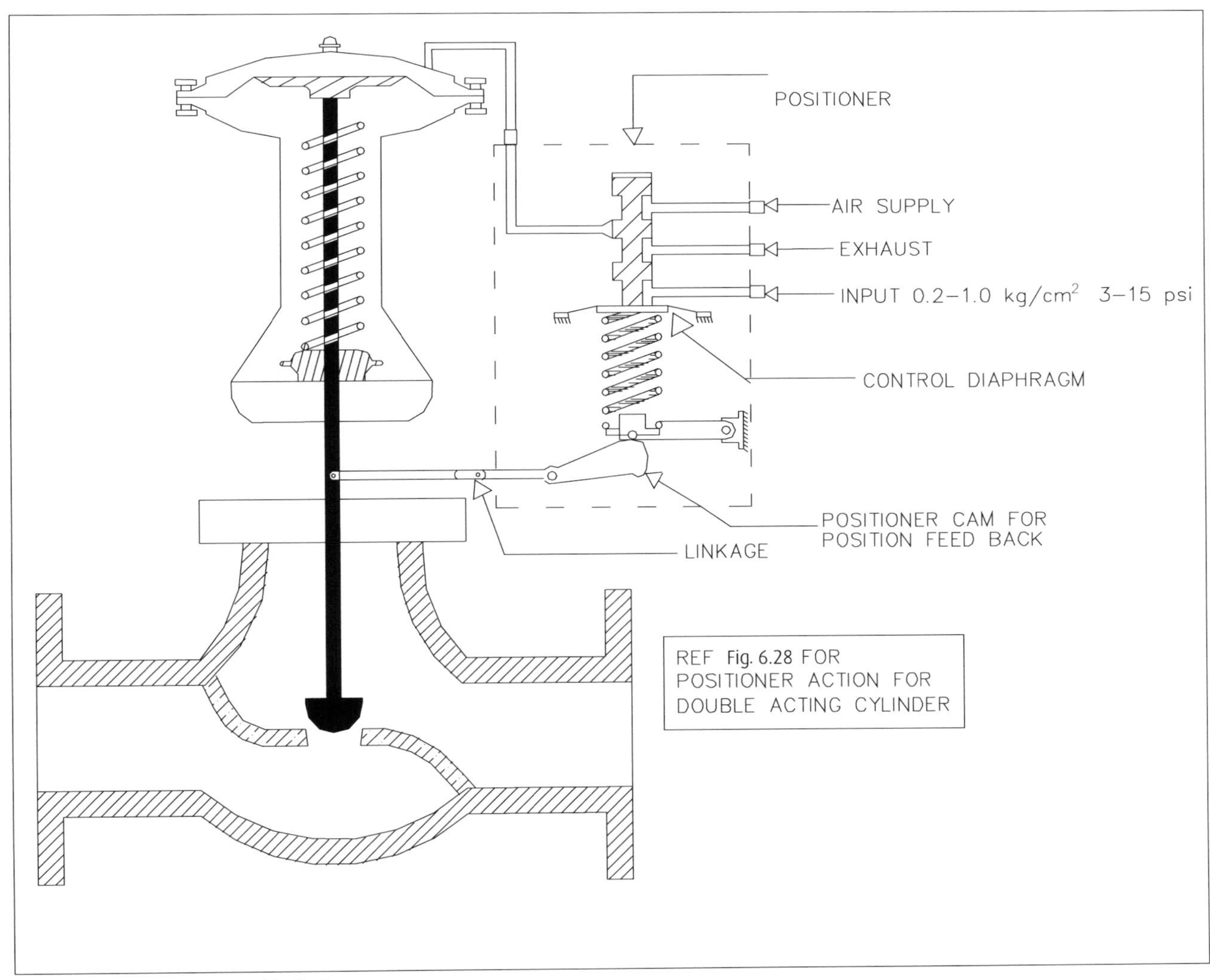

FIG. 6.36 Valve positioner with diaphragm actuator.

until the desired position (sensed by feedback cam) is reached. The operation will be similar when the valve is to be opened, the only difference, is the supply line to the valve is to be connected to the exhaust line; the main air supply line will remain isolated. Thus the displacement of the actuator diaphragm is minimum, but the valve position changes because of the change of the spring compression force. The operation is similar with the positioner for a double acting cylinder, shown in Fig. 6.28. When the control pneumatic signal changes, the diaphragm is pushed (or pulled up), and the pilot valve is also pushed down. The air supply gets connected to the upper part (B) of the cylinder (space above piston), whereas, in the space below (A) the piston gets connected to the exhaust. As a result, the power cylinder‘s pision starts moving down. As it moves down, the feedback spring is compressed through the cam. This compression of the spring will try to bring the diaphragm in its original position, by balancing the signal air pressure. So, until the diaphragm returns to its original size, the air will be pushed into the cylinder, that is, at the balanced condition, the air lines to the cylinder will blocked. Thus the accurate position is ensured by pushing more and more air on one side of the cylinder (Table 6.26).

TABLE 6.26 Pneumatic Positioner (DS)

Sl	Specifying Point	Standard/ Available Data	To Be Specified
1	Input	0.2–1.0 bar/3–15 psi/0.2–1.0 kg/cm^2	
2	Air supply	Manf. Std : >1.4 to <10 bar	
3	Split ranging	To specify range	If applicable
4	Accuracy/ hysteresis	<1%/<0.5% FSD	
5	Action/ characterization	Direct/reverse and Cam characteristic	To specify
6	Stroke length	15–100 mm std.	To specify
7	Materials	Case: die cast Al; flapper nozzle, bellow and cam: SS	To specify
8	Gage	40 mm dia. 3 gauges (Supp/Inst/OP)	
9	Environmental	(–)20–80°C 97% humidity	

5.1.6.2. Smart/electro pneumatic positioner (Fig. 6.37): A few big advantages that makes the smart positioner increasingly popular:

- Simple to use and reliable.
- A microprocessor-based, self-tuning, and self-calibrating controller with a PID configuration possible (helpful for valves in a special application). A noncontact-type position sensor is possible.
- Smart positioners have the advantage of low-cost digital communication through a two-wire system, reducing heavy cable cost.
- The actuator can be protected by the positioner against burn out when jammed.
- Remote calibration and configuration to change critical parameter to improve performance (e.g., HART, Fieldbus, etc.).
- In addition, to collect the position data, many other data, viz. Pr. Diff., flow, etc. can be collected from it.
- Diagnostic features such as total stroke, maximum travel speed, etc.
- Very easy to integrate the systems and FCE performance management is possible.
- With use of a wireless adapter (e.g., Smart Wireless THUM™ of Fisher) integrating with control system is possible at low cost.
- Any position locking is possible, The valve characterization may be customized.
- Because, it is self-tuning it is easier for commissioning.

1. Fig. 6.37 shows the basic block diagram of the smart positioner. The heart of the system is the mother board comprising the CPU and EE PROM which continuously run the set control algorithm. It consists of an I/O interface through which the controller input analog signal (if any 4–20 mA) and other digital input/output signals necessary for various controls and interlock are fed to the system. Normally, these positioners have a position sensing block to accept input from even noncontact-type position sensors. As discussed earlier, the I/P converter is built in to the system. The output of the I/P controller is fed through a pneumatic relay to boost the signal when necessary. The position sensor continuously monitors the pneumatic output for proper operation of the positioner. Normally, the smart positioner has a communication board to establish communication with external devices such as the HART communicator, PC programming, etc. for online calibration/programming. There will be on-board display drivers and optional boards for displays as well as optional features. It receives two power supplies such as an electric power supply (i.e., 24 VDC) and air supply. Not specifically shown, these controllers inside

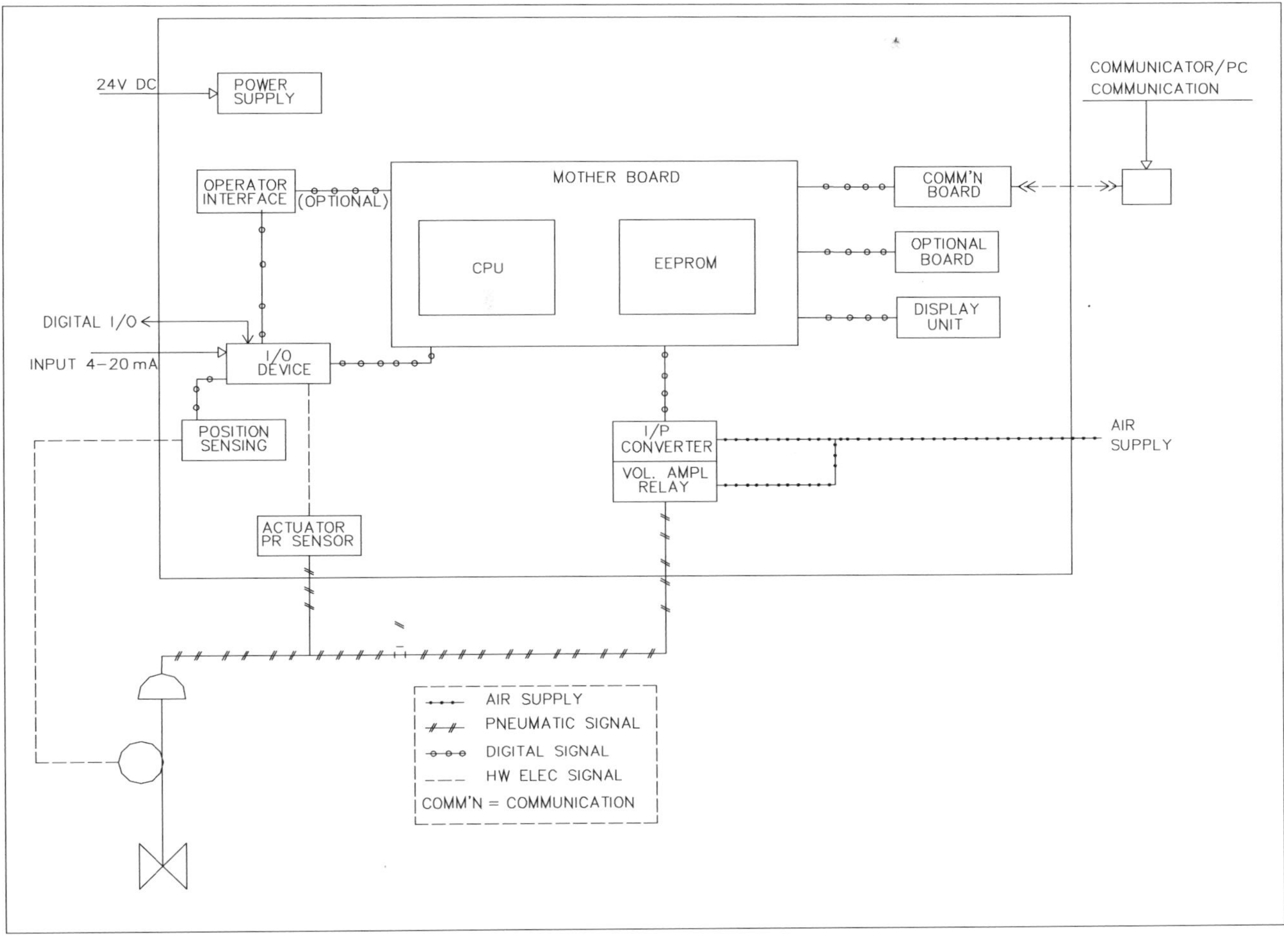

FIG. 6.37 Smart positioner block diagram.

the positioner have options for auto as well as local, manual operations. Important parameters of a smart positioner include:

- *Operating parameters*: These parameters need to be set and adjusted if necessary,
  - Signal range: Input signal (if any option) and split range signals (i.e., 4–12 or 12–20 mA).
  - Valve action: direct or reverse.
  - Flow characterization.
  - Travel limit: If other than 0%–100%.
  - Sensitivity limit (if necessary to adjust other than the factory setting within the specified limit of the manufacturer).
  - Speed of operation: time to travel 0%–100%
  - Digital input: (e.g., to hold at a fail lock condition), etc.
- *Adjusting parameter*: valve range (0%–100%), effective direction (air to close/open), display parameters and their ranges. Various control parameters needed for controls.
- *Monitoring parameters*: Performance monitoring parameter (to be set during the startup).
  - Alarm limit: Min./Max. position.
  - Leakage detection in the system.
  - Watchdog alarm.
  - Signal out of range <4 or >20mA.
  - Position out, timeout.
- Other monitoring and diagnostic conditions of the system shall include: Number of control actions performed, total stroke distance, maximum travel speed, stick slip, etc. to help the operator in assessing the service condition of the system. Also, it monitors performances of electronic circuits to be sure the overall performance of the system are in line with the predefined program.
- Communication: Each of these types of positioners has at least one communication port to set up remote operation and/or communication with the other systems such as a PC (through serial link e.g., RS 232/MODBUS), etc. A number of these can be configured to form a network

that is run by some special program (e.g., IBIS of ABB). For remote operation, and established system such as HART, Fieldbus, etc. can be used. Through this communication, new parameters can be set and may be put in to action immediately upon downloading the parameter value, etc.

- Wherever it is called for, the devices could be intrinsically safe, and/or FM, etc., certifications are also available from a standard manufacturer (Table 6.27).

5.1.7 Position Indicator

On the valve (FCE) itself, there are local, mechanical, and other types of local indicators to help commissioning and other personnel operating the valve (FCE) locally. In its basic form, it could be a small pointer attached to the valve shaft, which moves on a graduated scale (0%–100%) to give a local indication (similar to ones shown in Fig. 6.35). For the rotary actuator/valve, there may be a geared, shaft-driven mechanical device to give a valve position in mechanical digital reading. This could also be a small gauge-type device. There are local digital indicators available for electric actuators.

5.1.8 Position Transmitter

Two types of position transmitters are used in industries: standard conventional and smart type. Again, based on sensing it can be of two types; one is with physical contact, and the other is a noncontact type.

5.1.8.1. Standard electronic position transmitter: In this type a multiturn potentiometer is mechanically connected to the device sensing the mechanical position. This resistance change due to change in position is sensed in a bridge circuit excited by an AC supply. The change in resistance, is sometimes converted to a 4–20 mA DC signal in the signal conditioning unit of the transmitter. Normally, these instruments are provided with travel-limit sensing to generate high and low alarms in case travel exceeds the set positions (Table 6.28).

5.1.8.2. Smart Noncontact Position Transmitter: Noncontact position transmitters are used frequently with smart positioners. These are contact valve position transmitters with digital communication with, for example, HART protocol. These are highly accurate (<0.5%), and can be used with both linear as well as rotary actuators. The valve

TABLE 6.27 Electro Pneumatic Smart Positioner (DS)

Sl	Specifying Point	Standard/Available/*Indicative* Data	To Be Specified
1	Input	4–20 mA DC*/Digital inputs	*If no network
2	Air supply	0.5–7 bar (</= Actuator pressure)	Manf STD
3	Air consumption/ capacity/OP press	4–12 NL/min (based on air pressure)/4.5–10 Nm^3/h/0–6 bar (as per actuator)	Manf STD
4	Travel length/time/ speed/limit	10–100 mm/0%–100% travel time/0–200 Sec in each side/0%–100%	To specify
5	Valve action/ characterization	Direct or reverse/To specify FCE Characteristics	
6	Communication/link	HART/Fieldbus/Any link necessary	To specify
7	Accuracy/hysteresis	<1%/<0.5% FSD	
8	Sampling/sensitivity	10–20 ms/0.3% (typical)	
9	Environmental	(−)20–80°C 97% humidity	
10	Connections	Pneumatic 6mm or ¼″ and elect. ½″	To specify
11	Electrical classification	Agency approval certification	When applicable
12	Materials/enclosure	Case: Al; Elastomer: Nitrile/IP65	To specify
13	Accessories	Mounting kit, Pr. gage block, local indicator, AFR, PC and other adapter/ feedback module, wireless adapter, etc.	To specify as applicable
14	Mounting	Direct on actuator	To specify
15	Software and special feature	To specify if any special feature desired.	

TABLE 6.28 Standard Electronic Position Transmitter (DS)

Sl	Specifying Point	Standard/Available/*Indicative* Data	To Be Specified
1	Input	Single/dual multiturn potentiometer	
2	Signal output	4–20 mA DC with Hi. and Lo. alarm	
3	Power supply	24 VDC Std.	Manf. Std.
4	Span	Depends on application 9–110 mm travel	To specify
5	Overall accuracy	± 1% FSD	
6	Alarm contact and rating	2X COC/3 A@240 AC/0.3 A @220 VDC	To specify
7	Mounting:	On actuator	
8	Materials/enclosure	Case: Al/SS/steel mounting bracket/IP65	To specify
9	Environmental	(–)20–80°C 97% humidity	
10	Connections	Electrical. ½″ NPT	To specify
11	Electrical classification	Agency approval certification	When applicable
12	Electromagnetic comp.	EN 61326-1	
13	Accessories	Mounting kit, local	To specify
14	Special feature	To specify if any special feature desired.	

position sensing can be with a simple magnetic pickup avoiding mechanical contact arms, etc. These sensors could be a proximity type, LVDT or a capacitive type (Table 6.29). Even the Hall effect* type can be deployed. However, a proximity type with a self-diagnostic feature and availability in FM/intrinsic safe circuitry made them more popular (Fig. 6.38).

5.1.9 *Solenoid Valve (SV)*

Functions of Solenoid valve as an accessories to FCE shall include (Table 6.30):

- Interlock operation in a modulating control loop as shown in Fig. 6.39A. When interlock I operates, the FCE (valve) will go to a preset opening. Normally, when there is no interlock (logic0) SV is deenergized, and the FCE (valve) being fed with the signal from the I/P converter will open accordingly. Naturally, on the interlock being Logical 1, the solenoid valve will be energized, and the FCE (valve) will be fed with the preset signal to open the FCE (valve) to the preset value. In this circuit, SV energization action can be reversed by interchanging the ports. This is an example to illustrate an interlock operation. However, the same action can also be achieved electronically in the control loop. Refer to Fig. 6.39B for equivalence.
- The ON/OFF action can be achieved pneumatically by using a solenoid valve as shown in Fig. 6.39C for a double-acting cylinder (for a diaphragm actuator or single acting cylinder, a three-way valve would suffice).
- As discussed in Clause 4.5.6.3 of this chapter, the solenoid valve can be used to achieve fail lock condition due to electrical signal failure.
- In an electro-hydraulic circuit in turbine control there are a number of applications of the solenoid valve to achieve various trip and interlock operations. In those circuits, the port position changes depending on energization of the coil, so that the desired port gets connected to a different position. That is why solenoid valves are specified as 3/2 or 4/2, etc. to specify the number of port and position like numbers of pole and position for the electrical selector switch. This is discussed thoroughly in turbine control systems in the later part of the book.
- Solenoid valves may be single or double coil. In boiler BMS circuits, solenoid valves are generally double coiled, so in case of power failure, it does not suddenly change position.
- SS solenoid valves are generally used as FCE accessories. However, a Cu solenoid valve is also popular in instrumentation.

5.1.10 *Torque Switches*

This FCE accessory is essential for electrical actuators as to stop the motor at its end point when desired torque is reached (Table 6.31).

TABLE 6.29 Smart Noncontact-type Position Transmitter (DS)

Sl	Specifying Point	Standard/Available/*Indicative* Data	To Be Specified
1	Sensor type	Proximity/LVDT/Hall Effect	
2	Signal output	4–20 mA DC with HART/Foundation Fieldbus communication	
3	Power supply	24 VDC Std	Manf STD
4	Span	Depends on the application 5–110 mm or more travel	To specify
5	Accuracy/repeatability	<± 0.5%/0.4% FSD	
6	Electrical classification	Agency approval certification	As applicable
7	Mounting:	On actuator	
8	Materials/enclosure	Case: Al/SS/Steel mounting bracket/IP65	To specify
9	Environmental	(−)20–80°C 97% Humidity	
10	Connections	Electrical ½″ NPT	To specify
11	Turn on/update time	100 ms/25 ms	
12	Electromagnetic comp.	EN 61326-1	
13	Accessories	Mounting kit	To specify
14	Special feature	To specify if any special feature is desired	

***** *Hall effect: This is a semiconductor device. A transverse electric field perpendicular to both magnetic field as well as to the direction of current) will be generated in a semiconductor carrying an electric current , placed in a magnetic field perpendicular to the current.***

In Semiconductors holes and electrons are charge carriers, and collision of these charge carriers with impurities are in a straight path, in absence of magnetic field but in presence perpendicular magnetic filed the collision path deviates towards transverse direction due to Lorentz force.

Thus on account of Hall effect these charges move to one end in the transverse direction giving rise to electric potential across the device. Applying this effect many sensors have been developed some of them used in power plants are Position sensing, speed sensing(Turbine) etc. Detailed Hall Effects sensors have been discussed in Authors's book Entitled "Plant flow measurement and control Handbook" by Elsevier.

FIG. 6.38 Hall effect.

5.1.11 Volume Booster

As the name implies, it boosts the volume, hence, pressure in an enclosed space. Booster relay refers to volume boosting, pressure boosting, or both. The term booster relay is normally used when the input pressure is scaled up or down at the output as per requirement. The term volume (capacity) booster is more specifically used when the volume boosting of air with output and input pressure remains the same (that is with a ratio of 1:1) within the permissible deadband. This is basically a regulator where the input actually supplies the loading force. This works on the force balance principle. Again, force is equal to the pressure multiplied by the area, so by changing the ratio of the area at the input and output side, the required output pressures can be varied as the area ratio times input pressure. For example if the input side pressure (1 bar) is to be balanced by the output side force, and if the input diaphragm area ($6A$) is six times that of the output area (A), then input force $= 6A \times 1$ bar $= 6A$ unit of force. This force can be balanced by 6 bar pressure at the output as the output side area is A. So, by making the input side area less than the output side area, the pressure ratio can also be reduced.

Thus it is clear that booster relay can be used to change pressure level (Fig. 6.40C) and/or for bias. Application of volume boosting has been depicted in Fig. 6.40B. Here the same control signal is distributed to multiple users with 1:1 pressure ratio. For large actuators, volume boosters are used between positioners and actuators to improvise speed of response or is used, when positioner is not used. As shown in Fig. 6.40A, during a balanced condition, both supply and exhaust ports will be closed. Now when input pressure increases there will be downward pressure on the

TABLE 6.30 Solenoid Valve (*FCE Accessories*) (DS)

SL	Specifying Point	Standard/Available/*Indicative* Data	To Specify
1	Fluid	Air/Hydraulic fluid	To specify
2	Pressure rating	To specify (in case of pneumatic system > 10 bar)	
3	Size	¼″ or 6 mm generally for pneumatic application—to specify	
3	Differential pressure/flow/temperature	To specify (2–10 bar @800 L/m—typical value/30°C)	
4	No of coils	Single/double	
5	Coil voltage	Power supply available can be 240VAC/220VDC/24VDC	To specify
6	No of port and position	To specify depending on requirement say 3/2, etc.	To specify
7	Material body/core tube, plug, etc.	SS preferred	To specify
8	Connection type and size	¼″ Air side/electrical: ½″ NPT	To specify
9	Insulation class	Class H (for high temp)/F	
10	Environmental	(−)20–80°C 97% humidity	
11	Any other special feature	Electrical certification necessary	If applicable

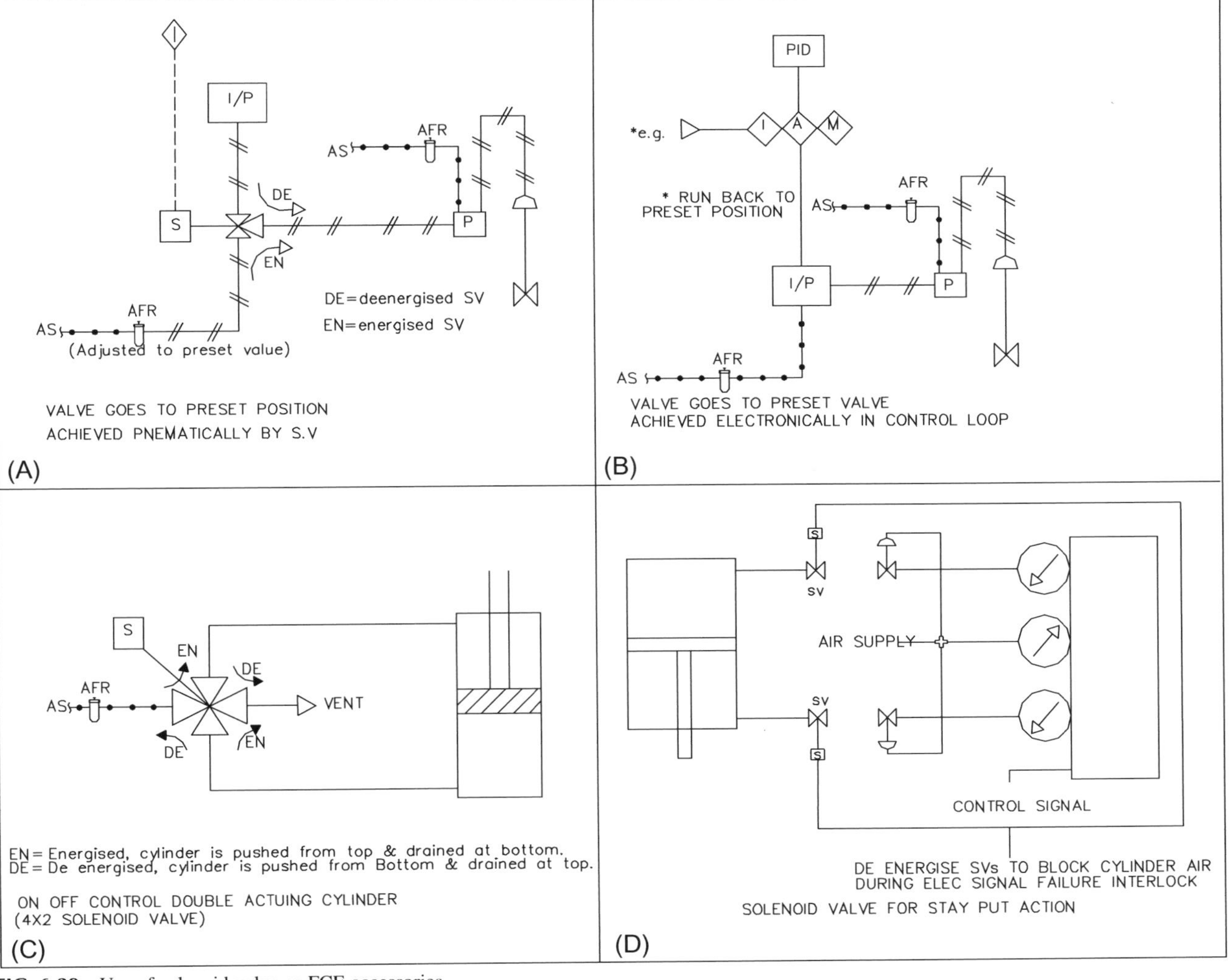

FIG. 6.39 Use of solenoid valve as FCE accessories.

TABLE 6.31 Torque Limit Switch (DS)

Sl	Specifying Point	Standard/Available Data	To Be Specified
1	Function	To stop the drive at torque limit	
2	Area classification	Agency approval certification	If applicable
3	Contact type	Single pole double throw, etc.	
4	Contact material	Gold alloy plated/self-cleaning type	
5	Operating torque	To specify based on actuator	
6	Encl. class	IP 67/NEMA 4X, etc.	To specify
7	Contact rating	AC 3 A @240 V 1 Ph 50 Hz 3 A DC 0.3 A@220 V	
8	Insulation resistance	>100 MΩ	
9	Life	>1 Million operation	
10	Mechanical resistance	High shock and vibration resistance	
11	Environmental	(−)20–50°C 97% humidity	
12	Termination	Suitable for 2.5 mm^2 cable	
13	Speed of operation	(1–360°/s typical)	
14	Environmental	(−)20–80°C 97% humidity	

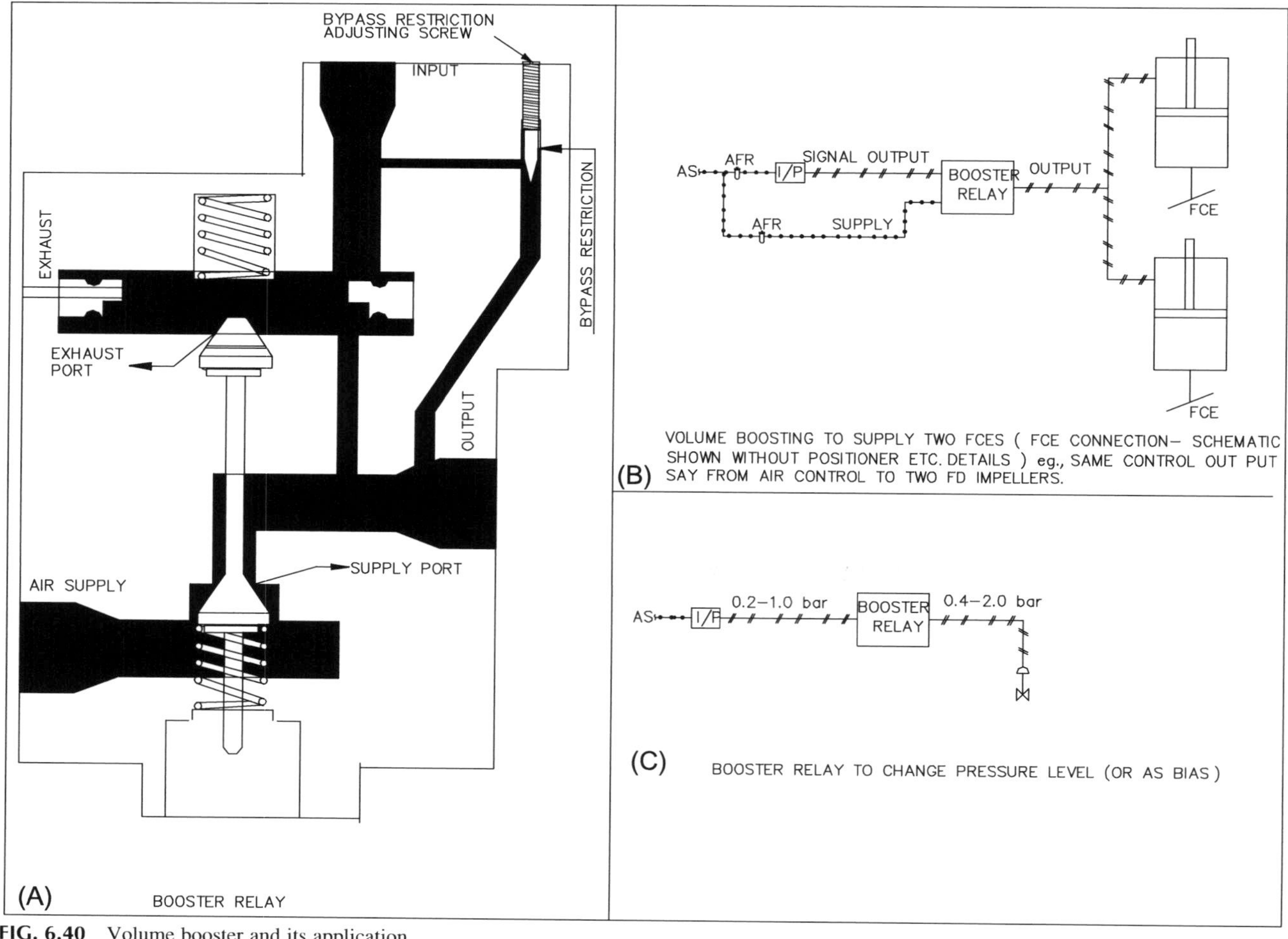

FIG. 6.40 Volume booster and its application.

TABLE 6.32 Booster Relay (DS)

Sl	Specifying Point	Standard/ Available Data	To Be Specified
1	Function	To boost volume/ vary pressure/to apply bias	
2	Input	0.2–1.4/0.2–3.5 bar (typical value)	If applicable
3	Supply pressure	2.1–4.0 bar (typical values)	
4	Gain ratio	To specify as per requirement	To specify
5	Consumption (typical)	200 lit/min	Manf. Std.
6	Capacity output	~150 lit/min	
7	Action	Proportional	
8	Connection	¼″ NPT	
9	Supply pressure effect	<3% FSD per 1 bar change in supply pressure	
10	Environmental	(−)20–50°C 97% humidity	

input side diaphragm to maintain close position of the exhaust path, and further force will push the valve downward allowing to air supply coming in the output chamber. Because air supply pressure is more, there will be air flow increase in the output chamber, hence, flow to the actuator. This will continue till the output side pressure acting on the diaphragm balances the force in the input side diaphragm (= input diaphragm area × input side pressure). When balance is reached, force on both diaphragms air supply line is closed. On the other hand if the input pressure falls the valve will go up and air from actuator will be exhausted with actuator supply line remaining closed, to bring down the output pressure. During this time, due to higher pressure in the outer diaphragm, it will port assembly up, and output side air will escape through exhaust port. From here it is clear that output side extra air volume is supplied by the relay. Also by varying input/output diaphragm area the output pressure can be varied with respect to input (Table 6.32).

BIBLIOGRAPHY

[1] Application and Selection of Control Valves: B.G. Liptak, A. Balint: Book-Process Control and Optimization Vol II (B.G. Liptak).

[2] Cavitation in Control Valves: Samson AG: Technical Information.

[3] Control Valve Characteristics: www.maintenanceresources.com/referencelibrary: Cashco Catalog.

[4] Control Valve Hand Book: Emerson Process Management: 4th edition (Fisher).

[5] Control Valve Selection and Sizing: K.L.M. Technology Group: Engineering Design Guidelines.

[6] Control Valve Technical Specification: G.Borden jr. & P.G. Friedmann: ISA 1998.

[7] Control Valves For Power Plants: MIL Controls Limited: Application Handbook.

[8] T. Bishop, M. Chapeaux, L. Jaffer, K. Nair, S. Patel, Ease Control Valve Selection, CEP Magazine, 2002.

[9] Guidelines for Selecting the Proper Valve Characteristic: Valve Magazine: Vol 15 no. 2.

[10] KCP&L Increases Plant Capacity by retrofitting New Control Valves1/30/2014D. Crawford, B. Tally, B. Beckmann, G. Knoch (KCP&L) & J. Wilson (fisher): Power Engineering Magazine.

[11] Key Design Component of Final Control Elements: ISA: Intech March/April 2010.

[12] Leakage Classification of Control valves: The Engineering Tool box, www.engineeringtoolbox.com/control-valve-leakage.

[13] Noise Control Manual: Dresser Flow Control: Masoneilan.

[14] Seat Leakage Classifications, www.maintenanceresources.com/referencelibrary, Cashco.

[15] Standard Design Criteria/Guideline for Main Plant Package (2 × 500 MW TPP), Central Electricity Authority, New Delhi, 2010.

[16] B. Fitzgerald, C. Liden, The Control Valve's Hidden Impact on the Bottom Line, Valave Magazine, 2003.

[17] Valve Sizing: Swagelok: Technical Bulletin.

[18] Valve Sizing and Selection Technical Reference: Warren Control.

[19] Valves Noise Calculations Prediction and Reduction: H.D. Baumann, J.B. Arnet, B.G. Lipatak, F.M. Cain: Process Control and Optimization, vol. II (B.G. Liptak).

[20] Ball Valve Options/Details: Autoclave Engineers: Catalog.

[21] Butterfly and rotary Process Valve: Tyco Flow control: Keystone Catalog.

[22] Condensed Catalog: Dresser Masoneilan.

[23] Cross-Reference of ASTM Material Specifications covering Cast and Forged Valves, Fitttings, Flanges and Union: Rare Energy: Materials Catalog.

[24] Diaphragm Valve: AquaMatic Catalog: Comercial Control Valve Pentair Water Treatment.

[25] Dirty Service Anticavitation Trim: Fisher Emerson Process Management: Product Bulletin, 2009.

[26] Drag Valve (BFP Recirculation): Control Component Inc.: Catalog.

[27] Drag Valve for Boiler Feedwater Control Application: Control Component Inc.: Catalog.

[28] GSI Series Weir Style Diaphragm Valves: Swaglok: Catalog.

[29] High Capacity Anti-Cavitation Solutions: The Dresser Masoneilan: Technical Write Up.

[30] Know Your Valve: S. Brame (Emerson Process Management): Hydrocarbon Engineering, 2005.

[31] Packing Selection Guidelines for Fisher Sliding Stem Valves: Fisher Emerson Process Management: Product Bulletin.

[32] Power Plant Applications: C. Sterud: Control Component Inc. Document.

[33] Replacement of BFP Recirculation Control Valve Solves a Maintenance Problem in Baldwin Unit 3: M.E. Liefer (Illinois Power) H. L. Miller and R.E. Katz (CCI): Control Component Inc. Document.

[34] Sunder's Diaphragm Valve: C.E. Gayler, B.G. Liptak: Process Control and Optimization Vol II (B.G. Liptak).
[35] The Ever Popular Gate Valve: G. Johnson: Internet Write Up.
[36] Types of Valves and Usage TLV. Co. Ltd: Dorot Control Valve: Internet Write Up.
[37] Valve Types: Butter Fly Valve: C.E. Gayler, B.G. Liptak, J.B. Arant: Process Control and optimization, vol. II (B.G. Liptak).
[38] Valve Types: Globe Valve: H.D. Baumann, J.B. Arnet, B.G. Lipatak, F.M. Cain, Process Control and optimization, Vol. II (B.G. Liptak).
[39] Combustion Fossil Power Systems: Combustion Engineering Inc.: Steam Generator Auxiliaries.
[40] R.C. Monroe, Consider Variable Pitch Fans, Hudson Product Corporation, USA.
[41] Energy Efficient Design of Auxiliaries in Fossil Fuel Power Plants: ABB Limited: Technical Paper.
[42] Fluidomat: www.fluidomat.com: Technical Paper.
[43] Hydro-Kinetic Drives…Hyadraulic Coupling: N.A. D'arcy: Internet Document.
[44] Hydrodynamic Couplings Principles | Features | Benefits: Voith Turbo I Hydrodynamic Couplings 5: Technical Catalog.
[45] Impellers With Airfoiled Blades: TLTTurbo.
[46] Increasing Efficiency of the Conventional Auxiliary Systems of Power Plants (Reduction of Life Cycle Cost by Operational Excellence): T. Schmager, P. Mannistö, P. Wikström, ABB Switzerland Ltd: Technical Paper.
[47] Inlet Outlet Dampers for Centrifugal Fans: GreenHeck: Catalog.
[48] Installation and Maintenance Manual for Scoop Coupling: Elecon Engineering Co. Ltd. India: Technical Manual.
[49] Variable Fill Fluid Coupling: Transfluid Transmissioni Industriali: Technical Catalog.
[50] Variable Speed Control: Hivectol HVI: Internet Paper.
[51] Variable Speed Drive Theory: L.M. Photonics Limited.
[52] Variable-Pitch Axial Flow Fans for Thermal Power Stations: L. Müller, Zweibrücken: www.troxtechnik.com/.
[53] Vortex Damper (Variable Inlet Vane Damper): Process Barron USA: Technical Write Up.
[54] A Precision Electric Actuator is Unaffected by Valve Stiction and Can Track Closed Loop Controller Demand Almost Perfectly Without Dead Time Lag, or Overshoot. K.C. Meyer, S. Kempf: Intech, December 2006.
[55] Actuator and Control Valve Selection: Prof. D.R. Yang: Korea University.
[56] Actuator Sizing Manual: Flow Serve: Worcester Controls.
[57] Control Valve Fail Safe Position: www.maintenanceresources.com/referencelibrary: Cashco Catalog.
[58] Electrical Actuator: Rotork Process Controls.
[59] Electrical Actuator for Industrial Process Controls: BECK: Technical Bulletin.
[60] Pneumatic and Electric Actuator: Elomatic (Emerson Process Management): Product Guide (DS).
[61] Pneumatic Actuators and Positioners/Mobile Type Air Cylinders: Rexroth Bosch Group: Catalog: Limitorque Actuation Systems.
[62] Valve Actuators and Control Systems: Limitorque Actuation Systems: Technical Catalog.
[63] Valve Automation: ABB Automation Inc. USA: Technical Bulletin.
[64] Air Filter Regulator: Worcester: Product Specification.
[65] Air Lock Relay: Rotex: Catalog.
[66] Air Lock: Yamatek Corporation: Product Specification.
[67] Air Supply Filter Regulator: ABB Limited: Technical Catalog.
[68] Analog Position Transmitter: Siemens AG: Catalog.
[69] Digital Position Transmitter: Westlock Controls: Tyco.
[70] Digital Positioners and I/P Signal Converters: ABB Instrumentation: Technical Bulletin.
[71] Electropneumatic Converters for Direct Current Signals: Samson AG: Product Specification.
[72] Electro-Pneumatic Positioner TZIDC: ABB Limited: Technical Catalog.
[73] Field type I/P Converter: Yamatek Corporation: Product Specification.
[74] Field View Digital Valve Controller: Fisher Emerson Process Management: Product Bulletin (Feb'10).
[75] Flame Proof Limit Switches: Yamatek Corporation: Specification.
[76] Handwheel Actuator: Fisher: Catalog.
[77] I/P Converter: Pneucon Valve Pvt Ltd: Product Specification.
[78] I/P Converter TEIP11ABB Instrumentation: Catalog.
[79] Low Torque Basic Switches: Omron: Catalog.
[80] Mechanical Position Indicators: Mission Inductries Inc.: Product Specification.
[81] Pneumatic Positioner: HNL Engineering Limited: Catalog.
[82] Position Transmitter: Fisher Emerson Process Management: Product Bulletin (Apr'09).
[83] Relays and Volume Boosters: Marsh Bellofarm: Product Catalog.
[84] Smart Position Transmitter: SMAR.
[85] Smart Valve Positioner: Yamatek Corporation: Product Specification.
[86] Smart Valve Positioner: Yamatek Corporation: Catalog.
[87] Smart Valve Positioner: Fisher (Emeson Process Management): Product Bulletin (Oct'09).
[88] Solenoid valve: Asco: Catalog.
[89] Valve Position Indicator: Proximity: Specification.
[90] Valve Position Monitoring and Smart Discrete Controls: J. Di Franco, Valve Magazine, 2007.
[91] Valve Position Transmitter: Cal Val Canada: Product Specification.
[92] Valve Positioner: www.maintenanceresources.com/referencelibrary: Cashco Catalog.
[93] VBL Volume Booster: Fisher: Product Bulletin (Jan'13).
[94] Volume Booster: Fisher: Product Bulletin (Oct'12).
[95] Volume Booster Relay: Robertshaw: Product Bulletin.

Chapter 7

Intelligent Control System

Chapter Outline

1 BASICS (DISCUSSION ON INTELLIGENT/SMART NETWORK SYSTEM)

1.1 Preamble

Intelligent/smart systems are normally encountered in power plants to control the process. **What is a system**? A system is part of the universe within certain domains in space and time. **What is an environment**?Outside the frontier of system is the environment. **What is an intelligent system**? Conceptually, an intelligent system learns how to act in a given environment to reach its objective. An intelligent system has a temporary objective that it has derived from the main objective. It has a few senses to gather knowledge about the given environment to reach those objectives. The system then stores these sense impressions as elementary concepts. Working on these concepts, it develops a new one, stores the relationship, and enriches the "concept." To continue with the internal information, it checks the inputs for updates and builds up the present situation. Also, it looks to its memory to find the set of rules, chooses the best suited (rule-based) for the application, and performs the action. The concept has now been extended further to integrate various sensors/devices with the help of internet IP addresses, which finally gave rise to the concept of the "Internet of Things (IoT)." These have been made possible on account of the advancement of embedded systems and their easy way of programming. When this IoT is applied in industrial controls, it is referred to as the "Industrial Internet of Things (IIoT)."

A smart transmitter keeps in its memory the family of characterization curves at various ambient temperatures. So, a smart transmitter can sense the environmental changes and can correct itself with the stored curve to offer better accuracy of measurement. An expert operator can establish relations among various parameters, so that in case of an upset in one parameter, based on knowledge, the operator adjusts directly some actuator and/or some other parameters to control the system precisely. Similarly, in an **expert system**, a few parameters and their relationships are stored, and based on a set of rules, the **expert system** regulates the system to achieve precision control.

1.1.1 Signals and Signal Processing (Basic Issues)

In industrial control systems, there are a few types of signals with classifications, as in Table 7.32.

There are many issues such as A/D conversion (ADC), D/A conversion (DAC), types details regarding Time division multiplexer (TDM) (viz.as sampling rate, accuracy, number of channels) input selection (such as common

Power Plant Instrumentation and Control Handbook. https://doi.org/10.1016/B978-0-12-819504-8.00007-X

ground, differential polarity issues) are general common issues one needs to be considered for smart digital controls. So, prior to going into the details, first these will be addressed in general.

1.1.1.1 General Considerations

Here, general considerations for I/O signal conditioning are discussed. Some issues are only some recommendations.

1. Spare Capacity: From a future expansion point of view, it is recommended to consider additional it is to consider ~10% fully wired module of each variant, terminated up to the final termination point (Marshalling Rack as applicable) of I/O capacity. At times, additional spare space with wiring for additional module insertion is also considered. However, with the introduction of the **HART** and the **fieldbus** and the system integration requirement of such issues discussed above has been reduced.
2. Signal Processing: Depending on the application, basic signal processing may include the following:
   - Optical/galvanic isolation as applicable.
   - Transmitter power supply with fuse and other protections.
   - Interrogation voltage (if different, suitable isolation).
   - Various computations.
   - Square rooting and flow computation (e.g., Pr./Temp. comp.).
   - Linearization of various signals (e.g., T/C).
   - Reference junction compensation.
   - Addition and deletion of input for scanning based on operator action.
   - Redundancy selection.
   - Short circuit/open circuit/noncoincidence (COC) fault detection.
   - Diode auctioneering for output redundancy.
   - Filtering for noise reduction/attenuation.
   - Filtering of digital signals (delay: 10–15 mS) prevents spurious action.
   - I/O data validation and reasonability check.
   - Online ADC gain drift check.
   - Self-surveillance and diagnostic (channel and module level).
3. Remote cabinet location: Cabling can be reduced by utilizing a remote I/O cabinet or remote terminal unit (RTU) and connecting the same to the system through various links/buses. In power plants, there are a few places where there is a cluster of measuring points. Some recommended places for such remote I/O locations could be: boiler platforms, boiler outside area (outside, fan area), mill area, BFP area, turbine area, SWAS room, switchgear room, and transformer area. Again, with the introduction of fieldbus/HART, such requirements are hardly felt, except at SCADA application.
4. Acquisition Timing: The timing for data acquisition, validation, and processing is guided by the systems under consideration, and the typical such values are: digital and SOE inputs the update times could be around 30–50 ms and 1 ms respectively and other signals updating @ 250 ms or better.

1.1.1.2 Critical Measurement

For critical measurements, modules—including power modules—may have facility to use the fault-tolerant mode discussed later.

1.1.1.3 Signal Grouping Philosophy

Based on design philosophy, there could be divisions of whole systems into groups, subgroups, subloops, and drive control devices (e.g., ATRS of Siemens/KWU). Depending on the functionalities of I/Os, these can be connected to the respective part (e.g., limit switch/MCC contacts at the drive module, but for input for a cluster of drives, at the subgroup level). There are variations in control system philosophies—for example, some prefer mixing input and output (I/O) in a single module to make a loop self-contained, whereas others believe in segregating I/Os with redundant signals in different modules. This grouping is not restricted to the I/O level alone, as it may be at the controller level also. A controller may take care of both the modulating as well as sequential controls for drives (e.g., FD fans, as in the Procontrol P13 system). However, there are systems where modulating controls are kept separate from interlock operations (e.g., the advanced control system for modulating control and PLC interface for interlock operation).

1.1.1.4 I/O Intelligence

With the introduction of embedded electronics, most of the major suppliers support intelligent I/O systems. Also, they use fieldbus/HART in place of hardwired input output.

1.1.1.5 Common Mode and Normal Noise Rejection

Usually, modules are designed to withstand a common voltage of **500 V** peak to peak/DC for a short period.

Typical common mode rejection ratio (CMRR) is in the tune of **100 dB** @50 Hz (or 60 Hz) with a source imbalance and a normal mode rejection ratio (NMRR) of **60** dB at 50/60 Hz (Fig. 7.1).

CMRR: It is a measure of the capability of a device to reject a signal that is common to both input leads; it is measured in decibels, the effectiveness to reject a common mode voltage. It is Ratio of A_v/A_{vcm} when expressed in dB. Here A_v and A_{vcm} stand for Differential and for common mode gain.
NMRR: It describes the ability of the instrument to reject a normal (differential) signal between the input leads. It is also expressed in dB. Discussed in Chapter 12

FIG. 7.1 Common and normal mode rejection ratio.

1.1.2 Network Basics

A number of computers connected together form a network. When this network is owned privately for a particular purpose, it is part of a local area network (LAN). A number of real-time systems (controllers) in plants form a LAN, for example, two or three PLCs may form a LAN or in a DCS, a number of control processors and work stations may form a LAN with a typical distance limitation such as **1–2 km**. If a completely integrated system in a power plant is considered, it can stretch over kilometers of LAN. When a LAN is developed through wireless means, it is known as a wireless LAN (**WLAN**). When this length limit goes beyond ~100 km, it will form a wide area network (**WAN**). Over the entire globe, various networks are connected via the Internet. With the development of embedded electronics and the Internet, this has become much easier and gives rise to the development of the IIoT. In power plants, there are different control systems (for SG/TG/offsites) from various manufacturers. However, it is necessary that these systems communicate with each other for data exchange and/or with central monitoring and control systems. So, some protocols need to be established so that these different systems can form an integrated control system.

What is a protocol? In connection with networking, a protocol can be conceived as an agreed-upon formally defined unambiguous convention for communication. So, it is a set of rules that the system hardware and software must follow in order to be recognized and understood by other systems in the network.

1.1.2.1 ISO-OSI Reference Protocol Model

The International Organization for Standardization (ISO) developed a reference model called the open system interconnection (OSI). An ISO OSI reference model consists of seven layers: **APPLICATION, PRESENTATION, SESSION, TRANSPORT, NETWORK, DATA LINK, AND PHYSICAL** (to remember more easily, use this phrasing: "**A**ll **P**eople **S**eem **T**o **N**eed **D**ata **P**rocessing"). The arrangement and data flow using this model is depicted in Fig. 7.2 with functional details in Table 7.1. As physical, data links, and network layers are more concerned with medium, these collectively are often referred to as the **media layer.** As the transport, session, presentation, and application layers are more connected with the hosts, they are often called **host layers**.

1. PHYSICAL LAYER (1): It is responsible for transmitting raw BIT over a communication channel, and to make sure if a node sends 1 receiving node gets ONE not ZERO. It defines the cable type, connector type, etc. This layer is concerned with signal, binary transmission and media. The issues are physical interface, bit rate, distance, etc.
2. DATA LINK LAYER (2): This layer develops a DATA FRAME from the raw data to convert raw data in a line without any transmission error to the network layer. Transmission speed is of great concern here. This layer ensures the buffering space, when there is a difference between the communication speed between the transmitter and the receiver. It has two sublayers: media access control (MAC) and logical link control (LLC). It is responsible for physical addressing and actual communication in a network. Here, the issues are: framing, addressing, header trailer bits, etc., for example, (say) Ethernet.
3. NETWORK LAYER (3): It is concerned mainly with path/route finding and the determination of logical addressing for data transfer. It regulates the subnet and looks for the path that will let packet data reach from the source to the destination. In a complex network with too many nodes, finding an easy sure route is a major task. Too many data packet transfers without a suitable route will result in a traffic jam, causing delays. In the meantime, more and more data will be coming. So, the network layer has to find an alternative route. Issues are: packet header and virtual path, for example, IP.
4. TRANSPORTATION LAYER (4): It is a layer for fragmentation and reassembly. It receives data from layer 3 and splits it into many data packets (to facilitate network layer action) and makes sure that these are correctly received at the other end. At the receiving end, this layer needs to reassemble the same. It uses message headers and control messages to make sure that the end user gets the messages in the order they were sent. It also decides the type of service to be provided to the session layer. It shields the upper layers from changes in hardware technologies. It ensures end-to-end connection. Issues are headers, error detection, reliable communication, etc.; for example, TCP/UDP.
5. SESSION LAYER (5): In a session layer, process-to-process communication takes place between the nodes in a network. Here, dialog control (tracking who is to

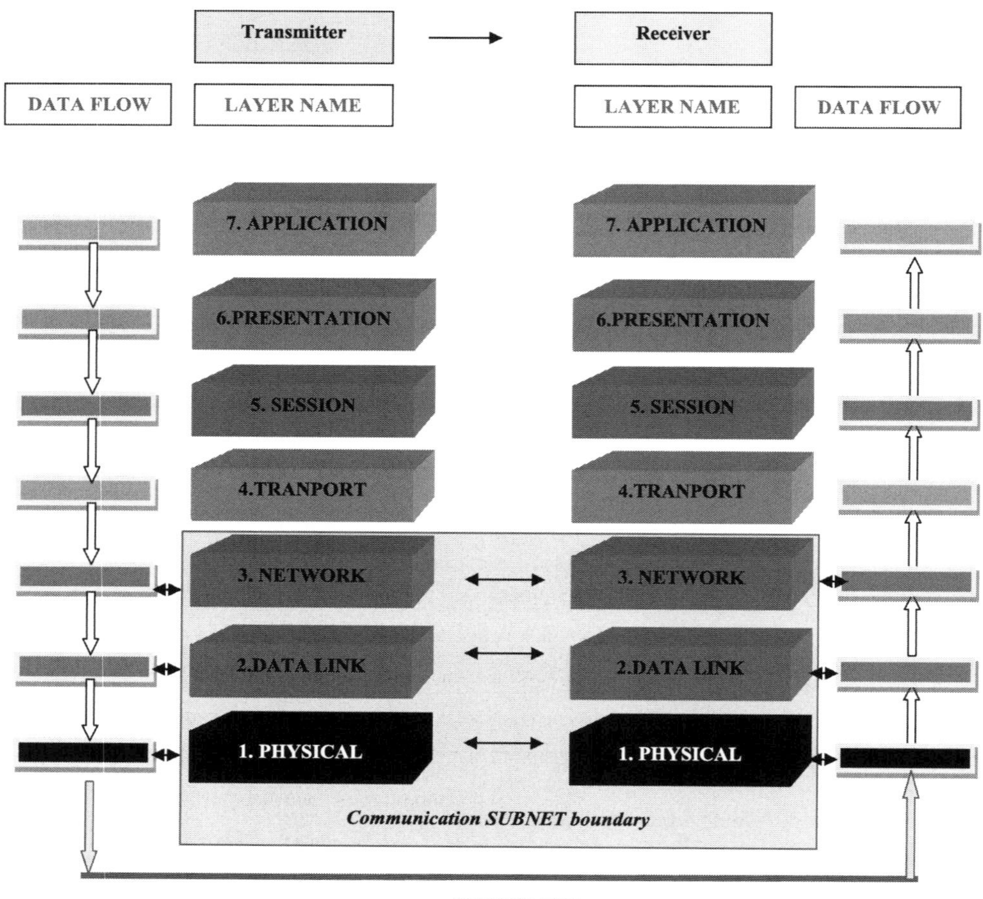

FIG. 7.2 ISO OSI model.

TABLE 7.1 Layer Functions

Layer No.	Layer Name	Main Function
1	Physical	Signal level and connection, media type (e.g., FO/wire)
2	Data link	Source/destination, data transmission and checksum
3	Network	Determination of data path in network
4	Transport	Error checking
5	Session	Opening/closing of communication path for a specific purpose
6	Presentation	Building block of data and encryption
7	Application	Data Meaning–purpose

speak and who is to listen) and TOKEN management to prevent two sets of parties from doing the same operation at the same time! Synchronization of check points takes place in this layer, that is, interhost communications. An example of this is NetBIOS (a protocol for some functions such as browsing and communication between servers).

6. PRESENTATION LAYER (6): When communication between the two nodes takes place, there shall be some defined syntax and semantics of information exchange. This layer looks after the same. The processors communicating may have different data representation, so someone has to define the same. It is the responsibility of this layer to carry out the same. Issues are conversions, encryption, and compression, for example, simple mail transfer protocol (SMTP) in email.
7. Application Layer (7): Email, FTP, etc., are examples of an application layer, that is, this layer is to offer service to the user (user interface layer). Issues are type of service.

1.1.2.2 TCP/IP Reference Model

TCP/IP is one popular reference model for radio and satellite network systems. It is a set of protocols used to communicate over the Internet. On account of its flexibility and wide functionality, it is in the control network also. In the suit of protocol TCP/IP, TCP is a reliable connection-oriented delivery service where data is transmitted in segments. **IP,** on the contrary, is responsible for routing and addressing packets between the hosts/nodes. This is connectionless as there is no established session used by shared devices. As delivery is not guaranteed, it is unreliable. TCP/IP layers are shown in Fig. 7.3. When compared with OSI, it is clear that some layers of OSI models have been clubbed together. Also, here the Internet layer is sometimes referred to as the transport layer. Four layers are discussed below:

1. Network Interface: This layer is sometimes referred to as the link layer, physical layer, etc. Basically when compared with OSI model comprise both physical and data link layer. This layer is transparent to the user. Basic **responsibilities** shall include error checking of incoming and outgoing data packets, data acknowledgement and resending of data if not acknowledged, and the computer network interface. Several support protocols and networks mainly include Ethernet, a network based on IEEE 802.3 and SLIP (serial line internet protocol), used to send data across a serial line. It places data packets in the data frame to transport across the network, CSLIP: Compressed SLIP and PPP: Point to Point Protocol is a serial line data encapsulation better than SLIP.
2. Internet Layer: It is created for a packet switching network layer (Eq. to Network layer of OSI model). IP address and address resolution take place within the Internet layer. It consists of the following protocols.
   - Internet Protocol (IP): The IP provides mechanism software to address and manage data.
   - Packets being sent to computers. There are several headers associated with the IP such as source IP address, destination IP address, identification (inform destination host whether to be passed on to the transport layer protocol TCP/UDP), and protocol check sum (to check the integrity of the data).
   - Address resolution protocol (ARP) to convert a logical address to a physical address. It is used to find specific address (hardware) from a specific IP number.
   - RARP: Reverse ARP is used to resolve an IP address without having own storage media.
   - ICMP: Internet control message protocol. This protocol is used to report the connection status back to the computer and to retry the connection. It also does error reporting.
   - IGMP: Internet group management protocol is used by a multicasting router for tracking group memberships.
3. Transport Layer: A transport layer is used for network connectivity, intended to see that data sent is received by the right node/machine for the right application. Error checking, flow control, verification of integrity, and completeness are functions of this layer, namely, TCP and UDP.

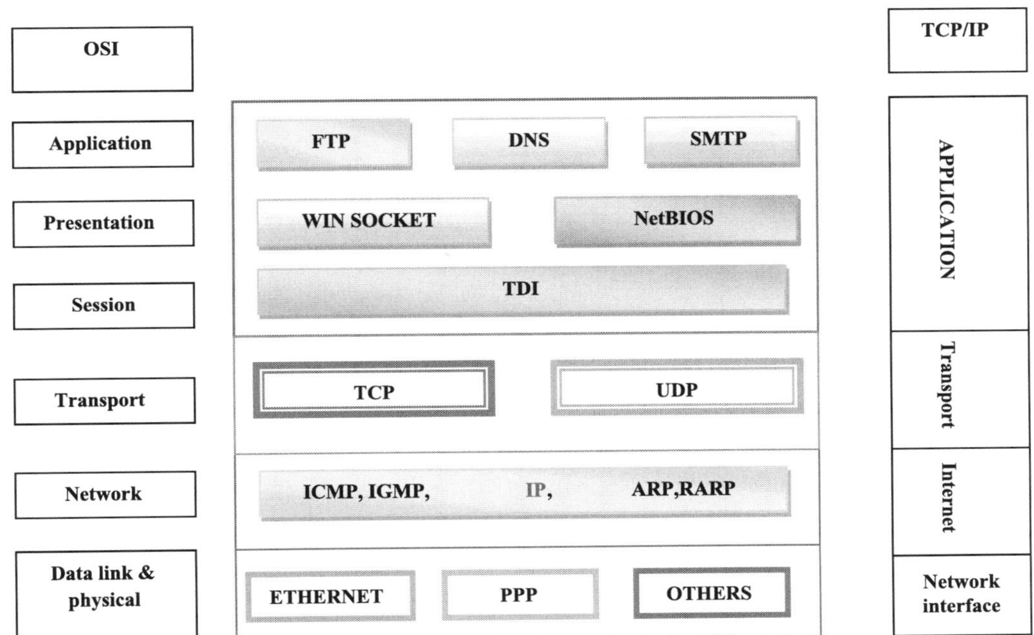

FIG. 7.3 TCP/IP REF model.

- Transmission control protocol (TCP) is a reliable connection-oriented delivery service where data is transmitted in segments. It is used for transport by some application. Reliability is achieved by assignment a sequence number to each segment sent. Several sequence numbers of TCP headers are **source port**, **destination port, sequence number** (sequence number of first byte of data), **TCP check sum,** etc. It is worth noting that there are **separate port numbers** for several applications, for example, *TCP port numbers 20 and 21 are assigned for file transfer protocol (**FTP**) data and control while 80 and 139 are for **HTTP** and **NetBIOS***. A TCP connection is done by three-way handshaking.
- User datagram protocol (UDP) is told by the application layer to which machine is supposed to transmit. It is fast but *not* a guaranteed transfer. Three headers of UDPs are source port, destination port, and UDP check sum. The port functions as a multiplexed message and gives an idea where the message is to be sent. For example, UDP ports 53 and 137 are used for domain name system (DNS) and NetBIOS, respectively.

4. Application Layer: This layer contains network applications and services that the user interfaces with in order to use the network (Internet). The application layer also has utilities such as printers. Some of these protocols here are FTP, HTTP, and SMTP.

1.1.2.3 Miscellaneous Network Devices

Major network connecting devices are:

- **Repeater** (physical layer): It extends a hardwired connection to copy bits from one network to another.
- **Bridge** (data link layer): Copies a data frame from one network to another (looking at header).
- **Router** (network level): Copies a packet from one network to another and makes decisions about a route.
- **Gateways** (network and above layers): As a simple router, data conversion, encapsulation, translation and encryption.

1.1.2.4 Common Link Connections

Link connections are physical connections used in computer networking. Some of these are also used in the fieldbus (for example, profibus uses RS 485). Other than a physical connection, there should be some protocol for data transfer, for example, a MODBUS protocol may be used with RS 232. Because the recommended standard (RS) is not truly maintained by manufacturers, later these were defined by the EIA.

1. RS (EIA) 232: This is used to send data over a short distance for point-to-point communication for serial transmission of data with a communication rate <20 kbps (a faster rate is also possible, e.g., MAX3225E). As shown in Fig. 7.4, "N" numbers SLCs to control system in typical, daisy chain connection. It has one serious disadvantage: with a break (as shown by a dotted break line) in the chain, all downstream devices will be disconnected. A typical connection of RS232 is shown in Fig. 7.5D. A few characteristics defined by the standard include: (1) Electrical characteristics: (voltage level, signaling rate, timing, slew rate, voltage withstand capability, short circuit behavior, and max load); pluggable connector and pin definition and function of each circuit in a connector; and an interface circuit for telecom applications. However, the standard does not define the external power supply, character encoding (say, ASCII), protocol, framing character, error checking methods, etc. Some of the standard data shall include but are not limited to the following:

- Binary 0: +5 to 15 VDC (Tx) and +3 to 13 VDC (Rx).
- Binary 1: (−)5 to 15 V (Tx) and (−)3 to 13 VDC (Rx).
- Slew rate is 4–30 V/μS.
- Load/output resistance: 3–7 KΩ/300 Ω.
- Start bit binary 0; stop bit binary 1.
- Data 5–8 bits.
- Parity even, odd, mark, or space.
- Leading and trailing idle bit: binary ones.
- Because there is no guideline for cable selection, one needs to decide on the basis of what is to be connected and what are the PIN connection (In RS 232 **9/25** D plug connectors are used) and gender of the plug. In some cases, four pins are also in use. The IBM PC AT connection is nine pin.

2. RS 422: It is a balanced voltage interface with a better data rate and distance when compared with RS232. Although defined mainly for communication between DTE and DCE, it also finds its use in the point-to-point interconnection of digital equipment. It defines the characteristic requirements of balanced line drivers and receivers. A typical baud rate of 100 K is achievable over a 1.2 km distance (normal maximum cable length),

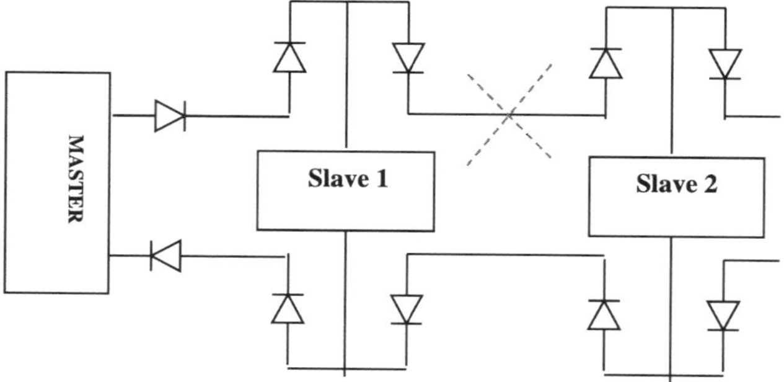

(Typical RS232 connection details shown in Fig 7.5d)

FIG. 7.4 Daisy chain connection for multiple slave connection.

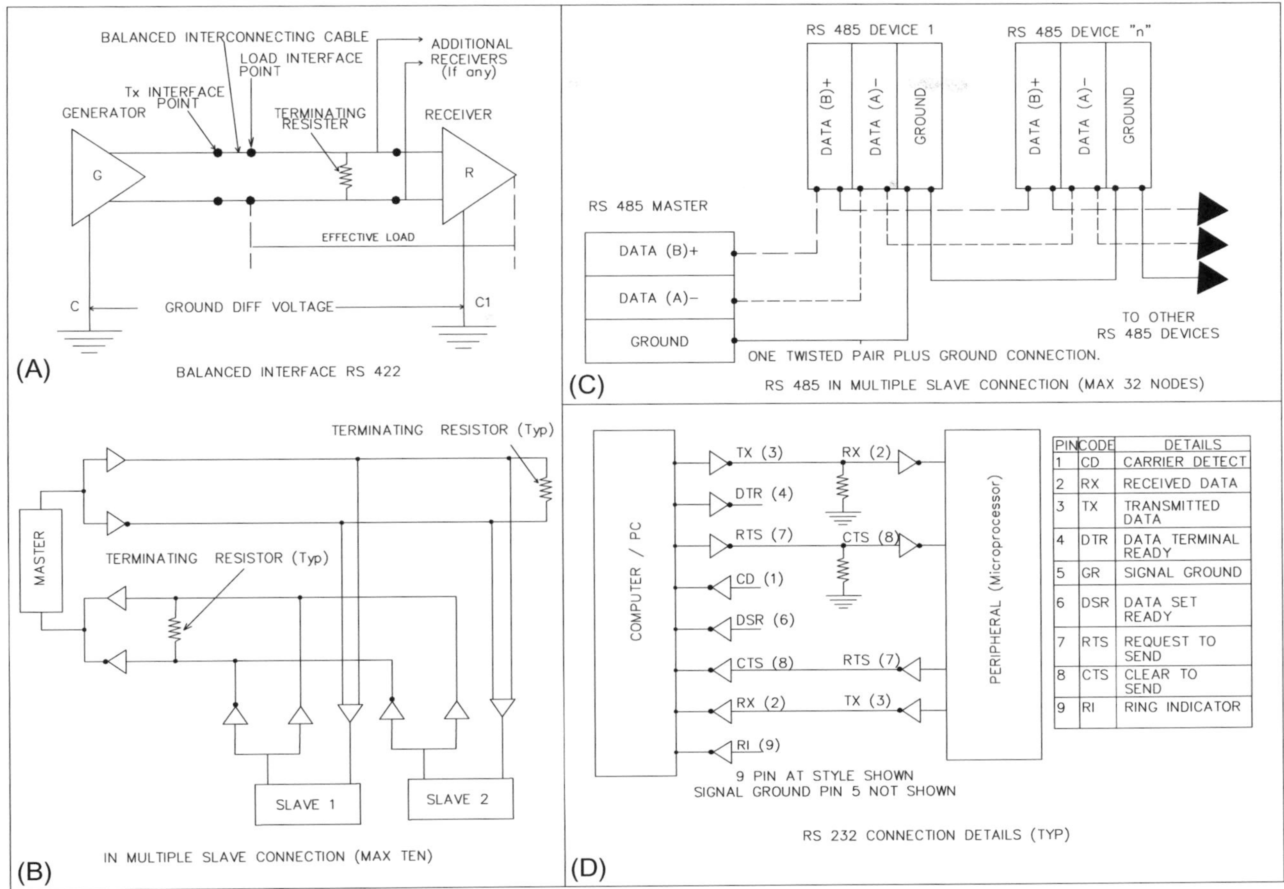

FIG. 7.5 Link connection details RS 232/422/485.

but the baud rate falls in logarithmic scale with an increase of distance (max 10 mbps is possible for 12 m typical for 24 AWG twisted pair cable). So, the length is decided based on the data rate necessary. 37 Pin in D SUB plug connectors are common. A typical RS 422 connection is shown in Fig. 7.5A. A single master can accommodate up to 10 slaves, as shown in Fig. 7.5B. The major characteristic features are: open O/P Voltage: + 10 Vdc (Max), max out current: 150 mA, CMV: @ RL 100 Ω: +3 V, loaded O/P voltage: + 2 Vdc (Min) (RL 100 Ω), differential receiver voltage: +10 V (+12 V capability), and input resistance of 4 KΩ.

3. RS (EIA) 485: On account of transmission over a twisted pair (half duplex—bidirectional communication but not simultaneous) cabling, RS485 is less costly and less sensitive to noise. It allows a bidirectional asynchronous serial transmitter over a long distance of 1.2 km at a baud rate of 100 kbps (in 10 m, max 35 mbps is possible). It can accommodate multiple nodes (up to 32) linear topologies in peer-to-peer communications, as depicted in Fig. 7.5C. There is one variant of RS485 in a four-wire connection in the master-slave mode and for duplex communication. It also uses a differential balanced line (three wires, even though the transmission is over two wires; third wire is ref voltage). However, the ground is important to avoid data loss. Cables are terminated with a resistor to minimize any noise effect. It has nine pins in a D SUB plug connecter. RS485 handles some software such as addressing, turn around delay, etc. RS485 is widely used in many protocols for electrical interface, for example, MODBUS or fieldbus: PROFIBUS. Some of the specification data are:
   - O/P voltage 1.5–6 VDC (±) max @load 100 Ω.
   - Short circuit current: ±250 mA per output.
   - Receiver sensitivity: (−) 7 < V < 12: ±200 mV.
   - Receiver input resistance: >12 KΩ.
   - Common mode voltage: ±3 V.

1.1.2.5 Network Transmission Techniques

Two major techniques are:

- **Broadcast** Mode: A short message called a packet is sent by a machine (node) over the communication channel shared/listened by all other machines (nodes)

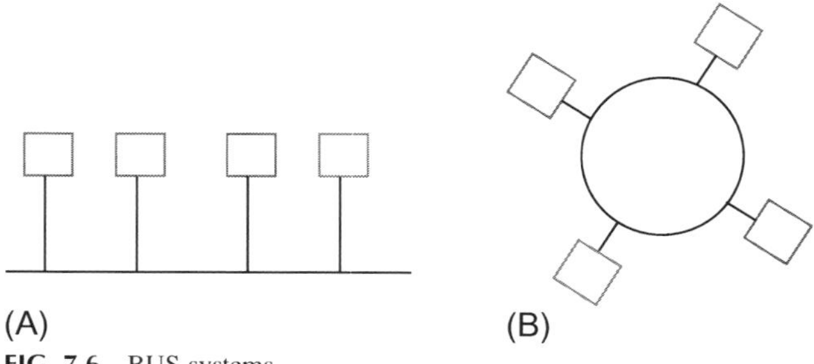

FIG. 7.6 BUS systems.

in the network. The packet contains an address, so the intended machine/node can check the address field and process the same while others ignore the message. So, with a specific code, the transmitting machine (node) can send the message to all desired destinations at a time. Multicasting has some differences with broadcasting; in multicasting, a subnet of machines (nodes) can be addressed. Any machine (node) in the network can become a member of the subnet to receive the group message (like group SMS or email). The broadcast may be bus or Ring, as shown in Fig. 7.6A and B with each box representing a node. On a round-robin basis, each channel may be allocated a time slot, on the basis of a time interval division. This type of allocation being static uncalled for time may be wasted. In a centralized dynamic allocation system, a central node will decide the time allocation to different nodes. In a decentralized dynamic system, a machine decides when it wishes to communicate, and to avoid any chaos, various standards have been developed to handle the same, for example, collision detection, etc.

- **Point-To-Point** Mode: In this mode, at any instant of time, there will be a transmission between a pair of machines (nodes). Naturally, a packet in this network may have to travel intermediate nodes. So, a routing algorithm plays a vital role.

As a general convention, it has been found that broadcast is used for a network with a limited geographical spreading whereas a larger network goes for point to point. IEEE has set forth a few standards; based on the same, various networks have been developed. Some of these are: Ethernet (CSMA/CD—IEEE 802.3), token bus (IEEE 802.4), token ring (IEEE802.5), wireless LAN (IEEE802.11), and wireless MAN (IEEE802.16).

1. Ethernet (IEEE 802.3): Ethernet (IEEE 802.3) uses Manchester (or differential Manchester) coding. This makes the bandwidth double. In an Ethernet, both half duplex and full duplex mode (simultaneous bidirectional communication) communication are possible. The data link layer (ref. clause no. 1.1.2.1.2) has two sublayers, as shown in Fig. 7.7.

So, one can conceive an Ethernet with three parts: (1) a physical layer (physical medium for communication; (2) a media access control (**MAC**) rule as a part of the data link layer embedded in the Ethernet allowing multiple machines to access the network; and (3) **LLC**, which stands for logical link control, and **SNAP** (subnetwork attachment points). **MAC has functions** such as data encapsulation (including frame assembly) before transmission, frame parsing (analyzing), and error detection after transmission. Media Access Control includes initiation of frame recovery after failure of transmission. The MAC—client sublayer, which is mainly related to the **LLC,** may be conceived as a logical link control (IEEE802.2), an interface between the Ethernet and the MAC of the upper layer; a bridge entity (LAN to LAN connections: Ethernet to Ethernet, or Ethernet to token ring); and Ethernet frame, a protocol structure for Ethernet IEEE802.3 (for MAC data frame and Ethernet data frame; see Figs. 7.8 and 7.9, respectively).

- o. **Preamble**: 7 byte (8 byte for DIX Ethernet): It is an alternate pattern of 0 and 1 7 bytes (101010…) to tell the receiving station a frame is coming and to provide a means to synchronize the frame reception portion of the receiving physical layer with 10 MHz square wave (5.6 μS).
- o. **SFD**: It stands for start of frame delimiter: one byte: Also, an alternate pattern of 0 and 1 with two consecutive 1s to indicate a neat bit is the left-most bit of the left-most byte.
- o. **Destination Address**: **6** byte long. The DA identifies the receiving station. Here, the first higher bit indicates **0 for ordinary address** and **1 for group addressing in MULTICAST** mode for multiple machine addressing, that is, to be received by all the stations in the group. Bit **46** distinguishes between a local address and a global address. If it is **0, it is a local address** decided by the LAN administrator and nobody outside the LAN is interested in it. If **46 bit** (adjacent to higher order bit—6 byte, meaning $6 \times 8 = 48$ bit $- 2 = 46$), **1 it stands for** the **global address** assigned by IEEE to have a unique address for each of the stations (in the order of 7×10^{13}).
- o. **Source Address**: 6 byte long SA identifies the sending station and is similar to the destination address.

FIG. 7.7 Data link layer and sub layer.

2	DATA LINK	LLC (or LLC +SNAP)
		MAC

Discussion on LLC(IEEE 802.2)

LLC and SNAP have been found in IEEE 802.2 standard

IEEE 802.2 is identical to all networks! It provides the interface between different protocols (TCP/IP etc.) and network types (such as Ethernet, Token ring etc.). LLC provides the following functions to upper network layer:

- ***TYPE 1 : Unacknowledged Connectionless (meaning no idea on what is going on)mode ---to Single destination, (point to point), multiple destination (multicasting) and to all destination (broadcasting)***
- ***TYPE 2 : Connection oriented Operational mode – a connection aware link between LLCs of end stations(using sequence numbering for frame receipt guarantee).***

LLC frame format consists of Destination Service Access Protocol (DSAP), Source Service ACCESS Protocol (SSAP), Control field, & Information Filed as shown below:.

DSAP Address	***SSAP Address***	***Control Field***	***Information Field***
8 bits	***8 bits***	***8 or 16 bits***	***N*8 bit***

DSAP: 1 I/G : 0 individual or 1 Group & 8 address bit

SSAP: 1 C/R : 0 Command or 1 Response & 8 address bit

Control has separate structure format to include fields like Information, Supervisory, & Un numbered. Information: LLC data or higher layer protocol

FIG. 7.8 LLC format.

Bytes ⟶	**7**	**1**	**6**	**6**	**2**	**0-1500**	**0-46**	**4**
Frame ⟶	**Preamble**	**SFD**	**Destination address**	**Source address**	**Length type**	**Data**	**PAD**	**Checksum**

FIG. 7.9 Ethernet frame.

Length Type: 2 bytes. It indicates the MAC data bytes in the data field or a frame-type ID. When the transmission speed increases, then the frame size must increase or the cable length must be decreased. The maximum length of the frame is mainly limited by the RAM size of the receiving end. There is also a limit in the minimum length of the frame (size fixed at 64 bytes from destination station to checksum, including both). This will be clear from Fig. 7.10, Bit collision. If the frame size is too small, then by the time first bit reaches the distant receiver "*y*" from "*x*," the frame transmission from *x* may be complete and at that time seeing no transmission, *y* may like to transmit and this will result in a collision. So, IEEE decided to have 64 bytes as a minimum.

- o. **Data Field**: It is a sequence of *n* bytes of any value (0–1500); it is specified with PAD.
- o. **PAD**: It is the field that is filled when the data is less than 46, so it is padded to make a **total frame 64** (hence may have 0–46 bytes}.
- o. **Check Sum**: This is 4 bytes. This sequence contains a 32-bit cyclic redundancy check (CRC), created by sending MAC and is recalculated by receiving MAC to check for the damage.

- **Physical Medium**: Multiple computers may be connected over a single channel in a multidrop way, as shown in Fig. 7.11.

Here, the common channel (e.g., coaxial cable, twisted pair cable, and fiberoptic cable) is being shared by all the drops (computers/nodes); and the logical topology is more important than the physical topology. In this method, each node is authorized to communicate over the bus/channel at any time but one at a time without any priority. Communication is carried out using CSMA/CD (carrier sense multiple access/collision detect). **CSMA/CD** is somewhat like in a meeting room where there are several speakers, and each wants to speak, meaning multiple access. To avoid noise in the room, each speaking one after another is like carrier sensing.

- □ Before transmission, each node verifies that there is no communication in the line.
- □ In case of two nodes that transmit simultaneously, there will be a collision (several data frames). In that case, both the machines send a jam signal to all the devices connected, preventing all from communication. A collision could be releasing its own signal (24 mA), or an error in the CRC. Also, all the nodes will interrupt its

FIG. 7.10 Bit collision (minimum frame length requirement).

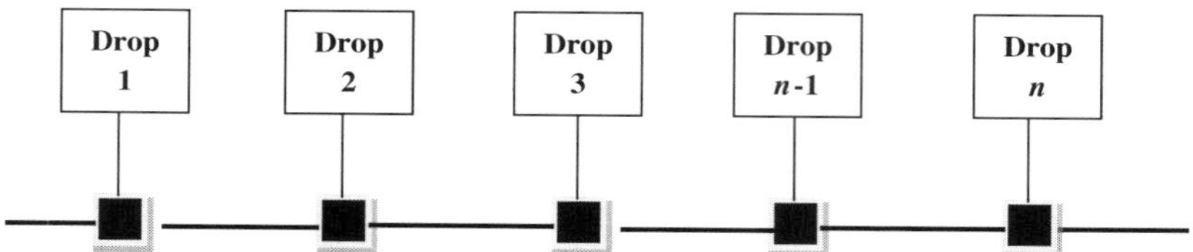

FIG. 7.11 Multidrop network.

communication and wait for a random period. After a lapse of that period, one may start checking the channel freeness to communicate. Obviously, there shall be a specified data packet size and a waiting time. Also, for one collision, there is time, if there is another collision, so machine shall wait for two waiting time and so on?

- **CSMA/CD** is applicable for a shared channel for a half-duplex mode of communication. Switching Ethernet is something different and could be full duplex.
- In Ethernet, names are specified by **xBASEy** where x represents communication speed in mbps and y stands for physical medium type. Refer to Table 7.2.
- **Switched Ethernet**: The physical topology of a switch is star but organized around switch very similar to gateway. There is another very similar device known as an Ethernet **HUB**, which is almost obsolete because of the falling cost of switches and limitations of the HUB. In an Ethernet switch, there will be 4–32 line cards, each with about eight connectors. These connectors are connected at the back plane for very high speed (> gigabits per seconds) communication utilizing own protocol. Because this is used for internal switching action and not related to the outside network, the protocol of the switch is immaterial, but the speed is important. Each of these connectors could be connected to a host machine, etc. Thus, it is possible to connect 32–256 nodes. A switch inspects the source and destination address message to draw up a table, which then allows it to know which machine is connected to which port. Knowing the recipient port switch, it will only transport the message on the appropriate port. Let one of the nodes connected to switch 1 wants to communicate, then switch will examine whether it belong to the same card, if so it will pass the same to it but If not, it will pass the same to the next card through hi speed back plane to the destination card. Now, what if more than one machine out of eight (say) wants to communicate? Then it will go for CSMA/CD. So, one card will have one **collision domain (eight only)**. Alternatively, it may have a suitable RAM to buffer the input data. When the buffer RAM is there, send and receive can go simultaneously in the full-duplex mode. In that case, there is no need for CSMA/CD. Fig. 7.12 may be referenced. When the Ethernet acts on full-duplex mode, there will be a separate set of connections for send and receive, that is, in duplex mode (say) 100BaseTx Ethernet, will have two pairs of cables. Thus, in a network with an Ethernet switch, there may be several communications going on simultaneously without any collision. In full-duplex mode in the same card, even transmitting and receiving may go on simultaneously. Therefore, it is not possible to predict which action is going on in a port.
  - Gigabit ethernet: In a gigabit Ethernet, as the name signifies, its speed is 1000 mbps minimum; this is achieved with Ethernet switches, as shown in Fig. 7.13.

Here, with respect to the switch and any node, there is only one sender so there is no collision. Therefore, CSMA/CD is not utilized. Also, when the switch is sending data, it is done through a separate line, hence full-duplex mode is possible. Except for the physical limitations of the medium, the switch does not have distance limitations, so it can be applied to a large area network. As of today, various technologies such as 10BaseT, 100BaseTx, 1000BaseT, or 1000baseTx use the same connector RJ45 to be connected to the same switch. This intermixing of various types poses a serious disruption problem that could be addressed by AUTO NEGOTIATION (N WAY). It is also covered in IEEE 802.3 1998. A reader with further interest may consult the same. Mostly in power plants, a network speed of 100 mbps is in use, but it may not be far off when higher speed will be demanded, so, the gigabit Ethernet has been touched upon.

Here, 10/1000 stands for mbps speed. 5/2/T/F stands for the cable type specified. BASE here stands for **base band**

TABLE 7.2 Ethernet Types

Name	Cable	MAX SEG (m)/ PORT (m)	Node/ SEG	Connector	Remarks
10BASE5	Thick coax	500	100	Vampire tap	Original now obsolete
10BASE2	Thin coax	200	30	BNC	Cheap (no hub)
10BASET	Twisted pair (CAT 3 or 5)	100	1024	RJ45	Easiest set up and cheapest. Common for Ethernet switch (or hub). Each node connected to each port. Use two different voltages (−)2.5, +2.5 v
10BASEF	Fiber optic	2000	1024		Long distance
100BaseFx	Fiber optic (type 62.5/ 125)	2 km			Fast Ethernet
100base Tx	2P twisted pair CAT5	100		RJ45	Known as fast Ethernet uses three different voltages (−) 1, 0, +1 V
1000BASE LX/SX	Fiber optic (multi/ mono)	550			Single (10 μ) or multiple (50, 62.5 μ) L = Long wave (∼1350 nm) S = short wave 860 nm
1000BASE CX	2p STP	25		GG45/RJ45	CAT7/Cat5e/Tough Cat5 STP
1000BASE T	4p UTP (CAT 5)	100		RJ45	(CAT 5) IEEE 802.3(40) 2006
10GB Base T	Cat7	100		GG45	

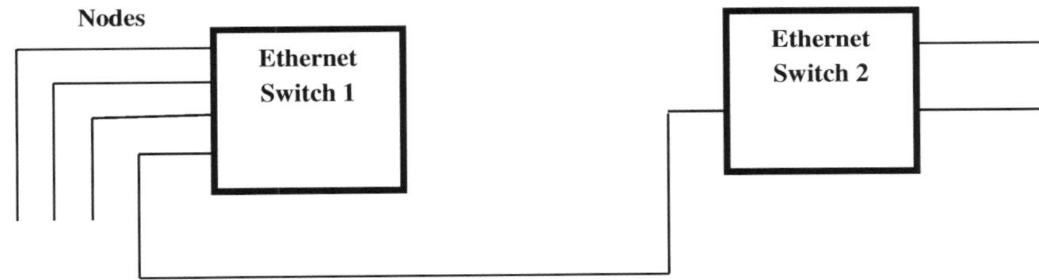

FIG. 7.12 Ethernet switch and network.

transmission. All 10/100/1000 Ethernet speeds are available in half as well as full-duplex configuration. The "X" after T and F stands for different standards, for example, both 10baseT and 100baseTx use only two pairs of Cat5 cable, yet they have different standards (10baseT—IEEE802.3 (14) 1990 and 100baseTx—IEEE802.3(24) 1995). 100 mbps and 1000 mbps Ethernets are covered by IEEE 802.3u and IEEE 802.3z, respectively. 40Gbase-CR4 and 100Gbase-Cr10 belonging to IEEE 802.3ba are a new Ethernet. System integrations are done based around this.

2. Token Passing (IEEE802.4 and IEEE802.5): When compared with CSMA/CD, the token passing system is deterministic, and has advantages such as higher throughput and priority setting. Based on topology, there types of token passing systems: (a) token passing bus (IEEE802.4) and (b) token passing ring (IEEE802.5). The following are the basic schemes of operation in token passing systems.

- Only one node can talk at a time.
- Node waits for a free token to communicate over the channel.
- Token circulates among the nodes until one wants to communicate.
- A station having no data to transmit passes the token to the neighbor.
- Token is held for a specified time.
- Node/device, when grabbing the token, takes up the following actions:

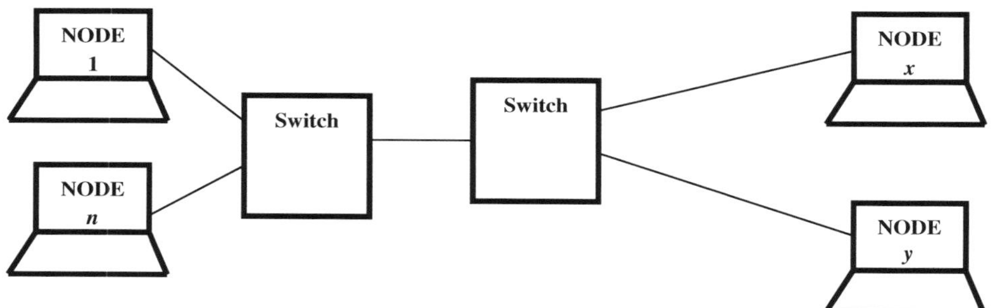

FIG. 7.13 Gigabit Ethernet.

- Sending node sets the token busy, adds information and a trailer packet.
- The entire message is sent over the channel for communication.
- Each node examines the header packet/frame to see if it belongs to it or to ignore it. The intended node copies the message set bit in the trailer field to acknowledge and sent back. The sender then checks that message received back to ensure that it was properly received. Then it frees the token in case it is not required.

□ Most popular Fieldbus like Foundation Fieldbus and Profibus utilize token passing media access.

□ It has a few advantages:
- High throughput at a higher network load (number. of nodes) when compared with CSMA/CD, but saturates after a certain number of nodes in the network.
- Deterministic.
- Priority setting (up to six).

□ It has a few disadvantages, such as:
- Complicated protocol and software.
- Higher bandwidth.
- Expensive cable and hardware.

▫ **Token Bus** (IEEE **802.4**): Fiber distributed data interface (**FDDI**) is 100 mbps LAN based on a token passing bus (IEEE 802.4). The token bus is a protocol suit for physical and data link layers having two subdivisions, MAC and LLC, with features such as:

- Physically, it is a bus network, but logically it has to be a ring network.
- Stations are organized in a doubly linked list.
- It uses the MAC address for distributed polling.
- It can broadcast, or may go for point to point also.
- Possible central station to manage and monitor the network.
- It has logical ring maintenance.
  - o. Initialization of ring, addition and deletion to and from ring.
  - o. Periodic token holder call to others to join if outside ring.
  - o. Maintenance of address of predecessor or and successor once the ring is formed (Fig. 7.14).

Where SD/ED = start/end delimiter, FC = frame control, FCS = frame check sequence Max from SD to ED is 8191 Bytes. Westinghouse WDPF uses a token bus system.

▫ **Token Ring** (IEEE **802.5**): In 1970, IBM first developed the system, which is quite compatible with the standard IEEE802.5. IEEE 802.5 specifies, data rate as 4–16 mbps, station per segment as 250 and signaling shall be base band token passing using differential Manchester encoding. As discussed above, when a node-possessing token wants to transmit information, it seizes the token, alters 1 bit of the token (which changes the token into a start of frame), appends the information, and sends the same to the next station/node. The next station/node sees and sends it to the next station. The information in this fashion circulates in the ring until it reaches the destination. The destination station copies the information and again circulates the token in the network until it reaches the sender. It then checks whether the information is copied. The token is then finally removed from the network. A token ring uses sophisticated means for priority management. A token ring has two fields, a PRIORITY FIELD and a RESERVATION FIELD, that control priority. Only a station/node with a priority value equal or higher than the priority value contained in the token can seize the token. A token ring deploys several means for fault management. A central station or any of the stations may be assigned

No. of bytes →	≥1	1	1	2 or 6	2 or 6	≥0	4	1
Frame →	Preamble	SD	FC	Destination address	Source address	LLC	FCS	ED

FIG. 7.14 Typical frame format of token bus.

to do the timing and ring maintenance management function. When a station fails, its token may be circulating in the ring, which prevents another from continuing communication in the network. Such fault management can be done by the station assigned for network maintenance. A typical IEEE802.5 and token ring are shown in Fig. 7.15, A typical frame format of token ring. SD/ED = start/end Delimiter SD Alters each station of arrival of a token. ED is the end of the token or data. AC = access control byte is for priority control (first three higher bits). FC = Frame control Byte states data or control. This bytes specifies the type of information. FCS = Frame Check Sequence filled in by source with a calculated value dependent on frame contents. The destination station recalculates the same to check if there is any damage. Frame Status: 1 byte data is for field terminating a command/data.

The frame status field includes the address-recognized indicator and the frame-copied indicator, as discussed above.

1.1.3 HART Protocol

A highway addressable remote transducer, or **HART**, is an open, bidirectional communication protocol utilizing the frequency shift keying (FSK) standard to exchange data between intelligent field devices with the host system. In HART, a 4–20 mADC signal can be sent simultaneously with digital information in a superimposed manner. Some of the systems may have the necessary OPC servers for system integration, for example, a HART OPC server ABB800xA system. Figs. 7.16 and 7.17 show the HART communication and data frame, respectively.

1.1.3.1 Hart Protocol Features

Major features of HART protocols in different headings are:

1. Technological advantage to support the user such as standard wiring of 4–20 mADC with superimposed digital data transmission, usefulness of intelligent devices for operational performance improvement, easy for commissioning, early warning for deviations for device and process, quick troubleshooting with identification and problem resolution, continuous validation of loop integrity and automation strategy, and a fieldbus communication gateway
2. System and plant availability features: Robust and accurate protocol for risk reduction for failure as well as avoidance of costly shutdowns, detection devices and connection problems in real time with early warning of deviations, and the ability to integrate devices and systems.
3. Advanced diagnostics and maintenance cost: Quick validation of the loop and configuration, quick reporting for any change in remote diagnostics (at the host), and the possibility of predictive maintenance and reduction of spare inventory.
4. Regulatory compliance and integrated safety level (SIL).
5. Enabled for record keeping and the possibility to test shutdown conditions.
6. Broader selection of products from multiple vendor support.

1.1.3.2 Characteristics of Hart Protocol

1. Technique: Bell 202 standard FSK, superimposition of digital signal with 4–20 mADC communication speed: 1200 bps.
2. Devices support: In loop supports up to 100. Based allowed load impedance it is 15 nos (typ). Powering provision and noise from each limits the number of devices in a loop.
3. Configuration: Master-slave mode: It can support two masters, for example, the host being the primary master and a handheld configurator can be the secondary master whereas intelligent devices, for example, Tx/control valves, could be the slaves, as shown in Fig. 7.16B.
4. Communication modes: Peer to peer, multiplexing type with one PC (say), connected via a multiplexer or a multidrop configuration, as shown in Fig. 7.16B.

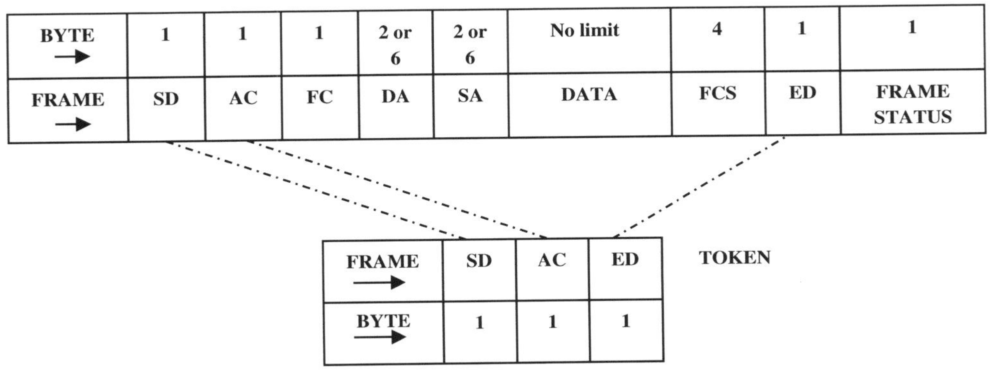

BYTE →
1
1
1
2 or 6
2 or 6
No limit
4
1
1
FRAME →
SD
AC
FC
DA
SA
DATA
FCS
ED
FRAME STATUS
FRAME →
SD
AC
ED
TOKEN
BYTE →
1
1
1

FIG. 7.15 Token ring data frame.

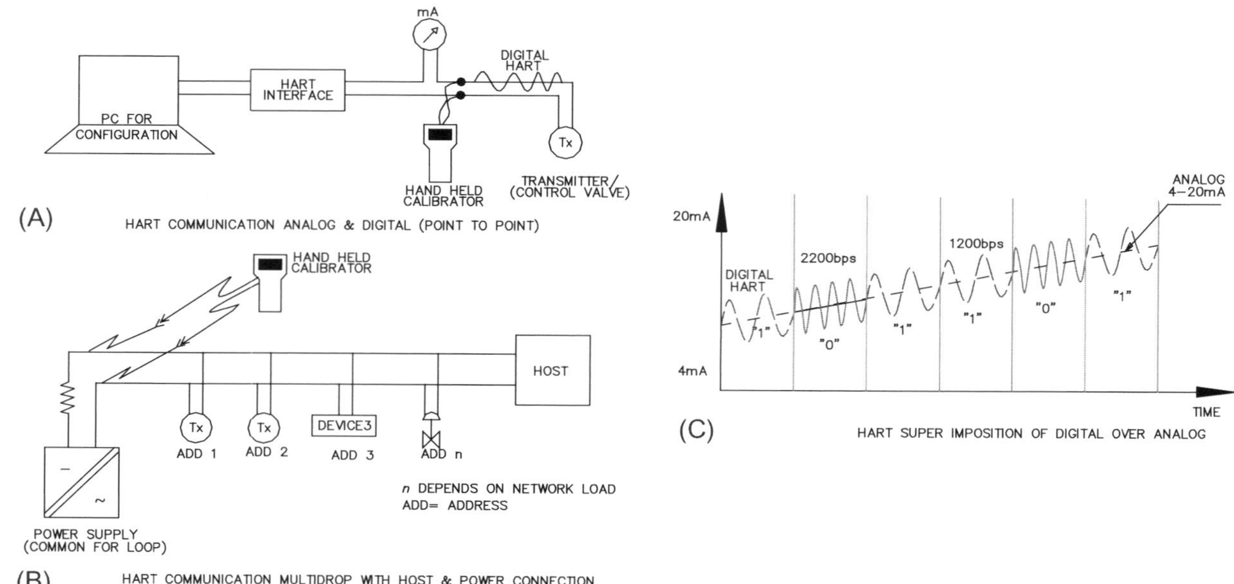

FIG. 7.16 HART communication.

Preamble	SD	AD	CD	BC	Status	Data	Parity

2 status byte

BC: byte count

CD: HART command byte

AD: Address

SD: Start byte

FIG. 7.17 Typical HART telegram timing (based on samson document).

5. Cable: A normal instrumentation twisted pair cable is sufficient. Depending on distance coverage, the specification may change, for example, up to 1500 m 0.2 mm^2 conductors with a common shield, whereas for a distance ~3000 m, the same cable but with individual shielding. For shorter distances, unshielded cables may be possible; however it is recommended to use screened cables.
6. Data Exchange: Status, diagnostics, calculated value, etc. Some of the generic parameters are: PV reading, analog output reading, secondary variable reading, change (tag, data, PV unit, range, damping, output transfer function, polling address), output trimming, loop checking data, zero and range change, modem action, etc.
7. Load Limit: 1100 Ohms, with provision of powering (Fig. 7.16B) the entire loop.
8. Noise immunity and error detection: Online devices can be added or removed without much degradation of performance. Noise immunity conforms to IEC 801-3 (radiated, radio-frequency, electromagnetic field immunity) and IEC 801-4 (electrical fast transient/burst immunity) Class 3. It can detect up to three corrupt bits in a telegram.
9. Hazardous application: IS compatible and exi.
10. Addressing: Special long format addressing format is used. During configuration mode in peer-to-peer mode, the tag and bus address are set.

1.1.3.3 Telegram

Typical structure is shown in Fig. 7.17 with timing as per Table 7.3 These are in UART character with a start, a stop, and parity bits.

1.1.3.4 OSI Model Versus HART

When compared with an OSI seven-layer model, it is found that layers 1, 2, and 7 are distinctively present. Network layers and transport layers are present but not so distinctive while session (5) and presentation (6) layers are missing, as shown in Fig. 7.19. It also compares wired and wireless HART.

TABLE 7.3 Typical HART Telegram Timings (Based on Samson Document)

Parameter	Calculation	Value
Byte/telegram		25 Message + 10 cont ch.
Telegram size	35 ch × 11	385 bits
Time per bit	1/1200	0.83 mS
Transaction time	385 × 0.83	0.32 S
User data rate %	(25(M) × 8)/385	52 (%)
Time/user data	0.32/25	13 mS

1. LAYER 1 PHYSICAL LAYER: Here, the transmission is in Bell 202 FSK with "0" being represented by a 2200 bps signal and "1" being represented by a 1200 bps signal (Fig. 7.16C).
2. LAYER 2 DATALINK LAYER: The function of this layer is to ensure communication from one device to another. The system consists of a MASTER (two numbers permitted) and a slave. A slave communicates only when it is called. For admitting two masters, the primary after each command gives a **pause** to allow for a secondary master to communicate. Here, one of the major problems is controlling noise when a new one is added, because each unit is a noise contributor.
3. LAYER 3 NETWORK LAYER and LAYER 4 TRANSPORT LAYER: Neither is prominent. The main functions of a network layer include routing and end-to-end security with service to the transport layer, whereas layer 4 ensures successful communication.
4. LAYER 7 APPLICATION LAYER: Various commands like universal (common to all devices), common to many devices, device family/generic command, device-specific command, response data type, and status report are the major functionalities of this layer. Some command and conformance relations are presented in Table 7.4.

1.1.3.5 Hart for System Integration

In integrated systems, HART devices can be connected via an intelligent HART-enabled I/O. Controller also and can be connected through multiplexer also. Integrated systems have a field management tool to configure and access data, for example, managing remote I/O and topology planning up to the field level from field devices. OPC servers are developed as a part of the system integration package (e.g., HART OPC for the ABB 800xA system). The OPC server is used to access specific field devices, utilizing various device libraries to configure, parameterize, operation and device communication. HART devices device-type managers are also available.

TABLE 7.4 Layer 7 Command Relations (HART)

Command Field Dev	Data Flow	Conformance for Master
Universal command	⟶	Read measured variable
Universal command	⟶	Read universal information
Common practice command	⟶	Write common practice parameter
Common practice command	⟶	Write selected parameter
Device SP. command	⟶	Read device specific information
Device SP. command	⟶	Read and write entire database

1.1.3.6 Hart Hand-Held Configurator

HART is an open communication protocol between the master and the field devices. This openness, interchangeability, and manufacturer independence is possible as long as communication is done using universal commands and common practice commands. To eliminate the problem with device-specific commands, a device description language (DDL) has been developed. Basically, a HART configurator consist of the following components:

- Microprocessors with associated memory (RAM temporary memory to execute the program and E^2PROM program and basic data).
- Value and parameter display and keyboard: For operation, calling data, etc.
- MODEM for communicate over the two wires, and RS 232 communication with the PC and other host devices. There are VDI guidelines toward the communication structure of this. Some of these features are:
  - o. All protocols defined, and commands are selectable and implementable.
  - o. Device description can be implemented.
  - o. User interface provides all extended communication, information, and control options.

General features and specifications for a universal HART HANDHELD CONFIGURATOR shall include but are not limited to those elaborated below:

1. Function: Clone/upload/download configuration, offline mode for developing/reviewing configurations.

2. Major keys: Quick menu access, HART menu, and home dedicated text edit.
3. Extensive store: >200 configurations, full alphanumeric keypad.
4. Edit function: From review/edit mode.
5. Programmable lock-out settings commission devices.
   - Step-by-step configuration.
   - Document as-left configuration.
6. Operations:
   - Reconfiguration for process changes.
   - Reconfiguration for batch runs.
   - Document as-found and as-left configurations.
   - Populate DMS database software.
   - Thumb key for UP/DOWN/SIDE selection.
7. Device maintenance:
   - Troubleshoot devices with HART.
   - View device error messages.
8. Device support: Normal available configurator with >600 devices support.
9. Onsite update: Via Internet
10. User interface: 10–3 line display, high-resolution pixel with backlit display.
11. Power: Lithium/alkaline battery backup.
12. Housing: Plastic case.
13. Certification: CE/ATEX, etc.
14. Firmware language: Multilanguage.
15. Major suppliers ABB, Emerson, YEL.

1.1.3.7 Wireless Hart Protocol

Wireless HART is one of the most preferred choices for process automation and system integration. In September 2007, the HART Communication Foundation developed and introduced the wireless HART, which has gained rapid acceptance in the industries. This has been developed for industrial wireless sensor networks following the international standards IEC 62591 and IEEE 802.15.4 (ref clause no. 1.1.5.3). Wireless HART needs to be reliable and secured even in the presence of interference. Mesh networking (fully connected, as in Fig. 7.26), channel hopping (changing frequency known to the transmitter/receiver), and time-synchronized messaging are the major technologies to get reliable communication. An effective power management mechanism, security and privacy for network communications, and coexistence with other wireless networks are very important.

1. **Wireless HART conceptual details**: As seen from the above discussions, HART is an open standard and vendor independent. Because of this, it is the world's most broadly supported protocol for the process industry. The control systems that do not accept HART information in digital form use external multiplexers. Wireless HART implements a wireless mesh communications network for process automation applications, maintaining compatibility with existing HART devices, commands, and tools. A wireless HART network includes three main elements: wireless field devices, gateways for communication of field devices and host applications, and a network manager (for configuring the network, managing message routes, and scheduling communications), which utilizes 2.4 GHz bandwidth as per standards such as IEC 62591 and, IEEE 802.15.4. It supports direct sequence spread spectrum technology and is time division multiple access (TDMA) synchronized.
2. **Benefits of Wireless HART**; Major features are enumerated below:
   - It is built on proven industrial standards to provide greater plant availability.
   - Well supported by multiple vendors to support interoperability with existing systems.
   - Fast, flexible, and easy system implementation and device commissioning with remote access to all devices for diagnostics, loop troubleshooting, and predictive maintenance.
   - Low cost (cost reduction on account of cable dispensation, reduction of labor for engineering, wiring, and associated costs) wireless access for a faster and easier process and additional information.
   - Large selection of devices and easy for system integration coupled with improved operation through regulatory compliance and elimination of manual data collection.
   - Extremely helpful for remote plant areas (tank farms, utilities, etc.).
   - Possible to instrument movable assets (railcars, rotating equipment, etc.)
   - Clear channel assessment tests available, hence it can coexist with other wireless networks. Also, it offers a self-configuring and self-healing network that can be easily extended to remote areas.
   - It offers a reliable, secured, and well-managed power economy for its operations.
3. **Working of Wireless HART** (Section 1.1.5.3): From the discussions in Sections 1.1.3.1–1.1.3.3, it is clear that the HART protocol is based on a request/reply communication mode, that is, HART devices transmit any information only on a request from the host to the device. However, the burst mode is an exception when the HART device can send a single piece of information continuously without repeated host requests in order to verify the analogue signal with digital value. Wireless HART utilizes the same principle, only the communication mode is without wire, that is, same protocol tools in wireless HART devices. Wireless HART is approved by international standard IEC62591/IEEE 802.15.4 as an open and free-to-use protocol duly supported by a large numbers of vendors. Wireless HART can be used on existing wired instruments to collect additional information and diagnostics. It provides a cost-effective,

simple, and reliable way to deploy new points of measurement and control without the wiring costs. In a wireless HART, there will be three basic components: a self-powered wireless device, an adapter for communicating with the host devices, and a network manager. Besides a protocol, wireless HART also provides system reliability, security, and power management. Based on a wireless mesh network communication protocol (fully connected in Fig. 7.26), the wireless HART protocol ensures compatibility between existing HART devices, commands, and tools. Fig. 7.18 may be referenced for wireless HART communication.

Through the wireless HART gateway, the host gathers data from instruments connected to the network in question. The wireless HART gateway converts data between the host and devices connected to the network. It incorporates a network manager, a security manager, and access point features. The security manager is an application embedded in the gateway to provide services to several wireless HART networks. The network manager is an application embedded in the gateway, and it is primarily responsible for distribution of network identity, configuring the network, scheduling communications between devices, managing message routes, and monitoring network health.

4. Characteristics of wireless HART: The basic characteristics of wireless HART are:
   - Real time: Supports time division multiple access (TDMA) with central network management.
   - Reliability: Channel hopping and channel black listing (restricting channel hopping to selected channels to protect a wireless service) in a mesh network. For reliability, it "hops" across channels to avoid interference to deliver high reliability in challenging radio environments, utilizing a 2.4 GHz license-free frequency band.
   - Coexistence with other wireless networks: Clear channel assessment tests, optimizes bandwidth and radio time, and time synchronization for on-time messaging.
   - Self-healing network: Adjusts communication paths to optimal with monitoring of paths for degradations to find alternate paths around obstructions. Mesh network and multiple access points. Built-in 99.9% end-to-end reliability [126].
   - Security: data integrity on MAC layer; data confidentiality on the network layer. In order to protect valuable information, it has robust, multitiered messaging with 128-bit encryption. There will be a unique encryption key for each message [126]. Data integrity and device authentication, rotation of encryption keys for the network. There will be multilevel security keys, indication of failed access attempts, reporting on authentication, and message integrity failure to make the system safe.
   - Power management: Truly speaking, this is not part of the wireless HART characteristics, yet it is a feature worth noting in the devices connected to a wireless HART network as it power optimizes at a network level. Also, as a part of the security features, there will be an adjustable transmit power level. Wireless devices do not have a wired power source, so they require minimal energy usage to extend the battery life or solar energy. However, there will be greater flexibility and lower installation costs [120].
   - Standards: This is built on international industrial standards such as IEEE 802.15.4 (250 kbps; with distance up to 250 m line of sight between devices) and IEC62591. It supports all phases of the plant lifecycle discussed in Chapter 13.

Comparison of wireless HART with wired HART and with the OSI layer model are detailed in Fig. 7.19. This shows that in the HART protocol, there are major changes in physical and data layers when wired and wireless protocols are compared. For the application layer, Table 7.4 may be referenced. Depending on applicability, there could be three types of commands in the application layer: universal commands, common practice commands, and wireless commands.

5. Components of wireless HART and functionalities: Major components are:
   - Gateway: The gateway provides a connection to the host network, that is, through this Ethernet/modbus network are accessed by wireless devices. It also provides the services of network and security managers and can support up to 80 devices. One or more host interfaces connect the gateway. Buffering and local storage for publishing data, event notification, and common commands time synchronization sourcing (i.e., processing queries from host applications) are common functions. Also, it provides the communication between GW and NM and host applications.
   - Network manager: It is the centralized brain and there can be a redundant network manager (only one active). Functionalities in the network manager include: location manager responsible for network topology, device configuration, and bandwidth; generation of network generation packets; maintaining reliable routing graphs; maintenance of communication schedule.
   - Security manager: It provides access control while also managing and distributing the encryption keys and keeping a list of devices in the network. Some consider this embedded system as part of the network manager.
   - Adapter: It is a device normally plugged into the existing HART-enabled device to pass information through a wireless HART to the host network. It could be battery powered or can be placed anywhere in the 4–20 mA loop and get power from it.

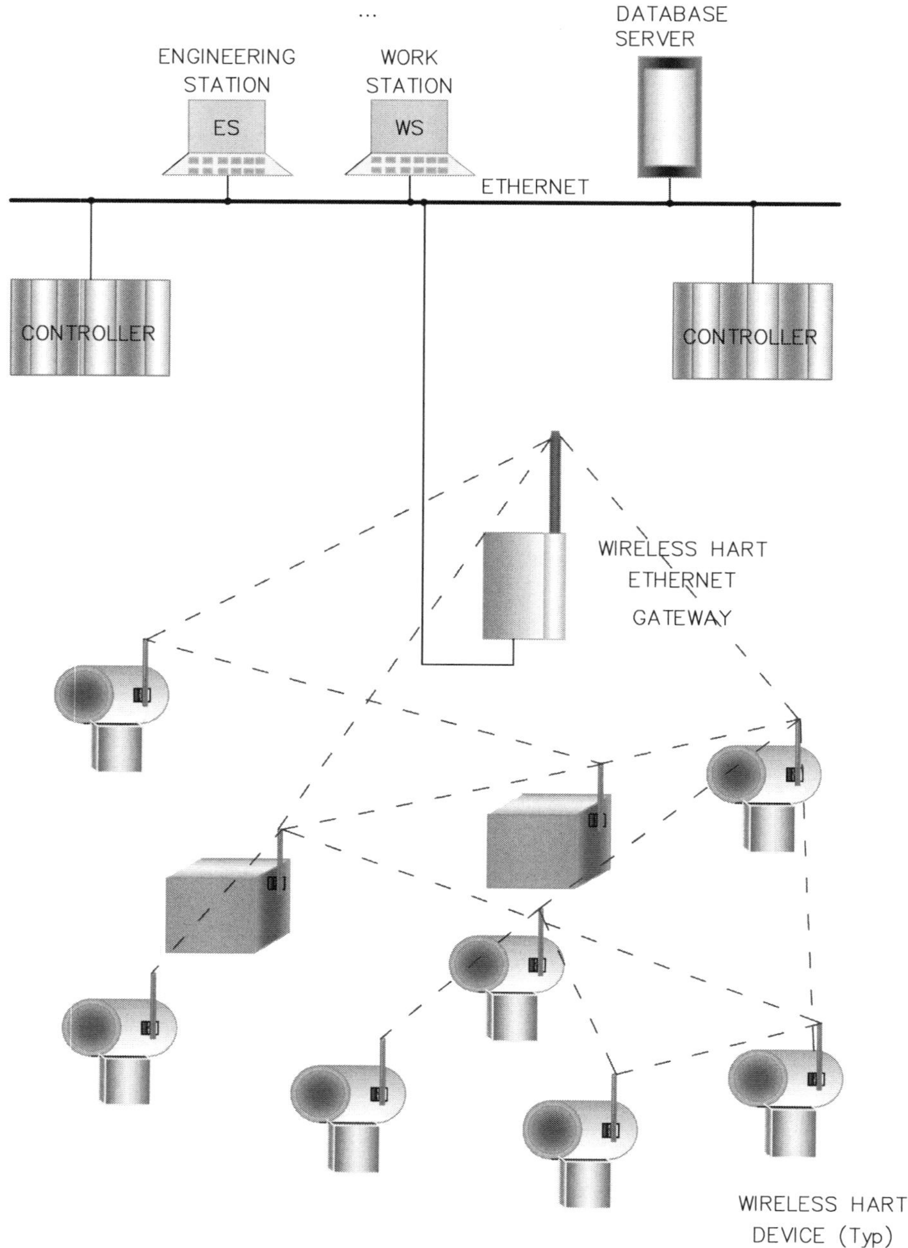

FIG. 7.18 Wireless HART.

- Repeater: It does not have any process connection of its own; it helps to route wireless HART messages from the wireless HART network to the host.

Apart from these, there will be host applications, devices, access points, and sniffers in wireless HART systems. With this, the discussions on wireless HART come to an end. Now let us look toward various bus systems, starting with MODBUS.

1.1.4 Modbus Protocol

In 1979, Modicon developed this protocol for communication among Modicon PLCs; later it became a de facto standard. It is basically an application-layer protocol for connecting devices (Fig. 7.20A). It works on a MASTER–SLAVE basis, that is, a transaction takes place upon a request from the MASTER (CLIENT), and the SLAVE(SERVER) responds, as shown in Fig. 7.20B. An exception response is sent for an error.

1.1.4.1 Transaction Methodology

The transaction methodology in MODBUS has two types; One is a transaction over the *MODBUS network* and the other is on a *different kind of network*. In the *former method*, the master has established a data format for the master's

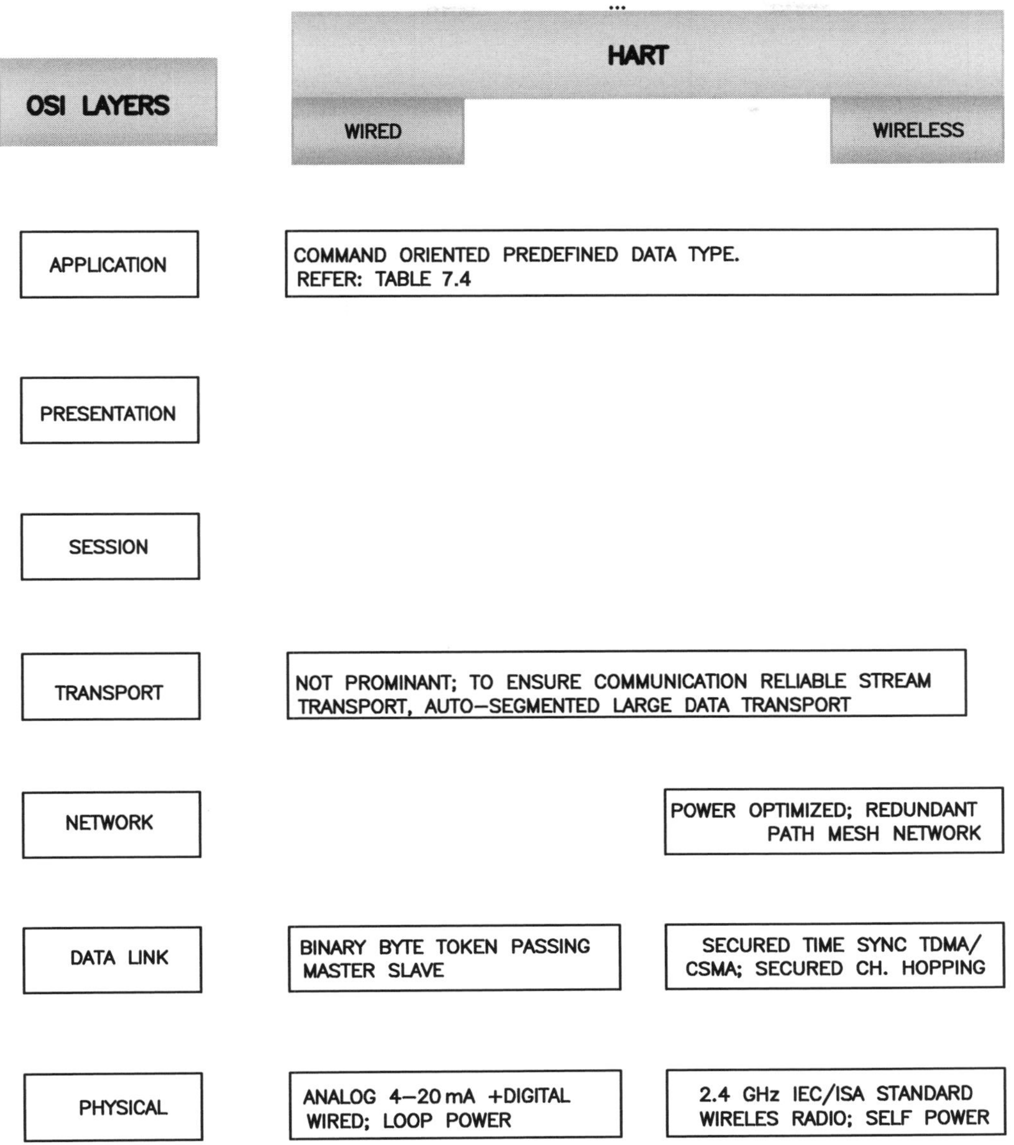

FIG. 7.19 HART and OSI layers.

query by placing the "device address," a "function code" defining the action, any "data" to be checked, and an "error" checking field. The slave message in the MODBUS protocol contains the field "confirming action," any "data" to be returned and an "error" checking field. For *different networks*, it uses a built-in port or a network adapter to communicate in peer-to-peer communication mode. In that case, the master initiates the request. Multiple internal paths are often provided for concurrent communications. As stated earlier, the master sends the query (or request) containing a few fields, as shown in Fig. 7.21.

QUERY: Function code tells the addressed slave the action to perform and any additional information, Data byte field information to slave regarding register to start and how many register to read, etc. The error check is for the slave to validate the data.

RESPONSE: The function code echoes the function code in the query. The data byte contains the data collected. In case of an error, the function code is modified to indicate an error.

MODBUS communication is based on a data packet called the protocol data unit (**PDU**) comprising the query and response as discussed above. Another PDU, known as the **exception response PDU**, indicates error where function code of query + 0X80 (error code) and a code specifying exception. The specification defines a number of functions, each with a number function code (1–127 and 128–255 for error code). New releases have function codes such as public, user defined, and reserved. MODBUS data types shall include read-only data types for discrete input/output (*each single bit*), input register, and holding register (*each 16 bit word*).

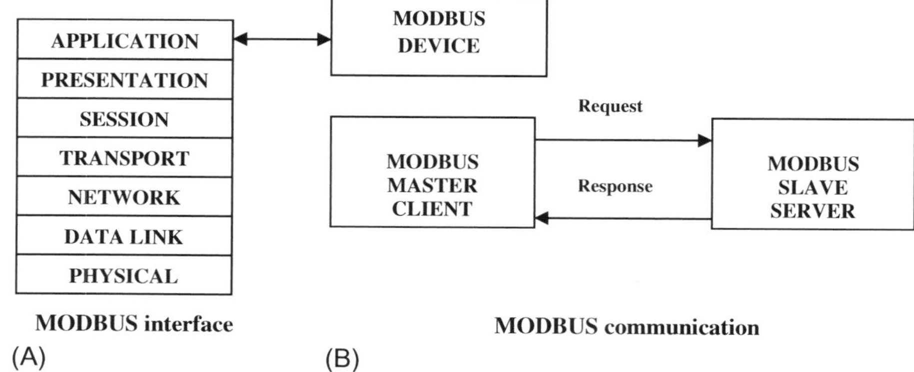

FIG. 7.20 MODBUS system.

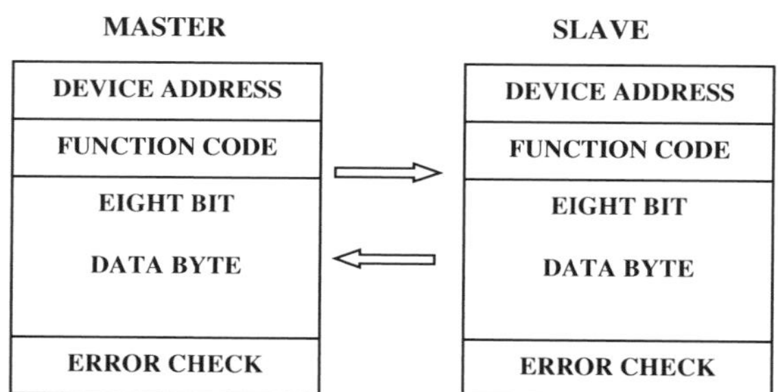

FIG. 7.21 Master slave query response with different fields.

1.1.4.2 Implementation Methods

MODBUS has been implemented in all physical media such as wire, fiberoptic, and radio. They are mainly implemented utilizing (1) SERIAL ASYNCHRONUS Master-Slave and (2) IP Master-Slave.

1. Serial Link: RS 232 (EIA232): Valid for **short distance (say, 15 m)** and **point-to-point** communication. RS 422: this link can be used for **bidirectional** communication over a comparatively **longer distance (<1.2 km)** in **point-to-point** mode, as shown in Fig. 7.22.

RS 485: As shown, Fig. 7.22A and B are both applicable for RS 485. In actual implementation, the protocol data unit (PDU) is transformed into the application data unit (ADU) by adding a header and error check sum having max256 byte, as shown in Fig. 7.23.

There exist two methods of serial communications: **ASCII and RTU**. The encoding difference between them has been given in Table 7.5 ASCII and RTU ENCODE.

2. IP based: A TCP/IP-based MODBUS implementation with multimaster system, bidirectional communication is possible. It uses a TCP/IP stack for communication and extends the PDU with an IP-specific header. The MODBUS TCP frame format is given in Table 7.6, TCP Format.

1.1.4.3 Discussions

The MODBUS Plus is proprietary and an extended version of MODBUS, which can be used in a 20 mA current loop for a distance of up to ~1000 m with a transmission speed of 19.2 kbps.

1.1.5 Fieldbus System

The main objective of a fieldbus in its simplest form is to enable the PLC/DCS to "control and instruct" the output device to change value and status, and also to "receive" information from the same or other devices. Although a fieldbus system replaces the massive cables (and wiring) with typically a set of two wires, it should be conceived also as an intelligent way of serial two-way communications with several norms and requirements.

1.1.5.1 Fieldbus Requirements

The fieldbus is a digital communication system for various field devices such as sensors, actuators, and field control systems. The fieldbus cable SHOULD *NOT* BE LOOKED AT AS "**JUST WIRE**." In an integrated system, it can be compared with one of the **main arteries** in the human body. There are a few advantages of a fieldbus that shall include but are not limited to the following:

- Huge reductions in cable, wiring, cable tray, marshalling cabinet, and junction boxes, along with a reduction in installation labor (hence, cost and complications).
- The fieldbus is reliable, cost effective, and deterministic. It offers greater flexibility in system design and layout for easier use, expansion, and modification.
- More extensive parameters and data exchange (calibration, service history configuration data, diagnostics, test information, device documentation, etc.) to and from the field devices at a much faster rate to facilitate speedier and easier commissioning, maintenance, and servicing.

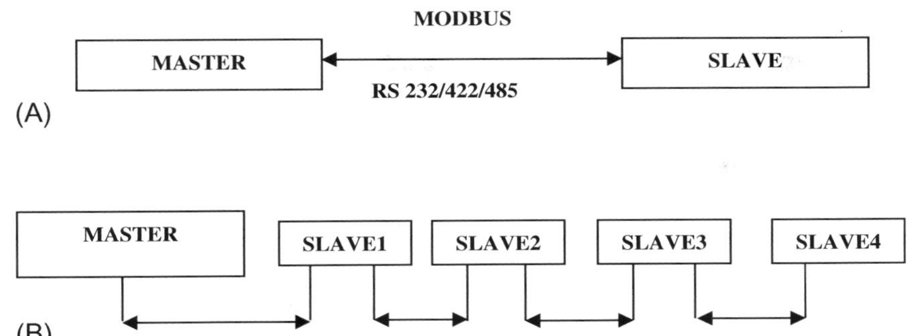

FIG. 7.22 (A) Point-to-point communication between master. (B) Multiple slave communication in RS 485.

HEADER 1 Byte	PDU	ERROR CHECK

FIG. 7.23 IP method of MODBUS implementation.

TABLE 7.5 ASCII and RTU ENCODE

	ASCII		RTU	
NAME	Length	Explanation	Length	Explanation
Coding system	NA	Hexadecimal, ASCII	NA	8 bit binary, hexadecimal
Start	1ch	Start with ":" colon (3A in ASCII)	3.5 idle	Min 3.5 ch time in silence
Address	2 ch	Station address	8 bit	Station address
Function	2 ch	Function code (say to read coil etc.)	8 bit	Function code (say to read coil, etc.)
Data	N ch	Data + variable length as per Message type.	N*8 bit	Data + variable length as per Message type
Check	2 ch	LRC error check	16 bit	CRC error check
End	2 ch	Carriage return etc.	3.5 idle	Min 3.5 Ch time in silence

- Openness and interface capability make it possible to integrate multiple products from different vendors in a system.

The IEC 61158-2 standard has been developed for the interconnection of various automation system components. Some of the salient features of IEC 1158-2 related to transmission technology are presented in Table 7.7.

The following are the basic transmission features noted from this standard combining with the field intrinsically safe concept (FISCO):

TABLE 7.6 TCP Format

Name	Length	Explanation
Transaction/ invocation identifier	2 bytes	Message synchronization between client and server
Protocol identifier	2 bytes	ZERO for MOD BUS/TCP
Length field	2 bytes	Number of remaining bytes
Unit identifier	1 byte	Slave address (255 if not used)
Function code	1 byte	Function code
Data byte	N bytes	Data as RESPONSE or COMMAND

- Various topologies such as linear, tree, and star topology are allowed with each segment with one power supply. No power feeding to a bus when a node is sending.
- Field devices are passive current sinks. The passive line terminates at both ends of the main bus line, and in a steady state, all field devices consume constant current.

1.1.5.2 Fieldbus Types

The fieldbus could be wire (or optical fiber) or it could be wireless. Also, there are several standard fieldbus systems available. Some of the popular types are: Profibus, Foundation Fieldbus, WorldFIP, DeviceNet, ArcNet, IEC/ISA SP50, and CAN open.

Again, out of so many fieldbus systems, Profibus, Foundation Fieldbus, etc., are more common in power plants. Apart from meeting the various requirements for fieldbus systems and for wireless communication, additional requirements come from their reliability and security. Therefore, a wireless fieldbus in general has been discussed.

TABLE 7.7 IEC 1158-2 Features

Features	Requirements	Features	Requirements
1. Data transmission	Digital Manchester Bit synchronized	5. Data security	Preamble, error free start and end delimiter
2. Speed	31.25 kbps voltage	6. Cable	Two-wire shielded twisted pair
3. Explosion protection	IS EExiaand ib EEx d/m/p/q	7. Number of stations	32/segment (Max. could be 126 address available)
4. Topology	Line/tree or combination	8. Repeaters	4

1.1.5.3 Wireless Fieldbus

As the name signifies, "wireless" communication is a way of communication in which there is no need of a hardwired physical medium (I use the term "hardwired physical medium" to exclude fiberoptics) to carry over the communication network. Before discussing the wireless fieldbus, it is better to gain a general idea about wireless communication, as some of the issues are applicable for wireless HART also.

1. General wireless communication: There are three ways of wireless communications: Radio frequency (RF) with a frequency range of 3 kHz to 300 GHz; Microwave, long-range communication with a frequency range of 300 MHz to 300 GHz; and infrared wave, short range of communication <430 THz (1 THz = 10^6 MHz). When looking at "The Wireless Landscape—Lots of Choices" mentioned in the standard, one would find that there are a number of technologies and standards to choose from for wireless communication. These are: **RFID**—active/passive, **Bluetooth**, **IEEE 802.15.4**, **Zigbee**, **Wifi** (802.11a/802.11b/802.11g/802), and **802.16**/ 2G/3G. In this connection, it is better to refer to the ISA 100 standard when dealing with various communication ways and means. The basic concept of the ISA100 standard is depicted in Fig. 7.24.

There are a few types of wireless networks possible, such as:

- **Wireless personal area networks (PAN)**: For interconnecting devices within a relatively small area, using Bluetooth, ZigBee, or Wi-Fi-Pan; this is limited in use in power plant controls.
- **Wireless local area network (LAN)**: Mostly used in power plants (already discussed).
- **Wireless wide area networks (WAN)**: These are wireless networks that typically cover large outdoor areas. Raw/intake water switch yard network integrations may utilize these types of networks.
- **Wireless metropolitan area networks (MAN)**: These types are wireless networks that connect several wireless LANs. These types may be used to connect several units in a network connection.
- **Mobile devices networks**: This is another kind of wireless networking device.

These details are stated here for the reader to conceive of various wireless networking types so that the wireless fieldbus can be interfaced.

IEEE 802.11—also known as Wi-Fi or wireless Ethernet–is a family of standards for wireless local area networks.

Various special wireless means of communications are illustrated in Fig. 7.25 for the reader to get some idea about RF communications. Similarly, there are a number of wireless topologies such as star, ring, tree, a fully connected line, and a bus, as shown in Fig. 7.26.

One of the major problems associated with wireless communication is communication security. There are a lot of risk operations that are controlled and monitored in wireless communication systems. The wireless medium is an open medium and without countermeasures, it is easy for an attacker to insert malicious packets or to simply jam the medium, this way challenging reliable and timely transmission. The recent trend to connect fieldbus systems to the Internet by means of gateways has led to research toward securing the gateway, but it is also required to protect a fieldbus against attacks from the inside, for example, by employing proper encryption and authentication schemes. Some of these are:

- **Integrity**: This is to ensure that the data are accurate, valid, and consistent. So, applicable integrity constraints and data validation rules must be satisfied before allowing the data to be accepted as authentic data in the network.
- **Encryption**: An algorithm to make it unreadable to outsiders except those with special knowledge. Encryption is a common method (initially used in military services) for secured communication and to shield data from piracy with privacy.
- **Authentication**: This is the act of confirming something as authentic, that is, that claims made are true [123].

There are a few ways that the wireless signals get disturbed. The major issues are:

- **Reflection**: It refers to the return of the signal by an obstacle (> wavelength) with unchanged composed frequency.

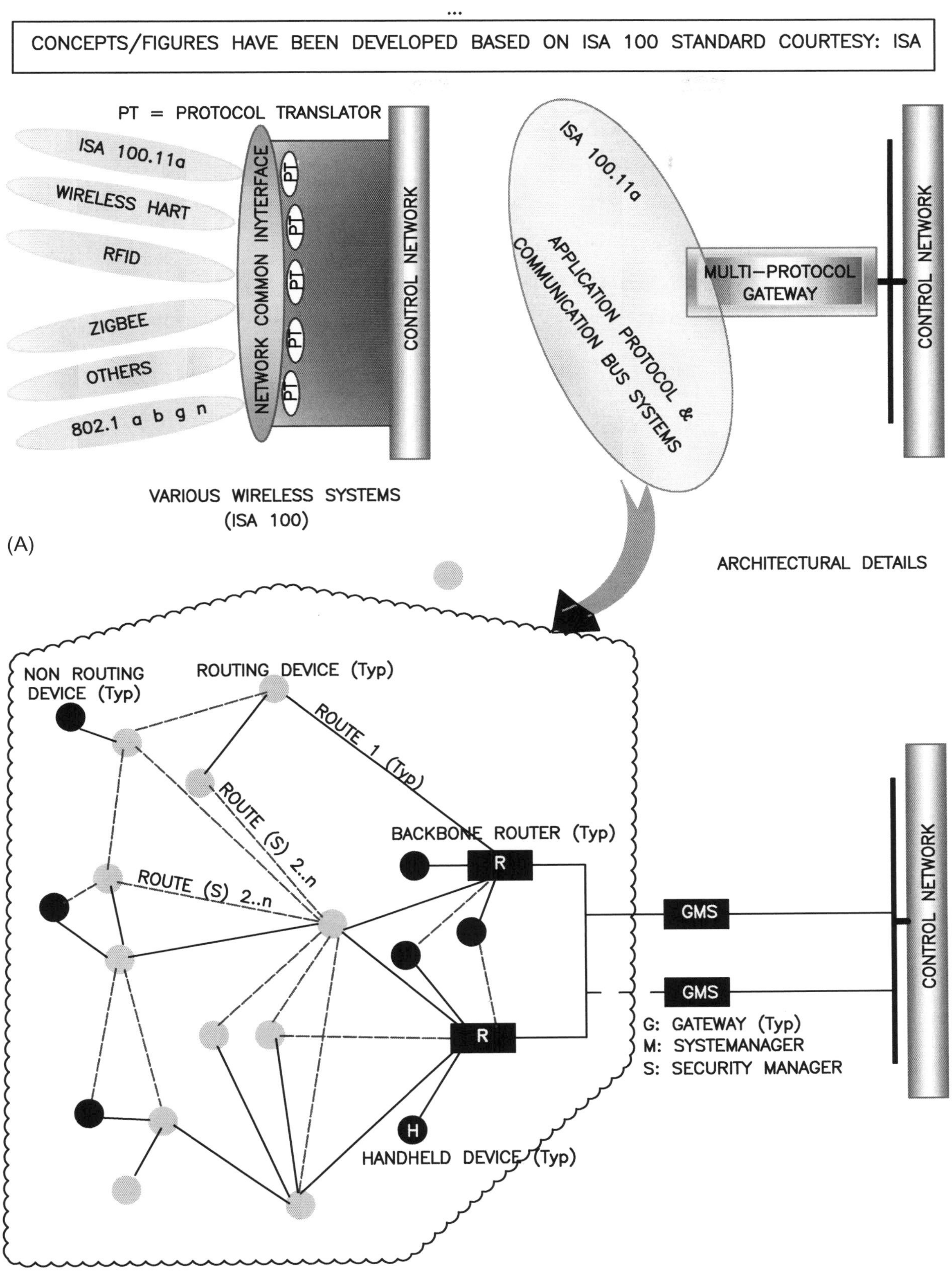

FIG. 7.24 Concept of ISA100.

RFID: **Radio-frequency identification (RFID) represents electromagnetic fields for identification and tracking tags containing stored information attached to objects. RFID systems mainly consists of smart label (Tag), RFID Reader and an antenna. There are two kinds of RFID viz. Passive RFID (with no internal power source—powered by electromagnetic energy transmitted) and Active RFID (with own battery power for broadcast own signal. (Example: Card in the car to pay Road toll)**

Bluetooth: **It is a short-range wireless communications technology with RF transceiver/the physical layer/medium which utilize unlicensed Industrial scientific medical (ISM) band centered at 2.4 gigahertz in star topology with frequency-hopping (repeatedly rapidly changing frequency). Bluetooth use a *piconet* (ad-hoc network) with one master and other slaves. In a piconet, the physical radio channel is shared by a group of devices that are synchronized to a common clock and frequency-hopping pattern.**

Zigbee: **This open global standard—IEEE 802.15.4, wireless technology have been developed to address the unique needs of low-cost, low-power wireless network such as machine to machine communication. It is suitable for high-level communication protocols used in personal area networks with small, low-power digital radios.**

**** For high data rate transfer wi fi PAN is used.***

FIG. 7.25 Concept on RFID, BLUETOOTH, ZigBEE, and WiFi PAN.

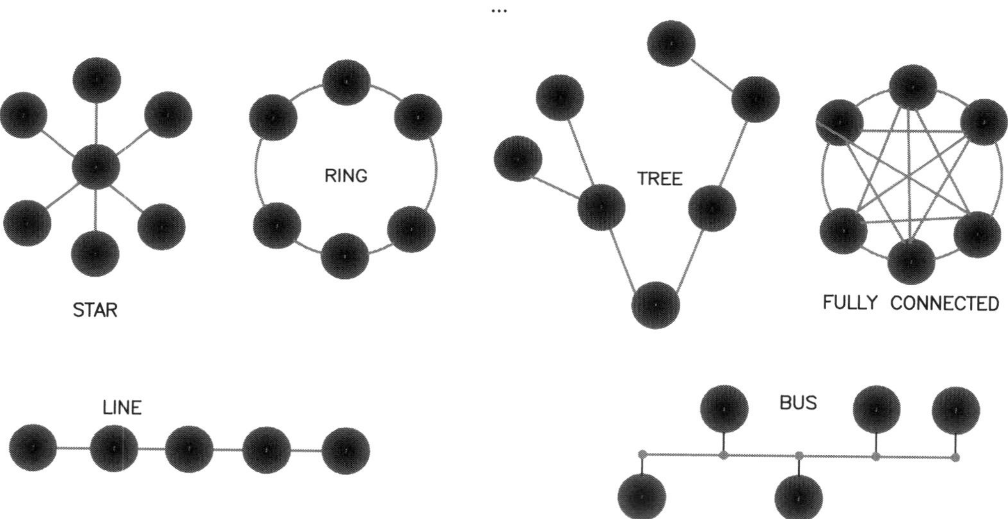

FIG. 7.26 Different wireless topologies.

- **Diffusion**: This refers to the change of the spatial distribution of a signal due to deviations in many directions by an obstacle (< wavelength).
- **Interference**: This may occur when two or more waves of the same frequency with different phase or direction of propagation overlap.

Frequency-hopping spread spectrum technology (FHSS) and direct-sequence spread spectrum technology (DSSS) are two techniques by which these signal disturbances may be reduced.

- **FHSS** uses a narrowband carrier that changes frequency in a pattern (known to both transmitter and receiver). Bluetooth and HART utilize this technique.
- **DSSS** transmitters spread the signal over a frequency band that is wider than required to accommodate the information signal by mapping each bit of data into a

redundant bit pattern of "chips" known as a chipping code. The longer the chipping code, the greater the probability that the original data can be recovered (and, of course, the more bandwidth required) [123]. Wifi and ZigBee utilize them quite often.

Radio signal path loss is very important in designing radio communication systems. Here, the major factors to be considered are: free space loss, absorption loss, diffraction multipath, terrain, atmosphere and building vegetation.

2. Wireless technology in field sensors:

Generally, the complete network of a plant can be conceived as three tier networks. At the top level/tier could be management information systems, which through gateway/Internet may be connected to the corporate office network. The middle-level/tier network is an area asset control network. The bottom-most tier may be a field or sensor network. A sensor network is mainly meant for system monitoring. So, wireless systems in the field play an auxiliary role to the existing control system. Some of the major issues related to wireless at this level become more important for process control, such as security, robustness, and power. In sensor wireless communications, there are two kinds of interference: Temporary interference in open air communication, perhaps due to a change in the weather, and permanent interference, which is because of reconfiguration due to the addition and deletion of sensors in the sensor network. Additionally, there could be disturbances on account of power usage and outage, including the contingency handling of a power outage, data transmission delay variance due to battery levels, etc., in the case of fieldbus power is supplied by the control system, but in case of wireless transmission, sensors and actuators could be powered by batteries or the net by solar cells. For battery-driven stations, energy is a scarce resource that should be used economically. Replacing batteries may be infeasible or can lead to machine downtime. Several mechanisms to conserve energy in protocols and applications have been developed in the context of wireless (sensor) networks. In the design of fieldbus protocols, however, the main concern was real-time communications, not energy efficiency [124]. Wireless local area networks (WLANs) succeeded in providing wireless network access at acceptable data rates. IEEE has set standards and specifications for data communications in a wireless environment. IEEE 802.11 is the standard for WLANs. In order to get some idea about wireless fieldbus communication in an instrument loop, Fig. 7.27 is shown for a deareator level control loop with redundant sensors and several ways of communications. Dotted (blue) lines show the radio communication possibilities.

3. Discussions on wireless PROFIBUS and foundation fieldbus: Wireless in a fieldbus acts like a cable for signal transmission from a technical point of view. Discussions on Profibus (Section 1.1.6) show that at the field level, it works on master-slave principles. Also, it is possible to have a wireless and a wired Profibus system, as typically shown in Fig. 7.28. Here, there is also a challenge with wireless transfer when the data needs to be transferred from the wireless Profibus to the cable interface in order that the data lies in the specified range of 9.6 kbps to a maximum 12 mbps. These are possible through suitable converters and gateways. Foundation fieldbus is one of the most popular fieldbuses used in industries. From discussions in Section 1.1.8, it will be noticed that in a Foundation fieldbus (FF), there are user layers, application layers, data link layers, and physical layers. Wireless foundation fieldbus systems use IEEE802.11. WFF can be hybrid with a wired FF. The hybrid is implemented on the interoperability of WFF and FF user applications and devices. With this, let us conclude the discussions on wireless communication and start the discussions on a safe fieldbus.

1.1.5.4 Safe Fieldbus

There is a very limited use of a safe fieldbus (GTs) in a power plant. Detailed discussions on a safe fieldbus are presented in Chapter 9 of reference [128]. The following are the basic transmission features in a field intrinsically safe concept (FISCO):

- Each segment has one power supply but no power is fed to the bus while a station is sending.
- Field devices are passive current sinks. In a steady state, all field devices consume constant current.
- Linear, tree, and star topology are allowed. Passive line termination at both ends of the main bus line.
- To increase reliability, redundant bus segments can be designed.

There are some other issues that shall include but are not limited to: [129]:

1. Safety aspects:
   - Self-diagnostic features of devices and real-time identification of dangerous failures.
   - Valve partial/full stroke testing is possible, with a reduced manual proof testing burden.
   - The safety fieldbus communication protocol needs to support the highest SIL of the SIS when it is certified. So far, SIL3 certification is available.
   - Devices may exist in a network in safety- and nonsafety-related applications, provided nonsafety-related devices are noninterfering. Fault tolerance is optional.
   - Enough security shall be there for inadvertent changes.

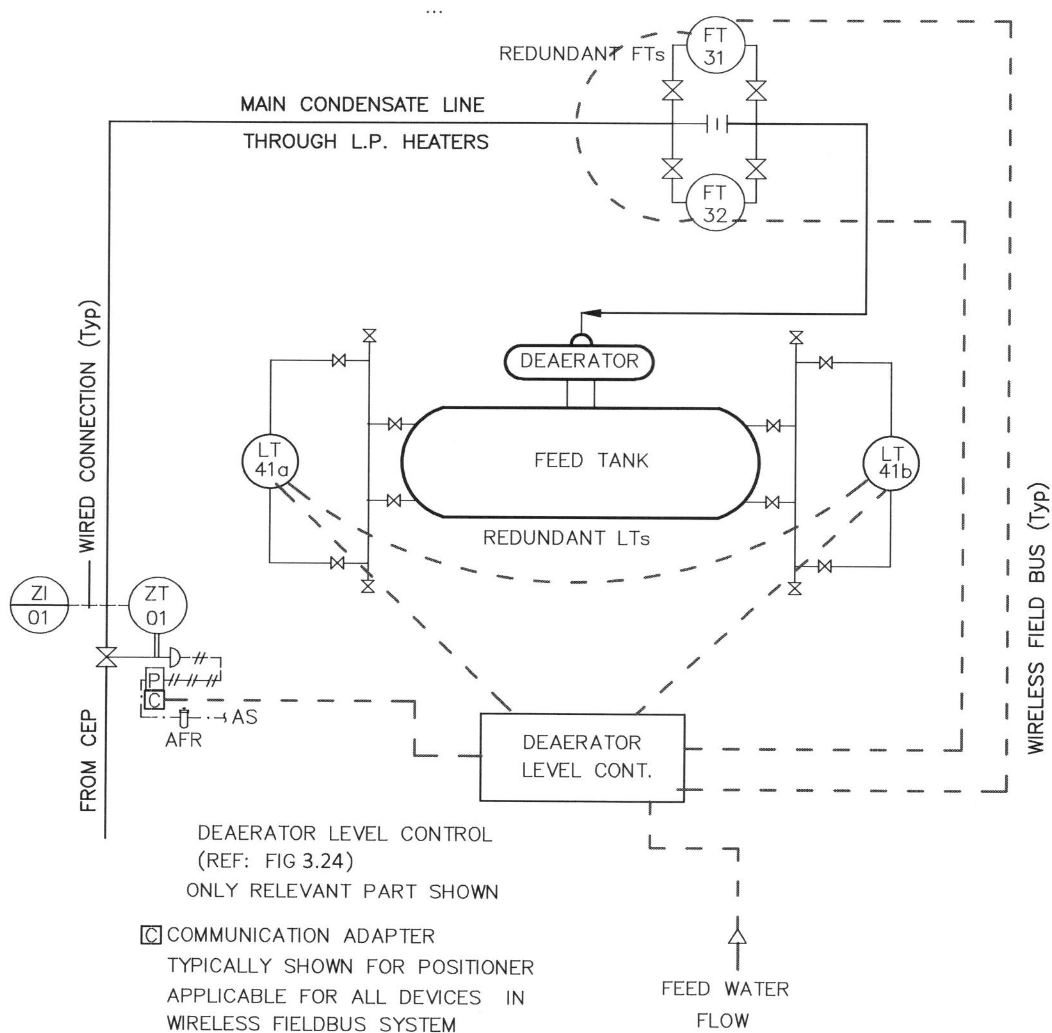

FIG. 7.27 Wireless fieldbus representation (loop).

2. Operability:
   - Programmable devices shall be interoperable and nonproprietary.
   - Trip on demand only and a reduction in nuisance trips.
   - Devices have self-diagnostics and the diagnostics shall be "transparent" to the user, with possible reporting.
   - Supports multidrop architecture, with online device replacement possibility.
   - Systems offer an asset management tool with flexibility in installation with testability of the system.

Safe fieldbus systems are required to meet the IEC 61508 standard for functional safety systems for specified SIL. PROFI SAFE and FF-SIF are examples of a safe fieldbus for Foundation fieldbus and Profibus, respectively. ISA TR 84.00.06 provides guidelines for fieldbus design. This is applicable for a fieldbus in general.

Now let us discuss various fieldbus types.

1.1.6 PROFIBUS

PROFIBUS is one of the most proven and popular fieldbus standards, introduced way back in 1989 in Germany in line with the DIN18245 standard. Now this has been incorporated in the international standard IEC 61158-2. It supports more than 15 million devices over a million networks worldwide having 2200 products.

1.1.6.1 PROFIBUS Family and Application Area

PROFIBUS systems consist of three compatible versions that can operate all together seamlessly.

- PROFI FMS (field message specification): This is high-end application level communication. It provides object-oriented transmission with structured data, loading, and control programs. The extension of other family members such as PROFIBUS DP and suitable integration of the same with Ethernet gave rise to PROFINET, which made FMS less important and popular.

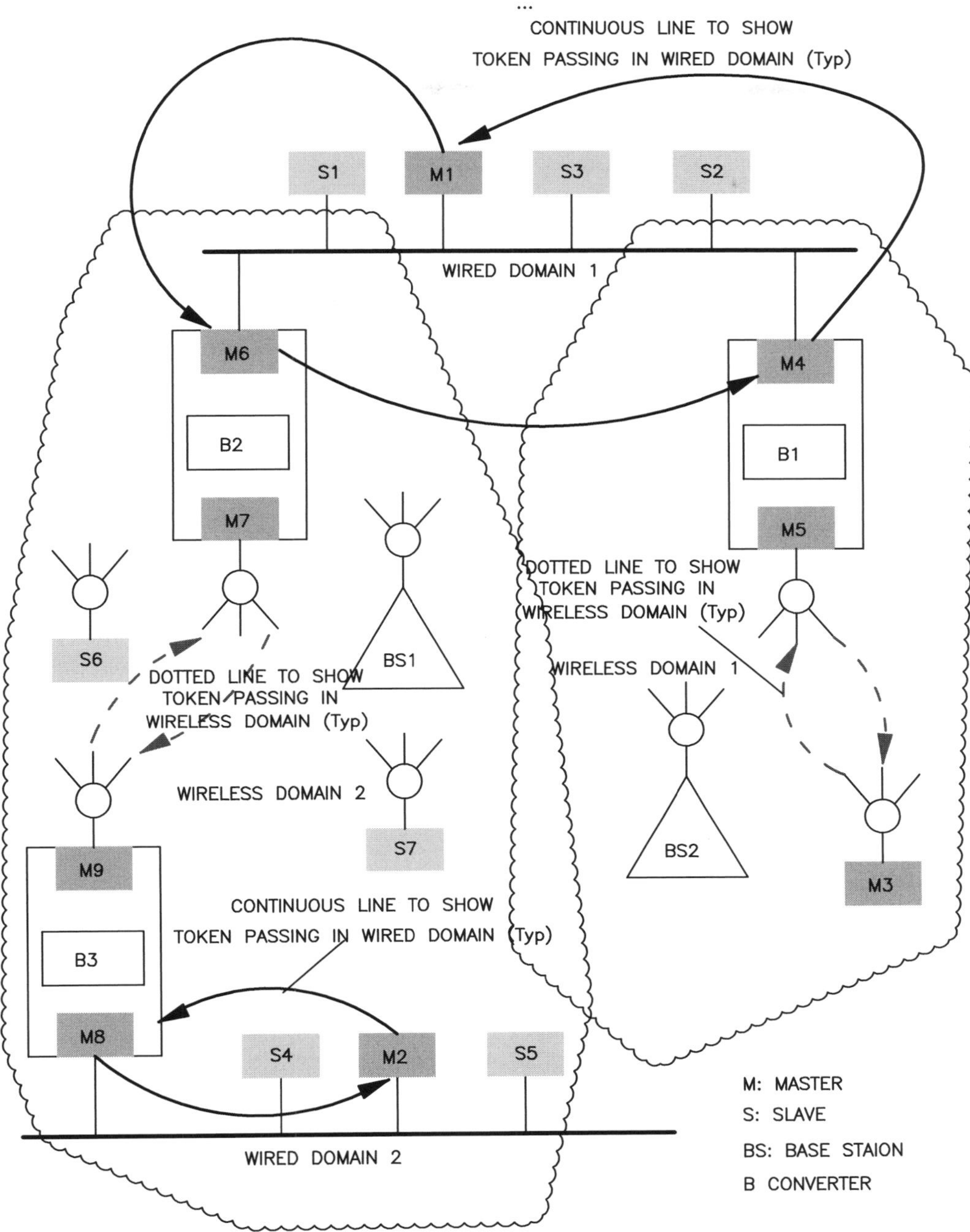

FIG. 7.28 Wireless and wired fieldbus system (Profibus).

- PROFIBUS DP (decentralized periphery): Low-cost field and controller level communication. The majority of PROFIBUS DEVICES belong to this category.
- PROFIBUS PA (process automation): This is a cost-effective two-wire connection for field devices (for example, T_x) to communicate.

For safety critical applications and IT applications, these are PROFISAFE and PROFINET, respectively. Table 7.8 provides applications of PROFIBUS.

1.1.6.2 PROFIBUS Structure (Fig. 7.29)

The two layers, the physical and fieldbus data link layer (FDL), in all three types of PROFIBUS are more or less identical. When compared with the ISO/OSI model, it is clear that layers 3–6 are missing in PROFIBUS where FDL handles transmission protocol, data security, and error detection. The original DP has been extended to include these functions, referred to as DPVs, which are found in the application layer. Various PROFIBUS variants are

TABLE 7.8 Profibus Applications

Application Area	Applicable Industries
Simple to large low cost distributed control and automation application	Factory automation, robotics
High speed time critical application	Process control and power industries
Complex communication tasks	Building automation, warehousing, and material handling
Hazardous application and process automation	Smart devices

TABLE 7.9 PROFI BUS Transmission Techniques

Tr. Technique	FMS/DP	PA
RS485	Twisted pair or with four wire with power supply speed: 9.6–1200 kbps	
Optical	High-speed communication with electrical isolation	Not applicable
IEC 61158-2	Not applicable	Shielded twisted pair cable @ 31.5 kbps. Supply and data on the same cable (IS possible)

shown in Figs. 7.48B and 7.49, where it is very clearly shown that PROFIBUS DP and PROFIBUS PA have been connected through couplers.

1.1.6.3 Physical Medium and Transmission

Three ways to realize transmission techniques in PROFIBUS are detailed in Table 7.9.

1.1.6.4 Topology and Communication

Out of 0–128 device numbering, **126 is reserved for a generic device** and **127 is used for a broadcast message**. But limit from IEC 61158-2 and RS 485, is 32 only per segment. The address setting means (HW and SW) are:

- Hardware means by local binary dip switch or rotary switch.
- Software setting over PROFIBUS.
- Special software and serial link by handheld device or control system.

The topology can be linear in RS 485 in a daisy chain. The circuits can include repeaters, couplers, and optical link modules (for transforming optical signals to electrical signals) as shown in Fig. 7.30. It follows the IEC 61158-2 standard.

1. SIGNAL COUPLER: The signal coupler provides the following features:
   - Electrical isolation between the safe and IS bus.
   - It is like a slave to the PROFI DP.
   - Baud rate adaption (PA; 31 kbps) is faster (1200 kbps) also.

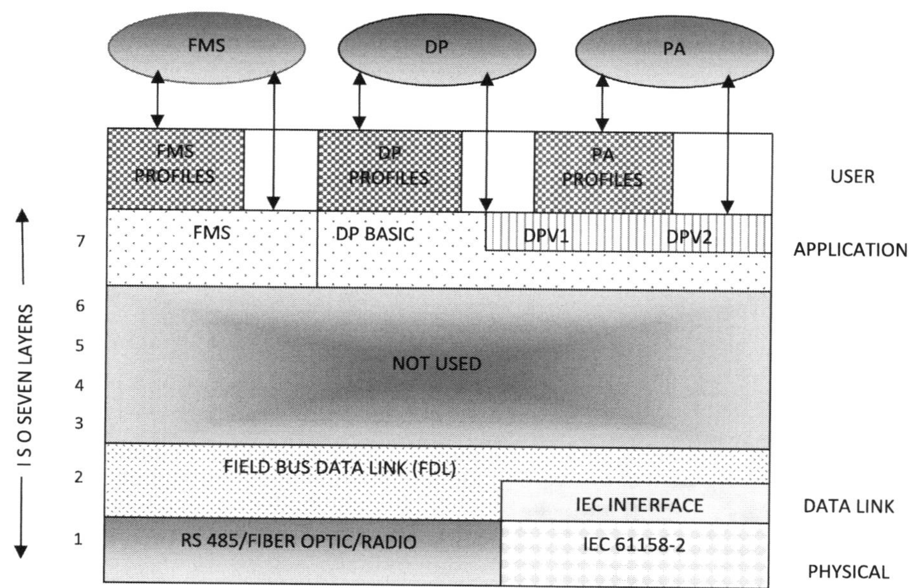

FIG. 7.29 PROFIBUS layer structure.

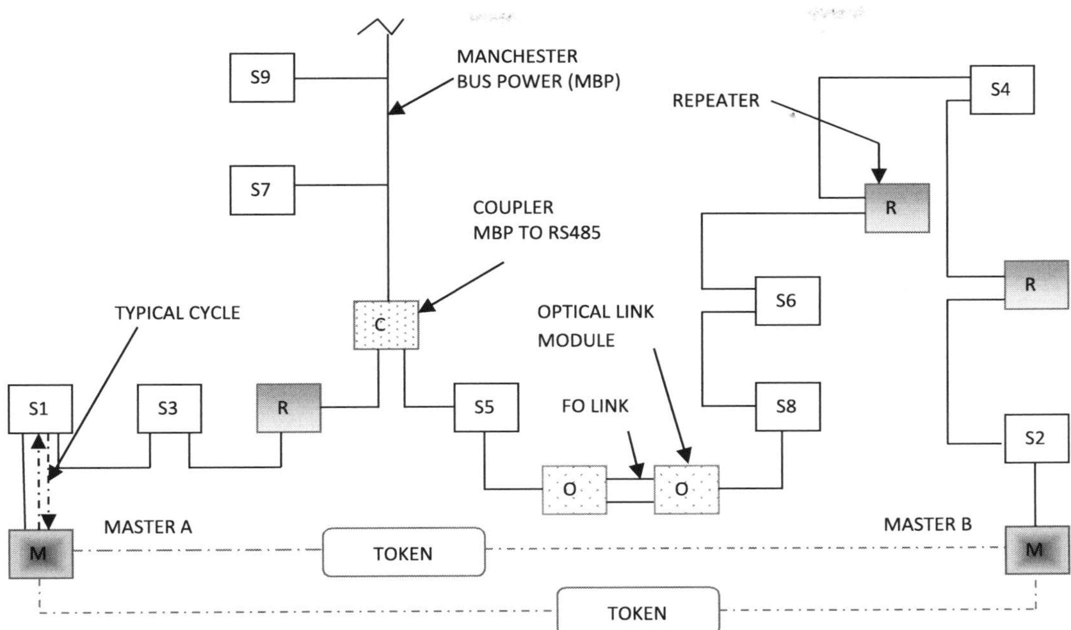

FIG. 7.30 PROFIBUS configuration and communication.

- Powering of PA bus.
- Adaption of the IEC 61158-2 technique for transmission in PA when the other side is RS 485.

Installed between the PROFIBUS DP and PA, the signal coupler is necessary as both systems have different technique for use of IS, transmission, and powering scheme.

2. PROFIBUS OPERATIONAL ASPECTS: PROFIBUS operates on a master-slave mode with even more than one master. Master stations control the network communication and the slave responds to the call from the master. Each of the masters can communicate with a number of slaves. The system is highly democratic, meaning that all the masters have equal rights while all the slaves connected to any master have equal rights. In case of a multimaster system, communication is regulated by the token passing method (ref clause no. 1.1.2.5.2). Having the token, the master takes over the control in the bus. A simplified method of communication and token passing is shown in Fig. 7.31, Token passing and communication (an arbitrarily chosen token from MA).

There again two classes of masters: Class I and Class 2; see Table 7.10.

3. Cyclic and Acyclic Communication: As shown in Figs. 7.30 and 7.31, the normal communication system in DP is cyclic. A master having a token (token holding time is calculated and specified by the user) talks to all its slaves in a cyclic manner, for example, M1 to S1 followed by S1 to M1 then to S3 (under control of M1), etc. After the communication is complete, it is passed to the next master (say, M2); when M2 does its communication M2 to S2 followed by S2 to M2 then to S4 (under control of M2), etc. Cyclic communication is transparent. When the master requires any information from a slave, it simply writes the same in the appropriate part of the memory of the master for the slave to retrieve the information. The slave on its part puts the required information again in the appropriate part of the memory of the master to be retrieved by the master. Slaves always monitor the bus, utilizing a watchdog timer to check the inactivity in the bus. In case any message is not received within a specified time, it senses an error in communication and goes to the failsafe condition while the master, during cyclic operation, checks the healthiness of the slave so that it responds to all the calls from the master. Acyclic operation is unique for the **DPM2** to access the network and read/write to **any slave during configuration and start up**.
4. Network Configuration and Start up: Like any other network, the PROFIBUS network also needs to be configured first. By the configuration, the master is notified about the basic characteristics of the slaves. This is done with the help of software supplied by the PROFIBUS master manufacturer. During configuration the token holding time is also set. The GSD files discussed in Section 1.6.5 help in such configurations. When the master device is initialized, the master (class 1)

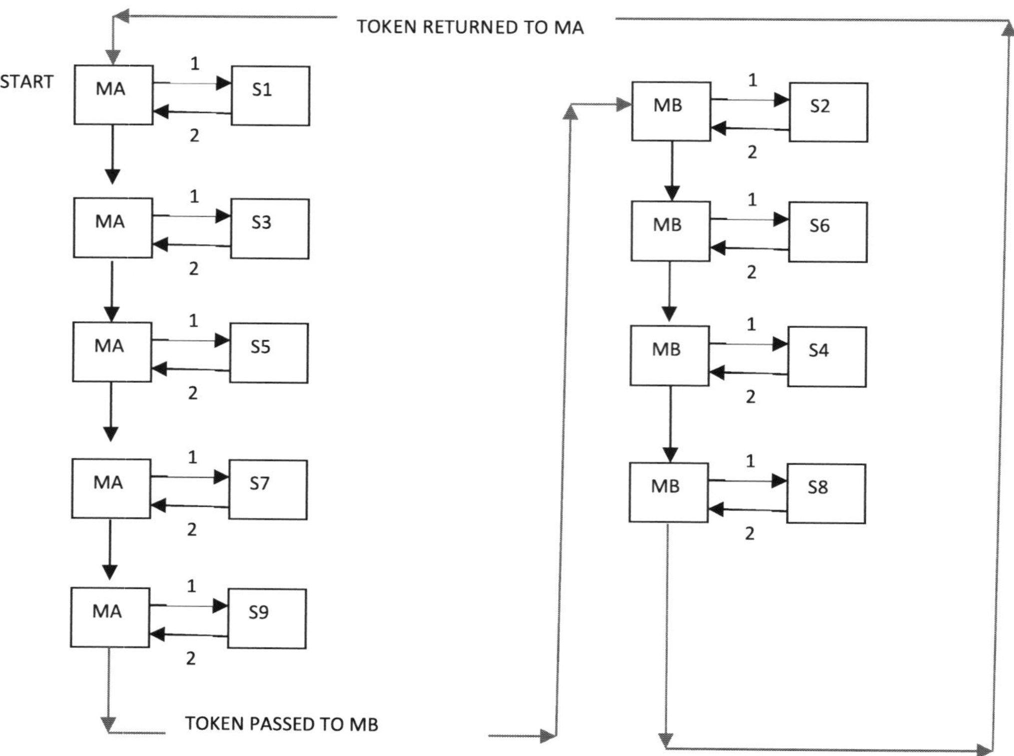

FIG. 7.31 Token passing and communication (PROFI).

TABLE 7.10 Master Classification (PROFI)

Functions	Class I Master	Class 2 Master
Designation	DPM1	DPM2
Processing and operation	Information processing to and from slave during normal operation	Engineering functions such as configuration tool, diagnostics, engineering, etc. It is used for operation and monitoring purposes. It is also used during start up
Authority	One master can write to the outputs of a particular slave	Can read and write to the outputs of any slave, even when it is under control of a master
Communication	It communicates with the slave in cyclic operation due to the DPV1 function of the extended DP	It uses acyclic communication for access

parameterizes, that is, sets the parameter to the slaves with the required settings and locks itself to the master. The next important operation during start up is to check configuration by the master to confirm that the slave is configured as per the requirement set forth by the master. At the final stage, a DIAGNOSTIC request is sent by the master (class 2) and this is to be responded to by the slave to satisfy the master that the slave has no problem.

1.1.6.5 GSD Files and Associated Parameters

In order to ease the configuration, all supporters of PROFIBUS standardized the so-called **"GENERAL STATION DESCRIPTION (GSD)"** to support interoperability. During the planning stage. GSD files are loaded to the master. The basic data supplied by GSD shall include but are not limited to:

- Device manufacturer and device identification number.
- Transmission rate and bus parameter.
- Number and format of the data cyclic communication.
- The device profile standardizes the PROFIBUS so that all device functions and parameters are standard, and it determines the access method. These include dynamic process value and status report. There are a few acyclic read-write functions such as operating and standard parameters. There is some manufacturer-specific

information that is only accessible if a class 2 Master has knowledge about them.

- Function block: It provides a uniform application interface for various blocks and their functions are required to be defined.
- There are two kinds of profiles: Class A and B. These are mainly to include the transducers block, which describes the coupling of signals to the process such as characteristic curve and sensor type. A Class A profile is limited to the absolute necessary information and B extends this to the available scope of device functions.
- To enable a class 2 master, the manufacturer specific feature and operating functions there are two different specifications for example, electronic device description (EDD) and field device tool (FDT)

1.1.6.6 PROFIBUS wiring

1. IEC 1158-2: IS requirement and cable as per specifications below (Table 7.11):
   - Design: Braided shielded* twisted pair.
   - Conductor: 0.8 mm^2 (AWG18).
   - Loop resistance: 44 Ω/km.
   - Impedance @ 31.25 kHz: 100 Ω ± 20%.
   - Wave attenuation: 3 dB/km.
   - Capacitance: 1 nF/km.
2. The RS 485 preferred connector could be nine-pin D SUB plug for the RS 485. Termination is necessary at both ends, preferably one with a master. The first and last segments can have 31 stations, but the rest shall be restricted to 30 stations. Wires are color coded and a shield may be connected to the body.
3. Optical Connections: Long distance, use of optical module including an optical link module (OLM). Line ring star configuration is possible.

1.1.6.7 PROFIBUS DP Features

The salient features of a PROFIBUS DP are presented below:

TABLE 7.11 PROFIBUS IS Parameter

Max Parameter	EEx ia IIC	EEx ia/ib IIB
Power	1.8	4.2
Current	110 mA	250 mA
Device	10 (depends on total current)	22 (depends on total current)

Supply current = Sum of Individual current + 9 mA (MCoding) + FDE fault (0–9mA).

1. Probable baud rate: 9.6/19.2/45.45/93.75/187.5 kbps and 1.5/3.0,6.0, 12 mbps.
2. Device support per segment: Max 32, including master.
3. Up to 244 bytes I/P andO/P per station.
4. Data can be read by the controlling and class 2 master.
5. Alarm acknowledgement.
6. Communication is cyclic and only one master can write to a slave for safety reasons.

1.1.6.8 PROFIBUS PA Features

1. Transmission technology: IEC 1158-2 (2W data and power).
2. Maintenance and diagnostics from available instruments.
3. Distance <1900 m. Max four repeaters.
4. Manchester coding without mean value @ ± 9 mA.
5. 126 addressable devices but up to 32 per line segment.
6. Remote DC voltage: Max 32.

1.1.6.9 Telegram

A Profibus telegram is presented in Fig. 7.32.

SD = start delimiter, SA = source address, DA = destination address, FC = frame control, DATA UNIT = data LE (LEr) = length byte (repeat), FCS = frame check sequence, ED = end delimiter.

1.1.7 PROFINET

One may argue that the PROFINET is an extension of P the ROFIBUS DP and its integration with the Ethernet gave rise to an open bus referred to as PROFINET. PROFIBUS International defined PROFINET as a universal standard that opens up new facilities such as (Table 7.12):

PROFINET, in accordance with IEC 61158 and 61784, is an open industrial Ethernet. It simultaneously supports TCP/IP Ethernet, OPC, and XML. In addition to PROFIBUS, it also supports other fieldbuses through proxy technology. For better availability, PROFINET can be configured in a ring using redundant cabling so, with the help of a media redundancy protocol (MRP), the changeover (time: 200 mS) can be made effective. Also, it supports an isochronous mode to synchronize acquisition with output, which is helpful for motor control flow measurement. Figs. 7.48 and 7.49 may be referred to for PROFINET CONFIGURATION.

1.1.7.1 Layer Features

Because it is basically extended from PROFIBUS, it has three well-defined layers, as indicated in Table 7.13

1.1.8 Foundation Fieldbus

According to the Fieldbus Foundation, "Foundation Field is an integrated total architecture for information Integration.

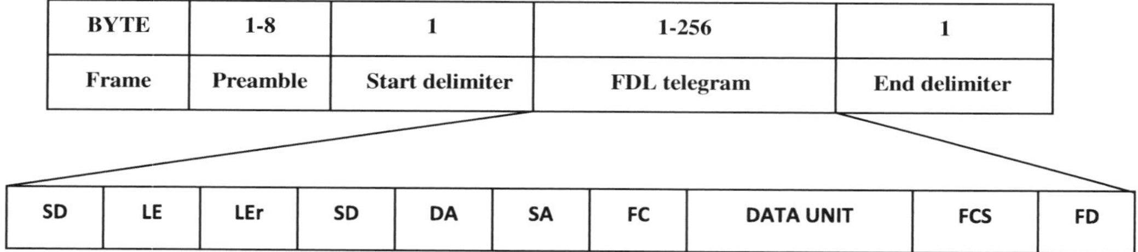

BYTE
1-8
1
1-256
1
Frame
Preamble
Start delimiter
FDL telegram
End delimiter
SD
LE
LEr
SD
DA
SA
FC
DATA UNIT
FCS
FD

FIG. 7.32 PROFIBUS telegram.

FOUNDATION fieldbus is an all-digital, serial, two-way communication system." It consists of H1, which interconnects the field equipment and high-speed Ethernet (HSE 100 mbps). A link device is an HSE device to connect one or more H1 Links. It has been incorporated in international standard IEC 61158-2.

1.1.8.1 Foundation Fieldbus Family and Application Area

As stated earlier, the Foundation fieldbus (FF) has types of bus systems:

- H1 bus based on IEC 61158-2 standard using filed devices such as sensors, transmitters, and I/Os.
- HSE bus: It is a high-speed Ethernet bus providing integration of high-speed controllers, subsystems via a link device, data servers, and workstations.

1.1.8.2 Foundation Fieldbus Benefits

A Foundation fieldbus offers the following benefits:

1. H1 bus:
   - Reduced wiring, marshalling, cabinets, equipment room size.
   - Common power supplies, IS.
   - More data/multivariables from each device.
   - Distortion-free digital communication enables control capability at the field level.
   - Better performance for self-diagnostic and communication capability.
   - For distributed control at the field, reduced hardware requirement.
   - Higher flexibility and sophistication.
   - More information for operation.
2. HSE bus: the benefits discussed in Section 1.1.8.2.1 are applicable to the HSE bus, but it also has the additional following benefits to offer:
   - High performance: Asset management functions such as diagnostics, calibration, and identification help to take proactive management actions.
   - Because the system is an open system, it is interoperable, enabling the user to mix various subsystems to form an integrated system.
   - The same functional blocks of H1 can be used here, eliminating the need for proprietary programming languages. The same language system can be used over the entire system.
   - Peer-to-peer communication enables communication between two devices without a central computer system. It is possible to bridge information between HI networks.
   - Because it is standard Ethernet, there is no need for any special tools and cables.

TABLE 7.12 PROFINET Facilities

IT integration	Utilization of industrial wireless LAN	Transfer of large volume of data
Real time	Distributed automation	Comprehensive diagnostics

TABLE 7.13 Profinet Layer Features

Physical Layer	Data Link	Application
Standard: IEC8802/IEEE 802.3		Standard: IEC 61158
Power and Communication: No	Link type: Token passing	Data transfer: PROFI DP
Speed: 100 and 1000 mbps	Error check: 16 bit CRC	Alert and trend: In field device
IS: No	Deterministic: Not really	
Distance: 100 m max	Comm. type master slave	Field control: No
Number of devices: Unlimited		

1.1.8.3 Fieldbus Model and ISO OSI Model

The FF specification is based on the layered communication model shown Fig. 7.33.

The Foundation fieldbus H1 consists of three layers: (1) a physical layer, (2) a data link layer, and (3) a user application layer, as shown in Fig. 7.33. As is shown here, **layers 3–6** are missing. In fact, the use application is made up of a function block and a device description (Section 1.8.10). It is directly based on a communication stack.

1. PHYSICAL LAYER: The H1 bus complies with the IEC 61158-2 physical layer. Upon receiving a message from the communication stack, the physical layer converts the same to a physical signal for transmission in the medium and vice versa. The conversion includes the additional removal of start/end delimiters, preambles, etc.
2. COMMUNICATION STACK: The communication stack comprises layers 2–7. The data link layer (DLL) regulates the message on to the bus through a deterministic link active scheduler (LAS-1.8.8). A field access scheduler (FAS) uses scheduled time restricted data and unscheduled data (on request, alarm) to provide services to the field message specification (FMS), which allows user applications to send messages to each other.
3. USER APPLICATON: It consists of a number of blocks as per the Foundation fieldbus. Each of these blocks represents different functions such a RESOURCE BLOCK, a TRANSDUCER BLOCK, and a FUNCTION BLOCK. While the first two blocks are used to configure various devices, the function block is used to develop a control strategy.
4. TELEGRAM: A complete telegram for a Foundation fieldbus is shown in Fig. 7.37.

1.1.8.4 H1 Bus Features

The H1 bus is quite similar to the PROFIBUS PA, with certain features:

- **Common power and IS** (where applicable) for the H1 bus; minimum voltage for communication: **9V**. H1 bus communication speed is **31.25 kbps** with Manchester encoded data; Standard: **IEC 61158-2. Token passing** is the communication mode deployed in the H1 bus**.** Basic units such as sensors, transmitters, and I/O devices can be connected to the H1 bus.
- There are two types of devices in the H1 bus: the **basic unit and** the **link master**. The link master can assume responsibility of the link active scheduler (clause 1.1.8.8). Max **32** devices per H1 bus in **line, tree, or star** topology. The maximum number of units in hazardous units are reduced on account of power supply limitations ($\ll$32). When a number of devices are connected to a common junction box, they form chicken foot, as shown in Fig. 7.35.
- Though daisy chains are allowed, it is preferable to connect devices, including the link master, through the JB. The connection between the devices and JB is called **a spur** whereas a JB to JB connection is termed a t**runk.**
- The maximum length of the H1 bus (sum of the trunk and spur) is **1900 m** without a repeater with a maximum possible spur length of 120 m (Fig. 7.34). Max four repeaters are allowed. So the maximum possible distance is **9500 m** (with **four** repeaters). A 1900 m connection length is possible with a shielded twisted pair cable with an 0.8 mm^2 conductor. The screen of the shielded cable shall be made ground at one end and both ends shall be properly terminated by resistors, as shown in Fig. 7.34.
- A standard bulk power supply is not fed instead it is fed through Filter (conditioner so as to avoid AC short circuit during data transmission (Fig.7.35A and B).

1.1.8.5 HSE Bus Features

HSE is based on standard Ethernet topology (ref clause 1.1.2.5.1) with a high speed of 100 mbps for communication.

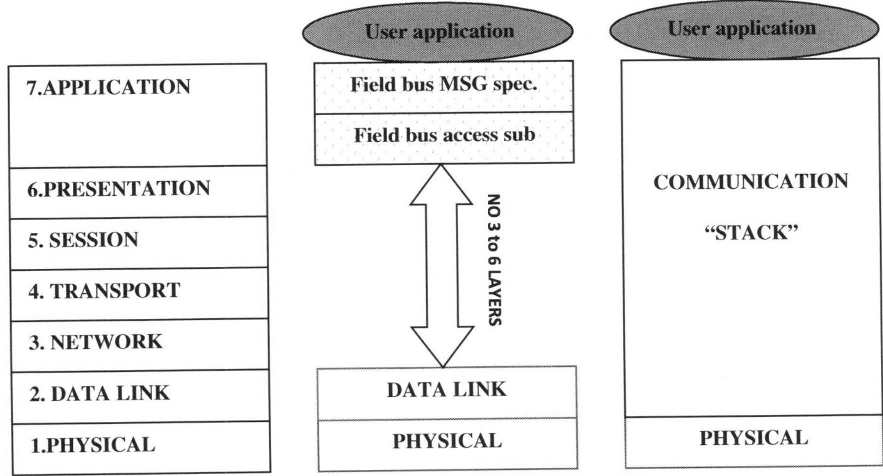

FIG. 7.33 Foundation Fieldbus model

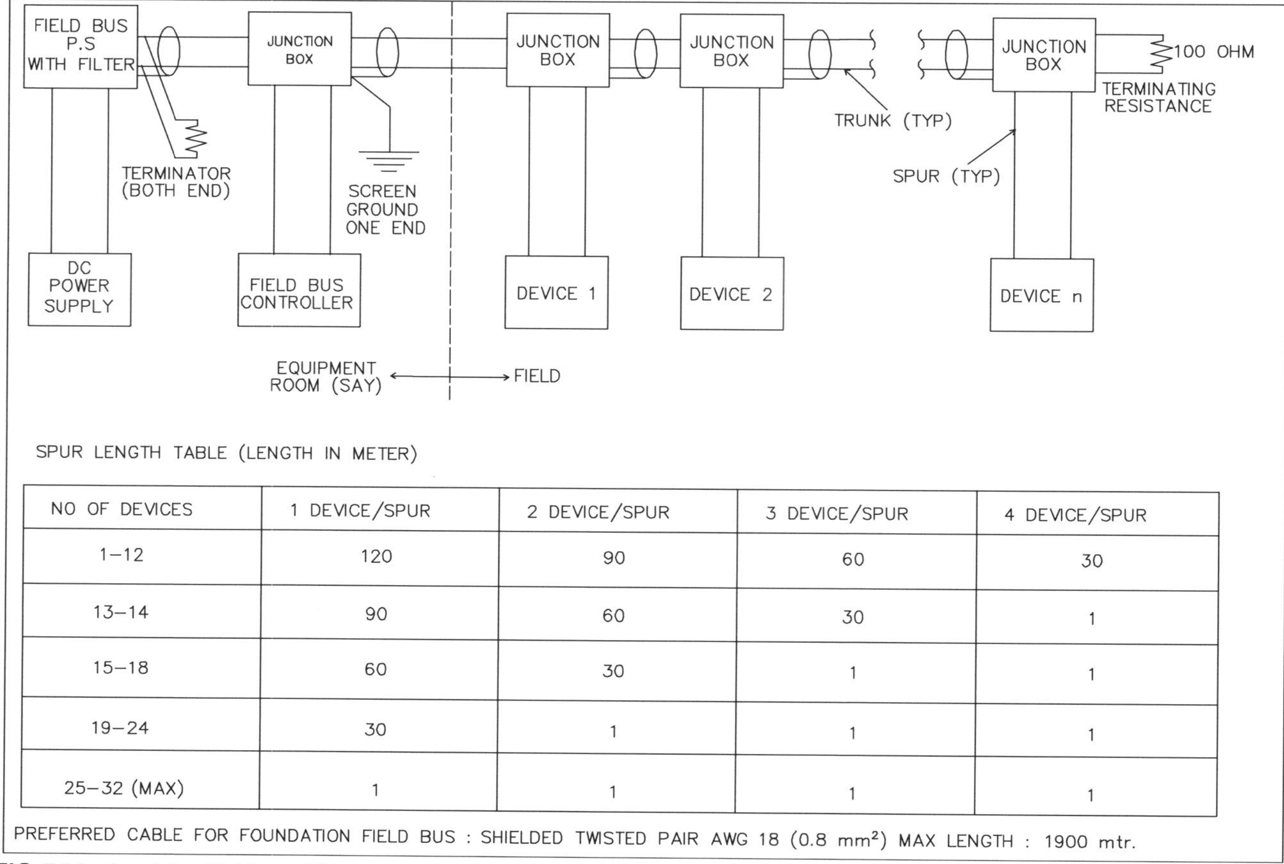

SPUR LENGTH TABLE (LENGTH IN METER)

NO OF DEVICES	1 DEVICE/SPUR	2 DEVICE/SPUR	3 DEVICE/SPUR	4 DEVICE/SPUR
1–12	120	90	60	30
13–14	90	60	30	1
15–18	60	30	1	1
19–24	30	1	1	1
25–32 (MAX)	1	1	1	1

PREFERRED CABLE FOR FOUNDATION FIELD BUS : SHIELDED TWISTED PAIR AWG 18 (0.8 mm^2) MAX LENGTH : 1900 mtr.

FIG. 7.34 Foundation Fieldbus cabling.

1.1.8.6 Linking Device

It is a device that connects one or more H1 buses to the HSE device/ network, and acts as a gateway for a field device access (FDA), field message specification (FMS), and system management(SM) services to the H1 bus. There is a difference in speed between the H1 bus and the HSE bus, so conversion of the data transfer rate and telegram need to be done at this device.

1.1.8.7 Gateway and Host Devices

A GATEWAY DEVICE is used to link the Foundation fieldbus to interface with other standard bus systems such as **Device Net**, **PROFIBUS**, etc. The **HOST DEVICES** are **OPC DA** devices so that non-HSE devices can communicate with the HSE network.

1.1.8.8 Communication Stack

In the H1 bus, there are a few services available, such as:

- Devices in the H1 bus can exchange data one with other.
- Devices are served in definite time.
- No two devices can access the bus simultaneously.
- Token management for scheduled communication (SC) and unscheduled communication (USC). The latter is executed in the break time between SCs.

All such services are controlled by the LAS responsible for control and scheduling of communication over the bus. Any link master can become an LAS. There can be more than one link master in a network and it is possible to operate the system with redundant LASs. If one LAS fails, the other link master will become LAS. LAS controls so many functions with the help of a series of commands it broadcasts over the bus, for example, it broadcasts a time distribution (**TD**) signal to synchronize various operation. There are basically two kinds of communications: scheduled communication (**SC**) and unscheduled communication (**USC**). Time-critical tasks such as process controls fall under SC. On the other hand, functions such as parameterization and diagnostics are USC. The major functions of LAS have been tabulated in Table 7.14. Many of these LAS functions are system management (SM) functions because LAS continuously polls the unassigned device, so it is possible to include any device at any time by a special command probe node (**PN**) and probe response (**PR**). Before any operation, LAS refers to the list to check any scheduled task.

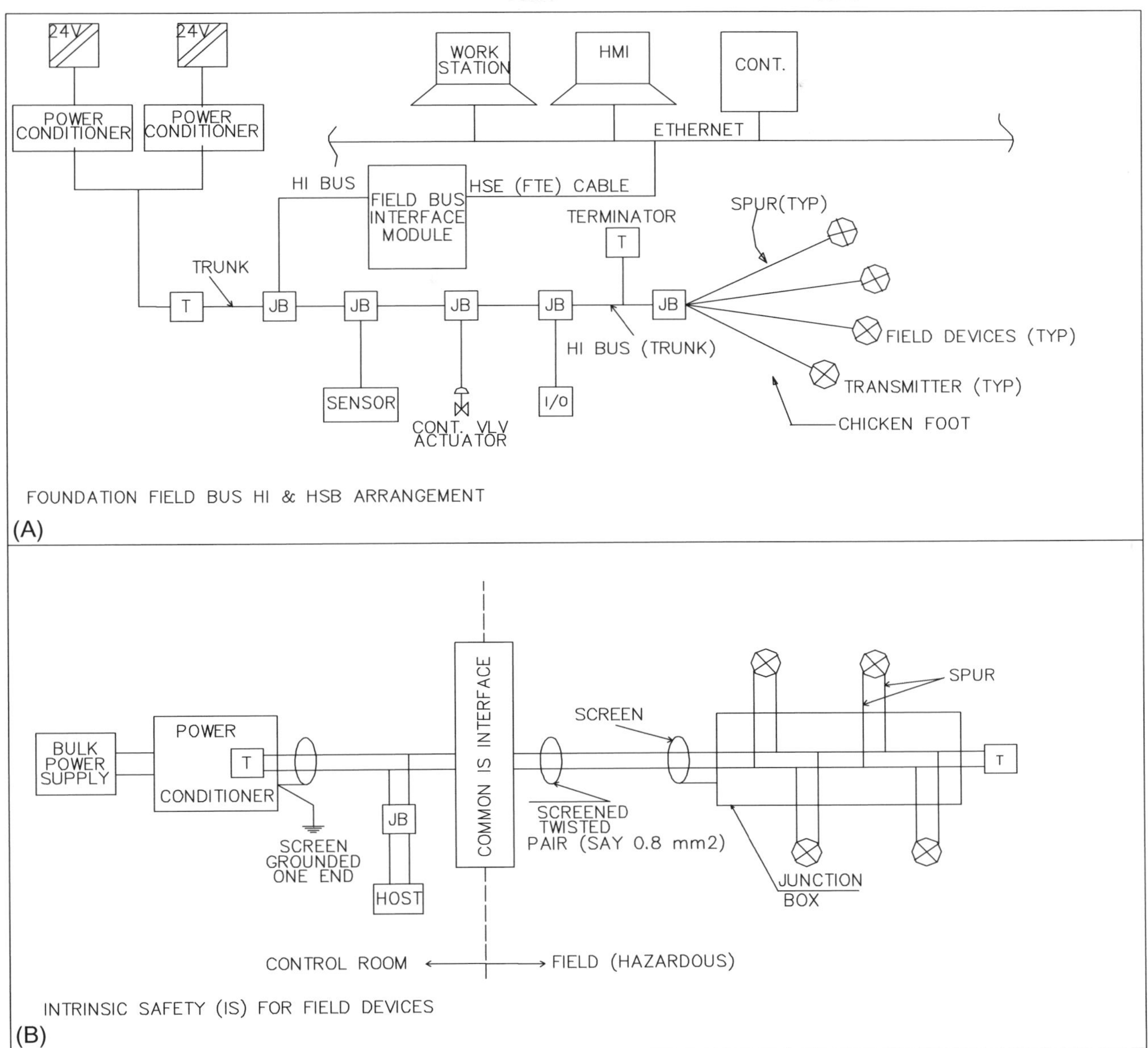

FIG. 7.35 Foundation Fieldbus configuration.

Depending on the device, there are separate schedule times for action, so that devices know precisely the task to be performed. In that case, it waits for the idle time to start the SC, as shown in Fig. 7.36. Here, **PT** stands for passing token. The LAS waits for a precise time to send compel data (**CD**) to activate the schedule operation.

1.1.8.9 Application Layer Functions

Field access sublayer (FAS) and field message specification (FMS) services are extremely important for functionality and performance of the fieldbus, but are invisible to the user.

1. FAS: It creates the virtual communication relationship (**VCR**) like a speed dialing system in a telephone. After configuration, such a VCR number is to be entered to communicate with another field device on the bus. There are different types of VCRs, and they are listed in Table 7.15.
2. FIELD MESSAGE SPECIFICATION: It is a standard message format used to send data across data sent over filed bus as "object description" where Index 0 stands for dictionary header, and 1–255 define standard data types. FMS also defines the virtual field device (VFD)

used to make the object or the device description (DD) of a field device as well as the associated data available to the network. These two are used to access from the remote.

1. USER APPLICATION: It is based on three types of blocks to represent different application functions: **resource block**, **transducer block,** and **function block.** Devices are developed based on the first two blocks whereas for control strategies, function blocks are used.
   - Resource block: Device characteristic for example, serial number, device name, and manufacturer.
   - Transducer block: Used to configure the device, read the sensor, and command the output value.
   - Function block: It is a control strategy builder input and out of function blocks are linked over the fieldbus. There may be many function blocks in a control strategy and these are scheduled. AI, AO, DI, DO manual loader, PID, bias gain, etc., are the standard ones. Control strategy building at the fieldbus is unique with the Foundation fieldbus.

TABLE 7.14 LAS Functions

Activity	Discussions	Remarks
Scheduled communication (SC)	(1) All time-critical tasks have a strict schedule created during configuration. (2) Compel data are sent to the device. Cyclic data are sent according to a list. (3) Each device has a different schedule based on its function and task	(1) Devices (say, a sensor) publish specific data upon receipt of compel data and it can be received by others (say, CV). (2) In SC, the point of time and sequence are well defined.
Unscheduled communication	(1) On request, device parameters and diagnostics data are sent. (2) LAS passes the token to all in the bus	(1) Sent between break of SC. (2) LAS grants permission to access the bus when the token is given to it. (3) Each device can send data until the time is expired or it finishes the data transmission
Live list	Maintains a live list to recognize the device sending data upon token passing	
Synchronization	All devices have exactly the same time as in DLL, due to time synchronization	
Redundancy	In case of more than one link master, if LAS fails the other takes over	

1.1.8.10 Device Description and Common File Format

For configuration and operation, the **actual physical**, **device description (DD), and common file format CFF)** are important. These DDs and CFFs are furnished by the device supplier or the host. Parameters and capabilities are defined in these files and used for offline configuration. DDs help the operation of devices from different suppliers to work on the same bus. CFFs are standard ASCII text files to describe the capabilities and functions of the device.

1.1.9 World FIP

This is another fieldbus available. Some of the basic features of the same are presented in Table 7.16. This is basically meant for factory automation.

1.1.10 OPC

This is another tool for system integration. What is OPC? OPC is open connectivity in industrial automation for interoperability supported by the creation and maintenance of open standards and specification. In the days of DOS-based systems, for printing in a printer one would have to write the driver of the printer. On the other hand, in Windows-based systems, printer support is built into the operating system (OS). The individual printer manufacturer gives a CD to put specific data into the printer support of the OS. The idea of an OPC is similar to that. OPC is a standardized interface for accessing process data.

Object **L**inking and **E**mbedding (OLE) COM (**C**omponent **O**bject **M**odel)/DCOM(**D**istributed **C**omponent **O**bject **M**odel) was developed by Microsoft. The same when applied in the area of process control, OPC (**OLE** for **p**rocess **c**ontrol), has been developed. So, OPC is based on the Microsoft COM/DCOM standard and has been expanded according to the requirements of the manufacturer when they need access to the data in the area of automation. The industrial automation product manufacturer can write OPC data access server and software and could become an OPC client. Typically, OPC clients are programs for the acquisition of operating data. OPC servers are provided for PLCs and fieldbus cards. An OPC server is an executable program that is started when the connection between the server and client is made. On account of DCOM characteristics, it is possible to access the OPC server even if it is another computer. Also, due to COM characteristics, languages such as C++, JAVA, and visual BASIC can be used.

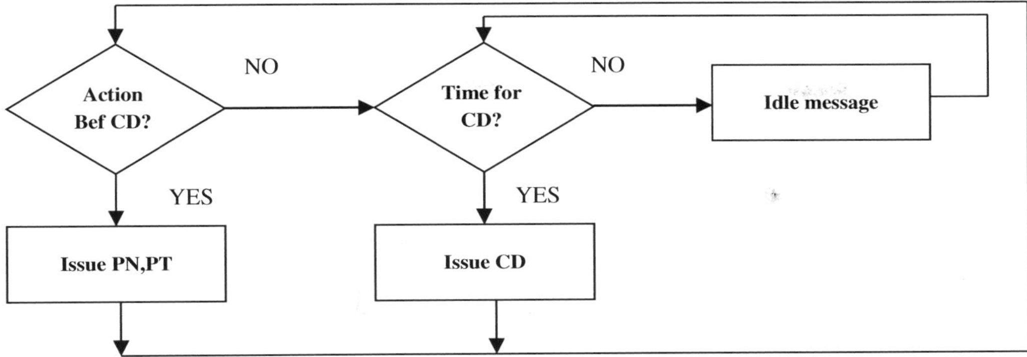

FIG. 7.36 Communication schedule.

TABLE 7.15 VCR Types

Cleint/Server	Report Distribution	Publisher/ Subscriber
Operator communication	Event notification, alarm, trend	Data publication
Changes for set point, device data remote diagnostics, display access	Operator alarm, trend to historian	Actual value or PID block to OP console

TABLE 7.16 World FIP

Features	Specification
MEDIA	TWISTED PAIR and FIBEROPTICS
SPEED	31.25 kbps/2.5 mbps (MAX)
ACCESS	BUS ARBITER
PHY. LAYER STD	IEC 1158-2/YES POWERING
COMMUNICATION TECHNIQUE	PRODUCER/CONSUMER—DETERMINISTIC
BUS POWER and "IS"	YES POWER and "IS" POSSIBLE
DISTANCE (m)	2000/4000 with REPEATER
DLL	SYNCHRONUS + MESSAGE
NUMBER OF NODES	256/NETWORK IN BUS TOPOLOGY
STANDARD	EN50170

Some of the features of OPC are listed below:

1. OPC DATA ACCESS (DA): To move real-time data from control systems such as PLC, DCS, RTU, etc., to the man-machine interface (MMI) or any other client.
2. ALARM AND EVENT (AE): Alarm and event notifications such as process alarms, operator's action, information messages, tracking, etc., are provided to the client on demand and upon event happening.
3. OPC BATCH PROCESSING: Special requirement; not so relevant for a power plant.
4. OPC DATA XCHANGE (DX): This is the most important part for interoperability in a multivendor environment, in the form of a data exchange from server to server and the client server. This service includes but is not limited to add remove, diagnostics, monitoring, and management systems across the Ethernet and fieldbus system, as shown in Fig. 7.38.
5. OPC HISTORICAL DATA ACCESS (HDA): This gives access to the stored data for trending. Historical data archiving has set rules.
6. OPC SECURITY: OPC specifies the **controlling** rules for client access to the server to safeguard the sensitive information as well as to prevent unauthorized modifications.
7. OPC XML DATA ACCESS (XML): XML is a language for encoding information in a machine-readable form that is mainly used in the Internet. XML DA provides a simple generalized flexible consistent rule for exposing plant data.
8. OPC UNIFIED ARCHITECTURE (UA): This is not based on Microsoft but a unified system provides standards for cross-platform capability such as Apple, Linux (JAVA), or Windows.
9. OPC Certification: Self-certification, interoperability workshop, and third-party testing are the ways to get the certification for an OPC compliant system.

1.1.11 Fault Tolerance (FT)

A fault-tolerant system continues to run in the event of a failure of some component (or due to ≥ 1 fault). Fault tolerance may mean graceful degradation of the system, but it continues to

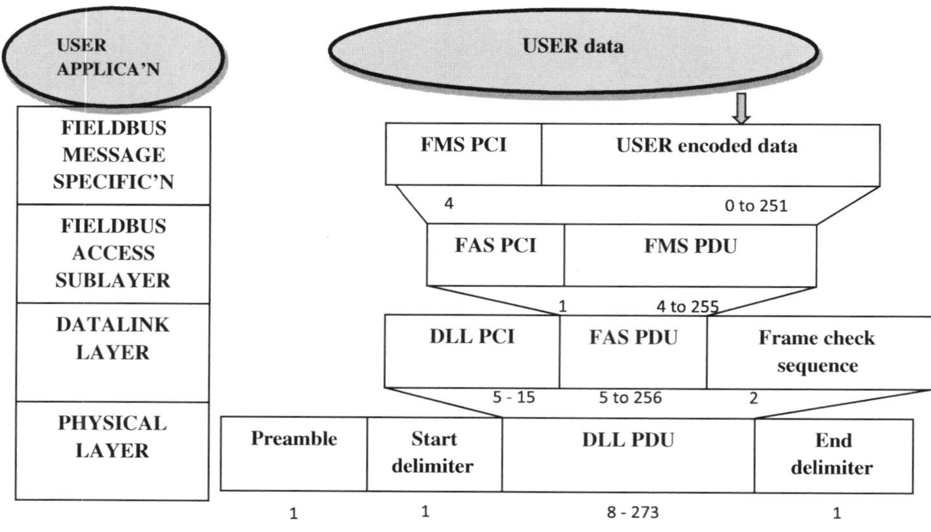

FIG. 7.37 F. Fieldbus telegram.

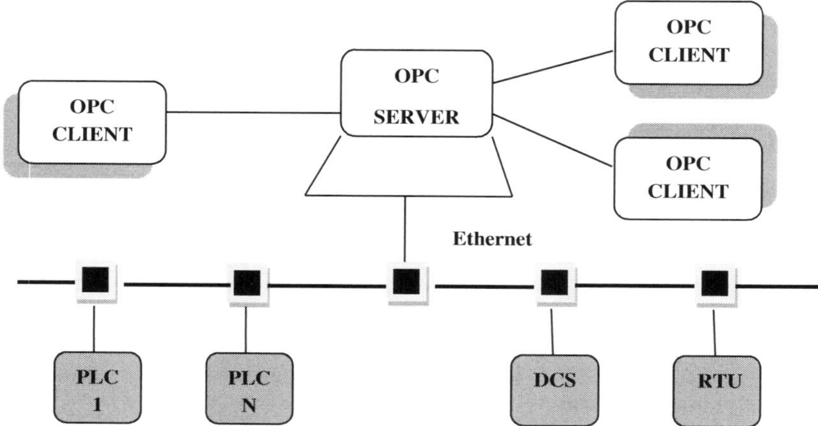

FIG. 7.38 OPC client server architecture.

run as a degraded system and the operating quality degradation is proportional to the severity of failure. Fault tolerance may not be restricted to one machine but to a system as well. In a TCP/IP packet, switching is continued even if the link is improper, resulting in a degradation in communication. It is important to distinguish among error, failure, and fault. The following definitions will be helpful [128].

- Error: An error can be defined as the deviation from correctness or accuracy. Because it is associated with certain values, it can be alternatively defined as the difference in the result of a computation and a correct result.
- Failure: Failure is the nonperformance of some action that is due or expected. The result of a fault could be failure. Error leads to failure or failure may be an effect of error. It could be in cause-effect mode with the fault. System reliability is directly related with the probability of failure or failure rate.
- Fault: Fault may be defined in various ways:
- A fault in a system is defined as some deviation from the expected behavior of the system, or a malfunction.
- Fault can be defined as an incorrect step, process, or data definition in general terms.

In terms of computing, fault can be defined as the deviation of one or more logic variables in computer hardware from a specified value.

- Fault tolerance: Fault tolerance is the property by which a system continues to operate (properly) in the event of the failure of (or one or more faults within) some of its components. The quality of operation of the system may continue without degradation or with a decrease in system performance. Tolerance stands for the capacity of endurance. Fault tolerance can be conceived as the capacity to endure faults in the system.

- Fault-tolerant design: A system that is designed with a fault-tolerance philosophy is basically a fault-tolerant design. So, a fault-tolerant design enables a system to continue its intended operation without interruption and/or at a reduced level, rather than failing completely, when some part of the system or component fails. Also in a fault-tolerant design, if the operating quality decreases at all, then the decrease is normally proportional to the severity of the failure; this is in contrast to a complete failure of the system for a nonfault-tolerant design. For further details, Section 1 of Chapter 11 of [128] may be referenced.

Fault-tolerant features are described in the following Table 7.17.

From the above table, one can infer that the major characteristics of a fault-tolerant system shall include features such as high availability (no single point failure), fault isolation of the failing component,fault containment and prevention of failure propagation, availability of reversion modes, and planned and unplanned service outages [128].

In **intelligent systems, the fault tolerance** needs to be treated a little differently. The fault may occur both in the software as well as the hardware. Out of them, it may be argued that the software faults are almost all are due to design failures such as **errors in interpretation** of the specification or **errors in implementation** of the algorithm. In software, fault tolerance can be conceived as a means to provide redundancy service to comply with the specification in spite of the fault. This is true when the error is not due to a design fault. This means that emphasis is put on the specification in defining whether it is a fault. And this puts pressure on the designer. Now, it is to be judged whether this is exception handling or fault tolerance. Therefore, fault tolerance in software is not a solution in itself. Software fault tolerance is **the ability for the software to detect and recover from a fault that has happened either in** the **software or hardware** (in which the software is running), in order to provide service as per the specification. Therefore, fault tolerance can be implemented by software, embedded in hardware, or a combination of both. When a control system provides satisfactory performance, even in the presence of variations from its original model without changes in the structure or parameter of the control loop, then the control system is **robust**. This may be due to any fault in original model could not be used instead diversified model has been used (without change in structure and parameter of the control loop). Now, if the control loop structure and parameter can be changed in the event of failure, then the control loop is said to be **reconfigurable**. Therefore a fault-tolerant control system can be characterized as robust and reconfigurable and/or a combination of both. A fault-tolerant system provides improvements in system RELIABILITY, SURVIVABILITY, and MAINTAINABILITY. To achieve a high degree of automation, the controller must be capable of additional performance requirements, for example, the additional task of fault tolerance meaning adaptation to the changes in the plant, environment, and control objective. To achieve reliability, it must detect failure, isolate it, and, if possible, contain the same. After hazard analysis (say) **HAZOP** (**HAZ**ordous **OP**eration) SIL (safety Instrumentation Level) of Various control systems are determined. Based on the same, the requirement of fault tolerance for the control systems may be determined. In order to increase the availability and reliability of systems, SIL for various control systems (for example, BMS/ATRS) is determined to make them reliable.

TABLE 7.17 Fault-Tolerance Characteristics and Implementation

FT Characteristics	FT Implementations Methods
No single point of repair	Replica
FA fault isolation	Redundancy
Fault limiting	Diversity
Available redundant or diverse route	Fault tolerance

1.1.12 Control Intelligence

What is artificial intelligence? It is the branch of control engineering (restricted to **control systems** only) to make intelligent systems, especially intelligent computer programs to solve control system problems. Both conventional as well artificial intelligence strive to harness the mathematical model and logical approach. Power plants keep on expanding in terms of size, geographical area, asset management, and penetration of new technologies. Naturally, with conventional control theory, the optimum performance may not be achievable. In view of nonlinearity coupled with plant dynamics, it is impossible to get a complete solution with conventional control theory based on linear theory. In power plant applications, expert systems and fuzzy logic have more application. A few application areas of expert systems and fuzzy logic in power systems are listed below:

- Power system operation—economic dispatch, load power flow, maintenance schedule.
- Power system planning, including generation expansion, reactive power planning.
- Power system control such as voltage control, load frequency control, etc.

- Power plant control such as control of boiler, turbine, including coordinate control.
- Power system automation heuristic.

The complexity of a dynamical system model and the increasing demand for closed-control loop performance for power plants necessitates the use of complex and sophisticated control systems. When there are significant variations in plant parameters, then the robust control loop may not be suitable, so the control system needs to be adaptive. In case of highly complex control systems, a self-learning/heuristic approach may be necessary.

1.1.12.1 Expert System

An expert system is an artificial intelligence (AI) computer program that achieves expert-level competence in solving problems in specific task areas by ringing a model of Knowledge about that specific task. So this is a KNOWLEDGE-BASED SYSTEM or EXPERT SYSTEM. In most cases, the knowledge base is built around the knowledge base used by a human expert. Naturally, each has a domain of operation called the task domain, which needs to be well defined. Some of the basic task definitions may be as given in Table 7.18:

TABLE 7.18 Some Expert Task Definitions

Task	Requirement
Interpretation	Complete analysis of data, correctly consistently
Monitoring	Alarm/event reporting
Diagnostic	Fault finding and analyzing
Prediction	Forecasting future based on knowledge in time
Planning	To define action for result to achieve goal
Design	Creating object and satisfying requirements specified

All the tasks defined above need to be cyclical and dynamic, which can reconfigure the control system in real time. Expert systems in control applications are normally rule-based systems having a database, rulebase, and an interpreter known as the INFERENCE ENGINE. An expert system draws the inference, based on the knowledge base. These inferences are expressed as parameters. These parameters are rule or external to the system. IF-THEN rules develops the premise whereas other parameters are connected by the logical AND-OR. Abnormal behaviors when detected are reported. One of the major disadvantages of the system is that it suffers from the bottleneck of adapting to a new situation. However, it has the following advantages:

- It is permanent and consistent.
- It can be transferred and reproduced.
- Easy documentation and development.

1.1.12.2 Fuzzy Logic System

Fuzzy logic developed by Zadeh is to address uncertainty and imprecision, which widely exist in engineering problems. From experience, it is found that an experienced operator can smoothly control a plant boiler by looking at displays and trends in recorder major parameters and adjusting a number of them. When the steam temperature is falling, the expert operator will control the spray valve at the same time he is to look at the feed water flow control. Similarly, when a plant is loading, by increasing the air flow there will be an effect on the furnace pressure. Like daily life, one has to precisely adjust various parameters in an up-and-down manner to get the optimum result. Fuzzy logic is based on that concept using generalized set theory. In a fuzzy set, the degree of association of an element continuously varies. The beauty of fuzzy logic is understood by a nonexpert also. Other advantages of fuzzy logic are that **constraints are softer** and it more **precisely** represents **operational constraints.** Coordinate control, steam temperature control, and feed water flow control are classical examples where fuzzy controls can be deployed. As discussed earlier, steam temperature control is coupled with the feed water flow control, but using a fuzzy control system, it can be decoupled by a necessary decoupling strategy such as a decoupling controller, gain scheduling, etc.

1.1.13 Firewall

A firewall is a means to trap inbound and outbound information packets, analyze them, and then either allow them to establish communication or discard them. A firewall is a subsystem of computer hardware and software that intercepts the data packet before allowing it in and out of the local area network/server. It is a kind of security used mainly for the networks in the public domain. The firewall is one kind security (or check point?), but control systems may have vendor-specific separate security systems also. General discussions on firewalls are presented in Appendix D.

1.1.14 Security in Control system

The control system evolved from a standalone system to an interconnected network that very much coexists with the IT network. Because these two environments, the IT and control system, are not the same, such integration at times introduces various security threats to the control network. Controllers with embedded firmware have less impact from the mobile code in the form of a virus. On the other hand, a complete integrated network gets easily affected with mobile codes as a

TABLE 7.19 Security Issues in Control System and IT

Security Issues	IT	Control Systems
Antivirus and mobile code	Commonly used	Almost impossible to use for vulnerability
Time-critical contents	Delay accepted generally	Real time no delay
Availability	Some delay/ interruption generally acceptable	Unacceptable has to be 24 × 7
Security awareness	Moderate in all sectors	Not much developed except physical means
Physical security	Secure (server room)	Remote unmanned—secured
Patch application	Regular interval	Rare, some cases vendor-specific

virus, as in an IT network. Complete integration of different networks such as corporate networks, WEB servers, etc., with control the system in the TCP/IP environment is not uncommon (and may be economical too). Also now, with the Internet of Things (IoT) (ref clause no. 1.1.15), the issues have become more vulnerable, that is, cybervulnerability for control systems. Traditional control system communications could be less secure (for isolated operation earlier) and the system administrator must be aware of the capabilities of the supplied security. In order to develop a proper security system for the control network, it is essential to have knowledge about different security threats, some of which are discussed below. A large network as a whole has its own security, but this may not be sufficient for control system vulnerability because most of these are mainly based on information technology (IT) security and may not be suited for control systems. Some of the unique features of the two systems are compared in Table 7.19

1.1.14.1 Types of Threats for Control System Security (CSS)

Today, various kinds of networks are connected to each other, including the control and operational networks for business interests. Also, open systems are the state of the art for a control network. This often makes the system vulnerable when proper security systems are implemented. Some of the probable sources of cyberattacks are discussed in Table 7.20

1.1.15 Internet of Things

From the name, it is clear that the term Internet of Things (IoT) stands to imply that there are things/device in the Internet. The Internet stands to represent a globally connected network, therefore an IoT would mean that there will be things/devices in the world connected through the Internet. In a broader sense, various things in the ecosystem and in the physical world could get connected with each other through a universal network Internet. Naturally, there has to be some internet protocol (IP) address so that the things in the world could be connected to the Internet for accessibility. This means that there should be some intelligent sensor so that it could be addressed. When action, decision, and automation are considered as the purpose of addressing the things, then it could give meaningful purpose for connecting the things/devices to the Internet. From an example, it will be clear. Suppose a person in India, traveling to the east coast of United States for a year, suddenly remembers that he forgot to switch off his air conditioner. If these things (AC machines) have suitable intelligent sensors, one would find it is meaningful to connect the same to the Internet so that it could be turned off through one's mobile telephone, that is, remote control. Items with RFIDs were in use for quite some time in infrastructure/logistics, for example, car intelligent cards to pay the toll tax for roads. Similarly, machine-to-machine (M2M) communications were also in service for quite some time. Now, these concepts have been extended to include billions of things pertaining to various sectors in the world that have been tried to be connected through the Internet to form the IoT, which is not only things nor a connection but a process. The Internet of Things is an ecosystem of inevitably related processes and other technologies from the perspective of a goal within a specific use case [127].

1.1.15.1 Definitions of Internet of Things

The IoT is significant because an object can be represented digitally so that the object is now connected to surrounding objects and database data. There are a lot of complexities around the "Internet of Things," but at the moment, it is better to stick to the basics. Depending on application/technological/perspectives, there will be different definitions for the IoT. In a generalized manner, the Internet of Things can be *conceived* as:

1. **General Definition:** A network of connected devices with a unique IP address with embedded technologies (or equipped with technologies) to enable them to send/receive data and communicate about the environment in which they reside and/or themselves so that action/decision/automation or the services for the devices can be initiated. The potential of the Internet of Things lies in the ways it is used to take advantage

TABLE 7.20 Control System Security (CSS) Attack Types

CSS Attacks	Discussions
Backdoor intruders	(1) Holes for intruding discovered by the attacker due to unintentional error, overlooking of system architecture perimeter design, or fixing boundary for the capabilities of the system. (2) In order to provide better connectivity many times the security perimeter of the control system is compromised, and/or the deployment of robust access technologies such as public facing services without proper security analysis of the control system. (3) For wireless communication there is a problem called the "residual effect." Once an attacker can discover an access point in wireless communication that have been made better connective without a proper security check, the attacker will gain the privilege to tamper or gain control in the network. (4) Many intrusion problems in an IT network can be managed by patch management. This is extremely difficult in CS, especially when there are in a remote location. (5) Many networks, such as networks for an electric company, need to share some data in the public domain but these are generated in CS, for example, electric power generation, frequency of operation, etc. In such cases, it is better that CS share the data with the business server (with a firewall), which may share the same with the web server through proper firewall access. This is because many times, the attacker collects important information from the public server. So the firewall must a separate public server from the internal network
Attack thru' Field devices	(1) Most of the field devices available are IT compatible (for example, transmitters) and support IoT (clause no. 1.1.15). Many modern devices have an embedded file server to provide operational and maintenance data. Operators and administrators have a secondary means for access. These were initially made for trusted network use only. In an integrated system, the attacker, by gaining control over these field devices, may create a tunnel path to the control system to get access to viable targets and investigate connected network security-related issues. (2) When the device is compromised, the attacker may gain control, even attack the master controller.
SQL data Injection Attack	(1) Database is an important aspect in CS, so naturally in the traditional use of this system, the security-associated computer and software of the related devices were looked over. (2) In an IT network, the database is mainly managed with structured query language (SQL). (3) When these two networks are connected, the attacker may gain access to the control data via the IT database, which may have sharing of the CS data. While making such an intrusion, the attacker may disregard the protection mechanism for the control system, and may gain control over the control database.
OPC attack	(1) OPC is extensively used in a control system for connectivity. (2) OLE, COM, and DCOM are used in a common computing environment. (3) Convergence of an isolated control system with a business network poses a big problem and the CS becomes vulnerable to the attacker. Use of patches, as in IT, is a big challenge in the control network.
Man in the Middle Attack MITM	(1) The address resolution protocol (ARP) helps maintain a route with mapping the network address to the physical address. (2) Attackers poison this ARP table and can force all network traffic, including control traffic, through the computer of the attacker. All resources on the network would talk to the attacker without knowledge. (3) In open architecture, data in transit is a major concern. It is MITM is exceptionally dangerous to the control system as it can stop the process, can capture, modify data, falsify information showing good data to HMI and wrong data to the field device.

of the insights from data, to take action/decision or to automate, to optimize, and, in more mature stages, transform processes (or business models) and industries. It is better to look at the various definition of IoT by various industrial giants.

2. **Technopedia** (https://www.techopedia.com/definition/28247/internet-of-things-iot): "The internet of things (IoT) is a computing concept that describes the idea of everyday physical objects being connected to the Internet and being able to identify themselves to other devices. The term is closely identified with RFID as the method of communication, although it also may include other sensor/wireless technologies, or QR codes."
3. **IBM** (https://www.ibm.com/blogs/internet-of-things/what-is-the-iot/): "…*T*he Internet of Things is the concept of connecting any device (so long as it has an on/off switch) to the Internet and to other connected devices. The IoT is a giant network of connected things and people–all of which collect and share data about the way they are used and about the environment around them."
4. **Forbes** (https://www.forbes.com/sites/jacobmorgan/2014/05/13/simple-explanation-internet-things-that-anyone-can-understand/#5e188c101d09): "The concept of basically connecting any device with an on and off switch to the Internet (and/or to each other)…The IoT

is a giant network of connected 'things' (which also includes people). The relationship will be between people-people, people-things, and things-things."

1.1.15.2 The Process and its Impact

Devices with built-in intelligent sensors and an IP address are connected to a platform referred to as the "IoT" to gather information, and integration of data from the different devices. Based on applications through IoT analytics, this information is applied and shared to address specific needs. The IoT is capable of pinpointing exactly what information is the most useful and which information can be ignored, so that useful action/decision/automation can be initiated to resolve the concerned problem. In this process, anything that can be connected will be connected through the Internet to ease the day-to-day life of the people as well as the ecosystem. In the Industrial IoT (IIoT), the Internet infrastructure is utilized with a goal to increase system/subsystem efficiency by accessing data/information from sensors, controls, and actuators to help automate the system and come up with optimum solutions to increase efficiency and assure safe operation with less downtime. In process industries, it automates in order to achieve enhanced productivity and safe distribution system operator tools utilizing digital controllers. In the energy sector, the following are some of the areas where IIoT can be deployed to optimize the process and get the most reliable, safe system with increased efficiency.

- Advanced metering infrastructure (AMI).
- SCADA (supervisory control and data acquisition).
- Smart inverters.
- Remote control operation of energy-consuming devices.
- Remote operation of various subsystems, for example, the raw water pump house.

1.1.15.3 Characteristics of IoT

The essential characteristics of a typical IoT are elaborated in Fig. 7.39. In the broader sense, there are seven main issues that can be put under the broad umbrella of IoT. These seven issues are elaborated upon as follows.

1. **Things:** This refers to any device/object that can be tagged. Naturally, these must have suitable sensors with IP addresses so that they can be tagged and referred to by others connected to the IoT. Anything that can be tagged or connected as such as it's designed to be connected. With the tremendous development of embedded electronics, it is possible to have a suitable sensing element. Such an object could be an industrial device or any tagged livestock with a suitable sensor (intelligent) attached to the device/object.
2. **Data/Information:** This stands to represent various data/information about the system that is of interest for analysis, action/decision making, or automation, for example, in order to automate a system, it is necessary to gather process information through sensing.
3. **Communication:** Through communication channels, various devices get connected for exchanging the required data/information and for analysis.
4. **Smart Analytics:** This stands to represent intelligent electronic technologies associated with the device as well as for its sensing and addressing capabilities. Today's embedded electronics make it possible to have application-specific intelligent sensing attached to the device. Such embedded electronics also help in gathering data analytics, including artificial intelligence. Usefulness and major developments actually have probably happened based on this property.
5. **Action:** Just gathering data would be meaningless unless some action is taken to properly utilize the gathered data. So, this can be considered the consequence of intelligence. Based on the same, actions/decisions/automation are initiated. This can be manual action (accepting an alternate route to avoid heavy traffic on the road), decision through discussions (climate change decisions), and automation such as in process automation of (say) starting a redundant pump automatically at a remote location when the running pump fails to develop pressure.
6. **Ecosystem:** This is the place of the Internet of Things from a perspective of other technologies, communities, goals, and the picture in which the Internet of Things fits [126].
7. **Connectivity:** This stands to represent the physical connection, devices, sensors, get connected: to an item, to each other, actuators, a process, and to "Internet" or another network. This enables communication between devices (M2M) and/or to the system.

These developments have been made possible for embedded electronics, which are application-specific and cheaper for individual application. Also, these embedded systems are efficient and low-power devices. On the IoT, one needs to remember that these devices need to be ON if real benefits are to be achieved. Because these are connected to the Internet, there will be cybersecurity threats also. Therefore, the security of the system is of utmost importance. Also, for IIoT implementation, a reliable secured Internet connection is essential. Also, for data analysis, the intelligence of the embedded system is essential. So technology up gradation to marry with needs is of too much importance. Two important issues in connection with the IoT are defined in Fig. 7.40. From the above discussions, it is clear that the hardware units in the IoT system can be categorized as:

- Sensors/actuators.
- Processing units.
- Storage units.
- Communication units.

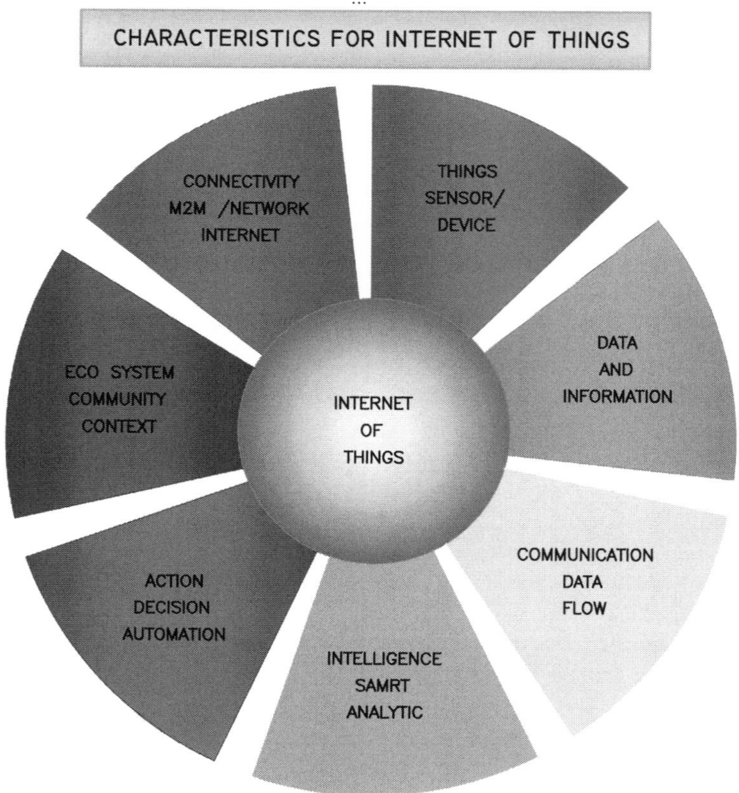

FIG. 7.39 IoT characteristics.

After identifying the hardware category, the necessary software components and protocols need to be developed and added to it so that it could be linked and a fully operation model can be developed. With this, the discussions on IoT comes to an end and also general discussions on intelligent control systems are concluded so that we can begin detailed discussions on PLC systems.

2 PROGRAMMABLE LOGIC CONTROL (PLC) SYSTEM

2.1 Introduction

A programmable control system is a digital computer deployed for automation and monitoring of electromechanical and/or electrochemical processes. Typical applications of a PLC in a power plant include the burner management system (BMS), the automatic turbine run-up system (ATRS), etc. On account of its versatility often people refer to them as "Cinderella of control". In modern times, it is very difficult to distinguish a PLC from a DCS. Normally, when the ratio of the digital signal to the analog signal is **high**, PLCs find their applications from an economy point of view. On the other hand, DCSs were in use for data acquisition systems (DAS), PIDs, and other complex control algorithms (that is, a digital signal to an analog signal is NOT **high**, but there is more complexity in the algorithm). Also, the response and reaction time in a DCS could be more when compared with a PLC. However ,in mid/large power plants, the DCS is more popular when compared with a PLC. The following advantages of a PLC over a DCS may be listed:

- Flexible, modular, and smaller physical size.
- Less-complicated application, with easier programming and reprogramming.
- Easier networking and interface with third-party devices.
- Easier expansion and system modification provisions.
- Low and moderate cost.

2.1.1 PLC Basics

In its basic form, a PLC may be conceived of as a combination of five basic units such as the power supply unit, I/O devices/systems, the central processing unit (CPU), memory (program/data), and the programming unit, connected via a bus in the configuration shown in Fig. 7.41. Apart from these, there may be a few other features to be supported, for example, an operator workstation (WS), an interface to third-party devices, and/or connection to the main plant control system. Naturally, in order to achieve the same, a **communication** module and **interface** devices

Cloud computing: **In order to deliver the services quickly for specific applications, Cloud computing in IoT provides a scalable way to manage all aspects of an IoT deployment. Thus, cloud serving helps to create powerful tools for IoT applications.**

Fog: **The fog extends the cloud to be closer to the things which produce and act on IoT data. Any device with computing, storage, and network connectivity can be considered as a fog node. These fog nodes, can be deployed anywhere with a network connection: on a factory floor, on top of a power pole etc. Industrial controllers are example of Fog node.**

FIG. 7.40 Fog and cloud computing in IoT.

would be necessary. All these will give rise to variations in the system configurations and intelligence.

1. Power Supply Unit (PSU): Generally, the PSU, supplying the desired quality (voltage, type, etc.) power to various modules, is a separate module, but it could be an integral part of each module. The general features of power supplies include suitable isolation and protection with necessary heat dissipation, which means a suitable quality and rating of power to drive all modules. There will be a separate power supply for remote I/Os. A common power supply may be 24 VDC; however, each manufacturer specifies the required power supply for its system, especially when higher (than 24 VDC) interrogation voltages are to avoid channel noise. The voltage of the PLC supply, by which the input contact status is sensed in the system, is the interrogation voltage.
2. 2 I/O system: Input (I) and Output (O) devices: Input devices are momentary or permanent digital signals from various contacts, such as process switches, electrical switches, push buttons, and/or registering inputs such as from a thumbwheel (BCD), etc. In addition to digital inputs, there are analog and pulse inputs also, for example, RTD/TC, 4–20 mA DC from process transmitters, and pulse input from flow meters. While choosing the required input module, one needs to decide a few things such as the number of inputs in one module, the interrogation voltage level (for larger distances, a higher interrogation voltage may give better noise immunity), and the type of input (whether in single or in differential mode). Another important aspect is to decide whether the input shall have any isolation circuitry such as optical/galvanic isolation, etc. Output devices can be digital output for a solenoid valve or MCC or analog output to drive the final control actuator. There may also be a registering type output to drive the panel meters. While choosing the required output module, one needs to decide a few things such as the number of outputs in one module, the output voltage level, the type of output, the allowable load, etc. The output could be potential free contact (contact rating and type is an important aspect here), wet output (for example, TTL), direct AC output, analog output, and/or special output for motion control and/or servo positioning control. A typical input-output interface is depicted in Fig. 7.42A–E.
3. Central processing unit (CPU) or control unit (CU): It is the heart of the entire system, performing all the tasks meant for the control system. The major tasks of the CPU are:
   - Scanning input execution of the program and allocating outputs.

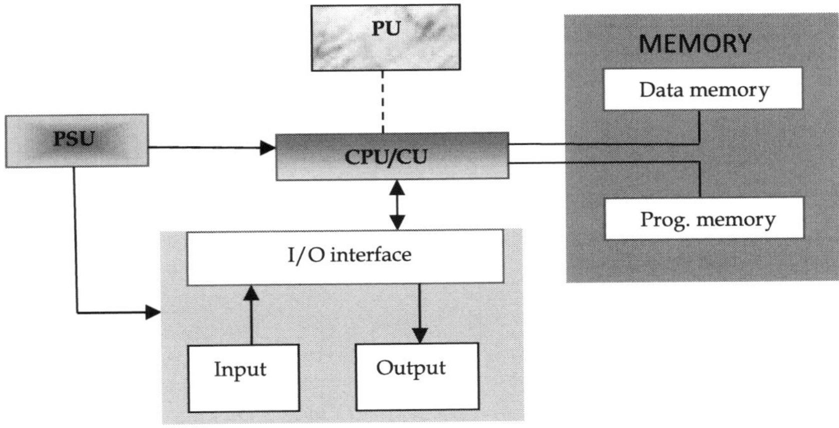

FIG. 7.41 Simple PLC configuration.

- Driving peripherals and communication with external devices.
- Traffic control on system data highway and diagnostics.

A microprocessor along with the main board memory forms the major intelligence unit of the system to carry out all these tasks in a faster and more flexible manner. The CPU internal program enables the PLC to execute control instructions embedded in the user's program. This is stored in nonvolatile memory or memory with a battery backup so that these are never lost. Housekeeping (such as checking the health of the devices, internal communication with other systems, etc.) is another important task the CPU performs.

When the systems starts after initialization, it performs four main steps (as given in Table 7.21) in a cyclic manner, as shown in Fig. 7.43.

Word size and scan time actually give an indication about the performance of a PLC. A word size from 8 bit to 32 bit and a scan time of 0.1 millisecond is typical for PLCs.

4. Memory: The memory unit plays a great role in the functioning of a PLC. One part of the PLC memory is used for the PLC's executive program, or operating system (OS), necessary to execute the user's application program, interruption handling. The other memory has two basic parts such as the program memory where the user's program shall reside and the data memory where the process data from the I/O shall be stored.

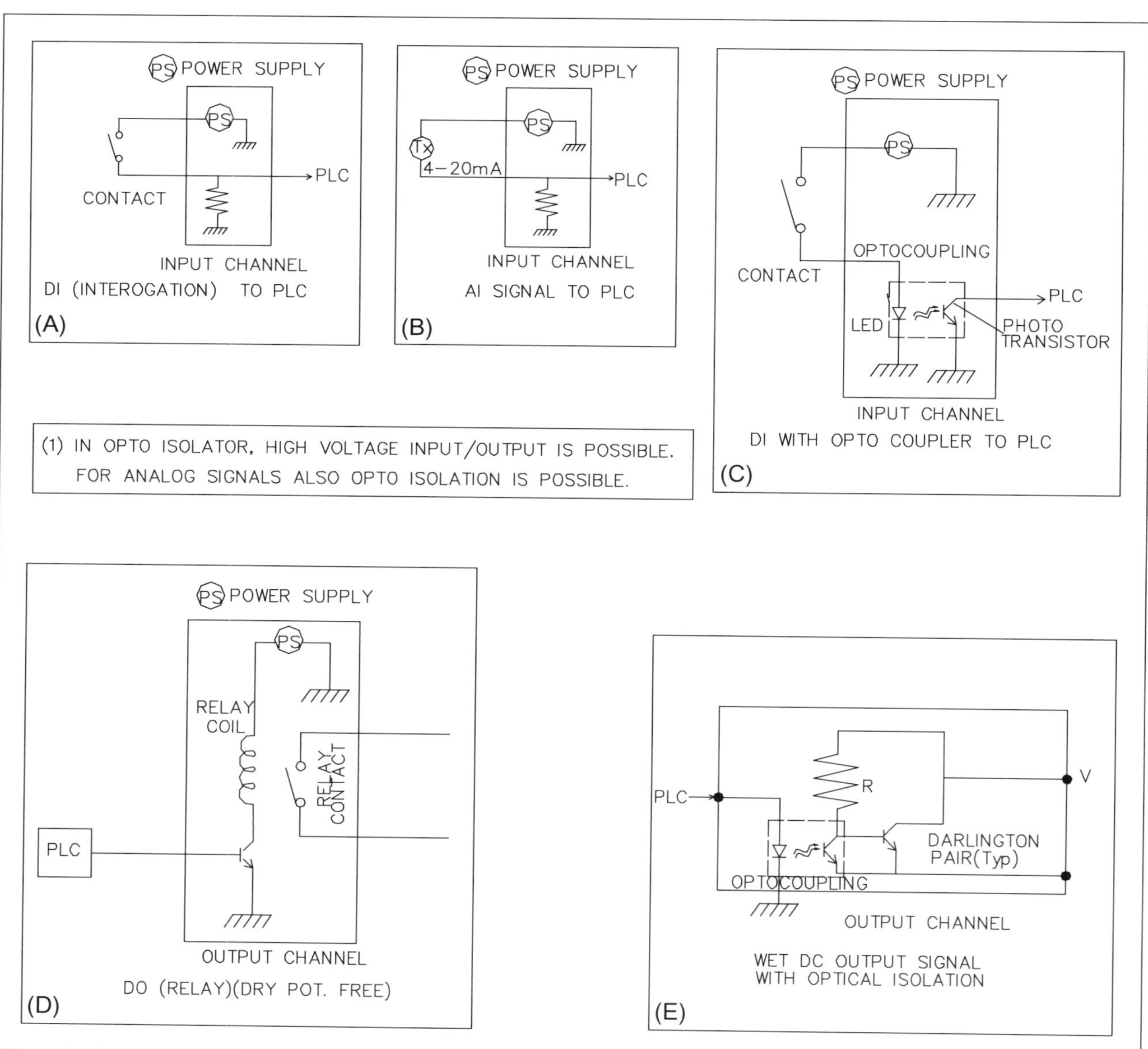

FIG. 7.42 PLC I/P output signals.

Volatile memories are backed up with a battery. The division of memory between data and program enhances the performance of the PLC. The size of the main memory is guided by the CPU selected. Integral RAM should be sufficient to store small programs, whereas for large PLC, the memory can be augmented to the tune of 64 MB. These may be backed up by a battery.

5. Programming unit: It is the interface between the user and the PLC during program development and loading. Starting from a small, calculator-sized handheld unit, a programming unit (PU) can also be a monitor-based system. Many manufacturers utilize a PC with a programming tool for program development. Program development can be online as well as offline.
6. Miscellaneous other units: In addition to the basic units discussed so far, there can be a few other units such as peripherals for operational aid such as color graphic monitors, printers, program aids such as PC-based systems with a license for special software, remote I/O, etc. Also, there could be specialized panels for the operator to operate the plant. Other operation aids shall include interface devices such RS 232, RS 422, etc., to link with other devices.
7. Program language: The normally used programming languages and methods shall include but are not limited to the following:
   - LADDER Diagram (LD): It is a graphical programming language to show the opening/closing of relays, counters, timers, flip flops, shift registers, etc., as shown in Fig. 7.44A.
   - Functional Block Diagram: As shown in Fig. 7.44B, this is a graphical language to depict the functional relationship among various functional blocks. This is very useful for control loop and logic representations.
   - Signal Flow Chart (Sequential Function Chart) (SFC): It is a special block representation with suitable direction graphics to elaborate the logical function. It consists of branches and nodes, as shown in Fig. 7.44C.
   - Instruction List (IL): Like an assembler language, it uses instruction sets to bring out PLC logic. Fig. 7.44D may be referenced.
   - Structured Text (ST): As shown in Fig. 7.44E, it is a high-level text language similar to PASCAL.
8. PLC Operational Aspect: The CPU tasks in Table 7.21 are discussed below.
   - Input Scan: The scanning of inputs takes place when the input data are stored in appropriate storage, provided the CPU is not in STOP mode and the I/Os are enabled. Scanning is skipped when an I/O is not enabled.
   - Program Execution: Immediately after the input scan, program execution or logical operation is implemented when the CPU is in RUN mode. In this part of the program, the application program will be executed according to the logic, and the output will be updated. Ladder logic shall be executed and or subroutines will be called if necessary in the program until the END of the program is reached.
   - Output Scan: Outputs are scanned immediately after the program execution. Outputs are updated. At this stage, if the application calls for it, the global data shall be updated also.
   - Program Communication: This is executed when the program device is connected and there is a service request from the programming device. Thus, communication with the programming device is established.
   - System Communication: When there is a service call from intelligent modules, this is executed. In case of time-consuming requests, such requests may be distributed in several cycles of programs.
   - Reconfiguration: In this part, the CPU checks the actual hardware configuration with the configured system. Empty modules and faulty modules shall not be scanned. If the CPU detects an earlier module that was not working but replaced the same with a good one, or an added module, it will reconfigure itself to include the same in the scan cycle. During this period, it recognizes changes in configuration.
   - Checksum: It is the last stage in each program cycle, calculating the check sum and comparing the same with the reference check sum. In case of a mismatch with respect to the reference value, the same will be flagged as a program failure.
   - It has the following functions being executed at the initial stage.
     o. Calculation of each program cycle time.
     o. Schedule start of the next program cycle.

TABLE 7.21 CPU Tasks

Step	Function	Explanation
1	INPUT SCAN	To scan all the inputs connected to P the LC and present the data to the CPU
2	PROGRAM SCAN	To execute the user program to perform PLC functions
3	OUTPUT SCAN	Present the result of the program to the output devices to energize and deenergize the output devices
4	HOUSEKEEPING	Internal diagnostics and communications

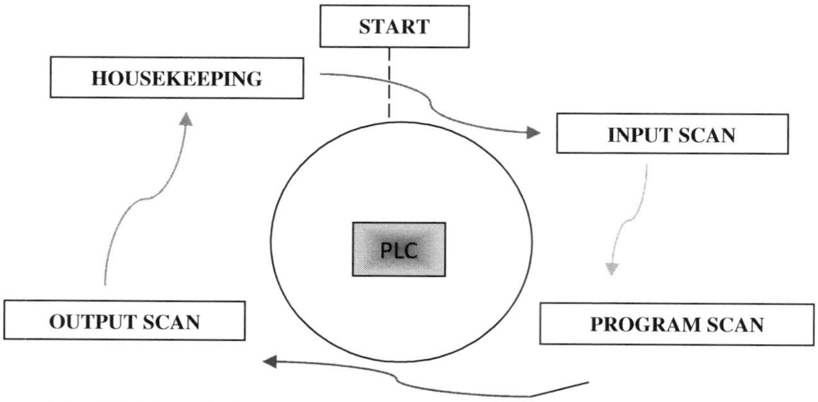

START
HOUSEKEEPING
INPUT SCAN
PLC
OUTPUT SCAN
PROGRAM SCAN

FIG. 7.43 PLC operating principles (CPU Operation).

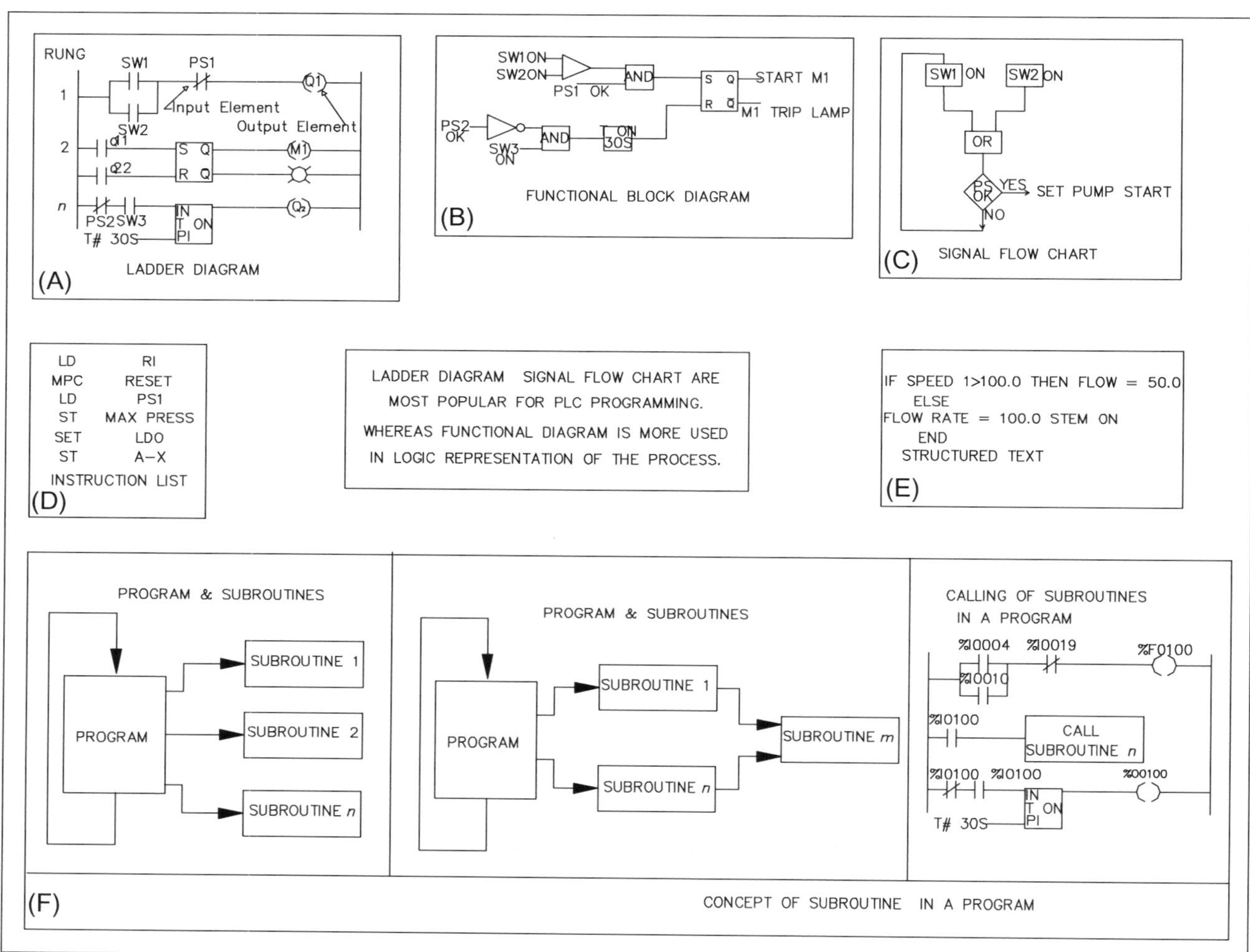

FIG. 7.44 Programming language types.

- o. Update fault table.
- o. Set/reset watchdog timer.
- o. Diagnostics.

9. The basic points for PLC selection in one application may be:

- PLC powering voltage type, value, and system current rating.
- Scan time and word processing capability.
- Available memory to cater to the application.
- Software and programming language.

- I/O capability, including analog signal handling capability.
- Remote I/O handling capability.
- Environmental requirement (especially for remote I/Os).
- Programming and connection of user with the PLC.
- Networking capabilities, fieldbus capabilities.
- Third-party device integration open process connect (OPC).

10. PLC in a Power Plant: It is not unlikely that various control systems supplied for a power plant may be from different vendors, for example, steam generator suppliers may include integral control systems for burner management systems (BMS), secondary air damper controls (SADC), soot blowing (SB) controls, high-pressure (HP) bypasses, etc. Similarly, a TG supplier may supply some integral controls systems such as an automatic turbine run-up system (ATRS), automatic turbine testing (ATT), turbine stress evaluation (TSE), etc. A third party may supply the main DCS. Also, there will be a number of OFFSITES whose suppliers may also supply different controls. It is necessary to connect different systems from different suppliers. Such integration requires compatibility in the physical connection, protocol, and networking speed (especially PLC with analog I/Os) compatibility. Fieldbus systems are one of the easier means for such integration with a reduction in cable cost. In subsequent discussions, we take up these issues.

2.1.2 Salient Features of PLC

The major features of the PLC shall include but are not limited to the following.

1. *Availability*: In order to avoid a catastrophic situation and costly loss, it is necessary that the system is available at all times. Some of the issues shall include:
   - Redundancy (controller, I/O system, etc.,), online change of configuration/system parts.
   - Efficient programming and scalable configuration: For redundancy, the system needs to be configured and designed in such a way that changeover is smooth and at the same time it is scalable in critical areas so that the system is cost-effective.
   - Diagnostics for early fault detection and fault tolerance for reliability, availability, maintainability, and survivability.
2. *Downtime Reduction*—Diagnostic: An automation system shall have an integrated self-diagnostic feature in all modules so as to have distributed diagnostic features throughout the plant. These may have features such as:
   - Module identification.
   - Time stamping.
   - Error message in the man-machine interface (MMI) with different degrees of detailing and status.
3. *Flexible I/Os for Various Plant Conditions*: Various kinds of I/O suitable for various conditions, including SIL (discussed later) rated I/O modules, IS I/O, etc., needed for various applications such as BMS and ATRS.
   - SIL rating—as applicable and the possibilities of intermixing of standard I/Os with a failsafe I/O to have a compact completely integrated automation system with international certification.
   - IS I/O **if** applicable, with necessary international certification.
   - I/O intelligence.
   - Configurable I/O modules.
   - Scalable I/O system.
4. *Security*: A seamless exchange of data and IT compatibility so as to access data from any location in an open system. Naturally, there shall be comprehensive security at all levels. Some of the applicable features may include:
   - IT standard security such as a micro**firewall** (to prevent all unauthorized access to the PLC devices without obstructing valid PLC control commands), a **virtual private network (VPN**, for authentication of communication between genuine partners to avoid unauthorized data modification and disclosure of system data and encryption), and **wired equivalent privacy (WEP**) in the wireless network as part of the IEEE 802.11 protocol for confidentiality.
   - Physical separation and accessibility at various levels that shall have autonomy.
   - System management for all these securities without interrupting the main operation.
   - Control system security: Also refer to Sections 1.13 and 1.14 of this chapter and Appendix D.
5. *Open Architecture*: It is preferred that the systems shall have open architecture so as to integrate various networks as well as go in for the industry standard fieldbus such as a Foundation fieldbus (FF), Profibus, etc. It is possible to integrate various other systems together such as drive controls, MCC, etc. IT enabled systems/IoT systems have facility for remote access and integration.
6. *Communication*: One complete integrated system comprises various levels such as field, controller, operator management, and the corporate level. So, such a system must therefore have a highly developed communication capability to transparently exchange data at all these levels. Some of these features shall include but are not limited to the following:
   - Combination bus system such as Profibus, FF.
   - IT compatibility/IoT or IIoT.
   - Integration of the office system.
   - Integrating communication of wireless LAN or remote access (for example, Zigbee).

7. *Engineering and Software*: The control system software shall be capable of meeting the specific needs of different control applications for the power plant, with special attention to safety and complexity of applications (with an EXPERT solution as applicable). The major key features are:
   - Blockwise modularity.
   - Data consistency.
   - Shared configuration for entire plant automation.
   - Open data interface (for example, GSD, OPC).
   - Support various protocols, for example, OPC UA/DA, HART, Profinet.
   - Standard programming language, for example, IEC 61131-3 compliant.
8. *Expandibility, Scalability, and Cost-Effectiveness*: A system with a modular flexible design can start with a single controller with a few modules, and then, depending on the application, can be built into a giant system. A large system can be integrated utilizing several controllers, field devices, and workstations—communicating over a fieldbus and Ethernet, that is, to have system flexibility coupled with the capability and power to integrate. In a scalable system, there will be a wide selection of CPUs and I/Os to meet the system requirements and system availability without incorporating unnecessary redundancy in the system. This means that common hardware and software (freely configurable modules) would be used for both small and large systems.

2.1.3 PLC Configurations

Some commonly used configurations are discussed below, starting with the simplest.

1. Basic PLC Configurations: In Fig. 7.45A, a PLC with on-chassis I/O cards and a power supply system (single or dual) is shown. Here, all I/Os are local to the PLC controller. Based on the selection of CPU, there will be a limit of such I/Os. Normally, 512/1024 I/Os are supported in this form basic PLC. Such an independent configuration may be suitable for control systems in a circulating water system (CW) and auxiliary cooling water (ACW) in the power plant. In case this needs to be connected to the main control system, then some modifications may be necessary. In Fig. 7.45B, a modified higher system is depicted. Here, in case of demand for additional I/Os, then a limited number (depends on the type of PLC) of I/Os can be added as extended I/Os (could be an extension module in the main chassis also). Sometimes in power plant applications such as a raw water pump house I/O and/or a pretreatment plant I/O connection to the DM plant PLC, it is desirable to have a remote I/O connected to the PLC. In these cases, all I/Os are terminated at the remote I/O cabinet/terminal board, which is connected to the PLC through a serial link. This could be done with the help of a fieldbus also. In this configuration, a data highway is shown to connect a computer. In this data highway, there could be workstations, programming station, and/or other PLCs of the same family as shown in Fig. 7.46A. The "DH +" of AB/Rockwell is an example of the same. In the data highway, other computer systems can be networked in the form of an Ethernet. Such standalone PLCs could be suitable for offsite plants such as water treatment plants, Coal handling plants, etc. Presently, people look for an integrated system with a main control system connection.
2. PLC Network—Redundancy and Safety Considerations: In Fig. 7.46A, where two PLCs each complete with I/Os power supply units in independent configurations have been connected via control network (for example, Ethernet). These I/Os to the PLC may be distributed over a few chassis and can be connected to the host/main PLC by fiberoptics (FO) and/or a fieldbus or they may be connected to the host/main in a dedicated way as shown. A common WS, PU, and/or any PC can be connected to the same network. Also, it is possible to share common input/outputs by the PLCs in the network. One thing to be noted here is that a dedicated communication processor has been deployed to reduce the load of the PLC main processor. The data highway may be redundant also to enhance system availability. Redundancies as options are available at various PLC sublevels such at I/Os, controllers, and communication subsystems.

   In Fig. 7.46B, two controllers, one the main controller with another as the hot standby, have been connected in the network. In mid- and large-size power plants, in most of the applications, people go for such redundant systems for all applications except for very noncritical services. A standby controller also shares the critical control information. Both are updated in a regular manner, that is, both share the same information over the bus but the standby controller outputs only when the main controller is out of service. In many systems, any of the two controllers can be set as the primary controller, and on failure of the same (being diagnosed by the system diagnosis program), the hot standby shall take over without a bounce (as both are updated simultaneously) to the process. Upon rectification of the primary controller, it may be put into service. In cases where a controller is not preassigned as the "primary," then this controller will start functioning as the standby. Otherwise, this controller will take over as the primary, and the working one will again be in the system as the standby. I/Os and the data

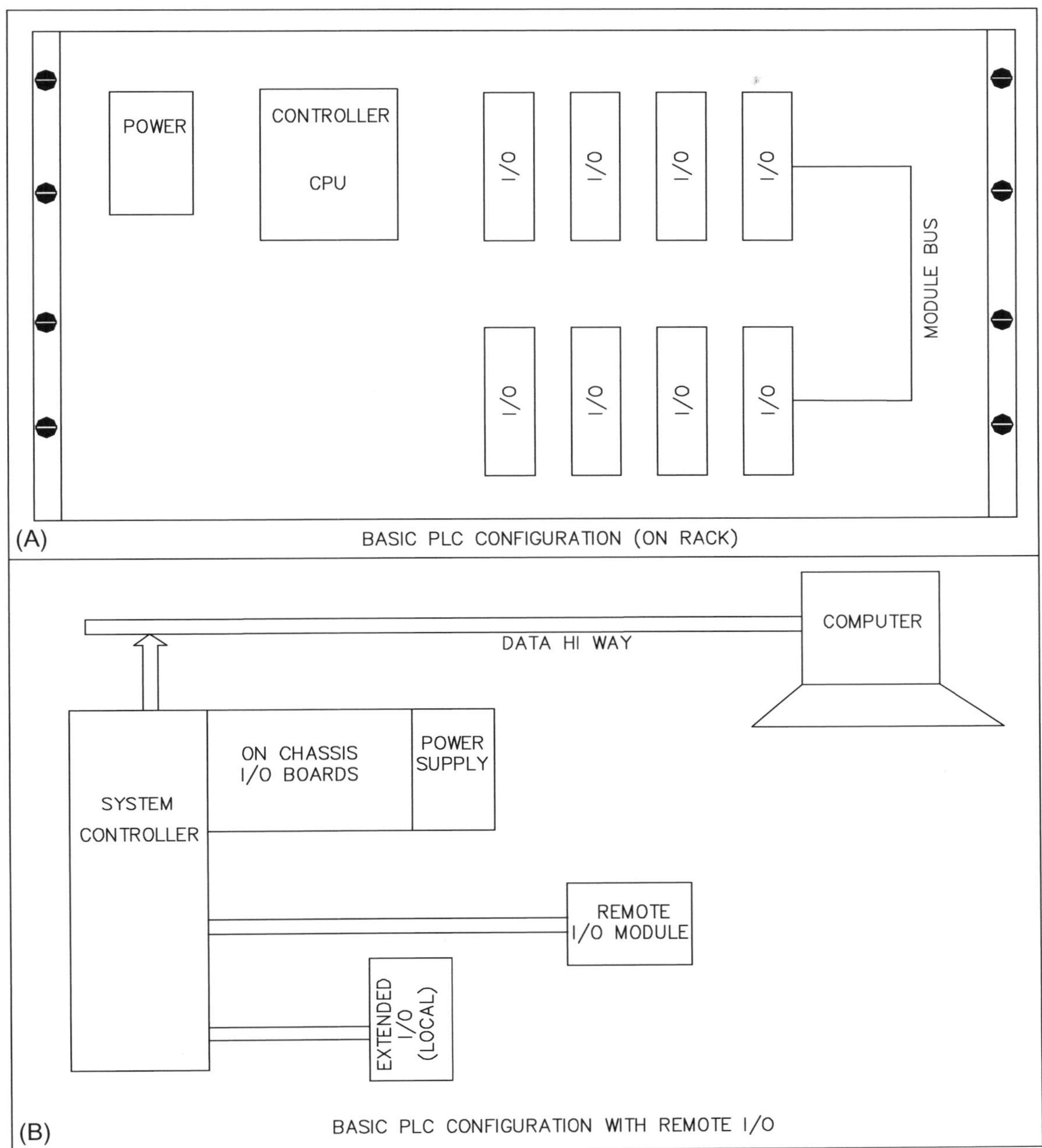

FIG. 7.45 Basic PLC configuration.

highway can also be redundant, as shown. Presently, it is possible to integrate various products from not only their own manufacturing range but also from other manufacturers. Another feature in this figure is that I/Os have built-in intelligence to be supported in the bus. Also, they can support a fieldbus (for example, a Profibus DP). In these I/O modules, parameterization, free configuration, etc., are possible. There are a few special intelligent modules available for the encoder, the high speed counter, the servo controller, servo positioning, etc. The **fault tolerance** (Section 1.1.11) of the PLC and I/O provides maximum availability of the system with single-point failure. With use of industrial-grade electronics suitable for a harsh environment, limited loading coupled with low power consumption and/or forced cooling make the system more reliable to be used in fault-tolerant mode. In a fault-tolerant system, redundancies are available at the controlled network, fieldbuses, CPU, I/O, etc., to achieve a bumpless transfer. For noise immunity, the FO may be used. In several

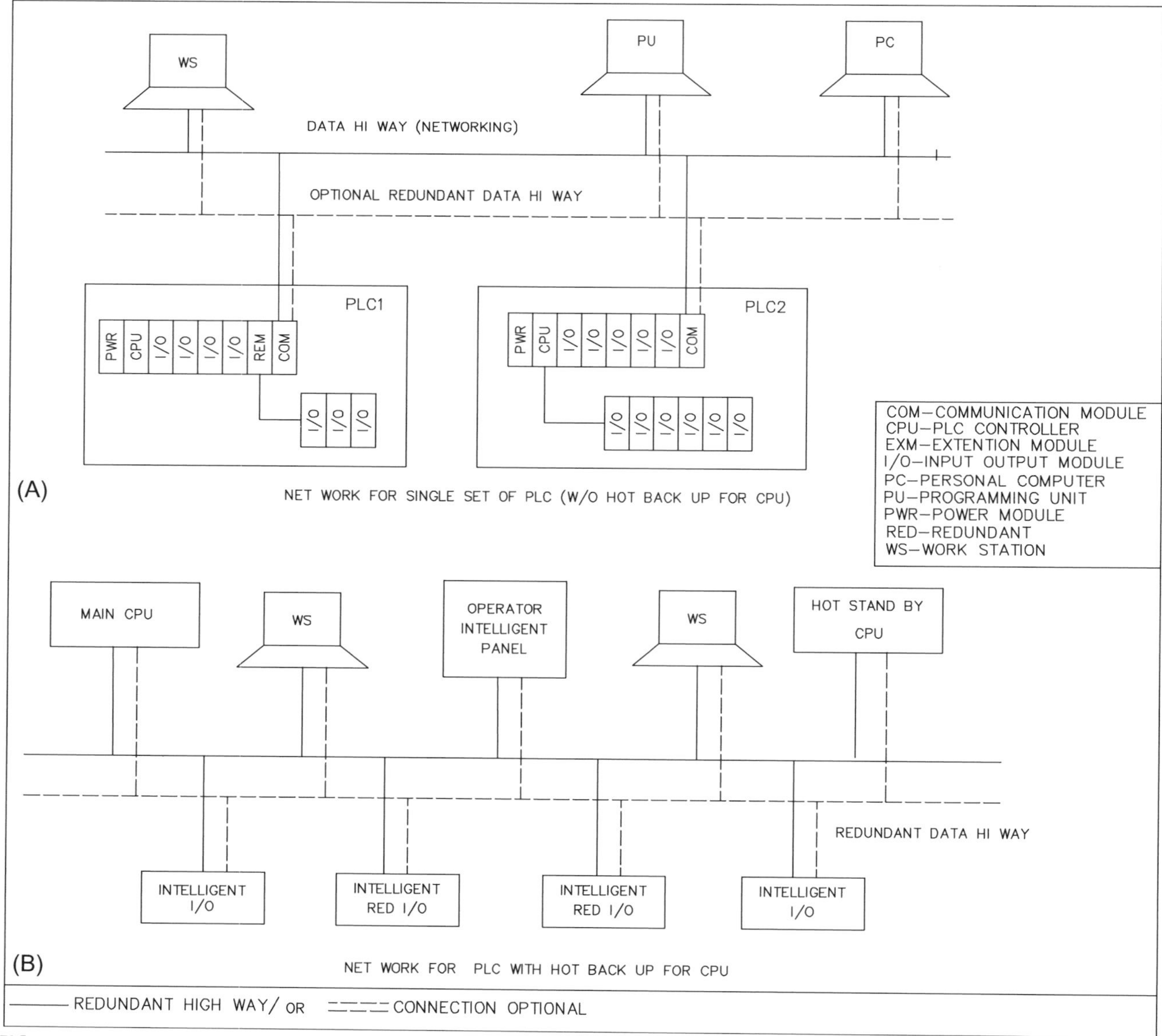

FIG. 7.46 PLC networking.

fault-tolerant PLCs such as the ABB AC880M control module, the application and hardware settings may be changed online during operation. FAULT-TOLERANT PLCs have international certification (for example, TUV). For safe systems, as applicable, the IEC 62061 SIL certification may be used. Fault-tolerant PLCs may be useful for BMS/ATRS/Digital EHG (DEHG) applications. For certain systems, designers go for triple modular redundant (TMR) PLCs. A TMR PLC increases safety, reliability, and system availability coupled with avoidance of nuisance tripping. In a TMR PLC, error detection is easier and the fault-tolerant situation is achieved. TMR is a specialized version of 2oo3 [128] with much more stringency in implementation, namely TMR logic is used in many governor controls of the turbine, GE gas turbine, or overpressure protection of the KWU turbine. In case of a complete TMR system, there are three main processors (MPs) as well as I/O processors.

Each MP operates in parallel with the other two. The I/O control processors manage the data exchange [128]. Triple I/O bus systems connect the trident systems. System 2oo3 vis-à-vis the TMR PLC system is depicted in Fig. 7.47A and B. The MPs use I/O data in the memory for the voting process. The features and advantages of TMR are [128]: High safety integrity, high availability, comprehensive diagnostics, low maintenance cost, and online repair. Also, programming is not very complex.

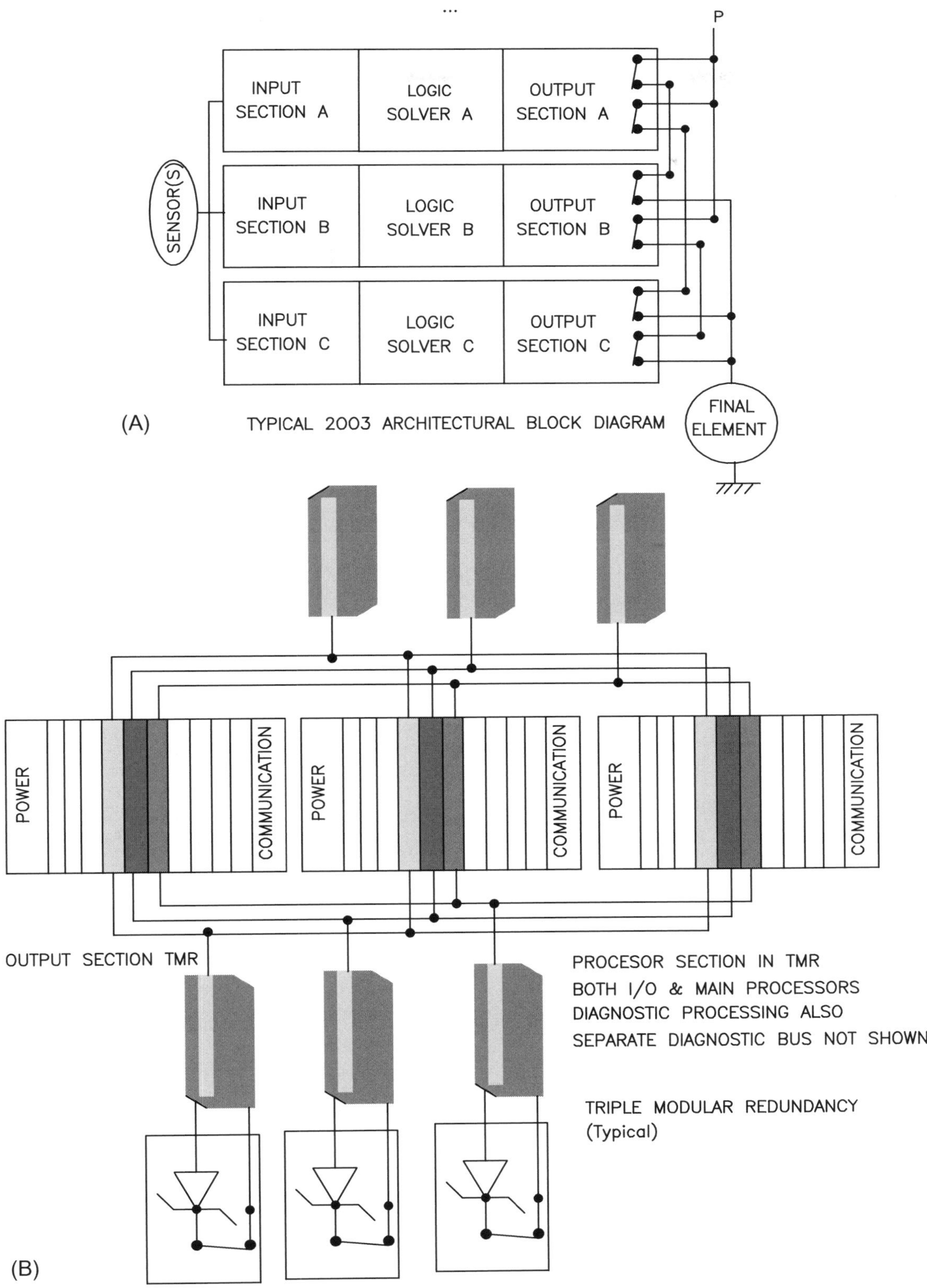

FIG. 7.47 S 2003 and TMR configuration.

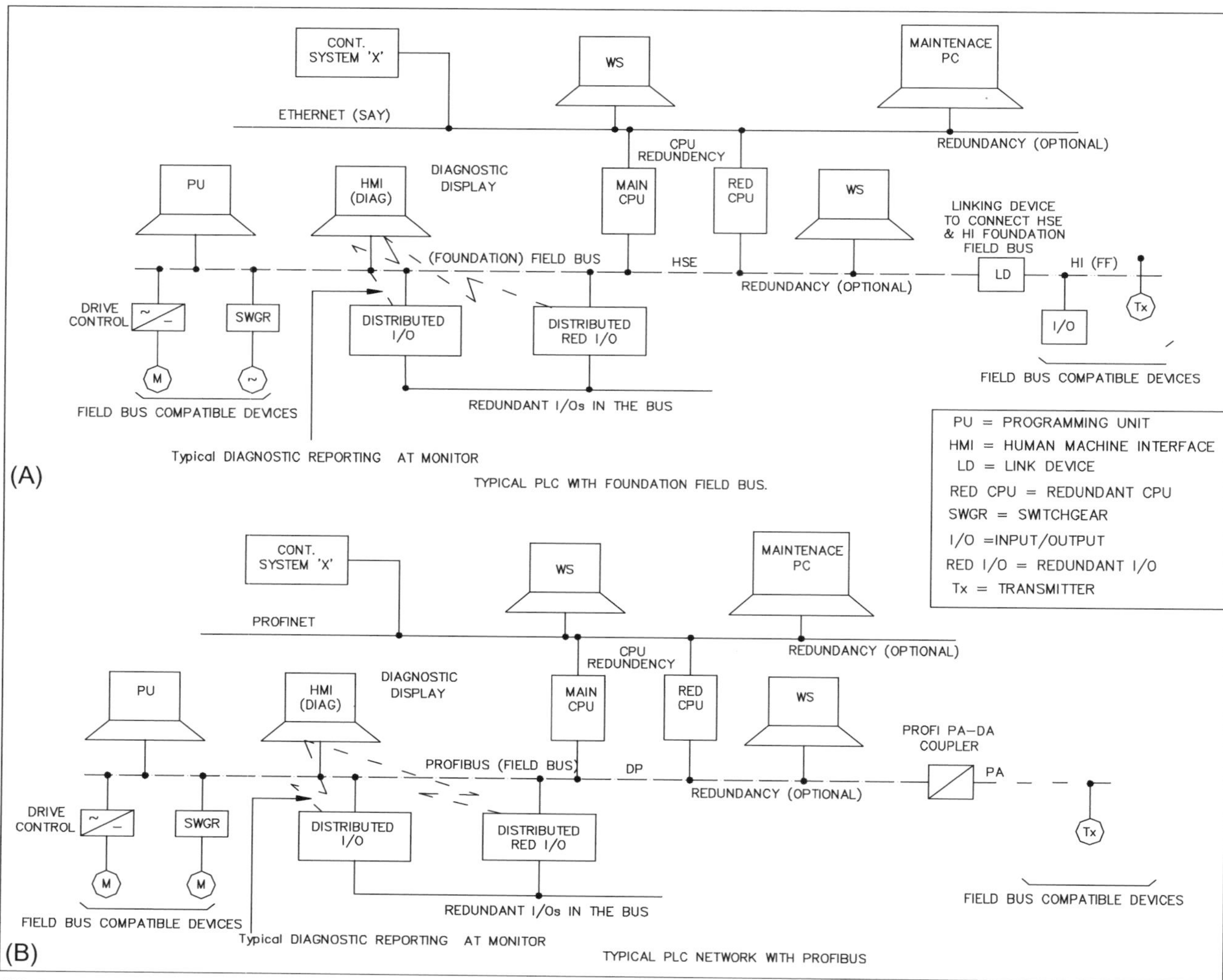

FIG. 7.48 PLC with fieldbus systems.

3. System Integration: In older times, PLCs were connected with systems with the help of interfaces such as the RS 422 to enable the main control systems to poll certain data/status from the PLC. Today, PLCs (better to call them intelligent systems) have become so open that parameters can be monitored from a location thousands of kilometers away. Also, now system integration is not restricted to a few vendors but almost all products can be integrated to the system. This has led to a huge cost reduction and freedom of choice for the products. This feature increase the lifecycle of the system as any off-the-shelf products can now be integrated to the system and the system can survive. Integration of a field device to the controller demands time and can be achieved with the help of fieldbuses such as PROFIBUS and the FOUNDATION FIELDBUS as well as supportive systems such as TCP/IP and OPC–OPC UA/DA. Almost all available intelligent systems are open to support these.

In Fig. 7.48A and B, such features are shown. Let us now look into the integration aspect of network.

- **FOUNDATION FIELDBUS INTEGRATION**: Fig. 7.48A shows systems integration by the Foundation fieldbus **HSE FF,** in which distributed I/Os, electrical drives, and switch gear/MCC have been integrated (with the controllers (CPUs) residing on the same **HSE FF**). Several other control systems (designated as "X"), PC-based systems, workstations, and MIS workstations can be connected to the PLC/intelligent controllers through a high-speed Ethernet. At the lower field level, field devices such as transmitters, sensors, other I/Os, etc., have also been integrated via H1FF. These two fieldbuses are connected by a linking device (LD).
- **PROFIBUS INTEGRATION**: Fig. 7.48B shows system integration by the Profibus **PROFI DP,** in which distributed I/OS, electrical drives, and switch gears/

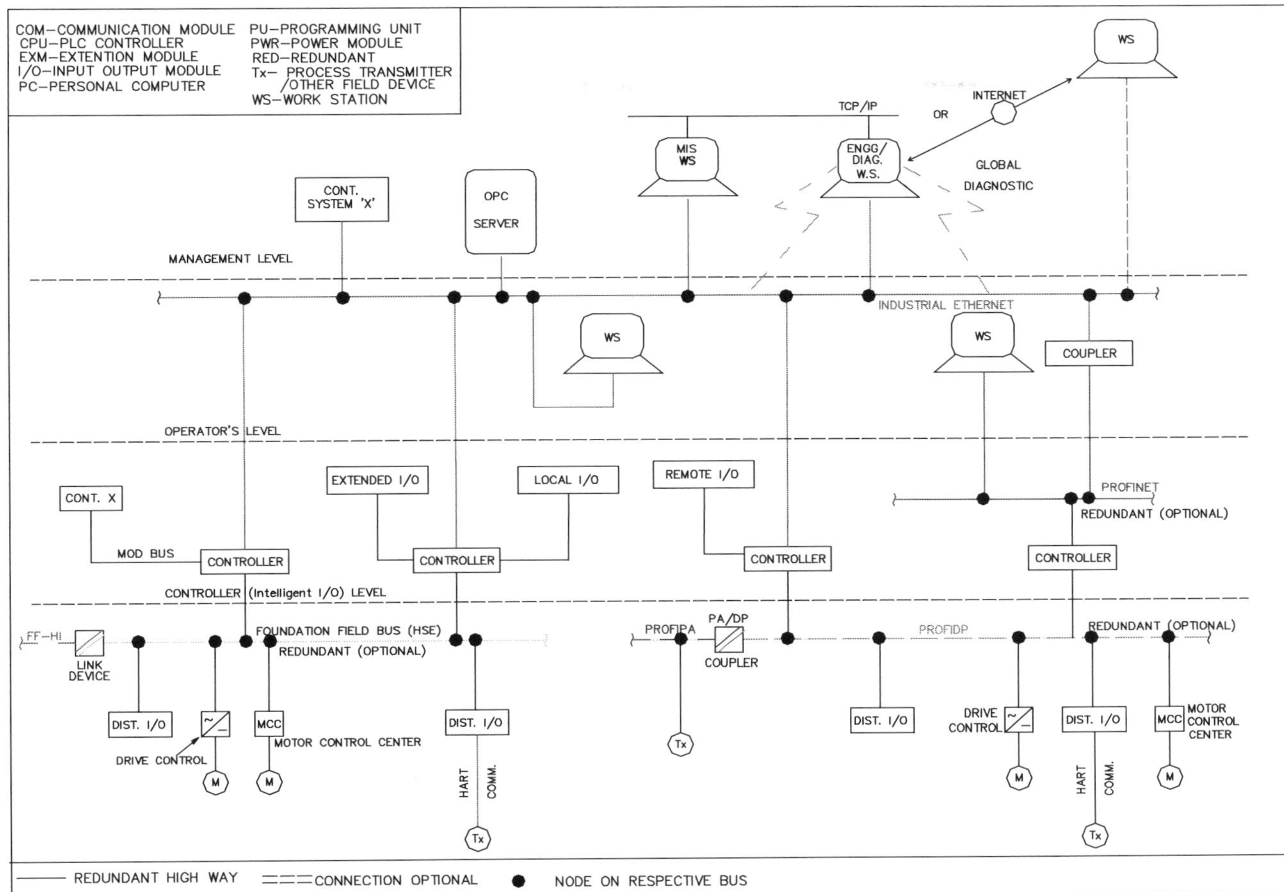

FIG. 7.49 Integrated plant control system.

MCCs have been integrated with controllers on the same **PROFI DP**. Several other control systems (designated as "X"), PC-based systems, workstations and MIS workstations can be connected to the PLC controllers through the PROFINET/industrial Ethernet. At the field level, field devices such as transmitters, sensors, other I/Os, etc., have also been integrated with the system via **PROFI PA,** which is connected to the **PROFI DP via** the **PA-DP** bus coupler.

Therefore, a seamless integration of systems is possible and by utilizing suitable object-oriented software, it is possible to exchange information for process monitoring and visualization. Therefore, a data exchange with other control systems and intelligent devices is possible with the help of PROFINET, industrial Ethernet, and message passing interface (MPI*). So, with integration, it is possible to develop a large communication network over a common bus to curtail additional costs for communication processors. In such connections, parameterization and optimization of field devices are possible. Also, it is possible to integrate industrial IT, which enables the systems to utilize the benefit of comprehensive diagnostics and troubleshooting from remote places.

A more comprehensive integrated system is shown in Fig. 7.49 with each subsystem having a different communication standard. This is typical system architecture to show how the systems can be integrated. The entire system architecture has been divided into four levels: the m**anagement** level, the o**perator** level, the **controller** level, and the **field** level. These level divisions shown here are somewhat arbitrary but functional division for design. Though **wireless WAN** has not been shown here, it can also be integrated.

Field level: Field devices with the Foundation fieldbus H1 have been integrated to the FF HSE by a link device, whereas the field devices that use PROFI PA have been integrated to the Fieldbus PROFI DP by the PA-DP coupler. Various I/O subsystems accepting HART have been utilized to put field devices with HART into the system. For integration of HART devices, a HART OPC server may be developed for the system (for example, 800xA of ABB). So at the field level, a few fieldbus systems have been well integrated at the controller level via Ethernet. It is worth noting that at this level, it is also possible to have wireless systems (such as wireless HART) for system integration.

Controller Level: In the controller level, there are separate controllers. These controllers may be two different scalable controllers or redundant controllers. At the controllers, the I/Os can be connected in a local/extended or as a remote I/O, as shown. Even these controllers may accept data/information by a link and MODBUS interface. There are separate controllers for two different fieldbus systems, for example, PROFI DP and FF HSE. These controllers in the fieldbus also connect/integrate the system to the Profinet or industrial Ethernet for operator/management interface. The Profinet and industrial Ethernet can be connected by a coupler or a switch.

Operator Level: At the operator level, there are various workstations that are connected to the Profinet and industrial Ethernet. There will be a seamless flow of information to the operator through the Profinet and industrial Ethernet (Fig. 7.50).

Management Level: Connected to the industrial Ethernet, there will be an MIS WS and an engineering/diagnostic WS. One server for the system has been connected. The server may include an OPC server for various third-party systems for seamless transfer of data and system management. Third-party systems have been represented by Cont. syst. "X". For IT compatibility (IEC 61131-3 standard), TCP/IP and Internet connections are shown so that the system can be accessed from a distant place. Also at many places, remote-control handheld units are used to adjust various settings with the help of Zigbee/wireless communications. Wireless HART, wireless fieldbus, and **IoT** can also be integrated in modern systems (also for remote operations). One has to be very careful about cyberattacks.

4. Diagnostics and Security: According to ISA, "Digital attackers are targeting organizations in the energy sector like never before" [ISA May 2018]. Naturally, with increased cyberattacks, it is very important to address cybersecurity so that the advantages of technological development can be utilized in a better way.

Mid- to large-sized PLCs integrate diagnostic functions to quickly detect and eliminate system faults to increase system availability. Controllers and I/Os should have a comprehensive self-diagnostic to quickly identify the fault. A fault LED in the front panel facilitates quick and easily module isolations. Also, the report can be generated from these diagnostic programs to the central maintenance monitor; a diagnostic alarm in the HMI/MMI (as shown in the figures) is quite common. As discussed earlier, the IoT/IIoT or industrial IT interface enables the system to utilize WEB server facilities for global diagnostics, remote operation, and troubleshooting. In a large network, there is a seamless transfer of data at various levels. However, it is not desirable that everybody has access to all this data as well as access to modify the data, so there are several security levels built into the systems for accessing and modifying data. Some data access is allowed for the plant operator (for example, control bias), but again, some data are only accessible by instrumentation technicians (for example, controller setting) and/or control strategy setting by Control engineer. Data access levels are arranged in a hierarchical pattern, as shown in Table 7.22.

2.1.4 Hardware and Configuration Considerations

Depending on the application in a power plant, there will be wide variations in I/O as well as controllers for PLC/intelligent systems. A coal-handing plant mainly handles digital I/Os in sequential operations in various subgroups, whereas

MPI: It Stands for Message Passing Interface, a standard for writing message passing program. This is an interface attempt for practical, portable, efficient and flexible standard for message passing. It is done on C++ /FORTAN language for shared and distributed memory. This is useful for global diagnostic feature also.

FIG. 7.50 Note on MPI.

TABLE 7.22 Security Levels

Security Level	Accessibility
Zero protection	Allow full access without password
Write protection	Read only access to the HMI/CPU or network communication W/O password. No access to CPU for writing and mode changing
Read/write protection	Same as above plus even reading data in the CPU is password protected

in DM plant, there will be a good mix of digital and analog I/Os. Sequential operation is also there, but something is different. The PLC in the BMS requires meeting the high safety requirements of the NFPA. Though the NPFA specifically does not call for SIL certification, in large plants people would go for a failsafe condition. Again, the PLC in the ATT/ATRS/BMS may have fewer I/Os when compared with the PLC for a total ash-handling plant, but they are more critical from safety and security point of view. So, they carry out the HAZOP (/FMEA) study to fix and recommend SIL-certified modules for major critical controls such as BMS, ATRS, protection systems, etc. The size and type of PLC applications in each of the above cases will be quite different. In this connection, it is worth noting that in the present era, it is necessary for all plants to have a safety lifecycle study in line with IEC 51611 (Chapter 14). In the market, most of the big manufacturers offer a wide range of products for the designer to choose from. A few features of various subsystems of PLC are:

1. I/O SUBSYSTEMS: The major considerations for an I/O subsystem shall include:
   - *Category*: Direct to the controller (local/extended or remote I/O), distributed I/O over a bus such as in the FF Bus, Profibus, control net, device net, and HART.
   - *I/O*: Digital I/O (signal level, output type, potential free, TTL, AC), analog I/O (types: 4–20 mADC, T/C, RTD, O/P: driving load impedance), pulse (voltage/frequency), registering I/O (thumbwheel), serial/parallel module, information transfer (for example, HART/FF/Profi). For analog inputs, whether single or differential.
   - *Power supply (AC/DC) and interrogation voltage*: To specify the voltage level, the quality of the power supply, and the interrogation voltage level if other than the power supply.
   - *I/O density*—Number of channels in each module. Normally, I/Os are available in 8/16/32 I/Os in one module. There are separate input and output modules, but mixed I/Os are not uncommon. Better to restrict to a maximum of 16 numbers in one module.
   - *System isolation*: Optical/galvanic isolation for I/Os.
   - *Connection medium*: Such as cable types, FO, wireless, etc.
   - *Module mounting and connectivity*—DIN mounting, cabling plug/FO termination, and adapter module bus isolation wherever applicable.
   - *Intelligence and special I/O*: Whether I/Os are intelligent, special modules for the encoder, high-speed counter, linear/servo positioning, motion control, etc.
   - *Event time stamping*: Critical I/Os with time-stamping facility.
   - *Module feature for safety*: Standard or failsafe module (as applicable).
   - *Diagnostic features*: Details of diagnostics built in, for example, watchdog timer, etc.
   - *General fault detections:* Such as wire breaking, short circuit, noncoincidence, etc.
   - *Remote I/O link*: A link to the controller by a serial link or a network, for example, Ethernet. In case of link type and associated protocol (for example, RS422/485 MODBUS protocol).
   - *Accuracy*: Depends mainly on ADC resolution and high accuracy is available.
   - *Flexibility:* I/Os shall be able to be selected and configured in flexible ways—in single redundant mode and/or via network and as a local/remote.
   - *Hot swap/standby*: Possibility to change/swap standby I/O modules online.
   - *Special features*: Whether IS type, single, dual, or TMR fault-tolerance type.
   - *Certification*: International certification for safe module (for example, Ex IS, if applicable).
2. CPU/CONTROL UNIT: Discussions on CPU requirements here are to supplement the short discussions on the CPU earlier. One needs to bear in mind that by selecting the CPU type, one basically selects the PLC type of the manufacturer, hence this helps in comparing PLC types. The data presented below are from **standard reputed manufacturers** and the data may not match with any particular system and standard. Also, all data may not be applicable simultaneously for any particular system.
   - *LED*: Indications (different colors) for run, fault, power, battery, data exchange etc.
   - *Clock frequency*: System-dependent available from 48 MHz to 133 GHz.
   - *Redundancy support:* Normally shall be available for CPU fallback.
   - *Memory:* RAM available from 2 to 64 MB.
   - *Cycle time:* <1 ms (Typical word processing time: 0.01–0.12 μS, whereas floating point processing time is higher at ~1 μS).
   - *Number of timers and counters:* 256–2048.
   - *Number of I/Os:* 1024 to 8192 bytes, sometimes limited by *the number* of modules in *the* I/O bus, say 96 modules. In case of redundancy, this will be limited further.
   - *Supportive communication interface/modules:* PROFIBUS DP (PA via coupler), FF BUS-HSE (H1 via a linking device), serial link like RS 232, RS 485, RS 422—if applicable. Apart from these, the manufacturer's own system/family dedicated bus for drive MCC etc. Fieldbuses discussed above may be supported as MASTER or MASTER-SLAVE. The number of channels for each Fieldbus or serial device may be single and/or dual along with cable redundancy, which could be a factor while comparing PLCs.

- *Open user communications, if any:* TCP/IP, ISO on TCP, and/or interface with standard systems such as OPC UA/DA support is also a key factor.
- *No application per controller/number of program per application:* Typically, 8/64.
- *Number of OPC servers per controller/number of clients per OPC server (if applicable):* one or two with a maximum of four. The frequency of OPC server updates can also be compared.
- *Number of Ethernet Channels (if applicable):* Two in the IEEE 802.3 interface in a twisted cable (STP) with an RJ 45 connector for a 10 mbps communication minimum (only for idea).
- *Network redundancy switching time*: Better than 1–2 s.
- *Certification:* International certification as applicable.
- *IP Class:* IP 20 minimum outdoor application if any may have higher IP.
- *Power supply/power consumption:* Normally 24 VDC, but depends on manufacturer (dissipation important for cooling arrangements).
- *Battery back type/internal back period:* Lithium etc. > 100 h for larger systems.
- *Environmental:* Temperature: (-20) to 50°C (storage higher range). Relative Humidity: < 95%.

3. MEMORY WITH A MANAGEMENT SCHEME: Memory management is important for functions such as user programs, maintaining system data, diagnostic buffers, clock timing, etc. It does not include BULK memory for a PLC-based automation system. The functional part of the basic memory management scheme presented may vary with different PLCs.

Storage space for the CPU for a user program, data, and configuration including system memory, diagnostic buffer, etc., can be divided as:

- LOAD MEMORY: This nonvolatile memory is either a memory card (**external**–battery backed-up **RAM** or **flash memory**) or integrated with the CPU (**internal** battery-backed **RAM**). External memory could be in the range of up to 64 MB. When a project is loaded, it is stored in the load memory first. It stores all blocks with a description and file format definition in the data block. The executive parts of the block will be less than the RAM size (Fig. 7.51).
- MAIN MEMORY: The main memory is integrated with the CPU. Main memory also has two parts. One is for the p**rogram code** or run relevant code; process input-output images are stored in this part for the code. The other part is for run time relevant **data** and data of the local data stack. By having separation in the main memory, it is possible to access both simultaneously (for example, S7-400 PLC). So, the main memory is used to store process-related modules, process images, local data etc. The main memory has battery-backed RAM, especially for the data part.
- WORKING MEMORY: This is volatile RAM, used to store some user elements by copying the same from the load memory for better performance.
- RETENTIVE MEMORY: This is nonvolatile RAM used to put some limited quantity of work memory values. In case of power failure, the CPU is designed to have hold-up time to store that data. It is used for markers, times counters, data blocks, bit memory etc.

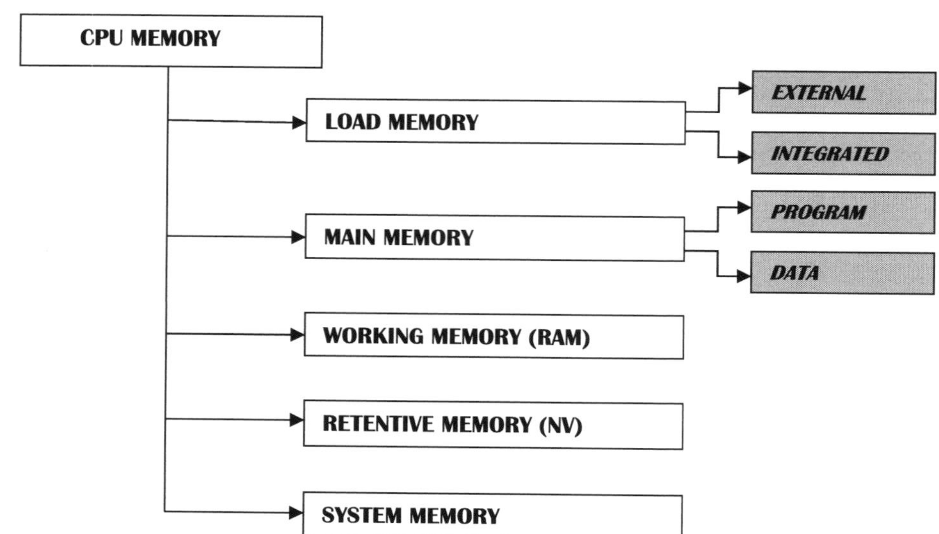

FIG. 7.51 Basic memory divisions (inspired by Siemens S7 PLC).

- SYSTEM MEMORY: System memory provides storage area for the marker (M), the timer (T), the counter (Z), and the local data stack. The CPU also supports a time of day clock and a diagnostic buffer. The entries are displayed in chronological order.

The CPU can understand whether the power is ON (after power down) and whether it is buffered or unbuffered. In the former case, the programmed section in the *main memory* goes to the working memory. Otherwise, for non-buffered ON, the executable parts are to be transferred from the *load memory*. The CPU assigns specialized memory areas and they have suitable code blocks suitably addressed, for example, for input (I), output (Q), bit memory (M), local stack (L), and data block (DB). Memory size and performance are very important (***The firmware is a basic program that PLC hardware can execute directly. It reads the program, written in a higher-level language, then executes the same. It includes OS, a real-time clock, and communication*).**

Memory Performance: Physical memory size minus the executing firmware gives the **memory size**. During start up, the amount of free memory in the controller is less on account of the empty project space. The rest of the memory can be utilized by the application program. To get an idea, a load-balancing memory diagram (***Courtesy AC 800M* systems of *ABB***) is shown in Fig. 7.52.

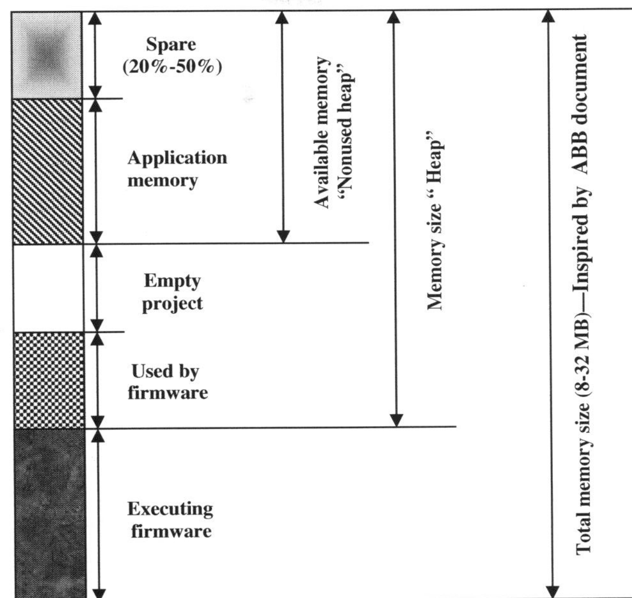

FIG. 7.52 Physical memory divisions in terms of use.

CPU Load% = (Total execution time/total time interval) × 100. Depending on the applications, suitable measures could be taken to keep this MAX 70% (typical 50%). This is important for the programs where communication is vital for the controller, as in a large distributed system. Timings for various functions have been indicated in Table 7.23.

The above data from a standard manufacturer have been presented to get an idea about timing.

4. EMBEDDED SYSTEM: An embedded system is simply a computer to provide a fixed specific task (for example, a mobile phone). The PLC features discussed above can be achieved by an embedded system meant for a specific purpose. Now, manufacturers come out with such solutions because system components residing within the control system make it possible to have greater flexibility and functionality (for example, ABB 800xA). A few features of embedded systems for intelligent control applications are as follows:

- Distributed intelligence and computing power: This means that control logic can be distributed at the place where it is needed most, and control logic can be distributed between several controllers exchanging measured and calculated values.
- Time-critical, real-time execution.
- Communication module: In the communication module, the protocol stack can be implemented in the hardware and the firmware to run in the associated embedded CPU.
- In the client server system, the software can be combined for operational flexibility.
- Controller level: Because the embedded system can withstand harsher environments, it is better suited at the controller, which may have to face harsh environments in the field.

TABLE 7.23 Some Typical Timings for Various Functions (for Idea Only)

Function	T (μS)	Function	T (μS)	Function	T (μS)
Single loop	1100	Override loop	3600	Dint function	0.31
Cascade loop	1900	Boolean function	01.5	Real to dint	1.13
F′ forward loop	1600	String function	13.45	Dint to real	1.17
Mid-range loop	1500	Real function	1.79	Real to time function	10.1

- In embedded systems, a number of components are reduced. Embedded systems can handle control logics and I/O scanning, so it is possible to modify the online logic change.
- The I/O system gives maximum flexibility in the embedded system. They also offer preprocessing such as I/O filtering, time stamping, and diagnostic features.
- System availability is very high and online replacement of modules is possible.
- The development charges of the embedded system are high, but once developed, the cost is reduced drastically and the system integration is smoother and easier.

Microcontrollers, field-programmable gate arrays (FPGA) along with associated software supports make it possible to have embedded electronics to easily carry out many high-end tasks and integrate various systems within a control system, each dedicated for a specific function. It is worth noting that there are differences between the microprocessor and microcontrollers (on a chip complete computer system). These are discussed at length in Appendix V of reference [**128**]. *FPGAs are specialized chips that are programmed to perform very specific functions in hardware. When a circuit board is manufactured containing an FPGA as part of the circuit, then the program during the manufacturing process can later be re-programmed to reflect any changes. Logics in FPGA are designed to be directly used by system designers, so FPGAs are not always optimized. FPGA architecture and layouts are very important for its performance. In this connection,* the *role of* hardware description language (HDL) is extremely important. *HDL is essential for electronic system/device design and development, especially in embedded controls. Interested readers may refer to Appendix V of reference [128].*

5. COMMUNICATION SYSTEMS AND INTERFACE DEVICES: Maximum data transparencies across various levels of automation are serious challenges. It is the responsibility of communication to ensure a seamless transfer of data at all network levels (including long-distance clients with a network via the Internet) as discussed earlier. The basic functions of communication systems can be classified as follows:
   - Adapt international standards so as to integrate other systems in the future.
   - Flexibility and Scalability: A combination of various communication options without affecting the system performance, including safety, security, and diagnostics.
   - Wireless communication as necessary, integrating routing functions for wide access.
   - Integration of office applications as applicable, and integrating the system through the Internet.

There are several methods for system interface and associated communication, for example, protocol support for HART devices, fieldbus, MODBUS (RTU), and OPCs (OPC UA/DA) for various device support, IoT, etc.

- OWN DATA HIGHWAY BUS (Fig. 7.46): All PLC manufacturers have their own system highway/module bus where various modules can be connected to communicate among them. A data highway plus (of Rockwell automation) a local network is an example of the same. The communication baud rate and way of transmission may be unique.
- REMOTE I/O (Fig. 7.46A): The power of a remote I/O hinges on the number of devices it can support via remote I/O. In its simplest form, a remote I/O helps in cutting wiring costs by connecting them to the system via serial link. These can also be directly hooked to the system bus—for example, Mitsubishi FX_{2N}-32DP-IF is a module to connect up to 256 I/Os (with 240 VAC supply and 24 V service voltage) to the Profibus with electrical isolation. A remote I/O can form a network with the help of the associated controller, as shown in Fig. 7.53. Typical speed of communication is 9.6–12,000 kbps. The bit rate depends on the device and network used as well as the distance. These remote I/Os can be placed in a normal chassis and an associated controller need to support remote I/O when used as link may also be referred to.
- SERIAL LINK (Fig. 7.54): A general configurable serial link shall include RS 232, RS 422(A), RS 423, RS 485, etc., which may be used to transfer data to various devices, for example, a printer and other devices in the ASCII or MODBUS/SNP protocol. Also, these may be used for connecting various control devices such as a single-loop controller. (A DM plant with a majority

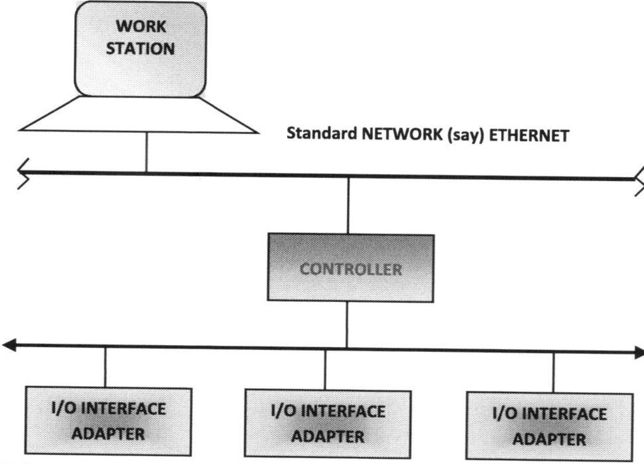

FIG. 7.53 Remote I/O link with associated network.

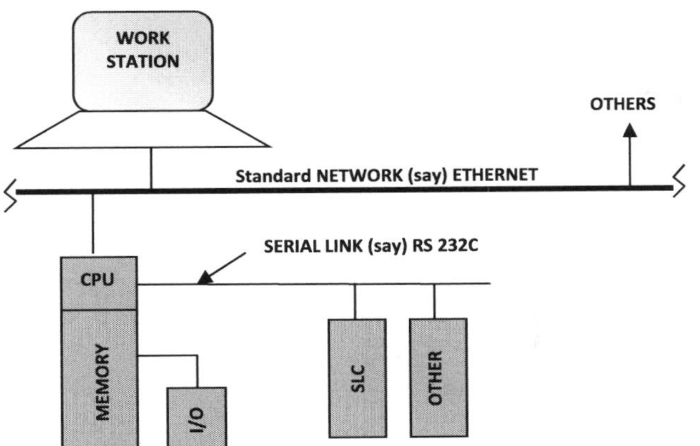

FIG. 7.54 Serial link and network connection.

of digital I/Os may have one SLC for degasser level control and the same SLC can be connected to a mid-size PLC via a serial link.) Some of the typical distance limits for these links are 15 M for RS 232 and 500 M for RS 485. A shielded cable with a suitable pin (D pin) may be utilized to communicate at around 9.6–19.2 kbps speed. The number of such ports varies from system to system.

- FOUNDATION FIELDBUS (Fig. 7.48A): As discussed earlier, many systems support full Foundation fieldbus devices for the PLC system. While the H1 bus is a slow bus supporting field devices @ 31.25 kbps, the HSE bus is a high-speed bus to support communication @ 100 mbps. The C1860 module of ABB (800xA) is an example of the same.
- PROFIBUS DP (Fig. 7.48B): As discussed earlier, PLC systems offer separate communication modules for Profibus DP master and slave. The master interface card distance is dependent on the baud rate. It could be as low as 9.6 kbps to 12 mbps. It allows comprehensive data processing as per the Profibus DPV1 standard. The number of slave units supported varies from system to system, and can be in the range of 64 to 128. Transmission is over bus network with a Profibus cable with a D connector. Message size vary between 244 bytes and 32 bytes per slave.
- PROFINET (Fig. 7.48B): There are CPUs with integral PROFINET interfaces (for example, Siemens S7-300/400). This is an equivalent of the industrial Ethernet with some features such as media redundancy protocol (MRP) to make a quick changeover (<200 mS). Like Profibus, it supports GSD files. Isochronous mode (synchronization of signal acquisition output by distributed output devices) of communication is helpful for closed-control loops.
- ETHERNET (Figs. 7.48 and 7.49): An Ethernet interface with 10BASE-T (10 mbps) is common to all standard manufacturers. It supports TCP/IP protocol (Ethernet IP) and SRTP MODBUS TCP (for example, GE Module IC693CMM321). In certain cases, a communication speed of 100 mbps (100base Tx) is also possible. CAT5 STP is a standard cable connection. In some systems such as the Rockwell PLC 5, an embedded web server provides WEB access to PLC diagnostics due to support of the **domain name server (DNS)** and the **simple network management protocol (SNMP)**.
- DEVICENET (Fig. 7.55): A communication protocol used for intelligent control devices to share information. This is used to connect low-level devices to the controller with diagnostic facilities and less start-up time. DeviceNet also has master/slave interface modules. These interface modules are used for read/write, configuration download, and device status monitoring. They have distance limitations, for example, for speed, 500 KPS is for 100 m whereas for 500 m it reduces to a quarter of the max speed. The master can support nearly 32–64 slave nodes. The message size is in the rage of 3 Kb in the master to 256 bytes in the slave.

CONTROLNET (Fig. 7.56): This open industrial protocol is used by a few PLCs, such as Rockwell's PLC 5. It provides high-speed access to the plant floor data, as it supports a number of controllers having their own I/O. Redundancy

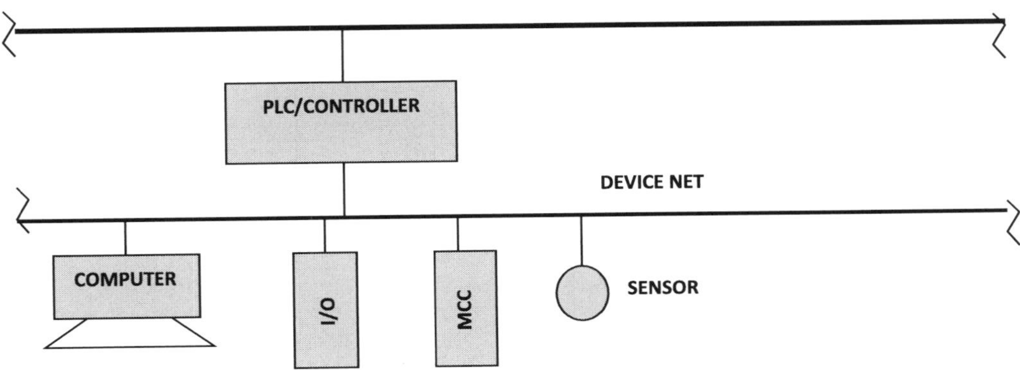

FIG. 7.55 Device net connection.

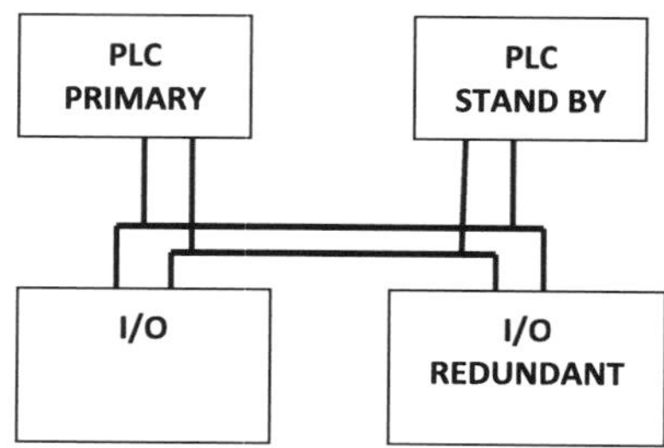

FIG. 7.56 Redundant control net.

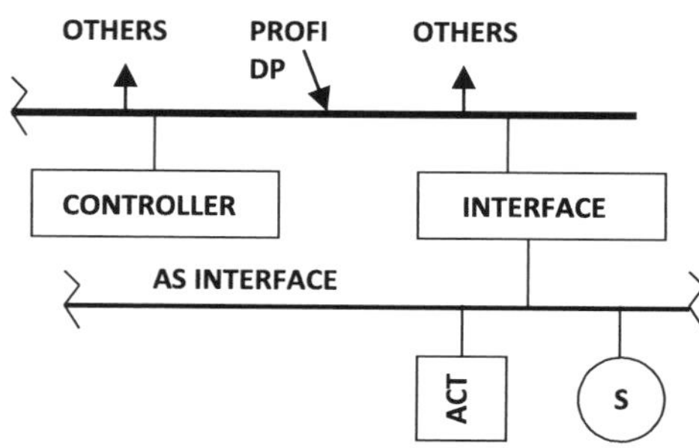

FIG. 7.57 As interface.

in the controller is possible with one as the primary. This is a deterministic system with a speed of 5 mbps connected by coaxial cable (RG6).

- AS INTERFACE (Fig. 7.57): This **a**ctuator **s**ensor interface is a protocol to connect field devices to the controllers. It defines the physical layer and the basic protocol, and it finds much use in a PLC. A communication speed of ~167 kbps for a distance around 100 M is possible. Many manufacturers (for example, Siemens SIMATIC- S, MitsubishiRX$_{2N}$-32ASI-M) utilize this interface to connect the field devices to any system bus via a PLC controller.

In addition to these, there are a few other networking systems such as (**C**ontroller **A**rea **N**etwork) CANopen (for example, Mitsubishi). Also now for remote operations, diagnostics people take the help of IoT or IIoT discussed earlier.

6. Man-Machine Interface (MMI): PC-based MMIs are quite common for PLCs. These can be in the client server also. This can be one simple standard/industrial PC with the latest operating system. MMI is explained later in this chapter.

2.1.5 Software Considerations

Various PLC suppliers offer software packages to reduce programming and set-up time (to a great extent) and to integrate the systems. These packages are also helpful for communication for data exchange and quick and direct data access. Programming is better done in a graphical environment in easy-to-use templates. Software should normally be compliant with IEC 61131-3 to meet the requirements of base level, conformity, and reusable definitions. PLC software development on Windows OS is quite common, for example, WINDOW CE v. 5.0 and WINCC for the Siemens PLC. These are only a few examples. Because there is faster progress in this area, it will be wise to select the latest but proven version prevalent at the time of the project (of course within the allocated budget, which should not be too constraining). Based on the main components and functional features expected of the software in a PLC (based on a standard PLC), the following are a few points that may be considered as some of the major issues:

- System Manager: Administrating all tools–Creation of a multifunctional communication subsystem, online help, project data storage on the CPU.
- Structured features of software programming.
- Configuration and parameterization of hardware, including limited configurations in run and Internet linking. Now, in some software during normal operation, it is possible to tell which ports are operating based on the IEEE802.1AB link layer discovery protocol (LLDP). Due consideration should be given for system integration with various systems such as fieldbus-driven controllers, accommodating various protocols and OPCs both wired and wireless versions, also integrating IoT devices as applicable.
- Program editing: It is an interface of the user program in a ladder diagram (may be higher languages such as FORTAN and PASCAL), statement list, or structured control language (SCL) as defined in IEC 61131-3 for complex algorithms. Graphical representations of the control logic are useful.
- Setting up data transfer of the bus and open command interface.
- Integrated system diagnostics: Reporting error, summary, and detailed diagnostics.
- Standard compliance.
- Special functions such as PID, etc.
- Creation of a program for fault tolerance and failsafe mode as applicable.
- Tool calling interface.
- Same basic set up for every engineering workplace.

1. STRUCTURED PROGRAMMING: Sometimes it is better to divide the complex and comprehensive program into smaller divisions that are self-content blocks. It is seen that these program blocks have definite

correlations and appear a number of times in the program. So any program can be implemented with this concept, and a third party accessing the program for any purpose can understand the system clearly. It allows several programs to run concurrently. As seen in Siemens Step 7 programming, some of these blocks are:

- Organization Block (OB): It is an interface between the operating system and the user program that is responsible for system start up of the job, interrupt handling, fault handling, etc.
- Function Block (FB): This is an actual program, and each has one assigned data block (DB) or it needs to be supplied with the current input values. FB displays can be suppressed for privacy.
- Data Block: It is the storage place for plant data.
- System function blocks: These are common functional blocks, and may be required repeatedly in the program, for example, clock functions, counter working hours, etc.

2. Program Editor: As the name implies, it can be used to edit programs in ladder diagram, statement list, and functional block diagram. At times, it is easier to understand and edit in graphical form for sequencing structure, selection of processing steps, synchronizing manual and auto operation, and diagnostic functions. Some useful commands in the structured control language are: Template block, template commands such as IF, THEN, ELSE, etc. Subroutines with GO TO commands are often useful.
3. Control and network Configuration: It is a part of the system software to implement the control strategies such as the PID loop, etc. The control algorithms are parameterized with some programming components and structuring data. Complex algorithms need to be resolved with the help of standard library functions and control structures and parameters. Also, establishing communications is a job for the system software. Graphical representation of individual station links helps to resolve the communication problem in a much easier way.
4. Diagnostics: It is also a prime function of system software to identify and navigate from the summary diagnostic display down to the component level diagnostics. Reporting of errors is a function of the system software.
5. Special Tools and functions: There are some special tools to meet the requirements such as high availability, fault tolerance, and powerful security and failsafe systems that may not necessarily deploy redundant systems. There are special tools for system integration and open system interface, for example, the tool calling interface.
6. Engineering and program Development: While selecting/evaluating a system, necessary emphasis may be placed on the following features for engineering and program development to get better performance from the system:
   - *User Friendliness*: It must be user friendly so as to optimize the program development work.
   - *Uniformity in Representation*: There shall be a common software framework on which all the shared functions can be performed. They shall also have common online representation in MMI so that they can be felt uniformly both by the operator as well as by the programmer.
   - Intelligence in Editor: The programming systems having built-in intelligence help to use the drag and drop feature during program development. Intelligence in the programming system helps to split the range screen to simultaneously edit a number of programs as well as data exchange in the system.
   - *Data Transparency*: Data shall be entered into the system only once, and this shall be transparent to all the editors in the questions so as to avoid any future mistakes and confusion. The system may have a centralized data management system in an object-oriented language to ensure better transparencies. A shared database must be consistent to the entire applications, including MMIs. Any modification at one place shall be updated at all relevant places simultaneously.
   - *System Library*: There will a number of program blocks, subblocks, face plates, etc., used many times in the program. These need to be kept in a secured library to use many times in the system. This helps in maintaining the consistency and quick response.
   - *Special Tools and Functions*: To ensure an open interface, safety, and security, some special tools may be built into the program development system.
   - *Program Development Device*: Laptops and PCs may be used, having features like ergonomic design, robustness, capable of eliminating the EMI effect with a USB port, and other connects.
7. Totally Integrated system (Software): A totally integrated system automation conforming to the industrial standard IEC 61131-3 with Internet access is shown in Fig. 7.58 to gain an idea (for example, ABB 800PEC).

2.2 PLC Types

Short discussions on PLC classifications with special reference to selection criteria are presented in this subsection.

2.2.1 Redundancy Criteria

One of the ways PLCs can be classified according to the redundancy at the I/O level and/or at the controller/CPU level.

1. I/O Redundancy: In case of redundancy at the I/O level, there will be two or more sets of inputs or outputs that may be distributed to two different chassis (for fire)/

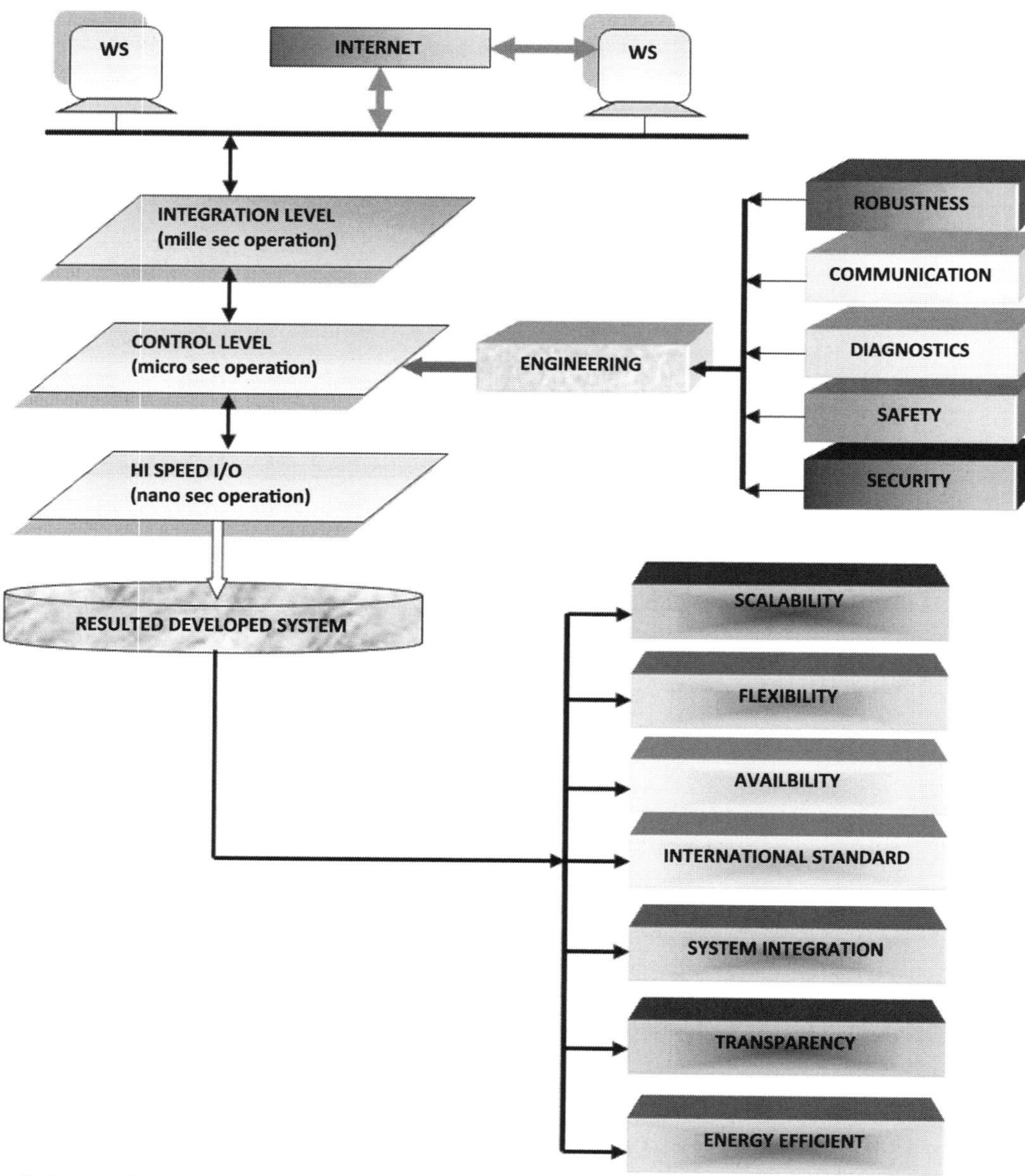

FIG. 7.58 Totally integrated system (total engineering).

modules. This is in case one module fails, the I/Os will be available through the other module with additional cost and cabling. In case of a remote I/O, such a distribution among two chassis may not be possible for the cost and complexity of the connection. How the two I/Os shall be handled in case of mismatch! (say) in one case there is snap of wire! The obvious answer is the PLC needs to monitor the health of each channel and report in case of such an event. Contact interrogation is another issue. For NO types of contacts, the same situation may happen when there is a wire snap as well as if the contact has not been made (that is, changed from NO to NC). Whereas in case of the NC type, that is, the power missing could be used to detect a cable snap, but problems are there also. For the NC type contact, the same situation will prevail when the cable is snapped or really contact changeover has taken place. However, the situation is little better with changeover contacts (COC). Where the contact forms a central terminal changes over to the other two contacts to form an NO or NC contact. In such cases, all three terminals, as shown in Fig. 7.59, are connected to the system and non-coincidence error is monitored (for example, ISKAMATIC B of the KWU-Siemens control system). In cases where there are voting circuits, then the same needs to be suitably taken care of in system

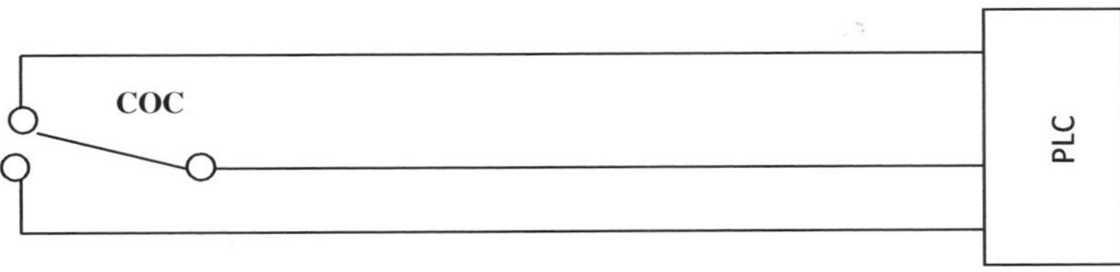

FIG. 7.59 Change over contact (noncoincidence error).

programming. A simple way to vote the two outputs to the actuators is to use a diode auctioneering circuit in the hardware.

2. CPU Redundancy Discussions: There can be redundancy in the CPU/controller, as shown in Fig. 7.46B. For bumpless transfer of control from one to another, Section 2.1.3.2 of this chapter may be referenced.
3. Communication Redundancy Discussions: As shown in Fig. 7.46A and B, the system buses may be redundant in nature to communicate data from various units over a redundant bus. Here, communication over the bus is an important aspect so that there is no mismatch between the two. Also, the physical bus may be routed through two different paths so that in case of damage to one, perhaps due to fire, the other can go on as it is. For two bus systems, the communication controllers shall be redundant, with suitable monitoring for uninterrupted communication. The situation is different for wireless systems.

2.2.2 I/O Connections

Based on the types of I/Os listed below, a PLC can be classified as:

- Directly (local I/O) (Fig. 7.46A) or an extension.
- Remote I/O (connected by a communication link) (Fig. 7.46A).
- Intelligent I/Os connected to the system over a system bus (Fig. 7.46B).
- Intelligent I/Os connected via a fieldbus (Figs. 7.48A and B).

2.2.3 System Integration

As seen earlier, from a small PLC, it is possible to develop a very large system. The capabilities of PLCs toward these are also a criteria classification.

2.2.4 Features

1. **General issues**: The importance and criticality of availability, diagnostics, safety, and security of the BMS/ATRS PLC are beyond question, when the same is compared (say) with the PLC in a raw water pump. It does not mean that the latter is not important, but it is less critical. This is because in case of nonavailability of the BMS/ATRS PLC, the whole plant will trip. Naturally, during selection, importance will be attached to these factors. Therefore, it could be concluded that one has to decide which factors are more important for the application, and thus various types of PLC subsystems with different distinguished features can be developed.
2. **Hardware and Software aspect**: The hardware and software features discussed in Sections 2.1.4–2.1.5 are important to classify the PLC types. Generalized specifications have been presented, and the designer needs to select the features required for the particular application so that the specification is neither overdone nor underspecified.

2.3 PLC (OLCS) IN BMS

2.3.1 Preamble

The focus of a burner management system (BMS) or a furnace safeguard supervision system (**FSSS**) is to ensure safe operation of the furnace and boiler. Naturally, the reliability and availability of BMS is very crucial to the control engineers in the power plant. **FSSS/BMS** is generally designed in accordance with the National Fire Protection Association (NFPA 8502) and other relevant standards. The functional details of BMS/FSSS are discussed in Chapter 8. The BMS/FSSS can be divided into subgroups such as the boiler purge igniter valve management, the main fuel management MFT, and the draft controls (if in BMS).

Implosion and explosion are two important factors for designing a BMS/FSSS. According to NFPA 8502, a furnace explosion can be caused due to the ignition of accumulated combustible mixtures within confined spaces of the furnace or associated boiler passes, ducts, and fans that convey gases to the stack. The degree of explosion is dependent on the proportion of air present at the time of ignition and the relative quantity of accumulative combustibles. The major causes for such an explosion could be (NFPA 8502 guideline) momentary loss of flame followed by delayed ignition, fuel leakage in an idle furnace ignited by any source, loss of flame of a burner while others are working, and complete flameout followed by an attempt to light up.

Similarly in line with NFPA 8502, implosion (very important, especially for large boilers) can be caused by any of the following causes:

o. Malfunction of equipment regulating the boiler gas flow, resulting in furnace exposure to excessive induced draft fan head capability.

o. Rapid decay in furnace temperature and pressure due to MFT.

Some key factors related to BMS *and* FSSS are given below:

Burner Management Part	Safeguard and Supervisory Part
• Burner oil and ignition process start/stop and sequence control of the oil burner	• Purging system
• Pulverized coal combustion control start/stop and sequence control of the coal burner	• Compliance with NFPA 8502 (NFPA85/86), etc.
• Pulverizer control such as preferential tripping of mills	• Emergency tripping MFT
• PA fan control	• Fuel leak test (low pressure)
	• Control of ID/FD fan as applicable

Generally, the basic design aspects of FSSS/BMS need to take care of the following criteria:

1. Inhibit start up in unsafe conditions (for example, permissive condition such as no MFT).
2. Protection against unsafe conditions (for example, trip burner at low oil pressure).
3. Provide operator with full status message (for example, status lamps).
4. Initiating safe condition for operation and shutdown interlock (for example, high/high or low/low furnace pressure or low/low feed flow* for the master fuel trip).
5. Reliable operation and avoiding spurious trips by a proper on/off sequence.
6. State-dependent interlocks.
7. Link to set point control.
8. Split of safety and control functions (as stated above).

*Also low/low drum level in case of drum boiler.

It is one of the duties of the BMS/FSSS to determine when to move from one state to another and what shall be the associated valve/damper position in each condition. The condition-based safety and sequential-based logic requirements are specified needs to be implemented in the system.

2.3.1.1 Some Features of BMS/FSSS

There are certain features generally considered essential for the BMS/FSSS, and of them, some are functional and some are related to failsafe conditions:

1. All major and large systems need to be failsafe, that is, any single fault in the sensor logic solver or actuator would not cause a loss of safety function. This is a similar requirement of safety instrumentation systems (SIS), to be discussed later. Therefore, people are also looking for SIL certifications for BMS/FSSS in many cases.
2. Major standards followed for BMS/FSSS in utility systems shall include:

NFPA 85 and NFPA 86.
EN 50156-1.
ISA S84.
FM7605—IEC 61508(/IEC61511).
VDE 0160/VDE 0116.
APP 556 AND API 14C (PETROLEUM).

3. A BMS/FSSS system should have self-monitoring features, and all these faults shall be brought to the notice of the operator in the right time so that suitable measures can be taken.
4. In many cases, especially in large utility stations, a major logic such as a master fuel relay (MFR) is implemented in the TMR logic, discussed in Section 1.1.11.
5. Functional criteria (see Chapter 8) shall include but are not limited to the following:
   - Any fuel firing can only be started after a purge sequence is complete, that is, following any trip/shutdown there shall be a purge sequence before any start to make sure that there is no stagnant sunburnt fuel after the burner trip has occurred.
   - Any start up can only be initiated after all required permissive conditions are fulfilled.
   - Continuous monitoring of the burner flame and trip in case of any undesired condition.
   - Continuous monitoring of boiler operation and trip MFR in case of any adverse situation.
   - Reliable MFT but spurious tripping shall be avoided (if necessary, defeating switches are used). Also, the use of an electromechanical MFT relay.
   - Continuous monitoring of start up, run, and shutdown logic sequence and initiate an alarm in case any fault is detected.
6. Equipment status and process alarm feedback in a clearly written message without ambiguity.
7. First out feature, SOE features, etc., for identification of the cause of a fault that is responsible for the failure. Also, identification of the fault up to the module level is possible.
8. System design shall be such that the system is modular in architecture.
9. It is desirable that the logic may be kept in EEPROM to combat a power failure situation.

10. Suitable redundant watchdog timer to detect system failure.
11. For dual firing, a smooth changeover of fuel at all operating conditions.
12. Use of redundant flame scanners for cleaning and replacement feature of the BMS PLC.

2.3.1.2 Standards Related to BMS/FSSS

NFPA 85 has been developed on the basis of all these six standards (8501–8506); hence, NFPA 85 in generalized form can be applied to BMS/FSSS. However, it should be noted that NFPA 85/86 defines "**what**" interlocks to be implemented but NOT "**how**" to implement. So, some means are necessary to ensure system safety (in the sense of a failure in a part of the system). There are many situations where it is desirable to obtain FM certification on a PLC for the BMS/FSSS. In that case, FM 7605-1999 may be referred to. The basic purpose of the standard is to evaluate the system with respect to "Performance requirements, making requirements, examination of manufacturing facilities, audit of quality assurance procedures and a follow up program." This standard calls for conformance to IEC 61508 (/61511). There are requirements for SIL-certified BMS/FSSS in some cases, especially for large units. There is an **ISA S84** BMS subcommittee that looks after the concept of a safety instrumented burner management system. Though detailed discussions are in Chapter 14, here also a short discussion on risk and risk avoidance is presented.

2.3.1.3 Risk

Apparently, it may seem that measures to reduce risk are simple, but in reality they are extremely complex because measures to reduce risk at various levels have totally different approaches. Fig. 7.60 shows risk reduction in a pressurized system, just as an example to get an idea.

In Fig. 7.60, it is seen that by taking measures at various levels, it may be possible to avoid accidents. However, they cannot be totally eliminated, meaning that there will always be a residual risk. It is clearer in Fig. 7.61, when a fire monitoring and fighting system is considered. By this, it can be reduced only as shown in Fig.7.61. **What is risk**? Risk can be defined as the probability of a hazardous event happening *multiplied* by the cost involved. It depends on the *country/ region*, *society*, *law*, *cost*, and *company policy*. The same risk may have different repercussions on different companies, for example, according to company policy one may accept tripping of a plant under one situation than going to a costly SIS with a higher SIL level. In this connection, Section 2.4 of Chapter 1 of [128] may be referenced by interested readers. Detailed discussions on the safety lifecycle and SIL requirements are detailed in Chapter 14. One of the major reasons for the need to develop standards IEC 61508/IEC 61511 (ANSI/ISA 84.00.01-2004) and a lifecycle was to devise some means to minimize the propagation of device failure through design, operating, inspection, and maintenance practices into the system.

2.3.1.4 Failure and Fault Classifications [128]

Failure and fault are two separate things.

1. Failure types: When analyzed, it has been found that there are three failure types spread over the entire lifespan of the device/system.
   - Early life failure: Early life failures are known as burnt-in or infant mortality failures that occur because of defects such as design defects, material defects, or errors in assembly/manufacturing in the components. Also, device handling during shipment,

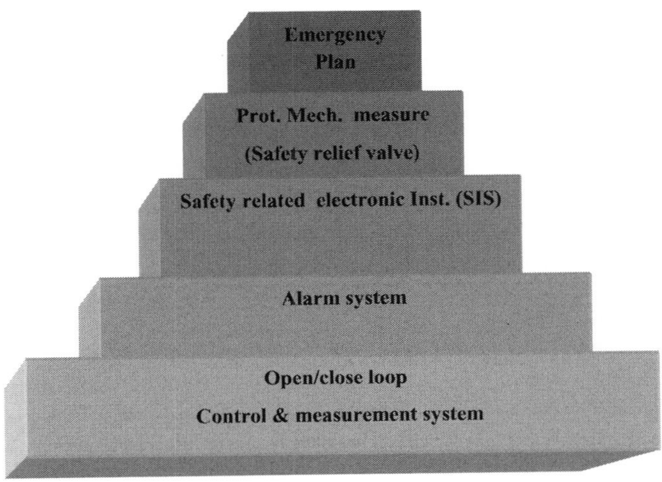
Emergency
Plan
Prot. Mech. measure
(Safety relief valve)
Safety related electronic Inst. (SIS)
Alarm system
Open/close loop
Control & measurement system
At the low level open/close loop to maintain pressure. Next level is alarm for the same. Above is safety related instrumentation. Last level is for mechanical pressure relief valve. Even it fails some emergency plan to avoid accident. Here in case of failure of lower level, higher level is at its limit to avoid damage. Each level has different approach to reduce risk. Also device in each level shall not be used as safety application of higher level i.e. separate instrument to be used for alarm and lower level.

FIG. 7.60 Risk reducing plan.

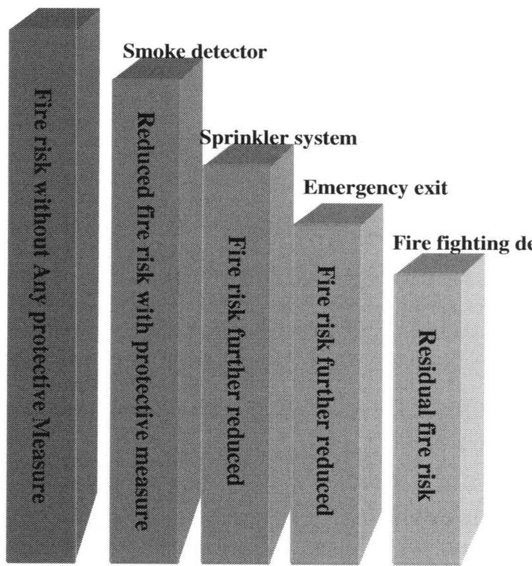
Smoke detector
Sprinkler system
Emergency exit
Fire fighting dev
Fire risk without Any protective Measure
Reduced fire risk with protective measure
Fire risk further reduced
Fire risk further reduced
Residual fire risk
From the example it is clear that each protective measure reduces risk level (Drawn not to scale). So in order to protect personnel and environment it is necessary to determine the risks involved and to take necessary protective measure to avoid risk to the extent possible. A Safe Instrumentation System (SIS) is one possibility to cope up with such situation.
International Standard IEC 61508 is applied safety related systems applicable for ELECTRICAL, ELECTRONIC EQUIPMENT AND PROGRAMMABLE ELECTRONIC EQUIPMENT. IEC61508 is a generalized one & IEC 61511 is applicable for process industries.

FIG. 7.61 Risk avoidance diagram.

unpacking, storage, transportation to the work site, and installation may be responsible for infant mortality failure.

- Normal/mid-life failure: This second failure is relatively constant and is referred to as a normal life period or "useful life." Normal or useful life failures mainly occur on account of random events with an increase of the stress on the device beyond its physical limits. Infant mortality failures are also possible at this stage.
- Third life failure or wear out: This third failure is the wear-out failure. This occurs at a later part of the lifecycle, meaning that it is dominant over the useful life failure rate, showing the end of the normal life of the device. Wear-out may be due to fatigue or depletion of materials. Electromechanical devices may have a gradual increase in failure rate, depending on how well these components are inspected and maintained. In cases of electronic/programmable devices, these rates of change may be very sharp due to the large number of aging components. The lifetime of an entire population of products often is graphically represented by a set of curves collectively called the **bathtub curve**, which is an important thing for failure analysis.

2. Fault classification [128]: All the faults are not the same but can broadly be classified as random failures or systematic failures. There is another kind of failure known as a common cause failure that can be a random or systematic failure–hence, it is defined separately.
   - *Random fault*: Random failure is characterized by the unpredictable failure of a device or component such as an electronic card. This kind of fault may not exist during delivery such as a short circuit fault, drift, etc. These are mainly hardware faults. The fault probability can be calculated.
   - *Systematic fault*: Systematic faults occur due to a combination of conditions resulting in a reproducible failure of the system, and are most often attributable to software issues in programmable safety systems. This failure may be the result of some error in the design, operation, or production process as well as installation and/or maintenance. Improper implementation of construction materials at any stage could be responsible for systematic failure also. It does exist at the time of delivery such as software fault, incorrect rating of measuring device, etc.
   - *Common cause fault:* According to IEC 61511, a CCF is defined as "a failure (that) is the result of one or more events, causing failures of two or more separate channels in a multiple channel system, leading to a system failure." As per IEC 61511(1), the lifecycle recommendations for the devices, systems, or protection layers need to be assessed for independence and the potential for CCF. So, understanding independence and common cause is very important, as at times they are interrelated. Any lack of independence will result in a potential for a CCF. Similarly, when a common cause is identified, it means there is a lack of independence. CCFs mainly occur due to external influences, and could cause more than one system component or layer of protection to fail. So, CCFs could develop serious threats to the reliability of the safety instrumentation system (SIS), and could be responsible for simultaneous failures of redundant components and safety

barriers. These can include faults due to electromagnetic interference (EMI), temperature, and environmental conditions. They have multiple effects and in redundant configurations, systematic faults become common cause faults.

2.3.1.5 NFPA 85 Vis-A-Vis IEC 61511

The standard IEC 61511 (2010) supplements the requirements of NFPA 85, as given in Table 7.24.

2.4 PLC (OLCS) in ATRS Turbine Protection and ATT

2.4.1 Preamble—A confession

There are various separate control subsystems for the turbine and these are very much interrelated (they may have a common hydraulic circuit) so, separate PLCs for the automatic turbine run-up system (ATRS), automatic turbine testing (ATT), or turbine protection may not be a practicable solution. For this reason, most turbine system suppliers prefer a composite control system such as an integrated DCS where functions such as ATRS, ATT, and turbine protection systems are part of the open-loop control system—OLCS functions whereas other functions like Electro Hydraulic Governor Control (EHG), Seal Steam Pressure Control etc. are Closed Loop Control System (CLCS). A turbine stress evaluator is a computing device that directly interacts with the ATRS and EHG. In this section, the basic requirements of the **OLCS/PLC** functions for a turbine control system shall be discussed. Although functionally all the turbine control systems are similar, there are variations in implementation of these systems by various manufacturers. It is not possible to address all such systems. Therefore, in the discussions below, one very popular and renowned turbine control system has been chosen from a globally reputed steam turbo generator (STG) manufacturer (courtesy: **Siemens KWU/BHEL**) as a model to detail the functionality and implementation of these systems.

2.4.2 OLCS for ATRS

The basic coverage and interface of the ATRS shall include but is not limited to the following subsystems:

- HP oil and lube oil system and barring gear.
- Condensate and evacuation system.
- Turbine drain, warm up, and starting system.
- Speed and load set point devices for EHG comprising:
  - o. Turbine rolling and autosynchronization.
  - o. Seal steam system interface.
  - o. LP bypass interface.

ATRS can be considered as a functional group in a turbine control system.

1. The ATRS functional group control (FGC) may be divided into subgroup controls (SGC), which executes commands to bring the associated sets of equipment to a defined status. Each of these subgroup controls can be divided into subloop control (SLC) and drive interface (DI) units. A subloop may be conceived as a watchdog to bring its associated equipment and/or parameters to the desired level. In a subloop, normally there may not be any sequential control logic. whereas the drive unit can be conceived as a software module to establish an interface between a remote-operated drive with plant commands (auto/manual) and provide feedback to the plant control system. The drive interface handles two-way communications with the SGC and SLC. As shown in Fig. 7.62A, the DI need not necessarily be a dedicated special device (as described here) but could be implemented with the help of an intelligent I/O card with the required number of I/Os. The ATRS consists of five (or maybe four when both control oil SGCs are combined) Though, HP oil systems for LP bypass and main control valves have been shown separately, but could be common also.
2. DIs are normally intelligent devices having a dedicated algorithm to implement the associated drive functions. There could be three drive/actuator types:

TABLE 7.24 NFPA Vis-A-Vis IEC 61511 (A Few Examples)

IEC 61511 Section Number	Requirements	Remarks
5.0	Management of functional safety activity	New to NFPA85
7.0	Demonstration of safety lifecycle	
8.0	Process hazard and risk analysis	
9.0 and 10.3.1	Determination of SIL for each SIS	
10.3.1	Requirement of proof test interval, mean time to repair	Not specific/in NFPA 85
11.9	SIF probability of failure	Not defined NFPA 85

- DI for unidirectional motors pertaining to various pumps, fans, etc. (for example, AOP).
- DI for a bidirectional motor for various valve actuators, for example, LO temp control.
- DI for solenoid valves, for example, a solenoid valve for the barring gear gate valve control.

SGCs and DIs shall have various signals with different nomenclature so that in design documents, these can be represented with suitable graphical symbols, for example, P = protection signal, R = release signal, A/M for auto/manual command, c = check back signal, T = operators guidance mode. It depends on the system supplier's convention.

3. Before a subgroup starts functioning, a number of preconditions, better known as "**operation release**," need to be fulfilled. Again, while in sequence, the logic system issues a number of commands and checks back that these criteria are fulfilled. If the check backs are not available, then it will not proceed to the next step. However, certain steps can be bypassed by the operator (if not related to protection), for example, the SGC turbine, with only a few release and step criteria. See Fig. 7.62B, **Example** Step Logic.
4. Timing and time monitoring are very important, as can be seen "**MO"**= monitoring Time shown with sequential step in Fig. 7.62B, "**Example** Step Logic". After issuing a command, the watchdog has to monitor that within the specified time check back and/or criteria for next step appear as input to the system. This is **monitoring time**, which could be defined as the time required to execute a command of any step and the time for the criteria for the next step to appear. During the system design stage, this is specified by the system supplier and during commissioning, it is set and/or adjusted. Under healthy conditions within this time criteria for next step should appear if not there is something wrong in the system so that alarm is to be initiated. **This is applicable for all sequential logic systems**. After the availability of all criteria, the subsequent steps are normally operated only after a waiting period to avoid any spurious actions. This ***waiting period*** after the appearance of criteria until starting the next step is called "waiting time."
5. Theoretically, if all steps and criteria are available, in a healthy situation, the system can proceed in auto to the final stage. In reality it may not be so due to sensor failure or plant exigency, and some steps need to be bypassed

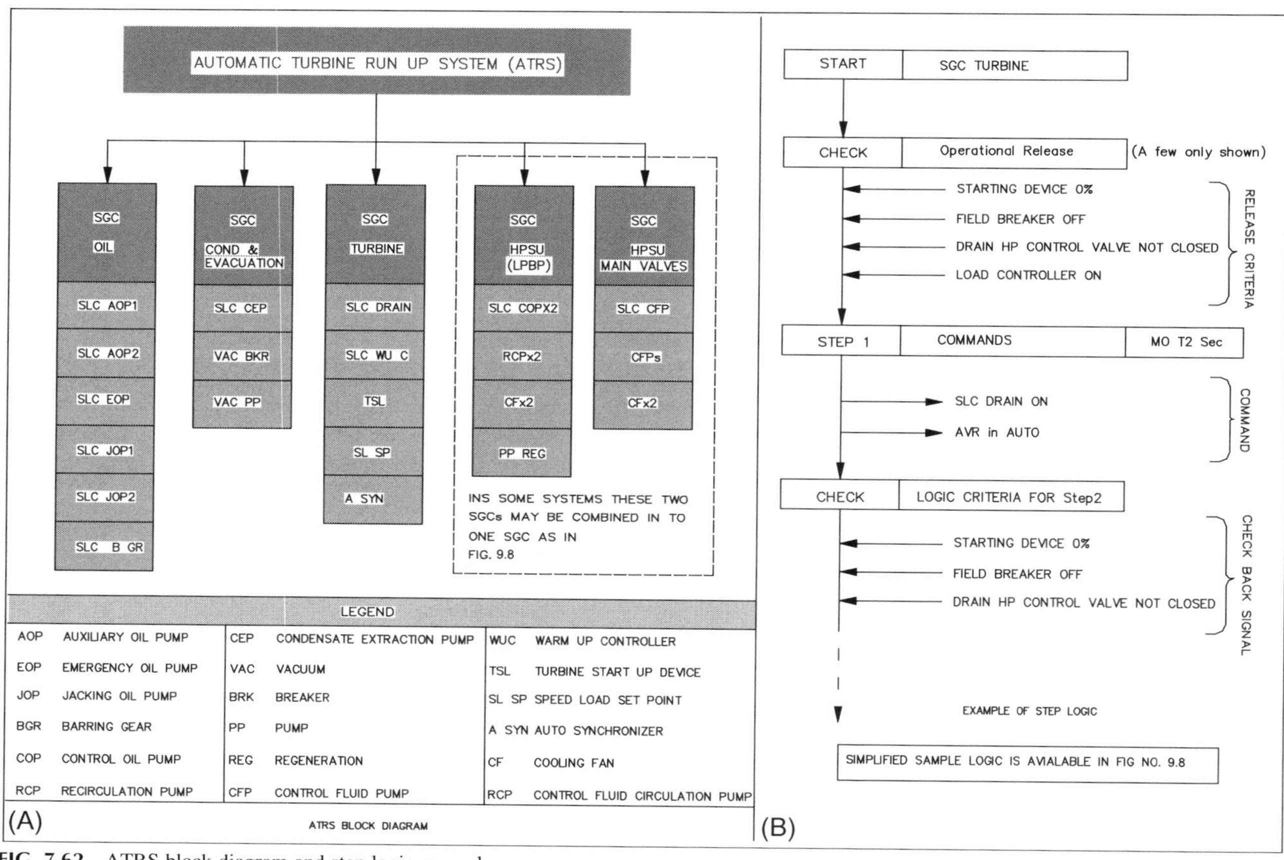

FIG. 7.62 ATRS block diagram and step logic example.

(not protection criteria). For these, there are three modes of operation: **auto**, **operator guide,** and **step by step** (semiauto mode). In ***auto mode,*** all steps will proceed based on the steps and criteria automatically in a sequential manner without manual intervention. In the ***operator guide mode,*** if some of the auto commands are blocked (that is, not forthcoming), looking at the signal, the operator may bypass some steps (not ***protection*** criteria) and issue a manual command so that the sequence can proceed. On the other hand, when there may be a failure of (say) the transmitters so that the check back criteria do not appear, then in ***step by step mode,*** the operator manually simulates the criteria so that the step can proceed. However, in any mode of operation, the ***protection can never*** be bypassed.

6. Brief functional aspect of ATRS are given in Fig. 7.62A (Also Chapter 9).
   - SGC OIL: It consists of a few SLCs such as SLC AOP 1 and 2 responsible for the operation of AOPs through a corresponding DI. Similarly, there will be an SLC for the emergency oil pump (DC) and the SLC jacking oil pumps. One must note that ***SLC On*** does not mean that the corresponding pump is ON. It is the system that is made ready so that, depending on the requirement, the necessary devices can be started/stopped such as SLC EOP on, which means that the EOP is to be kept ready so that in case of an AOP failure, the EOP can be started. There is also another SLC for the turning/barring gear gate valve.
   - SGC CONDENSATE and EVACUATION: This deals with condensate and evacuation systems such as SLC CEPs, vacuum breakers, vacuum pumps, and/or ejectors.
   - The SGC TURBINE handles not only the warm-up controllers and drains the SLC (by opening the drain valves pertaining to various control valves), but they also check the metal and steam temperature to admit steam based on the metal temperature conditions. This is very important for turbine start up. The SGC turbine interfaces with the governing system for speed and load set points (in conjunction with the turbine stress evaluator, TSE).
   - The SGC HPSU deals with control fluid pumps, cooling fans, etc., for HPSU for the main control valves and the main valves (Chapter 9, Section 4).

2.4.3 OLCS for Turbine Protection System

An intelligent turbine protection system with a redundant microcontroller/microprocessor is common. For large utility units, these are often implemented in the TMR mode at the I/O as well as at the controller level. GE turbine control system, KWU speed control. A turbine protection system is implemented in two of three voting logic having three independent tripping channels. Each of these channels shall have its own processor module, I/O, etc., in two of three mode. The output of the three channels shall utilizing two of three logic (may be implemented by an independent relay output for each channel), that is, effectively in TMR mode. Alternatively, the system can also be implemented with two channels with hot back-up processors. It is dependent on the purchaser's choice as well as the manufacturer's standard practice. In addition to the electronic protection system, protection through the hydraulic systems is also implemented. A turbine lockout relay and redundant trip solenoids are commonly used for turbine protection as the main output device. Tripping devices have a DC supply for operation so as to take care of power supply failures. Trip coils are continuously monitored for health and the operator is warned when necessary. Major electronic turbine protection shall include but is not limited to the following:

- Overspeed.
- Low vacuum.
- High exhaust temperature.
- High axial shift.
- High vibration.
- Low bearing oil pressure.
- High bearing temperature.
- High differential expansion.
- Remote manual trip.
- Generator protection.
- High/low extraction pressure.
- Low control oil pressure.

Tripping causes the opening of the drain path for the hydraulic fluid in the trip oil header (autostop emergency trip header). This causes the main stop and reheat stop valves to close very quickly (Chapter 9, Section 3). Normally, the scan time for these protection device sensors shall be very low, for example, 5 mS for overspeed protection. Autotesting of the major tripping devices, self-diagnostics, and changing to the preconfigured operating mode upon detection of a protection device that is "ON" are some of the features of OLCS for turbine protection systems.

2.4.4 OLCS for Automatic Turbine Testing System

This is a feature present in the OLCS in most modern utility turbine control systems. The ATT system allows online testing of turbine protective devices without hampering the normal operation, and keeps the protective devices alive for turbine protection. The major devices that fall under this category shall include:

- OP. and Cl. of HPSV.
- OP. and Cl. of HPCV.
- OP. and Cl. of IPSV.

- OP. and Cl. of IPCV.
- Overspeed.
- Low vacuum.
- L.O. Protection.
- El. Remote Trip.

The following features have been included in the OLCS for ATT:

1. Individual testing of each of the protective systems separately.
2. For turbine protection, an automatic SUBSTITUTE functional protective device shall be used.
3. Testing is possible ONLY IF the SUBSTITUTE DEVICE is healthy.
4. Each step in testing has a fixed execution time set by the manufacturer that needs to be monitored and, if exceeded, an alarm is to be issued.
5. Automatic resetting of a test program after a fault is detected.
6. Testing shall include the hydraulic circuit also and all alarms are to be initiated so that all portions are tested.
7. During testing, the trip oil may be shielded and instead the control oil may be used so that actual tripping can take place.

Here, the PLC/OLCS works as per the fixed program to be executed as per the turbine testing schedule to simulate the fault situation as per the manufacturer's recommendations.

2.5 PLC (OLCS) in Electrical System—SCADA

2.5.1 *Introduction*

Monitoring and control of the electrical parameters pertaining to the electrical facilities for the plant, including the generator and transformers (untt/station/generator transformer, part of the high-voltage system), may be a part of the main DCS or may be monitored by the PLC (SCADA) being integrated into the main DCS. Within power plants, all major high-voltage (HV) and low-voltage (LV) system parameters are monitored in the DCS. Normally, all **HV system parameters** (for example, MVA, MW, etc., of the unit transformer parameters, station transformer), **all generator and exciter parameters**, **all HV drive statuses,** and **analog parameters (for example, current, voltage** etc.), **along with the statuses of important LV drives** are taken into the DCS. In some cases, intelligent protection relays along with the associated logic may form a part of DCS/PLC also. Current/voltage transformers, transducers, contactors/relays, protection devices/relays, circuit breakers, battery backups (as needed), and communication channels along with the associated control system (PLC/DCS—OLCS) actually constitutes this electrical control system. **SCADA** is an important system that may be somewhat similar to PLC/DCS with some prominent differences. One of the major differences is its geographical spreading and control actions. The PLC/remote terminal unit (RTU) plays an important role in SCADA. SCADA of the present generation involves networking and sometimes even networking via the Internet. So, SCADA systems are very susceptible to cyberattack, making the entire system very vulnerable.

2.5.2 *What is SCADA? What are the Major Differences Between the Two Systems?*

SCADA is an acronym for **S**upervisory **C**ontrol and **D**ata **A**cquisition. Between the DCS and SCADA, some differences may be observed in **architecture**, **operation, communication,** safety, and **security.**

1. SCADA generally acts as the **set point control** supervision and coordinator and may not, truly speaking, be a real-time process/system control, whereas DCS takes part in real-time process control. In SCADA, real-time control is taken care of by the local controller (RTU or PLC).
2. While DCS is **process-driven** and functionally responsible for the **control** aspects of the system, SCADA is **event-based** and **monitoring**-oriented.
3. **Geographical spreading** is a common feature in SCADA and is mainly used in **WAN** (utilizing telecommunication channels) while DCS in general is restricted to local control via LAN. However, such a difference may not be truly applicable with the introduction of wireless fieldbus/HART and/or the IoT.
4. In DCS, MMIs have direct access to I/Os while in SCADA, PLC/RTUs run preprogrammed processes. So, in SCADA, MMIs connected with the server disseminate this data in a standard form acceptable to the operators, that is, in SCADA, the data are presented after being processed in the server/host.
5. While in newer generation SCADA being networked in WAN is very much vulnerable to Cyber attacks. Hence, in SCADA both safety and security are very important. Border router firewalls, etc., are some of the security measures of SCADA.
6. In SCADA, special protocols are in use. In SCADA, IEC 60870 and DNP3 are the major protocols.

2.5.3 *Major Components of SCADA*

The following system components are generally present in SCADA systems. However, there may be some variations according to the SCADA generation.

1. RTU/PLC—Field Interface Device: RTUs connected to the field devices, such as sensors/transmitters, local switches, and actuators (for example, a circuit breaker), convert the electronic signals from the sensors to digital

data–a language used to transmit over the communication channel to the supervisory system. RTUs also interface to convert the data from the supervisory system to the necessary electronic signal for the actuators. On account of band limitation, the automation of field devices is stored in local devices. On account of versatility, flexibility, and configurability, PLCs are also used as field device interfaces, like RTUs. Obviously, PLCs incorporate programmed intelligence in the form of logical procedures to be executed in the event of certain conditions. Communication modules connected with the PLC allow it communicate with remote devices. In its origin, SCADA was more connected with telemetry. In the present day, PLCs may replace both functions of RTUs and relays. In SCADA, from an overall control perspective, often it is necessary to influence the program within the PLC with use of a remote signal. This is "supervisory control" of SCADA, where the control strategy may get affected by the remote supervisory system over a communication channel. In switchyard SCADA, distributed PLCs/RTUs may be used to communicate with various intelligent protective relays.

2. Master Station or System Server: This is either a single computer or a network of processors used to provide MMI operator interface to SCADA. At this place, the information exchanged with the RTUs/PLCs may be processed for presenting to the operator to work with. The operator's MMI shall be connected to the server utilizing LAN/WAN. Also, this supervisory computer can acquire data, process the same, and send the supervisory commands. In contrast to older host computers (proprietary OS), today PCs and networking with standard servers (independent of vendors) are used so that it can connect even office-based applications in the network. However, this makes the system more vulnerable and calls for more protection.
3. Man-Machine Interface: In most cases, operator workstations (WS) are computer/PC terminals. In a client-server network, the WS acts as the client that requests and sends information to the servers. The operators' interface software is quite important for SCADA. An MMI package for SCADA includes the necessary software to provide trending, alarm handling and presentation, diagnostics, a logistic information maintenance schedule, etc., by accessing the system database.
4. Software Component: Software plays a very important role in SCADA, especially with the MMI package, which also contributes significantly to the cost. For a better-performing SCADA system, the software needs to be well defined, designed, checked, and tested. Proprietary software dependent on hardware puts more emphasis on the process concerned. In a multivendor environment, commercially off the shelf (COTS) are used. This puts the emphasis on a variety of instrumentations and equipment. Therefore, for software selection, a judicious decision is necessary. The major software components are:
   - PLC/RTU SOFTWARE: Local automation software for configuration and maintenance of control and monitoring applications.
   - SERVER(s) SOFTWARE: Necessary OS and application software to exchange data with RTUs/PLCs, including a graphical user interface (GUI) for historian, trending, and mimic functions.
   - MMI SOFTWARE: Normally, the same OS as in the servers is used to form suitable networking. Also, there will be application software to access the server in a network.
   - COMMUNICATION SOFTWARE: There shall be a suitable protocol when various systems are connected in a network. There shall be a suitable communication protocol drivers' resident in the server as well as at the PLC/RTU. It is the responsibilities of protocol drivers to prepare data for communication. Network management is another important aspect of communication software so that the network works without failure. Also, there will be a suitable means to monitor system communications. In case of any failure, a quick recovery is essential.

2.5.4 SCADA History and Architecture

There are three Generations of SCADA: first generation: *MONOLITHIC;* second generation: *DISTRIBUTED;* and third generation: *NETWORKED*.

A typical configuration of generations one and two is presented in Figs. 7.63 and 7.64, respectively. A third-generation SCADA system is shown in Fig. 7.65. This networked SCADA utilizes an open protocol to make the system vendor-independent ,and it could be distributed in the WAN also. It is possible to integrate third-party items and peripherals such as printers and monitors, and hardware bases may also be open standard servers. The major advantage in this generation is to utilize WAN protocols such as enhanced performance architecture (EPA) and TCP/IP communication between the systems. RTUs/PLCs are also open and they can communicate with masters utilizing an Ethernet (CSMA CD), as shown in Fig. 7.65.

From this architecture, it is clear that various intelligent I/O modules are connected to the master station via a standard system bus. These I/Os could be replaced by PLCs also. The master station also connects various intelligent electronic devices (**IED**s) and other devices via RS links and with the relevant standard protocols. The system may have more than one master connected via a router in a WAN. The operator station may be connected to the station bus based on protocol IEC 60870-5-101 for operator interface.

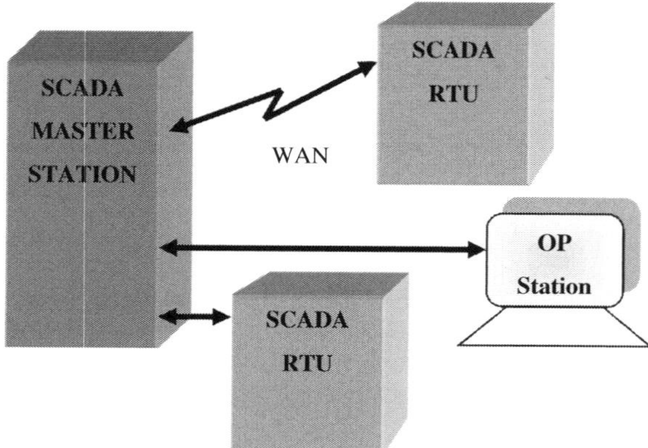

1st. generation: Monolithic: These are main frame based system using LAN, WAN as shown. These are stand alone SCADA. The Protocols generally developed by RTU manufacturer are proprietary in nature. Redundancies in these systems are two identical systems in Primary and Back up system. No other system data traffics were mingled here.

FIG. 7.63 1st generation (monolitic) SCADA.

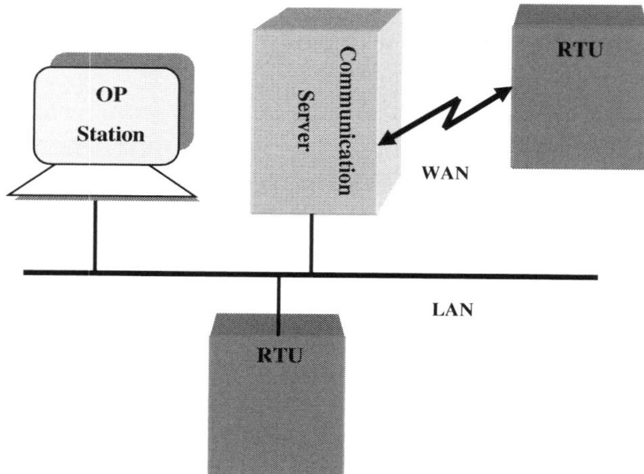

2nd. generation: Distributed: here the processing across multiple systems in LAN, each unit specific function. Communication processor is to communicate with RTUs. OP stations serve as MMI and some as Data base Server. Systems are connected using LAN protocols. WAN used to communicate with distant RTU. Obviously on account of distributions of functions the processing power is enhanced as well as more reliable and redundancy is easy to implement. LAN protocols were not open widely hence restricted to specific vendors.

FIG. 7.64 2nd generation (distributed).

2.5.5 SCADA Protocol

According to ISA, there are about 200 such real-time user layers and application protocols (proprietary and nonproprietary). Some of them include but are not limited to: Allen Bradley DF1, DH, and DH+, GE Fanuc, Siemens SINAUT, Modbus (Modbus/ModbusX), RTU/ASCII, Omron, Mitsubishi, and Westinghouse. In a generalized manner in SCADA, there are two sets of protocol. One is an RTU protocol consisting of a statement for RTU initiation and/or response. The other is for the master station initiation and response. Out of the many protocols available on the market, the IEC 60870-5 series and the distribution network protocol version 3–DNP3 are the most popular. The IEC is based on a three-layer EPA reference model for an RTU with relays and other IEDs (**Protocol outlines and terms are only discussed Ref: Std. book on SCADA).**

1. STANDARD IEC 60870: As shown in Fig. 7.66, EPA Layered Structure, the EPA has three layers. The IEC 60870-5 is very important for SCADA. The standard 101 provides a structure directly applicable for RTU and IEDs. At the data link layer, it specifies a multidrop (unbalanced) or point to point (balanced) (Section 1). The standard 101 defines the parameter support for multivendor devices. The following application functions can be utilized by a standard 101:
    - Station initialization.
    - Acquisition of events.
    - Cyclic data transmission.
    - Data acquisition by polling.
    - General interrogation.
    - Command transmission.
    - Clock synchronization.
    - Transmission of integrated total.
    - File transfer.
    - Parameter loading.
    - Testing procedure.

IEC 60870-5-104 uses TCP/IP in place of the EPA.

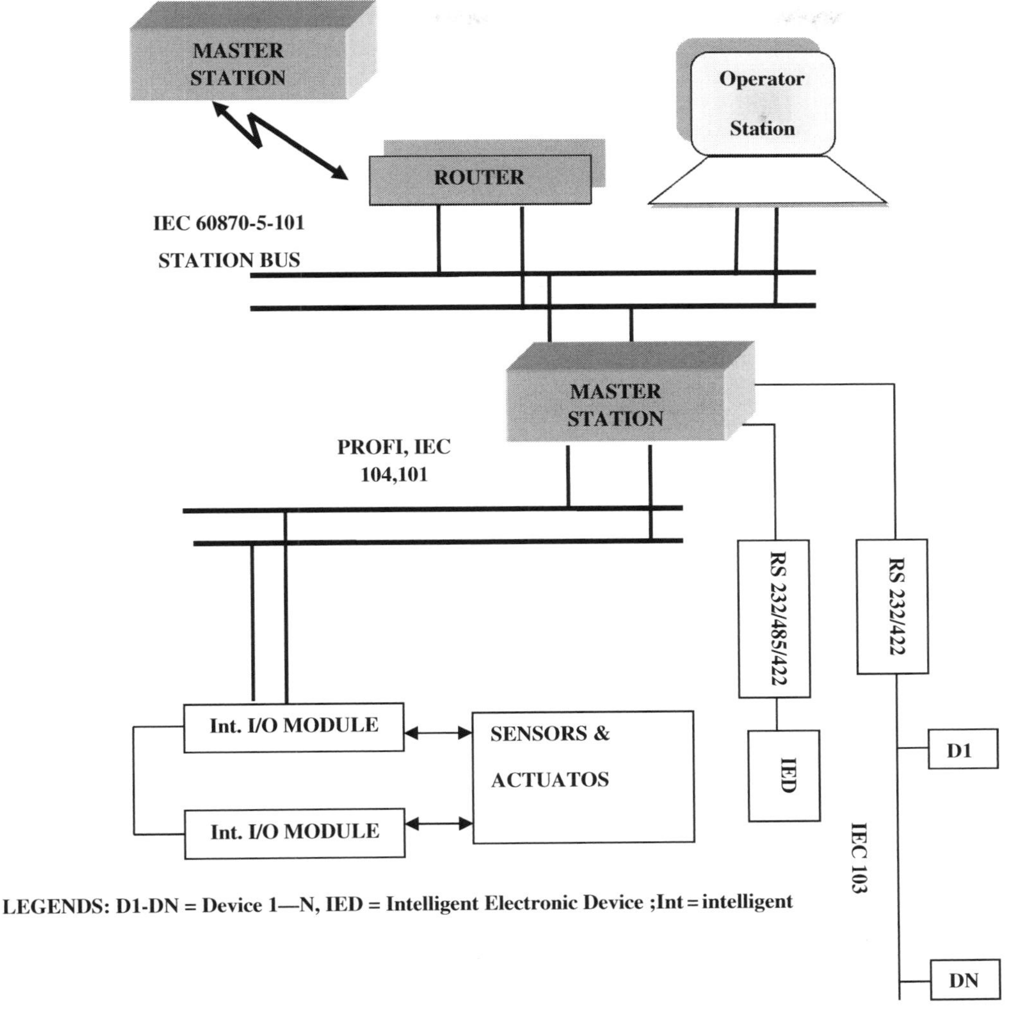

**EPA: Seven layer model adds significant over head for Bandwidth and processing power, so IEC came out with THREE LAYERED MODEL (Physical, Link and application Layer) with reduced overhead suitable model for SCADA. This is termed as ENHANCED PERFORMANCE ARCHITECTURE (EPA).*

FIG. 7.65 3rd generation (networked) SCADA.

APPLICATION LAYER (ISO LAYER 7)		IEC870-5-101/102/103
LINK INTERFACE		IEC870-5--5
DATA LINK LAYER ISO LAYER 2	LLC	
	MAC	
PHYSICAL INTERFACE		IEC870-5--1
PHYSICAL LAYER (ISO LAYER 1)		

EPA consists of three layered Structure:
- **Application Layer**
- **Data link Layer**
- **Physical Layer like RS232/485**

IEC 69870-5-
-1 Basic service requirement for physical & data link layer
-2 Selection of Link Transmission Procedure
-3 Rules for structuring application in Transmission frame
-4 Information data element specification
-5 Basic application function for telecontrol system

FIG. 7.66 EPA layered structure.

2. DISTRIBUTION NETWORK PROTOCOL (VERSION3)—DNP3: The DNP3 has been specifically developed for interdevice communications in SCADA such as RTU-master, RTU-IED, etc. Even though both have been developed from the EPA structure with some similarity, there are a number of *differences also* (for example, *NP3 always uses 16 bit addressing—source address and destination address—while IEC 60870 can use 8/16 bit and a source address is not mandatory) making it noncompatible. Also the DNP3 depends on the layer to confirm that the data transmitted is successfully received by the recipient. The IEC 60870-5 relies more on* the *security of the network*. On account of a pseudotransport layer, the DNP3 really has four layers: the *application layer*, the *pseudotransport layer, the data link layer, and the physical layer*, in place of the three-layer structure of the EPA. Some DNP3s are:
   - **Open Standard**: This can be used by vendors for RTU, IED, and master station/server.
   - **High Data Integrity**: Data and application layer data sent through a confirmed service.
   - **Flexible Structure**: Application layer is object-oriented with a structure that allows a range of implementation, retaining interoperability.
   - **Multiple Applications**: This shall include polled only, polled report by exception, unsolicited report by exception, and a mixture of the above. It can use several physical layers in the LAN and WAN.
   - **Minimized Overhead**: It can be used over an existing pair of wires.

The maximum data frame in DNP3 is 292 bytes, and out of this, 1 byte is used by the pseudotransport layer. DNP3 uses a CRC check to avoid noise susceptibilities and to confirm that the transmission is well received by the recipient. DNP3 supports various architectures such as: ONE to ONE (master-master), MULTIDROP, HIERARCHICAL, AND DATA CONCENTRATOR. Now, TCP/IP has been used to transport DNP3 via the Internet.

3. MODBUS: this is also almost an open standard used in SCADA communication. MODBUS X in a nonproprietary protocol.

2.5.6 SCADA Security

"*Open-based standards have made it easier for the industry to integrate various diverse systems together. It has also increased the risks of less technical personnel gaining access and control of these industrial networks*" (National Communication Systems bulletin). Therefore, in one hand integration is easier, but on the other hand, Internet connections make them very vulnerable to attacks. Major host attacks shall include but are not limited to the following:

- Use a denial of service (DoS)—shut down.
- Delete system file (downtime).
- Plant a Trojan horse and gain control.
- Modify logging (data loss).
- From log key stroke for user ID and password.
- Log sensitive data loss and loss in competition.
- Change data point to deceive operator.
- SCADA server as a launching point to defame.

There two **major** fear points: authorized access by a human or by an intention/accidental virus infection. To secure SCADA, the following are important steps:

- Patch host OS, application, and SCADA components (but may not be possible always).
- Control application communication between SCADA and other networks.
- Regulate application communication within SCADA.
- Regulate "who" and "what" to communicate in SCADA in the system.
- Close monitoring of the system and reacting immediately to any virus or other attack.

Several multiple detection mechanisms may include:

- Stateful application-aware firewall.
- Antivirus detection.
- Application control.
- Web filtering.
- VPN (Section 2, Chapter 7).
- Network anomaly and DoS prevention.
- Database protection.
- Wall application protection.
- Updating of antivirus and IPS signature database.
- Known SCADA exploits already in antivirus and IPS database.

A border router (main router in front of the main firewall of the organization), a proxy operating system, etc., are the major systems deployed in an industrial SCADA. For details, see the Standard SCADA book.

2.6 PLCS (OLCS) Offsite

2.6.1 Introduction

In power plant offsite controls, PLCs find quite a good number of applications. In many cases, these PLCs are **linked** to the main plant DCS (if the offsite controls are not **integrated** as one system with the main plant control). The boiler turbine and generator (BTG) are directly connected together for power generation. There are various material supplies necessary such as fuel (oil handling plant/coal handling plant), a water treatment plant (DM plant + pretreatment plant), treated cooling water, H_2 generation plant, etc., for the BTG to operate. Also plants such as ash-handling plants, effluent treatment plants, fire detection and protection systems, etc., are necessary for safe operation of a plant without pollution. So, the major offsite plants shall include:

• Coal handling	• Fuel oil handling	• Circulating water
• Ash handling	• Fire detection and protection	• Hydrogen generation
• Water treatment	• Equipment cooling water	• Effluent treatment

There will be a signal exchange from these plants to the main DCS, which is necessary to always keep the operator informed about the offsite plant condition/status so that the main control room operator can initiate any necessary action for safe and efficient operation of the plant, even during an emergency. This is more important where, in a station, there are several units operating with common offsite facilities. A few generalized features normally recommended for offsite PLCs are:

- PLC processors may be in hot standby mode, to include a fall-back feature.
- Redundancies in I/Os, with optical/galvanic isolations and expansion provisions, monitoring of power supply, contact bouncing, etc. (exception: smaller PLCs).
- Dual redundant communication channels to integrate various PLCs, as the case may be, so that they form a LAN with provisions for integration with the main plant DCS.
- Independent dual power supply scheme for each offsite PLC.
- There may be various modes of operations such as AUTO, SEMIAUTO, and OPERATORS GUIDE MODE (Section 2.4). These are often seen in case of water treatment and coal handling plants.
- In case of a power or other failure, the failsafe mode shall be chosen, and in case of sequence control, the last sequence prior to the failure shall be maintained.
- For offsite plants, a minimum of one operator station and one programming station (may be a special small operator interface for the stacker/reclaimer control cabin in place of a WS/monitor) shall be provided for the system. In case of multiple control locations, a suitable number of panels and operator interfaces shall be provided. Operator interface displays (alarm, mimic display, trend displays) and logs (on demand/event-based/periodic/alarm logs) are standard. For example, the mimic in a CHP/AHP is quite important for operation. Similarly, parameter trends in case of DM plant exchangers are very useful. In some cases, large video screens are helpful for a shared display. Like the main plant, offsite, also local PB stations/deinterlock switches (as applicable) shall interface with the PLC.

2.6.2 Coal Handling Plant (CHP) Control Systems

In a thermal power plant, a coal-handling plant is mostly the largest and costliest auxiliary plant. There are a number of subsystems in the plant such as

- Wagon tippler/track hopper control system.
- Conveyor control including crushers.
- Stacker/reclaimer.
- Travelling trippers.
- Dust extraction/suppression system.

All these systems are controlled by PLCs with remote I/Os. These PLCs shall normally have a provision for data exchange with the main DCS for status reporting and emergency operations from the unit control room. The major equipment in the CHP shall include a wagon tippler, a stacker/reclaimer, a paddle feeder, a number of belt conveyors/feeders, travelling trippers, a belt weigh scale, primary/secondary crushers, a magnetic separator, and a dust suppression system. A number of independent controls or PLCs may be deployed for subsystems such as for the **stacker cum reclaimer, the travelling tripper, the wagon tippler,** etc., so there may be a number of smaller PLCs and one main PLC. Also, there may be a number of remote I/Os, for example, at the MCC, for the main PLC of the CHP. These small PLCs and the main PLC may be integrated, forming a LAN. This LAN shall be connected to the DCS through a soft link. In larger plants, especially a station with a number of units, the CHP may be integrated with main DCS. SCADA is also introduced so that the coal washery/preparation system can form a part of the entire system. In the CHP, a few local panels are also deployed and may be connected with the PLCs. The majority of these are:

- Paddle feeder (up to the reclaimer) with dust suppression.
- Belt weighing system.
- Coal sampling unit.
- Electrical hoist.
- Magnetic separators.
- Sump pumps.
- Mobile tripper over bunkers (if applicable).

Out of the above, the belt weighing system, coal sampling system, and magnetic separator will interface with the associated PLC control system. All the systems require dust extraction and suppression systems (for example, bag filter control) with their own control and interface to the subsystem PLC. Alternatively, it may form a part of the subsystem PLC. Dust extraction/suppression for conveyors shall run when the conveyor systems are running while the same for the bunkers has to run around the clock.

1. LOCAL OPERATION: There will be a control room for the coal handling plant where from the plant shall be controlled from the Operators station for the following systems under main PLC. In some cases where there are independent PLCs, there will be a soft link for data exchange and control from the main PLC. In cases of systems with local panels, suitable arrangements for control changeover from each place shall be provided. A typical local control and operation could be conveyors, crushers, vibrating roller screens, paddle feeders, dust extraction and suppression systems, ventilation systems (group/individual), vibration and temperature monitoring of HT drives, magnetic separators, coal sampling system, a stacker reclaimer, a belt weighing system, and a traveling tripper.
2. There may be a number of flow paths the operator can select for feeding the coal. Therefore, separate flow path selection shall be made possible through mode selection at the PLC with the help of a few subroutines. Whenever such a path is selected, the PLC shall select the complete path so that the flow goes unhindered. Some of these paths are wagon tippler/track hopper to the bunker, wagon tippler/track hopper to crushed coal storage (also to the coal bunker), and a coal stock pile to the bunker for normal cases and emergencies.

2.6.3 Ash Handling Plant (AHP)

An ash handling system consists of some subsystems such as bottom ash (BA), fly ash (ESP/air heater), ash water, and ash disposal systems. The ash handling system may be dry (dry pneumatic conveying is more popular now on account of retention of original characteristics of fly ash and its utilization), semiautomatic wet, and a wet system. Similarly, the disposal system also may be wet or dry. I mentioned some of the subsystems because in case of a dry system, there is no need for an ash water system. For the purpose of control, PLC systems are used extensively. Normally, there will be one main PLC with operator stations in the main AHP control room (may be near the ash slurry pump house). There may be separate PLCs and local panels for a SILO system and one PLC for a water recovery system, as applicable. The local panels may have mimic displays in WS (or a conventional mimic panel) with a PLC interface. These PLCs shall form a LAN so that the operator from the main ash handling control room can oversee the entire system as well as take over control when necessary. Because the ash handling plant is very much related with boiler operations. this LAN is suitably connected to the main DCS for supervisory monitoring and control. Out of the various subsystems discussed, (dry) fly ash handling and disposal form the major part so far as I/Os are concerned. In each unit of a mid- to large-sized unit, the I/Os will be in the tune of >**2000** (when all switches are taken into account) with sequence and logic per network may in the tune of **1000**. The PLC shall be integrated through a dual redundant wire/fiberoptic communication channel with a video control panel where at all times the sequence shall be displayed and in case any sequence fails, then same shall be highlighted. The salient special features of these PLCs are:

- Suitable menus for set point and sequence control on MMIs.
- Remote control of ash dumping operation.
- Integration with DCS.
- Event and analog parameter trending.
- Manual/automatic sequence control and parameter set.

The time to empty each hopper, the hopper emptying sequence, etc., could form the basis for automatic operation, or it could be done on the basis of level sensing for better boiler operation. Now, in modern systems, system sequence optimization could be done with the help of an adaptive control or artificial intelligence through a learning process. There may be remote I/Os, on account of distance and economy, connected to the main PLC for interfacing with MCC and other systems.

2.6.4 Water Treatment Plant (WTP)

WTP including an effluent treatment plant: There are three different sections in a WTP: a pretreatment (PT) plant, a posttreatment or demineralized water (DM) plant, and a waste treatment or effluent treatment (ET) plant. The ratio of numbers of analog to digital measurement is near unity, and because at present there is hardly much difference between the DCS and PLC, so the selection of control could go either way. However on account of many sequential controls pertinent to each of the exchanger units in a DM plant and effluent plants (and cost also), the PLC is preferred as the control system. It has been found that in some cases, a number of PLCs are utilized for a total WTP plant. Also, some smaller units use PLCs with SLC (Section 4.3). A WTP encompasses everything from a raw water pump house to a DM storage tank, and the location of the pretreatment plants is far from the DM plant control room, where normally the main PLC and operator station are placed for control and operation. Sometimes it is better for pretreatment plants to have a separate PLC with remote I/Os for a raw water plant. Again, a chlorination plant is almost an independent plant and may be controlled by a micro PLC. Depending on plant size and configuration, the ET plant may be an elaborate system involving many I/Os, so a separate PLC with communication to the main system may be economical. Finally, all these PLCs may be integrated or linked to have relevant data available at the DM control room. Alternatively, these may be achieved by deploying remote I/Os and/or via a fieldbus to reduce cable cost. In a nutshell, there is a need to have an integrated LAN with a soft link or fieldbus. In terms of automation, there are differences in control systems deployed for PT, DM, and ET. In pretreatment plants, there will be a number of pumps, and their automatic selections and associated logics are major issues. Also, there will be a few measurements such as flow measurement by a Parshall flume. For a large power house, there may be a large number of pumps and valves, so these automatic selections are not very simple and a dedicated PLC may be a better option. Similarly, in a DM plant, there will be sequential operation of solenoid-operated valves for backwash and regeneration for the exchanger units. Regeneration can be done in various modes such as DP across an exchanger, time-based, or a conductivity-based system, so the PLC needs to take care of such selections. As in a DM plant, there are controls for chemical handling (for example, brine tank automation). The sequence control may be fully automatic, semiautomatic, or manual. A PLC in a DM plant needs to handle many analog parameters (for example, pH, Cond. analyzer) for monitoring and for modulating control (for example, degasser control) also. Therefore, the trending of parameters is important. In case of small DM plants, the use of a local dedicated sequencer is not uncommon. Selection of the IP rating for a PLC and local enclosure is quite important as the environment is very corrosive. In these plants, local operation of a few drives is common, for example, the clari-flocculator agitator, so local panels with operator interface and a link to the LAN may have to be considered.

2.6.5 Fuel Oil Handling System

This plant deals with the unloading and storage of fuel oil. Because FO is highly viscous, it is necessary to heat the same. Normally, this is done with the help of steam. The control operation of pumps and heaters (to maintain the desired temperature) and monitoring of various parameters such as the level are done with the help of a small PLC and a local panel with an operator station. These PLCs are connected to the main DCS via a the soft link/fieldbus integration method discussed earlier.

2.6.6 Circulating Water (CW) and Equipment Cooling Water System (ECW)

The two systems have different controlling philosophies. Also, location-wise, they are at different places. The ECW system is located in the main unit area whereas the same for the CW will be located outside the main unit area. In both systems, the PLCs are deployed with a soft link to the main DCS.

1. CW SYSTEM: As CW is directly related with the operation of the condenser, hence with the turbine, so in many cases the CW pumps along with associate valves form a part of the unit DCS may be extended to control this part also. However, in stations with a number of units, a separate PLC having full supervisory control at the unit control room could be deployed (a wise decision). In these cases, the CW PLC complete with the local operator stations used to control the pumps and valves with overridden control from Unit control room DCS which may be connected to PLC via soft link or with the help of fieldbus. Normally, an independent local control panel/operator station with control is deployed for CW chlorination/treatment plants. Such independent control may be connected with the main control system for supervision. The cooling tower water treatment automation may form a part of the CW PLC, or it could be achieved by a micro PLC with a link to the CW system PLC.
2. An ECW uses DM quality water to cool equipment in the unit. This water is cooled by raw water by a water-to-water heat exchanger. Because this is in the unit itself, so many times the control and operation of these pumps, valves, and associated logics are carried out as a part of the main DCS. Sometimes (especially when the system is common to, say, two units), separate small PLCs and local control panels with operator control and operating stations are provided. However, this may be connected to the main DCS of the plant.

2.6.7 Fire Detection and Protection System

PLCs here used in almost everywhere in fire system. Other than the PLC at the fire water pump house, other PLC/intelligent controls in the detection system are a little different from a conventional PLC. Normally, there is one operator station and programming station for each PLC located at the fire water pump house for operation of these pumps and valves. Also alarms from these systems shall be made available to various other locations such as the fire station, the fire panel, etc. Again, at present, the fire alarm panel and detector panel are also basically intelligent PLC-based systems. The addressable detectors interface with these panels. However, these alarm detectors and fire alarm panels have dedicated intelligent control systems that analyze analog addressable detector signals internally from an envelope of curves before really detecting a fire alarm. At the fire alarm panel, normally LCDs are provided to show the address of the detector (that is, the location of the fire and/or the manual call point). These fire alarms and detector panels may form a LAN with a PLC at the fire pump house PLC. In any case, these fire alarm panels and fire pump houses normally have a soft link to be connected to the DCS for polling various data as and when necessary.

3 ANNUNCIATION AND SEQUENCE OF EVENTE (SOE)

3.1 Preamble

In modern days, both annunciation and sequence of events are a part of the DCS, but these can be installed as independent systems also. An alarm is a part of the safety instrumentation system and an independent protection layer as per Annex F of IEC 61511-3:2003 (Chapter 14). The SOE recorder stores and records various events in sequence to study and analyze in a more precise manner to detect the root cause of the problem. After any trip and/or any major fault, it is of utmost importance to know the entire sequence of how it happened and what were the consequences. It is not always necessary that the first event is the root cause of the catastrophe. A simple and coarse example could be (say), the turbine tripped, the HP bypass did not come in time, and the boiler tripped. Here, even though the turbine trip occurred first, a plant having an HP bypass should begin operation in time. So for boiler tripping, the fault in the HP bypass was the culprit. There are systems available for SOE and annunciator systems in one system, independent of the DCS but suitably connected to it for the management information system.

3.2 Alarm Annunciator

An annunciator consists of multiple prewritten alarm display windows to draw the attraction of the operator. These alarm points operate through a set of logic and trouble contacts. Annunciation systems also include a shared power supply, an audible signal generator, a horn, etc. There are various sequences of operation implemented with the help of a logic system. A number of alarms may be grouped. There may be first out alarms to identify the first alarm occurrence in a group of multiple alarms. There will be different push buttons called "acknowledge," "rest," and "system (or lamp) test. The system/lamp test push buttons are used to test the alarm sequence (system test) or lamps (lamp test) in the annunciator. For the annunciator sequence, ISA Standard **ISA 18.1–1979 (R 2004)** may be referenced.

3.2.1 Alarm Designation Methods

According to ISA 18.1, these are designated as shown in Fig. 7.67. Option numbers have been listed in Table 7.25 (see Chapter 14). A number of alarms may be grouped together. As a result, one alarm may cause the generation of a number of alarms in the group or another group. In order to identify which occurred first, "first out" alarms are used.

3.2.2 Alarm Sequence

A typical alarm sequence has been presented in Table 7.26.

There could be quite a good number of alarm sequences; the same are available in **ISA 18.1** with a sequence diagram. To specify the annunciation alarm, the sequence needs to be mentioned.

SEQUENCE FEATURES:

1: ACK AND RESET PB. 2: ALARM AUDIBLE DEVICE. 3: LOCKIN OF MOMENTARY ALARM UNTIL ACKNOWLEDGED. 4: Audible DEVICE SILENCE AND FLASHING STOPS WHEN ACKNOWLEDGED. 5: AUTOMATIC RESET OF ACKNOWLEDGED ALARM INDICATIONS WHEN PROCESS CONDITIONS RETURN TO NORMAL. 6. OPERATIONAL TEST.

3.2.3 Optional Items

1. The annunciator is initiated by the trouble NO/NC contact. However, it is better to include one built-in switch by which at site this NO/NC contact can be altered that is, to Change NO to NC (vice versa) at alarm panel. Once this is specified, the supplier will include the necessary logic.
2. System test: Usually, the lamp burn out tests are carried out utilizing the LAMP TEST push button. However, optionally it is possible to include a SYSTEM TEST

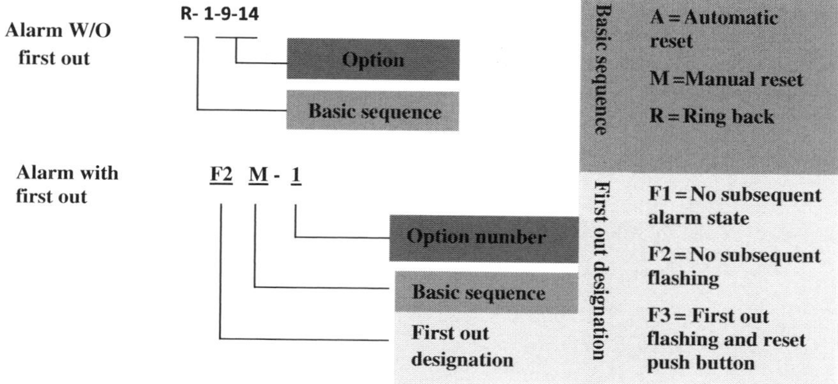

FIG. 7.67 Alarm designation method.

TABLE 7.25 Alarm Option Numbers

No.	Key Word	No.	Key Word	No.	Key Word
1	Silence push button	6	Not audible	11	Common ring back visual
2	Silence interlock	7	Automatic alarm silence	12	Automatic momentary ring back
3	First out reset interlock	8	Common ring back audible	13	Dim lamp monitor
4	No lock in	9	Automatic ring back silence	14	Lamp test
5	No flashing	10	No ring back audible		

TABLE 7.26 Alarm Sequence A Automatic Reset (ISA 18.1)

Line	Process Condition	Push Button Operation	Sequence State	Visual Display	Audible Device	Remarks
1	NORMAL		NORMAL	OFF	SILENT	
2	ABNORMAL		ALARM	FLASHING	AUDIBLE	LOCK IN
3A	ABNORMAL	ACKNOWLEDGE	ACKNOWLEDGE	ON	SILENT	MAINTAINED ALARM
3B	NORMAL	ACKNOWLEDGE	TO LINE 4			MOMENTARY ALARM
4	NORMAL		NORMAL	OFF	SILENT	AUTO RESET

push button so that the entire system can be tested, including the flasher, logic card, and audible devices.

3. Repeat auxiliary contacts: It is possible to include in the system the necessary auxiliary contacts for various purposes of interlock for external use.
4. Group alarms: It is possible to initiate a single window alarm as a group alarm. Depending on the initiating contact type NO/NC, these may be in parallel or in series. An HT motor winding temperature high alarm may be an example of this type.
5. First out: Here, alarms pertaining to a group (for example, BFP) may be part of a cluster of alarms. One alarm may cause the initiation of several other alarms. From all these alarms to distinguish the first alarm (cause), the first up is selected to detect the first alarm. When acknowledged, all other alarms will stop flashing except the first one. The first out alarm will be acknowledged only by a separate first out acknowledge push button pressed after pressing the normal acknowledge push button.

3.2.4 Annunciation Types

Alarm annunciation can be realized in various manners such as a CRT/monitor alarm or an independent annunciator with or without a link with the DCS (alternatively it could be part of the DCS/PLC). Here, the discussions have been restricted to independent annunciators (still prevalent in smaller units) that can be realized in relay logic, solid-state logic (for example, C MOS), and microprocessor-based logic with a serial link to the DCS. As per the location of the logic system pertinent to the annunciator, it can be classified into two types: integral and remote annunciator.

1. With the integral annunciator, there will be a rack containing the logic card, power supply, and display units. Display units are at the front panel door whereas the logic cards are inside the rack with terminal blocks at the rear of the rack (or at the two sides of the panel) for field and other terminations. In some integral annunciators, associated push buttons are part of the rack. However, normally the push buttons are kept at the control panel and connected to the system by external wires.
2. Remote annunciators have separate cabinets to house the logic cards whereas the display units are separate and mounted in the main panel, say. Here, push buttons are externally connected to the system. Such types are slowly becoming extinct.
3. Semigraphic annunciator: Semigraphic annunciator systems have been used since the 1960s. in this type of annunciation, the alarm points appear as separate

display points at various positions in the plant mimic and/or the semigraphic panel. These may be individual integral annunciators or the remote design. This type was normally used for process plants in the earlier days, but now are almost obsolete. However, a similar philosophy is implemented in workstations.

3.2.5 Alarm Display Types

Annunciator display units are generally arranged in a matrix form, that is, a number of rows and a number of columns. There are various such RXC combinations available from the manufacturer. Also, the individual window size and alarm display size have wide variations. Some typical sizes may be 50 mm high by 75 mm wide. The designer has to choose a suitable alarm display depending on the designer's control panel size and layout. The displays can again be various types. One of the simplest forms is a transparent window with an inscription and a back-lit translucent window. In the former case, the inscription is readable from outside even when there is no alarm. When the alarm light glows, it could be read from a long distance, as shown in Fig. 7.68A, Back-lit ordinary display. In the latter case, the inscription can only be read when the lamp glows, as shown in Fig. 7.68B. The lamps inside the windows could be LED or incandescent . Depending on the importance, the display lens colors could be white, yellow, red, etc.

3.2.6 Annunciator Specification

In Table 7.27, the standard features to be specified for the annunciator are presented with ISA 18.1 as the guideline. Also refer to Chapter 14.

3.3 Sequence of Events

Precise and continuous monitoring and recording of plant parameters is an absolute necessity. In modern times with highly sophisticated automatic control and protection, a single failure may result in a huge number of alarms, so operating personnel may be at a loss unless all such events are properly time-tagged. Open-loop control systems have scan times that are generally higher, so for precise analysis, a sequence of events recording is necessary to time tag such

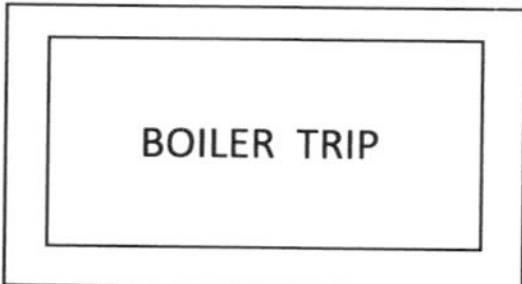

Back lighted Ordinary display

(A)

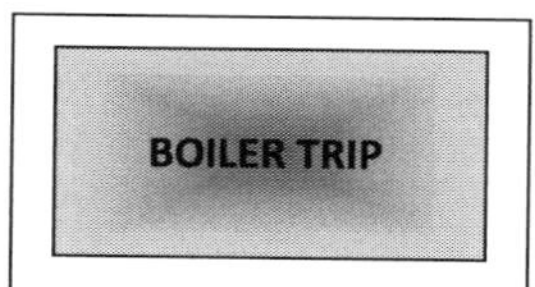

Back lighted Translucent, Readable when alarm

(B)

FIG. 7.68 Alarm window types.

inputs in 1 mS, so that the primary cause can be identified. In earlier plants (and even in smaller units still), there used to be a separate SOE/sequence of events recorder (SER). Now, in most cases this has become a part of the DCS/PLC.

3.3.1 SOE Features (Separate)

The SER normally has the following features:

1. A selected number of binary inputs in each of the subsystems such as:
   - SG flue gas.
   - SG fuel system.
   - SG pressure parts.
   - TG run-up system.
   - TG protection.
   - TG EHG system.
   - HP LP Bypass system.
   - CEP and Cond.
   - BFP Gr.
   - Misc. Others.
   - Reg. heater Gr.
   - Generator.
   - Exciter.
   - Unit elec. system.

All the binary inputs related to a process trip and pretrip conditions of the aforesaid subsystems are normally included in the SER. At least 100–120 points in a regenerative reheat power plant is a typical figure for a 200 MW plant. The exact quantity depends on the plant control philosophy and needs to be decided in the planning stage.

2. In many cases, SOE inputs are taken as a separate potential free contact for a hardwired connection to make the SOE an independent system; however, it could be a part of the main open-loop control system (OLCS) also. For use of common resources, SOE inputs may be connected to the DCS via link.
3. Like the alarm system, the SOE contacts normally also have the necessary optical/galvanic isolations and digital filters to avoid contact bounce.
4. Normally, the SOE records at least (say) a 100-point event for up to the next 3–5 min after the occurrence of any major disturbance, so that the analysis is proper and the sizing of the systems shall be done accordingly. Nearly 12,000–15,000 event records are typically available.
5. A standalone SER is available in high density packing Euro racks.
6. Independent SERs have interfacing facilities with the plant DCS, for example, RS 485 MODBUS protocol.
7. Independent SERs have separate self-diagnostic features and their own watchdog relays for output.
8. Standalone SERs are fully programmable and have a dedicated programming unit.

TABLE 7.27 Annunciation (DS)

Feature	To Be Specified	Remarks
Type	Integral/remote/semigraphic	
Sequence	ISA designation and sequence table/diagram	
Logic type	Relay, solid state, microprocessor-based (indicate soft link and protocol if any)	To specify
Total display	Actual display, active spare, dummy if any there quantity to be specified	
Power supply	Voltage, phase frequency, etc. interrogation voltage 24/48/110/220 V AC/DC standard	Also to specify back up power if applicable
Window size	Dimension of window (height, width, depth)	
Display area	Display width, height and RXC	
Display type	Ordinary or translucent type, color, etc.	
Cabinet details	IP class and cabinet mounting style.	For example, flush
Inscription details	Size and color each character and ch./line	And number of lines
Logic cabinet specification	Tentative acceptable size, enclosure class, wiring standard, etc.	For remote type if any
Filed contact type	NO/NC or site selectable	
Prefab cable	If any prefab cable connection is desired	
Contact discrimination and delay time	To specify first up discrimination time (1 ms, possible), also how much delay (say 3 ms*) contact can be considered as Authentic not spurious.	*If contact stays >3 ms it is nonspurious
Various control inputs	Number of push buttons with details of service	
Push button specification	Type, color for each service number of contacts and type rating	
Horn	Type, power supply, etc.	
Isolation feature	Optical/galvanic isolation as applicable	Optical isolation up to 2.5 kV available
Special alarm	Ground fault, power supply monitoring alarm (the type and details)	To specify
Type test	Associated type tests to be carried out	
Ambient condition	To specify temperature, RH	
Alarm test feature	Lamp test/system test	
EMI	To specify required standard	EN 61000-6; 2/4
Spares	Spares quantity for logic cards, flasher, lamp as per project philosophy	

3.3.2 SER as a Part of OLCS

When the SER is a part of the OLCS, generally it is then treated as a functional block in the program to collect the given numbers of channels in, say, 16 or 32 bits for each channel. Whenever it is embedded in a subroutine, then the sampling rate will be governed by subroutine execution time. The SER can be configured for pre- or posttriggered mode. The features stated in clause no. 3.3.1.1 are also applicable here.

3.3.3 SER Specification

General specifications for a standalone SER (Table 7.28):

4 INTEGRATED DCS-DDC MIS

4.1 Preamble (Explanation)

A DCS may be conceived as a **DDC MIS**. The term **DDC** (distributed digital control) means various control systems

TABLE 7.28 SER (DS)

Feature	To Be Specified	Remarks
Type	Standalone/part of OLCS	
Packing	Standard 19″ rack	
Capacity	Store at least 100 recorded data of all SOE point change state AND/OR recording at least for 3 min	About 12,000 to 15,000 recordings per unit
Cabinet details	IP class and mounting style in cabinet—flush, etc.	
Total display	Remote displays for back-lit lamp, LED, and/or monitor (CRT)	
Power supply	Voltage, phase frequency, interrogation voltage 24/48/110/220 V AC/DC standard	Also to specify back-up power if applicable
System processing	Built-in processor with necessary nonvolatile memory for setting as well as data	
Display character	About 60 characters with time stamp	In mS
Printer	Local printer if called for to specify printer types	
Character details	Size and color of each character	Also number of lines and characters per line
Filed contact type	NO/NC or site selectable	
Software	Details regarding software and software development	
Contact delay time	To specify delay time after which contact SOE can be initiated	To avoid contact bounce (say 3 ms)
Isolation feature	Optical/galvanic isolation as applicable	Optical isolation up to 2.5 kV available
Interface	To specify link and protocol for DCS	For example, RS 485 and Modbus
Aux. relay	Group/any relay output	
EMI compliance	To specify required standard	EN 61000-6; 2/4
Time synchronization	Pulse/GPS signal	
Spares	Spares quantity for various modules	As per project

functionally distributed (could be geographically). These controls could be an open-loop control system (OLCS) or a closed-loop control system (CLCS). An OLCS means a sequential control, interlock, and protection system where direct feedback may not be mandatory, and a CLCS is a feedback control system (analog/modulating control). **MIS** stands for management information system where all the tags shall be monitored and supervised such as the operator interface, calculations, alarm, etc. In the title, the word "***integrated***" *before "DCS" has been used to mean technological integration as well as functional integration.* Let us now look into the details on the issue as well as the features.

4.1.1 DCS Diagnostics Features

Fault detection and diagnostics are the keys to system stability and reliability. Whenever a fault occurs, it must be reported to the operating and maintenance personnel, at times in advance also. System diagnostics need to quickly analyze the fault by exploring various data. The diagnostic functions of the system must be able to pinpoint the cause of the fault as far as possible, locate the fault (physical location), and finally evaluate and analyze the system response due to this. Diagnostic software can monitor the faults at various levels, for example, network, network node, station unit, module, and I/O channel. In many systems, there is a separate fault diagnostic system, sometimes referred to as the **control diagnostic system** (namely, the Procontrol P14 of ABB). Most of the DCSs have intelligent I/O devices with self-diagnostic features. These diagnostic functions are active as long as it they are connected to the systems and they are well communicated over a bus to the other systems/system modules. It is possible to generate graphically the fault condition, location, etc., for the operator's displays. Now, with

intelligent field devices in conjunction with **fieldbus** systems, there have been *significant improvements* in system diagnostics and maintenance personnel are relieved to a great extent from pinpointing the fault. The maintenance types are slowly changing from reactive maintenance to preventive maintenance. The ability for a device to self-diagnose its health and integrity and put forward the information is too valuable to ignore, for example, process transmitters now detect plugged impulse lines/sensor drifts. Control valve diagnostics and the ability to generate valve signatures for online diagnostics allow many valve problems to be easily isolated and remedied without much cost associated with replacement. With a fieldbus system, it is possible to identify **failure**, **off specification**, **check function**, etc., for the field devices. In case of a fieldbus, it is possible to diagnose **function block diagnostics**, **resource block diagnostics**, and **transducer block diagnostics**, making it possible to spread the diagnostics at the **device level** (sensor degradation, dynamic error band, etc., in addition to internal diagnostics and process diagnostics), **loop level**, and **operation levels** (device and measurements identified by tags for fault detection) **identification**. Another good feature worth mentioning is that all these events are time-stamped at the source. Some systems such as "PRM with advance diagnostic application of YEL" have a diagnostic navigator, an advanced mechanism for fault detection. Systems such as the Invensys IA offer modules with fault tolerance at all levels. These are stated in this connection because, with all these intelligent device fault detections in conjunction with the diagnostic features of the DCS, it is possible to analyze the system as well as develop preventive maintenance features for the entire system.

4.1.2 DCS Security

For security aspects of the DCS, it is therefore essential that there shall be restrictions regarding access to various levels. Such variations normally encountered in power plants are indicated in Table 7.29.

Security threats in an integrated DCS come mainly from two sources: (1) human access at the plant location and (2) cyberaccess through the Internet. Many of the IT-enabled systems or IoT have protection and security for their servers and workstations against cyberattacks, which is a burning issue for integrated system. Invensys workstations are examples of the same. A secured network is distinguished from an unsecured one by putting an "s" after http, that is, https. In many systems, protection and security are provided for field device maintenance also. Some general ways to classify such access to the systems are described in Table 7.30. The security divisions are discussed in Table 7.29, Security Access Rights and Functions with password access.

4.1.3 DCS Redundancy and Fault Tolerance

Because DCS is the heart as well as the brain of the plant, there will be redundancies at various levels (field, I/Os, field I/O controller, I/O communication, controller, and main communication—maybe even for the network). At present, workstations (WS) have global data tag access, so during the failure of one WS, the operator can run the plant by accessing the same data from the other WS, provided he has security access for the data. Redundancies may be a 1:1 back up or two of three redundancies. When two controllers are working in tandem, then a few considerations need to be borne in mind, such as output from one has to be inhibited, and the output of the active controller has to be tracked by the back-up controller so that in case of failure of the active one, the hot standby has to fall back without any bump to the system. Also, another aspect one must keep in mind is that in case of a hot standby system, *either* some other processor has to check the healthiness of the controllers or each of them must also check the health of the other so that failures can be detected for changeover. A TMR system is a classical example of the same. In a redundant system, online upgradation is easier only suitable care should be taken to ensure, that but one needs to be careful not to compromise system safety. With reference to Sections 1.1.11 and 2.3.1.4, it is clear that fault tolerance and redundancy are not the same. So, in critical systems, fault-tolerant systems need to be deployed.

4.1.4 DCS Safety Aspects

There are two schools of thought regarding the integration of safety instrumentation with the basic plant control system (BPCS) with Safety Instrumentation system (SIS). Some feel that these two systems should be separate and should not be integrated into one network. Others allow them to be integrated as long as the SIS is an independent entity (no way depending on the BPCS). For further details, see Chapter 14; for example, the YEL Pro safe RS is an example of the same.

4.1.5 System Integration

A power plant is very much interconnected as well as interactive, so an integrated system is a good solution for plant control. This is not only restricted to the main plant but it may be necessary to integrate various offsites. Naturally, system integration ensures a seamless data exchange between these units and the offsite. In this connection, discussions in Section 2 may be referenced as these are applicable here also. So, the Ethernet and OPC foundation have provided DCS suppliers the ability to access and exchange information among the various networks and help the user to integrate the systems. Such integration provides improved operation and maintenance of the plant, better inventory, better management information systems, remote access for the manufacturer for troubleshooting, etc. This is thanks to international standard IEC 61158-2, which made it

TABLE 7.29 Security Access Rights and Functions

User	Allowed functions	Functions not allowed
Operator	• All displays, control of drives/devices • Changing of certain permitted set points and bias • Report printing in various forms permitted by the system	• Modification of controller parameter • Value pegging/forcing (**except** permitted ones—may be during sequence operations) • Modification of logics/loop/database/MMI SW • Taking off scan, changing any assignment for log and displays
Unit charge engineer	• All displays, control of drives/devices • Changing of certain permitted set points and bias • Report printing in various forms permitted by the system • Taking off scan, changing any assignment for log and displays	• Value pegging/forcing (**except** permitted ones—may be during sequence operations) • Modification of logics/loop/database/MMI SW
Shift/Station/ OandM in charge/Misc. Other users	• All displays, but no control of drives/devices • Report printing in various forms permitted by the system	• Control of drives and devices • Modification of controller parameter • Value pegging/forcing (**except** permitted ones if at all any) • Modification of logics/loop/database/MMI SW Taking off scan, changing any assignment for log and displays
CandI maintenance in charge	• All displays and printouts, control of drives/ devices. (normally during commissioning/plant modification) • Changing of certain permitted set points and bias • Value forcing and pegging • Modification of logics/loop/database	• Modification of MMI SW • Control of drives and devices after expiration of temporary password given to handle control function during commissioning and modifications
System administrator	All functions only in consultation with concerned personnel (may be joint access)	

TABLE 7.30 Security Access Category

Access	Functions	Remarks
Authentication	Password-protected user account for access to the system for control and making changes, for example, an operator may access to control and make a few changes in the set point or trending data collection interval whereas an engineer may have access to modify a part of the logic/loop	(1) Many times systems keep the record of such changes and they can be recalled later to identify who made the change and when (2) Most of the manufacturers have a minimum of three to four levels of access authentication
User	Like e-mail and other systems, a unique ID could be assigned along with a password	In case of an IT-enabled system, even users who are permitted can access through the Internet also
Classification	User access level can be determined plant-wise, and or function basis also	

possible to have some standardization of fieldbuses, which are an important means to integrate various systems.

4.1.6 Control Algorithms

Here, the concentration will be on the theoretical aspects of control systems, mainly feedback control systems, to refresh the memory.

1. BASIC PID ALGORITHM: A basic PID controller can be conceived, as shown in Fig. 7.69, Basic PID (Parallel Form), where an error is generated at the input of the controller where it is subject to mathematical treatment by the equation $O = K_p \times e + K_I \times \int e\, dt + K_D \times de/dt$. As the error is fed parallel to the proportional, integral, and derivative operators, it is a **parallel form**.

In most cases, the error is fed to the proportional controller, then from the same is fed to the other operators, as shown in Fig. 7.69, Basic PID noninteractive; even the value proportional to error is fed to the other operators.

K_C being a proportionality constant is dimensionless and expressed in percentage whereas T integral action time and TD are expressed in seconds as the unit of time. There is another term, proportional band, which is defined as PB $= (1/K_C) \times 100$. So as K_C increases, PB decreases. Generally, for stability, PBs are kept $>100\%$. As is known, a proportional controller introduces errors into the system whereas integral action integrates the error, so it provides zero errors at the steady state. For faster response, derivative actions are introduced into the system, but this in general is responsible for the generation of noise into the system.

There is another kind of basic PID control algorithm where the derivative is done on the proportional error prior to integrating the same. This is also known as an interacting PID controller (Fig. 7.70). A typical one is shown in Fig. 7.71. Here, the integration is done not only on the proportional error but also on the derivative of the proportional error. In case of persistence of error for a long time at the input on account of integral action, the output may grow to the highest possible value. This is **RESET** WINDUP, an important feature in a PID control system and in all modern controllers. **Antireset wind up** features are provided to avoid uncalled for overshooting and disturbances to the system. It is necessary that the control system be tuned for a proper response for a change in set point so that there is no output spike or bump due to derivative and proportional action. Also, it has to cope with the rate of change of the set point so that it is not too slow. Stability is another major goal of a control system. On account of delays, dead time, and high process gain, there may be instability and/or oscillation in the system. Therefore tuning of the control system is essential. What is tuning of the controller? By tuning the controller, one tries set the controller parameters such as gain, integral time, dead time, etc., so that the controller can effectively match the process dynamics. Conventionally the "Ziegler-Nichols" method is popular for controller tuning. In order to have a better system response, set points and measured variables are applied through filters, as shown in Fig. 7.72. This kind of approach is helpful to prevent sudden spikes. Sometimes, dead band timers are kept in the circuit to avoid a spurious signal affecting the control system.

2. SET POINT APPLICATIONS: As is clear from the above discussions, each of the terms such as P, I, and D acts on error. So while applying the set point to the controller, suitable measures in the controller need to be taken such that there will not be any spike or bump at the controller output due to a sudden change in the set point. There are contradicting factors: if it is set to take care of the rate of change of the set point, then the system response will be slow, whereas if it is set to reject disturbances, then there will be overshoot. So, a required softening effect along with suitable tuning is necessary.
3. VARIOUS CONTROL TYPES USED IN POWER PLANTS: Other than a PID control algorithm, there are a few other types used in power plant controls:
   - CASCADE CONTROLS: This is a multiloop structure where the outer loop has the set point (acts as the master controller) and the output of the same goes to the inner loop or slave loop as the set point to be compared with an intermediate parameter as a process variable. Here, a disturbance in the second variable can be corrected at the slave controller without affecting the primary variable. Also, the slave loop helps to reduce the process lag and this results in a faster system response. Steam temperature control is a typical example of this kind of loop.
   - FEEDFORWARD CONTROLS: Feedback controls have inherent system lag (between a changing of the manipulated variable and the effect of the change on the controlled variable). In a feed forward system, the major components of load are entered into a model to calculate the value of the manipulated variable to enhance the loop response. In combustion control, at many places steam pressure is used as

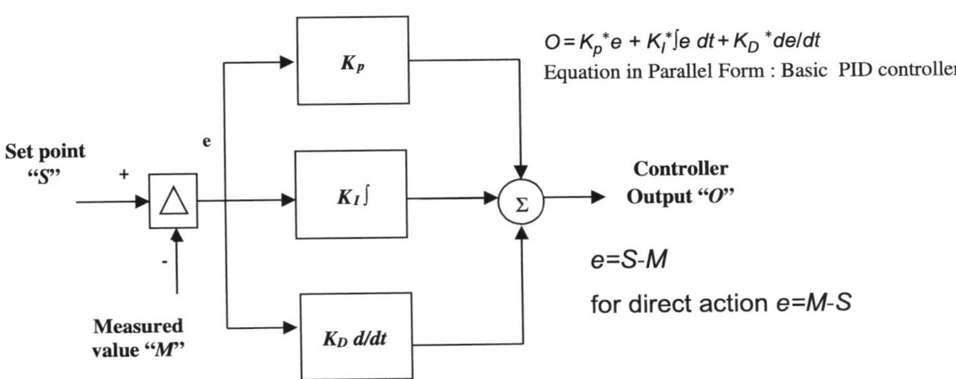

FIG. 7.69 Parallel form: basic PID controller.

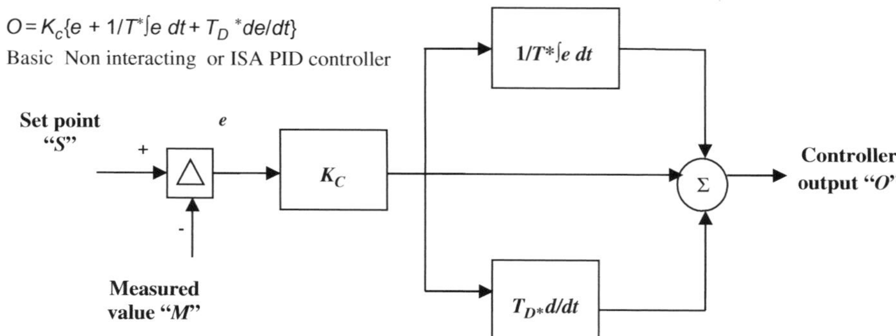

FIG. 7.70 Basic PID controller noninteractive.

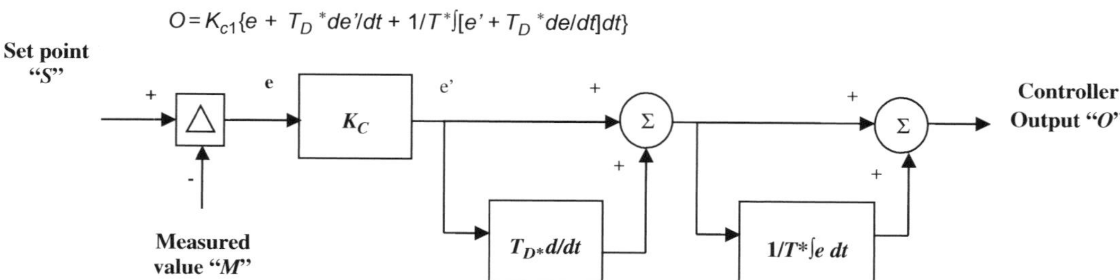

FIG. 7.71 Basic PID controller interactive type.

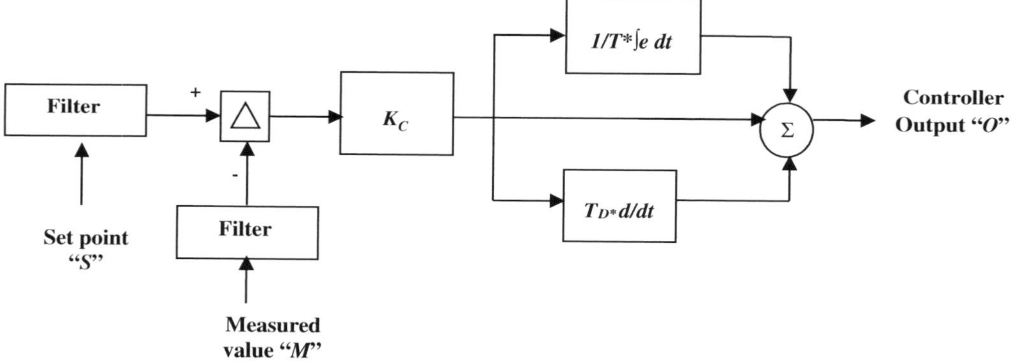

FIG. 7.72 PID controller with filter at set point and measured variable.

the main control variable whereas the steam flow/boiler load is used as the feed forward signal to make the loop responsive.

- RATIO CONTROL: It is an elementary feed forward control system. The advantage of this loop is that here, the control variable–the air: fuel flow ratio–can be kept constant. Perhaps not in the crude form but in a different fashion, the same type of control system is deployed for fuel flow control (with fuel type, the ratio may vary a little).

4. POSITION AND VELOCITY ALGORITHMS: The PID controller algorithm discussed so far produces the output to determine the position demand for the final control element. Thus, this type of control algorithm is known as a position algorithm. The control algorithms that determine the rate of change of the final control positions are known as velocity algorithms. These are effective in digital implementation in DCS. These are used for set point controllers for supervisory controls. In this system, the output is effective, keeping in mind the previous position so that it generates $\Delta O = O_n - O_{n-1}$. In a DCS, the previous position is stored in the memory. After the velocity algorithm is calculated, it is added to the previous position and is stored as the new "last position." For this, the transfer is effectively bumpless. With velocity algorithms, the reset action on the error is not effective, so it needs to be used as the override action.

5. AUTOTUNING AND ADAPTIVE CONTROL SYSTEM: There are two processes involved in the tuning of the controller; one is to learn the process dynamics while the other is to make the manipulated

**MFA: Model based systems are developed based on good knowledge of process dynamics and operational environment. In practice this is not always possible! MFA combine adaptive theory with intelligent controls (viz. Fuzzy logic/ artificial neural network) in learning control mode.*
Following are the basic features of MFA:

- *No precise quantitative process knowledge*
- *No Controller design for a specific process*
- *No Manual tuning of controller*
- *Stability criteria available to ensure closed control loop stability.*

The basic aim of these controllers is to produce an output to force the process variable to track the of the set point under variation in Set point, disturbances,& process dynamics. It does so through the process of learning and knowledge base

FIG. 7.73 Note on model free adaptive.

measured constant. Many DCS manufacturers use the AUTOTUNING method when the tuning parameters are continuously adjusted, depending on the process dynamics. They are used to inject excitations at the loop to sense the process response. These excitations could be very small, but enough to analyze the control variable in response to the change. Such an excitation injection could be at the gain, set, and/or the controller output, depending on the manufacturer's design. The aim is to understand the process dynamic. Once it is effectively done, necessary tuning parameters can be set until there is no further change in the system. In the majority of the advanced controllers, this are done with the help of mathematical models with prediction. There are other methods such as model free adaptive control. Model free adaptive (MFA*) control does not require precise quantitative knowledge of the process, no special control for any process, etc. MFAs could be single loop and/or multiloop types (Fig. 7.73).

A self-tuning controller is another option that is mainly applicable for a single-loop controller, but it could be a part of the DCS. Three basic components of self-tuning systems are the system **identifier** (the algorithm responsible for the process parameter estimation), the synthesizer (calculation of new controller parameters based on estimation algorithm result) and the control block (the main control algorithm whose parameters are evaluated by the synthesizer).

6. OPTIMIZATION: Like any other system, it is the aim of power plant control to maximize profit while maintaining the safety and production quality without degrading the equipment life. In a power plant, any economization is welcome as it saves energy, which is a scarce element. Normally, for optimization, independent variables are identified. These are fed to a black box, a mathematical model to calculate the performance parameters based on various constraints such as disturbances and intermediate variables.

4.1.7 Standard Library Functions

In any DCS, the control loops are functionally represented in terms of the algorithms in the DCS standard library. A list of such algorithms is presented in Table 7.31. The symbolic representation of some of these functional blocks has been in the control loops of this book. In this connection, discussions on IEC standards IEC 61131-3 and IEC61804, as discussed in Section 4.1.5, may be referenced.

4.2 DCS Configuration (OLCS/CLCS Discussions)

4.2.1 Introduction

In a power generation plant, sequential controls/interlocks and modulating controls are both interconnected and important. In a power plant, all subsystems are **connected** and the process is **continuous** and **fast responsive**, so the emphasis needs to be provided for both systems. In smaller plants, the PLC/SCADA could be a solution but in medium to large power plants, on account of a powerful **function library,** the DCS is the preferred solution. Most of the DCSs support **ISA S88** for interlock and have **embedded advanced control and safety** functions. Various subsystems of the DCS in a power plant are a closed-loop control system (CLCS) for MODULATING control, an open-loop control system (OLCS) for sequential control, interlock and protection and management information system/information system (MIS/IS) for monitoring and supervision, and interface functions, for example, monitoring, performance calculation, operator‘s interface, etc.

The basic system components of a DCS re shown in Fig. 7.74. There are two kinds of controls such as OLCS and CLCS (distinguished by the pattern inside) and there will be different components for IS/MIS functions. However, in reality in some control systems, such a distinction between two control systems may not be done in the DCS. Had that been done, then the OLCS could be replaced by a PLC (to curtail the cost). In fact, in a DCS such OLCS/CLCS are done very judiciously, as discussed in the following sections.

4.2.1.1 Control System Grouping

The entire plant can be divided into a unit level, a functional group level, a subgroup level, and a drive level. At the unit level, automatic start up or shut down commands can be issued. Functional levels are the levels where independent functions are performed: for example, the feed water system (a functional group). Within the functional group, there will

TABLE 7.31 Standard Library Functions

Type	Algorithm	Explanation	Remarks
Computation	Summing	Sum of inputs	
	Difference	Difference between inputs	Error Gen
	Median	Selection of mid value	Of inputs
	Average	Sum divided by number of inputs	
	Multiplier	Multiplying two/more inputs	
	Divider	Divided an input with other	
	Scale and bias	Scaling input + bias value	KX + C
	Function generator	Output as function of I/P	Functions to be defined
	High low selector	Selection of high or low of inputs	Max/min
	Dead time	Output after sp. time	
	Comparator	Compare with SP or with other input for sp difference	DO or switchover
	Comparator + CO	Same as above with CO	CO = Change over
	Lead/lag	Output increase or decrease allowed when compared	Between inputs
	Pulse generator	When input crosses limit, a pulse/trend of pulses	
	Differentiator	Calculation of rate of change	Of input
	Integrator	Integration	Of input
	Square root	Square root extractor	
	Flow computer	Pr. and temp compensation	$Q = \sqrt{(hP/T)}$
	Log function	Output as log of input	
	Exponential	Output as exp of input	
	Limiter/clamp	Output limiter	
	Limit value monitor	Generate DO when input crosses the set limit	
	Tr. Selection	Selection of transmitters on	
	One of two selection	One out of two	Chapter 3
	Two of three/ median	Selection of two of three inputs	Chapter 3
OLCS functions	AND/OR/NOT/ ExOR	OLCS functions of input(s)	Boolean logic
	Switch	Types of switching functions	Latch, unlatch
	Timer	Delay of specified time	On/off delay
	Counter	Counting of pulse, etc.	With DO
	Flip flop	Temporary memory RS/JK flip flop, etc.	
	Drive	Drive control with functions, various conditions, and logics	Trip, permissive, etc.
	Subgroup function	Logical subgroup function	Owner's choice
	Pulser	Generate mono shot spike or pulse when input reached	
	Changeover	Input changeover	May be based on third AI/ DI

TABLE 7.31 Standard Library Functions—cont'd

Type	Algorithm	Explanation	Remarks
CLCS functions	PID	PID functions with ratio, gap, antireset windup, etc.	
	Other PID functions	Feed forward/tracking/output limiting	
	Set point functions	Set point ramp, set point clamping	
	Auto manual	Switchover and tracking	
	Auto tuning	Auto/adaptive tuning	
	Others	To complete the list	

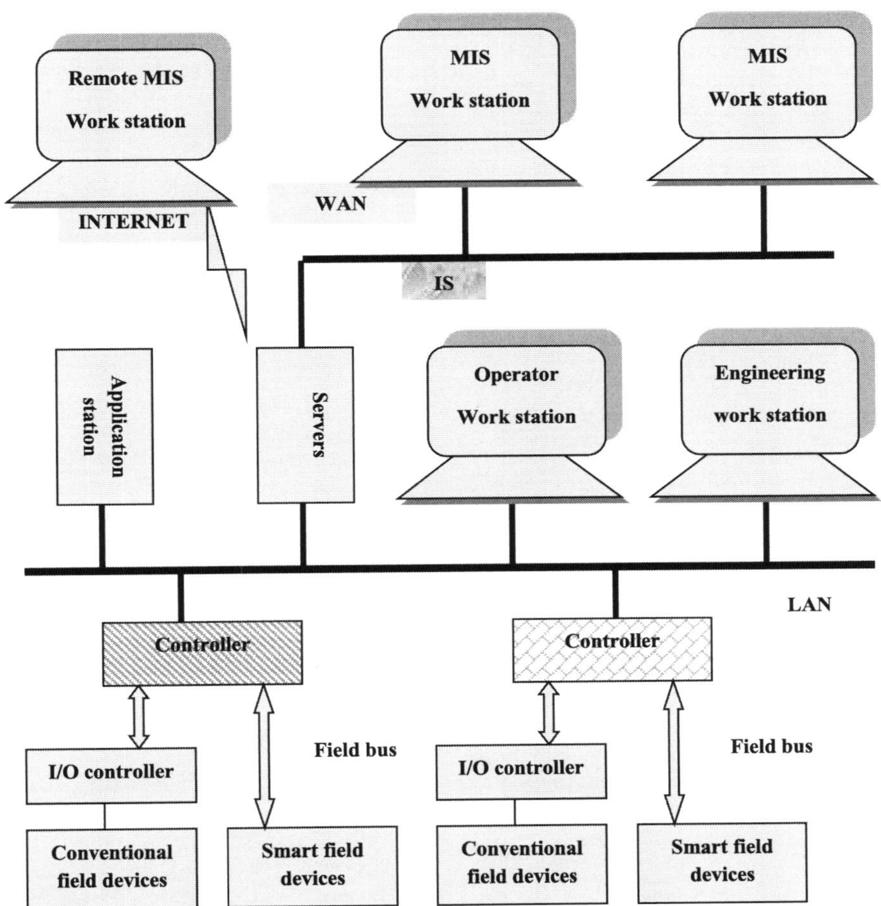

FIG. 7.74 Basic DCS components.

be subgroups such as the BFP subgroup comprising the BFP with its auxiliaries and associated valves. Drive level is the last level where the individual drives are addressed and controlled. Now, in case a particular functional group/subgroup is considered, for example, the BFP "A" subgroup, then there will be both OLCS and CLCS functions associated with it such as start up, shut down. and interlocks of BFP "A" will be OLCS functions whereas speed control of BFP "A" will be part of CLCS function (may be in the form of subloop discussed earlier in connection with turbine control). Normally, for each the functional groups/subgroups (**in power plants as the modulating control loops are so interrelated, many times it is very difficult really to distinguish between groups and subgroups for control functionality on account of functional overlapping. Therefore a judicious decision regarding such grouping is essential),** separate controllers complete with power supplies and I/Os are allocated to make the same independent so that the failure of one functional group/subgroup does not affect the other in terms of total control or its communication with the IS. So, the controller would mean its **associated I/O system, power supply, and communication system** also. Such groupings are mainly found in German and other European system designs. Some of the functional grouping criteria are discussed below:

1. GROUP CONSIDERATIONS: The grouping of drives depends on a number of factors such as the **system configuration** of the DCS control, **the numbers of I/Os each** controller can handle, the **communication speed** of the system, and the **update rate, meaning both the** process data update and the update over the network. As stated in clause no. 2.4.2.2 of this chapter, many of the drive-related I/Os can be segregated and put in a drive control card so that much of the I/O handling of the individual controller can be reduced. Also, because it is possible to manually control the individual drives, more equipment can be clubbed in the main controller (with drive control cards). This type of concept holds when the manufacturer divides the systems in such groups/subgroup and drive level controls. On the other hand, there are systems that do not go for such subdivisions. These systems have only I/O cards and a controller; some of the drive functions need to be done at the controller level. Naturally, the group division will be different. As each of the two systems discussed has some benefits and drawbacks, hence it is dependent on the manufacturer's control philosophy. Fieldbus systems drastically reduce the controller's timing for I/O handling. Also, some manufacturers use medium-sized multifunction controllers (with or without a back-up controller) to control several pieces of equipment (as in the PROCONTROL P14 of ABB), whereas some use a larger control processor with a HOT controller back up to control several pieces of equipment (for example, the CP of Foxboro Invensys). Naturally, the I/O handling capability, power, and functionality of these two types of controllers are not the same. In the former case, many advanced functions will be kept elsewhere. However, both types are workable and in service for many years, and it is purely the design philosophy of the manufacturer. With so many options available, the system designer has to choose the better one as per his plant control strategy and philosophy. However, the designer's main considerations shall be such that the **failure of** the **controller cannot threaten** the **complete loss of** the **control system (at** the **worst condition, it can affect a part only)**. It will be clear from an example. The system is chosen in such a way that the failure of one control system may affect one FD, one ID fan, and/or three mill systems (out of, say, six mill systems), but not the ENTIRE SYSTEM. It is a joint effort of the owner (and/or his engineering consultants) as well as the manufacturer to decide on the issue of controller assignment for various controls, based on available devices required to control the plant.
2. HOT BACK-UP CONTROLLERS: The hot back-up controllers should fall back on the failure of the active controller in such a way that it is procedureless and bumpless.
3. INTERRELATED CONTROLS: There are many controls that are interrelated, for example, the PA fan is connected mainly with the mill, so while selecting the controller for the PA fan, associated mills could be considered. If necessary, some of the signals may have to be repeated in two or more controllers, for example, steam flow and air flow are the signals that are required at many places (for example, at the steam pressure control, the steam temperature control, the drum level control, etc.) so these signals may be distributed in a hardwired I/O card or software at the I/O Bus level to a number of controllers.
4. ENVIRONMENTAL CONDITIONS: In power plants, in a majority of the cases, the electronics (except field devices) are placed in a controlled environment. However, there are cases, especially in RTUs, where electronics/control system components may be placed in harsh environments. So, while selecting the controls, due consideration shall be given to the possible worst environmental conditions and these should be mentioned clearly.
5. PROCESS REDUNDANCY: In a plant, if there are three ID fans deployed, then they are deployed for redundancy, so naturally the controller shall be such that a failure of one would not jeopardize the entire ID fan system. In cases where the drive control cards are used and these cards are separate and controllable from control room.
6. STREAM CONSIDERATION: In case several pieces of equipment are required to be clubbed together in a controller, then due consideration shall be given so that a failure of a single controller does not hamper the entire operation. Normally, FD fan A, air heater A, wind box A, etc., form a stream, so while grouping, such considerations shall be kept in mind so that in case of a failure of one common controller, the other stream can be kept running. In case cross-running of the equipment is possible, then due consideration must be given to the interconnecting valves/dampers. This will be clear from an example: If in a P&ID with the help of an interconnecting damper, it is possible to run (say) FD fan A with AH B in case of failure of AH A, then while selecting the controller, the interconnecting damper shall be suitably included in one or both (redundant) controllers.

4.2.1.2 OLCS FUNCTIONS

OLCS functions shall include SEQUENTIAL CONTROLS (ATRS), INTERLOCK, AND PROTECTION for various plant auxiliaries, valves, dampers, etc. For OLCS, a few features shall include but are not limited to the following:

1. OLCS HIERARCHY: As discussed above, the OLCS system can be conceived to be arranged in a hierarchical structure, such as unit level, group level (such as boilers, turbines, feed water systems, etc.), subgroup level such as the evacuation subgroup, and drive level. A hierarchical structure may be helpful for sequential control implementation. However, some system drive controls are implemented in I/O cards. It is purely the design philosophy of the manufacturer.
2. UNIT LEVEL START UP AND SHUT DOWN: It is possible nowadays to start up/shut down a complete plant from a single keystroke. Such start up or shut down is done at the unit level. While designing a control system of this kind, the safety aspect is given the highest priority at every stage so that the sequence may proceed properly.
3. GROUP LEVEL CONTROL: Such controls are quite common in most of the systems such as BMS, ATRS, ATT, etc., as discussed in Sections 2.3 and 2.4 of this chapter. At this group level, independent automatic start up and shutdown is possible. When in the unit level, all groups are started one by one from a single start/stop command. For group starting or shut down, each group needs separate command. All subgroups, subloops, and drive interfaces (as applicable) under each of these groups shall be started one by one under this single group command. It is interesting to note that these group commands need not be fully automatic, but can be semiautomatic also.
4. CHECK BACK: CLCS is a feedback control system whereas the sequential controls of OLCS have CHECK BACK (ref clause no. 4.2.1.7) signals for the sequence to proceed. It is like feedback in CLCS. After a command signal is issued, before proceeding, the system ensures that the previous command has been obeyed by all concerned within the allowed time gap. This is to ensure that there is no hindrance in the sequence. This is done by check back signals. If there is any constraint for the execution of the previous command, then the fault/constraint must be detected and removed (may be bypassed for noncritical check backs, even manually and/or by manual override). For example, in order to start a large fan, it is necessary that its lubrication pump runs so as to establish a flow of oil in the bearing. Now, if the command is issued to start the pump, but the oil pump does not start or cannot send oil, then the bearings will starve. Therefore, check back signals (that the lube pump is running and the oil pressure is OK) are essential for the sequence to proceed further. Now. if there is **no such check back** for a large fan, then one can imagine what would happen.
5. OPERATION RELEASE AND LOGIC CRITERIA: Before issuing any command, the safety and readiness of the concerned equipment and systems need to be checked. When a command is issued (even manually), it is the duty of the system to check whether all the conditions for the safety and readiness of the concerned equipment and systems are available before it can make the command effective. These criteria checked by the system at the start of the sequence for any subgroup/subloop/drive are normally referred to as **operation release.** For example, after a boiler shutdown, any fuel can be started only after purging is complete in the recent past (allowed time gap). In fact, if the fuel is not started within a specified time after the purge, another fresh purge cycle may have to be repeated before starting the first burner. Similarly, at each stage or for each interlock, such checks are done with the help of check back and other status signals referred to as **logic criteria,** which need to be satisfied (or bypassed) to proceed to the next logical step. **The interlock** will have the highest priority and can be executed at any stage.
6. TIMING **Safety**: See Section 2.4.2.4 of this chapter.
7. OPERATIONAL MODES: Normally, three modes of operation are seen in this type of OLCS: **AUTOMATIC, SEMIAUTOMATIC (step by step), and MANUAL (operators guidance)**. These are also discussed in clause no. 2.4.2.5 of this chapter (Fig. 7.75).

***Example to explain** (Fig no. 7.62B Example Step Logic): Start up Command for Turbine sub group (e.g. KWU- Siemens) turbine will have the Operation Releases. Some of them are: Starting device 0%, Position set point turbine controller < 0%, Field breaker OFF. Generator breaker OFF.*

Similarly in the same turbine start up example, For issuing command at one of the Subsequent step(3) shall have check backs as Logic Criteria before it can issue the command! Some of these criteria at that step would be Turbine speed > 15 rpm , Condenser pressure < 0.5Kg/cm2.,Any one CEP is ON------ this means that the only upon receipt of all operational released criteria available, start command shall be made effective. Similarly in subsequent step command shall be issued only after logic criteria are met. However in case of any safety condition/interlock (say), from Fire it will be made through as an emergent condition!

FIG. 7.75 Step logic explanation.

8. MONITORING and DISPLAY: In order to facilitate operation, it is customary to have the step number /step status, operational release, step criteria, and check back status displays made available to the operator. There are provisions for alarms for abnormal situations. All modes, loss of timing, etc., are made available to the operator. Also, it is important that access to these logic programs shall be restricted to the authorized persons. Certain steps/criteria bypasses are allowed, but this is by operator only when a faster action of operation is called for, but never at the cost of (human, equipment) overall safety and only authorized persons are allowed to make such actions. Also, a few other sensors monitoring like Non coincidence error, Open/Short circuit monitoring etc. are done.
9. COVERAGE: The OLCS discussed above is not only applicable to the main power plant sequence, interlock, and protection, but shall include all offsites as well. However, in these offsites, there may be a PLC integrated with the main DCS, but the criteria discussed in the OLCS could be utilized, depending on the applicability. The same thing is applicable for the electrical control system also.

4.2.1.3 CLCS Functions

These are modulating feedback controls deployed to achieve the desired parameter conditions. Some of the major features of CLCS could be:

1. CLCS CONTROL RANGE: CLCS is a control system that acts continuously, both in position and time, on control actuators (valves/dampers), mechanical equipment (hydraulic coupling), and electrical devices (VFD). Theoretically, a modulating control loop is to control the parameter over the entire range. In reality, such control is applicable for steady-state conditions ***mainly*** between >50% (60%) to ~100% MCR (maximum continuous rating) with variations in associated parameters within specified permissible limits. During transient stages and at lower load, it may not be possible to achieve CLCS regulation as per the performance specification (At the steady state, the loops are normally tested with STEP and RAMP functional signals).
2. CONTROL ALGORITHMS: Power plant CLCS loops are interrelated and require complex control algorithms. However, some of the basic forms of control algorithms may be standard **PID** (**P**roportional + **I**ntegral + **D**erivative), **PI**, or **P.** For the basic idea related to various control algorithms, Section 4 of this chapter may be referenced. Out of these, the PI controller is used in most of the applications. On account of noise, many engineers avoid **D** action. **PD** control in the governing control stages is common in many applications (for example, the KWU Siemens turbine). Because temperature is a slow-changing parameter and sensing is not fast, it is recommended to use PID controllers in temperature loops. Similarly, the I action in level controls may be chosen to a minimum. **Cascade** controls (superheater/reheater temperature control), **adaptive gain** controls (any loop, say, PA flow control) **state variable** controls (superheater temperature control), **feed forward** controls (steam flow in combustion control, or the difference between steam flow and feed water flow as the feed forward signal in the three-element drum level control), **ratio-bias** controls (fuel-air ratio control), **ON OFF** controls (DM makeup water control), etc., are common. Mathematical model-based controls are also applied in CLCS in power plants to get a better control loop response.
3. CONTROL TRANSFER and LOOP REACTION: In OLCS, because the devices and controls are discrete in nature, the transfer from auto to manual and vice versa is not much of a problem. However, in modulating controls, because the control actuator can assume any position prior to transfer from auto to manual or manual to auto, naturally in both cases, the control system outputs have to match. In case of a mismatch in output, there will be different inputs coming to the control actuator just before and after changeover. As a result, the control actuator may either close or open suddenly. On account of this, there may be a big change in the measuring parameter, that is, there will be a bump in the system. Therefore, it is necessary for the CLCS that both auto and manual outputs track each other so that in a transfer, the changeover is bumpless and procedureless. It expected that the control loop reaction time shall be restricted within **400–500 mS**, or even faster for some loops.
4. CLCS OUTPUT: Depending on the actuator used, the output of the CLCS shall be fed to the control actuator through various converters. This is because in most of the cases, the AO cannot directly interact with the actuator. So it is important that suitable measures are taken in the control actuator and/or in the converter so that in case of failure of signal from the CLCS, the actuators assume a safe position, as discussed in Chapter 6, Section 4.1.
5. CLCS OPERATION: The aim of all control systems is to achieve the maximum efficiency of the system/equipment under all conditions, provided the safety aspect is well taken care of and the permissible limit for the parameters is not exceeded. However, during the operation, if the permissible limits of the parameters are exceeded, then the CLCS needs to reorient itself so that the control loop performs in such a way that it ensures the safety and availability aspects of the control loop. For example, when there is load demand, the

coordinate control loop will load both the boiler and turbine in such a way as to get the maximum load output with high efficiency. Now, during this period, if there is a large disturbance as a follow up of a major upset or trip (of a major auxiliary), naturally at this point all the loops will be disturbed due to interactions. Also, the parameters may swing in the range out of the permissible limit. *At this point, the concerned CLCS loop shall first try to bring the parameters within the limits,* with the main aim to ensure the safety and availability of the system instead of looking for higher efficiency or load demand.

6. LOGIC CONTROL and TIME MONITORING for CLCS: There may be some interlock operation with some of the loops in the CLCS system. It is expected that these loops shall be performed within the same control block utilizing the logic functional block. If the succeeding heater level is high, then the higher heater will drain to the condenser. Such interlocks are normally implemented in the same control loop and not on separate blocks. Like OLCS, for the CLCS loop also, there shall be a time supervision facility so that if within a predefined monitoring time the output or feedback is not appearing, then the same shall be alerted to the operator and the loop may trip to manual mode.
7. CONDITIONAL OUTPUT: Under certain conditions, it is necessary that the CLCS control loop produces a preprogrammed output. This is mainly coming from the coordinate control loops discussed in Chapter 10. These may be RUN BACK, DIRECTIONAL BLOCKING, etc. These will be clear from various examples: From the load dispatch station (LDS), there may be a demand for a higher load from the unit. At this point, if there is a failure of one of major auxiliaries (say, one out of 2 X 60% FD fans has failed), then the boiler or unit loading will be decided by the capability of the running auxiliary. Firing, for example, will then run back to a max 60% (or less) load demand. Also, in a running plant in a similar situation of auxiliary failure, the CLCS loop **runs back** to the preprogram position. Again under such situations when any CLCS has reached maximum position (may be due to auxiliary failure or the full MCR is reached), then even if there is a further demand to increase, the loop will not allow a further increase. However, if there is lowering demand, then the same will be entertained. The situation will be similar in decreasing demand also. So, the loop is blocking a signal in one direction, hence it is called **directional blocking.** Thus it is seen that the RUN BACK comes mainly from a major auxiliary failure or not running whereas DIRECTION BLOCKING comes from auxiliary/system limitations.
8. FUNCTION LIBRARY: CLCS normally has a large functional library; see Section 4 of this chapter. The majority of suppliers support functional blocks, as per IEC 61131-3.

4.2.1.4 IS and MIS Functions

Apart from control functions, DCS functions such as alarm/SOE monitoring of various displays and logs are covered under the information system (IS). Various trend analyses, alarm SOEs and logs help operating personnel, technical personnel, and management personnel to assess the plant conditions both technically as well as commercially. So it could be argued that the information system is extremely valuable for running the plant as well as for plant performance analysis. When there is an IS system, what is the use for MIS? It is like plant overview display and detailed display; anyone will appreciate that both have separate requirements. The operator would prefer to view the overview display under normal conditions, and when required would go for detailed display. As the information in a plant is enormous, various information presentations have separate values to various personnel. For example, a shift engineer may be interested to note the trend of the parameter over the shift to check the health of the equipment, or the pre- and posttrip log so as to analyze the cause of the failure and how the plant was brought to operation. Again, a person at the helm of the organization located at a remote location would like to have all major plant information at his or her fingertips. So, it all depends on how the information is cooked and presented. In IS, all normal information is presented in different forms, whereas in MIS, this information is cooked and summarized before presentation in the desired format. MIS would include the overview, plant status, performance, and other calculations, including commercial calculations. Therefore, one can argue that MIS is a filtered IS with the filtering parameters set by the user subject to the user's accessibility and requirements. The major components of these systems are:

1. OPERATOR INTERFACE: The operator interface could be anything with which the operator interacts with the process presented before the operator (in this case) through a specialized intelligent network. The operator interface shall include but is not limited to call displays, alarm acknowledgements, various parameter modifications (as permitted), call logs, group parameter assignments, call for calculation results, including those for performance calculations, etc.
2. CALCULATIONS: Various calculations from simple calculations such as flow calculation to rate-of-change calculations to equipment efficiency to plant performance calculations.
3. ALARM ANALYSIS AND LOG: Display of various alarms, return to normal alarms, first up alarms, alarm logs, etc., with date and time stamping for future analysis.
4. LOGS: The various logs are: overview, demand, periodic, historical, alarm, SOE, etc.

5. SOE FUNCTIONS: SOE as predefined and/or following a trip; display and logging of these SOE points that could play a crucial role in the trip and controlled shutdown.
6. TRENDING: Trending of selected process parameters is a part of IS and provides a useful tool for operating personnel for analysis. Trending can be event-based or manually selected. Trending of parameters is also very useful during start up and/or (control) shut down, for example, trending of the metal temperature of the boiler and turbine during start up.
7. HISORY FUNCTION: Keeping the parameter history and retrieval of these parameters as and when required are also parts of the IS function.
8. SOME UTILITY FUNCTIONS: Some of the utility functions of IS are:
   - Assignment of any point from the database for display and log groups, their data collection intervals (1, 2, 5, 10, 30, 60, 120, 300 s), printing intervals, header information, selection of these points for their max, min, and average value, etc.
   - Modification of group, for Display and Log that is, inclusion and exclusion of tag from a group. However each of such change history can be available in DCS and can be called at any time to check when and by whom such change was initiated.
   - Other utility functions supported by the system such as MS office, MS projects, and other utility and office functions.

4.2.1.5 GRAPHICS

All the functions discussed so far need various kinds of graphics. There is the IEC 61131 standard general control graphics. IEC 61131-3 is basically for PLC-open for specifying syntax, semantics, and displays for various languages, for example, functional block diagram (FBD), instruction list, ladder diagram (LD), sequence function chart, and structured text. FBD, SFC, and LD have graphical representations. Programs are built from a number of different software elements written within the IEC defined language.

1. FUNCTION BLOCK: IEC 61131-3 defines the ADD/SQRT/PID/COS/SIN function block called a program organization unit (POU). Many manufacturers support the standard. It is possible to create a number of function blocks and use them multiple times in a program. Functional blocks used for both continuous and discrete functions have some defined rules for use. The standard defines an **independent function** that, when executed, would yield one or more results that can be represented as a function block. Some important properties are:
   - Each of these function blocks must have an identifier (multiple identifiers also).
   - Any value that is required for calculation is represented on the left side.
   - Function block I/Os shall have both values and status. Configuration values used in the block diagram and target need not be shown as a contained parameter.
   - IEC 61804—an international standard for process industry; there are function blocks to define the specific functionality of the device used in an automatic application. The device block defines the resource of the device, which contains information and functions about the device. All devices in IEC 61804 are expected to have a similar logical device structure and functional block structure, as is shown in Figs. 7.76 and 7.77.

   Some of the mode parameters used in function blocks are *interface direct, remote cascade input, contained variable, another function block cascade input variable, internal block algorithm, remote output in contained variable, another block track input variable,* etc.

   These are applicable to I/O as well as a source for set points, as the case may be. Based on the selected mode, the function block modes will be out of service (not evaluated), initialization manual mode, manual mode (set by the operator, not calculated), remote cascade mode, remote output mode, local override (set to track the tracked input parameter), automatic mode (local set value to generate primary output), or cascade mode (set point being supplied by another block).

   Fig. 7.78 shows the implementations of conceptual FB specifications in IEC 61804 in a product utilizing various fieldbuses/network systems.
   - All function blocks associated with the control will support a mode parameter. The values assigned to the permitted mode attribute are selected from the block designer.
   - There are 10 fundamental function blocks as adapted in fieldbus architecture. Also, there are as many as 19 advanced function blocks, and the status of the output parameter is evaluated on the basis of "GOOD," "UNCERTAIN," or "BAD" in fieldbus systems (Foundation fieldbus). Discrete input/output (DI, DO), analog input/output (AI, AO), PID/PI/I, P/PD, manual loader (ML), control selector (CS), bias/gain (BG), and ratio station (RA). Fig. 7.78 shows how conceptual FB specifications in IEC 61804 are implemented into a product utilizing various fieldbuses/network systems.
2. SEQUENTIAL FUNCTION CHART: IEC 61131-3 also defines SFC for the PLC. As shown in Fig. 7.79, SFC structure, at any time the state of logic evaluation is defined by:
   - STEP: In this, the POU with respect to input and output follows a defined rule.

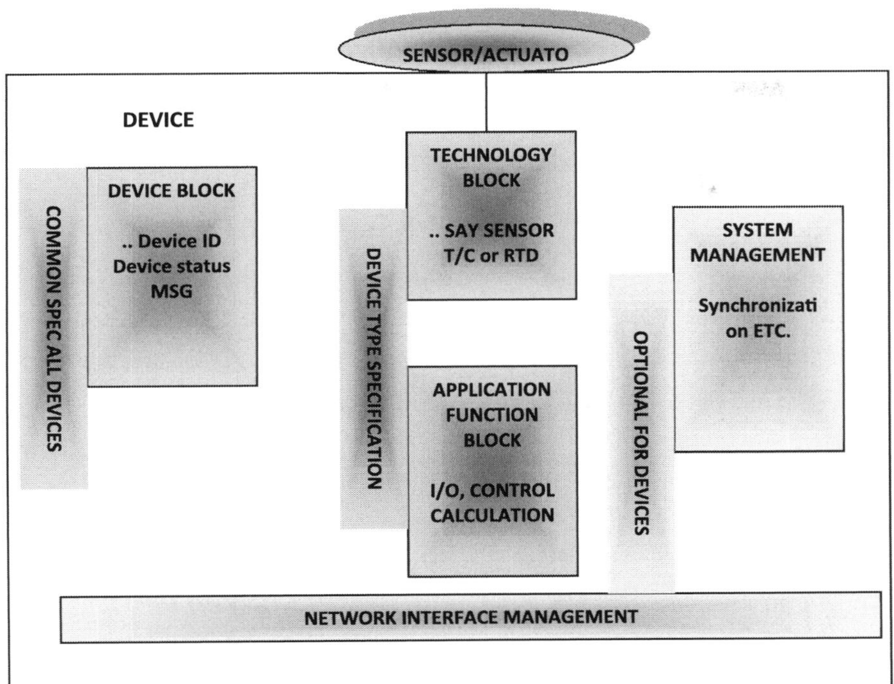

FIG. 7.76 IEC61804 typical device structure.

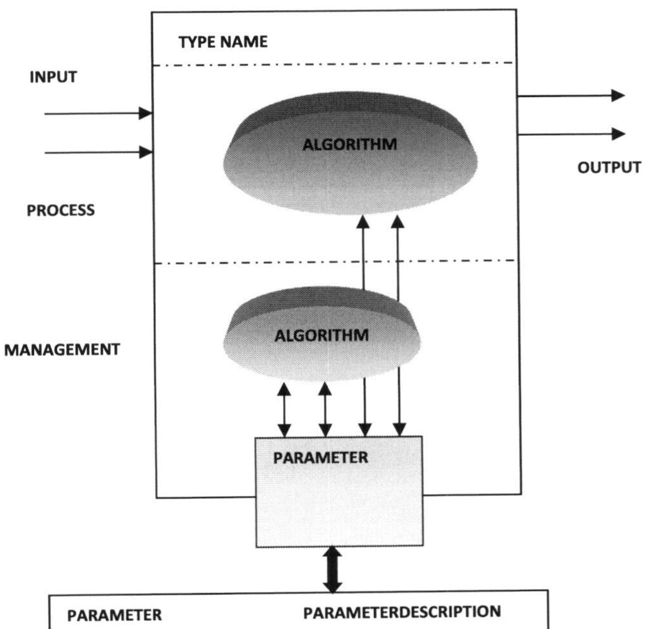

FIG 7.77 TYPICAL IEC61804 block overview.

- TRANSITION: It is the condition when the control passes one or more steps preceding the transition to one or more successor steps along a defined direction. Corresponding to each transition, there is a transition condition that needs to be true for transition triggering to occur.
- ACTION: It is used to assign results of an expression to a database item.

3. LADDER DIAGRAM: As shown in Fig. 7.44A, IEC 61131-3 defines the LADDER diagram with contacts (NO/NC), coils, and function blocks Like and OR. As a concluding discussion, it could be argued that IEC 61131-3 and IEC 61804-2 are the backbone in defining the graphical representation of control logics in modern control systems that utilize various fieldbus systems for integration. Although these are mainly defined for the PLC open, these are also stated here for some functions associated with DCS. In the limited volume of this book, it is not possible to discuss all the details; interested readers may refer to the associated standards mentioned above.

4.2.1.6 Input and Output Types

Whether it is a CLCS/OLCS/SI/MIS of the DCS, all functional subsystems require various types of hardware inputs and outputs. These are tabulated in Table 7.32 I/O, Types, showing the major I/O types in power plants.

4.2.1.7 OLCS Functions for Auxiliaries and Miscellaneous Systems

Three distinct parts in a main power plant (excluding offsite) control such as a boiler OLCS(FSSS or BMS/SB control), turbine OLCS (ATRS/ATT/Protection), and other

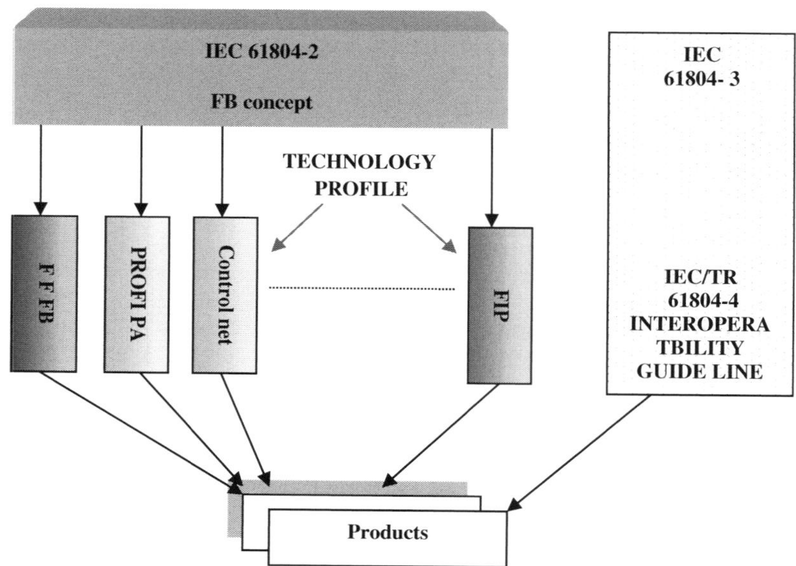

FIG. 7.78 IEC 61804 Conceptual FB mapped to specific communication system.

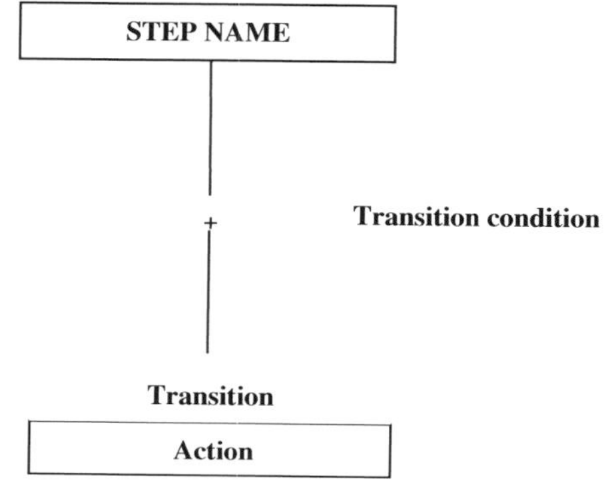

FIG. 7.79 SFC structure.

auxiliaries such as the controls of fans, pumps and regenerative system equipment.

Already, the boiler and turbine OLCS have been discussed earlier; here, the OLCS discussions will be restricted to other auxiliaries. As discussed in Section 4.2.1.2, there could be a hierarchical structure in some OLCSs. Fig. 7.80A shows the graphical representation of a group or subgroup functional control, which may be a subprogram (in software terms). There are two parts in this functional block: shut down/trip and run/operation. Each of these sections shall have the following signals with different significance (various signal definitions are, as per, ISA 77.22):

P: Protection signal: Used to avoid unsafe situations by taking corrective action. (namely, very low condenser level will initiate a shutdown program for the CEP group).

R: Release signal is permissive (when the condenser level is normal, then only the CEP can be initiated)—(ref clause no. 4.2.1.2.5).

A/M: for auto and manual command. In auto, it could be the standby pump selection.

C: Check back signal was already discussed above (clause no. 4.2.1.2.4). It is necessary for the sequence to proceed further.

ON or OFF is shown separately for groups and drives in Fig. 7.80. Drive control, step mode, and operator guide mode are in Section 4.2.1.2.7. Here also there will be the signals P, R, etc. with the same meaning as discussed. Here, the readers attention may be brought to the signal point B (block status discrepancy), which will come when a command is issued but for some reason not obeyed so that the operator can take corrective measure. For MOVs and SVs, the holding signal point H has been kept for normal operation. In Fig. 7.80C, two sets of logic sheets are shown. These are standard sheets used in various projects. In this particular example, it shows how a standby fan is selected (F100-32 in LHS) and how the same is applied to the drive control card (at the A input on the right side) for readers clear understanding only, how in a project drawing these are represented or shown. This drawing is meant to show how I/Os and flags (soft signal generated within the system as a result of some logical operation for use elsewhere) are indicated. The symbols used and the presentation methods are typically for example only; there could be other variations also. It could be in a LADDER diagram or SFC form also. Some important points for HT drives are:

- Normally the bearing and winding (motor) temperature are monitored (Fig. 3.38) and the associated drives are tripped (as protection commands OR for drive control) when they cross a set point. Also, for starting the drive

TABLE 7.32 I/O Types

I/O Type	Signal Type	Signal Value	Remarks
Analog input	High signal	0/4* to 20 mA DC	*Life zero is used mostly
	High signal	1–5 V DC	
	Resistance at 0°C	Pt 100 Ω, Cu 53 Ω, and other	ISA symbol L
	Thermocouple	B, J, K, R, S, T	ISA symbol for T/C
	mV	Depends on device	Analyzer output, rarely directly, connected to DCS
Analog output	High signal	0–10, 1–5 VDC, 4–20mA	Voltage signal generally used within control room
Discrete input	Binary Contact	24 VDC, 125/240 VAC (60/50 Hz)	Voltages are interrogation voltage
Discrete output	Binary contact	Same as above	Same as above
	Logic output	TTL, open collector	
Pulse I/O	Pulse	Depends on device	1–5 VDC common

it is ensured (release commands AND mode in drive control) that the temperature is within the allowable limit.

- Many of the large drives have their bearing vibrations (zero speed monitoring and/or other supervisory Instruments, similar to TSI applicable for turbine of the TDBFP) monitored in a similar fashion as stated above (for temperature measurements).
- Large fans/pumps are started at no load (Fig. 3.39) so that there will not be a high in-rush current (exception: starting the standby BFP/CEP in a running plant condition). For this, the inlet/outlet and valve/damper closed conditions are taken as the release signal (exception as stated above). Similarly, for drives with speed controls by a scoop tube (for example, ID fan/BFP), the minimum position of the scoop tube is taken as the release signal (exception: for automatic cutting in of the BFP, the scoop tube position of the standby BFP will follow the same of the running BFP to avoid bumps in the system).
- Normally, large drives have suitable oil pumps for supplying oil for bearing lubrication as well as supplying hydraulic fluid to other devices, for example, a scoop tube, an aerofoil/blade pitch, etc. Starting these oil pumps is one of the early preliminary conditions for starting the main drive. The following points also need to be satisfied:
- Wherever there is a lubrication pump with a drive, it needs to be started first (HT drives discussed above).
- After the LO pump has started, one needs to ensure that the oil pressure is ok for starting the main drive. If "LO pressure low" or "LO pumps do not run," the main drive will be tripped.
- There are a few drives (HT drive OR LT drive, namely the CEP or lube oil pumps of the main drives) for which there are standby selections. In those with suitable logic (similar to that in Fig. 7.80C), the same will be applied to the A point of the drive control.

There will be a wide variation of these controls based on system designs and the number of auxiliaries. So, here a brief discussion on the salient points will be touched upon so that readers can get an idea how to deal with the actual design (including ID FD fans when not covered in FSSS). The reader needs to note that only **major process** interlocks have been listed below; apart from these there could be some electrical release or protection signals also (for example, DC on or switchgear in remote position etc.). These are only typical.

1. **ID fan** (as applicable, for example, if a VFD is used, then the scoop tube condition is not applicable):
   a. Release Signal: Temperature and vibration signals, lubrication pump criteria, no load starting (both I/O dampers close), scoop tube in the minimum position (not for auto cut in), the entire flue gas path is clear, and no master fuel trip.
   b. Trip or Protective Signal: Temperature and vibration signals, LO pressure criterion, MFT due to loss of FD fans, both FD (and ID fans) were running and one FD fan trips, the selected ID fan will trip, and electrical protection (details not here).
   c. Action upon trip (as applicable): Bring the scoop tube to minimum, lube oil pumps in auto, close inlet/outlet damper (if permitted, see below). Open inlet and outlet dampers for natural cooling when both ID fans trip due to loss of FD fans (here in c says, if permitted). After a time gap, manually close the I/O dampers and trip the associated control loop to manual.

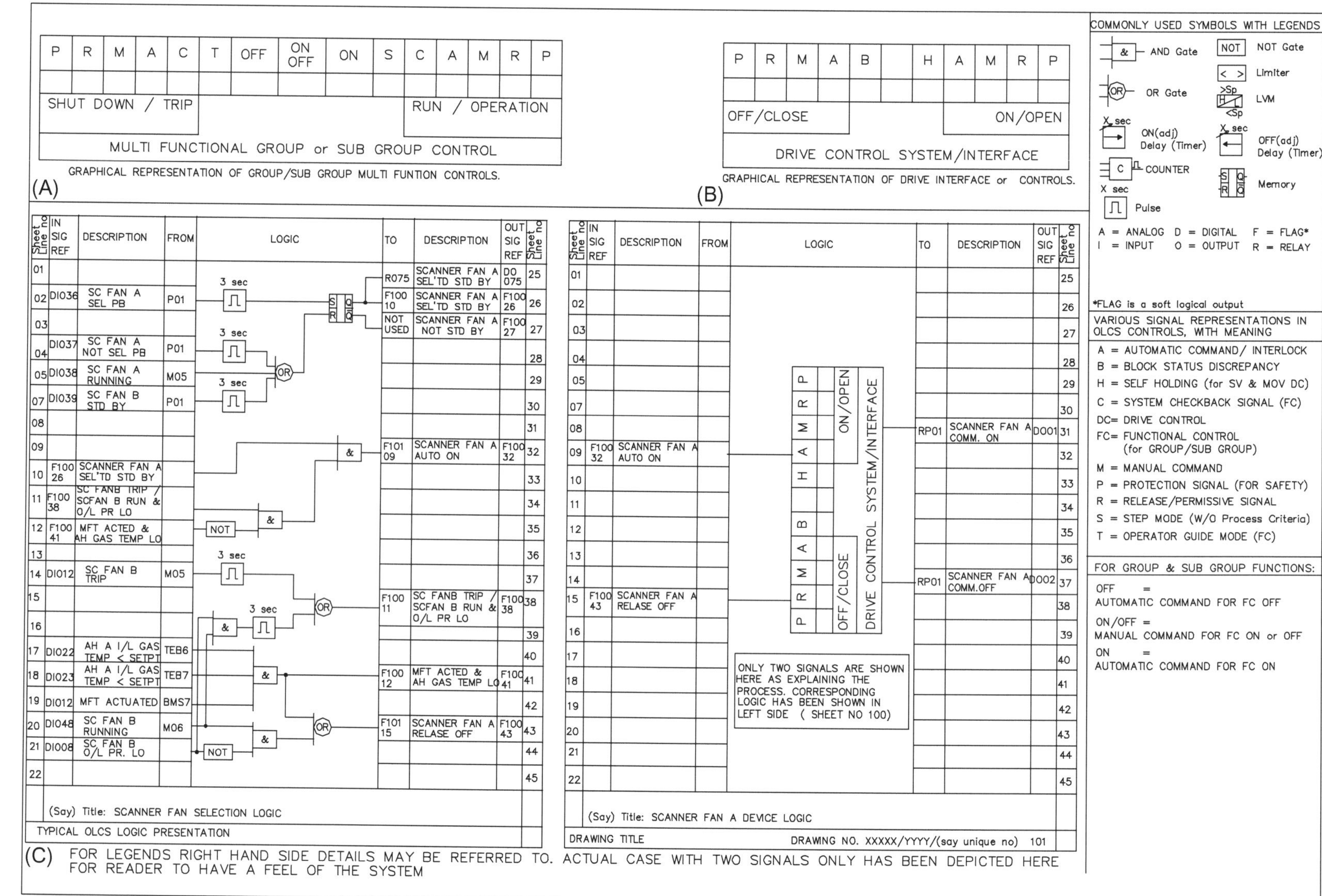

FIG. 7.80 Plant auxiliary interlock diagram (example).

   - **d.** Standby selection: No MFT, upon tripping of any running fan, if other permissive conditions discussed above and electrical BUS changeover logic are satisfied, the standby ID fan can be brought into service. In case there is cross connections of AH in the air flue gas path, necessary care shall be taken to check that the flue gas path is through before the standby fan is taken into service.
2. **FD fan** basic interlocks:
   - **a.** Release signal: Temperature and vibration monitoring, lubrication pump criteria, no load starting (O/L dampers), blade pitch or FCE minimum position, no master fuel trip, at least one ID fan running (to ensure the FD fan runs after ID fan running), and both ID fans are running (for second FD fan to start).
   - **b.** Trip or protective signal: Temperature and vibration signals, LO pressure criterion, MFT due to loss of ID fans, both ID (and FD) fans were running and one ID fan trips, (one) FD fan will trip, one ID fan and FD fans were running and ID fan trips, FD fan will trip, and electrical protection.
   - **c.** Action upon trip (as applicable): Bring blade pitch or FCE position to minimum, lube oil pumps in auto, close outlet damper (if permitted): Open outlet dampers for natural cooling, when both FD fans trip due to loss of ID fans (for this reason c if permitted). After a time gap, manually close the I/O dampers and trip control loop to manual.
3. **Boiler feed pump**: (See discussions above for normal and auto cut in).
   - **a.** Release signal: Temperature and vibration monitoring, lubrication pump criteria, No load starting (close position of O/L valve) for initial pump starting, scoop tube (as applicable) to its minimum position (30%), suction valve is open and pump is properly ventilated, recirculation valve open/in auto (at the time of BFP start), balance leak-off valve open, deaerator (feed tank) level normal (not low—for booster pump if exists), warm-up valve is open, BFP casing temperature is less than feed tank temperature, and no trip command exists.
   - **b.** Trip or protective signal: Temperature and vibration condition, lube oil pressure very low, working oil temperature very high, working oil pressure low, BFP O/L FW temperature high, deaerator level very low (for booster pump if exists), booster pump suction pressure low (booster pump if exists), booster pump discharge pressure low (booster pump if exists), BFP overload (discussed earlier), balance leak-off valve not full open, suction valve not open, and electrical protection.
   - **c.** Additional Trip conditions for **TDBFP**: TD exhaust steam pressure OR temperature hi hi, Live steam pressure hi hi, Gov/control Oil pressure lo lo, Axial Shift hi hi, Turbine Over Speed and Emergency trip from Control center.
   - **d.** Action upon trip (as applicable): Bring scoop tube (as applicable) to minimum position, lube oil pump in auto, warm-up valve open, recirculation valve open, discharge valve closed, and coasting down of turbine for TD BFP.
   - **e.** Standby selection: Auto BFP standby pump selection is extremely important and needs to be done almost immediately so that the boiler does not starve. This is moreso for large units, especially for OT boilers where the fuel: FW ratio is extremely important for safety and operation. A few pertinent points to be noted are:
     - **i.** Conditions for standby BFP selections have been elaborated in points iv–vi below. There could be two possibilities, as discussed in points ii and iii below.
     - **ii.** In cases where all the BFPs are MD BFPs, then any one can be selected as the standby, provided the suitable electrical bus selection is well taken care of (not applicable for 2X 100% BFPs). For standby selection, MDBFP is mentioned. Such selection details are shown in Fig. 7.81B.
     - **iii.** In case of a combination of 2X TDBFP + 1X MDBFP (or so), the MDBFP needs to be considered for initial start up and STANDBY. TDBFPs on account of the starting time of the turbine are not considered as standby. For standby selection, MDBFP is mentioned. Such selection details are shown in Fig. 7.81B.
     - **iv.** To ensure that the selected standby MDBFP shall have all permissive conditions stated above, the duly fulfilled and selected standby MDBFP has the O/L valve in the open position.
     - **v.** Standby MD BFP shall follow the speed master demand so that whenever it is started, the MDBFP straight away goes to the desired speed to meet the delivery demand.
4. **Condensate extraction pump**: Major interlock systems are:
   - **a.** Release signal: Temperature and vibration criteria (as applicable), lubrication pump criteria (as applicable), cooling water flow normal, no load starting (close position of O/L valve) for initial pump starting OR any CEP is running, suction valve is open and pump is properly ventilated, recirculation valve open and in auto, hot well level (Not low), and no trip command.
   - **b.** Trip or protection signal: Hot well level low low, Suction valve Close, pump running and discharge pressure low, electrical protection.

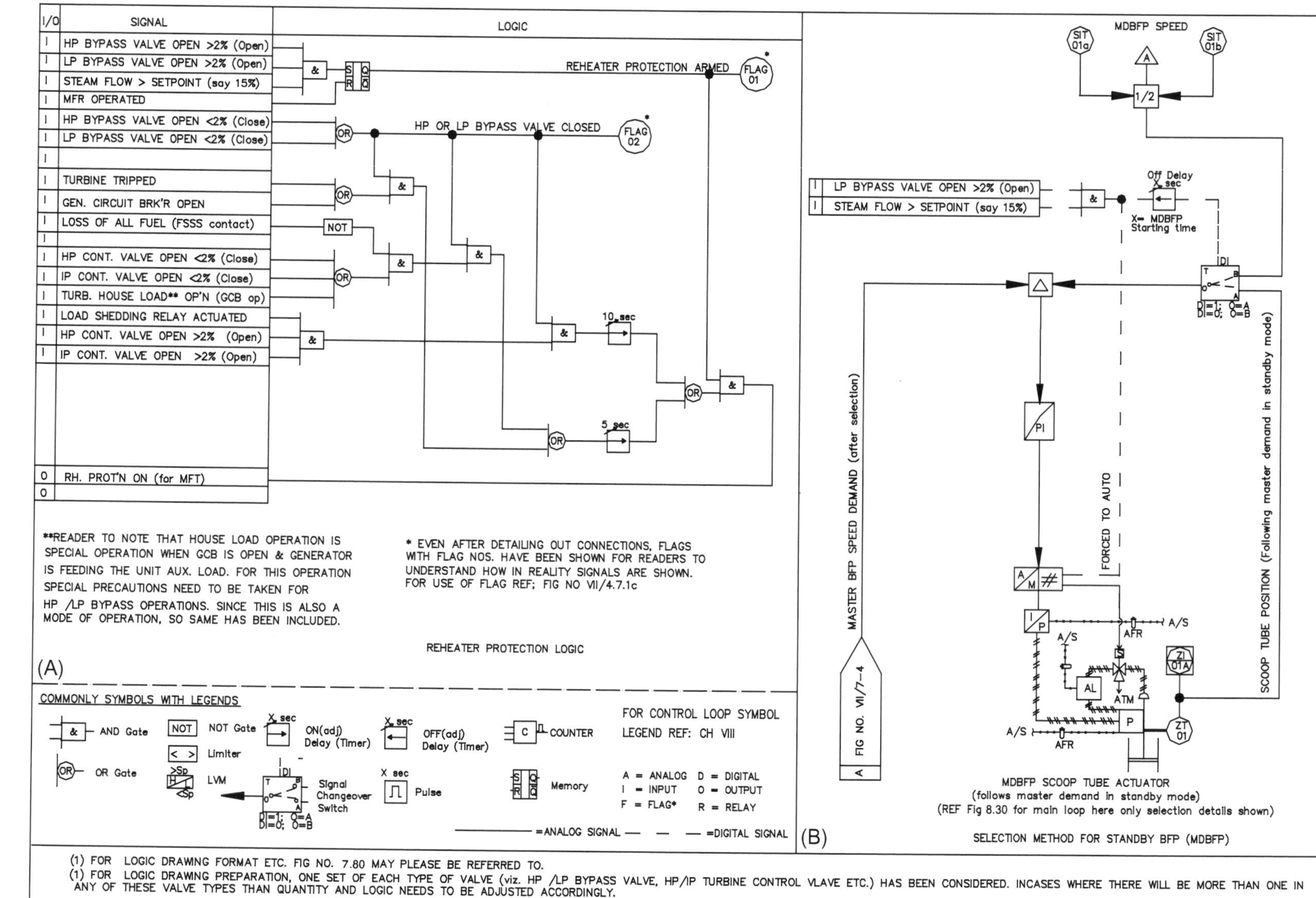

FIG. 7.81 Miscellaneous OLCS functions.

5. Reheater protection: The boiler will trip when any one or more of the conditions listed below occur. This is a part of FSSS (Chapter 8) as one condition. Here How such a condition occurs has been elaborated.

  a. Turbine tripped OR generator circuit breaker (GCB) open and HP OR LP bypass opening <2%, that is, closed. Then after a delay of 5 s, this signal will be generated.

  b. Turbine working (HP and IP control valve opening >2%) and load shedding relay actuated and HP OR LP bypass opening <2%, that is, closed. Then after a delay of 10 s, this signal will be generated.

  c. If turbine is not working (HP or IP control valve <2% opening, that is, during start up) OR turbine is in house load (parallel operation with HP/LP bypass) and the boiler is working (no contact loss of all fuel from FSSS) and HP OR LP bypass opening <2%, that is, closed. Then after a delay of 5 s, this signal will be generated (this may be during initial start of the unit with HP or LP bypass).

Arming of reheater protection: Once the HP and LP bypass are open 2% and steam has increased beyond a set value, then the reheater protection will come in to effect. Logic toward this has been elaborated in Fig. 7.81A.

4.3 Standalone Controller Unit (SCU) and its Integration With DCS

4.3.1 Preamble

Standalone controller units (SCU) basically refers to single loop or multiloop controllers. These are digital microprocessor/embedded electronics-based controllers packaged in one instrument, generally having sizes as per DIN standard sizes such as72 mmX144 (better 75 mm × 150 mm with bezel), 96 mm × 96 mm, 48 mm × 48 mm, etc., as shown in Fig. 7.82A and B. Some of these types of controllers are in use in smaller units for the main plant as well as for offsite plants of various unit sizes for the applications that are mainly relatively independent (or where the number of modulating loops is less), so that these can be integrated with the PLC for a complete control system. Even though these are standalone controllers, they are mostly integrated with the main DCS/PLCs. Actually, the reliability of the controller depends on the hardware selection, whereas the capability is a function of software, which is the main factor for the cost of the devices. In view of the following reasons, microprocessor-based digital controllers outperformed their analog counterparts in the present era.

- High accuracy to the tune of 0.075% (dependent mainly on A/D or D/A conversion).
- Flexibility in configuration and topology and display modes through a handheld calibrator or PC and integration of a number of controllers and integrating with PLC/DCS with easier communication and less hardware.
- Requirements of additional hardware and wiring are minimum (dependent on algorithm, and standard software in the system library) with the possibility of additional arithmetic functions such as totalizing, linearization, etc.

4.3.2 Standalone Controller Block Diagram

Typical architecture of a standalone controller is shown in Fig. 7.83.

As is seen, there are two basic parts: the basic control block and the supportive part. Basic control block consists of the main processor, RAM, ROM/EEPROM along with display electronics, communication ports, etc. Often, microcontrollers can be configured as controller block. At the input side it is supported by a signal conditioning unit (SCU) comprising a programmable gain controller, a signal conversion unit, a multiplexer and an A/D converter, whereas on the output side, there will be a D/A converter and relays (as applicable). As in the case of basic microprocessor systems, temporary data are stored in random access memory (RAM) while the main program is stored in long-life rechargeable battery-backed read only memory (ROM) of reasonable size or electrically erasable programmable ROM EEPROM. EEROM is basically a programmable ROM where during commissioning/at the factory, the basic programs are put. Operating parameters are entered into the system by the operator. These are battery backed up (in some cases with monitoring facilities). These controllers have a built-in standard library for control subblocks and arithmetic functions. During configuration, these functional blocks need to be linked by structuring data by the handheld unit and/or by the PC.

Some controllers have side panels for adjustments of gain, proportional band, integral action time, etc. (for example, the SLPC controller of YEL). Some manufacturers offer optional portable memory modules to have redundancy in the system parameters (for example, ABB MOD 30 series). Also, the operating parameter range, set points, etc., can be set from the front panel. Supporting SCU provides the set point, control variables, etc., to be fed via a programmable gain controller, multiplexer, and A/D converter. Most of these controllers support universal inputs (Fig. 7.84). Output support units have a D/A converter for analog output and other processing circuits to provide retransmission output, relay contact, etc., as applicable. The digital display may be a bar graph or digital readout type (or both, with an engineering unit) as shown in Fig. 7.82A and B. For controller parameters and/or set point adjustments, push buttons and keyboards are provided in the front with which the controller

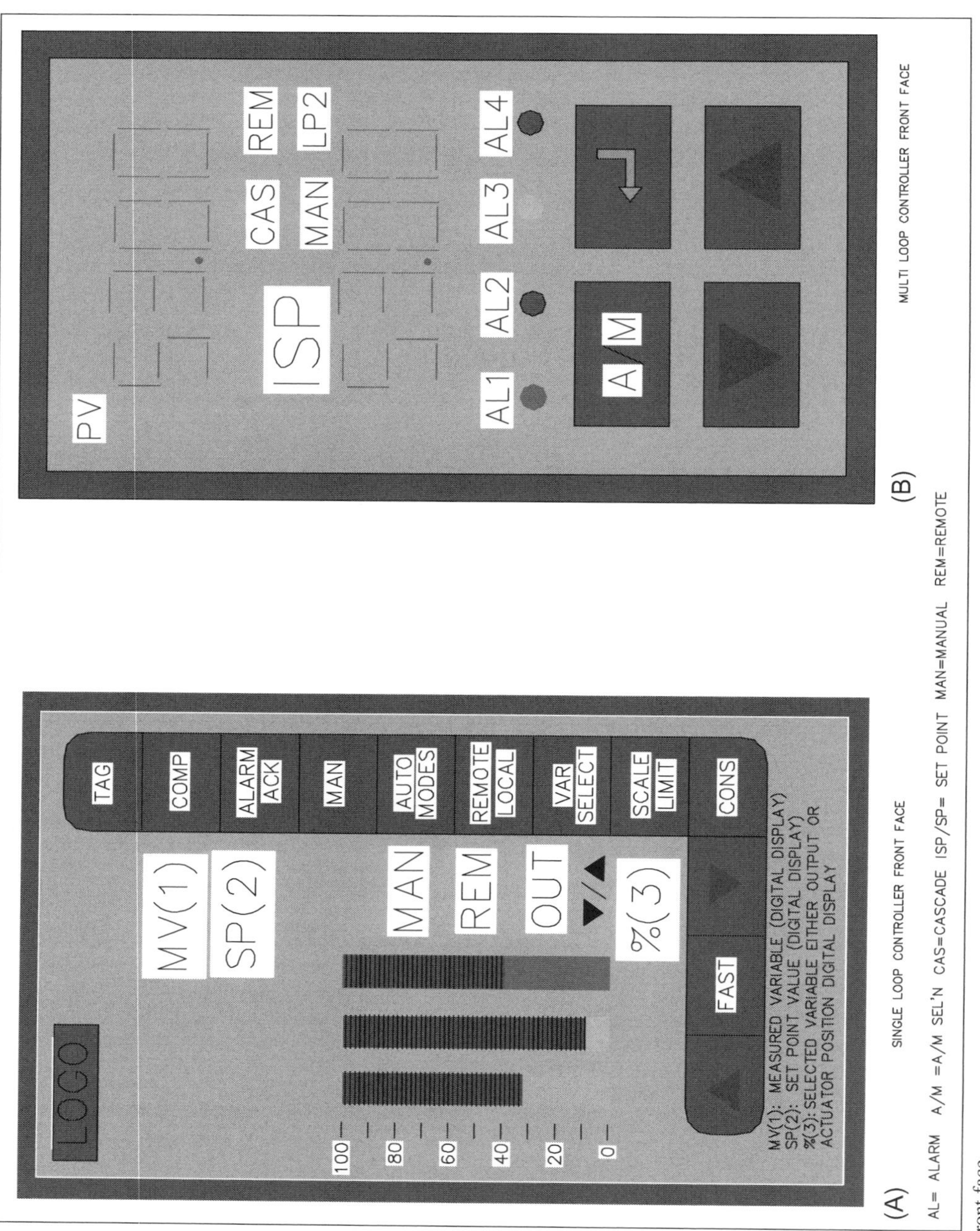

FIG. 7.82 Controller front face.

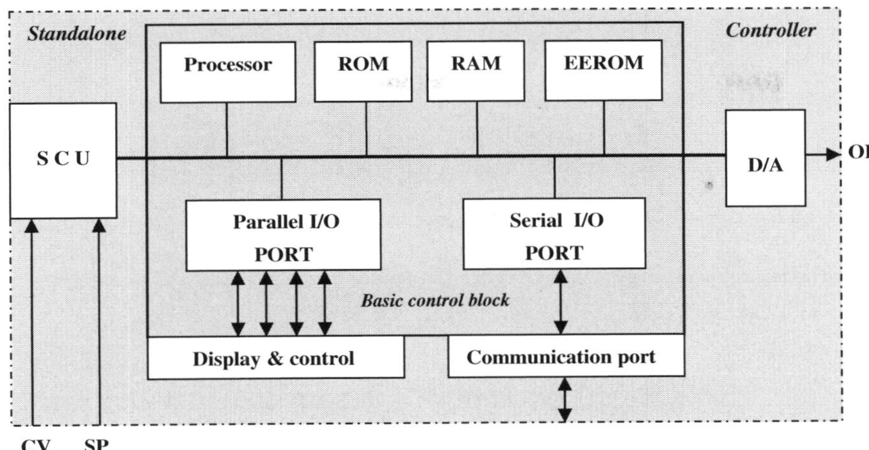

Legend: **SCU** = signal conditioning unit +A/D; **A/D**= analog to digital converter
D/A Digital to analog converter; **RAM** = Random access memory; **ROM** = Read only memory
EEROM Electrically erasable ROM **SP** = Set point; **OP** = Output; **CV** = Control variable

FIG. 7.83 Basic block diagram for standalone controller.

What is meant by universal input and how the same is handled? ***Universal inputs means that, at the same set controller input terminals, varieties of inputs such as 4-20 mADC, different many types of T/Cs , various kinds of RTD, mV etc. can be used. These calls for Programmable gain controllers and have capability to linearize these inputs as applicable. Also there shall be provision to provide reference junction compensation for T/Cs. For different class of input signals for the same input channel that there will be a number of terminals for each input.. Now advantage is that at a later date if input type is changed there will not be any difficulty to connect the same and run without any modification. Also it is seen that in many controllers*** **DI/DOs can be swapped** ***also.***

FIG. 7.84 Universal input/output.

may communicate in parallel mode. There will be a separate port to communicate with external PCs and other devices via a serial port. Serial communication may be carried out with the help of (say) an RS 232 in MODBUS protocol. Standalone controllers can support quite a good numbers of I/Os of different kind. This is done with the help of an extension board, as shown Fig. 7.85.

Here, in this controller based on a standard multiloop controller, there are a few parts such as the **bezel display board,** the **communication board,** the **extension board, and** the **main board;** all are in one packaged instrument. The extension board may be at the side, and the terminals at the back. Some of the features available from the standard manufacturer shall include but are not limited to the following:

- I/O linearization, math functions, standard library for control block, and auto tuning.
- Discrete/continuous logics, PID and advanced control functions with fewer instruments.
- Supports complex algorithms, diagnostics, and alarm management.
- Bump- and procedureless A/M transfer and digital communication over a bus.

4.3.3 Communication

Normally, these controllers communicate over RS 232/485 using MODBUS or another standard protocol. These controllers communicate with other instruments or systems of the same family over their own system bus, for example, ABB MOD 30 communicated with other instruments over an ICN bus to the DCS of the same family. Normally, manufacturers support the necessary software for communication with a PC for configuration.

In Figs. 7.86 and 7.87, the connections of the controllers are shown for communication with PLCs, DCS, and PCs as well as with a master controller, as necessary. This communication can be through the links using a standard protocol, and/or through the Ethernet. There may be a dedicated USB port connection for communication (for example, the YS series of YEL).

4.3.4 Typical Specifications for Standalone Controllers

See Table 7.33.

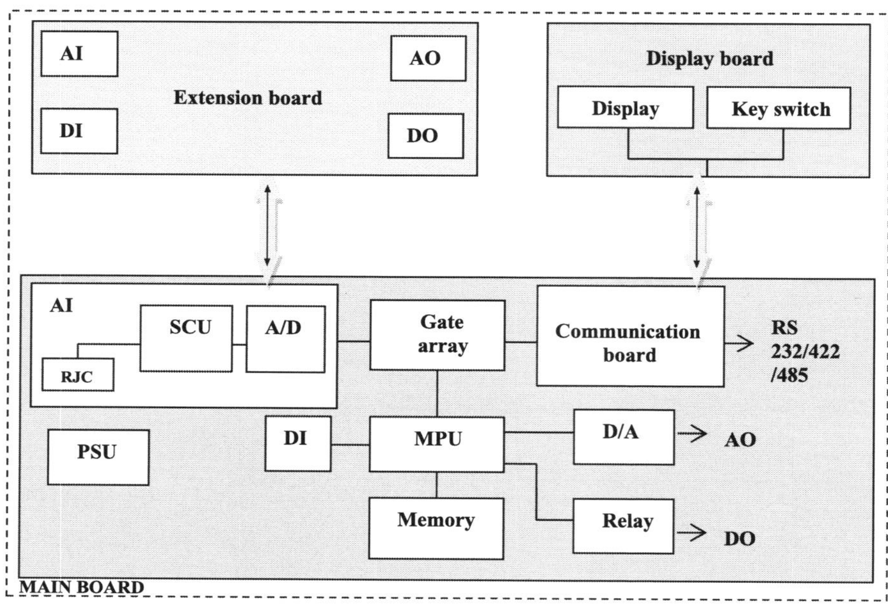

FIG. 7.85 Controller with extension board.

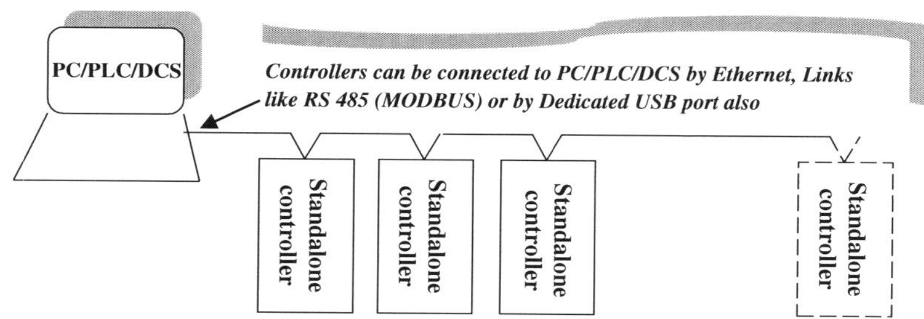

FIG. 7.86 Controllers connected to PC/PLC/DCS to communicate.

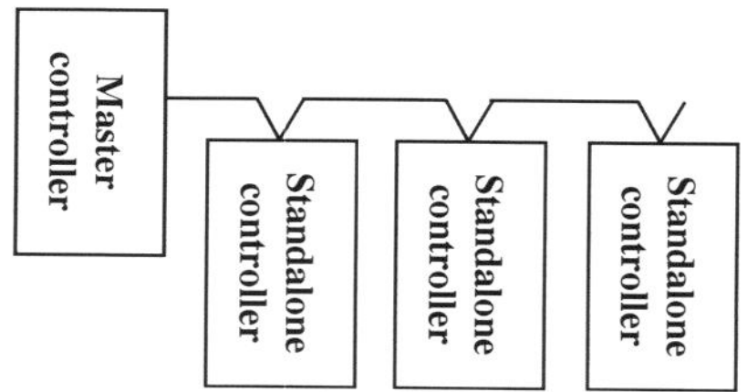

FIG. 7.87 Controllers in coordinated mode.

4.3.5 Support Functions

Standalone controllers provide lots of support functions such as:

- Multiloop functions.
- Standard logic functions.
- Linearization functions.
- Small PLC functions.
- Number of standard soft control function modules.
- Independent PID.
- Adaptive response.
- Dead time compensation.
- Cont. coordinator.
- Third-party support.
- Math unit.
- Cascade, feed forward control.
- Universal I/O support.
- Multipoint indicator/totalizer.
- DI/DO swapping.

5 MAN-MACHINE INTERFACE AND RECORDING

5.1 Introduction

In the industrial design field, the man (human)-machine interface is the place where people meet technology. This is the place where man **interacts** with the machine/process

TABLE 7.33 Standalone Controller (DS)

SL	Specifying Point	Standard/Available Data	To be Specified
1	Type	Single, multiloop, or others	Detailed in Sl.3
2	Functions	Controller/PLC/totalizer	To specify
3	Types of loops	Simple PI/PID/PD, cascade/remote/DDC, and/or totalizing	To be specified
4	I/O numbers and types	Number of AI/AO and DI/DO or others	*To specify
5	Universal inputs	Number of universal inputs for AI	To specify
6	Trans. power	Number of transmitters to be powered	To specify
7	Display type	Bar graph, digital back-lit LCD for curve generation	To be specified
8	Number of displays and functionality	Number and types of display with selection. Also LED/LCD and color	To be specified also
9	Control keys and functions	Number of control keys, for example, A/M selection, increment/decrement cont, set point, alarm acknowledge, mode selection	To be specified
10	Signal range	4–20 mA/RTD/mV for each I/O	As applicable
11	Rating	3A, @ 240 Vac DO contact	To specify
12	Sample rate	200 ms (only example)	To specify
13	Resolution	12/16 bit	
14	Set point limit	<(–)10% >110% etc.	
15	Diagnostic	Number and types of diagnostic alarms	To be specified
16	Process alarm	LL, L, H, HH with set point	To be specified
17	Audio alarm	Beep to be specified	
18	Communication	HW link with MODBUS Ethernet	To be specified
19	Configuration	Handheld tool, PC	To be specified
20	Power supply	1 PH, 240 VAC 50 Hz/110 VAC/60 Hz	To be specified
21	Location	Field, control room	To be specified
22	IP rating	IP 65/IP22	To be specified
23	Certification	International certification as applicable (for example, FM/CSA)	To mention standard ref.
24	Accuracy	+0.1% or better/0.05%*	*Repeatability
25	Software support	To specify for any third-party SW.	
26	Environmental condition	(–)20 to 55°C 0–95% RH	
27	Stability and drift	8 h/days in percentage	
28	Optional feature	Optional backup memory, etc.	
29	Special features	To be specified	
30	Size-bezel/cutout	To specify DIN, other standards	

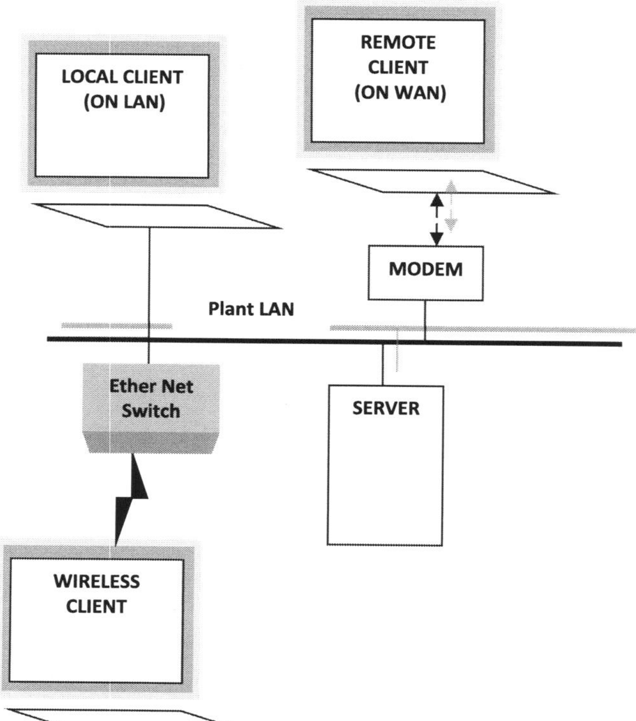

This is a client server network. Here there are a number of clients one is connected to LAN by wire another is wireless. There is a third client which is Remote client. This could be at a remote place and may be connected to the network (say, an Ethernet) with help of Dial up modem, or via Virtual Private Network (VPN). This makes the network IT compliant. In this type of operator interface the network can be accessed by the remote located operator for trouble shooting etc. this remote locations

FIG. 7.88 Client server network with remote client interface.

for effective analysis, control, and operation of the system. Machines operate in a way unnatural to human beings, so the MMI translates these machine actions into a format intelligible to human beings and vice versa. Starting from small switches, display device all in fact form the part of MMI. Now, in the DCS, high-resolution windows in the LED/LCD/plasma displays present various mimics, trends, diagnostics, and alarm displays. MMIs are not only now restricted to the operator's interaction with the process, but now thanks to the WAN Internet, remote man-machine access is also possible. Engineers can operate and troubleshoot DCS from a remote location, as shown in Fig. 7.88. Added to this, with the introduction of the IoT, even a smartphone can be an MMI for remote control.

IoT (IIoT)/WEB services help to precisely integrate systems for operation, control, and analysis of the process. Printers, recorders, etc., are the devices that generate a printable form of such displays. So these two devices go hand in hand.

5.1.1 Workstation (WS)

A workstation generally consists of a number of display devices such as LCD/LED monitors as well as some dedicated LCD/LED display panels (for example, the E300/200 from Mitsubishi Electric). LED displays are common in the modern era. Widescreen display (aspect ratio 4:3) are also available (for example, the Emerson Delta V). A workstation also includes a standard/custom made keyboard, pointing devices such as a mouse, a track ball, and touchscreen devices. In many places, multiple monitors/large screen monitors are supported by workstations. Some of the general features of workstations are:

- Access to a large numbers of tags (up to 3,00,000)
- User friendly, easy navigation through display
- Intuitive graphical configuration with modifiable custom displays
- Predefined faceplates/templates
- Clear overview for management decisions
- Sophisticated alarm handling and management
- Supervisory PCs and sophisticated engineering station
- Object-oriented programming and process optimization
- Secured multitasking operation
- Trending both in real time as well as historical data
- Third-party application and interface
- Multiplatform versatile set up and applications

1. WORKSTATION APPLICATION: Some of the applications of workstations are as follows:

• Overview display	This gives the bird's eye view of the entire plant so that concerned personnel can get an overall idea of the plant. In this, current alarms are suitably displayed to draw the operator's attention. Statuses of major drives and loops will be displayed suitably
• Group display	This display shows the operator a set of related parameters when arranged in a group defined by the user. Any alarm in the group shall be displayed suitably
• Alarm management	All alarms shall be time tagged, and various colors can be assigned to them based on user-defined priority and return to a normal message. There are several means of presentation of alarm management displays: such as a current alarm list (with the oldest pushed out of the list as first in, first out, etc. way), alarm history, list of unacknowledged alarms, acknowledged alarms, user-defined priority alarm, trip alarm, plant area alarm, event alarm, etc. So user-defined alarm filters can be used
• Event messaging	Event-based messaging can be sent to historians for data collection (say hi frequency data collection), and/or to the printing medium for event-based logs etc., with an event time tag, description, associated point tag, node number, etc.
• Mimic/process diagram	Generally, these are diagrams that present process information in a process flow pattern: process parameters, valve damper actuator status, alarm, etc., are presented in high-resolution graphic displays. These can be used to control, pan, zoom, or window any particular area. Depending on the software package, these may be developed utilizing a standard CAD format and may be downloaded to the system at a later date.
• Trending	The trending of parameters as selected by the user can provide the trend of real-time as well as historical data. It is possible to create trend groups to see the trending parameters of these related points in a group, for example, steam flow, feed water flow, and drum level could be a predefined trend to see how the drum level is affected with variations in the other two parameters. Zooming, panning, and windowing of a particular area for details are possible. Trending can be for X-Y and X-T, and could be in horizontal, vertical bar graphs, etc. In a historical trend average, the max/min value as per user choice could be selected over a defined time. This mainly depends on the memory space allocated and the sample rate
• Faceplate/logic Dia.	A controller faceplate display is very useful for the control application. Similarly, logic displays such as ladder diagrams are helpful for logical/sequential operation
• Point display	Normally, to access any particular tag to get the internal and external information about the same. When the cursor is taken and clicked on any window, it provides the point tag, point value, status, and other information through various navigational modes offered by the manufacturer as per the customer's choice
• Point information	This gives the complete database of the point, including point ID, node where it is connected, signal type, scan/alarm status, point value, etc.
• Point review	This is a special type of display to list the points having common characteristics as defined by the user, for example, superheater stage temperature. It shows the tag number, parameter name, process and signal range, limits (high/low), etc., for each of these parameters. Signal quality (poor, bad, good, etc.) time out, etc.
• Multiplatform	The multiplatform capability of the control system as well as the workstation must have a common comfortable work environment for the operator to access a variety of application, using multiplatforms for an integrated system
• Misc. other applications	Some other features of workstations shall include: **multilingual** displays, operator's pointing device interface, **diagnostic message display, multiwindow function** (eight to 10 such windows in a single display is common today), **user friendliness**, built-in **security functions** for access, etc.

2. WORKSTATION (WS)/OPERATOR CONSOLE TYPES: A monitor is an important part of a workstation in a DCS/PLC. The CRT belongs to the emissive type of display monitor that is almost obsolete now. Other discharge types of displays such as a light emitting diode (LED), a liquid crystal display (LCD), or a plasma display are extensively used as monitor displays. In view of the lower space requirements for LCD/LED types of displays, now it is possible to accommodate a number of such monitors being stacked one after another in a desktop arrangement. However, in certain compact designs, these may be placed one above the other with due consideration to ergonomics. Generally, there are two kinds of designs in use: the enclosed display console (Fig. 7.89A) and the open display console (Fig. 7.89B). In the open console, modern monitors are very slim with very good aesthetics (widescreens possible). For operational assistance, a touchscreen (finger/stylus), mouse,

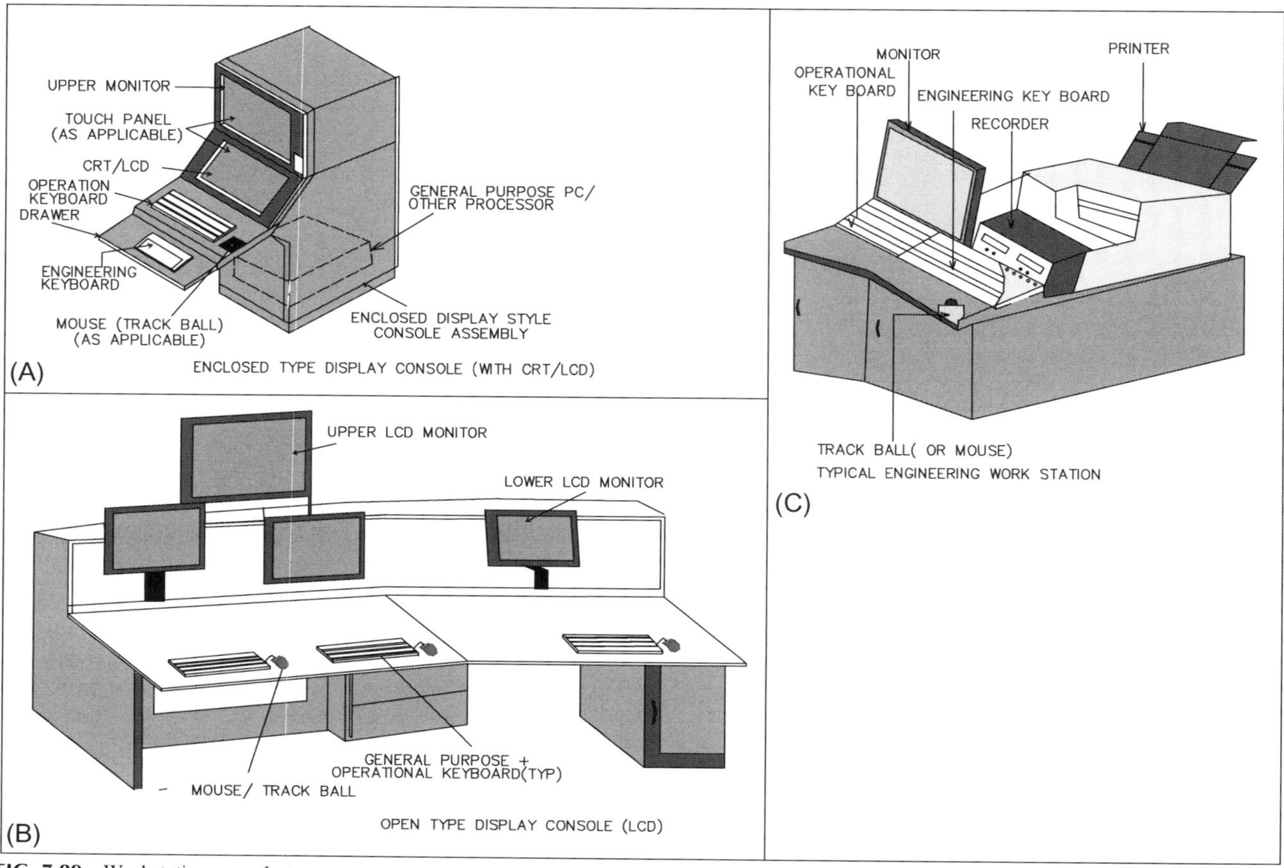

FIG. 7.89 Workstation console types.

and track balls are used. Based on the system design, any of them or a combination can be used. In DCS applications, in addition to the normal keyboard, a dedicated operational keyboard/stylus are also common as a very helpful device. All these workstations are nowadays intelligent, having own processing unit and hard disk drive (HDD). Normally, these consoles have small drawers, telephone jacks, etc. attached to them. Generally, these have a small door for attending the power and other connections (if not wireless) from the back. PCs associated with these workstations can be kept inside the console.

3. ENGINEERING WORKSTATION FEATURES: The main difference between an engineering workstation (EWS) and an operator's workstation (WS) lies with their purposes and functions. In the EWS, mainly developmental work is done so that the software can be exported to the control system without disrupting the operation of the system. Some of the features in a standard engineering station are:
   - Simultaneous working on control system, database, and graphics in a multiwindow environment.
   - Intuitive development system for generation of object-oriented graphics with elaborated and advanced software support so that various control strategies, complex graphics, point recording, etc., can be developed and maintained in the system under all conditions and including an upset situation. The software system should be comprehensive but user friendly so as to ease developmental work with suitable menu-driven systems. Some of the major functions of this are control strategy and graphic building, I/O device and historian configuration, grouping for trending parameters, security, and general and other application building tools. Also, open data integration should be possible.
   - Engineering workstation software needs to have dual functionality so that, if necessary, dynamic values could be accessed by an authorized person.
   - Offline configuration and export to the system is an important feature. Also, it may be possible to modify a part and update the system for running it on a modified program without much difficulty.
   - Embedded data system database.

- High security on access to various functional aspects of the engineering station.
- Other engineering functions such as import/export, upload/download, hardware modification, help functions, etc., with an embedded data system (database).

4. CHOICE BETWEEN CRT and LCD MONITOR: Each of them has some advantages and disadvantages. Comparisons between the two are presented in Table 7.34.
   - In a CRT [phased out], a high-voltage electromagnetic deflection and a low-voltage electrostatic field focus maintain the display quality in the P31/P1 screen of the CRT. There are X-Y linearity corrections for the X and Y data input. This correction of X and Y actually maintains the geometrical shape of the picture. Also, the information supplied to the CRT must be refreshed continuously (60/50 Hz) without flickering. The brightness control of the display is regulated by a potentiometer. The CRT needs high-voltage as well as low-voltage supply and power consumption is appreciable. Because electromagnetic and electrostatic fields are in operation, the performance is affected due to the EMI effect. Also, for accommodating electron scanning/deflection, some space at the backside is necessary.
   - In a thin-film transistor (TFT) LCD (cold cathode technology), separate tiny transistors form each of the pixels on the display. Because the transistors

TABLE 7.34 Comparison Between CRT and TFT LCD

TECH. Points	CRT	TFT LCD	Remarks
Visual search	Slow (with respect to LCD)	Faster (~20%)	Text search on display
Eye fixation	Longer and more	shorter and fewer	LCD ~10% less
Search error	More (~20%)	Less	
Geometrical distortion	Present, due to electron beam	active pixels in flat matrix number distortion	Electron beam is more tangential at the edges
Flicker	Flicker prone (as electron scanning)	Flicker free- not any issue	CRT display flickers needs refresh @ 50/60 Hz rate
Glare	Prone to glare *	Glare free	*Due to glass CRT
Brightness	Nonuniform	Uniform*	*Nearly two times CRT
Contrast	Better	Less (~equivalent available now)	At low light it is useful parameter to compare
Sharpness	Moderate to high	High	
Color purity	Excellent	Very good*	*Difference hardly noticeable/ perceptible
Viewing area	Space inefficient part area	Space efficient Full area	
Angle of view	Better	Less up to 160° available	Some consider this as better privacy
Size versus equivalent display	For eq. display Larger size	For eq. display Smaller size	Due to space efficient
Dead pixel	NA	Irreparable	
Aesthetics	Bigger	Slim and stylish	
Space	More at back	Less	20% space saving with LCD
Weight	More	Less (2–3 kg)/15″	
Flexible position	Moderate	Highly flexible	
Electrical energy	60–80 W	20–25 W	For (Typ) 15″ display
Heat emission	High	Minimum	Energy saving on AC
EMI	Affected	Not affected	LCDs available with speakers!
Response time	Better	12–16 ms	

are tiny, they require a tiny charge, hence redrawing a figure could be faster. This display is direct and resolution is not a problem. A typical 17" TFT display has nearly 1.3 million tiny TFTs, which obviously leaves a chance for malfunctioning of one or two in the display. When one or two transistors fails, there will not be any display at that pixel. This is **a dead pixel**, reader to note carefully that guarantee against dead pixel. This is critical for a TFT LCD. Manufacturers normally guarantee against failure of >11–12 dead pixels. Because a direct display takes place, the geometrical distortion is not any question. Also, power consumption is less. Although it has fewer viewing angles, the effective display area is more than that corresponding with a CRT display area. On account of the technological difference, it is not affected by electromagnetic interference (EMI).

An LED display is a flat-panel display using a light-emitting diode in place of cold cathode technology or a CCFL in LCD displays. It is preferred as it eliminates the drawbacks of backlighting (which badly impacts the image and decreases its sharpness and brightness). So, an LED is an improved version of the LCD to provide a vibrant color effect and better picture/image quality.

- DISPLAY AREA IN WORKSTATION MONITOR: Typically, a workstation display has a few divisions for operation and monitoring. Such divisions are generally decided by the system supplier **based on the system supplier's design**. Typical divisions (from a reputed manufacturer) are shown in Fig. 7.90. A few terms in this regard are listed below.
- Message banner (Fig. 7.90): This is to indicate the type of message with details for the same. This may be "overview," "alarm," "display," etc. It may also indicate the particular plant, for example, "mimic display of the boiler drum" to show that the mimic of the boiler drum is in the display. These are generally placed at the TOP OF THE DISPLAY. There may be another small dedicated display area to show the latest alarms of the area or of the plant (for example, a Centum VP System Message Banner on the top).
- Browser bar: On the left side of the main viewing area, there is a browser bar to call up various plant areas/various groups such as overview, trend, mimic, etc., as shown in Fig. 7.91 Browser bar and Frame details, so in the browser there may be several tabs. Based on the tab selected, the required display area/group can be called. This can display a list of operations and monitoring in a hierarchical structure in a tree-like fashion (similar to standard Windows Explorer). In some cases, additional tool boxes with required icons are also displayed for direct access.

Frame: In the frame, there are a number of tabs for various functional purposes. It is possible to call up multiple tabs from the frame. The number of such tabs is dependent on

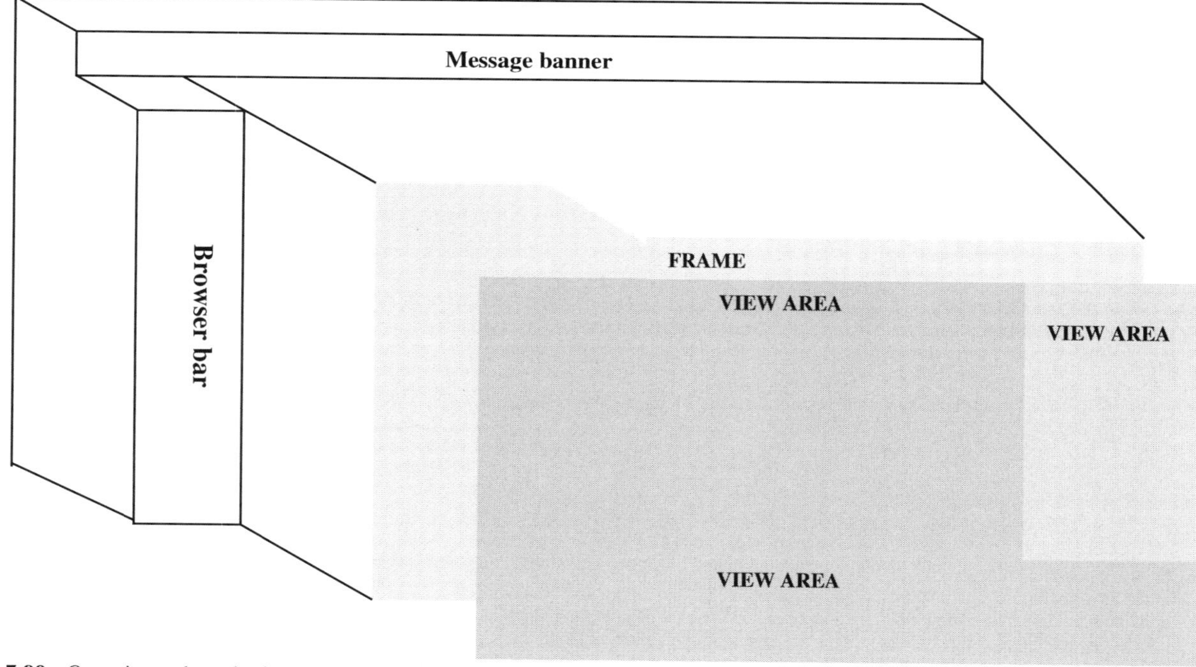

FIG. 7.90 Operation and monitoring area for workstation display.

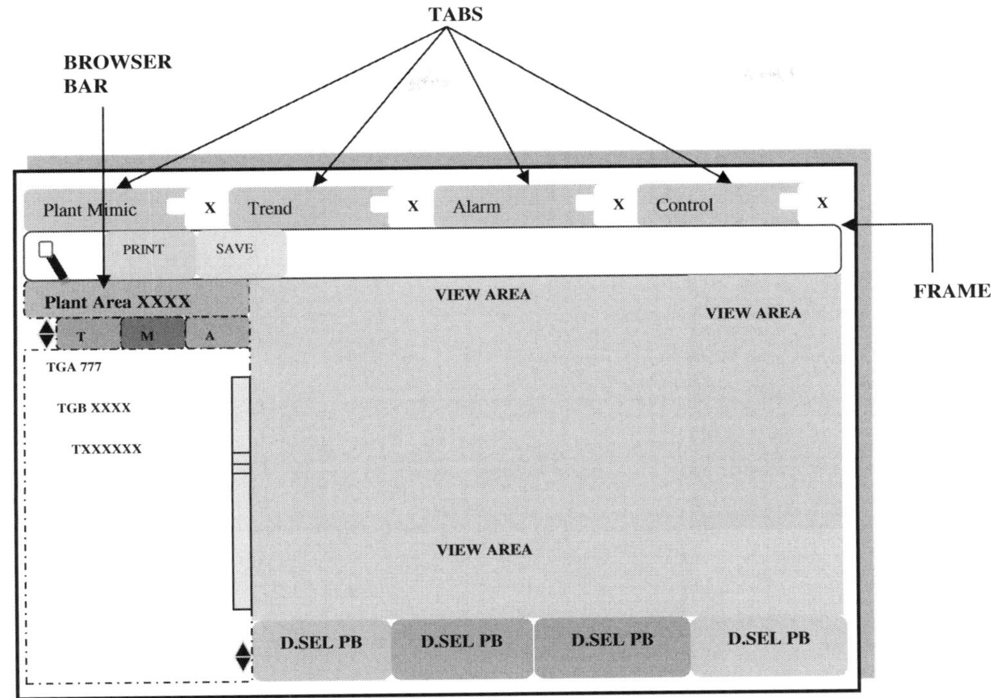

FIG. 7.91 Browser bar and frame details.

the number of views allowed in the display area. Fig. 7.91, Browser bar and Frame Details, may be referenced.

- View: Based on the system design, a number of types of views such as a normal plant view, a mimic view of an area, a controller faceplate (Fig. 7.92) of any of the control parameters in the plant view, a trend view of some parameters, and current alarms of the area in one window is possible. In some designs, at the bottom of the display there are some dedicated buttons for quick display changeover; some systems dedicate one or two bottom lines for plant alarm/trip/SOE point.
- Normally, it is possible to have a full-screen view of most of these display types. Also, it is possible to zoom to a specified area to have a closer look.

5. VARIOUS OPERATION AND MONITORING FUNCTIONS (Table 7.35)
6. CLASSIFICATION OF WORKSTATION (WS): Based on the hardware, workstations can be classified as **an intelligent** workstation where the system has its own controller, hard disk, and associated software. The other one could be a **dummy** terminal without its own disk. At the present time, all workstations are intelligent. Again, based on functions and use, the workstations can be broadly classified as in Table 7.36:

- CONSOLE/WORKSTATION HARDWARE DISCUSSIONS: The operator console may have a number of single or dual monitors of the CRT or TFT LCD type. Some generalized characteristic parameters about monitors and associated accessories in a console are discussed below:
  - Screen sizes: 15/17/19/21 inches are available but 19/21″ are the most popular ones.
  - Power supply: Normal 1 phase 240/110 V 50/60 Hz. The AC supply will do; however, the operating voltage inside the monitor may be as high as 15 kV in case of CRT. However, for a TFT such a high voltage is not required.
  - Refresh rate (CRT) 50/60 Hz.
  - Characters per line: ~128 in 75 lines in screen, with up to a 96 character set.
  - Aspect ratio: 4:3 or 16:9 for high definition.
  - CPUs: Normally reduced instruction set computer (RISC) processors having small, highly optimized sets of instructions are used. 64-bit processors are mainly used in larger systems.
  - Hard disc: ≫100 GB in IDE (integrated development environment–IBM PC)/SCSI (small computer systems interface type). However, for both the systems there is some minimum requirement of RAM that needs to be addressed; it is nearly 2 GB.

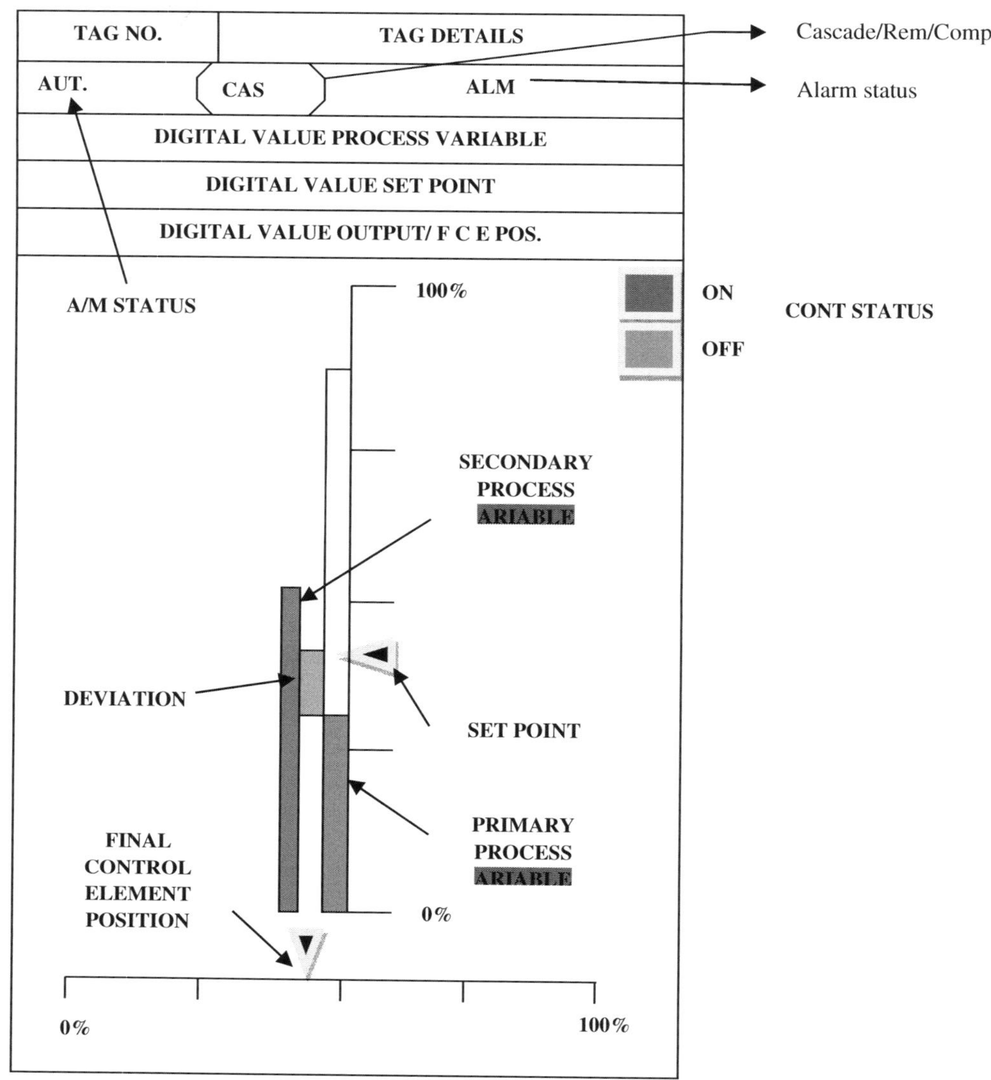

FIG. 7.92 Controller face plate details (Typ).

- The RAM size selection is quite critical. The requirements come from how fast the graphics need to be updated as a new display and the actual memory management system. Standard time for refreshing the most complex graphic is <1, so the RAM shall be chosen accordingly by the supplier. Therefore, the update rate needs to be specified. Also the RAM of the video systems shall be adequate.
- The power requirement is also important. As stated earlier, the power requirement and heat dissipation of the CRT is quite high compared to the TFT LCD/LED displays.
- Other hardware requirements are a graphical board, an Ethernet board, serial ports, and USB ports.
- Keyboard: Both a standard as well as a custom keyboard are in use. In most of the cases (may not be for a PC-based automation system), membrane keyboards with hermetically sealed switches are used. Inside may be keys based on the Hall effect and/or the Reed switch type.
- Mouse: It is used to move the cursor on the computer screen by sensing the movement of the wheel below it and/or by optical sensing. Latter ones can adjust the direction and speed of the cursor movement. Now, a LASER in place of the LED has been introduced in a mouse.
- Track ball: The track ball has a stationary part that is larger than the mouse, and has a ball in a socket. This allows the ball to move in all the two axes, causing the movement of the cursor similar to that in an electronic mouse. This has much operational convenience for control and operation of the plant.

TABLE 7.35 Work Station Operation and Monitoring Functions

Common Functions	Support Functions	Maintenance Function	Control Window	Monitoring Window
Browser bar	Remote operation	Status overview	Control loop	Mimics/graphics with attributes
Screen mode	Report	System alarm	Logic chart	Faceplate view
Window hierarchy	Security	Time setting	Sequence table	Tuning views
Plant hierarchy	Customization	Control system health status	Signal flow chart	Operator's guide mode
Print (scree)	Historical viewing and reporting	Workstation health status	Faceplate view	Alarm messaging
Short cut key	Printer			Message
Circulation/ selection	Alarm filtering			
Related build up	Multimonitor			
	Field control			
	Open interface			
	Messaging			

TABLE 7.36 Workstation Classification

Workstation	Funtionality
Operator WS	For monitoring control and operation of the plant
Engineering WS	Engineering and program development for example, a control loop and logic program
Historian WS	It performs acquisition of plant data for storage and retrieval for example, historical trends, alarms, event history
Application WS	Third-party interface, historian, database management function, etc.
Gateway WS	Protocol translator, gateway (for example, EPLCG of Honey well) to enable networking with different systems, for example, OLE of Microsoft
Portable WS	Special functions, for example, portable vibration monitor. could be plug-in/wireless

- **Touchscreen**: Less space, high accuracy, limited space requirements, time-critical application, and a simple interface male the touchscreen as a pointing device most popular. Because it is with the CRT/TFT LCD screen itself, there is no additional space. It also has a direct interface so this is in use in most of the applications (including those in cellular phones, video games, etc.). There are several technologies available to implement the design. These are discussed in Table 7.37.

It is important that the engineer takes into consideration a few factors such as response time, resolution, transmissivity, stylus type, reliability, etc. Because the image may be affected by any material in front of the image, transmissivity is important. The response time is important because the faster the sensing of the input, the better will be the design. A touchscreen needs to have a better MTBF (mean time between failure). Environmental factors are also important for durability, etc. A small comparison among the touchscreen techniques is presented in Table 7.38:

TABLE 7.37 Important Touchscreen Technologies

Resistive touchscreen	Resistive touchscreen consists of two film/polyester panels coated inside (facing each other) with electrically conductive materials. These two are kept separated by invisible spacer dots in (say) the air substrate. As shown, these are kept ahead of the CRT/TFT LCD screen. A small current goes through the screen. When pressed by a finger or any other object, the layers are pressed together and the panel behaves as a voltage divider, causing the flowing current to differ. This difference is registered as a switch closure/touchscreen event and sent to the controller to locate the path	CRT/LCD SCREEN; SPACER DOT; FILM with Inside CONDUCTIVE COATING; SUBSTRATE
Capacitive touchscreen	A capacitance touchscreen consists of a clear glass panel coated with a charge-storing coating of indium tin oxide (ITO*), as shown. On touching screen by a conductive stylus for example, finger, as shown by circle in the figure. A = a small charge is developed, the circuit at the corner measures the capacitance and the current, which is proportional to the distance from the corners. Their ratio gives the location. B = Location is calculated and sent to PC	CRT/LCD SCREEN; A; B; Clear glass; ITO* coating
Acoustic detection touchscreen	In this method, acoustic waves are sent across the glass surface because the glass surface is fixed. So, for a given sound speed, the sound reflections will be sequential. When a stylus or finger touches it, the sound is attenuated on account of the absorption of energy by the finger. This is detected by a detector to calculate the location. Some use piezoelectric transducers situated around the screen. When the finger touches the screen, a small vibration is created, then sensed by these transducers. For example, the Tyco International touchscreen system	
Infrared touchscreen	Infrared detection method depends on the interruption of IR beam. Infrared LEDs and an optomatrix comprising photodiodes are placed on opposite sides on the display, as shown. These LED beams cross each other in vertical and horizontal patterns. When the stylus or finger enters the grid to obstruct the beam, there will be an obstruction to a number of photodetectors, which send signals to the electronic board mounted behind the bezel. The X and Y coordinates are calculated, hence the position is detected and the signal sent to the processor	
Optical imaging touchscreen (camera)	In this method, a number of image sensors are placed at the corners. There will be an infrared light source in the field of the camera placed on the opposite corners. When a part of the screen is touched, there will be a shadow detected by a pair of cameras to pinpoint the exact grid and report to the processor	A; B
Near field imaging touchscreen	The sensor consists of glass coated with ITO. An excitation waveform to the conductive sensor is supplied and an electrostatic field is generated. When touched, the electric field is changed. The change is detected by the electronics	

7. CONSOLE/WORKSTATION SOFTWARE ASPECTS: Most of the standard workstations available are based on **Windows NT/Windows NT Embedded/Window 2000/ Window XP Pro/Window 7/10, or UNIX/Linux**, which is the platform for running various system software to support operational functional services such as operation and monitoring services (OMS), recording/reporting and calculating services (RCS), and data interface services (DIS). A standard manufacturer's MMI/WS software architecture based on this is shown in Fig. 7.93. Control processing services (CPS) is not part of a WS (shown for completeness). So software systems can be categorized as:

- Operating system: (As stated above).
- System software: This is the basic software for real-time data collection, process control message management, database management at the basic level, interfacing with other systems such as historians, application module, etc.
- Graphical user interface (GUI): A graphical window interface for display management.
- Application programming: Data-based management such as ACCESS, SQL, and a third-party interface.
- SOE, historian application, system maintenance support, etc.

TABLE 7.38 Comparison of Touchscreen Technologies

Technology	Resistive	Capacitive	Acoustic	Infrared	Optical	Near Field
Transmissivity	<78%–82%	<90%	<92%	~100%	>92%	<85%
Touch	Any thing	Conductive/finger	Finger/pen	Finger/pen	Finger	Finger
Ambient light	Unaffected	Unaffected	Unaffected	Consult Manuf'r	Consult Manuf'r	Unaffected
Response time	10–15 mS	<15 mS	~10 mS	<20 mS		~15 mS
Dust/dirt rating	Unaffected	Moderate	Moderate	Moderate	Moderate, (resistant to moisture)	Unaffeced
NEMA rating	4/12	4/12	12	4/12	12	4/12
Environment	0–50°C/95%	0–70°C/95%	0–50°C/95%	0–50°C/95%	0–70°C/95%	0–50°C/95%
Durability	3–5 years	<3 years	<5 years	<5 years	3–5 years	

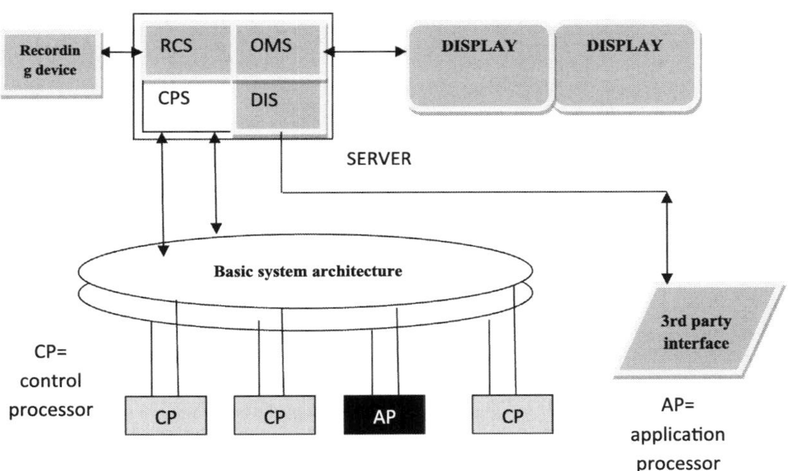

FIG. 7.93 Typical WS software architecture.

- Basic software support for 10/100 base T/F Ethernet, numbers of USB ports, com port including an RS232/485 link, keyboard, and other pointing devices.
- Software feature: This shall include but is not limited to:
  - o. Basic software support for 10/100 base T/F Ethernet, numbers of USB ports, com port including an RS232/485 link, keyboard, and other pointing devices.
  - o. Easy to use: User friendliness, intuitive menu/online help/multiwindow functioning/easy loading of applications/reliable run time applications.
  - o. Security feature for LOG in/out and also access control at various levels—multiple security.
  - o. Sharing and networking: Transfer of multiple interacting applications securely with LAN/WAN/remote node communication/sharing of data across workstations/support to transfer shop floor data for MIS presentation/support applications based on standard packages or software such as Excel, Access, C^{++}, VB, etc. Customized application software.
  - o. Openness to support OPC servers and other third-party software based on international standards.

Even though coordination efforts and testing time may be less when MMI software is procured from a DCS supplier, it is not mandatory. The following are a few points that may be considered for MMI software evaluation:

• System architecture	• Impleme-ntation easiness	• Scalability	• Integration facility
• System platform	• Operation-speed	• Perform-ance	• Provenness
• Administ-ration	• Multinet-work support	• Cost	

VMware in workstation systems is helpful in integrating various different systems with multiple operating systems. It is an operating system that sits directly on the hardware and is the interface between the hardware and the various operating systems. It expands the hardware, from the users point of view, to many different independent servers, all with their own processors and memory.

8. TYPICAL DISPLAY QUANTITY: In a typical DCS deployed for medium to large power plants, quantities are given in the following Table 7.39:

The process display as presented has some hierarchy in presentation for operational convenience. This has been presented in Table 7.40. This hierarchy is user-predefined. This may be clear from an example: For an operator in a boiler, the draft system may be the "AREA" whereas the air is a subsystem, or the FD fans may be the "PROOCESS UNIT" and in that there may be a "GROUP" of 8–16 parameters (for example, motor parameters).

5.1.2 Large Video Screen (LVS)

A giant video screen as in a sports stadium is being utilized in the control room as the man-machine interface (MMI) for control and monitoring. This would not only help to view the MMI operator's screen in a larger scale, but would provide a means to see the CCTV signal or any other camera signal also in a shared video. Live video for communication is also a possibility here. These are generally TFT LCD (or LED) displays having flicker-free uniform brightness display all over the screen with autobrightness adjustment facilities regarding the ambience. Normally, these are placed in a control room having the capability to absorb ambient light so that the videos can be seen without the need to darken the space (suitable light adjustment facility).

TABLE 7.39 Display Features

Attribute	Quantity	Remarks
Process diagram	15,000–25,000	Depends on detailing and number of auxiliaries
Number of windows/display	Min 8	Multiple window support
Dynamic field	Should not be any issue	
Dynamic field/diagram	As high as 200	Should be possible (analog + digital points)
Display update	<1 s	
Number of colors per display	256	Typical value

TABLE 7.40 Display Hierarchy

Display Type	Area	Process Unit	Group	Detail
Operating data	Overview	Sub system	Group	Detail
Operator graphics	Graphic displays as defined			
Trend (as defined) parameter can be MAX/MIN/AV, etc.	Area	Process unit	Group	
Hourly average			Group	
Sequence	Summary		Group Summary	Detail
Alarm	Summary	Process unit summary	Group alarm	
Help	Help display as required			
Message	Message summary			

There could be two or three such LVSs in the control room. Some of the technical features are:

1	Diagonal size (mm)	2000–2100 (80–84″)
2	Type	LCD/LED or any other proven technology
3	Resolution	1024 × 1024 pixels
4	Control unit	Similar to operator's workstation
5	Video signal from	MMI, compatible camera, or CCTV programmable
6	Switching unit	For selection of MMI, compatible camera and CCTV
7	Screen division	Multiple screens in selected area with provision to project any display on a particular screen/location following a predefined event
8	Interface and standard	PAL/NTSC/SECAM and others in VHS standard
9	Illumination level and viewing	predefined or preprogrammed as per the owner's choice/manufacturer's standard

5.1.3 Printers

Printers as an operator interface are important when hard copies of plant data or trends are necessary. So, in almost all cases, these printers are supplied with the DCS. Based on the technology, the printers can be classified into several categories: i**mpact printers and nonimpact printers**. Also, based on printing speed and types, these can be categorized as serial printers or line printers. Dot matrix printers, inkjet printers, and laser printers are mainly used with DCS. Only for large bulk printing would line printers be used, so these have limited use in a DCS.

1. DOT MATRIX PRINTER: This is an impact serial printer. Here, the character dots hit the ribbon against the paper. Character fonts are stored in the memory. The characters are formed by a matrix of dots. The printing head consists of a number of columns of needle hammers and there are six such horizontal columns that move across the paper. Based on the characters to be printed, these dots are fired in sequence. The character quality depends on the number of dots in the character matrix, that is, the more dots, the better the quality. A typical printing speed of 900 CPS is common. When compared with inkjet printers, the capital cost of these printers may be higher, but the printing cost and quality is lower (However, now deskjet printers with ink tank cost of printing is extremely low).
2. THERMAL MATRIX PRINTER: This is similar to a dot matrix printer, meaning that here also the printing head has 5 × 7 dots, and these dots are selectively heated to give character prints on the special paper on which the head moves. In DCS applications, in some cases these were in use, but they are not very popular these days.
3. INKJET PRINTERS: Inkjet printers are the most common type of printer. On account of a low initial cost, high-quality printing (per page printing cost may not be cheap), vivid color, and ease of use, these printers find huge application in a DCS. They work by propelling tiny droplets of liquid ink onto the plain paper. Based on this propulsion technique, there can be three types of inkjet printers:
   - THERMAL INKJET: This printer has a print cartridge with a series of tiny electrically heated chambers constructed by photolithography. The printer runs a pulse current through the heating elements, a steam of explosion occurs, and a bubble is formed that causes the ink droplet to be propelled onto the paper.
   - PIEZOELECTRIC INKJET: Here in place of heating, a piezoelectric crystal is used. So when the current pulse for printing is applied, the crystal changes the size and shape, forcing a droplet of ink onto the paper. This is quite expensive.
   - CONTINUOUS INKJET (Fig. 7.94A): This type of printer is used for the most part. A high-pressure pump directs liquid ink from a reservoir through a microscopic nozzle, creating a continuous stream of ink. There is a piezoelectric crystal that causes the liquid stream to breathe droplets of ink. This piezoelectric crystal is subject to an electrostatic field formed by deflection plates in horizontal and vertical directions to shift and drop at the desired point on the paper. These deflection plates form the character manipulate the dot structure. The charge droplets are separated by guard droplets to avoid electrostatic repulsion between two charged droplets. To allow the jet mechanism to work smoothly, a catch gutter is put in front of the last deflection plate. These printers can be black as well as colored ones. Currently, deskjet printers are available with an ink tank and the cost of printing is very low.
4. LASER JET PRINTER (Fig. 7.94B): Static electricity is the principle behind the working of a laserjet printer to act as temporary glue the major part is the revolving drum/cylinder type Photoreceptor. Initially, the photoreceptor is (say) positively charged by a corona wire or a charged roller. As the drum revolves, the printer sends a LASER beam across the surface to discharge certain points. In this way, the printer draws the letter and images as patterns of electric charge–an electrostatic image. The LASER is regulated based on the image. After the pattern is set, the printer coats the drum with positively charged

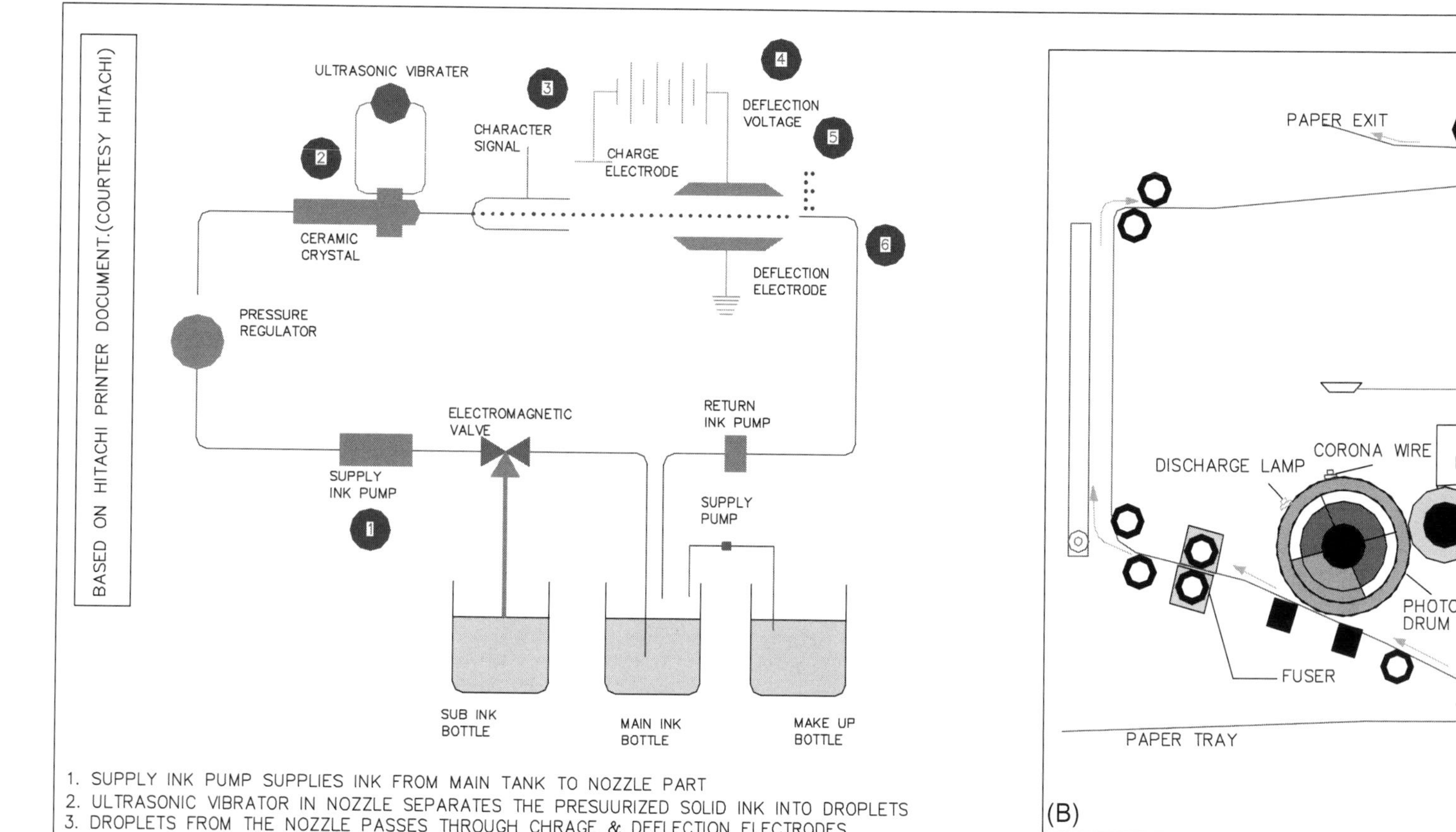
BASED ON HITACHI PRINTER DOCUMENT.(COURTESY HITACHI)
ULTRASONIC VIBRATER
CERAMIC CRYSTAL
CHARACTER SIGNAL
CHARGE ELECTRODE
DEFLECTION VOLTAGE
DEFLECTION ELECTRODE
PRESSURE REGULATOR
ELECTROMAGNETIC VALVE
RETURN INK PUMP
SUPPLY INK PUMP
SUPPLY PUMP
SUB INK BOTTLE
MAIN INK BOTTLE
MAKE UP BOTTLE
1. SUPPLY INK PUMP SUPPLIES INK FROM MAIN TANK TO NOZZLE PART
2. ULTRASONIC VIBRATOR IN NOZZLE SEPARATES THE PRESUURIZED SOLID INK INTO DROPLETS
3. DROPLETS FROM THE NOZZLE PASSES THROUGH CHRAGE & DEFLECTION ELECTRODES
4. EACH DROPLET PASSING THROUGH CHARGE ELECTRODE RECEIVES A DEFLECTION VOLTAGE WHICH VARIES BETWEEN DROPLETS AS PER CHARACTER SIGNAL. THESE CHARGE DROPLETS DEFLECT IN A A PREDETERMINED ARRAY.
5. AFTER DEFLECTION ELECTRODES, DROPLET CONTINUES TO TRAVEL IN PREDETERMINED PATTERN OUT OF HEAD TO SUBSTRATE.
6. DROPLETS NOT REQUIRED FOR PRINTING ARE RETRIEVED BY GUTTER & RECYCLED BACK TO INK BOTTLE.
(A)
PAPER EXIT
LASER UNIT
CORONA WIRE
TONER HOPPER
DISCHARGE LAMP
ROLLER
PHOTO ACCEPTER DRUM ASSEMBLY
FUSER
PAPER TRAY
(B)

FIG. 7.94 Inkjet and laser printing methods.

toner (a fine plastic powder that fuses at ~200°C to bond with paper). Because it is positively charged, it clings to the area of the drum discharged by the laser and not the positively charged background. With the pattern affixed, the receptor drum rolls over a sheet of paper that is moving along a belt below. Before the paper rolls below the drum, it is negatively charged by the corona wire or charged roller. This charge is stronger than the negative charge of the electrostatic image discussed, so that it pulls the power/toner back to the paper in the exact pattern. Laser printers can be black and/or a colored toner/cartridge. Because photocopying is done on the same principle, most of these printers can be used as a photocopier. The speed of printing depends on many factors and varies with the model. Typical speed 200/100 (color) is A4 pages per minute.

5. COMMENTS ON VARIOUS PRINTERS DISCUSSED: Dot matrix printers and inkjet printers take incoming spooled data for printing, so the printing process is slow, and they may stop printing when the printer is waiting for data. On the other hand, in these printers if the data required for printing is taken at a time and buffered, then the printing would be continuous. However, if it has to wait for such data, there may be the possibility of misalignment. Generally, in laser printers, a raster image processor scans the entire page line by line and stores the bitmap of the page in the raster memory before staring the printing procedure. In laser and dot matrix printers, A3 size printing is possible but a longer banner may not be possible. Various factors such as bandwidth, color, paper size for printing, and other facilities such as COPY, SCAN, FAX influence the price of a laser printer. In case of large-scale printing, the laser printing cost may be economical, even if the initial cost of the printer is high. Also, depending on the built-in processor/intelligence with the printer, it can be hooked directly with the network or with a server and/or any workstation/engineering station through a com port/USB port as well as in wireless mode.

5.1.4 Recorders

It is possible to take the print out of the screen to get the print out of any trend, X-Y plot, etc. However, in many plants, dedicated microprocessor-based recorders are used for selected parameters. Recorders are independent of DCS and could be the intelligent chart type and/or the chartless type. The advantage of utilizing microprocessor/microcontroller-based intelligent recorders is that these can be connected to the DCS through the RS 232/RS 485 link with say, the MODBUS protocol. Also, it is possible to connect the analog output of the DCS directly from the redundant output of the primary device to these recorders for a hardwired connection. It is choice of the designer/user. There are a few different kinds of recorders such as **strip chart pen** recorders (**1/2/3/6 pens**), **multipoint** recorders (3/6/12/24 **points**), **circular chart** recorders (1/2/3/6 pens), and **chartless** recorders (**none/6/12/24** inputs).

The number of pens/points: The number of pens/points stands to represent the input channel to the recorder. What is the difference between the pen and point? In case of a pen recorder, there will be a separate pen (each with a different colored ink), an independent measuring channel, and a signal processing unit for each of the input channels. Whereas in the case of a multipoint recorder, there will be only one common pen, measuring channel, and signal processing unit for all the input channels. Several input channel parameters are recorded by multiplexing/scanning the inputs sequentially at a predefined rate. Also, each channel shall be distinguished from the other, either by using a different colored ink and/or using separate dot symbols (say 1/2/3 print corresponding to Chapters 1–3, respectively).

Chart types and speed: Recording charts could be a strip chart, which is characterized by the uniform linear motion of the paper, either horizontally or vertically with measurement lines either rectilinear or curvilinear recordings or Circular chart with uniform rotation per unit time which normally one revolution per day circular, with 24 hrs divisions. In case of a chartless recorder, there is no chart and the trends are shown in LCD displays like the workstation of the DCS. However, in **ALL** recorders, the recording chart speed is variable, and many times it is required that the chart speed is made faster for better analysis of the parameter trends such as the boiler/turbine metal temperature during start up.

Enclosure rating: Mostly, recorders are available with IP55/NEMA four ratings.

Universal inputs: Earlier, there were divisions in recorders based on input measuring schemes such as bridge type, potentiometric type, etc. Now, almost all recorders available have a universal input.

Sizes: Almost all these recorders are available in standard din sizes, for example, 72 × 144, 144 × 144, and 288 × 288. The size of the recorder is dependent on the type of recorder and the number of pens/points.

Indications: Most of the recorders used are the indicating type, that is, the input parameters are indicated either by scale and pointer or by back-lit LCD, LED, or bar graph.

Power supply: 1 Ph, 110/240 VAC, 50/60 Hz (VA rating: type dependent).

Noise rejection: Normally ~120 dB @ 50/60 Hz is standard.

Memory protection: Memory protection in E^2PROM is available. Also, in a chartless recorder, USB ports are available for downloading some data, as required.

Advanced software functions: Advanced software functions such as math functions, for example, totalizing, + − X, etc., are available in these recorders.

Alarm function: If needed, such functions for the number of inputs are available, but may not be popularly used for recorders with DCS.

RS link: These recorders offer an RS link for serial communications.

1. CIRCULAR CHART RECORDER: As the name implies, the shape and movement of the chart is circular. In the chart, the time scale is graduated along the circumference and the parameter variation scale is in a radial manner. Some of the circular chart recorders have an event timer driven by the recorder's real-time clock to start/stop the recording on a preset timing. These are available in 1/2/3/4 pens. On each channel, it is possible in some recorders to record two or three events. A typical circular chart recorder is shown in Fig. 7.95A. It can also perform some mathematical and logical problems.
2. STRIP CHART RECORDER: In contrast to the circular chart, these recorders have the chart moving in linear motion, with the time scale either in the horizontal direction and the pen movement in the vertical direction, or vice versa. These are mainly used for 1/2/3 pen recording with 72 × 144 size recorders, and in these instruments, by opening the door, an extended view can be observed without hampering the recording. The reverse type, that is, pen movement in the horizontal direction, is shown in Fig. 7.95B, Strip Chart Recorder. These are used normally in multipoint/pen recordings with sizes such as 144 × 144 or above. The natural view is wider. Strip chart recorders have displays in the bar graph and/or digital displays. In the front, there may be switches to adjust speed, set alarms, etc.
3. CHARTLESS RECORDERS: This type of recorder, as the name implies, does not have any charts; these are LCD display very similar to a monitor display. These have powerful functional capability. These have internal flash memories for recording and logging. Also, they have a battery backup. These recorders are network compatible to be hooked up to the Ethernet network, as shown in Fig. 7.96. Its high mathematical capability makes it versatile. Similar to a workstation monitor, it

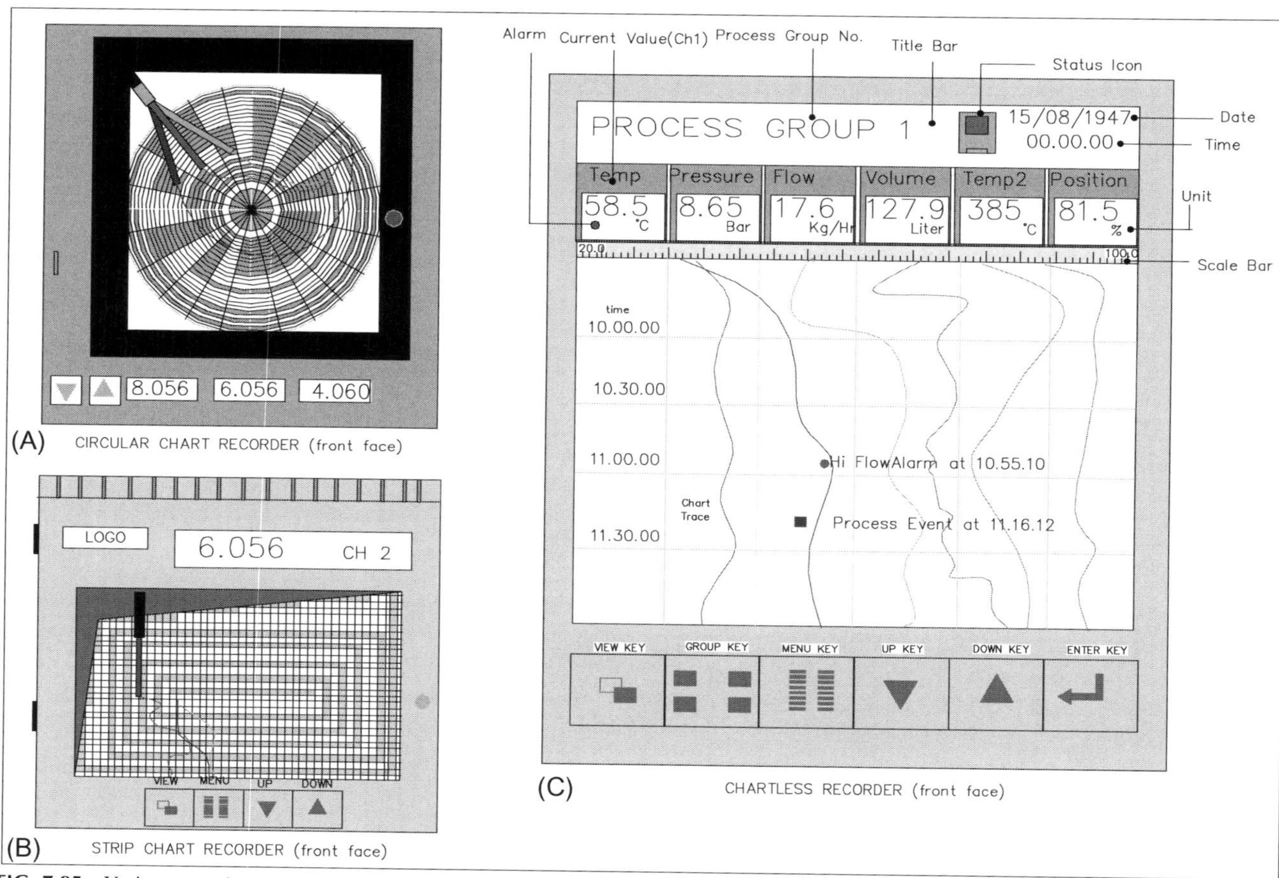

FIG. 7.95 Various recorder types.

can be used for historical trends, etc. Also, the data from these can be retrieved by a memory stick utilizing USB ports. Group display, menu, etc., push buttons are available in the front panels, as shown in Fig. 7.95C, Chartless Recorder. Some of these have embedded web servers.

4. X-Y RECORDER: There is yet another kind of recorder called an X-Y recorder that notes the variation of one parameter with respect to the other (say, boiler load versus air flow). There are two kinds of such plotters available. In one of them, the chart is stationary and the two signals are connected to two axes or, in case the chart is moving, one of the signals moves in the opposite direction while the other signal goes along a direction of 90 degrees. These plotting signals can be either analog or digital.
5. MISCELLANEOUS RECORDING SYSTEM: There are a few other recording systems such as a tape recording system, a storage oscilloscope, etc. All these devices are used to store data for future analysis.

5.2 Display Types

5.2.1 Introduction

Various kinds of displays play major roles in operation and management. The major display types are: **overview display**, **mimic display**, **group display**, **point display**, **alarm display**, **trend display**, **plant start-up and shut-down display**, **operator guide display**, **loop and logic display**, **controller faceplate display**, **X-T and X-Y display**, **bar chart,** etc. Each has a separate significance. The typical quantities of displays in a main power plant of medium/large units are enumerated in Table 7.41:

5.2.2 Overview Display (Typical)

An overview display frame normally gives a bird's eye view of the entire plant/subsystem to the operator. The design of the overview display normally is engineered jointly by the owner and manufacturer. Based on the owner's requirement for screen division (windows), and the content of each window, the manufacturer's engineer develops a system that can be well supported by the system resources available. Normally, from the overview displays, it is possible to quickly go to various detailed detail displays/trendings/control faceplates, etc. For a subsystem overview display, it should be possible to quickly switch to another subsystem display. Fig. 7.97 is an example of the overview of a boiler subsystem.

However here is possible to go quickly to see the status of other subsystems, e.g., Turbine/Generator sub system, with the help of dedicated illuminated soft key (shown in the left part). In overview display, there could be various windows in a shared screen to display, Mimic diagram, Trend display, Controller face plates as well as Alarm/SOE message, to apprise the operator of all important event status as well as control parameter variations. In each of these windows desired parameters can be chosen, in groups, etc. Normally date & times are shown at the Top/Bottom part, also it is possible to take print out of any desired display. To keep the operator apprised of the situation of the sub system, normally either at Top or at the bottom, three to four lines are dedicated to show latest alarm/SOE. In many displays such dedicated illuminated soft buttons are replaced by browser bars as shown in Fig 7.98.

From this browser bar, it is possible to view various loops/subloops, etc. One can also select the overview displays with the help of the browser bar. The browser bar enables the user to view the selected trend, mimic, etc., with

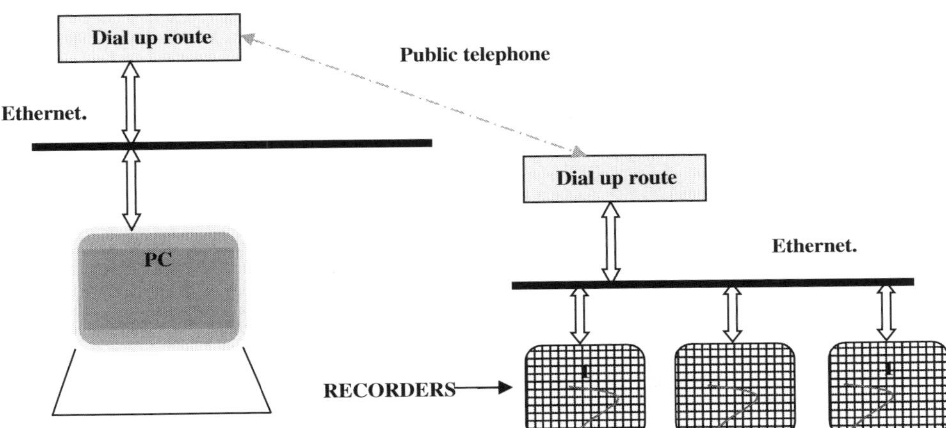

FIG. 7.96 Chartless recorder ethernet and web connection.

TABLE 7.41 Display Quantities (Typical for Mid/Large Unit Size TPS)

SL	Description	Quantity
1	Overview display	10 types
2	Control displays (group, sequence, loop, faceplate)	400–700
3	Mimic/P&ID with dynamic attributes	250–450
4	X-T trend (in groups/mix)	5 × 20–25 (75–180 total)
5	X-Y plot user selectable/historic	70–150
6	Bar chart (horizontal/vertical) π chart	75–130
7	Group display/individual displays	>100 (4–16 in each group)
8	Operator or start-up/shut-down guidance message	100–120 each
9	Status/system diagnostic message (manufacturer std)	As required
10	Shared window types of display	10–15
11	Alarm display type and LVS alarm	As required by owner
12	Misc. display types/qty	10/100

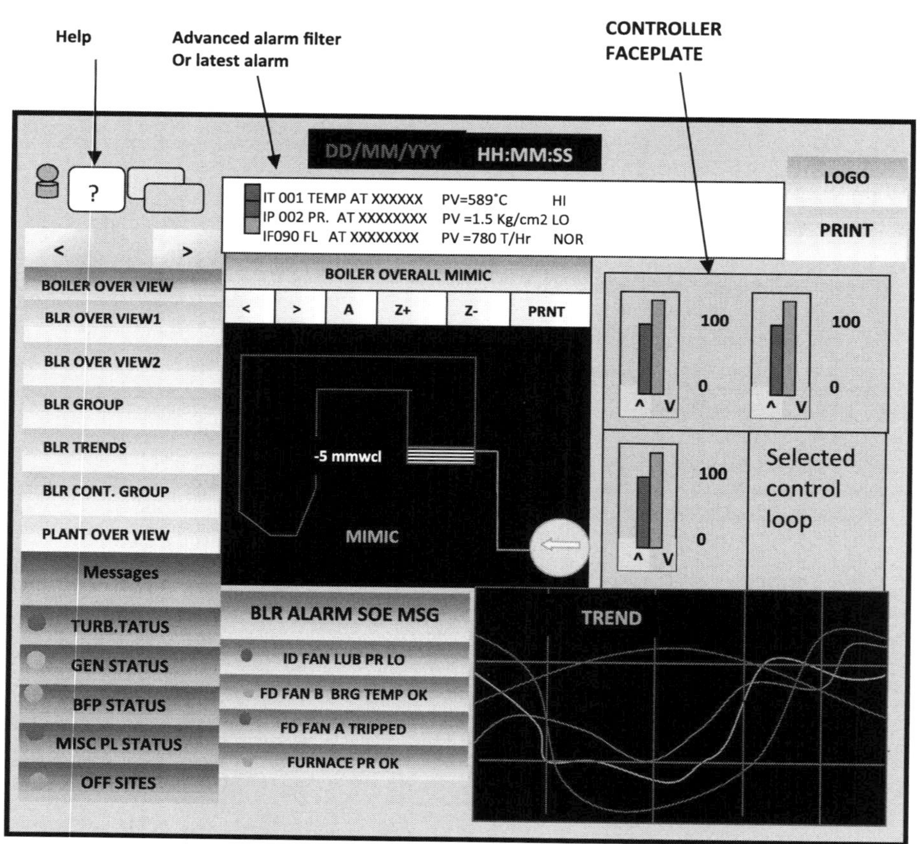

FIG. 7.97 System overview display (an example only).

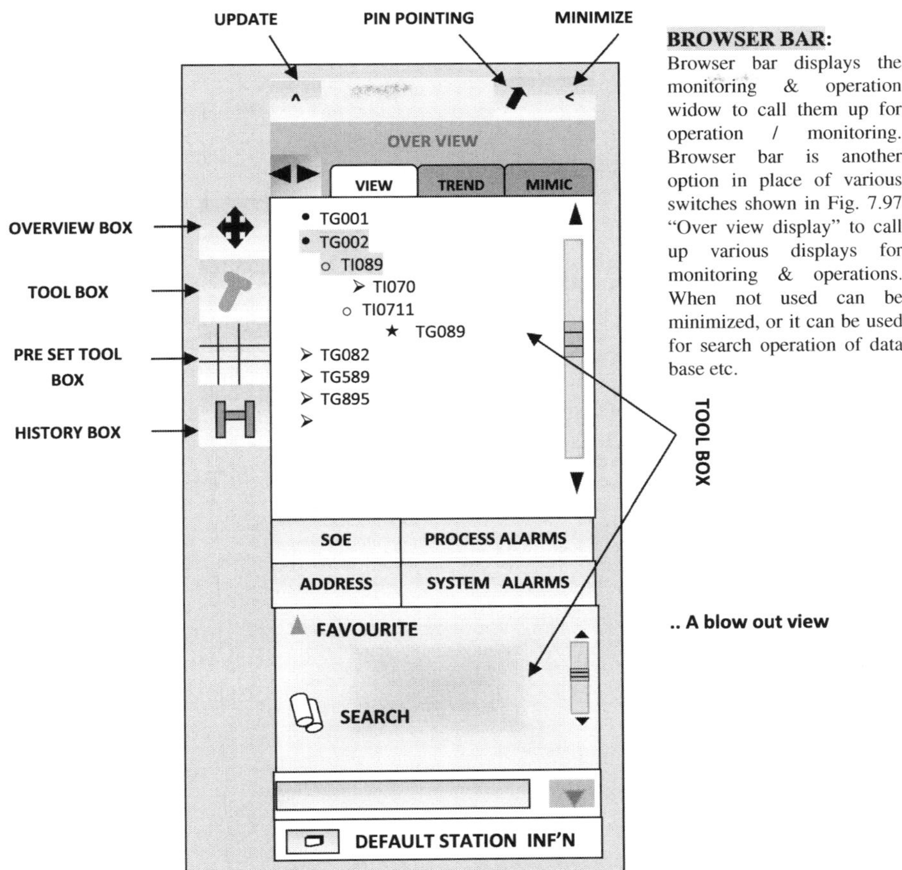
UPDATE
PIN POINTING
MINIMIZE
OVER VIEW
VIEW
TREND
MIMIC
OVERVIEW BOX
TOOL BOX
PRE SET TOOL BOX
HISTORY BOX
TG001
TG002
TI089
TI070
TI0711
TG089
TG082
TG589
TG895
SOE
PROCESS ALARMS
ADDRESS
SYSTEM ALARMS
FAVOURITE
SEARCH
DEFAULT STATION INF'N
TOOL BOX
BROWSER BAR:
Browser bar displays the monitoring & operation widow to call them up for operation / monitoring. Browser bar is another option in place of various switches shown in Fig. 7.97 "Over view display" to call up various displays for monitoring & operations. When not used can be minimized, or it can be used for search operation of data base etc.
.. A blow out view

FIG. 7.98 Browser bar details.

the help of the selected flag. There could be a toolbox or preselected tool boxes to carry out various operator functions as well as to go to the preselected display views (including history functions, search a tag).

5.2.3 Mimic/P&ID Display (Typical)

A mimic display is extremely helpful for plant operation. It displays the graphical form of the process within the purview of the particular mimic diagram. All the drives have dynamic attributes. This means that, by using the cursor, touch, or a stylus, the associated drives can be operated and/or its status updated for viewing. In a typical mimic diagram shown in Fig. 7.99A, whenever the illuminated soft button associated with AMMONIA/HYDRAZINE is touched, then the associated dosing control loop can be viewed. Similarly, by touching the sample line status button, it is possible to view the sampling point parameters. As shown in Fig. 7.99A, whenever FIC 11/21 is touched, it is possible to get the controller faceplate display as a pop-up window for control purposes. This is an example indicated as one of the possibilities available from the manufacturer for a plant. The basic design is engineered jointly by engineers from the owner and manufacturer. Normally, it is possible to add a fixed number of the owner's predefined library symbols in addition to the standard library symbols available. With the help of multiple window operations, it is possible to go to the trending of a few parameters in the mimic display. These mimic diagrams could be overview or individual. As shown in Fig. 7.99A, it is possible to highlight a particular loop mimic with different colors, and/or show a vessel level with shades. In most of the cases, mimic displays have provisions for ALARM/SOE lamps displayed for the drives in the mimic and/or associated/related drives/process.

5.2.4 Controller Faceplate Display

In this display, a number loops with associated drives shall be displayed along with related parameters such as measured process value, set point value, deviation value, drive position value, etc. These values can be depicted in the bar display as well as the digital value. In the faceplate, the remote/local and A/M selection status shall also be shown. The display configurations could be, say, eight loops at a time and/or a mix of a few loops along with an indication

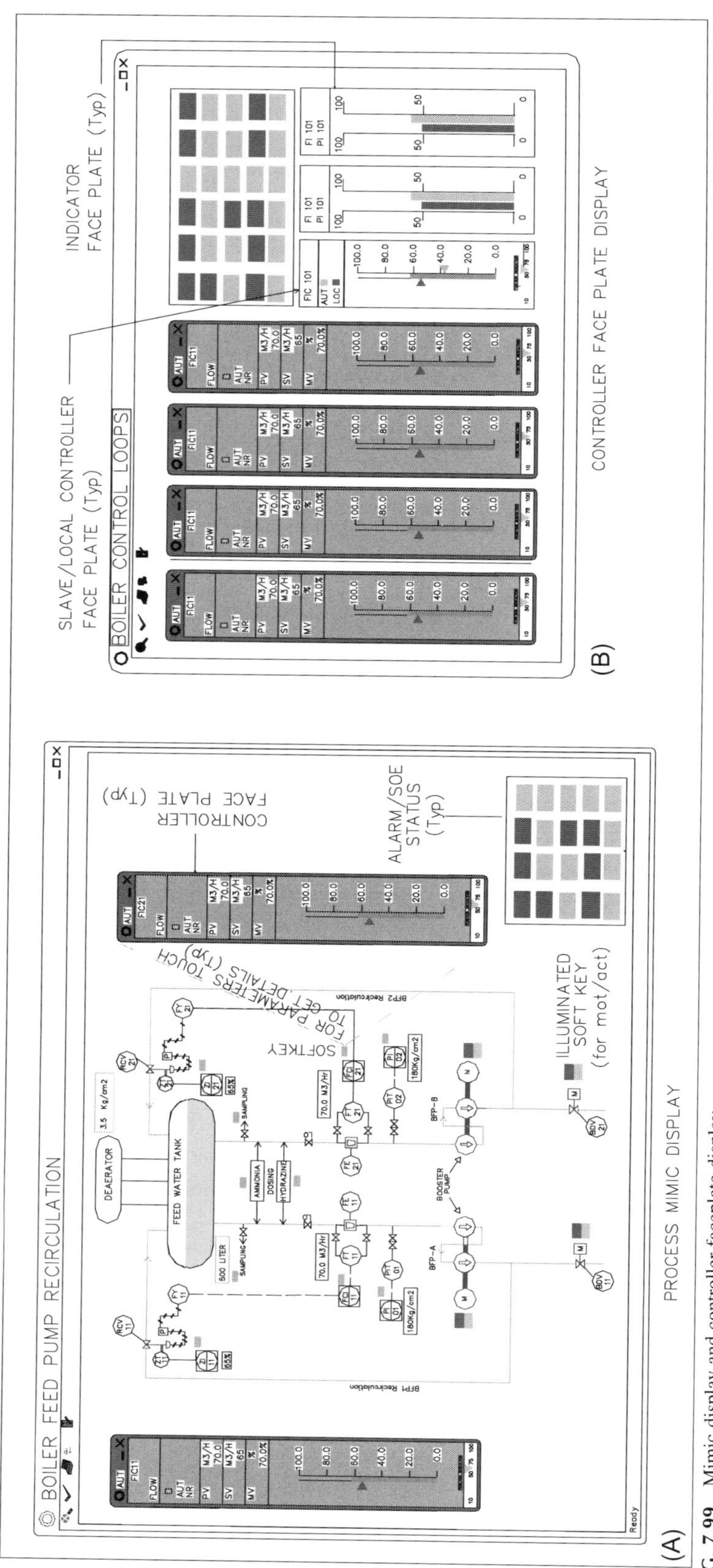

FIG. 7.99 Mimic display and controller faceplate display.

of a few related parameters, depending on the operator's choice as admitted by the manufacturer. As shown in Fig. 7.99B, it is possible to display (here four) numbers of the full control faceplate along with (here two) indications and a short controller faceplate of related parameters. This combination depends on system capability, but a combination of such displays can be decided jointly. By window sharing, it is possible to put forward the alarm/SOE status and to go other displays directly from these displays by the browser bar and/or by a set of dedicated soft buttons (Figs. 7.97 and 7.98).

5.1.4 Trend display: These are trends of analog parameters with respect to time, so these are X-T displays. There are selectable time spans for these displays.

While for current real-time parameters, these could be selected from 10/20/30 sec to 1, 2, 5, 10, 15, 20, 30, 60 min or it could be 2, 6, 8, 24,48, 72 h also. It is also possible to extend a trend to shift, daily, weekly, and monthly parameters. In the latter cases, the parameter values chosen could be average, max, or min values selectable by the operator. It is possible to get finer divisions in a time span for selected parameters and for event-based trending, for example, faster trending of furnace pressure following a disturbance in the system. Based on the available memory space, the sampling rate can be selected. Depending on the sampling frequency and time span, historically the parameters can also be trended. For a few parameters (for example, oil/coal/makeup water consumption, etc.) it is possible to get a trend on accumulated value over a period on shift, daily, weekly, or monthly basis. It is possible to scale a range as well as time trend display for detailed viewing. For a detailed view, it is possible to slide the view vertically as well as horizontally. These are shown in Fig. 7.100A. Normally, there are dedicated soft keys that allow the operating personnel to view any particular parameter in the desired way by squeezing and stretching the parameter curve and/or by zooming in/out. PAN and ZOOM for a trend display is quite common. Normally, there is a slider bar that, when put at a point, then a digital readout of all the parameters at the point of the slider (in time scale) is given in a space below (or above) the trend display. Although not shown, it is possible to go to other displays directly from these displays with the help of a browser bar and/or by a set of dedicated soft buttons (Figs. 7.97 and 7.98). On account of the presence of an object menu, it is possible to call up any trend display easily and immediately by simple clicking the point from any display.

5.2.5 Alarm Display

A typical alarm summary display is depicted in Fig. 7.100B. There may be a browser bar or dedicated soft switches to quickly switch to any other display. Alarm displays and SOE are helpful for operating personnel to analyze the fault as well as to get to know the health of the plant. The design of the alarm display generally comes from the plant control and operating philosophy of the owner and there could be a variety of such displays. In Fig. 7.100B, an alarm summary display includes mainly the information related to date, time, type of alarm (first up, ordinary, or return to normal), the actual alarm message, and the tag of the device

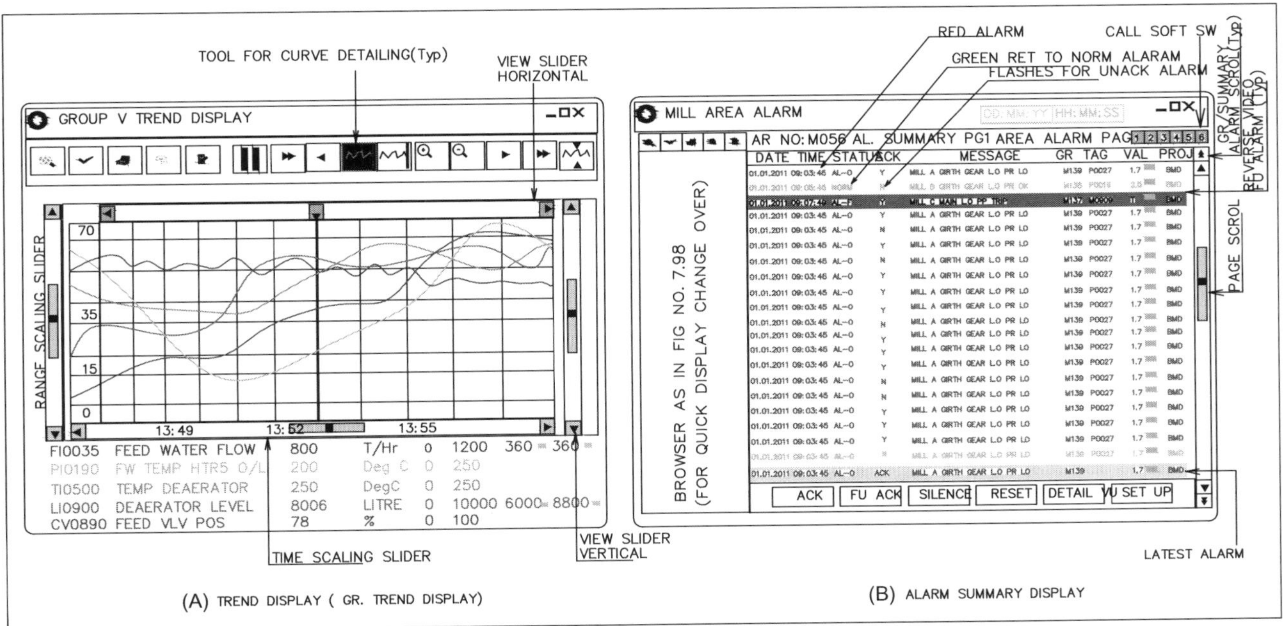

FIG. 7.100 Trend display and alarm display.

responsible. This also includes if the same is in any group (so that the operator can go to the group display to know the status of other points in the group) and area information. Normally, the alarm condition is depicted by a RED/Yellow color whereas return to normal is depicted by green color display. Ordinary and first alarms can be distinguished by color (say yellow and red) or can be distinguished by a reverse video, as shown. An alarm may appear from top or bottom. Sometimes, the latest one/two alarms are suitably distinguished. To distinguish between acknowledged and unacknowledged alarms, flashing may be attributed to the alarm line (for better ergonomics, only "Y/N" may be flashing). Some show the set point and parameter value in the display, but normally these can be displayed when a particular alarm value is clicked. When summary alarms are on display, details can be viewed as a pop-up window by clicking the detailed page associated with the alarm. Each of the alarms can be acknowledged/reset by selecting the alarm and clicking the relevant button. Silence is shown separately to indicate that these could be to stop the sound in case of a ring-back option, but it does not acknowledge the alarm. In case there are too many alarms in a display, it is possible to scroll, and in case of a summary alarm (area/group basis), other summary alarms can be directly viewed also.

5.2.6 X-Y Plot

In order to establish the correlation between the two variables, X-Y plots are done in the system. Normally, 4–8 locus lines are displayed on one screen. There can be preassigned such displays of 15–20 numbers in a DCS. Alphanumeric information shall be availed for these variables as well as for their chosen scale. PAN and ZOOM functions are common. Also, it is possible to superimpose the X-T plot of these variables on the X-Y plot. An XY plot using stored data is also quite common

5.2.7 Control System Display

Both the CLCS as well as the OLCS diagrams can be displayed.

5.2.7.1 CLCS Display

Under this display, the control loop schematic/loop drawing with associated attributes shall be displayed, as shown in Fig. 7.101A. There may be a browser bar or dedicated soft switches to quickly switch to any other display. As shown in Fig. 7.101A, a complete loop and or a combination of loops can be displayed with commonly used attributes. All these have been shown in one display to indicate the various possibilities of displays. The actual design needs to be firmed up jointly by the owner and the manufacturer. By using the loop A/M and local remote, the loop can be selected in the desired mode. Each of the input and output values is shown and it is possible to select the transmitter in case of multiple transmitters used for the measurement. The set point deviation parameters are also shown as a digital readout at the appropriate place. Valve position and controller outputs are also displayed. It is possible to look into the interlocks associated with the loop. Depending on applicability, the status of various devices is also displayed, for example, the status of the A/M, the solenoid valve, etc. It is worth noting that even if the controller tuning parameters can be viewed, they cannot be modified (unless authorized) in this display for safety. As a pop-up window Controller Faceplate/SFC, etc. by clicking concerned display.

5.2.7.2 OLCS Displays

OLCS displays are associated with interlock and sequential operations/sequence chain displays for groups and subgroups. These are used not only to trace the control sequence and to intervene (authorized personnel only), but also to change the mode of selection from auto to manual. In the entire chain and/or a sub block can be viewed with the status of each device along with step no, waiting and monitoring time. On the display, there may be a control plaque display enabling the user to know the operational step number and associated criteria as well as the missing criteria. A typical such tablet and plaque are shown in Fig. 7.102; for each step of the sequence control, all criteria status is by a RED or GREEN light to signify availability. In OP. Guide mode/ in step by step mode, operator can bypass the criteria (if the system permits). In case of a normal interlock, any input can be forced (if the system permits). All this is done with the help of plaque displays.

Normally, the OLCS display views either in SFC or a ladder diagram, as shown as a pop up in Fig. 7.101A. In many designs, various drives are sometimes generated, whether they are on-off or modulating separate drive displays. This could be a pop-up window in a control display or an independent display to show all the associated inputs and outputs pertinent to a drive (or a group of drives). There are many variations in displays, so the owner and manufacturer need to freeze the design based on the plant control philosophy and the manufacturer's recommendation.

5.2.8 Controller Tuning Display

It is possible to tune a particular controller from the DCS by authorized personnel from the control room using a DCS. A typical such controller tuning diagram is shown in Fig. 7.101B; there will be a number of dedicated soft keys to assists the controller tuning function. This is normally done by taking the loop into manual and freezing the output, or by doing it offline. After the controller is properly tuned, it can be put online. By changing various parameters of the controller tuning parameters such as the proportional band/ integral action time, etc., the control loop response curve can be observed in the display, as shown in Fig. 7.101B. The controller faceplate can be there as a pop-up display.

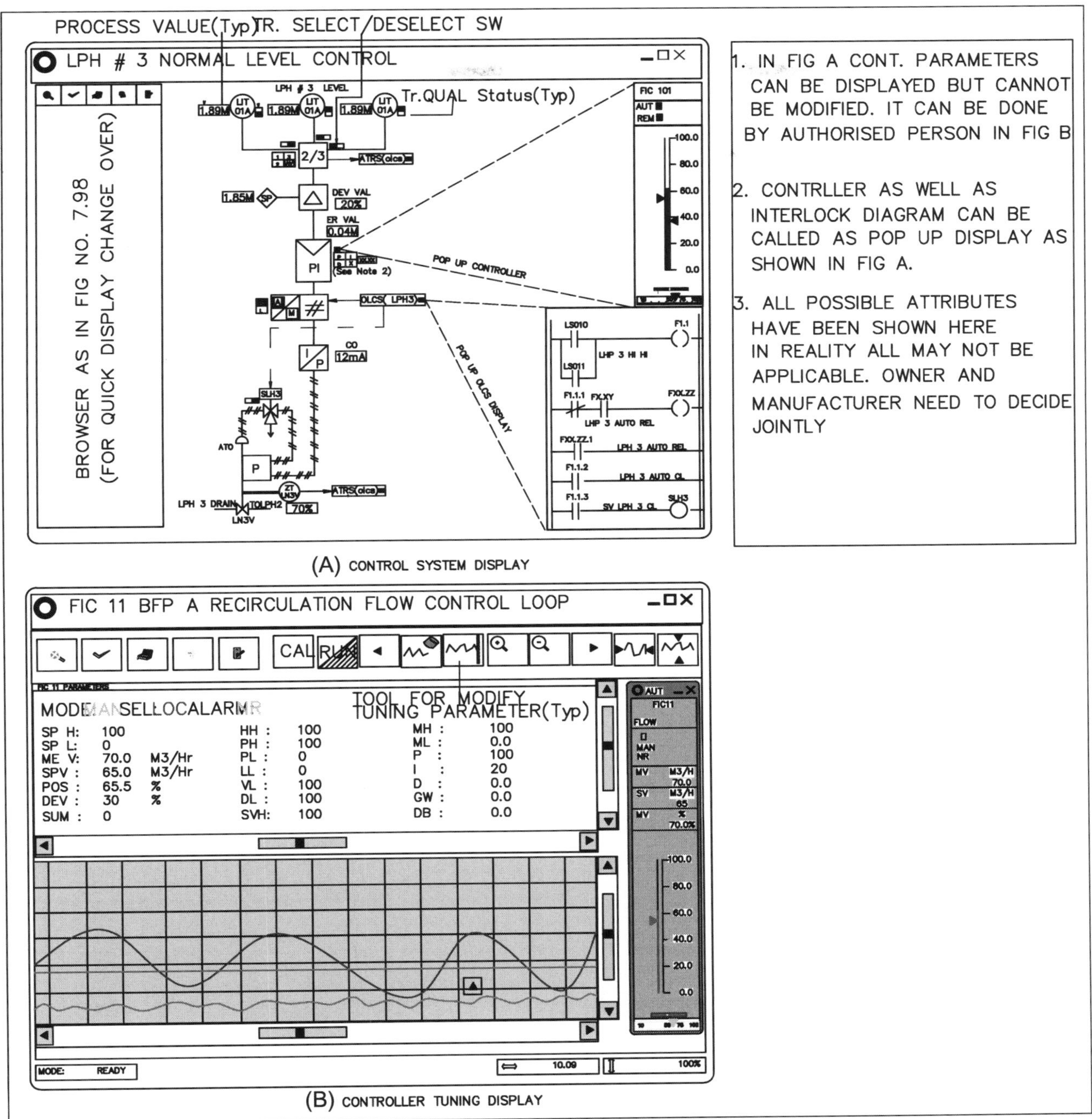

FIG. 7.101 Control system display and controller tuning display.

STEP NO	Criteria available
STEP DESIGNATION	Criteria missing
	Criteria available
	Criteria available
	Criteria available
	Criteria available

FIG. 7.102 OLCS display tabs and plaques.

Also, even during operation, the loop response can be observed. With the help of soft keys such as "CAL"/ "RUN," various mathematical algorithms can be built and the system can be run. Now, to support embedded control systems/microcontrollers, powerful mathematical tools such as Active X controls, MATLAB** (Fig. 7.103).

5.2.9 Miscellaneous Other Displays

There are various other kinds of displays such as:

5.2.9.1 Group Display

As discussed earlier, this display includes the information point ID, the value engineering unit, the set point current value, and the functional group. Presentation of this could be in various ways; these could be simple bar graph displays, as shown in Fig. 7.104. There could also be a group display in the form of a line message. Basically, group messages are helpful to compare related parameters in various ways to the comparative behavior of various parameters at different operating conditions. Group operations can be from group displays. In Fig. 7.99B, a group of predefined controllers has been depicted for performing the control tasks. Air flow is a parameter that is very much related with fuel flow/combustion control as well as steam temperature control. Again, steam flow is related to both combustion control as well as feed flow control, so naturally one may group all these controller faceplates in a group display for control operations. The designer needs to make such grouping according to one's choice, keeping parity with plant operation.

5.2.9.2 Point Display

It is possible to select a particular point from the mimic, group, alarm, etc., displays to get a detailed point display. In a point display, all database attribute for the particular point. It has been found that in many cases, even the terminal details of the particular point have been made available. This point display could be a calculated point also. By clicking the calculated point, the detailed calculation behind the same could also be displayed.

5.2.9.3 Bar Graph Display

Bar graph displays could be horizontal as well as vertical, and there could many such bar graphs in one display as long as the view is permitted. Many related parameters of commercial importance are also presented in terms of bar graphs and a π chart.

5.2.9.4 Large Video Screen

Large video screens are generally mounted on the tops of the wall the operator is facing so an alarm display in conventional annunciator, overlaying mimic would be better

***** MATLAB (Matrix Laboratory) is very useful tool for process control system. This is a fourth generation programming tool for numerical computation & analysis. This allows implementation of algorithms, graphic user interface. For programming it supports Languages like "C", C⁺⁺, FORTAN. Model based system design (such as - process model identification, stability assignment, process simulation, optimal /prediction control) embedded system etc. are well supported by this tool. It has predictive tool box, fuzzy tool box to make the designer's work much easier.***

FIG. 7.103 Note on MATLAB.

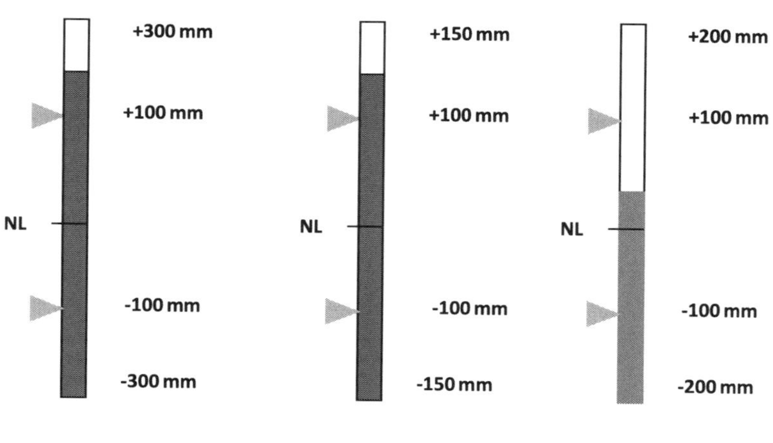

FIG. 7.104 Example of a group display (comparing Levels of main vessels).

proposition. Many also include special displays such as a CCTV view of the burner, a drum level display from an electronic water level indicator (for example, a Hydrastep). Also, a large-digit display of a selected parameter is also there so that people can view it from a long distance.

5.3 Log Types

Log types are very important for keeping track of the plant operation and performance. The format, point assignment, and generation of each type of log/report varies from project to project, and is finalized jointly by the owner's designer with the manufacturer.

Now, most of the log sheets are compatible in copying spreadsheets, for example, MS Office's Excel. Although most of the log/report functions are automatic, manual time tagged log/report generation is also possible from any display and/or type of log. In automatic log generation, it is possible for each log type dropping/inhibiting of any page/selectable Group/point in any log for a selectable period of time (or until further initiation). However, such an inhibiting action can be initiated only by an authorized person. Inhibiting details (such by whom, when, and when the same inhibition was cancelled) may be recorded. In a larger network with a number of printers, there may be a preassigned selection (primary and secondary) for each log type. If necessary, any particular log may be repeated at various printers. A DCS log/report can be categorized into three types, as shown in Fig. 7.105. These are **event-based**: When the log/report is generated after an event occurs; **time-based**: When the log/report is generated periodically after a time gap; and **on demand**: a manually generated log/report.

5.3.1 Event-Based Log/Report

These logs/reports shall be generated after an alarm or trip happens. An SOE is also an event-based log/report. Normally, dedicated printers are assigned for alarm/trip and SOE. During start up and shut down of a unit/subsystem, these logs/reports are generated. A control-related log as a predefined operator action such as a database modification (when authorized) is also an event-based log/report. It is also stored for future recall to see who did the change at what time and date.

5.3.1.1 Pre- and Posttrip Log/Report

Some predefined parameters (owner's choice, changeable at the site) to the tune of 200–300 tags (for medium and large plants) with 50–150 readings for each of them are logged in the pretrip and posttrip log. Naturally, suitable memory space shall be selected to collect these readings so that 50–150 prior readings at a predefined interval (normally onsite modifiable) just before the trip can be reported for the pretrip log. Similar will be for the posttrip log. A trip of major system/equipment (4–5) are assigned for such pre post trip for example, Boiler, turbine, generator, BFP, etc.

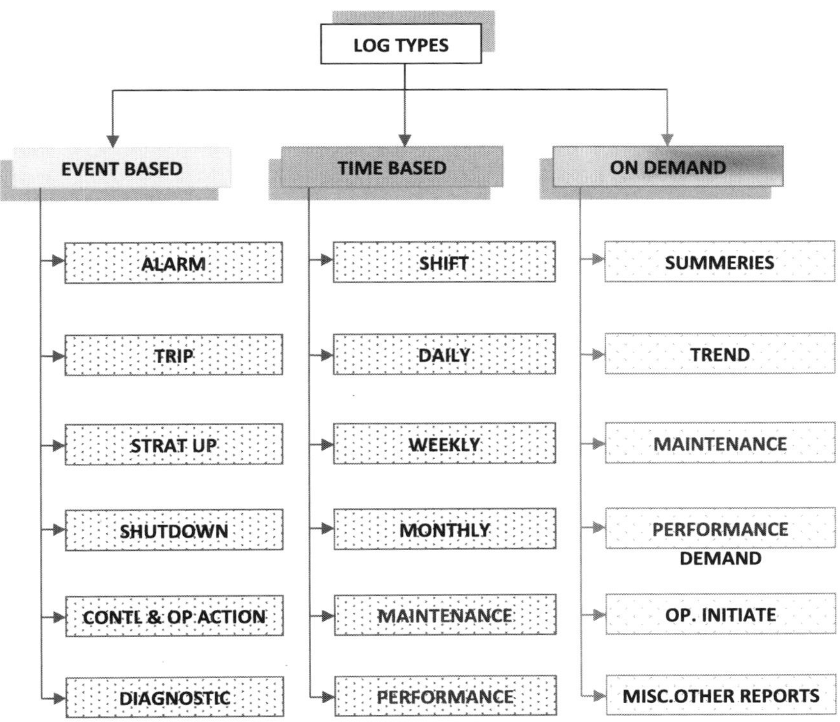

FIG. 7.105 Log/report types.

5.3.1.2 Start Up and Shut Down Log

During major equipment start up such as a boiler turbine, a number of parameters need to be monitored. Such parameters are also logged immediately after initiation of start up and control shut down of the equipment/system. These log types are also similar to the prepost trip logs discussed above. Because for the boiler as well as the turbine, the start up/shut down metal temperature plays a great role in indicating heat soaking. Therefore, the number of tags in these group logs/reports will be more, and this is decided by the owner as per the recommendation of the main system manufacturer. The DCS supplier sizes the system to accommodate the same because the heating surface and configuration vary greatly with the boiler/turbine design and unit size.

5.3.1.3 Control/Operator Action Log

These are also event-based logs. A control action such manual intervention, forcing or bypassing a criterion, etc., will be logged with the time tag and person who has done the same. Similarly, whenever any database attribute, controller parameter, etc., is changed, the same is logged/reported with time and date stamping along with the record of the person who has initiated the changes.

5.3.1.4 Diagnostic Log/Report

Any system faults that are diagnosed by the system are duly reported under the diagnostic logs. Generally, these logs/reports are not generated on any printer but are generated in specified printers especially for the system administrator. The exact format of such a log/report depends on the manufacturer, but general information such as the device ID up to the card level, the type of fault, the date and time of such a fault, any replacement suggestions, and time of fault rectification, etc.

5.3.2 Maintenance Log/Report

Normally, this is generated on demand. However, in certain cases the maintenance schedule is routinely produced on a time basis, that is, after a certain interval (for example, monthly). It gives the schedule of preventive maintenance and routine inspections for major equipment. Nearly 70–130 pieces of equipment (depends on unit size, and numbers of auxiliaries) are included in medium to large plants. The information such as equipment tag, description, redundancy, loss due to downtime, total running hours, current running hours prior to this schedule, lubrication schedule, etc. These log formats are designed as per the owner's choice.

5.3.3 Performance Log

This is the log of performance calculations. In most places, these are treated as "on demand logs/reports," but in many places, such calculations are done routinely and the log/reports are generated on time basis. A daily performance log is not uncommon. In most of the cases, such calculations are done on an ASME/other international standard and the results are printed in the standard format.

5.3.4 Time-Based Log/Report

Shift and daily logs/reports are time-activated logs/Reports that are generated at every time interval. However, such a time-activated log/report can be stopped by authorized persons when not necessary, for example, if for some reason the unit is in a long shutdown, then such logs can be deactivated. Normally, hourly records of 200–350 points (may be subdivided into groups) are included in each of these logs/reports. The exact format for each of the logs depends on the manufacturer. Generally, the information in this log/report shall include the date and time, the source ID, a description, a value with the unit, a set point if any, and the status, including alarm/trip status, group ID, and any other relevant information. There can be a monthly report of a few selected parameters, performance logs, etc., as defined by the user during the engineering stage.

5.3.5 On Demand Log

Any screen display can normally be taken as an on-demand log. Various summaries and trend logs are examples of on-demand logs/reports. Normally, 10–20 points in various groups (50–70) are included under a trend log. These trendings could be of historical value also. Various summary displays could be a summary of some functional groups/subgroups, an area summary, an alarm/event summary, an electrical drive summary, a cabinet-wise I/O summary as a process summary log/report. There could be a system summary log/report such as a scan summary, a summary of constants, a summary of substituted values, a fault diagnostic summary, etc.

5.3.6 Log Generation

This is a utility program normally all mid- to large-sized systems have. This facility makes it possible for the user to generate a set of logs/reports utilizing the database of the systems as well as some assigned constants, any offline manually entered value/parameter, etc. Each of the system suppliers has its own system rule to enter such parameters/constants manually for the generation of a log/report group. Normally, 8–16 such groups are permitted by various systems. In addition to **header information date and time**, the following facilities are normally included in the generation of logs/reports.

- Selection of a point tag from the database (both point tag and/or a calculated value in the system) for a particular log/report group/type. Selection of data collection intervals with various assigned collection frequencies such as 1, 2, 5, 10, 15, 30, 60, and 120 min. Even, if necessary, at a faster rate. For each log/report group separately.

- For a predefined point tag, say, 80–120 numbers arranged in different log/report groups, a faster rate of collection frequency on the order of 1, 2, 5, 10, 15, and 30 s is very helpful during start-up time.
- A free choice of initiating events for such data collection at the above-defined rates as well as use of calculated values for such logs/reports. Not only event-oriented, these logs/reports could be time-based, and or both time- and event-based (after event initiation for a particular period as defined by the user and permitted by the manufacturer). In case of time-based logs/reports, such printing intervals are user selectable.
- A free assignment of numbers of samples for each database, and a selection of average, max/min of these samples over a predefined period of time is possible.

Apart from the system suppliers' own automation software, various other third-party software is also available. These provide a development and runtime environment for process information management for HMI and SCADA applications, with advanced historian and optimization utility as well as powerful graphics. Normally, these are a resident of the MIS server and support DCS protocols for MIS utilization, direct log collections with tight integration with advanced historians, and/or controls.

5.4 Configuration and Communication

A distributed control system (DCS) initially built around a controller, I/Os, workstations, historians, and a configurator has undergone much evolution. This evolution process still continues. The major reason for such changes is the ever-increasing performance–price ratio. The evolution of communication technology and its supportive components in the area of DCS has changed the fundamental configuration of DCS to a great extent. LAN, WAN networking, and IoT (IIoT) contributed much to DCS communication and configurations. Ethernet TCP IP and OPC (Section 1, Chapter 7) made it easier to integrate third-party items into the DCS network. Fieldbus technologies such as Foundation fieldbus, PROFIbus (Section 1, Chapter 7), and IoT have brought about a large change in DCS communication and configurations. In power plants, other factors such as ever-increasing fuel costs, the need to reduce loss and increase energy savings, and statutory regulations to reduce environmental impact further accelerated system integration for smoother data exchange and communication to facilitate central monitoring as well as operation and maintenance from various remote places. These integrations also help in interfacing the DCS with a company's ERP projects.

As stated earlier, with the development of the fieldbus, much of the time-consuming data exchange and control has travelled to the field, giving rise to further functional and geographical distribution and the development of hybrid control systems. As indicated earlier, VMware makes it possible to have various OS intermingled so that the integration of the network can be further extended. Before discussions on networking, a brief discussion on the wire communication medium shall be taken up.

5.4.1 Communication Medium

Out of various kinds of cables used, the twisted pair cable, the coaxial cable, and the fiberoptic cable are the most popular. In twisted pair cables, depending on the communication speed allowed, there are several categories of cables such as CAT 3, 5, 5e, 6, and 7 that are also commonly used in networking. Fig. 7.106 shows the effects of speed with

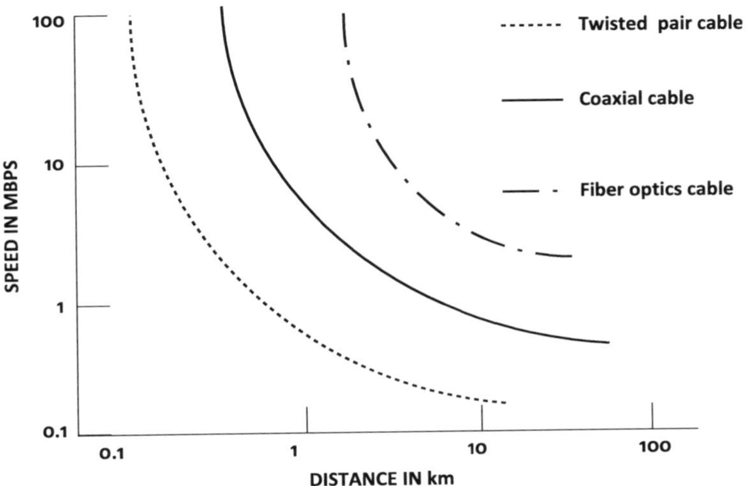

FIG. 7.106 Speed versus distance capabilities.

TABLE 7.42 Cat Cable (and Connector) Short Specification

Specification	CAT 5 UTP[a]	CAT6 UTP[a]	CAT5 STP[a]	CAT 7ScTP[a]
Conductor size Φ in mm	0.5	0.56	AWG24X7	27AWG
Pair (×2)	4 × 2			
Insulation	PVC/PE/FRLSOH			
Data rate (MHz)	100	250	600	600
Length (m)	100	100	70	100
LAN	100 Base Tx **5E-** 1000 BaseT	1000 BaseT		10 G Base
Attenuation dB/100 m	22	32		
IEC standard	IEC61156-5			
Connector	RJ45			GG45
ISO classification	D	E	D	F

[a]*UTP: unscreen twisted pair; STP/ScTP: screen twisted pair.*

distance for the various cables discussed above to gain an idea about variations among different types of cables.

5.4.1.1 Twisted Pair Cable Category (CAT CABLE)

Being moderately rugged in construction, easy to install and repair, moderately noise immune (when shielded), standardized, and with a low cost made this category of communication medium most popular. All Cat cables (mainly CAT 5, CAT6, and CAT 7), fall under the twisted pair cable. Table 7.42 gives short specifications (see Section 4.6.4, Chapter 12 and the manufacturer's catalog for details).

Fieldbus cables: Each of these fieldbuses has a defined data transmission rate and distance limitations (ref clause nos. 1.1.6 and 1.1.8) for the PROFIBUS and FOUNDATION FIELDBUS, respectively. For further details, see the manufacturer's catalog/data sheet and Table 7.43.

5.4.1.2 Coaxial Cable

Thin coaxial cable is used in 10BASE2 Ethernet, meaning a maximum distance of 200 m. Normally, AWG 14–18 solid copper with an approximate resistance of 100 Ohm/Km is used. These cables have attenuation of 14 dB @ 31.5 MHz/22–22 dB @ 100 MHz per 100 m. Braided copper wire is used as shield making with an overall diameter of ~6.75 mm 85% with shield. Black vinyl is used as the outer sheath. For connections, BNC connectors are used.

TABLE 7.43 Common Fieldbus Cable Short Specification and Comparison

Specification	PROFIBUS	FFD Bus
Conductor material	Annealed copper solid/stranded	
Size (Φ mm)	0.64–22 AWG	1.05 (AWG 18)–1.6 (AWG14)
Pairs	4 × 2	2 × 2, 1 × 4
Insulation	Foamed PE/PE/PVC	
Type	Twisted pair	
Shield and copper drain wire	Aluminum foil + metal side Cu drain wire (Φ 0.15 mm)	
Standards	IEC 60332 (Flame) and 61158	
Outer jacket	PVC/thermoplastic	
Impedance Ω/km	~110	100
Attenuation	~40 dB/km	<3 dB/km (Max 20 dB/km)

5.4.1.3 Fiberoptics

Single or multiple optical fibers are quite often used in a computer network as a communication medium. FO finds its application both in industrial Ethernet and a fieldbus. Four or six fibers with a core diameter of 62.5 μm and cladding of 125 μm are quite popular. Normally, 1000 MHz is the bandwidth with attenuation at ~1 dB/km. Snap connectors are mostly used. Connectors can be FC, LC, SC, ST, MTRJ, or V pin types. For further details, see Section 4.6.4 of Chapter 12 and the manufacturer's catalog/data sheet.

5.4.2 Control Network Structure

In its basic form, the control network for a DCS consists of three basic layers, as shown in Fig. 7.107. These layers are various I/Os connected to the controller via an I/O bus, a controller/control system, the operator's control console, and application processors. The data from the controller is presented to the operator via the data highway. On the data highway, there is one historian and application processor connected for carrying out miscellaneous functions such as trending, performance calculations, etc. There could be redundancy at various levels similar to what is shown in Fig. 7.46. As discussed earlier, there may be smart devices such as IEDs and transmitters connected to the system via a fieldbus, as shown in Fig. 7.108. Integration of various system buses into the systems is shown

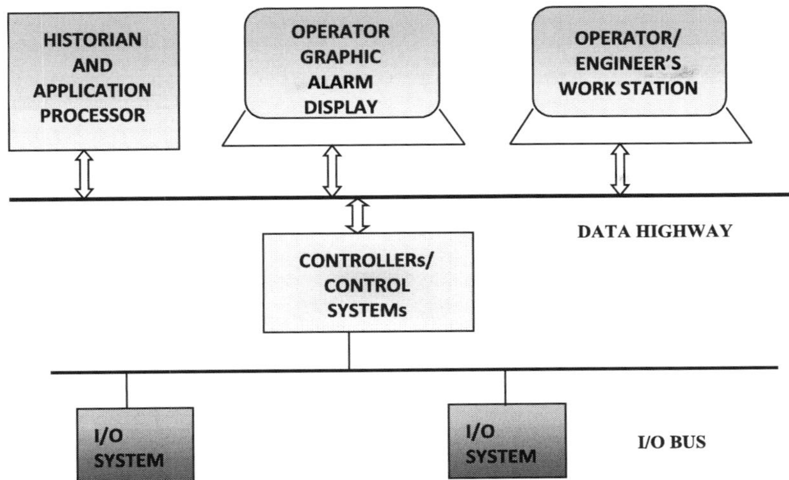
HISTORIAN AND APPLICATION PROCESSOR
OPERATOR GRAPHIC ALARM DISPLAY
OPERATOR/ ENGINEER'S WORK STATION
DATA HIGHWAY
CONTROLLERs/ CONTROL SYSTEMs
I/O SYSTEM
I/O SYSTEM
I/O BUS

FIG. 7.107 Basic functional structure of DCS.

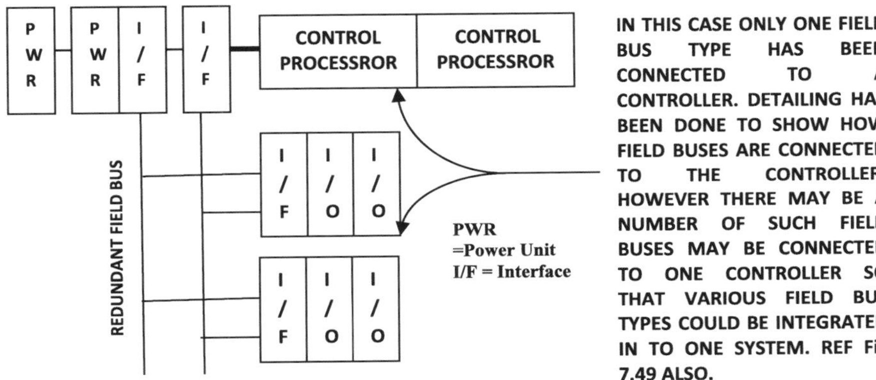
PWR
PWR
I/F
I/F
CONTROL PROCESSROR
CONTROL PROCESSROR
REDUNDANT FIELD BUS
I/F
I/O
I/O
I/F
I/O
I/O
PWR =Power Unit
I/F = Interface
IN THIS CASE ONLY ONE FIELD BUS TYPE HAS BEEN CONNECTED TO A CONTROLLER. DETAILING HAS BEEN DONE TO SHOW HOW FIELD BUSES ARE CONNECTED TO THE CONTROLLER. HOWEVER THERE MAY BE A NUMBER OF SUCH FIELD BUSES MAY BE CONNECTED TO ONE CONTROLLER SO THAT VARIOUS FIELD BUS TYPES COULD BE INTEGRATED IN TO ONE SYSTEM. REF Fig 7.49 ALSO.

FIG. 7.108 Redundant fieldbus, I/O with redundant controller.

in Fig. 7.109. In this case, the industrial Ethernet created at the top is responsible for control, monitoring, and supervision because all MMIs are at that level.

In the drawing, a number of controllers have been shown, and these could be OLCS/CLCS or an intermix depending on the manufacturer's system and the owner's choice for the particular application. These controller divisions could be on a process or area basis also. The configuration in Fig. 7.109 could be for one unit, and similar such units could be integrated to form a plant network with a common server and/or application processor. One such system to form a plant-level network is shown in Fig. 7.110. Common offsite/systems to the units may interface at that common level. For fewer units, the industrial Ethernet may be extended to form a single network, and a common system may be interfaced via one controller of the Ethernet. So there exist a number of possibilities. In Fig. 7.110, three-level structures of the DCS are shown. The first level pertains to the unit where there are three operator stations, one engineering station, and a performance server for one unit (say, unit 1). The second level is the plant level. At the second level, various units have been connected at a common network. Also, common systems and offsite control systems interface at this level so that the same can be utilized by the operator of various units. Also, at this level, a data exchange between the two units is possible (for example, at the start up time, the unit may check the auxiliary steam header condition of the other unit and open the interconnecting valve to take the supply from the other unit to start).

The third level is the group level, where information exchange with the central office is possible or load distribution programs, etc., can be run. Also, various MIS and enterprise resource planning (ERP) programs, run by the group, can interface at this level. So, for a large network, there are three layers of control and monitoring.

- MIS or group level layer: The top layer where all MIS services, ERP, interfaces, etc., are done through a remote network and/or the Internet (IoT/IIoT) so that from a central place, all managerial functions can be done, supervised, and/or advised. Also, units may get

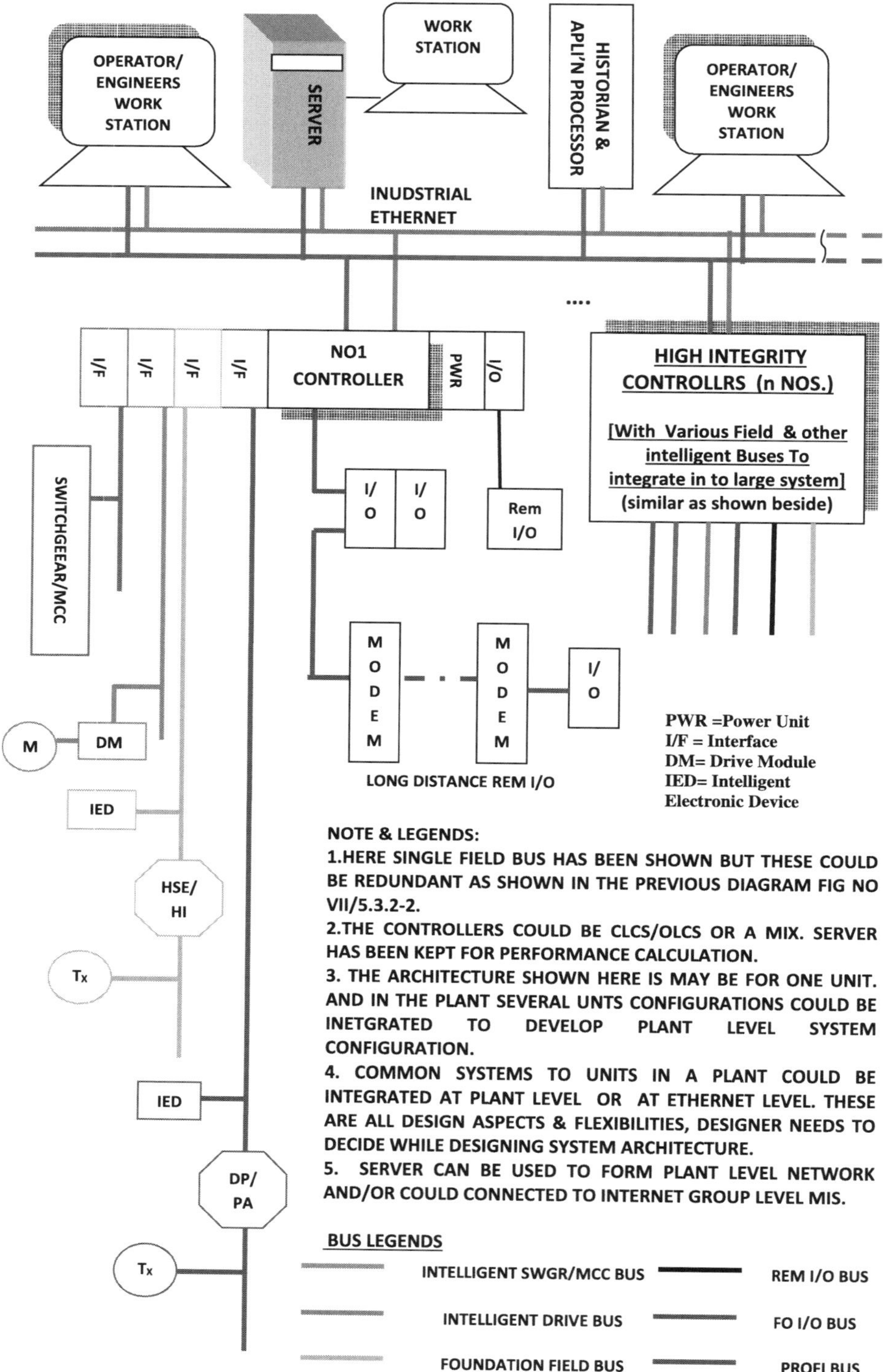

FIG. 7.109 Integrated DCS.

the load demand for the LDS through this level. This is the group/MIS level.

- Supervisory information system SIS or plant level layer: This is the plant level where the production management at the plant level is possible. Also, here resource sharing between the units is possible. Hence this layer may be termed the SIS layer.

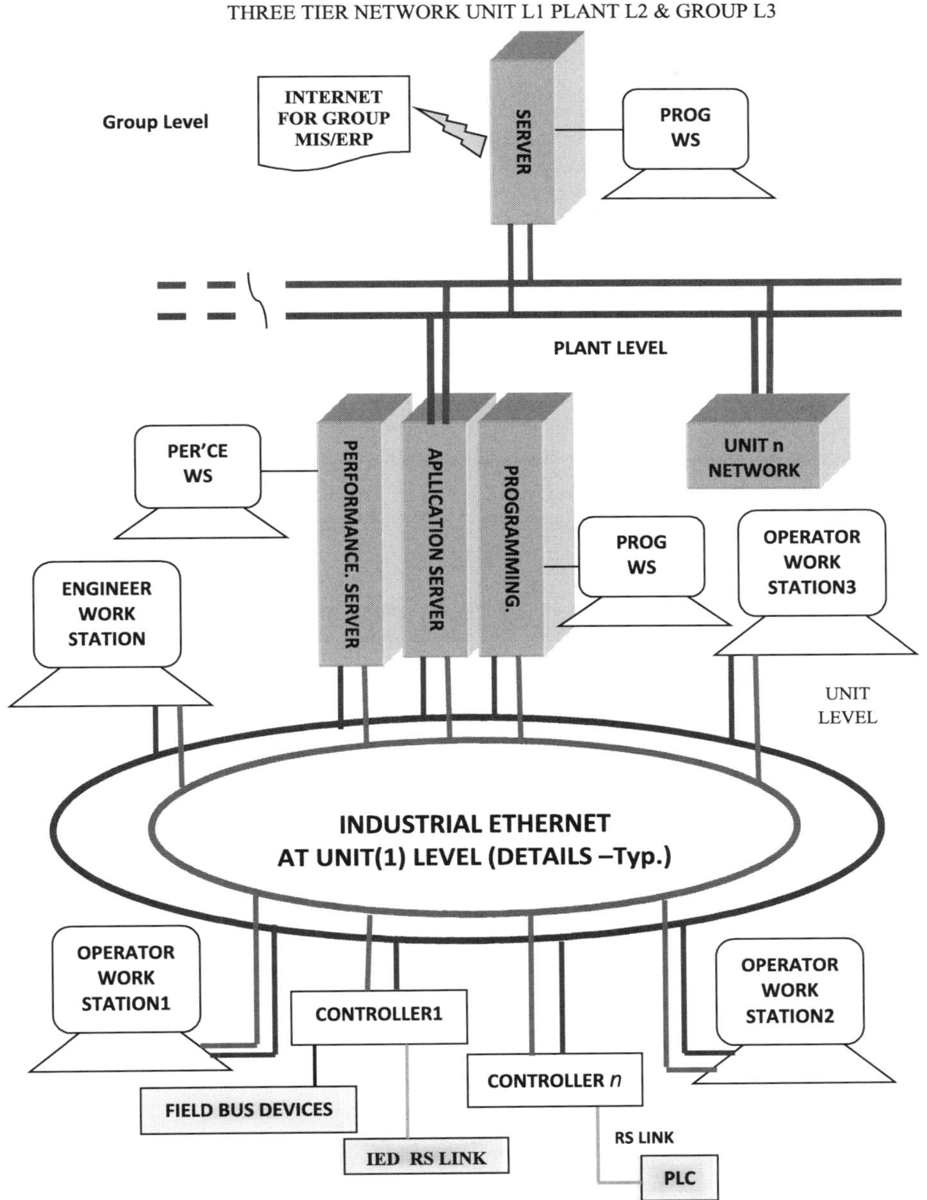

FIG. 7.110 Plant level network and interface to MIS.

- Control and monitoring of the unit layer: The bottom-most layer is for control and monitoring of individual units and/or local process monitoring and control.

5.4.3 System Configuration Aspects

There are many aspects such as physical configuration, control and monitoring strategy, safety and security, historian, system integration (fieldbus, IoT), and communication that must be taken into consideration when configuring the system or designing the system architecture for a DCS. Merely connecting various DCS components does not mean configuration. Actually configuration basically means the generation and exchange of a huge database of information, which enables the user to create or modify the control functions and download the same to the controllers from the programming unit (programming console/portable configurator plug in) via DCS. These configurations are not only restricted to the controller alone, but could be the operator's interface, display types, or even field devices such as an IED, field transmitters, etc. So, configuration is the strategy imparted to the system/unit so that it can operate in the desired way.

5.4.3.1 Physical Configuration

This involves connecting various nodes, devices, and cards by linking and structuring data. Many of the systems have the capability for autosensing, then configurations are easier, or systems sometime suggest such linking to make the task easier.

5.4.3.2 Control Strategy

A piece of information or data is useless unless it is known what it is and how it would help to achieve the target. In a DCS, provisions should be there to consistently reference and represent the information (tag). With the help of this tag and the structure of information, it is possible to apply the same in developing the control strategy. Normally, this information is independent of the basic physical device. To calculate how much HFO is consumed, one may call up the F0011 CALHFO, which will give an output as the difference of, say, two inputs from FT 0011A and 0011B. This tag is an independent transmitter, but only related to the tag. A modern DCS supports multiple languages to include various function blocks, sequencing charts, text ladder diagrams, signal flowcharts, etc., in line with international standards for control programming, IEC 61131. Most of the control strategies in a modern DCS support advanced safety functions. The quality of the DCS varies greatly with the quantity, type, and variety of the function/algorithm library supported by the system. In this connection, Table 7.29 may be referenced. Built-in subprograms/subroutines and function libraries make a project engineer's work not only much easier, but also enhances the power of the system in implementing advanced control strategies. A power plant is a fast-acting continuous process and the power plant control loops are very interactive. So, there is complexity in the control strategies for power plant control loops. Therefore, systems with plenty of functional libraries will be of much help. DCS systems should have online adjustment and upgradation features also. Time-tagged data for alarm and event management is very important for a power-plant DCS. As discussed earlier, all these must be displayed with proper attributes such as the state of the event (alarm, acknowledged, unacknowledged, priority, alarm value, set point, etc.). The operator's ability to suppress and filter alarms is a general feature. In this connection, ISA S50 and IEC 61804 for a fieldbus may be referenced. Alarm management, including the creation and management of active alarms, pre- and posttrip alarms, priority alarms, and SOE are important monitoring strategies. Update alerts and discrete alerts are also important aspects. Whenever a system is upgraded, the existing system information is necessary to update the operator to highlight the changes.

5.4.3.3 Safety and Security

The safety of the system comes from the ability to diagnose the problem, and in case of failure, to fall back into a system where the system can run safely. *System diagnostics* play a major role. Diagnostics cover hardware, controls, communication, and software. Normally I/Os and controllers have self-diagnostic features. In certain cases, there are fault LEDs at the module level; in some cases, a separate diagnostic reporting HMI is used, as shown in Figs. 7.48 and 7.49 (and Section 2.1.3.4 of this chapter). Now, in case of an actual failure, the system has to fall back on the redundant system. Apart from this, the safety lifecycle issue discussed in Chapter 14 also plays some role in deciding the DCS configuration and networking to give the safety system an independent entity.

Redundancy and fault-tolerant system: In order to tolerate failure and at the same time offer safety, redundancy is necessary. As discussed earlier, for fault tolerance, just redundancy may not work. Along with redundancy (with a suitable voting circuit), the use of devices with different technologies may have to be considered to avoid a common cause failure. In some critical cases, fault-tolerant systems are deployed, for example, the Mark V/VI control systems of GE or the speed protection of KWU, as discussed before. In case of a controller redundancy, one has to make sure that whenever such a fall back action takes place, there is no bump in the system. In communication, a redundancy system checks such failure by various means such as a CRC/check sum. In case of redundancy, it is easier to upgrade the system online, when precautions are taken to make sure that the system is not endangered due to online upgrading.

System Security, especially in the case of integrated SCADA, is extremely important. Identification of USER, AUTHENTICATION, AREA/SYSTEM wise security, etc. are extremely important aspect in a computer network. The security discussed in Table 7.22 is equally applicable here. Apart from that, each user must have a user account to work in the network with a suitable ID for a well-defined AREA/SYSTEM scope. There may be a number of levels of access in a network. There shall be at least a segregation of work and access nature among the operator, shift charge engineers, control engineers, programmers, etc. Normally, any modification done in the system is properly recorded with a user ID and a time of change, etc. (ref clause no. 4.1.2). Password protection is the most common means to restrict such access. Also, there may be restrictions in access to the system areawise, for example, a water treatment operator can have access to a network to know the status of the main plant condition, but cannot modify any data. For SCADA, additional security against cybercrime is important.

5.4.3.4 Historian

All DCSs have the ability to collect and store plant data and event records. In modern systems, this is an integral part of the system for generating various IS functions. All pre- and posttrip analyses and process failure analysis would be impossible without historians. So, its role in IS/MIS operations cannot be overestimated.

5.4.3.5 Integration

As already discussed and shown in Figs. 7.108–7.110 there will be a complete integration of various fieldbus systems and other systems. System integration discussed in clause no. 2.1.3.3 is also applicable here and clause no. 5.4.2 discussed above also portrays the benefit of system integration. A drastic reduction of cable cost and decentralization of many functions at the field instrument following the IEC1158-2 standard made fieldbus integration with the DCS smoother. The OPC-defined standard makes it possible to access information in the other control systems, making system integration easier.

5.4.3.6 Communication

Communication of the network is like physiological functions in the body. Now, industrial IT-compatible instruments/devices and/or IoT make it possible to have global access to the system (for example, remote fault finding and repair). On one hand, this aspect is very good and makes life easier; on the other hand, this calls for security threats also. Various functions of the communication system can be summarized as follows:

- Physical connection: Hardware connection of various nodes.
- Real-time data exchange: Exchange of real information across the network.
- Various modes of transfer: Both asynchronous as well synchronous data transfer and read-write operation among various subsystems and components.
- Message broadcast: Hosted messages are broadcast over all subsystems.
- Upload and download: Various configurations, when permitted, can be downloaded to various nodes. Also, data from the network could be uploaded for MIS.
- Autosensing and rearrangement: Whenever a new controller is put into the system and/or whenever a node is dropped, the communication system detects the same to rearrange the network for communication.
- Debugging and diagnostics: It has the ability to debug the system and report. Diagnostic reports are communicated over the network.
- Search function: Communication systems make it possible to pinpoint a node, component, and control strategy over the network.
- Online upgrade: Online DCS upgradability without disturbing the normal operation.
- Restarting strategy: COLD/WARM/HOT start strategies can be implemented.
- Alarm and event: Process, system, and device failure alarms and events are generated and communicated over the communication system to various parts of the network.
- Security: All access must be secured, otherwise the communication system will not allow any such access or communication.
- Time: To get a meaningful record, it is necessary that all devices and nodes in the network are time synchronized.
- Variety: There are varieties of networks and associated communications.

Thus, it is clear that the communication software has to work at its best to keep the process data fresh and to be used by various clients. A modern digital network can be conceived as shown in Fig. 7.111. In this architecture, there are basically three networks: the field network, the control network, and the information network. While the field network is mainly concerned with field devices and local controllers, the control network is the main control system doing all group control, performance calculations, alarms, and other monitoring with the operator interface. The third network is supervisory and information management. There are several network variations such as Ethernet, DeviceNet, Control Net, Profinet, FDDI, etc. Out of all these, the gigabit Ethernet is the most popular one. However, each has different protocols to manage the system. The aim of all these protocols is to ensure the network quality services, which depend highly on a predictive time delay and improved throughput of the network, the utilization and efficiency of the network, a medium access control, message scheduling, and minimizing lost data. The timing component is a vital issue in network communications. In a network, a time delay can be conceived as a device delay and a network delay. A device delay has two parts: a source device delay and a destination device delay. At the source, there two delays such as a preprocessing delay (T_{Preprop}) and a waiting delay (T_{wait}). In a destination delay, it is only a postprocessing delay (T_{Postprop}) while a network delay includes a total transmission delay (T_{Tx}) and a propagation delay. For a transmission rate of 3×10^8 m/s, the propagation delay can be ignored (Table 7.44).

$$\text{Total Delay}: T_{\mathrm{Delay}} = T_{\mathrm{Preprop}} + T_{\mathrm{wait}} + T_{\mathrm{Tx}} + T_{\mathrm{Postprop}}$$

So, in network communication, these timing considerations are important. The controller performance as normally evaluated by the integrated absolute error (IAE) or the integral time multiplied absolute error (ITAE) depends highly on the sampling period when a common transmission

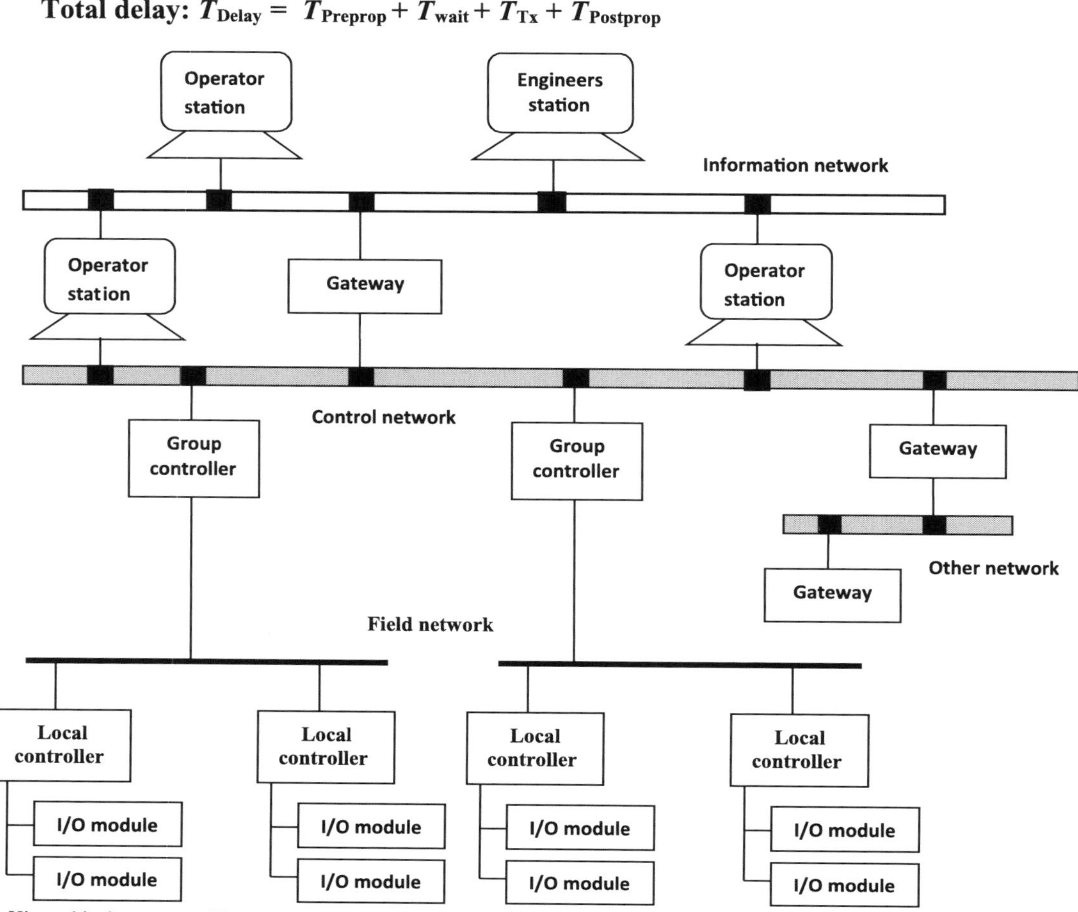

FIG. 7.111 Hierarchical system architecture.

medium is shared by the various devices. The sampling period influences the phase margin and **b**and**w**idth (BW), which is directly related to controller performance. On account of interactions of the network and control requirements, the selection of the best sampling period is really difficult, so there will be a compromise so that both services can be optimized. This is because a smaller sampling period is better for controller performance, whereas in a limited BW network, there are calls for high-frequency communication that may jeopardize network quality service. So one has to balance between the two.

Courtesy: Network design Consideration for Distributed Control Systems by Feng Li Lian, James Moyen, Dawn Tilbury.

6 MANAGEMENT INFORMATION SYSTEM

The target of all electric companies is to enhance production and work efficiency and reduce the overall cost. A management information system (MIS) that can be integrated with the plant DCS is an effort to meet those targets. In this chapter, a brief discussion on the same shall be presented to study MIS. For utility enterprises, a system controlling critical infrastructure for electricity generation, transmission, distribution, etc., can no longer be isolated. For substations, a remotely located pump house (for example, a raw water pump house) network-based access reduces operational costs by enabling remote monitoring and maintenance. Audit report generation, data gathering, etc., can be achieved by this to reduce the cost drastically. Now, if these are also integrated with supply and material management as well as account management, there will not only be a high reduction in cost, but also better control of the entire operation. Mid to large enterprises may have various data processing/ERP software projects such as SAP, etc. In this section, the DCS interfaces with these systems to have seamless connections to the upper level management systems, which shall be discussed. For a large network utilizing the Internet and other public systems, there may be cyberattack, which may mislead and jeopardize the entire operation. So network security, which shall mainly include assurances of the integrity of the network, a control system-

TABLE 7.44 Time Delay Significance

Terms	What Is It?	Remarks
Preprocessing delay	Time needed at source to acquire data from external environment and encoding the same into the data format of the network	This is dependent on the HW and SW characteristics of the device, may be varying in nature, and may be noticeable
Waiting delay	Waiting time in queue at the source/ sender's buffer	Depends on source node traffic and network traffic and may not be insignificant
Postprocessing delay	Time needed at the destination to decode the network data into physical data	
Transmission delay	Time required for transmission. T_{Tx} = message Size (*N*) XBit time + propagation delay (negligible)	Deterministic parameter. It depends on the data rate, message size, and node-to-node distance

specific security policy, securing remote access, and validation and authentication for every device and user on the network, is of prime importance.

6.1 MIS Characteristics

The MIS shall be economic and based on a network platform of Internet technology with high reliability, safety, and security.

For electric enterprises, the MIS shall have a few characteristics, listed below.

- Conformance to electric power industry standard.
- Tailored to unified reporting system standards for electric power companies.
- Easy to use and consistent user interface.
- Automatic production control with seamless connection with DCS gateway.
- Strict monitoring and control of environmental factors.
- Real time monitoring and control, including log operation.
- Automatic monitoring and control of material management, including fuel supplies.
- Effective maintenance management functionality.
- Accurate cost calculation and financial management.
- State-of-the-art technology and to support ERP and data processing systems.
- Future-oriented platform for process control ,especially for an electric enterprise.

Upper management programs interfacing with DCS are shown in Fig. 7.112.

6.2 MIS Position and Functionality

With respect to the complete network lies in between unit level and Co. group level as shown in Fig. 7.113.

The database at the server level constitutes a supervisory information system (IS), which is basically related to production monitoring and control. However, other upper management program data exchanges are done through MIS. Because IS is a subsystem of the MIS database, mainly for production monitoring and control, an overlap is shown in the drawing under reference.

Again, unit and plant performance calculations as well as optimization are also done using this database. Frankly speaking, this is also a part of IS. Because performance calculations and optimization are important subsystems of DCS, they need a fair bit of discussion (Section 6.3). The main application functions of MIS with IS are shown in Fig. 7.114.

Some of the IS and MIS Functions could be:

- Supervision functions at the unit/plant level perform various calculations, make online calculations, and display the results and/or analysis data for operator assistance.
- Plant performance calculation and optimization: Unit performance calculations and optimizations fall under this category and are discussed in Section 6.3. Here, the production people get to know the result of detailed analyses as well as the operator's guide. Automatic change of the set point is also possible in case the same is allowed.
- Production management: On a daily, weekly, or monthly basis, the heat and material balance calculations are carried out to see the periodic yield and utility consumption. Validated data are compared with the target to see the deviations and to try to minimize the same.
- Economy index analysis: Here, at any instant the plant performance such as boiler efficiency, turbine heat rate, etc., are calculated and thereby cost analysis can be done to pinpoint any deviations. From such a deviation, remedial measures can be suggested to minimize the cost. Variable costs on a daily basis can be calculated from validated and reconciled plant data.

MANAGEMENT INFORMATION SYSTEM

PRODUCTION MANAGEMENT	PRODUCT PLANNING, OPERATION, SAFETY,DCS GATEWAY
EQUIPMENT MANAGEMENT	ACCOUNTING, DEFECT, PROTECTION, WARNING,
SUPPLY MANAGEMENT	FUEL, STOCK, PURCHASES, MATERIALS
FINANCIAL MANAGEMENT	ACCOUNTS,PAYMENT,ASSETS, COST CALCULATION, ETC.
ASSISTANCE MANAGEMENT	HR,INSURANCE,ENVIRONMENTAL PROTECTION, ETC.
OFFICE AUTOMATION	OA, Archives, Etc.

FIG. 7.112 MIS profile.

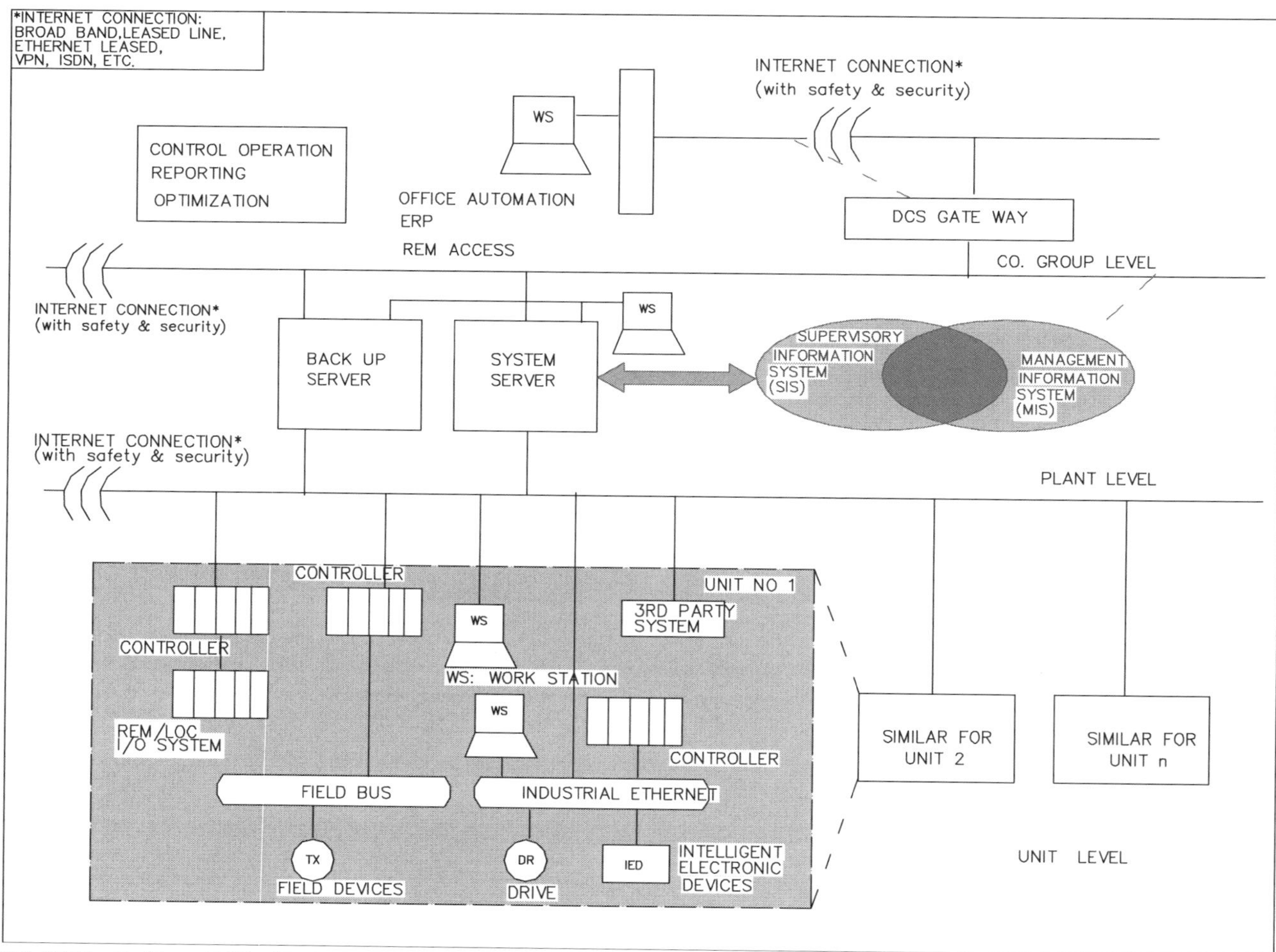

FIG. 7.113 Position of MIS with IS.

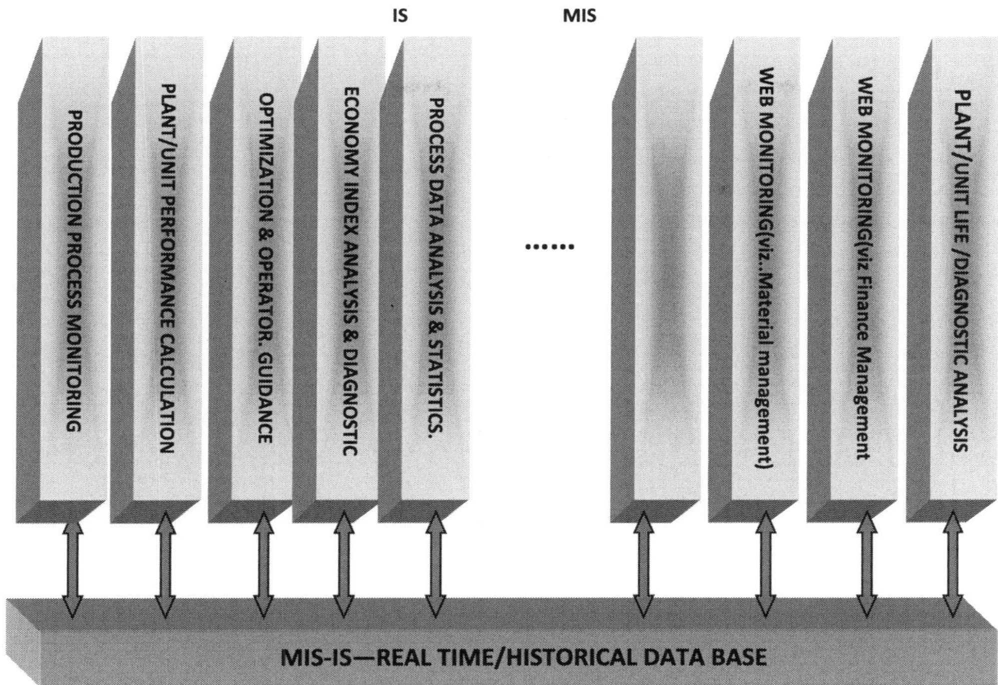

FIG. 7.114 Main application functionality (MIS).

- Equipment health monitoring and predictive maintenance: Major equipment auxiliary equipment data are monitored to detect early failure and predictive maintenance. Thermographic images, vibration analysis, and lubrication oil sample data are displayed (by transferring data from these systems). A condition-based maintenance coordinator module assesses the situation to suggest the status of acceptable/watch/marginal/unacceptable conditions of equipment with suggestions for maintenance.
- Maintenance scheduling: It suggest the operator based on reliability study the guide and schedule for maintenance from library of data for industry standard.
- Environment monitoring: In this module, the various acquired data are validated, and prealarm and alarm emission limits are generates with respect to the environment. It is possible to generate the necessary reports for public authorities.
- Material management: This manages all logistics and various arrival conditions. It is possible to link the material management functions from here. In most of the cases of large/mid-sized plants, the owners with a number of plants opt for ERP projects. Thus, it is possible to support the same data exchange from the MIS through the DCS gateway.
- WEB monitoring: MIS supports the exchange of data from MIS so that WEB monitoring of the plant from a central place is possible. Also, it supports the remote station (say, one substation, a remote pump house) from the central control room of the plant.

As stated above, plant performance calculations need special attention as they are directly related to plant equipment health. In the following section, the same will be dealt with.

6.3 Performance Calculation and Optimization

6.3.1 *Introduction*

It is quite possible that the power plant does not operate at the design conditions; naturally, there will be deviations from the designed plant efficiency and heat rate. The major influencing factors toward this could be:

- Variation in fuel properties.
- Variation in fuel calorific value.
- Process parameter variation.
- Ambient variations.
- Leakage factor.
- Boiler tube failure.
- Equipment degradation.
- Improper controller tuning/failure.

- Other mechanical problems.

Therefore, the performance calculation of the SG/TG at regular intervals is essential and system optimization needs to be carried out to get the best results. Also, other suitable actions such as regular soot blowing would lead to a savings of ~2.2%. Basic plant optimization consists of optimizations of major equipment and systems such as SG, TG, feed water heater, condenser, etc.

6.3.1.1 Calculations

In a power plant, there are various types of calculations such as:

- Class I calculation
- Class II calculation
- Other miscellaneous calculations

- Class **I** calculations: These are mainly equipment-protecting calculations for detection and alarming, and some of these are:
  - CRH steam approach to saturation temperature.
  - SH spray O/L approach to saturation temperature.
  - FW heater temperature. deviation from standard.
  - Drum saturation temperature rate of change.
  - Turbine rate of temperature rate of change.
  - Excess air deviation from standard.
  - Turbine steam-metal temperature difference.
- Class II calculations: These are performance calculations of equipment along with heat rate deviations, revenue calculations, tariffs, etc.:
  - Boiler efficiency.
  - Gross TG heat rate and deviation.
  - Gross unit heat rate and deviation.
  - Net unit heat rate.
  - Condenser performance.
  - Deaerator performance.
  - CEP/BFP performance.
  - Economizer performance.
  - AH performance.
  - ID/FD fan efficiency.
  - Unit availability.
  - HPT enthalpy drop efficiency.
  - IPT enthalpy drop efficiency.
  - LPT enthalpy drop efficiency (dry exhaust).
  - FW heater terminal temperature differential performance.
  - FW heater performance drain cooler approach.

In each of the above calculations, the deviation from standard shall also be calculated. The net unit heat rate versus load profile using the last value shall be interpolated at different load values.

- Other miscellaneous calculations: Apart from Class I and II calculations, there are a few other calculations (common for DCS) that shall include but are not limited to: Analog point sum/difference, average/max/min, selection of one analog point out of two/three, digital point status, duration of time at which an analog point is beyond a threshold, running point average of digital points, rate of change, time projection, variable alarm limit calculation, specific calculation needed for long term storage, and other calculations such as:
  - Running hours of rotating machine.
  - Number of start/stop devices.
  - Plant calculation.
  - SH/RH creep (stress).
  - Number of hot/warm/cold starts of TG.
  - Cumulative running hours of TG beyond threshold steam alarm.
  - Cumulative running hours of bearing beyond threshold alarm.

6.3.1.2 Standards

These performance calculations in a DCS are carried out in accordance with certain international standards. Various test codes such as **ASME PTC 4 for SG** and **ASME PTC 6 for TG** are the most popular. The following codes are widely used as **references,** as these are mostly used by the manufacturers at the testing facilities:

6.3.1.2.1 Steam Turbines

- DIN 1943: Thermal acceptance tests for steam turbines.
- BN EN 60953: Rules for steam turbine thermal acceptance tests.
- ASME PTC 6: Steam turbine performance test code.
- IEC 953: Rules for steam turbine thermal acceptance tests.

6.3.1.2.2 Gas Turbines

- DIN 4341 Acceptance rules for gas turbines.
- BS 3135 Specification for gas turbine acceptance test.
- ASME PTC 22 Gas turbine power plants—power test code.
- ISO 2314 Gas turbines—acceptance tests.
- ISO 2314 Acceptance tests for combined cycle power plants and amendments.

6.3.1.2.3 Steam Boilers

- ASME PTC 4.1 Steam generating units performance test code.

- ASME PTC 4.4 Gas turbine heat recovery steam generators performance test code.
- DIN 1942 Acceptance test for steam generators.

6.3.1.2.4 Reciprocating Engines

- IS: 10000 Part IV—1980: Method of tests for internal combustion engines.
- Declaration of power, efficiency, fuel consumption, and lubricating oil.
- Consumption.
- IS: 10000 Part VIII—1980: Method of tests for internal combustion engines.
- Performance tests.

6.3.1.3 Measuring Points

The measuring points and the accuracy of measuring instruments shall be the same as mentioned in the latest versions of the applicable standards. Today, very accurate instruments are easily available to cater to these requirements.

6.3.1.4 Data Presentation

Calculation results can be presented in the form of displays and/or printouts. A typical such printout is presented in Figs. 7.115 and 7.116. In the report, the synopsis of the calculation results of SG efficiency are presented with the turbine heat rate with different losses, courtesy of Kalki Communication Technologies (communication and computing technology meant primarily for energy sector) performance calculation presentation is an example! Different suppliers may have different formats but the information content will be more or less similar.

6.3.1.5 Optimization

Optimization packages may be a separate package that needs to be interfaced with the DCS software, or it may be supplied as a part of the system. The technological development of DCS in conjunction with software development made it possible to enhance the optimization package. Artificial intelligence (Section 1, Chapter 7) in the form of a neural network made it possible to frame a computer model to predict the best possible operations under given constraints. Most of these neural networks (**NN**) work in a similar fashion in terms of their complexities, applications, and interfaces with DCS. These performance packages are normally complete with performance analysis as well as diagnostic and optimization modules. **A** **g**eneric **NO**x **c**ontrol **i**ntelligent **s**ystem (**GNOCIS**) is an example of a similar package. Several optimization packages are gaining more and more popularity because of severe competition and pollution abatements. In the competitive market, when there is a rising trend for fuel prices and many restrictions on pollution, operating personnel have no other way to find the means to increase the efficiency of the plant. There are several versions available in the market. In some cases, the optimization package output is directly applied to the control system to automatically modify the set point, whereas in other case, these may be sent as guidance for the operator, who will evaluate the same from the perspective of overall gain. Mill selection is a classic example. Mill selection may affect the performance as well as emissions, so one has to evaluate the pros and cons before going for automatic selection and/or using the same as guidance only. Then, based on actual conditions, the operators can make a decision. Fig. 7.117 shows that the plant data are acquired with the help of the plant DCS, and it is then presented to the operator. Also, the same data is fed to the optimization module after passing through a prediction model. An optimization package in conjunction with the process model advises the plant operating personnel regarding the best possible action. If necessary, it can also generate the best set point at that time under the plant operating conditions.

The basic scope of optimization shall include but is not be limited to the following:

- The optimization package belongs to the overall plant production and automation process; even if it is a separate package, the same needs to be interfaced suitably.
- Building platform of unified process—real time and historian database.
- To act as a bridge between the production management and process control.
- Tapping the potential of the main and auxiliary equipment to realize optimization and reduce energy consumption.

Information highway, plant database, production information portal (process information development and application), equipment properties (characteristics), design data/parameters, power generation check list (instantaneous cost of production) with online economic analysis with solutions for cost reduction, use of neural networks, and equipment health monitoring are the data and tools for implementation of such optimization.

The actual implementation of the system is depicted in block diagram form in Fig. 7.118. The operator gets the display from the DCS as well as from the optimization package (as advice); it is up to the operator to select the same judiciously. Also, it is possible to automatically modify the closed loop set point. A brief discussion on the basic functionality of these optimization packages follows.

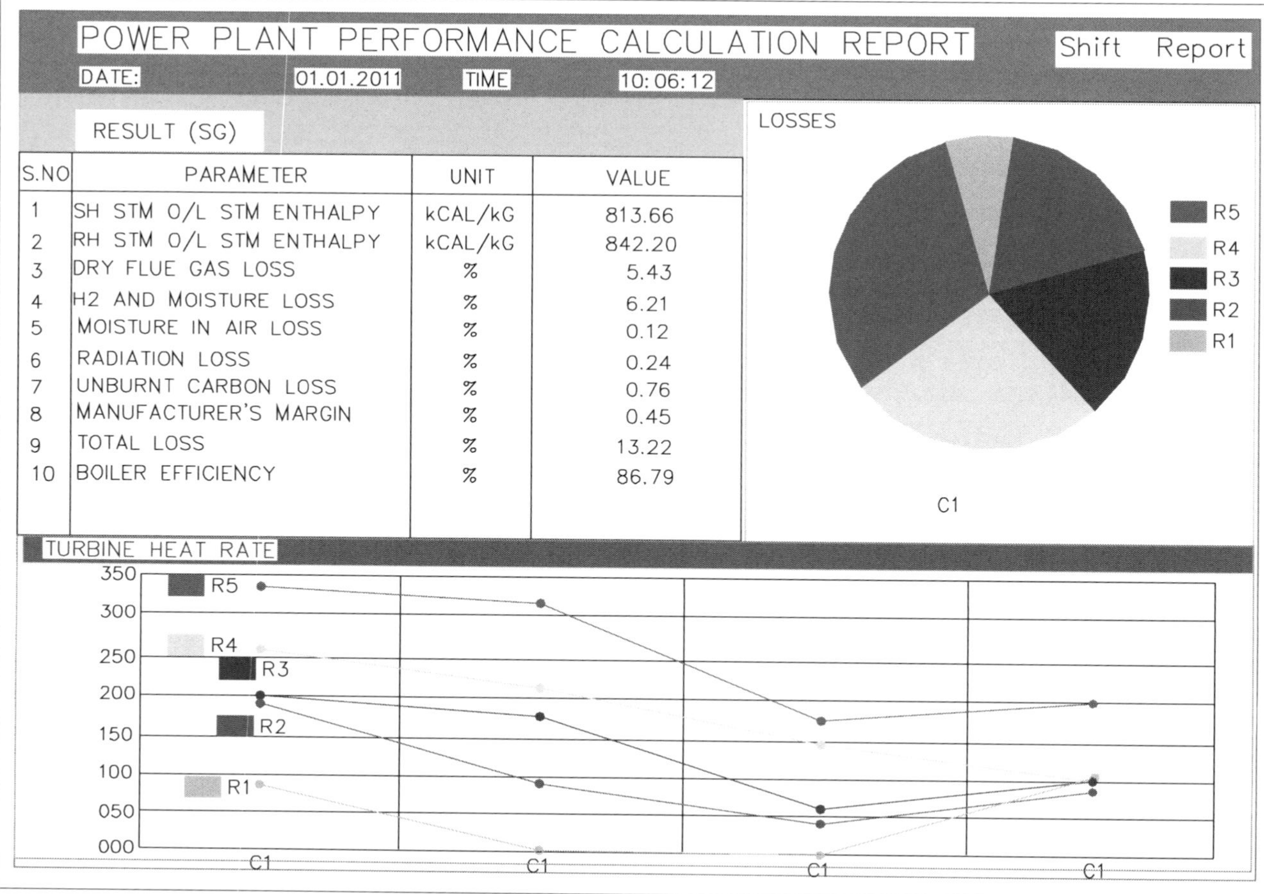

POWER PLANT PERFORMANCE CALCULATION REPORT Shift Report

DATE: 01.01.2011 TIME 10:06:12

RESULT (SG)

S.NO	PARAMETER	UNIT	VALUE
1	SH STM O/L STM ENTHALPY	kCAL/kG	813.66
2	RH STM O/L STM ENTHALPY	kCAL/kG	842.20
3	DRY FLUE GAS LOSS	%	5.43
4	H2 AND MOISTURE LOSS	%	6.21
5	MOISTURE IN AIR LOSS	%	0.12
6	RADIATION LOSS	%	0.24
7	UNBURNT CARBON LOSS	%	0.76
8	MANUFACTURER'S MARGIN	%	0.45
9	TOTAL LOSS	%	13.22
10	BOILER EFFICIENCY	%	86.79

FIG. 7.115 Performance calculation report (presentation) Type I data with curve and chart.

6.3.1.5.1 *Functioning of the Optimization Model*

Implementation of these optimization and prediction modules can be thought of as per the following stages:

- o. Initialization: During this stage, it gathers inputs from DCS and other data mentioned above (may be manual input), and does the validation checks before these can be transformed into model variables such as excess O_2, total fuel flow, and boiler efficiency.
- o. Transformed variable are used to RUN the model in two modes:
  - Straight prediction mode: The model is run to produce/calculate the predicted O/P.
  - Optimization mode: GOALS and LIMITS are fed and the model is run to optimally achieve the desired goal under the given limits.

However, in both modes the adjustment biases are different. Prediction biases can be updated but in the optimization mode, the biases are constant, so the overall impact on the entire system. So it is better to first go for the predicted mode to get the predicted output before correction, then model input, actual output, and predicted output are fed for gain bandwidth correction to update the models. A typical such system flow diagram, with "LIMIT" and "GOAL" settings, for various optimization packages is depicted in Fig. 7.119.

Fig. 7.119A shows the typical flow chart in connection with the CLCS loop while Fig. 7.119B shows the same for water chemistry management, discussed later in this section.

6.3.1.5.2 ***Boiler Optimization Package*** The major workload for the boiler optimization package shall include combustion control optimization to limit NO_x and CO_2 emissions. Under a rapid load-changing situation, there is a tendency for the emission levels to rise. NN models show appreciable improvement in this area with the help of real adaptive control to limit the emissions. For overall thermal efficiency, it is important to maintain the SH and HRH temperature at the specified value. Higher steam temperature not only degrades the boiler tube but also requires attemperation, which will reduce plant performance. An NN-based model can optimize and balance the burner tilt and

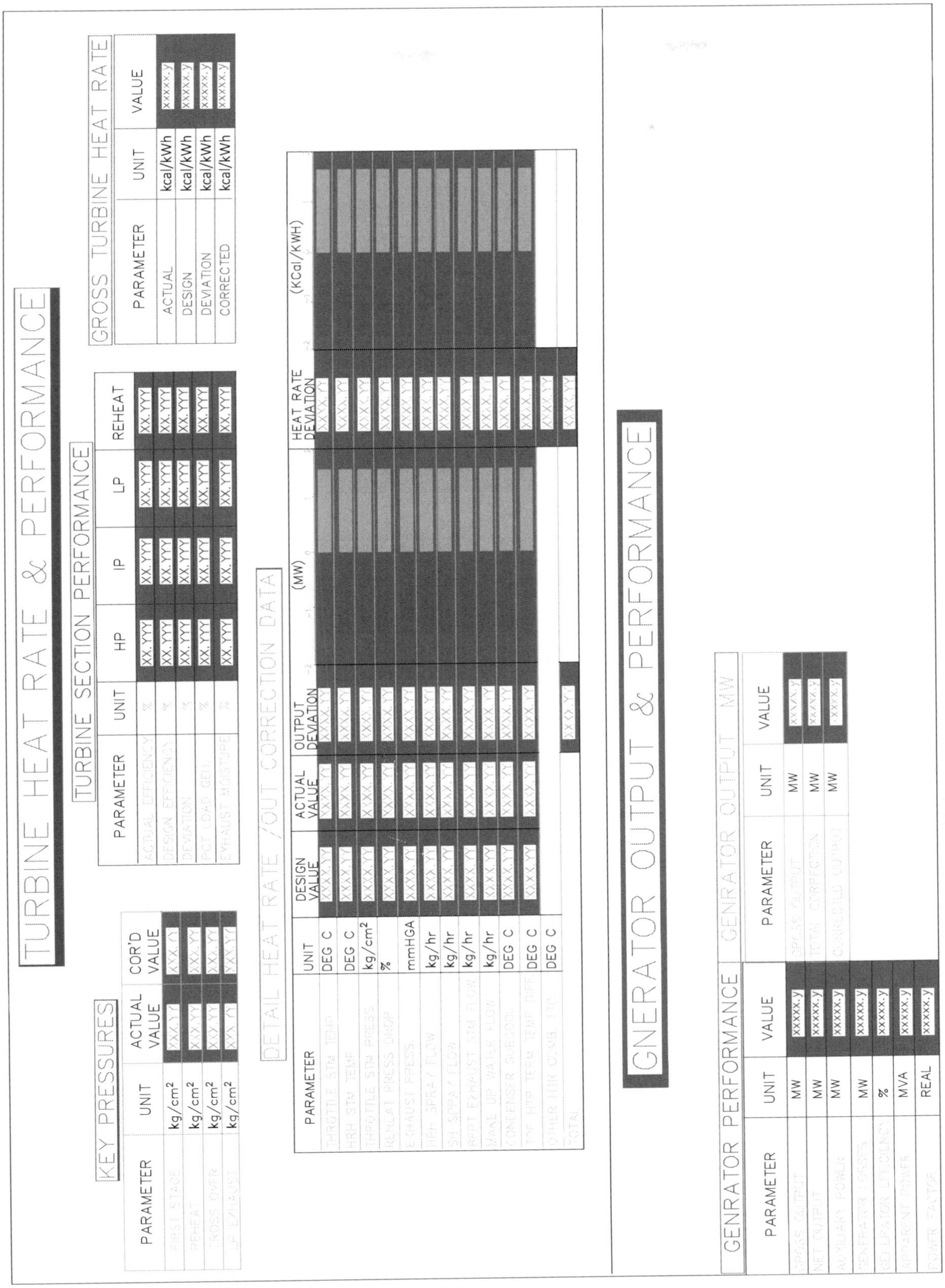

TURBINE HEAT RATE & PERFORMANCE

KEY PRESSURES

PARAMETER	UNIT	ACTUAL VALUE	COR'D VALUE
FIRST STAGE	kg/cm²	XXX.YY	XXX.YY
REHEAT	kg/cm²	XXX.YY	XXX.YY
CROSS OVER	kg/cm²	XXX.YY	XXX.YY
LP EXHAUST	kg/cm²	XXX.YY	XXX.YY

TURBINE SECTION PERFORMANCE

PARAMETER	UNIT	HP	IP	LP	REHEAT
ACTUAL EFFICIENCY	%	XX.YYY	XX.YYY	XX.YYY	XX.YYY
DESIGN EFFICIENCY	%	XX.YYY	XX.YYY	XX.YYY	XX.YYY
DEVIATION	%	XX.YYY	XX.YYY	XX.YYY	XX.YYY
PCT LOAD GEN.	%	XX.YYY	XX.YYY	XX.YYY	XX.YYY
EXHAUST MOISTURE	%	XX.YYY	XX.YYY	XX.YYY	XX.YYY

GROSS TURBINE HEAT RATE

PARAMETER	UNIT	VALUE
ACTUAL	kcal/kWh	xxxxx.y
DESIGN	kcal/kWh	xxxxx.y
DEVIATION	kcal/kWh	xxxxx.y
CORRECTED	kcal/kWh	xxxxx.y

DETAIL HEAT RATE /OUT CORRECTION DATA

PARAMETER	UNIT	DESIGN VALUE	ACTUAL VALUE	OUTPUT DEVIATION	(MW)	HEAT RATE DEVIATION	(KCal/KWH)
THROTTLE STM TEMP	DEG C	XXXX.YY	XXXX.YY	XXXX.YY		XXXX.YY	
HRH STM TEMP	DEG C	XXXX.YY	XXXX.YY	XXXX.YY		XXXX.YY	
THROTTLE STM PRESS.	kg/cm²	XXXX.YY	XXXX.YY	XXXX.YY		XXXX.YY	
REHEAT PRESS DROP	%	XXXX.YY	XXXX.YY	XXXX.YY		XXXX.YY	
EXHAUST PRESS.	mmHGA	XXXX.YY	XXXX.YY	XXXX.YY		XXXX.YY	
HRH SPRAY FLOW	kg/hr	XXXX.YY	XXXX.YY	XXXX.YY		XXXX.YY	
SH SPRAY FLOW	kg/hr	XXXX.YY	XXXX.YY	XXXX.YY		XXXX.YY	
RFPT EXHAUST STM FLOW	kg/hr	XXXX.YY	XXXX.YY	XXXX.YY		XXXX.YY	
MAKE UP WATER FLOW	kg/hr	XXXX.YY	XXXX.YY	XXXX.YY		XXXX.YY	
CONDENSER SUBCOOL	DEG C	XXXX.YY	XXXX.YY	XXXX.YY		XXXX.YY	
TOP HTP TERM TEMP DIFF.	DEG C	XXXX.YY	XXXX.YY	XXXX.YY		XXXX.YY	
OTHER HTR COMB. LTD.	DEG C					XXXX.YY	
TOTAL				XXXX.YY		XXXX.YY	

GNERATOR OUTPUT & PERFORMANCE

GENRATOR PERFORMANCE

PARAMETER	UNIT	VALUE
GROSS OUTPUT	MW	xxxxx.y
NET OUTPUT	MW	xxxxx.y
AUXILIARY POWER	MW	xxxxx.y
GENERATOR LOSSES	MW	xxxxx.y
GENERATOR EFFICIENCY	%	xxxxx.y
APPARENT POWER	MVA	xxxxx.y
POWER FACTOR	REAL	xxxxx.y

GENRATOR OUTPUT MW

PARAMETER	UNIT	VALUE
GROSS OUTPUT	MW	xxxxx.y
TOTAL CORRECTION	MW	xxxxx.y
CORRECTED OUTPUT	MW	xxxxx.y

FIG. 7.116 Performance calculation report (presentation) Type II with data.

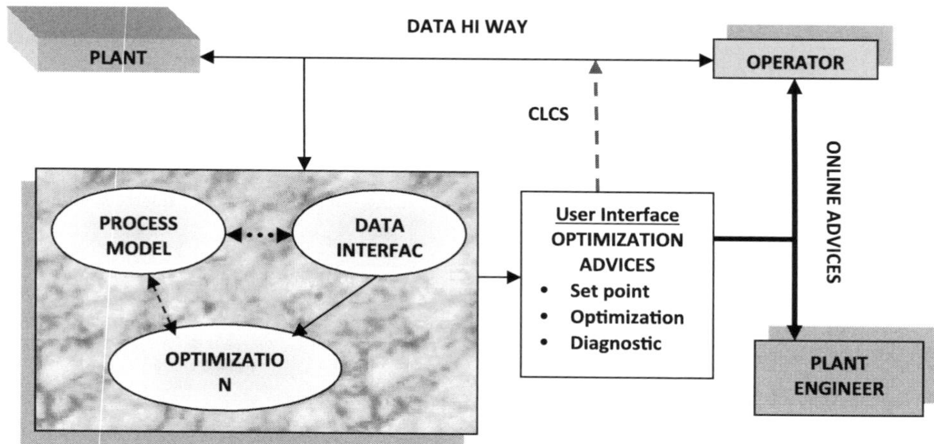

FIG. 7.117 Typical optimization package interface.

attemperation via real-time adaptive control. As stated earlier, GNOCIS is an example of this optimization package/module. As discussed, the main interface of the optimization package is to put forward a suitable display with all advice to clearly convey the recommended benefits and predictions. In this connection, the advanced monitoring and control of local combustion by OPTOCOM technology provides a new solution. Opticom technological details are explained with the help of Figs. 7.120 and 7.122. The operator, based on plant operating constraints, sets them to get the recommendation under the given sets of constraints. There are normally two sets of recommendations made available to the operator, along with the present plant conditions. In case of combustion optimization, ***the two*** possible recommendations could be: change the current mill operation settings either automatically or manually if the recommendations are accepted by the operator; or optimum mills in service and associated emission benefits. Based on the predicted benefit, the operator is to make the decision. A similar display shall also be available to the plant engineers, as shown in Fig. 7.117. Similarly, a closed loop can also be selected from the optimization workstation for set point controls. Energy savings by optimizing O_2 content in flue gas and flue gas temperature optimization would be extremely important for boiler optimization, as these reduce energy consumption greatly.

Another important boiler optimization package is the intelligent soot-blowing system (**ISBS**). It works on the basis of exhaust gas temperature and steam temperatures. Suitable models are developed to assess the soot formation in each of the heat transfer sections of the boiler. The ISBS gathers this data from the DCS and processes them to compare with the data from DCS historians to detect build up and allocate steam to remove the fly ash build up. The ash build up will jeopardize the heat transfer at the tube, so the steam conditions become poor. The ISBS is much more effective than the traditional soot-blowing operation/system based on the time lapsed. It has been found that with ISBS,

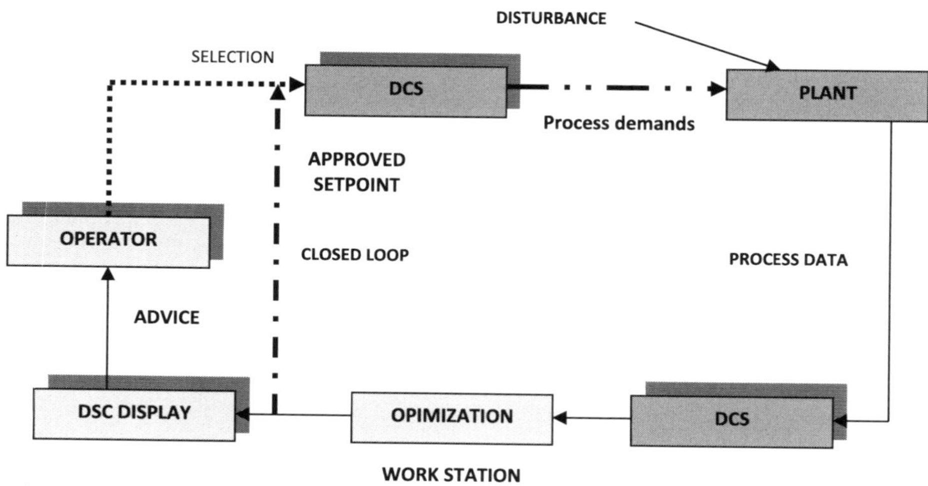

FIG. 7.118 Optimization implementation (typical).

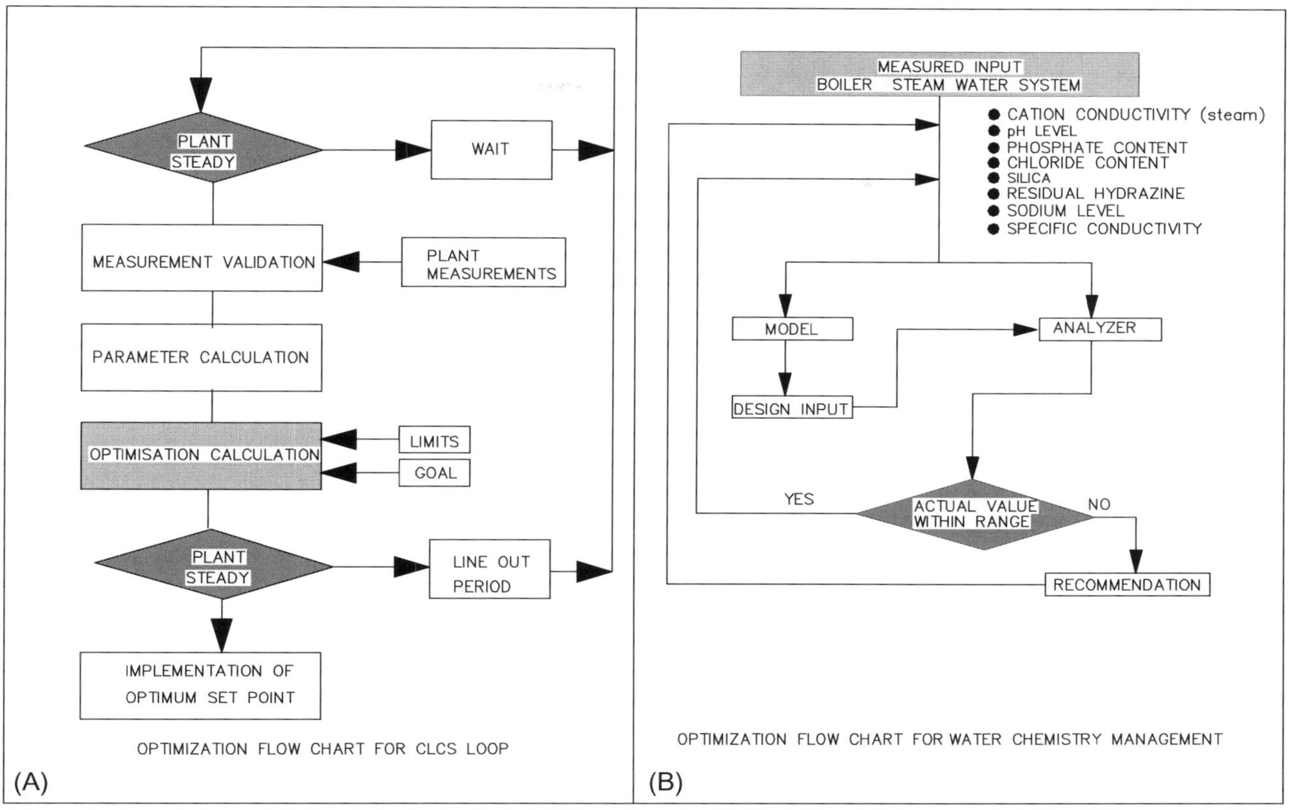

FIG. 7.119 Optimization of flow chart.

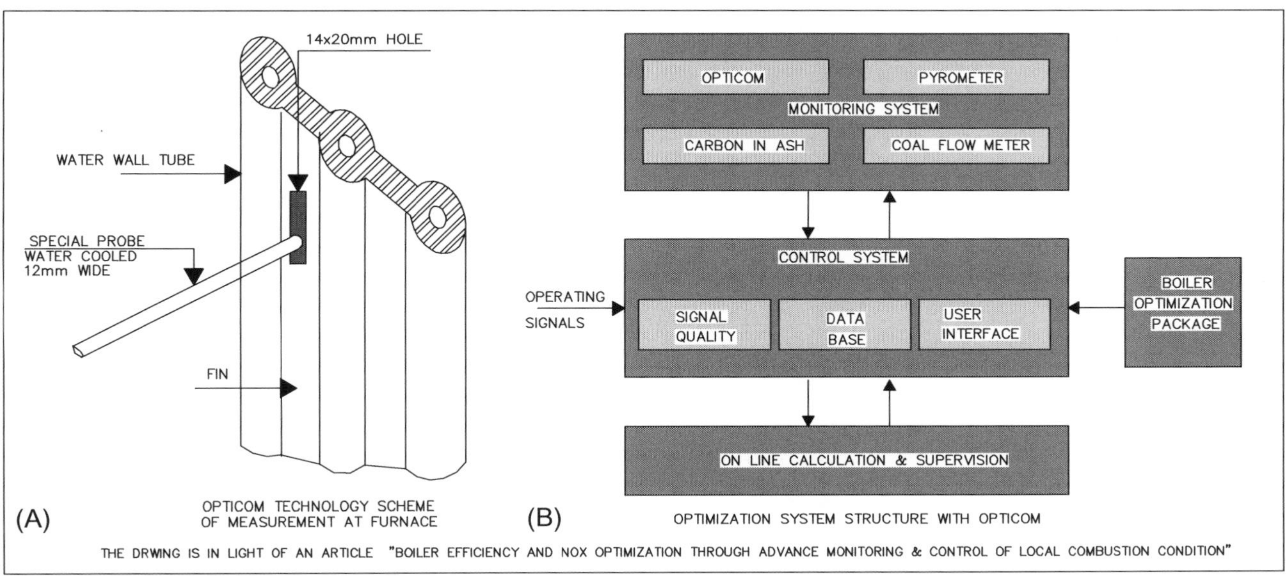

FIG. 7.120 OPTICOM and boiler optimization.

there is an improvement in boiler efficiency to the tune of ~07%–0.8%. It also gives guidance to operator section wise optimal soot blowing requirements, after assessment of fouling of heat transfer section. The advantages of ISBS are:

- Determination of boiler cleaning action dynamically.
- Guide existing soot-blowing system in real time to get the best result.
- Uses adaptive modeling and expert rules.
- Calculates the real-time heat transfer coefficient of each section.
- It initiates a strategic soot-blowing sequence to minimize the heat rate.

6.3.1.5.3 Steam Turbine Cycle Optimization Package The steam turbine cycle referred to here is for a utility station with TG and other regenerative heaters in a heat cycle. Cogeneration/combined cycle plant optimizations are dealt with separately. For turbine optimization, the steam parameters are very important, and these highly depend on the boiler, so a similar package may be more helpful. Temperature, pressure of MS, and HRH are major influencing parameters for steam turbine cycle optimization. Normally, the steam temperatures are maintained at a fixed set point (as suggested by the boiler and turbine manufacturer) by boiler controls. In case of OT boilers, steam temperature controls are extremely important because they greatly influence the combustion control. The MS pressure can be controlled by a fixed throttle pressure set point while the flow through the turbine is regulated as per the first-stage pressure. In the sliding pressure mode (within a band of load), the valve is in wide open condition and the pressure is regulated as per the load demand. The obvious advantages of the sliding pressure mode are reduced losses due to throttling, and higher steam temperature at a reduced load provides better thermal efficiency. All these parameters are highly dependent on the boiler operating conditions, so these parameters, which could be the outputs for the steam turbine cycle optimization package, may be set as targets for the boiler optimization package, as shown in Fig. 7.121.

In this case, the turbine cycle optimizer makes the target recommendations for MS pressure (PrMS), MS temperature (TMS), and HRH temperature (THR) as the target for the boiler optimizer. Full coordination and resources will be allocated to both the systems. On partial coordination, a direct link between the two systems may be broken, but still there will be coordination between the systems. As stated earlier, the two optimization packages will be similar, but both obviously would have different process models. In turbine cycle optimization, there may not be much CLCS optimization, but more will be OLCS recommendations. The main aims of the turbine optimizations with MS pressure will be:

- Determining the break points for constant pressure and sliding pressure.
- Determination of, for example, the optimal mode of operation of the turbine at different loads.
- Comparison of economy between constant pressure and sliding pressure operation.

6.3.1.5.4 Total Plant Optimization As discussed above, various optimization packages for SG and TG can be clubbed together in conjunction with several other software packages for the optimization of various other systems (for example, ESPert from EPRI); total plant optimization could be achieved. While on the subject, it is worth noting that water chemistry management is another important model. A typical such flow diagram is shown in Fig. 7.115B, Water chemistry management. This module helps keep the internal corrosion in check by continuously real time analyzer outputs with the adaptive model and recommends the operator's action for various dosing and other controls (for example, blow down control). So, this always tries to sense the root cause and suggest action. For total plant optimization, a single package can be developed, but it may be too complex to really achieve the desired target. Alternatively, there may be several optimization packages arranged in a hierarchical manner so that the top level will receive inputs from suboptimization packages for coordination to achieve the desired goal. Some of these suboptimization packages could be one regenerative heater, BFPs, BFPTs, etc. The latter approach is better in the sense that each of the models to be developed will have a reduced scope so that it can achieve the goal in a better way. Also, because each unit has fewer inputs, the chance of failure of the subsidiary optimizer due to input failure will be less. Even if it fails, the failure of one subsidiary optimizer will not jeopardize the entire system. Also, the testing of each module is easier. Also, each of the subsidiary optimization packages can be characterized for the unit very easily (namely, optimizing steam consumption in BFPTs and optimizing heat transfer areas for regenerative heaters by regulating heater level energy consumption can be reduced to get overall plant performance better). However there are problems also, for example, several suboptimizers may give conflicting control settings. Therefore, some compromises may have to be weighed, keeping in mind the overall objective. The overall objectives shall be to minimize the total unit cost, which must be related with the high-level plant variables such as boiler efficiency, NO_X, etc. The concept of composite performance analysis and optimization is explained and depicted as Fig. 7.123. This figure has been developed based on ideas from the Kalkitech performance and optimization (courtesy: Kalkitec, https://www.kalkitech.com/solutions/plant-performance-optimization/).

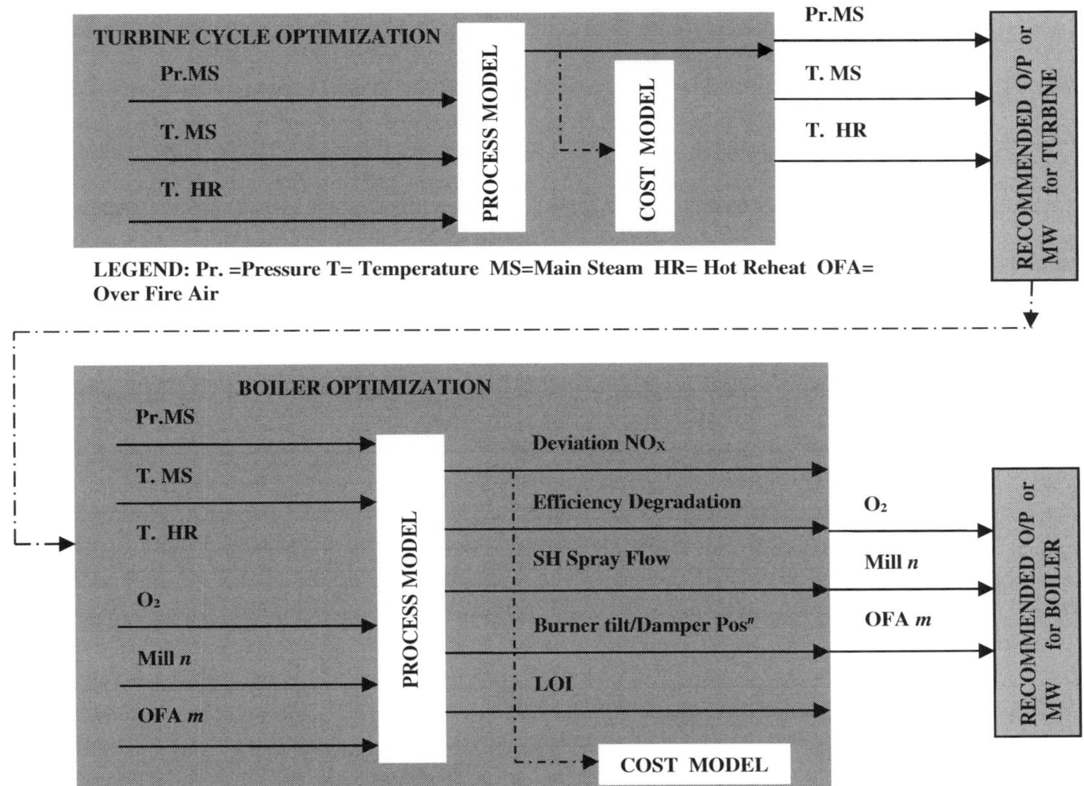

FIG. 7.121 Turbine cycle optimization (coordination with SG).

OPTICOM Technology: A new approach towards this is OPTICOM technology in which advanced monitoring and control of local combustion conditions are done to effective performance optimization. ****For power plant optimization the concept had been used by INRECO,AICIA& ENDESA Spain. The discussions below and the drawing Fig no. 7.120 has been prepared in light of an internet article " Boiler efficiency and NOx Optimisation through Advanced Monitoring and control of Local Combustion Conditions"(by Copado,Canadas,Gomez &pereze).**** *Major advantage of this system is direct combustion characterization in any part of the boiler or furnace. Allowing accurate assessment imbalance of air and fuel. It is capable of characterizing distribution of fuels with different properties within furnace. Optimization with OPTICOM has reduced heat rate by nearly 1% coupled with reduction of* NO_X *by 30%. Major features of this system shall include:*

- ★ *Collection and analysis of gas samples , on line local measurement of temperature gas concentration (* O_2*,CO* NO_X*,* SO_2*)*
- ★ *On line calculation and supervision of unit energy efficiency*
- ★ *Advisory and optimization for boiler efficiency and* NO_X *reduction process.*
- ★ *Measurement of type & stability of flames.*

OPTICOM technology essentially consists of a system that allows measurements to be taken in any area interior to the furnace near burners. In order to meet the goal these measurements are made through small openings in the fins which tubes making the water walls . In this way it is possible to take measurements at levels of each burner without much significant changes in structure. The probe is 10-12mm wide and capable of withstanding temperature 1400-1500°C. Sample collected is passed through heated filter to drive out fly ash and then conditioned to send the same to PLC which is used for measurement and feeding data to the control system. In this connection Fig. 7.120 may be referred to.

FIG. 7.122 Short note on OPTICOM technology for optimization.

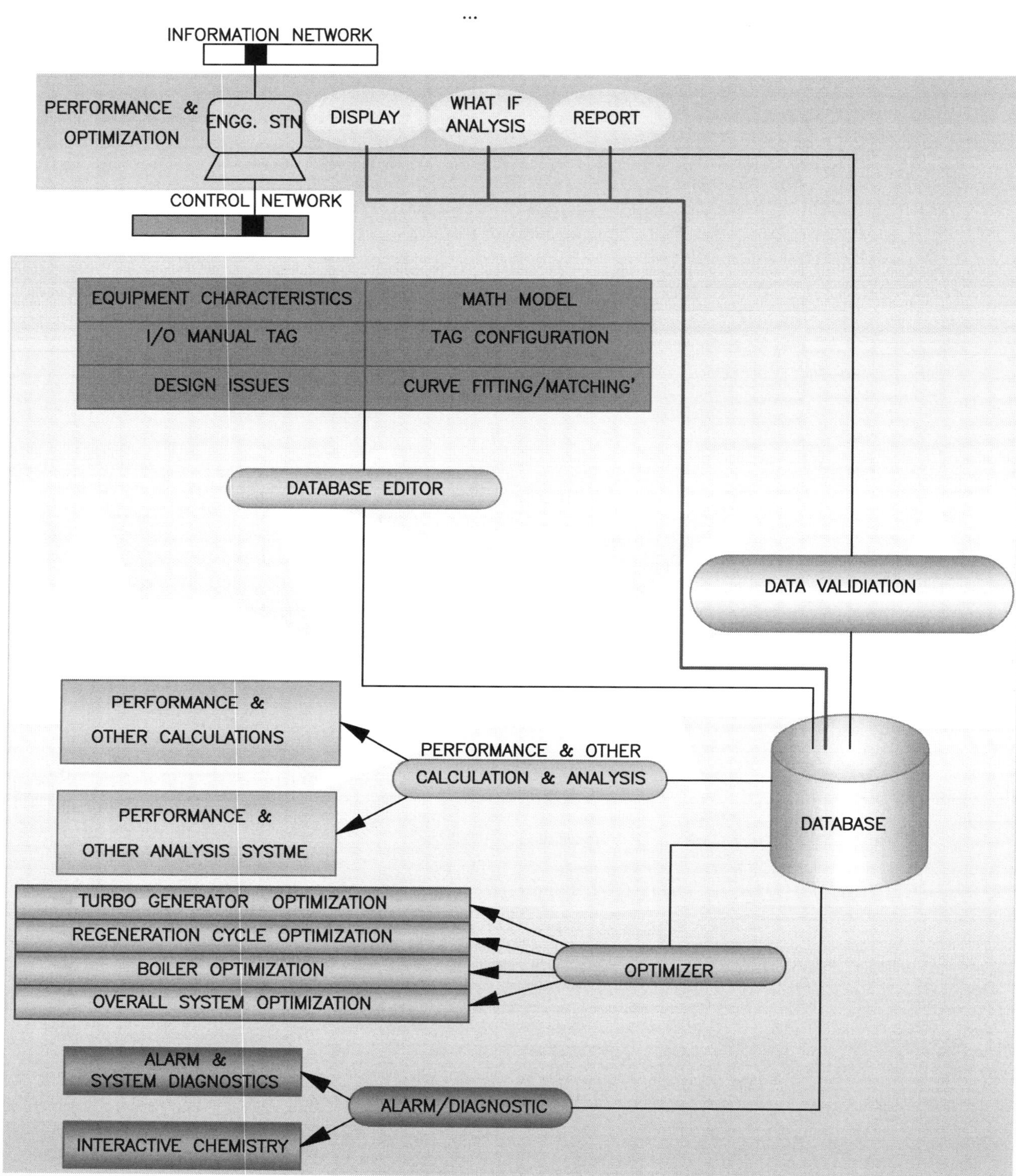

[DEVELOPED BWITH IDEA FROM KALKITECH. (COURTESY KALKITECH)]

FIG. 7.123 Concept of composite performance and optimization.

6.4 Gas Turbine Performance and Optimization

In case of a gas turbine, there are several factors, including the ambient condition, that affect the performance of the gas turbine. In order to get better performance, the aerodynamic design of the compressors and turbines has been much improved to minimize the conversion loss to attain higher efficiency. Another factor that affects the efficiency is GT operation at high-pressure and high-temperature conditions at the inlet. For sustained operation at high pressure and temperature, the durability of the GT components is seriously affected. In modern GTs, the process conditions are such that they almost reach their mechanical and thermal stress limits. So naturally, modern cooling systems and the selection of materials for the components are of immense importance. Ambient conditions such as air temperature and humidity affect the GT performance in two ways, that is, at the compressor inlet as well as the turbine outlet. The aerodynamic design tries to minimize conversion loss, and the GT gives optimum performance at a BASE LOAD range, beyond the same flow balance between GT and compressor varies and optimum performance is not achieved. Similarly, the fuel heat value affects the performance (Table 7.45).

Now, performance analysis helps to distinguish between a "natural" cause and machine degradation. Naturally, this can be done with the help of an engine performance model, combined with mechanical condition monitoring of the machine. So, various factors need to be properly addressed to get the various correction factors necessary to calculate the GT performance. GT manufacturers specify these correction factors for variations in inlet temperature, inlet pressure drop, coolant injection, and load variations. Depending on the applicability, there will be separate correction factors for heat rate and power output corresponding to each of these parameters. Once the operating data are collected, the performance analysis is carried out by comparison with the model with the help of software. The performance analysis will be different for a simple cycle and a combined cycle. There are several performance analysis products available on the market for example, EFFICIENCY MAP and BENTLY PERFORMANCE MAP. Traditionally, the controllers for GTs are tuned seasonally, but now there is software available that makes such repeat work unnecessary.

TABLE 7.45 Ambient Influence

Parameter	Compressor Inlet	Turbine Outlet
High temperature	So, lower density hence less mass flow	
High pressure		Less GT power O/P
Humidity	Higher specific volume, less mass flow in GT	

With this, the discussions on intelligent control systems come to an end as we look into the automatic controls of the boiler.

BIBLIOGRAPHY

[1] ABB 800XA Engineering Overview: ABB Limited.
[2] ABB 800xA high Integrity BMS Solutions: ABB Limited.
[3] ABB Your Partner Functional Safety: ABB Limited.
[4] AC 500 OPC Server: ABB Limited.
[5] AC 800M Control Software: ABB Limited.
[6] AC500-eCO: Your PLC From ABB: ABB Limited.
[7] An Intelligent Conveyor Control System for Coal-Handling Plant of TPP: Makarand Joshi.
[8] An Open Field Bus Comparison: MTL Document.
[9] Application Note: www.hachultra.com.
[10] Architecture for Secure SCADA and DCS Network: juniper Networks: White Paper 2010.
[11] Ash Handling for Power Plant: KOPAR OY Finland: Document.
[12] Automated System for Water Treatment Plant: Industrial Control Systems: www.ics-me.com.
[13] Automation of Coal Handling Plant: A. Dutta, A. Chowdhury, S. Karforma, S. Saha, S. Kundu, S. Neogi.
[14] Benefits of HART Protocol; SMAR.
[15] Boiler Efficiency and NOx Optimization through advanced monitoring and control of local combustion conditions: A. COPADO, F. RODRÍGUEZ,L. CAÑADAS, V. CORTÉS,P. GÓMEZ, E. PÉREZ-SANTOS: International Paper on July 2008 OPORTO (Portugal).
[16] Burner Management System Solution: Emerson Process Management: Delta V SIS.
[17] Centrum VP Document: Yokogawa Electric Limited.
[18] Centum CS Work Station:Yokogawa Electric Corporation.
[19] COAL-FIRED POWER PLANT HEAT RATE REDUCTIONS: Sargent Lundy USA: Jan. 22, 2009.
[20] Comander SR 100A:ABB Limited: Catalog.
[21] Communication Standards in Power Control:A.C West (Triangle microworks Inc.): ESAA 1999 Residential School in Power System Engineering.
[22] Compact HMI 800: ABB Limited: (V 4.1).
[23] Comparison Between CAT 5, 5e, 6 and 7: Discount Cables, USA.
[24] Comparison of Single Loop Controller With Delta V Multiloop Controller: Emerson Process Management: DELTA V White Paper November 2011.
[25] Computer Networks: Book: A.S. Tananbum and D.J. Witherall.
[26] Delta V Digital Automation System: Emerson Process Management.
[27] Demonstration of Advanced Wall-Fired Combustion Techniques for the Reduction of Nitrogen Oxide (NO_x) Emissions From Coal-Fired Boilers (500 MW):DOE Assessment: March 2004.
[28] Design Concept of Foundation Fieldbus Product: M. Yoshitsugu.
[29] Devicenet Automation: Lumberg Automation.
[30] Digitric 500: ABB Limited: Catalog.

[31] Dry System/modern controls upgrade Labadie AH: Shadduck W. Stillman: Union Electric Co.
[32] Effectively managing Power plant Information: Energy Tech magazine; September 2011.
[33] Effectively managing Power plant Information: Energy Tech magazine: September 2011.
[34] Ethernet.ph3: en.kiosk.net/contents/technologies.
[35] FAQ Sheet for S84/IEC 61511: www.primatech.com.
[36] Foundation Fieldbus: Samson Document.
[37] Foundation Fieldbus: www.instrumentation guide.com.
[38] Foundation Fieldbus Application Note: Fluke.
[39] Foundation System technology: Developer Training: Seminar Document.
[40] Fuzzy gain Scheduling PID Control for power plant wide range operation: R.G. Ramirez, K.Y.Lee: International Conference on intelligent system Application ISAP 2007.
[41] Hart Communications: Samson Document.
[42] Hart Protocol: www.hartcomm.org.
[43] Honeywell 620 Logic Controller: Honeywell: Catalog.
[44] Human Machine Interface: Mitsubishi Electric: Catalog.
[45] Industrial Systems 800xA AC 800M Control and I/O:AB B Limited.
[46] Industrial System 800 xA Device Management:AB B Limited: System Document.
[47] Integrated Thermal Power and Desalination Plant Optimization: P. Pechtl, M. Dieleman, M. Posch, B. Davari, M. Erbes, S. Schneeberger: GE Energy Services, Optimization Software.
[48] Intelligence-Based Hybrid Control for Power Plant: W. Wang, H.X. Li, and J. Zhang: IEEE Transaction on Control Systems Technology, vol. 10(2).
[49] Intelligent Failure Tolerant Control: R.F. Stengel: IEEE Control Systems.
[50] Intelligent systems: www.intelligent-systems.com.ar.
[51] Internet Document: http://technology.ezinemark.com.
[52] Introduction to HART: Emerson Process Management.
[53] Introduction to Intelligent Control System With High Degree of Autonomy: P.J. Antsaklis U of Notre Dame and K.M. Passino Ohio State University.
[54] Introduction to MODBUS: Contemporary Controls: Internet Document.
[55] Introduction Beyond Single Loop PID Control: ISA Philadelphia Chapter David B. Leach: Industrial Process optimization March 2003.
[56] Lecture Notes on Internet Protocol CSC343: Wake Forest University.
[57] Master Logic PLC: Honeywell: Catalog.
[58] ME 4012 Digital Turbine Control System D. Baran: Helmut Mauell GmbH.
[59] Memory and execution performance: ABB Limited.
[60] Mod30 1701RP controller: ABB Limited.
[61] MODBUS Protocol Ref. Guide: MODICON, Inc., Industrial Automation Systems.
[62] Model X500F SER: Ronan Engineering Company: Catalog.
[63] Network Design Consideration for Distributed Control Systems: F.L. Lian, J. Moyen, D.Tilbury.
[64] Network Modules: Mitsubishi Electric.
[65] New Generation Integrated Monitoring and Control System: S. Takahashi, S. Shimizu, T. Sekiai: Hitachi.
[66] Omron PLC-Based Power Plant Automatic Control for Pneumatic AH Equipment: Source Net Author: Admin: July 26, 2010.
[67] Operation and monitoring Function Centtum CS: Yokogawa Electric Corporation: September 2008 Document.
[68] Ovation Operator Work Station: Emerson Process Management.
[69] Planer Touch Screen Selection Guide: www.planartouch.com.
[70] Plant Performance Calculation Kalki Technologies India.
[71] PLC 5 Programmable Controllers: Rockwell Automation: Selection Guide 1785 and 1771.
[72] PLC-Based Controls for Ash Handling Plant: Diamon Power Inc. Document.
[73] Power Plant Automation: ABB Limited: Write Up.
[74] Preferred BMS Logic: Preferred Instruments USA.
[75] Preferred Utilities Manufacturing Corp—BMS Logic Discussions.
[76] Principles of LCD Display: Physics Group Project: May 2001.
[77] Process control and Optimization Vol II:by B.g. Liptak*:*P. M. B. SILVA GIRÃO.
[78] Process Control and Optimization Vol II by B.g. Liptak*: B. Kamali: Touch Screen Display.
[79] Process Control and Optimization Vol II by B.g. Liptak*: G.B. Sing, B.G. Liptak: Workstation Design.
[80] Process Control and Optimization Vol II by B.g. Liptak*: T.L. Belvins M. Nixon: *DCS Modern Control Graphics.
[81] Process Control and Optimization Vol II by B.g. Liptak*: T.L. Blevins M. Nixon: DCS Operator Graphics.
[82] Process Control and Optimization Vol II by B.g. Liptak*: R.R. Rhinehart, F.G. Shinskey and H.L. Wade: Control Modes.
[83] Process Control and Optimization Vol II by B.g. Liptak*: J.A. Moor, B.G. Liptak, T.L. Blevins and M. Nixon: DCS System Architecture.
[84] Procontrol P write up: ABB Limited.
[85] Profibus Cable: ABB Limited: Data Sheet 10/63-6.47.
[86] Profibus PA: Samson Document.
[87] Profibus Workshop: Siemens Limited: April 1999.
[88] S800L I/o Modules: ABB Limited.
[89] Safety in Process Instrumentation With SIL Rating: Siemens AG: 2007.
[90] Safety Instrumented BMS: M. Scott and B. Adler Applied Engineering Solutions.
[91] Safety Technologies Incorporated in the Safety Control Station: E. Toshiyuki K Shigehito.
[92] SCADA Applications in Thermal Power Plants: M.N. Lakhoua.
[93] SCADA Security: Challenges and Solution: Schneider Electric: Doc N: TBULM01012-32.
[94] Sceen Master 2000: ABB Limited: Catalog.
[95] Securing SCADA INFRASTRUCTURE: Fortinet CA USA: White Paper.
[96] Selecting and Using RS 232, RS422 and RS 485 Serial Data Standards: Maxim: December 2000.
[97] Sematic Controller Siemens AG: Catalog April 2000.
[98] Sematic Controller Software: Siemens AG: 2010.
[99] Sematic Modular controller: Siemens AG: 2010.
[100] Sequence of Event Recorder: Series 90/30/20 micro PLC: GE Document.
[101] Series 90-30 GE Intelligent Platforms Control Solutions: www.ge-ip.com.
[102] Sematic PC based Automation: Siemens AG: 2010.
[103] SLPC (YS 1700 1310): YEL: Catalog.
[104] Smart Process Plant Optimization and performance software Emerson Process Management: Document.
[105] Software Functional Design Specification: Honeywell: Project Document.

[106] Standard Tech. Specification For Main plant Package of Subcritical Thermal power plant (500MW and above): Central Electricity Authority New Delhi: New Delhi September 2008.

[107] Standard Design Criteria/Guideline for Balance of plant (2 × 500 MW TPP)Central Electricity Authority New Delhi: New Delhi 2010.

[108] Steam/Its Generation and Use: Book: Babcock and Wilcox.

[109] Steam Turbine Optimization: A.D. Gavrilos ABB/ETSI.

[110] Supervisory Control and Data Acquisition (SCADA) Systems: National Communication System: Technical Information Bulletin 04-1 (2004).

[111] Supervisory Information System at Power Plant Level (SIS) and Its Development: Y. Dong: Thermal Power Research Institute PR China.

[112] System 9000TS SER:RTK instrument: Document.

[113] Turbine ATRS NTPC: Korba Simulator.

[114] Turbine zSystems: EDC Singrauli India.

[115] Twisted Pair Cables: www.infocellar.com.

[116] What is OPC: OPC Foundation: OLE Wikipedia Also.

[117] Write up on ATRS BHEL-KWU Document.

[118] Write up on ATRS: BHEL Hardwar.

[119] Intelligent Sootblower Scheduling for Improved Boiler Operation: X. Cheng, R.W. Kephart, J.J. William: Westing House Process Control Inc.

[120] Industrial Ethernet Book; issue 48/39 Hirose Electric Europe BV; Internet document; http://www.iebmedia.com/index.php?id=5929andparentid=74andthemeid=255andhft=48andshowdetail=trueandbb=1andPHPSESSID=u8ogrsmgstv95133sa5mt8pbq1.

[121] Wireless HART; RP 400 Series; Smar; Catalog: RP400 www.smar.com/PDFs/catalogues/RP400CE.pdf.

[122] Wireless HART; HART Communication Protocol; www.hartcomm.org; Write up: http://www2.emersonprocess.com/siteadmincenter/PM%20Central%20Web%20Documents/HART%20Wireless%20Brochure%20v10%20Final.pdf.

[123] Fieldbus Why Wireless?; T. Kruidhof, L. Schotborg, and R. De vreede with Mr. Blankenstein; Rotterdam Main Port University of Applied Science—RMU; Maritime Symposium; Rotterdam; http://www.maritimesymposium-rotterdam.nl/uploads/Route/Wireless%20fieldbus%20system.pdf.

[124] Design and Implementation of Wireless Fieldbus for Networked Control Systems; D.K. Choi1, J. II Lee, D.S. Kim1, and W.C. Park; SICE-ICASE International Joint Conference 2006; Oct. 18–21, 2006, in Bexco, Busan, Korea; ftp://ftp.ucauca.edu.co/Facultades/FIET/DEIC/Materias/Redes%20Industriales/Articulos/Design%20and%20Implementation%20of%20Wireless%20Fieldbus%20for%20Networked%20Control.pdf.

[125] Enabling Inter-Domain Transactions in Bridge-Based Hybrid Wired/Wireless PROFIBUS Networks; L.L. Ferreira, E.Tover, and M. Alves; Conference: Emerging Technologies and Factory Automation, 2003. Proceedings. ETFA '03. IEEE Conference Volume: 1; October 2003; 10.1109/ETFA.2003.1247682; IEEE Xplore; https://www.researchgate.net/publication/4044841_Enabling_inter-domain_transactions_in_bridge-based_hybrid_wiredwireless_PROFIBUS_networks/figures?lo=1.

[126] Wireless HART: Applying Wireless Technology in Real-Time Industrial Process Control; S. Han; University of Connecticut; http://engr.uconn.edu/~song/classes/nes/WirelessHART.pdf.

[127] Bridging Digital, Physical and Human; What is the internet of things?; Internet document; iSCOOP; https://www.i-scoop.eu/internet-of-things/.

[128] Plant Hazard Analysis and Safety Instrumentation System; Swapan Basu; Elsevier B.V.; IChemE; 2016; https://www.elsevier.com/books/plant-hazard-analysis-and-safety-instrumentation-systems/unknown/978-0-12-803763-8; https://icheme.myshopify.com/products/plant-hazard-analysis-and-safety-instrumentation-systems-1st-edition.

[129] ARC White Paper, Foundation Fieldbus Safety Instrumented Functions Forge the Future of Process Safety, ARC Advisory Group, September 2008.

Chapter 8

Boiler Control System

Chapter Outline

1 BASIC CONTROL REQUIREMENT

1.1 Introduction

Modulating controls in a power plant is extremely important. It is always advisable to the designer to frame a basic control philosophy before developing a control loop. In the subsequent sections, brief discussions on the same are presented.

All the control loops discussed in the book shall cover the following points. Although this is stated in connection with boiler control, it is applicable to all the control loops discussed in Chapters 9–11 also.

- The objective of the control loop with relation to control parameters, manipulating variables, and final control elements (FCEs).
- Description of the loop/subloops.
- Automatic and manual operation, tripping to manual due to transmitter/sensor failure, as discussed in clause 1.2.
- Protection, interlock, and special features (If any).

Power Plant Instrumentation and Control Handbook. https://doi.org/10.1016/B978-0-12-819504-8.00008-1

1.2 Transmitter Selection

Most of the control loops in power plants are offered with redundancies in transmitters, as discussed in clause no. 1.7 of Chapter 3. Now, the control loops often trip to manual on account of failure of these transmitters in different ways. Any transmitter failure shall be alarmed in the operator's monitor. In the next subsections, the basic philosophy of tripping the control loop to manual under different conditions of transmitters is discussed, and this is applicable to all the control loops.

1.2.1 Control Loop to Manual in Case of One of Two Selection

While operating in auto, the loop shall trip to manual in any of the following events:

(i) Both transmitters fail (for example, out of range $4\,mA > O/P > 20\,mA$).
(ii) Both transmitters are healthy but the deviation between them is too high.
(iii) Operator has selected any one transmitter for control (for some reason) and it fails. Now, the operator may select the other transmitter (if healthy) and put the loop back to auto.
(iv) Average value of transmitter outputs selected and one transmitter fails (may be out of range) and the other could automatically be selected in auto, depending on the philosophy adapted.

1.2.2 Control Loop to Manual in Case of Two of Three Selection

While operating in auto, the loop shall trip to manual in any one of the following events:

(i) If the median or average is selected and two transmitters fail (e.g., out of range), the other could automatically be selected in auto, depending on the philosophy adapted.
(ii) The operator has selected any one transmitter for control (for some reason) and it fails. Now, the operator may select the average (or median) or any of the other two transmitters and put the loop back to auto, provided the other transmitter is healthy.
(iii) When the average/median of transmitters is selected and one transmitter fails (may be out of range), the average/median of the other two other could automatically be selected or the operator can select any of the two healthy transmitters manually, depending on philosophy adapted.

a. Utility functions associated with the control loops: Now, as control systems are implemented with state-of-the-art technology, a few utility functions can be implemented easily through system software. A few such conditions have been discussed, and would be followed as a general philosophy unless stated otherwise against any loop.
   i. Trip to Manual: Whenever the auto enable (release to auto) signal is absent, the loop will be forced to manual and cannot be put to auto by the operator until the auto enable signal is restored. An "auto enable" signal missing alarm may be generated in the diagnostic or any other monitor.

      In case any protection runback/forced open/close condition exists or appears in the A/M station, the operator shall not be able to reverse the command. The same will be situation in case of directional blocking due to run up or run down protection rising from the failure of the main auxiliary.
   ii. In certain control systems having three-tier password protection, it may be possible for the engineer to reverse a few of these commands, but in all such cases, the engineer's intervention will not only be alarmed but will be put on record with time stamping.
   iii. Generally, all output of the control loops is monitored against time. If the feedback of the control action does not reach the loop within the preset specified time, after the control command has been issued, or there is deviation between the command and control action, then the same may be alarmed at the monitor.
   iv. All set point reasonability are checked in the loop and in case of mismatch will be alarmed in the monitor.
   v. In case of changeover, the contact does not reach either end and an alarm shall be generated in the monitor as a noncoincidence error.

b. Position Signal: All the control (modulating) valves shall have a position transmitter for position indication at the DCS monitor. In some of the loops discussed in this book, no feedback from the FCE position is taken to balance the loop (instead, the feedback is taken through the process parameter—a general convention) for a better steady-state response.

The legend sheets are presented in Fig. 8.1 and are applicable to all control loop strategies depicted in the relevant chapters.

2 STEAM PRESSURE CONTROL WITH LOAD INDEX

General

For any steam-generation plant, steam pressure is the most vital parameter that indicates the state of balance between the supply and demand for steam, which, in other

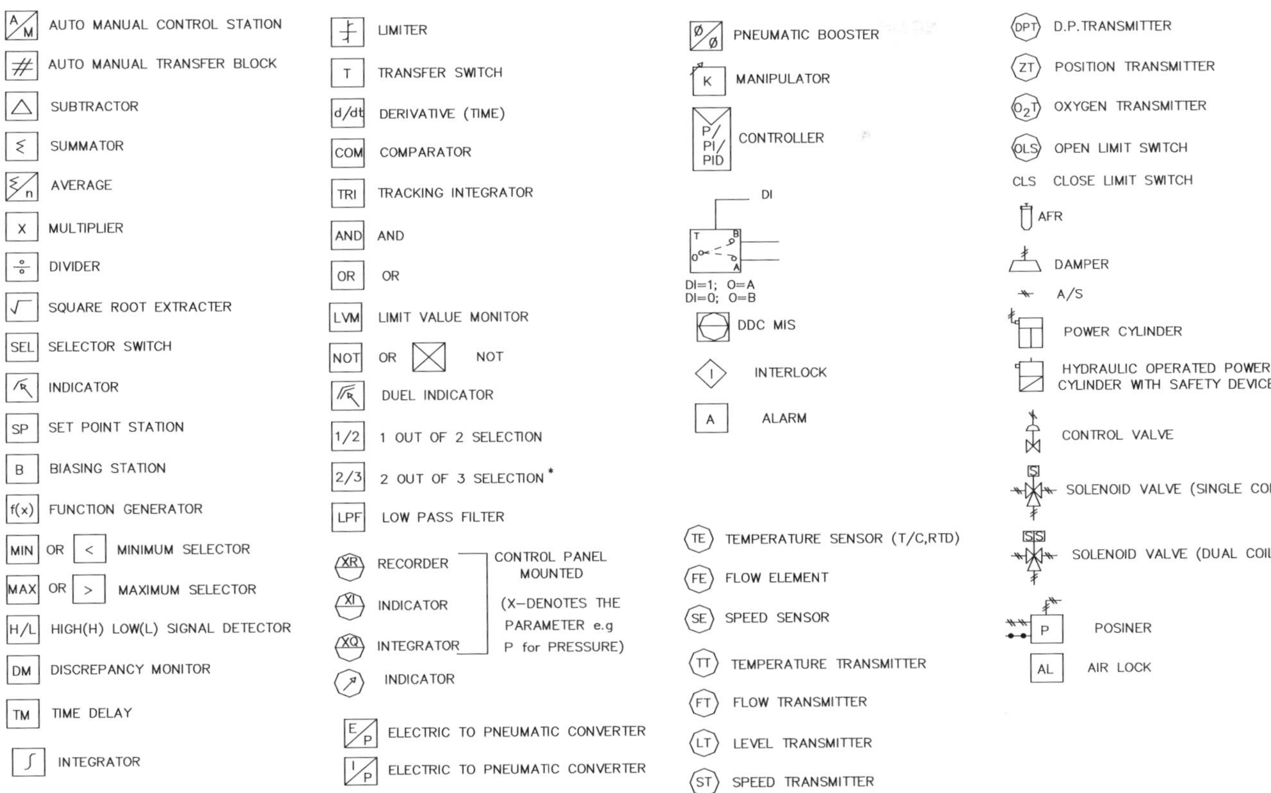

FIG. 8.1 Auto control loop system legend.

words, can be stated as the supply of heat input or fuel and the output as steam withdrawal from the system for various utilities like heating purpose or supplying to turbine for generation of power. In case supply exceeds demand, the pressure obviously increases and vice versa. The control loop strategies discussed in this chapter are for subcritical thermal power stations, that is, with a drum boiler.

2.1 Objective

The steam pressure control with load index as the feed forward signal is provided to maintain the main steam pressure in relation to combustion control. For a thermal power plant, it is the throttle steam pressure that is maintained at a fixed pressure or at a variable pressure for a sliding pressure control strategy.

2.2 Discussion

In the early low-pressure boilers stages with less development regarding superheaters, the measuring point of the steam pressure control was the drum itself. In a multiple steam generators interconnected, the common steam header pressure was the controlled parameter for each boiler with their individual bias setting to suit the prevailing site condition.

For the utility plants with individual turbines for any boiler, the very point of measurement for control purposes is the turbine inlet throttle pressure.

The boiler control system in general is to take care of the numerous aspects that cover the following requirements:

- **(i)** Integration with the coordinated control system (CCS) combining turbine and Generator for MW control.
- **(ii)** Combustion control with fuel and air control having a lead-lag control strategy.
- **(iii)** Burner management system (BMS).
- **(iv)** Drum or start-up separator (for a once-through SG plant) level control.
- **(v)** Superheater/reheater steam temperature control.
- **(vi)** Adequate display and data for the operator to continue in case of emergency and for the system engineers for analysis of any catastrophic situation.

During certain periods of time in the past, the normal practice was to provide separate and independent subsystem control for the boiler and turbine/generator with no or minimal coordination between them. Nowadays, the CCS has come up with a complete solution having the main and ultimate demand controller output issued simultaneously to each subsystem, ensuring more responsive control loops for better results.

Coordinating the entire plant with all the subsystems in tandem requires an intimate knowledge of these systems and the selection of the appropriate operating mode to make them work together. The main motivation behind the idea is to take care of the time lag and minimize the effect of interactions among the above subsystems. The time lag normally refers to the boiler, which has enormous thermal mass and is therefore relatively slow to respond to the requirement. Turbines, on the other hand, are very compact and almost immediately respond to any changes called for by the control system or operator's action.

The coordinated control loop, when operated under a *boiler follow mode* in which a load demand signal, as provided by the operator or load dispatch controls (LDC), controlled the position of the turbine governor or control valves. The throttle or system steam pressure changes and the boiler controller accordingly takes corrective action to control the input of fuel, combustion air, and feed water to maintain the system steam pressure at a predetermined fixed or variable level.

Alternatively, in a *turbine follow mode* system where the load demand is issued to the boiler whereby the fuel, air and water inputs were respectively controlled. As the steam pressure varies due to changes in heat input, the turbine controller adjusts the position of the governor valve to set the system steam pressure at a preset value. The main intention behind employing a turbine follow system is to maintain throttle pressure at all times, but the overall system performance is not regarded as a very efficient process. This is because the turbine follow mode is not capable of making maximum use of the stored energy in the boiler in order to maintain a constant pressure. A detailed discussion is included in clause no. 1.1 of Chapter 10.

However, the boiler control in relation to the total control system has been designed with the aim to cope with the ever-increasing power demand with good frequency control over the entire range. In addition to overall systems requirements, there must be a provision so as to extract the maximum energy from the boiler, having well-defined set limits within which the boiler is assured for efficient and safe operation.

The basic concept of a coordinated control system is to issue the load demand signals simultaneously to the boiler control system and to the turbine control system as a *feed forward signal. This is utilized mainly for minor or trim corrections in response to detection of errors in the throttle pressure, megawatt, and frequency adjustments*. The most important application of this feed forward signal is the use of a control philosophy to reduce interaction and to increase the possibilities of taking out the best dynamic response. Utilizing this signal is always advantageous as the same senses the load changes very fast and enables corrective action before any significant errors crop up. Feedback controllers may be used as a final or further trim at the appropriate level to make amends for minor deviations and adverse effects in getting a better steady-state response.

To be more specific, the throttle pressure is measured through redundant transmitters and compared with a fixed or variable set point, which provides the throttle pressure error. It is subsequently fed to a controller with integral action before being applied to the turbine control system. The throttle pressure error signal is utilized to readjust the load set point. On the other hand, the MW output delivered by the turbine is measured and compared with the MW set value to provide megawatt error. It is subsequently fed to a controller with integral action and is used along with the throttle pressure error signal before being applied to the boiler control system for the subsequent control of the air and fuel inputs to the boiler. The details are discussed in different sections of Chapter 10.

2.3 Control Loop Description

As indicated earlier, the boiler steam pressure control system is the basic system that is meant for maintaining the pressure, be it the main steam throttle pressure or the header pressure (multiboiler in a single-header configuration), by adjusting the fuel and air input in a bid to supply proper heat input as per the demand sensed by the change in the above steam pressure.

This control loop is actually a part of the boiler combustion control loop, and quite a few varieties of such loops to match the requirements of the individual boiler concerned have been developed over the years since inception. The final choice of configuration depends upon so many factors such as the type of fuels; the firing system; different subsystems such as the boiler, turbine, etc.; the pattern of load demands; and the layout of different subsystems and associated accessories.

Fig. 8.2 may be referred for a schematic representation of the idea for implementing the control loop in a typical 250 MW plant having a drum boiler with a down shot opposed fired (fossil/coal) furnace (Babcock design) with occasional oil support. The control loop may vary with the manufacturer's special design, but the basic idea would remain the same.

A similar loop with little difference in philosophy is shown in Fig. 8.11.

Here, the main steam pressure is measured with triple redundancy and sent to the coordinated control system after the necessary voting circuit.

The main steam pressure is compared with the set point, which may be a fixed set point or a variable set point depending on the power plant unit configuration and type of different subsystem such as the boiler, turbine, etc. The main steam pressure error signal is then fed to a controller with the output having built-in proportional, integral, and derivative action. Controller tuning is normally with an auto tuner, but the system engineers may tune the same to suit the particular site condition.

The controller output signal is then trimmed suitably through a function algorithm whose other input is the

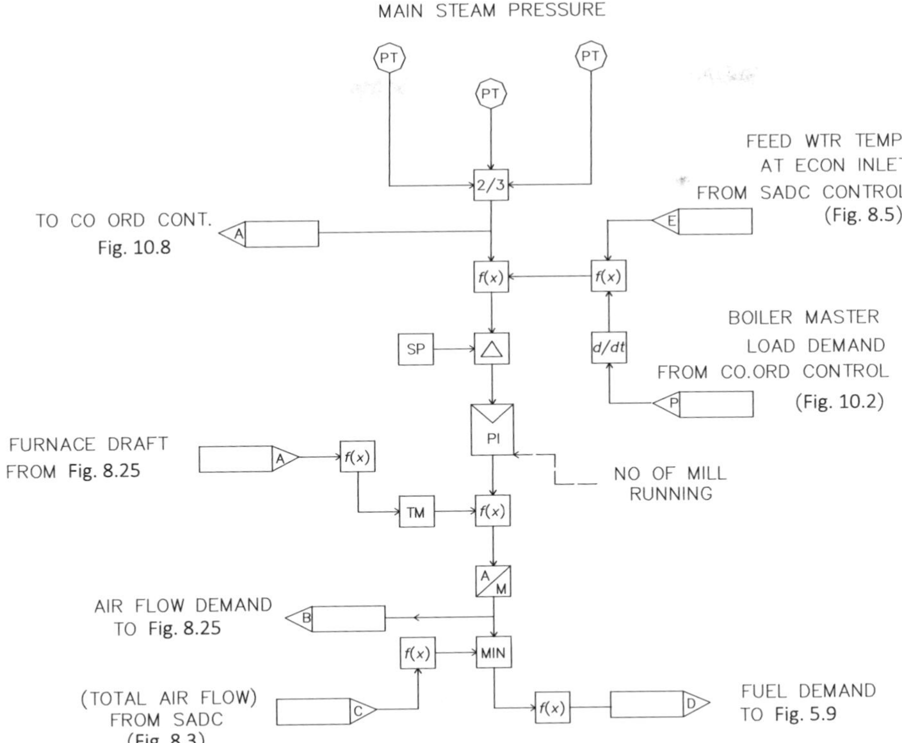

FIG. 8.2 Steam pressure control with load index.

characterized signal from the FW temperature at the economizer inlet and feed forward signal from the derivative of the boiler load demand signal generated in the coordinated control system (Chapter 10), as indicated in Fig. 10.2. The feed water temperature at the economizer inlet is also a part of the feed forward signal generated within the boiler control subsystem, meaning the enthalpy carried into the boiler by the feed water from the outside. When the temperature is less, the heat input demand would be more and vice versa.

Another trimming signal is taken from the furnace draft/pressure signal with sufficient redundancy before voting. The loading of the boiler corrects itself in accordance with the furnace's present running condition, that is, whether further loading or unloading is required to maintain the furnace pressure. With the furnace pressure always fluctuating, this signal is fed through a time delay and filter circuit to avoid oscillation in the controller output.

The trimmed output signal is the load demand for the air flow controller. The same signal is also utilized as the fuel flow demand but not before passing through a minimum selector with the total air flow as the other input, ensuring that fuel flow demand or actual flow can never be more than the air flow demand or actual flow.

The control philosophy in a tangential tilt burner, as shown in Fig. 8.11 may be different as utilized in another 250 MW power plant with TT burners. Here, the main steam pressure demand is compared with the pressure (fixed or sliding) set point generating the error signal, which is fed directly into a PI controller and the output of the controller is the desired pressure demand signal. In order to make the loop more responsive, the load index signal is added at the output of the controller. Because the loop is tuned with the load index signal at the output, any change in demand will affect the loop at the same time the derivative signal, which always contributes noise, has been eliminated. As shown, there could be a number of parameters to be considered as the load index signal:

- Steam flow, if measured (in higher MW plants, this may not be measured due to permanent loss in the flow element).
- First-stage pressure of the turbine added with high-pressure bypass (HPBP) flow (if partial HPBP is allowed).
- Generator load.

3 AIR FLOW CONTROL

General

Air flow control is a very vital part of the combustion control system, which generates the demand and ensures the adequate air flow required for complete combustion of fuels, as requisitioned by the combustion control system for keeping the steam pressure at its desired value.

3.1 Objective

The air flow control is provided to supply the total air flow to the exact quantity that satisfies the demand of the fuel firing rate considering the *stoichiometric ratio* (chemically correct or perfect ideal *air-fuel ratio*) and some excess air to guarantee complete combustion. The control loop discussed here actually computes the total required air flow in relation to the mill-wise air flow demand; when the relevant PA flow is subtracted from it, the balance amount is regarded as the secondary air demand required for stoichiometric combustion.

3.2 Discussion

Air flow control has been considered to be very important for any firing system. Inadequate air flow means the exhaust flue gas would contain the environment pollutant carbon monoxide due to incomplete combustion, which is not allowed by regulatory boards. Lack of air flow also means more fuel cost and less efficiency. For this reason, some amount of excess air is provided for the firing process to take place with the condition of as near to complete combustion as possible. On the other hand, too much excess air would not take part in the combustion process at all but would carry over the extra heat as simply waste, causing less efficiency toward the cost of equivalent fuel.

3.3 Control Loop Description

Figs. 8.3–8.6 may be referred to for a typical schematic representation of the control loops in a typical 250 MW plant with a drum boiler and fossil (coal)-fired furnace with occasional oil support (Babcock design), which may vary with the manufacturers' special design and fuel type.

3.3.1 Measurement of Different Parameters

3.3.1.1 Total Air Flow

Air flow is measured by different types of flow elements normally dictated by the boiler manufacturer. For plants having more than one forced draft fans whose suction or discharge points are selected for measurements are then summated to achieve total flow.

Triple redundant flow signal after necessary voting circuit is density compensated with redundant temperature transmitters before square rooted for getting linear signal. Pressure compensation is not required in this case. This signal is also utilized for fuel flow control and furnace draft control; binary signals derived from this signal may also be used for the BMS, as shown in Fig. 8.3.

3.3.1.2 Oxygen Percentage in Flue Gas

This measurement forms an important part in the air flow control philosophy so far as complete combustion and excess air presence is concerned. Fig. 8.5 indicates a

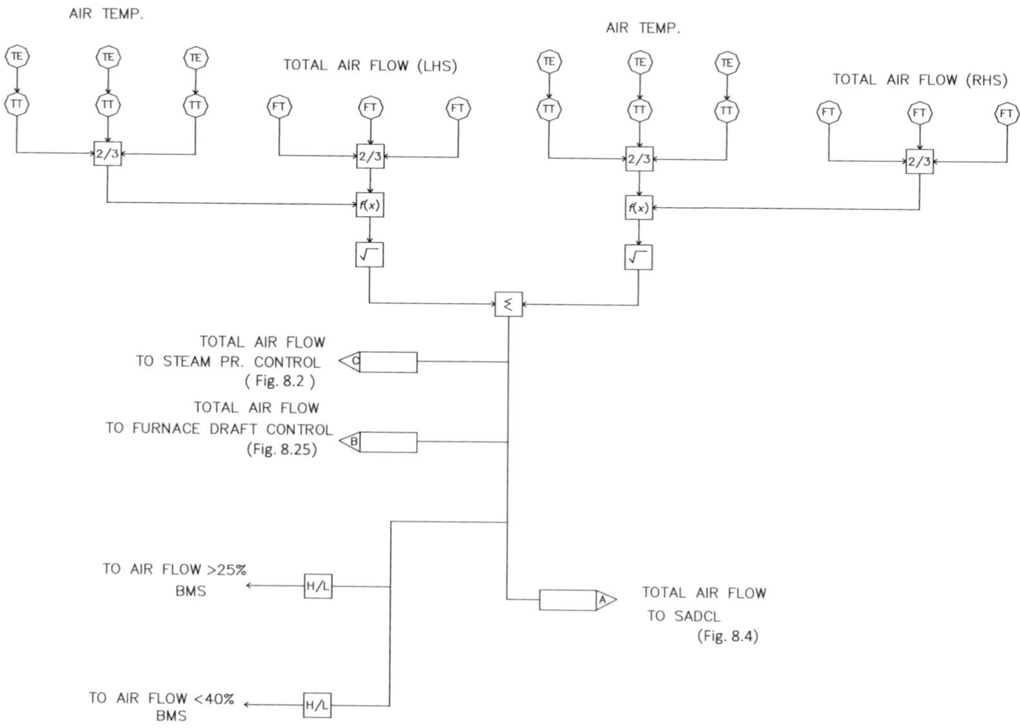

FIG. 8.3 Air flow control/SADC control 1.

triple-redundant measurement and voting circuit accordingly. The strategic tapping point of oxygen percentage measurement is also important and varies according to different schools of thought. The probable points may be at the economizer inlet, outlet, or the air heater inlet where the flue gas temperature would vary, and so the type of transmitters may also. Normally in situ or online instruments are preferred over sampling measurements for a faster response.

3.3.1.3 Measurement of Feed Water Temperature at Economizer Inlet

This measurement (Fig. 8.5) with sufficient redundancy and a voting circuit is utilized as a feed forward signal and a trimming factor for the steam flow signal. This characterized steam flow signal (from Fig. 8.27) forms the boiler load index.

3.3.1.4 Measurement of Secondary Air Flow

Fig. 8.4 envisages two numbers and secondary air dampers (SAD) catering to the combustion air required for mill-wise coal flow. The air flow in each line is measured with redundancy and a voting circuit and summated to achieve the total secondary air flow meant for the combustion air to each mill. Density compensation may be provided, depending upon the excursion of temperature.

The mill-wise secondary air flow is now corrected as per the oxygen trimming signal (oxygen percentage in flue gas) and becomes the process variable for the secondary air flow controller. The set value formation of this controller is dependent on the load.

3.3.2 Different Controls

3.3.2.1 Oxygen Trimming Controller

Fig. 8.5 may be referred to as the schematic representation of oxygen trimming in the control strategy. The combustion process needs excess air as discussed earlier; the residual oxygen present in the flue gas is an index of the same. This excess percentage of oxygen toward the optimum combustion condition is not a constant value but depends on the percentage of load. An approximate value is around 4% at boiler load <60% and at loads >60%, the presence of oxygen becomes gradually lesser and is about 1.5%–2% nearing 75–100% load. The graph of excess percentage of oxygen versus percentage of load is depicted in Fig. 8.5.

The main steam flow is taken (Fig. 8.27) as the boiler load with sufficient approximation (no leakage, no intermediate requirement, etc.) and corrected with the signal from the feed water temperature at the economizer inlet as the feed forward signal. After correction, the boiler load is converted to percentage of oxygen in the flue gas through the algorithm representing the graph indicated above. A bias is added to it as part of finer tuning or adjustment to suit site conditions to arrive at the desired set value of oxygen percentage. A selector switch makes provisions for a manual set point. A set point with oxygen percentage generates the error, which is connected to the controller with the P +I+D output option. To prevent overcorrection, the controller output is passed through a high/low limiter with hand set limit values corresponding to a particular boiler and sent for air flow control as well as supplying combustion air to all the mills' fuel control.

3.3.2.2 Secondary Air (SA) Flow Controller

As discussed above, there are two SAD provided in the air flow path to control and supply combustion air to each pulverizer/mill as per the command issued by the relevant controller.

Both the air flow demand and the characterized feeder speed signal as generated in Fig. 8.9 are passed through a maximum selector option so as to ensure higher air flow in both the increasing and decreasing load. The air flow demand as derived is the total combustion air flow required for each mill, including the PA flow as provided by separate PA fans. For that reason, the PA flow pertaining to each mill (Fig. 8.9) is subtracted from the above air flow demand to generate the actual air flow requirement or set point for the secondary air flow controller.

The controller output (Fig. 8.4) then passes through a maximum selector option with other signals related to minimum air flow, pulverized fuel position, oil position, etc. These other signals are analog but triggered from the binary signals initiated by the particular boiler operating conditions, which are:

(i) A dedicated controller is provided for maintaining the SA flow at 45% (typical value) when the requirement is initiated by the BMS related to the pulverized fuel in the operating condition.

(ii) A dedicated controller is provided for maintaining the SA flow at 10% (typical value) when the requirement is initiated by the BMS related to oil as fuel in the operating condition.

(iii) A dedicated controller is provided for maintaining the SA flow at 25% (typical value) when requirement is initiated by the BMS related to furnace safety conditions.

These typical values may vary from plant to plant; the ultimate strategy is to ensure an air-rich condition throughout the operation of the boiler subsystem without fire hazards and with optimum efficient output.

The final output signal from the above maximum selector then drives the respective SAD. Similar loops are envisaged for each pulverizer mill group, according to the boiler load.

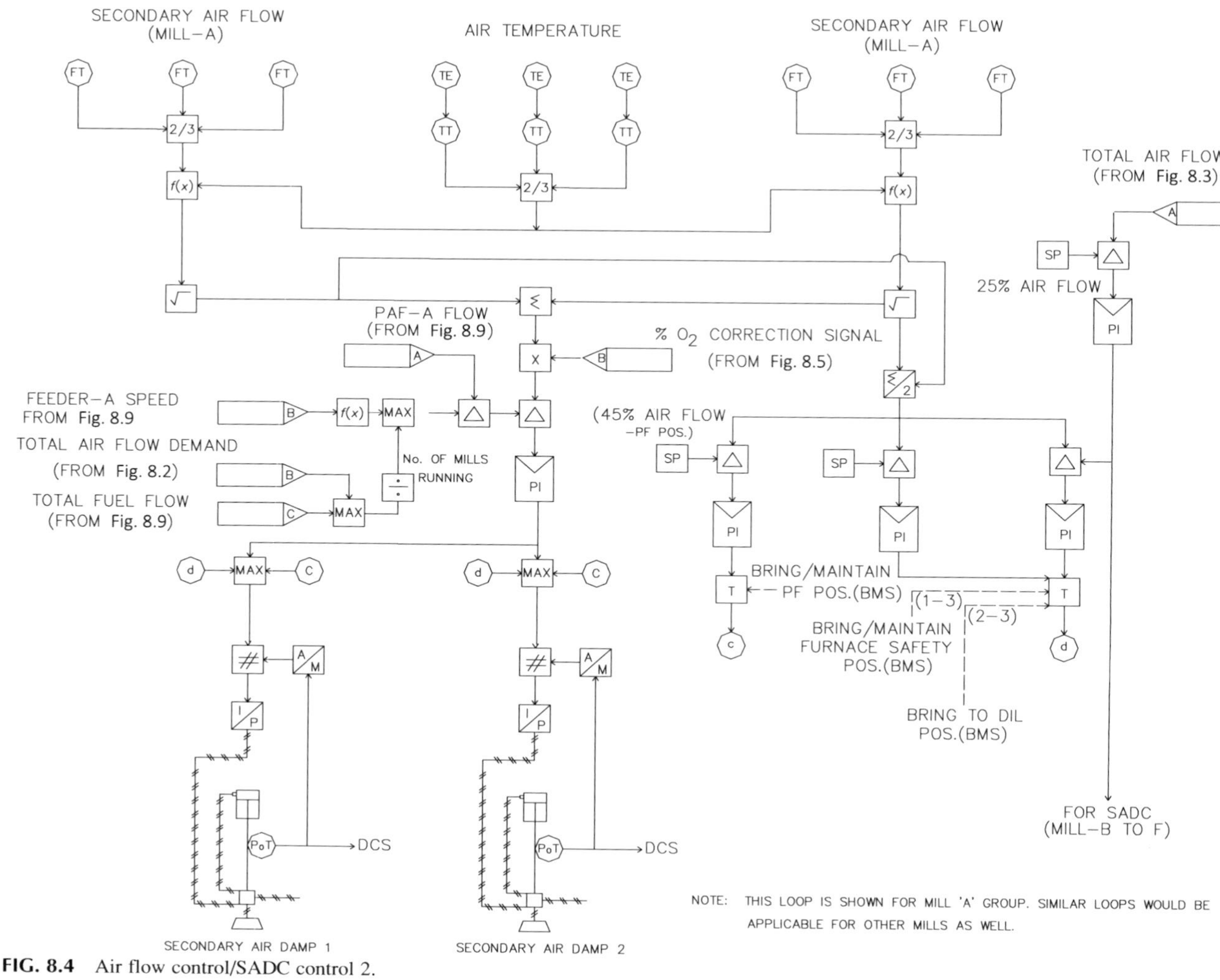

SECONDARY AIR FLOW (MILL–A)
AIR TEMPERATURE
SECONDARY AIR FLOW (MILL–A)
TOTAL AIR FLOW (FROM Fig. 8.3)
25% AIR FLOW
PAF–A FLOW (FROM Fig. 8.9)
% O2 CORRECTION SIGNAL (FROM Fig. 8.5)
FEEDER–A SPEED FROM Fig. 8.9
TOTAL AIR FLOW DEMAND (FROM Fig. 8.2)
TOTAL FUEL FLOW (FROM Fig. 8.9)
No. OF MILLS RUNNING
(45% AIR FLOW –PF POS.)
BRING/MAINTAIN PF POS.(BMS)
(1–3)
(2–3)
BRING/MAINTAIN FURNACE SAFETY POS.(BMS)
BRING TO DIL POS.(BMS)
FOR SADC (MILL–B TO F)
DCS
DCS
NOTE: THIS LOOP IS SHOWN FOR MILL 'A' GROUP. SIMILAR LOOPS WOULD BE APPLICABLE FOR OTHER MILLS AS WELL.
SECONDARY AIR DAMP 1
SECONDARY AIR DAMP 2

FIG. 8.4 Air flow control/SADC control 2.

F.W TEMP. AT ECO. INLET
FLUE GAS OXYGEN %
MAIN STEAM FLOW FROM DRUM LEVEL CONTROL (Fig. 8.27)
TO FUEL FLOW CONTROL (Fig. 8.2)
FOR SADC (MILL–B TOF)
% O2 CORRECTION SIGNAL TO AIR FLOW CONTROL (Fig. 8.4)

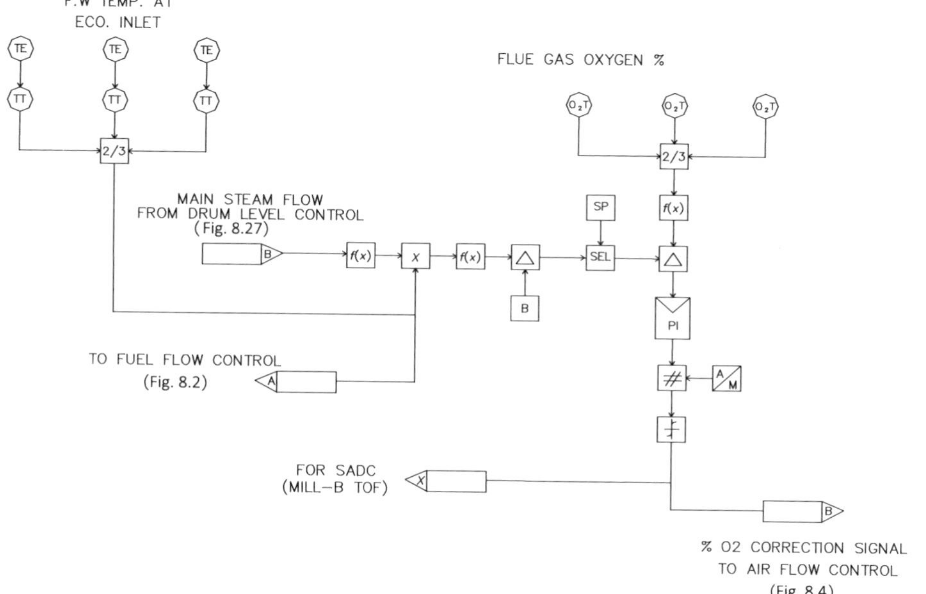

FIG. 8.5 Oxygen trimming in air flow control/SADC control.

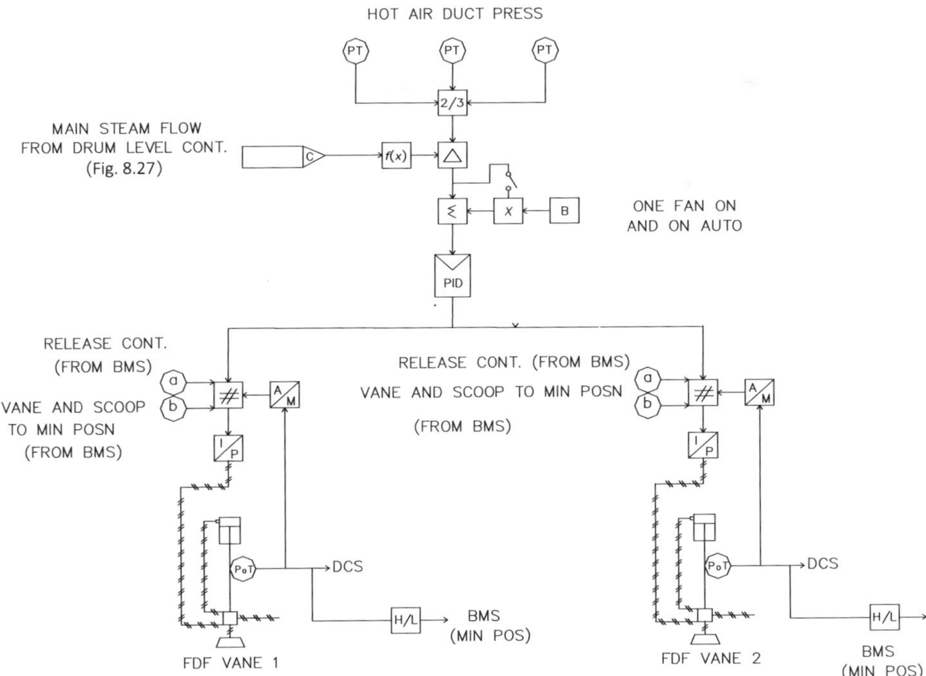

FIG. 8.6 Hot air duct pressure control (total air flow control).

3.3.2.3 Hot Air Duct Pressure Controller

Normally, two forced draft (FD) fans are provided with about 60% (typical) load-bearing capacity to supply the complete combustion air for all types of fuels. There are several FCEs that are deployed for controlling the air requirements, such as the FD fan vane, impeller, damper, etc. In this control loop philosophy, FD fans vanes are envisaged as the FCE and modulate to control the hot air duct pressure.

FD fans suck air from the atmosphere and heat it through air heaters before it is used for combustion. The hot air duct pressure normally decreases at a higher load and it becomes difficult to supply adequate air flow by the downstream control elements. This pressure is maintained at its desired value so that the downward SADs can control the air flow smoothly. The air pressure in the hot air duct is measured with a redundancy and voting circuit to form the process variable. The steam flow signal as generated from Fig. 8.27 for the drum level control is considered the load index and characterized to act as the desired pressure set value. Necessary arrangements are done to take care of increasing the demand signal in case only one fan is on and running on auto mode when the running fan would take more load, within its capacity, than that of the load shared by two fans.

3.3.3 Alternative Air Flow Control Concept

Boilers with tangential tilt (TT) burners (CE DESIGN) and independent FD fans and PA fans using cold PA systems. In Fig. 3.35, it is seen that the secondary/combustion air from the FD fan gets heated at the air preheater (there may be a SCAPH before APH), and goes directly to the wind box of the furnace and then to the combustion chamber. There are a number of FCEs in the path. All the FCEs located in the wind box are responsible for air distribution at different elevations of the boiler, whereas the FCE associated with the FD fans is responsible for controlling the total quantity of combustion air (CA)/secondary air (SA) for the furnace. As stated earlier, even though these loops are responsible for the SA/CA controls, they do so after computing the total air flow (i.e., including PA flow) so that at any condition, the stoichiometric ratio is maintained. For a typical TT boiler, the cross-section of any corner of the wind box is what is shown in Fig. 3.33. Here, it is clear that double-lettered elevations will have auxiliary air dampers, and some of them will have a fuel air damper, whereas all single-lettered elevations have a fuel air damper. A short list of the FCEs is presented in Table 8.1.

3.3.3.1 Secondary Air Damper Control

The SADC is a part of the furnace safeguard and supervisory (FSSS) system. The basic purpose of this control is to distribute the total quantity of SA among the various elevations so to ensure proper combustion. As stated earlier, these dampers can be categorized as:

- Fuel air damper—Those dampers that regulate the air surrounding a fuel compartment.

TABLE 8.1 FCEs for Combustion Air

FCE Type	FCE Location	Control	Remarks
Fan impeller *OR*	FD fan	To *control the total quantity of SA/CA* for the boiler. Main modulating control loop and part of combustion control	*Normally* Pneumatic Power Cylinder to regulate flow by restriction
Fan inlet damper *OR*			
Speed control by VFD (Associated with FD Fan motor)			Electrical type—to regulate flow by speed control of the fan, more energy efficient
Auxiliary air (AA) damper	Wind box at double-lettered elevations	To maintain furnace wind box differential pressure (DP) so as to regulate and *distribute* SA/CA to the corresponding elevation during operation	Furnace wind box DP control loop in secondary air damper control (SADC) system—a part of FSSS
Fuel air (FA) damper	Wind box at single-lettered elevations (as well as in oil elevations, when oil firing is taking place, AA becomes FA)	To regulate the damper opening in proportion to the feeder speed or coal flow to the corresponding elevation so as to regulate and *distribute* SA/CA to the corresponding elevation during operation	Fuel air damper control loop (proportionate to coal flow) in secondary air damper control (SADC) system—a part of FSSS

- Auxiliary air damper—Those dampers that regulate the air adjustment to the fuel compartment.

The main philosophy of control is that auxiliary air dampers in all elevations in a group are controlled in proportion to the fixed differential pressure between the wind box and the furnace, whereas the fuel air dampers of the four corners of each elevation are group-controlled proportionally to the rate of fuel fired in that elevation. During purging and at a lower load up to 30%, AA is regulated to maintain the wind box/furnace DP around 40–50 mm wcl. There is an exception: double-lettered elevations that have an oil burner (say AB, CD, EF, GH). When oil firing takes place at that elevation, the corresponding elevation damper will open at a fixed opening; otherwise they act as an AA. Normally, all fuel air dampers (FA) in single-lettered elevations are closed, but open after a small time delay (*tube mill will be different*) after the associated feeder starts, so as to ensure the availability of SA in the corresponding elevation.

3.3.3.2 Air Flow Control Loop (Figs. 8.7 and 8.8)

1. Discussions: As stated earlier, the alternative discussed now is a cold PA system, that is, the PA does not have its suction from the SA/CA as was the case discussed earlier. Therefore, to compute the total air flow, the PA fan flow discussed later needs to be added to the SA/CA flow.
2. Normally (as shown in Fig. 3.35) there are two independent lines from FD fans that go to the wind box, and the air flow is measured in these lines after air preheaters. Aerofoils, Pitot tube or Piccolo tube flow elements may be deployed. DP transmitters in two of three (as shown) or one of two redundancies are required, as these are very vital parameters.
3. Because these are hot SA/CAs, temperature measurements for density compensation are essential. Here also, the redundancy is shown in Fig. 8.7.
4. It is recommended that, in case of redundancy for both parameters, compensation should be done with two of three (one of two) output signals so that a better result will be obtainable.
5. Because FD fan pressure is low and there is not much chance of variation, pressure compensation has been avoided, s shown in Fig. 8.7.
6. Both side-computed SA/CA flows are added to arrive at the total SA/CA computation.
7. As discussed in clause no. 3.3.3.1.1, the PA flow is also added, as shown in Fig. 8.7. For a bowl mill, the individual mill PA flow computation, or for the Ball and tube mills, each side PA flow of each mill computation is done (to suit the possibilities of individual side operation of a mill) and added to constitute the total air flow to the furnace.
8. The exact quantity of total air flow corresponding to fuel is necessary to ensure optimum combustion (correct stoichiometric ratio maintained). Because incomplete combustion may give rise to CO accumulation, which is very dangerous and can cause an explosion, it is customary to have little excess air (clause no. 3.2). For this reason, lead/lag circuits are used. Here, as shown in Fig. 8.8, the actual total fuel

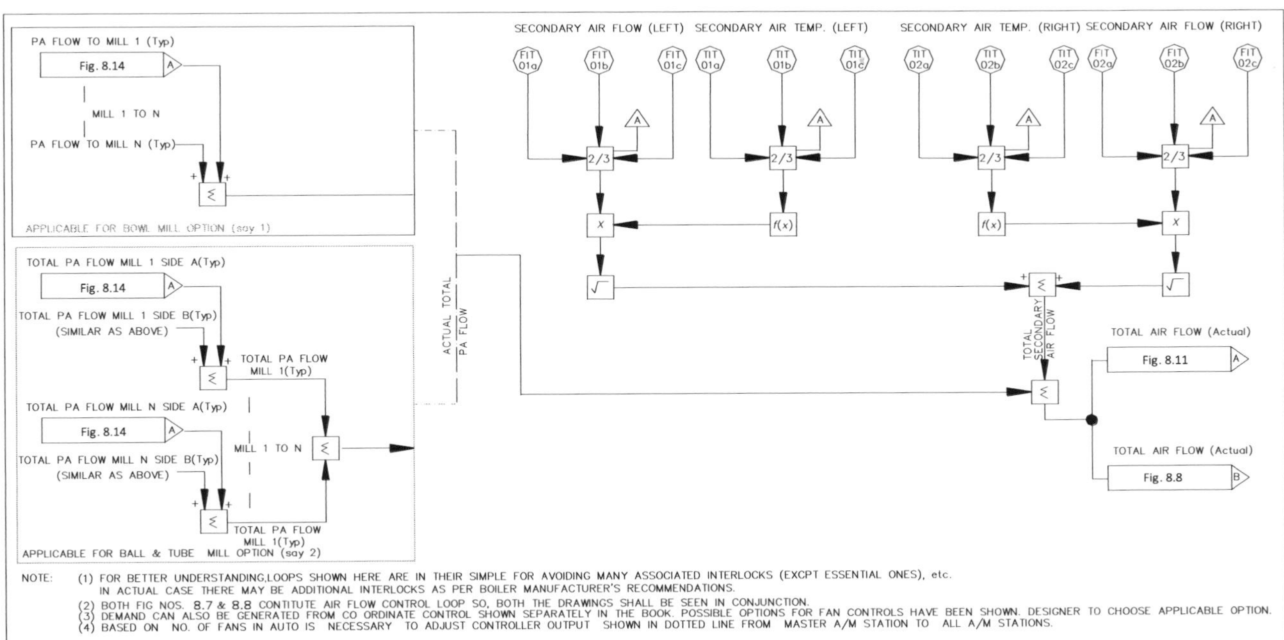

FIG. 8.7 Total air computation for secondary air flow control TT boiler.

flow is passed through a MAX selector where the air demand is also fed. When the fuel demand decreases, the air flow demand cannot decrease unless the actual fuel flow decreases (MAX circuit). Similarly, unless the air demand increases, the fuel flow cannot increase (MIN circuit as shown in Fig. 8.11).

9. In order to get the exact stoichiometric ratio as discussed in clause no. 3.2, the air demand is modified/trimmed with excess O_2 in the flue gas. There are two schools of thought: one believes this trimming should be done on the demand side (as done and as shown in Fig. 8.8) while the other thinks it is better to do the same in the actual measurement circuit, as discussed in clause no. 3.3.2.1. Now, corresponding to the operating load of the boiler, there will be a desired set value for O_2. So, the set point is given as a function of

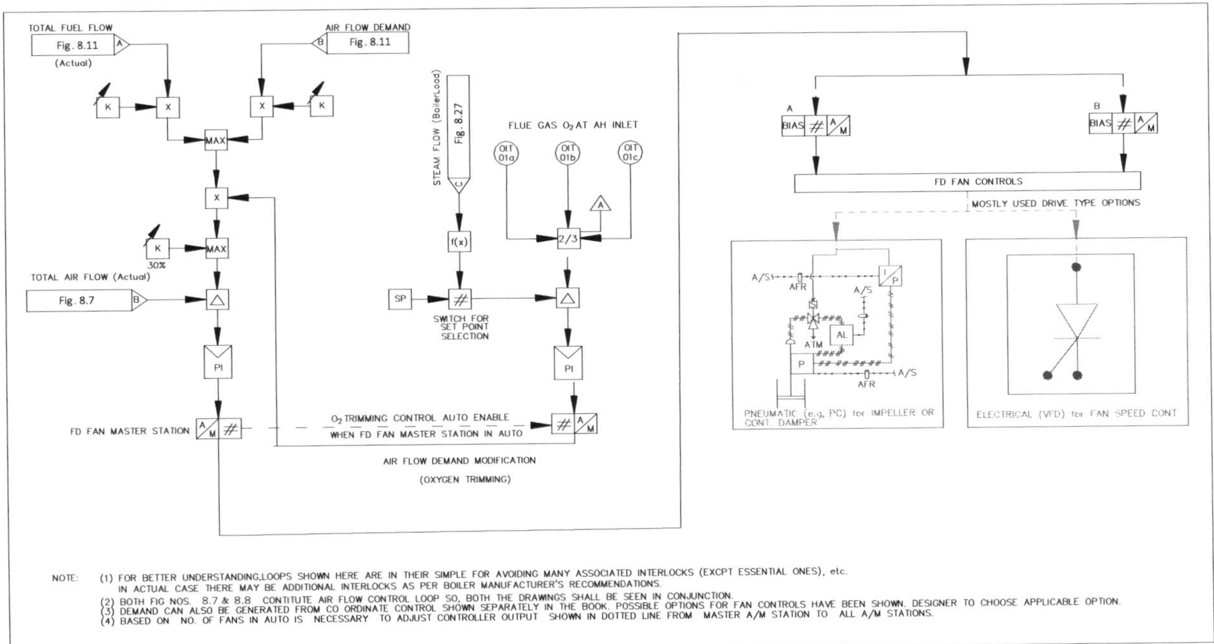

FIG. 8.8 Secondary air and oxygen trimming control TT boiler.

load, as shown in Fig. 8.8. At a lower load, this may be a fixed value, so two set points have been shown through a selector switch. The O_2 controller (PI) provides the trimming signal for demand modification. This modified demand signal is checked against the minimum value (30%) and allowed to enter the error generator.

10. Error generator output is fed to a PI controller to generate a demand signal for the FCE associated with the FD fans. There could be various options, as shown in Fig. 8.8 and given in Table 8.1.

4 FUEL FLOW CONTROL

General

A fuel flow control system is very important and a vital part of the combustion control system as air flow control system. On getting the boiler load demand signal as a command from the steam pressure controller with other influences, it generates the fuel flow demand after ensuring the necessary air flow is available for complete combustion of fuels to keep the steam pressure at its desired value.

4.1 Objective

The fuel flow control is provided to supply fuels in the exact quantity that would satisfy the demand of the fuel firing rate in close coordination with the air flow control system to ensure the *stoichiometric ratio* is maintained with some excess air flow throughout the combustion process. This control loop guarantees that the fuel flow must decrease prior to the air flow for a decreasing load demand, and the air flow must increase prior to the fuel flow for an increasing load demand so that an air-rich condition always prevails for safe and complete combustion.

4.2 Discussion

Air flow control is always associated with fuel flow control for various reasons. The increasing cost of fuels has ever been the center of consideration as the fuel quantity is always affected by incomplete combustion or too much excess air. Nevertheless, the percentage of carbon monoxide in the flue gas at the chimney outlet is also a serious matter of concern. The local/state pollution control boards are empowered to stop any (power) plant on allegations of polluting the environment.

Irrespective of mill type, the ultimate products, that is, powdered/pulverized fuel (PF), is transported by blowing pressurized and hot air called hot primary air through the mills up to the furnace. Separate fans performing this transportation duty are called primary air fans (PAF).

Manufacturers, plant design, and the layout of the steam generator unit decide the number of PAFs. There may be two or three centralized PAFs forming a common header catering to each mill, whereas the alternative design envisages an individual/dedicated fan for each mill. The control loop (Fig. 8.10) to be discussed here is with three centralized PA fans, but the control loop philosophy would remain almost the same with some minor modifications.

The suction of PA fans may be directly from the atmosphere or from the FD fans common discharge header. It is sent to a common trisector/separate primary air heater (AH), which is called the *cold PA system*. In the cold PA system, the PA discharge header is common and the major part of the PA fan discharge is sent to the PA heater for gaining heat. The rest acts as the cooling medium for controlling the PA temperature at a specific designed value with mill-wise hot and cold air control dampers.

When the PA fan suctions are connected to the common FD fans discharge header after the bisector AH, it is termed a *hot PA system*. In a hot PA system, a cold PA common header is formed before the AH.

Mill-wise or common PA fans are provided to transport fuel at a controlled temperature through the hot and cold air dampers (CADs). The large fossil-fired power stations mainly use coal or lignite as the main fuel. heavy fuel oil (LSHS) or light diesel oil (LDO) is also used as a supporting fuel. At the time of initial start-up, LDO is used to start the firing. However, in many plants, the use of LDO has been eliminated by high energy arc (HEA) igniters, even at the time of start-up, the HFO/LSHS can be used. Poor-quality or water-soaked (in the rainy season) PF also prompt the operator for oil support.

For heavy fuel oil (HFO) or an LSHS-fired boiler, it is necessary to keep these oils at a higher than ambient temperature to keep the viscosity at an acceptable limit. For this purpose, steam tracing or electrical heating is required for the supply pipe line up to the furnace ring header or the front/rear header and the return oil line also down to the pump suction.

4.3 Control Loop Description (Fig. 8.9)

The control loop may vary to some extent with the manufacturers' special design and fuel type, but the basic idea would remain the same; related control loops are presented for TT type boilers also. The control loop for ball tube mills is discussed in a separate section.

4.3.1 Measurement of Different Parameters (Fig. 8.9)

4.3.1.1 Primary Air Flow

The accurate PA flow measurement is very important, as it is utilized as the demand for coal/pulverized fuel (PF) flow in the control loop under consideration. Different types of flow elements are also used for measuring the PA flow, which is normally dictated by the boiler manufacturer as per their standard scope of supply.

Sufficient redundancy is provided (typically shown as triple redundant); after the necessary voting circuit, the output is density-compensated through temperature transmitters with a necessary redundancy and voting circuit and square rooted to provide a linear signal; pressure compensation is not required in this case. A computed signal is also utilized for air flow/SADC (Fig. 8.4) and for formation of the controller set point of the feeder speed controller.

4.3.1.2 Measurement of Feeder Speed

Accurate feeder speed signal being the measured/process variables is an essential part of fuel flow control strategy as this signal would enable the boiler load control to ensure the much needed supply of fuel flow. The bed height remains almost constant (by some external arrangement) and hence the feeder speed is equivalent to the quantity of material on the feeder. This measured signal, after the necessary redundancy (typically shown as triple redundant) and voting circuit, is sent to the speed controller.

4.3.1.3 Measurement of Differential Pressure Across the Mill

This measurement with sufficient redundancy and voting circuit is utilized as a means of feed forward signal and utilized as a trimming factor for feeder speed demand signal (Babcock design).

4.3.1.4 Measurement of Oil Consumption

One flow element in each supply and return oil line is provided. The difference between the two flow elements' signals is the actual oil consumption through the oil burners. As the requirement is for a short time, redundant transmitters have not been envisaged.

4.3.1.5 Measurement of Primary Air Pressure

This parameter with sufficient redundant measurements and a voting circuit is utilized as a measured/process variable for the controller.

4.3.1.6 Generation of Secondary Air Flow Demand

All the feeder speeds representing PF flows are summated to get the total solid fuels being fired; the actual oil flow is added to the PF flow with proper adjustment toward the calorific value to reach the actual heat input. This summated value is auctioned with air flow demand (Fig. 8.4) in a maximum selector output of which constitutes the all-important signal for the secondary air flow demand. As there are individual sets of SAD (Fig. 8.4) for each mill group, this same demand signal is utilized for all the mills.

4.3.2 Different Controllers

4.3.2.1 Primary Air Flow Controller

This controller takes care of the PA flow according to the PA flow demand from the master pressure controller (Fig. 8.2) in relation to the boiler load demand. This signal being the set value, the actual density-corrected PA flow acts as the measured/process variable. The PA dampers are provided for each mill group and are modulated as per the individual controller output (adaptive controller or P+I+D tuning facility).

4.3.2.2 Feeder Speed Controller

The feeder speed control loop (Fig. 8.9) is provided to vary the feeder speed, normally through the variable frequency drive (VFD) or any other suitable drive to meet the modified fuel demand. The fuel demand or set point is derived from the characterized PA flow. The primary air being the transporting medium for the solid fuel, the two flow signals have a particular relation or function of proportionality for a particular type of fuel. This relationship is exploited in the algorithm form where the PA flow constitutes the input and the output represents the equivalent feeder speed set point for the feeder speed controller. The actual feeder speed plays the role of a measured/process variable and the controller output (with P+I+D tuning facility) is further trimmed with a signal, which may be termed a feed forward signal.

Whenever the load increases, the air flow (both secondary and primary) has to increase first, followed by the fuel flow. If there is a delay between the two actions or a substantial difference, the DP across the mill would become low compared to the PA flow. This difference in values is obtained through a subtraction function and acts as the above-mentioned trimming signal of the controller output. This would enable increasing the feeder speed in advance to avoid a significant mismatch in complying with the actual fuel requirement. For a decreasing load, the DP across the mill would become high compared to the PA flow and would force a lowering of the feeder speed prior to the decreasing command coming out directly from the controller.

4.3.2.3 Primary Air Header Pressure Control (Fig. 8.10)

This control system applies to the common PA system and the pressure controller gets a fixed or variable (to load) set point selection via the selector switch, as shown in Fig. 8.10. For variable set points, there are some options available for the designer:

- In case of fall and tube mills, the maximum position of the damper of the PA flow through the mill out of both sides of all mills (i.e., the maximum coal flow condition) may be taken as the set point so that the PA fan regulates itself to cater to the maximum demand.

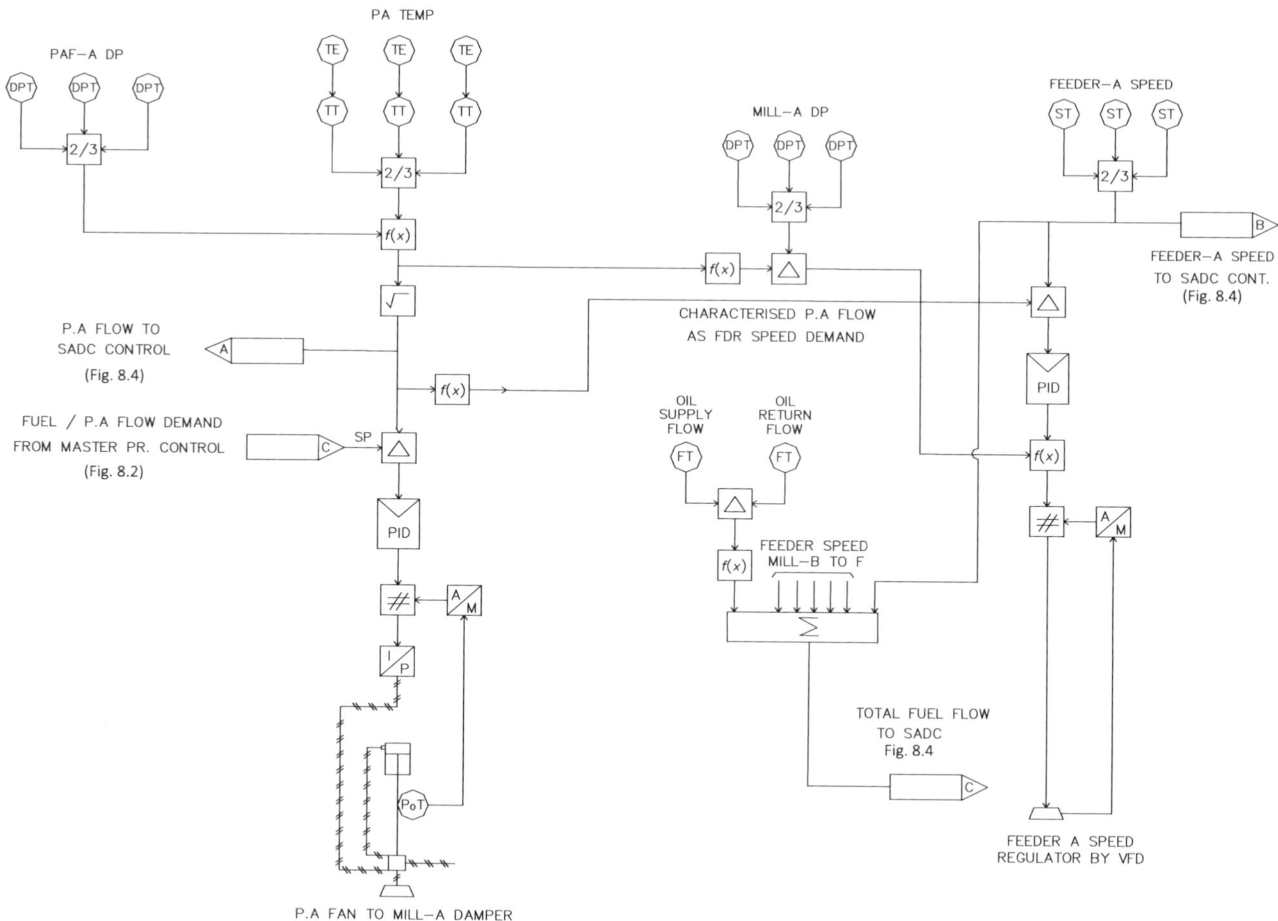

FIG. 8.9 Fuel flow control.

- Highest feeder speed condition for a bowl mill is mainly used in the TT burner boiler with a similar philosophy as discussed above.
- In some other cases, a base set point is added with the boiler load (the boiler load possible options are shown in Fig. 8.10) used as the set point.

The set point is compared with the actual PA header pressure to generate an error signal, which is fed to a P+I controller. The output of the controller regulates the working fan capacity by changing the fan speed through VFD or regulating the inlet vane/suction damper. Speed control is more efficient for obvious reasons such as lower loss. Controller output goes to the master A/M station so as to facilitate common manual operation. The final demand at the output of the A/M station (for both auto as well as common manual) is adjusted for gain depending on the number of PA fans in auto.

4.3.2.4 Coordination between Air Flow and Fuel Flow Controls

These two control loops are extremely interrelated. It is well known that the combustion process in the furnace of steam generators requires oxygen, which comes from the air in the atmosphere. To avoid incomplete combustion and inefficient operation of the plant, the air flow rate is very important, needing a close watch on its measurement and control along with the fuel flow.

The boiler load controller gets its set value from the main steam pressure at an appropriate point near the turbine inlet. The controller output gets properly trimmed from the feed forward signal from the coordinated controller output. The trimmed signal (Fig. 8.2) is then regarded as *demand for the secondary air flow*, provided this *boiler load demand is more than the fuel flow* (Fig. 8.9). It is then sent to each mill group for SA controls. This signal representing the total air flow, the PA flow is subtracted to get the mill-wise SA demand only, and the subsequent controller ensures that the SA dampers are acting accordingly.

The demand for PA flow, on the other hand, is generated from the above trimmed boiler load (from the master pressure control) signal and allowed to proceed further if this *boiler load demand is less than the total air flow* signal. This signal is then treated as the mill-wise PA flow set point. After the PA flow controller takes care of the PA flow, the feeder speed controller gets its set point from the characterized PA flow and regulates the fuel flow through the speed-changing device.

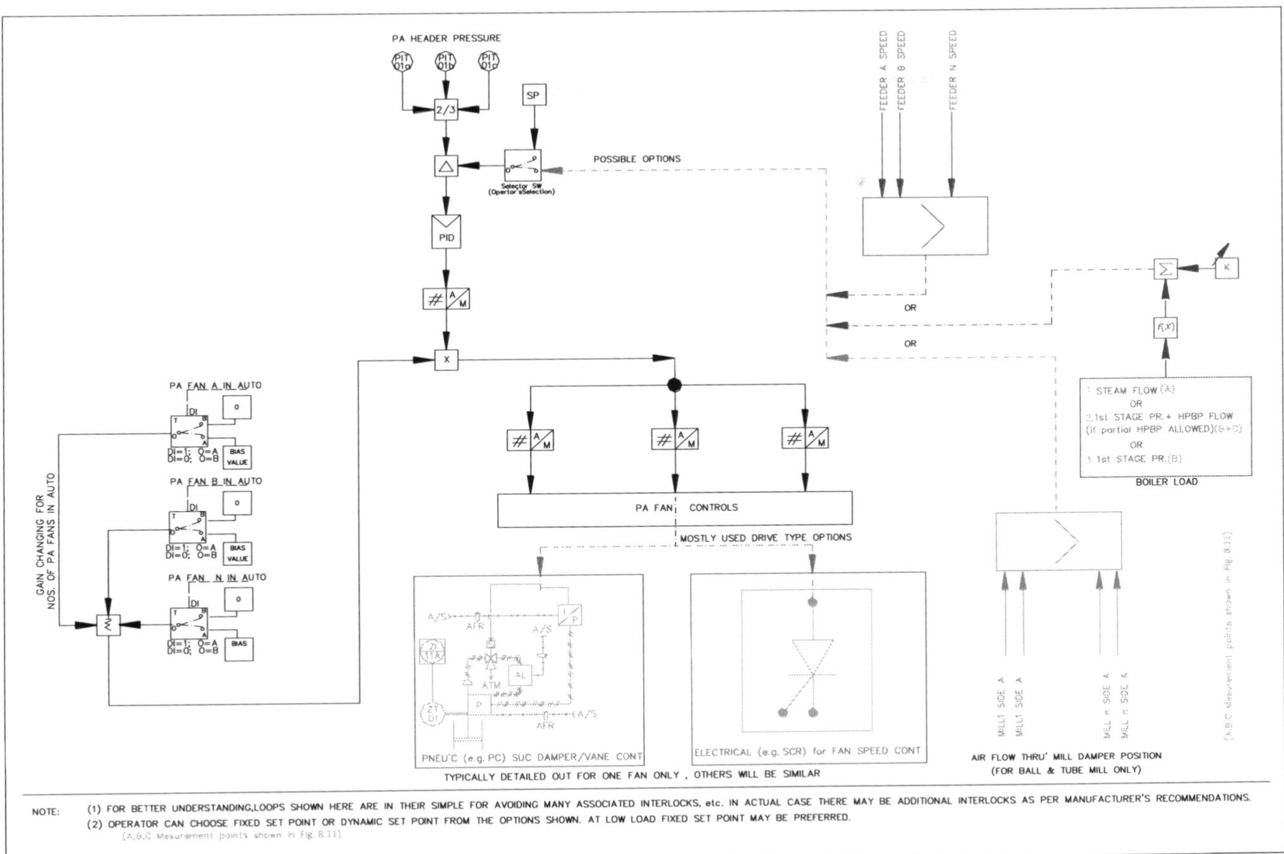

FIG. 8.10 PA header pressure (fuel flow control).

From the above discussion about the generation of demand for air (SA) flow and fuel flow (via PA flow), it may be noticed that whenever the *boiler load increases, the air (SA) flow increases first and then the PA/fuel flow increases* whereas in a *decreasing load, the PA/fuel flow decreases first and then the air (SA) flow decreases*. This unique feature is called the lead-lag relationship between the air and fuel flow. This particular idea is maintained throughout so as to prevent unsafe boiler operation and happens to be the essence of the control philosophy.

4.4 Fuel Flow Controls for Tangential Tilt Burner Boilers

Boilers with tangential tilt burners with bowl/ball and tube mills are discussed briefly, though the basic objective is the same as other boiler designs.

4.4.1 TT Boiler Master Demand Control (Fig. 8.11)

MS pressure is the determining factor for the generation of fuel demand signal in conventional loops without a coordinate control concept. For plants having coordinate control, the demand signal is generated in the coordinate control system, which is dealt with separately in Chapter 10 of this book. For direct control, a master pressure signal is generated from the turbine inlet pressure transmitters.

The manual (fixed) set point or sliding set point (common in an OT Boiler) is compared with the measured pressure signal obtained from the two of three redundant circuits, as shown in Fig. 8.11. Normally, a sliding pressure set point slides/changes with the load from >60% load (below which the auto loop is practically ineffective); to be precise, between 70% and 90% mainly. Beyond this range, the set point is fixed. In a load regime >90%, the pressure is kept fixed as a cushion to meet the eventualities. During sliding pressure operation, governor valves are kept wide open (less pressure drop across the control valve) so that the turbine inlet pressure slides as per load.

Supercritical/ultrasupercritical plants with OT boilers may go for a sliding pressure operation with the possibility of saving some energy. Thus, both options are shown in the drawing. The difference between the set point and measured value creates the error signal for the PI controller to generate the fuel demand signal. With the inertia of the boiler being higher in comparison to the system demand, the system load is taken as the feed forward. Because derivative signals are basically a source of high-frequency noise, the loop requires

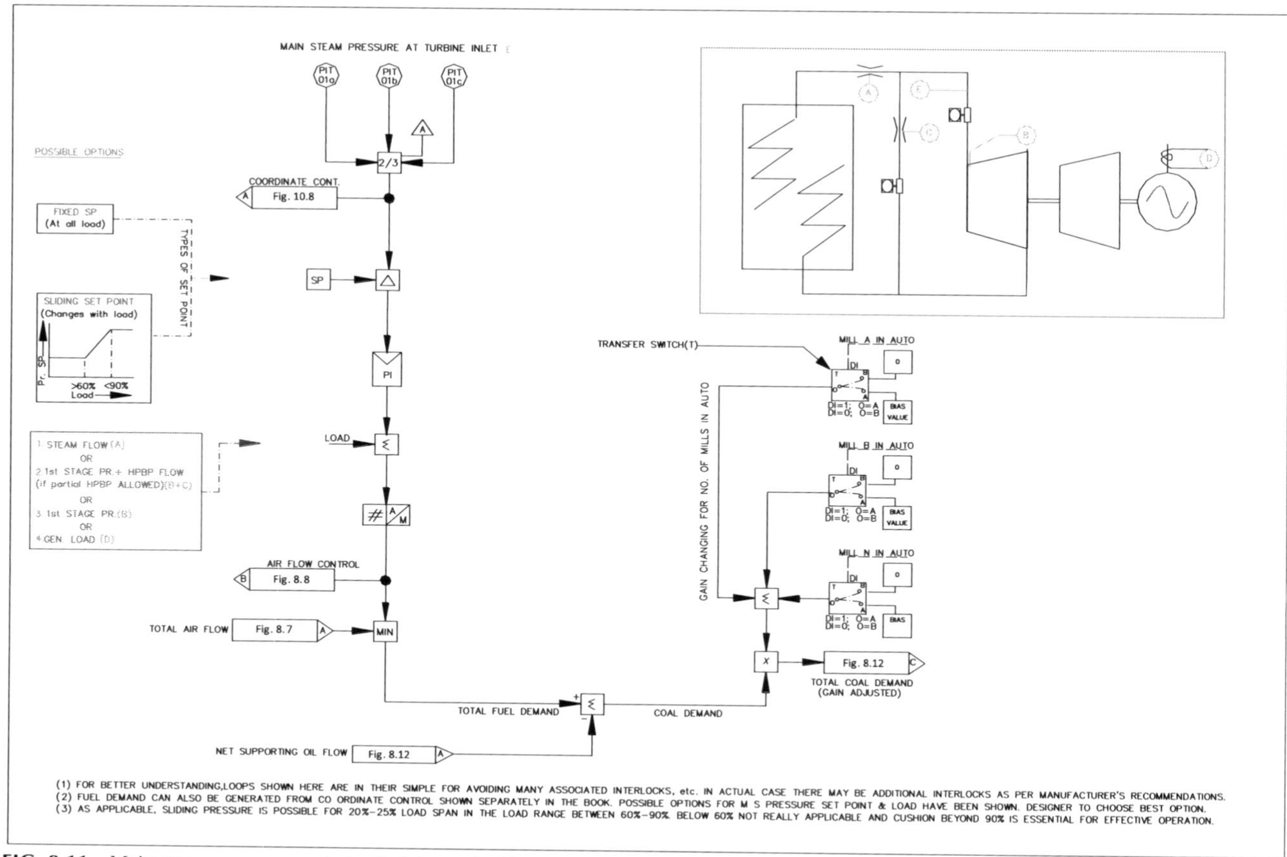

FIG. 8.11 Main steam pressure control—fuel demand—Tangential tilt boiler.

proper tuning with the boiler load as the feed forward is added to the demand from the master pressure controller to make the loop more responsive. A few options available to use as the load index are listed below:

- Steam flow (in larger boilers, steam flow may not be measured to avoid loss due to a permanent pressure drop).
- First-stage pressure/wheel chamber pressure, which is proportional to turbine load can be considered (with HP bypass flow where a partial bypass is envisaged in the system). In some places, a direct generator load is considered in place of first-stage pressure.

As discussed earlier, for actual fuel demand signal generation, it is necessary cross limiting fuel demand signal with Actual air flow, so that under all situations, air enriched condition prevails (i.e., for increasing load, air would increase first followed by fuel; for decreasing load, fuel is reduced first before air flow decreases). This is ensured by the MIN selector and the actual air flow. The fuel demand signal thus generated needs to gain correction for the number of mills in operation. The location of each of the measurement points is shown in Fig. 8.11. Generally, the fuel oil flow in the boiler is controlled by the oil pressure control, discussed separately. Out of the total fuel flow, there may be some contribution from the oil flow also. Oil flow measurements are shown in Fig. 8.12. The difference between the supply line flow and the return line flow of HFO gives the net HFO flow; this flow along with the LDO flow (insignificant in reality) with the necessary calorific value correction computes the total oil support of the fuel. This is subtracted from the total fuel demand to compute the net coal demand in Fig. 8.11.

4.4.2 TT Boiler Fuel (Coal) Control Loop for Bowl Mill (Fig. 8.12)

In bowl mills, which have very little holding capacity, the coal flow is computed by the feeder speed measured with the help of speed transmitters. To avoid a spurious signal, the feeder speed signal is passed through a low pass filter (LPF). A feeder speed signal is taken as valid only when the FSSS interlock permits, so that the preparatory stage flow signal is ignored. All feeder speed signals are summated and then compared with the net coal demand (Fig. 8.11) to generate the error signal for the PI controller whose output becomes the total coal demand signal at that instant. Immediately after the controller, the feeder master

A/M station provides the facility for common manual operation and the output (for both auto as well as common manual) is adjusted for gain depending on the numbers of feeders in auto, as shown in Fig. 8.12. Each of the feeder controls shall have its own A/M station to facilitate individual manual operation as well as adding bias.

The output of each of these A/M station drives the feeder speed variator with the help of a pneumatic power cylinder or may drive the SCR controller for the gravimetric feeder. Both options are shown in Fig. 8.12.

4.4.3 TT Boiler Fuel (Coal) Control Loop for Ball and Tube Mill (Figs. 8.13 and 8.14)

In order to understand the loop for this type of loop, it is better to understand the operation and flow diagram for this type of mill, as discussed in clause no. 5.3 of this chapter. In fact, discussions on this loop are more relevant in clause no. 5.3, but are discussed here so that the reader can get options for various kinds of fuel flow control loops. A short discussion on the operation principle of the mill is also included in clause no. 5.3.

Its worth noting that in this kind of mill, for proper performance of the mill, the coal-air ratio is very important—a typical such value could be 1.4–1.5. Typically, for a 100 t/h mill, the diameter is around >4.5 m and length is ~=7.2 m. This large-volume, slow-speed mill normally feeds two elevations in the TT boiler. Normally, a level inside the mill is maintained by controlling the feeder speed. It is to be noted that in this mill, instead of feeder speed, the PA flow is regulated to control the coal flow through the mill.

Now, the net total coal demand generated in Fig. 8.11 has to be balanced with the actual coal flow. For ball and tube mill feeder speed controls, a desired coal level in the mill and not coal flow directly. In the furnace, coal burns to release energy for the boiler so that steam is produced. Thus, the energy released by the boiler at any instant of time is proportional to the steam flow and to the total fuel/coal flow (ignoring HFO; it actually acts as supporting fuel for flame stabilizing). Therefore, the boiler load can be considered proportional to the actual coal flow. The boiler load can be any one from:

1. Steam flow from boiler.
2. Turbine wheel chamber pressure or generator load. However, in a plant with partial HPBP operation, the steam flow in the HPBP line needs to be summed up with any of the above parameters for boiler load computation.

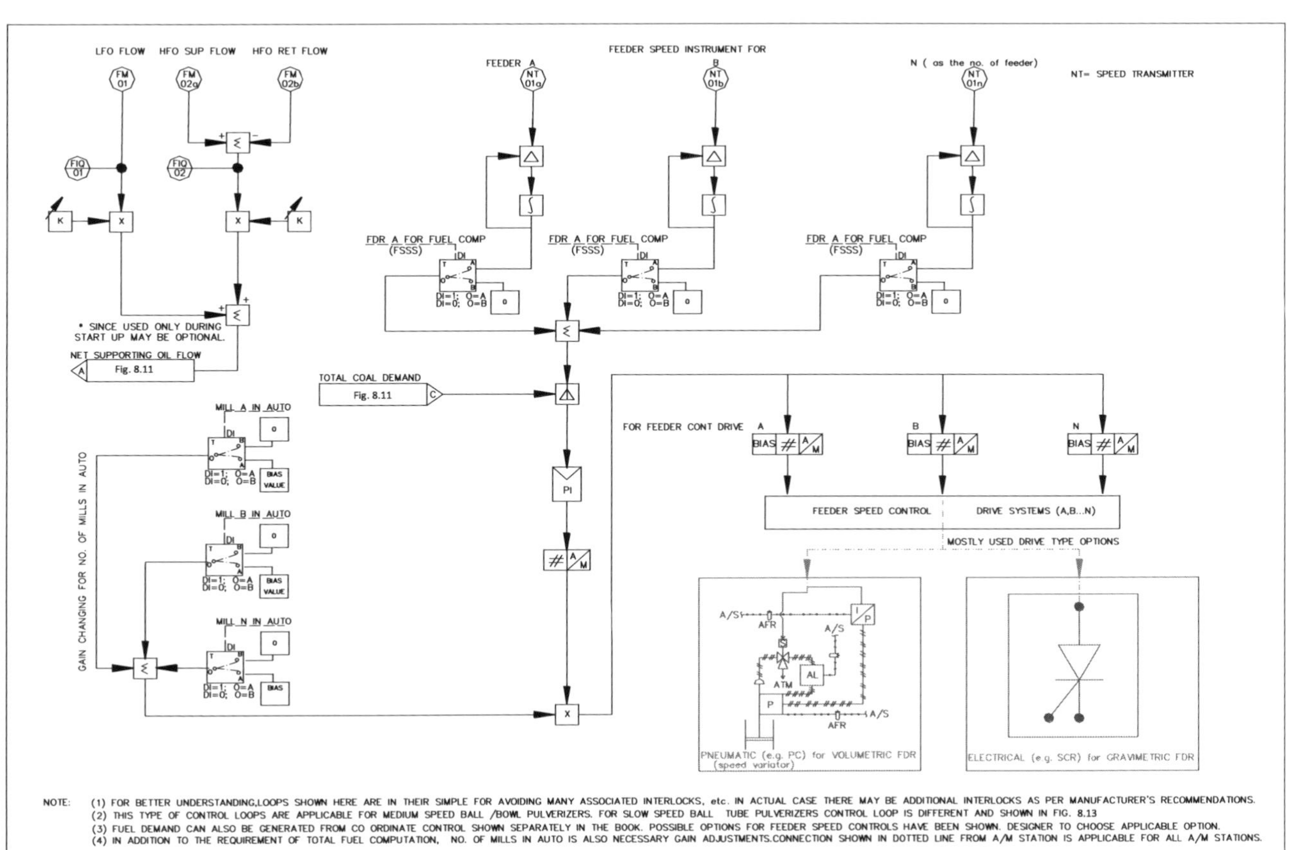

FIG. 8.12 Fuel flow control—tangential tilt boiler (bowl mill).

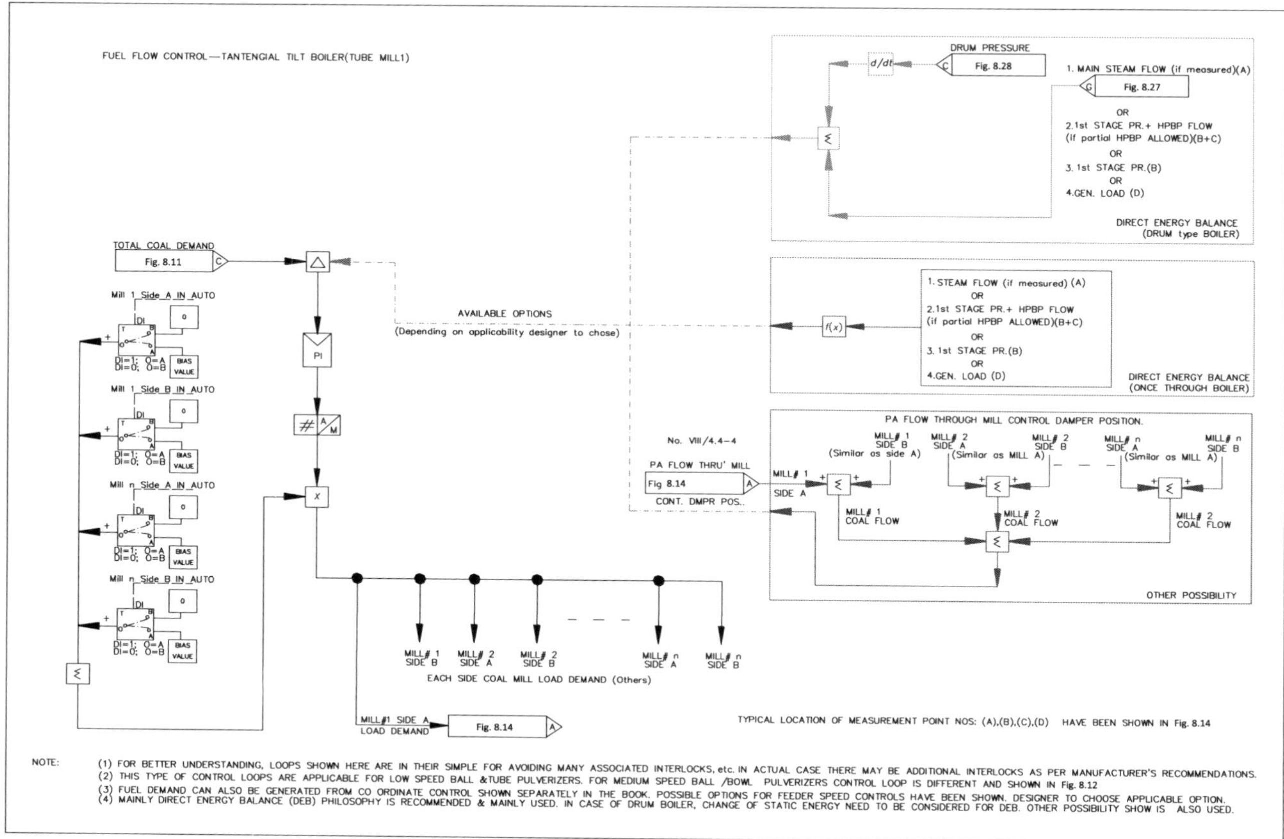

FIG. 8.13 Fuel flow control—tangential tilt boiler (tube mill).

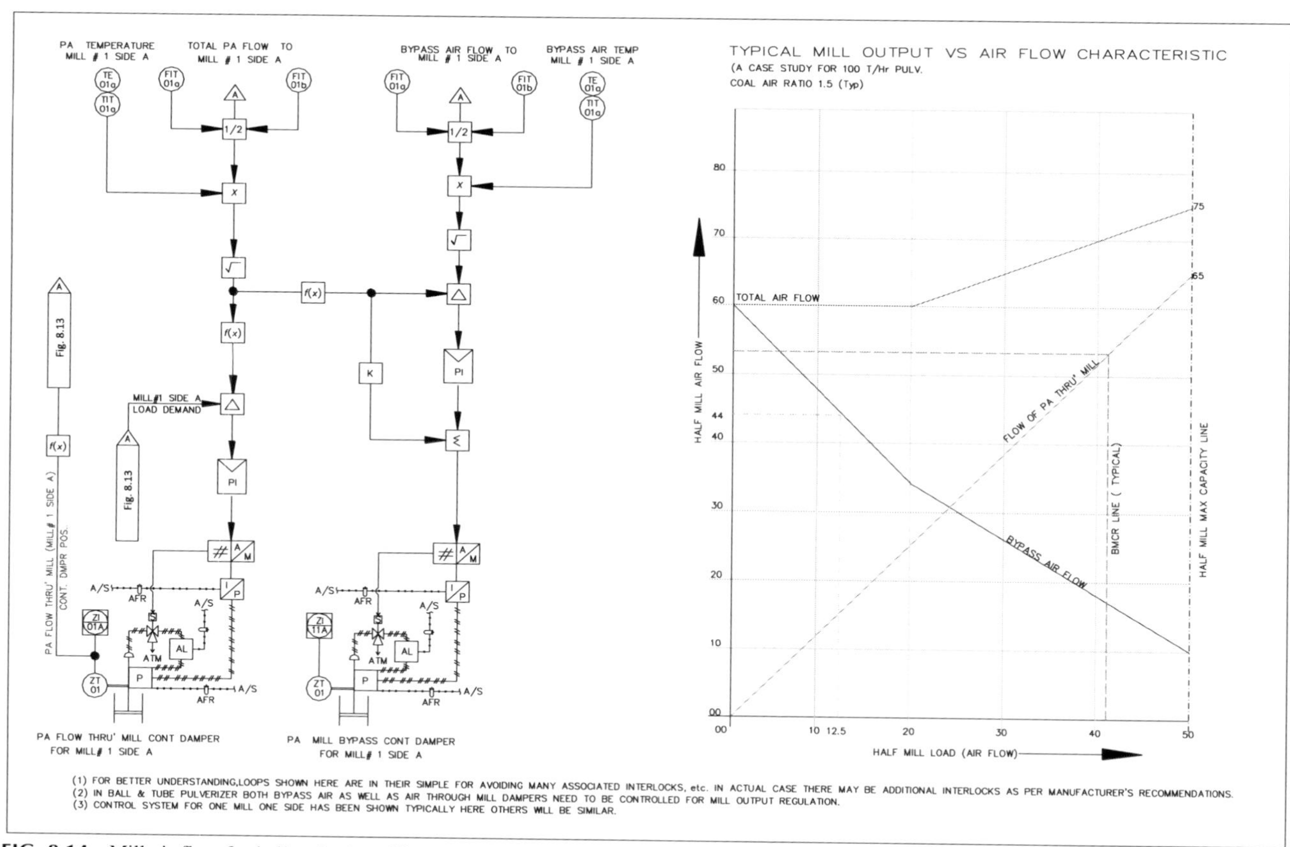

FIG. 8.14 Mill air flow for ball and tube mill.

All the parameters just listed are very true as far as the kinetic energy for OT boilers. For drum boilers, the drum acts as a source of static energy, and for any change in energy release from the boiler, this also participates in releasing energy. Therefore, the rate of change of the drum pressure needs to be summed up with the parameters discussed in 1 and 2 above. These are very distinctly shown in Fig. 8.13. For more precision, the fuel computation from the energy released signals and the energy supplied by the supporting fuel is subtracted from the total energy released.

As discussed earlier, in each side of the mill, the PA flow through the mill changes the coal flow and is functionally related to the coal flow in that side. The sum total of these signals forms a signal for total coal flow, as shown in Figs. 8.13 and 8.14.

The total coal flow, thus computed, is compared with the net coal demand to generate the error signal for the PI controller. Immediately after the controller, the master A/M station provides the facility of common manual operation also. The final coal demand at the output of the A/M station (for both auto as well as common manual) is adjusted for gain depending on the numbers of sides of mills in auto. Thus, the load demand (in other words, the position demand for the damper in the PA flow through the mill) for each side of each mill is generated, as shown in Fig. 8.13. It then goes to the individual damper through its own A/M station to facilitate individual manual operation as well as adding bias. The output of each of these A/M stations drives the PA flow through the mill damper with the help of a pneumatic power cylinder to attend the desired position.

4.4.4 Mill PA Flow For Ball and Tube Mill (Figs. 8.13 and 8.14)

PA flow controls are discussed in subsequent sections; however, to have a better idea, the PA flow loop for ball and tube mills in TT boilers are included here. The total PA flow and the bypass PA flow to each side of each mill are measured. The total PA flow and the bypass air flow to each mill have characterized variations as per mill load (each side). These are shown graphically in Fig. 8.14. Initially, each side bypass air flow is kept higher to dry and then take away the coal dust from the incoming coal and finally to mix with the PA flow through the mill to carry the coal to the boiler. Also, during the initial period, bypass air flow helps in gaining the velocity. As the PA flow through the mill increases with the mill load, the bypass air flow decreases as velocity boosting and heating is achieved by the PA flow through the mill. Normally, even at the highest load, the bypass air flow is kept open to, say, 10% so as to drive out the coal dust from the mixing box as well as have some drying effect on the incoming coal.

The PA flow at each side of the mill is measured and characterized for comparison with the mill load signal generated in Fig. 8.13. The error thus generated is fed to the PI controller PA FLOW THROUGH MILL damper position so that regulated quantity of coal goes out to the boiler.

As stated earlier, the bypass air flow maintains a particular characteristic with the total air flow, so the characterized total air flow signal is used as a set point for the bypass air flow damper control. The set point is compared with the actual bypass air flow to drive the P+I controller to generate the demand for the bypass air flow damper position. To make the loop more responsive, the function of total air flow control is added to the controller output as a feed forward signal.

5 COAL MILL CONTROL—MILL AIR FLOW CONTROL (FOR TT BOILER)

General

5.1 General

There are two main modulating control systems associated with mills: mill air flow control and mill temperature control. Mill temperature control was covered in clause no. 5.2 of this chapter. Mill air flow controls in opposed-fired and front-fired boilers based on Babcock designs are well covered in Section 4 of this chapter. Therefore, the mill air flow controls for the TT boiler (bowl mill) only are discussed in this section. Mill air flow controls for tube mills are covered in clause no. 4.4.3 of clause no. 4 and clause no. 5.3 of this chapter. Other control loops associated with tube mills are covered in clause no. 5.3.

PA (sourced by PA fans) before and after APH forms cold/tempering air (CA) and hot PA (HA), respectively. One set of these air lines is mixed before entering each mill through regulating dampers called hot air dampers (HADs) for hot PA and CADs for cold PA. As the mill load is increased by changing the feeder speed, naturally the PA flow through the mill has to be increased so that more coal is flown to the furnace; otherwise, there may be jamming. The mill outlet temperature is maintained within a temperature band, so that the coal is dried enough to be ignited instantaneously in the furnace. At the same time, it is not too high so that there is a chance of fire. Exactly for these reasons, both dampers are used. In older days, the controls of bowl mill cross-operation of HAD and CAD were in use to regulate both the air flow and mill temperature control. This cross operation showed poor performance due to excessive cross interference, and hence the necessity of two independent loops was felt. As such, it is very difficult to maintain the desired temperature at the mill outlet during cold weather or the rainy season due to wet coal. In modern philosophy, for PA flow change due to load change is catered to by HAD if due to this by chance temperature increases then only CAD is modulated to regulate Mill outlet temperature. To be specific, the PA flow to the mill is regulated by HAD and the temperature is regulated by CAD independently with protection interlocks.

5.1.1 Objective

The objective is to ensure the required quantity of PA flow through the mill so that the exact quantity of coal demand from the particular mill can be catered to at all varying load conditions without any mill jamming. The relative dispositions of these dampers and the flow element are shown in Fig. III/9.2–4, PA flow P and ID. The entire quantity of PA demand for the mill initially is catered to by hot PA to avoid low temperature. However, during summer and dry seasons, the mill outlet temperature may rise. The CAD then opens to increase the cold air flow, causing the PA flow to increase than demand (set point) from feeder speed controller; PA flow controller then corrects its output to lower HAD opening. Cross operation of the two dampers takes place through the process, but not in the loop. However, as a precautionary measure, an interlock has been incorporated to see that when the CAD is in manual or the temperature is too high, the HAD loop goes to manual so as to avoid any chances of a high mill outlet temperature.

5.1.2 Description (Fig. 8.15)

The PA flow to the mill is measured after mixing the HA and CA with the help of a Ventury/Pitot/Piccolo tube, redundant smart DPTs, and suitable temperature compensation arrangements, as shown.

The final set value is taken after the "MAX" selector having one manual set input and another being derived proportional value from feeder speed so that PA flow never becomes less than that required for corresponding feeder speed. In fact, to clear out any mill jamming, the operator can set a value other than that required for the corresponding feeder speed. The error signal generated by the difference between the set and measured values is fed to the controller. At the output of the controller, the feeder speed demand has been added as the feed forward signal, so that in case of any load change, the PA flow can act fast. Finally, the output of the controller through the A/M station goes to the FCE of the HAD to regulate the same.

5.1.3 Alarm and Interlock

1. Alarm:
   - **a.** The PA flow to the mill low is generated through the LVM. A spurious alarm during start-up is prevented by ANDing it with HAD/CAD in auto.
   - **b.** When CAD is in manual or high mill outlet temperature, an alarm is generated.
2. Interlock:
   - **a.** When CAD is in manual or high mill outlet temperature, this trips the loop operation to manual.
   - **b.** If the high temperature persists over a time period after operator action, the opening command to HAD is blocked.
   - **c.** Released to auto needs permission from FSSS/BMS.
   - **d.** CAD and HAD not in auto blocks input and output to/from the controller to avoid saturation.

5.2 Mill Temperature Control

General

Mill temperature control is an essential part so far as coal or pulverized fuel combustion is concerned. This is a vital control in consideration with the following points:

- **(i)** The coal while getting out of the pulverizer must be dry to avoid moisture (due to weather and other seasonal changes) vis-a-vis clogging up the coal pipe lines leading toward the furnace chamber.
- **(ii)** Coal types normally used in power plants are required to be preheated to make them more easily combustible.
- **(iii)** The appropriate extent of coal drying calls for lower pulverizer start-up costs.

Considering the above, the only way out is to maintain a fixed set point at a higher value capable of providing a sufficient margin to handle spikes in surface moisture.

5.2.1 Objective

The purpose of the mill temperature control is to supply fuel at an optimum temperature. By optimum, this means it should be higher than to maintain flowability and less than to avoid fire hazards. For this purpose, the transportation media, that is, the primary air inlet temperature to the mill, is considered the heating agent with a sufficient higher value of temperature and quantity so as to raise the mill outlet temperature, the measured/process variable of this control strategy.

5.2.2 Discussion

The mill outlet temperature is maintained at a set value by introducing preheated primary air (PA) into the mill's inlet. The supply of the right quantity of PA is equally important so as to ensure sufficient heat input, which would be exchanged through direct mixing with the solid but pulverized fuel. For obvious reasons, the amount of air would ultimately be proportional to the mill load in relation to the fuel or total air/main steam flow, as per the control strategy.

For each mill, one set of hot PA dampers and another set of cold PA dampers are provided as the FCEs with various types of operations to suit plant design.

In the early days, the hot PA and cold PA dampers were regulated in the same direction when there was a change in PA flow demand; on the other hand, they were supposed to move in the opposite direction to respond to the change in mill outlet temperature. According to some other schools of thought, the feeder speed controller output, after getting the command from the boiler fuel demand, is transmitted to the coal feeder speed changing system and to the position

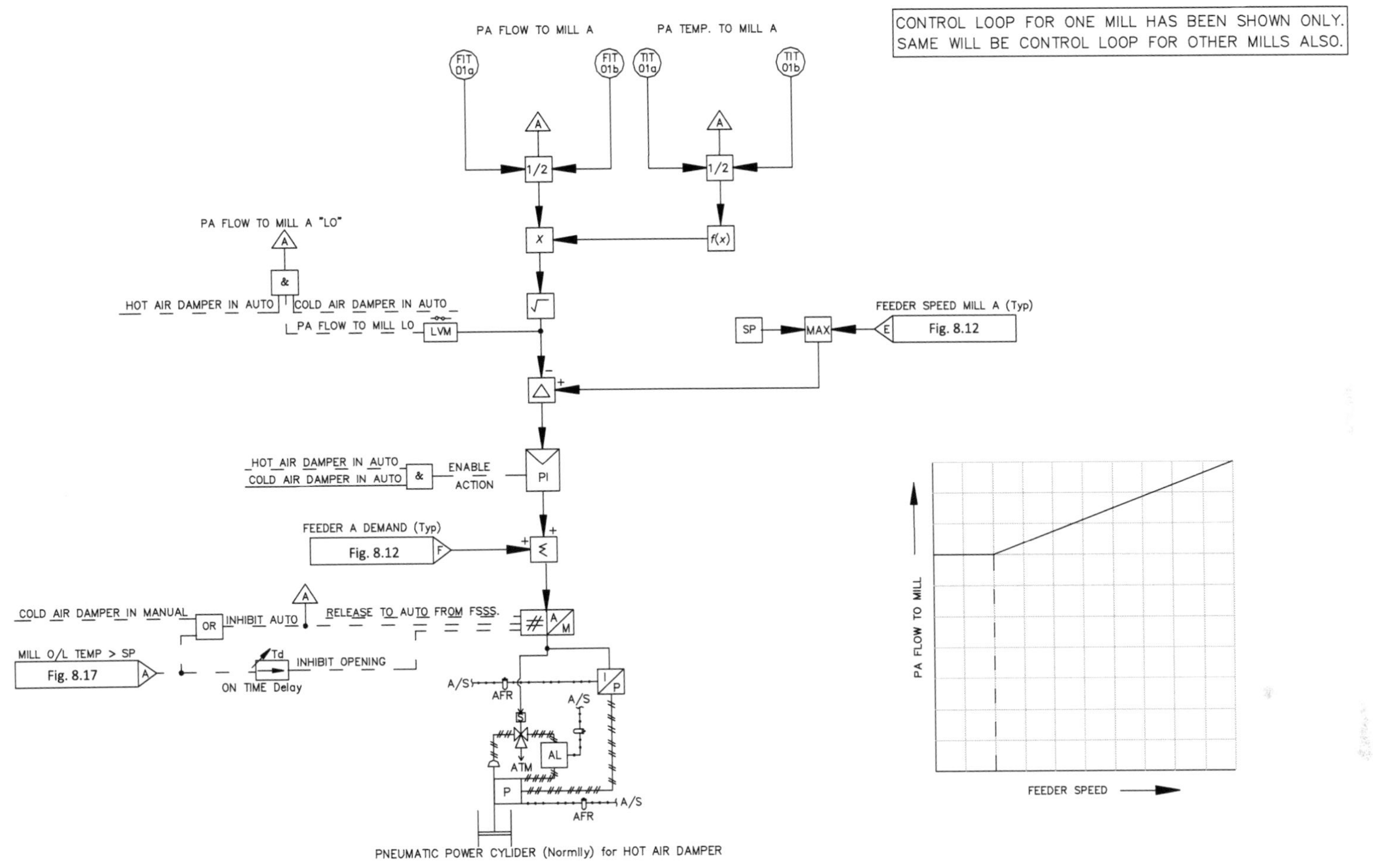

NOTE: (1) FOR BETTER UNDERSTANDING,LOOPS SHOWN HERE ARE IN THEIR SIMPLE FOR AVOIDING MANY ASSOCIATED INTERLOCKS, etc. IN ACTUAL CASE THERE MAY BE ADDITIONAL INTERLOCKS AS PER MANUFACTURER'S RECOMMENDATIONS.
(2) THIS TYPE OF CONTROL LOOPS ARE APPLICABLE FOR MEDIUM SPEED BOWL MILL .
(3) INTERNAL AUTO / MANUAL SIGNALS OF THE LOOP HAS NOT BEEN DETAILED OUT TO AVOID COMPLEXITY.
(4)TO GIVE AN IDEA TO THE READER, OPEN LOOP CONTROLS HAVE BEEN SHOWN IN DASHED LINE TO DISTINGUISH IT FROM MODULATING CONTROLS. THIS IS HOW IT COULD BE REPRESENTED IN REALITY FOR BETTER UNDERSATNDING

FIG. 8.15 Mill air flow control—for bowl mill TT boiler.

controller of the hot PA damper. This arrangement envisages simultaneous control actions in speed changing and repositioning of the HAD. The primary duty ascribed to this hot PA damper, according to this control strategy, is to maintain only the mill outlet temperature to a desired value and also to take care of the above feeder speed demand as a feed forward signal for damper position control. The function of the CAD position controller on the other hand is to control the total primary air flow to a value required to maintain the flow velocity for safe and smooth transport of the pulverized fuel to the furnace for combustion. The demand signal for the cold PA damper may be generated directly from the boiler load demand after the lead-lag route; the same signal also acts as a process trim on the feed forward control applied by the hot damper controller. There may be some modification where the HAD gets a feed forward signal from the PA flow controller also, and the CAD controller gets a feed forward signal from both the fuel and temperature controller.

Unfortunately, the control action was not so effective due to its sluggish response frittering away excessive time. This resulted in undesirable disturbances in the steam pressure whenever the boiler load or fuel quantity changes are called for.

The control loop considered for discussion in this section is of a typical 250 MW plant with a Babcock design where the PA flow to each mill is determined by a separate (refer to fuel flow control) damper after these two flow paths (hot and cold PA) are adjoined, that is, in the common line. The hot and cold PA damper set is dedicated to temperature control only. For no temperature change due to load change, these dampers would not respond to the new flow requirement. However, the PA fan inlet dampers would come into the picture to take care of the new load by maintaining the PA header pressure.

Hot and tempering air mixing requirements vary as the fuel moisture varies and as the primary air bias varies. A mill outlet temperature control system nowadays is so properly set up that its activity cannot even temporarily influence any impact in the PA flow.

5.2.3 *Control Loop Description (Fig. 8.16)*

The control loop philosophies are of various types as discussed in the above section and may vary with the manufacturer's special design and type of fuel, to some extent. As there is a separate section dedicated to the control loop for ball tube mills, this section will cover the control loop of other types of mills also.

5.2.3.1 Measurement of Mill Outlet Temperature

The temperature (measured variable) signals, after a sufficient (triple) redundancy and voting circuit, is transmitted to the controller.

5.2.3.2 Mill Outlet Temperature Control

Among the many control strategies to choose from, the selected loop as described in this subsection is a part of the total PA system that takes care of the temperature control, only without taking any notice of the other parameters and effects. As described elsewhere in this section, the mill-wise PA flow is hooked up with the fuel flow control by the PA to mill damper. Mill-wise hot air and CADs are provided to maintain the mill outlet temperature. The total impact of all the mill-wise PA flow and mill outlet temperature controls on the actual total PA flow is taken care of by the PAF inlet dampers by maintaining the PA header pressure. This control loop strategy is the most *noninteractive* between the PA flow and temperature, where both parameters are very important to assist the optimum transportation condition of the pulverized fuel.

The controller is set with a fixed desirable value for the mill outlet temperature and simultaneous action of HADs and CADs (CSAD) is contemplated in this control loop to maintain the temperature. The normal operating zone envisioned a full open condition of HAD and any temperature change is taken care of by the CAD. With HAD full open, the CAD modulates from the full closed position up to almost a 60% (typical value) opening. At 60% of the CAD position, if the mill outlet temperature increases, the controller output would call for a further opening signal to the CAD and simultaneously a decreasing signal to the HAD. This operating zone of simultaneous positioning of the two dampers is very limited. If the temperature still increases, there would be a certain point where the CAD would attain and remain in full open position and any further increase in temperature would be taken care of by the HAD. The situation would be just the reverse when the temperature starts decreasing to such a value when H the AD position would be increasing toward 60% (typical value) and then the CAD position starts decreasing. After a certain time, the HAD would again be at the full open condition while the CAD would modulate.

A further scope is envisioned in maintaining the mill outlet temperature if it goes beyond the capabilities of the above two dampers when the air temperature itself decreases at the outlet of the PA heater, due to low flue gas temperature resulting in poor heat exchange. The sole cause in that event being the flue gas heat content, some measures have been adopted to prevent it from falling across the heat exchangers, instead bypassing a part of the heat exchangers in the superheater/reheater section of the boiler. This is accomplished by providing another common temperature controller. The measured or process variables are the same mill outlet temperatures, but through a minimum selector and of those mills whose controllers are running on auto mode only. The set point is derived after putting a negative bias on the previously mentioned set point of the main mill outlet temperature control. This set point being lower, its controller output would not normally interfere with the main control

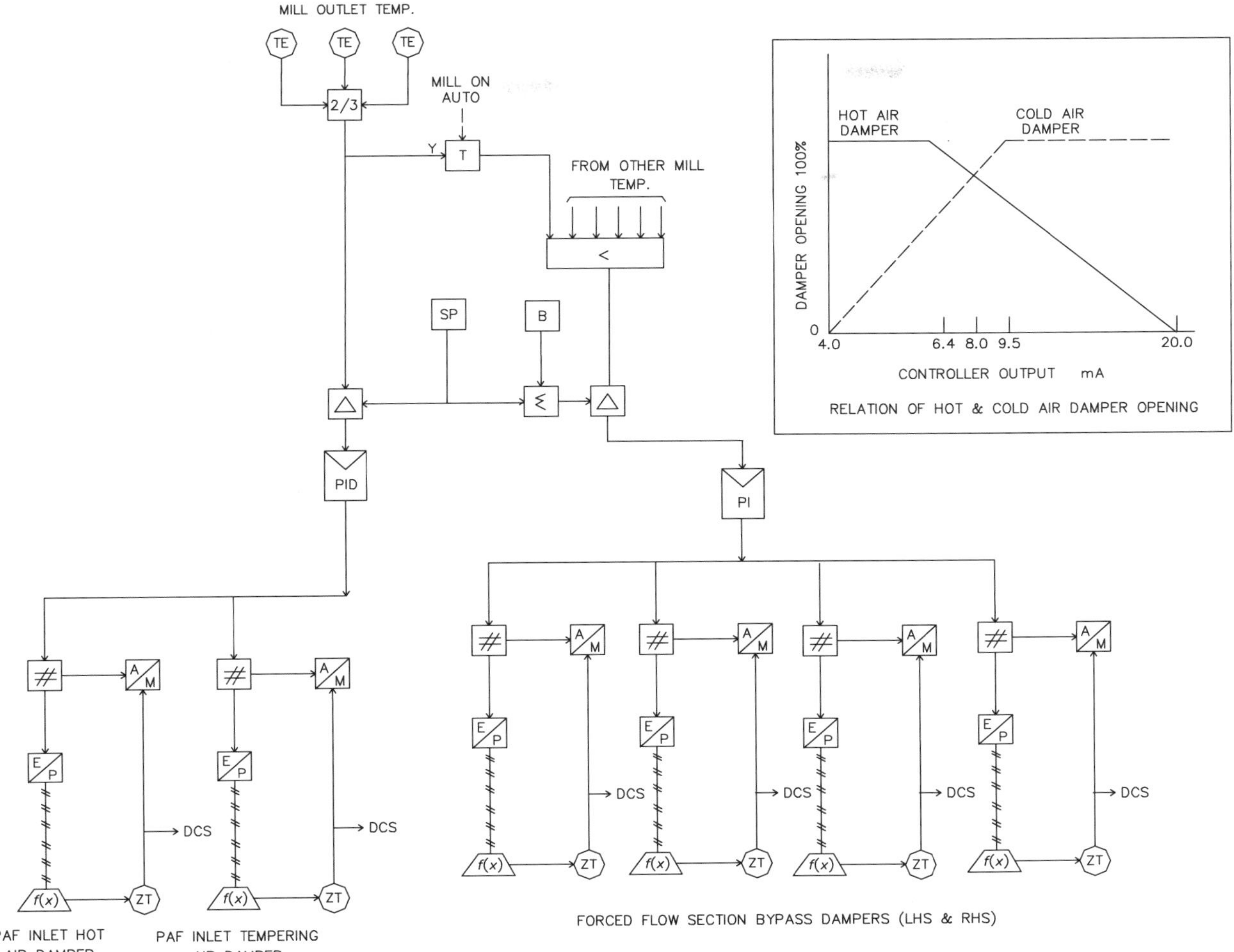

FIG. 8.16 Mill outlet temperature control.

system. Under the situation when the mill outlet temperature decreases, even by fully closing the CAD and fully opening the HAD, below the bias value, the second controller would take the charge of the affair and transmit the output command to the bypass dampers to take proportionate bypass position so as to allow a part of flue gas to flow bypassing the heat exchanger. The average flue gas temperature now being more than that without the bypass, it would then assist raising the mill outlet temperature with a higher PA temperature at the PA heater outlet.

5.2.4 Mill Outlet Temperature for TT Boiler (Fig. 8.17)

The control loop for the mill outlet temperature discussed here is mainly meant for TT boilers based on a CE design with a bowl mill. The control loop for the ball and tube mill is discussed separately in the next section. Fig. 8.17 may be read in conjunction with Fig. 8.15 and the associated P&ID Fig. 3.36.

1. Objective: To regulate the mill outlet temperature at the desired point so that the coal delivered from the mill is completely dry with the desired temperature. Also, in case of high mill outlet temperature, cold air is blown to avert a high-temperature fire hazard.
2. Discussions: Normally/initially entire requirement of PA flow necessary for a particular load at mill is tried to achieve through HAD enabling complete drying of the coal (especially during monsoon season) and to raise the mill temperature as desired. However, there may be eventualities during a hot/dry summer where the mill outlet temperature shoots up, which is an undesirable situation toward a fire hazard. Therefore, the CAD comes into operation whenever there is a need to bring down the mill temperature. Naturally, when this damper operates, that is, starts opening through process feedback, the HAD will close. So here also there is a cross operation of the two dampers but through a process. Additional arrangements of mill inerting systems with inert gases such as N_2, CO_2, etc., need

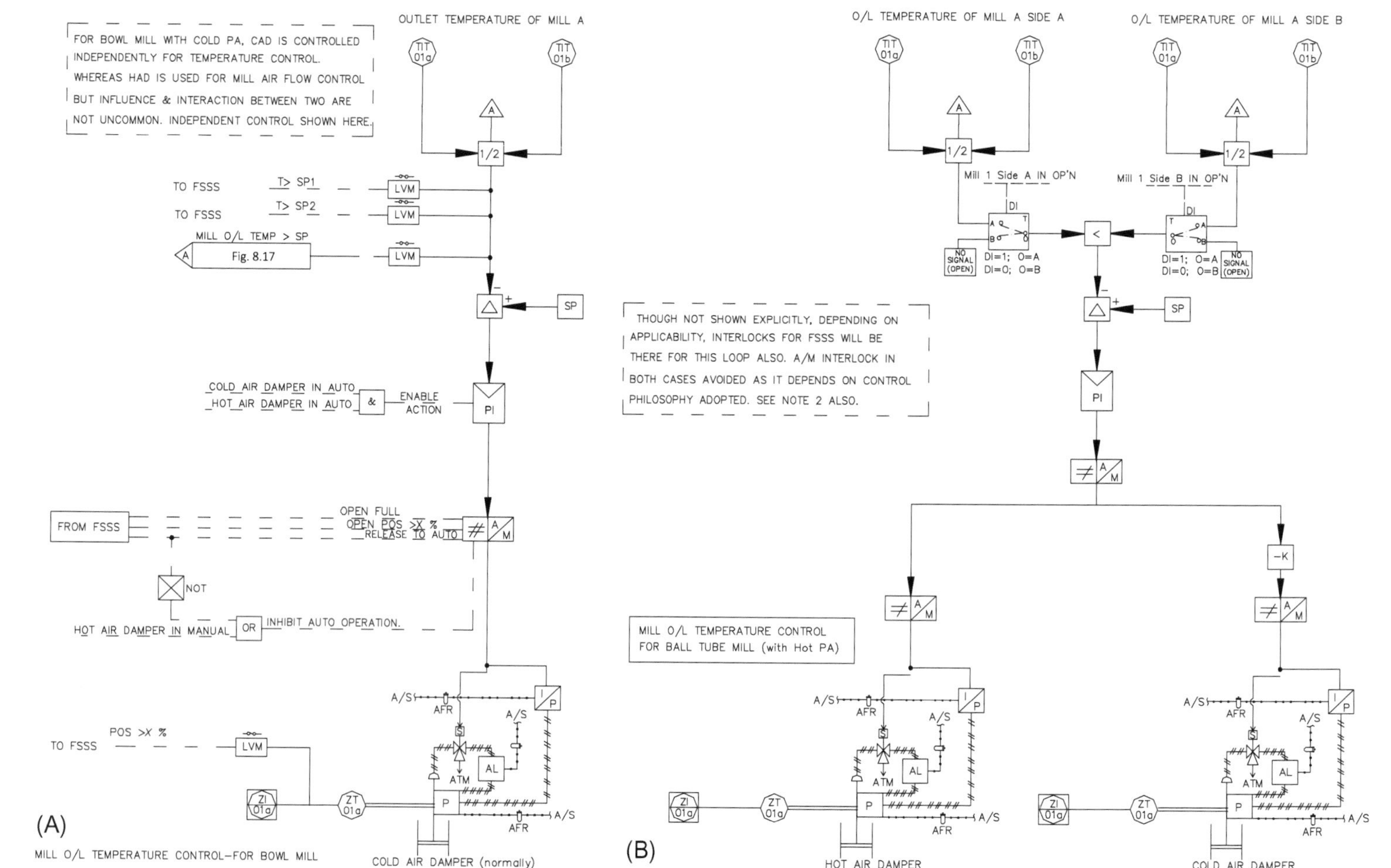

FIG. 8.17 Mill outlet temperature control for bowl mill and tube mill.

to be kept (reducing the air supply is also an aim). This is more important for ball and tube mills, especially when they are operating with one side only.

3. Descriptions: Mill outlet temperatures are measured by redundant temperature elements and transmitters are put in an error generator. The output of the error generator drives a PID controller. In general, with temperature being a sluggish parameter, it is always advisable to use PID controllers for better results. To prevent controller saturations, the controllers are put into service only when both the loops are in auto. The output of the controller through the I/P converters normally drives the pneumatic actuators meant for the CAD.
4. Interlock:
   a. As stated earlier, only when both HAD and CAD are in auto will the controller be put into operation.
   b. Because FSSS operations depend on mill temperature conditions, with the help of a limit value monitor (LVM), the necessary contact statuses have been shared with FSSS.
   c. The loop can be released to auto from the FSSS command.
   d. As protection, the full opening command as well as the $>x\%$ command for the mill CAD are issued from the FSSS so that there is sufficient cold air circulated.
   e. Whenever the auto release command from the FSSS is missing or the HAD is in manual, it is necessary to inhibit auto operation so that the operator puts all attention on the mill outlet temperature.
   f. That will be the check-back signal for FSSS from the loop for the damper position.

5.3 Mill Control (Ball-Tube Mill) (Fig. 8.18)

General

The ball tube mill is one of the many types of mills/pulverizers used for grinding different types of fuels such as coal or other minerals such as cement plant ingredients, etc.

A ball tube mill basically includes a large, hollow cylindrical drum that rotates along its axis, supported by a trunnion on each end. Wear-resistant liners are provided inside the drum, which contains different sizes of balls filled up to half of the drum. These balls are normally made of cast alloy steel or forged steel, and have sizes varying from 25 to 150mm or perhaps more for higher-capacity mills. The drum is made to rotate at a slower speed by standard high speed electric drives through a gear train consisting of a reduction gear box and large spur gear. The slow speed is normally available in the range of 13–35rpm, depending on the size of the drum and specifically the diameter of the drum. During drum rotation, all the balls move upward to about 60%–70% along the periphery and then cascade over each other with a tendency to move toward the center of the drum as a continuous process. This forces a mass of balls and higher-sized materials to mix with each other at the bottom of the drum, resulting in material size (around 25mm) reduced to small particles. The reasons behind this action may be attributed to the following forces:

(i) Impact of the falling heavy balls causes cracking of the materials.
(ii) Attrition of materials with each other and with liners.
(iii) Crushing of the heavy balls rolling over each other and liners with the materials in between.

Very high coal fineness is achievable in the range of 70–90μm at the outlet.

Raw materials are introduced from both ends of the drum through variable speed feeders so as to regulate the PF flow as required. An adequately heated PA is blown, through the hollow shaft of the screw feeder inside the pulverizing zone of the mill, for drying the materials; pulverized fuel (PF) is then made fine and light enough to become airborne and is vented out of the pulverizing zone from both sides of the drum to the combustion chamber. An external classifier is provided at both sides of the drum so as to allow only the fine particles to go outside. Due to the whirling action, larger particles experience more kinetic force and collide with the classifier wall, only to lose the force to zero. Then, this rejected material comes back downward for force due to gravity to the grinding zone with the raw feed. The new raw feed continuously enters through the ends of the drum to assume the already-vacated place of the departing airborne particles.

For this type of arrangement, a half-mill/one-side operation for partial loads is possible and is done in practice for a short time only, as this operation may result in high temperature and accumulation of pulverized fuel in the unused classifier side, causing an explosion.

The classifier should be so designed as to ensure the removal of fine feed particles from the grinding zone so that the production of extreme fines is eliminated and mill current in relation to power consumption is kept within limits. The relatively large quantity of raw feed and semi-ground feed mixture in the grinding zone may actually be regarded as the additional storage, which can cope with any unexpected and abrupt increase in the demand. However, it is to be noted that the electrical power consumed for a particular quantity of raw feed of this type of mill is very high, especially at the partial load.

There is the possibility of clogging at the feeding end of the pipe for the presence of moisture in the surface of raw feed, which may in turn reduce the capacity of the mill. It is therefore very important to reduce the moisture content in the raw feed. However, the arrangement of recirculating the heated and dried oversized rejects helps in reducing the moisture and thereby the clogging at the feed inlet. The hot air with powdered fuel flows out of the mill through the annular portion of the screw conveyer or feeder and the

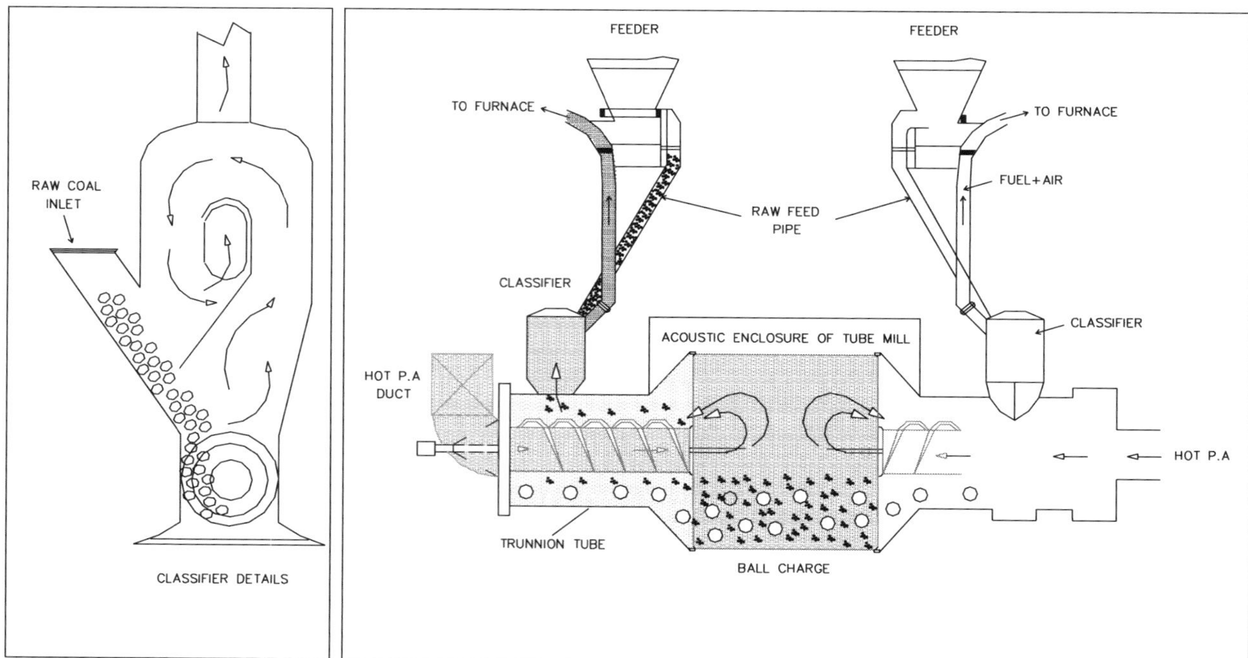

FIG. 8.18 Schematic diagram of ball-tube internal arrangements.

trunnion tube and thereby dries the raw feed while approaching the screw feeder. During the starting period, a portion of the hot PA is bypassed and sent through the raw feed to minimize the moisture in the raw feed. As the load increases, this bypass air is decreased and finally brought to zero above a particular load. However, according to another school of thought, the bypass damper is kept open to a minimum value, say 10% or even up to 100% mill load, so that even at a high load, the fine feed dust is not unnecessarily fed to the mill, increasing the mill current.

When the mill trips, a good quantity of heat trapped in the mill is lost. A good inerting arrangement of the mill may be necessary to eliminate fire hazards.

A trunnion seal is provided between the rotating drum and the related stationary components such as the raw feed piping system so that the material can move to the rotating drum end. The trunnion seal assists the passage of raw feed toward the rotating drum but, on the other hand, it prevents the escape of material into the surroundings. It is always desirable to form the best possible seal between the stationary and rotating components to thwart the tendency of the fine particles to escape. Replacing the trunnion seals is a cumbersome job that requires removing the inlet/outlet box from the foundation blocks under the trunnions and the downtime toward the same is significant. However, a sealing arrangement is provided by supplying pressurized air through separate seal air fans in an arrangement of common redundant fans or an individual fan.

Ball tube mills are not suitable for intermittent operation as the significant amount of heat energy stored in both the raw feed, particularly the coal and the balls, may lead to overheating and fires when the mill is kept idle. As per the operating principle and functional requirement of the mill, the noise level is significantly high and requires noise-suppressing arrangements.

5.3.1 Objective of the Control Strategies (Figs. 8.19–8.21)

There are a few control loops for the safe and efficient running of mills. Fuel flow and PA flow control details for ball and tube mills have already been presented in Section 4 of this chapter (Figs. 8.13 and 8.14).

The first and most important one is the *mill outlet temperature control*. It is provided to supply the fuels at an optimum temperature for various reasons, the same as the other mills described earlier. During start-up, the moisture and temperature are controlled by diverting a portion of the hot air between the feed pipe and classifier, and the control of this part is taken care of in the temperature control of the PA flow to the mill.

(i) In some plants, the *PA inlet pressure control loop* is provided to maintain the common PA fan discharge header pressure before the primary air heater and the take-off point of the cold PA header for the *cold PA system* (PA fan suction may be from the atmosphere or FD fan discharge). Mixing of hot PA (taken after the primary air heater) and cold PA is done for controlled temperature for each individual mill. Though not shown in the above drawing, the PA header pressure control loop assists the

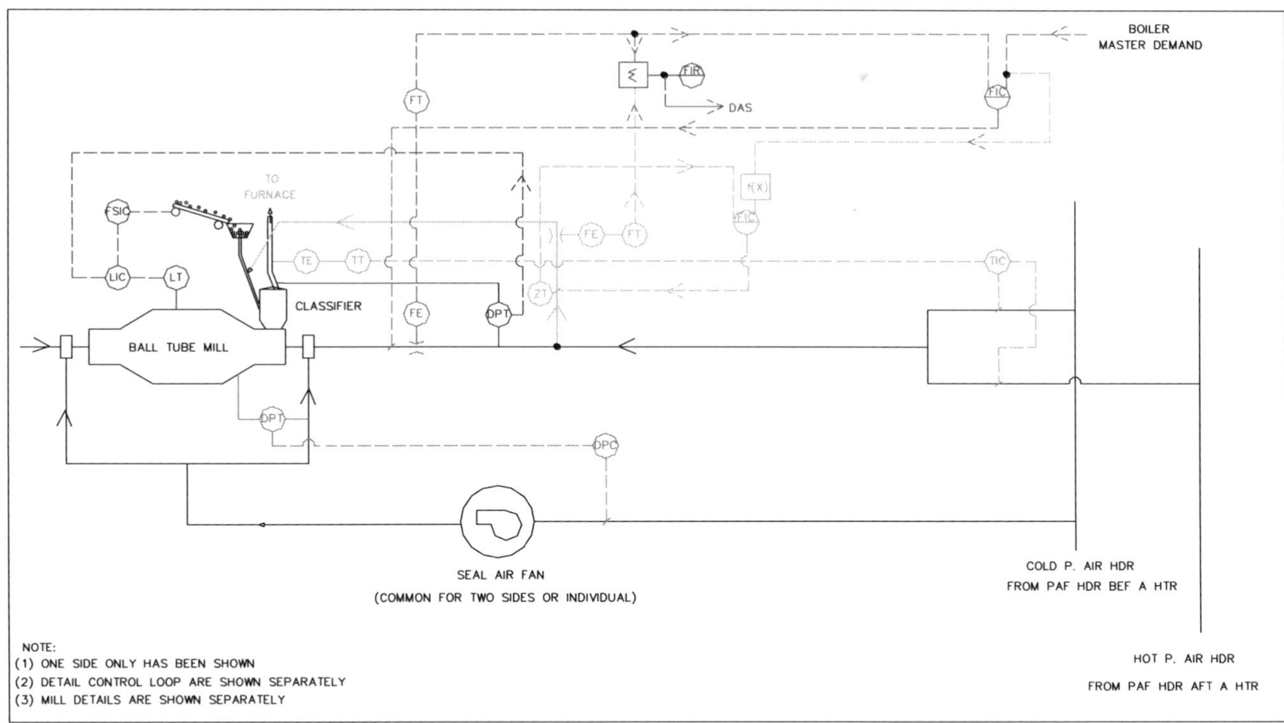

FIG. 8.19 Typical flow and control diagram of ball tube mill with common PA fan system.

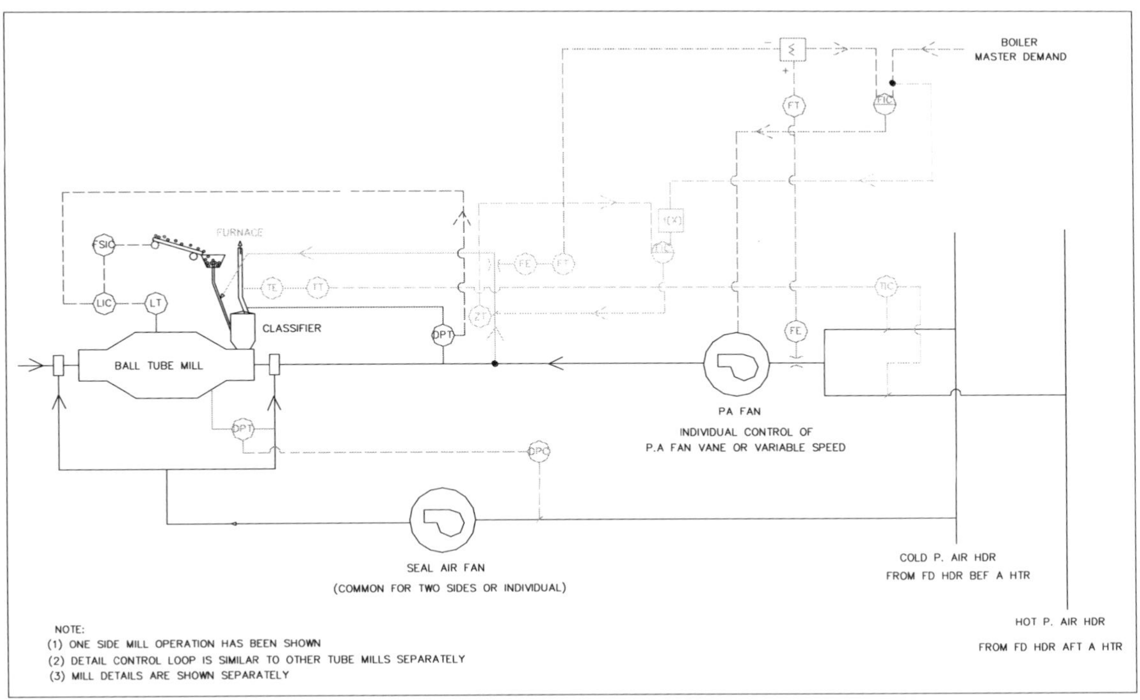

FIG. 8.20 Typical flow and control diagram of ball tube mill with individual PA fan system.

smooth and bumpless control of the actual PA flow to the mill accomplished by the PA to mill damper control. The PA pressure control may not be required when the mill-wise PA fan suction is taken from the temperature-controlled, individual hot PA header for a *hot PA system* (cold and hot PA headers are taken from the FD discharge before and after the common air heater, respectively, without a separate PA heater).

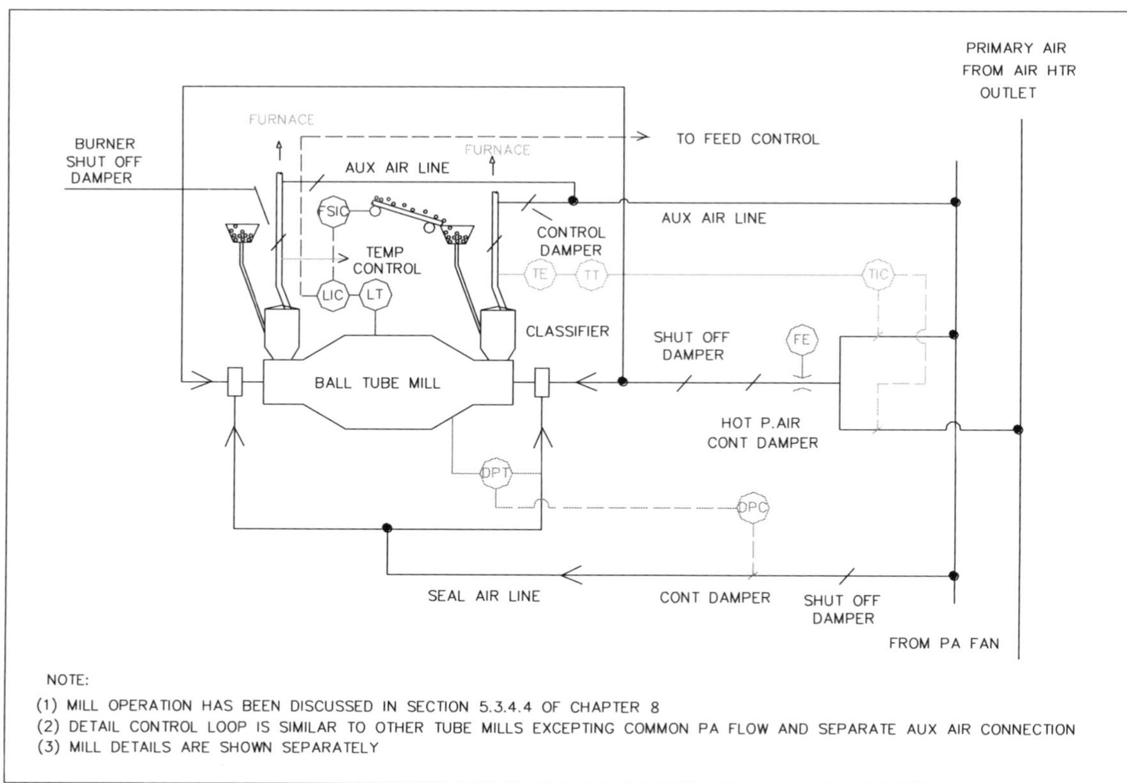

FIG. 8.21 Typical flow and control diagram of ball tube mill with common PA flow control.

(ii) *The mill drum level control loop* (Fig. 8.22) is meant for maintaining the level of the charges inside the drum.

This control is achieved by sensing the load on the mill by measuring the *differential pressure* across the mill or the characteristic *sound* produced by the mill at different feed levels. The set point is normally a fixed value and the controller output is used to regulate the speed of the solid fuel feeder, so as to maintain an appreciable fuel reserve through an assured level in varying load. On account of ambient sound, sonar transducers are used as corrective or confirmatory elements and not as a prime element to measure the level.

Mill load or fuel flow control is to follow the demand from the boiler master demand control signal and, unlike the bowl mill, this is achieved by regulating the quantity of the transporting agent, that is, the PA flow. The control strategies may be of different types as indicated below:

(a) Control of individual PA dampers with common PA fans.
(b) Control of individual PA fans through PA inlet vane/damper control.
(c) Control of individual PA fan speed through VFD.

The details are provided in clause no. 5.3.4.4 of this section and clause no. 4.4.3 of Section 4 of this chapter in conjunction with Fig. 8.13.

(iii) *Differential seal air pressure control is* achieved by regulating the seal air damper position through a controller with transmitters sensing the differential pressure across the mill air inlet and mill drum. Seal air to mill inlet pressure is varied to ensure some extra pressure to arrest the dusty air escaping the mill. A control loop for this purpose may be common for two sides of a mill or may be individual for each feeder, as dictated by the milling system.

5.3.2 Discussion

5.3.2.1 Mill Outlet Temperature Control

The basic mill temperature control loop has been discussed in clause no. 5.2 of this chapter. For ball tube mills, there is another consideration for addressing the problem of moisture in the raw feed before entrance to the mill inside. In fact, during start-up, the amount of hot PA is less and may not be sufficient to prepare the feed in the right condition for proper grinding and flowability.

This untoward situation is eliminated by increasing the temperature of the raw feed at the mill inlet by introducing hot PA into the classifier (mixing chamber), between the raw material feeder and the mill inlet through a bypass damper in the PA to mill line. This additional hot air

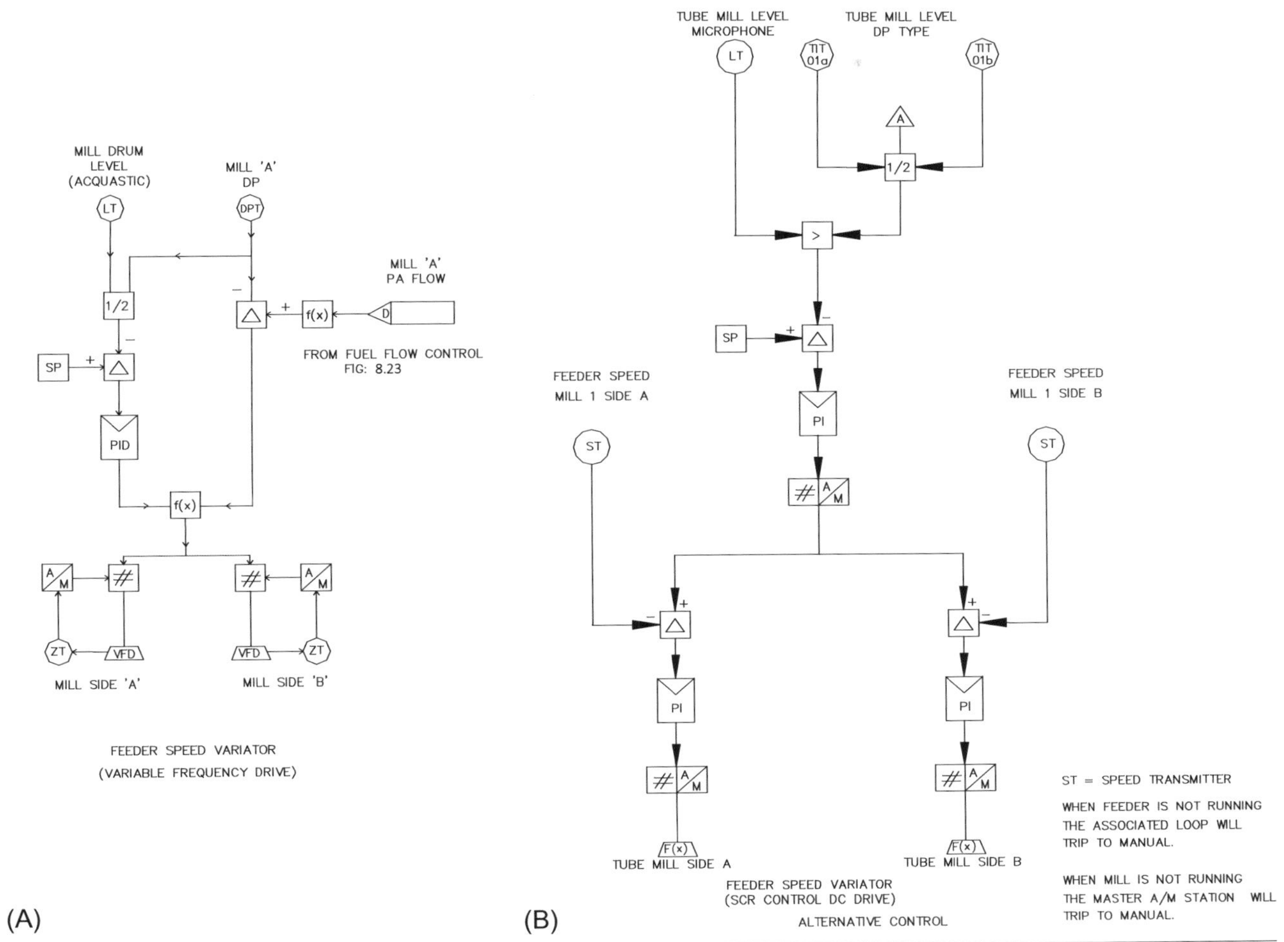

FIG. 8.22 Ball tube mill level control.

preheats the feed and reduces the moisture content. This arrangement is mainly operational during start-up, but may continue up to a certain load and even up to a full load, depending on the concerned mill design.

Although this appears to be a part of the mill temperature control, this initial control part is accommodated in the mill load control, that is, a part of the PA flow to the mill control, as depicted in Figs. 8.23 and 8.24.

5.3.3 Mill Drum Level Control Loop

As mentioned earlier, the mill drum level control loop is meant for maintaining the level of the charges in the side the drum. In this type of mill, the capacity of the raw feed storage is very high and is maintained at an average constant value by means of a level control of the charges, irrespective of actual fuel flow. The advantage of the availability of a constant amount of material at all loads is the reserve fuel, and this is highly favorable for the process control, especially when there is a sudden load increase. The extra amount of load is fed by the reserve capacity before the feeder response takes place. The shortfall of materials in that case is detected by the low level and the feeder then responds accordingly. However, there may be a feed forward signal from the PA flow to correct the feeder speed. This control is achieved by sensing the load on the mill by measuring *differential pressure* across the mill or the characteristic *sound*, which is also called the *electric eye*, produced by the mill at different levels.

5.3.3.1 Mill Drum Level Sound Detector

It has been observed that the sound or noise level of a mill, as generated during the normal range of operation, changes in accordance with the magnitude of the mill loading. Elaborate experiments have revealed that character of sound as well as its intensity are different, even when a mill is grinding a single material, under the operating condition of running wet or dry, in open or closed system without bearing any relation whether producing a coarse or fine

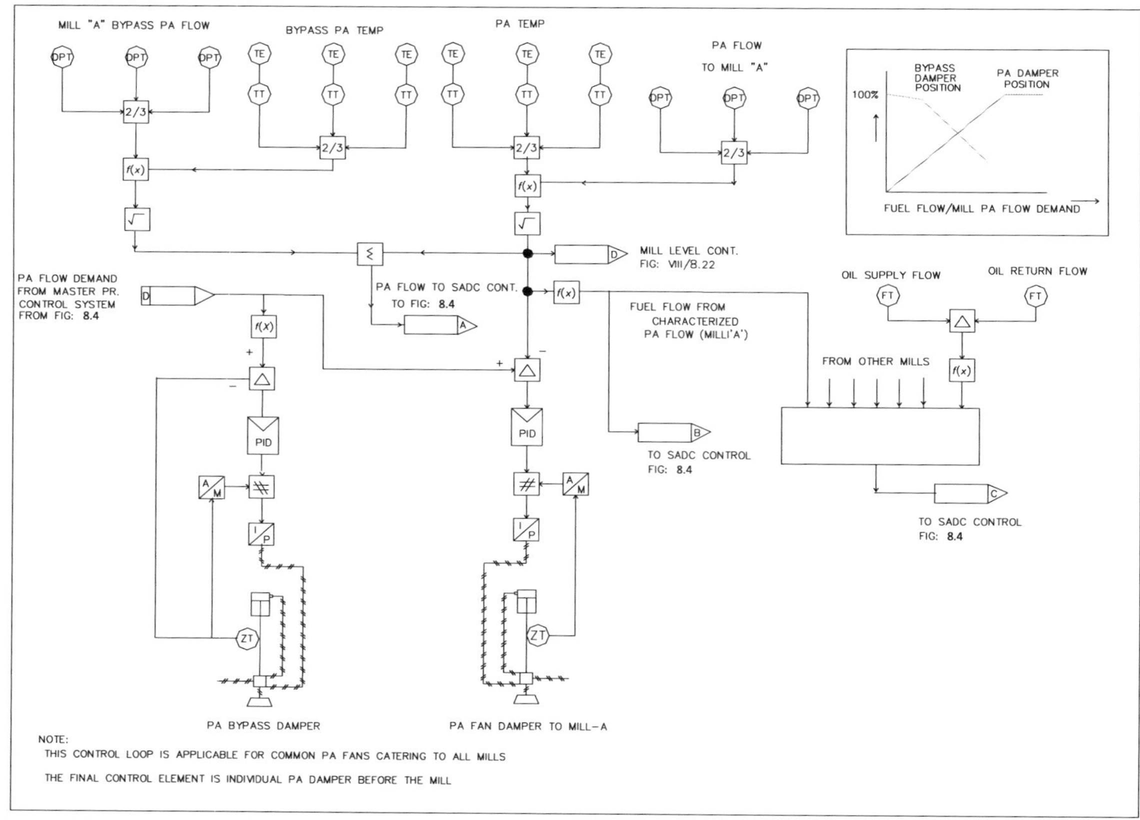

FIG. 8.23 Fuel flow control for ball tube mill (by PA Damper).

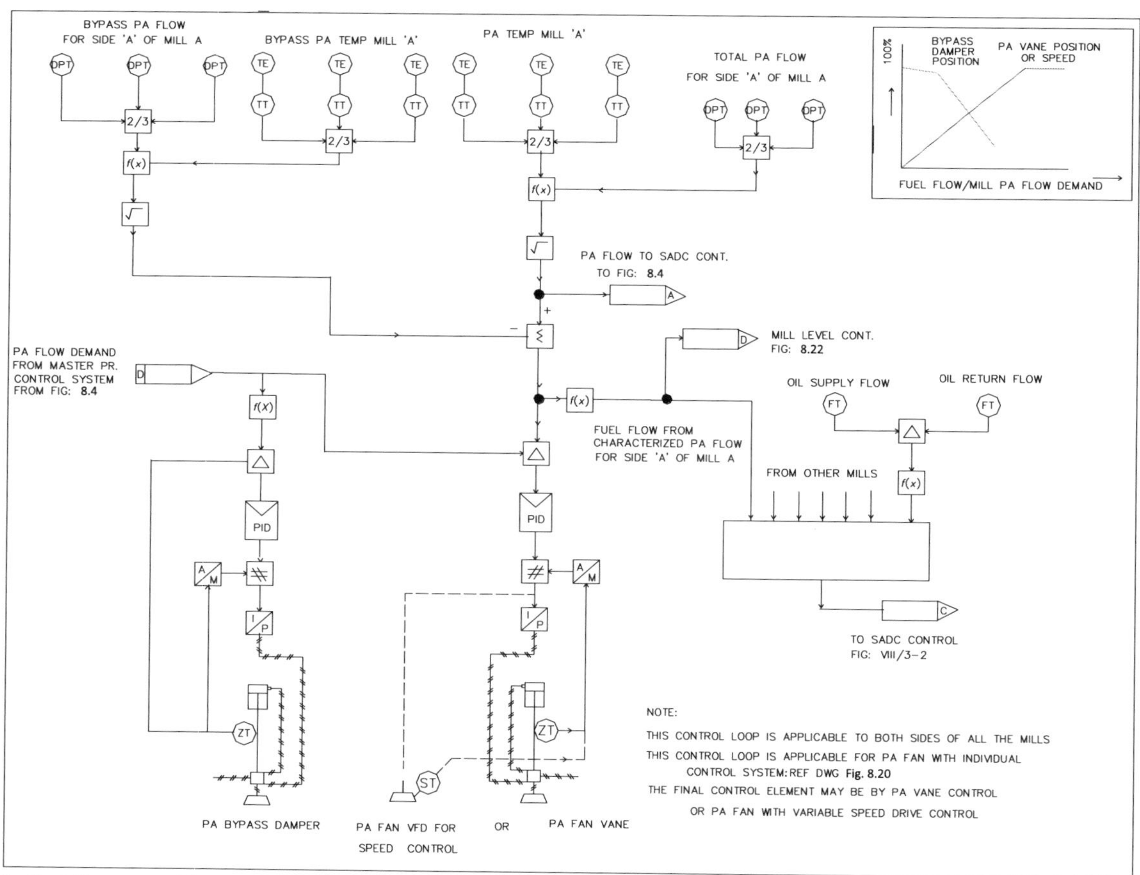

FIG. 8.24 Fuel flow control for ball tube mill (by Vane or speed control).

product or actual real time feed rate. For example, the sound intensity is going to be more as the drum charge level becomes less; on the other hand, as the charge becomes more, the sound intensity is going to be less and will turn out to be practically indistinct to the ears, that is, less than the exact sense of hearing of a human operator.

Automatic control of ball tube mills utilizes this sound property as the controlling variable. It has been proved to be a highly convenient, realistic, and dependable solution with economical sanction. By installing a microphone that is duly suitable for the range of frequency generated and proficiently utilizing the energy output of the detector to control the raw feed to mills and thereby maintaining the grinding efficiency at a its maximum level. The acoustic detection device is installed in a kind of parabolic reflector to permit the sensor to pick up the signal generated by only one mill at a time. The raw and low-level electrical signal output of the acoustic detector is then amplified and modified in an electronic transmitter to make it suitable for use in an automatic control loop. The detection system as a whole is calibrated to measure the optimum sound level predicting the mill charge level and the control loop strategy is built toward maintaining that level at a constant value by adjusting the speed of the raw feeder, allowing more or less material to the mill.

5.3.4 Control Loop Description

5.3.4.1 Mill Outlet Temperature Control

The control loop is similar to that discussed in Sections 5.2.3.1 and 5.2.3.2. The relevant drawings are Fig. 8.16.

5.3.4.2 PA Inlet Pressure Control Loop

As already discussed, this control loop is provided to maintain the PA header pressure before the mixing of hot and cold PA duly controlled for temperature. Fig. 8.10 is also applicable for this type of mill when the PA is common to all the mills. This control loop will not be applicable for individual PA fan systems where mixing is done before (hot PA system). For a control loop description, Section 4.3.2.3 of this chapter may be referred. Common PA fans are provided with suction normally from the atmosphere or maybe from the FD discharge header. The header pressure control system is done through different types of FCEs.

As the fuel/load control is solely done by the position adjustments of the PA damper near the mill, this control loop assists the smooth and bumpless control of the fuel flow transported by the PA flow to the mill as the upstream PA header pressure control takes the responsibility of providing an adequate quantity of air at any environmental condition without sacrificing the required downstream pressure.

5.3.4.3 Mill Drum Level Control Loop

Fig. 8.22 depicts the simple control loop. Any of the mill differential pressure transmitters or level (sound detector) transmitters is selected and the selected signal is connected to the controller as the process or measured variables against a fixed level set point. Sufficient redundancy in measurement may vary as per the plant's operating philosophy. The controller output is utilized for adjustment of feeder speed with the help of a VFD or SCR control for a gravimetric feeder/feeder speed variator.

At the higher load, the charge level inside the drum would decrease and the feeder speed should increase accordingly to replenish the material. For a decreasing load, the reverse action would take place. To take care of the sudden load change, the deviation between the characterized PA flow and the DP across the mill is used to modify the controller output to achieve the desired mill charge level.

5.3.4.4 Fuel Flow Control Loop

Mill load or fuel flow control is to follow the fuel demand from the boiler master demand control signal. and is achieved by regulating the quantity of PA that is the transporting agent only. Fig. 8.23, respectively, depict the functioning of the control loop, which is almost similar to that of other types of mills. For other mills, the fuel demand signal from the boiler master demand is first taken care of by the mill-wise PA flow control system if the demand is less than the prevailing air flow control system. The characterized PA flow is then construed as the feeder speed demand. The tube mill control system, on the other hand, envisages feeder speed control for maintaining the mill level control only and so the fuel flow control is achieved through control of the feeder-wise PA flow to the mill itself.

However, the feeder-wise PA flow as measured after redundant transmitter voting selection and density compensated through temperature correction is again characterized to get equivalent fuel flow. The total fuel flow is then computed by summating all the fuel (PF) flow of the running feeders and the supporting fuel (oil or gas), if any is being utilized at that time with proper weightage, taking consideration of their thermal or calorific value. The higher selection of this total equivalent fuel flow signal and the air flow demand signal from the boiler master demand (Fig. 8.2) is then taken as the actual air demand, just like other types of mills.

As already discussed in Section 5.2.3.2 (Mill outlet temperature control) of this chapter, there is another feeder-wise control system associated with fuel flow control, known as the bypass damper control. This feeder-wise damper is provided for each mill end for preheating the raw feed, which is an essential requirement during the start-up period. No process measurement signal is utilized in this subloop.

The same fuel demand from the boiler master demand (Fig. 8.2) is also taken as the set point for the position demand of the bypass damper as shown (Figs. 8.23 and 8.24) in the control strategy and the graphical representation of approximate positions of the two FCEs. The above-mentioned two position demands operate in the opposite direction. After being fully open for certain loads, ensuring the elimination of initial moisture, this bypass damper would begin to close gradually as the load increases.

There are two main types of fuel flow controls achieved through the proportionate PA flow only:

(i) Common PA fans with individual PA dampers.
(ii) Individual PA fans with vane or speed control.

Common PA fans with individual PA dampers (Figs. 8.19 and 8.23)

Here, the mill PA flow and bypass PA flows are combined to form a total mill-wise PA flow to the furnace. The boiler master demand acts as a set point here where the mill-wise PA flow is the measured value, as this air flow is only responsible for transporting the fuel to the furnace. The controller output is the demand signal for the individual PA damper. For bypass dampers, the boiler master demand is characterized to generate the set point while the actual position of this damper acts as the measured value for the controller, the output of which is the demand signal for the bypass damper.

For any load change, the two flows would readjust their positions to deliver the required PA flow. For a higher load, the bypass damper would tend to close to allow less flow for preheating the raw feed and the PA damper to the mill would open more to take care of the load demand.

Individual PA fans with vane or speed control (Figs. 8.20 and 8.24)

Here, the bypass PA flows need to be subtracted from the total mill-wise PA flow for the fuel flow control, and the total mill-wise PA flow to the furnace is required for air flow control. The reason is the FCE and the flow element are both located in the common primary air path to the individual mill. The boiler master demand acts as a set point here where the PA flow to the mill is the measured value. The controller output is the demand signal for the individual PA vane or the variable speed drive.

For the rest of the control loops, that is, for the bypass damper, the control strategy is the same as the previous one described above.

Mill-wise PA flow control common to both sides (Fig. 8.21).

This type of mill design vis-a-vis operation is somewhat different from other types as discussed earlier; it is mainly followed by manufacturers such as M/s Foster Wheelers Energy Corporation. Here, the boiler combustion control signal regulates the output of the mill by the primary air (PA) flow control damper placed in the common line to both the ends or sides. The predrying of the coal feed is done at the entry of each side before entering the drum, unlike what is done by the bypass PA damper in many types of tube mills. Another significant difference is the provision of an auxiliary air and purge air supply line taken from the cold PA for each side of the mill drum. The same is designed to the required minimum velocities of PA/fuel mixture for maintaining proper flowability inside the coal duct and preventing fuel settling during start-up or extreme low load operation. This feature extends the individual mill load range without encountering drifting or pulsating fuel flow to the burners. The other purpose is to purge the coal air line automatically when the burners are taken out of service.

The feed level control in the drum, the classifier outlet temperature control, and the seal air DP control are very similar to other types of mills, with the exception of the source of seal air. Here, the seal air supply is taken from the cold PA without any provision for a seal air fan.

6 FURNACE DRAFT CONTROL

General

The furnace draft control loop is very important in the boiler control system. The boiler has two distinct zones of functional operation: the firing zone and the heat exchanger zone. The firing zone, along with water walls (to insulate the heat content from around the side and top walls), is normally described as the furnace.

The pressure of the top portion of such a furnace is generally maintained at below atmospheric pressure that is, a −3 to −5 mm water column. This task is accomplished with the help of induced draft (ID) fans, which evacuate the products of combustion, that is, the combination of different gases from the furnace. By furnace draft control at such a typical pressure value implicates so many things that is quite surprising. The low but near to the atmospheric value ensures that *neither explosion nor implosion takes place*. It is known that an explosion in the furnace occurs when the rate of volume of flue gases or the products of combustion increase abruptly as a follow-up action of introducing a huge quantity of fuel and air. The situation causes instantaneous volume boosting at a much higher temperature as a result of exothermic reaction. With the furnace volume and the wall thickness being fixed and limited, the increased gas volume, if it goes beyond the design limit, may cause an explosion.

The near atmospheric pressure is not only safe from an explosion, but it also prevents the possibility of implosion, a situation that may arise due to a sudden collapse of flames, as discussed in Section 6.2 of this chapter. The advantages are safety against explosion/implosion, less wall thickness, fewer operation hazards and less of a possibility of leaking flame or hot flue gas, which means less heat loss and pollution.

The ID fans are normally supplied with inlet vanes for controlling the amount of volumetric flow handled by the fans. The control loop selected for the discussion of the furnace draft control in this section is a typical 250 MW thermal power plant with ID fans with an additional mechanism, that is, a hydraulic coupling along with a scoop tube control for varying the speed of the fan.

In fact, any one of the above items is sufficient to administer the control system as the FCE, but the present loop may be quite interesting as to how the things are taken care of by the two devices mainly for higher MW plants to achieve better and smoother controls.

6.1 Objective

The furnace pressure is maintained at a negative pressure and hence is popularly known as the furnace draft. The main objective of this control loop is to achieve the desired pressure at all loads by sucking the appropriate amount of flue gases, nonparticipating gases in the combustion process, and unburnt fuels (if any). The ID fans are provided to perform the duty with a necessary arrangement to have controllability, which acts as the FCE of this important loop. These control accessories are made to open (or increase) and close (or decrease) to suck the required amount of flue gases so that the furnace draft is maintained.

6.2 Discussion

Furnace draft control is a comparatively simple loop, considering the other control loops in a thermal power plant. However, with the furnace itself being a huge voluminous part of the boiler handling the burning of fuels, the safety implications are to be taken care of with due importance and significance. The applicable codes and standards must be followed to avoid any untoward situations while taking up the design aspects of the furnace-related equipment. *The National Fire Protection Association (NFPA) codes* provide the guidelines that one must read and follow before realization of anything regarding furnace application. The most relevant code is the NFPA 85: Boiler and Combustion Systems Hazards Code, which includes the protection measures to be taken to prevent fire or furnace explosion/implosion.

The NFPA also stipulates some additional interlocks for incorporation into the furnace draft control loop so that the required operating safety margins are ensured. For example, the interlocking signal from high and low furnace pressures is to be utilized to block the FCEs from further repositioning, which may further jeopardize the furnace condition.

Another important example of interlock, as per the NFPA codes, is taken from the master fuel trip (MFT) logic signal actuated from the dangerous condition that causes the emergency shutdown of the boiler. On the occurrence of such a sudden event, the flame within the furnace collapses violently, with all types of fuels completely and abruptly removed. This state of affairs invites a very dangerous situation such as an implosion; it can also be the reason for severe wear and tear of different boiler parts. To obviate those circumstances, the interlock allows the controller output to keep the FCEs to attain a predefined fixed position for a definite period of time and then releases the device back to normal operation. Normally, the ID fan suction head is more than the negative pressure the furnace structure is designed to withstand in the case of any eventuality.

The advantage of maintaining the furnace pressure at a constant value may be explained in another way. The Universal Law of Perfect Gas states that,

$$PV = MRT$$

where for any system, P stands for absolute pressure, V for volume, M for mass of the substance within the volume considered, R for the universal gas constant, and T for absolute temperature.

Considering the system as a whole, it is apparent that the volume V is also a constant value whereas the universal gas constant R is approximately constant, which means that $P = kMT$ (where k is a constant).

The above equation shows that the product of M and T can be made constant if the furnace pressure (or the draft) is kept constant. As the furnace temperature depends on the boiler load, or in other words, the thermal energy balance (heat input and output), the mass inside the furnace is automatically kept adjusted by the ID fans by sucking the required amount of flue gas. At the time of MFT (when the fuel is immediately cut), the temperature of the furnace drops rapidly with a subsequent drop in pressure. Instead of flue gas, there would only be hot air to be sucked by the ID fans. The drop in furnace pressure would cause the ID fans to suck less gas as instructed by the furnace draft controller and the mass M within the furnace would improve such that the product MT again equals the constant furnace pressure P. The amount of hot air pushed by the FD fans under this condition may be different for different boiler manufacturer, but ID fans finally take care of the mass availability within the furnace under any condition so far as implosion of the furnace is concerned.

The BMS also generates binary signals to enable the control system to run in automatic mode or the FCEs to attain a minimum position as per the demand of the prevailing situation.

6.3 Control Loop Description

See Figs. 8.25 and 8.26.

6.3.1 *Measurement of Different Various Parameters*

6.3.1.1 Furnace Pressure

The furnace pressure is measured by a differential transmitter with sufficient redundancy and at the appropriate location to get an average and representative measurement. For a larger furnace, there may be sets of transmitters at both the opposite walls for a better result. The span selection of a

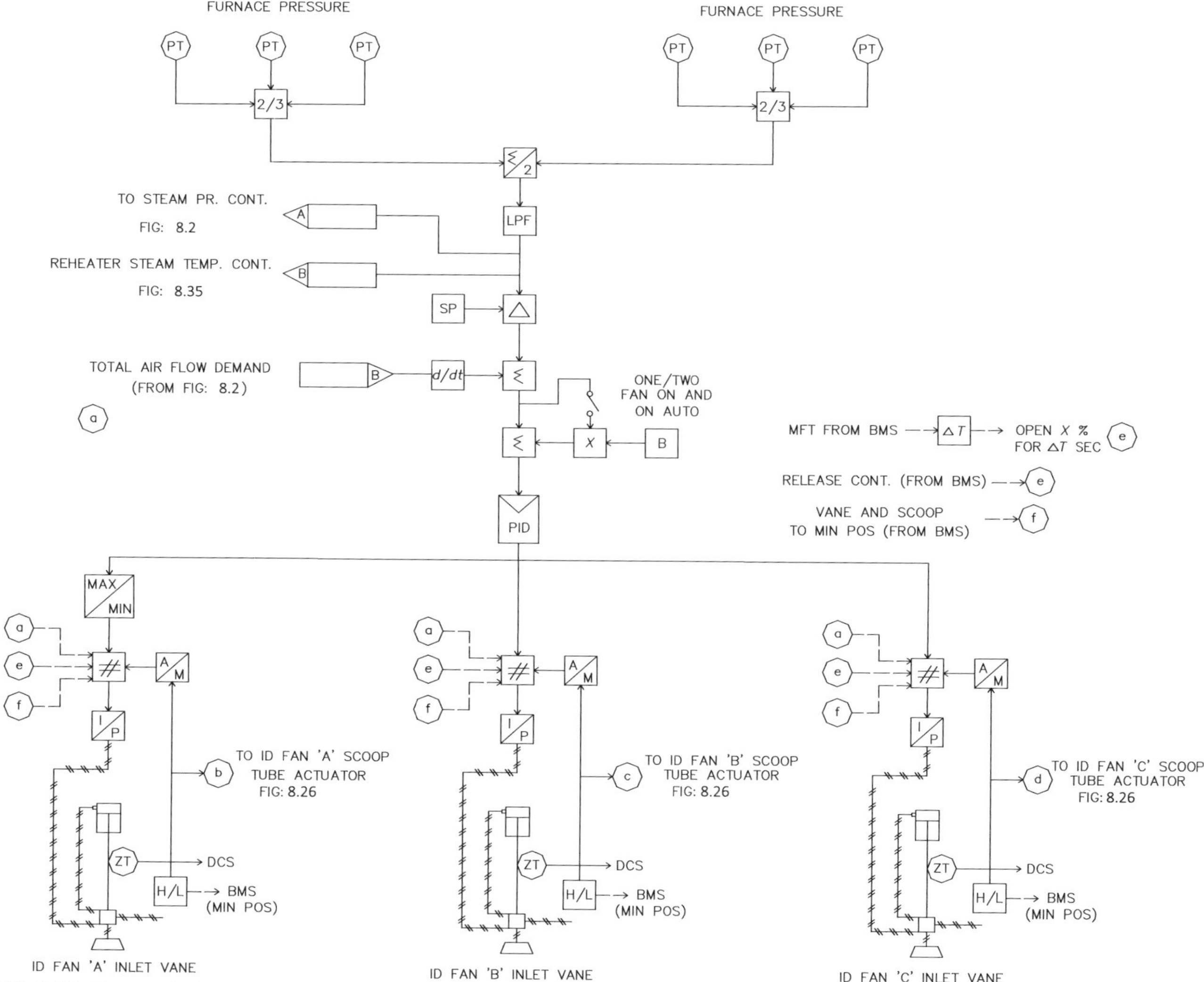

FIG. 8.25 Furnace draft control (inlet vane).

transmitter is very important; normally, these should not be too wide or narrow. Due to the fluctuating nature of the furnace pressure and dusty atmospheres, normally a special volume chamber is put for each transmitter for measurement purposes, as detailed in Fig. 12.13B. On account of the volume chamber, fluctuations are arrested and line choking is less as dust settles in the chamber.

6.3.1.2 ID Fan Inlet Vane Position

These parameters, being the position of the FCEs, are measured as a standard procedure and also used in the control strategy for the ID fan speed control through the hydraulic scoop tube control (or VFD).

6.3.2 Different Control

6.3.2.1 Furnace Draft

The furnace pressure signal is a very fast-changing parameter and may disturb the desired control action if connected to the controller directly. To circumvent the situation, a LPF algorithm is utilized to eliminate the high-frequency components of the furnace pressure transmitter output and allow only the signal with a comparatively low frequency. The LPF output signal is taken as the measured/process variable whereas the set value to the controller is a fixed one. The total air flow acts as a feed forward signal after passing through a derivative function and is summated to the error signal between the measured and set values. In an alternative (ISA recommended) loop, the FD fan impeller/vane control demand signal acts as the feed forward signal and is added at the output of the controller for a similar function.

Then, after the gain function is added to take care of the number of ID fans running on auto mode.

The final error signal then becomes the input to the PID controller. The controller output has to pass through a maximum/minimum limit so that the inlet vanes do not

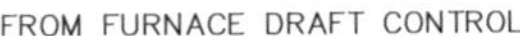

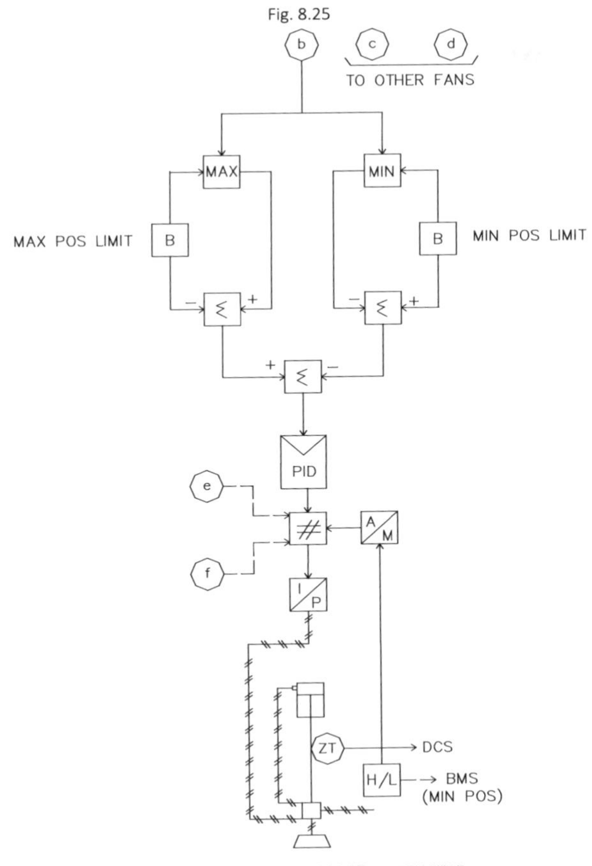

NOTE:
(1) SIGNAL C SHALL BE USED SIMILARLY FOR SCOOP TUBE OF IDF–B
(2) SIGNAL C SHALL BE USED SIMILARLY FOR SCOOP TUBE OF IDF–C

FIG. 8.26 Furnace draft control (scoop tube).

reach the end positions, causing an untoward situation for furnace safety. Regarding fine tuning of the furnace controller, one should avoid the use of a very fast integral. It is a fact that furnace pressure changes at a very fast rate but not instantaneously; therefore the process being compressible fluid, the capacitance effect needs to be considered which incorporates the size of the furnace, the huge ductworks between the furnace and fans, etc.

The inlet vanes assume the positions as per the controller output to maintain the furnace draft only, but they have got some other safety interlocked functions also. Those are indicated below as incorporated for a typical boiler, but they may be different to some extent as per the manufacturer's recommendation. The safety features are:

(i) The inlet vanes are forced to assume a certain preassigned open position (as dictated by the control strategy) for a certain time in case of MFT (other than FD or ID fans). This interlocking command enables the unburnt fuel, if any, to be evacuated to avert any explosive situation. For MFT due to both FD/ID tripping, vanes/dampers in the air and flue gas path would open slowly to full value to create a natural draft to ensure that the furnace does not go beyond design limits.

(ii) The BMS issues the release to automatic control command signal upon checking relevant conditions congenial to auto control operation.

(iii) The BMS also issues the commands to assume a minimum position at certain process conditions to prevent an implosion when the time period as described in point number (i) above is over. On a very negative furnace pressure signal also, the BMS binary signal output may be used as an interlocked condition that closes the ID inlet vane or decreases ID fan speed or blocks the increasing signal to prevent implosion.

There are other interlocked operations that also recommend blocking of the closing signal of the ID fan inlet vanes from the binary signal set to actuate on the preset high furnace pressure. The reverse is the case for low furnace pressure, upon occurrence of which the binary signal would block the opening signal of the ID fan inlet vanes. The settings of all these signals are ascertained by the boiler and fan supplier during the design stage dictated by safe operation of the plant.

The ID fans are equipped with another FCE called the *scoop tube actuator*, through the operation of which the speed of the ID fans can be varied and hence the suction capability can be enhanced or decreased. The scoop tube actuator, upon getting the modulating command, adjusts its position so as to allow more or less oil in a chamber to raise or lower the oil level. The oil chamber forms a part of the hydraulic coupling connected with the electric motor (Chapter 6, Section 3.1.4, Fig. 6.24) and the ID fan itself where the motor runs at a constant speed and the ID fan speed varies as per the oil level in relation tothe scoop tube position. In this control loop part (Fig. 8.26), the scoop tube actuators play a secondary role. Here, the scoop tube actuator is modulated so that the inlet vanes operate within certain maximum and minimum positions.

The vane position signal forms an input of a maximum selector, which has another input set manually to represent the maximum limit position of the inlet vane. There is one summator having its one input (with + sign) from the above maximum selector output and anther input (with − sign) from the above-mentioned vane maximum limit position. So long as the vane is operated within the maximum limit, this limit signal is selected as the output and the summator is getting both signals of the same magnitude but of opposite signs to cancel each other for a zero output. Similar is the case for the minimum position of the vane and that summator output is also zero when the vane is operated at a position more than the minimum one as set manually. The above two summator outputs again act as inputs to another summator whose output acts as an error signal to the scoop tube controller; it would also be zero for the inlet vanes being operated within the set limits. In that case, the controller output would not be changed and so the speed of the ID fans allow the inlet vane to operate freely as per the furnace draft excursion.

Now, the case is different for the situations when the vane crosses the maximum or minimum limit positions. For example, when the inlet vane position is more than the maximum limit, the excursion is positive and the difference in the signal forms the error signal (as the inlet vane position is more than the minimum position, the magnitude of the minus signal of the down most summator is zero). The controller output would be changed to attain the new position, which would enable the vane position to be brought back within limits. The normal operation envisages no change in ID fan speed but for adjustment and fine-tuning purposes, thereby providing a bias facility to adjust the optimum speed even at the entire range of operation. The similar interlocked operation is applicable as discussed in item (ii) and (iii) of the inlet vane above.

There could be other possibilities as well where the main furnace pressure control will be controlled by varying speed and the vane may be kept wide open. The vane control may only come into operation when the demand signal for the speed control is beyond a certain range. The philosophy is almost the reverse to what has been discussed. Because in this case and most of the time the vanes are kept wide open, the pressure loss will be less.

7 DRUM LEVEL CONTROL, FEED WATER CONTROL

General

The boiler drum level control system is provided for the conventional drum boilers applicable for subcritical power plants with coal/lignite as the pulverized fuel. The boiler drum in this case acts as a reserve vessel with feed water and saturated steam as the inlet outlet.

There are a number of control valves for drum level control with different types of operating ranges and for providing redundancy. To force the feed water into the drum, there are also a number of constant-speed feed water pumps or variable-speed feed water pumps. In case the feed water level in the drum becomes too high, water particles can be carried over by the main steam going to the turbine and may cause catastrophic damage. On the other hand, in case of a very low drum water level, the drum itself becomes overheated, which may also result in a catastrophic situation.

In general, the drum level control or the feed water flow control has been provided with two modes of automatic operation: single- and three-element control. The set point is the same for the drum level controls meant for both modes as set by the plant operator. For a single-element control system, the difference between the drum level measured/process variable and the drum level set point is the error signal. This is fed to the controller to take care of the rate of water being pumped into the drum by adjusting the position of the feed water flow control valve. The very name of the three-element control comes from the three process variables: *drum level, steam flow, and feed water flow*.

7.1 Objective (Fig. 8.28)

It is one of the most important control loops of the subcritical boiler category with the objective of maintaining the drum level at a fixed value for the reasons stated above.

7.2 Discussion

In a large-capacity thermal power plant, the normal range of operation is accomplished by the three-element control strategy that usually covers a large boiler load variation from 30% to 100%. The drum level at low boiler load is accomplished by a single-element control system with a low capacity control valve as the FCE.

There are generally two major control systems adopted for the three-element control strategy. One of them is to use the full capacity control valves ($2 \times 100\%$) as the FCEs with one 30% capacity valve or to use one 25%–100% valve with one 0%–100% bypass valve (say VRT or drag valves as discussed in Chapter 6).

The other control strategy utilizes the services of the variable speed boiler feed pumps (BFP) through the hydraulic coupling between the BF pump and the motor with the help of the scoop tube actuator and/or the turbine-driven boiler feed pump (TDBFP).

It has been observed through long experience that the three-element control system is a much more rugged and stable system than the single-element control. The advantage of using the three-element control strategy is that it can easily handle large and rapid load changes due to the fact that it is matching the mass balance between the main steam flow from the boiler and the feed water flow into it.

The reason for using a single-element control is rather than a compulsion because the accuracy is always under question at a low range of flow measurements. As discussed earlier, the main two constituents are the feed water flow and the main steam flow measurements are based on empirical formula developed by using a flow measuring element like a flow nozzle (or an orifice plate for comparatively smaller plant) where the differential pressure produced across the flow element is proportional to the square of flow rate. It is well known that at low (water and steam) flow rates where the relations between differential pressures and flow rates are different from what is expected. In other words, the three-element control at low loads introduces unreliability in the control action and the single-element control is used at low loads as a natural choice.

Normally, the operator starts the drum level control in manual mode and continues up to the full range of single-element control, but due to the simple nature of the control strategy, the control system can well be placed in automatic no sooner feed flow is established after any of the BFPs is started. As generally observed, the flow measurements are considered reliable with standard accuracy at about a reading of 20% or more. By keeping some cushion, when the flow figures are around 25%–30% of the measuring range, the drum level control can be transferred to the much-desired three-element control.

7.2.1 Drum Level Computation

The drum level is measured by differential pressure transmitters to be installed in line with Fig. 12.24A and C. It shows *the* temperature equalizing column has been used to ensure equal temperature vis-a-vis density in both limbs of *the* DPT. The level is measured by DPT to eliminate the incumbent pressure P_1 of *the* steam space at one limb. P_1 is subtracted from P_1+P (pressure due to water head) on the other limb.

Therefore, DP = pressure due to water head $P=h^*\rho^*g$.

For a particular place, g is always constant. So the water head P varies with h (level) and density, which in turn depends on temperature. In the drum, the incumbent pressure is due to the saturated steam pressure, and for saturated steam, each pressure corresponds to a particular temperature. Temperature is a sluggish parameter so saturation pressure is chosen to compensate/correct the density effect.

7.2.2 Shrinking and Swelling Phenomenon

The shrinking and swelling in the boiler drum is a very common phenomenon and must be taken care of in determining the three-element drum level control strategy of a subcritical boiler plant. Shrinking and swelling are caused due to a change in pressure in the drum, which also changes the water density. Rapid excursion of load vis-a-vis steam extraction causes a tremendous change in the boiler drum pressure, followed by the drum level. The hot water closer to the saturation point contains steam bubbles in the drum. During a rapid increase in load, a severe rise in level takes place because of an increase in the number and volume of the bubbles, which are the consequences of a drop in steam pressure (i.e., swelling/expansion of bubbles) as a result of load increase and also by the increase in steam generation from the greater firing rate to match the load increase. The double effect of lower pressure and higher temperature makes the situation worse. If the level in the drum becomes too high at this time, the water carryover into the superheater or the turbine maybe the worst consequence if the boiler is not made to trip from a very high water level.

Conversely, whenever there is a decrease in demand, the drum pressure increases and the firing rate changes, thus reducing or shrinking the volume of the bubbles (that is, the bubbles get smaller). An abrupt and significant loss in load could result in high drum pressure, causing shrinkage severe enough to trip the boiler on a low level. A more severe catastrophic situation followed by a boiler trip at high firing rates may create a condition similar to a furnace implosion. If the implosion is severe enough, the boiler walls are likely to be damaged due to the high vacuum in the furnace.

However, these increases/decreases in level are a transient phenomenon that is well taken care of by the control strategy.

It is true that the firing rate change affects the drum level, but not to the extent that may make it so severe as done by the abrupt change in steam extraction rate considered as the most significant cause of shrinking/swelling of the steam bubbles.

7.3 Control Loop Description

See Figs. 8.27 and 8.28.

7.3.1 Measurement of Different Parameters

7.3.1.1 Drum Level

Here, the level of the drum is measured with sufficient redundancy at both ends of the drum. This is necessary as there may be a difference in level in view of the long length of the drum and the dynamic nature of the process. The transmitters are not connected to the drum directly (due to the fact of reducing the number of nozzles on the drum itself) but to the temperature-compensated constant head unit (Fig. 12.21) connected to the drum stand pipe provided to facilitate multiple tapping points. After necessary voting, the selected raw level signal is compensated for pressure and the average value is taken for further action in the control loop.

7.3.1.2 Feed Water Flow

The feed water flow to the boiler drum is measured with sufficient redundancy and after the necessary voting, the selected raw level signal is compensated for temperature and the average value is taken for further action in the control loop. The attemperation water flow with due compensation for pressure and for temperature is added to get the total feed water flow, depending on the relative locations of the flow-measuring device and the spray line tapping.

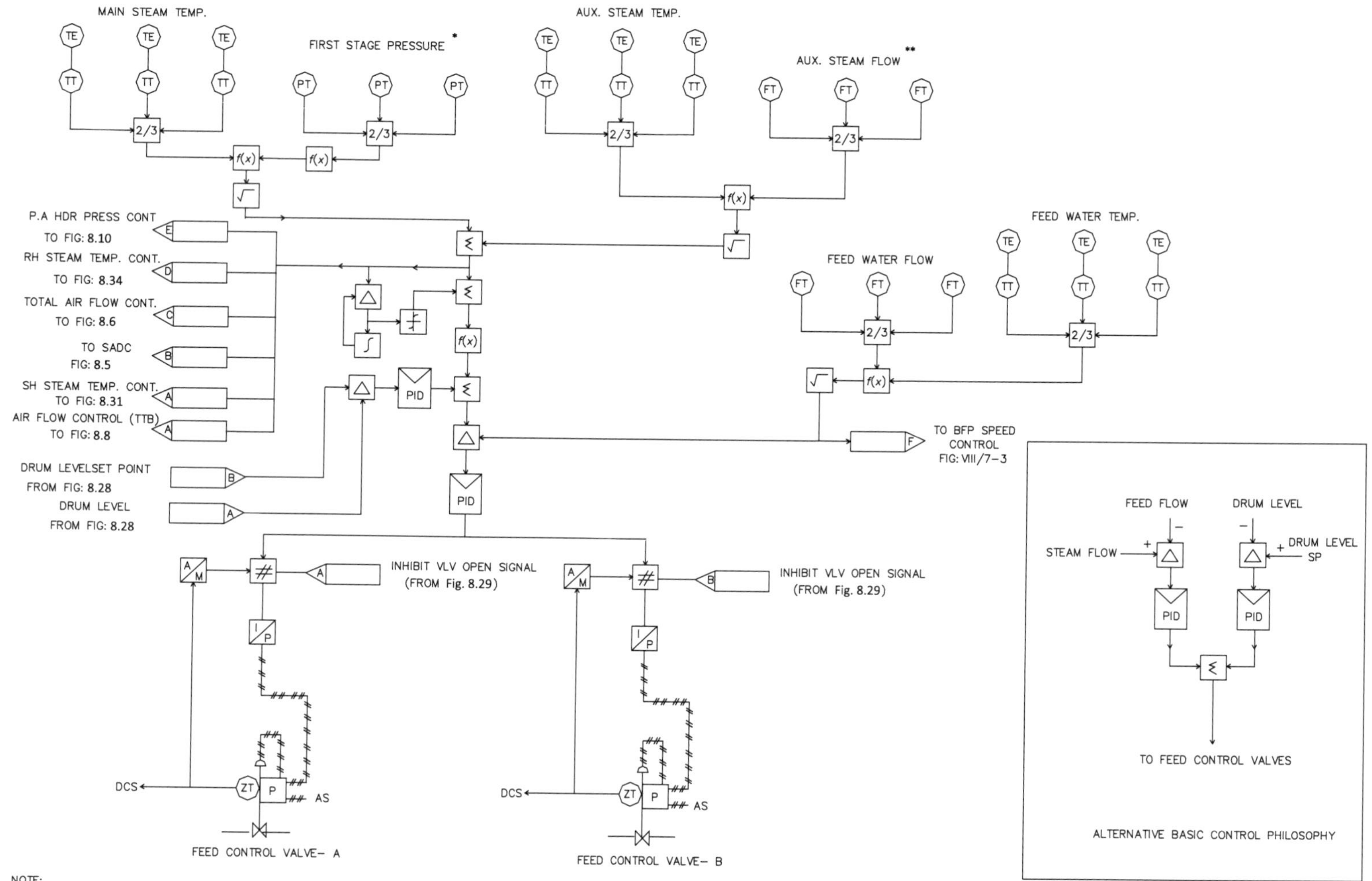

NOTE:

* MAIN STEAM FLOW IS CALCULATED FROM THE FIRST STAGE PRESSURE AT TURBINE INLET. IT MAY BE DONE BY FLOW MEASUREMENTALSO; IN THAT CASE PRESSURE COMPENSATION HAVE TO BE INCLUDED IN THE LOOP STRATEGY'

** ADDITION OF AUXILIARY STEAM FLOW IS NOT REQUIRED IF THE SOURCE IS TAKEN FROM CRH LINE.

FIG. 8.27 Drum level control.

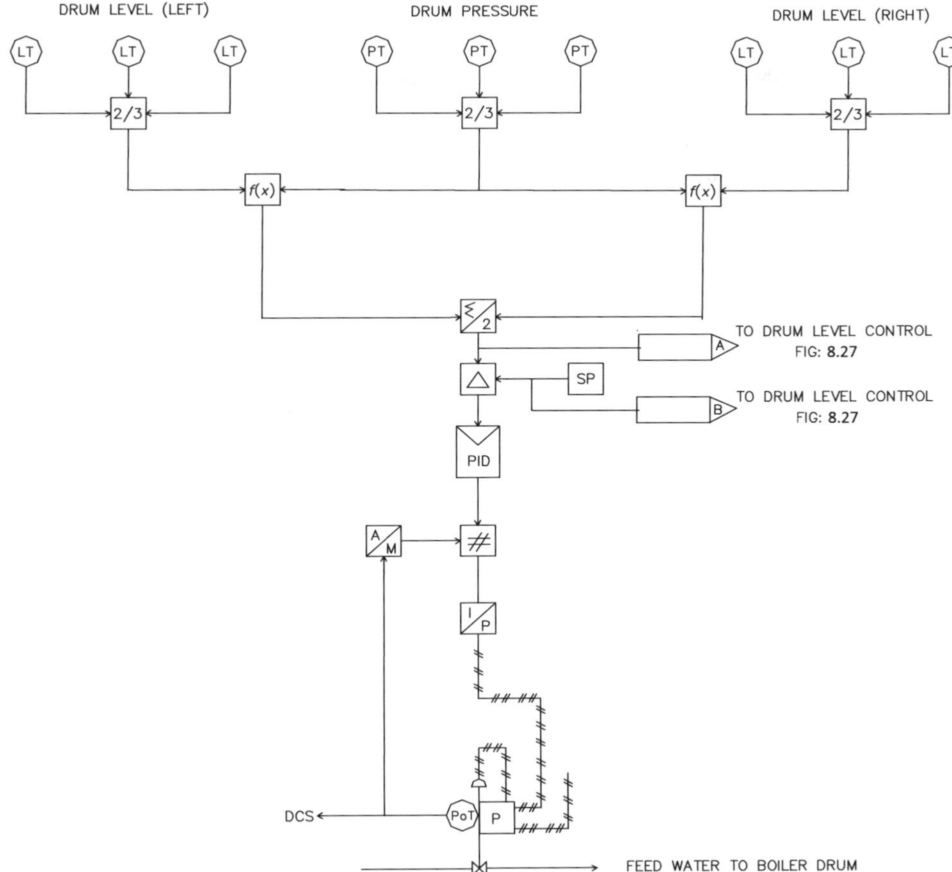

FIG. 8.28 Drum level control (low load).

7.3.1.3 Main Steam Flow

This flow measurement may be taken directly from the flow nozzle/transmitter combination or from the turbine first-stage pressure after suitable characterization through the first-stage pressure/steam flow relation algorithm. The measurement is done with sufficient redundancy and after necessary voting, the selected raw level signal is compensated for temperature. The other steam flow signals such as HP bypass steam flow and auxiliary steam flow, if any, are also measured and suitably compensated for pressure and temperature. They are then added to get the total steam flow value for further action in the control loop.

7.3.2 Different Control

7.3.2.1 Control Loop Description With Single-Element Control

As previously discussed, there are two stages of controlling the drum level, that is, single-element and three-element control. Although the three-element control is accomplished in two ways (as discussed later), there is no such option available for single-element control. The finally selected drum level signal acts as the process or measured variable and forms the error signal with the fixed drum level set point. The controller is with the PID algorithm for any adjustment for optimum tuning. A low-load control valve acts as a FCE.

In general, tuning of the single-element controller necessitates adjustments with large proportional and very little of integral gain settings.

7.3.2.2 Control Loop Description With Three-Element Control

7.3.2.2.1 Three-Element Control With Flow Control Valve as Final Control Element Drum level control normally uses the cascade control strategy, consisting of the formation of a master and a slave controller. The drum level signal, being the prime element of this control loop, forms the error signal with the fixed set point and is connected to the controller designated as the master one. The main steam flow is an indication of the rate at which water is being extracted from the drum. This measured or characterized (from first-stage pressure) main steam flow is taken as a feed forward signal and added to the output of the master controller.

An antiswelling and shrinkage circuit algorithm is customarily provided to take care of the problem, as discussed in Section 7.2.1 of this chapter. The main steam flow, before being added to the drum level (master) controller, is routed through this algorithm, which corrects the steam flow signal against a quick change. The amount of correction can be varied as required by adjusting the limiter function gradient and integral function integration rate. The correction would become zero with time when the steady-state condition is reached. For an abrupt increase in steam flow, a negative correction would apply to lessen the magnitude of the actually measured/calculated steam flow so that less water is requested for the time being to arrest the inflated drum water level (due to low drum steam pressure) from further increasing so as not to trip the boiler/MFT from a very high drum water level. The reverse is the action when there is a sudden load throw-off and the water supply is not lessened, as should have been the normal case, so that the boiler/MFT tripping can be avoided from a very low drum water level. Another method of incorporating the antishrinkage and swelling function as per another school of thought is to utilize the derivative signal of the drum pressure itself. For any quick change in load, the drum pressure ought to change as a result of energy imbalance. The rate of change with respect to time is the index of the load change rate and the signal thus derived is added to or subtracted from the steam flow (as per the tendency of pressure change, that is, increasing or decreasing) plus the drum level (master) controller output finally becomes the demand or set point for the feed water flow requirement. The difference between this set point and the measured feed water flow now becomes the error signal for another controller (slave) for feed water flow regulation. The controller output is then utilized to modulate the position of the redundant feed water flow control valves.

So far as the tuning of the three-element control system is concerned, it has some special requirements. Here, the feed water flow (slave) controller must be tuned with the much faster integral action than that of the drum level (master) controller. This philosophy is in fact applicable to all types of cascade control strategies.

There could be another simple way to achieve the same goal. Here, a demand is created by the steam flow feed flow error (in the ideal case, steam flow and feed flow should match, and only during changes will there be an error/difference between them). Thus demand created by a controller is added to the demand from the drum level controller. For steam flow, feed flow and drum level measurements with necessary corrections as stated in the main description are equally applicable here.

For the drum level control loop, the strategy is otherwise simple without many manual interlocks or control tracking. Though not specifically shown in the above drawings, the control loops may be forced to run in manual mode in case all pumps are not running or the drum level signal/feed water flow valve control output goes out of range, as may be deemed necessary by the appropriate authority.

Three-element control with scoop tube actuator for control of DP across feed valves (Fig. 8.29)

When the three-element control for the drum level is accomplished by the flow control valves, then another control loop is also envisaged for controlling the differential pressure (DP) across the feed control valves to enable them to operate smoothly. Controllability is also improved for control valves when subjected to a DP with a predictable range. In some plants, this DP is maintained at a fixed value by adjusting the BFP speed through hydraulic coupling vis-a-vis the scoop tube actuator. BFP speed control may be referenced for a schematic representation of the idea for implementing the control loop in a typical 250 MWTPS with a drum boiler.

In this control loop strategy, the DP across the feed valves is measured with sufficient redundancy, and after necessary voting, the selected raw DP signal is taken as a measured/process variable to form the error signal with the desired/set value of the DP, which is not a fixed value as in the other types of power plants, as stated earlier. This particular power plant has a variable set point for the DP after due characterization of the selected (high) position transmitters of the first-stage attemperation spray valves. All these position signals are passed through a high selection algorithm. There is another fixed input of the high selector that ensures the minimum value of this selector output as an 80%–90% (adjustable) position of the attemperation spray water valves. After characterization as per the predictable relation of the flow control valve DP with respect to the spray valve position, the output is again passed through a high/low selector that limits the derived set point from going beyond the maximum and minimum values of the DP across the flow control valves. The error signal thus obtained drives the PID controller whose output is utilized for positioning the scoop tube actuator for varying the speed of the BFPs.

The control loop strategy is made with a view to maintain the DP across the feed valves to such an extent that the upstream pressure of the flow control valve, which is incidentally the upstream pressure of all the spray control valves as well, slides to the minimum (for better efficiency) required value sufficient enough to push an adequate amount of spray water with the spray valves almost fully open. It is quite obvious that a wider opening of a control valve demands less DP across it with the flow being same. With the downstream pressure of the spray valves being dictated by the process, that is, the boiler load, the upstream pressure is now the minimum pressure required for the necessary flow of spray water. This minimum pressure calls for the minimum discharge pressure of the BFP required at that point of operation, thus minimizing the pumping loss.

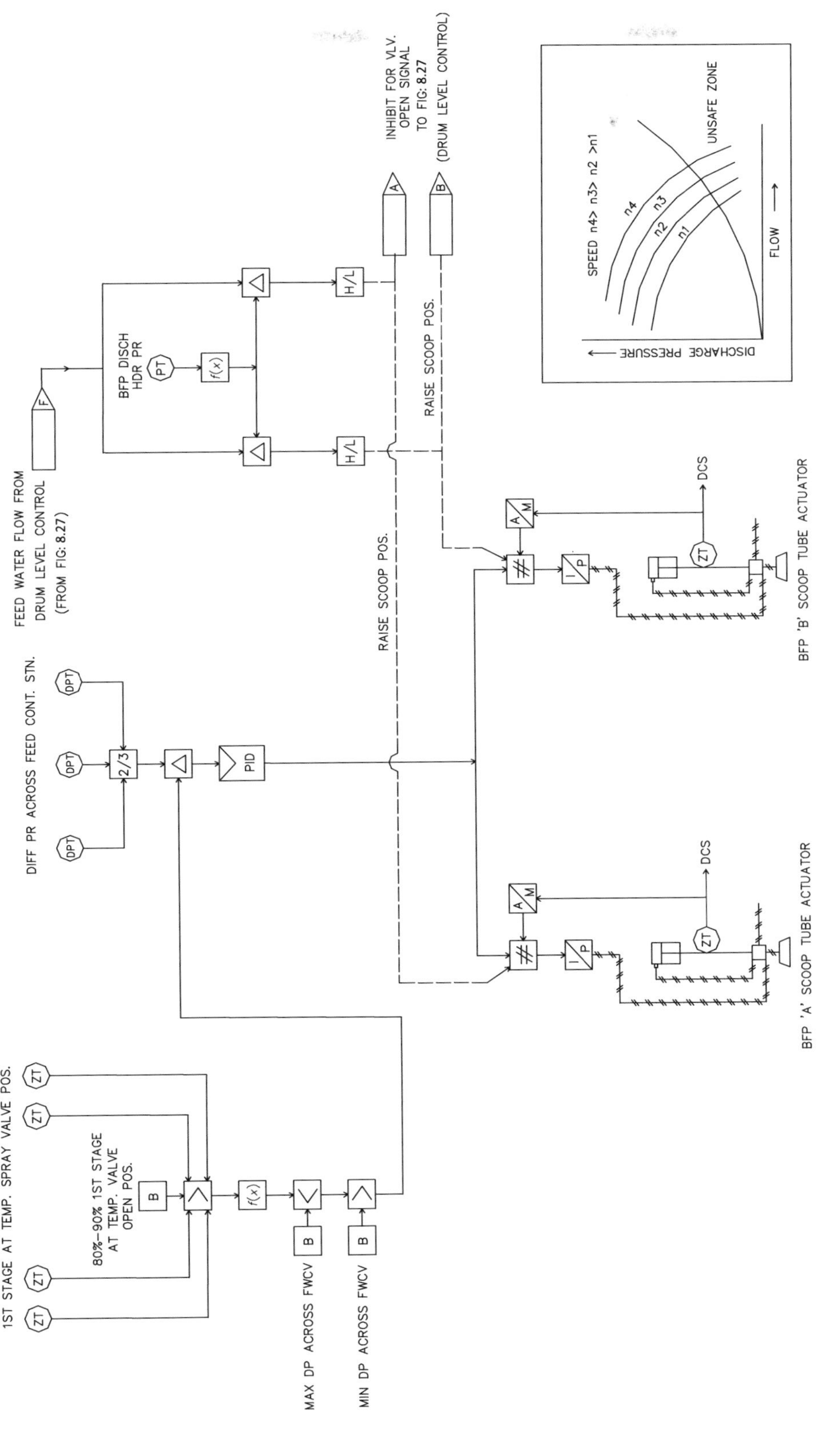

FIG. 8.29 BFP speed control (2 × 100% MDBFP).

The flow control valve readjusts its opening to suit this DP across it to maintain the feed water flow to maintain the drum level. This control strategy thus ensures a minimum but adequate upstream spray water pressure to keep the spray control valves operating in a satisfactory manner without running the BFPs at an unnecessarily high discharge pressure. For single-loop control strategy, the scoop tube actuators would remain in a fixed position, irrespective of whether the control mode is put on auto or manual.

With one BFP running, the second standby BFP (for $2 \times 100\%$ configuration) should have its scoop tube on auto follow-up mode so that, in the event of a tripping of the running pump, the second pump should start the forward feed to the boiler without much loss of time. The maximum allowable delay time in starting the second pump is dictated by the boiler drum storage capacity and the permissible time limit required to come to a very low drum level from a normal drum level. For $3 \times 50\%$ BFP configuration, the standby BFP would have its scoop tube actuator on auto follow-up mode to track the maximum position of the two running BFP's scoop tube actuators.

For TDBFPS, one motor-operated BFP is required as a fallback. The hydraulic coupling scoop tube always tracks the TDBFPs so that in case any TDBFP trips, the motorized BFP (with discharge valve open) can immediately meet the flow requirement. The BFP motor shall be rated in such a way for a loaded start.

The BFP running at safe pressure against the flow delivered at a particular point of time is detected for interlocked operation. Here, the characterized BFP discharge header pressure represents the limit of a safe discharge flow at the particular speed with which the pump is running. When the actual flow is more than the safe flow, binary signal is generated in the high/low signal detector (high in this case). This would force the scoop tube actuator to raise the speed of the BFP so that the operating point is shifted as to make the safe flow become more than the actual flow at a higher discharge pressure. An interlocked signal is also arranged at the same time to prevent the flow control valve from further opening, increasing the feed flow to the unsafe region of operation.

For a $3 \times 50\%$ BFP, individual pump suction flow is to be measured and compared to the safe flow from the characteristic curve at the corresponding discharge pressure.

7.3.2.2.2 Three-Element Drum Level Control by Pump Speed Variation (With Scoop Tube Actuator/ Turbine Driven BFP) (Fig. 8.30) The designers/manufacturers/customers of modern power plants prefer the feed water flow to be controlled by the variable speed feed pump with a scoop tube for the hydraulically coupled motor-driven BFP (MDBFP) and turbine-driven BFP (TDBFP). This arrangement requires a low load feed control valve for 0%–30% boiler load (single-element control), as previously discussed in Section 7.3.2.1 of this chapter. For a load higher than 30%, the usual control strategy, that is the drum level controller with the addition of total steam flow, duly incorporating the antishrinkage and swelling algorithm, becomes the set point of the feed water flow controller. The simplified schematic typically represents the control loop of a 500 MW power plant having one MDBFP and two TDBFPs with $3 \times 50\%$ pump configuration.

Here, parallel to the low-capacity FW control valve, that is, in the main FW line, there is another valve of the throttling cum isolation type that would be made (manually or automatically) to open slowly up to 100% at a boiler load of >30%. By this action, the low-capacity FW control valve would close gradually to 0% and then the FW is controlled by adjusting the speed of the pumps with the main line isolating valve fully open. The error between the feed water flow set point and the feed water flow measured/process variable is connected to the FW flow controller, but only after a manual or automatic changeover selection representing the boiler load has reached >30%.

As depicted in the above control strategy (Fig. 8.30), there are a number of controllers shown as PID in general, but the tuning of those controllers is very important. As discussed earlier, the speed controller may have a very fast integral action rate whereas the FW water controller may have a slower integral action rate with a drum level controller having a further slower integral action rate. However, all the whole tuning activities would depend on the plant condition, the quality/capability of the control systems, and the availability of equipment/accessories. To avoid a three-stage controller situation, in some control strategies the drum level controller is eliminated and the FW controller error signal is generated directly from the drum level set value minus the measured value plus steam flow with the suitable antishrinkage and swelling algorithm. Then, the FW flow signal is subtracted to get the error signal for the FW controller.

When the *boiler load is around 30%*, the main line valve is gradually opened and the low-load valve is gradually closed; it may be achieved automatically but manual selection often is a better option from a loop stability point of view.

DP mode error:

This signal is selected from the DP error across the low-load feed valve or the pump safe operation criteria through a maximum selection. The DP across the low-load feed valve is measured and the error with the DP set point forms one of the two inputs, which, if selected, controls the pump speed. This is, however, applicable to low-load operation only and the valve DP error signal is ignored at higher loads.

The other input that is the pump safe operation signal. The maximum of the measured flow signals of each individual pump (with sufficient redundancy and voting arrangement) passing through the characterizer (flow versus discharge pressure) is considered a safe discharge pressure

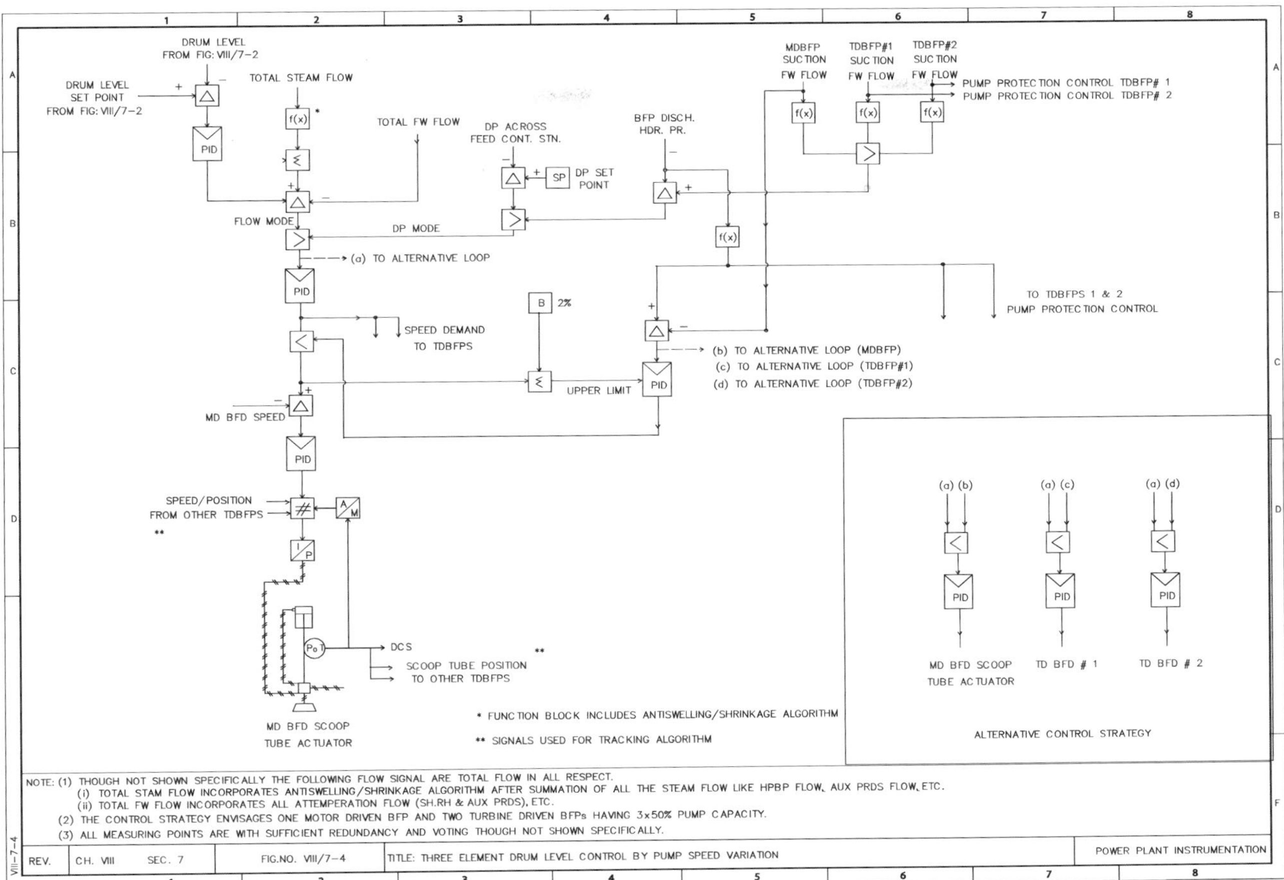

FIG. 8.30 Three-element drum level control by pump speed variation.

at any given flow and is input to a subtractor along with the actual BFP discharge pressure. This signal, if selected via the maximum selector, would be trying to maintain the safe discharge header pressure of the BFPs by adjusting the speed of the running pumps. This part of the control loop is known as DP mode.

The DP mode error signal is compared with the flow mode error signal in another maximum selector and acts as an input to the flow controller related to the boiler load. Another flow controller is provided to take care of the pump protection by bringing the individual suction flow to a safe value, according to the common discharge pressure. An individual pump's protection-related flow controller is free to act within the upper limit derived by adding a 2% (for example) signal value to the minimum selected signal of the two flow controllers stated above. Normally, the common boiler load-related controller would be selected as long as the discharge pressure is adequate enough to administer the required feed water flow.

The minimum selector output of flow controllers then constitutes the speed demand of the individual pump and is compared with the pump speed to determine the operating speed of the concerned pump.

As already indicated, that the MDBFP speed is varied by adjusting the scoop tube position of the hydraulic coupling, TDBFP speed is made to vary by the electrohydraulic governor control.

8 SUPERHEATER TEMPERATURE CONTROL

General

Superheated steam is principally used in power generation plants as the driving force for turbines. The application of the Rankine cycle in thermal power plants is available in most textbooks relating to the thermodynamic properties of steam. Very brief discussions are also in various sections of

Chapters 1 and 2, from which it is apparent that superheated steam is more thermally efficient than saturated steam when the question of driving a turbo generator arises.

As the extra heat causes the transmission of more energy than saturated steam, it is utilized for additional power generation from the available quantity of steam. Although superheated steam contains a large amount of heat energy, this energy is available from three sources, that is, the enthalpy of hot water, the enthalpy from evaporation (latent heat) of water to steam, and the enthalpy from the degree of superheat. The bulk of the energy is in the enthalpy of evaporation, and then from hot water and less energy from the superheated steam, which represents a comparatively smaller proportion of the total heat content.

Using superheated steam (meaning no condensation in the pipelines) has some added advantages that are very important, such as:

(a) It eliminates the problem of entering wet steam (with water droplets) inside the turbine, which causes increased friction, enhanced silica deposits, and pitting/erosion of turbine metals.

(b) It prevents water carryover in the saturated steam, which could cause steam formation and may give rise to overspeeding of the turbine, which is extremely dangerous.

(c) It permits higher steam velocities through the pipeline (up to 100 m/s), which automatically suggests that smaller pipelines can be used (provided that the pressure drop is not excessive).

8.1 Objective

The objective of this control (considered one of the most important) loop is to maintain a constant temperature of the boiler outlet steam or turbine inlet (fixed or sliding pressure) with the provision of variation due to transmission loss. The saturated steam from the boiler drum is superheated by adding heat through various superheaters (LHS and RHS). Before them are desuperheaters having water spray chambers. Hot feed water is sprayed through these spray chambers as per the command of the temperature controller, so that the steam temperature at the final superheater inlet goes down adequately so as to gain the required temperature inside the final superheater to maintain a constant temperature at the final superheater/boiler outlet.

8.2 Discussion

Superheated steam temperature control is a critical consideration for the efficient running of the steam generator vis-a-vis the turbo generator unit. Steam temperature needs to be stable to accomplish the highest turbine efficiency. This would also reduce the fatigue of the turbine metal, as discussed in Section 6 of Chapter 9.

As the steam temperature is controlled by spraying water with a comparatively colder temperature into the steam spray chamber, constant temperature control is a bit problematic as the temperature measurement itself is a sluggish phenomenon and the time delay arises out of the process itself. The spray chambers are generally located much before the superheater whose outlet temperature is to be measured and controlled. The changing process dynamics—including system gains, time constants, and delays—are also responsible for concern that varies as the turbine load changes. Considering all the aspects, it is apparent that the superheat steam temperature control has always been a critical subject for the efficient and most favorable operation of a power plant. The control strategy, in normal practice, employs cascade controls with controllers having PID algorithms at different levels to regulate the superheat temperature. The state-of-the-art technology has made modern controllers with self-tuning algorithms, enabling the controller block to adjust the controller parameters such as the proportional band, the integral time constant, and the derivative action rate. However, some instrumentation manufacturing houses and researchers feel that the self-tuning controllers may not be capable of tackling the situation alone against the tendency of changing gains and time constants of the controlled system during the disturbed condition.

New adaptive control strategies are now proposed for the improvement of superheat steam temperature control. The algorithm of these controllers is based on a cascade control system having PID functions, but self-tuning parameters are programmed to adjust themselves following the recursive least squares (RLS) method. As claimed by the inventors, the smart adaptive system can cope with the adverse disturbances much better than a conventional control function algorithm, with a faster and more accurate response as well as more effectiveness.

Another type of control algorithm, popularly known as a *predictive adaptive* controller, is also advantageous for maintaining the superheated steam temperature of large thermal power plants. Adaptive techniques combined with a predictive element enable the controller to incorporate continuous adjustments of parameter tuning by tracking the plant dynamic variations.

An advanced and new predictive adaptive controller algorithm uses a function series approximation technique called dynamic modeling technology (DMT), which is, as claimed, particularly designed to take care of the system under control with a long time delay and long time constants. The design is said to have incorporated the unique ability to build the models automatically while operating in a closed-loop configuration as well as controlling the transient responses that take place during boiler load changes. The model, or in other words the mathematical representation of the process response that normally takes expertise of a complex nature and detailed knowledge, need not be required here to be built as the DMT modeling method does the same thing by the use of advanced mathematical functions. For details, relevant textbooks may be referenced.

With superheated steam temperature control being a critical control loop, a different approach has also been made to address the problem with the help of a system that is known as *state variable control*. It is a unique set of variables that describes the state of a dynamical system at any point of time. In other words, the future behavior of a particular system can be predicted automatically to a close proximity if the data required by the state variable technique are provided judiciously. To be more precise, the state variables themselves represent the state of any type of system in general. In a thermodynamic system, the state variable input data are pressure, temperature, heat content, and related parameters of the system such as enthalpy, entropy, and internal energy. It is practically a special method of preparing a model of the concerned system employing a time-domain technique. Here, the actual physical system, that is, the thermodynamic representation of the boiler and turbine, is described by an ordinary differential equation of the n_{th} order. With the help of a state variable, a set of first-order differential equations is calculated and grouped, exploiting the use of a compact matrix notation that is depicted as a model and widely known as the state variable model. At any instant, the state equations receive the current inputs along with past states and bring out the relation to the current state and output of the system so as to enable the control system to generate the appropriate output.

8.3 Control Loop Description (Figs. 8.31 and 8.32)

All the methods other than the self-tuned and optimized cascade PID controllers are explored so that a reduction of steam temperature fluctuations around the set point is obtainable. However, the control loop strategy for superheater steam temperature control is discussed, incorporating the PID controllers for easy understanding of the control philosophy.

Figs. 8.31 and 8.32 are schematic representation for implementing the control loop in a typical 250 MWTPS having a drum boiler, two-stage spray water attemperation systems, and reheat steam temperature control through a gas recycling camper and a spray water attemperation.

8.3.1 Measurement of Different Parameters

The following measurements are done with triple redundancy and voting. The selected parameters are utilized in the control loop:

(i) First- and second-stage attemperator outlet temperature (left and right side).
(ii) Platen SH and final SH outlet temperature (left and right side).
(iii) Primary SH inlet pressure.

8.3.2 Control Loop Description

The steam path in the system for which the control strategy is described incorporates a primary SH, a first-stage attemperator, a Platen SH, a second-stage attemperator, and a final SH ad seriatim, irrespective of their physical location.

8.3.2.1 Control Loop Description for Two-Stage Attemperation

8.3.2.1.1 Control Loop Description for Second-Stage Attemperator Outlet Temperature Control Loop The ultimate measured/process variable is the *final SH outlet temperature*, which is compared with the constant set point (adjustable at low load) to form the error signal for connection to the master controller (or the outer controller) with PID algorithms. A derivative functional is normally provided in the temperature controller to compensate for the sensor's inherent sluggish behavior. The controller output now becomes the variable set point of the second-*stage attemperator outlet temperature controller* and forms the error signal along with the measured/process variable; it is connected to the controller. This PID controller may be termed the slave/inner controller and generates the output for the position adjustment of the second-stage attemperator spray valves. As already indicated, the second-stage attemperator is located (in the steam flow path) just before the final SH. The second-stage attemperator provides the necessary spray water injection while the main steam is flowing through it to adjust the attemperator outlet temperature at a value just needed to maintain the final SH outlet temperature at a constant value after the necessary temperature gain within the final SH.

The steam flow (in some plants, the air flow) is added to the controller output as the feed forward signal. The interlock shown from the steam flow $<25\%$ is provided to block the attemperation flow up to a 25% load.

8.3.2.1.2 Description for First-Stage Attemperator Outlet Temperature Control Loop Now comes the first-stage attemperator outlet temperature control. Actually, the first- and second-stage attemperator outlet temperature control loops are interconnected with each other in a cascade mode of control, justifying their interrelation within the system. The first-stage attemperator outlet temperature control loop is also configured in a master/slave mode whose master controller receives its variable set point from the final SH steam temperature master controller.

The first-stage attemperator, with its comparatively larger spray capacity, sets the temperature at the second-stage attemperator inlet or, in other words, the platen SH outlet such that the second-stage attemperator maintains,

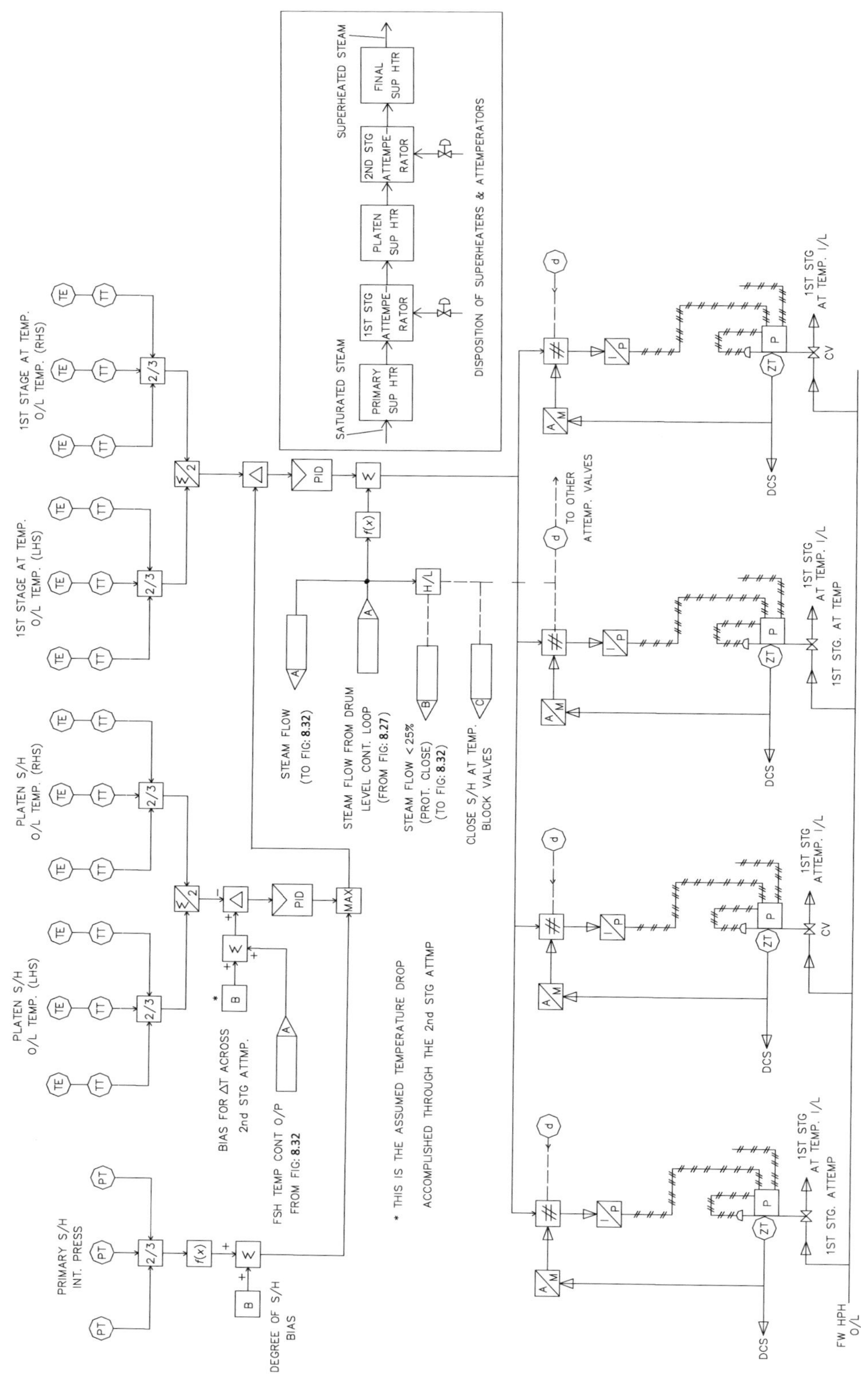

FIG. 8.31 Superheater steam temperature control.

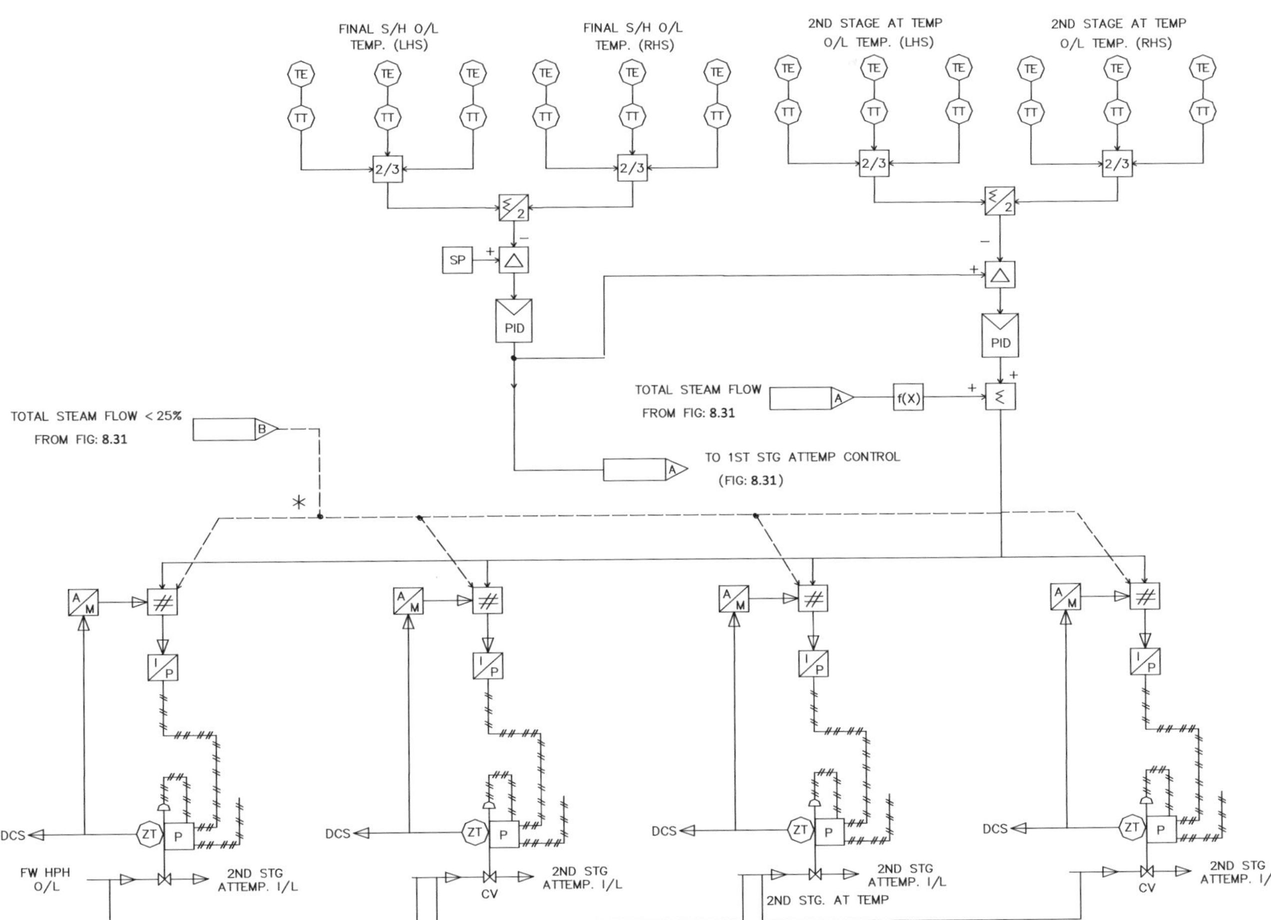

FIG. 8.32 Final superheater steam temperature control.

in the long run, the exact mean spray water flow while effectively controlling the final main steam temperature at the required value. Both the control loops are almost identical with the difference that the final SH steam temperature master controller (the outer controller) has a fixed set point whereas the first-stage attemperator control loop master controller has a variable set point, which is just required to attain a value as dictated by the final SH steam temperature master controller. However, as the final SH steam temperature master controller output is actually the variable set point or the desired temperature at the second-stage attemperator outlet, it cannot also be the set point for the first-stage attemperator control loop master controller (outer controller) meant for the second-stage attemperator inlet or, in other words, the platen SH outlet. Therefore, the differential temperature as envisioned by the designers is added to the desired second-stage attemperator outlet temperature (as derived from the final SH steam temperature master controller output) to get the anticipated variable set value of the second-stage attemperator inlet temperature.

The measured and selected values of the platen SH outlet temperatures of the left and right sides are then averaged and become the process/measured variables for the second part of this cascade control loop. The error signal is then connected to the first-stage attemperator control loop master controller (outer controller), the output of which becomes the variable set point for the controller assigned for the first-stage attemperator outlet temperature control but after a maximum selector.

Under certain operating conditions, there is the possibility of the control system demanding a first-stage attemperator outlet temperature below the saturation temperature, which is taken care of through the incorporation of this additional function so as to ultimately protect the platen superheater.

The primary SH inlet pressure (or drum pressure) is measured and the corresponding saturation temperature is calculated through a function generator.

To this signal is added (as a positive bias) an increment of around 20–35°C (meaning a certain degree of superheat), depending upon the operating pressure (the higher the pressure, the lower the permitted increment). The resulting summated signal, that is, the saturation temperature plus a certain degree of superheat, is treated as the minimum permitted first-stage attemperator outlet temperature and becomes another input of the above-mentioned maximum selector to ensure that the steam temperature does not come down near saturation at the first-stage attemperator outlet temperature after the application of spray water.

With this variable set point, the first-stage attemperator outlet temperature slave controller the output of the controller is added with a feed forward signal from the total steam flow as generated in the drum level control loop (Fig. 8.27) after appropriate characterization.

The output of the summator is now the position demand of the spray valves of the first-stage attemperator for maintaining its outlet temperature.

8.3.2.1.3 Safety Interlocks of the SH Temperature Control Loop The following safety interlocks are incorporated to protect different components of the thermal circuit. They are described as follows:

(i) The spray water isolation valves and block valves are interlocked to close automatically if the steam flow (or the equivalent boiler load) is equal to or <25% of the continuous maximum rating (CMR) to protect the superheater system from the effects of water droplets due to inefficient atomization of the spray water at low loads. Another interlocked operation is envisaged to release the signal to auto operation of the spray water modulating valves, unless the isolating valve is fully open; this logic prevents the possibility of the modulating valves being open due to a position demand request from the control loop when the availability of the spray water is uncertain.

(ii) The spray water modulating valves are interlocked to the closed position through a bias close signal from the steam flow <25% CMR when the isolating valve is fully closed; it also causes the controller output to trip to manual.

8.3.2.2 Control Loop Description for Boiler With Single-Stage Attemperation (TT Burner Boilers)

The description of the above control loop takes care of a typical power plant of 500 MW capacity having a steam drum, a firing system with a burner tilting arrangement, and provisions for both fixed and sliding pressure control systems.

1. Discussions: There are some differences in boiler design. For example, for TT burner boilers, the majority of the superheaters and reheaters are platen where the heat absorption is mainly by radiation, requiring more spray flow at lower loads. For this reason, during start-up and low-load, the set point is very critically chosen, as discussed in clause no. 8.3.2.2.3 below. For the same reason when the load is high, most of the heat is taken away by steam so the spray flow (mainly for SH steam) goes to the very minimum (even zero). The other type described in clause no. 8.3.2.1.2 has major sections of superheaters and reheaters as convection types, meaning more heat will be absorbed by the steam as the load increases with a tendency for the steam temperature to rise, hence a higher spray flow at a higher load. Also, burner tilting (mainly used for reheater control) will have an impact on heat radiation in superheater platen zones also. Therefore, the effect from the same will be taken care of here. Now, a boiler load index is the output of the boiler (in a drum boiler, the derivative of the drum pressure also contributes energy output, the *d/dt* of the drum pressure is added with the load index) and must match the fuel demand; otherwise the unbalanced energy will try to disturb the steam temperature. Such effects are used in the control strategy to make the sluggish temperature loop responsive.
2. *The measurements are* desuperheater (DSH) and final SH outlet temperature (left and right side).
3. *The control loop* is basically a cascade type and has a similar control strategy as that of the second-stage attemperator outlet temperature control part described above. The ultimate measured/process variable is the final SH outlet temperature, which is compared with the temperature set point to form the errors (Fig. 8.33).

Here, the set point is made variable, and is programmed as a function of the unit steam flow signal for both constant pressure control and sliding pressure control, but with separate characteristics. A manual set point is also provided and the output of a minimum selector passes the set point for generating the error signal, which is connected to the master controllers (outer loop controllers) with PID functional blocks (algorithms). A derivative function is usually provided for the temperature controllers to compensate for the sensors' inherent sluggishness.

When the difference between the left and right side temperatures at the final SH steam outlet header is within the limit, their average temperature acts as the measured/process variable for both the master controllers of the left and right side temperature control loops. In case of a high difference between the two sides, then the average signal is no longer considered as the measured/process variable. Instead, the individual the final SH steam outlet temperature becomes the measured/process variable for the respective control loops.

The controller output now becomes the variable set point of the desuperheater or attemperator outlet temperature controller and forms the error signal with the measured/process variable for the controller. This PID controller, which may be called the slave controller (inner controller), generates

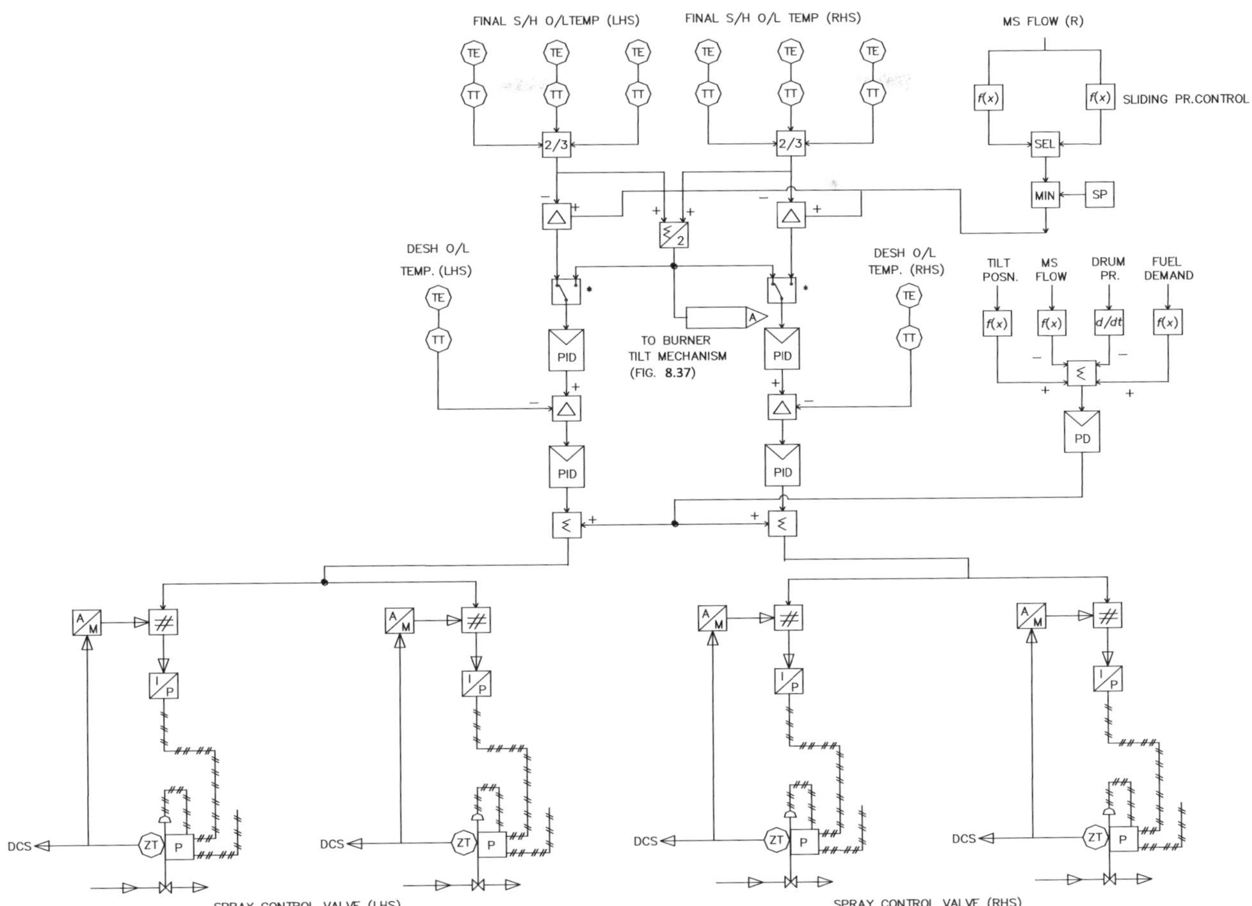

FIG. 8.33 Superheater steam temperature control (500 MW TPS) with burner tilt.

the output to position the desuperheater or attemperator spray control valve providing the necessary spray water flow injection to the main steam flowing through the desuperheater. It also adjusts its outlet temperature at a value just needed to maintain the final SH outlet temperature as desired after the necessary temperature gain within the final SH.

With reference to clause no. 8.3.2.2.1, the feed forward signals provided to make this control strategy more responsive are burner tilt position, main steam flow, drum pressure, and fuel demand.

The RH temperature, when controlled through the burner tilt operation, causes the SH temperature excursion, which is taken care of by the burner tilt position feed forward signal. The same logic applies to the fuel demand rate change. For instance, when fuel demand increases, it may cause overfiring; the spray valves are then open in advance to counter the effect due to overfiring.

The steam flow and drum pressure (derivative) signals are taken as the feed forward signals after the effects of overfiring. At that time, the steam flow and/or drum pressure would rise and is used as the index to recognize the tendency for the temperature to drop. This is thus countered through these signals by closing the spray valves.

All these signals are duly characterized to generate suitable output signals and are summated to make an input signal for a PD controller. The output is again added to the output of the slave controller (or the inner controller) to become the position demand for the individual spray valves.

9 REHEAT TEMPERATURE CONTROL

General

Exhaust steam from the high pressure turbine (HPT) is brought back to the boiler for making it again superheated (or reheated) steam at a lower pressure, but normally to the same temperature as the main steam temperature. The thermodynamic properties of steam and the application of the Rankine cycle in the thermal power plant suggests that superheated steam is more thermally efficient than saturated steam when the question of driving the turbo generator arises. That is the reason as to why the exhaust steam from the HPT is again reheated and the extra heat causes the transmission of extra energy. This is utilized for additional power generation from the available quantity of steam.

9.1 Objective

The objective of this control loop is to maintain a constant reheat steam temperature at the boiler outlet, meaning an almost constant temperature at the IP turbine inlet subject to variations due to transmission loss. The exhaust steam from the HPT is superheated by adding heat through various heat exchangers known as reheaters. Before the reheaters (left side and right side), there are arrangements for desuperheaters having water spray chambers in the CRH lines. Hot feed water is sprayed through these spray chambers as per the command of the temperature controller so that the steam temperature at the inlet of the reheaters goes down adequately so as to gain sufficient temperature, while passing through the reheaters, for maintaining constant temperature at the outlet of reheaters vis-a-vis boiler outlet. Desuperheaters are utilized as the last resort when all other FCEs reach their end positions.

9.2 Discussion

Similar to superheated steam temperature control, reheated steam temperature control is also a critical parameter for accomplishing the highest turbine/generator efficiency. This would also reduce the fatigue of the turbine metal, as discussed in Section 6 of Chapter 9.

It may be noted that *reheat steam*, having a low operating pressure, is subjected to different thermodynamic conditions than superheater steam. For the reheat system, problems may arise out of injecting colder spray water associated with unwarranted effects on the overall efficiency. The control system for reheater steam temperature thus normally avoids water spraying because long and uses a primary control system which would take care of the control of reheat steam by either adjusting through position control of *tilting burners* or by apportioning of *flue gas* flow through the superheater and reheater banks or recirculation of flue gas into the furnace from suitable take-off point. The provision of spray water injection is kept when the main FCE has reached the extreme position or the temperature has shot up to a particular predefined value when emergency spray is necessary. The system so far prevents the need for spray water and confines the use for fine tuning and emergency purposes only. In some control schemes, the system is used for controlling the furnace outlet temperature as well.

Therefore, there are different schools of thought for adapting the control philosophy (other than spray water control), depending on the type of boiler manufacturer. Basically, those are reheat temperature controls by:

(i) Control through gas recycling damper (Babcock boiler).
(ii) Control through gas dampers (main line) or bypass dampers (Babcock boiler).
(iii) Control through burner tilting mechanism (CE boiler).

9.2.1 Control Through Gas Recycling Damper

For this control, a damper at the chimney inlet is provided in a common flue path at the discharge of the ID fans. This arrangement ensure the generation of a back pressure with sufficient magnitude so as to force the flue gas from the ID fan discharge header to the furnace again through a separate duct connected between the main flue line and the furnace hopper. This recycled gas flow controlled by the dampers alters the furnace absorption and mass flow of the flue gas, which maintains the steam temperature within range at the reheater outlet. In this case, the size of the ID fan will be higher because it handles more flue gas.

There is another method of injecting flue gas into the furnace where the take-off point is much nearer to the furnace, which means the flue gas has a much higher temperature but much lower pressure. Recirculation fans are provided to generate pressure higher than the furnace. The gas recycling damper does the rest of the things.

Some added advantages of having a flue gas recirculation system for reheat temperature controls are:

(i) Total mass flow of the flue gas increases, enhancing more heat transfer to each heater bank.
(ii) Due to the admission of colder flue gas, the temperatures at different stages of the furnace/boiler would go down, though the overall enthalpy would increase and so does the heat transfer.
(iii) As the colder flue gas is injected near the burner, the resultant low temperature would help in the generation of less SO_x and NO_x compared to a system without a gas recirculation damper.

With the take-off point for the gas recirculation system being at a low pressure zone, gas recirculation fans are necessary to ascertain that the flue gas passage back to the furnace through a long way. By controlling the gas recirculation flow, the steam temperature of both the superheater and reheater may also be controlled. With the gas recirculation damper, the reverse flow of very hot flue gas must be arrested to protect the recirculation fans when they are not running.

9.2.2 Control Through Main Path Gas Dampers or Bypass Gas Dampers

In this arrangement, the quantity of flue gas passing through the reheater is varied by the gas bypass dampers to control the reheater temperature.

The gas dampers do the apportioning of the flue gas flow through both the superheater and the reheater banks by adjusting their position. Some arrangements allow two separate and dedicated sets of dampers in the flue gas path of both the superheater and reheater banks controlling the individual flue gas flow. With the dampers being located across the main path and controlling the total flue gas volume, the control system ensures that the dampers do not get a closed signal simultaneously, which may lead to constriction of the flue gas path. The resultant overpressurization may damage the boiler structure.

To obviate the untoward situation, the control philosophy suggests that when one particular set of dampers is fully opened, only then is the other set allowed to throttle.

Control through gas dampers is a slow response system and for large units, a sudden rise in temperature following a rise in furnace heat input may call for emergency spray to cope with the load demand.

9.2.3 Control Through Burner Tilt Mechanism

The steam temperature control system is genetically different in two types of furnaces, that is, one with a tangentially fired tilting burner and the other having fixed burners.

By this arrangement, the burning zone is raised or lowered by the burner tilting mechanism. The burners are corner-fired and focused tangentially to a common imaginary circle. When firing starts, there forms a huge fireball with swirling action. As all the burners in a corner are gagged to a common actuator and all the four corners get the same positioning signal, the fireball moves vertically upward or downward as the burner assemblies move up or down.

Such repositioning of burners vis-a-vis the fireball changes the pattern of heat transfer to the superheater and reheater banks and thus controls the steam temperature. The burners are so arranged that tilted up or lowered down and so the flame envelope in the furnace causes a good variation in the amount of radiation heat received by the reheater and superheater. In fact, the firing system in any type of furnace of a power plant boiler having reheater banks affects the steam temperature of both the superheater and the reheater. It is very difficult to control both steam temperatures by a single control system, as corrective action for one steam heater bank may have an adverse effect on the other bank, which calls for separate control strategies.

9.3 Control Loop Description

9.3.1 Control Loop Description With Gas Recycling Damper and Spray Water Attemperation

9.3.1.1 Measurement of Different Parameters

The measurements necessary with sufficient redundancy and voting include gas recycling flow and reheat attemperator flow, reheater outlet steam temperature (LHS/RHS), DP across the gas recycling damper, position of the gas recycling damper, position of the reheat steam outlet valves (for temperature balancing), main steam flow, and furnace pressure.

9.3.1.2 Control Loop Description

The FCEs in the system are the gas recycling damper, the stack inlet damper, the reheat attemperator valves, and the reheat steam outlet valves (for temperature balancing).

9.3.1.2.1 Function of Gas Recycling Damper (Fig. 8.34) The steam temperature is measured at both the outlet legs of the reheaters and averaged after selection and voting to form the ultimate measured/process variable for this control loop. The control strategy utilizes the recycled flue gas flow to the furnace hopper by throttling the single gas recycling damper receiving the controlled input command from a temperature/flow cascade control action.

The PID (temperature) controller employed for this purpose has a set point adjuster for setting the desired value of the reheat steam temperature control system. The controller output is then summated with another PID (flow) controller, the output of which signifies whether the requirement of gas recycling flow is satisfied. The gas recycling flow signal as measured forms the measured/process variable for this part of the control loop and the set point is derived from the main steam flow (as the load index), duly characterized to give the corresponding expected recycled flue gas flow. The (temperature) controller output trims the flow controller output to derive the required desired position of the gas recycling damper. This signal forms the set value of the position controller while the actual position of the damper is measured and connected to the position controller output, which is utilized to modulate the gas recycling damper.

9.3.1.2.2 Function of Stack Inlet Damper (Fig. 8.35) A secondary part of the reheater steam temperature control assists the main control loop to function properly. Its sole function is to maintain the differential pressure across the gas recycling damper at a constant value. The stack inlet damper acts as the FCE in this control loop.

The DP across the gas recycling damper is measured and after selection/voting, passes through a maximum/minimum limit to form the measured/process variable for a separate controller whose set point adjustment is made through a manual setter and id set at a fixed value. The controller output is utilized as the positioning command for the stack inlet damper, which modulates to maintain the DP across the gas recycling damper. It thereby ensures sufficient pressure head to force the flue gas from the ID fan discharge duct to the furnace chamber through the hoppers. Interlocked operation prevents a damper closing command when the furnace pressure is high.

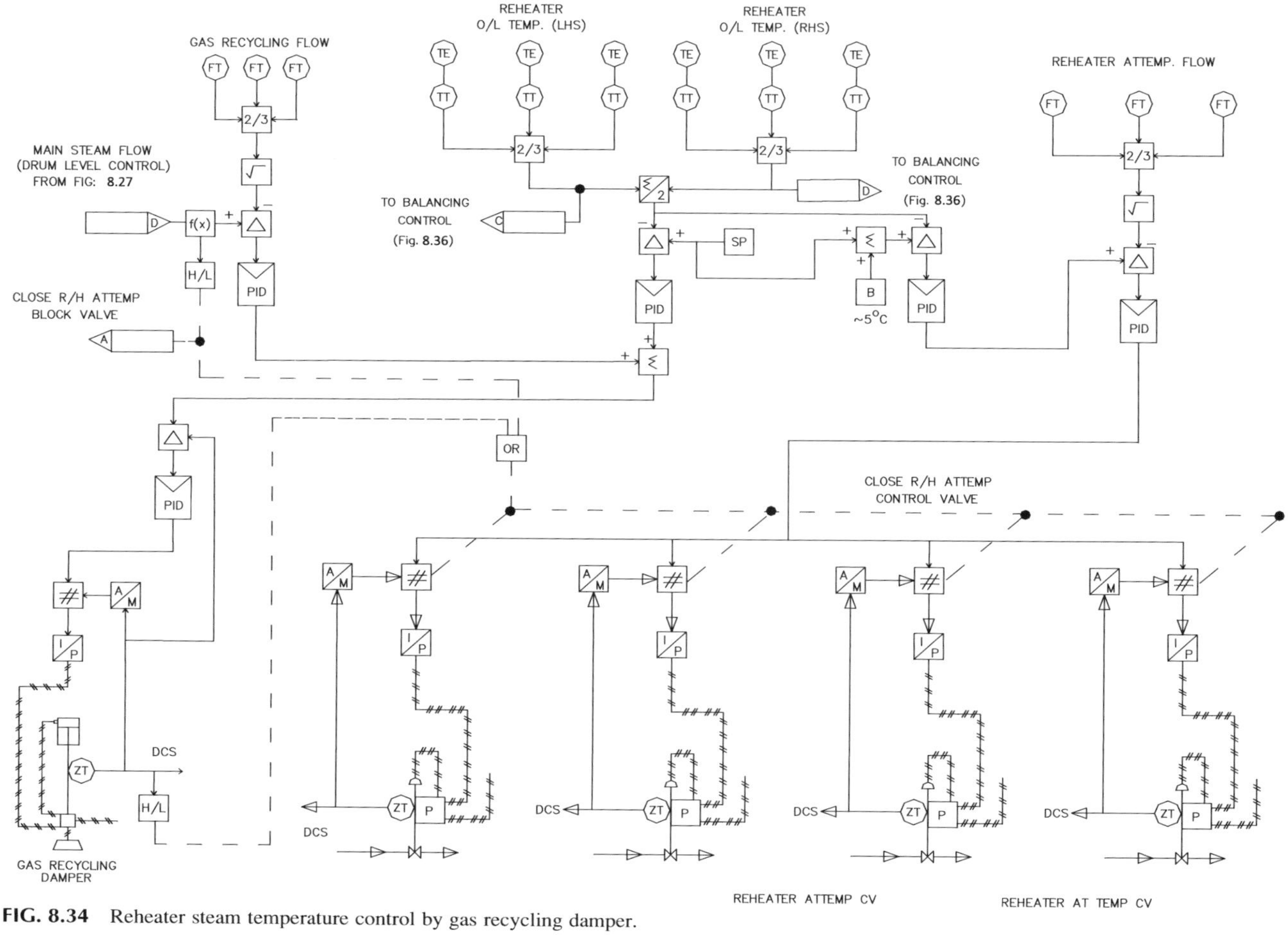

FIG. 8.34 Reheater steam temperature control by gas recycling damper.

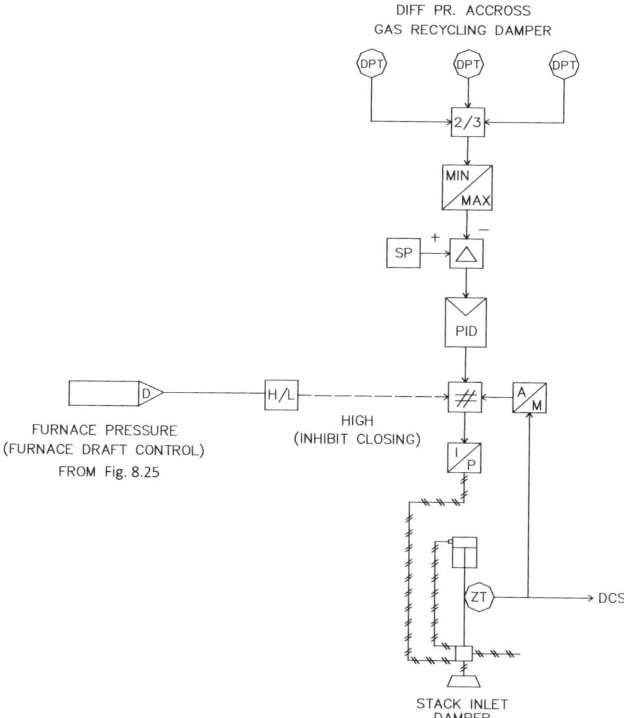

FIG. 8.35 Stack I/L Damper control (reheater steam temperature control by gas recycling damper.

9.3.1.2.3 Function of Reheat Steam Outlet Valves (for Temperature Balancing) (Fig. 8.36)

There are two butterfly valves located in the reheat steam line in each leg. The purpose of this control loop is to maintain equal steam temperature at both the LHS and RHS pipe lines. This is done by slightly restricting the steam flow through the low temperature leg and vis-a-vis increasing the steam flow through the high temperature leg.

The steam temperatures of both reheater outlet legs (LHS and RHS) are measured and averaged to form the common set value of two PID controllers, of whose measured/process variables are the individual reheater outlet steam temperatures. The outputs of these controllers are the position demands of the butterfly steam valves acting inversely so as to minimize the error. The purpose of the control strategy is not a fixed desired value, but to lessen the temperature difference between the two legs.

Another purpose of this loop is to ensure a minimum steam side pressure loss. For doing that, the control loop envisaged a separate PID controller whose fixed set value is the 100% position and the measured/process variable is the maximum selected position transmitter signal of the above two butterfly valves. The controller output is added to the main controller output to form the combined position demands of each valve. By this arrangement, the maximum

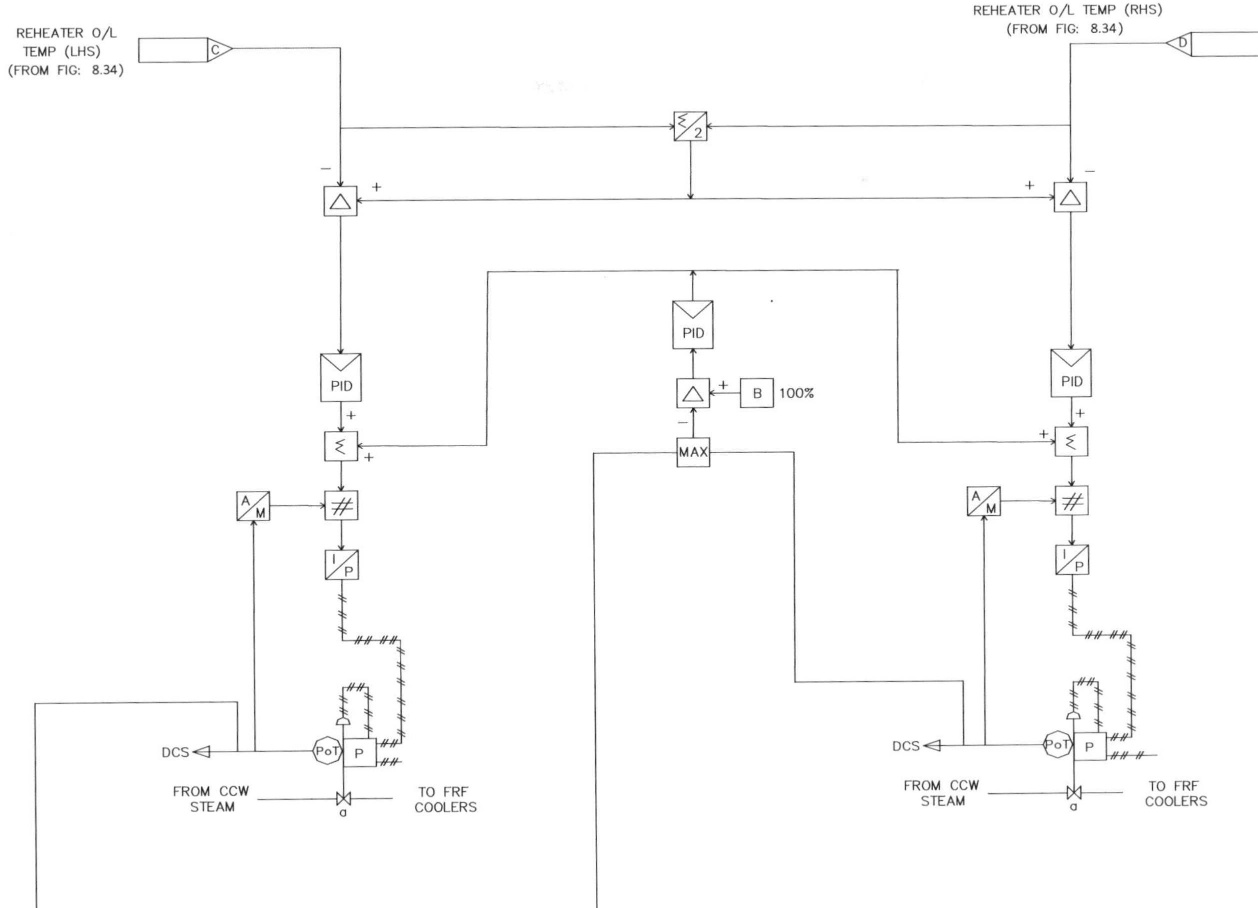

FIG. 8.36 Reheater outlet steam temperature control (by balancing control).

position selected valve would try to reach 100% and the other valve position would also increase by the amount, but the difference would remain the same. This is done by a minimum integration rate action and a high proportional band to allow a slow output signal until one valve reaches the 100% position without much disturbing the temperature balancing action.

As already discussed, there are spray water control systems for maintaining the reheat steam temperature in emergency cases after the gas recycling damper alone fails to bring down the temperature.

The normal control loop function is achieved by the gas recycling damper, but due to any reason, the temperature may shoot up. If that condition persist, then the spray control valve would come into action to save the reheater from burnout.

The dedicated controller for this purpose has the same temperature set value but added (biased) with some more degree (around 5–7°C) to prevent this loop from interfering with the normal loop. The spray control valves remain closed unless the temperature crosses that biased amount. The controller gets its measured/process variable from the average reheat steam temperature. The output becomes the RH attemperator water flow demand or set value for another controller, with the RH attemperator water flow signal measured directly as the measured/process variable. This controller output now becomes the position demand for the spray control valves.

An interlocked operation ensures that the gas recycling damper is automatically closed when the RH attemperator spray valve opens; at that time, the ID discharge damper is mechanically arranged with a limited stroke by using a mechanical stop or setting the linkage so that the full stroke of the actuator provides partial movement of the damper.

9.3.2 *Control Loop Description With Gas Dampers*

9.3.2.1 Control Loop Description With Gas Damper in Main Flue Gas Path (Fig. 8.37)

There are two types of arrangements for the boilers having reheater temperature control by gas dampers.

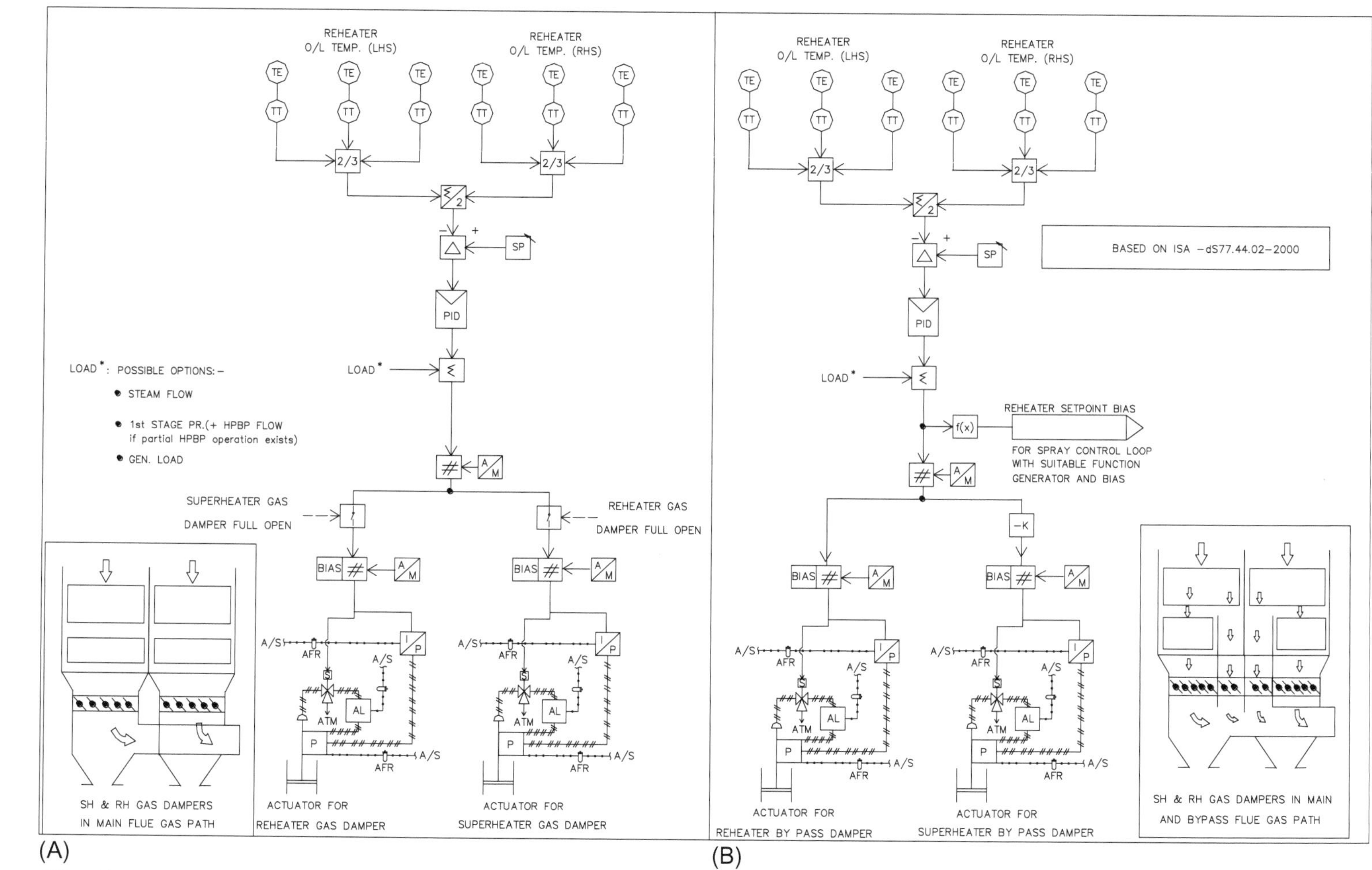

NOTE:
(1) FOR BETTER UNDERSTANDING,LOOPS SHOWN HERE ARE IN THEIR SIMPLE FOR AVOIDING MANY ASSOCIATED INTERLOCKS, etc. IN ACTUAL CASE THERE MAY BE ADDITIONAL INTERLOCKS AS PER MANUFACTURER'S RECOMMENDATIONS.
(2) THIS TYPE OF CONTROL LOOPS ARE APPLICABLE FOR MAINLY FOR BOILERS WHERE CONVECTION TYPE HEAT TRANSFERS ARE PREDOMINANT.
(3) INTERNAL AUTO / MANUAL SIGNALS OF THE LOOP HAS NOT BEEN DETAILED OUT TO AVOID COMPLEXITY.
(4)TO GIVE AN IDEA TO THE READER, OPEN LOOP CONTROLS HAVE BEEN SHOWN IN DASHED LINE TO DISTINGUISH IT FROM MODULATING CONTROLS. THIS IS HOW IT COULD BE REPRESENTED IN REALITY FOR BETTER UNDERSATNDING.

FIG. 8.37 Reheater temperature control with gas bypass dampers.

For some boilers, *gas dampers* are provided in the common flue gas path as the FCEs for reheater steam temperature, which, when positioned, acts as a three-way proportioning valve that takes the task of apportioning the flue gas flow through both the superheater and reheater banks. This means that the total flue gas flow would be uninterrupted with the ratio of gas flow through the SH, and the RH only would be readjusted.

In some boilers, two separate and dedicated sets of dampers are provided in the flue gas path of both the superheater and reheater banks controlling the individual flue gas flow. As these dampers are located across the main path, they control the total flue gas volume that may interrupt gas flow during auto control operation. In that situation, the flue gas path will be constricted so badly that the overpressurization may damage the structure itself, which is taken care of by not issuing a closing signal to both dampers simultaneously. To be on the safe side, one particular set of dampers throttles only when the other set of dampers is fully opened. Interlocked operation would also ensure that no gas dampers get a fully closed signal at any time during the entire range of operation.

Intermittent operation of emergency spray valves may be necessary here as the control through gas dampers involves a slow response system for large units. Whenever there is a possibility of sudden overheating accompanied by a rise in temperature following a rise in furnace heat input vis-a-vis a large load demand in a small span of time, the spray valves are expected to take care of the situation. The set point of the spray loop is also derived from the loop shown in Fig. 8.37B.

As shown in Fig. 8.37A, the measured variable (RH O/L temperature) is compared with the adjustable set point and the error thus created is sent to the PID controller. In order to make the loop more responsive, the load index is added at the output as the feed forward signal.

As the water spray is an emergency or secondary control system in case of reheat temperature control, the spray valves are normally shut unless the temperature at the reheat outlet reaches a predetermined value higher than the normal set value.

9.3.2.2 Control Loop Description With Bypass Gas Dampers

The control strategy is the same as the main line gas dampers, but here, bypass dampers in the SH and RH gas path are used to position in the opposite direction, as shown in Fig. 8.37B.

During start-up and at low load, the superheater bypass would be wide open and the reheater bypass would be at a minimum position. As the load increases, the reverse will be the operation. The aim of the loop is to maintain the DP across the damper at a nearly constant level by adjusting the loop parameters and relationships at all points of operations. The set point of the spray loop is also derived from the loop shown in Fig. 8.37B.

9.3.3 Control Loop Description Burner Tilting Arrangement

9.3.3.1 Measurement of Different Parameters

Measurements with sufficient redundancy and voting include burner tilt position, RH O/L steam temperature (LHS and RHS), RH desuperheater O/L steam temperature (LHS and RHS), and MS flow.

9.3.3.2 Control Loop Description

The FCEs are the burner tilting arrangement and the RH attemperation valves.

In the earlier stages, the burner tilting arrangement was utilized to control the reheat temperature only. Nowadays, the same is being utilized to control any of the superheat or reheat outlet temperatures, depending on the demand of the situation.

In larger units of boilers with TT burners, the area of platen superheaters is increased so the burner tilt controls can be used for temperature control of either the SH or RH, based on demand. As discussed in the previous section, in these cases the two loops (SH and RH temperature control) are interactive, so provisions are now made for selection through maximum demand—because the major heat transfer is not by convection.

The description of the control loop takes care of a typical power plant of 500 MW capacity having a steam drum, a firing system with a burner tilting arrangement, and provisions for a sliding pressure control system. Though this control loop needs a separate entity, it is included in the section of RH temperature control as many power plants of low/medium capacity still incorporate a burner tilting arrangement only for RH temperature control with the feed forward signal to the SH temperature control system.

9.3.3.2.1 Function of Burner Tilting Arrangement (Fig. 8.38) Here, the average control error signals from both the SH and RH controls loops are connected to the individual controller; one output, as selected through the maximum selector signifying the control signal related to the lower temperature out of the two, turns out to be the position demand of the burner tilting arrangement.

Interlocking signals are provided as per the following logic:

(i) When the steam flow vis-a-vis the boiler load is less than or equal to 25% MCR or at the MFT, the burner tilt would maintain its base position, that is, the horizontal position only.

(ii) In case any spray water valve opens fully, the burner tilt raise command would be inhibited for protecting both the SH and RH elements from overheating.

9.3.3.2.2 Function of Reheat Spray Control Valves (Fig. 8.38)

The measurements are RH desuperheater (DSH) O/L temperature (LHS and RHS), the final RH O/L temperature (LHS and RHS), the burner tilt position, and the MS flow.

The control loop is basically a cascade type and has a similar control strategy as that of the superheat temperature control part (Section 8). The ultimate measured/process variable is the RH outlet temperature (LHS/RHS), which is compared with the temperature set point to form the error. Here, the set point is made a variable one that is programmed as a function of the unit steam flow signal for both constant pressure control and sliding pressure control, but with separate characteristics. This error signal is connected to the master controllers (or the outer loop controllers) with PID functional blocks (algorithms). A derivative functional block is provided to compensate for the sensors' inherent sluggish behavior.

When the difference between the LHS/RHS temperatures at the final SH steam outlet header are within the limit, their average temperature acts as the measured/process variable for both the master controllers (or the outer loop controllers) of the LHS/RHS temperature control loops. In case of a high difference between the two sides, then an average signal is no more considered as the measured/process variable and instead the individual RH steam outlet temperature becomes the measured/process variable for their respective control loops.

The controller output now becomes the variable set point of the desuperheater or attemperator outlet temperature controller and forms the error signal along with the measured/process variable and is connected to the controller. The PID controller, also termed the slave/inner controller, generates the output for the position adjustment of the desuperheater or attemperator spray valves. Desuperheaters or attemperators are located (in the steam flow path) just before the reheater. The attemperator provides the necessary injection of spray water flow while the cold reheat steam is flowing through it to adjust the attemperator outlet temperature at a value just needed to maintain the reheater outlet temperature at the set value after the necessary temperature gain within the reheater.

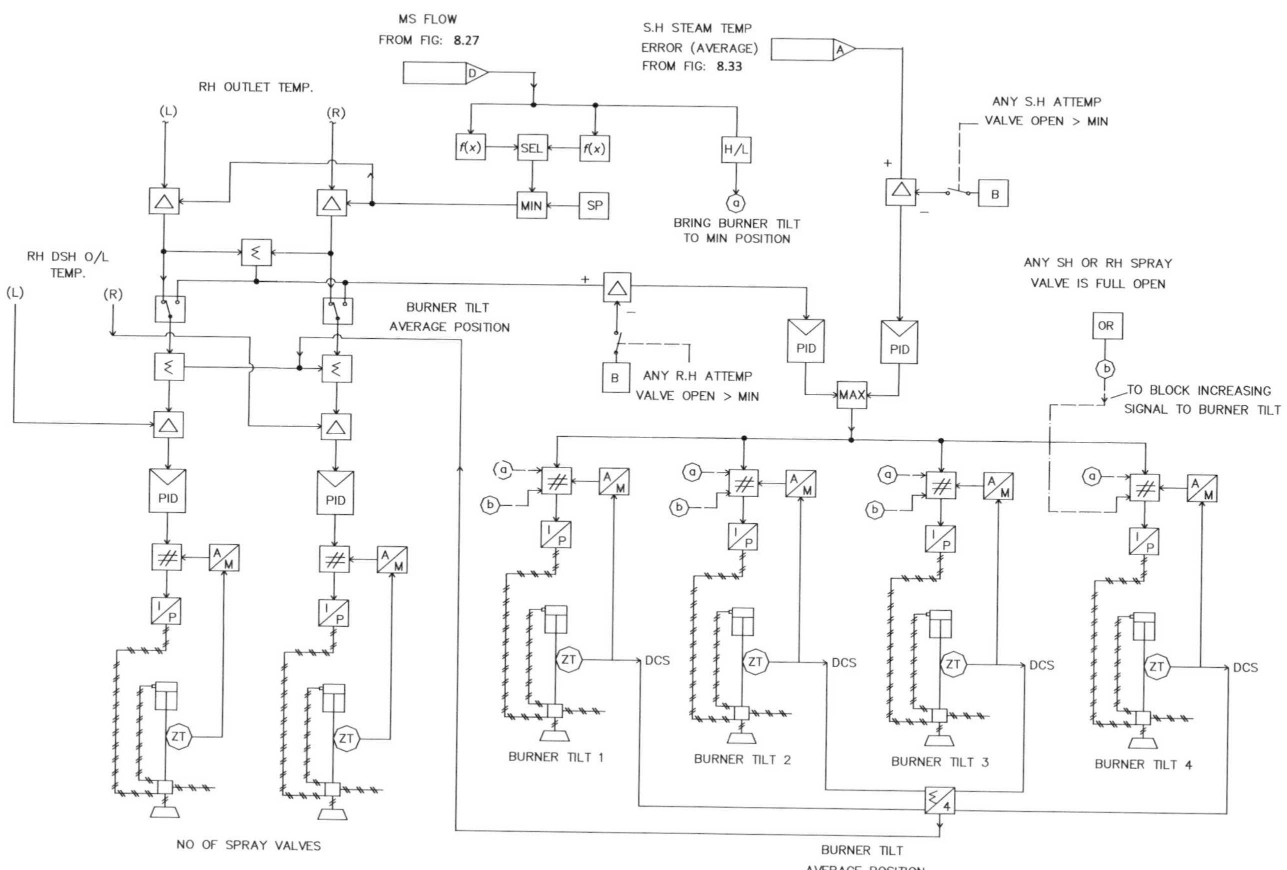

FIG. 8.38 Reheater (superheater) temperature control by burner tilt and spray water.

9.3.4 The Spray Water Valves in Reheater Control Loops: Common Features

There are certain common features regarding spray water valves that are applicable irrespective of the primary arrangement of reheat temperature controls. Those are indicated below:

(i) The spray water valves are generally provided at the inlet of the reheater to avoid any overheating, which may damage the reheater elements. Another reason for keeping the attemperation at the inlet is to also help avoid chances of water carryover.
(ii) All the primary control systems are designed to prevent any need for spray water in normal operation, confining such use for fine tuning, secondary control systems, and emergency purposes only. However, in large boilers (TT burner), as discussed here, the spray may be used as the primary control that also requires a minimum water flow at high load for the reasons already discussed in the previous section.
(iii) In some control schemes, the system is used for controlling the furnace outlet temperature as well.
(iv) All the FCEs utilized by the primary system are sluggish and any large step load increase may necessitate the action of spray water valves.
(v) In addition to boiler-specific interlocked operation, the spray valves must immediately be closed in case of turbine tripping. The reason behind the interlock is that the turbine tripping causes a collapse in the reheater flow. In that case, as per the recommendation of the Turbine Water Damage Prevention (TWDP) Act, cold spray water cannot enter the turbine anyway.

9.4 Other Reheat Steam Temperature Controls

This type of control is used for the compartmented boilers with mill bias. These are three-element controls such as reheat temperature, load demand, and total air flow with mill bias. Three-element controls are applicable mainly in the applications with rapid load changes as well as in variable steam and attemperation pressure. Here, the load demand is trimmed with reheat temperature control within the maximum allowable load demand. This load demand is compared with total air flow with mill bias. The controller decides the demand for excess air and mill bias. At lower loads, mill bias is increased first and then excess air bias, whereas at higher loads, it is just the reverse; the function generators are selected accordingly. The excess air is introduced in the bottom idle compartments. Details are available from ISA.

10 MISCELLANEOUS BOILER CONTROLS INCLUDING OVERFIRE AIR DAMPER

10.1 General

Many other control loops for the boiler unit, not yet described in specific sections separately, are indicated below, but there may be some other control loops not mentioned that are required for a particular type of make:

(i) Atomizing steam/air pressure control.
(ii) Air heater cold end or steam coil air preheater (SCAPH) temperature control.
(iii) Hot gas temperature control.
(iv) Continuous blowdown (CBD) tank level control.
(v) SCAPH drain tank level control.
(vi) Overfire air damper control.

10.2 Objective

(i) Atomizing steam/air pressure is controlled at a fixed value facilitating HFO/light fuel oil (LFO) atomization for proper burning of the oils.
(ii) The purpose of the air heater cold end or the SCAPH temperature control loop is for maintaining the average value of the cold end temperatures of the air heater, that is, at the air inlet and flue gas outlet at a fixed value above the dew point, which may vary from season to season and can be set in the controller accordingly.
(iii) Hot gas temperature is controlled to minimize the flue gas temperature imbalances between the primary and secondary air heater outlets. The temperatures of the two air heaters are compared and by positioning the gas dampers, the distribution of flue gas is readjusted in an effort to eliminate the difference between them.
(iv) Continuous blowdown (CBD) tank level control is provided to maintain a fixed level in the tank to facilitate the transfer of drum water blown down as unacceptable quality for the steam-water cycle.
(v) SCAPH drain tank level control is provided to maintain a fixed level in the drain tank, where the heating steam is condensed at the outlet of the SCAPH and collected, to facilitate the transfer of water for further use.
(vi) Overfire air damper control. The objective of this control loop is to minimize the NO_x level as per the directive of the local pollution control board. The FCEs are the OFA dampers for the tangentially fired burners. By proper positioning of these dampers at various elevations and locations, the desired level of NO_x can be maintained.

10.3 Discussion

(i) *Atomizing steam/air pressure control* is required for HFO or LFO atomization. The pressure of the atomizing media is always maintained at a fixed value, but the same may be more or less than the operating oil pressure, depending on the type of burner used.

(ii) *Air heater cold end or (SCAPH) temperature control* is provided for maintaining the average cold end temperature above the dew point so as to prevent condensation at the air heater surfaces and thereby prevent the resultant corrosion of materials. The flue gas contains sulfur oxides (SO_x) and moisture, which condenses when the temperature goes down at the flue gas outlet part of the air heaters (cold end of AH). This may happen from both ends, that is, when the incoming air temperature itself is low due to noncongenial ambient conditions where the air is sucked and/or the flue gas temperature is not capable enough of the raising air temperature. As a result, it may become as low as near the dew point.

To avoid that situation, this control loop is provided for a separate steam coil air preheater getting steam from a separate auxiliary steam source. Normally, the SCAPH is located in the bypass duct at the outlets of each FD fan for preheating the air. For a separate primary air heater system, SCAPH should also be provided in each PA fan bypass duct.

By this arrangement, the flow control valve provided at the SCAPH inlet of the steam line modulates as per the command from the controller and maintains the requisite temperature around 10°C, more than the acid dew point for flue gases.

It is to be noted that the service SCAPH is not normally required but for start-up, low load and may be under abnormal or due to climatic conditions.

(iii) *Hot gas temperature* control, as already described, is to minimize the gas temperature imbalances between the primary and secondary air heater outlets. The sustained differences in those would result in stratification of flue gases at the ID fan inlet, jeopardizing the unit efficiency. The two temperatures are compared and input into a controller whose output is the input of a function generator of each damper control loop. In auto operation, the controller for any temperature imbalance would tend to close the hotter side damper while the colder side damper would remain in the open position. Limit values are provided to restrict the damper movements between the preset maximum and minimum values. These dampers are to be made 100% manually open during start-up, and those positions are sent to the ID fan as start permissive.

(iv) *Continuous blowdown (CBD) tank level control and SCAPH drain tank level control*

The above two controls are provided to maintain the level only to facilitate the passage of the incoming media as and when required.

(v) *Overfire air damper control*

In certain furnaces, overfire air (OFA) dampers are provided to control the NO_x (NO_2 + NO) level in the product of combustion, that is, the flue gas. The NO_x level is also restrained through provision of the flue gas recirculation damper.

As discussed earlier, the production of NO_x in a thermal power plant depends on many factors, namely time, temperature, turbulence, stochiometric ratio, etc., which is described as thermal NO_x as airborne nitrogen (N_2) and oxygen (O_2) reacts with each other during the combustion process and can be kept under control with the help of OFA dampers. The other source of NO_x may be from the organically bound nitrogen compound in the fuel itself. NO_x formation and control loop strategies have been discussed in detail in Section 10.6 of this chapter.

10.4 Auxiliary Steam (BAS)

10.4.1 General

Auxiliary steam in a power plant plays a great role, especially during start-up of the boiler and turbine.

For smaller units without a bypass system, the boiler and turbine may have to be started simultaneously, so in those cases, separate auxiliary steam headers for the boiler and turbine known as boiler auxiliary steam (BAS) and turbine auxiliary steam (TAS) may be desirable.

Larger units generally consist of a common auxiliary steam header to supply auxiliary steam to the boiler and turbine. During start-up of the boiler, it is started with oil requiring BAS for FO heating/atomization, SCAPH (maybe)—thus, the requirement for auxiliary steam is high. Similarly, during turbine start-up, the requirement for auxiliary steam is as high as for deaerator pegging, the starting ejector, etc. In larger units, the turbine is generally started after the boiler is a little stabilized through the bypass system, suggesting that boiler start up means no requirement for auxiliary steam for the turbine. Similarly, the turbine start up means no auxiliary steam requirement for the boiler, as it is already stabilized. Thus the load from the common header is well distributed in the time sequence, hence it is better for larger units with a higher requirement for auxiliary steam having a common header in place of an individual BAS and TAS.

Whenever there are units in one plant (in the same geographical location), the auxiliary steam headers of each unit and the auxiliary boiler (if any) may be interconnected so that during the initial start up of one unit, the auxiliary steam may be supplied from another unit. Another aspect of the common auxiliary steam header is important. In larger units, the CRH pressure is high enough (>30 kg/cm^2) to form an auxiliary steam header instead of forming the same from the MS to minimize energy loss due to large pressure and temperature reductions. However, until the time the CRH line pressure is high enough, the source of auxiliary steam is MS, as shown in Fig. 3.15. Normally, auxiliary steam headers are formed with $P=10$ kg/cm^2 and $T=210$°C for medium plants and with $P=16$ kg/cm^2 and $T=230$°C for large plants. As discussed earlier, the requirement of for auxiliary steam (whether BAS or TAS) varies greatly during start-up and normal operations. At the time of start-up, the MS pressure (the source of BAS or TAS) may be lower than when the unit is stabilized. During this time, larger auxiliary steam flow is necessary, so at a lower pressure drop, a higher flowthrough control valve is required. On the contrary, when the system is stabilized, the control valve experiences a high pressure drop to allow a lower flow, which dictates the necessity to use separate control valves with higher and lower capacity, as shown in Figs. 3.12 and 8.40.

In this book, both BAS and TAS have been dealt with separately in Chapters 8 and 11, respectively. However in this section of Chapter 8, the common auxiliary steam header aspect has also been discussed, and is not repeated in Chapter 11.

10.4.1.1 Objective of Boiler Auxiliary Steam Control

The purpose of this control is to obtain a header of constant pressure and temperature, irrespective of changes in the boiler load.

10.4.1.2 Objective of Common Auxiliary Steam Header

The purpose of this control is to obtain a header of constant pressure and temperature, irrespective of changes in the load in the boiler and turbine as it is supplying auxiliary steam to both the boiler and turbine.

10.4.1.3 Discussions

The control system is basically for a pressure reducing and desuperheating station (PRDS) with two separate controls for maintaining the header pressure and temperature at constant value. In the available two types, one may be with the combined PRDS with the same control valve responsible for pressure reduction that has the desuperheater built in it, where water is controlled and sprayed using temperature control. In the other design, a pressure-reducing control valve is followed by a separate desuperheater where the water spray is controlled to maintain temperature. One typical combined PRDS is shown in Fig. 8.41.

10.4.1.4 Control Loop Description

In these loops, airlock relay and solenoid valves have been introduced to indicate that in case of failure of the air supply and electrical signal, the control valves may be locked in their previous position, that is, to attain the fail lock position.

10.4.1.5 Boiler Auxiliary Steam Control (Fig. 8.39)

This system consists of two separate loops:

1. Boiler auxiliary steam: Pressure control: Here the auxiliary steam header pressure is measured in one of two mode and compared with a fixed set point. The deviation thus created is used to drive a P+I controller. The output of the controller through the A/M station and I/P converter regulates the pressure-reducing valve to maintain constant pressure at the auxiliary steam header at all loads.
2. Boiler auxiliary steam: Temperature control: Steam temperature at the auxiliary steam header, that is, at the outlet of the pressure-reducing valve/desuperheater, is measured and compared with a fixed set point and the deviation thus created is fed to a P+I+D controller (as temperature is a slow-changing parameter). The output of the controller regulates the condensate flow to the desuperheater to control the temperature at the auxiliary steam header. In the figure under reference, a separate desuperheater has been shown, but it can be combined with a pressure-reducing station also, as shown in Fig. 8.41.

10.4.2 Boiler Auxiliary Steam Control (Fig. 8.40)

BAS high capacity and low capacity. As discussed in clause no. 10.4, there may be a necessity for high-capacity and low-capacity PRVs. The basic pressure and temperature control loop description is the same as discussed in clause no. 10.4.1.5. However, the loops may be operative in a split range, when the load is high then both the low and high capacity valves shall be operative. When the load reduces, the high-capacity valve shall start closing first. After that, the pressure shall be maintained by regulating the low-capacity valve. These high- and low-capacity control valves can be taken into service with the help of the interlock of the upstream isolating motorized valve from the loop. Until the low-capacity valve is 80% (~) open, the upstream isolating valve of the high-capacity control valve may be kept closed. Similarly, when the high-capacity control valve is around 20% open, the associated isolating valve may be closed and the low-capacity isolating valve may be opened. These two high- and low-capacity pressure-reducing valves are kept in parallel. The desuperheater is placed in series after these pressure-reducing valves. The temperature control, single and common to both the high- and low-capacity PRV, is the same as discussed in clause no. 10.4.1.5. However, if necessary, the temperature control for each high- and low-capacity PRV can be provided. In case of combined PRDS design, there shall be two such temperature control loops.

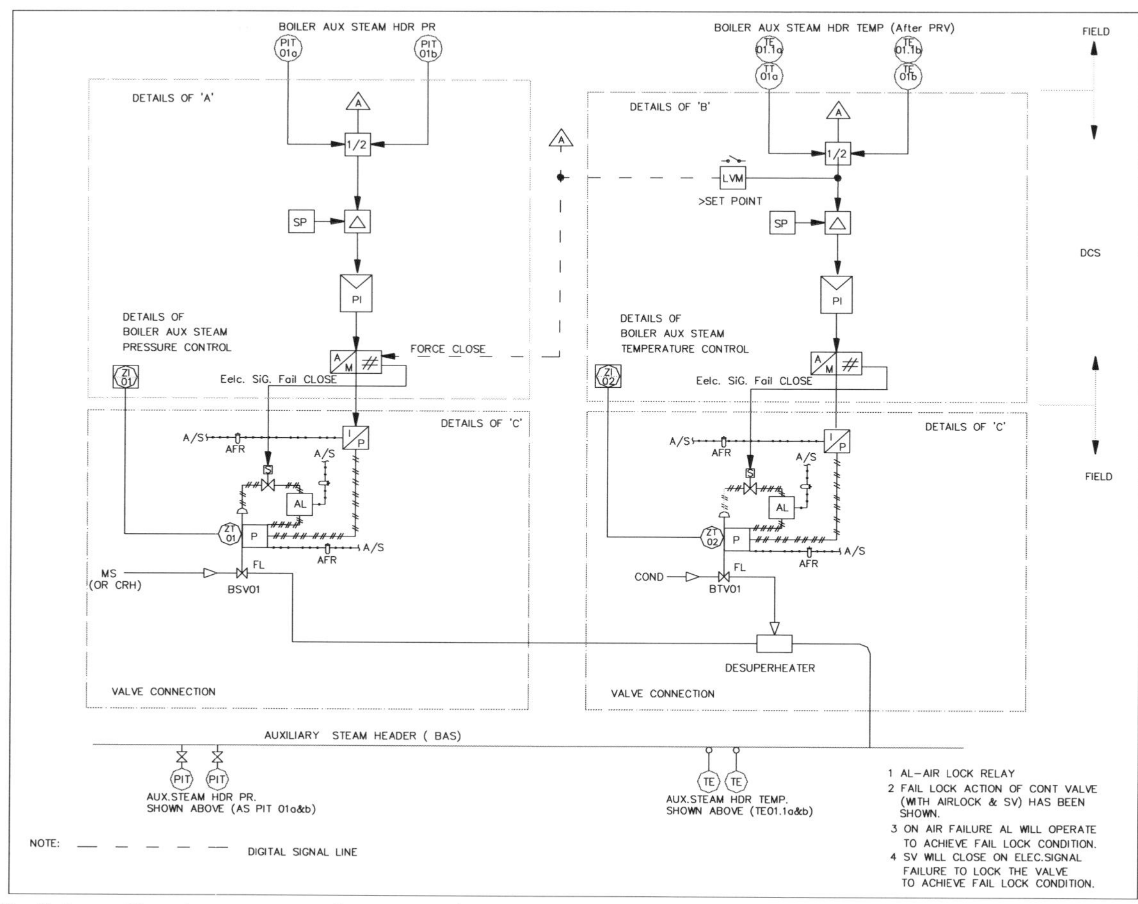

FIG. 8.39 Boiler auxiliary steam pressure and temp. control.

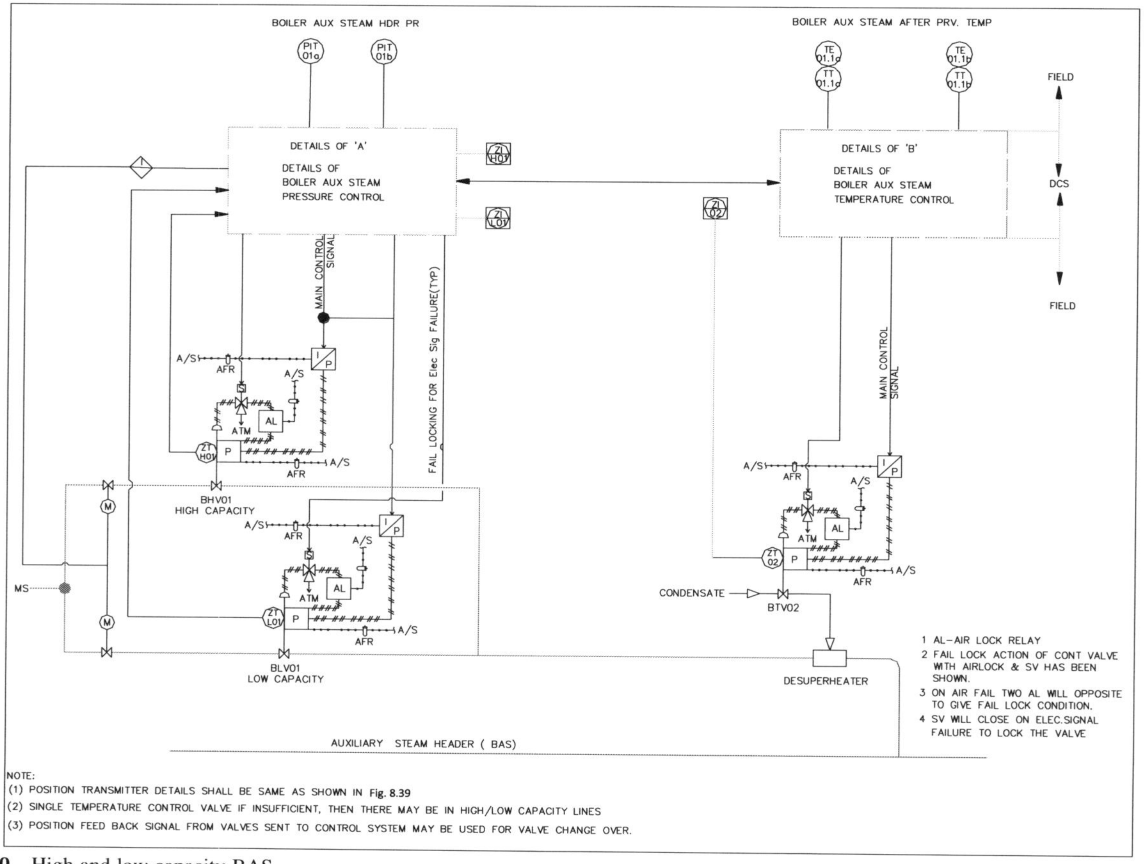

FIG. 8.40 High and low capacity BAS.

10.4.3 *Common Auxiliary Steam Control (Fig. 8.41)*

This is a header common for both the BAS and the TAS. Here, there may be three loops. One common pressure-control loop and one temperature-control loop for MS and CRH. There will be two pressure-reducing valves; one each for MS/CRH to the auxiliary steam line.

1. *Common auxiliary steam pressure control*: Here, the auxiliary steam header pressure is measured in one of two mode and compared with a fixed set point. The deviation thus created is used to drive a P+I controller. The output of the controller, through two sets of A/M stations and I/P converters, can regulate the pressure-reducing valves in the MS and CRH lines to maintain constant pressure at the common auxiliary steam header at all loads. Any one will be active based on selection through the interlock discussed later (clause no. 10.4.5). The CRH pressure, when developed, will source the auxiliary steam header, but initially the MS supplies the auxiliary steam. Whenever the CRH is selected, the PRV in the MS line will be closed following a ramp circuit.
2. *Common auxiliary steam temperature control*: There are two sets of steam temperature control loops to regulate the condensate flow to the desuperheater after PRV in each for the line from the MS and the CRH (These desuperheaters may be combined with PRVs also as shown in Fig. 8.41 for combined PRDS.) steam temperature at auxiliary steam header that is, at the outlet of the pressure-reducing valve and the desuperheater is measured and compared with a fixed set point; the deviation thus created is fed to a P+I+D controller as the temperature is a slow-changing parameter, hence Daction may be more effective. The output of the controller regulates the condensate flow to the desuperheater controlling the temperature at the auxiliary steam header. With suitable interlock, the condensate flow to the nonactive desuperheater has been prevented (discussed in clause no. 10.6).

10.4.4 *Control Loop Operation (Figs. 8.39–8.41)*

In this part, the auto and manual operation of the loop has been discussed.

1. This loop, as per clause no. 1.2.1 of Section 1 of Chapter 8, trips to manual on transmitter failure and/or selection. The operator needs to select the correct transmitter to continue with the loop.
2. Loop is operable in auto/manual position at any time.
3. There will be some additional signals for releasing control loops associated with (control valves CTV and CAV in Fig. 8.41) CRH PRV and the associated temperature control loop, as shown in Fig. 8.41. These loops can be released to auto only when the pressure in the CRH line is established. The control valve CTV in Fig. 8.41 has also been implicated as to prevent condensate flow when the associated pressure is not in service.

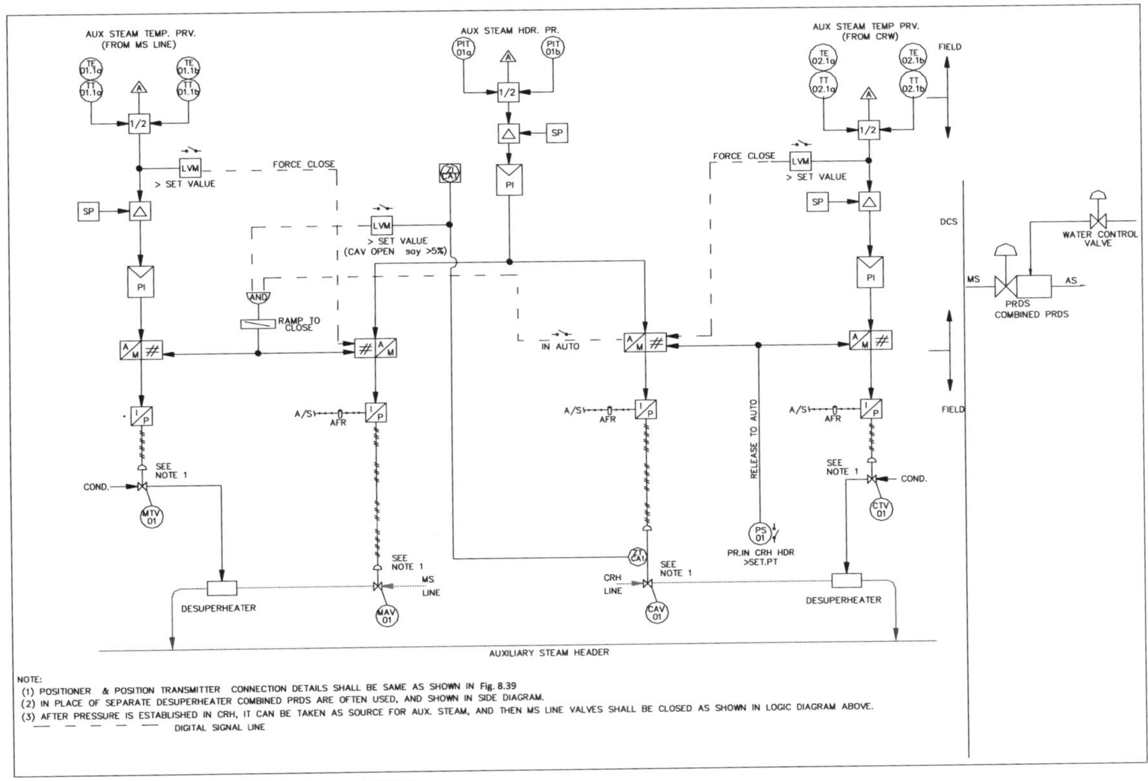

FIG. 8.41 Auxiliary steam header from MS and CRH.

10.4.5 Control Loop Interlock

There may be some specific interlocks for the loops:

1. BAS (Figs. 8.39 and 8.40): In case the temperature at the desuperheater (after PRV) outlet, that is, at the auxiliary steam header, is > set value, then the PRV (both high and low capacity) shall be forced closed. Also, in case of signal failure, the associated solenoid valve shall be closed to attain fail lock condition of the associated valve (fail lock condition).
2. Common Header Interlock:
   a. Associated PRV shall be closed when the temperature > set value sensed by LVM.
   b. Whenever the PRV in the CRH line (CAV) is in auto and open >5% (say), the PRV in the MS line and the associated temperature control valve shall be closed, as shown in Fig. 8.41.

Due consideration must be given while sizing the valves; there may be two sets of valves to cater to the requirements during start-up and normal operation.

Typical parameters for PRDS valves in auxiliary steam systems are as follows:

Some extreme conditions (typical parameters) for the PRV and temperature control valves are as follows in Table 8.2.

High pressure drop, high temperature, and high noise are very important in selecting the valves for these applications, especially for PRVs. Cage-guided low dB trim with multiple holes is used to minimize the noise effect, etc.

Use of auxiliary steam in boiler: Major uses of auxiliary steam include fuel oil heating, fuel oil atomization, and heating air in SCAPH.

10.5 Soot Blowing Steam PR and SCAPH Pressure Control

General

In a modern thermal power plant, there are steam requirements at different pressures and temperatures, depending on the nature and working parameters at the consumer ends. Soot blowing steam and steam coil air preheater (SCAPH) pressure control are examples of two such requirements.

TABLE 8.2 Typical Parameters for PRV and Temp Cont. Valve

Parameter	Pr. Red. Valve (PRV)	Temp. Cont. Valve
Inlet pressure	175/145 bar	190/165 bar
Outlet pressure	16/10 bar	16/10 bar
Temperature	540°C	160°C

10.5.1 Soot Blowing (SB) Steam Pressure Control

SB steam pressure control plays a vital role regarding the cleaning of the furnace and the air heater heat transfer surfaces.

In an SG plant, steam is readily available as and when the plant is operating. Suitable tapping point normally from auxiliary steam header whose process parameter (10–16 kg/cm^2 pressure, 210–250°C temperature) almost matches with the requirement of SB steam; exact requirement may be achieved through provision of separate pressure-reducing valves.

10.5.1.1 Objective

The objective of this control loop is to exactly maintain the steam pressure of the main SB header supplying all the furnace water wall and convection zone soot blowers. It is quite obvious that the MS and flue gas outlet temperatures are affected by the cleanliness of the heat-transferring surfaces. Soot blowing with the proper quality of steam as well as the proper sequence and frequency is the only measure to improve the overall efficiency and pollution control by reducing the unwanted accumulation of dirty substances.

10.5.1.2 Discussion

The experience entails that SB steam quality must be around 10–15 kg/cm^2 pressure and 50–60°C superheat with an aim to get sufficiently dry superheat steam to avoid moisture in the cleaning media. As the normal SB steam is sourced from the high-pressure steam already available, any reduction in the required pressure would raise the degree of superheat more than before. The source may be from MS, HR, or even from the auxiliary steam header. Normally, the source is so chosen that the temperature control is not necessary.

10.5.1.3 Control Loop Description (Fig. 8.42)

The SB steam control loop in a typical 250 MW plant with a drum boiler and fossil (coal)-fired furnace is shown. The control loop is a direct, single, and simple type.

10.5.1.4 Measurement of Parameters

Includes only SB steam pressure transmitters with sufficient redundancy and at the appropriate location to get an average and representative measurement.

10.5.1.5 Control Loop

The measured/process variable is compared with a fixed set value to form the error signal for the controller whose output is the position demand for a pressure-reducing valve having capacity dictated by the maximum number of blowers undergo simultaneous operation.

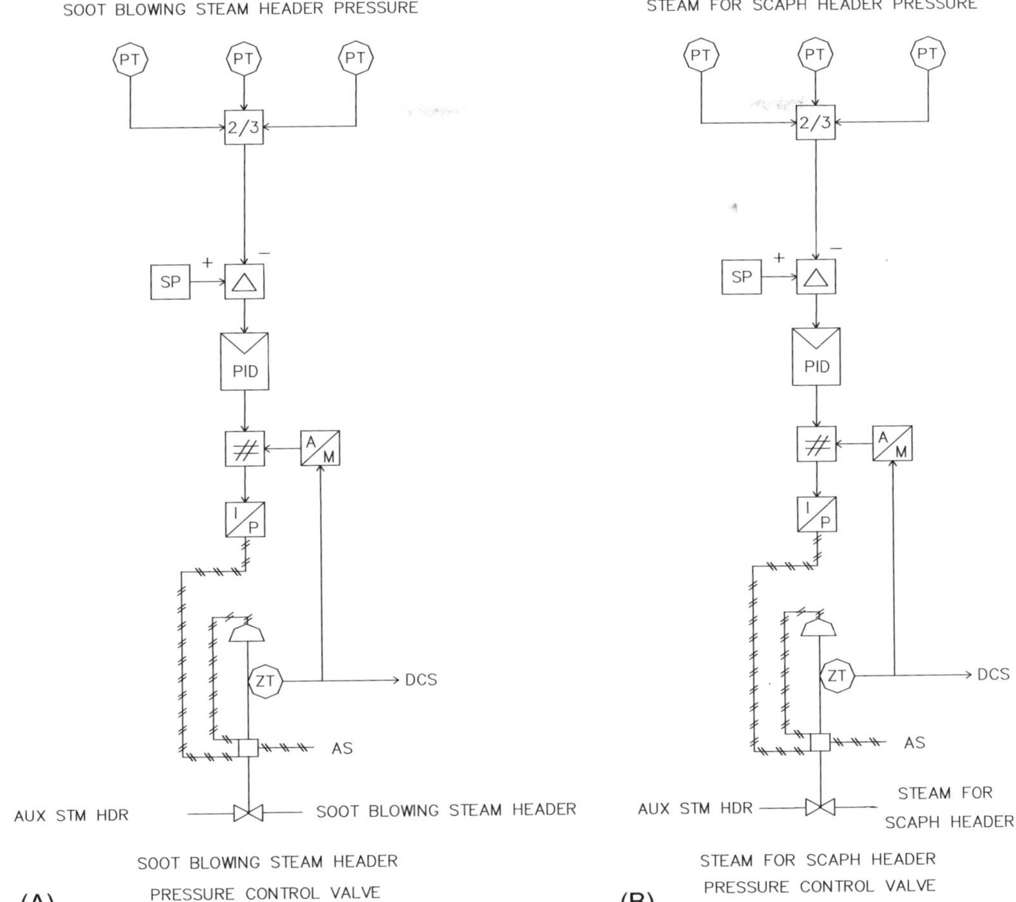

FIG. 8.42 Soot blowing and SCAPH steam pressure control.

There may be some individual or an array of soot blowers requiring different inlet pressures for their operation, which is taken care of with self pressure-reducing valves for the individual or group of blowers while supplying from the main SB steam supply header.

10.5.2 Steam Coil Air Preheater (SCAPH) Steam Pressure Control

SCAPH is one of the means for control of SO_x through process control loops. The control strategy adopted is to check the temperature fall below a certain value at specific points for avoiding deposits of condensation of sulfuric acid (H_2SO_4) and sulfurous acid (H_2SO_3) vapors on the exposed metal. The minimum average of the gas and air temperature is maintained, by which the corrosion is expected to be checked by not allowing the acid condensation as discussed above.

SCAPH is located just before the air heater so that the air temperature increases while entering the air heater. To implement this control, an external steam supply is required with separate control valves. The temperature/flow control valves provided need a steady and controlled steam pressure at the inlet so that the requisite and ultimate temperature control is achieved.

10.5.2.1 Objective

The objective of this control loop is to exactly maintain the SCAPH inlet steam pressure before the SCAPH temperature control valve.

10.5.2.2 Discussion

There may be more than one SCAPH for systems with separate heaters for secondary air and primary air. The requirement of steam supply pressure is normally the same and hence separate pressure-control valves may not be required. As the purpose is only raising the air temperature, the steam pressure may be around 5–7 kg/cm^2. The source is normally from the auxiliary steam header.

10.5.2.3 Control Loop Description

Fig. 8.42B may be referred to for a schematic representation of the idea for implementing the control loop in a typical

250 MW plant with a drum boiler and a fossil (coal)-fired furnace. The control loop is a direct, single, and simple type.

10.5.2.4 Measurement of Parameters

The SCAPH steam pressure is measured by a pressure transmitter with sufficient redundancy and at the appropriate location to get an average and representative measurement. There may be two different pressure-reducing stations if two types of air heaters (PA and SA) are provided with unequal steam pressure requirements.

10.5.2.4.1 Control Loop The measured/process variable is compared with a fixed set value, normally 10°C more than the acid dewpoint of the flue gas, to form the error signal. The controller output is the position demand signal of a pressure-reducing valve with the required capacity.

10.6 SO_x and NO_x Control

General

Different oxides of sulfur and nitrogen are termed SO_x and NO_x, and are generated as byproducts of combustion of different kinds of oils and coals in the furnace of the steam-generation plant. The fuels and atmospheric air contain these basic elements in various forms. After combustion, they are converted to their corresponding oxides with the presence of very high temperature, forming flue gas along with different oxides of carbon and other gases.

10.6.1 Objective

The main objective of all these controls is to reduce the sulfur and nitrogen oxides traced as byproducts of combustion. There are several methods of reducing the SO_x and NO_x levels in a thermal power plant. Some methods suggest a suitable control process by reducing the generation of those air pollutant gases at their very respective sources. Some methods, on the other hand, incorporate a chemical process for reducing those generated gases before discharge to the atmosphere. Such processes, however, do not take care about the reduction of those gas generation at source instead incorporate plants to convert them into harmless compounds following the local pollution control board's guideline.

10.6.2 Discussion

10.6.2.1 SO_x

As already discussed very briefly in Section 2.2 of Chapter 2, other aspects such as formation, aftereffects, and controls of SO_x, are incorporated in this section. Sulfur dioxide (SO_2) and trioxide (SO_3) are produced during the combustion process as byproducts and are present in the flue gas. They combine with moisture contained in the flue gas at a lower temperature and form sulfuric acid (H_2SO_4) and sulfurous acid (H_2SO_3). When the flue gas temperature falls below the H_2SO_4 dewpoint, droplets of condensed H_2SO_4 are formed on the metal surface of airheaters and ducts exposed to the flue gas. Under these conditions, corrosion occurs because of the presence of a thin film of acidic electrolytes over the surface, giving rise to localized and uniform corrosion, also known as acid dewpoint corrosion. It has been observed that long or extended shutdown of the unit may cause deposits of corrosion-aiding agents.

Too much excess air assists SO_x formation; on the contrary, if the stoichiometric ratio is less, then unburnt constituents cause corrosion at water wall surfaces. Later cases are somewhat eliminated when an ever-oxidizing atmosphere is maintained near the water walls mainly used for delayed mixing, as discussed later. Unlike NO_x, it cannot be minimized at the source but the condensation of sulfuric acid on the duct/air heater can be prevented by several methods.

10.6.2.2 NO_x

NO_x is not a compound/radical, but it represents a family of seven compounds/radicals and reacts in the atmosphere, forming ozone and acid rain. Out of these NO_x, NO_2 and is considered an important pollutant. Environmental concerns from NO_x include but are not limited to: ground-level ozone and acid rain formation, aquatic acidification, and deforestation. The Environmental Protection Agency (EPA) established national air quality standards that define the quality requirements of air with necessary safety margins. Electric utility plants contribute around 22% of NOx formation due to human activities (the other major source being transportation, ~56%).

Like sulfur, nitrogen (N_2) is also available from atmospheric air in quantity as well as from different kinds of fuel oils and coals being organic compounds that naturally contain N_2.

During the combustion process, only 5% of NO_2 is present against 95% of NO in the total available NO_x. The NO_x produced from the source of atmospheric air is often called thermal NO_x. Various test results and equations indicate that thermal NO_x conversion is dependent on the temperature and concentrations of both N_2 and O_2 and the time for which the combustion takes place. It is also observed that the same depends on both a fuel-rich flame front (diffusion flame where fuel and air are introduced separately and mixed through turbulence during combustion) and a fuel-lean flame front condition (premixed flame where premixed fuel and air are introduced to the furnace).

The other source of NO_x generation is called fuel NO_x, as the source is from fuels only. The test results show that this generation is dependent not on the temperature but on the availability of oxygen (O_2). O_2 reacts with the gaseous state of the fuel-bound nitrogen compounds such as NCH and NH_3 to generate NO in an air-rich combustion atmosphere only. On the contrary, these nitrogen compounds, being unstable, would simply reduce to N_2, a harmless gas, only under fuel-rich conditions. To summarize, the following design strategies are important,

(i) Lower thermal NO_x level can be achieved by using a low excess air operating strategy and a furnace designed for a lower gas temperature by introducing various methods such as burners with low turbulence diffusion flames, overfire air compartments, efficient water walls, and a flue gas recirculation system.

(ii) Lower fuel NO_x level design suggests the provision of controlling air flow for mixing with the fuel in the initial burning zone or combustion chamber.

(iii) Lowest NO_x level can be achieved if the coals used have the lowest fuel-nitrogen and lowest fuel-oxygen/nitrogen ratio.

NO_x generation thus can be reduced significantly and the methods, listed in Table 8.3 for ready reference, are typically adopted by the manufacturers to meet the stipulation of local/national pollution control boards and customer specifications.

Major approaches to reduce NO_x discharge into the atmosphere include either or both reducing the production of NO_x or mitigating the NOx already produced.

Out of these, the first may be less costly than the postproduction approach. Use of OFA in TT boilers is by far the most cost-effective way to control NO_x. Some of the major NO_x control methods have been tabulated (*based on EPA technical bulletin November 1999*) as Table 8.4, NO_x control methods with advantages and disadvantages. Apart from those listed in the table, there could be some other methods such as reduced air preheat, inject oxidant, nonthermal plasma, etc. Some of the abbreviations used in the table are BOOS (Burner Out Of Service), FGR (Flue Gas Recirculation), SCR (Selective Catalytic Reactor), SNCR (Selective Noncatalytic Reactor), and LEA (Less Excess Air).

TABLE 8.3 NO_x **Reduction Percentage by Different Methods**

Methods/ Systems	Type of NO_x Reduced	Reduction Possibility (~)	Remarks
Flue gas recirculation	Thermal NO_x	75%	Babcock design
Overfire air damper	Thermal NO_x	75%	CE/ Babcock design
Low NO_x burners	Thermal and fuel NO_x	25%	Applicable to any boiler
Staged burners	Thermal and fuel NO_x	70%	Applicable to any boiler
Air mixing and air flow control	Thermal and fuel NO_x	70%	Depends on boiler design

To combat NO_x pollution, three technologies are used:

- The primary technology is mostly concerned with air staging, that is, to reduce the availability of oxygen (less than the stoichiometric ratio) in the primary combustion zone and later balance the air downstream.
- The secondary technology is mainly concerned with fuel staging such as reburning, injection, and chemical reaction, such as SNCR (postproduction treatment).
- Posttreatment such as SCR.

  Apart from these, the use of oxygen in place of air utilizing an air separation plant is not uncommon. In IGCC, these have been discussed in the appendix. In ultrasupercritical boilers, in many system designs N_2 is separated from air and oxygen is used so that there will be only a remote chance of NO_x formation.

Following equipment/methods are utilized to curb the formation of NO_x:

1. Low NO_x burner (LNB) Fig. 8.47A–C: In an LNB, NO_x formation is limited by regulating the temperature profile and stoichiometric ratio.

The design features control the aerodynamic distribution and mixing pattern in the burner, always trying to delay complete mixing to the extent possible so that:

- Reduced oxygen in the primary flame region to limit formation of both thermal and fuel NO_x.
- Reduced flame temperature, hence less thermal NO_x formation.
- Reduced residence time due to distribution to cause less thermal NO_x formation.

Coal rank and volatile matter content have a direct impact on NO_x formation in LNB. Lower rank coals have a more volatile release, which inhibits NO_x formation near the burner on account of more conversion of nitrogen released in a fuel-rich environment near the burner into molecular nitrogen (N_2). On the contrary, LNB can lead to economic losses on account of more unburnt carbon (UBC). UBC can have an impact on tubes, electrostatic precipitator (EP) performance, and marketability of fly ash (ASTM C618 limits UBC to 6% in fly ash). However, this can be circumvented by using more fine coal and changing the air/fuel distribution. Coal with greater fineness will be better for less NO_x formation in LNB because less air will be required to carry it as well as less of a chance of UBC. On account

TABLE 8.4 Major NO_x Control Methods

Basic Principle	Technology	Description	Advantage	Disadvantage	Impact	Applicability
Reducing peak temperature	Low NO_x burner	Internal staged combustion	Low OP. cost compatible with FGR	High capital cost	Long flame, fan capacity, turn down stability	All fuel
	BOOS/OFA	Staged combustion	No cap. cost for BOOS, low cost for OFA	Higher air flow for CO High capital cost	Long flame, fan capacity, header pressure	All fuel, for BOOS multiple burners
	FGR	<30% flue gas recirculated with air, decreasing temperature	High NO_x reduction potential for low nitrogen fuels	Moderately high capital cost and operating cost affects heat transfer and system pressures	Fan capacity, furnace pressure, burner pressure drop, turndown stability	All fuels Low nitrogen fuels
	Air staging	Admit air in separated stages	Reduce peak combustion temperature	Extend combustion to a longer residence time at lower temperature	Adds ducts and dampers to control air furnace modification	All fuels
	Fuel Staging	Admit fuel in separated stages	Reduce peak combustion temperature	Extend combustion to a longer residence time at lower temperature	Adds ducts and dampers to control air Furnace modification	All fuels
Chemical reduction of NO_x	Fuel reburning	Inject fuel to react with NO_x	Moderate cost Moderate NO_x reduction	Extends residence time	Furnace temperature profile	All fuels (pulverized solid)
	SNCR (add-on technology) a. Urea b. Ammonia	Inject reagent to react with NO_x	a. Low capital cost. Moderate NO_x removal Nontoxic chemical b. Low operating cost. Moderate NO_x removal	a. Temperature dependent NO_x reduction less at lower loads b. Moderately high capital cost. Ammonia storage, handling, injection system	a. Furnace geometry Temperature profile b. Furnace geometry Temperature profile	All fuels
	SCR (add-on technology)	Catalyst located in the air flow, promotes reaction between ammonia and NO_x	High NO_x removal	Very high capital cost High operating cost Catalyst siting Increased pressure drop Possible water wash required	Space requirements Ammonia slip Hazardous waste disposal	All fuels

TABLE 8.4 Major NO_x Control Methods—cont'd

Basic Principle	Technology	Description	Advantage	Disadvantage	Impact	Applicability
Removal of N_2	Oxygen instead of air	Use oxygen to oxidize fuel	Moderate to high cost Intense combustion	Eliminates prompt NO_x Furnace alteration	Equipment to handle oxygen	All fuels
Reducing peak temperature	Combustion optimization	Change efficiency of primary combustion	Minimal cost	Extends residence time	Furnace temperature profile	Gas Liquid fuels
	Water/steam injection	Reduces flame temperature	Moderate capital cost NO_x reduction similar to FGR	Efficiency penalty Fan power higher	Flame stability Efficiency penalty	All fuels as low nitrogen fuels

of the longer flame of LNB, a deeper furnace may be called for to flame damaging water wall. In addition to this, there will be water wall wastage (corrosion due to unburnt sulfur/chlorides, etc. For such corrosion prediction, computational fluid dynamics (CFD) codes are available), especially for boilers with LNB and/or other means of external staging. The problem is acute in cases of supercritical boilers (more specifically using relatively high sulfur coal). For this reason part Offset auxiliary air is used discussed later. Low NO_x *Concentric Firing System* (LN*CFS*) improves coal air mixing. It forms an air circle outside the fire all created by fuel PA, as shown in Fig. 2.10. This is mainly developed to retrofit a TT furnace. It produces a stable flame front with an inner fuel-rich core fireball, which helps in converting bonded nitrogen to form an N_2 molecule. The air distribution philosophy behind a low NO_x burner is depicted in Fig. 8.47B. As shown in Fig. 8.47D, in a TT furnace, normally there will be a concentric firing system (CFS) between two adjacent low NO_x coal nozzles. Through the CFS, offset auxiliary air is introduced. The angle of these offset air nozzles that is, the yaw (horizontal) and tilt/pick (vertical), can be adjusted as shown in Fig. 8.47C. Normally, the tilt is automatically controlled as per RH (or SH) temperature control, that is, gagged with a burner tilt mechanism and the yaw is adjusted manually. Offset air is directed toward the furnace to reduce fouling and produce an oxidizing environment to minimize water wall wastage. LNCFS burners have nozzle attachments to allow coal to burn near the nozzle to minimize NO_x formation. Normally, above the top coal nozzle, there will be a CCOFA, as discussed below.

2. Overfire air (OFA): As indicated earlier, about 5%–20% of the total air required is diverted from the primary combustion zone and injected through some air port in the downstream of combustion for complete combustion and to reduce UBC. OFA in wall-fired boilers with LNB can help NO_x reduction by 10%–25%. In case of a TT furnace, OFA is an integral part of the design. In a TT furnace with OFA, an NO_x reduction between 20% and >60% is possible, depending on the initial NO_x level, but it is coupled with an increase in UBC and CO. However, the extent of this increase depends on the OFA design and coal property. There are two kinds of overfire air:
   - Closed coupled overfire air (CCOFA): In close proximity to the primary combustion zone, this is mainly responsible for reducing the residence time in a fuel-rich primary combustion zone and increasing the residence time in a fuel-lean burning zone.
   - Separate overfire air (SOFA): Located distinctly distant from the primary burning zone as well as the burnt out zone, it provides additional residence time in the primary combustion zone and less residence time in the burnt out zone.
   - The combination of these two helps in reducing NO_x and at the same time helps to reduce UCB.

   Dispositions of these dampers for TT furnaces have been shown in Figs. 8.47C and D. Just above the top coal nozzle, there is a CCOFA (like CFS) and above that there is another set of CCOFAs. In between the two CCOFAs, there is a port of CCOFA for tertiary air.

Separated by some distance, there will be SOFA dampers. Some of the design features in a TT furnace are:

- SOFA ports are above the wind box.
- Increased separation between auxiliary air and the fuel admission nozzle.
- Reduction of the control damper for secondary air flow.
- Reduction of secondary air flow while maintaining system pressure drop and injecting velocity for efficient mixing.
- Addition of separate tilt control of under bottom end air admission for CO control.
- In a TT furnace, especially in larger or supercritical boilers, there will be two sets of SOFAs. Lower SOFAs are essentially in the corners and the tilt control is gagged with the burner tilt control. There will be upper SOFA normally above the lower SOFA, and these are mounted on side walls, as shown in Fig. 8.47C. Normally, the upper SOFA dampers are connected with the automatic control loops. The upper SOFA may be regulated to control NO_x. Desired NO_x set point is compared with the actual value, then based on the error thus created, the controller issues the position demand for the associated damper. There have been loops where there may be MAX selection between the set point and the output of the function generator (having the boiler load as input), so that there will be a dynamic set point to take care of changes in load for the boiler. However, the yaw of each of these is normally adjusted manually. Reburning (Fig. 8.47E): Up to 25% of the total fuel heat input is provided by injecting secondary fuel (~90% of air required by the stoichiometry of the secondary fuel) above the main combustion zone to produce a slightly fuel-rich reburnt zone. This results in a hydrocarbon fragment that comes in contact with the incoming NO_x (from the upstream combustion zone), which, in a fuel-rich condition, forms hydrogen cyanide (HCN) and isocyanic acid (HNCO), and is finally converted to N_2 molecules. The furnace dimensions are very important here so as to achieve the desired goal, where sufficient residence time is essential. An increase in reburn heat input and a matching decrease in the main combustion heat input will decrease the stoichiometry in the reburn zone with an increase in NO_x reduction efficiency.

3. SNCR (Fig. 8.47F): This is basically post NO_x production control, where reagents such as urea and ammonia are injected into the furnace above the combustion zone so that the NO_x formed could be converted to N_2 molecules as per the reactions indicated below:

$$(NH_2)_2CO + NO + ½O \rightarrow 2H_2O + CO_2 + 2N_2$$

$$2NH_3 + 2NO + ½O_2 \rightarrow 2N_2 + 3H_2O$$

This injection can be done at the furnace where the temperature is above 1150°C. Urea is injected as an aqueous solution whereas NH_3 can be an anhydrous or an aqueous solution.

4. SCR: discussed separately below.
5. Miscellaneous other systems: Water/steam injection methods: By injecting water or steam into the flame, flame temperatures are reduced, thereby lowering thermal NO_x formation and the overall NOx levels. Water or steam injection can reduce NO_x up to 80% (when firing natural gas) and can result in lower reductions when firing oils. FGR, as discussed at length in the reheater temperature control loop, could be another method to limit thermal NO_x. For coal-fired units, this is good for RH steam temperature controls, but may not be a foolproof and effective method for NO_x reduction. Control loop optimization also helps in reducing NO_x.
6. Control loop considerations: As discussed earlier, the fuel-air ratio control by staging of combustions is a major way to combat NO_x production control. Reburning control is also a part of combustion control. In case of tangential tilt burners, the tilt controls of the CFS and CCOFA are done as a part of the reheat/superheat steam temperature control. As also said earlier, the yaw is usually manually adjusted. However, there may be separate auto control loops for SOFA, based on NOx measurements, so that in case of excess NO_x production, more air may be admitted through this port to reduce NO_x production by complete combustion.
7. Eight-corner method: There is another method to curb NO_x generation. The B and W eight-corner firing is an example of the same.

10.6.3 Description of Control Loops

The generation of SO_x cannot be avoided during the combustion of fuels, but can be minimized by proper selection of fuels where the user has that liberty to select as per his discretion. However, once generated, its detrimental effect on the duct/equipment is minimized by providing separate subplants, namely a steam coil air preheater (SCAPH) with its associated control loops. Concerned control loops are also provided for flue gas desulfurization (FGD) plants meant for eliminating postproduction SO_x by converting it to sulfites or sulfates.

10.6.3.1 Controls of SO_x

10.6.3.1.1 Controls of SO_x Through Main Plant Process Control Loops After various experiments, the relationship between the minimum metal temperature and the gas and air temperature has been established by which the corrosion of the air heater-exposed metal can be controlled. Normally, the average cold end metal temperature of the main air heaters throughout the year is kept at a minimum of 10°C above the acid dewpoint for flue gases.

The following methods are applied for the above controls:

(i) *Steam coil air preheater (SCAPH) is* located just before the air heater so that the air temperature increases while entering the air heater. Here, the average of the gas outlet and air inlet temperatures is taken as the measured variable and controlled against a calculated preset temperature value, which will not allow the acid condensation as discussed above and in clause no. 10.3(ii).

(ii) There is some arrangement made near the air heaters so that a portion of the total air is bypassed and the heat exchange takes place toward the amount of air that passes through the main line only. As the quantity through the air heater becomes less, the loss of temperature experienced by the flue gas is also less, which prevents the metal temperature from further going down.

The gas dampers at the inlet of the airheater and bypass line would modulate as per the controller output with the input, just like the SCAPH control loop. The controller output is so adjusted that the bypass damper would modulate only when the main line damper is fully open. During operation, if the bypass damper goes to the maximum position, then only the main line damper would start modulating. An interlock should be provided so that both dampers cannot operate simultaneously.

(iii) There is some other arrangement made near the air heaters so that a portion of the total hot air is diverted from the air heater outlet to the force draught fan (FD fan) inlet so that the air temperature increases while entering the air heater. The recirculated hot air flow quantity depends on the controller output with the same control strategy and inputs as described above. The amount of heat exchange that takes place would be almost the same (except the heat loss in the recirculation line and the additional heat required to raise the air temperature handled by the FD fan) as before toward more air passing through the air heater, but with a higher inlet temperature. The damper in the recirculation line modulates as per the controller output to maintain the average temperature of the flue gas outlet and the air heater inlet temperature.

10.6.3.1.2 Controls of SO_x Through a Separate Chemical Process Plant There are totally separate and special types of chemical process plants that are employed for reducing the SO_x level, and these are called *FGD plants*.

As SO_x or in particular SO_2 are acid gases, the removal process requires an alkaline scrubbing reagent (scrubber) or sorbent to react and convert them into a sulfur compound other than the gaseous form. There are broad classifications through which they can be identified such as wet or dry and regenerable or nonregenerable; those are enumerated below:

(a) Wet (nonregenerable) FGD processes include lime/limestone forced/natural oxidation, lime/limestone inhibited oxidation, lime and magnesium-lime, seawater process.

(b) Dry (nonregenerable) FGD processes include lime spray (dry and semidry), duct sorbent injection, furnace sorbent injection, gas phase oxidation/ammonia injection (both SO_x and NO_x).

(c) Regenerable FGD processes include sodium sulfite (wet), magnesium oxide (wet), sodium carbonate (wet), amine (wet), fluidized bed copper oxide process (both SOx and NO_x—dry), activated carbon (dry).

Here, regenerable means the process where the main scrubber is regenerated at the end of the reactions and reused. A small percentage is required to be injected into the system to compensate for the cyclical loss.

Scrubbing of sorbent-like limestone/lime is the most common method followed in FGD plants. It is a catalytic chemical process for the removal of SO_2 produced during combustion. In this system, a sulfur compound combines with a calcium-containing sorbent, generally lime (CaO) or limestone ($CaCO_3$), to form a slurry. The slurry after use can be treated as waste or a useful byproduct.

Though there are several processes indicated above, three types of FGD plants are normally available for large thermal power plants using chemical reagents such as a *wet scrubber (after ESP), a spray dry scrubber (before ESP) [dry and semidry], and a sorbent injector (furnace and duct).*

Wet Scrubber

This type of SO_2 removal system is the most efficient (>95%) with a gaseous and liquid phase reaction; SO_2 is transferred to the liquid under saturated condition. In general, it results in a liquid waste stream requiring wastewater treatment and a slurry as a byproduct, which needs a disposal system. As broadly classified, three main processes characterized by the use of absorbents dictate the scrubber design itself along with the generation of the waste and byproduct.

Theoretically, other types are also in use, but these may not have been manufactured recently. Those types include sodium-based or dual alkali-based systems. These two systems were developed to avoid the consequential fouling problem encountered from the lime (stone)-based wet scrubbing process. The other reason is the idea that the application of a higher reactivity of the sodium compound may affect higher SOx removal. These systems ultimately became unpopular due to higher costs, nonavailability, production of higher waste slurry, etc.

In *a lime/limestone-based wet scrubber*, flue gas is sprayed with an aqueous slurry of lime [CaO (dry lime) and $Ca(OH)_2$ (milk/slurry of lime)] or limestone ($CaCO_3$). Sulfur dioxide (SO_2) is removed during the series of chemical reactions between the SO_2 and the slurry to form calcium sulfate (gypsum) and sulfite.

$$\text{The reactions are} : CaO + SO_2 + H_2O = CaSO_3 + H_2O\ [\text{with dry lime}]$$

$$Ca(OH)_2 + SO_2 = CaSO_3 + H_2O\ [\text{with lime slurry}]$$

$$\text{and } CaCO_3 + SO_2 = CaSO_3 + CO_2\ [\text{with limestone slurry}]$$

In some plants, the cost of FGD installation is offset to some extent by converting the slurry into a useful byproduct such as gypsum [the remaining sulfite to $CaSO_4$]. This can be achieved with or without air added to the thickener or oxidation tank, which is referred to as natural or forced oxidation, respectively. The latter gives a better gypsum quality that finds wider application while the larger crystals produced render a simpler dewatering process.

To do this, compressed air is blown either in the scrubber process or in the following stage as a part of what is known as *forced oxidation*.

$$\text{The reaction is} : CaSO_3 + O_2 + H_2O \Longrightarrow CaSO_4, 2H_2O$$

The clean gas is then sent through the gas to gas heater (GGH). The GGH is provided to raise the temperature of the flue gas at the chimney inlet as it goes down during the process that takes place at the absorption tower. The incoming flue gas after the ESP is utilized to heat the outgoing flue gas after the absorption tower, as shown in Fig. 8.43.

The thermal performance of GGH is very significant, as the reheated gas outlet temperature must be kept above the minimum specified value to achieve adequate flue gas buoyancy and ensure plume dispersion. This is also required to get a sufficient chimney effect, which depends largely on the gas outlet temperature and the stack height. Finally, the reheated flue is discharged through the high stack. Efficiency of this type of plant is 99%. Generally, the system includes dewatering of gypsum for commercial purposes. The wastewater after proper neutralization is sent to an ash slurry sump or suitable place as applicable to suit the plant conditions.

The plants discussed here do have other auxiliary systems, a good amount of/control and is accomplished generally by dedicated PLC with communication link to main plant DCS for relevant data exchange. In Japan, it is common practice to utilize a $CaCO_3$ concentration analyzer to optimally control SO_2, as shown in Fig. 8.45. However, the ESP must be located before the FGD for a gypsum recovery system.

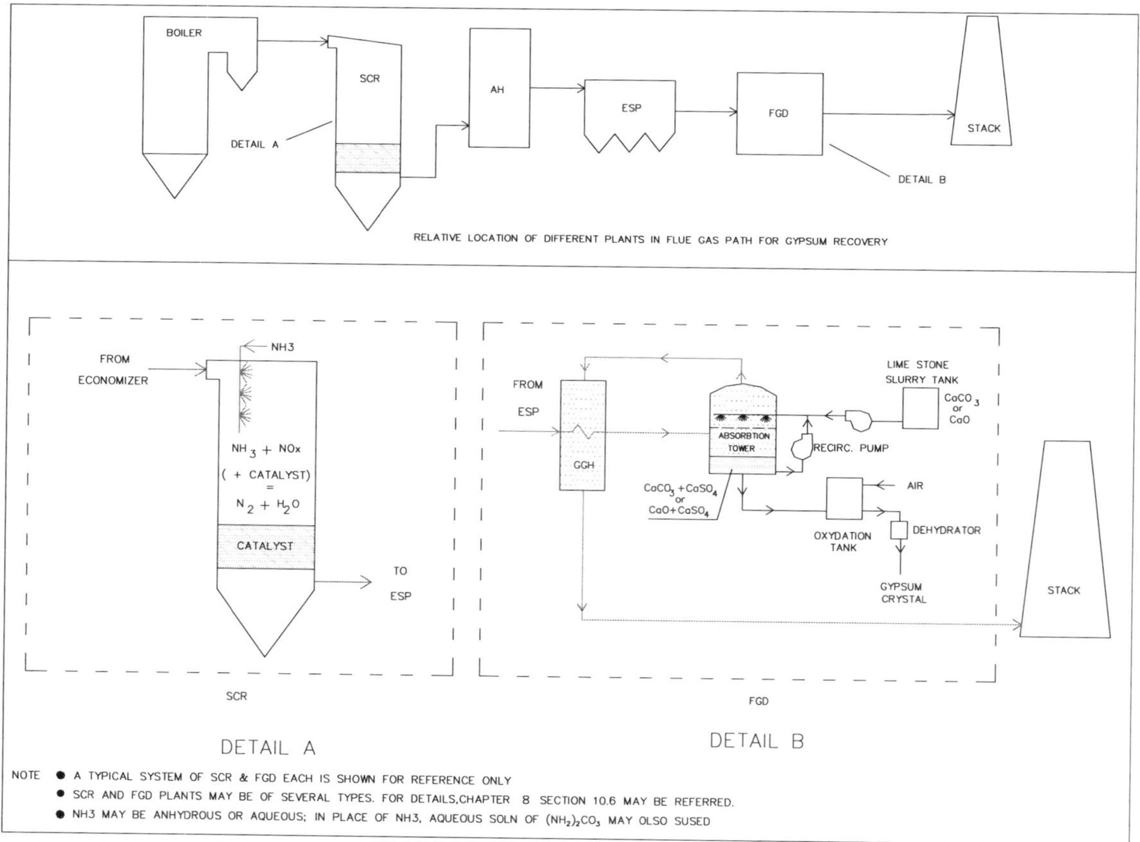

FIG. 8.43 Schematic diagram of SCR and FGD plants.

The other types including sodium-based or dual alkali-based systems are also in use, but may not have been manufactured much in recent days; they are very briefly discussed, mainly for academic interest.

(a) *A sodium-based wet scrubber* system uses sodium hydroxides (NaOH) or sodium carbonate (Na_2CO_3) as the absorbent and has a low liquid-to-gas ratio because of their high reactivity with respect to lime/limestone slurry absorbent. The reactions are:

$$2NaOH + SO_2 = Na_2SO_3 + H_2O \text{ and } Na_2CO_3 + SO_2 = Na_2SO_3 + CO_2$$

Further oxidation converts sodium sulfite (Na_2SO_3) into sodium sulfate (Na_2SO_4); both are highly soluble, meaning they need special disposal.

If sodium sulfite (Na_2SO_3) slurry is used in place of Na_2CO_3, the reaction would be:

$$Na_2SO_3 + SO_2 + H_2O \Longrightarrow 2NaHSO_3$$
$$\text{(sodium hydrogen sulfite solution)}$$

By heating this solution, the reaction reverses and Na_2SO_3 could be recovered rather than consumed. This process is called the *Wellman-Lord* process and was regarded as the most widely used wet and regenerative sorbent FGD treatment. The process is almost similar to the dual alkali process discussed later. Sodium sulfite (Na_2SO_3) acts here as the actual sorbent while the sodium hydrogen sulfite ($NaHSO_3$ acts as the sulfur-binding compound. However, the sulfur is not passed on to the calcium, but when heated, it is released again as SO_2, which is extracted as a concentrated mixture (approximately 85% SO_2) with water for further processing to sulfuric acid. In the process of scrubbing, sodium sulfite or sulfate is also formed as a byproduct. Due to some process loss in the sodium content in the form of Na_2SO_4, a small percentage of continuous make up through Na_2CO_3 (*soda*) or Na_2CO_3, $NaHCO_3$ (*trona*) is required to be added in the regeneration tank to balance this loss.

The reactions are as follows:

(i) Inside the scrubber:

$$Na_2SO_3 + SO_2 + H_2O \Longrightarrow 2NaHSO_3$$
$$\text{and } 2Na_2SO_3 + O_2 \Longrightarrow 2Na_2SO_4$$

(ii) Inside the regenerator:

$$2NaHSO_3 \Longrightarrow Na_2SO_3 + SO_2 + H_2O$$

(iii) In the make-up process:

$$Na_2CO_3 + SO_2 \Longrightarrow Na_2SO_3 + CO_2, Na_2CO_3, NaHCO_3$$
$$NaHSO_3 + NaHCO_3 \Longrightarrow Na_2SO_3 + H_2O + CO_2$$

Sulfuric acid formation reaction:

$$2SO_2 + O_2 + 2H_2O \Longrightarrow 2H_2SO_4$$

In some systems, magnesium hydroxide [$Mg(OH_2)$] is used as an absorbent and the reaction is:

$$Mg(OH_2) + SO_2 = MgSO_3 + H_2O$$

(b) *Dual alkali-based wet scrubber* system

This system utilizes sodium-based reagents such as sodium hydroxide (NaOH) or sodium carbonate/soda (Na_2CO_3) as the primary absorbent for SO_x treatment, similar to the Wellman-Lord process. Initially, when Na_2CO_3 (soda) or NaOH reacts with SO_2 in the absorption tower, the following reaction takes place:

$$Na_2CO_3 + SO_2 \Longrightarrow Na_2SO_3 + CO_2$$
$$\text{and } 2NaOH + SO_2 \Longrightarrow Na_2SO_3 + H_2O$$

After the formation of Na_2SO_3, it further reacts with SO_2 to form sodium hydrogen sulfite ($NaHSO_3$) and comes out of the absorption tower as a sulfur binding compound. Calcium-based reagents are then applied to this slurry for regeneration of the basic reagents, that is, (NaOH) or (Na_2CO_3).

This method was thought of due to the problem that arose for using a lime scrubbing system such as low solubility of lime and limestone in water and gypsum scaling problems. The solubility of sodium salts is much higher, which led to this concept based on a scrubbing liquor with $NaHCO_3$ and Na_2SO_3 as the SOx-binding compounds. The principle of the process is given in Fig. 8.44A.

The reactions at different places are given below:

(i) Inside the absorption tower:

$$Na_2SO_3 + SO_2 + H_2O \Longrightarrow 2NaHSO_3$$
$$\text{and } 2NaOH + SO_2 \Longrightarrow Na_2SO_3 + H_2O$$

(ii) Inside the precipitation tank:

$$2NaHSO_3 + CaO + H_2O \Longrightarrow Na_2SO_3 + CaSO_3, 2H_2O,$$
$$\text{and } NaHSO_3 + CaO \Longrightarrow CaSO_3, 2H_2O + NaOH$$

and

$$CaSO_3, 2H_2O + O_2 \Longrightarrow CaSO_4, 2H_2O \text{ (Gypsum)}$$

(iii) Inside the regeneration tank:

$$NaHSO_3 + Na_2CO_3 \Longrightarrow Na_2SO_3 + NaHCO_3$$
$$\text{and } NaHSO_3 + NaHCO_3 \Longrightarrow Na_2SO_3 + H_2O + CO_2$$

Due to some process loss in the sodium content in the form of $CaSO_3$ and $CaSO_4$ product, a small percentage of continuous make-up through Na_2CO_3 (*soda*) or Na_2CO_3, $NaHCO_3$ (*trona*) is required to be added in the regeneration tank.

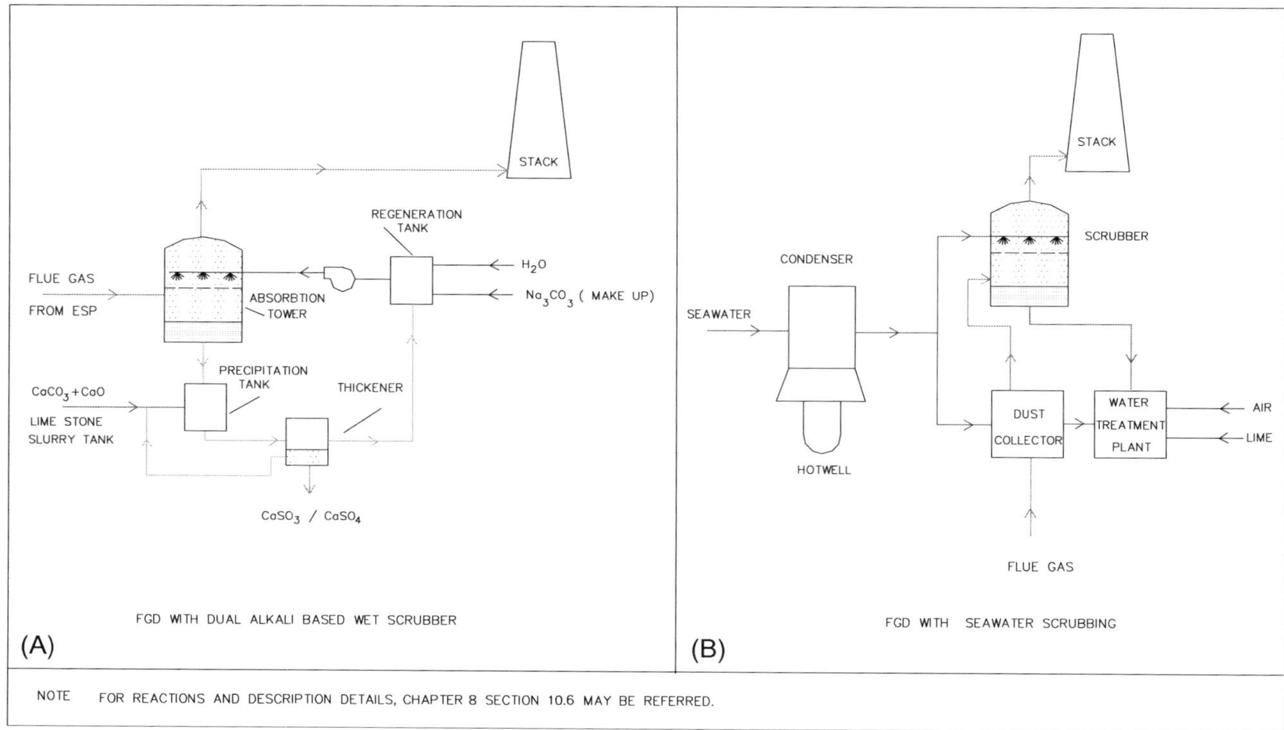

FIG. 8.44 Schematic diagram of FGD with dual alkali-based wet scrubber and seawater scrubbing.

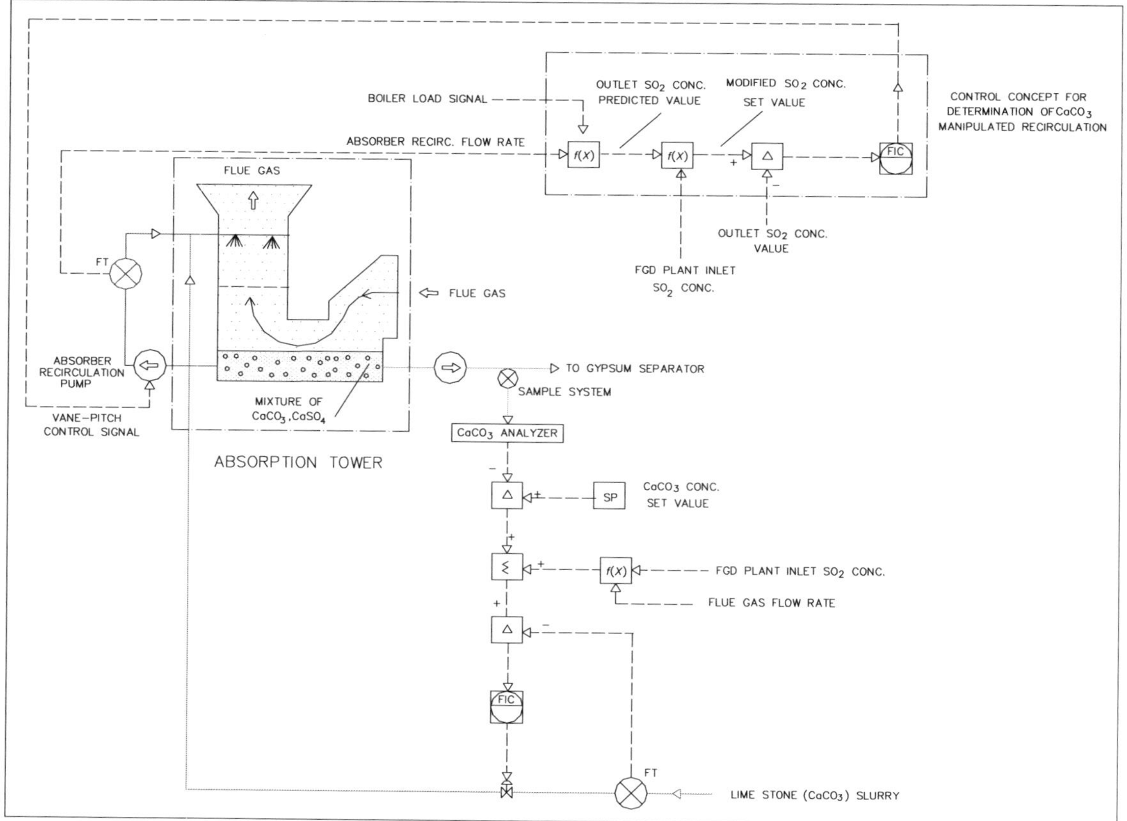

FIG. 8.45 Control loop schematic of SO_x control with limestone absorber and gypsum recovery.

Spray Dry Scrubber

The spray dry scrubber is made up of a semidry and a dry lime process. These processes have been developed as a competitive alternative to classical wet scrubber technology. The spray dry scrubber is next popular method with efficiency as good as that of a wet scrubber, if not better. The development in the process technology can render removal efficiencies of SO_x up to 98% some manufacturers claim. For a FGD plant with a spray dry scrubber, unlike a wet scrubber, the ESP/dedusting equipment is located after this plant, which is a major layout criterion. This process make use of water-based sorbent containing lime (CaO) or milk of lime. The lime-based reagent is injected into a reactor vessel in the form of milk of lime (in the case of a lime spray dryer or semidry process) or humidified powder (in case of dry processes). The atomized form of the reagent, when it comes in touch with the hot flue gases, becomes dry and then the reaction takes place between the hydrated lime and SO_x (mainly SO_2) in the flue gases. The solid reaction product is collected by downstream dedusting equipment (ESP in case of a thermal power plant or a baghouse filter for a sinter plant, for example) and part of it is recirculated. Recirculation of this reaction product is made through the lime slurry preparation tank to reduce the bulk consumption of lime and hence the running cost of the project.

The hydrated lime [$Ca(OH)_2$], when it reacts with SO_2, is converted into a mixture of calcium sulfate (gypsum) and sulfite. One of the many advantages of this process is that the requirement for a water treatment plant is eliminated.

The in-duct sorbent injection method is also available as a comparatively less costly approach; it can be located more upstream in the flue gas path. This system does not require any extra space or a reactor vessel and can be easily applied to older or existing plants without a desulfurization facility. This system is also described as part of the sorbent injector process at a lower temperature.

Additional activation of the sorbent may be achieved by spraying water to the flue gas at the downstream of the sorbent injection point where the sorbent is actually fed. The part of lime (CaO) that failed to make contact or react would now be converted to calcium hydroxide [$Ca(OH)_2$], which is more reactive to SO2, letting the calcium sulfite part be further oxidized to calcium sulfate.

The normal sulfur removal efficiencies of FGDs are on the order of 80%–85%. However, higher efficiencies can be attained if arrangements are made for water spraying after the sorbent injection point into the flue gas path (which reactivates the free sorbents in the flue gas), by spent sorbent recycling and by select optimized location of the sorbent injection point with respect to the temperature.

Sodium carbonate (Na_2CO_3) can also be used as a sorbent where sodium sulfate (Na_2SO_4) would be the byproduct, which is soluble and stipulates special handling. That problem, along with the availability and cost factor, makes it less popular as a sorbent.

Sorbent Injector

The third method is known as *a sorbent injector*, but it is less popular and offers more or less 60–65% efficiency. In this process, dry sorbent such as limestone or hydrated lime [$Ca(OH)_2$] is sprayed into the hot flue gases in the upper part of the furnace itself (high-temperature sorbent injector). Due to the reaction of sorbent and SO_2, gypsum is produced and captured later in a fabric filter or electrostatic precipitator (ESP), along with fly ash. In this system, the gypsum cannot be used for commercial purposes as it would be mixed with fly ash and react with SO_x and the calcium compound. The maximum temperature limit is around 1200°C and $CaSO_4$ is not stable above around 1250°C in typical flue gases from a typical coal-fired SG plant.

For the direct sorbent injection method, whether a low or a high temperature process, the efficiencies depend upon the approach temperature, the sorbent fineness, the injection point, and the recirculation of used sorbent.

Seawater Scrubbing

Other than scrubbing by alkaline chemical reagents, *seawater*, being a natural alkaline, can also be used as a sorbent, which absorbs SO_x/SO_2. Further oxidation by adding oxygen promotes the formation of SO_4^- ions and free H^+ ions. The H^+ ions then react with the carbonates present in the seawater to release CO_2 gas. The reactions are

$$SO_2 + H_2O + O_2 \Longrightarrow SO_4^- + H^+ \text{ and}$$

$$HCO_3 + H^+ \Longrightarrow CO_2 + H_2O$$

This type of FGD system can be suitable for coastal regions. Fig. 8.44B, Schematic Diagram of FGD with Seawater Scrubbing, may be referenced. A part of hot seawater from the condenser outlet is used to scrub SO_x/SO_2 along with fly ash particles. In the water treatment plant, the pH of the scrubbing liquid at the outlet of the FGD unit needs to be adjusted before the same is sent back to the sea. Lime may be used for adjusting the pH of discharge seawater.

10.6.3.2 Controls of NO_x

This control is very important as a part of the denitrification measure taken for the products of combustion, that is, flue gas.

10.6.3.2.1 Controls of NO_x Through Process Control Loops The presence of the NO_x level in the power plant can be reduced in many ways, as per the considerations set by different manufacturers. The main two basic ideas are:

Reduced production of NO_x through staging of air and/or fuel (discussed in this chapter, clause no. 10.6.3.2.2) and through separate chemical process plant.

10.6.3.2.2 Controls of NO_x Through Separate Chemical Process Plant There are several methods of denitrification of flue gas other than control measures taken at the source. This is important because the generation of NO_x cannot be fully avoided by protection at the source. NO_x removal from flue gas through a separate chemical process plant is achieved by the following methods:

(a) *Dry process* is more popular and needs discussion.
(b) *Wet process* is a complicated process requiring wastewater treatment and hence is seldom used.

Dry process may be of different types using the application of catalytic and noncatalytic reduction:

(i) *Selective catalytic reduction (SCR)* by spraying ammonia (*NH_3*) in the presence of a catalyst type depending on the flue gas temperature.
(ii) *Selective noncatalytic reduction (SNCR)* by spraying ammonia/urea without a catalytic presence requiring a high temperature on the order of 800°C/1000°C.
(iii) *Nonselective catalytic reduction (NSCR)* using the presence of multiple catalysts, namely *platinum (Pt)* with *methane (CH_4)* or *CO* or *H_2*.
(iv) *Catalytic cracking* using *platinum (Pt)* as the catalyst.

SCR, being the most widely used plant, is only discussed in this chapter. It is a means to convert NOx in the presence of various catalytic agents into nitrogen (N_2) and water (H_2O). Generally, SCR is located between the economizer and the air heater, and ammonia is injected before entering into a catalyst chamber. The approximate NO_x reduction possibility is about 90% by this method.

Aqueous or anhydrous ammonia, aqueous urea (also used as a reductant in the place of ammonia), etc., are added to the flue gas stream to remove NO_x. Other possible reductants include cynuric acid and ammonium sulfate.

Typical reactions are:

$$\begin{aligned}2NO_2 + 4NH_3 + O_2 &= 3N_2 + 6H_2O; 4NO + 4NH_3 + O_2 \\ &= 4N_2 + 6H_2O; NO + NO_2 + 2NH_3 \\ &= 2N_2 + 3H_2O\end{aligned}$$

Secondary reactions may be

$$2SO_2 + O_2 = 2SO_3;$$

$$\begin{aligned}2NH_3 + SO_3 + H_2O &= (NH_4)_2SO_4 \text{ or } NH_3 + SO_3 + H_2O \\ &= (NH_4)HSO_4\end{aligned}$$

CO_2 is formed while using urea (instead of ammonia) as per the reaction below:

$$4NO + 2(NH_2)_2CO + O_2 = 4N_2 + 4H_2O + 2CO_2$$

The ideal reaction has an optimal temperature range between 360°C and 450°C, but can operate at a further lower range from 230°C to 450°C requiring a longer reaction/residence time. The minimum effective temperature depends on the various fuels, gas constituents, and catalyst type/geometry.

10.6.3.2.3 Catalysts Used for SCR Operation SCR catalysts are manufactured from various ceramic materials used as a support or carrier, for example, titanium or aluminum oxides whereas active catalytic components are usually base metal oxides of vanadium, tungsten, etc., or zeolites. Precious metals such as platinum, etc., are also used as catalysts (clause no. 10.6.3.2.2 of this chapter).

Usage of base metal catalysts such as vanadium and tungsten is not suitable at higher temperatures because of the lack of high thermal durability; however, they are popular for their lower operating costs. Most of the industrial and utility boiler applications can well be covered by the temperature ranges offered by them. Zeolite oxides, on the contrary, are capable of operating at substantially higher temperature (continuous operating range at 630°C) than base metal catalysts with maximum short temperature withstanding capacity up to 850°C).

The shape/geometry of the catalysts is considered the influential factor for SCR system design. They are available in various forms/shapes such as granules, grids, honeycombs, plates, etc. The catalysts characteristically have their own advantages and disadvantages. For example, the honeycomb configurations are smaller than plates and less expensive, but they have higher pressure drops and are susceptible to plugging more often and more easily compared to plates. Plate catalysts are much larger in size and more expensive.

The instrumentation and controls involved here are not too many and can be integrated with the main DCS. In order to keep excess NH_3 (being harmful to human life) within limits, the NH_3 needs to be monitored and controlled. The catalyst bed loss or the differential pressure loss across the catalytic chamber also requires close monitoring.

10.6.4 Combined SO_x-NO_x Removal

10.6.4.1 SO_x-NO_x Removal by Fluidized Bed Copper Oxide Process (Fig. 8.46)

This process is an advanced technology that provides for the simultaneous control of SO_x and NO_x. The copper oxide process offers a number of advantages over conventional approaches to SO_x and NO_x control, as discussed above. The same are listed below:

(i) The combined removal of SO_x and NO_x is accomplished in a single reactor vessel.
(ii) The system is regenerative as far as copper oxide is concerned.
(iii) The process byproduct is a saleable form of sulfur (e.g., sulfuric acid).

Currently, this type of FGD is regarded as a potential type of NO_x control system also though during the initial period the aim was to eliminate SO_x only as that was considered as most important pollutant.

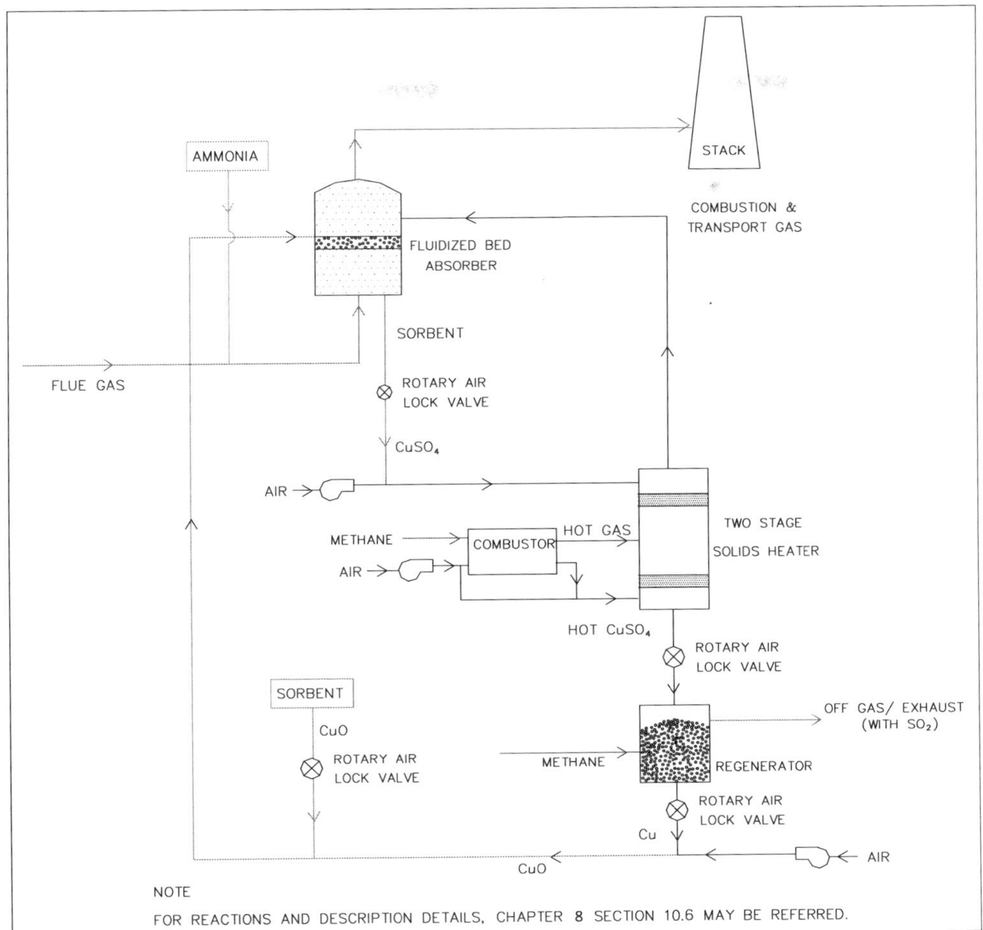

FIG. 8.46 Schematic diagram of SO_x NO_x control with fluidized bed copper oxide process.

There are three basic compartments envisaged in this system: the absorber, heater, and regenerator.

In the absorber, the sorbent is made fluidized by the flue gas injected from the bottom. The copper, as copper oxide, reacts with sulfur oxide to produce copper sulfate at about 400°C. The reactions are:

$$2CuO + 2SO_2 + O_2 \Longrightarrow 2CuSO_4 \text{ and } CuO + SO_3 \Longrightarrow (CuSO_4)$$

Ammonia (NH_3) is injected into the flue gas or premixed before it enters the absorber. In the absorber, it reacts with nitrogen oxides, with the copper sulfate acting as the catalyst. The reactions:

$$4NO + 4NH_3 + O_2 \Longrightarrow 4N_2 + 6H_2O$$
$$\text{and } 2NO_2 + 4NH_3 + O_2 \Longrightarrow 3N_2 + 6H_2O$$

These are exothermic reactions and the heat generated could be utilized through the flue gas in the air preheater located next to this plant.

This process then needs methane (CH_4). Part of it is burnt in a combustor and the hot gas is sent to the two-stage heater vessel. Copper sulfate sorbent is heated typically at about 500°C. Hot copper sulfate is then allowed to enter a moving bed regenerator from the top, whereas the methane is introduced at the bottom of the reactor. The copper sulfate when coming into contact with methane at ~500°C is reduced to Cu, releasing SO_2. There may be some copper oxide (CuO) contained in the sorbent entering the regenerator. This would react readily with the SO_2 in the exiting off-gas/exhaust gas to form copper sulfite ($CuSO_3$). To summarize, it may be stated that inside the regenerator, the sorbent may consist of copper sulfate, a small amount of copper oxide, and copper sulfite.

The reactions are

$$CuSO_4 + CH_4 \Longrightarrow Cu + SO_2 + CO_2 + H_2O,$$

$$CuSO_3 + CH_4 \Longrightarrow Cu + SO_2 + CO_2 + H_2O$$
$$\text{and } CuO + CH_4 \Longrightarrow Cu + CO_2 + H_2O$$

The concentrated SO_2 can be used to produce sulfuric acid, similar to the Wellman-Lord process. The Cu sorbent is transported back to the fluidized bed using air, during which it is oxidized to CuO as per the following reaction:

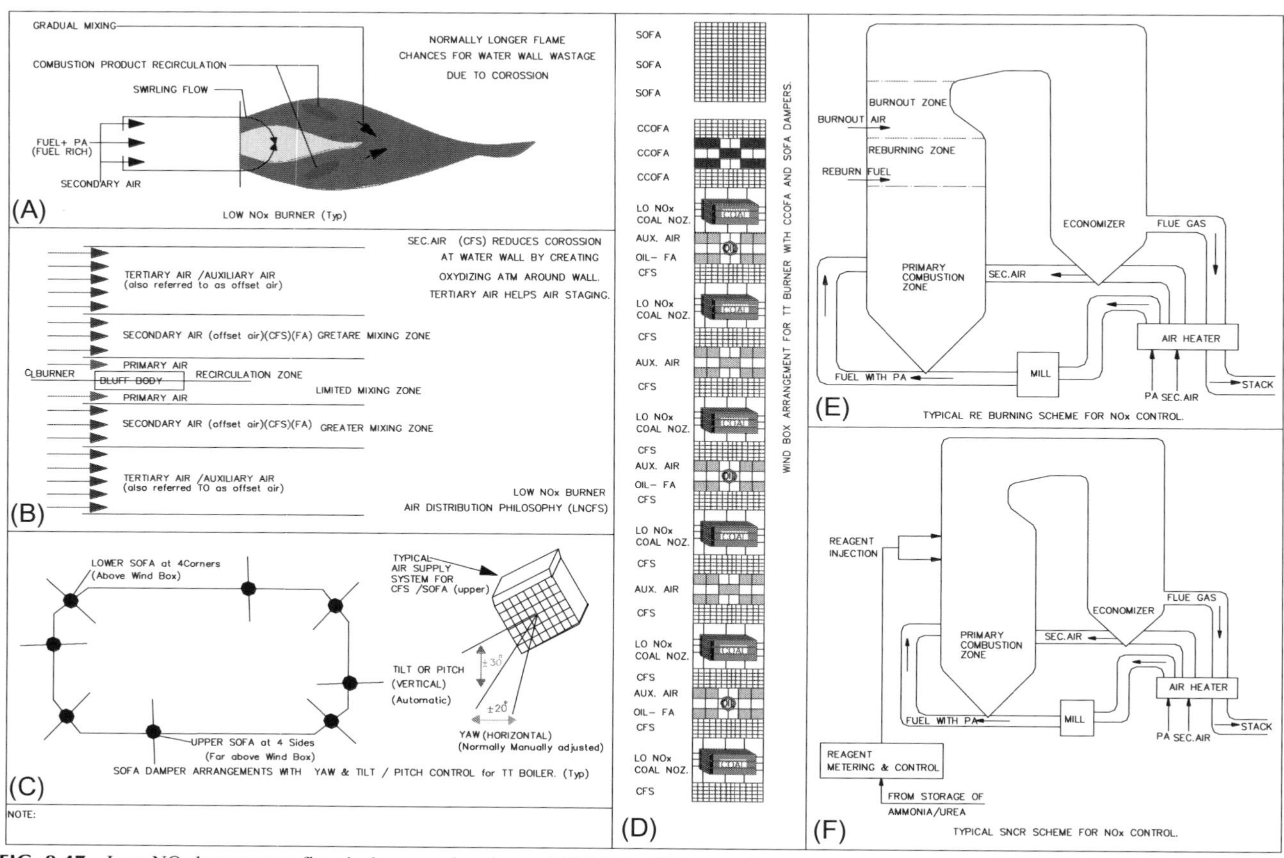

FIG. 8.47 Low NO_x burner over fire air damper reburning and SNCR for NO_x control.

$$2Cu + O_2 \Longrightarrow CuO \text{ and } 2CuSO_3 + O_2 \Longrightarrow 2CuSO_4$$

(Copper sulfite if any is also oxidized to the basic sorbent form).

The exhaust product gas/offgas, that is, SO_2 from the regenerator may also be utilized for recovery of sulfur in its elemental form in a separate place called the section plant. At this plant, the exhaust gas containing some methane is then converted as per the reaction:

$$2CH_4 + 3SO_2 \Longrightarrow S + 2H_2S + 2CO_2 + 2H_2O$$

$$\text{and } CH_4 + 2SO_2 \Longrightarrow 2S + CO_2 + 2H_2O$$

Hydrogen sulfide (H_2S) gas is also converted to yield elemental sulfur by the following reaction:

$$2\,2S + SO_2 \Longrightarrow 3S + 2(H_2O)$$

10.6.4.2 SO_x-NO_x Removal Through Gas phase Oxidation/Ammonia Injection

This is a new technology used for both denitrification and desulfurization of flue gas. It is a technology involving a radiation chemistry process where the physical effects of radiation are used to cause a desired reaction. In this process, an intense beam of electrons is fired into the flue gas with simultaneous injection of ammonia. This system has not become very popular due to the unavailability of more improved qualities of accessories such as an electron gun vis-a-vis an accelerator, etc., suitable for continuous operation in an industrial environment.

The application of the electron beam is meant for encouraging the oxidation process of sulfur dioxide to sulfur compounds. Ammonia injection makes it react with sulfur compounds to form ammonium sulfate. This byproduct is known for being used as a nitrogenous fertilizer after further processing to produce ammonium sulfate crystals through proper separation, drying, and screening before bagging or bulk loading. The ammonia injection also helps in lowering the presence of nitrogen oxides in the flue gas, as described above.

10.7 Fuel Oil Pressure Control

General

Fuel oil has been associated as a supporting fuel and may be of any type such as heavy furnace oil (HFO), heavy petroleum stock (HPS), low sulfur high stock (LSHS), etc. As already discussed in Section 4 of this chapter, it is normally used in any of the steam-generating plants for start up, low load operation, and coal flame stabilization at all loads if required. These oils need heating before being burned. For example, the flash points are 66°C for HFO and HPS and 75°C for LSHS. LDO firing, if provided, does not need preheating for burning.

The fuel oil is drawn from the HFO storage tank(s) by fuel oil pressurizing pumps and pumped through steam oil heaters. The oil, duly pressurized and heated, reaches near the boiler front in a ring main arrangement from which the exact requirement of oil is taken off by the oil burners. The extra oil not required for firing is returned back through the recirculation line to the storage tanks. The ring main oil pressure is maintained to the required set value by means of a control valve provided in a separate recirculation line near the pumps. Oil temperature is controlled by regulating steam flow to the fuel oil heaters.

The LDO firing system is provided to facilitate starting of units when no auxiliary steam or auxiliary boiler is available from an external source. The LDO firing system shall also be used for cold start-up of steam generators and for flushing of HFO lines.

It may be noted that the basic scheme for an LDO system is the same as discussed above for a fuel oil system, except that the suction for the LDO pressuring pumps is taken from the LDO tanks and there are no heating requirements.

10.7.1 Objective

There are two control loops and the objective of these control loops is to maintain the fuel oil header pressure at the pump discharge and at the burner front ring main where all the burners get their oil supply. It is very important that these header pressures are maintained at a fairly constant value and each burner gets more/less the same quantity of oil when the burner valves (open/close type) are fully open. Figs. 2.9 and 8.48A and B may be referenced in this regard.

10.7.2 Discussion

The very nature of fuel oil (HFO, HPS, or LSHS), being a highly viscous liquid, requires that the temperature must always be kept around a value necessary for maintaining sufficient fluid mobility. Another important requirement is to provide recirculation facility of the oil line from the burner front to the oil tank itself so that the oil viscosity is maintained through steam tracing or lagging with an electrical heating arrangement. This line is popularly called a long recirculation line, which covers the entire oil line while the unit is operating.

There is one recirculation line that enables oil back to the oil tank much before the burner front and after the fuel oil pumps/heaters. This line is known as the short recirculation line, and it covers a smaller part of the oil line while the unit is not operating, that is, on a long shut down. This valve remains 10% open even in a running unit to ensure fluid mobility.

The pressure-control valve is located in another recirculation line after the fuel oil pumps/heaters so as to maintain a fixed fuel oil discharge header pressure with the capacity as required when no burner is operating when it has to allow all the oil from the pump discharge back to the tank. The fuel oil pressure is around 15–25 kg/cm^2, depending on the type/size of the burners, the distance and height of the remotest burner, etc.

10.7.3 Control Loop Description (Fig. 8.48A)

The schematic representation depicts the control strategy as the direct, single, and simple type.

10.7.4 Control Loop (HFO/LFO Recirculation Control)

The measured/process variable is compared with a fixed set value to form the error signal. The controller output is the position demand signal of a control valve in the recirculation line. When there will be lees demand, the recirculation will open more and vice versa. These pressure transmitters are normally the sealed diaphragm type to avoid chocking or wax formation/deposition inside the sensor part. The impulse line up to the transmitter would, however, be steam traced or electrically heated.

10.7.5 Control Loop HFO/LFO Flow Control at Boiler Front (Fig. 8.48B)

The fuel oil pressure is measured by a pressure transmitter with sufficient redundancy and at the appropriate location near the burner front to get a representative measurement. The measuring point is important so that the highest and remotest burner gets sufficient pressure to cater to the requirement. The type and installation of the pressure transmitter are discussed above. The set point is sent through a "max" circuit to ensure a minimum header pressure at the boiler front. The error created by the set point and the actual pressure is sent to the PI controller to regulate the valves to cater to the HFO flow demand. When the opening of the valve is <2% (i.e., closed), signal is sent to FSSS to close the trip valve for positive isolation. The loop description and drawing are given for HFO only, but will be similar for LDO.

11 HP-LP BYPASS SYSTEM

General

There are situations that arise in a large power plant where the turbo generator (TG) sets are subject to undergoing extreme operating conditions such as a sudden load throw off, a full load trip out, or even a planned shutdown for a short span of time. In all conditions, the TG set is expected to run on the house load/low load or in case of a

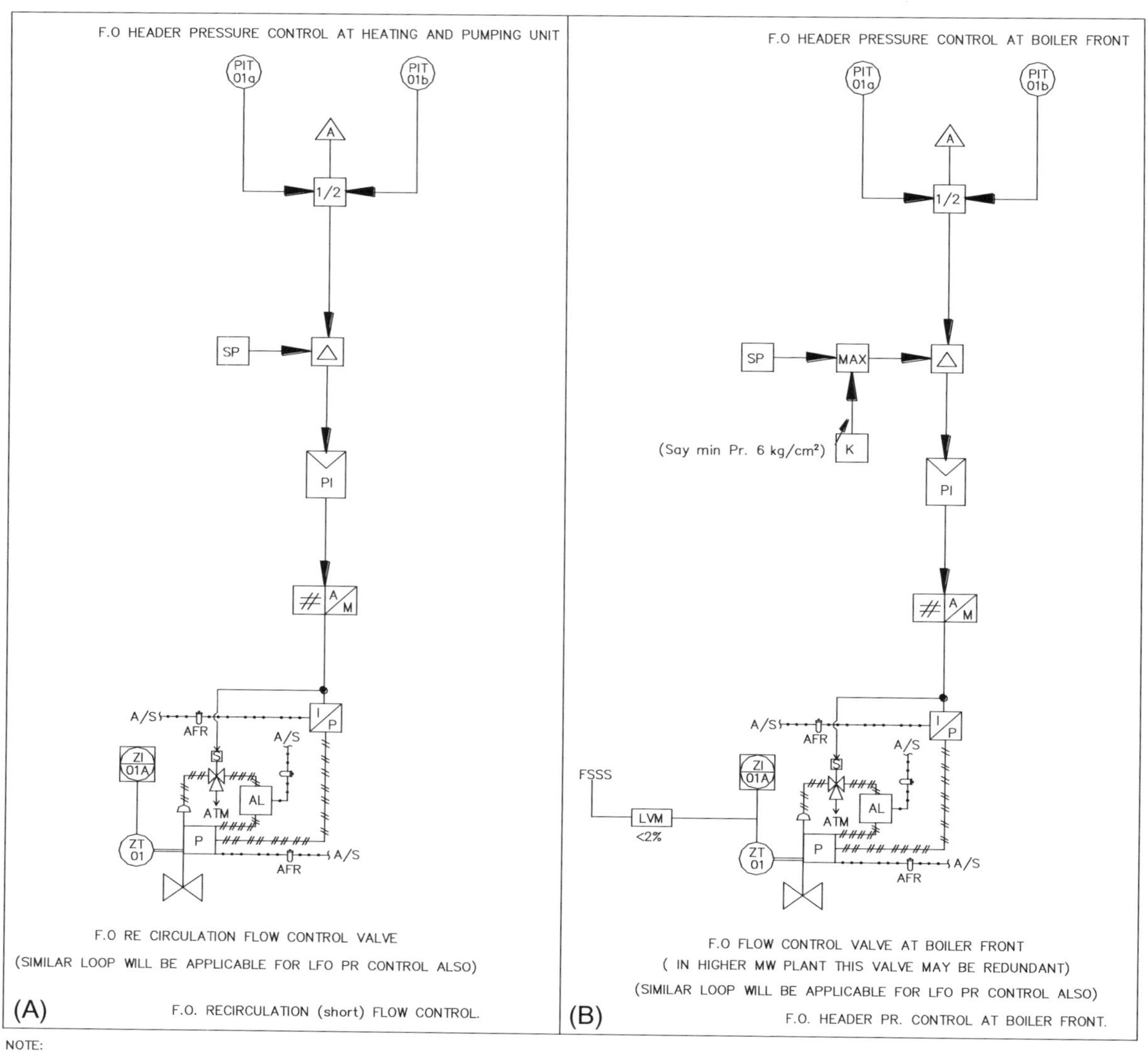

FIG. 8.48 Boiler oil pressure control.

turbine trip/shut down, the open generator circuit breaker, and to restart as quickly as possible so that revenue earning is continued. The design trend for the TG set has long been that the units must be suitable with the systems that enable them to run on house load operation. It may not be out of the way to mention that by house load operation, the unit can supply its own auxiliaries' electrical load and need not depend on the external power supply from the grid supplied by the other connected units. When there is a blackout, that is, a full load tripping has taken place from the entire grid failure, the house load operation facility keeps the boiler running and the TG set ready for further loading at the earliest. All these above-mentioned facilities are obtainable if the unit is designed for and supplied with an HP and LP bypass system having sufficient steam handling capacity of 60% or even 100%, depending on many influencing factors.

Many of the standard TG systems, HP and LP bypass is essential for its start-up and shutdown that is, TG cannot be started (say heat soaking) without HP and LP bypass. In some system designs, parallel operation of the bypass is also allowed.

The most important function of the steam turbine bypass system is practically to simulate the expansion and heat transfer process that normally takes place inside the operating steam turbine. In other words, the *HP-LP bypass system* provides an alternative route of the steam line from the boiler to the condenser. By doing so, the steam generation is almost uninterrupted or artificially loaded when there is no operating load (start-up), turbine trip, or a part load on the turbine. This system facilitates faster plant start-ups, that is, raising the steam parameters to suit the turbine requirements as needed by different kinds of start-ups. The most advantageous condition of the HP-LP Bypass system is to implement a hot start-up, which means a minimum life expenditure of the turbine components and auxiliaries with the overall advantage of the boiler, turbine, and generator's increased availability.

The turbine bypass system has manifold applications as it is suitable for handling almost all the possible start-up conditions such as cold start up, warm start-up, and hot restart as well as turbine shutdowns and disturbed situations such as load rejections and unit trips. In any case, bypass operation always calls for a loss of energy but this may be traded off by the time and money savings to meet the emergency and/or long start-up time of the boiler.

The HPBP system, during situations mentioned above, connects the main steam to the cold reheat line at matching pressure and temperature conditions through the HPBP valve, bypassing the HPT.

The LP bypass connects the hot reheat line to the condenser through the low-pressure bypass (LPBP) valve, bypassing the IPT and LPT with suitable steam conditioning.

In ultrasupercritical plants, there may be HP, IP, and LP bypasses, as there may be two stages of reheating and IPTs. However, all these systems operate in tandem.

The system is designed to incorporate the controls of the parameters such as pressure, temperature, and steam flow as they should have been at the turbine entry and exhaust points during the prevailing running condition of the turbine before the HP-LP bypass comes into action. It would have then been steered to the desired condition suitable for the purpose it is used for at that moment. Normally, the HP-LP bypass system does not directly interface with the turbine control and supervisory system, but is set to maintain all the above steam parameters by directly carrying out process measurements to tackle the situations for which it is deployed. The overall benefits that may be available by installing an HP-LP bypass system are summarized as follows:

- **(i)** The unit can be started even if there is a grid failure and consequent blackout.
- **(ii)** Safety valve pop-up situations are less/rare with an efficient HP-LP bypass system.
- **(iii)** The demineralized water requirement is less during start-up/trip-out conditions.
- **(iv)** Less rate of pressure rise in boiler under trip-out condition.
- **(v)** House load operation is practicable.

For medium and larger capacity HP-LP bypass systems, the steam flow measurement through the HPBP requires consideration for calculating the total steam flow and the corresponding feed water flow requirement to develop the control strategy toward the boiler control systems where parallel operation is allowed.

This is especially required where the steam flow to the turbine is measured by the characterized turbine first-stage pressure or if the flow sensor is installed after the bypass connection. The signal from the flow sensor at the HP bypass line or a characterized bypass valve position indication may be used to measure the bypass steam flow. On the other hand, if the steam flow is measured before the HP bypass connection, then separate HP bypass flow measurements can be eliminated.

The bypass system itself along with the control system is normally designed in a way that flow/pressure/temperature fluctuations are reduced to a possible minimum for avoiding inadvertent effects on the performance of other associated systems. The bypass system, especially the LPBP system design, must incorporate and ensure the proper conditioning of the bypassed steam while entering the condenser. It is extremely important that the steam conditioning is guaranteed to the level best against thermal or vibration damage to the sensitive heat transfer surfaces and structures within the condenser. European boiler codes, unlike American boiler codes, allow omitting the conventional safety valves and operating the system with the HP bypass valves only.

11.1 Objective

The objective of the control loops of the HP-LP bypass system is to maintain the steam pressure, temperature, and flow as demanded by the situation at suitable points.

11.1.1 High-Pressure Bypass System

The HP bypass system is designed to accomplish the following system requirements:

- **(a)** Boiler main steam pressure control.
- **(b)** Temperature and flow control of steam to match the CRH condition to cool the boiler reheater elements.
- **(c)** Steam flow control in the MS line to cool the final superheater in case of sliding pressure operation. For example, in an OT boiler, the steam temperature control is highly dependent on the feed water flow and firing rate. The effect of a firing rate reduction could take more while the feed rate has to be maintained to cool the superheater, resulting in more steam than that is required so the HPBP takes care of the extra steam.

11.1.2 Low-Pressure Bypass System

The basic function of an LP bypass valve is mainly for the protection of the condenser. The system is designed to condition reheat steam acceptable to the condenser. This obviously calls for a large valve (the FCE) due to the enormous increase in the specific volume of the steam to suit the subatmospheric pressure at the condenser.

The LP bypass system is designed to accomplish the following system requirements:

(a) Control the pressure of the steam after bypassing the HP turbine, or, in other words, the HRH steam pressure according to the loading of the turbine. The temperature and flow control of steam through the CRH line to cool the reheater elements. In cases, condenser pressure/temperature is high then LP bypass dumping is stopped/vented otherwise followed by firing rate reduction).

11.2 Discussion

Turbine bypass control valves have long been associated with demanding process conditions. In controlling the high-pressure/temperature steam of this critical service, the valves necessarily require a sophisticated valve design with powerful actuators that would enable full travel time in 2 s or less. They also are supposed to have the quality of a high degree of closure, normally with a leakage rating of Class V or better. Noise immunity is extremely important with stringent noise restrictions. For larger plants and supercritical plants, major suppliers recommend having in-body desuperheaters with a special design for better performance.

Regarding HPLP bypass valve actuators, hydraulic power units are preferred in Europe, whereas in the United States, pneumatic actuators are preferred due to their relatively low cost.

11.2.1 System Capacity

There are a number of possibilities and influencing factors to determine the capacity of the bypass system. The bypass system capacity to accomplish the previously mentioned functions is as follows:

(i) Bypass capacity of 15% only of maximum continuous rated (MCR) flow at valves wide open enables the unit for matching the steam-to-turbine metal temperatures with a reduction of the start-up time by about 30 min.

(ii) Bypass capacity of 40% of MCR flow at valves wide open can cope with the transient or disturbed conditions of minor nature is taken for granted to be enough. Larger capacity may also be considered for handling more critical situations.

(iii) Bypass capacity of 100% of MCR flow at valves wide open takes care of the situation with the boiler running at full load and the occurrence of any untoward situations such as a turbine or generator trip and would not let the safety valves blow.

(iv) The bypass systems with suitable capacity can bring down the steam generation in the boiler in 10 min or less to a house load capacity of approximately 10%–20% without creating excessive temperature gradients. In that case, some manufacturers recommend 60% HPBP and 100% LPBP systems.

The steam flow capacity of the bypass system is also influenced by a number of other factors:

(a) Condenser capacity, internal arrangements, and materials.
(b) Turbine rotor diameter.
(c) Start-up, shut down, and loading and unloading procedures and requirements.
(d) Safety considerations related to SH, RH, and condenser.
(e) Reheat pressure for turbine start.
(f) Number of warm starts, hot starts, and requirements for house load operation.
(g) Cost consideration.

11.2.2 Different Mode of Starts Through Bypass Systems

The following modes of operations are possible, which justifies the best use of bypass systems after making a proper size and design selection.

11.2.2.1 Cold Start

The HP-LP bypass system permits improved operating condition of the furnace, the primary and secondary superheaters and reheaters, and the main and reheater steam lines. The system as a whole improves the quality of the steam generated by the boiler before starting the turbine; this reduces the start-up times significantly. The turbine can be started from the turning gear and can reach the rated speed in 15–30 min, provided temperatures at different points of the turbine rotor are proper and following the turbine manufacturer's temperature gradients as per the thermal stress evaluator (TSE) guidelines. The bypass operation may take approximately 2.5–3.5 h. The steam flow through the superheater and reheater guaranties the tube cooling in an effective way and facilitates the boiler for going ahead with a higher furnace firing rate.

11.2.2.1.1 Role of HPBP System During Cold Start (Fig. 8.52) Step (i) During cold start, the bypass valve is kept open a crack or open up to the admissible minimum position to allow steam flow through the SH and RH banks as soon as the furnace starts firing.

Step (ii) After generation of sufficient steam capable of exerting the minimum required pressure, the bypass valve begins to open as required by the controller to maintain steam pressure to a certain value.
Step (iii) Valve position reaching at the preassigned value sufficient to initiate pressure build up through a ramp commences with more production of steam through more or less a fixed opening of the valve. This step allows more steam in pressure-building activity but with a gradient within the permissible limit by the system.
Step (iv) Steam pressure reaches the value desirable for preparing the turbine start-ups, and the bypass valve opens further to maintain the steam pressure at a constant value.
Step (v) After a time, the turbine is ready to be synchronized with a small load and the bypass valve would start closing no sooner than when the turbine takes further loading.
Step (vi) Bypass valve would be fully closed after a predetermined load is delivered by the TG set.

11.2.2.1.2 Role of LPBP System During Cold Start

During cold start, the LP bypass valve can be operated in two ways, as indicated below:

(i) The valve would remain closed to allow the reheat pressure to develop and then start opening to control the RH steam pressure as per the set point generated.
(ii) The valve would act similarly to the HPBP system to develop pressure in a controlled manner.

11.2.2.2 Warm Start

The benefits as available from the bypass systems as mentioned in the above Section 11.2.2.1 for cold start are also applicable to this mode. The only difference is the initial casing temperature of the HP turbine is usually above 100°C. However, the bypass systems enable the start-up operation to run in an efficient way through corelating the steam temperatures to the corresponding metal temperatures under all speed conditions.

11.2.2.3 Hot Start

Many of the advantages mentioned in earlier sections are applicable to this mode of operation as well. Sometimes the turbine-generator (TG) set experiences tripping out to save the set from disturbances of a minor nature. These tripping causes can be rectified with a comparatively low time span and then the unit becomes very much suitable for a hot restart. The bypass systems are designed to enable the unit start-up at the earliest possible time through close watch, and then by matching the steam temperature with the metal temperatures of the thick and heavy turbine parts. The incorporation of a hot start facility in the bypass systems enables the unit to shrug off the unnecessary cooling down and rewarming procedures.

11.2.2.4 Quick Start After Full Load (or Partial Load) Rejection

The duty of the bypass system calls for an immediate opening of the bypass control valves in the case of a partial- or full-load rejection. The bypass control systems open the bypass valves to the same degree as the turbine control valves were before being closed, that is, the occurrence of unacceptable conditions.

Protective systems should be provided to trip the boiler when the HP/LP bypass valves fail to open and when an insufficient cooling steam flow passes through the SH/RH. Adequate safety measures for condenser protection are essential during LP bypass operation.

11.3 Control Loop Description (Figs. 8.49 and 8.50)

The control schematics depict the strategy (a typical 250 MWTPS having a drum boiler) of the normal load operation and not the start-up, which is usually controlled manually.

11.3.1 Measurement of Different Parameters

The following parameters are measured with sufficient redundancy and voting before actual use in the control loops.

11.3.1.1 Measurements for HP Bypass Control Includes

MS header pressure near the turbine inlet, steam temperature at the HP bypass valve outlet for desuperheater spray water flow control to maintain the outlet steam temperature at the desired level, and desuperheater spray water pressure to bring down the take off feed water pressure to match the HRH/CRH steam pressure level.

11.3.1.2 Measurements for LP Bypass Control

(i) Turbine first-stage steam pressure.
(ii) Reheat outlet steam temperature.
(iii) Reheat outlet steam pressure.
(iv) LP bypass steam temperature at the desuperheater outlet.
(v) Position signal for both the LP bypass and spray control valve
(vi) LP spray water pressure.

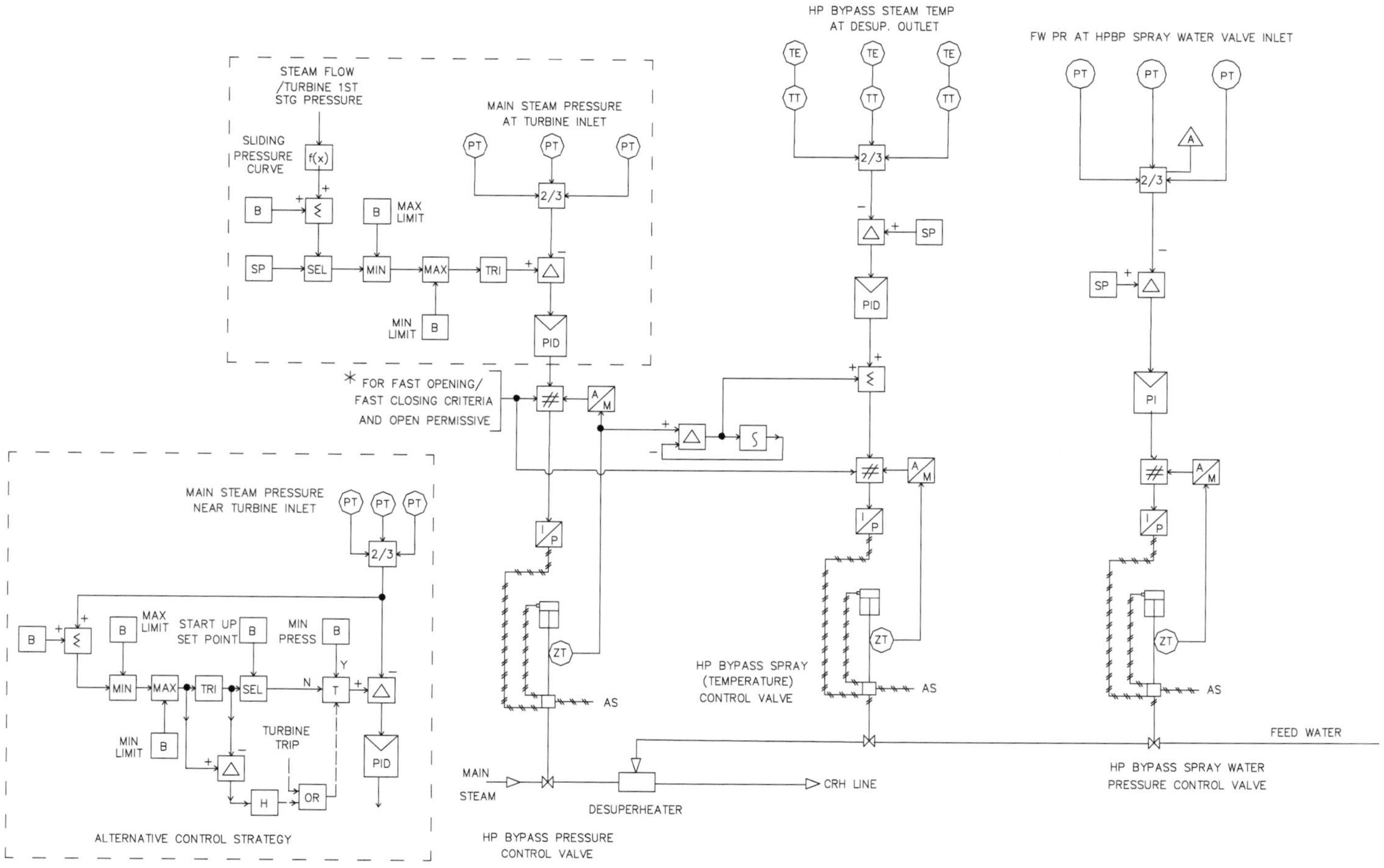

* FOR DETAILS FIG **8.51** MAY BE REFERRED

NOTE:
(1) PRESSURE CONTROL VALVE AND DESUPERHEATER MAY BE COMBINED IN ONE UNIT
(2) DESUPERHEATER MAY HAVE INTEGRAL SPRAY WATER VALVE
(3) CONTROL VALVE AND BLOCK VALVE MAY BE INCORPORATED IN A SINGLE BLOCK
(4) CONTROL VALVES AND BLOCK VALVE MAY BE SUPPLIED WITH HYDRAULIC ACTUATORS AS WELL

FIG. 8.49 HP bypass pressure and temperature control.

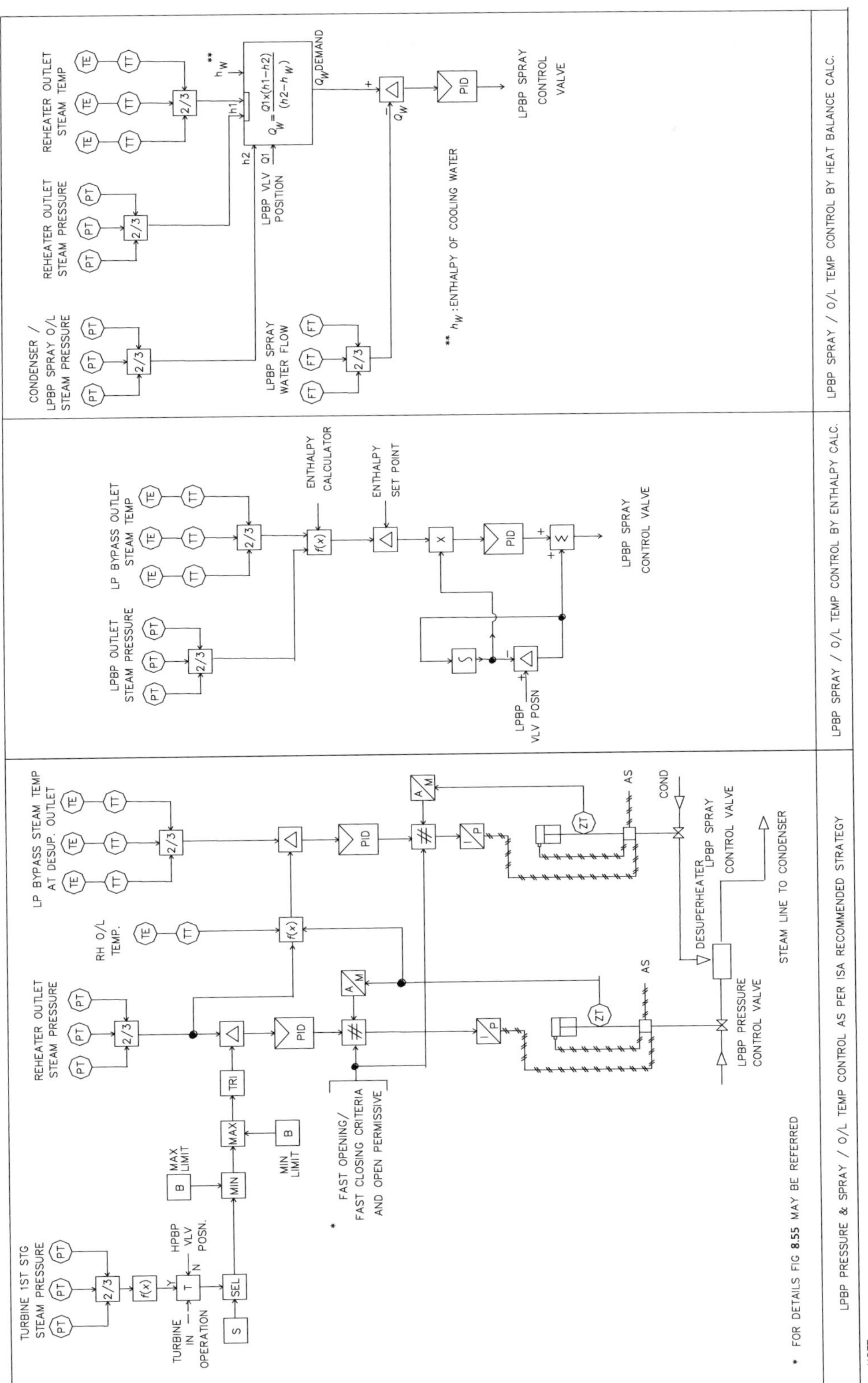

NOTE:
(1) PRESSURE CONTROL VALVE AND DESUPERHEATER MAY BE COMBINED IN ONE UNIT
(2) DESUPERHEATER MAY HAVE INTEGRAL SPRAY WATER VALVE
(3) CONTROL VALVE AND BLOCK VALVE MAY INCORPORATED IN A SINGLE BLOCK
(4) CONTROL VALVES AND BLOCK VALVE MAY BE SUPPLIED WITH HYDRAULIC ACTUATORS

FIG. 8.50 LP bypass control pressure and temperature control.

11.3.2 The Control Loop Strategy

The FCEs of these control strategies incorporate the following items:

(i) HP and LP bypass pressure-control valves.
(ii) HP and LP bypass spray water (temperature) control valves.
(iii) Spray water pressure control valve for HP bypass system.

11.3.2.1 HP Bypass Control

11.3.2.1.1 HPBP Pressure Control The control strategy is based on a single-element control concept. The selected signal of the main steam pressure near the turbine outlet becomes the measured/process variable. For a fixed main steam pressure control system, the HPBP set point is also a fixed value, which is normally higher than the master pressure control set point with a slight margin (~5.0 kg/cm^2). For a variable set point for a sliding pressure controlled system, the set point is derived as characterized from the load index, such as the main steam flow or the first-stage pressure, and then the bias for the margin is added. After the selection, the set point signal is passed through a maximum and a minimum selector. The steam pressure set point is thus limited within a minimum and maximum value and then passed through a tracking integrator (TRI) to have a ramped and smooth output to become the final set point so as to prevent process upsets. The maximum value of the integrating gradient is limited. In fact, the operating gradient is a very low value so that the instantaneous set value becomes lower than the main steam pressure in case of a sudden upset. For example, when there is a large load rejection, the steam pressure starts rising and if it becomes higher than the set point (with margin), the bypass valve opens as per the controller output command and tries to reduce the steam pressure. The set value for a variable pressure system would readjust its value as per the new load and the bypass valve would remain open up to a certain value until the steam pressure matches. The firing rate is then modified as per the load and the steam pressure is also readjusted so as to gradually close the bypass valve to bring back the normal operating condition.

There is another control loop strategy applicable to both fixed and variable pressure systems that is also shown in the above drawing as an alternative. Here, the main steam pressure itself is utilized as the set point after adding the bias for margin DP and passes through TRI to generate the final set point. The tracking integrator plays the vital role as described above.

The TRI would change its output at a very slow rate. Whenever there would be large load throw off, the MS pressure would shoot up at a very fast rate while the set point increased at a slower rate, which causes the HPBP valve to open to arrest the high pressure excursion. When the load and heat input stabilize, the bypass valve would again close. In a turbine trip, the bypass valve would open and the pressure control set point would be the minimum value, for example, the pressure required for a hot start-up. The rest of the system is the same as before.

11.3.2.1.2 HPBP Temperature Control (by Spray Water Valve) The control strategy is based on a very simple single-element control concept. The selected signal of steam temperature at the HP turbine bypass outlet becomes the measured/process variable with a fixed set point. The difference between the two signals form the error signal of the control loop and the controller (PID) output is supplemented with a feed forward signal from the HPBP valve position to get an advanced sense of the process condition. If measured separately, the bypass steam flow itself may be used as a feed forward signal for the temperature control system. Otherwise, the suitably characterized bypass valve position indexed by the steam pressure may be taken as an indicative signal of this steam flow. The combined signal is used to regulate the high-pressure desuperheater spray water control valve.

11.3.2.1.3 HPBP Spray Water Pressure Control The header pressure of the spray water should be maintained at a desired value. For this, there is a separate pressure control loop, as shown in Fig. 8.49. Here, the error generated by the set point and the measured variable (pressure at the spray water valve inlet) is fed to a controller, the output of which regulates the opening of the HPBP spray water pressure-control valve.

There are certain interlocked conditions other than the fast opening and fast closing criteria indicated in clause no. 11.3.2.3.1; they are indicated as follows:

- Valve is not permitted to open at the HPBP spray water pressure low.
- Valve is not permitted to open when the HPBP spray block valve is not fully open.

11.3.2.2 LP Bypass Control

11.3.2.2.1 LPBP Pressure Control The control strategy is based on a single-element control concept. The selected signal of the reheater outlet steam pressure at the turbine inlet (HRH) becomes the measured/process variable. The set point is of variable value, which is obtained from the characterized turbine first-stage pressure. The relation between the turbine first-stage pressure and the HP turbine exhaust, that is, the cold reheat (CRH) steam pressure, is exploited with some consideration of the line pressure loss in the reheater circuit determines the HRH set value.

When the turbine trips and/or during start-up, there will not be any first-stage pressure to act as the set point. So, the HPBP valve position (in case of more than one HPBP pressure control valve, the average valve positions) is taken as the set point through a selector circuit, as shown in the loop.

The difference between the two signals forms the error signal of the control loop and the controller (PID) output is used to regulate the high-pressure control valve.

11.3.2.2.2 LPBP Temperature Control (by Spray Water Valve) Efficient operation of any desuperheater calls for the injection of hot water, essentially to be at a temperature near the saturation temperature of the steam being cooled. It ensures that mainly the latent heat is extracted from the steam to evaporate the injected water. This design criteria assumes minimum suspension time experienced by the water particles in the steam path so as to make certain that the injection water is completely evaporated. Complete evaporation must be attained before the first pipe bend to avoid the impinging of water particles on the inside walls of the pipes. The desuperheater performance depends on the degree of atomization of the injected water and the mixing with the steam. Proper and speedy evaporation depends on the proper location and direction of the spray water jet as well.

The control strategy (Fig. 8.50A) for LP bypass temperature control is based on the ISA recommended simple single-element control concept. The selected signal of the steam temperature at the outlet of the LP turbine bypass becomes the measured/process variable. The set point is calculated essentially from the steam mass flow and steam condition. Normally, the steam mass flow online measurement is not done in this large diameter pipe, but in turn is calculated as a function of the bypass valve position duly characterized to the steam flow and the corresponding steam conditions. The difference between the two signals forms the error signal of the control loop and the controller (PID) output regulates the desuperheater spray water control valve lift position. This determines the requisite quantity of injection water flow that provides the guideline for the LPBP downstream steam condition.

An alternative loop (Fig. 8.50B) suggests making the calculated unit enthalpy from the LPBP valve outlet temperature and the pressure multiplied by the calculated steam mass flow as the measured variable. The set point is the manually adjustable enthalpy, as desired. This loop has been developed to take into consideration that steam conditions after the LP bypass desuperheater spray are very close to or at the saturation condition; the temperature after the desuperheater is not recommended to be used as a control signal. The feed forward signal is incorporated from changes in the LPBP valve position.

Another alternative loop (Fig. 8.50C) avoiding LPBP temperature measurement incorporates the heat balance method to determine the requisite quantity of spray or cooling water to maintain the enthalpy in relation tothe condenser outlet temperature. The heat loss by the steam is equal to the heat gained by the water when mixed to form a saturated condition. Heat lost by steam is the product of steam quantity and the difference between enthalpy at the LPBP inlet and the condenser inlet; heat gained by water is the product of water quantity and the difference between enthalpy at the condenser inlet and the spray water. The enthalpy is taken from the steam table at process temperature and pressure. The desired water quantity is the ratio of total heat lost by steam and enthalpy difference from the water side, which becomes the set point. Water quantity is the measured variable and the controller output regulates the spray water valve lift.

As has already been discussed, the LP bypass is basically provided for the protection of the condenser. As such, some condenser manufacturers stipulate the following guidelines in addition to the above control philosophy for start-up as well as intermittent and continuous control to maintain the LP bypass downstream steam temperature conditioning:

(i) Enthalpy restriction

Steam entering the condenser with enthalpy >670–680 kcal/kg is restricted; in the case of sudden high flow steam dumps, the enthalpy is recommended to be restricted to 660 kcal/kg. In certain cases, steam admission to the condenser with enthalpy >680 kcal/kg is also considered where the specific conditions of unit operation suggest it.

(ii) Pressure restriction

The maximum pressure of steam admission to the condenser is restricted to a limit value of 16 kg/cm^2 (g), mainly applicable for the dump condenser.

To accommodate the above guidelines, temperature and pressure at a suitable point are measured to calculate the enthalpy of the entering steam; this is considered as a measured/process variable. An appropriate set value as depicted above is provided for generating the control error. Also, a controller output is added to the conditioned steam flow derived from the LP bypass pressure control valve position to form the demand for the LPBP spray control valve position.

Interlocked conditions other than the fast opening/closing criteria (indicated in clause no. 11.3.2.3.2) are as follows:

The valve is not permitted to open for any of the following conditions.

(i) Insufficient desuperheater spray water pressure.
(ii) Block valve is not fully open.
(iii) High condenser pressure.
(iv) High condenser temperature.
(v) High condenser hot well level.

11.3.2.3 HP LP Bypass Interlocks (Fig. 8.51)

11.3.2.3.1 HP Bypass Interlock (Fig. 8.51A)

- *Fast opening criteria*: Under these severe conditions, it is required that the HPBP valve and the associated spray

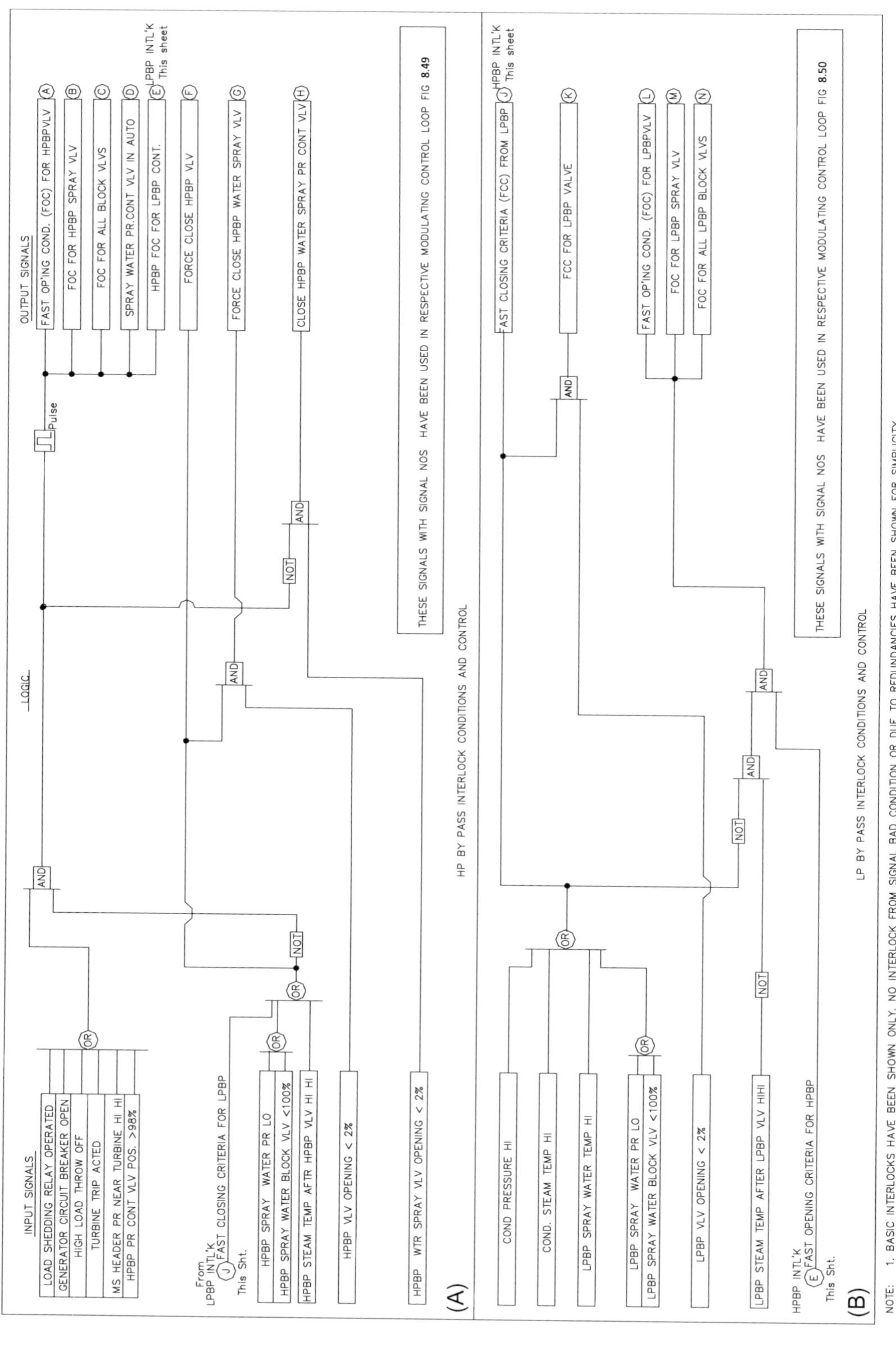

NOTE: 1. BASIC INTERLOCKS HAVE BEEN SHOWN ONLY. NO INTERLOCK FROM SIGNAL BAD CONDITION OR DUE TO REDUNDANCIES HAVE BEEN SHOWN FOR SIMPLICITY.

FIG. 8.51 HP LP bypass interlock conditions and control.

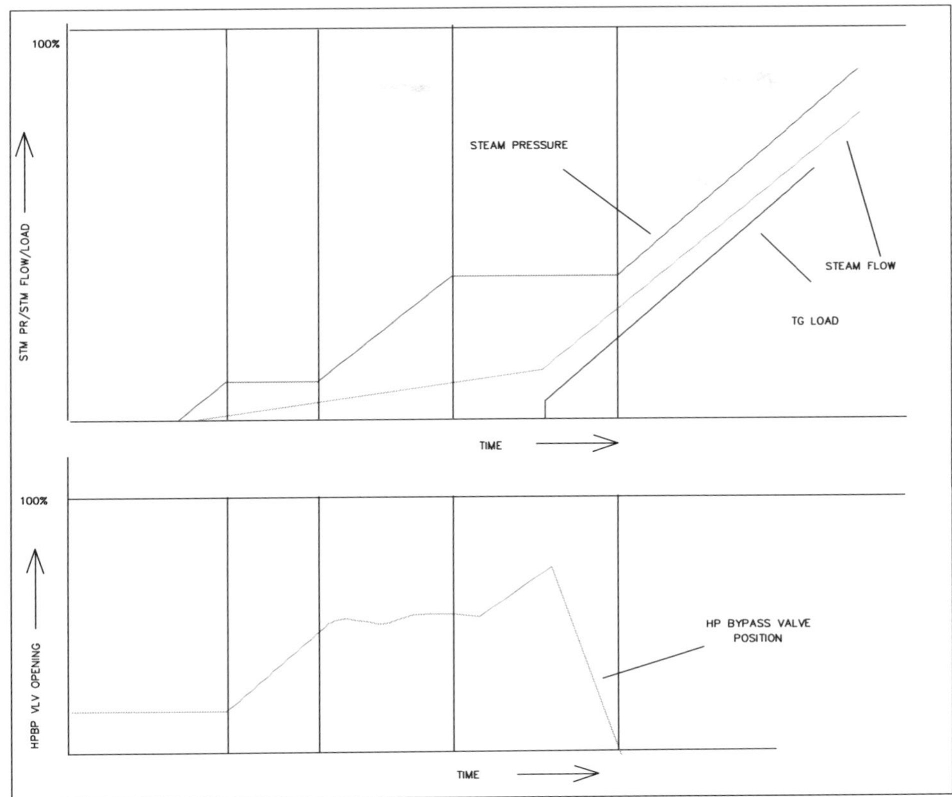

FIG. 8.52 Cold start-up: Graph of different parameters with respect to HPBP valve position.

valve and block valves shall open fully for a period to allow the HP bypass to dump the steam. During this time, the spray control pressure valve shall be auto so that suitable pressure is maintained. Suitable signals are sent to the LPBP system to enable its fast opening (Typically, these conditions are shown as signal output and the initiating criteria listed below as the input signal—applicable for all cases):

- o. Load shedding relay is operated.
- o. Generator circuit breaker is open.
- o. Turbine trip acted.
- o. High load dumping.
- o. MS pressure at/near turbine is very high.
- o. HPBP valve is almost open (to ensure full open).

- *Force/fast closing criteria*:
  - o. The PBP valve shall close under any of the following conditions:
    - Fast closing criteria from LPBP (discussed in clause no. 11.3.2.3.2).
    - HPBP spray water pressure low.
    - HPBP spray block valve not fully open.
    - Steam temperature after HPBP valve is The
  - o. The HPBP spray control valve shall close under all the above conditions, provided the associated HPBP valve is closed (<2% open).
  - o. The HPBP spray pressure control valve shall close when the HPBP spray control valve is closed, provided fast opening criteria do not exist.

11.3.2.3.2 LP Bypass Interlock (Fig. 8.51B)

- *Fast closing criteria*: It is necessary for condenser protection against conditions that would tend to increase condenser temperature.
  - o. The LPBP valve shall be closed under following condition:
    - Condenser temperature is very high.
    - Condenser pressure is very high.
    - LPBP spray water temperature is very high.
    - LPBP spray water pressure is low.
    - LPBP spray block valve is not fully open.
  - o. The LPBP spray valve shall be closed under the conditions stated above, provided the associated LPBP valve is closed (<2% open).
- *Force/fast closing criteria*:
  - o. The HPBP valve shall close under any of the following conditions:
    - Fast closing criteria from LPBP, as discussed in clause no. 11.3.2.3.2.
    - HPBP spray water pressure is low.

- HPBP spray water pressure is low.
- HPBP spray block valve is not fully open.
- HPBP valve outlet steam temperature is very high.

o. The HPBP spray control valve shall close under all the above conditions, provided the associated HPBP valve is closed (<2% open)

o. The HPBP spray pressure control valve shall close when the HPBP spray control valve is closed and no fast opening criteria exist.

- *Fast opening criteria*: Under these severe conditions, it is required that the LPBP shall open fast under fast opening conditions of the HPBP, provided there is no fast closing criteria or very high steam temperature after the LPBP valve. These will cause all LPBP pressure and spray valves and associated block valves to open.

12 BOILER OLCS: INTRODUCTION TO INTERLOCK AND PROTECTION OF BOILER BMS, SADC, SB CONTROL

General

The closed-loop control system (CLCS) always takes feedback from the process for which it is employed and readjusts the control output to get the desired process parameter. For more accurate or adaptive control, even the output signal is measured and feedback to the controller as another input of the control system. The CLCS is widely regarded as a regulatory part of the total control system and is accommodated in the auto control loops.

An *open-loop control* (OLC), on the contrary, does not utilize the same (feedback) and hence is also called a non-feedback control.

As the OLC does not normally utilize the feedback signal in computing the output, it also does not take care whether the desired condition has been achieved in the input. This also means that the control system is insensitive about the output without monitoring the processes status that it is controlling. As corollaries, it can be said that the open-loop system normally does not compensate for the disturbances in the system and at the same time cannot correct, unlike closed-loop control, any errors that might be generated while issuing the output.

However, for a sequential control system, the feedback signals are checked before proceeding to the next step.

The OLC is most suited for the following applications:

(i) Simple processes where feedback is not so critical or control within a moderate band is acceptable and cost toward the CLCS can be avoided.

(ii) A process of the nature where the control output mainly depends on the proper decision and assessment of the human operator and where automation may lead to wasting energies of different forms.

(iii) A process related to implementation of sequential or safety (protection)/interlocking logic systems where process measurement feedback signals are not required; for sequential logic, status feedback signals are required to proceed to the next step only.

In general, OLCS is the interlock/protection part of the total control system with a separate entity. It may also be noted that this system is designed to automate the required operator actions during an abnormal event or any change in operating status of the plant, for example, a trip, start-up, shutdown etc. All these above conditions necessitate the immediate and accurate handling/management of the situations, either through operator action or by some other means. It has been experienced and observed through experiments that the human response to emergencies is not beyond question. There is every possibility of erroneous operator action either toward making a judicious decision or timely intervention. These untoward situations can well be tackled if the actions are envisaged in logic form and then implemented through state-of-the-art technology, thereby avoiding damage to humans, the environment, and equipment. Such damage could result in more capital cost, loss of production due to a long shutdown time, and/ultimately loss of revenue; this justifies the application of a separate interlock/protection system.

The separation of the interlock/protection processing part from the regulatory control system has always been the center of discussion in control and instrumentation packages for thermal power plants. In a large power plant, the control systems are also large and complex as well as being developed with state-of-the-art technology that has sophisticated software and hardware. Though there are several signals required to be exchanged between the two systems, that is, between CLCS and OLCS, the same can be done through hardware or wet/soft signals without jeopardizing the resultant security and reliability. The logic must also differentiate between the safety protection part and the interlock part of the plant requirements based on the degree of possible hazards, including personnel injury as well as damage to the plant equipment and the environment. The degree of redundancy in the logic processing area is also determined by the above criteria for the selection of hardware/software.

12.1 Boiler OLCS

The boiler OLCS mainly comprises different sequential or safety (protection)/interlocking logics covering the following categories:

(i) The equipment, namely ID fans, FD fans, PA fans, boiler water circulating pumps, BF pumps (if in a boiler package), seal air fans, scanner air fans, mill/pulverizer motors, feeder motors, and lubricating oil pumps (LOPs) of all the drive motors needing lube oil without grease lubrication (covered in the BMS/furnace supervisory and safeguard system (FSSS) or a separate OLCS).

(ii) The on/off dampers and valves (motor-operated and/or solenoid-operated) related to air, oil, steam lines, etc. (Covered in BMS/FSSS).

(iii) Logic systems for boiler tripping, purging/firing, starting, shutdown, etc. (covered in BMS/FSSS).

12.1.1 OLCS for Lubricating Oil Pumps

The main drives such as ID fans, FD fans, PA fans, boiler water circulating pumps, BF pumps, etc., if equipped with forced lubricating oil and a control oil system (necessary to bring final control to the minimum necessary for starting a big fan) starting/tripping/shut down logic implementations of the oil pumps are taken care of in the OLCS.

12.1.1.1 Logics for LOPs

Normally the main drives have two LOPs. The starting logic incorporates that the first pump (as selected by the operator) would be started by the operator if LOP suction head/pressure is available. The LOP would continue to run to develop the lube oil pressure for the main drive. The second pump (as selected by the operator) would cut in automatically if the running LOP trips due to any reason or the LO pressure goes down below the minimum value. It would continue to run as long as the LO pressure goes up over a preset value.

The LOPs normally trip due to overload, low suction head/pressure, electrical circuit trouble, etc. It can be shut down manually if required by the operator.

12.1.2 OLCS for Main Drives

All the main drive motors, be it HT like 11/6.6 kV or LT like LT AC, related to the boiler part are taken care of in this part of the OLCS.

The typical list of drives mainly consists of ID fans (HT), FD fans (HT), PA fans (HT), mill/pulverizer motors (HT), feeder motors (LT), seal air fans (LT), scanner air fans (LT), BF pumps (HT), and boiler water-circulating pump (s), if any (HT).

There are in general two sets of interlocks/protection requirements for any drive motor: one from the drive or process side and the other from the motor side.

12.1.2.1 Logics for Drive Motors

The *start permissive* logics from the drive/process side are typically from:

(i) Process pressure/temperature/flow/levels are within acceptable limits.

(ii) Inlet/outlet dampers/valves, hydraulic coupling scoop tube, etc., open or closed as per the requirements from process/equipment side protection (no load starting, for example) purpose.

(iii) Running or tripping of any other system-related drives (to run a standby pump for example).

(iv) Drive bearing temperatures not high.

(v) Lube oil pressure is adequate.

(vi) Manual or automatic/sequential start (if any) command of selected drive.

Start permissive logics from the HT motor side are typically from:

(i) Winding temperatures are within acceptable limits.

(ii) Motor bearing temperatures are within acceptable limits.

The *tripping* logics from the drive/process side are typically from:

(i) Process pressure/temperature/flow/levels, etc., are beyond acceptable limits.

(ii) Some inlet/outlet dampers/valves are to open or close when the starting commands are issued for the main drives and/or any other conditions. If those inlet/outlet dampers/valves fail to execute those commands, then the trip signal is issued to that particular drive after some time delay.

(iii) Preferential tripping in case of a disturbed process condition.

(iv) Drive bearing temperatures are very high.

(v) Lube oil pressure is very low and/or pump running/trip status.

(vi) Manual or automatic sequential trip command.

Tripping logics of HT drive from process side are typically from:

(i) Winding temperatures are very high.

(ii) Motor bearing temperatures are very high.

12.1.3 On/Off Dampers and Valves

The *on (open) interlock* logics are typically from:

(i) Process-related interlock signals from high/low/normal pressure/temperature/flow/levels.

(ii) Running or tripping of any other system-related drives.

(iii) Main drive on or open command as per the process requirements.

(iv) Manual or automatic/sequential on (open) command of selected drive.

The *stop (close) interlock* logics are typically from:

(i) Process-related interlock signals from high/low/normal pressure/temperature/flow/levels.

(ii) Main drive stop or close command as per the process requirements.
(iii) Manual or automatic/sequential off (close) command of selected drive.

12.1.4 Logic Systems for Boiler Tripping, Purging, Starting, Shutdown

This part of the logic system is popularly known as the MFT logic part, and it is briefly discussed in Section 12.4 of this chapter. In many thermal power stations, in this particular section, MFT-related logics are developed in a dual redundant mode or even a triple redundant mode. The total inputs, outputs, and processing sections are redundant (dual or triple) to avoid any maloperation of processing units or faulty sensors.

12.2 OLCS in SADC

The SADC philosophy as discussed later in Section 12.5 of this chapter is different for the two major type of boilers. In both types of plants, the SAD are normally operated in a modulating mode through CLCS. In some phases of operation, however, they are subject to maintain some other set point or position as per the command issued by the BMS/FSSS.

12.2.1 Boiler With Fixed Burners and Flue Gas Recirculation Damper

The SADC-loop strategy was already discussed in Section 3 instead of Section 12.5 of this chapter. In this case, the OLCS in the BMS issues three distinct output commands requiring the SADC to take care of the same and modulate the dampers accordingly. The following modes of operation are recommended:

(i) *Furnace safety position mode*: BMS issues this command in case of any emergency situation (boiler tripping, for example) as envisaged during boiler operation. At this condition, the air flow is to be maintained at about 25% of the total air flow.
(ii) *Oil position mode*: BMS issues this command when the first set of oil burners is introduced after the purging operation is completed successfully. At this position, the minimum air flow, that is, about 10% the of secondary air flow of the corresponding oil burners only, is maintained so as to ensure an oil flame that may be extinguished in the presence of more air flow.
(iii) *Pulverized fuel (PF) position mode*: BMS issues this command during the introduction of the first set of PF burners. At this point of operation, the air flow of the corresponding PF burner is maintained at about 45% of the secondary air flow.

During normal operation, the main air flow signal from the auto control loop becomes dominant through the maximum selector.

12.2.2 Boiler With Tilting and Corner-Fired Burners and Overfire Air Damper

In this type of boiler, the auxiliary air dampers are modulated to maintain the differential pressure (DP) between the wind box (WB) and the furnace. Auxiliary air dampers are located in all two-lettered elevations, for example AA, AB, BC, CD, etc., in between the coal burners. Out of these, in some elevations a provision for oil firing (say, AB, CD, CE, etc.) is made. These dampers become fuel air dampers, maintaining a fixed position during oil firing at that particular elevation.

The only OLCS output from the FSSS is issued for adjusting the SADC modulating signal, that is, the case when the MFT trips. The set point for the (DP) between the wind box and the furnace would be typically around 150–200 mm of the water column.

12.3 OLCS in Soot Blower (SB) Control

Apart from the CLCS taking care of the pressure control, the SB control system also comprises the OLCS for selection of the blowers and duration of operation. As discussed in Section 12.6 of this chapter, there are many types of soot blowers and methods of operating them to suit the furnace, type of fuels, type of sensors, etc. In a nutshell, the following operations are possible for operating soot blowers individually or in a selected group as and when required.

12.3.1 Fixed Programmed Time-Based Operation

The OLC of the soot blowing system in this strategy is simply through step-by-step operation of the blower in a sequential manner. In the automatic mode of the system, the blowers operate one by one (or in a group) as logically recommended by the boiler manufacturer. They may be modified by the system/plant engineers to suit the site condition as experienced during operation. OLCs take care of the starting time of advancing, the time span of blowing, and the retracting and time gap between the end of one blowing and the starting of the next one of each particular blower (or group).

12.3.2 Temperature-Based Operation

The routine operations as discussed above may cause unwanted blowing and result in wear and tear of relatively cleaner tubes with wastage of an expensive soot blowing medium. On the contrary, the heavy soot-deposited tubes may starve the blowing at the right time and frequency.

This advanced technology enables a cleaning operation of the selected soot blower as per the decision taken by the system logic with the help of sensing the water tube wall temperatures associated with each soot blower. Temperature sensors welded to the tubes sense the actual surface temperature near each soot blower. The measured water tube wall temperature is compared with the corresponding temperature set point based on the saturation temperature of that point. If it become less than the set point, the associated soot blower is given a command to start the blowing operation. This is a need-based operation to clean the portions associated with the soot affected areas only.

Though effective for water tubes, this method was not suitable for areas where the fluid flowing through the pipes is a steam or steam-water mixture that has no further bearing on the saturation temperature. For supercritical boilers, where a mixture of water and steam is passed through the tubes above the supercritical pressure of water, there is no way of calculating the desired value to determine the metal temperature, which can be taken as an index of furnace dirtiness.

12.3.3 Heat Transfer/Heat Flux-Based Operation

The other strategy of the soot blowing procedure utilizes a more direct method of sensing. The local heat transfer rate is measured by employing a *heat flux meter* to assess the degree of soot deposition for selectively cleaning the tube walls or air heater ducts.

The sensing elements are located in each of the regions of the tube walls surrounding the soot blowers to detect the local heat transfer rate from the hot flue gas to the steam/water tubes in the furnace or heat exchanging elements (flue gas side) of the air heaters. The heat flux meter output is utilized for operation of the soot blower assembly. Details have been incorporated in Section 12.6.2.3 of this chapter.

Other control actions may be attained through the differential temperature.

12.3.4 Leak Test of Oil Lines

Leak tests of nozzle valves and trip valves are performed by a separate leak test valve in many plants before plant start-up; the logic diagram is shown in Fig. 8.53 for ready reference.

12.4 Burner Management System (BMS/FSSS)

General

The very name of the BMS suggests efficient management of the burners and all associated drives and accessories required by the firing system in the furnace. This system has become an indispensable and very important part of the steam generation domain, irrespective of the plant size. This management system includes safe starting operation and graceful emergency shutdown of the furnace vis-a-vis the steam generation package as a whole. A similar type of system offered by other major boiler manufacturers is known as the furnace safeguard supervision system (*FSSS*). Both names in this area are quite familiar and acceptable all over the world. Whatever the brand name of the product, the ultimate purpose of the system along with other normal functions is to act like a watchdog that is supposed to thwart any formation of an explosive fuel/air mixture and avoid implosion in the whole furnace area throughout the complete operating range. The BMS/FSSS is basically a dedicated fire protective system along with logic implemented for start-up, running, and shutdown. It is globally accepted that the guidelines of the National Fire Protection Association (NFPA 8502/8503) of the United States are to be followed as a standard.

NFPA defines the BMS as:

The control system dedicated to boiler furnace safety and operator assistance in the starting and stopping of fuel preparation and burning equipment and for preventing misoperation of and damage to fuel preparation and burning equipment.

BMS/FSSS, as it transpires, is required to perform logical or binary functions with quite a few numbers of inputs simultaneously and issue corresponding binary outputs, paying due attention to the protective aspects, that is, to avoid, well in advance, the firing hazards. There should not be any confusion regarding its role in the boiler operating system. The BMS/FSSS is a completely distinctively separate standalone system and not at all related in any way to the boiler related to normal automatic process controls such as steam pressure/temperature or boiler drum-level controls.

12.4.1 Process-Related Functions Expected From BMS

The design basis of BMS/FSSS is aimed to ensure the following jobs:

- **(i)** Satisfactory completion of the purge sequence before every starting of fuel firing to ensure that there is no unburnt fuels left over to cause secondary firing.
- **(ii)** Satisfactory compliance with each and every start permissive condition for start-up of individual fuel firing equipment.
- **(iii)** Flame (both oil as well as PF) monitoring when the fuel-firing equipment is in service.
- **(iv)** Sequential operation, including control of drives and monitoring of status feedback, during start up and shut down (cutting in/taking out of elevations, for example).

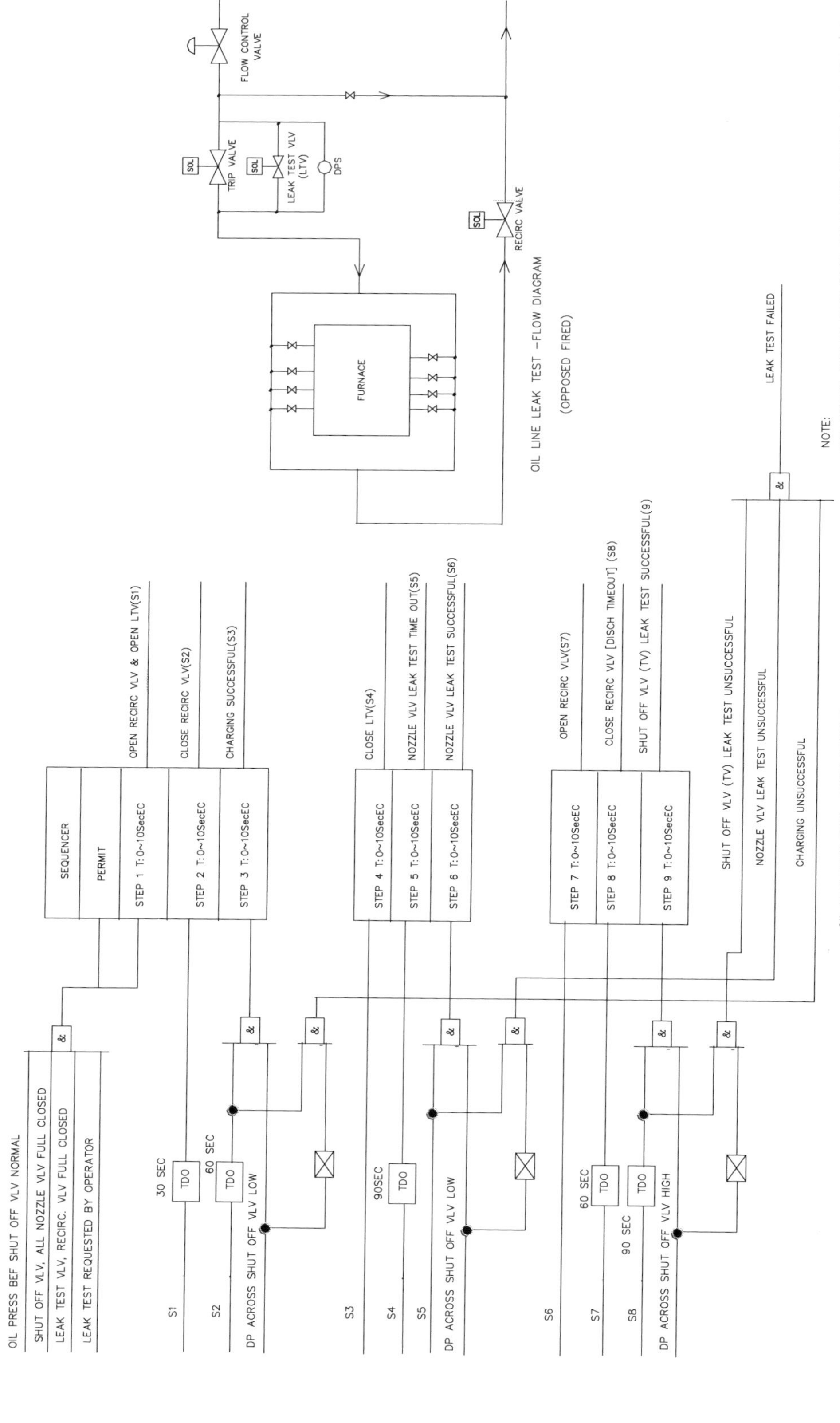

FIG. 8.53 Oil leak test—flow and logic diagram.

(v) Initiation of MFT through master fuel relay (MFR) upon adverse firing conditions beyond extreme limits that could be hazardous to both the equipment and personnel so as to stop/lose all sources of fuel, not only at the burner end but also upstream of the same (say, HOTV or mill trip).

(vi) Sensing and generating alarm conditions for annunciators or monitors and indicators to the operator's unit control board (UCB) to facilitate smooth manual operation if necessary.

(vii) Provision of all logic and safety interlocks, including tripping instructions compliant with NFPA 8502/8503 guidelines.

(viii) Inclusion of the first out feature in the system to identify the root cause of any trip, whether related to the burner or the boiler as a whole.

(ix) Provision of a complete BMS diagnostic or fault analysis system to immediately identify to the operator any system, subsystem, or module failure.

(x) To start/stop/trip burners/igniters on an individual basis in sequence and/or on a corner basis (TT boiler).

(xi) Boiler load-based operation of automatic cut in/cut off of burners/elevation. The operator may have the liberty to select the sequence of burner operation, for example, preferential tripping of mills in case of fixed wall burners.

(xii) Last but not least, provision of triple modular redundant (TMR) logic and/or a parallel back-up trip path by hardwared MFT or electromagnetic relay independent of processors and I/O modules.

With the advent of sophisticated microprocessor-based systems, BMS has long been implemented through the OLCS part of DCS or through programmable logic controllers (PLC) for more reliability, flexibility, and availability, which is discussed in detail in Section 2.2 of Chapter 7.

12.4.2 Hardware- and Software-Related Functions Expected From BMS

The following criteria should be available in BMS as a system:

(i) The system must be 100% foolproof, based on proven hardware and software.

(ii) The system would be provided with an automatic self-monitoring facility.

(iii) All modules must be of a failsafe design so that any single fault in any individual primary sensor, I/O channel, processor, etc., should not jeopardize safety functions.

(iv) All faults should be annunciated to the operator immediately with the first out facility. The BMS shall meet all applicable relevant safety requirements, including those stipulated in the latest editions of [DIN standard for electrical equipment for furnaces] Verband der Elektrotechnik or VDE 0116, Section 8.7, VDE 0160, NFPA 8502/8503.

(v) The BMS hardware may be configured in a triple-redundant, fault-tolerant arrangement as per the modern trend. Three independent sets of hardware (channel) comprising dedicated processors, I/O modules, and communication interfaces would receive the identical input signals. The complete logic part would be implemented independently in each channel. The primary sensors for protection logic should also be triple redundant.

12.4.3 BMS/FSSS Functional Configuration

The functional arrangement of the system can better be described in the following segregated areas such as:

(i) Logical functions as the heart of the system.

(ii) Sensors such as different process and position switches or manual commands.

(iii) Different drives, diaphragm/solenoid-operated valves, and actuators.

(iv) Operator's interface or a plaque/keyboard and monitor in the control board.

12.4.3.1 BMS/FSSS Logical and Functional Groups

The description of the starting, tripping, shutdown, or other interlocking logic signals is mainly incorporated for FSSS. The basic philosophy is more or less the same although, of course, major differences do exist in some areas where the equipment itself is different. Some of these are indicated below:

(i) *Firing system*:
- **(a)** BMS is built around fixed burners with front/opposed/downshot burners.
- **(b)** FSSS is built around tilting burners and a corner-fired system.

(ii) *Pulverizer system*:
- **(a)** BMS normally is based on pulverizers of the ball-mill type.
- **(b)** FSSS is based on bowl- and tube-type mills.

(iii) *Fuel oil recirculation system*:
- **(a)** BMS is meant for a long recirculation system only. A short recirculation valve is used for a long shutdown.
- **(b)** FSSS takes care of both long and short recirculation systems. There is an orifice across the recirculation valve (normally closed) that provides a constant return flow.

(iv) *Primary air system*:
- **(a)** BMS is built around both common PA fans with a header system and individual PA fans.
- **(b)** Same for FSSS, depending on the type of mill.

12.4.3.1.1 Leak Test Valves For some boilers, a leak test facility with valves is supplied for providing a test facility to check the passing through the HOTV and all the downward pipe lines and valves such as nozzle valves of oil burners, including HORV. A leak test valve is installed across the main oil shut off or trip valves (Fig. 8.53).

The procedure is described in a simplified way for getting some idea about the system. The following criteria must be fulfilled for starting the test:

(i) The oil pressure before the shut-off valve is adequate.
(ii) Shut-off valve is fully closed.
(iii) All nozzle valves of oil burners are fully closed.
(iv) Leak test valve is fully closed.
(v) Oil recirculation valve is closed.

After getting all of the above permissive signals, the test would start when the start command is issued by the operator. It is to be assured that the oil pressure control system is operating satisfactorily so as to maintain the required pressure of the oil header near the shut-off valve. Then the start command by the operator initiates the following activities:

(i) Oil recirculation valve is to open.
(ii) Leak test valve is to open.
(iii) Oil recirculation valve is then to close after 20–30 s if feedback from the above two signals is affirmative. This is done to flush the oil line before the actual testing is performed.
(iv) With the oil recirculation valve and oil nozzle valves closed and the leak test valves fully open (oil pressure control already in action), the oil line would be charged gradually through the leak test valve. The DP across the parallel combination of the oil shut-off valve and the leak test valve, which was initially high from the starting point, now shows a decreasing trend. The charging time is set at approximately 60 s so as to ensure complete charging. At the end of the charging time, the upstream and downstream pressures of the leak test valve should be the same with the DP showing very low. On the contrary, if the DP is still high, then the charging failure condition is acted and the leak test could not be carried out further.
(v) The leak test would then proceed to the next step with the above DP low permissive confirmation indicating that the oil line is in a fully charged condition. The leak test valve would then close automatically as the next step and the status is maintained for approximately 90 s. At the end of this time, if the DP is still found to be low, meaning no leakage in the downstream part of the oil line having burner nozzle valves and recirculation valve.
(vi) As a next step, the oil recirculation valve would then open automatically as a measure of checking the leakage through the main oil shut-off valve.
(vii) The oil recirculation valve is kept open for 60 s to drain all the oil in the pipeline and the DP should show high as the downstream pressure no longer exists. Then the oil recirculation valve is made to close fully.
(viii) Then the leak test valve is made fully closed for approximately another 90 s along with the oil recirculation and oil nozzle valves in the full closed position. At the end of those time periods, if the DP is not high, then the main oil shut-off valve appears to be leaking and the leak test failure condition is announced.

However, if the DP high condition can persist up to 90 s, the no leakage condition is achieved and the system may be declared as having passed the leak test.

12.4.3.1.2 Furnace Purge Logic Before every start-up of the furnace vis-a-vis the firing system of the boiler, it needs complete purging by blowing air through the furnace for a particular time as recommended by the manufacturer. This may be termed the prefiring purge and is applied strictly to make sure that all the residual unburnt fuels that might have accumulated are entirely eliminated. The air flow rate and the time period may vary, but normally a fairly accurate value may be taken as 30% of the total air for about 5–6 min. Some approximate amount of purged air is suggested to be close to nine times the furnace volume.

Purging is not only necessary for the prefiring purge, but is equally important to purge the furnace after every normal shutdown, termed the postfiring purge, to avoid possibility of explosion. Purging is also necessary for the oil burner valves at every time after burner closure, which is normally done by blowing atomizing media for a specified time period (also called scavenging); this is covered in the fuel oil-related logics.

During purging, the wind box dampers or air registers maintain a particular position along with other permissive conditions, allowing proper air velocities and travel through different routes consisting of radiant/convection/economizer zones. This complete purging system also enables the checking of proper operational functionalities of air dampers, burner valves, flame monitoring systems, etc., through ensuring permissive signals.

The furnace purge cycle normally needs the following conditions to be satisfied for eligible purge start initiation:

(i) Loss of AC/DC power for more than 2 s does not exist (to avoid any spurious tripping), meaning AC/DC power supplies are available [This is mainly done for FSSS with TT burners where equipment in elevation (e.g., nozzle solenoid valves) are normally run on AC supply whereas DC supplies are used for running equipment for common or unit, for example, HOTV would have a DC solenoid valve (may be double coiled)].

(ii) Drum level is not high (It is taken as absence (not) of drum level very high signal, which must exist for >10 s to avoid any spurious tripping).
(iii) Drum level not low (It is taken as absence (not) of drum level very low signal, which must exist for >10 s to avoid any spurious tripping).
(iv) Any BFP is on.
(v) Any induced draft (ID) Fan is on.
(vi) Any forced draft (FD) Fan is on.
(vii) All associated air and flue gas dampers are open.
(viii) All PAF are off.
(ix) All seal air fans are off.
(x) Air flow is adequate, for example, >30% and <40%.
(xi) Auxiliary air/SAD are proven modulating at the required position.
(xii) Windbox to furnace DP is proven satisfied (for TT boiler).
(xiii) Waterwall circulation is adequate.
(xiv) Furnace pressure not high.
(xv) Furnace pressure not low.
(xvi) All oil valves (trip and individual valves) are in the closed position.
(xvii) All pulverizers are off.
(xviii) All hot air gates are proven closed.
(xix) The flame scanners indicate no flame.

When all the above start permissive signals are available, the start purge command signal, if issued from the control board, initiates the purging process. It lasts for a specific period of time as set during commissioning (approximately 5 min). Upon expiration of the above counting period, it is taken a successful furnace purge cycle and then the system becomes ready for the next step of operation.

12.4.3.1.3 Fuel Oil Firing Logic When the LDO is used in a plant, it does not require temperature interlock and recirculation of LDO before purging. For HFO, both are necessary for maintaining the proper temperature vis-a-vis the viscosity of the same. So the recirculation-related logics are applicable to the HFO only. However, in some plant designs, the recirculation of LDO is also provided for better mobility and leak test facility.

The fuel oil trip valve (FOTV) and the fuel oil recirculation valve (FORV) are required to be opened manually after an MFT condition takes place and prior to the purge cycle being initiated for full-fledged boiler operation. This arrangement fulfills the condition of recirculating the hot oil near the burner area to raise the oil temperature as needed for oil firing.

Fuel Oil Recirculation (for HFO Only) Logic

The sequence of the oil recirculation procedure is as follows:

(i) FORV is opened first after MFT and before boiler restart.
(ii) FOTV can then be opened with all the following start permissive signals:
  (a) FORV is fully open.
  (b) All oil burner valves are fully closed.
  (c) MFT signal exists.
  (d) Open PB depressed.

After some period of recirculation, if the oil temperature and pressure are found adequate, both the FOTV and FORV valves must be closed to start the purge cycle.

Before the purge activities start, leak tests are performed as designed/provided by the manufacturer in some SG plants (discussed in Section 12.4.3.1.1 of this chapter).

FORV open logic: The valve can be opened by pressing the push button (PB) from the console if all the oil burner valves are closed.

FORV closed logic: The valve closes automatically when any oil burner valves are not closed OR by pressing the close push button.

Fuel Oil Trip Valve Logic

Fuel oil Trip Valve then can be opened again provided all of the following signals are available:

(i) Purge complete.
(ii) No MFT.
(iii) All oil burner valves are closed.
(iv) FO header temperature (for HFO only) is more than normal (~95°C).
(v) FO header pressure is more than normal (15–25 kg/cm^2).
(vi) Atomizing steam pressure is adequate, which may be more or less than the FO header pressure, depending on the burner design.
(vii) No close/trip command and open PB is depressed.

FOTV would trip automatically when any oil burner valve is not closed and if any of the following conditions occur:

(i) MFT signal exists.
(ii) FO header pressure is very low for more than the specified time.
(iii) FO temperature is very low for more than the specified time.
(iv) The atomizing steam pressure is low for more than the specified time.
(v) Loss of electrical power exists for more than 2 s.

Other than the above causes, the *FOTV* can be stopped manually by selecting the manual mode of operation and then by depressing the stop PB.

Fuel Oil Firing (Elevation or Mill Group-Wise) Logic

Fuel oil firing can now be initiated elevation-wise or mill group-wise as per the burners and firing systems supplied by the manufacturer. The following interlocks are checked for going ahead with fuel oil firing:

- **(i)** Fuel oil trip valve is fully open.
- **(ii)** Air flow is adequate (typically >30% and <40%).
- **(iii)** Burner nozzle tilts are at the horizontal position (for boilers with tilting burner nozzles only) [Items ii and iii above are applicable for first elevation starting only].
- **(iv)** No MFT.
- **(v)** Start oil firing command is available manually by the operator or from the automatic sequence logic (if provided).

The firing system is different for different manufacturers. As there are two distinct separate systems available worldwide, it is better to discuss separately the firing technique adopted by those two systems.

B and W boilers' firing systems are normally front-fired, opposed-fired, or downshot-fired, having individual igniters and oil guns for every coal burner distributed in two or more elevations/layers depending on the capacity of the power plant vis-a-vis the SG plant. Two such elevations for opposed-fired (may or may not be downshot burner) systems are typical for up to a 250 MW plant, whereas four such elevations for an opposed-fired system are typical for up to a 650 MW plant.

In CE boilers, the firing system is based on tangentially corner-fired nozzle burners spread over a number of elevations with in-between air supply compartments called fuel air, auxiliary air, etc. Oil burners are strategically located in some of the elevations in between the coal burners so that one oil burner may support two adjacent (top and bottom) coal burners. In this system, burners operate on a "pair" concept and not on an individual or single-unit basis.

A simple logic flow is described to have some idea how it works for TT boilers with FSSS and the same for the wall-fired boiler would follow.

If the permission for oil firing is available with all safety features taken care of, the following steps are taken manually or sequentially:

- **(a)** The oil elevation can be started in the "pairs" or "elevation" firing mode, selected automatically by the status of the PF feeders. The pairs mode is automatically selected when all feeders are off for more than 2 s. The elevation mode is automatically selected when any feeder is proven or an "auto start support ignition signal is established." Diagonally opposite corners, for example, one and three, are placed in service as one pair. Corner 1 is placed in service initially and after some time, Corner 3 is placed automatically in service. Then, Corners 2 and 4 are placed in service as one pair, similar to Corners 1 and 3.
- **(b)** The following start permissive signals would be checked before proceeding further:
  - **(i)** No MFT.
  - **(ii)** The auxiliary air dampers in the associated oil elevation are closed.
  - **(iii)** Associated oil gun is engaged.
  - **(iv)** Associated scavenger valve (meant for purging the oil burner that closes after firing) is closed.
  - **(v)** FO isolation valve is open.
  - **(vi)** Atomizing steam isolation valve is open.
- **(c)** The following events would take place sequentially for an oil burner (corner-wise) to fire:
  - **(i)** The associated oil igniter [may be HEA igniter/equivalent] advances to the firing position.
  - **(ii)** The steam atomizing solenoid operated valve opens.
  - **(iii)** The oil gun moves to the firing position.
  - **(iv)** The HFO solenoid-operated burner valve opens.
  - **(v)** The HEA igniter delivers spark energy to establish the oil flame in that corner.

If the flame is established (sensed by the associated oil flame scanner located strategically) after some time and all the commands are complied with, the igniter spark command stops and retracts to the normal position. The concerned oil corner is taken as in service. Other oil corners start in a similar way to make them in service. The oil elevation is taken as in service when a minimum three out of four corners are operating, enabling the next step, that is, proceeding with the starting of the PF firing.

Another simple logic flow is also described to have some idea how it works for B and W boilers as follows:

All the permissive signals for oil firing must be available with all safety features taken care of and then the following steps are taken manually or sequentially:

- **(a)** The firing would be started with LDO first to warm up the boiler to a certain stage such that the HFO firing operation gets sufficient thermal energy to start with. LDO firing arrangements are not provided for all HFO/coal elevations and are normally provided strategically just to warm up before HFO firing could start. For a particular HFO/coal elevation, any HFO burner can be selected manually or in a sequential manner automatically and started. When selected and the start command is available, the burner firing steps are sequentially performed. The following start permissive signals would be checked before proceeding further:
  - **(i)** No MFT.
  - **(ii)** HFO header pressure is normal and HOTV is open.
  - **(iii)** Associated scavenger valve is closed (meant for purging the oil burner that closes after firing).
  - **(iv)** FORV is open.

(v) Atomizing steam pressure is normal and the isolation valve is open.
(vi) Instrument air pressure is normal.

(b) The following events would take place sequentially for each oil burner to fire:

(i) The oil gun moves to the firing position.
(ii) The steam atomizing solenoid valve opens.
(iii) The associated oil igniter [may be HEA igniter/equivalent] advances to the firing position.
(iv) The HFO solenoid-operated burner valve opens; simultaneously, the HEA igniter delivers spark energy to establish the oil flame in that corner. The oil igniter then retracts from the firing position to the base position, that is, outside the furnace.

If the flame is established (sensed by the corresponding oil flame scanner) after some time and with all the commands complied with such as oil burner valve open, etc., the concerned oil burner is taken as in service or running. A similar procedure is repeated for starting other oil burners and making them in service. When the minimum requisite burners out of the total burners in that elevation are in service, the oil elevation as a whole is taken as in service. This signal would then enable the next step to proceed, that is, the starting of the PF firing.

The procedure of LDO firing is similar with a recirculation valve that may or may not be applicable and the atomizing and scavenging media may be air. The igniter may or may not be the same, depending on the situation.

12.4.3.1.4 Pulverized Fuel Firing Logic *Pulverizer Start Procedure*

The pulverizers and feeders can now be placed in auto mode of operation and after that, the pulverizers and feeders can be considered for starting after other permissive conditions are satisfied. Before the pulverizers/feeders, the *PA fans would be started* any time after the furnace is purged successfully.

Any start permissive condition signifying the availability of sufficient ignition energy is required for starting PF firing, which is different for every individual pulverizer/feeder set. In general, the following are the required conditions:

(a) Adjacent fuel oil elevation (upper or lower) in service (for corner-fired burners of CE boilers); if the opposite elevation is in service, that would also provide sufficient ignition energy for B and W boilers.
OR
(b) Adjacent PF elevation (upper or lower or opposite) is proven with the associated feeder speed running at greater than (for example) 50% and the boiler load is greater than around 30%.

Now, *any pulverizer can be started* after receiving all the start permissive conditions as stated under:

(i) Sufficient ignition energy for PF firing is available.
(ii) Air flow is adequate, for example, >30% and <40% (this signal is considered redundant after any feeder is proven on for >50 s).
(iii) Burner nozzle tilts are in the horizontal position (this signal is not required after any feeder is proven on for >50 s and is applicable to corner-fired boilers only).
(iv) No MFT.
(v) PA pressure is adequate (after the pulverizer is on, this permissive is no longer required).
(vi) All pulverizer discharge valves are open.
(vii) Pulverizer outlet temperature is less than the high value (around 95°C).
(viii) The cold air gate is open.
(ix) The tramp iron hopper (meant for taking out mill rejections) valve is open.
(x) The feeder inlet gate is open.
(xi) Pulverizer lube oil pressure is adequate for >5 min.
(xii) Lube oil sump level is adequate.
(xiii) Lube oil temperature is more than sufficient (around 30°C) [after the pulverizer is on, this permissive is not required].
(xiv) No mill trip.
(xv) Pulverizer seal air header to the mill under the bowl DP is adequate (about 200 mm wcl).
(xvi) Hot air gate is closed.
(xvii) All other permissive signals related to this particular type of pulverizer.

When the pulverizer is on, the following interlocked actions would be initiated:

(i) Open command to the hot air gate would be issued if the DP between the seal air to the mill and the mill is adequate.
(ii) Open CAD to 100% position signal to auto control system would be removed.
(iii) The pulverizer temperature control will now receive a release air and temperature control to auto signal.

The pulverizer can now be considered as in service and the next step of operation can begin *manually or automatically* by mode selection.

After receiving the start command, the pulverizer seal air valve would open first and the seal air header to the pulverizer under the bowl differential pressure would be established. After some time, when a DP of >200 mm (typical value) of wcl is sensed, then the pulverizer hot air gate can be opened and the *feeder can be started*. During this period, the closing operation would be locked temporarily for the pulverizer discharge/seal air valve and the associated auxiliary air dampers.

The pulverizer trips if any of the following conditions takes place:

- **(i)** After expiration of the time period set for establishing the DP across the pulverizer, if the same is not more than adequate (after starting), the pulverizer would be tripped.
- **(ii)** The pulverizer seal air header to the pulverizer under the bowl DP is very low (typically around 125 mm wcl) and remains after a specified time.
- **(iii)** Pulverizer start command given and not on after a specified time.
- **(iv)** Pulverizer on and lube oil pressure is low after a specified time.
- **(v)** Loss of ignition energy before it is self-sustaining.
- **(vi)** Any of the pulverizer discharge valves is not fully open.
- **(vii)** Hot PA duct, pressure very low.
- **(viii)** Both FD fans trip.
- **(ix)** Both PA fans trip.
- **(x)** If both the PA fans and the FD fans are running and any PA or any FD fan trips, then the pulverizers start tripping from the top elevations with some time gap in between tripping until half the pulverizers are available.

(xi) MFT.

Other than the above causes, the pulverizers can be stopped manually on manual mode of operation by the stop PB.

Feeder Starting and Associated Interlocks (Other Than Tube Mill)

The feeder can be placed in service if the following conditions are available:

- **(i)** Pulverizer ignition energy is available.
- **(ii)** The associated pulverizer remains on.
- **(iii)** The journal hydraulic pressure is satisfied.
- **(iv)** No MFT.
- **(v)** The pulverizer outlet temperature is more than sufficient (around 55°C).
- **(vi)** Feeder speed demand is at a minimum.
- **(vii)** Hot air gate is open (the feeder can be placed in service in manual operation without opening the hot air gate).

The *feeder can now be started* and if the feeder discharge chute is not plugged, after some time, the feeder start signal is taken as established and in service if the PF flow signal, that is, the coal on the belt signal, is received.

When the *feeder running* signal is established, the following actions take place, provided the corresponding pulverizer motor power is within limits and the differential pressure across the pulverizer bowl is not high:

- **(i)** Withdrawal of demand signal signifying drive feeder speed to minimum.
- **(ii)** All counting timers would be reset, signifying readiness for the next operation.

After the feeder is on after some time, the following events take place:

- **(i)** The total coal flow totalizing circuit takes this feeder into account after some time.
- **(ii)** The feeder speed control is released to auto after expiration of some more time period.
- **(iii)** Enables feeder speed to rise >50% and then if sufficient numbers of coal flame scanners sense flame in the corresponding elevation, the combined signals in and logic generate output after expiration of a comparatively long time period of about 3 min signifying a signal pulverizer ignition permissive.

The pulverizer ignition permissive confirms that the furnace flame has established and sufficient pulverizer ignition energy is available. At this point, the initial pulverizer start permissive signal to support PF firing is no more required as the flame is self-sustaining. However, it is recommended that a minimum of two pulverizer/feeder sets be established at >50% loading before the ignition support energy is withdrawn.

Normally, after the established running of the first pulverizer/feeder set, the adjacent pulverizer/feeder set is brought into service. The pulverizer/feeder set connected to a remote elevation from the running one is brought into service only in an emergency situation. Whatever the situation, standard good operating practice is that the second set be introduced and both feeders are loaded to >50% before ignition energy is removed.

Feeder Tripping and Associated Interlocks

The feeder is tripped automatically if any of the following conditions takes place:

- **(i)** No coal on belt.
- **(ii)** The pulverizer motor power is low.
- **(iii)** Discharge line plugged.
- **(iv)** MFT.
- **(v)** Associated pulverizer is off (may not be applicable for tube mill).
- **(vi)** Absence of ignition energy removed within specified time after feeder started.

Other than the above causes, the feeders can be stopped manually by selecting the manual mode of operation and then by depressing the stop PB.

Removing of feeders from service requires just the reverse procedure. The pulverizer ignition energy would again need to be reinstated when only two feeders are in service and operating at <50% loading. The fuel oil firing in the corresponding elevation is to be continued until the last pulverizer/feeder set is withdrawn.

For the manual shutdown of the feeder, that is, if the feeder goes off after it was on, the following sequential action would take place after expiration of some prejudged time period:

(i) HADs would be closed first if the hot air gate is not already closed for minimizing the chance of a pulverizer fire.
(ii) CADs to open 100%.
(iii) Pulverizer temperature control's air (flow?) and temperature control are transferred to manual.
(iv) Hot air gate is closed after some time.
(v) When the hot air gate is closed, the close command to the HADs is removed if he pulverizer outlet temperature remains <95°C.

Seal Air System and Operation of Fans Description

12.4.3.1.5 Master Fuel Trip Signals and Short Description

The BMS checks the health of the steam generation package as a whole and senses a boiler emergency trip signal or the MFT vis-a-vis closure of all fuel inputs immediately as well as on the occurrence of any of the following signals:

(i) Loss of all FD fans, all ID fans, all BFPs, scanner cooling air, all flames, all fuels, and airflow.
(ii) Very low deaerator level, drum level, and furnace pressure.
(iii) Insufficient water wall circulation.
(iv) Very high furnace pressure, drum level, and MS pressure (for supercritical boilers).
(v) Reheater protection operated.
(vi) Power supply failure to BMS/FSSS cabinets.
(vii) Emergency push button operated.

Short Description on Master Fuel Trip Signals

While the above signals are self-explanatory, short descriptions of insufficient water wall circulation.

and reheater protection operated need some elaboration.

(i) *Insufficient water wall circulation*

This type of tripping signal is applicable mainly for forced circulation water systems of the boiler having a separate boiler circulating water pump (BCWP). As the heat transfer design is not based on natural circulation, any sort of insufficient water circulation through the water wall would jeopardize heat transfer and cause the water wall itself to burn out due to loss of cooling media.

(ii) *Reheater protection operated*

This tripping signal is included to protect the reheater in situations when the boiler is working but there is no or low flow through the reheater. The conditions may be any of the following:

(a) Turbine trips/generator circuit breaker (GCB) opens while the HP/LP bypass remains closed for more than a specified time.
(b) Load shedding relay operates (heavy load throw off) with the turbine in working condition and with the HP/LP bypass closed for more than a specified time.
(c) HP or LP bypass system suddenly trips or closes while working during the house load operation, that is, the turbine and HP/LP BP systems are operating in parallel.
(d) HP or LP bypass system suddenly trips or closes while working during the start-up time, that is, when the turbine is bypassed and the boiler steam flow was channeled through the HP/LP BP systems.

12.4.3.2 Sensors Used in BMS/FSSS Functional Configuration

Different sensors are used in the *BMS/FSSS* as dictated by the system, such as different process switches, position switches, flame scanners, and associated electrical contacts or manual commands.

Flame sensors are conditioned in a separate standalone cabinet called a scanner cabinet. Sensors and scanners are of various types, namely ultraviolet, infrared, or even visible light. Because flame radiation has a wider range in visible light, nowadays, visible light scanners are more popular. Analog low signals sent by the sensors are not capable of generating binary contacts or analog signals for display. Therefore, they are transmitted to the BMS/FSSS cabinet via the scanner cabinet, where they are processed, conditioned, and converted to generate a binary signal.

These detectors are capable of automatic/manual checking of the optical path and have self-diagnostic features. These are available in flameproof and NEMA 4 × enclosures suitable for boiler applications.

12.4.3.3 Different Drives: Diaphragm/Solenoid-Operated Valves and Actuators

Several drives and actuators are connected to the BMS/FSSS for executing the commands as processed by the system. The drives include high-voltage (HV) unidirectional motors for FD fans, ID fans, PA fans, pulverizers, etc. Bidirectional drives include motorized valves. Some actuators may be solenoid-operated valves, pneumatic diaphragm, or hydraulic piston-operated valves (Solenoid valves are normally double coiled for maintaining a stay-put condition in case of wire snapping).

12.4.3.4 Operator's Interface or the Plaque/Keyboard, Monitor in the Control Board

This particular item is a combination of different facilities through which the operator can reach the remote drives or devices in case manual intervention is required. All the binary/analog process signals and actuator statuses are made available to the operator so as to enable him to make prompt and appropriate decisions in time of need.

The following facilities are normally available in the operator's interface:

(i) *Indicators*

This may be the analog type such as flame scanners or binary, for example, normal/abnormal process conditions through pressure/temperature/level/etc. switches, flame status in on/off form, actuator or bidirectional drives running/stopped/tripped open or close position, condition of unidirectional drives, etc.

(ii) *Command devices*

These command devices may be push buttons, selector switches, control switches, etc., required for manual operation or intervention as needed.

Normally, separate plaques/consoles are provided for hardware interfaces; a CRT monitor-based system with touchscreen facility or a separate keyboard are also quite common in modern plants.

12.4.4 BMS/FSSS: Manual Operation/Automation

BMS/FSSS was provided normally for achieving the basic safety interlocking features without intervening with the process control system. Through evolution, it also includes the start-up and normal operation/graceful shutdown of the SG plant. With the help of this system, the operator, through regular interaction and observation on the device status, may issue commands from the remotely located plaques/consoles or a special window in the monitor required for start-up and normal shut down of various devices and/or drives individually or in groups. The system is completely user-friendly and guides the operator to take proper action through the desired sequence. The sequence would not be allowed to proceed further until the requisite permissive signals are available and the proper mode of operation is selected. Once the start-up is done successfully, the boiler is ready for the loading/unloading process and protected by the BMS/FSSS from any hazardous situation. This type of operation is described as *manual* and the operator is required to take action at every step.

A more sophisticated and higher level of operation has also been envisioned, which is popularly called *automatic*. This type of system allows the operator to issue a single command that enables a group of devices/drives (subloop organized as per process requirement) to start-up or shut down in an appropriate sequential manner. Typically for boiler operations, complete automation is not required as per the load demand and pulverizer characteristics, which are brought into service one by one to suit the load requirement. Hence, the most straightforward way to divide the total devices/drives is to make it a pulverizer group-wise, which includes the igniter, the pulverizer, the feeder, etc., with manual intervention if needed. For example, the sequence of a particular pulverizer subloop operation may be:

(i) To arrange the ignition permissive signal.
(ii) If the ignition permissive is available, then the pulverizer motor to start.
(iii) Hot air gate to open.
(iv) Pulverizer outlet temperature control to be released in auto.
(v) Feeder to start.
(vi) Feeder speed control to be released in auto.
(vii) Windbox dampers or air register control to be released in auto.

12.4.5 Flame Monitoring System

As indicated earlier, the flame monitoring system is entirely a separate and standalone system used for monitoring each individual burner flame condition.

For tangentially corner-fired boilers, once the burners of a particular elevation are operated at $>30\%$ of their capacity, there is produced a *rotating fireball*. The ignition energy thus produced in one elevation can support the initial ignition energy required for a fresh introduction of fuel in another elevation. Effectively,the furnace is then considered a single burner and the flame detection is done on a statistical basis by multiple sensors. For a particular elevation, a single flame detector failing to indicate the flame of its adjacent burner is ignored and not shut off due to loss of flame condition. As indicated earlier, the statistical basis is applied by counting all the four corners' flame or no flame status and then determining the health of that elevation firing condition. F or example, if three of four flame detectors sense no flame, then only that elevation is considered as unsuitable for continuing further firing operation and is taken out of service.

In tangentially corner-fired boilers, the change in firing characteristics from start-up to fireball is gradually made by way of increasing the elevation fuel admission to about 30% loading capacity is based on hand on experience providing sufficient factor of safety.

For *other types of firing systems* such as front/opposite/downshot-fired boilers, the flame produced at all fuel firing rates demonstrates the individual burner characteristics, the same as exhibited by the tangentially corner-fired boilers at low firing rates up to 30% of individual capacity. This means that the type of flame produced does not have that energy capable of providing enough ignition permissive to burners at other locations. In these boilers, each PF burner requires a separate and individual ignition energy (from nearby oil burner firing). Therefore, the logic development should take care of that aspect for these boilers. BMS in many plants includes interlocks of auxiliaries such as ID/FD fans and associated dampers as well.

12.4.6 Seal Air Fan Control

Redundant seal air fans are provided to supply sealing air to pulverizers and the associated dampers and valves. SA fan suction is normally taken from the PA cold air duct and a differential pressure is available for sealing coal-laden dust from the pulverizers.

Any SAF can be started manually or automatically if selected and any PA fan run. The other SA fan would be on standby mode. The concerned discharge damper then receives a command to open. The standby fan would start automatically if the main fan does not start within a few seconds or if the DP is less than about 100 mm. The seal air fans are removed from service automatically when both PA fans are off for more than the specified time.

When the seal air fan is off, the associated seal air fan discharge damper is closed.

12.4.7 Scanner Air Fan Control

Redundant scanner air fans are provided to supply cooling air to the scanner. ScA fan suction is normally taken from the FD fan discharge and a DP is maintained. Any ScAF can be started manually or automatically if selected and the DP over the furnace is less than about 150 mm wcl; the concerned discharge damper then receives a command to open. The other ScAF would be OFF and on standby mode when its discharge damper receives a closing signal.

Standby fans would start automatically if the main fan does not start within a few seconds or if the differential pressure over the furnace is less than about 150 mm wcl. When both FD fans are off, the scanner air emergency damper opens and the cooling air is then supplied with atmospheric air. When any FD fan is ON, the scanner air emergency damper would be closed. Either scanner air fan can be removed from service manually if the scanner duct to furnace DP is >150 mm wcl. In case of loss of unit critical power for more than 2 s, both scanner air fans start and their outlet dampers open simultaneously.

If the scanner duct to furnace DP remains <150 mm wcl for >10 s,it is taken as loss of scanner cooling air for boiler tripping.

12.5 Secondary Air Damper Control

General

To accomplish the efficient burning of any type of fuel, the air supply plays a very vital role. The total air flow must take care of the stoichiometric ratio of the fuel being burnt and the optimum excess air vis-a-vis oxygen percentage, etc. The total air flow is normally provided to the furnace in the form of primary and secondary air flows. The low NO_x design is accomplished through a separate section of total air flow, which may be termed overfire air. There are specific purposes for the segregation of total air flow. For example, primary air contributes not only to the initial combustion air, but also for transporting the pulverized fuel. The contribution of secondary air flow is to provide additional air or oxygen to make sure of complete combustion of any unburnt hydrocarbon. The purpose of providing overfire air as already discussed above and in Section 10.6 of this chapter is the final part of combustion process to take place purposely delayed to minimize the generation of thermal NO_x level. By suitable design, normally the SAD are so adjusted that at the full open position, there is created an optimum swirl position that maintains the long flame and reduces the flame temperature.

For wall-fired boilers, secondary air (fuel air and tertiary air) combined together enters the individual burner. The proportioning of the fuel and tertiary air is done within the burner and also needs to be equal in all burners, but may change slightly due to plant conditions. This control strategy is depicted in Fig. 8.4.

12.5.1 Objective: Secondary Air Damper Control

The secondary air constitutes two parts: fuel air and tertiary/auxiliary air. Fuel air allows the higher hydrocarbon or char to burn, and the tertiary/auxiliary air is intended for ensuring the delayed combustion process purposely through a special design to reduce the NO_x level. The control strategies depend on the type of boiler as per the manufacturers own designs.

12.5.2 Discussion

The basic purpose of this system is to distribute secondary air according to the firing of various burners so as to ensure complete firing of fuel with minimum pollution.

12.5.2.1 Secondary Air Damper Control for Wall-Fired Boiler (With Flue Gas Recirculation System)

For wall-fired boilers, secondary air (fuel air and tertiary air) combined together enters the individual burner and the proportioning of fuel air and tertiary air is done within the burner.

The control strategy depicted in Fig. 8.4. The secondary air flow control system by the SAD was described in Section 3.3.2.2 of this chapter for a boiler with a flue gas recirculation system. As there are only two major parts of the total air flow, that is, primary and secondary air, their share of flow normally varies from 30% to 35% for the primary air and 65%–70% for the secondary air.

Individual secondary air flow is measured for each burner. This measured flow is utilized in the automatic

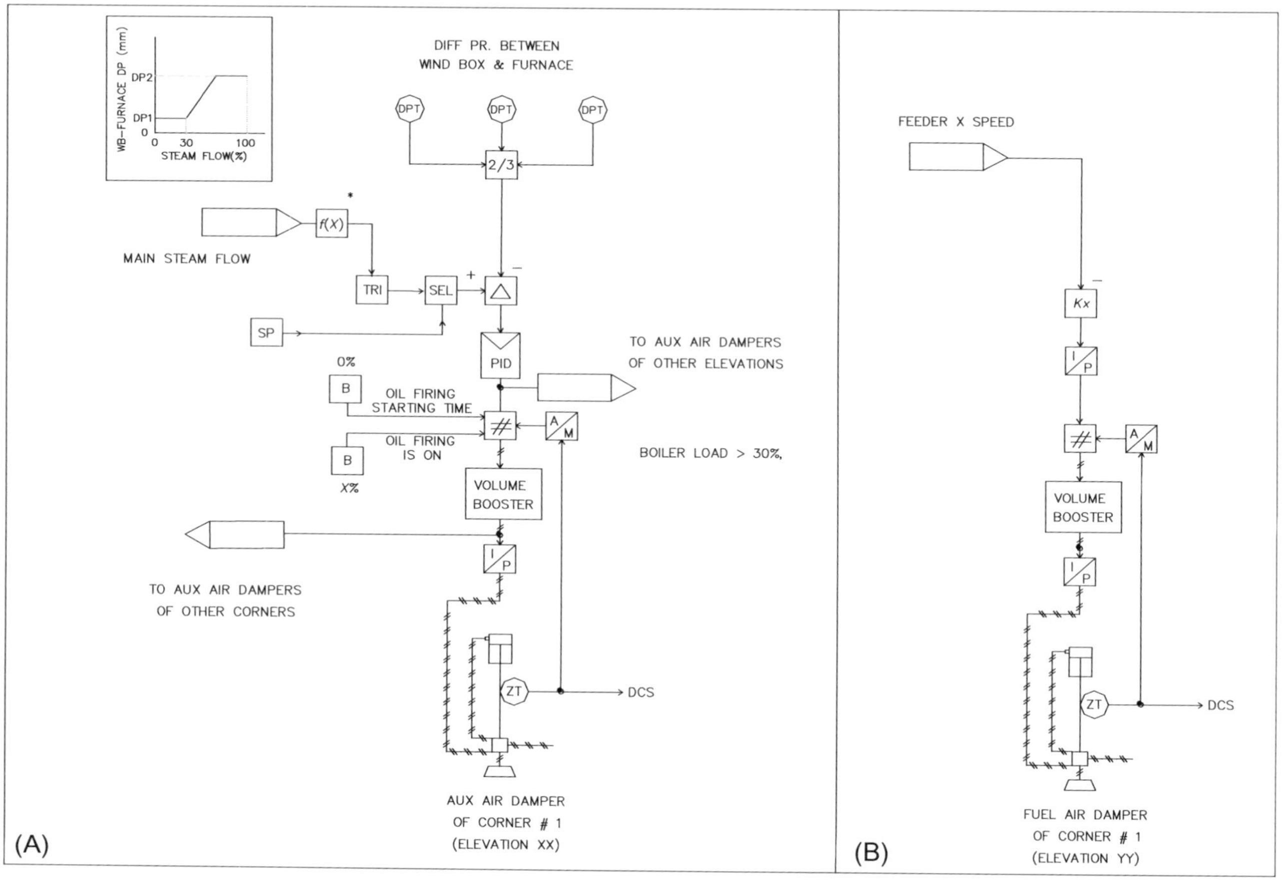

NOTE:
- AT BOILER LOAD > 30%, AUX AIR DAMPER ELEVATION NOT HAVING OIL NOZZLES WOULD CLOSE FROM TOP WHEN ADJACENT PULVERIZERS ARE OFF AND ANY OTHER FEEDER IS ON
- AT BOILER LOAD > 30%, AUX AIR DAMPER ELEVATION HAVING OIL NOZZLES WOULD CLOSE FROM TOP WHEN ADJACENT PULVERIZERS ARE OFF AND OIL BACK UP TRIP OCCURRED
- AFTER MFT, ALL AUX AIR & FUEL AIR ERS WOULD BE IN MANUAL AND OPEN TO 100%
- FUEL AIR DAMPERS OF A PARTICULAR ELEVATION WOULD MODULATE AS PER CORRESPONDING FEEDER SPEED / FUEL FLOW AFTER THE FEEDER IS PROVEN 'ON'

FIG. 8.54 Secondary air damper control (SADC) for TT boiler.

control loop as well as for the adjustment of equal secondary air flow for all the burners.

12.5.2.2 Secondary Air Damper Control for Tangentially Corner-Fired Boilers

The secondary air flow control system by the SAD for tangentially corner-fired boilers is somewhat different. This system is applicable for combustion in the tangentially fired furnace and the auxiliary air part of the secondary air flow is accomplished by the separate auxiliary air dampers, mainly for regulating the velocity and distribution of the air flow at the intermediate stage of combustion.

For this type of system, the individual secondary air flow is not measured like the former type of furnace. Here, the fuel air flow (provided for each coal nozzle), being a part of the total secondary air flow, is controlled by the load demand of the boiler through the individual demand of air flow generated from each pulverizer loading through feeder speed. As the concept of a fireball is just a swirling mass of huge flame produced by the burning of pulverized fuel (PF) and the sufficient air flow to support the combustion, the fuel air dampers act as FCEs and are placed in every fuel injection compartment to supply initial combustion air. The fuel nozzle discharge angle determines the size and rotational velocity of the fireball. Into this swirling mass of flame, the fuel air, a portion of the total secondary air, is injected with the quantity just required for stoichiometric conditions to enable the continuation of the combustion process with delayed mixing.

The remainder of the secondary air, that is, the auxiliary air, is directed with the velocity to form a layer of such air between the fireball and the inside walls of the furnace. This layer acts as a cushion over the fireball while rotating in the same direction. It also functions to resist the impingement of slag on the tubes with which the walls of the furnace are lined, and by this, soot blowing becomes easier. While doing this, the SADC control utilizes the wind box to furnace DP as the measured/process variable. This is maintained at a constant value at the initial stage, and after that, a variable set point up to a certain boiler load and ultimately a higher fixed value. This control is achieved through auxiliary air dampers placed below and over each fuel injection compartment with an aim to provide equal air flow through each compartment they are provided. The DP between the WB to the furnace is maintained at desired preset values by regulating the auxiliary air dampers in all double-lettered elevations. In some of them, oil nozzles are provided and those dampers would be kept at a fixed opening only when oil firing is taking place in that elevation.

12.5.2.3 Summery of Secondary Air Control System

To summarize the secondary air flow control system in a tangentially fired boiler with an overfire damper, it may be noted as:

(i) All the secondary air is fed through the wind boxes.
(ii) The wind box to furnace DP is maintained by a separate set of SAD called auxiliary air dampers.
(iii) A part of secondary air termed fuel air is controlled as per the requirement of fuel flow, that is, fuel air flow is proportional to fuel flow in a particular fuel injection compartment, which results in total fuel air flow being proportional to total fuel flow.

12.5.3 Control Loop Description

The control loops for the SADC system for boilers with flue gas recirculation systems were already described in Section 3.3.2.2 of this chapter. A control-loop description for the tangentially corner-fired boilers will be discussed.

12.5.3.1 Measurement of Parameters

The wind box to furnace DP is measured through suitable transmitters with sufficient redundancy and voting before actually being used in the control loops.

12.5.3.2 The Control Loop Strategy

The control strategy (Fig. 8.54A) for the SADC system is based on a simple, single-element control concept.

The selected signal of the wind box to furnace DP becomes the measured/process variable. The set point is normally a fixed value up to 30% of the boiler load. The set point begins to increase on a further increasing load through a function generator up to a certain load, where it then becomes fixed. However, the set point can be changed manually through selection. The controller output is used to regulate all the auxiliary air dampers adjacent to the coal elevation in tandem at a load >30%. At a load <30%, all these dampers modulate to the same signal.

Some auxiliary air damper elevations incorporate oil nozzles. Those dampers at a particular elevation would be closed during oil firing starting time and at a fixed opening when oil firing is proven on. Auxiliary air dampers not serving any adjacent coal/oil nozzles would be closed at boiler load >30%.

As shown in Fig. 8.54B, fuel air dampers are modulated in proportion to feeder speed/coal flow in the corresponding elevation.

12.6 Soot Blowing System

General

In steam-generating plants where an ash-bearing fossil fuel, such as coal, lignite, fuel oil, bagasse/refuse, or any other type of oil/byproduct, is burned, there has always been a perennial problem associated with the deposition of soot during the combustion process, and carried by the hot combustion products to the different heating elements in the furnace as well as in the air heaters. Deposition affects different heat transferring surfaces such as water walls, superheaters, reheaters, and economizer steam pipes covering both radiant and convection zones in the furnace chamber as well as the air heaters, which is called cold end deposition.

The fly ash carried along with flue gas combines with moisture and other harmful byproducts when it falls below the condensation temperature. Natural gas is comparatively a better fuel so far as this deposition generally called soot is concerned, as it is almost a clean fuel with negligible ash content. Oil has low ash content, which is water washable. The actual problem of soot deposition lies with the PF-fired boilers. The ash content and coal quality varies widely from country to country and even in different regions of a particular country; hence the form of soot also varies accordingly.

When the soot is deposited on the surfaces, it acts as an insulating media and encumbers proper heat transfer from the flue gas to the water/steam/air. The type of deposit or soot may be of dry and powdery substances or hard slag, depending on many factors such as temperature, percentage of ash content, percentage of sodium content in ash, etc. The most difficult soot deposits are caused by PF with high ash content, a high percentage of sodium content in ash, and a low ash fusion temperature, which also acts as an important factor; low value of which temperature means possibilities of more ashes would be stuck to the fluid tube outer surfaces and high deposition in turn.

As the practical behavior of the soot deposits is like a thermal insulation between the fluid inside all tubes of the furnace and the hot products of combustion, thereby reducing heat transfer, the more soot deposition on the furnace walls, the less effective the heat gained by the operating fluid, resulting in a steady decrement in fluid temperature. In other words, it may be said that the heat transfer from the hot flue gas to the fluid is very high and the temperature of the combustion products leaving the furnace is at a relatively low value when the furnace is in a very clean condition, as is normal in the initial stages of installation. However, over the course of time, the heat-exchanging surfaces in the furnace become coated with layers of soot deposition, and the heat transfer from the hot flue gas to the fluid inside the tubes is significantly reduced. As a consequence the temperature of the fluid decreases with the hot flue gas leaving the furnace is significantly increased. This phenomenon over a period of operation causes significant changes in the overall heat balance calculation and finally can pose serious problems for the operator in balancing the steam generation. The high flue gas outlet temperature has many associated problems, as already discussed, such as low overall efficiency, high NO_x generation, pollution control violations, etc. There may be some other problems such as the severe plugged condition of some sections and adequate flue gas flow may not be available or erosion may result due to high local gas velocity.

12.6.1 *Removal of Soot Depositions*

Periodic removal of soot deposits is a very important task so far as the efficiency of the unit is concerned. Due to the insulating nature of the soot, the overheating of the boiler tubes is very much natural as the temperature control loop would try to raise the temperature of the steam by allowing more energy input in order to maintain the final parameter.

To avoid this situation, the soot blowing system has been developed for selectively cleaning soot deposits from all possible areas. Therefore, it has become imperative for steam-generating plants to have a number of soot blowers physically installed at different locations in the walls of the furnace over the height of the combustion chamber, the superheater, reheater, economizer, etc. This is to intermittently but regularly clean the furnace and heat-exchanging walls in the steam and water tube surfaces carrying different types of fluid. Each soot blower, when selected manually or automatically, carries the spray assembly through advance and retract mechanisms and arranges to clean a region of the tube walls around it by blowing a fluid jet at a very high velocity.

For soot blowers, the operating blowing media may be anything such as steam, water, or compressed air through the spray nozzle head, which is made to pass through the furnace wall opening into the furnace chamber to eject the cleaning medium under pressure to hit directly against the surface area affected by the soot deposit. As already said, the operation of this type of advance and retract action is done only intermittently as and when necessary. Due to high velocity impingement and the thermal shock by the comparatively low temperature medium, the blowing medium causes blowing away of loose deposits and high impact loading on the hard and slag deposits. This action causes the same to fall from the tube surfaces, returning a relatively clean tube surface again being exposed for heat exchanging to a higher rate improving things as desired.

12.6.2 *Method of Removal of Soot Depositions*

The location of soot deposition on the different surfaces such as the superheater, reheater, economizer, etc., are not uniform. Even the deposition on furnace walls is not uniform over the height or width of the furnace walls. Depending on the flue gas flow rate/velocity and the travel path, some portion of the furnace experiences more deposition compared to other portions. Prediction of such high deposition zones is very complicated for a particular time period of operation. This is because the exact soot deposition profile may vary on the type of fuel and firing system, in addition to what has already been described.

12.6.2.1 Fixed Programmed Time-Based Operation

It has therefore become the normal practice to adopt the OLC strategy of the soot blowing system through step and sequential operation of the blowers in the well-defined time frame. Quick details are indicated in Section 12.3.1 of this chapter.

12.6.2.2 Temperature-Based Operation

This control strategy discussion above is based on certain presumptions and may not be entirely flawless. It may cause additional wear and tear and wastage of the soot blowing medium, whereas the tubes really needing soot blowing at the right time may not get the service in time.

Therefore some advanced method was sought to avoid such an unnecessary situation as per the experience and practical problems. A system strategy was evolved that enables cleaning the selected soot blower as per the decision taken by the system logic. A short discussion is incorporated in Section 12.3.1 of this chapter. Thermocouples welded to the tubes sense the actual surface temperature in the vicinity of each soot blower. The temperature set point is calculated from the saturation temperature of the water flowing through the water tubes as the representative of a soot-laden furnace condition at the particular. The online water tube wall temperatures, if less than the calculated set value, initiate the blowing operation of the associated soot blower. All the soot blowers thus take the need-based command for operation to clean the portions that are associated with the soot affected areas only, without disturbing the cleaner portion blowers as a routine job.

This method for obvious reasons is not effective for pipes having a steam or steam-water mixture, which has no further bearing on saturation temperature. At the same time, it is evident that the local fluid saturation temperature for the water line varies with elevation. It is therefore not a very easy task to determine an actual saturation temperature for obtaining the calculated preset temperature as an index of the degree of soot deposition in the furnace. For supercritical boilers, where a mixture of water and steam is passed through the tubes above the supercritical pressure of water, there is no way of calculating the metal temperature that can be taken as an index of furnace dirtiness. As already told, the metal temperature will vary significantly over the height of the water wall dependent on the local heat distribution and on the water-steam phase state of the mixture flowing through the tubes at that location, which is also next to impossible to determine.

Another method employs the measurement of the differential temperature of the metal tube and hot flue gas to initiate soot blowing at a higher value of differential temperature.

12.6.2.3 Heat Transfer/Heat Flux-Based Operation

There is another method of determining the strategy of soot blowing, which utilizes the more direct sensing system for selectively cleaning the tube walls of the furnace. The local heat transfer rate is measured by employing a *heat flux meter* (Fig. 8.55) to assess the degree of soot deposition and not by inferential measurement by the local wall temperature.

The sensing elements are located in each of the regions of the tube walls surrounding the soot blowers to detect the local heat transfer rate from the hot flue gas to the steam/water tubes in the furnace or heat-exchanging elements (flue gas side) of the air heaters.

The heat flux meter output signifying the local heat transfer rate is then compared with a manipulated set point that represents the corresponding lower (worst) heat transfer rate value vis-a-vis the maximum acceptable soot deposition a tube wall is covered with. The comparator generates an output whenever the sensed local heat transfer rate is less than the lower allowable set point. A logic processing unit utilizes the comparator output for automatic activation of the corresponding soot blower. As an alternative action, manual intervention is taken from the operator's place through an alarm/indication provided by the processing unit about this polluted tube/element condition.

On the other hand, another comparator may be provided for comparing the heat flux meter output to a manipulated upper heat transfer rate set point representing an acceptable clean condition of the tube walls for generating an output whenever the heat flux meter output is greater than the upper manual set value. When this comparator output is available, the alarm signal would be withdrawn and a healthy indication may take the place of a dirty tube condition. The corresponding soot blower also may be retracted manually or automatically by the logic function.

Nowadays, with PC-based systems utilizing software, the heat flux is computed and soot blowing is optimized. As stated earlier, excess soot blowing causes steam wastage, higher boiler maintenance, aggravated tube corrosion, less boiler efficiency, and higher stack opacity, which may not be permissible. The software with a dynamic model is available for the ideal heat transfer rate and cleanliness factor. The heat transfer rate, based on real-time parameters (temperature at various stages), is compared with the ideal value (model-based) to calculate the cleanliness factor. The other influencing factors included are fuel type, fuel flow, wind box-furnace DP, air parameters (flow, temperature, pressure), etc. The model is improvised in an interactive method with these factors to actually calculate the deviations. Then, soot blowing is optimized in one of the two modes:

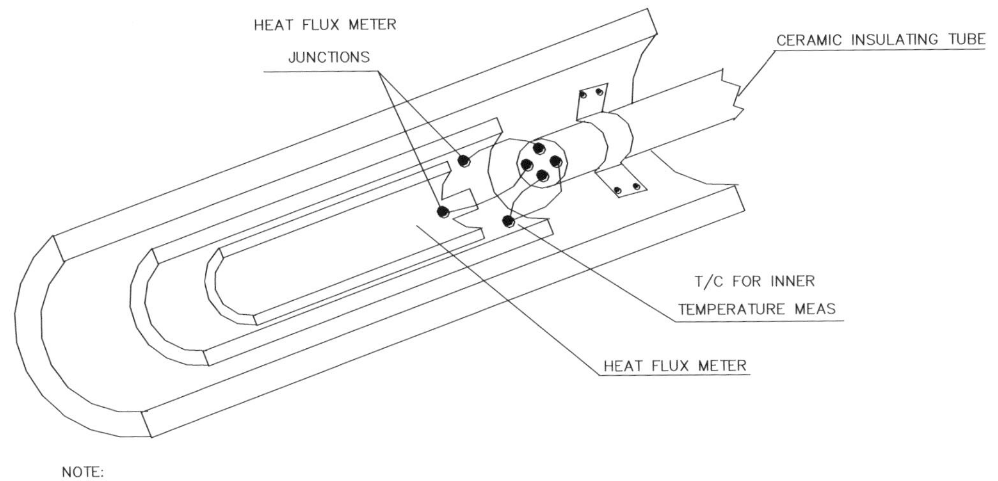

FIG. 8.55 Schematic diagram of heat flux meter.

(1) *Steam saving mode* to avoid excessive soot blowing, assuming that there is a limit of soot blowing frequency beyond which there will not be much improvement in reality.
(2) *Opacity mode*, when in all sections uniform blowing is initiated to reduce opacity. All these are nowadays done by utilizing artificial intelligence and neural networking ("Intelligent Soot Blower Scheduling for Improved Boiler Operation," by Xu Cheng, R.W. Kephart, and J.J. William).

12.6.2.3.1 Heat Flux Meter: Principle of Operation (Fig. 8.55) As already mentioned, the detector of the local heat transfer rate is basically a heat flux meter mounted directly to the outer side (the combustion chamber side of the furnace) of the tube walls. The transducer is attached to the heat transfer surface by a clamp, ceramic cement, or by any suitable bonding material. The transducer consists of a thermoelectrically dissimilar metal body (than that of the main fluid tubes) and has thermal and electrical connections with the tube metal. Electrical connections are taken from the outer and inner face of the metal body to the processing unit situated outside at a room or ambient temperature environment. The principle of operation of this heat flux meter is that there is created a small temperature difference (due to the thermal resistance of that metal body), which in turn generates an induced voltage or electromotive force (emf) across the transducer metal body whenever there is a flow of heat to or from a surface on which the transducer is placed.

The thermoelectric transducer generated signal is on the order of microvolts, but the measuring unit is self-powered without requiring any external excitation voltage. The thermal resistance introduced intentionally by the transducer for the purpose of producing a temperature difference vis-a-vis a potential difference across itself may easily be considered as negligible for all practical purposes.

The DC microvolt signal generated by the transducer is then treated and conditioned in a proper way to get analog (4–20 mA) and contact outputs for further application. The final output is proportional to the heat flux.

12.6.2.4 Type of Soot Blowers

12.6.2.4.1 Soot Blowers for Furnace Walls Generally, retractable blowers with short and single nozzles, popularly called *wall blowers*, are utilized to remove the soot deposition from the combustion chamber area. During forward movement, the blower goes inside the furnace wall by around 50 mm. The blower tip consists of a single nozzle with a high energy jet capable of spraying blowing media at supersonic speed. After the forward movement is completed, the blowing starts with a time lag. The blowing is also accompanied by rotation of the tip by 360° to clean the deposition up to 1500 mm of the radius. A very difficult type of soot deposition in the form of slag requires denser spacing as the effective cleaning radius becomes around 1000 mm. The frequency of operation depends on the build-up rate of soot deposition.

12.6.2.4.2 Soot Blowers for Radiation/Convection Area The radiation and convection area of the furnace includes heat-exchanging sections such as different superheaters/reheaters/economizers, etc. which are quite away from the furnace wall. The natural selection dictates long retractable blowers going inside the gaps of the tubes for a better approach to the tubes. It is considered the most effective way to clear out the soot deposition from the respective tube surfaces.

Generally, the available blowers of this type provide a pair of nozzles located diametrically opposite to each other. While in action, the blower moves forward, accompanied by a rotating motion as well. These two simultaneous motions make the blowing medium of a high-speed, high-energy jet form a resultant spray with a helical pattern for cleaning both the tube surfaces and the spaces in between the adjacent tube assemblies. Some manufacturers recommend the nozzles be provided at a slight angle opposite to each other (with respect to the perpendicular axis) to make them more effective. However, the overall effective cleaning radius may vary from 1200 to 3000 mm (approximately) for various reasons such as the flue gas temperature of a particular area selected for cleaning, the inherent ash characteristics of the fuel being used at that time, or the blowing media.

So far as the forward and reverse motions of long retractable blowers are concerned, the same may travel up to the mid way of the width of the furnace, thus covering the total furnace width with the help of the opposite end blowers. The supporting arrangement of long blowers must take care of the boiler overall expansion, which entails the front portion to be supported from the wall itself whereas the rear portion is supported by the boiler platform structure.

12.6.2.5 Blowing Media of Soot Blowers

In general, there are three theoretical media available for soot blowing. Out of that, water is normally not thought of in the recommended list mainly due to the thermal shock it may inflict to the main fluid tubes, causing fatigue and reducing tube life in turn. However, its use cannot be ruled out as it has some special uses in the case of soot deposition characterized with the tenacious and difficult nature of slag formation.

The other two blowing media—high pressure steam and air—are the natural choices. Both are equally effective for cleaning. The selection normally depends upon the availability and reliability, along with the cost factor.

Though the usage of *air as the blowing medium* is considered very efficient, its selection necessitates the installation of separate high-capacity compressors and associated pipeworks. Another disadvantage is the loss of the blowing medium in case of compressor failure. If redundant compressors are provided, that will increase the cost further. However, the most disadvantageous factor is the moisture content in compressed air.

The selection of *steam as the medium* necessitates the installation of pressure-reducing valve pipeworks, as steam would only be available no sooner than the steam generation starts. The pressure-reducing station enables getting steam at the valve outlet a dry quality of super heated steam as the pressure goes down with the steam temperature remains as before. The most significant advantage of steam blowing is that the kinetic energy content out of the nozzle jet in dry steam is almost 200% that of compressed air, which makes it a more effective medium for soot blowing.

12.6.2.6 Location and Operation of Soot Blowers

As already said, the type of soot deposition depends upon the fuel ash percentage content, the sodium percentage content in the ash, the ash fusion temperature, etc. The installation location and type of blower must also be decided as per the severity of the problem. Normally, the density of the soot blower location is more in the front and rare water walls of the furnace above the firing zone. There are also provided a set or two below the firing zone. For the convection zone, sets of blowers near and/or between the banks of platen SH, reheater, final SH, before the banks of primary SH and the economizer.

The sequence and also the frequency of operation are normally suggested by the boiler manufacturer based on which binary sequence control system is loaded, software-wise. However, the initial operating strategy though follow the loaded sequence, the accepted one may vary to suit the experience with the plant condition. The frequency normally followed is running one complete cycle in each working shift of the plant, which may also depend on ash content, deposition density in different zones, etc. The deposition study would be conducted so that some sets (maybe groups or singles) of blowers require more frequent operation than others. Nowadays, PC-based software is available to optimize such frequency of operation, as discussed earlier.

12.6.2.7 Soot Blowers for Air Heater

Scale or fine grain deposits are seen at the cold end of the air heaters when the flue gas temperature falls below the condensation temperature. At this point, the fly ash, moisture, and sulfur oxides (SO_x) combine to form the same on the heating surface of the cold end, which resists the heat transfer as an insulator.

Appropriate soot blowers clear out unwanted soot deposition from the heating surface and ensure that deposits are not subjected to further moisture. Many types of moisture ingress in the downside part of the flue gas path from the economizer or the steam/water tube leakages, water washing shut off valve leakage, or the unprotected FD fan inlets carrying rain water may enter the system. Above all these causes, the main and frequent source of moisture is the leakage of the steam soot blower system. A suitable quality of steam for soot blowing can eliminate that problem.

The AH soot blowers are generally installed at the cold end side where the depositions are observed. The gas outlet side is the natural selection spot, so as to avoid the entrance and mixing of fly ashes and driven away into windboxes. They are installed either in the gas outlet ductwork or as an integral part of the air heater.

12.6.2.7.1 Soot Blowers for Air Heater The number of soot blowers provided for air heaters suit different categories. These are, in general:

- **(i)** Retractable soot blowers are used for air heaters having large diameters of about 10 meters and above.
- **(ii)** Stationary and multinozzle soot blowers are used for small package air heaters.
- **(iii)** Soot blowers with nozzles mounted on a swinging arm are used for air heaters as a popular choice.

Retractable Soot Blowers for Air Heaters

This type of soot blower is the same as used for the radiation and convection area of the furnace, described in Section 12.6.2.4.2 of this chapter.

Swinging Soot Blowers for Air Heaters

These soot blowers are large and equipped with an electric power-driven source with a single- or double-nozzle arrangement. The nozzle assembly moves slowly over the face of the heat transfer surface. The nozzle arm simultaneously swings in an arc across the face of the heat transfer surface. The blowing cycle takes one pass across the heating face, either from the periphery of the rotor to the central rotor post or in the reverse direction.

12.6.2.7.2 Media of Soot Blowers for Air Heaters There are two widely accepted media available for air heater soot blowing. One of the blowing media is obviously the high-pressure *dry superheated steam* and the other is high-pressure (around 12 kg/cm^2) or *compressed air, with* both being equally applicable. The selection normally depends upon the availability and reliability along with the cost factor, as discussed in brief in Section 12.6.2.5 of this chapter.

For air heater service, saturated steam is applied sporadically instead of the moisture content in that quality of steam. However, as per experience, dry superheated steam is found to be the most effective medium of AH soot blowing than either compressed air or saturated steam. The temperature and pressure value of steam are required to be maintained around 70°C superheat and 15 kg/cm^2 for an efficient end result.

12.6.2.7.3 *Washing Medium for Air Heater Deposits*

Water as blowing media is not at all recommended for AH soot blowing, mainly due to the thermal shock. However, its use as a washing medium plays an important role. In case of failure of the soot blowing operation to remove tenacious soot formation, it has been experienced that a water washing cycle for the heating surface yields satisfactory results that would otherwise contribute a considerable amount of draft losses across the air heaters. The equipment normally utilized for this application is a standard one having stationary multiple nozzles and a medium velocity jet producer for different types of air heaters. Water washing may be done during plant shut down, with an isolated air heater portion or full plant running condition taking care of guidelines advised.

12.6.2.8 Pressure and Temperature Controls of Soot Blowing Media

It is extremely important that the blowing media must be supplied at the blowing nozzles with adequate pressure and temperature for efficient soot blowing. As has become apparent from the above discussion, steam is the most effective medium for soot blowing and maintaining its pressure and temperature at constant values is very important, as far as the dry superheated steam supply is concerned.

The control loops details are discussed separately in Section 10.5 of this chapter.

BIBLIOGRAPHY

[1] Cascade, Feed Forward and Boiler Level Control: By Allen D. Houtz (Consulting Engineer Automation System Group) Copyright © 2006 by Douglas J. Cooper. All Rights Reserved. http://www.controlguru.com/wp/p44.html.

[2] Innovative Boiler Master Design Improves System Response: By M/s George Keller, PE, Burns and Roe, and Bryan Baker and Russell J. Jones, E.ON U.S. POWER: Official Publication of 0'Electric Power 02/15/2007 http://www.powermag.com/innovative-boiler-master-design-improves-system-response/.

[3] Boiler-Tuning Basics, Part II, 2009: Tim Leopold, ABB Inc POWER: Official Publication of Electric Power, http://www.powermag.com/boiler-tuning-basics-part-ii/.

[4] Product Bulletin of M/s Foster Wheeler Energy Corporation (Doc. 5076T) on Pulverizer Operation and Control.

[5] YS170 Boiler Control Overview: By M/s Yokogawa corporation of America.

[6] Procidia Control Solutions Boiler Control Overview: By M/s Siemens Energy and Automation, Inc. (AD353–132 Rev. 1) March 2006.

[7] Turbine—Bypass Systems for better Plant Performances: Product Catalog of M/s Sulzer Valves (a product of CCI).

[8] Turbine Bypass Valves and their Application: By Adrian Croft: Technical Bulletins of WEIR Valves and Controls. dt 01/09/2005.

[9] Product Catalog on Turbine Bypass Systems of M/s Copes Vulcan (SPX Flow Control) 04/2008.

[10] Product Catalog on Blakeborough Desuperheating Equipment and Systems of M/s Weir Power and Industrial.

[11] Thermal Power Plants—vol. II—Fossil Fuel Fired Boiler Air and Gas Path: Chaplin R.A. http://www.eolss.net/Sample-Chapters/C08/E3-10-02-05.pdf. Copyright: Encyclopedia of Live Support Systems.

[12] US Patent 4,738,226 Apr. 19, 1988:By M/s Masmchl Kashlwamkl; Toshlkl Motai; Hisao Haneda, Japan.

[13] CCI Severe Service Applications in Fossil Power Plants: By M/s CCI, United States.

[14] Technical Publication on Coal Pulverizer Design Upgrades to Meet the Demands of Low NO_x: By: M/s Qingsheng Lin of Riley Power Inc. Presented at: Electric Power 2004 March 30–April 1, 2004, Baltimore, MD.

[15] Jerry Gilman, Special Section: Flow/Level; Boiler Control Systems Engineering, second ed.; 2010 Publisher (Book) ISBN-13: 9781936007202: ISA, Publication date: 7/1/2010.

[16] Boiler-Tuning Basics, Part I: Tim Leopold March 1, 2009 POWER: Official Publication of Electric Power http://www.powermag.com/boiler-tuning-basics-part-i/.

[17] Control of Desuperheating Process: By JOELW.KUNKLER (Senior applications specialist), Fisher products Severe Service Group. Copyright 2006: Valve Manufacturer's Association http://www.documentation.emersonprocess.com/groups/public/documents/articles_articlesreprints/ag365652.pdf.

[18] Steam Boiler With Gas Mixing Apparatus, Patent US 4738226 A: Inventor: Masamichi Kashiwazaki, Japan.

[19] Controlling SO_2 Emissions: A Review of Technologies: Ravi K. Srivastava (National Risk Management Research Laboratory EPA/600/R-00/093 November 2000, http://nepis.epa.gov/Adobe/PDF/P1007IQM.pdf.

[20] Performance Model of the Fluidized Bed Copper Oxide Process for SO_2/NO_x Control: By H. Christopher Frey: Paper 93–79.01, Proceedings of the 86th Annual Meeting. June 13–18, Denver, CO Copyright: 1993H.C. Frey http://www4.ncsu.edu/~frey/conf_pr/Frey93.pdf.

[21] Typical Installation Timelines for NOx Emissions Control Technologies on Industrial Sources: By M/s Institute of Clean Air Companies (United States); December 4, 2006 http://c.ymcdn.com/sites/ www.icac.com/resource/resmgr/ICAC_NOx_Control_Installatio.pdf.

[22] Control Technology Review: By M/s Golder Associates; (Section 4 of 063-7567) December 16, 2006.

[23] SNOXTM Flue Gas Cleaning Demonstration Project: By 'A DOE Assessment' (DOE/NETL-2000/1125), June 2000 http://www.netl.doe.gov/File%20Library/Research/Coal/major%20demonstrations/cctdp/Round2/SNOX2.pdf.

[24] Chapter 3 Sulfur: By M/s Zevenhoven and Kilpinen 6.1.2004 http://users.abo.fi/rzevenho/sulfur_1.PDF.

[25] Various documents of NTPC, India.

[26] Various documents of BHEL, India.

[27] DCS Integration for Intelligent Sootblowing: By M/s Seth Whitworth, XCEL Energy, Sandeep Shah, P.E., Clyde Bergemann, Inc., Huiying Zhuang, Clyde Bergemann, Inc. Published By Clyde Bergemann, Inc. May 2006.

[28] Boiler Soot-Blowing in Power Plants: Hank Van Ormer, Air Power, United States, http://www.airbestpractices.com/industries/power/boiler-soot-blowing-power-plants.

[29] U.S Patent no 4607961 Aug. 26, 1986: By M/s John R Wynnyckyj, Edward Rhodes. Canada.

[30] Product Catalog on Heat Flux Measurements By M/s International Thermal Instrument Company, Inc.

[31] Intelligent Sootblower Scheduling for Improved Boiler Operation: X. Cheng, R.W. Kephart, J.J. William: Westing House Process Control Inc.

[32] Putting Combustion Optimization to Work: Nancy Spring Sr.32. Editor: Power Engineering, May 2009.

[33] B and W NO_x Reduction System and Equipment at Moss Landing power station: B. Becker (B and W), D. Tonn (B and W), N. Stephenson (Cormetech United States), K Speer (Moss Landing United States).

[34] Reduction of NO_x Through Combustion Optimization: J. Cahill: Emerson Process Experts.

[35] Ultra Low NO_x Integrated System for Coal Fired Power Plants: G.H. Richards, C.Q. Maney, J.L. Marion, R. Lewis, C. Smith: ALSTOM power Inc.

[36] An Optimized Supercritical Oxygen Fired Pulverized Coal Power Plant for CO_2 Capture: A.H. Seltzer, Z. Fann: Foster Wheeler.

[37] Nitrogen oxides emission control options for coal fired electric utility boilers: R.K. Srivastav, R.E. Hall, S. Khan, K Culligan of US Environmental Protection Agency and B.W. Lani, US Department of Energy: J. Air Waste Manag. Assoc. 2005 (vol. 55), 1367–1388.

[38] Reducing NO_x emissions in tangentially fired boiler—a new approach: A. Kokkinos, D. Wasyluk, M. Brower Babcock, and WilCox and J.J. Barna 2000 Duke Power: ASME International Joint Power Generation Conference July 2000.

[39] Windbox arrangement: BHEL Role in Cleaner Environment: Power Gen India 2012.

[40] Comparison of NO_x Emission Reduction With Exclusive SOFA and Combination of SOFA and CCOFA on Tangential -Fired boilers: S.L. Tongmo Xu, S. Hui, H. Tan, Q. Zhou, H. Hu, School of energy and power engineering PR China: Internet Document.

Chapter 9

Turbo Generator Control System

Chapter Outline

1 INTRODUCTION

Modern thermal power plants are equipped with state-of-the-art technology instrumentation systems. Continuous advancement in science and technologies, especially in the field of electronics, with microprocessor-based hardware and custom software sparked the development of instrumentation systems, including both measurement and control. Turbo generator control systems are no exception. In the initial stage, the turbine is mainly controlled by a hydraulic system. As technology advanced, measurement systems became fully electronic and controlled by a combination of electronic and hydraulic systems. Today, the instrumentation package is equipped with an electronic system with hydraulic controls as a backup.

Unlike the boiler package, the turbine package is typically equipped with modern hardware, software, and control strategies for its equipment and systems. Modern turbine control systems include closed-loop control systems (CLCS) and open-loop control systems (OLCS) including automatic turbine testing systems (ATT) and the automatic turbine run-up system (ATRS).

The turbine control system, which includes both CLCS and OLCS, comprises five main subsystems or functional

Power Plant Instrumentation and Control Handbook. https://doi.org/10.1016/B978-0-12-819504-8.00009-3

groups categorized according to their functional requirement, and takes care of the entire operation of the turbine and generator. Those subsystems include the following:

1. Turbine high pressure (HP) control fluid subsystem for the low pressure bypass (LPBP) system.
2. Turbine HP control fluid subsystem for main valves.
3. Both systems (1 and 2) combined.
4. Turbine control oil subsystem
5. Condenser evacuation and gland seal subsystem
6. Turbine control subsystem (warmup, startup, loading, unloading, and shutdown)

In some plants, the generator exciter subsystem is also included for better coordination.

The subsystem controls have their own group control and various other subgroups, each containing similar or redundant drives. There are sufficient displays at every step and an intelligent alarm annunciation system to enable immediate action for detection as well as rectification of disturbing element(s). During the occurrence of fault conditions, if contributed by a part of the control system hardware/software, the subsystem control takes over so that the remaining part of the main plant can continue uninterrupted operation.

1.1 Subsystems or Functional Subgroups

As indicated above, the subgroups are functionally divided to achieve better control over the entire turbine system. The main items that particular subgroups control are discussed in the following clauses.

1.1.1 Turbine HP Control Fluid Subsystem

Whether combined or separate, subsystems have different standby liquid control systems (SLCs), drives for LPBP, and main valve systems to serve their assigned purposes, which are to supply HP power/control oil. This is done by using control oil pumps, recirculation pumps, cooling fans, pumps for regeneration circuits, etc.

1.1.2 Turbine Oil Subsystem

This subsystem takes care of the oil systems required for initial running of the turbine, and shutdown procedure, and includes SLCs and drives of auxiliary and emergency lube oil pumps, jacking oil pumps, turning gear gate valves, etc.

1.1.3 Condenser Evacuation and Gland Seal Subsystem

This subsystem includes SLCs for condensate extraction pumps and turbine drains as well, (although it is a part of the turbine control subgroup), and drives for vacuum pumps, air extraction valves, instrument/service air solenoid valves, vacuum breaker solenoid valves, etc.

1.1.4 Turbine Control Subsystem (Startup, Loading, Unloading, and Shutdown)

The heart of the turbo generator control system is the turbine control subsystem (startup, loading, unloading, and shutdown) comprised of SLCs for drains, warmup controller for generating the criteria for operation of drain valves, soaking of turbine metals, and holding/speeding of the turbine rotor. Also included are the startup device (TSL), speed and load setpoint devices of the turbine governing system, and the auto synchronizer.

1.2 Testing Process by ATT

In addition to the above subsystems, there are standalone ATTs that enable online testing of important drives or devices required for safe operation, and emergency tripping of the turbo generator. In other words, all the protective equipment required for turbo generator safety is tested automatically with uninterrupted normal operation, without jeopardizing the individual protection assignment bestowed on each device while the testing procedure is in progress.

The main function of the ATT system is to offer a failsafe operation of the protective drives or devices by checking and analyzing their readiness throughout the turbine operation regime. The system ensures that the abovementioned duty is done by performing the tests cyclically in the automatic mode of operation and manually at the operator's discretion. The essence of this particular standalone facility is to test all the protective devices sequentially in a well organized and systematic manner for verification of foolproof functioning of all the devices. The system is built to eliminate operator error, resulting from manual intervention, using the fully automatic test sequence. Individual tests for a particular device can be done by proper selection.

The standalone system for ATT includes the following items:

1. Turbine generator (TG) set over speed trip (mechanical).
2. Thrust-bearing trip (mechanical).
3. Remote trip devices.
4. Hydraulic and electrical low vacuum trip.
5. Opening and closing operation of control valves.
6. Opening and closing operation of emergency stop valves (ESVs).

Testing can be done for all of the items at any time, provided a protection device is substituted only after testing its readiness.

Testing can also be done manually at the operator's discretion or cyclically through an automatic program. For testing the turbine over speed trip, high-pressure oil is introduced via a shaft hole producing a simulated condition of over speed and actuation of trip bolts at TG normal speed, causing the hydraulic circuit oil to drain. The resultant fall in pressure is detected by the tester and sends a signal that

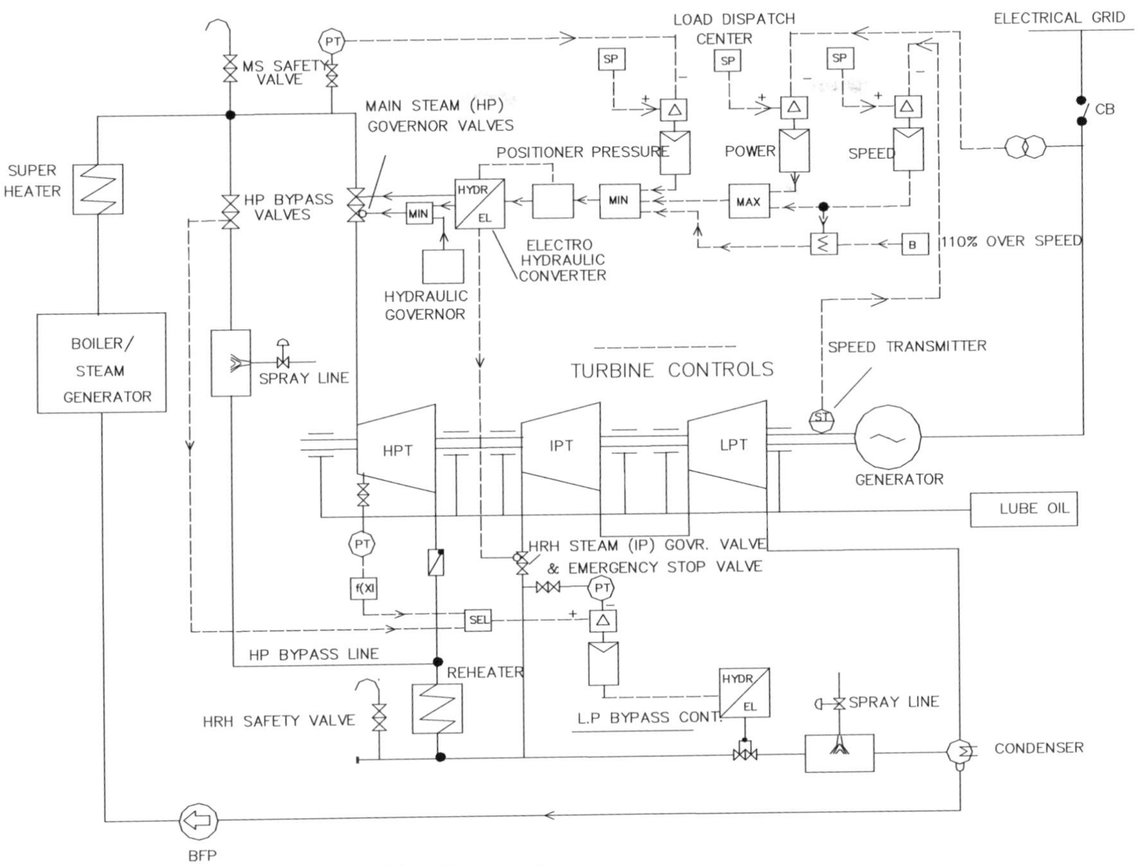

FIG. 9.1 Simplified flow diagram of steam cycle with turbine controls.

indicates a successful testing operation. The oil pressure, although low, will not be allowed to close the main steam control valves (MSCVs) and ESVs because they will be supplied with high-pressure oil by the test valve. Detailed discussions are included in Clause 5.

2 ELECTRO HYDRAULIC GOVERNOR CONTROL SYSTEM

2.1 General Introduction

The turbine control subsystem or functional group as mentioned earlier is a crucial part of automation of the TG set, which comprises the electro hydraulic governor (EHG) control system. This takes care of startup, shutdown, loading, and unloading of the TG set. With technological advancement, this control system transformed from a purely hydraulic system into an electro hydraulic system. However, the hydraulic governing system is not totally obsolete; it is used as a backup as shown in Figs. 9.1 and 9.4.

The advantage of an electro hydraulic system is that it uses electronic hardware and software for the measurements, processing, or signal conditioning and implementation of critical and complicated control strategies, which was not possible with a pure hydraulic system. The added values of an electronic system are the tremendous speed of dynamic response, flexibility for further modification at any stage, and interchanges of signals with other subsystems or functional groups without additional hardware. The advantage of the hydraulic system is undoubtedly the fast and vast actuating force required by the increasing size of the final control elements. The EHG control system as a whole provides smooth, reliable, and speedy functioning as required by the modern, large-capacity thermal power plants.

The EHG operates hand in hand with the ATRS and turbine stress evaluator (TSE), details of which are described in Clause 6. The TSE guides the EHG control system to generate the speed and megawatt power setpoint within the safe operating margin, and ultimately the position demand of the turbine governor valves, which are the hydraulically operated final control elements, through which speed or power can be maintained as the situation demands. The TSE collects various temperature measurements related to turbine parts and valves to calculate the safe margin for speed and generator load while considering allowable turbine metals stresses.

2.2 Task of the EHG

With the continuous development of TG control strategies, the following tasks are accomplished by the EHG:

1. Admission of steam and starting of turbine, raising speed gradually up to synchronizing speed, and then

maintaining the rated frequency of the generated power load in accordance with the droop (load-frequency) characteristics and associated provision for the proper adjustment of the same unit running on load.

2. Addressing the problem of low-frequency transients or speed excursion as well as permanent deviation taking place throughout the period of TG operation.
3. Generator load control under the guidance of TSE
4. Pressure control when the boiler is unsound but the turbine is in good condition, (initial pressure control in turbine follow mode).
5. Graceful switchover from load control to frequency control in case of an isolated grid condition.
6. Smooth and reliable control in case of a sudden disconnection of TG set from the grid.

2.3 Duty Assigned to the Electronic Part of EHG

As discussed previously, the duty assigned to the electronic part of the EHG is twofold:

1. Collection of measurement data, processing or conditioning of raw data, arranging of those data for transportation to other subsystems or functional groups as and when required, and sending them as output to the front desk indication/recording/data acquisition system.
2. Using the processed or conditioned data in various automatic control loops pertaining to EHG, which are TG set speed controller, generator power/load controller, steam pressure controller, and the valve position controller.

2.4 EHG Control Philosophy

As stated above, there are three basic control loops: speed control, power/load control, and pressure control. Another control is the valve position control, which actually receives the selected output out of the three controllers as its input (Fig. 9.1).

There are various modes of operation for governing control systems and these vary by manufacturer. Modes from one of the reputed manufacturers of EHG control systems are elaborated in Table 9.1.

2.4.1 Speed Control Loop

Fig. 9.2 is a ready reference for the speed control loop. The speed setpoint may be manually adjusted or may be an automatic signal generated elsewhere by some suitable method (e.g., "synchronizer"). It is passed through a tracking integrator whose output rises with a gradient and is finally equal to the input. The speed gradient is manually set and passes through a minimum selector whose other input is from the TSE margin, which is actually modified at every instant, depending on the condition of the turbine metal temperature.

TABLE 9.1 Operational Modes of EHG

SL	Operational Mode	Details
1	Startup mode	EHG operates by switching ON the supply and setting speed/load setpoint to zero, with hydraulic governor control in the upper end (as hydraulic controller and EHG are selected via MIN selector switch). After turbine is put in barring gear the actual speed is sensed, and the speed controller is in service until synchronization
2	Load control	After synchronization it is under load control either from load control center or various other means. The frequency change is selected by integral action of the load controller as per frequency droop[a] characteristic. Speed controller is in standby mode for load shedding
3	Load shedding	It is the sudden separation of the generator from the grid. At this point, the output of the load controller is immediately brought below that of the speed controller. Because of the max selector as shown in Fig. 9.1, the speed controller takes over the control and brings the speed to the reference setpoint. (See HP and LP bypass control also.)
4	Shutdown	At normal shutdown, the load controller is set to zero so that the speed controller can assume control and bring the system to safe condition

[a]*Frequency droop characteristic: The mismatch in power causes speed to change to bring the mismatch to zero, but there will changes in speed. The change in frequency due to the change in power is decided by regulation or droop. If 4% change in frequency/speed causes 100% change in power, then the droop characteristic is 4%, i.e., 0.04 per unit power. At 100% load frequency is 100%. If there is a drop in 50% power, then there will be increase in 2% frequency. Normally 4%–5% is the standard droop value (Fig. 9.2). There is another parameter called dead band (governor control), which is very important in governing the system. There is a minimum value of speed that cannot be picked by the sensing mechanism, so it may remain uncorrected. This minimum value is called the governor insensitivity or dead band. The governing system action depends on speed sensing. With change of time for the same turbine due to wear and tear, this value may change. Normally, dead band is expected to be 0.06% of the rated speed.*

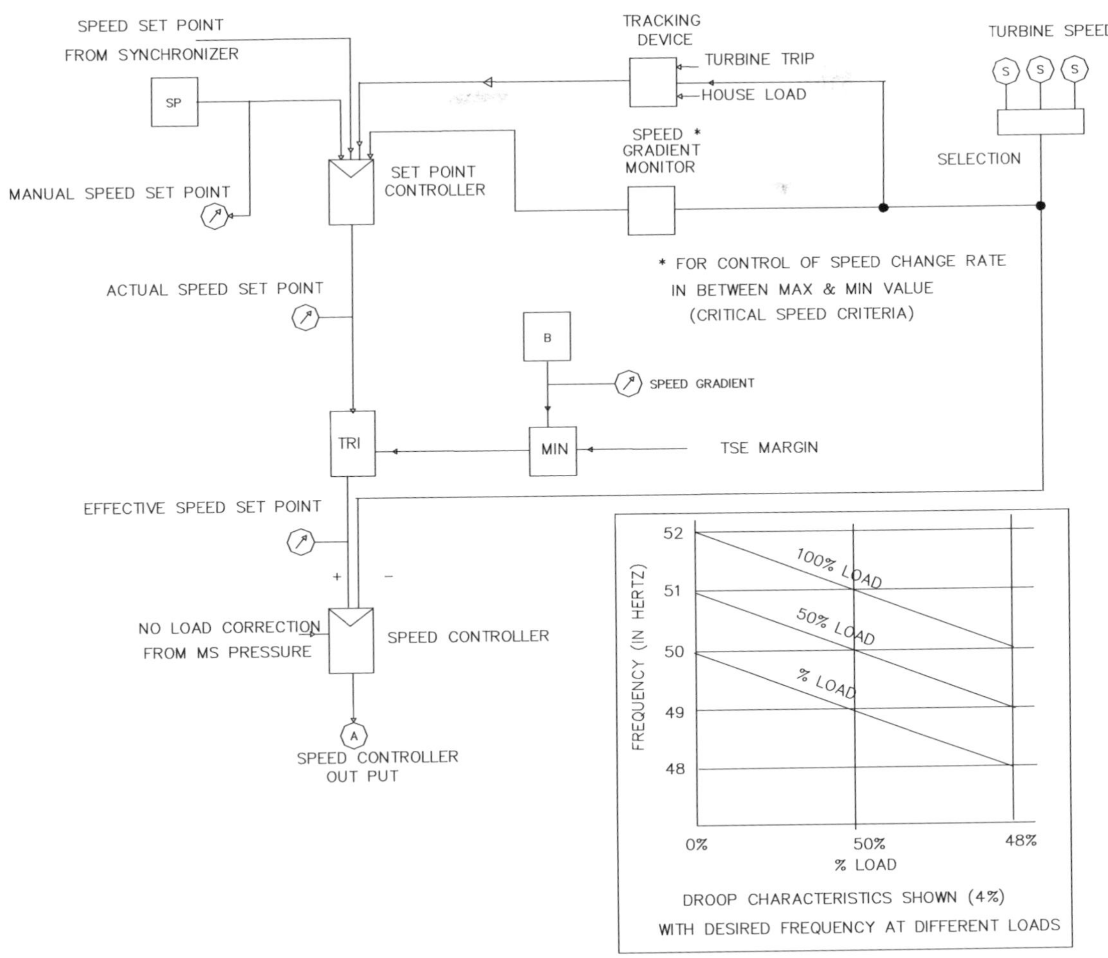

FIG. 9.2 Turbine speed controller block diagram.

This time-dependent speed set value at the time of startup is, therefore, the optimum speed by which the turbine can speed up. The automatic speed setter takes the TG set from barring speed to first-hold speed (~20% rated speed) for the soaking process for a specified time and temperature margin. Then, the speed setter raises the speed close to rated speed, and when the metal temperatures permit and generator voltage is matched through the automatic voltage regulator (AVR) with the grid voltage, the TG set is synchronized. The time-dependent set value generator, however, also takes care of the time when the TG set passes through critical speed (~2800 rpm). At speeds more than 2850 rpm, if the "difference between speed reference value and time-dependent set value exceeds a specified limit" and any of conditions take place, for instance, when TSE is not available or during startup, or reference speed is <50 rpm than actual speed during transition period from electro hydraulic controls (EHC) to hydraulic controller. If those or other conditions occur, the integrating action for the time-delayed setpoint would be blocked.

This optimum speed set value is then taken as the input of the speed controller whose measured value is the selected speed transmitter signal of an appropriate voting circuit. The integrator is ON only when the TSE has a healthy signal. If the TSE issues a warning signal, the integrator is disabled and can be enabled again after switching the TSE OFF. Then the speed setpoint will change as per the manual setter of the speed ramp value.

A speed gradient monitor is introduced to take input from the actual speed and calculate the real-time speed gradient and feedback to the setpoint generator. This does not allow the output to be less than the minimum set speed gradient. The speed gradient of the speed setpoint value passes between maximum and minimum speed gradient to avoid the critical speed criteria.

A tracking device is also introduced for follow-up action to take input from the actual speed. When the turbine trips, the speed setpoint follows the actual speed (minus ~60 rpm), and when the load controller is in action, it follows the actual speed plus 10% load proportionate value (within a band of 60–120 rpm). During follow-up action, the reference setpoint cannot be raised if the rate of change of speed is less than a preset limiting value. The speed controller action is P + D, that is, the output responds immediately with a derivative action after an error is sensed and then comes back to a value proportional to the error at a predetermined slope. At a steady-state condition, the speed controller output signal will always be proportional to the error value (that is, the difference between the set value and the measured variable). This arrangement is made because it is the best for the load sharing of various TG sets connected in a common grid. During the initial running of the TG set, the starting time constant of the TG set as a whole is the most important factor. If the TG set control is taken over by the power/load controller, the speed controller will still be operating in parallel as a standby and will activate when required.

The speed controller, with the help of other inputs, is used to take command of the TG set at the following times:

1. Startup of the TG set.
2. Synchronizing the generator with the grid.
3. Operation of TG set at a minimum load if required, for example, house load, huge load shedding, etc.
4. Operation of TG set throughout the 0%–100% range of load if required and dictated by extraordinary situations.
5. Taking charge of the valve lift control during over speeding when the value exceeds the limit
6. Smooth shutdown of the TG set.

For synchronizing the generator, the separate and standalone synchronizer provides the appropriate command to the speed controller input so that the TG set can attain the synchronizing speed at the proper time and with the proper speed as required by the machine (generator).

The TG set may be needed to operate at the minimum load delivered by the generator. This may happen during large load shedding, house load operation, or low load running, which is also called the idling mode of operation. Particularly in the case of load shedding, the speed of the TG set will tend to rise immediately above the synchronizing speed, but the speed controller will automatically take over the control and corrected output will bring back the TG set speed very close to the synchronizing speed. The speed controller temporarily takes over control of the TG set whenever rapid and/or a large change in speed is experienced.

2.4.2 Turbine Load Control Loop

Fig. 9.3 shows how the turbine load is controlled manually. The load or generator power setpoint may be manually adjusted or may be an automatic signal generated from the Load Dispatch Centre (LDC) or from the Coordinated Control System (CCS). The selected signal is passed through a high/low selector so that the demand does not go beyond the operating capability of a particular individual TG set. The LDC/CCS signal can be turned ON or OFF by the operator. IF the LDC signal is unhealthy, this component would automatically be switched OFF. In that event, the manual load setpoint would take control of the TG set. While the LDC/CCS signal is in command, manual load setpoint would act in standby mode continuously matching with the LDC/CCS signal through a tracking circuit so that bumpless transfer is achieved. The selected output is then passed through a tracking integrator whose output rises with a gradient to finally equal the input. Like the speed control loop, the load gradient can be manually set and passes through a minimum selector whose other input is from the TSE margin, which is actually modified at every instant depending on the condition of the turbine metal temperature.

Another component is added to this time-dependent integrated signal, which is a proportional fraction of the selected load setpoint. The total sum thus obtained is the P + I response, which behaves like a proportional output when the change is small and P + I when the change is significantly high. The proportional component of this load set value may be adjustable when the range is between 5% and 25%. It may be switched OFF when it is required to make the time-dependent load set value a purely ramp one. Here the integrator is only ON when the TSE has a healthy signal. If something goes wrong with the TSE, the integrator will be disabled and can be enabled again after switching the TSE OFF. If that happens, the load setpoint changes as per the manual setter of the load ramp value. The integrator would also be OFF when the load reference signal is raised and the pressure controller switches to limit pressure mode.

The final integrated setpoint signal continuously tracks the actual power/load until the main generator circuit breaker (GCB) is OFF, that is, before synchronization at the time the speed controller is in command and ensures bumpless and smooth transfer from speed controller to load controller during startup after the GCB is ON.

With this time-dependent load set value, another signal is now added that represents the frequency-related (droop-characterized) influence. This is a correction required on the load set value, as per the frequency–generator load/power delivered curve, that is typical for each generator. The total summation now is the optimum power/load demand, which takes care of the grid stability; the same signal is sent to the boiler control loop. The frequency influence proportional component is typically adjusted between 2% and 8% in steps of 0%–0.5%.The frequency influence component is likely to be automatically disconnected if any fault condition arises, or it can be switched OFF through manual intervention. However, with the introduction of the free governor mode of operation, all units in a grid that are run under governor control, without attaching any load restrictions, etc., to maintain grid frequency within a narrow band and droop characteristics, are restricted to within 3%–5%. In the approved grid code (in some countries), it is mandatory for each generator to have the

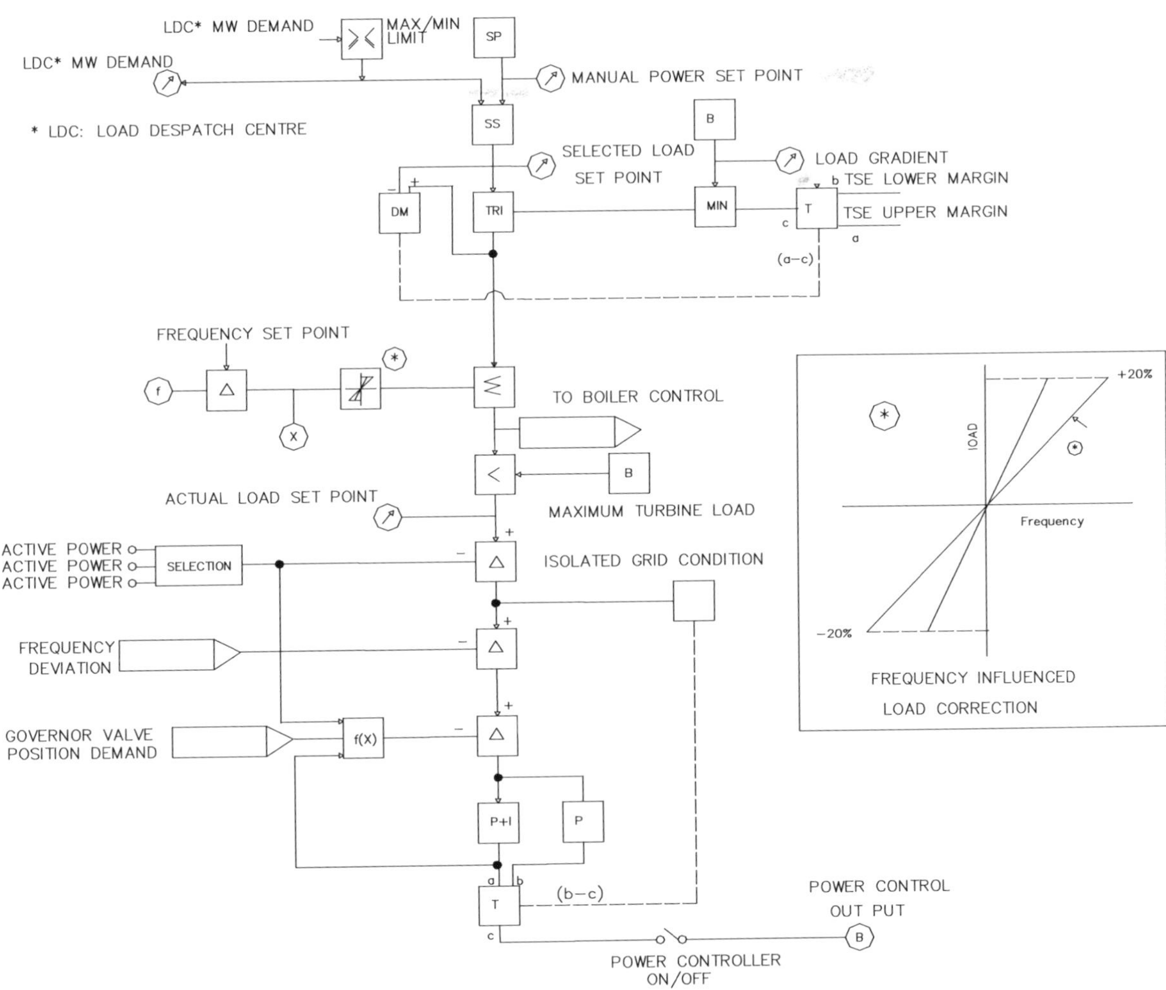

FIG. 9.3 Turbine load controller block diagram.

capability of 105% maximum continuous rating (MCR) capacity generation whenever the situation demands.

This power/load demand is allowed to be the set value for the controller if its value is less than the maximum permissible load. In other words, the above setpoint cannot be more than the manual TG set capability.

To avoid any problems in operation while in the maximum permissible load setting, and during TG operation with the load controller in action, a rate adjuster of maximum load limiter settings is provided for conservative operation of the TG set to avoid stressing the turbine thermally and mechanically. The maximum load limiter setting can be motor operated to handle an emergency shutdown of the turbine.

The load/power controller gets its measured value from the selected active power transmitter signal of an appropriate voting circuit. The potential transformers (PTs) of the generator output voltage, and the transformers of the generator output current are connected to the power transmitter measuring circuit.

The power/load controller action is P + I, that is, the output responds immediately with a proportional action for small changes integral for large changes, and then comes back to a final value at a predetermined slope. At a steady-state condition, the load/power controller output signal will always be the value to maintain the zero error between the set value and the measured variable value, which will allow the exact maintenance of the required power/load. The starting time constant of the TG set as a whole, is the most important factor to consider during the initial running period of the TG set for proper speed controller tuning, whereas the transfer function of the generator and power grid is the most important criterion for proper load controller tuning. In case the TG set control is taken over by the power/load controller, the speed controller still operates in parallel as on standby through suitable tracking circuits and activates when required. A gradient monitoring device/software is provided to block the power/load controller output in case the real-time gradient exceeds the preset maximum load gradient. This is done to arrest the tendency of the controller failure in the direction of 100% load.

When the GC is ON or OFF and load shedding has taken place, the speed controller takes the charge and load

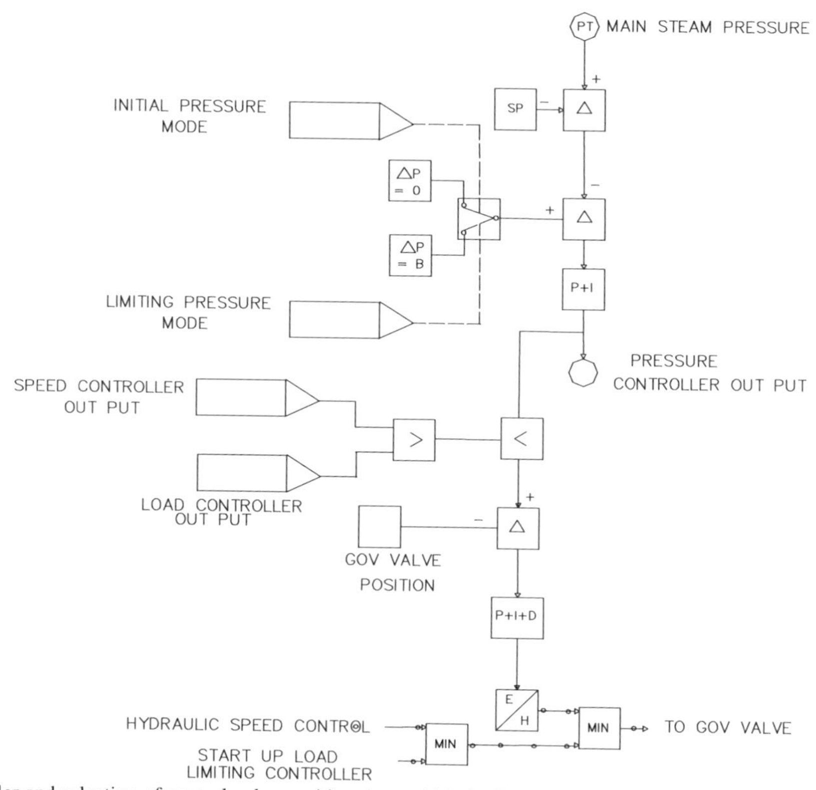

FIG. 9.4 Pressure controller and selection of control valve position demand block diagram.

controller output tracks the speed signal in the followup mode. If there is an isolated power grid condition when the load control is in action, the controller will automatically switch over to a P action output only with ~5% droop characteristics. The selected load controller output then constitutes an input signal rather than a setpoint of the valve lift controller through a series of selection procedures.

2.4.3 *Turbine Pressure Control Loop*

In this control loop configuration, both the measured variable and the set values are supplied by the boiler manufacturer. The controller output is set for P + I action, and operates the main steam valve lift controller via the same series of selection criteria (Fig. 9.4).

The turbine inlet steam pressure control loop is mainly operated with any one of the three operational modes selected from the control desk by the operator, which include the following:

1. Initial pressure control.
2. Limit pressure control.
3. Sliding or variable pressure control.

2.4.3.1 Initial Pressure Control

In this mode of control, the controller action is P + I, and the actual main steam pressure at turbine inlet is the measured variable, and the main steam pressure setpoint is issued from the CCS (furnished by the boiler manufacturer). Here the power/load delivered by the TG set is dictated by the boiler up to a certain load level. In other words, this is called a "turbine follow" mode, where the turbine is following the boiler running with some limitation beyond which it cannot deliver the load.

2.4.3.2 Limit Pressure Control

This mode of operation is envisaged when a condition arises where there is pressure deviation between the actual main steam pressure at the turbine inlet and the main steam pressure setpoint exceeds a predetermined value of ~10 kg/cm^2. With the help of this mode, the power/load controller takes care of the fast and small demands in load utilizing the boiler storage capacity. Here, the boiler is strong enough to deliver any power/load within its full capacity; this operational mode, though well regarded, is rare, because a healthy boiler with an appropriate automatic pressure control system, is typically able to maintain the main steam pressure within the desired low excursion.

2.4.3.3 Sliding or Variable Pressure Control

In this operational mode, the actual power/load signal, after necessary voting, is biased in such a way that the power/load controller output is modified to demand maximum MSCV opening. When the MSCV is fully opened, the turbine has no control on the power/load delivered by the TG set; rather the boiler determines the full power/load. Here the throttling losses across the MSCV are reduced to zero, thereby improving the overall efficiency, and the turbine control becomes less stringent because temperature variations occur less rapidly.

2.5 Selection of Controller Outputs

The response of the electronic turbine speed controller, turbine load controller, and turbine pressure controller is fast compared to the hydraulically operated MSCV. An interface called a valve lift controller or position controller has been introduced in between the selection unit of the final control element of these three controllers to provide the necessary thrust required by the actuator of the MSCV.

All three controllers running simultaneously are on standby mode, until one is selected for command. The controllers on standby mode continuously track their output with the output of the controller in command to provide a bumpless and procedureless transfer during changeover of the controllers.

There is a maximum selector provided between the turbine speed controller and the turbine load controller. After the output is chosen, it is taken to a minimum selector along with the pressure controller output. The final selected output of the controllers is then brought to the electro hydraulic converter for further selection with other hydraulic signals.

2.6 Selection of Hydraulic Signals

The output from the electro hydraulic converter is the normal commanding signal, but a purely hydraulic speed controller is provided as a backup. A separate hydraulic controller as a startup device is also provided to initiate the opening signal for the main steam stop valve (MSSV). The hydraulic signals of the hydraulic speed controller and startup devices pass through a minimum selector, and again, the minimum selected signal and the electro hydraulic converter signals are chosen to operate the MSCV. The startup device of the hydraulic controller output is kept high when MSSVs are opened. Hydraulic signals of the hydraulic speed controller output is also kept high so that hydraulic signals of the EHC take the command.

The electro hydraulic converter and hydraulic amplifier are connected to a set of followup piston valves to which trip oil is supplied through the restriction orifice and it flows in the secondary oil piping to control valves. The secondary oil pressure depends on the position of sleeves of the follow-up piston valves, which determines the amount of trip oil drainage.

2.7 Functional Description of Electro Hydraulic Converter

The electro hydraulic converter is not only a signal converter that has an electrical signal input and a hydraulic signal output, it is the interface between the selected controller's output and the MSCV actuator directly with suitable power-handling capacity. See Fig. 9.5 for the functional description.

The electro hydraulic converter constitutes items such as an electromagnetic plunger coil (item no. 1; getting signal from the selected controllers output), control sleeve (item no. 2; moving up and down as per control signal), pilot valve (item no. 3), spring (item no. 4), power piston (item no. 5), and position transmitter (item no. 12).

Fig. 9.5 (left side) is applicable for comparatively smaller units up to ~210 MW. The plunger coil's current value determines the vertical position of the control sleeve, which in turn dictates the drain cross-sectional area of control oil. The change in drain area determines the control oil pressure acting on the pilot valve, which will move upward or downward allowing power oil to enter one chamber of the power piston, while the other chamber of the piston opens the drain passage. The power piston continues to move as long as the pilot valve stops the action (discussed above). The balanced condition is restored by the position feedback system (item no. 7), which pushes or pulls the pilot valve along with the spring in the opposite direction so that the pilot valve blocks both the entry and drain passage to the power piston that has already moved to its new position. In the steady-state condition, the force balance is achieved through the force acting on the pilot valve (determined by the control oil pressure on the pilot valve multiplied by the actual effective area of the pilot valve) and the force offered by the compression spring.

The control oil pressure is a variable one depending upon two components; one is the output of the hydraulic speed controller entering the inlet (item no. 8). The other is the drain cross-sectional area between the spool of the pilot valve and position of the control sleeve as allowed by the plunger coil energized with the selected electronic controller.

It is, therefore, evident that the electro hydraulic converter plays such an important role because it provides the interface between the hydraulic and EHC at the inlet stage of the device with a built-in minimum selector. The TG set automatic control was developed with the idea that the leading control action is accomplished by the EHC, and the backup support is provided by the hydraulic speed controller.

Fig. 9.5 (right side) is applicable for larger units >210 MW. Here the "trip oil" is used to move the pilot valve spool at the bottom in place of "secondary oil," which represents the hydraulic governor output. Like the system discussed

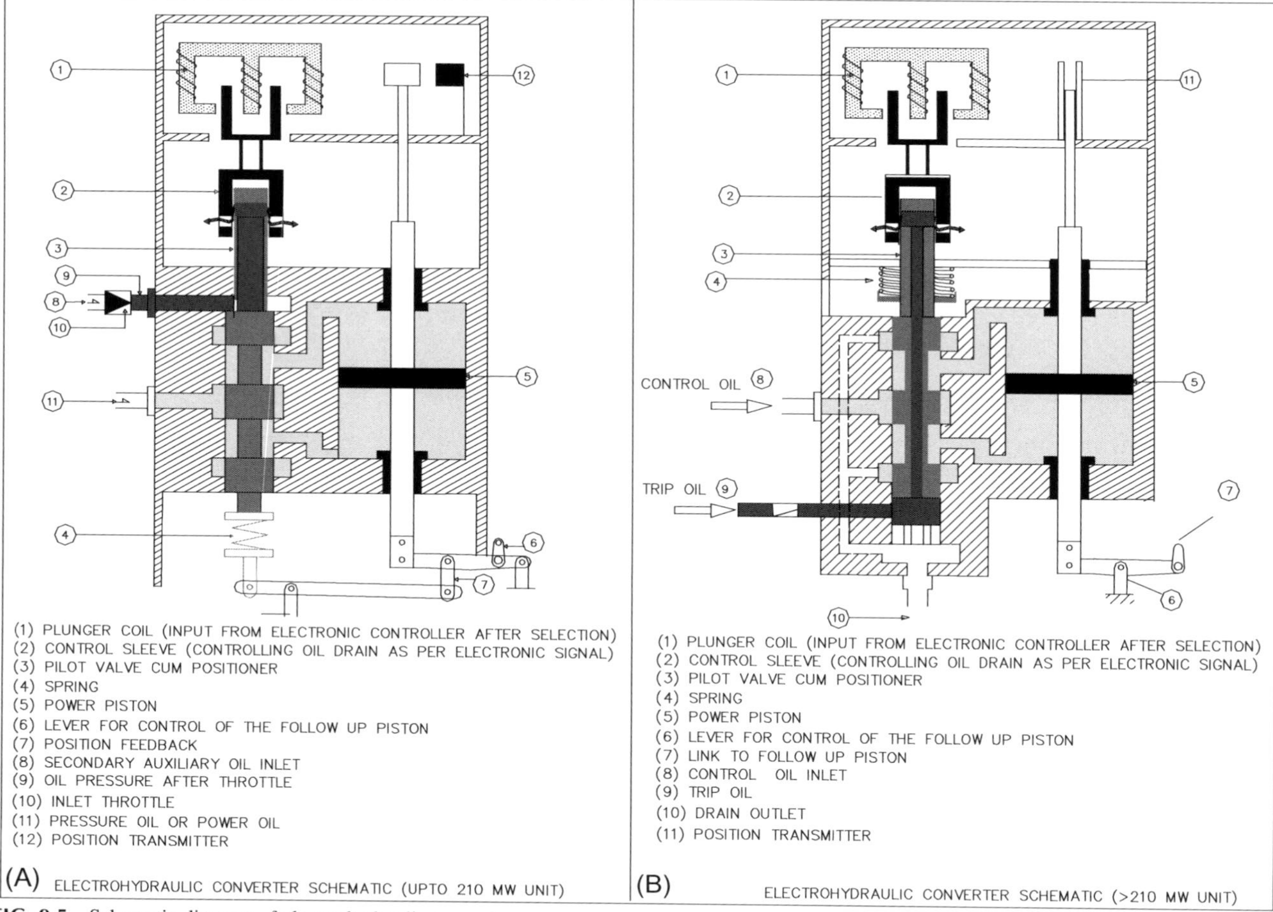

FIG. 9.5 Schematic diagram of electro hydraulic converter.

above, the pilot valve spool is at the central position under steady-state condition. In the deflected position, the control oil is admitted above or below the power piston, the motion is transmitted by a lever and cam shaft to sleeves of followup pistons to change the secondary oil pressure.

Another significant modification is the introduction of the electronic "admission controller" (PID), also referred to as position controller for proper positioning of the pilot valve spool. The EHC output serves as a setpoint, whereas the actual position of the power piston taken from two differential transformers acts as a measured variable, and the controller output operates the plunger coil and regulates the oil drains of the HP/intermediate-pressure check valves (IPCVs). The current to the plunger coil is increased to close the HP/IPCVs. The reference signal thus works in a reverse manner.

The drain port opening of the followup piston depends on the pilot spool valve of the EHC output and determines the related secondary oil pressure. The drain port openings of followup pistons of hydraulic governors depend on auxiliary secondary oil pressure and upstream auxiliary followup pistons (due to action/position of speeder gear), to determine its related secondary oil pressure. The followup piston valves constitute a minimum value gate for the output of both governors. This means the governor with the lower reference setpoint is effectively in control. This is also called hydraulic minimum selection of governors (applicable for bigger systems and that of the electro hydraulic converter).

2.7.1 *Starting of Control by EHC*

First, the electronic controller output signal is kept low, which is a high current signal (as per design), to go to the electro hydraulic converter plunger coil. This enables the control sleeve to move upward and, as per the strength of the signal, it will rest at any intermediate position. Further lowering of the electro hydraulic converter at its minimum value means the maximum current to the plunger coil, and by this the control sleeve, achieves the top position, and the drain cross-sectional area between control sleeve and the pilot valve becomes maximum. At this position, the oil pressure sent by the hydraulic speed controller is disturbed because of the wide drainage area. The hydraulic speed controller oil pressure is raised to a sufficiently high value so that EHC output, if it starts increasing or decreasing the current value, above the drain area will also decrease; opening of the MSCV will commence. The entire range of EHC output will then control the entire range of opening/closing of the MSCV.

2.7.2 Starting of Control by Hydraulic Controller

To achieve control by the hydraulic controller, EHC output is kept at maximum value, which means no current to the plunger coil. The control sleeves slip to the lowest position allowing no drain passage, and the oil pressure from the hydraulic controller acting on the pilot valve becomes maximum. With EHC output at maximum value, the hydraulic controller output may be varied to control the entire range of opening/closing of the MSCV. In this mode of operation, with total closing of the control oil drain, the movement of the pilot valve with a power piston is the function of the hydraulic controller output oil pressure.

2.8 Operational Mode of the Electro Hydraulic Converter

2.8.1 Start-Up and Synchronization

As described in Clause 2.6, the startup device of hydraulic controller output is kept high when MSSVs are opened, and hydraulic signal of the hydraulic speed controller output is also kept high at that time. The TG set is already rolling by the turning gear at a lower speed and sensed by the speed sensor. From this point, the speed setpoint is raised manually by using the electronic speed controller. By increasing the speed of the TG set, as per TSE guidance, and minimum and maximum speed gradient to take care of the critical speed or kink speed criteria of the whole set, the TG set is raised to the rated speed before synchronizing with the external power grid. The synchronization is done with the help of a separate synchronizing device dedicated to this purpose

2.8.2 On Load and House Load Operation

Immediately after synchronization of the TG set, the power/load controller takes the command of the TG set by increasing its output manually or automatically from the power grid controller or LDC. After the power/load controller is in control, the speed controller output signal is made to match with the power/load controller output as a standby controller. This standby mode is continued from full load operation up to house/station load operation. When there is a sudden operation of a load-shedding relay, the loaded TG set is abruptly separated from the power grid. The power/load controller output is reduced to a minimum because there is no demand from the grid, and the speed controller subsequently takes charge of the TG set to provide house/station load through the minimum selector. During the house/station load operation of the TG set, the speed controller almost maintains the rated speed because it is meant for maintaining the rated speed at house/station load during the startup of the TG set operation under load up to the beginning of full load operation.

2.8.3 Shut-Down Operation

In this mode of operation, the shutdown procedure of the TG set calls for a power/load controller output set to zero. From the minimum selector, the speed controller is in command and the power/load controller is switched OFF; in other words, it is disconnected from the grid. The speed set value then may be reduced by decreasing the house or station load gradually to zero. Finally, the speed of the TG set is gradually reduced to zero for initiating plant shutdown.

3 TURBINE PROTECTION SYSTEM

3.1 General Introduction

There are various conditions that may take place any moment during the operation of a thermal power plant that demand an immediate and emergency shutdown or "tripping" of the turbine. A few examples include low-bearing oil pressure causing serious malfunction and wear and tear of the turbine bearings; a very low/inadequate condenser vacuum may cause overheating of turbine blades located near the last row; excessive loss of metal or wearing of the thrust-bearing results in axial movement of the turbine causing misalignment of the rotating blades, which may cause catastrophic/excessive damage to the turbine internals; a loss of electrical load causing a dangerous over speed condition; or other critical conditions may occur necessitating emergency shutdown or tripping of the turbine instantaneously to protect it from an extremely dangerous operating condition or serious damage.

Turbine tripping actually means total and immediate shut off (typical time is 0.3 sec for a large turbine) of all steam valves responsible for steam admission to the turbine internal components, and a failure or delay in executing that action in the event any of the catastrophic conditions may cause extensive damage to various components of the plant. Fixing this damage may either be expensive or need prolonged repair time or both meaning high cost. Thus it is necessary that the protection system must be capable of responding at the fastest rate possible to every catastrophic condition.

Turbine protection, or to be more accurate, turbo generator protection, means that the turbine should always be given safe passage in the event of any untoward situation at any time during startup or shutdown and loading or

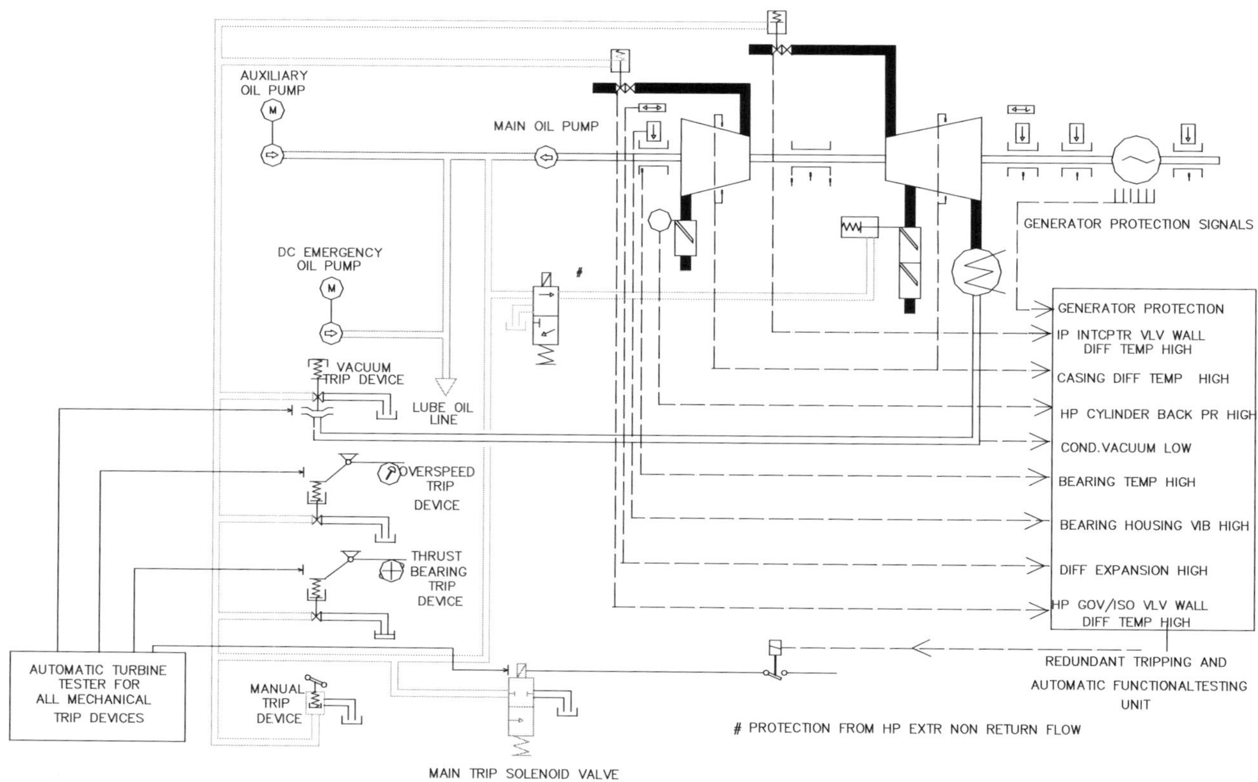

FIG. 9.6 Schematic diagram of TG protection system.

unloading conditions. The actions include tripping and post-tripping measures and limited occurrence of associated systems' failure.

The protective devices are designed and selected in such a way that whenever the turbine protection system senses a dangerous condition, it automatically reacts to stop the danger condition within the fastest time possible.

The mechanical and electronic trip mechanisms are the last line of defense for protecting the steam turbine. The devices deployed for protection implementation must be absolutely reliable and foolproof so that the duty imposed on them can be carried out in time of need only eliminating any possibility of spurious operation. One way to check their readiness is by using the ATT as described in Clause 5. The operating description of the protective devices is also described in Clause 5 as they are related to the ATT. It is immensely important to avoid spurious tripping and that the measurement of the ATT values are accurate along with the system processing them. If there is emergency tripping, the redundancy with proper design and number must be considered. Fig. 9.6 is an example of a turbo generator protection system.

Some protective devices are categorized as "passive protective devices." They receive signals from sensors of a "not so critical nature" and initiate prevention of further operation from startup to graceful shutdown. For example, if any of the start-permissive signals of a particular step goes wrong, the operation will not be allowed to proceed further. These are taken care of in the startup/shutdown procedure logic.

The protective devices, which receive signals from sensors responsible for turbine tripping signals, are popularly called "active protective devices." In case of an abnormal situation the active protective device, animated from the tripping signals, may initiate either opening/closing or starting/stopping of some areas to arrest the unwanted deviation, or may trip the whole turbo generator unit depending on the severity of the situation.

3.2 Turbine Tripping Input Signals

These following signals (typical only) will trip the whole turbo generator unit through active protective devices (Fig. 9.7):

1. TG set over speed (normally in duplicate for a hydraulic type; in case of electrical systems it is through a complex voting circuit, as discussed in Chapter 5).

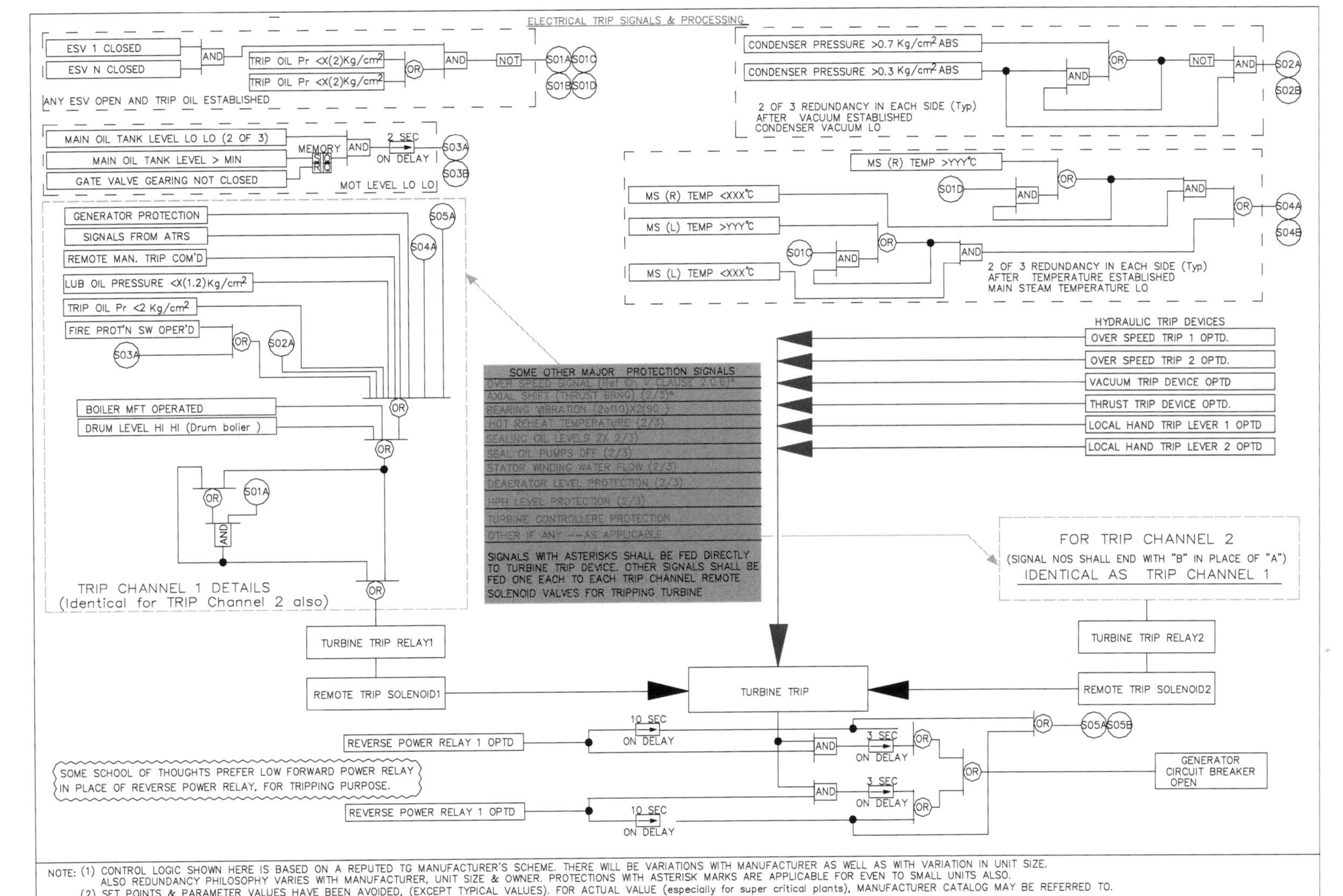

FIG. 9.7 Typical steam turbine protection scheme.

2. Thrust-bearing limit position (both hydraulic and electrical).
3. Very low condenser vacuum (both hydraulic and electrical).
4. Very low lube oil pressure.
5. Trip command from ATRS (during shutdown operation).
6. Very low main oil tank level.
7. Boiler tripped.
8. High bearing temperature.
9. High top and bottom casing temperature.
10. High HP wall (mid and inner) differential temperature.
11. High medium pressure wall (mid and inner) differential temperature.
12. High ESV wall (mid and inner) differential temperature.
13. Excessive differential expansion.
14. Excessive bearing housing vibration.
15. Very low steam temperature.
16. Emergency hand tripping [both from remote (electrical) as well as from local hand trip lever (hydraulic)].
17. Generator protection operated as listed below.

Tripping of equipment is carried out mainly in three classes, A, B, and C, as per their severity in spreading and damaging. Of these three, Class C is the only one related to generator protection so it is not detailed in this clause.

A. *Class A protection*: Simultaneous tripping of the turbine and generator take places if any one of the following signals (typical) occurs:
    a. Generator reverse power relay operated.
    b. Generator loss of excitation with under voltage.
    c. Generator stator earth fault.
    d. Generator rotor earth fault.
    e. Generator over voltage.
    f. Generator stator inter turn fault.
    g. Generator differential.
    h. Generator negative sequence.
    i. Generator distance backup.
    j. Generator pole slipping.
    k. Generator and generator transformer overall differential.
    l. Generator transformer high-voltage (HV) side over current.
    m. Generator transformer restricted earth fault.
    n. Generator transformer standby earth fault.
    o. Generator transformer fire protection.
    p. Generator transformer Buchholz relay.
    q. Unit transformer differential.
    r. Unit transformer instantaneous and inverse definite minimum time over current.
    s. Unit transformer low-voltage (LV)-side earth fault.
    t. Unit transformer fire protection.
    u. Unit transformer Buchholz relay.
B. *Class B protection*: Instantaneous tripping of turbine and generator; tripping (GCB opens) takes place after "low forward or reverse power relay operated" (with "'AND" logic) if any one of the following signals (typically) occur:
    a. Generator loss of excitation.
    b. Generator rotor single earth fault second stage.
    c. Generator over fluxing.
    d. Generator stator earth fault in the range of 0%–5%).
    e. Generator stator winding distillate flow.
    f. Very high generator distillate conductivity.
    g. Very high generator transformer winding temperature.
    h. Very high generator transformer oil temperature.
    i. Unit transformer LV-side voltage operated earth fault.
    j. Very high unit transformer oil temperature.
    k. Very high unit transformer winding temperature.

Out of the above tripping signals, the first three signals—TG set over speed, thrust-bearing limit position, and very low condenser vacuum—are normally received through mechanical/hydraulic, active, protective devices acting on the tripping circuit (hydraulic trip gear), in addition to parallel electronic/electrical tripping. Also, redundant over speed tripping sensors are taken through complex voting logic, as detailed in Chapter 5, Clause 2.1, for extra precautionary measures. An electronic system measures pulses from a toothed wheel on the turbine shaft and counts those pulses through a frequency to voltage converter.

Tripping signals 6–11 are all thermal or mechanical parameters measured through electrical transducers, and are processed in an electrical protection circuit supplied with 220 V DC, which is still regarded as the most reliable source backed by a battery array. If any signals from these parameters is activated, the trip solenoid valve(s) is actuated, which in turn drain(s) the power oil responsible for keeping the main governing valves and stop valves open.

The tripping systems for TG set over speed, thrust-bearing limit position, very low condenser vacuum, and common electrical and emergency hand tripping are described in detail in Clause 5, which also includes the corresponding trip testing systems.

Modern electronic trip systems are strongly preferred by some manufacturers who adopt highly reliable voting circuits for higher "tripping speed" units. The spring-loaded trip bolt/plunger may behave in an unstable manner at higher trip speeds and requires separate hydraulic relay valves, which use hydraulic fluid in long piping runs. The direct mechanical system requires extra shaft overhangs, which may be detrimental to vibration at higher speeds. However, even now manufacturers still use hydraulic systems as the last resort in case of power loss/wire snapping.

The main advantages of the electronic system are high accuracy and repeatability, fastest response, high flexibility from adjustability, higher reliability through voting logic, and a testing facility with or without running the machine. Many uncontrolled over speed events are the result of valves failing to close, even when the over speed trip device operates. Further, nearly all uncontrolled over speed failures are catastrophic, resulting in blade failures, shaft breakage, and retaining ring bursts. In case of turbine tripping, the following actions are required:

1. Immediate closing of the ESVs and HP control valves (HPCVs).
2. Immediate closing of the reheat inlet valves.
3. Immediate closing of the nonreturn valves in extraction lines.

3.3 Generator Protection

There is also a Class C type of generator protection (and also a Class E type envisaged by some manufacturers) and "manual trip," which are meant for generator tripping only. Class C protection is mainly for the faults beyond the generator transformer and HV circuit breaker and includes fault of the grid and bus bar, GCB pole discrepancies, and manual tripping. Class E protection is related to GCB failure.

3.4 Unit Protection

This is the coordinated and integrated protection of individual boiler, turbine, and generator. The provision of HP/LP bypass has added flexibility in unit operation by decoupling the boiler and turbine in case of turbine tripping. The inter-relationships are indicated below:

1. Boiler trip due to its own protection, and some other important tripping signals would cause tripping of both the turbine and generator; this is a Class U trip.
2. Turbine trip due to its own protection would cause tripping of the generator through low forward or reverse power relay, but no command to trip the boiler.
3. Generator trip under Classes A and B would act as already indicated above.
4. Class C generator trip would not cause turbine or boiler tripping; turbine may become stabilized due to the "load shedding relay/governing system." Boiler may cut back one or more mills and would dump extra steam to condenser via the HP/LP bypass.

4 ATRS

Modern power plants are becoming larger in size, especially with ultra-supercritical plants, to cope with increasing power demands, and at the same time, reduction of costs. Any delay in starting time creates less revenue and will never be accepted by modern day operation practice code. To reduce starting time (with shutdown as well) and eliminate human error, the ATRS was developed (Fig. 9.8).

The entire plant is grouped in relation to a particular functional requirement or subsystem. The main system identifies each subgroup with various subgroups consisting of drives or devices to respond in a timely manner according to the program logic (by hardware or software).

There may be some variations in the features of this system with different manufacturers, but the basic control philosophy and aim of the controls are the same. In view of this, an abridged discussion is presented here based on one of the most reputed and popular steam turbine manufacturers in Europe.

4.1 Subgroups Involved in Automatic Run Up

There are a number of subgroups connected to the automatic run up systems including:

1. Turbine HP control fluid subgroup for the LPBP system.
2. Turbine HP control fluid subgroup for main valves (Above two systems also may be combined).
3. Turbine control oil subgroup.
4. Condenser evacuation and gland seal subgroup.
5. Turbine control subgroup (warm up, startup, loading, unloading, and shutdown).

In some plants, the generator exciter subgroup is also included for better coordination.

The basic function of ATRS is to optimize the starting time. To accomplish that, several overlapping operations are performed. For example, admission of main steam into the steam chest and a period to warm up, maintain the exact TG set speed for efficient soaking and speeding up of TG shaft that runs through critical speeds, and shorten vibration run up.

4.2 TG system

As already described in several clauses in Chapters 2 and 3, the turbine of the modern large thermal power stations must include a high-pressure turbine (HPT) section fed with main steam from the boiler. HPT exhaust steam is called cold reheat (CRH) steam, because it gets reheated in the boiler and becomes hot reheat (HRH) steam. It is then readmitted into the intermediate pressure turbine (IPT) section and its exhaust is directly sent to a low-pressure turbine (LPT) section before finally exhausting to the condenser. Although the steam sections are different, the rotor shaft is common to them all with a series of (moving) blades arranged to rotate inside an axial bore, which passes centrally through the turbine casing with (fixed) blades. The rotation of the rotor shaft is achieved by passage of steam from a fixed blade to moving blades. The rotor of the generator, which is firmly fixed to the turbine rotor, normally is cooled by hydrogen

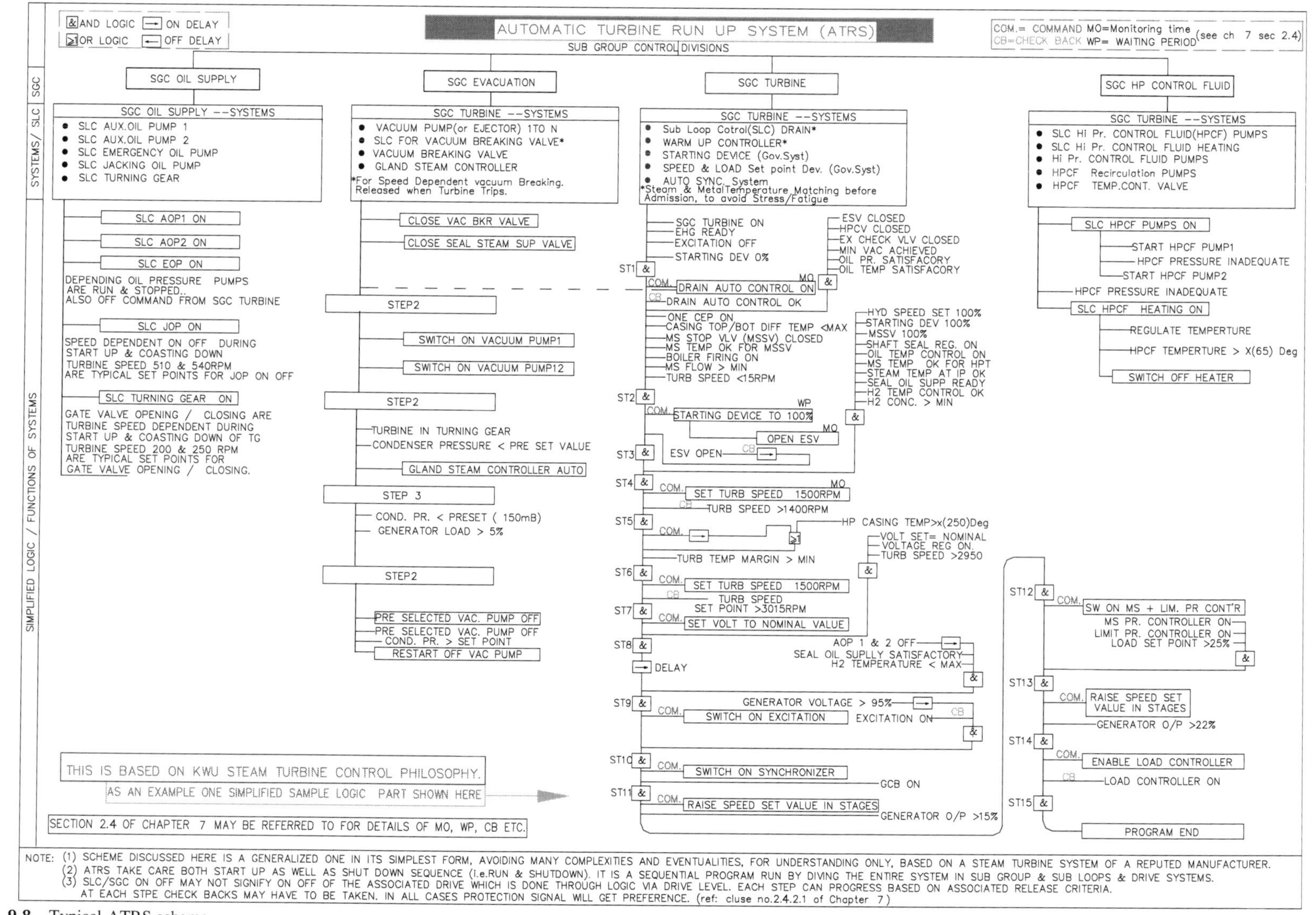

FIG. 9.8 Typical ATRS scheme.

gas. A steam chest on both the right and left sides of the turbine includes ESVs and HPCVs for controlling the steam to be admitted to the HPT.

4.3 Warmup Process

The huge and thick metal mass of the steam chest as well as the TG set rotor shaft must be brought to operating temperature as quickly as possible before the main steam at high pressure and high temperature from the boiler can be admitted into the steam chest. As steam with very high pressure and temperature is to flow through the turbine to ultimately generate electrical power output, it is extremely important the thermal stresses are kept under normal value during startup and normal operation. The startup procedure is more important as hot steam would stress the cold and thick turbine metals during initial entry of steam.

4.4 Startup and Thermal Stress

As steam with very high pressure and temperature passes through the turbine to receive power output, it is essential that the thermal stresses are kept under normal value during startup and normal operation. The startup procedure is very important as hot steam would stress the cold turbine metals during initial entry of steam, as detailed in Clause 6.

After many years of close observation, it was found that a set of particular initial steam temperatures with optimally regulated flow rates passed through for a well-judged duration of time, which would enable the shortest period of time to start up the turbine and maintain thermal stress within the permissive limit. This procedure of steam admission, especially in the primary stage, is necessary and important, and is guided by the TSE by imposing upper and lower margin limits for the startup operation. Any wrong step could lessen the life expectancy of the steam turbine and affect the startup time of the plant.

The startup procedure consists of cold startup and hot startup. Cold startup is most important because the temperature difference is at the maximum during the first admission of steam, and must be handled carefully to avoid reduction in plant life expectancy.

As the name implies, the cold startup of a TG set is applicable when the set is going to be started after a very long period of shutdown; the turbine case, chest, valves, and rotor are literally cold with temperatures from the ambient condition. The temperature rise (heating ramps) must be controlled (by controlling the steam flow rate and the quantity by experiencing minimum thermal stresses on the turbine metal, especially of thick components like the steam turbine shafts). It is also important to avoid high differential expansion between the rotor and the stator case, which could cause contact between them because of their different thermal inertias. The most critical part of the plant, as far as thermal stresses are concerned, is the turbine shaft in contact with the highest temperature steam, i.e., downstream of the first stage of the turbine. The startup control system must, therefore, be capable of keeping the controlled variables within the safe allowable limits. The "turbine control" subgroup contains a "subloop control" of "warming up main steam lines" and takes care of controlled heating of the pipes, components, etc., up to the inlet of HPCVs. Subgroup, "turbine control," then takes care of the rest of the controlled heating activities and the turbine load control afterward.

During the turbine startup, the steam flow and the heat transfer coefficient are very low. Therefore, the rotor temperatures increase slowly and when, after a certain period of time, the HP valves are opened more rapidly, the rotor surface temperature travels nearer to both the HPT inlet and first (impulse) stage outlet and steam temperature. At the end of the transient period, all temperatures are equal, because there is no steady state heat flow. After the heating ramp, the period is called heat soaking, which means the heating is continued for that period of time unless the desired temperature is achieved.

This is done to reduce the thermal stress peak by reducing the thermal shock before the start of the steam turbine load pickup. This can be achieved by allowing for a longer soak time for the turbine, i.e., once the turbine has reached full speed it is kept running without taking any load so that the steam flow can heat up the rotor a little bit more. The turbine startup is, therefore, initiated earlier than normal startup transient, and the turbine is kept idling for $\sim$1 h.

The load pickup phase is then started, and the rate of load change is steered to get a flat stress curve. After a desired load level is achieved, the rate of change is increased to the maximum allowable level. Here, two thermal stress peaks are observed and each of them contributes fatigue, and thereby, lifetime consumption. If the peak values are analyzed then, it can be seen that the first peak value may be slightly higher than the elastic limit, and thus correspond to a very low additional lifetime consumption. The second peak is higher than the first peak, however, still well below the previous standard startup value. The startup time is also reduced by considerable value from fuel savings.

When the heating is completed the turbine can be started under close supervision of the differential expansion measurements and when the unit can be synchronized (usually at speeds of 3000 rpm, which is called the synchronizing speed). After the unit is synchronized, then the TG can be loaded according to the related load ramp (MW/min). The conventional system available requires that the steam chest be sufficiently warm before transferring control of steam flow from the MSSV bypass valves to the governor valves. If the steam chest metal temperature is less than the steam saturation temperature at existing pressure, then the transfer is delayed to allow the steam chest to heat up more. The improved method suggested that the startup sequence can be squeezed by shortening the steam chest warmup process through lowering the governor valves to maintain rolling speed and opening the MSSV bypass valves to the full stroke, thereby overlapping the MSSV and governor valve HPCV control mode to partially pressurize the steam chest and allow more steam to flow in.

4.5 Sequential Operation of the Subgroups for ATRS

4.5.1 Turbine Oil Supply

This system provides pumping, bearing lubrication/cooling, jacking/lifting oil, turning gear operation, and vapor extraction of the system and control fluid. After getting the ON command, the following sequence starts:

1. The AC auxiliary oil pumps (AOPs) and DC emergency oil pumps (EOPs) start at a per oil pressure setting and stop at a per pressure setting when the shaft-driven main oil pump (MOP) starts functioning.
2. The temperature control loop for the turbine-bearing lube oil is activated and runs continuously as long as the turbine keeps running.
3. Jacking oil pump (JOP) inlet valve is opened during startup and the JOP is started manually, and then the SLC of all JOPs are turned ON. The valve closes and the JOP stops when the TG set attains a sufficient speed of >540 rpm (typically) to ensure adequate bearing lube oil. During turbine operation, the selected JOP(s) start (and valve opens) automatically if speed falls below 500 rpm (typically).
4. The hydraulic turbine mechanism for shaft-turning gear starts when its inlet valve opens. The preconditions required are hydrogen pressure, seal oil hydrogen DP is adequate, SLC is ON, and JOP is adequate at >130 bar (typically). During turbine operation, the turning gear starts (and valve opens) automatically if speed falls below 200 rpm (typically). The system will close automatically when the TG set begins to be driven by the steam, that is, the speed is typically >250 rpm.

When the system gets the shutdown command, the following actions are taken. The selected JOP(s) start (and valve opens) when speed falls below 500 rpm. The turning-gear system inlet valve will be open so that the system is ON immediately when the turbine speed comes down to ~ 200 rpm during coasting down. It will stop operating automatically when the casing temperature settings permit it to do so. AOPs will again start and stop as per their pressure setting and shutdown command.

4.5.2 Condensing System

The "condensing system" covers the drives and/or devices required for extraction of air and noncondensing particles from the condenser-associated systems, a vacuum breaker valve, and gland sealing system for the turbine shaft. The air-extraction function can be achieved by vacuum pumps or steam air ejectors. Usually two vacuum pumps run during startup and emergency conditions; only one runs during normal operation of the TG set. For a system with air ejectors, there is one "starting air ejector" for the startup purpose and running air ejectors for the normal operation.

Once the ON command is initiated, the sequence for systems with a vacuum pump and systems with air ejections starts like the following:

For a systems with vacuum pump:

1. Both vacuum pumps start and when the vacuum exceeds the preset value, one pump will be on the shutdown program as per selection with the other pump running. For a low vacuum condition, the second pump will start to maintain the vacuum and will stop again at the preset exceeded value of vacuum.
2. The associated cooling water supply valve opens whenever any vacuum pump starts.
3. The associated air-extraction valve opens whenever good running condition of the vacuum pump is established.
4. The vacuum breaker valve closes.
5. Warmup system starts for the gland steam line in association with the steam supply valve drain valve.
6. If the warmup condition is established and the vacuum has reached a preset value, the seal steam control valve will be released for gland steam startup pressure control and the gland steam cooler exhauster will be switched on.

For systems with air ejectors:

1. To start the air ejector, the auxiliary steam supply isolation valve opens with the startup pressure control valve crack open for the initial line warmup.
2. If warmup condition is established, the startup pressure control valve will be released to auto control.
3. The vacuum breaker valve closes.
4. When TG set load is established, the startup isolation valve will be closed, the selected normal ejector isolation valve will open, and the associated control valve will be released to auto control.

The rest of the sequence is the same as the vacuum pump system.

When the system gets a shutdown command, the following action is taken:

1. The vacuum breaker valve opens.
2. The gland seal leak off pressure control valve closes.
3. The main air ejector(s) stop or the vacuum pump stops, and associated isolation valves close.

4.5.3 Turbine Control

The turbine control subgroup covers the functional requirements of the automatic run up systems and the loading, and run down systems of the TG set as well. The run up system includes the warm-up function of the remaining part of the TG set and then controlled speeding up through different stages up to the synchronizing speed. The complete operation of the turbine control system is achieved through the TSE and the EHG system along with speed, generator output/load controller, and pressure controller (for special purposes). Once the ON command is initiated, the sequence follows these steps:

1. First, open the ESV with a hydraulically operated starting device. The ESV is the first step to allow the admission of live steam into the turbine. The following start-permissive signals are verified for availability:
   - **a.** Turbine is running (on turning/barring gear speed).
   - **b.** Differential temperatures between top and bottom section of both HPT and IPT casings are below the preset permissible values.
   - **c.** The condenser vacuum is well established above the preset value.
   - **d.** Condensate transfer subgroup and turbine oil supply subgroups are established.
   - **e.** MSSV must be closed.
   - **f.** Steam parameters are suitable before MSSV.
   - **g.** Steam generator with firing system in well-established condition.
   - **h.** Turbine drains are operating as desired (discussed later).
2. The speed setpoint is increased to warm-up speed (or the first hold speed for heat soaking) provided the following permissive signals are available (a time delay is introduced to make the system get ready):
   - **a.** Steam parameters are suitable before MSSV.
   - **b.** Steam parameters before ESV are suitable for admission to turbine.
   - **c.** Hydrogen (generator rotor coolant) temperature control is ready.
   - **d.** Turbine seal steam pressure control system is in operation.
   - **e.** Turbine oil temperature control system is in operation.
   - **f.** Generator seal oil supply is made available.
   - **g.** The starting device and speed changer are 100% positioned.

Upon availability of these start-permissive signals, the speed setpoint controller would raise, as guided by the TSE, the setpoint to "soaking up" speed for the speed controller, which will dictate the rate of steam flow by positioning the HPCV-required steam flow.

For a cold start, a certain amount of time delay is introduced after the TG set has reached its warmup speed, and then the speed setpoint controller will raise, as guided by the TSE, the setpoint to the rated speed when the turbine, with all its associated accessories, is adequately soaked with heat transmitted by the main steam.

As a next step, the generator's AVR is switched on when the TG MOP assumes the charge of total oil supply as the turbine is running at rated speed. The AVR is then set at its rated value and the generator is excited.

After receiving confirmation signals from the above commands, the generator is synchronized by the auto synchronizer device and switched on with the help of the GCB to the external power grid. No sooner than the generator is synchronized, a minimum amount of power is delivered to the grid. The speed set value is then raised in stages, and if generator output is more than a minimum value of ~5%, the main steam pressure controller is switched on to the limit pressure control.

When the system gets the shutdown command, either manually or automatically, and when a protective tripping action is initiated and closing command of ESV has already been issued, it will take the following actions:

1. The TG set is unloaded by lowering the load set value 20%–25%, and then it automatically switches OFF.
2. The remaining unloading action is accomplished by reducing the speed setpoint of the speed controller.
3. The GCB is opened by energizing the tripping circuit with the help of reverse power protection logic.
4. For the next restart, the necessary plant conditions are ascertained as a last step of the shutdown program.

4.5.3.1 Turbine Warmup and Drain Control

Turbine subgroup activates the warmup controller to ensure a smooth and safe warmup of the ESVs and HPCV turbine as permitted by the TSE. The automatic drain control forms a separate sub loop control with all the drains hooked up and does not follow any sequence. According to the operational philosophy, the drain valves can open or close any time depending on the parameter operating status of the plant without any bearing with the load, speed, or time. Depending on the relative importance and significance of various drain valves, the criterion (mainly temperatures of relevant points) for opening/closing of each individual drain valve is defined and suitable logic built in to ensure safe operation. After the warm up of the ESV and CV of the HPT, the ATRS emits steam into the turbine to roll it.

After sufficient warming up with other permissive criteria, the ATRS takes the turbine from the barring speed to the first hold speed (nearly 20% of the rated speed) with the acceleration controlled by the TSE/turbine stress control system to prevent components from overstressing. The ATRS ensures a definite hold for soaking of turbine internally at this speed for a specified period if the HP metal temperature is less than 250°C, otherwise the hold period is determined by the time it takes for temperature margins of the TSE to be within permissible limits.

Once the desired thermal soaking of the components and process requirements is complete, the turbine control raises the speed of the set to rated value with the guidance of the TSE by controlled acceleration, except for the time when the rotor crosses the critical speed region.

After attaining a speed near synchronous speed (~2950 rpm) with sufficient available temperature, the TG set is synchronized with the grid. For the saturated steam turbine, drain automatic control is accomplished by sensing level in the drain loops.

4.5.4 Control Oil Supply

With the advent of using very high pressure oil as control oil for operating the hydraulic actuators of LP bypass valves, ESVs, HPCVs, etc., a separate independent subgroup control has been envisaged. The type of oil used is a non-flammable fluid or a fire retardant fluid (FRF). The fluid known as "phosphate ester" is the FRF currently used in a majority of applications. This fluid provides excellent hydraulic capabilities and good fire-resistant characteristics. However, some typical problems were encountered in using phosphate ester over time including:

1. High water levels from:
   a. Moisture in the air breathed by the reservoir.
   b. Condensation formation inside the reservoir caused by rapid fluid temperature changes.
2. High acid levels in the fluid due to formation of phosphoric acids.
3. Gel and varnish formation throughout the system; gels can ultimately change to a varnish.
4. Low electrical resistivity of the fluid.

Recently some alternative fluids were introduced as replacements for the existing FRF fluids. These fluids are approved as FRFs by FM Approvals, which provides rigorous third-party testing, certification and loss prevention standards.

The replacement fluids do not create the harmful phosphoric acids like the existing fluids so they do not require the use of acid control filters. Replacement of the FRF fluids with these alternate fluids not only eliminates the problems and environmental hazards inherent with handling phosphate ester fluids, but it also eliminates the much larger problem of gel and varnish formation.

The replacement fluids lubricate better than the existing FRF fluids. If the fluid remains clean it helps prolong the life of most components in the hydraulic fluid system. The replacement fluids also have higher viscosity indices, which allow them to operate over a wider temperature range.

The basic components of the control oil supply subgroup include (2 × 100% or 3 × 50%) control oil FRF pumps, one circulating pump for fluid transferring to the filters, and temperature control systems with heaters and coolers for maintaining the requisite temperature. The FRF pumps are controlled by automatic start and stop (sub loop controller) logic to maintain the pressure and flow as required by the TG system.

The subgroup gets "startup operation" unit coordinated command when LPBP operation is about to start after the condenser vacuum attains more than the preset value, and/or ESV is about to get the open command. The sequence follows this order after the ON command is initiated:

1. The selected FRF pump will start.
2. For 3 × 50% FRF pump selection, the second pump as selected will start as per the unit load demand of maintaining pressure and flow.
3. The FRF control oil temperature control is released for auto operation.
4. The circulating pump will cut in to ensure continuous filtration no sooner than the temperature has attained the preset operating value.

When the subgroup gets the shutdown command, the shutting down process starts after the LPBP operation has completed the shutdown operation. When the system gets the shutdown command, it takes the following action:

1. The circulating pump stops.
2. The temperature controller is OFF.
3. The FRF pumps begin to stop as per selection.

4.5.5 HP Control Fluid Supply

This subgroup controls the HP control fluid (HPCF) pumps, HPCF heating, HPCF recirculation pumps, and HPCF temperature control valve. Both CF pumps get switched ON through the SLC. The standby pump starts if the running pump fails to develop the required pressure. The required CF pressure enables SLC heating ON and starts the CF circulation pump, which feeds the regeneration circuit. The temperature control valve is put into auto mode to regulate the temperature. CF temperature in the tank is maintained at ~56°C by the CF heater ON/OFF control. The heater switches OFF automatically if CF temperature exceeds 65°C.

In the shutdown program, all the oil pumps are switched OFF and an SLC CF pump is also turned OFF. SLC heating is kept on to maintain the temperature required for the next startup and can be manually switched OFF, if required.

5 ATT SYSTEM

5.1 Testing Process by ATT Systems

The ATT systems enable online testing of all the major drives (ESVs/HPCVs) or trip devices required for safe operation and emergency tripping of a TG. In other words, all the protective equipment required for TG safety is tested automatically with uninterrupted operation without jeopardizing the individual protection assignment bestowed on each protective device while running test procedures. The mechanical or electronic turbine trip mechanisms are the ultimate line of defense for protecting the steam turbine and must be available when they are needed. However, testing is permitted only under specified conditions, such as healthiness of a substitute device, no trip conditions exist for turbines, etc.

For turbine tripping, i.e., to close the ESVs (also HP/IPCVs through secondary oil), the trip oil is drained, meaning substantial quick reduction of trip oil pressure, which is achieved through any of these protection devices. Normally, ATT, if supported by the control system provided for the turbine, (not all manufacturers support ATT), extends into the trip oil piping network, and with the help of test oil, performs a similar function.

The main function of the ATT system is basically to ensure a fail-safe operation of the protective valves or devices by checking and analyzing their readiness throughout the turbine operation regime. The system ensures this duty by performing the tests periodically (Fig. 9.11C) in automatic mode of operation, and manually at the operator's discretion. The essence of this particular standalone facility is to test all the protective devices sequentially in a well-organized and systematic manner for verification of foolproof functioning of all the devices. The system is built to eliminate operator error, resulting from manual intervention, by using the fully automatic test sequence. Individual tests for a particular device can be done by proper selection. Due to limited scope, control philosophy of only one reputed manufacturer (KWU) is discussed so that the reader becomes familiar with the basic idea of this type of control.

5.2 Scope of ATT

ATT covers the following items:

1. TG set over speed trip (mechanical)
2. Thrust-bearing trip (mechanical)
3. Low vacuum trip (mechanical)
4. Electrical remote trip
5. Opening and closing operation of control valves
6. Opening and closing operation of ESVs.

Testing can be done for all the above items manually at the operator's discretion or periodically through the automatic program (Fig. 9.9).

During the test program, the trip oil pressure is reduced by the ATT for testing of each tripping cause so that the tripped signal is available at proper strategic locations. The HPCVs and ESVs, until they are tested, are not allowed to close during the testing of the protective devices. That's because they would still continue to be supplied with high-pressure oil via the trip test solenoid valve and the ATT changeover valve by the ATT system itself, so normal operation of the turbo generator is continued (Fig. 9.10).

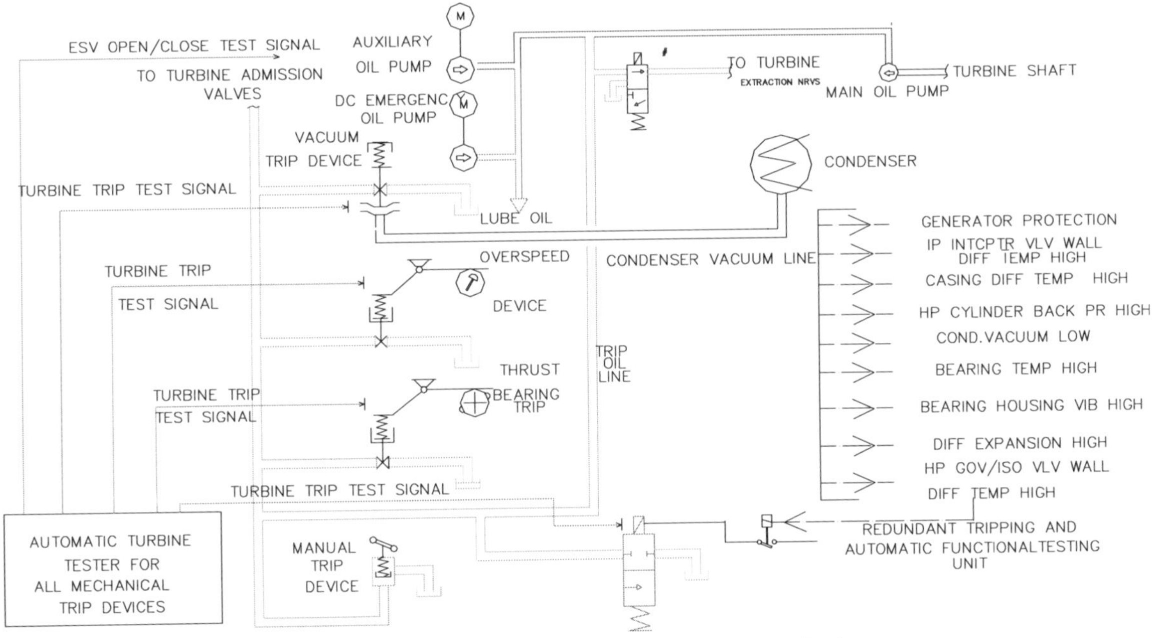

FIG. 9.9 Schematic diagram of ATT system.

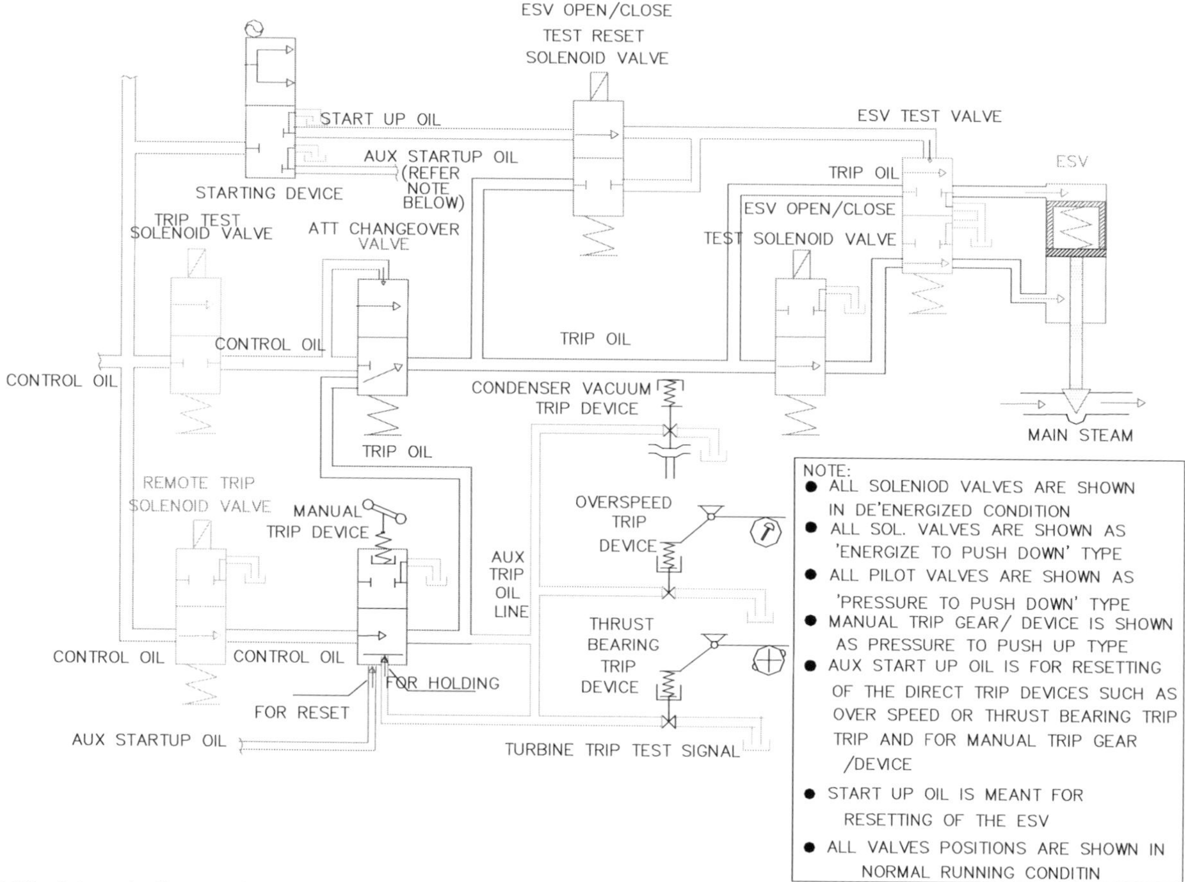

FIG. 9.10 Schematic diagram of ATT oil circuit.

All simulated electronic trip components, however, could be tested (through the remote electrical trip solenoid) more frequently, for example, once a month if they have no direct effect on the steam turbine operation.

5.2.1 TG Set Over Speed Trip (Mechanical and Electrical) and Testing

It is customary that the steam turbines are fitted with an emergency over speed tripping system in addition to normal operating speed governors. Uncontrolled over speed failures due to malfunctions or inappropriate control system of over speed protective devices are catastrophic, resulting in blade failures, shaft breakage and rotor winding retaining, ring bursts, etc.

5.2.1.1 Trip System

5.2.1.1.1 Direct Mechanical Type Over Speed Trip

The over speed measuring systems may be of any type, such as mechanical (fly wheel), hydraulic (positive displacement pump), or electromagnetic/electronic (Hall probe). In a thermal power plant, the hydraulic signal is used for the mechanical type of over speed tripping for faster response.

When the over speed trip mechanism is actuated (shaft speed exceeds the safe speed level, generally 10% more than the normal speed value), the TG set is tripped on an emergency shutdown basis by quickly draining the actuator/trip oil to close the ESVs. The system is independent of the TG set governor control system and forms one of the most important and last resorts for tripping criteria, as electromagnetic/electronic type (as a part of the remote electrical trip) tripping is also included in addition to direct mechanical tripping.

The over speed trip system consists of two over speed sensing bolt/plungers normally mounted on the frontend-bearing pedestal of the TG shaft (location may vary as per the manufacturer's design), speed setpoint adjustment screw, pawl, etc. The over speed sensing bolt/plungers are eccentric, that is, they have their center of gravity slightly away from the center and covered along with mounting. When the TG speed increases beyond the safe value, the centrifugal force of the unbalanced bolt/plunger overcomes a spring force at a preset trip speed and moves outward to strike the respective pawl from the trip lever. When struck, the trip lever releases the spring-loaded valve to drain the auxiliary trip oil and in turn drain the main trip oil, the pressure of which is no longer in a position to hold the ESVs. Different manufacturers use several methods like high-speed machines that use a weighted disk and a dished washer to accomplish the tripping action.

5.2.1.1.2 Electrical Type Over Speed Trip For the electrical/electronic trip, speed is sensed by the Hall probe technique or proximity type sensing and the subsequent electronic, and the relay circuit generates a switching contact when speed reaches beyond the setpoint. There is a trip solenoid valve along with a trip relay that energizes along with other electrical trips that act in parallel and drain the trip oil to trip the TG set.

5.2.1.2 Trip Testing System of Over speed Trip

5.2.1.2.1 Direct Mechanical Type Over Speed Trip Testing Testing of proper functioning of a turbine over speed trip device, although centered on the over speed trip mechanism, should be treated as a complete system. For functioning rather than simulating the trip condition, high-pressure oil is introduced by a hydraulic test signal transmitter (HTT) via a shaft hole producing a similar but artificial condition of actual over speed for actuation of trip bolts at TG normal speed. This causes the hydraulic circuit, that is, the trip oil, to drain. The resultant fall in trip oil pressure would be detected by suitable pressure switches and the signal sent to the turbine trip tester, which in turn sends the alarm indicating successful testing operation. All associated alarms and indicators are also provided for proper observation in different operational and management areas. After testing, resetting is done to move up the HTT spool and drain test oil.

5.2.1.2.2 Electrical Tripping Type Over Speed Trip Testing Some manufacturers provide an automatic online over speed test through the internal injection of a ramping frequency signal into the speed channels. This test verifies proper operation of the software, hardware, and hydraulic components of the trip circuit. Some manufacturers provide an automatic online over speed test by generating electrical relay contact at an over speed setpoint more than that of the mechanical trip setpoint, and club it along with other electrical trips to act on the remote trip solenoid.

5.2.2 TG Set Low Vacuum Trip (Mechanical and Electrical) and Testing

This parameter is also very important as far as turbine protection is concerned, which is the why both direct mechanical and electrical tripping are provided along with the provision of testing its functional operation.

5.2.2.1 Trip System for Low Vacuum Trip

5.2.2.1.1 Direct Mechanical Type Low Vacuum Trip The function of this protection device is to trip the turbine in case the condenser vacuum becomes lower than a very low preset value. The basic device is a spring diaphragm-operated sensing element with the upper side of the diaphragm connected to the condenser vacuum line and to the primary oil pressure through a piston with an isolated chamber. The lower side is connected to the atmospheric pressure with the spring acting against the force applied by the primary oil pressure. When the vacuum falls, that is, pressure increases on the upper side, the net downward force experienced by the diaphragm is more than the upward force offered by the combination of spring tension and atmospheric pressure. The result is the downward movement of the spring diaphragm assembly causing the trip oil to drain and ultimately to trip the turbine. The protection need not be bypassed during startup as the primary oil pressure builds up only when the TG set speed is more than the preset value.

5.2.2.1.2 Electrical Tripping Type of Low Vacuum Trip For the electrical tripping type, vacuum switches are used with sufficient redundancy and clubbed with other electrical tripping signals to operate the separate trip solenoid valves. The scheme/procedure of testing of this tripping system is similar to over speed tripping/testing, except the sensing of low vacuum, which is from an electro/mechanical type switch for the electrical part.

5.2.2.1.3 Direct Mechanical Type Low Vacuum Trip Testing Because the basic tripping device is a spring diaphragm with the upper side connected to the condenser vacuum line and the primary oil pressure, manipulation of any of the above parameters (depending on the manufacturer) to a higher value increases the downward force. The lower side is connected to the atmospheric pressure with the spring acting against the above combined and the forces do not change. The net downward force on the diaphragm will cause the downward movement of the spring diaphragm assembly by the simulated turbine trip, which is more than the upward force offered by the combination of spring and atmospheric pressure. A separate solenoid-operated test valve is provided to perform this test.

During testing, first the hydraulic circuit is established, and with the help of the HTT (test solenoid valve) discussed above, the trip device is disconnected from the condenser vacuum line and simultaneously the upper port is connected to the atmosphere via resistances provided by a long spiral tube and restricting orifice, which causes the trip oil to drain. After testing the HTT is deenergized and the vacuum trip gets reset automatically.

5.2.2.1.4 Electrical Tripping Type Low Vacuum Trip Testing This procedure is similar to that of over speed trip testing.

5.2.3 Thrust-Bearing/Axial Shift High Trip (Mechanical and Electrical) and Testing

Thrust-bearing trip (mechanical) and testing is another important parameter for the turbine protection for which a testing facility is provided in the ATT. Axial shift may be simply described as a rotor movement resulting from the wear and tear of the thrust pad. The difference in thrust created on the TG shaft by steam flowing through HP and combined IP/LP causes constant rubbing of the thrust pad in the thrust bearing by the high-speed TG shaft. The designed movement is kept at a minimum (typically 0.2–0.25 mm). The measuring/detection point is provided at a rotor anchor point where the thermal expansion of the rotor is zero so that the axial shift is not affected by thermal expansions of the casing/rotor.

5.2.3.1 Trip System of Thrust-Bearing/Axial Shift High Trip

5.2.3.1.1 Direct Mechanical Type Thrust-Bearing Trip Device This is a direct type of tripping provided for protection of the turbine against large amounts of wear and tear on the thrust-bearing pad and axial shift/movement. The physical arrangement is such that two pairs of cams are provided on the turbine shaft at an angle of 180° to each other. In each position of both pairs of cams, a pawl is positioned between them. The pawl touches any of the cams whenever the TG set shaft moves horizontally either forward or backward.

If the pawl comes into contact with any cam, the latch of the thrust-bearing trip device is released and actuates the trip part like the above tripping systems. Two pairs are provided diametrically apart to have redundancy in the protection system. The typical set value is between 0.6 and 0.75 mm. depending on the manufacturer of the turbine.

5.2.3.1.2 Electrical Tripping Type Thrust-Bearing Trip Device Redundant position pickups are provided with the standard turbo supervisory instrument (TSI) package to monitor the axial movement of the disk that is mounted on the rotor near (normally within 25 mm) the thrust-bearing collar. Movement of the disk may be taken as the axial movement of this collar (neglecting the linear thermal expansion between the two). If there is excessive movement of the disk from thrust-bearing wear (if it is more than the value set for a direct mechanical trip), relay contacts are generated in the TSI system, and are utilized for energizing the remote trip solenoid to initiate the turbine trip.

5.2.3.2 Thrust-Bearing Trip System Testing

5.2.3.2.1 Direct Mechanical Type Thrust-Bearing Trip Testing The test scheme/procedure is similar to the over speed trip test. The thrust-bearing trip function is simulated by the solenoid-operated HTT, which simulates the movement of the pawl. When the ATT-generated test signal is sent to activate the solenoid of the test valve, pressurized test oil is allowed to pass through it to a spring-loaded power cylinder, the shaft of which pushes the pawl in the direction of the action for actual tripping by a direct mechanical method. After testing, resetting is done when the HTT is deenergized, and auxiliary startup oil resets the device back to its normal position

5.2.3.2.2 Electrical Tripping Type Thrust- Bearing Trip Testing This procedure is similar to that of over speed trip testing.

5.2.4 Electrical Remote Trip Solenoid and Testing

All other trips, including generator tripping signals, are taken together as a part of the electrical remote trip and ATT. The remote electrical trip system includes electrical signals as well as mechanical process signals in the form of processed electrical contact or the process switch itself. These signals are discussed in Clause 3. All of these signals are tested individually, but through a common trip relay and a common remote trip solenoid valve when energized. The testing procedure is similar to that of earlier systems.

5.2.5 Testing of MSCVs (HP and IPCVs)

It is often desirable to test individual MSCVs or governor valves. To test them with minimum disturbance, the load must be reduced to a load that can be run by the other running governor valve(s) when a particular governor valve is closed. A valve test signal is generated in the ATT system for calculation of the valve position demand signals, which implements closures of the valve under test and appropriate proportionate increasing signal (from the EHC) for remaining valves to open further to maintain the load and stem flow. as demanded by the load/speed controller during the test.

For a system with many control valves, normally one control valve is tested. The corrected, individual valve position demand signals are generated at the EHC depending on the load. The valve test procedure is then followed by gradual closure of the governor valve under test, and at the same time opening of the others.

When a particular HPCV is closed, the associated ESV may be tested for closure so as not to disturb the load/speed control as dictated by the EHC. After the closing tests, as considered by the ATT, the further corrected test-related position demand signals enable gradual opening of the previously closed HPCVs, and proportional closing of the remaining HPCVs.

5.2.6 *Testing of ESV for HP and IP*

It is quite common for all types of valves to experience several constraints, such as deposits/buildup on the valve stem, different worn out components, poor stem condition, etc., which may pose problems with free and smooth operation of the valve, if not total control. The ON/OFF valves are affected the most due to the above effects, as the control valves normally modulate continuously and are less affected.

These valve operations should be checked on a regular basis. The frequency may vary and some manufacturers suggest that valves be tested more frequently to assure they do not get stuck in a particular position due to prolonged process demand. It is also a stringent requirement that each and every valve must be able to attain full shutoff condition to avoid excessive over speed. If the valves are not fully shut off, and the other devices do not work properly, the turbine may still excessively speed up beyond normal value, because the steam source is still present and uncontained despite the trip mechanisms actuating properly. The desired testing method demands checking of full-stroke excursion of valves, that is, the capability to operate through its entire range from fully open to closed position.

ESV test valves are provided for each ESV so that individual testing operations can be accomplished. ESVs are supplied with trip oil through this test valve, which provides port openings for both trip oil and startup oil. This startup oil is utilized for resetting of ESVs after the turbine tripping takes place. Handwheels are provided with each test valve to test the ESV operation from the field as well. There are separate and individual solenoid-operated ESV open/close tests and reset valves, which are operated through ATT for testing ESV opening and closing operations. The details are explained in Clause 5.3.1, point 2b. Upon getting a signal from the ATT, it allows control oil to act as auxiliary starting oil, which pushes down the plunger of the ESV test valve. This action enables the control oil to enter the upper part of the ESV power cylinder while the control oil for the lower part is guided to the drain. Fig. 9.10 depicts this operation. If the closing operation is tested on load condition, the TG set load must be brought down to a lower level such that sufficient steam can pass through the other valve to take care of the load. It is normal practice that a particular ESV may be tested for closing action after the associated HPCV is closed by the ATT command, so as not to disturb the load/speed control as dictated by the EHC.

5.3 Scope of ATT

5.3.1 *ATT Overview Based on KWU Steam Turbine Control Philosophy*

Listed below is the ATT overview (Fig. 9.11):

1. *Salient features of an ATT*: These are essential to the ATT so normal operation of the turbine is not affected and they include the following:
   a. Separate and individual testing of each protecting device and valves so that individual specialist functions/duty can be checked.
   b. Provide functional substitute so that during testing normal operation and protection is never affected.
   c. To ensure that substitute devices are healthy and the oil circuit is established so that during testing, normal operation and protection is never affected; for this reason before a main test, the preliminary test and hydraulic circuit is checked.
   d. After each test the resetting procedure must be followed so the system goes back to its original condition for readiness.
   e. During testing if the turbine trips or the test is not performed within a specified time, the test must be abandoned and solutions must be found.
2. *Test category and procedure*: There are two kinds.
   a. One is protective devices, such as mechanical over speed, thrust wear, low vacuum trips, and testing of remote solenoid valves. Each of them can be tested one at a time. The health of each substitute device must be tested and a hydraulic circuit must be established, i.e., control oil must be sent to the trip oil line during testing. During testing of the protective devices a substitute device is very important, for example, while testing the remote solenoid valve test trip, the solenoid through the changeover valve keeps the trip gear pressurized. Balance testing procedures were discussed previously and are depicted in Fig. 9.11B (left side).
   b. The other is the testing the control and stop valves (Fig. 9.10). Because this directly acts on main turbine load-controlling devices, some criteria must be considered such as the load must be greater than the preset value and EHG must be in operation so that normal load/pressure control of the turbine is not jeopardized due to ATT operation. Preset value stated above depends on MW rating of the TG and on turbine configuration, i.e., number and type of stop cum control valve, for example, a 500 MW turbine with a four sets of such valves. ATT is carried out at a load <160 MW. In any case, not more than one set of valves can be tested at a time. As stated earlier, with ATT initiation the positioner motor of

the control valve moves to lower the oil pressure beneath the servomotor to close the control valve. When it is fully closed, the open/close test solenoid valve is energized to drain the trip oil (through the ESV test valve) beneath the disk of the stop valve servomotor piston; the stop valve closes due to low oil pressure. After it is closed, the signal energizes another solenoid valve called the "test reset" valve, which allows trip oil to force the test valve spool to come down and supply oil to a place above, and press the servomotor piston and the piston starts sitting on the disk. As soon as it sits on the disk, the pressure increases and is sensed by a pressure switch indicating preloading completion of the ESV servomotor. Both of these solenoids are de-energized; the ESV test valve afterward moves up to admit trip oil beneath the disk, and the space above the piston disk is drained. Naturally the stop valve goes up to open. After it is opened, the command goes to the control valve motor position to move in reverse direction to open the control valve. This is depicted in Fig. 9.11B (right side).

3. Details of the testing and checking schedules of other important devices are detailed in Fig. 9.11C.

6 THERMAL STRESS EVALUATOR

Large-capacity thermal power plants are often required to start up and shutdown the turbine a continuously increasing number of times. In consideration of the plant's long life (~25–30 years), the turbine metal stress conditions must be maintained for the plant to reach its life expectancy. During each start and stop, some metal stress is experienced by the turbine, which reduces power plant life expectancy in proportion to the amount of stress. Therein lies the necessity of the TSE, which evaluates and guides the TG set to run at minimum stress. The methods and equipment for monitoring turbine stresses are considered essential for all large-capacity turbines.

The TSE calculates and monitors the operational heat stresses of the turbine during every operational phase. The goal is to achieve continuous optimization between material stresses on the turbine and flexibility in reacting to operational changes. The TSE makes it possible to conduct both qualitative and quantitative evaluation of the thermal stresses induced in the metal bodies during thermal shock. Detailed analysis of the TSE results in manufacturers developing materials with a high resistance to thermal shocks. There are a few methods used by different

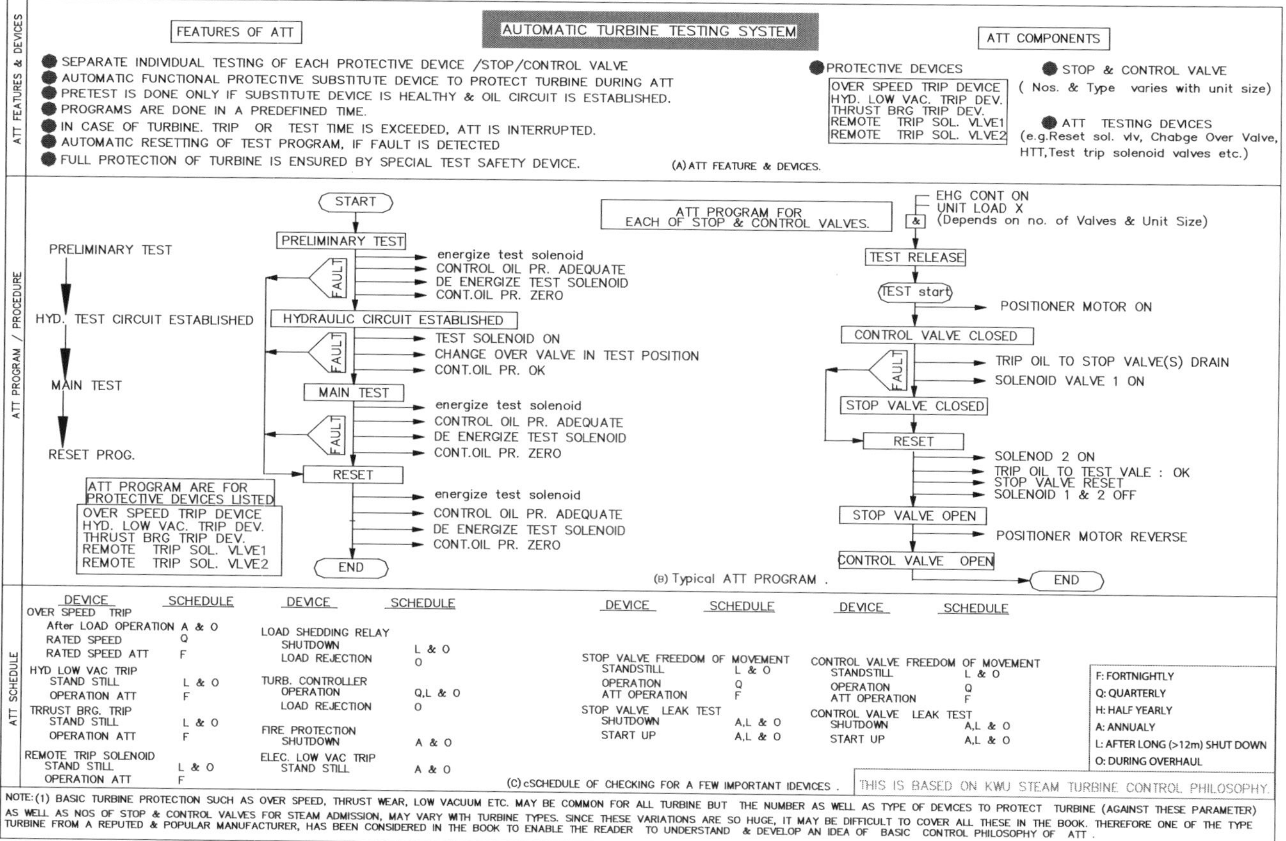

FIG. 9.11 Overview of ATT system (typical for a particular turbine type).

manufacturers to evaluate the thermal stresses at a given instant of thermal shock. The influence of the material anisotropy, that is, different properties at a given condition, on thermal stresses induced by thermal shocks is still unknown. However, simple expressions may be available using simulations to evaluate the thermal stresses when the material has anisotropies in physical properties, such as Young's modulus and a linear expansion coefficient.

6.1 General Causes of Turbine Stress

In rotating machines like a turbine operating in a high-temperature and high-pressure environment, repeated thermal shock causes damage and fracture. Such eventualities are big problems. There are many types of stress in metal bodies such as fatigue, creep, thermal stress, etc. Those type of stresses in turbine components are mainly caused by cyclic loading/unloading, steam temperature differences, and pressure and centrifugal forces(creep).The greater the applied stress value, the shorter is the material life. The damage is cumulative and materials do not recover, even when rested from cyclic loading and unloading.

The TSE helps to calculate and predict nonsteady-state thermal stresses to which the steam turbine is exposed during different stages of operation, especially during startup and shutdown, and as a result, changes in heat input to components during power operation are acquired and compared with the permitted limits. Such stresses mainly occur due to temperature difference between the exposed metal surface and the inner metal surface.

6.1.1 *Fatigue*

Material science is the progressive and localized structural damage that occurs when a material is subjected to repeated loading and unloading. The nominal maximum stress values are less than the ultimate tensile stress limit and may be below the yield stress limit of the material.

If the loads are above a certain threshold, microscopic cracks will begin to form at the surface. Eventually a crack will reach a critical size, and the structure will suddenly fracture. The shape of the structure also significantly affects the fatigue life. TSE determines the thermal stresses during operation and compares them with appropriate limits established on the basis of predictable operating conditions by means of appropriate low-cycle fatigue (typically less than 103 cycles) diagrams. TSE predicts various margins to limit rate of heating/cooling after evaluation of such fatigue and has now become a highly effective preventive maintenance feature for turbines.

6.1.1.1 Low-Cycle Fatigue

Low-cycle fatigue is the product of changes that take place in temperature and associated stresses whenever a dynamic change in TG operation is concerned such as start, shutdown, loading, unloading, etc. The TG set has certain components with a thick cross-sectional area, for example, HPT and IPT casing, rotor, HPCV and ESV parts, etc., which are most affected by low-cycle fatigue.

6.1.2 *Creep*

Creep is a phenomenon detected by dislocations on grain boundaries of materials made of steel or related products caused by high temperature associated with high centrifugal forces, which ultimately result in permanent deformation on the affected turbine component. It is quite natural that the rotating components such as rotors, HPT and IPT blades, center bore, serrations, blade roots, etc., suffer most due to creep phenomenon.

6.1.3 *Thermal Stress*

Clause 3.6 of Chapter 2 discusses thermal stress; however, some brief discussion is incorporated here for quick recapitulation. Whenever there is a difference of temperature between the metal surface and inner metal, or mid-metal, thermal stress develops. It will be quite significant when metal thickness is appreciable.

During startup and normal operation, steam with very high pressure and temperature passes through the turbine to become heated and deliver power output. The startup procedure is more important as hot steam can cause cold turbine metal stress during initial entry of the steam. During normal operation, it is not as critical as the main steam temperature; a suitable temperature controller is always kept near a fixed temperature (after about 60% load) where thermal shocks are not expected. It is, therefore, extremely important that the thermal stresses are kept under normal value during startup, whether it is a cold or a hot startup.

6.2 TSE

6.2.1 *Purpose*

Modern power plants are equipped with a TSE for the purpose of safe operation guidance under all possible types of running conditions. A TSE is an instrument cum device that measures the turbine stresses in thick materials caused mainly from casing temperature and temperature differentials compared with the permitted limits.

The TSE generates the speed and load upper and lower margins, which are incorporated in the turbine startup system, turbine automatic control systems, and are even used in the boiler control system while running on "coordinated control mode." The TSE guidance, popularly called, "TSE influence," dictates the speed or load gradients along which the set value changes. In other words, the purpose of TSE influence is to limit the acceleration and load ramp gradients in optimum mode as regarding both degree of component fatigue and economical operation.

6.2.2 *Measuring Temperature Points*

Some typical and important measuring temperature points for a 250 MW plant are indicated below, but the actual number and measuring points may vary with the size of the plant and even with continued development of the device.

1. ESV inner and middle wall metal temperature.
2. HPCV inner and middle wall metal temperature.
3. HP casing inner and middle wall metal temperature.
4. HP shaft surface and simulated middle wall metal temperature.
5. IP shaft surface and simulated middle wall metal temperature.

For items 1–3, the mid or inner wall metal temperatures are measured at exact position by making grooves with the help of a honing process from the top through the wall of the subject material. But for items 4 and 5, because it is a rotating object, the mid-metal temperatures are computed from the corresponding surface temperature, which is located at a representative temperature point of surface of the shaft.

Other parameters are measured/calculated in TSE for computing the lower and upper margins, such as:

- Effective power and reactive power.
- Steam before HPT and IPT pressure and temperature, also HPT mass steam flow.
- HPT exhaust steam pressure and temperature.
- HPT expansion section differential temperature.

During startup and shutdown, the ESV and HPCV inner and middle wall metal temperature sensors (thermocouples) are utilized by the TSE to calculate necessary margins to the controllers for controlling the steam admission rate and quantity.

For the remaining TG set operations, that is, under normal loading/unloading operations, HP casing, inner and middle wall metal temperature and HP and IP shaft surface, and middle wall metal temperature sensors (thermocouples) are utilized by the TSE to calculate necessary margins to the corresponding controller for controlling the steam admission rate.

6.2.3 *Principle of Operation*

For every channel, the differential temperature between mid-metal and inner metal is calculated and then compared with the permissible highest and lowest temperature differences, which are dependent on the mid-wall temperature and are derived through the function generator. The results, that is, the subtracted values, are the upper and lower temperature margins and are available for warmup, loading, unloading, etc. The maximum selected upper and lower temperature margins are then the representative values for increasing or decreasing steam flow during startup and then the rest of the on load operation. These values are indicated and utilized for further calculation. The upper load margin is added and lower load margin is subtracted from the actual load, and the borders of the load range, thus obtained, are displayed as maximum and minimum permissible load, respectively.

7 LPBP SYSTEM

The LPBP system enables live steam to bypass the LPTs. These sophisticated systems as offered by different manufacturers have been working satisfactorily with much needed operational flexibility. The LPBP system was invented basically to cope with several transient modes of operation to ensure the protection of the plant components during various phases of turbine operation, which is a startup and shutdown requirement in many plants.

With LPBP and high-pressure bypass (HPBP) systems functioning properly, the steam generator in fossil-fired utility boilers can operate on standalone mode as per its own system and process requirement, bypassing the turbine mainly during startup, shutdown, and load shedding or any other type of huge load disturbance. It is, therefore, necessary that the LPBP systems must be adequately designed and sized to meet the requirements of different operating modes as decided by both owner and manufacturer.

The system helps not only in reducing time toward startup and reloading of the TG set but also in enhancing the availability of the plant components and sizeable improvement on equipment life. The LPBP systems have been used in large, coal-fired thermal power plants as well as in combined-cycle power plants.

7.1 Purpose of LPBP System

Detailed discussions are in Chapter 8, Clause 11.

7.2 Components of a Turbine LPBP System

As mentioned earlier, the main objective of the LPBP (and HPBP) system is pressure and temperature conditioning of steam with suitable pressure reducing cum control valve, compatible desuperheaters, and spray water control valve. As the process demands all the valves to reduce the pressure in any way, they must be suitable without generation of noise and vibration (associated with damage of valve trim due to wear and tear) beyond the limit values as stipulated by controlling authorities. The LPBP system also controls the condenser protection by closing the associated valves in keeping with the system design demands. There are block valves in both steam and spray water lines to accomplish this job. The pressure control valve and desuperheater may be combined into one IP and LP turbine HRH bypass pressure control and desuperheating valve.

7.2.1 Isolation/Block Valves

The isolation/block valves upstream each of the LP bypass valves are provided to close the LPBP system for condenser protection as mentioned earlier. Isolation valves are installed according to regulatory body stipulation to ensure that leakage does not occur so that fluid can enter the condenser in LPBP closed condition. This function could be integrated in a composite control valve, provided the isolation valve's purpose is not compromised. The isolation valve, therefore, may be separate from the control valve, or a combined valve having control function with a quick closing function and suitable leakage class, as per the owner's specification/stipulation. This type of valve is normally hydraulically operated for achieving requisite very high actuating force and fast-closing criteria.

7.2.2 Pressure-Reducing and Control Valves

7.2.2.1 Sizing of Valves

Normally sizing of the LPBP system valves depends on the main system design as stipulated by the customer or consultant on their behalf as envisioned to take care of the individual plant requirements for a specific capacity. However, the range may go up 100% of the steam flow generated by the boiler as MCR. The pressure reducing/control valve and the desuperheater are available in combined configuration for less space requirement and other advantages like low noise and vibration design. Sizing of these valves actually depends on the quantity of steam flow to be bypassed to the condenser through the LPBP system, which in turn is dictated by a number of other variables such as condenser internals, startup, loading, unloading, and shutdown practices and requirements for the unit, plant safety aspects, and of course the economic aspects.

There are a number of considerations and factors that determine the size of the bypass system. The combined HPBP and LPBP system may be selected to take care of the following operating conditions:

1. Fifteen percent of MCR flow when the system is to match the steam-to-turbine metal temperatures, which may save half an hour during startup.
2. Forty percent (or more) of MCR flow when the system is to match the difference between the supplied steam flow and the steam utilized for power generation and other uses during any disturbed or transient conditions.
3. One hundred percent of MCR flow to keep the steam generation uninterrupted at full load in case of a turbine or generator trip without wasting energy and steam through operation of the safety valves.

7.2.2.2 Noise and Vibration of Valves

The LPBP system valves are used to reduce pressure; and hence, they are subject to generate a high noise level. The noise level as stipulated by different controlling authorities for the thermal power plant is 85 dBA from the source of noise. The standard method of abating noise is using acoustic insulation or providing back-pressure diffusers, downstream from the valve plug and near the outlet of the valve body. Some manufacturers' designs provide multiple zigzag channels in the specially invented outlet port of the valve that divide the fluid flow path into several discrete paths, which increases the frequency of the generated noise. The adjacent piping and structures can easily absorb this higher frequency noise to such an extent that noise abatement is ≥10 dBA as compared to conventional noise-reduction methods.

Some manufacturers' designs incorporate special types of contoured profiles so that wave interference is produced. Due to this design, a majority of the aerodynamic effects are canceled out resulting in reduction of noise and vibration. The injection of spray water for desuperheating within the body also causes noise attenuation in the combined pressure reduction and desuperheating system (PRDS) if applied for LPBP. Noise generated by the valve can thus be kept below 85 dBA throughout the entire operating range. A highly turbulent zone is created just near downstream of the valve seat of the specially designed and contoured cage. For highly efficient mixing of steam and desuperheating spray water, the ideal place for in-body injection is this turbulent zone. The noise is automatically attenuated as water injection takes place inside the valve body, thereby ensuring that this turbulent zone is removed from the valve surface and the source of the vibration is eliminated.

Excessive vibration in LPBP valves can damage pipe hangers, valve-trim materials, and shake accessories of different parts of the actuators causing high maintenance costs and unscheduled downtime leading to a loss of revenue.

7.2.3 Desuperheaters

In modern thermal power plant application normally a direct water spray desuperheating system is used. This type of desuperheater represents the vast majority of desuperheating applications. In water spray desuperheaters, superheated steam is passed through a section of pipe fitted with one or more spray nozzles. These inject a fine spray of cooling water into the superheated steam, which causes the water to be converted into steam, reducing the quantity of superheat by the temperature of the steam. The cooling water may be introduced into the superheated steam in a number of ways, consequently, there are a number of different types of water spray desuperheater. The design of water spray desuperheaters are affected by the following factors:

1. *Particle size*: The smaller the water particle size, the greater the ratio of surface area to mass, and the higher the rates of heat transfer. Because the water is directly injected into the moving superheated steam, the smaller

the particle size, the shorter the distance required for heat exchange to take place. The water is broken into small particles using either a mechanical device (such as a variable or fixed orifice nozzle) or steam-atomizing nozzles.

2. *Turbulence*: As the flow within the pipeline becomes more turbulent, the individual entrained water particles reside longer in the desuperheater, allowing for greater heat transfer. In addition, turbulence encourages the mixing of the cooling water and the superheated steam. Increased turbulence results in a shorter distance requirement for complete desuperheating to occur. The phenomenon of turbulence can be attributed to the following factors:
   a. *Pressure drop across the nozzle*: Subjecting the cooling water to a higher pressure drop will increase its velocity and induce greater turbulence.
   b. *Velocity*: By increasing the overall velocity of the water and steam mixture, the amount of turbulence is inherently increased. The increase in velocity is usually achieved by creating a restriction in the steam path, which further generates turbulence by vortex shedding. In addition to these high velocities, if poor piping design practices are used, the speed of the superheated steam could, in theory, approach Mach 1. At such speeds a number of problems occur (including the generation of shock waves). However, this would be far in excess of the velocities used in good piping design. Typical velocities of steam entering a desuperheater should be ~50–60 m/s.

In LPBP systems desuperheaters are very important where comparatively low-temperature water is sprayed over the hot steam to lower the mixture temperature as desired by the process. A thermally efficient design of desuperheater bodies can eliminate the stresses caused by various reasons such as thermal transients associated with cycling duty.

In any efficient type of desuperheater design, complete evaporation of the injected water is ensured to avoid water particles hitting the adjacent walls of the valve or downstream piping. A high degree of atomization of the injected water is responsible for accomplishing a superior class of mixing with the steam and resulting in quick evaporation enhancing desuperheater performance.

For an efficient desuperheater, proper location and direction of the spray water jet is extremely important so that complete evaporation is attained before the first pipe bend to stop any high-speed water particles impinging on the pipe walls causing erosion. The relative speed of steam flow to that of injection water flow determines the quality of spray water atomization. Full atomization is always followed by high water injection speed and injection of the desuperheating water into a one of turbulent, high-speed steam flow. Other factors, such as piping layout, selection of spray water valve, flow range, and selection and performance of the control system, influence the performance of the desuperheater as a whole.

A desuperheater may produce good atomization when there is multistage pressure reduction to ensure smooth downward pressure drops to eliminate any cavitation, flashing, trim erosion, noise, and vibration, and ultimately, to maintain exact temperature control at all steam flows as required by the plant operating condition. The ideal place for water injection into a high steam-flow turbulent zone for efficient desuperheating is downstream of the valve body just after the plug/seat/trim assembly. Appropriate selection of material, properly designed arrangement of the injection nozzles, and shape and hole pattern of the cage around the injection zone may give the best results for fully atomized spray water before the steam leaves the valve body. The efficient implementation of the whole process results in the shortest evaporation length.

There are various types of desuperheaters: single-point radial injection spray, multiple-point radial injection spray, single nozzle axial injection, multiple nozzle axial injection, etc. In the following clauses commonly used desuperheaters will be discussed.

7.2.3.1 Ring Type Desuperheater

Ring type desuperheaters are also suitable for LPBP applications. Here, a separate desuperheating chamber is provided downstream of the pressure-reduction area. In a ring-type desuperheater, the shape of the steam-flow steam causes the creation of turbulence inside the chamber. The steam velocity is increased by a contraction of the flow path and then allowed suddenly to expand. The injection and mixing take place in this turbulent region so that the shortest distance as desired for atomization is ensured for proper evaporation.

7.2.3.2 Steam-Assisted Desuperheaters

This kind of desuperheater is also another type of atomization outside of the valve body. The design is based on a combination of high-speed water injection into a high-velocity Venturi steam flow. Cooling liquid is introduced into the steam header through a stainless steel nozzle assembly, which uniquely divides one large jet of liquid into many small jets. Prior to entrance into the main header, each jet is bombarded by a higher pressure steam jet, creating a fine mist, which enters the stream flow without the need of a thermal liner inside the main header. The desuperheater thus reduces the size of the liquid particles so that the droplets can be quickly and efficiently evaporated.

Downstream temperatures can be controlled to within a small value higher than that of saturation. The thermal sleeve around the liquid tube ensures uniform expansion with the steam tube, thereby minimizing thermal stresses due to unequal elongation of the liquid and atomizing steam tubes. The cooling liquid is introduced through a series of small orifices, which are drilled circumferentially into the nozzle. Small slots are milled 90° to the drilled holes. Atomizing steam passes through each of the slots and blasts each of the cooling liquid jets. Liquid pressure of only a few bars above the steam header pressure is necessary to inject the cooling medium. Hardfacing (stellite) in the nozzle head minimizes erosion wear and tear.

Some designs introduce a compound swirl-type nozzle, which provides high injection speed while the atomizing steam provides a velocity that is virtually independent of the main steam flow that may be taken from a different source altogether. The spray pattern is centered in the piping to produce even temperature distribution at all flow conditions. For this type of desuperheater, the distance required for atomization is the shortest of all downstream desuperheaters. This design provides significantly better atomization than the ring-type and spring-loaded nozzle desuperheaters currently available in the industry.

7.2.3.3 Spring-Loaded Injection Nozzles

Spring-loaded nozzles are typically used for LP bypass applications by introducing cooling liquids directly into the hot fluid. Multiple nozzles, which can be opened or closed to vary the coolant flow and provide a higher turndown than single nozzle units and where the amount of coolant flow is dependent only on the ability to vary the pressure drop across the nozzle, are used. The atomizing principle of spring-loaded injection nozzles is based on high-speed injection. Due to the design of the spring-loaded nozzles, a sufficient injection pressure and injection speed exists at minimum load. The injection speed results, not only in good atomization, but also in a sufficient penetration of the spray water into the steam flow. The design contains an integral spring-loaded flow plug, which moves in response to a change in pressure differential between the coolant inlet pressure and the main header pressure. As the plug moves, nozzles are uncovered introducing coolant into the main header. The varying inlet coolant pressure is controlled by a separate water or coolant control valve. The coolant enters the unit and passes through the center of the spring spacer, spring, and plug. Unbalanced forces created by the coolant over line pressure differential cause the plug to move, exposing more or fewer flow nozzles. The coolant is atomized as it passes under pressure through the spray nozzles, which further assist in evaporation of the coolant.

In some designs, a spring-loaded nozzle incorporates a unique swirl pattern, which provides rotational energy to the spray water over the entire range of flow conditions. This ensures an even distribution, which produces good mixing of steam and spray water with extremely good rangeability.

7.2.4 Instrumentation and Control System

This system was discussed in Chapter 8, Clause 11.

8 TURBINE CONTROLS: SEAL STEAM PRESSURE CONTROL SYSTEM

8.1 Seal Steam Pressure Control: Background

The turbine shaft has to exit its casings to couple or connect with the unit that the turbines drive (generator, reduction gears, pumps, etc.). The seal steam system enables the turbine to be sealed where the shaft exits the casing—in effect keeping "air out and steam in." Sets of labyrinth packing are used along the turbine rotor where the rotor exits the turbine casing to maintain this pressure differential between the inside and outside. The labyrinths create many little chambers causing pressure drops along the shaft. The number of labyrinth sets depends greatly on the possible steam pressure in that area. Labyrinth packing alone will not stop the flow of steam from the turbine high-pressure part or prevent air into the turbine connected to the vacuum.

The gland sealing steam system provides low-pressure steam to the turbine gland in the final sets of labyrinth packing. This assists the labyrinth packing in sealing the turbine to prevent the entrance of air from reducing or destroying the vacuum in the associated condenser. Excess pressure (excess gland seal) also has to be removed and put into the gland steam condenser. The large, turbine gland sealing systems are basically self-sealing, in the sense at a load >40% HP steam coming out from the HPT side could be enough to seal the same on the LPT side. Up to nearly 40% load low-pressure auxiliary steam is supplied to the sealing header to seal HPT, IPT, and LPT glands. During this period the flow line to the condenser is closed with the help of a control valve. Above this load point the steam coming out of the HPT will be more than enough to seal the LPT gland and more, and the excess is sent to the condenser. At around this load point the auxiliary steam supply control valve is closed and the HPT and IPT glands (as applicable) leakage steam is routed through the leak control valve to other glands so that seal steam pressure can be maintained and excess steam will be sent to the gland steam cooler/condenser. During startup, vacuum building is the first step. It is necessary that some steam is supplied to the turbine glands from auxiliary steam via the seal steam supply control valve. With load building inside the turbine, HPT and IPT leakage will go to the gland steam supply header to increase the pressure, which will compel the supply valve to close. Normally the auxiliary steam header is maintained at a fixed pressure and temperature (at 10 or 16 kg/cm^2 and ~210° C). However, because this seal steam is required during startup and low load the seal steam pressure and temperature requirement may be much lower than the auxiliary steam header pressure and temperature. Therefore, it is necessary for the PRDS to be considered (not shown).

8.2 Control System

Because this control loop is connected with the startup of the turbine, it is often supplied as an integral part of the turbine instrumentation package by the turbine manufacturer. Also, during startup, trip, and/or large throw off, there will be a great deal of disturbance inside the turbine pressure, so the control system needs to be designed to cope with these eventualities, and must respond fast to avoid unnecessary leakage in the turbine. For this reason, the actuators for these valves are often the electro hydraulic type. Also, some special characteristics are incorporated in the loop at the proportional part to accommodate varying conditions of seal steam under different operating situation/eventualities.

8.3 Control Loop

See Fig. 9.12 for an example of seal steam pressure control.

8.3.1 *Objective*

Seal steam from the seal steam pressure header is used to seal turbine glands so that steam does not come out of the turbine, and air does not go into the turbine. In both cases, there will be loss of turbine efficiency. The main objective of this loop is to ensure constant pressure at the seal steam header so that there is no leakage through any of the turbine glands.

8.3.2 *Discussions*

As discussed earlier, at lower load auxiliary, the required amount of steam is to the header, which is used to seal the glands in HPT, IPT, and LPT. Rather than flow to the seal steam header, pressure is controlled by regulating the supply steam control valve. As the load increases (i.e., >40%) when the pressure inside the HPT and IPT increases, leakage from these glands occurs. The supply steam from the auxiliary steam may not be necessary and it will self seal, i.e., this leakage steam from HPT and IPT glands is utilized to seal the LPT gland and excess steam is regulated via leakage control valve.

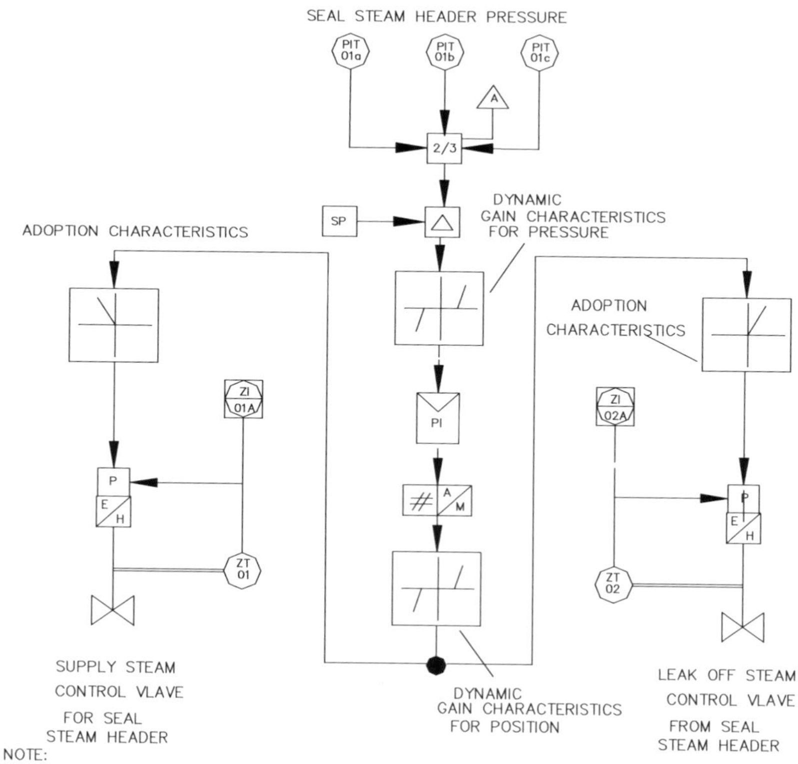

FIG. 9.12 Seal steam pressure control.

8.3.3 *Control Loop Description*

As discussed in the previous clause, there will be two control valves regulated to control the header pressure. The header pressure is measured by several pressured transmitters with due redundancies. It is important that redundancies are considered, but it depends on the manufacturer and owner. The measured value is compared with the setpoint, which could be set from a control room monitor and/or control plaque as applicable. The error thus created is fed into the controller. However, prior to sending the error signal to the controller it is passed through a dynamic gain characterizer. Apparently the loop is quite simple, but one has to bear in mind the eventualities.

During turbine trip, shutdown, quick loading, and large load throw off, there will be a huge fluctuation in steam pressure inside the turbine and these would take place at a very fast pace. To cope with such situations, the error signal thus created is passed through the dynamic gain characterizer so that when there is an error signal that is more than the preset value, the gain will change to a greater value. Output of the controller is fed to a proportional controller via auto/manual (A/M) station. So, position can be controlled manually as well. A dynamic gain characterizer is also kept here to meet the challenges discussed earlier.

Finally, the output is fed to the two control valves. As discussed earlier, the supply valve normally operates up to a preset (i.e., 40%) load, and after that it will close. Whereas, the leakage control valve comes in to the operation load more than the preset value (i.e., 40%), and at a lower load it remains closed. To utilize this characteristic to regulate two control valves from one output, each of them are given adoption-characteristic function generators to operate both valves in split range. Normally, the final actuator of these valves are hydraulic, so the electronic signal from the loop is fed via an electro hydraulic converter with a suitable positioner. In most cases, this control system forms a part of the integral instrumentation package of the turbine.

9 HYDROGEN SEAL OIL SYSTEM AND DIFFERENTIAL PRESSURE CONTROL

9.1 Hydrogen System

The brief description in Chapter 2, Clause 4.1, incorporates the requirement and basic process of hydrogen and seal oil system. Hydrogen leaking because of a dangerous explosive mixture with atmospheric oxygen is prevented by oil sealing supplied with a slightly higher oil pressure. A CLCS is provided to maintain seal oil to hydrogen differential pressure (DP) control along with hydrogen and seal oil temperature control.

9.1.1 *Moisture in Hydrogen Cooling System*

In its delivered form, whether from an external or in-house source, hydrogen gas has a dew point of approximately −50°C. The hydrogen dew point for this type of requirement is normally maintained at <0°C. Problems may arise when the generator is in shutdown condition, and hydrogen moisture levels increase beyond this point. The water particles deposit and cause the retaining rings of the rotor windings to crack if there is moisture on them. There are generators cooled by hydrogen with dew points of −30°C or lower depending on the climatic condition of the site.

9.2 Seal Oil System

Hydrogen is hazardous if it leaks and mixes with air with the possibility of an explosion. Due to the necessity of preventing leakage of hydrogen along the rotor shaft, a sealing system consisting of oil with pressure slightly more than the hydrogen pressure was developed. It receives makeup oil from the bearing lubricating oil system. The seal oil system not only prevents the escape of hydrogen from the generator casing, but also stops the ingress of air into this casing near the rotor shaft projected outside of the casing at both ends.

Seal oil pressure at the seal housing is ~0.25–0.4 kg/cm^2 higher than the generator coolant hydrogen gas pressure. The sealing arrangement is such that the seal oil flows in both directions, that is, the air side and the hydrogen side along the shaft. As the seal oil pressure is slightly more than hydrogen-side pressure, some oil flows toward the hydrogen (6–7 kg/cm^2) thus blocking the flow of hydrogen to the atmosphere along the shaft. Conversely, most of the seal oil flows toward the atmospheric pressure zone through the bearing housing. On the air side, the air-oil mixture combines with the bearing lubricating oil, while on the hydrogen side, some hydrogen gas mixes with the sealing oil. In some generator cooling systems, the bearing cooling oil and seal oil systems are different, whereas most of the manufacturers of generator cooling systems offer combined lubricating oil and hydrogen seal oil systems.

9.3 Description of Seal Oil Flow Diagram

The seal oil system supplies seal oil to the seal plates of the gas-tight TG set and all types of operations, irrespective of normal or emergency conditions. A DP between the seal oil and hydrogen gas is maintained to prevent hydrogen leakage (Fig. 9.13).

This system normally takes its suction, either from the governing or lubrication pump discharge, with standby emergency supply from AC and DC seal oil pumps. Alternatively, the suction is from the oil reservoir and supplies oil

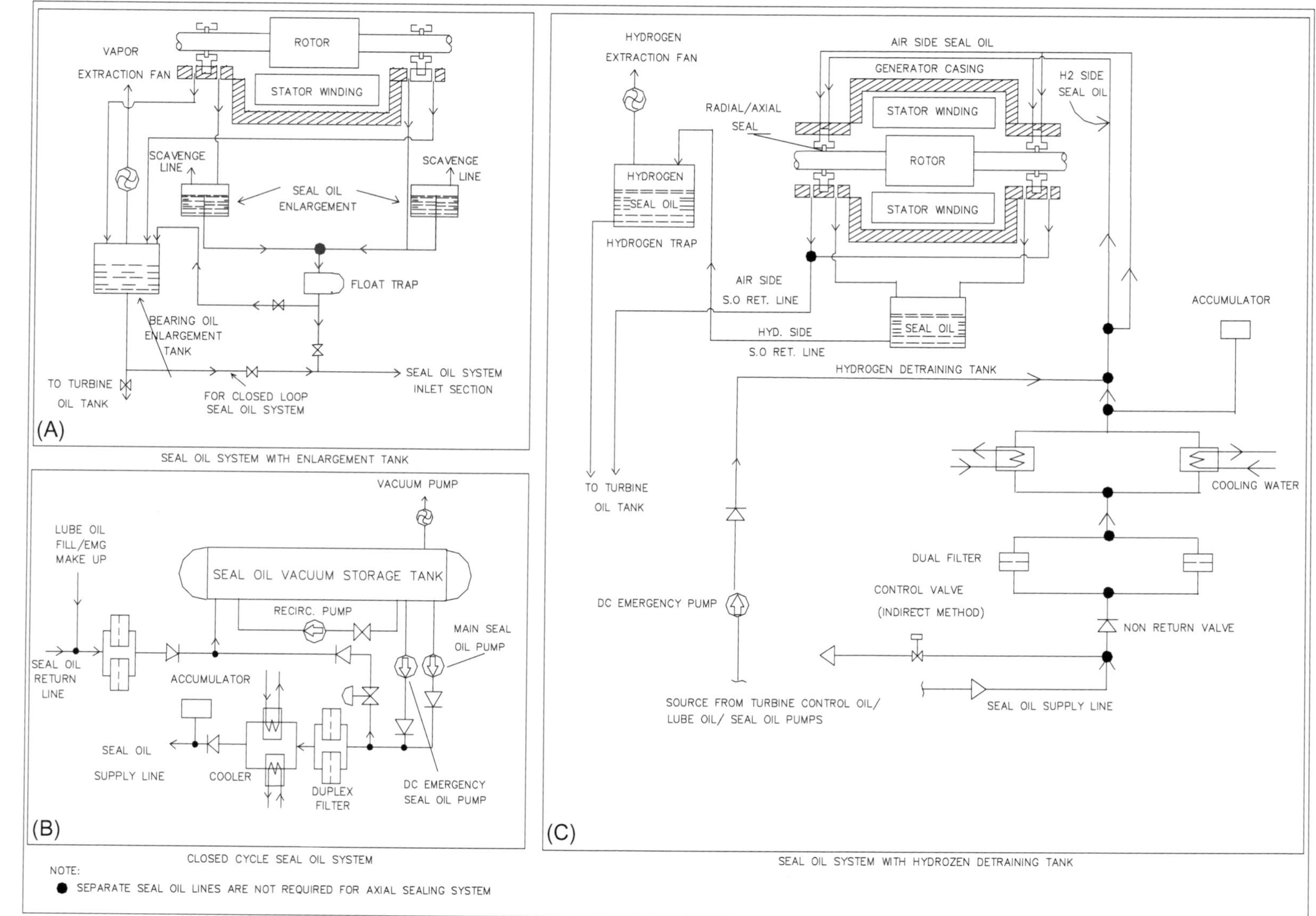

FIG. 9.13 Seal oil flow diagram.

to the seal plates with the help of separate seal oil pumps with sufficient redundancy. There is a recirculation line connected from seal oil pump discharge lines to the reservoir through a control valve, which modulates to maintain the DP between seal oil and hydrogen. The standby pump starts automatically when the pump discharge header pressure falls below a preset value through interlock with the pressure switch. There may be a flow switch instead that actuates the standby pump when the bypass flow falls below a preset limit. Many designers prefer the use of the flow-measuring device rather than DP sensors, because this flow-associated system can detect many types of malfunctions and anticipate the need for initiating standby pump operation before seal oil DP actually begins to drop.

In many thermal power plants the seal oil system is combined with the lubrication oil system of the turbine. The lubrication oil pump provides oil to the seal plates through a seal oil pressure-regulating valve. The backup provisions made for the lubrication system are also available for the oil supply of the sealing system, which makes the system simplified without requiring a separate source of seal oil, but necessitates a relatively high-pressure lubrication system. These systems typically operate with a pressure differential of $\sim$0.25–0.4 kg/cm^2. This is necessary to eliminate hydrogen contamination by air ingress in the seal oil, which is forced to flow along the shaft toward the hydrogen-filled generator.

9.3.1 Requirement of Accumulator

To cope with abrupt change or failure, the standby pump cannot be brought up to full pressure immediately and for maintaining uninterrupted seal oil flow an accumulator may be provided, which supplies seal oil for a few minutes when there is a sudden drop in seal oil pressure below the normal operating value. The pressure- or flow-measuring devices are excellent in detecting maloperation when the seal oil supply decreases at a relatively slow pace. The non-return or check valve in between the accumulator and the pump discharge header (after the recirculation line) prevents seal oil from flowing back and supplies the seal plate.

9.3.2 DP Control System

There are two types of control philosophies: indirect and direct. The indirect type has the control valve placed in the recirculation line, which modulates as per the output of the controller's measured variable—the DP between the seal oil and hydrogen against a fixed setpoint. The amount of excess oil necessary to maintain this DP returns to the reservoir (Fig. 9.14C and D).

The direct type envisages the control valve in the pump discharge line, which regulates the DP by directly allowing oil flow (Fig. 3.52) as per the controller output. In some double-injection types, separate sets of seal oil pumps are provided for the hydrogen-side and air-side sealing systems and DP is controlled only for the hydrogen side; one control system each for the exciter end and the turbine end is also provided. Seal oil and hydrogen temperature controls are shown and incorporated in Figs. 9.22 and 9.23, respectively.

9.3.3 Type of Pumps

There are three types of pumps available:

1. *Screw*: Generally used for the services where constant flow with high pressure is the requirement.
2. *Centrifugal*: Generally used where constant pressure is required for the system.
3. *Positive displacement gear type*: Recommended for constant flow services.

All of these pumps are supplied with an integral safety relief valve at the pump discharge side to return to the reservoir to avoid high-pressure buildup over and above the designed pressure. The pump discharge head is calculated on the basis of maximum generator hydrogen gas pressure plus the DP to be maintained between seal oil and gas, pressure drop of the system, and 10% spare capacity.

9.3.4 Requirement of Filter

Whatever the type of sealing arrangement, the effect of generator contamination is the same. Shaft-sealing system failure may take place when a particulate contaminant enters the seal oil system due to wear of the gland seals. This untoward situation lets hydrogen escape the generator and results in a forced outage with particulate contamination in the sealing process.

To circumvent particulate contamination, filtration of sealing oil is essential before it reaches the generator seals. Ensuring the proper mesh size of the filters prevents passage of particulate coarse enough to cause damage to associated materials. The international guideline as per ISO 4406 is generally followed to minimize erosion and maximize the component life of the seal oil system. Normally, the above guideline is expressed as "cleanliness level" and for the generator seal oil system it is recommended to be maintained at an ISO 14/12 cleanliness level. ISO 14/12 is now a standard recommended practice and it specifies using a 3–5 μg filter with a duplex construction and online changeover facility.

9.3.5 Requirement of Seal Oil Coolers

Note that function of seal oil is not only to arrest leakage of hydrogen gas from the generator but also to keep the seal metal at an acceptable temperature limit. During the sealing process, the seal oil takes the heat away from the sealing surface and experiences some rise in temperature that needs to be transferred to another cooling media to maintain the closed loop seal oil system. Redundant seal oil coolers

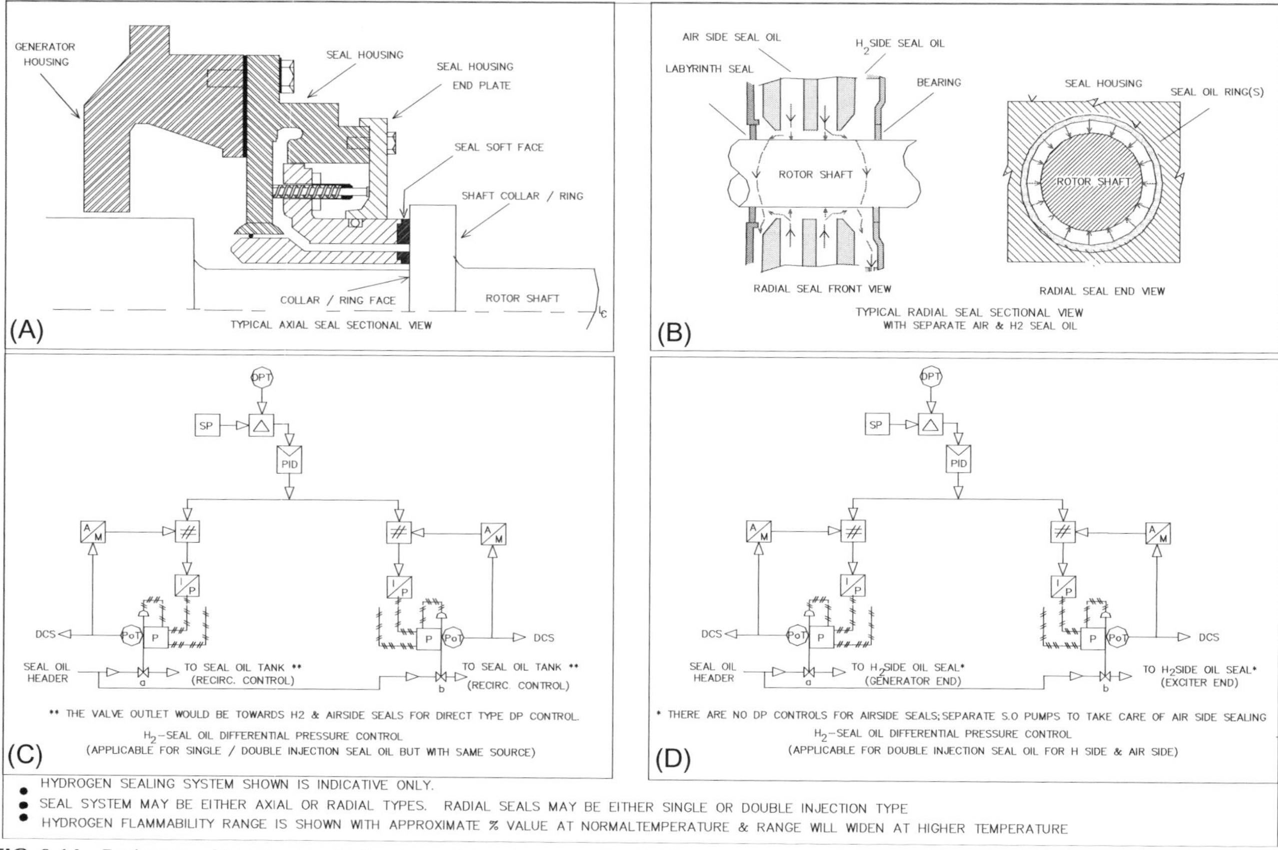

FIG. 9.14 Basic type of (axial and radial) seals used in hydrogen and seal oil systems.

are provided for that purpose and the cooling media is the water supplied from the clarified cooling water header.

9.3.6 *Basic Types of Seals*

Hydrogen seals are metal rings normally made of babbitt, bronze, brass, or maybe of carbon. They are fitted using precise geometry to maintain proper sealing over the total range of operating conditions. The seal system may consist of a single seal or a separate seal for hydrogen- and air-side leakage prevention. The seals are categorized into two basic sealing systems, which are widely used:

1. Axial ring shaft seal or thrust collar type
2. Radial oil seal or journal (ring) type

For the axial ring shaft seal (Fig. 9.14A) the seal ring floats on the shaft and does not rotate. The seal ring is held against the shaft ring face with the help of axial force exerted by the spring and partly by the seal oil pressure. The seal ring at the front of it is provided with a continuous soft metal face. The seal oil pressure forces itself to flow toward the seal face while the majority of the oil flows out and the rest flows in toward the rotor shaft. The pressurized oil is supplied at a particular pressure to both sealing sides to maintain required oil-film thickness formed in between the soft seal metal of axial thrust face and the rotating collar, which acts as lubricating oil and as a cooling agent for the heat generated from friction and to lessen the wear and tear caused by the same.

Radial oil seal (Fig. 9.14B) or journal-type seals are the most common type of seal. These seals are different, because they do not have any contact with the turbine shaft and thus require no soft metal seal face. The inner diameter of the seal is slightly larger than the outer diameter of the rotor shaft and is held tightly around the shaft with springs. The ring with the narrow clearance is placed between the sealing ring holder/bracket and the rotor shaft. High-pressure oil is injected from a set of holes/ports and grooves into a stationary seal ring surrounding the rotating shaft. High-pressure oil is thus spurted onto the shaft toward both sides of the seal. These seals are sometimes referred to as hydrogen gland seals. If only the ring is pressed without any clearance from free movement, the oil injection can be avoided. But in that case, friction between the ring and shaft develops and causes damage to both of them. The pressurized oil acts here as a cooling and sealing media, and lessens the DP across the seal (without seal oil the DP would have been the generator casing hydrogen pressure).

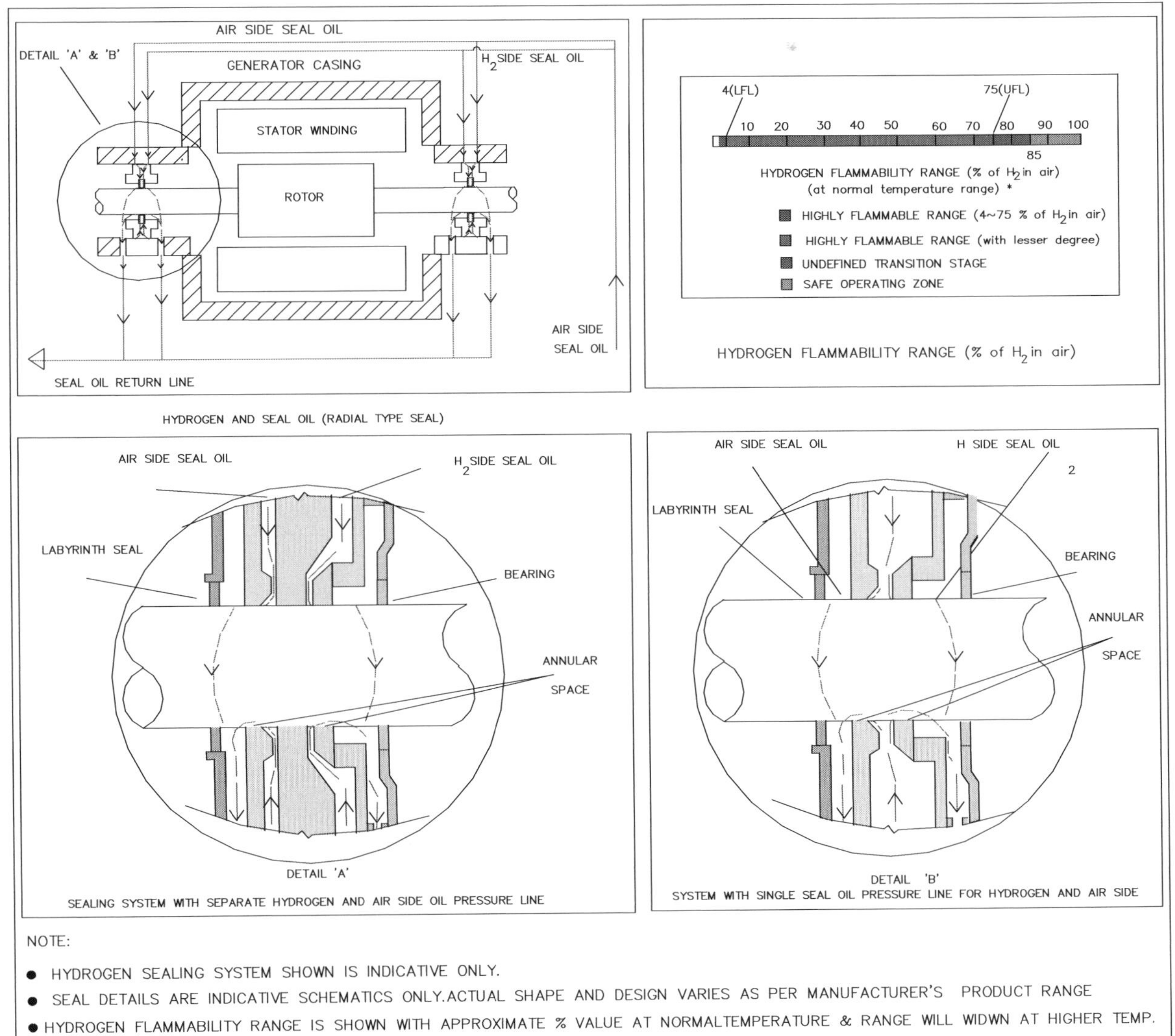

FIG. 9.15 Different types of radial seals used in seal oil lines and hydrogen flammability range.

The seal oil system may consist of both single or dual/multiple (for air and hydrogen side) pressurized oil lines. Fig. 9.15 incorporates the views of single seals and two separate hydrogen-and air-side seals. Two sets of holed rings are for the hydrogen side and the air side with a vacuum ring located in between these two rings to extract the oil from the seal area. These rings can accommodate both axial and radial shaft movements. Radial seals are used for both separate and common seal oil pressure lines as depicted in Fig. 9.15.

9.4 Hydrogen and Air Removal From Seal Oil

Other critical components of the seal oil system are included in the hydrogen and air removal systems, which supply clean and pure oil to the extent possible for sealing purposes. Examples are listed below:

1. *Degasification through spray nozzles*: Degasifying plants is a very common method of removing entrained hydrogen and other gases through "vacuum-treat" of the seal oil before applying it to the seals. The same is done directly in the main seal oil supply tank when separately provided. The seal oil is extracted right from the sealing location into this tank, which is maintained under vacuum. The oil reaches the tank through a spray nozzle and is converted into a fine mist that supports the removal of dissolved gases from the oil, and in turn, the system. In this type of plant, an additional recirculating pump is also provided to recirculate oil back to the supply tank through a series of spray nozzles for continuous gas removal (Fig. 9.13B).

2. *Removal through detraining tank*: A separate "detraining tank" is provided along with a "hydrogen trap" and "extraction fan" to accomplish removal of entrained hydrogen from the seal oil system. Due to the configuration of the seals and sealing arrangement, the majority of seal oil flows to the air-side drain of the seals and is led to the main/seal oil tank through bearing sumps, etc. A small portion of seal oil flows toward the hydrogen side and drains through to reach the detraining tank. The two sides of the tank are connected through pipelines of sufficient bore diameter to enable seal oil to drain and hydrogen to pressurize the tank. Two lines are used to keep the tank under pressure balance. The entrained hydrogen is almost separated from the oil, and the oil accumulated at the bottom of the tank is forced upward to the hydrogen trap by the generator hydrogen pressure. The trap is equipped with an extraction fan, which removes any remaining entrained hydrogen by venting it to the atmosphere at a safe place. The oil in the hydrogen trap, however, returns to the turbine oil tank through gravity action (Fig. 9.13C).
3. *Removal through scavenging via enlargement tank*: It is possible that air is entrained in the source of the seal oil, which may be released from the seal oil that flows into the generator-side hydrogen area causing reduced hydrogen purity if not removed. The removal process is popularly known as "scavenging." The hydrogen-side seal oil drain is arranged to accumulate in the enlargement tank whose oil level is maintained though overflow pipe. The tank may be a single type with one vertical separator or one for each side. The two overflow pipes are connected via loop seals to a float trap. This arrangement assists entrained hydrogen and mainly air bubbles to release and gather at the top of the tank. A small amount of gas (air and hydrogen) is vented from this seal oil enlargement tank where the entrained air is expected to be released from the seal oil to the atmosphere at a safe area. The seal drain enlargements are vented through the needle valves in the scavenging lines, and scavenging flowrates can be controlled according to the hydrogen purity reading. Because a small amount of gas (air and hydrogen) is continually vented to the atmosphere through the scavenging system, the pressure inside the generator decreases and must be replenished into the system in the generator casing from a source of high-purity hydrogen (Fig. 9.13A).

The air-side seal oil is also taken to a vessel known as a bearing drain enlargement tank with maintained oil level with mainly air on the top where it is released to the atmosphere through a vapor extractor.

The hydrogen-side seal oil drains after the float trap is sent to the bearing drain enlargement tank for normal plant operation, or sent to the seal oil recirculation line for standstill condition. For the common turbine oil tank, vapor extractors are provided to release the entrained air (mainly) to prevent air ingress in the generator casing through the generator-side seal oil as the only source.

In some closed loop seal oil systems, the hydrogen-side seal oil drain enlargement tank is also called a "H_2 detraining tank" without a scavenge connection line, whereas the air-side oil drain enlargement tank is called an "air detraining tank" with a vent connection to release entrained air and hydrogen. The major gas removal is done at the main seal oil tank as stated above.

9.5 Discussions on and Problems With Hydrogen and Seal Oil Systems

With aging and harsh process conditions, the geometry of various parts of the seal, such as rings, brackets, holding springs, etc., may change due to wear and tear or even by premature damage resulting in excessive leakage of seal oil. For example, dirty oil can also be responsible for wear of the seal rings due to abrasion. The application of the type of seal oil system depends on the type of shaft seal used in a particular generator. Axial and radial movement (although very small) and vibration of the generator rotor, drift in set value, or any other problem of the seal oil/hydrogen DP control system may be the source of increased seal oil flow.

In some generator cooling systems, the bearing cooling oil and seal oil systems are different; in that case, the air-oil mixture of the air-side, mixes with the bearing lubricating oil, while on the hydrogen side, the sealing oil drain, mixed with hydrogen gas, returns to the seal oil reservoir. The trouble with having individual seal oil systems is that somehow, at some point, the seal oil and bearing lube oil mix with each other, which causes problems. Most of the systems, however, provide combined lubricating oil and hydrogen seal oil systems to avoid this problem.

A great deal of care is taken to avoid mixing water with lubrication oil as it causes many problems. As per ASTM D95, the concerned regulatory authority, the limit value imposed on the presence of moisture in the lubricating oil is specified as 2000 ppm (0.2%) for a turbine/generator of a typical thermal power plant. The hydrogen dew point is affected with the amount of moisture present in the lubricating oil, so users prefer the moisture concentration maintained at a value well below 2000 ppm, if not 1000 ppm. There is always a chance of the contaminated lubricating oil mixing with the seal oil where in turn it can mix with the hydrogen gas.

In service, the hydrogen gas dew point usually is required to be maintained between −30 and 0°C depending on the seal oil system clearances and operation, size of the generator, and on the hydrogen gas pressure (higher gas pressure calls for lower dew point). It is obvious that generators operating with high dew point are very prone to insulation failure in the windings.

The moisture content in hydrogen is important because hydrogen gas is hygroscopic at about −50°C dew point, and looks like a dry sponge in the presence of oil with 1000 ppm water. Because of this, hydrogen may become physically "changed," for example, by an increase in stickiness or volume.

It is well known that one of the common sources of moisture contributing to the deterioration of hydrogen dew point is the presence of water in the hydrogen seal oil system. Experiment results indicate that <50 ppm of moisture in seal oil is obligatory to maintain hydrogen dew points below −10°C (provided other sources are negligible). Seal oil moisture >50 ppm increases the hydrogen dew point, and ultimately increases the dew point above acceptable limits and at point of seal oil moisture presence exceeds ~250 ppm of water.

Another important source of moisture in hydrogen gas is leakage from the cooling water provided by the hydrogen coolers. The dryers (and desiccants) used to remove moisture in the hydrogen gas path require regular and proper maintenance.

It is, therefore, critical that necessary actions are taken to limit the moisture level. The major source of water ingress in the TG set lubricating oil is the condensed water from inadequate performance of the steam sealing arrangement of the turbine. If the lube oil and seal oil use the same facility, this type of water contamination needs to controlled. It is a "must" situation to find the actual source of the seal oil moisture and attend to it to avoid destructive consequences. Routine analysis of moisture content in both lubricating oil and the hydrogen seal oil (whether common or separate) will ensure the correct levels are maintained, for example, in the range of 10 ppm (0.001%) for hydrogen seal oil. If the moisture content of hydrogen seal oil rises above 250 ppm, it is possible that the dew point of the hydrogen could shoot up beyond acceptable limits.

Increased hydrogen gas dew point in the generator cooling system affects the overall reliability of the generator and must be a priority. The formation of water vapor or moisture contamination inside a generator reduces the life of various generator components.

The reason moisture in hydrogen gas must be maintained at acceptable levels is the stator winding short circuit and failure due to stress corrosion cracking of retainer rings used to hold the rotor winding in place. If the retaining rings are not in place, the winding is flung from the rotor due to the centrifugal force caused by the rotor speed (typically 3000–3600 rpm depending on the country). The apprehension is quite legitimate as the repairing time may take 3–4 weeks at a minimum, which means a unit outage with costly repairs and loss of revenue. The worst-case scenario is that the failure of the retaining ring destroys the generator altogether.

The normal practice in a modern thermal power plant is to maintain an upper limit dew point at <0°C. Dew points higher than this might allow moisture in the hydrogen gas to condensate on the retaining rings when the unit is out of service.

In addition to dew point problem, water present in seal oil can be evaporated due to heat produced by the churning effect when it comes in contact with the high-speed turbine shaft. This water vapor, if it enters the generator shell, reduces the purity of the hydrogen gas. Additionally, the entrained air may also be released, if not properly taken care of, from the oil and mixed with the hydrogen, which further reduces hydrogen purity. Decreased hydrogen purity, as already discussed, results in increased windage loss with its consequent problems.

It is, therefore, very important that the concentration level of water and entrained air in the oil should be kept well below the limit. The seal oil reservoir, whether common with the turbine lube oil system or separate, needs suitable measures against water and air contamination. Strict stipulations call for the oil purifier to ensure 100% removal of free water and air, and up to 80% of the dissolved water and air present in the oil.

10 GENERATOR CONTROL SYSTEM

The generator control system normally consists of the following items:

1. Generator voltage control by AVR.
2. Deexcitation system of the rotor.
3. Generator protection system.
4. Instrumentation system.

Main systems pertinent to the generator are discussed in Chapter 2, Clause 4.2.

10.1 Generator Voltage Control by AVR

Power demand is increasing for large-capacity thermal power plants, and in light of modern lifestyle requirements, increasing population, etc., large-capacity thermal power plants are being developed at a faster rate. As a consequence to this increasing constant and huge power load, there is the possibility that stability of both power and voltage may deteriorate in the power system. This is a burning issue that needs to be addressed.

10.2 Effect of Excitation Control on Power System Stability

The rotor field excitation is controlled by the AVR. This control system is very effective during steady-state operation; however, during abrupt load changes or disturbances, there may be some negative influence on the corrective action taken toward damping of power swings. The reason for this is that the AVR always try to force field-current changes in the generator. The modern tendency is the introduction of an additional control loop to supplement the AVR action, which is supposed to suppress the untoward system swings, such as voltage oscillations. This supplementary control loop is popularly termed the power system stabilizer (PSS). It produces a correction signal at its output to compensate for the oscillations of all the electrical parameters.

The typical range of associated oscillations in frequencies without PSS might go up to ±3.0 Hz and cause problems in smooth transmission of power to the grid. The measured variables included in the PSS input lists include:

1. MW generation or active power.
2. Generator current.
3. Generator EMF.
4. Speed/frequency (real-time and set values).

Whenever a number of generators is connected in parallel, disturbances may arise from sharing power in between the generators. This needs quick suppression and a steady power must be maintained to return the power stability. Power system stability, in turn, is associated with voltage stability and ensures a constant voltage restoration. This can be maintained even when there are changes in load by taking care of the most severe operating conditions. The improvement of power system stability, however, needs the improvement of all other associated systems, such as by increasing the system voltage, availability of adequate power transmission lines, availability of sufficient numbers of capacitors to improve power factors to decrease reactive power, installation of the static volt ampere (VAR) compensator (SVC), if not already installed, etc., and finally, an improved method of generator excitation control system. Introduction of an improved control approach by digital control hardware and software makes it possible to haul out the capability of the generator to the maximum extent by improving the control algorithm.

10.3 AVR by Rotor Excitation Control

Excitation current needed for the generator is provided by the excitation system. The AVR is a vital component of that system along with the exciter or the power source, the measuring elements, the PSS, and the protection unit.

The excitation power source can be from the exciter, which is either a separate DC or AC generator. The exciter has its field winding (DC current) in the stator and armature winding in the rotor. In the case of an AC exciter generator, the three-phase AC is induced in the rotor winding, which is rectified using diode, thyristor, or transistor bridge installed in the rotor. However, for a brushless excitation system and with a pilot exciter, its armature in the stator and field is a permanent magnet. The main exciter, however, is the AC generator at the rotor. Different options and alternatives of excitation systems are depicted in Fig. 2.40.

Rotor excitation systems are, in general, broadly classified into three groups according to the power source used for excitation (IEEE, 2006):

1. *DC excitation systems*: Use DC generators to feed the field windings of the synchronous machine.
2. *AC excitation systems*: Use AC generators with the help of rotary or static rectifiers to feed the field winding of the generator.
3. *Static excitation systems*: Use transformers and rectifiers for conversion of AC to DC current for exciting the generator field winding.

Another general and broad classification of excitation system also exists, which categorizes by excitation power source. The two major classes:

1. *Separate excitation systems that are be static or brushless*: These systems are independent of disruptions and faults that occur in electric power systems, and can force excitation. Brushless systems are used for excitation of larger generators (power generation at ~600 MVA) and in flammable and explosive environments. Brushless systems consist of an AC generator, rotating diode bridge at the rotor, and field at the stator. When this system is equipped with a pilot exciter, it consists of another AC generator at the stator and realized with permanent magnet excitation at the rotor. Attempts to build a brushless system with a thyristor bridge were not successful because of problems with thyristor control reliability. The result of this problem is a significant disadvantage of these systems as well as the inability to provide generator deexcitation. Another disadvantage is slower response of the system, especially in case of low excitation (Fig. 9.16).
2. *Self-excitation systems*: The advantages of this system are simplicity and low costs. The thyristor or transistor bridge is supplied from generator terminals via transformer. The main disadvantage is that excitation supply voltage, and thereby excitation current, depends directly on generator output voltage. Brushless self-excitation systems also exist, although they are not used much.

Initially, the generator output voltage was controlled by a separate small generator or exciter machine coupled to the shaft of the generator. The field was installed on the stator with AVR controlling its input current. The exciter rotor acted as a DC generator, and output of the exciter is then controlled by AVR to deliver excitation for the main generator DC field through slip rings.

The above system introduced lag in buildup of magnetic fields in both the exciter and the main generator. Therefore,

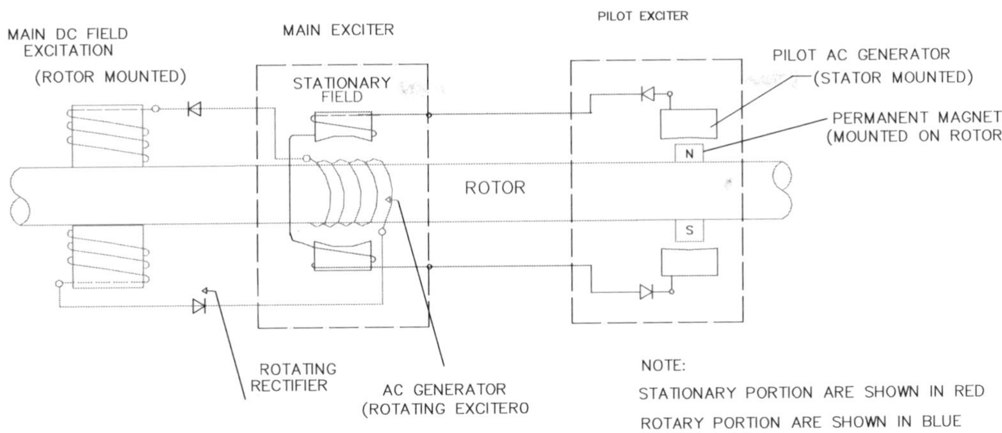

FIG. 9.16 Schematic diagram of brushless excitation system.

the idea of a self-excitation/shunt excitation system was developed. In this system, the separate exciter machine was eliminated and the power source was utilized directly from the generator output terminal with suitable controlled output rectifiers for the DC field excitation circuit. The advantage of this excitation system is that it can change the output voltage instantaneously to deliver the required current as needed to control the main generated voltage. Although the lag in the main generator field winding still persists, as dictated by its time constant, the availability of a higher voltage source to supply instant required field current reduces the lag.

In the self-excitation/shunt excitation system, generator output voltage is not available in the initial stage of starting up of the TG set. To take care of this situation, initially the excitation system was flashed by a brief application of DC from the station battery. This procedure helped to develop adequate field strength for generation of sufficient terminal voltage, which in turn could be fed back as a power source to start the normal excitation system. Some problems still exist in this system; as the machine starts at slow speeds, the excitation system needs to be on from the beginning. This type of field-flashing excitation is suitable for axial flow turbines where the turbine is already at a fairly high speed.

An alternative scheme was necessary to avoid all of these problems, which meant provision of another source of power during the TG startup time. The startup excitation would continue until the TG set is ready to produce prerequisite power to feed the self-excitation system. At this point, the excitation power source is switched over to a system connected to the generator output terminal. The power source of startup excitation may be from the station transformer or a diesel generator or a gas turbine, which is supposed to be available all the time.

The rectifiers available today are the thyristor-based bridge circuit with a digital voltage control system. Other important and necessary accessories include field circuit breakers, field discharge resistor, potential transformers, AC input circuit breakers, switch fuse unit, etc.

10.3.1 Brushless Excitation System for AVR

The AVR has slip rings, brushes, and commutators and is a bit cumbersome, hence the development of a brushless excitation system, which is widely used to provide DC to develop the rotor magnetic field for the main generator. The brushless excitation system consists of a main exciter and a pilot exciter. Fig. 9.16 shows an option for the arrangement and location of different accessories of the excitation system of a synchronous generator.

The pilot exciter includes a stationary armature winding in the form of an AC generator with rectifier and the DC magnetic field in the form of a permanent magnet, and is mounted on the same rotor shaft as the main generator of the TG set. The main exciter, on the other hand, includes a stationary DC magnetic field and an armature winding in the form of an AC generator with the rectifier on the same rotor shaft as the main generator of the TG set.

Whenever the TG rotor shaft rotates, an electromotive force (emf) develops across the generator or armature terminals of the pilot exciter due to the effect of the rotation of the magnetic field of the permanent magnet. The AC voltage thus produced is converted to DC voltage from DC by the rectifiers. This rectified DC output is then fed to the stationary field winding of the main exciter. As soon as this current flows through the field, emf develops across the rotating generator terminals of the main exciter due to the effect of relative motion of the magnetic field and exciter generator winding. This AC voltage is again converted to DC voltage from DC by the rectifiers. This rectified DC output is then fed to the rotating field winding on the rotor shaft of the main generator.

As the main exciter generator, its associated rectifiers and the main generator's field winding are all mounted on the rotor, and the connection among them does not require any sliding contacts in the form of slip rings, brushes, etc. Thus function of a brushless excitation system is achieved. The use of the brushless arrangement improves the reliability/availability and efficiency by lessening the losses. The maintenance problem is also reduced. Another kind of brushless excitation system is seen in Fig. 2.40.

10.3.2 Use of Thyristors/Semiconductor in a Brushless Excitation System

Normally thyristors are used in the circuit of rotating rectifiers of the main exciter instead of semiconductor diodes because of the following:

1. Higher current handling capacity and their suitability for the rotating machine application.
2. Thyristors are less sensitive to vibration, accelerative force, and extreme weather namely temperature.
3. Output currents can be controlled smoothly over a wide range for both normal operation, that is, forcing mode of operation and deexcitation mode of operation, which is also known as counter-excitation.

Many prefer power transistors to thyristors because of low (junction capacitance) dV/dT effect and better switching actions. Insulated gate-bipolar transistors (IGBTs) are popular because of their input MOSFET advantage, coupled with the bipolar advantage of bipolar junction transistors (BJTs). IGBT use in rectifiers as well as in AC drive control is very popular. Switching time of the IGBT may not be like MOSFET, but is faster than BJTs. ABB Unitrol is an example of excitation control by IGBT.

10.3.3 AVR Controller Output and Thyristors Gate Control in Brushless Excitation System

The digital or microprocessor-based control system takes the measured variable from the PT at the point of the generator output terminal, and checks the error value by subtracting it to the set value as desired. The controller output is in the form of pulses of equal magnitude, but their timing of appearance at the thyristor gate, that is, the firing or triggering gates, varies as per the controller output. Power transistor digital control systems issue current pulses to control the input base circuit, but for the IGBT, it is a voltage-controlled device.

10.3.4 Effect of VAR Control on AVR/Rotor Field Current

On many occasions, the rotor fails due to very high current flowing through it, which is needed to maintain generator terminal voltage. The highly mechanically stressed rotor insulation, when subject to excessive heat due to high rotor current, may fail at an earlier stage than the normal expected life. As the rotor repair work is time-consuming and costly, great effort is made to lower the rotor current to a value less than the limit, but still at a safe and stable one. By using a suitable method and hardware to control reactive VAR, the generator output current may be reduced with remarkable improvement in the power factor, which in turn, would require less generator emf to maintain the output terminal voltage compatible for connection to the grid.

10.4 Deexcitation System of the Rotor

Deexcitation is the withdrawal and suppression of excitation field current of the main generator when a major fault, namely a short circuit or excessive load current, takes place requiring immediate shutdown of the power-generating plant. A conventional deexcitation system requires shorting of the main generator's magnetic field windings on the rotor so that the flux becomes zero and eventually the generator terminal voltage also becomes zero.

The brushless deexcitation system, on the other hand, sends the current in the other direction for a short time by triggering the rotating rectifiers to be conductive only during the negative half-cycle of the main exciter generator or armature winding voltage waveform. As a result, during this part, the polarity of the generator field winding in the rotor becomes reversed. Due to this action, the resultant current through the main generator field winding decreases to zero at a fast rate, allowing the generator stator winding-induced voltage to become zero. The time taken for deexcitation is faster than the system by shorting the rotor terminals.

Another procedure for deexcitation is the static deexcitation system, which is the fastest deexcitation system and works by subjecting the rotor field winding with a reverse polarity voltage across its terminals with the help of an external source and through slip rings, etc.

10.5 Generator Protection System

All signals for a possible cause of tripping the generator and turbine as a protection measure are indicated in Clause 3.2. This is a standalone system provided by the TG supplier as a part of their standard scope. The system includes software and hardware with redundant measurements and multichannel measuring arrangements with an adequate periodical testing facility.

10.6 Instrumentation System

10.6.1 Process Instrumentation

The following instruments(typical) are supplied with the TG set as a minimum:

1. *Hydrogen purity in percentage*: Three indicators, hydrogen in air, hydrogen in carbon dioxide, and carbon dioxide in nitrogen, are provided.
   a. Hydrogen in air is meant for normal generator operation.
   b. Hydrogen in carbon dioxide indicator is useful when the generator is shutdown, and by slowly introducing carbon dioxide, hydrogen is driven out from the generator casing (as an intermediate measure to scavenge generator casing) for the preparation of a long shutdown.
   c. Carbon dioxide in nitrogen is meant for a long shutdown when nitrogen is introduced to drive out carbon dioxide which completely fills the generator casing and is capped thereafter.
2. Hydrogen pressure and seal oil/hydrogen DP.
3. Monitoring the seal oil system including vacuum treatment.
4. Monitoring and control of the hydrogen cooling system.
5. Monitoring and control of stator and rotor cooling system and primary/secondary water system.

10.6.2 Instrumentation System for Stator and Rotor

The required electrical instrumentation package is also included as a part of the standard scope of the TG supplier. The following typical measurements are provided:

1. Active power generation in MW, power imported and/or exported through different feeders.
2. Voltage in kV at the generator terminal, grid, and at any other points.
3. Current in amperes through generator bus and any other feeders.
4. Reactive power in mega VAR.
5. Frequency in Hz.

11 CONDENSER LEVEL AND DEAERATOR LEVEL CONTROL SYSTEM

Condenser level control and deaerator level controls are very interactive closed loop controls. The direct type of level control loop envisages the control valve after the condensate extraction pump (CEP) discharge to modulate to control the condenser hotwell level; the deaerator level control to modulate its own dump valve to a reserve tank for a high-level band; and to modulate the makeup valve from the reserve tank to the condenser in case of a low-level band. In contrast, the indirect type of level control envisages the control valve after CEP discharge to modulate to control the deaerator level; the condenser level control to modulate its own dump valve to the reserve tank for a high-level band; and to modulate the makeup valve from the reserve tank to the condenser in case of a low-level band.

11.1 Condenser Hotwell Level Control

A typical direct control loop (250 MW TPS) for a condenser hotwell level in its simplest form is depicted in Fig. 9.17. The main condensate control valve is placed in between the gland steam cooler and the drain cooler. The controller receives its measured variable input from the condenser hotwell level at a fixed setpoint. The controller output is connected to the abovementioned control valve, which opens more for a high level in the condenser hotwell and vice versa. In case this valve is fully open and the level still rises, then another control valve before this valve, connected to the reserve feedwater (RFW) tank, opens so that the level is maintained at the preset value. If these two valves are fully open and the condenser level is still high, then the control valve to the demineralized (DM) storage tank, whose connection is before the gland steam condenser, would start opening to dump the condensate. The corresponding deaerator level control loop is depicted in Fig. 9.19.

In another control loop configuration, there is a condensate minimum recirculation valve that opens and allows the requisite flow to go back to the condenser hotwell in a low-load and startup situation and maintain the condensate flow to maintain the level. The other configuration is the control valve in the main condensate line going to the deaerator via LPHs. As the load increases, the condenser level tends to increase and cause the main condensate valve to open gradually, and the CEP recirculation valve would close accordingly. After a certain time at a higher load, the CEP recirculation valve would be closed and the hotwell level maintained by the main condensate valve.

For some large-capacity thermal power plants (500 MW), the philosophy of using a control loop for the condenser hotwell level is different. A typical control loop (500 MW TPS) for the condenser hotwell level in its simplest form is depicted in Fig. 9.18.

Here there is a three-level domain with dead band in between each level. The normal level control in the condenser hotwell is controlled by the addition of makeup water by modulating the control valve in the makeup line from the DM plant to the condenser. The low-level control domain in the condenser is lower than the normal-level domain by a dead band and is controlled by a control valve located in the gravity makeup line to add condensate to the condenser. When this is still at a low level in the condenser and the valve is nearly completely open, the condensate transfer pump across the gravity makeup valve automatically starts to take care of the situation by supplying required water via an emergency makeup line. In the level rising situation, the gravity make-up control valve would close first, and the condensate transfer pump would stop when the level reaches normal.

In the eventuality of a high-level domain that is higher than the normal level domain by a dead band, the high-level control in the condenser hotwell is controlled by a control valve placed in the condensate dump line, and the excess

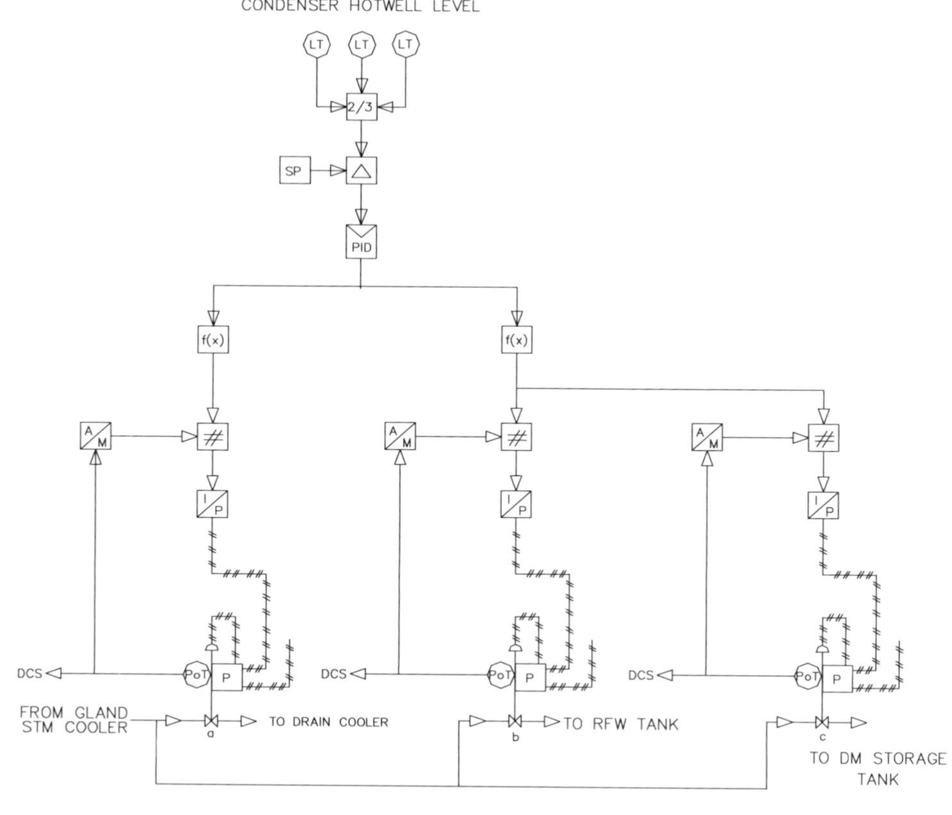

FIG. 9.17 Condenser hotwell level control loop A.

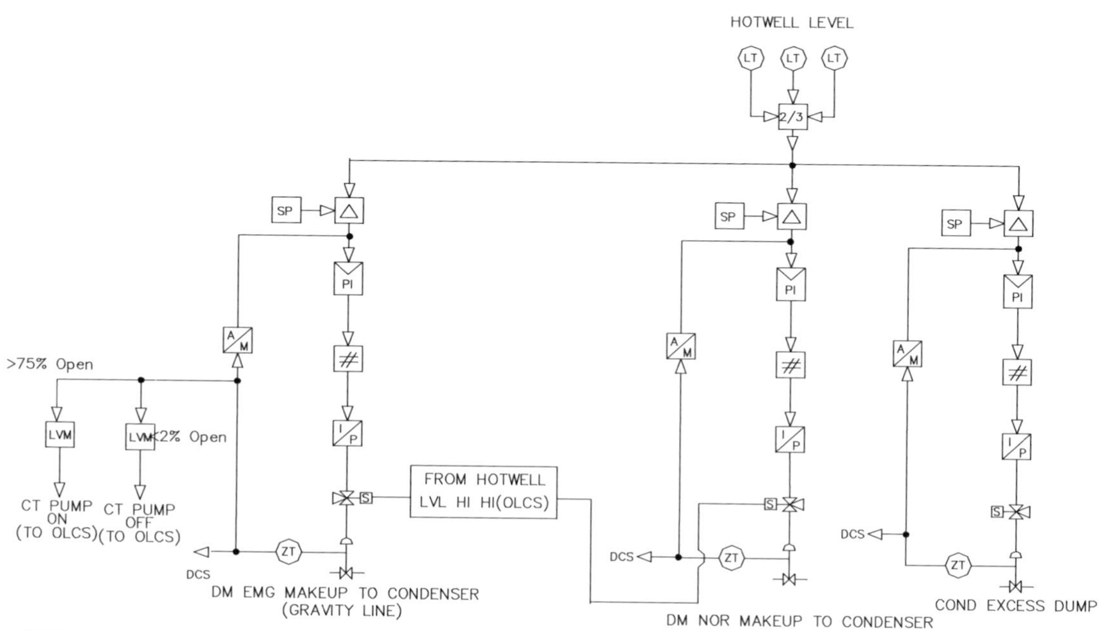

FIG. 9.18 Condenser hotwell level control loop B.

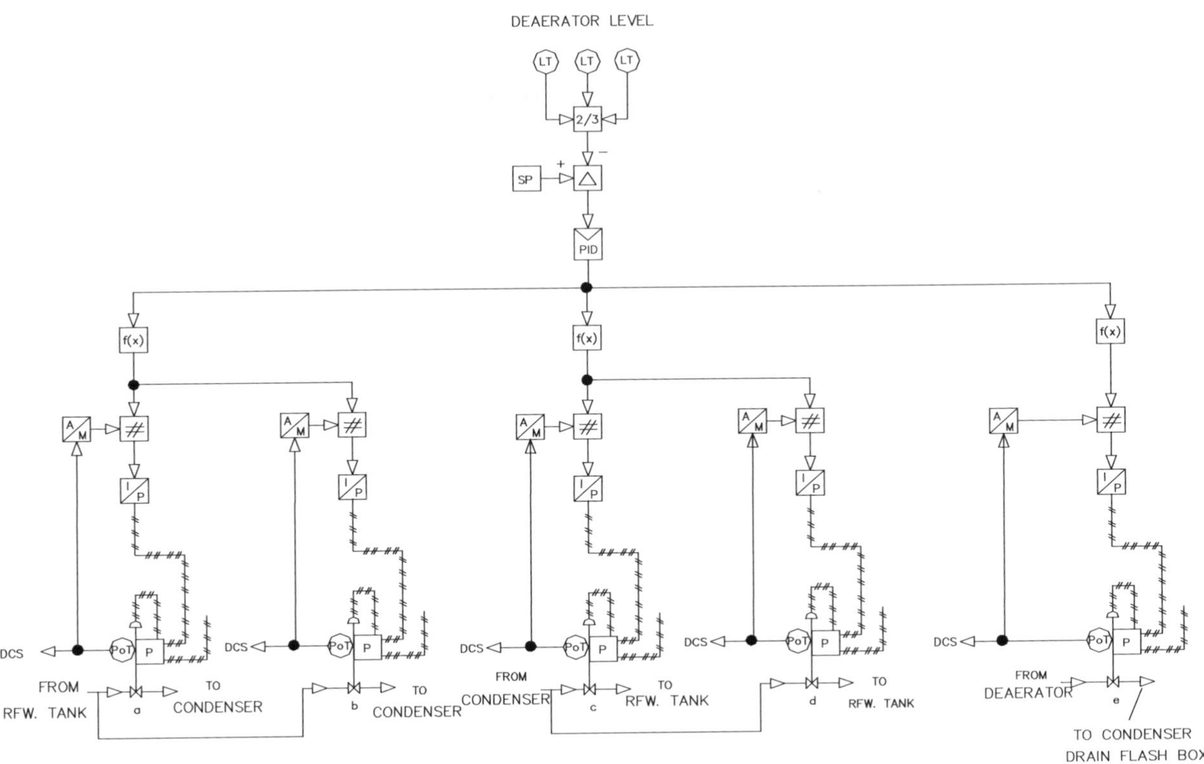

FIG. 9.19 Deaerator level control loop A.

condensate is dumped to the condensate storage tank. The corresponding deaerator control loop is depicted in Fig. 9.20.

To ensure CEP protection, flow control valves are provided in the recirculation line from CEP discharge to condenser. The flow control valve in each pump recirculation line modulates as per the output of the controller, which has its measured input signal as the condensate flow. The setpoint is fixed, which actually means the minimum flow is to be maintained. This loop is normally activated during startup and low-load application is detailed in Chapter 11, Clause 3.

11.2 Deaerator Level Control

A typical direct control loop for a deaerator level in its simplest form is depicted in Fig. 9.19 and matches with the control strategy of the condenser level control depicted in Fig. 9.17. In this control loop, there are two operating level domains where the separate control valves modulate to control the level of their respective assigned range. This may be achieved by a single control loop and split range control or by two sets of level transmitters and control loops.

For a high level in the deaerator, the control valve (condensate line to RFW tank) in the condensate line after the main condensate valve (as described in the condenser hotwell level control) modulates to maintain the deaerator level. In case of further high levels, the control valve in the line from deaerator to condenser drain flash box modulates to maintain the level after the (condensate line to RFW tank) control valve is fully open. When the level comes down from the high-level domain, this valve will remain closed. After some dead band (of level), if the level still decreases, then the control valve in the line from RFW tank to condenser modulates to allow more condensate to the condenser so that its level increases and the condenser level control valve opens more to force more water to deaerator.

The control loop philosophy for the 500 MW (or even 250 MW) TPS may be different from the above. A typical control loop for the deaerator level in its simplest form is depicted in Fig. 9.20 and matches with the control strategy of the condenser level control depicted in Fig. 9.18.

In this control loop philosophy, three element control systems are provided for a better response. The three elements:

1. Deaerator level.
2. Total outflow from deaerator such as feedwater flow plus total attemperation (SH + RH) flow.
3. Total inflow to the deaerator, such as HPH 5 drain flow, first emergency drain from HPH 6, condensate flow to deaerator, extraction steam flow to deaerator.

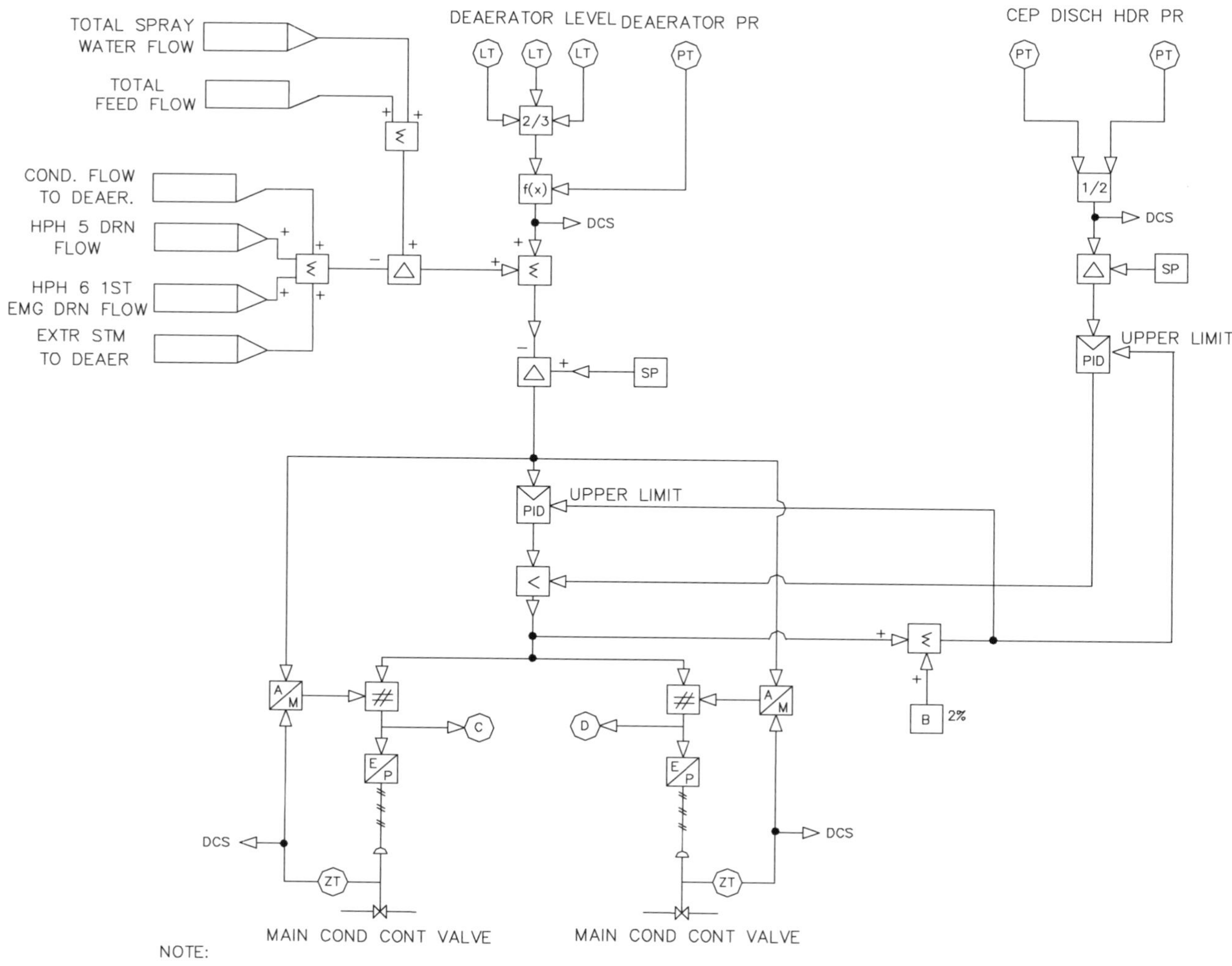

FIG. 9.20 Deaerator level control loop B.

The controller gets the deaerator level as a measured variable with a fixed setpoint and the difference between total outflow and inflow as the feed forward signal, so that the deaerator level is only minimally disturbed due to the excursion in demand. The controller output modulates the flow control valve(s) in the main condensate line. For the pump (CEP) safety from low discharge pressure, the controller output is taken through a minimum selector for maintaining minimum discharge pressure at the value above the CEP runout condition.

The interlocks are provided from the deaerator level, and in case the levels are very high as sensed by the level switches.

1. The pegging steam supply (from low-temperature auxiliary steam line or CRH line) valves along with their bypass valves will be closed.
2. The HPH 5 drain control valves will be closed.

12 VARIOUS TG OPTIONS AND MISCELLANEOUS TG CONTROLS

12.1 Steam Turbines

To stay ahead of competition, thermal power plants, like other commodities, need to have suitable options and features that match the demands or requirements of the ever-changing market. In general, steam turbines are applicable to both nuclear and thermal power plants. While being used in a combined cycle power plant (CCPP), steam turbines can be hooked up with the gas turbine in both multi-shaft and single-shaft configurations. In CCPP, the scheme is related to the number and type of installed gas turbines. In the single-shaft configuration, a gas turbine rotor shaft and a steam turbine rotor shaft are coupled together to drive the rotor of a single generator. For startup and shutdown operations, this configuration requires a switch gear to separate the steam turbine from the shaft train.

Multi-shaft configurations use independent gas turbine generators and steam turbine generators.

12.1.1 Factors influencing Options in Turbine Selection

Options of TG sets in a modern thermal power plant depend on many factors including:

1. Size of the plant.
2. Overall efficiency of the plant.
3. Lifetime of the plant.
4. Flexibility and reliability.
5. Minimum startup time.
6. Capital and running cost of the plant.
7. Optimized steam cycle path.
8. Reduction in downtime/ease of maintenance.
9. Minimum turbine manufacturing/delivery times.
10. Different main steam conditions for suitable material selection.
11. Ambient temperature and cooling system influence the size and design of the low-pressure end of the steam turbine.

To implement the above technological ideas and commercial aspects, turbine components, along with the associated equipment connected to it, are modified in design and/or with improved materials.

Different turbine stages, namely HPT, IPT, and LPT are accommodated in modules. In modern practice, there are, in general, three modules—HPT, IPT, and LPT—which are discussed in Chapter 2, Clause 3.

12.1.2 Selection of Modules

In addition to the traditional concept of accommodating each stage of turbine in a single module, other configurations, that is, more than one turbine stage in one module, are also available for subcritical steam power plants ranging from 50 to 700 MW output capacity requiring main steam temperatures up to 600°C at main steam pressures of 177 bar. The modular concept has evolved and is expected to reduce the number of alternative components, ensuring short lead times, moderate prices, and proven reliability.

12.1.2.1 IPT and LPT in Single Modules

For power plants with lower power output up to 210 MW with/without high back pressures, IPTs and LPTs are often put together in one module with axial exhaust. HPT is a single module with combined control and stop valves. The turbine casing structure normally follows the proven barrel-type design without horizontal flanges at the outer casing to ensure a homogenous distribution of the forces. The improved design, according to the manufacturer, provides the symmetrical expansion behavior of the turbine, and thus facilitates realization of small radial clearances between stationary and moving blades.

12.1.2.2 HPT and IPT in Single Modules

For medium-size power plants from 250 to 700 MW the selection is the other way around. In this option, LPT is provided in one module, whereas the other module comprises the combined HPT and IPT sections. Both modules are available with down, both sides, or a single side.

The combined HP/IP module is a double-shell design with horizontally split inner and outer casings including an admission section with stationary blade carriers. A thermal sleeve close to the center of the combined HP/IP turbine is provided for admission of the main steam. The reheat steam enters the IPT section of the same module, again close to the center of the machine. A suitable shaft gland seal with an abradable coating to improve the sealing efficiency separates the two HPT and IPT expansion sections.

12.1.3 Selection of Single-Flow or Double-Flow Turbines

IPTs and LPTs can be either a single- or double-flow type, which is mainly dictated by the capacity of the turbine. For a single-flow turbine, steam enters from one end in the axial direction and exits from the opposite end. The selection of the entry side depends on the design of thrust calculation and balancing requirement and axial shift limitations. For a double-flow turbine, the steam admission is provided at the center of a particular turbine and exits from both sides of the turbine.

12.1.4 Selection of Multiple Identical Modules

Very large capacity power plants comprise more than one identical IPT or LPT module to take care of the huge steam volume flow at the later stage. The nuclear power stations normally operate on low pressure comparative to fossil-fired pressures, hence they are provided with multiple identical IPTs or LPTs.

12.1.5 Selection of Governing System

There are multiple regulating valves for controlling the steam admission to the turbine at different loads and for receiving control signals from the EHC as already discussed in Clause 2. The final controller output distributed to these regulating valves are varied as per governing system philosophy. There are two types of governing systems: nozzle governing and throttle governing.

12.1.5.1 Nozzle Governing System

The nozzle governing system is the system where the regulating valves open one after another. In case of partial load condition, a load for a main steam pressure may be controlled by varying the nozzle area at a constant pressure. This type of control is widely used in the United States and Japan. Here the internal efficiency of the turbine is much more than the throttle governing system as the pressure drops across the regulating valves are not applicable.

12.1.5.2 Throttle Governing System

In the throttle governing system, which is popular in Western Europe, all the regulating valves open or close simultaneously. Here the load is controlled by varying the inlet pressure to the nozzle envisaged for operating at a constant nozzle area. The internal efficiency is lower than the nozzle governing system.

12.1.6 Selection of Constant Pressure or Sliding Pressure Controls

The turbine, as it gets high pressure and temperature steam from the boiler or the steam generator, may be a constant pressure or a variable (sliding) pressure at the normal controllable load zone (60%–100% load).

12.1.6.1 Constant Pressure Controls

For a constant pressure control system, the steam admission to the turbine is slowly increased to suit the turbine condition and load is increased up to a certain designed value by admitting steam with gradually higher pressure and temperature. Around this load, the steam condition at the turbine inlet is the same as in the full-unit load condition, and is kept at constant values by the boiler and turbine controls. Any increase in load is accomplished by increasing the steam flow through the turbine governor valves.

12.1.6.2 Sliding Pressure Controls

In this type of control operation, the steam condition is gradually increased. The temperature is brought up at the rated unit load value much earlier before the actual load reaches full load value, but the pressure is maintained by the steam generator at a value depending on the actual load. The relationship between the steam pressure and the unit load for any value of the unit load is predefined, and the turbine governing system takes care of this by full or wide opening of the regulating valves after a particular specified load. Before that, the nozzle or throttle governing system controls the load as per requirement.

Modified sliding pressure operation, however, suggests that the throttle valves operate slightly closed at 90%–95% load to utilize the reserve capacity of the valves and stored energy of the boiler can respond more quickly to rapid load change near full load. The throttling effect as temperature loss is nominal and cycling damage is disregarded.

12.1.7 Options Regarding Regulating Vales and ESVs

The steam admission to turbine is first through the ESVs with open/close function only. The actuator is hydraulic and suitably sized to meet the emergency closing feature. The regulating or governor valves are provided to maintain the steam flow/pressure to the turbine as desired by the EHC or hydraulic controller. These actuators are also hydraulic to meet fast-closing and huge power requirements for operating the valve for all situations.

Normally, these two types of valves provide different services for all the turbine stages—HPT, IPT, or LPT—(if any) are combined in a single casing and arranged at right angles to each other. The connection with the turbine is made through flanges. Separate valves are also provided by some manufacturers.

In general, the Venturi type valves have been used as turbine governing valves for quite a long time in large thermal power turbines to regulate the steam flow. The converging–diverging configurations of the valve passage restricts the total pressure loss at a low value, which is the main reason for their wide demand. However, to accommodate larger turbines there is a great demand for improved designs because of the complicated nature of the fluid, the basic mechanism causing valve vibration, and failure. The valve plug vibration is due to interactive force exerted by the fluid-induced excitation on the valve plug, and hydraulic forces acting on the plug at its balanced position causing plug vibrations in the lateral and vertical directions. There are several improved designs available by changing the plug, shape, etc. The flow patterns can be made much more symmetric, which reduces the intensity of steady forces and fluid plug interaction. For large power plants, there is not much choice in selection of these valves, but for smaller plants the other standard type valves can be used.

12.1.8 Options Regarding HPT/IPT and LPT Blading

Overall efficiency of steam turbine power plants depends on and is very much related to the turbine blading performance and efficiency. The construction of the blade path is normally designed from guidelines of the standard and proven flow elements such as airfoils, roots, grooves, shrouds, etc., with its own advantages and disadvantages. Other aspects regarding specific design boundary conditions such as aerodynamics, forces, materials, and temperatures are also considered for optimum blade design. However, the customer probably has no choice regarding the selection of blades but on the TG set has a choice of overall efficiency and performance only.

12.1.9 Options Regarding Type of Condensers

Condensers are a very important subsystem equipment/accessory. There are various types made to suit the turbine design, construction, connectivity, etc. Details are discussed in Chapter 2, Clause 3.5. There are so many options and types, and the selection depends on the considerations, such as minimum back pressure from maximum output, lower condensate sub cooling, enhanced condensate deaeration, minimum tube corrosion and erosion, minimum cooling water leakage, improved tube cleaning system, etc.

There may be an option for once-through cooling water supply from external and flowing water resources such as a river, sea water, etc., or in-house cooling tower arrangements.

12.1.10 Options Regarding Subcritical and Super Critical Steam Parameters

As modern technology and a suitable choice of super critical steam parameters with suitable design provides most efficient power plants with the lowest fuel cost and minimum emissions such as carbon dioxide, etc., there is a demand for super and ultra super critical steam parameters, especially when the capacity of the power plant is on the higher side.

The worldwide power demand, according to general forecasts, will be twice that of the present value. To cope with future demand, higher capacity, super critical thermal plants are the optimum and ultimate selection for fossil-fired steam-generating units with a view to reduced fuel consumption, reduced emissions, and maximum efficiency. The optimum (43%–46% efficiency) critical steam parameters are observed at ~270–285 bars and 600°C for HPT inlet and ~60 bars and 610°C–620°C for IPT inlet. The super critical steam turbine setup is influenced by the capacity of the plant, number of reheat cycles, plant atmospheric conditions, etc. The range of electrical power output generally covers from 600 to 1200 MW. In this category, normally the IPT and LPTs are double-flow design and the number of IPT and LPT modules is more than one, depending on the capacity of electrical power output of the plant.

12.1.11 Options Regarding Fossil-Fired or Nuclear Thermal Power Plants

Considering the limited resources of fossil-type fuels, there developed (around 1970) very large capacity nuclear thermal power plants up to 1500 MW and 27 kV generator voltage using radioactive elements as fuels. These plants are generally characterized by low steam temperature turbines with four-pole generators or 1500 rpm rotors with more than one IPT/LPT module of double-flow design, depending on the capacity of the plant.

12.2 TGs

Because cylindrical rotors for a high-speed generator were invented in 1901, its use as a TG set for steam turbine and gas turbine is now a part of the history. In 1912, Ludwig Roebel invented the continuously transposed stator bar, which made it possible to implement large-scale winding application. Around 1930, generators were designed with various pole options (i.e., two, four, and six poles) to match the speeds of the steam turbines. There was continuous development in stator winding insulation also, which started with some form of mica paper compounded by shellac varnish and later substituted by asphalt withstanding voltages up to 12 kV. The use of hydrogen as a generator coolant also was introduced around 1930. Today, TG sets ranging from 30 (and even less) to 1500 MW (or more) are available.

The main improvements for design upgrades to match the demands for higher capacity TGs are a more efficient cooling and insulation. As the future demand would always be for higher capacity units, the options would naturally be oriented around efficiency and stability with the availability of commercially viable technologies.

Around 1960, DM water was introduced first as a cooling media in the hollow conductor of stator winding and the insulation system was adapted to the use of mica tape with synthetic resin impregnation, which permitted even higher temperatures of ~150°C, and is still used today. By early 1960, the TG set unit ratings escalated to as high as 500 MW. By that time, there was tremendous improvement in the field of power-handling capacity of semiconductors, and close to the end of 1960, it was possible to change the then conventional DC machine excitation system to the static excitation system and even with a system with an AC exciter machine with rotating diodes, which totally eliminated the cumbersome system using slip rings, commutators, and brushes.

The early gas turbine models developed well before 1980 had a 25–40 MW capacity (low-capacity models were also available). Options for both 50 and 60 Hz with the same generator with a gear box wheel and pinion arrangement for frequency variation were presented as a standard. The gas turbine technology in early 1980 emerged with the introduction of 100 MW units and the same type of generators could be used for both gas turbines and steam turbines. The first 300 MW gas turbine was introduced in 1996 as a response to the growing demands from the user with a comparatively simple and robust air-cooling technology from some manufacturers.

12.2.1 Options for Power Output Ranges

12.2.1.1 Low Range Up to 150 MW

Low-range generators of ~150 MW TG sets are normally designed for low cost. The models used for gas turbines can easily be adapted for steam turbines. The maintenance of these types of generators claimed to be quite simple and requires only some recommended spare parts with a small storage space. All of these machines are made for easy transportation and mounting on site. Even the turbine manufacturers often can mount and couple the generators to the turbine themselves.

12.2.1.2 Medium Range Up to 500 MW

TG set generators with pressurized hydrogen as a coolant are filled up with a gauge pressure of $\sim$5.0–6.0 kg/cm^2. These machines are designed for both single-shaft and combined-cycle applications. The main theory of operation of the gas-cooled design is the same as that used for the air-cooled design. Normally the hydrogen-cooled types are used for high-capacity machines requiring more heat-dissipation facilities. They have also been used for both gas turbine and steam turbine applications.

Air-cooled TG set generators are also available from some manufacturers where the technological improvement is mainly made in the cooling arrangement, such as rotor axial cooling and winding indirect cooling using a special type of stator with a provision for multi-chamber air flow. These generators are distinguished by their simple design, which facilitates ease of operation and maintenance. By adopting suitable measures toward improved electrical design and cooling technology with increased air pressure inside the generator, the air-cooled generators also are increasing their marketshare by claiming high efficiency and usage up to 21 kV as a maximum limit.

12.2.1.3 Higher Range Up to 1500 MW

These thermal power plants, both fossil fired and nuclear, are all necessarily equipped with a hydrogen-cooling system for the rotor by axial flow of hydrogen through all conductors of a slot. Stator cooling is performed with direct flow of DM water through stainless steel tubes embedded in the hollow conductors of stator winding bars. Apparently the stator water cooling system has no upper limit as far as size of the TG. In contrast, the limiting factor is associated with the rotor winding. At 50 Hz, the rotor diameter of 1.25 m is the limit of mechanical stress. The rotor shaft dynamics calls for cautious and careful calculation in case active length extension goes beyond 8 m. However, experiments are still performed regarding multi-zone rotor cooling concepts with the possibility of higher hydrogen pressure with associated fan pressure in an effort to overcome the freeloading consequences, for example, stray flux.

12.2.2 Options for Cooling Media

As the stability, reliability, and performance of a generator depends greatly on the temperature rise of the stator and rotor windings, it is strongly believed that the cooling media should be air-cooled units instead of hydrogen-cooled units and substitution of hydrogen/water-cooling system by the hydrogen-cooling system only continues to budge the ratings upward. Some major benefits from an air-cooled TGs include elimination of the hydrogen generation/transportation/treatment system and seal oil system and less sealing, civil work, piping, maintenance, and better reliability. However, the maximum achievable capacity of air-cooled generators, proven by tests and experiments, is $\sim$400 MW so far.

12.2.3 Options for Insulation

Regarding options for insulation, important criteria are sensitivity to inconsistency in manufacturing, environmental compliance, and availability of the components from an alternative source. The generator windings insulation normally used today is based on the synthetic resin mica tape as discussed earlier, and the search for a better insulation is ongoing. Polymer insulations, as the experiments and test results indicate, might become a suitable option. Such new technology products are improved tape with high thermal conductivity using fillers, and higher mica content by denser roving carrier. The target for improving the insulation system to a higher thermal class (namely class 180) is in the close proximity of being introduced. The higher value of electrical field stress is also an added parameter allowing better heat transfer and more conductor in the slot. It is of utmost importance that both stator and rotor winding designs can accommodate their elongation due to thermal expansion.

12.2.4 Options for Stator Bar

For any new TG development, a refined stator bar improves efficiency to a great extent. The eddy currents and the circulating currents in the stator bars are very important factors and can represent $\sim$30% of the total losses in the stator bars. The modified stator bar, actually made in accordance with the design of a compact Roebel bar, comprising strands of minimum possible size reduces significant losses with practically no eddy currents; thus, the efficiency is enhanced. This design is beneficial for both directly and indirectly cooled stator bars.

12.2.5 Options for Superconducting Field Coil

In 1911 Heike Kamerlingh discovered that "superconductivity" of a conductor/certain materials is an exactly zero electrical resistance that takes place below a particular temperature called critical temperature. The electrical resistivity of a metallic conductor decreases as the temperature is lowered. However, in ordinary conductors, such as copper and silver, this decrease is limited by impurities and other defects. Even near absolute zero, copper shows some resistance. In 1986, experiments revealed that for some ceramic materials the critical temperatures are −183.15°C. From a practical perspective, even −183.15°C appears to be easy to reach with readily available liquid nitrogen having boiling point at −196.15°C, resulting in more experiments and applications. For TGs, field winding is the potential area where application of superconductors is considered for small generators only.

12.2.6 Options for Excitation of Field Coil

As discussed in Clause 10, there are different categories regarding options for excitation of field coils as listed below:

1. External static excitation system through slip ring, brush, commutators, etc.
2. DC generator/battery system through slip ring, brush, commutators, etc.
3. Rotating diode or thyristor rectifier circuit through main exciter with/without pilot exciter eliminating slip rings, etc.

AVR controls the generator terminal voltage by varying the excitation current. In case of "external static excitation" employing thyristors, the system is directly supplied by the three-phase voltage generated by the generator itself. There may arise a situation when the generator voltage drops to a certain value below the limits given by its AC feeding. There are some systems that provide slightly oversized transformer inputting from the generator, and supply the excitation system with a suitable voltage regulator. This keeps the control of the excitation current as required, even if the generator output voltages are at a low level. In some systems, there are adequately designed capacitor banks to maintain the said voltage, even if the main supply voltage experiences temporary heavy drops.

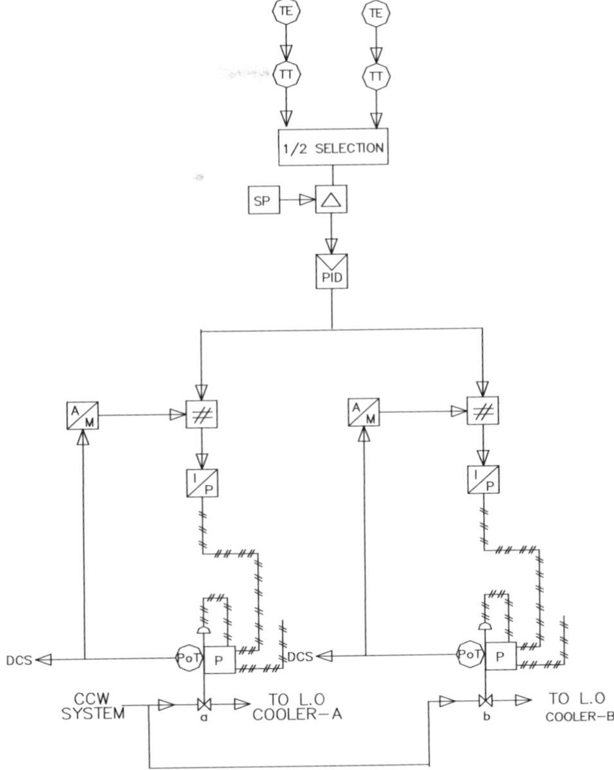

FIG. 9.21 TG lube oil temperature control.

12.3 Miscellaneous TG Control Loops

12.3.1 TG Lubrication Oil Temperature Control

1. *Objective*: As the name implies, the basic purpose of this control loop is to maintain the desired temperature of lubrication oil to ensure good flow ability and its proper action as coolant (Fig. 9.21).
2. *Discussions*: In some systems this control loop may be a part of the ATRS oil subgroup, because the oil goes to various turbine-bearing lubrication and other places, and is directly connected with startup and shutdown of the turbine. However, it could be a separate system in other turbines (especially for smaller units), so it is discussed separately.
3. *Loop description*: Lubrication oil temperature is measured in the header after the cooler with the help of the redundant temperature elements (e.g., Pt RTD) with a temperature transmitter, because this is an important parameter. The measured value is compared with the setpoint from the monitor (or setter in the plaque as applicable). Set point and measured value is compared to create an error value. The error thus created is fed to one PID controller with an A/M station. The output via converter is fed to each of the control valves to regulate coolant flow to the redundant LO coolers.

12.3.2 Generator Seal Oil Temperature Control

1. *Objective*: As the name implies, the basic purpose of this control loop is to maintain the desired temperature of seal oil to ensure good flow ability and proper action sealing medium (Fig. 9.22).
2. *Discussions*: As discussed earlier, there is a possibility of moisture in seal oil, which may deteriorate the performance of the sealing application, so it is necessary to maintain the desired temperature to circumvent such a situation.
3. *Loop descriptions*: Similar to item 3 in Clause 12.3.1.

12.3.3 Generator Hydrogen Temperature Control

1. *Objective*: Basic purpose of this control loop is to maintain the desired temperature of hydrogen to ensure good cooling (Fig. 9.23).
2. *Loop description*: Hydrogen temperature is measured in the line going to both the ends with the help of the temperature elements (e.g., Pt RTD) with a temperature transmitter. Because this is an important parameter, redundancy in measurement is recommended

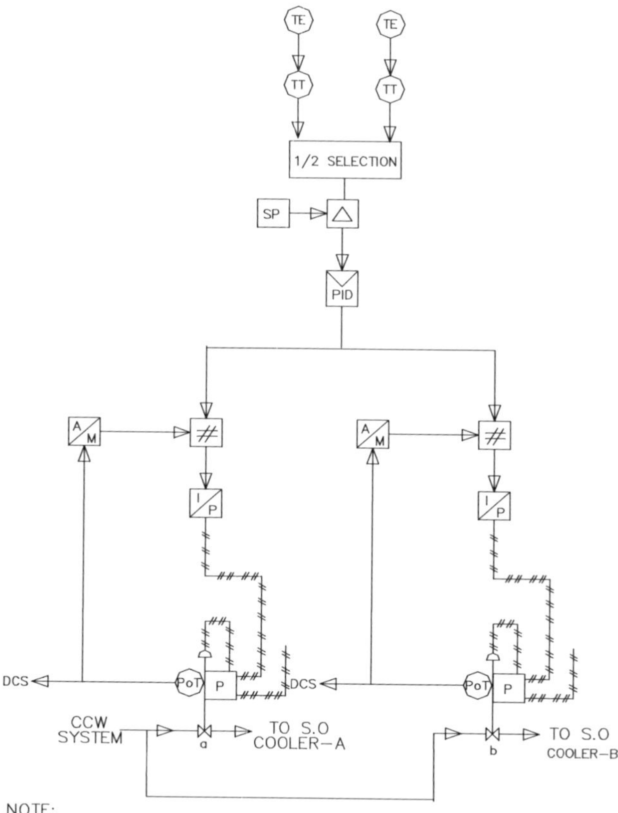

FIG. 9.22 Generator seal oil temperature control.

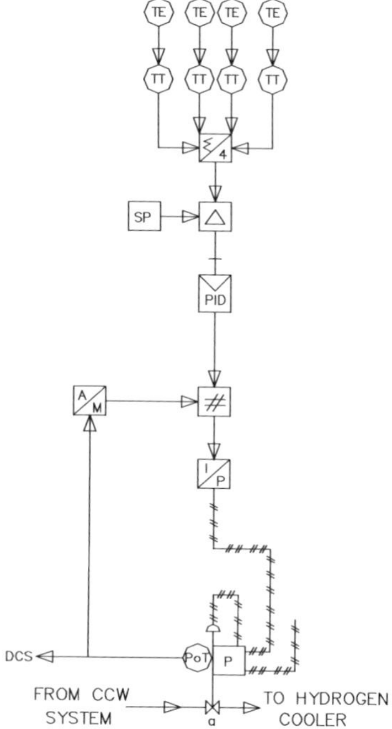

FIG. 9.23 Generator hydrogen temperature control.

so that the temperature on each side is considered and an average of all four is taken in the loop. The measured value is compared with the setpoint from monitor (or setter in the plaque as applicable). Setpoint and measured value is compared to create an error value. The error thus created is fed to one PID controller with an A/M station. The output via converter is fed to the control valve to regulate coolant flow to the hydrogen cooler.

12.3.4 Stator Coolant Flow Diagram and Temperature Control

1. *Objective*: The basic purpose of this control loop is to maintain the desired temperature of stator cooling water to ensure good cooling of the stator. The stator conductors are cooled by closed-circuit DM water. The stator holes through which water flows are quite narrow, and unimpeded flow through these holes is critical, because any overheating of the stator reduces generation and even leads to a catastrophic failure of the generator. Therefore, suitable control is envisioned maintaining the temperature of stator water.
2. *Discussion*: Stator winding of the TG is cooled by circulating DM water through the hollow conductor of stator winding. DM water (known as stator cooling water, SCW) is fed to the feed header mounted inside the generator casing, toward the turbine side, from there, Teflon lines are drawn to supply water to the conductors. This is done in a closed circuit. A pump (always with an auto standby pump) is used to drive the water through the filter, cooler, and the conductors and returned back to a hermetically sealed expansion tank, which is kept in vacuum (200–250 mmHg) and mounted above the generator. The temperature of stator cooling water is controlled by regulating the coolant supply to the heat exchanger. The maintenance quality of stator water is very important. For this reason, it is often supplied with standalone mixed bed units in a recirculation line. The conductivity measurement of stator water going to the generator is extremely important and is normally monitored. Also in some cases dissolved oxygen (DO) is often monitored. Because DM water passes through the stator conductors, which are quite narrow, any contamination or rise in temperature is very detrimental and often causes line blockage. A short flow diagram of typical stator cooling system is seen in Fig. 9.24A.
3. *Loop description*: SCW temperature is measured in the feed header with the help of the temperature elements (e.g., Pt RTD) with a temperature transmitter. Because this

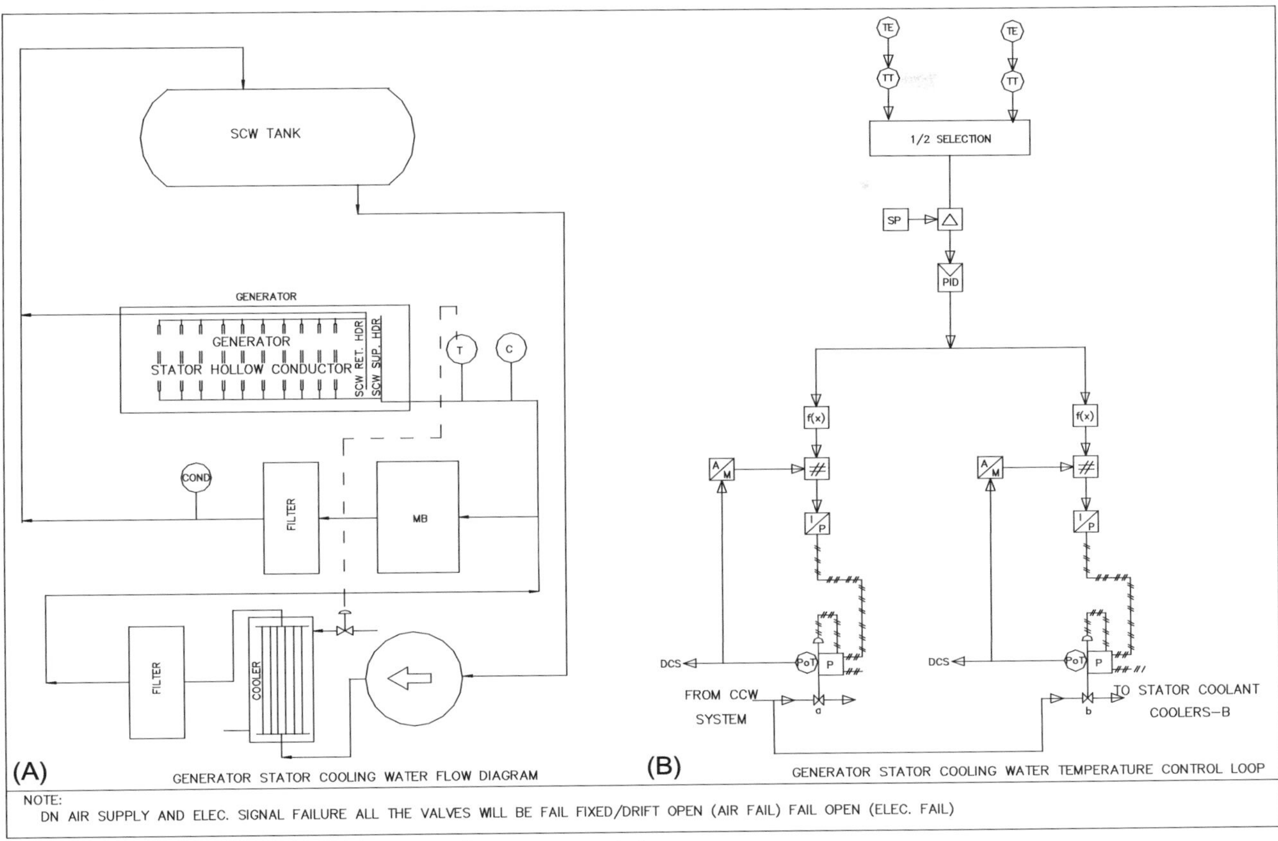

FIG. 9.24 Generator stator cooling water system and temperature control.

is an important parameter, redundancy in measurement is recommended, so in the loop 1 of 2 redundancy is shown. The measured value is compared with the setpoint from the monitor (or setter in the plaque as applicable). Setpoint and measured value is compared to create an error value. The error thus created is fed to one PID controller with an auto/manual station. The output via converter is fed to each of the control valve(s) to regulate coolant flow to the redundant SCW cooler(s).

BIBLIOGRAPHY

[1] Instrumentation and Control Systems for Steam Turbine Generators: By Wolfgang Kindermann and Hans Georg Schwarz. Germany. Seminar document, 1981.
[2] Instrumentation and Control for the Steam Turbine Generator Sets: By M/s KWU Germany. Seminar Document, 1981.
[3] Automatic Control of KWU Steam Turbine Generators: By Heribert Kurten Germany. Seminar Document, 1981.
[4] Electrohydraulical Controller (EHC) for Steam Turbines—Conventional Power Plants: By M/s KWU. Germany (Seminar Document—Write up). Doc no. V 793/8.1-8460/1/Pr/mt/engl 01.78.
[5] Electrohydraulic Steam Turbines Controller (EHC) (Iskamatic A): By M/s KWU. Germany (Seminar Document no. PSW 213/05.81 for Control Loop Strategies).
[6] Seal Oil System: Scavenging Type: BY M/s General Electric.
[7] Doc no. PI 30.23–2 (Excerpts from M/s Canteach CANDU). Electrical Equipment—Course PI30.2. Generator Auxiliary Systems, 1990 (ITPO.01).
[8] Application Update for the Power Generation market: By M/s Pall Corporation, http://www.hydrafab.com/pdf/AU-Seal%20Oil.pdf.
[9] Product Catalog of M/s Westinghouse Type Seal ring and Seal Oil System.
[10] Basic AC Electrical Generators (American Society of Power Engineers). Obtained by O&M Consulting Services Inc., http://www.asope.org/pdfs/AC_Electrical_Generators_ASOPE.pdf.
[11] 11th Annual Conference and Exhibition on Electric Power.
[12] Fuji Electric Review: Vol. 12 No. 5: By M/s Tatomi Iwamitsu & Ryoichi Kuramochi. udc 621.313.322-81, http://www.fujielectric.com/company/tech_archives/pdf/12-05/FER-12-05-129-1966.pdf.
[13] United States Patent no 4,152,636 dt May 1, 1979 by Dale I. Gorden, Munhall, PA.
[14] LPBP System—Product Bulletin from M/s Sulzer, USA.
[15] Turbine Bypass Systems (Vector Control Technology): By Nihon KOSO Co., LTD, Japan.
[16] Turbine Bypass Systems for greater plant performance-SULZAR valves a product of CCI.
[17] Overspeed turbine Trip: By Boyd Davis.
[18] Basic Electrical Theory: By Science and Reactor Fundamentals. Electrical CNSC Technical Training Group (Revision 1), 2003.

[19] Product Bulletin of M/s Whitby Hydro Energy Services Corporation: Engineering and Construction Services on Generator & Exciter Basics, http://www.whitbyhydro.on.ca/pdf/5.2.C_GenBasics.pdf.

[20] Workshop on Mechanical Systems (Power Cycle) in thermal power station: By M/s BHEL India: dtd, 1982.

[21] Steam Turbines Hydraulic Systems Problems and Solutions: By M/s Barry Sibul and Company. Copywrite 2012. Barry Sibul Company, 3720 S. Ocean Blvd., Suite 108, Highland Beach, FL 33487 Phone: 888-433-0003 FAX: 877-789-4133 Email: barry@sibul.com Web: www.sibul.com.

[22] Bringing to Light Lube Oil Moisture in Hydrogen Cooled Generators: By James R. Mickalec, AEP Pro Serv Inc. A Noria Publication on Machinery Lubrication. http://www.machinerylubrication.com/Read/132/oil-hydrogen-cooled-generators.

Chapter 10

Coordinated Control System

Chapter Outline

1 INTRODUCTION

The fundamental objective of a thermal power plant is for the boiler to supply the thermal energy that is required by the turbine, which then converts thermal energy to mechanical energy to match the electrical power delivered by the generator coupled with the turbine. There must be a balance among all these forms of power to ensure continuous supply of electricity. In a unit with a steam generator, a wellcoordinated control system with a control strategy is important to ensure a largely fluctuating power demand with an optimum rate of load change that results in a minimum amount of thermal stresses.

A modern power plant is a combination of several interacting control loops and subsystems. The control strategy for the main control loops must respond to a common control demand configuration, which determines individual set points and dictates the performance of the plant. The coordinated control system synchronizes the output of the boiler with the inherent slow response of the turbine and the faster response of the generator, to ensure optimum and stable performance for continuous load tracking of the load dispatch center (LDC) requirements (e.g., Free Governor Mode of Operation, FGMO), as well as the online load disturbances. Therefore, the system must be configured in such a way that the main subsystems—the boiler, the turbine, and the generator—would appear to act as a single entity.

The combined process of a boiler, a turbine, and a condenser is complex and nonlinear. Ultimately the objective of the control system is to maintain the steam pressure (constant or variable) at the turbine inlet so it can adjust the steam admission rate commensurate to the generator power output. With the help of the state-of-the art instrumentation and a control system, all the major subsystems (i.e., boiler, turbine, and generator) are controlled in tandem in different modes of operation to manage the loadbearing capacity of the subsystems, depending on the limitation of the subsystem's equipment/accessories.

1.1 Basic Operating Modes in Coordinated Control System

Modern, large thermal power plants connected to electrical grids can be described in three modes:

1. Coordinated control
2. Boiler follow
3. Turbine follow

Fig. 10.1A–C depicts various modes as indicated above. Fig. 10.1D and E depicts a coordination control system with direct energy balance (DEB) for drum-type boilers. The same DEB coordination control system for supercritical plants with a once-through boiler is seen in Fig. 10.11.

Power Plant Instrumentation and Control Handbook. https://doi.org/10.1016/B978-0-12-819504-8.00010-X

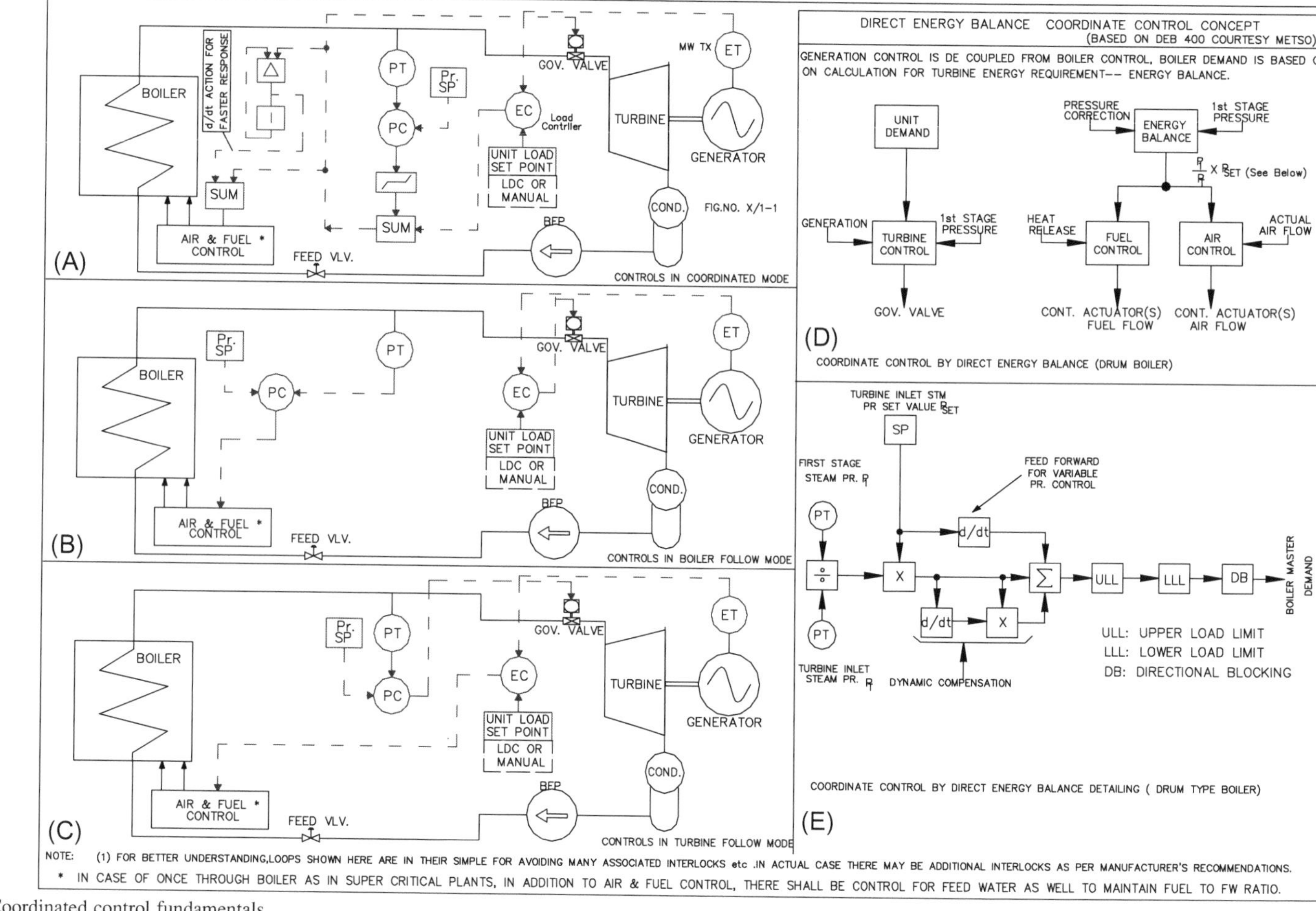

FIG. 10.1 Coordinated control fundamentals.

1.1.1 Coordinated Control Mode

Coordinated control mode is the most desired because in this mode of operation the boiler and turbine are capable of taking the full load in response to the load demands from the LDC without any contingency or limiting conditions. This mode ensures that any load demand will be processed suitably and the control output will be transmitted to both the boiler and turbine in a coordinated manner. This prevents a mismatch between their actions so the demand is managed without posing any constraint to the operating condition. In addition to the coordinated control mode of operation, both the boiler follow mode and turbine follow mode can be operated under the specific conditions of a particular plant. The coordinated control maintains a balance between the turbine and boiler control systems under steady-state conditions so the unit can generate the desired electrical load, and share its frequency control by increasing or decreasing the boiler and the turbine load demands. During transient operation, this system adjusts the correlation between the turbine and the boiler demands by utilizing the system's stored energy.

1.1.2 Boiler Follow Mode

In the boiler follow mode of operation, the turbine delivers the output as per its own control requirement, under the occurrence of a turbine load-limiting function, as per the unit load demand. The boiler follows the output delivered by the turbine by maintaining the main steam pressure at the turbine inlet at a fixed or variable value, depending on the requirement. In other words, the boiler follows the turbine by maintaining the steam pressure, and the load is delivered by the turbine governor valve position, and the steam pressure stays at the same value. This mode is automatically selected when a predefined turbine contingency occurs or by operator selection.

1.1.3 Turbine Follow Mode

In the turbine follow mode of operation, the boiler delivers the output as per its own control requirement, under the occurrence of a boiler load-limiting function, as per the unit load demand. The turbine follows the output delivered by the boiler by maintaining the main steam pressure at the turbine inlet at a fixed or variable value, depending on the requirement. This is done by providing the modified limit load demand to the turbine and boiler control. The turbine control receives an additional correction signal as required for maintaining the desired throttle pressure. This mode is automatically selected when a predefined boiler contingency occurs or by operator selection. There are three other auxiliary modes of operation that are applicable for temporary or short-term operation, but they are not really considered a part of the coordinate control system.

1.2 Startup Control Mode

When the unit is starting from rest, the operating mode is in the startup mode. In this mode, both of the basic operating modes (boiler follow mode and turbine follow mode) are possible because of the demand generated from the unit without referring to the LDC. There also may be an interlock, for example, whenever a generator breaker opens, the control is automatically transferred to the startup mode.

1.3 Base Control Mode

When the boiler is operated at a fixed load and the turbine delivers the corresponding mechanical output, the unit is in the base mode, which means the unit runs manually to set the fixed load.

1.4 Manual Control Mode

The control system of the plant goes automatically to manual mode if the boiler control station, which includes feedwater, fuel, and air control, is put on manual) or the turbine control station goes to manual, or by operator's action.

2 COORDINATE CONTROL MODE

As discussed in Clause 1, the basic objective of the coordinate control mode of operation in a thermal power plant is to receive the unit demand and generate the output accordingly by the boiler and turbine acting in tandem. In this mode of operation, the boiler follow and the turbine follow modes are not applicable. The boiler follow mode is possible in "master pressure control" mode in a system where a multiple number of boilers are connected to a common header. This will be discussed in a later clause.

2.1 Selection of Basic Modes

The selection of basic plant operating modes is generally at the operator's discretion unless there is some untoward situation that would command the unit to run on safe mode or manual mode. If the main "unit load controller" or the "coordinated load controller" operates on "auto mode," the operator can select any of the coordinated control, boiler follow, or turbine follow modes, depending on the existing conditions of the main subsystems. Whenever the main unit load controller operates on manual, all these modes would be deselected and trip to manual, then the manual demand may be routed to the boiler or turbine master controllers, as necessary.

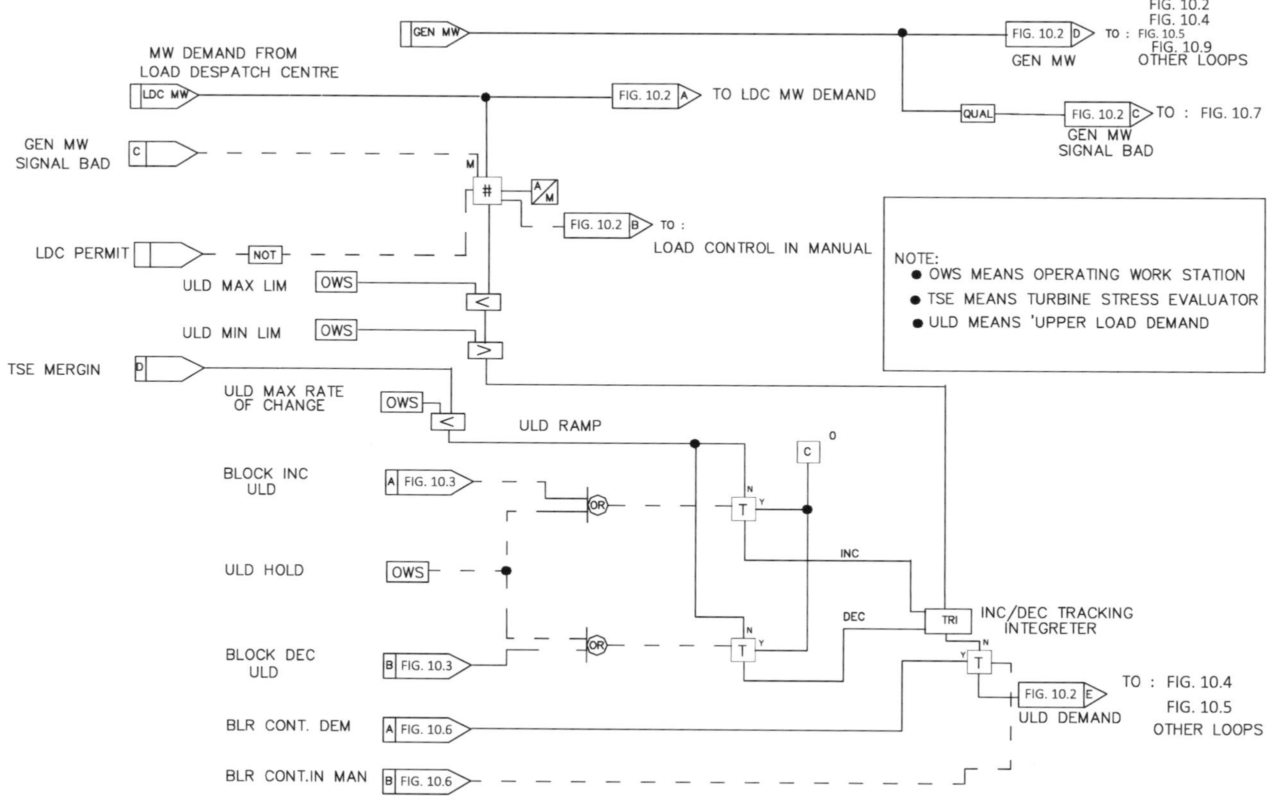

FIG. 10.2 Unit load demand control.

2.1.1 Generation of Unit Load Demand

In the coordinated control mode, the unit load demand is generated from the LDC as the particular unit would be connected to the electrical grid (Fig. 10.2). The LDC is a computer-controlled center that collects electrical data/parameters from the grid, such as voltage, frequency, individual feeder load, and generation of various generators connected to a particular grid. Depending on the health of each generating unit, demands of various feeders connected to it, and the software algorithm, LDC generates optimum demand pertinent to each unit connected to the system grid, which would then process the demand and pass on the relevant demand for proper execution by the boiler and turbine load control systems. Over a communication network, the LDC conveys the load demand to the particular unit in the form of electrical pulses, which are then integrated to generate demand.

2.1.2 Auto Manual Selection of Unit Load Demand

The demand from the LDC is taken to the auto manual station (AMS) and the operator may transfer it to manual position if the LDC signal becomes unavailable, corrupted, or not executable. The output from the AMS then passes through high and low limiters, which determine the limit of the maximum and minimum unit demand to be allowed as per the capacity of the unit or the load-bearing capability of the unit for a particular time or period of time.

2.1.3 Upper and Lower Limit Set of Unit Load Demand

These limiters are actually meant to provide manually set the maximum and minimum unit load for normal coordinated control mode; if there are other modes of operation, the online generator MW load with $\pm 10\%$, for example, may be used to set the limit values of unit load demand.

2.1.4 Upper and Lower Margin Set for Unit Load Demand Gradient

Any load demand varying within the two previously mentioned limits can only be made through a tracking integrator, which tracks the input and the output changes gradually in both directions at the ramp rate normally provided by the turbine "thermal stress evaluator" (TSE). This typically varies within an upper and lower rate margin set by the TSE. The TSE evaluates the online turbine load condition through the changing capability per unit time by sensing the turbine inner metal and mid metal temperatures at different strategic points, including governor valves, interceptor valves, etc. (Chapter 6, Clause 9). However, there may be a minimum selection of ramp rate, which also receives a manual bias input value. This is set during commissioning time or any other subsequent time to match the plant behavior and condition, or it may be an equipment limitation. Generally, the TSE dictates the unit load ramp rate under normal plant operating conditions. The minimum rate through manual bias plays a vital role during abnormal conditions for turbine protection and can be well maintained by the high-pressure/low-pressure bypass system or popping of the safety valves provided at appropriate points.

2.1.5 Blocking of Increase and Decrease of Unit Load Demand

Once the selected ramp rate is available at the output of the minimum selector, there are certain restrictions that do not allow the unit load demand to change, even if permission is made by the minimum selector. When one or more such restricting conditions take place, there are changeover arrangements that ignore the normal ramp rates, and zero ramp rates are enforced through various switching algorithms. This ensures that the tracking integrator stops integrating its output further in a particular direction(s) based on the existing plant conditions used to activate the transfer switch. The restricting conditions are categorized as blocking of increase ramp, blocking of decrease ramp, and "hold" of ramp rates.

2.1.5.1 Blocking of Decrease of Unit Load Demand

Occurrence of one or more of the following typical signals would activate the switching action of the ramp rate to zero so that any further decrease of unit load demand is blocked (Fig. 10.3):

1. very high mean steam pressure control deviation (in +ve direction) in coordinated control mode,
2. very high feedwater flow control deviation (in +ve direction),
3. very high fuel flow control deviation (in +ve direction),
4. very high total air flow control deviation (in +ve direction),
5. very high primary air pressure control deviation (in +ve direction).
6. Economizer outlet water temperature protection operated

Here, "high" signals mean the parameter is higher than the desired or set value.

2.1.5.2 Blocking of Increase of Unit Load Demand

Occurrence of one or more of the following typical signals would activate the switching action of the ramp rate to zero so that any further increase of unit load demand is blocked:

1. very high mean steam pressure control deviation (in −ve direction) in coordinated control mode,
2. very high feedwater flow control deviation (in −ve direction),
3. very high fuel flow control deviation (in −ve direction),
4. very high total air flow control deviation (in −ve direction),
5. very high primary air pressure control deviation (in −ve direction),
6. high demand in turbine follow control mode.

2.1.5.3 Blocking of Both Increase and Decrease (or Hold) of Unit Load Demand

Blocking this signal is normally issued by the operator from the workstation.

2.1.6 Generation of Ultimate Unit Load Demand, Boiler Master Demand, and Turbine Master Demand From Coordinated Control System

The tracking integrator output that maintains the restrictions described earlier constitutes the unit load demand signal. The frequency influence or the frequency-dependent component of MW demand is then added to the unit load demand to form the total unit load demand, that is, the set point for the controller of the coordinated control system. The generator MW signal acts as the measured/process variable signal and is compared with this set value and the output signal from the comparator. This forms the ultimate error signal of the controller of the coordinated control system. The controller output thus generated is the basic demand for both the turbine and boiler. The controller output may either be auto or manual, depending on the operator's action and after incorporating some pressure correction or pressure-related load demand. This is discussed separately in subsequent clauses of this chapter. "Master boiler demand" and "master turbine demand" signals are generated to manage the main steam pressure influence on the two main subsystems (Figs. 10.4 and 10.5).

2.1.6.1 Frequency-Influenced Load Demand for Coordinated Control System

The frequency influence signal is incorporated to take care of the droop characteristics (MW-frequency) on the coordinated control system (Fig. 10.3). The required dependence/relation of the load with the frequency according to the droop characteristics calculations is readily available by adding the supplementary load demand provided by the characterized droop unit. By measuring the speed through the frequency of the turbine generator set, the corresponding positive or negative MW corrections of the unit load demand signal is accomplished. However, there are maximum (approximately +10%) and minimum

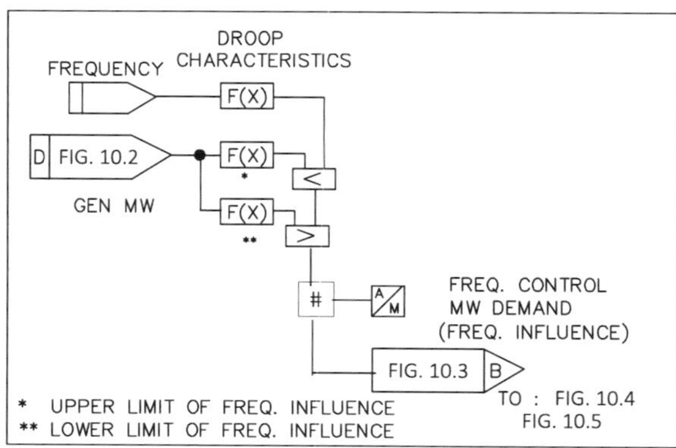

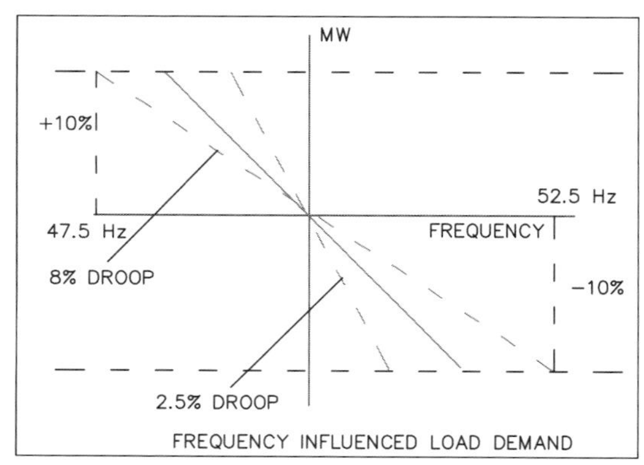

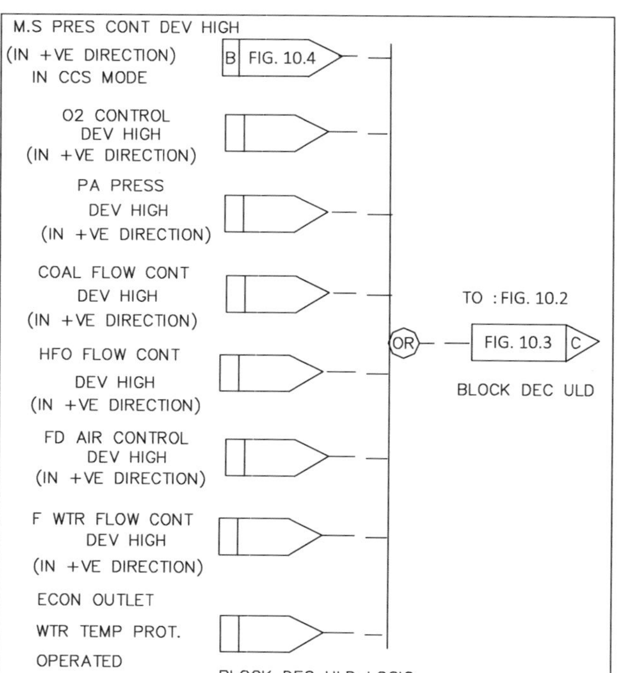

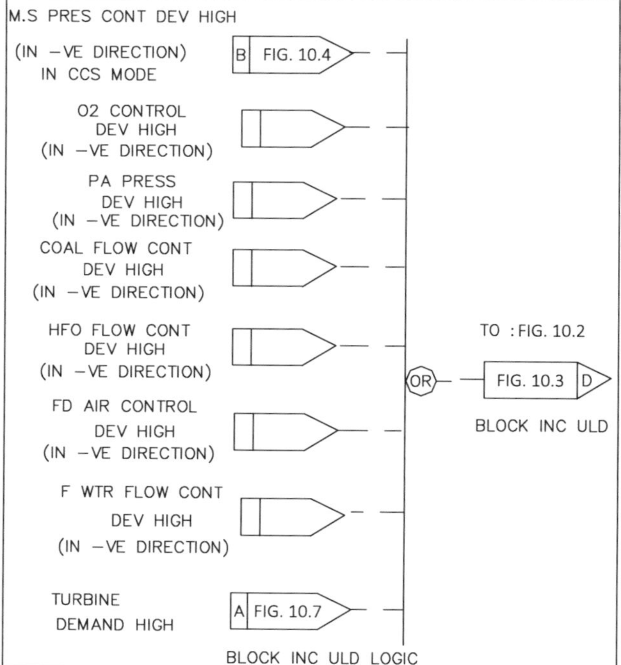

NOTE:
- OWS MEANS OPERATING WORK STATION
- TSE MEANS TURBINE STRESS EVALUATOR

FIG. 10.3 Frequency-influenced MW demand and directional blocking.

(approximately −10%) limits of this correction, which is also generated from the then-generation value denoting the allowable load changing requirement due to frequency excursion. These maximum and minimum values can be set from the workstation to suit the grid behavior.

Droop is the percent change in frequency causing unit generation to change by 100% of its capability. The previous proportional component of total load demand can be changed by adjusting the droop characteristic between 2.5% and 8% in steps of 0.5%. This is done by the operator/engineer through the bias, which multiplies/scales the basic frequency-dependent load component. The frequency influence may be connected or disconnected manually from the workstation or automatically if the frequency signal is not good.

2.1.7 Generation of Pressure Set Point for Coordinated Control System

For the unit with a variable or a sliding pressure control system, the main steam pressure at the turbine inlet has a definite relation with the unit electrical load (MW) generated (Fig. 10.4). The function generator algorithm is provided to get a variable main steam pressure with relation to MW generated. The pressure value thus available at the output is passed through a high/limiter and then through a tracking integrator output, which acts as a main steam pressure set point; an input to the error generator whose other input is the actual main steam pressure. The main steam pressure set value and the actual measured value are sent to the electrohydraulic controllers (EHC) and to the high-pressure bypass and the turbine load control

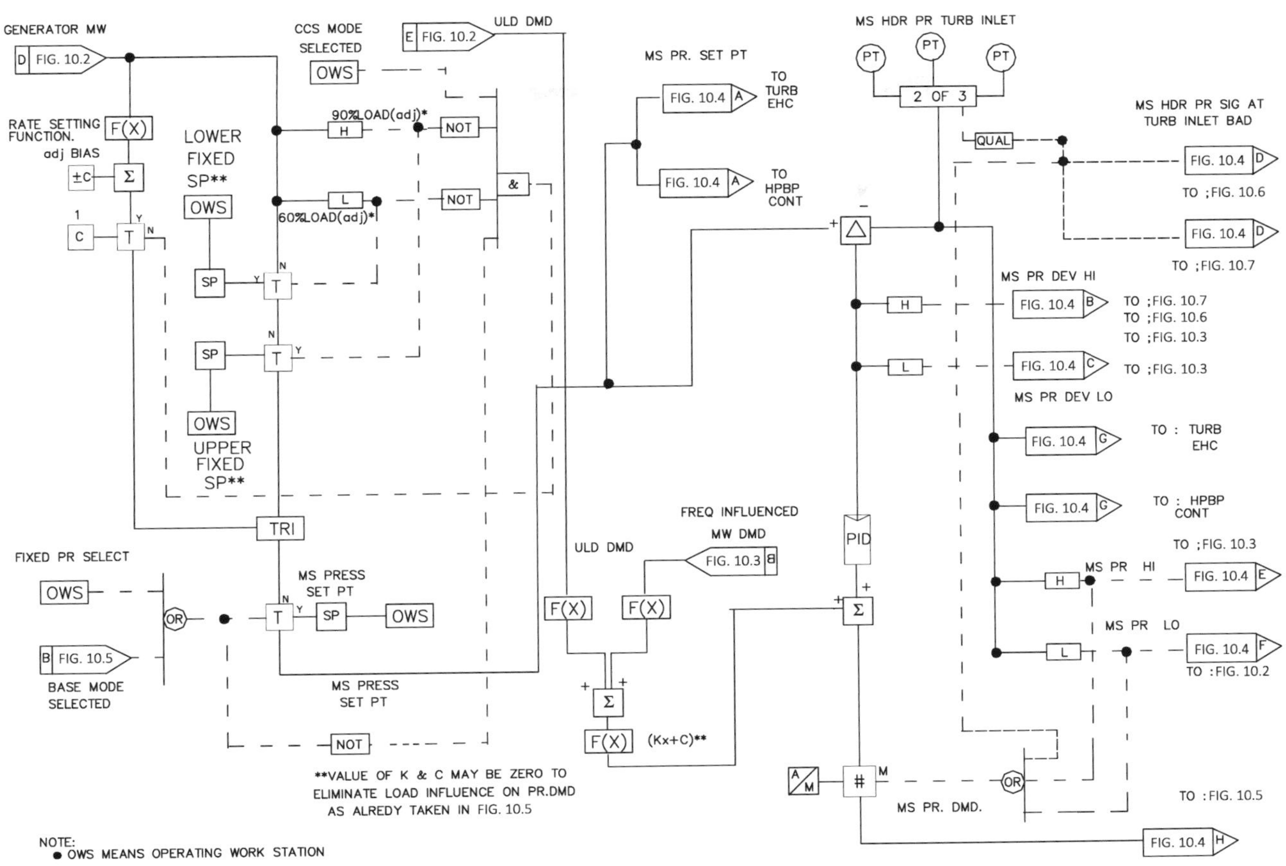

FIG. 10.4 Coordinated control: main steam pressure control.

systems for any correction required by those dedicated control systems or the turbine limit pressure control during any mismatch between the boiler and turbine controllers.

2.1.8 Generation of Main Steam Pressure Demand for Coordinated Control System

The electrical load (MW)/unit load-dependent main steam pressure signal via the function generator and a tracking integrator as discussed previously make an input to the pressure controller at a variable set point (Fig. 10.4). This controller is called "main steam pressure demand." The total unit load demand as discussed in Clause 2.1.6 and shown in Fig. 10.2 is, after passing through a function block and a high/low limiter, added as the influence of the load demand to the main steam pressure demand. The combined signal is the final demand signal for the main steam pressure control system, which would be routed to the boiler and turbine master controller.

Under certain conditions this main steam pressure control demand becomes predominant and acts on either the boiler master demand or turbine master demand (Fig. 10.5). The total unit load demand is compared with the actual generator load (as measured/process variable) to form the error signal of the main load controller of the coordinated control system. The output of this controller, in manual or auto mode, is made to pass through the high and low selector. That is, if the load controller demand output varies within the allowable plus or minus range (as set manually) of the main steam pressure controller demand, it would be selected as a boiler master demand and turbine master demand when the unit is run on the coordinated control mode. This is done to enable the system to generate electrical power as per the load demand ignoring the small allowable pressure excursions managed by both the boiler and turbine controls. When the pressure departure goes beyond the set values, pressure demand takes charge of the control to avoid unwanted fluctuation of the turbine inlet pressure. When this happens, neither the coordinated control mode nor the turbine following mode is selected, and the main steam pressure control demand becomes the boiler master demand. Conversely, when neither the coordinated control mode nor the boiler following mode is selected and the main load controller of the coordinated control system controller is on auto, the main steam pressure control demand becomes the turbine master demand.

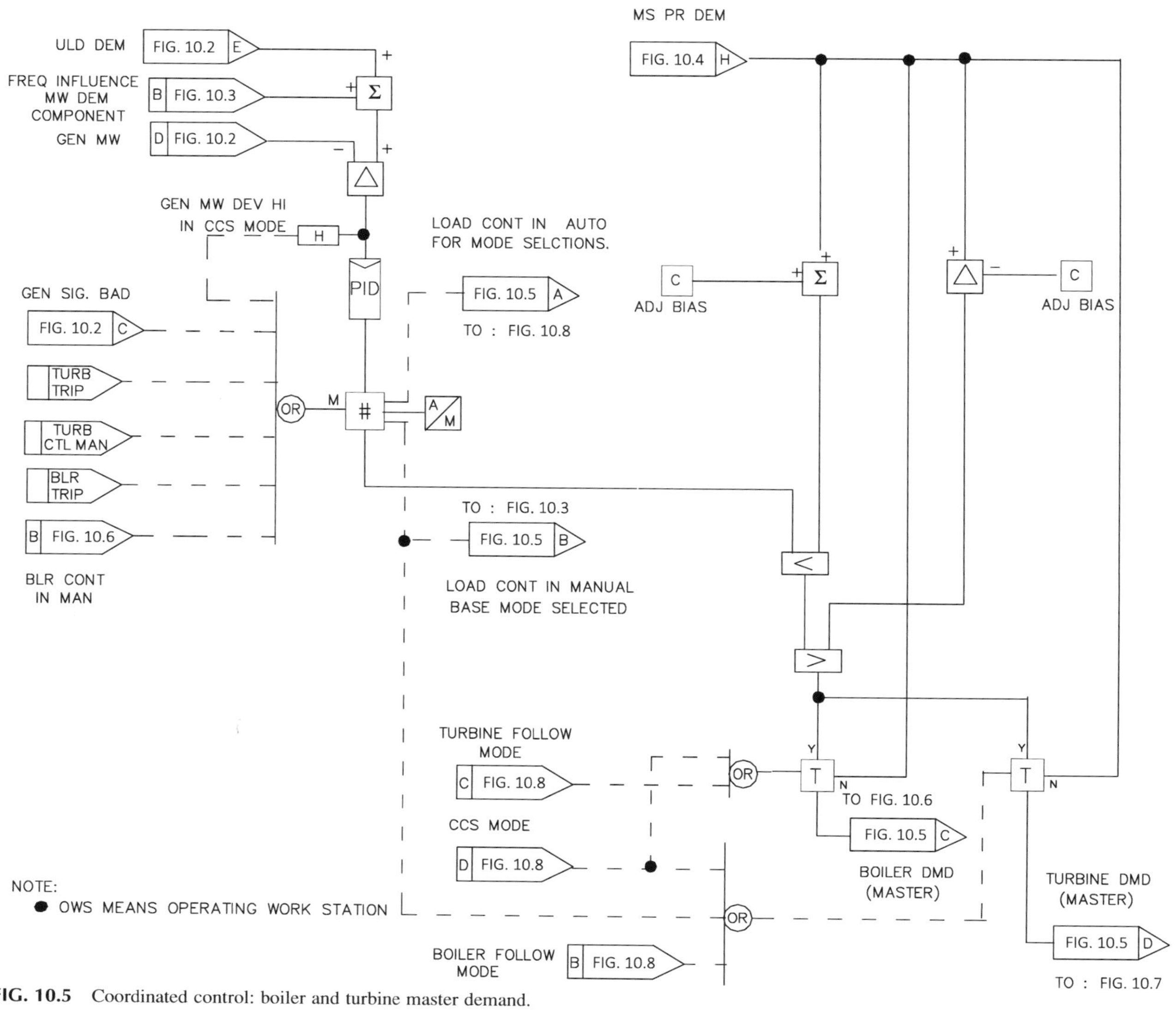

FIG. 10.5 Coordinated control: boiler and turbine master demand.

2.1.9 Generation of Boiler Load Demand

The boiler master demand, as generated in the coordinated control mode (Fig. 10.5) after the auto/manual station, is output through a tracking integrator and becomes the boiler load demand in normal running condition. In an emergency situation when the "run back" condition is enacted, the "run back target" value is considered the boiler load demand using a selector switch (Fig. 10.6).

The generation of boiler load demand becomes manual due to the following reasons:

1. A very high main steam pressure control deviation (in both direction) is in coordinated control mode.
2. Main steam pressure signal is bad.
3. Both forced draft fan (FDF) controls are in manual.
4. Both induced draft fan (IDF) controls are in manual.
5. Both primary air fan (PAF) controls are in manual.
6. Fuel control master is in manual.

2.1.10 Generation of Turbine Load Demand

The turbine master demand as generated in the coordinated control mode (Fig. 10.5) is output through an auto/manual station and becomes the turbine load demand in normal running conditions (Fig. 10.7).

The generation of the turbine load demand becomes manual due to following reasons:

1. High main steam pressure control deviation (in both direction) is in coordinated control mode.
2. Generator signal is bad in coordinated control mode.
3. Main steam pressure signal is bad in turbine follow mode.

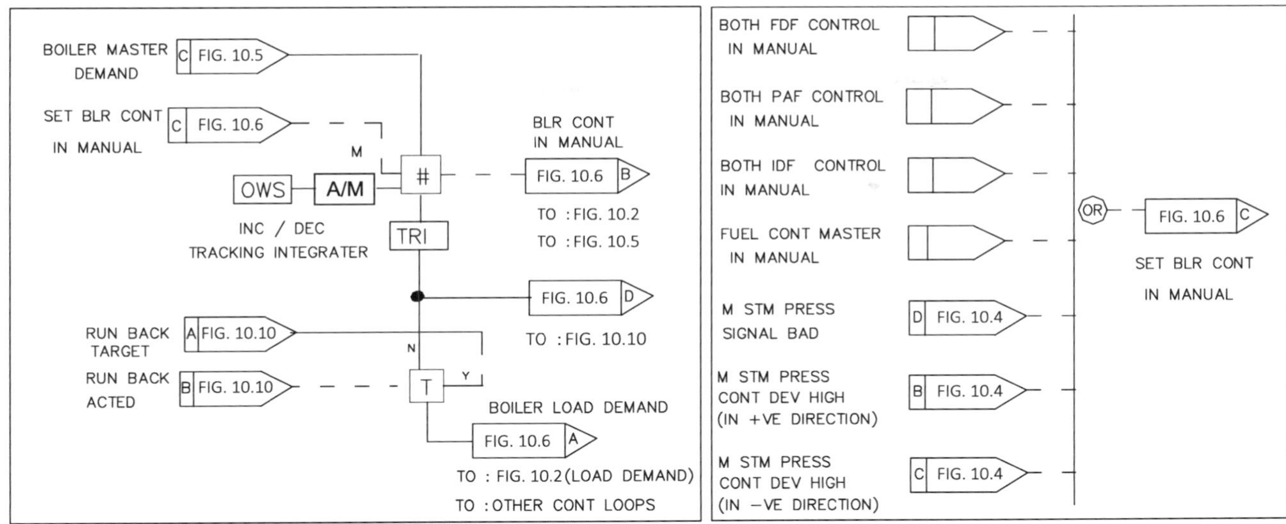

FIG. 10.6 Generation of boiler load demand.

4. High main steam pressure control deviation (in both direction) is in turbine follow mode.
5. Turbine EHC control is not in remote.

2.1.11 Selection of Control Mode in Coordinated Control System

The three control modes can be selected one at a time. Any particular mode of operation can be selected, and when the operator opts for a particular mode, the "unit load control in auto" signal comes on (Fig. 10.8). The boiler follow mode is also possible in master pressure control mode in a system where multiple boilers are connected to a common header. This will be discussed in a later clause.

3 TURBINE FOLLOW MODE

The turbine follow mode, as previously mentioned in Clause 1.1.3, is used when a boiler or steam generator is delivering the output as required by the unit load demand, the coordinated control system, or as per its own capability in manual mode. In this mode of operation, the unit load demand signal may be issued directly to the boiler control system when there is no contingency or limiting criteria, but there may be some other problem that poses a limitation for the unit to run on coordinated control mode. This occurs because the operation of the unit with the turbine follow mode is more stable than the boiler follow mode of operation, as per the study conducted by authorities to compare response characteristics under different modes of operation. During this mode of operation, the steam supplied by the boiler has to be accepted by the turbine at the desired pressure by positioning the governor valves as if the turbine is in auto mode and capable of producing full capacity output. This type of operation is normally performed when the unit is running on initial pressure control or during pressurizing of the system or limit pressure control.

3.1 Turbine Follow Mode: Steam Pressure Control Mode

In a turbine follow mode of operation (Fig. 10.8), the boiler output is at its maximum available capacity or as per the automatic load controller command. Here the turbine control system is forced to follow the boiler supplied load, whatever the value, and this mode of operation is applicable for the unit where difficulties are encountered in varying the boiler load. For any change in the electrical load, the steam flow will change and the steam pressure is expected to adjust accordingly. Since the boiler control system action is restricted due to equipment limitation, the steam pressure is now controlled at the desired value by the turbine control system. Fluctuations in the steam flow are expected because the frequency variation-related response in the turbine would now be canceled by the output of the pressure controller action.

If the boiler is supplying steam at a steady load, the function of steam pressure control in the turbine section is to maintain the steam flow admission rate so that the turbine's generator load matches the boiler load only. If there is any change in the thermal load supplied by the boiler, the pressure would change and the pressure controller action to maintain the pressure set value would change the turbine's generator load accordingly. This type of control philosophy is applicable for a once-through boiler where the boiler is dictating the steam flow delivery rate and the turbine adjusts its governing/control valve position to maintain the steam pressure at the turbine end.

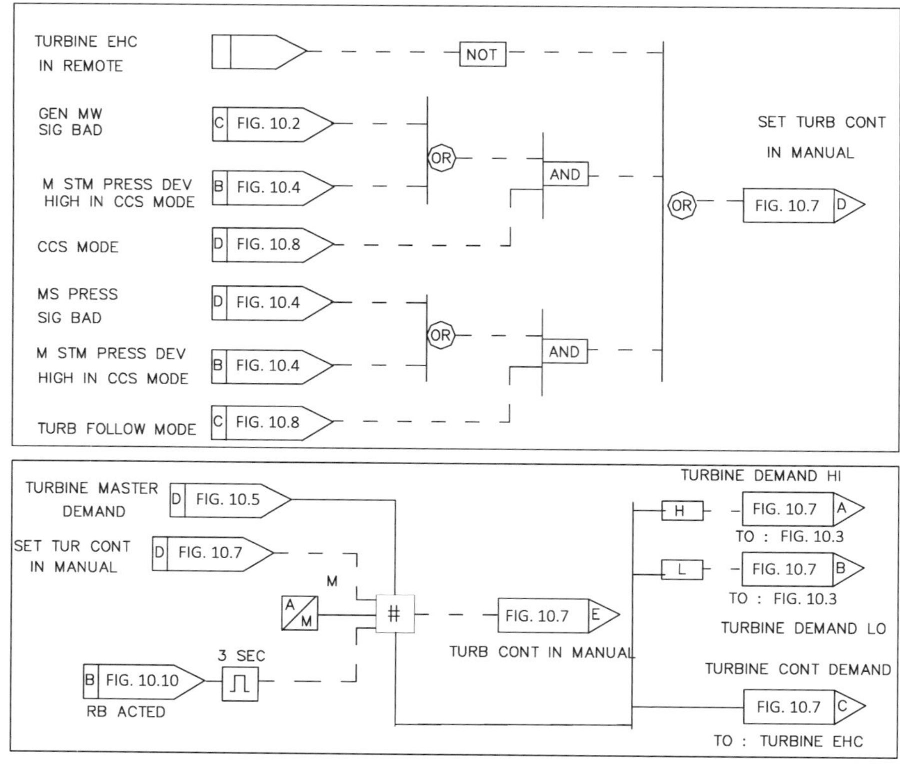

FIG. 10.7 Generation of turbine load demand.

3.2 Turbine Follow Mode under Coordinated Control System

The turbine follow mode in a coordinated control system means when there are load-limiting criteria in the boiler, such as nonavailability of adequate fuel, air, or feedwater support. This mode may be automatically entered by interlocked operation when such an emergency event occurs in a boiler, or it may be selected by the operator from the workstation.

As previously discussed, in this mode of operation, the unit load demand or boiler load demand is directly commanded to deliver steam flow rate; at the same time the turbine follows the above load demand accordingly so that the generator output is maintained at a desired level. The coordinated control system would provide necessary command/correction to the turbine control system to maintain a constant throttle or a variable pressure (Fig. 10.5).

4 BOILER FOLLOW MODE

The boiler follow mode, as previously mentioned in Clause 1.1.2, is used when only the turbine is controlling the load by positioning the governor valves. This mode ensures delivery of the required steam input to the turbine at the exact pressure (whether fixed or variable). The sole duty is accomplished by the steam generator plant through the creation of a suitable boiler master demand. This type of operation is possible when the unit is running under any of the following conditions:

1. When the turbine control system is in manual, the unit load demand may not be directly controlling the governor valve position, but at some value that may be fixed, for example, during the startup/pressurizing operation of steam.
2. Due to some constraint the turbine may operate within some restriction or in manual mode due to limitations of its own equipment or accessories; namely, outage of any heaters or CW/ACW pumps.
3. The turbine control is in auto mode delivering the load as per the unit load demand, which may be either in auto [LDC/automatic generation control (AGC)] or manual, and the boiler delivers the required steam at the required pressure without seeing the unit load demand.

4.1 Boiler Follow Mode: Master Steam Pressure Control Mode

This type of operation is applicable when a boiler maintains the steam pressure of a specified point only. The situation

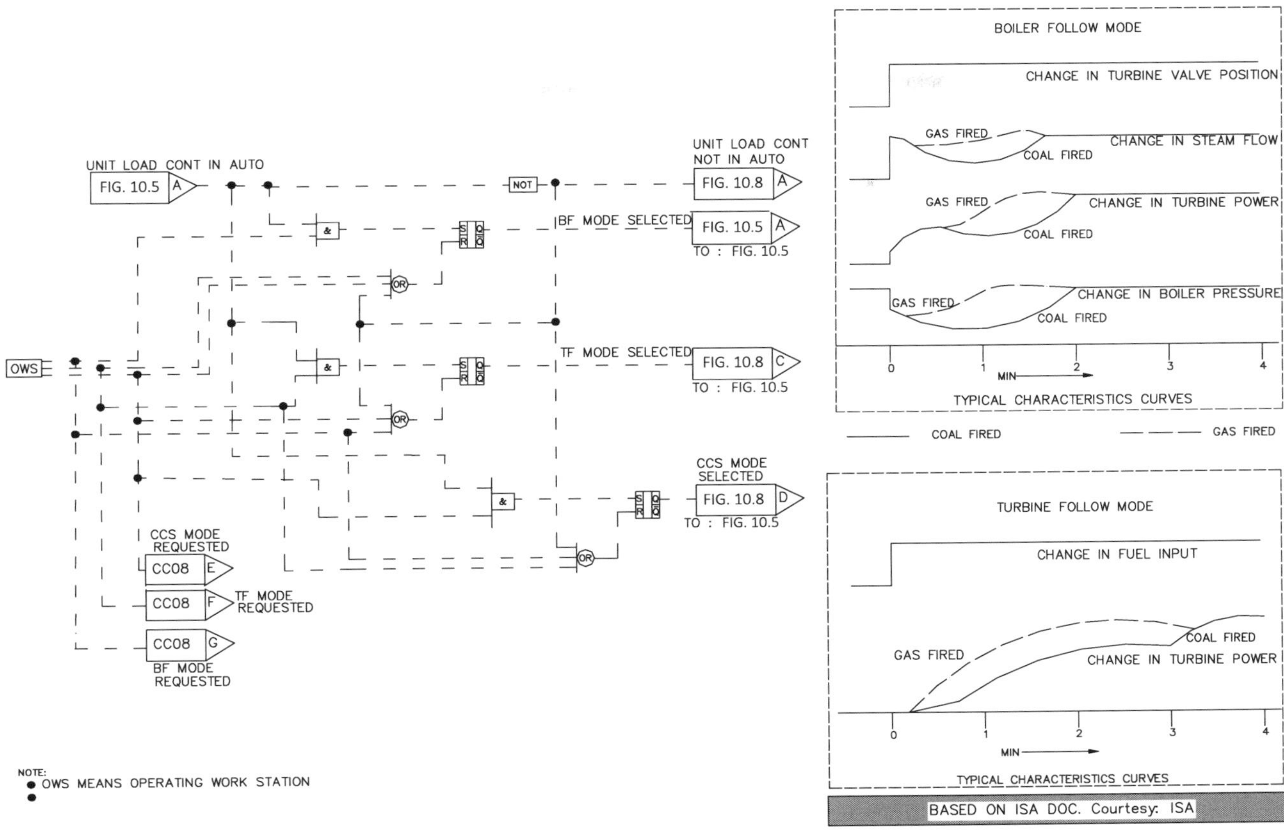

FIG. 10.8 Coordinated control: control mode selection.

may occur when a number of steam generators/boilers with different capacities are supplying steam to a common main header. There may be a single turbine or many turbines drawing steam from this header, according to the load demand decided by other systems. There may be steam extraction from the header or the different turbine stages to meet the process requirements. Fig. 10.4 is a schematic representation of this system.

Whatever the consumer network, the steam pressure of the header is supposed to change in response to the varying load demand. Each boiler connected to this main header has its own steam pressure controller with the same pressure control set value used to maintain the main header steam pressure according to its own capacity. This is called parallel operation of boilers. In this arrangement, the demand is mainly generated by the turbine through the governing/control valve modulating in response to the electrical load demand. Here, a boiler is just meeting the demand by supplying steam at a desired pressure or, in other words, is "following" the turbine demand. Since the control of the main header steam pressure is the only criterion for this type of operating mode, it is called "boiler following turbine" or "boiler following-master pressure control."

There must be a time delay for the controller to take corrective action as per the demand; the once-through type boilers are unable to cope with this type of requirement. This mode of operation can, therefore, be implemented only for the drum type boilers where some cushion is available for the energy stored in the drum.

4.2 Boiler Follow Mode Under Coordinated Control System

This control mode of operation is selected by the operator from the workstation, as previously discussed in Clause 2.1. In this mode of operation, the main steam pressure plays a vital role in the boiler side control while the turbine controls the load. Fig. 10.8 is a schematic representation of this system.

The boiler follow mode is part of a coordinated control system used when a turbine load-limiting function exists. This mode is automatically entered when a turbine contingency such as valve position limit or operator action occurs. The boiler must have fuel, water, and air on automatic to regulate the pressure. The turbine controls the MWs within its own limitation. The load limit is restricted as set by the adjustable bias over the main steam pressure (controller output) demand.

In the boiler follow mode, the output from the main load controller of the coordinated control system goes directly to the turbine control system as the turbine master demand by switching algorithms, whereas another switching provision at the same time enables the main steam pressure control demand to become the boiler master demand. During this mode of operation, the unit load demand is totally managed by the turbine governor valve position controls. By doing this, the main steam pressure at the governor valve inlet is liable to change. Unless the steam pressure is maintained at the desired value, the load control would become an impossible task. Hence the steam pressure is controlled in the boiler by controlling the fuel inputs.

5 RUN BACK SYSTEM

A run back system in a coordinated control system operation means the unit load demand comes from the unit load. As such, the load level decreases as dictated by the type of emergency situation. If the load is not already lower, then the new emergency load restriction is called a run back load. In some coordinated control systems, only the boiler demand is reduced to run back load and the turbine control is forced to manual.

The run back demand is a variable load value dependent on the type of emergency condition. In this condition, the load demand is gradually brought to a desired run back target set level as per the run back gradient, which is also dependent on the severity of the run back condition (Figs. 10.9 and 10.10).

The number of main drives, such as IDFs, FDFs, PAFs, and air preheaters (APHs), and the number of mill boiler feed pumps (BFPs), the condensate extraction pumps (CEPs), the circulating water pumps (CWPs), etc., determine the magnitude of the run back load. Here, the total number of drives (other than standby drives) is considered capable of delivering 100% load. For example, 3 × 50% BFPs means 100% equivalent load when any two BFPs are running (the other is a standby) and outage of two such pumps with only one pump running would mean the unit capacity would ~50% or the capacity of each pump. For 2 × 50% FDFs, there is 100% equivalent load and the outage of one such fan would mean the unit capacity would also be ~50% or the capacity of each fan.

If any of these drives fail, the equivalent load value, as shown in Fig. 10.9, would be <100% load. If the plant was running at a higher load (as delivered by the generator), then the new run back condition load, which has now become the run back target for that particular type of drive failure (FDF/IDF/BFP, etc.), may be selected as the emergency boiler load demand. When the generator MW demand signal is more than the generating capacity of any type of drives indicated above, the run back condition/action is now recognized for each type of drive. The low selector enables the minimum signal of all of the abovementioned run back targets to pass through as the input to the tracking integrator (TRI). The new load value would now be integrated gradually through the TRI with the gradient (rate of change) setter whose value also varies depending on the nature of the drive failure. As shown in Fig. 10.9, the run back rates are also set by providing suitable bias values applicable to each type of drive whenever an individual type of drives goes on run back condition. All the run back rates are input for maximum selection, and the output determines how fast the lowest run back load demand would be achieved to dictate the boiler load or unit demand.

This gradient would gradually bring down the boiler/unit load to an optimum condition to avoid any major bump in the system. The run back logic for mill tripping is different from that of one FDF/APH, etc. Normally, in medium to large units a minimum of one mill is kept as standby. Fig. 10.10 assumes M number of mills to run for 100% load out of a total of N number of mills. If Q number of mills running corresponds to the assigned run back target (~60%) and the number of mills running at any instant becomes $\leq Q$, the run back system is generated. Other run back inputs are from the burner management systems/the furnace safeguard supervision systems and/or "fuel control deviation high" logic. With ball and tube mills, each side could be considered equivalent to one other mill type.

While the boiler/unit load demand is brought to the run back (other than mill run back input) load, the mills are cut sequentially to tally with the new demand value. Failure of each type of drive determines the number of mills to be "ON" in run back condition, and the other mills are cut in a sequential manner. This sequence may be initiated from the top of the furnace up to the desired level, which is normally followed in TT boilers. Since the design of wall-fired boilers includes many types of firing systems—opposite firing, front firing, downshot firing, etc.—this preferential mill tripping depends on the guidelines indicated by the manufacturer.

Mill cutting is necessary for a run back condition as the reduced number of mills would then deliver near-full capacity output to maintain the new boiler load. This is done because a number of mills are not supposed to run at lower loads; this may jeopardize the stability of the mill. With this new scenario, the boiler control can run in automatic mode, but the turbine may be tripped to manual. The command of such an operation lasts for a short time, and the operator can take the turbine control to auto mode if the situation demands.

6 DISCUSSIONS AND EXPLANATIONS

In the 1950s, the United States developed the first application of the coordinated control system for a large once-through supercritical unit. Since its inception, the concept gained significant acceptance and recognition throughout the world for achieving engineering excellence. In view

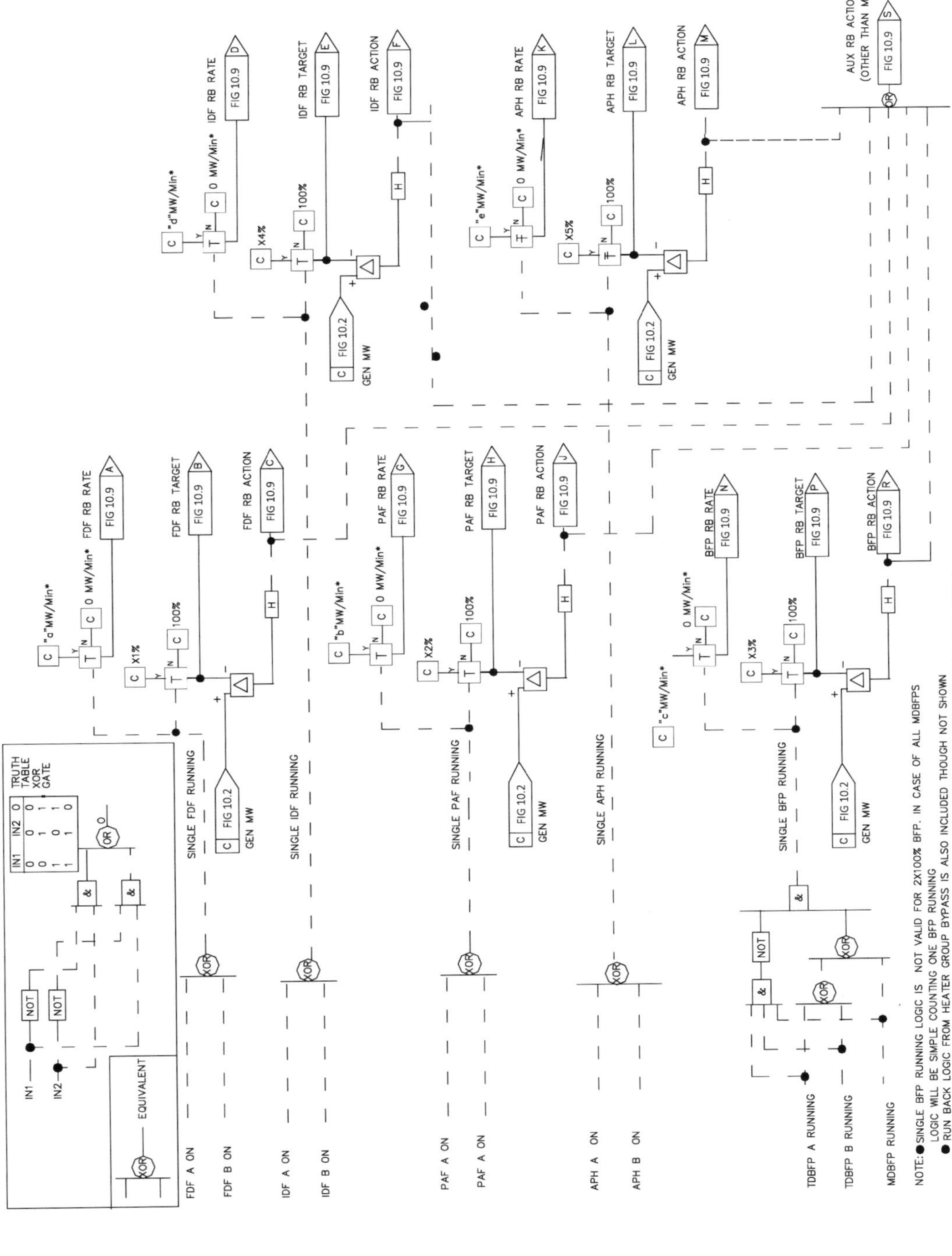

FIG. 10.9 Run back logic 1.

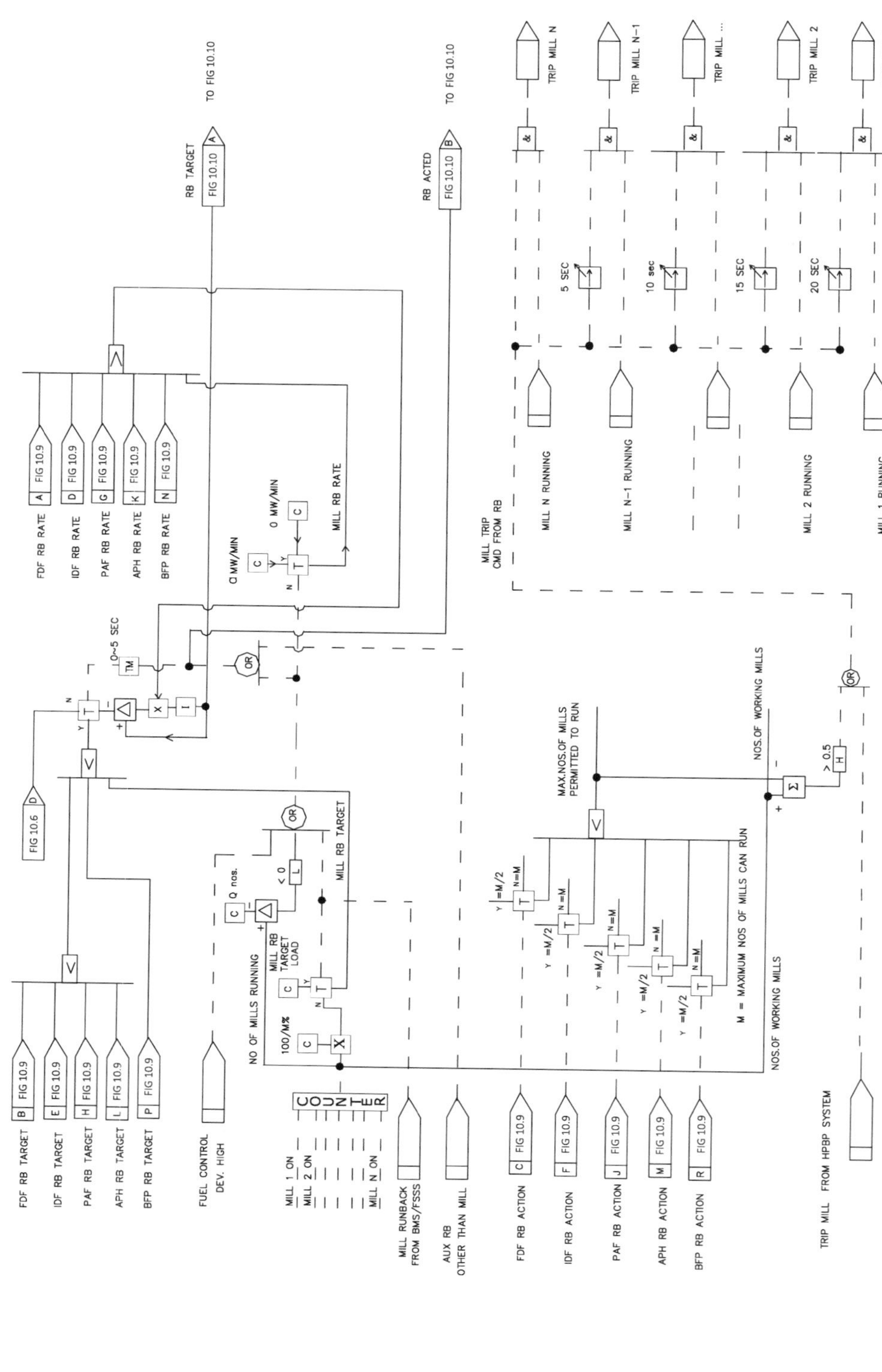

FIG. 10.10 Run back logic 2.

of the FGMO, as discussed in Chapter 9, and trouble-free operation of large modern thermal (fossil-fired) power stations, the coordinated control system continues to play a vital role.

Coordinated control implies a coordinated "single unit" operation of three components: a boiler with its associated accessories related to fuel, air, and feedwater; a turbine with related accessories; and a generator, including heaters, condenser, CWPs, CEPs, etc. Together these components meet the prime objective of the plant, that is, to supply adequate steam flow at the proper temperature and pressure for delivering electrical power generation according to demand. By this definition, the unit load demand is utilized as a requirement for the boiler thermal inputs as well as the electrical power generation through the turbine. The coordinated control system is generally used to get the boiler, the turbine, and the generator to work in tandem under all operating conditions such as a steady fixed load, during the changing of a load through ramp with a preprogrammed gradient in a normal application, and with different preprogrammed gradients under emergency conditions. The basic objective is to deliver maximum electric power, as per the demand initiated inhouse or by an external LDC or AGC under all conditions such as at steady load or during changed demand at the optimum rate, considering the process limitation such as the outage of vital drives or the capability of auxiliary equipment/systems. The constraints may be from any subsystem, such as the boiler, the turbine, and the generator, and the rate is programmed to be limited by the least responsive variable. The most common objectives for coordinated control systems are as follows:

1. Balance of electrical demand and generation; the MWs generated are matched with the unit load demand at the highest possible rate and in a linear trend.
2. Consequent to electrical balancing, control of the boiler-turbine balance in all situations, even under emergency conditions.
3. Subsequent to electrical balancing, safely and efficiently control the boiler inputs (fuel and air; feedwater as well for once-through boilers) to match energy input requirements. (However, this part of the three main objectives is maintained in the boiler control system.)
4. Safe and stable operation during limitations and emergencies such as during run back and directional blocking.
5. Simple operator interface to enable other modes of operation such as the boiler follow mode, the turbine follow mode, and the base mode, i.e., manual control of unit demand when the boiler will be responsible for load change and the turbine, and to follow and stabilize the throttle pressure.

Normally the load gradients are programmed to follow a linear path to reach the new target in spite of disturbances in the steam generator and/or the turbine, the main steam pressure excursion, and the nonlinear characteristics of turbine governor/control valves, etc. The energy balance between the thermal energy input to the steam generator with the corresponding electrical energy output from the generator is achieved by issuing a common demand to the boiler and turbine. This concept has been equally applicable for both drum and once-through units. The system must ensure that boiler and turbine controls are coordinated, even if the turbine is controlled independently, as per the recommended control philosophy, or as an outcome of emergency requirements.

The system uses a common unit demand for the boiler and turbine. When applied to the boiler, it is first subjected to the dynamic rate action to produce over firing on the increasing load and under firing on the decreasing load. This rate action, applied in proportion to the boiler load derived from the unit load, compensates for the inherent stored energy in the boiler, and is reflected in the turbine throttle pressure. This enables the boiler to act as quickly as possible to compensate for the stored energy already used by the faster operation of the turbine control through the turbine energy requirements corresponding to the unit load demand.

In DEB, another coordinated control strategy, the boiler control does not relate its coordinated action with the electrical generation; the boiler demand in this concept is calculated based only on the turbine energy requirements. Here (as per DEB 400, Metso) unit load demand acts on the governor valve to change the active power by varying steam flow, which changes the first stage pressure of the turbine, and thus the ratio of first stage pressure P_1 and throttle pressure P_T. This P_1/P_T is duly rationalized with the throttle pressure set point P_S and after it is dynamically compensated to act as the energy demand signal for the boiler. The boiler is decoupled from the electrical load change and connected to the system via energy balance utilizing direct coordination between the boiler and turbine. This calculation is called the energy balance signal and represents the turbine requirements for steam under all conditions. It responds as fast as the turbine valves move, even during emergency conditions or when the manual operation of the turbine is performed. Here the boiler demand minimizes the use and effect of the rate signal as the feedforward signal, which may provide more stable control on the load changes.

This coordinated control system is a better operating philosophy applicable to all boilers with different types of fuels and turbine combinations, such as large central stations with a single boiler tandem compound turbine-generators (fixed pressure and sliding pressure applications, multiple boilers connected on headers-cogeneration and combined heat and power, simultaneous burning of multiple fuels, etc.).

6.1 Demand From LDC

In all developed countries a large portion of power is generated by nuclear plants and combined cycle plants (gas turbine and HRSG/steam turbine), and most run on base load. Consequently, thermal power plants with fossil firing systems are utilized to operate on a wider range of generating and load-sharing conditions, such as load peaking up, low load operations, load cycling, etc. Even in India, there is a larger proportion of nuclear plants compared with hydel power plants. The following list contains a rough figure regarding the proportion of power generation in India:

1. Fossil-fuel fired thermal power plants: 75%–80%.
2. Hydel power plants: 15%–20%.
3. Nuclear power plants: 5%–7%.

This phenomenon caused a strong impetus to regulate all power plants from remote, centrally operated LDCs. To implement this, it is extremely important that the coordinated control system is responsive. The LDC is normally conceived as a global load center for the entire grid subdivided into regional grids. There are some conditions, e.g., a fluctuation in load, in which a particular region may not be controlled from an LDC. During those conditions, several units in that region can be operated from regional grids under AGC.

As previously discussed in Clause 1 and 2, the desired generation (MW) can also be set by the plant operators in addition to a signal from the LDC/AGC. Regardless, the response of the load generation controller should ensure a linear output at a maximum possible rate to avoid untoward bumps in the control systems. Although a competent boiler controller coordinates the combustion air and the corresponding fuel (and feed flow to maintain fuel to feedwater ratio in once-through boilers) and the turbine control includes an intelligent governor control system for MW control, only a well-coordinated system can ensure complete harmonized operation of the turbine and boiler. There are many problems associated with the turbine governors while the control action is passed on directly by the LDC/AGC.

Significant disturbing phenomena (2018) include the nonlinear response and the mechanical hysteresis affecting the governors to respond in a desirable way. These types of problematic elements need special provisions in the coordinated control system so that the response of the unit as a whole is commensurate with the load variation. If these types of anticipatory problems are not managed properly, the final action may be lacking in linear response and may consist of dead time and overshoot. These features may be incorporated in the coordinated control system and the turbine control to receive the command and act accordingly when put in a coordinated control mode and the turbine control is in a remote auto mode.

6.2 Boiler Turbine Balance

In the conventional coordinated control system, the controller output from deviation of unit demand and MW generated is issued to both the boiler and the turbine with the incorporation of signals appropriate to individual subsystems (for the boiler and turbine). For fine-tuning, this load signal is corrected by the throttle pressure deviation signal or limited by the throttle pressure controller ± bias values. In some coordinated control systems, the load demand for the boiler is not derived directly from the unit demand as described earlier, but through DEB.

6.2.1 Turbine Demand

For the turbine subsystem, the turbine load demand from the coordinated control system as generated is matched with the first stage pressure as a process feedback signal. The governor or the turbine control valves are set to move at a much faster rate to comply with generation demand. In this configuration, the system is able to achieve a response at the maximum possible rate to follow the changes in MW demand. The same is accomplished by consuming the energy already stored in the pipelines and the boiler steam superheaters to the greatest extent, thus fulfilling one of the main targets of the coordinated control system—intelligently managing the stored energy.

6.2.2 Boiler Demand

Here the boiler load demand is representing, under all conditions, the effective turbine valve opening through the turbine energy requirement from the boiler (Fig. 10.1D and E (by DEB) and Fig. 10.11). The effective turbine valve opening is simply the ratio of turbine first stage pressure (P_1) to the throttle pressure (P_T) and is well-accepted as the linear representation of the effective turbine valve opening. It appears it is independent of different disturbances caused during operation of the boiler subsystem, such as a change in the fuel quality or a limitation in the production capacity in a particular mill, etc. This signal, multiplied by the throttle pressure set value, constitutes a better load index, unlike steam flow or first stage pressure, which are commonly used as load signals and are apprehended to engender regenerative signal during boiler upsets. This signal acts similarly to a self-calibrating feedforward signal; hence it is recommended as a load index to the boiler control system. Because that signal is supposed to be the ideal feedforward signal from the user side, introduction of the same will initiate optimum firing rate by restraining the final demand through minimizing the intermediate over firing or under firing command that might be generated otherwise.

When load increases, the stored energy is withdrawn from the boiler and there is an equivalent drop in the throttle

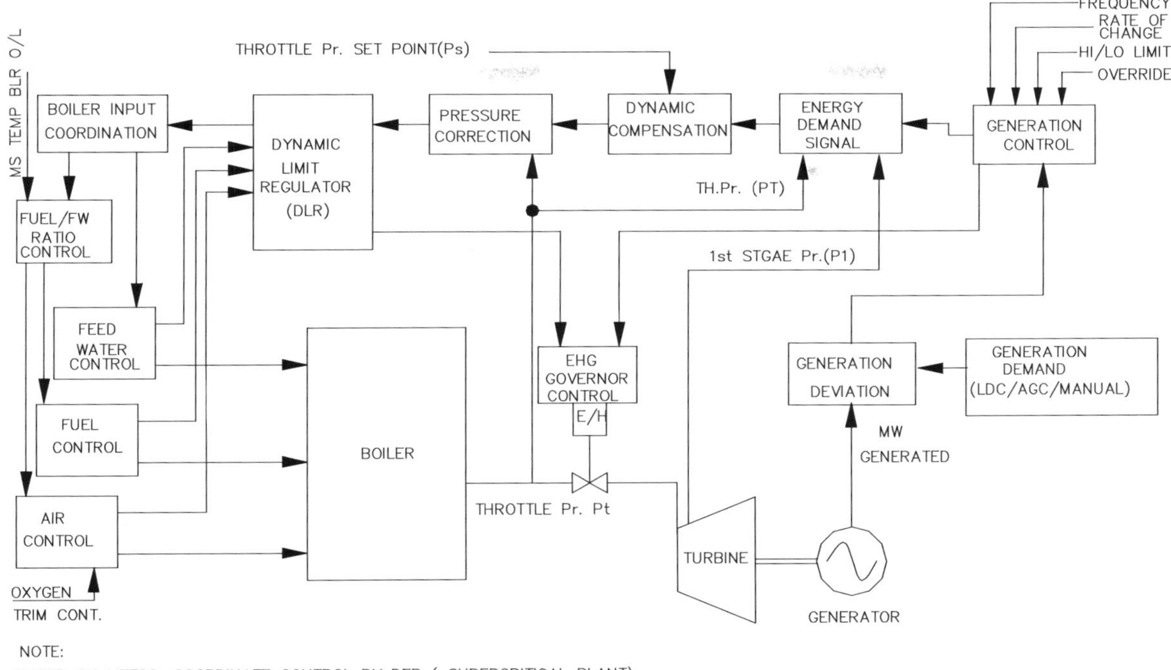

FIG. 10.11 Coordinated control-energy balance concept for supercritical boiler.

pressure. The tendency of the simple boiler control systems is to overcompensate with continuous integral control action, which may cause general instability. The desired controller action should have the self-tuning/calibrating capability as a measure of the energy balance signal with a proper combination of proportional and rate action to deliver the correct amount of the control output signal, thus the loss of stored energy in the boiler is compensated for as quickly as possible. As the pressure approaches the set point while the steady-state condition is attained, the controller output signal automatically decreases to a level where the energy input to the boiler equals the generation output. Overshoot and instability can thus be avoided and used to assist the immediate load change is restored to a new value appropriate with the new level. The derivative signal is appropriately scaled according to the preliminary demand (to avoid noise) before being added at the summator (Fig. 10.1E) to arrive at the final demand. The upper and lower load limiter and the directional blocking functional blocks are included to arrest any rapid change in the boiler demand.

Because there is no electrical generation (MW) signal in this energy balance concept, the controller output is assumed to provide for the "boiler-turbine coordination under all circumstances." This statement is valid even if it is not possible for the turbine control to be operated under automatic control mode; in other words, it is forced to operate in an emergency condition or put in manual control. The boiler demand signal is, as is apparent from the above description, always considered as the current energy requirements. The energy balance signal is an independent calculation, and the boiler control can now operate without the demand or feedback from the generation control.

6.3 The Concept of Heat Release

To establish energy balance between the boiler and turbine, the demand signal with an integrating type controller should be replaced with proportional and derivative or rate action since the former controller may cause instability. It is also recommended that the regenerative type of feedforward signal, such as steam flow, may also cause instability due to overshooting at the time of each load change. This problem may be more acute for the coordinated control system with the sliding pressure control mode.

According to this theory, the concept of "heat release" is utilized for calculation of an accurate measurement of the fuel or energy input to the boiler and not by the measurement of the steam flowing out from the boiler. This philosophy is most useful when there is a change of load or a change in fuel quality, which causes a change in the heating value of the fuel being burned. It is important that the calculation is accurate, otherwise the controller cannot maintain the desired air–fuel ratio control, which could ultimately result in untoward disturbances in the other control strategies, such as the drum level control and the steam temperature control. If a high amount of accuracy is not achieved, the rate and the magnitude of the load change must be restricted.

6.3.1 Measurement of Fuel Input

There are two conventional types of fuel measurement adopted in thermal power plants. They are utilized in the combustion controls and each of these measuring systems has its own limitations. The first type is the direct measurement of the fuel, and the second one is indirect measurement of the fuel based on the calculations and the associated functions related to other parameters.

Direct measurement of the fuel is a widely accepted and standard method, but it has practical limitations. The fuel flow can only be successfully measured in (natural) gas-fired steam generating plants. Measurement of solid fuels, such as pulverized coal, are extremely difficult to measure and are normally measured by "coal feeder speed" or air flow through the mill, which are also indirect methods.

The second traditional method of the fuel measurement is an empirical one using the measurement of boiler output, typically, steam flow or turbine first stage pressure. The primary limitation in using the boiler output measurement as a measure of energy input is that the output is proportional to the input only under steady-state conditions. A more accurate relationship between the energy input and output must also include changes in stored energy. The change in stored energy must be added to or subtracted from the boiler output to make it proportional to the energy input under both steady-state and dynamic conditions.

6.3.2 Energy Balance System

The design of a drum type boiler uses the working medium (water and steam mixture) at saturation pressure and temperature, and accumulates in the drum. Due to the load change, the drum pressure, the temperature, and the enthalpy also change. The majority of stored energy is in a fluid form. Because drum pressure is essentially linear with fluid enthalpy, it can be considered as a good index of stored energy.

The actual level of stored energy cannot be treated as a part of the load change, since it has no bearing on the fuel input. However, the changes in stored energy are of great importance since they are the result of changes in the fuel firing rate and/or the heating value of the fuel. The basic equation for boiler energy transfer is as follows:

$$\text{fuel input}(\text{kcal/H}) = \text{boiler output}(\text{kcal/H}) + \text{change in stored energy}(\text{kcal/H}) = KP_1 + K_1(dp_{DR}/dt)$$

where P_1 is the turbine first stage pressure or boiler steam flow

P_{DR} = drum pressure

From the above equations the rate of change of drum pressure ($d/dt\ P_{DR}$) is an indication of the boiler input and output balance similar to throttle pressure excursion and thermal energy balance between the boiler and turbine. In the steady-state condition, that is, when the drum pressure is not changing (the value of $d/dt\ P_{DR}$ is zero), the fuel input to the boiler equals the boiler output and, correspondingly, right at that moment, the energy level of the boiler and the turbine are perfectly balanced.

6.3.2.1 Limiting of Unit Load Demand According to Boiler Input Limitation

The coordinated control system should provide bumpless, trouble-free, safe, and economical boiler operation and not permit a significant imbalance between the boiler inputs by sensing several control errors. For example, it reduces the turbine requirements when a boiler input is limited, at the same time it reduces the boiler demand to match the limitation in the different types of boiler inputs. Positive action is to be taken to make certain that the system load does not exceed the productive ability of the least capable boiler input.

Basically, this part of the coordinated control system looks at the differences between the desired and the actual measurement/calculated values of all of the important process parameters, such as fuel, air, and feedwater control systems. If any of these process deviations exceeds their limit value, the rate of directional changes (increase or decrease) of the common unit load demand, that could have increased the deviation, is either blocked or held. In advanced control systems, when the deviation approaches its limit value, the rate of change of increase or decrease of the unit load demand is reduced as a precautionary measure. When any of these deviations reaches their limit value, these rate changes in the demand signal are blocked. If any of these deviations exceeds the limit value, the unit load demand signal is corrected at a rate proportional to the magnitude of the error until the balance is reestablished. However, if there is much difference between the unit load demand and the ultimate boiler load (reflected in high generator MW deviation), the turbine control is transferred to the turbine follow operation to follow the boiler in its limited condition.

The run back system is a coordinated control mode operation. A run back load occurs when the unit load needs to decrease to a lower load level as dictated by an emergency situation if it is not already lower than the new emergency load. Run back demand is a variable load value dependent on the type of emergency condition. In this condition the load demand is brought gradually to a desired run back level as per the run back gradient, which is also dependent on the severity of the run back condition (Figs. 10.2 and 10.5).

The "upper load limit" of a plant unit is normally set 10% above 100% load. As depicted in Fig. 10.2, the unit load demand is the output through a minimum selector that has its inputs as "demand from the LDC/AGC" and the abovementioned upper load limit. Then upper load demand is corrected by the "frequency-influenced" load value and sent to the boiler and turbine as the load demand after passing through the upper and lower limits set by the master pressure controller, as shown in Fig. 10.5

If any of the drives fails, the equivalent load value would be less than 100% load. If the plant was running at a higher load than the new run back condition load, the latter is now selected as the new boiler/unit load demand. When any drive failure-related signal is less than the generator output, the "run back condition" is recognized. The new load value is now to be achieved gradually through a limiter with a varying gradient setter whose value is also dependent on the nature of the drive failure.

As shown in Fig. 10.6, the number of drive-related equivalent loads of each type of drive is compared with the upper load limit. For any significant deviation, the comparator would trigger a switching action and allow the preloaded run back gradient signal, unique to the individual type of drives, to a selector where other such signals from different types of drives are also input. The selector has the capability to select the priority among all the types of drives connected to the run back system. This priority selected run back gradient signal is made through another switching action only when the run back condition takes place. This gradient enables the boiler load to be brought down gradually and smoothly in an optimum condition to avoid any major bump in the system.

As discussed previously, DEB can be applied to a supercritical plant with once-through boilers, based on the Metso documents (Fig. 10.11). Also, boiler control is decoupled from the main unit load deviation signal so that stored energy can be utilized quickly to meet the demand. The unit load deviation directly sends a signal to the generator control system, which also manages other influences such as frequency, high/low limit, rate of change, etc. The generator control directly affects the electrohydraulic governor control to change the first stage pressure (P_1). At the same time the energy demand signal is generated accounting for the P_1/P_T ratio duly normalized with the throttle pressure set point P_S at the dynamic compensation circuit. This signal after pressure correction is fed to the boiler input coordination input.

In once-through boilers there is no isolation between steam and water circuit via storage vessel boiler drum. Naturally, any change in parameter of steam directly affects the BFP, i.e., feed flow. So, in once-through boilers in supercritical plants, the ratio of fuel flow to feedwater is very important. Thus the boiler input coordinator sends a signal to the feedwater and fuel circuit via the ratio setter. For any combustion the fuel/air ratio needs to be maintained; therefore, via a ratio setter, a signal is sent to the air and fuel controller to manipulate the inputs to the boiler. During transit, the boiler-stored energy is utilized, so that unnecessary overshoot and undershoot could be avoided. There is another subblock, dynamic limit regulator, shown in Fig. 10.11, which is fed from boiler input control systems. These subsystems feed the status of the control systems, and their various auxiliaries, so that the run back and directional blocking signals can be generated to limit the demand for the boiler. These are mainly used during the failure of major auxiliaries, and/or when the capacity level is reached.

BIBLIOGRAPHY

[1] The Smart Grid begins with Efficient Generation; Energy Efficient Design of Auxiliary Systems in Fossil-Fuel Power Plants. A technology overview for design of drive power, electrical power and plant automation system: By M/s ABB, Inc. in collaboration with Rocky Mountain Institute, USA (ABB Energy Efficiency Handbook), http://www05.abb.com/global/scot/scot221.nsf/veritydisplay/5e627b842a63d389c1257b2f002c7e77/$file/Energy%20Efficiency%20for%20Power%20Plant%20Auxiliaries-V2_0.pdf.

[2] Modern Control and Actuating Systems on Steam Turbines by M/s VGB Power Tech 3/2005.

[3] D-E-B/400 Coordinated Boiler/Turbine Control System (Drum Units): By M/s Metso Automation, http://www.metso.com/automation/ep_prod.nsf/WebWID/WTB-041110-2256F-8E7F4?OpenDocument#.VBf0YZSSyt0.

[4] Controls for Once-Through Boilers (ISA Document): Excerpts from M/s F. Paul de Mello and other Authors. de Mello F.P Chapter IX (ISA ref book), http://www.isa.org/~powid/boiler_dynamics_controls_course/De%20Mello%20course%20notes/De%20Mello%20Chapter%209.pdf (Chapter IX).

[5] Boiler-Tuning Basics. Part II. By Tim Leopold, ABB Inc. Dtd.

Chapter 11

Balance of Plant Control System

Chapter Outline

1 BALANCE OF PLANT: INTRODUCTION

1.1 Introduction

The balance of plant (BOP) controls for a power plant are defined as the controls outside the main power plant control, such as controls pertinent to offsites like a coal handling plant, ash handling plant, etc. All these auxiliary and offsite plants have been covered in Chapters 2 and Chapter 7, Main Clause 2.5. However, various control systems of the main plant, which are not directly related to boiler-turbine control, have been grouped here as BOPs, so the BOP of main plants will be discussed in this chapter.

- *Boiler control system* (Chapter 8): The control systems directly related to boiler controls such as the furnace safeguard supervision system, secondary air damper control, combustion control, furnace pressure control, drum level (as applicable), steam temperature controls, high-pressure (HP) bypass, etc.
- *Turbo generator control system* (Chapter 9): This includes control systems directly related to turbine controls, such as electrohydraulic governor controls, automatic turbine run-up system (ATRS), automatic turbine tester, thermal stress evaluator, low-pressure (LP) bypass system, H_2 seal oil controls, condenser level control, etc.
- *Coordinated control system* (Chapter 10): Coordination in control between various systems.
- *BOP control system*: The BOP control loop is mainly related to equipment from regenerative systems. In cogeneration/combine cycle plants, all controls other than gas turbines may be called BOP controls, which are well covered in the various control systems discussed in this book.

Power Plant Instrumentation and Control Handbook. https://doi.org/10.1016/B978-0-12-819504-8.00011-1

The following control loops have been covered under BOP controls:

- Boiler feed pump (BFP) recirculation control
- Condensate extraction pump (CEP) recirculation control
- Gland steam condenser (GSC) control (minimum flow control-gland steam header pressure control covered in Chapter 9)
- Deaerator pressure control (deaerator level control covered in Chapter 9)
- HP and LP heater level controls (both normal and emergency)
- Turbine auxiliary steam system (TAS; ejector control)

There are a few common actions/monitoring such as auto/manual (A/M) changeover, final control element position monitoring in distributed control systems (DCSs), etc., which are not discussed separately in each loop as these discussions were covered in previous chapters (e.g., Chapter 8).

2 BFP RECIRCULATION CONTROL

2.1 General

When large pumps run, they create a great deal of heat from the churning effect of the impellers. So when the pump runs it is necessary to cool the pump impellers. During a higher load operation, fluid flowing through the pump acts as a coolant. When the pump just starts and/or when the pump is running at very low load, normal fluid flow quantity may not be sufficient to cool the impellers. Therefore a minimum quantity of fluid needs to flow through the pump, thus balance fluid is returned to the source. This is achieved by recirculation flow of the fluid through the pump. Recirculation is defined as returning the pump discharge flow to the point where the pump has suction so that a minimum flow requirement of the pump is ensured through the pump (even when there is low/no load demand to the system from the pump). For recirculation flow of a BFP (the pump has its suction from the deaerator), the pump entire/partial flow water from the BFP is returned to the deaerator during startup and at low load. Generally, the BFP handles the high flow at high differential pressure (DP); the pump sizes are large so even for small units such recirculation flow is necessary. The loop discussed here is applicable for each pump, so only one loop is illustrated.

2.1.1 Objective

The purpose of the BFP recirculation control loop is to ensure minimum flow through the BFP at all times.

2.1.2 Discussion

The BFP recirculation line is taken from the pump discharge ahead of the discharge valve and returned through the recirculation valve to a deaerator where the BFP has suction (Figs. 3.17 and 3.18). There are various BFP sizes, types, and unit sizes, and variations in controls. Generally, in power plants up to 200 MW (there are modulating types of controls for unit size <200 MW) an ON/OFF type of recirculation control is deployed. BFP recirculation control loops for unit sizes ≥500 MW are generally the modulating type. For smaller units a simple flow switch may be used to operate BFP recirculation control (Fig. 11.1A), whereas in 120 MW/200/210 MW units the suction flow of the BFP is measured and this signal is used to OPEN/CLOSE the BFP recirculation control valve (Fig. 11.1B). There will variations of temperature, so it is recommended to compensate the flow with feed water (FW) temperature at BFP suction.

2.2 Control Loop Description

The loop is applicable for each BFP (Figs. 11.1A and B and 11.2).

2.2.1 BFP Recirculation Control for Smaller Units

In this control loop, a flow switch (i.e., target type) is put at the suction of the pump (Fig. 11.1A). This loop will be functional when BFP is running. The flow switch is set at a flow rate slightly higher than the minimum flow requirement of the pump (to sense falling flow). As long as the flow through the pump is greater than flow setpoint, the remote control valve (RCV) will be closed. During startup or at low load of BFP, when flow through the BFP falls below the flow setpoint, a flow switch will operate to deenergize the solenoid valve. This means, [to have contact configuration: Normally open (NO) contact is open and normally closed (NC) contact is closed, meaning that if NO is connected then at low flow it will open the power to the solenoid valve. On the contrary, if NC is connected then at low flow, NC will bypass the solenoid and the power will be connected to return path via resistance. Both possibilities exist depending on circuit configuration. so, reader to take care of the same in the circuit. applicable in all such cases. Also refer to Fig. 11.3]. Consequently, air from the RCV will be vented, and the RCV will open. The contact configuration of the flow switch through selection of solenoid valve type is important. Normally the RCV is chosen as a fail-to-open type to ensure pump safety at all times (Fig. 11.3).

2.2.2 BFP Recirculation Control for Medium Units

The basic control loop philosophy was discussed earlier (Fig. 11.1A and B). The only difference is that here, flow is computed at the DCS with the help of a flow transmitter, which is more accurate and stable. In this control loop, a flow element and redundant transmitters are placed at the

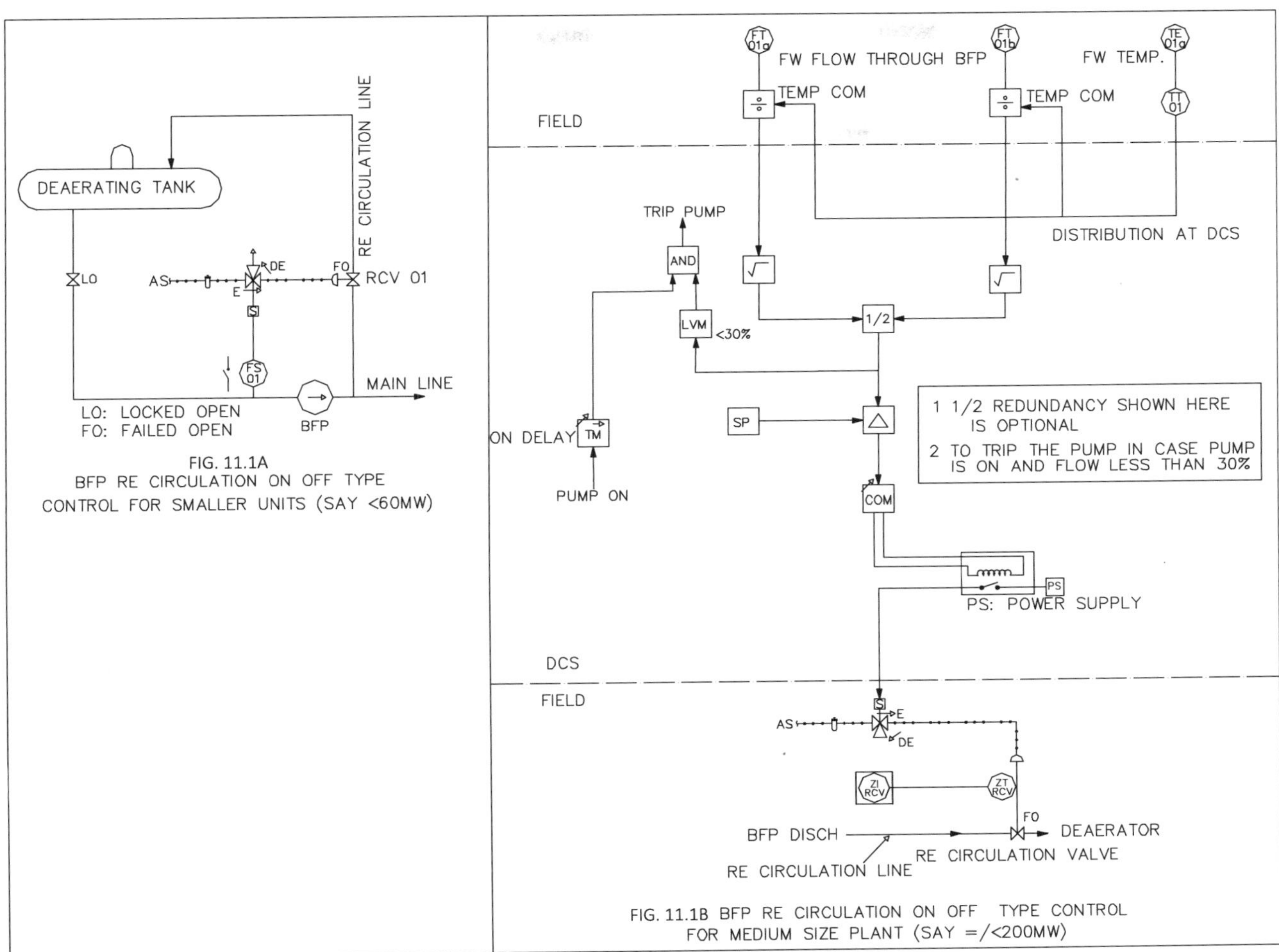

FIG. 11.1 BFP recirculation control.

suction of the pump. To take care of the variation in density, a temperature compensation is made in flow computation in DCS (Fig. 11.4).

Why in FW flow computation is temperature compensation given?

This is because, depending on the load, deaeration heating is done by auxiliary steam, cold reheat (CRH), and extraction steam, and in each case, the heating temperature is different, so the FW temperature is different. Also, when LP heaters come into service in combination, their FW temperature varies, so temperature compensation is necessary.

This flow is compared with the setpoint and fed to a controller in ON/OFF mode as a comparator. As long as the computed flow is less than the setpoint the output of the ON/OFF controller (COM) will be deenergized, and the solenoid valve will not get any power. As long as the solenoid valve is deenergized, air supply to the diaphragm of the RCV will be vented and it will open. When the flow setpoint is exceeded by the computed flow (with dead band), "COM" will generate output to close the NO contact and the solenoid valve will be energized to close the RCV. It can be configured in reverse, but as stated earlier, from a safety standpoint it is recommended to use air to close the valve so that during air failure the BFP recirculation valve is open. With reference to Clause 1.2.1 of Chapter 8, it may be noted that because there is no A/M station, if there is a transmitter failure alarm, the operator needs to manually select the correct transmitter for proper control loop operation.

2.2.3 *BFP Recirculation Control for Larger Units*

For plant sizes ≥500MW, modulating type control loops are deployed for BFP recirculation loops (irrespective of BFP types such as motor or turbine driven; Fig. 11.2). The flow computation will be similar as discussed in Clause 2.2.2. This flow is compared with the setpoint and fed to a P+I controller. The controller output is fed through an A/M station to an I/P converter, which will supply the pneumatic control signal to the positioner of the control valve. The control valve will throttle according to the signal from the controller.

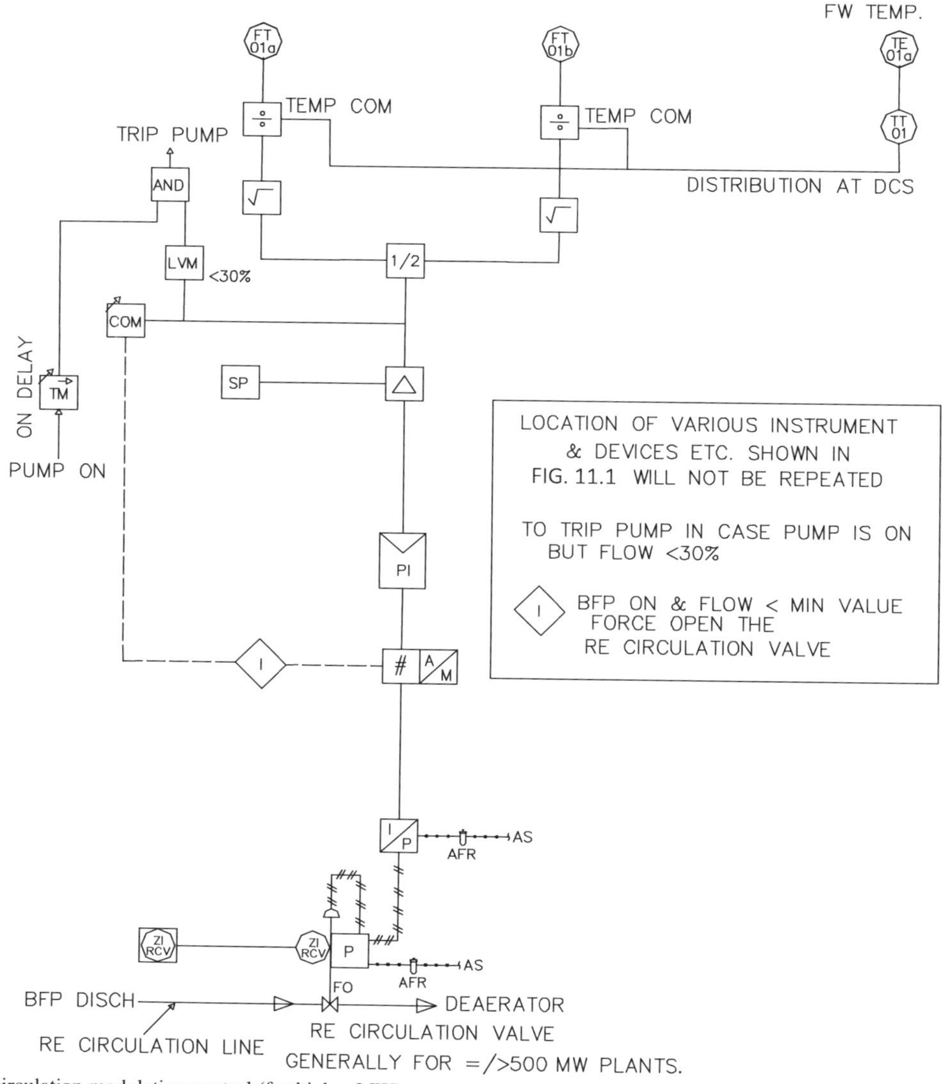

FIG. 11.2 BFP recirculation modulating control (for higher MW).

How NO & NC for a switch is decided? Generally the contact configuration decided based on their condition at NO FLOW NO POWER CONDITION. Now any switch when bought from market, has no flow nor it is energized, so closed contact at this condition is NC and Contact at Open condition is NO contact. Same philosophy is for electrical switches, relays also.

FIG. 11.3 NO or NC contact configuration.

****Since Computed flow Q= √(hP/T) where h= DP head, P= static Pressure & T =Temperature.***
When representing them in the loop Square rooting to be done after Pressure and / or Temperature compensation — not immediately after DP transmitter O/P. This is control loop representation tip!! Water is incompressible, fluid so Pressure compensation is not applicable.

FIG. 11.4 Flow computation.

2.3 Control Loop Operation

The loop operation in small- and medium-size units was discussed in Clauses 2.2.1–2.2.2 (Fig. 11.2). As discussed in Chapter 8, Clause 1.2.1, this control loop may be tripped to manual in case of transmitter failure and/or by selection. The operator needs to ensure the correct transmitter selection to continue with the loop.

2.4 Control Loop Interlock

To make the loop safe and avoid operator error in manual, an interlock can be incorporated to open the valve in case the BFP is ON and flow is less than the set value. Generally, it is set a little higher than the actual setpoint to avoid tripping the pump (Fig. 11.2). To save the pump, an interlock is incorporated so that when the pump is ON (time delay to avoid spurious tripping) and flow is less than the set value sensed by the limit value monitor (LVM; setting a little below recirculation set value), then the pump will be tripped (Figs. 11.1B and 11.2).

2.5 Control Valve Considerations

For medium-size plants, typical parameters for the RCV will be I/P pressure: 200–240 bar; O/P pressure: 8–10/12; bar temperature: ~200°C with a drop in the range of ~200–230 bar; and outlet temperature very near saturation temperature. Because of this, there will be a high possibility of flashing and cavitation. As a result trim durability will be at stake due to high DP at low lift. Therefore it is recommended that for medium and large-size plants drag, VRT, and multistep multistage RCVs may be deployed.

3 CEP RECIRCULATION CONTROL

3.1 General

Basic recirculation control, as discussed in main Clause 2.1, is also applicable for CEP. However, in smaller units the CEP recirculation may not be applicable because of the small pump size (because of low flow and lower DP), but for medium to large-size units it is applicable. With CEP recirculation, the tapping for the recirculation line may be done after the GSC, to ensure minimum flow through the CEPs and through the GSC. This is also possible for 2 × 100 CEPs, and there may be a single CEP recirculation line and a valve for two sets of CEPs. However, for the 3 × 50 CEP configuration this will not be suitable. There are variations in CEP recirculation loops and these are discussed in following clauses.

3.1.1 Objective

The purpose of the CEP recirculation control loop is to ensure minimum flow through the CEP at all times. In some cases (with 2 × 100 CEP configurations), the same control loop is deployed to ensure minimum flow through the GSC.

3.1.2 Discussion

The CEP recirculation line is taken from the pump discharge after the discharge valve and returned through the recirculation valve to the condenser hotwell (from where the CEP takes its suction) as shown in condensate P&ID (Figs. 3.22 and 3.24). As stated earlier, for smaller units an auto control loop may not be necessary for CEP recirculation control. Generally, for larger units (≥500 MW) independent direct CEP recirculation control loops are deployed. For 2 × 100 CEPs, a common control loop is used for both CEP and GSC minimum flow recirculation control (Fig. 11.6A) for medium-size units. In these cases, the RCV is placed after the GSC. A few variations in CEP recirculation control loops are found in medium-sized units where the recirculation controls work with the condenser hotwell level control also. Here, the hotwell level signal is related with the control parameter for regulate CEP and GSC minimum recirculation controls (Fig. 11.6B). However, these loops are an indirect way of achieving the goal and have inherent limitations that are discussed in a later clause.

3.2 Control Loop Description

The control loop description given in this clause is related to the modulating part only. Associated interlocks for the loops are discussed separately in Clause 3.4 (Figs. 11.5–11.8).

3.2.1 CEP Recirculation Controls Multiple CEPs (i.e., 3 × 50%)

In Fig. 11.5A, two CEPs are shown with independent recirculation. Control loop details for both CEPs are identical, so only one will be discussed. Fig. 11.5B illustrates individual CEP condensate flow, which is computed in DCS (for condensate flow computation, temperature compensation is not essential as flow is measured at CEP discharge where there may not be many variations in temperature because there is no heat exchanger in between and the temperature is almost the same as condenser) by measuring the flow by orifice (or flow nozzle) and flow transmitter in one of two modes. This flow is compared with the setpoint (for example, 30% flow capacity of each CEP) and the deviation thus created is fed to the P+I controller, which generates the position demand of the recirculation valve. The position demand is fed to the valve through the I/P converter and positioner. As long as

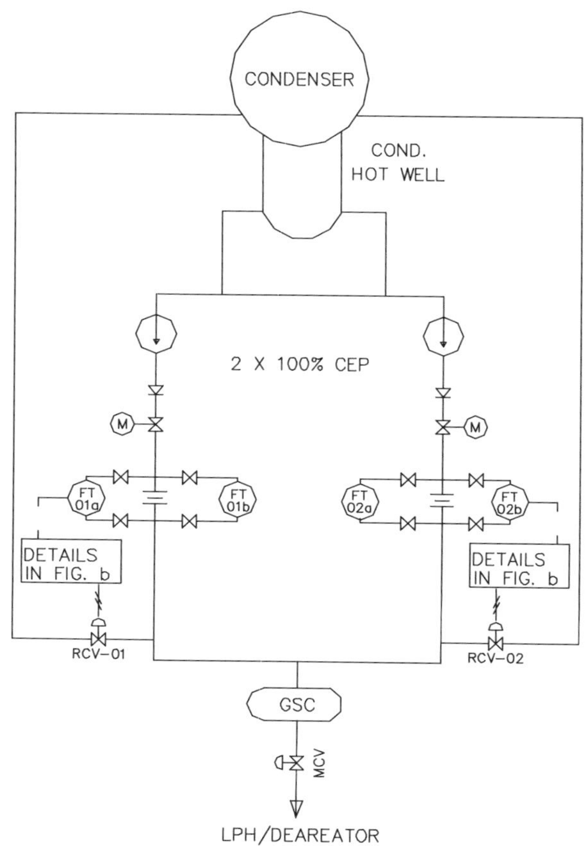

FIG. 11.5A INDEPENDENT RECIRCULATION FOR 2X100% CEPs

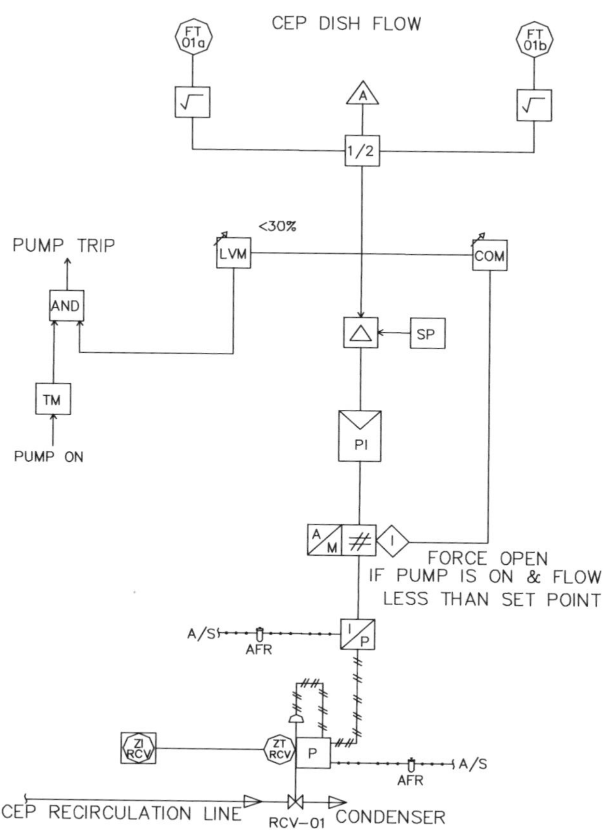

FIG. 11.5B INDEPENDENT CEP RECIRCULATION FLOW CONTROL FOR ONE PUMP OTHER (S) WILL BE IDENTICAL

FIG. 11.5 Independent CEP recirculation control loop.

the flow is less than the setpoint the recirculation valve will throttle to return the condensate to the surface condenser. When the set value is exceeded the valve will start closing.

3.2.2 CEP Recirculation Control Double (i.e., 2 × 100%) CEPs—Direct Control Option

Fig. 11.7 shows that the control loop is almost the same as that discussed in Clause 3.2.1 with the exception that the flow measurement is not done for individual CEPs; one set of measurements is done for both CEPs as well as the GSC. Here it is noted that in this loop no outlet (including main system flow) for condensate is upstream the GSC, so the flow measurement just ahead of the GSC gives the net flow through the GSC (as well as through the CEP), which can be computed. Consumers (other than main system flow to low-pressure heaters, LPHs) of condensate, such as the de-superheating flow to auxiliary steam control system, LP bypass cooling, etc., may be taken from anywhere after the GSC. Because of the 2 × 100% CEP configuration, there is one advantage: one set of loops will suffice for CEPs and for the GSC.

3.2.3 CEP Recirculation Control for 2 × 100% CEPs—Indirect Control Option

Fig. 11.8 shows that there are actually two loops in split range control mode, and two control valves, the RCV (works to return the condensate to the condenser) and the main condensate valve (MCV; which controls main condensate flow to LPHs, deaerators, etc.), operate in opposite way. When the MCV opens from 0% to 100% with a 4–20 mADC signal, the RCV closes from 100% to 0% (at which point the RCV is fully closed depends on which open position of the MCV maintains 30% flow so that minimum flow through the CEP and GSC is achieved) as shown in Fig. 11.8B. In this loop, the hotwell level is the controlling parameter, as the level increases (it may with load as more and more steam is condensed or more drain), i.e., setpoint is lower than the actual level. Deviation thus computed will be fed to the P+I controller, which will cause an increase in controller output and the MCV (fed by an I/P converter and positioner) will start opening. Because the RCV is in opposite mode, which was fully open at initial

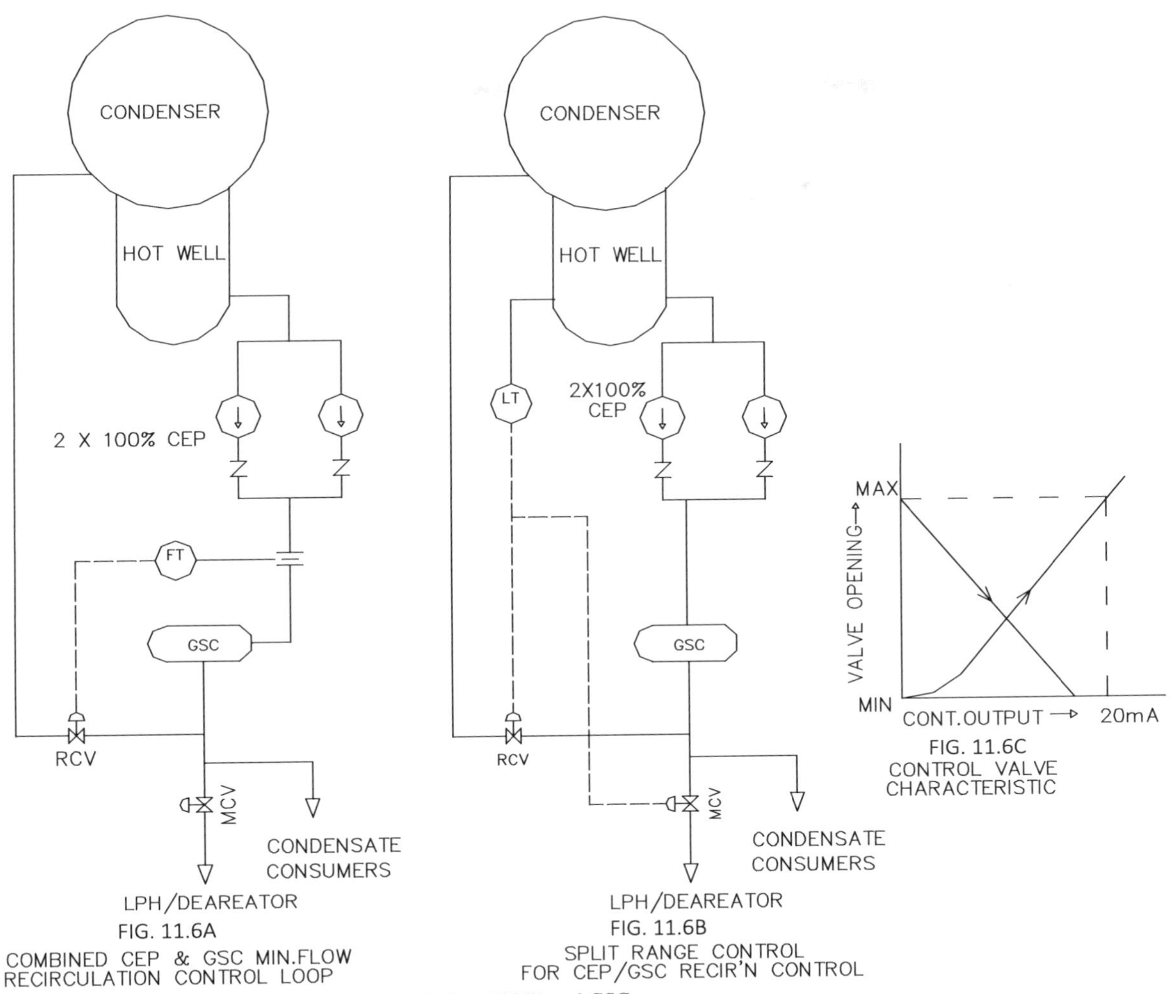

FIG. 11.6 Minimum recirculation flow control for CEP (2 × 100%) and GSC.

position, it will start closing; as flow through the CEP (hence GSC) increases, requirement of recirculation deceases and the RCV starts closing. So by modulating these two valves the hotwell level is controlled. During startup and low load, and when condensate consumers (for example, desuperheating condensate for auxiliary steam or LP bypass tempering water flow) will be drawing condensate, the level would also tend to fall (because of less steam condensation at low load) and the MCV may be closed and the RCV will be opened, which may be uncalled for—making the system less energy efficient. Also, in such an operation the main condensate flow will decrease due to closing of the MCV by control action, resulting in a decreasing deaerator level. This will cause makeup in the condenser and the condenser level will tend to rise and the MCV will open slowly. This is an indirect way of controlling the hotwell level, which is different from that discussed in Chapter 9, Clause 11.

3.3 Control Loop Operation

All these loops, as per Chapter 8, Clause 1.2.1, may trip to manual in case of transmitter failure and/or selection. The operator needs to select the correct transmitter to continue with the loop. In Figs. 11.5 and 11.7 it is clear that the loop may be switched to manual by the operator; however, in such cases suitable measures, as discussed in Clause 3.4, will be taken to keep the pump from running dry. In Fig. 11.8, it is seen that each of the two control valves can be switched to manual. However, the matter is trivial in this case, because if the loop is switched to manual there is no direct parameter to ensure minimum flow through the CEP and GSC. Instead, this can be done indirectly by looking into the hotwell level parameter. Also, the loop has inherent limitation, as discussed earlier. In addition to this there is the chance of operator error (in manual mode as there is no direct parameter for controlling the RCV)

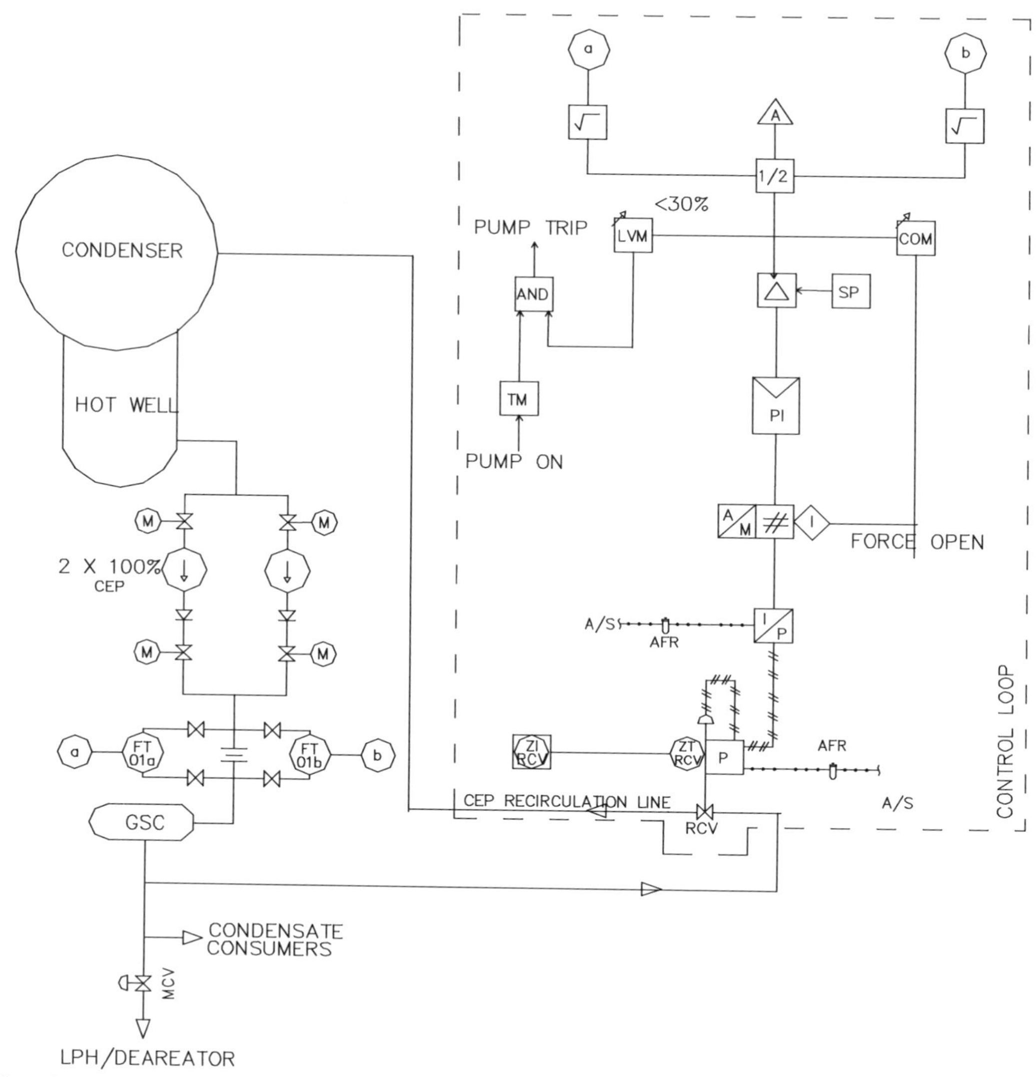

FIG. 11.7 Independent CEP recirculation control loop (2 × 100%).

and the RCV may fail to open, and the situation may not be congenial. The way to circumvent this problem is to provide one common A/M station also after the controller to operate both the valves from there (thus any operator error can be avoided).

3.4 Control Loop Interlock

The following interlocks may be incorporated to save equipment (Figs. 11.5 and 11.7). To make the loops safe and avoid operator error, an interlock can be incorporated to open RCV in case the CEP is ON and flow is less than the set value (set a slightly higher than actual set value to take preventive action) if in manual. To trip the pump if flow is less than the set value (30%), after the pump is on for some time (adjustable), is a last resort.

3.5 Control Valve Considerations

The recirculation valve has to absorb full DP, i.e., pump discharge pressure at the inlet and condenser pressure at the outlet. Also the temperature is very near saturation, so for such a drop, there is the chance of cavitation and flashing. To prevent seat damage these valves should have good leakage. Some typical parameters are DP (=40–50 bar − 0.1 bar at condenser) approximately equal to 50 bar at 40°C. Because of high erosion due to cavity, materials like 17–4 PH and 440C steel should be used. Multistage design is a good solution.

4 GSC MINIMUM FLOW CONTROL

4.1 General

The basic purpose of this loop is to ensure minimum flow through the GSC so it is cooled. This loop as a separate

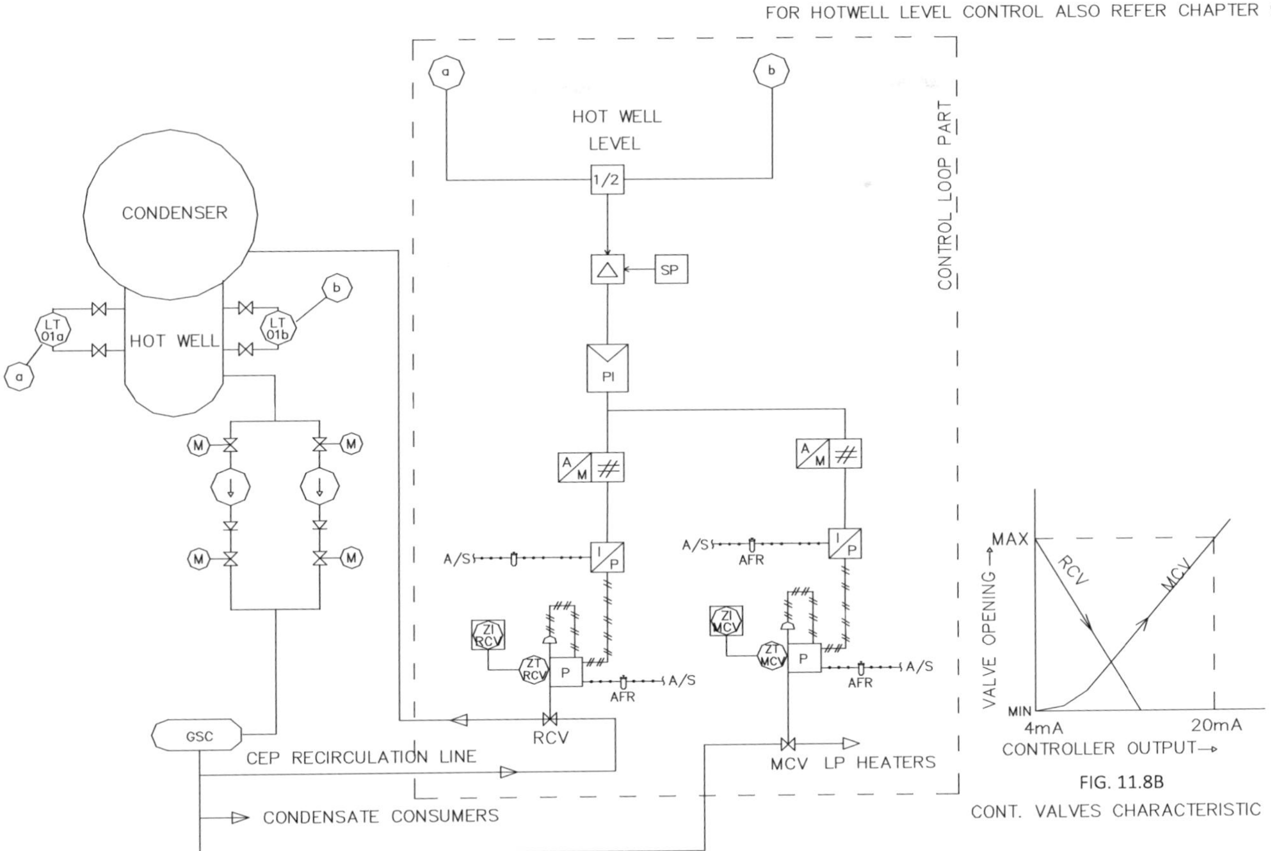

FIG. 11.8 CEP and GSC recirculation control loop (indirect operation).

entity is mainly applicable for medium and large-size power plants where there are 3 × 50% CEPs (for 2 × 100% CEPs this may not be applicable). In this type of configuration, generally condensate consumers are placed after the GSC, and GSC flow measurement is done immediately before GSC so that it can manage those consumers in addition to normal condensate flow to LPHs.

4.1.1 Objective

The purpose of the GSC control loop is to ensure minimum flow through the GSC at all times.

4.2 Control Loop Description

4.2.1 GSC Control

Condensate flow through GSC is measured immediately before the GSC with the help of an orifice plate and flow transmitter (flow compensation may not be applicable really) similar to Fig. 11.9. The flow is computed at the DCS. The computed flow is compared with the setpoint. Deviation thus created is fed to one P+I controller. The control loop is provided with an A/M station so that the valve can be regulated manually. The output of the A/M station is fed to the control valve through an I/P converter and positioner. Position of the control valve can be monitored at the DCS.

4.3 Control Loop Operation

This loop, as in Chapter 8, Clause 1.2.1, may trip to manual if there is transmitter failure and/or selection. The operator needs to select the correct transmitter to continue with the loop.

4.4 Control Loop Interlock

There may not be any specific interlock for the loop.

4.5 Control Valve Considerations

These are similar to those discussed in Clause 3.5.

5 DEAERATOR (PRESSURE) CONTROL

5.1 General

Why deaerator pressure control? Any liquid at saturation temperature has zero solubility of gas. So to get rid of

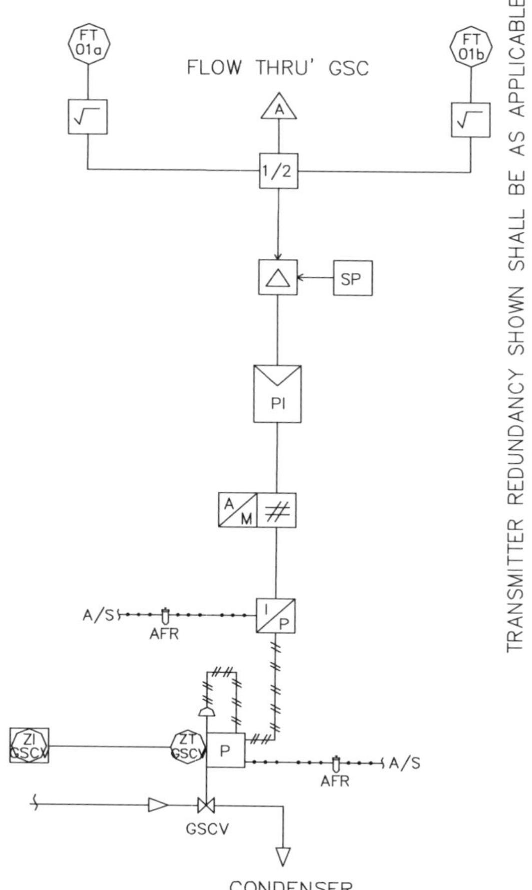

FIG. 11.9 GSC minimum flow control.

soluble gases, the condensate needs to be kept at its saturation temperature. Condensate contains dissolved O_2, CO_2, etc. Dissolved oxygen is the major cause of boiler corrosion and carbon dioxide makes the water acidic; therefore, it is necessary to eliminate them. This elimination is done chemically (as discussed in Chapter 5) and physically (mechanical). In the deaerator these dissolved gases are eliminated by heating with the help of pegging steam from the auxiliary steam and CHR steam. Also, at higher load when extraction steam is available, the heating steam is supplied from extraction steam of the turbine. The deaerator acts as an FW heater. The incoming condensate is sprayed in a fine mist (droplets) in an area surrounded by steam drawn from various sources like auxiliary steam, CRH, and extraction steam. In a droplet condition, the heat transfer area for water is larger when kept in intimate contact with the steam. This water is then passed over trays to form very thin sheets of water for gases to escape (vented out of the deaerator into the atmosphere). It is to be noted that various pressure values indicated below are arbitrary typical values only for mid- large-size plants.

5.1.1 Objective

The purpose of the deaerator pressure control loop is to maintain required pressure (because pegging pressures are different at different conditions) in the deaerator, during startup, and at low load. At normal or higher load, when extraction steam is available, the heating and deaeration are done as per extraction steam pressure, according to turbine load.

5.1.2 Discussion

Minimum pegging steam pressure is normally ~1.5 bar. As the boiler load increases, it is desired to have higher temperature FW, so, with a change in system load (until extraction pressure is available), the pegging steam pressure (when CRH is established, i.e., CRH pressure is sufficient) is also varied for energy efficiency. The initial (during cold start) pegging steam is supplied at 1.5 bar (may be at 0.8 bar for some smaller units) from auxiliary steam. At low load and/or during bypass operation, hot startup, etc., the pegging steam may be supplied from CRH or auxiliary steam (if CRH is not available at suitable pressure) at ~3.5 bar.

With the power plant load at approximately 55%, and extraction steam is available, the heating steam is supplied from extraction steam; pressure varies with load and is uncontrolled (pressure variation directly as per turbine load). The extraction steam (supplied to the deaerator) pressure typically goes up to 8 bar for medium- to large-size units. Another important aspect in designing the loop is rate of change of pressure in the deaerator. During sudden load throw off the pressure in the deaerator decays at a faster rate, which is not healthy for deaerator operation, so necessary measures are taken so that the high rate of change of pressure in the deaerator is arrested.

5.2 Control Loop Description

5.2.1 Deaerator Pressure Control (Simplified Version)

As discussed earlier, this loop consists of two distinct controls (Fig. 11.10). One is to control the auxiliary steam control valve (ASV 02) to keep the deaerator pressure at a fixed minimum value of 1.5 bar, pegging the deaerator at the set pressure during startup conditions. Another loop comes into action when pressure in the CRH line is established, and the deaerator is maintained at 3.5 bar. When the CRH line is brought in to service the auxiliary steam valve in the first loop closes automatically (it has a lower setpoint) as well as by interlock.

1. *Control loop for auxiliary steam valve*: Set pressure is 1.5 bar compared with the actual pressure to generate the deviation. The deviation thus generated will drive the controller. Setpoint range is 0–8 bar so that the operator can set it to 3.5 bar during hot startup.

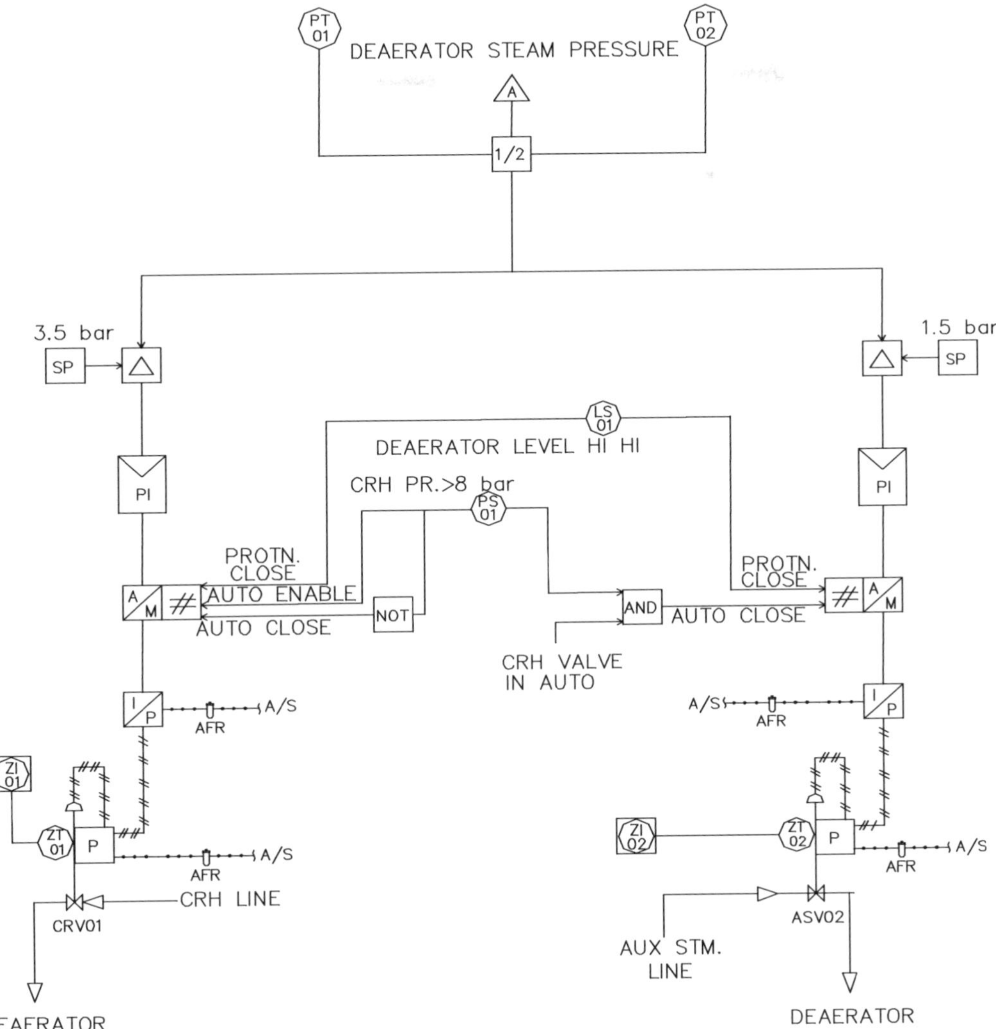

FIG. 11.10 Deaerator pressure control loop (simplified).

2. *Control loop for CRH valve*: Set pressure is 3.5 bar compared with the actual pressure to generate the deviation. The deviation thus generated will drive the controller in the same way as discussed above. This loop can operate only when the steam pressure in the CRH line is greater than the set value (sensed by a pressure switch), typically 8 bar.
3. *This loop has a serious disadvantage*: When a huge load is shaded by turbine, then the pressure decay in the deaerator takes place quickly. When this happens, the loop has no protection for such fast decay and immediately goes to either 3.5 bar or 1.5 bar, depending on the condition of pressure at the CRH line at that time. This is not desirable and it is detrimental to deaerator; hence, the modified loop as shown in Fig. 11.11, is considered where such faster depressurization as well as pressurization needs to be prevented.

5.2.2 Deaerator Pressure Control (Normally in Use)

The deaerator pressure control loop (Fig. 11.11) also has two distinct controls. One loop is to control the auxiliary steam control valve (ASV 02) to keep the deaerator pressure at a fixed minimum value. The other loop is activated when pressure in the CRH line is established. At this point the CRH valve is on auto and the deaerator is maintained at the desired set value. The same deviation station has been used, where two setpoints may be fed via suitable interlock. When the CRH is brought into service the auxiliary steam control valve is closed by interlock and automatic closure, as discussed in Clause 5.2.1, and will not be effective. The CRH valve will be brought into service after pressure in the CRH line attains the desired value.

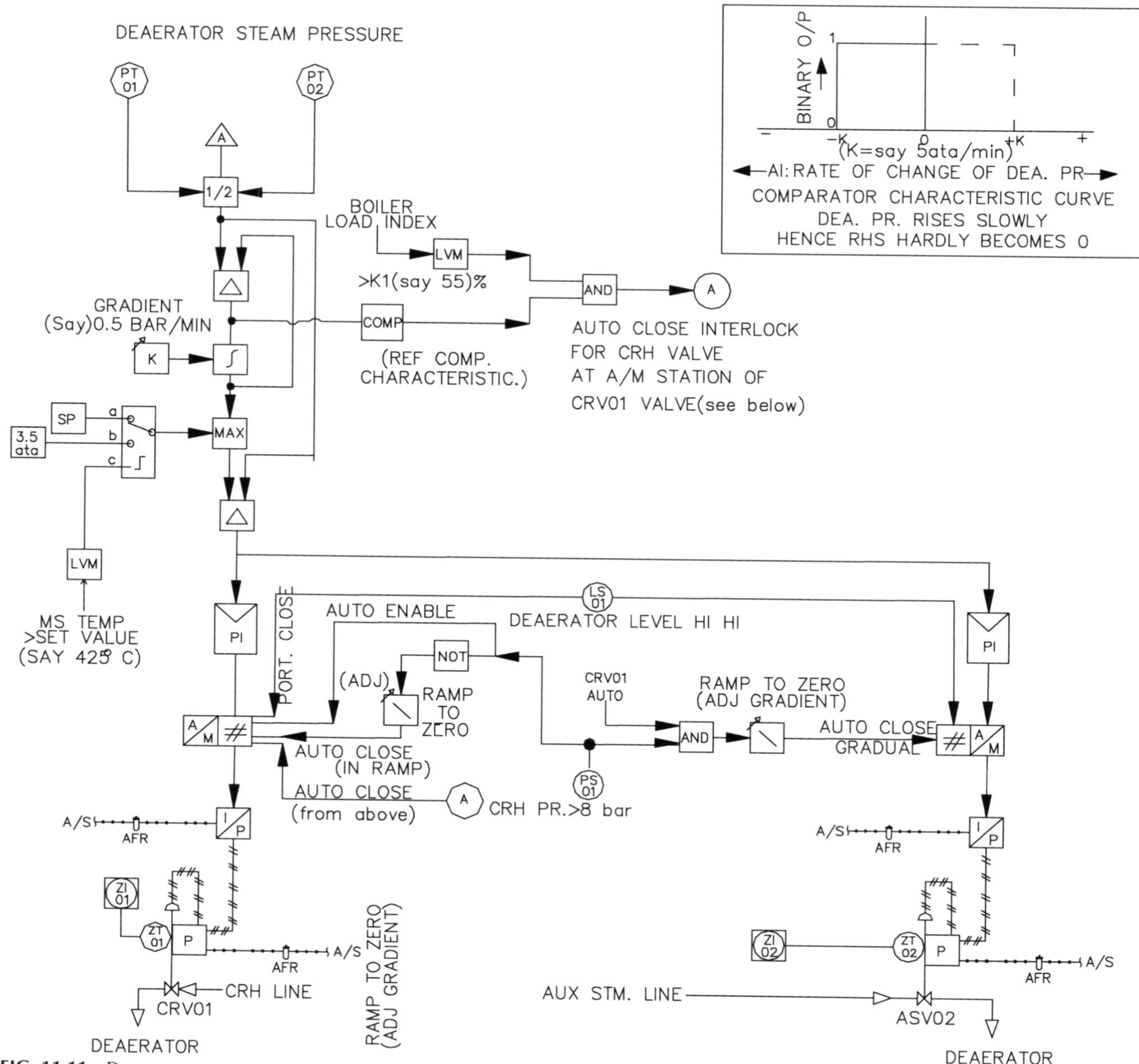

FIG. 11.11 Deaerator pressure control loop (normally in use).

1. *Setpoint establishment*: Normally the operator puts the setpoint to 1.5 bar during cold startup, whereas during low load, hot startup, or at the time of large load throw off, etc., the setpoint is 3.5 bar. The setpoint is generated based on the integrator circuit used. The gradient of integration is adjustable (for example, 0.5 bar/min). The measured value after integration (low-pass filter to filter out sudden changes) is one of the setpoints (mainly helpful at high load conditions), along with fixed setpoints through a max select circuit. During large load throw off, if the deaerator was running at higher pressure, the actual pressure decays very quickly. Because the integrator in the circuit generates the setpoint, such large changes will not be allowed; instead the same change will appear as a slow changed setpoint (as that time actual pressure prior to large load throw off was higher than the fixed setpoint). So slight decay from the original high pressure will appear as the setpoint, but the actual value is less, and the steam control valve will open to arrest the sudden decay. (This could be made clear from this hypothetical example: Near the full load of the turbine extraction pressure was 7.5 bar, and there was a sudden load throw off and pressure decays to 6.5 bar in 1 min. Now, looking at the integrator circuit, the

output of the integrator just prior to load throw of was 7.5, then in 1 min actual pressure decays to 6.5 bar, but because the integrator has a decay rate of 0.5 bar/min, the integrator output will appear as 7 bar in 1 min, and because of the MAX selector this will be selected as the setpoint until the time this value is higher than the fixed set value. In this example, because the actual pressure is 5 bar, the differential is that one of the valves will operate to arrest the decay rate.) Similarly, if someone wants to increase the setpoint the same will appear as a slow change in setpoint. There is an interlock operation in the setpoint generation circuit. If main steam temperature is >425°C then 3.5 bar will appear, otherwise the operator's selected set will be connected to the max selector. This is done so that during cold startup the operator can set a 1.5 bar setpoint through the same circuit. In the loop, the same setpoint circuit has been utilized in both the loops through the interlock.

2. Based on the setpoint generated above, a deviation from actual value will be created. This deviation signal is fed to the respective controllers for the auxiliary steam valve as well as the control valve in the CRH line. Depending on the interlock discussed below, one of the two valves will be selected for maintaining the pressure in the deaerator. Another advantage of the loop is that the unselected valve, through the interlock, will be closed at a desired rate as shown by the ramp line (shown from last value to zero; the gradient of the ramp should be adjustable).

5.3 Control Loop Operation

In this clause auto and manual operation of the loop will be discussed (Figs. 11.10 and 11.11). This loop, as per Chapter 8, Clause 1.2.1, may trip to manual if there is transmitter failure and/or selection. The operator needs to select the correct transmitter to continue with the loop. At any time the loop is operable in the A/M position. The control valve in the CRH line will be enabled only if pressure in the CRH line >8 bar, for example.

5.4 Control Loop Interlock

There may be some specific interlock for the loops. During auto close interlock, irrespective of the auto or manual operation selected, the valve will close. This closure is not instantaneous; the gradient of closing command is gradual from its last position to zero position (close position), which is adjustable in DCS.

1. Control valve in CRH will have auto close whenever pressure in the CRH line is *not* greater than the set value.
2. Control valve in auxiliary steam line will auto close when pressure in the CRH line is more than the set value (for example, 8 bar) *and* the control valve in the CRH is on.
3. When the boiler maximum continuous rating (BMCR) is >55% (or any value when extraction steam is available for deaerator heating) *and* rate of change of the deaerator pressure is within the preset limit, the CRH valve (CRV 01) will be closed in auto through the A/M station. This is applicable for the loop in Figs. 11.10 and 11.11 and to ensure that the CRH valve is closed when extraction steam is established (see below).
4. During protection close interlock, irrespective of the operation selected, the valve will close. This closure is instantaneous and immediate. Both the control valve in the CRH line and control valve in the auxiliary steam line will close in case of very high level in the deaerator.

5.5 Fixed Setpoint Interlock

When there is a main steam temperature set value (i.e., >425°C), then a digital "1" signal will be generated to select the fixed set value (i.e., 3.5 bar) for the CRH. This is done through an interlock (Fig. 11.11). If the main steam temperature is less than the set value (i.e., >425°C) the fixed set value will be whatever the operator selects—cold startup set value or higher set value for hot startup or low load operation.

As long as extraction steam is available the CRH valve will remain closed. Without interlock valve closure may not be achievable if the rate of change of deaerator pressure is very slow. To ensure the CRH valve is closed when extraction steam is available, closing interlock with comparator has been incorporated. Also if there is a sudden load throw off/unit trip when the rate of change of pressure is high, then this interlock will be ineffective (to note the comparator characteristics) so that the CRH can be activated.

6 LP HEATER LEVEL CONTROL

6.1 General Discussion on Heater Drain

ASME TDP-1 2006 provides effective recommendations to prevent water induction damage in the turbine. TDP-1 offers guidance on how to identify systems that have potential for water ingress to the turbine. It also provides guidance for automatic disposal of water. Heater level control is extremely important as per these recommendations. If the heater level rises there may be water ingression to the turbine (of course, this is also prevented by extraction line swing check valves and power-assisted nonreturn valves), through extraction lines. Also, if the heater level is not properly maintained then there will be improper heat transfer. Heater drain system design is more complex because of the two-phase flow phenomena.

The heater numbering system as used in the book is elaborated in Fig. 11.12. Normally each FW heater will drain to the preceding heater with lower extraction steam supply, for example, HPH 5 drains to the deaerator (Heater 4). LPH 1 (located at the condenser neck) may directly drain to the condenser through a loop seal or a control valve. The emergency (high level) drain flow path goes directly to the condenser from heater through a separate header and control valve. Emergency drain valves normally operate when normal drain valves fail to maintain the heater level and/or normal draining equipment (downstream equipment in normal draining circuit) is not available (for example., if tubes rupture in LPH 2, it is not available so, LPH 3 cannot normally drain in LPH 2). Improper design of the system would result in loss of plant efficiency. The two-phase flow at the downstream of the control valve would result in high erosion, so material with higher thickness needs to be chosen. Normal and emergency drain flow does not necessarily match. Therefore normal drain control valves should be designed to cover from minimum load through full load of the plant, whereas the emergency drain flow control valve should be designed with minimum flow during startup and full load plus emergency situation (Fig. 11.13).

6.1.1 Objective

The purpose of LPH 2 (or LPH 3) level control loops is to maintain the LPH 2 (or LPH 3) level. Under normal conditions, LPH 2 (or LPH 3) level is maintained by draining water from LPH 2 (or LPH 3) to LPH 1 (or LPH 2). However, in case of emergency level LPH 2 (or LPH 3) is drained to the condenser/drain flash tank to cope with the situation.

6.1.2 Discussion

It is clear from the previous discussion that these level controls are introduced to the system as per TWDPS requirements. According to ASME TDP-1 2006, "No single failure of equipment, device, signal, or loss of electrical power should result in water or cold steam entering the turbine." To meet these requirements, several redundancies have been incorporated in the loop—e.g., redundancy in transmitters, closing of control valve to the following heater, and opening of emergency drain control valve of the same heater—directly through the solenoid valve (level switch) and through the DCS interlock.

6.2 Control Loop Description

Both LPH 2 and LPH 3 have two, nearly identical loops: One for normal level control and the other for high level control (emergency drain). The loops in each heater have separate setpoints. Because the loops are almost identical, only one will be discussed at length. In level measurement, three level transmitters (each covering an entire level range) are considered in two of three modes. With vertical heaters, the entire range may be too large, so separate sets of level transmitters (in one of two modes) may be chosen for normal- and high-level control to have better control range. The level in the heater is important for the turbine system to close the extraction (power-operated) check valve to prevent ingress of steam/water to the turbine. This voted signal is used in ATRS. Also, when the turbine suddenly trips, extraction pressure falls, at that temperature, water may convert to steam and enter the turbine through the extraction line. As a result, power-operated check valves may be closed during high heater level and during the turbine trip. In the loops, the position signal of each of the drain valves (normal and emergency) is monitored. The signals for <2% open of normal drain (to recognize that the normal drain valve is closed) and >98% open for emergency drain (to recognize that the emergency drain valve is open) are also used for an interlock in the open loop control system (OLCS; Figs. 11.14 and 11.15).

6.2.1 LPH 2 (or LPH 3) Level Control

1. The measured value is compared with the normal set value of the heater. Deviation thus created is fed to the P+I controller. Output of the controller is used to throttle the normal level control valve to the preceding heater, for example, the drain control valve to LPH 1 (or to LPH 2 for LPH 3).

Normal convention to number the heaters and extraction lines from ONE (1) at the lowest pressure extraction at LPT and accordingly name that heater as LPH 1. The heater & extraction line numbering increase as we go along the Low pressure to High pressure (turbine) i.e. for a system with six regenerative heaters lowest pressure extraction (no 1) steam will send steam to LPH1 and highest pressure (e.g. CRH no 6) will send steam to HPH6.

FIG. 11.12 Extraction and heater numbering system.

Physical Location of heaters plays some important role in heater draining system. LPH 1 has steam from Ex1 steam line having very marginal pressure difference WRT condenser so it is located at the neck of the condenser. Also pressure gradient between heaters are not much so it is better to have greatest elevation difference so that there will be proper drain flow.

FIG. 11.13 Heater location and its importance.

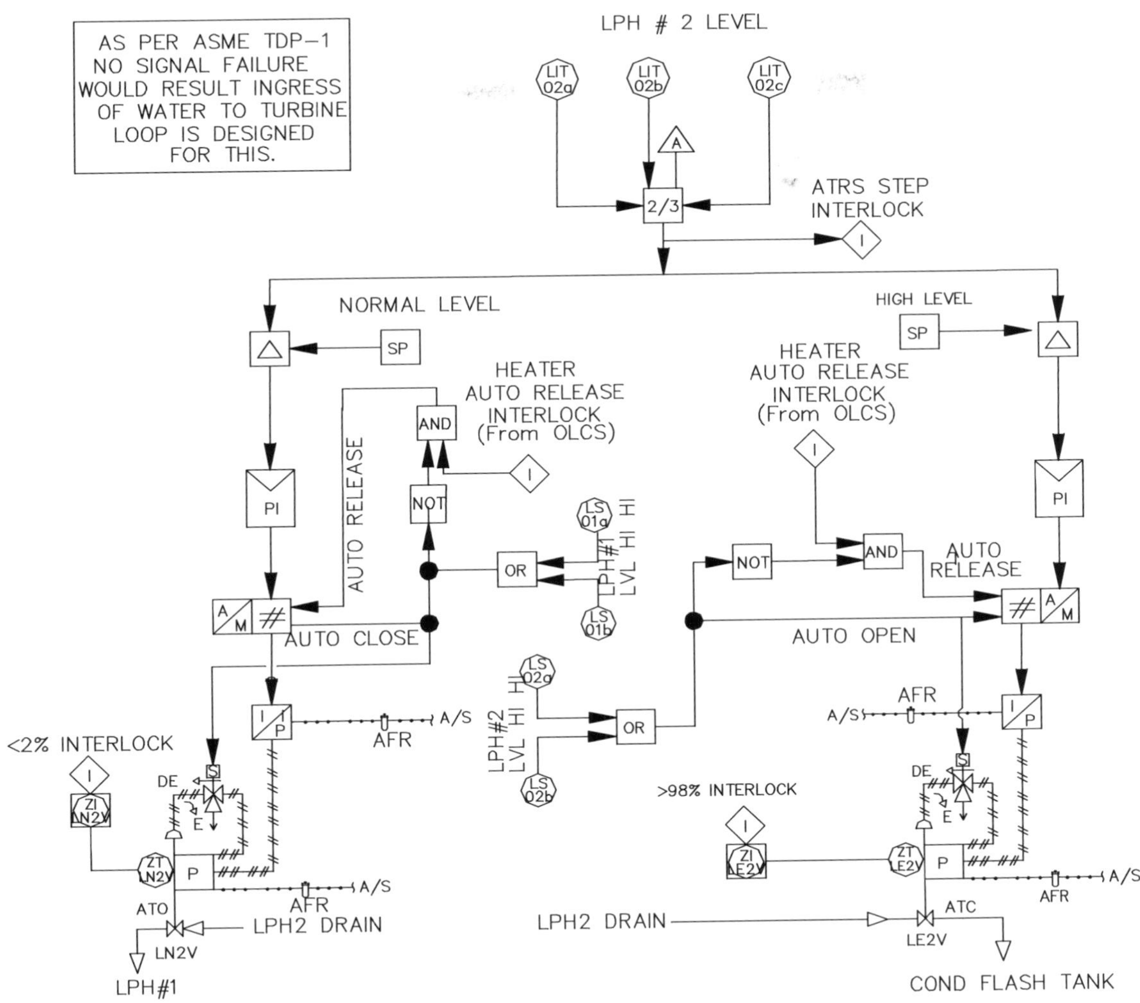

FIG. 11.14 LPH 2 level control.

2. After opening the normal level control valve, and the level in the heater cannot be maintained and the level is elevated (for example, due to heater tube rupture), once the high level set value is reached automatically, the other the high level set value is reached automatically, the other emergency level control loop activates. If the downstream heater in the heater drain circuit is not available, then the emergency drain control loop also activates. Similarly, based on the high set value and the actual level, the emergency control valve will throttle to drain LPH 2 (or LPH 3) to the condenser via the LP flash tank.

6.3 Control Loop Operation

In this clause, A/M operation of the loop will be discussed. This loop, as per Chapter 8, Clause 1.2.1, may trip to manual in case of a transmitter failure and/or selection. The operator needs to select the correct transmitter to continue with the loop. At any time, the loop is operable in the A/M position. The LPH 2 (or LPH 3) normal drain valve loop can be released to auto if the level in LPH 1 (or LPH 2) is not greater than the hi hi value, and the interlock system is fine, based on plant condition for interlock, and protection of the plant from the OLCS. This is shown as "heater auto release interlock" in Figs. 11.14 and 11.15. LPH 2 (or LPH 3) emergency drain valve loops can be released to auto if the level in LPH 2 (or LPH 3) is *not* greater than hi hi value, and the interlock system is okay, based on plant condition for interlock and protection of the plant from OLCS. If the LPH 2 (or LPH 3) level is greater than hi hi value, a direct open emergency drain valve command will be issued to bring the level down quickly.

6.4 Control Loop Interlock

There may be some specific interlock for the loops. LPH 2 (or LPH 3) interlocks are used for the following:

1. During a hi hi level in LPH 1 (or LPH 2) the normal drain control valve from LPH 2 (or LPH 3) will be

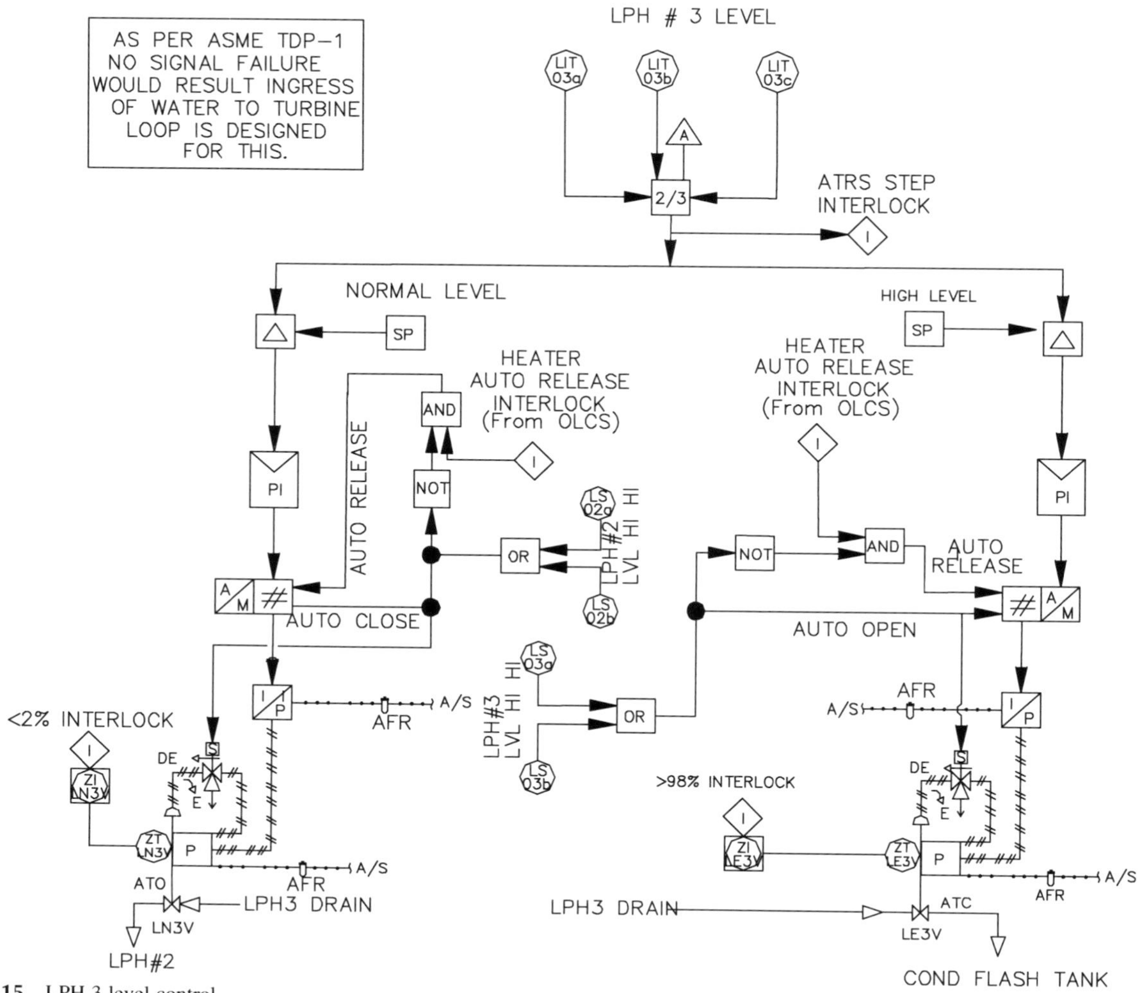

FIG. 11.15 LPH 3 level control.

closed via interlock via DCS and directly to the control valve (ATO type) with the help of the solenoid valve. This circumvents the failure and/or cable snap from DCS, so the system does not suffer.

2. The emergency drain valve (ATC type) of LPH 2 (or LPH 3) will be fully opened during a high level in LPH 2 (or LPH 3). The opening of the valve is done directly and by DCS.

6.5 Control Valve Consideration

As discussed earlier, there is a major problem in the heater drain control valve sizing due to the flashing liquid condition. Based on inlet and outlet valve pressure, control valves are selected a size larger to reduce velocity in the valve and lessen the chance of erosion. When a larger valve size is selected and operated continuously at low opening the valve may poorly perform and have less integrity. Sometimes it is better to use a higher body size with reduced trim. A few tips on selection:

- *Trim size*: 70%–80% opening to give full flow (is using an emergency drain, consider tube rupture).
- *Control valve*: Must meet minimum flow at slightly >15% opening.
- *Materials*: Hardened materials such as 400 series SS for valve trim to lessen erosion, for which 2–3mm erosion allowance should be considered; chrome Moly Steel is a good choice—ASTM A217 Gr C5/WC9 are also chosen for this service.
- *Control valve body*: Use erosion-resistant materials.

7 HP HEATER LEVEL CONTROL

7.1 General

The basic philosophy behind HP heater level control has been discussed in Clause 6.1. In ≥500 MW units, there may be two sets of heaters in parallel, for example, HPH 5A and HPH 5B with HPH 6A and HPH 6B instead of a single HPH 5 and HPH 6.

7.1.1 Objective

The purpose of HPH 5 (or HPH 6) level control loops is to sustain the level in the heater. Under normal conditions, the HPH 5 (or HPH 6) level is maintained by draining HPH 5 (or HPH 6) to the deaerator (or HPH 5; cascade drain). However, during an emergency, the level HPH 5 (or HPH 6) is drained to the condenser/HP drain flash tank to cope with the situation.

7.1.2 Discussion

To meet these requirements of ASME TDP-1 2006, several redundancies are incorporated in the loop, as discussed in Clause 6.1.2.

7.2 Control Loop Description

HPH 5 and HPH 6 each have two loops: one for normal level control and one for high level control (emergency drain). Both loops in each heater have separate setpoints. Because both loops are almost identical, only one will be discussed at length. For level measurement, three level transmitters (each covering an entire level range) are considered in two of three modes. With vertical heaters where the entire range may be too large, separate sets of level transmitters (in one of two modes) may be chosen for normal and high level controls for better control range. See Clause 6.2 for discussions on ATRS and valve position signal.

7.2.1 HPH 5 (or HPH 6) Level Control

1. The measured value is compared with the normal set value of the heater. Deviation thus created is fed to the P+I controller. Output of the controller is used to throttle the normal level drain control valve to the deaerator (or HPH 5).
2. When opening the normal level control valve and the level in the heater rises and cannot be maintained, (heater tube rupture), once the high level is reached, the other emergency level control loop is activated. Also, if the downstream heater is not available then the emergency control loop will activate.

Similarly, based on the high set value, the emergency control valve will throttle to drain HPH 5 (or HPH 6) to the condenser by HP Flash tank (HPH 6 may drain to the deaerator in an extreme case to condenser via flash tank; Figs. 11.16 and 11.17).

7.3 Control Loop Operation

In this clause, A/M operation of the loop will be discussed. The loop in Chapter 8, Clause 1.2.1, may trip to manual during transmitter failure and/or selection. The operator should select the correct transmitter for control. The loop is operable in A/M position. The HPH 5 normal drain valve loop can be released to auto if the level in the deaerator is not greater than the high value, and is released from the OLCS (i.e., overall system interlock system based on overall plant condition for interlock and protection of the plant). The HPH 5 emergency drain valve loop can be released to auto if the level in the HPH 5 is not greater than the hi hi value, and is released from the OLCS. The HPH 6 normal drain valve loop can be released to auto if the level in HPH 5 is not greater than hi hi value and is released from the OLCS. The HPH 6 emergency drain valve loop can be released to auto if the level in HPH 6 is not greater than the high value, and is released to from the OLCS.

7.4 Control Loop Interlock

Listed below are some specific interlocks for the loops.

7.4.1 HPH 5 Interlocks

1. When there is a high level in the deaerator (heater 4), the normal drain control valve from HPH 5 is closed by interlock by DCS and directly to the control valve (auto type) with the help of the solenoid valve. This is done to circumvent the failure or snap of the cable from DCS.
2. The emergency drain valve (ATC type) of HPH 5 is fully opened when there is a high level in the HPH 5. The opening of the valve is done directly by DCS.

7.4.2 HPH 6 Interlocks

1. During HI HI level in HPH 5, the normal drain control valve from HPH 6 is closed with interlock by DCS and directly to the control valve (ATO type) with the help of the solenoid valve. This circumvents the failure or snap of the cable from DCS.
2. Emergency drain valve (ATC type) of HPH 6 is opened fully in case of a high level in HPH 6. The opening of the valve is done directly or via DCS, as discussed previously. Here in case of emergency, drain will go to condenser. In some cases ist high level drain could be to deaerator, then final emergency drain to condenser as shown in the drawing.

7.5 Control Valve Consideration

See Clause 6.5.

8 EJECTOR CONTROL AND TAS

8.1 General

Auxiliary steam in a power plant plays a vital role during startup of the boiler and turbine. In many plants with

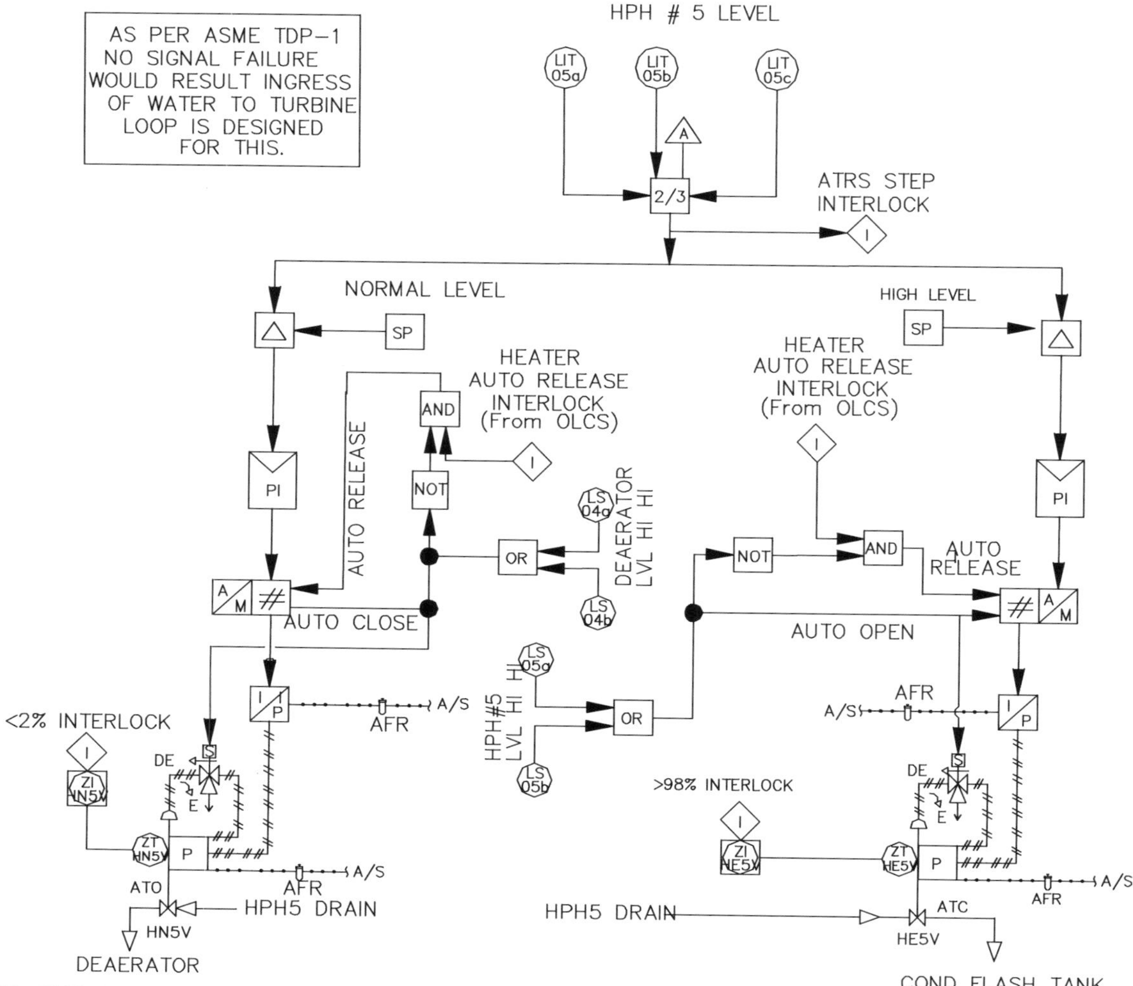

FIG. 11.16 HPH 5 level control.

smaller- and medium-sized units, there may be separate auxiliary steam headers for the boiler and turbine. They are a boiler auxiliary steam (BAS) and a turbine auxiliary steam (TAS). Bigger plants may have a common auxiliary steam header. A general discussion on auxiliary steam controls is in Chapter 3, Main Clause 5 and Chapter 8, Main Clause 10.4.

8.1.1 Ejector Control

The exhaust steam of turbine is condensed in the condenser maintained at vacuum, so that the steam pressure drop between the inlet and exhaust of the turbine is increased, resulting in more active power available at the turbine output to drive the generator. For water-cooled surface condensers, the vacuum is maintained by either a vacuum pump or a steam ejector. These ejectors use steam as the motive fluid to remove any noncondensable gases that may be present in the surface condenser. In the steam ejector, pressurized steam is passed through an ejecting nozzle, and immediately after the nozzle steam is allowed to expand to create a vacuum. TAS is utilized for this purpose, and by reducing the TAS pressure with the help of one pressure-reducing valve (PRV) steam is supplied to the ejector. It is a simple pressure-reducing control (Fig. 11.20).

8.1.2 Objective

The purpose of TAS control is to obtain a header of desired constant pressure and constant temperature, irrespective of changes in load in turbine.

8.1.3 Discussion

See Chapter 8, Clause 10.4.1.3 for further discussion.

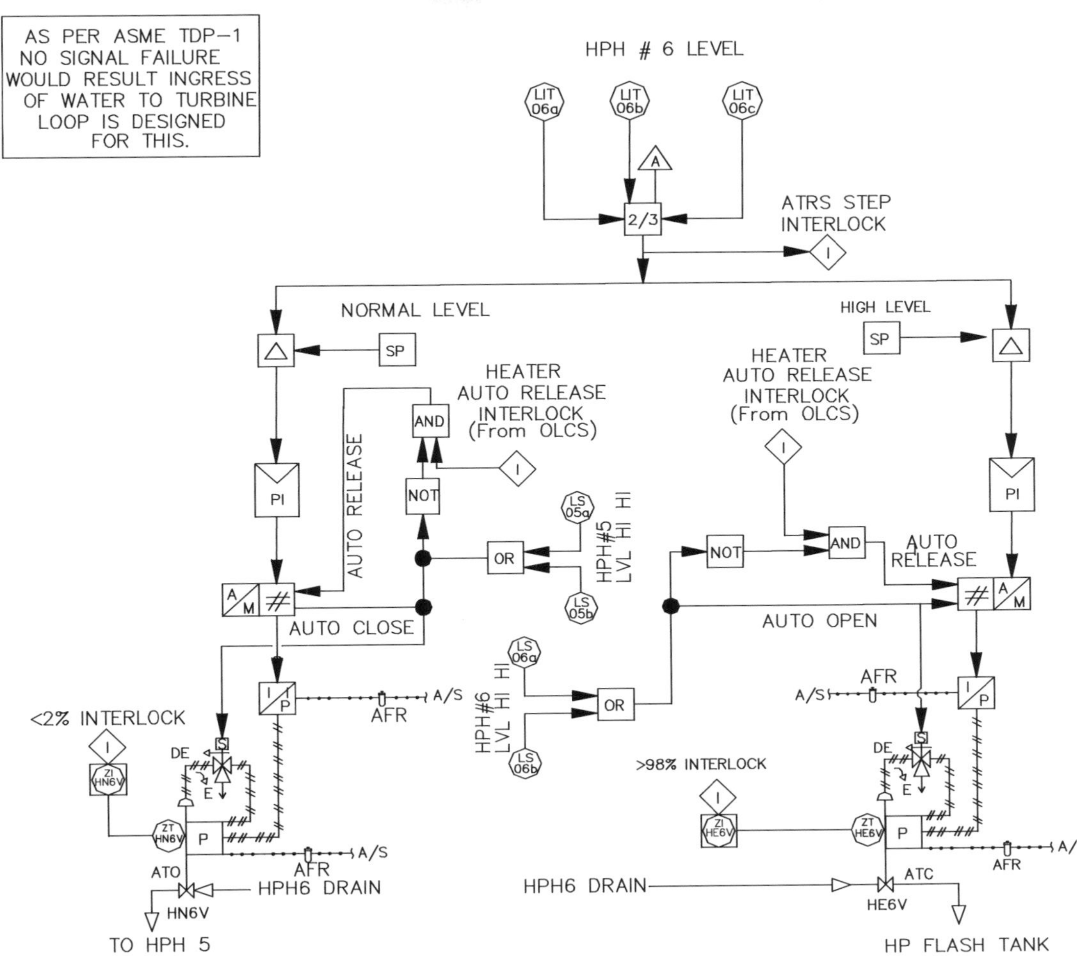

FIG. 11.17 HPH 6 level control.

8.2 Control Loop Description

The control loops involved are pressure control and temperature control. (In these loops both air lock relays and solenoid valves have been introduced to indicate the connection detail in a practical loop.) During air supply failure and electrical signal failure control valves may be locked in their previous positions (i.e., they stay put; Figs. 11.18 and 11.19).

8.2.1 TAS Control

There are two control loops (Fig. 11.17):

1. *TAS pressure control*: TAS header pressure (downstream of PRV) is measured in one of two modes and compared with a desired setpoint. Deviation thus created is used to drive a P+I controller. The output of the controller through A/M station and I/P converter regulates the PRV to maintain constant pressure at the auxiliary steam header at all loads.
2. *TAS temperature control*: Steam temperature at the auxiliary steam header (i.e., downstream of de-superheater) is measured and compared with the fixed setpoint. Deviation thus created is fed to a P+I controller (may be P+I+D also, as temperature is a slow changing parameter P+I+D may be more effective). The output of the controller regulates the condensate flow to the desuperheater to control the temperature at the auxiliary steam header. There may be separate desuperheaters or a combined pressure-reducing and desuperheating system (PRDS) may be deployed.

8.2.2 High & Low Capacity TAS

As discussed in Chapter 8, depending on load requirement, it may be necessary to have separate high and low capacity PRDS systems. These loops may operate in split range as well. When the load is high and even with full opening of the low capacity valve, pressure cannot

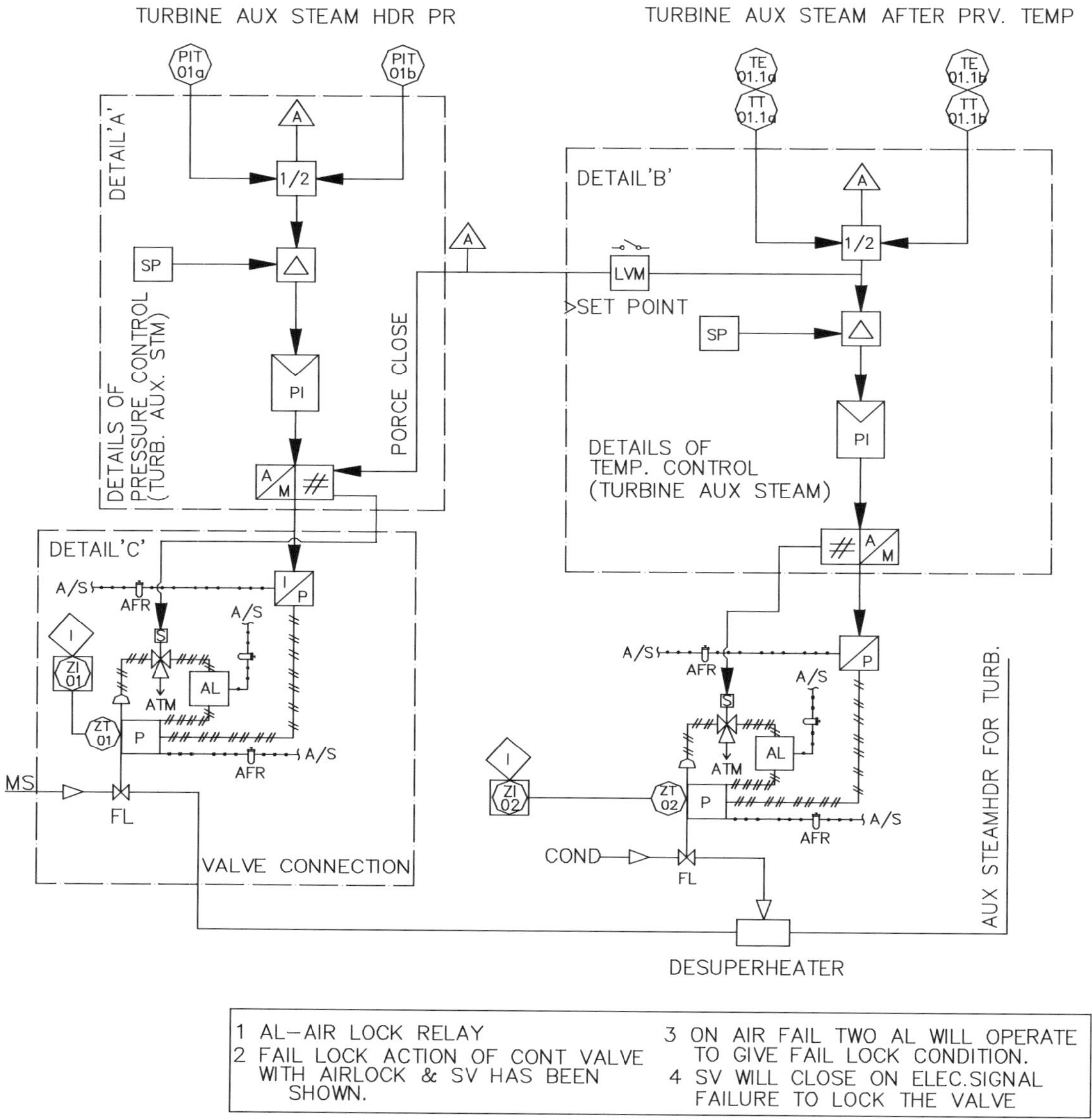

FIG. 11.18 Turbine auxiliary steam pressure and temperature control.

be maintained and the high capacity valve will start to open. During a transition period, low- and high-capacity valves may be operative, so there may be some overlap in split-range operation. When the load reduces, the high capacity valve will start to close first, then the low capacity PRDS will start to operate (Fig. 11.19). For detailed discussions, see Chapter 8, Clause 10.4.2.

8.3 Control Loop Operation

In this clause A/M operation of the loop will be discussed. This loop (Chapter 8, Clause 1.2.1) may trip to manual during transmitter failure and/or selection. The operator should select the correct transmitter. The loop is operable in the A/M position.

8.4 Control Loop Interlock

There may be specific interlocks for the loops. If the temperature at the outlet of the desuperheater (after PRV; i.e., at auxiliary steam header) is greater than the set value. then the PRV (both high- and low-capacity) will be closed. If there is a signal failure, the associated solenoid valve will be closed from the DCS to secure fail lock condition of the associated valve (Figs.11.18 and 11.19). If the temperature is greater than the set value sensed by LVM, the associated PRV will be closed.

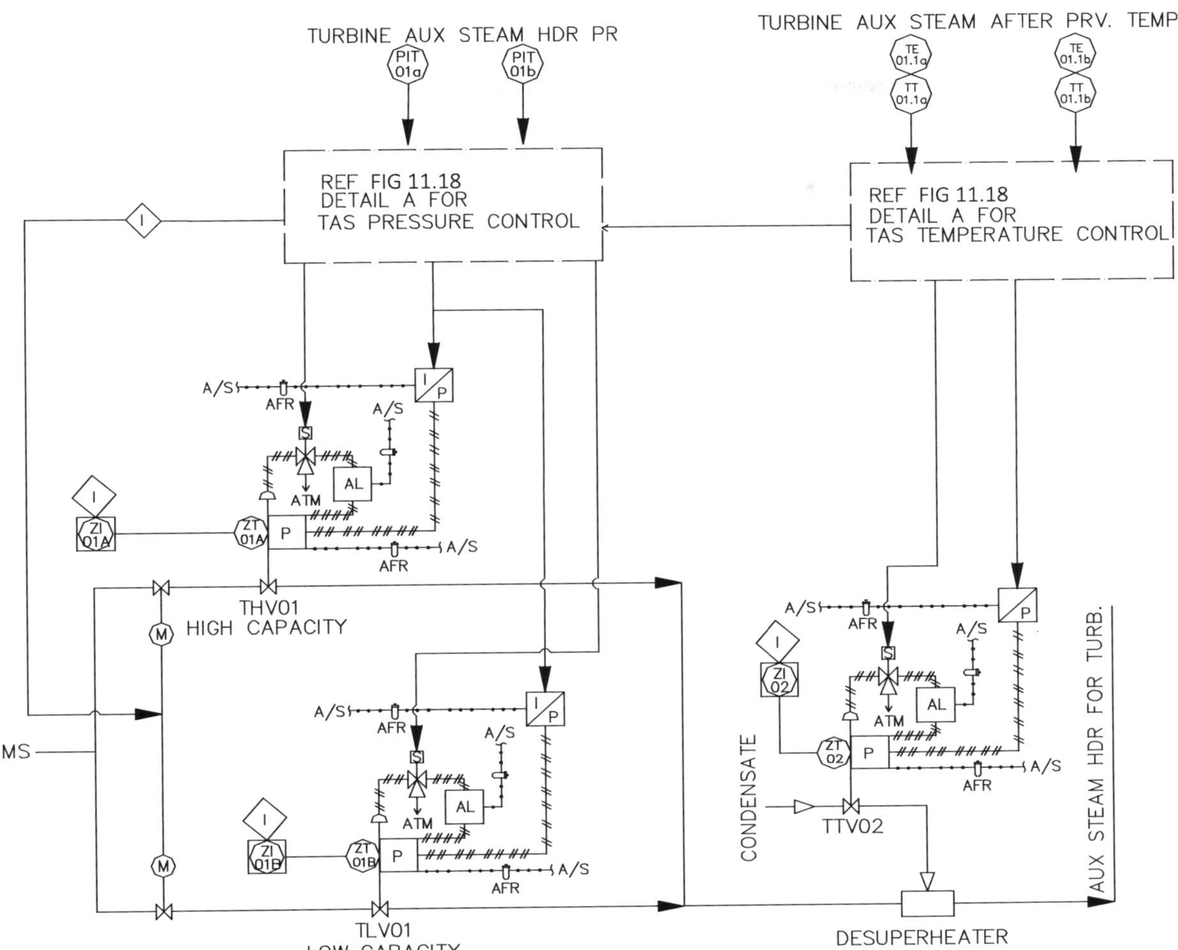

FIG. 11.19 High and low capacity TAS.

8.5 Control Valve Consideration

As discussed earlier, there may be two sets of valves to meet plant requirements: during startup, and normal operation. Therefore proper attention is given to sizing these valves. Typical parameters for PRDS valves in auxiliary steam systems are shown in tabular form:

Some extreme conditions (typical parameters) for PRVs and temperature control valves include:

Parameters	PRV	Temperature Control Valve
Inlet pressure	175(/145) bar	190(/165) bar
Outlet pressure	16(10) bar	16(10) bar
Temperature	540°C	160°C

When high pressure drops, and high temperature and high noise occur, selecting values for these applications is very important, especially for PRVs. Cage-guided low dB trim with multiple holes are used to minimize noise effect.

8.6 Use of Auxiliary Steam in Turbine

Major uses of auxiliary steam on the turbine side are as follows:

- Deaerator pegging
- Steam ejector
- Gland steam pressure control
- BFP drive (if BFP turbine is used)

8.7 Ejector Control

Auxiliary steam at 16 bar may not be directly fed to the ejector. Normally pressure is reduced before it is fed to the ejector as shown in Fig. 11.20. The loop is similar to the pressure-reducing loop discussed in Clause 8.2.1.

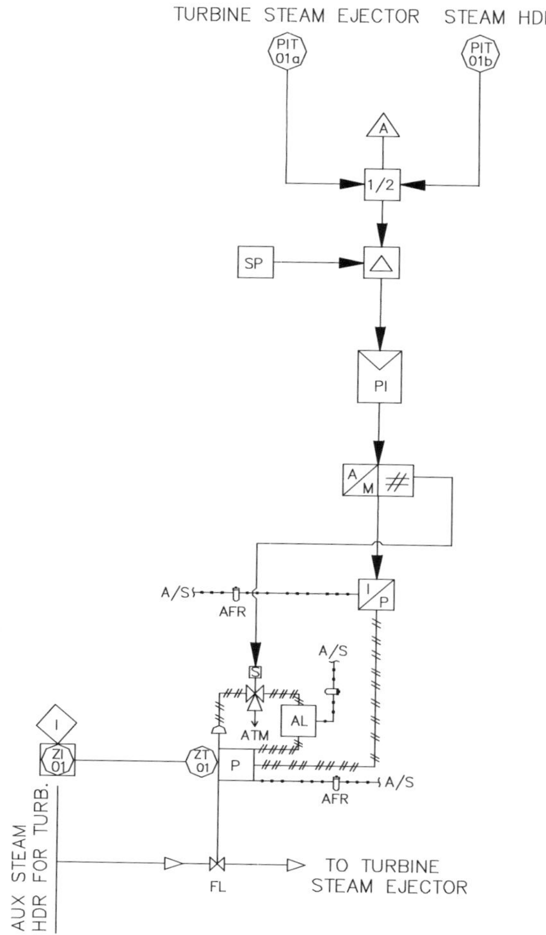

FIG. 11.20 Turbine ejector steam pressure control.

BIBLIOGRAPHY

[1] Cavitation in Control Valves: Samson AG: Technical information.

[2] Control Valve Handbook: Emerson Process Management: Fourth Edition (Fisher).

[3] Standard Design Criteria/ Guideline for Main Plant Package (2 × 500 MW TPP). Central Electricity Authority, New Delhi.

[4] Drag Valve (BFP Recirculation): Control component Inc.: Catalog Drag Valve for boiler.

[5] Feedwater Control Application: Control component Inc.: Catalog.

[6] Drag Valve for boiler Feedwater Control Application: Control component Inc.: Catalog.

Chapter 12

Installation Practices

Chapter Outline

1 INTRODUCTION

The performance of a plant greatly depends on the performance of the control and instrumentation system (C&I system). A high-quality instrument may not perform well if the mounting and installation, including cabling, etc., are improper. Different types of instruments have different types of installation requirements. Because ISA practices are very common, in the book efforts have been made to follow ISA guidelines. As a consequence, many references have been made to ASME/ANSI standards. In the book, SI systems have been followed. However, in some cases it was necessary to use only FPS because of the standard associated with such a specification: for example, ANSI flange rating shall be only in the pound class or the NPT/BS thread shall be only in inches. However, equivalences have been shown in many cases. In this connection, the latest revision of "American National Standard for Metric Practice" ANSI/IEEE Std. 268 may be referenced. Field Instrument installation practices have been divided mainly into two parts: (1) *mechanical installation*, and (2) *electrical installation*. Under mechanical installation, instrument mounting, source connection details, installation of an impulse line, and other accessories, auxiliaries, and supporting details have been discussed. On the other hand, cable, cabling, etc., have been covered under electrical installation.

Some of the standards/common practices, mainly in line with ISA guideline 1994, are enumerated below:

1. **Process tap**: When the pressure $\leq$62.5 kg/cm^2 and the temperature $\leq$426°C, a 15 nb pipe shall be used. If the pressure or temperature exceeds the above limits, then a 20 nb pipe is recommended.. The material for the process pipe, fittings, and root valves shall be a material compatible with the process fluid.
2. **Impulse line**: For subcritical steam services, it shall be designed at the process design pressure at the *saturation temperature*. For other services, the line shall be designed at the process design pressure and temperature.
3. **Impulse line ID and thickness**: To avoid plugging and to have sufficient mechanical strength, a minimum ID >9.14 mm and a thickness of 1.25 mm are normally accepted. Only by calculation, on technical reasons, may deviations be considered. The latest revision of ANSI B31.1 Section 104 may be referred to for such calculations.
4. **Commonly used tubing materials**:
   a. *SS tubing*: Noncorrosive: ASTM A213 GR TP 304 (for temperatures up to 537°C), ASTM A213 GRTP 316 (for temperatures up to 648°C), and ASTM A213 GR TP 316L (for temperatures up to 450°C) or better. In larger plants, where pressure and temperatures ratings are high, ASTM A213 Gr. TP 316L and 316H are recommended.
   b. *Carbon steel*: Noncorrosive: ASTMA210 GR A1 (temperatures up to 412°C).

Power Plant Instrumentation and Control Handbook. https://doi.org/10.1016/B978-0-12-819504-8.00012-3

   c. *Copper tube*: ASTM B88 or B75 or annealed soft temper (maximum temperature 205°C) mainly used for air service and gas services.
5. **Valves and fittings**: Blowdown valves shall be the gradual opening type and shall be suitable for design pressure and temperature. Fitting materials shall be suitable for the application, and shall be compatible with the tubes and pipes. Any mercury (Hg) or Hg compound, materials with lower melting points such as lead, arsenic, antimony, bismuth, cadmium, zinc, etc., shall not be used.
6. **Impulse line preparation**: Before laying the impulse line, the same needs to be prepared as per Clause no. 3.1.11 of this chapter.
7. **Slope of impulse line**: Normally, 80 mm/m is standard practice. If it is not possible to maintain this slope, then the minimum shall not be <20 mm/m. For liquid services or level measurements, it shall be downward, whereas for air and gas services, it shall be upward from the source point to the instrument.
8. **Line drain/vent**: A high point vent and a low point drain or a combination shall be used to purge the unwanted gas/liquid.
9. **PREFERRED relative position of source point and instrument**: These are preferred locations with distinct advantages, but are *not* mandatory.
   a. *Steam and water service*: The instrument should preferably be below the tapping point to avoid any entrapped gas bubbles. This will also help in getting a lower temperature at the instrument end. Also, this arrangement avoids the additional need for a high-pressure vent.
   b. *Vacuum service*: The instrument shall be located above the source point so that the line is self-draining; this will avoid any errors in measurement.
   c. *Air and gas services*: The instrument shall be located above the source point—here also self draining of condensed moisture.
10. Preferred location of tapping point on pipe and allied systems.
11. **Refer** Fig. 12.1: Theoretically, in a circular pipe, cross-section tapping can be done at any position in the plane, but there are certain preferences in locating the tapping points based on the services as detailed.
    a. *For steam, air, and gas services*: The preferred tapping point location could be the top point, but it can be at any other position (upper half) with 45 degrees on both sides of the vertical line, as shown in Fig. 12.1A–C, to avoid any liquid ingress.
    b. *For liquid services*: The preferred tapping point locations are either side of the horizontal line, but can be taken at any other position (lower half) with 45 degrees on both sides of the horizontal line, as shown in Fig. 12.1D–F, to avoid any entrapped gas. However, it should not be at the bottom or any bottom positions beyond those discussed above to avoid the ingress of pipe dirt.
    c. *Temperature sensors* (elements) shall be placed in flowing streams and not on stagnant fluid to get a representative reading; normally, it should be directed in such a way as to oppose the flow to create turbulence.
    d. *Purging connections*: In many cases (as applicable), these are used to prevent the plugging from freezing and corroding. These purge fluids must compatible with the process fluid. In any case, care shall be taken to see that purging does not introduce any errors into the system. The quantity of gas/liquid purge shall be regulated. Purge fluids are normally independent of process fluid so that it is available even when the process fluid is not operating. Normally, a purging system shall have a check and shut-off valve.
12. **Source point**: The tapping socket, nipple, and root valve (first set of valves at the tapping point) mainly constitute the source point of most instruments (an exception is the temperature tapping point when there is no nipple or root valve). The terminal point for a piping engineer is the socket with a plug for the temperature element, and the root valve for tapping points pertinent to other measurements. Based on the application, the root valves could be of the socket welded/flanged/threaded type. These will be discussed later, but it is important that these are *well defined* so that there is no coordination problem in the future. In case of insulation in the pipe, the root valve shall be out of the insulation (at least the operating handle shall be accessible without opening insulation). Normally, in a power plant where any line pressure > **40 kg/cm^2** (especially for steam service), **two root valves** are used. In a lower-pressure application, a single root valve may suffice. In case of flue gas service, which is in draft, *no* root valve is necessary (in fact, an improper root valve is a problem in those cases).
13. **Mounting location of instruments**: The mounting of instruments is done on the basis of functional, operational, and maintenance facilities. In power plants, to facilitate maintenance, a number of instruments (mainly transmitters and switches) in nearby locations are grouped together and kept in a separate enclosure referred to as a local instrument enclosure (LIE, closed type) or a local instrument rack (LIR, open type). This helps in many ways, such as: (i) extra protection for the instruments, (ii) instead of separate drain lines, a common drain line from the enclosure or rack, and (iii) convenience for maintenance personnel. The major issues here are:
    a. The field instruments are to be suitably located to minimize the impulse pipe length.

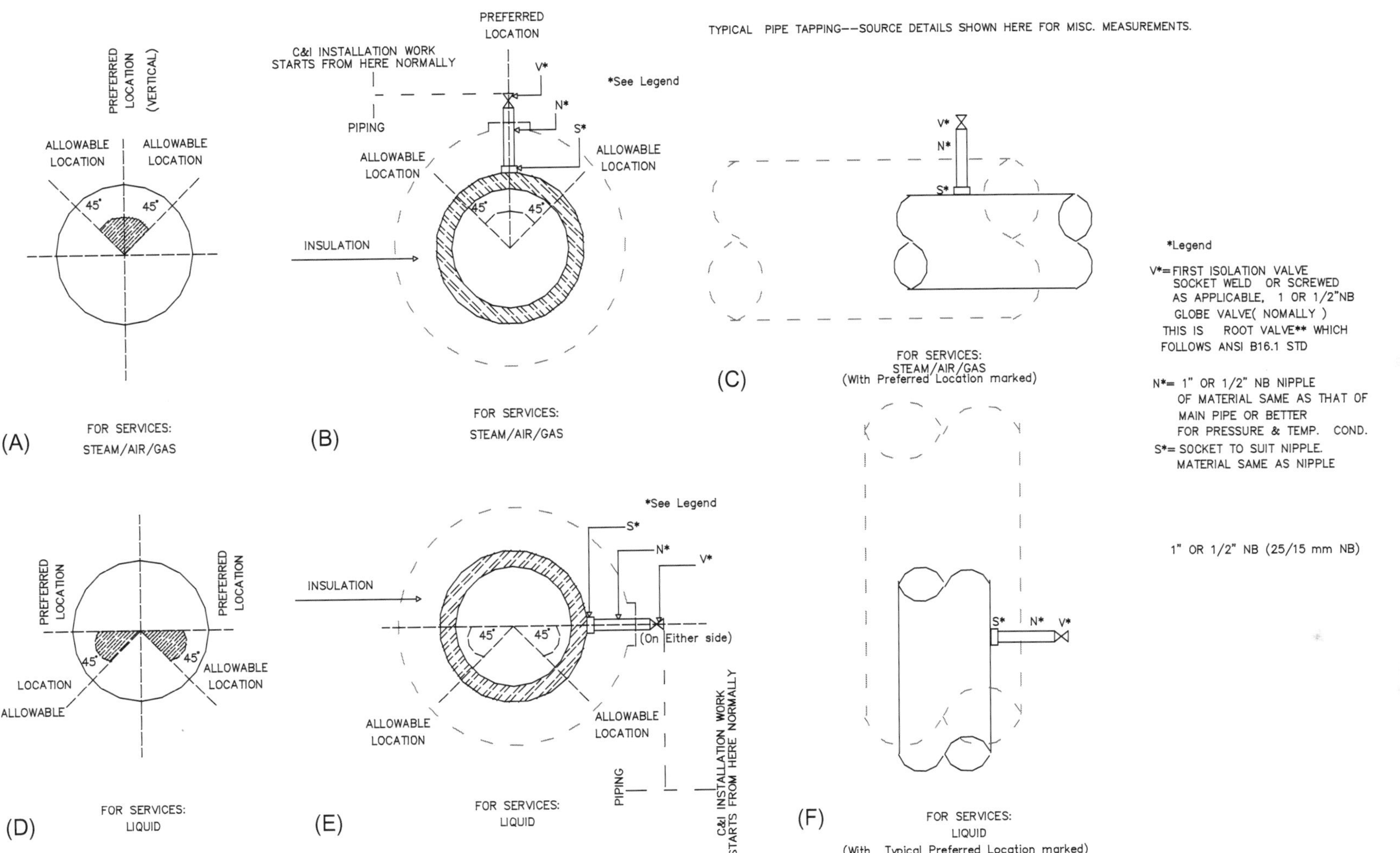

FIG. 12.1 Tapping point location (source point).

b. All instruments shall be easily accessible for calibration and maintenance from grates, platforms, or ladders. This is extremely important for instruments intended for safety/emergencies. In a place where there is a chance of flooding, the instruments are to be located at least 1–1.2 m above ground.

c. Normally, while routing the impulse piping/tubing for height clearance, the following guidelines from ISA may be referenced: within a structure—8 ft (2.44 m); within a yard—10 ft (3.05 m); over a secondary road—14 ft (4.27 m); and over a railroad and main building—22 ft (56.71 m).

d. In case an instrument is to be located outside the handrail of the platform, then it must reach 300 mm from the handrail horizontally, <1.5 m above the platform, and not >0.5 m from the fixed ladder (for example, instruments in the boiler platform).

e. Local instruments such as pressure and temperature gauges, etc., shall be readable from the platform or walkway. The drum level gauge (DP/optical) drum pressure gauges at the boiler operating platform shall be located in such a way that these are easily accessible and readable from a long distance.

f. Local controller/receiver instruments shall be located and readable in the vicinity of the control element.

g. Instruments and instrument racks/enclosures shall not obstruct the walkway.

h. From a maintenance point of view, clearance shall be provided for the removal of covers and cases/openings of doors of enclosures. Also, enough clearance (space) for lifting equipment shall be kept free.

14. General power supplies and signal types: power supplies and signal types have been indicated below.

a. See Table 12.1.

Hydraulic supplies for various actuators on the turbine or boiler side normally have self-contained hydraulic power unit (HPU).

TABLE 12.1 Supplies (Normal)

Electrical supplies	LER: 230/110 VAC 1PH 50/60 Hz
Instrument (field) supplies	24 VDC
Field panel	230/110 VAC 1PH 50/60 Hz or 24 VDC
Main ring pneumatic supplies	7 Bar (g) Max 10 Bar (g)
Instrument air supplies (normal)	1.4 Bar (g)

b. *Signal types*: In addition to digital protocols and fieldbus signals, other types are indicated in Table 7.32, and hence are not repeated again.

15. Instrument and tube/pipe support considerations:

a. Local instruments such as pressure/temperature gauges can be supported by piping on the main process pipe/equipment.

b. Instruments shall be supported in LIE/LIR, and on any other support meant for support purposes only, but shall not be supported from the process pipe to avoid getting affected by heat and vibrations from that pipe. A separate 50 (2″) NB Scheme 40 galvanized pipe mounting with a suitable support and stanchion is a common practice. Pipe stands shall be securely anchored and grouted.

c. Pipe stands directly embedded in concrete are not acceptable.

d. The weld area, drills, and threads on galvanized steel shall be prepared and sprayed with cold galvanizing coating.

e. The instrument tube/impulse pipe shall be supported at regular intervals to maintain structural stability and to prevent strains on instrument, equipment, or piping.

f. In all cases, suitable measures need to be taken to accommodate thermal expansion due to variations in ambient temperature.

g. All tubing/cables may be supported in steel trays.

16. Environmental protection: Some basic environmental protections that are commonly considered are enumerated below.

a. At all points, the ambient temperature for electronics must be kept within the operating range recommended by the manufacturer with some safety factor/margin.

b. Normally, hard enclosures are acceptable for instruments, but in case of space constraints, a soft type may be chosen if permitted.

c. Normally, process lines enter the enclosure of the instrument from the side or bottom, but never from the top to minimize the impulse tube length as well as for protection of electrical installations due to leakage of the impulse tube. These are mainly for local gauge panels/racks. However, in cases of **transmitter enclosure or SWAS sample conditioning panel,** the impulse/sample lines normally enter the enclosure from the top. In all these cases, suitable separation is kept between the electrical points and mechanical instruments. For this reason, the **junction box in** the **transmitter enclosure is kept at the backside**, away from the instrument mounting of the transmitter enclosure. Also, in an SWAS, there are normally separate sections for sample conditioning and the SWAS analyzer panel.

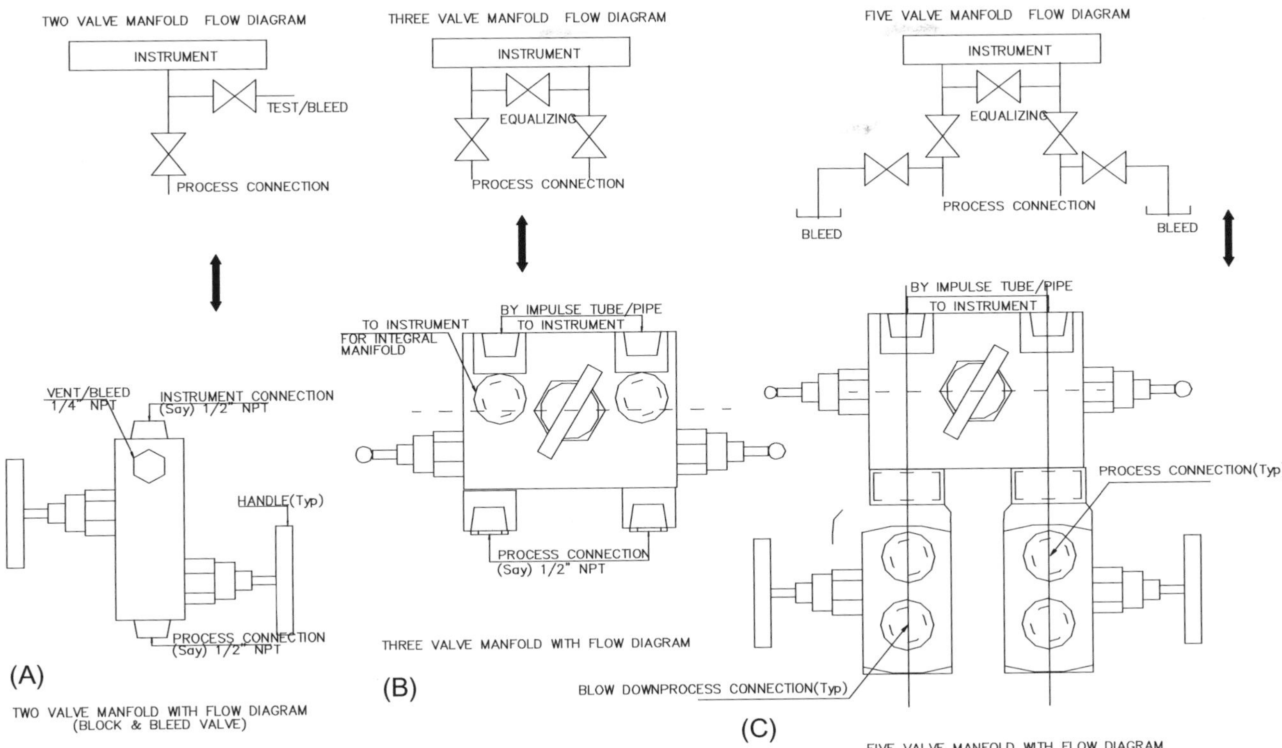

FIG. 12.2 Source point—valve manifold.

- **d.** Local instrument enclosures shall be normally wall mounted of pipe/structure supportive. In case of a bigger enclosure (a transmitter enclosure), it could be self-supportive with the proper grouting.
- **e.** Material for the metallic enclosure shall be suitably chosen so that it is corrosion resistant. IP 65/66 (NEMA 4/4 ×) is normally chosen for outdoor installation.
- **f.** The space inside the enclosure shall be sufficient to make maintenance easier.
- **g.** To prevent freezing, insulating protection shall be provided for the instruments as well as associated accessories such as a valve manifold.

17. Valve manifold: the valve manifold is a very important accessory for installation. Two valve manifolds are used mainly for pressure instruments. Three valve manifolds are used for DP Instruments. Five valve manifolds are also used for DP measurements where there is a provision for line drain. Detailed drawings of valve manifolds along with a flow diagram are shown in Fig. 12.2. Valve manifolds used with instruments are generally the forged type and as far as possible, they shall be mounted integral with the instruments, especially for DP instruments. The valves are normally tested for 1.5 times the maximum process pressure they must withstand. A block design of the manifolds is also used.

1.1 A Source Point for Pressure Tapping (Figs. 12.3 and 12.4)

As shown in the figure under reference, a socket/boss shall be welded to the pipe as a source point.

1. Pressure/DP tapping on pipe: (Fig. 12.3): Normally, a socket of length and diameter of 50 mm is butt-welded to the pipe to accommodate a nipple (a normal pipe of same or better material).Nipple is placed at small groove and welded with socket. 250 mm could be an optimum size for a 25/15 NB nipple). The thickness of the nipple depends on the application (mainly on the pressure rating), as detailed in Fig. 12.3. A root valve is welded (or threaded for low-pressure applications) to the nipple, as shown in Fig. 12.1B–E.
2. Pressure/DP tapping on gas duct/furnace wall: Because in gas services there are chances for dust, an upward pipe of a suitable diameter and length (at 45 degrees with the wall of the duct or furnace) is welded with the wall of the duct. A straight path of this longer pipe is kept for purging and poking, whereas at the side, another pipe is welded at 45 degrees with the longer pipe for pressure and DP connections. As seen in Fig. 12.4, the orientation is such that the dust will be self-draining to prevent the impulse tube getting plugged due to dust. A port for poking is kept closed by a cap with a gasket

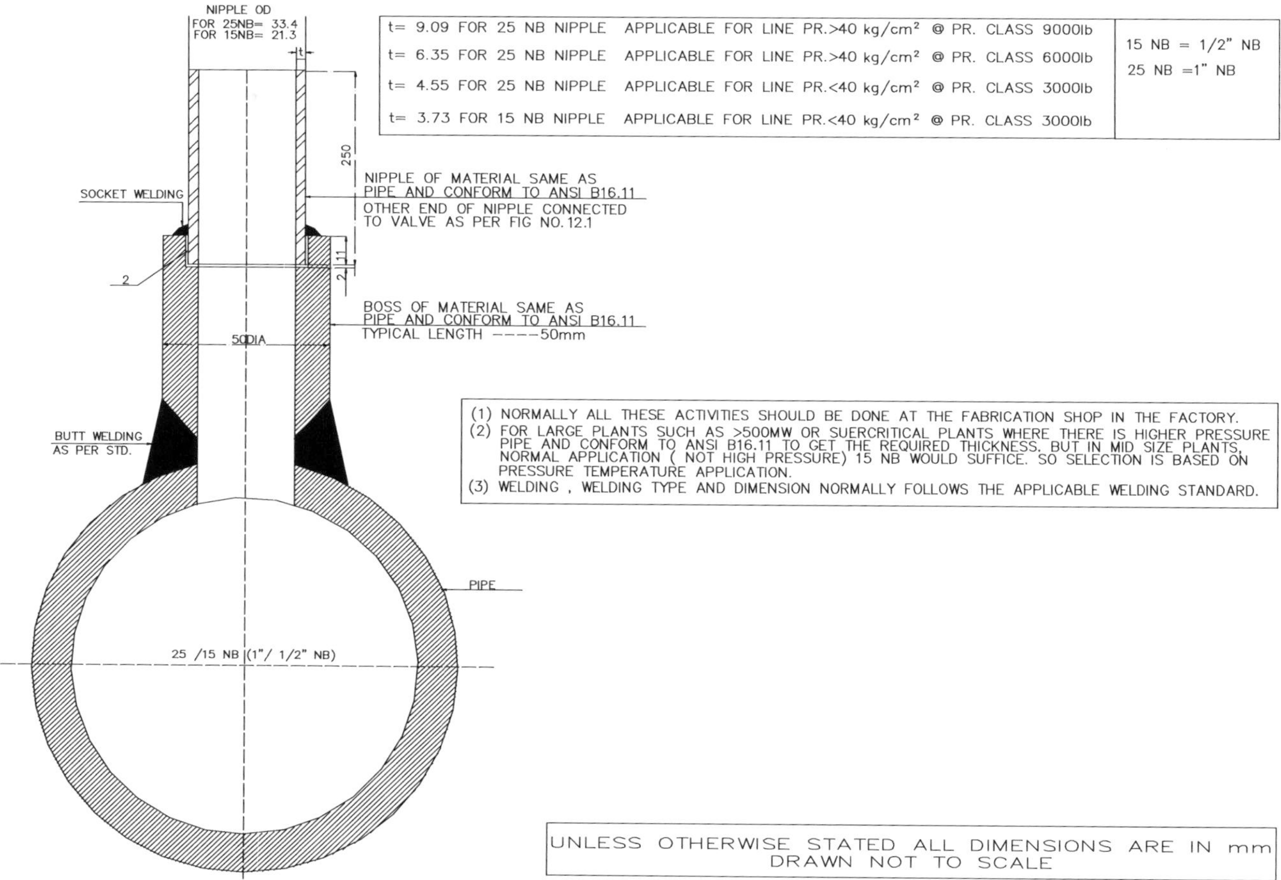

FIG. 12.3 Source connection for Pr. DP type instruments.

and chain so as to avoid leakage. As shown in the drawing under reference, normally 40 mm NB pipes are chosen for such application.

1.2 Source Point for Temperature Tapping (Figs. 12.5–12.8)

Here, the boss/socket shall be similar to the same as that for pressure tapping. As per ASME PTC 19.3, the thermowell shall be immersed inside the pipe at least 1/3 of the pipe ID so that the representative temperature can be monitored. Again, there are some limitations regarding the length of the thermowell. If it is too small, it is very difficult to fix or maintain. If it is too large, then there may be the possibility for breakage due to the flowing fluid velocity if it is not properly supported. Although it is **not** a rule, empirically people consider a thermowell length of 150 mm or half the pipe ID (±13 mm), whichever is higher within the normal lower and upper limit of length as 75 and 150 mm. However, for a gas duct, it is more and the upper limit may be 1500 mm as a good solution in selecting the thermowell length. Details regarding immersion length, insertion length, extension neck, etc., for thermocouples have already been discussed in Chapter 4, Clause 3.1, so they are not repeated here. Unless otherwise stated, material for the boss shall be suitable for the application. In most of the cases, the thermowell material shall be 316 SS, or better as per the data sheet.

A few installation details of the thermowell are discussed below:

1. Refer to Figs. 12.5A, B and 12.6: Inclined installations are mainly adapted for a size of 75–100 mm NB pipelines so that a suitable length of the thermowell can be chosen. It is seen that the thermowell is mounted at an angle of 45 degrees with a vertical so that the thermowell opposes the flow, as shown in the figure under reference (flows vertically upward). In case the pipe size <75 mm NB, then an expander can also be used for such an installation in a vertical or inclined plane to get a suitable thermowell length. For a very small size pipe such as <50 mm NB, the installation shown in Fig. 12.5B can be adopted. Here, the pipe junction has been considered for installation of the thermowell. As shown, a suitable expander has been used to accommodate the thermowell,

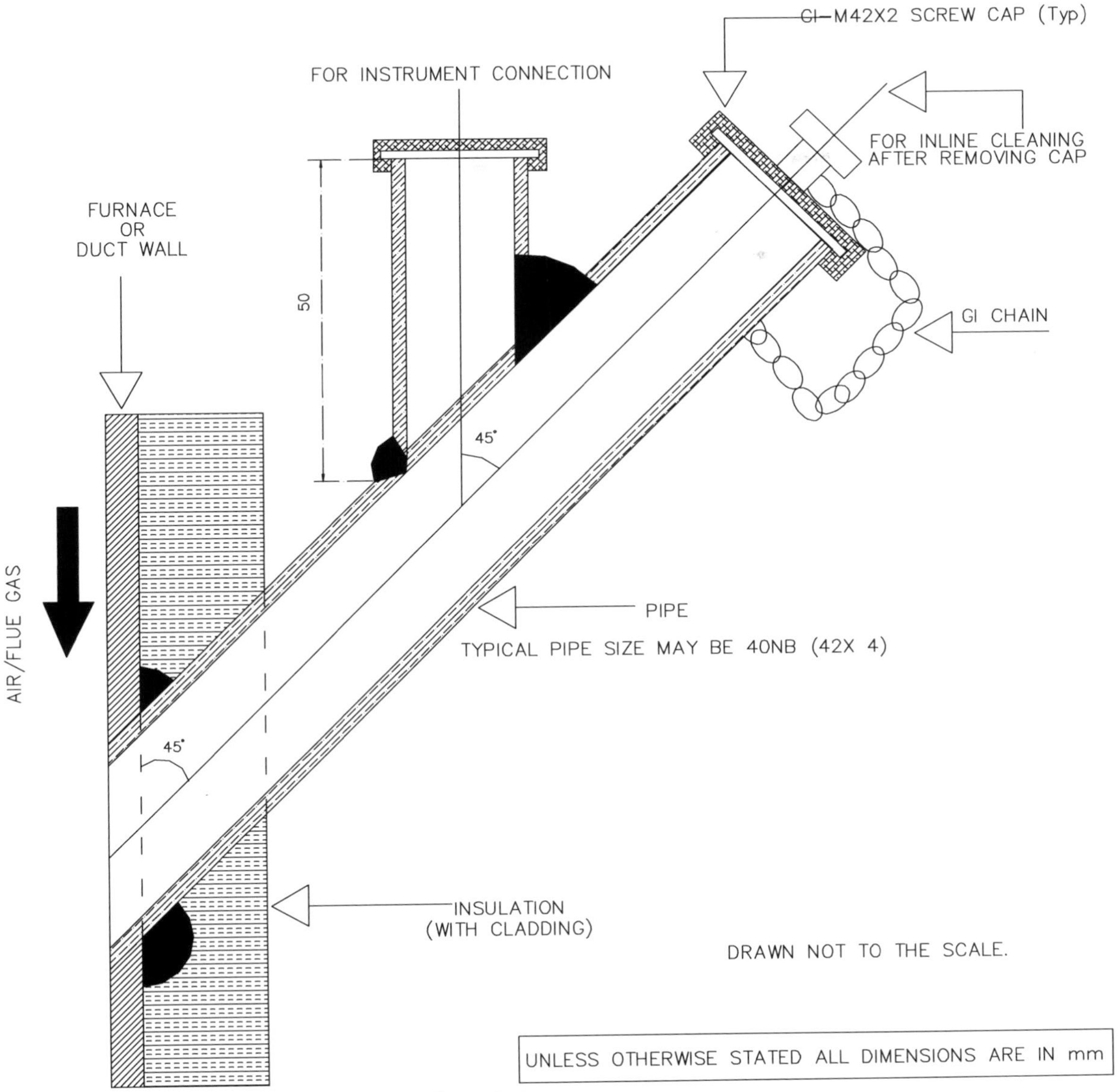

GI M42X2 SCREW CAP (Typ)
FOR INSTRUMENT CONNECTION
FOR INLINE CLEANING
AFTER REMOVING CAP
FURNACE
OR
DUCT WALL
50
GI CHAIN
45°
AIR/FLUE GAS
PIPE
TYPICAL PIPE SIZE MAY BE 40NB (42X 4)
45°
INSULATION
(WITH CLADDING)
DRAWN NOT TO THE SCALE.
UNLESS OTHERWISE STATED ALL DIMENSIONS ARE IN mm

FIG. 12.4 Source connection for Pr. DP type instruments in gas duct.

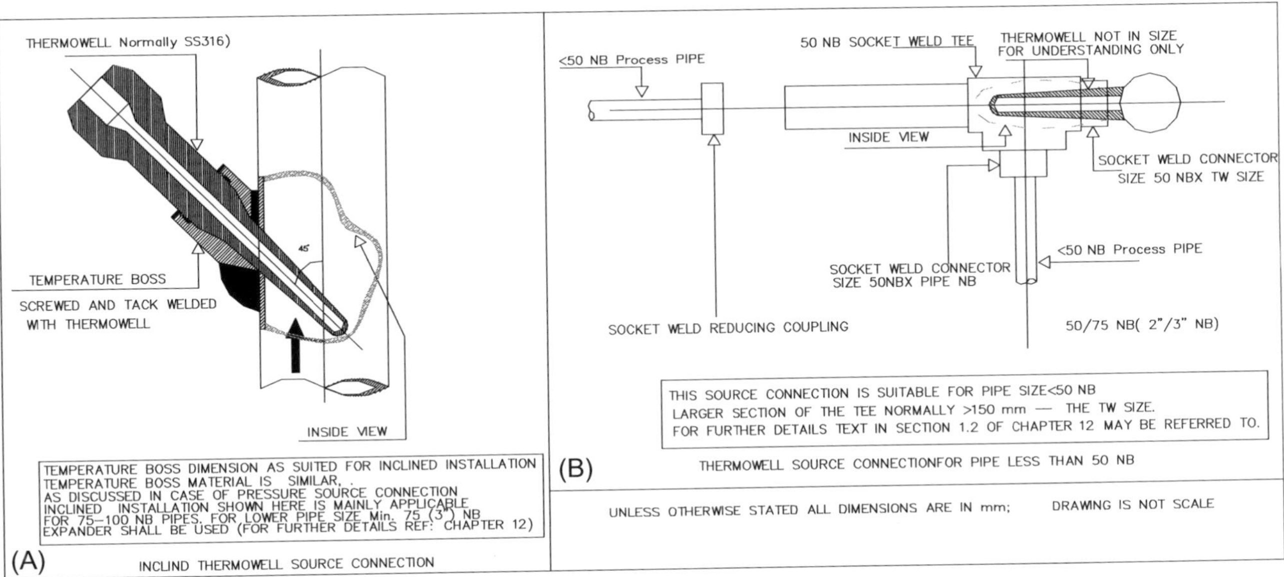

FIG. 12.5 Source connection for temperature element (1).

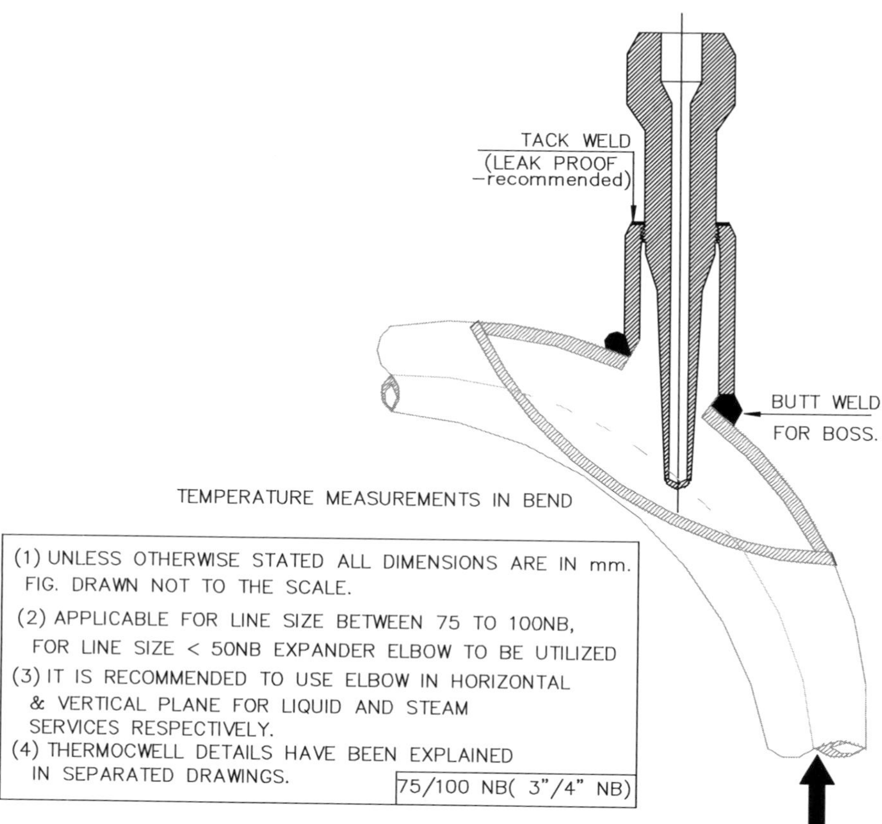

TACK WELD
(LEAK PROOF
—recommended)
BUTT WELD
FOR BOSS.
TEMPERATURE MEASUREMENTS IN BEND
(1) UNLESS OTHERWISE STATED ALL DIMENSIONS ARE IN mm.
FIG. DRAWN NOT TO THE SCALE.
(2) APPLICABLE FOR LINE SIZE BETWEEN 75 TO 100NB,
FOR LINE SIZE < 50NB EXPANDER ELBOW TO BE UTILIZED
(3) IT IS RECOMMENDED TO USE ELBOW IN HORIZONTAL
& VERTICAL PLANE FOR LIQUID AND STEAM
SERVICES RESPECTIVELY.
(4) THERMOCWELL DETAILS HAVE BEEN EXPLAINED
IN SEPARATED DRAWINGS.
75/100 NB(3"/4" NB)
PROCESS FLOW TO CREATE MORE TURBULENCE

FIG. 12.6 Source connection for temperature element (2).

and in such a case, a 150mm thermowell is inserted. In case suitable bends are available in the process pipeline, then the same can be looked into in cases of line sizes between 75 and 100mm NB, as shown in Fig. 12.6.

2. Fig. 12.7A and B: Major thrust has been put in these drawings to show the mounting details of the thermowell on the pipes for power plant applications; our discussions shall be for power plants only. There are two distinct cases. One case is where the process pressure $>40\,kg/cm^2$: Here, the thermowell is welded to the boss/socket in addition to tightening the thermowell with its external threads with the internal threads of the boss/socket. Naturally, surface preparation in the boss is necessary to weld the thermowell with the boss to make the system able to withstand high pressure. Normally, in these cases shop welding is done for the boss with the pipe, as shown in Fig. 12.7A. In cases where the process pressure $<40\,kg/cm^2$, a connection between the thermowell and boss could be the only thread, as discussed above and shown in Fig. 12.7B. However, in addition to the thread connection, tack welding is a good engineering practice for leakage. In most of the cases (exception: very low pressure services), the thermocouples shall be drawn from solid bar stock for better mechanical stability and strength. In power plants, many of the pipelines where the thermowell is to be mounted are insulated, so a suitable extension length (ISA symbol L) shall be considered, as shown in the drawings. The internal bore of the thermowell normally follows ASME PTC19.3. Normally, a 50mm diameter boss is selected the greatest diameter of the thermowell shall be matching with internal bore of the boss which is normally 40mm so that with external threads it could be accommodated in the boss. Normally, the lowest tip thickness of the thermowell is around 4.5–4.8mm.
3. Fig. 12.8: A slip-on flange connection of the thermowell is very suitable for gas duct installations. One pipe (of size say, 42.4×4mm) is welded with the wall of the duct. On the other end of the pipe, there will be a flange to accommodate the flange of the thermowell. The length of this pipe shall be enough so that it goes past the insulation for fitting the thermowell. A typical pipe length of 150mm outside the insulation has been shown to accommodate/adjust various thermowell lengths with the help of the *slip-on flange* discussed above. The slip-on flange as per ASME 19.3 has been shown (but it could also be other types that match). A thermowell with matching flange shall be inserted through the pipe to the gas duct.

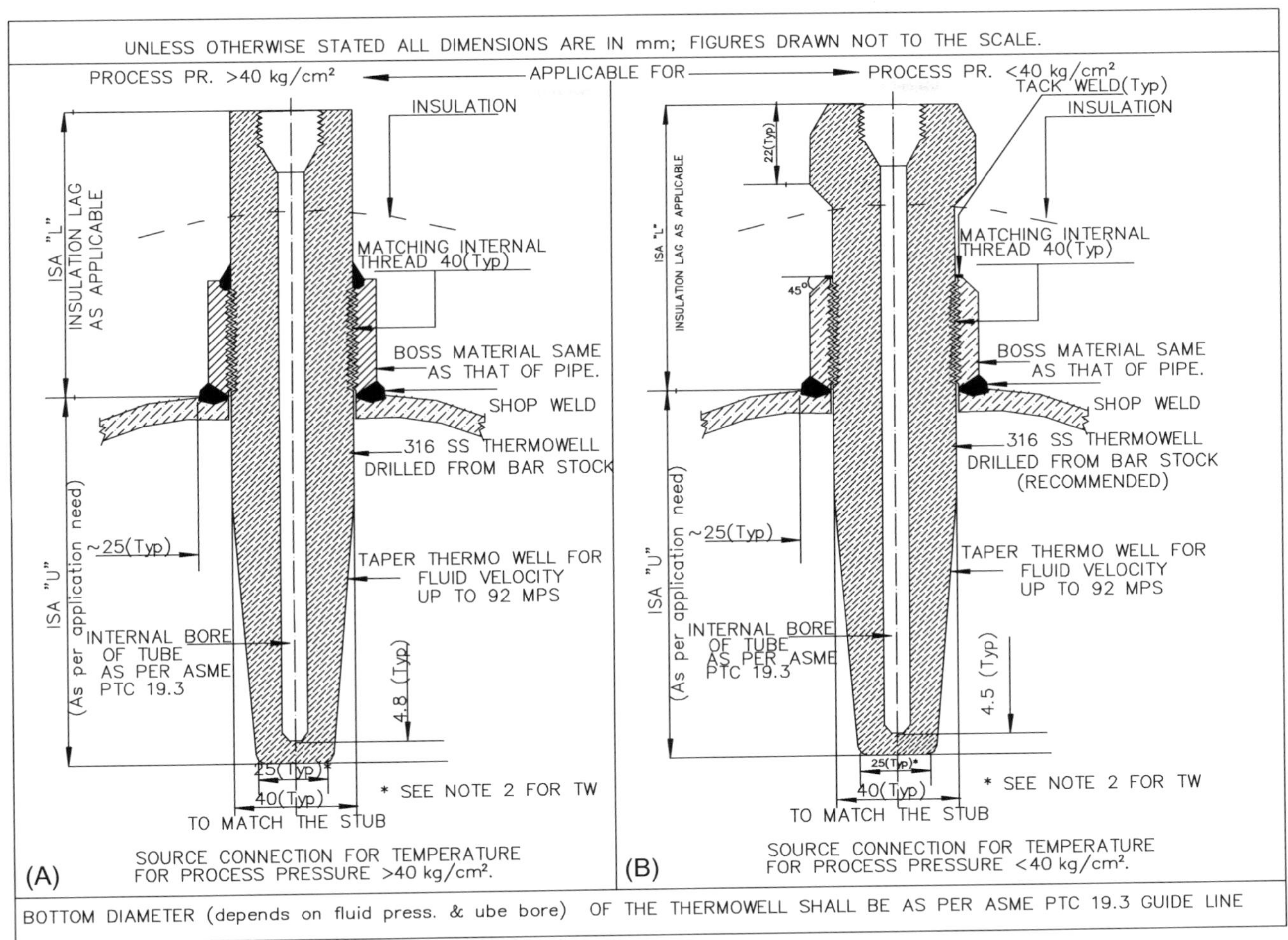

FIG. 12.7 Source connection for temperature element (3).

1.3 Source Connection for Flow Measurements

Out of variety of flow meters used in a power plant but major important flow measurements are based on differential pressure (DP) measurement type, so primarily discussions shall be on the same. In this method, impulse lines are used to measure the DP due to restrictions such as orifice plate, flow nozzle, Venturi tube, etc. There are a few problems (listed below) associated with impulse lines, for example, damping of pressure signals/response, blockage, leakage at the coupling, temperature (hence density) difference, condensate in the line for gas flow measurement, and gas bubble in the impulse line for liquid flow measurement.

1. Issues related with impulse line
   - **a.** Impulse line length and diameter: In order to have a better response, a narrower diameter of the impulse line is necessary, but if it is too narrow, there will be a blocking problem. A capillary effect error and the need to vent the bubble will be problematic. It is recommended that the minimum diameter shall be 10mm. With a variation in length, there will be a variation in diameter also. The length and diameter of the impulse line as per ISO/CD 2186:2004 has been given in Table 12.2.
   - **b.** The location of the DP instrument with respect to the primary sensor has a bearing on installation; the same is discussed in Clause no. 1.9 above.
   - **c.** Location of tapping point and orientation: Discussed in Clause no. 1.10 above.
   - **d.** Routing and slope. For proper functioning as well as accurate measurement, it is recommended that there be a suitable length of the impulse line and a suitable slope is maintained. Clause nos. 1.12c and 1.7 above may be referenced.
   - **e.** Temperature effect: Both ambient temperature and a difference in temperature between the two limbs of the impulse line have direct bearings on the accuracy of measurement. During installation, it should not happen that one limb of the impulse line is exposed to sunlight while the other is in shade. Similarly, both impulse lines shall be taken close to each other so that

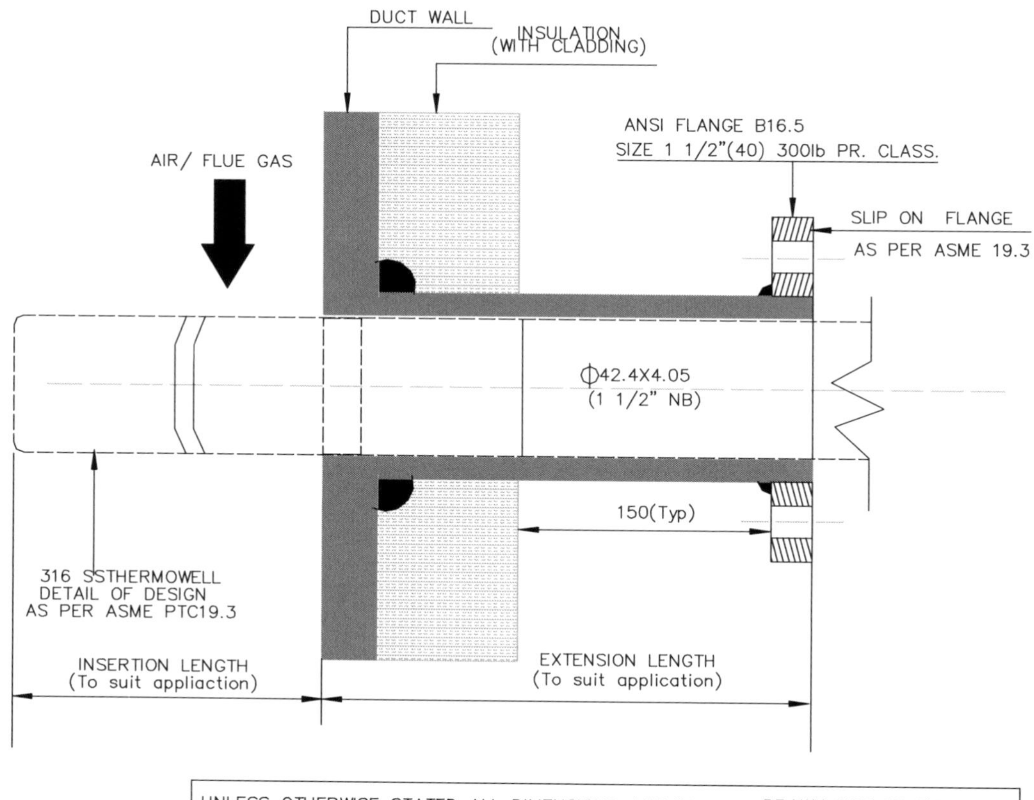

FIG. 12.8 Source connection for temperature element in gas duct.

TABLE 12.2 Impulse Line Diameter (Min) and Length

Dia for Flow Type	Line Length (0–16 M)	Line Length (16–45 M)
Steam/water/dry air/gas	7–9 mm	10 mm
Wet air/gas	13 mm	13 mm
Oil (low/medium viscosity)	13 mm	19 mm
Dirty fluid	25 mm	25 mm

both limbs are mostly at the same temperature (Fig. 12.9A). This is important because in case of a variation in temperature between the two sides of the impulse line, there will be a variation in density in the fluid in those two limbs. Therefore, it affects the DP {$DP = h^*(\rho_1 - \rho_2)^*g$, so a variation of ρ will affect the reading, even for the same h}.

f. Fluid in Impulse pipe: For steam and condensate services, the line is filled with water prior to the start of measurement. In most of the cases, condensate pots are used to ensure that the steam does not reach the instruments.

g. Proper connections: There shall be suitable isolation, drain/vent valves, and connections for measurement accuracy. Also, blocking and leakage at the coupling shall be avoided.

2. Flow nozzle: Flow nozzle connections have been shown in Fig. 12.9A and B. A horizontal installation flow nozzle is preferred but it can be used in vertical downward installations that are also used for steam (wet). It is not conventional (because of removal difficulty) to place a flow nozzle between two flanges. Normally, it is welded with spool pieces. As shown in Fig. 12.9A, in order to keep the temperature at two limbs the same, impulse lines shall be run close to each other. In vertical installations, normally two limbs are first raised vertically before taking a horizontal run. Horizontal installation has been depicted in Fig. 12.9B with D and D/2 tappings, which are very common for flow nozzles in power plant applications. As shown in Fig. 12.9B, the straight length of the spool piece length is 8D. For further details, see Clause 4.1 of Chapter 4.

3. Orifice plate: A square edge orifice plate with a drain and vent hole has been shown in Fig. 12.10A. Orifice plates are mounted between the two flanges. There are a number of accessories, for example, a carrier ring, to mount the orifice plate. Corner tap, flange tap, and D D/2 taps are

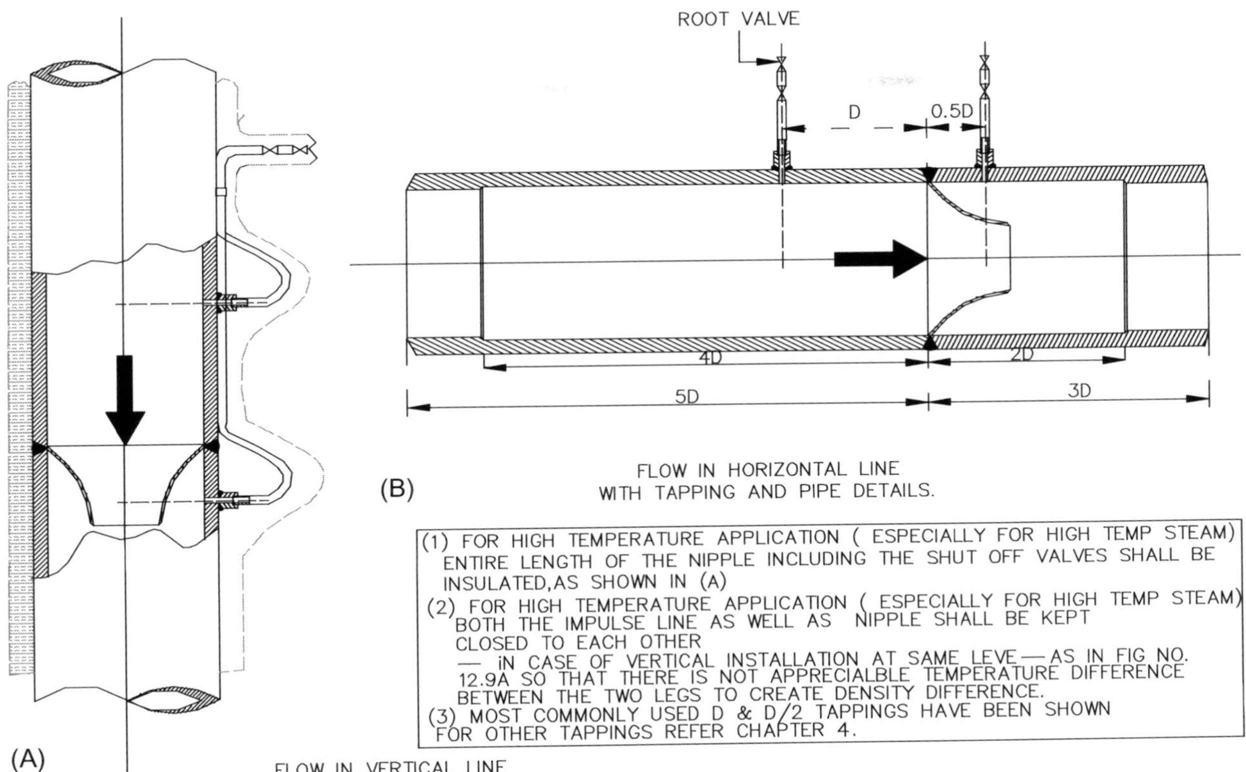

ROOT VALVE
D
0.5D
4D
2D
5D
3D
(B)
FLOW IN HORIZONTAL LINE
WITH TAPPING AND PIPE DETAILS.
(1) FOR HIGH TEMPERATURE APPLICATION (ESPECIALLY FOR HIGH TEMP STEAM) ENTIRE LENGTH OF THE NIPPLE INCLUDING THE SHUT OFF VALVES SHALL BE INSULATED,AS SHOWN IN (A)
(2) FOR HIGH TEMPERATURE APPLICATION (ESPECIALLY FOR HIGH TEMP STEAM) BOTH THE IMPULSE LINE AS WELL AS NIPPLE SHALL BE KEPT CLOSED TO EACH OTHER
— iN CASE OF VERTICAL INSTALLATION AT SAME LEVE—AS IN FIG NO. 12.9A SO THAT THERE IS NOT APPRECIALBLE TEMPERATURE DIFFERENCE BETWEEN THE TWO LEGS TO CREATE DENSITY DIFFERENCE.
(3) MOST COMMONLY USED D & D/2 TAPPINGS HAVE BEEN SHOWN FOR OTHER TAPPINGS REFER CHAPTER 4.
(A)
FLOW IN VERTICAL LINE
HIGH TEMPERATURE FLUID(STEAM AT HIGH TEMP)
INSULATION SHOWN AND NIPPLE AT ONE LEVEL SHOWN
(TO AVOID DENSITY DIFFERENCE BETWEEN TWO LIMBS
— IMPULSE LINES SHALL ALSO BE CLOSE TO EACH OTHER)
FOR PROPER INSTALLATION GASKET (NOT SHOWN) SHALL BE USED FOR . TIGHTENING THE FLAGE AND TO MAKE LEAK PROOF INSTALLATION
UNLESS OTHERWISE STATED ALL DIMENSIONS ARE IN mm. ; DRAWN NOT SCALE
THREE PAIRS OF TAPPINGS ARE NOT UNCOMMON , AND MAY BE AT 120 deg APART

FIG. 12.9 Source connection for flow elements (1).

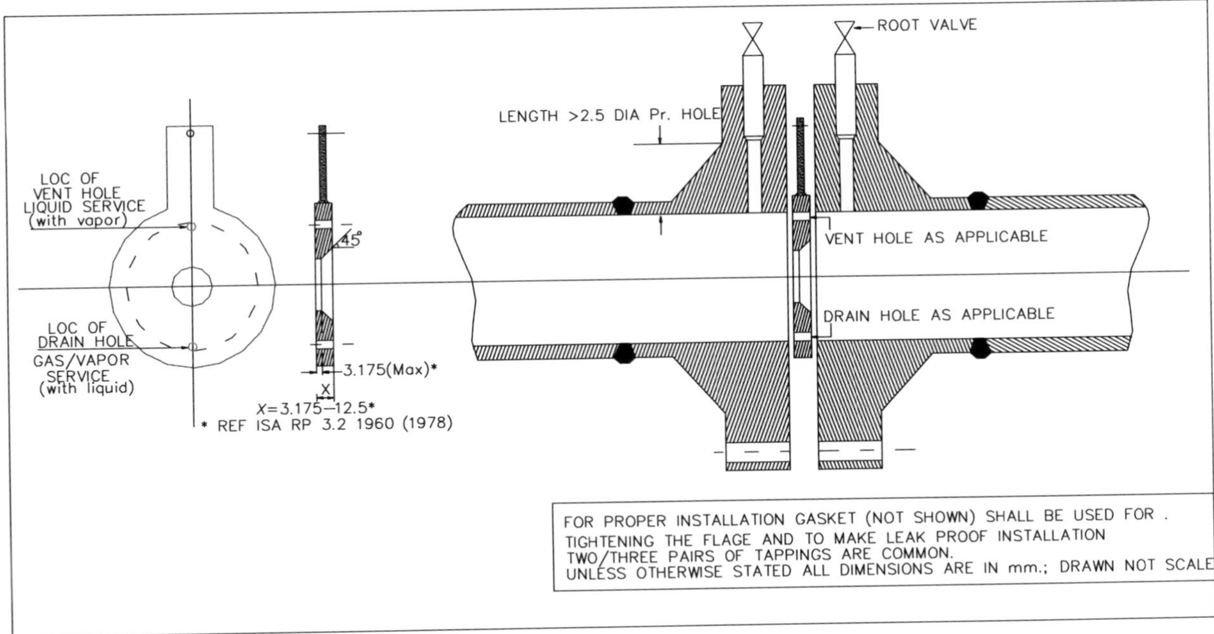

FIG. 12.10 Source connection for flow elements (2).

the various kinds of tapping available, as detailed in Clause 4.1 of Chapter 4. Fig. 12.10B shows an orifice plate with a weld neck flange and flange tappings. In the drawing, both drain and vent holes have been shown but only one shall be used.

1.4 Source Point Connection for Level Measurement

Basically, the tapping point types are similar to those for pressure measurements, only the sizes will be different, as shown in Fig. 12.11.

However, one has to keep in mind that in case there is tapping at the bottom of the tank, then the nozzle shall be raised 150 mm above the bottom of the tank to avoid dirt. Also, level instrument flange connections have been depicted in Fig. 12.11C.

2 PIPE VALVE FITTING (PVF) SPECIFICATION RATINGS

In most of the cases, all materials in power plants follow the latest editions of ANSI B31.1 and ANSI B16.11 for power piping, flanges, fittings, etc. There are other international standards also. In this section, various sizes, ratings, and material specifications for pipes, valves, and fittings shall be discussed briefly. All impulse lines (could be pipe/tube) shall be properly specified. Pipes are specified either by OD X thickness or as a nominal bore. Specifying a pipe by nominal more (NB) is the most common. Size, specified by NB, is actually the diameter of an imaginary circle (may be anywhere between the OD and ID). However, as long as NB is specified, the OD of the pipe is specified (because corresponding to each NB, there is a specific pipe OD). The advantage is that, for the same NB, there can be pipes of varied thicknesses. So, the thickness of the pipe is specified by the pipe schedule, which also determines the capability of the pipe to withstand pressure. This is clear from Table 12.3. Tubes are specified by OD X thickness. Impulse lines could be impulse pipe, impulse tube, or a combination of the two. From a commercial point of view, on the field side, impulse pipes are quite common. In those cases, CS alloy steel pipes could be used. When instruments are placed in a local instrument enclosure (LIE), it is customary (for manufacturing uniformity) to use a stainless steel impulse tube inside the enclosure. Bulkhead fittings are used to bring the impulse line into the enclosure.

1. **Pipe dimensions**: As discussed above, there will be variations in dimensions of impulse pipes based on NB size

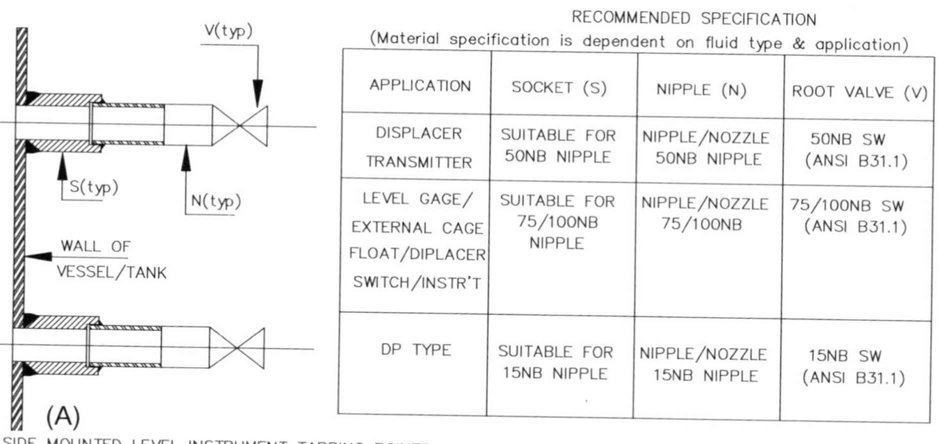

RECOMMENDED SPECIFICATION
(Material specification is dependent on fluid type & application)

APPLICATION	SOCKET (S)	NIPPLE (N)	ROOT VALVE (V)
DISPLACER TRANSMITTER	SUITABLE FOR 50NB NIPPLE	NIPPLE/NOZZLE 50NB NIPPLE	50NB SW (ANSI B31.1)
LEVEL GAGE/ EXTERNAL CAGE FLOAT/DIPLACER SWITCH/INSTR'T	SUITABLE FOR 75/100NB NIPPLE	NIPPLE/NOZZLE 75/100NB	75/100NB SW (ANSI B31.1)
DP TYPE	SUITABLE FOR 15NB NIPPLE	NIPPLE/NOZZLE 15NB NIPPLE	15NB SW (ANSI B31.1)

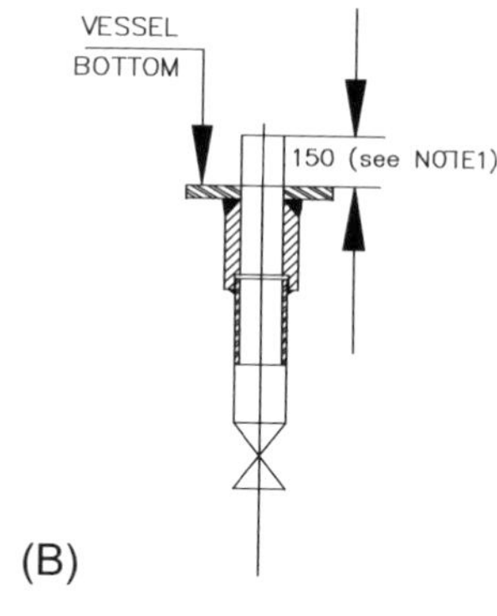

NB mm	NB Inch
15	1/2
20	3/4
25	1
40	1 1/2
50	2
75	3
100	4

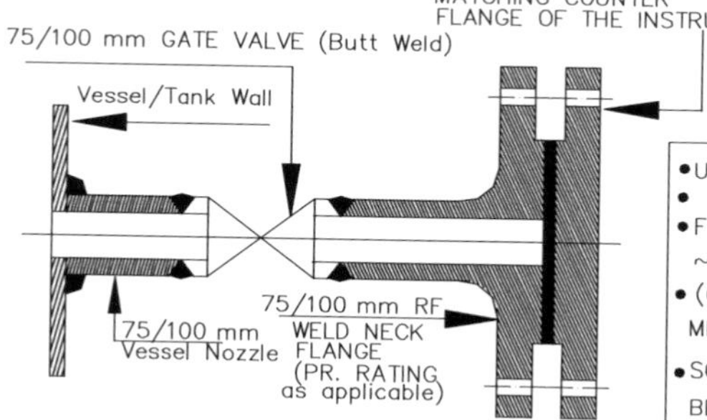

- UNLESS OTHERWISE STATED ALL DIMENSIONS ARE IN mm.
- DRAWN NOT TO SCALE
- FOR BOTTOM POINT INSTALLATION NIPPLE EXTENDED ~ 150 mm TO GET RID OF DIRT. AT THE BOTTOM
- (C) IS MAINLY APPLICABLE FOR LEVEL MEASUREMENT OF VISCOUS FLUID–MAINLY
- SOURCE CONNECTION POINT FOR LEVEL MEASUREMENT SHALL BE AT LOCATED SUITABLY SO THAT THERE IS NOT MUCH INTERFERENCE/TURBULENECE OF INLET/OUTLET.

FIG. 12.11 Source connection for level instruments (1).

TABLE 12.3 Pipe Dimension

Pipe NB (in.)	OD (in.)	Identification (Sch.)			Thickness (in.)
		Iron Pipe Size	Steel Pipe Sch.	SS Sch.	
1/4	0.405	–	–	10S	0.065
		STD	40	40S	0.088
		XS	80	80S	0.119
3/8	0.675	–	–	10S	0.065
		STD	40	40S	0.091
		XS	80	80S	0.126
1/2	0.84	–	–	5S	0.065
		–	–	10S	0.083
		STD	40	40S	0.109
		XS	80	80S	0.147
		–	160	–	0.187
		XXS	–	–	0.294
3/4	1.050	–	–	5S	0.065
		–	–	10S	0.083
		STD	40	40S	0.113
		XS	80	80S	0.154
		–	160	–	0.219
		XXS	–	–	0.308
1	1.315	–	–	5S	0.065
		–	–	10S	0.109
		STD	40	40S	0.133
		XS	80	80S	0.179
		–	160	–	0.250
		XXS	–	–	0.358
1 1/4	1.660	–	–	5S	0.065
		–	–	10S	0.109
		STD	40	40S	0.140
		XS	80	80S	0.191
		–	160	–	0.250
		XXS	–	–	0.382

and schedule. Important parameters required for instrumentation from ANSI B36.10 for selected sizes have been presented in Table 12.3.

2. **Impulse line valve and fitting materials and ratings**: Various impulse lines and valve materials have been enumerated below. All fittings are forged steel as per ANSI B16.11. The materials for forged tube fittings are ASTM A182 Gr. 316H and A182 Gr. 316L for high pressure and temperature applications and other applications, respectively. Bar stock tube fittings can be 316SS. Valves are considered to be the forged type. All materials discussed are forged/bar stock types. Service-wise, a list has been placed based on large power plants with higher temperature and pressure ratings. In medium to lower MW power plants, the stringency may be less, and accordingly a lower grade may be selected. The data given is on an approximate basis; the actual choice shall be based on design pressure and temperature conditions.

a. Service: **MS, Ms up to pressure reducing valves in AS and HPBP line.**
   i. Impulse line: ASTM A335 Gr. P91 Sch: XXS (1/2″)
   ii. Valve Body, flange, and fittings: ASTM A182 Gr. F-91
   iii. Valve stem materials: ASTM A182 Gr. F-6a
   iv. Valve pressure class: 3000 lb.
   v. Fitting: ASTM A182 Gr. 316H
   vi. Ratings of fitting/pipe: 9000 lb.

b. Service: **HRH, HRH up to pressure reducing valves LPBP line.**
   i. Impulse line: ASTM A335 Gr. P91 Sch: 160 (1/2″)
   ii. Valve body, flange, and fittings: ASTM A182 Gr. F-91
   iii. Valve stem materials: ASTM A182 Gr. F-6a
   iv. Valve pressure class: 1500 lb.
   v. Fitting: ASTM A182 Gr. 316H
   vi. Ratings of fitting/pipe: 3000 lb.

c. Service: **Downstream of auxiliary steam PRV up to drain manifold.**
   i. Impulse line: ASTM A335 Gr. P91 Sch: 160 (1/2″)
   ii. Valve body, flange, and fittings: ASTM A182 Gr. F-91
   iii. Valve stem materials: ASTM A182 Gr. F-6a
   iv. Valve pressure class: 1500 lb.
   v. Fitting: ASTM A182 Gr. 316L
   vi. Ratings of fitting/pipe: 3000 lb.

d. Service: **BFP disch./SH and RH attemperation/ spray lines.**
   i. Impulse line: ASTM A106Gr. C Sch: 160 (1/2″)
   ii. Valve body, flange, and fittings: ASTM A105
   iii. Valve stem materials: ASTM A182 Gr. F-6a
   iv. Valve pressure class: 2500 lb. (RH attemperation. line valve 1500 lb)
   v. Fitting: ASTM A182 Gr. 316L
   vi. Ratings of fitting/pipe: 6000 lb.

e. Service: **CRH up to tee off for HPBP/Ex 5 to HPH**
   i. Impulse line: ASTM A335Gr. P22 Sch: 80 (1/2″)

ii. Valve body, flange, and fittings: ASTM A182 Gr. F-22
iii. Valve stem materials: ASTM A182 Gr. F-6a
iv. Valve pressure class: 900 lb.
v. Fitting: ASTM A182 Gr. 316L
vi. Ratings of fitting/pipe: 3000 lb.

f. Service: **CRH after tee off for HPBP**
   i. Impulse line: ASTM A106Gr. C Sch: 80 (1/2″)
   ii. Valve body, flange, and fittings: ASTM A105
   iii. Valve stem materials: ASTM A182 Gr. F-6a
   iv. Valve pressure class: 900 lb.
   v. Fitting: ASTM A182 Gr. 316L
   vi. Ratings of fitting/pipe: 3000 lb.

g. Service: **Other lines in FW/cond line**
   i. Impulse line: ASTM A106Gr. B Sch: 80 (1/2″)
   ii. Valve body, flange, and fittings: ASTM A105
   iii. Valve stem materials: ASTM A182 Gr. F-6a
   iv. Valve pressure class: 900 lb.
   v. Fitting: ASTM A182 Gr. 316L
   vi. Ratings of fitting/pipe: 3000 lb.

h. Service: **Air flue gas line In Furnace pass**
   i. Impulse line: ASTM A335Gr. P22 Sch: 80 (3/4″)
   ii. Valve body, flange, and fittings: ASTM A 182 Gr.F22
   iii. Valve stem materials: ASTM A182 Gr. F-6a
   iv. Valve pressure class: 900 lb.
   v. Fitting: ASTM A182 Gr. 316H
   vi. Valve pressure class: 900 lb.
   vii. Ratings of fitting/pipe: 3000 lb.

i. Service: **Air flue gas line Outside run**
   i. Impulse line: ASTM A106Gr. C Sch: 80 (3/4″)
   ii. Valve body, flange, and fittings: ASTM A105
   iii. Valve stem materials: ASTM A182 Gr. F-6a
   iv. Valve pressure class: 900 lb.
   v. Fitting: ASTM A182 Gr. 316L
   vi. Ratings of fitting/pipe: 3000 lb.

3. **Air Line**: All instrument air supply lines shall be provided to each pneumatic instrument/valve/consumer, etc., through an air filter regulator (AFR). These connections shall be done through a copper tube ASTM B88/B75. Minimum thickness for such a copper tube shall be 1.6 mm, and normally it shall be coated with PVC. For purging/cleaning purposes, the service air line shall be used for various enclosures. Normally, the AFRs will have a 5-μm filter and shall be suitable to withstand the pressure of 10 Bar. These AFRs normally have 50 mm pressure gauges and a bottom drain/blowdown port. Normally for air lines, the instrumentation installation point starts from the isolation ½″ valve in the main air header or subheader, within 2–3 m from the instruments/pneumatic devices. All internal pneumatic distribution shall be done in instrumentation scope. For example, for a control valve, one air line point with an isolation valve may be provided by others from there air lines for I/P converter, Valve positioner may have to be arranged in instrumentation scope. Also, interconnections by copper tubes from the positioner to the control valve are done under instrumentation scope. For such air line distribution, a ½″ hot-dipped, galvanized mild steel pipe may be used. Fittings could be ASTM A 234Gr.WPB.

3 MECHANICAL INSTALLATION OF INSTRUMENTS

3.1 General

Right from the planning stage or design stage, it is recommended to properly plan the complete installation of the instrumentation systems for the entire plant, so that there is uniform interchangeability and a lower inventory is required during normal running of the plant. As discussed earlier, there could be a number of packages under which the instruments will be supplied and installed. Naturally, if not planned earlier, there will be the possibility that there may not be any standardization of installation as well as many different kinds of fittings and accessories (hence more inventory). Therefore it is better to attach instrumentation installation practices and specifications as a part of the specification for all the packages so that there will be uniformity. The impulse line is an important item in the mechanical installation of instruments. A few common points one needs to decide before going any further on installation diagrams are:

1. There could be tubes and pipes. Both can be used for the same purpose. Generally, tubes are preferred to pipes in small diameter applications, that is, an impulse line. One major difference between tubes and pipes is that a tube is not threaded (*as the wall thickness is too thin to hold a thread*) at the end to form a connection; instead, tube fittings are used. The major advantages of tubes over pipes are as follows:
   - **Bending quality**: is very nice—stronger with a thin wall, making it easy for bending.
   - **Stronger**: Because there is no thread at the end, it is stronger with no weak section.
   - **Less turbulence**: Smooth bends hence a streamlined flow.
   - **Economy in space and weight**: Better bending, smaller OD, and less weight for fittings.
   - **Flexibility**: Less rigid, hence immune to the transmission of vibrations between the connections when compared with a pipe.
   - **Fewer fittings**: Fewer fittings are necessary, for example, no bent is necessary. Also all fittings act as union.
   - **Less leakage**: When properly fitted, there will be no leakage. It is seen that a high-pressure tube may burst, but the fittings are properly fitted and hardly leak.

- **Good look**: On account of good consistency, a smoother contour, and fewer fittings, it looks better, especially when a number of tubes are placed parallel.
- **Cleaner fabrication**: As there is no thread, there is no chance of foreign material nor a need for sealing.
- **Easier assembly and disassembly.**
- **Less maintenance.**

However, as tubes are costlier when compared to pipes, from a cost consideration many may prefer an impulse pipe. Whatever it may be, the same philosophy need to be adopted for the entire plant unless there is a genuine reason behind any exceptions. As discussed in previous sections, the material selection is important for the impulse line. Keeping this in view, the entire impulse pipe has been considered in the field side, up to the LIE. Inside the enclosure, an SS tube has been considered so that it can cover almost all applications. This is not sacrosanct, but has instead been chosen to economize the cost and to limit the variation in designing the LIE.

2. Another important aspect of the installation diagram is the demarcation between the scope division between the mechanical and instrument engineering section. This scope division is normally done in such a way that the progress of installation work of one branch does not affect another. These instrument devices are installed in place only after mechanical installation and testing activities (hydraulic testing of power and process piping, steam blowing, etc.) are more or less over. These are more so for in-line devices such as various flow elements (Venturi/nozzle/orifice), Control Valves, Online Flow meter, etc. In fact, for all in-line devices, matching connection details (say, matching flanges), end-to-end dimensions, etc., are necessary before actual installation of these instrumentation devices in the pipe. Mechanical contractors use matching spool pipe to complete various mechanical activities discussed above, and later the actual instrumentation devices are installed. Here, a **good coordination** between the mechanical and instrument disciplines is extremely important. For example, when a Venturi is to be installed in the air duct, *precise dimension* matching is absolutely necessary. The same is the case for a steam flow nozzle in a critical pipe, a piece of which (the spool piece) needs *to be sent to the flow element manufacturer's* works for positioning the flow nozzle. Normally, the terminal point of the instrumentation terminal point is the root valve, that is, the instrumentation work starts with suitable fittings (or welding at the root valve). Mechanical scope shall include the stub, nipple (matching with application), and root valve. Root valve sizes may be 1″/3/4″, etc., so, any reducer as necessary (after root valve) shall be in the instrumentation scope. Now, in cases where there is no root valve, for example, as in flue gas, normally the Y pipe that is welded to the flue gas duct/furnace wall is kept complete with a necessary screwed cap (may be fitted with a chain), as shown in Fig. 12.4. The inclined Y pipe is provided so that from the straight part, a pocking rod can be inserted as necessary to remove clogging. In case of temperature element installations, the stub is left with the blanking plate so that the mechanical activities can be completed (without the actual thermowell). Then the thermowell is inserted so as to avoid damage due to mechanical activity. All these discussions are based on standard practices only and there could be some exceptions that need to be sorted out by the owner through proper coordination and mutual agreement between the engineers of different disciplines.
3. The source connection details discussed in Clause 1 are very important and need to be prepared and circulated with all mechanical and other packages for implementation so that at the time of installation, there is no mismatch. Normally, a ½″ globe valve with socket weld connections is used as the root valve. This is for *guidance* only; another type as per the owner's choice could also be used. However, consistency should be maintained throughout the plant.
4. Each owner adopts specific standards regarding the number of root valves and the same needs to be followed during design. As discussed earlier, normally double root valves are recommended for a process pressure $>40\,kg/cm^2$ and in steam services (may be at lower than $40\,kg/cm^2$). Also, in case of drain of high-pressure lines where double root valves have been used, double drain valves need to be considered for line draining. All these are done as preventive action against leakage.
5. All impulse lines shall be duly coated with color as per the prevalent standard applicable, but such coatings must not be flammable or combustible. All impulse lines shall have free movement to take care of thermal expansion, etc. For preventing steam from reaching the instruments, pig tail/siphon, CHU, etc., are used. In some installations, in place of CHU, a temperature equalization column is used.
6. Bulkhead fittings shown in installation diagrams are the points through which impulse lines enter the LIE. Bulkhead fittings have been chosen for both end tube connections, but they can also be chosen as the **interface between** the **pipe and tube** (to eliminate pipe to tube union). For example, a bulkhead male or a bulkhead female in place of a bulkhead union used in the drawings. All fittings and tubes **inside** the **LIE have been chosen as SS**, so that these may be used for all applications. However, other suitable materials as discussed in Clause 2 of this chapter may also be chosen.

Because the valve manifold is normally inside the LIE, it has been chosen to also be SS. To avoid more fittings, etc., transmitters with integral valve manifolds have been recommended. These are *recommendations only*.

7. In the installation diagram discussed later, it will be found that in all cases, line drains have been used, so instead of a five-valve manifold, a three-valve manifold would suffice.
8. In most of the cases, a main instrument air header or major subheader is provided at two or three places in the boiler, turbine area, offsite, and other places by the mechanical group. From these headers, branch distribution lines are drawn under the instrumentation scope of work. Normally, a 2″ GI pipe is used as the main subheader with a 1″ GI pipe as the branch lines. Normally, corresponding to each group of consumption points, there shall be one dedicated tapping point in the air line complete with a suitable isolation valve so that from there, a pneumatic line can be drawn. In addition to these, there may be purge lines for various measurements in the mill and flue gas system. Also, there could be purge air lines for the LIE.
9. Purging process connections are used when there is a chance of the line getting plugged due to dust and/or freezing. Normally, a medium of fluid is to be used, which does not interfere with the process medium. Air purging (as a part of cleaning the line) in case of flue gas, mill, and a fly ash system is quite common in power plants. Care shall be taken to see that this purging does not introduce errors in measurement. In case of a freezing problem, the purging line shall be properly traced. The purging connections meant cleaning is for intermittent purging and normally arranged with flexible pipes/hose and quick connect fittings. But there are certain measurements such as DP across the mill or the level in fly ash bins where continuous air purging is necessary to carry out the measurement. In such cases, rotameters and DP regulators are used to purge the required quantity of fluid. Normally, purge connections are kept near the process connection or near the instrument for long runs of the impulse line. Normally, check valves and shut off valves are used for purge connections.
10. Heat tracing may be required for impulse lines/instruments for highly viscous services (namely, HFO), or vapors that have the chance to be frozen. Each of these applications needs to be studied to determine the type and degree of heat tracing necessary. The types of heat tracing can be categorized as: self-limiting electric heat tracing up to 120°C, steam tracing $<100°C$, and constant resistance electric tracing for temperature-controlled applications.
11. Impulse line preparation: Impulse lines are laid on a flat smooth surface and shall be straightened by hand without pulling or heating to keep the inside diameter unchanged. Only the required length should be uncoiled as repeated operation may bring about kinks or damage. Bending tools or hacksaws/tube cutters with a sharp edge shall be used for bending and cutting of the tube. All impulse line cuts are squared off with fine files, etc. Minimum bending radius for the tube shall be 2¼ OD (for tube $OD < 12.5\,mm$) and three times (for tube $OD > 12.5\,mm$).
12. Impulse line joinings and fittings installation is done as per ANSI B31.1.
13. Welding procedure and performance qualifications are normally done as per ANSI B31.1, Section IX, at the time of qualification. Filler materials, etc., to be used are also as per ANSI B31.1, Section II.
14. For pipe threads, tape sealants could be used in clean air, water (liquid), and steam services up to 230°C. Other sealants can be used as per the manufacturer's recommendations. These are not applicable for tubes.
15. Hydrostatic and Pneumatic Testing: Normally, the ANSI B31.1 standard is referred to as the testing method for hydrostatic and pneumatic testing procedures for impulse lines. All impulse lines, after the erection work is over, are tested for the ability to withstand 1.5 times the maximum pressure. During such tests, the instruments are kept isolated with the help of an instrument shut-off valve. Similarly, the air lines are tested by the pressure-decay method, as per ANSI B 31.1.
16. The salient point of each of the installation diagrams has been discussed below, as the diagrams are self-explanatory and need not be repeated. Also for Bill of material relevant drawing need to be consulted as these are not again repeated in write up.

3.2 Pressure Instrument Installations (Figs. 12.12–12.17)

The entire pressure instrument installations have been categorized as a local pressure instrument, for example, local gauges/switches, pressure instruments, namely pressure switches, and transmitters mounted in the LIE/LIR and DP instruments (LIE/LIR).

3.2.1 *Pressure Gauge Installation (Fig. 12.12)*

1. Steam service—the instrument above the source point (Fig. 12.12A):
   a. This is for steam service, hence the tapping has been shown from the top of the pipe.
   b. A pipe union has been used to make pipe fixing easier.

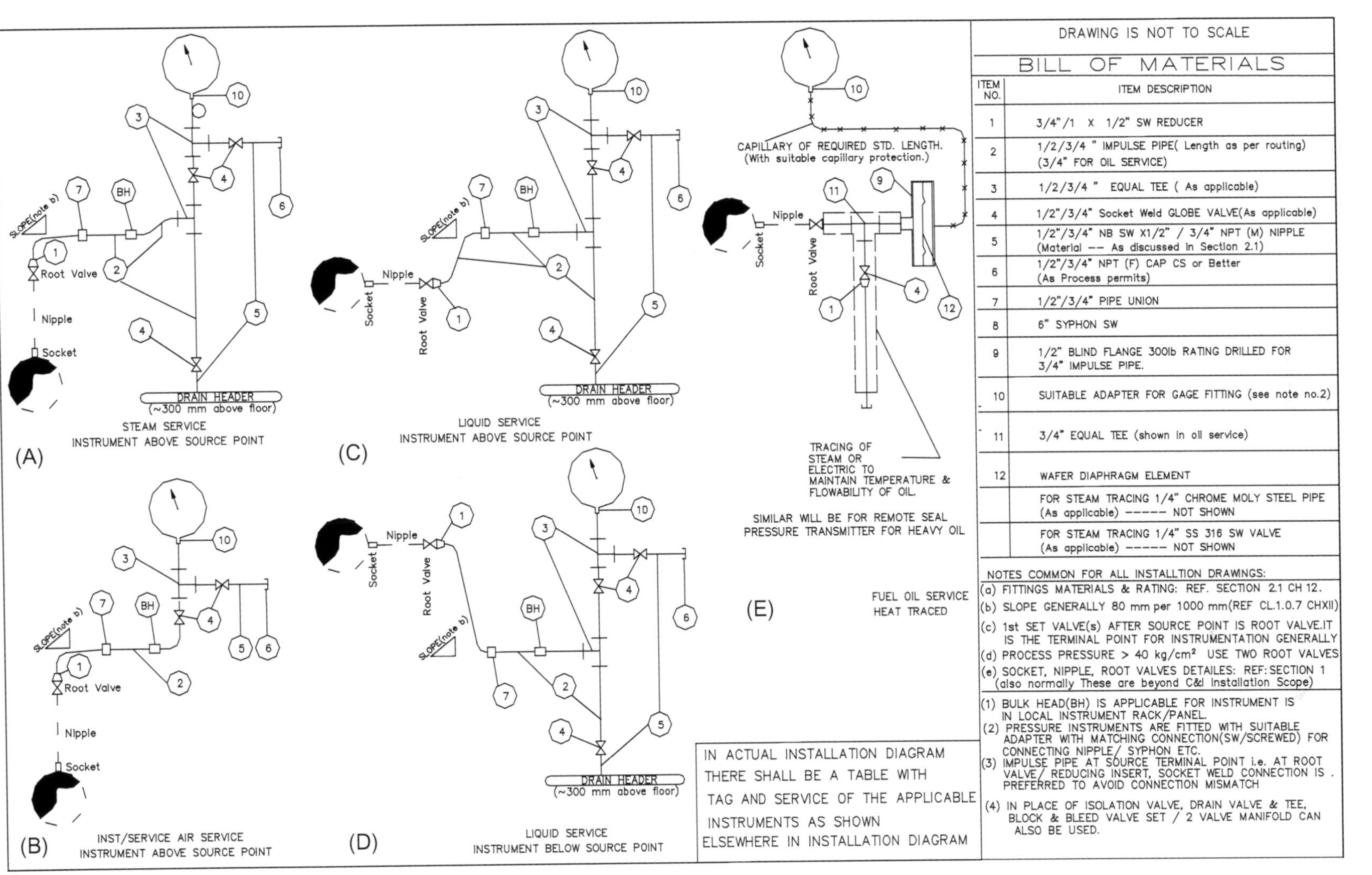

FIG. 12.12 Local pressure instrument installation 1.

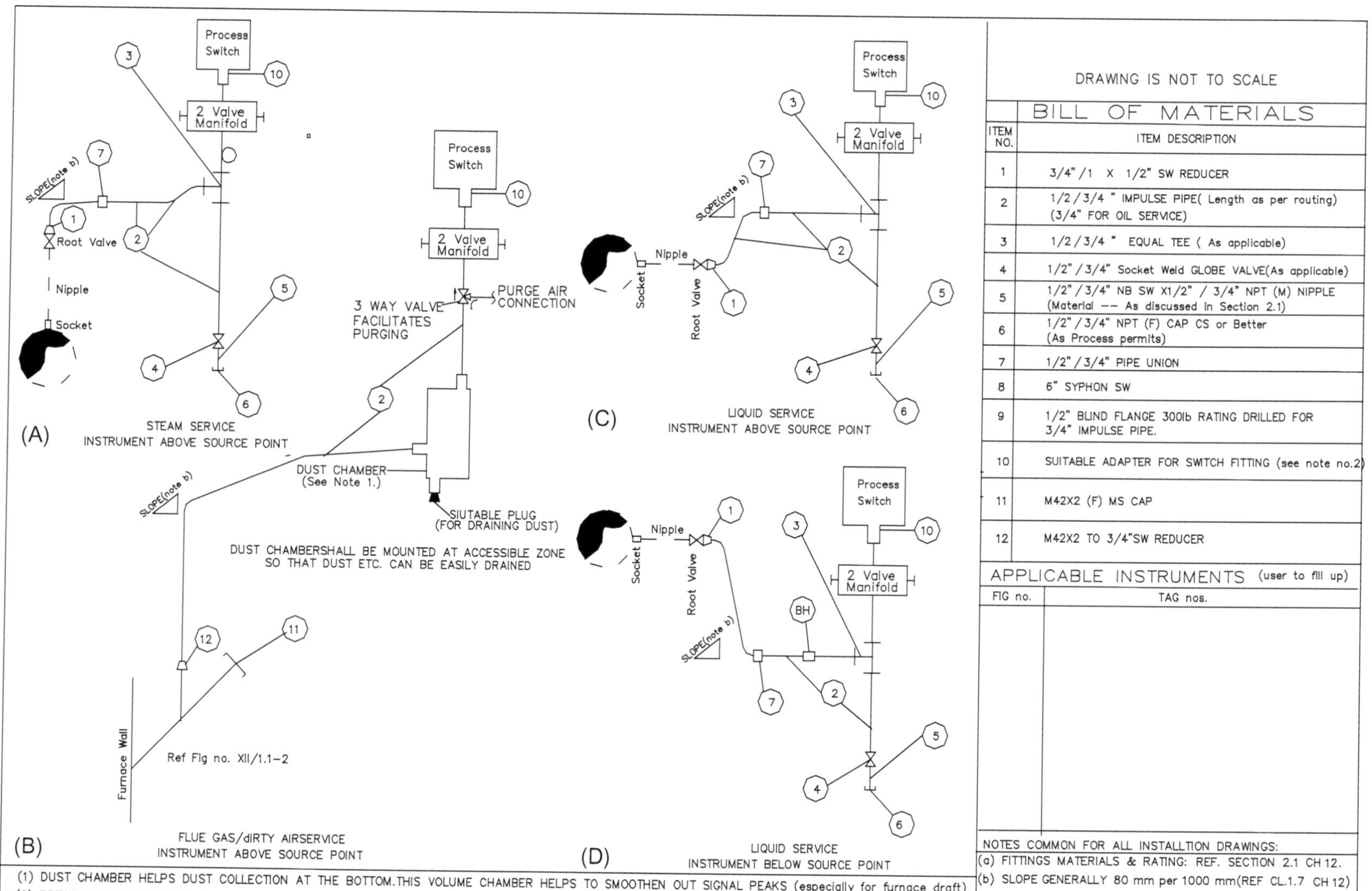

FIG. 12.13 Local pressure instrument installation 2 (Pr. switch).

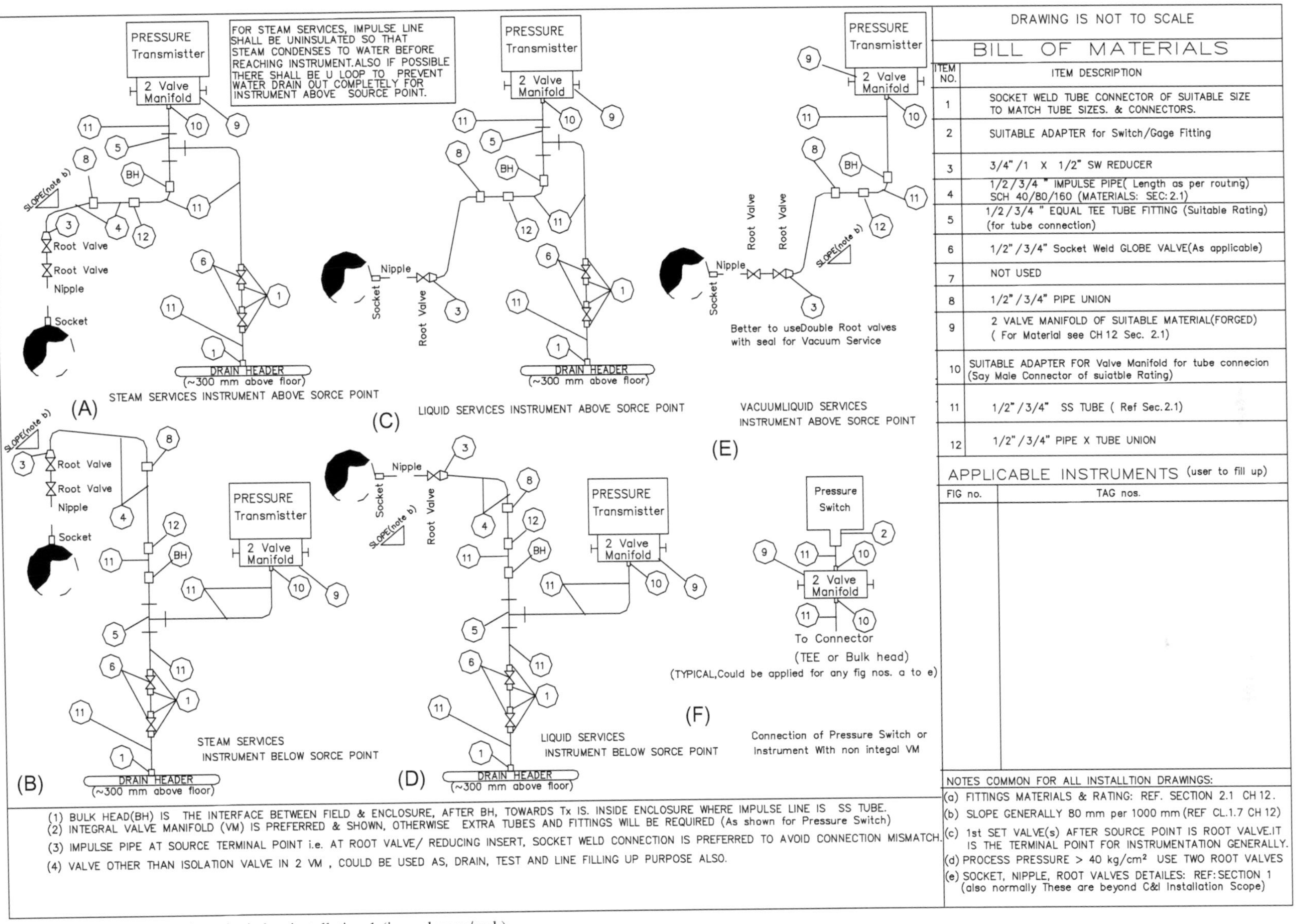

FIG. 12.14 Pressure transmitters/switches installation 1 (in enclosure/rack).

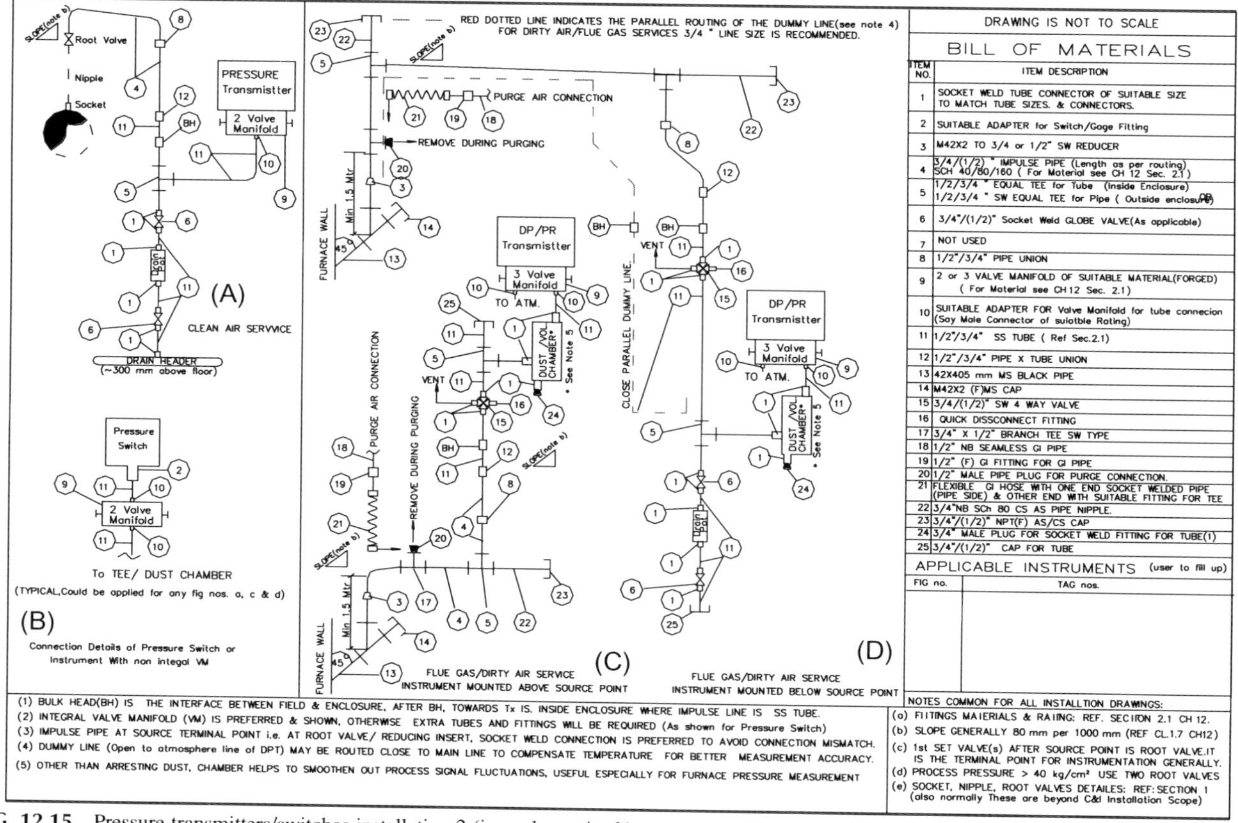

DRAWING IS NOT TO SCALE
BILL OF MATERIALS
ITEM NO. ITEM DESCRIPTION
1 SOCKET WELD TUBE CONNECTOR OF SUITABLE SIZE TO MATCH TUBE SIZES. & CONNECTORS.
2 SUITABLE ADAPTER for Switch/Gage Fitting
3 M42X2 TO 3/4 or 1/2" SW REDUCER
4 3/4/(1/2) " IMPULSE PIPE (Length as per routing) SCH 40/80/160 (For Material see CH 12 Sec. 2.1)
5 1/2/3/4 " EQUAL TEE for Tube (Inside Enclosure) 1/2/3/4 " SW EQUAL TEE for Pipe (Outside enclosure)
6 3/4"/(1/2)" Socket Weld GLOBE VALVE(As applicable)
7 NOT USED
8 1/2"/3/4" PIPE UNION
9 2 or 3 VALVE MANIFOLD OF SUITABLE MATERIAL(FORGED) (For Material see CH 12 Sec. 2.1)
10 SUITABLE ADAPTER FOR Valve Manifold for tube connecion (Say Male Connector of suiatble Rating)
11 1/2"/3/4" SS TUBE (Ref Sec.2.1)
12 1/2"/3/4" PIPE X TUBE UNION
13 42X405 mm MS BLACK PIPE
14 M42X2 (F)MS CAP
15 3/4/(1/2)" SW 4 WAY VALVE
16 QUICK DISSCONNECT FITTING
17 3/4" X 1/2" BRANCH TEE SW TYPE
18 1/2" NB SEAMLESS GI PIPE
19 1/2" (F) GI FITTING FOR GI PIPE
20 1/2" MALE PIPE PLUG FOR PURGE CONNECTION.
21 FLEXIBLE GI HOSE WITH ONE END SOCKET WELDED PIPE (PIPE SIDE) & OTHER END WITH SUITABLE FITTING FOR TEE
22 3/4"NB SCh 80 CS AS PIPE NIPPLE.
23 3/4"/(1/2)" NPT(F) AS/CS CAP
24 3/4" MALE PLUG FOR SOCKET WELD FITTING FOR TUBE(1)
25 3/4"/(1/2)" CAP FOR TUBE
APPLICABLE INSTRUMENTS (user to fill up)
FIG no. TAG nos.
NOTES COMMON FOR ALL INSTALLTION DRAWINGS:
(a) FITTINGS MATERIALS & RATING: REF. SECTION 2.1 CH 12.
(b) SLOPE GENERALLY 80 mm per 1000 mm (REF CL.1.7 CH12)
(c) 1st SET VALVE(s) AFTER SOURCE POINT IS ROOT VALVE.IT IS THE TERMINAL POINT FOR INSTRUMENTATION GENERALLY.
(d) PROCESS PRESSURE > 40 kg/cm² USE TWO ROOT VALVES
(e) SOCKET, NIPPLE, ROOT VALVES DETAILES: REF:SECTION 1 (also normally These are beyond C&I Installation Scope)
Root Valve
Nipple
Socket
PRESSURE Transmistter
2 Valve Manifold
(A)
CLEAN AIR SERVICE
DRAIN HEADER (~300 mm above floor)
Pressure Switch
2 Valve Manifold
To TEE/ DUST CHAMBER
(TYPICAL,Could be applied for any fig nos. a, c & d)
(B)
Connection Details of Pressure Switch or Instrument With non integal VM
RED DOTTED LINE INDICATES THE PARALLEL ROUTING OF THE DUMMY LINE(see note 4) FOR DIRTY AIR/FLUE GAS SERVICES 3/4 " LINE SIZE IS RECOMMENDED.
PURGE AIR CONNECTION
REMOVE DURING PURGING
FURNACE WALL
Min 1.5 Mtr
DP/PR Transmitter
3 Valve Manifold
TO ATM.
DUST /VOL CHAMBER*
* See Note 5
VENT
PURGE AIR CONNECTION
REMOVE DURING PURGING
CLOSE PARALLEL DUMMY LINE
FLUE GAS/DIRTY AIR SERVICE INSTRUMENT MOUNTED ABOVE SOURCE POINT
(C)
FLUE GAS/DIRTY AIR SERVICE INSTRUMENT MOUNTED BELOW SOURCE POINT
(D)
(1) BULK HEAD(BH) IS THE INTERFACE BETWEEN FIELD & ENCLOSURE, AFTER BH, TOWARDS Tx IS. INSIDE ENCLOSURE WHERE IMPULSE LINE IS SS TUBE.
(2) INTEGRAL VALVE MANIFOLD (VM) IS PREFERRED & SHOWN, OTHERWISE EXTRA TUBES AND FITTINGS WILL BE REQUIRED (As shown for Pressure Switch)
(3) IMPULSE PIPE AT SOURCE TERMINAL POINT i.e. AT ROOT VALVE/ REDUCING INSERT, SOCKET WELD CONNECTION IS PREFERRED TO AVOID CONNECTION MISMATCH.
(4) DUMMY LINE (Open to atmosphere line of DPT) MAY BE ROUTED CLOSE TO MAIN LINE TO COMPENSATE TEMPERATURE FOR BETTER MEASUREMENT ACCURACY.
(5) OTHER THAN ARRESTING DUST, CHAMBER HELPS TO SMOOTHEN OUT PROCESS SIGNAL FLUCTUATIONS, USEFUL ESPECIALLY FOR FURNACE PRESSURE MEASUREMENT

FIG. 12.15 Pressure transmitters/switches installation 2 (in enclosure/rack).

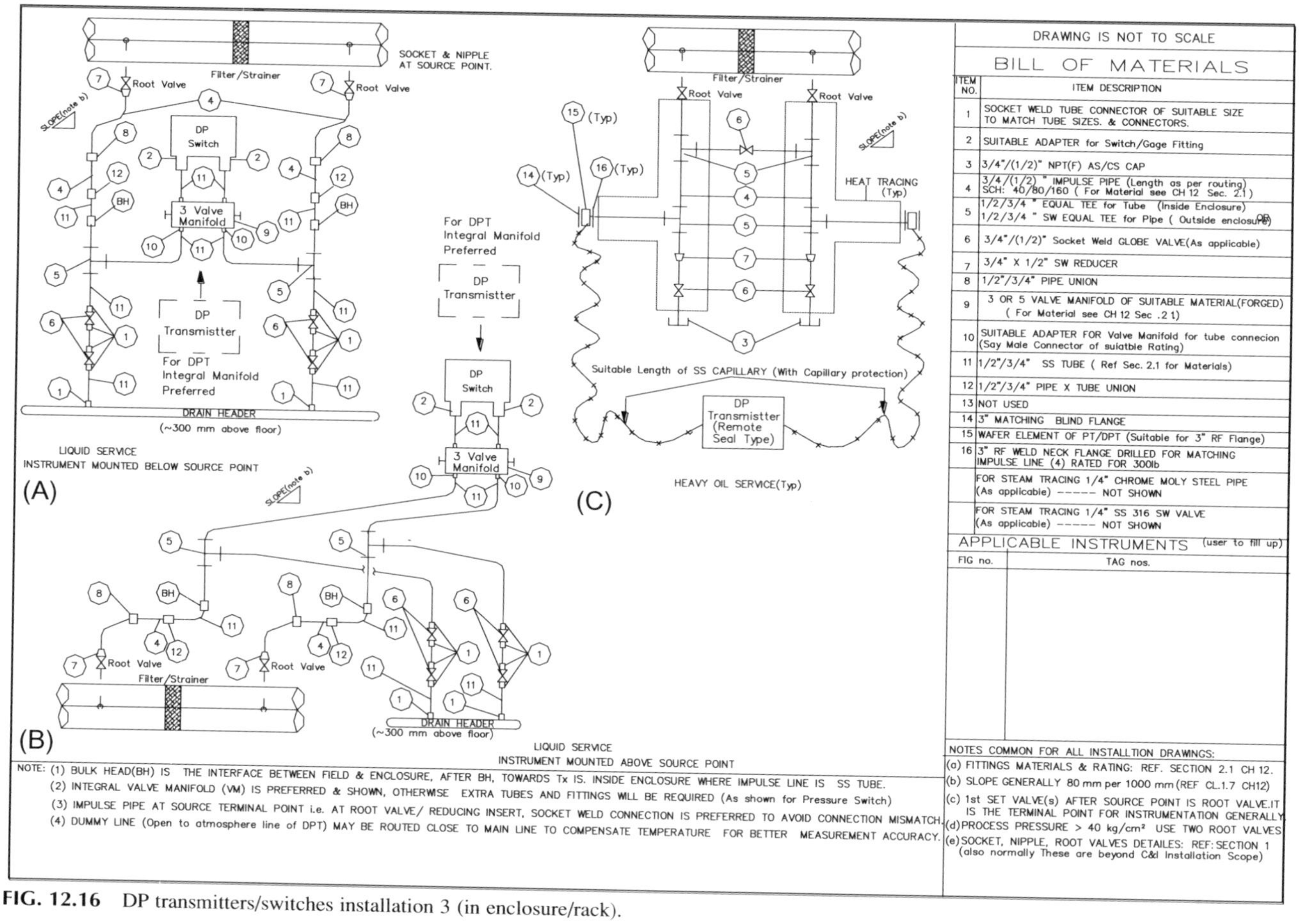

DRAWING IS NOT TO SCALE
BILL OF MATERIALS
ITEM NO. ITEM DESCRIPTION
1 SOCKET WELD TUBE CONNECTOR OF SUITABLE SIZE TO MATCH TUBE SIZES. & CONNECTORS.
2 SUITABLE ADAPTER for Switch/Gage Fitting
3 3/4"/(1/2)" NPT(F) AS/CS CAP
4 3/4/(1/2) " IMPULSE PIPE (Length as per routing) SCH: 40/80/160 (For Material see CH 12 Sec. 2.1)
5 1/2/3/4 " EQUAL TEE for Tube (Inside Enclosure) 1/2/3/4 " SW EQUAL TEE for Pipe (Outside enclosure)
6 3/4"/(1/2)" Socket Weld GLOBE VALVE(As applicable)
7 3/4" X 1/2" SW REDUCER
8 1/2"/3/4" PIPE UNION
9 3 OR 5 VALVE MANIFOLD OF SUITABLE MATERIAL(FORGED) (For Material see CH 12 Sec .2 1)
10 SUITABLE ADAPTER FOR Valve Manifold for tube connecion (Say Male Connector of suiatble Rating)
11 1/2"/3/4" SS TUBE (Ref Sec. 2.1 for Materials)
12 1/2"/3/4" PIPE X TUBE UNION
13 NOT USED
14 3" MATCHING BLIND FLANGE
15 WAFER ELEMENT OF PT/DPT (Suitable for 3" RF Flange)
16 3" RF WELD NECK FLANGE DRILLED FOR MATCHING IMPULSE LINE (4) RATED FOR 300lb
FOR STEAM TRACING 1/4" CHROME MOLY STEEL PIPE (As applicable) ----- NOT SHOWN
FOR STEAM TRACING 1/4" SS 316 SW VALVE (As applicable) ----- NOT SHOWN
APPLICABLE INSTRUMENTS (user to fill up)
FIG no. TAG nos.
NOTES COMMON FOR ALL INSTALLTION DRAWINGS:
(a) FITTINGS MATERIALS & RATING: REF. SECTION 2.1 CH 12.
(b) SLOPE GENERALLY 80 mm per 1000 mm (REF CL.1.7 CH12)
(c) 1st SET VALVE(s) AFTER SOURCE POINT IS ROOT VALVE.IT IS THE TERMINAL POINT FOR INSTRUMENTATION GENERALLY
(d) PROCESS PRESSURE > 40 kg/cm² USE TWO ROOT VALVES
(e) SOCKET, NIPPLE, ROOT VALVES DETAILES: REF:SECTION 1 (also normally These are beyond C&I Installation Scope)
SOCKET & NIPPLE AT SOURCE POINT.
Filter/Strainer
Root Valve
DP Switch
3 Valve Manifold
DP Transmitter
For DPT Integral Manifold Preferred
DRAIN HEADER (~300 mm above floor)
LIQUID SERVICE INSTRUMENT MOUNTED BELOW SOURCE POINT
(A)
For DPT Integral Manifold Preferred
DP Transmitter
DP Switch
3 Valve Manifold
HEAT TRACING (Typ)
Suitable Length of SS CAPILLARY (With Capillary protection)
DP Transmitter (Remote Seal Type)
HEAVY OIL SERVICE(Typ)
(C)
(B)
LIQUID SERVICE INSTRUMENT MOUNTED ABOVE SOURCE POINT
NOTE: (1) BULK HEAD(BH) IS THE INTERFACE BETWEEN FIELD & ENCLOSURE, AFTER BH, TOWARDS Tx IS. INSIDE ENCLOSURE WHERE IMPULSE LINE IS SS TUBE.
(2) INTEGRAL VALVE MANIFOLD (VM) IS PREFERRED & SHOWN, OTHERWISE EXTRA TUBES AND FITTINGS WILL BE REQUIRED (As shown for Pressure Switch)
(3) IMPULSE PIPE AT SOURCE TERMINAL POINT i.e. AT ROOT VALVE/ REDUCING INSERT, SOCKET WELD CONNECTION IS PREFERRED TO AVOID CONNECTION MISMATCH.
(4) DUMMY LINE (Open to atmosphere line of DPT) MAY BE ROUTED CLOSE TO MAIN LINE TO COMPENSATE TEMPERATURE FOR BETTER MEASUREMENT ACCURACY.

FIG. 12.16 DP transmitters/switches installation 3 (in enclosure/rack).

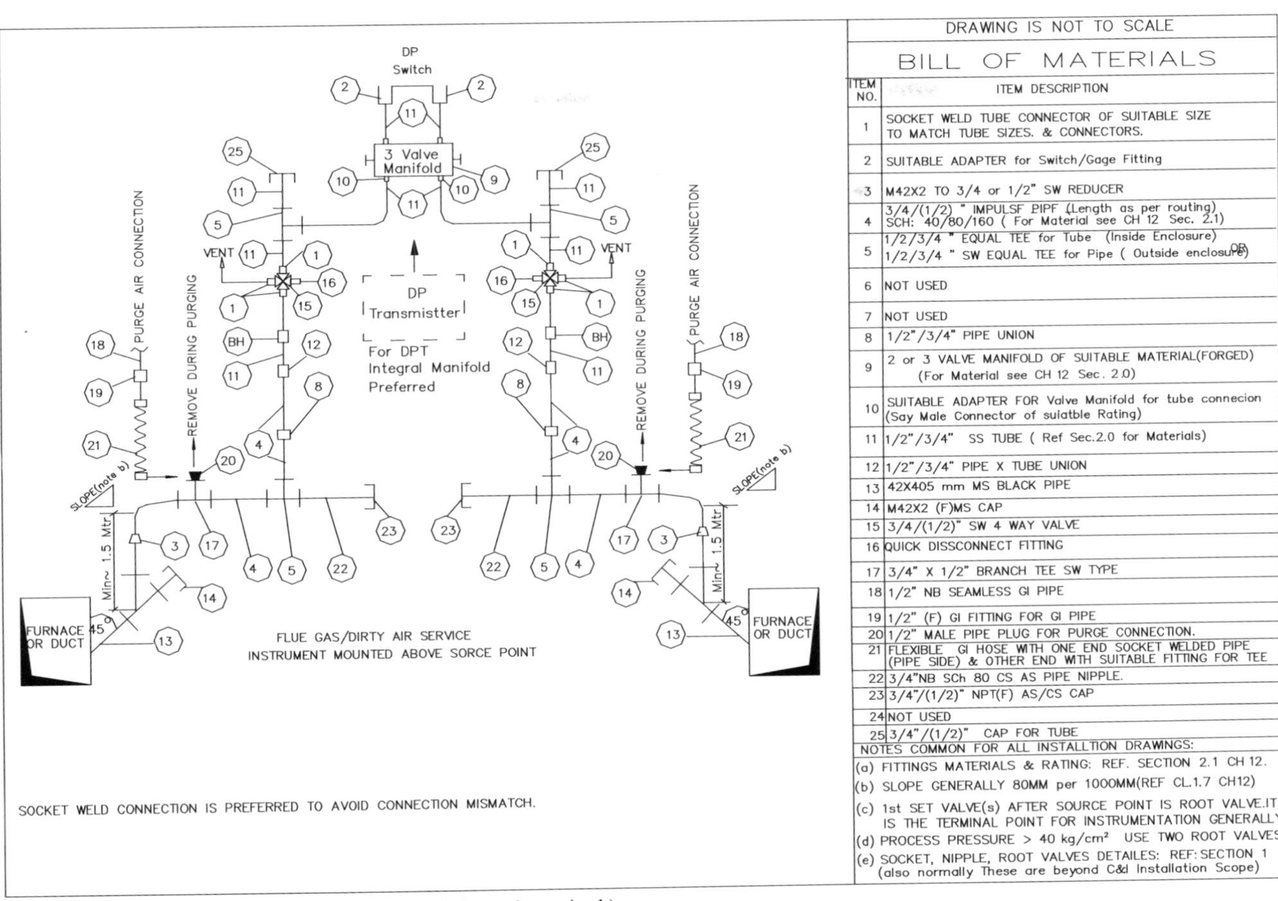

FIG. 12.17 DP transmitters/switches installation 4 (in enclosure/rack).

c. A bulkhead fitting is shown for local gauge board installation possibilities.

d. Attention is drawn toward the two valves connected to one TEE just ahead of the siphon. The purposes of the two valves are: one is for isolation (to isolate the line from the gauge) and the other one is for testing of the gauge, draining from the gauge, etc. Now, in place of these two valves, a block and bleed valve in one block or a two-valve manifold can be used. A simple block and bleed valve can be a better choice for pressure gauges. This is applicable for all the pressure gauge installations discussed below.

e. There is a line drain connection shown with a valve. This is required when the gauge is first put into service (may be after a long period, the same is not in operation), the line can be drained to clear dirty water. In case of a pressure gauge mounted in a local gauge board/panel, then the line draining can be done at a common header to avoid flooding of the floor.

f. Because the distance between the tapping and the instrument may not be long and the instrument is taken above the tapping point, a pigtail or a syphon is a must so as to ensure that the steam does not reach the instrument at all times.

2. Air service—instrument above the source point (Fig. 12.12B):

a. This is for air service, hence the tapping has been shown from the top of the pipe.

b. Clause nos. 3.2.1b–d are applicable here also.

3. Liquid service—instrument *above/below* the source point (Fig. 12.12C and D):

a. This is for liquid service, hence the tapping has been shown from the side of the pipe.

b. Clause nos. 3.2.1b–e are applicable here also.

4. Fuel oil (FO) service—instrument is with a diaphragm seal (Fig. 12.12E):

a. This is for liquid service, hence the tapping has been shown from the side of the pipe.

b. For heavy fuel oil service for high viscosity, the entire line, including the line drain, needs to have heat tracing so that the flowability is maintained.

c. A diaphragm seal with a capillary filled with inert liquid (silicone oil) has been used so that the FO does not come in contact with the pressure gauge physically and the gauge is always responsive and not clogged due to the FO. This inert liquid actually transmits the FO pressure to the gauge.

3.2.2 Pressure Switch (Local) Installation (Fig. 12.13)

1. Steam service—instrument above the source point (Fig. 12.13A): Similar as in Clause no. 3.2.1.1 above.
2. Flue gas/dirty air services: instrument above the source point (Fig. 12.13B):
   a. As indicated earlier, here tapping is given from the top part of the inclined Y pipe so that through the straight part, pocking action can be initiated as necessary. This end is normally kept closed with the help of the screwed cap, as shown.
   b. A dust chamber has been shown; this is helpful in two ways. First, it helps dust to be collected in it. Second, it also acts as a volume chamber that (like the capacitor in an electrical circuit) helps to dampen the fluctuation in the measurement, i.e., acts as a low-pass filter). This is very helpful in arresting the fluctuations in furnace pressure measurement, which is very fluctuating in nature. From time to time, dust collected at the chamber can be cleaned from the bottom.
   c. A three-way valve with a purge air connection has been kept to purge the line by operation of the three-way valve and isolating the instrument. The dust can be returned to the system with the help of purge air.
3. Liquid service—instrument above and below the source point (Fig. 12.13C and D):
   a. Same as Clause no. 3.2.1.3a.
   b. Clause nos. 3.2.1b–e are applicable here also.

3.2.3 Pressure Transmitter/Switch (in LIE/LIR) Installation 1 (Fig. 12.14)

It is common in power plants that the transmitters/switches are grouped together and placed in the LIE/LIR. However, in many offsite plants, the transmitters are *not* grouped but are installed individually. In the installation diagram referred to above for transmitters/switches in LIE/LIR has been shown, if they are **not** in LIR/LIE also installation will be same with *probable changes* like: Bulkhead may not be applicable, Drain Header may not be there instead drain line may be capped, but while draining care should be taken to ensure that the place is not flooded.

Item 1 (SW tube connector), 11(impulse tube) may not be applicable if impulse pipe is used and equal TEE could be SW* type when used with pipe (*SW = Socket Weld).

1. Steam service—instrument above and below the source point: (Fig. 12.14A and B):
   a. This is for steam service, hence the tapping has been taken and shown from the top of the pipe.
   b. Pipe union has been used so that pipe fixing is easier.
   c. Bulkhead fitting is shown for LIE/LIR mounting.
   d. Attention is drawn toward item 12 meant for a pipe-to-tube interface, which is used to have a tube connection just prior to the LIE/LIR so that all tube fittings could be used inside the LIE/LIR, including the bulkhead. However, this can also be avoided by having bulkhead male/bulkhead female connector. In these diagrams, the tube bulkhead union has been used.
   e. All fittings inside the LIE/LIR are SS tube fittings. Because of this, SW tube connectors have been used to have same SW valves (which can be used in pipes directly) in all cases and to have less inventory.
   f. Integral valve manifolds have been shown for the transmitter (as transmitter connections are generally universal and a matching manifold is easily available), but it could be separate, as shown for switches in Fig. 12.14F. In that case, extra fittings will be necessary for mounting of the manifold with instruments.
   g. Connector to the valve manifold could be a male tube connector also; fittings for the switch shall have to be matching with the switch connection available (but this could be made matching while specifying a **standard** switch connection for the plant).
   h. Drain header in the LIE/LIR shall be common, as shown in Fig. 12.15.
   i. In case of transmitters when the distance of the tapping point to the instruments is not very short, then a pigtail siphon could be avoided. However, if the transmitter is located above or near the source point, as a safety measure prior to transmitter connection, one U loop in the impulse tube could be done also.
   j. It is preferred that in steam services and liquid services, transmitters are kept below the source point, as in Fig. 12.14B (refer Clause no. 1.9a).
2. Liquid service—instrument above and below the source point (Fig. 12.14C and D):
   a. This is for liquid service, hence the tapping has been taken and shown from the side of the pipe.
   b. Clause nos. 3.2.3.1b–h above are applicable here also.
   c. Same as Clause no. 3.2.3.1j above.
3. Vacuum service—instrument above the source point (Fig. 12.14E):
   a. This is for vacuum service, hence the tapping has been taken and shown from the top of the pipe.
   b. As a safety measure against leakage, double root valves could be used as these valves must be highly leakproof (may be high-pressure quality).

c. Clause nos. 3.2.3.1b–h above are applicable here also.
d. Same as Clause no. 3.2.3.1j above.

3.2.4 Pressure Transmitter/Switch (in LIE/LIR) Installation 2 (Fig. 12.15)

Here, the basic things are similar to those discussed in Clause no. 3.2.3 above. Discussions on pressure measurements in clear air, flue gas, and dirty air are presented.

1. Pressure measurement in clean air (Fig. 12.15A):
   a. This is for vacuum service, hence the tapping should be from the top of the pipe.
   b. Clause nos. 3.2.3.1b–e and 3.2.3.1g, h above are applicable here also.
   c. Integral valve manifolds have been shown for a transmitter (as transmitter connections are generally universal and a matching manifold is easily available), but it could be separate as for corresponding switches, as shown in Fig. 12.15B. In that case, extra fittings will be necessary.
   d. Here, the line drain is collected through a drain pot having one valve each upstream and downstream of the drain pot, so that the water can be trapped at the drain pot for intermittent draining.
2. Pressure measurement in flue gas and dirty air (Fig. 12.15C and D):
   a. This is for gas service, hence the tapping should be from the top of the pipe.
   b. Clause nos. 3.2.3.1b–e and 3.2.3.1h above are applicable here also.
   c. Same as Clause no. 3.2.4.1c above.
   d. When the instrument is mounted below the tapping point (as in Fig. 12.15D), the line drain is collected through a drain pot having one valve each upstream and downstream of the drain pot, so that the dust collected can be drained intermittently.
   e. As discussed earlier, by removing cap 14, the line can be cleaned by a pocking rod.
   f. In order to purge the line thoroughly, a purge air connection at the field side with quick connect fittings has been kept at the first equal TEE division point (normally fitted with a plug). At the time of purging, the plug is removed and the air connection is fitted in.
   g. Also, instrument tapping has been shown to be taken after a straight length via another TEE, so that if necessary, this length can also be cleaned by a pocking rod. This is kept closed by cap 23.
   h. Inside the LIE/LIR, provisions have been kept with the help of four-way valve for air purging.
   i. As discussed in Clause no. 3.2.2.2b above, a dust chamber has been shown (mainly for flue gas measurement) to facilitate dust elimination and reduce fluctuations in measurement.
   j. Another interesting feature in this installation diagram is the dummy line (shown by a *dotted line*). In air and flue gas, pressure measurement is done by the DPT, keeping one port exposed to the atmosphere. The question is if the length of the impulse line is long, naturally the length of the gas column will also be long. On account of fluctuations in atmospheric temperature, the gas column pressure will also vary (even if the actual pressure to be measured does not vary) due to changes in density. To compensate for the same, a dummy line connected to the unused port of the transmitter is routed along with the main line so that both limbs have the same length of gas column at the same temperature. The dummy line is shown for fig. d but applicable for Fig. 12.15C and D also.

3.2.5 DP Transmitter/Switch (in LIE/LIR) Installation 3 (Fig. 12.16)

These are mainly used for DP across a strainer/filter and/or DP in a flue gas duct.

1. DP measurement for liquid services (Fig. 12.16A and B):
   a. This is for liquid service for DP measurement, hence two sets of tappings have been taken and shown from the side of the pipe.
   b. Clause nos. 3.2.3.1b–e and 3.2.1.3g, h above are applicable here also.
   c. Integral valve manifolds have been shown for the transmitter (as transmitter connections are generally universal and a matching manifold is easily available), but it could be separate for corresponding switches, as shown in associated figures. In those cases, extra fittings will be necessary.
2. DP measurement for heavy oil services (Fig. 12.12C)
   a. This is for liquid service for DP measurement, hence two sets of tappings have been taken and shown from the side of the pipe.
   b. For heavy fuel oil service for high viscosity, the entire line, including the line drain, needs to have heat tracing so that flowability is maintained.
   c. Remote diaphragm seal with capillary filled with inert liquid (say, silicone oil) has been used so that the FO does not come in contact with the pressure gauge physically and the gauge is always responsive and not clogged due to FO.
   d. Both remote seals (may be wafer type) are placed against a flange in each line, with the help of blind matching flanges, as shown.
   e. Capillary length shall be long enough to meet the requirement of installation. If there is extra length, it must be coiled but should not kink at any place. The capillary may have necessary capillary protection.

   f. It is recommended that remote seals (wafer) are suitable for the 3″ RF flange, hence matching flanges shall be selected.
   g. Even though line drains have been shown, precautions shall be taken when draining the line with high viscous slippery fluid (not to drain on the floor).

3.2.6 DP Transmitter/Switch (in LIE/LIR) Installation 4 (Fig. 12.17)

These are mainly used for DP across a heat exchanger in the flue gas path.

1. DP measurement for flue gas/dirty air service (Fig. 12.17).
   a. This is for gas service for DP measurements, hence two sets of tappings have been taken and shown from the top/upper part of the duct.
   b. Same as Clause no. 3.2.5.1c.
   c. Clause nos. 3.2.3.1b–e and 3.2.1.3g above are applicable here also.
   d. Clause nos. 3.2.4.2e–h are also applicable here.

3.3 Flow (DP Type) Instrument Installations (Figs. 12.18–12.20)

The flow measurements considered here under the installation diagram are all based on the restriction type, that is, with DPT. With other online flow measurements such as a magnetic flow meter, a mass flow meter, etc., one has to ensure matching flanges at both ends so that the meters can be installed, as discussed in Clause no. 3.1.2 above. Flow element types with applications are: **flow nozzle** *for steam and feed water* flow measurements (Fig. 12.18), an **orifice plate** for *condensate flow measurement* (Fig. 12.19), and a **Venturi** for *air flow* measurement (Fig. 12.20). In all these flow measuring system installations, instruments above as well as below the source points have been shown. Detailing has been done for flow elements in a horizontal pipe. Even when the flow elements are not in horizontal pipe installations, the details will be more or less the same, so they are not repeated. When the flow element is in a vertical pipe for a **steam flow,** *lower tapping shall be brought up to the upper tapping point* and then two lines shall be brought close to each other to the instrument. Similarly for **liquid flow** service, the *upper tapping shall be brought up to the lower tapping point* and then the two lines shall be brought close to each other to the instrument.

3.3.1 Flow Metering Installation for Flow Nozzle (Fig. 12.18) Steam and FW Service (Fig. 12.18A and B)

1. For steam service, the tapping has been shown from the top of the pipe while for liquid services, it shall be from the side, as shown separately.
2. Clause nos. 3.2.3.1b–h and 3.2.3.1j above are also applicable here.
3. A three-port **condensate pot** for **steam** application is essential when the instrument is located *above* the source point, and better to use even when the instrument is located below the source point.
4. In a majority of the cases, D and D/2 tappings are used in these applications.
5. On account of high-pressure applications, flow nozzles are used in most of the cases wherever flow elements are deployed.

3.3.2 Flow Metering Installation for Orifice Plate (Refer Fig. 12.19): Condensate Service (Fig. 12.19A and B)

1. For condensate service, the tapping has been taken and shown from the side.
2. Clause nos. 3.2.3.1b–h and 3.2.3.1j above are also applicable here.
3. A weld-neck flange orifice plate has been shown here.
4. In a majority of the cases, flange tappings are used in these applications.
5. Because this normally is not high pressure, a weld-neck orifice would be used.

3.3.3 Flow Metering Installation for Venturi (Fig. 12.20): Air Service (Fig. 12.20A and B)

1. For air service, the tapping has been taken and shown from the top.
2. Clause nos. 3.2.3.1b–h and 3.2.3.1j above are also applicable here.
3. Root valves may not be necessary, hence they are not shown in the drawing.
4. When the instrument is mounted below the tapping point (as in Fig. 12.20B), the line drain is collected through a drain pot having one valve each upstream and downstream of the drain pot, so that the dust collected can be drained intermittently.
5. Provisions have been made inside the LIE/LIR for purging the line with air with the help of a four-way valve.
6. Venturi forms a part of the duct.

3.4 Level Instrument Installations (Figs. 12.21–12.23)

Various level instrument installations have been discussed. These instruments include DP instruments with and/or without a suitable seal, local level gauges (gauge glass) and float/displacer instruments, etc. The following level instrument installations have been considered:

- Closed/pressurized vessel level measurement with a DP transmitter (Fig. 12.21).

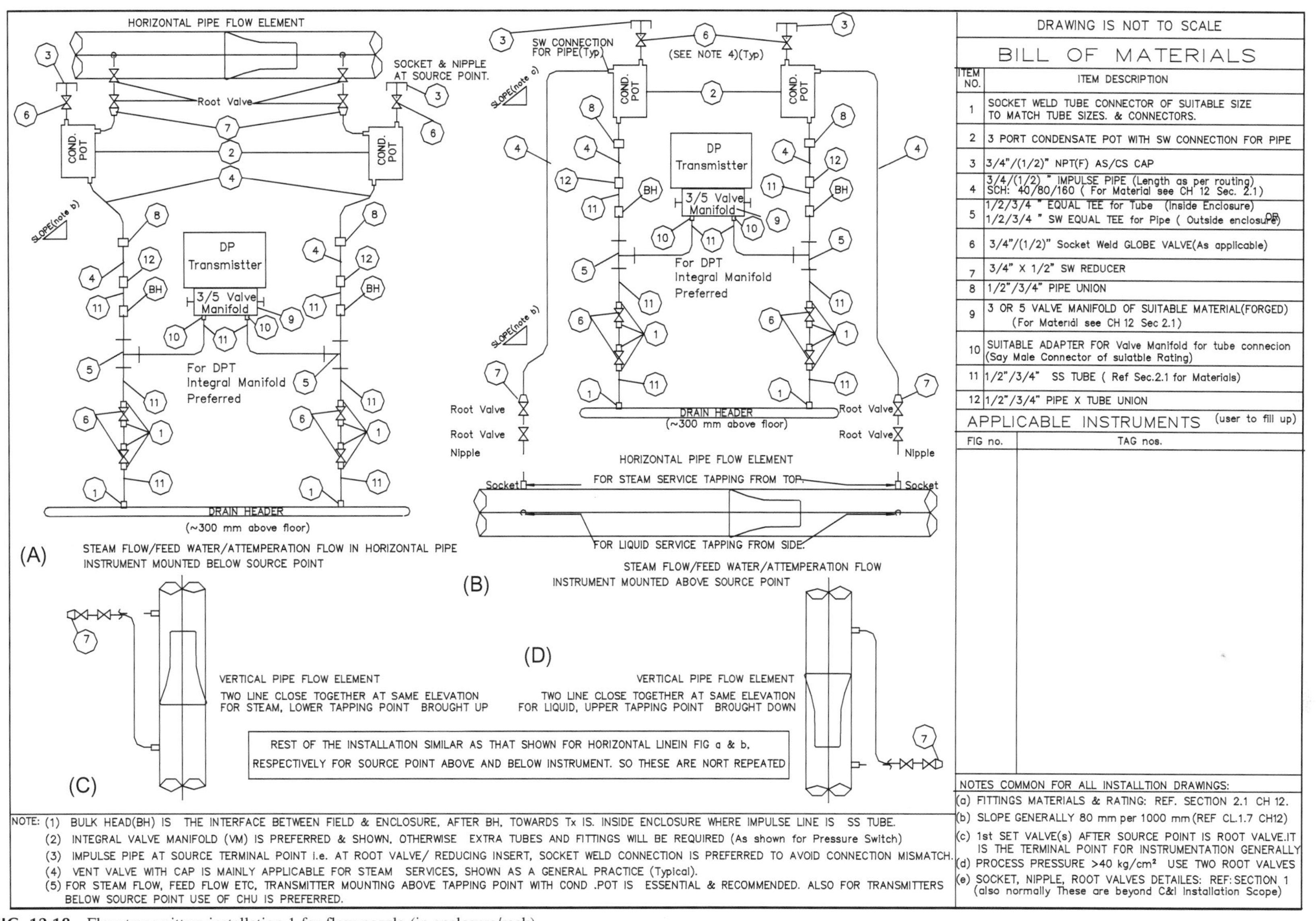

FIG. 12.18 Flow transmitters installation 1 for flow nozzle (in enclosure/rack).

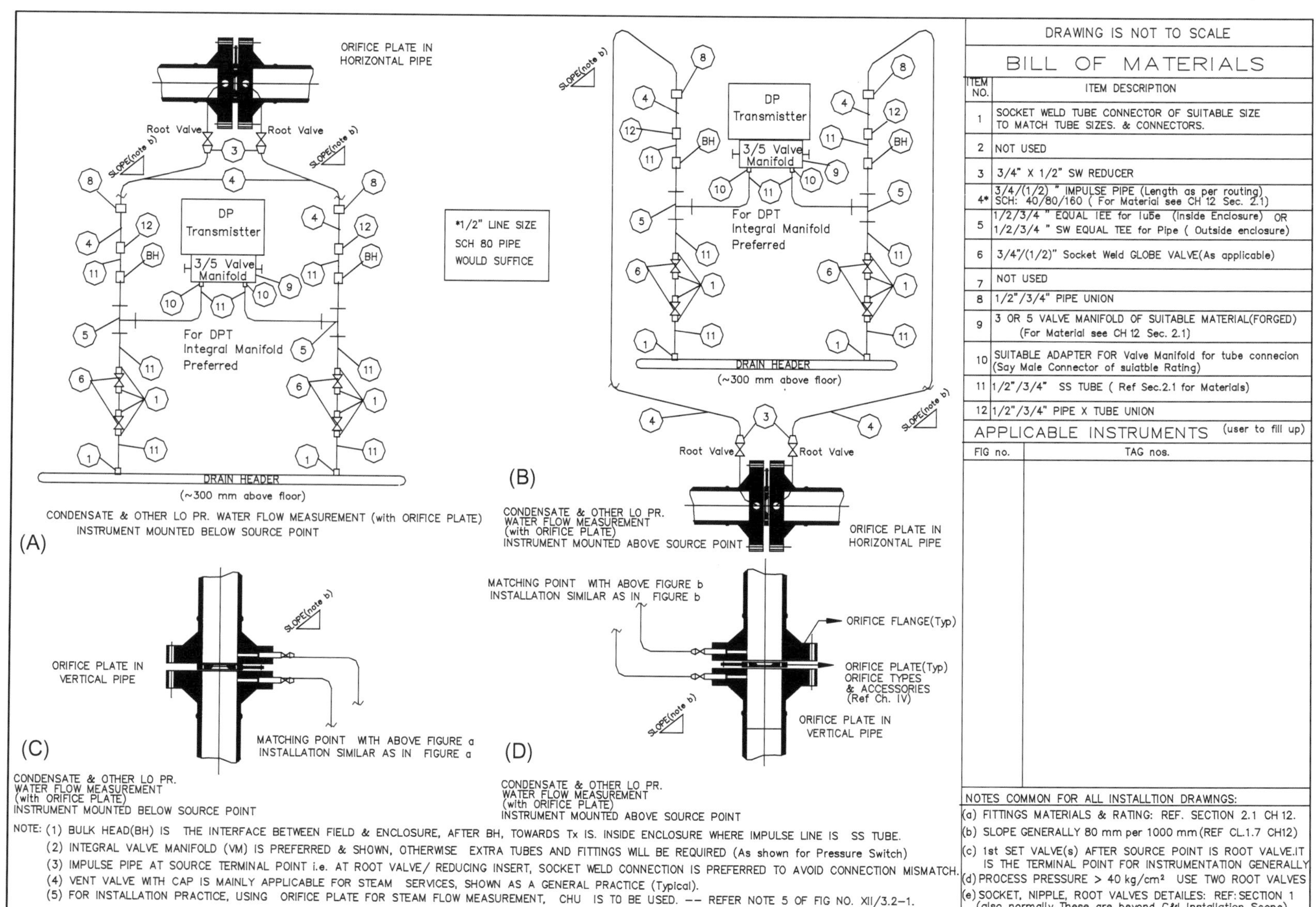

FIG. 12.19 Flow transmitters installation 2 for orifice plate (in enclosure/rack).

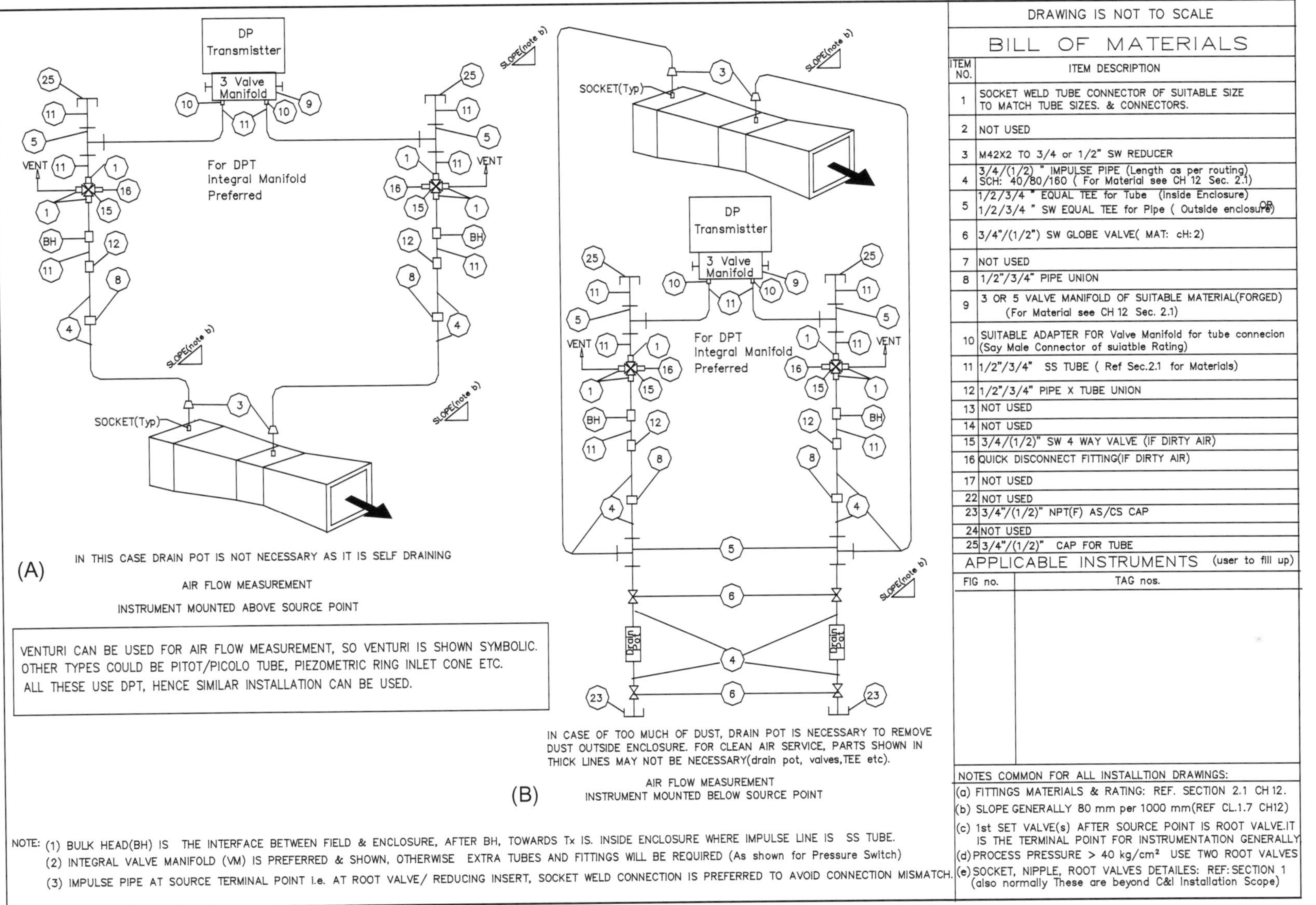

FIG. 12.20 Flow transmitters (in enclosure/rack) installation 3.

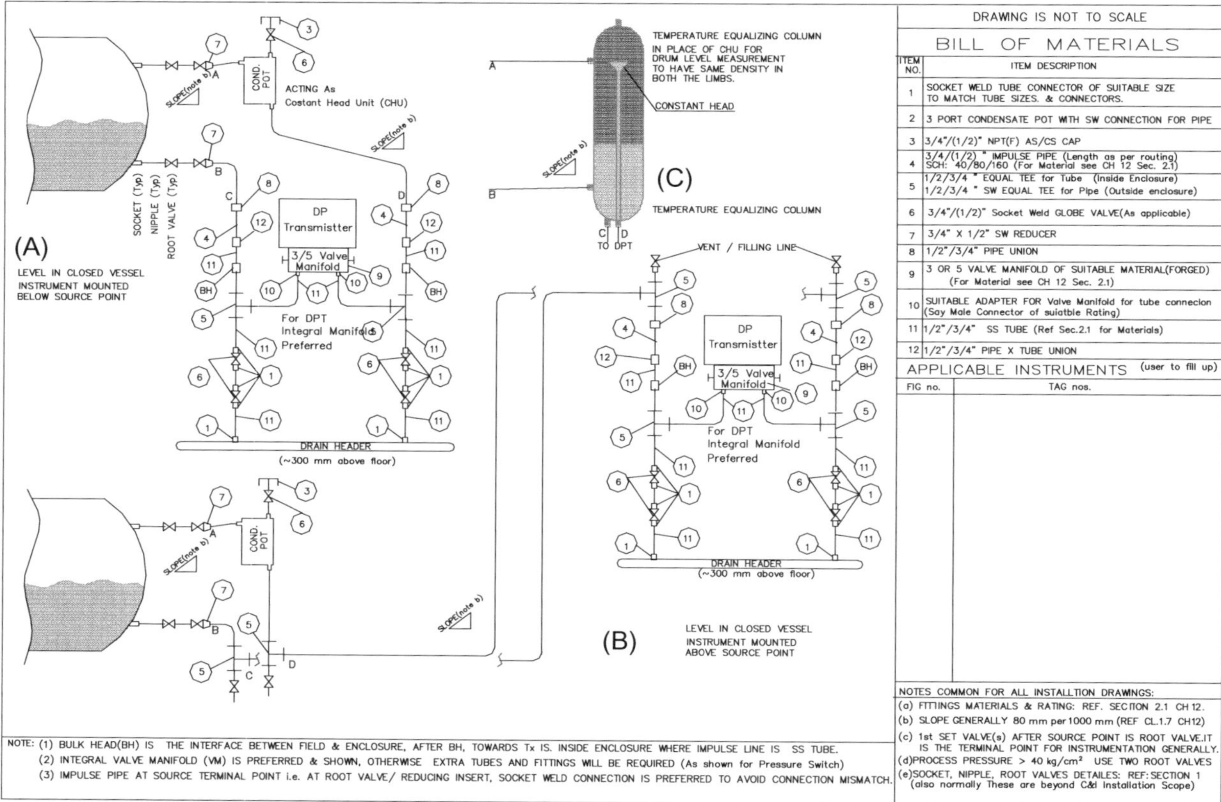

FIG. 12.21 Level measurement in closed chamber (noncorrosive/viscous).

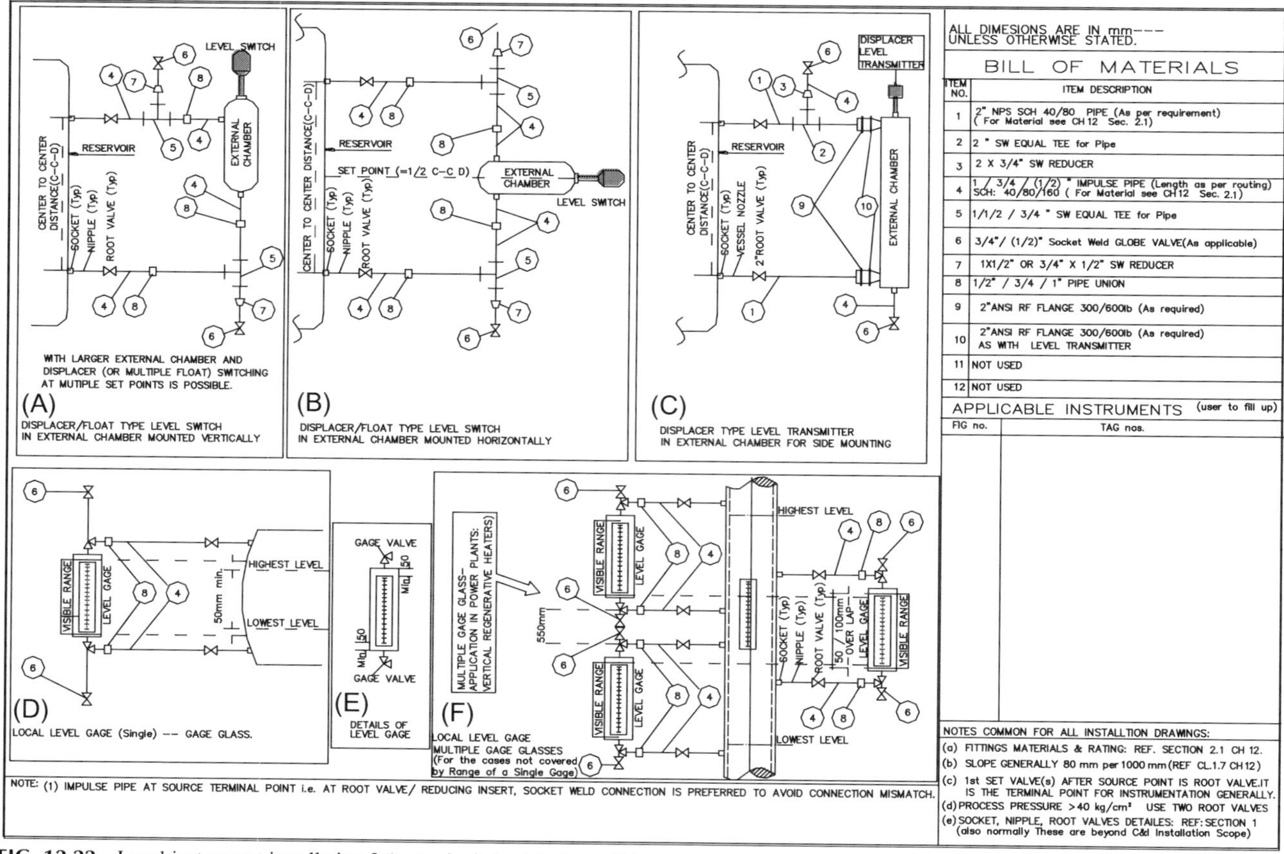

FIG. 12.22 Level instrument installation 2 (gauge/switches/displacer Tx).

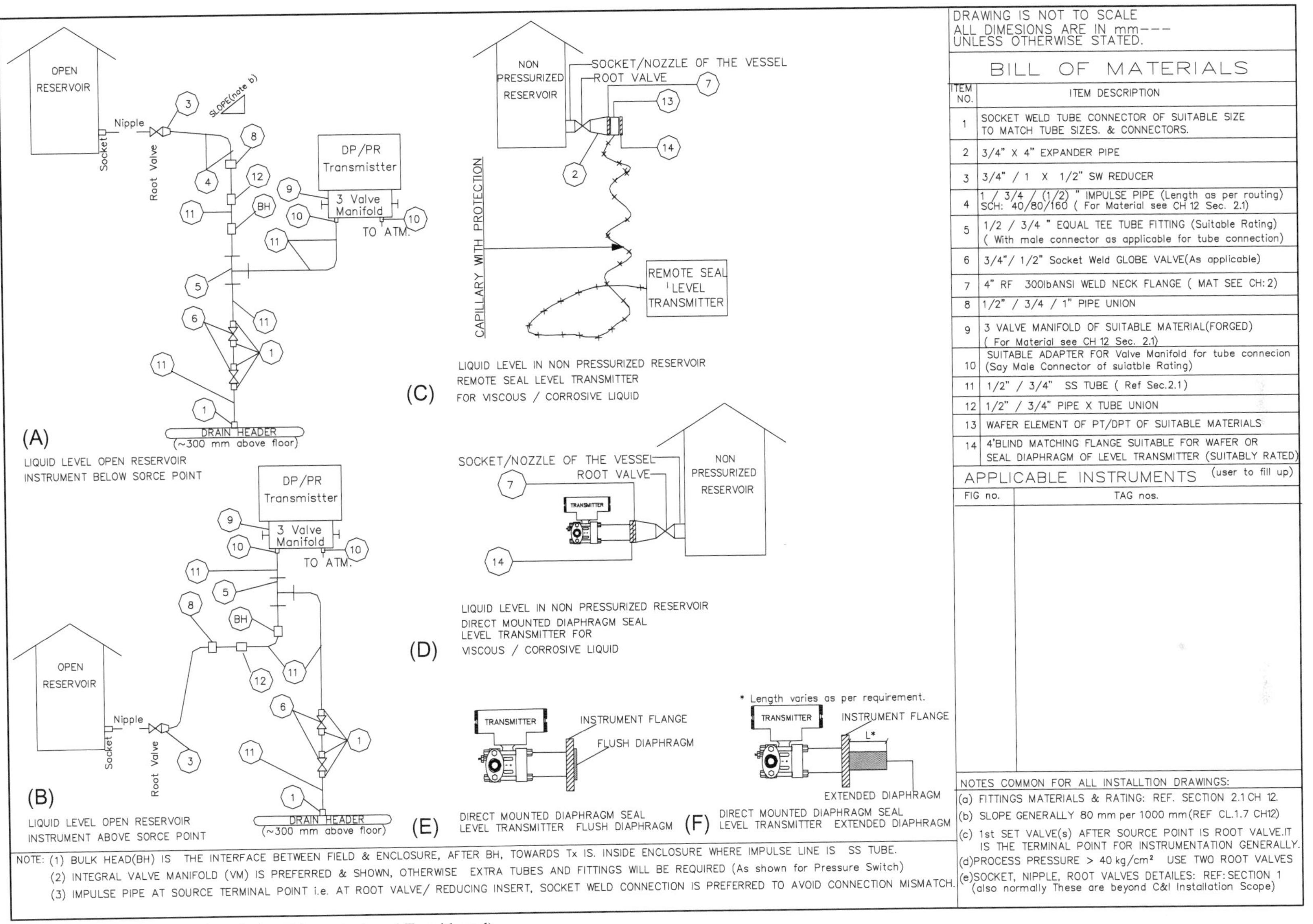

FIG. 12.23 Level instrument installation 3 (open tank and Tx with seal).

- Open/nonpressurized vessel level measurement with displacer/float instruments and level gauges (gauge glass) (Fig. 12.22).
- Open/nonpressurized vessel level measurement with DP transmitter with/without an instrument seal (Fig. 12.23).

3.4.1 Level Instrument Installation for Closed/ Pressurized Vessel (Fig. 12.21)

Drum level (for a drum boiler) and deaerator level measurements are examples of such categories of level measurements. The vessel is pressurized with steam at the top of the vessel with pressure P. There are two tappings: one (T_1) at the top above the maximum possible water level and the other (T_2) is from the lower side, well below the allowable low level. These two tappings are connected by an impulse line to the two sides of the DP transmitter. The side connected to T_1 will have pressure $=P$ whereas the other tapping will have pressure $=P+h\rho g$. Where $h=$ level in the vessel, $\rho=$ density of the medium, and $g=$ acceleration due to gravity. Therefore, it is seen that a difference in pressure is proportional to h in case ρg does not vary. Now g at a place is constant and if there is no variation in ρ, then the level can be measured by the DP method. If there is a chance of variation in ρ, as in the case of a drum level, necessary compensation is provided (detailed in boiler level control, Chapter 8). Another point worth noting is that $T_2-T_1=h\rho g$. So far this is ok. The connection from the steam part is not directly fed to the transmitter, instead passing through the constant head unit (CHU) so that there is no possibility of steam reaching the transmitter. Also, in one side a constant head is maintained and at the other end, head varies with level. Normally, these transmitters are located below the tapping point, naturally tapping from top T_1 will have more head than that of T_2. As the level rises, the difference will be narrower, hence in normal cases the DPT output will be less. But in reality it is found that the *O/P* is proportional to the level. How is this done? This is normally done by a full span suppression detailed in Chapter 4 and/or the standard book on instrumentation may be referenced. As stated earlier, the CHU has to maintain constant head, but if more steam in the space above the point of CHU (where T_1 is connected) condenses, then it will try to increase the level. This is not possible because the CHU connected to the vessel will downward tilt toward the vessel, so that the spillover liquid goes back to the vessel to maintain constant head. There is another important point: if the two legs of the DPT are not taken close to each other and exposed, there could be a difference in density ρ in the two limbs and there may be an inaccuracy in the reading. For drum level measurements, instead of a condensate pot, a temperature equalizing column, as shown in Fig. 12.21C, may be used. Here, the central limb with a cup at the top acts as the constant head unit. The top connection is at the steam space. When the steam comes inside the temperature equalizing column, it condenses. If more steam condenses, then the cup will spill and would try to increase level outside. Because this part is connected to the water part of the vessel, then it will ultimately go to the main vessel due to the downward slope toward the vessel. Here, because the water limb in the constant head part i.e., the central limb) and the water in the main part are at the same temperature, there is the same temperature in both limbs. In most of the cases, a temperature equalizing column is used.

1. Clause nos. 3.2.3.1b–h and 3.2.3.1j above are also applicable here.
2. A three-port condensate pot is used as the constant head unit.
3. It is recommended as well as in practice that most of these level instruments are located below the tapping point. However, if at any case when instruments are to be located above the source point, then in those cases the elevation rise in both the limbs should be made the same, as shown in Fig. 12.21B, to avoid inaccuracy in the level reading

3.4.2 Open/Nonpressurized Vessel Level Measurement With Displacer/Float Instruments and Level Gauges (Gauge Glass) (Fig. 12.22)

For instrument details on gauge glass, displacers and float instruments have been discussed in Chapter 4 and are not repeated here. Because the vessel is an open type, a single tapping would suffice but two tappings are taken for the mechanical stability of the installation. The lower limb connection of the external chamber is always chosen at a point that is lower than the required set point so as to ensure that the level in the external chamber tallies with the same in the vessel. Center-to-center distance of tappings is chosen in such a way that the level set point falls within that at the same time the instrument installation is stable. However, the center-to-center distance of tapping is of great importance for side-mounted level switches (* 2b of the figure under reference).

A few common points worth noting are as follows:

- External chamber for level switches/transmitters not only helps with easier maintenance of the instruments, but also allows the use of isolation valves so that the entire system can be isolated from the process.
- Normally, the vent and drain connections shown help entrapped bubbles to escape as well as drain the line whenever required, especially when the instrument has not been in use for a long time. Normally, for level transmitters, the connections are flanged, as shown in Fig. 12.22C.
- The drawing under reference is self explanatory so only a few pertinent points are presented.

1. Fig. 12.22A for a top-mounted level switch.
   - **a.** External chamber level and that in the vessel are the same within the range.
   - **b.** Single or multiple floats can be inserted to get switching actions at varied levels. However, all such set points shall be within the length of the external chamber, that is, the length of the external chamber to be chosen to cover the entire range of set points.
   - **c.** Displacer instruments can also be used for switching actions.
2. Fig. 12.22B for a side-mounted level switch.
   - **a.** Basic installation is similar as above, only here these types of installations are applicable for single set point float instruments.
   - **b.** In these installations, the tapping points are so chosen that the set point lies at the middle of the center-to-center distance of the tappings.
3. Fig. 12.22C for top-mounted level transmitter.
   - **a.** Tappings for the external chambers shall be so chosen that it covers the entire range of level for the vessel.
   - **b.** For transmitters, only one external chamber will have a single displacer instrument mounted from the top.
4. Fig. 12.22D and E for level gauge glass.
   - **a.** Normally, gauge glasses are supplied with the gauge valve check valve, as shown in the drawing under reference.
   - **b.** In many cases, these gauge glasses are supplied and erected as a part of the vessel.
   - **c.** Vent and drain valves have been provided to allow venting and draining.
   - **d.** The drawing shows a single gauge glass for the vessel, so naturally the range must cover the highest and lowest levels for the vessel.
5. Fig. 12.22E and F for multiple level gauge glasses.
   - **a.** In cases where the height of the vessel is too long (vertical FW heaters), multiple gauge glasses are used with suitable overlap.
   - **b.** Same as Clause nos. 3.4.2.4a–c.
   - **c.** Vent and drain valves have been provided to allow venting and draining.

3.4.3 Open/Nonpressurized Vessel Level Measurement With DP Transmitter With/Without an Instrument Seal (Fig. 12.23)

Because this is a nonpressurized tank (may be with the roof also), hence the level can be easily measured by a simple pressure transmitter (or rather by a DP transmitter having one leg open to the atmosphere), depending on the desired span of measurement. There could be two kinds of situations, one where for noncorrosive nonviscous liquid ordinary Pr./DP transmitter can be used whereas for corrosive (say Strong Acid Cat Ion tank for DM Plant—by other types of instrument such as capacitance are used more) or Viscous fluid such as Heavy fuel oil, etc. Pr./DP transmitters are used with suitable seals.

1. Transmitters without seal: As discussed earlier, the transmitter could be a normal pressure or DPT, and the installation details will be the same discussed in Clause no. 3.2.5.1 above and as shown in Fig. 12.23A and B (in case of DPT, one end shall be kept open to the atmosphere).
2. Transmitter with seal: In these types of measurements, the tanks are provided with a suitable nozzle fitted with the necessary root valve of sizes ¾″ to 1″. Normally, in order to have better response and accuracy, transmitter seals with min. 3″ diameters are chosen. So, after the root valve, a suitable expander complete with 3″ or 4″ flange (RF type preferable) of required rating (300 lb) is used.
   - **a.** For a remote wafer seal, the wafer is placed between the expander flange and another matching blind flange of the required pressure rating. Pressure sensed by the remote seal is sent to the transmitter via a capillary filled with inert fluid. Capillary length is to be decided based on the requirements. All capillaries must have suitable mechanical protection. The transmitter may be located at a remote location, as shown in Fig. 12.23C.
   - **b.** Other types of seals could be flush or extended. In either case, instruments come with seals and flanges that directly fit into the flange of the expander, as shown in Fig. 12.23E and F, for flush or extended seals, respectively.

3.5 Transmitter Mounting and Transmitter Enclosure

There are a number of ways the transmitter can be mounted. Sometimes, these are mounted on a stanchion or local pipe stand, etc. in power stations, normally the transmitters are grouped and put in a local open transmitter rack or enclosure, depending on whether the area is a closed space or an open area, respectively. Transmitters in the turbine hall may be placed in an open rack, the same as inside the turbine building, whereas the transmitters in the boiler platform or boiler area, which are basically open spaces, could be enclosures. However, open racks with a canopy are also seen in many boiler installations. It is recommended to have a transmitter enclosure in an open space so that the same is safe. In the subsequent sections, brief discussions about transmitter mounting details and transmitter enclosure details are presented. The discussions presented are *not* sacrosanct, but recommendations only. From a technoeconomic point of view, there could be some other solutions also

3.5.1 Transmitter Mounting

Almost all the present transmitters come with a mounting bracket. In the present discussions, a mounting bracket for a″ pipe mounting has been considered and shown on Fig. 12.24A and B. The transmitters are fixed with the mounting bracket, which in turn is clamped with a 2″ pipe. These transmitters can be fixed on horizontal as well as vertical pipes, depending on the convenience of installation. When a single transmitter is mounted on a pipe, it may be convenient and easy to mount on a vertical pipe, whereas it is convenient to mount the same on a horizontal pipe in case of a transmitter enclosure and/or transmitter rack. Based on the density of transmitters in one row of a transmitter enclosure and the design of the same, process entry could be from the side and from the front also. There is another factor: in many transmitters, the mounting in any plane is not permissible (even though in most of the current smart transmitters, such things may not be applicable). Fig. 12.24A and B shows these two types of pipe entries (and mounting on horizontal and vertical pipes). It is preferred to have the transmitter mounted in such a way that the local display/indicator of the transmitter is faced toward the front, so that the reading of the transmitter can be seen standing in front of the transmitter rack/enclosure.

3.5.2 Transmitter Enclosure/Rack

As stated earlier that the rack could be open and/or closed depending on the owner's design, naturally discussions here have been restricted to the structural part of the system. These open and closed enclosures are often referred to as local instrument racks (LIR) and local instrument enclosures (LIE) also. In both the cases, a suitable junction box is located for cable grouping, especially where the fieldbus or protocols are not used, so that a number of transmitters can be connected by a set of two wires. However, in case of transmitters connected by a fieldbus, the number of cables would be less and a small terminal box may suffice.

1. Normally, these are all a freestanding type constructed with a minimum 3″ angle or channel frame. In case of an open rack, there shall be a canopy to protect the instruments from water/rain as well as from falling objects in the plant. Normally, the canopies are extended both in the front as well as in the back side. These are made with a 2.5–3.5 mm thick sheet.
2. In case of an open rack, there are at least four vertical members that may be welded on a 3 or 6 mm thick horizontal plate at the four ends with reinforcing steel members, as typically shown in detail "A" in Fig. 12.24 for grouting.
3. In case of a closed transmitter enclosure, like any standard panel, this shall be mounted with the help of a 75 or 100 mm channel by grouting (similar to what has been shown in detail "A" in Fig. 12.24), or the channel may be welded to the metallic insert at the place. The enclosure is placed on the channel discussed above. In case of transmitter enclosure, an antivibration mounting or pad may be used also.
4. Closed transmitter enclosures have sheet metal work with a cold rolled, cold annealed (CRCA) sheet of thickness 1.6–2.0 mm. There may be a door at the front to see the meter reading, checking, etc. Also, there will be a door at the backside for maintenance purposes as well as for accessing the junction box discussed later.
5. The layout inside the LIR/LIE shall be done (as shown in Fig. 12.24C and D) in such a way so that the blowdown lines and impulse lines are easily accessible from the front. An integral valve manifold may give more space to work with inside the LIR/LIE.
6. There shall be a common drain line to which all drain lines from the transmitters would be connected. In many installations, it is necessary to have provisions for purge air. Suitable provisions for the same need to be arranged.
7. The transmitters are mounted on 2″ pipe, which in turn is supported on an angle/channel with the help of suitable clamps, as detailed in Fig. 12.24D.
8. A suitable junction box of required terminals is generally mounted at the back side of the LIR/LIE. Normally, these junction boxes should have an IP55 protection rating. Cable from the Junction box may be taken out with the help of a suitable pipe, through the cable gland at the gland plate at the bottom of the junction box, as detailed in Fig. 12.24D. In case of an LIE, the cable gland may be at the bottom of the LIE. As discussed earlier, fewer cables will be required if transmitters are used in smart mode, that is, a number of transmitters connected via a single cable. In the junction box, there may be provisions for power connection also.

4 ELECTRICAL INSTALLATION OF INSTRUMENTS

4.1 General

Proper functioning of DCS and instrumentation depends on a number of factors such as seamless system integration, selection of proper control strategy, reliability of system components as well as proper installation of electrical systems including proper cable connections, grounding, etc. Electrical installation of DCS and instrumentation mainly consists of electrical grounding, cabling, cable tray and conduits, junction boxes, marshaling rack, and selection of cable and specifications.

In power plants, there is equipment with different voltage levels. During electrical installation, suitable care must be taken so that there is no interferences. There will

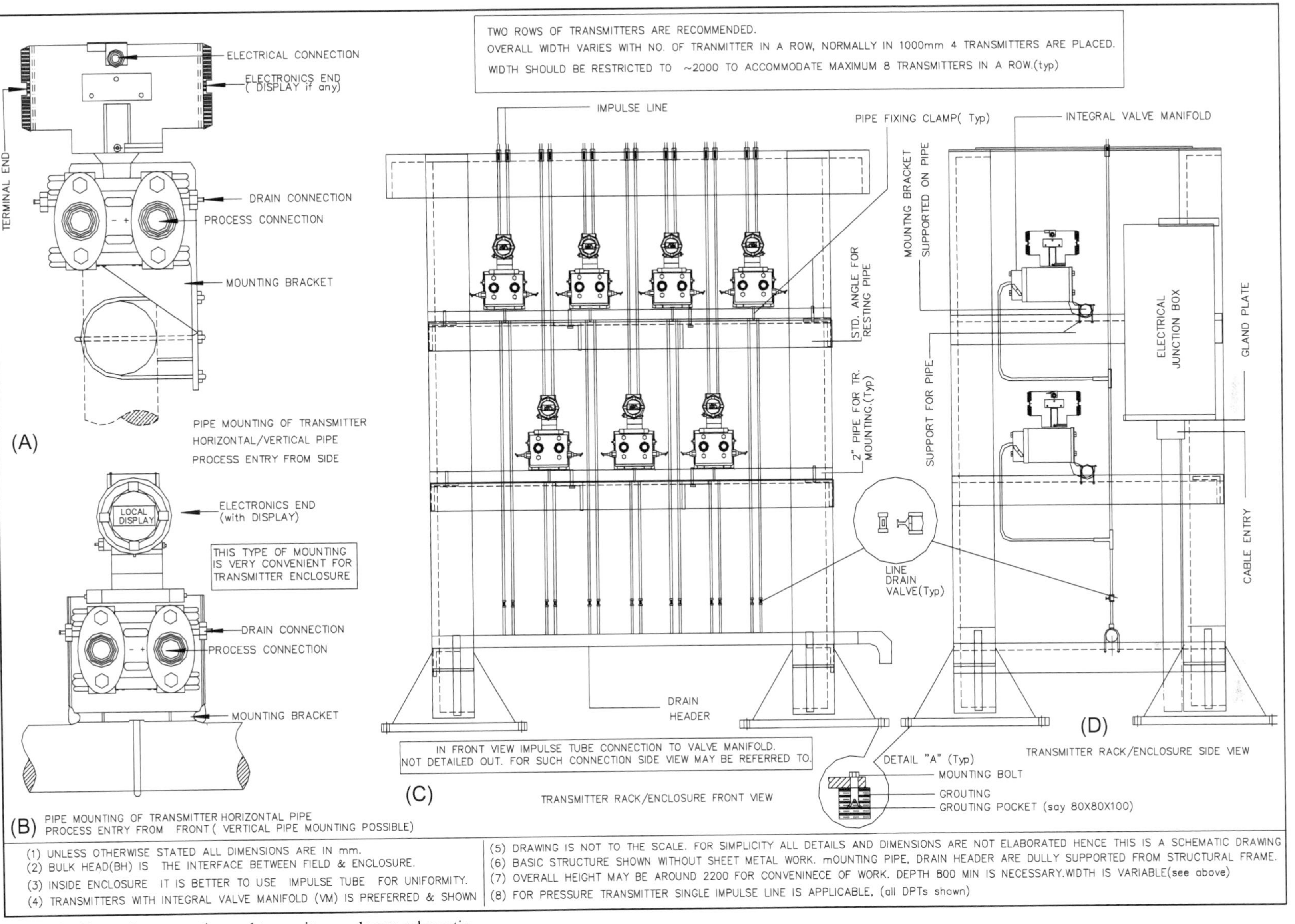
TWO ROWS OF TRANSMITTERS ARE RECOMMENDED.
OVERALL WIDTH VARIES WITH NO. OF TRANMITTER IN A ROW, NORMALLY IN 1000mm 4 TRANSMITTERS ARE PLACED.
WIDTH SHOULD BE RESTRICTED TO ~2000 TO ACCOMMODATE MAXIMUM 8 TRANSMITTERS IN A ROW.(typ)
ELECTRICAL CONNECTION
ELECTRONICS END (DISPLAY if any)
TERMINAL END
DRAIN CONNECTION
PROCESS CONNECTION
MOUNTING BRACKET
(A)
PIPE MOUNTING OF TRANSMITTER HORIZONTAL/VERTICAL PIPE PROCESS ENTRY FROM SIDE
LOCAL DISPLAY
ELECTRONICS END (with DISPLAY)
THIS TYPE OF MOUNTING IS VERY CONVENIENT FOR TRANSMITTER ENCLOSURE
DRAIN CONNECTION
PROCESS CONNECTION
MOUNTING BRACKET
(B) PIPE MOUNTING OF TRANSMITTER HORIZONTAL PIPE PROCESS ENTRY FROM FRONT (VERTICAL PIPE MOUNTING POSSIBLE)
IMPULSE LINE
PIPE FIXING CLAMP(Typ)
STD. ANGLE FOR RESTING PIPE
2" PIPE FOR TR. MOUNTING.(Typ)
MOUNTNG BRACKET SUPPORTED ON PIPE
SUPPORT FOR PIPE
LINE DRAIN VALVE(Typ)
DRAIN HEADER
IN FRONT VIEW IMPULSE TUBE CONNECTION TO VALVE MANIFOLD. NOT DETAILED OUT. FOR SUCH CONNECTION SIDE VIEW MAY BE REFERRED TO.
(C)
TRANSMITTER RACK/ENCLOSURE FRONT VIEW
INTEGRAL VALVE MANIFOLD
ELECTRICAL JUNCTION BOX
GLAND PLATE
CABLE ENTRY
(D)
TRANSMITTER RACK/ENCLOSURE SIDE VIEW
DETAIL "A" (Typ)
MOUNTING BOLT
GROUTING
GROUTING POCKET (say 80X80X100)
(1) UNLESS OTHERWISE STATED ALL DIMENSIONS ARE IN mm.
(2) BULK HEAD(BH) IS THE INTERFACE BETWEEN FIELD & ENCLOSURE.
(3) INSIDE ENCLOSURE IT IS BETTER TO USE IMPULSE TUBE FOR UNIFORMITY.
(4) TRANSMITTERS WITH INTEGRAL VALVE MANIFOLD (VM) IS PREFERRED & SHOWN
(5) DRAWING IS NOT TO THE SCALE. FOR SIMPLICITY ALL DETAILS AND DIMENSIONS ARE NOT ELABORATED HENCE THIS IS A SCHEMATIC DRAWING
(6) BASIC STRUCTURE SHOWN WITHOUT SHEET METAL WORK. mOUNTING PIPE, DRAIN HEADER ARE DULLY SUPPORTED FROM STRUCTURAL FRAME.
(7) OVERALL HEIGHT MAY BE AROUND 2200 FOR CONVENINECE OF WORK. DEPTH 800 MIN IS NECESSARY.WIDTH IS VARIABLE(see above)
(8) FOR PRESSURE TRANSMITTER SINGLE IMPULSE LINE IS APPLICABLE, (all DPTs shown)

FIG. 12.24 Transmitter mounting and transmitter enclosure schematic.

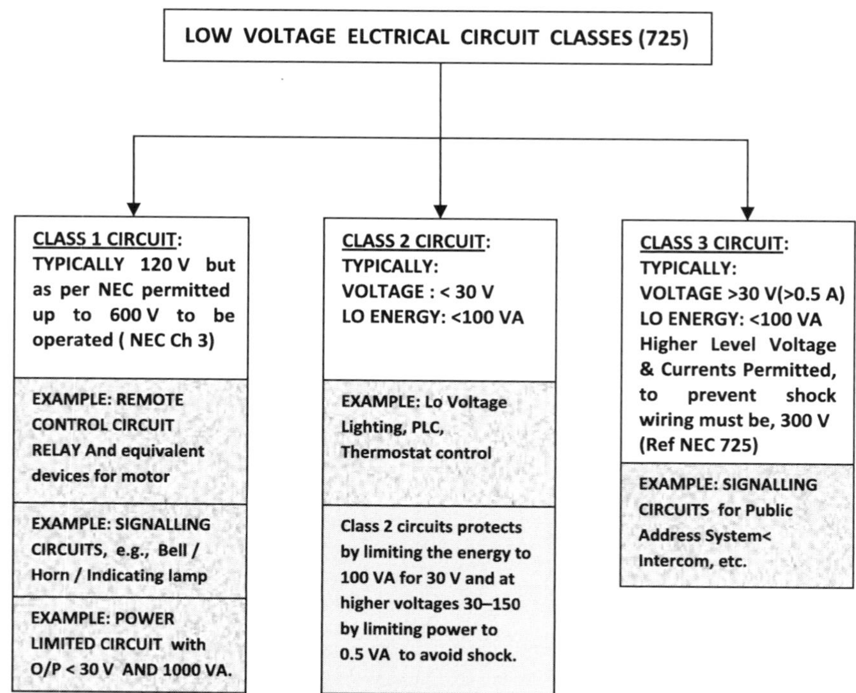

FIG. 12.25 Electrical circuit classification in line with NEC (low voltage).

be high voltage (HV) at the switch yard and switchgears (for large motors). These **HV** systems in the plant may be, for example, 11, 6.6, 3.3 kV, etc. There will be medium voltages (**MV**) like 415 V for MCC, etc. Apart from these, there will be low voltages (**LV**) like 240 V, 110/120 V, and 24 VDC. Even though control and instrumentation as well as SCADA are involved in monitoring and control of HV and MV systems, instrumentation cables are generally installed separately from other cables. Instrumentation and control (I and C), is more concerned with 24 VDC, but there may be power supply systems from the other LV systems also, for example, 240 VAC or 120 VAC control voltages. One needs to go for best practices in I and C cabling, its health, and complete integrity of the system as per certain standards such as NEC/NFPA 70, IEEE 518, 422, etc. Prior to moving to these standards, it is important to throw some light on the classification of circuits as per 725 of NEC, which is meant for remote-control, signal and power limiting devices.

4.1.1 Circuit Classifications: Short Discussions Have Been Presented on Article 725 of NEC Titled

"Class 1, Class 2, and Class 3 Remote Control, Signaling, and Power-Limited Circuits." On account of a limited load and power source, there are some changes from the general rule. Class 1, Class 2, Class 3 remote control, signaling, and power limited circuits have been dealt with in Article 725. Remote control, signaling, and power limited circuits are quite different from power and lighting, and as per NEC their intermixing is restricted. According to Article 725 of NEC, the low-voltage circuits can be classified, as shown in Fig. 12.25 Electrical Circuit Classification, which is self explanatory.

Circuit Classification for Instrumentation Applications. Common circuits used in instrumentations have been listed below with probable circuit classification in line with NEC and associated factors:

- 4–20 mADC signal derived from ~24 VDC power supplies, typically has a fuse of rating 0.1–0.25 A. Depending on the listed power supplies and circuit card, these may normally be classified as Class 2 or Class 3 circuits.
- 0–10 or 0–1 V normally falls under Class 2, provided the listed power source has been used; otherwise these may be class 1.
- Discrete signal 24 VDC with fuse rated <0.25 A is the same as that discussed above.
- Thermocouples, bimetallic junctions, etc., are considered under Class 2.

Digital signals under 25 V, RS 422, RS 232, etc., derived mainly from information technology equipment are generally Class 2, unless not listed.

- 120 VAC or 24 VDC used to operated relays, solenoid valves, etc., could be Class 2 or 3 provided they have listed power supplies; otherwise Class 1 circuit shall be applicable. Generally 120 VAC falls in Class 1. A 120 VAC discrete signal derived from a branch circuit with a fuse <1 A falls under Class 1.

240 VAC circuits as well as remote control and signaling circuits can also be classified as per the logic explained above.

4.1.2 Signal Noise and Interferences

In comparison to older instrumentation system, modern instrumentation systems based on digital signals are very sensitive to any noise and signal distortion due to noise. Therefore, it is needless to emphasize modern wiring techniques and better qualities of cables to achieve a better solution. Noise coupling methods (based on the IEEE Guide for Instrumentation and Control Equipment Grounding in Generating Stations Standard 1050) could be *conductive* (common impedance), *capacitive* (electric field), *inductive* (magnetic field), or *radiative* (electromagnetic).

1. Conductive (Common Impedance) Coupling: When a common junction/wiring is shared by two circuits, such as a common return path, then this noise or interference occurs. This could be intentional for grounding or could be due to undesired leakage between the two circuits. The resulting ground potential can couple into the signal path, as shown in Fig. 12.26A. So, the current in one circuit may appear as noise in the other circuit, and such noise voltage depends on the value of **common impedance**. These may manifest in the plant on account of different ground potential at various locations in the plant. Many times, it may be futile to short out this potential. These may occur even if the

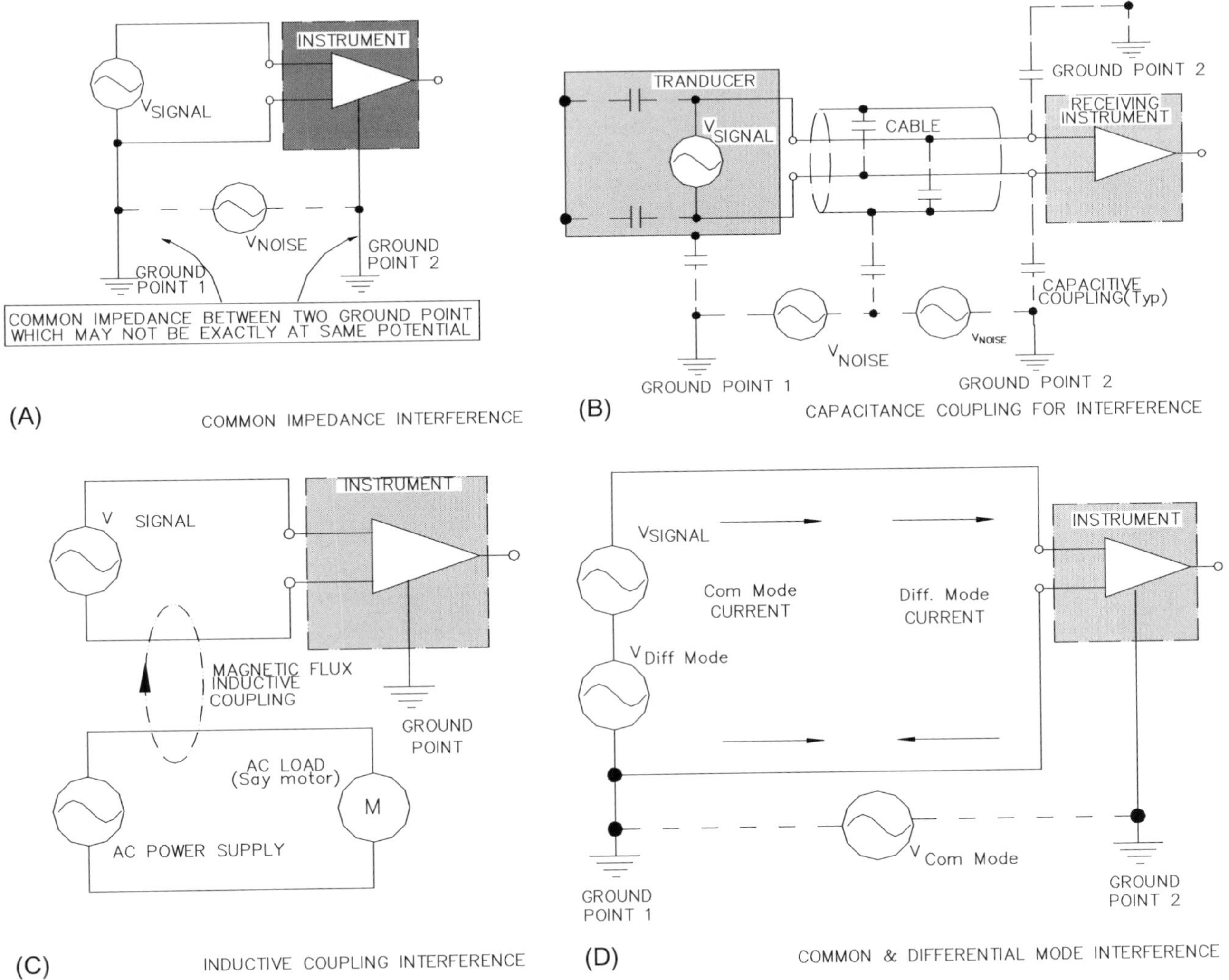

FIG. 12.26 Various noise interferences

TABLE 12.4 Interference Modes (Fig. 12.26D)

Characters	Common Mode	Differential Mode
Introduction	Introduced into the signal channel from outside, having at least one terminal not a legitimate part of the circuit. Chassis is always a terminal unless isolated.	Introduced into the channel through the same path as the legitimate circuit. Other than the main signal path, no current path exists.
Cause	Caused due to potential difference between ground points, and/or electromagnetic pickup or other electrical coupling.	May be due to conversion of common mode current to differential mode noise.
Effects	It indirectly acts on receiver and signal error can be considered as common mode interference to differential mode noise. In a two-wire system, usually each wire has the same magnitude but is opposite in phase.	Normally has frequency characteristic different from the desired signal. In a two-wire system, usually each wire has the same magnitude but is opposite in phase.

Suitable RC filter, Zenners, Varistors can be used to minimize interference.

receiving instrument has high common mode rejection (CMR), when the shield is not correctly terminated or terminated at two points, giving rise to a ground loop. In a grounded thermocouple, this is quite prevalent.

2. Capacitive Coupling: In all portions, such as between the insulated conductors, there is a capacitance. Any voltage change regardless of location will try to push a current through the various possible capacitances for coupling in the circuits, as shown in Fig. 12.26B (as $I = C\ dV/dT$). So, the length of two parallel wirings will increase the noise due to capacitance coupling. High impedance circuits are more susceptible to it.
3. Inductive Coupling: There could be many closed loops having mutual inductance, which is proportional to the area enclosed by the closed loop, as shown in Fig. 12.26C. There is a transformer effect and it is equally applicable for a DC circuit whenever their current is interrupted or it changes periodically ($E = M\ dI/dT$). This coupling is an increasing function of the length of the coupled conductors and a decreasing function to the distance between them (Noise source: dI/dT noise medium: mutual inductance coupled noise: voltage).
4. Radiative (Electromagnetic) Coupling: When the circuit/loop route within the electromagnetic radiation profile of interfering source. Normally these electromagnetic fields couple signal voltage in the form of common mode voltage. This coupling is proportional to the loop area and frequency. Even though these frequencies may be much higher than the control circuits, they may yet cause harmonic interferences.
5. Interference Modes: Common mode, differential mode, and cross are a few modes of signal interference. Out of these, the first two modes are very much interrelated and their characteristics have been detailed in Table 12.4, and the associated drawing to be referred to is Fig. 12.26D.
   - Cross Talk: It is another means of interference. When AC signals or pulsating DC signals are transmitted over a multipaired cable, then there is a tendency of signals to be superimposed on the signals carried by the other pair of cables. This is because of the effects of both inductive and capacitive coupling, and it is proportional to the frequency of signals. The impedance change will cause a proportional effect on capacitive coupling and an inversely proportional one on inductive coupling. So, to minimize the cross-talk effect, one has to assess the predominant coupling and act accordingly.
6. Cable considerations: In the following sections, a brief discussions of the type of cable and their dressings and terminations is explained. Some tips for various signal types have been elaborated in Table 12.5.
7. Considerations to effectively eliminate electrical noise: In a line with standard IEEE 518, the following methods may be adopted to effectively eliminate electrical noise:
   - To maintain physical separation between the electrical noise source and sensitive equipment and between electrical noise-bearing wires and sensitive signal wires.
   - Proper shielding and grounding practices with use of the appropriate filter.
   - Twisted pair wiring should be used in critical signal circuits and in noise-producing circuits to minimize magnetic interference.
8. Shielding: It is a generalized term used as a barrier between an emitter and susceptor. Shield effectiveness (SE) is measured in decibels (dB) and 20–80 dB is normal effective shielding. Only short discussions have been presented here; for details, see the latest edition of IEEE 1050. Most of the electronic devices are housed in steel cabinets that provide some degree of shield, but their SE depends on a number of factors such as seam,

TABLE 12.5 Signals and Cable Types—A Guideline

Points	Analog/Pulse/Digital (Lo) Signals	Discrete (Hi) Signal
Signal types	4–20mADC, thermocouple, RTD, strain gauge, digital/network communication signal, pulses,	Process switches, limit switches, relay contacts, etc. with higher interrogation voltages.
Conductor	Twisted pair/triad	Twisted pair
Screening	Individually shielded pairs/triads. Screen coverage 100%	Overall shield—100% coverage is preferred.
Grounding	At field drain wire/screen to be cut and taped, grounding at marshalling end at control (equipment) room. The screen wire/drain wires shall be properly terminated at the specified ground terminal board. Drain wire/screen of one loop must not be shorted at the marshalling rack/junction box.	

cable penetration, apertures, etc., as any opening in the enclosure provides a highly effective coupling. Similarly, when electronic systems are packaged in nonconductive material (ABS), their cases should be treated with conductive materials to provide shielding. A cable shield, as already discussed, has a very important role to play in diminishing the noise effect on the signal. In most of the cases, only overall or individual pair shields are used because the capacitance per unit length is greatly increased when both are used. For digital signals, in some cases an overall shield will suffice. Common shields used in the cables are copper braid and an aluminized mylar tape with a (copper) drain wire. Metallic ducts/conduits also act as a shield. A double-grounding arrangement of aluminum/copper tapes gives very poor performance. So, **it is advisable to ground the shield conductors at one point only** for single-point grounding **and** the **shield envelope should have an insulated jacket to prevent multiple grounds.** Some shielding tips in line with IEEE 518 standards are:

- The shield shall never be left floating, as it enhances the electrostatic coupling and for digital logic signals normally has shield grounded at power supply end.
- Low-level signal sources requiring a differential amplifier shall have shield grounding as per equipment manufacturer's recommendation.
- Multipair conductors used with thermocouples should have individual insulated shields so as to maintain the particular thermocouple ground potential.

9. Other Methods of Noise Minimizing Techniques: There are a few other means and methods to minimize noise. These could be the use of isolation transformers/neutralizing transformers/differential amplifiers, etc.

10. Classification of Wiring as per Noise Susceptibility Level (*NSL*): According to noise susceptibility, IEEE 518 classified wiring has four classes of wiring:

- Level 1 High susceptibility analog signal <50 V and digital signal <16 V mainly include: fieldbus, 4–20 mA with or without a transmitter protocol—HART, RTD/T/C, millivolt/pulses, signals with digital hardware, discrete I/O, and wiring connected to a sensitive analog signal, for example, strain gauge, phone lines.
- Level 2 Medium susceptibility analog signal >50 V and switching circuits for example, common returns to medium susceptible equipment, AI (tachometer), DI/O DC/AC.
- Level 3 Low susceptibility: switching signal >50 V AI >50 V with current <20A and AC feeder <20A, fused control bus 50–250 VDC, indicating lamp >50 V and thyristor field exciter electrical systems.
- Level 4 AC and DC Buses of 0–1000 V with current 20–800 A: Thyristor AC power input and DC output and other electrical systems (not discussed here).

As per IEEE 518 standard, within the level, conditions may exist that require specific cables and regrouping is not allowed. For class code and other details, see the latest edition of IEEE 518; also see Clause no. 4.3.

4.2 Grounding/Earthing

During the discussions, the terms "grounding" and "earthing" have been used as synonyms because both these terms are used in various standards. Improper grounding would bring about disastrous results in safety as well as plant operation due to distortions.

4.2.1 Grounding Philosophy

As discussed earlier, grounding has two objectives: to ensure safety and to minimize the effect of undesirable electrical coupling. So, the main points for grounding philosophy may be as follows (IEEE std. 1050-1989):

1. To maintain a safe voltage across the entire plant and to minimize high-voltage transients and lighting surges. Also, to minimize a shock effect that may appear on noncurrent carrying equipment, as the resulting cabinet voltage may appear due to insulation failure and to provide low impedance path for leakage for static charge development, etc.
2. To minimize the effect of the electrical noise existent in an industrial environment.
3. To minimize any disturbing effects within the control equipment and any propagation effects that may result in excessive ground current as well as any fault currents that may occur.
4. To minimize noise interference in the instrumentation system by providing a common reference point.

4.2.2 Grounding/Earthing Types

In the instrumentation system, there could be three types of grounding: dirty grounding, mainly for 120, 240, and 480 VAC power grounds. Apart from these there will be structural grounding such as cabinet ground, or any structure grounded to make the same safe from lightning, etc. So these are protective groundings.

On the other hand, DC grounds of 24 VDC used in PLC/DCS, having normally one common reference point, are referred to as clean grounding or reference grounding. Intrinsic safe circuits also have grounding and these are called IS grounding. Normally, IS grounds are treated separately and these are done with a blue cable. All these are shown in detail in Fig. 12.27A. Clean grounding is shown with dark green whereas dirty grounding is shown with light green in the drawing under reference. There are three approaches toward grounding: single point grounding, multipoint grounding, and floating grounding (IEEE Standard 1050) (Fig. 12.28).

1. Single Point Grounding (Fig. 12.27B and C): In single point grounding, circulation of the ground current is eliminated. This is quite effective for signal frequency <300 kHz. However, this may not be effective at high frequency. At a ground cable length of >0.15 times the signal wavelength, the cable is no longer low impedance. In both the variations shown in Fig. 12.27B and C, the signal is grounded at one point; only in case of cabinets well apart, as in Fig. 12.27C, are electrical grounds locally floating.
2. Multiple Grounding: As shown in Fig. 12.27D and E, there are two variations, one due to a high-frequency signal (say >300 kHz) or one when long ground cable say 200 M (IEEE the 1050). These types of circuit constructions are easy but rather difficult to maintain. Of course, these may give rise to multiple ground loops. However, instead of direct grounding, if bypass capacitance is used at one shield end (say, at the source end), it is possible that the cable shield appears grounded at the high frequency (for capacitance), whereas at the low frequency, the capacitance carries hardly any current. Therefore, it would appear as a single point ground.
3. A floating ground, as shown in Fig. 12.27F, is used to isolate circuits or equipment from a common ground plane. It is implemented by interconnecting the systems, yet isolating them from the common ground plane. The system is good, but may have a serious problem when there is static charge accumulation, which may be destructive while producing high noise.

4.2.3 Electrical Earthing in Instrumentation (Fig. 12.29B and C)

As discussed earlier, protective earthing of power supplies, cabinet earthing, etc., are electrical earthing done with the objective to prevent hazardous potential from developing between the adjacent equipment grounds to protect personnel and equipment hazard.

- Enclosures have a suitable AC ground bus for termination of the equipment ground cable.
- To connect only one multistranded ground cable between the equipment and the ground.
- All individual chassis in an enclosure shall have an intentional ground connection.
- The armor of all field cables should be terminated at the cable glands through which earthing connections are made to the cabinet earth bus.
- For a metallic junction box, an earth stud bolt should be connected to the nearest steel structure. Otherwise, the earth stud bolt will be located at the metal gland plate, which has direct metallic contact with the junction box.
- Each bus bar (25/40 mm × 6 mm) inside the cabinet will be connected to a grounding dispatcher by cable (usually 35 mm^2 copper cable with green and yellow stripes). This earth dispatcher connects all bus bars to the grounding loop.

4.2.4 Instrumentation Earthing

As per IEEE standard 1050-1989, "The purpose of signal ground is to reference all control signals within a system to a single point supposed to be at earth potential." In any enclosure, there shall be no more than one reference point. As a general principle of instrumentation earthing, all individual or overall shields/screens of cables shall be isolated from electrical earthing and terminated at a different bus bar, which may have a final earthing at a suitable place in the building that may be common with electrical earthing. This instrument earth is called the reference earth/reference point in the instrumentation loop. A proper earthing hierarchy is helpful in eliminating unwanted signals. Often, it has been found that most of the field noise problems are

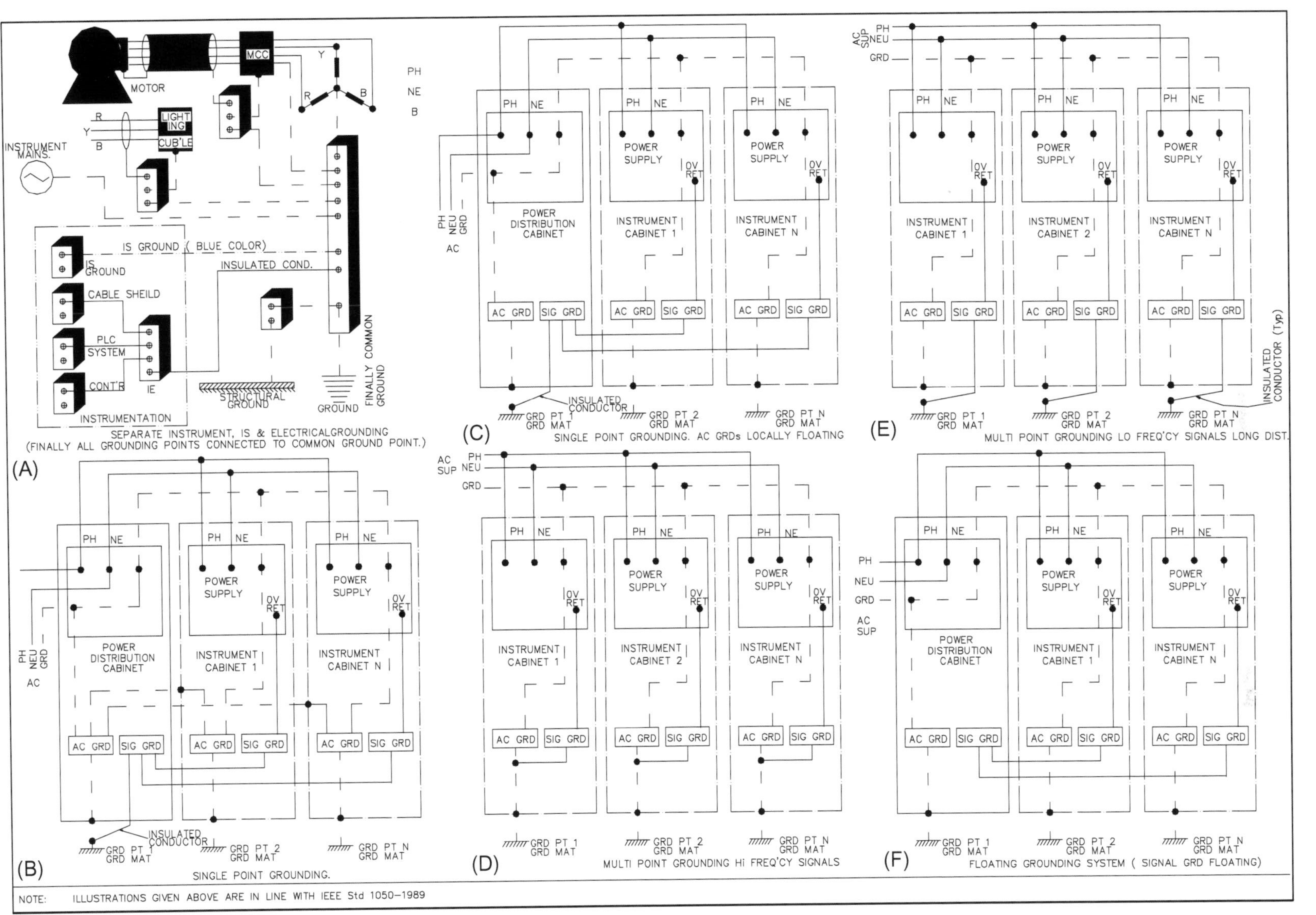

FIG. 12.27 Various grounding systems.

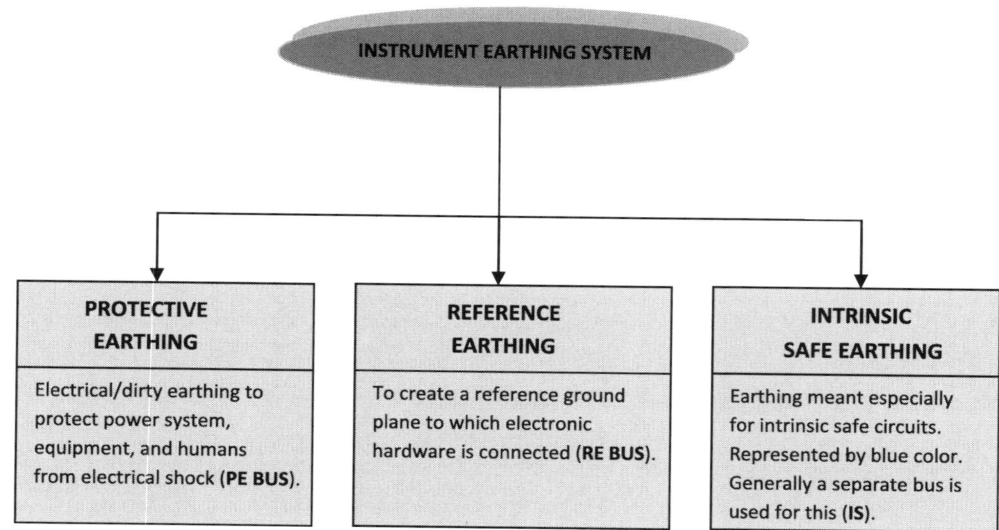

FIG. 12.28 Instrument earthing types.

associated with poor grounding. The signal is properly guarded only when the shield is connected to true zero potential.

There are two possibilities for the input signal. One is the grounded signal and the other is the floated signal. In the former case, the correct point of the shield grounding would be the place where the signal is grounded. If the source ground and output ground differ in potential, then there will be a common mode rejection problem. The correct point of shield grounding is the point where the signal is grounded. In fact, for a grounded signal, the signal ground should be where the transducer is located. Use of a differential amplifier for the grounded source gives better results. As a thumb rule, the shield should be grounded at the signal source point and the receiving end may be left floating.

A floated signal may be necessary for safety and performance. The proper pathway from the input circuit to the output ground is provided by the shield and ground resistance at the input. In practice, it is extremely difficult to achieve this.

1. Field Transmitter Grounding: The majority of the transmitter manufacturers with output 4–20 mADC output with or without HART, even transmitters with a fieldbus, recommend local grounding of their product through the grounding terminal they provide. This is because modern electronics are vulnerable to surge current. Often, these boards are provided with integral surge protection, so it becomes essential to give a path for the surge current from the surge protection circuit. Normally, there are two ground points in the transmitter: internal ground and external ground (mainly protective grounding). These are internally connected suitably via a filter, etc. For a nonisolated system, grounding of the transmitter with a pipe structure will be sufficient. Otherwise, a proven local ground with min 12 AWG ground wire G, locally has to be provided.
2. Signal Cable Shield Grounding: Various suggestive methods for cable shield grounding are discussed below and these should be read in conjunction with the illustrative drawings referenced.
   - **(a)** Transducer cable (Fig. 12.29A): The individual shield (drain wire) of a single pair (/2C) cable shall be terminated at the earth or ground terminal block inside the transducer/instrument enclosure.
   - **(b)** The individual screen cable of a single pair (/2C) from the instrument to the junction box cable in the junction box shall be terminated at a terminal block inside the junction box.
   - **(c)** Referring to Fig. 12.29A, the individual screen cable of a multipair (/M Core) from the junction box shall also be terminated at the terminal block, matching with the individual shield of the single pair (/2c) cable discussed in B above. But, the overall shield from a multipair cable from the junction shall be terminated at the terminal block or bus bar (no matching criterion) for final grounding through a marshaling/system cabinet.
   - **(d)** All individual shields/overall shields from a multipair (/M core) cable shall be terminated into their respective instrument earth bus at the marshaling/system cabinet (Fig. 12.29A).
   - **(e)** An individual shield from a digital single pair cable for a digital signal going to the junction box shall be terminated at the terminal block and shorted, then connected to the bus bar. Also, the overall shield from the multipair (/M core) cable going

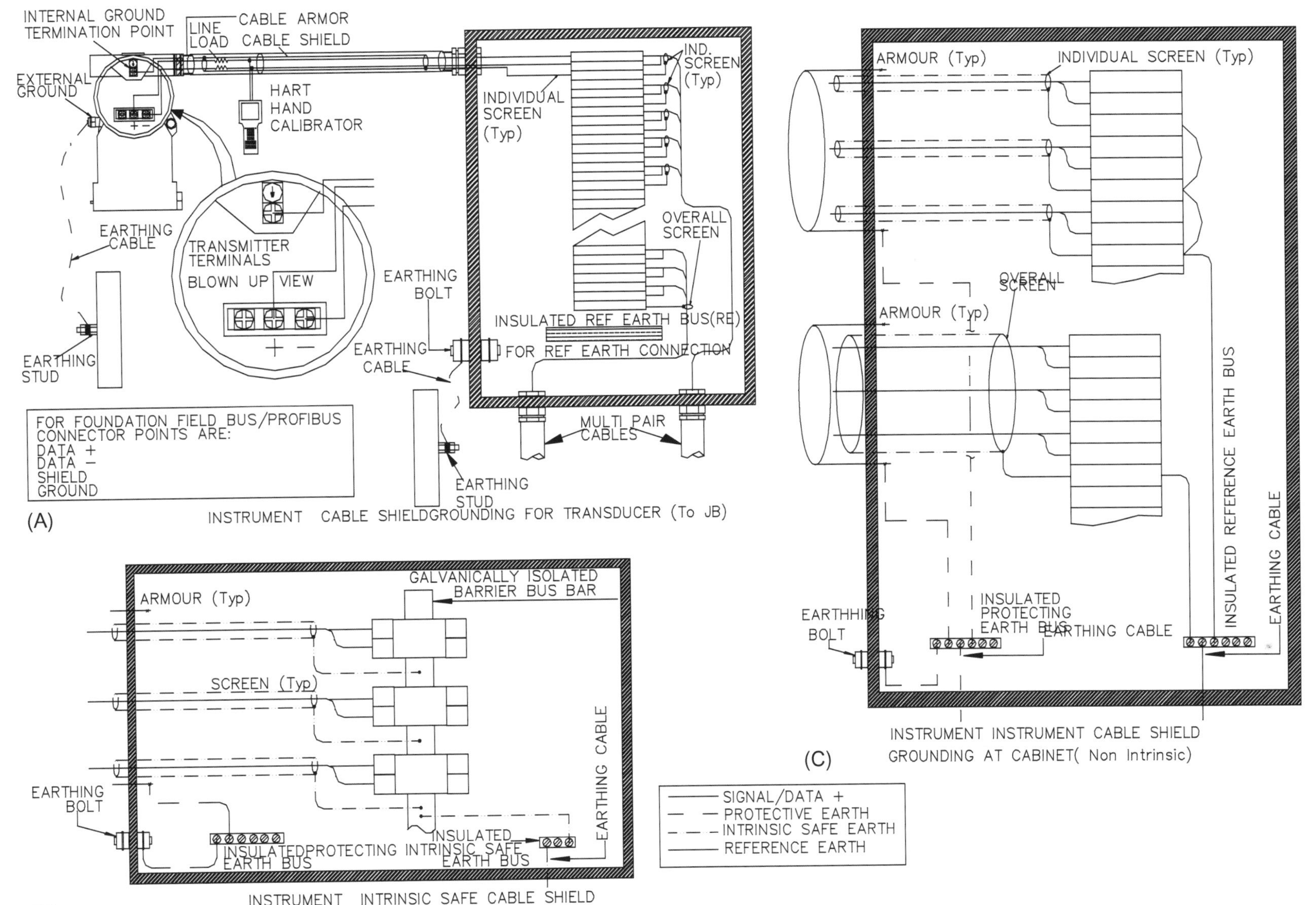

FIG. 12.29 Instrument cable shield grounding methods.

inside the junction shall be terminated at the bus bar.

(f) The instrument bus bar will be connected to a ground dispatcher by a 25 mm^2 green yellow-striped copper cable.

(g) From field instruments all the way up to the marshaling/system cabinet, the shield drain wire should be treated and terminated similar to the signal wire.

(h) The exposed part of the drain wire within the junction box and the marshalling/system cabinet should be inside the insertion jacket to avoid short circuiting of various loops.

(i) All spare pairs and triads should always be terminated at both ends, but grounded at the marshaling/system cabinets.

(j) In the central control room (CCR) or the control equipment room (CER), the marshalling/system cabinet must be equipped with both AC and DC or reference bus bars. The DC bus bar should be isolated from the cabinet structure. All shields and DC common wires are connected here. It is important to ensure that these are grounded at one point. All DC/shield buses in all cabinets shall be consolidated as the master DC ground bus of that building. Then, this may be connected to the plant's overall grid. Fig. 12.29C may be referenced.

(k) Directly to a building column or electrically conducting mass determined to be at true earth from, the termination point of equipment ground reference and Ground reference itself are some of the methods for connecting common reference point with ground.

(l) Grounding of Centralized System: This is similar to what is shown in Fig. 12.27B. Generally, there will be one power supply system (common source) with power distribution done through MCBs, etc. Each cabinet shall have a separate equipment signal ground. Out of many cabinets one should act as a concentrator for both signal and equipment ground.

(m) Grounding of Distributed System: When individual control systems are separated from each other (may be common for a large network with integration of subsystems). Some of the design conditions in line with IEEE 1050-1989 may be: an effort may be made to have a common power supply source (power transformer), each of the cabinets/clusters will have local protective ground, signal between the systems shall be properly coupled (transformer/DC/optical) having CMR ground voltage under fault and one of the system cabinet may be considered as master where both ground types of points are collected. A remote station's signal ground may be left floating.

(n) Instrument grounding connections, especially with a thermocouple, are shown in Fig. 12.30B and C. For a thermocouple, the negative leg is connected to the earth. For floated, such grounding is done at the computer end via a screen and not locally, as in a grounded system.

(o) Intrinsic safety circuit grounding principles are not much different from what has been discussed above. Normally, blue cables are used for these. Fig. 12.29B may be referred to in this connection. Overall, the screens of these cables in cabinets are connected to the insulated IS bus bars, whereas the same for an individual screen goes to the IS bus bar via a galvanic isolation circuit. Normally, the resistance between the grounding dispatcher and the IS bars is restricted to 0.5 Ohms. In a power plant, there is not much application of intrinsic safety circuits, as in the oil sector.

(p) Other pertinent issues could be: Unused conductors and shields should be terminated at both ends. In some cases, it will be useful to ground the spare wiring at the marshaling/system cabinet. The power supply leads shall be kept closely spaced for canceling the coupling due to a magnetic field and care shall be taken as per the manufacturer's recommendations for circuits with MOSFET, which is susceptible to static charges (Fig. 12.31).

4.3 Instrument Cables and Wiring Practices

A broad classification of various kinds of cables used in instrumentation is presented below. Wiring practices based on ISA, IEEE, and PIP are also discussed in this section.

4.3.1 Typical Cable Type

Cable types and uses have been presented in line with PIP standards (Figs. 12.32 and 12.33).

1. Instrument Cable Classification (Table 12.6):
2. See Table 12.7.

4.3.2 Cabling and Wiring Practices

Generally, the installation of wiring and low-voltage instrumentation cables needs to comply with international standards such as NEC requirements. A few points in this connection could be:

1. The routing of cables to the control room shall preferably be through the overhead cable trays and routing cable trays through areas with fire hazards shall be avoided.
2. Cables that are laid directly under a raised floor shall obtain the proper approval of NEC article 354 (a nonmetallic smooth-walled conduit with a conductor or cables inside). AC power wiring (120 VAC or higher) routed in the raised floor shall utilize conduits, metal-enclosed

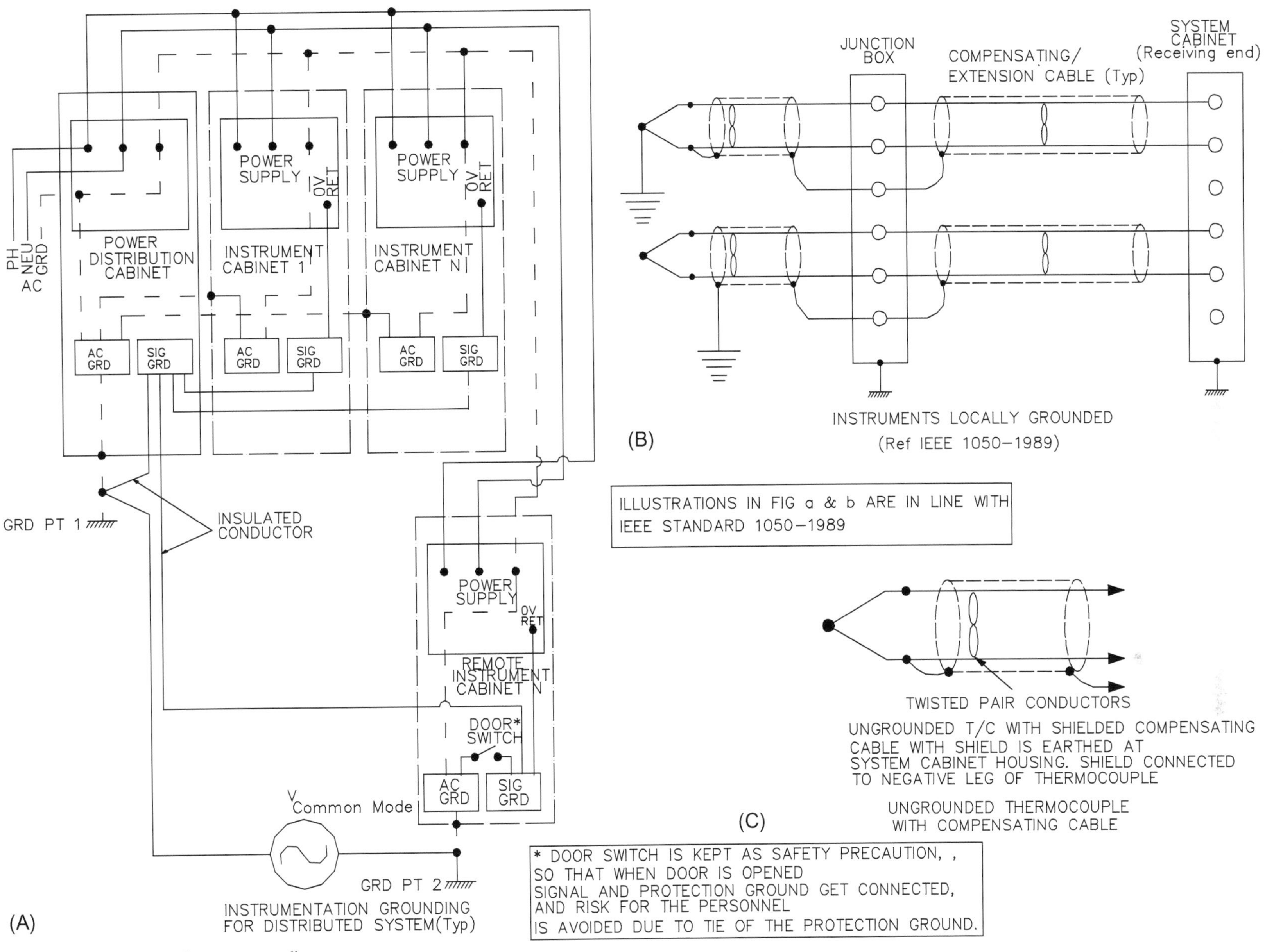

FIG. 12.30 Instruments and systems grounding.

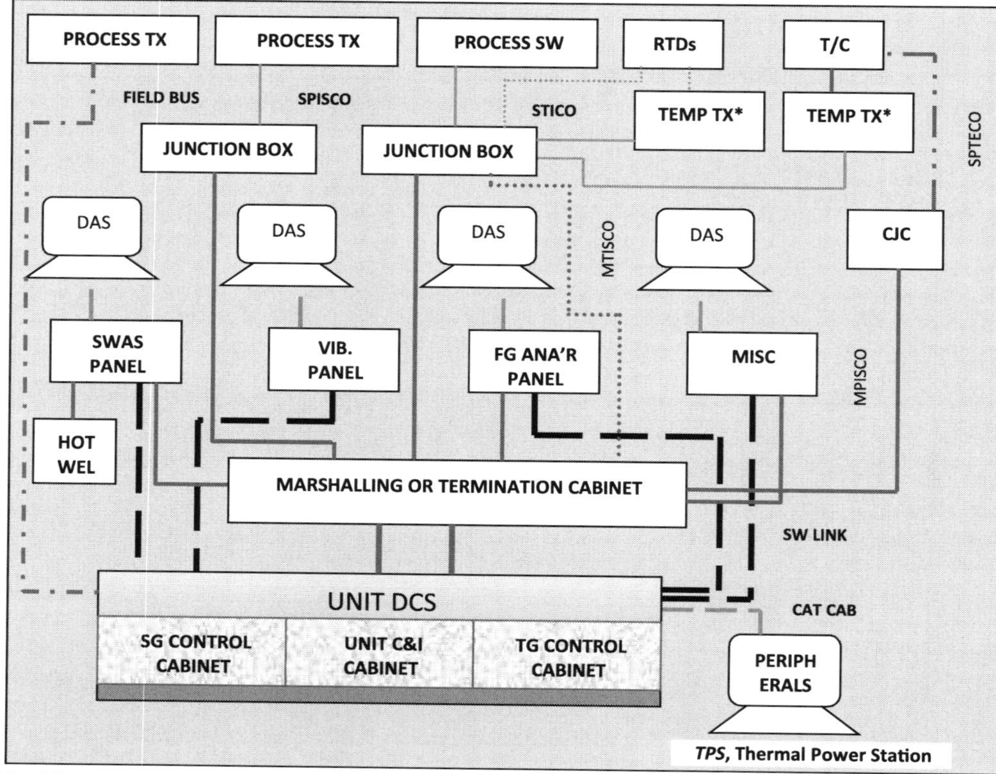

FIG. 12.31 C&I cabling diagram in TPS (Typ).

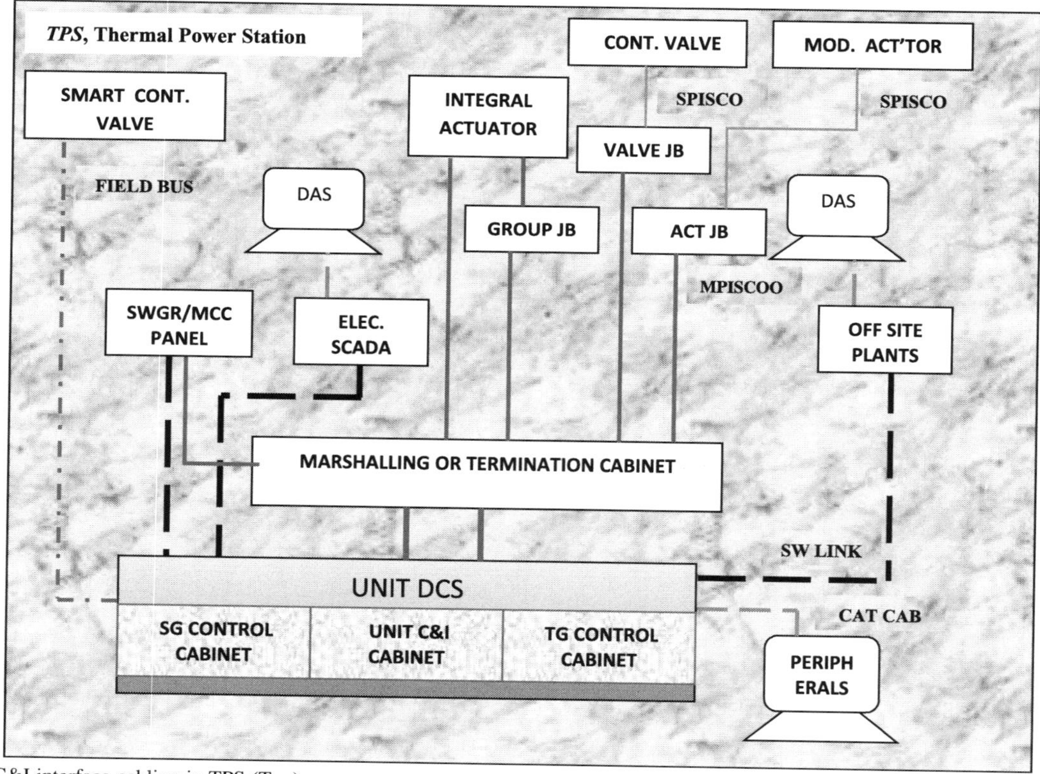

FIG. 12.32 C&I interface cabling in TPS (Typ).

ITC: A type of cable introduced after 1996, covered in NEC art 727. Earlier in case of low power Instruments, 600 V grade cables were used to meet the waveform and power limit of NEC. These were disadvantageous because of cost and size (diameter). ITCs are rated for 300 V insulation, with conductor size ranging from 22 (current limit <3 A not 5 A) to 12AWG. Conductors are copper and/thermocouple.

FIG. 12.33 Note on ITC.

TABLE 12.6 Instrument Cable Classification and Specification

Cable\Name	Abbreviation	Minimum Specification
Single Pair Instrument Signal Cable with Overall Shield	SPISCO	Pair of two copper conductor 16 AWG, overall shield
Single Pair Thermocouple Extension Cable with Overall Shield	SPTECO	Each pair of two solid alloy T/C extension wires with color code as per ISA MC 96.1 with overall shield.
Single Triad Instrument Signal Cable with Overall Shield	STISCO	Three copper conductors of min 16 AWG with overall shield.
Multi Pair Instrument Signal Cable with Overall Shield	MPISCO	Multipair (say 12/24 pair) with overall shield. Each pair two copper conductors of min 20 AWG.
Multipair Thermocouple Extension Cable Individual Shield	MPTECI	Multipair (say 12/24 pair). Each pair of two solid alloy T/C extension wires with color code as per ISA MC 96.1 with overall shield minimum conductor size 20 AWG
Multipair Instrument Signal Cable Individually shielded Pair with Overall Shield	MPISCI	Multipair (say 12/24 pair) with overall shield. Each pair two copper conductors of min 20 AWG with individual shield insulated from other pair shield.
Multitriad Instrument Signal Cable with Overall Shield	MTISCO	Multitriad (say 12/24 pair) with overall shield. Each triad copper conductor of min 20 AWG.
Multitriad Instrument Signal Cable Individually Shielded Triads with Overall Shield	MTISCI	Same as above with individual shield insulated from other shield.

raceways, or cable trays with adequate separation from the signal cable.

3. For a safety instrumentation system (SIS), ISA S84.01 needs to be followed. If the electronic equipment room (process control room alone is not classified) is classified as the information technology room, then NEC Art 645 is to be followed.
4. Splitting multiple conductor cables among multiple devices/boxes in the field is not permitted.
5. Instruments/electronics located in the field in a large open area or cable run >500 m should have lightning protection devices in line with NEC art 280.
6. Devices may be protected by a metal oxide varister or diode to protect the same back EMF (AC/DC, respectively).
7. All the pairs/triads must have a suitable tag plat at both ends. Cables should have suitable markings and defined color codes.

4.4 Segregation/Separation Requirements

As discussed in Clause no. 4.1.2.10, there will be different noise susceptibility levels (*NSL*), and accordingly, there will be a separation between different NSL wiring as in IEEE 518 and PIP standards.

4.4.1 Class Codes

According to IEEE standard 518, the class codes are: **A**: analog input/output, **B**: Pulse Input, **C**: contact/interrupt input, **D**: decimal switch inputs, **E**: output data lines, **F**: display/contact output, **G**: logic input buffer, **S**: special (level) for special spacing (see std), and **U**: HV potential unfused >600 VDC.

4.4.2 Tray and Conduit Spacing

The spacing between trays and conduits presented below in tabular form is a summary of standards IEEE 518-1982 and PIP PCCEL001. The following points need to be considered while referring to the tables below:

1. All distances are in **inches.** For the top and bottom tray, it is the distance between the top of one tray to the bottom of the other tray while for side trays, these are between the sides of adjacent trays. And for conduits, these are outside the surface of the conduits.
2. When unlike signal levels cross, they should cross at 90 degrees (if not at all possible, a grounded metal barrier should be placed at the cross-over of the two levels).
3. Trays of all levels are based on metal trays solidly grounded with good ground.

TABLE 12.7 Wire and Cable Requirements for Instrument Circuits (PIP PCCEL001)

1. NEC class circuits (notes 1, 2)			
Instrument class	**Circuit example**	**Conduit wiring**	**Tray cable**
A. 120VAC or less	Switches, solenoids, relays	600V PVC insulated (THWN) 14 AWG (min)	600V TC insulated
B. NEC article 727 circuits, conduits, and cable tray installations (notes 1, 9)			
B1. Instrument circuit class and type	**Circuit example**	**Single pair or triads**	**Multipair or triad cable**
B2. 24VDC or less	4–20mADC, RTDs, weigh cell, solenoids, alarms, switches, relays, secondary motor control, digital process transmitter	300V shielded twisted pair or traids Cable types: SPISCO, STISCO (Notes 2, 4, 8)	300V ITC* PVC insulated pairs or triads overall shield. Cable types: MPISCO, MTISCO (Notes 2, 3, 8)
B3. Frequency or pulse train digital communication (Note 5)	Speed, vibration, turbine meter	300V shielded twisted pair or triads Cable types: SPISCO,STISCO (Notes 2, 8)	300V insulated individually shielded pairs or triads 300V insulated Cable types: MPISCO,MTISCO (Notes 2, 4, 8)
B4. Thermocouple measurement (Notes 7)	Thermocouple	300V ITC 16 AWG Individually shielded twisted pair. Cable types: SPTECO (Notes 2, 8)	300V ITC, 20 AWG individually shielded twisted pairs, overall shield cable types: MPTECO (Overall shield) (Notes 2, 8)
C. Data highway and other high speed circuits			
Data highway (Notes 6, 8)	EIA-422A data highway/high speed communication	Follow cable and system manuf.'s recommendations	Follow cable and system manuf.'s recommendations

Notes (Based on PIP PCCEL001):
1. Type A circuits shall be separated from all other types of instrument circuits by conduits, tray dividers, or separate trays.
2. The conductor sizes given are the minimum wire size. The actual shall be as per load and voltage drop. Use of parallel wires to meet the requirements is not allowed.
3. Higher DC levels require owner's approval.
4. Instrument tray cable (ITC*) can only used in power limiting circuits, with $V < 150$ V and current <5 A.
5. Vibration signals shall be run in a galvanized steel flexible conduit and armor from the probe to the transducer.
6. Maximum separation between redundant highways shall be obtained within the operating plant, and use of a single tray for redundant highways is not acceptable.
7. Type B2 and B4 can be routed in the same conduits and trays.
8. Difference between manufacturer's recommendation and these should be resolved mutually.
9. Intrinsically safe wiring should be separated from other wiring by a physical barrier.

4. Trays containing level 1 and 2 wiring should have a solid bottom and proper tray cover to provide complete shielding. Covers connected to side rails must be positive and continuous to avoid a high reluctance air gap.
5. The tray and conduit of Level 1 and 2—as defined in Clause no. 4.1.2.10 should not be run parallel to any high power enclosure. In cases where the recommended spacing cannot be maintained, parallel runs >5 ft should be avoided.
6. Level 3 and 4 cables may be run in the same tray with a suitable barrier between them.
7. All these are not applicable for nonmetallic conduits.
8. It is important to note that zero separation distance between signals of the same level does not mean different signals of the same level can be used in the same cable. Even if the signals are in the same NSL, the cables shall be separate for separate signal types (Tables 12.8–12.10).

Now, in practice in power plants, it is often required that the same tray/raceway be used for different types of cables, including power (HT) cables, as it is not feasible to run a separate instrumentation (shielded) cable at every place in the plant. In this regard, IEEE standard 422 meant for cabling in power generating stations may be referenced. In vertically stacked trays, various cables may be taken with minimum separations, such as unit HT (11/6.6/3.3 kV) power cable tray: ~ 36″ (914 mm); medium voltage (415 V) cable tray: ~24″ (610 mm); and control supply cable trays: ~12″ (305 mm).

TABLE 12.8 Tray Spacing (in inches)—NSL-IEEE518

Level	1	2	3	3S[a]	4	4S[a]
1	0 (Note 1)	1-(Note 2)	6	6	26	26
2	1-(Note 2)	0 (Note 1)	6	6	18	26
3	6	6	0 (Note 1)	0 (Note 1)	6 (Note 3)	12
3S[a]	6	6	0 (Note 1)	0 (Note 1)	8	18
4	26	18	6 (Note 3)	8	0 (Note 1)	0 (Note 1)
4S[a]	26	26	12	18	0 (Note 1)	0 (Note 1)

[a]*See Clause 4.3.1.*
Note: *1. For zero distance (see Clause 4.4.2, step 8). 2. If a separate tray is impractical,* a common tray with a grounded steel barrier *can be used for two signal levels. 3. Junction/pull boxes, unlike levels, shall be* separated by ground barriers.

TABLE 12.9 Conduit Spacing (in inches)—NSL-IEEE518

Level	1	2	3	3S[a]	4	4S[a]
1	0 (Note 1)	1	3	3	12	12
2	1	0 (Note 1)	3	3	9	12
3	3	3	0 (Note 1)	0 (Note 1)	0 (Note 1)	6
3S[a]	3	3	0 (Note 1)	0 (Note 1)	6	9
4	12	9	0 (Note 1)	6	0 (Note 1)	0 (Note 1)
4S[a]	12	12	6	9	0 (Note 1)	0 (Note 1)

[a]*See Clause 4.3.1.*
Note: *1. For zero distance (see Clause 4.4.2, step 8).*

TABLE 12.10 Tray-Conduit Spacing (in inches) NSL-IEEE518

Level	1	2	3	3S[a]	4	4S[a]
1	0 (Note 1)	1	4	4	18	18
2	1	0 (Note 1)	4	4	12	18
3	4	4	0 (Note 1)	0 (Note 1)	0 (Note 1)	8
3S[a]	4	4	0 (Note 1)	0 (Note 1)	6	12
4	18	12	0 (Note 1)	6	0 (Note 1)	0 (Note 1)
4S[a]	18	18	8	12	0 (Note 1)	0 (Note 1)

[a]*See Clause 4.3.1.*
Note: *1. For zero distance (see Clause 4.4.2, step 8).*

In most of the cases, the bottommost one or two trays (as applicable) may be left for instrumentation cabling. All these are subject to the owner's approval.

4.5 Cable Installation and Termination

In this part, a brief discussion shall be provided on installation and terminations

4.5.1 Cable Installation Points

The cables from redundant devices and redundant cables shall be routed in different paths, as per the recommendation of the IEEE 422 standard. Also, cables of one unit shall be segregated from those of the other unit. Also, cable routing shall be done in such a manner that in case of fire, at least half the critical drives of the unit and the interunit auxiliaries are safe (not affected due to a single fire). A few other points to be considered may be:

1. Cables shall be laid strictly in line with the cable schedule prepared.
2. All cables must have identification tags at both terminating ends, both sides of the walls or floor crossing, at regular intervals in the tray or trench, etc., as the case may be.
3. All trays and subtrays must be numbered at regular intervals.
4. No jointing of cables for <250 m.
5. Buried cables shall be protected by a concrete slab with route markings at a 20 m gap and at the bends.
6. Road-crossing cables need to be pass through high density polyethylene (HDPE) buried in PCC and suitable clearance shall be maintained with other LV and MV electrical cables as per the owner's approval (not <1 ft).
7. All cables shall be neatly dressed and clamped/tied in the trays. Fiberoptic (FO) cables should be inside and could be placed inside cable trays whenever feasible.
8. Whenever FO cables are to be buried, these must be inside 2″ GI/rodent-proof HDPE and put in a trench ~1.5 m deep that is covered with brick and sand. For canal crossing, FO cables in 2″ GI/rodent-proof HDPE shall be put inside HUME pipes.
9. During installation, the cable manufacturer's recommendation on cable pulling tension and cable bends must be followed to avoid damaging the cable.

4.5.2 Termination: A Few Points on Terminations are Enumerated Below

1. A few accessories need to be used for safe cabling such as surge suppressers, opto isolators, FO card cages, etc. This also includes sets of terminating kits such as a crimping tool or a Max Termi (a special type of termination used for better contacts that is used mainly in German systems) tool.
2. Termination of twisted pair cables apparently is easy but some experience is needed to do the job. Cutting of the proper length of cable for termination is very important so as to have an adequate margin for connecting connectors, leaving excess cable at the termination point. For cable termination in the cabinet (draw-out type), sufficient cable shall be coiled and kept inside.
3. For cables such as UTP, the amount of cable to be kept exposed for RJ45 connectors is to be ensured as typically around 1″. Overtwisting, etc., should be avoided. After installation, initial termination testing is a part of the installer's job.
4. FO cables in networking are becoming important, so the installer should be well trained regarding the fundamentals of FO cable installations. In this case, workmanship is extremely important for the performance of the network.

4.6 Brief Specification of Some Basic Items

In this part, a short specification of a few items is presented.

4.6.1 Conduit

Conduits are normally used for cables between field instruments to a local junction box and/or the cases where the cables are fewer. Basic requirements could be as follows:

1. Type: rigid or flexible.
2. Materials: A hot-dipped galvanized mild steel/stainless steel inner part coated with enamel lacker/zinc chromate. Also, needs a flame-retardant property.
3. Sizes (mm) 12/16/20/32/40; heavy duty.
4. Fitting: suitable fittings to get IP rating.

4.6.2 Junction Box

1. Number of ways: 12/24/36/48/72/96/with spare termination: 20%.
2. Material: Stainless steel, thick fiberglass-reinforced polyester (FRP).
3. Grounding stud: To be provided.
4. Type: Screw fitted for door.
5. Terminal block: Suitable rail-mounted terminal block suitable for cable type.
6. Protection class: IP 55/NEMA 4.
7. External color: Owner's choice; internal color is normally brilliant white.
8. Gland plate: Minimum 4 mm thick gland plate detachable, for punches for cable glands.

4.6.3 Cable Glands

Cable glands provide environmental protection by sealing the outer sheathe, earth continuity, sealing against other ingress and hold the cable level against pulling force. Gland size varies with cable size.

1. Type: Single/double compression; normal or hazardous application.
2. Size and Materials: Brass, nickel-plated brass, stainless steel, aluminum.
3. Accessories,: Gland reducer, lock nut, serrated washer, etc.
4. Selection: Done based on type and material of armor.

4.6.4 Instrumentation Cable

Instrumentation cables can be various types and categories, so it is very difficult to cover all types. So a generalized specification have been covered. The cable covered is mainly twisted pair cables. There exist IEC standards.

1. See Table 12.11.
2. See Table 12.12.
3. LAN cable: There could be a variety of LAN cables used in automation and computer networking. These are CAT 5e type UTP/FTP/SFTP with a PVC/LSOH sheathe. Similar will be for CAT 6 and CAT 7 cables. These depend on the network types deployed, as discussed in Table 7.3. Actually, there will be a difference

TABLE 12.11 Instrument Copper Cable (DS)

Sl.	Specifying Point	Standard/Available Data	Remarks
1	Voltage grade	225/300/600 V application based	
2	Standards	ASTM D 2843/2863, BSBS5308, IEEE 383, IEC 60227/60228/60754, IS 1554/8784/10810, SS 4241475, VDE:/0207 part 4 and 5/0815, etc.	
3	Temperature	70°C/85°C for HR cable ratings	
4	Tolerance	As per owner's approval	
5	Conductor: (electrolytic)	A high conductive annealed bare copper stranded (7 × 0.3 mm) conductor with cross-section: 0.5 mm^{2}*	*Other sizes as per application, for example: 1.0/1.5 mm^{2}.
6	Insulation	PVC of thickness 0.25–0.35 mm	
7	Inner sheathe	PVC	
8	Armor	Galvanized steel wire	As applicable.
9	Outer sheathe	Extruded PVC with FRLS property >1.8 mm thick with resistance to water. and fungus. Termite and rodent attack proof	
10	Pairs	1, 2, 4, 8, 12, 24, 48	As required.
11	Twisting	50 mm maximum lay	
12	Shielding	Individual/overall (Table 12.7)	As per application and owners' approval.
13	Shield type	Polyester tape—aluminum mylar tape (28 μm) with tinned copper drain wire (7X20AWG—0.51 mm^{2}) in metallic side of the tape.	Normally 50% overlap.
14	Fillers	Nonhygroscopic flame retardant	As applicable.
15	FRLS properties (see relevant std. mentioned above)	Oxygen index 29%	As applicable.
		Temperature index >250°C	
		Acid generation <20%	
		Smoke density <60%	
16	Electrical parameter	Mutual cap. per km: bet cond. at 1 kHz: 100—120 nF Pair to pair: 115 nF Unbalanced capacitance @ 1 kHz: 1 nF	Rough values as per manufacturer's value.
		Insulation resistance: >100 MΩ	
		Conductor resistance/km: 36 Ω	
		Ch. impedance: <320 Ω @ 1 kHz	
		Crosstalk @ 1 kHz: 60 dB	
		Attenuation: 1.2 dB/km max	
17	Accessories	As required like marker, jointer, etc.	
18	Color code of pairs	VDE standard may be followed	
19	Markings	FRLS, pair number, circles, etc., as per standard.	
20	Tolerance	As per relevant standard	Owner's approval.

TABLE 12.12 Instrument Fiberoptic Cable (DS)

Sl.	Specifying Point	Standard/Available Data	Remarks
1	Type and construction	Multimode tight buffered color-coded fiber/dielectric and interlock armor. Each fiber is protected in a minicable and armid yarn/plastic is applied either inside the minicable or within the inner jacket.	
2	Application	Indoor outdoor LAN, aerial, self-supporting or direct buried cable.	
3	Standards	IEC60092/60331/60332/60754/60793/60794	
4	Number of fibers	4/6/8/12/36 as per application	Owner's choice.
5	Core and cladding diameter	62.5 and 125 μm	
6	Primary coating	0.274 μm	
7	Filler and rip chord	Strengthening material filler with optional rip cord.	
8	Inner sheathe	Elastomer jacket	
9	Outer sheathe	Halogen-free, low-smoke flame retardant (PVC/LSZH)	
10	Wave length	850/1300 nm (Typ)	
11	Attenuation	3.5/1.5 dB/km for optical wave length of 850/1300 nm, respectively.	
12	Testing/color coding	IEC 60793	
13	Mechanical properties	Tensile strength, impact, etc., as per IEC 60794-1	
14	Bending radius	10 OD/20 OD during service and installation, respectively.	Manufacturer's recommended.
15	Testing	IEC 60793-1	
16	Temperature	70°C	
17	Connectors and accessories	Patch and splice mount/wall mount as applicable, connectors, etc.	

TABLE 12.13 LAN Cable (DS)

Sl.	Specifying Point	Standard/Available Data	Remarks
1	Type	CAT (UTP/FTP/SFTP)	
2	Standard	EIA/TIA 568B IEC11801/60332/60754/61034 ISO 11801	
3	Conductor dia (mm)	0.5 (0.64 mm 2 × 7 wire)	
4	Insulation	Solid PE thickness 0.2 mm	
5	Sheathing	PVC/LSOH/LSFROH	
6	Impedance	Typical characteristic impedance: 120 Ω	
7	Temperature	70°C Max	
8	Screen	For FTP/SFTP Al foil overall with tinned copper drain wire. SFTP will have copper braid wire.	
	Flame propagation/smoke/ halogen acid test, etc.	As per IEC listed above	
	Attenuation typical	1000 Base T @ 1 MHz—2 dB/100 m for CAT 5 cable whereas same for CAT 6 cable will be @4 MHz—3.8 dB/100 m	

What is thermocouple wire? **Wire used in a thermocouple (T/C) from point of sensing to the point of cold junction/where the signal is measured. Different T/Cs use a different mixture of metals.**

What is Limit of error? What is the difference between standard and special limit? **Limit of error generally refers to the accuracy of wire. Special limit has the same features as the standard limit, with added features with better accuracy. The accuracy of measurements vary with the thermocouple types.**

What is the difference between thermocouple grade/compensating cable and extension wire? **Thermocouple wires are used to make the sensing point of the thermocouple. Extension wire is used to extend the a thermocouple signal to the reading instrument. The extension wire has lower ambient temperature limit. Thermocouple cable may be used to replace extension, wire but extensionwire cannot be used as a sensing wire. Compensating and extension wire are generally represented by putting "C" and "X" after the thermocouple type, respectively. For example, SC represents compensating cable for type S T/C and SX represents extension wire for S type T/C. Both have different materials of construction as indicated in the table. Naturally for thermocouples with noble metals extension wire are less costly. Thus economy is also a factor when deciding to have two types of cables for these types of cables.**

What is the maximum length of thermocouple Wire? **As a rule of thumb, the thermocouple wire with thickness 20AWG or higher can be used for a length of <100 ft. There are really many factors to consider when limiting distances. Two main factors may be loop resistance and noise interference. Since different T/Cs have different materials loop resistance depend on the same as well as the AWG of the conductor. Normally it is recommended to keep the total loop resistance below 100 ohms. For noise reduction, shield/wire braids may be used to pick up.**

FIG. 12.34 Notes on thermocouple and T/C wires

in performance between the CAT 5/6/7 cables and the WRT different network and frequency. Standard manufacturer's specifications may be consulted for details (Table 12.13).

4. Thermocouple cables: Out of different kinds of thermocouple types, mainly the ISA type K, R, and S are in use in power plants. In some cases, type J is used (such as mill applications). During discussions, concentration shall be paid on these three types. There are two kinds of cables used for thermocouples: One is a compensating cable and the other is cheaper variety extension cables. These are especially so for costly thermocouple materials such as ISA type R and S in power plants (ISA Std) (Fig. 12.34, Tables 12.14 and 12.15)

TABLE 12.14 Compensating and Extension Wire With Error Limit Color Code

Property	ISA Type J	ISA Type K	ISA Type R	ISA Type S
Temperature range	0–760°C	0–1260°C	0–1600°C	
Error limit (std)	±2.2°C (±0.75%)	±2.2°C (±0.75%)	1.5°C for 0–600°C; 0.0025XT for 600–1600°C ±0.057 mV	
Error limit (sp)	±1.1°C (±0.4%)	±1.1°C (±0.4%)		
T/C material +ve	Iron*	Chromel (Ni Cr)	Pt 13% Rh	Pt 10% Rh
T/C material −ve	Constantan (Cu/Ni)	Alumel (NiAl#)	Pt	Pt
Comp. cable +ve	Iron*	Fe*/Cu**	Cu (IEC)	Cu(IEC)
Comp. cable −ve	Constantan (Cu/Ni)	CuNi	Cu Ni(IEC)	Cu Ni(IEC)
Ext'n. cable +ve	Iron*	Chromel (Ni Cr)	Cu (ANSI)	Cu (ANSI)
Ext'n. cable +ve	Constantan (Cu/Ni)	Alumel (NiAl#)	Cu Ni (ANSI)	Cu Ni (ANSI)
In. ANSI color +/−	White/red	Yellow/red	Black/red	
In.IEC color +/−	Black/white	Green/white	Orange/white$	

Continued

TABLE 12.14 Compensating and Extension Wire With Error Limit Color Code—cont'd

Property	ISA Type J	ISA Type K	ISA Type R	ISA Type S
Jac. color ANSI	Black	Yellow	Green	
Jac. color IEC	Black	Green	Orange	
Remarks	*Magnetic	As per IEC *KCA **KCB #Magnetic	$Not well defined	

TABLE 12.15 Compensating Cable/Extension Wire (DS)

Sl.	Specifying Point	Standard/Available Data	Remarks
1	Voltage grade (V)	300	
2	Standards	ANSI MC96.1 DIN 45722/IEC 584 IEEE 383, 70,000 BTU/UL 1581 NEC Article 725 PLTC NEC Article 510: Class 1, Division 2 UL Listed Subject 13	As applicable.
3	Temp. range	0–100/200°C (surrounding temp.)	
4	Conductor:	20 AWG solid alloy/0.5mm^2	
5	Insulation	Color-coded PVC	Refer Table 12.14 for color codes.
6	Insul'n thickness	0.38 mm	Jacket thickness as per of pairs.
7	Color sheathe	Color-coded PVC	Refer Table 12.14 for color codes.
8	Armor	Galvanized steel wire	As applicable.
9	Outer sheathe	Extruded PVC with FRLS property >1.8 mm thick with resistance to water. and fungus. Termite and rodent attack proof	
10	Pairs	1, 2, 4, 8, 12	As applicable.
11	Twisting	50 mm maximum lay	
12	Shielding	Individual/overall (Table 12.7)	As per application and owners' approval.
13	Shield type	Polyester tape—aluminum mylar tape (28 μm) with tinned copper drain wire (7X20AWG-0.51 mm^2) in metallic side of the tape.	Normally 50% overlap.
14	FRLS properties (see relevant std. mentioned above)	Oxygen index 0.29% Temp index >250°C Acid generation <20% Smoke density < 60%	As applicable.
15	Electrical parameter	Mutual cap. per km: bet cond. at 1 kHz: 200 nF (Typ) Insulation resistance: >100 MΩ Crosstalk @ 1 kHz: 60 dB	Manufacturer's data sheet may be consulted for exact details as it varies with types of cable.
16	Markings	FRLS, pair number, circles, etc., as per standard.	
17	Tolerance	As per relevant standard	Owner's approval.

4.6.5 *Types of Cable Terminations: There are a Number of Cable Termination Types Shown Below*

1. Plain screwed type with lug.
2. Plug-in Connector: These may be used in cases of junction boxes/instruments, etc.
3. Cage Clamp Type as may be found in a marshaling rack, system cabinet, CJC box.
4. Prefab cable: Annunciator, console to system cabinet, etc.
5. Manufacturer's standards: such as in PA system, wire wrap, Max Termi type as found in TG cabinets, etc.
6. With cable connectors such as RJ 45 for CAT cables, etc.
7. Opto cable connectors.

BIBLIOGRAPHY

[1] 2 Valve Manifold, Instrument, www.swagelok.com, Product Catalog.
[2] 3 Valve Manifold, Instrument, www.swagelok.com, Product Catalog.
[3] 5 Valve Manifold, Instrument, www.swagelok.com, Product Catalog.
[4] Valve Manifold, Instrument, Sabre Valve, Product Catalog.
[5] Fitting, Tubing and Nipples (HP)1/30/2014 Autoclave Engineers, Product Catalog.
[6] M.J.R. Harris, J.M. McNaught, Impulse Lines for Differential-Pressure Flow Meters—Best Practice Guide, NEL, Glasgow, UK, September 2005
[7] Installation Guidelines for C and I, Laying of Impulse Pipe, BHEL India Write up.
[8] Instrument Installation, Telisman Energy Inc., General Specification.
[9] Instrumentation Tubing and Their Connections, Instrumentation-tubing.blogspot.in/2012.
[10] Instruments Installation Details, Copper Cameron Valves United States, Product Bulletin.
[11] Local Instrument Enclosures and Rack, Pyrotech, Technical Catalog.
[12] Misc. Fittings and Accessories, Hex Valves (Richard Industries Valve group), Technical Product Buletin.
[13] Polaris Engineering Standard, Polaris Engineering, Internet Document.
[14] Standard Design Criteria/Guideline for Main Plant Package (2X500 MW TPP), Central Electricity Authority, New Delhi, 2010.
[15] Valve Manifold, Baxcell Instrument Valves and Fittings Pvt. Ltd.
[16] W.H. Howe, L.D. Dinapoli, J.B. Arant, in: B.G. Liptak (Ed.), Venturi Tube, Flow Tubes and Flow Nozzles, Instrument Engineers' Handbook.
[17] S.M. Al Abeedjah, Best Practices for Process Instrumentation Cabling, Intech, March 2008.
[18] D. Munch, T. Reinert, Best Practices for Terminating Fiber Optic Cabling, Ideal Industries Inc., Electrical Construction and Maintenance, April 2004.
[19] D. Payerle, S. Lindhart, Best Practices for Terminating Fiber Optic Cabling, IDEAL Industries, Inc., Electrical Construction and Maintenance, September 2005.
[20] K.Y. Lung, Considerations for Instrument Grounding, Agilent Technologies Tiwan, Application Note.
[21] Electrical Connection HART Version, ABB 2600 T Transmitter, ABB Catalog.
[22] Extension and Compensating Cable, ABB Limited, Product Catalog.
[23] R. Morrison, Grounding and Shielding, President INSTRUM, CALEX.
[24] Grounding and Wiring of Protection and Control Equipment, ABB Limited, Technical Document, February 2002.
[25] Instrumentation Tray Cables, Wire Wisdom, Internet Document, February 1997.
[26] Power Separation Guidelines, Data Network Resource, http://www.rhyshaden.com/pwrsep.htm (Internet Document).
[27] Process Industry Practices Process Control, PIP PCCEL001, December 1999.
[28] Proper Grounding of Instrumentations and control Systems in Hazardous Locations, J Zullo MTL Americas, Explosion Protection and Hazardous Locations Conference 2009—IDC Technologies.
[29] Thermocouple Cables PVC Insulation for Fossil Power Plant, Nexans Cables, Prodcut Catalog.
[30] XLPE Instrument Cables, Caledonian Cables UK, Product Catalog.

Chapter 13

Advanced Ultrasupercritical Thermal Power Plant and Associated Auxiliaries

Chapter Outline

1 OVERVIEW OF ELECTRICAL POWER GENERATION SCENARIO

In the present era, the electric power generation sector is experiencing gradual changes. Renewable power generation, utilizing mainly wind and solar energies, is growing considerably and contributing a sizeable chunk of the share of total electrical power produced in different countries.

For example, the share of renewable power generation in European countries varies from 10% to 25% of total generation. As per the US Energy Information Administration, the published figures show that renewable energy sources

Power Plant Instrumentation and Control Handbook. https://doi.org/10.1016/B978-0-12-819504-8.00013-5

accounted for 19.35% of US electricity in the first quarter of this year.

However, at present the renewable electricity generation (excluding hydro) is estimated to account for nearly 8% of global electricity generation. This figure still indicates the inevitability of electrical power generation through the burning of fossil fuels as well as nuclear or hydro power plants. There are a variety of opportunities for thermal generation to make the most of its potential as a vital resource of the future energy system through technological advancements.

So, as it appears, the requirement for thermal generation with the unique feature of adjusting power output as per demand is not going to decrease appreciably in the coming decades. The growing role in balancing the supply and demand of electrical generation with a stable capacity such as thermal power generation that is less affected by the weather will continue to be a necessity in the near future. Maintaining system stability is a very important feature, including stability of the transmission line that faces risks, particularly when the system generation loss or network fault takes place suddenly and/or unexpectedly. In these critical situations, the fundamental role of thermal generation is to provide all important system services, that is, stabilizing the power network through an inertial response or fast frequency power recovery.

It is also expected that thermal generation will be able to adapt to the demands of the changing situation and achieve new capabilities through performing necessary requirements such as faster start ups and shutdowns, higher ramping rates, more frequent changes in output, and the availability of a lower minimum generation.

The competitiveness in the present scenario also demands an improvement of thermal generation's environmental performance (that is, reducing the emissions of CO_2 and atmospheric pollutants), efficiency, and flexibility. Worldwide environmental regulations stress that thermal power generation must improve its environmental performance considering the factors such as cost, security of supply, flexibility, and competitiveness. The power world expects today that thermal electricity generation will be highly efficient and operating with maximum flexibility, a long lifetime, and cost competitiveness. In line with the demands for CO_2 reduction, the machine efficiencies are continuously increased.

Thermal generating plants have significantly reduced their emissions of NO_x, SO_x, and particulates over the past 20–30 years. Further reductions can be expected from the implementation of the Industrial Emissions Directive. It is predicted that emissions will be influenced by the changing operational regimes of thermal plants, and this needs to be taken into consideration by matching legally binding emission levels.

Carbon capture and storage (CCS) systems could enable thermal generation to become increasingly CO_2-free. A CCS system is a proven technology, but the full CCS chain from capture to storage still needs to be demonstrated on an industrial scale to ensure that the system is technically and financially feasible before its full-fledged deployment.

It is true that thermal generating plants using coal as fuel will continue to contribute the largest share of total electrical power generation capacity in the near future. Consequently, these power plants are also responsible for major contributions to CO_2 emissions worldwide. Improving their efficiency is an important way to cut global CO_2 emissions. The worldwide average efficiency of fossil fuel-fired generation was 36% in 2011 and is expected to improve to 40% in 2035. Similarly, in the European Union, the efficiency of coal-fired generation will increase from 38% to 44% and subcritical plants are expected to be almost entirely phased out by 2035. Increasing the net efficiency of coal-fired power plants is a major requirement for the intelligent and cost-effective use of CCS technologies. The most promising strategies to achieve this are based on boosting the steam temperature and pressure in new AUSC power plants through proper selection of construction materials and clean coal technologies based on oxy-combustion and cofiring technologies, among others.

The priority grid access granted to renewables means that fossil-fuel power plants will increasingly move from a base load operation to a load-following one. The flexibility of power plants therefore requires special attention in future research.

Other fields requiring special research attention are fuel pretreatment, fuel flexibility, emissions reductions (SO_2, NO_x, and dust), and polygeneration.

The European Union's Large Combustion Plant Directive (LCPD) (Directive 2001/80/EC) sets absolute emission limit values for SO_2, NO_x, and dust for individual new installations (coal or oil/gas-fired plants). Scope of the fuel pretreatments can improve power plant efficiency. Diversifying fuels can have a remarkable impact on efficiency while minimizing the effects of fluctuations in fuel price and availability. Using the best available technologies can increase efficiency. One of the most effective ways to increase the electrical efficiency of coal-fired power plants is to increase the temperature and pressure of the steam used to drive the turbines. Historically, changes in steam conditions have followed the development of improved steels. Current ultrasupercritical (USC) plants run with superheater temperatures up to 604°C, for a maximum electrical efficiency of 47%. The new target is advanced ultrasupercritical (AUSC) operation with superheater temperatures up to 700° C and efficiencies up to 50%. Combined-cycle plants using natural gas, the use of biomass in pulverized coal power plants, and integrated gasification combined cycle (IGCC) plants, which turn coal into gas, can all potentially reduce emissions even further, especially with carbon capture1. Further into the future, IGCCs with hybrid fuel cells could possibly reach 60% efficiency with zero emissions. For both new coal plants and retrofits, combustion using pure oxygen rather than air seems a promising way to reduce the cost of carbon capture.

Advanced fossil fuel power generation from coal is another option, yielding not only electricity and heat but also chemical feedstocks and alternative fuels for transport.

1.1 Some important predictions are:

(i) The global share of fossil fuels in total power generation is expected to fall from 68% in 2011 to 57% in 2035.

(ii) Global coal-fired generation will increase from 9140 $\times$ 10^6 MWh in 2011 to 12310 $\times$ 10^6 MWh in 2035, despite coal's share of total generation falling from 41% to 35%.

(iii) The average efficiency of coal-fired generation worldwide will improve from 36% to 40% in the same period as old plants are retired and replaced by newer technology.

(iv) In the EU, the efficiency of coal-fired generation will increase from 38% to 44% by 2035 as subcritical plants are almost entirely phased out.

(v) Gas-fired generation will rise from 4847 TWh in 2011 to 8310 TWh in 2035, maintaining a constant 22% share of total generation. In the EU, low electricity demand growth, support for renewables, high gas prices, and low CO_2 prices will stifle new gas-fired generation before 2020. After 2020, gas-fired generation will increase as inefficient coal capacity is retired, CO_2 prices rise, and more system flexibility is needed to match the growth in renewables. Compared to coal, gas-fired generation is cheaper and quicker to build, more flexible to run, and has lower emissions.

2 SHARE OF SUPERCRITICAL AND ULTRASUPERCRITICAL PLANTS IN GLOBAL THERMAL POWER PRODUCTION

Several new ultrasupercritical pulverized coal plants have cropped up in recent years around the world. The International Energy Agency's (IEA's) December 2012 report "Technology Roadmap: High-Efficiency, Low-Emissions Coal-Fired Power Generation" finds that in 2011, about 50% of new coal-fired power plants used mostly supercritical and ultrasupercritical coal units, a share that doubled in the past decade. Ultrasupercritical plants are already operational in Japan, South Korea, Italy, Germany, and China (Fig. 13.1). China is notably spearheading the ultrasupercritical

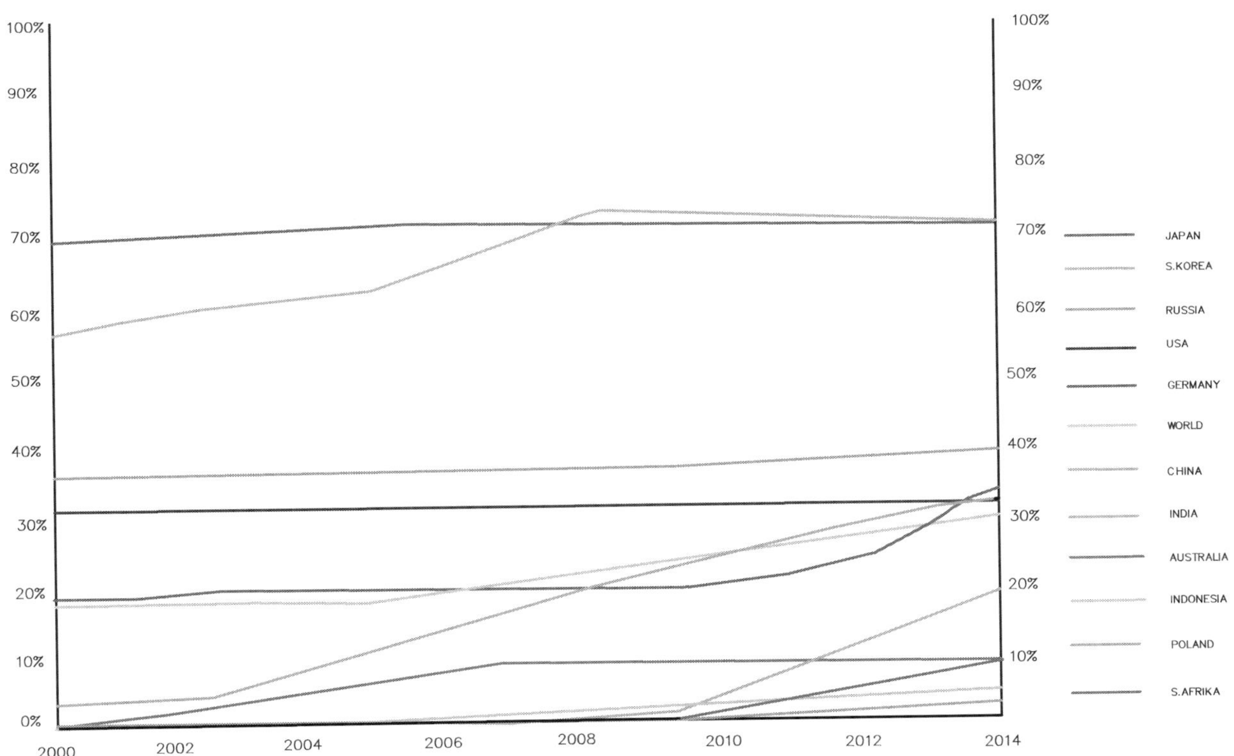

FIG. 13.1 Share of super/ultrasuper critical plants.

revolution: As of 2011, it had 116 GW of 600-MWe ultrasupercritical units and 39 GW of 1000-MWe ultrasupercritical units in operation out of a total coal-fired fleet of 734 GW.

2.1 Leading the World

The share of supercritical and ultrasupercritical capacity is highest in a handful of countries such as Japan and South Korea, where supercritical technology was adopted before 2000, the International Energy Agency says in a new report. Driven by fuel supply concerns and immense power needs, China's share of supercritical and ultrasupercritical capacity and India's share of supercritical capacity have also grown tremendously [Source: Technology Roadmap: High-Efficiency, Low-Emissions Coal-Fired Power Generation © OECD/IEA 2012, Fig. 5, p.11].

In the rest of the world, at least 75% of the world's operating coal power units use subcritical technologies, and most are more than 25 years old and less than 300 MW, the agency noted. Despite the significant fuel efficiency gains afforded by supercritical and ultrasupercritical technologies, the technologies' widespread adoption is constrained by growing water shortages, a need for flexible generation to balance renewables, and costs, the IEA says.

Nevertheless, AUSC plants aren't too far in the future, the agency speculates. Though progress is needed in development and deployment of AUSC plants operating at 700°C and above, most groups involved have announced plans to advance a full-scale demonstration within a decade (between 2020 and 2025). China/India have already started their own programs, funded by both the public and private sectors, the IEA notes.

(i) India's first 800 MW coal-fired power plant based on a more efficient AUSC thermal power technology is expected to be operational by 2018.

(ii) The United State's first such facility—Southwestern Electric Power's (SWEPCO's) 600-MW John W. Turk, Jr. Power Plant—was switched on in Arkansas. The project commissioned in December 2012 operates above supercritical pressure and at steam temperatures above 593°C, allowing it to employ a more efficient steam cycle that tamps down fuel consumption by 13% compared to a subcritical boiler. It also reduces reagent consumption, solid waste, water use, and operating costs.

3 DEVELOPMENT IN POWER GENERATION

Being the major player in total electrical power generation, modern superthermal power plants are now equipped with AUSC technology with reduced fuel and higher efficiencies to take care of the greenhouse effect and the increasing cost of appropriate fuel. Years of relentless research have made it possible through improved design, materials, and equipment. The major areas of improvement are the boiler, the turbine, and the boiler feed pump and its drive while the basic equipment/subsystems are more or less the same, although with more sophistication and technological development. Another major subsystem, the regenerative heating cycle, also has undergone several modifications associated with the improved boiler and turbine.

There are some other subsystems or offsite packages such as a water treatment plant (DM plant and pretreatment plant), CHP, AHP, an oil-handling plant, CW and ACW systems, a condensate polishing unit that are almost the same as before, except for capacity enhancement (Table 13.1).

3.1 The following are the major areas where conceptual and technological improvements have taken place but

TABLE 13.1 Steam Conditions, Materials, Efficiency, and Coal Consumption for Four Generations of Coal Plants (Nicol, 2013)

	Superheater Conditions	Materials for High Temperature Components	Efficiency (%,) LHV, Hard Coal	Coal Consumption (g/kwh)
Subcritical	≤540°C <22.1 MPa	Low-alloy CMn Mo ferric steels	<35	≥380
SC	540–580°C 22.1–25 MPa	Low-alloy CrMo steels, 9%–12% Cr martensitic steels	35–40	380–340
USC	580–620°C 22–25 MPa	Improved 9%–12% Cr martensitic/austenitic steels	40–45	340–320
AUSC	AUSC 700–725°C 25–35 MPa	Advanced 10%–12% Cr steels and nickel alloys	45–52	320–290

Source: THEMATIC RESEARCH SUMMARY, Advanced fossil fuel power generation.

research and development work is still going on to make thermal generation more attractive and acceptable.

(i) Fuel pretreatment and fuel flexibility.
(ii) Advanced firing systems, such as oxyfuel combustion.
(iii) Polygeneration.
(iv) Steam conditions and materials of construction.
(v) Conversion efficiency.
(vi) Emissions.

3.1.1 Fuel Pretreatment and Fuel Flexibility

Circulating fluidized bed (CFB) combustion, with its advantages of fuel flexibility, heat integration and low emissions, is well suited to using pure oxygen instead of air. CFB with Oxyfuel combustion will allow the use of low-quality domestic coal and biomass with CCS, thus improving energy security while cutting emissions. A flexi-burn CFB aims to develop and demonstrate a power plant concept that allows flexible high-efficiency air-firing of fossil fuels with biomass, and oxygen-firing with carbon capture—the latter providing the potential for an almost 100% reduction in CO_2 emissions. However, it is expensive. The project combines the intrinsic advantages of CFB combustion (fuel flexibility and low emissions) with oxygen-firing for CCS. CFB technology appears to be ideally suited to oxygen firing, and it will allow CCS to be used with lower-quality indigenous coals and biomass. This in turn will address the need for security of supply, reduced dependency on imported coal, and lower CO_2 emissions.

The expectation from this type of project is the demonstration of high-efficiency, utility-scale power generation with CCS, burning a large variety of indigenous and imported coals starting from lignite to anthracite with the facility of cofiring biomass. The main novelty is the combination of the latest advances in CFB boiler design with a supercritical once-through steam cycle, an air separation unit (ASU) to produce oxygen, and a CO_2 capture unit. The project has developed new simulation tools to support the FLEXI BURN CFB concept. A demonstration at 30 MW scale proved that an oxygen-fired CFB boiler can be operated in a reliable, controllable, and safe manner. Based on this and other tests at scales up to 30 MW as well as field measurements at a 460 MWe coal-fired plant in Lagisza, Poland, the project team developed the concept to commercial scale. Flexi-burn CFB is a first-generation CFB plant that is capable of operation with either air or oxygen. Second-generation oxyfuel CFB plants will use only oxygen for combustion, which aims significantly to reduce (by around 50%) the overall efficiency penalty imposed by CO_2 capture. Some projects are aimed to integrate the cogasification of coal, biomass, and wastes with processes for CO_2 separation and capture. Both fluidized bed and entrained-flow gasification processes were considered because of their suitability for different feedstocks. Separate work packages, including tailoring gasification schemes for integration with CO_2 separation, and developing materials for gas cleaning, char upgrading, and CO_2 separation have been envisaged. A 500 kW fixed-bed reactor has been designed, built, and operated within an existing IGCC power plant in Puertollano, Spain.

As an alternative, a new technology called *chemical looping combustion* (CLC), which is not yet in a realistic proposition for practical usefulness, is available at a much lower cost. In CLC, the oxygen required to burn the fuel is produced internally by oxidation and reduction reactions; this is discussed later in this section.

CLC in large-scale plants, currently under investigation, is expected to open new possibilities for using multiple fuels and reducing the energy penalty for CO_2 capture. More details are in Section 3.1.2.2.

3.1.2 Advanced Firing Systems

For both new coal plants and retrofits, oxyfuel combustion is a promising option that increases the concentration of CO_2 in the flue gas and so reduces the cost of CO_2 capture. However, it normally requires the flue gas to be recycled around the furnace, increasing the risk of corrosion. High oxygen concentrations in oxyfuel combustion save energy by reducing the flue gas recirculation rate, allow the use of smaller CO_2 boilers, bring new opportunities for using waste heat, and improve system flexibility.

Oxyfuel combustion is an example of an advanced firing system. Here, the process of burning a fuel uses pure oxygen instead of air as the primary oxidant. Because the nitrogen component of air is not heated, fuel consumption is reduced and higher flame temperatures are possible.

Highly dedicated research teams in many countries are working to use an oxygen-enriched gas mix instead of air, which is currently being done in firing fossil-fueled power plants. Nitrogen is removed almost entirely from the inlet air taken from the atmosphere. Approximately 95% oxygen is available at the outlet stream. Pure oxygen firing results in a very high flame temperature and hence the mixture is diluted through mixing with recycled flue gas (RFG) or staged combustion.

The RFG can also be used to carry fuel into the boiler and ensure adequate convective heat transfer to all boiler areas. In the absence of nitrogen, oxyfuel combustion produces approximately 75% less flue gas than air-fueled combustion. The exhaust gas thus produced consists primarily of CO_2 and a lesser percentage of H_2O of around 10%. The justification for using oxyfuel is to produce a CO_2-rich flue gas ready for sequestration. Oxyfuel combustion has some other significant advantages over traditional air-fired plants. Some important issues:

(i) The mass and volume of the flue gas are reduced by approximately 75% and so accordingly are the size of the flue gas treatment equipment and the heat lost in the flue gas.

(ii) Nitrogen oxide production is greatly reduced as nitrogen is freed from air and so the flue gas is primarily CO_2, suitable for sequestration as the concentration of pollutants in the flue gas is higher, making separation easier.

(iii) Most of the flue gases are condensable, which makes compression separation possible.

(iv) Condensation heat can be captured and reused rather than lost in the flue gas. However, it is to be admitted that this method costs more than a traditional air-fired plant. The separation of oxygen process needs lots of energy, nearly 15% of production by a coal-fired power station.

Oxyfuel combustion is often combined with staged combustion for nitrogen oxide reduction because pure oxygen can stabilize the combustion characteristics of a flame.

For both new pulverized coal combustion plants and retrofits, combustion using pure oxygen rather than air seems a promising option. The expectation is that this will minimize the cost of carbon capture because the flue gas from oxyfuel combustion contains around 90% CO_2, compared to 9%–14% CO_2 in the case of conventional combustion using air. Globally, the Callide Oxyfuel Project in Australia is a world-leading demonstration of how carbon capture technology can be applied to an existing coal-fired power station. The Callide Oxyfuel Project, one of a handful of low-emission coal projects in the world to move beyond the concept stage, is now in the demonstration phase. Combustion in an atmosphere of O_2 with RFG leads to higher concentrations of CO, SO_x, HCl, and fine particles, so the corrosion potential is higher. It is necessary to determine the critical corrosion parameters during oxy-coal combustion, investigate both high- and low-temperature corrosion, and characterize the fly ash produced and its deposition on plant elements. The impact of varying combustion parameters on the flue gas composition and their further effect on the corrosion process were studied and evaluated for use in developing advanced process layouts and boiler design strategies to increase efficiency and reduce maintenance costs while paving the way for full-scale deployment of coal firing in oxyfuel power stations. Various test projects are undertaking systematic and focused applied RD&D, involving both experimental studies and combustion modeling, to resolve existing technical uncertainties and barriers that inhibit commercial deployment of the technology. Attention is being paid to generating design rules and methods that can be employed to scale up results from pilot and laboratory studies to retrofit oxy/coal/FG recirculation systems (existing boilers) and new plants. These projects will therefore enable full-scale early demonstration plants to be designed with greater confidence, and will improve assessments of the commercial risks and opportunities. However, this requires large quantities of oxygen to be extracted from air, for example, a 500 MW power plant would need around 16,000 tons/day of oxygen. Conventionally, this would be obtained by liquefying and distilling air at temperatures down to—195°C—a surprisingly energy-efficient process in modern air separation plants with a high rate of heat recovery. Membrane separation of gases, however, can be even more energy-efficient when high purity is not required. One of the main drawbacks of CCS is the additional energy it uses for operation. This energy penalty reduces the power plant's efficiency, increasing both the cost of electricity and the use of resources. Reducing the efficiency penalty associated with CCS is therefore a major challenge for the years to come. The objective of another ongoing project is to demonstrate the concept of second-generation oxyfuel combustion that halves the overall efficiency penalty of CO_2 capture in power plants, from approximately 12 to 6 percentage points. This is done by using higher oxygen concentrations in oxyfuel combustion, reducing the flue gas recirculation rate and the energy penalty. The use of higher oxygen concentrations has other important advantages: smaller boilers, lower capital/operating costs, more opportunities to exploit high-temperature waste heat and process integration, smaller flue gas volumes, and improved system flexibility. Depending on the demonstration test results, the study will then focus on the best use of the heat and electrical energy available from different parts of the process to reduce the overall energy penalty of CCS.

For Oxyfuel combustion, the system involves three major components: (i) an ASU for oxygen production, (ii) a boiler suitable for oxy-combustion (fuel conversion/combustion unit), and (iii) a CO_2 purification and compression unit.

All these associated components, along with some possible design options, are depicted in Fig. 13.2A (incorporated in Fig. 13.2). Different systems can be configured with these components, resulting in different energetic and economic performances.

Two possible configurations of Oxyfuel combustion systems are either low- or high-flame temperature boiler designs. In low-temperature designs, the flame temperatures are similar to that of air-fired combustion, that is, around 1650°C. On the other hand, when flame temperatures exceed 2480°C, it falls under the advanced high-temperature design. Recycling of flue gas vis-a-vis the combustion products is necessary for low-temperature designs, whether it be for new or retrofitted boilers to lower the flame temperature to make it similar to the heat transfer characteristics of air-fired boilers. The highlight of high-temperature designs is to use increased radiant heat transfer in new power plant applications to reduce the size and capital cost of the boiler.

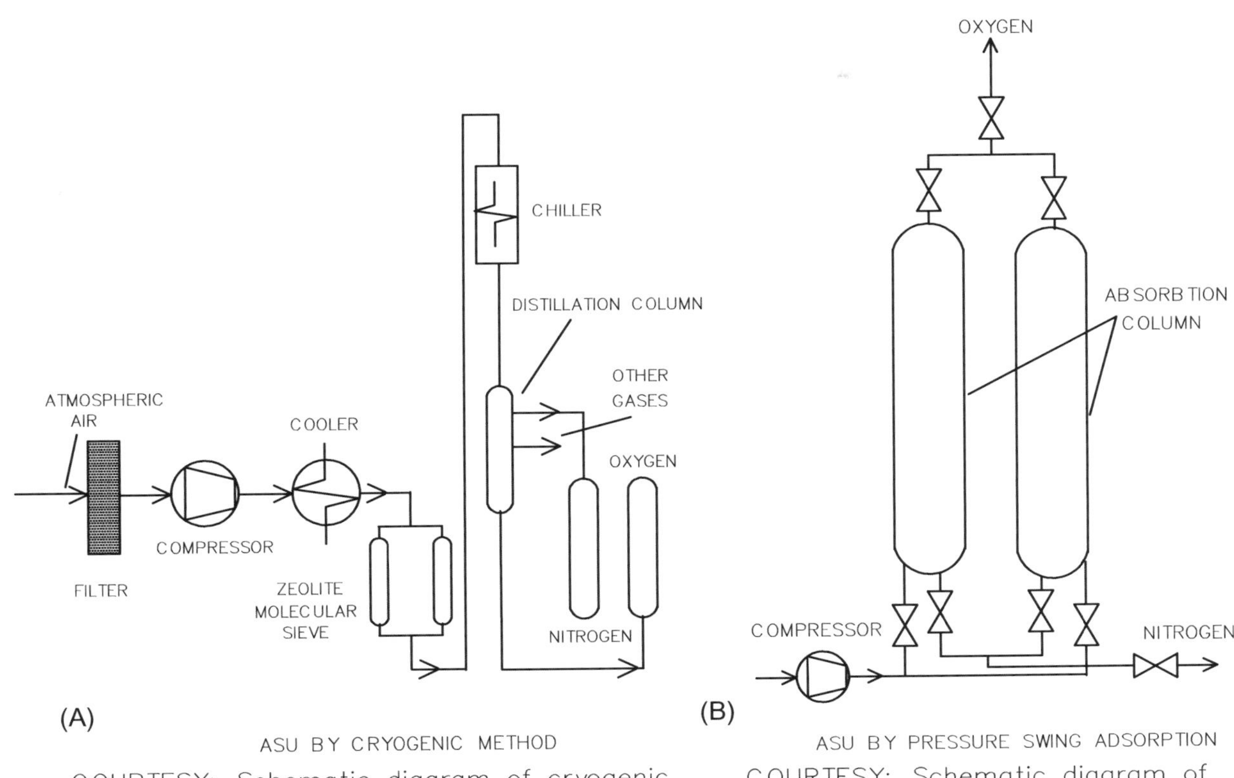

FIG. 13.2 Oxy fuel combustion general configuration.

3.1.2.1 Air Separation Unit

So far, the high-purity oxygen is generated via cryogenic distillation and pressure swing adsorption (PSA). However, cryogenic distillation is costly and energy intensive while PSA has limited purity of less than 95%. Both of these techniques possess the production capability of 20–300 tons of oxygen per day and oxygen purity of around 95%.

3.1.2.1.1 Air Separation Unit Using Cryogenic Process Nowadays, oxyfuel combustion systems comprise a cryogenic process to supply oxygen; atmospheric-pressure combustion for fuel conversion; considerable flue gas recycling; conventional pollution control technologies for SO_x, NO_x, mercury, and particulates; and mechanical CO_2 compression. As the costs of today's Oxyfuel combustion technologies are too high, a research program is developing advanced technologies to reduce the costs along with the efforts focused on developing pressurized (at high temperatures as well) oxy-combustion power generation systems and improving the performance of ancillary system components.

Cryogenic distillation, or the cryogenic liquefaction process, is similar to conventional air distillation. The ambient air will be drawn and compressed by a multistage air compressor and purified by an air filter to remove the impurities. Then, the temperature of the compressed air will be reduced to remove carbon dioxide, trace hydrocarbons, and water vapor prior to liquefaction. The liquefied air will be transferred into the distillation column where the nitrogen will be extracted from the top of the column due to its relatively lower boiling point compared to oxygen, which will be removed from the bottom of the column. The system in its simplest form is depicted in Fig. 13.3A, ASU by Cryogenic Distillation and PSA Method. The excessive feed gas in the column will be recirculated to the distillation column for several stages for further purification until the desired concentration of oxygen is achieved.

Cryogenic distillation has the advantages of high daily gas production volume (>100 tons/day) and excellent oxygen purity (>99%). To date, well-known global gas producers such as Air Products and Linde have commissioned more than 5000 oxygen product plants in the world.

3.1.2.1.2 Air Separation Unit Using Pressure Swing Adsorption PSA is a noncryogenic air separation process that is commonly used in commercial practice. This process as depicted in Fig. 13.3B, ASU by PSA method, involves the

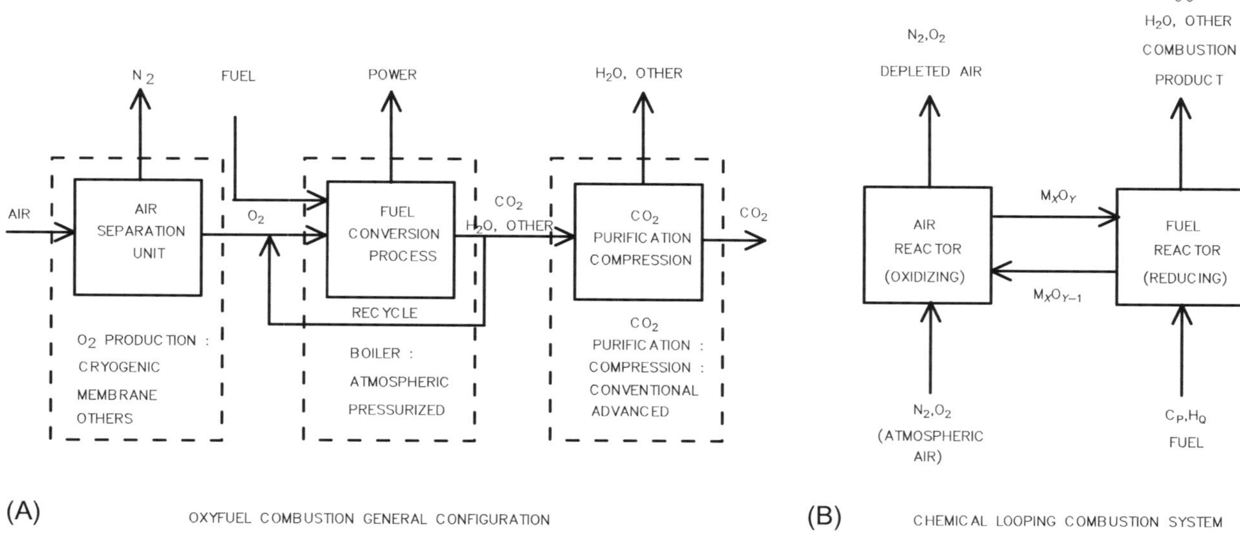

FIG. 13.3 ASU by cryogenic distillation and PSA method.

adsorption of the gas by adsorbents such as zeolite and silica in a high-pressure gas column.

In the PSA process, the air is drawn from the ambient atmosphere and compressed into high-pressure gas. The gas will be transferred into a column that is filled with the desired adsorbent materials, depending on the required gas. The system will be pressurized for a predetermined period and depressurized to atmospheric pressure, where the low sorbing gas will slowly leave the column first, followed by the other gases. If the adsorption process occurs under vacuum conditions instead of a pressurized environment, the process will be called vacuum swing adsorption (VSA). Generally, there are two or more adsorbent columns in the PSA process to avoid system downtime during the pressurized and depressurized processes. The PSA is appropriate to be utilized at a relatively lower daily production volume of 20–100 tons of oxygen and an oxygen purity $\geq$90%. Until now, M/s Praxair has pushed the production limit to 218 tons of oxygen per day with the purity up to 95% by integrating the PSA and VSA into one process, namely vacuum pressure swing adsorption (VPSA).

3.1.2.1.3 Air Separation Unit Using Membrane Technology Separation of air to produce enriched nitrogen and oxygen has been of great significance to the chemical industry. A membrane-based gas permeation process has gained due importance over conventional methods such as cryogenic distillation, solid adsorption, and solvent absorption because it is economical, compact in size, modular in configuration, and has the capacity to offer low specific power consumption. Typical applications of the membrane-produced nitrogen include blanketing, purging, inerting, and underbalanced drilling. Also, the oxygen-enriched air is mainly used for combustion enhancement in furnaces, medical respiration, and undersea breathing. Most membranes used in air separation are made of solid polymers.

The small difference between the kinetic diameters of nitrogen (3.64 Å) and oxygen (3.46 Å) makes their separation very difficult by porous polymeric membranes with pore sizes normally larger than 5 nm. Thus, air separation is achieved via the nonporous polymeric membrane where the solution-diffusion is the separation mechanism.

Due to the inherently low selectivity (the state-of-the-art O_2/N_2 separation factor: 6–9), the polymer membranes are actually used to produce oxygen-enriched air with an oxygen purity less than 50%. To be precise, it is believed that membrane technology is able to cater to small oxygen gas production volume at the range of 10–25 tons/day with an oxygen purity of 25%–40%. Until now, there have been no commercially feasible membranes that have high permeability and selectivity for large-scale commercial gas production. For more purified oxygen, a multistage operation has to be adopted but with more energy cost. To make a possible breakthrough, the mixed O_2-ion and electronic ceramic membranes have attracted more interest due to a 35% reduction of production costs to produce high-purity O_2 compared to the conventional distillation method.

Membrane units do not require a phase change and have relatively small footprints, being smaller than conventional systems. The lack of mechanical complexity in membrane systems is another advantage while their modularity allows an easy scale-up and results in a significant flexibility. Membrane devices for gas separation usually operate under continuous steady-state conditions. A strong push for technology to produce economically high-performance membranes was given by the development of the innovative concept of high-flux asymmetric membranes (Loeb and Sourirajan, 1962), initially for reverse osmosis and then adapted to gas separation. In addition, the implementation of new preparation methods (hollow fiber spinning) allowed a significant reduction in the effective thickness of the membrane selective layer, with a consequent increase of permeation rate and membrane surface area inside the membrane module. Gases dissolve and diffuse into polymeric films under a pressure gradient, generating a mass transport through the film. It is a pressure-driven operation where gas species are separated by "dense" membranes in virtue of differences in solubility and diffusivity. Nonpermeating molecules that remain at the feed-stream side leave the membrane unit as retentate stream. The right selection of the polymer determines the ultimate performance of the gas separation module.

Gas separation such as O_2/N_2 separation is a pressure-driven process where the driving force is induced by the difference in pressure between the downstream and upstream sides. The membrane used in the O_2/N_2 gas separation process is generally a nonporous layer so as to arrest the severe leakage of gas. Currently, membrane technology is still not commercially popular in O_2/N_2 separation and other gas separation applications.

3.1.2.2 Chemical Looping Combustion Systems

The projects with CLC aim to demonstrate the technical, economic, and environmental feasibility of high-temperature, high-pressure, packed-bed CLCs in large power plants.

A CLC uses two or more reactions to oxidize hydrocarbon-based fuels. In its simplest form, a chemical species (normally a metal) is first oxidized in air to form an oxide. This oxide is then reduced using a hydrocarbon in a second reaction. In recent years, interest has been shown in CLC as a carbon capture technique. Carbon capture is facilitated by CLC because the two redox reactions generate two intrinsically separated flue gas streams: a stream from the oxidizer, consisting of atmospheric N_2 and residual O_2, but sensibly free of CO_2; and a stream from the reducer containing CO_2 and H_2O with very little diluent N_2. The oxidizer exit gas can be discharged to the atmosphere with minimal CO_2 pollution. The reducer exit gas contains almost all the CO_2 generated by the system, so the CLC can be said to exhibit inherent carbon capture. Water vapor can easily be removed from the reducer flue gas via condensation, leading to a stream of almost pure CO_2. This gives the CLC clear benefits when compared with competing carbon capture technologies, as the latter generally involve a significant energy penalty associated with either postcombustion scrubbing systems or the work input required for air separation plants.

Packed-bed CLC reactor technology opens up the possibility of using multiple fuels such as coal, petroleum coke, and biomass in the same power plant. It will deliver power cost effectively with a reduced energy penalty for CO_2 avoidance compared to currently available techniques. The DemoCLOCK project has already developed suitable materials to serve as oxygen carriers in the CLC process. Simulation is now under way to design the test reactor, which will operate at up to 1100°C and 22 bar. This combination of parameters cannot be handled by a single-walled vessel, so the reactor incorporates a thick layer of internal insulation to reduce the skin temperature.

Chemical looping systems produce O_2 internally within the process, eliminating the large capital and operating/energy costs associated with O_2 generation. CLC is considered a "transformational" technology with the potential to meet program cost and performance goals.

In CLC systems, O_2 is introduced to the system via oxidation-reduction cycling of an O_2 carrier, as depicted in Fig. 13.2B, CLC System (incorporated in Fig. 13.2). The O_2 carrier is usually a solid, metal-based compound. It may be in the form of a single metal oxide, such as an oxide of copper, nickel, or iron, or a metal oxide supported on a high-surface-area substrate (e.g., alumina or silica), which does not take part in the reactions. For a typical CLC process, combustion is split into separate reduction and oxidation reactions in multiple reactors. The metal oxide supplies oxygen for combustion and is reduced by the fuel in the fuel reactor, which is operated at an elevated temperature.

This reaction can be exothermic or endothermic, depending on the fuel and the oxygen carrier. The combustion product from the fuel reactor is a highly concentrated CO_2 and H_2O stream that can be purified, compressed, and sent to storage for beneficial use. The reduced O_2 carrier is then sent to the air reactor, also operated at an elevated temperature, where it is regenerated to its oxidized state. The air reactor produces a hot spent/depleted air stream, which is used to produce steam to drive a turbine, generating power. Then the O_2 carrier is returned to the fuel reactor, restarting the reduction-oxidation cycle.

Current CLC R&D efforts are focused on developing and refining oxygen carriers with aspects such as acceptable cost, sufficient oxygen capacity, durability, developing effective/sustainable solid circulation/separation techniques, improving reactor design, effective heat recovery/integration, and overall system design/optimization.

3.1.3 Polygeneration

Polygeneration is a new idea in the power industry and refers to any process that transforms fuels into multiple energy forms or products. The simplest example is a CHP plant that produces cooling as well as heat and power. However, polygeneration also refers to more complex plants that yield chemical products as well as heat and power. Polygeneration system technology is similar to chemical plant technologies where very large gas-to-liquids (GTL) and coal-to-liquids (CTL) plants have been running for several years. The fuel could be coal, oil, gas, biomass, or mixtures of these. Polygeneration related to thermal power plants is based on coal or biomass. The process starts with a gasifier that converts fuel into a mixture of carbon monoxide, hydrogen, and steam; this is popularly called synthesis gas, or syngas. At the next step, carbon monoxide and steam after further reaction turn into CO_2 and more hydrogen, which is known as the water-gas shift reaction. The CO_2 is removed while the hydrogen can be burned in a gas turbine to generate electricity. Alternatively, carbon monoxide and hydrogen may be allowed to react to turn into methane or heavier hydrocarbons (synthetic gasoline and diesel) as per Fischer-Tropf (F-T) reactions; this is referred to as methanation. Methane can be used as a substitute natural gas (SNG) or as a building block for a huge range of chemical products, starting with methanol. Converting coal into so-called syngas may assist in reducing the global dependence on oil and natural gas resources. An innovative coal-to-SNG technology based on steam gasification is currently in development for plants in the 50–500 MW range. A follow-up project is targeting a demonstration of the complete process chain with coal/lignite. Products of polygeneration could include liquid transport fuels, hydrogen, fertilizers, and chemical feedstocks.

The CO_2-free SNG project aims to scale up technologies for gasification, methanation of the resulting syngas, and CO_2 sequestration, originally developed for biomass conversion, and to evaluate their technical and economical potential for use with coal. The necessary gasification and methanation tests will take place at an existing gasification plant. The new process includes a substantially simplified gas cleaning system based on carbonate scrubbing that will substantially increase process efficiency compared to current state-of-the-art systems. The single-stage pressurized water/carbonate scrubbing process simultaneously removes CO_2, sulfur, and tar. Methanation tests showed full conversion of CO in a compact honeycomb methanation reactor, with a very high selectivity for methane. A commercially available and economically feasible catalyst was identified for large-scale application. One 5 MW coal/lignite-to-SNG plant was simulated with parallel production of electric power and came out with 3.20 MW SNG plus net electricity production of 254.5 kW, which means the efficiency reached as high as 69.1%. The SNG produced meets the standard specification and so could be injected into the gas grid. A chemical scrubber was identified as the best option for removing CO_2 from the raw SNG. It is expected that larger SNG plants based on this technology should be profitable. Thermodynamic evaluation of plants rated at $\geq$50 MW showed ~63% simulated SNG efficiency. Based on detailed cost estimates including sensitivity analysis, it was concluded that the 50 MW plant could be an economically viable investment.

3.1.4 Steam Conditions and Materials of Construction

The increase of steam temperature and suitable candidate materials for associated pipeworks and equipment are the all-important research programs going on all over the world. Flexibility of power plants Coal-fired power plants are increasingly required to balance power grids by compensating for the variable electricity supply from renewable energy sources. This needs high flexibility in terms of the ability to withstand frequent start ups and load changes, and to provide frequency control. Designs are now being developed to give future plants such flexibility.

3.1.4.1 Economics of AUSC Technology

Though AUSC technology has already been in development for more than 20 years, it still requires an extensive materials research program. The main outstanding issue is the fabrication of large components. This research is expected to take more than 10 years, with substantial costs and technical risks. Consortia of utilities, manufacturers, and research establishments are combining their resources to overcome these barriers.

The intelligent and cost-effective use of CCS requires new strategies to increase the net efficiency of coal-fired power plants. The most promising strategies are: (i) increase the working steam temperature and pressure in new USC power plants (350–370 bar, 700–720°C minimum); (ii) promote clean coal technologies based (for example) on oxy-combustion and cofiring technologies, which means continuously increasing the percentage of biomass in the mixture with coal (to reduce CO_2 capture losses as well as the amount of CO_2 to be captured and stored); and (iii) the intelligent and cost-effective use of CCS.

Nationwise detailed research programs are discussed elaborately in Section 4 of this chapter.

TABLE 13.2 Emission Limit Values (ELVs) for Existing and New Coal Plants, According to the LCPD

1		SO_2	NO_x	PM 10
2	Coal plant (>500 MW) old	400 mg/Nm^3 (1300 mg/kWh)	From Jan. 1 2016: 200 mg/Nm^3 (650 mg/kWh)	
3	Coal plant (>500 MW) new	200 mg/Nm^3 (650 mg/kWh)	200 mg/Nm^3 (650 mg/kWh)	
4	Coal plant (>300 MW) old	Linear reduction from 1200 mg/Nm^3 (300 MW) to 400 mg/Nm^3 (500 MW) (from 3600 to 1300 mg/kWh)	500 mg/Nm^3 (1620 mg/kWh)	50 mg/Nm3 (160 mg/kWh)
5	Coal plant (>300 MW) new	200 mg/Nm^3 (650 mg/kWh)	200 mg/Nm^3 (650 mg/kWh)	30 mg/Nm^3 (100 mg/kWh)

Source: THEMATIC RESEARCH SUMMARY, Advanced fossil fuel power generation.

3.1.5 Conversion efficiency

In the future, direct electrochemical conversion of coal to electricity via fuel cells could offer a significant increase in efficiency and easier CO_2 capture. High-pressure CFB gasifiers can be run with high-ash coals, but the efficiency and reliability need improvement.

Direct electrochemical conversion of coal to electricity offers very significant increases in efficiency, with consequent reductions in CO_2 emissions and new possibilities when sequestrating CO_2. Direct carbon fuel cells (DCFCs) can exceed the performance of commercial MCFCs (molten carbonate fuel cells), and have given their name to the DCFC project, which seeks to apply DCFC technologies to coal conversion. The projects will focus on developing coal DCFCs, increasing scale, improving cell design, seeking new active and conductive structures and surface-promoted catalysts, addressing durability, investigating coal sources, and optimizing coal processing for this application. The project will conclude by making recommendations for a commercial-scale demonstrator. The OPTIMASH project aims to optimize the efficiency and reliability of gasifiers fueled with coals containing high proportions of ash. High-pressure CFB gasifiers are the target technology. The objective of the project is to develop a pilot gasifier capable of producing syngas at a pressure of 10 bars and a flow rate suitable for a 1 MW combustion plant. Indian high-ash coals are the main target of the project, but to ensure fuel flexibility, the project will also study Turkish high-ash coals. The project will allow the global efficiency of the gasification technology to be optimized for high-ash coal by minimizing steam use, optimizing particle size versus residence time, developing strategies to avoid particle agglomeration, investigating corrosion risks, increasing fuel flexibility, developing efficient ash disposal systems, and testing technologies for gas cooling, tar, and fly ash removal.

Direct conversion of coal to electricity via fuel cells has been discussed in Section 11 of this chapter.

3.1.6 Emission limits as per Large Combustion Plant Directive 2001/80/EC (LCPD) are given in Table 13.2.

Cofiring biomass in coal plants and IGCC plants in which coal or other solid fuels is converted into combustible gases have a relatively high emissions reduction potential, especially when used with carbon capture.

The gas turbines in the next generation of IGCC power plants may be able to burn undiluted hydrogen-rich syngas, with benefits for fuel flexibility and carbon capture. Innovative SCR-deNO_x catalysts for coal, biomass, and cocombustion are in development. These will improve the removal of NO_x and mercury while avoiding the formation of SO_2 from sulfur in the fuel.

The overall objective of the H_2-IGCC project is to provide and demonstrate technical solutions to allow the use of state-of-the-art gas turbines (GTs) in the next generation of IGCC plants. The goal is to enable combustion of undiluted hydrogen-rich syngas with low NO_x emissions, high fuel flexibility, high efficiency, and high reliability. The challenge is to operate a GT on hydrogen-rich syngas in a stable and controllable manner, with emissions and processes similar to current state-of-the-art GTs running on natural gas. The project aims to tackle this challenge as well as that of fuel flexibility by enabling the burning of back-up fuels, such as natural gas, without harming reliability. H2-IGCC is divided into four subprojects:

- **(i)** *Combustion*: Demonstrate the use of (undiluted) high-hydrogen syngas in typical natural gas combustion systems, with minimal modifications, so as to conserve the ability to burn a variety of fuels; demonstrate the safe use of undiluted high-hydrogen syngas in lean premixed combustion at comparatively low emission levels.
- **(ii)** *Materials*: Demonstrate cost-effective material/coating technologies to tackle a component's life-limiting problems of overheating/hot corrosion resulting from higher temperatures and residual contaminants in the syngas; validate material performance data, life prediction, and monitoring methods

applicable to industrial implementation in advanced IGCC plants.

(iii) *Turbomachinery*: Deliver a compressor design with a stability margin enabling the switch between fuels without compromising its efficiency; deliver a turbine design and cooling system capable of coping with the resulting heat transfer environment dominated by water vapor; verify designs using a large-scale virtual testing environment to meet industrial standards.

(iv) *System analysis*: Provide a detailed system analysis that generates realistic technical and economic results for future IGCC plants based on GTs. Various studies have shown that selective catalytic reduction (SCR) enhances the oxidation of elemental mercury in the flue gas, and so improves mercury retention in both the wet flue gas desulfurization and the dry adsorption processes. The devCat project addresses the control of NO_x, SO_3, and mercury (Hg) emissions from power plants. The goal is to study and develop special deNO_x catalysts for SCR for coal, biofuels, and cocombustion, targeting efficient NO_x reduction and mercury oxidation while also ensuring low conversion of SO_2^- to SO_3. Existing catalyst technology will be further developed with innovative designs for biofuel applications, including the use of nanotechnology. The project will work on minimizing investment and operating costs, avoiding deactivation when biofuels are burned, and devising new ways to regenerate deactivated catalysts.

4 INTERNATIONAL DEVELOPMENTS

Electricity is the most versatile energy carrier in the modern world and its growth has outpaced that of any other similar kind, leading to ever-increasing shares in the overall scenario.

Compared to oil and natural gas reserves, coal deposits are abundant and dispersed broadly in many countries throughout the world. In terms of supply, it is the most stable and most economical of all fossil fuels. Many countries around the world rely on coal as a primary power generation fuel source, including countries with high energy consumption. For example, China and the United States use coal for ~80% and 50%, respectively, of the power they generate. On a global basis, coal is the major source of energy, accounting for ~40% of all electrical power generated.

Going forward, the number of coal-fired thermal power plants is projected to grow further and generation will continue to be a crucially important energy source for meeting the ever-increasing worldwide demand, especially from the population in developing countries. At the same time, the combustion of coal and other fossil fuels generates CO_2, a greenhouse gas, and CO_2 from coal-fired thermal power plants accounts for roughly 30% of the world's energy-derived CO_2 emissions. Because rising demand for energy in such emerging countries as China and India is projected to increase those countries' coal use by a substantial margin, reducing CO_2 emissions from coal-fired thermal power generation has become an international issue and therefore deserves particular attention. This issue is reflected in the ongoing development of low-carbon technologies for power generation along with the adaptation of CCS technology. It is also fairly representative for technologies that are important in terms of their potential capacity to contribute to a low-carbon world economy.

Modern development incorporates a variety of next-generation technologies and is making advances in combined combustion by utilizing biomass fuels. With respect to long-term initiatives, manufacturers and researchers are seeking the practical application of oxygen-blown coal gasification technology, which is expected to provide the next generation of coal-fired thermal power generation. Establishing this revolutionary technology and applying it together with an integrated coal gasification combined cycle (IGCC) system and an integrated coal gasification fuel cell combined cycle (IGFC) system (see Section 11 of this chapter) will dramatically increase the generating efficiency and make it possible to substantially reduce CO_2 emissions in order to realize zero emissions. In addition, development of AUSC technologies will further enhance the high efficiency of the state-of-the-art USC technology at the present time. Ultimately, the main object of development must strive to realize innovative, zero-emission, coal-fired thermal power by combining IGCC and IGFC systems with CO_2 capture and storage (CCS) technologies.

The integrated coal gasification fuel cell combined cycle (IGFC) adds fuel cells to both gas and steam turbines for a triply integrated power generation configuration.

With its intrinsic advantages of fuel flexibility and low emissions, CFB technology appears to be ideally suited to oxygen-fired combustion. CFBs will allow the use of indigenous coals and biomass with CCS, thus addressing the need both for security of supply and reduced dependence on imported coal. High-temperature, high-pressure CLC in packed beds for large-scale plants is currently under investigation. Compared to the currently available techniques, packed-bed reactor technology is expected to open new possibilities in using multiple fuels (such as coal, petroleum coke, and biomass), and to deliver power cost effectively with a lower energy penalty for CO_2 avoidance. High-pressure CFB gasifiers have the advantage that they can be fuelled with high-ash coals. However, the efficiency and reliability of such gasifiers still need to be optimized. The focus, among other things, is on developing strategies to avoid particle agglomeration, investigating corrosion risks, increasing fuel flexibility, and developing efficient ash disposal systems.

Direct electrochemical conversion of coal to electricity offers a theoretically significant increase in efficiency, with consequent reductions in CO_2 emissions plus easier CO_2 sequestration. Such a system is, however, a long way from fruition. Current activities are directed toward developing combined coal-fuel cell conversion technology with a focus on increasing scale, improving cell design/durability, and finding new catalysts.

The ability to burn undiluted hydrogen-rich syngas in the next generation of IGCC plants will benefit the downstream carbon sequestration process. Work is ongoing to provide and demonstrate technical solutions to allow the use of state-of-the-art, highly efficient, and reliable gas turbines with low NO_x emissions and high fuel flexibility. Attention has been directed, among other things, toward using (undiluted) high-hydrogen syngas in typical natural gas combustion systems with minimal modifications; demonstrating cost-effective material and coating technologies to overcome problems that limit component life; designing compressors with a stability margin to allow switching between fuels without compromising efficiency; and generating realistic technoeconomic results for future gas turbine-based IGCC plants.

For advanced firing systems for both new PC combustion plants and retrofits to existing plants, oxyfuel combustion seems a promising option. The use of pure oxygen rather than air is expected to minimize the cost of carbon capture because the resulting flue gas contains around 90% CO_2. Combustion in an atmosphere of oxygen plus RFG leads to higher concentrations of contaminants and corrosion potential. Therefore, critical corrosion parameters need special attention during oxy-coal combustion. Higher oxygen concentrations in oxyfuel combustion reduce the flue gas recirculation rate and the energy penalty attached to CCS. They also enable the use of smaller boilers with lower capital and operating costs, make it easier to take advantage of high-temperature waste heat and process integration, reduce flue gas volumes, and improve system flexibility. However, to achieve this objective, it is necessary to demonstrate and analyze the effect of high O_2 concentrations on combustion performance, materials performance, fuel flexibility, controllability of solid recirculation within the boiler, and the effects on the CCS.

In a thermal power station, plant efficiency increases while fuel consumption and flue gas emissions decrease when steam parameters increase, but the maximum steam temperature is limited by the materials of the wetted parts for their service lifetimes without failure. China, India, Japan, the EU and the United States all have material research programs aiming for the next generation of increased steam temperatures and efficiency, known as AUSC technology. It is well known that the boiler side superheater (SH)/reheater (RH) metal temperatures will be more than that of steam [e.g., it may be 740°C (steam-side) and 770°C (gas side) to have 700°C] and the candidate metals selected for AUSC technology are nickel alloys and new high-temperature steels.

Nickel alloys have high creep resistance, fire-side corrosion resistance, and weldability above 630°C but low thermal conductivity. Also, some nickel alloys have limited steam-side oxidation resistance. They cannot be easily cast or forged for large rotors or thick-section components. As nickel alloys are expensive, their use is minimized by only using them in the hottest components of the boiler and steam turbine to maintain economic favorability. An alternative plant configuration to shorten the nickel alloy pipework is an attractive prospect that is elaborately discussed in Section 9 of this chapter. New austenitic steels use aluminum to form a protective alumina oxide for improved oxidation resistance up to 900°C. They are estimated to have creep rupture strength of over 100 MPa for 100,000 hours at 660°C. It is said that these alloys could have lower coefficients of thermal expansion than that of traditional austenitic steels.

Martensitic steels are now well understood, and interesting concepts in Europe and Japan to increase creep and steam-side oxidation resistance are being researched. These concepts include Z-phase strengthening, boron modification, fine tuning the composition with the carbon-to-nitrogen ratio, optimized heat treatment, niobium-free steels, chromium limitation of steels, and low carbon steels. Unfortunately, attempts to increase steam-side oxidation resistance while maintaining creep resistance have not yet been found. Low alloy ferritic steel has been seen as obsolete for the last 30 years. However, research has seen a paradigm shift toward the development of fully ferritic >15% chromium steels (without martensitic transformation) as they have sufficient steam-side oxidation resistance up to 650°C and do not require expensive postweld heat treatment (PWHT).

Coatings (carbide/metallic spraying) can be applied on the inside diameter surface for protection against steam-side oxidation and/or on the outer diameter surface against fire-side corrosion, that is, to combat any type of chemical attack. When used, it will allow the use of lower-cost materials and require a radical change in the component design and manufacturing process, depending on the method of coating, which requires strict process control; otherwise, the coating may not adhere. However, coatings are not currently used in PCC plants and boiler manufacturers have avoided coatings as they are said to have failed to work in practice. The developments and status of these major material research programs are stated below.

4.1 Development in the European Union (EU)

The EU has a deregulated power market and tight emission standards. The subsidized renewable energy and extraction of gas from shale formations could potentially become significant energy sources. The new PCC plants in Europe are only required with the closure of old PCC plant or nuclear power although Europe has been leading the coal technology and still running some of the most efficient and environmentally clean PCC plant globally.

TABLE 13.3 EU Research Program

Starting Year	Program Name-Span (Years)	Program Name-Span (Years)	Program Name-Span (Years)	Program Name-Span (Years)
1998	COST 501-3 years	COST 522-5 years	AD700-1A-3 years	AD700-1B-6 years
1999	MARCKO DE 2-4 years	KOMET 650-1 years		
2002	AD700-2A-4 years	AD700-2B-9 years		
2004	COST 536-4 years	MARCKO 700-4 years	ETR: AD700-3-4 years	COMTES 700: AD700-3-7 years
2006	NRWPP 700: AD700-4-1 years			
2007	E.ON 50 +: AD700-4-3 years			
2008	HWT1:PHASE 1-4 years	CRESTA-4 years		
2010	IMPACT-1 year			
2011	COMTES +: HWT II 3 years	COMTES +: ENCIO-6 years	Next Gen Power-3 years	MACPLUS - 4 years
2012	HWT1:PHASE 2-3 years			
2017	FSDP: build-4 years			
2021	FSDP: Operation-5 years			

Source: "Gantt chart of the European Research Program" by Kyle Nicol.

High-temperature materials used in modern USC PCC plants have been developed largely in Europe and research continues to develop advanced USC plants with optimized efficiency and economy with a clean environment. The EU research program/projects consist of a consortium of participants, including component manufacturers, major utilities, and research establishments in Europe carefully selected to meet the needs of the program in terms of competencies and knowledge of one or more aspects of the supply chain (Table 13.3).

Above, Table 13.3 shows different high-temperature material research programs in Europe with plans up to 2026. A brief description of the functional activities of the programs is indicated below:

(i) *COST*: The European Cooperation in the Field of Science and Technology (COST) program aimed to developed new steels that have slightly improved performance on material grades 91/92; it began in 1980 and ended in 2008. COST 501, COST 522, and COST 536 developed some new steels, such as grade 911 for 625°C piping/tubing and alloys FB2/CB2 for 610°C rotors.

(ii) *KOMET 650*: Known as Power station options: developments in materials and measurement techniques and tests under operating parameters at 650°C (a wholly German joint research program) and a correlating efficiency of >47% with 15 individual projects in the areas of materials, measurement techniques, and modeling. Small-scale tests were undertaken in the laboratory. Large-scale tests were conducted at four steam loops placed in the Westfalen PCC plant. Each steam loop tested 10 different materials up to temperatures of 650°C and pressures of 19 MPa.

COST and KOMET 650 concluded that steels have a practical limit of 630°C and that nickel alloys are suitable above 630°C. This prompted the EU utilities and component manufacturers to start the Advanced 700°C (AD700) Material Research Program in 1998. AD700 aimed to achieve 700°C/37.5 MPa SH steam parameters using new nickel alloys. It was initially divided into four phases over a period of 20 years from 1998 to 2018.

(iii) *AD700*

(a) AD700-1 (Phase 1): The aims were to confirm the technical and economic feasibility of the concept (part A), finished in December 2001and investigate material property requirements and then plan a material development program (part B) (COMTES700, 2013). Single steam cycles with parameters of 700/72 0/35 reaching 50.7% and double reheat steam cycles with parameters of 700/720/35 reaching 52% on coastal sites. Part

A also developed new concepts for power plant configuration to minimize the use of expensive materials. M/s Siemens has developed a horizontal boiler, known as the compact design. In that design, a 550 MW unit is 32 m shorter than a two-pass boiler or 60 m shorter than a tower boiler and the steam turbine plinth is raised to 30 m (and not the standard 16 m). The compact boiler reduces the amount of expensive high-temperature pipework by 80% (Figs. 13.21–13.23). Another arrangement is to have the boiler and steam turbine in line. These configurations have been depicted in Section 8 of this chapter (Fig. 13.7). Alstom and Hitachi Power Europe have also designed AUSC boilers. Part B was finished in 2004; the material testing proposed Sanicro 25 and Inconel 740 as the nickel alloys for use in an AUSC plant.

(b) AD700-2 (Phase 2) started in 2002 and ended in 2007; the preparatory work for a component test facility (CTF) was completed in Phase 3. Large-scale components were designed for the chosen host plant. AD700-2 chose candidate materials for component testing in Phase 3. Part of the characterization that finished in 2011 included long-term creep tests of martensitic steel (P92, H1F28, NF12), austenitic steel (Alloy 174, SAVE 25), and nickel alloys (Alloy 4020, Inconel 740, Nimonic 263).

(c) AD700-3 (Phase 3): The aim of AD700-3 was to demonstrate the novel manufacturing concepts and performance of new materials in operational boilers of large-scale components, that is, required to minimize the technical risk involved in a full-scale demonstration plant (FSDP). Phase 3 was split into three subprojects: the COMponent TESt facility for 700°C (COMTES700), the turbine control valve (TCV) project, and the Esbjerg test rig (ETR).

(d) AD700-4 (Phase 4): This phase was to preengineer a FSDP in order to assess the economic and technical viability of AUSC technology and make a decision on whether to go ahead with the build. The first three phases produced a notable research spin-off technology called the Master Cycle modifying the regenerative FW preheating system, adding an extra 1.5% net efficiency (Fig. 13.8).

(iv) COMTES 700 installed a CTF in Gelsenkirchen (Scholven PCC plant, Germany) comprising mainly a parallel main steam line with an optional bypass to the reheater. This steam line is further superheated through an evaporator and superheater to 705°C and 22.6 MPa flowing at 12 kg/s, which passes into a header, a pipe with an HP bypass valve, and an HP turbine control valve (TCV). Finally, the steam is spray cooled and added to the main steam again. Although the maximum steam pressure proposed for AUSC plant designs is higher at 35 MPa, the steam loop could not exceed the 22 MPa limit of the steam turbine. Components were fabricated from the alloys, for example, T24, HCM12, TP310N, HR3C, Alloy174/617/617m, and Inconel 740 for monitoring/inspection over the test duration.

Testing at Scholven PCC plant provided the first opportunity (2004 and 2009) to investigate the effects of high-temperature gradients (creep/fatigue properties) for 24,000 h at 700°C on nickel alloys. Repair welds of thick-section (50 mm) Alloy 617B steam pipe developed cracks in the heat-affected zone (HAZ) along the grain boundaries, caused by a phenomenon called stress relaxation cracking (SRC). An investigation showed that Alloy 617B hardens with time due to precipitate formation around the grain boundaries. However, optimized welding procedures and PWHT (Alloy 617B was exposed at 980°C for 3 h) did well to avoid SRC. By 2009, valuable operational experience had been gained regarding manufacturing, bending, and welding of the new materials, including testing of nickel alloys and operational behavior (flue gas corrosion and steam oxidation) in high temperatures, determination/evaluation of residual service life, data collection, etc. A special working group evaluated the results and suggested improved component design, updated boiler codes (design of components from verified alloys), problem identification, and further research requirements.

(v) TCV: The CTF included testing an HP turbine control valve (HP TCV), which was manufactured by Alstom and Siemens. The complex configuration makes HP TCVs critical components in the steam turbine with regard to safety and optimal performance. They are sensitive to thermal stresses and are susceptible to fatigue damage as they are exposed to the highest steam parameters and largest thermal gradients of any steam turbine component. So, their testing in the CTF represented other steam turbine components.

(vi) The Esbjerg test rig (ETR) consisted of a single loop of SH tubes placed in a boiler of the Esbjerg PCC plant (this was the fourth steam loop installed). The ETR operated up to maximum steam parameters of 720°C and 27 MPa. The steam loop was manufactured by Alstom Power Boilers (Stuttgart) and operated from 2004 to 2008. The materials TP347HFG, S304Hcu, TP310N, HR3C, Sanicro 25, HR6W, and Inconel 740 (VGB, 2013b) were assessed for resistance to fire-side corrosion and steam-side oxidation.

TABLE 13.4 Candidate Materials for the EU Program for FSDP

Component	Candidate Material
Header inlet superheater 1, 2, 3, and 4	P92, P92, Alloy 617, and Alloy 617m
Header inlet reheater 1.1 and 1.2	13CrMo4-5 and Alloy 617m
Header outlet superheater 1, 2, 3, and 4	P92, Alloy 617, Alloy 617m, and Alloy 263
Header outlet reheater 1.1 and 1.2	Alloy 617m
Superheater 1	T92
Superheater 2	Alloy 617m, Alloy 174
Superheater 3	Alloy 617m, Alloy 174, HR3C
Superheater 4	Inconel 740
Reheater 1.1	Alloy 617m, HR3C, S304, T91, 10° CrMo9.10
Reheater 1.2	Alloy 617m
Casing, outer casing	Cast steel (9%–10% Cr)
Casing, inner casing	Alloy 625 (cast), welded with 9%–10% martensitic steel
Valves casing	Alloy 625 (cast)
Valves weld-on ends	Alloy 617m
Blades	Martensitic steel, Nimonic80, Waspalloy
Rotor HP and IP	Alloy 617 welded with 2% chromium, 10% chromium steel

Source: Kyle Nicol "Status of advanced ultrasupercritical pulverized coal technology."

(vii) NRWPP700 is the preengineering study at the North Rhine-Westphalia Power Plant at 700°C from October 2006 to 2008 in three stages.

Stage 1 was for the technical and economic feasibility of a demonstration unit. A boiler with steam parameters of 705/720/36.5 and a steam turbine generated a total of 500 MWe net including the auxiliary load. Research projects carried out the production of an HPBP valve casing (Alloy 617), an HP pipe (by extrusion process from Alloy 617), a qualification of SH material (Inconel 740), and thick-section SH pipes (Alloy 617), processing and relaxation of new nickel alloys under modified cyclic load, components design and investigation of development possibilities for high-temperature sensors and nondestructive investigations of different welds to produce steam turbine rotors fabricated from nickel alloys. The candidate alloys are shown in Table 13.4. It is important to note that Alloy 263, Inconel 740, and Alloy 174 are still in the development stage. The membrane wall is SH 1 in this design.

Stage 2—This stage transferred the technical and economic findings of Stage 1 to a commercial 1000 MWe unit to test for economic feasibility against a 1000 MWe USC plant. It was proved technically possible to scale up the boiler and steam turbine. For highest steam turbine efficiency, a conventional configuration was chosen with a single flow HPT, a double flow IPT, and two double flow LPTs. Using an inland state-of-the-art USC plant (600/610/28) with 45% efficiency as a starting point, the following efficiency improvements have been estimated for a new AUSC PCC plant as +5.2% points −3% for increased steam temperature and pressure, 0.2% for increased feed water temperature, 0.7% for seawater cooling, and 1.9% toward miscellaneous improvements.

Stage 2 identified the need for the following additional research projects: short- and long-term properties and production/manufacturing of nickel alloys; an improved version of P92 (to reduce the amount of nickel alloy), and further qualifying works on austenitic material Sanicro 25.

Stage 3—Commercial plant planning (1000 MWe, lignite coal with predrying). This stage transferred the technical and economic findings of Stage 2 to a commercial predrying lignite-fired 1000 MWe unit test for economic feasibility (against a 1000 MWe USC plant) running at 650°C, 675°C, and 705°C with lignite predrying was assessed reaching efficiencies of 51.6%, 52.06% and 52.63% respectively (net, LHV). However, differences in temperatures, pressures, and dimensions required some redesigning.

(viii) E.ON 50+: NRWPP700 did the first task of preengineering after which E.ON took over the FSDP build in a project called E.ON 50+. In 2008, it was decided that a 500 MW FSDP would be built at Wilhelmshaven (Germany). The electricity produced from a demonstration unit is likely to be more expensive than the projected market value, as the unit is smaller and it is a prototype. Commercial AUSC PCC units were likely to be built in the 1000–1100 MW range to utilize economies of scale, but this was postponed in 2010 due to technical problems (such as cracking of thick-section components). However, research continues in order to solve these technical problems and knowledge from all former/current projects for consolidation in the FSDP being planned to be built from 2017 to 2021 and operated from 2021 to 2026.

(ix) MARCKO DE2 (Material realization for a CO_2 power plant) started in April 1999 and ended in March 2003. It helped qualify Alloy 617 for outlet headers and SH tubes in AUSC boilers. Alstom Power Boilers in Stuttgart successfully fabricated a steam header from Alloy 617m via gas tungsten arc welding (GTAW) and submerged arc welding (SAW) procedures. GTAW narrow gap orbital welding has been shown to be the weld procedure of choice for martensitic steels and nickel alloy tube-to-tube welds in PCC plants. The process is fully automated and produces efficient and reliable welds. Less filler material is required due to efficient use and the narrow gaps reduce the amount required. The automated process is quicker than the manual process. Automated welding is reproducible and ensures a constant level of quality.

(x) The MARCKO 700 (Material qualification for the 700/720°C power plant) project, which began in August 2004 and ended in June 2008, was headed by Professor Karl Maile of the Stuttgart Materials Testing Institute. It helped qualify materials T24, 12CrCoMo, and Alloy 617 for use in tubes/components; and T91, T92, and VM12 for use in membrane walls. MARCKO 700 started long-term creep rupture tests of components and welds.

(xi) HWT: The project Hochtemperatur Werkstoff Teststrecke or high-temperature material test track is titled "Investigation of the long-term service behavior of tubes for future high-efficiency power plants." HWT I installed four test loops in a commercial boiler, unit 6 at the GKM PCC plant near Mannheim (Germany). This boiler has a peak temperature of 1260°C and the test loop produced steam at 725°C and 16 MPa at a flow rate of 0.32 kg/s. Forty-three different alloys were used; there were three martensitic steels and nine austenitic steels in the first loop, 11 austenitic and three nickel alloys in the second loop, three austenitic and seven nickel in the third loop, and finally seven nickel alloys in the last loop. All aspects of fire-side corrosion, steam-side oxidation, dissimilar metal welds (DMW), and similar welds with various fabrication methods and designs will be assessed. Additionally, there are two external loops undergoing creep rupture tests with live steam at 630–725°C testing 10 austenitic steels/nickel alloys. Information gathered from the privately funded HWT1 (2008–2015) has been used to qualify new SH materials.

(xii) COMTES+: This was set up in 2011 to further qualify candidate AUSC materials. Collectively, the AD700 programs as well as the KOMET and MARCKO projects have successfully demonstrated the assessed manufacturability of critical components from nickel alloys at higher steam parameters and 650°C steels developed to significantly improve the economics of AUSC technology. However, problems remain: the cracking of thick-section nickel alloys must be overcome, manufacturing costs must be reduced, fire-side corrosion under biomass and waste cofiring must be assessed, and performance under static operation to severe cyclic operation must be guaranteed. The issues tackled in AD700 continued under the names of COMTES+ (ENCIO and HWT II), NextGenPower, and MACPLUS. COMTES+ will reuse components from COMTES700; results from COMTES+ were expected by 2017 and generated all the knowledge necessary to design, build, and operate an FSDP.

(xiii) HWT II: Hochtemperatur Werkstoff Teststrecke (HWT) II primarily investigates the operation and failure behavior of thick-section components under base load and cyclic operation. The work was completed in a test loop at GKM PCC plant (Germany). HWT II started in January 2011 and ended by December 2014. Steam extracted from unit 6 at 530°C was heated to 725°C in SHs fabricated from materials P92, DMV310, A263, A740, and A617B. The SH pipes A740 and A617B were welded to a header fabricated from A617B and A263, respectively.

A bypass valve (the body is forged from Alloy 617B) is installed after the header to release steam from the test loop in case of an emergency. The HP bypass supplied by HORA was originally manufactured for the E.ON 50+ demonstration plant. To avoid cracking of the HP bypass valve, the valve has been redesigned and PWHT is used. The valve is of the flow to open design with 2.75 kg/s of steam flow. The bypass piping includes a pressure-reducing and desuperheating system (PRDS) for simulating reheater inlet conditions with start-up and shut-down procedures and definitive cycling conditions. The steam (725°C) pipe is extruded from A617B or A263. Operation of the test loop started in October 2012.

(xiv) ENCIO: The European Network for Component Integration and Optimization (ENCIO) project started in 2011 and is expected to finish in 2017–18. The aim of the ENCIO project is to demonstrate and qualify fabrication, welding, behavior, erection, and repair concepts for up to 140 mm thick-section nickel alloy components for a long lifetime. The work included test loops in a new test facility operating at 700°C, 17.7 MPa, and a 5 kg/s flow rate of steam. The host plant is the Andrea Palladio Fusina PCC plant (Venice, Italy). The steam cycle is arranged so that steam at 705°C is sent to the four independently operated test loops, which investigate the following items:

Test loop 1: Development of a pipe repair concept for aged materials: This loop simulated a pipe repair situation. This task was investigated to characterize the aged materials, followed by repair welds that are used to optimize welding procedure specifications.

Test loop 2: Lifetime monitoring for components at 700°C and hot isostatic pressing (HIP) of components. This loop was used for online measuring and to develop a monitoring concept of the creep behavior of nickel alloy pipes (Alloy 617B). This loop was used to test components, such as T-pieces, valve bodies, and turbine components that have been fabricated using HIP. HIP is a commercial manufacturing process where materials are placed in an inert atmosphere in a pressure containment vessel for a certain amount of time. The material undergoes creep, plastic deformation, and diffusion bonding, which eliminates internal voids and microporosity. This increases the density of the material, which improves the material properties such as workability and fatigue resistance.

Test loop 3: New materials for thick-section components: This loop explored possible improvements in the weldability of Alloy 617 by means of improved melting processes to reduce the amount of impurities within the ingot.

Test loop 4: Testing of turbine cast components and welds: This loop aims to prove full-scale welds between thick-section Alloy 617 OCC and Alloy 625 cast (steam turbine).

(xv) NextGenPower is a project that aimed to demonstrate the fabrication and utilization of nickel alloys and material coatings for use in boilers, pipeworks, and steam turbines in AUSC PCC plants. The NextGenPower project began in May 2010 and ended in April 2014. The project was divided into four main categories:

(a) Boiler and pipework (Doosan and Babcock): Using coated steels in the boiler and heat exchangers could be a cheaper alternative than using nickel alloys. This project assessed the application method, welding compatibility (thus the ability to be repaired or replaced), and the performance of steel coatings in fire-side and steam-side corrosion. Mechanical tests were proposed to quantify the life expectancy of coated steels. These tests include creep, low cycle fatigue (with dwell, notched relaxation), cyclic notched tests, slow strain rate tensile tests, fracture toughness, and hardness. Tests were conducted selectively upon the parent metal, the weld metal, the cross weld, the longitudinal weld, and the service-simulated material. Projects include fire-side corrosion of the membrane walls/superheaters, steam-side oxidation in the boiler, and creep and fatigue of tubes, pipeworks, and steam turbines.

(b) Steam turbine (Skoda): This project demonstrated the capability to cast, forge, and weld nickel alloys for rotors, casings, and valve chests. A program of mechanical tests was carried out to verify component properties. Nondestructive inspection techniques were optimized and qualified. For nickel alloys, steam-side oxidation is negligible and the mechanical properties were modeled on the basis of the material composition.

(c) Integration: The main objective of this is to integrate the work done as indicated above, including operating conditions and environments, modeling, case studies, and dissemination as well as making sure that at a later stage, the results can be integrated into the power plant.

(xvi) MACPLUS (material component performance driven solutions for long-term efficiency increase in ultrasupercritical power plants) assessed six aspects of AUSC technology via laboratory testing, industrial-scale test loops, and computer models.

MACPLUS started in January 2011 and ended in June 2015. The total MACPLUS is split into projects such as an investigation of new refractory materials/new ferritic and martensitic steels (for use in headers and pipework to avoid premature cracking of welds) and a means to understand the operational behavior of austenitic steels and nickel alloys in SHs using laboratory specimens and materials from previous test loops, such as the AD700-3. Material modifications via fabrication and composition will be assessed along with long-term behavior extrapolation via modeling. An SH tube and welded joints of thick-section pipes of modified nickel alloy were fabricated and tested to demonstrate improved performance, assessment of fabrication, and the application processes of multilayer and multimaterial boiler tubes for application in aggressive fire-side conditions. The project also assessed high-temperature components in a steam turbine using the latest generation high alloy steels for use at 620–670°C and nickel alloys at 670–720°C to be completed via metallurgical-thermomechanical modeling and manufacturing.

(xvii) E.ON New Build and Technology aimed to develop advanced design and testing criteria for high-temperature component development/integration/standardization into an AUSC PCC plant built on existing testing methods as well as to identify future development needs.

(xviii) IMPACT (Innovative Materials Design and Monitoring of Power Plant to Accommodate Carbon

Capture) based in the United Kingdom is a joint venture that ran from 2010 to 2013. There were three objectives within the project: the development of advanced welded MARBN steels for power plants, improving the design for welded components to reduce premature cracking, and improving the strain and materials monitoring to allow high-temperature operation. The proportion of boron in steel has been previously altered with inconsistent results. Fujio Abe of Japan found that alloying boron with nitrogen in certain amounts achieves high creep rupture resistance and MARBN steel (MARtensite plus boron plus nitrogen) was discovered. The IMPACT project has optimized the composition of MARBN steel and short-term tests have shown that MARBN, at up to 675°C, has 20%–40% higher creep rupture resistance than P92; this further increases by 10%–15% with heat treatment at 1200°C. Potentially, MARBN has 55% higher creep rupture resistance than P92. However, MARBN steel shows a substantial weld strength reduction and is susceptible to cracking in the HAZ. The proportions of the alloying elements in MARBN need not be exact but a wide range is acceptable for similar properties. In May 2012, Goodwin Steel Castings (UK) produced an eight-ton melt of MARBN, from which ingots and castings were fabricated and welded. These components are now in long-term tests.

(xix) Z Ultra (Z-phase strengthened steels for ultrasupercritical power plants) started in February 2013 and aimed to improve the creep rupture strength of martensitic 12% chromium steel by increasing the amount of Z phase precipitates for operation with 650°C steam. Numerous steels, filler materials, and welding processes were assessed. Atomic-scale microstructural modeling and investigations were employed to hasten research, improve understanding, and provide design tools and lifetime estimation methods for the operation of future power plants. The project coordinator is Fraunhofer IWM (Germany) and it involves 10 EU and some non-EU partners.

(xx) Rafako (Poland) conducted a research project that aimed to calculate the highest possible steam parameters using commercially available materials. These materials must meet the applicable European standards and therefore represent a boiler that is commercially available. First, the critical boiler components were identified. Second, the global steel market was analyzed. Third, the most suitable materials were selected for critical components. Finally, the highest possible steam parameters for the selected materials were calculated, regarding strength only, using the TRD-EN program to achieve a 200,000 h operating time. Corrosion/oxidation had been assumed not to become a problem within the 200,000 h operating time. Some elements were recalculated by the ANSYS program (finite elements method). The result is a boiler with steam parameters of 653/672/30 and 48% efficiency (LHV, net, 25MJ/kg coal heating value, 94% boiler efficiency, 4.5 kPa condenser pressure) using materials of only 15% austenitic steel and 5% nickel-based alloy. A problem with this study, however, was that a turbine operating at these steam parameters was not commercially available.

4.2 Development in the United States

The United States has huge coal reserves and has helped pioneer the use of new high-temperature materials in PCC plants. The program on AUSC power plant materials in the United States started in 2001. The research program was split into two consortia, the major US boiler manufacturers (Alstom, Babcock and Wilcox, Foster Wheeler, Riley Power, GE Energy) and the US steam turbine manufacturers (Alstom, GE energy, Siemens). The Oak Ridge National Laboratory and the National Energy Technology Laboratory supported both consortia. The research program was managed by the Energy Industries of Ohio with the Electrical Power Research Institute serving as the program's technical lead.

The boiler side of the program was split into two phases. Phase I was split into eight tasks. Phase II extends and enhances the tasks in Phase I. The boiler side of the program assessed alloys in both air/pure oxygen-fired boilers for cyclic operation, material coatings, and effects on fire-side corrosion with high-sulfur indigenous coals. The turbine side of the program was split into two phases; Phase I selected candidate materials and Phase II was for testing these materials in six tasks, as shown in Table 13.5.

TABLE 13.5 US Research Program

Starting Year	Program Name (Years)
2001	Materials R&D Program—15 years
2012	Component Test Facility Design—1 year
2013	Component Test Facility Build— 1 year
2014	Component Test Facility Operation—3 years
2015	Demonstration Plant-Design and Permit—2 years
2017	Demonstration Plant-Build—4 years
2021	Demonstration Plant-Operation—4 years

Source: "Gantt chart of the US research program" by Kyle Nicol.

4.2.1 The National Energy Technology Laboratory

The National Energy Technology Laboratory aimed for a 760°C maximum temperature and a 35 MPa maximum pressure, with efficiencies of 45%–47% (net, HHV) and a corresponding drop in carbon dioxide emissions of 15%–22%. Alstom estimated a 7% efficiency point increase toward the AUSC steam parameters. Alstom argued that a 760°C maximum temperature should be reached as opposed to 700°C for three reasons. First, Ni alloys can reach this temperature. Second, the cost of the precipitation-strengthened NI alloy needed for 760°C is the same as the solution-strengthened alloy for 700°C. Finally, a conventional PCC plant configuration can exploit temperatures of 760°C.

4.2.1.1 Task 1: Conceptual Design

Alstom has completed designs for a 550 and 1100 MW tower AUSC boiler and steam turbine set based on a proven conventional configuration and experience. In late 2012, Alstom designed a 1000 MW two-pass AUSC boiler and assessed the possibility of placing the HP turbine higher to shorten the steam pipe, leaving the IPT/LPT at ground level.

Babcock and Wilcox (Band W) have a modified AUSC tower boiler design that is based on proven USC boilers. The modified tower design combines features of both a tower boiler and two-pass designs; the advantages include minimized steam-side oxidation and fire-side corrosion, improved reheat control, lower overall height, and shorter pipework. B&W state that they have a boiler ready for MS and RH temperatures of 700°C and 730°C, respectively. B&W are working with steam turbine manufacturer Toshiba on alternative steam cycles to minimize piping length for air/oxyfuel boilers with and without the addition of CCS. Factors under consideration included single/double reheat, boiler capacity, steam turbine configuration, base load and cyclic operation (starting conditions, turndown cycling, FW pump drive, sliding pressure, pure or throttle reserve, pressure shelf at minimum load, and rate of load change), and the application of carbon capture (condensate heat exchangers for oxy-combustion heat recovery and turbine steam extraction for postcombustion solvent regeneration). B&W are confident that they would be able to manufacture a boiler capable of handling steam at 760°C. The following tasks of conceptual design, material properties, steam-side oxidation, fire-side corrosion, welding, fabricability, coatings, design data, and rules (including code interface) were undertaken for review, research, and observation. The necessary information (except conceptual design) has been collected, including material costs for revisiting optimization of the boiler design.

4.2.1.2 Task 2: Economic Analysis

For more attractive economics, the material research program in the United States uses higher-strength materials to achieve temperatures higher than 700°C. An example of this concept can be demonstrated with a comparison between pipework manufactured of materials Alloy 617 and Inconel 740H. A single reheat 750 MW boiler running at steam parameters 704/704°C and 732/760°C was used as a reference. The delivered cost for Inconel 740H is around 50% of the cost of alloy 617. Welding Inconel 740H is estimated to be less than half the cost of welding Alloy 617. The combined effect of these results demonstrates that the utilization of 740H against Alloy 617 in AUSC pipework would result in a significant cost savings with regard to both capital and maintenance. Using historic prices for nickel, all studies show that the cost of electricity (COE) from AUSC technology is more expensive by 1.5%–13% than USC technology, unless there is a carbon tax. If CCS were deployed, the cost of electricity from AUSC technology would be cheaper than USC, for both air- and oxygen-fired combustion.

4.2.1.3 Task 3: Mechanical Properties

Mechanical testing is extensive and is split among the US boiler manufacturers and national laboratories. All data from mechanical testing are accumulated in a long-term material property database. B&W are using finite element software (3D-FEA) to analyze the stresses of thick-section components. Mechanical testing has found that Inconel 740H has a higher allowable stress than other code-approved nickel alloys, for example, Alloy 617/Alloy 230.

Therefore, Inconel 740H requires thinner walls for a given stress than weaker alloys, which reduces the quantity of material usage and the amount of welding required while putting less load on the structural supports, thus reducing time, cost, and risk. Additionally, thinner pipes are less susceptible to cracking with cyclic operation and Inconel 740H has a large forging window (its low flow stress allows for long extrusions of pipes/tubes with a range of widths). Candidate materials have been selected through basic mechanical tests, including creep rupture testing to 10,000 h. Full-scale pressurized creep tests on candidate alloys are now being performed on heat treated cold-bent tubes. This will help determine whether cold strain affects creep rupture strength. This data will be used as the basis to set rational cold-work limits for the guidelines in the ASME BPVC.

4.2.1.4 Task 4: Steam-Side Oxidation

Initially the materials, both coated and noncoated, were screened by laboratory tests. Long-term testing is under way to improve the understanding of long-term oxidation kinetics and exfoliation. In 2012, B&W performed steam-side oxidation tests of pieces (coated pieces also) at 620°C/650°C/750°C/800°C to 10,000 h. Results have concluded the following (McCauley, 2012):

(i) Austenitic steels show good resistance to oxidation and exfoliation up to 700°C.
(ii) Several outstanding ferritic and austenitic steels were identified.
(iii) Nickel alloys show the least oxidation and an absence of exfoliation.
(iv) Parabolic kinetics have been exhibited, allowing the prediction of oxidation rates.

4.2.1.5 Task 5: Fire-Side Corrosion

This task will evaluate the long-term corrosion resistance of materials, cladding, and coatings under a variety of fire-side conditions. Initially, alloys were screened in laboratory tests that simulate a membrane wall and SH/RH conditions using a range of coals at different temperatures for 1000 h. The candidate materials have since been tested in steam loops in commercial-sized boilers. A steam loop was installed at the Niles PCC plant (United States) with high-sulfur coal. AUSC steam parameters (760°C) were achieved by throttling up the steam flow. The steam loop replicates a membrane wall and SH/RH components. Tubes are monitored during planned outages by means of inside and outside diameter measurements for corrosion/oxidation wastage, and photographs were taken to record the surface condition. Steam loop field testing would validate laboratory corrosion/oxidation testing and provide knowledge on the reliability of component fabrication. The conditions at the membrane walls were similar to those at the SH tubes and therefore testing of the membrane wall materials in the steam loops was not undertaken.

Results from the steam loop at the Niles PCC plant show that 9%–12% chromium martensitic steel may not have sufficient fire-side corrosion resistance when firing high-sulfur coals; weld overlays and coatings may be required. Alloys for oxy-combustion were tested at the laboratory stage only. To date, no major difference has been found in boiler materials required for oxygen-fired boilers to those for air-fired boilers. In 2012, an additional steam loop manufactured from candidate materials was placed in another commercial boiler. The field tests ran for 12–18 months, operating with steam up to 760°C firing high-sulfur coals, and included the use of air-cooled probes. The materials tested were Super304H, HR3C, HR6W, Haynes 230, Haynes 230 with an Amstar thermal spray, Inconel 617, Inconel 617 with EN33/EN622 laser cladding, Haynes 282, and Inconel 740. Results show that resistance to fire-side corrosion for nickel alloys is a function of the chromium level; it decreases rapidly as chromium increases from 22% to 27% and then levels out. Inconel 740 showed greater fire-side corrosion resistance than Haynes 230 and CCA617. Results have shown parabolic kinetics, allowing the prediction of corrosion rates. Material testing in Oxyfuel environments has shown less fire-side corrosion attack than for the air-fired boilers. Overlays and coatings will be necessary on membrane walls in units burning high sulfur coals, as the 9%–12% chromium martensitic steels do not have adequate fire-side corrosion resistance. External/internal coatings with application methods have been identified.

4.2.1.6 Task 6: Welding

For a number of materials, including Haynes 230, the weld metal is the weakest link. The weld usually fails at a shorter time interval or a lower stress level than the parent alloy. Welding procedures, filler material chemistry, and PWHT are under investigation to improve the weld strength. Welding of tubes is a well-understood and developed process; however, welding of thick-section plates is difficult and requires gas-shielded metal arc welding (GMAW) (Viswanathan, 2008b). B&W is researching the weld lifecycle, and this will allow simultaneous testing of various material conditions at a given temperature, including stress-relieved, as-welded, the effect of filler metal, and stress. Continuing work includes DMW, weld metal chemistry, and effects on the weld strength reduction factor. B&W has had success with the following welds:

(i) Welding of tubes, pipes, and a thick section of Inconel 740.
(ii) Welding of thick-section stainless steels and Alloy 282.
(iii) Welding of small diameter alloys and narrow groove welding.
(iv) Circumferential pipe weld for thick sections (75 mm) made of Inconel 740H.
(v) Butt welding procedures using 5–7.6 cm qualified thick-section components. Alloys include Haynes 230 via pulsed GMAW using a matching filler and Inconel 740 via a hot wire GTAW using a matching filler.

By late 2012, the program had developed/qualified successful and repeatable welds for numerous alloy combinations, sizes, and configurations. Ongoing work includes better weld performance and simulating weld repair activities for Ni alloy boiler components.

4.2.1.7 Task 7: Fabricability

The program has subjected alloys to common fabrication processes and produced prototype assemblies for machining, bending and swaging are being qualified.

(i) *Forming*: Press forming of headers/pipes, bending of plates (CCA617)/tubes, and swaging of tubes.
(ii) *Machining*: Weld grooves for header and pipe; longitudinal and circumferential seams; socket weld

grooves for tube to header joints; and held grooves for tube circumferential seam.

(iii) *Welding*: SAW for headers and longitudinal/circumferential pipe seams, shielded metal arc welding (SMAW) and GTAW for tube to header socket joints, tube/tube joints, and DMW (for example, between CCA 617 and Super304H and T91 tubing). B&W fabricated full-scale membrane wall panels from T23 and T92 and have trialled PWHT field repairs. The following fabrications methods have been successful: hot bending trials, machining trials, and tube swaging trials. Controlled cold strain, recrystallization, and precipitation studies have been completed for all candidate materials showing successful results (Inconel 740, Alloy 230, CCA617, S304H, HR6W, and SAVE12). A mock-up header using Inconel 740 and hot-wire, narrow-groove GTAW was manufactured. A 9.7 ton ingot of Inconel 740 (750 mm OD) was produced. Fabrication studies have been successfully completed for the production of large boiler components of all alloys.

4.2.1.8 Task 8: Material Standards

Inconel 740/740H was found to have sufficient high creep rupture strength of more than 100 MPa for 100,000 h, fire-side corrosion resistance of 2 mm per 200,000 h, and weldability for operation at 760°C (approved by ASME in September 2011).

4.2.2 Steam Turbine

The steam turbine part of the program is split into phases. Phase 1 screens alloys for further investigation in Phase 2. Phase 2 is split into the tasks, for example, rotor/disc testing (large-scale forgings), blade/air foil alloy testing, valve internal alloy testing, rotor alloy welding and characterization, cast casing alloy testing, and casing welding and repair.

Phase 1 found the following materials for steam turbine application: Haynes 282, Alloy 617, Alloy 263, Sanicro 25, Inconel 740, and Alloy 625. Haynes 282 appears the most attractive for steam turbine rotors. Phase 2 is under way.

Alstom has used its conventional and proven steam turbine design for the AUSC steam parameters, with the application of nickel alloys and austenitic steels in the highest temperature regions. Nickel alloy and ferritic steel are used to forge rotors and casings. There is no change in standard procedures and design rules for the safe introduction of a higher-temperature steam turbine. However, new fabrication techniques are required for new materials. One challenge is the DMW when welding ferritic steel to nickel alloy for the HP/IP steam turbines. To simulate the rotor DMW, Alstom has welded a full-scale test block of Alloy 617/Alloy 625 to ferritic steel. However, the stability in operation of the DMW remains unproven. Operational flexibility is ensured by using a shrink-ring inner casing design that has minimal thermal stresses during cyclic operation. This ensures the capability for rapid start-up and shut-down load changes. Research on castings identified key materials for large turbine and valve components, including Haynes 282. The first ingot of triple-melted Haynes 282 was produced in 2012 and was subjected to microstructural analysis. This work was undertaken by the Foundry Research Institute in Poland. However, work continues on the scaling up of castings and forgings for turbine components.

4.2.3 Candidate Materials

Table 13.6 summarizes material selection for the critical components of a 1000 MW two-pass boiler and steam turbine by Alstom.

TABLE 13.6 Candidate Materials for the US Program—Alstom 1000 MW Two-Pass Boiler

Component	Candidate Material
Economizer	Carbon steel
Membrane walls	T23, T92
Superheater panels	S304H, Inconel 617, Inconel 740
Superheater platens	347 HFG, Inconel 617, Inconel 740
Superheater finish third	Inconel 740
Superheater finish in	Inconel 740
Superheater finish out	Inconel 740
Reheat low temp 1	T23, P91
Reheat low temp 2	S304H
Reheat pendants	S304H
Reheat platens	S304H, HR120, Inconel 617, 230
Valves	Haynes 282
Blades	Haynes 282, Alloys 617, Alloy 263, Sanicro 25, Inconel 740 and Alloy 625, austenitic steel, martensitic steel
Casing	Haynes 282, Alloys 617, Alloy 263, Sanicro 25, Inconel 740, Alloy 625, ferritic steel
Rotor	Alloy 617, Alloy 625, ferritic steel

Source: Kyle Nicol "Status of advanced ultrasupercritical pulverized coal technology."

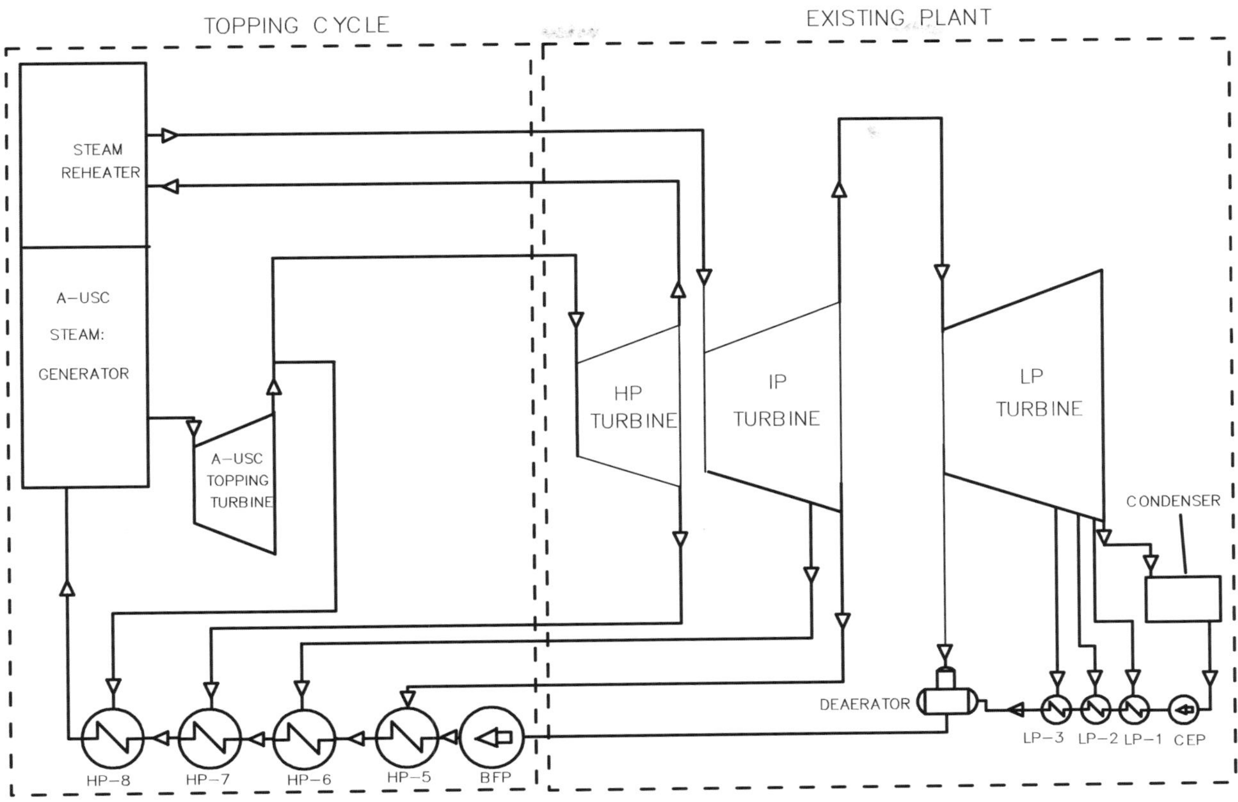

FIG. 13.4 Concept of A-USC with topping cycle in existing power plant.

4.2.4 *Component Test Facility and Full-Scale Demonstration Plant*

The US program was aiming for an operational CTF at the start of 2014, followed by the design, construction, and operation of an FSDP in order to guarantee operational characteristics and economics. A supply chain for advanced material components was felt necessary. The US FSDP will consist of a single unit (possibly 600 MW) expected to be operational in 2021 with a probable SH steam temperature of 700°C. However, this temperature will increase, providing the materials are capable and those temperatures offer an economic advantage.

4.2.5 *Topping Cycle*

Aspects of the AUSC technology could be used in existing subcritical PCC plants with the addition of a topping cycle. Steam at 680°C would pass from the SHs to the topping TG set before passing through the existing three-stage steam turbine; this can raise the efficiency by 3–3½ percentage points. The concept is depicted in Fig. 13.4.

4.3 Development in Japan

In Japan, energy is expensive due to scarce indigenous reserves. Japan has a high population density, creating relatively high amounts of pollutant emissions per unit area that adversely affect the environment and public health. To counter these two factors, Japan uses fossil fuels efficiently and applies substantial pollutant mitigation technologies, enabling them to operate one of the most efficient/cleanest fleets of PCC plants in the world.

R&D on superalloys for use in AUSC technology started in Japan in 2000. The work was headed by the Electrical Power Development Center (EPDC) and was strongly supported through the Ministry of Trade and Industry (Sven, 2001). Prior to 2008, the National Institute for Materials Science (NIMS) completed a feasibility study of AUSC technology with a 700°C SH steam temperature. The study showed that efficiencies can reach 46%–48% (net, HHV, bituminous coal), which is economically and environmentally favorable. That technical viability looks promising and it is technically feasible to retrofit AUSC technology to older PCC plants.

In March 2008, AUSC technology was selected by the Japanese government's Cool Earth-Innovative Technology Program toward reducing CO_2 emissions. The program has four committees: the lead AUSC technology development committee, and three subcommittees for the boilers, valves, and steam turbines. Twelve companies and institutions are participating in the program and are organized into their working topics:

(i) *Boiler*: ABB, Bailey Japan, Central Research Institute of Electric Power Industry,
(ii) Ishikawajima-Harima Heavy Industries, MHI, NIMS, and Sumitomo.
(iii) *Valves*: Fuji Electric (lead), Hitachi, Mitsubishi Heavy Industries, Toshiba, ABB Bailey Japan (bypass valves), Okano Valve, and Toa Valve (safety and control valves)
(iv) *Steam turbine*: Babcock-Hitachi, MHI, and Toshiba.

Observers include many Japanese universities, the Japan Steel Works and the Ministry of Economy, Trade, and Industry. Table 13.7 shows that the Japanese AUSC research program started in August 2008. It has detailed plans for nine years until 2017 and rough plans until 2026, aiming to have an operational demonstration plant by 2021.

4.3.1 Chemical and Mechanical Tests

The first part of the Japanese research program will identify new materials through mechanical and chemical tests. Mechanical tests include fatigue, tensile, long-term creep rupture, bending, welding, and nondestructive testing. Chemical tests include steam-side oxidation and fire-side corrosion, which are explained below.

4.3.2 Fire-Side Corrosion

A material corrosion database is being compiled from materials testing in laboratories, steam loops, and investigation of operating power plants. The amount of corrosion can be extrapolated from the data about the corrosion rate of the material, the flue gas composition, the metal and gas temperatures, and the ash deposition via computational fluid dynamics (CFD) analysis. Material protection methods such as weld overlays and chromizing have been tested. Laboratory testing has shown that nickel and nickel/iron-based alloys have similar fire-side corrosion resistance to conventional austenitic steels. Fire-side corrosion and protection methods with biomass cofiring were undertaken for the testing of new materials.

TABLE 13.7 Japan Research Program

Starting Year	Program Name Span (Years)	Program Name Span (Years)	Program Name Span (Years)
2000	Prior R&D (8 years)		
2008	System design (3.5 years)	Boiler: material development (4 years)	Boiler: long-term creep rupture test (8 years)
2008	Turbine: material development (3.5 years)	Turbine: long-term creep rupture test (8.5 years)	Turbine: fabrication process (3.5 years)
2008	Valve: material testing (1.5 years)	Boiler: fabrication process (3.5 years)	Turbine: cooling, seals, and some mechanical tests (3.5 years
2010	Valve: trial manufacture (3.0 years)		
2012	Boiler component test: facility design (1.0 years)	Turbine rotor test: test facility design (1.0 years)	Turbine rotor test: test rotor production (3.0 years)
2013	Boiler component test: component design (1.0 years)	Boiler test: component prodduction and installation (2.0 years)	Turbine rotor test: test facility build (1.5 years)
2015	Boiler component test: (2.0 years)	Turbine rotor test: rotating test (2.0 years)	
2017	Demonstration plant: design and build (4 years)		
2021	Demonstration plant: operation (5 years)		

Source: "Gantt chart of the Japanese research program" (Saito, 2012; Fukuda, 2012a, b; Imano, 2012; Sven, 2001) by Kyle Nicol.

4.3.3 *Steam-Side Oxidation*

A materials oxidation database was compiled from materials testing in laboratories and the investigation of operating plants. The amount of oxidation can then be extrapolated from the oxidation rate of the material and the temperatures of the steam and metal. Materials protection methods will be investigated, such as shot blasting for austenitic steels and the varying of silicon content for new 9%–12% chromium martensitic steels. Ni alloys show slight internal oxidation but without affecting creep resistance. Shot blasting has proved to be effective for increasing the oxidation resistance of austenitic steels. Scale stability and protective coatings of new materials need longer-term testing.

4.3.4 *New Materials*

By 2012, the Japanese program had already developed the following new materials specifically for AUSC technology: FENIX700, USC141, and USC800MOD.

4.3.4.1 FENIX700

FENIX700 is a low-cost nickel/iron alloy developed for steam turbine rotors. FENIX700 has a high creep rupture strength of more than 100 MPa at 700°C for 100,000 h. It is typically one-third cheaper than nickel alloys. A 12.5 tons rotor has been forged.

4.3.4.2 USC141

USC141 is a high-strength Ni alloy developed for tubes, steam turbine blades, nozzles, and bolts. USC141 has a low thermal expansion and a high creep rupture strength (>180 MPa at 700°C for 100,000 h). A solution-treated version of USC141 is in qualification for use in boiler tubes.

4.3.4.3 USC800MOD

Hitachi developed USC800MOD, a nickel alloy capable of operating at up to 800°C steam temperatures; Imano (2013) investigated its properties and manufacturability. Results estimate that the creep rupture strength is more than 100 MPa at 800°C for 100,000 h. Trial fabrication of a three-ton forging and extrusion of tubes has been successful.

4.3.5 *Steels for 650°C USC Technology*

The Electrical Power Development Center lead a program called STX-21, which contains a subproject called "advanced ferritic heat resistant steels for 650°C USC steam boilers." That project focuses on the continued development of 9%–12% chromium martensitic steels for 650°C steam parameters (the same principal aim of COST 536). This project is scheduled to last for 15 years. The first five-year phase involved a total of 35 researchers at the National Institute for Materials Science in Tsukuba and has made significant progress with modifying martensitic steels via boron additions.

4.3.6 *Component Test Facility and Full-Scale Demonstration Plant*

Toward the end of 2012, CTF and FSDP components were prepared with component testing that started in mid-2014 and finished by 2017. The SH panels, large OD pipes, safety valves, and an HPBP valve were tested in steam loops installed in a commercial boiler. Three rotors made of candidate materials were tested at actual speed in a rotor test rig at 700°C. The results verified the predicted material life assessment and the effectiveness of maintenance procedures, such as repair welding. In 2012, four years into the program, candidate materials were selected for components through preliminary testing. Common issues with all nickel alloys are that they are expensive, segregate easily during manufacture, and have a low tensile strength with a high linear coefficient of expansion. There are plans in Japan to design, construct, and build a 600 MW FSDP sometime between 2017–2021, with operation and feedback between 2021 and 2026.

4.4 Development in China

In China, the total generating capacity is forecast to reach 2380 GW in 2030 (1056 GW in 2011), of which thermal power would be 1270 GW (765 GW in 2011). The country is upgrading its entire fleet of coal-fired power plants. They would replace the old, small, inefficient, and dirty units with new large, efficient, and clean units in order to sustain resources, minimize pollution, and produce lower cost, reliable power. They would then maintain some of the cleanest and most efficient PCC plants. The Waigaoqiaw PCC plant achieved a peak efficiency of 46.5% (net, LHV). The Guodian Taizhou PCC plant was built with two 1000 MW double reheat USC units (600//610/610/31) aiming for an electrical efficiency of 47.6% (LHV, net). In July 2010, the National Energy Administration (NEA) launched the National 700°C USC Coal-Fired Power Generation Technology Innovation Consortium, whose principal aim was to develop and commercialize AUSC technology. The NEA and 18 other members were involved in the program.

4.4.1 *The Research Program*

The research program has five main areas of study: identifying the critical high-temperature components, material research, developing programme manufacturing capability for new alloys, operating a component testing facility, and an FSDP. The program has a council that meets once a year to make the significant decisions. A committee made up of experts arranges a workshop every six months to organize, guide, and inspect the research projects, review achievements, propose new research, and thus develop the technical roadmap. The technical committee is subdivided into four groups: system and engineering solutions, boilers, turbines, and the material group. Once a new project is proposed, the members capable of completing the work apply for government funding (NEA or the Ministry of Science and Technology). There is a secretariat that is responsible for dealing with daily operations and arranging a meeting every three months.

4.4.2 Ministry of Science and Technology

The Ministry of Science and Technology set up a project called "Research on advanced boilers key pipes of USC thermal power units." The project is coordinated by CISRI along with nine partners. This project consists of three subprojects:

(i) Manufacturing the boilers key pipes of a 600–700°C USC unit: This subtopic focuses on selecting materials for the temperature range of 600–700°C in the demonstration plant and the assessment of material such as P92, P122, G115/G112, CCA617, and Inconel 740H.

(ii) Research on alloy GH2984 and its technology process used in 700°C USC thermal power units: Developed in China, GH2984 is a low-cost nickel/iron-based alloy that has been used in SH/RH tubes in Chinese PCC plants since the early 1990s. This subtopic aims to optimize its constituents for application in tubes/headers and rotors operating at 700°C.

4.4.3 Material Special Workgroup (MSW)

The MSW has agreed on four criteria that make up their goal. The first criterion is establishing which materials and components should be developed in China. Second, those selected materials will be optimized. The third criterion is to assess materials worldwide for application in Chinese PCC plants. The fourth criterion is to ensure that these materials are economically competitive. The MSW is funded by the NEA and MST. Table 13.8 shows a list of the candidate alloys for an FSDP.

4.4.4 NEA

In 2012, projects funded by the NEA were segregated into six distinct subtopics:

(i) *Research on overall design proposal*: This includes parameters, capacity, thermal system, and the layout scheme. In 2012, the chosen unit capacity was 600 MW and 700/720/35 steam parameters with efficiencies above 50% with the viability of a double reheat system.

(ii) *Research on key materials*: Material characteristics, nondestructive, and weld testing. The program is actively seeking international cooperation on key materials.

TABLE 13.8 Candidate Materials for the Chinese FSDP

Component	Candidate Materials
Water cooling tubes	T91, HCM12
Pipe and header	P91, P92, G115/G112, GH2984G, CCA617CN
Tube and pipe	GH2984
Tube (superheater and reheater)	T91, T92, NF709R, Sanicro25, GH2984G, Inconel 740HM

Source: "Status of advanced ultrasupercritical pulverized coal technology" by Kyle Nicol.

(iii) *Research proposal on key technology of the boiler*: This topic would develop details of the boiler such as performance, grid layout, frame design, process design of steam/water/flue gas/air, expansion system, piping thrust, heating surface arrangement of all levels, NO_2 removal, and preheater system selection. A draft boiler design has been completed.

(iv) *Research on turbine key technology*: This project develops fabrication techniques and refines the design of steam turbine components.

(v) Construction and operation of a test platform for key components (CTF-700).

(vi) *Feasibility study on constructing an FSDP*: This includes site selection, general layout, capacity, steam parameters, detailed design, and economic evaluation of an FSDP. In 2013, the NEA decided that the FSDP would be 600 MW with a SH steam at 700°C.

4.4.5 CTF-700 and FSDP

The aims are to test the performance and reliability of full-scale components over a long period of time in order to verify their use and therefore limit the risk involved in building an FSDP. The Clean Energy Research Institute, which is part of the Huaneng Group, is undertaking this project and plans to build a CTF to operate at 700°C (CTF-700).

The CTF-700 would be capable of reaching 725°C and 35 MPa with 3 kg/s of steam flow. It would test an evaporator panel, SH, header, a large diameter pipe, and valves. It was installed in Unit 2 of the Huaneng Nanjing PCC plant (two-pass, 300 MW, 540°C, and 25 MPa superheat steam). Materials tested were those developed in China and possibly foreign materials from other programs. The tests will run for >100,000 h (approximately 11 years) in order to verify performance. The Chinese research program aims to have an operational 660 MW FSDP in 2021, after only 11 years from the program start date.

4.4.6 Future Projects

In 2012, there were nine additional 700°C material research topics that were being prepared to request support from national scientific research funds:

(i) Boiler membrane wall technique.
(ii) Components of the boiler superheater and reheater.
(iii) Research on boiler header.
(iv) High-temperature steam pipes and fittings.
(v) Manufacturing technology of turbine HP-IP rotor.
(vi) Manufacturing technology of the high-temperature cylinder valve housing.
(vii) Turbine high temperature blades/fasteners, valve core wear-resisting components.
(viii) High-temperature turbine forgings/castings.

4.4.7 Plant Configuration

The cross compound at the high/low position arrangement (CCHLPA) of the steam turbine is where the HPT/IPTs are both mounted at the boiler steam header level and the LPT remains at ground level (Fig. 13.14). The CCHLPA reduces the pipework length, cost, and pressure drop. A 1350 MWe unit with double RH/CCHLPA and steam parameters (600/620/610/30) is estimated to reach 48.92% net efficiency.

4.5 Development in Russia

The prominence of coal power in Russia has been declining since the 1990s. Russia has one of the largest reserves of coal in the world. However, there is a partnership between the private/public sectors that has set up a research program for cleaner and more efficient power from coal called Foundation: Energy Without Borders. The partners are the Ministry of Energy, the Institutes of Development, the Ministry of Education and Science, and the Ministry of Industry and Trade. Within the Foundation: Energy Without Borders program (Portfolio 1, Action 1), AUSC is a key technology aiming at 51%–53% net efficiency with steam parameters 700/720/35. Research in Russia has previously tested steels for use up to 650°C. Inter RAO are researching steels for use in AUSC technology. Austenitic steels 12Cr18H12T and 10Cr16H16B2MBR are being developed for heating surfaces and martensitic steels 10X9MFB, 10Cr9B2MFBR, 10Cr9K3B2MFBR, and 12Cr10M1V1FBR are currently in development for the pipes and rotors. The All-Russian Thermal Engineering Institute envisaged a diagram of a possible CTF and the program was open for collaboration, both on a bilateral and a multilateral basis.

4.6 Development in India

Coal is India's largest fossil energy reserve, at 293 Gt. The country saw an annual consumption in 2012 of 0.5% Gt and the power from coal is expected to rise to 320 GW in 2031. The Indian government has a program called "the AUSC project" under the ninth National Mission on Clean Coal Technologies, according to the National Action Plan for Climate Change.

The Indian Material Research Program was initiated in 2008. The program is being funded and undertaken by a joint venture between the Indira Gandhi Center for Atomic Research (IGCAR); Bharat Heavy Electrical (BHEL), the equipment manufacturer; and the power-generating company, National Thermal Power Corporation (NTPC). IGCAR would develop the materials; Misra Dhatu Nigam and BHEL would design, manufacture, and commission the equipment; and NTPC would construct the test loop and demonstration unit. IGCAR would design and develop new 70% nickel alloys capable of 710°C and 35 MPa superheater steam parameters, raising the net efficiency from 38%–40% to 46% (LHV) and reducing the carbon dioxide emissions by 15%–20%. IGCAR has experience with new nickel alloys through its prototype fast breeder nuclear reactor project. Materials include Super 304H, Inconel 617, Haynes 230, T92, and T91. BHEL would manufacture and commission the demonstration unit for NTPC to operate.

Components for the USC plant in India are often purchased from other countries, though indigenous commercialization of AUSC technology has a large potential market. In 2013, BHEL submitted the project design memorandum for an 800 MW AUSC boiler, including technical and economic details, to the office of the Principal Scientific Adviser of the government of India. The Welding Research Institute (WRI) at BHEL is studying new welding techniques such as GTAW. The Hyderabad-based public sector Misra Dhatu Nigam would help fabricate the new Ni alloys. New high-temperature components would be tested in a CTF at the NTPC 210 MW unit at the Dadri PCC plant in Uttar Pradesh. Providing the successful completion of a CTF, operation of a demonstration unit is planned to start in 2018 and is likely to be based at the NTPC Dadri complex.

5 LEAKAGES AND LOSSES IN THERMAL POWER PLANT CAUSING REDUCTION IN EFFICIENCY

Electrical efficiency is lost to the auxiliary load and internal losses. The auxiliary load is the power required by electrical devices to operate the PCC plant. Significant auxiliary loads include the coal pulverizing mills, motor-driven boiler feed water pumps (BFP), forced and induced draught fans, particulate control devices, and pumps in flue gas desulfurisation units. Internal losses are heat losses due to friction in moving components, such as motor-driven BFPs, forced

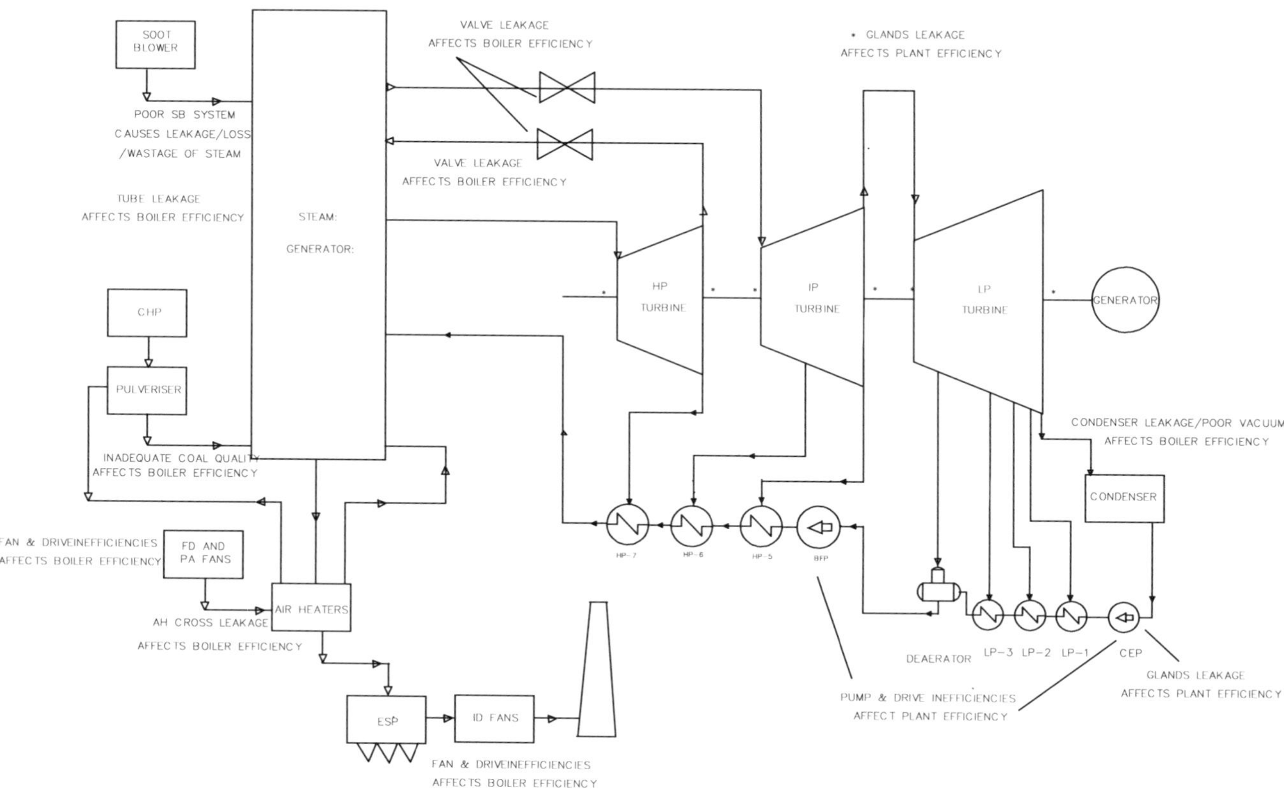

FIG. 13.5 Sources of leakage/losses affecting efficiency plants.

and induced draught fans, the electricity generator, and turbine bearings. For a unit operating with superheat steam temperature at 600°C, efficiency losses to the auxiliary load and component inefficiencies equate to around 10% points of electrical efficiency. Electrical and mechanical improvements in components have reached an electrical efficiency point of diminishing returns, and component losses can no longer be significantly reduced. However, components need cooling and this heat can be used in the water/steam cycle to raise electrical efficiency by ~0.1% point.

Therefore, achieving high efficiency with low emission is top priority while designing, installing, and operating a thermal power plant. A new plant may not experience leakages but as time passes, leakages at different locations take place around various equipment throughout the thermal plants, affecting the efficiency badly (Fig. 13.5).

5.1 Boiler Tube

Boiler tube leakage is one of the most important outage causes, resulting in a loss of efficiency in a thermal power plant. Sometimes, thermal power plants run for a few hours with undetected tube leaks due to confusion with temporary outages/absence of relevant monitoring systems. The furnace walls containing high-pressure water/steam are subject to fire-side corrosion and thermal stress smelt, which can even cause an explosion vis-a-vis a total catastrophe of the boiler, apart from the shutdown of power plants as well as loss of fuel, maintenance time, and cost. Approximately 60% of boiler outages are due to tube leakages that call for special awareness to avoid secondary damages to the furnace refractory and pressure parts such as water wall tubes, heaters, and superheater/reheater tubes. Details on tube failures are discussed in Chapter 2, Section 2.10.

5.2 Flue Gas Path

The long flue gas path comprising the furnace, economizer, air heater, flue gas ducts, ESP casing, expansion bellow, etc., is always susceptible to leakage. This circuit is maintained at a negative pressure and consequently, comparatively cold atmospheric air is soaked through these leakages, which subsequently increases the flue gas volume and hence, the ID fan power drawing increases. Leakages can be found by measuring the O_2 level at various locations of the flue gas path. If the O_2 level is more than that of the boiler furnace, it indicates there is an air ingress into the flue gas path.

5.2.1 Air Preheater (APH)

It is well known that APH leakage is a major factor in the loss of boiler efficiency that necessitates the needs to employ improved design of, say, the radial, axial, and rotor posts as well as the circumferential seals usually installed (for example) on rotary regenerative APHs.

APHs are one of the vital components of any power plant due to the well-known heat transfer and efficiency-related benefits. Being a significant component of the overall plant efficiency calculation, APHs deliver upward of approximately 10%–12% of the heat transfer in the boiler process at only a nominal percent of the investment. For every 20°C decrease in the gas outlet temperature of the air heater, the boiler efficiency rises about 1%, with inherent fuel consumption reductions. It has been observed worldwide that leakage rates of more than 30% are not uncommon while rates of 15%–20% are typical. However, with properly designed, installed, and maintained seals, it is possible to keep the APH leakage rate ≤10%. Degradation of seals or partial failure of seals means higher air-to-gas leakage, increased fuel and fan power consumption, and lower heat transfer. It also affects the downstream air pollution control (APC) equipment because of higher gas flow rates and pressure drops. Reduced APH leakage is vital for proper operation of the boiler with improved efficiency. One step is the single train component design while another step is to install an improved APH sealing system with reliable sensors.

The Burmeister and Wain Energy A/S (BWE) APH has a number of upper and lower radial seals. Each radial seal consists of two plates hinged together. Each radial seal control consists of one actuator, one distance sensor, and one mechanical safety device.

The control of the sealing system is integrated in a local control panel with a PLC. The gap distance between the radial seal plate and the APH rotor as measured by the distance sensors is compared with the desired distance stored in the controller (PLC), and action is taken to correct any deviation. Each radial seal is controlled individually.

BWE has developed a very reliable sensor for the APH sealing system. The sensor can be installed in a hot flue gas condition and does not need any external cooling system. The advance sealing arrangement with a control system ensures less leakage because it is extremely efficient.

5.2.2 Steam Sealing System Around Turbine

The functions of the turbine glands and seals are two-fold.

(i) Prevention/reduction of steam leakage between the rotating and stationary components of the turbines if the steam pressure is > atmospheric.

(ii) Prevention/reduction of air ingress between the rotating and stationary components of the turbines when steam pressure is < atmospheric. The last few stages in the low-pressure (LP) turbines are normally under vacuum.

The steam leakage or air ingress could occur where the shaft is extended through the turbine end walls to the atmosphere and is always associated with power loss. Modern steam turbines use labyrinth glands to restrict steam and air leakage. As an important and major component of thermal power plants, the highly efficient turbines are essential to cope with the modern age where the machine efficiencies are continuously increased. It is therefore necessary to enhance the efficiency of individual turbine cylinders and sections by minimizing steam leakages/energy losses through the end seals, and the central seals between sections for the integrated cylinders. The HPT gland sealing leakages are always issues of concern that affect the efficiency; that is the reason that gland-sealing technologies are frequently the focus of R&D activities. Various concepts were and are still developed and proven to work in real plant operations in low tolerances so as to keep the efficiency figure to its closest design value.

5.2.3 Turbine Gland Seals and Nonbladed Areas

The (shaft) seals comprising the end seals and the central seals between sections for the integrated cylinders are mainly designed in line with the same approaches as those applied to other types of gland seals. However, at present, in order to trim the leakage, the design has been significantly modernized. For example, labyrinth and once-through gland seals are now available with very attractive features by means of retractable packing offering zero or controlled radial clearances. For this type of seal, the radial clearances increase when the turbine is at a standstill or running with a low steam flow. They decrease under normal operating conditions with the desired load. For retractable packing, the springs push the packing segments outward toward the stator component (the diaphragm/casing ring). This technology has now been proved to be a cost-effective and reliable measure for preventing the wear of seals and improving steam turbine efficiency (conventional fixed packing seals with the flat springs press down the packing segments toward the shaft). The retractable packing seals are extensively applied at many turbines manufactured by ALSTOM, GE, Westinghouse, Hitachi, Toshiba, and others. Changeover to retractable packing at many plants completely solved the rubbing problem at start up without sacrificing turbine efficiency.

Retractable seals are quite effective in the HP/IP sections but are not always suitable for the LP stages because the available pressure differences across the packing segments may be inadequate to retract them. For the LP stages, the use of so-called "sensitized" seals, which rely on steam pressure to displace, in contrast to retractable seals. While the turbine is working under load, such seals are kept in the operating position as sensitized coil springs hold the packing segments in place. Retractable packing systems were additionally improved with the inclusion of a built-in brush bristle sandwiched between two solid faceplates. The brush material is a Haynes 25 cobalt-based super-alloy. Several thousand extremely fine diameter bristles, with wire diameter in the range of 0.1–0.15 mm (4–6 mil), are packed together, forming a hedge against the leakage steam path. The bristles are inclined radially in the direction of the shaft rotation, commonly by about 45 degrees, to prevent them from picking up on the rotor. The back plate provides stiffness to the brush pack and prevents it from being deflected downstream by the steam pressure across the seal. During operation, the aerodynamic forces, due to the leakage flow and well-known bristle "blow-down" effect, make the bristles move down and close the bristle tip clearance, further reducing the leakage flow. Because retractable brush seal packing operates at zero clearance when it is "closed," steam leakage is limited to the flow that can percolate its way through the tight maze created by the brush seal's bristle pack. In doing so, the steam leakage flow through the seal dramatically reduces and remains almost invariable in the operation process, even if the turbine works in a cycling manner.

Another seal design concept, called "leaf seal," is under consideration by MHI. Such a seal comprises a number of thin metal plates ("leaves") inclined in the circumferential direction so that their tips are kept in a noncontact state with a negligibly small clearance when the rotor is rotating. This is provided by a lifting force produced due to a hydrodynamic pressure effect acting between the leaf and rotor. The tip of the leaf is lifted by a balance of the pushing force due to pre-pressure of the setting, the hydrodynamic lifting force, and additional lifting force due to the pressure difference across the seal. The result is that both the seal and rotor are prevented from wear and the durability of the seal is increased when the turbine is running. This distinguishes the leaf seal from contact-type brush seals. In addition, because the seal itself is in the shape of a plate with axial width, it has a higher rigidity in the direction of the steam pressure difference and the sealing function can be kept to higher differential pressure values compared with brush seals.

The considered advanced that seal technologies, such as retractable, brush, and abradable seals, have already been used in power plant practice, are not only responsible for reduction of the steam leakages through the seals but also designed to avoid/minimize the energy drops in the non-bladed steam path areas by reducing their aerodynamic resistance and diminishing pressure drops. This primarily concerns steam admission and outlet zones, bleeding chambers and channels connecting them to the steam main-stream, and all the areas where the steam flow turns.

Siemens has optimized the HP steam admission section (907-MW Boxberg Q steam turbine) through development of a helical configuration that yielded a more uniform distribution over the entire stationary blade ring area. New diffuser geometry was developed for the HP/IP exhaust hoods specifically to counteract backflow and vortex formation caused by flow separation, making it possible to reduce pressure drops and energy losses due to internal circulation flows. Special attention was paid to excluding any excessive aerodynamic resistances/vortex formation sources at all the steam path turns, between the main steam flow and bleeding chambers, and so on.

Turbine efficiency gains due to such specific individual measures are counted in fractions and if taken together, they can raise the final turbine efficiency quite remarkably. Modern steam turbines demand improvement on all these "small things," along with efficient steam path design.

5.3 Vacuum Sealing System Around Condenser

The high-pressure region of a turbine is working at pressures as high as 365 bar while the low-pressure region operates at a pressure of 0.03–0.07 bar. Turbine back pressure is a key parameter in steam-to-power efficiency. The typical design back pressure for a system is around 0.08 bar. If this vacuum level deteriorates due to air in-leakage, leaky valves, or condenser circulating water (CW)/cooling tower problems, the efficiency of the entire system can decline rapidly. If even small amounts of noncondensable gas (air) accumulate, they would inhibit heat transfer in the condenser and adversely affect performance. Large amounts can virtually block the condensation process, which of course would cause a substantial rise in back pressure. To prevent excessive accumulation by removing the noncondensable gases, most power plants use steam jet air ejectors and/or liquid ring vacuum pumps. However, reduction of this air in-leakage means fewer requirements for ejection steam or electrical power for vacuum pumps. Also, water leakage in the condenser would raise the levels of dissolved O_2 in feed water and cause increased corrosion and deterioration of the boiler and feed systems.

It is understood that effective cleaning and testing strategies will maximize megawatt output while minimizing condenser-related outages during normal operating cycles, permitting an accurate calculation of return on investment. Combined efforts of cleaning, leak detection, and testing are required to achieve maximum condenser performance. Many plants have an established cleaning regimen, usually annually, as well as an eddy current testing regimen that could take place up to every few years, depending on the condenser's age and condition. However, many of the leak-detection programs occur on an as-needed basis. By combined proactive cleaning, leak detection, and eddy current testing, the total performance of the condenser and its components will be improved. Condensers are designed with air removal systems to handle a certain amount of air in-leakage and keep the unit running at peak efficiency. When leakage exceeds air removal system capability, the condenser efficiency is adversely affected and would certainly indicate the problem by increased plant heat rate and condenser back pressure.

Air in-leakage can be related to the shell, rupture discs, shaft seals, man ways, vacuum pumps, flanges, and one or more of the many bolt holes in the equipment and test probe penetrations. With technology development nowadays, more sensors are tapped into the system than ever before and there is a potential for a leak for every penetration. Similarly, there are many sources of water leakage, for example, water box flanges, hot well components, through-wall penetrations, tube-to-tube sheet joints, faulty/temporary tube plugs left in too long that loosened over time, or permanent plugs if put in incorrectly. Moreover, condenser tube fouling may result in corrosion and finally tube failure. Some of the old leak detection methods still used today, especially in an emergency, include smoke, shaving cream, plastic wrap, and dependency on sight and sound. Some of these, however, are often inaccurate and certainly not easy to replicate. It is wise to rely today on tracer gases, helium, and sulfur hexafluoride (SF6). The choice of the most appropriate tracer gas for the site-specific conditions is important because a less ideal tracer may cost time and money.

5.4 Hydrogen Sealing and Cooing System

An electric generator produces large amounts of heat that must be removed to maintain efficiency. Hydrogen is the coolant for generator rotor winding that high capacity generators use (~70% of all electric power generators over 60 MW worldwide), H_2 for its low density, high specific heat and thermal conductivity, high availability and relatively inexpensiveness that make it a superior coolant for this application. In addition, H_2 also takes part in the removal of heat from the surface of hot generator winding insulation. The winding insulation, besides having good dielectric strength, has excellent thermal conductivity to ensure the fast removal of heat.

In its pure state, H_2 is fine, but a mixture of hydrogen in the air (4%–74%) can be explosive. Therefore, hydrogen leakage continues to be a safety concern, and maintaining efficiency at high value at power plants. Power plants using hydrogen-cooled generators must maintain the recommended H_2 purity/pressure in the generator casing for efficiency, safety, and equipment reliability through constant monitoring. There is a high probability that H_2 would leak because it is difficult to seal for its lightness and pressurized sealing system. All H_2-cooled generators do leak, but an acceptable leakage rate must be adhered to.

Monitoring hydrogen purity is important for two reasons. The first is efficiency: The purer the hydrogen, the more efficient the generator because hydrogen has less windage loss than air. The second reason is the explosive mixture ratio—the purer, the safer. The possible hydrogen leakage sources may be attributed to failing of hydrogen seals, including seal springs, improperly installed hydrogen seals, excessive seal oil pressure (the usual culprit), dirt/debris in the seals, or the brass ring/shaft (sometimes) gets scored.

5.4.1 *Stator Cooling Water (SCW) System*

For a large generator stator, huge heat originates due to losses and is removed by circulating cold water through the hollow generator stator windings to maintain the design efficiency. An SCW system is used for the same thing through flow and other parameter measurements and control, including water chemistry. Removal of stator-originated heat is also aided by hydrogen from the surface of hot winding insulation. An SCW system is a closed-loop system that cools the copper stator bars. The holes though which the water flows are narrow. Unimpeded flow through all stator bar openings is critical to generator operation to avoid overheating of the stator bars, which can result in reduced generating capacity or even catastrophic failure of the generator.

SCW systems must contain only deionized water. The cooling water after removing heat from the stator coolers passes through heat exchangers to release the heat. The water is then passed through a mixed-bed polisher that removes any soluble ionic contaminants that enter the water. The stator ion exchange resins often act as a filter for particulates in the water, though some systems have a separate filter. The ion exchange resin will eventually become exhausted, but it is common that the pressure differential across the resin bed (created by accumulated particulates) may require that the resins be replaced before the ion exchange capacity is reached.

The stator winding bars are typically copper; recent research on the chemistry of copper in oxidizing and reducing conditions reveals that dissolved oxygen (DO) and pH play a major role in determining the copper corrosion product formation rate and transportation rate through the SCW system. However, it is important that the major cause of problems in SCW systems has not been corrosion in isolation but, rather, deposit accumulation in critical areas. Copper oxides are released from one area of the stator coolers and deposited in another, which is determined by the amount of DO in the system and particularly variations in that oxygen concentration. Copper forms cuprous oxide (Cu_2O) under reducing (low-DO) conditions and cupric oxide (CuO) at high DO condition. Both these oxides are stable and create a passive oxide layer on the stator bars. A slightly alkaline pH increases the stability of the oxide layer.

5.4.2 *Regimes for Stator Cooling Water System*

The recommended operating regimes for the system are categorized by their levels of dissolved oxygen (DO) and pH. Of the four options indicated below, three are generally recommended. Each of the options has pros and cons that must be balanced against the particular plant suitability and each of them can be found in operating power plants, for example, low-DO, neutral pH option, high-DO, Neutral pH option, Low-DO, High pH option.

5.4.2.1 Low-DO, Neutral pH Option

This treatment option is found in about 50% of the stator cooling systems worldwide. The water is fully oxygenated when the system is first filled. As the water circulates, it reacts with the copper in the system and a thin layer of passive cuprous oxide is formed. The O_2 is consumed, and the DO gradually approaches zero. The DO is likely to remain at <20 ppb as long as no water is added to the system. O_2 scavengers or reducing agents are not commonly added to stator cooling systems. The main idea of this system is to keep dissolved O_2 out or at the minimum possible value.

The corrosion or oxidation rate of copper in a system containing 400 ppb of DO is more than five times its rate at low levels of O_2 under neutral conditions.

Every time makeup water is added to the stator system, it receives a little shot of dissolved oxygen, which may be more due to significant leaks in the stator cooling water system. The stator cooling water can swing back and forth between low and high oxygen conditions. The effect of transient oxygen spikes is most profound on low DO and neutral pH systems and release oxides into the system. The DO level may be increased considerably due to air ingress which can be arrested to a minimum by putting a nitrogen cap on the stator cooling water storage tank. This system sweeps out the oxygen that permeates the membrane and is capable of reducing the makeup water DO to about 3 ppb. In parallel, carbon dioxide can enter the system and is absorbed into the water via the makeup water or air ingress. It drops the pH to acidic levels, thereby increasing the corrosion rate of copper through the formation of bicarbonate and carbonate in the water. The mixed bed ion exchanger/polisher gets exhausted and if the same is not changed timely, the released carbonate can form insoluble copper carbonate in the stator. Some thermal power stations (for example, the Tarong Power Station in Queensland, Australia) employ O_2 removal systems through a series of three gas transfer membranes to prevent DO contamination resulting from makeup water.

5.4.2.2 Low DO, Higher pH Option

The corrosive response during O_2 transitions can be decreased considerably by making the pH of the stator water at 8–9. This can be done through adding controlled amounts of sodium hydroxide to the water. Initially, sodium will be neutralized for it will be exchanged with H_2 on the cation resin of the mixed-bed ion exchanger/polisher. If the caustic continues to be added, sodium leakage from the resins will make the water gradually increase and maintain a desired pH.

In another method, a sodium exchange polisher on a side stream is added to control the amount of water that passes through the sodium exchanger to achieve the desired pH. Alternatively, the mixed bed ion exchanger/polisher will totally replace all strong base anion resins; however, this will obviously be unable to remove soluble copper. During a shutdown or major turbine outage, stator water can become oxygenated, which may cause deterioration of the stator cooling system shortly after the unit comes back in service.

5.4.2.3 High DO, Neutral pH Option

This system occupies almost 40% of the water-based stator cooling systems worldwide. In this operating system, CuO is formed and deposited on the copper. It will create a passive layer on the metal and tightly adhere to the surface. This layer tends to be thicker than the Cu_2O formed under low-oxygen conditions. Because the dissolved oxygen will be depleted by its reaction with copper, it may need to add air to the system to maintain sufficient DO. Because the system is operating continuously under high (>2 ppm) levels, this chemical process is impervious to additions of DO. However, it may still be susceptible to low-pH corrosion from carbon dioxide and carbonates, thus producing a need to be removed by the mixed-bed ion exchanger/polisher.

H_2 leaks into the system can replace the DO and cause a low DO condition. Low/high DO transients in the system help in causing oxides to be released.

5.4.2.4 High-O_2, High-pH Option

This system is not recommended because it increases the possibility of clip corrosion.

5.4.3 Monitoring Stator Water

Table 13.9 indicates the necessary monitoring parameters for an efficient SCW system.

The make-up water daily consumption is also indicative information. Measurement of the DO in SCW offers valuable information about tiny H_2 leakages in the stator, which may be undetectable through the H_2 gas trap or consumption pattern, but it leads to lowering of the DO content of SCW.

The Electric Power Research Institute (EPRI) suggests a change of resin at any condition such as: (a) conductivity of 0.5 μS/Cm, (b) every 18–20 months, and (c) pressure drop across the ion exchanger/polisher resin exceeds 1.0 kg/cm^2.

TABLE 13.9 Parameters Necessary for Efficient SCW System

Parameters	Range	Remarks
Conductivity	0.5–2.0 μS/cm	10.0 μS/Cm (max) High conductivity indicates plugging
Dissolved oxygen (DO)/pH	<20 ppb/8.5–9 (1) <20 ppb/7 (2) >2000 ppb/7 (3)	(1) Low DO/high pH (2) Low DO/neutral pH (3) High DO/neutral pH
Electrochemical potential (ECP)	<223 mv (low O_2 regime) >315 mv (high O_2 regime)	>266 mv (short time action level) <305 mv (short time action level)
Hot/cold water temperature	60°C/35°C	(90°C/40°C max.) high temperature indicates plugging
Coil/bar temperature	95°C	High temperature indicates plugging
Total coil flow		Low flow indicates plugging
Individual bar low		Low flow indicates plugging
Water pressure at inlet to conductor	3.5 kg/cm^2	
Hydrogen leakage		Calculated/inferred from trend of reduction of DO value
ΔP coil inlet/outlet	1.5 kg/cm^2	High ΔP indicates plugging
ΔP gas/water	1–2 kg/cm^2	High value may lead to H_2 leakage
ΔP across ion exchanger/polisher	0.5 kg/cm^2	High ΔP indicates plugging

Source: System understanding, diagnosing, monitoring and addressing leakage and pluggage in water-cooled generator stator windings by A.K. Sahay, CC-OS, B. Manjul, Kahalgaon, V.K. Garg, Badarpur.

A laboratory analysis of the copper concentration in the ion exchanger/polisher once every two weeks is a good indication toward the performance of the resin or filter.

5.4.4 *Electrochemical Potential*

A research project developed by EPRI has concluded that online ECP monitoring as a significant parameters for copper release and assessing overall SCW system chemistry.

It shows that a change in the oxidation state of copper (+1 to +2 and vice versa) induces stresses, which are a driving factor for particulate release and deposition.

The ranges of ECP as established by the EPRI for two operating regimes are as follows:

Parameter	EC (normal operating values)	EC (short-term action level)
Low oxygen regime	<223 mv	>266 mv
High oxygen regime	>315 mv	<305 mv

Essentially, a summative parameter that describes chemical reactivity better than any single parameter, even oxygen, and is governed by oxidizing and reducing agents in water, pH, temperature, and other ionic impurities. The normal efficient operation of the stator cooling water system is affected adversely by leakage and plugging of the water flow path.

There are a number of possible/probable causes for the development of leaks in a liquid-cooled stator winding. During normal operation, the stator winding is subjected to thermal shocks, cyclic operation, corrosion, and mechanical/electrical vibrations. These normal operational stresses can lead to the development of leaks in a liquid-cooled stator winding. The possible sources of leakages in an SCW system include stator hydraulic components and connections, leaks in the stator bar, leaks in the clip window braze, leaks through porosity, leaks in the clip-to-strand connection, leaks through serrated spacers, leaks initiated by mechanical causes, erosion in hollow conductors, and end-winding vibrations.

Stator bar insulation failures due to these/or other reasons may occur and cause extensive collateral damage to the generator stator core and field. This can cause a forced outage for several months, depending on the extent of the damage. Leaks in braze/weld joints typically fall between 10-5 and 10-2 std. cc/s. Leaks below 10-7 std. cc/s will usually be plugged by the water itself.

As the stator winding is subjected to various kinds of thermal shocks as indicated earlier, the potential for damage and methods of leak detection are very different. Water-cooled stator windings have thousands of brazed joints that are potential sources of leaks.

Leaks in stator hydraulic components and connection leaks in the generator stator hydraulic system may originate at any of the following hydraulic components: copper tubing, pipes, piping connections, elbows, fittings, sleeve joints, tube extensions, connection sleeves, Teflon hoses and hose fittings, liquid-cooled series loop brazes, connection rings, inlet/outlet headers, nipple connections, liquid-cooled high voltage bushings and connections, and "P" bar liquid connections and hoses, if applicable.

Small leaks in these locations typically will not result in winding damage so long as the hydrogen gas pressure is maintained above the stator cooling water pressure. However, capillary action can cause water to leak from a bar even though the gas pressure is higher than the cooling water pressure. Larger leaks can be detected by online testing methods and vacuum/pressure decay tests. Even extremely small water leaks can be detrimental if allowed to persist. This is particularly the case if the SCW system is left in operation during outages when the generator is degassed. Under these circumstances, the pressure differential provides the greatest drive that forces water into the groundwall insulation.

Addressing the plugging issues: The Electric Power Research Institute guidelines suggest the measurements as depicted in Table 13.9 to detect plugging or flow restrictions that are an indication of the quantity of Cu released from the system.

Cleaning method: This is usually done by ethylenediaminetetraacetic acid (*EDTA*), as it removes only copper oxide and does not react with Cu metal. But the limitation of this method is that if the strand is severely choked, this gentle organic acid may not be good enough and in that case, cleaning is to be done by 9% H_2SO_4.

Prevention of pluggage: (a) online ECP and DO monitoring, (b) offline Cu trend analysis from spent ion exchanger/polisher resins, (c) for high oxygen regime: a large diameter vent to provide air access to the storage tank and forced aeration to the water tank, and (d) During shut down when stator water system is drained then it needs to be dried out with Nitrogen (Grade-1) quality.

5.5 Valve Leakage and Losses

5.5.1 Valve Leakage

A valve leakage results in unscheduled shutdowns and frequent equipment repair or replacement as well as wasted fuel/process liquids. Valve leakages are almost preventable by using properly designed valve externals, internals, and seals that may permit zero leakage but not to 100%. Most leaks are deeply secreted somewhere well inside the equipment components and piping networks around the vicinity of a power plant. For instance, drain valves remain open during start up and shut down, and gate and globe valves can experience rapid erosion and wear due to primary sealing components being in the flow path of the high-pressure steam. These leakages gradually eat away the performance, for example, the efficiency and increased costs, and may pose a critical problem by exposing plant personnel and the environment to potentially dangerous conditions.

However, "zero leakage" does not always mean zero leakage; the usual definition actually means acceptably slow leakage internally with no visible external leakage. But research is still in progress for applying the best available technology and adopting new standards to reach zero leakage, both internally and externally.

Internal leaks are entirely different from visible external leakage, for example, the isolation valve on the HPBP line between the boiler and the steam turbine that redirects steam to the condenser. Any leakage through that valve lowers the generator output while increasing fuel consumption, showing up as a gradually increasing heat rate and accompanying emissions remediation expenses to produce the same amount of electricity.

External leaks appear as drips and puddles that collect on plant floors or equipment for liquid services, but for high-pressure/temperature water or steam services, the leaking fluid would exude from inside the valve with a hissing sound as the cloud of steam escapes to the atmosphere. However, both are risky and can be dangerous to working personnel and visitors. The loss of flow media in a closed-loop system has a negative impact on the efficiency of that system. External leakage can also result in lower operating pressure, which can lead to vibrations that can cause excessive stress on system components.

Internal leakage takes place around the valve seat or through the closure member when the valve is closed. Internal leakage is normally contained within the piping system. Such leakage can be external when a valve is used to dead-end a pipe. Unintentional internal valve leakage can result in an inefficient system, impacting the system parameters. In block/isolation services, this may result in maloperation of the equipment as it was intended and consequently, a loss of fluid as well. Typical leak areas include the packing/top flange sealing and the piston/seat.

Control valves of several types and quantities are required for thermal power plants (discussed in previous chapters) with their inlet/outlet isolation valves; all of them are prone to leakage and contribute to overall leakage. Isolation valves for very high-temperature/pressure services need a seal to be made by the metal components of the valve to avoid the limitation of thermoplastic seals.

However, the fluids involved may contain some solid content or abrasive materials that erode the seal surface, thereby producing leakage paths. Over the course of time, even minor leaks can grow into major ones. Overall, losses from a single leaky valve may add up to millions of heat energy never reaching the turbines. Low/zero leakage valves can typically improve a plant's heat rate performance from 1% to 2% to as high as 3%.

Triple- and double-offset butterfly valves (Chapter 6) give the highest leak-tight characteristics of all types of valves. Of all butterfly valve designs, the triple-offset type offers the best sealing and longest life while delivering bidirectional gas tightness. The use of triple/double-offset valves in turbines/boilers/other PCC applications is not uncommon.

5.5.2 *Reduction of Losses in Control Valves*

Existing stop and control valves of steam turbines for thermal power plants with substantially high main steam conditions are frequently characterized with very great steam velocities in the narrowest section of the diffuser-type seat (that is, practically at the turbine entrance) of approximately 100–140 m/s. As a result, they have a pressure drop of about 4%–7% with a corresponding enthalpy drop for the turbine of about 7–10 kJ/kg. Some special investigations of this problem are brought to the concept of profiled valves with a cup featuring a perforated surface as an alternative to plate-type valves. The proposed solutions can be illustrated by the example of the unit of the main stop and control valves for the 360 MW steam turbine 18K360 of Zamech (Poland) with a steam pressure of 18 MPa. The unit comprises the stop valve and control valve, arranged on both sides of the common saddle. Following are the distinguishing features:

(1) Profiled valve heads, whose surfaces, together with the surfaces of the inlet and outlet parts of the saddle, form an axially symmetrical annular duct.
(2) Perforation on the valve head surfaces for damping possible flow pulsations.
(3) A special system for additional loading of the valve under conditions of high lifting, when the tapered bushing obstructs the access of steam to the unloading valve.
(4) A confuser saddle instead of a diffuser saddle, which is entirely unconventional for control valves, ensures stable operation under any operating conditions.

5.5.3 *Reduction of Aerodynamic Resistance*

In order to reduce the aerodynamic resistance of the fully opened valves, the maximum steam velocity in the saddle throat was reduced from the initial value of 150 m/s for the old-type valves to 70 m/s by increasing the fit diameters of the valves. As a result, the total resistance of the considered unit did not exceed 1.5% of the inlet steam pressure. Such a solution became possible due to the application of an unconventional system for counterbalancing the axial thrust on the control valve stem when its internal cavity is connected to the confuser saddle duct, not through the central unloading orifice but through the holes perforated on the streamlined surface of the valve head. Thus, at small openings of the valve, its deep unloading from the acting steam pressure forces is ensured, and there arises the possibility to increase the fit valve diameters but not increase the servodrive power. The single-seat angle valves of ABB were also optimized in terms of power drops, flow losses, and sensitivity to vibration. The design of their spindle seals diminishes the steam leakage amount along the stem and reduces the required leakage piping to a minimum. The diffusers after the main and intercept control valves also act as the inlet ducts to the inner casings. The turbine design in these HP and IP areas significantly reduces the cost and duration of overhauls by allowing easy dismantling and reassembling. The gland seal pipes of the valves are fitted directly to the valve body and do not need to be removed for the overhauls. The same to a great degree can be said about the combined stop-and-control-valve units of Siemens. Their number and size are set according to the mass flow rate to diminish the maximum steam velocity and hence the pressure drop and energy losses.

5.5.4 *Valve Stem Sealing*

Proper sealing of the valve stems minimizes the steam leakages along the valve stems. Steam leakages through the seals of the HP valve stems are commonly quite remarkable, especially for supercritical-pressure steam turbines, and result in noticeable energy losses. In addition, the outside steam given to the seals and flowing along the valve stem causes an axial temperature unevenness in the valve steam chests, with a resulting increase of unsteady thermal stresses in them. The presence of these seals also makes the valve steam-chests cool faster when the turbine is stopped. The temperature differences, arising between the turbine's HP cylinder and valve steam chests in the cooling-down process, create certain difficulties at subsequent start ups. All these problems vanish if the valve stem's seal is replaced with a hermetic assembly without steam flow along the stem. This task was successfully solved by specialists of the All-Russia Thermal Engineering Research Institute (VTI), together with LMZ, with the use of a liquid-metallic seal (LMS). Its behavior was experimentally tested even under superelevated steam conditions. Due to a low friction coefficient (less than 0.05), the LMS does not hamper the valve motion freedom. After thorough bench tests and long-term field tests at the HP control valves on an actual supercritical-pressure 300 MW steam turbine in service since 1987, such LMSs were installed, in particular, at all the HP control valves of six 300 MW turbines of LMZ at the Konakovo power plant and four similar turbines of the Lukoml power plant. The LMSs have been successfully operated without additional inspections and maintenance between the overhauls (every 6 years). No forced outages have taken place because of these seals. According to the power plant data, the resulting increase in the turbine efficiency makes up about 0.2%. It is obvious that with the increase of the main steam conditions, the effect of applying the LMSs rises.

6 EVOLUTION TOWARD ADVANCED ULTRASUPERCRITICAL THERMAL POWER PLANT BOILERS

6.1 Modification and Developments in Modern Efficient Boilers

The modern boiler in a thermal power plant is becoming mandatory, both from an economic point of view and as a positive step toward the reduction of the greenhouse effect and particulate emissions by the enhancement of efficiency. The best available technique (BAT) is becoming a future political demand, stating the achievable emission limits/efficiency that will become a benchmark for the power industry. Using improved technology, it is expected that boilers can be reliably operated with 600/610°C steam parameters using well-known materials while plant efficiencies of 46–48% (LHV, EN) can be achieved.

The boiler technology is heading toward achieving 50% Rankine cycle efficiency. Various intricate points related with technological improvements in a component design have been adopted by leading manufacturers/researchers toward achieving this goal.

6.1.1 The increase in the cycle efficiency in modern power plants is mainly achieved by:

- **(i)** Increasing boiler efficiency by optimizing the combustion process to maximize the energy release from coal, as already discussed in Section 4 of this chapter.
- **(ii)** Increasing the steam parameters with a suitable selection of metals. This issue has been discussed in detail in Sections 6.7.2 & 7.1.3 of this chapter.
- **(iii)** Redesigning/optimization of the steam cycle such as the inclusion of a double reheat module and a tuning turbine/back pressure extraction steam turbine (BEST) to maximize the energy conversion to power, as discussed in Section 8 of this chapter.

The technological steps in plant design are inevitably shifting steam from subcritical to supercritical, ultrasupercritical, and AUSC parameters toward optimization in heat rates with enhanced boiler efficiency and reduced emissions (CO_2 and NO_x) with special design features such as Fuel flexibility, effective design for APH, optimization of excess air/control of air-coal ratio, compact design for the furnace/MS piping to fit elevated parameters, judicious selection of single/double flue gas train, etc.

In modern power stations, flexibility of operation is mostly accomplished by the Benson technology in conjunction with a well-proven, reliable, and most-effective operational mode such as the sliding pressure control mode (Chapter 10).

6.2 Different Features of Modern Boilers

Different types of boilers are nowadays available to suit different types of fuel, turbines, steam cycle requirements, types of firing, etc. Some of the main features are described below.

6.2.1 Tower Type and Two-Pass Type

Two principal options are adopted for the section of the boiler after the furnace. Either a two-pass arrangement (Figs. 13.18 and 13.19) with a vestibule cage and downpass (Figs. 13.16 and 13.17) may be used, or a tower concept with the SH/RH/ECON sections above the furnace (Fig. 13.15) may be used. The walls and roof of the boiler may consist of tubes forming part of the evaporator or SH circuits. Inside these are mounted banks of tubes acting as the principal SH/RH/ECON circuits. The configurations of all possible boiler types (discussed in Section 9) though the basic types are tower type and the two-pass type.

Two-pass boilers of close coupled and divided second pass have been the preferred boiler design almost from the very beginning of thermal power plant installation; they are still available in the range from 80 to 640 MWe. The majority of boiler installations worldwide are of the two-pass design. From the last quarter of the 20th century, the necessity of a more effective boiler design with enhanced fuel flexibility, boiler efficiency, and load change rate was considered and subsequently introduced. The tower-type single-pass design is one such option that finds its application in cases where the fuel used is highly erosive in nature; it is adopted for high ash coal. As the flue gases travel through the SH/RH and economizer (ECON) pressure parts section without any change in direction, the erosion rate in these types of boilers is lower. In the case of the once-through (OT) design, the tower type is the preferred choice and is used widely. The supercritical units normally go with the tower-type construction due to the fact that it can easily adapt to the spiral wall construction. All the horizontal heat transfer sections of SH/RH and ECON in the top pass are designed as drainable sections.

The tower-type design makes it easy to arrange a *split ECON*, which is divided into two separate heating surfaces located before and after the SCR plant in the flue gas stream. The feed water is first led to the ECO located between the SCR plant and APH and from here, it is led to the ECO in the boiler top. The total heating surface of the two economizers is sized to maintain the same heat absorption as the basic design with a single ECON. The splitting of the heating surface is chosen to ensure the best possible operating conditions for the SCR plant over the entire boiler operating range (315–400°C). The split ECON will lead to a high reduction of the boiler steel structure and a reduction of the membrane walls by about 1.5 m.

However, there are certain disadvantages with the tower design of boilers.

Because the entire horizontal heat transfer surfaces have to penetrate though the furnace water wall tubes in the upper portion, the sealing arrangement requires a careful design and erection. These areas are vulnerable points of air ingress into the boiler if it is a balanced draft or leakage points if it is a pressurized unit. Due to air ingress in these points, if it happens, the pressure parts get eroded near the penetration points and can cause leakage. This type of heat transfer surface arrangement does not allow parallel erection of pressure parts. Any pressure part failure in the SH/RH and a part of ECON is cumbersome to attend and leads to higher downtime. The very tall structure of the tower design requires extra care and checking to be done if the unit is to be put in a high wind load/cyclone-prone area. These types of boilers have a higher amount of ash settlement on the heat transfer surface and so they need a higher frequency of soot blowing.

The advantages of the two-pass boiler are the lower total height of the boiler and the easy erection of the boiler top, including headers, heating surfaces, and boiler ceiling. However, the design of the boiler top of a two-pass boiler is quite complicated. The temperature difference between the first-pass and vestibule/second-pass membrane wall will often lead to cracks after some years of operation. An additional problem in the two-pass boiler is when the flue gas is leaving the first pass and entering into the second pass, the flue gas particles will be concentrated close to the rear wall of the second pass. This results in the erosion of the SH banks and normally calls for erosion shields. The two-pass boiler has some geometrical limitations that make it difficult to optimize the boiler pressure part design. The pitch of the first hanging superheater banks SH-1 needs to have a mutual distance (400–800 mm) to avoid blocking the slag.

The tower design has a number of advantages such as reduction of area/space/weight of the boiler, a fully drainable pressure part, no extraction of fly ash, a uniform flue gas temperature profile, and easy installation of the selective catalyst reactor (SCR).

6.2.1.1 Reduction of Area or Space Covered by the Boiler

The footprint or, in other words, the space covered by the tower, will be smaller than that of the two-pass boiler. The tower provides a very compact boiler design, including accommodation of SCR/APH. Although the total height is increased, due to the reduced footprint, the boiler steel structure has also been reduced consequently.

6.2.1.2 Reduction of Weight of Boiler Pressure Part

For a tower boiler, the heating surface is fully effective because the flue gas flow is perpendicular to the bank heating surfaces, facilitating efficient heat transfer. However, for the two-pass boiler, the heat absorption efficiency of the hanging superheaters is reduced. So the tower boiler pressure part's weight is less compared to a two-pass boiler.

6.2.1.3 Fully Drainable Pressure Parts

Fully drainable pressure parts are accessible as all heating surfaces are horizontally arranged. During start up, the condensate can be drained easily, facilitating faster heating of the pipelines and faster rising of the steam temperature. For short overhauls, dry preservation can easily be used. Steam-side oxidation and increased formation of magnetite (Fe_3O_4) layers inside the superheater tubes is critical in USC boilers. The risk of blocking the superheaters by exfoliated magnetite is considerably reduced in tower boilers with horizontally arranged superheaters.

6.2.1.4 No Extraction of Fly Ash

With the arrangement of the flue gas path from the boiler and downward through the SCR/APH, there is no need for additional/intermediate extraction of fly ash (as needed for the two-pass boiler, where the flue gas is changing from a downward to an upward flow below the ECO). When the flue gas is leaving the APH, it proceeds horizontally and the hoppers in the flue gas duct below the APH can be designed only to collect water from APH washing and not for extraction of fly ash, which will end up in the filter.

6.2.1.5 Uniform Flue Gas Temperature Profile

The tower design in combination with tangential firing results in a very uniform flue gas profile entering the first SH. Temperature peaks in the flue gas path are avoided, resulting in fewer temperature imbalances in SHs. This is very important for materials operated at their limit in the creep range.

6.2.1.6 Easy Installation of Selective Catalyst Reactor

The arrangement of the boiler and APH makes it natural and easy to install an SCR. Ammonia injection will be installed just after the boiler outlet. The flue gas duct with possible installation of a static mixer ensures that there will be a uniform distribution at the SCR inlet. Flue gas velocity through a typical SCR is approximately 5–6 m/s, but higher velocities (because of air-to-gas leakage) will decrease residence time and therefore affect ammonia injection rates and slip. In flue gas desulfurization systems, a lower residence time can affect lime or limestone injection rates and thus the SO_2 removal efficiency.

TABLE 13.10 Comparison Table Tower Type versus Two-Pass Boilers

Sl	Tower Type	Two Pass
1	Uniform flue gas profile and reduced temperature peak in pressure part	Uneven flue gas flow profile and high ash concentration on second pass rear wall
2	Excellent RH temperature characteristic	Cold built in conditions for final RH, effective heating surfaces
3	No ineffective (dead) areas	Partly ineffective heating surfaces
4	Reduced footprint, increased height	Reduced height, enlarged footprint
5	Low pressure loss due to high number of parallel tubes in superheater banks	Higher pressure loss due to limitation in heating surface design
6	Easy installation of SCR	Difficult installation of SCR and increased duct work
7	Smooth membrane wall temperature increase	Thermo stress and cracks in membrane wall between first pass and second pass
8	No extraction of fly ash, extraction of fly ash below ECO fully drainable superheaters, fast start up	Risk of blocking the hanging superheaters by exfoliated magnetite
9	Simple boiler suspension, penthouse not required	Complicated boiler suspension

Source: ADVANCED BOILER DESIGN by Flemming Skovgaard Nielsen, Carsten Søgaard, BURMEISTER and WAIN ENERGY A/S (BWE) October 2014.

The boiler ceiling of the tower is uncooled. The boiler suspension is very simple and does not require any penthouse. It has a very simple design of membrane walls with a smooth increase in temperature. On the tower boiler, the first SH banks SH-1, which are typically arranged just above the final RH, can be designed with a smaller pitch (100–200 mm). The number of parallel tubes can be higher and so the pressure loss will be smaller.

Construction time of the tower boiler is longer than the two-pass boiler, but it is expected that a better performance in the total lifetime of more than 35 years would be available from the former. Table 13.10 shows a comparison between the two types.

6.2.2 Types of Water/Membrane Wall

Supercritical, USC, and AUSC plant boilers are necessarily of the OT type, which is more flexible with respect to heat absorption because the evaporation fully takes place in the furnace and superheating starts already in the membrane wall. The problems of erosion and corrosion are minimized in OT boilers as two phased mixtures do not exist. The OT boilers are divided into two basic types of furnace design as far as water wall or heat absorption membranes are concerned.

6.2.2.1 Vertical Tubes

Here, the units are designed with vertical tubes like the construction techniques employed in natural circulation boilers. Following the practice of US boiler construction, making use of a partial membrane to minimize the amount of refractory/skin casing. In order to obtain the necessary mass flux within the tubes to provide adequate heat transfer/tube cooling, a multiple pass arrangement has been adopted. Because the units are meant only for operation in the supercritical regime, this design strategy was successful.

6.2.2.2 Spiral/Meandering Tubes

The Benson technology boilers, because they were originally designed for operation in the subcritical regime, made use of the meandering tube arrangement where the tubes passed completely around the furnace enclosure as a means of obtaining more uniform heat absorption from tube to tube. This construction was necessary for the subcritical OT design as a means to minimize the temperature difference between tubes. The meandering tube design used the refractory/skin casing construction. This type of design is offered by Siemens KWU (owners of the *Benson boiler* patent). The basic differences are indicated below in Table 13.11.

All OT boilers incorporate relatively small bore evaporator tubes (~25 mm), which are generally arranged in a spiral/vertical fashion to form the furnace envelope.

As the water is evaporated to high quality in an OT boiler, it is important to guard against dryout occurring in high heat flux zones or to take other precautions against the phenomenon associated with burnout. A useful solution promotes the use of rifled-bore tubing, which, by creating centrifugal forces, causes more of the liquid phase to remain in contact with the tube wall, thus delaying dryout to higher qualities and/or enabling lower water velocities to be adopted.

6.2.3 Tangential (T) Firing—Front/Opposed Firing

In a T-fired furnace cross-section, it is obvious that there are no dead corners whereas front/opposed fired boilers will have dead corners at the outer burners close to the side walls.

BWE (Denmark) designs a T-firing system for coal combustion with air excess down to λ 1.15. The BWE combustion system for T-firing will have over burner air

TABLE 13.11 Comparison Table for Spiral Versus Vertical Water Wall

Sl	Spiral Water Wall	Vertical Water Wall
1	Spiral furnace system applicable for all size units	Vertical furnace wall system limited to larger capacity units
2	Benefits of averaging of lateral heat absorption variation (each tube forms a part of each furnace wall)	Lower water wall pressure drop, thereby reducing the required feed pump power
3	Simplified inlet header arrangement	Traditional furnace water wall support system
4	Large number of operating units	Elimination of intermediate furnace wall transition header
5	Use of smooth bore tubing throughout entire furnace wall system	Less welding in the lower furnace wall system
6	One material utilized throughout entire water wall system	Easier to identify and repair tube leaks
7	No individual tube orifices…less maintenance and plugging potential	Less complicated windbox openings

Source: Presentation on Supercritical Boilers by BHEL on Aug. 28, 2012.

(OBA) injected just above each burner. The OBA is part of the air staging and contributes to a reduction of NO_x formation. It also ensures sufficient O_2 content along the membrane wall and in this way, protects the membrane walls against CO corrosion. No CO corrosion is reported on T-fired boilers with a BWE system.

The use of circular burners provides an annular airside protection against CO corrosion of the furnace walls, contrary to traditional jet burners in tangentially fired systems. This protection is enhanced by the use of the OBA nozzles, which maintain a layer with higher stoichiometry close to the walls.

The T-firing concept results in a very uniform flue gas temperature profile at the furnace outlet. The temperature imbalance in the first heating surfaces caused by the flue gas profile is thus reduced significantly. In the USC boiler design, this is essential because the materials operate in the creep range, where allowable stress drops fast when the temperature increases. Front, opposed, or even worst box boilers will have high temperature peaks in the profile of flue gas temperature at the furnace outlet, resulting in temperature peaks in the SH banks. By T-firing, it is possible to operate with longer flames and without swirls in the tertiary air sectors of the burner. Front and especially opposed-fired boilers are very sensitive to variations in coal composition and the related shape of the flame. Often, it is required with a heavy swirl in order to reduce the flame length. In opposed-fired boilers, the flames will meet at the middle of the furnace and generate NO_x.

6.2.4 *Double Reheat Cycle*

One of the obvious initiatives to increase plant efficiency is to introduce the double RH cycle. This will increase plant efficiency by approximately 1%. When boilers are designed with double RH, it is extremely important to make accurate boiler calculations and to balance the heat absorption between the two RH parts. Even with the increased cost for the associated equipment and accessories toward additional RH/turbine modules, pipework, and valve arrangements, it may be commercially attractive to design a thermal power plant with a double RH cycle—a well-proven technology to be used in an effort toward 50% efficiency (LHV). Development/modification in double RH steam cycle/parameters is discussed in detail in Section 8 of this chapter.

6.2.5 *Single Train-Double Train*

There is a tendency to design new utility boilers in many countries with a single-train flue gas path and mono components. For years, some boiler plant manufacturers have design implemented with only one APH and very high availability (as per operation experience) and because the risk of failure is reduced significantly. All critical components such as bearings can be designed for a high lifetime (200,000 h for APH typically for bearing). The single-train mono components concept is used in up to 800 MW plants and is recommended for modern boiler design. Single-train mono components require one FD/PA fan, one/two ID fans, one APH (with two drives), and no dampers in the flue gas path where the equipment/duct system is arranged as a single flue gas path.

6.2.6 *Advantages and Disadvantages of Single Train*

The main advantages of the single-train design are: Lower cost due to a reduced number of components; lower APH leakage rate (reduced from 8% to 6%), less power consumption, SCR design 100% (not 2 × 60%), no temperature imbalances, and no dampers in the main flue gas path. The disadvantages are a higher starting current on the FD/PA fan and a higher load on single steel columns in the APH/SCR area.

Together with the single APH concept, it is suggested to have two 100% main drives; they can be in operation simultaneously. In case one drive fails, the other will take over. On top of one of the drives, an AC emergency drive will be placed and connected to the emergency diesel generator. The layout of a single train: The single APH is typically arranged at elevation +35 m. A single center column is positioned at the center line supporting the main APH rotor bearing. Due to the large diameter of the APH, the casing is supported on beams outside the footprint of the boiler itself.

A double-train component require two FD fans, two/more (for cold PA) PA fans/APH, 2–3 ID fans, and ~16. dampers in the flue gas path. For the double-train solution, the equipment and the duct system are arranged as two parallel flue gas paths. The components are usually arranged so as to make it possible to run the boiler at 60% load with one of the FD fans out of operation.

6.3 Optimized Combustion System

The complete optimized combustion system with a coal firing arrangement basically consists of low-NO_x burners where three coaxial burner air outlets are primary air, secondary air (SA), and tertiary air (TA). The other two corner-wise air flow arrangements are OBA and over fire air (OFA); the air nozzles are above the burner zone. The combustion system may, however, be applicable to different types of fuel such as cofiring coal/biodust/oil/gas, as have been commissioned in many plants worldwide.

6.3.1 *Tangential Firing System*

BWE has developed a technique using circular burners and air staging. By inclusion of OFA, this system is extended to in-furnace air staging, resulting in reduced NO_x formation. The complete BWE tangential firing system with coal is used in the latest USC boilers in Denmark for a fuel range including coal, oil, gas, and biomass. The principle of air staging means applying just enough air to make the combustion stable but not enough to allow the N_2 to be oxidized to NO and NO_2. The NO_x formation is governed by local furnace parameters such as gas temperature and composition. In low-NO_x burner designs, the combustion air is controlled so that a staged mixing of fuel and air takes place. The result is a long flame where zones with simultaneous high temperature and high air-fuel ratio are avoided. By implementing low-NO_x burners alone, the NO_x level can be reduced by 40%–50% compared to traditional "high NO_x" coal burners. The combustion system is composed of five air flows.

For in-furnace air staging, the overall air-fuel ratio in the burner zone is reduced to just above stoichiometry (~1.05), which results in an even better NO_x performance. Finally, the principle of air staging is extended to the whole furnace by use of over fire air (OFA). With both low- NO_x burners and OFA, the NO_x level is typically reduced by 60%–70% compared to a high NO_x combustion installation.

6.3.2 *Twin Vortex (Tangential) Firing*

M/s MHI has designed a twin vortex (tangential) firing system where two similar vortices are produced in every level; those are equipped with fuel burners. The system is depicted in Fig. 13.6 with the relative position of different headers, pipes, and sprayers. For a *firing control system*, they use MACT (Mitsubishi Advanced Combustion Technology) with advanced pollution minimum (APM) burners with 60% of total combustion air and the rest at 40% through additional air ports for low NO_x and stable combustion. Here, the *reheat temperature* control is achieved by three types of control element/gears such as by the use of:

a. *Gas-biasing dampers* in the second pass as the primary control system.
b. An additional tilting mechanism for burners as a secondary control.
c. Interstage attemperation as an emergency control that may come into operation when a predetermined RH temperature exceeds the set value. The SH *temperature* is controlled solely by the spray control of the three-stage SH through the use of interstage attemperation.

6.4 Fuel Flexibility and Cocombustion

The multifuel concept has been installed in many boiler plant combustion systems, and the burners are suitably designed for firing different types of fuel. Standard PF burners are designed with an inner gas and simultaneously an inner HFO/LFO oil lance. The combustion system can be designed for 100% coal/oil/gas firing. In this way, the fuels can be changed quickly; it is also possible to operate with different fuels on different burner levels. For example, the Avedøreværket Unit 2, a 400 MW multifuel USC boiler by M/s BWE, has four burner levels, each designed for coal, oil, and gas.

Another concept, the cocombustion of biomass (straw, wood chips/pellets) and coal in utility boilers, is a proven technology. A heat input of 10%–15% on biomass is possible and it is still possible to use the fly ash for cement and other products, for example, bricks, fertilizers, road material, etc. Cocombustion of biomass is carried out through the PF coal burner by a center lance/inner tube injecting fine particles of biomass/refused derived fuel (RDF) into the furnace. Furthermore, the normal pulverized fuel system can be used for biomass firing. Wood pellets are simply grinded in the mills and combusted via traditional PF coal burners.

6.5 Controlled Safety Pressure Relief Systems (CSPRS)

Preferably, hydraulically operated safety valves are used nowadays to protect the high-pressure parts as well as the

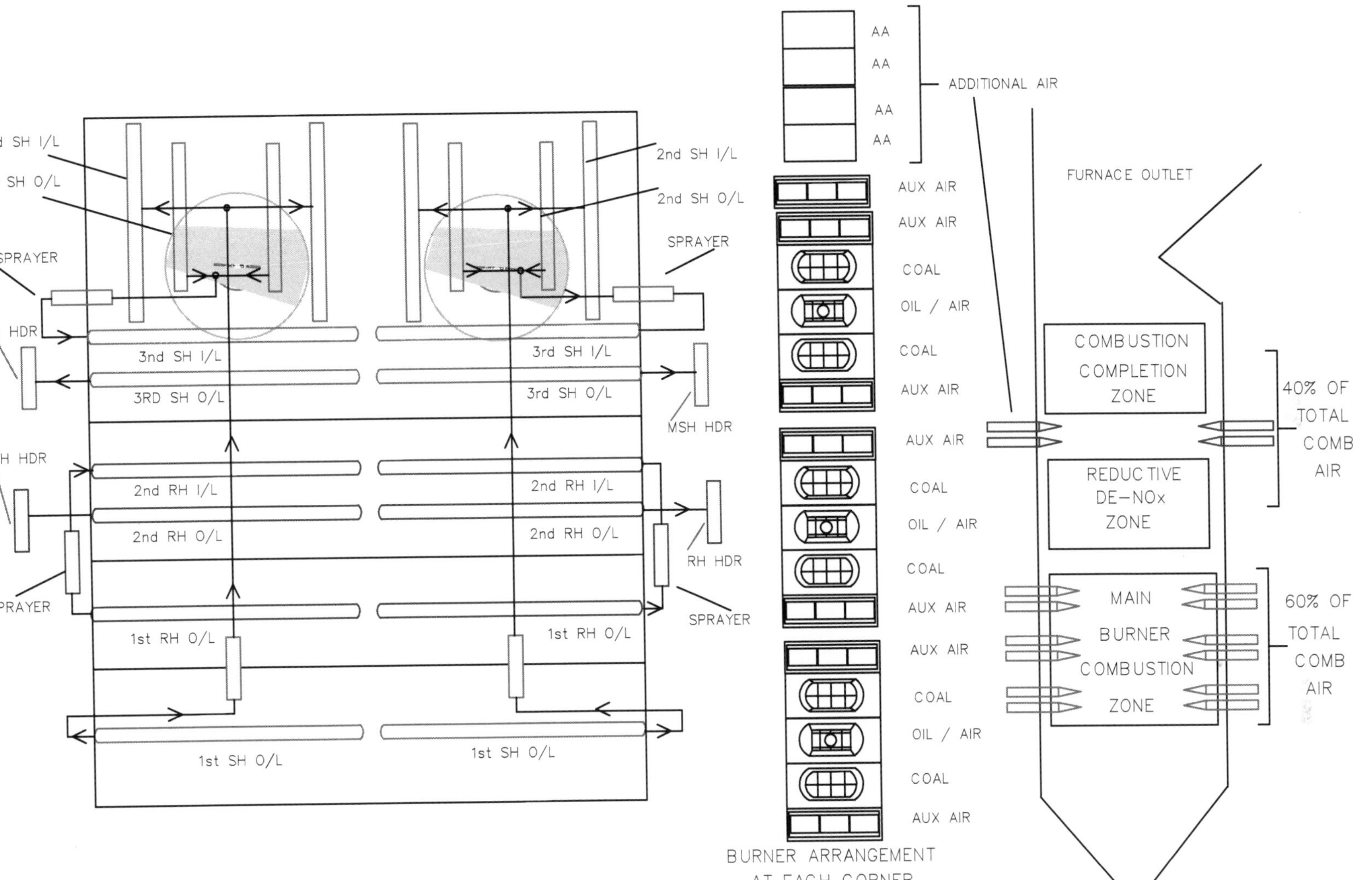

FIG. 13.6 Arrangement of twin vortex (tangential) firing system by MHI.

TABLE 13.12 Comparison Table CSPRS Versus Traditional System[a]

Sl	Controlled Pressure Relief	Traditional Safety Valve
HP side	100% HP combined bypass units with safety function Hydraulic spring-loaded (two valves)	80% HP bypass valves (two valves) 23% HP vent valves (two valves) Shut-off valves in front of HP bypass (two valves) 103% HP spring-loaded safety valves (two valves)
IP side	116.5% RH safety valve Hydraulic spring-loaded (two valves)	100% RH spring-loaded safety valves (two valves) 40% RH vent valves (two valves)

[a]*All the figures are approximate and may differ from plant to plant.*

Source: ADVANCED BOILER DESIGN by Flemming Skovgaard Nielsen, Carsten Søgaard, BURMEISTER and WAIN ENERGY A/S (BWE) October 2014.

reheater in power stations. In comparison to the use of spring-loaded safety valves, the number of installed components can be reduced drastically by the use of a combined bypass and safety valve arrangement (Table 13.12).

A modern boiler is typically equipped with valves as indicated in the above table 100%. Hydraulic spring-loaded bypass units with safety functions (two valves) for the HP/MS pressure line and 100% + 16.5% hydraulic spring-loaded safety valves (two valves) for the IP or RH steam pressure line. With this design, there is always sufficient flow through the RH. RH safety valves (RHSV) will be designed for 100% HP flow plus a spray water flow of, for example, 16.5%. The span between normal operation pressure at MCR 103% and the set pressure of the safety valves can be used for overload operation. By having combined HPBPs and HPSVs, the whole span can be used.

It is normally foreseen that the LP bypass valves at the turbine will have a capacity of 60%–70%. At turbine trip, the RHSVs will open. The boiler load will be reduced to 60%–70% load. The RHSVs will then be closed and the full RH flow will pass through the LPBP valves. The HPBP and the RHSVs will be designed for three types of operation: safety relief (opening time <2 s), quick opening (closing time 5 s), and control (opening/closing time 15–20 s).

The safety relief control unit will open the valves at the set pressure as if they were simple spring-loaded safety valves. The hydraulic relief system is in line with EN Code 12952-10.

The "quick opening function" relieves the pressure at low load and part load when the sliding pressure is exceeded by a certain margin. In this way, the quick opening function avoids stagnation in the steam flow, essential for the sufficient cooling of SHs and RHs.

The "control" operation mode is used for pressure and flow control during start-up conditions. By continuous boiler operation after a turbine trip, the HP bypass valves are controlled so that the boiler pressure is controlled as per the sliding pressure curve.

The safety valve design described is in line with the references of BWE and in line with the design of new USC boilers in Northern Europe. The pipe layout will be simplified due to a reduced number of branches of safety/bypass/vent valves with less piping.

The HP bypass valves are connected closely to the main steam line and in this way, kept at sufficient temperature. The two HPBP valves will always operate in parallel, avoiding temperature imbalances between the LHS/RHS of the boiler outlet. The RH safety valves will be equipped with exhaust pipes/silencers requiring only a small bore heating line.

6.6 Increasing Overall Plant Efficiency by Increased Steam Parameter

The next step in modern boiler design is to increase the steam parameter further. Studies performed in the late 1990s showed that an increase of steam temperatures to 650°C is not possible with the known martensitic steels (P91 and P92) for headers and steam piping. Instead, the use of Ni alloy materials such as alloys 174, 263, 617, and 740 is required. Because these materials have very good creep properties (creep strength does not limit the design stress at temperatures below 650°C), it is therefore natural to take a large step up to 700°C steam temperature. It has been considered that single steam cycles with parameters 700/72 0/35 reaching 50.7% and double reheat steam cycles with parameters 700/720/35 reaching 52% on coastal sites is possible.

6.7 Materials for Advanced Ultrasupercritical Boiler Selection of Metals

At ultrasupercritical plant boilers, the materials after the evaporation stage are usually in the creep range due to the elevated steam parameter. Any increase in temperature, however small, will effect a remarkable diminution of the allowable stress, which can be prevented by taking suitable measures such as reducing the temperature imbalances in the boiler vis-a-vis the furnace. Modern boiler design makes this possible by taking a number of well-proven measures, namely a uniform flue gas profile, intermediate outlet headers, crossover of steam from the left to the right boiler side, and in some cases, the extraction of MS via four outlet headers (instead of the usual two). With ongoing programs on superior metallurgy in the near future, thermal power plant industries are fully geared up to maximize the bottom line of plant operation with a cleaner and greener environment.

6.7.1 *The First Generation*

The first generation of supercritical power plants was faced with a variety of problems in design, operations, and material-related problems that resulted in reduced availability and reliability. For example, EDDYSTONE STATION, the 325 MW GENERATING UNIT 1 in Philadelphia was installed around 1960 with the 3600 rpm tandem compound turbine shaft that consisted of the newly introduced SP (superpressure) element and the combined VHP/HP (very high pressure/high pressure) element and their associated electric generator system. The SP element received steam at 35 MPa and 650°C. Due to drastically increased maintenance costs, the power industry had to revert back to subcritical units.

6.7.2 *International Development Projects*

AUSC boilers with an increased steam temperature of 700°C are considered to be the key technology for the realization of a lower environmental load and the increased efficiency of power-generating facilities. Development projects have been ongoing for years in many countries worldwide, which is discussed in detail in Section 4 of this chapter.

6.7.2.1 Boiler Material Selection

The 700°C class AUSC boiler is a promising technology for high-efficiency coal-fired thermal power generation. For AUSC boilers, water enters at 320°C and steam leaves at around 480°C, with the metal temperatures around 540°C. This demands that the creep strength be much greater for the SH/RH tubing. For the secondary SH, the material will be austenitic stainless steel for supercritical boilers, advanced austenitic stainless steel for USC, and Ni alloys for AUSC. For RH temperatures designed higher than the SH temperature, the grades of material used are slightly higher than for the SH. T22 has sufficient strength but the fire-side corrosion resistance and steam-side oxidation resistance is very low for AUSC conditions. Steam-side oxidation is a major issue that can be reduced by efficient water treatment techniques, but beyond a particular operating condition, the water treatment will not be effective. Suggestions have been made to use T91/T92 for the water wall, which has strength and corrosion resistance but requires careful heat treatment. Inconel 617 satisfies all the requirements but its cost is a serious drawback. High fire-side corrosion is required by all the materials used in AUSC boilers.

It is also expected that an AUSC boiler can be retrofitted to an existing power-generating facility in most cases and is useful as a replacement for a growing number of aging coal-fired boilers. The reasons behind that idea is that the basic configurations remain unchanged and there are no drastic changes required in the peripheral equipment of a power station. The only main difference is that the final steam temperature of AUSC boilers is 100°C higher than that of conventional coal-fired boilers. It is envisioned that CO_2 emissions from 1000 MW coal-fired power plants can be reduced by approximately 12% from existing ultrasupercritical (USC) power generation. In an AUSC plant, the overall efficiency is also expected to improve by nearly 10% (as a relative value) from that of the USC, which indicates the possibility of maintaining high efficiency when combined with carbon dioxide capture and storage (CCS) systems in the future. The ultra high-temperature materials that are developed can be applied to not only coal-fired power generation, but also the rest of the cycles (steam turbines and waste heat recovery boilers) of gas turbine combined cycle plants with ever-increasing temperatures.

To realize such boilers, it is critical to develop and validate the creep-enhanced Ni-based alloys and advanced ferritic steels, both of which can withstand temperatures substantially higher than conventional boilers. In addition to the long-term creep rupture strength and manufacturing techniques, the steam oxidation properties and the coal ash-induced high-temperature corrosion properties are also essential requirements.

The components can be categorized into three sections: (i) heat transfer tube panels, (ii) SHs/RHs, and (iii) thick section components and steam pipes.

6.7.2.2 The Heat Transfer Tube Panels

Three newly developed alloy steels have been selected for furnace panels: a high alloyed 12% Cr tube HCM12 developed by M/s Sumimoto Metal Industries (SMI) and Mitsubishi Heavy Industries (MHI) with all adequate features. Moreover, due to the duplex microstructure of nearly 30% ferrite and 70% tempered martensite, welding is possible without preheat and PHPT; other low alloyed $2^1/_2$% Cr tube HCM2S developed by SMI/MHI and 7CrMoVTiB1010 developed by Mannesmann also with all adequate features and no PHPT.

6.7.2.3 Superheaters/Reheaters Tube

High-temperature (creep) strength properties and corrosion resistance are the basic two characteristics required for metals or metallic compounds being used in AUSC plants. The other two are low cycle fatigue strength and easy machinability.

High-temperature (creep) strength properties or creep causes permanent deformation of materials due to high mechanical stresses if subjected to heating for a long time (increases and effects of creep deformation generally become noticeable at ~35% of the melting point). This also prevents normal operation when it goes beyond the limit. The rate of deformation is a function of physical properties/ exposure time temperature/applied structural load and is "time-dependent" occurs gradually upon the application of stress/strain accumulates as a result of long-term stress

6.7.2.4 Resistance to Steam Oxidation and High-Temperature Corrosion

Two other important characteristics are necessary for selecting AUSC materials: resistance to steam oxidation and high-temperature corrosion. A steam oxidation scale is formed on the inner surface of high-temperature heat transfer tubes and piping as a result of the reaction with superheated steam. When heat transfer is prevented by scale formation, the metal temperature increases, accelerating creep damage. The deposition of exfoliated scale may result in blocked tubes/pipes or steam turbine erosion. When used in AUSC plants, the Ni alloys are supposed to be exposed to temperatures of ≥700°C, that is, higher than conventional steel. As these alloys contain more added elements than the currently used austenitic steels, many unclear points of their behavior–including the growth rate and properties of scale–are still unknown. Ni-based alloys with superior heat resistance are intended to be used in heat transfer tubes and MS/RH steam pipes, all of which are subject to steam temperatures of ≥700°C. The possibility of using less-expensive advanced ferritic steels for the major piping with steam temperatures of 650°C or below has also been considered to improve the economy by decreasing the amount of expensive Ni-based alloys whenever they are unnecessary. To realize AUSC boilers, therefore, it is important to design/examine Ni-based alloys and advanced ferritic steels suitable for use at temperatures above 600°C for the pressure parts as boiler component which must be suitable and evaluated for the creep strength and corrosion resistance.

Most modern nickel-based alloys have been developed from a Ni20%Cr alloy in order to achieve suitable creep properties. For example, MHIi uses six types of Ni-based alloys: HR6W, HR35, and Alloy 141 as well as Alloy 617, Alloy 263, and Alloy 740 (another newly developed alloy steel INCONEL 740 alloy based on alloy 263 which is a nickel, chromium, cobalt alloy and is age hardenable by precipitation of gamma prime but also benefits from solid solution hardening), the properties of which were well assessed in the United States and Europe as A-USC candidate materials. Of these, HR6W, HR35, and Alloy 617 have superior workability and high-temperature strength; they are also considered especially suitable for use in thick large-diameter pipes. For heat transfer tubes, in addition to these three alloys, Alloy 263, Alloy 740, and Alloy 141 are also included as candidate materials. Their high strength is enhanced by the increased precipitation of the γ' phase (Ni3 (Al, Ti) intermetallic). Three types of advanced ferritic steels—high B-9Cr steel, low C-9Cr steel, and SAVE12AD—were developed and designed by adding Co and/or B to the conventional material (modified 9Cr-1Mo steel) in order to improve the high-temperature strength and prevent deterioration of the creep strength at the heat affected zone.

6.7.2.5 Thick Section Components and Steam Pipes

For thick section components and steam pipes, two possible materials are:

(i) An improved version of ferritic steel or martensitic 9–12% Cr steel alloy is desirable to increase the present temperature range to 650°C. At the moment, it does not seem possible to obtain high creep strength and high oxidation resistance properties in the ferritic steel family.
(ii) An Ni superalloy with a 100,000 h rupturing strength at 700°C is necessary for construction of MS lines and an outlet header with proper wall thickness. Nimonic 263 alloy or an improved version of Inconel 617 may meet the demands of the MS lines and outlet header. Inconel 740 (Ni, Cr, Co alloy) based on Alloy 263 has been found to be age hardenable by the precipitation of gamma prime and solid solution hardening. The ash corrosion resistance is also found to be quite appreciable.

Composition of Inconel 740 is: C (0.03)/Cr (25)/Mo (0.5)/ Co (20.0)/Al (0.9)/Ti (1.8)/Nb (2)/Mn (0.3)/Fe (0.7), Si (0.5), Ni (balance).

The manufacturing process includes the following:

(i) A welding technique vis-a-vis residual stress analysis for assessment of reheat cracking sensitivity. The initial creep strain corresponds to the magnitude of residual tension. If a small initial strain can yield a crack in a material and it happens within a short while, the material is considered highly sensitive to reheat cracking.
(ii) Bending techniques, including cold/hot bending with a proper heat/thermal treatment.

6.8 Changes in Technology, Parameters, and Controls

There has been a gradual but huge modification in the area of technology, parameters, and controls in the journey toward highly efficient and low pollutant AUSC thermal power plants. B&W, in their technical paper BR-1896, “Steam Generator for 700C TO 760C AUSC Design and Plant Arrangement: What Stays the Same and What Needs to Change,” indicated clearly the gradual changes in three development stages. Table 13.13 incorporates the essence of that technical paper.

TABLE 13.13 Comparison Table of Basic Technology, Parameters, and Controls Used in SC, USC, and A-USC Plants

Feature	Supercritical	Ultrasupercritical	Advanced Ultrasupercritical
Steam pressure and temperature	24 MPa, 540–565°C	~25.5 MPa, 600–650°C	~30 MPa, 700C–760°C
Pressure control	Constant pressure	Variable pressure	Variable pressure
Load control	Unit load master feed forward to firing feed water flow	Firing rate demands some use of frequency control (condensate throttling)	Firing rate demands some use of frequency control (condensate throttling)
Steam temperature control	Feed water: firing rate ratio, low single stage spray attemperation	Multiple stage spray attemperation versus load program	Multiple stage spray attemperation versus load program
Feed water control	Feed forward with unit load master	Furnace enthalpy diff pick up versus load and trim with 1st stage attemp. diff. temperature	Furnace enthalpy differential pick up versus load and trim with first stage attemp. diff. temperature
Water treatment chemistry	All volatile treatment (AVT) with full CPU. Some use OWT	Oxygenated water treatment (OWT) with full CPU. AVT used in early operation	Oxygenated water treatment (OWT) with full CPU. AVT used in early operation
Turbine bypass	5%–10% on mainsteam	40%–100% HP and LP	40%–100% HP and LP
Start up system	Division and throttling valves between furnace and SH, 10 MPa flash tank with steam and water drain recovery system	Vertical separator and water collection tank (WCT) with boiler circulation pump. WCT level control valve to condenser for flow up to 7%	Vertical separator (VS) and water collection tank (WCT) with boiler circulation pump (BCP). WCT level control valve to condenser for flow up to 7%
Start up operation	Pump minimum furnace circulation flow 25%–33%. SH shut-off and throttling vlvs closed. Bypass vlvs throttle water to flash tank and returns steam to SH. When up-stream enthalpy high enough to pressurize the SH, throttling division valve opened on a program to raise SH to 24.1 MPa (OT operation)	Pump minimum furnace circulation flow (30%–35%) with a minimum FW flow of 7% and use of the BCP controlling WCT level. FW flow meets minimum as BCP handles less and less water drains. BCP is shut off above Benson Load point (30%–35% load). OT operation FW control listed above becomes highest demand	Pump minimum furnace circulation flow (40%–45%) with a minimum FW flow of 7% and use of the BCP controlling WCT level. FW flow meets minimum as BCP handles less and less water drains. BCP is shut off above Benson Load point, about 40%–45% load. OT operation FW control listed above becomes highest demand
Arrangement configuration	Two pass and tower	Two pass	Two pass, tower, and modified tower
Piping material	P22	P92	740H nickel
Furnace enclosure material	T-2 smooth and multilead ribbed, vertical multipass UP-UP furnace with 1st to 2nd pass full or partial mix and 2nd to 3rd pass full or partial mix	T-12 smooth and multilead ribbed spiral lower furnace (vertical tube lower furnace needs T-23 or high flow per foot perimeter), vertical upper furnace above transition	T-12 T-22 T-23 T-91 T-92 smooth and multilead ribbed lower furnace (spiral or vertical based on flow per foot of perimeter), vertical upper furnace above transition
Superheater material	T-22, 304H	Previous column plus T-91, T-92, 347HFG, 310HCbN	Previous column plus super 304H, 230, 740H

Source: Technology Comparison of B&W Supercritical, Ultrasupercritical, and Advanced Ultrasupercritical Steam Generators

7 MODERN AUSC PLANT TURBINES AND DEVELOPMENTS

For most of these manufacturers, the range of the rated output for large steam turbines designed and produced for fossil-fuel power plants stretches to 800–1200 MW, even though the MHI and ALSTOM power turbosystems mark a readiness to deliver steam turbines with a single capacity of up to 1400 and 1800 MW, respectively.

7.1 Rise of Steam Turbine Output and Efficiency With Steam Parameters

7.1.1 Steam Turbines With Subcritical Main Steam Pressure of the 1960s–1980s

Beginning from the early 1950s and up to the early 1990s, the increase of MS conditions, especially MS pressure, was the basic and most productive way for raising the efficiency of the then newly designed power steam turbines.

The LMZ 200-MW RH steam turbine (1958) may be taken (as claimed) as the starting point for tracing the steam turbine efficiency rise. Around 1960, a great number of these turbines of several modifications were installed at power plants of the then USSR and some European and Asian (including India) countries. Those turbines attained a gross efficiency of 44.7% with a subcritical MS pressure of 12.8 MPa as well as MS/RH steam temperatures of 540°C with 210 MW output.

7.1.2 Supercritical and USC Pressure Steam Turbines of the 1960s–1980s

Mass commercial operation of supercritical power plants with MS pressure (>21.4 MPa) and the same MS/RH steam temperatures of about 535–545°C also commenced in the early 1960s. There was a clear rise in efficiency (46.7%–47.2%) and a decrease in heat rate (7710–7626 kJ/kWh) achieved in LMZ supercritical-pressure steam turbines with MS pressure up to 23.5 MPa and the same MS/RH steam temperatures of 540°C delivering rated output in the range of 300–1200 MW compared to the above mentioned data of 8045 kJ/kWh and 44.7% for the subcritical-pressure 200-MW turbine of the same producer. It is understandable that the increase in the turbine output and a resulting increase in the blading length as well as the use of longer last stage blades (LSB) providing a decreased energy loss with the exit velocity, as well as some improvements in general design solutions, also played a certain role in this efficiency rise.

Supercritical MS pressure made it possible to reach the record-breaking values of the single capacity for a fossil-fired steam-turbine power unit became equal to 1300 MW with the first such unit was put into commercial operation in the United States (1972–73, at Cumberland and Amos). Subsequently, nine such 1300-MW units at US power plants increased to nine in stages until 1991. These units, equipped with cross-compound (CC)/double-shaft steam turbines of ABB, have remained the largest among the fossil-fuel power units up to now. The largest single-shaft high-speed steam turbine of 1200 MW output and a maximum output of 1380 MW, was developed by LMZ and commissioned in 1979 (the Kostroma power plant, Russia).

The rated steam conditions of the above-mentioned 1300-MW units installed at the US power plants are mainly 24.1 MPa, 538/538°C. These units were preceded and followed by other supercritical-pressure steam-turbine units with approximately the same steam conditions, but fewer single capacities were put into operation at US power plants, for example: the Widows Creek Unit 7 (500 MW, 1960), Bull Run Unit 1 (900 MW, 1966), and Paradise Unit 3 (1100 MW, 1969).

According to the North American Electric Reliability Council (NERC), even for the first years of commercial operation, the reliability indices (equivalent availability factor and forced outage rate) for supercritical coal-fired plants in the United States with an output of 400–799 MW—excepting the early, very first "supercriticals"—were quite comparable and even better than those for similar subcritical units. For years, supercritical units have retained their leading positions among the United States' most efficient steam-turbine power plants. In the list of top US coal-fired power plants of 2001, 10 of 20 power plants with the highest efficiency are completely/partially furnished with supercritical units, even after operating for many years. For example, the operating plant heat consumption of the Bull Run power unit (the most efficient in the United States for many years) with 889 MW output and steam conditions of 24.1 MPa, 538/538°C, heat rate 8861 kJ/kWh corresponding to an efficiency of 40.63%, even though this unit was commissioned in 1967 and is still running. The total number of supercritical-pressure steam-turbine units commissioned at US power plants until 1991 was 155. Thirteen of them, launched between 1967 and 1972, were double-reheat with the steam temperatures equal to 538/538/538°C or 538/552/566°C. The greatest number of such units were installed in the former Soviet Union: 180 with 300 MW turbines and 15 with 500 MW turbines. Fourteen with a considerable number of supercritical steam-turbine units were launched in Japan; some units of this class were also constructed and implemented in Germany, Italy, Denmark, the United Kingdom, China, the Netherlands, Finland, and other countries. The relatively low prices of natural gas and the impressive developments of the 1990s in gas turbine (GT) technologies provoked a boom in power-generating facilities with the use of simple-cycle GT and combined-cycle units that featured pretty high efficiency, low capital expenditures, and lower gas emissions.

In the early 1960s, the United States also pioneered the development and mastering of pilot commercial USC steam-turbine plants, that is, with an MS pressure of

30 MPa and above. Most of these units were also designed with elevated MS temperatures of 593°C and more. In addition, the increased MS pressure required increasing the RH steam temperature or using double-reheat to avoid inadmissibly high wetness at the LP exhaust. The most well-known of these pilot USC units were the 125 MW Philo Unit 6 with a GE turbine and steam parameters of 31 MPa and 621/565/538°C, and the 325 MW Eddystone Unit 2 with a Westinghouse turbine of steam conditions of 34.5 MPa and 650/565/565°C. Some experimental power units were operated in Germany, for example, the Hattingen Units 2 and 4 of a 107 MW output, with a low supercritical MS pressure of 22.1 MPa but elevated MS temperature of 600°C; they were preceded by a few experimental steam turbines of a small capacity with advanced MS temperatures of up to 650°C. In 1967, a unit with a 100 MW back-pressure turbine of KhTGZ of steam conditions 29.4 MPa and 650/565°C was implemented at the Kashira power plant (Moscow, USSR); the outlet of this turbine was fed to three existing older 50 MW condensing turbines of medium steam conditions.

Very high temperatures required the use of austenitic steels but the disadvantages of such steel is that it is ill-welded with common Cr-steels and is more vulnerable to unsteady thermal stresses at the transient operating conditions. This was mainly responsible for the temporary abandonment of the use of advanced steam temperatures in steam-turbine plants. In this case, the USC steam pressure unavoidably required the use of double-reheat to get an acceptable wetness degree for the LP LSB. In 1989–91, two USC double-reheat 700 MW 3600-rpm units with steam conditions of 31 MPa and 566/566/566°C were put into operation at Kawagoe (Japan). The Toshiba turbine consisted of an integrated super HP and HP cylinder, a double-flow IP one, and two double-flow LP cylinders. Most of the turbine features, with the exception of the integrated SHP-HP cylinder, are very similar to those of the preceding supercritical-pressure 700 MW units for Japanese power plants. Field tests of the Kawagoe Unit 1's turbine completely confirmed its expected high performance. The turbine's gross (thermal) efficiency was found equal to 48%, as compared to 46.1% for conventional supercritical 700 MW turbines of the same producer. The unit's net efficiency at the rated operating conditions amounted to 41.9%. Despite the USC steam conditions, the Kawagoe units were designed to be capable of operating in a cycling mode.

7.1.3 Materials for Advanced Ultrasupercritical Turbine Components

The main components related to the turbine as a whole are the casing/shell, the steam cylinder valve bodies and bolting, turbine rotors and discs, vanes, and blades.

7.1.3.1 Casings/Shells

The casings of steam turbines are typically large structures with complex shapes that provide the pressure configuration for the steam turbine. The design of the turbine includes an inner casing or cylinder enclosing the hot gas path, so that the main steam from the steam generator first flows into the steam chest, through the inner cylinder, over the vanes/blades, and then returns through the annulus between the inner cylinder and the outer casing before being sent to the reheater. For higher temperatures, Ni-based alloys will be required with adequate strengthening. The candidate alloys chosen for evaluation in the AD700 program goals included both Fe-based superalloys and Ni alloys 155, 230, 263, 617, 625, 706, 718, 901 and Waspaloy39 a modified version of CCA617, and a new alloy, Inconel 740, appear to meet the strength and creep-rupture criteria for the 760°C goal of the U.S. USC steam program with an extensive data generation effort for wrought versions. The major requirements for the Ni alloys are suitability for operation at 760°C with enough creep rupture strength, the ability to cast them into the required size/shape (and to inspect for defects), the ability to perform initial fabrication welding on cast/wrought forms/DMW), and the ability to make repair welds on aged material. The effort required is considerable and involves the development of rupture, creep, and rupture ductility relationships for these materials.

7.1.3.2 Bolting

The major requirements for bolting materials are: (i) high resistance to stress relaxation (aging characteristics) at temperatures that can rise to the steam temperature (max) experienced by the casing itself, (ii) thermal expansion characteristics compatible with the bolted structure, and (iii) low notch sensitivity, that is, the minimum concentration of stress around a point.

Long-term creep data are available for a number of alloys, including U-700, U-710, and U720 variants as well as Nimonic alloys 105 and 115. The choice of materials for bolting becomes somewhat clear-cut and appears to have no significant manufacturing issues. These alloys are available as bar stock and are suitable for rolling/grinding to shape. Similar requirements exist for GTs, although there may be some scale-up issues to be addressed.

7.1.3.3 Rotors/Discs

The HP rotor/discs experience the highest steam conditions, so that an Ni alloy will be required for temperatures >620–700°C. However, it can be noted that the size of the blades is significantly less depending on the overall steam turbine design. The IP rotors experience steam at the maximum system temperature around 620–760°C, but at much less pressure. This phenomenon suggests that the strength requirement may be relieved compared to the HP rotor but the issue of oxidation in the steam path has to be addressed. The manufacture of large Ni alloy components such as a turbine rotor has been recognized as a key issue for a 620–760°C power plant.

As far as metal-to-metal joining is concerned, all the alloys can be fusion welded. The only issue of concern is the welding of dissimilar metals where the nickel-based alloys are required to be welded to chromium steel. In order to reduce alloy costs, some components may be manufactured by welding the nickel alloy to steel so that the nickel alloy is only used for the parts of the component subjected to the highest temperatures. However, all such welds need to be rigorously investigated and qualified. Turbine casings/valve bodies are cast components and therefore it will be necessary to examine the properties of cast versions of the candidate alloys. The higher operating temperatures of a 620–760°C power plant mean that turbine components, for example, shaft seal/retaining springs/erosion shields and wear-resistant valve stem bushes, will require Ni-based alloys; candidate materials include the Inconel alloys 617, 625, and the new 740 as well as Haynes 230. Except for 740, these alloys are approved by the ASME Boiler and Pressure Vessel Code so that a significant design database exists for them, although this does not include fatigue data.

7.1.3.4 Blading

The choice of blading material will depend on the rotor temperature/size/shape/thermal expansion characteristics of the material used for the blades, which will be designed using CFD modeling. The current supercritical steam plants with comparatively low operating temperatures use vanes and blades made from 12 Cr ferritic steels such as type 422, or proprietary alloys of similar composition. For higher temperatures, a wide choice of wrought Ni-based alloys is available with a substantial design database from their application in GTs. For operation with steam at 760°C, it is considered likely that materials new to steam use will be necessary for many stages in the HP turbine and maybe even in the IP turbine. For rotors and discs, modern secondary steel-making practices enable large rotors to be produced from the Cr-Mo-V and 12 Cr alloys used up to 620°C, and the European programs have explored the capabilities for Ni-based alloys. The IP rotor is a critical item because of the size, but it is expected that the turbine can be designed so that this component can use ferritic steels. For the airfoils, complex shapes are involved to achieve maximum efficiency. A range of wrought (and cast) Ni-based alloys suitable for use in the higher-temperature HP turbine is available.

Finally, there exists a major need to demonstrate that the available materials can be made into actual components that work as intended, and to obtain property data for design purposes and service life prediction. Research has been ongoing in the field of powder metallurgy for producing the components for AUSC plants such that the disadvantages associated with welding and forging can be reduced. The production of large, complex components can be done using the near net shape technology. Components thus produced will require only minimal machining and the weldability is also enhanced due to the homogeneity of the microstructure. The feasibility study of the powder metallurgy process was carried out. It was concluded that the parts produced had acceptable metallurgical and mechanical properties compared to the existing cast/forged products. Table 13.14 indicates some of the recently developed materials for AUSC plants.

TABLE 13.14 Turbine Material Selection FOR AUSC Plants

Component	Up to 700°C	Up to 760°C
Casings (shells, valves, nozzles)	CCA617, Inconel 625, Nimonic 263	CCA 617, Inconel 740
Bolting	Nimonic 105, Nimonic 115, Waspaloy	Nimonic 105, Nimonic 115, U700, U710, U720
Rotors/discs	Inconel 625, Inconel 740, Haynes 230	CCA 617, Inconel 740
Nozzles/ blades	Wrought Ni-based	Wrought Ni-based
Piping	CCA 617	Inconel 740

Source: Newer Materials for Supercritical Power Plant Components—A Manufacturability Study by Anish Nair, S. Kumanan, NIT, Tiruchirappalli.

7.2 USC Boiler-Turbine Load Cycling Capability

New steam generator designs are expected to be capable of variable pressure operation. The purpose of variable/sliding pressure is to not change the temperature of the steam turbine high-pressure components with load cycling. A constant pressure steam generator can maintain the turbine inlet temperature. However, the turbine control valve will throttle the steam at partial loads and thereby lower the metal temperature during load cycling. Variable pressure operation means the turbine inlet valves operate nearly wide open and do not throttle the steam, resulting in maintaining a high full-load steam temperature and metal temperatures of the turbine and SH outlet headers and piping at partial loads. Lower-temperature intermediate components of the steam generator will cross some isotherms during load cycling while the higher-temperature outlet components do not cross very many isotherms until shutdown.

Variable pressure is an important operating mode capability due to the increase to a higher steam temperature with AUSC and the situation with electric grids needing units to load cycle. The load change rate of response will be slower than a constant pressure with some throttle pressure reserve. Variable pressure needs to use additional methods to improve the load change response. A technique called frequency control/condensate throttling is used to quickly start the load change without serious overfiring from the delayed control action.

Condensate throttling is applied in case of a high decrease in grid frequency. If the grid frequency requires a high power demand and the unit is in modified sliding pressure operation, the turbine control opens the governor valves to raise the load by using the steam storage capacity of the boiler (except if the valves are fully open already). Simultaneously, the main condensate control valve is throttled to a calculated position, allowing a reduced condensate mass flow flowing through the LP feed water heaters. Considering a certain time response, the extraction steam mass flows of the LP feed water heater and the deaerator/feed water tank are reduced. The surplus steam that remains in the turbine generates additional power. The resulting load increase depends on the amount of preset throttling of the turbine governor valves, the main condensate control valve, and the actual unit load. By means of additional fast-acting valves in the above-mentioned extraction steam lines, the response time behavior can be optimized.

This condensate throttling serves as a compensator for the transient time behavior of the boiler. The accumulated condensate is stored in the condenser hot well or a separate reserve condensate tank. Parallel to the above-mentioned measures, the firing rate of the boiler is increased to meet the load requirements. The feeding of the boiler is continued and increased, so the level of the feed water storage tank is decreased accordingly.

During this time, the condensate flow reduction is gradually released and has reached a steady-state condition again. Refilling of the feed water storage tank is initiated by releasing the condensate control valve to control the level of the hot well or the condensate collecting tank. The maximum allowable condensate mass flow is monitored and the refill flow is limited to a maximum value. Due to the increased condensate flow through the LP feed water heaters and into the deaerator, the steam extraction from the affected turbine extractions is increased. The generator output reduces correspondingly. To compensate for this influence, the slow coal pulverizers can then meet the firing demand for the load change without the catch-up delay creating a severe temperature overshoot and damage. The boiler-firing rate is increased but the firing rate control should not be too fast. However, the maximum allowable SH outlet flow is limited to 100% BMCR. The water storage design of the hot well/deareator is increased to handle the condensate flow transient.

Another very rapid response system for primary frequency control that can be supplied with USC steam turbines is HP stage bypass. The system is working in Siemens 600–1200 MW turbines equipped with sliding pressure and full arc admission over the load range. It permits additional HP steam to be admitted to the HP turbine some stages after the first blade row when the bypass valve is opened, also to give full arc admission at that stage. The system is normally designed to give a short-term 5% increase in power. It is, however, possible to design such systems for an additional 10% or even 15% increase in power (Almstedt et al., 2007; Wechsung et al., 2012). Normal operation at 100% MCR is achieved with the stage bypass valve closed. Efficiency falls upon opening it, but providing frequency control is a necessity to stabilize the grid. In any case, the HP stage bypass system is the most efficient means available for achieving rapid load increase (1% per second) because throttling losses are at a minimum at both points (Quinkertz et al., 2008). The stage bypass system is available for use within the whole load range. In contrast, throttling is normally operated at around a fairly high load.

8 REDESIGNING/OPTIMIZATION OF STEAM CYCLE

A single steam RH loop was established in the early 1950s. Contemporary developments in high-temperature materials allowed OT boilers to operate with supercritical steam conditions.

A double RH steam cycle was introduced with the early supercritical boilers in the late 1950s. The double RH cycle added another steam RH loop requiring an additional IP turbine module. This increases electrical efficiency by 1%–2% points.

8.1 The first supercritical coal-fired double RH unit in the world, the 104 MWe thermal power plant, was commissioned in 1956 in the Hüls AG chemical factory, unit 1, (Marl, Germany) with steam conditions of 600°C/30MPa (HP), 560°C/10.7 MPa (IP 1), and 510°C (IP2).

8.2 Philo unit 6 was the first supercritical double-RH unit commissioned in 1957 in southeastern Ohio. Owned by American Electric Power (AEP), the 120 MWe unit operated at steam conditions 621°C/31MPa (HP), 565°C/11 MPa (for IP 1), and 538°C (for IP2) that reached to 40% electrical efficiency. The boiler and turbine were manufactured respectively by B&W and GE. Philo unit 6 was put on hold in 1975 and decommissioned in 1979. The unit was demolished in1983 after 103,110 operating hours because it was not economical to upgrade the plant to comply with the requirements of the Clean Air Act (Weitzel, 2012a; ASME, 2003a).

8.4 Eddystone 325 MWe unit 1 was the first ultrasupercritical (USC) thermal power plant, and it was commissioned in 1959 in Pennsylvania. It was unit 1 of the

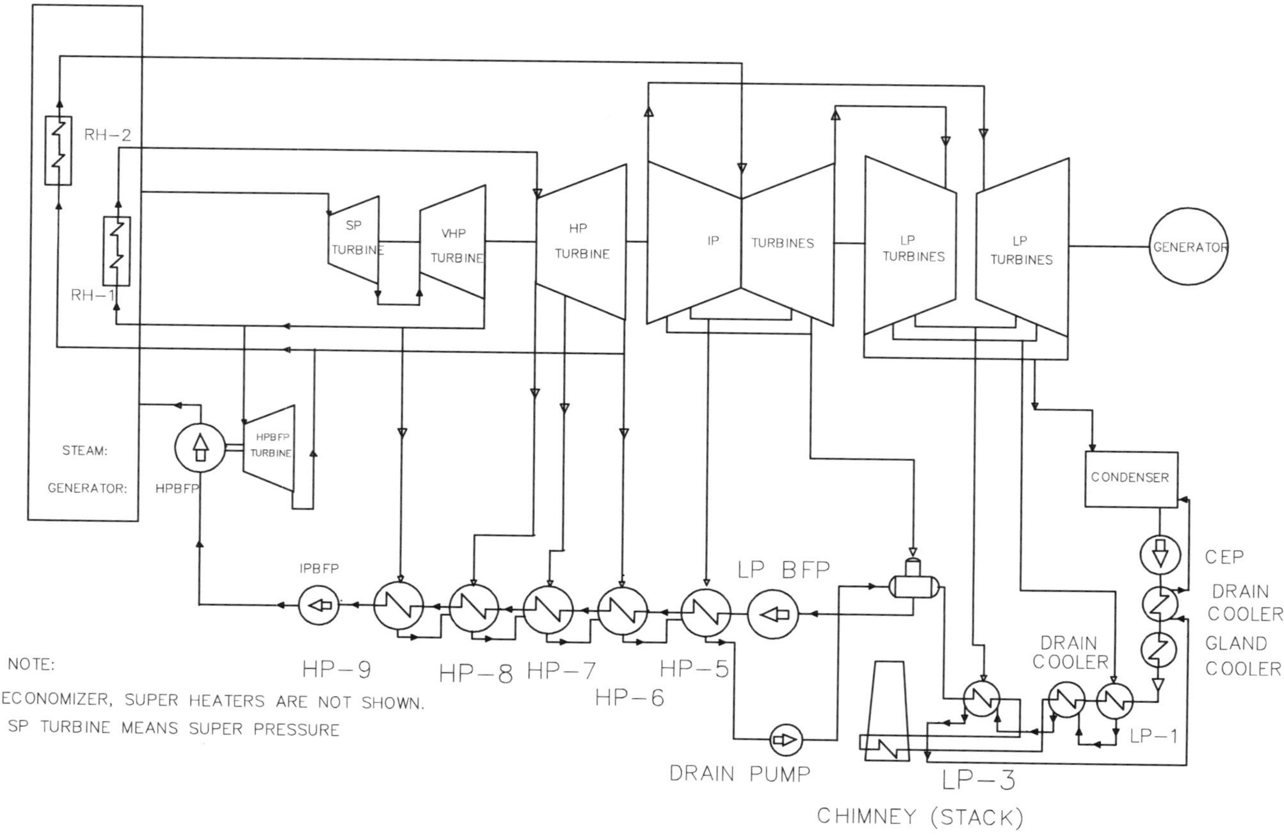

FIG. 13.7 Steam water cycle with super pressure concept.

Philadelphia Electric Company, later PECO, and then Exelon. ABB Combustion Engineering manufactured the two-pass boiler, avoiding the necessity for spray desuperheating and providing constant RH temperature at most load levels (see the single line diagram of the steam-water cycle, Fig. 13.7 of the Eddystone 325 MWe unit 1).

The cross-compound turbine, manufactured by Westinghouse Electric, was rated at 325 MWe and had steam conditions of 649°C/34.5 MPa (super pressure or SP), 565°C/11 MPa (for VHP), and 565°C (for HP). Soon after, several instances of rotor vibration occurred due to seal rubbing damaging the nested seals at the shaft ends, resulting in steam leaks. Nested seals were replaced with conventional spring-backed labyrinth seals, which reduced the SH steam temperature from 649°C to 621°C. To add to the woes, excessive fire-side corrosion of the SH/RH surfaces then reduced the SH steam parameters to 610°C and 33 MPa (ASME, 2003b).

In the start-up procedure of this double-RH unit, a minimum of 30% of the design water/steam flow was required through the boiler at a throttle pressure of 24 kPa. The turbine was bypassed until the steam reached a certain temperature, at which point the throttle and hot RH inlet control valves were partially opened. The gradual increase in control valves opening up to 95%–100% then raised the electric output to 260 MWe with a throttle pressure of 24 MPa. This start-up procedure required simultaneous operation of the boiler, turbine, and bypass systems. The load was further increased by raising the operating pressure from 24 to 34.5 MPa. This is probably the first sliding pressure mode operation reducing pumping power and elimination of control valves' throttling pressure drop thus increasing electrical efficiency (ASME, 2003b).

In the 1970s and 1980s, the performance further deteriorated with the use of lower-quality coals and additional APC systems. During a start up in 1993, the VHP module overheated, which caused some of the rotating blades to fail. The unit was derated to 230 MW until 1995 when a new 3600 rpm rotor was installed. The boiler stop valves cracked during operation; this was solved with a new valve using different materials (ASME, 2003b).

The ultrapure feed water caused leaching of the alloying elements from the steam side of the high-temperature materials. Copper oxide deposits were found on the turbine blades. Further deposits were prevented with a condensate polishing filter fitted to the end of the LP heater, and substituting copper tube heaters with ones made of steel (ASME, 2003b).

8.1 Master Cycle With Double Reheat

DONG Energy (China) has devised and patented a modified double-RH cycle, called the Master Cycle. It solves the SH problem and modifies the steam cycle to increase plant efficiency and decrease capital costs (Kjaer and Drinhaus, 2010a, b; MPS, 2008).

The AUSC plant SH steam conditions are very high and so when the turbine bleeds for feed water preheating, it also becomes too hot (>300°C), which is known as the "superheat problem." It is detrimental to both double-RH and single-RH AUSC units as the SH problem would be significant whenever the plant is operating with superheated steam over 700°. Costly desuperheaters were used to tackle that situation, that is, spraying over the bleed steam to bring down the temperature to utilize for regenerative heating. The superheating problem has been left unattended for too long and it is the Achilles heel of double-RH cycles. However, a simple modification of the conventional double-RH cycle, named the Master Cycle (MC), solves the superheating problem. Here, some portion of the HP turbine exhaust, that is, the first cold RH steam, is fed to an additional small turbine. This is known as the tuning turbine (T turbine), which might replace the feed pump turbine (FPT). The new T turbine is shown in Fig. 13.8 with all the expansion lines. When the new tapping points for regenerative heaters are compared with the tapping points of a conventional double-RH cycle, it is evident that the superheating problem is reduced dramatically.

The Master Cycle works in the following way, along with the steam and water cycles.

8.1.1 *Steam Cycle*

The steam flow is reduced by approximately 15%–30% through the RH tubing in the boiler, the associated piping, and the IP module, as the same amount of bleed steam is shifted to the T turbine. In other words, the Master Cycle moves the heat uptake from the two RHs to the furnace walls and the SHs, meaning significant size reductions in RH tubing/piping and the IP turbine itself.

It is quite obvious that the T turbine bleed steam temperatures are not as high as before. The lower cost steels for bleed tubing as well as the stop/check valves can be used,

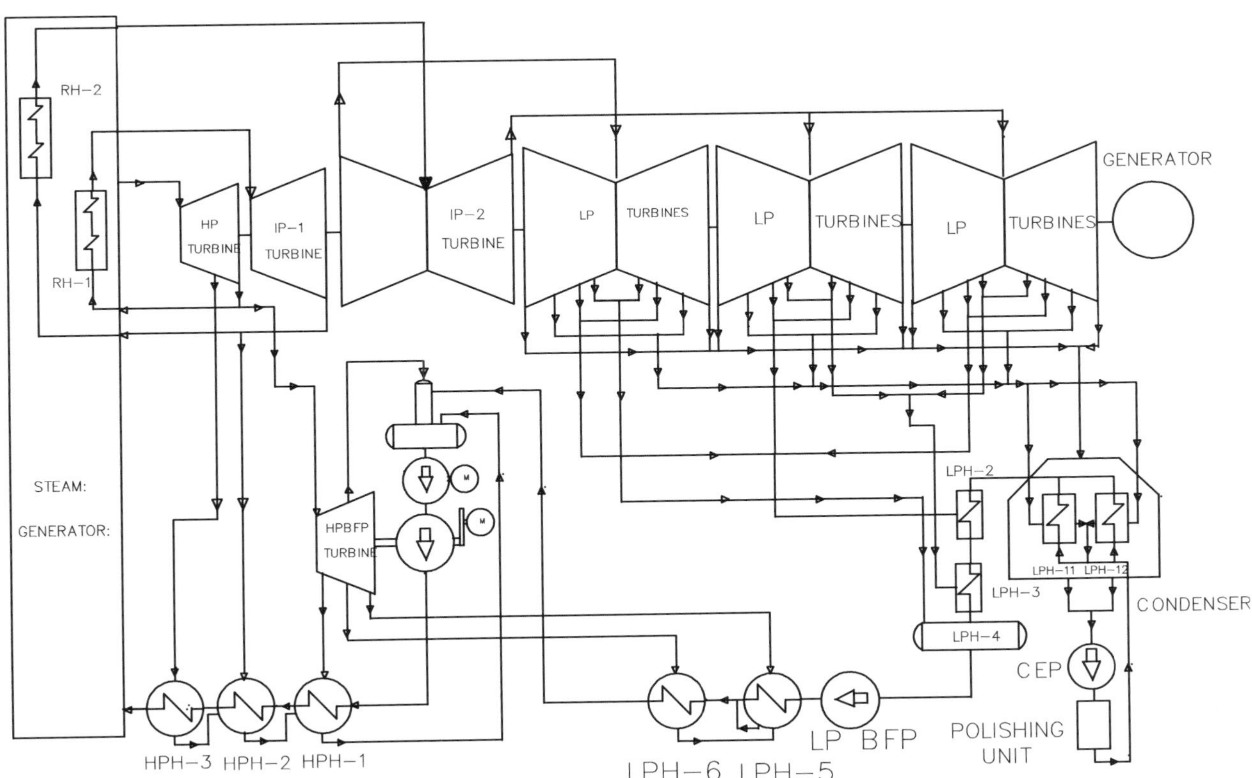

FIG. 13.8 Master (steam water) cycle with double reheat.

which means a reduction in capital costs by reducing the need for high-temperature material. The final feed water can still reach an optimum 330°C with preheating in spite of maintaining the performance of large, state-of-the-art single-line rotating air preheaters. The components altered from a conventional double-RH cycle are the T Turbine/turbine-driven BFP, the HPH 1/LPH 5/6, the deaerator/associated piping/tubing, accessories, etc. The mechanical output from a T turbine drives the BFP, as the T turbine is similar in size to a condensing turbine. A variable speed balancing motor can be used to guarantee the correct speed of the BFP at high loads; it can be used as a generator at low loads. For a 900 MWe USC Master Cycle, the balancing motor is sized at 4–5 MWe.

The enthalpy drop of the T-turbine is relatively large, meaning it should be a high-speed turbine to improve blading efficiency. It seems that a speed range of 5000–5500 rpm is ideal to both the T-turbine and the feed pump. As the T turbine bleed steam ends in the HPH 1, LPH 1/2, the complicated condenser and CW systems with conventional condensing turbine-driven BFP, and the associated building structure, are avoided. This configuration is already operational in Avedøre unit 1.

The T-turbine is operating with the inlet valves wide open and a control stage is not needed. The T-turbine steam path is unconventional and so the first stage experiences a relatively high volume flow compared to a conventional FPT. However, as steam is extracted for regenerative heaters, the blades get shorter and the final stage group only experiences the steam flow for one heater. This also means that a major problem in relation to T turbine design is not limiting the length of the last-stage rotating blades, as for a conventional FPT, but getting sufficiently long and effective turbine blades. The T turbine could be used with its own generator, but is unfavorable for a number of thermodynamic and practical reasons (Kjaer and Drinhaus, 2010a, b; MPS, 2008).

8.2 Echelon Cycle With Double Reheat

Echelon Cycle, a variation of the Master Cycle, has been developed by the Shanghai Turbine Works. It is also a method of regenerative preheating that reduces the amount of high-cost materials and capital costs for both single- and double-reheat turbines. The Echelon Cycle uses an additional smaller turbine called the back pressure extraction steam turbine (BEST), which takes some steam from the exhaust of the HP or VHP module to drive the boiler feed water pump as well as supplies steam for regenerative feedwater preheating. With this system, the extraction steam temperature is reduced sharply because the high-temperature extraction steam is moved from the main turbine and the steam source of the small extraction turbine is from the cold reheater. So the highest extraction steam temperature will not exceed 500°C, and the high-temperature risk of the heat recovery system will be eliminated completely. The Echelon Cycle (Fig. 13.9) reduces the amount of expensive high-temperature materials used in regenerative preheating (Yang et al., 2014).

8.3 Double Reheat With Outer Coolers

At the North China Electric Power University in Beijing (China), a variety of studies on coal-fired double-RH cycles by the researchers at the National Thermal Power Engineering and Technology Research Center (NTPETRC) have come out with some modified cycles. One of those efficiency-enhancing designs is the double RH with outer coolers. Outer steam coolers are widely known as feed water desuperheaters (or less frequently, a topping desuperheater) and lots of units employ them.

Outer steam coolers use an additional heat exchanger to heat high-temperature feed water before heating the low-temperature feed water. This means that the superheated steam bleed is cooled by up to 100°C before heating the low-temperature feed water, which uses the heat more efficiently. Outer steam coolers would be required on steam bleeds 2 and 5, as depicted in Fig. 13.10.

A simulation test at full load results in an increase of electrical efficiency by 0.16% points to 46.99% compared to a reference 1000 MWe double-RH unit with the steam conditions 600/610/610/30 and eight stages of steam bleeds, which reaches 46.83% electrical efficiency. (The reference double-RH unit has the SH problem with steam bleeds 2 and 5, which are both first-stage IP module bleeds. Bleeds 2 and 5 can reach 562.6°C and 521.7°C, respectively.) For a simulation test at half load, the increase in electrical efficiency rises to 0.19% points. The half load is assessed because the SH steam bleed temperature remains constant in sliding pressure units, which means that the SH problem has more of an effect on unit performance at lower loads. The payback period may be 7.2 years for the additional outer steam coolers over the reference double-RH plant.

8.4 Double RH With Regenerative Heating (Reference Fig. 13.11)

This method is similar to the Master Cycle discussed earlier in Section 8.1 of this chapter. For the simulation test at full load, the regenerative turbine increases electrical efficiency by 0.67% points to 47.5% compared to the reference double-reheat unit, as indicated in Section 8.3 of this chapter. At half load, the increase in electrical efficiency rises to 0.79% points to 47.62%. The payback period may be 3.8 years for the additional regenerative heating arrangement over the reference double-RH plant, that is, almost half the payback period for the method using outer steam coolers.

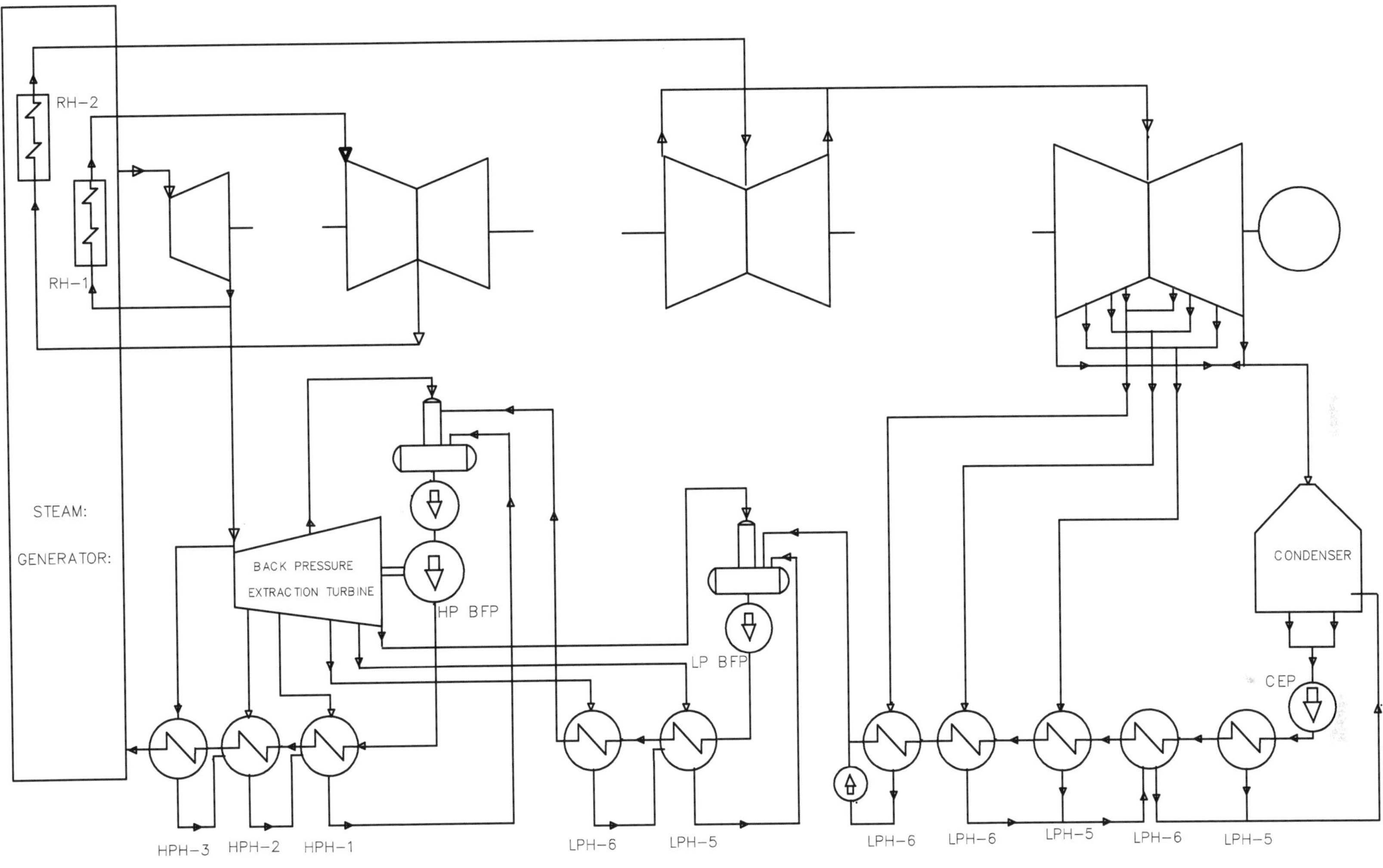

FIG. 13.9 Echelon cycle with double reheat.

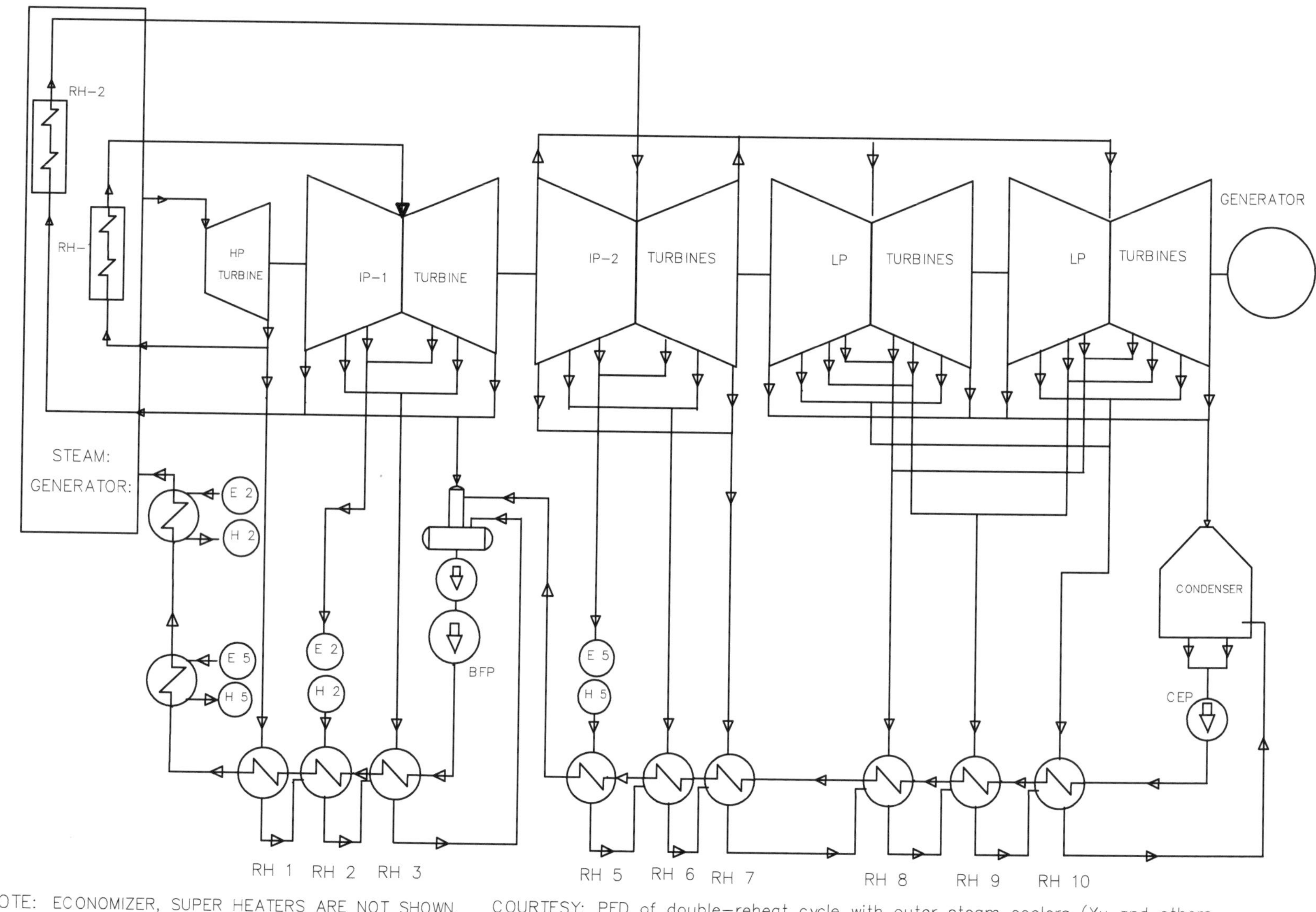

FIG. 13.10 Double reheat with outer coolers.

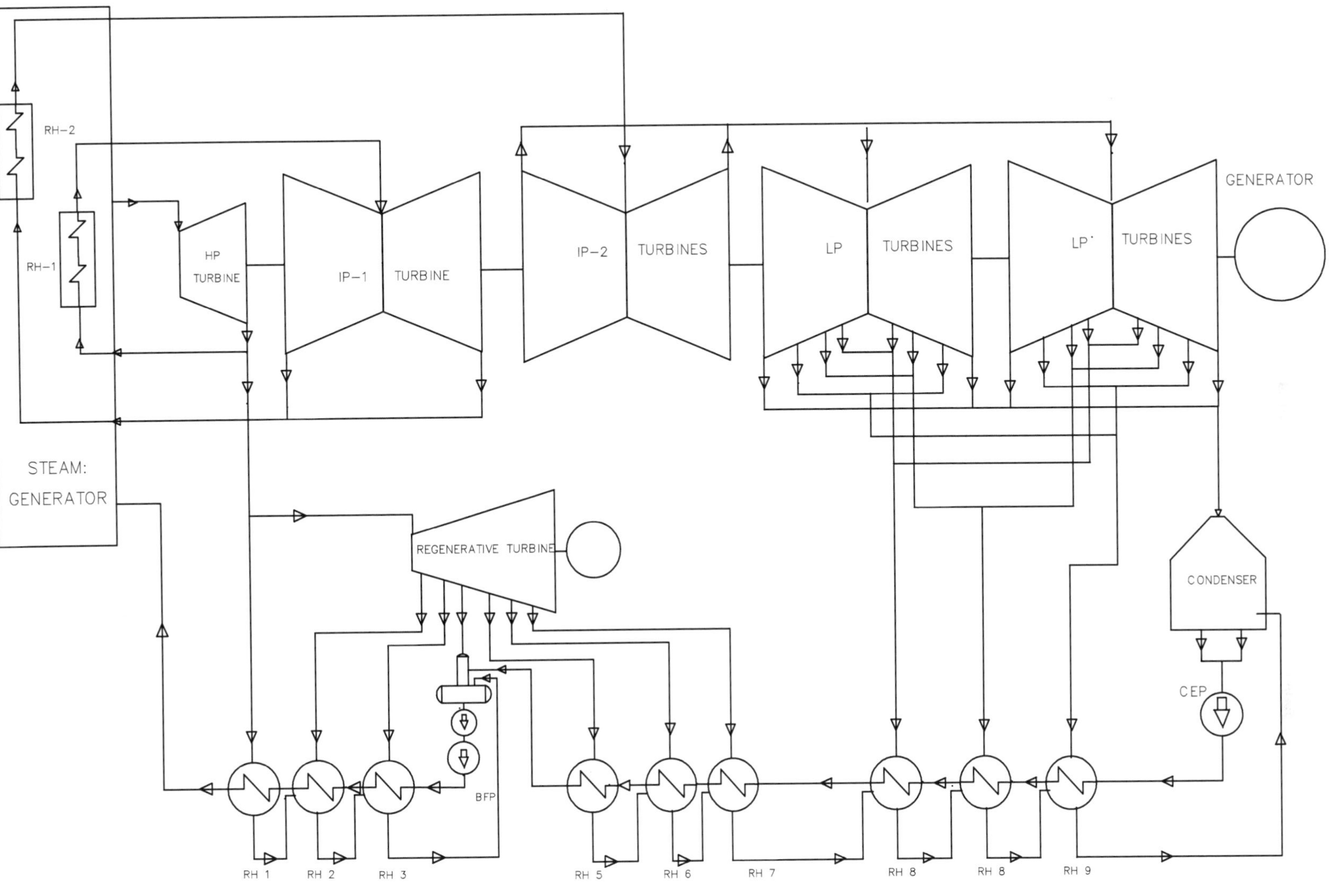

FIG. 13.11 Double reheat with regenerative heating.

9 BOILER AND TURBINE CONFIGURATION

It is obvious that the cost of the Ni alloy components will be a strong economic incentive for changes in how the boiler and turbines are configured and arranged in the plant layout. To be more precise, the nickel price is high enough that new arrangements and the mutual position of boiler and turbine must be a consideration. Researchers throughout the world are on the lookout for new plant concepts that will suggest what will certainly change, what will probably change, and what should remain unchanged in the plant and boiler design. These considerations are really vital when moving from a 600°C current state-of-the-art design to a plant design with a 700°C boiler, turbine, and other auxiliary equipment layout.

The operation and start up of the 700°C plant will be similar in control methods and techniques to a 600°C plant. Due to arrangement features, the steam temperature control range and the OT minimum circulation flow will be slightly different.

Efficiency improvement with 700°C plus operation over a 600°C power plant results in a 12% (~) reduction in fuel consumption/CO_2 emissions. The reduced flue gas weight per MW generated reduces cleanup costs for the lower SO_x/NO_x and particulate emissions.

Due to the high cost of nickel steam piping, new concepts have been proposed for the arrangement of both the boiler and turbine in the plant. Siemens AG has proposed a compact designed horizontal boiler. China has some interest in the two trains of turbine concept with the HP and IP sections at high elevation of the steam outlet headers and another IP section (of the double reheat concept) with the LP sections in the present turbine hall elevation and the condenser on the ground level. Horizontal tube banks could provide a means to better handle the exfoliation of steam-side oxidation where distribution and removal can be improved. The higher feed water temperature from the AUSC turbine cycles makes it difficult to achieve a low boiler outlet gas temperature entering the air heater, so flue gas cooling using a condensate in between the LPHs may be needed. The soot blowing (SB) and other auxiliary steam requirements could be met by a lower-pressure boiler. An SB with compressed air would save high pressure pure water. However, everything must be understood and gain acceptance through proof of performance and the industry must remain economically viable backed up by advancing technical development.

9.1 Configuration of a Tower Boiler With Master Cycle

As mentioned in Section 8.1.1, the HP exhaust/cold RH steam line is shifted to the tuning turbine, there is 15%–30% less steam flow in the RH piping resulting in reduction of about 6% amount of RH tubing in the boiler thus decreasing capital costs of the boiler (Fig. 13.12). From bottom to top, the order is as follows: furnace screen, which includes the primary SH (SH1); the final SH (SH3); the final primary RH (RH1.2); the final secondary RH (RH2.2); the secondary SH (SH2); the first section of the secondary RH (RH2.1); the first section of the primary RH (RH1.1); the final section of the ECON; the SCR; the first section of the ECON; and finally the rotating APH (Kjaer and Drinhaus, 2010a, b).

9.2 Compact Design Boiler

The compact design boiler has been developed by Siemens to reduce the steam pipework. These are Benson-type boilers employing a horizontal firing system while the turbine plinth is elevated to a higher altitude to minimize the amount of expensive high-temperature pipework by up to 80%. For a 550 MW unit, the compact design boiler is 31 meters tall, a two-pass boiler is 63 meters tall, and a tower boiler is 91 meters tall. This concept has not yet been demonstrated for a coal-fired boiler. However, a subcritical compact design boiler was installed as an HRSG at the Cottam Development Center (UK) test facility. Another arrangement is to have the boiler/turbine inline.

9.3 Partially Underground (U/G) Tower Type Boiler

Double-RH thermal power plants have nearly 450 m of pipework between the boiler and turbine, manufactured from high-cost materials, against 150 m of such pipework for a single-RH plant. Therefore, alternative boiler-turbine configurations could reduce the length of the pipework and thus dramatically reduce capital costs. Researchers at the NTPETRC of the North China Electric Power University in Beijing proposed a novel partially underground tower boiler design to decrease the capital costs of double-RH units. This configuration as depicted in Fig. 13.13 is intended to reduce the boiler height by 70 m and the steam piping length by 160 m.

Compared to a typical 1000 MWe double-RH design, this submerged boiler increases electrical efficiency by 0.1% points. The high efficiency and lower investment cost lower the cost of electricity. Further investigations revealed that the economic advantages of this partially U/G boiler become more favorable with increasing coal prices and high capital costs, especially for AUSC double-RH units, which employ high-cost nickel alloys.

9.4 Cross Compound at High/Low Position Arrangement (CCHLPA)

In line with the view stated in Section 9.3, this configuration is also intended to reduce the costly pipework. The plant engineers at the Waigaoqiao 3 thermal power plant have designed a turbine configuration with the aim to shorten

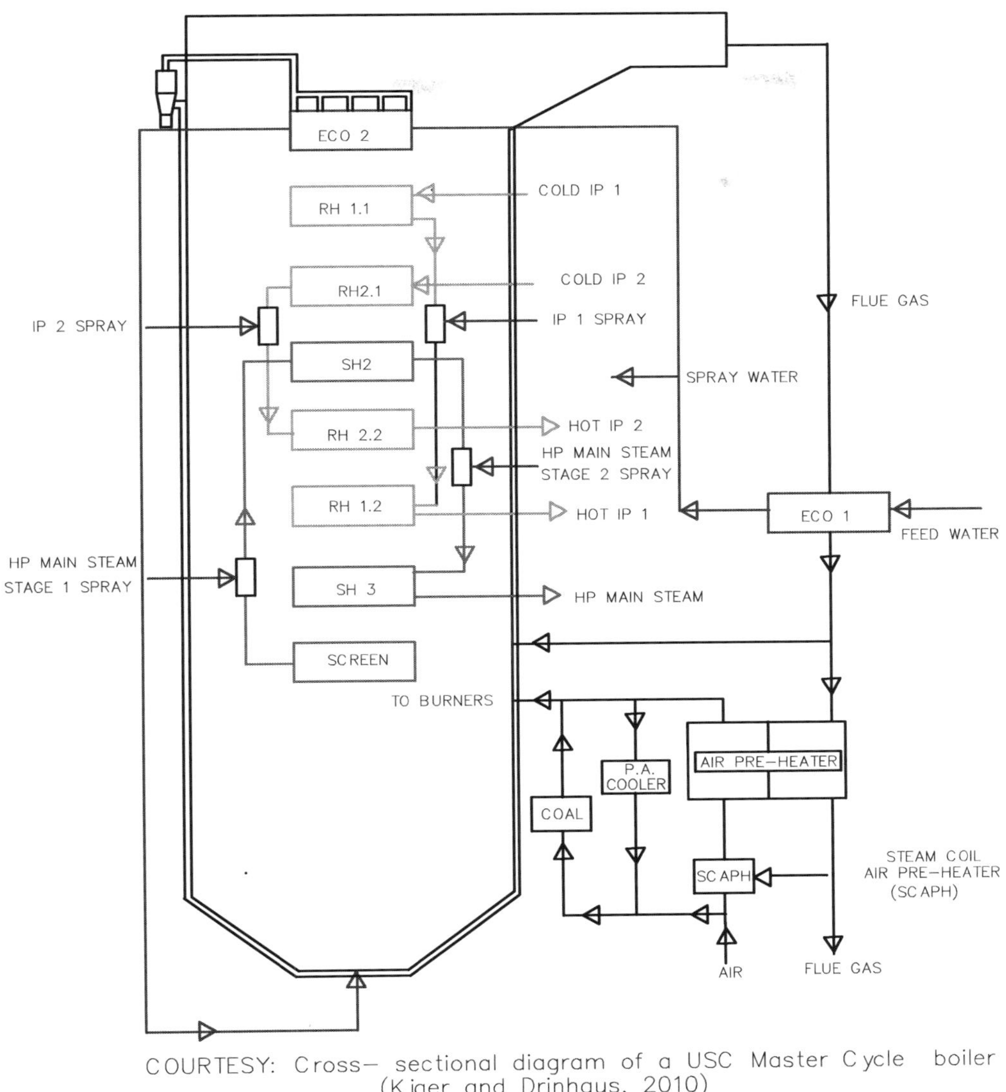

FIG. 13.12 Boiler layout in an AUSC plant with master cycle.

the pipe length in comparison to the standard turbine hall layout. This redesign is known as the cross compound at high/low position arrangement (CCHLPA). In this arrangement, the turbine layout is done in a cross compound mode, that is, the turbine modules are split into two trains (Fig. 13.14). Here, the VHP, HP, and one IP module constitute the first train; this is mounted at around the 80–85 m level, the same as that of the boiler steam outlet headers. The remaining two IP modules/LP modules are arranged in the second train located at nearly 17 m elevation, that is, at the conventional turbine hall. Shorter pipework reduces not only cost but also the pressure drop and temperature loss of steam, which slightly increases efficiency.

9.5 B&W AUSC Conceptual Design—Modified Tower (Fig. 13.15)

A modified tower design was developed for a Toshiba steam turbine of 898 MW, 30 MPa 700°C/730°C. The design fuel is an Indian coal. Featured is the use of stringer supports for a horizontal surface. All bank heating surfaces in the gas upflow pass must be on a multiple of the interval spacing of the stringers. The material of the stringers is T-92 supporting nickel SH/RH tubing. The down pass is also stringer-supported and is typical of current practice. A series or parallel down pass may be used when the RH temperature control range is limited or wide, respectively.

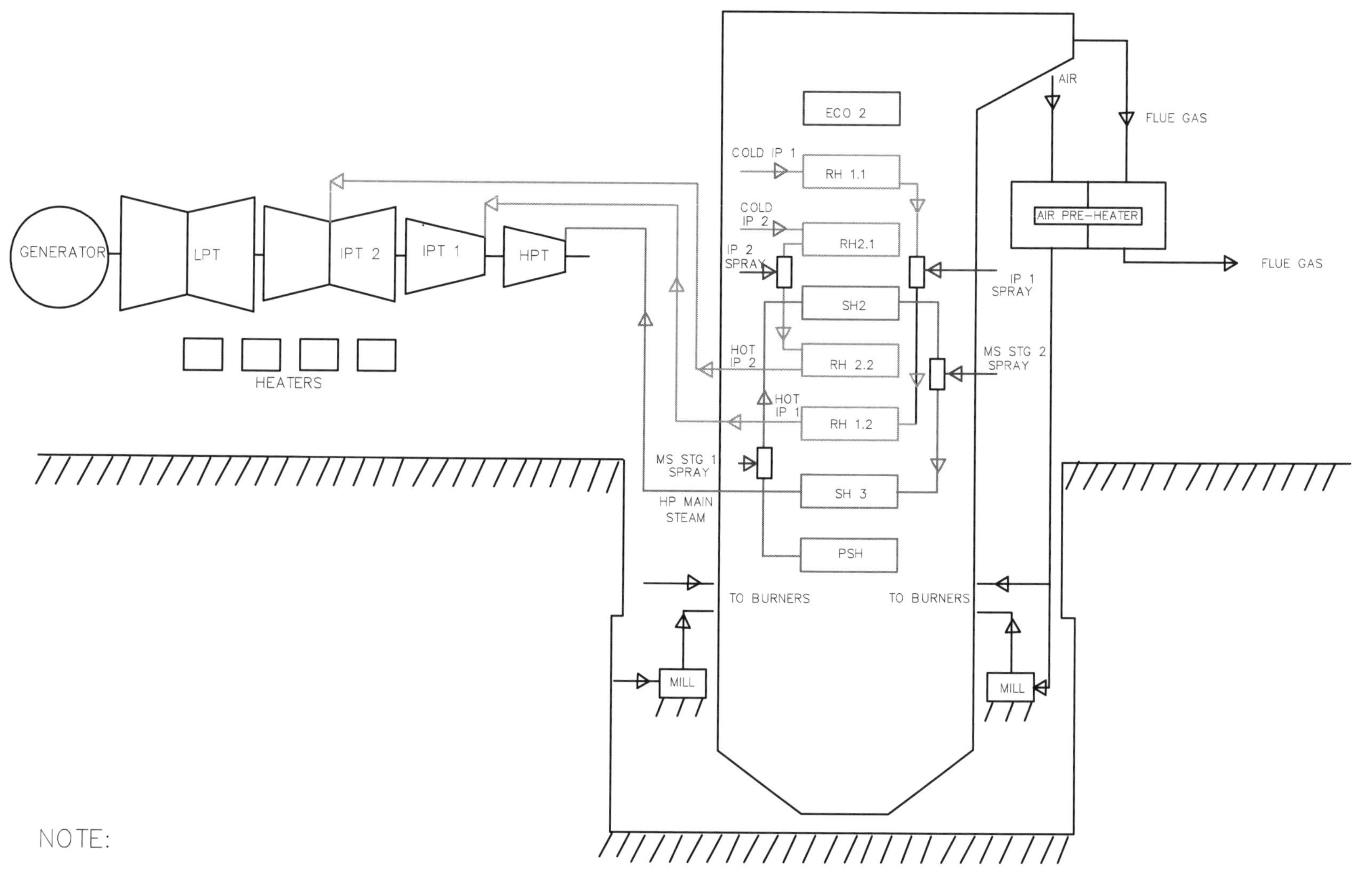

COURTESY: Cross-section of partislly underground tower type boiler (Xu and others, 2014)

FIG. 13.13 Partially underground tower type boiler.

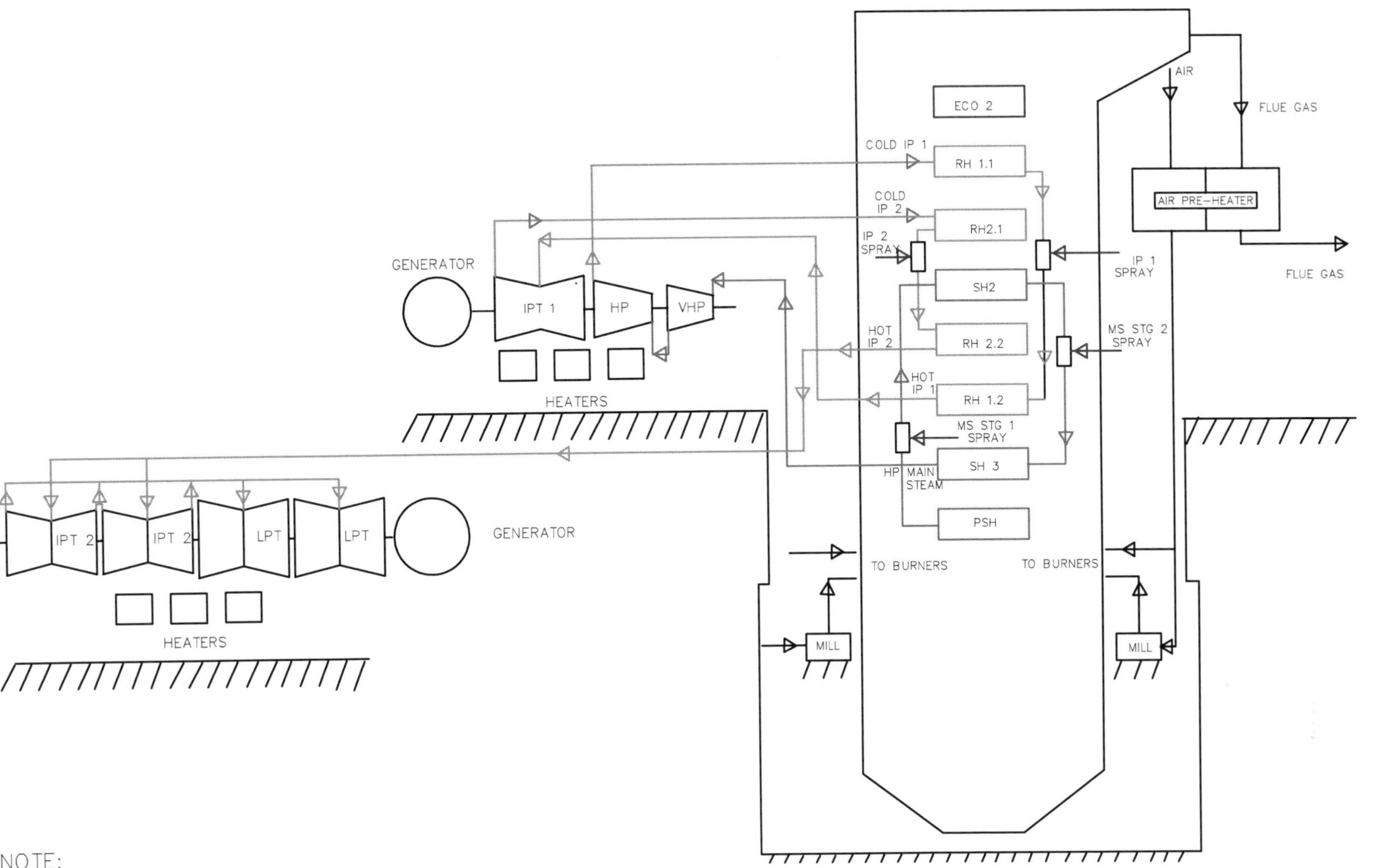

FIG. 13.14 Cross compound at high/low position arrangement.

The high-ash coal requires much lower gas velocity limits, so the convective heat transfer degrades quickly with reduced load. Gas recirculation and gas biasing may be employed to meet performance requirements. An LP boiler placed after the ECON reduces the temperature of exit gas and provides auxiliary steam and not extracting very high condition steam. This modified tower arrangement puts the outlet headers closer to the steam turbine, although, due to the high-ash fuel, the furnace height is much higher. By locating the base of the boiler below ground, the nickel steam leads are shorter and offer significant savings. The feed water heaters and deareator building bay may also be moved from between the boiler-turbine to along one side of the boiler.

9.6 Downdraft Inverted Tower Type A-USC Boiler

Considering the shortening of the MS/RH piping at a very high working temperature level, different development projects were carried out, even back in the mid-20th century. For example, the boiler "PK-37" can be mentioned in the first arrangement solution. It was developed in the 1960s at the Machine Engineering Plant in Podolsk and made as an inverted conventional arch boiler (Fig. 13.15). The first USC boiler of a 100 MW power unit (PK-37) was commissioned at the Kashirskaya State District Power Plant (GRES) as early as 1962. Presently, the Alstom Power Company and B&W work on a similar project of a power

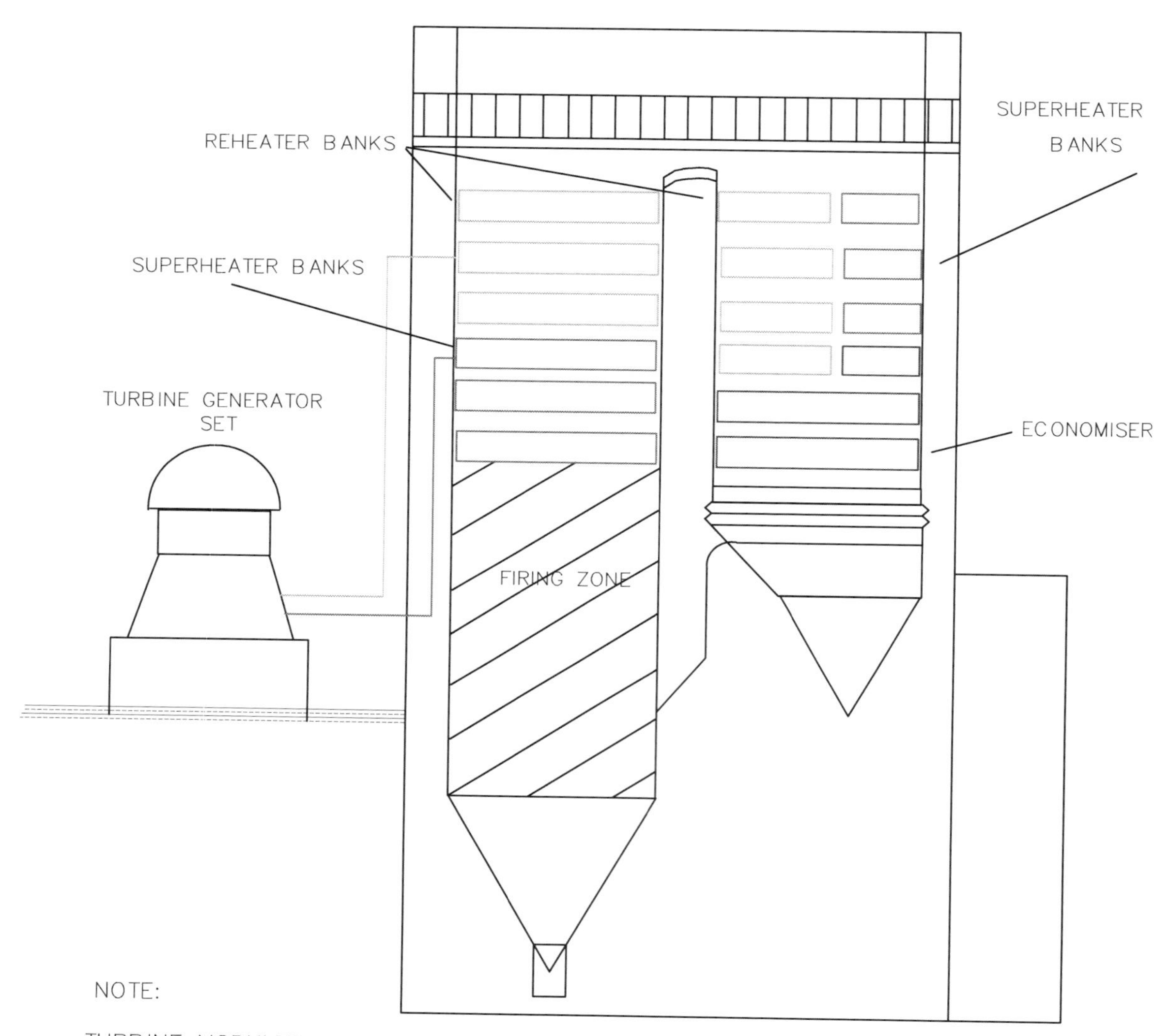

FIG. 13.15 Modified tower type A-USC boiler.

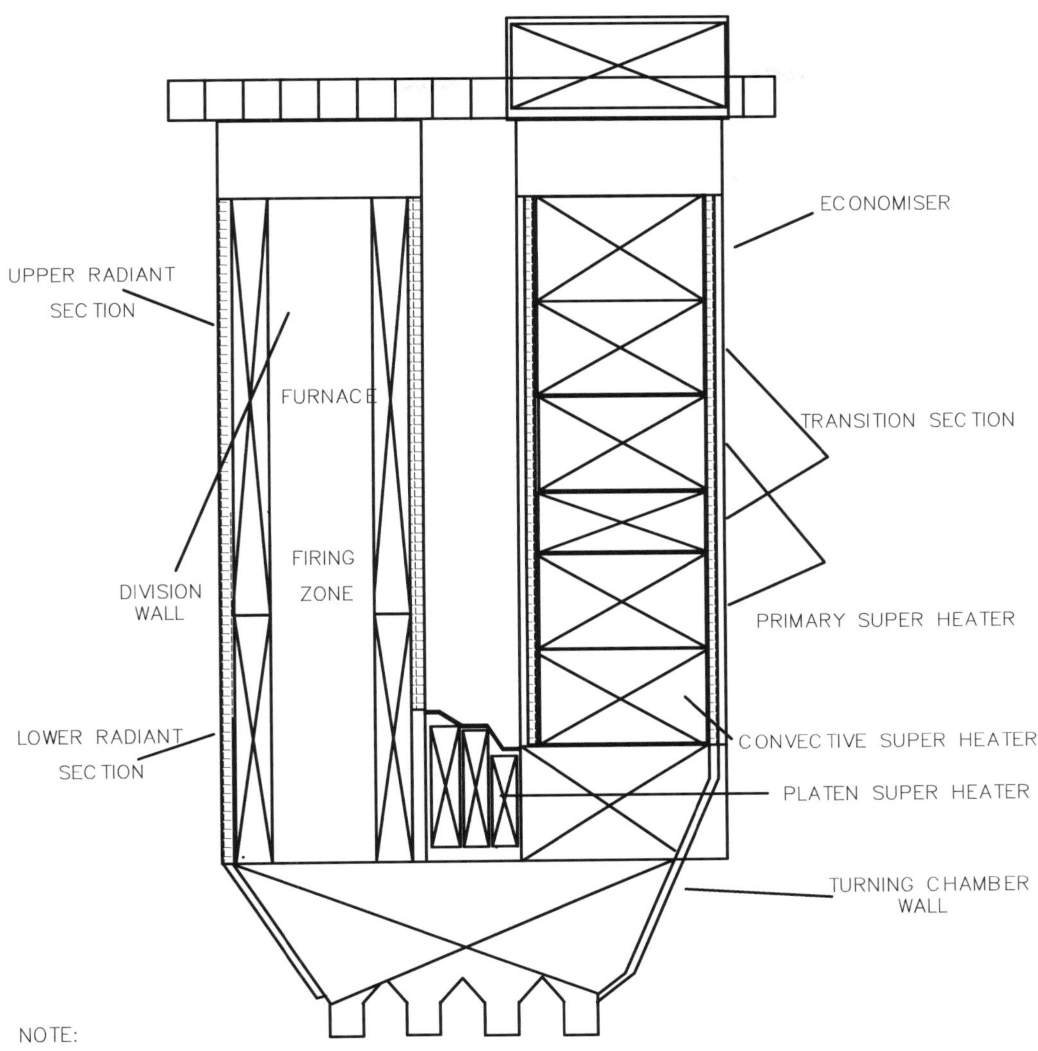

FIG. 13.16 Downdraft inverted arch USC boiler.

plant. Although these versions provide considerable length reduction of high-temperature pipelines, they have some disadvantages that make their maintenance difficult. This is due to the fact that the ash hopper is in the area of the primary SH, and the successive passage of two corners by combustion products will inevitably deflect current lines. This potentially throws them in the ash hopper, which can lead to heat loss and extra difficulties in providing a bottom ash-handling system. An additional point to emphasize is that an inverted conventional arch boiler has burner units located in the upper part of the boiler plant, which complicates the supply of pulverized coal. However, in this case, ash and slag removal is also more complicated than for a standard design. In order to enable ash removal effectively from the boiler gas duct, the lower furnace wall shall have a slope of 50–60 degrees, requiring a high-elevation furnace and terminal headers that lead to an increasing length of high-temperature pipelines (Fig. 13.16).

B&W has considered a conceptual design for the AUSC boiler (SG) that employs a downdraft inverted tower design in order to shorten the length and thereby reduce the expenditure toward the costly Ni-based pipeworks. The proposed design assumes the height of the SG unit to be about 49 m against ~137 m for the conventional configuration. The downdraft inverted tower design is estimated to be a 36% cost reduction for the entire SG island scope. The design concept envisages two simultaneous construction paths, as depicted in Fig. 13.17.

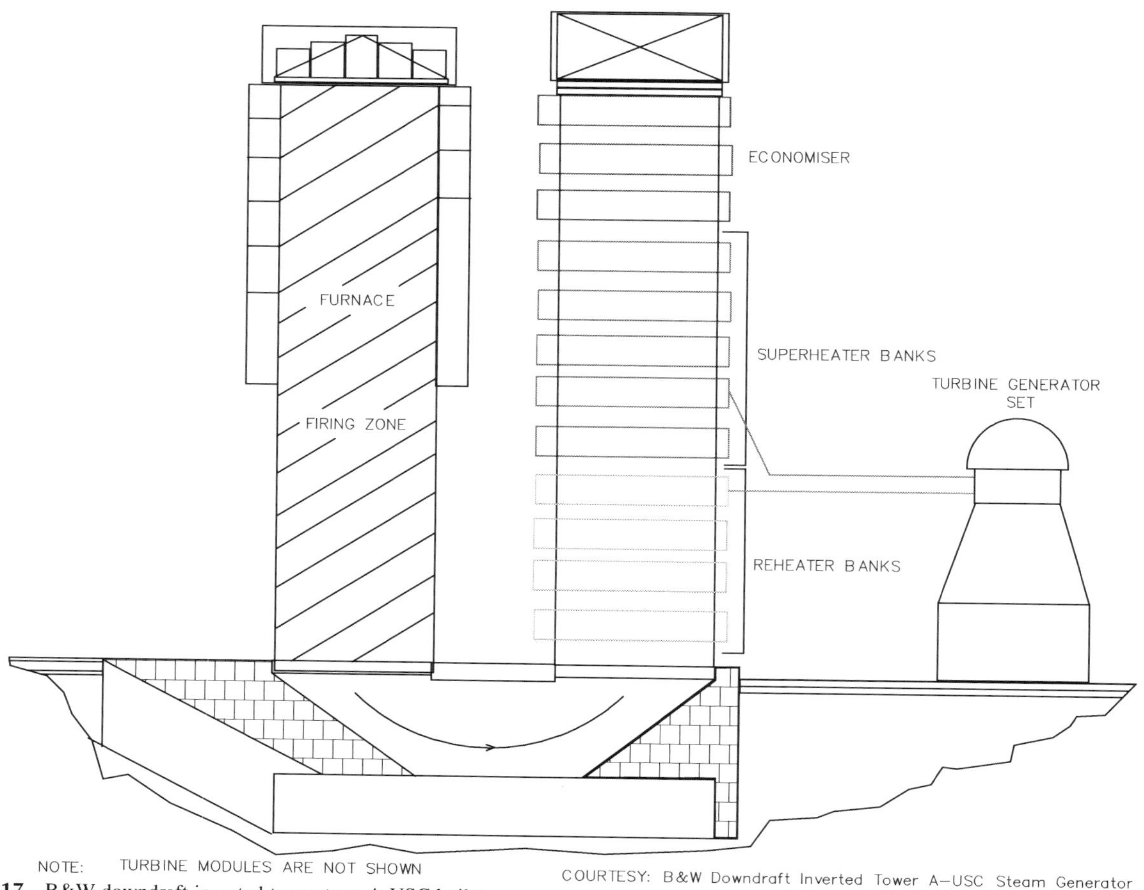

FIG. 13.17 B&W downdraft inverted tower type A-USC boiler.

The conceptual downdraft inverted tower AUSC boiler includes a downdraft furnace enclosure, a hopper tunnel, and a convection pass enclosure, with the hopper tunnel joining the downdraft furnace enclosure and the updraft convection pass enclosure. Flue gas passes down through the downdraft furnace enclosure through the refractory insulated hopper tunnel and up through the convection pass enclosure. This structure permits the high temperature steam lines to the steam turbine located at the base of the boiler rather than at the boiler top, as with conventional boilers. This results in a significant reduction in Ni-based steam piping length/cost. An alternative furnace hopper tunnel could be constructed above ground with tube panels and a refractory.

9.7 Two-Pass AUSC Boiler

B&W has developed a 700–760°C AUSC SG with features and characteristics that are the same as the 600°C USC SG. With higher steam temperature, there is more superheating than steam generation/evaporation duty, and the gas-to-steam temperature difference is lower so more convection bank heating surface is needed and the superheating duty of the enclosure is greater. The design requirements of the fuel are setting the constraints on the flue gas side, which are the same for the SC/USC/AUSC furnaces. The furnace enclosure material is currently selected to allow operation to 454°C to less than 510°C average outlet fluid temperature. This is about +66°C for the enclosure and +93°C for the final SH over the temperature for USC. AUSC produces a lower heat rate and lower CO_2 emissions, starting at about a 12% improvement over 600°C.

9.7.1 First B&W Conceptual AUSC Design—Air Fired With Gas Biasing

The first B&W air-fired AUSC thermal power plant of 750 MW net at 34.5 MPa 735°C/760°C was based on bituminous coal. The conceptual design was developed to provide the material quantity and service temperature requirements for early cost analysis in the materials development program. The SG was the OT type of system with the most typical arrangement available in US utility service; it is depicted in Fig. 13.18.

This system is called a two-pass or B&W Carolina type. The top supported pendant heating surface is not governed

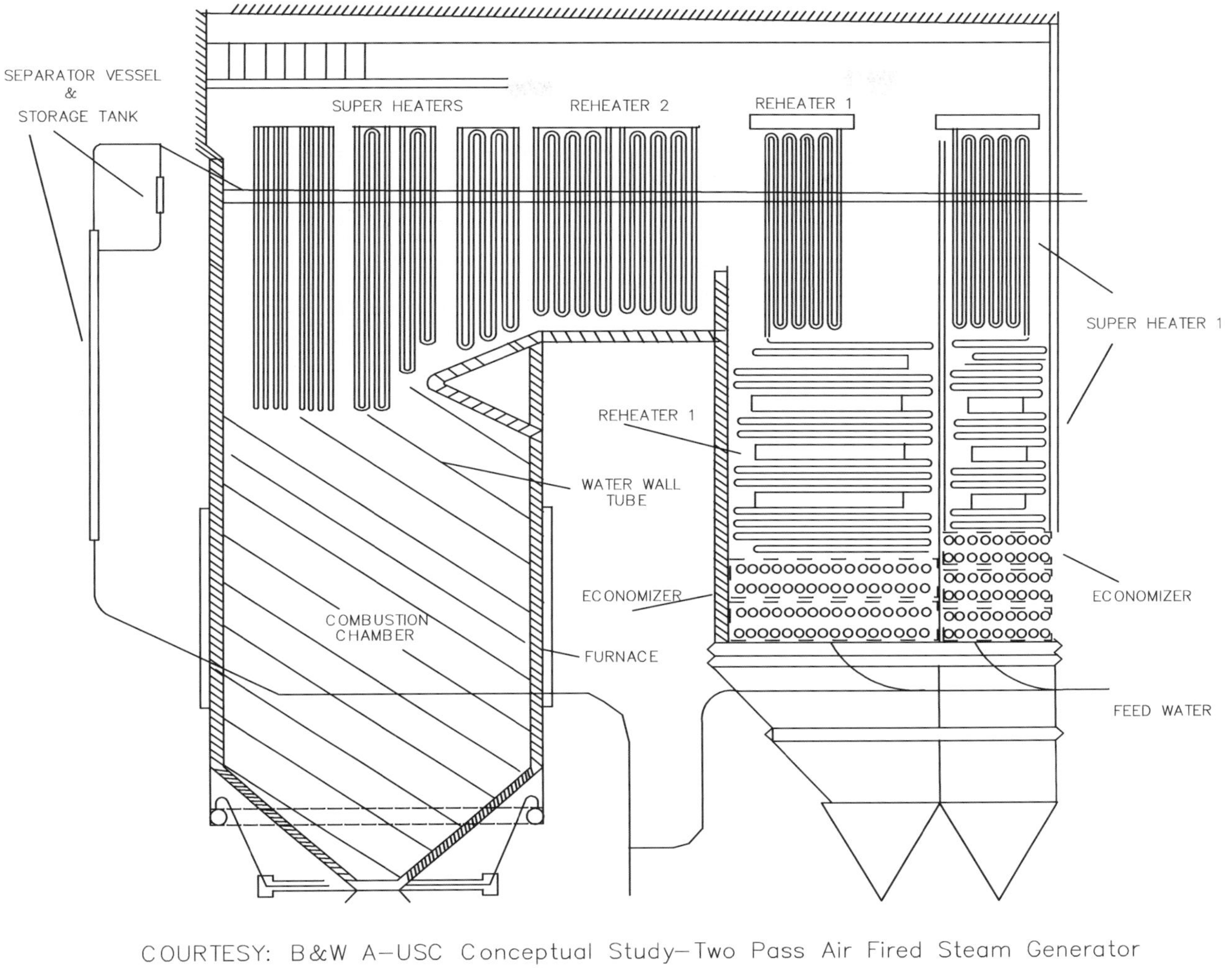

COURTESY: B&W A–USC Conceptual Study–Two Pass Air Fired Steam Generator

FIG. 13.18 First B&W conceptual two pass A-USC—air fired boiler with gas biasing.

by stringer supports and hence may be located on any spacing interval. The down pass flow of the flue gas path is baffled to provide the scope for gas biasing used for controlling the RH steam temperature through a wider load range. The gas-side operation is set by the fuel design limitations. For the design fuel, the gas velocity limits are high so the convection heat transfer is more effective. A lower temperature difference between the gas and steam called for increased the heating surface requirements. The steam outlet terminals are at the structure top.

9.7.2 B&W Conceptual A-USC Design—Oxygen Fired Series Back Pass

A similar B&W conceptual design was developed using O_2 combustion in place of air firing to provide the conditions, materials quantity, and service temperature requirements for a comparison to the earlier air-fired design. Again, the steam conditions produced 750 MW net at 34.5 MPa and 735°C/760°C with bituminous coal as the design fuel. The SG was of the OT type, selected similar to a two-pass air-fired arrangement but with an in-series down pass (Fig. 13.19).

O_2 combustion uses about 95% O_2 replacing air and a flue gas recycling system for increasing the CO_2 concentration. Cleaning and compression of the flue gas would prepare the CO_2 for pipeline transportation and deep well sequestration. The type of fuel being used determines the type of flue gas recirculation system employed, that is, cold/cool or warm. Contemplation of hot flue gas recirculation is not at all considered because of the concentration of the sulfur and chloride contained in the fuel itself. Gas from the ECON outlet was used in the early OT supercritical units. Gas recirculation (GR) dilutes the furnace gas

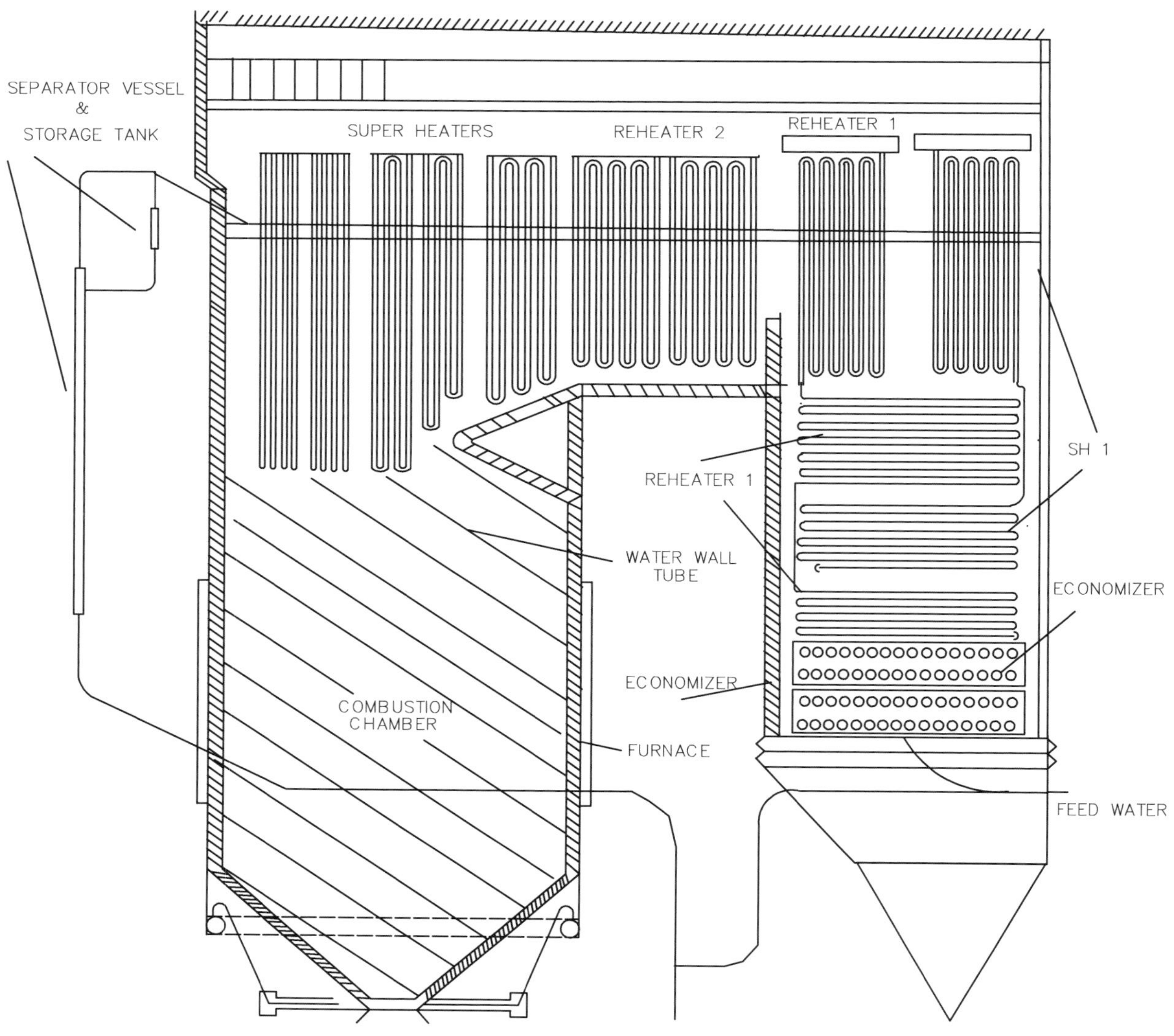

FIG. 13.19 B&W conceptual A-USC design—oxygen fired series back pass.

temperature and provides adequate gas flow rates that ensure excellent effectiveness of the convection heating surface. The RH steam temperature control with GR is also effective and within the operating parameters of the two constraints. The required heating surface in the down pass can be less than the two-pass, parallel-path, or gas-biasing design. The furnace absorption and convection pass absorption is enhanced by the higher CO_2 due to the higher density and specific heat.

9.7.3 Boiler by MHI: Air Fired ,Two Pass With Gas Biasing (Fig. 13.20)

The MHI-designed air-fired AUSC thermal power plant of 600–1000 MW net at 27.05 MPa, 568°C/5.84 MPa, and 596°C was based on subbituminous coal with high spontaneous combustibility and low caloric value. The SG is of the OT type with the most typical arrangement available for a two-pass utility boiler. The furnace is provided with low NO_x burners (eight burners in a twin vortex firing system in each elevation). This ensures highly efficient combustion with better mixing of fuel/air and a long flame path by effective use of the furnace. Combustion stability is mutually supported by adjacent burners. Low NO_x emission is also aided through proper design. Less furnace slagging is expected due to uniform combustion without a peak heat flux though sufficient wall deslaggers (68) as well as long (26) and half (10) retractable soot blowers. The rifled tube vertical water walls offer a simple structure and low pressure drop across them plus easy removal of slag. The

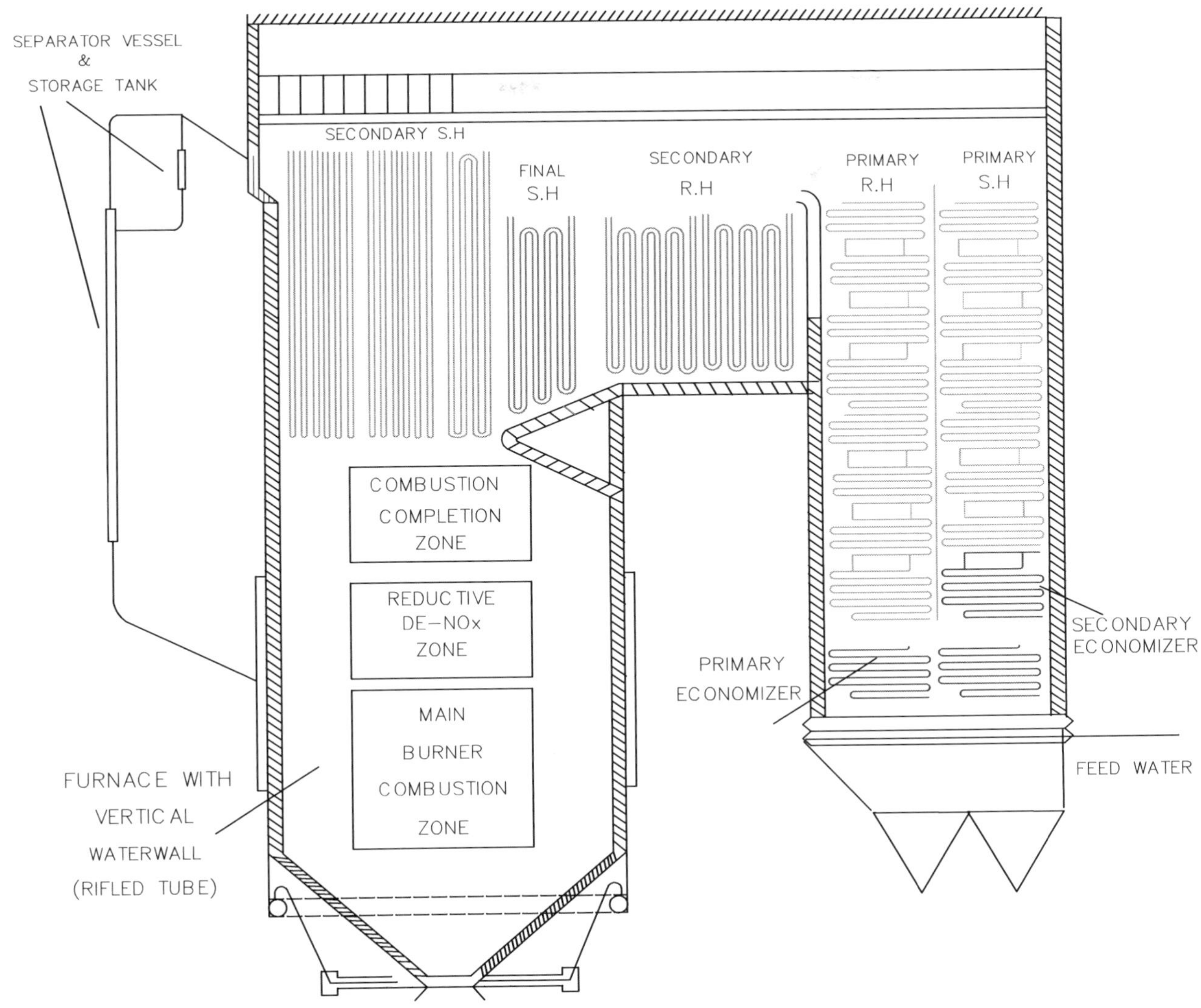

FIG. 13.20 Two pass boiler by MHI.

down pass flow of the flue gas path is baffled to provide the scope for gas biasing used for controlling the RH steam temperature through a wider load range.

The gas side operation is set by the fuel design limitations. For the design fuel, the gas velocity limits are high so the convection heat transfer is more effective. The lower differential temperature between the gas and steam called for increased heating surface requirements. The steam outlet terminals are at the very top of the structure. For more details on the firing system, see Fig. 13.6 in Section 6.3.2 of this chapter.

9.8 Horizontal (Vertical Tube) Boiler

A horizontal vertical tube boiler with a horizontally oriented furnace chamber (Benson boiler) may serve as an example of another approach implementation, as depicted in Fig. 13.21A.

Originally, the given solution was engineered for CCPP by Siemens, but then the project was reoriented to building a power plant with steam parameters equal to 35 MPa and 700°C/720°C.

In a horizontal boiler arrangement, the problem encountered in an inverted conventional arch boiler with burners located in the upper part of the boiler (Section 9.6) complicating the supply of pulverized coal was overcome. In this regard, the detailed research/engineering study of the given version was conducted on a horizontal steam boiler with USC steam parameters having a nominal capacity of 1000 MW. Combustion products flowing over the heating surface is ordinary. Flue gases downstream of the furnace successively pass over the platen SH, the primary SH, the

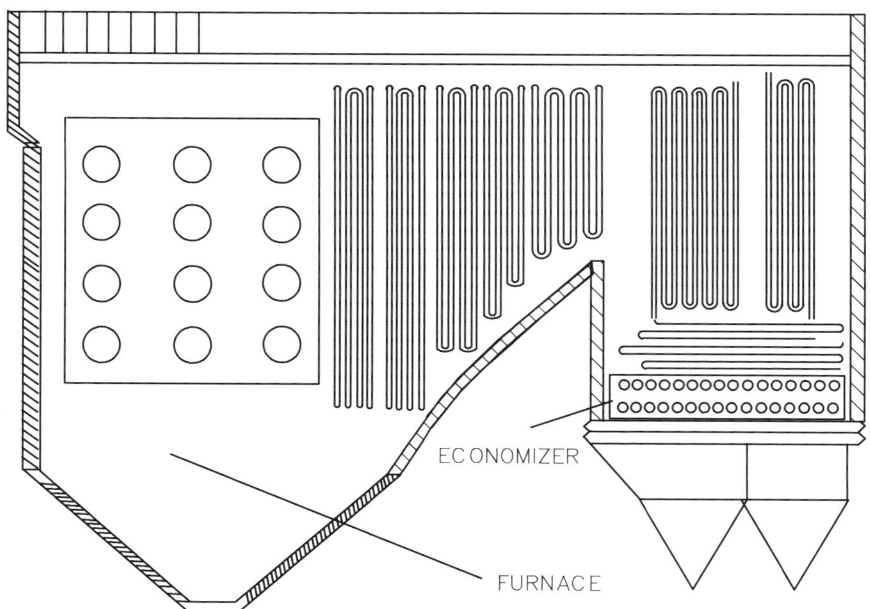

COURTESY: The sectional drawing of horizontal Benson boiler arrangement (ISSN 0973–4562 Volume 11, Number 18 (2016) pp 9297–9306)

(A)

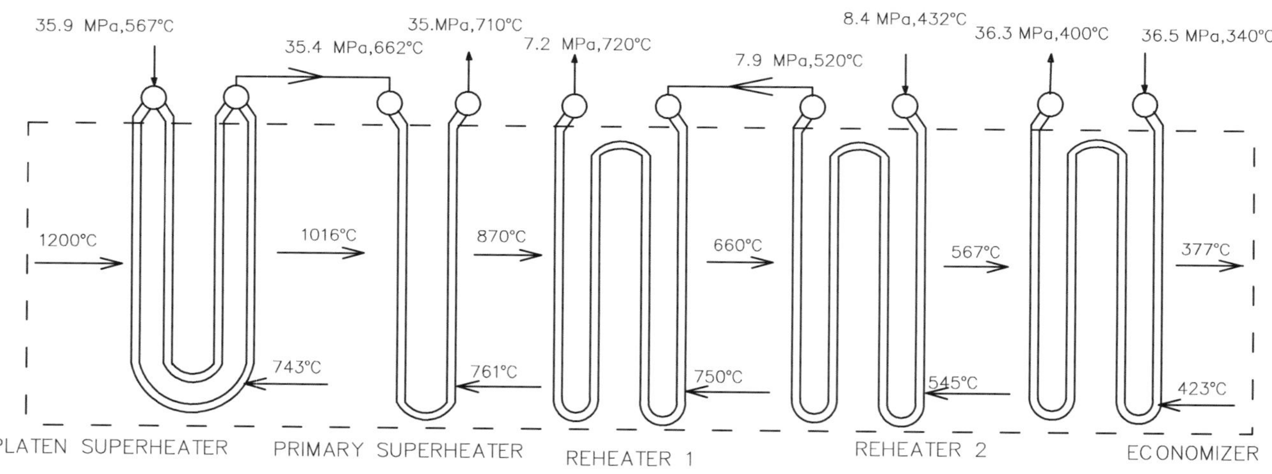

COURTESY:– Principal organization scheme of gas circuit of horizontal boiler (ISSN 0973–4562 Volume 11, Number 18 (2016) pp 9297–9306)

(B)

FIG. 13.21 Horizontal (vertical tube) benson boiler arrangement.

second/first of the RH and ECON, and the regenerative APH. The flue gas flow path with the indication of pressure part element temperatures is shown in Fig. 13.21B (included in Fig. 13.21).

Compared to a classic arrangement, the developed boiler unit has a distinctive feature of slag removal arrangement, with ash hoppers located along the full length of the furnace chamber, what exerts a significant impact on flow structure in boiler furnace. In the course of boiler unit design research, three tasks concerning furnace volume were completed: the optimal position in terms of flow structure and the angle of slope for burner units were determined; new construction of ash hoppers was presented, providing a minimum draft pressure drop; and a unique design of the gas duct turn after the furnace chamber was worked out, providing a minimum velocity profile variation after the turn to

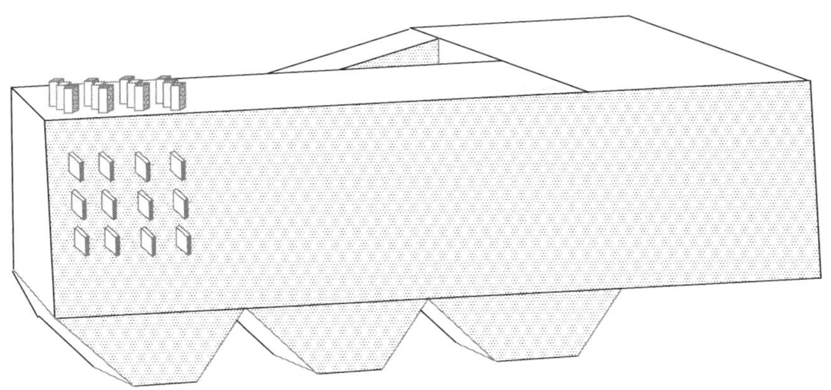

(A) HORIZONTAL (VERTICAL TUBE) BENSON BOILER ARRANGEMENT

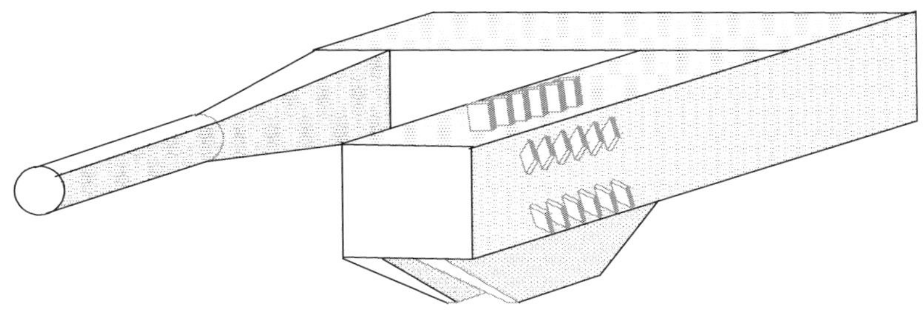

(B) HORIZONTAL (VERTICAL TUBE) BENSON BOILER ARRANGEMENT

BURNER ARRANGEMENTS—HORIZONTAL BOILER

COURTESY: 3D models of horizontal boiler versions (ISSN 0973–4562 Volume 11, Number 18 (2016) pp 9297–9306)

(C)

FIG. 13.22 Horizontal boiler 3D layouts.

reduce nonuniform heat flow distribution over the ECON heating surface. In order to tackle the given issues, three-dimensional (3D) modeling technologies (the software application ANSYS CFX) were applied.

In terms of a numerical study, different structural variations of the horizontal boiler were examined. The following parameters varied: turn shape, furnace chamber dimensions, burner unit location, and ash hopper shape. Fig. 13.22A and B shows two versions of a horizontal boiler with different shapes of the ash hopper and the burner locations.

Experimental modeling of the combustion process in the furnace is very expensive. Instead, a numerical investigation of such a complicated process requires validation experiments. Preassessment of furnace aerodynamic efficiency could be done by aerodynamic venting of the flow path without firing. Calculations showed that the inline arrangement of ash hoppers along the furnace chamber allows an alignment of vortex formation, achieving its stability. Taking into account the heat-recovery surfaces, the total length of an SG unit reaches 85.5 m; the construction area required for its installation is nearly twice the size of a classic arrangement. Therefore, terminal headers reach 34.5 m and let the high temperature pipeline length be reduced by almost 50%. Apart from that, there is the possibility of terminal headers leading to the lower part of the tube banks of the convective heat-transfer surfaces, so that they reach the level of 16 m. Here, the total length of the MS/RH piping is shortened almost four times compared to the classic design

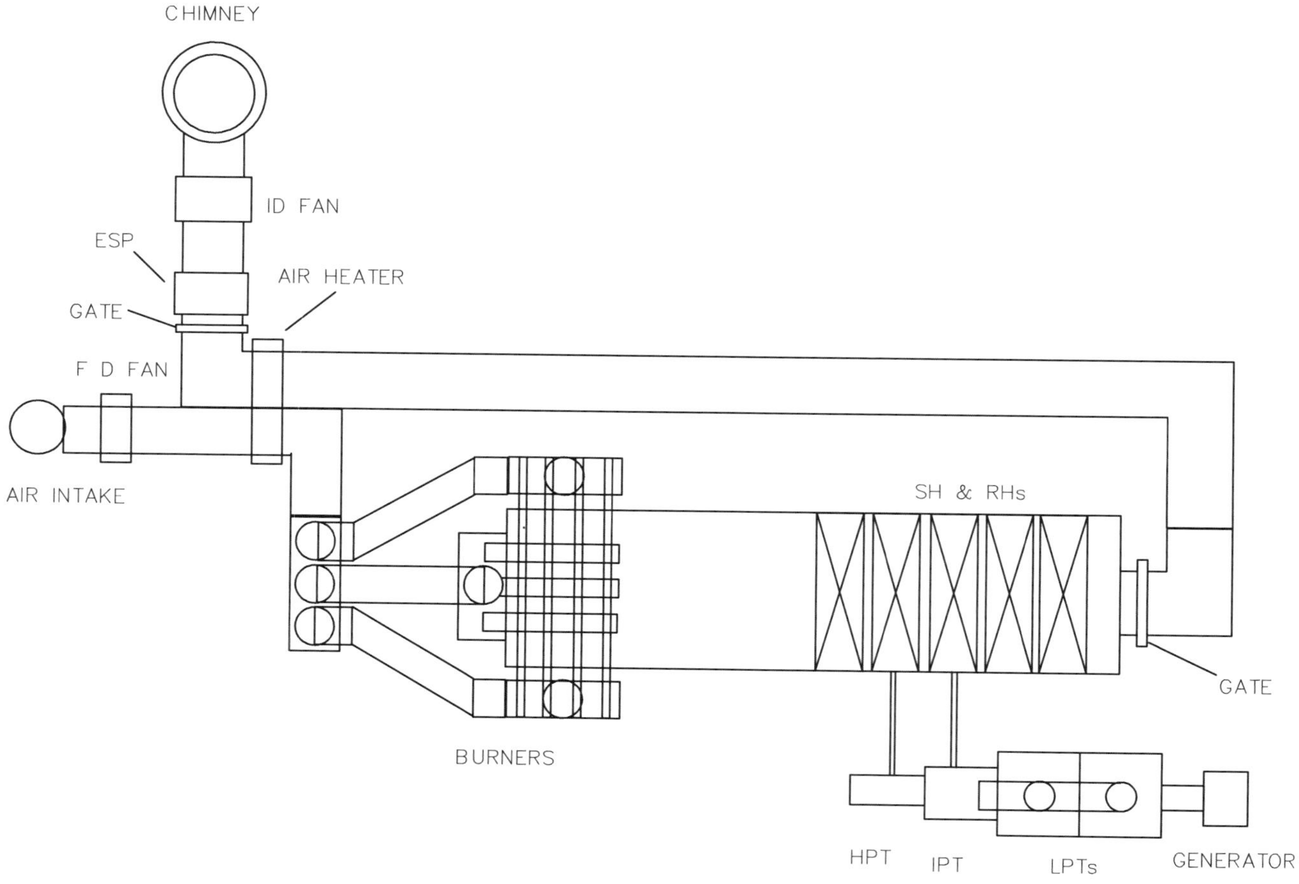

FIG. 13.23 Horizontal (vertical tube) boiler and TG plant layout.

of an SG unit. Thus, a horizontal SG unit makes it possible to reduce the share of capital costs, which falls on high-temperature pipelines, to 25% of the initial share, providing a cost reduction of the whole power plant by 10%–12%. Three-dimensional aerodynamic and one-dimensional heat and hydraulic calculations showed the perspective of the given design arrangement. The boiler efficiency reached 93.1%. The developed basic mutual arrangement of the SG/turbine is shown in Fig. 13.23.

The turbine unit is to be located near the boiler SH surfaces, thereby reducing the length of the MS piping from 150 to 70 m (~less than 50%) compared to a tower layout.

When passing to the horizontal arrangement, the flow structure in the boiler changes significantly. Hence, for the given version, an extensive study based on 3D modeling was conducted on furnace chamber aerodynamics. A burner unit arrangement and an ash hopper design were developed that affected aerodynamics greatly. Heat and aerodynamic calculations proved the high efficiency of the solutions proposed. The efficiency of both boiler unit versions (M-type and horizontal) exceeds 93%.

9.9 M-Type Boiler Arrangement (Fig. 13.24)

It is reasonable to suppose that a T-shaped arrangement (discussed later) is the most favorable priced solution among all inverted arrangements considered. Because of the burning of a large amount of coal (including low-ash coal), the problem of erosive wear of the heating surface becomes of key importance. However, it is obvious that a T-shaped arrangement requires major optimization.

An M-type boiler unit is the result of a T-shaped boiler design review.

Applying an inverted furnace chamber assumes a top-bottom flow of flue gases. Burners/nozzles are located in the upper third of the furnace and the flue gas outlet is in the lower third part. Such a design allows locating the MS/RH steam terminal headers significantly lower than

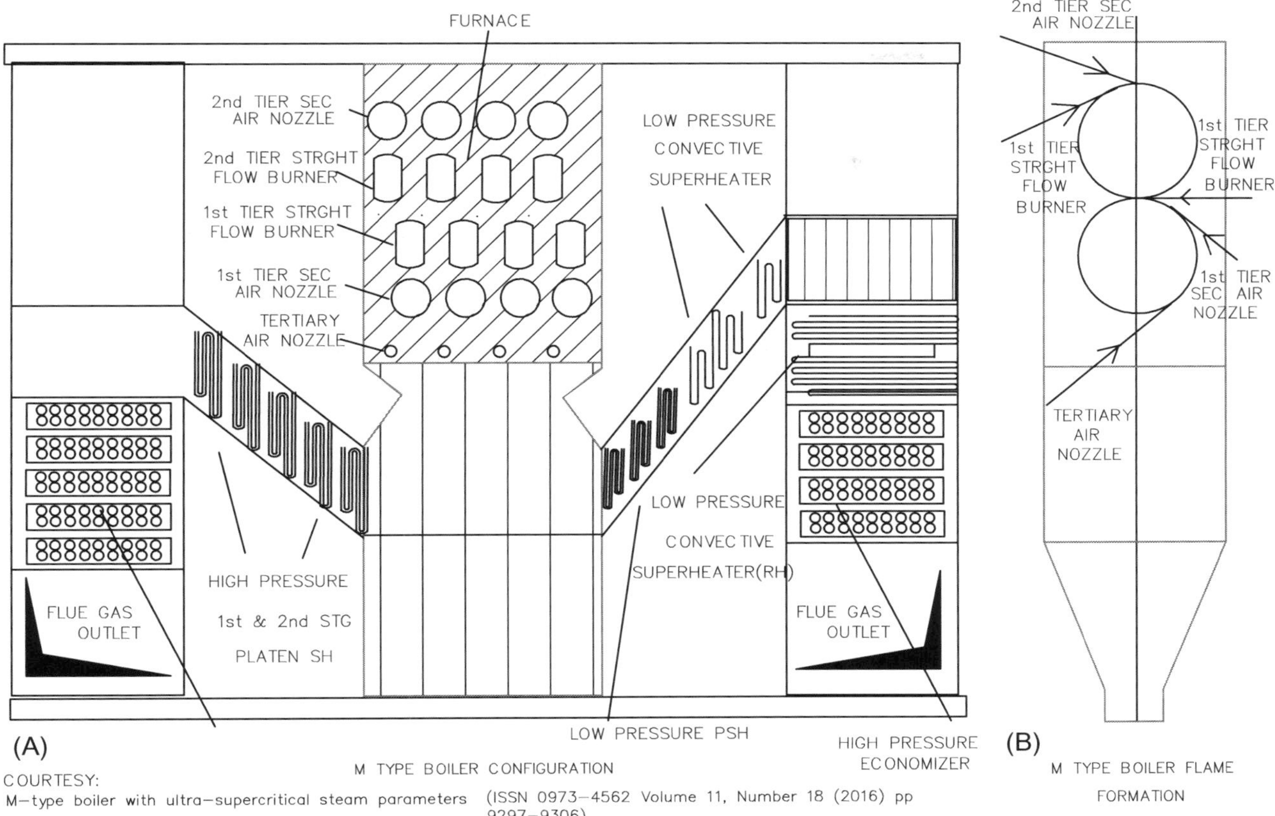

FIG. 13.24 M-type boiler arrangement.

standard boilers because of the lower location of the sloping gas ducts. The primary/secondary SHs are spaced between two opposite sloping gas ducts, providing for terminal header locations nearly on the same level.

The furnace is common with two sides separated and dedicated to completely different high-pressure and low-pressure heat exchangers. The height of the boiler enables the turbine to be installed near the boiler along the rear furnace wall, down and convection passes. The closeness of the SH/RH terminal headers with the turbine is obvious, resulting in the reduction of the length of steam pipelines and the total metal consumption of a boiler. Instead of locating the SH/RH outlet headers in the upper part, their designed location is chosen underneath the sloping gas ducts. By this arrangement, the installation level of the terminal headers comes down from 70 to 20 m. The whole layout and combined effect facilitate an almost 65%–70% total reduction of steam pipeline length compared to the inverted conventional arch arrangement of a power plant of the same capacity.

The burners are located at the top of the furnace with the flue gas output at its bottom, reassuring greater fuel particle residence time in the furnace chamber. This is a necessary precondition toward complete combustion and an increased degree of fuel burn-up that results in the reduction of combustible content in slag and flue ash or less combustible loss.

It may be referred to the advantages of M-type SG plant that its application allows to reduce fly ash with flue gases approximately by 15% due to their separation upstream of gas port and further flow through ash hopper, followed by reduction of fly ash wear of the convective heating surfaces.

Compared to inverted conventional arch and tower-type arrangements, the installation area of platen SHs is expanded as well.

The furnace chamber dimensions of an M-type boiler in a horizontal section are 14,640 × 26,840 mm. Pulverized coal combustion takes place within the vertical vortex system as depicted in sixteen pulverized-coal burners with an opposite-fired shifted arrangement are located at two levels. Secondary air nozzles (16 pieces) are also located at two levels: the nozzles of the second level are installed downward, and the nozzles of the first level are located opposite and upward. Secondary air nozzles (8 pieces) with an opposite shifted arrangement are located one level upward. Burners and nozzles are located in such a way that every set of burners and nozzles (two pulverized-coal burners, two secondary air nozzles, and a tertiary air nozzle) would create two opposite vertical vortices, as depicted in

Fig. 13.24B, M-type Boiler Flame Formation (included in Fig. 13.24). Pulverized-coal burners and air nozzles are installed on the front and rear walls. Pulverized-coal combustion takes place in a vertical-horizontal tangential flame. Eight pairs of vertical vortices are formed on all fronts.

The boiler is designed for bottom-ash removal: at an elevation of 11.15 m, the front/rear walls taper off at a 52-degree angle, forming hopper work points. At the furnace chamber bottom, the flue gases are divided into two streams, either of which turns to the sloping gas duct and contributes to the additional separation of ashes, reducing fly ash. On sidewalls before the turn to the sloping gas ducts, the furnace walls form aerodynamic noses, providing a more proportional sweep of the HP/LP superheaters.

Divided into two streams at the furnace chamber outlet, the flue gases pass through the sloping gas ducts. In the left sloping gas duct in the direction of the gases, the first and second stages of the high-pressure platen SH (live steam) are located inline. In the right sloping gas duct in the direction of gases, the low-pressure platen SH (RH steam, second stage) and low-pressure convective SH (third stage) are located inline. Downstream of the sloping duct, gases enter convective passes. In the left convective pass in the direction of the gases, five ECON tube banks are located. In the right convective pass in the direction of the gases, the regulated stage of the low-pressure convective SH (first stage) and the four ECON tube banks are located. Stringed tubes of ECON and outlet tubes are located in the turning chamber. Due to a different number of ECON tube banks at the LHS/RHS of the down-taking duct, identical heat absorption can be achieved in both gas ducts. Then, the flue gases pass three APHs in a parallel way.

Membrane walls are designed to be gas tight. The walls of the sloping gas ducts, the turning chamber, the convective pass, and the ECON tube banks are screened with membrane panels. According to calculations, the boiler gross efficiency is 93.07%, and the estimated fuel consumption is 91.13 kg/s. The efficiency of both boiler unit versions (M-type and horizontal) exceeds 93%.

9.10 T-Type Boiler Arrangements (Fig. 13.25)

Various designs of T-type boiler arrangements are contemplated by the researchers in pursuit of the reduction of costly high-temperature pipelines. Similar to M-type boilers, it has two furnace outlets. However, the pressure part elements, for example, SH/RH, are arranged in both the wings in a symmetrical configuration, unlike M-type boilers. The SH/RH pipelines of both the wings are combined externally to form a single SH and cold/hot RH lines. The turbine modules are located in the close vicinity and the advantage of a low height boiler brings into being the combined effect of reduced pipe lengths.

9.10.1 A horizontal T-type boiler is shown in Fig. 13.25A, horizontal T-type boiler arrangement. The approximate pipe length is estimated to be:

$$\text{Main steam pipe length} \Longrightarrow 70\,\text{m}, \text{reheat steam pipe length} \Longrightarrow 85\,\text{m}$$

9.10.2 A vertical standard T-type boiler is shown in Fig. XIII/9.10-1B, vertical T-type boiler arrangement. The approximate pipe length is estimated to be:

$$\text{Main steam pipe length} \Longrightarrow 106\,\text{m}, \text{reheat steam pipe length} \Longrightarrow 82\,\text{m}$$

9.10.3 A vertical Inverted T-type boiler is shown in Fig. 13.25C, vertical inverted T-type boiler arrangement. The approximate pipe length is estimated to be:

$$\text{Main steam pipe length} \Longrightarrow 37\,\text{m}, \text{reheat steam pipe length} \Longrightarrow 50\,\text{m}$$

10 VARIOUS TURBINE CONTROL SYSTEMS

In Chapter 10 of this book, the turbine control systems are described, mainly on the Siemens KWU make turbine. Although the control philosophy is somewhat similar, certainly there are changes, for example, a constant turbine inlet pressure is popular in the United States whereas the sliding or variable pressure control is preferred in Europe. Here in this section, the control philosophy of some other reputed world-class manufacturers will be discussed very briefly to have some idea about their products.

10.1 Thermal Stress Controlled Loading of Steam Turbine Generators

In a steam turbine generator, the unit load reference signal sets the turbine load in accordance with various limiting bands so as to ascertain safety and to uplift the life expectancy of the high-temperature rotating machine. However, it may be noted that the turbine tripping, testing, servo mechanism systems, and turbosupervisory systems are almost similar and hence are not included in this chapter.

It was the normal practice that the operator had to reduce the mismatch at a safe rate during the acceleration phase of the start up as well as determine the allowable rates of change of metal temperatures during the loading of the turbine generator (TG) set. In an effort to minimize start-up time without damaging the turbine, various techniques such as acceleration control, load control, etc., were utilized, including heat soaking periods on the turning gear to reduce the initial mismatch. The initial operation in the less efficient full-arc (all the GCVs operating in tandem) steam admission mode were used to achieve uniform warming

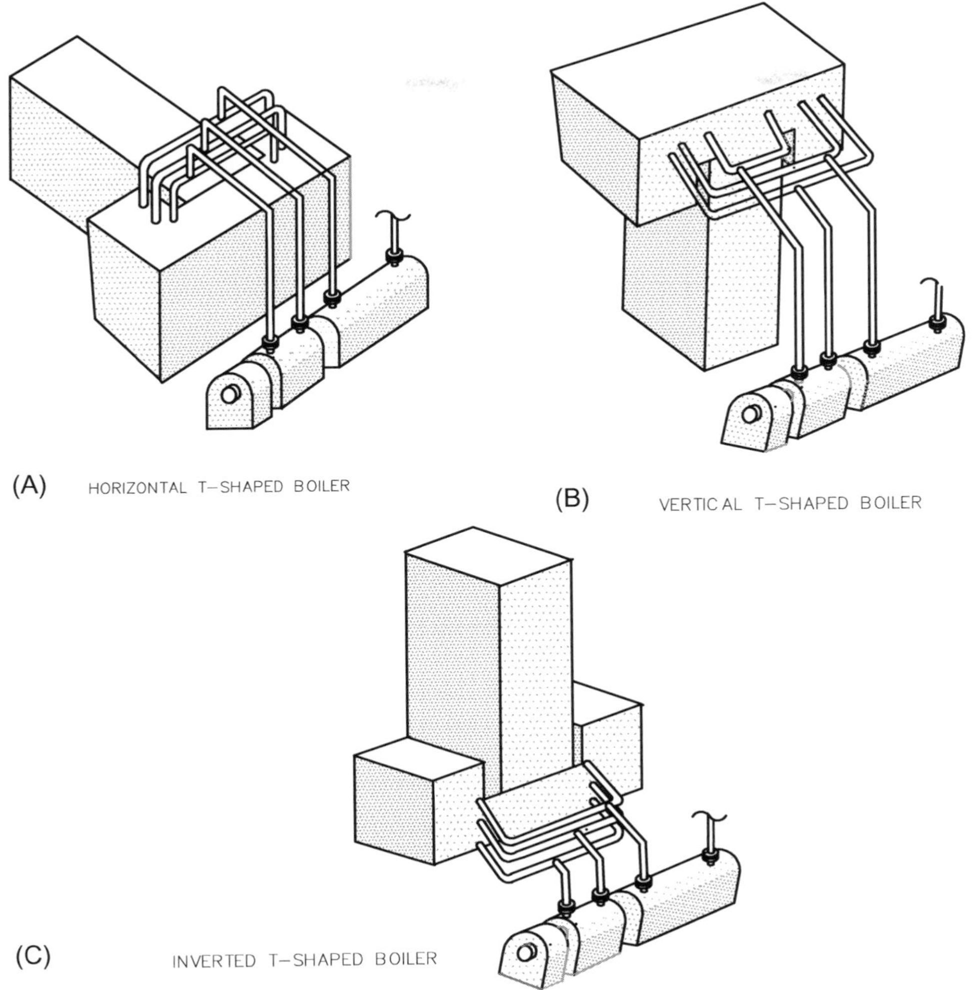

FIG. 13.25 Various types of T-type boiler arrangements.

of the high-pressure turbine inlet parts. These systems were usually based on ideal boiler conditions. Because turbine start ups can take several hours, systems that will reduce these times as well as allow for fluctuations in the steam parameters from the boiler are of great value.

A sophisticated approach to start up and loading control by means of continuously calculating the rotor surface and bore stresses from speed and temperature measurements, and then loading at the maximum permissible stress, is envisioned in this type of turbine configuration with a matching control philosophy. It is highly recommended that this system will be useful for achieving rapid or even faster start up, and loading would be possible through better thermal stress distribution among various parts of the different turbine sections relative to their designed capabilities.

10.1.1 Full Arc and Partial Arc Mode of Steam Admission

At the high-pressure turbine (HPT) inlet, four governing control valve (GCV)s are provided to manipulate steam admission uniformly through all the four individually dedicated nozzle arc sets disposed around the first stage turbine inlet; this is known as *full arc admission*. On the other hand, when the GCVs are manipulated to admit steam in sequence in a thermodynamicallly more efficient mode to one nozzle arc at a time, the configuration is known *as partial arc admission*.

The entire steam admission system to the turbine is regulated by providing a means for transferring between full arc and partial arc mode through the mode reference signal.

10.1.2 The Stress Evaluator

The stress evaluator determines the mode reference signal output so as to minimize the thermal stress differences between the HPT/IPT rotors with the ultimate aim to reduce the turbine loading time (without overstressing the rotors/other turbine parts) to make a rapid load change, whether loading/unloading of steam turbine generators (TG). In recent years, as the trend is toward larger units, it results in higher thermal stresses for any given temperature transient and obviously the start up and loading of a large TG set has become of more concern. The factors contributing to thermal stresses during start up are:

(i) Mismatch between the temperature of the admitted steam and the metal temperature and (ii) the degree of mismatch between the above two, which depends upon the past operating history, that is, whether the turbine is involved in a cold start or a hot start. However, the mismatch is essentially corrected during the acceleration phase of the start up.

At load condition, the steam flow is high enough through the turbine without any substantial mismatch and the metal temperature will follow the steam temperatures closely. Stress may develop for a rapid load change beyond the permissible rate of change, which is not allowed by the stress-controlled controller. Metal temperature control vis-a-vis thermal stresses is based primarily on analytical/statistical correlations between stress levels and expected rotor life. According to GE, faster start up/loading is possible through better thermal stress distribution among various parts of different turbine sections relative to their designed capabilities. Stress signals of HPTs/IPTs are calculated and utilized in a load/mode controller suitably to make the complete control system.

10.1.3 The Control System

The control system envisages one HP turbine, one IP turbine, and two LP turbines, all arranged in tandem. The number and arrangement of additional LPTs/IPTs as well as the number and arrangement of generators are not important to an understanding of the control system. Steam flows from the boiler through MSSVs with built-in bypass valves and then through the GCVs. Each of the latter is connected to a different nozzle arc distributed symmetrically along the HPT inlet chest supplying first stage or HP rotor blades. The admission of steam is controlled through an MSSV servo mechanism and four GCV servo mechanisms operating the respective valve, as dictated by the control system. The servo mechanisms are of the electrohydraulic (EHC) type, driving high-pressure hydraulic rams in response to electrical signals issued by the controller. The servo mechanisms receive input from the load control unit, which provides as its output a suitable valve positioning signal corresponding to a desired rate of steam flow. The remainder of the primary control loop includes a speed controller receiving redundant speed signals in the form of digital pulses from a toothed wheel on the turbine shaft. All the above-mentioned valves are manipulated in such a way so as to either admit steam uniformly (*full arc admission*, Fig. 13.26) or else in sequence (*partial arc admission*, Fig. 13.27). A transfer device is provided for switching between full arc and partial arc mode as well as to indicate the degree of transfer that has taken place.

One type of transfer is to enable the MSSV integral bypass valve to throttle the steam during full arc admission for initial warm up and shut down for a long period. Another type of transfer enables all GCVs to throttle in unison during full arc admission with, of course, MSSV and its integral bypass valve in the full open condition.

The load/mode controller uses a first-stage HPT/IPT shell and other temperatures from redundant sensors to calculate both the rotor surface stress and the rotor bore stress at the T/C insertion location. Because the rotor is assumed to be at rated speed, only the thermal stresses are affected by load changes and need be considered. A state-of-the-art means for calculating the rotor stresses in a stress-controlled start-up system is fully utilized as the main parameter for the load/mode control unit.

A sophisticated approach to start up and loading control by means of continuously calculating the rotor surface/bore stresses from the speed and temperature measurements and then loading at a maximum permissible stress is almost similar to other manufacturers. However, here one of the outputs from the load/mode controller is a load reference signal, indicating a desired rate of steam flow change to produce the load as required by the main load controller.

The total load control is achieved by means of a so-called *load/mode controller* block, which provides a load rate of rise (or load reference) setting signal and also a mode signal for adjusting the steam admission mode. The stress evaluator calculates stress signals representing the maximum allowable stress, one of which is responsive both to the load and mode. The other signal is responsive primarily to load in accordance with the higher of these stresses while the mode is adjusted according to the difference between these stresses so as to minimize the time required to change the load.

The mode reference signal supplied to the full/partial arc transfer device indicates a desired relative point between full arc and partial arc steam admissions to the first stage.

In order to relate the stresses in the different rotors (having different materials with different allowable stresses) to a common ground for comparison, the stresses are normalized by dividing the actual stress by the allowable stress for each particular material and location as part of the computation. By convention, stresses resulting from an increasing temperature are calculated as positive quantities, and stresses due to a decreasing temperature are calculated as negative.

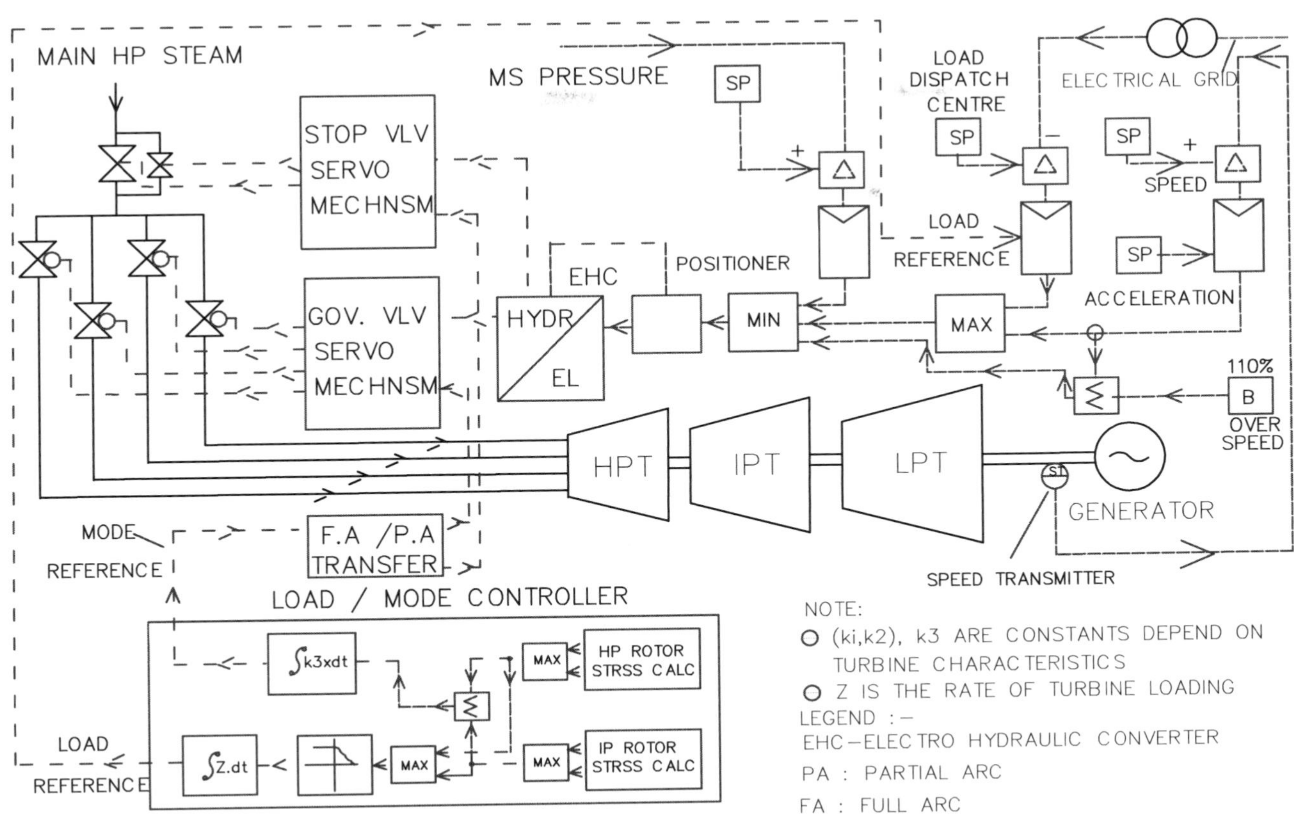

FIG. 13.26 Simplified schematic of stress controlled turbine control system.

The maximum absolute value of all the rotor surfaces and bore stresses of HP and IP rotors is applied for further calculation represented by function and integration. The output for load increasing use positive stress values and for control of load reduction, the control action is changed to pass the highest negative stress and function is changed to produce a negative rate. A suitable switch on the control console allows the operator to select a load increase or reduction as required.

The HPT/IPT rotor stresses are input to a summation unit with the appropriate sign convention, and the output, indicating a difference signal, is applied to a device for calculation through an integration unit, producing a mode reference signal. This signal is applied for transferring the mode between full arc and partial arc and indicating the relative point at which the valves are to be positioned. Absolute value of the mode reference signal is unimportant and the convention selected here is that mode reference varies between –1.0 (partial arc) to 1.0 (full arc).

In Fig. 13.27, the upper two figures show the two extreme positions between the full arc and partial arc with the MSSV bypass valve in action; the lower two figures show the partial arc condition wherein the MSSV and its bypass are open. The first GCV is wide open, admitting steam to the nozzle arc, while the second GCV is partially open, admitting reduced steam flow to the nozzle arc. The third and fourth GCVs are closed so that the corresponding nozzle arcs are blocked. These GCV positions are shown arbitrarily for the turbine running at a load zone between 25% and 50% (for example).

It may be particularly noted that the total steam flow (hence load) is substantially the same, even if the admission mode changes from one mode to another mode.

It is also a fact that the first-stage temperature on full arc admission is much higher than that on partial arc admission at the same load. Taking advantage of this phenomenon, the control system contemplates continuous controlling steam admission between full and partial arc or at any intermediate point during transient operation in order to control the first-stage temperature to optimize turbine stress conditions. During constant load operation, control is gradually returned to the more efficient partial arc admission through a small bias arranged suitably in the control loop.

The allowable rate of load increase Z, expressed in %/minute, is calculated at a function unit after the max unit in the load reference calculation path and is indicated by the slope of the function in the corresponding function unit itself. The integration units in both the load and mode

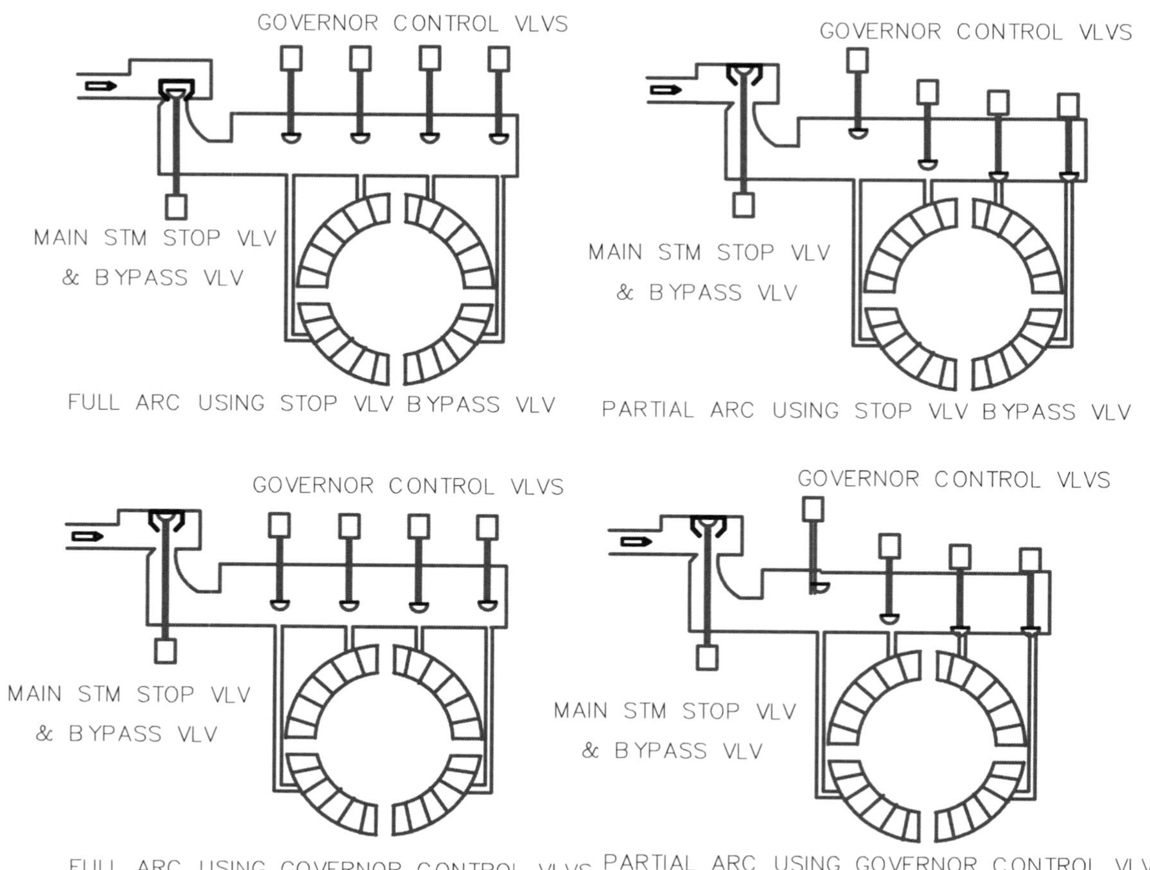

FIG. 13.27 Concept of partial and full arc admission in turbine control.

reference path are coupled with the effect of a delay of one minute between each calculation.

Provisions have been made for the operator to interrupt a load change by specifying a hold condition. In this case, the admission mode is removed from stress control by setting the HPT/LPT stress difference equal to zero, and is slowly moved by the above-mentioned bias to the most advantageous position to subsequently resume the load change. This procedure serves to continue the change in the first-stage temperature during hold and permits a greater loading rate at the resumption of loading.

In a steady-state condition when neither an increase nor a decrease of load is taking place, the admission mode control remains operative in case of a load change called for by the speed governor if a system frequency disturbance occurs. In this case, the small bias mentioned above enables control to return to partial arc admission in the steady state.

Here, the admission mode is continuously controlled between full arc and partial arc, that is, the mode control is transferred back and forth between the two. Stresses in the high-pressure rotor have no effect on the loading rate during intervals of time when the reheat rotor stress is greater and determining the loading rate. Hence, the admission mode reference may be moved by the controller to increase the rate of temperature change and therefore the stress in the HP rotor.

As stated before, the maximum stress (of the four stress values) at any one time is acting to set the loading rate for the turbine. During intervals of time in which the HP rotor stress is greater, the loading rate is dependent entirely on the HP rotor stress. The overall rate of loading can be increased by moving the admission mode reference so as to reduce the rate of first-stage temperature change, hence, stress in the HP rotor. Because the total temperature change for both rotors depends only on the initial and final load points, the overall temperature change is the same for any method of control.

It may be noted that the admission mode is used to control the temperature of the high-pressure rotor independently of the reheat rotor while the rate of loading is affecting either or both rotor temperatures. Therefore, the time required to perform the change in load can be less.

Stated in another way, the stress signals calculated for both the HP rotor and the IP rotor are responsive to temperature changes, which depend upon load changes. However, the HP stress signal is also dependent upon the mode of steam admission while the reheat stress signal is not. The turbine is loaded according to the higher of these two signals. When the HP signal is highest, the mode is adjusted toward partial arc to reduce HP stress (despite the fact that the load is being added in accordance with HP stress). When the IP signal is highest, it is determining the additions of load. However, the HP rotor temperature is not being increased as fast as it could within allowable stress limits. Therefore, even though the mode adjustment cannot affect the IP stress directly, it is adjusted toward full arc in order to increase the rate of temperature change for the HP rotor.

To summarize, the load control means an arrangement to position the GCVs to admit a desired total steam flow to both the turbines. Admission mode transfer means an arrangement to adjust the relative openings among said valves to the nozzle arcs to effect changes in first-stage temperature independently of total steam flow. A stress evaluator calculates the first and second signals for the HP and IP turbines, respectively, which simulate allowable thermal stresses. The first signal is substantially responsive to setting the mode transfer means and the second signal is substantially unresponsive thereto.

(a) Adjustment of the load change rate control is related to the higher of the HP and IP signals.
(b) Mode transfer is related to the difference between the HP and IP signals so as to reduce the same.

10.1.4 Steam Turbine Control with Megawatt Feedback and Pressure Influence

After synchronous speed is attained, the electrical output of the generator is connected to the load and loading toward full output can begin. While ramping up the loading, the turbine may require a greater steam flow than the boiler is capable of supplying, resulting in a fall of the steam pressure.

One type of control system programs the loading of the TG set by comparing a generator megawatt demand with the actual generator output to produce an error signal. This error signal is used, either directly or after further processing, to control the GCVs feeding steam admitted to the turbine. As noted above, during the start up and loading process, less than the full operational supply of steam is available from the boiler. Accordingly, it is possible that steam pressure perturbations may occur during loading. If a sharp downward swing occurs in the steam pressure, the control system sharply opens the steam valves to compensate by attempting to increase the supply of steam to the turbine. However, because a limited supply of steam is available, rapidly opening the steam valves instead of increasing the steam fed to the turbine further reduces the available steam pressure. The reduced steam pressure causes the error signal to open the steam valves even wider, which, due to the limited steam supply available, further reduces the steam pressure and acts as positive feedback to increase the instability of the system.

10.1.4.1 Objects

GE provides a TG control system to overcome the positive feedback state by modulating the MW error signal relative to a negative slope in the steam pressure for regulating turbine steam admission. Signals representing the TG set electrical output and the boiler steam pressure are fed to the turbine load/governor control system. A control signal for controlling GCVs feeding steam to the turbine is gain controlled between 1 and 0 in the presence of a negative rate of change of steam pressure less than a predetermined value. At other less negative values and all positive values of rate of change of steam pressure, the control signal is unaffected. A linear relationship is employed for controlling the gain for increasingly negative values of steam pressure rate change. At an extreme negative value, the gain is controlled to zero, whereby the control signal remains unchanged.

10.1.4.2 Description of the Control Philosophy

The main idea of this part of the control system is to create a gain control means responsive to a negative value of the rate of MS pressure change less than a predetermined value for reducing the load reference error, whereby a response of the steam turbine generator is rendered relatively unresponsive to steam-pressure reductions exceeding the predetermined value. According to this system, there is a method for controlling a steam turbine generator using the megawatt measured value and a set value for producing a load reference error responsible for controlling the amount of steam fed to the turbine. The signal representing the negative rate of change of MS pressure is utilized to trim or make the load reference error relatively unresponsive. This trimming is active when the rate of change signal indicating a negative rate of change is less than a predetermined value up to a further negative value, beyond which the response of the steam turbine generator is rendered entirely unresponsive to steam-pressure reductions.

Referring to Fig. 13.28 the actual MW output of a generator forms the measured signal after sufficient redundancy is compared with the set point from the monitor/setter in the plaque as applicable. Set point and measured value are compared to create an error value "*e*".

The error thus created is fed to a divider as the first input. The steam pressure *P* after sufficient redundancy forms the second input of the divider, which divides the error signal by

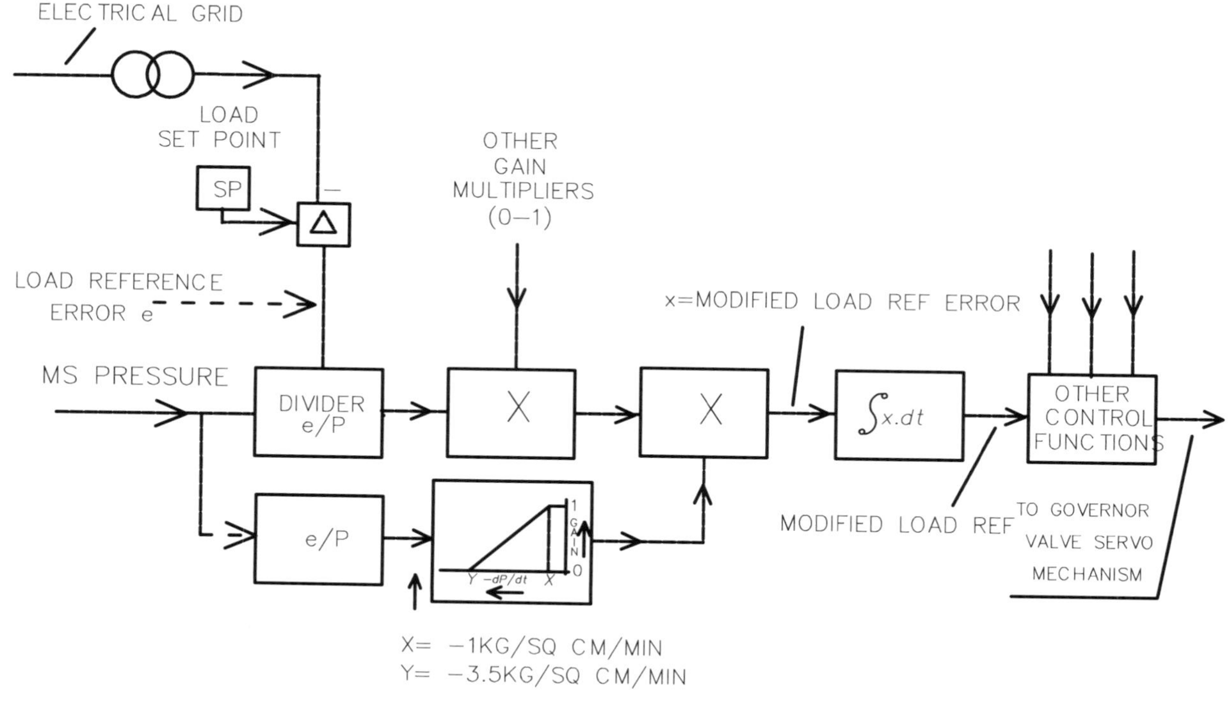

FIG. 13.28 Pressure influenced turbine control.

the steam pressure signal and produces the output as *e*/*P*, representing the MW load reference error signal. That is then input into a dynamic gain controller whose gain control signal is generated elsewhere from some other influences and varies between zero and one. The output of this gain controller is applied as an input to a second dynamic gain controller. The steam pressure signal is also fed to the input of a differentiator, which calculates the steam pressure rate change with time. The derivative thus formed is applied as an input to a gain function generator, which is responsive to negative values of the rate of change of steam pressure having a value less than a predetermined value.

In the normal range of steam pressure derivatives at or greater than the predetermined value (e.g., -1 kg/cm^2/m) and all positive values, the gain control signal applied to the gain controlled amplifier remains 1, meaning the load reference error signal will be unaffected by the pressure variation.

Between the predetermined value (-1 kg/cm^2/m) and an extreme negative value (-3.5 kg/cm^2/m) of the derivative, the signal produced by the gain function generator progressively reduces the gain from 1 to 0 along a linearly sloping curve. This means the load reference error signal will be trimmed according to the severity of the rapid pressure change and the lesser effects to the GCVs.

When the of rate of change of the steam pressure goes beyond the extreme positive value (-3.5 kg/cm^2/m), the gain function generator output becomes 0, which means that the load reference error signal will be turned to 0 and the position of the GCVs would not change during this situation.

Then, the output of the gain function generator is applied to the input of a gain controller. The output of the gain controller becomes an input of an integrator whose output is an input to other control functions of the main turbine governor control system furnishing the control signal to the GCVs.

10.2 Improved Turbine Control by Supervisory Controller and EHC

Recently, the scarcity and high cost of energy has fostered the development of larger, more refined, and more efficient TGs, for which the electrical utilities have sought a means to ensure the ability to start/stop/change loads, etc., in response to changing load demands in the most flexible and economical manner. This has led to the development of highly refined supervisory C&I systems, but it has also made the duty of the operator more demanding by requiring that he absorb and process an increasing amount of information as he further directs control of the TG set.

10.2.1 Computer-Aided Electrohydraulic Control (EHC) (Fig. 13.29)

In this system, a conventional, well-tested, highly reliable analog EHC system having direct control of the TG operation along with a dedicated and interactive supervisory

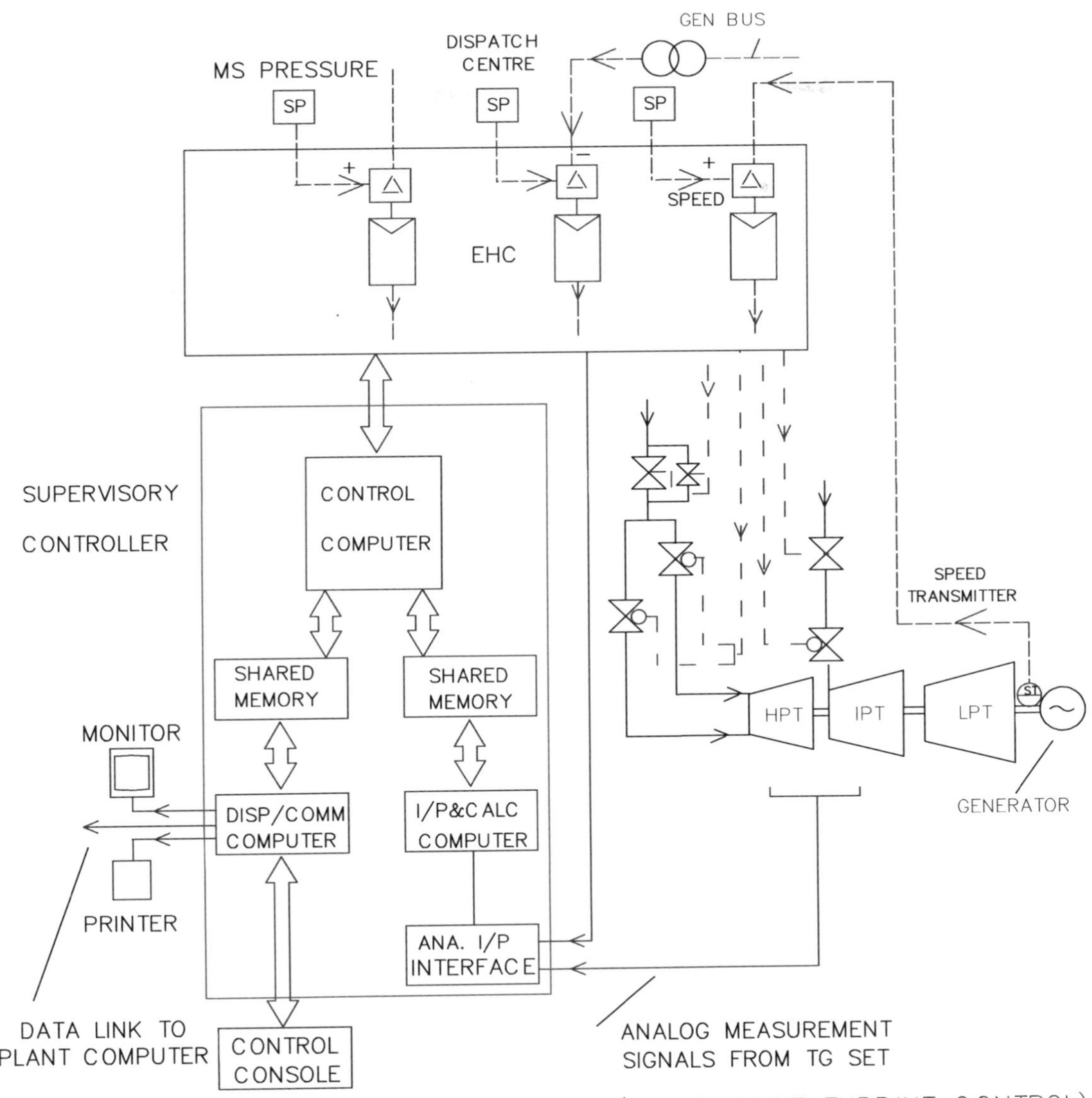

FIG. 13.29 Computer aided EHC for turbine control.

control system capable of optimally and automatically starting/loading/unloading comprising a hierarchy of computer subsystems are there to provide C&I capabilities during all operating phases of the TG. The computer subsystems are programmed for coordinated interaction and communication with each other and each microcomputer subsystem is programmed and configured to handle a separate group of control responsibilities, forming a distributed control system. The system architecture showing the computer control system, the EHC, and the turbine steam admission has been depicted in Fig. 13.29.

The computer hierarchy includes an input and calculations unit having the means to interface with the analog input data sources and sensors, which report on various operating parameters of the TG, boiler, the balance of the plant, and turbine metals from which thermal and mechanical stress and other derived quantities are calculated. A separate communications and display unit is provided to interface with a plant computer and with an operator control panel and other display and readout devices whereby operating personnel may interact with the control system and a control computer, standing at the top of the hierarchy, for receiving information from the other computers, for making decisions based on that information and, through input/output ports, for providing the EHC system with directions for optimal control of the TG within its thermal/mechanical limitations. This system aims to provide a cheaper alternative to a large mainframe computer for TG control by arranging a computer-based DCS dedicated to supervisory control. The improved supervisory and protective capabilities in the integrated dedicated control system has the provision for various operating modes such as a monitor mode, a supervisory control mode, and a subloop control mode whereby a large plant computer, requiring minimal programming, can direct TG operations and receive reports regarding its progress.

10.2.2 Control Computer and EHC Communication (Fig. 13.30)

Special interfacing networks have been envisaged to couple the computer subsystem to external operating elements associated with that particular computer (e.g., to interface with the electrohydraulic control system or to take in measured analog data).

The supervisory controller of this system provides multiple operating modes as indicated earlier. These include a monitor mode wherein the operating personnel are guided through all phases of TG operation by announcements and directions that appear on a MONITOR or other readout device and in which the operator causes advancement from one turbine operating phase to another; a control mode where the operating decisions are automatically made and the turbine is advanced through all operating phases with minimum operator interaction; a remote automatic mode where TG control is turned over to a centralized automated dispatch system (ADS) or a coordinated boiler-turbine control system (CBC) once the turbine has reached a basic target load and where ADS or CBC operates by interacting with the controller; and a plant computer control mode wherein the control system functions as a subsystem in an overall plant control scheme so that a very minimal, straightforward program for the plant computer is required to achieve TG control.

The operator can judiciously direct the EHC system by causing the turbine to proceed through a logical operating sequence while omitting steps not needed under the prevailing conditions. For example, to effect a start up of the turbine, steps are included for rotor prewarming and for chest warming, followed by a step to prepare for roll off, which includes a validation check of calculations made and a determination that the available steam parameters

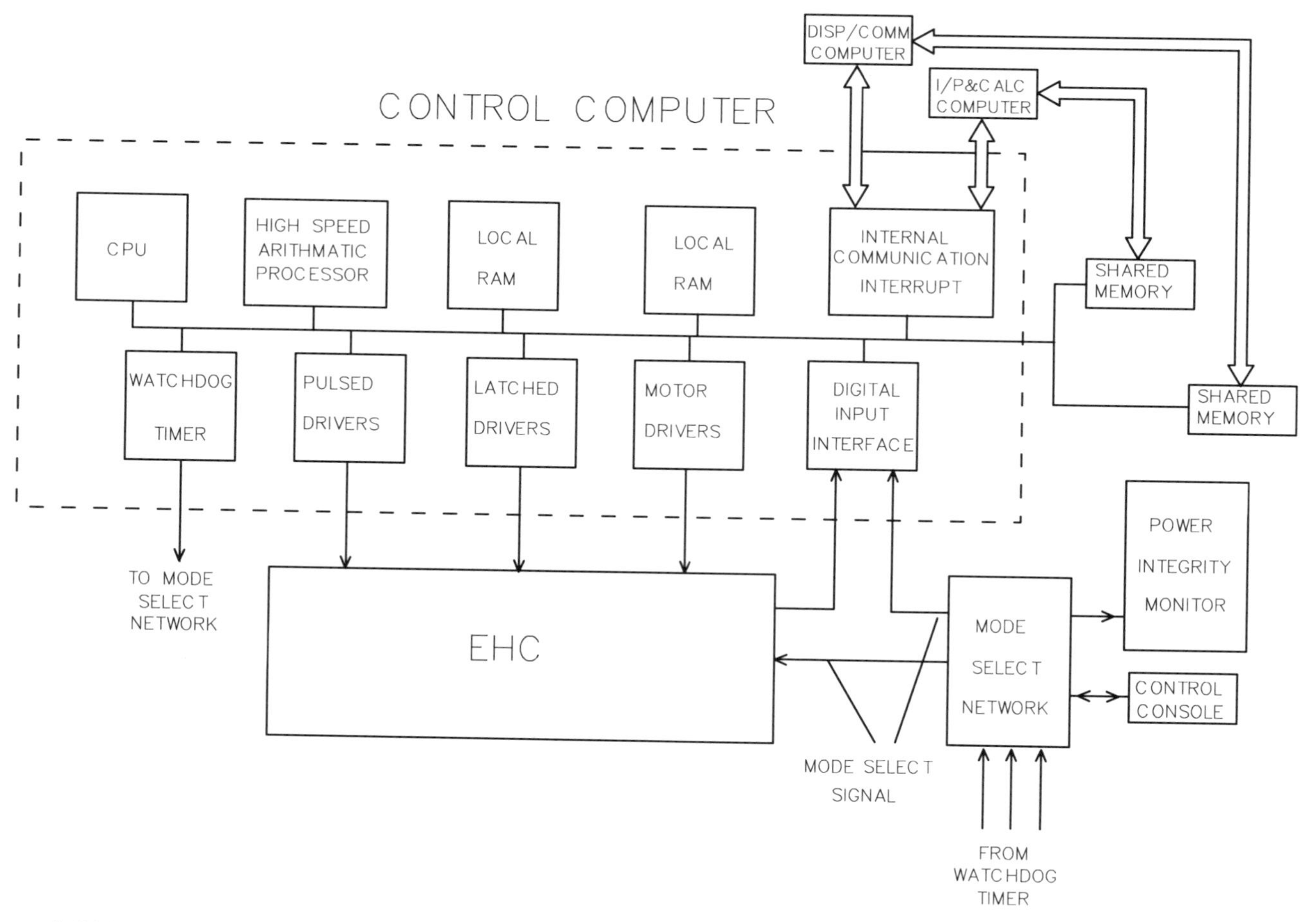

FIG. 13.30 TG control computer and EHC communication.

are in satisfactory condition. The progress of these and other steps is monitored by posting appropriate information to the operator through the MONITOR display. Once preparation for roll off is complete, the turbine is rolled free of the turbine gear (a motor-gear drive arrangement to turn the rotor during prewarming) and a first target rotor speed and an acceleration rate to reach the speed are selected. When the first preselected speed has been reached, the controller determines whether the turbine speed may be further increased or whether to hold speed until sufficient warming and reduction in turbine stresses have taken place. In any case, the controller directs the operation by selecting optimal speed levels and acceleration rates while maintaining acceptable levels of stress to turbine components until a speed is reached at which the TG can be synchronized to supply electrical power at the required line frequency. Other TG functions controlled or monitored by the computer-based supervisory control system include application of the generator field; initiation of synchronization of the generated power frequency to the line or power grid frequency; loading and unloading to and from a target power load; turbine admission mode selection whereby partial arc or full arc admission of steam is selected as a function of turbine operating conditions to provide the most efficient operation; and turbine stress analysis and control. The computers are programmed to take in data pertaining to TG operation, to process that data, to decide how the turbine should be made to respond, and to either automatically direct the EHC system or to provide proper information to an operator so that he can manually direct the EHC system.

The electrical power generating plant includes a TG set that is advantageously controlled by a dedicated microcomputer-based control system, according to the present invention. In the power plant, the boiler supplies high-pressure, high-temperature steam to drive the HPT, IPT, and LPT. The TGs are coupled in tandem to each other. Steam to the HPT is initially admitted through the MSSV and subsequently through a set of governor control valves GCV1 and GCV2. Although two control valves are mentioned for the purpose of describing the process, a plurality of stop and control valves is commonly used, with the control valves arranged circumferentially in a well-known manner in nozzle arcs around the inlet to the HPT, as depicted earlier. Such an arrangement of control valves effectively provides admission of steam to the HPT in either the partial arc mode of operation w the herein steam is admitted through fewer than all the control valves (GCV1and 2) or in the full arc mode wherein steam is admitted simultaneously through all the GCVs.

The speed and the amount of load of the turbine are dependent upon the quantity and condition of the steam admitted to the turbines through MSSVs, GCVs, and the intercept valves (ICVs for IPT). Speed and load control of the turbine generally are provided by the EHC. An EHC system is preferably of the type disclosed in US patent 3,097,488 to Eggenberger et al., a disclosure incorporated herein by reference thereto. It is an analog, feedback controller adapted to receive input information regarding turbine operation as from the speed transducer, the electrical load transducer, and by appropriately positioning the control valves of the GCV 1 and 2 in conjunction with MSSVs and ICVs to maintain turbine operation at desired load.

The EHC system is capable of standalone control of the turbine according to operator guidance in consideration of operating conditions and safety limits, and provides a means for steam admission mode selection and protective measures against such abnormal conditions as turbine overspeed, excessive temperature, and vibration.

A dedicated supervisory controller is provided to interact with the EHC system and give direction thereto for optimal turbine-generator performance under all operating conditions and during all operating phases. Supervisory control information thus given to the EHC system is determined by continuous measurements of TG operating parameters and a database of information related to other nonsensed TG parameters. The supervisory controller comprises a hierarchy of computer subsystems, including a control computer having interfacing capabilities with the EHC system; a display and communications computer unit; and an input and calculations computer unit. The distribution of functions between computers may be referred to herein as providing DCS. The control computer is the basic decision-making computer in the hierarchy, communicating with the display and communications computer and with the input and calculations computer through shared memory (dual port RAM) units. An analog input interface unit is a subsystem to provide signal conditioning, isolation, and analog-to-digital conversion for analog signals of turbine-generator operating parameters. Analog signals are adequately redundant and are obtained by direct measurements on the turbine or secondarily through the EHC system.

10.2.3 The Input and Calculations Computer (Fig. 13.31)

The input and calculations computer reads the input signals after they have been converted to digital format, validates the input signals by comparing them to maximum and minimum acceptable values and to companion input values, and converts the input signals to engineering units. Data thus taken in is retained until updated by the subsequent acquisition of data, and is supplied, as requested, to operating programs and subprograms, either within the input and calculations computer or the control computer.

The input and calculations computer also provides a means for calculating thermal and mechanical stresses to

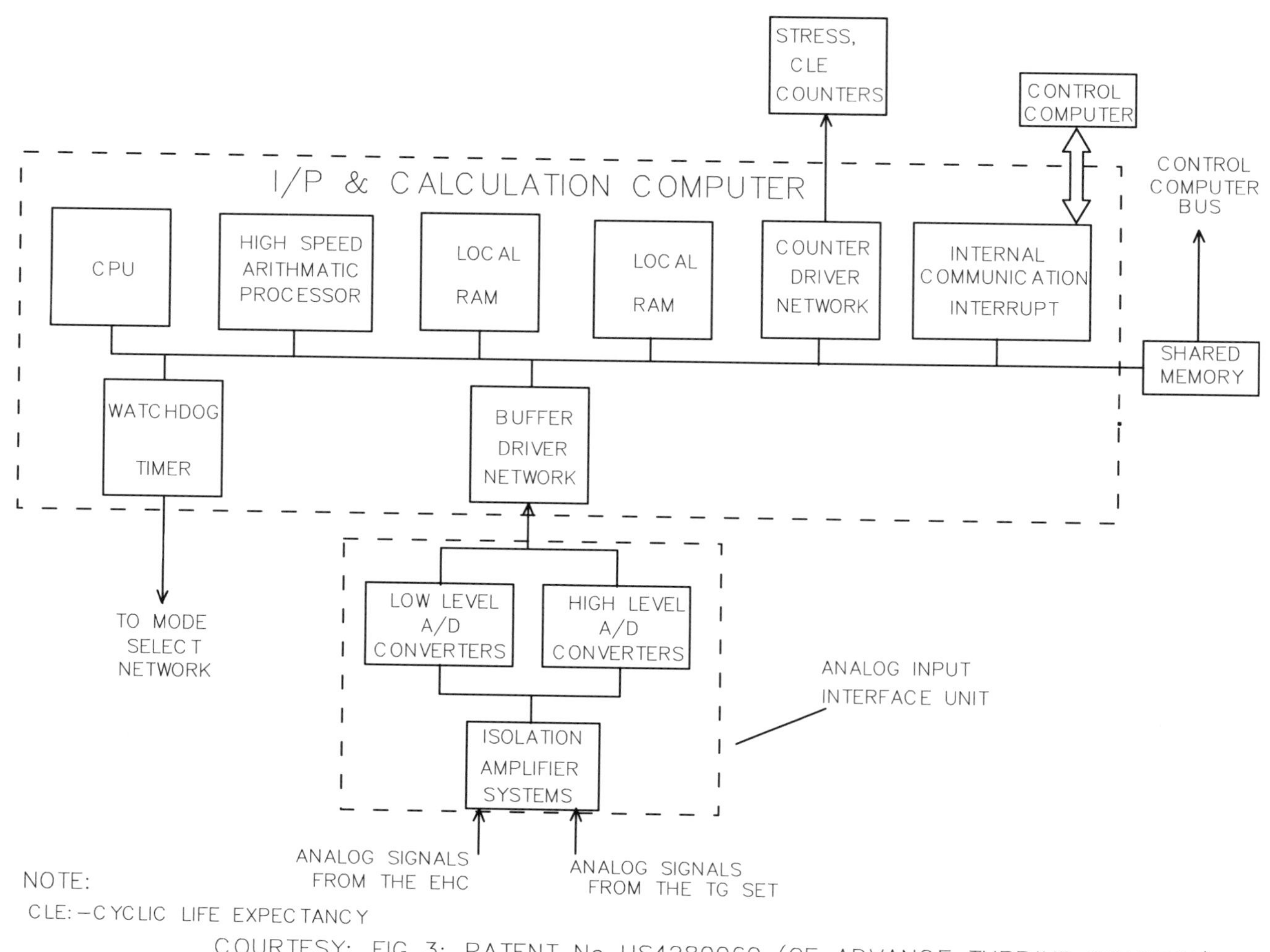

FIG. 13.31 TG control input and calculation computer system architecture.

turbine components, such as the turbine rotor and shell, and for supplying this derived information to the control computer. Based on the determined stress levels, the control computer provides direction to the EHC system, which has direct control of the turbine, so that stress is minimized. Stress is determined according to the teaching of Zwicky, Jr. in US patent 3,446,224, and according to subsequent improvements in the art, including the teachings and methods of US patent 4,046,002 to Murphy et al. and US patent 4,104,908 to Timo et al. The useful life of a turbine component part is affected by the unavoidable cyclic stresses that occur as a result of the cyclic heating/cooling/centrifugal loading that occur during start up/load changes/shutdowns, and sudden changes in steam conditions, the input and calculations computer determines the amount of life expended during these stress cycles for predetermined turbine parts. The values may be expressed as a percentage of life expended for the stress cycle and are referred to as cyclic life expenditure, or CLE. The life expended for each stress cycle is accumulated to provide an output indicative of CLE for the particular turbine part (e.g., the rotor) according to the part's physical properties and geometry. That information is stored within the permanent memory of the input-calculations computer. The CLE is displayed to operating personnel by display devices interfaced to the input and calculations computer.

Furthermore, the input and calculations computer takes into account the turbine rotor material of construction and the behavioral characteristics of that material above and below the fracture appearance transition temperature (FATT), which is the boundary temperature between brittle and ductile behavior of the rotor material. Material at lower temperatures is relatively more brittle whereas at higher temperatures, the ductility is increased. Certain stress levels occurring below the transition temperature may be undesirable while the same stress levels above the transition temperature may be acceptable. Hence, the transition temperature divides a stress versus temperature plot into brittle and ductile regions, which are further divided into zones of potential risk of permanent rotor damage. The input-calculations computer provides for a comparison between the instantaneous/actual rotor stress and allowable rotor stress, and accumulates data in separate counter registers, respectively scoring incidents in the brittle and ductile regions. Both the stress determining and calculating methods are programmed into the input/calculations computer and are made according to the teachings of the above US patents.

The input-calculations computer also includes a watchdog timer, a counter driver network, and buffer/driver network. The watchdog timer monitors the performance of the computer and in the event of failure, provides a signal indicative thereof so the control system can automatically be put into a safe operating mode (monitor mode). This computer is periodically put through a test according to its programming, and unless satisfactory results are received by the watchdog timer before a preselected time-out period expires, the failure mode is selected. The counter driver network is an interface network accepting digital data relative to CLE events and to high-stress events categorized with respect to the FATT. It transfers that data to stress and CLE counters and these high-stress events and fractional life expenditures are accumulated and displayed. The input-calculations computer determines the stress data in accord with the program, operating upon sensor infomation brought in from the TG through an analog interface. The counter driver network preferably comprises a buffer and shift register.

10.2.4 Display and Communications Computer (Fig. 13.32)

The display-communications computer is an input/output subsystem interfaced to an operator control panel. The functions are vividly clear from the figure and include these features: (i) It allows the operator to interact with the control

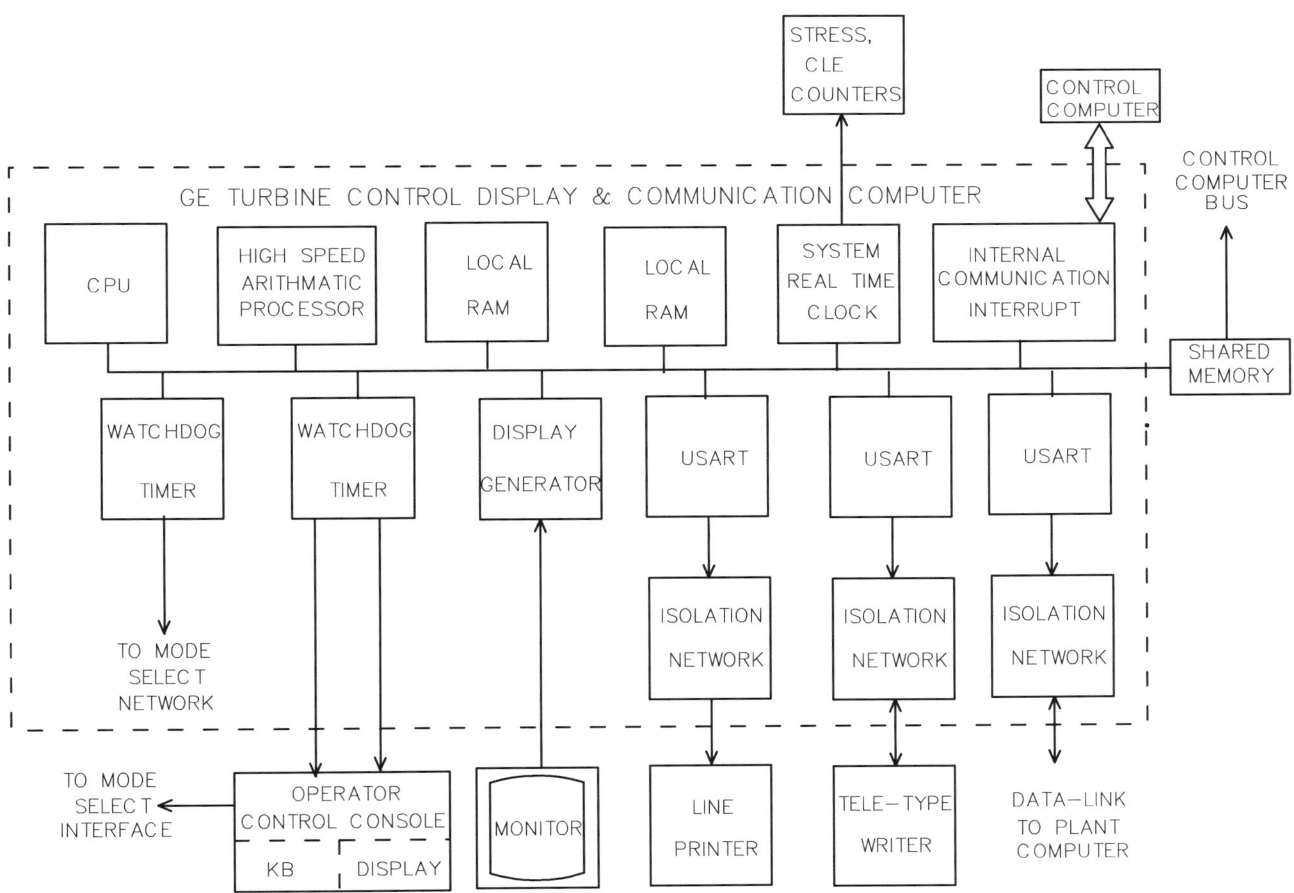

FIG. 13.32 TG control display and communications computer system architecture.

system, (ii) it enables the printer units to provide a permanent record of data and messages printed out from the control system, (iii) it provides displays in display units, which presents messages/requests to the operator, and (iv) it provides a data link to a plant computer whereby, in one operating mode of the controller, the plant computer provides input commands to and receives progress reports from the control system. In this mode, the plant computer uses the control system as a subsystem in the overall plant control. However, it is to be noted that the plant computer is not programmed to duplicate the functions of the control system.

10.2.4.1 The Analog Input Interface

The analog input interface provides isolation and signal conditioning and accepts the analog input signals pertaining to TG operation. The analog signals are the fundamental pieces of information upon which the control system operates to determine further, derived information or control parameters according to which the TG can best be operated. The analog input signals may be obtained directly, in which case the sensing devices, such as T/Cs and RTDs, are connected directly to the input interface. Alternatively, the analog signals may be obtained indirectly, and the analog signals brought to the input interface via the EHC system. Analog input signals include:

(i) Temperature measurement for control valves inner/outer surfaces, steam crossover chamber, RH bowl, high-pressure shell, lube oil, and main steam. (ii) Pressure measurement for RH, MS chest and turbine rotor speed, generator speed, control valve position, load set motor load level, and admission mode select motor in admission mode.

In the plant computer mode of operation, the supervisory controller is used in conjunction with a large, external mainframe computer. In the computer mode, the controller either supplies data to the plant computer regarding turbine operation, or receives from the plant computer inputs that operating personnel would otherwise supply. Examples of such inputs include target loads, allowable CLE, and operating holds.

10.2.5 Different Operating Modes of TG Operation

TG operation can be performed from various places, for example, the turbine control computer console, the EHC console, the plant computer console ,or the load dispatch system.

The area covered in TG subsystems is indicated below.

10.2.5.1 Operating Modes of the Supervisory Controller

This modes of the controller include:

- **(i)** Monitor mode: EHC takes the charge of controlling TG operation.
- **(ii)** Control mode: Supervisory controller takes charge of TG operation.
- **(iii)** Remote automatic mode: Control command shall be from the load dispatch system or the coordinated control system.
- **(iv)** Plant computer mode: In this mode of operation, the supervisory controller is used in conjunction with a large, external mainframe computer. In the computer mode, the controller either supplies data to the plant computer regarding turbine operation, or receives from the plant computer inputs that operating personnel would otherwise supply. Examples of such inputs include target loads, allowable cyclic life expenditures, and operating holds. The exchange of information between the supervisory controller and the plant computer is solely via data link.

These modes are a result, principally, of the programmed coordination of the separate microcomputers of the supervisory controller, but certain hardwired items, including the mode selector and the mode selection switches of the control panel, are necessary for implementation.

10.2.5.2 Operating Modes of the EHC

The EHC is of the type having the following modes of operation: (i) manual mode, (ii) supervisory remote mode, (iii) remote load control mode for load control by an automatic dispatching system (ADS) or a coordinated boiler control system (CBC), and (iv) Standby mode.

Mode selection in the EHC is compatible with mode selection in the supervisory controller and the selection of incompatible modes is inhibited. The interrelation between the supervisory controller and the EHC is depicted in Fig. 13.30.

Because the EHC has direct control of the TG, the activation of a particular mode (as by operator selection) within the EHC prevails over mode selection in the supervisory controller. For example, changing the EHC mode from remote to manual forces the mode of the supervisory controller to change from a control mode to a monitor mode. It may be noted that signals indicative of mode or status of the EHC system are generated within the EHC system and presented to the supervisory controller through a digital input interface unit. The control computer handles the status of signals as programmed and places the supervisory control system into a mode that is compatible with the EHC system.

The selection of an overall operating mode ordinarily begins by making a selection on the EHC controller. With the EHC system in any of the manual/remote load control/standby modes, the control is conventional and the only mode available to the supervisory controller is the monitor mode. In this mode, the supervisory system will guide an operator through all phases of turbine operation, providing information on turbine operating conditions, alarming those conditions that become abnormal, and generally providing the operator with information so that he can set the EHC controller for the most efficient and economical TG performance.

With the supervisory controller in any of its remaining modes, that is, the control/remote auto/plant computer mode, all modes of the EHC system except the supervisory remote mode are inhibited. In the control mode (selectable through the control panel switch), the supervisory controller assumes control of the TG set so that only minimal intervention is required from an operator in automatically starting and loading/unloading to and from a so-called target load. Following synchronization of the generator to the power line, and having reached the target load, the turbine load control can be turned over to a centralized load dispatch system such as ADS or CBC. Alternatively, inputs can be accepted from a plant computer to provide coordination of the controlled TG with all other plant equipment, including other TG sets.

10.2.6 Working Procedure of Control Computer through EHC

As shown in Fig. 13.30, the control computer directs/controls the EHC system through pulsed/latched/motor drivers, each of which contains the required number of circuits to provide a complete set of output signals as necessary for the EHC system.

10.2.6.1 Function of Drivers

The pulsed drivers provide output pulses of sufficient power/time duration to cause the operation (e.g., increasing/decreasing/latching) of devices such as relays located within the EHC system to increase/decrease set points, for example, for turbine speed/acceleration rate control.

Latched drivers provide outputs for on/off operation of devices within the EHC system such as indicator lamps, which require a sustained application of power.

Motor drivers provide outputs for driving set point motors within the EHC system, such as those for setting the turbine load or for selecting the steam admission mode.

10.2.6.2 Function of Digital signals

The control computer is apprised of the operating status through the digital input interface. The status of the EHC system includes a remote control mode so that supervisory control from the control computer can be effected. The digital input interface also accepts digital signals from the mode selector through which the operating mode of the supervisory control system is effected. The mode selector accepts signals (from each watchdog timer of the system) that are indicative of the corresponding computer's status. In the event of a computer malfunction detected by any one of the system's watchdog timers, the mode selector directs the EHC system, the control computer, and the entire system into a safe operating mode. The mode selector is interfaced to the control computer bus through a digital input interface and is also in two-way communication with the operator control console so that the operating personnel can effect the operating mode changes and be apprised and announced of those changes mandated by the mode selector.

10.2.6.3 Function of Power Integrity Monitor

The power integrity monitor provides continuous system power supply monitoring (of any impending source failure) and alerts the mode selector to respond by sending signals to the EHC system and supervisory controller through the input interface to force both into safe operating modes.

10.2.6.4 Function of Control Console (Fig. 13.33)

The control system structure has multiple operating modes that are coordinated with various operating modes provided on the associated feedback control system, such as the EHC, for example.

10.2.6.5 Keyboard and Display

The operator interacts with the supervisory control system through a control console that includes an alphanumeric display by which the operator commands and other data entered through the keyboard for program control are displayed and corrected (if necessary) prior to entry into the supervisory control system. The keyboard includes a cancel switch by which a displayed quantity may be cancelled prior to being entered. It also includes a manual override switch to allow a program hold (which may be related to a turbine operating parameter) to be overridden and an enter switch to transfer displayed values into the supervisory control system. Affirmation by the operator that the turbine is properly conditioned to be accelerated is expressed through a continue switch. In essence, the continue switch provides an override of a halt built into the turbine start-up routine to prevent the turbine from being accelerated off, turning gear without operator acknowledgement.

The control panel further comprises a bank of indicator-selector switches that allow one of the various operating modes of the supervisory controller to be manually chosen, a push button for a lamp test for all other indicator lamps on the console, and a malfunction indicator (if any within the supervisory control system). For selection of a target load and a loading rate to reach the selected target load, a target load switch and a rate limit switch are provided. These switches alert the supervisory controller that either a target load or a rate limit, as appropriate, is to be selected. The selection is then made through the keyboard, display unit, and enter switch.

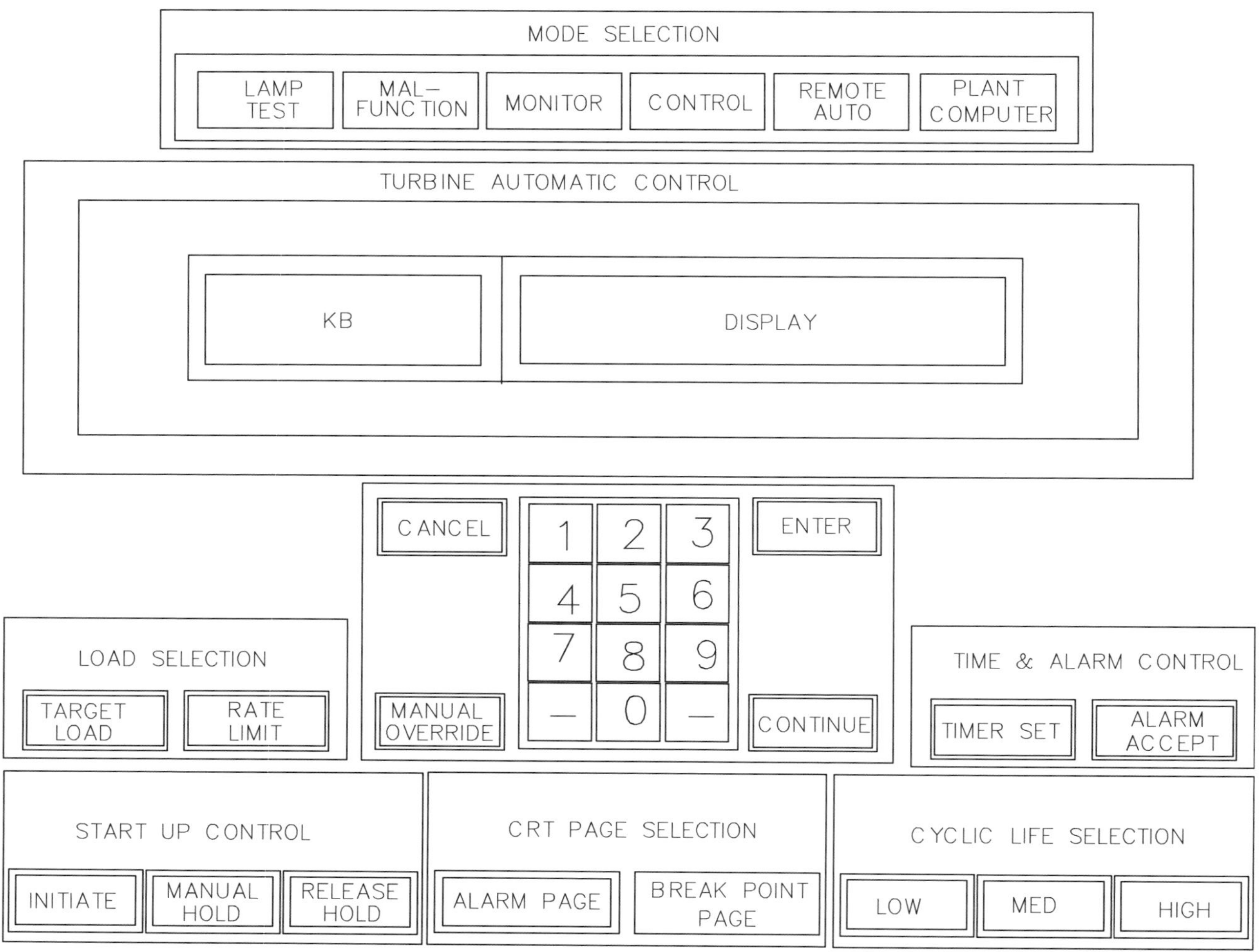

FIG. 13.33 Turbine control console layout.

10.2.6.5.1 Start-Up Controls Start-up controls include the initiate switch, which is used to initiate a turbine start-up sequence, the manual hold switch to impose a hold on the turbine start up, and the release hold switch.

10.2.6.5.2 Monitor Page Selectors Monitor page selectors include an alarm page and a break point page switch. These switches provide for changing the "page" of information displayed on the monitor. The alarm page switch is actuated to bring to the monitor screen a listing of parameters being monitored for alarm purposes and to show the status of those parameters. The alarm page may be changed to show a different set of alarm parameters by continued actuation of the alarm page switch/keyboard/display unit/enter switch. The break point page switch, on the other hand, changes the monitor display so that information is presented that pertains to a particular operating phase of the TG set, that is preparation for roll off.

10.2.6.5.3 Cyclic Life Expectancy (CLE) Selectors Switches for selecting the allowable expenditure of turbine rotor life during nonsteady state operating phases of the turbine include low/medium/high selector switches. This set of switches provides for manual selection of stress limits that may be imposed on the turbine during operation in which cyclic stress will occur, for example, during a turbine start up.

10.2.6.5.4 Time and Alarm Control Time and alarm control switches include a time set switch and an alarm acknowledge button. The time set switch sets the time frame of the supervisory control system to synchronize with the actual time of day so that the data reporting time from the controller is accurately made. The alarm acknowledge switch allows the operator to recognize the alarm to the controller.

10.2.6.5.5 Mode Selectors The mode selector bank comprises a monitor switch, a control switch, a remote auto switch, and a plant computer switch. These switches allow the controller operating mode to be manually selected and indicated. This control console is preferably located near the control console of the EHC system so that an operator is in close touch with both the control system and the EHC system.

10.2.7 Automatic Turbine Start-Up Procedure

10.2.7.1 The Prewarming Step

A start-up sequence is initiated from the operator control panel by initiating a dedicated switch. The control system proceeds then in logically arranged steps, beginning with rotor prewarming. During the rotor prewarming step, the supervisory controller determines the turbine rotor bore temperature at three locations, announces these temperatures to the operator, and indicates whether rotor prewarming is required before turbine roll off can take place. The prewarming step is essentially a manual operation with the controller providing guidance. Its purpose is to assure that the rotor bore material has enough ductility for the centrifugal stresses that occur as the rotor accelerates. Minimum temperatures at three locations within the turbine must be reached and the rotor shell sufficiently warmed before the start up can proceed further. Progress of rotor prewarming and other phases of the start up are monitored and described on the monitor.

10.2.7.2 The Chest Warming Step

Next, a determination is made as to whether chest warming is required. If so, the operator is advised by an appropriate message on the monitor. Once the rotor is up to a safe operating temperature, a step is provided for determining whether the steam chest of the GCVs requires a delay for warming and pressurization to take place. Then the operator is alerted by messages in the appropriate step, followed by the necessary delay step. A chest warming consists of two phases: control valve chest pressurization and heat soaking. In the pressurization phase, a determination is made, based on the differential temperature (dT) between the steam and the valve chest outer wall, as to whether the valve chest can be pressurized quickly or slowly; the pressurization proceeds at a rate that prevents excessive dT. When the chest pressure reaches 85% of the MS pressure, the heat soaking phase of chest warming begins. In this phase, a gradual warming at pressure is allowed until the dT between the MS and valve chest outer wall has decayed sufficiently to prevent excessive dT during turbine acceleration. With these conditions satisfied, the control valve chest is ready for turbine roll off and is announced upon satisfactory chest warming.

10.2.7.3 The Roll-Off Step

When satisfactory chest warming and rotor prewarming have been achieved, the next step is preparation for roll off. However, either the plant computer or the operator may, at any point, impose a hold on the start-up procedure. The operator-imposed hold is by a manual hold switch located on the control panel. The hold is removed by a release hold switch. As preparation for roll off begins, the operator will be requested by the controller to select an allowable level of CLE for that particular start up. The operator selection of CLE is expressed through high, medium, and low cyclic life selection through individual switches. Preparation of the turbine for roll off includes checks for (i) validity of the calculations made, (ii) assuring that all control equipment is properly set for automatic start up, (iii) the generator field is adequately warmed, (iv) the boiler steam is of satisfactory condition/ steam enthalpy is sufficient, (v) there are no remaining excessive temperature mismatches within the turbine, and (vi) operator overrides exist.

Once these conditions are satisfied and the control system is free of unacceptable alarms, the turbine is ready for the acceleration step, which provides control of the speed and acceleration rate of the TG in accordance with the prewarming requirements and thermal stress level limitations. Acceleration rates are dictated by thermal stress levels at the surface of the HPT rotor. Interim speed holds are provided before reaching the target speed so that heat soaks can be used to reduce rotor thermal stress. Such speed holds are also dictated by steam to metal temperature mismatches to limit thermal stresses from anticipated changes in heat-transfer coefficients. Stresses and bore temperatures are calculated by subprograms executed by the input-calculations computer and the results are provided to the control computer in performing the acceleration step.

If the results of these checks are satisfactory, the TG rotor is rolled free of the turning gear by increasing the steam admission and a first target speed and acceleration rate are dictated to the EHC system by the supervisory controller. A determination is automatically made, after reaching first target speed, as to whether to proceed to a second higher target speed or to hold momentarily for sufficient warming of the turbine and to ensure that stress reduction has occurred. Then, intermediate target speeds and acceleration rates are selected and set until synchronous speed has been reached.

10.2.7.4 The Synchronization Step

Once the turbine has reached at least 98% of its target speed at a time prior to reaching synchronous speed, the generator field excitation system is notified and applied in the next operation in the start-up subprogram. The excitation system returns a signal upon achieving the matching between the generator output and the power distribution system voltage to which the generator is tied so that the start-up task can proceed to the synchronization step. This step provides further checks of the turbine speed and makes determinations as to whether the line speed matching apparatus (in the EHC system) and the automatic synchronizer are in service. If these items are not in service, messages are given to the operator and he may override holds in the subprogram that occur in such cases. When the turbine is at line speed, a supervisory controller announces to the operator that synchronous conditions are achieved and holds until synchronization has been achieved by the operator or by an automatic synchronizer activated by the supervisory controller.

10.2.7.5 The Loading of the Turbine

With the TG synchronized, the control computer of the hierarchy directs its attention to the first step of the loading subprogram that is to determine whether the turbine is minimally loaded above a minimum load, for example, 2% of the rated load by increasing the load setting through a load set motor in the EHC system. The time for which the load set motor is to run and therefore how much the load is to be increased is first calculated in the first step. The calculated time is a function of steam pressure, minimum load, and load set motor speed. Following the increase to a minimum load, there is a program step for determining whether the turbine start up is under hot or cold conditions. For a cold start, there is a delay prior to selecting a load reference value in the next program step. For hot start conditions, the load reference is set to a minimum load value (2%) in the next step. The load reference in either case is used in calculations to determine the time duration for pulsing the above-mentioned load set motor. Just after the minimum turbine loading, further loading action is either held there or to advance toward a higher target load at an optimum rate as determined by turbine temperatures/rotor stresses. The target load and maximum allowable loading rate are selected by the operator through the target load switch and the rate limit switch.

Having set the load reference for either a hot or cold start up, there follows a step to calculate the optimal loading rate for the turbine. This step is actually a subprogram that provides a loading rate such that turbine rotor stresses are maintained within limits. The rate of change of stress and of steam temperature as well as their instantaneous values are used in the calculation. This permits faster and more uniform loading of the TG. The subroutine of this step also includes the calculation of an initial loading rate, which is used only during the first part of a start up to avoid inappropriately high calculated rates due to initially low rotor stresses. The calculated loading rate is used in the next step to determine whether the present load is acceptably close to the target load. The criterion is that the present load be within a small percentage of the target load. If the criterion is satisfied, the operator is notified by message and the loading subprogram is complete. On the other hand, if the present load is not sufficiently close to the target load, the program checks for various holds in the next step and either holds as desired or proceeds to calculate a new loading rate. Examples of holds that may occur include operator-imposed holds are for (i) generator warming, (ii) valve chest wall temperature differences, (iii) low rotor bore temperature, (iv) excessive rotor expansion, and (v) excessive main steam pressure.

Appropriate load hold messages are provided to the operator at this step. After the holds are taken care of, with a newly calculated loading rate, there is next provided a step to calculate a time during which the load set motor (located in the EHC system) will be pulsed to a new load setting. The time calculated and the speed of the motor are determinative of the new load setting. The calculation of running time is based on the calculated optimal loading rate from the previous step of the subprogram and on the load reference as set in previous steps. As part of the operation to determine a motor run time period, the load reference is increased by a fraction of the calculated loading rate and a new load reference is used in the next calculations. With the calculated time set, the next operational step is to pulse the load set motor for that time. The program then returns through the first loading rate calculation step and the decisional step, being routed there to repeat the above steps until the target load is attained with sufficient exactness.

10.2.8 Admission Mode: Full Arc or Partial Arc

The procedure has been described earlier in Section 10.1.

Once the target load is attained, the operator can turn the load control of the turbine over to a central load dispatch system/coordinated boiler control system by switching to the remote automatic mode where the supervisory controller remains as a monitor and retains control of the steam admission mode and other control parameters to arrest turbine overstressing.

10.2.9 System Architecture of Supervisory Controller (Fig. 13.34)

The supervisory controller is a complete system in charge of TG control and supervision. Functions carried out and software impregnated in the individual computer are shown in Fig. 13.34.

11 FUEL CELLS AND INTEGRATED GASIFIER FUEL CELLS

Today's state-of-the-art ultrasupercritical (USC) coal-fired power generating units can achieve a net energy efficiency of around 45% (LHV). Intensive R&D activities are ongoing to develop AUSC technology that uses steam temperatures of up to 700°C and it is expected to achieve an energy efficiency of around 50% (LHV, net). However, the maximum efficiency of a Rankine cycle is restricted by the Second Law of Thermodynamics. The actual

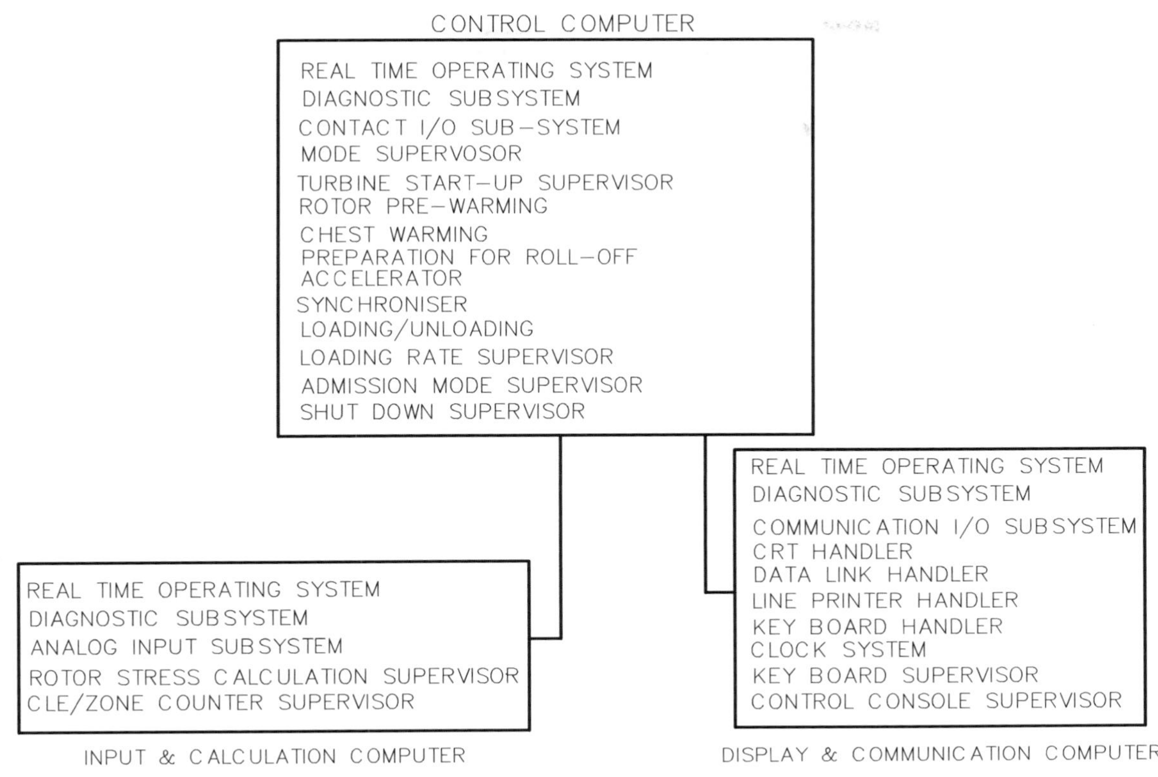

FIG. 13.34 Turbine supervisory controller software architecture.

Rankine cycle efficiency is limited to below the Carnot efficiency. Given the present status of the technologies, any further substantial increase in efficiency will be difficult and at high cost. So, the researchers have turned to the development of alternative systems, that is, unconventional power-generation concepts.

Some of the studies focus on improving the basic power cycle, for example the integrated gasification fuel cell (IGFC) concepts, the chemical looping concept, and a renewed look at MHD or magnetohydrodynamic and indirect coal combustion GT cycles. Other analyses seek to replace the working fluid with one that reduces parasitic losses intrinsic to using water. Systems based on a supercritical CO_2 Brayton cycle have been the subject of a number of studies and R&D. Bottoming/topping cycles are being studied as a means to extract additional energy from the process.

Fuel cells are electrochemical devices that convert chemical energy in fuels into electrical energy (and heat) directly. Also, because combustion is avoided, the emission of pollutants from fuel cells is minimal. Therefore, fuel cells can produce power with high efficiency and low environmental impact. As such, fuel cells are considered one of the most promising technological solutions for sustainable power generation.

11.1 Principle of Operation and Basic Structure of Fuel Cells

Fuel cells are electrochemical energy conversion devices that convert the chemical energy in a fuel and oxidant directly to electricity without direct combustion. They can be thought of as "gas batteries" where the electrochemically active materials are gases that can be ducted to the electrodes from outside the battery case. The reaction products are also gases and can be removed similarly. A fuel cell can be "discharged" continuously to produce electricity so long as the reactants are supplied and the products removed.

The fuel cell has many of the same features as a battery. The power production takes place at a constant temperature, so it is not constrained to the theoretical upper limit for heat engines (Carnot cycle efficiency). Therefore, it potentially can be much more efficient than combustion-based systems. The electrochemical reactions do not involve direct combustion, so thermal NO_x production is negligible. Reactants are consumed exactly in proportion to the electric energy output, so the efficiency remains high even when the level of power production is reduced.

In practice, fuel cell system efficiencies remain limited by the unavoidable energy losses and inefficiencies inherent in most engineered systems. Continued R&D is aimed at reducing these losses to improve overall efficiency. However, with the use of combined heat and power (CHP) with SOFC, a total cogeneration efficiency of about 90% is achievable. An attraction of fuel cell systems is that natural gas or coal-derived fuel gas both make suitable fuels for running a fuel cell Fuel cells as an energy conversion system stem primarily from the fact that they offer the highest efficiency and lowest emissions than that of any known fossil-fuel power generation technology.

Typically, a fuel cell consists of three main components: an anode, a cathode, and an electrolyte that is in contact with the anode and cathode on either side. The basic structure of a fuel cell is very simple: a fuel and an oxidant (often O_2 from air), supplied from external sources, are introduced to the anode and cathode side, respectively. The motive force of the operation is the chemical potential gradient of ions across the electrolyte. Fuel cell electrolytes are electronically insulating but ionically conducting, allowing certain types of ions to transport through them.

There are different kinds of fuel cells discussed later. Mainly, solid oxide fuel cells (SOFC) are used for large power and heat generation.

11.1.1 Fuel Cell through Migration of Positive Ions (Fig. 13.35)

One type of fuel cell envisages positive ions passing through the electrolyte from the anode to the cathode.

Hydrogen in fuel is channeled through a field flow plate to the anode on one side of the fuel cell and oxidant (O_2/air) is channeled to the cathode on the other side of the fuel cell. At the anode, the platinum catalyst causes the H_2 to split into positive hydrogen ions (protons) and negatively charged electrons. A polymer electrolyte membrane (PEM) allows only the positively charged ions to pass through it to the cathode. Negatively charged electrons are to pass through an external circuit (load) to the cathode creating an electric current. At the cathode, the positively charged H_2 ions and negatively charged electrons combine with oxygen to form water, which comes out from the cell.

11.1.2 Fuel Cell Through Migration of Negative Ions

The other version is the transmission of negatively charged ions, which depends on the selection of the electrolyte, anode, cathode, and catalyst utilized in them. The electrochemical reduction of the O_2 takes place at the cathode to form oxide ions (O_2^-) that migrate through the electrolyte, to the anode, and oxidize the fuel (H_2 in this case). This releases water, heat, and electrons that flow around an external circuit and do useful work (Fig. 13.35). For practical application, fuel cells would be connected in a series of cells to obtain higher voltage/power. When single cells are stacked together to two more cell components of interconnect and sealant are required.

A variety of fuel cells differ from each other regarding power density and efficiency but the fundamental difference is the electrolyte used in the cells. There are five main fuel cell types: (i) a polymer electrolyte membrane fuel cell (PEMFC), (ii) an alkaline fuel cell (AFC), (iii) a phosphoric acid fuel cell (PAFC), (iv) a molten carbonate fuel cell (MCFC), and (v) a SOFC.

The type of fuel cell and the operating temperature range are primarily related to the electrolyte material, which broadly dictates the operating temperature range of the fuel

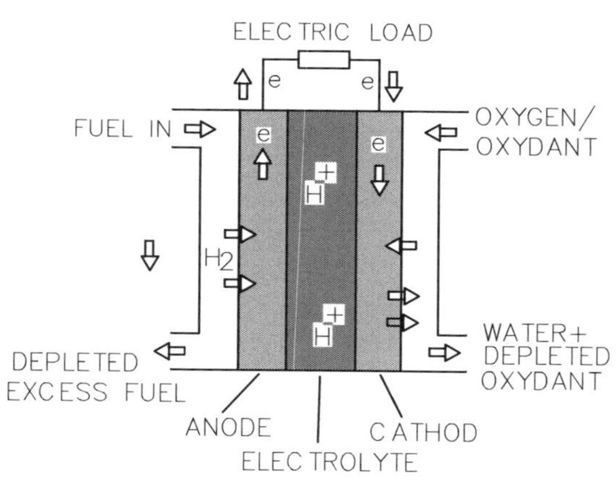

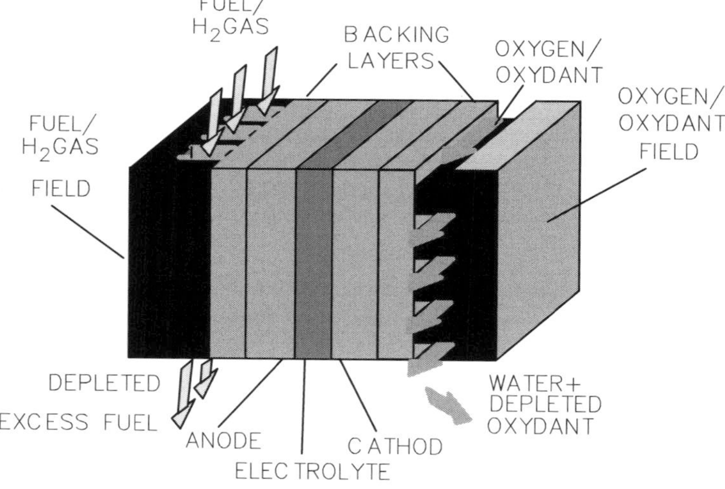

FIG. 13.35 Fuel cell operation and schematic diagram.

cell. The operating temperature and useful life of a fuel cell determine the physiochemical and thermomechanical properties of materials used in the cell components (electrode, electrolyte, interconnect, current collector). Aqueous electrolytes are limited to ≤200°C because of high vapor pressure and rapid degradation at higher temperatures.

AFC/PAFC/PEMFC are considered low-temperature fuel cells, whereas the MCFC and SOFC are high-temperature fuel cells. The operating temperature also plays an important role in deciding the degree of fuel processing required. In low-temperature fuel cells, all the fuel must be converted to hydrogen prior to entering the fuel cell. In addition, the anode catalyst in low-temperature fuel cells (mainly platinum) is strongly poisoned by CO. In high-temperature fuel cells, CO and even CH_4 can be internally converted to hydrogen or even directly oxidized electrochemically.

11.2 Fuel Cell Developments

The potential market for fuel cells is large, including base-load stationary power plants operating on coal or natural gas. Another opportunity exists in repowering older existing plants with high-temperature fuel cells. MCFCs and SOFCs coupled with coal gasifiers have the best attributes to compete for the large base-load market. A lot of effort has been made and the work is still ongoing to develop high-efficiency, low-emission, coal-fueled power systems using MCFCs/SOFCs. Recently, progress has also been made in the development of direct-carbon fuel cells (DCFC) that convert the chemical energy in carbon directly into electricity without the need for gasification.

11.2.1 *Solid Oxide Fuel Cells Operating Principles (Fig. 13.36A)*

The basic building block is an oxide-ion conducting ceramic electrolyte, operating at an even higher temperature (600–1000°C). SOFCs work on a solid nonporous metal oxide electrolyte

that is usually Y_2O_3-stablilized ZrO_2. The anode is typically a porous Ni-ZrO_2 cermet and the cathode is commonly a porous strontium-doped lanthanum manganite ($LaMnO_3$).

H_2 is normally used as fuel, but carbon monoxide (CO) can also be used as the fuel together with hydrogen. H2 and/or CO react with O_2– at the anode, releasing electrons, H_2O,

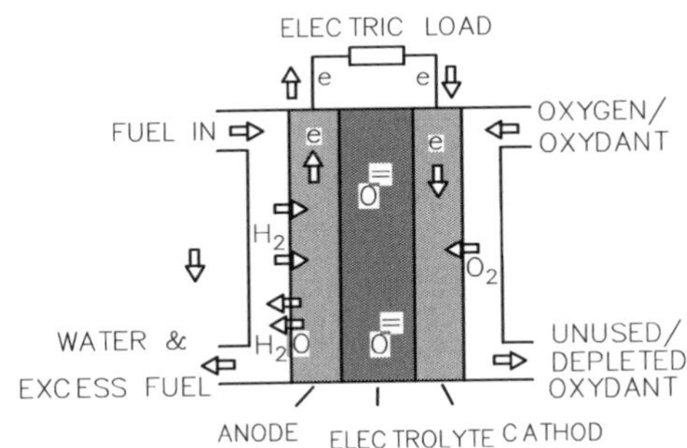

(A) SOFC OPERATION & SCHEMATIC DIAGRAM

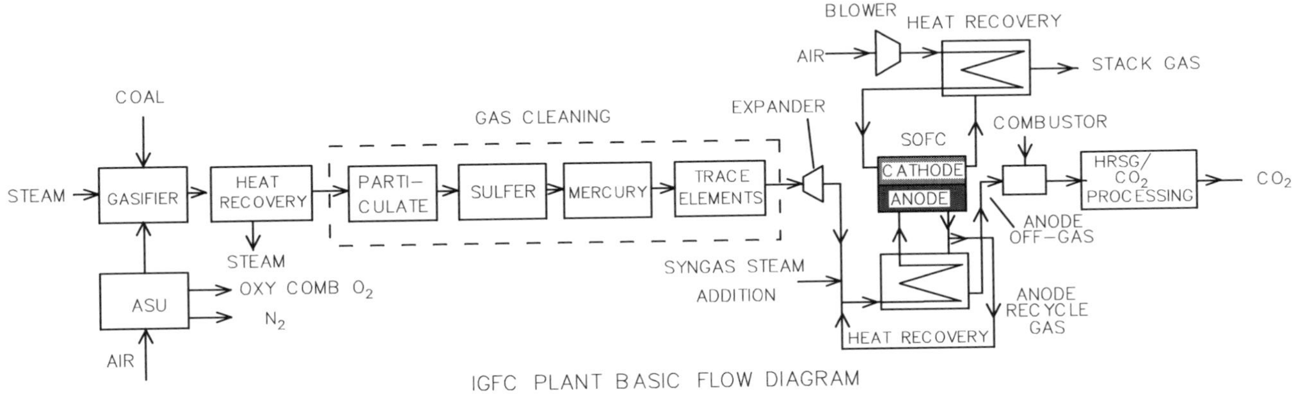

(B)

FIG. 13.36 Simple SOFC operation and schematic diagram.

and heat. The high operating temperature of SOFCs enables the direct oxidation of methane (CH_4). Consequently, the direct use of hydrocarbon gas instead of hydrogen or carbon monoxide is possible.

Advantages may be classified as follows:

- **(i)** High efficiency: The high operating temperature allows use of most of the waste heat for cogeneration or in bottoming cycles. Efficiencies ranging from around 40% (simple cycle small systems) to more than 50% (hybrid systems) are available.
- **(ii)** A system efficiency of 60% (HHV), including CO_2 capture >97%, may also be achieved with an advanced catalytic gasifier and pressurised SOFC.
- **(iii)** Carbon capture is easy because the anode (fuel) and cathode (air) streams are separated by the electrolyte. All carbon enters the SOFC with the fuel on the anode side and exits in the anode off-gas as CO_2. The residual fuel in the anode off-gas (approximately 10%–15%) can be combusted in oxygen, producing a stream that contains only H_2O and CO_2. Condensing out the H2O leaves an exhaust stream that contains mainly CO_2 ready for compression and storage.
- **(iv)** Fuel flexibility is optimum as SOFCs can operate on H2 and hydrocarbon fuels, including coal-derived syngas and natural gas.
- **(v)** A variety of shapes are available because the electrolyte is solid. The cell can be cast into different shapes such as tubular, planar, or monolithic.
- **(vi)** Minimum corrosion problems in the cell are achieved for solid ceramic construction of the unit cell. Solid electrolyte also avoids electrolyte movement or flooding in the electrodes.
- **(vii)** Future cost reduction is possible, the potential for which makes SOFCs attractive.

Disadvantages

- **(i)** High temperature of the SOFC has its drawbacks, such as causing severe constraints on materials selection and fabrication difficulties. Thermal expansion mismatching among materials and sealing between cells is difficult in the flat-plate configurations.
- **(ii)** Corrosion of metal stack components is a challenge and limits stack-level power density (though significantly higher than in PAFC/MCFC), thermal cycling, and stack life.
- **(iii)** Expensive, as the key fabrication issue is the use of costly chemical vapor deposition to fabricate these tubular components (Current research is focusing on simpler planar systems that show promise for less-expensive fabrication techniques.)

Recent developments

Early on, the limited conductivity of solid electrolytes required cell operation at around 1000°C. However, the high temperature imposes some limitations to SOFC, especially to the materials used. Currently, yttrium-stabilized zirconia (YSZ) is the most commonly used electrolyte for SOFC. The recent development in fabrication processes allows YSZ membranes to be produced as thin films (~10 μm) on porous electrode structures. These thin-film membranes improve the performance and reduce operating temperatures to 650–850°C, leading to the development of a compact and high-performance SOFC that utilizes relatively low-cost construction materials. Electrolytes made of other materials, such as scandium-doped zirconia, gadolinium-doped ceria, or cerium gadolinium oxide, are found to have higher reactivity at even lower temperatures but their applications are limited due to the availability and price of Sc and Gd as well as some technical problems. Improvements in electrolyte/cathode materials and designs increased the cell power density by 36%. An individual cell's active power-generating area has increased by more than a factor of 5 and the stack size has increased by a factor of 25 in recent years. Module stacks rated at ~25 kWe have been tested for more than 1500 h and a voltage degradation of less than 1%/1000 h has been observed (US DOE, 2013). M/s Fuel Cell Energy (FCE) developed a 5 MW plant in 2015.

11.2.2 Operating Principle of Molten Carbonate Fuel Cell (MCFCs)

MCFCs, using a molten alkali-metal carbonate electrolyte, operate at a temperature (~650°C) where rejected heat can be used efficiently in a steam or gas bottoming cycle and conventional materials still can be used for the balance of the plant (BOP) equipment. Efficiency ≥60% may be achieved in larger sizes, with natural gas used as fuel.

MCFCs use carbonate salts of alkali metals suspended in a porous ceramic matrix as the electrolyte. The high cell operating temperature keeps the alkali carbonates in a highly conductive molten salt form. The higher temperature makes the cell less prone to CO poisoning than lower temperature systems. Therefore, MCFC systems can operate on a diverse range of fuels, including coal-derived syngas, methane, or natural gas.

The electrode reactions for MCFC are:
at the cathode:

$$2CO_2 + O2 + 4e^- \rightarrow 2CO_3^{2-} \tag{13.1}$$

at the anode:

$$2H_2 + 2CO_3^{2-} \rightarrow 2H_2O + 2CO_2 + 4e^- \tag{13.2}$$

and overall:

$$2H_2 + O_2 \rightarrow 2H_2O$$

At the high operating temperature, the cell reactions proceed vigorously, and the nickel in the anode catalyzes the reaction between CO and steam, producing hydrogen and CO_2. In other words, CO can be used in MCFCs as a fuel. Natural gas needs to be steam reformed in the presence of a suitable catalyst to convert it into an H_2-enriched gas mixture by the reaction: $CH_4 + H_2O \rightarrow 3H_2 + CO$. The CO_2 generated at the anode is recycled to the cathode where it is consumed. This requires additional equipment to either transfer the CO_2 from the anode exit gas to the cathode inlet gas or produce CO_2 by combustion of the anode exhaust gas and mix this with the cathode inlet gas.

Advantages are indicated below:

- **(i)** No expensive electrocatalysts are needed as the Ni electrodes provide sufficient activity.
- **(ii)** Fuel flexibility as both CO and certain hydrocarbons are fuels for the MCFC simplifying the balance of plant (BOP).
- **(iii)** Improving system efficiency up to the low 50s.
- **(iv)** The high temperature waste heat allows the use of a bottoming cycle to further boost the system efficiency to >60%.

Disadvantages are the following

Some difficulties with MCFC technology put it at a disadvantage compared to SOFC.

- **(i)** Complexity of working with a liquid electrolyte rather than a solid.
- **(ii)** Chemical reaction inside the cell indicates exhaustion of carbonate ions from the electrolyte are used up in the reactions at the anode, making it necessary to compensate (usually by recycling the anode exhaust), representing additional BOP components.
- **(iii)** Higher temperatures means material/mechanical stability and stack life.

Recent developments

The performance of single cells has improved considerably in the past few decades. The power density of a single cell has increased from about 10 to >150 mW/cm^2, and the cell area has been scaled up by 50%. Stack performance improvement has been achieved in the areas of cell conversion efficiency, thermal management, thermal cycle capability, and high-power operation. Developments in advanced materials have resulted in extended stack life (>40,000 h) and reduced product cost. The cost of material for a bipolar plate has been lowered by a factor of 7 and advanced corrosion-resistant cathode current collectors have reduced corrosion (by a factor of 2) and electrolyte loss. The stack temperature distribution has been improved significantly, allowing 20% higher power operation of full-size stacks. This latest improved design is being incorporated into full-size stacks. The full-size stack capacity has steadily increased. Today, a single cell stack can produce up to 2.8 MW electrical power, has a stack life of five years, and is 9000 cm^2 in area. Work is ongoing at FCE to further improve the cell technology to increase the output to 3 MWe and increase the stack life to seven years.

The development of an internal reforming MCFC system eliminates the need for a separate fuel processor for reforming carbonaceous fuels. It integrates a reformer within a cell stack so that the heat generated by the cell reactions can be effectively used as the heat of reaction for fuel reforming. There has been a significant increase in the number of MCFC systems installed (from 2010 to 2011, the megawatts of MCFCs shipped annually increased almost six times), clearly indicating the commercialization of this technology.

11.2.3 Direct Carbon Fuel Cells

In a DCFC, the overall cell reaction is based on the electrochemical oxidation of C to CO_2. This reaction proceeds via mechanisms that vary with cell design and on electrolytes that are under development, which include solid oxide, molten carbonate, and molten salt. Depending on the electrolyte, O_2, CO_3, or OH ions participate in the oxidation-reduction reaction.

11.2.3.1 Molten Salt DCFC

The molten salt DCFC uses molten hydroxide (NaOH/KOH) at 500–650°C as the electrolyte in a metallic container acting as a cathode. Air is purged into the molten salt at the bottom of the container to supply O_2 at the cathode. Fuel is fed to the cell in the form of a rod made from graphite or coal-derived carbon dipped into the electrolyte. This fuel rod also acts as an anode of the cell and hence it runs as a battery and not as a fuel cell.

11.2.3.2 Molten Carbonate DCFC

Here, the fuel cell uses molten carbonates at 750–800°C as the electrolyte and fine particles of carbon dispersed into the electrolyte as the fuel. Mixed molten carbonates of lithium, potassium, and/or sodium are used due to high carbonate conductivity and good stability in the presence of CO_2. The ionic species that carry the charge between the electrodes are the carbonate ions (CO_3^{2-}).

The major technical issues related to this type of fuel cell are high cathode polarization losses, corrosion of metal clad bipolar plates, and scaling up. Furthermore, the fuel related issues include lack of a suitable fuel delivery system for long-term continuous operation, poor understanding of the relationship between carbon structure and its chemical and electrochemical reactivity, and electrolyte tolerance to high percentages of contaminants such as sulfur and ash.

Molten carbonate systems offer the most attractive near-term opportunities for utility applications. For the long-term, there is some risk that the MCFC manufactured cost will not decrease to the levels needed for widespread use in distributed generation and in coal-based IGFC systems.

For applications with coal, contamination of the fuel cell by trace coal constituents is the primary area of concern. Substances such as chlorides, sulfides, arsenic, alkali metals, zinc, cadmium, lead, and mercury vapors are capable of poisoning/reducing the fuel cell performance High-temperature purification systems can reduce some trace contaminants to very low levels. The current US fuel cell program is focused on the use of natural gas, though an IGFC system running on coal-derived fuel gas is envisioned by the DOE as a logical next step. Such systems offer the highest efficiency and lowest emissions of coal-based systems, but their cost currently is high.

11.2.3.3 Solid Oxide DCFC

This type of fuel cell uses O_2 ion (O_2^-) conducting ceramic (typically YSZ) as the electrolyte, similar to that in SOFC, and operates in a temperature range of 800–1000°C. This type of DCFCs has three subcategories of differing in materials and design of the anode and method of fuel delivery to the electrode/electrolyte interface: (i) carbon mixed with a molten metal, (ii) carbon mixed with a molten salt including molten carbonate, and (iii) solid carbon as fuel in a fluidized bed reactor.

11.2.3.3.1 Liquid Metal Anode SOFC Here, molten metal is used as the anode (resides in a layer between the fuel chamber and the solid electrolyte) and solid carbon as the fuel carrier. O_2^- ions react electrochemically with the liquid metal, generating metal oxide, which is the active species for the oxidation of C, producing CO_2. However, the exact mechanism occurring and the species involved in the liquid metal anode (LMA) media are not well defined and depend upon the metal used.

The molten metal blocks direct contact of the electrolyte with gaseous impurities and hence reduces electrolyte degradation. Furthermore, the fuel contaminants can become a fuel source themselves as they undergo electrochemical oxidation (Toleuova et al., 2013).

11.2.3.3.2 Solid Carbon in Molten Salt Here, a circulating liquid-molten salt/carbonate containing C fuel as the anode and O_2-ion conducting ceramic as the electrolyte. In one configuration, the cell employs a cathode-supported tubular cell geometry. Air is supplied via a concentric tube to the cathode consisting of a metal current collector and strontium-doped $LaMnO_3$ (LSM) as the catalyst layer. Circulating molten salt/carbonates mixed with C fuel particles are supplied to the anode, which also has a metal mesh/coil current collector. Various types of fuels such as biomass/coal/coke/tar have been tested on this cell. This type of fuel cell is a hybrid between molten carbonate and SOFC with similar materials issues (corrosion of Ni anode/other cell components, and stability of the YSZ electrolyte in molten carbonate environments) (Badwal and Giddey, 2010; Jain et al., 2008).

11.2.3.3.3 Solid Carbon as Fuel in a Fluidized Bed Reactor This technology is based on the direct electrochemical reaction between the solid carbon at the anode and oxygen ions (O_2^-) being transported through the ceramic electrolyte membrane from the cathode to the anode. The anode is in direct contact with the C particles, typically using a fluidized bed reactor. In the fluidized bed reactor, fine particles of carbon fuel are suspended by blowing in a nonreactive gas such as CO_2 through the bottom of the reactor for continuous fuel feed to the anode/electrolyte interface. A collection of unit cells is arrayed along the reactor. Mostly, the developmental work on this technology has so far been concentrated on button cells consisting of a ceramic electrolyte disk with a nickel-based anode and an LSM-based cathode. The major technical issues apart from those associated with SOFC are the solid fuel delivery to the anode/electrolyte interface, and a lack of understanding of carbon oxidation reaction mechanisms at the interface (Badwal and Giddey, 2010; Gur and Huggins, 1995).

11.3 Fuel Cell Power Systems

The fuel flexibility of MCFC and SOFC allows the syngas produced by coal gasification to be used to fuel the fuel cells. In addition, the high cell operating temperatures offer the best opportunity for thermal integration with the coal gasification systems. Various fuel cell power system configurations that can potentially achieve high energy efficiency and excellent environmental performance have been proposed and investigated.

11.3.1 Integrated Gasification Fuel Cell Systems

Fuel cells integrate readily with coal gasifiers. Such IGFC systems are potentially the most efficient and least polluting method to generate electricity from coal. The IGFC power plant is similar to an IGCC power plant, but with the gas turbine power island replaced with a fuel cell power island. Various IGFC power plant design concepts have been developed, generally consisting of three main parts: a gasification island, fuel cleaning/processing, and a power island. The power system configuration varies depending on the choice of technologies. Given the number of technologies available for gasification, syngas cleaning and processing, fuel cell systems, and waste heat recovery, a number of IGFC plant configurations have been proposed and studied.

In a recent study, Newby and Keairns (2011) analyzed four IGFC plant configurations. All the plants are designed for coal-fed base load operation with a net plant capacity of 500 MWe, use conventional dry syngas cleaning and polishing technology, apply advanced planar SOFC technology with separate anode and cathode off-gas streams, and incorporate anode off-gas oxy-combustion for nearly complete carbon capture (Fig. 13.36B, IGFC Plant Basic Flow Diagram (included in Fig. 13.36)). This can be described as follows:

(i) Plant 1 (baseline design): Like an IGCC plant, this consists of the coal receiving and storage area, the ASU, the gasification area, the gas cleaning area, the power island, and the CO_2 dehydration and compression area. An oxygen-blown, entrained-flow gasifier is selected. The power island consists of a syngas expander, the atmospheric-pressure SOFC unit with DC-AC inverters, an anode off-gas oxy-combustor, an HRSG, and a steam bottoming cycle.

(ii) Plant 2: Essentially the same as Plant 1, except a pressurized SOFC is utilized.

(iii) Plant 3: A catalytic gasifier is used to produce a syngas with higher concentrations of methane; the rest is like Plant 1 (Fig. 13.37A, Simplified Flow Diagram of Catalytic IGFC with Atmospheric SOFC, included in Fig. 13.37).

(iv) Plant 4: A catalytic gasifier is used; the rest is the same as in Plant 2 as depicted in Fig. 13.37B, simplified flow diagram of catalytic IGFC with pressurized SOFC, included in Fig. 13.37.

It has been experienced from this study that compared with conventional bituminous-coal-fired power plants with 90% CO_2 capture, the IGFC plants could achieve higher electrical efficiency with >98%–99% CO_2 removal. The emissions of other air pollutants from the IGFC power plants were also lower. The study also indicates that the introduction of a pressurized SOFC results in a substantial increase in the net plant efficiencies. However, using a pressurized SOFC provides little or no cost benefit over atmospheric pressure SOFC plants. The researchers claimed that an IGFC using an advanced catalytic coal gasifier and atmospheric-pressure SOFC would provide the greatest benefits, with the cost of electricity projected to be significantly lower than IGCC, PCC, and NGCC (natural gas turbine combined cycle), all with CCS.

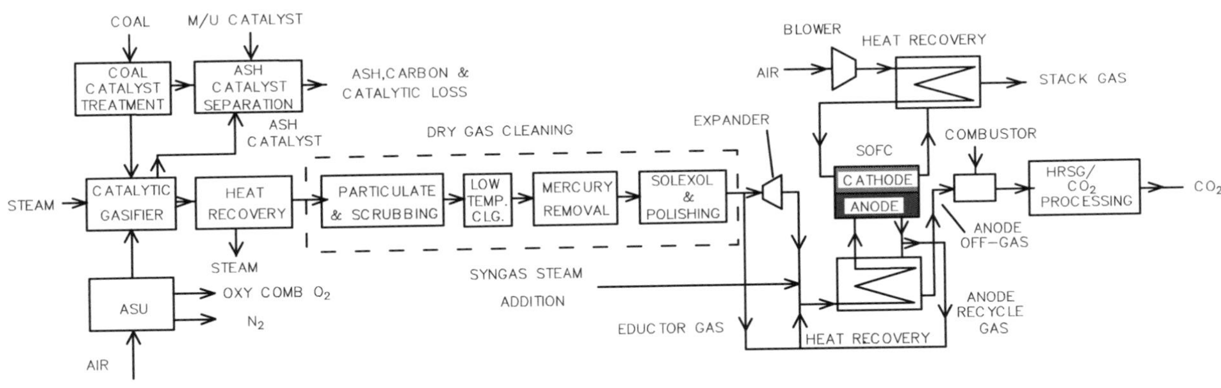

(A) SIMPLIFIED FLOW DIAGRAM OF CATALYTIC IGFC WITH ATMOSPHERIC SOFC

COURTESY: Simplified flow diagram of SOFC IGFC plant—atmospheric pressure SOFC IGFC plant (Gerdes and others, 2009)

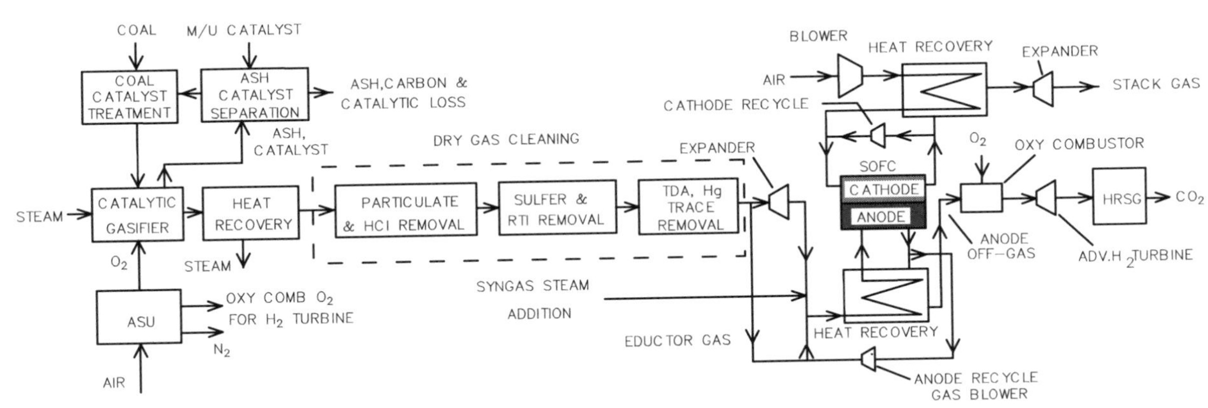

(B) SIMPLIFIED FLOW DIAGRAM OF CATALYTIC IGFC WITH PRSSURIZED SOFC

COURTESY: Simplified flow diagram of SOFC IGFC plant—pressurised SOFC IGFC plant (Gerdes and others, 2009)

FIG. 13.37 Flow diagram of catalytic IGFC plant with SOFC.

11.3.2 Typical Power Summary of a 550 MW IGFC Power Plant

The approximate scenarios of an IGFC plant using SOFC are given in Table 13.15.

The future development of coal-based IGFC systems will depend on the success of current gas-based technology and on the resolution of key technical issues, particularly the types and levels of contaminants in coal-derived fuel gas that must be controlled.

TABLE 13.15 Power Summary of a IGFC using SOFC

Sl No.	Power Generation	IGFC With Atmospheric-Pressure SOFC	IGFC With Pressurized-SOFC	Remark
A		**MWe (Approx)**	**MWe (Approx)**	
1	SOFC power	551	542	
2	Syngas expander power	36	7	
3	Steam turbine power	113	130	
	Total power generation	700.4	679	
B	**Auxiliary load consumption**	**MWe (Approx)**	**MWe (Approx)**	**Remark**
1	Coal handling/crushing	2.7	1.8	
2	Sour water recycle slurry pumps	0.157	0.125	
3	Ash handling	1	0.8	
4	ASU air compressor/aux	40.5	44	
5	Oxygen compressor	13	12.5	
6	CO_2 compressor	51	15.5	
7	N_2 compressor	0.64	0.5	
8	Anode recycle compressor	4.25		
9	Claus tail gas recycle compressor	1.0	0.8	
10	BFP	1.8	2.0	
11	Condensate pump	0.12	0.138	
12	Syngas recycle compressor	0.43	0.343	
13	Quench water pump	0.46	0.356	
14	CWP	2.0	2.2	
15	Ground water pump	0.44	0.35	
16	Cooling tower fans	1.4	1.0	
17	Selexol auxiliary power	2.8	2.24	
18	Cathode air compressors	10.3	37	
19	Cathode recycle blower	11.4		
20	Transformer losses	2.6	2.5	
21	BOP	2.7		
22	Claus/TGTU auxiliaries	0.164	0.131	
23	Miscellaneous loads	0.15	10.5	

TABLE 13.15 Power Summary of a IGFC using SOFC—cont'd

Sl No.	Power Generation	IGFC With Atmospheric-Pressure SOFC	IGFC With Pressurized-SOFC	Remark
24	Total auxiliary power consumption	150.4	129	
C	Net power output	550	550	
D	Net plant efficiency (HHV)	40%	50.1%	
E	Net plant heat rate kJ/kWh	8.993	7.187	
F	Raw water consumption m^3/h	6.4	4.5	

Source: Exhibit 3-10 Case 1-1 and Exhibit 3-29 Case 2-1 Plant Performance Summary (100 Percent Load); ANALYSIS OF INTEGRATED GASIFICATION FUEL CELL PLANT CONFIGURATIONS; NETL February 25, 2011.

BIBLIOGRAPHY

[1] ADVANCED BOILER DESIGN by Flemming Skovgaard Nielsen, Carsten Søgaard, October 2014 http://www.bwe.dk/download/articles_pdf/advanced_boiler_design.pdf.

[2] Modern boiler design by Flemming Skovgaard Nielsen, BWE Paolo Danesi, BWE M.V. Radhakrishnan, BWE Energy India January 2012 http://www.bwe.dk/download/articles_pdf/modernboilerdesign2012.pdf.

[3] Steam Boilers' Advanced Constructive Solutions for the Ultra-supercritical Power Plants Nikolay Rogalev, Vadim Prokhorov, Andrey Rogalev, Ivan Komarov and Vladimir Kindra National Research University "Moscow Power Engineering Institute" Krasnokazarmennaya str. 14, 111250, Moscow, Russian Federation. www.ripublication.com/ijaer16/ijaerv11n18_10.pdf.

[4] Article written by: Dr V T Sathyanathan, edited by: Lamar Stonecypher, updated: 9/24/2010 https://www.brighthubengineering.com/power-plants/59662-difference-between-tower-types-and-two-pass-boilers/.

[5] Improvement of Pulverized Coal Combustion Technology for Power Generation http://criepi.denken.or.jp/en/energy/research/pdf/Improvement.pdf.

[6] Application and development prospects of double-reheat coal-fired power units by Kyle Nicol https://www.usea.org/sites/default/files/media/Application%20and%20development%20prospects%20of%20double-reheat%20coal-fired%20power%20units%20-%20ccc255.pdf.

[7] Status of advanced ultra-supercritical pulverised coal technology by Kyle Nicol (CCC/229 ISBN 978-92-9029-549-5) December 2013 www.usea.org/sites/default/files/122013_Status%20of%20advanced%20ultra-supercritical%20pulverised%20coal%20technology_ccc229.pdf.

[8] ASME: A modified double reheat cycle; Excerpt from Energy-Tech Magazine https://www.energy-tech.com/turbines_generators/article_f20ce7b1-a9f6-56bd-a05b-919eb3b0644f.html.

[9] CHEMICAL ENGINEERING TRANSACTIONS VOL. 32, 2013 by Paola Bernardo, Gabriele Clarizia* Istituto di Ricerca per la Tecnologia delle Membrane, ITM-CNR, Via P. Bucci, cubo 17/C, 87030 Arcavacata di Rende, CS, Italy g.clarizia@itm.cnr.it https://pdfs.semanticscholar.org/af85/934fa668a312d3e71c284ced478c12edfd16.pdf.

[10] Development of Materials for Use in A-USC Boilers in Mitsubishi Heavy Industries Technical Review Vol. 52 No. 4 (December 2015) By Nobuhiko Saito, Nobuyoshi Komai, Yasuo Sumiyoshi, Yasuhiro Takei, Masaki Kitamura and Tsuyoshi Tokairin. http://www.mhi.co.jp/technology/review/pdf/e524/e524027.pdf.

[11] THEMATIC RESEARCH SUMMARY Advanced fossil fuel power generation (© European Union 2014) https://setis.ec.europa.eu/energy-research/sites/default/files/library/ERKC_%20TRS_Advanced_fossil_fuel_power_generation.pdf.

[12] First U.S. Ultrasupercritical Power Plant in Operation by Sonal Patel on 02/01/2013 (Power Magazine).

[13] Leak Detection "Ins" and "Outs" by Barry Van Name http://www.power-eng.com/articles/print/volume-120/issue-2/features/leak-detection-ins-and-outs.html.

[14] Air preheater leaks: Mind the gap by *Pavan Kumar Ravulaparthy* http://www.powerengineeringint.com/articles/print/volume-22/issue-2/features/air-preheater-leaks-mind-the-gap.html.

[15] Upgrading and efficiency improvement in coal-fired power plants by Colin Henderson CCC/221 ISBN 978-92-9029-541-9, August 2013 copyright © IEA Clean Coal Centre http://www.wbpdcl.co.in/files/PPT_on_EERM/ccc221coalpowerR_M.pdf.

[16] Hydrogen Cools Well, but Safety is Crucial By Nancy Spring, Senior Editor http://www.power-eng.com/articles/print/volume-113/issue-6/features/hydrogen-cools-well-but-safety-is-crucial.html.

[17] Forgotten water: Stator cooling water chemistry 12/15/2007 By David G. Daniels, M and M Engineering http://www.powermag.com/forgotten-water-stator-cooling-water-chemistry/?pagenum=1.

[18] Newer Materials for Supercritical Power Plant Components–A Manufacturability Study Anish Nair1,*, S.Kumanan2 1 Production Engineering, NIT, Tiruchirappalli, 620015, anishn@live.com 2 Abstract Production Engineering, NIT, Tiruchirappalli, 620015, kumanan@nitt.edu.

[19] Excerpt from Wikipedia on Oxy Combustion https://en.wikipedia.org/wiki/Oxy-fuel_combustion_process.

[20] Oxy Combustion by NETL India https://www.netl.doe.gov/research/coal/energy-systems/advanced-combustion/oxy-combustion.

[21] Chemical-looping-combustion by NETL India https://www.netl.doe.gov/research/coal/energy-systems/advanced-combustion/clc.

[22] Integrated Gasification Fuel Cell (IGFC) Integrated Gasification Fuel Cell (IGFC) Systems 11th Annual SECA Workshop July

2010, Pittsburgh, PA Dale L. Keairns and Richard A. Newby Booz Allen Hamilton www.netl.doe.gov/file%20library/events/2010/seca/presentations/Keairns_Presentation.pdf.

[23] Thermal stress controlled loading of steam turbine-generators By James H Moore Jr (Inventor), Current Assignee: General Electric Co https://patents.google.com/patent/US3561216A/en.

[24] Steam turbine control with megawatt feedback (Patent by GE) Inventor Jens Kure-Jensen, Bernd A. K. Westphal, Thane M. Drummond https://www.google.co.in/patents/US4853552.

[25] Dedicated microcomputer-based control system for steam turbine-generators Inventor Jens Kure-Jensen, Richard S. Gordon, Charles L. Devlin, Frederick C. Krings. https://patents.google.com/patent/US4280060.

[26] High-efficiency power generation–review of alternative systems by Qian Zhu. March 2015 © IEA https://www.usea.org/sites/default/files/032015_High-efficiency%20power%20generation%20-%20review%20of%20alternative%20systems_ccc247.pdf.

[27] RECENT PROGRESS OF OXYGEN/NITROGEN SEPARATION USING MEMBRANE TECHNOLOGY by K. C. CHONG*, S. O. LAI*, H. S. THIAM, H. C. TEOH, S. L. HENG laiso@utar.edu.my http://jestec.taylors.edu.my/Vol%2011%20issue%207%20July%202016/11_7_8.pdf.

[28] Technical Paper BR-1925 on Component Test Facility (ComTest) Phase 1 Engineering for 760C (1400F) Advanced Ultra-Supercritical (A-USC) Steam Generator Development By P.S. Weitzel, Babcock and Wilcox, Barberton, Ohio, U.S.A www.babcock.com/products/-/media/317b6dd371ed479783e6a4e8411189cb.ashx.

[29] Technical Paper BR-1896 on A Steam Generator for 700C TO 760C Advanced Ultra-Supercritical Design and Plant Arrangement: What Stays the Same and What Needs to Change By P.S. Weitzel, Babcock and Wilcox,Barberton, Ohio, U.S.A http://www.babcock.com/products/-/media/3d7d9be9f78e4b6189355a80b7ed499f.ashx.

[30] Lünen—State-of-the-Art Ultrasupercritical Steam Power Plant Under Construction. By Dr. Frank Cziesla Siemens AG, Energy Sector, Dr. Jürgen Bewerunge Trianel Kohlekraftwerk Lünen GmbH and Co.KG, Andreas Senzel Siemens AG, Energy Sector http://m.energy.siemens.com/US/pool/hq/power-generation/power-plants/steam-power-plant-solutions/coal-fired-power-plants/Luenen.pdf.

[31] Material Development for boilers and steam turbines at 700°C by Rudolf Blum-Elsam Engineering A/S Fredericia, Denmark, Rod W. Vanstone- Alstom Power Ltd. Rugby, U K https://www.phase-trans.msm.cam.ac.uk/2005/LINK/103.pdf.

[32] Development of Materials for Use in A-USC Boilers By NOBUHIKO SAITO, NOBUYOSHI KOMAI, YASUO SUMIYOSHI, YASUHIRO TAKEI, MASAKI KITAMURA and TSUYOSHI TOKAIRIN.

[33] Increasing the flexibility of coal-fired power plants By Colin Henderson www.usea.org/sites/default/files/092014_Increasing%20the%20flexibility%20of%20coal-fired%20power%20plants_ccc242.pdf.

[34] ASME: A modified double reheat cycle https://www.energy-tech.com/turbines_generators/article_f20ce7b1-a9f6-56bd-a05b-919eb3b0644f.html.

[35] Description of Twin Firing Furnace by By LandT MHPS Boilers Pvt. Ltd http://www.lntmhps.com/products/boilers/.

[36] Appendix D Advanced Cycles A Final Phase IV Report Prepared by Electric Power Research Institute, June 2014 www.canadiancleanpowercoalition.com/files/1314/2056/8847/Advanced_Cycles_-_Phase_IV.pdf.

[37] Understanding, diagnosing, monitoring and addressing leakage and pluggage in water-cooled generator stator windings By A.K. Sahay, CC-OS, B. Manjul, Kahalgaon V.K. Garg, Badarpur http://indianpowerstations.org/Presentations%20Made%20at%20IPS-2012/Day-2%20at%20PMI,NTPC,%20NOIDA,UP/Nalanda%20Hall/Session-11%20Chemistry%20for%20Competitive%20Edge/Paper%202%20Stator%20Water%20System.pdf.

[38] Understanding, Diagnosing and Repairing Leaks in Water-Cooled Generator Stator Windings By Joseph A. Worden, Jorge M. Mundulas, GE Power Systems, Schenectady, NY www.ge.com/content/dam/gepower-pgdp/global/en_US/.

Chapter 14

Plant Safety Lifecycle and Safety Integrated Level

Chapter Outline

The author thanks the International Electrotechnical Commission (IEC) for permission to reproduce Information from its International Standards. All such extracts are copyright of IEC, Geneva, Switzerland. All rights reserved. Further information on the IEC is available from www.iec.ch. IEC has no responsibility for the placement and context in which the extracts and contents are reproduced by the author, nor is IEC in any way responsible for the other content or accuracy therein.

In addition the quotation from IEC Standards should include the following footnotes:

IEC 61508-1 ed.2.0 "Copyright © 2010 IEC Geneva, Switzerland. www.iec.ch"

IEC 61511-1 ed.1.0 "Copyright © 2003 IEC Geneva, Switzerland. www.iec.ch"

1 PREAMBLE

The earlier importance of systematic safety lifecycle studies was only restricted to a few industries, such as nuclear plants and the oil and gas industries, mainly. However, with the introduction of IEC standards IEC 61508 and IEC 61511, the importance of the safety lifecycle for any plant operation has been properly felt. A safety lifecycle study not only provides solutions for hazardous environmental conditions, but it also takes care of various kinds of risks that are associated with each and every plant. The standard creates the platform to think that plant safety is an independent entity. Safety must not be mixed with basic plant control systems (BPCS) in the form of plant protection, and is to be considered independently. So, there shall be a separate safety

Power Plant Instrumentation and Control Handbook. https://doi.org/10.1016/B978-0-12-819504-8.00014-7

instrumentation system (SIS). As a consequence of this, there shall be a safety integrated level (SIL) for the associated instrumentation and control systems. This chapter has been developed with the intent to give an idea to the reader about safety lifecycles, which have become part and parcel of any plant or industry as a safeguard against loss of property, personnel, and environment. This chapter has mainly been developed from the author's first book titled, "**Plant Hazard Analysis and Safety Instrumentation Systems**" that was published by AP-Elsevier and approved by IChemE UK (courtesy: Elsevier), which deals with a detailed analysis of plant hazard analysis and suggested safety instrumentation as a safeguard against hazards. The interested reader may go through the book from Elsevier. In this limited space, starting with hazards, risks, and risk analysis, short discussions pertinent to the safety lifecycle as per IEC 61508 and IEC 61511 are presented. Also, the chapter covers brief discussions on alarm philosophy as per international standards, as well as on electrical area classification, enclosure classes, and safety integrated levels (SIL) for various systems. Coal-fired power plants are comparatively less hazardous when compared with gas-fired plants or gas turbines. However, the importance of a plant safety lifecycle study and the assessment of the safety integrated level for various instrumentation and control systems (e.g., BMS) cannot be overestimated. As a part of a plant safety lifecycle study and SIL assessment, it is necessary to carry out a hazard analysis (e.g., HAZOP, most popular for power plants), which helps in identifying various hazards during operation that otherwise could have been missed.

In a plant, safety can be met not only by electrical systems but also by mechanical systems, for example, spring-operated pressure safety valves. Now, one should bear in mind that these standards meant for a plant lifecycle study are applicable for electrical, electronics, and programmable electronic (**E/E/PE**) devices and systems **only**. For PEs, both hardware and software should be taken into account. In addition to IEC standards, there are similar standards from the International Society of Automation (ISA), ISA 84.00.00. However, these standards are practically the same as those discussed in IEC standards. In fact, many of the ISA standards in this regard directly follow the IEC standards. Although there are similarities between the two standards IEC 61508 and 61511, the purpose of the two is different. IEC61508 has been developed for manufacturing units, that is, how a safety lifecycle during manufacturing is to be followed so that there will be fewer chances of equipment failure. On the other hand, IEC 61511 is meant for the process plant owners, that is, for process plants (operations). There are seven parts (parts 1–7) of IEC 61508 and at the time of this book, these are available in ED2: 2010. On the other hand, there are three parts of IEC61511 (parts 1–3) and ED2: 2016 at the time of the book. For a safety lifecycle study and arriving at any SIL required for the system, it is necessary to first identify the probable hazards/risks for the system. Then through risk analysis, the required control measures are undertaken and/or SIL assessment is done. With this preamble in mind, it is time to look into some of the focus areas of the standards to get to know the plant safety lifecycle, then slowly discuss other issues, including SIL. With the advent of a number of standards such as EEMUA 191, ISA 18.2, and IEC 62682, the focus on alarm systems in modern plants has been changed. So, along with alarm systems (in the perspective of an independent protection layer, IPL) other SIS components such as field sensors (FS), final control elements (FCE), and logic solvers (LS) are also covered. Enclosure protection is an important issue and it is guided by IP ratings/NEMA ratings. At the end of the chapter, the IP ratings of enclosures along with their equivalence in NEMA is presented so that the reader can get an equivalence.

2 GENERAL DISCUSSIONS

Assets are normally acquired against a lot of effort, toil, and monetary cost. People always wish to protect these. Unfortunately, this is not always possible on account of hazards in various forms. Until recently in the process industry, people would incorporate necessary safety measures in the form of protections under basic process control systems (BPCSs). In the arena of industrial hazards and risk analysis, a "system" is defined as a subject of risk assessment, which includes mainly the process, product, facility, and environmental and logical groups. So, the safety associated with it needs to be treated separately and **independently** from the BPCS. Sometimes, people incorporate redundancy in the system design as fall back action. This is not always true, as is the case with common cause failure. After 1995, people felt the need to integrate safety systems with BPCS without compromising the **functional independence** between the two to get the best-secured industrial systems. In order to treat these independently in a standardized manner, several international standards such as IEC 61508, IEC 61511, and ISA 84 evolved [1]. These standards have been developed with the aim that at a process upset or system or equipment failure, the designed system would allow the process safety to be managed in a systematic way following a risk-based management system. Safety instrumented systems (SISs) play a great role in mitigating technical risks in industrial plants. An SIS consists of a well-engineered hardware and software control system used to monitor the condition of the plant within the operating limits. When any risk condition arises, it triggers an alarm and will take the entire system to a safe condition to mitigate all kinds of risks as far as possible.

A safety lifecycle, according to the IEC standard, can be considered as a cyclic process or a closed loop comprising in cyclic fashion of **identify-analyze-design-verify** and is comparable with "**plan, do, check, act**" of ISO 31000.

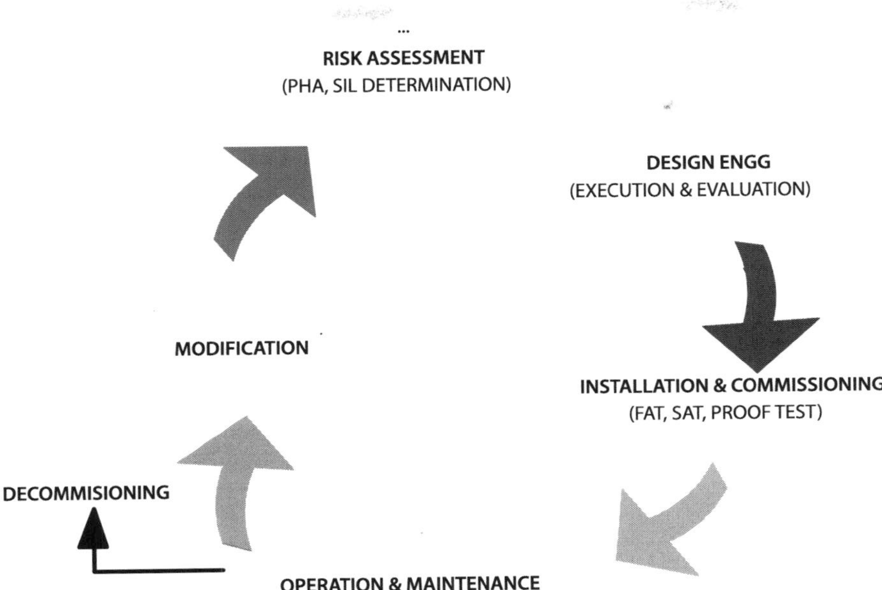

FIG. 14.1 Safety lifecycle of SIS.

In view of this, the SIS lifecycle could be conceived as what is shown in Fig. 14.1.

Unless protected, a system runs at a risk. This is inherent risk prior to any action taken to change the consequence. Designers aim to bring the system within, or in fact below, that risk limit by incorporating various protection measures for mitigation. Even after such protections, the little risk left is often referred to as residual risk. Some protections come from other technological means, but the major protections come through the interface of the BPCS with a safety system to develop a safety instrumentation system (SIS). Readers should not confuse the operational interlock and protections of BPCS with SIS. As stated earlier, SIS (E/E/PE) should be separate and independent of BPCS (even though these are integrated in a common network). The concept has been clarified through Fig. 14.2.

When there is risk in the system, there will be some hazards associated with the process. So, prior to moving on to other discussions, it is important to clarify all these details so that when these are referenced at later stages, the reader would have an understanding.

2.1 Definitions and Explanation of a Few Related Terms

1. **Hazard:** The term hazard has been defined by many agencies in different ways based on their terms of reference. These are detailed in [1] so interested readers may refer to the same.
   - General definition: A hazard can be considered as a state with a set of conditions of a system, which together with other conditions in the environment or in the environment of the system will lead to an accident. So, a hazard can be any biological, chemical, mechanical, environmental, or physical agent that has the potential to cause harm or damage to humans, other organisms, plant machinery, assets, or the environment in the absence of its control.
   - ISO/IEC Definition: As per ISO/IEC 51 or IEC 61508, a hazard is defined as, "the potential source of harm." In IEC 61508, harm has been defined as physical injury or damage to the health of people, either directly or indirectly as a result of damage to property or to the environment.
2. **Hazard Analysis:** Hazard analysis uncovers the hazards that exist in the workplace (here, it is power plants), focusing on the system or project. By hazard analysis, risk-based decisions are taken to develop the means to quantify, track, and develop mitigation means as well as control hazards, develop follow-up actions, verify effectiveness, and communicate. Refer to Section 2.3 for this.
3. **Risk:** According to ISO/IEC guide 51/IEC 61508, risk is, "the combination of probability of occurrence of harm and the severity of that harm." From here, it transpires that risks refers to the likelihood that a hazard can cause actual damage.
4. **Basic process/plant control system (BPCS):** According to IEC 61511, "BPCS is a key layer of protection (that) responds to input signals from the process, its associated equipment, other programmable systems, and/or (the) operator and generates output signals, causing the process and its associated equipment to operate in the desired manner but which does not perform any safety instrumented functions with a claimed **SIL 1**."
5. **Safety instrumented system (SIS)**: An SIS is designed to prevent or mitigate a hazardous event by taking the process to a safe state whenever a predefined or predetermined condition occurs to the system. It is a

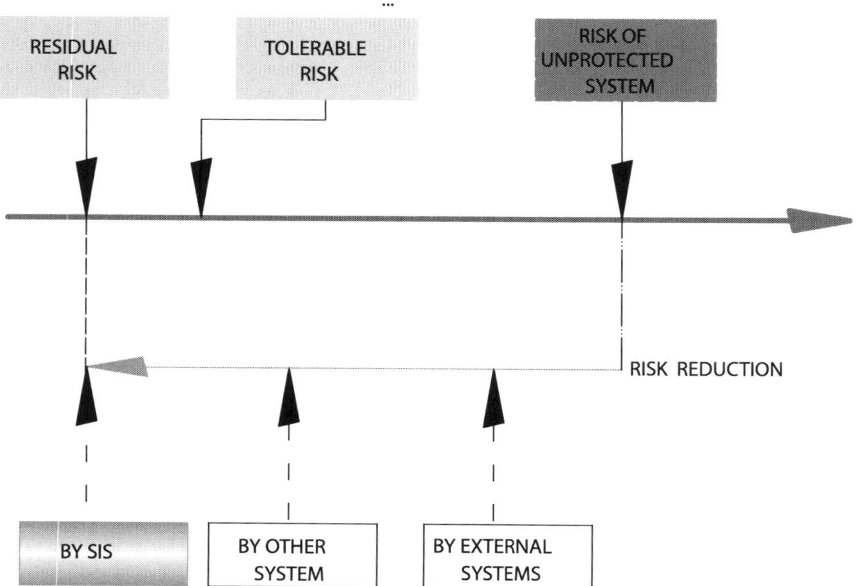

FIG. 14.2 Risk reduction by SIS.

combination of sensors, logic solvers, and final control elements. In modern days, these are in PEs, consisting of hardware and software. It is worth noting that the SIS should not be mixed with the BPCS, and the SIS should have a separate entity independent of BPCS, even if both systems are integrated through networking. There are schools of thought that do not even allow the integration of SIS and BPCS.

6. **Safety instrumented function (SIF) and Safety Instrumented system (SIS):** An SIF consists of sensors, logic solvers, and a final control element combination (independent of the BPCS). The SIF takes the system or process into a safe zone in the event of a hazardous situation/event, which is determined by predefined conditions for the process. SIF should not be confused with SIS, even though the standard uses the term SIS and SIF somewhat interchangeably in places. However, according to Clause 3.2.68 of IEC 61511–1:2003, the safety function is defined as a, "function to be implemented by an SIS, other technology safety related system or external risk, reduction facilities, which is intended to achieve or maintain a safe state for the process, with respect to a specific hazardous event." Corresponding to each hazard, there shall be one specific SIF, for example, the closure of the main fuel in the event of loss of flame is one SIF. In another example, if there is very low feed water flow or very low drum water level, then through the automatic protection interlock, the MFR will be tripped. From here, one can conclude that an **SIS** consists of a number of **SIFs** (functional parts) that are included for each set of hazards. For details, see Figs. 14.23 and 14.24 along with discussions in Section 5.
7. **Functional Safety:** According to ISA, "the ability of SIS or other means of risk reduction to carry out the actions necessary to achieve or to maintain a safe state for the process and its associated equipment." Also, functional safety in SIS highly depends on the proper functioning of sensors, logic solvers, and the FCE so that a reduced risk level can be achieved.
8. **Safety integrity level (SIL):** It is a measure of performance of an SIS. It is determined by the probability of failure on demand (PFD) for the SIF (SIS). There are four SIL levels represented by number: SIL 1, 2, 3, and 4. The higher the SIL number, the better the performance and the lower the probability of the failure on demand (PFD) value. However, with an increase in the SIL number, the cost and complexity of the system increases, but the risk level decreases. It is worth noting that there could be an individual component PFD but not SIL. SIL is only given to a system (SIS). SIL certification can be issued by the company (self-certification allowed) or another competent authority to indicate that appropriate procedures, analysis, and calculations have been followed and are compatible for use in the appropriate SIL level.
9. **Probability of failure on demand (PFD):** It is the probability that SIF/SIS fails to perform its intended safety function during a potentially dangerous condition. PFDavg is normally used in calculations when they are regularly inspected and tested.
10. **Some other associated terms:** A few terms associated with hazard and risk analysis are:
    - Accident: It is an undesired, unplanned (may not always be unexpected) event that will result in a

specified level of loss (in terms of health, property, production, etc.).
- Mishap: It stands for bad luck, misfortune, etc. In terms of industry, it could be an accident, which is associated with uncontrolled release of energy and toxic material exposure.
- Near miss/incident: It is normally used in good sense, meaning an event occurred but it involved very minor or no loss (in terms of health, property, production, etc.).
- Safety: Freedom (or nearly freedom) from accident or loss.

2.2 Discussions on BPCS and SIS

As indicated earlier, an SIS also consists of sensors, final control elements, and logic solvers. Typical layout has been shown in Fig. 14.3. In this diagram, the user interface and interface with BPCS have been shown through a communication bus. Here, it is interesting to note that there could be a separate BPCS and SIS, but these two could be **integrated** as long as they meet the requirements of standards such as IEC 61508/61511 or ANSI/ISA 84.

Functional safety is very important for the safety lifecycle and a brief idea about the same along with failure categories is given below.

1. **Functional safety concept:** Basically, functional safety stands on the following concepts:
   - All processes/systems have inherent hazards that cannot be brought to zero value.
   - All processes/systems have a tolerable failure rate without causing any harm for the system. Also, this failure rate is specific to the system in question.
   - For all processes/systems, these failure rates can be categorized in terms of SIL.

Brief discussions on functional safety are later in this chapter.

2. **Failure Category:** One of the major reasons for the need to develop the IEC 61508/IEC 61511 (ANSI/ISA 84.00.01-2004) standards was to devise some means so as to minimize the propagation of device failure

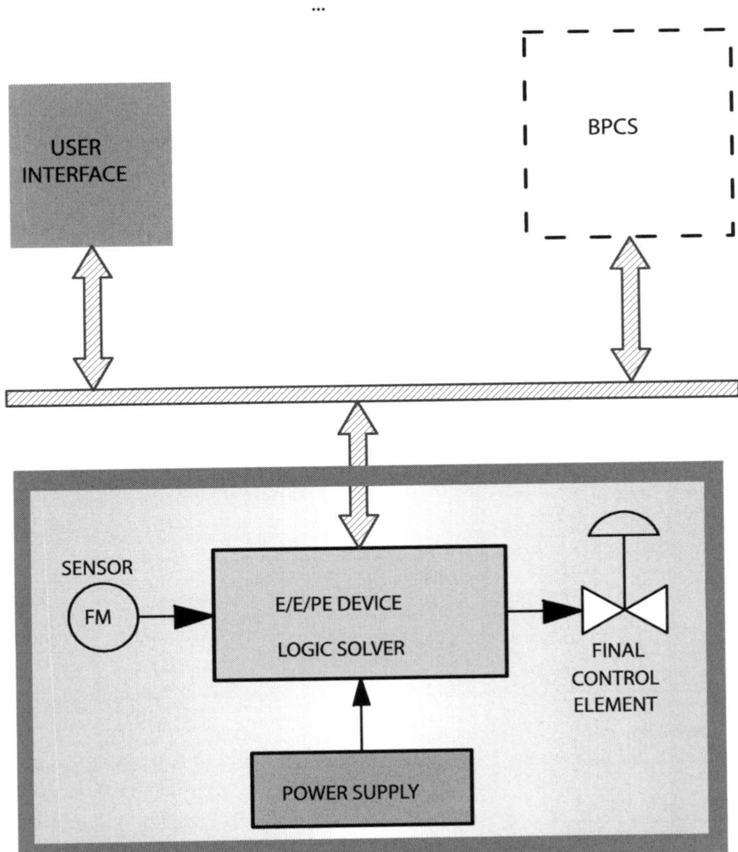

FIG. 14.3 SIS boundary and layout.

through design, operation, inspection, and maintenance practices into the system. Failure categories in a functionally safe system are:

- Random failure: These are uncontrolled, unnoticed failures, sometimes inherent with the process. They cannot be reduced in a systematic way, instead only with proper attention given for early detection to reduce the amount of loss (e.g., unprecedented grid collapse).
- Systematic failure: The following accounts for systematic failure and it is possible to reduce the same. These may come from shortcomings in system design, implementation, or manufacturing defects or from not following proper statutes, standards, or good engineering practices. These can be reduced through analysis and remedial measures.

Brief discussions on fault and failure types are given in Section 2.5. Safety functions may come as a result of hazard analysis, for implementation as SIS. Let the discussions start with hazard analysis issues so as to understand with a few more details.

2.3 Hazard Analysis Issues

For any hazard analysis, it is always necessary to identify and record all possible hazards in the plant as well as at the workplace. In order to carry out hazard analysis and/or for hazard identification, it is better to form a team comprising both experienced and fresh people so that experienced people familiar with the job will help in hazard identification and fresh eyes may throw light on new hazards. Hazard analysis can be conceived as a preventive approach to identify any potential hazards, which should be realistic and relevant to the facility. Also, hazards must be prevented, eliminated, or reduced. This means that in a modern plant, it is necessary to detect the hazard early and there should be suitable safety measures to mitigate/control it. The major areas of such hazard analysis mainly include [1]:

1. All aspects of work, including accident or incident/near-miss record.
2. Include all nonroutine activities (e.g., maintenance, repair).
3. Include people "offsite" (not regular) and foreseeable unusual conditions.
4. Include assessment groups for different levels of risk.

2.3.1 Hazard Study Issues and Scope

It is normal practice to identify the hazard, find control measures, and carry out follow-up action. In order to control the hazard, one has to look for safety interfaces also. In the following subsections hazard study and scope

1. Hazard study: The following points need to form a part of basic hazard study/analysis:
   - Hazardous component identification.
   - Possible malfunction of equipment or system including software.
   - Safety interface including software.
   - Operating condition and environmental constraints, if any.
   - Available support equipment/system.
   - Operating procedure, regular test, maintenance diagnostic features.
   - Safety-related equipment, safeguards.
   - Possible alternate approach and emergency procedures.
2. Scope of Hazard analysis: The scope of the PHA shall include but not be limited to the following:
   - Equipment in the process and any failures, including electrical, control, and instrumentation (EC&I) system [hardware (HW) and software (SW)].
   - Facility assessment of explosion/toxicity, human error and failure factors.
   - Environmental impact on the process and hazards caused by the same.
   - Identification of previous incidents that had likely potential to be catastrophic.
   - Hazards inherent to the process as well as hazards caused by reactions.
   - Consequences of failure of engineering and administrative controls, etc.
   - Engineering and administrative controls pertinent to hazards and their interrelationship.
   - Qualitative evaluation of failure of safety controls on assets, humans, and the environment.
   - Steps required to correct or avoid deviation.

2.3.2 Industrial Hazards

In industries and process plants, there are many sources of hazards. Listed below are some of the sources of hazards encountered in industry:

1. Mechanical moving parts.
2. Fire/explosion.
3. Sources and propagation of stored energy in the form of chemical, mechanical, and electrical.
4. Noise of different forms.
5. Error due to erosion in pipe.
6. Toxic and corrosive liquids and gases.
7. Human error (e.g., operating error).
8. Software error (design error).
9. Cyberattack and loss of network security.
10. Nuclear radiation (nuclear plants).
11. Biological hazards (bacterial growth).

These days, with the integration of various networks—including the integration of control networks—with a company's main servers as well as ERPs, a network security threat is a very big hazard. Some of the major security threats come on account of:

- Control networking.
- Standardization and open system.
- Unsecured remote connections.
- Public information to hackers.

2.3.3 Aim of Plant Hazard Analysis

In the world, there is no system that is totally free from hazards. So, there will be a risk of damage. Therefore, people will always try to eliminate risk as far as possible with the help of hazard analysis. The basic aims of hazard analysis are:

1. Identify the potential hazard and analyze the magnitude and likelihood of the hazard.
2. Identify equipment, instrument failure, process upset, and human error that could manifest a potential hazard and have a detrimental effect such as loss of asset and/or a negative environmental impact, for example, emissions from the boiler.
3. Identify and evaluate existing protection systems against risk consequences.
4. Document all potential hazards with recommended mitigation strategies, which when compared with the existing protection, may require additional independent protection.
5. Statutory requirements; Clause 4.2 of NFPA 654 (2013) is a classic example: "Design of fire and explosion safety provisions shall be based on process analysis of the facility, the process, and associated fire and explosion hazards." It also says, "PHA shall be documented and to be used for life of the process."

2.3.4 Major Hazard Analysis Steps

With reference to Fig. 14.4, the following are the major plant hazard analysis steps:

1. Study and preparation of PHA (some cases preceded by preliminary hazard analysis).
2. Formation of a team with a team leader (experienced in PHA and one plant expert).
3. Scope and boundary definition in accordance with system demand description. This is very important in the sense that without this, the entire hazard analysis process could be a huge one and effective control will be a serious problem. Major description issues shall include:
   - Specifications/detailed design requirements.
   - Operational details.
   - General compliance to standards and other statutory requirements.
   - Applicable standard human convention factors (e.g., lamp color, etc.).
   - Accident experience, failure reports, and similar plant data.
4. Hazard identification: Irrespective of the process chosen, this is necessary because by itemizing the hazards, analysts can imagine and develop a picture of the complexity of the plant and the breadth of safety analysis required. Hazard identification gathers information about:
   - Characteristics of the hazard, including inherent nature, multiplicity, etc.
   - Form and some quantification of the hazard.
   - Where and when in the project/facility it is present.
   - How this hazard could result in an undesirable event or a chain reaction.
5. Collect and collate all up-to-date relevant information to support the PHA.
6. Selection of appropriate PHA techniques out of those shown in Fig. 14.5.
7. Agreement for schedule of work, also in terms of time.
8. Conduct regular meetings and documentation.
9. Depending on applicability, conduct a functional analysis.
10. Develop a preliminary hazard list, identify contributory hazards, initiators, etc.
    - Equipment failure because of hardware, software, and human factors.
    - Operation and maintenance (O&M) condition.
    - Man-machine interface (interaction procedure).
    - All other systems and procedures.
11. Develop a safety system baseline, recommendations, and requirements for mitigating hazards.
12. Detailed documents of PHA result (risk assessment discussed separately).

Hazard analysis culminates in implementation [including management of change (MOC) if called for] and follow up.

2.3.5 Hazard Analysis Types and Classification and HAZOP Outline

Broadly, hazard analysis methods can be classified as qualitative, guided word, and quantitative. Major hazard analysis methods with categorization have been depicted in Fig. 14.5. Proper evaluation of various plant hazard analysis techniques is important before selecting the same for the plant. In case of power plants, the HAZOP hazard analysis technique is very popular. Also, various quantitative techniques could also be applied.

HAZOP analysis has been described with the help of a number of tables and figures. These tables and figures are

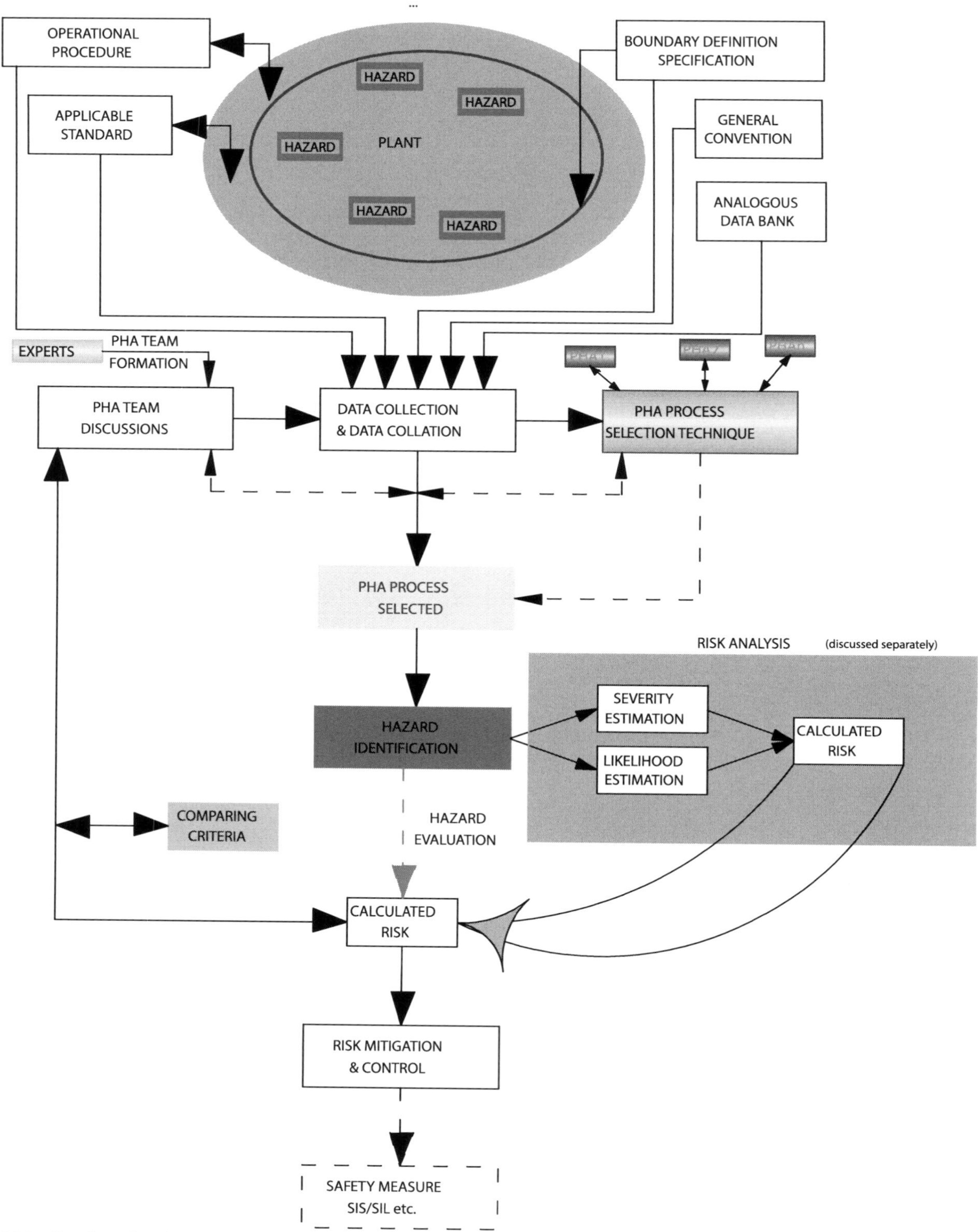

FIG. 14.4 Plant hazard analysis stages.

self-explanatory, and hence are not elaborated upon further to limit the volume of the book. Interested readers may refer to the author's book in reference [1]. Table 14.1 shows the basic steps and functions of a HAZOP study. It is worth noting that normally, a HAZOP study is carried out with the help of a team of personnel from different disciplines (with one instrumentation and control engineer). Normally, a team consists of a team leader, one scribe, and other

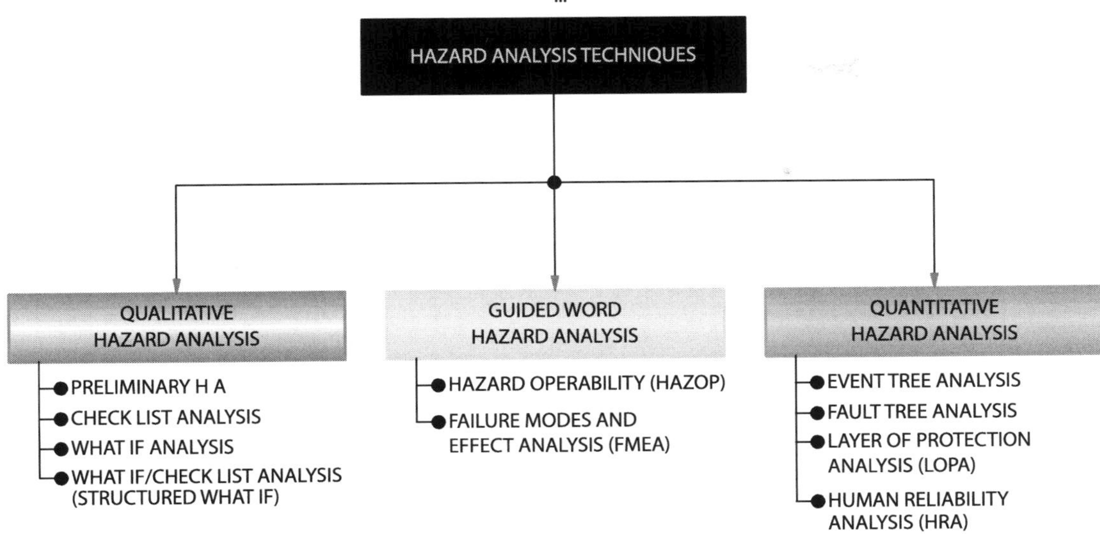

FIG. 14.5 Various PHA techniques.

TABLE 14.1 HAZOP Steps and Functions

Step	Explanation
Intention	Process designer to highlight outlines about one section/P&ID. General scope and intention discussed. Relevant part highlighted with dotted line. Process designer to explain the part and general discussion on the same.
Deviation	Line-by-line study commences with team leader choosing the relevant guide word. Deviations with potentiality for hazards are noted. Team leader goes through all relevant guide words one by one, and when all guide words are exhausted, the next line is chosen and this line is highlighted firm, meaning that its work is done. When all lines in this way are complete, that is, all are highlighted firm, additional words may be chosen to check the entire P&ID.
Cause	Cause for each of the deviations is identified
Consequence	For each of these deviations, a consequence (combination of likelihood of occurrence and severity) is identified through creative and brainstorming discussions. Consequences that warrant action are recorded.
Safeguard	Existing safeguards are evaluated during meetings and new control measures, if any, are prescribed.
Corrective action	When warranted, a detailed QRA or reliability analysis may be undertaken for complex systems at a later time.
Note	The purpose of this study is to identify hazards that require solutions and NOT the proper solution.

members totaling around 7–8 personnel. There will be several stages of analysis such as preparation, evaluation, and documentation. A total of 3–12 weeks in the different stages mentioned above will be necessary for a complex plant. Starting with a flow diagram, P&ID, etc., there will be requirements of a number of engineering documents to complete the study.

From Figs. 14.5 and 14.6, it could be noted that this analysis is carried out with a few guided words; such guided words have been indicated in Tables 14.2–14.3.

Additional guided words used in HAZOP are indicated in Table 14.3.

Hazards can be applied to power plants at various stages of the project. HAZOP analysis applicable at various stages of the project is indicated in Fig. 14.6.

In order to understand HAZOP, a process overview of HAZOP analysis is presented in Fig. 14.7.

With this short discussion on HAZOP at an end, look into the consequences/impacts of hazards and ALARP systems, which is a good tool for hazard analysis.

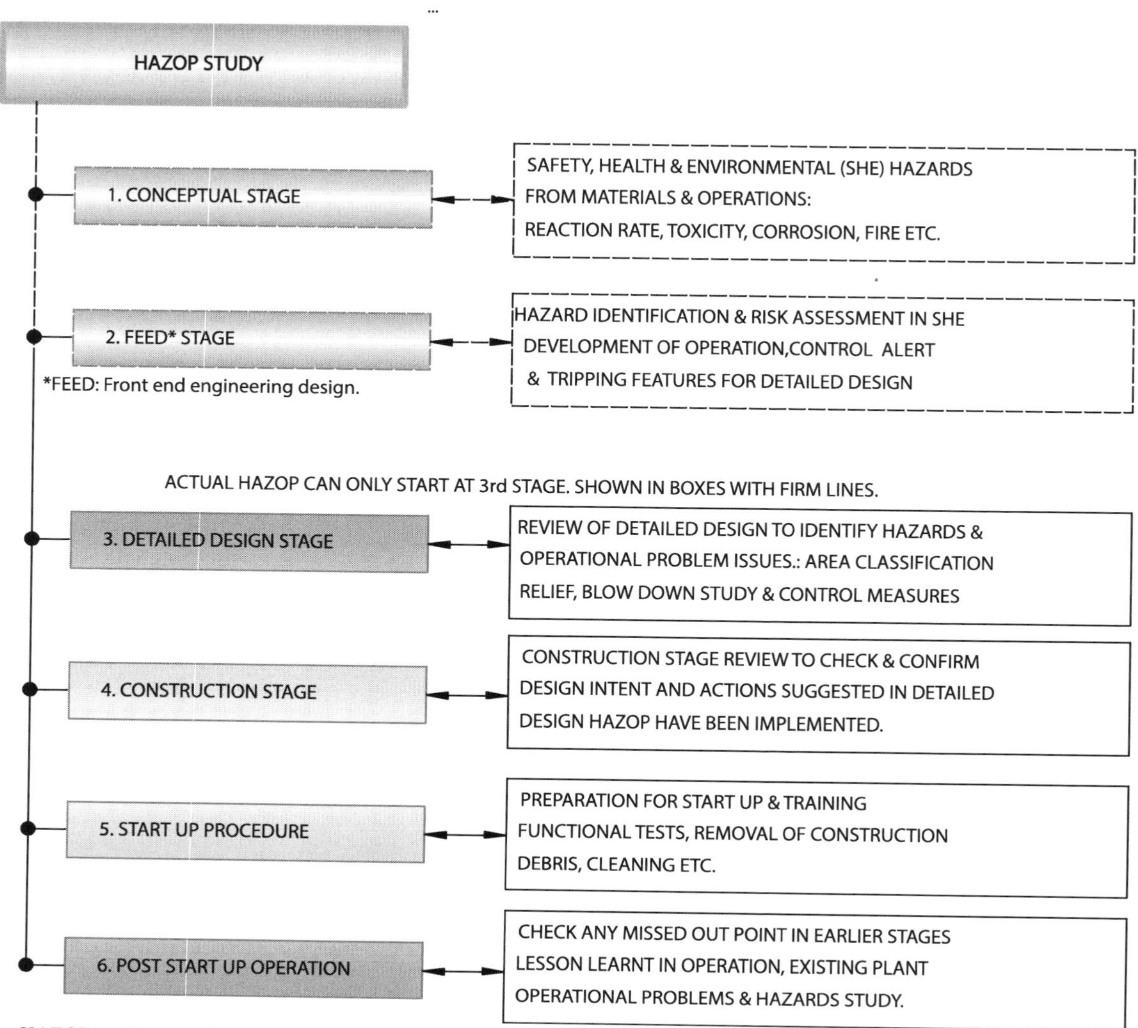

FIG. 14.6 HAZOP studies at various stages.

2.4 Consequence, Vulnerability, and ALARP

Consequence, vulnerability, speed of onset, and ALARP are also very important for risk analysis, and they will be discussed later.

Speed of onset refers to the time a risk event tends to manifest itself, that is, the time lapse between the event occurring and when one can feel the effect.

In this section, these will be discussed briefly. Let the discussion start with consequence/impact of hazard and risks.

2.4.1 Consequence (Impact) and Consequence Assessment

Hazards/risks always have direct impact on assets, humans, and environments. In order to reduce such impacts, hazard analysis is carried out. Therefore, consequence/impact is an important aspect in hazard/risk analysis vis-a-vis the safety lifecycle study. According to the DHS Risk Lexicon, 2010 edition, consequence (or impact) is the effect of an incident, event, or occurrence. The nature of consequence can vary widely and depends on the entity. Consequences include (but are not limited to) loss of life, injuries, economic impacts, psychological consequences, environmental degradation, and inability to execute essential missions, direct or indirect [see DHS Risk Lexicon, 2010 edition page (p. 10)]. From the discussions in previous sections, it is clear that the consequence is dependent on the likelihood of occurrence and severity. From the subsequent discussions, it will be made clear that consequence is dependent on the likelihood of occurrence and the severity. Therefore, one may argue that the consequence spectrum represents the risk scenarios of several events with different severities and each of these is associated with probabilities of occurrence.

Consequence analyses/assessments are carried out to estimate the extent to which casualties or damage may occur

TABLE 14.2 Generally used Guided Words With Meaning

Guided Word	General Meaning	Remarks
No (not/none)	Negation (of intent)	No forward flow
More (higher)	Quantitative increase	More of any physical parameter
Less (Lower)	Quantitative decrease	Less of any physical parameter
As well as (more than)	Quantitative increase additional activity	Design/operating intent achieved along with additional thing.
Part of	Quantitative decrease	Only part of intent achieved
Reverse	Opposite of intention	Reverse reaction/flow
Other than	Complete substitution/miscellaneous	Not original intention achieved some different thing happened—alternative mode of operation

TABLE 14.3 Additional Guided Words With Meaning

Guided Word	Meaning	Application
Early	Relative to clock time	Timing before intention
Late	Relative to clock time	Timing after intention
Before	Sequence order	The step (before) is effected out of sequence
After	Sequence order	The step (after) is effected out of sequence
Faster	Different (earlier) from timing intention	Faster reaction
Slower	Different (later) from timing intention	Slower reaction
Where else	Other location	Flow/transfer/source/destination

Possible parameters with which guided words could be associated are: flow, pressure, temperature, level, viscosity, mixing, stirring, transfer, reaction, composition, addition, separation, time, speed, phase, particle size, measure, control, pH, sequence, start, stop, signal, operate, maintain, communication, and service.

as a consequence of an accident/unexpected event. This will be clear from an example in the context of power plant operation. Sheering of one bearing in a turbine will not only make the turbine inoperative, it can make the entire unit shut down or cause some accident to the nearby operator. Therefore, in consequence analysis, all consequences are identified, not only the immediate ones but the whole chain of events triggered by one accident or unexpected event. Though not mandatory, normally consequence assessments are carried out by expert agencies, often by a team of senior personnel chosen from the enterprise. Consequence assessment software packages are also available on the market. Suitable software is available for this. This software normally is a little costly and (often) proprietary in nature. ALOHA is a common one available on the market. Consequence assessment criteria exclusively depend on the enterprise.

2.4.2 *Vulnerability*

Vulnerability and exposures in conjunction with hazards give rise to risk (see Fig. 14.8).

The vulnerability is basically susceptibility. Vulnerability would have at a specified point in the future to a specific hazard as a result of realizing all its current adaptive capacity through anticipatory adaptation.. As mentioned before, vulnerability refers to the susceptibility of a risk event entity in terms of its preparedness, exposure, and adaptability. Vulnerability is very closely related with network security in programmable electronic (PE) systems mainly related to a logic solver in the SIS, to be discussed later. From Fig. 14.8, it is clear that the higher the vulnerability, the higher the risk or consequence should the event happen. Vulnerability analysis is also carried out to assess and rate the severity and exposure. For more exposure, there would be higher vulnerability. Vulnerability analysis could also determine whether current safeguards are sufficient or need to be augmented with additional safeguards. The capability to anticipate, respond, adapt, and prevent along with available options are the major criteria for vulnerability analysis. For vulnerability analysis, some ratings are assigned to vulnerability, which with their meanings have been presented in Table 14.4.

2.4.3 *ALARP Principles*

ALARP stands for "**as low as reasonably practicable**." The basic principle of ALARP is well understood from Fig. 14.9, which as such is self-explanatory.

The triangle is a representative of increasing cumulative risk to which a person is exposed. "Reasonably practicable" is a narrower range than physically possible [5]. The entire region between intolerable risk (upper tolerability limit) and negligible risk level (lower tolerability limit) is the ALARP region. At the ALARP level, the trouble, time and cost of

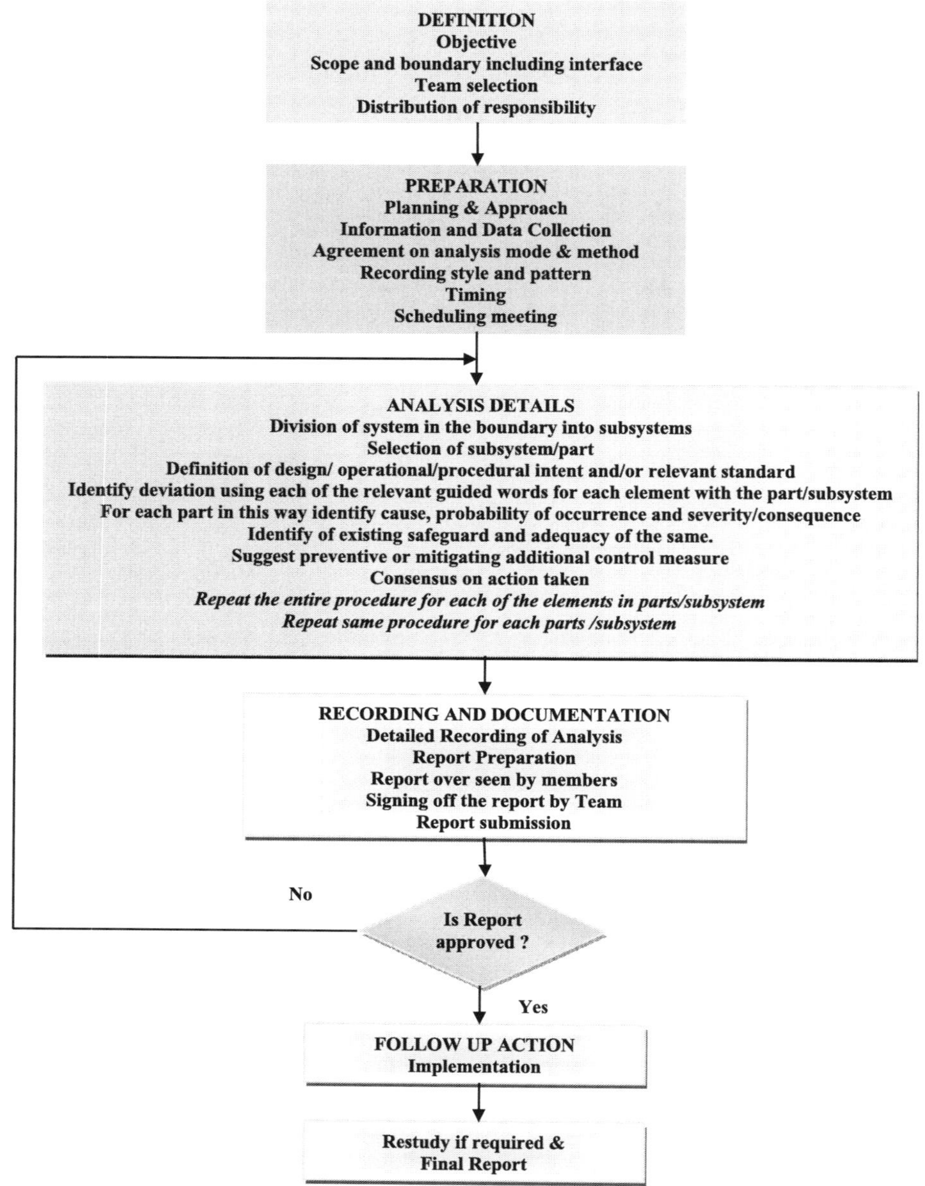

FIG. 14.7 Overall view of HAZOP study [1].

further reduction is not worthy, that is, unreasonably high. However, as long as the risk is even just above the ALARP level, it is necessary to reduce the risk to bring it to the ALARP level. The utility of ALARP is easily understood from Table 14.5 given for power plants.

From the examples, it is clear that at various stages, it has some different things to offer. However, it is necessary that the ALARP shall be periodically reviewed because with time and technology, there may be changes in the ALARP level. Now let us look at a few other relevant points on ALARP.

1. **ALARP Point:** The ALARP concept is taken from the British health and safety system (Act 1974). Though cost-benefit studies are done to arrive at the ALARP point, it is a subjective method as it requires duty holders and others to exercise their judgment very carefully. An ALARP point is like a break-even point where risk level meets resources and effort. This has been clarified through Fig. 14.10. In the figure, one can see that as one moves on the X-axis on the right side, the risk level decreases but the cost/effort shoots up a sharp line.
2. **ALARP Process**: Briefly, the entire process can be divided into the following steps for each risk:
   - Assessment of damage due to risk.
   - Find the methods for risk reduction (brainstorming).

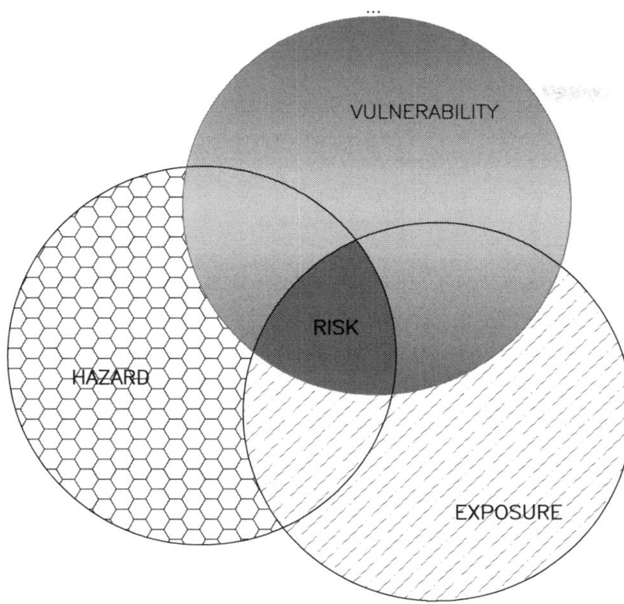

FIG. 14.8 Relation between risk and vulnerability.

- Checks and balance of the risk vis-a-vis sacrifice. If the risk and sacrifice are disproportionate, then the same may be curtailed; if not, then it may be added into the list for risk reduction methods. If there is no potentially feasible method, it means it is ALARP.
- Now, for cost-benefit analysis (CBA), the cost and benefit of each of the methods listed are found accurately.
- The next stage is the implementation and reassessment stage, when it is time to check to see if there is any risk in the ALARP region. If no, it is ALARP; if yes, then the process is repeated.
- Other issues associated with the ALARP process are ALARP demonstrations, ALARP assessments, and good practices.

2.5 Fault and Failure Discussions

From Section 1.1.11 of Chapter 7 and Subsection 2.2.2 of this chapter, we have gathered a fair idea about error, fault, fault tolerance, failure, and failure categories. Here, discussions on different kinds of fault failure categories along with their implications on system design shall be covered.

2.5.1 Discussions on Fault Types and Fault Tolerant Characteristics

The discussions presented here supplement the discussions in Section 1.1.11 of Chapter 7. Here, various fault types and fault tolerance characteristics are briefly discussed.

TABLE 14.4 Vulnerability Ratings

Class (Rating)	Vulnerability for:
Very High (5)	Totally lacking in planning
	Lack of addressing risk both at the corporate and process levels
	Risk response not implemented
	No emergency/contingency planning
High (4)	Planning for strategic risks performed only
	Low corporate and process level/capability to address risk
	Part response implemented without achieving objective
	Some emergency/contingency planning only in place
Moderate (3)	Sensitivity analysis for planning performed
	Medium corporate and process capability to address risk
	Response implemented, partial objective achieved.
	Most emergency/contingency plans in place not rehearsed
Low (2)	Most options, including strategic options for planning, defined
	Medium to high corporate and process level to address risk
	Response implemented, objective achieved—except extremes
	Emergency/contingency in place+some rehearsal
Very low (1)	Real options in place and strategic planning optimized
	High corporate and process level to address risk
	Redundant emergency plans in place with testing of critical ones
	Emergency/contingency plans in place with rehearsal

1. **Fault Type:** Faults may be due to human error, hardware failure, software failure, and/or design problems. Depending on timing, faults can be mainly three types:
   - Intermittent fault: The fault that occurs for some time, then vanishes and reappears again. These faults are mainly due to hardware. Many times, the system does not function well when a part of it is hot, but if it is allowed to cool down, it again starts functioning well.

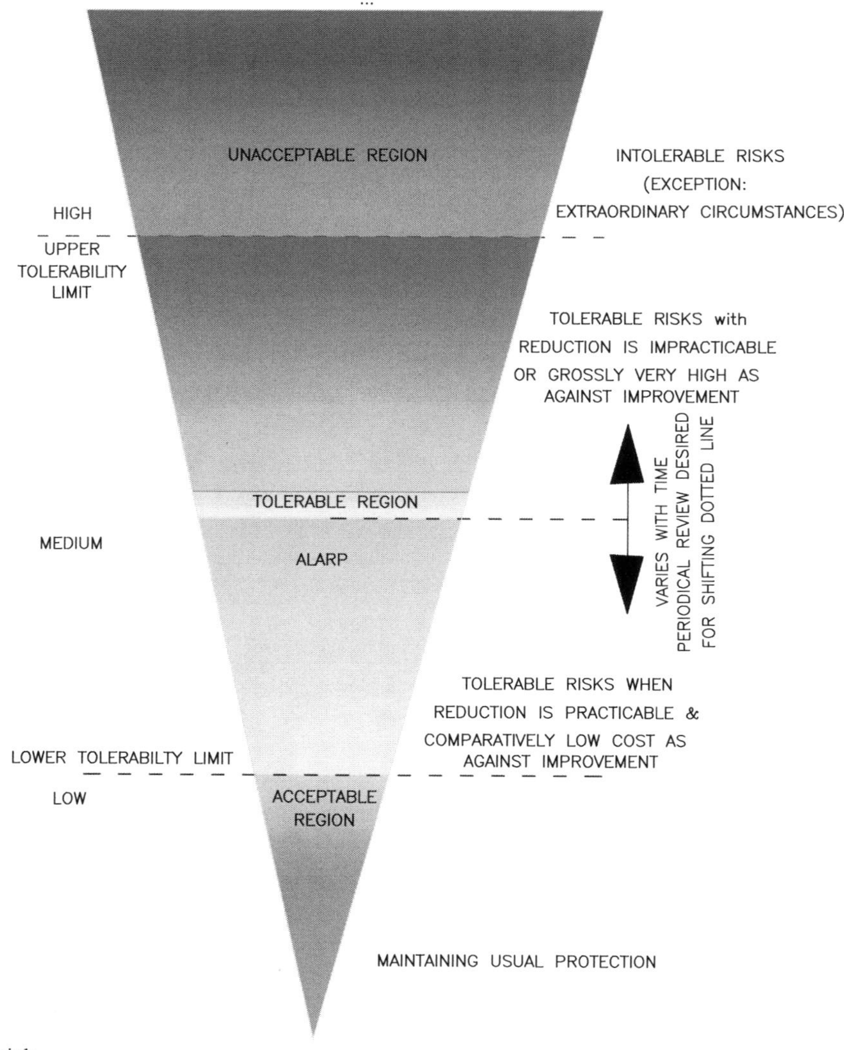

FIG. 14.9 ALARP principle.

TABLE 14.5 ALARP Power Plant Application

Stages	Fossil Fuel Power Plant
Conceptual	More options to choose from various alternatives such as subcritical or supercritical technology/combined cycle plant or IGCC
FEED/basic engineering	Alternative layout for accommodation of common control tower for multiple units, or, say, alternative ash handling system technology
Detail engineering	Alternatives for safety system incorporation, integration of control systems
Operation	Collecting feedback for safety improvement, frequent failure causes (e.g., introduction of drum level trips from, say, the hydrastep control, etc.)

Also, the system could malfunction due to one loose connection. All these are examples of intermittent fault.

- Permanent fault: It is a persistent fault perhaps due to the failure of a component/subsystem, for example, an input short circuit.
- Transient fault: This type of fault occurs once, then disappears.

Again, depending on the nature of the fault, it can be two types: silent or Byzantine.

- Silent fault: Here, the output stops, that is, no output due to failure.
- Byzantine fault: Here, the system produces output but it gives incorrect results. Obviously, it is difficult to tackle Byzantine faults. For details on these, the book reference [1] may be consulted.

2. **Fault tolerance characteristics:** From Section 1.1.11 of Chapter 7, it has been noted that a fault-tolerant

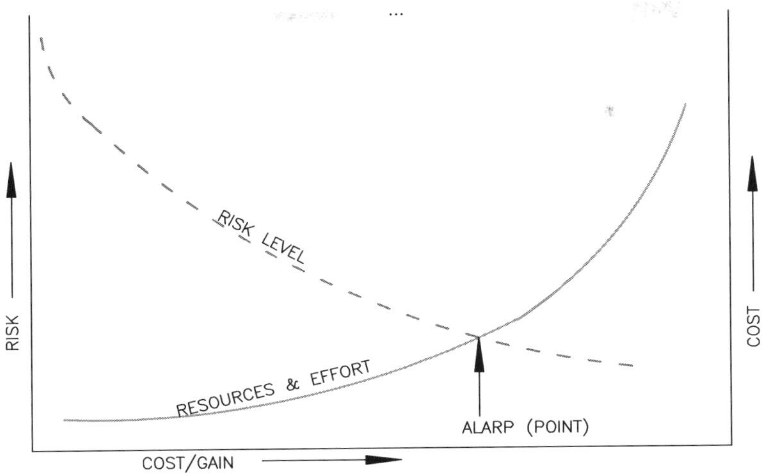

FIG. 14.10 ALARP point.

design enables a system to continue its intended operation, perhaps at a reduced level, rather than a complete breakdown of the system. System fault tolerance can be resilient (e.g., mask the process failure by replication), for example, for a two of three sensor configuration or a redundant network communication with a one-link failure. Some are such that there may not be any degradation, for example, a triple modular redundant (TMR) system in sensors, logic elements, and final control elements (FCE). Fault tolerance is directly connected with dependability, which consists of "availability" (ready for use immediately), "reliability" (how much the system can run without failure), "safety" (how safe is the failure? What are the results, that is, nothing serious happens for one failure?), and "maintainability" (ease of repair). Fault tolerance caters to both system availability and safety, but two issues are independent. Table 14.6 gives fault-tolerance characteristics. The reader should compare this table with Table 7.19. There is a similarity between the two but they are not exactly the same.

2.5.2 *Discussions on Failure Categories and Bath Tub Curve*

Fault categories discussed in Section 2.2 are further elaborated here. It is common that control systems operate within a range of control, beyond which it is not possible to keep control systems to take action to avoid failure. This will be clear from an example: when there is a catastrophe in a grid, it is almost impossible to keep the governor control of any connected set within a given range. In such situations, there may be a failure in one part, leading to the high possibility that this will propagate and give rise to a hazardous situation. It is needless to say that the SIS must

TABLE 14.6 Fault Tolerance Characteristics [1]

Name	Explanation	Example
No single point of failure/repair	System continues its operation uninterrupted for a single failure and/or during repair for that.	With an uninterrupted power supply (UPS) of suitable capacity connected to a computer, the computer continues its operation with main supply failure, or input supply fuse failure and repair for the same.
Fault Isolation for the failing component	Ability of the system to isolate itself from the failed component and continue its operation. So, necessary fault detection and isolating devices must be provided.	It is quite common to isolate a part of the grid in case of massive grid failure. PLC I/Os with galvanic/ photovoltaic isolation do a similar function.
Fault containment to prevent propagation of failure	It is possible in some cases, that on account of failure of one subsystem, the fault may propagate. So, suitable measures shall be built in to prevent such propagation.	Firewall in network in the classical example of the same.
Availability of recovery	A process by which failure shall be recovered.	It is possible in two ways: one is forward recovery when the system will be taken to a new correct state but not the last correct state. In back recovery, it is brought back to the last correct state. These are often found in network communications.

address all such situations right from the beginning so as arrest hazard propagation. This will lead to fewer consequences for human life, the environment, and assets. Two types of failures discussed in Subsection 2.2.2 set two benchmarks for SIF design: (1) Minimum performance against random failures, and (2) The rigor of protective management to reduce the potential for systematic errors. Major faults for two types of failure categories are indicated in Table 14.7.

Apart from random failure and systematic failure, there is another kind of failure popularly known as common cause failure (CCF). This could be either random or systematic, and is responsible for the failure of multiple devices, systems, etc. On account of CCF, redundant devices for SIF protection purposes would be effective when these devices use different technologies and/or are from different sources so that there is no chances of common cause failure. Such kinds of failures are often found when the device specifications are violated, in multiple devices (when sourced from the same device) simultaneous failures can occur (and redundant instruments may fail). In ISA TR84.00.02-2002 Part 1, common cause failure has been categorized under physical failure. However, in the note, the CCF of systematic failure has been mentioned (Fig. 14.11).

The following issues are responsible for the performance and failure probability of SIF:

- Random failure rate of devices.
- Design parameters.
- Systematic failure.
- Common cause failure (random failure of multiple devices).

1. **Random Failure**: As per ISA TR84.00.02-2002 Part 1, "a failure is classified as physical when some physical stress (or combination of stressors) exceeds the capability of the installed components." Random failures normally exhibit the following characteristic features:
   - Physical stress due to unusual/abnormal process conditions, corrosion, or material deterioration.
   - Adverse environmental conditions.
   - Occurrence at any time during life.
   - May not be detectable during accepting stage.
   - No particular pattern for occurrence.

TABLE 14.7 Faults for Failure Categories

Random Failure	Systematic Failure
Stress, aging	Systematic fault, design error, hardware/ software error, error in management of change (MOC*), human interaction error, installation error.

   Random failures can be detected by diagnostics internal to the devices, externally configured diagnostics, and proof tests. Device redundancy is the most common form to back each other up if a failure in one device occurs (provided there is no CCF).
2. **Systematic faults:** Systematic failure is **a** reproducible type of failure of the system. Mostly, this occurs due to error in design, operation, the production process, installation, and/or maintenance as well as software issues in programmable safety systems. This can also occur due to improper implementation of MOC at any stage. Manufacturing errors can be addressed by diversity, and this may increase the SIF complexity. The following errors are mainly responsible for systematic failure:
   - Specification: Inadequate or incorrect design or software/design.
   - Manufacturing: Manufacturing defects (HW/SW error) contributed during manufacturing.
   - Implementation: Installation/programming error, interface issues, and/or not following standards.
   - O and M: Inspection/testing flaws.

   Common cause failure: As per IEC 61511, CCF is defined as, "a failure (that) is the result of one or more events, causing failures of two or more separate channels in a multiple channel system, leading to a system failure." Understanding independence and common cause is very important and they may be interrelated. Any lack of independence will result in the potential for a CCF. Similarly, when a common cause is identified, it means there is a lack of independence. CCFs mainly occur due to external influences, and could cause more than one system component or layer of protection to fail. On account of CCFs, there may be a reduction in SIF-SIS performance. CCFs are treated in different ways and strategies for coping with the situation may be different, depending on the nature of the failure (e.g., systematic or random). There are two types of CCFs: single failure, multiple failure, or a single event to cause multiple failures. CCFs may be introduced at the design stage (improper understanding of failure/software development, large variations in process/environmental conditions), improper selection or hardware deficiency, improper procedure/operation, Human error, inadequate testing/maintenance, etc.
3. **Bath tub curve:** System failures are well understood by Fig. 14.12 (self-explanatory).

   The major issues related to bathtub curves are infant mortality and wearing out. Avoidance issues are:
   - **Infant mortality**: Defects elimination, appropriate and adequate specifications, sufficient design tolerance, sufficient component derating, early stress testing, and design evaluation.

Management of Change (MOC*): MOC is a part of operation section of safety lifecycle and last part of active part of lifecycle. It normally refers to modification of SIS to ensure that even after the change safety integrity of SIS is maintained. Starting from initiation it passes through a number of stages (7–8 stages) to end with implementation and closing of the change proposal.

FIG. 14.11 Management of change (MOC).

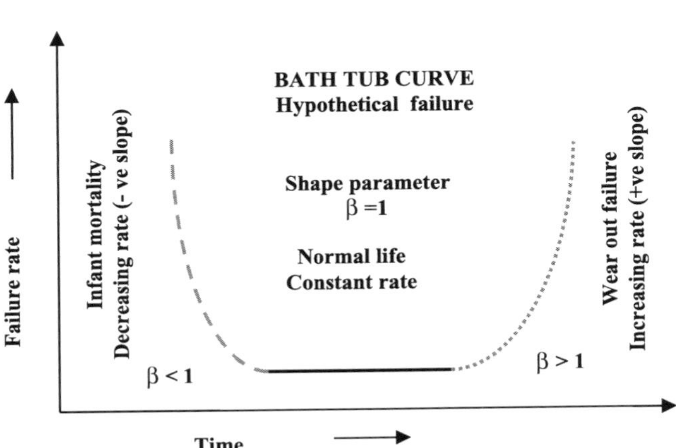

FIG. 14.12 Bathtub curve for failure rate.

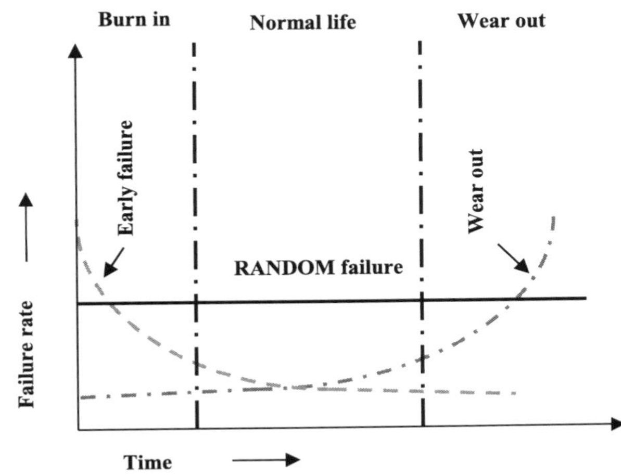

FIG. 14.13 Typical component failure curves.

- **Wearing out:** Electronic items wear out after a reasonable period of operation. In a mechanical assembly, it is often seen that a particular component fails much sooner than expected, but when replaced, the assembly can a last longer period.

Individual units may have different failure rates; some may fail during the early stage while some may survive through a normal life or even during the wearing-out stage. Failure categories of different individual units are shown in Fig. 14.13 (to show variations).

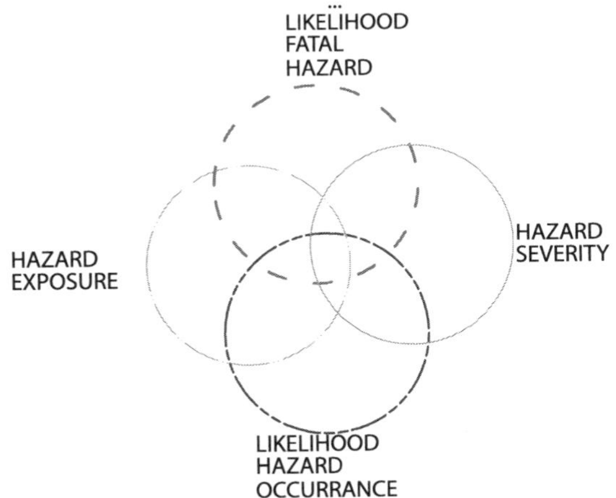

FIG. 14.14 Risk component combination.

3 DISCUSSIONS ON RISK AND RISK ASSESSMENTS

Prior to taking up risk analysis issues, it is better to address some pertinent issues associated with risks and risk analysis. There are two kinds of risks. The first is raw (inherent) risk, which represents the risk before taking any control or mitigation action. The second category is residual risk, that is, the risk that could be faced after putting in place controls or mitigation actions (Fig. 14.2). It has been found that risk is a combination of likelihood of hazard, severity, exposure time, etc. Risk components are therefore severity, the latency (or exposure) period, the likelihood of hazard occurrence, and the likelihood that the hazard leads to an accident. Fig. 14.14 shows the four factors and their combination, which will generate a new risk. Factors affecting risk components are: hazard complexity, exposure, energy, automation, scale, new versus old technology replacements, new technology, and pace of change.

3.1 Risk Frequency

This defines the likelihood of the risk, that is, it stands for the probability of risk. These are categorized as: very likely: (at least once in six months), likely: (at least once a year), unlikely: maybe once in lifetime, and very unlikely: Maybe 1%.

Typical examples are shown here. Risk frequency data and release data are available in HSE (UK), OREDA, and OGP publications also.

3.2 Severity

Severity is loosely used to indicate the impact of risk, that is, consequence. These are slightly harmful (e.g., superficial cuts, minor cuts, etc.), harmful (e.g., burns, serious pains, minor fractures), and extremely harmful (e.g., major fractures, amputation). There are some other ways to categorize also. Typical categorizations could be:

1. Minor: Minor system damage without causing injury.
2. Major: Low-level exposure to personnel, activates public alarm.
3. Critical: Minor injury to personnel, fire, or release of chemicals to the environment.
4. Catastrophic: Major injury, death, big leakage (e.g., the Bhopal gas leak).

3.3 Risk Level (Based on Action and Time)

The levels of risk are often categorized based on the potential. These are as follows:

1. Very low: These risks are acceptable and may not need any action.
2. Low: No control may be necessary unless these are available at low cost.
3. Medium: Suitable considerations shall be there to see if the risk can be lowered, wherever applicable, to a tolerable level within a defined time limit. However, due considerations shall be given for the additional cost for risk reduction. Whenever the risk is associated with harmful consequences, proper maintenance of risk reduction controls is essential.
4. High: Good amount of effort is applied to reduce risk on an urgent basis within a defined time frame. It is essential to give due consideration toward the choice between suspending or restricting the activity or to apply interim control measures until the main risk reduction control is implemented. Whenever the risk is associated with a harmful consequence, it is necessary to make sure that risk reduction controls are properly maintained.
5. Very high: Unacceptable. Substantial improvements in risk reduction control measures are necessary to reduce the risk to an acceptable level. Activities need to be halted until risk reduction control is implemented. Otherwise, work shall remain prohibited.

Risk associated with very harmful consequences needs risk assessment and analysis. The above categorizations are qualitative in nature. For quantitative calculations, one may need to take the help of probability and associated software, which is also available from various agencies for different applications. Interrelations among these factors are depicted in Fig. 14.14.

3.4 Control Measure and Risk Target

From hazard analysis, it transpires that there will be a risk target and some control measures are taken to bring the risk to an acceptable limit. Here it will be discussed.

1. **Control Measure**: In hazard analysis, one part is to recommend control measures to reduce risks. There are several control measure categories possible; they are shown in Fig. 14.15. In control measures, the preferred actions are to climb up the hill, as shown in Fig. 14.15 (based on health and safety authority, www.hsa.ie/eng). These control measure categories are discussed below [1].
   - Elimination of hazard: Total elimination is not always possible, but in some cases it is. For example, if unleaded petrol is used, attendants are not facing the hazard of lead contamination.
   - Substitute: Sometimes, to eliminate a hazard in a particular material or system, for example, a vulnerable thing can be substituted by other for example, on account of health hazard, DEHA is substituted for hydrazine to scavenge oxygen from boiler water.
   - Isolating: Isolating the hazard is achieved by restricting hazard propagation in plants and equipment. For example, in a hazardous area, process

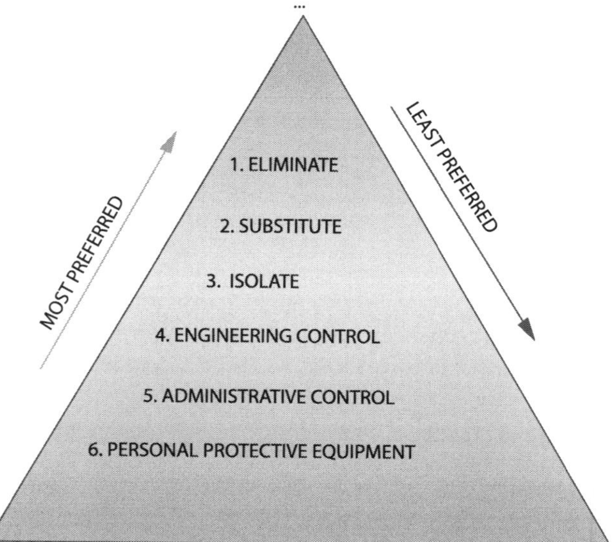

FIG. 14.15 Control measures for risk control.

transmitters use a flame-proof enclosure or IS circuit so that the hazard does not reach the electrical circuit or sufficient energy does not reach the hazardous areas.
- Engineering control: By redesigning the process and putting in barrier Machinery guard is an example.
- Administrative control: Adaptation of safe control practices and procedures through training. Personal protective equipment: Gloves, helmet, etc. are used for this.

2. **Risk Target**: A risk target is a measure that expresses the consequence of a risk in relevant terms of the project and the organization concerned. Value is a function of risk and return. So, any decision is associated with a value, that is, it either increases, preserves, or erodes value. This is well depicted in Fig. 14.16. This is self-explanatory. Enterprise would always like to set the risk target in sweet spot depicted to get the best possible value i.e., it accepts some target risk [6]

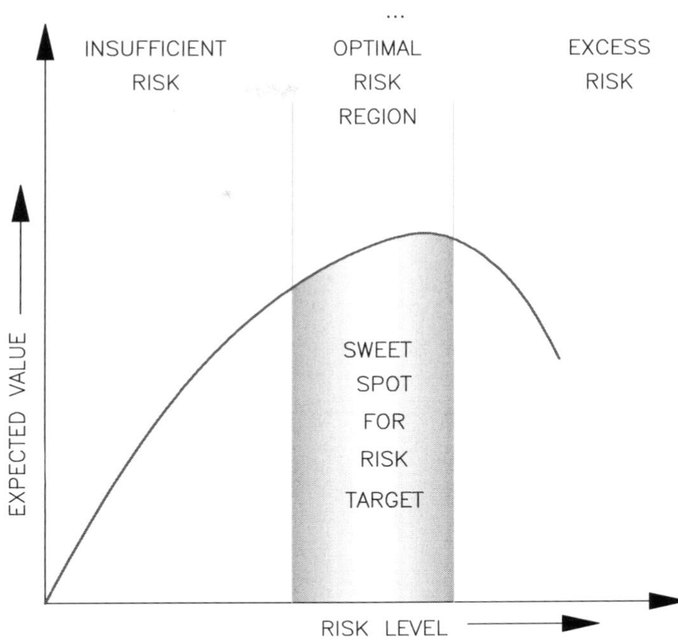

FIG. 14.16 Risk target [6].

3.5 Risk Analysis and Assessment [7]

For plant hazard analysis, there will be always a risk target, which is a measure that expresses the consequence of a risk in relevant terms of the project and organization concerned. In order to get the measure, it is necessary to go for risk analysis and risk assessment.

1. **Risk Analysis:** As per the latest version of IEC/ISO 31010 (IEC 60300-3-9), risk analysis is the "systematic use of available information to identify hazards and to estimate the risk to individuals, populations, property, or the environment." So, essentially risk analysis finds, organizes, and categorizes sets of risks.
2. **Risk assessment:** Risk assessment will be clarified from the following activities:
   - Identification of hazard and analysis and evaluation of risk.
   - Find an appropriate way to control and mitigate hazards.
   - The main aim of risk assessment is to remove hazards, or reduce the risk level by adapting necessary control measures to move toward safety.
3. **Risk Assessment procedure:** The following points are major issues considered part of a risk-assessment procedure:
   - Hazard identification and evaluation of risk; likelihood, severity, and level of risk.
   - Standard operating conditions.
   - Emergency situation (nonstandard operation).
   - Review of all associated information.
   - Actual and potential exposure of personnel (latency, frequency, intensity).
   - Environmental impact.
   - Design engineering control and documentation.

Risk register, risk matrix, etc., are tools for risk analysis and assessment.

3.5.1 Risk Register

A risk register is basically a record of identified risks for a project, as shown in Fig. 14.17.

The major characteristic features of a risk register have been listed below:

1. Short description of each risk along with associated consequences.
2. Factors influencing the likelihood and impact.
3. Grading of risks, for example, low, medium, high, extreme, etc.
4. Risk acceptability.
5. Existing and proposed actions for risk mitigation.
6. Key risk indicator (KRI) and upward reporting factor.

3.5.2 Risk Matrix

A risk matrix may be considered as a quantitative or semi-quantitative tool for qualitative hazard analysis. It is very important to develop a risk matrix design very precisely so that there will not be a false sense of security after the risk matrix is done. Simply, if the likelihood or impact of any risk is not properly defined, then as a result of wrong calculations, any particular risk may be considered in the low risk level, but in reality it is not so. In that case, one may be happy to note that it is low level, hence a false sense of security.

RISK IDENTIFICATION									QUALITATIVE RISK ASSESSMENT			RISK RESPONSE ACTION			RISK MONITORING & CONTROL	
SERIAL NO.	RISK ID.	STATUS	CATEGORY	EVENT DETAILS	CAUSE DETAILS	EFFECT DETAILS	+ OR − IMPACT	PRIMARY OBJECTIVE	LIKELIHOOD	CONSEQUENCE	RISK MATRIX (L: LOW) (M: MEDIUM) (H:HIGH) (E:EXTREME)	STRATEGY	ACTION DETAILS	RESPONSIBLE PERSON	INTERVAL	STATUS REVIEW COMMENTS DATE
01	PRF 001	ACTIVE	EXTERNAL	PROJECT FUND PART LOAN NOT RELEASED IN TIME	BUDGET CONSTRAINT ALLOCATION CHANGED CURTAIL IN INTERIOR	PROJECT DELAY	(−) IMPACT	TIMING	HIGH	VERY HIGH	LIKELIHOOD 1 2 3 4 5 H E L M 1 2 3 4 5 CONSEQUENCE X	MITIGATE	PHASING IN PLANNED ALLOTMENT, HENCE WORK PHASED OUT	CHIEF PLANNING ENGINEER AND FINANCE OFFICER	MONTHLY	BANK FOLLOW UP WITH GOVT. LICENCE FINAL DEADLINE DATE DD.MM.YYYY
02	PSC 001	ACTIVE	CONSTRUCTION	UNIDENTIFIED UTILITY TRANSPORTATION COST	ADDITIONAL CABLE COST AND LAYING COST	ADDITIONAL PROJECT COST IMPACT	(−) IMPACT	COST	MEDIUM	MEDIUM	LIKELIHOOD 1 2 3 4 5 H E L M 1 2 3 4 5 CONSEQUENCE X	MITIGATE	PURSUE WITH UTILITY COMPANY FOR ALTERNATE ROUTE	ELECTRICAL ENGINEER	MONTHLY	LETTER TO UTILITY COMPANY DD.MM.YYYY

FIG. 14.17 Risk register (typical).

There are several standard guidelines and published risk matrices, but at the beginning one has to decide the intent for which it is to be developed. Table 14.8 is an example of a risk matrix available from the Center for Chemical Plant Safety (CCPS). Here, risk levels are described as I,II, III, and IV.

The following tables, which are self explanatory, show various features of risk matrices. They also show how a risk matrix could be qualitative as well as quantitative. Table 14.9 indicates the risk levels.

Likelihood and consequence ranges are explained by Tables 14.10 and 14.11.

As stated earlier, risk matrices could be qualitative, semiquantitative, or quantitative. A typical quantitative risk matrix is shown in Table 14.12.

Here, it is to be noted that both frequency and consequences are quantified. However, frequency and consequence can be qualitative also, namely:

Qualitative frequency terms could be:

- Frequent.
- Probable.
- Occasional.
- Remote.
- Improbable.
- Incredible.

Similarly, a qualitative consequence could be:

- Catastrophic.
- Critical.
- Marginal.
- Negligible.

TABLE 14.8 Risk Matrix

Frequency	Consequence			
	1	2	3	4
4	IV	II	I	I
3	IV	III	II	I
2	IV	IV	III	II
1	IV	IV	IV	III

TABLE 14.9 Risk Level

Risk Level	Category	Description
I	Unacceptable	Should be mitigated by engineering and/or administrative control to a risk level III or less, within a specified period (say six months).
II	Undesirable	Should be mitigated by engineering and/or administrative control to risk level III or less, within a specified period (say 12 months).
III	Acceptable with controls	Should be verified that procedures and controls are in place.
IV	Acceptable	No mitigation required.

TABLE 14.10 Likelihood Ranges Based on the Levels of Protection

Likelihood Range	Quantitative Frequency Criteria (Typical)
Level 4	Initiating event or failure (e.g., leakage/rupture)
Level 3	One-level protection (e.g., pipe leakage, overload)
Level 2	Two-level protection (e.g., electric actuator uprooting)
Level 1	Three-level protection (for example, vessel failure)

TABLE 14.11 Consequence Range

Consequence Range	Quantitative Safety Consequence Criteria
4	Onsite/offsite: Potential for multiple life threatening injuries or fatalities
	Environmental: Uncontained release with potential for major environmental impact
	Property: (including plant): Plant damage value in excess of $100 million
3	Onsite/offsite: Potential for single life threatening injury or fatalities
	Environmental: Uncontained release with potential for moderate environmental impact
	Property (including plant): Plant damage value in the range of $10–$100 million
2	Onsite/offsite: Potential for an injury that requires medical attention
	Environmental: Uncontained release with potential for minor environmental impact
	Property (including plant): Plant damage value in the range of $1–10 million.
1	Onsite: Potential for injuries that require only first aid Offsite: Noise or odor.
	Environmental: Contained release with local impact only
	Property (including plant): Plant damage value in the range of $100,000 to $1 million.

TABLE 14.12 Quantitative Risk Matrix

	Consequence			
Probability	**$1000**	**$10,000**	**$1,000,000**	**$10,000,000**
Every month	Medium	High	High	High
Every year	Low	Medium	High	High
Once in ten years	Negligible	Low	Medium	Medium
Once in hundred year	Negligible	Negligible	Low	Low

Now let us look at Table 14.12, and note that when either the frequency or consequence set of values is replaced by a qualitative set of values mentioned above, then the matrix would be semiquantitative. Similarly, when both the frequency and consequence set of values are replaced by a qualitative set of values, the risk matrix would be qualitative.

With this idea on risk analysis, let us look at what is needed by standard IEC 61508 or 561511 for the safety lifecycle.

3.6 Risk Graph

In the preceding section, the risk matrix was discussed. A risk graph is similarly another way to categorize risk. This also takes into account exposure to risk and it can be used to determine the SIL in process industries (especially when there are many hazards to analyze). The discussion presented here is generalized, hence uncalibrated. However, for each application, one needs to calibrate the same, that is, a project-specific calibration. In this approach, there are four sets of parameters: C, F, P and W. Each is arranged as a column, as shown in Fig. 14.18. All these parameters "describe the nature of the hazardous situation when safety instrumented systems fail or are not available. One parameter is chosen from each of four sets and the selected parameters are then combined to decide the safety integrity level allocated to the safety instrumented functions" (Annex D1of IS/IEC 61511-3:2003). Based on consequence, the first hazard is divided; these are Ca through Cd. Such categorization is based on injury/fatality, as shown in Fig. 14.18. Then, exposure time is under consideration, that is, how much time a person is exposed to the hazard represented by F as shown. These are rare exposures or frequent exposures. The hazard event is tested for probability (or likelihood; represented by P as shown) of avoidance. Both the exposure rate and probability of avoidance are combined with all four consequences as shown. Now, based on

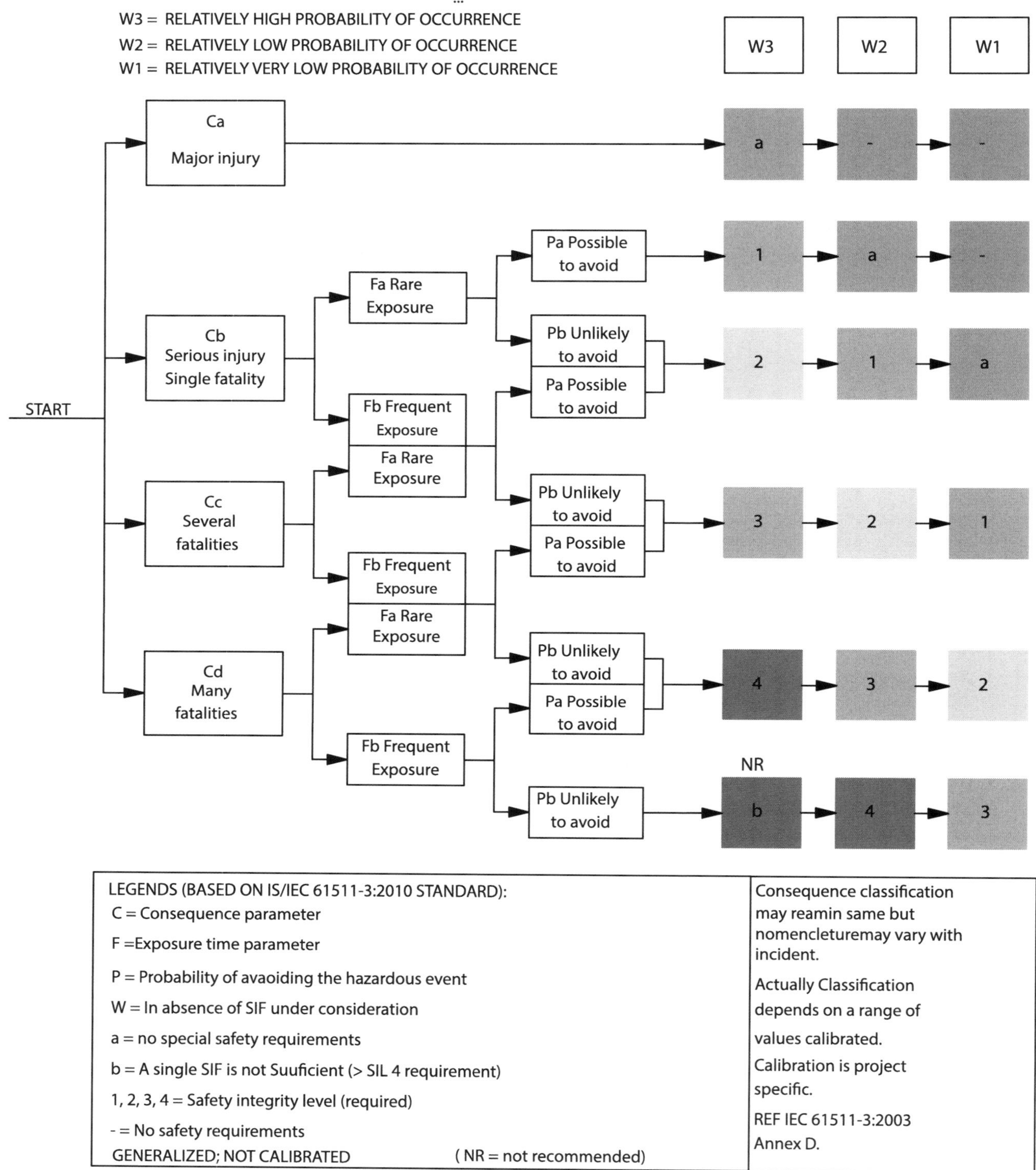

FIG. 14.18 Risk graph.

frequency of occurrence (represented by W as shown), the SIL is decided. Compare the same with the risk matrix, where consequence and frequency of occurrence were the main considerations only, but the SIL was not really assigned from there. The reader's attention is drawn to the top right side SIL box to the bottom left SIL box. The bottom left box is b, meaning not a recommended operation, that is, no single SIF is sufficient (should be avoided). On the other hand, the top right side says no safety requirement. This is a semiquantitative approach. The reader may refer to Annex D of (IS)/IEC 61511-3:2003 for further details and for calibration.

4 SAFETY LIFECYCLE

The safety lifecycle for any system is based on IEC61508, IEC 61511, and ISA 84.00.00. Because ISA standards have equivalence with IEC standards and, in many cases, the IEC standards have also been mentioned, this is not discussed separately. As already indicated in the preamble, the purposes of these two standards are different so naturally the approaches are also different. One thing common to all these is that each of them has three stages: analysis, implementation/realization, and operation. The basic idea about the safety lifecycle can be conceived as shown in Fig. 14.19. In addition, there will be planning and management sections. All stages verification are very important. For a detailed explanation and interpretation of the entire process, [1] may be referenced. Conceptually, the safety lifecycle can be represented as a cyclic system [3]. This is shown in Fig. 14.19.

4.1 Safety Lifecycle Approach

Through hazard analysis, calculated risk in a process is identified prior to suggesting control measures to reduce the same. The gap between the calculated risk in a process and tolerable risk is normally met by changes in process design, other risk-mitigation methods, and safety instrumentation systems. For instrumentation engineering, in order to offer functional safety, a major concern is safety instrumentation systems. With the publication of safety standards for manufacturing and process industries, the users become more knowledgeable about safety issues. They are also focusing SISs to satisfy their needs with more cost-effective ways through integration with control systems, scalable architectures, and less frequent proof testing. A protective system needs to address the overall health of safety loops in an integrated safety solution from sensor to actuator, including logic solvers. Having accepted that no system is completely immune to failures, one needs to take the necessary measures to ensure that even in case of failure, it should provide a safe condition. During my working tenure, I found that many do not like to put much importance to this issue, in design of may systems and/ plants (e.g., fossil fuel power plant designed even in 2007), because the concerned plant does not handle with explosives or toxic materials. This is not a correct approach. The important issue here is that all enterprises, especially industrial ones, must adapt a safety lifecycle in their system. This will provide ultimate benefits to the system in the long run.

4.2 Failure Types and Safety Lifecycle

For random failure (Ref. 1.5.1.1), a probabilistic performance-based approach could be one way to address the same for E/E/PE systems and the safety integrity level (SIL) could be considered for such purposes. As noted earlier, systematic failures (Ref. 1.5.1.2) are normally connected with design failures or software errors, including incorrect specifications. A safety lifecycle is adapted for systematic failures. So safety standards meant for E/E/PEs take care of both. SISs, discussed later, are developed to prevent or mitigate hazardous events to protect people or the environment or to prevent damage to the process equipment. In this connection, another important issue is SIL discussed later, is a discrete level for specifying the safety integrity requirements of safety functions, but it is not a measure of risk.

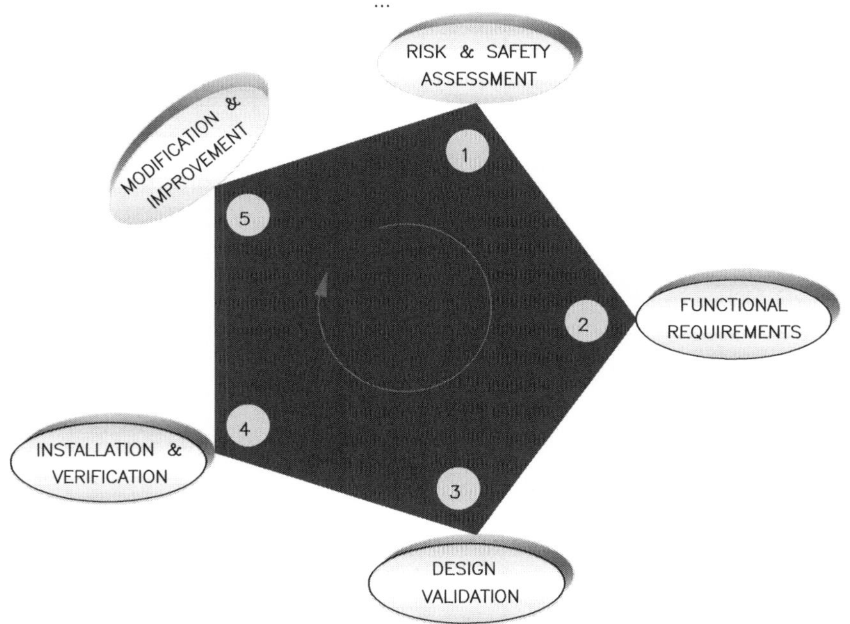

FIG. 14.19 Concept of safety lifecycle.

Safety standards give guidance on best practices and offer recommendations without absolving the users' responsibility for safety. These standards deal with technical issues as well as the planning, documentation, and assessment of all activities required to manage safety throughout the entire life of a system.

4.3 Safety Lifecycle Stages

The safety lifecycle is an important issue in safety instrumentation, and can be categorized into three broad areas. The first is the analysis phase. In this phase, the identification of hazards and hazardous events, the likelihood of these hazardous events, and any potential consequences are handled. Also in this phase, the availability of a layer of protection as well as the need for any SISs and the allocated SIL are covered. The second phase is realization, focusing on the design and fabrication of the SIS. The final phase is operation, which covers startup, operation, maintenance, modification, and eventual decommissioning of the SIS. These phases encompass the entire lifecycle process of the safety system from concept through decommissioning.

With this concept in mind, safety lifecycle discussions are presented below. With reference to Fig. 14.20, which depicts the safety lifecycle as per standard IEC 61508 reproduced here with the permission of IEC (Ref At the beginning of the chapter—e mail dt 20th July'2018) and duly acknowledged at the beginning of the chapter.

The differences between the two standards are indicated in Table 14.13.

It is also pertinent to have an idea about the responsibilities of the end user and manufacturer. This has been also included in the standard and is reflected in Table 14.14.

In the following section, both discussions are elaborated upon.

4.4 IEC 61508 Safety Lifecycle

The following issues are the main pillars on which this safety standard is standing:

- System lifecycle.
- System subdivisions.
- Functional safety.
- Risk reduction.
- SIL.
- ALARP.

In the standards, there are two types of subsystems (Clause 7.4.3.1.2 P2 of standard), which are indicated in Tables 14.15 and 14.16 (Tables 1 and 3 of IEC 61508-2).

As indicated above, there will be three stages; we will start with analysis.

4.4.1 Analysis Part of IEC 61508

The following stages are under the analysis part.

Concept: Understanding of the equipment under control and its environment (physical and legal) to determine hazard sources and hazard information; hazard interaction with other equipment.

Overall scope definitions: Here, the system boundary and hazard scopes are defined.

Hazard and risk analysis: Here, the list of hazards and events are prepared with a sequence. It is followed by finding the likelihood and consequence (Section 3.5).

Safety reallocation: This includes overall safety requirements and safety reallocation.

4.4.2 Analysis Part of IEC 61508

This section deals with technology and architecture selections. Here, the major issues are:

- Perform reliability and safety evaluations to determine if you met your target SIL requirement.
- Conceptual design of SIS.
- Detailed design of SIS.
- System development includes detailed design and engineering, installation planning, installation, and commissioning, including acceptance tests.

Here, it is worth noting that there are two parts associated with the PE system for realization/implementation. These are hardware (HW) and software (SW) implementation parts, including the specification of safety requirements and the safety integrity for each HW and SW part. There should be proper means to validate the planning, design development, and integration of each HW and SW part. Complete system/overall validation. This stage also includes the installation and commissioning of the entire system meant for the safety system.

4.4.3 Operation Part of IEC 61508

This really starts with design validation through operation and maintenance to check whether the system really solves the safety issue. Necessary modifications, including overall modification and retrofitting as applicable, are to be carried out to verify proper implementation, that is, a review of safety lifecycle activities, and ensure that all steps were carried out and documentation is in place [2]. The final step is decommissioning or disposal.

The system has been described here very briefly. It is recommended to refer to Section 4 of Chapter 6 of [1] for gathering detailed ideas on the same. Now, it is time to concentrate on IEC 61511.

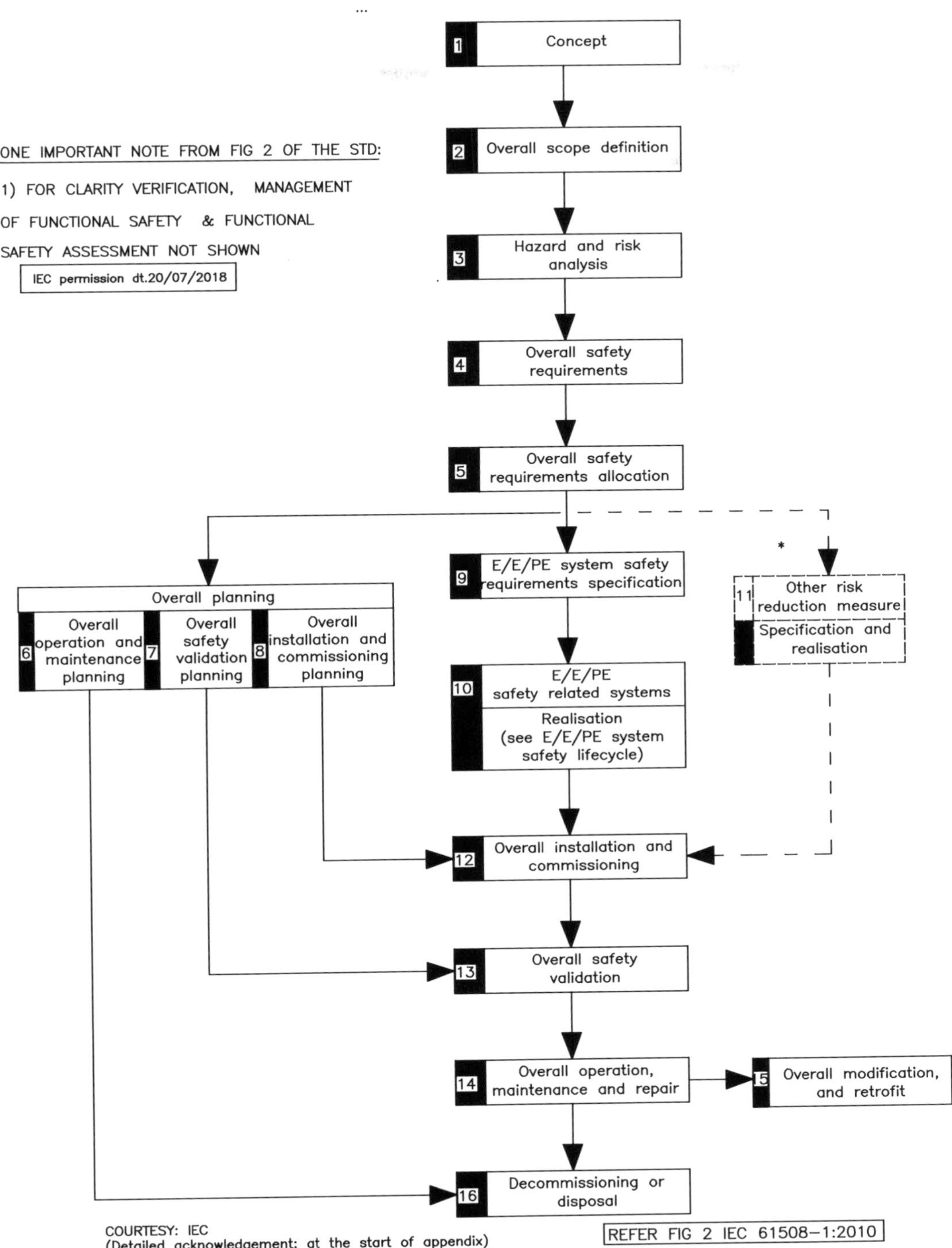

FIG. 14.20 Safety lifecycle phase IEC 61508 [Refer Acknowledgement for permission (dt. 20.07.2018) by IEC].

4.5 IEC 61511 Safety Lifecycle

Let the discussion start with Fig. 14.21, which depicts the safety lifecycle described in standard IEC 61511, reproduced here as per permission from IEC (Ref: E-mail dated July 20, 2018) duly acknowledged at the beginning of the chapter.

Here, it is noted that each phase number is given in a small box in the left bottom-most corner of each box in Fig. 14.21. Also, applicable clause numbers are indicated in each of the main boxes (with the exception of Clause 1). There are three parts of IEC 61511. Some detailing of Part 1 has been elaborated in Table 14.17. Parts 2 and 3

TABLE 14.13 Differences Between IEC 61508 and IEC 61511

Issues	IEC 61508	IEC 61511
Nature	Generic safety standard	Specific safety standard
Industry	Wide range of industries	Process industries only
Divisions	Technical and other requirements	Mainly analysis, realization, and operations
Coverage	Safety-related systems including external risk-reduction systems	Mainly on safety instrumentation systems
Focus	Major focus on supplier community (that is, manufacturing)	Primary focus on system designer, integrator, user, and end user

TABLE 14.14 Phase-Wise Responsibility of the End User and Supplier (Manufacturer)

Phase	Responsibility	Activities
Pre design Phases 1–5 and 9	End user	Hazard identification, specify requirements, and setting up target SIL
Design and Installation Phase 6–8, 10–13	Manufacturer/ supplier	Development as per requirements and get target SIL
O and M Phases 14–16	End user	Operation, maintenance, modification, and SIL maintenance

TABLE 14.15 Definition/Requirements of Type A and B Subsystems

Conditions	Type A (All Necessary)	Type B (Any One)
Failure mode	Well defined for constituent components	For at least one constituent, component not well defined
Faulty subsystem	Behavior completely determinable	Cannot be completely determined
Dependable failure data	Sufficient failure data available from field	Insufficient failure data available from field

TABLE 14.16 Hardware Safety Integrity: Architectural Constraints

Safe Failure Fraction	Type A (as per Table 14.15) Hardware Fault Tolerance[a] 0	1	2	Type B (as per Table 14.15) Hardware Fault Tolerance[a] 0	1	2
<60%	SIL 1	SIL 2	SIL 3	Not allowed	SIL 1	SIL 2
60% to <90%	SIL 2	SIL 3	SIL 4	SIL 1	SIL 2	SIL 3
90% to <99%	SIL 3	SIL 4	SIL 4	SIL 2	SIL 3	SIL 4
>99%	SIL 3	SIL 4	SIL 4	SIL 3	SIL 4	SIL 4

[a] *A hardware fault tolerance of* N *means that* N + 1 *faults could cause a loss of the safety function.*

of the standard, "Guidelines for the application of Part 1" and "Guidelines for determination of the required SIL," respectively, are not really relevant for the table. Clause 19, belonging to Part 1, is for information requirements, and Annex A of Part 1 is for differences with IEC 61508. Basically, it is the management of functional safety and safety assessment with safety lifecycle structure and planning. The system is completed with proper verification. Here also there are three stages: analysis, implementation, and operation.

4.5.1 Analysis Part of IEC 61511

The basic structure of analysis consists of;

1. Hazard and Risk Assessment (Clause 8).
2. Allocation of a safety function protection layer (exists around the BPCS in different forms as additional protection for the system) (Clause 9).
3. Specification of SIS as safety requirement.

Frankly speaking, Subsections 3.2.1.1 and 3.2.1.2 are not part of the IEC standard but external things to interface (refer to Clause 8 and 9 of the standard).

4.5.2 Implementation Part of IEC 61511

Basic structure of Analysis part consists of;

Design and engineering of SIS.

Design and development of other means of risk reduction (Clause 9 not part of standard).

Installation, commissioning, and validation.

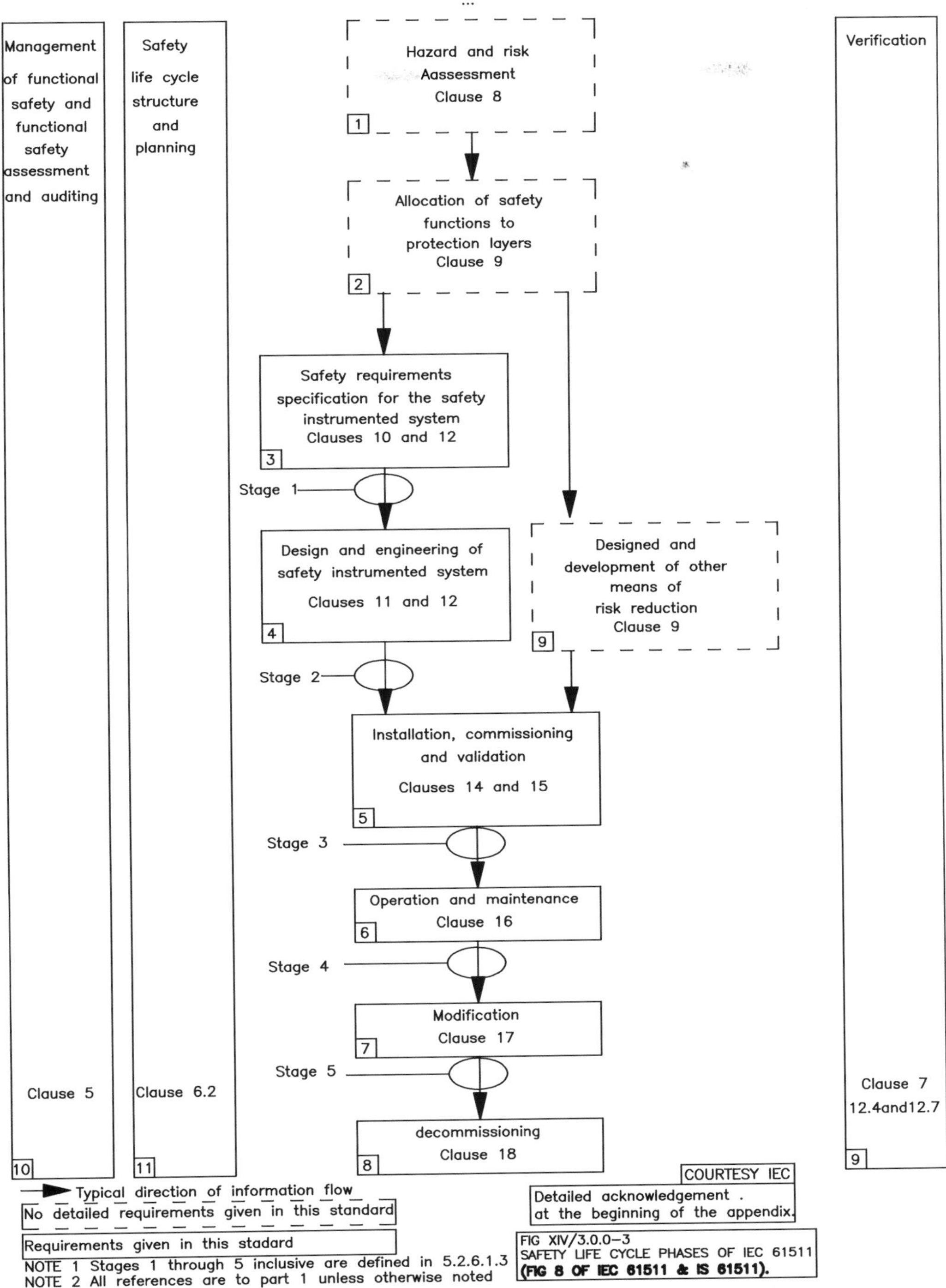

FIG. 14.21 Safety lifecycle phase IEC 61511 [Refer Acknowledgement for permission (dt.20.07.2018) by IEC].

4.5.3 Operation Part of IEC 61511

This stage starts after erection and commissioning are over. Here, the stages are:

Operation and maintenance.
Modification (as required).
Decommissioning.

Throughout all stages, there will be a verification that the end user needs to verify to establish the safety system.

Basic requirements for the lifecycle have been established. However, so far flow metering is concerned main issue is around SIS and SIL for the same. Naturally, without discussions of SIS and SIL, the discussions will be

TABLE 14.17 Elaboration of Framework of IEC 61511 (Based on Fig. 14.1 of Standard)

Cl	Ph	Heading	Activity	TR/SP	PC	Responsibility
2		Reference		SP		
3		Definition abbreviation		SP		
4		Conformance		SP		
5	10	Management of functional safety		SP		ALL
6	11	Safety lifecycle requirements—structure and planning		SP		ALL
7	9	Verification		SP		ALL
8	1	Hazard and risk assessment (concept scope)	Hazard identification and requirement specification	TE	A	End users and operators
9	2	Allocation of safety functions to protection layers		TE	A	
10 12	3	Safety requirement for safety instrumentation system (SIS) Clause 12 also	Configuration, Meeting requirements	TE	A	Engineering and equipment supplier
11 12	4	Design and Engineering SIS (**also 12**)		TE	R	
13,14,15	5	Installation, commissioning and validation (**also 13 and 15**)		TE	R	
16	6	Operation and maintenance	Operate maintain modify		O	End users and operators
17	7	Modification			O	
18	8	Decommissioning			O	

Clause no. 1 mainly refers to the scope of the standard.
Used abbreviations: Cl, clause no; Ph, phase; TR, technical requirement; SP, support part; PC, phase category; A, analysis; R, realization; O, operation.

incomplete. So, the discussions will be completed with the discussions on SIS and SIL in the next section.

4.5.4 Summery

From the above discussions, it is clear that in each of the stages, there will be some functions to be performed, which are summarized below:

1. **Analyzing phase:** The following are the major steps:
   - Experiment design.
   - Hazard identification.
   - Risk assessment.
   - Comparison to risk tolerance criteria.
   - Risk reduction allocation.
   - Safety function definition.
   - Safety function specification.
   - Reliability. Verification
2. **Implementation phase:** The following are the major steps:
   - Equipment design.
   - Software configuration.
   - Equipment build (61508).
   - Factory acceptance testing (61508).
   - Construction/installation.
   - Site acceptance testing.
   - Validation.
   - Training.
   - Prestartup safety review.
3. **Operation phase:** The following are the major steps:
   - Operation.
   - Training.
   - Proof testing.
   - Inspection.
   - Maintenance.
   - Management of change.
   - Decommissioning.

4.5.5 IEC 61511 Discussions on Management of Functional Safety

In line with IEC 61508, short details on various phases of IEC 61511 are described above. Because IEC 61511 is very much applicable for power plant applications, the same has been discussed further here. Let the discussion start with management of functional safety (Clause 5). Identification of management activities and ensuring meeting the requirements of functional safety are the main objective of the section. Major issues in this connection are:

1. General:
   - The policy and strategy for safety identification.
   - Evaluation of achievement.
   - Communication in the organization.
   - Assurance of usage of safety instruments.
2. Organization and resources:
   - Persons, departmental organizations.
   - Competence to carry out the job.
3. Risk evaluation and management.
4. Implementing and monitoring.
   - Hazard analysis and risk assessment.
   - Assessment and auditing.
   - Verification.
   - Validation.
   - Postincident/accident activities.
   - Procedure for quality management.
   - Evaluation of SIS performance.
5. Assessment auditing and revision.
6. SIS configuration management.

4.5.6 Standard IEC 61511 and Safety Lifecycle Requirements

Phase definition and establishment of requirements for safety lifecycle activities are two of the major objectives, which also include planning and technical activities, to ensure that SIS meets the safety requirements. The major issues are listed in Table 14.18.

4.5.7 Some Revisions in Edition 2 of The Standard (61511)

In this section, some of the changes in the second edition of standard 61511 are elaborated upon below.

1. SIS design and engineering
   In terms of hardware fault tolerance (HFT) requirements, the revised requirements are now based only on the type of device (A or B type, Section 4.4) and the target SIL. The safe failure fraction (SFF) concept has been abandoned in IEC 61511 Edition 2. The new requirements for architectural constraints are presented in Table 14.19 below.
2. Systematic Capability: A "systematic capability" concept has been included within IEC 61511 Edition 2 and has been further aligned to IEC 61508 Edition 2

TABLE 14.18 SIS Lifecycle Overview—Part (IEC 61511-1)

Box	Title	Objective	Clause Ref
1	Hazard and risk assessment	Determination of sequence of event leading to hazard, hazard events with risk involvement with the same, risk reduction and safety functions for risk reduction.	8
2	Allocation of safety functions to protection layers	Allocation of safety function to protection layers and each SIF and associated SIL.	9
3	SIS safety requirement specification	To specify requirements to each SIS in terms of required SIF with associated safety integrity to achieve functional safety.	10
4	SIS design engineering	SIS design to meet SIF and associated safety integrity.	11, 12.4
5	SIS installation, commissioning, and validation	SIS integration and testing. Validation of SIS in all respects of requirements in terms of SIF and associated safety integrity.	12.3, 14, 15
6	SIS operation and maintenance	Maintenance of functional safety during operation and maintenance	16
7	SIS modifications	Correcting, enhancing, and adapting SIS to ensure that the required SIL is achieved and maintained.	17
8	Decommissioning	Reviewing and ensuring that the appropriate SIF is maintained.	18
9	SIS verification	Testing and evaluation of output at a given phase with respect to products and standards for the given input.	7, 12.7
10	SIS functional safety assessment	Investigation and determining that functional safety is achieved by SIS	5

TABLE 14.19 SIL and HFT

SIL	Minimum Required HFT
1 (Any mode)	0
2 (Low demand)	0
2 (High and continuous demand)	1
3 (Any mode)	1
4 (Any mode)	2

requirements. The main intent of the "prior use" evaluation has also been better expressed.

3. Application Program: In Edition 2 application software has been replaced by application programs.

4.5.8 Sort Details About Safety Requirement Specification (SRS)

In connection with power plant applications, safety requirement specifications should include the following at a minimum. In this connection, Clause 10 of IEC 61511-1 may be referenced.

- Functional details of instrumentation for functional safety.
- Issues related to common cause failures.
- Defining the safety state of the process for safety instrumentation and concurrent safety states.
- Demand and demand rate for safety instrumentation.
- Proof test interval.
- Response time to bring to a safe state.
- SIL and mode of operation (on demand/continuous).
- Safety instrument measurement details with trip points.
- Process relation of input/output and process logic requirements.
- Output action and criteria for successful operations.
- Failure mode and required response.
- SIS and BPCS and other interfaces.
- Manual shutdown.
- Application software details.
- Inclusion of override, inhibition requirements, etc.

We now end this brief discussion on standards for the safety lifecycle to move on to discussions on safety instrumentation systems.

5 SIF SIL AND SIS

By this time, it is clear that safety instrumentation is independent of BPCS and is used to take the system into a safe state. Fig. 14.2 shows various methods of risk reduction in a common figure to include all risk reduction methods. Here, SIS is the main concern to us. An SIS is one of the most commonly used and engineered safeguard systems offering good flexibility to the designers. It is better to address the first barrier, then to SIFs. Safety functions are a type of barrier. Barrier functions, specifically safety barriers, are planned for prevention, regulation, and mitigation of undesired events. This will be clear from an example of a fire situation with several safety barriers, for example, a smoke detection system, a heat/thermal detection system, an exit signal/exit gate, a manual call/alarm, smoke control—air handling, a fire door/fire wall, a sprinkler system, and an evacuation/emergency plan. Out of these, the first four are detection and alerting safety barriers, whereas the last four are protection-type safety barriers that can be classified into active/passive barriers, proactive/reactive, high-demand/low-demand, and technical, human, or organizational. All these can also be seen from the examples of fire (e.g., detection parts are proactive whereas protection parts are reactive). In fact, the safety barriers can be categorized according to their safety influence. Safety barriers need to be evaluated before selection. These safety barriers actually are provided as layers of protection, something similar to several layers in an onion (each layer is an independent protection layer). As per layers of protection analysis (LOPA), safety barriers or independent protection layers (IPL) can be credited with risk reduction if they are:

- Effective in preventing the consequence.
- Independent of the initiating event.
- Independent of other credited independent protection layers (IPLs).
- Auditable.

Safety instrumentation systems actually act as one of the protecting layers. For further details on LOPA, see Section 4 of Chapter 5 of the book in reference [1].

According to Clause 3.2.68 of IEC 61511-1:2003, a safety function is defined as a, "function to be implemented by an SIS, another technology safety-related system, or external risk reduction facilities, which is intended to achieve or maintain a safe state for the process, with respect to a specific hazardous event." Again, according to IEC 61511-1:2003, Clause 3.2.71, "safety instrumented function (SIF): a safety function with a specified safety integrity level (that) is necessary to achieve functional safety and which can be either a safety instrumented protection function or a safety instrumented control function." So, there are basically two types of SIFs: a safety instrumented control function (Fig. 14.22) in continuous mode or a safety instrumented protection function operating in the demand mode. An SIS explanation and the interrelation between SIS and SIF in line with IEC 61511-1:2003 is depicted in Fig. 14.22 for better understanding.

Let's examine two modes for a boiler/furnace fire rate that is controlled continuously as per demand to make sure that the furnace environment is always air-enriched and the air-fuel ratio is maintained. This is a safety instrumented

According to IEC 61511 *Safety instrumented control function* stands for safety instrumented function with a specified SIL operating in continuous mode which is necessary to prevent a hazardous condition from arising and/or to mitigate its consequences. *Safety instrumented control system* is instrumented system used to implement one or more safety instrumented control functions. Also *Safety instrumented system (S1S)* is instrumented system used to implement one or more safety instrumented functions (SIF). An S1S is composed of any combination of sensor (s), logic solver (s), and final elements(s)

FIG. 14.22 SIS details as per IEC 61511-1:2003.

control function, and could be a part of a basic plant control system (BPCS) also (**or** can be an independently cooked signal put in a control loop as an inhibiting signal). On the other hand, flame out (total) in a boiler or furnace could cause the release of toxic and inflammable gases such as CO. This is a specific hazard, and there is a demand on the safety instrumented function part of a system through which the automatic protective interlock will trip or close the main fuel valve to stop any inflow of fuel to the furnace and bring the system to a safe state. This action will initiate the master fuel relay (MFR) trip. Another important issue here is that corresponding to each hazard, there shall be one specific SIF, for example, the closure of the main fuel in the event of loss of flame is one SIF. If there is a very low feed water flow or a very low drum water level, then through the automatic protection interlock, the MFR will be tripped. From here, one can conclude that the SIS consists of a number of SIFs that are included for each set of hazards. As far as MFR is concerned, all separate SIFs have the same result (tripping of MFR), which in turn operates different actuators to bring the system to a safe state. By definition, each SIF needs to have one SIL and it depends on the amount of risk reduction by the SIF for the particular hazard. The SIL is selected based on the risk posed by the hazard. From previous discussions, it is well known that risk is constituted by the consequence and frequency of occurrence of the hazard. Therefore, there will be different SILs for different SIFs. Obviously, the highest SIL will result in a safer SIS. Now, there could be some situations when an SIF may have multiple initiating cause scenarios. In such cases, the **highest** SIL corresponding to the scenarios will be used. Refer to Fig. 14.23 for the interrelation among them (Here, the safer state has been shown for the SIF or SIL.)

As has been discussed earlier, there could be several safety instrumented functions (SFI) in any safety instrumented system (SIS) comprising sensors, logic solvers, and final control elements. In this connection, Fig. 14.24 is also used to study SIS, SIF, and SIL together.

1. **Safety instrumented system (SIS) Explanation:** SIS is meant to prevent, control, or mitigate hazardous events and take the process to a safe state when predetermined conditions are violated. An SIS can be one or more SIFs, which are composed of a combination of sensors, logic solvers, and final elements. An SIS or an SIF is extremely important when there is no other noninstrumented way of adequately eliminating or mitigating risks.
2. **Safety Integrity Level (SIL) Explanation:** SIL basically represents, "To what extent can a device or the process be expected to perform safely? And, in the event of a failure, to what extent the process be expected to go to the safe state in the event of a failure!" Therefore, an SIL gives a measure of safety risk or risk reduction to a tolerable limit for a given process. The IEC 61508 standard also specifies the measures such as "fault avoidance" (systematic faults) and "fault control" (systematic and random faults) to be taken in the design of safety functions consisting of sensors, logic solvers, and final elements. From IEC 61508, one knows that SIL depends highly on two major factors: hardware failure tolerance (HFT) and safe failure fraction (SFF).
3. **Hardware failure tolerance (HFT):** This is the ability of hardware to continue to perform a specified safety function in the presence of faults or errors. HFT of N means that $N+1$ faults will cause a loss of safety function for the unit.
4. **Safe failure fraction (SFF):** This is the ratio of the average failure rates, safe plus dangerous detected failure, and safe plus dangerous failure (IEC 61508). SFF will be well understood with the help of Fig. 14.25 in conjunction with Table 14.20.

   Now let us look into SIL more closely.
5. **Probability of failure on demand (PFD):** PFD_{avg}: Probability of failure on demand is the probability of a functional unit or system failing to respond to a demand for action arising out of a potentially hazardous

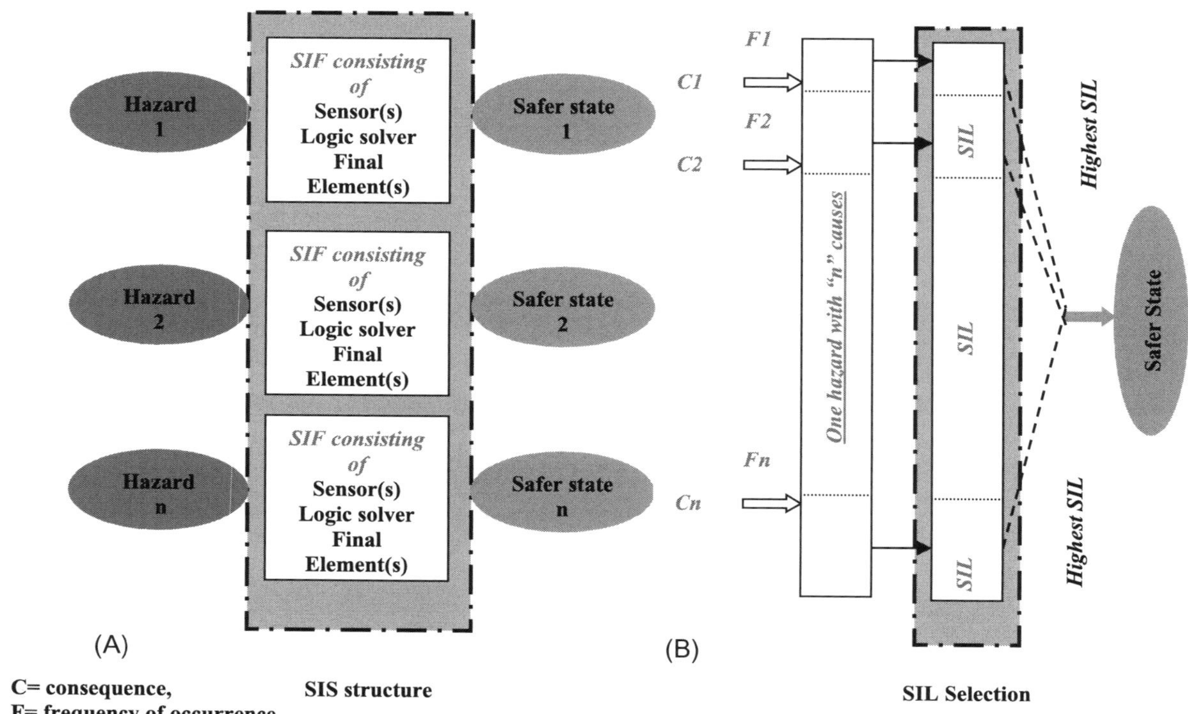

FIG. 14.23 Relationship of SIS, SIF and SIL [1].

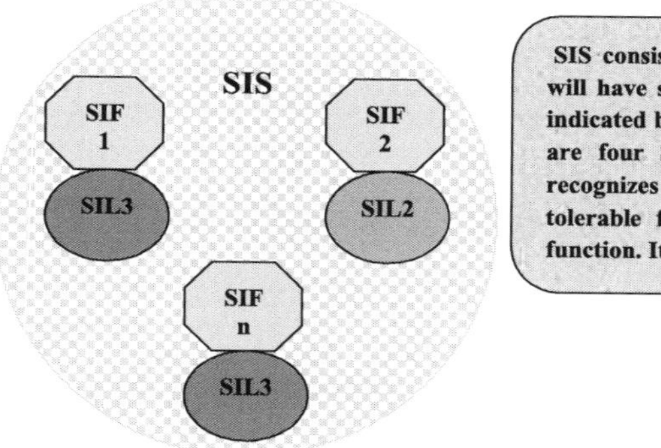

FIG. 14.24 SIS SIF and SIL relations [1].

condition. In other words, a device will fail to perform its specified safety function when it is asked to do so. The probability average PFD_{avg} is used in calculations for system reliability. Probability of failure is directly related with availability.

6. **Availability:** Availability is defined as the probability that the equipment will perform its task.

Now let us look into SIL more closely.

5.1 Safety Integrity Level (SIL) Discussions

As per IEC 61511, each SIF shall have an associated SIL, which is a measure of safety system performance and is related to the probability of failure on demand (PFD) for the associated SIF. There are four defined SILs: SIL 1, 2, 3, and 4. The higher the SIL number, the lower the PFD for the safety system, indicating a better system performance. Also, it has been found that the higher the SIL number, the higher the cost and complexity of the system. SIL is applicable and calculated for the entire SIF system, but not on individual products or components. Reliability and availability of SIF due to SIL are achieved by design, installation, and testing. SIL is also dependent on architectural constraints.

5.1.1 SIL Categories

The following four points on SIL are worth noting:

1. SIL 0/none: lowest risk.
2. SIL 1: 95% of the SIFs.

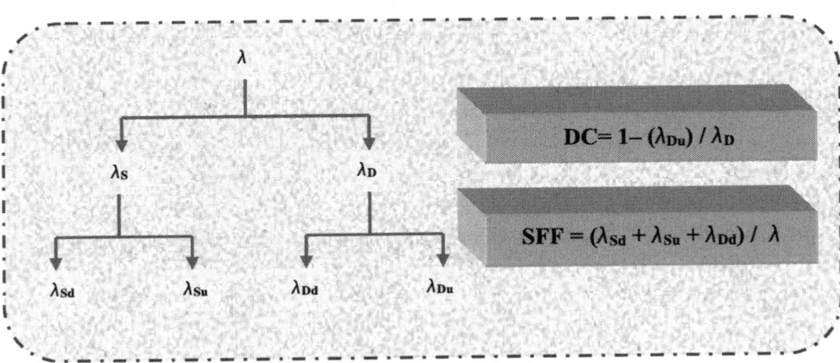

FIG. 14.25 Failure rate SFF and DC with their relationship.

TABLE 14.20 Lists of Symbols for Failure Rate and Related Terms

	Symbols Used to Represent		
Various Terms	Detected	Undetected	Others
Safe failure (λ_S)	λ_{Sd}	λ_{Su}	NA
Dangerous failure (λ_D)	λ_{Dd}	λ_{Du}	NA
Safe failure fraction	NA	NA	SFF
Diagnostic coverage	NA	NA	DC

TABLE 14.21 SIL and PFD (IEC 61508)

Safety Integrated Level (SIL)	Mode of Operation	
	On Demand	Continuous
4	$\geq 10^{-5}$ to $< 10^{-4}$	$\geq 10^{-9}$ to $< 10^{-8}$
3	$\geq 10^{-4}$ to $< 10^{-3}$	$\geq 10^{-8}$ to $< 10^{-7}$
2	$\geq 10^{-3}$ to $< 10^{-2}$	$\geq 10^{-7}$ to $< 10^{-6}$
1	$\geq 10^{-2}$ to $< 10^{-1}$	$\geq 10^{-6}$ to $< 10^{-5}$

3. SIL 2: 5% of SIFs.
4. SIL 3: <1% (offshore platforms/nuclear).
5. SIL 4: highest risk (nuclear industry).

Summarily SIS is meant to prevent, control, or mitigate hazardous events and take the process to a safe state when predetermined conditions are violated. An SIS can be one or more SIFs, which are composed of a combination of sensors, logic solvers, and final elements. SIS or SIF is extremely important when there is no other noninstrumented way of adequately eliminating or mitigating process risks.

TABLE 14.22 Interrelationship Among SIL, PFD_{avg}, Availability, and Consequence

SIL	PFD_{avg}	Availability (%)	Consequence (Fatality)
SIL1	10^{-2} to 10^{-1}	90–<99	Minor onsite injury
SIL2	10^{-3} to 10^{-2}	99–<99.9	Major onsite injury or fatality
SIL3	10^{-4} to 10^{-3}	99.9	Multiple onsite fatalities
SIL4	10^{-5} to 10^{-4}	>99.99	Fatality in the community

SIL 1: lowest SIL; SIL4: highest defined SIL.

5.1.2 *SIL PFD and Availability Interrelations*

The interrelation among SIL PFDs is given in Tables 14.21 and 14.22.

5.2 SIL Determination Techniques

The SIL is determined for the entire loop pertinent to one SIF with the SIL/PFD value specified by the manufacturer. Out of many plant hazard analysis (PHA) techniques, plant hazard analysis (quantitative) techniques such as fault tree, event tree, or layer of protection analysis (LOPA) can be used to determine the SIL. However, prior to that, let us look into the major steps involved to determine the SIL for SIF:

- To carry out the PHA of the process by selecting any suitable method of PHA.
- Reveal the study result with cause, consequence, recommended safety, etc.
- Identification of hazards and existing safeguards.

- Identification and initiating cause and development of a hazardous scenario.
- Quantification efforts for hazard frequency and safeguard reliability.

The following techniques can be utilized to determine the SIL of SIF:

1. Direct by calculation.
2. SIL determination by fault tree.
3. SIL determination by the safety matrix method.
4. SIL determination by risk graph.
5. LOPA for SIL determination.
6. SIL determination by comparison.

The relation between the SIL and associated factors is depicted in Fig. 14.26A; it is like a fire triangle. Another important issue is that because the sensors and final elements are further exposed (in the field) to physical and chemical loading and parameter variations, failure rates are high for the sensors and final elements. Hence a larger share, as shown in the pie chart in Fig. 14.26B, is accountable for the failure of sensors and final elements.

Now let us discuss a few methods of SIL determination. IEC 61511:3 provides guidelines for SIL determination. In this connection, Table 14.23 may be referenced.

SIL by risk graph and by the SIL matrix method are discussed here. For further details, see Chapter 8 of book reference [1]. SIL determination by a risk matrix is shown with the help of Fig. 14.27, where the determination method has been depicted.

SIL determination by risk graph has already been shown in Fig. 14.18 Risk Graph. If one recalls the discussions there, it has been stated that the risk graph is not calibrated and it is to be calibrated for a specific project. Brief guidelines for the calibration method are illustrated in Table 14.24. These are only guidelines in line with standard IEC 61511:3-2003.

Various SIL determination methods are elaborated upon in Table 14.25 as guidance.

All hazard analysis methods are not relevant for power plant applications. LOPA is a good method applicable for major process plants. In order to have conceptual details about LOPA—a quantitative method of hazard analysis and ways of providing an independent protection layer (IPL)—Fig. 14.28 has been presented.

After gathering some knowledge on hazard and risk analysis, SIF, SIS, and SIL (with determination), it is necessary to look into details about safety requirements of various components (exception: the safe fieldbus discussed in Section 1.1.5 of Chapter 7) to complete the discussion on the safety lifecycle. If we look at Fig. 14.28, it is seen that alarm is the first as well as a very important layer applicable for all plants, including power plants. So let us have some details about alarm management, covering alarm philosophy and the alarm lifecycle.

6 ALARM SYSTEM

As alarm system is considered as an independent protection layer (IPL). It is important to have a good idea about the alarm system to complete the discussions of the process safety lifecycle as well as for SIL. While on the subject, a short discussion on the alarm lifecycle is also important. From ISA technical reports, one gets that an alarm is an audible and/or visible means of indication to the operator about equipment malfunction, process deviation, or abnormal condition that requires a response. As per Annex F of IEC 61511-3:2003, an alarm is an independent protection layer. Unlike safety interlocks where actions are automatic, in case of alarm, the action part lies with the operator (manual action) in accordance with a response procedure to stop hazardous event propagation. When an operator's response to an alarm is included in a layer of protection analysis, operator reliability in responding to the alarm should also be considered in making claims for safety alarms. Operator process knowledge and the capability to take correct timely action are part of the behavior of the alarm system as a whole, and need to be **validated** using tests, drills, or simulated environments. Here, the design, management, and performance greatly influence operator action as well as the level of actual risk reduction. For example, if the operator is flooded with alarms and a number of them are spurious, the operator is highly disturbed. This makes it a high probability that even a very knowledgeable operator might miss some important issues. The result of poor alarm management will affect the SIL, and as a consequence the **SIF**. In this regard, the standard ANSI/ISA-18.2, "Management of Alarm Systems for the Process Industries" (ISA-18.2), may be referenced. The basic intent of this standard is to improve the safety of the process through the alarm system. Ineffective alarm systems are often found as contributing factors to accidents. **Alarm rationalization** involves comparing potential alarms with various criteria set in the philosophy document and generating an alarm database. All alarm points are not rationalized, therefore it is necessary to identify the potential alarms for the rationalization process. As alarm response is really a function of man-machine interface and not specific to the process, so it does not vary with variations in process. In order to treat the alarm system as an SIF consisting of sensors, logic solvers, and final control elements, one needs to consider the alarm system, operator, and action together. Therefore, the basic parts of the alarm systems are:

1. Physical sensor,
2. Logic solver to compare the process value with the set point,
3. HMI for audiovisual signal,
4. Operator to detect, diagnose, and respond,

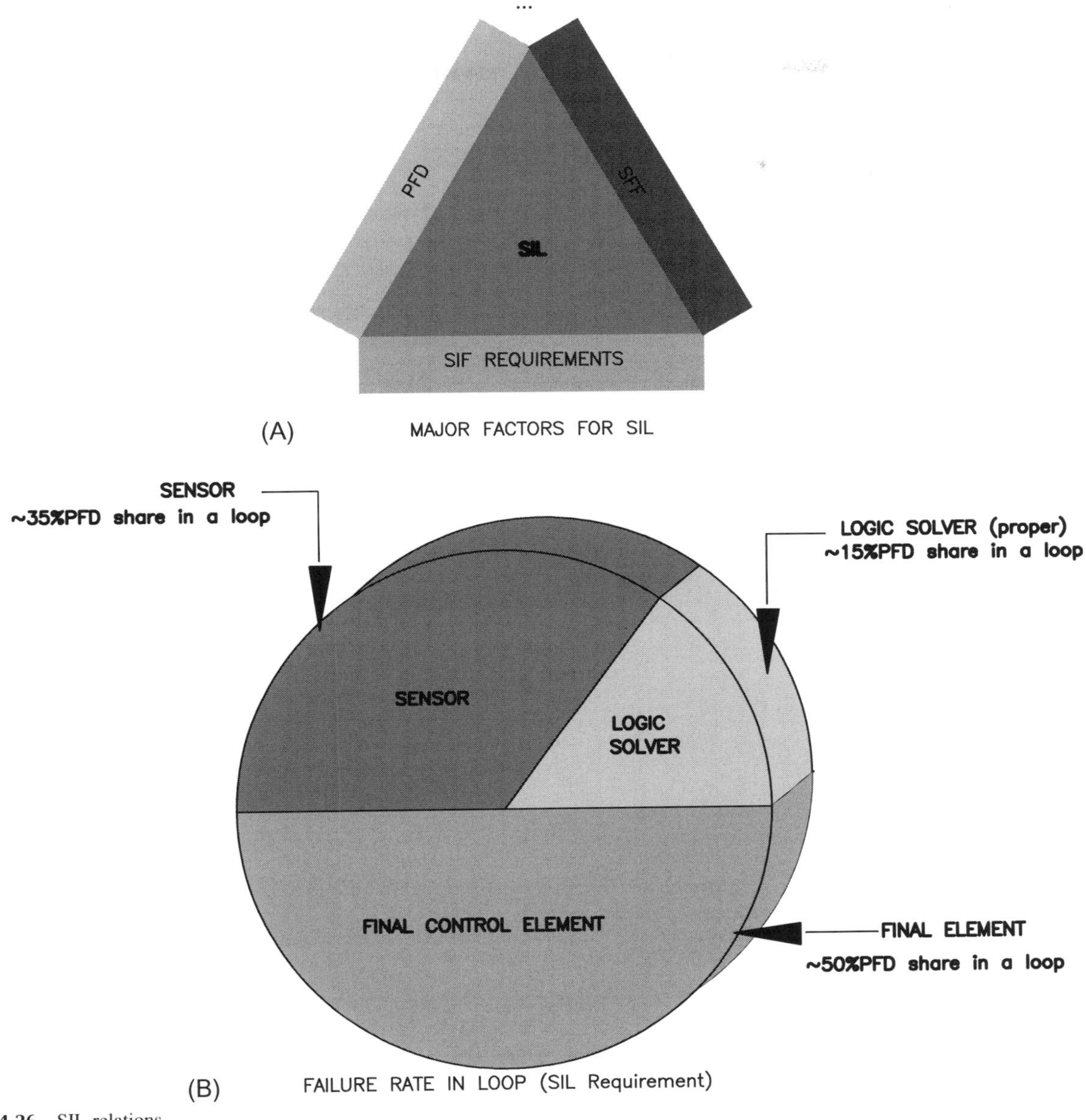

FIG. 14.26 SIL relations.

5. Logic solver to receive the response via HMI.
6. Communication of operator response through HMI command.
7. Logic solver to interpret the same, and obey the command.

An alarm provides vital support to the operators with an early warning about plant abnormalities and/or situations that need their attention so as to prevent, control, and mitigate many effects of abnormal situations. So, as far as facility operations are concerned, it is personally felt that a pretrip alarm is of more significance for an operator's action than the actual trip alarm. This is because of the fact that after pretrip alarm, the operator can take certain measures to prevent tripping—hence, downtime. However, in the case of a trip alarm, the operator is to take care of posttrip actions only and downtime may not be saved. This will be clear from one BMS action: When there is a high furnace pressure pretrip alarm, the operator can regulate the air/fuel flow to prevent the boiler from tripping. However, in the case of a very high furnace pressure, the boiler will trip and the operator needs to see that posttrip actions are suitably taken but down time could not be prevented. This suggests that for operating personnel, pretrip conditions need more attention, even if the priority assignment may be lower. So, alarm management is a major issue to deal with. What is alarm management really?

TABLE 14.23 SIL Determination Methods as Per IEC

IEC Annex	IEC Ref:	Heading	Method
Annex B	IEC 61511-3:2003	Semiquantitative method	Fault tree
Annex C		The safety layer matrix method	Safety layer matrix
Annex D		Determination of the required SIL semiqualitative method: Calibrated risk graph	Risk graph
Annex E		Determination of the required SIL, qualitative method: risk graph	Risk graph
Annex F		Layer of protection analysis	LOPA

6.1 BASICS of Alarm Management

Basic alarm management mainly encompasses the following:

1. Development of alarm philosophy documentation.
2. Definitions of nuisance alarms and standing alarms.
3. Definitions for alarm priorities based on impact and reaction time.
4. Definition of procedures for management of changes to the alarm system.
5. Alarm rationalization and implementation.
6. Creation of an alarm response procedure.
7. Design of an advanced alarm system such as dynamic alarming.
8. Establishment of benchmarking for alarm system performance.
9. Alarm system performance monitoring and assessment.
10. Auditing alarm system practices.
11. Training procedures for technical personnel.
12. Creation of reporting procedures and record keeping.

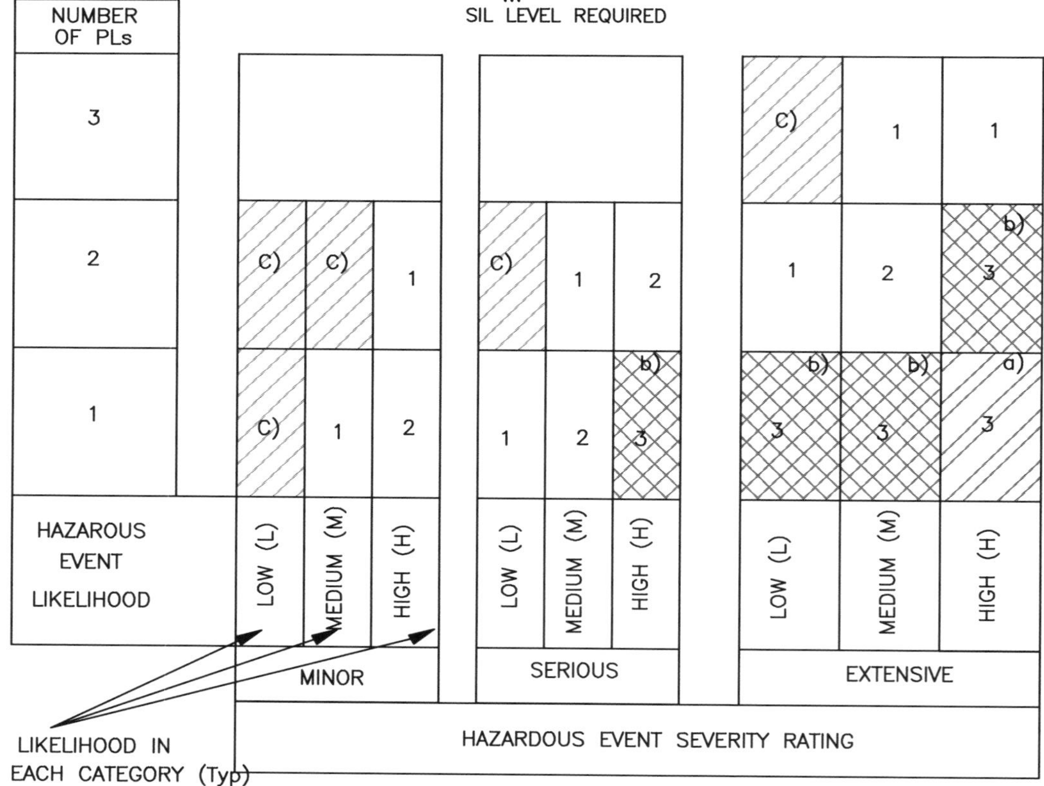

FIG. 14.27 Risk matrix for SIL determination.

TABLE 14.24 Guidelines for Calibration Parameter (Basis: IEC61511-3:2003)

Risk Parameter	Symbol	Classification	Comments
Consequence (C)	C_A	Minor injury	Injury and death, normal healing time
Consequence (C)	C_B	0.01 to 0.1	Injury and death, normal healing time
Consequence (C)	C_C	0.1 to 1.0	Injury and death, normal healing time
Consequence (C)	C_D	>1.0	Injury and death, normal healing time
Occupancy (F)	F_A	<0.1	Rare occupancy
Occupancy (F)	F_B	>0.1	Frequent and permanent occupancy
Probability (P)	P_A	>90%	Good probability of avoidance
Probability (P)	P_B	<10%	Poor probability of avoidance
Demand (W)	W_1	0.1D/year	Demand rate
Demand (W)	W_2	0.1–1D/year	Demand rate
Demand (W)	W_3	1–10D/year	W >10D/y—higher integrity

TABLE 14.25 Uses of SIL Determination Methods

Purpose	Safety Matrix	Risk Graph	LOPA	FTA
Initial analysis	X	X	X	NR
Detailed analysis	NR	NR	X	X
Multiple causes and protections	NR	NR	X	X
Dependencies	NR	NR	NR	X
SIL/PFD_{avg}	SIL	SIL	PFD_{avg}	PFD_{avg}
Human factor	NR	NR	X	X
SIL level	1	1	1 and 2	>1

6.2 Alarm Management Benefits

1. Reduction in unplanned downtime.
2. Prevention of damage to equipment and system.
3. Reduction in avoidable maintenance of equipment.
4. Increased productivity.
5. Production optimization and increased throughput.
6. Savings on SIL requirements.
7. Energy savings.
8. Reduction in penalties due to regulatory incidents.

6.3 Standards for Alarm Systems

Details about an alarm system with a lifecycle are discussed in EEMUA 191 and standards such as ISA 18.2:2009 and IEC 62682:2014 also deal with alarm systems. The major standards are:

- Standard EEMAUA in 2013 includes guidance for HMI management techniques, alarm configuration, and appendices covering alarm suppression, geographically distributed processes, intelligent fault detection, requirements for alarms, and supporting checklists.
- Standard ISA 18.2:2009 is not really about hardware or software. It is about the work processes of people. ISA-18.2 is a comprehensive standard developed per stringent methods based on openness, balancing of interests, due process, and consensus [8]. An alarm definition as per ISA-18.2 is well depicted in Fig. 14.29.

6.4 Basics of Alarm Systems

From ISA-18.2, one gets an alarm definition as "an audible and/or visible means of *indicating* to the *operator* an equipment malfunction, process deviation, or *abnormal condition requiring a response*."

6.4.1 *Alarm Definition and Dead Bands*

The alarm definition as per ISA-18.2 is well depicted in Fig. 14.29.

Dead bands in alarm systems are very fundamental for the interpretation of abnormal conditions and the same is depicted in Fig. 14.30. It shows set/reset points and dead bands.

6.4.2 *Alarm Objectives*

When one looks a little deeper into the alarm system, one finds that causes (expected or unexpected), consequence (direct and ultimate), corrective action, and acknowledgements are important for developing alarm objectives. When there is no ultimate consequence and no corrective action is required, then it is not an alarm (even this conforms to the definition elaborated above). However, when it is an alarm, there shall be some objective associated with the same.

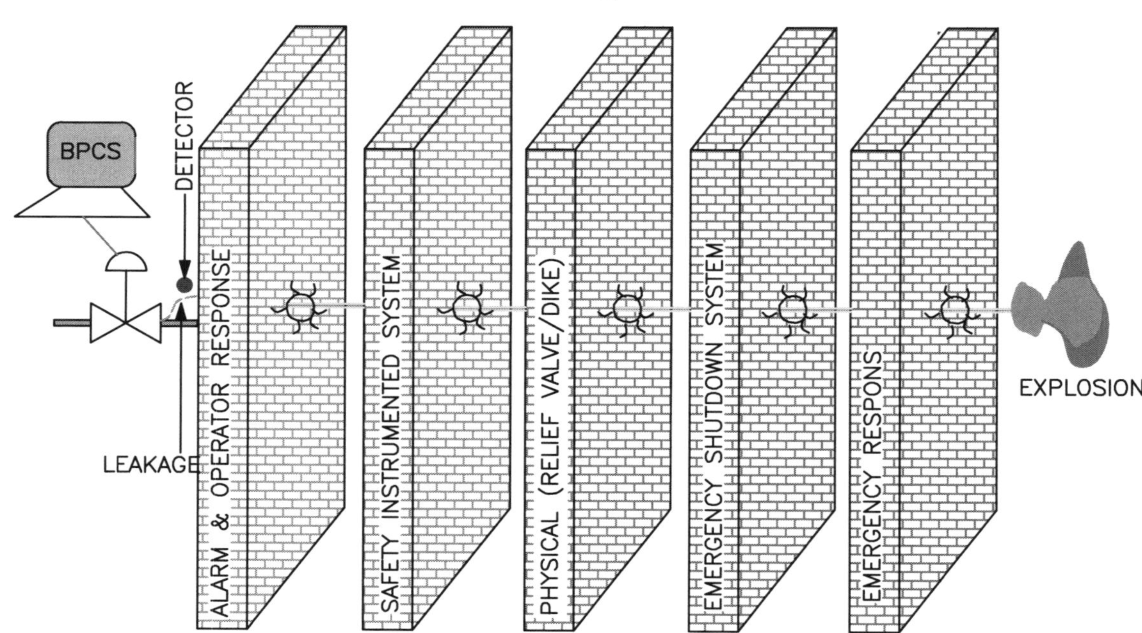

FIG. 14.28 Example of layer of protection.

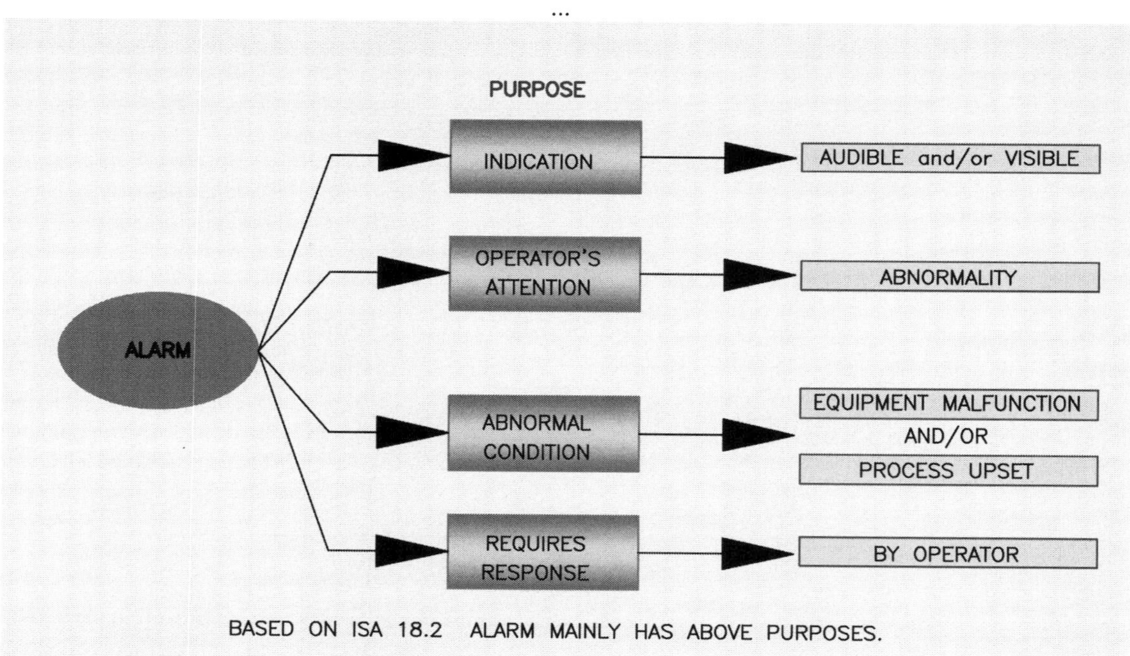

FIG. 14.29 ISA alarm definition.

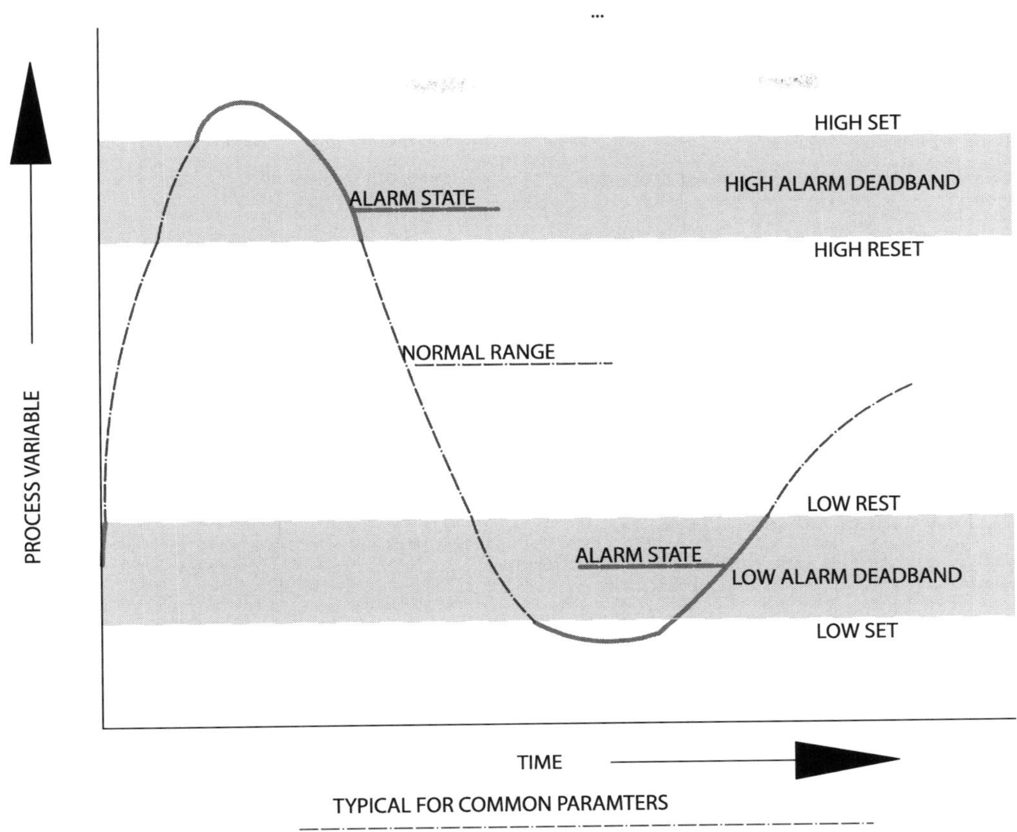

TYPICAL FOR COMMON PARAMTERS

PARAMETER TYPE	DEAD BAND % RANGE	DELAY (ON/OFF) SECONDS
PRESSURE	2	15
TEMPERATURE	1	60
FLOW	5	15
LEVEL	5	60

FIG. 14.30 Typical process alarm with dead bands.

Therefore, the basic objective of an alarm system normally shall include:

1. Safety aspect: Clear identification of safety to the people, property, and environment.
2. Usability: it is usable when it meets the following criteria:
   - Relevant for the user.
   - Defined response for the alarm.
   - Rate of appearance must be commensurate with the ability of the operator to handle.
3. Performance: Evaluation during the design and commissioning stages to ensure that the alarm is effective at all operational conditions and stages of the facilities. Important issues are:
   - Defined response.
   - Adequate time to respond.
   - Relevant, unique, and useful.
   - Well defined and prioritized and, if applicable, grouped.
4. Return on investment (ROI): High-standard, well-engineered alarm systems to achieve the desired goal to get a return on investment.

An alarm is a function of HMI and not of the process, hence it is better to look into the key features of various guidelines (e.g., EEMUA) and standards (e.g., ISA 18.2 and IEC 62682).

6.5 Alarm Features and Performance Indicators

A survey by Automation World revealed that 68% of respondents indicated that an alarm overload negatively affects their ability to properly operate the process (From Alarm Management, 31st Annual European AIChE Seminar, April 23, 2015). EEMUA 191 is only a guideline, meaning a set of guidance and recommendations. ISA 18.2 and IEC 62682 (which has adapted the majority of ISA 18.2, including the alarm lifecycle (defining work process should be in place)) [1] are major standards.

6.5.1 EEMUA: Features and Performance Indicators [1]

1. The following are the features in line with EEMUA:
   - Alarm philosophy and design principles.
   - Details on alarm design and risk assessment.
   - Alarm system specification.
   - Design implementation issues.
   - Details about field sensors and displays.
   - Performance assessment and alarm management.
   - Performance matrices and questionnaires.
   - Separate section for alerts (e.g., autostarting of stand by pump).
   - Alarm suppression hazard study.
   - Risk assessment and priority determination.
   - Separate section for bath process.
2. The following are the performance indicators in line with EEMUA:
   - Average alarm rate.
   - Maximum alarm rate (following an upset).
   - Percentage of time alarm rates outside the acceptable target rate.
   - EEMUA also recommends that for a control system:
   - Configured alarms > 1000 with a standing alarm < 10 and a shelved alarm (other than maintenance shelved alarms) < 30.

The following are definitions of the terms used in the performance indicators of EEMUA.

- **Standing Alarm**: An alarm that is in an active alarm state for a significant period of time, for example, 4–8 h [courtesy: ABB: Alarm Management; Meet the experts, September 2011] is referred to as a standby alarm.
- **Stale alarm:** When an alarm goes into the alarm state but never returns to a normal state within 24 hours. This is similar to a standing alarm.
- **Shelved alarm**: A shelved alarm is an alarm that results on account of alarm suppression techniques or mechanisms initiated by the operator to temporarily suppress a set of alarms until an underlying problem can be corrected. Shelved alarms must be presented to the operator.

Tables 14.26 and 14.27 are important issues pertinent to EEMUA guidelines:

TABLE 14.26 Average Alarm Rate Steady-State Operation (IEEMU191 Guidelines)

Long Term Average Alarm Rate Steady-state operation	Alarm per Operator		Acceptability
	Number per Hour	Numbers per 10 Minutes	
More than 1 per minute	>60	>10	Very likely to be unacceptable
1 in 2 minutes	30	5	Likely to be overdemanding
1 in 5 minutes	12	2	Manageable
Less than 1 per 10 minutes	<6	<1	Very likely to be acceptable

TABLE 14.27 Average Alarm Rate Following an Upset (IEEMU191 Guidelines)

Number of Alarms Displayed in 10 Minutes Following a Major Plant Upset	Acceptability
More than 100	Definitely excessive and very likely abandonment of the system
20–100	Hard to cope with
Less than 10	Should be manageable, but if each requires complex action by an operator, then it is difficult

6.5.2 ISA: Features and Performance Indicator [1]

The following are the major features as per ISA18.2. The performance indicator has been indicated in Table 14.28.

- Alarm system lifecycle.
- Standard IEC 62682 developed based on ISA.
- Clear performance indicators.
- Detailed alarm philosophy document.
- Alarm identification and rationalization.
- System requirement specifications.
- Alarm and HMI design.
- Implementation, operation, and maintenance.

TABLE 14.28 Performance Indicator as per ISA

Metric	Target Value	
Annunciated Alarms per Time	Target Value Likely to be Acceptable	Target value max. Manageable
Per 10 minutes per operating position	~1	~2
% of hours containing > 30 alarms	<1%	
% of 10 min containing > 5 alarms	<1%	
Max number of alarms in 10 min period	10 or less	
% of time alarm system in flood condition[1]	<1%	
% contribution on the top 10 most frequent alarms on alarm load	1–5% max with action plan to address	
Quantity of chattering[2] and fleeting[3] alarms	0 action plan if it occurs	
Stale alarm* (*ref cl. 4.1.2)	< 5% on any day action plans to address	
Priority distribution	3 priorities: ~80% low, ~15% medium, ~5% high or 4 priorities ~80% low,~15% medium, ~5% high< 1% highest	
Unauthorized alarm suppression	0 outside controlled approved methodology	
Improper alarm attribute change	0 outside approved methodology or MOC	

Notes: (1) Alarm flooding condition is that condition when 10 or more alarms are received by an operator within 10 minutes. (2) Chattering alarm is on then off and on again in a short period of time (e.g., <1 min). (3) Fleeting alarms turn on and off very quickly, but do not necessarily repeat—mainly occur due to contact chattering. In chattering and fleeting alarms suitable time delay at input will give better results.

- Monitoring and assessment.
- Management of change handling and auditing.

6.6 Alarm System Discussions

6.6.1 *Operator Response to Alarm*

In case of an alarm, there are two sets of logic solving issues. The first is the electronic system, which gets the signal from the sensor and decides the alarm and alarm type The other one is the human brain, which actually initiates action. So, human intelligence also acts as a logic solver to initiate the action. It is seen that in an HMI alarm response, both

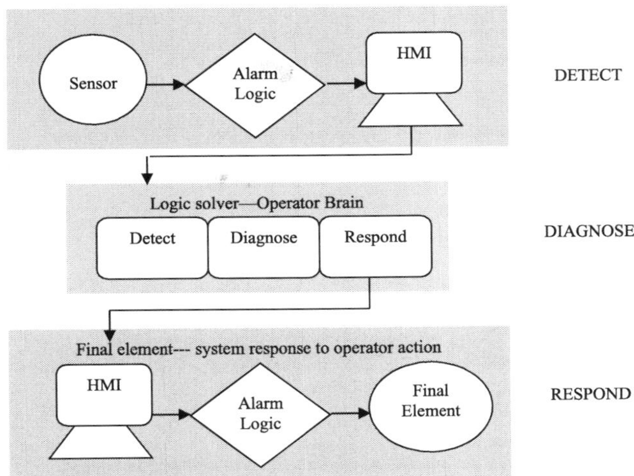

FIG. 14.31 Model of operator response to alarm.

detection and action are done through the HMI unit with a human interface, as shown in Fig. 14.31. So, an alarm system is highly dependent on the human factor, which is extremely vulnerable under stress for several reasons. These include quick decision making in an emergency, making multiple decisions in a short time, suffering from too much or not enough confidence, boredom from the job, not well-defined or conflicting data or conditions (confusion), and lack of training (especially in new technology).

6.6.2 *Process Safe Time*

Now, with the help of Fig. 14.32, an attempt will be made to see how process safe time is determined. When an untoward incident occurs after a small gap of time (sensor detection and alarm logic reaction), an alarm appears with the help of a sensor, alarm logic, and HMI. After the appearance of an alarm in HMI, there will be some time for the operator's detection, diagnosis, and response. Now, after the operator responds, actual thing that happen will be reaction time, which comprises of process dead band and system reaction time. A minimum sum total of operator action time and reaction time will be necessary for any corrective action to happen. So, if a hazardous event occurs at T = 0, after this the system will be safe.

6.7 Alarm Philosophy, Rationalization, and Lifecycle Discussions

In order to study an alarm lifecycle, it is important to have some prior discussions on two other issues: alarm philosophy and alarm rationalization, which are highly connected with a lifecycle study for alarm systems. Let the discussion start with the alarm philosophy defined in ISA 18.2: 2009 as well as IEC 62682:2014.

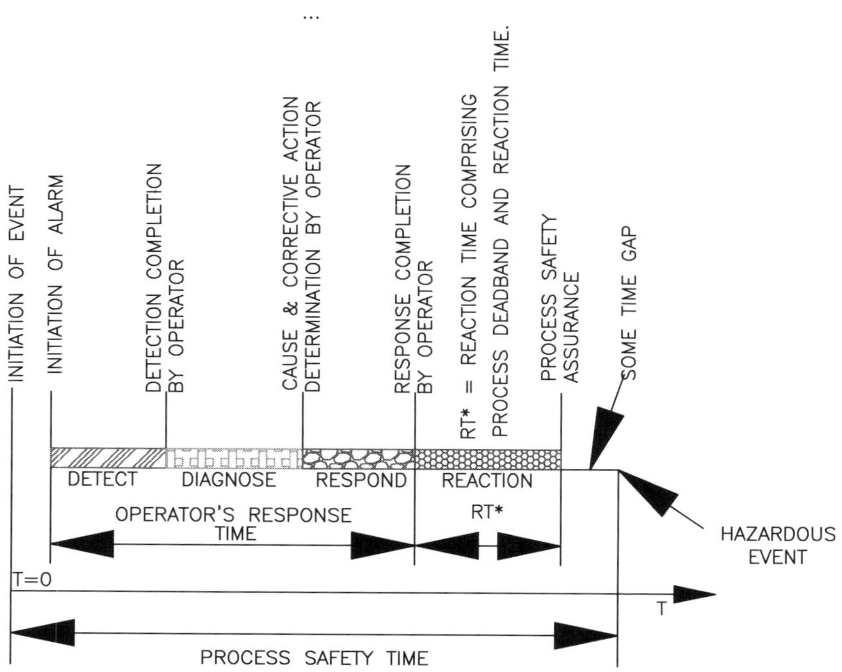

FIG. 14.32 Process safe time determination.

6.7.1 Alarm Philosophy

ISA 18.2 gives the basic content/topic to be covered in the philosophy document, as depicted in Table 14.29 (*the associated* ***subection numbers of the standard*** *are given in the table*).

This document provides guidance for a consistent approach to alarm management and defined activities of the alarm management lifecycle. It provides the following benefits.

1. Optimally designed alarm system to:
   - Cater to user needs.
   - Meet ergonomic requirements.
   - Define the relevant user's role.
   - Understand easily.
2. Alarm design consistency.
3. Risk management goals/objectives.
4. Good engineering practices (Table 14.30).
5. Effective alarm rationalization.
6. Efficient operator response to alarms.
7. Prioritization of alarms.
8. Identification, justification, and classification of alarms.
9. Alarm conditions and attribute specification.
10. Design of alarms.
11. Implementation and testing of alarms.
12. Commissioning of alarms.
13. Training, maintenance, and testing.
14. Operation and use of alarms.
15. Maintenance and testing of alarms during service.
16. MOC for alarm systems.
17. Performance monitoring and assessment of alarm systems.
18. Auditing of alarm system performance.

Table 14.31 Provides some details about design details and operator action.

6.7.2 General Discussions on Alarm and Alarm Philosophy

Here, a few pertinent issues for alarm systems and alarm philosophy are discussed.

1. Alarm state as per ISA has already been discussed in Chapter 7 so is not repeated.
2. **Pertinent definitions of alarm philosophy**: A few types of alarms defined in the alarm philosophy are defined below:
   - **Shelving:** This alarm suppression technique/mechanism is used by the operator to manually suppress an alarm temporarily during the operation process for a definite period of time for operational facility (e.g., a low flow alarm for a nonoperating pump). Alarm shelving can be done for short-term maintenance purposes also. Visibility and accountability of such alarms is very important for safe operation.
   - **Suppression by design**: Any mechanism within the alarm system preventing an alarm from the operator's visibility based on plant state or other condition; an advanced design.

TABLE 14.29 Requirements of Alarm Philosophy Content (ISA 18.2/IEC62682)

Content	Required	Recommended
Purpose of alarm system	6.2.2	
Definitions	6.2.3	
References	6.2.4	
Roles and responsibilities for alarm management	6.2.5	
Alarm design principles	6.2.6	
Rationalization	6.2.7	
Alarm class definition	6.2.8	
Highly managed alarms (or site equivalent)		6.2.9
HMI design guidance	6.2.10	
Prioritization method	6.2.11	
Alarm set point determination		6.2.12
Alarm system performance monitoring	6.2.13	
Alarm system maintenance	6.2.14	
Testing of alarms	6.2.15	
Approved advanced alarm management techniques		6.2.16
Alarm documentation	6.2.17	
Implementation guidance	6.2.18	
Management of change	6.2.19	
Training	6.2.20	
Alarm history preservation	6.2.21	
Related site procedures		6.2.22
Special alarm design considerations		6.2.23
Alarm system audit	6.2.24	

- **Out of Service:** The state of the alarm during which the indication is suppressed manually (maybe) for long maintenance/shutdown.
- **Dynamic alarming:** Dynamic alarming techniques are used for the elimination of alarm floods by automatic suppression of redundant and consequential alarms resulting from anticipated equipment malfunction or process abnormalities.

TABLE 14.30 Good Engineering Practices for Alarm

Issue	Characteristics
Relevance	Not spurious or of low operational value
Unique	No duplication
Timing	Not too long before response needed, not too late to do anything
Prioritize	Indicates the importance of the operator to deal with the problem
Understanding	Clear and easy to understand message
Diagnostics	Identifying the problem that has happened
Advisory	Indicative of the action to be taken
Focus	Drawing attention to the most important issues

TABLE 14.31 Design Details and Operator Action for Alarm Systems

Item	Description
Design Issues	Context sensitivity, functional aspect (alert, action guide), Following development guidelines, consequential effect, time to respond, rationalization, prioritization, redundancy of equipment, CCF issues, defined and authorized management of change (MOC).
Alarm HMI	State, priority and type, information message, sound system, acknowledgement, summary, alarm suppression and shelving and other relevant functional details, for example, alarm tree, active alarm list, history messages, acknowledged alarm list.
Operator action	Fault identification, coordination, maintenance depending on plant states such as normal condition, start up, upset/emergency shutdown, or controlled shutdown. Great role during emergencies. Operator training is an important issue

- **First out alarm:** This has already been discussed in Chapter 7.
- **Conditional alarming techniques:** This technique helps in eliminating alarms based on operating conditions, for example, a low pressure alarm for pump header pressure will not be active until once before header pressure is established.

6.7.3 *Alarm Rationalization*

Alarm philosophy and alarm rationalization are the main issues pertinent to the alarm lifecycle (Fig. 14.33).

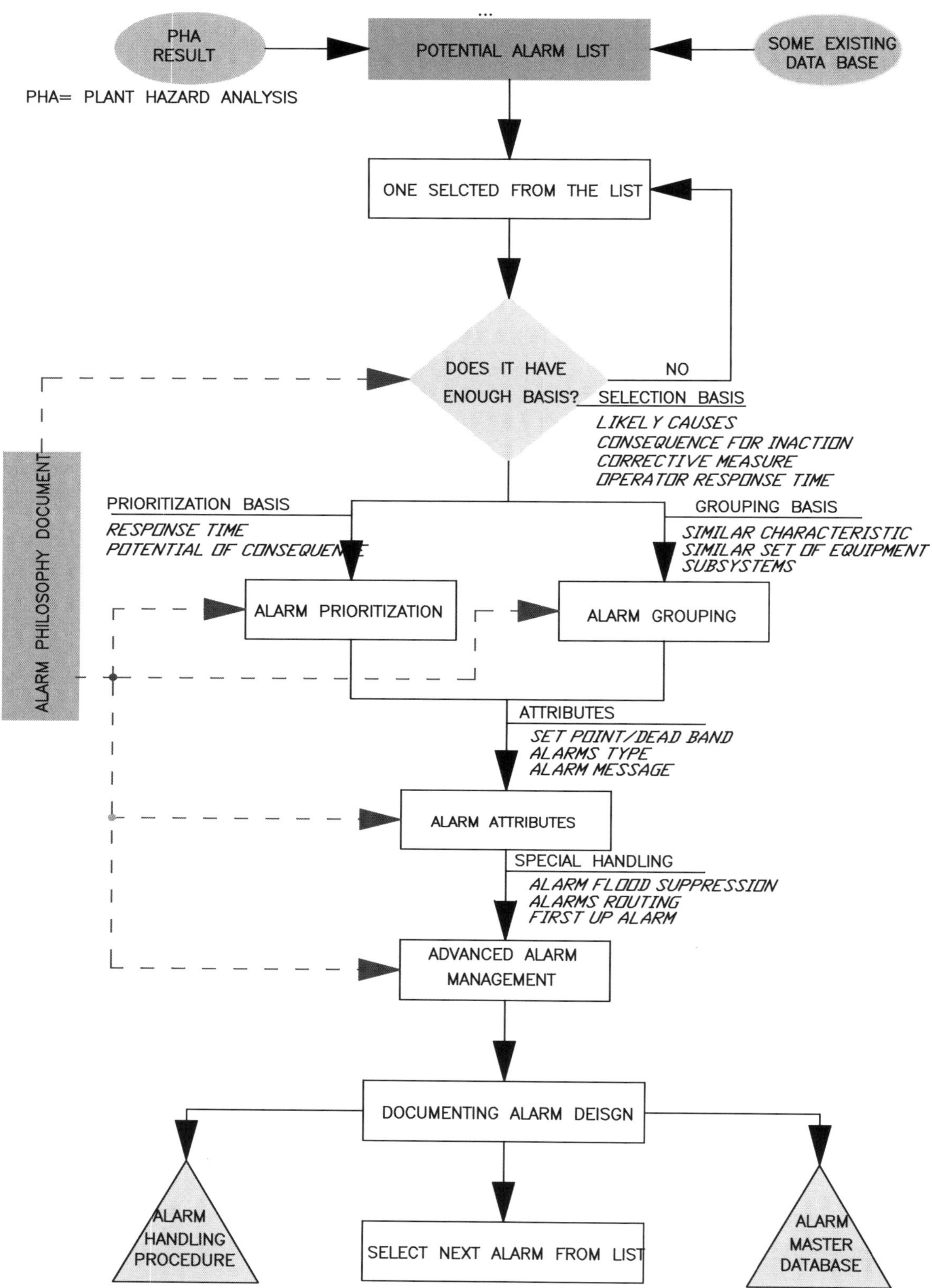

FIG. 14.33 Alarm rationalization.

Start with a review of potential alarms against the criteria set in the philosophy document to make sure that these are necessary and appear in the master alarm database (MADB).

1. From the database, one alarm is selected and tested for a validity and basis test along with objective analysis. The conditions are analyzed such as abnormal conditions requiring operator action. Other major factors here are consequence, corrective action, confirmation, and operator response time.
2. Based on the response time and potential consequences, the selected alarm is prioritized.
3. Based on the facility, alarms may be grouped as per location, functionality, etc.
4. In the next stages, the attributes are assigned, including a special one with special handling. Until this stage, all activities are performed as per the alarm philosophy guideline.
5. Alarm design, including response procedures and a master alarm database, is documented.
6. Next, another alarm is chosen and the same procedure is repeated.
7. Alarm tree view: The process alarms in a hierarchical manner for the operator to browse the alarm status configuration.
8. Active alarm list: Triggered but not cleared alarms in tabular form.
9. Acknowledged alarm list and message history.

6.7.4 Alarm Lifecycle Discussions

Operating personnel frequently come across the following undesired issues:

- Alarm overload/alarm floods.
- Nuisance alarms.
- Chattering alarms.
- Standing/stale alarms.
- Redundant alarms.
- Alarms with no response/with wrong priority.

One of the milestone activities of ISA 18.2: 2009 and IEC62682:2014 is the alarm lifecycle presented in Fig. 14.34, Alarm Management lifecycle with three categories of parts.

These are (1) an alarm comprising A, B, C, and I, mainly for the master alarm database and MOC, (2) D, E, F, and G for alarm suppression and dead band, etc., and (3) H for alarm analytics and metrics.

Each of the blocks shown in the figure are discussed below:

1. Philosophy: The alarm lifecycle starts with the development of an alarm philosophy, which gives guidance for alarm management practices for the facility in all lifecycle stages. This has been discussed at length above.
2. Identification: Though not defined in ISA 18.2, this stage involves a review of documents (e.g., P and ID reviews, PHA reports, etc.) to ensure that the results are useful for rationalization. It is helpful for documenting causes, consequences, and the time to respond for each identified alarm. For existing systems, existing alarms will be included.
3. Rationalization Stage: Each potential alarm is tested against the criteria in the alarm philosophy to justify that it meets the requirements of being an alarm. Details on alarm rationalization are covered in Section 6.7.3.
4. Detailed Design: At this stage, an alarm needs to cater to the requirements in the philosophy and rationalization documents for basic alarm design, setting up of attributes (set point, dead band, etc.), conducive operator graphics design, elimination of bad actors, and advanced alarm design.

   So, this stage is for an accurate translation of requirements into an alarm. Poor design or an incomplete design will end up with problematic issues for alarm management.
5. Implementation: In this stage, the alarms are put into operation with operator training, testing, and commissioning. Testing and training will be applicable for new systems, new alarm additions, and any change in process design. This also includes the activation of alarm designs in the running system.
6. Operation: At the operation stage, an alarm performs its function of the drawing attention of the operator to the presence of an abnormal situation. Key activities shall include:
   - Alarm management by system tools.
   - Accessing information at the rationalization stage.
   - Operator awareness of alarm.
   - Avoidance of alarm before it could happen.
   - Ensuring expected operator action.
7. Maintenance: In this stage, recommended procedures to remove an alarm from service and return an alarm to service are done. The major issue here may be ensuring the visible accountability for a restored critical alarm to active service.
8. Monitoring and Assessment: This stage is meant to gather data from the operation and maintenance stages and compare alarm system performance matching that in the philosophy document. In order to respond effectively, an operator should be presented with no more than one to two alarms every 10 minutes. So, the major issue is the rate alarms are presented to the operator. Another key activity at this stage is to identify "nuisance," "chattering," "fleeting," or "stale" alarms. So, at this stage, suitable analytics and facility-

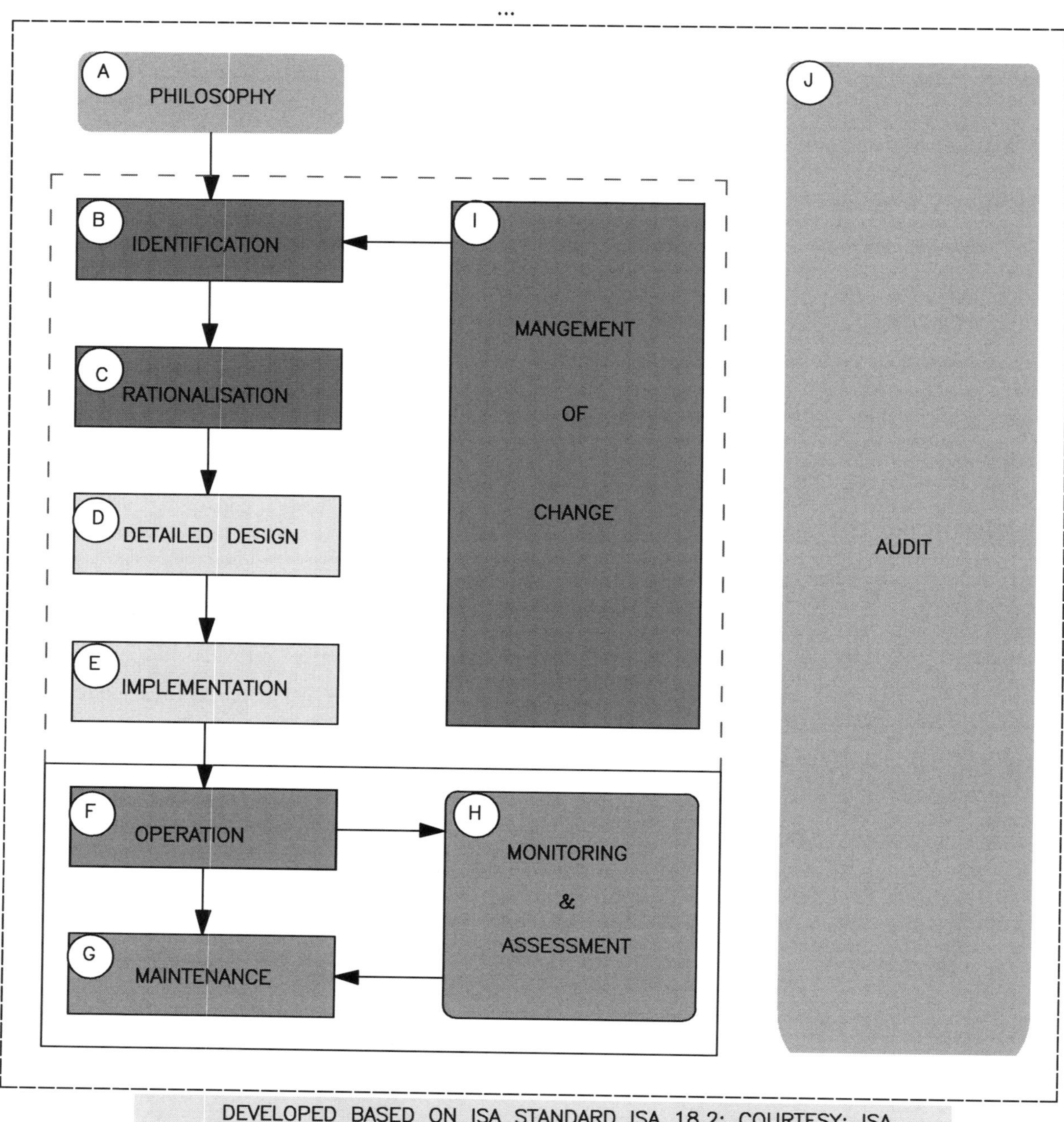

FIG. 14.34 Alarm management lifecycle.

specific metrics are tailored to analyze raw alarms. This is used for key performance indicators and reporting.

9. Management of Change (MOC): The major activity of the stage involves authorization for all changes to the alarm system, including the addition/deletion of alarms and changes to alarms. Upon approval (by the appropriate authority), the modified alarm is processed through the stages of rationalization, detailed design, and implementation, and documentation updating is necessary.
10. Audit: The audit stage is mainly meant for periodic review of the work processes and the performance of the alarm system with a goal to maintain the integrity of the alarm system throughout its lifecycle and to identify areas of improvement. The audit process may call for updating the philosophy document to reflect any changes carried out during the audit.

7 SIS—FIELD DEVICES

An SIS is a collection of one or more SIFs, each of which has one or more components designed to execute specific safety-related tasks to bring industrial processes to a safe state in the event of hazardous conditions. At the component level, there will be interfaces of responsibilities of the end user and manufacturer, as discussed in Table 14.14. An important issue one should keep in mind is that the safety data provided by manufacturers for their devices, with due validation by a third certifying authority, are laboratory results that need to be further verified in the field. According to the OREDA-97 report, 92% of these failures were due to the nonperformance of their safety function upon demand by the sensors and final elements. In this section, we will discuss SIF components in the field, that is, sensors and final control elements.

7.1 SIS—Field Sensors

Basic issues pertinent to field sensors are briefly enumerated below. For details, the reader may consult Chapter 9 of the book referenced as [128].

7.1.1 Selection of Sensors and Sensor Types

In modern SIFs, continuous process transmitters in conjunction with comparator circuits may be used for hazard detection in place of process switches. A "healthy" transmitter shows a changing output signal through diagnostics. The purpose of diagnostics is to detect **covert** failures as **overt** (detected) failures. In contrast to that, process switches provide no indication of "healthy" operation. So, when a process switch needs to change states at the set point, it might also fail and be incapable of registering a dangerous situation—hence process switches suffer from covert (hidden) failure.

Factors influencing selecting sensors: It is important to look into the pertinent issues, listed below, which influence the selection of sensors for various measurements:

1. Use of transmitters (not process switches) with built-in diagnostics, redundancy, and availability.
2. Common cause strength and diversity for CCF elimination.
3. High ambient temperature may be a common cause of failure. The following issues may be considered:
   - Robustness: Robust transmitters (protection of electronics).
   - Installation: Install away from the hot point.
   - Diagnostics: Suitable software to predict failure due to high temperature.
4. Factors influencing performance:
   - Transmitter performance: In addition to overall accuracy, designers need to take care of many other data such as orientation (e.g., loss of performance due to orientation change, that is, positioning in a tilted condition), ambient temperature variations, high static pressure, drifts (temperature/voltage), noise pick up, etc.
   - Process interfaces: Various issues such as impulse line plugging or coating (error in reading for thermowell or magnetic flow meter). In smart instruments, it is possible to detect such changes and eliminate errors in reading.
   - Sensor robustness quality control: Use of suitable material and ability to withstand sudden changes.
5. Installation and maintenance issues.

7.1.2 Redundancy and Voting

In SIS applications, the importance of reliability and availability cannot be overestimated. In such cases, redundant sensors may be deployed in one of two, one of three, or two of three (or two of four, typical for nuclear plants). Out of these, two of three is a very good and common choice but sensor voting in two of three architecture demands special attention. High, low, average, and median are the typical selection criteria for "voting" modules, already discussed in Chapter 3.

7.1.3 Device Diagnostics and Impact

Apart from monitoring performances, smart instruments are capable of predicting a great deal of information about what's going on with a process, well beyond the specific variable measurement or control. Some smart instruments are able to detect deterioration in a key component or realize sensor drift. Table 14.32 shows diagnostic prediction in some common measurements.

7.1.4 Safety and Reliability Data

Safety and reliability data from internationally reputed manufacturers have been presented to note the type of data and the range of data normally associated with safety sensors:

- Safety accuracy: 1%–2%.
- Response time: 750 ms (max).
- Damping set to 0 s.
- Updates: Every 50 ms.

Reliability data in terms of failure rate have been presented in tabular forms in Tables 14.33–14.35.

7.2 SIS—Final Control Element

According to the published reports and shown in Fig. 14.35, final elements account for nearly 50% of SIS failure. FCEs consist of control or shutdown valves/dampers, actuators, and associated accessories such as solenoid valves, positioners, etc. Therefore, the FCE selection is extremely

TABLE 14.32 Diagnostic Prediction of Measurement Problem (Typical)

Measurement	Diagnostic Issue	Measurement	Diagnostic Issue
Pressure	Plugging of impulse line	Vortex flow	Change in medium
Pressure/Level	Diaphragm leakage	Coriolis flow	Slug flow, tube coating
Temperature Sensor	Coating in thermowell, drift, sensor open/short	Magnetic flow	Faulty electrode/ ground, process noise, tube coating
Differential pressure	Process noise, zero drift	pH	Faulty electrode, cleaning

TABLE 14.33 Typical Failure Data-1 as per IEC 61508 (FIT: Failure in Time)

Device	λ	λ_S	λ_{Dd}	λ_{Du}	SFF
Transmitter: XX ...	366 FIT	2150 FIT	1160 FIT	3680 FIT	90.06%

important to improve the reliability and availability and reduce the risks of the associated SIF.

7.2.1 Valve as Final Control Element in SIS

In the majority of cases, valves are used as final elements. In SIS designs, valves are found in the following ways in applications as final elements:

- Single control valve for control and on/off valve.
- Single valve (may be control type) as on/off valve.
- Control valve as redundant final control element.

From Chapter 6, it has been noted that safe position of final control elements to make a safe state is very important. In many cases, solenoid valves play an important role in cases of emergency shutdown. Now, for all such cases, there will be a fail-safe condition of these valves, namely fail open or fail close and fail lock (the last position is mainly for modulating valves). In fail lock condition, regardless of the valve's "natural" failsafe state, the system makes the valve lock positioner air inside the valve actuator. However, on account of leakage, this position cannot be held permanently over a very long period.

Similar conditions may be achieved by a digital positioner. So, with the help of a digital positioner, it is possible

TABLE 14.34 Typical Failure Data-2 as per IEC 61508

Transmitter Type	Meas. Range	SFF	PFD_{avg}	$\lambda_{Dd}+\lambda_S$	λ_{Du}
XXX	10 mBar	75%	$8.54*10^{-4}$	614FIT	195FIT

TABLE 14.35 Typical Proof Test Results

$T_{[Proof]} = 1$ year	$T_{[Proof]} = 5$ years	$T_{[Proof]} = 10$ years
$PFD_{avg} = 8.54E-04$	$PFD_{avg} = 4.26E-03$	$PFD_{avg} = 8.5E-03$

to use a control valve as a shutdown valve; it can also be used as a redundant valve to avoid the CCF effect. However, in shutdown, valves have a high shutoff capability (bubble tight leakage). Various final control elements are indicated in Fig. 14.35. From the above figure, it is clear that the digital positioner plays a vital role in SIS; it also helps in testing.

7.2.2 Testing Method of Final Control Elements

Traditional methods of testing can highlight the movement but do not provide any internal valve diagnostic data such as valve friction, process build up, pressure status, valve torque, etc. Typically, SIS valves are subject to the following conditions:

- Increased packing friction.
- Seizing of shaft in bearing.
- Bearing corrosion.
- Fluid build up on shaft.
- Fracture valve shaft/stem.
- Spring issue: Broken or permanent setting.
- Linkage breakaway friction.
- Air exhaust slow or blocked.
- Spring return actuator dented.
- Increased valve breakaway friction.
- Bending of shaft/stem.
- Increased friction of the closure element in the seal.

Now, let us look into two testing methods: partial stroke testing and full stroke testing.

7.2.3 Partial Stroke Test

With a partial stroke test (PST), it is possible to improve safety, as failures are detected at an earlier stage than when only proof testing is conducted. It is possible to maintain the same SIL rating, even with longer proof test intervals. Partial stroke testing can be of two types: manual PST or automatic computerized PST with digital positioners.

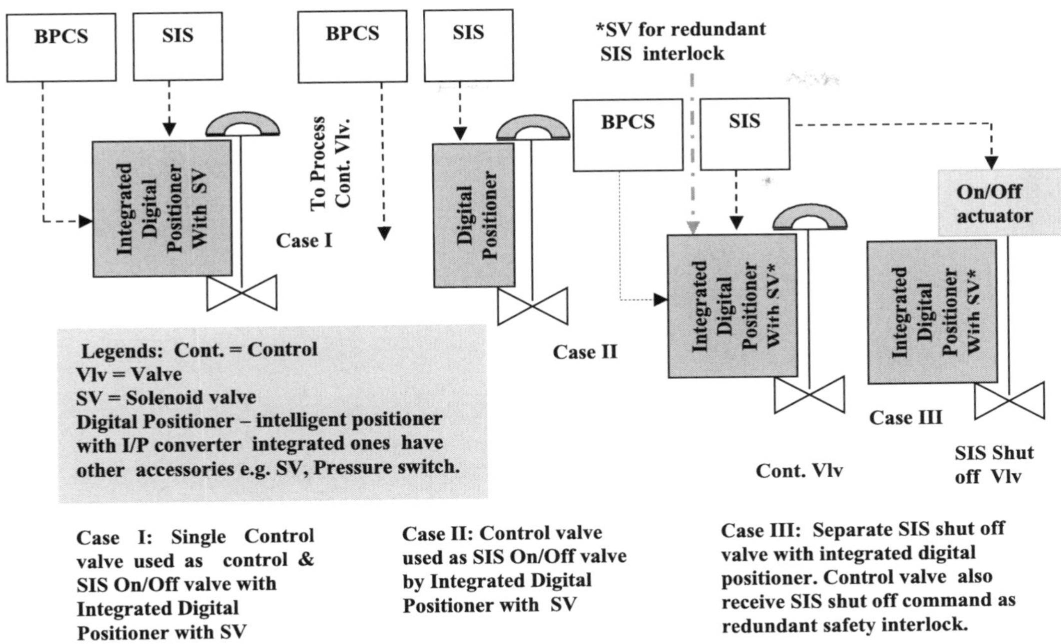
BPCS
SIS
Integrated Digital Positioner With SV
Case I
BPCS
SIS
To Process Cont. Vlv.
Digital Positioner
Case II
*SV for redundant SIS interlock
BPCS
SIS
On/Off actuator
Integrated Digital Positioner With SV*
Integrated Digital Positioner With SV*
Case III
Cont. Vlv
SIS Shut off Vlv
Legends: Cont. = Control
Vlv = Valve
SV = Solenoid valve
Digital Positioner – intelligent positioner with I/P converter integrated ones have other accessories e.g. SV, Pressure switch.
Case I: Single Control valve used as control & SIS On/Off valve with Integrated Digital Positioner with SV
Case II: Control valve used as SIS On/Off valve by Integrated Digital Positioner with SV
Case III: Separate SIS shut off valve with integrated digital positioner. Control valve also receive SIS shut off command as redundant safety interlock.

FIG. 14.35 Various valve configurations in SIS.

The modern trend is to opt for the second choice. People have realized the need for a digital positioner so, for detailed analysis, automatic computer-programmed PSTs are preferred. Partial stroke testing can be done while the plant is running. In this connection, the following points are worth noting:

- Time for actuator pressure to settle to a defined value before partial stroke.
- Ramp time to limit the speed of movement.
- Maximum time for test.
- Maximum permissible valve position.
- Overshooting above position test is terminated.
- Minimum permissible air pressure for actuator [9].

7.2.4 Full Stroke Test

SIFs in an SIS are tested at regular intervals to prove their full functioning. A full stroke test (FST) involves the valve being closed under supervision and then the test result being documented. Failure types of partial stroke tests and full stroke tests are enumerated in Table 14.36, Failure types tested at PST and FST.

7.2.5 Diagnostic Feature in Partial Stroke Testing

A few features involved that make a digital positioner better when compared with the same with solenoid valves are:

- When the actuator is activated by the solenoid valve, extra air pressure is applied to overcome the force in the return spring. However, this can give rise to dead time during closing, and it could be a critical level in some cases.
- In case of a digital positioner, the same may not be needed and it is capable of maintaining the opening. This shortens the closing time.
- Also, digital positioners have diagnostic programs that are capable of small changes around the set point to detect the friction force changes.

These are elaborated upon in Table 14.37.

Smart digital positioners can provide the necessary data for predictive diagnostics to improve system availability. These are also capable of reducing spurious trips during testing as well as maintaining safety availability, should a "safety demand" arise during testing.

8 SIS—LOGIC SOLVER

From reports and as shown in Fig. 14.26, it has been found that a maximum of 15% of failures is due to the logic solver. So, proper care must be taken to ensure that the logic solver is always available. The logic solver is the major impetuous for proper operation of the plant by taking the corrective actions. From the discussions in Section 2 of Chapter 7, it is clear that a good logic solver in a safety application may be characterized by:

- Engineering.
- Communication.
- Diagnostics.

TABLE 14.36 Failure Types Tested at PST and FST [10]

Failure Type	Failure Mode	PST	FST
Valve packing is seized partial or full stroke	Valve fails to close (or open)	X	X
Valve packing is tight partial or full stroke	Valve is slow to move to closed or open position	X	X
Air line to actuator crimped	Valve is slow to move to closed or open position	X	X
Air line to actuator blocked	Valve fails to move to closed or open position	X	X
Valve stem sticks	Valve fails to close (or open)	X	X
Valve seat is scarred	Valve fails to seal off		X
Valve seat contains debris	Valve fails to seal off		X
Valve seat plugged due to deposition or polymerization	Valve fails to seal off		X

TABLE 14.37 Fault Detection by Diagnostics [11]

Fault	Diagnostic Means
Valve friction	Step response and hysteresis tests
Valve leakage	Packing chamber by pressure switch. Also change in friction
Valve shut off change	Zero shifts
Set point deviation	Raw data
Packing degradation	Travel counter
Actuator spring	Breakage by travel or pressure characteristics
Unstable performance	Travel sensing

- Safety.
- Security.
- Robustness.

However, in this section, the same will be viewed from an SIS point of view. Major points for discussion will be around safe PLC and integration issues.

8.1 General Discussions

As such, a logic solver is not an off-the-shelf item like sensors. Depending on applications (especially for SIF/SIS), these are configured as per the need of the system. It is therefore important to select a system best suited for SIS application.

8.1.1 Types of Logic Solvers in SIS

For a simple system, a single-loop controller may suffice, but for very large and complex facilities, an integrated approach could be a better solution. With IIoT in service, it has been found that not only are these integrated, but they are integrated with IT systems. However, in case of integration through the Internet, there are always security threats from malware or unauthorized access that can give rise to issues such as espionage, password stealing, sabotage, etc.

1. Single-loop approach: At times, in smaller and less complex systems, a discrete logic device for each loop could be a better solution, so that the complications and expense of programmability can be avoided.
2. Safe PLC approach: A safe PLC (or DCS) is very much similar to a conventional PLC (or DCS), except that it contains additional integrated safety functions to control safety functions and to bring the automated process to a safe state. Safe PLCs (or DCSs) provide flexibility in programming as well as a familiar, easy-to-use environment for the programming personnel. Summarily, any safety designed around a standard PLC (or DCS) will cost more time, engineering effort, I/O HW, wiring support, etc. Safe PLCs have redundant processors (may be dual cross wired), flash memory, RAM, etc., with continuously monitoring watchdog and detection systems. In this approach, BPCS and SIS use different hardware and separate engineering tools. The two systems are connected by a gateway for data exchange, as shown in Fig. 14.36.
3. Integrated approach: At a plant location, an owner needs to meet the lifecycle activities and thoroughly document the same. In addition to these, the maximum availability of the system is ensured also. If for the timing, controversial issue of BPCS-SIS integration is kept aside, then naturally to meet all these needs, a separate or nonintegrated approach for safety system alone may not be cost-effective not only for two separate sets but also for additional cost due to safe PLC (or DCS). Also, fieldbus systems make system integration much easier, so for complex systems, an integrated approach is taken. This

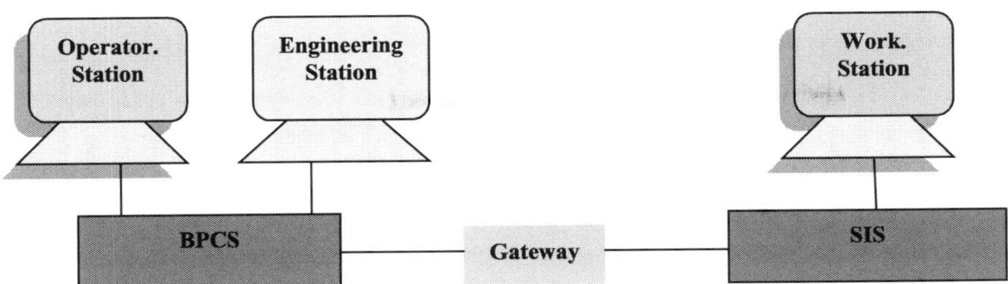

FIG. 14.36 BPCS-SIS interface approach.

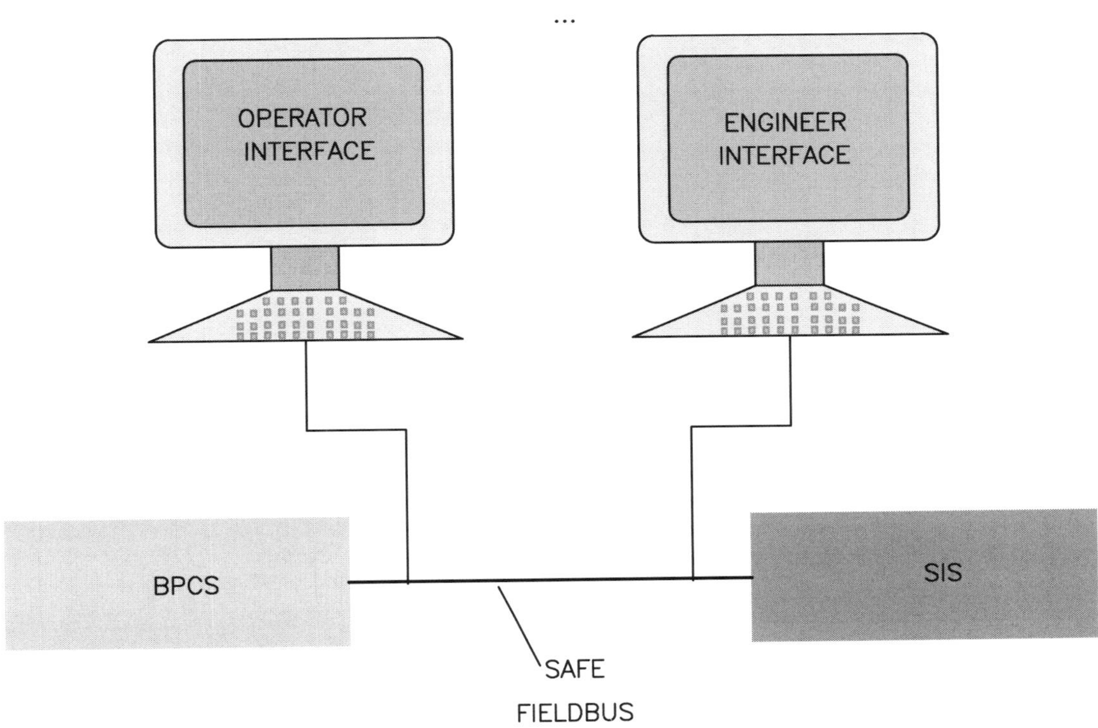

FIG. 14.37 Integrated BPCS and SIS via fieldbus.

is also found to be cost effective. Through judicious selection of the system hardware, software, integration policy, and a proven track record, it is possible to develop a system without compromising system safety. As the owner/end user will be finally responsible for facility operation, it is up to the judgment of the end user to finally make the call.

In an integrated approach, there are two approaches:

Integrated: BPCS-SIS is based on different sets of hardware, but has a uniform communication system and a common engineering tool, as shown in Fig. 14.37. Common: BPCS and SIS are combined in the process control system, using a common **platform** (not common) of hardware (controller, fieldbus, I/O). However, standard and safety-related programs are executed in parallel and independent of each other. A typical configuration is shown in Fig. 14.38.

8.1.2 Discussions on a Separate and Integrated BPCS-SIS Network

Clearly, there are two schools of thought regarding a separate and integrated BPCS-SIS network. Before going into this controversy, it is better to make a few points that are well stated. In both cases (separate or integrated), SIS

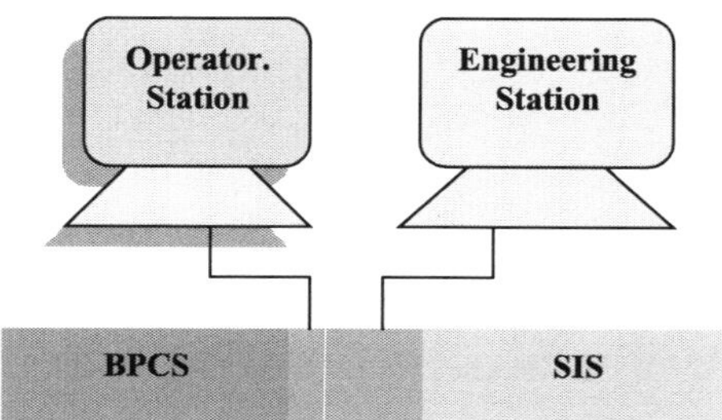

FIG. 14.38 Common BPCS-SIS.

and BPCS have functionally separate controls and protection functions, and they are not mingled. Here the question is whether they will be in a same platform/network or not. Also standards are silent on the issue. Manufacturers may advocate for integration but it is better for the end user to utilize his or her own judgment in deciding the same. This is because the end user is ultimately responsible.

1. Advantages of **separate** SIS and BPCS: Major positive points in favor of separate SIS and BPCS shall include, but are not limited to, the following:
   - Independent failure: Less chance for simultaneous failure.
   - Security: Prevents changes in BPCS to cause changes in SIS; here in modern integrated systems, they are kept separate even though the same platform is used.
   - Different requirements for controllers: SIS is designed to fail in a safe way, whereas in BPCS controllers, the main focus is on system availability.
   - SIS has extensive diagnostics, special software error checking, protected data storage, and fault tolerance. In BPCS, the same may not be required to that extent.
2. Advantages of an **integrated** system: Some advantages of an integrated system have been discussed already. These mainly shall include, but are not limited to, the following:
   - One common engineering tool and approach for BPCS and SIS.
   - Less engineering effort for common systems.
   - Less chance of error when compared to two different systems.
   - Common controller platform.
   - Common data mapping.
   - Direct and seamless communication between SIS and BPCS.
   - Automatic integration of safety-related alarm and messages with time stamping.
   - Possible flexible modular redundancy approach.
   - Redundancy in I/O is independent of that in CPUs.
   - On account of cost-effective design, higher redundancy and fault-tolerant design can be utilized.
   - Effective higher security and access protection possible for a common platform.
   - Based on SIF requirement, redundancy is selected.
   - Safety is not a function of redundancy.
   - Utilization of a safe fieldbus system.
   - Reduction in space, hardware, wiring, assembly, installation engineering, and overhead, hence significant cost savings for the complete lifecycle of the plant.
   - Possibility for increased flexibility and fault tolerance.
   - Cost effective and a financial benefit.

8.2 Logic Solver Types and SIL

The standard also includes fault-tolerant tables to show what configurations of logic solvers and field devices are suitable for different integrity levels. So, this is applicable for SIS field devices discussed in Section 7. For this, Table 14.38 may referenced.

TABLE 14.38 Technology and Configurations to Meet SIL [45]

SIL Level	Sensors	Logic Solver	Final Element
SIL 1	Switches, transmitters	Relay, safety solid state, PLCs, safety PLCs	Standard valves
SIL 2	Fault-tolerant sensors, transmitters with comparison, safe transmitters	Relay, safety solid state, and safety PLCs	Fault-tolerant valves, simplex valves with partial stroking
SIL 3	Fault-tolerant transmitters	Relay, safety solid state, and redundant safety PLCs	Fault-tolerant valves with partial stroking

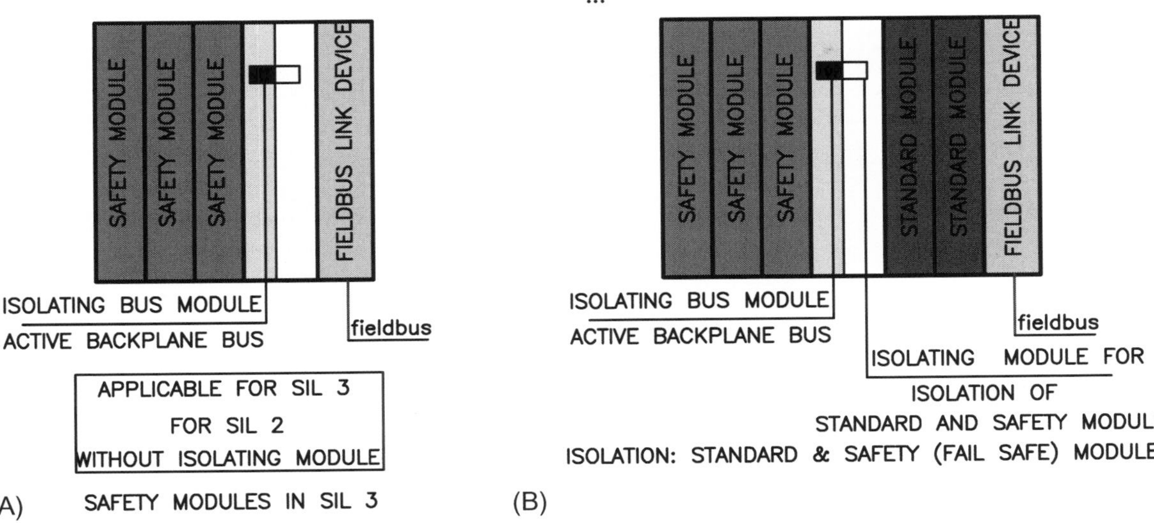

FIG. 14.39 Module disposition.

Typical module dispositions of modules in SIL2 and SIL3 logic solvers and fieldbus connections have been detailed in Fig. 14.39 (courtesy: Siemens) to understand how in reality BPCS and SIS are handled.

8.3 Controller Requirements and Redundancy

1. **Type:** The programs of BPCS and SIS must be functionally separate so that faults in BPCS applications have no effect on safety-related applications and vice versa. Special tasks with very short response times can also be implemented [14]. For safety applications, controllers and I/O modules need to individually certified by a third party; they must also comply with SIL 2/SIL 3 (as the case may be) as per IEC 61508. For safety-related applications, a few restrictions are followed such as: Mutual influencing is prevented, the BPCS program must be separated, and data exchange is done by special conversion blocks. Also, in certain systems, the safety functions are executed twice in different processor sections of one CPU through redundant, multichannel command processing.
2. **Redundancy:** In order to increase availability, redundant subsystems may be deployed. The redundant controllers may work in a one of two principle (or dual cross wired) comprised of two subsystems of identical design. They could be kept electrically isolated from one another, and are synchronized over fiberoptic cables.

9 ENCLOSURE PROTECTION RATINGS

Enclosure protection ratings are very important, so various protections by IP ratings and their equivalence in NEMA are presented in Figs. 14.40 and 14.41.

With this, the discussions on SIS components as well as plant lifecycles and SIL come to an end. This also marks the end of the main write up of the book, with appendices to follow. It is hoped that the information provided in the book is useful to readers. Any queries/comments on the book are most welcome.

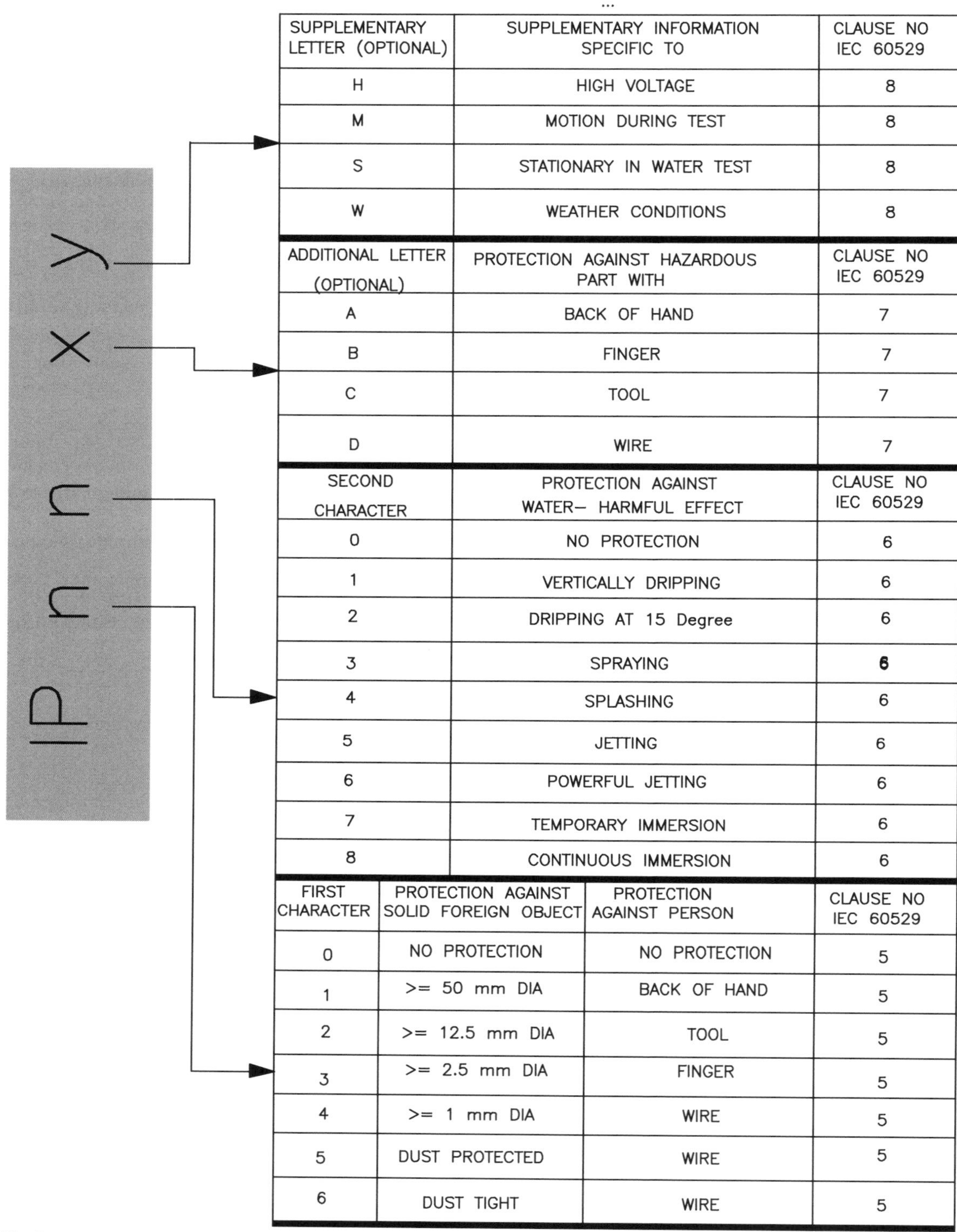

SUPPLEMENTARY LETTER (OPTIONAL)	SUPPLEMENTARY INFORMATION SPECIFIC TO	CLAUSE NO IEC 60529
H	HIGH VOLTAGE	8
M	MOTION DURING TEST	8
S	STATIONARY IN WATER TEST	8
W	WEATHER CONDITIONS	8

ADDITIONAL LETTER (OPTIONAL)	PROTECTION AGAINST HAZARDOUS PART WITH	CLAUSE NO IEC 60529
A	BACK OF HAND	7
B	FINGER	7
C	TOOL	7
D	WIRE	7

SECOND CHARACTER	PROTECTION AGAINST WATER– HARMFUL EFFECT	CLAUSE NO IEC 60529
0	NO PROTECTION	6
1	VERTICALLY DRIPPING	6
2	DRIPPING AT 15 Degree	6
3	SPRAYING	6
4	SPLASHING	6
5	JETTING	6
6	POWERFUL JETTING	6
7	TEMPORARY IMMERSION	6
8	CONTINUOUS IMMERSION	6

FIRST CHARACTER	PROTECTION AGAINST SOLID FOREIGN OBJECT	PROTECTION AGAINST PERSON	CLAUSE NO IEC 60529
0	NO PROTECTION	NO PROTECTION	5
1	>= 50 mm DIA	BACK OF HAND	5
2	>= 12.5 mm DIA	TOOL	5
3	>= 2.5 mm DIA	FINGER	5
4	>= 1 mm DIA	WIRE	5
5	DUST PROTECTED	WIRE	5
6	DUST TIGHT	WIRE	5

FIG. 14.40 Ingress protection code.

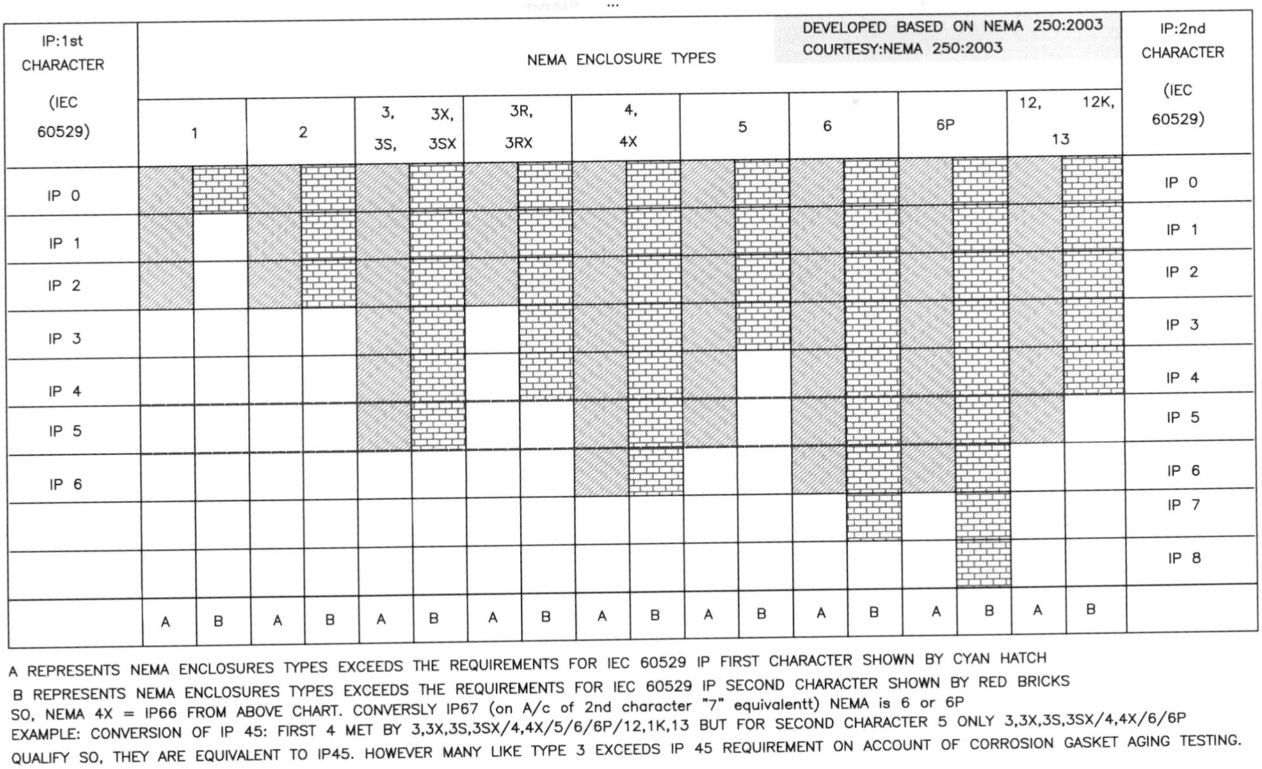

IP:1st CHARACTER (IEC 60529)

NEMA ENCLOSURE TYPES

DEVELOPED BASED ON NEMA 250:2003
COURTESY:NEMA 250:2003

IP:2nd CHARACTER (IEC 60529)

1 | 2 | 3, 3S, | 3X, 3SX | 3R, 3RX | 4, 4X | 5 | 6 | 6P | 12, 12K, 13

IP 0 | IP 1 | IP 2 | IP 3 | IP 4 | IP 5 | IP 6

IP 0 | IP 1 | IP 2 | IP 3 | IP 4 | IP 5 | IP 6 | IP 7 | IP 8

A B A B A B A B A B A B A B A B A B

A REPRESENTS NEMA ENCLOSURES TYPES EXCEEDS THE REQUIREMENTS FOR IEC 60529 IP FIRST CHARACTER SHOWN BY CYAN HATCH
B REPRESENTS NEMA ENCLOSURES TYPES EXCEEDS THE REQUIREMENTS FOR IEC 60529 IP SECOND CHARACTER SHOWN BY RED BRICKS
SO, NEMA 4X = IP66 FROM ABOVE CHART. CONVERSLY IP67 (on A/c of 2nd character "7" equivalentt) NEMA is 6 or 6P
EXAMPLE: CONVERSION OF IP 45: FIRST 4 MET BY 3,3X,3S,3SX/4,4X/5/6/6P/12,1K,13 BUT FOR SECOND CHARACTER 5 ONLY 3,3X,3S,3SX/4,4X/6/6P
QUALIFY SO, THEY ARE EQUIVALENT TO IP45. HOWEVER MANY LIKE TYPE 3 EXCEEDS IP 45 REQUIREMENT ON ACCOUNT OF CORROSION GASKET AGING TESTING.

FIG. 14.41 Conversion of NEMA to IP rating.

BIBLIOGRAPHY

[1] S. Basu, Plant Hazard Analysis and Safety Instrumentation Systems, Elsevier; IChemE, 2016. https://icheme.myshopify.com/products/plant-hazard-analysis-and-safety-instrumentation-systems-1st-edition.

[2] USPAS, Controlling Risks Safety Lifecycle, http://uspas.fnal.gov/materials/12UTA/06_lifecycle.pdf, January 2012.

[3] A. Richardson, A Culture of Safety; Flow Control, https://www.flowcontrolnetwork.com/a-culture-of-safety/, October 2011.

[4] J.L. Bergstrom, Safety Instrumented Systems (SIS) and Safety Lifecycle, Process Engineering Associates, LLC, September 2009.

[5] Commission for Energy Regulation, ALARP Guidance a Part of Petroleum Safety Frame Work, (2013) (Guidance Document CER/13/282; CER; Nov. 29, 2013).

[6] Patchin Curtis, Mark Carey Risk Assessment in Practice; COSCO; Deloitte and Touche LLP; https://www2.deloitte.com/content/dam/Deloitte/global/Documents/Governance-Risk-Compliance/dttl-grc-riskassessmentinpractice.pdf

[7] Swapan Basu Plant Flow Measurement and Control Handbook; Elsevier B.V.; https://www.elsevier.com/books/plant-flow-measurement-and-control-handbook/basu/978-0-12-812437-6.

[8] Understanding and Applying the ANSI/ISA 18.2 Alarm Management Standard; PAS Understanding of ISA 18.2; https://www.isa.org/standards.

[9] T. Karte, I.J. Kiesbauer, A.G. Samson, Partial stroke testing for final elements, in: Conference Proceedings of Petroleum and Chemical Industry Conference (PCIC) Europe, 2005.

[10] B.G. Liptak, Instrument Engineers' Handbook: Process Control, Chapter 6.10; Emergency Partial-Stroke Testing of Block Valves; A.S. Summers.

[11] T. Karte, J. Kiesbauer, Smart Valve Positioners and Their Use in Safety Instrumented Systems, (2009) Samson W023040en.pdf; Industrial valve.

Appendix A

Process and Mechanical Standard Table

A.1 STANDARD PREFIXES

TABLE A.1 Conversion Tables of Units

Coded Prefix p	Meaning of Prefixed Unit	Multiply By	Coded Prefix	Meaning of Prefixed Unit	Multiply By
da	deka	10	d	deci	10^{-1}
h	hecto	100	c	centi	10^{-2}
k	kilo	1000	m	milli	10^{-3}
M	mega	10^{6}	μ	micro	10^{-6}
G	giga	10^{9}	n	nano	10^{-9}
T	tera	10^{12}	p	pico	10^{-12}
P	peta	10^{15}	f	femto	10^{-15}
E	exa	10^{18}	a	atto	10^{-18}
Z	zeta	10^{21}	z	zepto	10^{-21}
Y	yotta	10^{24}	y	yocto	10^{-24}

A.2 STANDARD UNITS

TABLE A.2 Standard Units for Temperature

Celsius	C	K – 273.15	(F – 32) × 5/9	(R – 491.67) × 5/9
Rankine	R	K × 9/5	F + 459.67	C × 9/5 + 491.67
Fahrenheit	F	R – 459.67	C × 9/5 +32	K × 9/5 – 459.67
Kelvin	K	C + 273.15	R × 5/9	(F + 459.67) × 5/9

TABLE A.3 Standard Units for Pressure

Bar	bar	10^5 N/m^2	Atmospheres	atm	1.01325×10^5 N/m^2
Pascal	Pa	1 N/m^2	Inches of mercury	inHg	3.387 kPa
Pounds per square inch	psi	1 lb/in^2	Millimeters of mercury	mmHg	0.1333 kPa
Pounds per square foot	psf	1 lb/ft^2	Torr	torr	1.333224 Pa
Kilo psi	ksi	1000.0 psi			

TABLE A.4 Standard Units for Volume

Liter	L	1/1000.0 m^3	Barrel (petroleum)	barrel	158.9873 L
Gallon	gal	3.785412 L	Fluid ounce	fl oz	29.57353 ml
Pint (U.S. liquid)	pint	1/8. gal	Tablespoon	tbl	1/32. pint
Quart (U.S. liquid)	qt	2 pint	Teaspoon	tsp	1/3. tbl
Pint (U.S. dry)	dpint	0.5506105 L	Cup	cup	16. tbl
Quart (U.S. dry)	dqt	2 dpint			

TABLE A.5 Standard Units for Mass

Gram	g	0.001 kg	Short Ton	ton	2000 lbm
Pound mass	lbm	0.45359237 kg	Long Ton	ton_l	2240 lbm
Tonne	T	1000 kg	Ounce	oz	28.34952 g
Tonne	T	2205 lb	Grain	gr	64.79891 mg
Troy pound	lbt	0.3732417 kg	Hundredweight	cwt	112 lbs/4 quarter
Carat (metric)	carat	0.2 g	Pennyweight	dwt	1.55174 g

TABLE A.6 Standard Units for Distance

Foot	ft	0.3048 m	Astronomical unit	AU/au	1.49598×10^{11} m
Inch	in	1.0/12.0 ft	Angstrom	Ang	$10-^{10}$ m
Mile	mile	5280.0 ft	Angstrom	\\AA	1 Ang
League	league	3 mile	furlong	furlong	220 yd
milliinch	mil	0.001 in.	fathom	fathom	6 ft
Parsec	pc	3.085678×10^6 m	Rod	rd	16.5 ft
Meter	m	1.094 yd	Meter	m	10^6 μm
Meter	m	3.281ft	Nautical mile/h	knot	6080 ft/s 1.853 km/s
Meter	m	39.37 in.			

TABLE A.7 Standard Units for Area

Square mile	640 acres	259 hectares	Square meter	1.196 sq. yds	10.76 sq. ft
Acre	4840 sq. yds	0.405 hectares	Square kilometer	100 hectares	0.3861 sq. mile
Square yds	9 sq. ft	0.836 sq. m	Hectares (ha)	10^4 sq. m	2.471 acres

TABLE A.8 Standard Units for Force/Weight

Newton	N	1 kg m/s^2	Poundal	poundal	1 lbm ft/s^2
Dyne	dyn	$10-^5$ N	Kilopound	kip	1000 lbf
Pound force	lb	lbm G	Kilogram force	kgf	kg G
Pound force	lbf	lbm G			

TABLE A.9 Standard Units for Energy

Joule	J	1 N m	British Thermal Unit	Btu	107.6 m kg
British Thermal Unit	BTU/Btu	1055.056 J	British Thermal Unit	Btu	0.2520 kcal
Calorie	cal	4.1868 J	Erg	erg	10^{-7} J
Calorie	Cal	4.1868 kJ	Ton of TNT	TNT	4.184×10^9 J
British Thermal Unit	Btu	778 ft-lb	Electron volt	eV	1.602177×10^{-12} erg

TABLE A.10 Standard Units for Power

Watt	W	1 J/s	Horse power	hp	550 ft lb/s
Kilowatt	kW	738 ft-lb/s	Horse power	hp	746 W
Kilowatt	kW	102 m.kg/s	Horse power	hp	76 m kg/s

TABLE A.11 Standard Units for Density

Density	Cu ft/lb	0.0624 Cu m/kg	Concentration	lb/Cu ft	16.02 kg/Cu m
Density	Cu m/kg	16.02 Cu ft/lb			

TABLE A.12 Standard Units for Electromagnetism

Coulomb	Co	1 Amp s	Oersted	Oe	79.57747 Amp turn/m
Faraday	faraday	96,485.31 Co	Webber	Wb	V s
Farad	farad	Co/V	Tesla	Tesla	Wb/m^2
Stokes	stokes	10^{-4} m^2/s	Henry	H	Wb/A

TABLE A.13 Standard Units for Viscosity

Dynamic	pascal s	10 Poise (P)	Kinematic	m^2/s	10^4 Stokes (St)
Dynamic	pascal s	10 dyne s/cm^2	Kinematic	1 cm^2/s	1 cSt

A.3 FLANGE DIMENSIONS

They are determined by the pipe size and the pressure class required for the application. Most of these dimensions have been standardized and published as ASME, MSS, API, or other standardization organization specifications. ASME/ANSI B16.5 provides dimensions and tolerances for flanges in pipe sizes from 1/2″ through 24″ and in classes ranging from 150 through 2500.

TABLE A.14 Flange Dimensions: Class 150

Nominal Pipe Size (NPS; in.)	Diameter of Flange (in.)	Number of Bolts	Diameter of Bolts (in.)	Bolt Circle (in.)
1/4	3-3/8	4	1/2	2-1/4
1/2	3-1/2	4	1/2	2-3/8
3/4	3-7/8	4	1/2	2-3/4
1	4-1/4	4	1/2	3-1/8
1-1/4	4-5/8	4	1/2	3-1/2
1-1/2	5	4	1/2	3-7/8
2	6	4	5/8	4-3/4
2-1/2	7	4	5/8	5-1/2
3	7-1/2	4	5/8	6
3-1/2	8-1/2	8	5/8	7
4	9	8	5/8	7-1/2
5	10	8	3/4	8-1/2
6	11	8	3/4	9-1/2
8	13-1/2	8	3/4	11-3/4
10	16	12	7/8	14-1/4
12	19	12	7/8	17
14	21	12	1	18-3/4
16	23-1/2	16	1	21-1/4
18	25	16	1-1/8	22-3/4
20	27-1/2	20	1-1/8	25
24	32	20	1-1/4	29-1/2

TABLE A.15 Flange Dimensions: Class 300

Nominal Pipe Size (NPS; in.)	Diameter of Flange (in.)	Number of Bolts	Diameter of Bolts (in.)	Bolt Circle (in.)
1/4	3-3/8	4	½	2-1/4
1/2	3-3/4	4	½	2-5/8
3/4	4-5/8	4	5/8	3-1/4
1	4-7/8	4	5/8	3-1/2
1-1/4	5-1/4	4	5/8	3-7/8
1-1/2	6-1/8	4	¾	4-1/2
2	6-1/2	8	5/8	5
2-1/2	7-1/2	8	¾	5-7/8
3	8-1/4	8	¾	6-5/8
3-1/2	9	8	3/4	7-1/4
4	10	8	3/4	7-7/8
5	11	8	3/4	9-1/4
6	12-1/2	12	3/4	10-5/8
8	15	12	7/8	13
10	17-1/2	16	1	15-1/4
12	20-1/2	16	1-1/8	17-3/4
14	23	20	1-1/8	20-1/4
16	25-1/2	20	1-1/4	22-1/2
18	28	24	1-1/4	24-3/4
20	30-1/2	24	1-1/4	27
24	36	24	1-1/2	32

TABLE A.16 Flange Dimensions: Class 400

Nominal Pipe Size (NPS; in.)	Diameter of Flange (in.)	Number of Bolts	Diameter of Bolts (in.)	Bolt Circle (in.)
1/4	3-3/8	4	1/2	2-1/4
1/2	3-3/4	4	1/2	2-5/8
3/4	4-5/8	4	5/8	3-1/4
1	4-7/8	4	5/8	3-1/2
1-1/4	5-1/4	4	5/8	3-7/8
1-1/2	6-1/8	4	3/4	4-1/2
2	6-1/2	8	5/8	5
2-1/2	7-1/2	8	3/4	5-7/8
3	8-1/4	8	3/4	6-5/8
3-1/2	9	8	7/8	7-1/4

Continued

TABLE A.16 Flange Dimensions: Class 400—cont'd

Nominal Pipe Size (NPS; in.)	Diameter of Flange (in.)	Number of Bolts	Diameter of Bolts (in.)	Bolt Circle (in.)
4	10	8	7/8	7-7/8
5	11	8	7/8	9-1/4
6	12-1/2	12	7/8	10-5/8
8	15	12	1	13
10	17-1/2	16	1-1/8	15-1/4
12	20-1/2	16	1-1/4	17-3/4
14	23	20	1-1/4	20-1/4
16	25-1/2	20	1-3/8	22-1/2
18	28	24	1-3/8	24-3/4
20	30-1/2	24	1-1/2	27
24	36	24	1-3/4	32

TABLE A.17 Flange Dimensions: Class 600

Nominal Pipe Size (NPS; in.)	Diameter of Flange (in.)	Number of Bolts	Diameter of Bolts (in.)	Bolt Circle (in.)
1/4	3-3/8	4	1/2	2-1/4
1/2	3-3/4	4	1/2	2-5/8
3/4	4-5/8	4	5/8	3-1/4
1	4-7/8	4	5/8	3-1/2
1-1/4	5-1/4	4	5/8	3-7/8
1-1/2	6-1/8	4	3/4	4-1/2
2	6-1/2	8	5/8	5
2-1/2	7-1/2	8	3/4	5-7/8
3	8-1/4	8	3/4	6-5/8
3-1/2	9	8	7/8	7-1/4
4	10-3/4	8	7/8	8-1/2
5	13	8	1	10-1/2
6	14	12	1	11-1/2
8	16-1/2	12	1-1/8	13-3/4
10	20	16	1-1/4	17
12	22	20	1-1/4	19-1/4
14	23-3/4	20	1-3/8	20-3/4
16	27	20	1-1/2	23-3/4
18	29-1/4	20	1-5/8	25-3/4
20	32	24	1-5/8	28-1/2
24	37	24	1-7/8	33

TABLE A.18 Flange Dimensions: Class 900

Nominal Pipe Size (NPS; in.)	Diameter of Flange (in.)	Number of Bolts	Diameter of Bolts (in.)	Bolt Circle (in.)
1/2	4-3/4	4	3/4	3-1/4
3/4	5-1/8	4	3/4	3-1/2
1	5-7/8	4	7/8	4
1-1/4	6-1/4	4	7/8	4-3/8
1-1/2	7	4	1	4-7/8
2	8-1/2	8	7/8	6-1/2
2-1/2	9-5/8	8	1	7-1/2
3	9-1/2	8	7/8	7-1/2
4	11-1/2	8	1-1/8	9-1/4
5	13-3/4	8	1-1/4	11
6	15	12	1-1/8	12-1/2
8	18-1/2	12	1-3/8	15-1/2
10	21-1/2	16	1-3/8	18-1/2
12	24	20	1-3/8	21
14	25-1/4	20	1-1/2	22
16	27-3/4	20	1-5/8	24-1/2
18	31	20	1-7/8	27
20	33-3/4	20	2	29-1/2
24	41	20	2-1/2	35-1/2

TABLE A.19 Flange Dimensions: Class 1500

Nominal Pipe Size (NPS; in.)	Diameter of Flange (in.)	Number of Bolts	Diameter of Bolts (in.)	Bolt Circle (in.)
1/2	4-3/4	4	3/4	3-1/4
3/4	5-1/8	4	3/4	3-1/2
1	5-7/8	4	7/8	4
1-1/4	6-1/4	4	7/8	4-3/8
1-1/2	7	4	1	4-7/8
2	8-1/2	8	7/8	6-1/2
2-1/2	9-5/8	8	1	7-1/2
3	10-1/2	8	1-1/8	8
4	12-1/4	8	1-1/4	9-1/2
5	14-3/4	8	1-1/2	11-1/2
6	15-1/2	12	1-3/8	12-1/2
8	19	12	1-5/8	15-1/2
10	23	12	1-7/8	19

Continued

TABLE A.19 Flange Dimensions: Class 1500—cont'd

Nominal Pipe Size (NPS; in.)	Diameter of Flange (in.)	Number of Bolts	Diameter of Bolts (in.)	Bolt Circle (in.)
12	26-1/2	16	2	22-1/2
14	29-1/2	16	2-1/4	25
16	32-1/2	16	2-1/2	27-3/4
18	36	16	2-3/4	30-1/2
20	38-3/4	16	3	32-3/4
24	46	16	3-1/2	39

TABLE A.20 Flange Dimensions: Class 2500

Nominal Pipe Size (NPS; in.)	Diameter of Flange (in.)	Number of Bolts	Diameter of Bolts (in.)	Bolt Circle (in.)
1/2	5-1/4	4	3/4	3-1/2
3/4	5-1/2	4	3/4	3-3/4
1	6-1/4	4	7/8	4-1/4
1-1/4	7-1/4	4	1	5-1/8
1-1/2	8	4	1-1/8	5-3/4
2	9-1/4	8	1	6-3/4
2-1/2	10-1/2	8	1-1/8	7-3/4
3	12	8	1-1/4	9
4	14	8	1-1/2	10-3/4
5	16-1/2	8	1-3/4	12-3/4
6	19	8	2	14-1/2
8	21-3/4	12	2	17-1/4
10	26-1/2	12	2-1/2	21-1/4
12	30	12	2-3/4	24-3/8

A.4 VISCOSITY

Dynamic viscosity of liquid water at different temperatures up to the normal boiling point is listed below.

TABLE A.21 Dynamic Viscosity of Liquid Water

Temperature in °C	10	20	30	40	50	60	70	80	90	100
Viscosity in mPa s	1.308	1.002	0.7978	0.6531	0.5471	0.4668	0.4044	0.3550	0.3150	0.2822

Dynamic viscosities of some gases at 1 kg/cm^2 are listed below.

TABLE A.22 Dynamic Viscosity of Gas

	Air	Hydrogen	Helium	Argon	Xenon	Carbon dioxide	Methane	Ethane
Temperature in °C	0/27	0/27	27	27	0/27	27	27	27
Viscosity in μPa s	17.4/ 18.6	8.4/9	20	22.9	21.2/ 23.2	15	11.2	9.5

TABLE A.23 Dynamic Viscosity of Differential Liquids

Acetone	0.306	Mercury	1.526
Benzene	0.604	Methanol	0.544
Castor oil	985	Nitrobenzene	1.863
Corn syrup	1380.6	Nitrogen (liquid) at −196°C	0.158
Ethanol	1.074	Propanol	1.945
Ethylene glycol	16.1	Olive oil	81
Glycerol (at 20°C)	1200	Pitch	2.3×10^{11}
HFO	2022	Sulfuric acid	24.2

Viscosity (in cP) of liquids (at 25°C unless otherwise specified); 1 cP (centipoise) = 1 mPa s. HFO, heavy fuel oil.

TABLE A.24 Dynamic Viscosity of Other Liquids

Blood (37°C)	$(3–4) \times 10^{-3}$	Molten chocolate	45–130
Honey	2–10	Ketchup	50–100
Molasses	5–10	Lard	≈100
Molten glass	10–1000	peanut butter	≈250
Chocolate syrup	10–25		

Viscosity (in Pa s) of various fluids (at 25°C unless otherwise specified).

A.5 DYNAMIC VISCOSITY

Definition: If a fluid with a viscosity of 1 Pa s is placed between two plates, and one plate is pushed sideways with a shear force of 1 Pa, it moves a distance equal to the thickness of the layer between the plates in 1 s.
Symbol: Greek letter mu (μ); http://en.wikipedia.org/wiki/Viscosity-cite_note-3 used by mechanical and chemical engineers; Greek letter: Eta (η); used by chemists, physicists.
SI unit: pascal (Pa), Pascal second (Pa s) or Newton second/sq. meter (N s/m^2) or kg/(m s).
Cgs unit: poise (P; named after Jean Louis Marie Poiseuille) or centipoise (cP).
Relations: 1 Pa s = 10 P = 1000 cP.

A.6 KINEMATIC VISCOSITY

Definition: Ratio of viscosity and density or μ/ρ.
Symbol: Greek letter nu (v).
SI unit: m^2/s (no special name). The SI unit of ρ is kg/m^3.
Cgs unit: Stokes (St; named after Sir George Gabriel Stokes) or centistokes (cSt).
Relations: 1 m^2/s = 10,000 St = 10^6 cSt or 1 mm^2/s = 1 cSt. Water at 20°C has a kinematic viscosity of about 1 cSt.

Sometimes the ratio of the inertial force (resistant to change or motion) to viscous (heavy and gluey) force is required for calculation purposes, for example, Reynolds number ($Re = \mu L/v$).

A.7 EFFECT OF TEMPERATURE ON VISCOSITY

The temperature and viscosity of liquids are inversely proportional. To be more specific, when temperature increases, viscosity decreases. For liquid services, kinematic viscosity of liquid is defined at a set temperature (for example, 40°C for oil as per the ISO standard). The curve is linear with viscosity values in the logarithmic scale (vertical axis) and temperature values in linear scale (horizontal axis). Each viscosity-temperature characteristic line is plotted according to their viscosity index, which is actually the gradient of the line. The greater the value of the viscosity index, the smaller the change in viscosity for a given change in temperature and vice versa. A sample characteristic curve is shown in Fig. A.1.

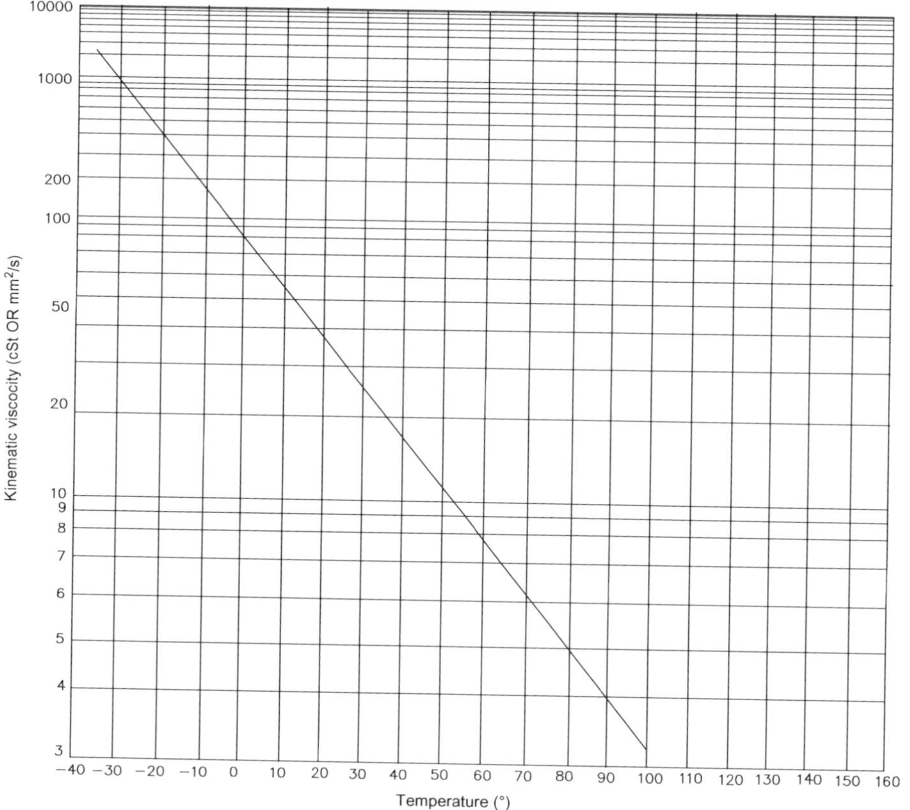

FIG. A.1 Typical kinematic viscosity-temperature graph of a liquid.

TABLE A.25 Combustion Equations and Heat Release

Combustible	Molecular Weight	Reaction	Heat Release in Btu/lb
Carbon	12	$C + O_2 => CO_2$	14,100
Hydrogen	2	$2H_2 + O_2 => 2H_2O$	61,000
Methane	16	$CH_4 + 2O_2 => CO_2 + 2H_2O$	23,900
Ethane	30	$2C_2H_6 + 7O_2 => 4CO_2 + 6H_2O$	22,300
Propane	44	$C_3H_8 + 5O_2 => 3CO_2 + 4H_2O$	21,500
Butane	58	$2C_4H_{10} + 13O_2 => 8CO_2 + 10H_2O$	21,300
Pentane	72	$C_5H_{12} + 8O_2 => 5CO_2 + 6H_2O$	22,000
Sulfur	32	$S + O_2 => SO_2$	4000
Hydrogen Sulfide	34	$2H_2S + 3O_2 => 2SO_2 + 2H_2O$	7100

The viscosity of gases, on the other hand, increases as temperature increases and is approximately proportional to the square root of the temperature. The viscosity-temperature characteristic line is plotted with both parameters in linear scale, and the graph indicates a hyperbolic curve.

BIBLIOGRAPHY

[1] ASME/ANSI B16.5 for flange details.

[2] Viscosity of Hydraulic Oil: By Martin Cuthbert MEng (Hons), Webtec Products Ltd, UK.

[3] Documents on Reynold's Number: By National Aeronautics and Space Administration, Glen Research Centre.

Appendix B

Electrical Data and Tables

TABLE B.1 Stranded Annealed Copper Conductor Size With Electrical Resistance

Size (AWG)	Diameter of Each Strand	Cross-Sectional Area (in mm^2; 7 Strands)	Resistance in ohm/km at 20°C (as per EN 50288-7[a])	Resistance in ohm/km at 20°C (as per UL 13[b])
24	0.579	0.205	–	91.1
22	0.729	0.327	–	57.6
–	0.9	0.5	36.8	–
20	0.919	0.517	–	35.8
–	1.11	0.75	25	–
18	1.16	0.82	–	22.8
–	1.29	1.0	18.5	–
16	1.46	1.3	–	14.2
–	1.59	1.5	12.3	–
14	1.85	2.1	–	8.94
–	2.01	2.5	7.56	–
12	2.32	3.3	–	5.63
–	2.58	4.0	4.7	–

Multipair Instrumentation Cables: Size, Resistance, and Other Physical Data.
[a]*European standard.*
[b]*Underwriters Laboratory Inc. (United States).*

TABLE B.2 Multicore Stranded Annealed Copper Conductor Size With Physical Data and Current Rating (Conforming IStrun-1 694)

Nominal Area (in mm)	Number/ Nominal Diameter of Each Strand	Core Diameter (in mm)	Nominal Insulation Thickness (in mm)	Nominal Sheath Thickness (in mm)			Overall Diameter (in mm)			Rated Current (in amp)
				2 Core	3 Core	4 Core	2 Core	3 Core	4 Core	
0.5	16/0.2	2.2	0.6	0.9	0.9	0.9	6.3	6.7	7.3	4
0.75	24/0.2	2.45	0.6	0.9	0.9	0.9	6.8	7.2	8.0	7
1.0	32/0.2	2.6	0.6	0.9	0.9	0.9	7.1	7.5	8.3	12
1.5	30/0.25	2.9	0.6	0.9	0.9	1.0	7.6	8.2	9.3	15

Continued

TABLE B.2 Multicore Stranded Annealed Copper Conductor Size With Physical Data and Current Rating (Conforming IStrun-1 694)—cont'd

2.5	50/0.25	3.6	0.7	1.0	1.0	1.0	9.0	9.6	10.6	20
4.0	56/0.3	4.3	0.8	1.0	1.0	1.0	10.4	11.4	12.6	27

Collective Screened Pair Cables as Recommended by EN 50288-7.

All dimensions are approximate.

Individual and Collective Screened Pair Cables as Recommended by EN 50288-7.

TABLE B.3 Seven Strands of 0.30 mm Diameter of Each: Conductor With 0.5 mm^2 Cross Section

Number of Pairs	Weight (kg/km)	Diameter Over Bedding (mm)	O/D (mm)	Conductor Resistance (ohms/km)
1	175.00	5.2	9.5	36.00
2	250.00	7.6	12.0	36.00
4	315.00	8.8	13.5	36.00
8	440.00	11.3	16.0	36.00
12	565.00	13.5	18.0	36.00
24	1000.00	18.3	24.0	36.00

TABLE B.4 Seven Strands of 0.37 mm Diameter of Each: Conductor With 0.75 mm^2 Cross Section

Number of Pairs	Weight (kg/km)	Diameter Over Bedding (mm)	O/D (mm)	Conductor Resistance (ohms/km)
1	195.00	5.6	10.0	24.50
2	250.00	7.6	12.0	36.00
4	370.00	10.9	14.5	24.50
8	535.00	12.8	17.5	24.50
12	670.00	15.1	20.0	24.50
24	1230.00	20.8	27.0	24.50

TABLE B.5 Seven Strands of 0.44 mm Diameter of Each: Conductor With 1.3 mm^2 Cross Section

Number of Pairs	Weight (kg/km)	Diameter Over Bedding (mm)	O/D (mm)	Conductor Resistance (ohms/km)
1	240.00	6.8	11.0	14.20
2	385.00	10.4	15.0	14.20
4	480.00	12.0	16.5	14.20
8	720.00	15.7	20.5	14.20
12	1095.00	18.9	24.5	14.20
24	1780.00	25.9	32.0	14.20

TABLE B.6 Seven Strands of 0.3 mm Diameter of Each: Conductor With 0.5 mm^2 Cross Section

Number of Pairs	Weight (kg/km)	Diameter Over Bedding (mm)	O/D (mm)	Conductor Resistance (ohms/km)
2	310.00	8.7	13.5	36.00
4	375.00	10.2	15.0	36.00
8	540.00	13.1	18.0	36.00
12	685.00	15.7	20.5	36.00
24	1255.00	21.5	27.5	36.00

TABLE B.7 Seven Strands of 0.37 mm Diameter of Each: Conductor With 0.75 mm^2 Cross Section

Number of Pairs	Weight (kg/km)	Diameter Over Bedding (mm)	O/D (mm)	Conductor Resistance (ohms/km)
2	345.00	9.7	14.5	24.50
4	430.00	11.2	16.0	24.50
8	620.00	14.4	19.0	24.50
12	930.00	17.4	23.0	24.50
24	1495.00	24.0	30.0	24.50

TABLE B.8 Seven Strands of 0.44 mm Diameter of Each: Conductor With 1.3 mm^2 Cross Section

Number of Pairs	Weight (kg/km)	Diameter Over Bedding (mm)	O/D (mm)	Conductor Resistance (ohms/km)
2	435.00	11.4	16.0	14.20
4	550.00	13.4	18.0	14.20
8	970.00	17.6	23.0	14.20
12	1250.00	21.1	27.0	14.20
24	2265.00	29.2	36.5	14.20

TABLE B.9 Temperature Versus Resistance Table (PRT 100)

Deg °C	Ohm	Deg °C	Ohm	Deg °C	Ohm	Deg °C	Ohm
0	100	41	115.93	82	131.66	123	147.19
1	100.39	42	116.31	83	132.04	124	147.57
2	100.78	43	116.7	84	132.42	125	147.94
3	101.17	44	117.08	85	132.8	126	148.32
4	101.56	45	117.47	86	133.18	127	148.7
5	101.95	46	117.85	87	133.56	128	149.07
6	102.34	47	118.24	88	133.94	129	149.45
7	102.73	48	118.62	89	134.32	130	149.82
8	103.12	49	119.01	90	134.7	131	150.2
9	103.51	50	119.4	91	135.08	132	150.57
10	103.9	51	119.78	92	135.46	133	150.95
11	104.29	52	120.16	93	135.84	134	151.33
12	104.68	53	120.55	94	136.22	135	151.7
13	105.07	54	120.93	95	136.6	136	152.08
14	105.46	55	121.32	96	136.98	137	152.45
15	105.85	56	121.7	97	137.36	138	152.83
16	106.24	57	122.09	98	137.74	139	153.2
17	106.63	58	122.47	99	138.12	140	153.58
18	107.02	59	122.86	100	138.5	141	153.95
19	107.4	60	123.24	101	138.88	142	154.32
20	107.79	61	123.62	102	139.26	143	154.7
21	108.18	62	124.01	103	139.64	144	155.07
22	108.57	63	124.39	104	140.02	145	155.45
23	108.96	64	124.77	105	140.39	146	155.82
24	109.35	65	125.16	106	140.77	147	156.19
25	109.73	66	125.54	107	141.15	148	156.57
26	110.12	67	125.92	108	141.53	149	156.94
27	110.51	68	126.31	109	141.91	150	157.31
28	110.9	69	126.69	110	142.29	151	157.69
29	111.28	70	127.07	111	142.66	152	158.06
30	111.67	71	127.45	112	143.04	153	158.43
31	112.06	72	127.84	113	143.42	154	158.81
32	112.45	73	128.22	114	143.8	155	159.18
33	112.83	74	128.6	115	144.17	156	159.55
34	113.22	75	128.98	116	144.55	157	159.93
35	113.61	76	129.37	117	144.93	158	160.3
36	113.99	77	129.75	118	145.31	159	160.67
37	114.38	78	130.13	119	145.68	160	161.04
38	114.77	79	130.51	120	146.06	161	161.42
39	115.15	80	130.89	121	146.44	162	161.79
40	115.54	81	131.27	122	146.81	163	162.16

Continued

TABLE B.9 Temperature Versus Resistance Table (PRT 100)—cont'd

Deg °C	Ohm	Deg °C	Ohm	Deg °C	Ohm
164	162.53	205	177.68	246	192.63
165	162.9	206	178.05	247	192.99
166	163.27	207	178.41	248	193.35
167	163.65	208	178.78	249	193.71
168	164.02	209	179.14	250	194.07
169	164.39	210	179.51	251	194.44
170	164.76	211	179.88	252	194.8
171	165.13	212	180.25	253	195.16
172	165.5	213	180.61	254	195.52
173	165.87	214	180.97	255	195.88
174	166.24	215	181.34	256	196.24
175	166.61	216	181.71	257	196.6
176	166.98	217	182.07	258	196.96
177	167.35	218	182.44	259	197.33
178	167.72	219	182.8	260	197.69
179	168.09	220	183.17	261	198.05
180	168.46	221	183.53	262	198.41
181	168.83	222	183.9	263	198.77
182	169.2	223	184.26	264	199.13
183	169.57	224	184.63	265	199.49
184	169.94	225	184.99	266	199.85
185	170.31	226	185.36	267	200.21
186	170.68	227	185.72	268	200.57
187	171.05	228	186.09	269	200.93
188	171.42	229	186.45	270	201.29
189	171.79	230	186.82	271	201.65
190	172.16	231	187.18	272	202.01
191	172.53	232	187.54	273	202.36
192	172.9	233	187.91	274	202.72
193	173.26	234	188.27	275	203.08
194	173.63	235	188.63	276	203.44
195	174	236	189	277	203.8
196	174.37	237	189.36	278	204.16
197	174.74	238	189.72	279	204.52
198	175.1	239	190.09	280	204.88
199	175.47	240	190.45	281	205.23
200	175.84	241	190.81	282	205.59
201	176.21	242	191.18	283	205.95
202	176.57	243	191.54	284	206.31
203	176.94	244	191.9	285	206.67
204	177.31	245	192.26	286	207.02

Continued

TABLE B.9 Temperature Versus Resistance Table (PRT 100)—cont'd

Deg °C	Ohm	Deg °C	Ohm	Deg °C	Ohm
287	207.38	328	221.94	369	236.31
288	207.74	329	222.29	370	236.65
289	208.1	330	222.65	371	237
290	208.45	331	223	372	237.35
291	208.81	332	223.35	373	237.7
292	209.17	333	223.7	374	238.04
293	209.52	334	224.06	375	238.39
294	209.88	335	224.41	376	238.74
295	210.24	336	224.76	377	239.09
296	210.59	337	225.11	378	239.43
297	210.95	338	225.46	379	239.78
298	211.31	339	225.81	380	240.13
299	211.66	340	226.17	381	240.47
300	212.02	341	226.52	382	240.82
301	212.37	342	226.87	383	241.17
302	212.73	343	227.22	384	241.51
303	213.09	344	227.57	385	241.86
304	213.44	345	227.92	386	242.2
305	213.8	346	228.27	387	242.54
306	214.15	347	228.62	388	242.89
307	214.51	348	228.97	389	243.24
308	214.86	349	229.32	390	243.59
309	215.22	350	229.67	391	243.93
310	215.57	351	230.02	392	244.28
311	215.93	352	230.37	393	244.62
312	216.28	353	230.72	394	244.97
313	216.64	354	231.07	395	245.31
314	216.99	355	231.42	396	245.66
315	217.35	356	231.77	397	246
316	217.7	357	232.12	398	246.35
317	218.05	358	232.47	399	246.69
318	218.41	359	232.82	400	247.04
319	218.76	360	233.17	401	247.38
320	219.12	361	233.52	402	247.73
321	219.47	362	233.87	403	248.07
322	219.82	363	234.22	404	248.41
323	220.18	364	234.56	405	248.76
324	220.53	365	234.91	406	249.1
325	220.88	366	235.26	407	249.45
326	221.24	367	235.61	408	249.79
327	221.59	368	235.96	409	250.13

Continued

TABLE B.9 Temperature Versus Resistance Table (PRT 100)—cont'd

Deg °C	Ohm	Deg °C	Ohm	Deg °C	Ohm
410	250.48	451	264.45	Negative temperature (–)	
411	250.82	452	264.79		
412	251.16	453	265.13	200	18.49
413	251.5	454	265.47	199	18.93
414	251.85	455	265.8	198	19.36
415	252.19	456	266.14	197	19.79
416	252.53	457	266.48	196	20.22
417	252.88	458	266.82	195	20.65
418	253.22	459	267.15	194	21.08
419	253.56	460	267.49	193	21.51
420	253.9	461	267.83	192	21.94
421	254.24	462	268.17	191	22.37
422	254.59	463	268.5	190	22.8
423	254.93	464	268.84	189	23.23
424	255.27	465	269.18	188	23.66
425	255.61	466	269.51	187	24.09
426	255.95	467	269.85	186	24.52
427	256.29	468	270.19	185	24.94
428	256.64	469	270.52	184	25.37
429	256.98	470	270.86	183	25.8
430	257.32	471	271.2	182	26.23
431	257.66	472	271.53	181	26.65
432	258	473	271.87	180	27.08
433	258.34	474	272.2	179	27.5
434	258.68	475	272.54	178	27.93
435	259.02	476	272.88	177	28.35
436	259.36	477	273.21	176	28.78
437	259.7	478	273.55	175	29.2
438	260.04	479	273.88	174	29.63
439	260.38			173	30.05
440	260.72			172	30.47
441	261.06			171	30.9
442	261.4			170	31.32
443	261.74			169	31.74
444	262.08			168	32.16
445	262.42			167	32.59
446	262.76			166	33.01
447	263.1			165	33.43
448	263.43			164	33.85
449	263.77			163	34.27
450	264.11			162	34.69

Continued

TABLE B.9 Temperature Versus Resistance Table (PRT 100)—cont'd

Deg °C	Ohm	Deg °C	Ohm	Deg °C	Ohm	Deg °C	Ohm
Negative temperature (–)		Negative temperature (–)		Negative temperature (–)		Negative temperature (–)	
167	32.59	128	48.82	89	64.7	50	80.31
166	33.01	127	49.23	88	65.11	49	80.7
165	33.43	126	49.64	87	65.51	48	81.1
164	33.85	125	50.06	86	65.91	47	81.5
163	34.27	124	50.47	85	66.31	46	81.89
162	34.69	123	50.88	84	66.72	45	82.29
161	35.11	122	51.29	83	67.12	44	82.69
160	35.53	121	51.7	82	67.52	43	83.08
159	35.95	120	52.11	81	67.92	42	83.48
158	36.37	119	52.52	80	68.33	41	83.88
157	36.79	118	52.92	79	68.73	40	84.27
156	37.21	117	53.33	78	69.13	39	84.67
155	37.63	116	53.74	77	69.53	38	85.06
154	38.04	115	54.15	76	69.93	37	85.46
153	38.46	114	54.56	75	70.33	36	85.85
152	38.88	113	54.97	74	70.73	35	86.25
151	39.3	112	55.38	73	71.13	34	86.64
150	39.71	111	55.78	72	71.53	33	87.04
149	40.13	110	56.19	71	71.93	32	87.43
148	40.55	109	56.6	70	72.33	31	87.83
147	40.96	108	57	69	72.73	30	88.22
146	41.38	107	57.41	68	73.13	29	88.62
145	41.79	106	57.82	67	73.53	28	89.01
144	42.21	105	58.22	66	73.93	27	89.4
143	42.63	104	58.63	65	74.33	26	89.8
142	43.04	103	59.04	64	74.73	25	90.19
141	43.45	102	59.44	63	75.13	24	90.59
140	43.87	101	59.85	62	75.53	23	90.98
139	44.28	100	60.25	61	75.93	22	91.37
138	44.7	99	60.66	60	76.33	21	91.77
137	45.11	98	61.06	59	76.73	20	92.16
136	45.52	97	61.47	58	77.13	19	92.55
135	45.94	96	61.87	57	77.52	18	92.95
134	46.35	95	62.28	56	77.92	17	93.34
133	46.76	94	62.68	55	78.32	16	93.73
132	47.18	93	63.09	54	78.72	15	94.12
131	47.59	92	63.49	53	79.11	14	94.52
130	48	91	63.9	52	79.51	13	94.91

Continued

TABLE B.9 Temperature Versus Resistance Table (PRT 100)—cont'd

Deg °C	Ohm	Deg °C	Ohm	Deg °C	Ohm	Deg °C	Ohm
Negative temperature (–)		Negative temperature (–)		Negative temperature (–)		Negative temperature (–)	
129	48.41	90	64.3	51	79.91	12	95.3
						11	95.69
						10	96.09
						9	96.48
						8	96.87
PRT 100 is the most commonly used resistance temperature						7	97.26
Detector (RTD) and hence the chart is reproduced						6	97.65
						5	98.04
Thermocouple elements used are of various types, and for that reason are not included in this appendix due to space						4	98.44
						3	98.83
Constraint						2	99.22
						1	99.61
						0	100

BIBLIOGRAPHY

[1] IS:1554(PT-I)-1988—Indian Standard.

[2] BS: PAS 5308-1:2009 (Part-1)—British Standard.

[3] BS: PAS 5308-2:2009 (Part-2)—British Standard.

[4] BS EN 50288-7:2005.

[5] RTD Temperature vs. Resistance Table as per DIN 43 760.

Appendix C

International Society of Automation, (ISA), Standard, Materials, Human Engineering, and Control Room

ISA standards and other material data, human engineering, and control room concepts outline.

C.1 ISA STANDARDS

Presented below is a list of important and relevant ISA standards and technical papers.

TABLE C.1 List of Important ISA Standards

Publication Number		Title
Main Ref.	Sub Ref.	
1-55617-531-0		Standards Library for Measurements and Controls (ISA)
12.00.01	− 01.01, 02.01	Electrical Apparatus for Use in Hazardous Locations
12.10		Area Classification in Hazardous (Classified) Dust Location
12.16.01	−16.01, 22.01, 23.01, 25.01, 26.01	Electrical Apparatus for Use in Hazardous Locations (ISA)
18.1		Annunciator Sequence and Specifications (ISA)
20		Process, Measurements, Instruments, Primary Elements, Valves (ISA)
26		Dynamic Response Testing of Process Control Instrumentation
37.1		Electrical Transducer Nomenclature and Terminology (ISA)
37.3		Strain Gauge Pressure Transducers (ISA)
37.6		Potentiometric Pressure Transducers (ISA)
37.8		Specifications and Test of Strain Gauge Force Transducers (ISA)
37.10		Piezoelectric Pressure and Sound Pressure Transducers (ISA)
37.12		Potentiometric Displacement Transducer
5.1		Instrumentation Symbol and Identifications (ISA)
5.2		Binary Logic Diagram for Process Operations (ISA)
5.3		Graphic Symbol for Distributed controls/Shared Display System (ISA)
5.4		Instrumentation Loop Diagrams (ISA)

Continued

TABLE C.1 List of Important ISA Standards—cont'd

Publication Number		Title
Main Ref.	**Sub Ref.**	
5.5		Graphic Symbols for Process Display (ISA)
50.02.2 TR 50.02-9	–.–. 2,3,4,5,6	Fieldbus Standard for Use in Industrial Control Systems (ISA)
51.1		Process Instrumentation Terminology
71.01	–. 01,02,03,04	Environmental Conditions Process Measurements, Control Systems (ISA)
75.01.01		Flow Equations for Sizing (CVs) (ISA)
75.02		CV Capacity Test Procedure (ISA)
75.03		Integral Flanged Globe Style CV bodies (ISA)
75.04		Dimension for Flangeless CV (ISA)
75.05		CV Terminology (ISA)
75.11		Inherent Flow Characteristics, Rangeability of CV (ISA)
75.12		Socket Weld, Screw End Globe Style CV (ISA)
75.13		Method of Evaluating Performance of Positioner (ISA)
75.14	–.14,15	Dimensions of Butt Weld End Globe Style CV (ISA)
75.16		Dimensions of Flanged Globe Style CV (ISA)
75.17		CV Aerodynamic Noise Production
75.19		Hydrostatic Testing of CV
75.22		Dimension of Flanged Globe Style Angle CV (ISA)
77.13.01		FFPP Steam Turbine Bypass
77.20		FFPP Simulator Functional requirements (ISA)
77.41		FFPP Boiler Combustion controls (ISA)
77.42.01		FFPP Feedwater Control Systems (ISA)
77.43		FFPP Unit/Plant Development (ISA)
77.44		FFPP Steam Temperature Control Systems (ISA)
77.70		FFPP Instrument Piping Installations (ISA)
82.02.01, 82.03		Safety Standard for Electrical and Electronic Equipment (ISA)
82.02.02, 82.02.04		Safety Standard for Electrical Equipment (ISA)
92.02.01-1, 92.02.02		Carbon Monoxide Detection Instruments (ISA)
RA8425		Standards Library for Automation and Control
REWIC01		Programmable Logic Controllers in Safety-Related System
REWIC02		Achieving Safety in Distributed Systems (ISA)
RP2.1		Manometer Tables (ISA)
RP42.1		Nomenclature for Instruments Tube Fittings (ISA)
RP60.3		Human Engineering for CC (ISA)
RP60.4		Documentation for CC
RP60.6		Name Plate, Labels, Tags for CC

TABLE C.1 List of Important ISA Standards—cont'd

Publication Number		Title
Main Ref.	Sub Ref.	
RP60.8		Electrical Guide for CC
RP60.9		Pipe Guide for CC
TR77.60.04		FFPP Human Machine interface

CC, control centers, CV, control valve, FFP, fossil fuel power plant.

C.2 MATERIAL COMPOSITIONS

Presented below is a list of important and relevant materials with the associated composition normally used in power plants.

C.2.1 Carbon and Ceramics

TABLE C.2 Carbon and Ceramics Composition Details

Material Type	Materials	Description
Carbon	Karbate (carbon)	Impervious carbon
	Karbate (graphite)	Impervious graphite
Ceramics	Pfaudler glass lining	Glass-lined metallic equipment (steel)
	Plate glass lining	Polished plate glass, flat or bent
	Pyrex	Glass type

C.2.2 Plastics and Rubber

TABLE C.3 Plastics and Rubber Composition Details

Material Type	Materials	Description
Plastic	Koroseal	Plasticized polyvinyl chloride
	Polythene	Polyethylene
	Teflon	Polymerized tetrafluoroethylene
	Tygon	Synthetic compound
Rubber	Ace hard rubber	Vulcanized rubber
	Butyl	Solid copolymer of isobutylene and isoprene
	Neoprene	Polymer of chloroprene
	Nitrile Rubber or NBR	Synthetic rubber copolymer of acrylonitrile and butadiene

C.2.3 Metals

All elements are indicated in their chemical formula and the amount in percentage if not in traces. This composition is a broad spectrum. For actual composition, relevant standards such as ASTM and DIN should be referenced.

TABLE C.4 Commonly Used Metal Alloy Composition Details

Brass	Cu 60–65, Zn 35–40, Pb 0.5–3
Brass red	Cu 85, Zn 15
Bronze (comm)	Cu 90, Zn 10
Bronze phosphor	Cu 88, Zn 4, Sn 4, Pb 4
Bronze phosphor 10%	Cu 89.5–90, Sn, P 10–10.5
Hastelloy A	Ni 17–21, Mo 17–21, Fe
Hastelloy B	Ni 24–32, Mo 3–7, Fe 0.02–0.12, C
Hastelloy C	Ni 14–19, Mo 4–8, Fe 0.04–0.15, C 12–16, Cr 3–5.5, W
Hastelloy D	Ni 8–11, Si 2–5, Cu 1, Al
Monel	Ni 67,Cu 30, Fe 0.15, C
Nickel	Ni 99.4, Mn 0.2, Cu 0.15, Fe 0.05, Si
Nickel-silver A/B	A: Cu 65, Ni 18, Zn 17 B: Cu 55, Ni 18, Zn 27
Platinum	Pt 99.99
Silver	Ag 99.9
Stainless steel (SS) 301	Fe 16–18, Cr 6–8, Ni 0.08–0.15, C
SS302	Fe 17–19, Cr 8–10, Ni 0.08–0.15, C
SS303	Fe 17–19, Cr 8–10, Ni 0.15 (max), C 0.07 (min), P, S, Se 0.6
SS304	Fe1 8–20, Cr 8–11, Ni 0.08 (max), C 2 (max), Mn
SS310	Fe 24–26, Cr 19–22, Ni 0.25 (max), C
SS316	Fe 16–18, Cr 10–14, Ni 0.1 (max), C 1.75–2.75, Mo
SS317	Fe 16–18, Cr 10–14, Ni 0.1 (max), C 1.75–2.75, Mo
SS321	Fe 17–19, Cr 8–11, Ni, Ti 5 × C
SS347	Fe 17–19, Cr 9–12, Ni, Cb, 10 × C (min)
SS403	Fe 11.5–13, Cr 0.15 (max), C
SS410	Fe 11.5–13.5, Cr 0.15 (max), C
SS416	Fe 12–14,Cr 0.15 (max), C 0.07 (min), P, S, Se 0.6 (max), Mo
SS430	Fe 14–18, Cr 0.12 (max), C
SS446	Fe 23–27, Cr 0.35 (max), C 0.25 (max), N
Steel	Normal carbon steel C < 0.2, Mn 1.5 (max), Si 0.1%–0.2%, Cr 0.4, S 0.06–0.3, P 0.04
Tantalum	>99.9 Ta
Worthite	Fe 20, Cr 24, Ni 0.07 (max), C 3.25, Si, 3, Mo 1.75, Cu 0.5, Mn

C.2.4 Service Temperature Limit of a Number of Commonly Used Materials

TABLE C.5 Temperature Limits for Commonly Used Materials

ASTM Designation/ Trade Name	Description	Temperature Range (°C)
EPDM	Ethylene propylene terpolymer	−40 to 135
NBR	Nitrile	−54 to 82
NR	Natural rubber	−29 to 93
PTFE	Polytetrafluoroethylene Polytetrafluoroethylene (glass/carbon filled)	−73 to 204 −73 to 232
Flexible graphite, Grafoil		−185 to 540

C.2.5 Designation of Commonly Used High Nickel Alloys

TABLE C.6 Casting Designation of Commonly Used Alloys

Casting Designation	Wrought Trade Name	Generic Designation
CF3		304L
CF8		304
CF3M		316L
CF8M		316
CG8M		317
CW2M	New Hastelloy C	Alloy C276
CXW2M	Hastelloy C22	Alloy C22
CW6MC	Inconel 625	Alloy 625
CY40	Inconel 600	Alloy 600
CZ100	Nickel 200	Alloy 200
LCB		LCB
LCC		LCC
N7M	Hastelloy B2	Alloy B2
WCB		WCB
WCC		WCC

C.3 ERGONOMIC CONTROL ROOM DESIGN CONCEPTS

Human engineering based on ISA RP 60.3 and ergonomic design of control rooms based on ISO 11064 (various parts) are discussed. In Fig. C.1, reachability of humans along with control room design basics are depicted to get an idea about control room design. Today's superior ergonomic design makes it possible to have several large screens at the back (behind the workstation console), screens in the front, live video, and process control graphics, etc. For a better control room environment, remote graphics make the control room computer free. In modern control rooms, operators control local temperature, lighting, etc. Also, large screen units and workstation consoles are motorized so that height as well as distance between them can be adjusted for better viewing. Modern control room concepts (based on ABB EOW-Technical Specification) are shown in Fig. C.2. In some control systems, even dedicated iPads are used as control devices (e.g., 800 × of ABB).

C.3.1 Human Body Dimensional Details

The data given below are based on ISA RP60.3. Only major relevant data are presented, and are based on normal humans (male). Unless otherwise stated, data are in meters.

TABLE C.7 Human Body Dimensional Details (Human Engineering)

Human Position	Dimensional Element	5th Percentile	95th Percentile
Standing on the floor	Vertical reach (hand up)	1.9	2.2
	Eye to floor	1.5	1.7
	Side arm reach from center of body on either side of body.	0.7	0.8
	Forward arm reach (in front)	0.7	0.8
	Chest circumference	0.87	1.1
	Elbow to floor	1.0	1.2
(In body)	Head height	–	0.255
	Chin to eye		0.125
	Arm swing (aft)		40°

Continued

TABLE C.7 Human Body Dimensional Details (Human Engineering)—cont'd

Human Position	Dimensional Element	5th Percentile	95th Percentile
Sitting on a chair placed on the floor	Sitting height to floor	1.3	1.4
	Eye distance from floor (standard chair)	1.2	1.3
	Standard chair height from floor	0.45	0.45
	Vertical reach	1.1	1.3
	Top of head to seat (sitting level)	0.84	0.95
	Eye level to seat level	0.7	0.83
	Thigh clearance (above seat)	0.12	.162
	Forearm length	0.34	0.405
	Seat length	0.37	0.54

C.3.2 Ergonomic Considerations for Control Room Concepts

In the past, basic human factors were a functional grouping of instruments for operational facilities. However, in the modern control room there are so many ergonomic factors to be considered that the design is foolproof and helpful to the operators. The human factor, the machine (hardware and software), the work environment, and the control (operation and management) need to be in harmony and integrated during all phases of the design process. In line with ISO 11064, the basic steps for such a design are depicted in Fig. C.1C. In the following clause, other factors are discussed.

C.3.2.1 Layout

- The control room layout must take into account both 5th and 95th percentile data.
- Layout should be done after task analysis and both link and hierarchical task analysis.
- Emergency exit must be considered for 99th percentile.

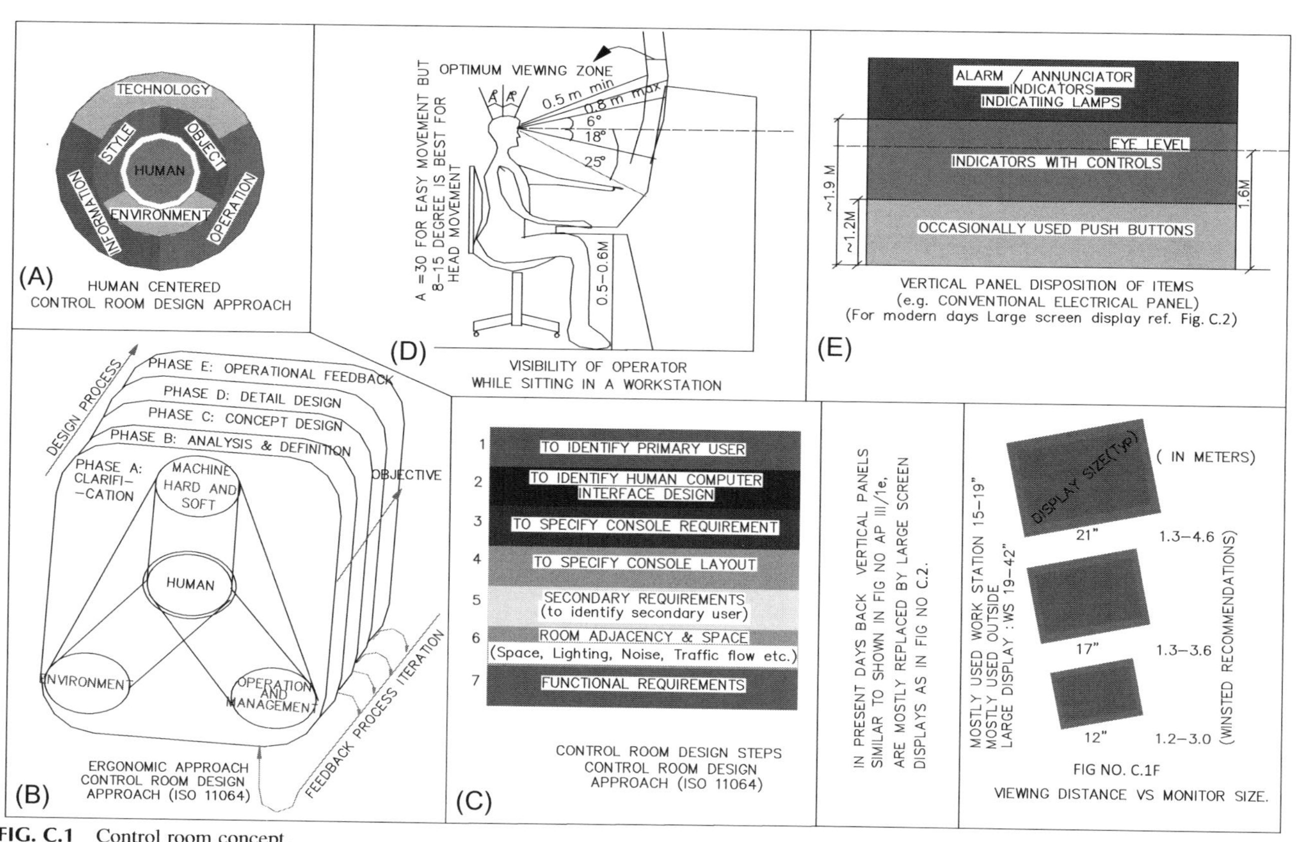

FIG. C.1 Control room concept.

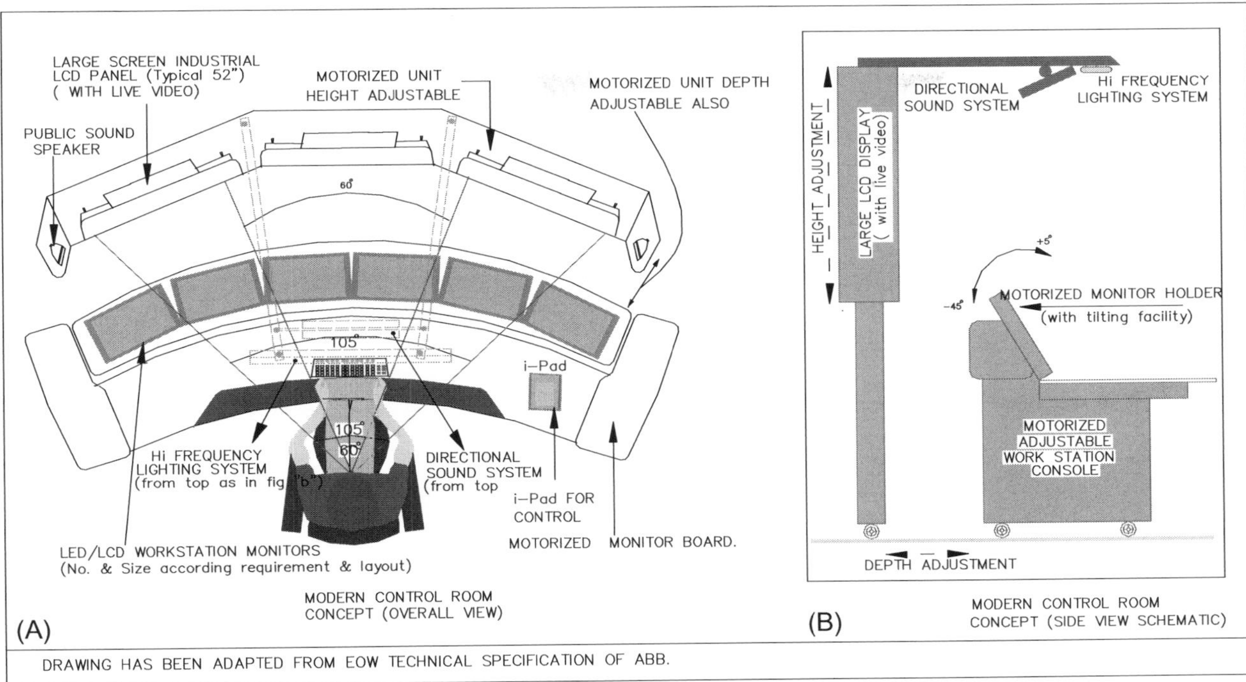

FIG. C.2 Modern control room concepts.

- Adequate space and suitable use of space needs to be designed, while taking into consideration traffic flow, etc. Flow from general circulation areas should be discouraged.
- Line of sight and communication means are important factors in the layout of the control room.
- Allocation of responsibility and the requirements of supervision during high as well as low staffing periods is essential.
- While creating the layout the designer must consider a 300–700 mm distance between intimate zones (other's).

C.3.2.2 Temperature and Air Flow

Temperature varies little from place to place, but a comfortable range is 24 ± 2°C (some places may demand lower than this, for example, 18–20°C), and air flow should be 0.11–0.15 m/s with suitable makeup. The current trend is for the operator to have his own comfort control.

C.3.2.3 Lighting

Suitable lighting is extremely important to avoid maladies like headaches and eye pain. The displays should be glare free, with no reflecting surfaces, including veiling reflection on the video display units. All illumination should be flicker free. Normally 500–800 lux is recommended. The current trend is to use a high-frequency illumination system as shown in Fig. C.2.

C.3.2.4 Sound Level and Alarm

Sound level in the control room should never be >85 dB. Noise level should never cause a hindrance to the operators' attention to warnings and decision making. It has been found that blinking of lights and pulsating sound draws attention. During emergencies this will be a hindrance, making it difficult for the operator to take action. In modern plants, hundreds of alarms may occur when a unit trips, but many of them may be repetitive, so there should be some grouping and group acknowledgment of the alarms if possible. Alarm management is very important for this reason in modern units.

C.3.2.5 HCI Features and Alarm Management

See Chapter 7, Clause 5.

C.3.2.6 Coding

Coding by color, symbol, sound, shape, inverse video, etc., is quite common. However, these codes should follow international standard and should be consistent throughout the plant.

C.3.2.7 Text and Labels

These must also follow international standards and be consistent throughout the plant. (For further reference, see ISA RP 60.4 & 60.6.)

BIBLIOGRAPHY

[1] Standard Design Criteria/Guideline for Main Plant Package (2 × 500 MW TPP):Central Electricity Authority: New Delhi.
[2] B.G. Liptak, Process control and optimization, in: B.G. Liptak (Ed.), Composition of Metallic and Other Materials, vol. II, CRS Press/Taylor & Francis, 2006.
[3] B.G. Liptak, Process control and optimization. In: B.G. Liptak (Ed.), Human Engineering, vol. II, CRS Press/Taylor & Francis.
[4] EOW-x2 A superior complete workstation for System 800xA: ABB Limited: Technical Specification.
[5] Ergonomics—Control Room Design: AngloAmerical AA Standard.
[6] T. Naito, N. Takano, E. Inamura, A. Hadji, Control Room Design for Efficient Plant Operation, (2011). Yokogawa Technical Report.
[7] Planning and Design of a Control Room: The Winsted Corporation USA: Tech. Write Up.
[8] Control Room Design, http://www.hse.gov.uk/comah/sragtech/techmeascontrol.htm.
[9] System 800xA Extended Operator Workplace EOW-x: ABB Limited.

Appendix D

Network Control and Communication

D.1 NETWORK CONTROL AND COMMUNICATIONS

The following information related to computer networks is a supplement to discussions in Chapter 7. The subtopics include:

A brief discussion is presented below related to computer networks meant to supplement the discussions presented in Chapter 7. The following subtopics are covered:

- Use of fiber optics in the network.
- Electromagnetic spectrum mainly used in computer networks.
- Firewall protection of network (network security).
- Fault Tolerant Ethernet (FTE; network availability).

D.2 FIBER OPTIC AND ASSOCIATED NETWORKS

Since there is no electromagnetic interference, fiber optic networks are becoming increasingly popular. Basic fiber optic cross sections are shown in Fig. D.1A, and a typical fiber optic network is presented in Fig. D.1B. Normally, there are two taps: in one tap there is a light source such as an LED acting as a transmitting end, and the other is a receiving tap with a photo diode. There will be another interface that acts as a converter to convert the light signal to a full-strength electrical one (i.e. boosting the weak signal). At times, in place of an active repeater, passive star connections are used. As a primary light source both an LED and semiconductor laser can be used.

D.2.1 LED Versus Semiconductor Laser

A comparative table between an LED and a semiconductor laser two is presented in Table D.1.

D.2.2 Optical Fiber in a Gigabit Ethernet

Gigabit Ethernet supports both copper and fiber cabling. A gigabit Ethernet with fiber cable has a light source that has to be turned on and off in 1 ns. This is impossible for an LED so a semiconductor laser is used. Gigabit Ethernet cabling is presented in Table D.2.

D.3 ELECTROMAGNETIC SPECTRUM

Any computer commuter network depends on an electrical or optical signal transmission to carry out communication within or outside the network. Therefore the electromagnetic spectrum plays a vital role in computer network communication. Fig. D.1C illustrates a wide variation of wavelength vis-á-vis frequency. Out of the entire spectrum, electromagnetic waves with frequency in the range of 10^4 to 10^{16} are more important for computer network communication. Networks with communication in the frequency range of 10^4 to 10^8 twisted pair cables UTP/STP can also be used. Coaxial cables in the range of 10^5 to 10^9 can be used, whereas optical fibers are used in the frequency range of 10^{12} to 10^{14}. However, for wireless communication duly modulated data can be sent over the medium via radio communication.

D.4 LARGE INTEGRATED COMPUTER NETWORK

Robust connectivity technology, economy, and business competition are forcing industries to integrate various kinds of networks together. Integration also helps provide a quick and rational flow of information. As a byproduct, there can be problems associated with unauthorized access and protective measures must be put in place. The danger with a control system was discussed in Chapter 7, Clause I. A typical large network comprising Control System LAN with a remote control system, a corporate LAN, a business LAN, and Internet are depicted in Fig. D.1D. Networks also need a backup control system in case of a major problem at a remote place. In large networks and in the public domain, security vulnerability is a problem. This vulnerability is addressed by connecting the servers/network to a set of subsystem "firewalls." A firewall only

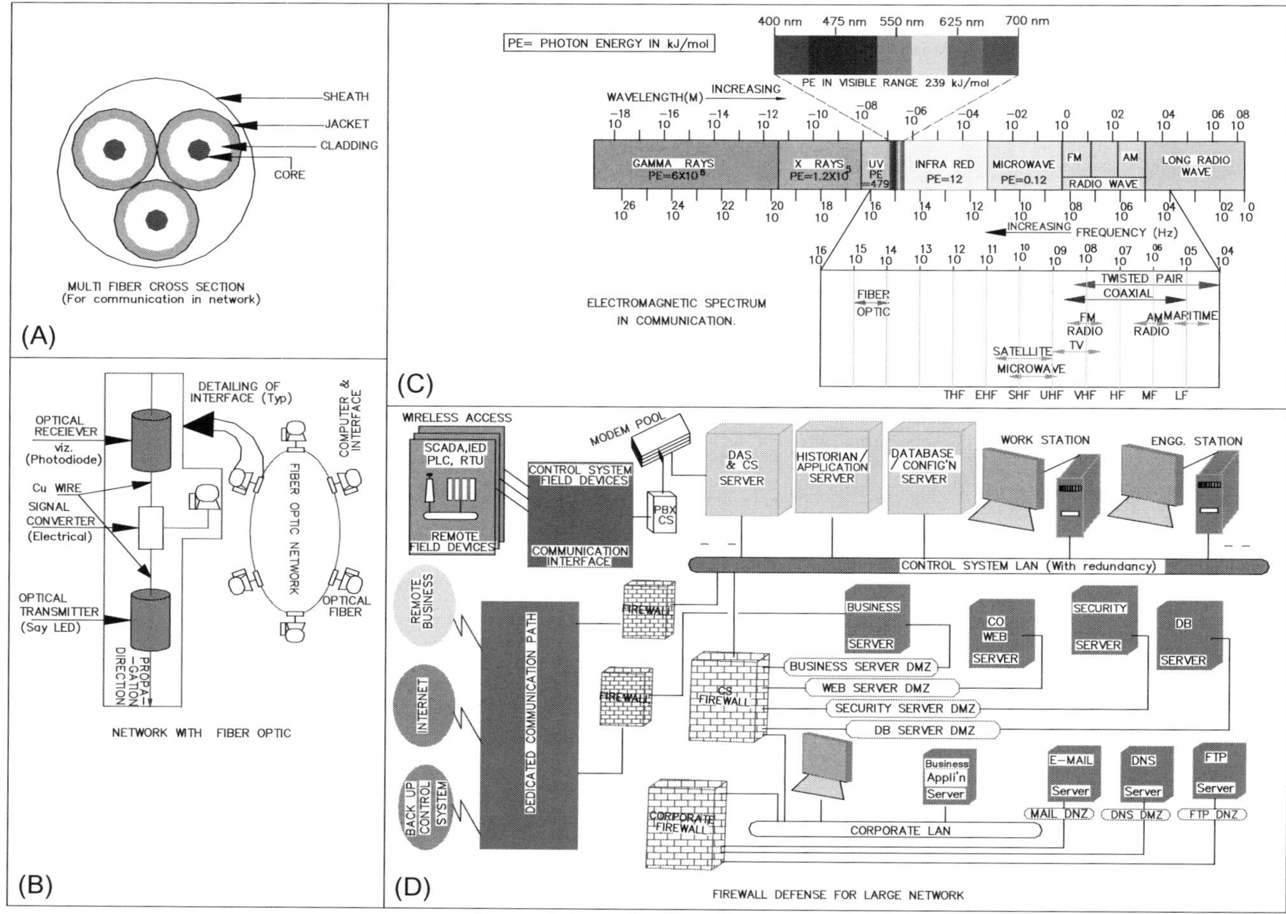

FIG. D.1 Communication and networking.

TABLE D.1 LED Versus Semiconductor Laser

Item	LED	Semiconductor Laser
Data	Low data rate	High data rate
Fiber type	Multimode	Single or multimode
Distance	Short	Long
Cost	Low	Expensive
Lifetime	Longer life	Shorter life
Temperature stability	Not much	Substantially sensitive

TABLE D.2 Gigabit Ethernet Cable Types

Name	Cable	Segment (m)	Advantage
1000Base-SX	Fiber optics	550	Multimode (50 and 62.5 μm)
1000Base-LX	Fiber optics	5000	Multimode (50 and 62.5 m) or single (10 μm)
1000Base-CX	2P–STP	25	Shielded twisted pair
1000Base-T	4P–UTP	100	Standard Cat 5 UTP

allows data that can be accessed from the outside (as permitted in the network design). An external client sends a request to the system through the firewall. If the request meets the set rule, only then will it be serviced by the related server, otherwise the request will be denied. Any request from the public domain must be verified by two firewalls. For example, if any request for a control system is permitted, then that request is sent to the control network. Also, any control system and business server data exchange (from the same network system) will be done through the firewall. These are a few precautionary measures to restrict access to the control/business server/corporate network. Since there will be huge data flow in a large network, separate servers are dedicated for such security and authentication services. However, it must be kept in mind that for control systems, there should be a separate security system provided by the vendor as a lone firewall, which is mainly developed for IT/Internet application. This was discussed in Chapter 7.

D.5 FIREWALLS

Firewalls protect computer networks from unauthorized access. They may be hardware (HW), software (SW), or a combination of both. Apart from the broadband router, a network firewall is a proxy server acting intermediately between internal and external networks by receiving and selectively blocking data at the boundary. It also helps to hide the LAN addresses from outside access to avoid Address Resolution Protocol-ARP poisoning. See Fig. D.2.

D.5.1 Types of Firewalls

Firewalls can be classified in the following ways.

D.5.1.1 HW and SW Firewall Classification

Table D.3 lists the advantages and disadvantages of HW versus SW firewalls.

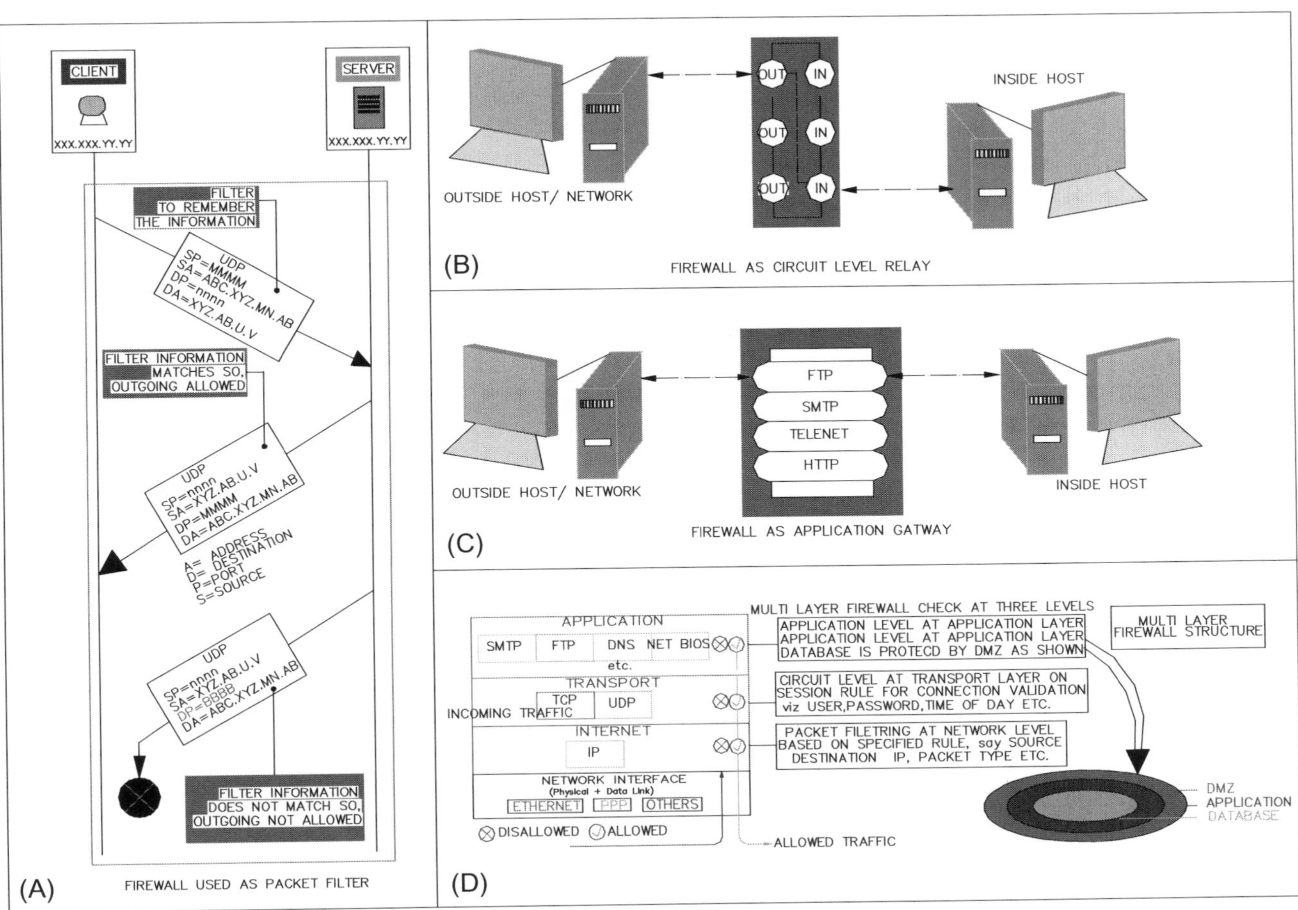

FIG. D.2 Firewall functional details.

TABLE D.3 HW Versus SW Firewall

Comparison	HW Firewall	SW Firewall
Advantages	No operating system, so immune to viruses, generally Faster, better performance Single duty, so very effective	Less expensive Free software available Can be implemented in an existing network Low administrative cost
Disadvantages	On single failure system, may collapse Proprietary, prior knowledge may be necessary High cost for maintenance as well as for installation	Vulnerable to malicious attack Operating-system dependent, so possibility of getting affected by virus Requires additional host resources like CPU/memory, etc. Low performances

D.5.1.2 Other Classifications of Firewall

The following information is how firewalls are classified. The functionality of firewalls will be discussed below.

D.5.1.2.1 Packet-Filtering Firewall

A packet-filtering firewall is classified as listed below:

- *Stateless packet-filtering firewall*: Also known as static IP filtering, this firewall is inexpensive with a high-throughput firewall. It is included with router configuration software or with most open-source operating systems. These are highly vulnerable.
- *IP packet-filtering firewall*: Every packet is handled on an individual basis. Previously forwarded packets belonging to a connection have no bearing on the filter's decision to forward or drop the packet.
- *Stateful packet-filter firewall*: It is a pure packet-filtering environment.

D.5.1.2.2 Application Gateway/Proxies Firewall

A proxy acts as an intelligent intermediary between hosts on the internal network and hosts on the external network. These are expensive, but secure.

D.5.2 Firewall Functionality

A true firewall is HW and SW that intercepts the data between external (Internet) and internal networks (private computer/network). Based on functionality, a firewall can be divided into the following categories (Table D.4) with the functionality of each of them.

D.5.3 Demilitarized Zone

A demilitarized zone (DMZ) isolates the host, which is accessible from outside the internal server. As shown clearly in Fig. D.1D, the business server has access from the outside and has a DMZ in the CS firewall. Basically, the DMZ is the outward facing level of the application. It is a subnetwork that resides between the known/trusted internal network and external network, providing services to the outside without allowing direct access. Some system components of the DMZ include:

- Public-facing server.
- Public-facing File Transfer Protocol (FTP) server.
- E-mail gateway.
- Public-facing Domain Name Server (DNS).
- Traffic management and security server.
- Streaming video.

D.5.4 General Discussion on Firewalls

Firewall policy is extremely important. No firewall can protect a network against inadequate or mismanaged firewall policies, so planning and implementation is vital. Any weakness in the policy, and failure to implement the policy, will result in failure of the firewall. If an IP Virtual Private Network (VPN) is used, then the placement of the VPN, with respect to the firewall, needs to be considered. Some of the policy decisions may be based on the following:

- Internal and external access and their extent.
- Remote user access.
- Virus protection and avoidance.
- Encryption requirement, if any.
- Program usage.

D.6 FAULT TOLERANT ETHERNET—AN APPROACH

Modern plant control systems are highly distributed and communication between the nodes becomes extremely critical. So, for satisfactory operation of the system, it is essential that these critical components are redundant and fault tolerant (as discussed in Chapter 7, Clause 1.0).Fault tolerance of a network varies with system architecture design (see Fig. D.3 for a typical Fault Tolerant

TABLE D.4 Firewall Types According to Functionality

Type Name	Feature
Packet filter (Fig. D.2A and D)	• First line of defense (Fig. D.2A). • Internet and other digital network data travel in packet of limited sizes. It consists of data, ACK, request or command, protocol information, source, destination IP address, port error checking code, etc. • Filtering consists of examining incoming and outgoing packets compared with a set of rules for allowing and disallowing the transmission or acceptance. • It is fast because it does not check any data in the packet except the IP header. It works in the network layer (Internet) of the OSI model. · • IP address can be spoofed.
Circuit relay/gateway (Fig. D.2B and D)	• It is one step above the packet filter and commonly known as "Stateful Packet Inspection." It checks the validation of the connection between the two ends (in addition to packet filtering operation) based on the following: 1. Source destination IP address/port number. 2. Time of the day. 3. Protocol. • User and password. • It operates on the transport layer. Stateful inspection makes the decision about connection based on the data stated above.
Application gateway (application proxies) (Fig. D.2C and D)	• It acts as proxy for the application at the application layer of OSI. • This firewall authorizes each packet for each protocol differently. • Following specific rules it may allow some commands to a server, but not others or limit access to certain types based on the authenticated user. • Setup is complex; every client program needs to be set up. Also, each protocol must have a proxy. • True proxy is safer.
NAT/PAT	• Firewalls using Network Address Translation (NAT) or Port Address Translation (PAT) hide the network using NAT/PAT. • In NAT, there will be a single IP address used for the entire network. • Disadvantage is that it cannot properly pass protocols containing the IP address in the data portion.

Ethernet). While in some cases it is better to continue operation, sometimes it is better to go for a safe state and not continue operation. Critical controls such as BMS and EHG systems in any situation have to run so that equipment and/or human safety is ensured. Cost is also another factor when designing such a system. Redundancy normally is achieved by duplicating the components; apparently it is not that difficult, but in reality unless there is some diversity built in the redundancy becomes meaningless. If two communication channels are used and routed in the same path, redundancy may not be effective because of accidental breakage of the communication cable. Also, there is the probability of similar failure due to external causes. Another problem associated with such duplication is when and how to change over. Before further discussions about FTE, a brief history of various developments is outlined below.

IEEE 802.1D Spanning Tree Protocol (STP) is aimed at media redundancy. It supports various kinds of Ethernet mesh rings or a combination thereof, avoids looping problems in Ethernet connections, and uses Bridges to connect Ethernet LANs, which does the forwarding and filtering operation. The basic idea for an STP is to get rid of the looping associated with ordinary bridge operation and to choose tree topology. The shortest bridge is chosen as the root.

Another standard, IEEE 802.1w Rapid Spanning Tree Protocol (RSTP), has been created for faster recovery time (1 s) from topology changes. RSTP provides faster recovery by monitoring link status of each port and then generating topology change after a link status change. Based on the standard Ethernet device, manufacturers are developing proprietary protocols to minimize recovery time. To get a better idea about these standards, consult the relevant standards and/or related standard material available on the Internet.

Another aspect to consider is the failure of electronics, so switches are also duplicated in many cases. A complete redundant network including redundant devices, port, switches, etc., would provide the safest solution. However, in such cases, apart from cost, the recovery time is in the range of 10–30 s for complete recovery, which depends on the complexity of the network. FTE developed by Honeywell could be a good solution. In the following clauses, Fault Tolerant Network based on Honeywell PKS is discussed.

D.6.1 What is FTE?

The FTE connects a group of nodes typically associated with communication paths between them, so the network can tolerate all single faults and many multiple faults. FTE can rapidly detect faults and, in case of communication failure, the switchover time is around 1 s. FTE uses commercial off-the-shelf equipment, but with increased system availability.

D.6.2 Some Benefits of Fault Tolerant Ethernet

- *Rapid response*: In conventional Ethernet, there are two separate networks with each node (server) connected to both networks. The switchover time, in case of communication failure, is ~30 s. FTE employs single network and does not require a server, so changeover time is less.
- *Possible communication path*: FTE provides more communication path possibilities than the Dual Ethernet Networks, as shown in Table D.5.
- *Full redundancy in single network*: A conventional Ethernet network with redundancy usually has two independent Ethernets with different performance and configuration. Whereas, in an FTE single Ethernet there is no such problem, and at the same time it provides multipath capabilities in its unique topology.

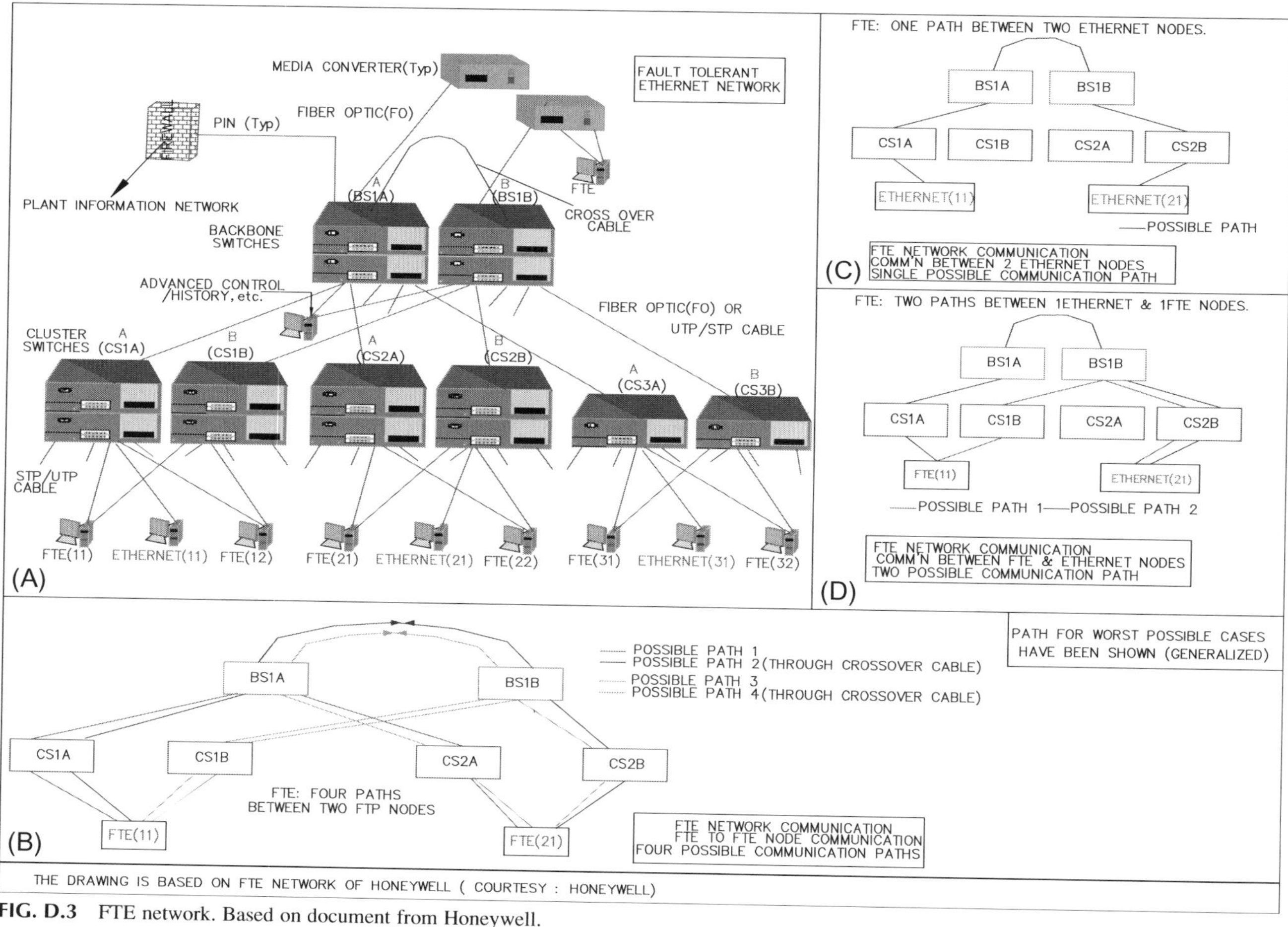

FIG. D.3 FTE network. Based on document from Honeywell.

TABLE D.5 Comparison Between Dual Ethernet and FTE

Connectivity	Supporting DWG Reference	DUAL Ethernet	FTE
Number of networks	Fig. D.3A	2	1
DCN to DCN	Fig. D.3B[a]	2	4
DCN to SCN	Fig. D.3D[a]	1	2
SCN to SCN in same tree	Similar to Fig. D.3C	1	1
SCN to SCN in different trees	Fig. D.3C[a]	0	1

DCN, dual connected nodes; SCN, single connected nodes.
[a]*Each possible path(s) shown by shaded lines in the drawing.*

D.6.3 FTE Topology

Two parallel trees of switches and cabling "A" and "B" are linked at the top to form one Fault Tolerant Network. Each FTE node has two ports that connect to a switch in each tree. In contrast, Ethernet nodes can connect to either if the switches are A or B. There may be one or more levels of switches, and there can be multiple pairs of switches in each level. In Fig. D.3A, these have been designated as "cluster" and "backbone" switches. In the Honeywell FTE Node there are FTE software and Network interface controllers—one for each tree—and these networks are expandable. There can be 511 FTE nodes forming a community, within which it can talk as long as there is a possible path. Similarly, a 511 non-FTE Ethernet can also be accommodated. As shown in Fig. D.3A, a firewall or router can be connected between an FTE and other networks.

D.6.4 Major Components of the FTE

- *Cable*: Copper STP/CAT 5 for connections mainly in cluster switches used normally up to 100 m. Fiber optics (62.5/125) are used for connections up to 2 km and mainly used in backbone switches.
- *Media converter*: Used for flexibility in networking for converting between Cu cable and fiber optic cable.
- *FTE software*: Proprietary in nature, meant for multiple communication paths between nodes and faster switching.
- *Switches*: 10/1000 Mbps ports with run-time diagnostics and remote diagnostics available in Cu/fiber optic models.
- *Network interface controllers*: Two single-port or one dual-port card.

BIBLIOGRAPHY

[1] Computer Networks: Book: by A.S. Tananbum & D.J. Witherall.
[2] Standard Design Criteria/Guideline for Main Plant Pacakge (2 × 500 MW TPP). Central Electricity Authority: New Delhi.
[3] Firewall Presentation, http://www.cs.northwestern.edu/%7Eychen/classes/mitp-458/firewalls.ppt Internet Document.
[4] Firewall Architecture: Nextstep Broadband: White Paper.
[5] Internet Firewall FAQ: P.D. Robertson, M. Curtin & M.J. Ranum: Internet Document.
[6] Firewalls—Overview and best Practices: Decipher Information Systems: White Paper.
[7] D. Kuipers, M. Fabro (Eds.), Control Systems Cyber Security: Defense in Depth Strategies, Idaho National Laboratory, 2006.
[8] NETWORK ARCHITECTURE STANDARD: California Technology Agency: 19-Sept'2012.
[9] Fault tolerant Ethernet (FTE) Specification and Technical Data: Honeywell Industry Solution.
[10] Rapid Spanning Tree in Industrial Network: M. Galea Ruggedcom Inc.: Internet Document.
[11] The ABCs of Spanning Tree Protocol: Contemporary Controls.
[12] Ethernet Fault Tolerance and Redundancy. (Mar'2007) Emerson Process Management.

Appendix E

Supercritical/Ultra-Supercritical Power Plants

Depending on the operating condition, steam power plants can be divided into subcritical, supercritical, and ultra-supercritical categories. The criticality is based on the critical pressure point for steam, i.e., 221 MPa and 374°C. Power plants operating at steam pressure below critical pressure are subcritical power plants, and plants operating with a steam pressure >221 MPa and steam temperature >374° are supercritical power plants. The supercritical stage is a thermodynamic expression describing the state of the substance; here. it is water and steam, where there is no distinction between the liquid [water and gaseous (steam)] phases. In the subcritical stage latent heat is necessary to convert water to steam; at the supercritical state this is not necessary. In ultra-supercritical plants, the steam is operated at higher pressure and temperature (for example, 300 bar and 630°C). Why go supercritical or ultra-supercritical? The answer is clear from the information related in Table E.1.

This table shows a clear increase in efficiency from subcritical to supercritical, and finally to ultra-supercritical. Also, there is a substantial reduction of CO_2 (as well as SO_2) starting from subcritical to ultra-supercritical. The efficiency mainly increases with temperature and not from pressure when a plant switches from supercritical to ultra-supercritical. However, there is an increase in per kW capital expenditure and auxiliary power consumption of the unit as it goes from a subcritical and finally to an ultra-supercritical plant. A cycle efficiency increase from 30% to 50% results in a 30% CO_2 reduction. Water treatment, requirement of condensate polishing unit (CPU), etc., are critical in these types of plants. Another major task in supercritical and ultra-supercritical technology implementation is the material selection for critical parts. However, initially discussions will be about the most critical item, i.e., supercritical steam generation.

The following points are significant to understanding of supercritical technology:

- When water is heated above critical pressure at a constant value, temperature is never constant.
- In the whole system, two phases never exist. There is no distinction between water and steam; hence, density of the two phases is the same.

TABLE E.1 Comparison Among Typical Subcritical/SC and USC Plants

Plant Types/Parameter	Subcritical	SC	USC
Steam pressure (MPa)	16.5	29.0	36.5
Steam temperature (°C)	538/540	580	700
Auxiliary consumption (%)	4–6	5–7	6–8
Efficiency (%)	<40	42–45	>48
CO_2 (g/KWh)	855	780	710
SO_2 (g/KWh)	2.4	2.2	2.0

Based on POWER ASIA 2004 Supercritical and Ultra-Supercritical Plant.

- The actual location of transition from liquid to steam is different under different conditions; hence, sliding pressure is possible.
- For effective heat transfer under different load and pressure, a boiler is capable of optimizing liquid and gas amounts.

E.1 POWER PLANTS WITH SUPERCRITCAL AND ULTRA-SUPERCRITICAL STEAM GENERATORS

These types of steam generating (SG) plants operate on pressure and temperature more than the critical point (that is. 220.6 bar and 374°C), which results in no boiling of water and it directly transforms to steam. Because of this, the boiler is not applicable for supercritical (SC) or ultra-supercritical (USC) SG units. The subcritical SG units normally operate at 150–170 bar pressure and are called boilers, while the SC SG units operate at a pressure ranging from 230 to 265 bar with a temperature ~600°C. The USC SG units operate at pressures around 300 bar and 620–640° C or more for better efficiency.

E.1.1 Historical Developments of SC and USC Thermal Power Plants

For a chart of historical developments of SC and USC thermal power plants, see Table E.2. Thermal power plants operate on the Rankine cycle and their efficiency depends on the operating pressure. Higher operating pressure and temperature means higher efficiency; hence, the modern trend is to install power plants in the supercritical and ultra-supercritical range. Higher efficiency means lower coal consumption per MWh and subsequently lower emissions of pollutants such as CO2; these advantages force designers to an ever-increasing operating point, which may increase to 385 bars and 750°C. The approximate efficiency values are expected to go beyond 50% and even may touch the magic value of 52% by the year 2020 (Table E.3).

The main constraint is the quest for materials with a resistance to all degenerating factors like corrosion, stress,

TABLE E.2 Historical Developments toward the SC and USC Thermal Power Plants

Year	Steam Pressure (bar)	Steam Temperature (°C)	Efficiency	Reheat Cycle
1957	310	612	35%	Single
1960	322	612	35%–40%	Single
1965–1975	242	566	35%–40%	Double
2000	265	600C	43%	Single/double
Currently	320	620	45%	Single/double

All data indicated above are approximate.

TABLE E.3 Comparative Features of Subcritical, Supercritical/Ultra-Supercritical Units

	Power Plant With Subcritical SG Plant	Power Plant With SC or USC Plant
Efficiency	35% (maximum)	45% +
Installation cost	Less than SC or USC plant	Slightly higher than subcritical plant
Operating cost	Same as SC or USC plant	Same as subcritical SG plant
Fuel cost	More than SC or USC plant/per MW generation	Less than subcritical SG plant/per MW generation
CO_2 emission	High/per MW generation	Low/per MW generation
Additional equipment	Boiler drum (for drum type boiler) Boiler recirculating pump, separator, storage vessel, 100% CPU (for OT SG Plant)	Boiler recirculating pump, separator, storage vessel, 100% CPU (for OT SG plant)
Load change ramp	3–4%/min	7–8%/min

All data indicated above are approximate.

stress corrosion, high-temperature creep, oxidation, cyclic and thermal fatigue, etc. Austenitic steels or Ni-based super alloys cope with modern demands.

However, there were initial problems regarding the operation of SC and USC power plants before that were addressed appropriately and emerged as the state-of-the-art technology, such as:

- Complicated startup process.
- Lower operational flexibility.
- Unstable availability and reliability.
- Higher maintenance costs.
- Turbine-governing valve erosion.
- Thermal stress in turbine casing and governing valve body.
- Erosion due to impingement of solid particle on turbine blade.

The turbine is also prone to water induction along with the main steam system due to the once-through (OT) process than drum type plants. The OT configuration also demands a high quality of feedwater because there is no drum in which the blowdown can be made or chemicals can be injected to maintain the feedwater quality. The CPU with a 100% full flow capacity along with oxygenated treatment is extremely necessary for SC/USC plants.

It is well established that higher steam temperatures mean greater efficiency, but it also results in higher bleed steam temperatures being used for the regenerative feedwater heaters. This phenomenon reduces efficiency for the regenerative feed heater area and eats up a portion of the overall efficiency gain. The idea of replacing a single-reheat by a double-reheat cycle is twofold. It helps to extract more work done by the turbine and lowers the degree of superheat of the bleed steam thus arresting the probable efficiency loss. Another important advantage of the double-reheat cycle is the substantial reduction of steam temperature and pressure at the low-pressure (LP) turbine inlet because the maximum allowable temperature is limited by the rotor and crossover hood materials. For example, a plant with a single-reheat cycle can go up to 60–70 bar reheat pressure whereas with the double-reheat cycle can go up to 95–100 bar for first reheat pressure and 25–35 bar for the second reheat pressure, which means lower LP turbine (LPT) inlet pressure accordingly. The double-reheater system clearly offers lower generating costs and less corrosion at the last stage. The implementation of this, however, depends on the cost and complexity of the additional equipment, piping, etc.

There are other options. For example, double LPT and double condensers provide more volume-handling capability, so more work gets done. Possible configurations with both double reheaters and LPTs are seen in Fig. E.7B. There are two ways the circulating (cooling) water can be fed to the condenser. First, using a parallel path, meaning the same quantity of cooling water is fed to the two condensers. The other is a series path, which means the circulating water (CW) is first fed to one condenser then the outlet of the same is fed to the other. In the former, the drop in temperature will be the same in both the condensers, but, in the latter, the temperature will drop more in the first condenser than the second. For the same mass flow of water (Fig. E.6C) the average condenser pressure will be better in the series path. The higher the condenser pressure drop, the higher is the efficiency; hence, the series circuit is the preferred method.

Another important aspect of a SC/USC power plant is the selection of final feedwater temperature, which normally varies from 315 to 330°C, depending on a single- or double-reheat cycle.

E.2 SUPERCRITICAL AND ULTRA-SUPERCRITICAL STEAM GENERATORS

E.2.1 Temperature-Entropy Relationship of SC/USC SG Plants

Fig. E.7A shows the typical behavior path, which is conspicuously different from that of a subcritical boiler. Here the path does not follow the saturation line in normal operating range except in startup, shutdown, and abnormal situations like abrupt load shedding/runback.

E.2.2 Constant and Variable Pressure Operation

SC/USC plants must operate with an OT boiler configuration, and either with a constant pressure or with a variable pressure, depending on the load. At the initial stages of OT, SG plants for SC units were and still are operated in many regions/countries to maintain constant throttle pressure. Unit load demand in that case constitutes the demand set value for both the feedwater flow and the firing rate. The ratio of firing rate to feedwater flow is utilized for the steam temperature control strategy. During normal operation, the fluid within the furnace is always maintained at SC pressure to avoid subcritical two-phase flow, which leads to high temperature, and subsequently to tube failures.

In contrast, SG plants with a Benson design enable the unit to run in a variable pressure mode operation. In this design, two-phase flow is allowed to take place up to the critical point at a predefined load permitted by the design criteria. After this point (Benson point), the process turns to a single-phase flow condition achieving a full load as per demand. In the precritical point operation, a minimum recirculation flow is maintained by the boiler circulation pump (BCP) along with a (or pair of) vertical steam separator(s) and a storage vessel.

At the Benson point or the load at which the steam separator(s) and storage vessel run too dry to permit initial firing, the BCP is shut off. The boiler feed pump (BFP) is then controlled so the feedwater flow will meet the demanded furnace enthalpy pick-up function (from the economizer outlet to the primary superheater inlet) in an OT operation. With this method, the relative proportion of a steam generator surface serving as an evaporator and as a superheater is controlled as a function of load and is stabilized. The final steam temperature control range meets a setpoint from ~50% to 100% load. The reheat steam temperature control range meets a setpoint from ~60% to 100% load.

The constant pressure type SG plants are generally supplied with vertically arranged tubes and are popular in the United States. The tube layout inside the furnace is relatively simple, and hence, significantly less costly and easier to fabricate, assemble, and construct.

On the other hand, the variable pressure type is generally designed with spiral tubes wrapped around the furnace, and is used more extensively in Europe. The spiral tube design demands more complicated, costly fabrication, and construction than vertically arranged tubes. This type of SG plant is more advantageous than one with constant pressure because it is efficient, even at lower loads. Also, with its circulation pump, the plant can also be cycled in on-and-off mode, as required in an expeditious way, thus enabling the unit to match the grid demand more effectively and more quickly.

E.2.3 Control and Monitoring of Parameters in an SG Plant

SG plants have no energy reserves. The complete plant may be conceived of as a bunch of heated tubes with high-pressure water and steam at the outlet. This simple configuration shows that the output is a function of feedwater flow and the amount of heat imparted to the tubes. The heat content per unit mass or enthalpy of the outlet steam depends only on the ratio of heat supplied to the feedwater flow. When the pressure is constant at a particular load, the temperature of the outlet steam becomes dependent only on the enthalpy of steam from the heat input to feedflow ratio. This means if feedflow is increased without increasing the heat input, the temperature of the outlet steam would decrease and vice versa. In an SC/USC power plant, the output pressure and flow are maintained by the feedwater pump and the main steam temperature is maintained by the heat input to the feedflow ratio, in other words, the heat supplied to the system. The desuperheating water spray system is mainly used for emergency or for a transient requirement. The reheat temperature control philosophy is the same as a conventional boiler, in that it is through redistribution of heat or by emergency spray valves.

In the absence of reserve energy, the OT SG plant needs continuous monitoring, and the control system, quickly and accurately, enables the feedwater flow and firing rate to deliver the steam to the turbine as required by the generator according to the unit or grid demand. The distributed control system (DCS) plays a more vital role in these types of plants than the drum-type boilers. To avoid any delays originating from the inertia factors in the process, predominantly in the pulverized fuel circuit, the judicious location of transducers and the speed of control action of the temperature measuring system are extremely important.

The SG plant load demand signal, as generated from the DCS according to coordinated control mode selection, is passed on to the dedicated control system of feedwater flow/pressure, heat input from fuel, and air flow to act as their strategic set values as a function of the load signal for rearranging the final control element position. Some typical automatic control loop strategies are depicted in Figs. E.1–E.7.

In some control strategies, the set values of the feedwater flow are determined directly from the measurement of the steam conditions at different locations in the steam cycles. This is assumed to eliminate uncertainty related to any untoward disturbances in the process, as it is expected that the measurement of steam conditions affirms exactly how much water it needs independently from other systems.

E.2.4 Water Circulation and Startup Procedure

The startup procedure of this type of boiler is a little different from subcritical boilers. As stated earlier, there is no boiler drum in this type of boiler like in the once-through (OT) type. However, there are separator vessels and separator storage tanks (SSTs), which are important during startup. The relative position of these vessels and tanks in a flow diagram is depicted in Fig. E.8, and they play an important role during startup. Also, the quality of feedwater required for its operation is significant. It is better to have a clear idea about the various modes of operation for these types of boilers. There are basically five modes of operations:

- Filling up the boiler
- Cleaning up the boiler
- Wet-mode operation of the boiler (up to 30% load)
- Dry-mode operation of the boiler (>30% load)

E.2.4.1 Filling Up

When the water system of the boiler is empty (economizer to separator), then is it filled with feedflow at around 10% turbine maximum continuous rating (TMCR). When the

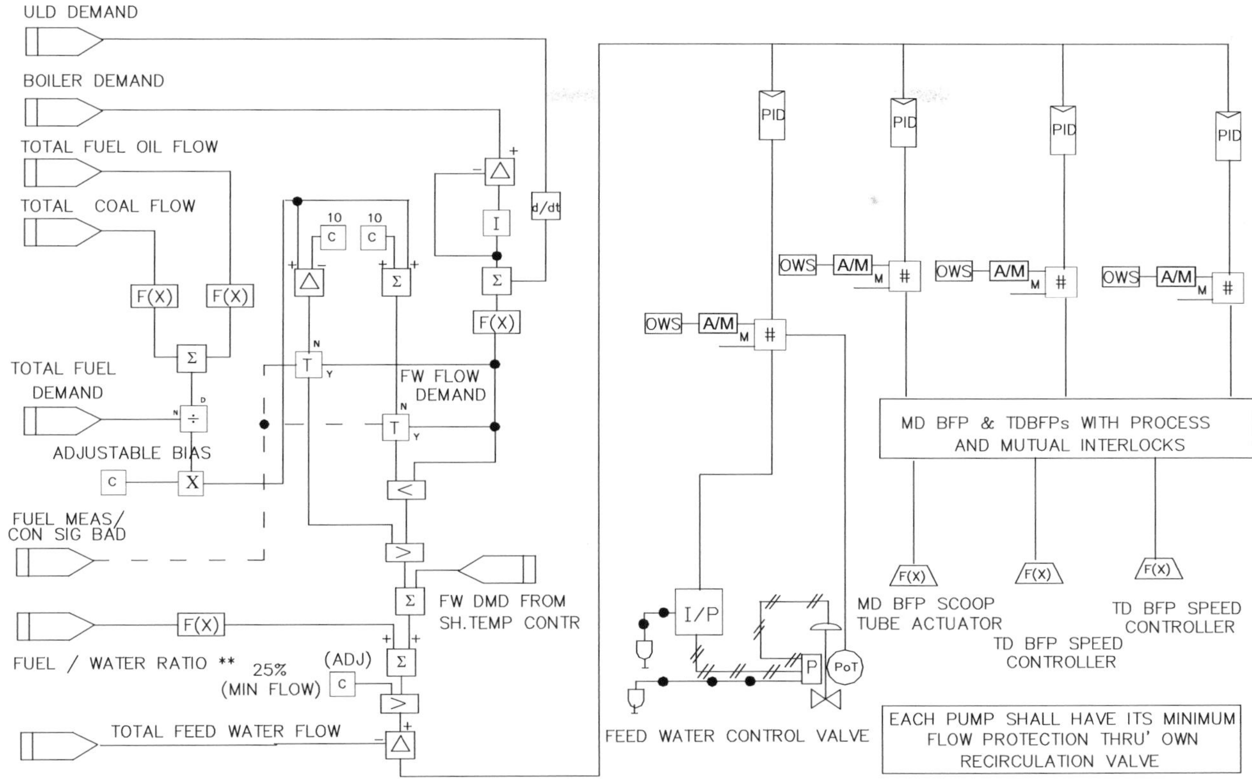

** FUEL / WATER RATIO SIGNAL IS GENERATED AT FIG E.4

NOTE

- TYPICAL CONTROL STRATEGIES ARE SHOWN. PROCESS & MUTUAL INTERLOCKS ARE EXCLUDED FOR SIMPLICITY.
- NUMBER OF SENSING ELEMENTS WOULD BE ACCORDING TO REDUNDANCY LEVEL ENVISAGED.
- NUMBER OF DRIVES & FINAL CONTROL ELEMENTS WOULD BE ACCORDING TO PLANT OPERATION PHILOSOPHY
- NO. OF FEED PUMPS MAY VARY. MDBFP IS MAINLY PROVIDED FOR START UP AND EMERGENCY SITUATION.

FIG. E.1 Boiler load/feedwater control.

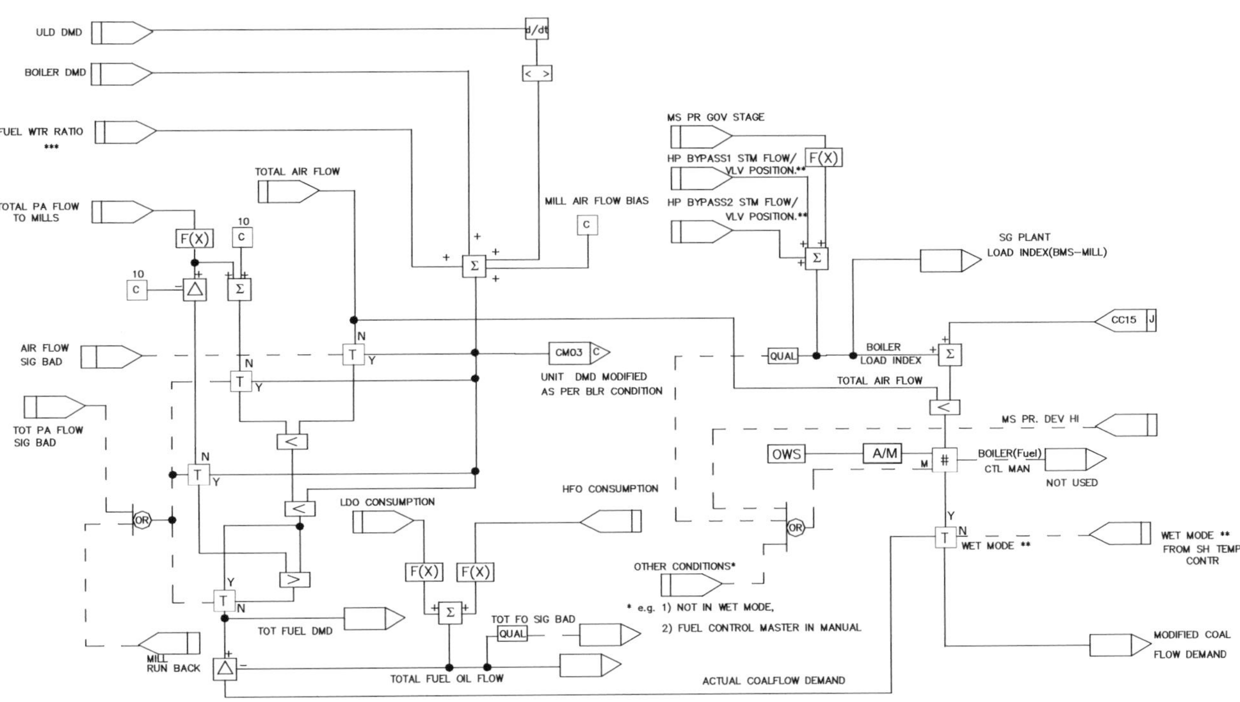

** WET MODE SIGNIFIES SATURATED STEAM AT SG PLANT OUTLET

*** FUEL / WATER RATIO SIGNAL IS GENERATED AT FIG E.4

NOTE

- TYPICAL CONTROL STRATEGIES ARE SHOWN. PROCESS & MUTUAL INTERLOCKS ARE EXCLUDED FOR SIMPLICITY.
- NUMBER OF SENSING ELEMENTS WOULD BE ACCORDING TO REDUNDANCY LEVEL ENVISAGED.
- NUMBER OF DRIVES & FINAL CONTROL ELEMENTS WOULD BE ACCORDING TO PLANT OPERATION PHILOSOPHY.

FIG. E.2 SG plant load/fuel control.

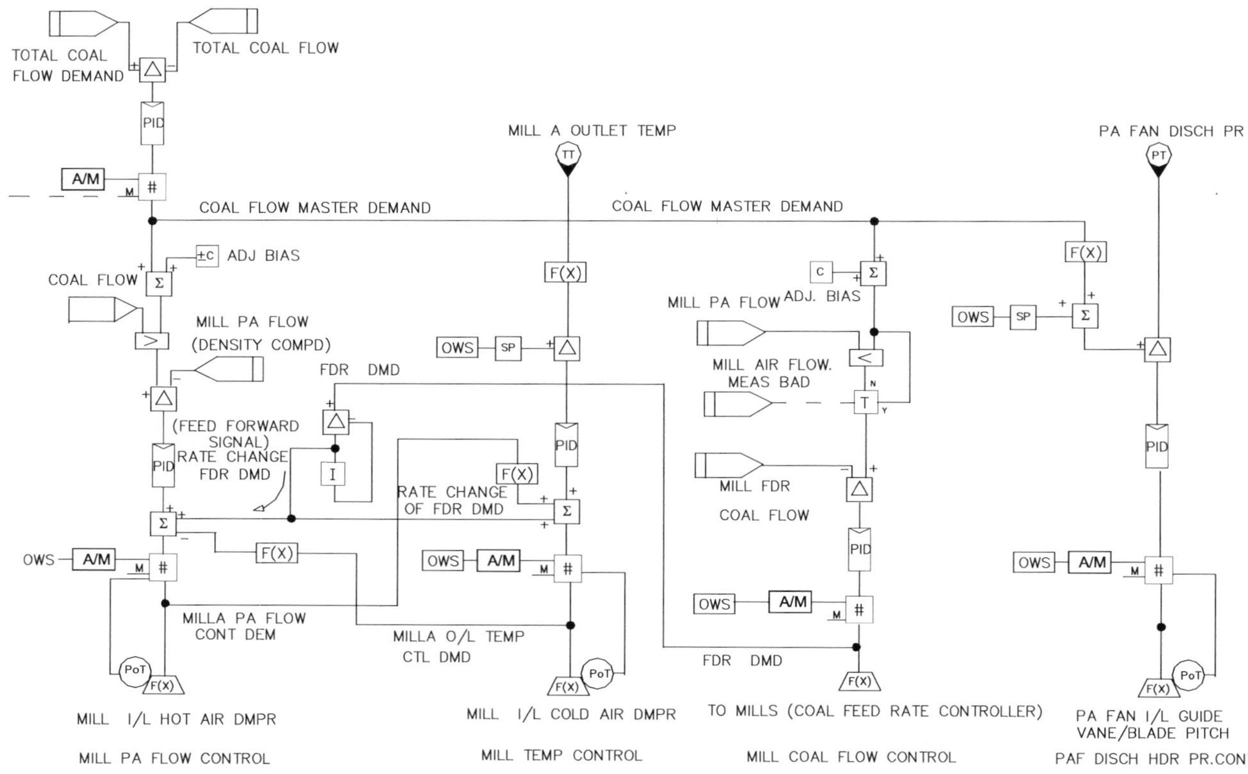

NOTE
- TYPICAL CONTROL STRATEGIES ARE SHOWN. PROCESS & MUTUAL INTERLOCKS ARE EXCLUDED FOR SIMPLICITY.
- NUMBER OF SENSING ELEMENTS WOULD BE ACCORDING TO REDUNDANCY LEVEL ENVISAGED.
- NUMBER OF DRIVES & FINAL CONTROL ELEMENTS WOULD BE ACCORDING TO PLANT OPERATION PHILOSOPHY.

FIG. E.3 PA flow-press/mill temperature/fuel flow control.

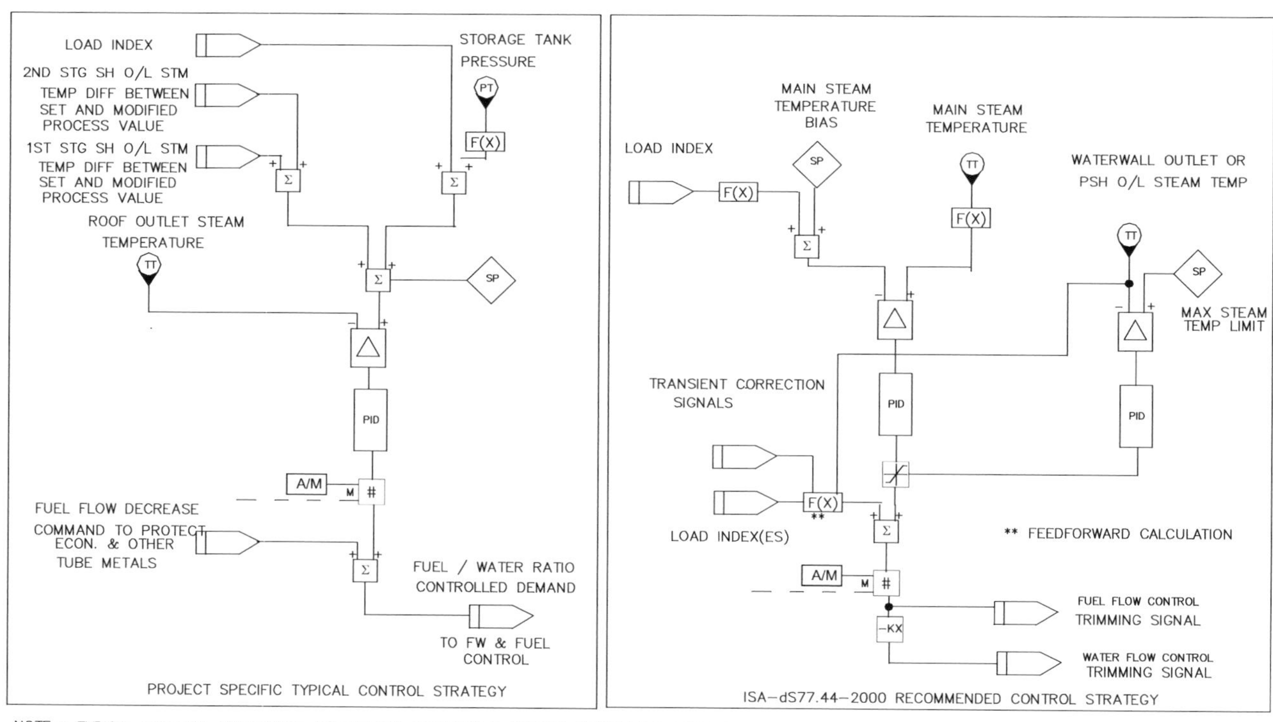

NOTE
- TYPICAL CONTROL STRATEGIES ARE SHOWN. PROCESS & MUTUAL INTERLOCKS ARE EXCLUDED FOR SIMPLICITY; WRITE UP IN APPENDIX V MAY BE REFERRED.
- NUMBER OF SENSING ELEMENTS WOULD BE ACCORDING TO REDUNDANCY LEVEL ENVISAGED.
- NUMBER OF DRIVES & FINAL CONTROL ELEMENTS WOULD BE ACCORDING TO PLANT OPERATION PHILOSOPHY.

FIG. E.4 Fuel/water ratio control.

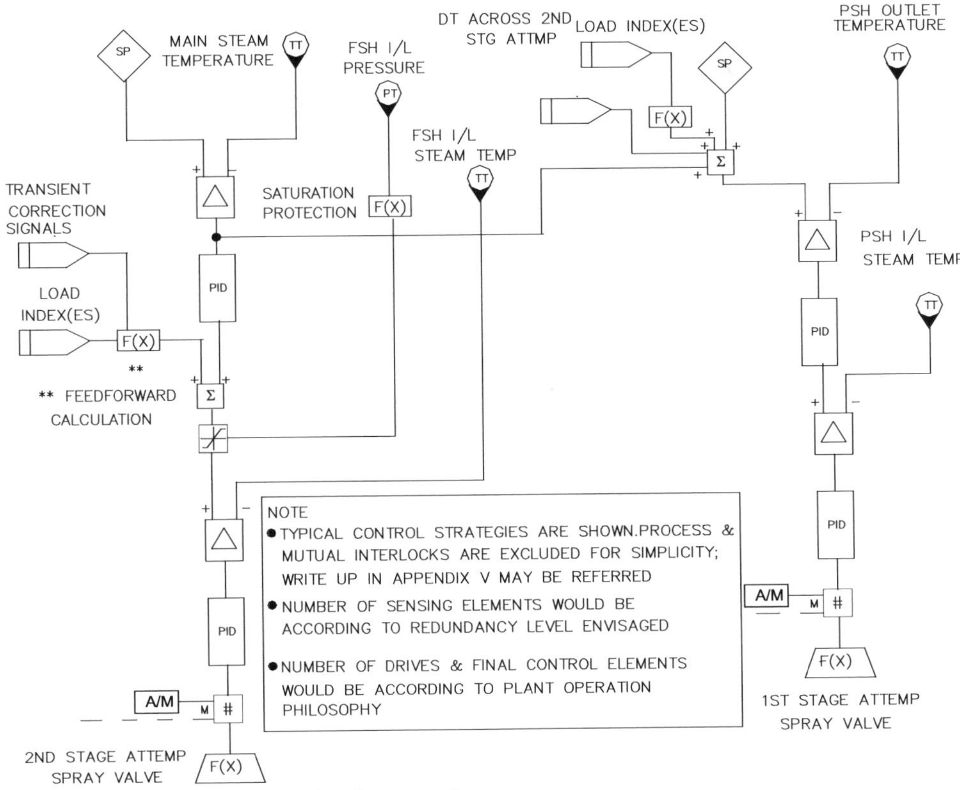

FIG. E.5 Main steam temperature control (attemperation flow control).

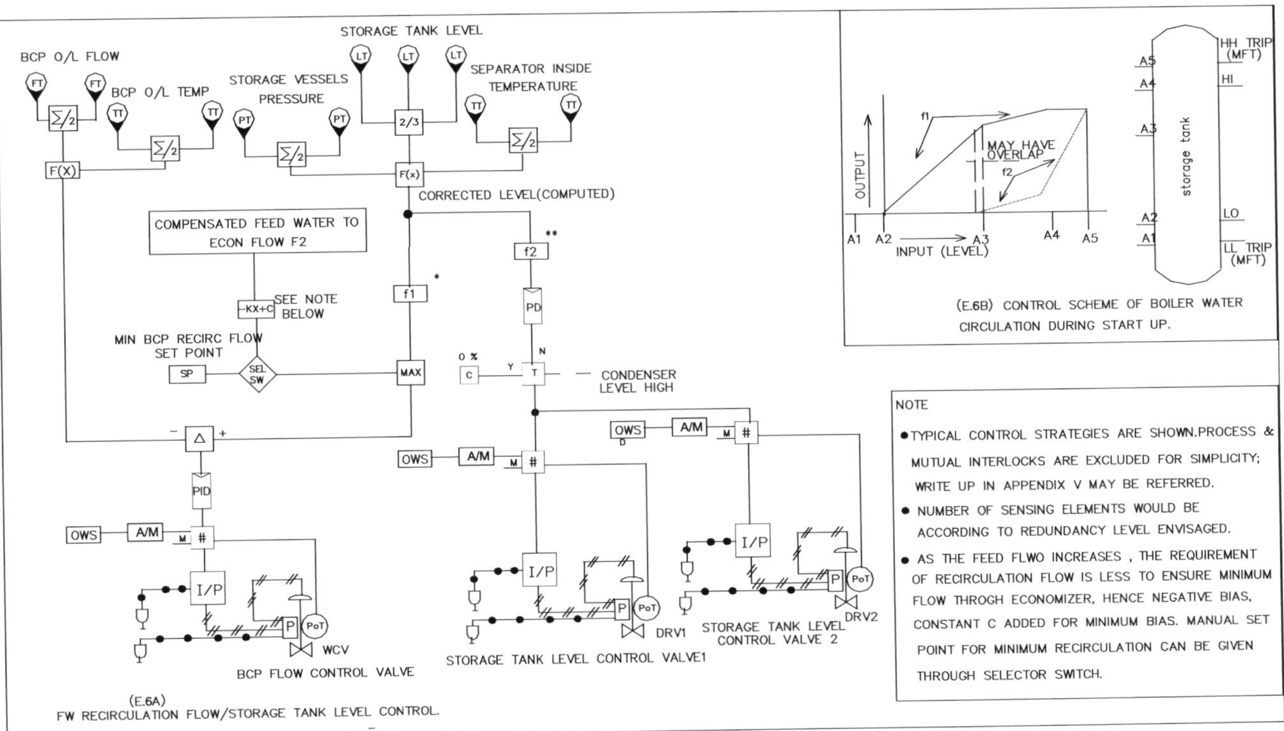

FIG. E.6 Feedwater recirculation and steam SST level control.

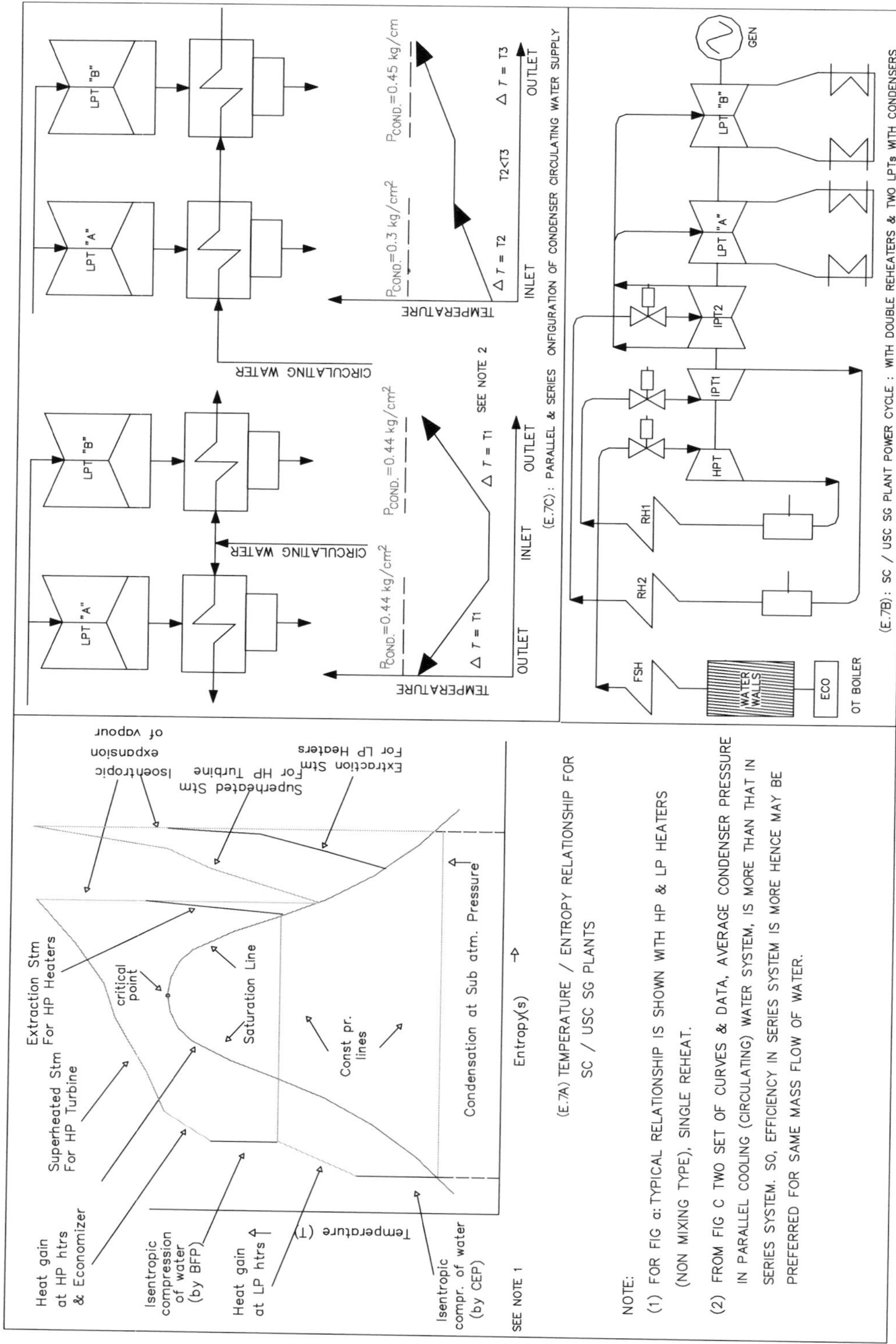

FIG. E.7 Process cycle and components for SC and USC plants.

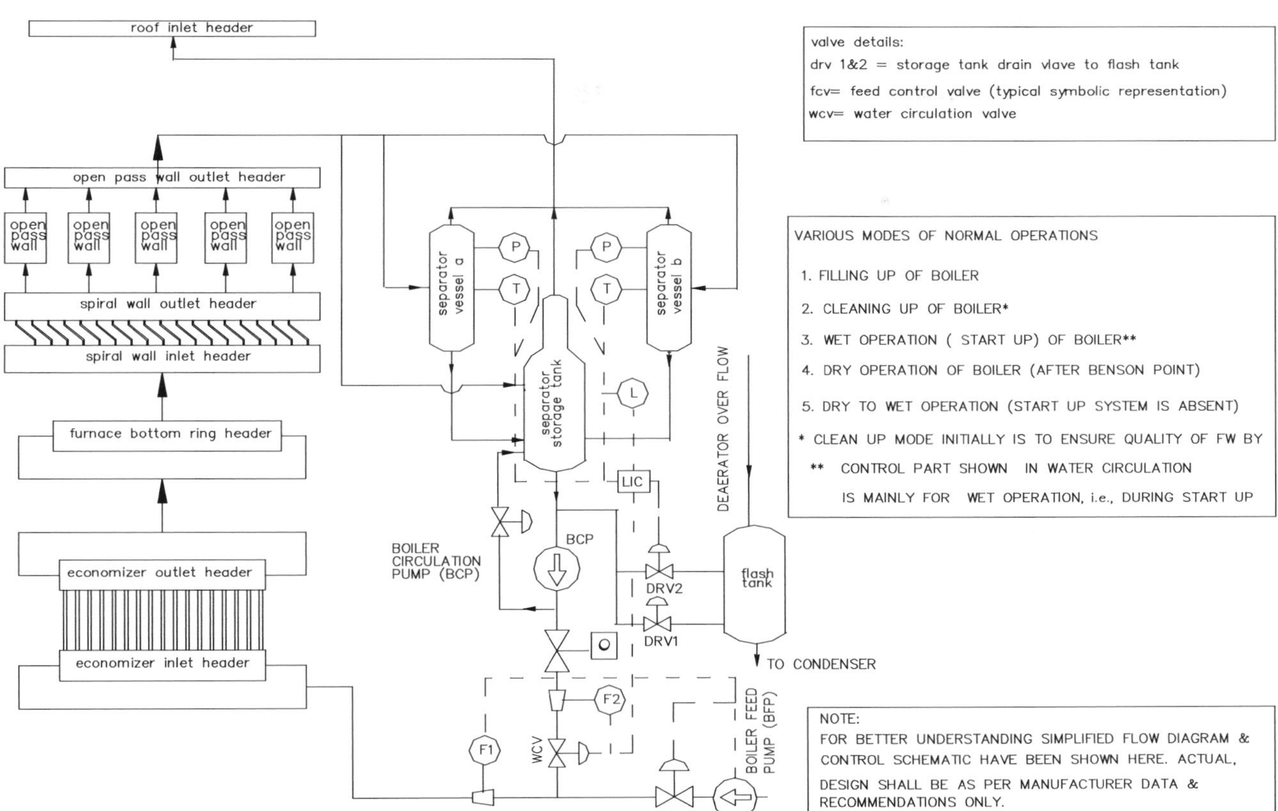

FIG. E.8 Boiler water circulation (also startup control scheme).

level reaches the filling setpoint, the drain valve DRV (Fig. E.8) is opened slowly and kept open for a certain time to check that the level is maintained. Then feedflow is increased, up to 30%. The water level is maintained by regulating the drain valve for a certain time until it is full. For simplicity, the number of drain valves is kept to one, but there may be more.

E.2.4.2 Boiler Clean Up

As discussed earlier, the feedwater quality is important for OT SC boilers. When at the deaerator outlet and separator outlet, if feedwater quality is not right, then it is necessary to flush water from the system into the flash tank. During this time, ~10% of the TMCR value of feedwater is sent; it flows through the economizer and evaporator and is discharged to the flash tank via the drain valve DRV (Fig. E.8), and then finally to the condenser. From the condenser, it flows through the CPU to remove any impurities. The process continues until the quality is achieved.

E.2.4.3 Wet Mode Operation

This mode of operation is done during initial boiler light up until 30% MCR value (Fig. E.8). While in this operation, the Benson point is reached. Throughout this time, the economizer minimum flow is ensured by the BCP, which is started during lightup, and following the beginning of the BFP and cleanup operation. The water circulation valve (WCV) is the only control value used in the discharge of the BCP. The WCV is responsible for maintaining the required flow through the economizer. This valve also helps to maintain the level in the SST. The opening demand of the WCV is mainly determined by the economizer water flow, but it is done in such a way as to help maintain the level of the SST. As shown in Fig. E.8 (simplified for understanding; the actual system may be a little different, but the basic philosophy is similar) there are certain level values in the SST level as determined by the SG manufacturer. Here levels A1 and A5 (Fig. E.6B) are designated as very low and very high levels, respectively, for the master fuel trip (MFT; only during the startup period like the dry mode of operation there will be no level since the MFT is not applicable), whereas A2 and A4 are low and high level, respectively. Normally during initial startup the level is maintained between A2 and A3. Setpoint for flow through economizer can be derived from the following:

- Manual set value
- Minimum set value is derived from feedflow. Here, (−) KX has been chosen because as feedflow increases there

will be less CW flow to meet the minimum flow requirement through the economizer. This and the above setpoints can be selected by the operator.

- From the level in the SST level, as the level increases, there should be more circulation flow to maintain the level in the storage tank.

These set values are fed to the system through the "MAX" selector so that minimum flow through the economizer is ensured via CW flow. Two function generators (f1 and f2) are illustrated in Fig. E.6. Function generator (f1) characteristics are blue. Between A2 and A3 the flow setpoint is proportional to the level value to circulate more water for steaming to maintain the level. After A3 up to A4 it is also linear but with lower slope, because near A3 DRV comes in to operation to maintain the level. At A4 (f1) it reaches maximum. When A5 is reached, the setpoint is zero to indicate that the valve is closed following MFT during startup. Function generator (f1) is shown after A2 as BCP should start after the level is above the low level; hence, the WCV cannot open prior to that. For generating the circulation setpoint from the storage tank level, the actual level is measured and duly compensated for by operating the temperature and pressure to get a corrected level signal computation. Actual circulation flow is measured and compared with the setpoints discussed. The error thus generated will be fed to the PI controller to regulate the WCV. When the water level in the SST is high, the DRV is operated by the function generator (f2) to send some water from the storage tank to the flash tank. This valve DRV is important for other operations as discussed in Clauses 2.4.1–2. The characteristics are shown in Fig. E.6B by f2. There are different philosophies for circulation based on manufacturer's recommendations, for example, NTPC Sipat uses the separator level control parameter to regulate the feedwater flow control valve (BFP speed), whereas minimum feedwater through the economizer is controlled by the WCV.

E.2.4.4 Dry Mode of Operation

At 30% TMCR the separator should be completely dry with no level, and at this point (referred to as Benson point) the boiler changes from wet mode to dry mode. The BCP is shut and the boiler runs in OT mode. Economizer water flow is controlled by feed pump speed and the feed control valve (FCV; bypass valve full open). The boiler startup system is isolated and platen and final superheating spray comes on. During transition from FCV (30%) to bypass or 100% FCV there may be fluctuation of pressure, which when stabilized, the FCV (100%) opens. In dry mode the steam temperature control loop and feedwater control loop are interactive. These are discussed in the following clauses.

E.2.5 Chemistry of OT SC Boiler

Like other boilers, water in an SC boiler needs to be treated. Also, there is a different dosing system to control the chemistry of water and steam. Various treatments such as oxygenated treatment and all volatile treatment (AVT; as discussed in Chapter 5, Clause 4.0) are also applicable for SC OT boilers. For OT SC boilers, because of the absence of a drum, the quality requirements are stringent. Some of these requirements, set by the Electric Power Research Institute (EPRI), are listed in Table E.4. For feedwater both treatment and associated commissioning values are detailed using AVT-O and AVT-R symbols.

For SC plants and subcritical plants with a CPU, oxygenated treatments are performed for long-term protection of the pretreatment system by creating a hematite layer. Also, CPU operating time is due to low condensate ammonia and iron transfer. Normally this is done after the startup period is over. For oxygenated treatment good quality make-up water and condensate feedwater is necessary, so a CPU is essential. As shown in Fig. 5.48, the normally high-purity oxygenated water is dosed at the CPU outlet and at the deaerator outlet. O_2, hydrogen peroxide, or air can be used as an oxidant with an O_2 level at ~30–15 ppb for OT boilers (30–50 ppb for drum type boilers). Based on dissolved oxygen and flow of condensate/feedwater the dosing is regulated with the help of the flow control valve. Cat conductivity and pH monitoring is extremely important. As shown in Fig. 5.47, there is a relationship between pH and conductivity and it is crucial for OT SC boilers.

E.2.5.1 Control and Monitoring of Parameters in a Turbine Generating Plant

The turbine and generator control part in a turbine generating plant is similar to that of conventional power plants. They mainly differ in the mechanical and metallurgical aspect, for example, the double-reheat casing (as discussed above), two nos. double flow LP casings, blade shaping and provision of diffusers, etc. The exhaust diffuser at the LPT outlet stage provides for the necessary recovery of the exhaust steam pressure, which enables the outlet pressure of the last stage blading to become lower than the condenser pressure.

E.2.5.2 Control Loop Strategies

E.2.5.2.1 Steam Temperature Control

Due to the very configuration of an OT SG plant, final steam temperature is affected by the ratio of firing rate to feedwater flow. At wet mode and/or at lower load it is necessary that fuel and feedwater ratio is maintained at a point so that there is no chance of over fire. In feedwater

TABLE E.4 Target Value of Feedwater, Main Steam, and Recommendation

FW/MS/RH	Parameter	Unit	Target Value	Remarks
FW	Ammonia	ppm	0.02–0.07	
FW	Iron	ppb	≤2	(AVT-O/AVT-R: <2[a])
FW	Dissolved oxygen at Econ I/L	ppb	30–150	(AVT O: <10[a]; AVT-R: <5[a])
FW	pH		8.8–8.5	(AVT-O/AVT-R: <9.2–9.6[a])
FW	CAT conductivity	μS/cm	≤0.15	(AVT-O/AVT-R: <0.2[a])
MS and RH	CAT conductivity	μS/cm	≤0.15 (<0.45[a])	Action needed if it is 0.3
MS and RH	Silica	ppb	≤10 (<40[a])	Action needed if it is 20
MS and RH	Sodium	ppb	≤2 (<12[a])	Action needed if it is 4
MS and RH	Chlorides	ppb	≤2 (<12[a])	Action needed if it is 4
MS and RH	Sulfate	ppb	≤2 (<12[a])	Action needed if it is 4
MS and RH	TOC	ppb	≤100 (<100[a])	Action needed if it is >100

MS, main steam; RH, reheat; FW, feedwater.
[a]*Parameters in parentheses are the guideline of parameters during startup and commissioning.*

control strategy, minimum feedflow at all conditions is always ensured.

The control strategy changes either the firing rate or feedwater flow or both to control the steam temperature. In a typical OT SG plant, unlike a subcritical boiler, the boiler master demand (BMD) signal is sent simultaneously to the fuel demand and the feedwater demand. The boiler demand is fed to both the fuel flow and feedwater flow control system in parallel as the BMD is commensurate with the MW output, which is the final product of the fuel inputs and feedwater flow. In certain boilers feedwater flow and/or fuel flow change is allowed to take part as per the BMD as long as the BMD is within the specified range of the desired feedwater/fuel ratio added with approximately ±10% swing.

A separate temperature controller loop (Fig. E.4A) is provided as a ratio controller whose output is then constituted as the trimming signal for any of the fuel flow and feedwater flow control (with a scaling factor) or both depending on the plant configuration, manufacturer's/designer's recommendation, process variable response pattern, process constraints, etc.

Analysis of the individual effect of this ratio control on fuel flow control for final steam temperature control illustrates an adjustment between the control of temperature and the control of load. On the other hand, the individual effect of this ratio control on feedwater flow control for final steam temperature control depicts an adjustment between control of temperature and steam outlet pressure. This type of control may eventually try to over fire or under fire the firing rate in response to load changes. The immediate response may create an erroneous steady-state fuel demand causing departure of temperature and pressure from their respective setpoints. It may be advantageous to utilize the ratio factor for both the fuel and water demand in an opposite direction to establish overall system stability because its effect on one process would offset the other.

However, note that due to the process safety requirement, which ensures minimum feedwater flow through economizer/furnace tubes, the firing rate and feedwater flow rate are released for adjustment to control steam temperature when after load means more than safe water flow.

Although it is useful to utilize the fuel/water ratio in steam temperature control, the inherent time delay in the water circuit makes it somewhat slow in response even if the fuel flow trimming signal is much faster than its water flow signal trimming counterpart.

To circumvent the time delay problem, attemperation through spray water valves at different stages is also provided to control the main steam temperature. The philosophy behind provision of both systems is that spray attemperation would take immediate action to control initial temperature departures while the correction through the proper fuel/water ratio would be achieved from a steady-state component of the steam temperature control loop.

The traditional way to control steam temperature through multiple stages of spray attemperation is similar for subcritical boilers. The steam temperature control is equipped to take care of the faster transients to avoid the time delay (depending on the SG plant loads) of the water at the entry of the economizer, and attaining the desired superheater temperature at the SG plant outlet. Without the transient correction approach, the tube metals can experience severe damage due to the miscalculation in actual spray water from the water flow requirement demanded to combat the transit time delay.

E.2.5.2.2 Fuel/Water Ratio Control

It was discussed above that for an OT SG plant it is extremely important to utilize this derived parameter. How it is derived, however, depends on the manufacturer's recommendation. This ratio determines all the final values of the main steam at the SG plant outlet.

Because of OT configuration, all feedwater entering the SG plant becomes steam and therefore the throttle pressure is directly related to feedwater flow, which is controlled by the BFPs. Since all the feedwater is sent to the SG plant as per the unit master from the BMD, it must be evaporated through heat input by adjusting the firing rate control system. The superheater spray flow is taken out from the total feedwater flow. Any change in superheat spray flow would not influence a permanent adjustment in outlet main steam temperature since it does not change the ratio of firing rate to feedwater flow.

The control strategies of main steam temperature control with the help of the fuel/water ratio and spray attemperation are depicted in Fig. E.4A.

E.2.5.3 Short Description of the Control Strategies

E.2.5.3.1 Fuel/Water Ratio Control

Fuel/water ratio control is derived from the output of a temperature controller. Some manufacturers recommend an operator setpoint summed up with differential temperatures of both the two stages superheater temperature controllers' input, load index (as the feed forward signal), and the characterized value of storage tank pressure (to take care of the saturation problem), whereas the "roof outlet temperature" is the measured or process variable. Corrections toward protection of the economizer and waterwall tubes are done by adding them to the controller output. The final output is then suitably processed to form a trimming signal for either of the fuel and water flow control systems or both.

ISA-dS77.44-2000 recommends the main steam temperature as the measured or process variable against the operator setpoint summed up with the load index (as the feed forward signal). The controller output is passed through a limiter to take care of the overheating of the waterwall tubes or platen superheater (PSH) tubes. The transient corrections from load index/indices and other signals are then added to the abovementioned signal output from the limiter to avoid under or overheating. The final output is then suitably processed to form a trimming signal for the fuel and water flow control system.

In other designs, mainly that of a TT furnace, differential temperature across the primary desuperheater is controlled by feedflow. Differential temperature for the final desuperheater is adjusted with the difference between the rate of change of main steam flow and that of the fuel rate flow (e.g., as seen in the Sipat Super Thermal Power Project of NTPC, India).

E.2.5.3.2 Main Steam Temperature Control (Attemperation Flow Control)

This is typically accomplished by traditional control strategies, but with some features added to take care of the OT configuration. Generally, two-stage attemperation systems are provided with independent or coupled systems. The first-stage attemperator controls the steam outlet temperature of the primary or PSH). The second-stage attemperator is used to control the final superheater steam outlet temperature. Both control systems operate in cascade mode as provided for subcritical boilers.

To take care of the transient conditions, feed forward signals are normally added to the primary controller to improve the control performance, and are usually applied only on the final outlet temperature control strategy. The actual provision of feed forward signal sources and their characterization, however, depends on many factors. They should be applied as a minimum but not be limited to be sourced from the load index/indices, heat distribution signals like gas damper position or fuel nozzle tilt positions, fuel flow, airflow, etc.

The individual spray attemperator demand is distributed between both stages of attemperation. The individual spray demands are developed from the spray distribution and coordination control strategy as determined from the boiler manufacturer's recommendations and specifications. Both valves participate during transient boiler operation to minimize temperature excursions. The first spray demand protects the primary superheater from going into saturation at the inlet and from high temperature at the outlet. The control system must coordinate the first and second spray demands to minimize correction from the final steam temperature controller.

E.2.5.3.3 Feedwater Recirculation Flow Control

Fig. E.6 shows a typical control strategy in its simplest form. This control strategy is developed to ensure minimum feedwater flow through the economizer, waterwall tubes, and the steam separator with the help of the BCP and associated valve for various functions, which include:

1. To protect the economizer and waterwall during the low load.
2. To minimize the BFP discharge flow and achieve heat recovery during startup.

The SST level is controlled for the BCP protection (to ensure NPSH) by regulating the recirculation flow valve position. Here the PI(D) controller gets the SST level as the variable setpoint and the recirculation flow (density compensated) as the process or measured variable. When the valve is positioned at full closed status or recirculation flow is less than a predetermined minimum value, a negative deviation of ~5%–10% is imposed on the controller through a switching action to ensure sustained full closed signal to the valve.

The other interlocks, although not shown in the abovementioned control strategy, are normally provided as follows:

1. To avoid sudden decrease in SST level:
   a. During the start of the BCP, the upper limit of valve opening command is set at 20% for a limited duration.
   b. During a low load operation, the rate change of the valve position command is curbed if the SST level changes (mainly decreasing) too fast.
2. To avoid overflow, the valve receives a 100% opening command when the recirculation flow is more than a predetermined high value.
3. An increase bias may be added to the setpoint after BCP starts to raise expeditiously the economizer inlet temperature. By this action, more hot water from the economizer outlet mixes with the economizer inlet, thus raising the temperature after mixing. The bias may be an adjustable fixed value or economizer inlet temperature with proper function generator.

E.2.5.3.4 Steam SST Level Control

Fig. E.6 shows a typical control strategy in its simplest form. During water filling, startup, and cleanup operations, the level of this tank is maintained by draining the water to the flash tank. The drain valve opening is proportional (through the rate adjuster) to level, for example, the valve fully opens at 4000 mm and fully closes at 3000 mm. At normal load there would not be any water at the storage tank and the valve would remain fully closed. The interlocks include:

1. At a very high alarm level, high rate of change is automatically selected to arrest overflow.
2. Gain compensation of the drain valve opening signal may be incorporated to take care of the change in specific volume of water passing through the drain valve.
3. When level is decreasing at a high rate, a negative bias is added to the level signal to arrest the fast decreasing rate of level. For a fast opening signal, the rate change is restricted toward the opening path.
4. The drain valve would receive closing signal at a high condenser level.

BIBLIOGRAPHY

[1] PF-Fired Supercritical Boilers. (March 2002) Operational Issues and Coal Quality Impacts. Technical Notes 20: By M/s B.J.P. Buhre, R. Gupta, S. Richardson, A. Sharma, C. Spero, T. Wall. Dtd.

[2] Latest Development in Supercritical Steam Technology. Powergen Asia 2008; Kualalampur-Malaysia. Dtd. By Miro R. Susta (IMTE AG, Power Consulting Engineers. Switzerland).

[3] Ultra Super Critical Pressure Coal Fired Boiler—State of the Art Technology Applications: By Yoshio Shimogori (Babcock-Hitachi K.K.).

[4] Tube Vertical. (2001) Variable Pressure Furnace for Supercritical Steam Boiler. POWER-GEN International: Las Vegas Nevada, U.S.A. Dtd. December 11-13, 2001: By M/s D.K. McDonald (The Babcock & Wilcox Company, USA) & S. S. Kim, U.S. Department of Energy, National Energy Technology Laboratory, USA).

[5] FWPG Benson VT Boiler—Process and Operational Description (Product Bulletin TP_PC_04_02.DOC): By M/s Foster Wheeler Power Group, Inc.

[6] Design Factors and Water Chemistry Practices—Supercritical Power Cycles. (September 8–11, 2008) By M/s Frank Gabrielli, Alstom-Windsor, USA & Horst Schwevers. Alstom, Germany (PREPRINT-ICPWS XV, Berlin) Dtd.

[7] Steam Generator for Advanced Ultra-Supercritical Power Plants 700 to 760C; (Technical Paper BR-1852;ASME 2011 Power Conference). By P.S. Weitzel, Babcock & Wilcox Power Generation Group, Inc Dtd.

[8] Design Features of Advanced Ultrasupercritical Plants. Part III By Paul S. Weitzel, Babcock & Wilcox Power Generation Group, Inc. Dtd.

[9] Controlling Chemistry During Start Up and Commissioning of Once—Through Supercritical Boilers. (December 2008) K. Kirschenheiter, M. Chuk, C. Layman, K. Sinha-Bechtel Corporation.

[10] ISA Recommended Control Loop: ISA-ds77.44.02-2000.

Appendix F

Integrated Gasifier and Combined Cycle Plant (Pollution Control)

F.1 INTRODUCTION

Integrated gasification combined cycle (IGCC) technology will have great demand in the near future to cater to large energy needs. It offers a clean, efficient option for the production of electricity from coal and other low-cost fuels. Like the circulating fluidized bed combustion boiler (CFBC; Fig. F.4), in this technology a variety of fuels such as coal, biomass petroleum coke/resid (residual after distillation), and waste can be used as fuel. In IGCC technology the gas turbine combine cycle enables high IGCC efficiencies, while the gasification block cleanly converts coal to fuel for the gas turbine. Apart from the core gasification technology and the power block, there are a variety of auxiliary technologies to complement the core technology in its applications. This enables synthesized gas (syngas) conversion into a variety of liquid fuels, gases, and chemicals. The important environmental benefits of IGCC include the following:

- Gasification allows both gas turbine generator (GTG) and steam turbine generator (STG) to produce energy.
- IGCC efficiency can be up to ~45% higher than supercritical power plants.
- Pollutants like S, Hg, N_2, etc., are removed prior to combustion [unlike in pulverized coal (PC) boilers].
- Because of the inherent clean technology, it is more environmentally friendly than other technology.

F.1.1 What is Gasification?

Gasification is a thermochemical process (not combustion) at an elevated temperature and pressure in reducing (dearth of oxygen) atmosphere used to convert feedstock mixed with steam into a clean syngas (no toxic product) and other byproducts. The major items needed for this process are fuel and oxidant to produce syngas, which is finally cleaned. The basic block diagram of a typical IGCC is presented in Fig. F.1A. From a pollution standpoint, the natural gas combined cycle plants provide better results when compared with various other types of technologies, but IGCC has better results when compared with PC/supercritical/ultra-supercritical plants (Table F.1).

As mentioned, IGCC is very environmentally friendly. In the subsequent clauses, the basic process along with the advanced control systems deployed for this type of plant will be discussed. IGCC is basically an integration of various systems and not a single process, so naturally, the various systems will have different needs.

F.2 BASIC PROCESS

The gasification unit is the main part of the IGCC system. High-purity oxygen at a high pressure of $\sim 70\ kg/cm^2$ reacts with fuel at a temperature of nearly 1200°C to produce syngas. This syngas is cooled and cleaned and used as fuel in a gas turbine system in the combined cycle process. Also, there are a number of byproducts available such as sulfur/sulfuric acid, as part of the integrated process. This clean syngas also provides an energy source for various other applications, e.g., H_2 for fuel cells. As seen in Fig. F.1A, the basic building blocks consist of a gasification unit, a cleanup unit, a combined cycle power block, other utility/byproduct units, and an oxidant supply system including an air separation unit (ASU). The ASU, which separates oxygen from air cryogenically, plays an important role by supplying oxygen for the gasifier and nitrogen in an increasing mass flow in the power block (combined cycle plant). Various supporting units consist of the following:

1. *Coal storage and feed preparation*: This consists of a coal-handling plant, which provides the means for receiving, unloading, storing, reclaiming, and conveying coal for a storage facility. Mainly, solid material handling equipment is used in PC power plants. For gasification preparation, fuel is necessary. In a gasifier, the

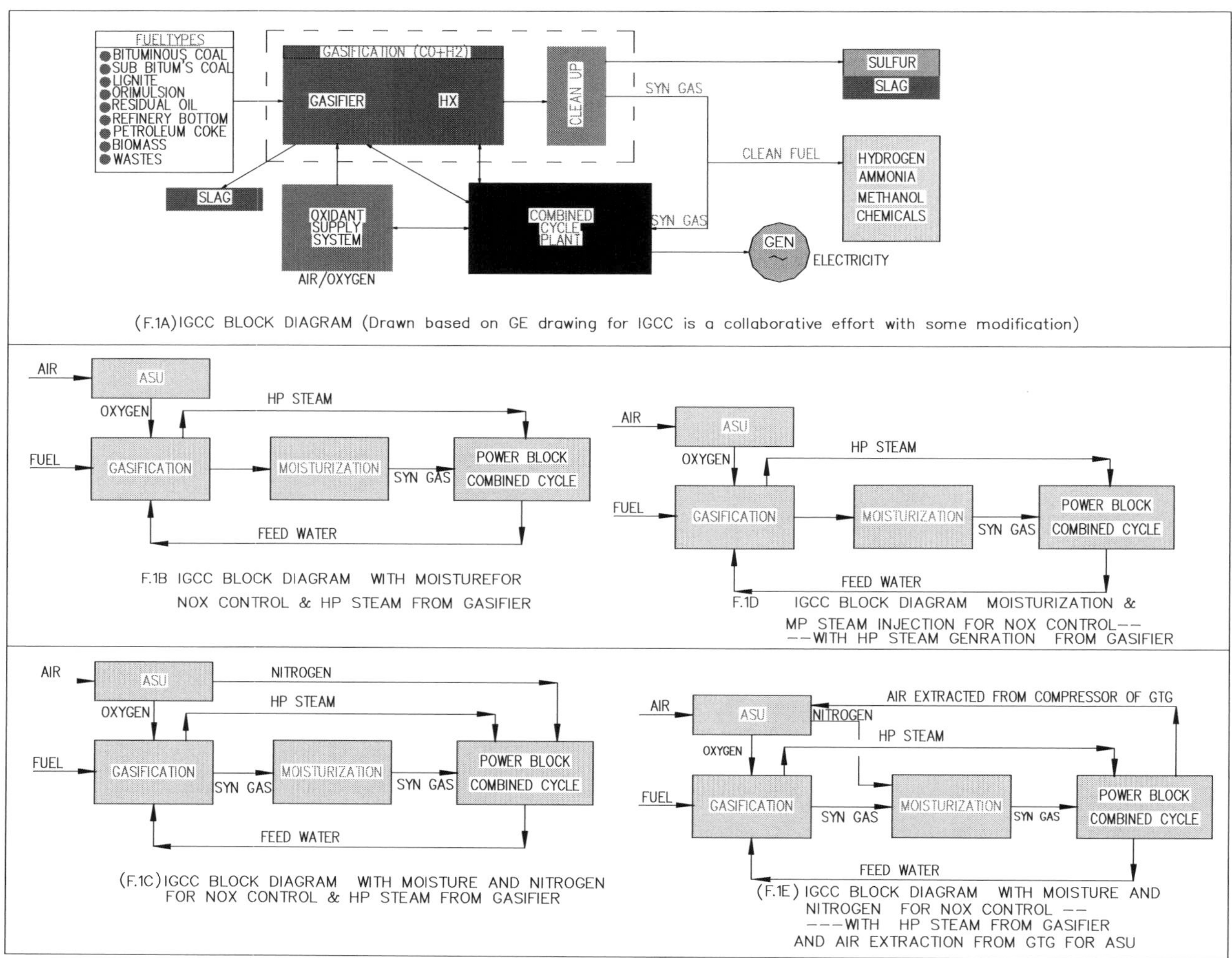

FIG. F.1 IGCC block diagram and variations in types.

TABLE F.1 Comparison of Pollutants Between IGCC and Super Critical/Ultra Super Critical Boilers

Criteria	Ultra/Supercritical PC (Coal)	IGCC (Coal)
NO_X (kg/GJ)	0.05	0.03
SO_X (kg/GJ)	0.03	0.15
CO (kg/GJ)	0.02	<0.005
CO_2 (kg/GJ)	100	96
PM (kg/GJ)	0.01	<0.005
Mercury (%)	30–90	>90

coal is pumped in slurry form or can be sent in dry form via a lock hopper system.

a. The coal slurry is prepared by wet grinding of wet coal in a rod mill. The coal needs to be ground to a specified particle size and distributed to maintain proper slurry solid concentration (60%–65% by weight). Prepared slurry is stored in an agitated tank. Normally recycled water is used here. The disadvantage is that the requirement of water may absorb some heat.

b. The Shell and Siemens design uses a dry coal feed system, with the help of a locked hopper, which consists of one top and one bottom valve (one opening at a time). During operation, the top valve (with the bottom valve closed) is opened to receive the coal fines. After the top valve is closed, the hopper is pressurized by N_2 from the ASU, at a pressure a little above the gasifier pressure, and then the bottom valve opens (with the top valve closed) for discharging to the gasifier. After this operation, the bottom valve is closed and the hopper is depressurized before the top valve is opened again.

2. *ASU*: The gasification process requires an oxidant such as air/oxygen. To minimize the size of the gasifier, high-purity, pressurized oxygen is used most often. However, it accounts for nearly 15% of the total gasification plant cost and utilizes most of the power. ASUs working on cryogenic distillation are utilized in most cases to obtain 95% oxygen purity. In cryogenic distillation, air is liquefied at an extremely low temperature (−185°C). Atmospheric air is compressed in multiple stages with intercooling and again cooling by chilled water. Then the air enters the "cold box" with a distillation column of many stages with an argon column for better purification. After removing other constituents, oxygen and nitrogen are warmed by a heat exchanger. Oxygen is stored for continuous gasifier operation. Nitrogen is compressed and may be used in a GTG or as a byproduct.
3. *Syngas cooling*: The raw syngas leaving the gasifier is >1500°C. Thus gas can be cooled by direct quenching, or heat recovery can be done utilizing a radiant, convective heat exchanger. Typically, there are three stages of heat recovery.
   a. Raw syngas is first cooled in a high-temperature, radiant heat exchanger to produce high-pressure (HP) steam at approximately >575° C for the power block. Hot syngas flows through the central waterwalls and HP steam is generated in the tube (>110 kg/cm^2).
   b. In the next stage, it is cooled by a convection-type heat exchanger, which produces medium pressure steam. It is then water scrubbed to remove fine soluble particles, such as ammonia, chloride, etc.
   c. In the last stage, it is cooled to ~35°C to get low-pressure steam in the low-temperature gas cooler (LTGC).
4. *Syngas cleanup*: To meet emission regulations and to protect downstream equipment, it is necessary to cleanup and condition the gas to get rid of fly ash and other fine particles, sulfur, ammonia, and chloride traces of heavy metals such as Hg and CO_2. Depending on the application, the cleaning process may vary as well as adjusting the H_2 to CO ratio requirements for the downstream system. Typical subsystems may include:
   a. *Scrubber, filter, and cyclone*: As discussed above, raw syngas is quenched and scrubbed with water in a trayed column to remove fine char and ash particulates in the slurry-fed gasifier. For the dry gasifier, cyclones and candle filters are removed and recycled back into the system, and finally, water quenched and scrubbed. With water scrubbing fine particles, ammonia, chlorides, traces of H_2S, and other contaminants are removed. Spent water is sent for sour treatment, whereby decantation fine particles are further removed.
   b. *Water gas shift (WGS)*: Most of the sulfur in coal is converted to H_2S, but depending on gasification temperature, ~3%–10% sulfur is converted to COS. Approximately 99% of COS is first converted to H_2S by syngas through the catalytic hydrolysis reactor. For a higher H_2 to CO ratio, syngas is passed through a sulfur-tolerant shift catalyst to convert CO to CO_2 and water into H_2. In some cases, steam injection is done for this ratio adjustment.
   c. *Acid gas removal (AGR)*: This is used for removal of H_2S and CO_2 followed by sulfur recovery for sulfuric acid plants. Removal of H_2S and CO_2 can be done by a physical process that is needed when syngas with <1 ppmv sulfur. This is not only costly, but it requires more power. However, for power generation where up to 30 ppmv sulfur is acceptable, chemical solvent processes such as methyl diethanolamine (MDEA) and sulfinol are used. This process operates slightly above atmospheric temperature (unlike the cryogenic physical process), so power requirement is less.
   d. *Activated carbon (not shown) filter to capture Hg*: Cooled syngas from LTGC is passed through a sulfide-activated carbon bed to recover ~90% of Hg and other heavy metal contaminants. Since there is sulfur in activated carbon, it is normally placed ahead of AGR to eliminate the possibility of sulfur slippage.
5. *Power block–combined cycle (see* Chapter 3, *Clause 10)*: Clean syngas is utilized in GTG for power generation. The exhaust gas from the GTG is used in heat recovery steam generators (HRSGs) for the generation of steam that is finally used to run STGs for the generation of power. A combined cycle plant is supported by a number of utility processing facilities such as steam condensation collection facilities (Fig. F.2), distribution, pumping, and treatment units. The gasification unit heat is recovered through the heat exchanger where feedwater comes from the combined cycle and is heated to get HP steam. This steam is later superheated in the HRSG (Fig. F.2) to be utilized in STG for power generation (Fig. F.1A).
6. *Other uses for syngas*: Along with producing power, syngas can be converted into a variety of liquid fuels, gases, and chemicals with the help of various conversion methods such as Fischer-Tropsch synthesis.
7. *Gasification process*: Coal gasification is quite a complex process, but for instrumentation, such detailing may not be necessary and can be conceived as a few reactions given in Table F.2. In the

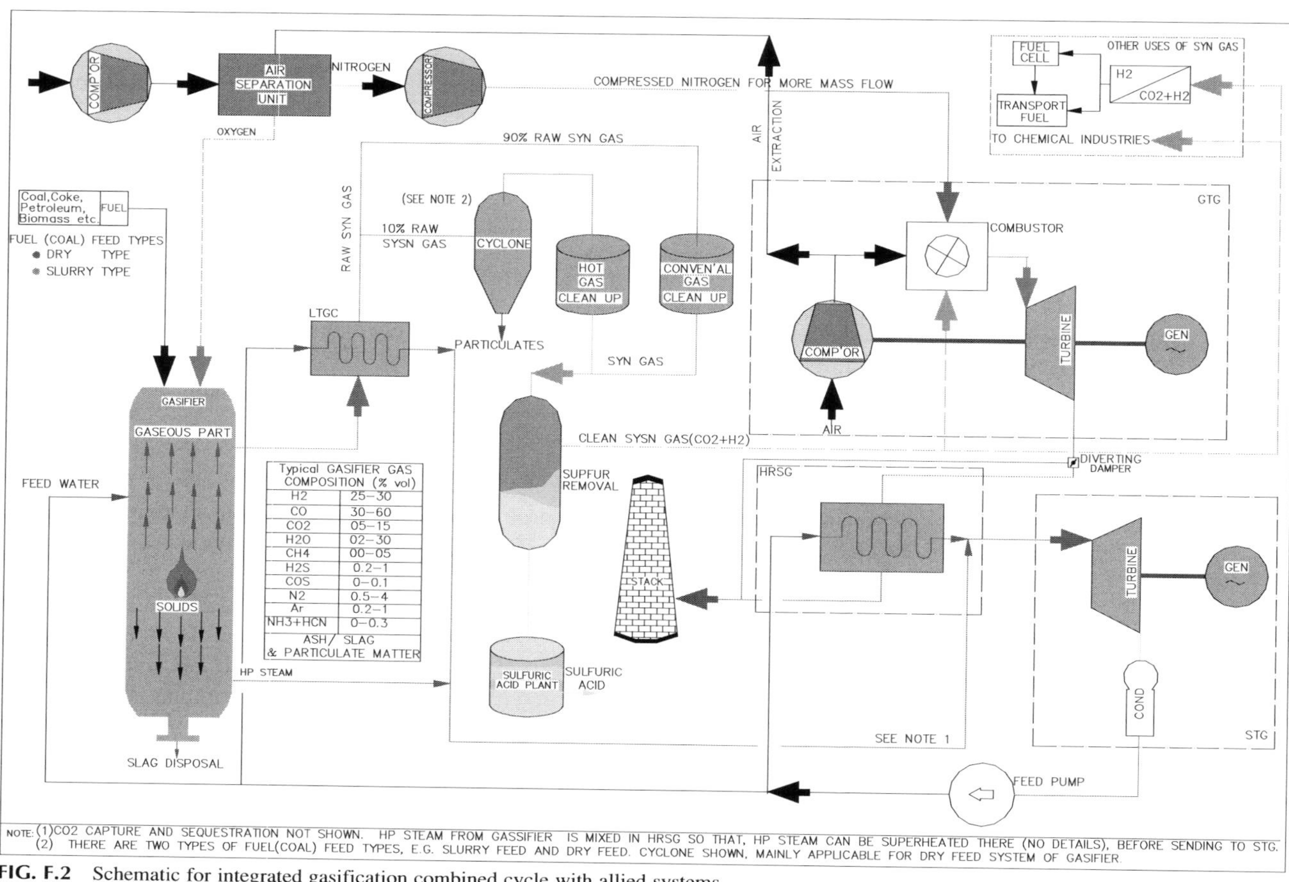

FIG. F.2 Schematic for integrated gasification combined cycle with allied systems.

gasification process, syngas is produced with a mixture of gas comprised mostly of CO and H_2, along with other constituents, such as CO_2, sulfur compound COS, H_2S, etc., as per the details given in Table F.3, when coal, steam, and oxygen are input in the system. The data show that a wide range of syngas composition is obtainable with variance in gasification types, feedstock, and operation parameters. In gasification most of the fuel's sulfur is converted to H_2S and, to a lesser degree, COS. Nitrogen in the fuel is converted mostly to NH_3, and a small amount of HCN and chlorine is converted to HCl. Gasification reactions are reversible, and the reactions and their conversions are subjected to thermodynamic equilibrium and reaction kinetics. Thermodynamic modeling is a very useful tool in calculating the amount of steam and oxygen required, finding the composition of syngas produced, and optimizing the process efficiency. It's possible to produce syngas with low methane content, using a high temperature with excess amount of steam (stoichiometric requirement). At a very high temperature, syngas production needs more oxygen, which would again lower system efficiency (because of the power for ASU). A few alternatives for gasification via system integrations are depicted in Figs F.1B–E. The design uses moisturization for NO_X control as shown in Fig. F.1B. These systems also use radiant and convective heat exchangers/syngas cooler along with an economizer in the feedwater line for heat recovery. The HP steam from the cooler is superheated by HRSG in the power block to be utilized in

TABLE F.2 Gasification Process

Reaction Type	Reaction
Gasification with O_2	$2C + O_2 \leftrightarrow 2CO$
Combustion with O_2	$C + O_2 \leftrightarrow CO_2$
Gasification with CO_2	$C + CO_2 \leftrightarrow 2CO$
Gasification with steam	$C + H_2O \leftrightarrow CO + H_2$
Gasification with H_2	$C + 2H_2 \leftrightarrow CH_4$
WGS water gas shift	$CO + H_2O \leftrightarrow H_2 + CO_2$
Methanation	$CO + 3H_2 \leftrightarrow CH_4 + H_2O$

TABLE F.3 Composition of Typical Gasifier Gas (in % vol)

H_2	CO	CO_2	H_2O	CH_4	H_2S	COS	N_2	Ar	NH_3 + HCN	Ash/Slag/ Particulate Matter
25–30	30–60	5–15	2–30	0–5	0.2–1	0–0.1	0.5–4	0.2–1	0–0.3	

the STG. In Fig. F.1C, nitrogen from the ASU has been utilized in GTG to boost the output with lower water consumption, but at the cost of using auxiliary power. Here, the design uses moisturization for NO_X control. Fig. F.1D is similar but there are ways to control NO_X by moisturization followed by medium pressure steam injection. In Fig. F.1E, moisturization of syngas and N_2 control NO_X.

8. *Design considerations*: A few points related to system design are listed below so the various subsystems could be integrated properly since integration plays a major role in IGCC.

a. *ASU air extraction and GTG integration*: Syngas has a very low caloric value. This can add significant mass flow to the GTG when compared with the performance-firing natural gas, with which most GTGs are optimized. Therefore when syngas is fired, there is extra capacity in the compressor, so extra mass flow will boost output to the limit firing temperature, the GT expander flow, and expander blade materials. Similarly, it is economical to use all the N_2 available from the ASU. N_2 is a function of coal feedstock, gasifier performance, etc. Adding N_2 and moisture to the maximum limit is called power augmentation, and this results in significant air extraction from the compressor for the ASU.

b. As noted before, moisture, N2, and CO_2 can be considered as diluents for NO_X.

c. Proper design integration involves a trade-off between overall plant efficiency, capital cost, operability, availability, and proper selection of equipment, especially the GTG on which the entire integration hinges. Considerations regarding this are presented below:

i. *Design Points*:
- Feedstock and its flexibility. This in turn determines the performance of the gasifier and composition of the syngas.
- Ambient condition and site elevation necessary for GTG.
- GTG design and operating domain.
- Emission limit, standards, and wastewater discharge guideline as applicable.

ii. *Integration considerations*:
- Gas turbine air extraction to the ASU
- NO_X control strategy
- Gas turbine power augmentation
- Heat recovery integration
- Steam generation conditions
- CO production

F.3 CONTROL SYSTEM

As IGCC is an integrated system, it is necessary that a coordinated control approach is deployed using the control strategy shown in Fig. F.3A. In the control strategy load, setpoint as well as total power generation could also be fed into the system so that a coordinated control system can integrate various generation units to initiate the actual requirement for the power block of a particular subsystem. The control demand sends signal simultaneously to both GTG fuel control, i.e., syngas flow, as well as to the ASU so that oxygen flow to the gasifier is regulated immediately. When there is a change in demand to the GTG syngas flow, there will be change in pressure of the syngas header pressure. Through controllers this regulates the fuel and oxygen flow to the gasifier so that syngas demand is met and the syngas header pressure is restored. This is also sent to the coordinate controller as feedback to balance the new demand. Gas turbine demand is given as a feedforward signal to make the system more responsive. Like the coordination control loop in PC boilers, a frequency signal is introduced.

Fig. F.3B shows typical system architecture for IGCC. Here, a number of controllers are hosted on a workstation (s) that has an OPC server. Other workstations are kept for data collection and implementation. There are a number of controllers to regulate operation of the various subsystems such as a gasifier, cooler, and a conditioning unit. The IGCC control network could be connected to other plant LANs via a firewall to limit access to the system. For details on functionality of firewalls in a network, refer to Appendix D (Fig. F.4).

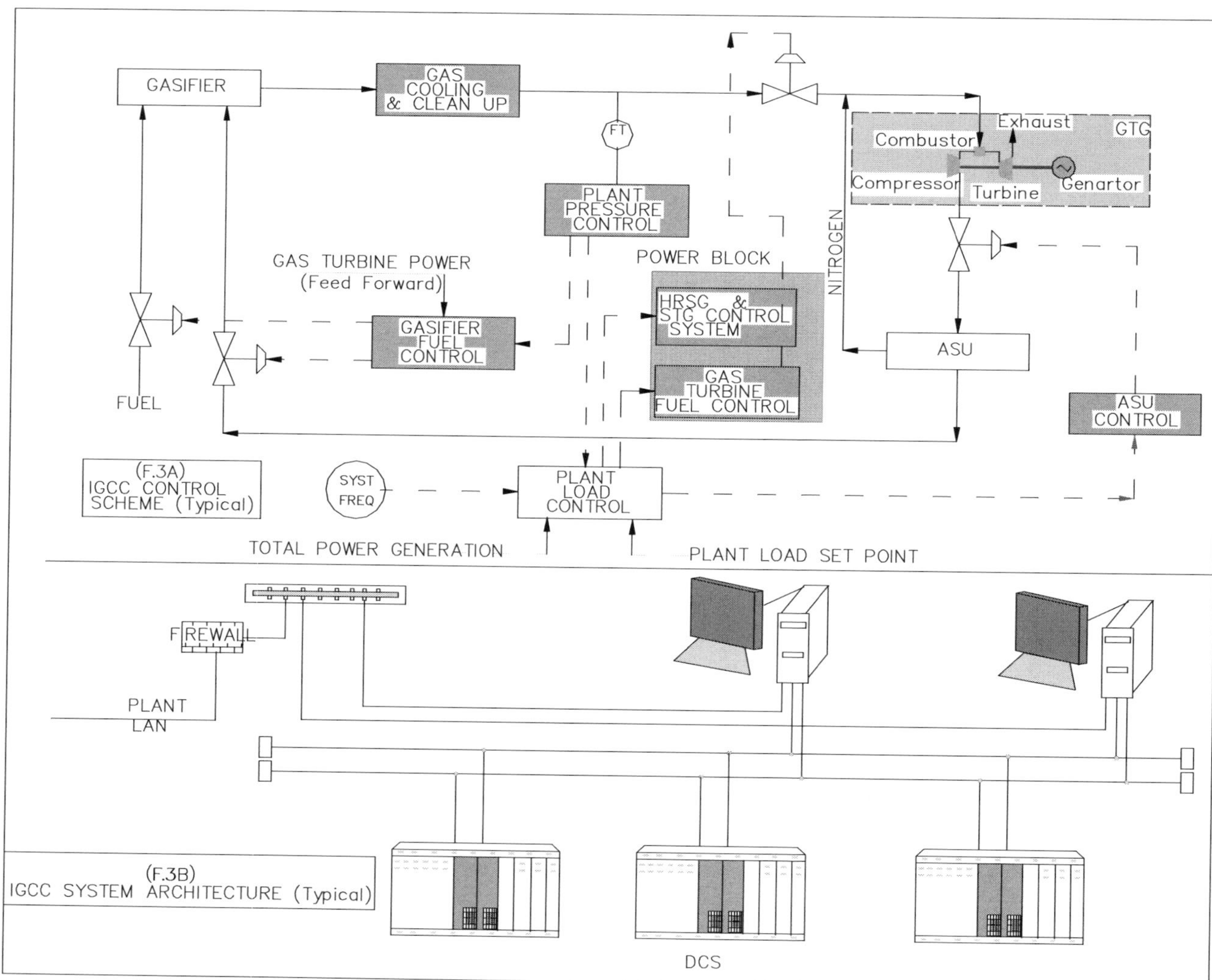

FIG. F.3 IGCC control scheme and system architecture.

**CFBC:*
"When an evenly distributed air or gas is passed upward through a finely divided bed of solid particles such as sand supported on a fine mesh, the particles are undisturbed at low velocity. As air velocity is gradually increased, a stage is reached when the individual particles are suspended in the air stream—the bed is called "fluidized."
With further increase in air velocity, there is bubble formation, vigorous turbulence, rapid mixing, and formation of dense defined bed surface. The bed of solid particles exhibits the properties of a boiling liquid and assumes the appearance of a fluid—"bubbling fluidized bed."
At higher velocities, bubbles disappear and particles are blown out of the bed. Therefore, some particles have to be recirculated to maintain a stable system—"circulating fluidized bed." The fluidized bed combustion (FBC) takes place at ~ 840 to 95°C. Since this temperature is below the ash fusion temperature, melting of ash and associated problems are avoided.
A CFBC is a good choice if the following conditions are met:
- *Capacity of boiler is large to medium.*
- *Sulfur emission and NO_X control is important.*
- *The boiler is required to fire low-grade fuel or fuel with highly fluctuating fuel quality.*

Major performance features of the circulating bed system are as follows:
(a) It has a high processing capacity because of the high gas velocity through the system.
(b) The temperature of ~87°C is reasonably constant throughout the process because of the high turbulence and circulation of solids. The low combustion temperature also results in minimal NO_X formation.
(c) Sulfur present in the fuel is retained in the circulating solids in the form of calcium sulfate and removed in solid form. The use of limestone or dolomite sorbents allows a higher sulfur retention rate, and limestone requirements have been demonstrated to be substantially less than with bubbling bed combustor.
(d) The combustion air is supplied at 1.5–2 psig rather than 3–5 psig as required by bubbling bed combustors.
(e) It has high combustion efficiency.
(f) It has a better turndown ratio than bubbling bed systems.
(g) Erosion of the heat transfer surface in the combustion chamber is reduced, since the surface is parallel to the flow. In a bubbling bed system, the surface generally is perpendicular to the flow. "

FIG. F.4 Note on CFBC.

BIBLIOGRAPHY

[1] GE Gas turbine Performance Characteristics: GE Power Systems:GER 3567H(10/0).

[2] An Introduction to CO_2 Separation and Capture Technologies. H. Herzog MIT Energy Laboratory.

[3] Advanced Process Control At an Integrated Gasification Combined Cycle Plant. M. Abela (Isab Energy Service), N. Bonavita, R. Martini (ABB PS&SS SpA).

[4] Integrated Gasification Combined Cycle GE IGCC Technology. (Dec'2003). J. Tobin, S.K. Bae GE Power Systems: APEC Clean Fossil Energy Seminar.

[5] IGCC—The Challenges of Integration. (Jun'2005) R. F. Geosits and Lee A. Schmoe—Bechtel Corporation: GT2005.

[6] IGCC experience and further developments to meet CCS Market Needs. (Sept'2009). J. Krag Siemens AG Energy Sector Fossil Power Generation division.

[7] The Tampa Electric Integrated Gasification Combined Cycle Project: US Department of Energy and Tampa Electric Company:A Joint report (Internet Document).

Appendix G

A Few Operational Features of the Unit

G.1 INTRODUCTION

Basic measurements and control aspects pertinent to thermal power plants have been covered in the main body of this book. In addition, there are a few other operational features where instrumentation and control engineering are either directly or indirectly related. A brief discussion on the same is presented here. In Chapters 7–9 and 11 various controls pertinent to boiler turbine regenerative systems and BOP systems (both closed loop and open loop control systems) have been discussed at length. With this knowledge, a short discussion is presented in this appendix on the following topics:

- Unit protection system.
- House load operation.
- Bus transfer system (BTS).

Out of these three systems, the first two directly impact the control systems as previously discussed in the main body of this book, whereas a BTS indirectly effects the control systems. For abbreviation definitions, refer to Figs. G.1 and G.3.

G.2 UNIT PROTECTION SYSTEM

As the name suggests, the discussion in this clause is about the protection of the unit, but it does not mean that in all cases entire units (i.e., all subsystems of the units) will be tripped. Since the degree of damage and/or abnormal situations varies, so does the type of protection. Common protections are offered against the following:

- Personal damage.
- Equipment/system damage.
- Safety operation mode for boiler turbine and generator, by limiting certain actions and even tripping equipment and/or a system.

It is clear there are abnormal situation classifications since there are different classes of trips. Internationally, these classes are termed as Class A, Class B, and so forth. The most popular classes of trips for mechanical systems are shown in Fig. G.1A.

G.2.1 Classification of Trips

Trips are mainly classified as Class U, Class A, Class B, Class C, Class D (also Class E—electrical system not included), etc. These classes are discussed in the following clauses.

G.2.1.1 Class U Trip

This stands for tripping of the unit or tripping of the boiler, turbine, and generator modules including regenerative cycle equipment (Fig. G.2). [This may not be applicable for cogeneration plants with a combination of boilers and turbo generator(s); TGs]. Major electrical fault (Class A electrical fault), extremely high condenser pressure, or damage of turbine bearing may fall under this category. In utility units without high-pressure (HP)/low-pressure (LP) bypass, when the TG trips, the boiler has to trip because there is no flow path left. The situation is totally different if the unit has an HP/LP bypass system. In this case, people may question the need for tripping the boiler even if the (TG) trips. The query is valid, because in that case, there is no Class U tripping in a true sense. It may be called Class A tripping. In reality, this may not be true, because HP/LP bypass can be run for a certain amount of time, when there is the probability of getting back the unit within a certain time, to avoid a long starting time (inertia) of the boiler. Faults are not easily repairable and it is questionable whether the HP/LP bypass should be run and energy wasted. The repair time is long for major electrical faults or damage to the turbine bearing, etc. Also, if the condenser pressure is high, HP bypass cannot be run. It has been found that many units' boilers are directly or indirectly tripped from high condenser pressure conditions.

All of these causes effectively call for an entire unit tripping. There are two schools of thought. One idea is that the boiler should be tripped directly, instead of attempting

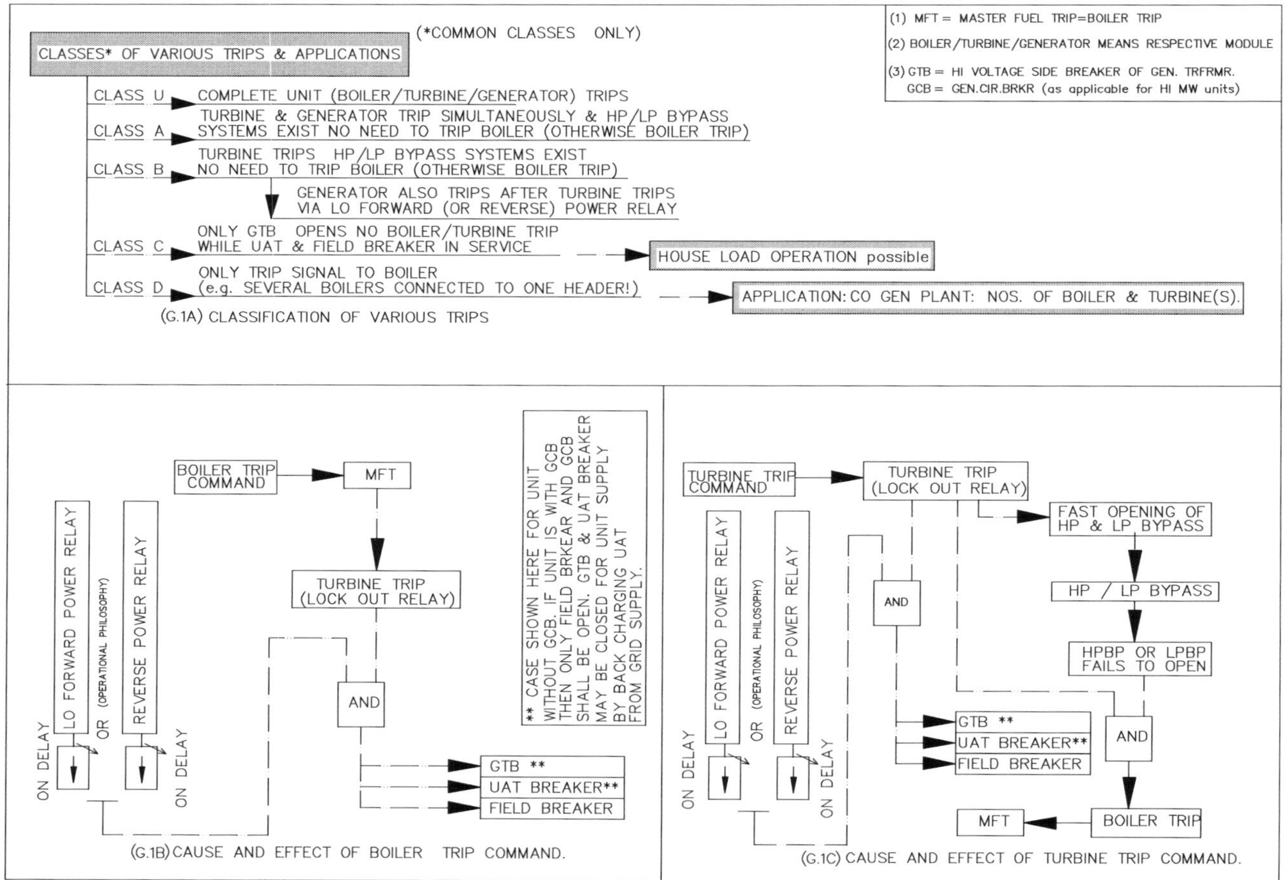

FIG. G.1 Classification cause and effect of various trips, Scheme 1.

to start the HP/LP bypass. This gives rise to Class U tripping. The other idea is that the boiler is not directly tripped; instead it may be tripped indirectly and/or manually. In the latter case, Class U tripping has no significance.

G.2.1.2 Class A Tripping

In this situation both the turbine and generator trip. This is mainly caused by electrical trips (Chapter 9, Clause 3.1). In utility units without HP/LP bypass, the boiler will have to be tripped (same as Class U). In units with HP/LP bypass, the boiler may not be tripped; instead HP/LP bypass operates by fast-opening criteria. If either the HP or LP bypass fails to open, for reheater protection (Fig. 7.81A), the boiler will trip.

G.2.1.3 Class B Tripping

It is similar to Class A tripping; the only difference is that the generator will trip after the turbine trips from low forward power or reverse power relay with a time delay (Chapter 9, Clause 3.1). What is the significance of this? The generator is run until the motive power is not totally exhausted so that when the generator trips, due to the remaining steam inside the turbine, it does not cause over speeding of the turbine. Time delay varies from machine to machine, and use of low forward power or reverse power relay varies with manufacturers. Some believe that generators should be tripped when motive power is lost, just before setting the motor action of the generator. On the other hand, some manufacturers believe that a very short period of motoring action of the generator is necessary to ensure that there is no motive power left to eliminate any possibility of overspeed (e.g., Siemens KWU machines).

G.2.1.4 Class C Tripping

In this type of tripping, the generator connection to the grid may be cut off by tripping the generator transformer high-voltage (HV) circuit breaker (GTB), without tripping the breaker for the unit auxiliary transformer (UAT) and

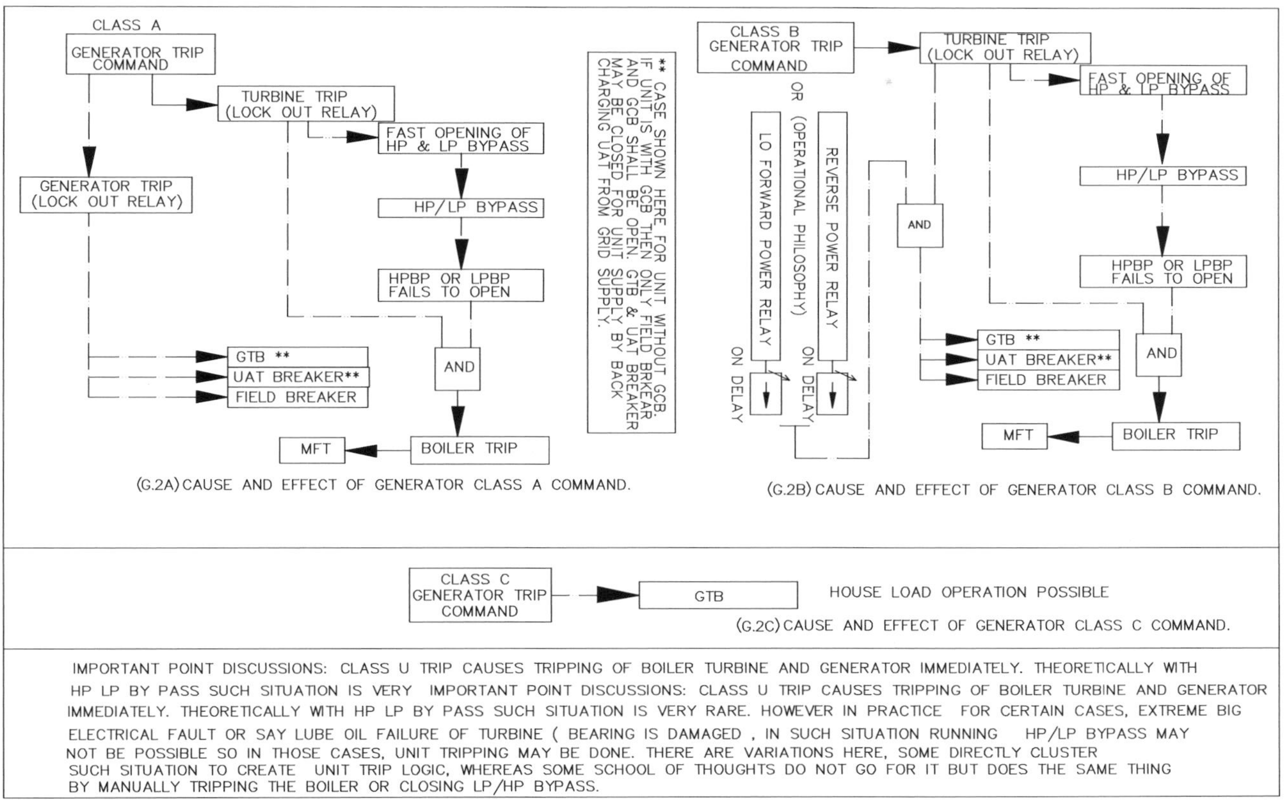

FIG. G.2 Classification cause and effect of various trips, Scheme 2.

generator field breaker, so that it can supply power to the unit board-house load operation.

G.2.1.5 Class D Tripping

This is tripping of the boiler, and is more significant for cogeneration plants.

G.2.2 Cause and Effect of Various Types of Tripping

In this clause, various causes and actions for boiler turbine and generator tripping are presented. Causes for boiler turbine tripping have been discussed at length in Chapters 8 and 9, and reheater protection was detailed in Fig. 7.81A.

G.2.2.1 Cause and Effect of Boiler Trip

Various boiler trip commands are discussed in Chapter 8 and detailed in Fig. 7.81A (for reheat protection). Some commands trip the master fuel relay to trip the master fuel trip to cut all sources of fuel (i.e., cutting off all energy) to the boiler (see Figs. G.2B and G.3C to understand the meaning of ** in Fig. G.2B). Since there is no steam supply, the turbine needs to be tripped via the turbine lockout relay (tripping solenoid), and consequently, the same generator needs to be tripped. However, the generator is not tripped immediately, but tripped after it is ensured that the motive power is lost as detailed in Clause 2.1.3. As a result GTB and the UAT breaker and field breaker will be opened. In larger units, where there is a generator circuit breaker (GCB), the GCB is opened along with the field breaker and the UAT breaker and the GTB can remain closed to the supply power from the unit board by back-charging the UAT. This possibility is discussed to a greater extent in Clause 4.0.

G.2.2.2 Cause and Effect of Turbine Trip

Various turbine trip commands are discussed in Chapter 9, such as the trip turbine lockout relay (trip solenoid valve) to trip the turbine immediately. Units with an HP/LP bypass will issue a fast-opening command. Tripping of the generator from the turbine trip will be similar to what was discussed in the previous clause. However, if either HP or LP bypass fails to open, then for reheater protection (as detailed in Fig. 7.81A) the boiler will trip (Fig. G.1C).

G.2.2.3 Cause and Effect of Generator Trip

There are classifications of generator tripping: Class A, B, and C. Of these, Class C trips are mainly concerned with faults beyond the generator transformer HV side. In this trip, house load operation is possible as long as the

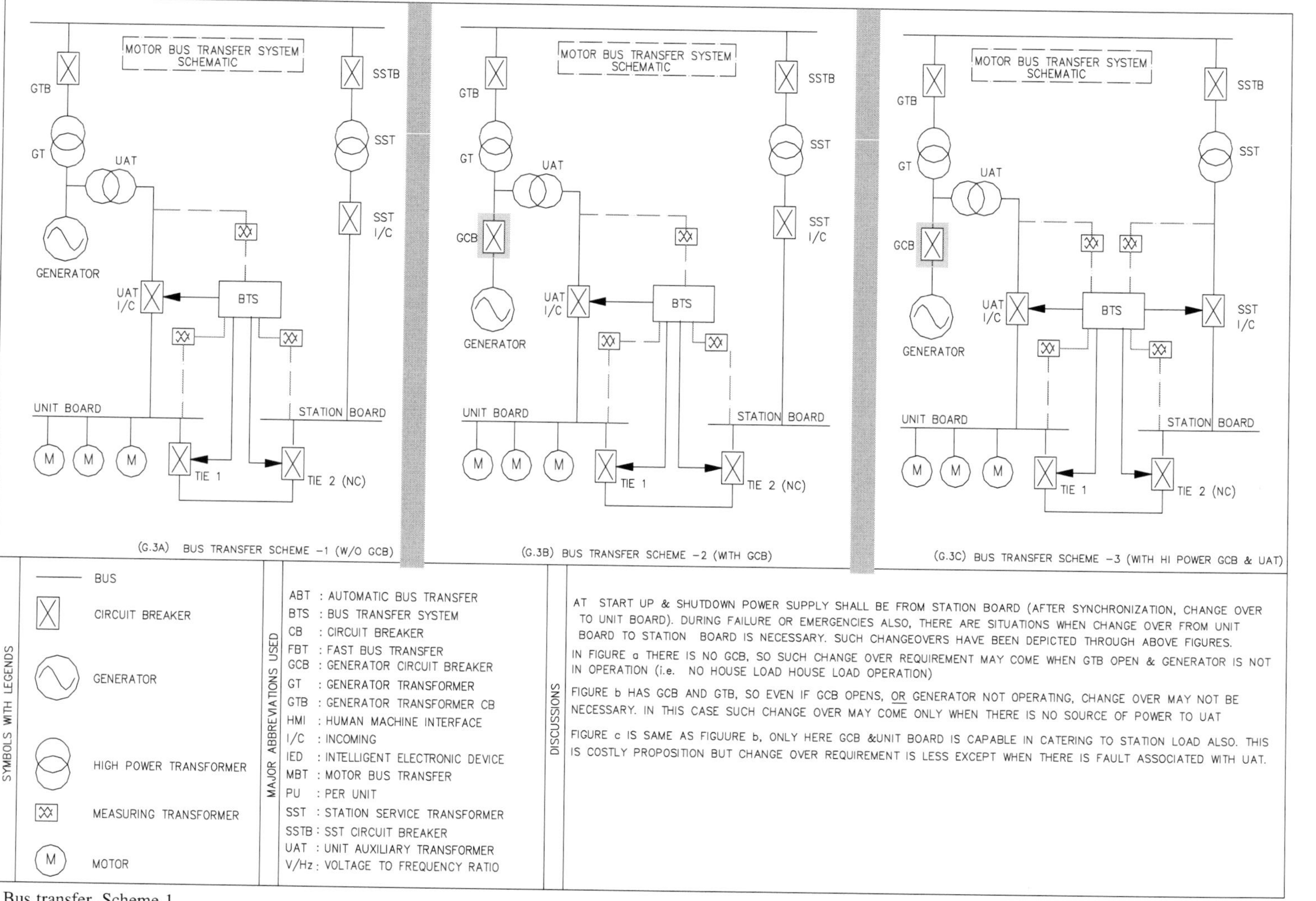

FIG. G.3 Bus transfer, Scheme 1.

generator system is healthy. Two major Class C tripping causes include: "manual tripping of GTB when system fails to take power" or "bus bar failure." Major tripping causes for Class A and B tripping were listed in Chapter 9, Clause 3.1 (also see Fig. G.2A–C).

As shown in Fig. G.2A, a Class A trip will immediately trip the turbine and generator. Tripping of the boiler depends on the presence of HP/LP bypass. In case the system does not have an HP/LP bypass the boiler will trip, otherwise there will be fast opening of the HP/LP bypass. However, if either the HP or LP Bypass fails to open (and turbine lock-out relay operated) then the boiler has to be tripped for reheater protection (Fig. 7.81A). Class B and C tripping of the generator are similar to what has been previously discussed.

G.3 HOUSE LOAD OPERATION

A unit will be in house load operation when it is disconnected from the grid and feeding the power for its own auxiliaries. Class C tripping of a generator is a good example of this. If an external fault beyond GT occurs, grid frequency is beyond the permissible operating limit, or Class C tripping has taken place, then house load operation is possible.

G.3.1 Electrical Side Changes

As per the discussions on unit protection, a fault beyond GT causes generator tripping via Class A tripping. So, if the unit is to be run in house load operation, then the wipeout condition needs to be initiated. This is done so that external faults, such as backup over current protection, would cause the GTB to open and the generator will not trip. Again, if the unit is to be isolated from the grid when grid frequency is beyond permissible value, then from a separate bus bar PT (Potential Transformer) is to be used to check grid frequency variation, and the same signal should be used for tripping of the GTB. On the other hand, generator PT should be used to check frequency variation, even during house load operation.

G.3.2 Boiler Effect

When the GTB opens naturally there will be a sudden surge in the boiler from 100% to HP bypass flow. Upon closing the turbine valve and quick opening of the HP bypass (as well as LP bypass), there will be a surge and safety pop-up may occur. Experienced manual handling may be able to stop this from happening. While reducing the boiler load, care should be taken to see that the main steam temperature does not go below the permissible value (for example, 490°C).

G.3.3 HP Bypass

Quick-opening command is used for HP bypass, and after it is open, operation with bias in the HP bypass pressure setpoint may be removed for better operation of the system.

G.3.4 Run Back for Coordinate Control

Boiler runback command may be issued to reduce the fuel to the desired point. During this operation, the runback rate and runback target may have to be adjusted for house load operation.

G.3.5 LP Bypass Control

When the control valve in the turbine attains position for house load condition, the first stage pressure will be reduced. This means a reduced set point for LP bypass pressure control. At this point, the reheater pressure may rise, so LP bypass will open to maintain the reheater pressure.

G.3.6 Turbine Control

On the turbine side, to get the GTB to open, the two governor controls will select speed as a controlling parameter, i.e., speed will assume control in the electrohydraulic governors (EHG). Also, in house load the work done is less in the turbine; hence, cold reheat pressure and HP exhaust temperature may rise. The trimming device has to operate to increase flow through the HP turbine to reduce HP exhaust temperature.

G.3.7 Protection

For house load operation, the following must be available:

- Turbine in EHG mode of control not in hydraulic control mode.
- All turbine protections are ON and working.
- Functioning of load shedding relay.
- Functioning of trimming device to regulate flow through high-pressure turbine.
- TSE is available.
- Coordinate control system is operational.
- AVR in auto.
- All major auxiliaries are available.

G.4 BUS TRANSFER SYSTEM (BTS)

Bus transfer is the practice of transferring electrical load from one bus to an alternative source of power and other industrial plants during emergency or control transfer. The bus transfer has to happen at a high speed in a secure manner so there is no adverse, economic, or other effect on the plant operation. Often the connected loads are induction motors of various ratings naturally technical analysis for probable adverse effect has to be assessed beforehand. This clause starts with a brief discussion about how various loads are connected to various boards/bus systems in thermal power plants and what bus transfer really means.

G.4.1 Bus Configuration

Normally thermal power plants have at least two sources of power for auxiliaries associated with the generating unit.

G.4.1.1 Two-Breaker (Main Tie) System

A typical unit with a generator and connected loads is shown in Fig. G.3A. The generator is normally connected to the grid via a high-power, step-up transformer called Generator Transformer GT and a GTB. In Fig. G.3A, there are two sources of power to the feed unit board motor loads: one is from the generating unit (from a point upstream of the GT) itself via a UAT, and the other is from a station service transformer (SST). Normally, unit auxiliary loads, such as a boiler feed pump, induced draft/forced draft fans, mills, etc., are connected at the unit board and are supplied via UAT, the UAT I/C is closed, and the tie (1) is open. Start-up and shut-down power, when the main the generator is offline, is provided from an alternative source grid, via SST through the tie (Tie 1 and Tie 2 are closed and UAT I/C is open).

When the unit is synchronized with the grid, the unit board is transferred to the UAT. This is called station to unit transfer. During unit tripping conditions, e.g., a generator trip, a turbine trip, etc., along with UAT/GT trip situations (on differential, winding temperature, etc.; partly discussed above), it is required that the unit board is transferred from UAT to SST. This transfer is called unit to station transfer. Such transfers can be manual or automatic. During start-up and controlled shutdown, manual transfer may be possible; however, with the automatic bus transfer system (ABTS) a minimum dead time is preferred (during which the motor is deenergized) for the motor. In this method of transfer, there are two ties, Tie 1 and Tie 2. Tie 2 is chosen as normally closed (NC) so that in the event of breaker failure logic if both UAT I/C and Tie 1 fail to open then dangerous back-feeding to the generator (permanent paralleling) is avoided.

Many (large) units have GCBs in addition to GTBs, as is seen in Fig. G.3B. This system has one important advantage: There is less chance of bus transfer, because if the unit trips (GCB open) the GT can back-charge UAT to keep the unit board alive without BTS. However, under conditions such as GT/UAT trip, an automatic transfer is still required to ensure supply to the motor board. However, in both cases discussed so far, house load operation is possible (with some electrical side changes discussed in Clause 3.1). Fig. G.3B and C are the same; the only difference is that in Fig. G.3B UAT is sized to cater to the load of unit + station board load. This has additional capital cost, but the ABTS system can be a three-breaker (or main tie main) scheme, which is normally used in industrial system. This kind of scheme has been seen in one 250 MW unit.

G.4.1.2 Three-Breaker Scheme (Main Tie Main)

This scheme, as shown in Fig. G.4A, is more popular in industrial plants, but with a UAT it is highly rated; a similar scheme also might be used in thermal power plants. There are two sources, a main source and alternative source, each capable of catering to both motor buses. These two buses are connected via a tie. If the tie is the NC type, then the entire motor bus comprised of Bus 1 and Bus 2 is supplied by either of the two sources. On the other hand, for the normally open type tie, Bus 1 is fed by the main source and Bus 2 is fed by an alternative source. If one source fails, the source supplies power to that failed motor bus via closure of the tie.

G.4.1.3 Cogeneration Plant Islanding Operation

In Fig. G.4B a typical configuration is shown for a process cum power plant (e.g., alumina plant) with a cogeneration unit. Critical plant loads are normally supplied via a cogeneration unit in an islanded operation to avoid fluctuations in an unstable grid. Noncritical loads are supplied from the grid. Grid supply acts as a backup for the critical loads so that failure from the islanded supply will be delivered from the grid via the tie.

G.4.2 Considerations for Bus Transfer

The following parameters need attention during bus transfer:

- Coasting down duration of bus voltage during open circuit.
- Electrical/mechanical stress on the connected motors with load.
- Bus transfer system blocking during a short circuit.

When an induction motor is disconnected from the supply, then self-generated voltage, known as residual voltage, appears across the terminal. When a bus is disconnected, the motor in the bus starts decelerating and the residual voltage deteriorates due to decaying trapped air gap flux in the motor. Also, the frequency of residual voltage continuously drops. Decay time is governed by the open circuit time constant of the rotor. The larger the size of the motor, the longer is the time for voltage to decay. Also, the higher the load on the motor, the faster is the frequency decay.

This has a direct impact on how fast the phase angle will change. So, low inertia load will result in change of the phase angle since frequency decay will be faster thus creating a slip in the frequency between motor bus and new source. (Note: Fans in thermal power plants have higher inertia.) Another important factor is the V/Hz ratio. According to IEEE/ANSI Standard C50.41-2000, for fast motor BTS, resultant V/Hz at the instant of transfer is <1.33 per unit. With increase in phase angle (assuming voltage at motor bus and source are the same), V/Hz will increase. Also as the voltage difference between two sources increases the V/Hz increases. Due consideration is needed for transient effects, including:

- Magnitude of residual voltage between the motor and the bus.
- Phase angle between motor bus residual voltage and alternative source.
- Resultant V/Hz across the motor at the time of transfer.

The term "across the motor" is used because each motor will have a different residual voltage and a different impact, so the aggregate is normally considered. The worst-case scenario is when at the time of transfer there is little voltage decay the motors are in phase opposition. At this point, the impact will create havoc, i.e., twice the normal voltage will be applied creating 9–15 times inrush current to the motor. Force will be proportional to the inrush current so the extent of damage is possible.

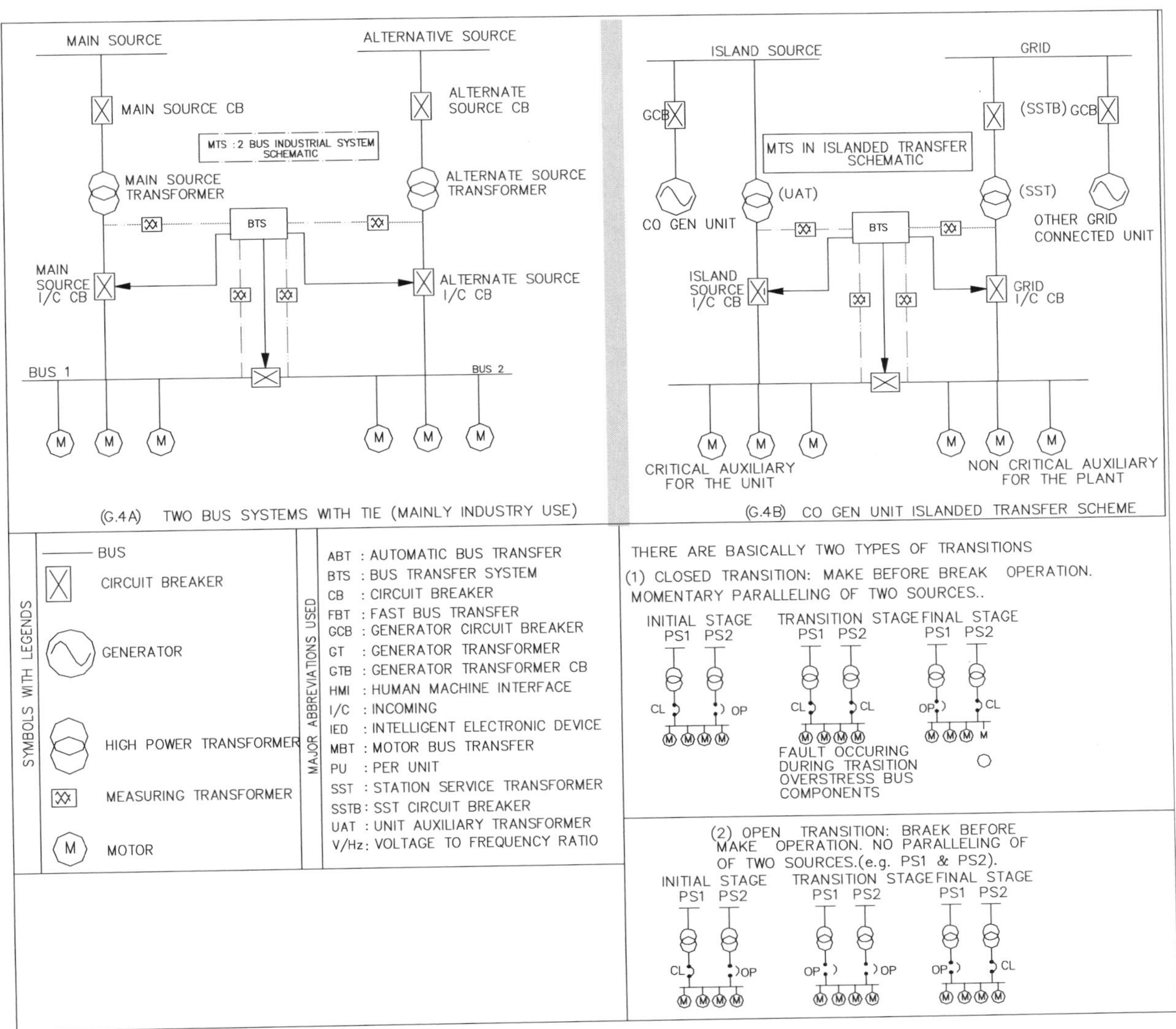

FIG. G.4 Bus transfer, Scheme 2.

G.4.3 Various Methods for Bus Transfer

There are two methods of bus transfer:

1. *Close transition*: Make before break or momentary paralleling of sources.
2. *Open transition*: Break before make without paralleling of sources.

Both methods are shown in Fig. G.4 (Discussions).

G.4.3.1 Closed Transition

A new source is connected to the motor bus when an old source trips to transfer the bus without interruption. However, it is necessary to evaluate the voltage difference frequency as well as phase angle before making this transition. This has the advantage that the process is not interrupted, and it is simple to implement with a synchronizing check relay. The major problem is if there is a fault during transition, bus components are overstressed and cannot be used when the main source loss is due to a fault.

G.4.3.2 Open Transition

In this method, unlike the previous method, an alternate source breaker closes only after the original/other breaker is open, leaving the motor bus without a source for a moment, as shown in Fig. G.4 (Discussions). There are three main types of open transition:

1. *Fast transfer*: This is used to minimize dead time of the motor bus. The alternative source breaker is closed in the fast transfer method when the phase angle of the motor bus and the new source is within, or moves in to the phase-angle limit within the fast transfer enable time. One cycle phase-angle response is necessary, along with a check back to ensure that the bus voltage level is within upper and lower limits before closing the breaker. Major advantages of these systems include: High-speed transfer, transient torque less because of quick transfer, minimum interruption of the process, no parallel operation, and no breaker fault effect.
2. *In-phase transfer*: Takes place when the phase angle of the residual voltage of the motor bus is in phase with the new source. For this, calculation of *df/dt* may be necessary. This can be done by a phase-angle relay capable of predicting the in-phase condition in advance of the new source breaker closing time, similar to a synchronizer. This is faster than the residual voltage method of transfer. This method minimizes V/Hz ratio. The new source breaker will be closed by predicting movement through zero phase coincidence between the motor bus and the new source during transfer. It is an excellent back-up for fast transfer. However, this may pose serious problems when the bus transfer is initiated at a time the main source has a fault. Load shedding may be required for in-phase transfer; however, it is safe enough when the fast transfer is blocked.
3. *Residual voltage transfer*: Auxiliary loads are transferred in this method when motor bus residual voltage reaches a value that could cause damage to the motors. Typically, this is ~25%–30% rated voltage or when the motor bus falls to 0.33 pu (no matter what the phase angle; it could be much less than the 1.33 required by standard). This is a simpler, risk-free transfer method, and is usually too slow to cause plant interruptions.

G.4.3.3 Timing Diagram

A typical timing diagram for an open type transfer is shown in Fig. G.5A. Before enabling any transfer, the old source breaker must be opened. There are two methods for this: simultaneous and sequential.

G.4.4 New Transfer Technology

Today, with the introduction of intelligent electronic devices (IEDs; see Chapter 7) that can communicate with other IEDs, systems are simplified and more effective. In the past, a large amount of wiring and huge spaces were necessary to implement systems with relay logics. Now IEDs can be connected via the PLC, which in turn can be directly connected to the plant distributed control system for decentralized control and central monitoring.

G.4.5 Electrical Power Distribution

An electrical single line with a typical electrical power distribution scheme is presented in Fig. G.6. In this scheme, how power is drawn from the transmission system and distributed within the power plant is illustrated. From an overall economic point of view, in some large units there are various voltage-level bus systems, e.g., 11, 6.6, ad 3.3 kV, and low-voltage (LV) systems of 0.415 kV, which are adjusted according to the country. In midsize plants, normally one HV and one LV (i.e., 6.6 and 0.415 kV) bus system is kept. Typically, the loads are distributed over two sets of boards.

There could be several changeover options, as seen in Fig. G.6. From the HV bus LV buses are created with the help of a step-down transformer. This scheme is presented to show the power distribution in thermal power plants and to demonstrate how to study the key single-line diagram. The notes in Fig. G.6 will show how electrical equipment is specified and how tag numbers typically are allocated. In this scheme, separate GCBs are shown, but in many midsize units, these may be missing.

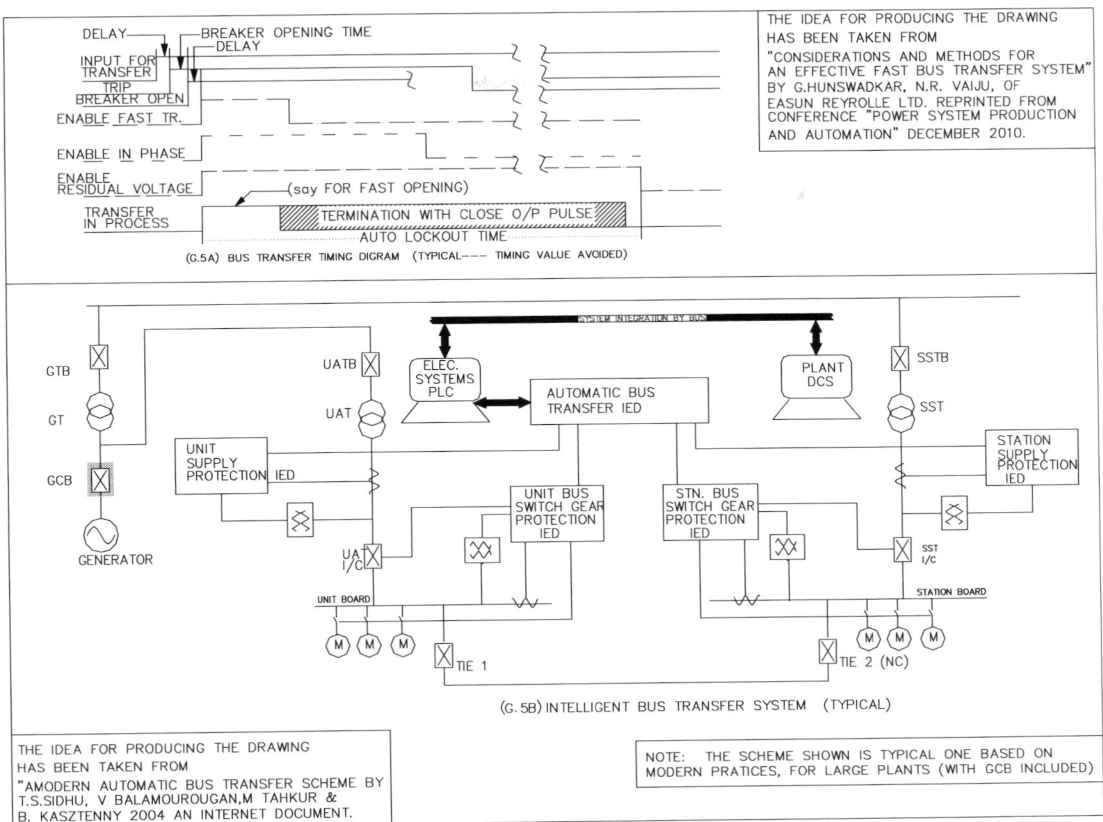

FIG. G.5 Bus transfer, technical details.

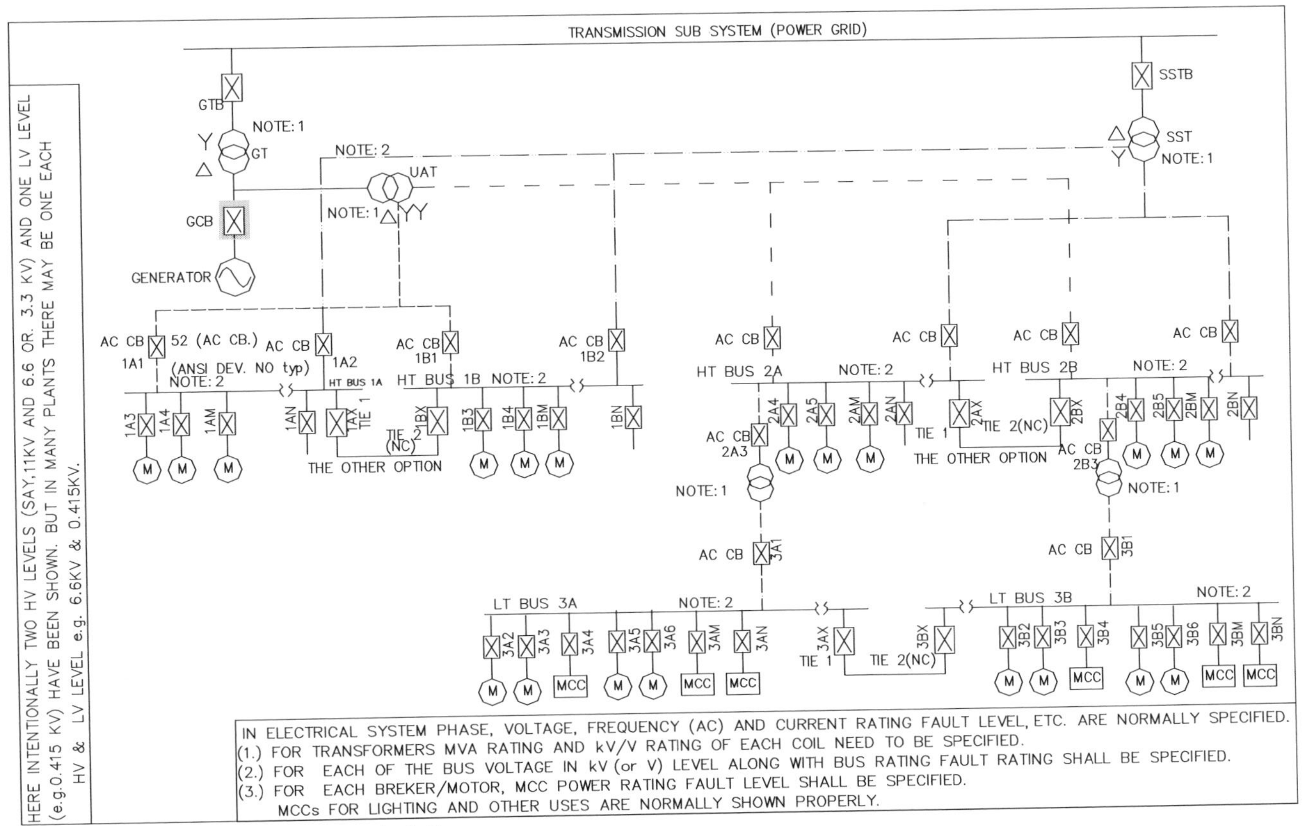

FIG. G.6 Electrical power distribution, single line.

BIBLIOGRAPHY

[1] Workshop on Mechanical systems (power cycle) in Thermal Power Station BHEL New Delhi.
[2] A modern Automatic Bus transfer scheme by T.S. Sandhu, V. Balamourougan, M. Thakur, & B. Kasztenny (Internet document).
[3] Considerations and Methods for an effective Bus Transfer System by G. Hunswadkar, N.R. Viju. (December 2010) Easun Reyrolle Ltd. Reprinted in from conference. Power System Production and Automation. New Delhi.
[4] Fast transfer Systems. (July 2008). A SystemSolution Apprach by A. Raje, A. Raje & A chowdhary Member IEEE.

Index

Note: Page numbers followed by *f* indicate figures and *t* indicate tables.

B

C

D

G

N

Printed and bound by CPI Group (UK) Ltd, Croydon, CR0 4YY
13/12/2024
01805846-0001